FAO Fisheries Series
No. 66
FAO Statistics Series
No. 180

Collection FAO:
Pêches N° 66
Collection FAO:
Statistiques N° 180

Colección FAO:
Pesca N° 66
Colección FAO:
Estadística N° 180

F

FAO yearbook
annuaire
anuario

Fishery statistics
Capture production

Statistiques des pêches
Captures

Estadísticas de pesca
Capturas

Vol. 94/1

2002

FOOD AND AGRICULTURE ORGANIZATION OF THE UNITED NATIONS
Rome, 2004

ORGANISATION DES NATIONS UNIES POUR L'ALIMENTATION ET L'AGRICULTURE
Rome, 2004

ORGANIZACIÓN DE LAS NACIONES UNIDAS PARA LA AGRICULTURA Y LA ALIMENTACIÓN
Roma, 2004

Prepared by the Fishery Information, Data and Statistics Unit of the Fisheries Department, FAO, on the basis of information available as of 31 December 2003.

Information from this publication may be quoted if reference is made to the source.

FAO provides time series of fishery statistical data in computer readable form to users of the *FAO Yearbook of Fishery statistics - Capture production.* The time series are given as annual data for countries or area/species groups/major fishing areas for a period of years starting with the year 1950 up to the latest year published in the Yearbook.

Inquiries about the technical content of this publication should be addressed to:

The Senior Fishery Statistician
Fishery Information, Data and Statistics Unit
Fisheries Department
FAO
Viale delle Terme di Caracalla
00100 Rome,Italy

E-mail: FIDI-Inquiries@fao.org

Requests for copies of this publication should be sent to the sales agents listed at the back, or to:

Sales and Marketing Group
FAO
Viale delle Terme di Caracalla
00100 Rome, Italy

Information on world fisheries and aquaculture can be obtained from the Internet site: **www.fao.org/fi**

Préparé par l'Unité de l'information, des données et des statistiques sur les pêches, Département des pêches, FAO, sur la base des renseignements disponibles au 31 décembre 2003.

Il est possible de reproduire les articles et renseignements contenus dans cette publication sous réserve d'en indiquer la source.

La FAO met en vente des séries de données statistiques publiées dans l'*Annuaire statistique des pêches – Captures.* Les séries chronologiques donnent des chiffres annuels par pays ou zones/groupes d'espèces/zones de pêche principales, qui portent sur une période allant de l'année 1950 à la dernière année publiée dans l'Annuaire.

Adresser toutes demandes d'ordre technique concernant cette publication au:

Statisticien principal des pêches
Unité de l'information, des données et des statistiques sur les pêches
Département des pêches
FAO
Viale delle Terme di Caracalla
00100 Rome, Italie

Courriel: FIDI-Inquiries@fao.org

Adresser les commandes de la publication soit aux dépositaires (liste en dernière page), soit au:

Groupe des ventes et de la commercialisation
FAO
Viale delle Terme di Caracalla
00100 Rome, Italie

Des renseignements sur la pêche et l'aquaculture mondiales peuvent être obtenus en consultant le site Internet : **www.fao.org/fi**

Preparado por la Dependencia de Información, Datos y Estadísticas de Pesca, Departamento de Pesca, FAO, teniendo en cuenta los datos disponibles hasta el 31 de diciembre de 2003.

Las informaciones contenidas en esta publicación se pueden reproducir si se cita su origen.

La FAO ofrece series de datos estadísticos para su lectura en computadoras a las personas que utilizan el *Anuario Estadístico de Pesca – Capturas.* Las series cronológicas se facilitan como datos anuales por países o áreas/grupos de especies/principales áreas de pesca para el período que inicia con el año 1950 y termina con el último año publicado en el Anuario.

La correspondencia concerniente al contenido técnico de esta publicación debe dirigirse al:

Estadístico Superior de Pesca
Dependencia de Información, Datos y Estadísticas de Pesca
Departamento de Pesca
FAO
Viale delle Terme di Caracalla
00100 Roma, Italia

Correo electrónico: FIDI-Inquiries@fao.org

Los pedidos de esta publicación deben ser dirigidos a los puntos de ventas que figuran en la última página, o al:

Grupo de Ventas y Comercialización
FAO
Viale delle Terme di Caracalla
00100 Roma, Italia

La información sobre la pesca y la acuicultura mundiales se puede obtener en la página de Internet: **www.fao.org/fi**

FAO Fishery Information, Data and Statistics Unit/Unité de l'information, des données et des statistiques sur les pêches/Dependencia de Información, Datos y Estadísticas de Pesca.
Capture production 2002.
Captures 2002.
Capturas 2002.
FAO yearbook. Fishery statistics. Capture production/Annuaire FAO. Statistiques des pêches. Captures/Anuario FAO. Estadísticas de pesca. Capturas.
Vol. 94/1. Rome/Roma, FAO. 2004. 642 p.

Designations of countries or areas	Désignations des pays ou zones	Denominaciones de los países o áreas
The designations employed and the presentation of material in this information product do not imply the expression of any opinion whatsoever on the part of the Food and Organization of the United Nations concerning the legal or development status of any country, territory, city or area or of its authorities, or concerning the delimitation of its frontiers or boundaries.	Les appellations employées dans ce produit d'information et la présentation des données qui y figurent n'impliquent de la part de l'Organisation des Nations Unies pour l'alimentation et l'agriculture aucune prise de position quant au statut juridique ou au stade de développement des pays, territoires, villes ou zones ou de leurs autorités, ni quant au tracé de leurs frontières ou limites.	Las denominaciones empleadas en este producto informativo y la forma en que aparecen presentados los datos que contiene no implican, de parte de la Organización de las Naciones Unidas para la Agricultura y la Alimentación, juicio alguno sobre la condición jurídica o nivel de desarrollo de países, territorios, ciudades o zonas, o de sus autoridades, ni respecto de la delimitación de sus fronteras o límites.

ISBN 92-5-005139-5

Table of Contents / Table des matières / Tabla de materias

	Standard symbols	Signes conventionnels	Símbolos convencionales
...	Data not available; unobtainable; data not separately available but included in another category	Données non disponibles; données que l'on n'a pas pu obtenir; données non disponibles séparément, mais comprises dans une autre catégorie	No hay datos; no se han podido obtener datos; datos no disponibles por separado pero incluidos en otra partida estadística
—	None; magnitude known to be nil or zero	Néant; quantité que l'on sait égale à zéro	Ninguna; cantidad que se sabe es nula o cero
0	More than zero but less than half the unit used	Quantité supérieure à zéro, mais inférieure à la moitié de l'unité utilisée	Más de cero pero inferior a la mitad de la unidad empleada
F	FAO estimate from available sources of information or calculation based on specific assumptions	Estimations de la FAO d'après les sources d'informations disponibles ou calculées sur la base de suppositions spécifiques	Estimación de la FAO partiendo de fuentes de información disponibles o calculada sobre la base de suposiciones específicas
t	tonnes (=1000 kg)	tonnes (=1000 kg)	toneladas (=1000 kg)
kg	kilograms	kilogrammes	kilogramos
no	number	nombre	número
nei	not elsewhere included	**nca** non compris ailleurs	**nep** no especificado en otra partida
...A	FAO English name of the species item not ascertainable	Le nom anglais utilisé par la FAO pour la catégorie d'espèces n'est pas vérifiable	Se desconoce el nombre utilizado por la FAO en inglés para la partida de especies
...B	FAO French name of the species item not ascertainable	Le nom français utilisé par la FAO pour la catégorie d'espèces n'est pas vérifiable	Se desconoce el nombre utilizado por la FAO en francés para la partida de especies
...C	FAO Spanish name of the species item not ascertainable	Le nom espagnol utilisé par la FAO pour la catégorie d'espèces n'est pas vérifiable	Se desconoce el nombre utilizado por la FAO en español para la partida de especies
S	Summation of catches	Somme des captures	Suma de las capturas

Introduction

1. This Volume of the **Yearbook of Fishery Statistics** presents the annual statistics, for a varying series of recent years ending in 2002, on a worldwide basis, on nominal catches of fish, crustaceans, molluscs and other aquatic animals, residues and plants, taken for all purposes (commercial, industrial, recreational and subsistence) by all types and classes of fishing units (fishermen, vessels, gear, etc.) operating both in inland, fresh and brackish water areas, and in inshore, offshore and high seas fishing areas. Beginning with Volume 82 statistics for mariculture, aquaculture and other kinds of fish farming, are excluded from all national, regional and global totals. **Beginning with Volume 90/1, the names and species composition of some groups of the FAO International Standard Statistical Classification of Aquatic Animals and Plants (ISSCAAP) have been revised. See paragraph 3 of the NOTES ON SPECIES ITEMS.**

2. Despite the importance of recreational fishing regarding some species and for certain countries, figures include recreational catches only where available.

3. The annual period used is the calendar year (1 January-31 December), with the exceptions of capture data in the Antarctic fishing areas and for some countries, mentioned in the NOTES ON INDIVIDUAL COUNTRIES OR AREAS, for which a split-year is used. Since Volume 94/1, the new fishing season (1 December–30 November) of the Commission for the Conservation of Antarctic Marine Living Resources (CCAMLR) has been adopted. Split-year data are shown under the calendar year in which the split-year ends.

4. Catches are expressed in tonnes, except those for whales, seals and crocodiles which are given in numbers and corals, pearls and sponges which are given in kilograms. See paragraph 11 of the NOTES ON COUNTRIES OR AREAS.

5. Beginning with Volume 48 data on aquatic mammals (tables B-61, B-62 and B-63 expressed in numbers and B-64) and data on aquatic plants (A-6, B-91, B-92, B-93 and B-94) are excluded from all national, regional and global totals. For practice prior to Volume 48, see notes in respective Yearbook. Beginning with Volume 60, data on corals, expressed in kilograms given in table B-82 are also excluded from all national, regional and global totals. Beginning with Volume 62 data on pearls, expressed in kilograms in table B-81 and sponges, expressed in kilograms in table B-83 are likewise excluded. Beginning with Volume 66 data on crocodiles expressed in numbers in table B-73 are excluded from all national, regional and global totals.

6. FAO has completed the separation of the aquaculture time series from the capture production time series dating back to 1950. This separation was based on national reporting, when available, and other sources of historical information. The entire global time series for aquaculture production can be downloaded using the FishStat Plus software available at: http://www.fao.org/fi/statist/FISOFT/FISHPLUS.asp

7. Where necessary the data for 1950-2001 published in the preceding volumes of the Yearbook of Fishery Statistics have been revised. Where figures in this volume differ from those previously published, the amended data represent the most recent version. Some statistics provided to FAO by national offices, in particular those for 2002, are provisional and may be amended in future volumes, and in other FAO publications.

Introduction

Dans ce volume de l'**Annuaire statistique des pêches** figurent, pour diverses périodes récentes prenant fin en 2002, les statistiques établies sur une base mondiale des captures nominales de poissons, crustacés, mollusques et autres animaux aquatiques, résidus et plantes aquatiques, effectuées à toutes fins (commerciales, industrielles, récréatives et de subsistance) par tous les types et catégories d'unités de pêche (pêcheurs, bateaux, engins, etc.) opérant tant dans les eaux continentales, douces et saumâtres, que dans les zones de pêche du littoral, du large et de la haute mer. A partir du Volume 82, les statistiques ayant trait à la mariculture, à l'aquaculture et à d'autres types de pisciculture, sont exclues de toutes les données au niveau national, régional et global. **A partir de Volume 90/1, les dénominations et la composition par espèces de quelques groupes de la Classification statistique internationale type des animaux et des plantes aquatiques (CSITAPA) de la FAO, ont subi des révisions. Voir le paragraphe 3 des NOTES SUR LES CATEGORIES D'ESPECES.**

Malgré la grande incidence de la pêche récréative sur certaines espèces et pour certains pays, les chiffres indiqués comprennent les captures de la pêche récréative, seulement lorsque de telles données sont disponibles.

La période annuelle utilisée est l'année civile (1er janvier-31 décembre), sauf pour les données relatives aux captures effectuées dans les zones maritimes de l'Antarctique et pour quelques pays, mentionnés dans les NOTES SUR DIVERS PAYS OU ZONES, pour lesquels on utilise l'année fractionnée. A partir du volume 94/1 on utilise la nouvelle période de pêche (1er décembre–30 novembre) de la Commission pour la conservation de la faune et la flore marines de l'Antarctique (CCAMLR). Les données relatives aux années fractionnées figurent sous l'année civile durant laquelle se termine l'année fractionnée.

Les captures débarquées sont exprimées en tonnes, à l'exception des données relatives aux baleines, phoques et crocodiles qui sont indiquées numériquement et les données sur les coraux, les perles et les éponges qui sont indiquées en kilogrammes. Voir le paragraphe 11 des NOTES SUR LES PAYS OU ZONES.

A partir du volume 48, les données sur les mammifères aquatiques (tableaux B-61, B-62 et B-63 exprimées en nombres et B-64) et les données sur les plantes aquatiques (A-6, B-91, B-92, B-93 et B-94) sont exclues de tous les totaux au niveau national, régional et global. Pour la méthode suivie avant le volume 48, se reporter aux notes de l'Annuaire en question. A partir du volume 60, les données sur les coraux exprimées en kilogrammes indiquées au tableau B-82 sont également exclues de tous les totaux au niveau national, régional et global. A partir du volume 62, les données sur les perles exprimées en kilogrammes indiquées au tableau B-81, et sur les éponges exprimées en kilogrammes indiquées au tableau B-83 sont, de même, exclues. A partir du volume 66, les données sur les crocodiles exprimées en nombre au tableau B-73 sont exclues de tous les totaux au niveau national, régional et global.

La FAO a complété la séparation des séries de production en aquaculture et capture à partir de 1950. Cette séparation a été basée sur des reportages nationaux, si disponibles, et d' autres sources d' information historique. La série mondiale complète pour la production d'aquaculture peut être téléchargée en utilisant le logiciel FishStat Plus disponible à: http://www.fao.org/fi/statist/FISOFT/FISHPLUS.asp

Le cas échéant, les données relatives aux années 1950-2001 publiées dans les volumes précédents de l'Annuaire statistique des pêches ont été révisées. Lorsque les chiffres indiqués dans le présent volume diffèrent de ceux déjà publiés, les données modifiées représentent la version la plus récente. Certaines statistiques communiquées à la FAO par les services nationaux sont provisoires, notamment pour 2002, et pourront être modifiées dans les futurs volumes ainsi que dans d'autres publications de la FAO.

Introducción

En el presente volumen del **Anuario Estadístico de Pesca** se presentan las estadísticas mundiales, para diversas series de los últimos años que terminan en 2002, de las capturas nominales de peces, crustáceos, moluscos y demás animales, residuos y plantas acuáticos, hechas con cualquier fin (comercial, industrial, recreativo y de subsistencia), por unidades de pesca de todos los tipos y categorías (pescadores, barcos, artes, etc.) en aguas continentales, dulces y salobres y en áreas de pesca de bajura, media altura o altura. A partir del volumen 82 las estadísticas correspondientes a maricultura, acuicultura y otros tipos de pisicultura, están excluidas de las cantidades a nivel nacional, regional y global. **A partir de volumen 90/1, las denominaciones y la composición por especies de algunos de los grupos de especies de la Clasificación Estadística Internacional Uniforme de los Animales y Plantas Acuáticos (CEIUAPA) de la FAO fueron revisados. Véase el párrafo 3 en las NOTAS SOBRE LAS PARTIDAS DE ESPECIES.**

A pesar de la importancia que la pesca recreativa tiene para algunas especies de pescado y para ciertos países, las cifras incluyen las capturas de la pesca recreativa solamente cuando se hallan disponibles.

El período anual utilizado es el año civil (1 de enero-31 de diciembre), excepto en el caso de las capturas efectuadas en las áreas marítimas del Antártico y para algunos países, mencionados en las NOTAS SOBRE LOS DISTINTOS PAISES O AREAS, para los que se utiliza el año emergente. Desde el volumen 94/1 se utiliza el nuevo período de pesca (1 de diciembre–30 de noviembre) de la Comisión para la Conservación de los Recursos Vivos Marinos Antárticos (CCAMLR). Los datos correspondientes a los años emergentes se incluyen en el año civil en que termina el año emergente.

Las capturas se expresan en toneladas, excepto en el caso de ballenas, focas y cocodrilos que se dan en número y en el caso de corales, perlas y esponjas que se dan en kilogramos. Véase el párrafo 11 en las NOTAS SOBRE LOS PAISES O AREAS.

A partir del volumen 48 los datos relativos a los mamíferos acuáticos (cuadros B-61, B-62 y B-63 expresados en números y B-64) y los datos relativos a las plantas acuáticas (A-6, B-91, B-92, B-93 y B-94) están excluidos de las cantidades totales a nivel nacional, regional y global. Con respecto a la práctica seguida antes del volumen 48, véanse las notas en el Anuario respectivo. A partir del volumen 60 los datos relativos a los corales expresados en kilogramos que aparecen en el cuadro B-82 están también todos excluidos de las cantidades totales a nivel nacional, regional y global. A partir del volumen 62 los datos relativos a las perlas expresados en kilogramos que aparecen en el cuadro B-81 y a las esponjas expresados en kilogramos que aparecen en el cuadro B-83 están igualmente excluidos. A partir del volumen 66 los datos relativos a los cocodrilos, expresados en números, en el cuadro B-73 están excluidos de las cantidades totales a nivel nacional, regional y global.

La FAO ha completado la separación de las series cronológicas de acuicultura y de producción de captura a partir de 1950. Esta separación se basó en los informes nacionales, cuando eran disponibles, y en otras fuentes de información histórica. La serie cronológica mundial completa para la producción de acuicultura puede ser descargada utilizando el software de estadística FishStat Plus disponible en: http://www.fao.org/fi/statist/FISOFT/FISHPLUS.asp

Siempre que ha sido necesario se han revisado los datos correspondientes a 1950-2001 publicados en volúmenes anteriores del Anuario Estadístico de Pesca. Cuando las cifras que aparecen en este volumen difieren de las publicadas anteriormente, los nuevos datos representan la última versión disponible. Algunas estadísticas facilitadas a la FAO por las oficinas nacionales, en particular las correspondientes a 2002, son provisionales y podrán modificarse en volúmenes futuros y en otras publicaciones de la FAO.

Introduction

8. National focal points for fishery statistics, in particular those of countries fishing in more than one major fishing area, report their annual catches to various fishery commissions, as well as to FAO. To eliminate duplication in requests to these national offices, FAO cooperates with regional fishery bodies, particularly through the Coordinating Working Party on Fishery Statistics (CWP), to standardize reporting forms, procedures, definitions, classifications and other related documentation. This system reduces discrepancies between the figures appearing in the Yearbook of Fishery Statistics and those published in the bulletins issued by the commissions. Some discrepancies may still exist, but effort is constantly being made to eliminate them.

9. For the time being, the flag of the vessel is used to assign its nationality unless the wording of chartering and joint operation contracts indicates otherwise.

10. To facilitate use of the Yearbook, a series of notes and lists are included.

11. As usual, government officers and staff of international organizations have made possible the timely publication of this Yearbook by their prompt attention to our requests, and the care they devoted to checking material submitted to them. FAO expresses its thanks to them and welcomes the support of national and international organizations, and of interested individuals, in improving the scope and accuracy of the Yearbook.

12. Great care is taken by FAO in ensuring as far as possible the quality of the data presented in this Yearbook, supplementing data reported by countries with information from other sources, where available, including regional fishery bodies, field projects and independent surveys, specialist literature and fishery-independent sources. However, the accuracy and reliability of world aggregations of fishery statistics ultimately depends upon the quality of national data sources, collection methods, periodicity of updating and reporting. Fishery data quality is known to be very uneven among countries. Although improvements to data quality are made by FAO on a continuing basis, it is clear that many more can be made. Any input from data users in this regard will be most welcome.

Introduction

Les centres nationaux des statistiques des pêches, notamment ceux des pays exploitant plus d'une principale zone de pêche, déclarent leurs captures annuelles aux diverses commissions des pêches, de même qu'à la FAO. Pour éviter que les demandes adressées à ces services nationaux ne fassent double emploi, la FAO coopère avec les organes régionaux des pêches, en particulier par l'entremise du Groupe de travail chargé de coordonner les statistiques des pêches (CWP), pour normaliser les formules de déclaration, les procédures, les définitions, les classifications et toute la documentation connexe. Ce système permet de réduire les divergences entre les chiffres figurant dans l'Annuaire statistique des pêches et ceux des bulletins publiés par les commissions. Certaines de ces divergences demeurent, mais on s'efforce toujours de les éliminer.

Pour le moment, le pavillon du navire est utilisé pour déterminer sa nationalité, à moins que le libellé des contrats d'affrètement et d'opérations conjointes ne l'indique autrement.

Pour faciliter l'emploi de l'Annuaire, on a introduit une série de notes et de listes.

Comme de coutume, c'est grâce à la rapidité et au soin avec lesquels les fonctionnaires des services publics et le personnel des organisations internationales ont répondu à nos demandes et vérifié les matériaux qui leur étaient soumis, que cet Annuaire a pu être publié en temps voulu. La FAO tient à les remercier et se félicite de l'aide que lui apportent les organisations tant nationales qu'internationales et les personnes intéressées, afin d'améliorer l'étendue et la précision de l'Annuaire.

La FAO essaie autant que possible, avec le plus grand soin, d'assurer la qualité des données présentées dans cet Annuaire, en complétant les données communiquées par les pays par des informations provenant d'autres sources telles que: organes régionaux des pêches, projets de terrain, enquêtes indépendantes, documents rédigés par des spécialistes et sources extérieures au secteur des pêches. Néanmoins, la précision et la fiabilité des agrégats mondiaux des statistiques des pêches dépendent en dernière analyse de la qualité des sources nationales de données, des méthodes de collecte, de la périodicité de leur mise à jour et de leur présentation. La qualité des données halieutiques peut être fort inégale d'un pays à l'autre. Même si la FAO s'engage à améliorer constamment la qualité des données, il ne fait pas de doute que nombre d'autres améliorations peuvent être apportées. Toute contribution des utilisateurs des données à cet égard sera la bienvenue.

Introducción

Los centros nacionales de estadísticas pesqueras, en particular los de aquellos países que pescan en más de una de las áreas principales de pesca, comunican sus datos sobre capturas anuales a las diversas comisiones de pesca y a la FAO. Para eliminar duplicaciones en la solicitud de datos a esas oficinas nacionales, la FAO colabora con los órganos regionales de pesca, particularmente mediante el Grupo Coordinador de Trabajo sobre Estadísticas de Pesca (CWP), para uniformar los formularios de comunicación de datos, y los procedimientos, definiciones, clasificaciones, etc. Esto permite además reducir las discrepancias entre las cifras que aparacen en el Anuario Estadístico de Pesca y las publicadas en los boletines preparados por las diversas comisiones. Pueden subsistir aún algunas discrepancias, pero siempre se intenta eliminarlas.

Por el momento, se entiende que el pabellón del barco determina su nacionalidad, a menos que en los contratos de fletes o de operación conjunta se indique otra cosa.

Para facilitar la consulta del Anuario se incluyen una serie de notas y listas.

Igual que en ocasiones anteriores, la pronta publicación de este Anuario ha sido posible gracias a la colaboración de funcionarios estatales y de organizaciones internacionales que han respondido con prontitud a nuestras peticiones y han controlado con atención el material que se les ha presentado. La FAO desea manifestar a todos ellos su reconocimiento y agradece ya desde ahora la ayuda que organizaciones nacionales e internacionales o personas privadas le presten para mejorar el alcance y exactitud de este Anuario.

La FAO presta gran atención a asegurar en la medida de lo posible la buena calidad de los datos que se presentan en este Anuario, complementando los datos comunicados por los países con información de otras procedencias, cuando se dispone, tales como los órganos regionales de pesca, los projectos de campo y encuestas independientes, bibliografía especializada y fuentes que no dependen de la pesca. Sin embargo, la exactitud y fiabilidad del acopio mundial de estadísticas de pesca se basan fundamentalmente en la calidad de las fuentes de datos nacionales, en los métodos de recopilación de los mismos, en la periodicidad de la actualización y la comunicación. Como se sabe, la calidad de los datos pesqueros nacionales puede variar mucho de un país a otro. Aunque la FAO está mejorando constantemente la calidad de los datos, es evidente que se puede mejorar todavía mucho. Cualquier aportación de los usuarios de los datos a este respecto será muy bien recibida.

Notes and lists

Notes et listes

Notas y listas

General notes

1. This volume of the **Yearbook of Fishery Statistics** presents, for the most recent series of calendar years and split-years, annual statistics on NOMINAL CATCHES of freshwater, brackishwater and marine species of fish, crustaceans, molluscs and other aquatic animals and plants, killed, caught, trapped or collected for all commercial, industrial, recreational and subsistence purposes.

2. In view of the importance of recreational fishing regarding some stocks and for certain countries, and the difficulty of distinguishing in many cases between recreational and subsistence fishing (and in accordance with the recommendation of the 16th Session of the Coordinating Working Party on Fishery Statistics - CWP, Madrid, Spain, 20-25 March 1995), data should cover recreational fisheries.

3. The NOMINAL CATCHES concept refers to the landings converted to a *live weight* basis; the closely related concept LANDINGS refers to the quantities on a *landed weight* basis. In many fisheries the landed quantities (LANDINGS) are identical to the quantities caught (NOMINAL CATCHES).

4. There are many instances where the catches on board fishing vessels or factory ships are gutted, eviscerated, filleted, salted, dried, etc., or reduced to meals, oil, etc. The data on the LANDINGS of such species and products require conversion by accurate yield rates (conversion factors) to establish the live weight equivalents (nominal catches) at the time of their capture. See diagram on the following pages.

5. Many national statistical publications do not use the terms "landings" and "catches" with their precise meanings as described above. In such publications the term "catches" is used sometimes to refer to quantities on a landed weight basis, i.e. "landings" which might consist of gutted, eviscerated and filleted fish, as well as of meals and oils. However, only where the "primary production" (this phrase "primary production" is used in its economic and not in its biological sense) is landed whole is it correct to describe such landed quantities as "catches".

6. In some national statistics the following terms are in common use to refer to NOMINAL CATCHES, i.e. landings on a live weight basis:

   a) *landings on a round, fresh basis;*

   b) *landings on a round, whole basis;*

   c) *landings on an ex-water weight basis.*

7. For the "primary production" data on seaweeds, pearls, shells, corals, sponges, etc., it is preferable to use the term PRODUCTION as being more appropriate than NOMINAL CATCH. The term CATCH is also more suitable than NOMINAL CATCH in the case of whales and seals, where the "primary production" data for these species are not expressed in weight units but in numbers.

8. Data concerning the nominal catch of fish included within group 36 (tunas, bonitos and billfishes) are generally reviewed in collaboration with the regional agency concerned with tuna statistics (i.e. ICCAT, IOTC, SPC and IATTC). Due to differences in the date by which these agencies require data to be submitted, figures for the most recent year are often subject to significant revision.

Notes générales

1. Dans ce volume de l'**Annuaire statistique des pêches** figurent, pour les plus récentes séries d'années civiles et d'années fractionnées, les statistiques annuelles des CAPTURES NOMINALES de poissons, crustacés, mollusques et autres animaux et plantes aquatiques d'eau douce, d'eau saumâtre et d'eau marine, tués, capturés, piégés ou ramassés à des fins commerciales, industrielles, récréative et de subsistance.

2. Etant donné que la pêche sportive a une grande incidence sur certains stocks et pour certains pays, qu'il est souvent difficile d'établir une distinction entre celle-ci et la pêche de subsistance (et en conformité avec la recommandation de la seizième session du Groupe de travail chargé de coordonner les statistiques de pêches - CWP, Madrid, Espagne, 20-25 mars 1995), les données doivent concerner la pêche sportive.

3. L'expression CAPTURES NOMINALES désigne l'équivalent en *poids vif* des quantités débarquées; l'expression étroitement apparentée QUANTITES DEBARQUEES désigne le *poids mis à terre*. Dans de nombreuses pêcheries, les mises à terre (QUANTITES DEBARQUEES) sont identiques aux quantités capturées (CAPTURES NOMINALES).

4. Dans de nombreux cas, les captures sont vidées, éviscérées, filetées, salées, séchées, etc., ou réduites, en farine, en huile, etc., à bord des bateaux de pêche ou de navires-usines. Les données sur les QUANTITES DEBARQUEES de la sorte, à savoir après traitement, doivent être converties à l'aide de taux de rendement précis (coefficients de conversion) afin de déterminer les équivalents en poids vif (captures nominales) au moment de la prise. Voir le diagramme aux pages suivantes.

5. De nombreuses publications statistiques nationales n'emploient pas les expressions «quantités débarquées» et «captures» avec le sens précis indiqué ci-dessus. Le terme «captures» y sert parfois à désigner les quantités sur la base du poids mis à terre, c.-à-d. les «quantités débarquées» qui peuvent être des poissons vidés, éviscérés et filetés, ainsi que des farines et des huiles. Toutefois, c'est seulement lorsque la «production primaire» (au sens économique et non au sens biologique) est débarquée à l'état entier qu'il est correct de parler de «captures» pour ces mises à terre.

6. Dans certaines statistiques nationales, on utilise couramment les expressions suivantes pour désigner les CAPTURES NOMINALES, c.-à.-d. l'équivalent en poids vif des quantités débarquées:

   a) *quantités débarquées sur la base du poisson entier, frais;*

   b) *quantités débarquées sur la base du poids du poisson entier;*

   c) *quantités débarquées sur la base du poids du poisson à sa sortie de l'eau.*

7. Pour les données sur la «production primaire» des algues, perles, coquillages, coraux, éponges, etc., il est préférable d'utiliser le terme PRODUCTION, qui est plus approprié que l'expression CAPTURES NOMINALES. Par ailleurs, le terme CAPTURES convient mieux que l'expression CAPTURES NOMINALES dans les cas des baleines et des phoques où la «production primaire» ne s'exprime pas en unités pondérales mais en nombres.

8. Les données concernant les captures nominales de poissons appartenant au groupe 36 (thonidés, bonites et marlins) sont en général révisées en collaboration avec l'organisation régionale chargée des statistiques des thonidés (CICTA, CTOI, CPS et CITT). Comme les dates pour lesquelles ces organisations demandent les données ne correspondent pas toujours, les chiffres pour l'année la plus récente sont souvent sujets à des révisions significatives.

Notas generales

1. En el presente volumen del **Anuario Estadístico de Pesca** se presentan, para las más recientes series de años civiles y emergentes, estadísticas anuales de las CAPTURAS NOMINALES de especies de aguas dulces, salobres y marinas, de peces, crustáceos, moluscos y otros animales y plantas acuáticos, matados, capturados, entrampados o cobrados para todo fin de carácter comercial, industrial, pesca de recreo o de subsistencia.

2. Dada la importancia que la pesca de recreo tiene para algunas poblaciones de pescado y para ciertos países, y la dificultad de distinguir en muchos casos entre dicha pesca y la de subsistencia (y en conformidad con la recomendación de la decimosexta sesión del Grupo Coordinador de Trabajo sobre Estadísticas de Pesca - CWP, Madrid, España, 20-25 de marzo de 1995), los datos deberán incluir la pesca de recreo.

3. El concepto de CAPTURAS NOMINALES se refiere a los desembarques expresados en su *peso en vivo*; el concepto DESEMBARQUES íntimamente relacionado, se refiere al *peso descargado*. En muchas pesquerías las cantidades desembarcadas (DESEMBARQUES), son idénticas a las cantidades capturadas (CAPTURAS NOMINALES).

4. En muchos casos, las capturas a bordo de las embarcaciones de pesca o de los buques factoría se destripan, evisceran, filetean, salan, secan, etc., o se reducen en harina, en aceite, etc. Los datos sobre los DESEMBARQUES de estas especies y productos hay que convertirlos, mediante índices precisos de rendimiento (factores de conversión) para fijar su equivalente en peso en vivo (capturas nominales) en el momento de la captura. Véase el diagrama en las páginas siguientes.

5. Muchas publicaciones estadísticas nacionales no utilizan los términos «desembarques» y «capturas» con la significación precisa que se describe anteriormente. En tales publicaciones, el término «captura» se emplea algunas veces para referirse a cantidades basadas en el peso desembarcado, es decir «desembarques» que pueden consistir en pescado destripado, eviscerado y fileteado, así como en harinas y aceites. Sin embargo, sólo cuando la «producción primaria» (esta frase «producción primaria» se utiliza en su sentido económico y no en el biológico) se desembarca entera es correcto describir estas cantidades desembarcadas como «capturas».

6. En algunas estadísticas nacionales son de uso común las siguientes expresiones para referirse a las CAPTURAS NOMINALES, es decir, los desembarques expresados según su peso en vivo:

   a) *desembarques de pescado fresco, entero;*

   b) *desembarques de pescado entero;*

   c) *desembarques según el peso del pescado al sacarlo del agua.*

7. Para los datos de «producción primaria» de algas marinas, perlas, mariscos, corales, esponjas, etc., es preferible usar el término PRODUCCION, ya que es más apropiado que la expresión CAPTURA NOMINAL. El término CAPTURA también es más adecuado que la expresión CAPTURA NOMINAL en el caso de ballenas y focas donde los datos de «producción primaria» para estas especies no se expresan en unidades de peso sino en números.

8. Los datos relativos a la captura nominal de pescado incluido en el grupo 36 (atún, bonitos y agujas) se revisan generalmente en colaboración con los organismos regionales que se encargan de las estadísticas atuneras (CICAA, CAOI, SCP y CIAT). Debido a las diferentes fechas en que esos organismos necesitan recibir los datos, es frecuente que las cifras correspondientes a los años más recientes se sometan a una considerable revisión.

CATCH CONCEPTS: DIAGRAMMATIC PRESENTATION

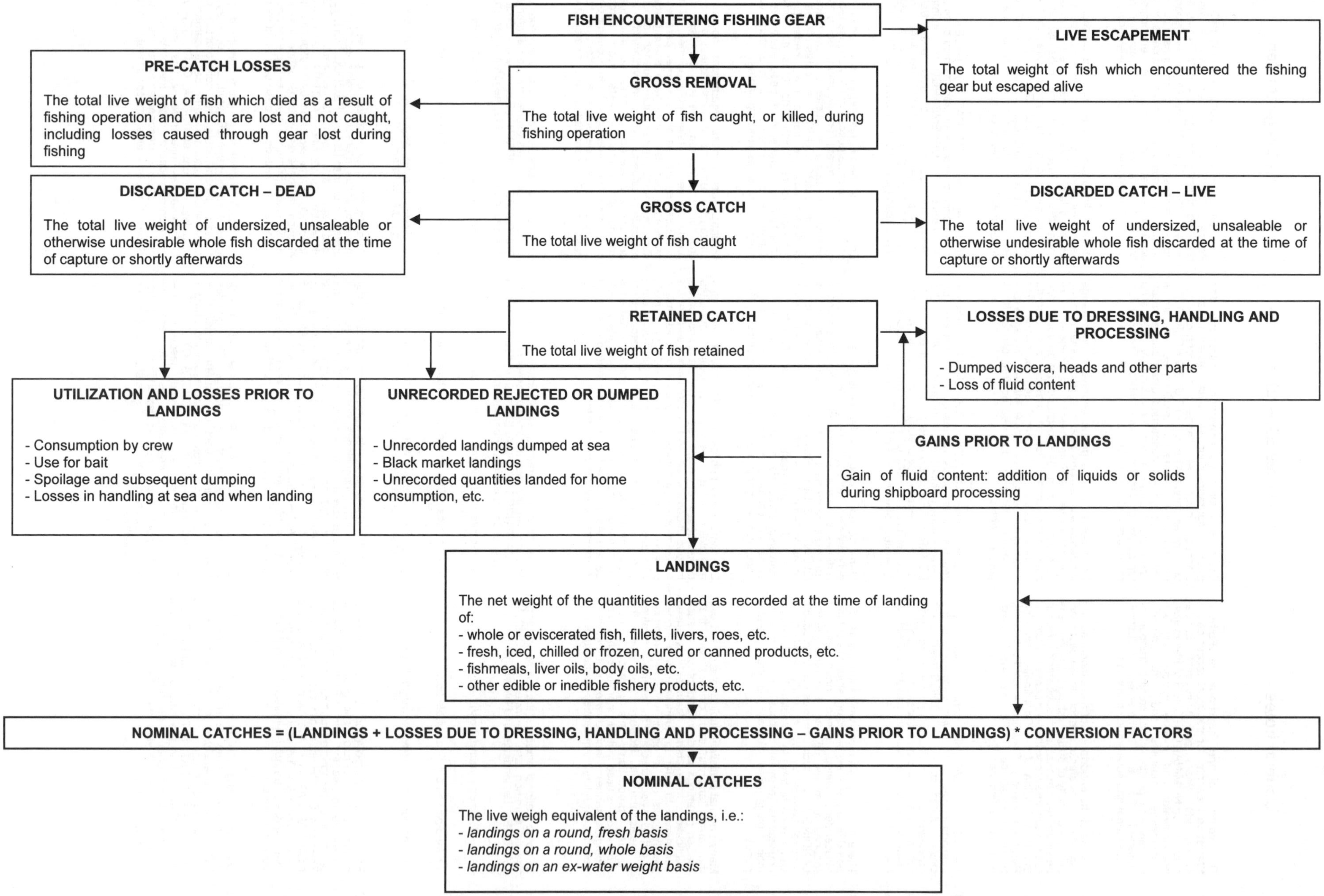

CONCEPTS RELATIFS AUX CAPTURES: DIAGRAMME EXPLICATIF

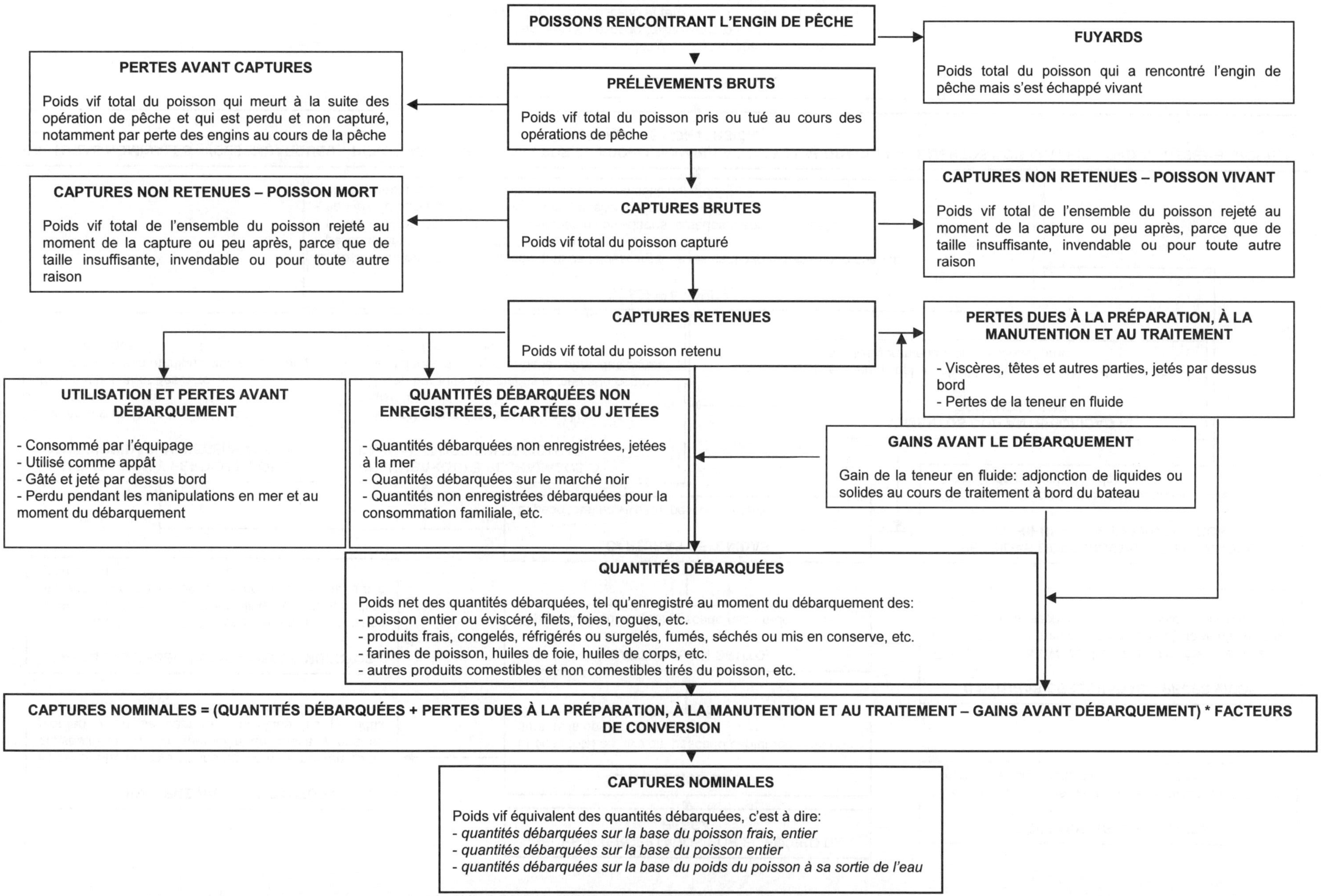
POISSONS RENCONTRANT L'ENGIN DE PÊCHE
FUYARDS
Poids total du poisson qui a rencontré l'engin de pêche mais s'est échappé vivant
PERTES AVANT CAPTURES
Poids vif total du poisson qui meurt à la suite des opération de pêche et qui est perdu et non capturé, notamment par perte des engins au cours de la pêche
PRÉLÈVEMENTS BRUTS
Poids vif total du poisson pris ou tué au cours des opérations de pêche
CAPTURES NON RETENUES – POISSON MORT
Poids vif total de l'ensemble du poisson rejeté au moment de la capture ou peu après, parce que de taille insuffisante, invendable ou pour toute autre raison
CAPTURES BRUTES
Poids vif total du poisson capturé
CAPTURES NON RETENUES – POISSON VIVANT
Poids vif total de l'ensemble du poisson rejeté au moment de la capture ou peu après, parce que de taille insuffisante, invendable ou pour toute autre raison
CAPTURES RETENUES
Poids vif total du poisson retenu
PERTES DUES À LA PRÉPARATION, À LA MANUTENTION ET AU TRAITEMENT
- Viscères, têtes et autres parties, jetés par dessus bord
- Pertes de la teneur en fluide
UTILISATION ET PERTES AVANT DÉBARQUEMENT
- Consommé par l'équipage
- Utilisé comme appât
- Gâté et jeté par dessus bord
- Perdu pendant les manipulations en mer et au moment du débarquement
QUANTITÉS DÉBARQUÉES NON ENREGISTRÉES, ÉCARTÉES OU JETÉES
- Quantités débarquées non enregistrées, jetées à la mer
- Quantités débarquées sur le marché noir
- Quantités non enregistrées débarquées pour la consommation familiale, etc.
GAINS AVANT LE DÉBARQUEMENT
Gain de la teneur en fluide: adjonction de liquides ou solides au cours de traitement à bord du bateau
QUANTITÉS DÉBARQUÉES
Poids net des quantités débarquées, tel qu'enregistré au moment du débarquement des:
- poisson entier ou éviscéré, filets, foies, rogues, etc.
- produits frais, congelés, réfrigérés ou surgelés, fumés, séchés ou mis en conserve, etc.
- farines de poisson, huiles de foie, huiles de corps, etc.
- autres produits comestibles et non comestibles tirés du poisson, etc.
CAPTURES NOMINALES = (QUANTITÉS DÉBARQUÉES + PERTES DUES À LA PRÉPARATION, À LA MANUTENTION ET AU TRAITEMENT – GAINS AVANT DÉBARQUEMENT) * FACTEURS DE CONVERSION
CAPTURES NOMINALES
Poids vif équivalent des quantités débarquées, c'est à dire:
- quantités débarquées sur la base du poisson frais, entier
- quantités débarquées sur la base du poisson entier
- quantités débarquées sur la base du poids du poisson à sa sortie de l'eau

CONCEPTOS DE CAPTURA: PRESENTACIÓN DIAGRAMÁTICA

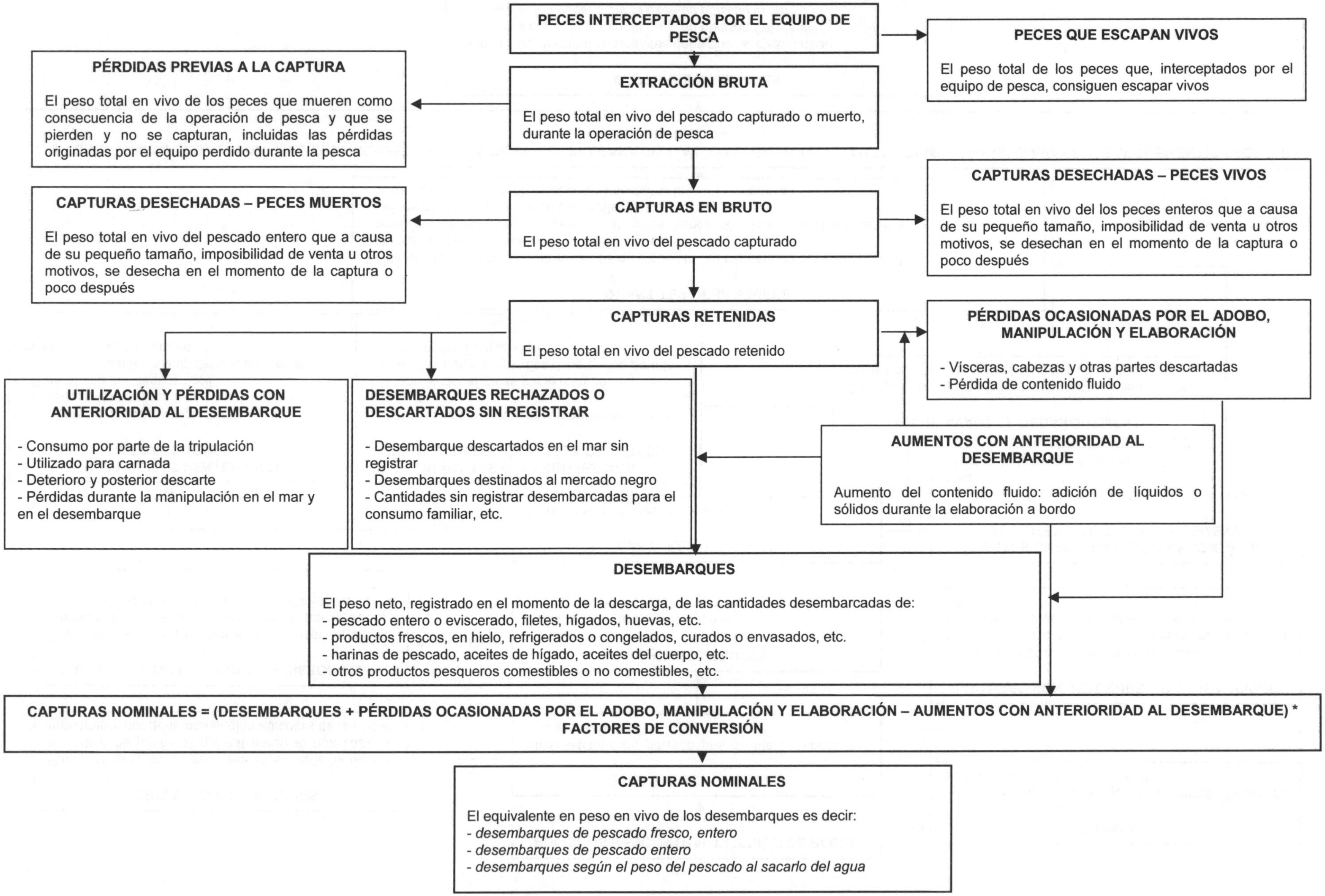
PECES INTERCEPTADOS POR EL EQUIPO DE PESCA
PECES QUE ESCAPAN VIVOS
El peso total de los peces que, interceptados por el equipo de pesca, consiguen escapar vivos
PÉRDIDAS PREVIAS A LA CAPTURA
El peso total en vivo de los peces que mueren como consecuencia de la operación de pesca y que se pierden y no se capturan, incluidas las pérdidas originadas por el equipo perdido durante la pesca
EXTRACCIÓN BRUTA
El peso total en vivo del pescado capturado o muerto, durante la operación de pesca
CAPTURAS DESECHADAS – PECES MUERTOS
El peso total en vivo del pescado entero que a causa de su pequeño tamaño, imposibilidad de venta u otros motivos, se desecha en el momento de la captura o poco después
CAPTURAS EN BRUTO
El peso total en vivo del pescado capturado
CAPTURAS DESECHADAS – PECES VIVOS
El peso total en vivo del los peces enteros que a causa de su pequeño tamaño, imposibilidad de venta u otros motivos, se desechan en el momento de la captura o poco después
CAPTURAS RETENIDAS
El peso total en vivo del pescado retenido
PÉRDIDAS OCASIONADAS POR EL ADOBO, MANIPULACIÓN Y ELABORACIÓN
- Vísceras, cabezas y otras partes descartadas
- Pérdida de contenido fluido
UTILIZACIÓN Y PÉRDIDAS CON ANTERIORIDAD AL DESEMBARQUE
- Consumo por parte de la tripulación
- Utilizado para carnada
- Deterioro y posterior descarte
- Pérdidas durante la manipulación en el mar y en el desembarque
DESEMBARQUES RECHAZADOS O DESCARTADOS SIN REGISTRAR
- Desembarque descartados en el mar sin registrar
- Desembarques destinados al mercado negro
- Cantidades sin registrar desembarcadas para el consumo familiar, etc.
AUMENTOS CON ANTERIORIDAD AL DESEMBARQUE
Aumento del contenido fluido: adición de líquidos o sólidos durante la elaboración a bordo
DESEMBARQUES
El peso neto, registrado en el momento de la descarga, de las cantidades desembarcadas de:
- pescado entero o eviscerado, filetes, hígados, huevas, etc.
- productos frescos, en hielo, refrigerados o congelados, curados o envasados, etc.
- harinas de pescado, aceites de hígado, aceites del cuerpo, etc.
- otros productos pesqueros comestibles o no comestibles, etc.
CAPTURAS NOMINALES = (DESEMBARQUES + PÉRDIDAS OCASIONADAS POR EL ADOBO, MANIPULACIÓN Y ELABORACIÓN – AUMENTOS CON ANTERIORIDAD AL DESEMBARQUE) * FACTORES DE CONVERSIÓN
CAPTURAS NOMINALES
El equivalente en peso en vivo de los desembarques es decir:
- desembarques de pescado fresco, entero
- desembarques de pescado entero
- desembarques según el peso del pescado al sacarlo del agua

Notes on species items

1. Data available world-wide on nominal catches of aquatic animals and plants, taken in inland and marine waters, for all kinds of commercial, industrial, recreational and subsistence purposes, are at present broken down at either the species, genus, family or higher taxonomic levels into 1347 statistical categories called *species items*.

2. These 1347 *species items* are arranged by FAO within the 50 *groups of species* constituting the nine *divisions* of the FAO "International Standard Statistical Classification of Aquatic Animals and Plants" (ISSCAAP). For these *groups of species* and *divisions* see the ISSCAAP list on the following page. The two-digit group codes shown in this list are reflected in the table numbers in Section B.

3. Following a recommendation of the 19th Session of the Coordinating Working Party on Fishery Statistics - CWP (Nouméa, New Caledonia, 10-13 July 2001) the names and composition of former groups 33, 34 and 37 of the FAO International Standard Statistical Classification of Aquatic Animals and Plants (ISSCAAP) were revised. The species items of the former group 33 "Redfishes, basses, congers" were classified as coastal or demersal fishes and accordingly assigned to the new groups 33 "Miscellaneous coastal fishes" and 34 "Miscellaneous demersal fishes". The species formerly included in group 34 "Jacks, mullets, sauries" were moved to group 37, which was renamed "Miscellaneous pelagic fishes".

4. Each *species item* is identified by means of the following descriptors (subject to constant review and improvement):

   a) FAO English name;

   b) FAO French name;

   c) FAO Spanish name;

   d) scientific name (at the species, genus, family or higher taxonomic levels);

   e) taxonomic code; and

   f) inter-agency 3-alpha code.

5. The taxonomic code descriptors are taken from FAO's **Aquatic Sciences and Fisheries Information System**. The following illustrates the meaning assigned to the different digits of this code:

Main grouping	Order or other taxonomic level	Family	Genus	Species
1,	75	(04)	003,	01

6. In all tables, except A-1(e) (where selected species items of significant commercial importance are listed in the order of 2002 catch size), the species items are arranged in order of their taxonomic codes within their respective ISSCAAP groups.

7. Several countries still report their catches by large groups of species. In these circumstances the catch data presented by individual species items are likely to be underestimated. Therefore, when examining the statistics for a particular species, it should be noted that an unknown proportion of the catches for that species might have been reported by the national office under the generic, family or order name of the species, or even more roughly as, for example, miscellaneous fishes. Consequently, **species items totals frequently underestimate the real catch of the individual species**.

Notes sur les catégories d'espèces

1. Les statistiques disponibles sur une base mondiale des captures nominales d'animaux et de plantes aquatiques effectuées dans les eaux continentales et dans les eaux marines à toutes fins commerciales, industrielles, récréatives et de subsistance sont, à l'heure actuelle, ventilées au niveau de l'espèce, du genre, de la famille ou des niveaux taxonomiques supérieurs dans 1347 catégories statistiques qui sont appelées des *catégories d'espèces*.

2. Ces 1347 *catégories d'espèces* sont rassemblées par la FAO en cinquante *groupes d'espèces* constituant les neuf *divisions* de la 'Classification statistique internationale type des animaux et des plantes aquatiques' (CSITAPA) de la FAO. Pour ces *groupes d'espèces* et ces *divisions*, on se reportera à la liste de la CSITAPA figurant à la page suivante. Les codes de groupe à deux chiffres indiqués sur la liste se retrouvent dans les numéros des tableaux de la section B.

3. Sur la base d'une recommandation de la 19ème session du Groupe de travail chargé de coordonner les statistiques des pêches - CWP (Nouméa, Nouvelle-Calédonie, 10-13 juillet 2001), la dénomination et la composition des anciens groupes 33, 34 et 37 de la Classification statistique internationale type des animaux et des plantes aquatiques (CSITAPA) de la FAO ont été révisées. Les catégories d'espèces de l'ancien groupe 33 "Rascasses, perches de mer, congres" ont été classifiées comme espèces côtières ou démersales et réparties sur cette base entre les nouveaux groupes 33 "Poissons côtiers divers" et 34 "Poissons démersaux divers". Les espèces auparavant incluses dans le groupe 34 "Chinchards, mulets, balaous" ont été classées dans le groupe 37 rebaptisé "Poissons pélagiques divers".

4. Chaque *catégorie d'espèces* est identifiée au moyen des descripteurs suivants (sous réserve de révision et d'amélioration constantes):

   a) nom anglais utilisé par la FAO;

   b) nom français utilisé par la FAO;

   c) nom espagnol utilisé par la FAO;

   d) nom scientifique (de l'espèce, du genre, de la famille ou des niveaux taxonomiques supérieurs);

   e) code taxonomique; et

   f) code interinstitutions alpha-3.

5. Les descripteurs de code taxonomique proviennent du **Système d'information sur les sciences aquatiques et la pêche** de la FAO. On trouvera ci-après la signification des divers chiffres constituant ce code taxonomique.

Groupe principal	Ordre ou autre niveau taxonomique	Famille	Genre	Espèce
1,	75	(04)	003,	01

6. Dans tous les tableaux, à l'exception de A-1(e) (où certaines catégories d'espèces ayant une importance commerciale notable sont énumérées selon le chiffre des captures de 2002), les catégories d'espèces sont disposées selon l'ordre du code taxonomique dans leurs groupes CSITAPA, respectifs.

7. Plusieurs pays continuent à communiquer leurs captures par grands groupes d'espèces. Dans ces circonstances les données des captures par catégories d'espèces individuelles sont donc probablement sous-estimées. Par conséquent, lorsqu'on examine les statistiques d'une espèce donnée, il convient de noter qu'une proportion inconnue des captures de cette espèce peut avoir été consignée par l'organisme national sous le nom du genre, de la famille ou de l'ordre de l'espèce, ou même, plus approximativement dans la catégorie, par exemple, poissons divers. C'est pourquoi **les totaux correspondant aux espèces données sous-estiment souvent la capture réelle de chacune des espèces**.

Notas sobre las partidas de especies

1. Los datos procedentes de todo el mundo sobre las capturas nominales de animales y plantas acuáticos capturados en aguas continentales y marinas con fines comerciales, industriales, de recreo o de subsistencia se desglosan en la actualidad en unas 1347 categorías estadísticas, denominadas *partidas de especies*, que corresponden a los niveles taxonómicos de especie, género, familia o superiores.

2. Utilizando esas 1347 *partidas de especies*, la FAO desglosa los cincuenta *grupos de especies* que constituyen las nueve *divisiones* de la Clasificación Estadística Internacional Uniforme de los Animales y Plantas Acuáticos (CEIUAPA) de la FAO. Estos *grupos de especies* y las nueve *divisiones* indicadas pueden verse en la lista de la CEIUAPA que aparece en la página siguiente. Los códigos de dos dígitos que en esa lista identifican a cada uno de los grupos se utilizan luego en los cuadros de la Sección B.

3. Sobre la base de una recomendación de la 19a sesión del Grupo Coordinador de Trabajo sobre Estadísticas de Pesca - CWP (Noumea, Nueva Caledonia, 10-13 de julio 2001) fue revisada la denominación y la composición de los antiguos grupos 33, 34, y 37 de la Clasificación Estadística Internacional Uniforme de los Animales y Plantas Acuáticos (CEIUAPA) de la FAO. Las partidas de especies del antiguo grupo 33 "Gallinetas, lubinas, congrios" fueron clasificadas como especies costeras o demersales y, de acuerdo con esto, distribuidas entre los nuevos grupos: 33 "Peces costeros diversos" y 34 "Peces demersales diversos". Las partidas de especies anteriormente comprendidas en el grupo 34 "Jureles, lisas, papardas" se incluyeron en el grupo 37, que pasó a denominarse "Peces pelágicos diversos".

4. Cada una de las *partidas de especies*, se identifica con los siguientes descriptores (que se revisan y mejoran continuamente):

   a) nombre inglés utilizado por la FAO;

   b) nombre francés utilizado por la FAO;

   c) nombre español utilizado por la FAO;

   d) nombre científico (de la especie, género, familia o niveles taxonómicos superiores);

   e) código taxonómico; e

   f) código interinstitucional alfa-3.

5. Los descriptores del código taxonómico se han tomado del **Sistema de información de las ciencias acuáticas y la pesca** de la FAO. El cuadro siguiente ilustra el significado de los distintos dígitos de dicho código:

Grupo principal	Orden u otro nivel taxonómico	Familia	Género	Especie
1,	75	(04)	003,	01

6. En todos los cuadros, excepto en el cuadro A-1(e) (en el cual algunas especies de gran importancia comercial aparecen ordenadas según el volumen de sus capturas en 2002), las partidas de especies aparecen por orden de su código taxonómico dentro de los respectivos grupos de la CEIUAPA.

7. Varios países siguen informando de sus capturas por grandes grupos de especies. En estos casos, es probable que aparezcan subestimados los datos sobre la captura de las distintas categorías de especies. Por consiguiente, al examinar las estadísticas de una especie particular, debe tenerse en cuenta que una proporción desconocida de las capturas de esa especie puede haber sido ya comunicada por la oficina nacional con el nombre genérico, el de la familia o el del orden de la especie, o incluso más aproximativamente como, por ejemplo, peces diversos. Por consiguiente, **los totales de las categorías de especies subestiman frecuentemente la captura efectiva de la especie individual**.

Code / Code / Código	ISSCAAP DIVISION Group of species	CSITAPA DIVISION Groupe d'espèces	CEIUAPA DIVISIÓN Grupo de especies
1	Freshwater fishes	Poissons d'eau douce	Peces de agua dulce
11	Carps, barbels and other cyprinids	Carpes, barbeaux et autres cyprinidés	Carpas, barbos y otros ciprínidos
12	Tilapias and other cichlids	Tilapias et autres cichlidés	Tilapias y otros cíclidos
13	Miscellaneous freshwater fishes	Poissons d'eau douce divers	Peces de agua dulce diversos
2	Diadromous fishes	Poissons diadromes	Peces diádromos
21	Sturgeons, paddlefishes	Esturgeons, spatules	Esturiones, sollos
22	River eels	Anguilles	Anguilas
23	Salmons, trouts, smelts	Saumons, truites, éperlans	Salmones, truchas, eperlanos
24	Shads	Aloses	Sábalos
25	Miscellaneous diadromous fishes	Poissons diadromes divers	Peces diádromos diversos
3	Marine fishes	Poissons marins	Peces marinos
31	Flounders, halibuts, soles	Flets, flétans, soles	Platijas, halibuts, lenguados
32	Cods, hakes, haddocks	Morues, merlus, églefins	Bacalaos, merluzas, eglefinos
33	Miscellaneous coastal fishes	Poissons côtiers divers	Peces costeros diversos
34	Miscellaneous demersal fishes	Poissons démersaux divers	Peces demersales diversos
35	Herrings, sardines, anchovies	Harengs, sardines, anchois	Arenques, sardinas, anchoas
36	Tunas, bonitos, billfishes	Thons, pélamides, marlins	Atunes, bonitos, agujas
37	Miscellaneous pelagic fishes	Poissons pélagiques divers	Peces pelágicos diversos
38	Sharks, rays, chimaeras	Squales, raies, chimères	Tiburones, rayas, quimeras
39	Marine fishes not identified	Poissons marins non identifiés	Peces marinos no identificados
4	Crustaceans	Crustacés	Crustáceos
41	Freshwater crustaceans	Crustacés d'eau douce	Crustáceos de agua dulce
42	Crabs, sea-spiders	Crabes, araignées de mer	Cangrejos, centollas
43	Lobsters, spiny-rock lobsters	Homards, langoustes	Bogavantes, langostas
44	King crabs, squat-lobsters	Crabes royaux, galatées	Cangrejos reales, galateidos
45	Shrimps, prawns	Crevettes	Gambas, camarones
46	Krill, planktonic crustaceans	Krill, crustacés planctoniques	Krill, crustáceos planctónicos
47	Miscellaneous marine crustaceans	Crustacés marins divers	Crustáceos marinos diversos
5	Molluscs	Mollusques	Moluscos
51	Freshwater molluscs	Mollusques d'eau douce	Moluscos de agua dulce
52	Abalones, winkles, conchs	Ormeaux, bigorneaux, strombes	Orejas de mar, bígaros, estrombos
53	Oysters	Huîtres	Ostras
54	Mussels	Moules	Mejillones
55	Scallops, pectens	Coquilles St-Jacques	Vieiras
56	Clams, cockles, arkshells	Clams, coques, arches	Almejas, berberechos, arcas
57	Squids, cuttlefishes, octopuses	Encornets, seiches, poulpes	Calamares, jibias, pulpos
58	Miscellaneous marine molluscs	Mollusques marins divers	Moluscos marinos diversos
* 6	Whales, seals and other aquatic mammals	Baleines, phoques et autres mammifères aquatiques	Ballenas, focas y otros mamíferos acuáticos
* 61	Blue-whales, fin-whales	Baleines bleues, rorquals communs	Ballenas azules, rorcuales
* 62	Sperm-whales, pilot-whales	Cachalots, globicéphales	Cachalotes, calderones
* 63	Eared seals, hair seals, walruses	Otaries, phoques, morses	Lobos marinos, focas, morsas
* 64	Miscellaneous aquatic mammals	Mammifères aquatiques divers	Mamíferos acuáticos diversos
7	Miscellaneous aquatic animals	Animaux aquatiques divers	Animales acuáticos diversos
71	Frogs and other amphibians	Grenouilles et autres amphibies	Ranas y otros anfibios
72	Turtles	Tortues	Tortugas
* 73	Crocodiles and alligators	Crocodiles et alligators	Cocodrilos y aligatores
74	Sea-squirts and other tunicates	Ascidiens et autres tuniciers	Ascidias y otros tunicados
75	Horseshoe crabs and other arachnoids	Limules et autres arachnoïdés	Límulos y otros arácnidos
76	Sea-urchins and other echinoderms	Oursins et autres échinodermes	Erizos de mar y otros equinodermos
77	Miscellaneous aquatic invertebrates	Invertébrés aquatiques divers	Invertebrados acuáticos diversos
* 8	Miscellaneous aquatic animal products	Produits divers d'animaux aquatiques	Diversos productos de animales acuáticos
* 81	Pearls, mother-of-pearl, shells	Perles, nacres, coquilles	Perlas, madreperlas, conchas
* 82	Corals	Coraux	Corales
* 83	Sponges	Eponges	Esponjas
* 9	Aquatic plants	Plantes aquatiques	Plantas acuáticas
* 91	Brown seaweeds	Algues brunes	Algas pardas
* 92	Red seaweeds	Algues rouges	Algas rojas
* 93	Green seaweeds	Algues vertes	Algas verdes
* 94	Miscellaneous aquatic plants	Plantes aquatiques diverses	Diversas plantas acuáticas

ISSCAAP = International Standard Statistical Classification of Aquatic Animals and Plants

CSITAPA = Classification statistique internationale type des animaux et des plantes aquatiques

CEIUAPA = Clasificación Estadística Internacional Uniforme de los Animales y Plantas Acuáticos

* See paragraph 5 of the INTRODUCTION

* Voir le paragraphe 5 de l'INTRODUCTION

* Véase al párrafo 5 en la INTRODUCCIÓN

Systematic list of aquatic organisms	Liste systématique des organismes aquatiques	Lista sistemática de los organismos acuáticos

1 PISCES

1,02 PETROMYZONTIFORMES

1,02(01) PETROMYZONTIDAE
1,02(01)XXX,XX *Petromyzontidae*
1,02(01)001,01 *Petromyzon marinus*
1,02(01)002,01 *Lampetra fluviatilis*

1,05 HEXANCHIFORMES

1,05(02) HEXANCHIDAE
1,05(02)002,01 *Hexanchus griseus*
1,05(02)005,02 *Notorynchus cepedianus*

1,06 LAMNIFORMES

1,06(01) CETORHINIDAE
1,06(01)003,01 *Cetorhinus maximus*

1,06(02) ODONTASPIDIDAE
1,06(02)005,01 *Carcharias taurus*

1,06(06) ALOPIIDAE
1,06(06)006,XX *Alopias spp*
1,06(06)006,01 *Alopias vulpinus*
1,06(06)006,03 *Alopias superciliosus*

1,06(08) LAMNIDAE
1,06(08)002,XX *Isurus spp*
1,06(08)002,01 *Isurus oxyrinchus*
1,06(08)002,03 *Isurus paucus*
1,06(08)003,01 *Lamna nasus*
1,06(08)007,01 *Carcharodon carcharias*

1,07 ORECTOLOBIFORMES

1,07(03) GINGLYMOSTOMATIDAE
1,07(03)009,01 *Ginglymostoma cirratum*

1,08 CARCHARHINIFORMES

1,08(01) SCYLIORHINIDAE
1,08(01)001,04 *Galeus melastomus*
1,08(01)003,XX *Scyliorhinus spp*
1,08(01)003,01 *Scyliorhinus canicula*
1,08(01)003,02 *Scyliorhinus stellaris*
1,08(01)030,04 *Halaelurus canescens*

1,08(02) CARCHARHINIDAE
1,08(02)XXX,XX *Carcharhinidae*
1,08(02)004,01 *Prionace glauca*
1,08(02)010,01 *Carcharhinus plumbeus*
1,08(02)010,03 *Carcharhinus limbatus*
1,08(02)010,16 *Carcharhinus obscurus*
1,08(02)010,17 *Carcharhinus falciformis*
1,08(02)010,20 *Carcharhinus brachyurus*
1,08(02)017,03 *Galeocerdo cuvier*

1,08(03) SPHYRNIDAE
1,08(03)XXX,XX *Sphyrnidae*
1,08(03)005,01 *Sphyrna zygaena*
1,08(03)005,06 *Sphyrna lewini*

1,08(04) TRIAKIDAE
1,08(04)007,XX *Mustelus spp*
1,08(04)007,03 *Mustelus canis*
1,08(04)007,07 *Mustelus henlei*
1,08(04)007,09 *Mustelus lenticulatus*
1,08(04)007,12 *Mustelus schmitti*
1,08(04)007,13 *Mustelus mustelus*
1,08(04)011,03 *Galeorhinus galeus*

1,09 SQUALIFORMES

1,09(01) SQUALIDAE
1,09(01)XXX,XX *Squalidae*
1,09(01)XXX,XX *Squalidae, Scyliorhinidae*
1,09(01)002,01 *Somniosus microcephalus*
1,09(01)002,02 *Somniosus rostratus*
1,09(01)002,03 *Somniosus pacificus*
1,09(01)007,04 *Squalus acanthias*
1,09(01)008,01 *Centrophorus granulosus*
1,09(01)008,03 *Centrophorus squamosus*
1,09(01)010,XX *Etmopterus spp*
1,09(01)014,01 *Deania calcea*
1,09(01)016,01 *Centroscymnus coelolepis*
1,09(01)016,02 *Centroscymnus crepidater*
1,09(01)018,01 *Dalatias licha*
1,09(01)019,01 *Centroscyllium fabricii*

1,09(02) PRISTIOPHORIDAE
1,09(02)004,XX *Pristiophorus spp*

1,09(03) SQUATINIDAE
1,09(03)XXX,XX *Squatinidae*
1,09(03)004,01 *Squatina squatina*
1,09(03)004,04 *Squatina argentina*

1,09(05) OXYNOTIDAE
1,09(05)006,01 *Oxynotus centrina*
1,09(05)006,02 *Oxynotus paradoxus*

1,10 RAJIFORMES

1,10(01) RHINOBATIDAE
1,10(01)XXX,XX *Rhinobatidae*
1,10(01)005,09 *Rhinobatos percellens*
1,10(01)005,10 *Rhinobatos planiceps*

1,10(02) PRISTIDAE
1,10(02)XXX,XX *Pristidae*

1,10(04) RAJIDAE
1,10(04)XXX,XX *Rajidae*
1,10(04)001,XX *Raja spp*
1,10(04)001,01 *Raja batis*
1,10(04)001,02 *Raja clavata*
1,10(04)001,03 *Raja radiata*
1,10(04)001,04 *Raja montagui*
1,10(04)001,06 *Raja circularis*
1,10(04)001,07 *Raja fullonica*
1,10(04)001,09 *Raja microocellata*
1,10(04)001,10 *Raja naevus*
1,10(04)001,11 *Raja oxyrinchus*
1,10(04)001,32 *Raja georgiana*
1,10(04)002,XX *Bathyraja spp*
1,10(04)002,01 *Bathyraja eatonii*
1,10(04)002,03 *Bathyraja murrayi*
1,10(04)002,16 *Bathyraja meridionalis*

1,10(05) DASYATIDAE
1,10(05)003,XX *Dasyatis spp*
1,10(05)003,01 *Dasyatis akajei*
1,10(05)003,26 *Dasyatis pastinaca*

1,10(07) MYLIOBATIDAE
1,10(07)XXX,XX *Myliobatidae*

1,10(08) MOBULIDAE
1,10(08)XXX,XX *Mobulidae*

1,11 TORPEDINIFORMES

1,11(01) TORPEDINIDAE
1,11(01)002,XX *Torpedo spp*

1,12 CHIMAERIFORMES

1,12(01) CHIMAERIDAE
1,12(01)003,01 *Chimaera monstrosa*
1,12(01)004,XX *Hydrolagus spp*
1,12(01)004,01 *Hydrolagus colliei*
1,12(01)004,11 *Hydrolagus novaezealandiae*

1,12(02) RHINOCHIMAERIDAE
1,12(02)001,01 *Rhinochimaera atlantica*

1,12(03) CALLORHINCHIDAE
1,12(03)001,XX *Callorhinchus spp*
1,12(03)001,01 *Callorhinchus milii*
1,12(03)001,03 *Callorhinchus capensis*

1,16 POLYPTERIFORMES

1,16(02) PROTOPTERIDAE
1,16(02)002,XX *Protopterus spp*

1,17 ACIPENSERIFORMES

1,17(01) ACIPENSERIDAE
1,17(01)XXX,XX *Acipenseridae*
1,17(01)001,02 *Acipenser gueldenstaedtii*
1,17(01)001,04 *Acipenser ruthenus*
1,17(01)001,05 *Acipenser stellatus*
1,17(01)001,09 *Acipenser transmontanus*
1,17(01)001,15 *Acipenser medirostris*
1,17(01)005,01 *Huso huso*

1,17(02) POLYODONTIDAE
1,17(02)002,01 *Polyodon spathula*

1,21 CLUPEIFORMES

1,21(05) CLUPEIDAE
1,21(05)XXX,XX *Stolothrissa, Limnothrissa spp*
1,21(05)001,05 *Clupea harengus*
1,21(05)001,07 *Clupea pallasii*
1,21(05)011,XX *Alosa spp*
1,21(05)011,02 *Alosa pontica*
1,21(05)011,03 *Alosa sapidissima*
1,21(05)011,04 *Alosa alosa*
1,21(05)011,05 *Alosa fallax*
1,21(05)011,06 *Alosa pseudoharengus*
1,21(05)011,07 *Alosa aestivalis*
1,21(05)011,08 *Alosa mediocris*
1,21(05)012,XX *Sardinella spp*
1,21(05)012,03 *Sardinella gibbosa*
1,21(05)012,04 *Sardinella longiceps*
1,21(05)012,10 *Sardinella aurita*
1,21(05)012,17 *Sardinella maderensis*
1,21(05)012,22 *Sardinella zunasi*
1,21(05)012,23 *Sardinella lemuru*
1,21(05)012,24 *Sardinella brasiliensis*
1,21(05)013,01 *Sardinops melanostictus*
1,21(05)013,02 *Sardinops caeruleus*
1,21(05)013,03 *Sardinops sagax*
1,21(05)013,05 *Sardinops ocellatus*
1,21(05)013,09 *Sardinops neopilchardus*
1,21(05)014,XX *Caspialosa spp*
1,21(05)018,01 *Dorosoma cepedianum*
1,21(05)023,01 *Anodontostoma chacunda*
1,21(05)024,01 *Brevoortia aurea*
1,21(05)024,02 *Brevoortia pectinata*
1,21(05)024,03 *Brevoortia tyrannus*
1,21(05)024,04 *Brevoortia patronus*
1,21(05)029,01 *Dussumieria acuta*
1,21(05)029,02 *Dussumieria elopsoides*
1,21(05)030,02 *Ethmalosa fimbriata*
1,21(05)031,01 *Etrumeus teres*
1,21(05)031,04 *Etrumeus whiteheadi*
1,21(05)033,XX *Harengula spp*
1,21(05)034,05 *Hilsa kelee*
1,21(05)038,01 *Tenualosa ilisha*
1,21(05)038,04 *Tenualosa toli*
1,21(05)042,01 *Opisthonema libertate*
1,21(05)042,02 *Opisthonema oglinum*
1,21(05)049,01 *Spratelloides gracilis*
1,21(05)058,03 *Clupanodon thrissa*
1,21(05)059,01 *Clupeonella cultriventris*
1,21(05)060,01 *Konosirus punctatus*
1,21(05)064,01 *Sardina pilchardus*
1,21(05)066,01 *Sprattus sprattus*
1,21(05)066,02 *Sprattus fuegensis*
1,21(05)067,02 *Ethmidium maculatum*
1,21(05)072,01 *Herklotsichthys quadrimaculat.*
1,21(05)078,01 *Strangomera bentincki*

1,21(06) ENGRAULIDAE
1,21(06)XXX,XX *Engraulidae*
1,21(06)002,01 *Engraulis encrasicolus*
1,21(06)002,02 *Engraulis japonicus*
1,21(06)002,06 *Engraulis anchoita*
1,21(06)002,07 *Engraulis mordax*
1,21(06)002,08 *Engraulis ringens*
1,21(06)002,12 *Engraulis capensis*
1,21(06)015,01 *Cetengraulis edentulus*
1,21(06)015,03 *Cetengraulis mysticetus*
1,21(06)020,11 *Anchoa hepsetus*
1,21(06)038,04 *Lycengraulis grossidens*
1,21(06)050,XX *Stolephorus spp*

1,21(11) CHIROCENTRIDAE
1,21(11)002,XX *Chirocentrus spp*
1,21(11)002,01 *Chirocentrus dorab*

1,21(12) PRISTIGASTERIDAE
1,21(12)001,03 *Ilisha elongata*
1,21(12)001,12 *Ilisha africana*
1,21(12)003,03 *Pellona ditchela*

1,22 GONORYNCHIFORMES

1,22(02) CHANIDAE
1,22(02)001,01 *Chanos chanos*

1,23 SALMONIFORMES

1,23(01) SALMONIDAE
1,23(01)004,XX *Salmo spp*
1,23(01)004,01 *Salmo salar*
1,23(01)004,02 *Salmo trutta*
1,23(01)009,XX *Oncorhynchus spp*
1,23(01)009,02 *Oncorhynchus gorbuscha*
1,23(01)009,03 *Oncorhynchus keta*
1,23(01)009,05 *Oncorhynchus masou*
1,23(01)009,06 *Oncorhynchus nerka*
1,23(01)009,07 *Oncorhynchus tshawytscha*
1,23(01)009,08 *Oncorhynchus kisutch*
1,23(01)009,09 *Oncorhynchus mykiss*
1,23(01)010,XX *Salvelinus spp*
1,23(01)010,02 *Salvelinus fontinalis*
1,23(01)010,05 *Salvelinus alpinus*
1,23(01)010,07 *Salvelinus namaycush*
1,23(01)020,02 *Hucho hucho*

1,23(02) THYMALLIDAE
1,23(02)005,01 *Thymallus thymallus*

1,23(03) PLECOGLOSSIDAE
1,23(03)016,01 *Plecoglossus altivelis*

1,23(04) OSMERIDAE
1,23(04)XXX,XX *Osmerus spp, Hypomesus spp*
1,23(04)002,01 *Mallotus villosus*
1,23(04)003,01 *Osmerus eperlanus*
1,23(04)003,03 *Osmerus mordax*
1,23(04)008,01 *Hypomesus olidus*
1,23(04)008,02 *Hypomesus pretiosus*
1,23(04)012,01 *Thaleichthys pacificus*

1,23(05) ARGENTINIDAE
1,23(05)015,XX *Argentina spp*
1,23(05)029,01 *Glossanodon semifasciatus*

1,23(12) COREGONIDAE
1,23(12)001,XX *Coregonus spp*
1,23(12)001,01 *Coregonus albula*
1,23(12)001,02 *Coregonus lavaretus*
1,23(12)001,03 *Coregonus oxyrinchus*
1,23(12)001,12 *Coregonus clupeaformis*
1,23(12)001,13 *Coregonus artedi*

1,23(14) ALEPOCEPHALIDAE
1,23(14)007,01 *Alepocephalus bairdii*

1,24 ESOCIFORMES

1,24(03) ESOCIDAE
1,24(03)001,01 *Esox lucius*

1,25 STOMIIFORMES

1,25(02) STERNOPTYCHIDAE
1,25(02)030,01 *Maurolicus muelleri*

1,28 OSTEOGLOSSIFORMES

1,28(01) OSTEOGLOSSIDAE
1,28(01)001,01 *Arapaima gigas*
1,28(01)004,XX *Heterotis spp*

1,28(02) NOTOPTERIDAE
1,28(02)003,XX *Notopterus spp*

1,28(06) MORMYRIDAE
1,28(06)XXX,XX *Mormyridae*

1,28(07) GYMNARCHIDAE
1,28(07)001,01 *Gymnarchus niloticus*

1,29 ELOPIFORMES

1,29(01) ELOPIDAE
1,29(01)003,02 *Elops saurus*
1,29(01)003,04 *Elops lacerta*

1,29(02) MEGALOPIDAE
1,29(02)004,01 *Megalops atlanticus*
1,29(02)004,02 *Megalops cyprinoides*

1,30 ALBULIFORMES

1,30(01) ALBULIDAE
1,30(01)005,01 *Albula vulpes*
1,30(01)006,01 *Pterothrissus belloci*

1,31 AULOPIFORMES

1,31(12) CHLOROPHTHALMIDAE
1,31(12)XXX,XX *Chlorophthalmidae*

Systematic list of aquatic organisms	Liste systématique des organismes aquatiques	Lista sistemática de los organismos acuáticos

1,31(16)	SYNODONTIDAE	
1,31(16)XXX,XX		*Synodontidae*
1,31(16)001,02		*Harpadon nehereus*
1,31(16)068,01		*Saurida tumbil*
1,31(16)068,04		*Saurida undosquamis*
1,32	**MYCTOPHIFORMES**	
1,32(08)	MYCTOPHIDAE	
1,32(08)XXX,XX		*Myctophidae*
1,32(08)017,01		*Lampanyctodes hectoris*
1,32(08)017,15		*Lampanyctus achirus*
1,32(08)030,02		*Electrona carlsbergi*
1,32(08)032,01		*Gymnoscopelus nicholsi*
1,38	**CHARACIFORMES**	
1,38(01)	CHARACIDAE	
1,38(01)XXX,XX		*Characidae*
1,38(01)046,05		*Colossoma macropomum*
1,38(01)076,01		*Piaractus brachypomus*
1,38(12)	CURIMATIDAE	
1,38(12)001,XX		*Prochilodus spp*
1,38(12)001,20		*Prochilodus reticulatus*
1,40	**CYPRINIFORMES**	
1,40(01)	CATOSTOMIDAE	
1,40(01)XXX,XX		*Catostomidae*
1,40(01)011,XX		*Ictiobus spp*
1,40(02)	CYPRINIDAE	
1,40(02)XXX,XX		*Cyprinidae*
1,40(02)001,XX		*Abramis spp*
1,40(02)001,02		*Abramis brama*
1,40(02)002,01		*Cyprinus carpio*
1,40(02)007,01		*Tinca tinca*
1,40(02)012,01		*Alburnus alburnus*
1,40(02)013,01		*Barbus barbus*
1,40(02)014,01		*Chondrostoma nasus*
1,40(02)016,01		*Carassius carassius*
1,40(02)016,02		*Carassius auratus*
1,40(02)018,XX		*Rutilus spp*
1,40(02)018,01		*Rutilus rutilus*
1,40(02)018,02		*Rutilus frisii*
1,40(02)019,01		*Scardinius erythrophthalmus*
1,40(02)020,XX		*Leuciscus spp*
1,40(02)020,01		*Leuciscus idus*
1,40(02)020,05		*Leuciscus leuciscus*
1,40(02)020,08		*Leuciscus cephalus*
1,40(02)024,XX		*Labeo spp*
1,40(02)025,02		*Cirrhinus molitorella*
1,40(02)035,01		*Ctenopharyngodon idellus*
1,40(02)043,01		*Hypophthalmichthys molitrix*
1,40(02)043,02		*Hypophthalmichthys nobilis*
1,40(02)070,01		*Rastrineobola argentea*
1,40(02)072,02		*Acanthobrama terraesanctae*
1,40(02)097,01		*Vimba vimba*
1,40(02)100,01		*Pelecus cultratus*
1,40(02)115,01		*Aspius aspius*
1,40(02)132,01		*Leptobarbus hoeveni*
1,40(02)144,01		*Mylopharyngodon piceus*
1,40(02)151,01		*Megalobrama amblycephala*
1,40(02)161,XX		*Puntius spp*
1,40(02)161,02		*Puntius javanicus*
1,41	**SILURIFORMES**	
1,41(02)	ARIIDAE	
1,41(02)XXX,XX		*Ariidae*
1,41(02)042,08		*Galeichthys feliceps*
1,41(06)	PLOTOSIDAE	
1,41(06)064,XX		*Plotosus spp*
1,41(07)	SILURIDAE	
1,41(07)031,01		*Silurus glanis*
1,41(07)050,XX		*Kryptopterus spp*
1,41(08)	BAGRIDAE	
1,41(08)078,XX		*Chrysichthys spp*
1,41(08)078,02		*Chrysichthys nigrodigitatus*
1,41(08)111,XX		*Bagrus spp*
1,41(10)	ICTALURIDAE	
1,41(10)002,XX		*Ictalurus spp*
1,41(10)004,07		*Ameiurus nebulosus*
1,41(18)	CLARIIDAE	
1,41(18)030,XX		*Clarias spp*
1,41(18)030,01		*Clarias batrachus*
1,41(18)030,03		*Clarias gariepinus*
1,41(18)030,07		*Clarias anguillaris*
1,41(30)	PANGASIIDAE	
1,41(30)002,XX		*Pangasius spp*
1,41(32)	MOCHOKIDAE	
1,41(32)008,XX		*Synodontis spp*
1,43	**ANGUILLIFORMES**	
1,43(02)	ANGUILLIDAE	
1,43(02)002,XX		*Anguilla spp*
1,43(02)002,01		*Anguilla anguilla*
1,43(02)002,04		*Anguilla japonica*
1,43(02)002,06		*Anguilla rostrata*
1,43(02)002,07		*Anguilla australis*
1,43(06)	MURAENIDAE	
1,43(06)XXX,XX		*Muraenidae*
1,43(09)	MURAENESOCIDAE	
1,43(09)011,XX		*Muraenesox spp*
1,43(09)011,02		*Muraenesox cinereus*
1,43(13)	CONGRIDAE	
1,43(13)XXX,XX		*Congridae*
1,43(13)001,01		*Conger conger*
1,43(13)001,02		*Conger orbignyanus*
1,43(13)001,04		*Conger oceanicus*
1,43(13)001,05		*Conger myriaster*
1,47	**BELONIFORMES**	
1,47(01)	BELONIDAE	
1,47(01)XXX,XX		*Belonidae*
1,47(01)001,01		*Belone belone*
1,47(01)013,XX		*Tylosurus spp*
1,47(02)	SCOMBERESOCIDAE	
1,47(02)002,01		*Scomberesox saurus*
1,47(02)007,01		*Cololabis saira*
1,47(03)	HEMIRAMPHIDAE	
1,47(03)003,11		*Hyporhamphus sajori*
1,47(03)004,XX		*Hemiramphus spp*
1,47(03)004,06		*Hemiramphus brasiliensis*
1,47(04)	EXOCOETIDAE	
1,47(04)XXX,XX		*Exocoetidae*
1,47(04)010,02		*Cypselurus agoo*
1,48	**GADIFORMES**	
1,48(01)	MURAENOLEPIDIDAE	
1,48(01)001,XX		*Muraenolepis spp*
1,48(01)001,01		*Muraenolepis microps*
1,48(02)	MORIDAE	
1,48(02)XXX,XX		*Moridae*
1,48(02)010,01		*Mora moro*
1,48(02)014,01		*Pseudophycis bachus*
1,48(02)030,01		*Antimora rostrata*
1,48(02)040,01		*Salilota australis*
1,48(03)	BREGMACEROTIDAE	
1,48(03)018,01		*Bregmaceros mcclellandi*
1,48(04)	GADIDAE	
1,48(04)001,01		*Brosme brosme*
1,48(04)002,02		*Gadus morhua*
1,48(04)002,11		*Gadus macrocephalus*
1,48(04)002,12		*Gadus ogac*
1,48(04)003,01		*Lota lota*
1,48(04)005,01		*Molva molva*
1,48(04)005,02		*Molva dypterygia*
1,48(04)006,01		*Phycis blennoides*
1,48(04)008,01		*Urophycis brasiliensis*
1,48(04)008,02		*Urophycis chuss*
1,48(04)008,03		*Urophycis tenuis*
1,48(04)010,01		*Melanogrammus aeglefinus*
1,48(04)012,01		*Eleginus navaga*
1,48(04)012,02		*Eleginus gracilis*
1,48(04)013,01		*Microgadus proximus*
1,48(04)013,02		*Microgadus tomcod*
1,48(04)015,01		*Pollachius virens*
1,48(04)015,02		*Pollachius pollachius*
1,48(04)016,01		*Theragra chalcogramma*
1,48(04)019,01		*Boreogadus saida*

Systematic list of aquatic organisms	Liste systématique des organismes aquatiques	Lista sistemática de los organismos acuáticos

1,48(04)028,XX		*Gaidropsarus spp*
1,48(04)032,01		*Trisopterus esmarkii*
1,48(04)032,02		*Trisopterus minutus*
1,48(04)032,03		*Trisopterus luscus*
1,48(04)033,01		*Micromesistius poutassou*
1,48(04)033,02		*Micromesistius australis*
1,48(04)034,01		*Merlangius merlangus*
1,48(05)	MERLUCCIIDAE	
1,48(05)004,XX		*Merluccius spp*
1,48(05)004,XX		*Merluccius capensis,M.paradox.*
1,48(05)004,01		*Merluccius merluccius*
1,48(05)004,02		*Merluccius senegalensis*
1,48(05)004,03		*Merluccius australis*
1,48(05)004,04		*Merluccius bilinearis*
1,48(05)004,05		*Merluccius gayi*
1,48(05)004,06		*Merluccius hubbsi*
1,48(05)004,07		*Merluccius productus*
1,48(05)004,08		*Merluccius polli*
1,48(05)004,12		*Merluccius albidus*
1,48(05)004,14		*Merluccius angustimanus*
1,48(05)004,19		*Merluccius capensis*
1,48(05)017,XX		*Macruronus spp*
1,48(05)017,01		*Macruronus magellanicus*
1,48(05)017,02		*Macruronus novaezelandiae*
1,48(06)	MACROURIDAE	
1,48(06)XXX,XX		*Macrouridae*
1,48(06)001,XX		*Macrourus spp*
1,48(06)001,03		*Macrourus berglax*
1,48(06)001,04		*Macrourus whitsoni*
1,48(06)001,05		*Macrourus carinatus*
1,48(06)001,06		*Macrourus holotrachys*
1,48(06)004,XX		*Coryphaenoides spp*
1,48(06)004,01		*Coryphaenoides rupestris*
1,48(06)030,01		*Lepidorhynchus denticulatus*
1,50	**GASTEROSTEIFORMES**	
1,50(01)	GASTEROSTEIDAE	
1,50(01)001,01		*Gasterosteus aculeatus*
1,50(05)	HYPOPTYCHIDAE	
1,50(05)009,01		*Hypoptychus dybowskii*
1,51	**SYNGNATHIFORMES**	
1,51(03)	MACRORAMPHOSIDAE	
1,51(03)004,01		*Macroramphosus scolopax*
1,51(03)021,01		*Centriscops humerosus*
1,52	**LAMPRIFORMES**	
1,52(01)	LAMPRIDAE	
1,52(01)001,02		*Lampris guttatus*
1,52(01)001,03		*Lampris immaculatus*
1,52(04)	TRACHIPTERIDAE	
1,52(04)002,XX		*Trachipterus spp*
1,52(05)	REGALECIDAE	
1,52(05)001,01		*Regalecus glesne*
1,58	**OPHIDIIFORMES**	
1,58(02)	OPHIDIIDAE	
1,58(02)XXX,XX		*Ophidiidae*
1,58(02)001,XX		*Genypterus spp*
1,58(02)001,01		*Genypterus blacodes*
1,58(02)001,02		*Genypterus chilensis*
1,58(02)001,03		*Genypterus maculatus*
1,58(02)001,05		*Genypterus capensis*
1,58(02)005,02		*Brotula barbata*
1,61	**BERYCIFORMES**	
1,61(02)	BERYCIDAE	
1,61(02)003,XX		*Beryx spp*
1,61(02)012,01		*Centroberyx affinis*
1,61(05)	TRACHICHTHYIDAE	
1,61(05)XXX,XX		*Trachichthyidae*
1,61(05)002,02		*Hoplostethus atlanticus*
1,61(11)	HOLOCENTRIDAE	
1,61(11)XXX,XX		*Holocentridae*
1,62	**ZEIFORMES**	
1,62(01)	ZEIDAE	
1,62(01)XXX,XX		*Zeidae*
1,62(01)001,01		*Zeus faber*
1,62(01)004,01		*Zenopsis conchifer*
1,62(01)004,02		*Zenopsis nebulosus*
1,62(03)	CAPROIDAE	
1,62(03)XXX,XX		*Caproidae*
1,62(04)	OREOSOMATIDAE	
1,62(04)XXX,XX		*Oreosomatidae*
1,63	**ATHERINIFORMES**	
1,63(02)	ATHERINIDAE	
1,63(02)XXX,XX		*Atherinidae*
1,63(02)002,01		*Odontesthes regia*
1,63(02)003,01		*Atherina boyeri*
1,63(02)013,02		*Menidia menidia*
1,65	**MUGILIFORMES**	
1,65(01)	MUGILIDAE	
1,65(01)XXX,XX		*Mugilidae*
1,65(01)001,02		*Mugil cephalus*
1,65(01)001,12		*Mugil liza*
1,65(01)001,28		*Mugil soiuy*
1,65(01)006,02		*Agonostomus monticola*
1,65(01)010,01		*Joturus pichardi*
1,65(01)012,06		*Liza saliens*
1,68	**SYNBRANCHIFORMES**	
1,68(02)	SYNBRANCHIDAE	
1,68(02)001,01		*Monopterus albus*
1,70	**PERCOIDEI**	
1,70(00)	CAESIONIDAE	
1,70(00)112,XX		*Caesio spp*
1,70(01)	CENTROPOMIDAE	
1,70(01)025,XX		*Centropomus spp*
1,70(01)025,10		*Centropomus undecimalis*
1,70(01)167,XX		*Lates spp*
1,70(01)167,01		*Lates calcarifer*
1,70(01)167,07		*Lates niloticus*
1,70(02)	SERRANIDAE	
1,70(02)XXX,XX		*Serranidae*
1,70(02)040,XX		*Mycteroperca spp*
1,70(02)040,01		*Mycteroperca bonaci*
1,70(02)040,04		*Mycteroperca microlepis*
1,70(02)040,06		*Mycteroperca phenax*
1,70(02)040,09		*Mycteroperca venenosa*
1,70(02)040,10		*Mycteroperca xenarcha*
1,70(02)042,XX		*Epinephelus spp*
1,70(02)042,01		*Epinephelus marginatus*
1,70(02)042,02		*Epinephelus aeneus*
1,70(02)042,19		*Epinephelus tauvina*
1,70(02)042,20		*Epinephelus striatus*
1,70(02)042,22		*Epinephelus analogus*
1,70(02)042,24		*Epinephelus flavolimbatus*
1,70(02)042,25		*Epinephelus guttatus*
1,70(02)042,28		*Epinephelus morio*
1,70(02)042,30		*Epinephelus niveatus*
1,70(02)042,34		*Epinephelus goreensis*
1,70(02)042,36		*Epinephelus nigritus*
1,70(02)072,01		*Lepidoperca pulchella*
1,70(02)081,02		*Centropristis striata*
1,70(02)259,01		*Acanthistius brasilianus*
1,70(02)405,05		*Paralabrax humeralis*
1,70(04)	TERAPONTIDAE	
1,70(04)089,XX		*Terapon spp*
1,70(05)	POLYPRIONIDAE	
1,70(05)058,01		*Polyprion americanus*
1,70(05)058,02		*Polyprion oxygeneios*
1,70(05)220,02		*Stereolepis gigas*
1,70(06)	MORONIDAE	
1,70(06)006,02		*Morone saxatilis*
1,70(06)006,03		*Morone americana*
1,70(06)006,04		*Morone chrysops*
1,70(06)345,XX		*Dicentrarchus spp*
1,70(06)345,01		*Dicentrarchus punctatus*
1,70(06)345,03		*Dicentrarchus labrax*
1,70(08)	PERCICHTHYIDAE	
1,70(08)297,01		*Lateolabrax japonicus*
1,70(10)	CENTRARCHIDAE	

Systematic list of aquatic organisms	Liste systématique des organismes aquatiques	Lista sistemática de los organismos acuáticos

1,70(10)014,02 *Micropterus salmoides*

1,70(11) PRIACANTHIDAE
1,70(11)026,XX *Priacanthus spp*
1,70(11)026,05 *Priacanthus macracanthus*

1,70(12) APOGONIDAE
1,70(12)XXX,XX *Apogonidae*

1,70(13) ACROPOMATIDAE
1,70(13)007,01 *Synagrops japonicus*

1,70(14) PERCIDAE
1,70(14)002,01 *Perca fluviatilis*
1,70(14)002,02 *Perca flavescens*
1,70(14)007,01 *Percarina demidoffi*
1,70(14)015,XX *Stizostedion spp*
1,70(14)015,01 *Stizostedion vitreum*
1,70(14)015,02 *Stizostedion canadense*
1,70(14)015,03 *Stizostedion lucioperca*
1,70(14)059,01 *Gymnocephalus cernuus*

1,70(15) SILLAGINIDAE
1,70(15)XXX,XX *Sillaginidae*

1,70(16) BRANCHIOSTEGIDAE
1,70(16)XXX,XX *Branchiostegidae*
1,70(16)001,03 *Caulolatilus princeps*
1,70(16)001,04 *Caulolatilus chrysops*
1,70(16)400,02 *Lopholatilus chamaeleonticeps*

1,70(19) LACTARIIDAE
1,70(19)165,02 *Lactarius lactarius*

1,70(20) POMATOMIDAE
1,70(20)213,01 *Pomatomus saltatrix*

1,70(22) RACHYCENTRIDAE
1,70(22)221,01 *Rachycentron canadum*

1,70(23) CARANGIDAE
1,70(23)XXX,XX *Carangidae*
1,70(23)004,XX *Trachurus spp*
1,70(23)004,01 *Trachurus trachurus*
1,70(23)004,02 *Trachurus picturatus*
1,70(23)004,03 *Trachurus japonicus*
1,70(23)004,05 *Trachurus murphyi*
1,70(23)004,06 *Trachurus symmetricus*
1,70(23)004,08 *Trachurus mediterraneus*
1,70(23)004,11 *Trachurus lathami*
1,70(23)004,13 *Trachurus capensis*
1,70(23)004,14 *Trachurus trecae*
1,70(23)004,15 *Trachurus declivis*
1,70(23)011,27 *Pseudocaranx dentex*
1,70(23)043,XX *Decapterus spp*
1,70(23)043,07 *Decapterus maruadsi*
1,70(23)043,08 *Decapterus russelli*
1,70(23)044,XX *Caranx spp*
1,70(23)044,26 *Caranx crysos*
1,70(23)044,29 *Caranx hippos*
1,70(23)044,31 *Caranx ruber*
1,70(23)044,42 *Caranx rhonchus*
1,70(23)046,04 *Selene setapinnis*
1,70(23)046,05 *Selene dorsalis*
1,70(23)047,XX *Trachinotus spp*
1,70(23)047,01 *Trachinotus blochii*
1,70(23)047,03 *Trachinotus carolinus*
1,70(23)048,XX *Seriola spp*
1,70(23)048,01 *Seriola dumerili*
1,70(23)048,02 *Seriola quinqueradiata*
1,70(23)048,06 *Seriola lalandi*
1,70(23)072,02 *Lichia amia*
1,70(23)090,03 *Alectis alexandrinus*
1,70(23)099,01 *Parastromateus niger*
1,70(23)134,01 *Elagatis bipinnulata*
1,70(23)151,01 *Gnathanodon speciosus*
1,70(23)179,01 *Megalaspis cordyla*
1,70(23)231,XX *Scomberoides spp*
1,70(23)268,01 *Chloroscombrus chrysurus*
1,70(23)268,02 *Chloroscombrus orqueta*
1,70(23)283,01 *Parona signata*
1,70(23)291,01 *Selar crumenophthalmus*
1,70(23)422,01 *Selaroides leptolepis*
1,70(23)425,01 *Seriolina nigrofasciata*

1,70(26) MENIDAE
1,70(26)327,01 *Mene maculata*

1,70(27) BRAMIDAE
1,70(27)XXX,XX *Bramidae*
1,70(27)003,01 *Brama brama*

1,70(28) CORYPHAENIDAE
1,70(28)071,01 *Coryphaena hippurus*

1,70(29) ARRIPIDAE
1,70(29)051,01 *Arripis georgianus*
1,70(29)051,02 *Arripis trutta*

1,70(30) EMMELICHTHYIDAE
1,70(30)XXX,XX *Emmelichthyidae*
1,70(30)010,01 *Emmelichthys nitidus*

1,70(32) LUTJANIDAE
1,70(32)XXX,XX *Lutjanidae*
1,70(32)027,XX *Lutjanus spp*
1,70(32)027,02 *Lutjanus argentimaculatus*
1,70(32)027,20 *Lutjanus argentiventris*
1,70(32)027,22 *Lutjanus purpureus*
1,70(32)027,25 *Lutjanus campechanus*
1,70(32)027,33 *Lutjanus synagris*
1,70(32)028,01 *Ocyurus chrysurus*
1,70(32)225,01 *Rhomboplites aurorubens*

1,70(33) NEMIPTERIDAE
1,70(33)XXX,XX *Nemipteridae*
1,70(33)184,XX *Nemipterus spp*
1,70(33)184,05 *Nemipterus virgatus*
1,70(33)230,XX *Scolopsis spp*

1,70(34) LOBOTIDAE
1,70(34)029,01 *Lobotes surinamensis*

1,70(35) LEIOGNATHIDAE
1,70(35)XXX,XX *Leiognathidae*
1,70(35)169,XX *Leiognathus spp*

1,70(36) HAEMULIDAE
1,70(36)XXX,XX *Haemulidae (=Pomadasyidae)*
1,70(36)032,01 *Orthopristis chrysoptera*
1,70(36)163,01 *Isacia conceptionis*
1,70(36)207,05 *Plectorhinchus mediterraneus*
1,70(36)209,04 *Pomadasys argenteus*
1,70(36)209,18 *Pomadasys incisus*
1,70(36)209,19 *Pomadasys jubelini*
1,70(36)263,03 *Brachydeuterus auritus*
1,70(36)395,01 *Conodon nobilis*

1,70(37) SCIAENIDAE
1,70(37)XXX,XX *Sciaenidae*
1,70(37)005,01 *Totoaba macdonaldi*
1,70(37)009,02 *Sciaena gilberti*
1,70(37)009,21 *Sciaena umbra*
1,70(37)011,01 *Aplodinotus grunniens*
1,70(37)016,XX *Cynoscion spp*
1,70(37)016,04 *Cynoscion analis*
1,70(37)016,09 *Cynoscion nebulosus*
1,70(37)016,14 *Cynoscion regalis*
1,70(37)016,17 *Cynoscion striatus*
1,70(37)038,XX *Micropogonias spp*
1,70(37)038,02 *Micropogonias furnieri*
1,70(37)038,04 *Micropogonias undulatus*
1,70(37)039,XX *Menticirrhus spp*
1,70(37)039,07 *Menticirrhus saxatilis*
1,70(37)039,08 *Menticirrhus littoralis*
1,70(37)070,01 *Umbrina cirrosa*
1,70(37)070,03 *Umbrina canosai*
1,70(37)070,09 *Umbrina canariensis*
1,70(37)096,01 *Macrodon ancylodon*
1,70(37)106,01 *Argyrosomus regius*
1,70(37)106,06 *Argyrosomus hololepidotus*
1,70(37)108,01 *Atractoscion aequidens*
1,70(37)108,03 *Atractoscion nobilis*
1,70(37)147,01 *Genyonemus lineatus*
1,70(37)166,07 *Pteroscion peli*
1,70(37)186,03 *Otolithes ruber*
1,70(37)194,02 *Paralonchurus peruanus*
1,70(37)210,02 *Pogonias cromis*
1,70(37)298,05 *Nibea mitsukurii*
1,70(37)303,04 *Larimichthys croceus*
1,70(37)303,05 *Larimichthys polyactis*
1,70(37)363,01 *Leiostomus xanthurus*
1,70(37)411,01 *Sciaenops ocellatus*
1,70(37)457,XX *Pseudotolithus spp*
1,70(37)457,01 *Pseudotolithus brachygnathus*
1,70(37)457,02 *Pseudotolithus senegalensis*
1,70(37)457,05 *Pseudotolithus elongatus*
1,70(37)553,01 *Atrobucca nibe*
1,70(37)561,01 *Pennahia argentata*

1,70(38) LETHRINIDAE
1,70(38)XXX,XX *Lethrinidae*

Systematic list of aquatic organisms	Liste systématique des organismes aquatiques	Lista sistemática de los organismos acuáticos

Code	Family	Species
1,70(38)155,XX		*Gymnocranius spp*
1,70(38)172,04		*Lethrinus atlanticus*
1,70(39)	SPARIDAE	
1,70(39)XXX,XX		*Sparidae*
1,70(39)008,XX		*Pagellus spp*
1,70(39)008,01		*Pagellus bogaraveo*
1,70(39)008,02		*Pagellus erythrinus*
1,70(39)008,03		*Pagellus acarne*
1,70(39)008,07		*Pagellus bellottii*
1,70(39)033,XX		*Diplodus spp*
1,70(39)033,01		*Diplodus argenteus*
1,70(39)033,03		*Diplodus sargus*
1,70(39)034,XX		*Calamus spp*
1,70(39)060,XX		*Dentex spp*
1,70(39)060,02		*Dentex macrophthalmus*
1,70(39)060,04		*Dentex canariensis*
1,70(39)060,06		*Dentex dentex*
1,70(39)060,10		*Dentex angolensis*
1,70(39)060,11		*Dentex congoensis*
1,70(39)063,02		*Spondyliosoma cantharus*
1,70(39)076,01		*Oblada melanura*
1,70(39)102,01		*Archosargus probatocephalus*
1,70(39)102,02		*Archosargus rhomboidalis*
1,70(39)105,01		*Argyrops spinifer*
1,70(39)107,01		*Argyrozona argyrozona*
1,70(39)118,02		*Cheimerius nufar*
1,70(39)191,XX		*Pagrus spp*
1,70(39)191,01		*Pagrus caeruleostictus*
1,70(39)191,03		*Pagrus pagrus*
1,70(39)191,15		*Pagrus auratus*
1,70(39)203,01		*Petrus rupestris*
1,70(39)220,01		*Pterogymnus laniarius*
1,70(39)224,01		*Rhabdosargus globiceps*
1,70(39)235,08		*Sparus aurata*
1,70(39)261,01		*Boops boops*
1,70(39)264,XX		*Chrysoblephus spp*
1,70(39)277,XX		*Lithognathus spp*
1,70(39)277,01		*Lithognathus lithognathus*
1,70(39)277,02		*Lithognathus mormyrus*
1,70(39)293,01		*Sarpa salpa*
1,70(39)330,04		*Acanthopagrus schlegeli*
1,70(39)330,05		*Acanthopagrus latus*
1,70(39)330,06		*Acanthopagrus bifasciatus*
1,70(39)353,01		*Stenotomus chrysops*
1,70(40)	CENTRACANTHIDAE	
1,70(40)075,XX		*Spicara spp*
1,70(40)075,01		*Spicara maena*
1,70(41)	MULLIDAE	
1,70(41)XXX,XX		*Mullidae*
1,70(41)007,XX		*Mullus spp*
1,70(41)007,01		*Mullus surmuletus*
1,70(41)007,02		*Mullus barbatus*
1,70(41)007,03		*Mullus argentinae*
1,70(41)251,XX		*Upeneus spp*
1,70(41)499,03		*Pseudupeneus prayensis*
1,70(42)	ECHENEIDAE	
1,70(42)XXX,XX		*Echeneidae*
1,70(46)	GERREIDAE	
1,70(46)XXX,XX		*Gerreidae*
1,70(46)036,XX		*Gerres spp*
1,70(47)	KYPHOSIDAE	
1,70(47)003,02		*Girella tricuspidata*
1,70(47)003,03		*Girella nigricans*
1,70(47)035,XX		*Kyphosus spp*
1,70(47)035,01		*Kyphosus cinerascens*
1,70(50)	DREPANIDAE	
1,70(50)132,01		*Drepane punctata*
1,70(50)132,02		*Drepane africana*
1,70(57)	PENTACEROTIDAE	
1,70(57)004,02		*Paristiopterus labiosus*
1,70(57)007,01		*Pseudopentaceros richardsoni*
1,70(59)	CICHLIDAE	
1,70(59)XXX,XX		*Cichlidae*
1,70(59)051,XX		*Oreochromis (=Tilapia) spp*
1,70(59)051,01		*Oreochromis mossambicus*
1,70(59)051,02		*Oreochromis niloticus*
1,70(59)051,03		*Oreochromis aureus*
1,70(59)052,01		*Sarotherodon galilaeus*
1,70(59)053,06		*Cichlasoma managuense*
1,70(59)054,04		*Cichla ocellaris*
1,70(59)313,XX		*Astronotus spp*
1,70(59)316,XX		*Haplochromis spp*
1,70(60)	CEPOLIDAE	
1,70(60)083,01		*Cepola rubescens*
1,70(63)	LABRIDAE	
1,70(63)XXX,XX		*Labridae*
1,70(63)005,01		*Labrus bergylta*
1,70(63)243,01		*Tautoga onitis*
1,70(63)244,02		*Tautogolabrus adspersus*
1,70(63)331,02		*Semicossyphus pulcher*
1,70(65)	SCARIDAE	
1,70(65)XXX,XX		*Scaridae*
1,70(65)055,06		*Sparisoma cretense*
1,70(66)	POMACANTHIDAE	
1,70(66)XXX,XX		*Pomacanthidae*
1,70(70)	CHEILODACTYLIDAE	
1,70(70)270,02		*Cheilodactylus bergi*
1,70(70)270,03		*Cheilodactylus variegatus*
1,70(70)305,XX		*Nemadactylus spp*
1,70(70)305,04		*Nemadactylus macropterus*
1,70(71)	LATRIDAE	
1,70(71)XXX,XX		*Latridae*
1,70(77)	POLYNEMIDAE	
1,70(77)XXX,XX		*Polynemidae*
1,70(77)001,09		*Polydactylus quadrifilis*
1,70(77)002,01		*Eleutheronema tetradactylum*
1,70(77)003,02		*Galeoides decadactylus*
1,70(77)004,01		*Pentanemus quinquarius*
1,70(92)	NOTOTHENIIDAE	
1,70(92)XXX,XX		*Nototheniidae*
1,70(92)007,02		*Eleginops maclovinus*
1,70(92)015,XX		*Dissostichus spp*
1,70(92)015,01		*Dissostichus mawsoni*
1,70(92)015,02		*Dissostichus eleginoides*
1,70(92)019,02		*Notothenia rossii*
1,70(92)019,03		*Notothenia gibberifrons*
1,70(92)019,04		*Notothenia neglecta*
1,70(92)019,06		*Notothenia squamifrons*
1,70(92)019,10		*Notothenia kempi*
1,70(92)019,11		*Notothenia acuta*
1,70(92)430,01		*Nototheniops larseni*
1,70(92)430,02		*Nototheniops nudifrons*
1,70(92)430,03		*Nototheniops mizops*
1,70(92)440,01		*Patagonotothen brevicauda*
1,70(92)440,02		*Patagonotothen ramsayi*
1,70(92)448,XX		*Trematomus spp*
1,70(92)448,01		*Trematomus eulepidotus*
1,70(92)529,01		*Pagothenia hansoni*
1,70(92)535,01		*Pleuragramma antarcticum*
1,70(93)	BATHYDRACONIDAE	
1,70(93)452,01		*Parachaenichthys georgianus*
1,70(94)	CHANNICHTHYIDAE	
1,70(94)XXX,XX		*Channichthyidae*
1,70(94)009,01		*Neopagetopsis ionah*
1,70(94)416,01		*Chaenocephalus aceratus*
1,70(94)417,01		*Champsocephalus gunnari*
1,70(94)418,01		*Pseudochaenichthys georgianus*
1,70(94)466,01		*Chionodraco rastrospinosus*
1,70(94)466,02		*Chionodraco hamatus*
1,70(94)470,01		*Channichthys rhinoceratus*
1,70(94)480,01		*Chaenodraco wilsoni*
1,70(95)	AMBASSIDAE	
1,70(95)XXX,XX		*Ambassidae*
1,70(96)	EPIGONIDAE	
1,70(96)373,XX		*Epigonus spp*
1,70(96)373,01		*Epigonus telescopus*
1,70(97)	HARPAGIFERIDAE	
1,70(97)001,01		*Pogonophryne permitini*
1,71	**ZOARCOIDEI**	
1,71(02)	ANARHICHADIDAE	
1,71(02)001,XX		*Anarhichas spp*
1,71(02)001,01		*Anarhichas lupus*
1,71(02)001,03		*Anarhichas minor*
1,71(15)	ZOARCIDAE	
1,71(15)004,01		*Zoarces viviparus*
1,71(15)012,02		*Macrozoarces americanus*
1,71(15)024,XX		*Lycodes spp*

Systematic list of aquatic organisms	Liste systématique des organismes aquatiques	Lista sistemática de los organismos acuáticos

1,72 TRACHINOIDEI

1,72(04) AMMODYTIDAE
1,72(04)002,XX *Ammodytes spp*
1,72(04)002,04 *Ammodytes personatus*

1,72(08) PERCOPHIDAE
1,72(08)202,01 *Percophis brasilianus*

1,72(09) PINGUIPEDIDAE
1,72(09)006,02 *Pseudopercis semifasciata*
1,72(09)196,02 *Parapercis colias*

1,72(12) TRACHINIDAE
1,72(12)010,02 *Trachinus draco*

1,72(13) URANOSCOPIDAE
1,72(13)352,02 *Uranoscopus scaber*
1,72(13)484,01 *Kathetostoma giganteum*

1,72(72) TRICHODONTIDAE
1,72(72)103,01 *Arctoscopus japonicus*

1,73 GOBIOIDEI

1,73(20) ELEOTRIDAE
1,73(20)XXX,XX *Eleotridae*

1,73(21) GOBIIDAE
1,73(21)XXX,XX *Gobiidae*
1,73(21)XXX,XX *Gobiidae*
1,73(21)003,XX *Gobius spp*
1,73(21)003,03 *Gobius niger*

1,74 ACANTHUROIDEI

1,74(02) ACANTHURIDAE
1,74(02)XXX,XX *Acanthuridae*

1,74(05) EPHIPPIDAE
1,74(05)XXX,XX *Ephippidae*
1,74(05)206,XX *Platax spp*

1,74(06) SCATOPHAGIDAE
1,74(06)330,XX *Scatophagus spp*

1,74(07) SIGANIDAE
1,74(07)001,XX *Siganus spp*

1,75 SCOMBROIDEI

1,75(01) SCOMBRIDAE
1,75(01)XXX,XX *Scombridae*
1,75(01)001,01 *Sarda sarda*
1,75(01)001,02 *Sarda orientalis*
1,75(01)001,04 *Sarda chiliensis*
1,75(01)002,XX *Scomber spp*
1,75(01)002,01 *Scomber japonicus*
1,75(01)002,05 *Scomber scombrus*
1,75(01)002,07 *Scomber australasicus*
1,75(01)006,01 *Orcynopsis unicolor*
1,75(01)010,01 *Acanthocybium solandri*
1,75(01)012,02 *Gymnosarda unicolor*
1,75(01)014,XX *Rastrelliger spp*
1,75(01)014,01 *Rastrelliger brachysoma*
1,75(01)014,03 *Rastrelliger kanagurta*
1,75(01)015,XX *Scomberomorus spp*
1,75(01)015,03 *Scomberomorus commerson*
1,75(01)015,04 *Scomberomorus guttatus*
1,75(01)015,05 *Scomberomorus lineolatus*
1,75(01)015,06 *Scomberomorus cavalla*
1,75(01)015,07 *Scomberomorus maculatus*
1,75(01)015,08 *Scomberomorus regalis*
1,75(01)015,09 *Scomberomorus sierra*
1,75(01)015,11 *Scomberomorus tritor*
1,75(01)015,12 *Scomberomorus niphonius*
1,75(01)015,15 *Scomberomorus brasiliensis*
1,75(01)023,XX *Auxis thazard, A.rochei*
1,75(01)023,01 *Auxis thazard*
1,75(01)024,01 *Euthynnus alletteratus*
1,75(01)024,04 *Euthynnus lineatus*
1,75(01)024,06 *Euthynnus affinis*
1,75(01)025,01 *Katsuwonus pelamis*
1,75(01)026,01 *Thunnus thynnus*
1,75(01)026,02 *Thunnus orientalis*
1,75(01)026,03 *Thunnus tonggol*
1,75(01)026,04 *Thunnus atlanticus*
1,75(01)026,05 *Thunnus alalunga*
1,75(01)026,08 *Thunnus maccoyii*
1,75(01)026,10 *Thunnus albacares*
1,75(01)026,12 *Thunnus obesus*

1,75(03) ISTIOPHORIDAE
1,75(03)XXX,XX *Istiophoridae*
1,75(03)004,02 *Istiophorus platypterus*
1,75(03)004,04 *Istiophorus albicans*
1,75(03)005,05 *Makaira nigricans*
1,75(03)005,07 *Makaira indica*
1,75(03)009,03 *Tetrapturus audax*
1,75(03)009,04 *Tetrapturus albidus*
1,75(03)009,05 *Tetrapturus angustirostris*
1,75(03)009,06 *Tetrapturus pfluegeri*

1,75(04) XIPHIIDAE
1,75(04)003,01 *Xiphias gladius*

1,75(05) GEMPYLIDAE
1,75(05)001,01 *Thyrsites atun*
1,75(05)002,01 *Thyrsitops lepidopoides*
1,75(05)005,01 *Lepidocybium flavobrunneum*
1,75(05)007,01 *Ruvettus pretiosus*
1,75(05)009,01 *Rexea solandri*
1,75(05)017,01 *Promethichthys prometheus*

1,75(06) TRICHIURIDAE
1,75(06)XXX,XX *Trichiuridae*
1,75(06)002,XX *Benthodesmus spp*
1,75(06)003,02 *Trichiurus lepturus*
1,75(06)006,01 *Lepidopus caudatus*
1,75(06)012,01 *Aphanopus carbo*

1,76 STROMATEOIDEI, ANABANTOIDEI

1,76(03) STROMATEIDAE
1,76(03)XXX,XX *Stromateidae*
1,76(03)004,01 *Stromateus fiatola*
1,76(03)009,XX *Pampus spp*
1,76(03)009,01 *Pampus argenteus*
1,76(03)011,XX *Peprilus spp*
1,76(03)011,01 *Peprilus alepidotus*
1,76(03)011,04 *Peprilus triacanthus*
1,76(03)011,05 *Peprilus simillimus*

1,76(04) ARIOMMATIDAE
1,76(04)012,01 *Ariomma indica*

1,76(05) ANABANTIDAE
1,76(05)002,01 *Anabas testudineus*

1,76(08) CENTROLOPHIDAE
1,76(08)XXX,XX *Centrolophidae*
1,76(08)010,XX *Seriolella spp*
1,76(08)010,01 *Seriolella porosa*
1,76(08)010,02 *Seriolella brama*
1,76(08)010,04 *Seriolella punctata*
1,76(08)010,05 *Seriolella caerulea*
1,76(08)015,02 *Hyperoglyphe antarctica*
1,76(08)015,03 *Hyperoglyphe bythites*
1,76(08)020,01 *Psenopsis anomala*

1,76(09) OSPHRONEMIDAE
1,76(09)007,01 *Osphronemus goramy*

1,76(10) BELONTIIDAE
1,76(10)013,02 *Trichogaster pectoralis*

1,76(11) HELOSTOMATIDAE
1,76(11)006,01 *Helostoma temminckii*

1,77 OTHER PERCIFORMES

1,77(10) SPHYRAENIDAE
1,77(10)001,XX *Sphyraena spp*

1,77(19) CHANNIDAE
1,77(19)001,XX *Channa spp*
1,77(19)001,01 *Channa argus*
1,77(19)001,03 *Channa striata*
1,77(19)001,04 *Channa micropeltes*

1,78 SCORPAENIFORMES

1,78(01) SCORPAENIDAE
1,78(01)XXX,XX *Scorpaenidae*
1,78(01)001,XX *Sebastes spp*
1,78(01)001,01 *Sebastes marinus*
1,78(01)001,03 *Sebastes entomelas*
1,78(01)001,04 *Sebastes flavidus*
1,78(01)001,09 *Sebastes alutus*
1,78(01)001,10 *Sebastes paucispinis*
1,78(01)001,12 *Sebastes mentella*

1,78(01)001,15 *Sebastes pinniger*
1,78(01)001,16 *Sebastes goodei*
1,78(01)001,32 *Sebastes melanops*
1,78(01)009,02 *Scorpaena scrofa*
1,78(01)017,03 *Helicolenus dactylopterus*
1,78(01)079,01 *Sebastolobus alascanus*

1,78(02) TRIGLIDAE
1,78(02)XXX,XX *Triglidae*
1,78(02)003,01 *Chelidonichthys kumu*
1,78(02)003,02 *Chelidonichthys lucerna*
1,78(02)003,03 *Chelidonichthys cuculus*
1,78(02)003,04 *Chelidonichthys lastoviza*
1,78(02)003,07 *Chelidonichthys capensis*
1,78(02)020,XX *Prionotus spp*
1,78(02)025,01 *Pterygotrigla polyommata*
1,78(02)025,02 *Pterygotrigla picta*
1,78(02)070,01 *Eutrigla gurnardus*

1,78(07) HEXAGRAMMIDAE
1,78(07)005,01 *Ophiodon elongatus*
1,78(07)014,02 *Pleurogrammus azonus*
1,78(07)014,03 *Pleurogrammus monopterygius*

1,78(08) ANOPLOPOMATIDAE
1,78(08)004,01 *Anoplopoma fimbria*

1,78(09) PLATYCEPHALIDAE
1,78(09)XXX,XX *Platycephalidae*
1,78(09)018,01 *Platycephalus indicus*

1,78(11) CONGIOPODIDAE
1,78(11)030,01 *Zanclorhynchus spinifer*

1,78(13) COTTIDAE
1,78(13)XXX,XX *Cottidae*
1,78(13)012,XX *Myoxocephalus spp*
1,78(13)030,01 *Scorpaenichthys marmoratus*

1,78(16) NORMANICHTHYIDAE
1,78(16)022,05 *Normanichthys crockeri*

1,78(20) CYCLOPTERIDAE
1,78(20)003,01 *Cyclopterus lumpus*

1,83 PLEURONECTIFORMES

1,83(01) BOTHIDAE
1,83(01)XXX,XX *Bothidae*

1,83(02) PLEURONECTIDAE
1,83(02)002,01 *Hippoglossus hippoglossus*
1,83(02)002,02 *Hippoglossus stenolepis*
1,83(02)004,05 *Pleuronectes platessa*
1,83(02)004,08 *Pleuronectes quadrituberculat.*
1,83(02)004,15 *Pleuronectes vetulus*
1,83(02)005,01 *Reinhardtius hippoglossoides*
1,83(02)008,01 *Atheresthes stomias*
1,83(02)008,02 *Atheresthes evermanni*
1,83(02)010,01 *Eopsetta jordani*
1,83(02)011,02 *Glyptocephalus cynoglossus*
1,83(02)011,03 *Glyptocephalus zachirus*
1,83(02)014,01 *Hippoglossoides platessoides*
1,83(02)014,04 *Hippoglossoides elassodon*
1,83(02)024,02 *Limanda aspera*
1,83(02)024,04 *Limanda ferruginea*
1,83(02)024,05 *Limanda limanda*
1,83(02)043,01 *Lepidopsetta bilineata*
1,83(02)045,03 *Microstomus pacificus*
1,83(02)045,04 *Microstomus kitt*
1,83(02)048,02 *Platichthys flesus*
1,83(02)049,01 *Psettichthys melanostictus*
1,83(02)050,01 *Pseudopleuronectes herzenst.*
1,83(02)050,03 *Pseudopleuronectes americanus*
1,83(02)053,XX *Rhombosolea spp*
1,83(02)059,02 *Pleuronichthys decurrens*

1,83(03) SOLEIDAE
1,83(03)XXX,XX *Soleidae*
1,83(03)007,01 *Solea solea*
1,83(03)007,04 *Solea lascaris*
1,83(03)017,01 *Dicologlossa cuneata*
1,83(03)057,XX *Austroglossus spp*
1,83(03)057,01 *Austroglossus microlepis*
1,83(03)057,02 *Austroglossus pectoralis*
1,83(03)081,XX *Microchirus spp*

1,83(04) CYNOGLOSSIDAE
1,83(04)XXX,XX *Cynoglossidae*

1,83(05) SCOPHTHALMIDAE
1,83(05)XXX,XX *Scophthalmidae*
1,83(05)003,XX *Lepidorhombus spp*
1,83(05)003,01 *Lepidorhombus whiffiagonis*
1,83(05)064,01 *Scophthalmus rhombus*
1,83(05)064,02 *Scophthalmus aquosus*
1,83(05)092,01 *Psetta maxima*

1,83(07) PSETTODIDAE
1,83(07)001,01 *Psettodes erumei*
1,83(07)001,02 *Psettodes belcheri*

1,83(08) PARALICHTHYIDAE
1,83(08)009,02 *Citharichthys sordidus*
1,83(08)046,XX *Paralichthys spp*
1,83(08)046,01 *Paralichthys olivaceus*
1,83(08)046,03 *Paralichthys californicus*
1,83(08)046,06 *Paralichthys dentatus*

1,83(10) ACHIROPSETTIDAE
1,83(10)002,01 *Mancopsetta maculata*

1,90 TETRAODONTIFORMES

1,90(01) OSTRACIIDAE
1,90(01)XXX,XX *Ostraciidae*

1,90(02) TETRAODONTIDAE
1,90(02)XXX,XX *Tetraodontidae*
1,90(02)006,XX *Sphoeroides spp*
1,90(02)006,12 *Sphoeroides maculatus*
1,90(02)011,01 *Takifugu vermicularis*

1,90(08) MOLIDAE
1,90(08)002,01 *Mola mola*

1,90(09) MONACANTHIDAE
1,90(09)XXX,XX *Monacanthidae*
1,90(09)004,XX *Cantherhines(=Navodon) spp*
1,90(09)010,01 *Stephanolepis cirrhifer*
1,90(09)018,01 *Parika scaber*

1,90(10) BALISTIDAE
1,90(10)XXX,XX *Balistidae*
1,90(10)002,01 *Balistes carolinensis*

1,93 BATRACHOIDIFORMES

1,93(01) BATRACHOIDIDAE
1,93(01)XXX,XX *Batrachoididae*

1,95 LOPHIIFORMES

1,95(01) LOPHIIDAE
1,95(01)XXX,XX *Lophiidae*
1,95(01)001,01 *Lophius piscatorius*
1,95(01)001,02 *Lophius budegassa*
1,95(01)001,03 *Lophius americanus*
1,95(01)001,05 *Lophius vaillanti*
1,95(01)001,06 *Lophius vomerinus*

1,99 PISCES MISCELLANEA

2 CRUSTACEA

2,02 ANOSTRACA

2,02(02) ARTEMIIDAE
2,02(02)001,01 *Artemia salina*

2,13 THORACICA

2,13(01) SCALPELLIDAE
2,13(01)005,01 *Mitella pollicipes*

2,13(02) LEPADIDAE
2,13(02)007,XX *Lepas spp*

2,13(03) BALANIDAE
2,13(03)015,01 *Megabalanus psittacus*

2,25 STOMATOPODA

2,25(01) SQUILLIDAE
2,25(01)XXX,XX *Squillidae*
2,25(01)001,02 *Squilla mantis*

2,26 EUPHAUSIACEA

2,26(01) EUPHAUSIIDAE

2,26(01)005,XX *Euphausia spp*
2,26(01)005,01 *Euphausia superba*
2,26(01)011,01 *Meganyctiphanes norvegica*

2,28 NATANTIA

2,28(01) PENAEIDAE
2,28(01)XXX,XX *Xiphopenaeus,Trachypenaeus spp*
2,28(01)001,XX *Penaeus spp*
2,28(01)001,01 *Penaeus aztecus*
2,28(01)001,03 *Penaeus merguiensis*
2,28(01)001,04 *Penaeus californiensis*
2,28(01)001,05 *Penaeus duorarum*
2,28(01)001,09 *Penaeus japonicus*
2,28(01)001,10 *Penaeus stylirostris*
2,28(01)001,11 *Penaeus vannamei*
2,28(01)001,12 *Penaeus monodon*
2,28(01)001,16 *Penaeus chinensis*
2,28(01)001,17 *Penaeus kerathurus*
2,28(01)001,19 *Penaeus brasiliensis*
2,28(01)001,20 *Penaeus semisulcatus*
2,28(01)001,22 *Penaeus setiferus*
2,28(01)001,23 *Penaeus brevirostris*
2,28(01)001,25 *Penaeus indicus*
2,28(01)001,28 *Penaeus latisulcatus*
2,28(01)001,29 *Penaeus occidentalis*
2,28(01)001,30 *Penaeus penicillatus*
2,28(01)001,31 *Penaeus notialis*
2,28(01)001,32 *Penaeus paulensis*
2,28(01)016,XX *Metapenaeus spp*
2,28(01)016,01 *Metapenaeus monoceros*
2,28(01)016,06 *Metapenaeus endeavouri*
2,28(01)016,07 *Metapenaeus joyneri*
2,28(01)017,01 *Parapenaeus longirostris*
2,28(01)019,XX *Parapenaeopsis spp*
2,28(01)019,02 *Parapenaeopsis atlantica*
2,28(01)022,01 *Xiphopenaeus kroyeri*
2,28(01)022,02 *Xiphopenaeus riveti*
2,28(01)043,02 *Trachypenaeus curvirostris*
2,28(01)067,01 *Artemesia longinaris*

2,28(02) ARISTAEIDAE
2,28(02)XXX,XX *Aristeidae*
2,28(02)021,01 *Plesiopenaeus edwardsianus*
2,28(02)031,01 *Aristeus antennatus*
2,28(02)031,02 *Aristeus varidens*

2,28(04) PANDALIDAE
2,28(04)XXX,XX *Pandalus spp, Pandalopsis spp*
2,28(04)002,XX *Pandalus spp*
2,28(04)002,01 *Pandalus hypsinotus*
2,28(04)002,03 *Pandalus borealis*
2,28(04)002,05 *Pandalus montagui*
2,28(04)002,07 *Pandalus goniurus*
2,28(04)002,08 *Pandalus kessleri*
2,28(04)005,01 *Heterocarpus reedi*
2,28(04)005,08 *Heterocarpus vicarius*
2,28(04)037,02 *Pandalopsis japonica*

2,28(07) SERGESTIDAE
2,28(07)XXX,XX *Sergestidae*
2,28(07)009,03 *Acetes japonicus*

2,28(12) PALAEMONIDAE
2,28(12)XXX,XX *Palaemonidae*
2,28(12)XXX,XX *Palaemonidae*
2,28(12)003,02 *Nematopalaemon schmitti*
2,28(12)018,06 *Palaemon longirostris*
2,28(12)018,10 *Palaemon serratus*
2,28(12)023,XX *Macrobrachium spp*
2,28(12)023,07 *Macrobrachium rosenbergii*

2,28(23) CRANGONIDAE
2,28(23)XXX,XX *Crangonidae*
2,28(23)003,03 *Crangon crangon*
2,28(23)006,XX *Sclerocrangon spp*

2,28(28) SICYONIIDAE
2,28(28)028,01 *Sicyonia brevirostris*
2,28(28)028,06 *Sicyonia ingentis*

2,28(29) SOLENOCERIDAE
2,28(29)006,01 *Pleoticus muelleri*
2,28(29)006,03 *Pleoticus robustus*
2,28(29)072,01 *Solenocera agassizii*
2,28(29)073,XX *Haliporoides spp*
2,28(29)073,01 *Haliporoides triarthrus*
2,28(29)073,02 *Haliporoides sibogae*
2,28(29)073,03 *Haliporoides diomedeae*

2,29 REPTANTIA

2,29(01) PALINURIDAE
2,29(01)XXX,XX *Palinuridae*
2,29(01)001,XX *Panulirus spp*
2,29(01)001,01 *Panulirus longipes*
2,29(01)001,08 *Panulirus argus*
2,29(01)001,15 *Panulirus cygnus*
2,29(01)001,16 *Panulirus gracilis*
2,29(01)002,01 *Jasus lalandii*
2,29(01)002,02 *Jasus frontalis*
2,29(01)002,03 *Jasus verreauxi*
2,29(01)002,05 *Jasus tristani*
2,29(01)002,06 *Jasus edwardsii*
2,29(01)002,07 *Jasus novaehollandiae*
2,29(01)002,08 *Jasus paulensis*
2,29(01)008,XX *Palinurus spp*
2,29(01)008,02 *Palinurus mauritanicus*
2,29(01)008,04 *Palinurus elephas*
2,29(01)008,05 *Palinurus delagoae*
2,29(01)008,06 *Palinurus gilchristi*

2,29(03) ASTACIDAE
2,29(03)027,02 *Astacus astacus*
2,29(03)076,01 *Pacifastacus leniusculus*
2,29(03)139,01 *Austropotamobius pallipes*

2,29(04) CAMBARIDAE
2,29(04)001,01 *Procambarus clarkii*

2,29(05) PARASTACIDAE
2,29(05)XXX,XX *Parastacidae*
2,29(05)158,01 *Euastacus armatus*

2,29(15) SCYLLARIDAE
2,29(15)XXX,XX *Scyllaridae*
2,29(15)001,04 *Ibacus ciliatus*
2,29(15)005,01 *Thenus orientalis*

2,29(42) NEPHROPIDAE
2,29(42)005,XX *Metanephrops spp*
2,29(42)005,02 *Metanephrops mozambicus*
2,29(42)005,03 *Metanephrops andamanicus*
2,29(42)005,04 *Metanephrops challengeri*
2,29(42)006,02 *Nephrops norvegicus*
2,29(42)007,01 *Homarus americanus*
2,29(42)007,18 *Homarus gammarus*

2,29(49) UPOGEBIIDAE
2,29(49)001,03 *Upogebia pugettensis*

2,29(59) CALLIANASSIDAE
2,29(59)001,XX *Callianassa spp*

2,30 ANOMURA

2,30(19) GALATHEIDAE
2,30(19)XXX,XX *Galatheidae*
2,30(19)001,01 *Pleuroncodes planipes*
2,30(19)001,02 *Pleuroncodes monodon*
2,30(19)003,01 *Cervimunida johni*

2,30(20) LITHODIDAE
2,30(20)XXX,XX *Lithodidae*
2,30(20)018,XX *Paralithodes spp*
2,30(20)018,01 *Paralithodes camtschaticus*
2,30(20)018,02 *Paralithodes platypus*
2,30(20)018,03 *Paralithodes brevipes*
2,30(20)070,XX *Lithodes spp*
2,30(20)070,01 *Lithodes antarcticus*
2,30(20)070,02 *Lithodes murrayi*
2,30(20)070,05 *Lithodes aequispina*
2,30(20)123,01 *Paralomis granulosa*
2,30(20)123,02 *Paralomis aculeata*
2,30(20)123,03 *Paralomis spinosissima*
2,30(20)123,04 *Paralomis formosa*

2,31 BRACHYURA

2,31(09) CANCRIDAE
2,31(09)010,03 *Cancer irroratus*
2,31(09)010,04 *Cancer magister*
2,31(09)010,06 *Cancer pagurus*
2,31(09)010,07 *Cancer borealis*
2,31(09)010,08 *Cancer productus*

2,31(10) XANTHIDAE
2,31(10)050,01 *Menippe mercenaria*

2,31(11) PORTUNIDAE
2,31(11)004,XX *Portunus spp*
2,31(11)004,01 *Portunus pelagicus*

2,31(11)004,04 *Portunus trituberculatus*
2,31(11)012,02 *Callinectes sapidus*
2,31(11)012,05 *Callinectes danae*
2,31(11)090,01 *Carcinus maenas*
2,31(11)090,02 *Carcinus aestuarii*
2,31(11)140,01 *Scylla serrata*

2,31(21) MAJIDAE
2,31(21)001,02 *Mithrax armatus*
2,31(21)005,01 *Maja squinado*
2,31(21)145,XX *Chionoecetes spp*
2,31(21)145,01 *Chionoecetes opilio*

2,31(23) ATELECYCLIDAE
2,31(23)001,01 *Erimacrus isenbeckii*

2,31(43) GERYONIDAE
2,31(43)088,XX *Geryon spp*
2,31(43)088,01 *Geryon quinquedens*

2,99 CRUSTACEA MISCELLANEA

3 MOLLUSCA

3,07 GASTROPODA

3,07(01) LITTORINIDAE
3,07(01)001,XX *Littorina spp*
3,07(01)001,01 *Littorina littorea*

3,07(02) MURICIDAE
3,07(02)002,XX *Murex spp*
3,07(02)018,XX *Rapana spp*
3,07(02)023,01 *Concholepas concholepas*

3,07(03) HALIOTIDAE
3,07(03)001,XX *Haliotis spp*
3,07(03)001,09 *Haliotis gigantea*
3,07(03)001,11 *Haliotis midae*
3,07(03)001,13 *Haliotis rubra*
3,07(03)001,14 *Haliotis tuberculata*

3,07(04) TROCHIDAE
3,07(04)006,XX *Ex Trochus spp*

3,07(05) TURBINIDAE
3,07(05)002,XX *Ex Turbo spp*
3,07(05)002,02 *Turbo cornutus*

3,07(06) STROMBIDAE
3,07(06)002,XX *Strombus spp*

3,07(08) BUCCINIDAE
3,07(08)001,01 *Buccinum undatum*

3,07(09) MELONGENIDAE
3,07(09)003,XX *Busycon spp*

3,07(42) VOLUTIDAE
3,07(42)004,XX *Cymbium spp*
3,07(42)007,01 *Zidona dufresnei*

3,16 BIVALVIA

3,16(04) ARCIDAE
3,16(04)001,XX *Arca spp*
3,16(04)005,08 *Scapharca subcrenata*
3,16(04)071,XX *Anadara spp*
3,16(04)071,01 *Anadara granosa*

3,16(05) UNIONIDAE
3,16(05)XXX,XX *Ex Unionidae*

3,16(06) PTERIIDAE
3,16(06)006,XX *Ex Pinctada spp*

3,16(07) OSTREIDAE
3,16(07)002,03 *Ostrea chilensis*
3,16(07)002,05 *Ostrea edulis*
3,16(07)002,20 *Ostrea lutaria*
3,16(07)002,25 *Ostrea conchaphila*
3,16(07)008,XX *Crassostrea spp*
3,16(07)008,01 *Crassostrea gigas*
3,16(07)008,02 *Crassostrea rhizophorae*
3,16(07)008,03 *Crassostrea virginica*
3,16(07)008,11 *Crassostrea iredalei*

3,16(08) PECTINIDAE
3,16(08)XXX,XX *Pectinidae*
3,16(08)001,05 *Aequipecten opercularis*
3,16(08)003,09 *Pecten maximus*
3,16(08)003,11 *Pecten jacobaeus*
3,16(08)003,13 *Pecten novaezelandiae*
3,16(08)012,01 *Zygochlamis delicatula*
3,16(08)012,02 *Zygochlamis patagonica*
3,16(08)014,04 *Placopecten magellanicus*
3,16(08)030,01 *Argopecten gibbus*
3,16(08)030,02 *Argopecten irradians*
3,16(08)030,03 *Argopecten purpuratus*
3,16(08)030,04 *Argopecten ventricosus*
3,16(08)036,03 *Chlamys islandica*
3,16(08)066,01 *Patinopecten caurinus*
3,16(08)066,07 *Patinopecten yessoensis*

3,16(09) ARCTICIDAE
3,16(09)045,01 *Arctica islandica*

3,16(10) MYTILIDAE
3,16(10)XXX,XX *Mytilidae*
3,16(10)001,01 *Mytilus coruscus*
3,16(10)001,03 *Mytilus chilensis*
3,16(10)001,05 *Mytilus edulis*
3,16(10)001,08 *Mytilus platensis*
3,16(10)001,12 *Mytilus galloprovincialis*
3,16(10)001,17 *Mytilus planulatus*
3,16(10)026,01 *Choromytilus chorus*
3,16(10)028,XX *Modiolus spp*
3,16(10)032,01 *Perna perna*
3,16(10)032,02 *Perna viridis*
3,16(10)032,03 *Perna canaliculus*
3,16(10)038,01 *Aulacomya ater*

3,16(11) VENERIDAE
3,16(11)XXX,XX *Veneridae*
3,16(11)001,05 *Chamelea gallina*
3,16(11)003,01 *Venerupis pullastra*
3,16(11)007,04 *Chione stutchburyi*
3,16(11)017,XX *Meretrix spp*
3,16(11)017,01 *Meretrix lusoria*
3,16(11)020,01 *Ruditapes decussatus*
3,16(11)020,02 *Ruditapes philippinarum*
3,16(11)024,01 *Tawera gayi*
3,16(11)025,02 *Tivela mactroides*
3,16(11)037,02 *Saxidomus giganteus*
3,16(11)041,XX *Paphia spp*
3,16(11)055,02 *Protothaca staminea*
3,16(11)055,03 *Protothaca thaca*
3,16(11)075,01 *Mercenaria mercenaria*

3,16(12) MACTRIDAE
3,16(12)001,02 *Pseudocardium sybillae*
3,16(12)005,XX *Tresus spp*
3,16(12)020,01 *Spisula solidissima*
3,16(12)020,02 *Spisula polynyma*
3,16(12)020,05 *Spisula solida*
3,16(12)040,XX *Mulinia spp*

3,16(15) DONACIDAE
3,16(15)002,XX *Donax spp*

3,16(16) SOLENIDAE
3,16(16)003,XX *Solen spp*
3,16(16)005,02 *Ensis directus*
3,16(16)007,01 *Siliqua patula*

3,16(17) MYIDAE
3,16(17)006,01 *Mya arenaria*

3,16(18) HIATELLIDAE
3,16(18)089,01 *Panopea abrupta*

3,16(21) CORBICULIDAE
3,16(21)025,02 *Corbicula japonica*

3,16(23) CARDIIDAE
3,16(23)XXX,XX *Cardiidae*
3,16(23)002,03 *Cerastoderma edule*
3,16(23)004,02 *Clinocardium nuttallii*

3,16(24) MESODESMATIDAE
3,16(24)002,01 *Paphies australis*
3,16(24)039,01 *Mesodesma donacium*

3,16(39) SEMELIDAE
3,16(39)001,03 *Semele solida*

3,21 CEPHALOPODA

3,21(02) SEPIIDAE
3,21(02)XXX,XX *Sepiidae, Sepiolidae*

3,21(02)002,02 *Sepia officinalis*

3,21(04) LOLIGINIDAE
3,21(04)001,XX *Loligo spp*
3,21(04)001,02 *Loligo gahi*
3,21(04)001,03 *Loligo opalescens*
3,21(04)001,05 *Loligo pealei*
3,21(04)001,12 *Loligo reynaudi*

3,21(05) OMMASTREPHIDAE
3,21(05)XXX,XX *Loliginidae, Ommastrephidae*
3,21(05)003,01 *Ommastrephes bartrami*
3,21(05)010,01 *Illex illecebrosus*
3,21(05)010,02 *Illex coindetii*
3,21(05)010,03 *Illex argentinus*
3,21(05)023,01 *Dosidicus gigas*
3,21(05)058,01 *Todarodes sagittatus*
3,21(05)058,03 *Todarodes pacificus*
3,21(05)059,01 *Nototodarus sloani*
3,21(05)060,01 *Martialia hyadesi*

3,21(09) OCTOPODIDAE
3,21(09)XXX,XX *Octopodidae*
3,21(09)001,XX *Pareledone spp*
3,21(09)005,07 *Octopus vulgaris*
3,21(09)024,XX *Eledone spp*

3,99 MOLLUSCA MISCELLANEA

4 MAMMALIA

4,06 PINNIPEDIA

4,06(01) OTARIIDAE
4,06(01)001,01 *Eumetopias jubatus*
4,06(01)002,01 *Callorhinus ursinus*
4,06(01)006,01 *Arctocephalus australis*
4,06(01)006,03 *Arctocephalus pusillus*
4,06(01)016,01 *Otaria byronia*

4,06(02) ODOBENIDAE
4,06(02)004,01 *Odobenus rosmarus*

4,06(03) PHOCIDAE
4,06(03)005,01 *Phoca groenlandica*
4,06(03)005,02 *Phoca vitulina*
4,06(03)005,03 *Phoca hispida*
4,06(03)005,04 *Phoca fasciata*
4,06(03)005,05 *Phoca caspica*
4,06(03)005,06 *Phoca sibirica*
4,06(03)005,07 *Phoca largha*
4,06(03)008,01 *Erignathus barbatus*
4,06(03)010,01 *Cystophora cristata*
4,06(03)011,01 *Halichoerus grypus*

4,14 SIRENIA

4,14(02) TRICHECHIDAE
4,14(02)003,02 *Trichechus inunguis*

4,22 ODONTOCETI

4,22(02) ZIPHIIDAE
4,22(02)005,01 *Hyperoodon ampullatus*
4,22(02)019,02 *Berardius bairdii*

4,22(03) PHYSETERIDAE
4,22(03)001,01 *Physeter catodon*

4,22(04) DELPHINIDAE
4,22(04)004,02 *Globicephala melas*
4,22(04)004,03 *Globicephala macrorhynchus*
4,22(04)006,01 *Delphinus delphis*
4,22(04)022,01 *Orcinus orca*
4,22(04)029,01 *Tursiops truncatus*

4,22(05) PHOCOENIDAE
4,22(05)002,01 *Phocoena phocoena*

4,22(06) MONODONTIDAE
4,22(06)014,01 *Delphinapterus leucas*
4,22(06)018,01 *Monodon monoceros*

4,23 MYSTICETI

4,23(02) BALAENOPTERIDAE
4,23(02)001,01 *Balaenoptera acutorostrata*
4,23(02)001,02 *Balaenoptera edeni*
4,23(02)001,03 *Balaenoptera borealis*
4,23(02)001,04 *Balaenoptera musculus*
4,23(02)001,06 *Balaenoptera physalus*
4,23(02)003,01 *Megaptera novaeangliae*

4,23(03) BALAENIDAE
4,23(03)001,01 *Eubalaena glacialis*
4,23(03)002,01 *Balaena mysticetus*

4,23(04) ESCHRICHTIIDAE
4,23(04)001,01 *Eschrichtius robustus*

4,99 MAMMALIA MISCELLANEA

5 AMPHIBIA, REPTILIA

5,12 ANURA

5,12(01) RANIDAE
5,12(01)001,XX *Rana spp*

5,31 TESTUDINES

5,31(06) EMYDIDAE
5,31(06)021,XX *Malaclemys spp*

5,31(07) CHELONIIDAE
5,31(07)005,02 *Chelonia mydas*
5,31(07)017,01 *Eretmochelys imbricata*
5,31(07)018,01 *Caretta caretta*

5,31(11) TRIONYCHIDAE
5,31(11)024,01 *Trionyx sinensis*

5,36 CROCODILIA

5,36(01) CROCODYLIDAE
5,36(01)XXX,XX *Crocodylidae*
5,36(01)001,03 *Caiman crocodilus*
5,36(01)002,01 *Alligator mississippiensis*
5,36(01)003,01 *Crocodylus porosus*
5,36(01)003,02 *Crocodylus siamensis*
5,36(01)003,03 *Crocodylus johnstoni*
5,36(01)003,04 *Crocodylus niloticus*
5,36(01)003,05 *Crocodylus novaeguineae*
5,36(01)003,06 *Crocodylus rhombifer*
5,36(01)003,07 *Crocodylus moreletii*
5,36(01)003,08 *Crocodylus acutus*
5,36(01)008,01 *Paleosuchus palpebrosus*
5,36(01)008,02 *Paleosuchus trigonatus*

6 INVERTEBRATA AQUATICA

6,15 DEMOSPONGIAE

6,15(01) SPONGIDAE
6,15(01)XXX,XX *Spongidae*

6,18 SCYPHOZOA

6,18(41) RHIZOSTOMIDAE
6,18(41)007,XX *Rhopilema spp*

6,19 ANTHOZOA

6,19(01) CORALLIIDAE
6,19(01)003,01 *Corallium rubrum*
6,19(01)003,02 *Corallium japonicum*
6,19(01)003,03 *Corallium elatius*
6,19(01)003,04 *Corallium konojoi*
6,19(01)003,05 *Corallium secundum*
6,19(01)003,07 *Corallium sp. nov.*

6,49 POLYCHAETA

6,56 XIPHOSURA

6,56(01) LIMULIDAE
6,56(01)001,02 *Limulus polyphemus*

6,89 ECHINODERMATA

6,91 ASTEROIDEA

6,91(05) ASTERIIDAE
6,91(05)001,01 *Asterias rubens*

6,93 ECHINOIDEA

Systematic list of aquatic organisms	Liste systématique des organismes aquatiques	Lista sistemática de los organismos acuáticos

6,93(02) STRONGYLOCENTROTIDAE
6,93(02)004,XX *Strongylocentrotus spp*

6,93(04) ECHINIDAE
6,93(04)007,01 *Paracentrotus lividus*
6,93(04)014,01 *Echinus esculentus*
6,93(04)017,01 *Loxechinus albus*

6,94 HOLOTHURIOIDEA

6,94(14) STICHOPODIDAE
6,94(14)004,01 *Stichopus japonicus*

6,96 ASCIDIACEA

6,96(09) PYURIDAE
6,96(09)005,01 *Pyura stolonifera*
6,96(09)005,02 *Pyura chilensis*
6,96(09)037,01 *Microcosmus sulcatus*

6,97 THALIACEA

6,97(01) SALPIDAE
6,97(01)XXX,XX *Salpidae*

6,99 INVERTEBRATA AQUATICA MISCELL.

7 PLANTAE AQUATICAE

7,11 CYANOPHYCEAE

7,11(01) OSCILLATORIACEAE
7,11(01)001,XX *Spirulina spp*

7,41 CHLOROPHYCEAE

7,41(08) ULVACEAE
7,41(08)002,04 *Ulva pertusa*

7,71 PHAEOPHYCEAE

7,71(02) LAMINARIACEAE
7,71(02)002,01 *Laminaria digitata*
7,71(02)002,02 *Laminaria japonica*
7,71(02)002,04 *Laminaria hyperborea*

7,71(04) ALARIACEAE
7,71(04)003,01 *Undaria pinnatifida*

7,71(05) LESSONIACEAE
7,71(05)001,XX *Lessonia spp*
7,71(05)001,01 *Lessonia nigrescens*
7,71(05)001,02 *Lessonia trabeculata*
7,71(05)002,XX *Macrocystis spp*

7,71(06) FUCACEAE
7,71(06)006,01 *Ascophyllum nodosum*

7,71(08) DURVILLAEACEAE
7,71(08)001,01 *Durvillaea antartica*

7,71(09) CYSTOSEIRACEAE
7,71(09)001,01 *Cystoseira barbata*

7,87 RHODOPHYCEAE

7,87(02) PHYLLOPHORACEAE
7,87(02)001,XX *Phyllophora spp*
7,87(02)002,04 *Gymnogongrus furcellatus*

7,87(12) GRACILARIACEAE
7,87(12)004,XX *Gracilaria spp*

7,87(16) GIGARTINACEAE
7,87(16)001,04 *Chondrus crispus*
7,87(16)002,XX *Gigartina spp*
7,87(16)002,03 *Gigartina skottsbergii*
7,87(16)003,XX *Iridaea spp*

7,87(20) BANGIACEAE
7,87(20)002,XX *Porphyra spp*
7,87(20)002,08 *Porphyra tenera*

7,87(22) RHODOMELACEAE
7,87(22)002,02 *Digenea simplex*

7,87(26) GELIDIACEAE
7,87(26)002,XX *Gelidium spp*

7,87(27) KALLYMENIACEAE
7,87(27)001,01 *Callophyllis variegata*

7,92 ANGIOSPERMAE

7,92(02) ZOSTERACEAE
7,92(02)001,01 *Zostera marina*

7,92(04) CYPERACEAE
7,92(04)001,XX *Scirpus spp*

7,99 PLANTAE AQUATICAE MISCELLANEA

See paragraph 4 of the NOTES ON SPECIES ITEMS

Voir le paragraphe 4 des NOTES SUR LES CATÉGORIES D'ESPÈCES

Véase al párrafo 4 en las NOTAS SOBRE LAS PARTIDAS DE ESPECIES

Notes on major fishing areas

1. The 26 *major fishing areas*, internationally established for fishery statistical purposes, consist of:

   a) seven major inland fishing areas, covering the inland waters of the continents;

   b) nineteen major marine fishing areas, covering the waters of the Atlantic, Indian, Pacific and Southern Oceans with their adjacent seas.

2. The list of these 26 areas appears on the preceding page facing the world map on which the boundaries of the various major fishing areas are shown together with their identifying two-digit codes. These two-digit area codes are reflected in the table numbers in Section C.

3. The major fishing areas, inland and marine, are identified in all tables by two kinds of descriptors: their FAO names and/or two-digit codes. Space restrictions require the use in column headings of the word *fishing area* when referring to *major fishing areas*.

4. Breakdown of catch statistics by subareas, divisions, subdivisions, etc., within these major fishing areas, are not given in this volume. These details appear in statistical bulletins issued regularly by various regional fishery organizations, councils, commissions, committees, etc., e.g. ICES, NAFO, ICCAT, GFCM, CECAF, etc.

5. Changes have been made in some of the boundaries between major fishing areas. The catch statistics for all areas and all years in this volume have been adjusted in accordance with the boundaries as now shown in the world map.

   The fishing area 07 ("Former USSR area – Inland waters") referred to the area that was formerly the Union of Soviet Socialist Republics. Since 1988, information for each new independent Republic is shown separately with the exception of capture data in the Antarctic fishing areas for which separate data begin in 1991. The new independent Republics are: Armenia, Azerbaijan, Georgia, Kazakhstan, Kyrgyzstan, Tajikistan, Turkmenistan, Uzbekistan (statistics now assigned to the fishing area "Asia – Inland waters") and Belarus, Estonia, Latvia, Lithuania, Republic of Moldova, Russian Federation, Ukraine (assigned to the fishing area "Europe – Inland waters").

   Since 1981, fishing area 41 includes the waters lying between 55° and 60° South latitude and between 50° and 60° West longitude which during the 1970-80 period were included in the fishing area 48.

   Division of catches between fishing areas 57 and 71 was estimated following boundary change in the Strait of Malacca in 1986.

   Beginning with the 1989 data, catches have been adjusted in accordance with the boundary change between fishing areas 87 and 81; and between 87 and 77.

   Boundary between fishing areas 51 and 57 was modified starting with the 1999 data. As a consequence data for Sri Lanka have been moved from area 51 to area 57.

   Boundary between fishing areas 57 and 71 in the Australian-Indonesian region was modified starting with the 2001 data.

6. Freshwater species are usually recorded as caught in inland waters. Small quantities of several freshwater species are caught regularly in parts of some seas with low salinities; these catches are included in the statistics of the appropriate marine area. Similarly, the catches of diadromous (anadromous and catadromous) species are shown in either the marine or inland area where caught.

Notes sur les principales zones de pêche

Les 26 *principales zones de pêche* établies au plan international à des fins statistiques, comprennent:

a) sept principales zones de pêche continentales, couvrant les eaux continentales des continents;

b) dix-neuf principales zones de pêche maritimes, couvrant les eaux des océans Atlantique, Indien, Pacifique et Austral et leurs mers limitrophes.

La liste de ces 26 zones figure à la page précédente en regard de la carte mondiale sur laquelle on a porté les limites des principales zones de pêche avec les codes à deux chiffres qui les identifient. Ces codes à deux chiffres se retrouvent dans les numéros des tableaux figurant à la Section C.

Les principales zones de pêche, tant continentales que maritimes, sont identifiées dans tous les tableaux par deux types de descripteurs: leur nom FAO et/ou leur code à deux chiffres. Faute d'espace, il a fallu utiliser le terme *zone de pêche* dans les têtes de colonne pour indiquer les *principales zones de pêche*.

Les ventilations des statistiques des captures par sous-zones, divisions, sous-divisions, etc., au sein de ces principales zones de pêche ne sont pas données dans le présent volume. Ces détails figurent dans les bulletins statistiques qui sont publiés périodiquement par les divers organismes régionaux de pêche, conseils, commissions ou comités des pêches, par exemple, CIEM, NAFO, CICTA, CGPM, COPACE, etc.

Des modifications ont été apportées à certaines des limites tracées des principales zones de pêche. Les statistiques des captures relatives à toutes les zones de pêche et à toutes les années incluses dans le présent volume ont été ajustées de manière à correspondre aux limites qui sont maintenant indiquées sur la carte mondiale.

La zone de pêche 07 (Zone de l'ex-URSS – Eaux continentales) correspondait au territoire de l'ancienne Union des Républiques socialistes soviétiques. Depuis 1988 les renseignements sont donnés séparément pour chaque république indépendante sauf pour les données relatives aux captures effectuées dans les zones maritimes de l'Antarctique pour lesquelles les données séparées commencent en 1991. Les nouvelles Républiques indépendantes sont: Arménie, Azerbaïdjan, Géorgie, Kazakhstan, Kirghizistan, Tadjikistan, Turkménistan, Ouzbékistan (statistiques maintenant attribuées à la zone de pêche Asie - Eaux continentales) et Bélarus, Estonie, Lettonie, Lituanie, République de Moldova, Fédération de Russie, Ukraine (statistiques attribuées à la zone de pêche "Europe - Eaux continentales").

Depuis 1981 la zone de pêche 41 comprend les eaux situées entre 55° et 60° de latitude sud et entre 50° et 60° de longitude ouest qui durant la période 1970-80 étaient incluses dans la zone de pêche 48.

La répartition des captures entre les zones de pêche 57 et 71 était estimée après changement apporté en 1986 à la limite relative au détroit de Malacca.

À partir des données de 1989 les captures ont été ajustées de manière à correspondre à la modification apportée aux limites entre les zones de pêche 87 et 81, et entre les zones 87 et 77.

La limite entre les zones de pêche 51 et 57 a été modifiée à partir des données de 1999. En conséquence les données pour le Sri Lanka ont été portées de la zone de pêche 51 à la zone de pêche 57.

La limite entre les zones de pêche 57 et 71 dans la région australien-indonésienne a été modifiée à partir des données de 2001.

Les poissons d'eau douce sont d'ordinaire enregistrés comme captures faites dans les eaux continentales. De petites quantités de poissons d'eau douce de plusieurs espèces sont couramment capturées dans certaines mers dont la salinité est faible; ces captures figurent dans les statistiques relatives à la zone maritime appropriée. De même, les captures de poissons d'espèces diadromes (anadromes et catadromes) apparaissent dans les statistiques ayant trait soit aux zones maritimes, soit aux zones continentales où elles ont été effectuées.

Notas sobre las áreas principales de pesca

Las 26 *áreas principales de pesca* establecidas internacionalmente con fines estadísticos se dividen en:

a) siete áreas principales de pesca continental, que cubren las aguas continentales de los continentes;

b) diecinueve áreas principales de pesca marítimas, que cubren las aguas de los océanos Atlántico, Índico, Pacífico y Austral así como las de sus mares adyacentes.

La lista de estas 26 áreas de pesca puede verse en la página opuesta a la carta mundial, en la que se indican los límites de las distintas áreas principales de pesca con los códigos de dos dígitos correspondientes a cada área de pesca. Estos dos dígitos se utilizan luego en los números de los cuadros de la Sección C.

Las áreas principales de pesca, continentales y marinas, se identifican en todos los cuadros con dos tipos de descriptores: su nombre FAO y/o código de dos dígitos. Por razones de espacio, en los encabezamientos de las columnas se utiliza el término *área de pesca* para designar las *áreas principales de pesca*.

En el presente volumen no se desglosan las estadísticas de captura por subáreas, divisiones, subdivisiones, etc., dentro de las áreas principales de pesca. Esos desgloses pueden verse en los boletines estadísticos publicados regularmente por las distintas organizaciones regionales de pesca, consejos, comisiones, comités, etc., como el CIEM, la NAFO, la CICAA, el CGPM, el CPACO, etc.

Los límites de algunas de las áreas principales de pesca han sido modificados en los años pasados. Las estadísticas de captura de todas las áreas y años que figuran en este volumen se han ajustado en consecuencia, teniendo en cuenta los límites que ahora aparecen en la carta mundial.

El área de pesca 07 (Área de la ex URSS – Aguas continentales) correspondía a la antigua Unión de Repúblicas Socialistas Soviéticas. A partir de 1988 la información de cada una de las nuevas repúblicas independientes se expone por separado excepto en el caso de las capturas efectuadas en las áreas marítimas del Antártico para las cuales los datos separados comienzan en 1991. Las nuevas repúblicas independientes de la ex URSS son: Armenia, Azerbaiyán, Georgia, Kazajstán, Kirguistán, Tayikistán, Turkmenistán, Uzbekistán (los datos han sido atribuidos al área de pesca Asia - Aguas continentales) y Belarús, Estonia, Letonia, Lituania, República de Moldova, Federación de Rusia, Ucrania (datos atribuidos al área de pesca "Europa – Aguas continentales").

Desde 1981 la área de pesca 41 incluye las aguas comprendidas entre 55° y 60° de latitud Sur y entre 50° y 60° de longitud Oeste que durante el período 1970-80 eran incluidas en la área de pesca 48.

La distribución de las capturas entre las áreas de pesca 57 y 71 fue una estimación debida a la modificación hecha en 1986 al límite relativo al estrecho de Malaca.

A partir de datos de 1989 las capturas se han ajustado teniendo en cuenta las modificaciones hechas a los límites entre las áreas de pesca 87 y 81, y entre las áreas 87 y 77.

El límite entre las áreas de pesca 51 y 57 ha sido modificado a partir de datos de 1999. Consecuentemente se cambiaron los datos por Sri Lanka desde la área de pesca 51 a la 57.

El límite entre las áreas de pesca 57 y 71 en la región australiana-indonesia ha sido modificado a partir de datos de 2001.

Las especies de agua dulce se consignan de ordinario considerando que se han capturado en aguas continentales. Pequeñas cantidades de varias especies de agua dulce se capturan regularmente en algunas zonas marítimas de poca salinidad; dichas capturas se incluyen en las estadísticas correspondientes al área marítima en cuestión. De igual forma, las capturas de especies diadromas (anadromas y catadromas) se incluyen en el área marítima o continental en que se han capturadas.

List of major fishing areas

Liste des principales zones de pêche

Lista de las áreas principales de pesca

Code Code Código	Major fishing areas	Principales zones de pêche	Áreas principales de pesca	Km²	%
	INLAND WATERS	EAUX CONTINENTALES	AGUAS CONTINENTALES	...	...
01	Africa - Inland waters	Afrique - Eaux continentales	Africa - Aguas continentales	...	...
02	America, North - Inland waters	Amérique du Nord - Eaux continentales	América del Norte - Aguas continentales	...	...
03	America, South - Inland waters	Amérique du Sud - Eaux continentales	América del Sur - Aguas continentales	...	...
04	Asia -Inland waters	Asie - Eaux continentales	Asia - Aguas continentales	...	...
05	Europe - Inland waters	Europe - Eaux continentales	Europa - Aguas continentales	...	...
06	Oceania - Inland waters	Océanie - Eaux continentales	Oceanía - Aguas continentales	...	...
07	(Former USSR area – Inland waters)	(Zone de l'ex-URSS – Eaux continentales)	(Área de la ex URSS – Aguas continentales)	...	...
08	Antarctica - Inland waters	Antarctique - Eaux continentales	Antártida - Aguas continentales	...	...
	MARINE AREAS	ZONES MARITIMES	ÁREAS MARITIMAS	360 900 000	100.0
	Atlantic Ocean and adjacent seas	*Océan Atlantique et mers limitrophes*	*Océano Atlántico y mares adyacentes*		
18	Arctic Sea	Mer Arctique	Mar Artico	9 300 000	2.6
21	Atlantic, Northwest	Atlantique, nord-ouest	Atlántico, noroeste	6 300 000	1.7
27	Atlantic, Northeast	Atlantique, nord-est	Atlántico, nordeste	14 400 000	4.0
31	Atlantic, Western Central	Atlantique, centre-ouest	Atlántico, centro-occidental	14 500 000	4.0
34	Atlantic, Eastern Central	Atlantique, centre-est	Atlántico, centro-oriental	14 100 000	3.9
37	Mediterranean and Black Sea	Méditerranée et mer Noire	Mediterráneo y mar Negro	3 000 000	0.8
41	Atlantic, Southwest	Atlantique, sud-ouest	Atlántico, sudoccidental	17 500 000	4.8
47	Atlantic, Southeast	Atlantique, sud-est	Atlántico, sudoriental	18 300 000	5.1
	Indian Ocean	*Océan Indien*	*Océano Índico*		
51	Indian Ocean, Western	Océan Indien, ouest	Océano Índico, occidental	29 300 000	8.1
57	Indian Ocean, Eastern	Océan Indien, est	Océano Índico, oriental	31 100 000	8.6
	Pacific Ocean	*Océan Pacifique*	*Océano Pacífico*		
61	Pacific, Northwest	Pacifique, nord-ouest	Pacífico, noroeste	21 500 000	6.0
67	Pacific, Northeast	Pacifique, nord-est	Pacífico, nordeste	7 600 000	2.1
71	Pacific, Western Central	Pacifique, centre-ouest	Pacífico, centro-occidental	33 300 000	9.2
77	Pacific, Eastern Central	Pacifique, centre-est	Pacífico, centro-oriental	48 100 000	13.3
81	Pacific, Southwest	Pacifique, sud-ouest	Pacífico, sudoccidental	27 700 000	7.7
87	Pacific, Southeast	Pacifique, sud-est	Pacífico, sudoriental	30 800 000	8.5
	Southern Ocean	*Océan Austral*	*Océano Austral*		
‣ 48	Atlantic, Antarctic	Atlantique, Antarctique	Atlántico, Antártico	11 800 000	3.3
‣ 58	Indian Ocean, Antarctic	Océan Indien, Antarctique	Océano Índico, Antártico	12 700 000	3.5
‣ 88	Pacific, Antarctic	Pacifique, Antarctique	Pacífico, Antártico	9 600 000	2.7

‣ See paragraph 3 of the INTRODUCTION

See paragraph 5 of the NOTES ON MAJOR FISHING AREAS for the changes of boundaries between some major fishing areas.

‣ Voir le paragraphe 3 de l'INTRODUCTION

Voir le paragraphe 5 de NOTES SUR LES PRINCIPALES ZONES DE PÊCHE par les modifications apportées à certaines des limites tracées des principales zones de pêche.

‣ Véase el párrafo 3 en la INTRODUCCIÓN

Véase el párrafo 5 en las NOTAS SOBRE LOS DISTINTOS PAÍSES O ÁREAS por las modificaciones de los límites de algunas de las áreas principales de pesca.

List of countries or areas / Liste des pays ou zones / Lista de países o áreas

A	B	C	D	E	F	G	H
Afghanistan	AFG	AF	004	4	Afghanistan	Afghanistan	Afganistán
Albania	ALB	AL	008	5	Albania	Albanie	Albania
Algeria	DZA	DZ	012	1	Algeria	Algérie	Argelia
Amer Samoa	ASM	AS	016	6	American Samoa	Samoa américaines	Samoa Americana
Andorra	AND	AD	020	5	Andorra	Andorre	Andorra
Angola	AGO	AO	024	1	Angola	Angola	Angola
Anguilla	AIA	AI	660	2	Anguilla	Anguilla	Anguila
Antigua Barb	ATG	AG	028	2	Antigua and Barbuda	Antigua-et-Barbuda	Antigua y Barbuda
Argentina	ARG	AR	032	3	Argentina	Argentine	Argentina
Armenia	ARM	AM	051	4	Armenia	Arménie	Armenia
Aruba	ABW	AW	533	2	Aruba	Aruba	Aruba
Australia	AUS	AU	036	6	Australia	Australie	Australia
Austria	AUT	AT	040	5	Austria	Autriche	Austria
Azerbaijan	AZE	AZ	031	4	Azerbaijan	Azerbaïdjan	Azerbaiyán
Bahamas	BHS	BS	044	2	Bahamas	Bahamas	Bahamas
Bahrain	BHR	BH	048	4	Bahrain	Bahreïn	Bahrein
Bangladesh	BGD	BD	050	4	Bangladesh	Bangladesh	Bangladesh
Barbados	BRB	BB	052	2	Barbados	Barbade	Barbados
Belarus	BLR	BY	112	5	Belarus	Bélarus	Belarús
Belgium	BEL	BE	056	5	Belgium	Belgique	Bélgica
Belize	BLZ	BZ	084	2	Belize	Belize	Belice
Benin	BEN	BJ	204	1	Benin	Bénin	Benin
Bermuda	BMU	BM	060	2	Bermuda	Bermudes	Bermudas
Bhutan	BTN	BT	064	4	Bhutan	Bhoutan	Bhután
Bolivia	BOL	BO	068	3	Bolivia	Bolivie	Bolivia
Bosnia Herzg	BIH	BA	070	5	Bosnia and Herzegovina	Bosnie-Herzégovine	Bosnia y Herzegovina
Botswana	BWA	BW	072	1	Botswana	Botswana	Botswana
Brazil	BRA	BR	076	3	Brazil	Brésil	Brasil
Br Ind Oc Tr	IOT	IO	086	1	British Indian Ocean Ter	Terr. brit. océan Indien	Ter. brit. océano Indico
Br Virgin Is	VGB	VG	092	2	British Virgin Islands	Iles Vierges britanniq.	Islas Vírgenes Britán.
Brunei Darsm	BRN	BN	096	4	Brunei Darussalam	Brunéi Darussalam	Brunei Darussalam
Bulgaria	BGR	BG	100	5	Bulgaria	Bulgarie	Bulgaria
Burkina Faso	BFA	BF	854	1	Burkina Faso	Burkina Faso	Burkina Faso
Burundi	BDI	BI	108	1	Burundi	Burundi	Burundi
Cambodia	KHM	KH	116	4	Cambodia	Cambodge	Camboya
Cameroon	CMR	CM	120	1	Cameroon	Cameroun	Camerún
Canada	CAN	CA	124	2	Canada	Canada	Canadá
Cape Verde	CPV	CV	132	1	Cape Verde	Cap-Vert	Cabo Verde
Cayman Is	CYM	KY	136	2	Cayman Islands	Iles Caïmanes	Islas Caimán
Cent Afr Rep	CAF	CF	140	1	Central African Republic	Rép. centrafricaine	República Centroafricana
Chad	TCD	TD	148	1	Chad	Tchad	Chad
Channel Is	...	..	830	5	Channel Islands	Iles Anglo-Normandes	Islas Anglonormandas
Chile	CHL	CL	152	3	Chile	Chili	Chile
China	CHN	CN	156	4	China	Chine	China
China,H.Kong	HKG	HK	344	4	China, Hong Kong SAR	Chine, RAS de Hong-Kong	China, RAE de Hong Kong
China, Macao	MAC	MO	446	4	China, Macao SAR	Chine, RAS de Macao	China, RAE de Macao
China,Taiwan	TWN	TW	158	4	Taiwan Province of China	Prov. chinoise de Taïwan	Prov. china de Taiwán
Christmas Is	CXR	CX	162	6	Christmas Island	Ile Christmas	Isla Christmas
Cocos Is	CCK	CC	166	6	Cocos (Keeling) Islands	Iles des Cocos (Keeling)	Islas Cocos (Keeling)
Colombia	COL	CO	170	3	Colombia	Colombie	Colombia
Comoros	COM	KM	174	1	Comoros	Comores	Comoras
Congo Dem R	COD	CD	180	1	Congo, Dem. Rep. of the	Rép. dém. du Congo	Rep. Dem. del Congo
Congo Rep	COG	CG	178	1	Congo, Republic of	République du Congo	República del Congo
Cook Is	COK	CK	184	6	Cook Islands	Iles Cook	Islas Cook
Costa Rica	CRI	CR	188	2	Costa Rica	Costa Rica	Costa Rica
Côte dIvoire	CIV	CI	384	1	Côte d'Ivoire	Côte d'Ivoire	Côte d'Ivoire
Croatia	HRV	HR	191	5	Croatia	Croatie	Croacia
Cuba	CUB	CU	192	2	Cuba	Cuba	Cuba
Cyprus	CYP	CY	196	4	Cyprus	Chypre	Chipre
Czech Rep	CZE	CZ	203	5	Czech Republic	République tchèque	República Checa
Denmark	DNK	DK	208	5	Denmark	Danemark	Dinamarca
Djibouti	DJI	DJ	262	1	Djibouti	Djibouti	Djibouti
Dominica	DMA	DM	212	2	Dominica	Dominique	Dominica
Dominican Rp	DOM	DO	214	2	Dominican Republic	République dominicaine	República Dominicana
Ecuador	ECU	EC	218	3	Ecuador	Equateur	Ecuador
Egypt	EGY	EG	818	1	Egypt	Egypte	Egipto
El Salvador	SLV	SV	222	2	El Salvador	El Salvador	El Salvador
Eq Guinea	GNQ	GQ	226	1	Equatorial Guinea	Guinée équatoriale	Guinea Ecuatorial
Eritrea	ERI	ER	232	1	Eritrea	Erythrée	Eritrea
Estonia	EST	EE	233	5	Estonia	Estonie	Estonia
Ethiopia	ETH	ET	231	1	Ethiopia	Ethiopie	Etiopía
Faeroe Is	FRO	FO	234	5	Faeroe Islands	Iles Féroé	Islas Feroe
Falkland Is	FLK	FK	238	3	Falkland Is.(Malvinas)	Iles Falkland(Malvinas)	Islas Malvinas(Falkland)
Fiji Islands	FJI	FJ	242	6	Fiji Islands	Iles Fidji	Islas Fiji
Finland	FIN	FI	246	5	Finland	Finlande	Finlandia
France	FRA	FR	250	5	France	France	Francia
Fr Guiana	GUF	GF	254	3	French Guiana	Guyane française	Guayana Francesa
Fr Polynesia	PYF	PF	258	6	French Polynesia	Polynésie française	Polinesia Francesa
Fr South Tr	ATF	TF	260	1	French Southern Terr	Terres australes fr.	Tierras Australes Fr.
Gabon	GAB	GA	266	1	Gabon	Gabon	Gabón

List of countries or areas | Liste des pays ou zones | Lista de países o áreas

A	B	C	D	E	F	G	H
Gambia	GMB	GM	270	1	Gambia	Gambie	Gambia
Georgia	GEO	GE	268	4	Georgia	Géorgie	Georgia
Germany	DEU	DE	276	5	Germany	Allemagne	Alemania
Ghana	GHA	GH	288	1	Ghana	Ghana	Ghana
Gibraltar	GIB	GI	292	5	Gibraltar	Gibraltar	Gibraltar
Greece	GRC	GR	300	5	Greece	Grèce	Grecia
Greenland	GRL	GL	304	2	Greenland	Groenland	Groenlandia
Grenada	GRD	GD	308	2	Grenada	Grenade	Granada
Guadeloupe	GLP	GP	312	2	Guadeloupe	Guadeloupe	Guadalupe
Guam	GUM	GU	316	6	Guam	Guam	Guam
Guatemala	GTM	GT	320	2	Guatemala	Guatemala	Guatemala
Guinea	GIN	GN	324	1	Guinea	Guinée	Guinea
GuineaBissau	GNB	GW	624	1	Guinea-Bissau	Guinée-Bissau	Guinea-Bissau
Guyana	GUY	GY	328	3	Guyana	Guyana	Guyana
Haiti	HTI	HT	332	2	Haiti	Haïti	Haití
Honduras	HND	HN	340	2	Honduras	Honduras	Honduras
Hungary	HUN	HU	348	5	Hungary	Hongrie	Hungría
Iceland	ISL	IS	352	5	Iceland	Islande	Islandia
India	IND	IN	356	4	India	Inde	India
Indonesia	IDN	ID	360	4	Indonesia	Indonésie	Indonesia
Iran	IRN	IR	364	4	Iran (Islamic Rep. of)	Iran (Rép. islamique d')	Irán (Rep. Islámica del)
Iraq	IRQ	IQ	368	4	Iraq	Iraq	Iraq
Ireland	IRL	IE	372	5	Ireland	Irlande	Irlanda
Isle of Man	IMY	IM	833	5	Isle of Man	Ile de Man	Isla de Man
Israel	ISR	IL	376	4	Israel	Israël	Israel
Italy	ITA	IT	380	5	Italy	Italie	Italia
Jamaica	JAM	JM	388	2	Jamaica	Jamaïque	Jamaica
Japan	JPN	JP	392	4	Japan	Japon	Japón
Jordan	JOR	JO	400	4	Jordan	Jordanie	Jordania
Kazakhstan	KAZ	KZ	398	4	Kazakhstan	Kazakhstan	Kazajstán
Kenya	KEN	KE	404	1	Kenya	Kenya	Kenya
Kiribati	KIR	KI	296	6	Kiribati	Kiribati	Kiribati
Korea D P Rp	PRK	KP	408	4	Korea, Dem. People's Rep	Rép. pop. dém. de Corée	Rep. Pop. Dem. de Corea
Korea Rep	KOR	KR	410	4	Korea, Republic of	République de Corée	República de Corea
Kuwait	KWT	KW	414	4	Kuwait	Koweït	Kuwait
Kyrgyzstan	KGZ	KG	417	4	Kyrgyzstan	Kirghizistan	Kirguistán
Laos	LAO	LA	418	4	Lao People's Dem. Rep.	Rép. dém. pop. lao	Rep. Dem. Pop. Lao
Latvia	LVA	LV	428	5	Latvia	Lettonie	Letonia
Lebanon	LBN	LB	422	4	Lebanon	Liban	Líbano
Lesotho	LSO	LS	426	1	Lesotho	Lesotho	Lesotho
Liberia	LBR	LR	430	1	Liberia	Libéria	Liberia
Libya	LBY	LY	434	1	Libyan Arab Jamahiriya	Jamahiriya arabe libyen.	Jamahiriya Arabe Libia
Liechtensten	LIE	LI	438	5	Liechtenstein	Liechtenstein	Liechtenstein
Lithuania	LTU	LT	440	5	Lithuania	Lituanie	Lituania
Luxembourg	LUX	LU	442	5	Luxembourg	Luxembourg	Luxemburgo
Macedonia	MKD	MK	807	5	Macedonia, Fmr Yug Rp of	Ex-Rép.youg.de Macédoine	Ex Rep.Yug. de Macedonia
Madagascar	MDG	MG	450	1	Madagascar	Madagascar	Madagascar
Malawi	MWI	MW	454	1	Malawi	Malawi	Malawi
Malaysia	MYS	MY	458	4	Malaysia	Malaisie	Malasia
Maldives	MDV	MV	462	4	Maldives	Maldives	Maldivas
Mali	MLI	ML	466	1	Mali	Mali	Malí
Malta	MLT	MT	470	5	Malta	Malte	Malta
Marshall Is	MHL	MH	584	6	Marshall Islands	Iles Marshall	Islas Marshall
Martinique	MTQ	MQ	474	2	Martinique	Martinique	Martinica
Mauritania	MRT	MR	478	1	Mauritania	Mauritanie	Mauritania
Mauritius	MUS	MU	480	1	Mauritius	Maurice	Mauricio
Mayotte	MYT	YT	175	1	Mayotte	Mayotte	Mayotte
Mexico	MEX	MX	484	2	Mexico	Mexique	México
Micronesia	FSM	FM	583	6	Micronesia,Fed.States of	Micronésie(Etats féd.de)	Micronesia(Estados Fed.)
Moldova Rep	MDA	MD	498	5	Moldova, Republic of	République de Moldova	República de Moldova
Monaco	MCO	MC	492	5	Monaco	Monaco	Mónaco
Mongolia	MNG	MN	496	4	Mongolia	Mongolie	Mongolia
Montserrat	MSR	MS	500	2	Montserrat	Montserrat	Montserrat
Morocco	MAR	MA	504	1	Morocco	Maroc	Marruecos
Mozambique	MOZ	MZ	508	1	Mozambique	Mozambique	Mozambique
Myanmar	MMR	MM	104	4	Myanmar	Myanmar	Myanmar
Namibia	NAM	NA	516	1	Namibia	Namibie	Namibia
Nauru	NRU	NR	520	6	Nauru	Nauru	Nauru
Nepal	NPL	NP	524	4	Nepal	Népal	Nepal
Netherlands	NLD	NL	528	5	Netherlands	Pays-Bas	Países Bajos
NethAntilles	ANT	AN	530	2	Netherlands Antilles	Antilles néerlandaises	Antillas Neerlandesas
NewCaledonia	NCL	NC	540	6	New Caledonia	Nouvelle-Calédonie	Nueva Caledonia
New Zealand	NZL	NZ	554	6	New Zealand	Nouvelle-Zélande	Nueva Zelandia
Nicaragua	NIC	NI	558	2	Nicaragua	Nicaragua	Nicaragua
Niger	NER	NE	562	1	Niger	Niger	Níger
Nigeria	NGA	NG	566	1	Nigeria	Nigéria	Nigeria
Niue	NIU	NU	570	6	Niue	Nioué	Niue
Norfolk Is	NFK	NF	574	6	Norfolk Island	Ile Norfolk	Isla Norfolk

List of countries or areas | Liste des pays ou zones | Lista de países o áreas

A	B	C	D	E	F	G	H
N Marianas	MNP	MP	580	6	Northern Mariana Is.	Iles Mariannes septentr.	Islas Marianas Septent.
Norway	NOR	NO	578	5	Norway	Norvège	Noruega
Oman	OMN	OM	512	4	Oman	Oman	Omán
Pakistan	PAK	PK	586	4	Pakistan	Pakistan	Pakistán
Palau	PLW	PW	585	6	Palau	Palaos	Palau
Palest, O.T.	PSE	PS	275	4	Palestine, Occupied Tr.	Palestine, terr.occupés	Palestina, Terri.Ocupado
Panama	PAN	PA	591	2	Panama	Panama	Panamá
Papua N Guin	PNG	PG	598	6	Papua New Guinea	Papouasie-Nlle-Guinée	Papua Nueva Guinea
Paraguay	PRY	PY	600	3	Paraguay	Paraguay	Paraguay
Peru	PER	PE	604	3	Peru	Pérou	Perú
Philippines	PHL	PH	608	4	Philippines	Philippines	Filipinas
Pitcairn Is	PCN	PN	612	6	Pitcairn Islands	Iles Pitcairn	Islas Pitcairn
Poland	POL	PL	616	5	Poland	Pologne	Polonia
Portugal	PRT	PT	620	5	Portugal	Portugal	Portugal
Puerto Rico	PRI	PR	630	2	Puerto Rico	Porto Rico	Puerto Rico
Qatar	QAT	QA	634	4	Qatar	Qatar	Qatar
Réunion	REU	RE	638	1	Réunion	Réunion	Reunión
Romania	ROU	RO	642	5	Romania	Roumanie	Rumania
Russian Fed	RUS	RU	643	5	Russian Federation	Fédération de Russie	Federación de Rusia
Rwanda	RWA	RW	646	1	Rwanda	Rwanda	Rwanda
St Helena	SHN	SH	654	1	Saint Helena	Sainte-Hélène	Santa Elena
St Kitts Nev	KNA	KN	659	2	Saint Kitts and Nevis	Saint-Kitts-et-Nevis	Saint Kitts y Nevis
St Lucia	LCA	LC	662	2	Saint Lucia	Sainte-Lucie	Santa Lucía
St Pier Mq	SPM	PM	666	2	St. Pierre and Miquelon	Saint-Pierre-et-Miquelon	San Pedro y Miquelón
St Vincent	VCT	VC	670	2	Saint Vincent/Grenadines	Saint-Vincent/Grenadines	San Vicente/Grenadinas
Samoa	WSM	WS	882	6	Samoa	Samoa	Samoa
Sao Tome Prn	STP	ST	678	1	Sao Tome and Principe	Sao Tomé-et-Principe	Santo Tomé y Príncipe
Saudi Arabia	SAU	SA	682	4	Saudi Arabia	Arabie saoudite	Arabia Saudita
Senegal	SEN	SN	686	1	Senegal	Sénégal	Senegal
Serbia-Monte	SCG	SC	891	5	Serbia and Montenegro	Serbie-et-Monténégro	Serbia y Montenegro
Seychelles	SYC	SC	690	1	Seychelles	Seychelles	Seychelles
Sierra Leone	SLE	SL	694	1	Sierra Leone	Sierra Leone	Sierra Leona
Singapore	SGP	SG	702	4	Singapore	Singapour	Singapur
Slovakia	SVK	SK	703	5	Slovakia	Slovaquie	Eslovaquia
Slovenia	SVN	SI	705	5	Slovenia	Slovénie	Eslovenia
Solomon Is	SLB	SB	090	6	Solomon Islands	Iles Salomon	Islas Salomón
Somalia	SOM	SO	706	1	Somalia	Somalie	Somalia
South Africa	ZAF	ZA	710	1	South Africa	Afrique du Sud	Sudáfrica
Spain	ESP	ES	724	5	Spain	Espagne	España
Sri Lanka	LKA	LK	144	4	Sri Lanka	Sri Lanka	Sri Lanka
Sudan	SDN	SD	736	1	Sudan	Soudan	Sudán
Suriname	SUR	SR	740	3	Suriname	Suriname	Suriname
Svalbard Is	SJM	SJ	744	5	Svalbard and Jan Mayen	Iles Svalbard/Jan Mayen	Islas Svalbard/Jan Mayen
Swaziland	SWZ	SZ	748	1	Swaziland	Swaziland	Swazilandia
Sweden	SWE	SE	752	5	Sweden	Suède	Suecia
Switzerland	CHE	CH	756	5	Switzerland	Suisse	Suiza
Syria	SYR	SY	760	4	Syrian Arab Republic	Rép. arabe syrienne	República Arabe Siria
Tajikistan	TJK	TJ	762	4	Tajikistan	Tadjikistan	Tayikistán
Tanzania	TZA	TZ	834	1	Tanzania, United Rep. of	Rép.-Unie de Tanzanie	Rep. Unida de Tanzanía
Thailand	THA	TH	764	4	Thailand	Thaïlande	Tailandia
Timor-Leste	TLS	TL	626	4	Timor-Leste	Timor-Leste	Timor-Leste
Togo	TGO	TG	768	1	Togo	Togo	Togo
Tokelau	TKL	TK	772	6	Tokelau	Tokélaou	Tokelau
Tonga	TON	TO	776	6	Tonga	Tonga	Tonga
Trinidad Tob	TTO	TT	780	2	Trinidad and Tobago	Trinité-et-Tobago	Trinidad y Tabago
Tunisia	TUN	TN	788	1	Tunisia	Tunisie	Túnez
Turkey	TUR	TR	792	4	Turkey	Turquie	Turquía
Turkmenistan	TKM	TM	795	4	Turkmenistan	Turkménistan	Turkmenistán
Turks Caicos	TCA	TC	796	2	Turks and Caicos Is.	Iles Turques et Caïques	Islas Turcas y Caicos
Tuvalu	TUV	TV	798	6	Tuvalu	Tuvalu	Tuvalu
Uganda	UGA	UG	800	1	Uganda	Ouganda	Uganda
Ukraine	UKR	UA	804	5	Ukraine	Ukraine	Ucrania
Untd Arab Em	ARE	AE	784	4	United Arab Emirates	Emirats arabes unis	Emiratos Arabes Unidos
UK	GBR	GB	826	5	United Kingdom	Royaume-Uni	Reino Unido
USA	USA	US	840	2	United States of America	Etats-Unis d'Amérique	EstadosUnidos de América
US Minor Is	UMI	UM	581	6	US Minor Outlying Is.	Iles Mineures EloignésEU	Is Menores PeriféricasEU
US Virgin Is	VIR	VI	850	2	US Virgin Islands	Iles Vierges américaines	Islas Vírgenes de los EU
Uruguay	URY	UY	858	3	Uruguay	Uruguay	Uruguay
Uzbekistan	UZB	UZ	860	4	Uzbekistan	Ouzbékistan	Uzbekistán
Vanuatu	VUT	VU	548	6	Vanuatu	Vanuatu	Vanuatu
Venezuela	VEN	VE	862	3	Venezuela	Venezuela	Venezuela
Viet Nam	VNM	VN	704	4	Viet Nam	Viet Nam	Viet Nam
Wallis Fut I	WLF	WF	876	6	Wallis and Futuna Is.	Iles Wallis-et-Futuna	Islas Wallis y Futuna
Westn Sahara	ESH	EH	732	1	Western Sahara	Sahara occidental	Sahara Occidental
Yemen	YEM	YE	887	4	Yemen	Yémen	Yemen
Zambia	ZMB	ZM	894	1	Zambia	Zambie	Zambia

List of countries or areas | Liste des pays ou zones | Lista de países o áreas

A	B	C	D	E	F	G	H
Zimbabwe	ZWE	ZW	716	1	Zimbabwe	Zimbabwe	Zimbabwe
Other nei	...	..	896	9	Other nei	Autres nca	Otros nep

Column A = FAO multilingual country or area code (maximum 12 characters) used for statistical purposes.

Column B = Alpha-3 country or area code: ISO (International Organization for Standardization).

Column C = Alpha-2 country or area code: ISO (International Organization for Standardization).

Column D = Three-digit UN numerical country or area code as published in the "STANDARD COUNTRY OR AREA CODES FOR STATISTICAL USE" Statistical Papers, Series M., No. 49, Rev. 4; United Nations, New York, 2003.

Column E = Continents:
- 1 = Africa
- 2 = America, North
- 3 = America, South
- 4 = Asia
- 5 = Europe
- 6 = Oceania
- 9 = Other nei

Column F = Country or area names in English (maximum 24 characters).

Column G = Country or area names in French (maximum 24 characters).

Column H = Country or area names in Spanish (maximum 24 characters).

Colonne A = Noms de pays ou zone en code multilingue FAO (n'excédant pas 12 lettres) utilisés à des fins statistiques.

Colonne B = Code Alpha-3 des pays ou zones: ISO (Organisation internationale de normalisation).

Colonne C = Code Alpha-2 des pays ou zones: ISO (Organisation internationale de normalisation).

Colonne D = Code numérique à trois chiffres pour les pays ou zones des Nations Unies, tel que publié dans "STANDARD COUNTRY OR AREA CODES FOR STATISTICAL USE" (Code type pour les pays ou les zones des Nations Unies à des fins statistiques), Statistical Papers, Series M., No. 49, Rev. 4; United Nations, New York, 2003.

Colonne E = Continents:
- 1 = Afrique
- 2 = Amérique du Nord
- 3 = Amérique du Sud
- 4 = Asie
- 5 = Europe
- 6 = Océanie
- 9 = Autres nca

Colonne F = Noms de pays ou zone en anglais (n'excédant pas 24 espaces typographiques).

Colonne G = Noms des pays ou zone en français (n'excédant pas 24 espaces typographiques).

Colonne H = Noms de pays ou zone en espagnol (n'excédant pas 24 espaces typographiques).

Columna A = Nombres multilingües de los países o áreas utilizados por la FAO para fines estadísticos (máximo, 12 espacios).

Columna B = Código Alfa-3 de los países o áreas: ISO (Organización Internacional de Normalización).

Columna C = Código Alfa-2 de los países o áreas: ISO (Organización Internacional de Normalización).

Columna D = Código numérico de tres dígitos atribuido al país o área en cuestión por las Naciones Unidas, tal como aparece en la publicación "STANDARD COUNTRY OR AREA CODES FOR STATISTICAL USE" (Código estándar por los países o áreas de las Naciones Unidas para fines estadísticos), Statistical Papers, Series M., No. 49. Rev. 4; United Nations, New York, 2003.

Columna E = Continentes:
- 1 = África
- 2 = América del Norte
- 3 = América del Sur
- 4 = Asia
- 5 = Europa
- 6 = Oceanía
- 9 = Otros nep

Columna F = Nombres de los países o áreas en inglés (máximo, 24 espacios).

Columna G = Nombres de los países o áreas en francés (máximo, 24 espacios).

Columna H = Nombres de los países o áreas en español (máximo, 24 espacios).

Notes on countries or areas

1. The designations employed and the presentation of material in this publication do not imply the expression of any opinion whatsoever on the part of the Food and Agriculture Organization of the United Nations concerning the legal or development status of any country, territory, city or area or of its authorities, or concerning the delimitation of its frontiers or boundaries.

2. The term *country or area* as used in the stubs and the column headings of the tables also covers territories, cities, land areas, as well as provinces, districts, enclaves, exclaves and other parts of territories or combinations of countries or areas such as economic or customs unions.

3. In all tables a country or area entry is designated by a multilingual code of not more than 12 characters. This code is keyed in the LIST OF COUNTRIES OR AREAS to the other descriptors for each country or area entry - the names in English, French and Spanish, the ISO's 3-alpha and 2-alpha codes, the UN's three-digit numerical code, etc.

4. Country or area names and designations are subject to nationally announced changes. Name changes announced after 31 December 2002 have not necessarily been incorporated in this volume but will be reflected in future ones.

5. The flag of the vessel performing the essential part of the operation catching the fish should be considered the paramount indication of the nationality assigned to the catch data.

   The flag State of the vessel performing the essential part of the fishing operation should be responsible for the provision of catch and landing data.

   Where a foreign flag vessel is fishing in the waters under the national jurisdiction of another State, the flag State of the vessel should have at all times the responsibility to provide relevant catch and landing data. The only exceptions to this shall be:

   (a) where the vessel undertakes fishing under a charter agreement or arrangement to augment the local fishing fleet, and the vessel has become for all practical purposes a local fishing vessel of the host country;

   (b) where the vessel undertakes fishing pursuant to a joint venture or similar arrangement in waters under the national jurisdiction of another State and the vessel is operating for all practical purposes as a local vessel, or its operation has become, or is intended to become, an integral part of the economy of the host country.

   In any situation where there is uncertainty as to the application of these criteria, any agreement, charter, joint venture or other similar arrangement should contain a provision setting out clearly the responsibility for reporting catch and landing data, which should be reported to the flag State and, where relevant, to any coastal State in whose waters fishing operations are to take place or competent subregional, regional or global fisheries organization or arrangement.

6. National data cover all quantities caught by fishing craft flying the flag of the reporting country and landed not only in the domestic harbours of the reporting country but also in foreign harbours. National catch excludes quantities caught by foreign fishing craft and landed in domestic ports.

Notes sur les pays ou zones

1. Les appellations employées dans cette publication et la présentation des données qui y figurent n'impliquent de la part de l'Organisation des Nations Unies pour l'alimentation et l'agriculture aucune prise de position quant au statut juridique ou au stade de développement des pays, territoires, villes ou zones, ou de leurs autorités, ni quant au tracé de leurs frontières ou limites.

2. Le terme *pays ou zone* employé dans les talons et les têtes de colonne des tableaux, doit s'entendre également des territoires, villes, zones terrestres, ainsi que des provinces, districts, enclaves, parties détachées d'un Etat, autres parties de territoires ou regroupements de pays ou zones, tels qu'unions économiques ou douanières.

3. Dans tous les tableaux, les entrées des pays ou zones sont désignées par un code multilingue n'excédant pas douze espaces typographiques. Dans la LISTE DES PAYS OU ZONES ce code est associé aux autres descripteurs, c'est-à-dire aux noms en anglais, français et espagnol, aux codes ISO alpha-3 et alpha-2, au code numérique à trois chiffres des Nations Unies, etc.

4. La liste des appellations des pays ou zones est sujette de temps à autre à des modifications annoncées par les pays ou zones. Les modifications de noms survenues après le 31 décembre 2002 n'ont pas toujours été incorporées dans le présent volume, mais il en sera tenu compte dans les futurs volumes.

5. Le pavillon du navire effectuant la partie essentielle de l'opération de pêche devrait être considéré comme le facteur déterminant du pays ou zone auquel sont attribuées les données des captures.

   L'État du pavillon du navire effectuant la partie essentielle de l'opération de pêche devrait fournir les données sur les captures et les débarquements.

   Lorsqu'un navire battant pavillon étranger pêche dans des eaux placées sous la juridiction nationale d'un autre Etat, l'Etat du pavillon du navire devrait fournir à tout moment les données pertinentes sur les captures et les débarquements. Les seules exceptions à ce principe sont les suivantes:

   (a) lorsque le navire pêche au titre d'un contrat ou d'un arrangement d'affrètement visant à compléter la flottille de pêche locale et que le navire est devenu à toutes fins utiles un navire de pêche local du pays d'accueil;

   (b) lorsque le navire pêche dans le cadre d'une opération conjointe ou d'arrangements similaires dans des eaux placées sous la juridiction nationale d'un autre État et que le navire opère à toutes fins utiles en tant que navire local ou que ses activités font, ou sont appelées à faire, partie intégrante de l'économie du pays d'accueil.

   Dans toute situation où il existe une incertitude quant à l'application de ces critères, l'accord, le contrat, l'opération conjointe ou tout autre arrangement similaire devrait contenir une disposition fixant clairement les responsabilités en matière de notification des données sur les captures et les débarquements. Celles-ci devraient être notifiées à l'État du pavillon et, le cas échéant, à tout État côtier dans les eaux duquel les opérations de pêche ont lieu ou à tout organisme ou accord de pêche sous-régional, régional ou mondial compétent.

6. Sont comprises dans les données nationales toutes les quantités capturées par les bateaux de pêche battant le pavillon du pays déclarant et débarquées non seulement dans les ports du pays déclarant, mais aussi dans des ports étrangers. Ne sont pas comprises dans les captures nationales, les quantités capturées par des bateaux étrangers et débarquées dans des ports nationaux.

Notas sobre los países o áreas

1. Las denominaciones empleadas en esta publicación y la forma en que aparecen presentados los datos que contiene no implican, da parte de la Organización de las Naciones Unidas para la Agricultura y la Alimentación, juicio alguno sobre la condición jurídica o nivel de desarrollo de países, territorios, ciudades o zonas, o de sus autoridades, ni respecto de la delimitación de sus fronteras o límites.

2. En los encabezamientos de las columnas o renglones de los cuadros, el término *país o área* puede referirse también a territorios, ciudades, zonas terrestres, provincias, distritos, enclaves, exclaves, otras partes de territorio o grupos de países o áreas, como uniones económicas o aduaneras.

3. En todos los cuadros, los países o áreas se indican con un código multilingue que no utiliza más de doce espacios. En la LISTA DE PAÍSES O ÁREAS dicho código aparece acompañado de otros descriptores - los nombres en inglés, francés y español, los códigos ISO alfa-3 y alfa-2, y el código numérico de tres dígitos de las Naciones Unidas, etc.

4. Los países o áreas anuncian en algunas ocasiones cambios de sus nombres y denominaciones. En el presente volumen no se han incorporado todos los cambios de nombres anunciados después del 31 de diciembre de 2002, pero se tendrán presentes en volúmenes futuros.

5. El pabellón de la embarcación que efectúa la majoría de las operaciones de pesca debería considerarse como indicación decisiva para establecer a qué país o área hay que asignar los datos de captura.

   El Estado del pabellón del barco que efectúa la parte principal de la operación pesquera debería ser el responsable de facilitar datos sobre capturas y desembarques.

   Cuando un barco de pabellón extranjero faene en aguas sometidas a la jurisdicción nacional de otro Estado, el Estado del pabellón de ese barco debería tener en todo momento la responsabilidad de facilitar datos pertinentes sobre capturas y desembarques. Las únicas excepciones a esto serán:

   (a) cuando el barco faene con arreglo a un acuerdo de fletamento o como medida para aumentar la flota pesquera local, y el barco se ha convertido a todos los efectos prácticos en un barco pesquero local del país hospedante;

   (b) cuando el barco emprenda las operaciones pesqueras en virtud de un acuerdo de empresa mixta o análogo en aguas sometidas a la jurisdicción nacional de otro Estado y faene a todos los efectos prácticos como un barco local, o sus operaciones se hayan convertido, o se tenga intención de que se conviertan, en parte integrante de la economía del país hospedante.

   En cualquier situación en que haya incertidumbre respecto de la aplicación de estos criterios, cualquier acuerdo, fletamento, empresa mixta u otro acuerdo similar debería contener una disposición por la que se establezca claramente la responsabilidad de la presentación de datos sobre capturas y desembarques que deberían notificarse al Estado del pabellón y, según proceda, a cualquier Estado ribereño en cuyas aguas se vayan a realizar operaciones pesqueras o a cualquier organización o acuerdo pesquero subregional, regional o mundial competente.

6. Los datos nacionales incluyen todas las cantidades capturadas por embarcaciones pesqueras que enarbolan el pabellón del país que comunica los datos y desembarcadas en los puertos del país en cuestión o en puertos extranjeros. Las capturas nacionales no incluyen las cantidades capturadas por embarcaciones pesqueras extranjeras y desembarcadas en puertos del país al que se refieren los datos.

Notes on countries or areas

7. Final data have been provided by many national offices; others submitted provisional figures only. Whenever national offices failed to report their annual catch statistics in time for publication, FAO has estimated the quantities to reflect more realistic catch data (F designated figures).

8. The NOTES ON INDIVIDUAL COUNTRIES OR AREAS list the few national exceptions to the standards described above and also to those standards specified in other NOTES in this Yearbook.

9. Users are reminded that, although data are shown to the nearest tonne, they are not necessarily of this degree of accuracy.

10. Countries included in the low-income food-deficit countries (LIFDCs) grouping are those classified (i) by the World Bank as low-income in terms of GNP per capita, and (ii) by FAO as having a trade deficit for food in terms of calorie content. Countries which have formally objected to being included in the grouping are not included.

Notes sur les pays ou zones

De nombreux services nationaux ont communiqué des données définitives; d'autres n'ont fourni que des chiffres préliminaires. Chaque fois que les services nationaux n'ont pu déclarer leurs statistiques de captures annuelles en temps utile pour la publication, la FAO procédé à des estimations pour arriver à des chiffres de captures plus proches de la réalité (chiffres désignés par F).

Les quelques exceptions nationales à ces normes ainsi qu'aux normes indiquées dans les autres NOTES figurant dans le présent Annuaire sont indiquées dans les NOTES SUR DIVERS PAYS OU ZONES.

Il est rappelé aux utilisateurs que les chiffres sont arrondis à la tonne la plus proche, mais qu'ils n'ont pas nécessairement un tel degré de précision.

Les pays inclus dans le groupe pays à faible revenu et à déficit vivrier (PFRDV) sont ceux classifiés (I) par la Banque mondiale comme ayant un revenu par habitant bas en termes de PIB, et (ii) par la FAO comme ayant un déficit commercial alimentaire en termes de contenu en calories. Les pays qui se sont formellement opposés à leur inclusion dans ce groupe n'y figurent pas.

Notas sobre los países o áreas

Muchas oficinas nacionales han facilitado datos difinitivos; otras, en cambio, sólo han presentado cifras provisionales. Cuando las oficinas nacionales no han comunicado a tiempo sus estadísticas nacionales de captura, la FAO ha estimado las cantidades para que reflejaran más objetivamente los datos de captura (cifras designadas F).

En las NOTAS SOBRE LOS DISTINTOS PAÍSES O ÁREAS pueden verse algunas excepciones nacionales a estas normas; véanse también las que aparecen en otras NOTAS de este Anuario.

Se recuerda a los usarios que, si bien los datos están redondeados a la tonelada más próxima, no tienen necesariamente ese grado de precisión.

Los países incluidos en el grupo países de bajos ingresos y con déficit de alimentos (PBIDA) son aquellos clasificados (i) por el Banco Mundial como teniendo una renta per cápita baja en términos de PNL, y (ii) por la FAO como teniendo un déficit comercial alimentario en términos de contenido en calorías. Los países que se han formalmente opuesto a su inclusión en este grupo no están incluidos.

A - Summaries

A - Résumés

A - Resúmenes

A-1 (a)

Fish, crustaceans, molluscs, etc — **World capture production**
Poissons, crustacés, mollusques, etc — **Captures mondiales**
Peces, crustáceos, moluscos, etc — **Capturas mundiales**

Division, fishing area, continent **Division, zone de pêche, continent** **División, área de pesca, continente**	**1996** **t**	**1997** **t**	**1998** **t**	**1999** **t**	**2000** **t**	**2001** **t**	**2002** **t**
World total **Total mondial** **Total mundial**	**93 846 643**	**94 298 586**	**87 672 034**	**93 774 052**	**95 502 004**	**92 862 087**	**93 190 654**
Capture production in inland waters ***Captures dans les eaux continentales*** ***Capturas en aguas continentales***	***7 433 689***	***7 575 487***	***8 067 057***	***8 527 706***	***8 730 632***	***8 698 092***	***8 738 167***
By ISSCAAP divisions	*Par divisions de la CSITAPA*				*Por divisiones de la CEIUAPA*		
1 Freshwater fishes Poissons d'eau douce Peces de agua dulce	5 882 020	5 964 376	6 232 049	6 814 558	6 803 720	6 904 600	6 750 536
2 Diadromous fishes Poissons diadromes Peces diádromos	476 171	452 464	511 363	564 821	644 906	406 592	348 418
3 Marine fishes Poissons marins Peces marinos	60 122	69 232	65 794	66 207	76 773	85 893	76 419
4 Crustaceans Crustacés Crustáceos	448 022	574 190	671 700	524 803	604 832	667 695	924 269
5 Molluscs Mollusques Moluscos	562 514	508 781	581 126	552 829	595 755	629 035	634 312
6 Whales, seals and other aquatic mammals Baleines, phoques et autres mammifères aquatiques Ballenas, focas y otros mamíferos acuáticos	*	*	*	*	*	*	*
7 Miscellaneous aquatic animals Animaux aquatiques divers Animales acuáticos diversos	4 840	6 444	5 025	4 488	4 646	4 277	4 213
8 Miscellaneous aquatic animal products Produits divers d'animaux aquatiques Diversos productos de animales acuáticos	*	*	*	*	*	*	*
9 Aquatic plants Plantes aquatiques Plantas acuáticas	*	*	*	*	*	*	*
By inland fishing areas/ continents	*Par zones de pêche continentales/continents*				*Por áreas de pesca continentales/continentes*		
01 Africa - Inland waters Afrique - Eaux continentales Africa - Aguas continentales	1 856 887	1 927 214	2 003 833	2 019 268	2 052 981	2 051 183	2 092 924
02 America, North - Inland waters Amérique du Nord - Eaux continentales América del Norte - Aguas continentales	214 628	210 597	195 771	186 608	189 254	174 959	170 614
03 America, South - Inland waters Amérique du Sud - Eaux continentales América del Sur - Aguas continentales	344 197	332 798	334 638	347 938	345 276	368 803	377 313
04 Asia - Inland waters Asie - Eaux continentales Asia - Aguas continentales	4 597 426	4 702 743	5 095 470	5 502 187	5 690 272	5 734 686	5 722 141
05 Europe - Inland waters Europe - Eaux continentales Europa - Aguas continentales	402 228	382 615	416 770	450 679	431 835	347 242	354 270
06 Oceania - Inland waters Océanie - Eaux continentales Oceanía - Aguas continentales	18 323	19 520	20 575	21 026	21 014	21 219	20 905
Capture production in marine fishing areas ***Captures dans les zones de pêche maritimes*** ***Capturas en áreas de pesca marítimas***	***86 412 954***	***86 723 099***	***79 604 977***	***85 246 346***	***86 771 372***	***84 163 995***	***84 452 487***
By ISSCAAP divisions	*Par divisions de la CSITAPA*				*Por divisiones de la CEIUAPA*		
1 Freshwater fishes Poissons d'eau douce Peces de agua dulce	34 119	33 825	32 222	32 793	31 957	33 275	28 132
2 Diadromous fishes Poissons diadromes Peces diádromos	1 277 100	1 192 904	1 175 234	1 223 469	1 119 089	1 236 433	1 141 310
3 Marine fishes Poissons marins Peces marinos	73 448 603	72 669 615	65 802 684	70 448 384	71 789 721	69 603 377	70 177 288
4 Crustaceans Crustacés Crustáceos	4 982 219	5 202 875	5 594 187	5 757 380	5 910 698	5 801 518	5 781 432
5 Molluscs Mollusques Moluscos	6 173 240	6 963 416	6 357 448	7 114 923	7 261 396	6 931 864	6 793 067
6 Whales, seals and other aquatic mammals Baleines, phoques et autres mammifères aquatiques Ballenas, focas y otros mamíferos acuáticos	*	*	*	*	*	*	*
7 Miscellaneous aquatic animals Animaux aquatiques divers Animales acuáticos diversos	497 673	660 464	643 202	669 397	658 511	557 528	531 258
8 Miscellaneous aquatic animal products Produits divers d'animaux aquatiques Diversos productos de animales acuáticos	*	*	*	*	*	*	*
9 Aquatic plants Plantes aquatiques Plantas acuáticas	*	*	*	*	*	*	*

A-1 (a)

Fish, crustaceans, molluscs, etc — World capture production
Poissons, crustacés, mollusques, etc — Captures mondiales
Peces, crustáceos, moluscos, etc — Capturas mundiales

Division, fishing area, continent / Division, zone de pêche, continent / División, área de pesca, continente	1996 t	1997 t	1998 t	1999 t	2000 t	2001 t	2002 t
By marine fishing areas / Par zones de pêche maritimes / Por áreas de pesca marítimas							
21 Atlantic, Northwest / Atlantique, nord-ouest / Atlántico, noroeste	2 069 186	2 053 310	1 965 230	2 035 542	2 082 554	2 240 365	2 245 008
27 Atlantic, Northeast / Atlantique, nord-est / Atlántico, nordeste	11 066 416	11 766 088	10 986 318	10 516 374	11 014 721	11 143 204	11 048 962
31 Atlantic, Western Central / Atlantique, centre-ouest / Atlántico, centro-occidental	1 720 699	1 821 208	1 782 450	1 791 133	1 802 775	1 686 404	1 764 352
34 Atlantic, Eastern Central / Atlantique, centre-est / Atlántico, centro-oriental	3 572 444	3 635 172	3 803 544	3 700 408	3 649 655	3 929 630	3 373 623
37 Mediterranean and Black Sea / Méditerranée et mer Noire / Mediterráneo y Mar Negro	1 531 975	1 444 025	1 401 557	1 533 687	1 505 217	1 570 335	1 550 099
41 Atlantic, Southwest / Atlantique, sud-ouest / Atlántico, sudoccidental	2 479 862	2 756 890	2 379 280	2 622 527	2 369 739	2 287 502	2 089 660
47 Atlantic, Southeast / Atlantique, sud-est / Atlántico, sudoriental	1 325 437	1 355 366	1 537 315	1 528 601	1 633 902	1 648 084	1 701 440
48 Atlantic, Antarctic / Atlantique, Antarctique / Atlántico, Antártico	95 088	79 600	93 268	105 914	123 562	109 257	134 595
51 Indian Ocean, Western / Océan Indien, ouest / Océano Indico, occidental	3 897 309	3 983 306	3 797 772	4 067 446	3 995 996	3 981 292	4 243 330
57 Indian Ocean, Eastern / Océan Indien, est / Océano Indico, oriental	4 190 529	4 459 036	4 700 229	4 647 945	4 785 689	4 877 380	5 100 261
58 Indian Ocean, Antarctic / Océan Indien, Antarctique / Océano Indico, Antártico	5 689	8 867	10 140	9 775	12 017	10 169	8 004
61 Pacific, Northwest / Pacifique, nord-ouest / Pacífico, noroeste	23 542 610	24 686 259	24 815 317	24 130 564	23 204 042	22 550 874	21 436 229
67 Pacific, Northeast / Pacifique, nord-est / Pacífico, nordeste	2 833 342	2 766 092	2 742 459	2 551 689	2 477 803	2 759 090	2 702 885
71 Pacific, Western Central / Pacifique, centre-ouest / Pacífico, centro-occidental	8 730 620	8 966 502	9 285 155	9 598 268	9 874 399	10 103 215	10 510 202
77 Pacific, Eastern Central / Pacifique, centre-est / Pacífico, centro-oriental	1 619 642	1 704 354	1 407 976	1 446 226	1 719 304	1 860 373	2 037 267
81 Pacific, Southwest / Pacifique, sud-ouest / Pacífico, sudoccidental	663 750	839 819	862 699	783 595	714 513	752 661	739 868
87 Pacific, Southeast / Pacifique, sud-est / Pacífico, sudoriental	17 068 356	14 397 205	8 034 212	14 176 309	15 804 615	12 653 427	13 765 143
88 Pacific, Antarctic / Pacifique, Antarctique / Pacífico, Antártico	-	0	56	343	869	733	1 559
By continents / Par continents / Por continentes							
1 Africa / Afrique / Africa	3 743 156	4 042 588	4 144 658	4 351 125	4 575 163	4 903 000	4 706 303
2 America, North / Amérique du Nord / América del Norte	7 724 879	7 830 230	7 357 332	7 437 466	7 607 570	7 967 884	8 072 096
3 America, South / Amérique du Sud / América del Sur	19 445 089	16 898 107	10 368 256	16 186 887	17 647 134	14 560 804	15 644 088
4 Asia / Asie / Asia	37 404 978	39 202 032	39 709 194	40 275 778	39 931 639	39 760 539	39 773 158
5 Europe / Europe / Europa	17 087 777	17 529 588	16 693 187	15 637 580	15 738 457	15 634 136	14 809 361
6 Oceania / Océanie / Oceanía	832 952	1 066 841	1 149 366	1 143 567	1 044 548	1 086 480	1 109 887
9 Other nei / Autres nca / Otros nep	174 123	153 713	182 984	213 943	226 861	251 152	337 594

* See paragraph 5 of the INTRODUCTION. * Voir le paragraphe 5 de l'INTRODUCTION. * Véase el párrafo 5 en la INTRODUCCIÓN.

A-1 (b)

Fish, crustaceans, molluscs, etc — **Capture production by groupings of major fishing areas**
Poissons, crustacés, mollusques, etc — **Captures par groupes de principales zones de pêche**
Peces, crustáceos, moluscos, etc — **Capturas por grupos de áreas principales de pesca**

Group of major fishing areas Groupe de principales zones de pêche Grupo de áreas principales de pesca		1996 t	1997 t	1998 t	1999 t	2000 t	2001 t	2002 t
World total **Total mondial** **Total mundial**		**93 846 643**	**94 298 586**	**87 672 034**	**93 774 052**	**95 502 004**	**92 862 087**	**93 190 654**
Inland waters ***Eaux continentales*** ***Aguas continentales***		***7 433 689***	***7 575 487***	***8 067 057***	***8 527 706***	***8 730 632***	***8 698 092***	***8 738 167***
Marine areas ***Zones maritimes*** ***Areas marítimas***		***86 412 954***	***86 723 099***	***79 604 977***	***85 246 346***	***86 771 372***	***84 163 995***	***84 452 487***
Atlantic Ocean Ocean Atlantique Oceano Atlantico	18+21+27+31+34+37+41+47	23 766 019	24 832 059	23 855 694	23 728 272	24 058 563	24 505 524	23 773 144
Indian Ocean Ocean Indien Ocean Indico	51+57	8 087 838	8 442 342	8 498 001	8 715 391	8 781 685	8 858 672	9 343 591
Pacific Ocean Ocean Pacifique Oceano Pacifico	61+67+71+77+81+87	54 458 320	53 360 231	47 147 818	52 686 651	53 794 676	50 679 640	51 191 594
Southern Ocean Océan Austral Océano Austral	48+58+88	100 777	88 467	103 464	116 032	136 448	120 159	144 158
Atlantic, Northern Atlantique du Nord Atlantico del Norte	18+21+27	13 135 602	13 819 398	12 951 548	12 551 916	13 097 275	13 383 569	13 293 970
Atlantic, Central Atlantique central Atlantico central	31+34+37	6 825 118	6 900 405	6 987 551	7 025 228	6 957 647	7 186 369	6 688 074
Atlantic, Southern Atlantique du Sud Atlantico del Sur	41+47+48	3 900 387	4 191 856	4 009 863	4 257 042	4 127 203	4 044 843	3 925 695
Indian Ocean Ocean Indien Oceano Indico	51+57+58	8 093 527	8 451 209	8 508 141	8 725 166	8 793 702	8 868 841	9 351 595
Pacific, Northern Pacifique du Nord Pacifico del Norte	61+67	26 375 952	27 452 351	27 557 776	26 682 253	25 681 845	25 309 964	24 139 114
Pacific, Central Pacifique central Pacifico central	71+77	10 350 262	10 670 856	10 693 131	11 044 494	11 593 703	11 963 588	12 547 469
Pacific, Southern Pacifique du Sud Pacifico del Sur	81+87+88	17 732 106	15 237 024	8 896 967	14 960 247	16 519 997	13 406 821	14 506 570
Northern regions Regions septentrionales Regiones del Norte	18+21+27+61+67	39 511 554	41 271 749	40 509 324	39 234 169	38 779 120	38 693 533	37 433 084
Central regions Regions centrales Regiones centrales	31+34+37+51+57+71+77	25 263 218	26 013 603	26 178 683	26 785 113	27 333 035	28 008 629	28 579 134
Southern regions Regions meridionales Regiones del Sur	41+47+48+58+81+87+88	21 638 182	19 437 747	12 916 970	19 227 064	20 659 217	17 461 833	18 440 269

A-1 (c)

Fish, crustaceans, molluscs, etc — Capture production by principal producers in 2002
Poissons, crustacés, mollusques, etc — Captures par principaux producteurs en 2002
Peces, crustáceos, moluscos, etc — Capturas por productores principales en 2002

Country or area / Pays ou zone / País o área	1993 t	1994 t	1995 t	1996 t	1997 t	1998 t	1999 t	2000 t	2001 t	2002 t
China (1)	9 351 437	10 866 836	12 562 706	14 182 107	15 722 344	17 229 927	17 240 032	16 987 325	16 529 389	16 553 144
Peru	9 004 777	11 999 217	8 937 342	9 515 048	7 869 871	4 338 437	8 428 601	10 658 620	7 986 103	8 766 991
USA	5 522 917	5 535 324	5 224 858	5 001 191	4 983 440	4 708 980	4 749 646	4 717 638	4 944 336	4 937 305
Indonesia	3 089 787	3 329 174	3 532 653	3 604 795	3 857 660	3 961 004	4 044 543	4 120 126	4 273 662	4 505 474
Japan	7 252 278	6 615 893	5 969 517	5 931 872	5 928 334	5 303 555	5 188 948	4 984 813	4 713 006	4 443 000
Chile	5 949 565	7 720 578	7 434 180	6 690 665	5 811 550	3 265 293	5 050 180	4 299 942	3 797 140	4 271 475
India	3 062 712	3 257 607	3 265 240	3 447 954	3 523 448	3 373 492	3 472 150	3 666 428	3 777 092	3 770 912
Russian Fed	4 368 067	3 703 972	4 311 809	4 675 738	4 661 853	4 454 759	4 141 158	3 973 535	3 628 459	3 232 295
Thailand	2 927 689	3 015 196	3 031 074	3 013 961	2 902 898	2 930 354	2 952 308	2 997 394	2 932 374	2 921 216
Norway	2 415 131	2 366 119	2 524 111	2 648 457	2 863 059	2 861 223	2 627 534	2 699 421	2 687 303	2 743 184
Iceland	1 715 581	1 556 962	1 612 548	2 060 168	2 205 944	1 681 951	1 736 267	1 982 524	1 980 715	2 129 655
Philippines	1 834 323	1 845 335	1 860 701	1 783 601	1 805 806	1 833 380	1 872 818	1 896 638	1 949 026	2 030 542
Korea Rep	2 257 192	2 357 891	2 319 917	2 413 785	2 204 202	2 027 932	2 118 521	1 824 995	1 990 722	1 668 979
Viet Nam	932 143	1 025 909	1 084 939	1 223 644	1 276 325	1 293 954	1 386 300	1 450 590	1 490 303	1 508 000 F
Mexico	1 102 932	1 191 646	1 329 340	1 464 188	1 489 112	1 179 860	1 205 603	1 315 581	1 398 592	1 450 654
Denmark	1 614 289	1 873 316	1 999 033	1 681 517	1 826 852	1 557 335	1 405 005	1 534 089	1 510 461	1 442 068
Myanmar	739 702	746 241	751 232	601 788	780 295	830 117	919 410	1 069 726	1 166 868	1 312 642
Malaysia	1 049 321	1 067 650	1 112 375	1 130 372	1 172 922	1 153 719	1 251 768	1 289 245	1 234 733	1 275 555
Bangladesh	764 824	770 790	792 389	814 787	829 426	839 141	959 215	1 004 264	1 068 417	1 103 855
China,Taiwan	1 133 676	967 189	1 010 201	967 483	1 038 048	1 091 768	1 099 715	1 093 889	1 005 199	1 042 756
Canada	1 140 499	1 026 900	853 682	909 511	975 548	1 013 929	1 027 855	1 012 099	1 052 543	1 013 997
Argentina	931 303	951 387	1 170 126	1 291 355	1 400 162	1 164 829	1 078 313	908 979	931 734	944 346
Morocco	624 511	754 327	848 951	642 886	791 906	710 436	745 431	896 620	1 083 953	894 957
Spain	1 085 959 F	1 094 524 F	1 181 324 F	1 174 683	1 205 401	1 263 016	1 172 995	1 045 616	1 092 525	882 633
Brazil	717 090 F	740 100 F	706 708	715 482	744 585	706 789	703 941	766 846	806 672	822 159
South Africa	562 374	523 449	575 547	441 370	514 238	559 679	587 933	643 526	750 099	766 284
UK	860 229	878 018	909 963	867 686	890 936	923 219	840 719	747 571	741 045	689 919
Namibia	790 332	649 199	569 833	517 828	513 216	607 831	580 084	589 904	547 498	624 891
France	623 409	622 568	610 841	564 694	574 076	549 545	593 479	634 819	605 629	620 078
Pakistan	608 344	537 277	525 849	537 489	589 795	596 980	654 530	614 069	600 798	599 104
Turkey	548 758	589 803	633 970	527 826	459 153	487 200	573 824	503 348	527 730	566 682
New Zealand	426 668	449 660	556 635	425 564	612 424	640 236	599 466	549 318	561 940	559 289
Venezuela	395 861	438 218	500 817	496 115	470 126	503 494	399 809	356 835	418 207	515 384
Faeroe Is	246 294	237 708	287 796	304 587	329 145	364 497	358 362	454 399	524 837	490 000 F
Nigeria	238 409	267 059	349 482	337 993	387 923	463 024	455 628	441 377	452 146	481 056
Netherlands	461 756	420 053	438 092	410 798	451 799	536 626	514 611	495 754	518 162	464 035
Egypt	302 829	311 600	335 300	340 434	371 332	406 204	422 665	384 314	428 651	425 170
Cambodia	101 023	95 018	103 000	94 710	102 800	107 900	269 156	284 368	428 200	406 182
Senegal	382 212	352 421	354 617	411 759	457 366	403 872	412 125	402 047	403 202	375 824
Ghana	372 619	335 437	352 844	476 733	447 088	442 641	492 776	452 070	447 681	371 178
Other nei	138 834	134 742	297 280	174 123	153 713	182 984	213 943	226 861	251 152	337 594
Iran	322 006	308 101	341 383	351 725	342 287	367 212	387 200	384 000	336 450	324 853
Tanzania	331 267	288 399	359 800	323 921	356 960	348 000	310 509	332 779	335 900	323 530
Ecuador	286 633	330 248	505 395	702 974	548 988	310 022	497 872	592 547	586 570	318 540
Panama	184 821	185 147	203 490	150 027	162 223	202 598	120 872	219 166	256 408 F	305 081
Sri Lanka	215 400	218 400	229 171	228 945	235 099	263 330	276 080	293 410	279 640	298 260
Sweden	341 897	386 814	404 572	370 881	357 406	410 886	351 254	338 534	311 816	294 963
Ireland	278 774	294 154	388 303	332 233	292 809	325 020	283 358	276 292	356 430	282 331
Italy	397 533	398 730	396 791	365 899	343 693	306 096	282 790	302 149	310 397	269 846
Ukraine	299 404	272 259	387 716	407 305	368 800	462 196	408 711	391 831	360 914	265 599
Angola	126 200	132 413	122 781	137 815	146 304	163 149	175 799	238 351	252 518	260 797
Germany	253 010	230 161	239 890	236 411	259 352	266 622	238 925	205 689	211 282	224 451
Poland	395 406	438 032	429 372	342 793	348 913	242 010	235 724	217 682	225 065	222 441
Uganda	219 814	213 129	208 789	195 088	218 026	220 628	226 097	219 356	220 726	221 898
Congo Dem R	196 789	155 897	158 627	163 010	162 211	178 041	208 448	209 300 F	214 600 F	220 000 F
Korea D P Rp	397 398 F	371 961	327 083	253 125	236 462	228 000 F	220 000 F	212 850 F	206 500	205 000 F
Greenland	116 650	117 417	128 890	116 018	120 596	128 542	160 253	159 711	158 485	202 828
Portugal	287 802	263 065	260 253	259 846	221 879	224 171	208 407	187 521	191 017	200 037
58 countries / 58 pays / 58 países	***82 298 460***	***87 826 640***	***87 921 128***	***89 454 763***	***89 918 129***	***83 197 635***	***89 139 666***	***90 758 691***	***88 053 022***	***88 400 919***
Other countries / Autres pays / Otros países	***4 254 390***	***4 255 412***	***4 459 033***	***4 391 880***	***4 380 457***	***4 474 399***	***4 634 386***	***4 743 313***	***4 809 065***	***4 789 735***
World total / Total mondial / Total mundial	**86 552 850**	**92 082 052**	**92 380 161**	**93 846 643**	**94 298 586**	**87 672 034**	**93 774 052**	**95 502 004**	**92 862 087**	**93 190 654**

These countries or areas are those with capture production of 200 000 tonnes or more in 2002.

(1) See note on China in NOTES ON INDIVIDUAL COUNTRIES OR AREAS.

Ces pays ou zones sont ceux dont les captures ont été de 200 000 tonnes ou plus en 2002.

(1) Voir la note sur la Chine dans les NOTES SUR DIVERS PAYS OU ZONES.

Estos países o áreas son referentes a los que totalizan unas capturas de 200 000 toneladas o más en 2002.

(1) Véase la nota sobre China en las NOTAS SOBRE LOS DISTINTOS PAÍSES O ÁREAS.

A-1 (d)

Fish, crustaceans, molluscs, etc — Capture production by groups of species
Poissons, crustacés, mollusques, etc — Captures par groupes d'espèces
Peces, crustáceos, moluscos, etc — Capturas por grupos de especies

Species group Groupe d'espèces Grupo de especies	1996 t	1997 t	1998 t	1999 t	2000 t	2001 t	2002 t
11 Carps, barbels and other cyprinids Carpes, barbeaux et autres cyprinidés Carpas, barbos y otros ciprínidos	663 465	608 078	605 024	613 623	570 965	548 894	592 962
12 Tilapias and other cichlids Tilapias et autres cichlidés Tilapias y otros cíclidos	560 516	561 796	597 782	637 413	686 066	688 101	682 639
13 Miscellaneous freshwater fishes Poissons d'eau douce divers Peces de agua dulce diversos	4 692 158	4 828 327	5 061 465	5 596 315	5 578 646	5 700 880	5 503 067
21 Sturgeons, paddlefishes Esturgeons, spatules Esturiones, sollos	4 713	4 441	3 777	2 849	2 603	2 269	1 859
22 River eels Anguilles Anguilas	17 119	14 724	12 299	12 489	16 437	12 856	14 038
23 Salmons, trouts, smelts Saumons, truites, éperlans Salmones, truchas, eperlanos	1 032 524	924 572	889 982	913 309	805 139	891 042	806 998
24 Shads Aloses Sábalos	635 738	636 789	701 166	783 722	856 488	661 200	585 303
25 Miscellaneous diadromous fishes Poissons diadromes divers Peces diádromos diversos	63 177	64 842	79 373	75 921	83 328	75 658	81 530
31 Flounders, halibuts, soles Flets, flétans, soles Platijas, halibuts, lenguados	945 886	1 030 607	935 111	956 111	1 008 471	948 651	918 840
32 Cods, hakes, haddocks Morues, merlus, églefins Bacalaos, merluzas, eglefinos	10 784 248	10 373 733	10 334 525	9 333 765	8 673 042	9 244 845	8 392 479
33 Miscellaneous coastal fishes Poissons côtiers divers Peces costeros diversos	5 938 912	6 452 786	6 151 293	6 060 207	6 039 972	6 199 635	6 171 124
34 Miscellaneous demersal fishes Poissons démersaux divers Peces demersales diversos	2 690 372	2 740 042	2 969 434	2 943 038	3 023 335	3 005 719	3 021 049
35 Herrings, sardines, anchovies Harengs, sardines, anchois Arenques, sardinas, anchoas	22 385 424	21 730 998	16 664 950	22 637 492	24 892 060	20 628 706	22 472 563
36 Tunas, bonitos, billfishes Thons, pélamides, marlins Atunes, bonitos, agujas	4 853 169	5 186 555	5 758 303	5 972 769	5 828 375	5 722 174	6 088 337
37 Miscellaneous pelagic fishes Poissons pélagiques divers Peces pelágicos diversos	14 332 473	13 789 709	11 214 197	10 684 205	10 645 490	12 249 788	11 639 160
38 Sharks, rays, chimaeras Squales, raies, chimères Tiburones, rayas, quimeras	815 533	825 970	817 903	833 397	857 849	819 327	818 542
39 Marine fishes not identified Poissons marins non identifiés Peces marinos no identificados	10 762 708	10 608 447	11 022 762	11 093 607	10 897 900	10 870 425	10 731 613
41 Freshwater crustaceans Crustacés d'eau douce Crustáceos de agua dulce	410 241	526 777	645 056	498 817	568 469	628 083	818 993
42 Crabs, sea-spiders Crabes, araignées de mer Cangrejos, centollas	999 674	991 140	1 075 725	1 059 348	1 091 717	1 091 958	1 134 205
43 Lobsters, spiny-rock lobsters Homards, langoustes Bogavantes, langostas	212 914	233 346	217 075	228 539	227 800	222 052	222 132
44 King crabs, squat-lobsters Crabes royaux, galatées Cangrejos reales, galateidos	75 694	70 686	79 224	77 641	68 011	46 326	41 910
45 Shrimps, prawns Crevettes Gambas, camarones	2 559 141	2 633 493	2 755 710	3 026 265	3 073 108	2 949 714	2 979 336

A-1 (d)

Fish, crustaceans, molluscs, etc — Capture production by groups of species
Poissons, crustacés, mollusques, etc — Captures par groupes d'espèces
Peces, crustáceos, moluscos, etc — Capturas por grupos de especies

Species group Groupe d'espèces Grupo de especies	1996 t	1997 t	1998 t	1999 t	2000 t	2001 t	2002 t
46 Krill, planktonic crustaceans Krill, crustacés planctoniques Krill, crustáceos planctónicos	91 156	75 653	90 441	101 957	114 430	104 219	125 987
47 Miscellaneous marine crustaceans Crustacés marins divers Crustáceos marinos diversos	1 081 421	1 245 970	1 402 656	1 289 616	1 371 995	1 426 861	1 383 138
51 Freshwater molluscs Mollusques d'eau douce Moluscos de agua dulce	561 185	507 773	580 047	552 452	595 286	628 205	633 551
52 Abalones, winkles, conchs Ormeaux, bigorneaux, strombes Orejas de mar, bígaros, estrombos	141 740	139 241	115 663	120 050	119 148	123 398	110 740
53 Oysters Huîtres Ostras	187 924	184 587	160 085	157 507	249 647	198 161	186 699
54 Mussels Moules Mejillones	203 439	239 939	237 747	219 183	276 276	275 676	264 101
55 Scallops, pectens Coquilles St-Jacques Vieiras	535 166	532 891	554 767	612 702	660 700	702 737	741 516
56 Clams, cockles, arkshells Clams, coques, arches Almejas, berberechos, arcas	919 100	814 048	838 118	841 658	798 339	824 201	825 651
57 Squids, cuttlefishes, octopuses Encornets, seiches, poulpes Calamares, jibias, pulpos	3 149 719	3 456 667	2 857 605	3 598 201	3 660 404	3 307 969	3 173 272
58 Miscellaneous marine molluscs Mollusques marins divers Moluscos marinos diversos	1 037 481	1 597 051	1 594 542	1 565 999	1 497 351	1 500 552	1 491 849
71 Frogs and other amphibians Grenouilles et autres amphibies Ranas y otros anfibios	2 955	3 622	3 009	1 807	2 328	2 486	2 258
72 Turtles Tortues Tortugas	1 103	977	1 182	1 243	1 010	819	1 369
74 Sea-squirts and other tunicates Ascidiens et autres tuniciers Ascidias y otros tunicados	21 331	5 976	3 443	3 905	3 858	2 427	2 320
75 Horseshoe crabs and other arachnoids Limules et autres arachnoïdés Límulos y otros arácnidos	1 598	2 607	3 252	2 397	1 696	1 299	1 387
76 Sea-urchins and other echinoderms Oursins et autres échinodermes Erizos de mar y otros equinodermos	131 361	122 550	110 304	121 574	122 378	107 156	124 995
77 Miscellaneous aquatic invertebrates Invertébrés aquatiques divers Invertebrados acuáticos diversos	344 165	531 176	527 037	542 959	531 887	447 618	403 142
World total Total mondial Total mundial	**93 846 643**	**94 298 586**	**87 672 034**	**93 774 052**	**95 502 004**	**92 862 087**	**93 190 654**

A-1 (e)

Fish, crustaceans, molluscs, etc — **Capture production by principal species in 2002**
Poissons, crustacés, mollusques, etc — **Captures par principales espèces en 2002**
Peces, crustáceos, moluscos, etc — **Capturas por especies principales en 2002**

3-alpha code Code alpha-3 Código alfa-3	English name Nom anglais Nombre inglés	Scientific name Nom scientifique Nombre científico	1998 t	1999 t	2000 t	2001 t	2002 t
VET	Anchoveta(=Peruvian anchovy)	*Engraulis ringens*	1 729 064	8 723 265	11 276 357	7 213 077	9 702 614
ALK	Alaska pollock(=Walleye poll.)	*Theragra chalcogramma*	4 054 223	3 277 286	2 938 230	3 144 465	2 654 854
SKJ	Skipjack tuna	*Katsuwonus pelamis*	1 895 381	1 977 381	1 962 706	1 814 500	2 030 648
CAP	Capelin	*Mallotus villosus*	982 312	904 045	1 484 818	1 670 906	1 961 724
HER	Atlantic herring	*Clupea harengus*	2 421 462	2 411 408	2 381 011	1 952 975	1 872 013
JAN	Japanese anchovy	*Engraulis japonicus*	2 093 888	1 820 259	1 725 685	1 836 502	1 853 936
CJM	Chilean jack mackerel	*Trachurus murphyi*	2 025 758	1 423 447	1 540 494	2 508 834	1 750 078
WHB	Blue whiting(=Poutassou)	*Micromesistius poutassou*	1 185 003	1 321 195	1 472 105	1 823 305	1 603 263
MAS	Chub mackerel	*Scomber japonicus*	1 924 691	1 950 387	1 473 060	1 812 712	1 470 673
LHT	Largehead hairtail	*Trichiurus lepturus*	1 436 303	1 416 885	1 478 033	1 475 127	1 452 209
YFT	Yellowfin tuna	*Thunnus albacares*	1 212 869	1 215 822	1 182 782	1 330 165	1 341 319
PIL	European pilchard(=Sardine)	*Sardina pilchardus*	949 776	907 612	939 335	1 134 206	1 089 836
COD	Atlantic cod	*Gadus morhua*	1 211 067	1 094 077	940 346	943 667	890 358
MAC	Atlantic mackerel	*Scomber scombrus*	666 964	618 014	685 575	710 578	769 068
CPI	California pilchard	*Sardinops caeruleus*	381 898	405 402	546 079	685 497	722 071
ANE	European anchovy	*Engraulis encrasicolus*	506 844	631 978	626 338	667 326	661 326
SPR	European sprat	*Sprattus sprattus*	696 228	684 189	658 412	647 417	620 020
AKS	Akiami paste shrimp	*Acetes japonicus*	587 376	598 602	639 219	577 497	585 647
MHG	Gulf menhaden	*Brevoortia patronus*	497 461	694 242	591 434	528 506	582 497
NPH	Japanese Spanish mackerel	*Scomberomorus niphonius*	551 780	595 103	539 094	522 756	553 652
SQA	Argentine shortfin squid	*Illex argentinus*	693 542	1 144 998	930 839	743 440	511 087
SQJ	Japanese flying squid	*Todarodes pacificus*	378 605	497 887	570 427	528 523	504 438
SAA	Round sardinella	*Sardinella aurita*	509 079	411 645	362 218	436 689	444 934
BET	Bigeye tuna	*Thunnus obesus*	418 128	429 982	430 679	408 696	430 289
IOS	Indian oil sardine	*Sardinella longiceps*	247 065	208 898	402 936	438 737	409 894
HKP	Argentine hake	*Merluccius hubbsi*	527 229	372 167	244 988	305 672	409 488
GIS	Jumbo flying squid	*Dosidicus gigas*	27 466	134 773	182 399	223 784	406 356
HMC	Cape horse mackerel	*Trachurus capensis*	482 086	398 598	423 607	354 046	386 361
POK	Saithe(=Pollock)	*Pollachius virens*	329 793	340 351	312 570	325 750	381 335
PRA	Northern prawn	*Pandalus borealis*	317 693	338 394	369 688	345 811	375 878
GAZ	Gazami crab	*Portunus trituberculatus*	283 971	284 851	351 051	350 168	369 247
CKI	Araucanian herring	*Strangomera bentincki*	317 564	782 142	722 522	324 617	347 368
SAP	Pacific saury	*Cololabis saira*	180 973	187 898	306 069	376 173	335 473
PCO	Pacific cod	*Gadus macrocephalus*	411 371	402 245	370 912	330 884	332 692
HEP	Pacific herring	*Clupea pallasii*	508 627	471 307	457 197	406 938	320 298
CHU	Chum(=Keta=Dog)salmon	*Oncorhynchus keta*	311 965	281 260	276 355	307 662	311 275
JSC	Yesso scallop	*Patinopecten yessoensis*	294 211	305 510	310 104	293 268	310 303
THG	Golden threadfin bream	*Nemipterus virgatus*	266 383	250 591	296 319	290 920	309 392
SCA	American sea scallop	*Placopecten magellanicus*	99 432	131 962	196 993	254 196	280 572
CRY	Yellow croaker	*Larimichthys polyactis*	207 772	257 758	302 127	254 166	272 140
DPC	Daggertooth pike conger	*Muraenesox cinereus*	258 172	253 978	237 743	259 625	271 350
PIN	Pink(=Humpback)salmon	*Oncorhynchus gorbuscha*	371 552	386 928	285 338	360 973	266 907
HAD	Haddock	*Melanogrammus aeglefinus*	283 689	249 451	212 821	228 821	266 501
PIA	Southern African pilchard	*Sardinops ocellatus*	196 581	175 969	161 448	200 100	264 886
NIP	Nile perch	*Lates niloticus*	339 183	306 282	302 905	282 245	257 272
ANC	Southern African anchovy	*Engraulis capensis*	110 296	180 954	267 986	289 323	254 643
TLN	Nile tilapia	*Oreochromis niloticus*	229 684	227 342	245 988	249 441	253 871
GRM	Patagonian grenadier	*Macruronus magellanicus*	473 633	447 063	233 774	296 830	248 442
ALB	Albacore	*Thunnus alalunga*	229 936	254 843	218 333	232 193	238 281
JAP	Japanese pilchard	*Sardinops melanostictus*	295 788	515 477	305 767	339 377	236 602
TRV	Southern rough shrimp	*Trachypenaeus curvirostris*	179 544	403 027	312 968	247 940	234 102
JJM	Japanese jack mackerel	*Trachurus japonicus*	340 565	227 290	271 501	235 874	229 539
GRN	Blue grenadier	*Macruronus novaezelandiae*	324 750	281 105	283 706	257 801	221 315
HIL	Hilsa shad	*Tenualosa ilisha*	207 269	215 991	220 372	230 221	221 107
COM	Narrow-barred Spanish mackerel	*Scomberomorus commerson*	171 957	183 218	186 595	177 337	216 742
HOM	Atlantic horse mackerel	*Trachurus trachurus*	350 298	321 195	231 059	250 026	216 622
GIT	Giant tiger prawn	*Penaeus monodon*	240 301	264 218	253 003	208 617	214 242
ATK	Okhotsk atka mackerel	*Pleurogrammus azonus*	293 271	214 269	223 456	214 592	213 933
MHA	Atlantic menhaden	*Brevoortia tyrannus*	276 230	208 000	207 122	261 401	211 574
VEP	Pacific anchoveta	*Cetengraulis mysticetus*	180 705	70 358	120 943	227 783	203 704
SAG	Goldstripe sardinella	*Sardinella gibbosa*	174 691	162 710	172 219	185 912	191 300
BOA	Bonga shad	*Ethmalosa fimbriata*	175 931	184 036	163 359	182 440	188 517
RAG	Indian mackerel	*Rastrelliger kanagurta*	294 607	296 905	184 810	166 605	185 649
CLB	Atlantic surf clam	*Spisula solidissima*	131 700	142 370	165 765	166 310	176 322
RUS	Indian scad	*Decapterus russelli*	145 747	162 437	181 363	163 798	176 269
SCD	Blue swimming crab	*Portunus pelagicus*	138 123	147 767	158 223	173 727	174 646
KAW	Kawakawa	*Euthynnus affinis*	157 699	167 267	170 554	163 439	168 348
PHA	South Pacific hake	*Merluccius gayi*	162 518	140 910	193 504	246 265	162 290
CLQ	Ocean quahog	*Arctica islandica*	157 282	147 933	124 131	150 260	162 212
OYA	American cupped oyster	*Crassostrea virginica*	135 222	132 207	221 553	175 042	159 168
MUS	Blue mussel	*Mytilus edulis*	137 268	130 020	147 972	163 911	156 341
71 species 71 espèces 71 especies			***40 987 495***	***47 505 336***	***49 567 394***	***47 288 387***	***48 259 323***
Other species Autres espèces Otras especies			***46 684 539***	***46 268 716***	***45 934 610***	***45 573 700***	***44 931 331***
World total Total mondial Total mundial			**87 672 034**	**93 774 052**	**95 502 004**	**92 862 087**	**93 190 654**

These selected species are those with capture production of 150 000 tonnes or more in 2002.

Ces espèces sont celles dont les captures ont été de 150 000 tonnes ou plus en 2002.

Estas especies se refieren a las que totalizan unas capturas de 150 000 toneladas o más en 2002.

A-2

Fish, crustaceans, molluscs, etc	Capture production by countries or areas	All fishing areas
Poissons, crustacés, mollusques, etc	Captures par pays ou zones	Toutes les zones de pêche
Peces, crustáceos, moluscos, etc	Capturas por países o áreas	Todas las áreas de pesca

Country or area Pays ou zone País o área	1993 t	1994 t	1995 t	1996 t	1997 t	1998 t	1999 t	2000 t	2001 t	2002 t
Afghanistan	1 200 F	1 300 F	1 300 F	1 300 F	1 250 F	1 200 F	1 200 F	1 000 F	800 F	900 F
Albania	2 500 F	2 100 F	1 379	2 125	1 013	2 683	2 745	3 320	3 310	3 955
Algeria	101 895	135 402	105 872	81 989	91 580	92 346	102 396	113 156	133 623	134 320
Amer Samoa	27	111	152	210	431	586	518	830	3 663	6 963
Andorra	0	0	0	0	0	0	0	0	0	0
Angola	126 200	132 413	122 781	137 815	146 304	163 149	175 799	238 351	252 518	260 797
Anguilla	330	333	150 F	200 F	250 F	250 F	250 F	250 F	250 F	250 F
Antigua Barb	1 097	1 145	1 610	1 463	1 665	1 708	1 660	1 754	1 824	2 374
Argentina	931 303	951 387	1 170 126	1 291 355	1 400 162	1 164 829	1 078 313	908 979	931 734	944 346
Armenia	1 850	1 033	821	580	580	698	1 144	1 133	866	465
Aruba	260	260	140	160	205	182	175	163	163	163 F
Australia	228 533	202 480	204 561	200 249	196 674	203 202	211 176	190 090	194 411	194 178
Austria	420	388	404	450	465	451	432	439	362	350
Azerbaijan	21 733	18 901	10 545	6 702	5 161	4 760	20 866	18 797	10 893	11 334
Bahamas	10 073	10 311	9 944	10 197	10 439	10 124	10 473	11 070	9 290	9 300 F
Bahrain	8 958	7 628	9 389	12 940	10 050	9 849	10 620	11 718	11 230	11 204
Bangladesh	764 824	770 790	792 389	814 787	829 426	839 141	959 215	1 004 264	1 068 417	1 103 855
Barbados	3 214	2 818	3 581	3 512	2 809	3 644	3 250	3 100	2 676	2 500
Belarus	2 993	786	715	821	499	457	514	553	943	5 575
Belgium	36 089	34 254	35 599	30 823	30 500	30 835	29 876	29 800	30 209	29 028
Belize	3 528	7 670	9 657	12 038	24 710	26 309	51 583	51 540	14 892	24 753
Benin	39 221	39 923	44 379	42 175	43 785	42 139	40 436	32 324	38 415	40 663
Bermuda	404	394	449	465	461	466	453	286	315	393
Bhutan	320 F	310 F	310 F	300 F	300 F	300 F	300 F	300 F	300 F	300 F
Bolivia	5 518	5 353	5 692	5 988	6 038	6 055	6 052	6 106	5 940	5 800 F
Bosnia Herzg	2 500 F	2 500 F	2 500 F	2 500 F	2 500 F	2 500 F	2 500 F	2 500 F	2 500 F	2 500 F
Botswana	600 F	400 F	200 F	81	160	191	157	166	118	139
Brazil	717 090 F	740 100 F	706 708	715 482	744 585	706 789	703 941	766 846	806 672	822 159
Br Ind Oc Tr	0	0	0	0	0	0	0	0	0	0
Br Virgin Is	343	470	532	506	105	116	115	43	50 F	50 F
Brunei Darsm	1 728	4 445	4 719	7 405	4 521	5 049	3 186	2 487	1 492	2 058
Bulgaria	13 658	6 585 F	8 012	8 854	11 237	18 946	10 556	6 998	6 420	15 007
Burkina Faso	7 000	8 000	8 000	8 000	8 000	8 335	7 600	8 500	8 500 F	8 500 F
Burundi	17 000 F	22 000 F	21 101	3 041	20 296	13 426	9 199	17 315	8 964	9 000 F
Cambodia	101 023	95 018	103 000	94 710	102 800	107 900	269 156	284 368	428 200	406 182
Cameroon	65 257	79 000 F	94 131 F	98 400 F	102 000 F	106 800 F	110 000 F	112 109	111 031	120 135
Canada	1 140 499	1 026 900	853 682	909 511	975 548	1 013 929	1 027 855	1 012 099	1 052 543	1 013 997
Cape Verde	7 000	8 256	8 495	9 155	9 705	9 424	10 360	10 586	9 653	8 000 F
Cayman Is	445	125	125	110	125	125	125	125	125	125
Cent Afr Rep	13 250 F	13 500 F	13 750 F	14 000 F	14 250 F	14 500 F	15 000	15 000 F	15 000 F	15 000 F
Chad	87 300	80 000	90 000	100 000	85 000	84 000	84 000 F	84 000 F	84 000 F	84 000 F
Channel Is	2 854 F	2 783	2 949 F	4 346	4 238	4 117	3 601	3 589	3 927	3 449
Chile	5 949 565	7 720 578	7 434 180	6 690 665	5 811 550	3 265 293	5 050 180	4 299 942	3 797 140	4 271 475
China	9 351 437	10 866 836	12 562 706	14 182 107	15 722 344	17 229 927	17 240 032	16 987 325	16 529 389	16 553 144
China,H.Kong	217 544	211 010	194 999	183 856	186 000	180 000	127 780	157 012	173 972	169 790
China, Macao	1 898	1 890	1 604	1 418	1 500 F	1 500 F	1 500 F	1 500 F	1 500 F	1 500 F
China,Taiwan	1 133 676	967 189	1 010 201	967 483	1 038 048	1 091 768	1 099 715	1 093 889	1 005 199	1 042 756
Christmas Is	0	0	0	0	0	0	0	0	0	0
Cocos Is	0	0	0	0	0	0	0	0	0	0
Colombia	123 497	97 799	120 699	125 829 F	122 918 F	122 908 F	122 995 F	129 644	130 000 F	135 000 F
Comoros	11 645	12 976	13 000 F	12 700 F	12 500 F	12 500 F	12 000	13 200	12 180	12 200 F
Congo Dem R	196 789	155 897	158 627	163 010	162 211	178 041	208 448	209 300 F	214 600 F	220 000 F
Congo Rep	46 748	42 664	45 776	45 473	38 082	42 955 F	43 509	44 729	42 240 F	43 000 F
Cook Is	985 F	933 F	1 090	900 F	820 F	770 F	750 F	720 F	700 F	1 768 F
Costa Rica	16 170	17 132	17 437	24 327	26 669	24 757	28 218	35 398	34 733	32 938
Côte dIvoire	76 972	73 978	70 189	69 168	64 169	69 572	74 365	75 772	73 556	79 743 F
Croatia	26 626	17 477	16 265	18 233	17 035	21 938	18 900	21 062	18 489	21 230
Cuba	84 460	77 634	79 967	84 208	76 694	60 770	69 135	61 061	44 730	33 838
Cyprus	10 016	9 427	9 320	12 526	24 819	19 295	39 638	67 482	81 071	1 978 F
Czech Rep	3 185	3 955	3 929	3 524	3 321	3 952	4 190	4 654	4 646	4 983
Denmark	1 614 289	1 873 316	1 999 033	1 681 517	1 826 852	1 557 335	1 405 005	1 534 089	1 510 461	1 442 068
Djibouti	300 F	320 F	350 F	350 F	350 F	350 F	350 F	350 F	350 F	350 F
Dominica	794	882	950	1 030	1 079	1 212	1 200 F	1 200 F	1 200 F	1 217
Dominican Rp	12 857	22 832	17 874	12 894	14 535	10 109	8 537	11 029	13 217	18 339
Ecuador	286 633	330 248	505 395	702 974	548 988	310 022	497 872	592 547	586 570	318 540
Egypt	302 829	311 600	335 300	340 434	371 332	406 204	422 665	384 314	428 651	425 170
El Salvador	12 451	13 559	15 072	14 433	11 897	12 658	16 525	9 590	19 010	34 455
Eq Guinea	3 507	5 069	2 306	5 040	6 090	6 005	7 001	3 634	3 500 F	3 500 F
Eritrea	475	2 706	3 559	3 252	1 038	1 629	6 891	12 612	8 820	7 832
Estonia	147 185	123 680	132 028	108 446	123 613	118 714	111 793	113 146	105 167	101 452
Ethiopia	4 175	5 285	6 325	8 770	10 370	14 000	15 858	15 681	15 390	12 300
Faeroe Is	246 294	237 708	287 796	304 587	329 145	364 497	358 362	454 399	524 837	490 000 F
Falkland Is	1 974	5 914	27 190	31 540	17 113	43 616	39 164	62 928	59 824	35 656
Fiji Islands	27 817	28 964 F	28 836	25 029	27 755	28 158	36 713	40 000 F	42 972	42 800 F
Finland	156 294	164 269	167 484	179 077	180 185	171 681	160 560	156 501	150 096	144 808
France	623 409	622 568	610 841	564 694	574 076	549 545	593 479	634 819	605 629	620 078
Fr Guiana	6 931	7 819	8 089	7 377 F	6 602	6 709	6 271 F	5 237 F	5 194 F	5 568 F
Fr Polynesia	8 075	8 175	9 020	9 910	11 670	12 473	12 336	13 899	15 404	15 543
Fr South Tr	460	524	519 F	437 F	375 F	388 F	425 F	272 F	263 F	414 F
Gabon	31 789 F	31 015 F	40 437 F	46 113	43 584	53 609	51 143	47 470	42 871	40 875
Gambia	21 308	22 781	23 752	31 601	32 254	29 002	30 000	29 016	34 527	45 769
Georgia	18 240 F	7 413 F	3 560 F	2 453	2 583	3 001	1 413	2 200	1 830	2 537
Germany	253 010	230 161	239 890	236 411	259 352	266 622	238 925	205 689	211 282	224 451
Ghana	372 619	335 437	352 844	476 733	447 088	442 641	492 776	452 070	447 681	371 178
Gibraltar	0	0	0	0	0	0	0	0	0	0
Greece	159 098	181 122	151 788	149 435	157 088	108 580	118 771	99 280	94 388	88 983 F
Greenland	116 650	117 417	128 890	116 018	120 596	128 542	160 253	159 711	158 485	202 828
Grenada	2 103	1 599	1 499	1 577	1 530	1 837	1 658	1 701	2 250	2 171
Guadeloupe	8 600	8 800	9 500	9 570	10 480	9 084	9 114	10 100	10 000 F	10 100 F
Guam	431	435	185	121	158	253	223	275	278	231

A-2

Fish, crustaceans, molluscs, etc — Capture production by countries or areas — All fishing areas
Poissons, crustacés, mollusques, etc — Captures par pays ou zones — Toutes les zones de pêche
Peces, crustáceos, moluscos, etc — Capturas por países o áreas — Todas las áreas de pesca

Country or area Pays ou zone País o área	1993 t	1994 t	1995 t	1996 t	1997 t	1998 t	1999 t	2000 t	2001 t	2002 t
Guatemala	7 582	7 211	8 253	7 653	6 896	10 847	21 018	39 203	22 400 F	24 164
Guinea	60 600 F	63 800 F	67 860	63 360	62 441	69 764	87 314	91 513	105 227	104 000 F
GuineaBissau	5 350 F	6 000 F	6 328	7 000 F	7 250 F	6 000 F	5 000 F	5 000 F	5 000 F	5 000 F
Guyana	44 123	46 367	47 900 F	48 583	53 998	52 840	53 844	48 887	53 405	48 017
Haiti	5 150 F	5 500 F	5 517 F	5 245 F	5 301 F	5 259 F	5 000 F	5 000 F	5 000 F	5 000 F
Honduras	19 577	15 445	22 917	21 550	20 811	15 994	21 229	14 507	12 563	11 402
Hungary	7 886	8 307	7 314	7 606	7 406	7 265	7 514	7 101	6 638	6 750
Iceland	1 715 581	1 556 962	1 612 548	2 060 168	2 205 944	1 681 951	1 736 267	1 982 524	1 980 715	2 129 655
India	3 062 712	3 257 607	3 265 240	3 447 954	3 523 448	3 373 492	3 472 150	3 666 428	3 777 092	3 770 912
Indonesia	3 089 787	3 329 174	3 532 653	3 604 795	3 857 660	3 961 004	4 044 543	4 120 126	4 273 662	4 505 474
Iran	322 006	308 101	341 383	351 725	342 287	367 212	387 200	384 000	336 450	324 853
Iraq	19 941	25 147	28 208	30 737	31 302	22 574	22 423	20 767	16 500 F	13 000 F
Ireland	278 774	294 154	388 303	332 233	292 809	325 020	283 358	276 292	356 430	282 331
Isle of Man	4 850	3 571	3 734	3 537	4 289	2 214	2 609	3 552	3 112	3 127
Israel	5 125	4 061	4 941	5 229	5 204	6 300	5 884	5 818	5 024	4 880
Italy	397 533	398 730	396 791	365 899	343 693	306 096	282 790	302 149	310 397	269 846
Jamaica	23 000 F	25 277 F	24 232 F	31 029 F	20 035	17 610	17 387	5 676	5 700 F	5 650 F
Japan	7 252 278	6 615 893	5 969 517	5 931 872	5 928 334	5 303 555	5 188 948	4 984 813	4 713 006	4 443 000
Jordan	395	410	425	440	450	470	510	550	520	526
Kazakhstan	56 700	46 433	48 402	44 273	31 826	25 000 F	36 170	36 620	21 654	24 910
Kenya	182 052	202 577	192 706	180 988	161 054	172 592	205 287	215 106	164 151	144 512
Kiribati	27 072	27 753	30 306	31 871	30 052	35 304	52 741	25 563	32 375	31 001
Korea D P Rp	397 398 F	371 961	327 083	253 125	236 462	228 000 F	220 000 F	212 850 F	206 500	205 000 F
Korea Rep	2 257 192	2 357 891	2 319 917	2 413 785	2 204 202	2 027 932	2 118 521	1 824 995	1 990 722	1 668 979
Kuwait	8 466	7 752	8 616	8 255	7 826	7 799	6 271	6 000 F	5 846	5 900 F
Kyrgyzstan	127	131	185	160 F	120 F	80 F	48	52	57	48
Laos	19 500 F	23 800	27 370	23 000 F	18 857	19 642	30 041	29 250	31 000 F	33 440
Latvia	141 973	138 105	149 194	142 644	105 682	102 331	125 389	136 403	128 176	113 677
Lebanon	2 020	2 225	4 085	4 135	3 655	3 520	3 560	3 666	3 670	3 970
Lesotho	22 F	22 F	26 F	28 F	30 F	30 F	30	32	24	24 F
Liberia	7 782	7 721	8 829	8 308	8 580	10 830	15 472	11 726	11 286	11 500 F
Libya	31 085 F	33 500 F	34 400 F	32 976 F	31 877 F	32 911 F	32 850 F	33 387 F	33 239 F	33 666 F
Liechtensten	0	0	0	0	0	0	0	0	0	0
Lithuania	117 171	49 162	57 368	88 514	44 002	66 578	72 962	78 988	151 831	150 146
Luxembourg	0	0	0	0	0	0	0	0	0	0
Macedonia	164	196	208	78	130	131	135	208	128	148
Madagascar	114 261	116 431	115 653	114 475	116 391	126 395	129 630	132 093	135 583	141 284
Malawi	67 951	58 579	53 664	63 569	56 340	41 111	45 392	43 000 F	40 619	41 329
Malaysia	1 049 321	1 067 650	1 112 375	1 130 372	1 172 922	1 153 719	1 251 768	1 289 245	1 234 733	1 275 555
Maldives	100 852	118 168	119 048	120 508	116 257	128 968	133 547	132 427	126 190	160 981
Mali	64 300	62 850	132 900	111 910	99 550	98 000	98 536	109 870	100 000 F	100 000 F
Malta	838	2 356	4 635	9 197	1 036	1 180	1 244	1 074	893	1 004
Marshall Is	471 F	393 F	381 F	2 772	400 F	500 F	500 F	8 060	35 544	38 742
Martinique	5 853	5 800 F	5 300	3 500 F	5 500	5 500	6 000	6 310	6 200	6 200 F
Mauritania	59 452 F	51 746 F	53 147 F	60 324 F	57 756 F	61 660 F	76 026 F	80 849 F	84 881 F	78 902 F
Mauritius	20 579	18 145	16 395	11 869	14 025	12 093	12 205	9 615	10 985	10 706
Mayotte	500 F	600 F	1 033 F	1 553	2 101 F	3 172 F	3 130	3 030 F	11 051	4 299
Mexico	1 102 932	1 191 646	1 329 340	1 464 188	1 489 112	1 179 860	1 205 603	1 315 581	1 398 592	1 450 654
Micronesia	17 888 F	22 926 F	7 629 F	9 069 F	10 127 F	16 581 F	12 843 F	23 884 F	19 653 F	20 387 F
Moldova Rep	630	708	709	603	569	491	309	344	387	565
Monaco	3 F	3 F	3 F	3 F	3 F	3 F	3 F	3 F	3 F	3 F
Mongolia	165	184	158	221	180	311	524	425	117	129
Montserrat	58	62	48	38	45	46	50 F	50 F	50 F	46
Morocco	624 511	754 327	848 951	642 886	791 906	710 436	745 431	896 620	1 083 953	894 957
Mozambique	30 195	27 456	26 913	34 915	39 703	36 677	36 175	37 729	30 074	36 462
Myanmar	739 702	746 241	751 232	601 788	780 295	830 117	919 410	1 069 726	1 166 868	1 312 642
Namibia	790 332	649 199	569 833	517 828	513 216	607 831	580 084	589 904	547 498	624 891
Nauru	500	500	400 F	300 F	250 F	200 F	150 F	100 F	50 F	21
Nepal	7 418	7 340	11 230	11 230	11 230	12 000	12 752	16 700	16 700	17 900
Netherlands	461 756	420 053	438 092	410 798	451 799	536 626	514 611	495 754	518 162	464 035
NethAntilles	1 200 F	1 100 F	1 020 F	13 762 F	19 532 F	19 754 F	19 431 F	19 882 F	22 805 F	12 901 F
NewCaledonia	2 670	3 185	2 644	2 998	2 438	3 105	3 152	3 386	3 309	3 417 F
New Zealand	426 668	449 660	556 635	425 564	612 424	640 236	599 466	549 318	561 940	559 289
Nicaragua	8 169	10 034	10 995	15 442	16 176	19 892	23 909	28 008	22 799	23 670
Niger	2 162	2 516	3 616	4 156	6 328	7 013	11 000	16 250	20 800	23 520
Nigeria	238 409	267 059	349 482	337 993	387 923	463 024	455 628	441 377	452 146	481 056
Niue	120 F	150 F	150 F	200 F	200 F	200 F	200 F	200 F	200 F	200 F
Norfolk Is	0	0	0	0	0	0	0	0	0	0
N Marianas	136	176	192	225	250	235	193	189	197	198
Norway	2 415 131	2 366 119	2 524 111	2 648 457	2 863 059	2 861 223	2 627 534	2 699 421	2 687 303	2 743 184
Oman	105 772	118 572	139 861	121 618	118 995	106 171	108 809	120 421	129 907	142 670
Pakistan	608 344	537 277	525 849	537 489	589 795	596 980	654 530	614 069	600 798	599 104
Palau	1 104	1 001	1 025	999	904	951	960	1 047	1 075	1 008 F
Palest, O.T.	...	...	1 229	2 493	3 791	3 625	3 600 F	3 000 F	2 500 F	2 378
Panama	184 821	185 147	203 490	150 027	162 223	202 598	120 872	219 166	256 408 F	305 081
Papua N Guin	25 741 F	26 219 F	39 077 F	37 823 F	46 185 F	78 316 F	65 770 F	96 407 F	122 483 F	148 700 F
Paraguay	19 000 F	20 000 F	21 000 F	22 000	28 000	25 000 F	25 000 F	25 000 F	25 000 F	25 000 F
Peru	9 004 777	11 999 217	8 937 342	9 515 048	7 869 871	4 338 437	8 428 601	10 658 620	7 986 103	8 766 991
Philippines	1 834 323	1 845 335	1 860 701	1 783 601	1 805 806	1 833 380	1 872 818	1 896 638	1 949 026	2 030 542
Pitcairn Is	8 F	8 F	8 F	8 F	8 F	8 F	8 F	8 F	8 F	8 F
Poland	395 406	438 032	429 372	342 793	348 913	242 010	235 724	217 682	225 065	222 441
Portugal	287 802	263 065	260 253	259 846	221 879	224 171	208 407	187 521	191 017	200 037
Puerto Rico	1 877	2 275	3 173	2 701	3 187	3 006	3 020	4 154	3 794	2 529
Qatar	6 994	5 086	4 271	4 739	5 032	5 279	4 207	7 142	8 606	6 880
Réunion	1 679	2 531	2 500	3 607	4 288	4 579	4 043	4 082	3 635	3 700
Romania	13 819	22 251	49 275	18 259	8 446	9 061	7 843	7 372	7 637	6 989
Russian Fed	4 368 067	3 703 972	4 311 809	4 675 738	4 661 853	4 454 759	4 141 158	3 973 535	3 628 459	3 232 295
Rwanda	3 500 F	3 400 F	3 300 F	2 952	4 428	6 641	6 433	6 726	6 828	7 000 F
St Helena	726	702	915	819	897	1 060	632	718	866	598
St Kitts Nev	250 F	212	192	352	216	407	471	469	555	355

A-2 Fish, crustaceans, molluscs, etc — Capture production by countries or areas — All fishing areas
Poissons, crustacés, mollusques, etc — Captures par pays ou zones — Toutes les zones de pêche
Peces, crustáceos, moluscos, etc — Capturas por países o áreas — Todas las áreas de pesca

Country or area Pays ou zone País o área	1993 t	1994 t	1995 t	1996 t	1997 t	1998 t	1999 t	2000 t	2001 t	2002 t
St Lucia	1 336	1 252	1 188	1 274	1 308	1 589	1 718	1 855	1 983	1 637
St Pier Mq	282	294	317	747	3 571	6 108	5 892	6 485	3 802	3 654
St Vincent	1 945	3 840	1 005	921 F	6 093	33 973	17 759	27 694	52 485	43 879
Samoa	673	1 108	2 519	2 727	7 041	7 547	10 204	13 004	12 966 F	12 392 F
Sao Tome Prn	2 334	3 391	3 565	3 980	3 338	3 477	3 756	3 500 F	3 500 F	3 500 F
Saudi Arabia	48 021	54 612	45 609	47 698	49 314	51 206	46 618	49 761	49 167	55 330
Senegal	382 212	352 421	354 617	411 759	457 366	403 872	412 125	402 047	403 202	375 824
Serbia-Monte	1 693	1 589	1 438	1 612	1 439	1 279	1 251	1 096	986	1 391
Seychelles	5 178	4 469	4 008	4 707	14 043	23 886	34 223	32 770	53 534	63 209
Sierra Leone	66 288	62 439	64 870	67 304 F	72 628	63 065	59 407	74 730	75 210	82 990
Singapore	9 304	11 301	10 102	9 943	9 250	7 733	6 489	5 371	3 342	2 769
Slovakia	1 185	1 627	1 950	1 414	1 364	1 361	1 396	1 368	1 531	1 746
Slovenia	2 284	2 346	2 167	2 367	2 367	2 228	2 027	1 856	1 827	1 686
Solomon Is	42 970 F	46 904 F	62 019 F	50 990 F	64 005 F	61 332 F	58 428 F	24 926 F	30 277 F	31 003 F
Somalia	27 750 F	29 850 F	27 950 F	26 050 F	24 150 F	22 250 F	20 250 F	20 200 F	20 000 F	18 000 F
South Africa	562 374	523 449	575 547	441 370	514 238	559 679	587 933	643 526	750 099	766 284
Spain	1 085 959 F	1 094 524 F	1 181 324 F	1 174 683	1 205 401	1 263 016	1 172 995	1 045 616	1 092 525	882 633
Sri Lanka	215 400	218 400	229 171	228 945	235 099	263 330	276 080	293 410	279 640	298 260
Sudan	40 000	44 000	44 000	45 000	47 000	49 500	49 500	53 000 F	58 000	58 000
Suriname	9 500	14 465	13 000 F	13 000 F	14 000 F	16 195	16 200	17 500 F	18 915	18 700 F
Svalbard Is	0	0	0	0	0	0	0	0	0	0
Swaziland	68 F	65 F	60 F	60 F	65 F	70 F	70 F	70 F	70 F	70 F
Sweden	341 897	386 814	404 572	370 881	357 406	410 886	351 254	338 534	311 816	294 963
Switzerland	1 822	1 481	1 588	1 841	1 859	1 809	1 840	1 659	1 715	1 544
Syria	4 554	5 520	5 782	5 773	6 131	7 097	7 938	6 572	8 291	9 178
Tajikistan	253	127	100	40 F	75 F	100 F	80 F	78 F	137	181
Tanzania	331 267	288 399	359 800	323 921	356 960	348 000	310 509	332 779	335 900	323 530
Thailand	2 927 689	3 015 196	3 031 074	3 013 961	2 902 898	2 930 354	2 952 308	2 997 394	2 932 374	2 921 216
Timor-Leste	...	...	...	...	...	...	408 F	362	356	350 F
Togo	16 964	13 052	12 201	15 098	14 290	16 655	22 924	22 277	23 163	20 946
Tokelau	200 F	200 F	200 F	200 F	200 F	200 F	200 F	200 F	200 F	200 F
Tonga	2 375	2 499	2 597	2 915	2 871	4 076	4 221	3 760	4 673	4 804
Trinidad Tob	8 998	14 046	11 500 F	9 435	11 283	9 175	8 826	8 651	10 707	12 539
Tunisia	82 914	85 610	83 355	83 734	87 012	88 075	93 186	95 550	98 482	96 685
Turkey	548 758	589 803	633 970	527 826	459 153	487 200	573 824	503 348	527 730	566 682
Turkmenistan	16 080	15 140	9 740	9 014	8 179	7 014	9 058	12 228	12 749	12 812
Turks Caicos	6 255	5 963 F	7 824	7 833	5 951	6 383	5 401	5 975	6 611	5 953
Tuvalu	1 460	561	399	400 F	500 F	500 F	500 F	500 F	500 F	500 F
Uganda	219 814	213 129	208 789	195 088	218 026	220 628	226 097	219 356	220 726	221 898
Ukraine	299 404	272 259	387 716	407 305	368 800	462 196	408 711	391 831	360 914	265 599
Untd Arab Em	99 600	108 600	105 884	107 000	114 358	114 739	117 607	105 456	112 561	97 574
UK	860 229	878 018	909 963	867 686	890 936	923 219	840 719	747 571	741 045	689 919
USA	5 522 917	5 535 324	5 224 858	5 001 191	4 983 440	4 708 980	4 749 646	4 717 638	4 944 336	4 937 305
US Minor Is	0	0	0	0	0	0	0	0	0	0
US Virgin Is	650 F	550 F	470 F	400 F	350 F	300 F	263	300 F	300 F	300 F
Uruguay	118 815	120 732	126 446	123 330	136 954	140 707	106 583	113 339	104 903	108 765
Uzbekistan	4 358	3 095	3 611	1 494	3 075	2 799	2 871	3 387	4 070	2 009
Vanuatu	40 515 F	47 190 F	57 646 F	45 615 F	70 822 F	74 908 F	93 041 F	68 896 F	24 521 F	17 139 F
Venezuela	395 861	438 218	500 817	496 115	470 126	503 494	399 809	356 835	418 207	515 384
Viet Nam	932 143	1 025 909	1 084 939	1 223 644	1 276 325	1 293 954	1 386 300	1 450 590	1 490 303	1 508 000 F
Wallis Fut I	150	193	170	180	176	300	300 F	300 F	300 F	300 F
Westn Sahara	0	0	0	0	0	0	0	0	0	0
Yemen	82 356	81 885	107 970	104 955	115 600	127 620	124 385	114 751	142 198	159 262
Zambia	65 768	70 057	70 546	66 332	65 923	69 938	67 327	66 671	65 000 F	65 000 F
Zimbabwe	21 230	20 219	16 463	16 387	18 156	16 371	12 410	13 114	13 000 F	13 000 F
Other nei	138 834	134 742	297 280	174 123	153 713	182 984	213 943	226 861	251 152	337 594
World total Total mondial Total mundial	**86 552 850**	**92 082 052**	**92 380 161**	**93 846 643**	**94 298 586**	**87 672 034**	**93 774 052**	**95 502 004**	**92 862 087**	**93 190 654**
World excl. China Monde excl. Chine Mundo excl. China	**77 201 413**	**81 215 216**	**79 817 455**	**79 664 536**	**78 576 242**	**70 442 107**	**76 534 020**	**78 514 679**	**76 332 698**	**76 637 510**

A-3

Fish, crustaceans, molluscs, etc	Capture production by countries or areas	Inland waters
Poissons, crustacés, mollusques, etc	Captures par pays ou zones	Eaux continentales
Peces, crustáceos, moluscos, etc	Capturas por países o áreas	Aguas continentales

Country or area Pays ou zone País o área	1993 t	1994 t	1995 t	1996 t	1997 t	1998 t	1999 t	2000 t	2001 t	2002 t
Afghanistan	1 200 F	1 300 F	1 300 F	1 300 F	1 250 F	1 200 F	1 200 F	1 000 F	800 F	900 F
Albania	850 F	700 F	219	317	180	823	814	955	1 373	1 167
Algeria	0	0	0	0	0	2	0	0	0	0
Amer Samoa	0	0	0	0	0	0	0	0	0	-
Andorra	0	0	0	0	0	0	0	0	0	0
Angola	7 000	7 000	6 000	6 000	6 000	6 000	6 000 F	6 000 F	6 000 F	6 000 F
Anguilla	0	0	0	0	0	0	0	0	0	0
Antigua Barb	0	0	0	0	0	0	0	0	0	0
Argentina	11 800	12 785	17 191	19 189	22 735	23 197	27 558	30 418	33 757	31 233
Armenia	1 850	1 033	821	580	580	698	1 144	1 133	866	465
Aruba	0	0	0	0	0	0	-	-	-	-
Australia	3 011	584	821	705	612	593	694	666 F	663 F	646 F
Austria	420	388	404	450	465	451	432	439	362	350
Azerbaijan	21 733	18 901	10 545	6 702	5 161	4 760	20 866	18 797	10 893	11 334
Bahamas	0	0	0	0	0	0	0	0	0	0
Bahrain	0	0	0	0	0	0	0	-	-	-
Bangladesh	452 109	517 746	527 739	535 617	534 285	538 689	649 418	670 465	688 920	688 435
Barbados	0	0	0	0	0	0	0	0	-	-
Belarus	2 993	786	715	821	499	457	514	553	943	5 575
Belgium	511	511	511	511	511	511	536	511	511	511
Belize	1	1	0	0	0	0	0	0	0	-
Benin	32 805	32 707	37 449	34 193	32 871	31 778	31 894	26 400	30 000	29 993
Bermuda	0	0	0	0	0	0	0	0	0	-
Bhutan	320 F	310 F	310 F	300 F	300 F	300 F	300 F	300 F	300 F	300 F
Bolivia	5 518	5 353	5 692	5 988	6 038	6 055	6 052	6 106	5 940	5 800 F
Bosnia Herzg	2 500 F	2 500 F	2 500 F	2 500 F	2 500 F	2 500 F	2 500 F	2 500 F	2 500 F	2 500 F
Botswana	600 F	400 F	200 F	81	160	191	157	166	118	139
Brazil	186 990 F	191 485 F	193 042	193 309	178 871	174 190	185 471	199 159	217 275	217 750
Br Ind Oc Tr	0	0	0	0	0	0	0	0	0	0
Br Virgin Is	0	0	0	0	0	0	0	0	0	0
Brunei Darsm	25	4	7	15	17	35	26	23	16	14
Bulgaria	1 675	995 F	762	1 127	1 881	2 336	2 475	861	1 540	1 452
Burkina Faso	7 000	8 000	8 000	8 000	8 000	8 335	7 600	8 500	8 500 F	8 500 F
Burundi	17 000 F	22 000 F	21 101	3 041	20 296	13 426	9 199	17 315	8 964	9 000 F
Cambodia	67 900	65 000	72 500	63 510	73 000	75 700	231 000	245 600	385 000	360 300
Cameroon	23 000	27 000 F	30 000 F	35 000 F	40 000 F	45 000 F	50 000 F	55 000	52 500 F	65 000
Canada	36 327	36 333	38 756	38 295	38 798	40 744	40 587	40 667	38 140	39 999
Cape Verde	0	0	0	0	0	0	0	0	0	0
Cayman Is	0	0	0	0	0	0	0	0	0	0
Cent Afr Rep	13 250 F	13 500 F	13 750 F	14 000 F	14 250 F	14 500 F	15 000	15 000 F	15 000 F	15 000 F
Chad	87 300	80 000	90 000	100 000	85 000	84 000	84 000 F	84 000 F	84 000 F	84 000 F
Channel Is	0	0	0	0	0	0	0	0	0	0
Chile	7	5	...	...	...	4	...	...	...	...
China	1 182 390	1 327 785	1 607 385	1 762 860	1 886 967	2 280 244	2 285 364	2 233 230	2 149 932	2 247 926
China,H.Kong	0	0	0	0	0	0	0	0	0	0
China, Macao	0	0	0	0	0	0	0	0	0	0
China,Taiwan	1 216	1 059	874	407	403	449	561	549	591	599
Christmas Is	0	0	0	0	0	0	0	0	0	0
Cocos Is	0	0	0	0	0	0	0	0	0	0
Colombia	30 538	34 983	23 524	23 061	20 610	21 673	28 788	24 854	25 000 F	25 000 F
Comoros	0	0	0	0	0	0	0	0	0	0
Congo Dem R	192 589	152 117	154 751	159 037	158 367	174 087	204 503	205 000 F	210 000 F	215 000 F
Congo Rep	27 850	24 752	26 811	25 873	18 987	25 455	25 268	20 204 F	20 200 F	20 500 F
Cook Is	0	0	10	0	0	0	0	0	0	0
Costa Rica	710	840	900	1 090	840	1 000 F	1 000 F	1 000 F	1 000 F	1 000 F
Côte dIvoire	13 477	15 604	11 335	11 562	12 032	12 501	10 656	10 502	10 630	23 843
Croatia	284	340	364	434	408	10 F	10	17	34	25
Cuba	9 886	9 537	8 270	9 744	7 603	5 154	4 591	3 005	2 007	2 034
Cyprus	...	5	65	64	70	70	70	78	70 F	60 F
Czech Rep	3 185	3 955	3 929	3 524	3 321	3 952	4 190	4 654	4 646	4 983
Denmark	337	243	264	196	232	349	206	183	99	77
Djibouti	0	0	0	0	0	0	0	0	0	0
Dominica	0	0	0	0	0	0	0	0	0	0
Dominican Rp	2 037	3 774	2 106	288	1 067	1 095	733	187	1 158	2 363
Ecuador	372	300	300	300	400	400	400	400	400	400 F
Egypt	207 473	217 700	244 300	240 900	261 167	281 141	250 318	253 470	295 422	292 645
El Salvador	4 461	3 819	4 324	2 967	2 808	2 443	2 654	2 831	1 692	2 663
Eq Guinea	600	700	450	900	850	970	1 101 F	1 076	1 000 F	1 000 F
Eritrea	0	0	0	0	0	0	0	0	0	-
Estonia	2 411	1 909	2 366	2 361	2 439	3 878	3 108	3 190	2 461	4 578
Ethiopia	4 175	5 285	6 325	8 770	10 370	14 000	15 858	15 681	15 390	12 300
Faeroe Is	0	0	0	0	0	0	0	0	0	-
Falkland Is	-	-	-	1	1	1	1	1	1	1
Fiji Islands	2 907	2 964 F	3 586	3 034	4 325	5 111	5 622	5 700 F	5 921	5 800 F
Finland	51 522	47 895	48 436	47 618	47 618	36 813	36 813	34 840	34 820	34 782
France	4 400	4 450	4 500	4 540	4 540	4 500	2 134 F	2 131 F	2 130 F	2 130 F
Fr Guiana	0	0	0	0	0	0	0	0	0	0
Fr Polynesia	0	0	0	0	53	53	53	53	53	53
Fr South Tr	0	0	0	0	0	0	0	0	0	0
Gabon	3 500 F	4 500 F	7 648 F	9 433	9 441	9 000	10 000	10 417	9 850	9 400
Gambia	2 400	2 400	2 500	2 500	2 500	2 500	2 500 F	2 500 F	2 500 F	2 500 F
Georgia	549	16	90	6	1	4	17	22	8	10
Germany	10 837	10 908	22 987	22 987	22 916	22 916	22 868	22 868	22 818	22 798
Ghana	52 000	54 700	60 000	73 580	70 000	74 500	74 500	74 500	74 500	74 500
Greece	2 960	3 452	3 606	2 903	2 601	2 818	3 280	3 433	3 181	3 000
Greenland	0	0	0	0	0	0	0	0	0	0
Grenada	0	0	0	0	0	0	0	0	0	0
Guadeloupe	0	0	0	0	0	0	0	0	0	0
Guam	0	0	0	0	-	6	-	-	-	-
Guatemala	4 228	3 776	4 025	4 000	5 121	6 523	6 976	7 301	7 300 F	7 300 F

A-3 Fish, crustaceans, molluscs, etc — Capture production by countries or areas — Inland waters
Poissons, crustacés, mollusques, etc — Captures par pays ou zones — Eaux continentales
Peces, crustáceos, moluscos, etc — Capturas por países o áreas — Aguas continentales

Country or area Pays ou zone País o área	1993 t	1994 t	1995 t	1996 t	1997 t	1998 t	1999 t	2000 t	2001 t	2002 t
Guinea	4 600	3 800	3 100	2 780	3 600	4 000	4 000	4 000	4 000	4 000 F
GuineaBissau	250 F	250 F	250 F	250 F	250 F	200 F	200 F	200 F	200 F	200 F
Guyana	800	800	700 F	800	625	625	603	800	800	800
Haiti	600 F	500 F	500 F	500 F	500 F	500 F	500 F	500 F	500 F	500 F
Honduras	86	92	127	98	126	119	102	61	111	102 F
Hungary	7 886	8 307	7 314	7 606	7 406	7 265	7 514	7 101	6 638	6 750
Iceland	907	698	739	608	404	416	370	176	160	160 F
India	575 905	552 874	608 378	633 425	641 775	692 439	696 083	905 700	975 403	813 755
Indonesia	308 649	336 141	329 710	335 707	304 258	288 666	327 627	318 334	310 240	316 030
Iran	75 021	89 157	88 800	109 286	103 795	140 263	143 400	123 500	73 645	55 853
Iraq	17 808	20 926	22 955	19 049	20 519	9 111	9 330	8 378	8 000 F	8 000 F
Ireland	2 929	3 604	3 761	3 806	3 804	3 976	801	881	902	789
Isle of Man	0	0	0	0	0	0	0	0	0	0
Israel	1 813	1 478	1 214	1 845	1 476	2 164	2 145	1 852	1 286	1 568
Italy	9 515	9 921	10 035	6 764	6 690	4 667	5 436	4 565	5 527	4 242
Jamaica	450 F	450 F	450 F	450 F	450 F	450 F	450 F	450 F	450 F	450 F
Japan	91 032	92 219	91 455	93 501	85 612	78 822	71 270	70 612	61 354	60 843 F
Jordan	350	350	350	350	350	350	350	400	350	350
Kazakhstan	56 700	46 433	48 402	44 273	31 826	25 000 F	36 170	36 620	21 654	24 910
Kenya	176 435	198 805	187 241	174 692	154 955	165 992	198 653	210 343	156 763	137 792
Kiribati	0	0	0	0	0	0	0	0	0	-
Korea D P Rp	23 000 F	20 000	20 000	20 000	20 000	16 000 F	12 000 F	8 000 F	4 928	5 000 F
Korea Rep	12 263	10 492	9 646	8 034	6 934	6 845	6 316	7 141	5 971	5 690
Kuwait	0	0	0	0	0	0	0	0	0	0
Kyrgyzstan	127	131	185	160 F	120 F	80 F	48	52	57	48
Laos	19 500 F	23 800	27 370	23 000 F	18 857	19 642	30 041	29 250	31 000 F	33 440
Latvia	553	495	514	536	544	501	610	612	581	581
Lebanon	20	20	20	20	20	20	20	20	20	297
Lesotho	22 F	22 F	26 F	28 F	30 F	30 F	30	32	24	24 F
Liberia	4 000	4 000	4 000	4 000	4 000	4 000	4 000	4 000	4 000	4 000 F
Libya	0	0	0	0	0	0	0	0	0	0
Liechtensten	0	0	0	0	0	0	0	0	0	0
Lithuania	1 146	1 187	1 260	1 295	1 713	1 737	1 715	1 912	1 854	758
Luxembourg	0	0	0	0	0	0	0	0	0	0
Macedonia	164	196	208	78	130	131	135	208	128	148
Madagascar	30 000	30 000	30 000	30 000	30 000	30 000	30 000	30 000	30 000	30 000
Malawi	67 951	58 579	53 664	63 569	56 340	41 111	45 392	43 000 F	40 619	41 329
Malaysia	1 971	2 064	3 939	3 683	3 949	4 626	3 366	3 549	3 446	3 450
Maldives	0	0	0	0	0	0	0	0	0	0
Mali	64 300	62 850	132 900	111 910	99 550	98 000	98 536	109 870	100 000 F	100 000 F
Malta	0	0	0	0	0	0	0	0	0	-
Marshall Is	0	0	0	0	0	0	0	0	0	0
Martinique	3	0	0	0	0	0	0	0	0	0
Mauritania	5 000 F	5 000 F	5 000 F	5 000 F	5 000 F	5 000 F	5 000 F	5 000 F	5 000 F	5 000 F
Mauritius	3	0	0	0	0	0	0	0	0	0
Mexico	111 178	111 125	122 020	122 501	113 552	100 335	91 462	106 817	91 952	82 648
Micronesia	4 F	4 F	5 F	5 F	5 F	5 F	5 F	5 F	5 F	5 F
Moldova Rep	630	708	709	603	569	491	309	344	387	565
Mongolia	165	184	158	221	180	311	524	425	117	129
Montserrat	0	0	0	0	0	0	0	0	0	0
Morocco	1 617	1 750	1 500	1 500	2 100	1 703	2 163	1 608	1 660	2 112
Mozambique	4 689	4 925	5 173	7 510	11 668	8 994	9 052	11 813	5 284	12 156
Myanmar	142 065	146 365	148 347	146 494	149 069	149 279	159 746	189 708	235 376	304 529
Namibia	1 200	1 200	1 200	1 200	1 500	1 500	1 500	1 500	1 500	1 500
Nepal	7 418	7 340	11 230	11 230	11 230	12 000	12 752	16 700	16 700	17 900
Netherlands	1 601	2 446	4 107	2 157	2 293	1 547	2 303	2 250 F	2 200 F	2 578
NethAntilles	0	0	0	0	0	0	0	0	0	0
NewCaledonia	0	0	0	0	0	0	0	0	0	0
New Zealand	1 160	1 100	1 115	1 079	1 025	1 307	1 152	1 089	1 076	900
Nicaragua	547	824	538	1 142	1 293	1 256	1 120	1 076	1 051	1 592
Niger	2 162	2 516	3 616	4 156	6 328	7 013	11 000	16 250	20 800	23 520
Nigeria	95 627	103 800	117 903	89 521	93 644	139 020	139 393	132 315	154 175	187 233
Niue	0	0	0	0	0	0	0	0	0	0
Norfolk Is	0	0	0	0	0	0	0	0	0	0
N Marianas	0	0	1	0	0	0	0	0	0	-
Norway	435	432	413	338	439	507	514	578	570 F	570 F
Oman	0	0	0	0	0	0	0	0	0	0
Pakistan	109 185	118 703	121 405	142 092	167 530	163 524	179 865	176 468	180 100	181 000
Palau	0	0	0	0	0	0	0	0	0	0
Palest, O.T.	...	...	0	0	0	0	0	0	0	0
Panama	28	285	130	80	91	23	20	20	20 F	20 F
Papua N Guin	13 500 F	13 500 F	13 500 F	13 500 F	13 500 F	13 500 F	13 500 F	13 500 F	13 500 F	13 500 F
Paraguay	19 000 F	20 000 F	21 000 F	22 000	28 000	25 000 F	25 000 F	25 000 F	25 000 F	25 000 F
Peru	38 290	48 837	50 789	28 890	32 221	35 327	36 223	32 297	35 653	29 966
Philippines	210 775	204 325	186 006	177 355	159 353	146 004	146 234	151 753	135 845	130 881
Pitcairn Is	0	0	0	0	0	0	0	0	0	0
Poland	31 391	27 500	24 889	22 037	13 832	13 236	13 875	17 543	17 789	17 947
Portugal	3	4	2	0	0	0	0	0	1	12
Puerto Rico	0	0	0	0	0	0	0	0	0	-
Qatar	0	0	0	0	0	0	0	0	0	0
Réunion	0	0	0	0	0	0	0	0	0	0
Romania	8 562	10 598	9 048	6 145	4 574	4 630	5 336	4 896	5 206	4 867
Russian Fed	216 866	217 950	212 874	233 272	227 091	271 311	307 823	292 368	206 430	208 522
Rwanda	3 500 F	3 400 F	3 300 F	2 952	4 428	6 641	6 433	6 726	6 828	7 000 F
St Helena	0	0	0	0	-	-	-	-	-	-
St Kitts Nev	0	0	0	0	0	0	0	0	0	-
St Lucia	0	0	0	0	0	0	0	0	0	0
St Pier Mq	0	0	0	0	0	0	0	0	0	0
St Vincent	0	0	0	2	1	0	0	0	0	0
Samoa	0	0	0	0	0	0	0	1	1 F	1 F

A-3 Fish, crustaceans, molluscs, etc / Poissons, crustacés, mollusques, etc / Peces, crustáceos, moluscos, etc

Capture production by countries or areas / Captures par pays ou zones / Capturas por países o áreas

Inland waters / Eaux continentales / Aguas continentales

Country or area Pays ou zone País o área	1993 t	1994 t	1995 t	1996 t	1997 t	1998 t	1999 t	2000 t	2001 t	2002 t
Sao Tome Prn	0	0	0	0	0	0	0	0	0	0
Saudi Arabia	0	0	0	0	0	0	0	0	0	0
Senegal	27 650	30 000 F	31 000 F	23 000 F	31 000 F	21 000 F	34 000 F	22 450	20 000 F	20 000 F
Serbia-Monte	1 416 F	1 329 F	1 069	1 232	1 063	866	828	672	570	937
Seychelles	0	0	0	0	0	0	0	0	0	0
Sierra Leone	14 000	15 000	15 000	14 500 F	14 500	14 190	14 480	14 000	14 000	14 000
Singapore	25	23	0	0	0	0	0	0	0	0
Slovakia	1 185	1 627	1 950	1 414	1 364	1 361	1 396	1 368	1 531	1 746
Slovenia	297	339	316	289	302	269	243	226	206	226
Solomon Is	0	0	0	0	0	0	0	0	0	0
Somalia	250 F	250 F	250 F	250 F	250 F	250 F	250 F	200 F	200 F	150 F
South Africa	832	800	800	850	850	900	900 F	900 F	900 F	900 F
Spain	9 215	6 284	8 869	8 710 F	8 710 F	8 710 F	8 710 F	8 710 F	8 710 F	8 710 F
Sri Lanka	15 000	9 500	15 000	22 250	27 250	29 900	31 450	36 700	29 870	28 130
Sudan	37 500	40 000	40 000	40 500	42 000	44 000	44 000	48 000 F	53 000	53 000
Suriname	187	138	140 F	150 F	200 F	200 F	200 F	200 F	200	200 F
Swaziland	68 F	65 F	60 F	60 F	65 F	70 F	70 F	70 F	70 F	70 F
Sweden	2 273	2 254	1 934	1 810	2 011	1 559	1 478	1 459	1 234	1 436
Switzerland	1 822	1 481	1 588	1 841	1 859	1 809	1 840	1 659	1 715	1 544
Syria	2 535	3 570	3 832	3 103	3 557	4 347	5 338	3 991	5 969	6 355
Tajikistan	253	127	100	40 F	75 F	100 F	80 F	78 F	137	181
Tanzania	294 582	247 614	317 029	262 276	306 750	300 000	260 020	280 000	283 000	273 850
Thailand	175 140	196 397	186 665	205 903	203 671	200 715	206 840	201 405	202 400	205 500
Timor-Leste	...	...	...	...	...	...	0	0	0	0
Togo	6 000	5 000	4 998	5 000	5 000	5 000	5 000	5 000	5 000	5 000
Tokelau	0	0	0	0	0	0	0	0	0	0
Tonga	1	0	0	0	0	0	0	0	0	0
Trinidad Tob	0	0	0	0	0	0	0	0	0	0
Tunisia	400	243	440	706	1 010	896	808	832	860	870
Turkey	44 801	45 067	47 976	49 600	50 460	54 500	50 190	42 824	43 323	43 938
Turkmenistan	16 080	15 140	9 740	9 014	8 179	7 014	9 058	12 228	12 749	12 812
Turks Caicos	0	0	0	0	0	0	0	0	0	0
Tuvalu	0	0	0	0	0	0	0	0	0	0
Uganda	219 814	213 129	208 789	195 088	218 026	220 628	226 097	219 356	220 726	221 898
Ukraine	13 189	14 786	6 847	9 468	6 215	4 898	4 728	4 429	4 343	3 823
Untd Arab Em	0	0	0	0	0	0	0	0	0	0
UK	1 909	2 191	2 146	1 930	1 491	4 569	4 835	2 743	3 142	3 431
USA	54 377	42 648	36 688	33 471	38 347	36 129	36 413	25 339	29 578	29 943
US Virgin Is	0	0	0	0	0	0	0	0	0	0
Uruguay	621	966	849	598	2 216	1 931	2 423	2 302	451	387
Uzbekistan	4 358	3 095	3 611	1 494	3 075	2 799	2 871	3 387	4 070	2 009
Vanuatu	0	0	0	0	0	0	0	0	0	0
Venezuela	28 251	35 412	54 175	49 911	40 881	46 035	35 219	23 739	24 326	40 776
Viet Nam	146 839	79 587	94 689	164 936	177 589	138 800	169 107	170 000 F	133 280 F	149 200 F
Wallis Fut I	0	0	0	0	0	0	0	0	0	0
Zambia	65 768	70 057	70 546	66 332	65 923	69 938	67 327	66 671	65 000 F	65 000 F
Zimbabwe	21 230	20 219	16 463	16 387	18 156	16 371	12 410	13 114	13 000 F	13 000 F
World total Total mondial Total mundial	**6 596 214**	**6 726 895**	**7 275 116**	**7 433 689**	**7 575 487**	**8 067 057**	**8 527 706**	**8 730 632**	**8 698 092**	**8 738 167**
World excl. China Monde excl. Chine Mundo excl. China	**5 413 824**	**5 399 110**	**5 667 731**	**5 670 829**	**5 688 520**	**5 786 813**	**6 242 342**	**6 497 402**	**6 548 160**	**6 490 241**

A-4

Fish, crustaceans, molluscs, etc	Capture production by countries or areas	Marine fishing areas
Poissons, crustacés, mollusques, etc	Captures par pays ou zones	Zones de pêche maritimes
Peces, crustáceos, moluscos, etc	Capturas por países o áreas	Areas de pesca marítimas

Country or area Pays ou zone País o área	1993 t	1994 t	1995 t	1996 t	1997 t	1998 t	1999 t	2000 t	2001 t	2002 t
Albania	1 650 F	1 400 F	1 160	1 808	833	1 860	1 931	2 365	1 937	2 788
Algeria	101 895	135 402	105 872	81 989	91 580	92 344	102 396	113 156	133 623	134 320
Amer Samoa	27	111	152	210	431	586	518	830	3 663	6 963
Angola	119 200	125 413	116 781	131 815	140 304	157 149	169 799	232 351	246 518	254 797
Anguilla	330	333	150 F	200 F	250 F	250 F	250 F	250 F	250 F	250 F
Antigua Barb	1 097	1 145	1 610	1 463	1 665	1 708	1 660	1 754	1 824	2 374
Argentina	919 503	938 602	1 152 935	1 272 166	1 377 427	1 141 632	1 050 755	878 561	897 977	913 113
Aruba	260	260	140	160	205	182	175	163	163	163 F
Australia	225 522 F	201 896 F	203 740 F	199 544 F	196 062	202 609	210 482	189 424	193 748	193 532
Bahamas	10 073	10 311	9 944	10 197	10 439	10 124	10 473	11 070	9 290	9 300 F
Bahrain	8 958	7 628	9 389	12 940	10 050	9 849	10 620	11 718	11 230	11 204
Bangladesh	312 715	253 044	264 650	279 170	295 141	300 452	309 797	333 799	379 497	415 420
Barbados	3 214	2 818	3 581	3 512	2 809	3 644	3 250	3 100	2 676	2 500
Belgium	35 578	33 743	35 088	30 312	29 989	30 324	29 340	29 289	29 698	28 517
Belize	3 527	7 669	9 657	12 038	24 710	26 309	51 583	51 540	14 892	24 753
Benin	6 416	7 216	6 930	7 982	10 914	10 361	8 542	5 924	8 415	10 670
Bermuda	404	394	449	465	461	466	453	286	315	393
Bosnia Herzg	0	0	0	0	0	0	0	0	0	0
Brazil	530 100 F	548 615 F	513 666	522 173	565 714	532 599	518 470	567 687	589 397	604 409
Br Ind Oc Tr	0	0	0	0	0	0	0	0	0	0
Br Virgin Is	343	470	532	506	105	116	115	43	50 F	50 F
Brunei Darsm	1 703	4 441	4 712	7 390	4 504	5 014	3 160	2 464	1 476	2 044
Bulgaria	11 983	5 590 F	7 250	7 727	9 356	16 610	8 081	6 137	4 880	13 555
Cambodia	33 123	30 018	30 500	31 200	29 800	32 200	38 156	38 768	43 200	45 882
Cameroon	42 257	52 000 F	64 131	63 400	62 000 F	61 800 F	60 000 F	57 109	58 531	55 135
Canada	1 104 172	990 567	814 926	871 216	936 750	973 185	987 268	971 432	1 014 403	973 998
Cape Verde	7 000	8 256	8 495	9 155	9 705	9 424	10 360	10 586	9 653	8 000 F
Cayman Is	445	125	125	110	125	125	125	125	125	125
Channel Is	2 854 F	2 783	2 949 F	4 346	4 238	4 117	3 601	3 589	3 927	3 449
Chile	5 949 558	7 720 573	7 434 180	6 690 665	5 811 550	3 265 289	5 050 180	4 299 942	3 797 140	4 271 475
China	8 169 047	9 539 051	10 955 321	12 419 247	13 835 377	14 949 683	14 954 668	14 754 095	14 379 457	14 305 218
China,H.Kong	217 544	211 010	194 999	183 856	186 000	180 000	127 780	157 012	173 972	169 790
China, Macao	1 898	1 890	1 604	1 418	1 500 F	1 500 F	1 500 F	1 500 F	1 500 F	1 500 F
China,Taiwan	1 132 460 F	966 130 F	1 009 327 F	967 076	1 037 645	1 091 319	1 099 154	1 093 340	1 004 608	1 042 157
Christmas Is	0	0	0	0	0	0	0	0	0	0
Cocos Is	0	0	0	0	0	0	0	0	0	0
Colombia	92 959	62 816	97 175	102 768 F	102 308 F	101 235 F	94 207 F	104 790	105 000 F	110 000 F
Comoros	11 645	12 976	13 000 F	12 700 F	12 500 F	12 500 F	12 000	13 200	12 180	12 200 F
Congo Dem R	4 200	3 780	3 876	3 973	3 844	3 954	3 945	4 300 F	4 600 F	5 000 F
Congo Rep	18 898	17 912	18 965	19 600	19 095	17 500 F	18 241	24 525	22 040 F	22 500 F
Cook Is	985 F	933 F	1 080	900 F	820 F	770 F	750 F	720 F	700 F	1 768 F
Costa Rica	15 460	16 292	16 537	23 237	25 829	23 757	27 218	34 398	33 733	31 938
Côte dIvoire	63 495	58 374	58 854	57 606	52 137	57 071	63 709	65 270	62 926	55 900 F
Croatia	26 342	17 137	15 901	17 799	16 627	21 928	18 890	21 045	18 455	21 205
Cuba	74 574	68 097	71 697	74 464	69 091	55 616	64 544	58 056	42 723	31 804
Cyprus	10 016	9 422	9 255	12 462	24 749	19 225	39 568	67 404	81 001	1 918
Denmark	1 613 952	1 873 073	1 998 769	1 681 321	1 826 620	1 556 986	1 404 799	1 533 906	1 510 362	1 441 991
Djibouti	300 F	320 F	350 F	350 F	350 F	350 F	350 F	350 F	350 F	350 F
Dominica	794	882	950	1 030	1 079	1 212	1 200 F	1 200 F	1 200 F	1 217
Dominican Rp	10 820	19 058	15 768	12 606	13 468	9 014	7 804	10 842	12 059	15 976
Ecuador	286 261	329 948	505 095	702 674	548 588	309 622	497 472	592 147	586 170	318 140
Egypt	95 356	93 900	91 000	99 534	110 165	125 063	172 347	130 844	133 229	132 525
El Salvador	7 990	9 740	10 748	11 466	9 089	10 215	13 871	6 759	17 318	31 792
Eq Guinea	2 907	4 369	1 856	4 140	5 240	5 035	5 900 F	2 558	2 500 F	2 500 F
Eritrea	475	2 706	3 559	3 252	1 038	1 629	6 891	12 612	8 820	7 832
Estonia	144 774	121 771	129 662	106 085	121 174	114 836	108 685	109 956	102 706	96 874
Faeroe Is	246 294	237 708	287 796	304 587	329 145	364 497	358 362	454 399	524 837	490 000 F
Falkland Is	1 974	5 914	27 190	31 539	17 112	43 615	39 163	62 927	59 823	35 655
Fiji Islands	24 910	26 000 F	25 250	21 995	23 430	23 047	31 091	34 300 F	37 051	37 000 F
Finland	104 772	116 374	119 048	131 459	132 567	134 868	123 747	121 661	115 276	110 026
France	619 009	618 118	606 341	560 154	569 536	545 045	591 345	632 688	603 499	617 948
Fr Guiana	6 931	7 819	8 089	7 377 F	6 602	6 709	6 271 F	5 237 F	5 194 F	5 568 F
Fr Polynesia	8 075	8 175	9 020	9 910	11 617	12 420	12 283	13 846	15 351	15 490
Fr South Tr	460	524	519 F	437 F	375 F	388 F	425 F	272 F	263 F	414 F
Gabon	28 289	26 515	32 789	36 680	34 143	44 609	41 143	37 053	33 021	31 475
Gambia	18 908	20 381	21 252	29 101	29 754	26 502	27 500	26 516	32 027	43 269
Georgia	17 691 F	7 397 F	3 470 F	2 447	2 582	2 997	1 396	2 178	1 822	2 527
Germany	242 173	219 253	216 903	213 424	236 436	243 706	216 057	182 821	188 464	201 653
Ghana	320 619	280 737	292 844	403 153	377 088	368 141	418 276	377 570	373 181	296 678
Gibraltar	0	0	0	0	0	0	0	0	0	0
Greece	156 138	177 670	148 182	146 532	154 487	105 762	115 491	95 847	91 207	85 983 F
Greenland	116 650	117 417	128 890	116 018	120 596	128 542	160 253	159 711	158 485	202 828
Grenada	2 103	1 599	1 499	1 577	1 530	1 837	1 658	1 701	2 250	2 171
Guadeloupe	8 600	8 800	9 500	9 570	10 480	9 084	9 114	10 100	10 000 F	10 100 F
Guam	431	435	185	121	158	247	223	275	278	231
Guatemala	3 354	3 435	4 228	3 653	1 775	4 324	14 042	31 902	15 100 F	16 864 F
Guinea	56 000 F	60 000 F	64 760	60 580	58 841	65 764	83 314	87 513	101 227	100 000 F
GuineaBissau	5 100 F	5 750 F	6 078 F	6 750 F	7 000 F	5 800 F	4 800 F	4 800 F	4 800 F	4 800 F
Guyana	43 323	45 567	47 200 F	47 783	53 373	52 215	53 241	48 087	52 605	47 217
Haiti	4 550 F	5 000 F	5 017 F	4 745 F	4 801 F	4 759 F	4 500 F	4 500 F	4 500 F	4 500 F
Honduras	19 491	15 353	22 790	21 452	20 685 F	15 875 F	21 127 F	14 446 F	12 452 F	11 300 F
Iceland	1 714 674	1 556 264	1 611 809	2 059 560	2 205 540	1 681 535	1 735 897	1 982 348	1 980 555	2 129 495
India	2 486 807	2 704 733	2 656 862	2 814 529	2 881 673	2 681 053	2 776 067	2 760 728	2 801 689	2 957 157
Indonesia	2 781 138	2 993 033	3 202 943	3 269 088	3 553 402	3 672 338	3 716 916	3 801 792	3 963 422	4 189 444
Iran	246 985	218 944	252 583	242 439	238 492	226 949	243 800	260 500	262 805	269 000
Iraq	2 133	4 221	5 253	11 688	10 783	13 463	13 093	12 389	8 500 F	5 000 F
Ireland	275 845	290 550	384 542	328 427	289 005	321 044	282 557	275 411	355 528	281 542
Isle of Man	4 850	3 571	3 734	3 537	4 289	2 214	2 609	3 552	3 112	3 127
Israel	3 312	2 583	3 727	3 384	3 728	4 136	3 739	3 966	3 738	3 312
Italy	388 018	388 809	386 756	359 135	337 003	301 429	277 354	297 584	304 870	265 604

A-4

Fish, crustaceans, molluscs, etc	Capture production by countries or areas	Marine fishing areas
Poissons, crustacés, mollusques, etc	Captures par pays ou zones	Zones de pêche maritimes
Peces, crustáceos, moluscos, etc	Capturas por países o áreas	Areas de pesca marítimas

Country or area Pays ou zone País o área	1993 t	1994 t	1995 t	1996 t	1997 t	1998 t	1999 t	2000 t	2001 t	2002 t
Jamaica	22 550 F	24 827 F	23 782 F	30 579 F	19 585	17 160	16 937	5 226	5 250 F	5 200 F
Japan	7 161 246	6 523 674	5 878 062	5 838 371	5 842 722	5 224 733	5 117 678	4 914 201	4 651 652	4 382 157 F
Jordan	45	60	75	90	100	120	160	150	170	176
Kenya	5 617	3 772	5 465	6 296	6 099	6 600	6 634	4 763	7 388	6 720
Kiribati	27 072	27 753	30 306	31 871	30 052	35 304	52 741	25 563	32 375	31 001
Korea D P Rp	374 398 F	351 961	307 083	233 125	216 462	212 000 F	208 000 F	204 850 F	201 572	200 000 F
Korea Rep	2 244 929	2 347 399	2 310 271	2 405 751	2 197 268	2 021 087	2 112 205	1 817 854	1 984 751	1 663 289
Kuwait	8 466	7 752	8 616	8 255	7 826	7 799	6 271	6 000 F	5 846	5 900 F
Latvia	141 420	137 610	148 680	142 108	105 138	101 830	124 779	135 791	127 595	113 096
Lebanon	2 000	2 205	4 065	4 115	3 635	3 500	3 540	3 646	3 650	3 673
Liberia	3 782	3 721	4 829	4 308	4 580	6 830	11 472	7 726	7 286	7 500 F
Libya	31 085 F	33 500 F	34 400 F	32 976 F	31 877 F	32 911 F	32 850 F	33 387 F	33 239 F	33 666 F
Lithuania	116 025	47 975	56 108	87 219	42 289	64 841	71 247	77 076	149 977	149 388
Madagascar	84 261	86 431	85 653	84 475	86 391	96 395	99 630	102 093	105 583	111 284
Malaysia	1 047 350	1 065 586	1 108 436	1 126 689	1 168 973	1 149 093	1 248 402	1 285 696	1 231 287	1 272 105
Maldives	100 852	118 168	119 048	120 508	116 257	128 968	133 547	132 427	126 190	160 981
Malta	838	2 356	4 635	9 197	1 036	1 180	1 244	1 074	893	1 004
Marshall Is	471 F	393 F	381 F	2 772	400 F	500 F	500 F	8 060	35 544	38 742
Martinique	5 850	5 800 F	5 300	3 500 F	5 500	5 500	6 000	6 310	6 200	6 200 F
Mauritania	54 452 F	46 746 F	48 147 F	55 324 F	52 756 F	56 660 F	71 026 F	75 849 F	79 881 F	73 902 F
Mauritius	20 576	18 145	16 395	11 869	14 025	12 093	12 205	9 615	10 985	10 706
Mayotte	500 F	600 F	1 033 F	1 553	2 101 F	3 172 F	3 130	3 030 F	11 051	4 299
Mexico	991 754	1 080 521	1 207 320	1 341 687	1 375 560	1 079 525	1 114 141	1 208 764	1 306 640	1 368 006
Micronesia	17 884 F	22 922 F	7 624 F	9 064 F	10 122 F	16 576 F	12 838 F	23 879 F	19 648 F	20 382 F
Monaco	3 F	3 F	3 F	3 F	3 F	3 F	3 F	3 F	3 F	3 F
Montserrat	58	62	48	38	45	46	50 F	50 F	50 F	46
Morocco	622 894	752 577	847 451	641 386	789 806	708 733	743 268	895 012	1 082 293	892 845
Mozambique	25 506	22 531	21 740	27 405	28 035	27 683	27 123	25 916	24 790	24 306
Myanmar	597 637	599 876	602 885	455 294	631 226	680 838	759 664	880 018	931 492	1 008 113
Namibia	789 132	647 999	568 633	516 628	511 716	606 331	578 584	588 404	545 998	623 391
Nauru	500	500	400 F	300 F	250 F	200 F	150 F	100 F	50 F	21
Netherlands	460 155	417 607	433 985	408 641	449 506	535 079	512 308	493 504	515 962	461 457
NethAntilles	1 200 F	1 100 F	1 020 F	13 762 F	19 532 F	19 754 F	19 431 F	19 882 F	22 805 F	12 901 F
NewCaledonia	2 670	3 185	2 644	2 998	2 438	3 105	3 152	3 386	3 309	3 417 F
New Zealand	425 508	448 560	555 520	424 485	611 399	638 929	598 314	548 229	560 864	558 389
Nicaragua	7 622	9 210	10 457	14 300	14 883	18 636	22 789	26 932	21 748	22 078
Nigeria	142 782	163 259	231 579	248 472	294 279	324 004	316 235	309 062	297 971	293 823
Niue	120 F	150 F	150 F	200 F	200 F	200 F	200 F	200 F	200 F	200 F
Norfolk Is	0	0	0	0	0	0	0	0	0	0
N Marianas	136	176	191	225	250	235	193	189	197	198
Norway	2 414 696	2 365 687	2 523 698	2 648 119	2 862 620	2 860 716	2 627 020	2 698 843	2 686 733	2 742 614
Oman	105 772	118 572	139 861	121 618	118 995	106 171	108 809	120 421	129 907	142 670
Pakistan	499 159	418 574	404 444	395 397	422 265	433 456	474 665	437 601	420 698	418 104
Palau	1 104	1 001	1 025	999	904	951	960	1 047	1 075	1 008 F
Palest, O.T.	...	...	1 229	2 493	3 791	3 625	3 600 F	3 000 F	2 500 F	2 378
Panama	184 793	184 862	203 360	149 947	162 132	202 575	120 852	219 146	256 388	305 061
Papua N Guin	12 241 F	12 719 F	25 577 F	24 323 F	32 685 F	64 816 F	52 270 F	82 907 F	108 983 F	135 200 F
Peru	8 966 487	11 950 380	8 886 553	9 486 158	7 837 650	4 303 110	8 392 378	10 626 323	7 950 450	8 737 025
Philippines	1 623 548	1 641 010	1 674 695	1 606 246	1 646 453	1 687 376	1 726 584	1 744 885	1 813 181	1 899 661
Pitcairn Is	8 F	8 F	8 F	8 F	8 F	8 F	8 F	8 F	8 F	8 F
Poland	364 015	410 532	404 483	320 756	335 081	228 774	221 849	200 139	207 276	204 494
Portugal	287 799	263 061	260 251	259 846	221 879	224 171	208 407	187 521	191 016	200 025
Puerto Rico	1 877	2 275	3 173	2 701	3 187	3 006	3 020	4 154	3 794	2 529
Qatar	6 994	5 086	4 271	4 739	5 032	5 279	4 207	7 142	8 606	6 880
Réunion	1 679	2 531	2 500	3 607	4 288	4 579	4 043	4 082	3 635	3 700
Romania	5 257	11 653	40 227	12 114	3 872	4 431	2 507	2 476	2 431	2 122
Russian Fed	4 151 201	3 486 022	4 098 935	4 442 466	4 434 762	4 183 448	3 833 335	3 681 167	3 422 029	3 023 773
St Helena	726	702	915	819	897	1 060	632	718	866	598
St Kitts Nev	250 F	212	192	352	216	407	471	469	555	355
St Lucia	1 336	1 252	1 188	1 274	1 308	1 589	1 718	1 855	1 983	1 637
St Pier Mq	282	294	317	747	3 571	6 108	5 892	6 485	3 802	3 654
St Vincent	1 945	3 840	1 005	919 F	6 092	33 973	17 759	27 694	52 485	43 879
Samoa	673	1 108	2 519	2 727	7 041	7 547	10 204	13 003	12 965 F	12 391 F
Sao Tome Prn	2 334	3 391	3 565	3 980	3 338	3 477	3 756	3 500 F	3 500 F	3 500 F
Saudi Arabia	48 021	54 612	45 609	47 698	49 314	51 206	46 618	49 761	49 167	55 330
Senegal	354 562	322 421	323 617	388 759	426 366	382 872	378 125	379 597	383 202	355 824
Serbia-Monte	277	260	369	380	376	413	423	424	416	454
Seychelles	5 178	4 469	4 008	4 707	14 043	23 886	34 223	32 770	53 534	63 209
Sierra Leone	52 288	47 439	49 870	52 804 F	58 128	48 875	44 927	60 730	61 210	68 990
Singapore	9 279	11 278	10 102	9 943	9 250	7 733	6 489	5 371	3 342	2 769
Slovenia	1 987	2 007	1 851	2 078	2 065	1 959	1 784	1 630	1 621	1 460
Solomon Is	42 970 F	46 904 F	62 019 F	50 990 F	64 005 F	61 332 F	58 428 F	24 926 F	30 277 F	31 003 F
Somalia	27 500 F	29 600	27 700 F	25 800 F	23 900 F	22 000 F	20 000 F	20 000 F	19 800 F	17 850 F
South Africa	561 542	522 649	574 747	440 520	513 388	558 779	587 033 F	642 626 F	749 199	765 384
Spain	1 076 744 F	1 088 240 F	1 172 455 F	1 165 973	1 196 691	1 254 306	1 164 285	1 036 906	1 083 815	873 923
Sri Lanka	200 400	208 900	214 171	206 695	207 849	233 430	244 630	256 710	249 770	270 130
Sudan	2 500	4 000	4 000	4 500	5 000	5 500	5 500	5 000 F	5 000	5 000
Suriname	9 313	14 327	12 860 F	12 850 F	13 800 F	15 995	16 000	17 300 F	18 715	18 500 F
Svalbard Is	0	0	0	0	0	0	0	0	0	0
Sweden	339 624	384 560	402 638	369 071	355 395	409 327	349 776	337 075	310 582	293 527
Syria	2 019	1 950	1 950	2 670	2 574	2 750	2 600	2 581	2 322	2 823
Tanzania	36 685	40 785	42 771	61 645	50 210	48 000	50 489	52 779	52 900	49 680
Thailand	2 752 549	2 818 799	2 844 409	2 808 058	2 699 227	2 729 639	2 745 468	2 795 989	2 729 974	2 715 716
Timor-Leste	...	...	...	...	...	...	408 F	362	356	350 F
Togo	10 964	8 052	7 203	10 098	9 290	11 655	17 924	17 277	18 163	15 946
Tokelau	200 F	200 F	200 F	200 F	200 F	200 F	200 F	200 F	200 F	200 F
Tonga	2 374	2 499	2 597	2 915	2 871	4 076	4 221	3 760	4 673	4 804
Trinidad Tob	8 998	14 046	11 500 F	9 435	11 283	9 175	8 826	8 651	10 707	12 539
Tunisia	82 514	85 367	82 915	83 028	86 002	87 179	92 378	94 718	97 622	95 815
Turkey	503 957	544 736	585 994	478 226	408 693	432 700	523 634	460 524	484 407	522 744

A-4

Fish, crustaceans, molluscs, etc — **Capture production by countries or areas** — **Marine fishing areas**
Poissons, crustacés, mollusques, etc — **Captures par pays ou zones** — **Zones de pêche maritimes**
Peces, crustáceos, moluscos, etc — **Capturas por países o áreas** — **Areas de pesca marítimas**

Country or area Pays ou zone País o área	**1993** **t**	**1994** **t**	**1995** **t**	**1996** **t**	**1997** **t**	**1998** **t**	**1999** **t**	**2000** **t**	**2001** **t**	**2002** **t**
Turks Caicos	6 255	5 963 F	7 824	7 833	5 951	6 383	5 401	5 975	6 611	5 953
Tuvalu	1 460	561	399	400 F	500 F	500 F	500 F	500 F	500 F	500 F
Ukraine	286 215	257 473	380 869	397 837	362 585	457 298	403 983	387 402	356 571	261 776
Untd Arab Em	99 600	108 600	105 884	107 000	114 358	114 739	117 607	105 456	112 561	97 574
UK	858 320	875 827	907 817	865 756	889 445	918 650	835 884	744 828	737 903	686 488
USA	5 468 540	5 492 676	5 188 170	4 967 720	4 945 093	4 672 851	4 713 233	4 692 299	4 914 758	4 907 362
US Minor Is	0	0	0	0	0	0	0	0	0	0
US Virgin Is	650 F	550 F	470 F	400 F	350 F	300 F	263	300 F	300 F	300 F
Uruguay	118 194	119 766	125 597	122 732	134 738	138 776	104 160	111 037	104 452	108 378
Vanuatu	40 515 F	47 190 F	57 646 F	45 615 F	70 822 F	74 908 F	93 041 F	68 896 F	24 521 F	17 139 F
Venezuela	367 610	402 806	446 642	446 204	429 245	457 459	364 590	333 096	393 881	474 608
Viet Nam	785 304	946 322	990 250	1 058 708	1 098 736	1 155 154	1 217 193	1 280 590 F	1 357 023 F	1 358 800 F
Wallis Fut I	150	193	170	180	176	300	300 F	300 F	300 F	300 F
Westn Sahara	0	0	0	0	0	0	0	0	0	0
Yemen	82 356	81 885	107 970	104 955	115 600	127 620	124 385	114 751	142 198	159 262
Other nei	138 834	134 742	297 280	174 123	153 713	182 984	213 943	226 861	251 152	337 594
World total Total mondial Total mundial	**79 956 636**	**85 355 157**	**85 105 045**	**86 412 954**	**86 723 099**	**79 604 977**	**85 246 346**	**86 771 372**	**84 163 995**	**84 452 487**
World excl. China Monde excl. Chine Mundo excl. China	**71 787 589**	**75 816 106**	**74 149 724**	**73 993 707**	**72 887 722**	**64 655 294**	**70 291 678**	**72 017 277**	**69 784 538**	**70 147 269**

A-5

Fish, crustaceans, molluscs, etc — **Capture production by low-income food-deficit countries (LIFDCs)**
Poissons, crustacés, mollusques, etc — **Captures par pays à faible revenu et à déficit vivrier (PFRDV)**
Peces, crustáceos, moluscos, etc — **Capturas por países de bajos ingresos y con déficit de alimentos (PBIDA)**

Country or area Pays ou zone País o área	Population in 2002** Population en 2002** Población en 2002** '000	1996 t	1997 t	1998 t	1999 t	2000 t	2001 t	2002 t
Africa Afrique Africa	*720 678*	*4 356 573*	*4 634 250*	*4 652 300*	*4 855 576*	*5 041 384*	*5 254 919*	*5 006 441*
Angola	*13 184*	137 815	146 304	163 149	175 799	238 351	252 518	260 797
Benin	*6 558*	42 175	43 785	42 139	40 436	32 324	38 415	40 663
Burkina Faso	*12 624*	8 000	8 000	8 335	7 600	8 500	8 500 F	8 500 F
Burundi	*6 602*	3 041	20 296	13 426	9 199	17 315	8 964	9 000 F
Cameroon	*15 729*	98 400 F	102 000 F	106 800 F	110 000 F	112 109	111 031	120 135
Cape Verde	*454*	9 155	9 705	9 424	10 360	10 586	9 653	8 000 F
Cent Afr Rep	*3 819*	14 000 F	14 250 F	14 500 F	15 000	15 000 F	15 000 F	15 000 F
Chad	*8 348*	100 000	85 000	84 000	84 000 F	84 000 F	84 000 F	84 000 F
Comoros	*747*	12 700 F	12 500 F	12 500 F	12 000	13 200	12 180	12 200 F
Congo Dem R	*51 201*	163 010	162 211	178 041	208 448	209 300 F	214 600 F	220 000 F
Congo Rep	*3 633*	45 473	38 082	42 955 F	43 509	44 729	42 240 F	43 000 F
Côte dIvoire	*16 365*	69 168	64 169	69 572	74 365	75 772	73 556	79 743 F
Djibouti	*693*	350 F	350 F	350 F	350 F	350 F	350 F	350 F
Egypt	*70 507*	340 434	371 332	406 204	422 665	384 314	428 651	425 170
Eq Guinea	*481*	5 040	6 090	6 005	7 001	3 634	3 500 F	3 500 F
Eritrea	*3 991*	3 252	1 038	1 629	6 891	12 612	8 820	7 832
Ethiopia	*68 961*	8 770	10 370	14 000	15 858	15 681	15 390	12 300
Gambia	*1 388*	31 601	32 254	29 002	30 000	29 016	34 527	45 769
Ghana	*20 471*	476 733	447 088	442 641	492 776	452 070	447 681	371 178
Guinea	*8 359*	63 360	62 441	69 764	87 314	91 513	105 227	104 000 F
GuineaBissau	*1 449*	7 000 F	7 250 F	6 000 F	5 000 F	5 000 F	5 000 F	5 000 F
Kenya	*31 540*	180 988	161 054	172 592	205 287	215 106	164 151	144 512
Lesotho	*1 800*	28 F	30 F	30 F	30	32	24	24 F
Liberia	*3 239*	8 308	8 580	10 830	15 472	11 726	11 286	11 500 F
Madagascar	*16 916*	114 475	116 391	126 395	129 630	132 093	135 583	141 284
Malawi	*11 871*	63 569	56 340	41 111	45 392	43 000 F	40 619	41 329
Mali	*12 623*	111 910	99 550	98 000	98 536	109 870	100 000 F	100 000 F
Mauritania	*2 807*	60 324 F	57 756 F	61 660 F	76 026 F	80 849 F	84 881 F	78 902 F
Morocco	*30 072*	642 886	791 906	710 436	745 431	896 620	1 083 953	894 957
Mozambique	*18 537*	34 915	39 703	36 677	36 175	37 729	30 074	36 462
Niger	*11 544*	4 156	6 328	7 013	11 000	16 250	20 800	23 520
Nigeria	*120 911*	337 993	387 923	463 024	455 628	441 377	452 146	481 056
Rwanda	*8 272*	2 952	4 428	6 641	6 433	6 726	6 828	7 000 F
Sao Tome Prn	*157*	3 980	3 338	3 477	3 756	3 500 F	3 500 F	3 500 F
Senegal	*9 855*	411 759	457 366	403 872	412 125	402 047	403 202	375 824
Sierra Leone	*4 764*	67 304 F	72 628	63 065	59 407	74 730	75 210	82 990
Somalia	*9 480*	26 050 F	24 150 F	22 250 F	20 250 F	20 200 F	20 000 F	18 000 F
Sudan	*32 878*	45 000	47 000	49 500	49 500	53 000 F	58 000	58 000
Swaziland	*1 069*	60 F	65 F	70 F	70 F	70 F	70 F	70 F
Tanzania	*36 276*	323 921	356 960	348 000	310 509	332 779	335 900	323 530
Togo	*4 801*	15 098	14 290	16 655	22 924	22 277	23 163	20 946
Uganda	*25 004*	195 088	218 026	220 628	226 097	219 356	220 726	221 898
Zambia	*10 698*	66 332	65 923	69 938	67 327	66 671	65 000 F	65 000 F
America, North Amérique du Nord América del Norte	*31 605*	*126 445*	*118 982*	*101 915*	*119 273*	*108 576*	*85 092*	*73 910*
Cuba	*11 271*	84 208	76 694	60 770	69 135	61 061	44 730	33 838
Haiti	*8 218*	5 245 F	5 301 F	5 259 F	5 000 F	5 000 F	5 000 F	5 000 F
Honduras	*6 781*	21 550	20 811	15 994	21 229	14 507	12 563	11 402
Nicaragua	*5 335*	15 442	16 176	19 892	23 909	28 008	22 799	23 670
America, South Amérique du Sud América del Sur	*12 810*	*702 974*	*548 988*	*310 022*	*497 872*	*592 547*	*586 570*	*318 540*
Ecuador	*12 810*	702 974	548 988	310 022	497 872	592 547	586 570	318 540
Asia Asie Asia	*3 144 283*	*25 265 980*	*27 223 720*	*28 771 318*	*29 377 114*	*29 439 545*	*29 485 422*	*29 896 949*
Afghanistan	*22 930*	1 300 F	1 250 F	1 200 F	1 200 F	1 000 F	800 F	900 F
Armenia	*3 072*	580	580	698	1 144	1 133	866	465
Azerbaijan	*8 297*	6 702	5 161	4 760	20 866	18 797	10 893	11 334
Bangladesh	*143 809*	814 787	829 426	839 141	959 215	1 004 264	1 068 417	1 103 855
Bhutan	*2 190*	300 F	300 F	300 F	300 F	300 F	300 F	300 F
Cambodia	*13 810*	94 710	102 800	107 900	269 156	284 368	428 200	406 182
China	*1 272 403*	14 182 107	15 722 344	17 229 927	17 240 032	16 987 325	16 529 389	16 553 144
Georgia	*5 177*	2 453	2 583	3 001	1 413	2 200	1 830	2 537
India	*1 049 549*	3 447 954	3 523 448	3 373 492	3 472 150	3 666 428	3 777 092	3 770 912
Indonesia	*217 131*	3 604 795	3 857 660	3 961 004	4 044 543	4 120 126	4 273 662	4 505 474
Iraq	*24 510*	30 737	31 302	22 574	22 423	20 767	16 500 F	13 000 F
Korea D P Rp	*22 541*	253 125	236 462	228 000 F	220 000 F	212 850 F	206 500	205 000 F
Kyrgyzstan	*5 067*	160 F	120 F	80 F	48	52	57	48
Laos	*5 529*	23 000 F	18 857	19 642	30 041	29 250	31 000 F	33 440
Maldives	*309*	120 508	116 257	128 968	133 547	132 427	126 190	160 981
Mongolia	*2 559*	221	180	311	524	425	117	129
Nepal	*24 609*	11 230	11 230	12 000	12 752	16 700	16 700	17 900
Pakistan	*149 911*	537 489	589 795	596 980	654 530	614 069	600 798	599 104
Philippines	*78 580*	1 783 601	1 805 806	1 833 380	1 872 818	1 896 638	1 949 026	2 030 542
Sri Lanka	*18 910*	228 945	235 099	263 330	276 080	293 410	279 640	298 260
Syria	*17 381*	5 773	6 131	7 097	7 938	6 572	8 291	9 178
Tajikistan	*6 195*	40 F	75 F	100 F	80 F	78 F	137	181
Turkmenistan	*4 794*	9 014	8 179	7 014	9 058	12 228	12 749	12 812
Uzbekistan	*25 705*	1 494	3 075	2 799	2 871	3 387	4 070	2 009
Yemen	*19 315*	104 955	115 600	127 620	124 385	114 751	142 198	159 262

A-5 Fish, crustaceans, molluscs, etc — Capture production by low-income food-deficit countries (LIFDCs)
Poissons, crustacés, mollusques, etc — Captures par pays à faible revenu et à déficit vivrier (PFRDV)
Peces, crustáceos, moluscos, etc — Capturas por países de bajos ingresos y con déficit de alimentos (PBIDA)

Country or area Pays ou zone País o área	Population in 2002** Population en 2002** Población en 2002** '000	1996 t	1997 t	1998 t	1999 t	2000 t	2001 t	2002 t
Europe Europe Europa	*19 253*	*5 524*	*4 142*	*5 771*	*5 894*	*6 581*	*6 881*	*12 178*
Albania	*3 141*	2 125	1 013	2 683	2 745	3 320	3 310	3 955
Belarus	*9 940*	821	499	457	514	553	943	5 575
Bosnia Herzg	*4 126*	2 500 F	2 500 F	2 500 F	2 500 F	2 500 F	2 500 F	2 500 F
Macedonia	*2 046*	78	130	131	135	208	128	148
Oceania Océanie Oceanía	*6 529*	*169 426*	*218 605*	*257 907*	*280 684*	*229 296*	*223 122*	*240 735*
Kiribati	*87*	31 871	30 052	35 304	52 741	25 563	32 375	31 001
Papua N Guin	*5 586*	37 823 F	46 185 F	78 316 F	65 770 F	96 407 F	122 483 F	148 700 F
Samoa	*176*	2 727	7 041	7 547	10 204	13 004	12 966 F	12 392 F
Solomon Is	*463*	50 990 F	64 005 F	61 332 F	58 428 F	24 926 F	30 277 F	31 003 F
Tuvalu	*10*	400 F	500 F	500 F	500 F	500 F	500 F	500 F
Vanuatu	*207*	45 615 F	70 822 F	74 908 F	93 041 F	68 896 F	24 521 F	17 139 F
LIFDCs PFRDV PBIDA	***3 935 158***	***30 626 922***	***32 748 687***	***34 099 233***	***35 136 413***	***35 417 929***	***35 642 006***	***35 548 753***
Other countries Autres pays Otros países	***2 289 820***	***63 219 721***	***61 549 899***	***53 572 801***	***58 637 639***	***60 084 075***	***57 220 081***	***57 641 901***
World total Total mondial Total mundial	***6 224 978***	**93 846 643**	**94 298 586**	**87 672 034**	**93 774 052**	**95 502 004**	**92 862 087**	**93 190 654**

** Population (mid-2002) estimated by the Population Division of the United Nations, New York.

** Les données relatives à la population (mi-2002) ont été estimées par la Division de la Population des Nations Unies, New York.

** Los datos relativos a la población (mitad-2002) están estimados por la División de Población de las Naciones Unidas, Nueva York.

A-6

Seaweeds and other aquatic plants
Algues et autres plantes aquatiques
Algas y otras plantas acuáticas

Capture production by countries or areas
Captures par pays ou zones
Capturas por países o áreas

All fishing areas
Toutes les zones de pêche
Todas las áreas de pesca

Country or area Pays ou zone País o área	1993 t	1994 t	1995 t	1996 t	1997 t	1998 t	1999 t	2000 t	2001 t	2002 t
Argentina	2 418	3 139	2 403	2 500	900	900 F	900 F	3	...	...
Australia	18 275	17 519	22 316	18 570	21 152	20 811	20 774	13 650	13 547	9 916
Canada	21 560	30 214	26 671	24 332	34 696	27 472	30 660	34 507	14 932	16 751
Chile	107 109	116 755	250 038	216 815	178 839	197 495	230 203	247 376	234 253	244 020
China	124 250	151 000	150 000	151 630	184 270	170 520	215 550	204 290	266 870	297 160
China,H.Kong	1	0	0	0	0	0	0	0	0	0
China,Taiwan	206	212	96	106	124	87	123	125	91	148
Cook Is	40 F	40 F	40 F	45 F	50 F	50 F	50 F	50 F	50 F	50 F
Estonia	0	411	548	163	2 444	2 880	1 319	201	325	912
Fiji Islands	376	400 F	400 F	320 F	400 F	480	536	520 F	480 F	480 F
France	60 333	79 527	75 550	84 372	75 502	67 032	70 996	68 232	66 283	80 366
Iceland	11 303	13 982	11 841	14 411	19 505	17 770	18 147	17 501	20 367	14 962
India	87 500 F	90 000 F	92 500 F	95 000 F	97 500 F	100 000 F	100 000 F	100 000 F	100 000 F	90 000 F
Indonesia	118 395	110 438	111 575	161 544	125 979	47 515	23 152	42 712	34 450	35 770
Ireland	29 923	33 417	35 456	33 360	35 386	36 100 F	36 100 F	36 100 F	36 100 F	36 100 F
Italy	...	600	500	1 850	1 950	2 000	2 000 F	2 000 F	2 000 F	2 000 F
Japan	167 604	137 722	151 042	154 078	149 910	116 892	120 877	119 030	122 145	127 897
Korea Rep	22 096	26 899	22 373	22 916	23 234	12 599	12 895	13 023	14 933	14 800 F
Latvia	1	0	0	0	0	0	0	0	0	0
Madagascar	423	702	787	787	1 000	2 510	1 933	5 792	5 045	2 909
Mexico	60 600	36 777	49 334	34 598	42 067	12 456	32 090	33 555	46 927	30 124
Morocco	7 108	5 357	7 858	7 625	8 094	7 049	5 920	6 080	10 015	7 919
Namibia	226	175	799	936	851	897	660	829	800	500
New Zealand	0	0	16	148	34 283	7 831	733	0	1	14
Norway	169 606	185 065	185 033	173 160	191 681	179 762	178 542	192 426	175 210	182 641
Peru	243	170	413	307	155	1 650	1 733	1 312	5 505	6 176
Philippines	1 144	1 062	919	884	494	417	433	413	447	684
Portugal	3 000 F	2 686	2 816	1 703	1 743	968	1 949	1 224	1 198	...
Russian Fed	6 753	7 661	10 546	18 667	26 441	30 279	25 715	53 653	27 541	55 387
Senegal	50 F	50 F	50 F	50 F	50 F	100 F	150 F	200 F	223	232
South Africa	7 960	4 376	6 252	8 500 F	11 500 F	14 400 F	17 300 F	20 511	32 138	32 000 F
Spain	13 939 F	14 291 F	14 392 F	14 871 F	14 885 F	13 940	13 818	14 214	14 011	234
Tanzania	3 600	3 500 F	3 500 F	3 500 F	3 610	4 000	4 500	5 000	5 000 F	4 500
Timor-Leste	...	...	...	...	...	...	1	1	1	-
Ukraine	424	4 238	3 172	1 190	331	314	160	26	...	147
UK	4 769	5 683	7 400	8 400	8 100	...	-	-	-	-
USA	85 108	116 841	73 999	59 295	70 729	25 689	79 236	42 058	37 165	47 183
World total Total mondial Total mundial	**1 136 343**	**1 200 909**	**1 320 635**	**1 316 633**	**1 367 855**	**1 122 865**	**1 249 155**	**1 276 614**	**1 288 053**	**1 341 982**

The data in this A-6 table and in tables B-91, B-92, B-93 and B-94 refer to the annual production of seaweeds and other aquatic plants (expressed in metric tons and on a wet-weight basis) through harvesting of wild stocks.

See paragraph 5 of the INTRODUCTION.

Les données présentées dans les tableaux A-6 et B-91, B-92, B-93 et B-94 se rapportent à la production annuelle d'algues marines et autres plantes aquatiques (exprimée en tonnes et en poids vert) fournie par la récolte des espèces sauvages.

Voir le paragraphe 5 de l'INTRODUCTION.

Los datos de este cuadro A-6 y de los cuadros B-91, B-92, B-93 y B-94 se refieren a la producción anual de algas marinas y otras plantas acuáticas (expresada en toneladas métricas y peso en húmedo) obtenidas mediante la recolección de plantas silvestres.

Véase el párrafo 5 en la INTRODUCCIÓN.

B - Capture production: by species groups

B - Captures: par groupes d'espèces

B - Capturas: por grupos de especies

B-00 (a)

Fish, crustaceans, molluscs, etc — **Capture production by species groups and fishing areas**
Poissons, crustacés, mollusques, etc — **Captures par groupes d'espèces et zones de pêche**
Peces, crustáceos, moluscos, etc — **Capturas por grupos de especies y áreas de pesca**

Species group Groupe d'espèces Grupo de especies	Fishing area Zone de pêche Area de pesca	1996 t	1997 t	1998 t	1999 t	2000 t	2001 t	2002 t
11 Carps, barbels and other cyprinids	01	81 926	86 704	78 960	96 396	97 962	98 307	92 378
Carpes, barbeaux et autres cyprinidés	02	28 105	24 924	21 241	20 971	22 094	21 943	19 457
Carpas, barbos y otros ciprínidos	03	502	213	265	307	1 745	390	392
	04	429 436	376 916	377 296	382 408	336 298	310 843	355 909
	05	117 076	111 636	120 232	106 265	103 828	108 379	114 431
	06	0	28	2	3	4	6	7
	27	5 834	6 674	6 311	6 574	8 172	8 399	9 975
	37	476	882	617	643	854	575	375
	61	110	101	100	56	8	52	38
	Group total	663 465	608 078	605 024	613 623	570 965	548 894	592 962
12 Tilapias and other cichlids	01	366 526	361 520	394 271	423 730	467 477	475 290	472 135
Tilapias et autres cichlidés	02	86 191	84 500	72 068	69 657	75 289	66 090	61 990
Tilapias y otros cíclidos	03	16 173	15 460	15 370	14 242	16 226	22 123	22 194
	04	86 676	95 384	112 104	125 424	122 422	119 686	121 397
	06	2 347	2 350	2 604	2 600	2 590	2 588	2 590
	34	2 603	2 582	1 365	1 760	2 062	2 324	2 333
	Group total	560 516	561 796	597 782	637 413	686 066	688 101	682 639
13 Miscellaneous freshwater fishes	01	1 376 824	1 443 241	1 492 828	1 462 412	1 441 072	1 425 036	1 473 314
Poissons d'eau douce divers	02	51 261	49 959	51 319	51 767	58 297	47 613	51 295
Peces de agua dulce diversos	03	322 890	312 455	315 347	328 254	323 038	344 523	352 681
	04	2 791 199	2 883 899	3 067 956	3 625 592	3 611 382	3 751 883	3 494 506
	05	115 531	105 834	101 028	95 408	114 859	100 874	106 866
	06	9 357	9 353	9 158	9 122	9 137	9 026	8 994
	27	14 971	17 105	17 928	17 062	11 133	11 466	12 099
	31	7 303	3 656	2 870	4 130	6 772	6 937	-
	37	2 822	2 825	3 031	2 568	2 956	3 522	3 312
	Group total	4 692 158	4 828 327	5 061 465	5 596 315	5 578 646	5 700 880	5 503 067
21 Sturgeons, paddlefishes	02	526	615	525	420	281	173	172
Esturgeons, spatules	04	2 179	1 733	1 531	1 320	1 288	1 231	915
Esturiones, sollos	05	1 337	1 365	1 222	856	704	633	526
	21	86	91	69	9	3	4	-
	27	0	0	1	2	-	-	-
	37	584	637	429	242	85	33	39
	67	1	0	0	0	242	195	207
	Group total	4 713	4 441	3 777	2 849	2 603	2 269	1 859
22 River eels	01	637	985	801	959	2 164	2 129	2 002
Anguilles	02	730	796	739	770	942	648	269
Anguilas	04	5 795	2 089	2 064	2 255	5 687	2 907	4 728
	05	3 787	3 194	3 133	3 092	3 101	2 483	2 326
	06	931	810	1 286	1 209	1 151	1 244	1 036
	21	551	455	496	353	395	220	494
	27	2 959	4 327	2 290	2 622	1 982	2 098	1 990
	31	35	19	9	2	1	0	11
	34	0	1	4	-	-	-	-
	37	962	1 257	917	682	464	602	648
	57	208	203	160	129	133	175	167
	81	524	588	400	416	417	350	367
	Group total	17 119	14 724	12 299	12 489	16 437	12 856	14 038
23 Salmons, trouts, smelts	01	11	16	24	29	31	67	112
Saumons, truites, éperlans	02	32 374	27 972	29 200	25 267	22 515	22 669	19 563
Salmones, truchas, eperlanos	03	463	1 208	733	527	566	658	978
	04	38 676	35 498	33 261	28 219	29 420	24 738	24 659
	05	64 252	69 832	66 252	83 333	82 718	80 962	88 611
	21	1 301	1 458	1 374	428	1 070	405	355
	27	14 553	13 283	11 199	10 262	14 507	9 324	10 355
	37	18	22	26	-	-	0	-
	61	455 779	471 418	428 221	380 303	350 771	401 629	371 815
	67	423 012	301 117	318 745	382 944	300 917	349 383	288 126
	77	2 084	2 747	944	1 996	2 623	1 206	2 424
	81	1	1	3	1	1	1	0
	Group total	1 032 524	924 572	889 982	913 309	805 139	891 042	806 998
24 Shads	01	290	253	359	270	284	258	238
Aloses	02	3 187	3 941	3 825	5 199	3 070	4 228	4 462
Sábalos	03	642	455	567	523	407	340	313
	04	220 802	212 274	239 546	250 449	366 137	210 293	159 638
	05	96 744	86 878	120 291	156 813	120 766	47 425	33 658
	06	480	480	480	480	480	480	480
	21	6 415	6 887	8 281	7 367	5 919	4 007	6 239
	27	122	113	58	59	140	103	158
	31	343	246	280	117	288	171	286
	34	11 585	8 810	8 113	4 608	7 489	5 633	8 819
	37	3 862	3 439	6 391	13 550	14 141	30 727	30 601

B-00 (a)

Fish, crustaceans, molluscs, etc — **Capture production by species groups and fishing areas**
Poissons, crustacés, mollusques, etc — **Captures par groupes d'espèces et zones de pêche**
Peces, crustáceos, moluscos, etc — **Capturas por grupos de especies y áreas de pesca**

Species group Groupe d'espèces Grupo de especies	Fishing area Zone de pêche Area de pesca	1996 t	1997 t	1998 t	1999 t	2000 t	2001 t	2002 t
	51	5 547	5 265	4 607	4 548	4 736	4 784	6 039
	57	197 740	192 141	180 806	190 310	194 226	209 540	206 628
	61	75 549	106 577	116 818	137 932	126 545	127 786	112 677
	67	15	34	62	-	105	178	66
	71	12 415	8 996	10 682	11 497	11 755	15 247	15 001
	77	-	0	-	-	-	-	-
	Group total	635 738	636 789	701 166	783 722	856 488	661 200	585 303
25 Miscellaneous diadromous fishes Poissons diadromes divers Peces diádromos diversos	02	1 040	1 219	1 132	1 155	1 327	1 371	1 126
	04	369	34	3 000	83	163	201	329
	05	569	467	1 042	1 243	1 354	1 104	1 927
	06	350	350	350	350	350	350	350
	21	2 144	2 785	3 032	3 000	3 123	2 934	2 847
	27	67	135	101	40	571	658	818
	31	9	12	14	4	12	15	17
	37	-	-	-	-	-	8	4
	51	395	334	335	341	125	134	159
	57	13 989	13 790	23 061	15 685	15 982	15 665	16 535
	61	18	9	26	47	32	81	73
	71	44 227	45 706	47 280	53 973	60 282	53 137	57 344
	77	0	1	0	0	7	...	1
	Group total	63 177	64 842	79 373	75 921	83 328	75 658	81 530
31 Flounders, halibuts, soles Flets, flétans, soles Platijas, halibuts, lenguados	01	...	547	381	763	2 490	2 134	1 858
	21	84 379	91 352	92 171	113 787	118 636	122 063	116 383
	27	318 478	310 917	298 034	317 559	300 255	303 387	276 423
	31	2 527	1 546	1 891	1 673	1 765	1 695	1 922
	34	35 323	38 371	46 234	41 604	50 540	49 910	44 910
	37	13 340	11 718	11 059	10 469	11 167	10 874	8 695
	41	10 356	11 976	10 902	8 701	8 689	8 095	8 047
	47	1 996	2 671	2 960	3 069	2 928	4 021	5 320
	48	-	-	-	-	0	-	-
	51	20 730	25 299	20 676	15 878	27 872	14 647	18 916
	57	30 824	33 443	28 204	26 965	26 879	26 200	28 228
	61	181 394	186 917	197 870	210 735	225 682	203 918	188 575
	67	217 113	273 510	198 371	174 192	203 808	173 733	189 849
	71	14 922	24 987	16 791	20 281	18 157	19 443	21 484
	77	9 229	8 937	4 853	6 470	6 284	4 823	4 887
	81	4 514	8 005	4 278	3 554	2 954	3 269	3 025
	87	761	411	436	411	365	439	318
	Group total	945 886	1 030 607	935 111	956 111	1 008 471	948 651	918 840
32 Cods, hakes, haddocks Morues, merlus, églefins Bacalaos, merluzas, eglefinos	02	23	24	25	25	25	-	-
	05	98	163	177	22	11	55	127
	21	120 222	127 256	140 007	143 127	138 976	148 393	140 651
	27	3 191 871	3 135 608	3 256 629	3 291 268	3 330 963	3 607 974	3 406 129
	31	...	4	2	2	6	3	4
	34	31 447	25 650	25 150	25 212	19 545	25 435	25 462
	37	90 708	72 072	75 371	64 641	71 692	64 917	47 992
	41	860 118	815 951	779 466	618 103	508 894	547 832	615 675
	47	287 275	261 732	303 451	309 702	309 995	324 936	309 613
	48	25	21	18	13	3	6	9
	51	1 251	1 348	1 456	748	1 472	2 436	2 372
	57	2 716	2 864	5 890	7 566	9 143	7 440	9 130
	58	-	13	76	153	376	206	386
	61	3 529 239	3 516 455	3 030 625	2 463 168	1 959 990	1 885 369	1 307 857
	67	1 665 540	1 669 564	1 714 162	1 511 756	1 631 326	1 830 568	1 882 188
	71	24	21	33	765	50	60	42
	77	138	1 273	806	379	718	1 051	1 421
	81	251 354	351 951	435 214	389 941	350 209	339 181	299 470
	87	752 199	391 763	565 958	507 147	339 571	458 922	343 785
	88	-	-	9	27	77	61	166
	Group total	10 784 248	10 373 733	10 334 525	9 333 765	8 673 042	9 244 845	8 392 479
33 Miscellaneous coastal fishes Poissons côtiers divers Peces costeros diversos	01	15 875	18 986	20 939	20 750	24 658	28 359	28 718
	02	...	...	...	...	...	220	213
	04	37 088	41 467	35 911	35 721	39 120	41 978	31 976
	05	52	109	219	451	360	494	447
	06	1 850	1 850	1 850	1 850	1 850	1 850	1 850
	21	20 492	20 055	19 826	28 628	19 522	20 343	21 158
	27	880 800	1 267 534	1 065 549	789 791	768 549	939 102	929 214
	31	164 649	157 162	159 994	130 630	155 314	150 992	137 757
	34	263 289	291 576	267 289	292 312	277 854	287 190	283 954
	37	183 355	170 147	173 388	176 952	174 397	161 557	160 344
	41	226 233	234 505	213 701	207 297	236 185	251 775	248 419
	47	23 740	19 664	52 202	36 793	47 795	61 670	73 764
	48	-	0	1	10	6	4	16
	51	821 440	856 296	777 742	848 740	805 850	798 578	821 758
	57	488 324	485 137	511 989	511 983	511 301	517 302	506 572
	58	15	5	6	11	0	0	1
	61	1 572 287	1 623 465	1 635 578	1 614 745	1 724 308	1 640 243	1 591 033
	67	92 432	62 763	54 050	54 531	47 878	59 881	40 847

B-00 (a)

Fish, crustaceans, molluscs, etc — **Capture production by species groups and fishing areas**
Poissons, crustacés, mollusques, etc — **Captures par groupes d'espèces et zones de pêche**
Peces, crustáceos, moluscos, etc — **Capturas por grupos de especies y áreas de pesca**

Species group Groupe d'espèces Grupo de especies	Fishing area Zone de pêche Area de pesca	1996 t	1997 t	1998 t	1999 t	2000 t	2001 t	2002 t
	71	1 026 052	1 076 116	1 030 488	1 159 252	1 076 667	1 100 750	1 149 324
	77	43 004	46 049	44 181	57 328	46 348	60 739	63 251
	81	25 463	26 477	23 120	27 367	26 536	25 896	26 830
	87	52 472	53 423	63 270	65 065	55 474	50 711	53 678
	88	-	-	0	0	0	1	0
	Group total	5 938 912	6 452 786	6 151 293	6 060 207	6 039 972	6 199 635	6 171 124
34 Miscellaneous demersal fishes	04	490	447	270	494	402	620	337
Poissons démersaux divers	21	56 985	59 856	66 934	66 707	73 940	81 218	91 644
Peces demersales diversos	27	435 011	418 836	425 131	393 488	420 190	422 671	393 672
	31	12 416	14 408	19 486	17 505	15 203	21 535	19 132
	34	94 725	76 143	77 018	50 478	38 726	40 055	38 351
	37	21 004	23 071	18 757	17 683	17 558	19 894	20 118
	41	58 259	59 688	74 873	66 381	55 017	74 027	68 142
	47	58 742	65 833	73 371	53 312	51 583	55 853	63 594
	48	3 602	3 812	3 208	3 915	9 120	5 016	8 422
	51	90 794	189 577	118 150	166 673	168 664	155 063	194 034
	57	68 457	72 513	94 098	93 470	84 293	79 175	77 957
	58	5 661	8 824	10 030	9 573	11 542	9 842	7 270
	61	1 469 582	1 394 197	1 655 751	1 637 181	1 696 006	1 712 536	1 719 785
	67	107 235	111 003	88 790	97 110	86 495	79 323	76 243
	71	43 313	47 778	48 101	55 292	55 184	53 714	53 763
	77	9 816	9 547	7 942	5 446	4 517	4 418	3 983
	81	128 844	159 270	157 473	155 554	154 231	144 720	142 461
	87	25 436	25 239	30 009	52 479	79 913	45 377	40 773
	88	-	0	42	297	751	662	1 368
	Group total	2 690 372	2 740 042	2 969 434	2 943 038	3 023 335	3 005 719	3 021 049
35 Herrings, sardines, anchovies	01	1 900	1 850	1 800	1 798	1 500	1 700	1 700
Harengs, sardines, anchois	21	575 240	606 757	521 264	472 439	460 612	543 766	439 255
Arenques, sardinas, anchoas	27	2 875 382	3 089 362	3 013 492	2 995 072	2 890 562	2 427 552	2 350 717
	31	679 660	758 862	734 996	865 288	713 794	641 554	783 083
	34	1 753 261	1 828 435	1 869 042	1 695 482	1 748 123	1 926 048	1 740 654
	37	674 739	642 385	625 771	781 261	722 642	796 135	810 153
	41	139 838	161 501	115 964	59 628	58 851	89 073	82 478
	47	233 659	328 352	419 566	495 703	582 167	604 524	614 005
	51	373 255	433 033	415 773	407 553	546 667	577 895	586 653
	57	435 808	497 817	490 752	393 681	394 122	431 941	408 334
	61	1 990 976	2 556 096	2 932 317	2 856 880	2 534 514	2 614 151	2 443 271
	67	71 607	84 451	73 503	68 758	74 750	87 058	100 795
	71	1 014 955	1 070 578	1 051 747	999 289	1 028 734	1 094 228	1 036 418
	77	562 780	621 114	544 447	485 986	705 994	844 121	908 267
	81	525	829	956	908	1 368	1 470	1 145
	87	11 001 839	9 049 576	3 853 560	10 057 766	12 427 660	7 947 490	10 165 635
	Group total	22 385 424	21 730 998	16 664 950	22 637 492	24 892 060	20 628 706	22 472 563
36 Tunas, bonitos, billfishes	21	10 721	7 665	7 842	10 785	8 399	8 750	7 284
Thons, pélamides, marlins	27	47 620	44 300	40 738	41 155	40 581	30 237	29 246
Atunes, bonitos, agujas	31	88 413	88 404	93 736	85 069	87 792	103 718	84 069
	34	346 623	322 931	351 789	356 802	304 768	338 374	279 141
	37	80 682	75 137	81 243	74 811	71 572	70 006	59 238
	41	72 769	72 144	63 213	64 577	65 990	65 189	67 295
	47	48 631	38 599	50 823	51 926	51 299	48 500	50 565
	51	796 802	800 918	783 481	913 946	918 054	864 343	996 906
	57	440 357	475 267	508 308	512 704	472 477	424 030	448 125
	61	567 298	748 063	924 890	961 585	917 339	848 314	855 844
	67	10 063	15 888	11 654	4 710	11 220	12 162	12 205
	71	1 699 445	1 759 571	2 102 575	2 050 157	2 101 913	2 046 013	2 281 419
	77	372 757	416 664	422 196	393 477	383 416	455 509	558 360
	81	37 555	36 171	42 007	29 628	37 050	37 703	39 970
	87	233 433	284 833	273 808	421 437	356 505	369 326	318 670
	Group total	4 853 169	5 186 555	5 758 303	5 972 769	5 828 375	5 722 174	6 088 337
37 Miscellaneous pelagic fishes	01	1 326	1 229	1 150	1 180	3 182	5 154	5 941
Poissons pélagiques divers	02	...	...	372	487	392	558	379
Peces pelágicos diversos	03	850	850	850	850	850	760	750
	04	562	1 600	1 500	1 455	1 583	1 685	1 733
	05	8	110	350	361	350	326	390
	21	77 942	66 326	77 935	58 871	49 971	65 508	88 109
	27	2 547 671	2 630 715	2 000 933	1 857 952	2 404 692	2 622 398	2 903 351
	31	35 373	36 900	38 351	32 601	33 106	33 744	31 092
	34	439 833	444 084	513 196	435 006	570 112	477 161	477 244
	37	116 031	110 585	100 465	97 217	106 059	117 976	119 937
	41	40 265	42 416	39 028	31 276	41 725	34 979	45 976
	47	554 647	507 630	534 727	486 932	501 730	414 020	440 596
	51	555 851	430 288	437 632	451 621	289 597	276 419	322 533
	57	543 336	537 456	584 052	560 548	507 071	515 504	529 577
	58	-	-	-	-	-	0	0
	61	3 267 951	3 162 978	2 674 518	2 462 646	2 469 381	2 667 858	2 587 752
	67	388	206	1 593	251	302	537	148
	71	1 394 094	1 477 321	1 513 491	1 616 062	1 653 087	1 709 494	1 789 831
	77	28 341	48 042	51 052	85 841	85 380	48 199	33 747

B-00 (a)

Fish, crustaceans, molluscs, etc — Capture production by species groups and fishing areas
Poissons, crustacés, mollusques, etc — Captures par groupes d'espèces et zones de pêche
Peces, crustáceos, moluscos, etc — Capturas por grupos de especies y áreas de pesca

Species group Groupe d'espèces Grupo de especies		Fishing area Zone de pêche Area de pesca	1996 t	1997 t	1998 t	1999 t	2000 t	2001 t	2002 t
		81	57 373	63 536	62 766	77 222	55 824	57 466	61 219
		87	4 670 631	4 227 437	2 580 236	2 425 826	1 871 096	3 200 042	2 198 855
		Group total	14 332 473	13 789 709	11 214 197	10 684 205	10 645 490	12 249 788	11 639 160
38	Sharks, rays, chimaeras	21	56 131	48 219	51 296	49 186	49 360	39 716	35 639
	Squales, raies, chimères	27	74 119	106 720	94 286	98 654	104 071	107 084	82 758
	Tiburones, rayas, quimeras	31	31 629	33 108	30 760	27 069	23 679	22 064	24 836
		34	35 840	68 091	52 464	51 721	52 223	51 528	49 094
		37	16 047	16 334	14 886	12 157	12 514	11 395	10 863
		41	56 595	58 696	57 896	61 308	60 197	62 855	57 721
		47	3 058	7 781	6 531	8 334	10 688	15 221	13 826
		48	45	32	8	17	0	13	0
		51	177 750	118 146	120 715	119 038	126 347	121 676	141 293
		57	117 205	111 844	127 229	121 470	119 552	116 425	116 615
		58	0	11	26	38	96	121	347
		61	31 201	37 365	34 383	57 887	56 079	35 970	39 671
		67	8 700	6 496	5 142	9 340	8 628	8 313	10 420
		71	138 975	140 049	143 301	139 951	146 527	148 968	151 084
		77	38 188	35 128	38 346	39 191	43 727	37 032	34 990
		81	17 075	24 720	18 773	22 584	20 172	22 452	24 149
		87	12 975	13 230	21 856	15 433	23 948	18 487	25 211
		88	-	-	5	19	41	7	25
		Group total	815 533	825 970	817 903	833 397	857 849	819 327	818 542
39	Marine fishes not identified	21	14 493	14 376	5 848	7 487	8 173	4 483	4 516
	Poissons marins non identifiés	27	90 327	74 377	82 501	49 251	72 227	36 178	36 964
	Peces marinos no identificados	31	222 149	252 532	213 099	175 905	164 058	152 559	162 579
		34	294 759	301 696	289 463	405 072	285 493	433 548	249 344
		37	105 579	88 163	93 606	72 160	81 048	68 265	66 403
		41	158 570	165 118	157 430	181 165	187 841	165 941	176 164
		47	90 506	106 528	59 888	60 929	47 629	91 020	99 275
		48	-	-	7	0	-	0	1
		51	610 747	661 111	632 249	662 681	633 757	740 547	699 449
		57	1 350 880	1 503 736	1 693 158	1 660 958	1 882 441	1 999 007	2 231 001
		58	-	1	0	0	0	-	-
		61	5 147 881	4 853 220	4 967 320	5 010 112	4 689 171	4 204 409	4 122 073
		67	106 820	106 601	97 853	98 191	31 003	74 345	4 864
		71	2 224 261	2 138 367	2 137 686	2 280 509	2 356 699	2 527 456	2 566 893
		77	213 310	181 446	159 623	114 666	126 461	135 485	129 559
		81	49 338	49 134	28 774	10 319	9 015	39 741	42 723
		87	83 088	112 041	404 257	304 202	322 884	197 441	139 805
		88	-	-	0	-	-	-	0
		Group total	10 762 708	10 608 447	11 022 762	11 093 607	10 897 900	10 870 425	10 731 613
41	Freshwater crustaceans	01	8 042	8 423	8 995	7 801	9 140	9 498	10 931
	Crustacés d'eau douce	02	10 174	14 177	13 430	9 681	3 835	7 885	10 310
	Crustáceos de agua dulce	03	2 677	2 157	1 506	3 235	2 437	...	...
		04	386 236	498 702	617 997	474 888	548 927	605 818	792 853
		05	2 774	2 989	2 783	2 800	3 758	4 507	4 481
		06	338	329	345	412	372	375	418
		Group total	410 241	526 777	645 056	498 817	568 469	628 083	818 993
42	Crabs, sea-spiders	04	363	267	44	47	77	150	219
	Crabes, araignées de mer	21	142 599	150 626	147 582	172 414	166 103	173 491	182 638
	Cangrejos, centollas	27	42 033	58 913	53 020	51 037	53 379	56 441	55 784
		31	57 004	69 237	68 121	52 584	60 703	49 241	58 834
		34	2 331	4 114	6 558	5 586	5 231	6 717	6 886
		37	2 195	2 104	2 520	4 491	2 891	2 594	3 485
		41	13 845	16 668	15 285	16 091	17 991	16 318	16 888
		47	5 329	3 415	5 639	4 855	6 875	5 251	5 300
		51	7 652	9 814	11 083	11 839	20 952	15 697	15 977
		57	23 284	24 528	31 378	31 429	46 864	46 902	49 325
		61	488 600	431 585	442 492	437 240	500 701	500 522	507 295
		67	65 038	73 923	131 368	102 383	34 839	33 594	39 353
		71	126 790	127 540	144 405	141 990	151 625	162 730	170 387
		77	15 145	12 657	8 665	7 818	13 402	12 700	10 556
		81	1 094	1 120	1 138	1 035	732	652	528
		87	6 372	4 629	6 427	18 509	9 352	8 958	10 750
		Group total	999 674	991 140	1 075 725	1 059 348	1 091 717	1 091 958	1 134 205
43	Lobsters, spiny-rock lobsters	21	71 866	78 146	77 155	83 105	83 062	83 803	82 422
	Homards, langoustes	27	54 509	58 122	55 760	59 818	55 202	55 364	56 017
	Bogavantes, langostas	31	30 412	29 219	28 118	31 709	31 249	24 704	29 121
		34	2 174	9 378	4 010	2 821	2 886	2 982	1 154
		37	7 455	7 203	4 927	4 623	4 023	4 133	3 739
		41	8 026	7 502	6 002	6 334	6 469	7 135	7 016
		47	3 032	3 117	3 348	2 931	2 690	3 452	3 989
		51	3 452	3 315	3 634	3 187	2 918	2 818	3 137
		57	16 337	16 285	16 599	19 058	20 855	18 202	15 784
		61	2 221	1 724	1 794	1 842	3 316	3 531	2 787

B-00 (a)

Fish, crustaceans, molluscs, etc — Capture production by species groups and fishing areas
Poissons, crustacés, mollusques, etc — Captures par groupes d'espèces et zones de pêche
Peces, crustáceos, moluscos, etc — Capturas por grupos de especies y áreas de pesca

Species group Groupe d'espèces Grupo de especies	Fishing area Zone de pêche Area de pesca	1996 t	1997 t	1998 t	1999 t	2000 t	2001 t	2002 t
	67	...	...	...	25	27	18	18
	71	6 717	10 085	8 498	6 339	7 208	8 385	9 928
	77	2 618	2 731	2 504	2 305	3 573	3 629	3 322
	81	3 953	6 418	3 980	3 873	3 976	3 762	3 618
	87	142	101	746	569	346	134	80
	Group total	212 914	233 346	217 075	228 539	227 800	222 052	222 132
44 King crabs, squat-lobsters	27	175	177	313	471	476	814	1 802
Crabes royaux, galatées	41	201	414	457	271	369	310	517
Cangrejos reales, galateidos	48	214	1	1	2	2	15	111
	58	-	0	0	0	3	0	0
	61	48 254	39 019	41 217	45 646	36 020	23 244	17 059
	67	9 526	8 177	10 941	7 675	6 848	7 281	7 616
	77	164	-	-	-	1	1 265	3 157
	87	17 160	22 898	26 295	23 576	24 292	13 397	11 648
	88	-	-	0	-	-	-	0
	Group total	75 694	70 686	79 224	77 641	68 011	46 326	41 910
45 Shrimps, prawns	01	3 000	2 850	2 700	2 574	2 100	2 400	2 400
Crevettes	04	34 418	44 296	23 900	23 365	34 186	37 062	102 657
Gambas, camarones	21	200 651	174 549	207 259	243 054	269 324	263 745	292 213
	27	167 166	191 183	176 695	173 335	172 670	142 743	149 372
	31	163 062	170 023	181 549	173 843	202 638	209 006	186 447
	34	57 810	53 206	83 505	70 383	59 356	57 907	56 130
	37	25 384	30 279	27 525	29 407	35 293	29 057	31 259
	41	45 681	47 343	58 911	48 673	76 986	109 959	92 848
	47	5 915	5 569	16 230	5 333	10 421	13 329	5 844
	51	297 846	272 896	336 556	322 311	311 223	291 178	281 543
	57	217 697	231 575	182 546	234 839	241 260	232 937	216 201
	61	776 922	834 742	932 180	1 150 109	1 119 120	991 000	981 353
	67	24 422	24 331	11 471	18 086	20 381	22 952	29 973
	71	435 422	440 527	414 268	438 591	442 815	474 998	484 354
	77	73 054	72 952	66 723	68 316	59 171	53 902	50 819
	81	1 824	1 949	1 684	2 410	2 842	2 861	2 206
	87	28 867	35 223	32 008	21 636	13 322	14 678	13 717
	Group total	2 559 141	2 633 493	2 755 710	3 026 265	3 073 108	2 949 714	2 979 336
46 Krill, planktonic crustaceans	27	-	-	88	-	-	36	-
Krill, crustacés planctoniques	41	-	-	74	-	4	-	-
Krill, crustáceos planctónicos	47	-	-	254	-	-	-	-
	48	91 150	75 653	90 025	101 957	114 426	104 183	125 987
	58	6	-	-	-	-	-	-
	Group total	91 156	75 653	90 441	101 957	114 430	104 219	125 987
47 Miscellaneous marine crustaceans	21	2	-	1	5	6	1	1
Crustacés marins divers	27	1 621	1 783	1 389	1 276	1 067	1 672	2 585
Crustáceos marinos diversos	31	366	351	364	168	181	186	163
	34	1 034	1 317	1 743	875	249	536	139
	37	9 804	8 827	7 873	9 927	10 428	10 024	8 740
	41	313	849	2 161	6 293	227	1 317	247
	47	2	-	-	-	-	1 062	1 653
	51	15 600	12 367	12 797	11 460	11 251	27 066	31 290
	57	40 568	45 940	52 272	54 445	36 398	41 365	43 329
	61	1 007 693	1 165 004	1 314 828	1 198 681	1 304 743	1 335 243	1 281 908
	67	...	...	...	...	49	63	58
	71	2 751	8 275	7 359	4 498	5 215	5 147	5 843
	77	631	515	1 079	1 237	1 247	1 376	5 257
	81	156	157	92	64	307	293	273
	87	880	585	698	687	627	1 510	1 652
	Group total	1 081 421	1 245 970	1 402 656	1 289 616	1 371 995	1 426 861	1 383 138
51 Freshwater molluscs	01	530	610	625	606	921	851	1 197
Mollusques d'eau douce	02	240	122	185	240	565	1 029	1 052
Moluscos de agua dulce	04	557 745	503 071	574 737	546 606	588 720	620 951	625 462
	05	-	-	-	-	-	-	451
	06	2 670	3 970	4 500	5 000	5 080	5 300	5 180
	61	-	-	-	-	-	74	209
	Group total	561 185	507 773	580 047	552 452	595 286	628 205	633 551
52 Abalones, winkles, conchs	21	4 513	3 668	3 149	5 721	4 567	6 339	6 221
Ormeaux, bigorneaux, strombes	27	29 369	33 733	18 337	22 335	33 739	33 900	30 357
Orejas de mar, bígaros, estrombos	31	39 257	33 711	31 933	30 197	24 153	25 322	20 732
	34	8 235	6 598	9 849	7 290	6 637	9 659	8 571
	37	4 518	5 682	4 829	5 545	5 716	4 675	1 739
	41	558	1 322	1 010	683	1 621	1 519	1 314
	47	735	330	524	481	490	527	516
	51	43	40	40	29	45	51	50
	57	5 081	4 905	4 920	5 297	5 207	5 304	5 845

B-00 (a)

Fish, crustaceans, molluscs, etc — **Capture production by species groups and fishing areas**

Poissons, crustacés, mollusques, etc — **Captures par groupes d'espèces et zones de pêche**

Peces, crustáceos, moluscos, etc — **Capturas por grupos de especies y áreas de pesca**

Species group Groupe d'espèces Grupo de especies	Fishing area Zone de pêche Area de pesca	1996 t	1997 t	1998 t	1999 t	2000 t	2001 t	2002 t
	61	23 352	24 084	25 540	22 257	20 249	19 385	20 550
	67	6 369	...	...	...	-	4	7
	71	448	183	348	282	241	288	292
	77	4 309	3 465	2 374	2 108	2 673	3 230	3 368
	81	1 364	1 515	1 627	1 493	1 590	1 369	1 371
	87	13 589	20 005	11 183	16 332	12 220	11 826	9 807
	Group total	141 740	139 241	115 663	120 050	119 148	123 398	110 740
53 Oysters Huîtres Ostras	21	39 569	49 099	33 329	47 106	26 475	18 571	13 236
	27	1 400	1 852	1 434	881	683	899	1 387
	31	113 560	99 330	106 853	89 056	198 341	161 729	148 186
	34	100	109	89	125	101	151	120
	37	2 246	1 863	1 204	952	319	157	188
	41	873	828	744	1 547	884	895	876
	47	...	...	...	-	-	-	1
	51	32	16	9	8	2	1	1
	57	-	-	-	-	350	298	180
	61	18 259	17 210	9 905	11 610	15 943	10 092	16 679
	67	4 989	6 187	11	6	14	20	19
	71	1 887	1 830	2 118	2 120	2 010	992	1 697
	77	3 071	4 079	3 383	3 028	3 745	3 317	3 301
	81	1 933	2 174	1 000	1 057	766	832	816
	87	5	10	6	11	14	207	12
	Group total	187 924	184 587	160 085	157 507	249 647	198 161	186 699
54 Mussels Moules Mejillones	04	4	-	-	-	-	-	-
	21	17 585	17 228	15 051	15 651	21 098	18 353	23 631
	27	101 062	118 949	122 237	114 376	128 707	146 696	133 725
	31	223	295	3 802	451	316	1 081	1 272
	34	10	6	7	...	401	102	8
	37	39 046	53 310	38 899	44 453	46 143	46 092	51 820
	41	1 234	1 548	1 606	1 381	1 215	2 540	2 161
	57	71	-	-	-	21	20	20
	61	3 028	7 309	7 934	8 265	6 952	4 955	4 479
	67	455	...	...	1	675	448	251
	71	20 216	18 017	17 454	6 554	41 246	22 782	23 282
	77	583	2 206	2 230	832	643	823	782
	81	454	1	665	2 978	4 468	2 271	1 725
	87	19 468	21 070	27 862	24 241	24 391	29 513	20 945
	Group total	203 439	239 939	237 747	219 183	276 276	275 676	264 101
55 Scallops, pectens Coquilles St-Jacques Vieiras	21	122 704	116 169	108 136	135 115	201 288	256 929	283 771
	27	55 636	71 666	73 299	73 406	74 769	81 361	70 411
	31	684	581	834	223	256	253	391
	37	93	137	65	80	588	154	476
	41	36 952	39 817	28 441	42 700	37 404	42 598	51 181
	57	6 872	6 093	5 543	7 622	8 317	4 421	2 525
	61	276 406	266 957	294 211	305 510	310 104	293 268	310 303
	67	2 372	2	3 228	2 642	2 012	1 052	1 264
	71	7 276	3 488	6 392	5 403	4 601	5 901	4 334
	77	17 290	2 320	2 726	1 864	6 287	3 100	9 875
	81	5 209	19 050	4 684	6 281	2 912	7 015	4 526
	87	3 672	6 611	27 208	31 856	12 162	6 685	2 459
	Group total	535 166	532 891	554 767	612 702	660 700	702 737	741 516
56 Clams, cockles, arkshells Clams, coques, arches Almejas, berberechos, arcas	04	844	956	914	304	303	792	850
	21	384 556	345 924	315 694	325 053	329 462	347 200	372 074
	27	58 500	57 086	116 751	94 255	65 960	49 247	50 815
	31	39 590	48 875	38 107	45 883	55 716	54 439	53 151
	34	54	69	139	148	120	112	171
	37	52 633	39 346	36 511	43 249	47 403	44 929	38 920
	41	126	196	664	2 443	279	297	1 075
	47	0	-	-	-	-	-	36
	51	57	8	19	26	30	11	19
	57	61 448	63 188	34 444	42 944	41 498	37 540	37 651
	61	181 775	144 085	156 515	145 198	132 789	119 870	116 738
	67	7 111	5 415	8 262	6 777	6 517	6 658	8 984
	71	72 085	57 595	74 531	83 453	67 778	104 017	108 471
	77	7 265	6 792	5 770	6 913	7 143	5 848	4 386
	81	1 274	1 120	1 984	2 090	2 548	2 574	2 374
	87	51 782	43 393	47 813	42 922	40 793	50 667	29 936
	Group total	919 100	814 048	838 118	841 658	798 339	824 201	825 651
57 Squids, cuttlefishes, octopuses Encornets, seiches, poulpes Calamares, jibias, pulpos	21	38 908	48 733	44 724	26 767	26 402	18 409	19 890
	27	53 371	50 983	49 948	53 067	58 619	44 474	51 988
	31	31 119	20 750	24 135	20 113	24 384	23 232	18 114
	34	190 620	150 352	183 729	250 402	216 428	213 558	96 870
	37	59 266	60 139	54 281	50 692	53 877	53 858	52 419
	41	747 875	1 016 557	750 044	1 192 738	1 000 427	803 766	543 286
	47	8 170	4 145	7 801	8 301	7 506	4 654	9 331

B-00 (a)

Fish, crustaceans, molluscs, etc — **Capture production by species groups and fishing areas**
Poissons, crustacés, mollusques, etc — **Captures par groupes d'espèces et zones de pêche**
Peces, crustáceos, moluscos, etc — **Capturas por grupos de especies y áreas de pesca**

Species group Groupe d'espèces Grupo de especies	Fishing area Zone de pêche Area de pesca	1996 t	1997 t	1998 t	1999 t	2000 t	2001 t	2002 t
	48	52	81	-	-	0	20	49
	51	94 868	148 579	112 432	121 305	119 166	82 139	115 430
	57	104 647	109 443	115 729	111 790	127 253	119 986	125 264
	58	-	1	-	-	-	-	0
	61	1 165 463	1 154 074	1 033 618	1 117 839	1 221 972	1 137 512	1 130 859
	67	685	2 787	2 293	3 132	748	2 997	842
	71	343 913	360 035	366 124	364 644	437 671	403 546	433 191
	77	201 442	213 080	30 387	150 569	202 531	166 988	189 990
	81	73 299	84 833	71 006	44 064	35 801	57 831	80 183
	87	36 021	32 095	11 354	82 778	127 619	174 999	305 566
	88	-	-	-	-	-	-	0
	Group total	3 149 719	3 456 667	2 857 605	3 598 201	3 660 404	3 307 969	3 173 272
58 Miscellaneous marine molluscs Mollusques marins divers Moluscos marinos diversos	04	481	52	165	73	166	112	120
	21	2 394	600	168	1 920	1 469	1 384	3 210
	27	878	1 023	1 302	663	392	424	196
	31	673	1 873	2 921	6 706	2 716	1 913	2 971
	34	746	1 489	12 599	2 258	1 189	385	3 423
	37	18 120	15 480	15 120	13 389	10 245	15 401	17 769
	41	1 175	1 851	1 408	4 937	2 474	1 082	2 334
	47	-	-	-	-	-	-	2 000
	51	15 939	7 363	5 015	3 747	5 034	3 580	2 613
	57	381	263	844	8 396	1 940	1 941	1 440
	61	938 870	1 502 215	1 488 978	1 457 426	1 410 260	1 405 909	1 381 743
	67	1 873	4 593	2 879	1 850	1 904	1 649	2 094
	71	53 322	54 772	59 749	59 051	53 179	61 483	60 783
	77	1 307	1 857	1 462	1 742	3 816	2 950	2 983
	81	342	173	230	111	54	61	90
	87	980	3 447	1 702	3 730	2 513	2 278	8 080
	Group total	1 037 481	1 597 051	1 594 542	1 565 999	1 497 351	1 500 552	1 491 849
71 Frogs and other amphibians Grenouilles et autres amphibies Ranas y otros anfibios	02	422	2 034	1 201	337	338	361	146
	03	0	0	0	0	7	9	5
	04	2 533	1 550	1 767	1 435	1 957	2 116	2 078
	05	...	38	41	35	26	-	29
	Group total	2 955	3 622	3 009	1 807	2 328	2 486	2 258
72 Turtles Tortues Tortugas	02	18	17	26	15	53	52	27
	03	0	0	0	0	0	-	-
	04	36	48	27	1	3	42	40
	31	227	86	136	31	44	34	42
	34	42	164	189	463	122	315	845
	41	0	0	0	0	0	...	...
	57	353	248	337	273	372	45	71
	71	417	403	452	448	384	319	332
	77	0	0	3	1	21	...	...
	87	10	11	12	11	11	12	12
	Group total	1 103	977	1 182	1 243	1 010	819	1 369
74 Sea-squirts and other tunicates Ascidiens et autres tuniciers Ascidias y otros tunicados	37	28	22	22	30	30	76	78
	51	-	-	-	-	95	-	-
	58	7	-	-	-	-	-	-
	61	16 747	2 780	891	1 171	1 443	1 053	1 025
	87	4 549	3 174	2 530	2 704	2 290	1 298	1 217
	Group total	21 331	5 976	3 443	3 905	3 858	2 427	2 320
75 Horseshoe crabs and other arachnoids Limules et autres arachnoïdés Erizos de mar y otros equinodermos	21	1 598	2 606	3 182	2 238	1 557	1 071	1 257
	31	-	1	70	159	139	228	130
	Group total	1 598	2 607	3 252	2 397	1 696	1 299	1 387
76 Sea-urchins and other echinoderms Oursins et autres échinodermes Erizos de mar y otros equinodermos	21	15 139	12 201	13 126	14 298	13 440	8 778	9 363
	27	911	612	564	643	463	478	650
	31	10	15	15	15	10	49	10
	37	69	55	60	76	204	104	166
	51	7 258	7 293	3 371	1 767	2 139	2 229	3 158
	57	419	610	833	859	1 048	787	920
	61	27 020	28 009	25 353	24 156	24 161	22 753	25 229
	67	7 576	9 044	8 081	7 329	7 115	6 678	6 448
	71	8 408	7 683	6 857	6 123	6 505	6 780	7 185
	77	12 244	10 401	6 221	8 682	9 328	8 626	8 556
	81	282	627	845	645	718	862	778
	87	52 025	46 000	44 978	56 981	57 247	49 030	62 532
	88	-	-	-	-	-	2	0
	Group total	131 361	122 550	110 304	121 574	122 378	107 156	124 995

B-00 (a)

Fish, crustaceans, molluscs, etc — **Capture production by species groups and fishing areas**
Poissons, crustacés, mollusques, etc — **Captures par groupes d'espèces et zones de pêche**
Peces, crustáceos, moluscos, etc — **Capturas por grupos de especies y áreas de pesca**

Species group Groupe d'espèces Grupo de especies	Fishing area Zone de pêche Area de pesca	1996 t	1997 t	1998 t	1999 t	2000 t	2001 t	2002 t
77 Miscellaneous aquatic invertebrates Invertébrés aquatiques divers Invertebrados acuáticos diversos	02	337	297	483	617	231	119	153
	04	1 494	2 460	1 480	2 048	2 031	1 578	1 735
	21	-	223	299	921	202	481	507
	27	0	-	-	-	-	24	4
	31	6	2	4	0	139	-	-
	37	909	904	1 764	1 737	908	2 595	579
	47	-	-	-	-	106	44	2 212
	48	-	-	-	-	5	-	0
	57	23 527	29 747	7 077	35 524	36 686	25 228	22 837
	58	0	12	2	-	-	-	-
	61	278 735	410 601	441 444	410 337	346 443	340 147	297 549
	71	38 285	86 579	74 425	91 744	144 866	77 337	77 520
	77	872	351	59	31	248	36	25
	81	-	-	-	-	22	29	21
	Group total	344 165	531 176	527 037	542 959	531 887	447 618	403 142
World total Total mondial Total mundial		**93 846 643**	**94 298 586**	**87 672 034**	**93 774 052**	**95 502 004**	**92 862 087**	**93 190 654**

B-00 (b)

Fish, crustaceans, molluscs, etc — **Capture production by species items**
Poissons, crustacés, mollusques, etc — **Captures par catégories d'espèces**
Peces, crustáceos, moluscos, etc — **Capturas por partidas de especies**

3-alpha code Code alpha-3 Código alfa-3	Scientific name Nom scientifique Nombre científico	Species group Groupe d'espèces Grupo de especies	1996 t	1997 t	1998 t	1999 t	2000 t	2001 t	2002 t
BUF	*Ictiobus spp*	11	793	991	959	697	1 280	1 569	1 558
CTM	*Catostomidae*	11	542	550	464	465	231	187	173
FBM	*Abramis brama*	11	59 110	50 388	51 460	46 470	47 222	51 239	52 632
FBR	*Abramis spp*	11	2 426	1 507	2 427	2 680	2 414	2 594	2 678
FCP	*Cyprinus carpio*	11	79 592	69 865	75 779	70 751	65 633	61 727	66 884
FTE	*Tinca tinca*	11	1 157	1 879	1 211	1 419	2 336	2 475	2 809
ALR	*Alburnus alburnus*	11	168	130	240	268	247	545	291
PTB	*Barbus barbus*	11	186	239	215	176	118	119	83
HON	*Chondrostoma nasus*	11	24	26	28	28	32	27	32
FCC	*Carassius carassius*	11	8 584	7 691	7 096	6 539	6 406	6 321	5 889
CGO	*Carassius auratus*	11	3 454	1 708	1 943	2 531	2 379	2 472	2 362
FRO	*Rutilus rutilus*	11	7 807	7 197	6 698	6 717	6 929	7 378	7 943
RFR	*Rutilus frisii*	11	9 210	2 320	6 624	6 905	10 120	7 199	6 433
FRX	*Rutilus spp*	11	32 707	25 069	23 876	18 093	21 540	21 928	24 166
SRE	*Scardinius erythrophthalmus*	11	39	113	192	175	147	102	637
FID	*Leuciscus idus*	11	3 612	3 660	3 799	3 095	2 943	2 811	3 039
FIE	*Leuciscus leuciscus*	11	220	255	305	180	104	91	73
LUH	*Leuciscus cephalus*	11	...	92	166	167	70	35	36
LEW	*Leuciscus spp*	11	47	39	32	38	33	31	24
RHI	*Labeo spp*	11	3 456	2 362	3 375	3 840	3 672	3 609	2 506
MUC	*Cirrhinus molitorella*	11	3	4	7	7	6	6	9
FCG	*Ctenopharyngodon idellus*	11	20 206	21 615	9 911	5 881	16 047	19 164	21 567
SVC	*Hypophthalmichthys molitrix*	11	24 480	22 527	15 100	18 058	11 279	6 860	5 990
BIC	*Hypophthalmichthys nobilis*	11	2 356	2 036	3 008	4 251	4 826	778	1 220
ENA	*Rastrineobola argentea*	11	49 670	40 315	42 336	48 816	49 618	41 384	35 455
AHT	*Acanthobrama terraesanctae*	11	1 155	626	1 171	1 048	1 052	811	1 170
VIV	*Vimba vimba*	11	270	390	331	361	279	269	324
FSC	*Pelecus cultratus*	11	498	403	1 780	1 826	1 235	1 590	1 316
ASU	*Aspius aspius*	11	580	416	353	416	542	642	1 282
FCH	*Leptobarbus hoeveni*	11	6 892	5 836	3 241	4 608	3 382	2 362	2 270
BKC	*Mylopharyngodon piceus*	11	15	16	22	27	34	36	25
WUB	*Megalobrama amblycephala*	11	0	0	-	-	-	-	3
FJB	*Puntius javanicus*	11	19 622	19 469	20 189	17 939	17 791	17 425	17 840
FAB	*Puntius spp*	11	39 003	37 272	56 480	56 774	53 430	55 049	55 940
FCY	*Cyprinidae*	11	285 581	281 072	264 206	282 377	237 588	230 059	268 303
TLM	*Oreochromis mossambicus*	12	19 253	20 025	20 175	23 482	22 141	22 520	23 290
TLN	*Oreochromis niloticus*	12	207 777	187 928	229 684	227 342	245 988	249 441	253 871
OEA	*Oreochromis aureus*	12	9 704	7 654	5 210	4 630	3 007	1 981	2 020
TLP	*Oreochromis (=Tilapia) spp*	12	281 916	302 170	299 248	321 518	352 884	347 229	337 420
SAR	*Sarotherodon galilaeus*	12	350	462	391	405	262	110	93
CHL	*Cichlasoma managuense*	12	566	367	344	351	324	410	318
CLO	*Cichla ocellaris*	12	65	35	12	7	4	4	4
AST	*Astronotus spp*	12	308	177	141	251	188	183	...
CIM	*Haplochromis spp*	12	10 472	13 053	13 077	10 731	11 699	9 695	8 759
CIX	*Cichlidae*	12	30 105	29 925	29 500	48 696	49 569	56 528	56 864
FLU	*Protopterus spp*	13	7 653	12 313	11 091	11 428	8 827	9 911	7 337
DAG	*Stolothrissa, Limnothrissa spp*	13	70 277	100 636	87 869	78 245	81 992	73 322	80 216
FPI	*Esox lucius*	13	25 823	24 014	24 045	23 285	25 311	24 474	26 339
ARP	*Arapaima gigas*	13	457	465	210	338	273	204	...
HTT	*Heterotis spp*	13	4 451	5 716	8 385	9 872	9 514	10 127	8 382
FKN	*Notopterus spp*	13	3 890	3 456	3 743	3 878	4 294	3 842	3 819
OEY	*Mormyridae*	13	...	4 481	18 781	16 030	7 691	24 509	12 567
OGN	*Gymnarchus niloticus*	13	1 108	3 952	7 685	5 887	...	...	9 097
CSM	*Colossoma macropomum*	13	4 991	3 600	3 014	3 921	5 928	4 693	4 068
CSD	*Piaractus brachypomus*	13	397	...	166	648	324	487	...
CHA	*Characidae*	13	144 295	128 789	140 913	125 605	121 684	70 540	64 063
PLR	*Prochilodus reticulatus*	13	7 448	2 007	6 117	9 768	10 942	11 220	10 133
PRL	*Prochilodus spp*	13	18 740	10 927	15 670	25 320	27 313	26 343	32 261
SOM	*Silurus glanis*	13	8 434	9 392	11 411	9 266	9 396	8 573	8 070
CAG	*Kryptopterus spp*	13	17 560	15 938	13 943	15 927	13 110	13 522	14 980
CSR	*Chrysichthys nigrodigitatus*	13	8 505	5 814	15 538	10 679	7 311	12 413	5 678
CST	*Chrysichthys spp*	13	7 596	6 513	7 100	5 592	7 800	7 397	7 397
CAN	*Bagrus spp*	13	9 865	12 458	13 386	16 239	11 495	11 574	8 001
CAF	*Ictalurus spp*	13	9 745	11 822	9 524	13 710	11 567	11 126	9 027
ITE	*Ameiurus nebulosus*	13	1	5	12	14	10	8	7
CLZ	*Clarias gariepinus*	13	20 601	24 703	21 956	27 220	36 520	39 867	37 190
CLN	*Clarias anguillaris*	13	20 370	22 433	21 618	21 511	31 506	39 535	39 208
CTO	*Clarias spp*	13	36 379	36 769	54 160	48 456	56 787	49 965	45 710
PGZ	*Pangasius spp*	13	541	522	917	1 061	1 274	1 100	1 200
CSY	*Synodontis spp*	13	9 267	11 560	13 384	12 856	12 571	13 737	12 827
FSI	*Siluroidei*	13	95 574	100 229	167 437	178 915	170 234	167 099	177 331
FBU	*Lota lota*	13	3 334	3 354	3 437	3 107	3 070	2 944	3 026
FLT	*Monopterus albus*	13	-	-	-	-	162	100	100
NIP	*Lates niloticus*	13	312 017	329 244	339 183	306 282	302 905	282 245	257 272
PEX	*Lates spp*	13	1 205	2 378	3 925	1 523	3 070	1 657	1 660
MPS	*Micropterus salmoides*	13	1 014	1 058	684	784	903	693	1 191
FPE	*Perca fluviatilis*	13	25 594	27 095	24 357	23 066	21 263	21 463	24 264
FPY	*Perca flavescens*	13	2 151	2 695	2 639	2 421	2 414	3 164	4 307
PND	*Percarina demidoffi*	13	-	-	-	-	-	-	18
STV	*Stizostedion vitreum*	13	7 736	7 755	8 212	8 784	9 164	6 959	7 021
SZC	*Stizostedion canadense*	13	1	...	0	...	...	...	...
FPP	*Stizostedion lucioperca*	13	26 259	18 022	18 708	17 166	15 996	15 534	17 343
ACC	*Gymnocephalus cernuus*	13	18	10	9	56	99	65	32
AGR	*Aplodinotus grunniens*	13	368	549	573	470	577	506	502
FGB	*Eleotridae*	13	2 959	3 140	2 928	3 093	2 975	4 454	4 930
FGX	*Gobiidae*	13	4 670	5 350	4 863	4 961	5 370	5 196	6 875
FPC	*Anabas testudineus*	13	3 905	3 754	4 637	6 340	6 730	5 800	6 300
FGS	*Trichogaster pectoralis*	13	30 792	21 728	22 422	23 776	22 456	21 357	20 940

B-00 (b)

Fish, crustaceans, molluscs, etc — **Capture production by species items**
Poissons, crustacés, mollusques, etc — **Captures par catégories d'espèces**
Peces, crustáceos, moluscos, etc — **Capturas por partidas de especies**

3-alpha code Code alpha-3 Código alfa-3	Scientific name Nom scientifique Nombre científico	Species group Groupe d'espèces Grupo de especies	1996 t	1997 t	1998 t	1999 t	2000 t	2001 t	2002 t
FGO	*Helostoma temminckii*	13	12 611	18 376	16 598	23 320	18 771	19 429	20 140
CNA	*Channa argus*	13	169	120	...	28	...	...	...
FSS	*Channa striata*	13	30 966	28 646	21 520	23 784	26 839	24 998	25 988
FIS	*Channa micropeltes*	13	11 614	10 117	8 253	8 787	8 149	6 357	6 440
FSN	*Channa spp*	13	128 350	136 916	188 042	88 915	95 235	93 221	80 513
FRF	*Osteichthyes*	13	3 552 457	3 649 506	3 712 400	4 373 981	4 358 824	4 549 150	4 389 297
APG	*Acipenser gueldenstaedtii*	21	614	861	771	409	292	277	234
APR	*Acipenser ruthenus*	21	36	17	11	38	15	13	17
APE	*Acipenser stellatus*	21	699	502	355	262	204	196	153
APN	*Acipenser transmontanus*	21	-	-	-	-	206	185	200
AAM	*Acipenser medirostris*	21	-	-	-	-	36	10	7
HUH	*Huso huso*	21	135	172	129	85	95	68	68
STU	*Acipenseridae*	21	3 229	2 889	2 511	2 055	1 755	1 520	1 180
ELE	*Anguilla anguilla*	22	8 687	10 164	7 445	7 454	7 887	7 434	7 113
ELJ	*Anguilla japonica*	22	1 014	916	904	830	765	677	610
ELA	*Anguilla rostrata*	22	1 316	1 270	1 244	1 125	1 338	868	774
ELU	*Anguilla australis*	22	621	523	453	425	470	512	497
ELX	*Anguilla spp*	22	5 481	1 851	2 253	2 655	5 977	3 365	5 044
SAL	*Salmo salar*	23	6 493	5 579	4 744	4 107	4 700	4 769	4 292
TRS	*Salmo trutta*	23	4 963	4 677	4 896	4 354	3 372	3 844	3 382
TRO	*Salmo spp*	23	3 183	3 186	2 879	2 990	3 033	3 008	3 432
PIN	*Oncorhynchus gorbuscha*	23	294 915	318 717	371 552	386 928	285 338	360 973	266 907
CHU	*Oncorhynchus keta*	23	411 395	347 560	311 965	281 260	276 355	307 662	311 275
CHE	*Oncorhynchus masou*	23	2 552	1 808	2 577	1 989	1 814	1 610	1 882
SOC	*Oncorhynchus nerka*	23	188 584	132 075	78 972	130 128	124 782	108 618	103 297
CHI	*Oncorhynchus tshawytscha*	23	10 509	13 000	9 840	8 735	8 437	8 771	13 909
COH	*Oncorhynchus kisutch*	23	28 201	13 191	19 386	15 449	18 035	20 006	20 750
TRR	*Oncorhynchus mykiss*	23	4 391	3 370	6 031	5 761	3 614	3 758	4 463
ORC	*Oncorhynchus spp*	23	-	-	-	-	50	204	6
SVF	*Salvelinus fontinalis*	23	3	4	4	4	5	7	11
ACH	*Salvelinus alpinus*	23	96	130	125	67	76	56	56
LAT	*Salvelinus namaycush*	23	975	1 109	1 054	985	1 129	1 130	878
CHR	*Salvelinus spp*	23	92	109	129	139	149	134	119
HUC	*Hucho hucho*	23	1	1	1	1	1	1	1
TLV	*Thymallus thymallus*	23	52	43	38	41	37	39	38
PCA	*Plecoglossus altivelis*	23	12 732	12 624	11 386	11 380	11 172	11 148	10 663
SME	*Osmerus eperlanus*	23	7 915	7 541	5 752	5 639	9 837	4 334	5 682
SMR	*Osmerus mordax*	23	2 429	2 862	3 334	2 024	1 977	1 242	1 942
PSM	*Hypomesus olidus*	23	3 981	5 964	6 471	5 691	3 254	4 241	3 263
SUS	*Hypomesus pretiosus*	23	1	-	-	-	-	-	-
EUL	*Thaleichthys pacificus*	23	34	27	6	8	13	143	329
SMX	*Osmerus spp, Hypomesus spp*	23	2 127	5 262	3 921	3 567	5 008	4 410	5 266
FVE	*Coregonus albula*	23	7 675	7 648	6 177	6 048	6 145	6 087	6 349
PLN	*Coregonus lavaretus*	23	5 742	5 639	5 495	5 313	5 126	4 819	4 782
HOU	*Coregonus oxyrinchus*	23	63	30	61	35	31	9	11
WHL	*Coregonus clupeaformis*	23	14 393	14 218	14 277	13 683	14 227	14 108	12 862
CIS	*Coregonus artedi*	23	1 145	1 223	1 048	804	836	870	788
WHF	*Coregonus spp*	23	12 770	13 097	13 181	12 976	12 576	12 173	15 718
SLX	*Salmonoidei*	23	5 112	3 878	4 680	3 203	4 010	2 868	4 645
SHC	*Alosa pontica*	24	925	983	831	117	228	308	619
SHA	*Alosa sapidissima*	24	1 869	1 527	2 048	1 259	1 280	1 652	742
ASD	*Alosa alosa*	24	59	56	2	...	38	45	94
TSD	*Alosa fallax*	24	2	1	...	...	12	16	22
ALE	*Alosa pseudoharengus*	24	6 731	7 517	9 062	8 355	7 135	5 061	7 489
SHH	*Alosa mediocris*	24	88	75	47	62	51	90	42
SHZ	*Alosa spp*	24	2 342	1 932	2 844	2 721	2 195	3 055	3 347
ASP	*Caspialosa spp*	24	58 957	62 746	88 366	99 714	79 274	45 408	26 253
SHG	*Dorosoma cepedianum*	24	1 272	1 989	1 291	3 007	977	1 781	2 780
CHG	*Anodontostoma chacunda*	24	6 752	4 546	3 850	5 488	5 156	5 383	5 045
HIX	*Hilsa kelee*	24	100 591	95 677	105 281	89 354	221 899	113 378	88 183
HIL	*Tenualosa ilisha*	24	227 308	216 065	207 269	215 991	220 372	230 221	221 107
TOL	*Tenualosa toli*	24	2 131	3 011	3 714	3 505	2 645	5 257	5 820
DAS	*Clupanodon thrissa*	24	4 933	13 846	11 360	9 538	6 387	9 194	4 748
CLA	*Clupeonella cultriventris*	24	118 857	106 151	136 557	197 860	164 322	97 406	83 471
DOD	*Konosirus punctatus*	24	18 647	14 850	20 787	17 770	12 335	17 210	13 401
ANR	*Lycengraulis grossidens*	24	90	100	100	120	120	140	130
EIL	*Ilisha elongata*	24	51 969	77 881	84 671	110 624	107 823	101 382	94 528
ILI	*Ilisha africana*	24	11 585	8 797	8 073	4 608	7 467	5 571	8 812
PEO	*Pellona ditchela*	24	16 077	14 176	10 763	10 014	13 544	13 885	13 516
DCX	*Clupeoidei*	24	4 553	4 863	4 250	3 615	3 228	4 757	5 154
LAU	*Petromyzon marinus*	25	34	20	17	38	36	32	50
LAR	*Lampetra fluviatilis*	25	141	87	95	129	163	118	106
LAS	*Petromyzontidae*	25	76	33	39	73	45	54	120
MIL	*Chanos chanos*	25	4 181	839	3 313	562	2 734	705	860
GTA	*Gasterosteus aculeatus*	25	385	462	992	1 047	1 681	1 566	2 473
GIP	*Lates calcarifer*	25	55 167	59 385	70 739	69 917	74 207	68 863	73 931
STB	*Morone saxatilis*	25	2 153	2 802	3 046	3 004	3 135	2 949	2 864
PEW	*Morone americana*	25	995	1 164	966	1 042	1 221	1 226	1 022
ROY	*Morone chrysops*	25	45	50	166	109	106	145	104
LEF	*Bothidae*	31	62	29	39	433	648	564	499
HAL	*Hippoglossus hippoglossus*	31	3 775	4 006	3 346	3 498	3 576	4 366	4 048
HAP	*Hippoglossus stenolepis*	31	28 110	39 291	41 609	43 620	40 848	40 161	43 724
PLE	*Pleuronectes platessa*	31	117 699	121 421	103 586	113 421	113 239	110 838	98 646
RFE	*Pleuronectes vetulus*	31	...	...	425	1 100	1 140	1 395	1 476
GHL	*Reinhardtius hippoglossoides*	31	108 053	97 779	87 301	117 459	113 880	112 787	110 640

B-00 (b) Fish, crustaceans, molluscs, etc / Poissons, crustacés, mollusques, etc / Peces, crustáceos, moluscos, etc

Capture production by species items / Captures par catégories d'espèces / Capturas por partidas de especies

3-alpha code Code alpha-3 Código alfa-3	Scientific name Nom scientifique Nombre científico	Species group Groupe d'espèces Grupo de especies	1996 t	1997 t	1998 t	1999 t	2000 t	2001 t	2002 t
ARF	*Atheresthes stomias*	31	10 681	6 320	8 280	12 083	18 774	14 401	16 611
KAF	*Atheresthes evermanni*	31	9 739	8 476	9 567	10 743	23 473	19 049	17 609
EOJ	*Eopsetta jordani*	31	...	1 937	1 464	1 480	1 870	1 822	1 793
WIT	*Glyptocephalus cynoglossus*	31	15 048	17 830	19 226	18 981	18 805	20 052	19 458
GLZ	*Glyptocephalus zachirus*	31	...	...	289	544	541	571	600
PLA	*Hippoglossoides platessoides*	31	16 036	17 230	19 237	21 645	16 805	18 486	15 660
FTS	*Hippoglossoides elassodon*	31	11 392	12 867	18 886	14 318	16 266	16 092	13 174
YES	*Limanda aspera*	31	101 354	149 302	80 500	56 830	69 971	54 918	63 625
YEL	*Limanda ferruginea*	31	3 840	5 266	9 604	13 960	20 971	24 273	18 948
DAB	*Limanda limanda*	31	17 967	17 509	22 134	20 917	16 599	17 172	16 179
ROS	*Lepidopsetta bilineata*	31	26 202	32 752	15 199	17 192	27 517	24 213	29 271
MIP	*Microstomus pacificus*	31	...	12 342	9 992	10 558	9 412	7 441	6 653
LEM	*Microstomus kitt*	31	12 078	12 283	14 378	14 203	13 974	15 461	10 346
FLE	*Platichthys flesus*	31	10 997	12 045	16 069	12 470	14 584	15 344	15 056
YSE	*Psettichthys melanostictus*	31	...	...	34	605	79	129	187
YFL	*Pseudopleuronectes herzenst.*	31	18 066	18 079	20 135	19 569	15 423	14 503	13 816
FLW	*Pseudopleuronectes americanus*	31	8 155	8 313	6 904	6 417	8 185	9 187	7 665
FSA	*Rhombosolea spp*	31	251	278	28	57	1	37	204
NYD	*Pleuronichthys decurrens*	31	-	-	-	-	1	5	4
SOL	*Solea solea*	31	43 338	36 682	44 378	44 078	48 846	48 558	41 168
SOS	*Solea lascaris*	31	246	288	305	222	297	339	335
CET	*Dicologlossa cuneata*	31	885	656	488	713	805	700	703
SOW	*Austroglossus microlepis*	31	409	517	393	463	571	589	644
SOE	*Austroglossus pectoralis*	31	909	837	859	768	800	844	702
SOA	*Austroglossus spp*	31	480	1 197	773	926	965	650	492
THS	*Microchirus spp*	31	50	57	77	71	71	80	139
SOX	*Soleidae*	31	12 606	13 552	16 050	9 960	6 242	5 106	4 228
TOX	*Cynoglossidae*	31	28 387	29 018	28 598	30 502	36 915	34 740	32 751
MEG	*Lepidorhombus whiffiagonis*	31	15 866	15 243	14 020	12 668	13 441	12 915	10 856
LEZ	*Lepidorhombus spp*	31	6 405	7 006	7 446	6 969	8 361	8 161	6 277
BLL	*Scophthalmus rhombus*	31	3 221	2 764	2 548	2 535	2 983	3 068	2 784
FLD	*Scophthalmus aquosus*	31	46	48	521	166	268	177	98
TUR	*Psetta maxima*	31	8 881	7 445	7 527	7 591	9 181	9 085	6 781
SCF	*Scophthalmidae*	31	1 377	964	528	478	643	622	482
HAI	*Psettodes erumei*	31	15 571	23 309	13 626	15 267	14 179	16 625	18 809
SOT	*Psettodes belcheri*	31	0	2	0	-	1	2	2
IYO	*Citharichthys sordidus*	31	-	-	-	-	150	-	-
BAH	*Paralichthys olivaceus*	31	10 628	9 953	9 617	8 877	9 179	8 436	8 502
YSF	*Paralichthys californicus*	31	...	...	523	620	390	393	428
FLS	*Paralichthys dentatus*	31	5 425	4 775	6 348	5 795	6 063	6 554	7 415
BAX	*Paralichthys spp*	31	11 428	11 605	9 780	7 472	7 257	6 251	5 405
MMM	*Mancopsetta maculata*	31	-	-	-	-	0	-	-
FLX	*Pleuronectiformes*	31	260 223	269 334	262 474	267 867	274 281	241 489	243 947
MOY	*Muraenolepis microps*	32	-	-	-	4	5	0	0
MRL	*Muraenolepis spp*	32	-	-	0	1	2	3	5
RIB	*Mora moro*	32	694	1 410	1 324	1 122	1 358	1 244	1 391
NEC	*Pseudophycis bachus*	32	10 596	11 087	16 495	12 555	5 365	4 530	4 443
ANT	*Antimora rostrata*	32	18	2	35	31	45	7	4
SAO	*Salilota australis*	32	8 071	6 741	11 806	15 888	14 915	6 193	4 471
MOR	*Moridae*	32	-	567	11 190	31 362	39 446	32 742	29 432
UNC	*Bregmaceros mcclellandi*	32	1 287	1 351	2 587	2 095	1 629	2 527	2 480
USK	*Brosme brosme*	32	30 296	24 745	30 986	34 748	32 531	28 524	27 126
COD	*Gadus morhua*	32	1 340 022	1 375 079	1 211 067	1 094 077	940 346	943 667	890 358
PCO	*Gadus macrocephalus*	32	429 475	443 895	411 371	402 245	370 912	330 884	332 692
GRC	*Gadus ogac*	32	2 121	1 729	1 697	1 903	931	1 133	977
LIN	*Molva molva*	32	54 619	50 988	60 826	53 949	43 320	36 969	40 247
BLI	*Molva dypterygia*	32	9 788	11 751	12 768	15 699	16 146	19 366	12 088
GFB	*Phycis blennoides*	32	5 056	8 339	7 807	6 535	6 575	5 695	5 872
HKU	*Urophycis brasiliensis*	32	3 152	2 570	6 081	2 842	3 498	9 076	9 606
HKR	*Urophycis chuss*	32	2 904	2 850	2 803	3 228	3 365	4 008	5 443
HKW	*Urophycis tenuis*	32	8 303	6 830	5 959	6 378	8 158	8 324	12 801
HAD	*Melanogrammus aeglefinus*	32	361 929	334 105	283 689	249 451	212 821	228 821	266 501
COW	*Eleginus navaga*	32	659	1 152	1 185	480	674	1 166	962
SAF	*Eleginus gracilis*	32	21 110	27 803	40 426	47 032	35 763	33 753	32 591
MGX	*Microgadus proximus*	32	...	...	3	1	0	0	-
TOM	*Microgadus tomcod*	32	140	90	119	42	55	57	1
POK	*Pollachius virens*	32	355 319	319 307	329 793	340 351	312 570	325 750	381 335
POL	*Pollachius pollachius*	32	12 988	12 164	10 952	10 379	12 059	11 937	12 634
ALK	*Theragra chalcogramma*	32	4 548 585	4 486 510	4 054 223	3 277 286	2 938 230	3 144 465	2 654 854
POC	*Boreogadus saida*	32	20 788	6 826	3 592	22 005	40 743	39 445	37 194
ROL	*Gaidropsarus spp*	32	...	...	...	24	43	20	39
NOP	*Trisopterus esmarkii*	32	275 667	211 394	97 478	112 613	204 845	93 405	106 924
POD	*Trisopterus minutus*	32	603	428	428	637	890	755	860
BIB	*Trisopterus luscus*	32	14 881	14 213	12 762	13 300	18 751	18 203	17 508
WHB	*Micromesistius poutassou*	32	632 471	712 279	1 185 003	1 321 195	1 472 105	1 823 305	1 603 263
POS	*Micromesistius australis*	32	156 154	166 205	201 363	185 310	152 830	156 086	146 240
WHG	*Merlangius merlangus*	32	98 595	88 163	75 081	76 115	75 154	59 923	53 081
HKE	*Merluccius merluccius*	32	92 332	76 958	66 088	70 285	71 241	56 766	66 187
HKM	*Merluccius senegalensis*	32	16 485	16 430	17 396	18 187	13 515	17 925	7 516
HKN	*Merluccius australis*	32	37 022	39 047	43 200	47 560	52 478	48 736	42 566
HKS	*Merluccius bilinearis*	32	43 399	33 565	31 892	27 568	25 756	33 000	24 293
PHA	*Merluccius gayi*	32	323 470	265 573	162 518	140 910	193 504	246 265	162 290
HKP	*Merluccius hubbsi*	32	681 999	648 301	527 229	372 167	244 988	305 672	409 488
NHA	*Merluccius productus*	32	195 406	227 705	227 754	217 000	206 720	172 903	130 793
HKB	*Merluccius polli*	32	39	0	38	147	44	139	2 222
HOF	*Merluccius albidus*	32	32	...	5	12	5	2	6
MRG	*Merluccius angustimanus*	32	...	267	165	143	250	391	380
HKK	*Merluccius capensis*	32	832	777	906	2 060	658	1 863	3 225

B-00 (b)

Fish, crustaceans, molluscs, etc — **Capture production by species items**
Poissons, crustacés, mollusques, etc — **Captures par catégories d'espèces**
Peces, crustáceos, moluscos, etc — **Capturas por partidas de especies**

3-alpha code Code alpha-3 Código alfa-3	Scientific name Nom scientifique Nombre científico	Species group Groupe d'espèces Grupo de especies	1996 t	1997 t	1998 t	1999 t	2000 t	2001 t	2002 t
HKX	*Merluccius spp*	32	2 075	1 126	2 064	1 679	2 011	1 885	1 793
HKC	*Merluccius capensis,M.paradox.*	32	286 443	260 955	302 542	307 642	309 337	323 073	306 387
GRM	*Macruronus magellanicus*	32	434 797	123 678	473 633	447 063	233 774	296 830	248 442
GRN	*Macruronus novaezelandiae*	32	201 017	285 580	324 750	281 105	283 706	257 801	221 315
RHG	*Macrourus berglax*	32	1 080	4 545	7 170	8 158	8 574	1 912	1 538
WGR	*Macrourus whitsoni*	32	-	-	-	1	9	48	158
MCC	*Macrourus carinatus*	32	-	-	-	20	65	-	0
MCH	*Macrourus holotrachys*	32	-	1	-	-	-	-	-
GRV	*Macrourus spp*	32	926	1 072	3 066	2 753	10 854	3 424	6 437
RNG	*Coryphaenoides rupestris*	32	14 196	19 907	23 260	18 223	30 968	53 669	28 021
CVY	*Coryphaenoides spp*	32	-	-	-	-	-	-	0
LDE	*Lepidorhynchus denticulatus*	32	670	2 361	4 627	3 678	3 833	4 783	5 349
RTX	*Macrouridae*	32	1 889	6 241	4 912	4 034	3 428	5 631	15 006
GAD	*Gadiformes*	32	43 848	29 071	20 411	22 487	11 244	40 144	15 644
LAD	*Elops saurus*	33	143	746	980	1 978	141	480	675
CEC	*Elops lacerta*	33	379	735	1 546	1 822	755	558	1 118
TAR	*Megalops atlanticus*	33	515	763	604	331	349	1 185	1 237
TAI	*Megalops cyprinoides*	33	1 096	1 158	1 058	1 433	1 284	1 469	1 572
BOF	*Albula vulpes*	33	383	1 558	1 674	370	249	526	210
BUC	*Harpadon nehereus*	33	196 431	226 491	194 188	194 307	177 753	184 676	143 949
LIG	*Saurida tumbil*	33	14 751	10 874	10 931	10 791	11 203	8 342	7 172
LIB	*Saurida undosquamis*	33	333	321	286	103	846	814	89
LIX	*Synodontidae*	33	124 080	135 924	139 055	137 330	132 602	128 794	132 659
CAX	*Ariidae*	33	289 052	311 811	297 831	323 391	317 886	340 523	347 305
CAE	*Plotosus spp*	33	2 644	2 210	2 860	2 380	3 533	3 013	3 152
MUI	*Muraenidae*	33	710	268	165	335	358	354	359
KLA	*Hypoptychus dybowskii*	33	6 613	8 832	4 801	4 806	-	462	7
HCZ	*Holocentridae*	33	121	92	101	90	90	109	313
MUF	*Mugil cephalus*	33	32 299	30 699	24 256	27 307	29 284	26 705	23 927
MUB	*Mugil liza*	33	3 287	1 550	2 876	2 855	3 186	2 537	2 294
MYZ	*Mugil soiuy*	33	1 039	2 718	3 674	5 364	5 575	2 810	2 601
AJW	*Agonostomus monticola*	33	...	...	...	...	...	...	28
MUA	*Joturus pichardi*	33	659	339	374	404	497	504	417
LZS	*Liza saliens*	33	3	2	2	2	0	10	25
MUL	*Mugilidae*	33	301 711	344 067	370 659	381 401	386 432	396 783	374 270
FUS	*Caesio spp*	33	48 183	57 080	51 423	54 251	48 235	52 992	55 163
SNO	*Centropomus undecimalis*	33	5 132	4 867	5 012	5 119	5 034	5 903	7 388
ROB	*Centropomus spp*	33	4 058	3 792	4 925	7 830	8 349	7 966	6 736
MAB	*Mycteroperca bonaci*	33	...	...	...	...	9	7	6
MKM	*Mycteroperca microlepis*	33	...	...	...	...	15	10	13
MKH	*Mycteroperca phenax*	33	...	...	1	14	27	17	26
MKV	*Mycteroperca venenosa*	33	...	...	...	...	3	1	3
GBS	*Mycteroperca xenarcha*	33	180	125	155	82	80	210	200
GPB	*Mycteroperca spp*	33	423	584	518	628	536	1 040	1 347
GPD	*Epinephelus marginatus*	33	1 715	1 426	1 105	409	317	503	525
GPW	*Epinephelus aeneus*	33	682	603	561	633	683	655	730
GPN	*Epinephelus striatus*	33	466	598	571	438	257	325	322
GPS	*Epinephelus analogus*	33	43	48	32	39	30	28	28
EEL	*Epinephelus flavolimbatus*	33	...	15	18	...	73	36	67
EEU	*Epinephelus guttatus*	33	60	42	52	72	67	88	159
GPR	*Epinephelus morio*	33	618	525	292	797	1 424	128	86
EFV	*Epinephelus niveatus*	33	...	...	...	...	6	4	9
EEG	*Epinephelus goreensis*	33	196	63	76	290	277	62	64
ELG	*Epinephelus nigritus*	33	16	23	23	-	45	44	50
GPX	*Epinephelus spp*	33	125 471	137 329	150 939	159 018	168 713	166 226	169 024
LDP	*Lepidoperca pulchella*	33	...	...	193	32	...	46	97
BSB	*Centropristis striata*	33	1 978	1 590	1 531	1 717	1 516	1 667	1 907
BSZ	*Acanthistius brasilianus*	33	10 799	9 164	5 258	5 914	4 222	4 890	3 624
BAP	*Paralabrax humeralis*	33	4 954	2 789	2 554	3 278	4 373	2 011	1 522
BSX	*Serranidae*	33	48 123	48 294	53 679	55 124	66 191	72 445	71 655
THO	*Terapon spp*	33	-	-	1	2	0	1	2
TEJ	*Stereolepis gigas*	33	1	1	3	2	2	3	2
SPU	*Dicentrarchus punctatus*	33	1 273	798	963	782	379	460	841
BSS	*Dicentrarchus labrax*	33	8 263	7 051	6 686	7 648	8 513	9 118	10 159
BSE	*Dicentrarchus spp*	33	5 243	5 788	7 230	5 505	7 157	5 953	5 593
BAJ	*Lateolabrax japonicus*	33	9 512	10 558	12 180	12 167	10 538	12 256	12 216
BIR	*Priacanthus macracanthus*	33	7 386	4 824	3 670	2 756	3 986	3 079	2 142
BIG	*Priacanthus spp*	33	91 272	87 311	89 071	90 814	85 572	91 147	91 995
APO	*Apogonidae*	33	211	180	60	365	60	344	413
WHS	*Sillaginidae*	33	19 649	21 190	20 535	24 014	23 525	22 996	23 962
MOO	*Mene maculata*	33	5 540	12 842	12 539	14 132	13 214	14 857	18 060
RUF	*Arripis georgianus*	33	1 302	1 287	1 008	1 066	1 143	992	860
ASA	*Arripis trutta*	33	8 165	7 550	6 847	8 306	8 506	7 668	8 799
RES	*Lutjanus argentimaculatus*	33	13 488	13 242	9 354	16 129	14 442	13 944	14 262
HUS	*Lutjanus argentiventris*	33	4 960	3 163	3 490	3 076	3 472	3 468	3 813
SNC	*Lutjanus purpureus*	33	6 439	7 490	7 191	10 796	8 215	8 111	8 079
SNR	*Lutjanus campechanus*	33	6 346	6 264	5 166	5 412	4 841	4 869	4 617
SNL	*Lutjanus synagris*	33	3 682	3 807	4 250	3 384	3 578	4 041	3 536
SNA	*Lutjanus spp*	33	73 384	84 765	86 541	92 149	83 653	90 333	91 288
SNY	*Ocyurus chrysurus*	33	7 332	7 436	6 145	6 967	6 470	4 529	4 568
RPU	*Rhomboplites aurorubens*	33	...	678	480	731	911	1 057	980
SNX	*Lutjanidae*	33	74 725	79 686	85 389	86 347	84 676	94 406	92 235
THG	*Nemipterus virgatus*	33	242 261	263 399	266 383	250 591	296 319	290 920	309 392
THB	*Nemipterus spp*	33	205 683	198 987	221 903	225 200	229 367	222 410	246 017
MOB	*Scolopsis spp*	33	1 342	1 294	1 473	1 090	1 795	1 852	1 654
THD	*Nemipteridae*	33	2 750	4 703	3 469	4 066	2 917	4 506	6 565
POY	*Leiognathus spp*	33	2 690	2 500	3 211	3 180	2 573	2 395	2 434
PON	*Leiognathidae*	33	198 581	218 807	186 246	215 875	186 865	205 463	214 736

B-00 (b)

Fish, crustaceans, molluscs, etc — Capture production by species items
Poissons, crustacés, mollusques, etc — Captures par catégories d'espèces
Peces, crustáceos, moluscos, etc — Capturas por partidas de especies

3-alpha code Code alpha-3 Código alfa-3	Scientific name Nom scientifique Nombre científico	Species group Groupe d'espèces Grupo de especies	1996 t	1997 t	1998 t	1999 t	2000 t	2001 t	2002 t
PIG	*Orthopristis chrysoptera*	33	-	1	1	-	0	0	0
GRP	*Isacia conceptionis*	33	2 048	2 034	2 133	2 947	3 293	3 321	5 655
GBR	*Plectorhinchus mediterraneus*	33	1 869	904	459	890	297	271	622
GRL	*Pomadasys argenteus*	33	1 644	1 156	2 847	2 235	1 972	1 792	1 621
BGR	*Pomadasys incisus*	33	1	5	-	-	0	0	-
BUR	*Pomadasys jubelini*	33	1 439	1 535	1 253	1 038	1 017	1 418	1 435
GRB	*Brachydeuterus auritus*	33	20 033	29 274	18 853	22 385	21 939	21 167	21 716
BRG	*Conodon nobilis*	33	172	118	114	84	39	53	82
GRX	*Haemulidae (=Pomadasyidae)*	33	75 707	74 229	74 277	74 121	77 392	85 591	78 280
IAG	*Sciaena gilberti*	33	9 099	3 565	6 154	6 442	4 744	4 328	6 051
CBM	*Sciaena umbra*	33	430	222	240	186	128	171	156
WEP	*Cynoscion analis*	33	7 887	5 797	11 187	8 929	6 339	4 438	3 487
SWF	*Cynoscion nebulosus*	33	3 696	3 875	6 865	6 944	6 486	4 951	6 127
STG	*Cynoscion regalis*	33	3 261	3 317	3 821	3 146	2 438	2 273	2 162
WKS	*Cynoscion striatus*	33	31 641	39 318	32 394	19 588	22 873	22 734	20 370
WKX	*Cynoscion spp*	33	43 883	40 593	44 830	51 244	58 914	19 077	18 544
CKM	*Micropogonias furnieri*	33	78 698	76 857	64 502	45 117	61 501	77 915	75 540
CKA	*Micropogonias undulatus*	33	9 049	12 434	11 522	12 180	12 138	13 018	11 762
CRX	*Micropogonias spp*	33	6 062	5 852	6 709	5 391	4 830	4 790	4 212
KGF	*Menticirrhus saxatilis*	33	49	21	17	13	28	39	8
KGG	*Menticirrhus littoralis*	33	1 070	1 443	2 061	1 822	961	690	786
KIX	*Menticirrhus spp*	33	1 144	1 303	1 402	1 062	1 352	1 957	2 047
COB	*Umbrina cirrosa*	33	796	545	429	469	382	450	378
CKY	*Umbrina canosai*	33	13 018	6 658	5 284	7 660	9 221	16 256	17 796
UCA	*Umbrina canariensis*	33	11	10	10	839	9	2 218	20
WKK	*Macrodon ancylodon*	33	7 266	7 625	9 229	5 595	6 642	4 139	6 261
MGR	*Argyrosomus regius*	33	3 435	3 905	3 605	4 923	4 095	4 030	5 543
KOB	*Argyrosomus hololepidotus*	33	1 497	1 126	1 297	1 442	1 486	4 325	1 191
AWE	*Atractoscion aequidens*	33	381	519	632	459	412	415	326
WEW	*Atractoscion nobilis*	33	46	28	71	112	101	124	194
KIC	*Genyonemus lineatus*	33	242	167	64	74	105	137	97
DRS	*Pteroscion peli*	33	1 879	1 291	1 190	1 260	1 126	1 159	1 128
LKR	*Otolithes ruber*	33	-	-	13	10	...	...	...
PDR	*Paralonchurus peruanus*	33	4 263	2 737	4 363	6 063	5 729	4 167	1 886
BDM	*Pogonias cromis*	33	505	1 755	3 297	1 181	2 876	3 059	2 950
HOC	*Nibea mitsukurii*	33	1 940	1 177	1 285	1 566	1 999	2 156	1 148
LYC	*Larimichthys croceus*	33	102 297	71 515	71 229	66 076	123 274	76 715	83 412
CRY	*Larimichthys polyactis*	33	280 036	167 010	207 772	257 758	302 127	254 166	272 140
SPT	*Leiostomus xanthurus*	33	2 554	3 073	3 359	2 599	3 141	3 091	2 487
RDM	*Sciaenops ocellatus*	33	1	2	3	6	6	3	4
CKL	*Pseudotolithus brachygnathus*	33	715	618	354	318	811	1 083	855
PSS	*Pseudotolithus senegalensis*	33	2 727	2 781	4 659	4 837	3 949	4 800	3 905
PSE	*Pseudotolithus elongatus*	33	12 662	11 497	9 461	6 855	13 405	10 932	12 066
CKW	*Pseudotolithus spp*	33	29 349	23 762	32 890	32 669	31 520	33 262	47 771
CRL	*Atrobucca nibe*	33	297	288	237	223	450	644	535
CRV	*Pennahia argentata*	33	8 599	6 085	5 089	4 940	4 180	3 796	3 394
CDX	*Sciaenidae*	33	527 546	565 757	526 451	613 004	508 441	506 365	527 683
LBR	*Gymnocranius spp*	33	228	192	168	128	145	160	150
LTN	*Lethrinus atlanticus*	33	231	334	380	...	...	...	...
EMP	*Lethrinidae*	33	81 537	80 745	80 905	86 139	95 610	100 838	100 837
SBR	*Pagellus bogaraveo*	33	2 259	4 508	2 957	2 550	1 876	1 324	1 694
PAC	*Pagellus erythrinus*	33	4 716	5 745	5 077	4 933	5 772	5 055	5 008
SBA	*Pagellus acarne*	33	1 305	1 462	1 680	1 368	1 607	1 420	1 402
PAR	*Pagellus bellottii*	33	12 001	13 133	13 997	18 053	7 708	7 929	7 651
PAX	*Pagellus spp*	33	6 638	4 594	4 432	6 472	5 460	5 188	4 790
DIG	*Diplodus argenteus*	33	39	84	15	5	2	2	1
SWA	*Diplodus sargus*	33	463	681	468	695	759	580	1 085
SRG	*Diplodus spp*	33	8 010	7 376	6 597	9 026	8 965	8 528	8 174
PRG	*Calamus spp*	33	1 052	1 541	553	1 266	587	534	449
DEL	*Dentex macrophthalmus*	33	2 442	2 733	3 580	5 479	4 868	2 448	1 899
DEN	*Dentex canariensis*	33	-	-	-	-	20	-	-
DEC	*Dentex dentex*	33	3 293	1 715	1 662	1 451	1 388	1 119	1 294
DEA	*Dentex angolensis*	33	763	490	1 416	1 767	504	838	990
DNC	*Dentex congoensis*	33	102	119	392	1 272	350	571	445
DEX	*Dentex spp*	33	8 054	6 400	13 863	13 599	17 596	19 668	13 243
BRB	*Spondyliosoma cantharus*	33	5 147	5 148	4 640	5 128	5 487	5 318	5 861
SBS	*Oblada melanura*	33	1 040	881	862	781	980	660	645
SPH	*Archosargus probatocephalus*	33	1 656	1 743	1 366	1 183	1 528	1 214	1 111
AVB	*Archosargus rhomboidalis*	33	...	...	...	...	...	...	53
KBR	*Argyrops spinifer*	33	2 784	3 013	2 992	3 015	3 803	4 575	5 124
SLF	*Argyrozona argyrozona*	33	883	780	505	541	500	287	249
SLD	*Cheimerius nufar*	33	73	62	0	29	25	25	46
RPG	*Pagrus pagrus*	33	6 978	6 771	6 691	9 299	7 338	5 318	5 975
GSU	*Pagrus auratus*	33	29 114	29 618	29 014	31 465	31 402	33 067	35 296
SBP	*Pagrus spp*	33	5 346	6 522	6 439	4 026	5 301	10 385	10 702
RER	*Petrus rupestris*	33	28	35	...	22	10	7	6
PGA	*Pterogymnus laniarius*	33	1 200	1 323	1 042	1 349	9 503	6 956	3 798
WSN	*Rhabdosargus globiceps*	33	237	165	296	335	300	169	107
SBG	*Sparus aurata*	33	5 843	5 564	5 806	7 479	8 301	9 552	9 389
BOG	*Boops boops*	33	30 408	27 054	29 456	28 219	29 139	26 001	27 896
RSX	*Chrysoblephus spp*	33	268	224	154	149	70	78	124
SNW	*Lithognathus lithognathus*	33	7	8	9	2	...	...	...
SSB	*Lithognathus mormyrus*	33	968	1 228	2 509	1 999	2 230	3 047	3 961
STW	*Lithognathus spp*	33	6	136	2	116	56	20	4
SLM	*Sarpa salpa*	33	1 941	2 421	2 237	2 304	2 056	2 129	2 241
MLM	*Acanthopagrus schlegeli*	33	207	277	404	604	717	878	784
YWF	*Acanthopagrus latus*	33	234	249	280	464	350	584	402
AAB	*Acanthopagrus bifasciatus*	33	...	...	...	...	...	1 586	332
SCP	*Stenotomus chrysops*	33	2 952	2 196	1 893	1 676	1 206	1 845	3 312

B-00 (b) Fish, crustaceans, molluscs, etc / Poissons, crustacés, mollusques, etc / Peces, crustáceos, moluscos, etc

Capture production by species items / Captures par catégories d'espèces / Capturas por partidas de especies

3-alpha code Code alpha-3 Código alfa-3	Scientific name Nom scientifique Nombre científico	Species group Groupe d'espèces Grupo de especies	1996 t	1997 t	1998 t	1999 t	2000 t	2001 t	2002 t
SBX	*Sparidae*	33	133 684	149 316	144 847	163 586	185 622	205 950	199 143
BPI	*Spicara maena*	33	285	505	395	419	255	381	593
PIC	*Spicara spp*	33	12 319	12 271	13 210	9 367	8 227	8 953	7 213
MUR	*Mullus surmuletus*	33	15 207	14 233	13 176	12 765	15 149	14 477	13 202
MUT	*Mullus barbatus*	33	5 331	4 531	4 878	5 618	4 433	5 667	5 159
MWU	*Mullus argentinae*	33	75	73	99	229	498	225	78
MUX	*Mullus spp*	33	20 564	16 181	15 001	18 842	19 307	17 127	15 523
GOX	*Upeneus spp*	33	49 504	52 579	51 738	55 362	51 657	61 716	63 716
GOA	*Pseudupeneus prayensis*	33	7 126	3 323	2 560	2 426	2 391	3 682	4 450
MUM	*Mullidae*	33	25 368	17 753	15 833	16 842	17 757	21 442	32 475
MOJ	*Gerres spp*	33	9 500	8 190	7 642	8 118	8 553	10 035	8 555
GDJ	*Gerreidae*	33	16 597	9 857	7 432	6 219	5 540	5 737	4 684
GIY	*Girella tricuspidata*	33	561	588	576	579	635	584	548
GIQ	*Girella nigricans*	33	-	-	-	-	3	2	1
KYC	*Kyphosus cinerascens*	33	-	8	-	1	-	-	17
KYP	*Kyphosus spp*	33	3	2	2	1	1	9	54
SPS	*Drepane punctata*	33	939	757	612	976	792	824	927
SIC	*Drepane africana*	33	2 774	3 790	2 268	1 545	2 856	2 790	2 650
USB	*Labrus bergylta*	33	-	-	-	2	1	-	-
TAU	*Tautoga onitis*	33	119	116	116	95	111	138	160
CUN	*Tautogolabrus adspersus*	33	1	1	3	4	3	9	13
YFH	*Semicossyphus pulcher*	33	...	...	118	58	78	68	52
WRA	*Labridae*	33	21 983	22 662	19 486	18 748	18 716	19 091	11 990
PRR	*Sparisoma cretense*	33	-	-	-	56	89	162	153
PWT	*Scaridae*	33	1 021	1 362	1 283	1 836	1 321	1 582	2 226
ANW	*Pomacanthidae*	33	0	0	7	17	1	3	37
TGA	*Polydactylus quadrifilis*	33	7 359	10 705	10 023	10 770	12 237	12 447	15 493
FOT	*Eleutheronema tetradactylum*	33	3 673	4 614	4 217	2 414	6 980	1 770	1 296
GAL	*Galeoides decadactylus*	33	17 768	18 385	12 890	10 764	13 948	16 943	13 683
PET	*Pentanemus quinquarius*	33	3 005	3 356	4 430	4 162	3 612	3 947	1 439
THF	*Polynemidae*	33	51 837	50 769	53 842	53 098	53 406	52 593	55 300
BLP	*Eleginops maclovinus*	33	439	194	1 314	2 220	1 919	403	337
NOR	*Notothenia rossii*	33	-	1	-	1	0	0	5
NOG	*Notothenia gibberifrons*	33	-	-	-	5	1	2	1
NON	*Notothenia neglecta*	33	-	-	-	0	-	2	-
NOS	*Notothenia squamifrons*	33	15	4	3	15	368	329	330
NOK	*Notothenia kempi*	33	-	-	-	-	-	-	0
NOL	*Nototheniops larseni*	33	-	-	-	-	0	-	0
NOD	*Nototheniops nudifrons*	33	-	-	-	-	0	0	-
TRT	*Trematomus spp*	33	-	-	-	0	-	0	-
TRH	*Pagothenia hansoni*	33	-	-	-	-	0	-	-
ANS	*Pleuragramma antarcticum*	33	-	-	-	-	-	0	-
NOX	*Nototheniidae*	33	16	34	7	226	3	198	232
AIB	*Ambassidae*	33	3 075	2 594	2 186	1 733	1 575	1 876	1 986
PRC	*Percoidei*	33	155 971	110 833	95 569	136 617	117 120	102 757	129 345
ELP	*Zoarces viviparus*	33	147	97	54	43	28	37	43
OPT	*Macrozoarces americanus*	33	41	15	17	18	19	18	12
PAS	*Ammodytes personatus*	33	115 766	108 666	90 688	82 918	66 129	92 967	72 901
SAN	*Ammodytes spp*	33	858 452	1 242 837	1 039 873	762 248	740 852	912 436	904 930
FLA	*Percophis brasilianus*	33	9 231	11 887	10 191	7 236	7 862	7 664	4 936
UPR	*Pseudopercis semifasciata*	33	3 438	2 483	2 222	2 679	1 905	1 887	1 641
NEB	*Parapercis colias*	33	2 115	2 227	2 313	2 286	2 130	2 441	2 376
WEG	*Trachinus draco*	33	265	337	395	308	331	1 065	1 040
GBN	*Gobius niger*	33	-	-	-	-	1	1	1
GPA	*Gobiidae*	33	27 350	27 646	51 521	41 788	43 596	40 146	37 116
SUR	*Acanthuridae*	33	4 902	7 659	8 235	6 247	5 749	6 969	7 846
BAT	*Platax spp*	33	1 584	2 975	2 793	2 877	2 598	2 830	3 227
SPA	*Ephippidae*	33	80	61	87	309	359	406	260
SCT	*Scatophagus spp*	33	4 204	2 152	2 496	2 541	2 698	3 217	3 235
SPI	*Siganus spp*	33	30 023	27 479	29 261	30 424	33 039	34 881	31 642
CLI	*Ophiodon elongatus*	33	4 754	3 589	2 559	3 154	3 214	2 698	2 925
ATK	*Pleurogrammus azonus*	33	213 363	249 356	293 271	214 269	223 456	214 592	213 933
HUM	*Pleurogrammus monopterygius*	33	88 030	59 538	51 579	51 436	44 592	57 096	37 759
FLI	*Platycephalus indicus*	33	2 900	4 196	2 857	2 248	2 310	1 699	1 199
FLH	*Platycephalidae*	33	4 389	4 576	3 305	4 772	4 834	4 169	4 791
SMQ	*Scorpaenichthys marmoratus*	33	...	...	168	173	148	119	97
SWU	*Cottidae*	33	-	2	5	3	64	75	161
NRC	*Normanichthys crockeri*	33	3 924	20 426	236	4 843	853	223	4 371
BXF	*Ostraciidae*	33	...	1	1	55	68	118	80
PUF	*Sphoeroides maculatus*	33	11	18	17	37	36	35	57
PUP	*Takifugu vermicularis*	33	9 708	7 471	3 897	4 787	-	-	3 110
PUX	*Tetraodontidae*	33	8 512	7 510	8 536	9 666	14 668	12 249	9 027
FLF	*Cantherhines(=Navodon) spp*	33	211 059	297 227	236 189	241 209	225 403	204 270	159 983
FIL	*Stephanolepis cirrhifer*	33	1 772	16 318	9 364	2 999	-	-	933
PKB	*Parika scaber*	33	442	1 095	312	738	1 279	1 142	1 012
FFX	*Monacanthidae*	33	...	...	...	...	...	...	15
TRG	*Balistes carolinensis*	33	38	53	59	36	50	66	101
TRI	*Balistidae*	33	4 070	8 531	6 047	5 923	3 543	3 571	3 477
TFD	*Batrachoididae*	33	...	1	6	74	112	102	108
ARG	*Argentina spp*	34	24 700	27 431	46 509	31 629	28 630	49 036	37 169
DES	*Glossanodon semifasciatus*	34	8 134	7 431	7 142	6 312	5 970	5 414	4 926
ALC	*Alepocephalus bairdii*	34	0	0	0	0	12	616	259
LAN	*Lampanyctodes hectoris*	34	33	243	6 553	0	...	...	...
LAC	*Lampanyctus achirus*	34	0	-	-	-	-	-	-
GYN	*Gymnoscopelus nicholsi*	34	-	-	-	-	0	-	0
LXX	*Myctophidae*	34	0	0	0	5	67	335	37
DPC	*Muraenesox cinereus*	34	189 864	201 111	258 172	253 978	237 743	259 625	271 350
PCX	*Muraenesox spp*	34	19 286	20 507	19 365	24 172	19 427	20 258	16 943
COE	*Conger conger*	34	18 216	17 600	17 660	16 616	15 295	14 578	14 487

B-00 (b)

Fish, crustaceans, molluscs, etc — Capture production by species items
Poissons, crustacés, mollusques, etc — Captures par catégories d'espèces
Peces, crustáceos, moluscos, etc — Capturas por partidas de especies

3-alpha code Code alpha-3 Código alfa-3	Scientific name Nom scientifique Nombre científico	Species group Groupe d'espèces Grupo de especies	1996 t	1997 t	1998 t	1999 t	2000 t	2001 t	2002 t
COS	*Conger orbignyanus*	34	89	138	189	180	122	586	655
COA	*Conger oceanicus*	34	29	17	48	43	49	40	61
ELS	*Conger myriaster*	34	17 314	19 136	11 913	10 160	8 304	7 676	17 210
COX	*Congridae*	34	15 448	14 784	13 308	11 769	11 861	11 734	14 003
SNS	*Macroramphosus scolopax*	34	-	89	443	-	-	44	0
CUQ	*Centriscops humerosus*	34	1	2	11	39	...	52	24
CUS	*Genypterus blacodes*	34	43 643	55 223	59 402	55 118	49 254	51 756	47 512
CUC	*Genypterus chilensis*	34	982	745	584	415	608	730	213
CUB	*Genypterus maculatus*	34	1 343	1 661	2 753	1 943	3 542	3 889	1 558
KCP	*Genypterus capensis*	34	6 807	5 955	5 594	7 820	7 922	11 460	13 418
CEX	*Genypterus spp*	34	1 121	439	425	196	614	552	1 029
BRD	*Brotula barbata*	34	811	1 842	2 213	3 862	1 720	1 926	1 390
OPH	*Ophidiidae*	34	294	355	489	557	525	693	752
ALF	*Beryx spp*	34	8 109	8 026	6 516	5 638	9 469	9 529	12 947
CXF	*Centroberyx affinis*	34	1 407	1 551	1 984	1 689	1 583	1 415	1 275
ORY	*Hoplostethus atlanticus*	34	48 988	44 545	42 420	36 924	27 972	26 183	30 731
TRC	*Trachichthyidae*	34	833	1 052	40	807	1 489	2 346	1 528
JOD	*Zeus faber*	34	6 161	5 615	6 517	7 698	8 627	8 345	9 079
JOS	*Zenopsis conchifer*	34	36	12	55	22	28	62	66
ZNE	*Zenopsis nebulosus*	34	352	429	544	378	315	257	416
ZEX	*Zeidae*	34	288	641	754	763	778	687	1 086
BOR	*Caproidae*	34	-	-	5	-	-	127	91
ORD	*Oreosomatidae*	34	18 799	22 038	21 102	22 690	22 960	24 413	18 075
WRF	*Polyprion americanus*	34	1 292	857	659	776	617	475	508
WHA	*Polyprion oxygeneios*	34	1 178	1 687	1 597	1 555	1 504	1 589	1 612
ULP	*Caulolatilus princeps*	34	...	...	535	627	1 077	984	929
CKZ	*Caulolatilus chrysops*	34	-	-	-	0	6	806	857
TIL	*Lopholatilus chamaeleonticeps*	34	1 463	1 890	1 419	616	688	285	295
TIS	*Branchiostegidae*	34	10 188	9 185	8 907	8 831	9 481	11 296	12 223
EMM	*Emmelichthys nitidus*	34	2 061	1 826	2 188	2 959	2 875	1 918	2 825
EMT	*Emmelichthyidae*	34	624	438	653	452	582	526	1 821
LOB	*Lobotes surinamensis*	34	-	-	-	-	1	1	1
SWH	*Paristiopterus labiosus*	34	19	27	75	6	9	3	9
EDR	*Pseudopentaceros richardsoni*	34	305	53	156	121	127	19	63
CBC	*Cepola rubescens*	34	-	-	-	-	-	-	17
CTA	*Cheilodactylus bergi*	34	3 142	4 893	11 508	3 260	1 432	1 367	498
HAW	*Cheilodactylus variegatus*	34	283	462	140	288	380	306	360
TAK	*Nemadactylus macropterus*	34	4 366	5 441	5 239	5 589	5 739	6 129	6 149
MOW	*Nemadactylus spp*	34	1 277	1 620	1 287	1 412	1 215	1 289	1 146
TRU	*Latridae*	34	747	776	601	735	624	595	571
TOA	*Dissostichus mawsoni*	34	-	-	42	296	751	626	1 354
TOP	*Dissostichus eleginoides*	34	32 263	30 062	35 641	34 835	40 138	40 254	37 904
TOT	*Dissostichus spp*	34	-	-	-	-	-	572	-
NOT	*Patagonotothen brevicauda*	34	-	-	-	3	0	-	9
PAT	*Patagonotothen ramsayi*	34	-	-	-	-	-	-	10
PGE	*Parachaenichthys georgianus*	34	-	-	-	-	0	-	-
SSI	*Chaenocephalus aceratus*	34	-	-	-	1	0	1	5
ANI	*Champsocephalus gunnari*	34	5	227	134	268	4 201	2 033	3 633
SGI	*Pseudochaenichthys georgianus*	34	-	-	-	3	0	6	6
KIF	*Chionodraco rastrospinosus*	34	-	1	-	1	-	1	-
LIC	*Channichthys rhinoceratus*	34	-	9	6	2	2	1	3
WIC	*Chaenodraco wilsoni*	34	-	-	-	0	-	11	-
ICX	*Channichthyidae*	34	-	-	0	0	0	2	2
EPI	*Epigonus telescopus*	34	3 085	4 515	2 894	2 873	4 343	2 132	2 874
CDL	*Epigonus spp*	34	513	1 727	5 284	2 999	5 792	4 648	1 598
PGR	*Pogonophryne permitini*	34	-	-	-	-	0	0	0
CAA	*Anarhichas lupus*	34	27 004	30 961	37 587	39 573	40 600	38 948	36 229
CAS	*Anarhichas minor*	34	1 109	1 180	1 599	1 545	2 987	3 257	3 522
CAT	*Anarhichas spp*	34	8 695	15 066	18 474	8 638	6 300	15 117	5 837
ELZ	*Lycodes spp*	34	18	-	2	1	28	48	62
UUC	*Uranoscopus scaber*	34	...	...	...	15	50	46	104
STZ	*Kathetostoma giganteum*	34	2 123	3 991	2 196	3 370	3 638	4 233	3 272
JAS	*Arctoscopus japonicus*	34	9 220	8 403	8 285	9 064	8 223	10 039	12 630
SNK	*Thyrsites atun*	34	32 829	39 627	45 914	41 693	40 235	41 421	41 124
WSM	*Thyrsitops lepidopoides*	34	1 232	309	285	136	-	-	-
LEC	*Lepidocybium flavobrunneum*	34	0	1	55	57	125	134	137
OIL	*Ruvettus pretiosus*	34	2 687	2 661	3 110	2 739	2 697	3 805	4 455
GEM	*Rexea solandri*	34	4 344	2 253	1 899	1 360	1 696	1 396	995
PRP	*Promethichthys prometheus*	34	-	-	-	-	-	6	7
BEH	*Benthodesmus spp*	34	-	-	-	-	-	1	80
LHT	*Trichiurus lepturus*	34	1 283 139	1 206 353	1 436 303	1 416 885	1 478 033	1 475 127	1 452 209
SFS	*Lepidopus caudatus*	34	17 397	13 945	16 503	11 991	4 730	8 989	10 021
BSF	*Aphanopus carbo*	34	13 658	9 930	9 678	9 398	12 387	14 834	14 650
CUT	*Trichiuridae*	34	119 436	219 561	150 705	190 272	206 611	166 033	206 201
DRI	*Ariomma indica*	34	120	79	49	35	40	45	40
SEO	*Seriolella porosa*	34	2 042	2 593	2 462	3 322	3 542	3 990	5 368
SEM	*Seriolella brama*	34	1 760	4 108	3 101	3 881	4 259	4 101	4 174
SEP	*Seriolella punctata*	34	4 926	11 253	10 993	9 029	11 218	11 268	9 175
SEU	*Seriolella caerulea*	34	1 467	2 432	2 296	2 366	2 407	1 962	1 975
BSP	*Seriolella spp*	34	10 382	8 922	6 890	8 619	8 982	11 394	9 830
BWA	*Hyperoglyphe antarctica*	34	2 432	2 974	2 630	2 755	2 793	2 954	3 001
HGY	*Hyperoglyphe bythites*	34	...	...	...	2	7	5	6
BUP	*Psenopsis anomala*	34	13 290	11 777	13 734	10 871	10 721	10 942	9 899
CEN	*Centrolophidae*	34	459	654	597	1 495	987	719	856
REG	*Sebastes marinus*	34	11 034	11 413	7 968	52 592	75 556	67 044	68 929
WRO	*Sebastes entomelas*	34	...	7 751	4 897	4 243	3 604	2 609	442
YRO	*Sebastes flavidus*	34	...	2 744	3 602	3 468	3 170	2 077	1 206
OPP	*Sebastes alutus*	34	31 523	28 224	27 264	29 717	25 631	25 001	28 716
SBC	*Sebastes paucispinis*	34	...	724	596	197	27	33	22

B-00 (b)

Fish, crustaceans, molluscs, etc — **Capture production by species items**
Poissons, crustacés, mollusques, etc — **Captures par catégories d'espèces**
Peces, crustáceos, moluscos, etc — **Capturas por partidas de especies**

3-alpha code Code alpha-3 Código alfa-3	Scientific name Nom scientifique Nombre científico	Species group Groupe d'espèces Grupo de especies	1996 t	1997 t	1998 t	1999 t	2000 t	2001 t	2002 t
REB	*Sebastes mentella*	34	4 842	5 234	4 619	25 043	76 177	98 661	93 109
SPG	*Sebastes pinniger*	34	...	1 262	1 313	772	60	49	54
SGO	*Sebastes goodei*	34	...	1 850	1 273	918	445	618	161
RMG	*Sebastes melanops*	34	...	...	...	...	109	148	127
RED	*Sebastes spp*	34	273 211	237 912	247 159	177 281	119 779	97 395	106 254
RSE	*Scorpaena scrofa*	34	...	...	...	...	1	2	5
BRF	*Helicolenus dactylopterus*	34	5 581	7 715	6 932	7 873	7 305	5 343	3 311
SJU	*Sebastolobus alascanus*	34	...	...	...	...	331	268	295
SCO	*Scorpaenidae*	34	74 451	69 824	50 844	51 086	48 351	45 931	41 904
KUG	*Chelidonichthys kumu*	34	2 395	3 061	2 972	2 242	3 154	4 250	4 085
GUU	*Chelidonichthys lucerna*	34	5	4	5	22	2 472	1 217	2 770
GUR	*Chelidonichthys cuculus*	34	214	153	188	319	4 860	7 293	6 165
CTZ	*Chelidonichthys lastoviza*	34	...	...	...	...	56	51	68
GUC	*Chelidonichthys capensis*	34	497	642	839	578	686	609	696
SRA	*Prionotus spp*	34	812	550	889	1 187	1 863	356	418
BEG	*Pterygotrigla polyommata*	34	58	136	92	94	120	149	140
JGU	*Pterygotrigla picta*	34	87	71	87	74	55	65	49
GUG	*Eutrigla gurnardus*	34	1 514	1 319	1 216	1 497	1 918	1 668	1 264
GUX	*Triglidae*	34	17 906	15 825	15 774	40 365	64 370	15 039	11 484
SAB	*Anoplopoma fimbria*	34	30 764	28 006	24 232	26 471	25 351	22 948	20 055
ZSP	*Zanclorhynchus spinifer*	34	-	0	-	-	-	-	-
LUM	*Cyclopterus lumpus*	34	11 340	15 551	7 138	9 823	7 081	9 494	12 775
MON	*Lophius piscatorius*	34	69 104	72 930	60 156	54 680	55 405	54 777	50 150
ANK	*Lophius budegassa*	34	1 021	735	1 156	664	383	525	880
ANG	*Lophius americanus*	34	25 826	29 580	27 844	26 320	22 075	25 178	26 756
MVA	*Lophius vaillanti*	34	...	...	...	...	7	5	5
MVO	*Lophius vomerinus*	34	15 433	17 222	24 457	21 783	21 402	21 996	23 721
ANF	*Lophiidae*	34	12 014	5 444	4 512	4 200	6 867	14 745	15 957
DPX	*Perciformes*	34	45 072	50 813	54 388	50 486	44 898	57 122	67 645
HER	*Clupea harengus*	35	2 328 688	2 533 909	2 421 462	2 411 408	2 381 011	1 952 975	1 872 013
HEP	*Clupea pallasii*	35	257 870	437 897	508 627	471 307	457 197	406 938	320 298
SAG	*Sardinella gibbosa*	35	157 104	156 914	174 691	162 710	172 219	185 912	191 300
IOS	*Sardinella longiceps*	35	223 355	289 142	247 065	208 898	402 936	438 737	409 894
SAA	*Sardinella aurita*	35	554 692	461 627	509 079	411 645	362 218	436 689	444 934
SAE	*Sardinella maderensis*	35	125 701	139 827	131 970	129 066	140 943	128 945	134 435
JSS	*Sardinella zunasi*	35	10 663	5 593	1 973	6 674	4 603	766	796
SAM	*Sardinella lemuru*	35	88 590	138 636	153 965	89 286	88 744	103 710	108 370
BSR	*Sardinella brasiliensis*	35	97 093	117 642	82 283	25 518	17 053	39 039	27 894
SIX	*Sardinella spp*	35	912 328	1 013 043	1 125 520	1 069 477	1 032 808	914 401	783 645
JAP	*Sardinops melanostictus*	35	430 837	417 939	295 788	515 477	305 767	339 377	236 602
CPI	*Sardinops caeruleus*	35	450 596	498 653	381 898	405 402	546 079	685 497	722 071
CHP	*Sardinops sagax*	35	1 493 936	722 807	937 269	442 790	338 131	135 712	27 953
PIA	*Sardinops ocellatus*	35	106 381	144 681	196 581	175 969	161 448	200 100	264 886
SRP	*Sardinops neopilchardus*	35	169	385	519	894	1 253	1 399	992
MHS	*Brevoortia aurea*	35	6 294	2 396	2 936	2 202	1 123	344	490
MHP	*Brevoortia pectinata*	35	511	1 112	519	291	337	368	141
MHA	*Brevoortia tyrannus*	35	304 665	322 239	276 230	208 000	207 122	261 401	211 574
MHG	*Brevoortia patronus*	35	491 612	597 565	497 461	694 242	591 434	528 506	582 497
RAS	*Dussumieria acuta*	35	41 347	38 045	42 960	48 265	36 754	37 303	29 601
RAL	*Dussumieria elopsoides*	35	48	19	17	10	15	17	15
BOA	*Ethmalosa fimbriata*	35	144 538	179 630	175 931	184 036	163 359	182 440	188 517
RRH	*Etrumeus teres*	35	85 563	58 362	58 569	36 837	29 407	32 861	27 908
WRR	*Etrumeus whiteheadi*	35	67 773	97 279	57 669	59 032	38 877	56 762	64 450
SAS	*Harengula spp*	35	839	1 023	708	946	1 715	2 279	1 075
THP	*Opisthonema libertate*	35	73 558	69 411	90 002	60 999	84 051	48 919	59 126
THA	*Opisthonema oglinum*	35	11 082	18 912	22 674	26 418	27 029	11 085	7 703
SRH	*Spratelloides gracilis*	35	657	785	896	561	669	505	1 063
PIL	*Sardina pilchardus*	35	996 334	998 734	949 776	907 612	939 335	1 134 206	1 089 836
SPR	*Sprattus sprattus*	35	671 616	700 239	696 228	684 189	658 412	647 417	620 020
FAS	*Sprattus fuegensis*	35	0	0	29	17	97	4	1
MES	*Ethmidium maculatum*	35	9 792	13 241	40 845	29 436	23 991	14 016	27 337
HES	*Herklotsichthys quadrimaculat.*	35	140	120	22	20	20	...	20
CKI	*Strangomera bentincki*	35	446 669	441 154	317 564	782 142	722 522	324 617	347 368
ANE	*Engraulis encrasicolus*	35	527 610	502 390	506 844	631 978	626 338	667 326	661 326
JAN	*Engraulis japonicus*	35	1 254 487	1 666 503	2 093 888	1 820 259	1 725 685	1 836 502	1 853 936
ANA	*Engraulis anchoita*	35	21 023	25 211	13 417	13 025	12 164	13 133	16 203
NPA	*Engraulis mordax*	35	14 103	7 925	2 335	11 137	19 460	19 677	9 027
VET	*Engraulis ringens*	35	8 863 714	7 685 098	1 729 064	8 723 265	11 276 357	7 213 077	9 702 614
ANC	*Engraulis capensis*	35	41 792	62 640	110 296	180 954	267 986	289 323	254 643
AVA	*Cetengraulis edentulus*	35	8	0	119	0	0	2	0
VEP	*Cetengraulis mysticetus*	35	112 528	195 630	180 705	70 358	120 943	227 783	203 704
ENP	*Anchoa hepsetus*	35	...	1	16	8	0	0	0
STO	*Stolephorus spp*	35	267 262	296 356	279 854	274 846	278 834	312 851	302 003
ANX	*Engraulidae*	35	328 256	288 793	964 051	235 545	242 158	389 474	254 893
DOB	*Chirocentrus dorab*	35	11 873	12 392	14 432	17 326	16 994	14 760	14 814
DOS	*Chirocentrus spp*	35	46 855	56 304	45 340	50 265	49 883	51 441	57 515
CLU	*Clupeoidei*	35	304 872	312 794	324 863	356 750	316 579	340 110	337 060
BON	*Sarda sarda*	36	26 390	30 999	47 445	37 629	28 958	30 937	26 571
BIP	*Sarda orientalis*	36	370	498	162	134	95	287	378
BEP	*Sarda chiliensis*	36	23 927	18 929	7 239	3 187	972	1 471	1 158
BOP	*Orcynopsis unicolor*	36	2 136	476	224	861	1 199	1 053	1 046
WAH	*Acanthocybium solandri*	36	2 303	2 947	2 643	3 202	2 774	3 238	3 202
DOT	*Gymnosarda unicolor*	36	625	489	470	426	451	647	789
COM	*Scomberomorus commerson*	36	157 862	168 665	171 957	183 218	186 595	177 337	216 742
GUT	*Scomberomorus guttatus*	36	41 881	40 346	50 829	55 241	44 089	46 146	57 243
STS	*Scomberomorus lineolatus*	36	96	901	107	135	55	38	431
KGM	*Scomberomorus cavalla*	36	14 155	15 551	12 396	14 556	11 625	11 569	14 610

B-00 (b)

Fish, crustaceans, molluscs, etc — **Capture production by species items**
Poissons, crustacés, mollusques, etc — **Captures par catégories d'espèces**
Peces, crustáceos, moluscos, etc — **Capturas por partidas de especies**

3-alpha code Code alpha-3 Código alfa-3	Scientific name Nom scientifique Nombre científico	Species group Groupe d'espèces Grupo de especies	1996 t	1997 t	1998 t	1999 t	2000 t	2001 t	2002 t
SSM	*Scomberomorus maculatus*	36	12 594	9 084	8 840	9 744	7 367	7 236	7 698
CER	*Scomberomorus regalis*	36	307	481	441	125	190	147	...
SIE	*Scomberomorus sierra*	36	7 154	6 813	6 651	8 657	9 141	7 678	6 707
MAW	*Scomberomorus tritor*	36	1 296	1 377	1 456	703	834	963	1 186
NPH	*Scomberomorus niphonius*	36	301 356	365 585	551 780	595 103	539 094	522 756	553 652
BRS	*Scomberomorus brasiliensis*	36	8 435	8 065	7 923	5 674	6 643	6 313	8 321
KGX	*Scomberomorus spp*	36	41 659	42 972	42 745	42 820	38 638	38 934	34 943
FRI	*Auxis thazard*	36	910	1 368	1 339	1 650	1 413	1 409	896
FRZ	*Auxis thazard, A.rochei*	36	187 478	217 724	201 712	263 606	223 848	216 881	275 375
LTA	*Euthynnus alletteratus*	36	9 948	11 495	11 403	10 351	11 221	8 146	11 448
BKJ	*Euthynnus lineatus*	36	614	129	600	140	360	1 865	1 304
KAW	*Euthynnus affinis*	36	143 585	167 206	157 699	167 267	170 554	163 439	168 348
SKJ	*Katsuwonus pelamis*	36	1 570 873	1 594 931	1 895 381	1 977 381	1 962 706	1 814 500	2 030 648
BFT	*Thunnus thynnus*	36	52 581	48 710	41 414	35 296	35 957	36 385	35 140
PBF	*Thunnus orientalis*	36	16 098	10 954	7 593	16 845	16 185	9 143	9 095
LOT	*Thunnus tonggol*	36	104 316	99 962	101 822	105 909	136 036	118 310	102 184
BLF	*Thunnus atlanticus*	36	3 572	2 261	2 809	2 917	2 319	3 830	4 607
ALB	*Thunnus alalunga*	36	201 492	219 425	229 936	254 843	218 333	232 193	238 281
SBF	*Thunnus maccoyii*	36	16 336	15 350	17 671	19 828	15 940	17 152	15 218
YFT	*Thunnus albacares*	36	1 065 273	1 180 131	1 212 869	1 215 822	1 182 782	1 330 165	1 341 319
BET	*Thunnus obesus*	36	378 227	408 203	418 128	429 982	430 679	408 696	430 289
SFA	*Istiophorus platypterus*	36	13 646	16 157	17 426	17 000	16 896	14 880	15 813
SAI	*Istiophorus albicans*	36	2 259	1 870	1 814	1 869	2 370	2 281	2 774
BUM	*Makaira nigricans*	36	28 963	27 118	31 064	30 155	30 174	29 761	29 417
BLM	*Makaira indica*	36	2 447	2 240	3 229	3 020	2 599	2 639	2 515
MLS	*Tetrapturus audax*	36	13 163	12 294	14 536	10 502	9 740	8 360	6 617
WHM	*Tetrapturus albidus*	36	1 172	1 034	1 029	962	1 180	463	575
SSP	*Tetrapturus angustirostris*	36	44	20	5	8	5	5	56
SPF	*Tetrapturus pfluegeri*	36	61	69	90	133	136	91	73
BIL	*Istiophoridae*	36	30 388	30 549	38 276	38 992	26 729	26 168	22 626
SWO	*Xiphias gladius*	36	89 404	101 973	108 301	98 923	111 350	101 929	106 749
TUX	*Scombroidei*	36	277 773	301 204	328 849	307 953	340 143	316 733	302 293
CAP	*Mallotus villosus*	37	1 527 422	1 603 338	982 312	904 045	1 484 818	1 670 906	1 961 724
GAR	*Belone belone*	37	3 027	3 453	3 203	2 771	2 147	2 507	3 341
NED	*Tylosurus spp*	37	36 501	40 184	39 549	40 129	43 204	39 425	40 797
BEN	*Belonidae*	37	1 228	431	397	481	721	738	579
SAU	*Scomberesox saurus*	37	484	814	1 368	1 154	481	5 171	2 081
SAP	*Cololabis saira*	37	276 111	388 643	180 973	187 898	306 069	376 173	335 473
HAJ	*Hyporhamphus sajori*	37	990	1 193	1 160	913	956	613	670
BAL	*Hemiramphus brasiliensis*	37	494	442	1 500	864	1 036	3 383	3 688
HAX	*Hemiramphus spp*	37	11 885	10 305	9 468	10 174	12 842	11 834	13 296
JFL	*Cypselurus agoo*	37	8 501	7 486	8 933	6 738	9 615	8 286	6 909
FLY	*Exocoetidae*	37	42 392	55 480	69 731	363 026	104 720	63 040	65 977
LAG	*Lampris guttatus*	37	172	126	369	542	466	538	915
LAI	*Lampris immaculatus*	37	-	-	-	-	-	-	0
TRP	*Trachipterus spp*	37	49	60	74	94	87	128	80
REL	*Regalecus glesne*	37	10	64	60	34	20	1	0
ODR	*Odontesthes regia*	37	690	494	560	3 414	1 357	833	1 396
ATB	*Atherina boyeri*	37	39	123	375	381	434	423	489
SSA	*Menidia menidia*	37	170	259	255	583	319	661	150
SIL	*Atherinidae*	37	15 982	17 152	13 755	19 098	22 880	23 074	27 433
TRF	*Lactarius lactarius*	37	7 345	8 709	10 321	7 743	6 736	7 708	5 903
BLU	*Pomatomus saltatrix*	37	23 265	19 105	17 326	13 950	16 362	27 012	44 548
CBA	*Rachycentron canadum*	37	5 072	5 583	5 549	6 634	8 557	7 600	8 017
HOM	*Trachurus trachurus*	37	474 625	455 109	350 298	321 195	231 059	250 026	216 622
JJM	*Trachurus japonicus*	37	349 166	350 619	340 565	227 290	271 501	235 874	229 539
CJM	*Trachurus murphyi*	37	4 378 843	3 597 117	2 025 758	1 423 447	1 540 494	2 508 834	1 750 078
PJM	*Trachurus symmetricus*	37	2 176	1 160	1 793	1 126	1 316	3 839	1 026
HMM	*Trachurus mediterraneus*	37	21 002	17 065	15 122	12 898	19 111	19 308	23 047
RSC	*Trachurus lathami*	37	976	601	328	495	107	345	455
HMC	*Trachurus capensis*	37	470 372	407 482	482 086	398 598	423 607	354 046	386 361
HMZ	*Trachurus trecae*	37	86 118	94 493	57 816	92 476	70 651	49 307	46 015
HMG	*Trachurus declivis*	37	15 441	10 656	9 379	15 545	12 665	7 729	6 303
JAX	*Trachurus spp*	37	247 950	236 646	321 010	315 284	367 090	355 132	320 455
TRZ	*Pseudocaranx dentex*	37	3 833	3 813	4 166	4 671	4 021	3 531	3 776
RSA	*Decapterus maruadsi*	37	62 378	69 764	71 168	67 689	40 803	47 460	49 586
RUS	*Decapterus russelli*	37	145 320	150 027	145 747	162 437	181 363	163 798	176 269
SDX	*Decapterus spp*	37	1 132 890	1 042 486	1 079 873	1 032 812	1 034 891	1 112 780	1 175 061
RUB	*Caranx crysos*	37	816	923	851	823	830	905	989
CVJ	*Caranx hippos*	37	3 059	2 529	3 432	4 591	2 292	3 564	3 424
CXR	*Caranx ruber*	37	-	-	-	3	5	7	13
HMY	*Caranx rhonchus*	37	3 301	3 344	2 864	2 710	3 935	281	1 745
TRE	*Caranx spp*	37	168 747	169 096	170 014	184 139	163 330	165 808	159 631
MOA	*Selene setapinnis*	37	3 445	4 024	3 850	3 043	2 362	4 551	2 695
LUK	*Selene dorsalis*	37	1 015	832	755	772	1 236	1 936	1 649
POO	*Trachinotus blochii*	37	-	-	4	0	31	-	-
POM	*Trachinotus carolinus*	37	44	297	321	207	242	181	159
POX	*Trachinotus spp*	37	9 238	5 218	5 366	6 499	2 372	1 782	2 455
AMB	*Seriola dumerili*	37	1 057	724	1 315	1 847	2 033	2 130	2 094
AMJ	*Seriola quinqueradiata*	37	246	...	...	...	...	...	...
YTC	*Seriola lalandi*	37	1 042	961	1 517	885	971	542	447
AMX	*Seriola spp*	37	60 974	63 417	80 195	69 133	97 491	105 570	91 929
LEE	*Lichia amia*	37	2 795	3 577	3 793	4 788	1 777	1 222	2 730
ALA	*Alectis alexandrinus*	37	160	742	1 023	563	502	862	611
POB	*Parastromateus niger*	37	44 676	48 464	44 524	47 590	46 346	56 676	60 725
RRU	*Elagatis bipinnulata*	37	12 402	13 147	18 940	15 874	14 923	15 039	16 845
GLT	*Gnathanodon speciosus*	37	441	471	536	489	1 125	1 286	2 644
HAS	*Megalaspis cordyla*	37	59 552	70 059	77 525	80 113	78 517	77 685	89 418

B-00 (b)

Fish, crustaceans, molluscs, etc — Capture production by species items
Poissons, crustacés, mollusques, etc — Captures par catégories d'espèces
Peces, crustáceos, moluscos, etc — Capturas por partidas de especies

3-alpha code Code alpha-3 Código alfa-3	Scientific name Nom scientifique Nombre científico	Species group Groupe d'espèces Grupo de especies	1996 t	1997 t	1998 t	1999 t	2000 t	2001 t	2002 t
QUE	*Scomberoides spp*	37	25 067	22 364	25 282	27 630	24 588	28 161	28 167
BUA	*Chloroscombrus chrysurus*	37	4 534	6 186	12 485	12 250	9 810	17 255	20 367
HSO	*Chloroscombrus orqueta*	37	1 706	952	565	1 409	...	1 008	1 100
PAO	*Parona signata*	37	2 364	2 236	2 209	2 070	1 854	1 551	1 314
BIS	*Selar crumenophthalmus*	37	72 630	80 881	93 486	99 057	105 493	111 912	132 692
TRY	*Selaroides leptolepis*	37	31 031	35 139	35 852	44 537	46 888	42 892	46 326
RNJ	*Seriolina nigrofasciata*	37	7 269	7 096	6 042	6 045	5 415	6 238	6 218
CGX	*Carangidae*	37	293 707	296 225	293 145	301 819	281 079	292 138	327 912
POA	*Brama brama*	37	12 631	10 439	10 061	10 192	10 036	17 727	6 467
BRZ	*Bramidae*	37	-	-	-	-	2	5	225
DOL	*Coryphaena hippurus*	37	24 623	30 701	44 170	31 307	38 172	42 920	39 491
ECN	*Echeneidae*	37	-	-	3	1	12	...	...
MAS	*Scomber japonicus*	37	2 177 781	2 427 389	1 924 691	1 950 387	1 473 060	1 812 712	1 470 673
MAC	*Scomber scombrus*	37	559 258	555 621	666 964	618 014	685 575	710 578	769 068
MAA	*Scomber australasicus*	37	2 994	8 777	7 260	15 874	12 108	11 801	15 136
MAZ	*Scomber spp*	37	13 169	11 878	10 879	8 964	8 916	9 547	9 083
RAB	*Rastrelliger brachysoma*	37	25 224	22 978	23 350	25 713	26 771	28 091	32 657
RAG	*Rastrelliger kanagurta*	37	399 939	303 946	294 607	296 905	184 810	166 605	185 649
RAX	*Rastrelliger spp*	37	425 521	427 233	458 361	477 519	458 497	465 408	460 255
MAX	*Scombridae*	37	51 342	60 706	63 435	49 331	44 263	51 290	46 843
BLB	*Stromateus fiatola*	37	240	323	292	636	320	4 432	1 547
SIP	*Pampus argenteus*	37	37 452	31 369	30 252	32 246	32 525	39 656	43 770
XPO	*Pampus spp*	37	220 364	242 547	303 024	337 919	338 848	352 493	386 317
HVF	*Peprilus alepidotus*	37	14	15	13	16	23	18	22
BTG	*Peprilus spp*	37	1 889	568	1 175	1 004	809	742	872
BUX	*Stromateidae*	37	61 130	61 324	59 923	59 896	45 523	46 645	53 158
BAR	*Sphyraena spp*	37	83 742	84 886	97 487	109 084	91 821	95 769	102 280
MOX	*Mola mola*	37	...	...	...	12	...	15	1
PPX	*Perciformes*	37	70 923	69 623	80 012	93 670	75 446	90 089	103 288
SBL	*Hexanchus griseus*	38	-	-	-	-	-	1	7
NTC	*Notorynchus cepedianus*	38	...	...	2	3	4	5	4
BSK	*Cetorhinus maximus*	38	1 984	1 169	192	210	389	287	180
CCT	*Carcharias taurus*	38	-	-	-	-	1	-	-
ALV	*Alopias vulpinus*	38	20	67	393	495	650	614	464
BTH	*Alopias superciliosus*	38	...	149	125	5	5	2	...
THR	*Alopias spp*	38	...	34	55	66	...	...	...
SMA	*Isurus oxyrinchus*	38	494	2 140	2 377	1 215	2 140	2 170	4 875
LMA	*Isurus paucus*	38	-	1	1	-	4	3	1
MAK	*Isurus spp*	38	...	92	38	...	116	47	117
POR	*Lamna nasus*	38	1 496	1 917	2 221	2 690	2 865	2 135	1 008
WSH	*Carcharodon carcharias*	38	-	-	-	-	2	0	-
GNC	*Ginglymostoma cirratum*	38	-	-	-	-	407	89	17
SHO	*Galeus melastomus*	38	...	...	...	...	...	...	288
SYC	*Scyliorhinus canicula*	38	5 144	5 613	5 740	5 818	6 182	7 072	6 471
SYT	*Scyliorhinus stellaris*	38	306	378	258	274	274	264	195
SCL	*Scyliorhinus spp*	38	54	78	51	275	525	508	331
HAN	*Halaelurus canescens*	38	-	-	-	-	-	-	0
BSH	*Prionace glauca*	38	1 281	6 811	3 496	4 615	8 401	9 517	20 370
CCP	*Carcharhinus plumbeus*	38	-	-	-	-	41	24	28
CCL	*Carcharhinus limbatus*	38	3	9	10	11	601	521	11
DUS	*Carcharhinus obscurus*	38	-	7	0	...	80	0	3
FAL	*Carcharhinus falciformis*	38	23 048	16 876	22 996	23 059	16 050	16 960	21 405
BRO	*Carcharhinus brachyurus*	38	...	...	15	14	25	38	38
TIG	*Galeocerdo cuvier*	38	-	-	-	-	-	1	13
RSK	*Carcharhinidae*	38	52 477	44 502	47 760	41 816	38 195	35 608	38 628
SPZ	*Sphyrna zygaena*	38	10	223	109	19	35	27	40
SPL	*Sphyrna lewini*	38	37	180	10	40	48	517	808
SPY	*Sphyrnidae*	38	...	998	1 028	147	1 957	1 951	1 540
CTI	*Mustelus canis*	38	...	...	...	...	334	321	493
CTK	*Mustelus henlei*	38	...	...	3	5	3	4	3
MTL	*Mustelus lenticulatus*	38	1 350	3 464	1 707	1 662	1 643	1 563	1 403
SDP	*Mustelus schmitti*	38	10 456	10 130	13 422	12 274	8 157	10 766	8 140
SMD	*Mustelus mustelus*	38	...	...	...	...	15	...	58
SDV	*Mustelus spp*	38	15 206	11 533	15 368	10 536	12 135	12 825	15 199
GAG	*Galeorhinus galeus*	38	3 594	3 478	3 654	4 259	4 324	4 108	4 181
GSK	*Somniosus microcephalus*	38	61	73	87	51	45	58	57
SOR	*Somniosus rostratus*	38	...	...	...	...	...	...	2
SON	*Somniosus pacificus*	38	-	-	-	1	-	-	3
DGS	*Squalus acanthias*	38	23 622	23 980	20 382	25 569	31 731	28 886	28 208
GUP	*Centrophorus granulosus*	38	...	...	...	73	54	93	161
GUQ	*Centrophorus squamosus*	38	53	58	133	452	506	538	1 203
SHL	*Etmopterus spp*	38	0	2	...	573	...	4	127
DCA	*Deania calcea*	38	...	...	36	17	46	117	188
CYO	*Centroscymnus coelolepis*	38	336	280	232	717	1 154	2 535	2 986
CYP	*Centroscymnus crepidater*	38	-	-	3	-	-	-	12
SCK	*Dalatias licha*	38	175	352	434	373	628	564	560
CFB	*Centroscyllium fabricii*	38	4	0	-	-	271	271	28
DGX	*Squalidae*	38	37 373	30 562	29 331	22 793	9 613	9 259	6 383
DGH	*Squalidae, Scyliorhinidae*	38	2 022	2 083	2 011	2 113	3 032	2 700	3 058
PWS	*Pristiophorus spp*	38	...	...	...	...	270	423	371
AGN	*Squatina squatina*	38	18	34	44	25	20	22	16
SUG	*Squatina argentina*	38	4 281	4 410	4 311	3 368	3 123	3 339	2 288
ASK	*Squatinidae*	38	2 143	400	392	465	595	618	692
OXY	*Oxynotus centrina*	38	...	...	...	81	33	63	86
OXN	*Oxynotus paradoxus*	38	...	...	...	...	...	...	1
SHX	*Squaliformes*	38	3 423	1 252	887	1 673	710	876	1 047
GUD	*Rhinobatos percellens*	38	404	...	...	...	...	...	...
GUF	*Rhinobatos planiceps*	38	460	333	344	95	2 624	1 060	822
GTF	*Rhinobatidae*	38	1 535	1 550	1 882	1 955	4 230	3 811	3 162

B-00 (b)

Fish, crustaceans, molluscs, etc — **Capture production by species items**
Poissons, crustacés, mollusques, etc — **Captures par catégories d'espèces**
Peces, crustáceos, moluscos, etc — **Capturas por partidas de especies**

3-alpha code Code alpha-3 Código alfa-3	Scientific name Nom scientifique Nombre científico	Species group Groupe d'espèces Grupo de especies	1996 t	1997 t	1998 t	1999 t	2000 t	2001 t	2002 t
SAW	*Pristidae*	38	...	48	...	41	42	...	...
RJB	*Raja batis*	38	508	441	411	558	866	817	559
RJC	*Raja clavata*	38	1 773	1 588	1 367	1 366	1 277	1 296	1 257
RJR	*Raja radiata*	38	1 493	1 431	1 252	996	1 076	1 211	1 781
RJM	*Raja montagui*	38	977	1 163	1 179	1 260	1 341	1 563	1 451
RJI	*Raja circularis*	38	438	438	410	435	369	330	301
RJF	*Raja fullonica*	38	65	55	50	86	65	105	103
RJE	*Raja microocellata*	38	-	-	1	11	-	-	-
RJN	*Raja naevus*	38	4 077	4 721	4 015	3 638	3 064	2 885	2 735
RJO	*Raja oxyrinchus*	38	346	311	327	194	140	89	211
SRR	*Raja georgiana*	38	-	-	-	11	36	7	24
SKA	*Raja spp*	38	47 768	55 330	58 826	63 094	64 787	60 472	52 614
BEA	*Bathyraja eatonii*	38	-	-	-	1	5	0	1
BMU	*Bathyraja murrayi*	38	-	-	-	-	0	-	-
BYE	*Bathyraja meridionalis*	38	-	-	-	-	0	-	-
BHY	*Bathyraja spp*	38	-	-	-	1	0	-	-
RAJ	*Rajidae*	38	-	-	-	6	-	-	-
WST	*Dasyatis akajei*	38	4 029	3 959	4 329	4 407	5 388	4 312	4 512
JDP	*Dasyatis pastinaca*	38	...	...	-	-	4	11	-
STI	*Dasyatis spp*	38	1	2	5	6	10	7	10
EAG	*Myliobatidae*	38	0	1	1	15	12	16	21
MAN	*Mobulidae*	38	...	...	342	802	931	106	110
SRX	*Rajiformes*	38	160 646	171 369	166 695	178 623	177 990	170 929	173 881
TOE	*Torpedo spp*	38	16	18	19	34	32	43	34
CMO	*Chimaera monstrosa*	38	21	15	32	12	15	123	69
RAT	*Hydrolagus colliei*	38	-	-	-	-	-	-	2
CYV	*Hydrolagus novaezealandiae*	38	1 614	2 064	1 956	1 975	1 819	1 572	2 055
HYD	*Hydrolagus spp*	38	...	0	36	491	1 548	3 019	2 535
RCT	*Rhinochimaera atlantica*	38	-	-	-	-	-	2	1
CHB	*Callorhinchus milii*	38	595	962	972	1 274	1 310	1 294	1 188
CHM	*Callorhinchus capensis*	38	366	484	482	356	380	405	422
ELF	*Callorhinchus spp*	38	2 265	2 151	3 186	2 609	1 993	1 576	880
HOL	*Chimaeriformes*	38	50	5	5	21	40	76	103
SKH	*Selachimorpha(Pleurotremata)*	38	7 694	36 309	25 294	26 480	33 236	33 219	19 876
SKX	*Elasmobranchii*	38	387 921	367 887	365 171	374 692	395 780	370 657	374 026
GRO	*Osteichthyes*	39	19 633	10 218	11 238	7 661	10 399	3 188	2 360
PEL	*Osteichthyes*	39	25 717	8 375	15 772	4 483	2 978	312	4 731
FIN	*Osteichthyes*	39	41 461	49 272	55 357	40 657	62 309	33 694	30 758
MZZ	*Osteichthyes*	39	10 675 897	10 540 582	10 940 395	11 040 806	10 822 214	10 833 231	10 693 764
PRF	*Macrobrachium rosenbergii*	41	6 140	5 228	5 400	5 966	5 475	5 204	5 256
PPF	*Macrobrachium spp*	41	7 093	5 872	4 875	7 528	6 020	3 230	3 169
PPZ	*Palaemonidae*	41	14 692	13 865	10 681	11 556	12 520	12 317	13 013
AAS	*Astacus astacus*	41	...	10	10	10	...	...	...
PCL	*Pacifastacus leniusculus*	41	-	-	-	10	81	80	50
AUP	*Austropotamobius pallipes*	41	0	0	0	0	0	0	0
RCW	*Procambarus clarkii*	41	2 513	2 524	2 519	2 521	2 522	2 503	2 514
AYA	*Euastacus armatus*	41	-	-	-	66	24	23	69
CJF	*Parastacidae*	41	6	6	14	6	0	1	1
AYS	*Astacidae, Cambaridae*	41	6 971	12 130	11 909	7 084	2 214	6 507	9 275
FCX	*Crustacea*	41	372 826	487 142	609 648	464 070	539 613	598 218	785 646
CRK	*Cancer irroratus*	42	4 341	7 530	7 696	7 033	9 516	8 294	8 331
DUN	*Cancer magister*	42	34 503	21 251	18 487	19 024	19 941	22 210	26 275
CRE	*Cancer pagurus*	42	29 185	38 011	40 635	39 341	42 764	44 143	43 886
CRJ	*Cancer borealis*	42	334	745	1 255	1 549	1 114	1 245	1 190
ROC	*Cancer productus*	42	...	...	574	359	494	537	552
STC	*Menippe mercenaria*	42	3 675	3 354	3 347	2 363	3 197	3 332	3 422
SCD	*Portunus pelagicus*	42	139 950	138 733	138 123	147 767	158 223	173 727	174 646
GAZ	*Portunus trituberculatus*	42	303 170	252 502	283 971	284 851	351 051	350 168	369 247
CRS	*Portunus spp*	42	2 761	4 790	4 262	9 147	9 348	9 524	5 662
CRB	*Callinectes sapidus*	42	115 552	121 643	115 996	105 237	92 659	79 755	89 079
CRZ	*Callinectes danae*	42	2 020	2 600	3 014	1 626	1 597	1 062	1 310
CRG	*Carcinus maenas*	42	837	951	995	913	889	1 144	1 258
CMR	*Carcinus aestuarii*	42	65	44	66	44	30	37	74
MUD	*Scylla serrata*	42	17 059	13 998	13 566	16 271	17 581	19 340	20 536
MXT	*Mithrax armatus*	42	-	-	114	110	54	35	22
SCR	*Maja squinado*	42	5 705	6 193	5 919	5 828	6 376	7 518	6 280
CRQ	*Chionoecetes opilio*	42	66 831	74 341	77 517	98 633	104 252	110 044	117 187
PCR	*Chionoecetes spp*	42	57 001	85 914	139 755	110 115	60 190	55 999	44 749
HBZ	*Erimacrus isenbeckii*	42	789	612	409	440	198	162	117
CRR	*Geryon quinquedens*	42	1 978	3 528	2 743	3 382	8 391	6 720	4 590
GER	*Geryon spp*	42	7 370	4 884	5 734	5 829	7 491	5 244	6 087
CRA	*Brachyura*	42	206 548	209 516	211 547	199 486	196 361	191 718	209 705
LOJ	*Panulirus longipes*	43	1 106	1 082	1 098	1 166	1 716	1 924	1 782
SLC	*Panulirus argus*	43	38 438	36 721	34 115	38 038	37 713	31 835	36 133
LOA	*Panulirus cygnus*	43	9 902	9 896	10 400	13 065	14 605	11 353	9 050
NUG	*Panulirus gracilis*	43	207	221	791	678	805	765	405
SLV	*Panulirus spp*	43	9 472	11 282	10 047	10 286	11 203	12 687	14 401
LBC	*Jasus lalandii*	43	1 767	1 879	2 076	2 097	2 058	1 974	3 029
LOF	*Jasus frontalis*	43	36	32	21	22	17	21	9
LOG	*Jasus verreauxi*	43	161	158	124	123	152	114	111
LBT	*Jasus tristani*	43	327	321	376	336	316	425	301
LOR	*Jasus edwardsii*	43	3 121	5 009	2 707	2 818	2 789	2 551	2 485
JSN	*Jasus novaehollandiae*	43	4 856	4 888	4 615	4 655	4 756	4 677	4 386
JSP	*Jasus paulensis*	43	357	295	308	345	192	183	334
PSL	*Palinurus mauritanicus*	43	1	0	25	11	9	5	1
SLO	*Palinurus elephas*	43	435	475	267	228	196	253	222

B-00 (b)

Fish, crustaceans, molluscs, etc — Capture production by species items
Poissons, crustacés, mollusques, etc — Captures par catégories d'espèces
Peces, crustáceos, moluscos, etc — Capturas por partidas de especies

3-alpha code Code alpha-3 Código alfa-3	Scientific name Nom scientifique Nombre científico	Species group Groupe d'espèces Grupo de especies	1996 t	1997 t	1998 t	1999 t	2000 t	2001 t	2002 t
SLN	*Palinurus delagoae*	43	10	10	6	7	8	10	7
SLS	*Palinurus gilchristi*	43	918	892	864	429	305	1 053	651
CRW	*Palinurus spp*	43	1 967	9 143	3 984	1 876	2 550	2 732	1 112
VLO	*Palinuridae*	43	332	233	239	204	228	213	276
IBC	*Ibacus ciliatus*	43	1 115	642	696	676	1 600	1 607	1 005
THQ	*Thenus orientalis*	43	3 015	2 962	3 538	1 797	2 312	2 540	2 464
LOS	*Scyllaridae*	43	1 098	968	1 095	128	111	164	150
NEM	*Metanephrops mozambicus*	43	132	156	192	152	180	141	130
MEC	*Metanephrops challengeri*	43	670	1 093	989	925	1 034	1 093	1 020
MWF	*Metanephrops spp*	43	-	-	-	-	39	105	97
NEP	*Nephrops norvegicus*	43	58 990	61 596	57 379	61 770	56 628	56 310	56 513
LBA	*Homarus americanus*	43	71 866	78 146	77 155	83 105	83 062	83 803	82 422
LBE	*Homarus gammarus*	43	2 590	3 219	2 933	3 285	2 600	2 781	2 900
UOP	*Upogebia pugettensis*	43	...	...	...	...	10	4	4
CZP	*Callianassa spp*	43	-	-	-	25	17	14	14
LOX	*Reptantia*	43	25	2 027	1 035	292	589	715	718
LQL	*Pleuroncodes planipes*	44	164	-	-	-	-	1 263	3 156
PQG	*Pleuroncodes monodon*	44	7 726	8 939	12 602	12 710	11 129	1 754	2 499
CZJ	*Cervimunida johni*	44	6 402	10 322	9 426	7 273	5 069	2 178	925
LOQ	*Galatheidae*	44	105	106	81	130	352	85	808
KCD	*Paralithodes camtschaticus*	44	34 370	23 333	32 718	37 413	28 932	16 794	11 409
KCI	*Paralithodes platypus*	44	8 762	10 268	4 508	5 455	5 233	4 500	4 598
KCY	*Paralithodes brevipes*	44	204	418	194	256	347	254	350
KCS	*Paralithodes spp*	44	9 848	8 331	11 073	7 792	6 938	7 464	7 746
KCR	*Lithodes antarcticus*	44	1 959	2 573	3 222	2 425	3 006	3 025	3 254
KCM	*Lithodes murrayi*	44	-	-	1	0	0	0	0
KAQ	*Lithodes aequispina*	44	4 666	4 917	3 897	2 746	1 797	2 245	2 291
KCZ	*Lithodes spp*	44	-	-	-	0	-	-	-
PAG	*Paralomis granulosa*	44	1 274	1 478	1 502	1 439	5 203	6 745	4 761
KCU	*Paralomis aculeata*	44	-	-	0	0	0	0	-
KCV	*Paralomis spinosissima*	44	214	0	-	0	0	4	55
KCF	*Paralomis formosa*	44	-	-	-	2	2	11	56
KCX	*Lithodidae*	44	-	1	0	0	3	4	2
ABS	*Penaeus aztecus*	45	55 369	47 836	50 722	61 206	63 817	68 869	58 191
PBA	*Penaeus merguiensis*	45	72 746	71 809	79 674	82 053	85 387	85 029	87 219
YPS	*Penaeus californiensis*	45	345	186	321	250	158	142	87
APS	*Penaeus duorarum*	45	15 510	10 872	12 086	7 504	7 177	8 627	9 382
KUP	*Penaeus japonicus*	45	5 606	6 995	5 113	6 096	10 193	6 084	3 041
PNV	*Penaeus vannamei*	45	5 735	5 046	4 277	2 182	1 619	2 710	3 376
GIT	*Penaeus monodon*	45	177 054	178 139	240 301	264 218	253 003	208 617	214 242
FLP	*Penaeus chinensis*	45	56 534	71 317	79 595	70 725	85 545	97 583	99 690
TGS	*Penaeus kerathurus*	45	4 542	6 078	5 498	5 966	7 700	4 952	4 321
PNB	*Penaeus brasiliensis*	45	8 743	10 758	7 796	9 092	10 728	5 010	5 903
TIP	*Penaeus semisulcatus*	45	3 469	3 517	3 019	2 821	2 298	2 703	2 609
PST	*Penaeus setiferus*	45	28 808	32 841	39 799	44 633	52 593	40 724	43 186
CSP	*Penaeus brevirostris*	45	2 136	2 776	2 090	2 462	2 604	2 903	2 072
WKP	*Penaeus latisulcatus*	45	3 350	3 629	3 386	4 145	3 492	3 298	3 292
WWP	*Penaeus occidentalis*	45	1 158	1 505	1 261	2 706	2 020	1 259	1 245
REP	*Penaeus penicillatus*	45	5 020	2 473	647	316	308	312	509
SOP	*Penaeus notialis*	45	18 371	22 233	19 172	32 721	25 250	23 659	17 276
PPS	*Penaeus paulensis*	45	-	177	13	12	56	23	7
PEN	*Penaeus spp*	45	282 433	299 937	259 772	230 991	231 760	237 410	224 271
MPN	*Metapenaeus monoceros*	45	-	-	-	-	-	757	1 607
ENS	*Metapenaeus endeavouri*	45	2 400	2 339	2 930	2 691	3 631	3 525	3 146
SHI	*Metapenaeus joyneri*	45	2 211	1 976	3 651	3 633	2 621	2 385	1 736
MET	*Metapenaeus spp*	45	48 322	57 307	63 831	56 911	63 840	61 275	65 479
DPS	*Parapenaeus longirostris*	45	16 201	19 262	22 789	19 076	23 499	25 466	15 823
NPP	*Parapenaeopsis spp*	45	14 047	16 722	14 689	12 889	11 945	11 576	10 222
BOB	*Xiphopenaeus kroyeri*	45	29 074	36 721	30 358	28 612	36 730	44 615	37 434
TIT	*Xiphopenaeus riveti*	45	2 683	2 830	1 970	1 752	2 000	2 341	2 300
TRV	*Trachypenaeus curvirostris*	45	167 723	180 255	179 544	403 027	312 968	247 940	234 102
ASH	*Artemesia longinaris*	45	263	166	146	37	39	283	298
BOS	*Xiphopenaeus,Trachypenaeus spp*	45	12 156	7 354	9 881	6 457	4 720	4 955	2 777
SSH	*Plesiopenaeus edwardsianus*	45	352	479	337	605	54	39	34
ARA	*Aristeus antennatus*	45	1 428	1 359	1 672	1 526	2 173	2 391	2 108
ARV	*Aristeus varidens*	45	1 578	1 323	2 808	1 827	3 024	3 405	2 046
ARI	*Aristeidae*	45	2 258	2 406	1 231	2 128	4 463	1 833	1 768
PYX	*Pandalus hypsinotus*	45	...	467	388	288	275	359	231
PRA	*Pandalus borealis*	45	285 421	280 316	317 693	338 394	369 688	345 811	375 878
AES	*Pandalus montagui*	45	...	...	...	...	697	609	206
DUJ	*Pandalus goniurus*	45	...	-	1 199	330	1 200	247	32
DUK	*Pandalus kessleri*	45	...	123	55	97	75	94	81
PAN	*Pandalus spp*	45	28 015	28 649	30 463	30 649	35 640	31 308	33 632
CHS	*Heterocarpus reedi*	45	10 535	10 239	7 301	7 951	5 448	4 863	4 112
HUV	*Heterocarpus vicarius*	45	-	-	40	37	20	131	0
NDJ	*Pandalopsis japonica*	45	...	12	86	35	11	36	22
PSH	*Pandalus spp, Pandalopsis spp*	45	16 737	20 419	6 890	14 651	16 530	19 049	26 907
AKS	*Acetes japonicus*	45	460 871	495 680	587 376	598 602	639 219	577 497	585 647
SHS	*Sergestidae*	45	61 611	48 290	43 427	43 402	34 327	43 690	37 422
NLC	*Nematopalaemon schmitti*	45	-	-	-	-	1 464	1 382	1 400
PIQ	*Palaemon longirostris*	45	...	...	...	10	19	26	18
CPR	*Palaemon serratus*	45	583	611	509	534	520	532	514
PAL	*Palaemonidae*	45	400	373	510	589	513	281	215
CSH	*Crangon crangon*	45	32 444	37 922	30 683	37 133	33 461	32 150	33 012
CVL	*Sclerocrangon spp*	45	...	38	59	82	20	45	41
CRN	*Crangonidae*	45	-	31	41	-	42	45	38
RSH	*Sicyonia brevirostris*	45	10 549	1 789	4 409	1 826	3 254	2 909	890

B-00 (b)

Fish, crustaceans, molluscs, etc — **Capture production by species items**
Poissons, crustacés, mollusques, etc — **Captures par catégories d'espèces**
Peces, crustáceos, moluscos, etc — **Capturas por partidas de especies**

3-alpha code Code alpha-3 Código alfa-3	Scientific name Nom scientifique Nombre científico	Species group Groupe d'espèces Grupo de especies	1996 t	1997 t	1998 t	1999 t	2000 t	2001 t	2002 t
YII	*Sicyonia ingentis*	45	...	...	185	630	756	165	215
LAA	*Pleoticus muelleri*	45	9 874	6 479	23 203	15 928	37 150	78 866	51 412
RRS	*Pleoticus robustus*	45	198	209	195	286	391	305	441
SOK	*Solenocera agassizii*	45	...	...	...	...	...	686	680
KNS	*Haliporoides triarthrus*	45	1 771	1 510	1 882	1 611	1 766	1 738	1 441
HJD	*Haliporoides diomedeae*	45	15	32	29	135	169	309	482
DCP	*Natantia*	45	592 492	591 681	548 708	561 795	573 038	593 881	663 568
KRI	*Euphausia superba*	46	91 156	75 653	90 353	101 957	114 430	104 183	125 987
KRX	*Euphausia spp*	46	0	-	-	-	-	-	-
NKR	*Meganyctiphanes norvegica*	46	-	-	88	-	-	36	-
AMS	*Artemia salina*	47	...	...	513	691	561	538	378
GOO	*Lepas spp*	47	2	-	-	-	-	-	-
MBZ	*Megabalanus psittacus*	47	879	579	683	620	620	685	199
MTS	*Squilla mantis*	47	5 446	4 507	3 680	6 023	6 331	6 606	6 231
SQY	*Squillidae*	47	695	2 133	2 219	2 068	2 141	2 612	3 344
SVX	*Stomatopoda*	47	181	176	459	871	881	479	464
CRU	*Crustacea*	47	1 074 218	1 238 575	1 395 102	1 279 343	1 361 461	1 415 941	1 372 522
CMJ	*Corbicula japonica*	51	27 314	22 209	20 609	21 015	19 295	17 369	17 988
MOF	*Mollusca*	51	533 871	485 564	559 438	531 437	575 991	610 836	615 563
PEE	*Littorina littorea*	52	2 946	3 299	2 753	3 144	2 780	2 781	2 315
PER	*Littorina spp*	52	1 989	3 153	2 311	1 716	1 378	1 506	1 360
MUE	*Murex spp*	52	1 304	1 264	2 584	1 290	1 580	2 132	1 677
RPN	*Rapana spp*	52	4 349	5 494	4 671	4 419	4 713	3 753	851
SNE	*Concholepas concholepas*	52	5 269	7 520	3 394	4 583	2 524	1 372	2 308
ABG	*Haliotis gigantea*	52	1 941	2 218	2 269	2 109	2 146	1 982	2 223
ABP	*Haliotis midae*	52	735	330	524	481	490	527	516
ABR	*Haliotis rubra*	52	5 425	5 240	5 247	5 620	5 532	5 609	6 126
HLT	*Haliotis tuberculata*	52	62	75	36	37	64	62	34
ABX	*Haliotis spp*	52	2 909	2 649	2 470	2 144	2 241	1 993	2 004
TOS	*Turbo cornutus*	52	17 480	19 010	21 748	18 397	17 120	16 515	16 697
CON	*Strombus spp*	52	40 921	34 947	32 396	30 727	25 565	27 474	23 304
WHE	*Buccinum undatum*	52	24 723	27 659	13 768	17 794	29 818	30 146	27 643
WHX	*Busycon spp*	52	5 487	4 379	3 643	6 181	4 853	6 078	5 288
CXY	*Cymbium spp*	52	6 648	5 166	4 678	5 737	4 989	5 422	4 347
ZDF	*Zidona dufresnei*	52	558	1 322	1 010	683	1 621	1 519	1 314
GAS	*Gastropoda*	52	18 994	15 516	12 161	14 988	11 734	14 527	12 733
OCH	*Ostrea chilensis*	53	-	5	1	6	9	202	7
OYF	*Ostrea edulis*	53	3 456	3 598	2 565	1 776	880	959	1 560
DRY	*Ostrea lutaria*	53	1 931	2 172	1 000	995	766	832	816
OYH	*Ostrea conchaphila*	53	...	...	...	6	7	8	11
OYG	*Crassostrea gigas*	53	25 160	25 692	12 101	12 272	17 983	11 240	17 781
OYM	*Crassostrea rhizophorae*	53	4 589	3 722	4 896	3 906	3 228	5 244	2 235
OYA	*Crassostrea virginica*	53	148 540	144 707	135 222	132 207	221 553	175 042	159 168
CSI	*Crassostrea iredalei*	53	291	152	89	95	79	96	97
OYC	*Crassostrea spp*	53	3 957	4 539	4 211	6 244	5 142	4 538	5 024
MUK	*Mytilus coruscus*	54	2 191	3 211	1 469	1 414	1 133	1 085	946
MYC	*Mytilus chilensis*	54	5 715	4 724	4 900	4 344	5 237	6 758	1 416
MUS	*Mytilus edulis*	54	117 550	136 147	137 268	130 020	147 972	163 911	156 341
MSR	*Mytilus platensis*	54	370	354	375	474	412	523	440
MSM	*Mytilus galloprovincialis*	54	39 046	53 310	38 899	44 453	46 143	46 092	51 820
MYA	*Mytilus planulatus*	54	75	1	1	1	1	1	0
CHC	*Choromytilus chorus*	54	323	266	127	155	217	166	91
MOD	*Modiolus spp*	54	20	30	20	7	96	22	32
MSL	*Perna perna*	54	223	295	3 802	451	316	1 081	1 272
MSV	*Perna viridis*	54	20 220	18 017	17 454	6 554	41 173	22 782	23 282
MSC	*Aulacomya ater*	54	13 428	16 078	22 831	19 738	18 933	22 584	19 433
MSX	*Mytilidae*	54	4 278	7 506	10 601	11 572	14 643	10 671	9 028
QSC	*Aequipecten opercularis*	55	7 183	11 443	14 486	15 746	14 891	20 464	16 731
SCE	*Pecten maximus*	55	24 098	26 660	25 172	17 277	18 204	30 590	32 843
SJA	*Pecten jacobaeus*	55	52	95	50	68	570	150	470
SCZ	*Pecten novaezelandiae*	55	5 080	18 848	4 592	6 152	2 912	6 792	4 408
ZYE	*Zygochlamis delicatula*	55	124	201	91	128	0	222	118
ZYP	*Zygochlamis patagonica*	55	36 952	39 817	28 441	42 700	37 404	42 598	51 181
SCA	*Placopecten magellanicus*	55	109 382	102 028	99 432	131 962	196 993	254 196	280 572
SCB	*Argopecten irradians*	55	230	452	690	216	154	30	135
SCQ	*Argopecten purpuratus*	55	2 095	4 013	23 546	30 141	11 830	6 544	2 086
SCH	*Argopecten ventricosus*	55	17 290	2 320	2 726	1 864	6 287	3 100	9 875
ISC	*Chlamys islandica*	55	22 752	24 673	18 946	12 018	13 485	9 469	8 663
SCG	*Patinopecten caurinus*	55	2 372	2	3 228	2 642	2 012	1 052	1 264
JSC	*Patinopecten yessoensis*	55	276 406	266 957	294 211	305 510	310 104	293 268	310 303
SCX	*Pectinidae*	55	31 150	35 382	39 156	46 278	45 854	34 262	22 867
ARK	*Arca spp*	56	35 401	42 891	32 748	42 103	48 293	45 061	45 656
MCL	*Scapharca subcrenata*	56	16 328	14 133	10 120	10 413	7 308	4 899	9 828
BLC	*Anadara granosa*	56	995	493	12 555	8 260	5 156	1 971	2 409
BLS	*Anadara spp*	56	47 233	43 872	34 395	36 904	37 448	67 195	70 923
CLQ	*Arctica islandica*	56	180 114	168 150	157 282	147 933	124 131	150 260	162 212
SVE	*Chamelea gallina*	56	50 501	38 794	36 798	44 605	49 822	48 213	40 529
CTS	*Venerupis pullastra*	56	2 826	4 058	5 166	4 969	2 388	2 420	2 821
KNU	*Chione stutchburyi*	56	815	541	1 325	1 396	1 789	1 748	1 687
HCJ	*Meretrix lusoria*	56	4 394	4 133	6 147	4 110	2 973	2 289	3 004
HCX	*Meretrix spp*	56	13 481	14 027	17 146	14 767	14 177	13 375	14 600
CTG	*Ruditapes decussatus*	56	1 469	1 288	1 943	1 899	1 487	1 317	1 720
CLJ	*Ruditapes philippinarum*	56	56 095	56 514	51 392	57 051	56 540	51 026	49 252

B-00 (b)

Fish, crustaceans, molluscs, etc — **Capture production by species items**
Poissons, crustacés, mollusques, etc — **Captures par catégories d'espèces**
Peces, crustáceos, moluscos, etc — **Capturas por partidas de especies**

3-alpha code Code alpha-3 Código alfa-3	Scientific name Nom scientifique Nombre científico	Species group Groupe d'espèces Grupo de especies	1996 t	1997 t	1998 t	1999 t	2000 t	2001 t	2002 t
TWG	*Tawera gayi*	56	...	...	...	...	1	291	271
TVM	*Tivela mactroides*	56	126	196	664	1 684	273	280	835
BCL	*Saxidomus giganteus*	56	1 431	1 293	1 726	1 019	1 113	1 159	1 428
NCL	*Paphia spp*	56	53 131	35 968	49 866	70 160	49 227	54 770	55 228
PTS	*Protothaca staminea*	56	-	-	-	-	-	102	146
TCL	*Protothaca thaca*	56	20 016	12 475	24 254	16 429	16 303	26 483	5 360
CLH	*Mercenaria mercenaria*	56	22 964	6 287	1 234	2 536	15 150	11 793	19 005
CLV	*Veneridae*	56	6 389	5 293	4 968	5 915	6 843	6 115	4 205
HCL	*Pseudocardium sybillae*	56	14 791	14 720	11 104	16 428	13 033	12 323	12 325
TQZ	*Tresus spp*	56	-	-	-	-	1	2	9
CLB	*Spisula solidissima*	56	155 584	141 421	131 700	142 370	165 765	166 310	176 322
CLT	*Spisula polynyma*	56	25 612	27 365	26 008	26 722	22 985	20 273	19 979
ULO	*Spisula solida*	56	...	...	...	765	1 335	1 440	3 030
MUN	*Mulinia spp*	56	999	2 757	2 549	1 536	1 491	1 699	7 034
DON	*Donax spp*	56	0	0	0	490	495	628	463
RAZ	*Solen spp*	56	1 873	581	632	701	468	608	404
CLR	*Ensis directus*	56	-	14	49	64	99	36	90
RAP	*Siliqua patula*	56	-	-	-	-	14	61	146
CLS	*Mya arenaria*	56	5 824	6 777	8 131	7 793	8 248	10 069	9 251
GEC	*Panopea abrupta*	56	1 768	3 719	4 027	4 139	3 821	3 894	4 767
COC	*Cerastoderma edule*	56	37 089	38 132	86 750	70 400	47 304	23 764	18 902
KCL	*Clinocardium nuttallii*	56	-	22	-	18	49	67	80
COZ	*Cardiidae*	56	2 583	1 587	1 078	112	411	2 002	810
AFQ	*Paphies australis*	56	702	1 295	1 496	1 489	1 713	1 875	1 560
CLM	*Mesodesma donacium*	56	7 204	7 831	7 042	1 728	1 259	1 396	1 388
TUW	*Semele solida*	56	4 418	2 199	1 900	2 071	4 212	3 054	2 162
CLX	*Bivalvia*	56	146 944	115 222	105 923	92 679	85 214	83 933	75 810
CTC	*Sepia officinalis*	57	10 909	13 319	11 966	13 865	12 816	13 148	15 199
CTL	*Sepiidae, Sepiolidae*	57	370 497	456 107	429 547	439 357	486 810	527 915	493 435
SQP	*Loligo gahi*	57	68 504	21 711	51 751	42 505	67 016	57 730	24 976
SQO	*Loligo opalescens*	57	...	70 915	2 709	90 662	117 711	85 829	72 879
SQL	*Loligo pealei*	57	12 490	16 161	18 879	18 749	16 942	14 211	16 684
CHO	*Loligo reynaudi*	57	7 549	3 696	6 670	7 169	6 000	3 373	7 406
SQC	*Loligo spp*	57	216 651	227 602	218 114	199 843	219 179	225 958	237 964
OFJ	*Ommastrephes bartrami*	57	...	49 870	54 951	36 076	47 368	23 870	22 483
SQI	*Illex illecebrosus*	57	28 971	34 837	26 588	9 883	11 222	5 698	5 525
SQM	*Illex coindetii*	57	395	411	216	338	402	250	527
SQA	*Illex argentinus*	57	656 481	980 300	693 542	1 144 998	930 839	743 440	511 087
GIS	*Dosidicus gigas*	57	142 186	162 504	27 466	134 773	182 399	223 784	406 356
SQE	*Todarodes sagittatus*	57	5 875	5 571	6 639	4 890	4 909	4 261	5 197
SQJ	*Todarodes pacificus*	57	715 908	603 367	378 605	497 887	570 427	528 523	504 438
TSQ	*Nototodarus sloani*	57	53 699	64 602	55 570	31 358	25 603	45 119	62 234
SQS	*Martialia hyadesi*	57	3 845	8 429	2	27	33	4	...
SQU	*Loliginidae, Ommastrephidae*	57	363 638	237 928	341 747	362 903	364 161	293 354	311 450
PRD	*Pareledone spp*	57	-	-	-	-	0	-	-
OCC	*Octopus vulgaris*	57	72 936	60 313	72 465	56 055	50 478	53 103	41 619
OCM	*Eledone spp*	57	1 939	2 258	1 758	1 433	1 990	2 357	2 187
OCT	*Octopodidae*	57	230 773	210 844	245 385	295 452	259 330	263 709	206 401
CEP	*Cephalopoda*	57	186 473	225 922	213 035	209 978	284 769	192 333	225 225
MOL	*Mollusca*	58	1 037 481	1 597 051	1 594 542	1 565 999	1 497 351	1 500 552	1 491 849
FRG	*Rana spp*	71	2 955	3 622	3 009	1 807	2 328	2 486	2 258
TTG	*Malaclemys spp*	72	-	-	0	-	0	-	-
TUG	*Chelonia mydas*	72	63	28	22	15	17	15	16
TTH	*Eretmochelys imbricata*	72	23	19	12	12	18	11	17
TTL	*Caretta caretta*	72	10	7	5	5	9	7	8
TUL	*Testudinata*	72	54	65	53	16	56	94	67
TTX	*Testudinata*	72	953	858	1 090	1 195	910	692	1 261
SSE	*Pyura chilensis*	74	4 549	3 174	2 530	2 704	2 290	1 298	1 217
SSG	*Microcosmus sulcatus*	74	28	22	22	30	30	76	78
SSX	*Ascidiacea*	74	16 747	2 780	891	1 171	1 538	1 053	1 025
SPX	*Salpidae*	74	7	-	-	-	-	-	-
HSC	*Limulus polyphemus*	75	1 598	2 607	3 252	2 397	1 696	1 299	1 387
ECH	*Echinodermata*	76	7 054	6 521	4 040	5 702	6 936	7 035	7 129
STF	*Asteroidea*	76	4	0	13	3	6	11	40
URC	*Strongylocentrotus spp*	76	45 583	45 724	39 344	39 656	36 745	33 097	33 038
URM	*Paracentrotus lividus*	76	63	48	59	84	198	101	164
URS	*Echinus esculentus*	76	425	25	1	13	1	5	0
UCH	*Loxechinus albus*	76	51 437	45 560	44 843	55 654	54 096	46 794	60 166
CUJ	*Stichopus japonicus*	76	9 205	9 377	8 391	7 866	8 376	8 129	8 092
CUX	*Holothurioidea*	76	17 590	15 295	13 613	12 596	16 020	11 984	16 366
JEL	*Rhopilema spp*	77	326 850	516 749	513 401	529 401	515 203	432 886	389 697
WOR	*Polychaeta*	77	100	347	408	422	257	543	559
INV	*Invertebrata*	77	17 215	14 080	13 228	13 136	16 427	14 189	12 886
World total Total mondial Total mundial			**93 846 643**	**94 298 586**	**87 672 034**	**93 774 052**	**95 502 004**	**92 862 087**	**93 190 654**

B-11 Carps, barbels and other cyprinids / Carpes, barbeaux et autres cyprinidés / Carpas, barbos y otros ciprínidos

Capture production by species, fishing areas and countries or areas
Captures par espèces, zones de pêche et pays ou zones
Capturas por especies, áreas de pesca y países o áreas

Species, Fishing area Espèce, Zone de pêche Especie, Area de pesca	1993 t	1994 t	1995 t	1996 t	1997 t	1998 t	1999 t	2000 t	2001 t	2002 t
Buffalofishes nei	**Poissons-taureaux nca**		**...C**		***Ictiobus spp***				**1,40(01)011,XX**	**BUF**
02 USA	982	896	480	793	991	959	697	1 280	1 569	1 558
02 Fishing area total	*982*	*896*	*480*	*793*	*991*	*959*	*697*	*1 280*	*1 569*	*1 558*
Species total	*982*	*896*	*480*	*793*	*991*	*959*	*697*	*1 280*	*1 569*	*1 558*
Suckers nei	**Cyprins sucets nca**		**Chupadores nep**		***Catostomidae***				**1,40(01)XXX,XX**	**CTM**
02 USA	872	292	662	542	550	464	465	231	187	173
02 Fishing area total	*872*	*292*	*662*	*542*	*550*	*464*	*465*	*231*	*187*	*173*
Species total	*872*	*292*	*662*	*542*	*550*	*464*	*465*	*231*	*187*	*173*
Freshwater bream	**Brème d'eau douce**		**Brema común**		***Abramis brama***				**1,40(02)001,02**	**FBM**
04 Azerbaijan	412	309	219	402	346	314	52	55	127	112
Georgia	18	11	27	3	1	2	1	3	1	2
Kazakhstan	20 420	19 302	20 520	18 770 F	11 000 F	9 800 F	11 000 F	12 000 F	12 630	14 110
Kyrgyzstan	20	19	11	9 F	7 F	5 F	3 F	3 F	4	3
Tajikistan	123	82	58	...	...	...	...	...	37	25
Turkey	...	...	...	...	...	...	259	200	151	198
Turkmenistan	170	158	137	75	100	142	147	153	126	107
Uzbekistan	598	579	474	220 F	289	387	353	335	540	72
04 Fishing area total	*21 761*	*20 460*	*21 446*	*19 479 F*	*11 743 F*	*10 650 F*	*11 815 F*	*12 749 F*	*13 616*	*14 629*
05 Belarus	464	201	145	244	182	130	98	27	198	403
Belgium	60	60	60	60	60	60	60	60	60	60
Bulgaria	...	70 F	74	91	90	82	71	25	4	6
Czech Rep	202	265	286	247	232	253	297	261	247	243
Denmark	130	90	85	64	93	141	79	47	28	26
Estonia	318	152	155	167	222	240	181	194	318	385
Finland	3 449	3 266	3 239	2 334	2 334	1 590	1 590	1 660	1 660	1 660
Latvia	233	145	156	151	172	135	158	155	152	141
Lithuania	276	324	420	397	448	454	466	467	470	86
Netherlands	229	30	...	75	65	399	355	350 F	300 F	282
Poland	2 400	2 400	1 043	3 053	1 498	1 013	1 291	1 883	1 859	1 797
Romania	468	1 090	1 604	827	328	951	1 052	936	800	911
Russian Fed	27 441	28 154	29 142	29 050	29 507	31 715	25 360	23 794	27 264	27 670
Slovenia	9	18	16	10	11	11	10	11	10	9
Sweden	7	3	13	24	16	20	12	-	-	-
Switzerland	70	59	22	24	20	16	16	13	9	10
Ukraine	2 033	1 922	908	986	810	721	832	873	942	502
UK	1	1	2	1	9	8	8	2	0	2
05 Fishing area total	*37 790*	*38 250 F*	*37 370*	*37 805*	*36 097*	*37 939*	*31 936*	*30 758 F*	*34 321 F*	*34 193*
27 Denmark	4	3	2	2	1	3	1	1	0	0
Estonia	6	6	9	8	8	7	13	10	10	16
Finland	1 216	1 010	986	809	941	1 264	1 255	886	987	949
Latvia	33	70	81	91	69	56	77	63	94	111
Lithuania	9	-	-	-	-	-	-	-	-	426
Poland	453	469	462	569	869	1 044	1 030	1 062	717	829
Russian Fed	-	-	-	-	-	-	-	1 351	1 382	1 427
Sweden	2	3	1	1	2	2	8	6	4	5
27 Fishing area total	*1 723*	*1 561*	*1 541*	*1 480*	*1 890*	*2 376*	*2 384*	*3 379*	*3 194*	*3 763*
37 Russian Fed	1 360	387	618	324	636	474	327	328	108	47
Ukraine	9	9	4	22	22	21	8	8	-	-
37 Fishing area total	*1 369*	*396*	*622*	*346*	*658*	*495*	*335*	*336*	*108*	*47*
Species total	*62 643*	*60 667 F*	*60 979*	*59 110 F*	*50 388 F*	*51 460 F*	*46 470 F*	*47 222 F*	*51 239 F*	*52 632*
Freshwater breams nei	**Brèmes d'eau douce nca**		**Bremas nep**		***Abramis spp***				**1,40(02)001,XX**	**FBR**
04 Iran	17	29	5	3	7	20	9	20	10	38
04 Fishing area total	*17*	*29*	*5*	*3*	*7*	*20*	*9*	*20*	*10*	*38*
05 Lithuania	-	5	4	7	10	12	11	13	24	23
Poland	50	50	267	373	...	...	235	490	400	398
Romania	-	-	-	-	-	3	1	-	-	7
Russian Fed	2 246	1 236	1 720	1 664	1 046	2 058	1 951	1 593	1 751	1 886
Slovakia	79	102	141	111	102	99	98	94	95	103
Slovenia	-	5	5	5	5	5	4	0	1	-
Ukraine	55	484	53	101	62	86	122	80	137	107
05 Fishing area total	*2 430*	*1 882*	*2 190*	*2 261*	*1 225*	*2 263*	*2 422*	*2 270*	*2 408*	*2 524*
27 Germany	53	196	181	162	275	144	249	124	174	107
Lithuania	-	-	-	-	-	-	-	-	2	9
27 Fishing area total	*53*	*196*	*181*	*162*	*275*	*144*	*249*	*124*	*176*	*116*
Species total	*2 500*	*2 107*	*2 376*	*2 426*	*1 507*	*2 427*	*2 680*	*2 414*	*2 594*	*2 678*
Common carp	**Carpe commune**		**Carpa**		***Cyprinus carpio***				**1,40(02)002,01**	**FCP**
01 Ethiopia	...	...	74	94	27	62	71	75	74	79
Kenya	279	305	360	334	216	48	52	47	49	50
Lesotho	15 F	15 F	16 F	16 F	18 F	18 F	18 F	18 F	8	8 F
01 Fishing area total	*294 F*	*320 F*	*450 F*	*444 F*	*261 F*	*128 F*	*141 F*	*140 F*	*131*	*137 F*

B-11 Carps, barbels and other cyprinids / Carpes, barbeaux et autres cyprinidés / Carpas, barbos y otros ciprínidos

Capture production by species, fishing areas and countries or areas / Captures par espèces, zones de pêche et pays ou zones / Capturas por especies, áreas de pesca y países o áreas

Species, Fishing area Espèce, Zone de pêche Especie, Area de pesca	1993 t	1994 t	1995 t	1996 t	1997 t	1998 t	1999 t	2000 t	2001 t	2002 t
02 Canada	222	649	619	687	543	354	741	516	506	878
Dominican Rp	1 579	2 881	597	62	109	27	180	...	397	705
Mexico	18 171	16 187	22 677	21 237	16 787	14 345	10 945	15 300	14 700	13 706
USA	1 106	824	852	1 034	1 076	1 072	1 103	724	704	656
02 Fishing area total	*21 078*	*20 541*	*24 745*	*23 020*	*18 515*	*15 798*	*12 969*	*16 540*	*16 307*	*15 945*
03 Chile	-	-	-	-	-	4	-	-	-	-
Venezuela	0	4	544	218	1	3	5	1 390	0	0
03 Fishing area total	*0*	*4*	*544*	*218*	*1*	*7*	*5*	*1 390*	*0*	*0*
04 Armenia	1	0	1	91	32	14	12	9	7	11
Azerbaijan	1 446	803	370	84	49	87	92	93	51	38
China,Taiwan	104	54	44	26	19	22	45	49	50	54
Georgia	499	0	49	0	0	2	11	12	5	5
Indonesia	5 287	5 490	5 613	7 081	6 644	7 082	7 127	7 035	8 228	8 730
Iraq	7 703	5 394	5 691	5 538	4 163	2 336	...	...	...	...
Israel	108	147	89	35	38	150	165	189	97	98
Japan	5 338	4 968	4 896	4 771	4 607	4 477	4 259	4 079	3 558	3 359
Kazakhstan	431	537	477	440 F	320 F	230 F	500 F	650 F	710	1 371
Korea Rep	-	1 469	1 684	1 979	842	874	438	...	...	...
Kyrgyzstan	-	-	36	31 F	23 F	15 F	9 F	10 F	11	9
Nepal	25	...	...	...	...	...	...	...	...	...
Tajikistan	...	...	...	...	...	...	48	59	24	51
Thailand	8 692	8 200	10 144	7 420	7 418	11 508	13 689	6 961	6 800	7 100
Turkey	16 035	15 900	17 081	15 631	16 000	20 000	17 797	14 137	12 265	12 965
Turkmenistan	342	774	484	115	154	140	150	144	93	75
Uzbekistan	771	605	864	193 F	843	804	826	617	906	148
04 Fishing area total	*46 782*	*44 341*	*47 523*	*43 435 F*	*41 152 F*	*47 741 F*	*45 168 F*	*34 044 F*	*32 805*	*34 014*
05 Albania	20 F	30 F	34	45	38	230	216	230	300	260
Belarus	59	18	39	39	12	8	5	17	15	3 845
Belgium	30	30	30	30	30	30	30	30	30	30
Bulgaria	-	-	19	16	281	251	302	143	880	521
Croatia	72	97	96	143	126	1 F	1	3	1	2
Czech Rep	2 284	2 832	2 919	2 522	2 312	2 899	3 006	3 558	3 560	3 909
Denmark	1	1	2	1	1	1	0	0	0	0
Germany	66	386	386	386	386	386	386	386	386	386
Greece	...	...	...	208	256	279	247	220	198	263
Hungary	3 103	3 265	2 856	2 717	2 255	3 373	3 279	3 212	2 470	2 787
Latvia	-	5	5	3	3	6	5	3	5	5
Lithuania	52	41	27	31	14	13	12	16	16	18
Macedonia	9	13	32	10	9	...	...	25	6	22
Moldova Rep	515	469	472	408	349	280	178	192	212	274
Netherlands	0	-	-	-	-	-	-	-	-	-
Poland	-	-	-	77	82	78	37	45	50	54
Portugal	0	0	0	0	0	0	0	0	0	-
Romania	562	560	577	441	173	147	310	458	566	486
Russian Fed	1 491	2 637	752	3 933	2 713	3 212	3 507	4 007	2 698	2 614
Slovakia	612	779	1 063	778	746	778	822	854	967	1 166
Slovenia	96	97	89	86	90	94	78	71	75	83
Switzerland	-	-	1	1	2	1	1	1	1	2
Ukraine	149	232	153	598	27	33	41	44	39	40
05 Fishing area total	*9 121 F*	*11 492 F*	*9 552*	*12 473*	*9 905*	*12 100 F*	*12 463*	*13 515*	*12 475*	*16 767*
06 New Zealand	-	-	0	0	28	2	3	4	6	7
06 Fishing area total	*-*	*-*	*0*	*0*	*28*	*2*	*3*	*4*	*6*	*7*
27 Denmark	0	0	0	-	0	0	0	-	-	10
Germany	-	-	-	-	-	-	1	-	1	1
27 Fishing area total	*0*	*0*	*0*	*-*	*0*	*0*	*1*	*-*	*1*	*11*
37 Russian Fed	29	5	2	2	3	3	1	-	1	-
Ukraine	-	-	-	-	-	-	-	-	1	3
37 Fishing area total	*29*	*5*	*2*	*2*	*3*	*3*	*1*	*-*	*2*	*3*
Species total	*77 304 F*	*76 703 F*	*82 816 F*	*79 592 F*	*69 865 F*	*75 779 F*	*70 751 F*	*65 633 F*	*61 727*	*66 884 F*
Tench	**Tanche**		**Tenca**			*Tinca tinca*		**1,40(02)007,01**		**FTE**
04 Azerbaijan	5	0	0	2	0	0	0	0	-	-
Kazakhstan	-	-	-	-	-	-	-	-	-	183
Turkey	...	...	...	...	...	...	...	690	778	800
04 Fishing area total	*5*	*0*	*0*	*2*	*0*	*0*	*0*	*690*	*778*	*983*
05 Belarus	23	11	13	8	4	3	90	41	7	65
Belgium	15	15	15	15	15	15	15	15	15	15
Bulgaria	...	...	...	...	...	...	...	...	2	3
Croatia	1	1	2	3	0	-	-	-	-	-
Czech Rep	36	29	30	23	21	29	30	27	24	24
Denmark	0	0	0	0	0	0	0	1	0	0
Germany	33	36	36	36	36	36	36	36	36	36
Hungary	-	-	-	-	5	-	-	-	-	-
Latvia	5	16	15	15	24	22	41	29	35	40
Lithuania	6	9	10	9	12	9	11	13	15	16
Poland	100	100	70	97	101	91	102	160	113	141
Romania	-	-	-	-	-	2	7	4	31	19
Russian Fed	879	1 501	1 300	936	1 647	990	1 071	1 307	1 409	1 456
Slovakia	5	5	8	8	9	8	8	8	5	5
Slovenia	2	1	2	1	2	2	3	2	1	1

B-11 Carps, barbels and other cyprinids / Carpes, barbeaux et autres cyprinidés / Carpas, barbos y otros ciprínidos

Capture production by species, fishing areas and countries or areas
Captures par espèces, zones de pêche et pays ou zones
Capturas por especies, áreas de pesca y países o áreas

Species, Fishing area Espèce, Zone de pêche Especie, Area de pesca	1993 t	1994 t	1995 t	1996 t	1997 t	1998 t	1999 t	2000 t	2001 t	2002 t
Sweden	0	0	0	0	0	1	1	-	-	-
Switzerland	3	3	3	4	3	3	3	3	3	4
Ukraine	...	11	...	...	...	...	...	...	...	...
05 Fishing area total	*1 108*	*1 738*	*1 504*	*1 155*	*1 879*	*1 211*	*1 418*	*1 646*	*1 696*	*1 825*
27 Denmark	0	0	0	0	-	0	0	-	0	0
Germany	-	-	-	-	-	-	1	-	1	1
27 Fishing area total	*0*	*0*	*0*	*0*	-	*0*	*1*	-	*1*	*1*
Species total	*1 113*	*1 738*	*1 504*	*1 157*	*1 879*	*1 211*	*1 419*	*2 336*	*2 475*	*2 809*
Bleak	**Ablette**		**Alburno**		***Alburnus alburnus***			**1,40(02)012,01**		**ALR**
05 Albania	...	...	151	162	68	149	160	190	478	234
Bulgaria	-	-	2	6	62	67	92	24	3	19
Lithuania	-	-	-	-	-	-	-	3	4	6
Romania	-	-	-	-	-	24	16	30	60	27
Slovenia	-	-	-	-	-	-	-	-	-	3
Ukraine	...	12	-	-	-	-	-	-	-	2
05 Fishing area total	...	*12*	*153*	*168*	*130*	*240*	*268*	*247*	*545*	*291*
Species total	...	*12*	*153*	*168*	*130*	*240*	*268*	*247*	*545*	*291*
Barbel	**Barbeau fluviatile**		**Barbo común**		***Barbus barbus***			**1,40(02)013,01**		**PTB**
05 Bulgaria	...	...	92	120	142	113	93	43	3	4
Hungary	39	66	32	37	64	46	50	30	52	41
Romania	-	-	-	-	-	32	11	17	34	12
Slovakia	16	20	24	19	21	15	10	17	19	15
Slovenia	16	15	13	10	12	9	12	11	11	11
05 Fishing area total	*71*	*101*	*161*	*186*	*239*	*215*	*176*	*118*	*119*	*83*
Species total	*71*	*101*	*161*	*186*	*239*	*215*	*176*	*118*	*119*	*83*
Common nase	**Nase commun**		**Condrostoma común**		***Chondrostoma nasus***			**1,40(02)014,01**		**HON**
05 Bulgaria	...	...	...	...	3	3	4	8	7	17
Slovakia	33	32	31	24	23	25	24	24	20	15
05 Fishing area total	*33*	*32*	*31*	*24*	*26*	*28*	*28*	*32*	*27*	*32*
Species total	*33*	*32*	*31*	*24*	*26*	*28*	*28*	*32*	*27*	*32*
Crucian carp	**Carassin(=Cyprin)**		**Carpín**		***Carassius carassius***			**1,40(02)016,01**		**FCC**
01 Ethiopia	-	-	47	61	191	88	101	110	108	103
01 Fishing area total	-	-	*47*	*61*	*191*	*88*	*101*	*110*	*108*	*103*
04 Armenia	25	10	2	28	23	42	26	38	32	54
Azerbaijan	...	...	...	...	...	...	6	4	17	9
China,Taiwan	35	23	14	5	5	5	28	35	37	26
Japan	4 921	4 402	4 286	4 205	4 008	3 881	3 493	3 423	2 948	2 706
Kazakhstan	4 500	4 000	4 317	3 950 F	2 840 F	2 060 F	1 900 F	1 800 F	1 707	1 611
Tajikistan	...	...	...	...	...	...	...	...	...	17
Turkmenistan	41	75	63	85	1	225	233	228	154	141
Uzbekistan	...	...	...	...	...	...	...	...	331	38
04 Fishing area total	*9 522*	*8 510*	*8 682*	*8 273 F*	*6 877 F*	*6 213 F*	*5 686 F*	*5 528 F*	*5 226*	*4 602*
05 Albania	10 F	9 F	...	...	...	77	62	65	326	260
Belarus	384	60	64	69	35	106	138	154	188	239
Bulgaria	...	...	...	...	328	392	360	179	138	247
Latvia	-	5	7	7	11	10	14	15	24	36
Moldova Rep	61	171	164	128	166	159	104	132	127	183
Poland	30	30	60	46	49	47	46	92	84	111
Portugal	3	4	2	0	0	0	0	0	0	0
Romania	-	-	-	-	33	2	23	128	97	87
Slovakia	2	-	-	-	-	2	1	0	...	...
Slovenia	-	-	-	-	-	0	1	0	0	2
05 Fishing area total	*490 F*	*279 F*	*297*	*250*	*622*	*795*	*749*	*765*	*984*	*1 165*
27 Denmark	0	0	-	-	0	-	0	0	0	0
Lithuania	-	-	-	-	-	-	-	-	-	15
Poland	-	-	-	-	1	0	3	3	3	4
27 Fishing area total	*0*	*0*	-	-	*1*	*0*	*3*	*3*	*3*	*19*
Species total	*10 012 F*	*8 789 F*	*9 026*	*8 584 F*	*7 691 F*	*7 096 F*	*6 539 F*	*6 406 F*	*6 321*	*5 889*
Goldfish	**Poisson rouge(=Cyprin doré)**		**Pez rojo**		***Carassius auratus***			**1,40(02)016,02**		**CGO**
02 USA	...	...	5	5	10	10	7	9	10	12
02 Fishing area total	...	...	*5*	*5*	*10*	*10*	*7*	*9*	*10*	*12*
04 Kyrgyzstan	1	3	2	2 F	1 F	1 F	1 F	1 F	2	1
Uzbekistan	...	...	361	170 F	...	...	...	...	...	...
04 Fishing area total	*1*	*3*	*363*	*172 F*	*1 F*	*1 F*	*1 F*	*1 F*	*2*	*1*
05 Czech Rep	30	49	39	49	40	40	43	35	37	33
Greece	...	...	...	530	296	235	548	448	415	350
Lithuania	26	41	41	30	30	38	32	45	38	33
Romania	2 135	2 130	2 320	1 954	817	1 113	1 199	1 212	1 149	1 097

B-11 Carps, barbels and other cyprinids — Capture production by species, fishing areas and countries or areas
Carpes, barbeaux et autres cyprinidés — Captures par espèces, zones de pêche et pays ou zones
Carpas, barbos y otros ciprínidos — Capturas por especies, áreas de pesca y países o áreas

Species, Fishing area Espèce, Zone de pêche Especie, Area de pesca	1993 t	1994 t	1995 t	1996 t	1997 t	1998 t	1999 t	2000 t	2001 t	2002 t
Russian Fed	219	92	-	-	-	-	-	-	-	-
Slovakia	48	60	90	76	53	54	62	0	73	63
Slovenia	-	-	-	-	-	0	1	1	1	-
Ukraine	737	783	704	638	461	452	638	628	747	773
05 Fishing area total	*3 195*	*3 155*	*3 194*	*3 277*	*1 697*	*1 932*	*2 523*	*2 369*	*2 460*	*2 349*
Species total	*3 196*	*3 158*	*3 562*	*3 454 F*	*1 708 F*	*1 943 F*	*2 531 F*	*2 379 F*	*2 472*	*2 362*
Roach	**Gardon**		**Rutilo**			***Rutilus rutilus***			**1,40(02)018,01**	**FRO**
04 Azerbaijan	63	53	19	74	89	62	81	8	64	39
Turkmenistan	111	110	88	71	69	48	39	41	1	2
Uzbekistan	247	88	200	90 F	379	392	613	1 035	1 300	1 457
04 Fishing area total	*421*	*251*	*307*	*235 F*	*537*	*502*	*733*	*1 084*	*1 365*	*1 498*
05 Belgium	160	160	160	160	160	160	150	160	160	160
Bulgaria	...	...	...	...	...	10	40	...	...	14
Denmark	84	24	44	30	27	64	28	35	19	12
Estonia	50	70	92	209	150	128	167	234	231	464
Latvia	59	42	47	36	52	45	53	60	49	47
Lithuania	353	318	314	431	593	645	647	635	643	229
Poland	2 000	2 000	1 050	2 185	792	593	653	947	837	800
Slovenia	2	4	2	2	2	2	2	3	4	4
Switzerland	350	312	169	146	164	168	154	136	137	159
Ukraine	4 467	4 513	717	798	795	1 140	666	449	519	314
UK	-	-	-	-	4	5	5	-	-	-
05 Fishing area total	*7 525*	*7 443*	*2 595*	*3 997*	*2 739*	*2 960*	*2 565*	*2 659*	*2 599*	*2 203*
27 Denmark	36	32	26	13	11	22	11	8	7	7
Estonia	212	188	240	293	342	321	157	244	272	303
Finland	1 983	1 385	1 439	2 172	2 271	1 465	1 465	1 412	1 459	1 501
Germany	-	430	337	176	242	347	462	269	320	347
Latvia	15	16	7	10	12	11	14	10	11	12
Lithuania	...	...	...	...	...	...	...	...	...	517
Poland	675	645	937	910	1 042	1 069	1 309	1 241	1 338	1 544
Sweden	0	0	1	0	1	-	-	-	-	-
27 Fishing area total	*2 921*	*2 696*	*2 987*	*3 574*	*3 921*	*3 235*	*3 418*	*3 184*	*3 407*	*4 231*
37 Ukraine	...	6	8	1	0	1	1	2	7	11
37 Fishing area total	*...*	*6*	*8*	*1*	*0*	*1*	*1*	*2*	*7*	*11*
Species total	*10 867*	*10 396*	*5 897*	*7 807 F*	*7 197*	*6 698*	*6 717*	*6 929*	*7 378*	*7 943*
Kutum	**...B**		**...C**			***Rutilus frisii***			**1,40(02)018,02**	**RFR**
04 Azerbaijan	...	...	...	...	...	...	...	...	...	16
Iran	12 727	9 277	8 435	9 210	2 320	6 624	6 905	10 120	7 199	6 417
04 Fishing area total	*12 727*	*9 277*	*8 435*	*9 210*	*2 320*	*6 624*	*6 905*	*10 120*	*7 199*	*6 433*
Species total	*12 727*	*9 277*	*8 435*	*9 210*	*2 320*	*6 624*	*6 905*	*10 120*	*7 199*	*6 433*
Roaches nei	**Gardons nca**		**Rutilos nep**			***Rutilus spp***			**1,40(02)018,XX**	**FRX**
04 Iran	714	1 366	1 178	878	203	607	626	1 515	1 316	787
Kazakhstan	3 702	2 251	2 650	2 420 F	1 740 F	1 260 F	1 000 F	850 F	579	1 682
04 Fishing area total	*4 416*	*3 617*	*3 828*	*3 298 F*	*1 943 F*	*1 867 F*	*1 626 F*	*2 365 F*	*1 895*	*2 469*
05 Albania	6 F	4 F	-	-	-	-	-	-	-	-
Belarus	417	62	83	95	42	24	19	9	33	59
Finland	5 793	6 850	6 915	6 775	6 775	5 362	5 362	3 999	3 999	3 999
Greece	...	...	...	164	81	252	309	345	324	334
Netherlands	10	20	54	100	123	107	100	90 F	80 F	64
Romania	670	1 295	693	301	125	292	234	174	278	276
Russian Fed	25 557	22 180	19 630	21 888	15 864	15 950	10 355	13 973	14 545	16 300
Slovenia	10	10	8	6	8	9	10	11	7	8
Sweden	2	2	4	0	1	-	-	-	-	-
05 Fishing area total	*32 465 F*	*30 423 F*	*27 387*	*29 329*	*23 019*	*21 996*	*16 389*	*18 601 F*	*19 266 F*	*21 040*
27 Russian Fed	-	-	-	-	-	-	-	501	653	585
27 Fishing area total	*-*	*-*	*-*	*-*	*-*	*-*	*-*	*501*	*653*	*585*
37 Russian Fed	102	325	250	80	107	13	78	73	114	72
37 Fishing area total	*102*	*325*	*250*	*80*	*107*	*13*	*78*	*73*	*114*	*72*
Species total	*36 983 F*	*34 365 F*	*31 465*	*32 707 F*	*25 069 F*	*23 876 F*	*18 093 F*	*21 540 F*	*21 928 F*	*24 166*
Rudd	**Rotengle**		**Escardinio**			***Scardinius erythrophthalmus***			**1,40(02)019,01**	**SRE**
04 Kazakhstan	...	...	...	...	...	...	...	...	...	485
04 Fishing area total	*...*	*...*	*...*	*...*	*...*	*...*	*...*	*...*	*...*	*485*
05 Bulgaria	...	...	...	...	...	71	90	39	3	45
Denmark	-	-	-	-	-	-	0	2	1	0
Greece	...	...	...	3	70	69	30	50	45	50
Lithuania	11	16	17	14	18	14	13	17	20	23
Slovenia	-	6	7	7	6	4	4	1	0	0
Ukraine	21	15	5	15	19	34	38	38	33	32
05 Fishing area total	*32*	*37*	*29*	*39*	*113*	*192*	*175*	*147*	*102*	*150*

B-11 Carps, barbels and other cyprinids — Capture production by species, fishing areas and countries or areas
Carpes, barbeaux et autres cyprinidés — Captures par espèces, zones de pêche et pays ou zones
Carpas, barbos y otros ciprínidos — Capturas por especies, áreas de pesca y países o áreas

Species, Fishing area Espèce, Zone de pêche Especie, Area de pesca	1993 t	1994 t	1995 t	1996 t	1997 t	1998 t	1999 t	2000 t	2001 t	2002 t
27 Lithuania	...	...	...	...	...	...	...	...	...	2
27 Fishing area total	...	...	...	...	...	...	...	...	...	2
Species total	*32*	*37*	*29*	*39*	*113*	*192*	*175*	*147*	*102*	*637*
Orfe(=Ide)	**Ide mélanote**		**Cachuelo**		***Leuciscus idus***				**1,40(02)020,01**	**FID**
05 Albania	8 F	5 F	-	-	-	-	-	-	-	-
Belarus	45	22	18	19	11	14	11	5	18	26
Bulgaria	-	-	7	8	10	12	14	...	0	1
Czech Rep	34	42	31	27	0	0	0	0	0	1
Estonia	3	1	1	1	2	1	2	4	4	4
Finland	311	442	442	229	229	158	158	181	181	181
Latvia	7	1	2	-	1	1	1	-	1	1
Lithuania	1	1	3	1	1	-	-	1	1	1
Poland	20	20	0	0	-	-	-	9	7	7
Romania	-	-	-	-	1	5	1	-	-	-
Russian Fed	2 560	2 093	1 977	2 914	3 037	3 280	2 604	2 378	2 257	2 492
Slovenia	22	25	19	17	17	16	14	15	14	14
Ukraine	...	6	...	...	...	...	...	...	...	...
05 Fishing area total	*3 011 F*	*2 658 F*	*2 500*	*3 216*	*3 309*	*3 487*	*2 805*	*2 593*	*2 483*	*2 728*
27 Estonia	187	165	98	131	88	69	50	61	36	26
Finland	356	302	309	263	262	242	238	287	290	284
Latvia	5	7	4	2	1	1	2	2	2	1
27 Fishing area total	*548*	*474*	*411*	*396*	*351*	*312*	*290*	*350*	*328*	*311*
Species total	*3 559 F*	*3 132 F*	*2 911*	*3 612*	*3 660*	*3 799*	*3 095*	*2 943*	*2 811*	*3 039*
Common dace	**Vandoise**		**Leucisco**		***Leuciscus leuciscus***				**1,40(02)020,05**	**FIE**
04 Turkey	242	215	223	215	250	300	176	104	91	73
04 Fishing area total	*242*	*215*	*223*	*215*	*250*	*300*	*176*	*104*	*91*	*73*
05 Slovenia	-	5	5	5	5	5	4	0	0	0
05 Fishing area total	*-*	*5*	*5*	*5*	*5*	*5*	*4*	*0*	*0*	*0*
Species total	*242*	*220*	*228*	*220*	*255*	*305*	*180*	*104*	*91*	*73*
Chub	**...B**		**...C**		***Leuciscus cephalus***				**1,40(02)020,08**	**LUH**
05 Bulgaria	...	...	...	...	92	121	150	43	4	17
Lithuania	-	-	-	-	-	-	-	3	3	10
Romania	-	-	-	-	-	45	17	24	28	9
05 Fishing area total	*...*	*...*	*...*	*...*	*92*	*166*	*167*	*70*	*35*	*36*
Species total	*...*	*...*	*...*	*...*	*92*	*166*	*167*	*70*	*35*	*36*
Chubs nei	**...B**		**...C**		***Leuciscus spp***				**1,40(02)020,XX**	**LEW**
05 Slovakia	46	81	63	47	39	32	38	33	31	24
05 Fishing area total	*46*	*81*	*63*	*47*	*39*	*32*	*38*	*33*	*31*	*24*
Species total	*46*	*81*	*63*	*47*	*39*	*32*	*38*	*33*	*31*	*24*
Rhinofishes nei	**Labéos nca**		**Labeos nep**		***Labeo spp***				**1,40(02)024,XX**	**RHI**
01 Ethiopia	...	...	...	1 994	2 007	3 168	3 621	3 451	3 387	2 303
Kenya	175	143	605	1 462	355	207	219	221	222	203
01 Fishing area total	*175*	*143*	*605*	*3 456*	*2 362*	*3 375*	*3 840*	*3 672*	*3 609*	*2 506*
Species total	*175*	*143*	*605*	*3 456*	*2 362*	*3 375*	*3 840*	*3 672*	*3 609*	*2 506*
Mud carp	**Carpe de vase**		**Carpa de fango**		***Cirrhinus molitorella***				**1,40(02)025,02**	**MUC**
04 China,Taiwan	6	6	5	3	4	7	7	6	6	9
04 Fishing area total	*6*	*6*	*5*	*3*	*4*	*7*	*7*	*6*	*6*	*9*
Species total	*6*	*6*	*5*	*3*	*4*	*7*	*7*	*6*	*6*	*9*
Grass carp(=White amur)	**Carpe herbivore(=chinoise)**		**Carpa china**		***Ctenopharyngodon idellus***				**1,40(02)035,01**	**FCG**
01 Egypt	...	...	10 000	15 343	16 553	4 233	707	12 826	17 918	19 528
01 Fishing area total	*...*	*...*	*10 000*	*15 343*	*16 553*	*4 233*	*707*	*12 826*	*17 918*	*19 528*
02 USA	8	8	19	10	15	13	11	17	31	31
02 Fishing area total	*8*	*8*	*19*	*10*	*15*	*13*	*11*	*17*	*31*	*31*
04 China,Taiwan	179	109	90	79	83	81	95	104	132	160
Iran	3 935	3 727	4 392	4 205	4 517	5 039	4 600	2 500	580	1 301
Kazakhstan	1	-	2	2 F	1 F	1 F	1 F	-	-	-
Nepal	6	...	...	...	...	...	...	...	...	...
Uzbekistan	500	13	19	10 F	30	92	7	17	13	1
04 Fishing area total	*4 621*	*3 849*	*4 503*	*4 296 F*	*4 631 F*	*5 213 F*	*4 703 F*	*2 621*	*725*	*1 462*
05 Albania	1 F	1 F	0	1	0	3	5	45	10	14
Bulgaria	-	-	3	8	8	17	20	12	18	23
Czech Rep	47	44	51	47	49	53	70	60	60	69
Germany	-	10	10	10	10	10	5	5	5	3

B-11 Carps, barbels and other cyprinids — Capture production by species, fishing areas and countries or areas
Carpes, barbeaux et autres cyprinidés — Captures par espèces, zones de pêche et pays ou zones
Carpas, barbos y otros ciprínidos — Capturas por especies, áreas de pesca y países o áreas

Species, Fishing area Espèce, Zone de pêche Especie, Area de pesca		1993 t	1994 t	1995 t	1996 t	1997 t	1998 t	1999 t	2000 t	2001 t	2002 t
	Hungary	343	381	366	346	305	301	318	356	309	400
	Poland	2	0	3	4	11	4	2	4	4	5
	Romania	397	360	170	124	12	46	21	83	73	10
	Slovakia	14	19	23	15	21	16	17	15	9	19
	Slovenia	-	-	-	-	-	2	2	3	2	2
	Ukraine	-	3	1	2	-	-	-	0	-	1
05	*Fishing area total*	*804 F*	*818 F*	*627*	*557*	*416*	*452*	*460*	*583*	*490*	*546*
Species total		*5 433 F*	*4 675 F*	*15 149*	*20 206 F*	*21 615 F*	*9 911 F*	*5 881 F*	*16 047*	*19 164*	*21 567*
Silver carp		**Carpe argentée**		**Carpa plateada**		***Hypophthalmichthys molitrix***				**1,40(02)043,01**	**SVC**
04	China,Taiwan	59	49	37	29	35	34	11	9	11	18
	Iran	4 225	8 246	16 965	17 510	15 825	11 629	14 400	8 750	3 680	4 295
	Israel	103	181	40	40	17	11	9	32	16	30
	Kazakhstan	362	272	267	240 F	170 F	120 F	70 F	30 F	-	-
	Kyrgyzstan	-	-	28	24 F	18 F	12 F	7 F	8 F	19	17
	Nepal	42	...	...	...	...	...	...	...	...	...
	Tajikistan	...	...	...	...	...	...	...	...	...	16
	Uzbekistan	916	402	1 003	481 F	893	249	322	586	544	75
04	*Fishing area total*	*5 707*	*9 150*	*18 340*	*18 324 F*	*16 958 F*	*12 055 F*	*14 819 F*	*9 415 F*	*4 270*	*4 451*
05	Albania	100 F	70 F	11	52	33	104	130	140	101	129
	Bulgaria	...	250 F	415	488	471	553	488	42	403	85
	Germany	-	76	76	76	76	76	39	39	39	23
	Hungary	1 354	1 136	521	862	1 483	731	676	365	997	525
	Poland	6	0	198	211	180	106	136	185	120	198
	Romania	1 779	1 708	1 950	1 272	1 940	428	1 308	634	644	411
	Russian Fed	...	...	...	...	...	31	11	8	11	8
	Slovakia	2	-	-	-	-	4	5	7	8	8
	Slovenia	-	6	7	7	6	4	4	-	0	0
	Ukraine	2 896	3 083	2 448	3 187	1 380	1 008	442	444	267	152
05	*Fishing area total*	*6 137 F*	*6 329 F*	*5 626*	*6 155*	*5 569*	*3 045*	*3 239*	*1 864*	*2 590*	*1 539*
37	Ukraine	-	-	2	1	-	-	-	-	-	-
37	*Fishing area total*	-	-	*2*	*1*	-	-	-	-	-	-
Species total		*11 844 F*	*15 479 F*	*23 968*	*24 480 F*	*22 527 F*	*15 100 F*	*18 058 F*	*11 279 F*	*6 860*	*5 990*
Bighead carp		**Carpe à grosse tête**		**Carpa cabezona**		***Hypophthalmichthys nobilis***				**1,40(02)043,02**	**BIC**
04	China,Taiwan	244	147	105	67	72	99	241	204	202	183
	Iran	819	866	1 414	1 751	1 565	2 522	3 600	4 250	245	520
	Nepal	25	...	...	...	...	...	...	...	...	...
04	*Fishing area total*	*1 088*	*1 013*	*1 519*	*1 818*	*1 637*	*2 621*	*3 841*	*4 454*	*447*	*703*
05	Bulgaria	...	...	...	...	...	...	...	...	...	223
	Croatia	-	-	1	-	-	-	-	-	-	-
	Czech Rep	4	3	13	3	5	6	8	10	12	12
	Germany	-	12	12	12	12	12	6	6	6	4
	Hungary	291	45	127	74	83	-	-	-	-	-
	Romania	1 047	1 103	700	449	299	369	396	356	313	278
05	*Fishing area total*	*1 342*	*1 163*	*853*	*538*	*399*	*387*	*410*	*372*	*331*	*517*
Species total		*2 430*	*2 176*	*2 372*	*2 356*	*2 036*	*3 008*	*4 251*	*4 826*	*778*	*1 220*
Silver cyprinid		**...B**		**...C**		***Rastrineobola argentea***				**1,40(02)070,01**	**ENA**
01	Kenya	42 505	69 134	56 827	49 670	40 315	42 336	48 816	49 618	41 384	35 455
01	*Fishing area total*	*42 505*	*69 134*	*56 827*	*49 670*	*40 315*	*42 336*	*48 816*	*49 618*	*41 384*	*35 455*
Species total		*42 505*	*69 134*	*56 827*	*49 670*	*40 315*	*42 336*	*48 816*	*49 618*	*41 384*	*35 455*
Kinneret bleak		**...B**		**...C**		***Acanthobrama terraesanctae***				**1,40(02)072,02**	**AHT**
04	Israel	967	300	434	1 155	626	1 171	1 048	1 052	811	1 170
04	*Fishing area total*	*967*	*300*	*434*	*1 155*	*626*	*1 171*	*1 048*	*1 052*	*811*	*1 170*
Species total		*967*	*300*	*434*	*1 155*	*626*	*1 171*	*1 048*	*1 052*	*811*	*1 170*
Vimba bream		**...B**		**...C**		***Vimba vimba***				**1,40(02)097,01**	**VIV**
05	Bulgaria	...	...	...	...	84	15	50	4	1	2
	Estonia	0	0	0	1	0	0	0	0	0	0
	Latvia	-	-	1	8	14	13	13	19	20	15
	Lithuania	-	1	3	2	3	3	11	48	40	-
	Poland	...	...	...	...	...	...	...	5	2	1
	Romania	-	-	-	-	26	10	31	3	2	2
	Slovakia	28	30	40	20	25	27	14	11	10	11
	Slovenia	-	-	-	-	-	1	1	2	2	1
	Ukraine	29	16	24	25	10	18	13	13	20	15
05	*Fishing area total*	*57*	*47*	*68*	*56*	*162*	*87*	*133*	*105*	*97*	*47*
27	Estonia	118	104	188	164	185	165	122	101	83	115
	Latvia	49	39	54	50	43	79	106	73	89	84
	Lithuania	...	...	...	...	...	...	...	...	...	78
27	*Fishing area total*	*167*	*143*	*242*	*214*	*228*	*244*	*228*	*174*	*172*	*277*
Species total		*224*	*190*	*310*	*270*	*390*	*331*	*361*	*279*	*269*	*324*

B-11

Carps, barbels and other cyprinids — **Capture production by species, fishing areas and countries or areas**
Carpes, barbeaux et autres cyprinidés — **Captures par espèces, zones de pêche et pays ou zones**
Carpas, barbos y otros ciprínidos — **Capturas por especies, áreas de pesca y países o áreas**

Species, Fishing area Espèce, Zone de pêche Especie, Area de pesca	1993 t	1994 t	1995 t	1996 t	1997 t	1998 t	1999 t	2000 t	2001 t	2002 t
Sichel	**...B**		**Peleco**		***Pelecus cultratus***				**1,40(02)100,01**	**FSC**
04 Kazakhstan	-	-	27	25 F	20 F	15 F	10 F	10 F	8	-
Tajikistan	3	1	0	...	...	...	...	...	8	4
Turkmenistan	24	20	16	13	1	0	0	0	-	-
04 Fishing area total	*27*	*21*	*43*	*38 F*	*21 F*	*15 F*	*10 F*	*10 F*	*16*	*4*
05 Lithuania	-	-	-	-	-	4	3	8	12	-
Russian Fed	721	871	435	412	256	1 645	1 571	550	1 022	638
Ukraine	...	108	37	2	13	11	14	7	7	4
05 Fishing area total	*721*	*979*	*472*	*414*	*269*	*1 660*	*1 588*	*565*	*1 041*	*642*
27 Lithuania	...	...	...	...	...	...	...	...	...	4
Russian Fed	-	-	-	-	-	-	-	384	348	519
27 Fishing area total	...	...	...	...	...	...	...	*384*	*348*	*523*
37 Russian Fed	88	53	54	39	86	79	215	267	178	140
Ukraine	-	-	-	7	27	26	13	9	7	7
37 Fishing area total	*88*	*53*	*54*	*46*	*113*	*105*	*228*	*276*	*185*	*147*
Species total	*836*	*1 053*	*569*	*498 F*	*403 F*	*1 780 F*	*1 826 F*	*1 235 F*	*1 590*	*1 316*
Asp	**Aspe**		**Aspio**		***Aspius aspius***				**1,40(02)115,01**	**ASU**
04 Azerbaijan	31	14	5	9	6	4	2	1	5	4
Kazakhstan	313	314	415	380 F	270 F	195 F	250 F	350 F	476	1 026
Tajikistan	11	8	6	...	...	...	...	...	6	5
Turkmenistan	8	7	6	10	6	12	8	11	16	21
Uzbekistan	...	...	44	20 F	...	...	...	...	...	...
04 Fishing area total	*363*	*343*	*476*	*419 F*	*282 F*	*211 F*	*260 F*	*362 F*	*503*	*1 056*
05 Bulgaria	-	-	3	4	5	7	8	9	7	1
Czech Rep	-	-	-	10	15	16	16	13	17	18
Hungary	39	52	20	22	44	38	42	38	21	20
Lithuania	0	5	3	4	6	9	6	6	5	2
Poland	...	...	...	...	...	...	...	6	5	6
Romania	-	-	-	-	-	11	7	17	4	7
Russian Fed	51	...	55	84	36	45	53	51	56	133
Slovakia	8	16	13	12	9	9	9	8	8	9
Slovenia	1	-	-	-	-	0	-	0	0	-
Ukraine	20	16	18	25	19	7	15	26	7	7
05 Fishing area total	*119*	*89*	*112*	*161*	*134*	*142*	*156*	*174*	*130*	*203*
27 Lithuania	...	...	...	...	...	...	...	...	...	8
Russian Fed	-	-	-	-	-	-	-	6	9	15
27 Fishing area total	...	...	...	...	...	...	...	*6*	*9*	*23*
Species total	*482*	*432*	*588*	*580 F*	*416 F*	*353 F*	*416 F*	*542 F*	*642*	*1 282*
Hoven's carp	**Barbus d'Hoven**		**Barbo de Hoven**		***Leptobarbus hoeveni***				**1,40(02)132,01**	**FCH**
04 Indonesia	4 606	5 376	5 454	6 892	5 836	3 241	4 608	3 382	2 362	2 270
04 Fishing area total	*4 606*	*5 376*	*5 454*	*6 892*	*5 836*	*3 241*	*4 608*	*3 382*	*2 362*	*2 270*
Species total	*4 606*	*5 376*	*5 454*	*6 892*	*5 836*	*3 241*	*4 608*	*3 382*	*2 362*	*2 270*
Black carp	**Carpe noire**		**Carpa negra**		***Mylopharyngodon piceus***				**1,40(02)144,01**	**BKC**
04 China,Taiwan	22	20	20	15	16	22	27	34	36	25
04 Fishing area total	*22*	*20*	*20*	*15*	*16*	*22*	*27*	*34*	*36*	*25*
Species total	*22*	*20*	*20*	*15*	*16*	*22*	*27*	*34*	*36*	*25*
Wuchang bream	**Carpe de Wuchang**		**Carpa de Wuchang**		***Megalobrama amblycephala***				**1,40(02)151,01**	**WUB**
05 Albania	1 F	1 F	0	0	0	-	-	-	-	3
05 Fishing area total	*1 F*	*1 F*	*0*	*0*	*0*	-	-	-	-	*3*
Species total	*1 F*	*1 F*	*0*	*0*	*0*	-	-	-	-	*3*
Java barb	**Barbeau de Java**		**Barbo de Java**		***Puntius javanicus***				**1,40(02)161,02**	**FJB**
04 Indonesia	15 027	19 084	18 102	19 622	19 469	20 189	17 939	17 791	17 425	17 840
04 Fishing area total	*15 027*	*19 084*	*18 102*	*19 622*	*19 469*	*20 189*	*17 939*	*17 791*	*17 425*	*17 840*
Species total	*15 027*	*19 084*	*18 102*	*19 622*	*19 469*	*20 189*	*17 939*	*17 791*	*17 425*	*17 840*
Asian barbs nei	**Barbeaux d'Asie nca**		**Barbos de Asia nep**		***Puntius spp***				**1,40(02)161,XX**	**FAB**
04 Indonesia	10 523	12 113	12 344	13 253	11 976	12 131	11 263	12 406	11 649	11 840
Thailand	23 061	25 567	22 468	25 750	25 296	44 349	45 511	41 024	43 400	44 100
04 Fishing area total	*33 584*	*37 680*	*34 812*	*39 003*	*37 272*	*56 480*	*56 774*	*53 430*	*55 049*	*55 940*
Species total	*33 584*	*37 680*	*34 812*	*39 003*	*37 272*	*56 480*	*56 774*	*53 430*	*55 049*	*55 940*
Cyprinids nei	**Cyprinidés nca**		**Ciprínidos nep**		***Cyprinidae***				**1,40(02)XXX,XX**	**FCY**
01 Egypt	928	2 154	1 059	1 261	1 386	2 117	4 515	1 232	5 040	4 385

B-11 Carps, barbels and other cyprinids — Capture production by species, fishing areas and countries or areas
Carpes, barbeaux et autres cyprinidés — Captures par espèces, zones de pêche et pays ou zones
Carpas, barbos y otros ciprínidos — Capturas por especies, áreas de pesca y países o áreas

Species, Fishing area Espèce, Zone de pêche Especie, Area de pesca	1993 t	1994 t	1995 t	1996 t	1997 t	1998 t	1999 t	2000 t	2001 t	2002 t
Ethiopia	-	-	362	387	639	768	878	860	843	759
Kenya	136	127	118	0	0	0	0	0	0	0
Madagascar	5 100	4 313	4 123	4 000	4 000	4 000	4 000	4 000	4 000	4 000
Malawi	6 775	7 265	538	505	522	299	8 302	7 500 F	6 291	6 401
Morocco	500	1 000	900	800	800	900	1 200	1 000	900	1 100
Nigeria	3 293	3 021	4 476	4 460	6 199	5 006	6 296	4 823	5 901	6 004
Uganda	1 149	2 872	2 948	1 539	13 476	15 710	17 600	12 181	12 182	12 000
01 Fishing area total	*17 881*	*20 752*	*14 524*	*12 952*	*27 022*	*28 800*	*42 791*	*31 596 F*	*35 157*	*34 649*
02 Mexico	3 086	3 255	1 739	3 732	4 838	3 982	6 805	3 988	3 806	1 711
USA	3	10	4	3	5	15	17	29	33	27
02 Fishing area total	*3 089*	*3 265*	*1 743*	*3 735*	*4 843*	*3 997*	*6 822*	*4 017*	*3 839*	*1 738*
03 Brazil	100 F	100 F	119	284	212	258	302	355	390	392
03 Fishing area total	*100 F*	*100 F*	*119*	*284*	*212*	*258*	*302*	*355*	*390*	*392*
04 Armenia	312	46	42	33	7	37	21	19	15	9
China,H.Kong	0	0	0	0	0	0	0	0	0	0
India	258 502	314 939	339 828	220 994	194 608	167 186	177 394	152 350	148 622	183 785
Iran	7 748	8 001	7 802	12 364	12 870	19 573	10 800	10 000	2 680	3 645
Iraq	4 836	10 213	9 981	7 694	4 782	1 703	5 723	4 013	3 800 F	3 800 F
Israel	48	52	49	12	15	64	80	60	70	...
Korea Rep	6 262	4 626	3 311	1 010	2 221	1 977	2 294	...	...	...
Kyrgyzstan	49	30	39	35 F	26 F	18 F	11 F	12 F	8	7
Laos	2 500 F	3 200 F	4 000 F	3 500 F	2 800 F	3 000 F	4 500 F	4 400 F	4 650 F	5 000 F
Lebanon	10	10	10	10	10	10	10	10	10	25
Philippines	7 117	5 568	8 880	6 497	5 717	6 453	4 677	5 032	5 562	9 127
Tajikistan	-	-	-	-	-	-	-	-	31	27
Turkey	1 303	1 499	1 535	1 380	1 900	1 800	406	699	626	240
Uzbekistan	350	585	0	-	378	332	337	441	132	89
04 Fishing area total	*289 037 F*	*348 769 F*	*375 477 F*	*253 529 F*	*225 334 F*	*202 153 F*	*206 253 F*	*177 036 F*	*166 206 F*	*205 754 F*
05 Belarus	221	122	105	92	61	47	27	19	77	195
Belgium	50	50	50	50	50	50	50	50	50	50
Bulgaria	...	...	...	...	1	251	250	102	-	14
Finland	193	277	277	15	15	-	-	-	-	-
Germany	1 156	1 653	1 653	1 653	1 653	1 653	1 653	1 653	1 653	1 653
Greece	...	...	...	173	37	38	44	40	35	32
Hungary	1 342	1 557	1 744	1 731	1 730	1 510	1 666	1 710	2 155	2 418
Italy	2 450	2 628	2 540	1 146	2 378	1 155	1 900	725	1 821	799
Latvia	-	30	36	34	42	26	27	-	-	-
Lithuania	5	1	3	1	2	-	-	-	-	0
Poland	0	0	0	0	395	330	-	2	2	3
Romania	274	337	153	112	16	396	126	39	298	282
Russian Fed	16 754	9 306	11 887	9 852	17 067	23 353	20 331	19 906	18 008	20 012
Slovakia	51	71	64	8	44	37	38	53	12	29
Slovenia	60	71	59	54	40	42	35	35	25	24
Switzerland	4	4	34	42	20	10	6	8	13	13
05 Fishing area total	*22 560*	*16 107*	*18 605*	*14 963*	*23 551*	*28 898*	*26 153*	*24 342*	*24 149*	*25 524*
27 Finland	-	-	-	8	8	-	-	-	-	-
Lithuania	-	-	-	-	-	-	-	-	19	-
Russian Fed	-	-	-	-	-	-	-	67	88	113
Sweden	0	1	0	0	0	-	-	-	-	-
27 Fishing area total	*0*	*1*	*0*	*8*	*8*	-	-	*67*	*107*	*113*
37 Russian Fed	5	-	-	-	1	-	-	167	159	95
37 Fishing area total	*5*	-	-	-	*1*	-	-	*167*	*159*	*95*
61 Russian Fed	-	55	61	110	101	100	56	8	52	38
61 Fishing area total	...	*55*	*61*	*110*	*101*	*100*	*56*	*8*	*52*	*38*
Species total	*332 672 F*	*389 049 F*	*410 529 F*	*285 581 F*	*281 072 F*	*264 206 F*	*282 377 F*	*237 588 F*	*230 059 F*	*268 303 F*
Group total	***673 996***	***756 801***	***780 522***	***663 465***	***608 078***	***605 024***	***613 623***	***570 965***	***548 894***	***592 962***

B-12 Tilapias and other cichlids / Tilapias et autres cichlidés / Tilapias y otros cíclidos

Capture production by species, fishing areas and countries or areas
Captures par espèces, zones de pêche et pays ou zones
Capturas por especies, áreas de pesca y países o áreas

Species, Fishing area Espèce, Zone de pêche Especie, Area de pesca	1993 t	1994 t	1995 t	1996 t	1997 t	1998 t	1999 t	2000 t	2001 t	2002 t
Mozambique tilapia	**Tilapia du Mozambique**		**Tilapia del Mozambique**		***Oreochromis mossambicus***			**1,70(59)051,01**		**TLM**
04 Indonesia	15 849	15 337	13 293	16 943	17 715	17 865	21 172	19 831	20 210	20 980
04 Fishing area total	*15 849*	*15 337*	*13 293*	*16 943*	*17 715*	*17 865*	*21 172*	*19 831*	*20 210*	*20 980*
06 Papua N Guin	2 310 F	2 310 F	2 310	2 310 F	2 310 F	2 310 F	2 310 F	2 310 F	2 310 F	2 310 F
06 Fishing area total	*2 310 F*	*2 310 F*	*2 310*	*2 310 F*	*2 310 F*	*2 310 F*	*2 310 F*	*2 310 F*	*2 310 F*	*2 310 F*
Species total	*18 159 F*	*17 647 F*	*15 603*	*19 253 F*	*20 025 F*	*20 175 F*	*23 482 F*	*22 141 F*	*22 520 F*	*23 290 F*
Nile tilapia	**Tilapia du Nil**		**Tilapia del Nilo**		***Oreochromis niloticus***			**1,70(59)051,02**		**TLN**
01 Burundi	30 F	50 F	50	50	50	50	50	120	120	120 F
Egypt	97 679	95 819	122 207	125 307	130 992	128 446	112 811	131 276	145 291	138 450
Kenya	12 196	11 319	11 827	10 765	13 953	14 652	17 524	19 347	7 292	16 252
Mali	19 290	18 855	39 842	30 450	1 719	26 675	26 821	32 961	30 000 F	30 000 F
Rwanda	...	...	...	...	...	2 233	2 640	2 646	2 650	2 750 F
Sudan	...	...	10 000	10 500	11 000	16 000	16 000	18 000 F	20 000	20 000
01 Fishing area total	*129 195 F*	*126 043 F*	*183 926*	*177 072*	*157 714*	*188 056*	*175 846*	*204 350 F*	*205 353 F*	*207 572 F*
02 El Salvador	2 801	2 323	2 494	1 265	1 297	1 017	1 216	1 171	560	1 169
Jamaica	150 F	150 F	150 F	150 F	150 F	150 F	150 F	150 F	150 F	150 F
02 Fishing area total	*2 951 F*	*2 473 F*	*2 644 F*	*1 415 F*	*1 447 F*	*1 167 F*	*1 366 F*	*1 321 F*	*710 F*	*1 319 F*
04 Thailand	53 903	63 400	55 746	29 253	28 727	40 173	49 840	40 037	43 100	44 700
04 Fishing area total	*53 903*	*63 400*	*55 746*	*29 253*	*28 727*	*40 173*	*49 840*	*40 037*	*43 100*	*44 700*
06 Fiji Islands	7	8 F	6	37	40	288	290	280 F	278	280 F
06 Fishing area total	*7*	*8 F*	*6*	*37*	*40*	*288*	*290*	*280 F*	*278*	*280 F*
Species total	*186 056 F*	*191 924 F*	*242 322 F*	*207 777 F*	*187 928 F*	*229 684 F*	*227 342 F*	*245 988 F*	*249 441 F*	*253 871 F*
Blue tilapia	**...B**		**...C**		***Oreochromis aureus***			**1,70(59)051,03**		**OEA**
02 Cuba	9 348	9 483	8 199	9 672	7 553	5 122	4 564	2 940	1 948	2 002
02 Fishing area total	*9 348*	*9 483*	*8 199*	*9 672*	*7 553*	*5 122*	*4 564*	*2 940*	*1 948*	*2 002*
04 Israel	160	228	69	32	101	88	66	67	33	18
04 Fishing area total	*160*	*228*	*69*	*32*	*101*	*88*	*66*	*67*	*33*	*18*
Species total	*9 508*	*9 711*	*8 268*	*9 704*	*7 654*	*5 210*	*4 630*	*3 007*	*1 981*	*2 020*
Tilapias nei	**Tilapias nca**		**Tilapias nep**		***Oreochromis (=Tilapia) spp***			**1,70(59)051,XX**		**TLP**
01 Benin	10 200 F	10 235	11 713	11 500 F	11 500 F	11 500 F	11 648	9 600 F	10 900 F	10 900 F
Botswana	300 F	200 F	100 F	48	80	88	93	92	88	101
Ethiopia	4 175	5 285	3 703	5 066	6 076	7 952	9 088	7 000	6 870	4 603
Gabon	1 500 F	1 500 F	2 600 F	3 022	3 630	3 500	3 800	3 800	3 122	3 478
Gambia	1 000	1 000	1 050	1 050	1 050	1 050	1 050 F	1 050 F	1 050 F	1 050 F
Kenya	7 886	6 354	7 568	7 635	19 329	19 732	18 203	19 853	19 121	17 418
Malawi	10 219	6 635	3 863	5 080	4 472	5 104	6 808	6 200 F	5 154	5 244
Mauritius	3	0	0	0	0	0	0	0	0	0
Nigeria	8 003	7 874	9 060	10 074	12 614	16 300	19 662	13 402	18 332	17 218
Senegal	...	...	...	...	...	...	...	9 133	8 100 F	8 100 F
Tanzania	20 213	19 610	25 869	35 160	25 100	24 000	38 000	40 000	45 000	43 000
Togo	5 000	3 500	3 500	3 500	3 500	3 500	3 500	3 500	3 500	3 500
Uganda	74 280	80 176	83 223	75 027	81 379	78 500	84 540	96 468	96 172	98 000
Zimbabwe	370	440	300	320	523	412	420	830	800 F	800 F
01 Fishing area total	*143 149 F*	*142 809 F*	*152 549 F*	*157 482 F*	*169 253 F*	*171 638 F*	*196 812 F*	*210 928 F*	*218 209 F*	*213 412 F*
02 Dominican Rp	328	659	1 188	100	207	77	439	...	529	1 175
Mexico	75 835	75 024	74 646	74 354	74 814	65 178	59 343	68 772	60 336	54 901
Panama	-	-	10	15	56	11	13	16	16 F	16 F
USA	1 732	1 098	2 717	4	0	0	2 651	1 040	1 277	1 752
02 Fishing area total	*77 895*	*76 781*	*78 561*	*74 473*	*75 077*	*65 266*	*62 446*	*69 828*	*62 158 F*	*57 844 F*
03 Brazil	8 400 F	7 700 F	6 234	7 300	6 927	7 444	6 616	7 893	8 329	8 248
03 Fishing area total	*8 400 F*	*7 700 F*	*6 234*	*7 300*	*6 927*	*7 444*	*6 616*	*7 893*	*8 329*	*8 248*
04 China,Taiwan	427	607	498	145	146	152	86	79	98	97
Philippines	19 071	17 649	21 244	17 663	20 935	23 477	25 278	28 874	28 881	30 586
Sri Lanka	15 000	9 500	15 000	22 250	27 250	29 900	28 520	33 220	27 230	24 900
04 Fishing area total	*34 498*	*27 756*	*36 742*	*40 058*	*48 331*	*53 529*	*53 884*	*62 173*	*56 209*	*55 583*
06 Guam	-	-	-	-	-	6	-	-	-	-
06 Fishing area total	-	-	-	-	-	*6*	-	-	-	-
34 Gabon	...	...	...	598	520	19	378	421	301	191
Gambia	75	5	94	198	85	23	20	12	27	67
Senegal	1 024	2 803	2 382	1 807	1 977	1 323	1 362	1 629	1 996	2 075
34 Fishing area total	*1 099*	*2 808*	*2 476*	*2 603*	*2 582*	*1 365*	*1 760*	*2 062*	*2 324*	*2 333*
Species total	*265 041 F*	*257 854 F*	*276 562 F*	*281 916 F*	*302 170 F*	*299 248 F*	*321 518 F*	*352 884 F*	*347 229 F*	*337 420 F*
Mango tilapia	**...B**		**...C**		***Sarotherodon galilaeus***			**1,70(59)052,01**		**SAR**
04 Israel	215	362	316	350	462	391	405	262	110	93
04 Fishing area total	*215*	*362*	*316*	*350*	*462*	*391*	*405*	*262*	*110*	*93*

B-12 Tilapias and other cichlids / Tilapias et autres cichlidés / Tilapias y otros cíclidos

Capture production by species, fishing areas and countries or areas / Captures par espèces, zones de pêche et pays ou zones / Capturas por especies, áreas de pesca y países o áreas

Species, Fishing area Espèce, Zone de pêche Especie, Area de pesca	1993 t	1994 t	1995 t	1996 t	1997 t	1998 t	1999 t	2000 t	2001 t	2002 t
Species total	*215*	*362*	*316*	*350*	*462*	*391*	*405*	*262*	*110*	*93*
Jaguar guapote	**...B**		**Guapote tigre**		***Cichlasoma managuense***				**1,70(59)053,06**	**CHL**
02 El Salvador	464	382	608	566	367	344	351	324	410	318
02 Fishing area total	*464*	*382*	*608*	*566*	*367*	*344*	*351*	*324*	*410*	*318*
Species total	*464*	*382*	*608*	*566*	*367*	*344*	*351*	*324*	*410*	*318*
Peacock cichlid	**...B**		**Sargento**		***Cichla ocellaris***				**1,70(59)054,04**	**CLO**
02 Panama	28	285	120	65	35	12	7	4	4 F	4 F
02 Fishing area total	*28*	*285*	*120*	*65*	*35*	*12*	*7*	*4*	*4 F*	*4 F*
Species total	*28*	*285*	*120*	*65*	*35*	*12*	*7*	*4*	*4 F*	*4 F*
Velvety cichlids	**...B**		**Acarahuazu**		***Astronotus spp***				**1,70(59)313,XX**	**AST**
03 Peru	...	...	320	308	177	141	251	188	183	...
03 Fishing area total	*...*	*...*	*320*	*308*	*177*	*141*	*251*	*188*	*183*	*...*
Species total	*...*	*...*	*320*	*308*	*177*	*141*	*251*	*188*	*183*	*...*
Mouthbrooding cichlids	**...B**		**...C**		***Haplochromis spp***				**1,70(59)316,XX**	**CIM**
01 Kenya	3 506	4 196	4 822	3 914	2 453	2 577	2 731	2 699	1 195	1 259
Tanzania	9 122	13 204	11 960	6 558	10 600	10 500	8 000	9 000	8 500	7 500
01 Fishing area total	*12 628*	*17 400*	*16 782*	*10 472*	*13 053*	*13 077*	*10 731*	*11 699*	*9 695*	*8 759*
Species total	*12 628*	*17 400*	*16 782*	*10 472*	*13 053*	*13 077*	*10 731*	*11 699*	*9 695*	*8 759*
Cichlids nei	**Cichlidés nca**		**Cíclidos nep**		***Cichlidae***				**1,70(59)XXX,XX**	**CIX**
01 Madagascar	21 100	21 500	21 600	21 500	21 500	21 500	21 500	21 500	21 500	21 500
Malawi	...	...	...	...	...	...	18 841	19 000 F	20 533	20 892
01 Fishing area total	*21 100*	*21 500*	*21 600*	*21 500*	*21 500*	*21 500*	*40 341*	*40 500 F*	*42 033*	*42 392*
02 Guatemala	-	2	5	0	21	157	173	192	200 F	200 F
Nicaragua	...	...	...	...	...	...	750	680	660	303
02 Fishing area total	*...*	*2*	*5*	*0*	*21*	*157*	*923*	*872*	*860 F*	*503 F*
03 Brazil	15 100 F	15 450 F	12 045	8 565	8 356	7 785	7 375	8 145	13 611	13 946
03 Fishing area total	*15 100 F*	*15 450 F*	*12 045*	*8 565*	*8 356*	*7 785*	*7 375*	*8 145*	*13 611*	*13 946*
04 Israel	97	90	64	40	48	58	57	52	24	23
04 Fishing area total	*97*	*90*	*64*	*40*	*48*	*58*	*57*	*52*	*24*	*23*
Species total	*36 297 F*	*37 042 F*	*33 714*	*30 105*	*29 925*	*29 500*	*48 696*	*49 569 F*	*56 528 F*	*56 864 F*
Group total	***528 396***	***532 607***	***594 615***	***560 516***	***561 796***	***597 782***	***637 413***	***686 066***	***688 101***	***682 639***

B-13 Miscellaneous freshwater fishes / Poissons d'eau douce divers / Peces de agua dulce diversos

Capture production by species, fishing areas and countries or areas / Captures par espèces, zones de pêche et pays ou zones / Capturas por especies, áreas de pesca y países o áreas

Species, Fishing area Espèce, Zone de pêche Especie, Area de pesca	1993 t	1994 t	1995 t	1996 t	1997 t	1998 t	1999 t	2000 t	2001 t	2002 t
African lungfishes	**Protoptères d'Afrique**		**Protopteros africanos**		***Protopterus spp***			**1,16(02)002,XX**		**FLU**
01 Benin	100 F	127	121	110 F	105 F	100 F	100 F	100 F	110 F	110 F
Kenya	146	398	408	164	1 704	1 717	1 600	1 608	1 822	1 707
Nigeria	798	436	678	1 284	3 950	1 679	536	1 364	2 183	520
Uganda	3 949	7 868	7 724	6 095	6 554	7 595	9 192	5 755	5 796	5 000
01 Fishing area total	*4 993 F*	*8 829*	*8 931*	*7 653 F*	*12 313 F*	*11 091 F*	*11 428 F*	*8 827 F*	*9 911 F*	*7 337 F*
Species total	*4 993 F*	*8 829*	*8 931*	*7 653 F*	*12 313 F*	*11 091 F*	*11 428 F*	*8 827 F*	*9 911 F*	*7 337 F*
Dagaas	**Dagaas**		**Dagaas**		***Stolothrissa, Limnothrissa spp***			**1,21(05)XXX,XX**		**DAG**
01 Burundi	14 100 F	18 200 F	18 163	1 508	17 868	8 646	7 030	10 839	6 138	6 160 F
Mozambique	689	925	3 173	5 574	9 921	7 313	9 052	11 813	5 284	12 156
Tanzania	44 582	51 467	50 360	40 179	48 000	46 800	42 000	40 000	43 000	43 000
Zambia	9 722	8 910	8 674	7 593	7 813	9 822	8 955	8 863	8 500 F	8 500 F
Zimbabwe	19 958	19 232	15 280	15 423	17 034	15 288	11 208	10 477	10 400 F	10 400 F
01 Fishing area total	*89 051 F*	*98 734 F*	*95 650*	*70 277*	*100 636*	*87 869*	*78 245*	*81 992*	*73 322 F*	*80 216 F*
Species total	*89 051 F*	*98 734 F*	*95 650*	*70 277*	*100 636*	*87 869*	*78 245*	*81 992*	*73 322 F*	*80 216 F*
Northern pike	**Brochet du Nord**		**Lucio**		***Esox lucius***			**1,24(03)001,01**		**FPI**
01 Algeria	-	-	-	-	-	2	...	...	...	...
01 Fishing area total	-	-	-	-	-	*2*	...	...	...	...
02 Canada	3 424	1 787	2 496	2 406	2 334	2 568	2 615	2 808	2 464	2 684
02 Fishing area total	*3 424*	*1 787*	*2 496*	*2 406*	*2 334*	*2 568*	*2 615*	*2 808*	*2 464*	*2 684*
04 Azerbaijan	82	27	15	7	23	18	21	28	25	11
Kazakhstan	733	564	637	580 F	420 F	300 F	550 F	700 F	819	920
Turkey	304	406	453	225	350	200	276	224	192	217
Uzbekistan	282	122	36	20 F	7	10	60	23	24	1
04 Fishing area total	*1 401*	*1 119*	*1 141*	*832 F*	*800 F*	*528 F*	*907 F*	*975 F*	*1 060*	*1 149*
05 Belarus	468	42	29	27	11	9	12	162	171	312
Belgium	20	20	20	20	20	20	20	20	20	20
Bulgaria	-	-	0	0	19	41	74	14	2	6
Croatia	23	22	30	30	30	1 F	1	1	1	1
Czech Rep	133	173	82	163	159	168	183	180	176	172
Denmark	7	8	6	5	7	11	8	5	7	6
Estonia	103	47	42	97	82	114	133	156	180	199
Finland	13 334	9 704	9 681	9 763	9 763	7 621	7 621	7 887	7 887	7 887
Germany	480	205	205	205	205	205	205	205	205	205
Greece	...	...	...	12	9	5	8	10	9	10
Hungary	165	190	37	46	203	158	241	280	191	190
Ireland	1 493	2 000	2 000	2 000	2 000	2 000	...	...	...	...
Latvia	30	41	55	56	47	55	72	72	73	88
Lithuania	62	78	78	66	71	56	62	71	68	66
Moldova Rep	39	45	56	53	39	38	25	19	36	62
Netherlands	1	-	-	-	-	-	-	-	-	-
Norway	...	...	...	...	13	7	...	...	...	...
Poland	350	350	245	1 076	280	226	262	363	302	345
Portugal	0	0	0	0	0	-	-	0	0	-
Romania	8	10	11	8	4	40	47	47	95	83
Russian Fed	5 338	4 367	4 938	5 688	4 646	5 567	6 118	8 843	8 427	9 741
Serbia-Monte	...	...	37	16	33	25	19	22	16	18
Slovakia	89	193	134	103	85	68	69	76	73	64
Slovenia	9	12	11	11	12	9	10	11	9	9
Sweden	168	158	154	156	149	139	177	175	145	168
Switzerland	35	38	60	36	34	33	34	48	48	46
Ukraine	10	136	8	26	23	15	18	15	14	5
UK	5	7	4	5	2	4	4	6	5	4
05 Fishing area total	*22 370*	*17 846*	*17 923*	*19 668*	*17 946*	*16 635 F*	*15 423*	*18 688*	*18 160*	*19 707*
27 Denmark	6	6	5	6	7	5	3	4	3	4
Estonia	65	37	30	42	23	17	19	21	19	19
Finland	4 202	2 175	2 150	2 582	2 614	4 028	4 042	2 562	2 541	2 571
Germany	-	135	112	83	101	99	131	117	104	104
Lithuania	...	...	...	...	...	...	...	...	...	8
Poland	1	0	0	0	-	-	-	-	9	9
Russian Fed	-	-	-	-	-	-	-	21	17	18
Sweden	279	272	240	204	189	163	145	115	97	66
27 Fishing area total	*4 553*	*2 625*	*2 537*	*2 917*	*2 934*	*4 312*	*4 340*	*2 840*	*2 790*	*2 799*
Species total	*31 748*	*23 377*	*24 097*	*25 823 F*	*24 014 F*	*24 045 F*	*23 285 F*	*25 311 F*	*24 474*	*26 339*
Arapaima	**...B**		**Paiche**		***Arapaima gigas***			**1,28(01)001,01**		**ARP**
03 Peru	...	...	420	457	465	210	338	273	204	...
03 Fishing area total	...	...	*420*	*457*	*465*	*210*	*338*	*273*	*204*	...
Species total	...	...	*420*	*457*	*465*	*210*	*338*	*273*	*204*	...
Bonytongues nei	**...B**		**...C**		***Heterotis spp***			**1,28(01)004,XX**		**HTT**
01 Benin	400 F	349	742	700 F	650 F	600 F	600 F	500 F	550 F	550 F
Mali	643	629	1 328	453	329	397	399	1 099	1 000 F	1 000 F
Nigeria	4 375	2 897	4 832	3 298	4 737	7 388	8 873	7 915	8 577	6 832

B-13

Miscellaneous freshwater fishes — Capture production by species, fishing areas and countries or areas
Poissons d'eau douce divers — Captures par espèces, zones de pêche et pays ou zones
Peces de agua dulce diversos — Capturas por especies, áreas de pesca y países o áreas

Species, Fishing area Espèce, Zone de pêche Especie, Area de pesca	1993 t	1994 t	1995 t	1996 t	1997 t	1998 t	1999 t	2000 t	2001 t	2002 t
01 Fishing area total	*5 418 F*	*3 875*	*6 902*	*4 451 F*	*5 716 F*	*8 385 F*	*9 872 F*	*9 514 F*	*10 127 F*	*8 382 F*
Species total	*5 418 F*	*3 875*	*6 902*	*4 451 F*	*5 716 F*	*8 385 F*	*9 872 F*	*9 514 F*	*10 127 F*	*8 382 F*
Knifefishes	**...B**		**...C**		***Notopterus spp***			**1,28(02)003,XX**		**FKN**
01 Nigeria	...	703	383	10	1	144	33	39	22	19
01 Fishing area total	*...*	*703*	*383*	*10*	*1*	*144*	*33*	*39*	*22*	*19*
04 Indonesia	4 256	3 866	4 790	3 880	3 455	3 599	3 845	4 255	3 820	3 800
04 Fishing area total	*4 256*	*3 866*	*4 790*	*3 880*	*3 455*	*3 599*	*3 845*	*4 255*	*3 820*	*3 800*
Species total	*4 256*	*4 569*	*5 173*	*3 890*	*3 456*	*3 743*	*3 878*	*4 294*	*3 842*	*3 819*
Elephantsnout fishes nei	**...B**		**...C**		***Mormyridae***			**1,28(06)XXX,XX**		**OEY**
01 Mali	...	...	...	...	...	...	...	7 691	7 000 F	7 000 F
Nigeria	...	...	6 439	...	4 481	18 781	16 030	...	17 509	5 567
01 Fishing area total	*...*	*...*	*6 439*	*...*	*4 481*	*18 781*	*16 030*	*7 691*	*24 509 F*	*12 567 F*
Species total	*...*	*...*	*6 439*	*...*	*4 481*	*18 781*	*16 030*	*7 691*	*24 509 F*	*12 567 F*
Aba	**...B**		**...C**		***Gymnarchus niloticus***			**1,28(07)001,01**		**OGN**
01 Nigeria	...	...	3 179	1 108	3 952	7 685	5 887	...	...	9 097
01 Fishing area total	*...*	*...*	*3 179*	*1 108*	*3 952*	*7 685*	*5 887*	*...*	*...*	*9 097*
Species total	*...*	*...*	*3 179*	*1 108*	*3 952*	*7 685*	*5 887*	*...*	*...*	*9 097*
Cachama	**...B**		**Cachama**		***Colossoma macropomum***			**1,38(01)046,05**		**CSM**
03 Brazil	...	...	11 379	4 298	2 838	2 760	2 905	4 965	3 789	4 068
Peru	...	...	...	693	762	254	1 016	963	904	...
03 Fishing area total	*...*	*...*	*11 379*	*4 991*	*3 600*	*3 014*	*3 921*	*5 928*	*4 693*	*4 068*
Species total	*...*	*...*	*11 379*	*4 991*	*3 600*	*3 014*	*3 921*	*5 928*	*4 693*	*4 068*
Pirapatinga	**...B**		**Cachama blanca**		***Piaractus brachypomus***			**1,38(01)076,01**		**CSD**
03 Peru	...	...	...	397	...	166	648	324	487	...
03 Fishing area total	*...*	*...*	*...*	*397*	*...*	*166*	*648*	*324*	*487*	*...*
Species total	*...*	*...*	*...*	*397*	*...*	*166*	*648*	*324*	*487*	*...*
Characins nei	**Characinidés nca**		**Carácidos nep**		***Characidae***			**1,38(01)XXX,XX**		**CHA**
01 Mali	3 215	3 142	6 650	9 504	6 969	8 316	8 362	5 493	5 000 F	5 000 F
Nigeria	4 309	7 169	8 802	7 141	7 646	19 835	11 883	12 500	10 274	6 989
Uganda	10 033	10 000	10 400	12 160	11 718	11 744	13 165	10 331	10 331	7 100
01 Fishing area total	*17 557*	*20 311*	*25 852*	*28 805*	*26 333*	*39 895*	*33 410*	*28 324*	*25 605 F*	*19 089 F*
03 Argentina	8 000	8 700 F	11 700 F	13 000 F	15 500 F	16 000 F	2 800 F	3 000 F	3 400 F	3 100 F
Brazil	83 000 F	81 000 F	81 433	80 598	65 273	60 983	65 149	67 297	22 850	22 000
Colombia	12 834	8 832	5 698	5 504	3 730	6 427	8 667	6 600 F	6 600 F	6 600 F
Paraguay	7 000 F	7 350 F	7 700 F	8 000 F	10 000 F	9 000 F	9 000 F	9 000 F	9 000 F	9 000 F
Uruguay	439	872	763	338	1 924	1 687	1 762	1 690	107	260
Venezuela	4 878	5 827	8 778	8 050	6 029	6 921	4 817	5 773	2 978	4 014
03 Fishing area total	*116 151 F*	*112 581 F*	*116 072 F*	*115 490 F*	*102 456 F*	*101 018 F*	*92 195 F*	*93 360 F*	*44 935 F*	*44 974 F*
Species total	*133 708 F*	*132 892 F*	*141 924 F*	*144 295 F*	*128 789 F*	*140 913 F*	*125 605 F*	*121 684 F*	*70 540 F*	*64 063 F*
Netted prochilod	**Prochilode réticulé**		**Boquichico reticulado**		***Prochilodus reticulatus***			**1,38(12)001,20**		**PLR**
03 Peru	...	...	5 088	7 448	2 007	6 117	9 768	10 942	11 220	10 133
03 Fishing area total	*...*	*...*	*5 088*	*7 448*	*2 007*	*6 117*	*9 768*	*10 942*	*11 220*	*10 133*
Species total	*...*	*...*	*5 088*	*7 448*	*2 007*	*6 117*	*9 768*	*10 942*	*11 220*	*10 133*
Prochilods nei	**Prochilodes nca**		**Sábalos sudamericanos nep**		***Prochilodus spp***			**1,38(12)001,XX**		**PRL**
03 Argentina	...	...	...	...	...	...	16 000 F	17 700 F	19 600 F	18 200 F
Venezuela	9 767	16 907	20 430	18 740	10 927	15 670	9 320	9 613	6 743	14 061
03 Fishing area total	*9 767*	*16 907*	*20 430*	*18 740*	*10 927*	*15 670*	*25 320 F*	*27 313 F*	*26 343 F*	*32 261 F*
Species total	*9 767*	*16 907*	*20 430*	*18 740*	*10 927*	*15 670*	*25 320 F*	*27 313 F*	*26 343 F*	*32 261 F*
Wels(=Som)catfish	**Silure glane**		**Siluro**		***Silurus glanis***			**1,41(07)031,01**		**SOM**
04 Azerbaijan	63	19	6	11	9	8	8	9	8	4
Georgia	1	0	2	-	-	-	-	-	-	-
Kazakhstan	1 855	1 496	1 513	1 380 F	990 F	700 F	750 F	750 F	780	608
Tajikistan	15	7	12	...	...	...	...	...	10	12
Turkey	723	857	896	705	1 000	1 000	958	1 019	813	987
Turkmenistan	42	54	33	42	7	8	9	7	3	2
Uzbekistan	13	4	20	10 F	9	14	16	8	16	0
04 Fishing area total	*2 712*	*2 437*	*2 482*	*2 148 F*	*2 015 F*	*1 730 F*	*1 741 F*	*1 793 F*	*1 630*	*1 613*
05 Bulgaria	18	20 F	30	27	34	44	106	59	4	34
Croatia	10	20	19	24	20	1 F	1	1	1	2

B-13 Miscellaneous freshwater fishes / Poissons d'eau douce divers / Peces de agua dulce diversos

Capture production by species, fishing areas and countries or areas
Captures par espèces, zones de pêche et pays ou zones
Capturas por especies, áreas de pesca y países o áreas

Species, Fishing area Espèce, Zone de pêche Especie, Area de pesca	1993 t	1994 t	1995 t	1996 t	1997 t	1998 t	1999 t	2000 t	2001 t	2002 t
Czech Rep	49	38	42	36	44	49	52	53	57	61
Germany	55	12	12	12	12	12	12	12	12	12
Greece	...	...	...	11	13	20	14	15	18	20
Hungary	126	153	205	201	121	113	145	104	120	134
Latvia	...	1	1	1	2	1	2	-	-	-
Lithuania	-	0	0	0	-	-	-	0	0	0
Poland	6	0	0	0	0	1	1	7	4	4
Romania	66	113	72	22	9	58	87	73	116	120
Russian Fed	3 923	3 915	4 590	5 864	7 026	9 305	7 031	7 199	6 540	5 978
Serbia-Monte	...	...	113	60	66	49	47	47	33	52
Slovakia	14	20	26	21	21	20	20	22	28	29
Slovenia	5	6	6	4	7	7	5	7	6	8
Switzerland	1	1	1	1	1	0	1	1	1	1
Ukraine	0	59	1	2	1	1	1	3	3	2
05 Fishing area total	*4 273*	*4 358 F*	*5 118*	*6 286*	*7 377*	*9 681 F*	*7 525*	*7 603*	*6 943*	*6 457*
Species total	*6 985*	*6 795 F*	*7 600*	*8 434 F*	*9 392 F*	*11 411 F*	*9 266 F*	*9 396 F*	*8 573*	*8 070*
Glass catfishes	**...B**		**...C**		***Kryptopterus spp***				**1,41(07)050,XX**	**CAG**
04 Indonesia	16 466	15 602	15 283	17 560	15 938	13 943	15 927	13 110	13 522	14 980
04 Fishing area total	*16 466*	*15 602*	*15 283*	*17 560*	*15 938*	*13 943*	*15 927*	*13 110*	*13 522*	*14 980*
Species total	*16 466*	*15 602*	*15 283*	*17 560*	*15 938*	*13 943*	*15 927*	*13 110*	*13 522*	*14 980*
Bagrid catfish	**...B**		**...C**		***Chrysichthys nigrodigitatus***				**1,41(08)078,02**	**CSR**
01 Nigeria	-	2 000	4 000	8 505	5 814	15 538	10 679	7 311	12 413	5 678
01 Fishing area total	-	*2 000*	*4 000*	*8 505*	*5 814*	*15 538*	*10 679*	*7 311*	*12 413*	*5 678*
Species total	-	*2 000*	*4 000*	*8 505*	*5 814*	*15 538*	*10 679*	*7 311*	*12 413*	*5 678*
Black catfishes nei	**...B**		**...C**		***Chrysichthys spp***				**1,41(08)078,XX**	**CST**
01 Benin	1 100 F	1 143	1 289	1 300 F	1 400 F	1 500 F	1 542	1 300 F	1 500 F	1 500 F
Gabon	1 000 F	1 500 F	2 000 F	2 124	1 131	1 950	380	887	797	797
Mali	2 572	2 514	5 320	4 172	3 982	3 650	3 670	4 395	4 000 F	4 000 F
Senegal	...	...	...	...	...	...	...	1 218	1 100 F	1 100 F
01 Fishing area total	*4 672 F*	*5 157 F*	*8 609 F*	*7 596 F*	*6 513 F*	*7 100 F*	*5 592*	*7 800 F*	*7 397 F*	*7 397 F*
Species total	*4 672 F*	*5 157 F*	*8 609 F*	*7 596 F*	*6 513 F*	*7 100 F*	*5 592*	*7 800 F*	*7 397 F*	*7 397 F*
Naked catfishes	**...B**		**...C**		***Bagrus spp***				**1,41(08)111,XX**	**CAN**
01 Ethiopia	...	...	...	...	...	94	107	120	119	98
Kenya	34	37	127	157	158	107	147	161	178	183
Nigeria	8 903	5 944	6 531	3 594	4 726	5 463	5 421	4 539	4 902	920
Tanzania	1 954	1 801	2 900	1 136	2 800	2 500	2 000	2 300	2 000	2 000
Uganda	10 110	4 681	4 879	4 978	4 774	5 222	8 564	4 375	4 375	4 800
01 Fishing area total	*21 001*	*12 463*	*14 437*	*9 865*	*12 458*	*13 386*	*16 239*	*11 495*	*11 574*	*8 001*
Species total	*21 001*	*12 463*	*14 437*	*9 865*	*12 458*	*13 386*	*16 239*	*11 495*	*11 574*	*8 001*
Catfishes nei	**Barbottes nca**		**Bagres nep**		***Ictalurus spp***				**1,41(10)002,XX**	**CAF**
02 Canada	39	43	37	37	29	27	35	19	344	338
El Salvador	205	178	250	240	230	213	208	142	169	125
Mexico	3 815	3 165	4 423	5 358	4 691	4 027	4 232	3 845	3 135	2 263
USA	5 463	5 262	5 264	4 110	6 872	5 257	9 235	7 561	7 478	6 301
02 Fishing area total	*9 522*	*8 648*	*9 974*	*9 745*	*11 822*	*9 524*	*13 710*	*11 567*	*11 126*	*9 027*
Species total	*9 522*	*8 648*	*9 974*	*9 745*	*11 822*	*9 524*	*13 710*	*11 567*	*11 126*	*9 027*
Brown bullhead	**...B**		**...C**		***Ameiurus nebulosus***				**1,41(10)004,07**	**ITE**
06 New Zealand	0	0	1	1	5	12	14	10	8	7
06 Fishing area total	*0*	*0*	*1*	*1*	*5*	*12*	*14*	*10*	*8*	*7*
Species total	*0*	*0*	*1*	*1*	*5*	*12*	*14*	*10*	*8*	*7*
North African catfish	**Poisson-chat nord-africain**		**Pez-gato**		***Clarias gariepinus***				**1,41(18)030,03**	**CLZ**
01 Ethiopia	...	...	643	812	1 200	1 677	1 917	4 000	3 926	4 318
Lesotho	2 F	2 F	2 F	2 F	2 F	2 F	2 F	2 F	2	2 F
Mali	16 075	15 712	33 210	17 129	14 933	15 009	15 091	27 468	25 000 F	25 000 F
Nigeria	-	-	-	2 658	8 568	5 268	9 994	4 474	10 419	7 375
01 Fishing area total	*16 077 F*	*15 714 F*	*33 855 F*	*20 601 F*	*24 703 F*	*21 956 F*	*27 004 F*	*35 944 F*	*39 347 F*	*36 695 F*
04 Turkey	759	...	...	...	...	...	216	576	520	495
04 Fishing area total	*759*	...	...	...	...	...	*216*	*576*	*520*	*495*
Species total	*16 836 F*	*15 714 F*	*33 855 F*	*20 601 F*	*24 703 F*	*21 956 F*	*27 220 F*	*36 520 F*	*39 867 F*	*37 190 F*
Mudfish	**...B**		**...C**		***Clarias anguillaris***				**1,41(18)030,07**	**CLN**
01 Egypt	23 680	19 895	19 608	20 370	22 433	21 618	21 511	31 506	39 535	39 208
01 Fishing area total	*23 680*	*19 895*	*19 608*	*20 370*	*22 433*	*21 618*	*21 511*	*31 506*	*39 535*	*39 208*
Species total	*23 680*	*19 895*	*19 608*	*20 370*	*22 433*	*21 618*	*21 511*	*31 506*	*39 535*	*39 208*

B-13 Miscellaneous freshwater fishes / Poissons d'eau douce divers / Peces de agua dulce diversos

Capture production by species, fishing areas and countries or areas / Captures par espèces, zones de pêche et pays ou zones / Capturas por especies, áreas de pesca y países o áreas

Species, Fishing area Espèce, Zone de pêche Especie, Area de pesca	1993 t	1994 t	1995 t	1996 t	1997 t	1998 t	1999 t	2000 t	2001 t	2002 t
Torpedo-shaped catfishes nei	**...B**		**...C**		***Clarias spp***				**1,41(18)030,XX**	**CTO**
01 Benin	1 600 F	1 399	2 325	2 100 F	1 900 F	1 700 F	1 537	1 300 F	1 500 F	1 500 F
Kenya	254	581	574	339	1 724	1 532	1 570	1 592	1 611	1 581
Malawi	...	...	...	4 556	...	3 935	8 471	7 600 F	6 367	6 478
Nigeria	13 292	13 585	20 745	8 480	10 202	17 220	10 963	14 012	10 925	7 640
Tanzania	4 531	1 613	7 920	2 107	7 750	7 500	2 500	2 000	2 500	2 000
Uganda	3 616	2 380	2 284	1 916	2 130	2 170	2 833	2 987	2 987	2 500
01 Fishing area total	*23 293 F*	*19 558*	*33 848*	*19 498 F*	*23 706 F*	*34 057 F*	*27 874*	*29 491 F*	*25 890 F*	*21 699 F*
04 China,Taiwan	-	-	-	-	-	6	-	-	-	-
Indonesia	7 603	7 485	9 911	8 385	7 257	7 535	6 413	5 496	7 609	7 110
Philippines	6 574	2 100	2 669	2 696	2 396	1 628	2 058	2 200	2 366	2 601
Thailand	8 137	7 100	8 080	5 800	3 410	10 934	12 111	19 600	14 100	14 300
04 Fishing area total	*22 314*	*16 685*	*20 660*	*16 881*	*13 063*	*20 103*	*20 582*	*27 296*	*24 075*	*24 011*
Species total	*45 607 F*	*36 243*	*54 508*	*36 379 F*	*36 769 F*	*54 160 F*	*48 456*	*56 787 F*	*49 965 F*	*45 710 F*
Pangas catfishes nei	**...B**		**...C**		***Pangasius spp***				**1,41(30)002,XX**	**PGZ**
04 Thailand	1 110	1 311	1 000	541	522	917	1 061	1 274	1 100	1 200
04 Fishing area total	*1 110*	*1 311*	*1 000*	*541*	*522*	*917*	*1 061*	*1 274*	*1 100*	*1 200*
Species total	*1 110*	*1 311*	*1 000*	*541*	*522*	*917*	*1 061*	*1 274*	*1 100*	*1 200*
Upsidedown catfishes	**...B**		**...C**		***Synodontis spp***				**1,41(32)008,XX**	**CSY**
01 Benin	130 F	21	125	120 F	110 F	110 F	110 F	100 F	110 F	110 F
Egypt	...	...	...	...	...	...	...	2 678	3 937	5 520
Kenya	25	37	28	24	408	418	506	443	48	142
Mali	1 929	1 886	3 990	4 657	5 973	4 075	4 097	3 296	3 000 F	3 000 F
Nigeria	3 781	10 909	9 709	4 466	5 069	8 781	8 143	6 054	6 642	4 055
01 Fishing area total	*5 865 F*	*12 853*	*13 852*	*9 267 F*	*11 560 F*	*13 384 F*	*12 856 F*	*12 571 F*	*13 737 F*	*12 827 F*
Species total	*5 865 F*	*12 853*	*13 852*	*9 267 F*	*11 560 F*	*13 384 F*	*12 856 F*	*12 571 F*	*13 737 F*	*12 827 F*
Freshwater siluroids nei	**Silurides d'eau douce nca**		**Siluroideos de agua dulce nep**		***Siluroidei***				**1,41(XX)XXX,XX**	**FSI**
01 Togo	500	500	500	500	500	500	500	500	500	500
Uganda	1 365	1 500	1 600	120	868	887	1 000	...	...	...
01 Fishing area total	*1 865*	*2 000*	*2 100*	*620*	*1 368*	*1 387*	*1 500*	*500*	*500*	*500*
03 Argentina	1 800	1 900 F	2 500 F	2 800 F	3 000 F	3 000 F	3 500 F	3 500 F	4 100 F	3 800 F
Brazil	43 900 F	44 900 F	40 048	47 302	52 601	49 866	57 014	58 474	58 748	59 641
Colombia	473	6 414	8 718	1 392	2 648	9 215	10 544	10 454 F	10 600 F	10 600 F
Paraguay	8 500 F	9 000 F	9 500 F	10 000 F	13 000 F	12 000 F	12 000 F	12 000 F	12 000 F	12 000 F
Uruguay	32	21	36	72	252	203	319	293	81	83
Venezuela	11 216	9 890	19 497	18 143	15 248	14 833	7 470	6 226	8 967	8 675
03 Fishing area total	*65 921 F*	*72 125 F*	*80 299 F*	*79 709 F*	*86 749 F*	*89 117 F*	*90 847 F*	*90 947 F*	*94 496 F*	*94 799 F*
04 India	...	...	...	...	...	63 724	71 152	64 179	57 999	67 621
Indonesia	11 941	13 451	13 215	13 508	10 117	12 466	13 854	13 630	13 203	13 490
Iran	670	28	5	22	6	32	24	25	1	21
Iraq	500	1 338	1 416	1 436	1 738	711	1 402	953	900 F	900 F
Korea Rep	...	-	-	279	251	...	136	...	...	...
04 Fishing area total	*13 111*	*14 817*	*14 636*	*15 245*	*12 112*	*76 933*	*86 568*	*78 787*	*72 103 F*	*82 032 F*
Species total	*80 897 F*	*88 942 F*	*97 035 F*	*95 574 F*	*100 229 F*	*167 437 F*	*178 915 F*	*170 234 F*	*167 099 F*	*177 331 F*
Burbot	**Lotte de rivière**		**Lota**		***Lota lota***				**1,48(04)003,01**	**FBU**
02 USA	15	16	15	14	9	28	13	20	9	10
02 Fishing area total	*15*	*16*	*15*	*14*	*9*	*28*	*13*	*20*	*9*	*10*
05 Belarus	28	16	14	17	13	18	20	7	29	38
Bulgaria	-	-	-	-	-	-	-	1	2	5
Czech Rep	2	-	-	-	-	-	-	-	-	-
Denmark	0	0	0	0	0	0	0	0	0	0
Estonia	16	8	29	33	27	18	52	38	37	41
Finland	1 765	1 384	1 384	1 227	1 227	679	679	831	831	831
Germany	-	2	2	2	2	2	2	2	2	2
Lithuania	4	3	5	5	14	10	13	13	9	3
Norway	...	...	...	...	1	1	...	...	...	...
Poland	0	0	0	0	-	-	-	0	-	-
Russian Fed	1 746	881	993	1 531	1 533	2 121	1 763	1 765	1 628	1 683
Slovakia	1	2	3	1	3	4	4	2	4	4
Sweden	83	112	93	66	56	61	59	-	-	-
Switzerland	12	18	11	6	9	6	7	8	8	10
05 Fishing area total	*3 657*	*2 426*	*2 534*	*2 888*	*2 885*	*2 920*	*2 599*	*2 667*	*2 550*	*2 617*
27 Estonia	1	3	2	3	4	3	1	2	1	1
Finland	724	517	465	420	436	468	475	345	337	330
Germany	-	-	-	-	1	1	1	1	2	2
Lithuania	...	...	...	...	...	...	...	...	...	12
Poland	-	-	-	-	9	10	12	18	26	38
Russian Fed	-	-	-	-	-	-	-	13	16	14
Sweden	9	12	11	9	10	7	6	4	3	2
27 Fishing area total	*734*	*532*	*478*	*432*	*460*	*489*	*495*	*383*	*385*	*399*

B-13 Miscellaneous freshwater fishes / Poissons d'eau douce divers / Peces de agua dulce diversos

Capture production by species, fishing areas and countries or areas / Captures par espèces, zones de pêche et pays ou zones / Capturas por especies, áreas de pesca y países o áreas

Species, Fishing area Espèce, Zone de pêche Especie, Area de pesca	1993 t	1994 t	1995 t	1996 t	1997 t	1998 t	1999 t	2000 t	2001 t	2002 t
Species total	*4 406*	*2 974*	*3 027*	*3 334*	*3 354*	*3 437*	*3 107*	*3 070*	*2 944*	*3 026*
Lai	**...B**		**...C**		***Monopterus albus***				**1,68(02)001,01**	**FLT**
04 Thailand	-	-	-	-	-	-	-	162	100	100
04 Fishing area total	-	-	-	-	-	-	-	*162*	*100*	*100*
Species total	-	-	-	-	-	-	-	*162*	*100*	*100*
Nile perch	**Perche du Nil**		**Perca del Nilo**		***Lates niloticus***				**1,70(01)167,07**	**NIP**
01 Egypt	...	...	2 495	2 730	2 551	3 255	2 572	3 278	5 328	5 775
Ethiopia	...	...	908	270	230	191	75	65	63	37
Kenya	100 037	104 102	102 546	97 145	73 555	76 663	103 014	109 815	78 534	58 432
Mali	3 858	3 771	7 980	5 712	4 978	5 024	5 052	6 592	6 000 F	6 000 F
Nigeria	9 062	6 118	4 993	3 746	4 224	5 250	6 366	4 447	6 139	3 030
Senegal	...	...	...	...	...	...	...	1 451	1 300 F	1 300 F
Tanzania	156 401	123 557	155 860	121 161	152 000	150 000	100 000	90 000	96 000	92 000
Uganda	95 005	101 208	92 722	81 253	91 706	98 800	89 203	87 257	88 881	90 698
01 Fishing area total	*364 363*	*338 756*	*367 504*	*312 017*	*329 244*	*339 183*	*306 282*	*302 905*	*282 245 F*	*257 272 F*
Species total	*364 363*	*338 756*	*367 504*	*312 017*	*329 244*	*339 183*	*306 282*	*302 905*	*282 245 F*	*257 272 F*
Freshwater perches nei	**Perches d'eau douce nca**		**Percas de agua dulce nep**		***Lates spp***				**1,70(01)167,XX**	**PEX**
01 Burundi	2 000 F	2 650 F	2 888	1 205	2 378	3 925	1 523	3 070	1 657	1 660 F
01 Fishing area total	*2 000 F*	*2 650 F*	*2 888*	*1 205*	*2 378*	*3 925*	*1 523*	*3 070*	*1 657*	*1 660 F*
Species total	*2 000 F*	*2 650 F*	*2 888*	*1 205*	*2 378*	*3 925*	*1 523*	*3 070*	*1 657*	*1 660 F*
Largemouth black bass	**...B**		**Perca atruchada**		***Micropterus salmoides***				**1,70(10)014,02**	**MPS**
02 Dominican Rp	...	...	...	...	...	...	...	...	...	425
Mexico	...	996	1 212	1 014	1 058	684	784	903	693	766
02 Fishing area total	...	*996*	*1 212*	*1 014*	*1 058*	*684*	*784*	*903*	*693*	*1 191*
Species total	...	*996*	*1 212*	*1 014*	*1 058*	*684*	*784*	*903*	*693*	*1 191*
European perch	**Perche européenne**		**Perca**		***Perca fluviatilis***				**1,70(14)002,01**	**FPE**
04 Kazakhstan	817	728	736	670 F	480 F	350 F	450 F	500 F	547	422
04 Fishing area total	*817*	*728*	*736*	*670 F*	*480 F*	*350 F*	*450 F*	*500 F*	*547*	*422*
05 Belarus	215	98	93	101	75	49	38	34	86	105
Belgium	25	25	25	25	25	25	15	25	25	25
Bulgaria	...	...	...	...	63	81	102	26	6	23
Czech Rep	39	38	31	33	34	36	37	34	34	30
Denmark	37	32	27	21	16	35	35	24	13	7
Estonia	525	672	623	634	887	815	680	561	300	246
Finland	13 754	13 043	13 028	12 634	12 634	9 470	9 470	9 599	9 599	9 599
Germany	277	215	215	215	215	215	215	215	215	215
Greece	...	...	...	24	15	19	24	15	23	20
Ireland	200	...	...	...	...	200	...	...	...	...
Latvia	16	20	27	22	29	34	34	34	34	36
Lithuania	92	115	114	83	114	104	116	115	114	84
Netherlands	340	136	219	376	336	155	177	170 F	150 F	131
Norway	...	...	...	...	9	3	...	...	...	...
Poland	120	120	224	601	179	163	186	297	245	267
Romania	-	-	39	38	8	40	60	28	42	32
Russian Fed	2 221	1 976	2 234	2 628	3 527	4 459	3 455	3 726	3 824	6 649
Slovakia	18	30	32	18	11	13	13	13	14	13
Slovenia	1	2	2	1	1	1	0	0	1	1
Sweden	220	211	216	173	278	251	171	150	135	169
Switzerland	460	396	373	475	364	422	491	395	262	288
Ukraine	125	288	319	126	80	107	79	155	74	64
05 Fishing area total	*18 685*	*17 417*	*17 841*	*18 228*	*18 900*	*16 697*	*15 398*	*15 616 F*	*15 196 F*	*18 004*
27 Denmark	75	72	63	42	67	65	78	76	60	72
Estonia	940	551	384	396	315	237	296	280	386	578
Finland	6 143	3 670	3 848	5 013	5 226	5 251	5 224	3 775	3 796	3 877
Germany	536	672	733	549	567	607	499	286	326	263
Latvia	54	34	37	34	27	21	53	24	49	59
Lithuania	...	...	...	...	...	...	...	...	1	48
Poland	404	744	728	568	1 405	1 009	938	625	874	790
Russian Fed	-	-	-	-	-	-	-	-	161	69
Sweden	112	145	127	94	108	120	130	81	66	81
27 Fishing area total	*8 264*	*5 888*	*5 920*	*6 696*	*7 715*	*7 310*	*7 218*	*5 147*	*5 719*	*5 837*
37 Russian Fed	-	-	-	-	-	-	-	-	1	1
37 Fishing area total	-	-	-	-	-	-	-	-	*1*	*1*
Species total	*27 766*	*24 033*	*24 497*	*25 594 F*	*27 095 F*	*24 357 F*	*23 066 F*	*21 263 F*	*21 463 F*	*24 264*
American yellow perch	**Perchaude**		**Perca canadiense**		***Perca flavescens***				**1,70(14)002,02**	**FPY**
02 Canada	...	1 439	1 282	1 440	2 073	2 083	1 884	1 847	2 524	3 622
USA	1 412	1 387	1 246	711	622	556	537	567	640	685
02 Fishing area total	*1 412*	*2 826*	*2 528*	*2 151*	*2 695*	*2 639*	*2 421*	*2 414*	*3 164*	*4 307*

B-13 Miscellaneous freshwater fishes / Poissons d'eau douce divers / Peces de agua dulce diversos

Capture production by species, fishing areas and countries or areas / Captures par espèces, zones de pêche et pays ou zones / Capturas por especies, áreas de pesca y países o áreas

Species, Fishing area Espèce, Zone de pêche Especie, Area de pesca	1993 t	1994 t	1995 t	1996 t	1997 t	1998 t	1999 t	2000 t	2001 t	2002 t
Species total	*1 412*	*2 826*	*2 528*	*2 151*	*2 695*	*2 639*	*2 421*	*2 414*	*3 164*	*4 307*
Percarina	**Percarina**		**Percarina**		***Percarina demidoffi***			**1,70(14)007,01**		**PND**
37 Ukraine	-	-	-	-	-	-	-	-	-	18
37 Fishing area total	-	-	-	-	-	-	-	-	-	*18*
Species total	-	-	-	-	-	-	-	-	-	*18*
Walleye	**Sandre américain**		**Lucioperca americana**		***Stizostedion vitreum***			**1,70(14)015,01**		**STV**
02 Canada	8 160	7 543	7 832	7 718	7 751	8 206	8 782	9 160	6 949	7 007
USA	8	22	17	18	4	6	2	4	10	14
02 Fishing area total	*8 168*	*7 565*	*7 849*	*7 736*	*7 755*	*8 212*	*8 784*	*9 164*	*6 959*	*7 021*
Species total	*8 168*	*7 565*	*7 849*	*7 736*	*7 755*	*8 212*	*8 784*	*9 164*	*6 959*	*7 021*
Sauger	**Sandre canadien**		**Lucioperca canadiense**		***Stizostedion canadense***			**1,70(14)015,02**		**SZC**
02 USA	3	1	4	1	-	0	-	-	-	-
02 Fishing area total	*3*	*1*	*4*	*1*	...	*0*	...	...	...	...
Species total	*3*	*1*	*4*	*1*	...	*0*	...	...	...	...
Pike-perch	**Sandre**		**Lucioperca**		***Stizostedion lucioperca***			**1,70(14)015,03**		**FPP**
04 Azerbaijan	33	27	19	22	11	7	5	5	19	35
Iran	16	95	10	6	4	105	19	20	26	32
Kazakhstan	5 983	5 436	6 089	5 570 F	4 000 F	2 900 F	2 500 F	2 000 F	1 628	1 718
Kyrgyzstan	15	17	7	6 F	5 F	3 F	2 F	2 F	1	2
Tajikistan	87	22	20	...	...	...	...	...	21	24
Turkey	5 683	5 181	5 877	8 042	1 500	3 000	1 906	1 633	1 644	1 850
Turkmenistan	69	78	47	30	28	109	87	96	42	34
Uzbekistan	379	161	282	130 F	117	175	118	127	136	30
04 Fishing area total	*12 265*	*11 017*	*12 351*	*13 806 F*	*5 665 F*	*6 299 F*	*4 637 F*	*3 883 F*	*3 517*	*3 725*
05 Albania	25 F	20 F	5	10	4	-	-	-	45	30
Belarus	67	8	9	6	3	3	10	17	24	36
Belgium	15	15	15	15	15	15	10	15	15	15
Bulgaria	29	30 F	22	26	45	54	68	27	5	18
Croatia	13	12	12	12	14	1 F	1	1	1	1
Czech Rep	101	164	86	130	157	125	130	134	139	144
Denmark	12	10	13	11	39	38	13	19	5	7
Estonia	554	460	393	393	280	750	655	652	484	938
Finland	1 454	978	994	1 128	1 128	985	985	759	759	759
Germany	582	251	251	251	251	251	251	251	251	251
Hungary	212	243	226	224	199	156	169	200	196	190
Latvia	-	17	22	19	20	21	33	35	38	44
Lithuania	73	55	44	56	65	51	60	78	87	11
Moldova Rep	5	4	6	5	4	1	2	1	12	19
Netherlands	44	185	79	100	89	61	104	160 F	250 F	295
Norway	...	...	...	...	5	3	...	...	...	...
Poland	230	230	208	289	150	128	136	203	162	214
Romania	361	859	353	94	30	78	154	155	92	118
Russian Fed	4 975	3 212	3 271	3 994	4 082	3 161	3 644	3 863	3 427	4 036
Slovakia	52	89	94	76	70	65	64	56	62	79
Slovenia	1	5	5	5	5	5	4	5	4	4
Sweden	356	354	298	279	311	287	307	291	241	371
Switzerland	6	10	13	10	6	7	6	5	9	9
Ukraine	409	291	159	210	152	163	94	58	93	247
UK	1	3	1	1	0	0	2	1	1	-
05 Fishing area total	*9 577 F*	*7 505 F*	*6 579*	*7 344*	*7 124*	*6 409 F*	*6 902*	*6 986 F*	*6 402 F*	*7 836*
27 Denmark	1	0	1	1	12	15	2	9	4	2
Estonia	458	167	264	333	180	141	116	25	33	39
Finland	1 192	1 294	1 352	1 348	1 502	2 256	2 203	1 065	1 027	1 222
Germany	719	327	319	295	309	240	273	257	243	273
Latvia	48	33	35	54	20	17	25	13	21	34
Lithuania	1	...	...	...	...	...	...	...	28	162
Netherlands	-	-	-	-	-	-	-	1	4	7
Poland	242	266	232	223	330	253	401	343	316	295
Russian Fed	-	-	-	-	-	-	-	422	409	428
Sweden	89	71	53	33	55	47	39	36	26	27
27 Fishing area total	*2 750*	*2 158*	*2 256*	*2 287*	*2 408*	*2 969*	*3 059*	*2 171*	*2 111*	*2 489*
37 Russian Fed	546	866	950	2 221	2 022	2 229	1 676	1 843	1 875	1 577
Ukraine	100	217	353	601	803	802	892	1 113	1 629	1 716
37 Fishing area total	*646*	*1 083*	*1 303*	*2 822*	*2 825*	*3 031*	*2 568*	*2 956*	*3 504*	*3 293*
Species total	*25 238 F*	*21 763 F*	*22 489*	*26 259 F*	*18 022 F*	*18 708 F*	*17 166 F*	*15 996 F*	*15 534 F*	*17 343*
Walleyes nei	**Sandres nca**		**Luciopercas nep**		***Stizostedion spp***			**1,70(14)015,XX**		**STP**
05 Ukraine	...	3	-	-	-	-	-	-	-	-
05 Fishing area total	...	*3*	-	-	-	-	-	-	-	-
Species total	...	*3*	-	-	-	-	-	-	-	-

B-13 Miscellaneous freshwater fishes / Poissons d'eau douce divers / Peces de agua dulce diversos

Capture production by species, fishing areas and countries or areas / Captures par espèces, zones de pêche et pays ou zones / Capturas por especies, áreas de pesca y países o áreas

Species, Fishing area Espèce, Zone de pêche Especie, Area de pesca	1993 t	1994 t	1995 t	1996 t	1997 t	1998 t	1999 t	2000 t	2001 t	2002 t
Ruffe	**Grémille**		**Acerina**		***Gymnocephalus cernuus***				**1,70(14)059,01**	**ACC**
05 Denmark	2	5	15	13	5	9	3	2	1	1
Latvia	-	5	4	5	5	-	1	-	-	-
Lithuania	-	-	-	-	-	-	52	97	64	0
05 Fishing area total	*2*	*10*	*19*	*18*	*10*	*9*	*56*	*99*	*65*	*1*
27 Lithuania	...	...	...	...	...	...	...	...	...	31
27 Fishing area total	...	...	...	...	...	...	...	...	...	*31*
Species total	*2*	*10*	*19*	*18*	*10*	*9*	*56*	*99*	*65*	*32*
Freshwater drum	**Malachigan**		**...C**		***Aplodinotus grunniens***				**1,70(37)011,01**	**AGR**
02 USA	495	611	470	368	549	573	470	577	506	502
02 Fishing area total	*495*	*611*	*470*	*368*	*549*	*573*	*470*	*577*	*506*	*502*
Species total	*495*	*611*	*470*	*368*	*549*	*573*	*470*	*577*	*506*	*502*
Gudgeons, sleepers nei	**Gudgeons, dormeurs nca**		**Durmientes nep**		***Eleotridae***				**1,73(20)XXX,XX**	**FGB**
04 Indonesia	1 080	1 207	1 514	1 109	1 290	1 078	1 243	1 125	2 604	3 080
04 Fishing area total	*1 080*	*1 207*	*1 514*	*1 109*	*1 290*	*1 078*	*1 243*	*1 125*	*2 604*	*3 080*
06 Papua N Guin	1 850 F	1 850 F	1 850	1 850 F	1 850 F	1 850 F	1 850 F	1 850 F	1 850 F	1 850 F
06 Fishing area total	*1 850 F*	*1 850 F*	*1 850*	*1 850 F*	*1 850 F*	*1 850 F*	*1 850 F*	*1 850 F*	*1 850 F*	*1 850 F*
Species total	*2 930 F*	*3 057 F*	*3 364*	*2 959 F*	*3 140 F*	*2 928 F*	*3 093 F*	*2 975 F*	*4 454 F*	*4 930 F*
Freshwater gobies nei	**Gobies d'eau douce nca**		**Góbidos de agua dulce nep**		***Gobiidae***				**1,73(21)XXX,XX**	**FGX**
01 Benin	1 100 F	1 302	914	900 F	850 F	800 F	800 F	700 F	800 F	800 F
01 Fishing area total	*1 100 F*	*1 302*	*914*	*900 F*	*850 F*	*800 F*	*800 F*	*700 F*	*800 F*	*800 F*
04 Philippines	9 160	4 466	3 431	3 585	4 300	3 803	4 027	4 563	4 280	5 920
Turkey	189	230	262	185	200	200	118	107	116	85
04 Fishing area total	*9 349*	*4 696*	*3 693*	*3 770*	*4 500*	*4 003*	*4 145*	*4 670*	*4 396*	*6 005*
05 Ukraine	-	...	...	...	...	60	16	...	...	70
05 Fishing area total	-	...	...	...	...	*60*	*16*	...	...	*70*
Species total	*10 449 F*	*5 998*	*4 607*	*4 670 F*	*5 350 F*	*4 863 F*	*4 961 F*	*5 370 F*	*5 196 F*	*6 875 F*
Climbing perch	**Anabas**		**Perca trepadora**		***Anabas testudineus***				**1,76(05)002,01**	**FPC**
04 Thailand	8 132	3 956	6 651	3 905	3 754	4 637	6 340	6 730	5 800	6 300
04 Fishing area total	*8 132*	*3 956*	*6 651*	*3 905*	*3 754*	*4 637*	*6 340*	*6 730*	*5 800*	*6 300*
Species total	*8 132*	*3 956*	*6 651*	*3 905*	*3 754*	*4 637*	*6 340*	*6 730*	*5 800*	*6 300*
Snakeskin gourami	**Gourami peau de serpent**		**Gurami piel de serpiente**		***Trichogaster pectoralis***				**1,76(10)013,02**	**FGS**
04 Indonesia	26 911	23 587	24 904	30 407	21 375	20 936	23 265	21 733	20 857	20 440
Thailand	751	200	186	385	353	1 486	511	723	500	500
04 Fishing area total	*27 662*	*23 787*	*25 090*	*30 792*	*21 728*	*22 422*	*23 776*	*22 456*	*21 357*	*20 940*
Species total	*27 662*	*23 787*	*25 090*	*30 792*	*21 728*	*22 422*	*23 776*	*22 456*	*21 357*	*20 940*
Kissing gourami	**Gourami embrasseur**		**Gurami besador**		***Helostoma temminckii***				**1,76(11)006,01**	**FGO**
04 Indonesia	18 954	16 675	19 166	12 611	18 376	16 598	23 320	18 771	19 429	20 140
04 Fishing area total	*18 954*	*16 675*	*19 166*	*12 611*	*18 376*	*16 598*	*23 320*	*18 771*	*19 429*	*20 140*
Species total	*18 954*	*16 675*	*19 166*	*12 611*	*18 376*	*16 598*	*23 320*	*18 771*	*19 429*	*20 140*
Snakehead	**Poisson tête de serpent**		**Cabeza de serpiente**		***Channa argus***				**1,77(19)001,01**	**CNA**
04 Korea Rep	-	-	-	169	120	...	28	...	...	...
04 Fishing area total	-	-	-	*169*	*120*	...	*28*	...	...	...
Species total	-	-	-	*169*	*120*	...	*28*	...	...	...
Striped snakehead	**Tête de serpent strié**		**Cabeza de serpiente cabrío**		***Channa striata***				**1,77(19)001,03**	**FSS**
04 Philippines	13 104	5 619	6 018	5 457	4 547	4 856	5 789	6 386	6 698	7 388
Thailand	18 591	21 400	21 810	25 509	24 099	16 664	17 995	20 453	18 300	18 600
04 Fishing area total	*31 695*	*27 019*	*27 828*	*30 966*	*28 646*	*21 520*	*23 784*	*26 839*	*24 998*	*25 988*
Species total	*31 695*	*27 019*	*27 828*	*30 966*	*28 646*	*21 520*	*23 784*	*26 839*	*24 998*	*25 988*
Indonesian snakehead	**Tête de serpent d'Indonésie**		**Cabeza de serpiente rojo**		***Channa micropeltes***				**1,77(19)001,04**	**FIS**
04 Indonesia	7 908	13 236	9 021	11 614	10 117	8 253	8 787	8 149	6 357	6 440
04 Fishing area total	*7 908*	*13 236*	*9 021*	*11 614*	*10 117*	*8 253*	*8 787*	*8 149*	*6 357*	*6 440*
Species total	*7 908*	*13 236*	*9 021*	*11 614*	*10 117*	*8 253*	*8 787*	*8 149*	*6 357*	*6 440*

B-13 Miscellaneous freshwater fishes / Poissons d'eau douce divers / Peces de agua dulce diversos

Capture production by species, fishing areas and countries or areas / Captures par espèces, zones de pêche et pays ou zones / Capturas por especies, áreas de pesca y países o áreas

Species, Fishing area Espèce, Zone de pêche Especie, Area de pesca	1993 t	1994 t	1995 t	1996 t	1997 t	1998 t	1999 t	2000 t	2001 t	2002 t
Snakeheads(=Murrels) nei	**Poissons tête de serpent nca**		**Cabezas de serpiente nep**		***Channa spp***				**1,77(19)001,XX**	**FSN**
01 Nigeria	1 073	185	368	921	86	2 589	2 038	2 990	2 951	3 313
01 Fishing area total	*1 073*	*185*	*368*	*921*	*86*	*2 589*	*2 038*	*2 990*	*2 951*	*3 313*
04 India	75 127	90 762	93 933	94 944	105 496	159 065	50 349	58 669	58 856	45 584
Indonesia	31 407	31 634	31 940	32 365	31 326	26 108	36 309	33 368	31 274	31 580
Kazakhstan	12	9	10	10 F	7 F	5 F	10 F	10 F	12	8
Turkmenistan	0	0	0	0	1	0	0	0	1	-
Uzbekistan	-	-	242	110 F	...	275	209	198	127	28
04 Fishing area total	*106 546*	*122 405*	*126 125*	*127 429 F*	*136 830 F*	*185 453 F*	*86 877 F*	*92 245 F*	*90 270*	*77 200*
Species total	*107 619*	*122 590*	*126 493*	*128 350 F*	*136 916 F*	*188 042 F*	*88 915 F*	*95 235 F*	*93 221*	*80 513*
Freshwater fishes nei	**Poissons d'eau douce nca**		**Peces de agua dulce nep**		***Osteichthyes***				**1,99(XX)XXX,XX**	**FRF**
01 Algeria	0	0	0	0	0	0	0	0	0	0
Angola	7 000	7 000	6 000	6 000	6 000	6 000	6 000 F	6 000 F	6 000 F	6 000 F
Benin	7 595 F	7 464	9 552	6 563 F	5 096 F	3 838 F	3 733 F	3 000 F	3 440 F	3 433 F
Botswana	300 F	200 F	100 F	33	80	103	64	74	30	38
Br Ind Oc Tr	0	0	0	0	0	0	0	0	0	0
Burkina Faso	7 000	8 000	8 000	8 000	8 000	8 335	7 600	8 500	8 500 F	8 500 F
Burundi	870 F	1 100 F	...	278	...	805	596	3 286	1 049	1 060 F
Cameroon	23 000	27 000 F	30 000 F	35 000 F	40 000 F	45 000 F	50 000 F	55 000	52 500 F	65 000
Cape Verde	0	0	0	0	0	0	0	0	0	0
Cent Afr Rep	13 250 F	13 500 F	13 750 F	14 000 F	14 250 F	14 500 F	15 000 F	15 000 F	15 000 F	15 000 F
Chad	87 300	80 000	90 000	100 000	85 000	84 000	84 000 F	84 000 F	84 000 F	84 000 F
Comoros	0	0	0	0	0	0	0	0	0	0
Congo Dem R	192 589	152 117	154 751	159 037	158 367	174 087	204 503	205 000 F	210 000 F	215 000 F
Congo Rep	27 850	24 752	26 811	25 873	18 987	25 455	25 268	20 204 F	20 200 F	20 500 F
Côte dIvoire	12 477	14 604	10 835	11 162 F	11 732 F	12 301 F	10 556 F	10 475	10 500	22 000
Djibouti	0	0	0	0	0	0	0	0	0	0
Egypt	65 814	80 906	62 895	55 738	63 547	95 924	83 715	34 196	37 282	37 957
Eq Guinea	600	700	450	900	850	970	1 101 F	1 076	1 000 F	1 000 F
Eritrea	0	0	0	0	0	0	0	0	0	-
Ethiopia	...	...	588	86	-	-	-	-	-	-
Fr South Tr	0	0	0	0	0	0	0	0	0	0
Gabon	1 000 F	1 500 F	3 018	4 251	4 636	3 500	5 814	5 726	5 922	5 121
Gambia	1 400	1 400	1 450	1 450	1 450	1 450	1 450 F	1 450 F	1 450 F	1 450 F
Ghana	52 000	54 700	60 000	73 580	70 000	74 500	74 500	74 500	74 500	74 500
Guinea	4 600	3 800	3 100	2 780	3 600	4 000	4 000	4 000	4 000	4 000 F
GuineaBissau	250 F	250 F	250 F	250 F	250 F	200 F	200 F	200 F	200 F	200 F
Kenya	9 103	2 015	1 215	3 059	745	5 960	4 221	4 886	5 237	4 984
Lesotho	5 F	5 F	8 F	10 F	10 F	10 F	10 F	12 F	14	14 F
Liberia	4 000	4 000	4 000	4 000	4 000	4 000	4 000	4 000	4 000	4 000 F
Libya	0	0	0	0	0	0	0	0	0	0
Madagascar	3 800	4 187	4 277	4 500	4 500	4 500	4 500	4 500	4 500	4 500
Malawi	50 957	44 679	49 263	53 428	51 346	31 773	2 970	2 700 F	2 274	2 314
Mali	16 718	16 341	34 580	39 833	60 667	34 854	35 044	20 875	19 000 F	19 000 F
Mauritania	5 000 F	5 000 F	5 000 F	5 000 F	5 000 F	5 000 F	5 000 F	5 000 F	5 000 F	5 000 F
Morocco	1 000	600	500	600	900	500	700	500	600	800
Mozambique	4 000	4 000	2 000	1 936	1 747	1 681	...	...	...	...
Namibia	1 200	1 200	1 200	1 200	1 500	1 500	1 500	1 500	1 500	1 500
Niger	2 162	2 516	3 616	4 156	6 328	7 013	11 000	16 250	20 800	23 520
Nigeria	38 738	42 959	33 708	29 776	11 375	2 093	16 589	48 445	36 986	102 976
Réunion	0	0	0	0	0	0	0	0	0	0
Rwanda	3 500 F	3 400 F	3 300 F	2 952	4 428	4 408	3 793	4 080	4 178	4 250 F
St Helena	0	0	0	0	-	-	-	-	-	-
Sao Tome Prn	0	0	0	0	0	0	0	0	0	0
Senegal	27 650	30 000 F	31 000 F	23 000 F	31 000 F	21 000 F	34 000 F	10 648	9 500 F	9 500 F
Seychelles	0	0	0	0	0	0	0	0	0	0
Sierra Leone	14 000	15 000	15 000	14 500 F	14 500	14 190	14 480	14 000	14 000	14 000
Somalia	250 F	250 F	250 F	250 F	250 F	250 F	250 F	200 F	200 F	150 F
South Africa	832	800	800	850	850	900	900 F	900 F	900 F	900 F
Sudan	37 500	40 000	30 000	30 000	31 000	28 000	28 000	30 000 F	33 000	33 000
Swaziland	68 F	65 F	60 F	60 F	65 F	70 F	70 F	70 F	70 F	70 F
Tanzania	57 779	36 362	62 160	55 975	60 500	58 700	67 520	96 700	86 000	84 350
Togo	500	1 000	998	1 000	1 000	1 000	1 000	1 000	1 000	1 000
Tunisia	400	243	440	706	1 010	896	808	832	860	870
Uganda	20 307	2 444	3 009	12 000	5 421	-	-	2	2	1 800
Zambia	56 046	61 147	61 872	58 739	58 110	60 116	58 372	57 808	56 500 F	56 500 F
Zimbabwe	902	547	883	644	599	671	782	1 807	1 800 F	1 800 F
01 Fishing area total	*861 312 F*	*797 753 F*	*830 689 F*	*853 155 F*	*848 696 F*	*844 053 F*	*873 609 F*	*858 402 F*	*843 494 F*	*941 557 F*
02 Anguilla	0	0	0	0	0	0	0	0	0	0
Antigua Barb	0	0	0	0	0	0	0	0	0	0
Aruba	0	0	0	0	0	0	-	-	-	-
Bahamas	0	0	0	0	0	0	0	0	0	0
Barbados	0	0	0	0	0	0	0	0	-	-
Belize	1	1	0	0	0	0	0	0	0	-
Bermuda	0	0	0	0	0	0	0	0	0	-
Br Virgin Is	0	0	0	0	0	0	0	0	0	0
Canada	5 809	6 707	8 445	9 649	8 089	8 914	9 317	10 756	8 247	10 568
Cayman Is	0	0	0	0	0	0	0	0	0	0
Costa Rica	710	840	900	1 090	840	1 000 F	1 000 F	1 000 F	1 000 F	1 000 F
Cuba	486	...	...	...	...	...	...	...	...	...
Dominica	0	0	0	0	0	0	0	0	0	0
Dominican Rp	130	234	315	105	751	990	96	179	183	1
El Salvador	776	932	967	891	912	863	867	1 177	535	1 022
Greenland	0	0	0	0	0	0	0	0	0	0
Grenada	0	0	0	0	0	0	0	0	0	0

B-13 Miscellaneous freshwater fishes / Poissons d'eau douce divers / Peces de agua dulce diversos

Capture production by species, fishing areas and countries or areas / Captures par espèces, zones de pêche et pays ou zones / Capturas por especies, áreas de pesca y países o áreas

Species, Fishing area Espèce, Zone de pêche Especie, Area de pesca	1993 t	1994 t	1995 t	1996 t	1997 t	1998 t	1999 t	2000 t	2001 t	2002 t
Guadeloupe	0	0	0	0	0	0	0	0	0	0
Guatemala	4 228	3 774	4 020	4 000	5 100	6 366	6 803	7 109	7 100 F	7 100 F
Haiti	600 F	500 F	500 F	500 F	500 F	500 F	500 F	500 F	500 F	500 F
Honduras	86	92	127	98	126	119	102	61	111	102 F
Jamaica	300 F	300 F	300 F	300 F	300 F	300 F	300 F	300 F	300 F	300 F
Martinique	0	0	0	0	0	0	0	0	0	0
Mexico	4 850	8 166	10 724	9 766	5 185	6 482	3 284	8 716	3 911	4 280
Montserrat	0	0	0	0	0	0	0	0	0	0
NethAntilles	0	0	0	0	0	0	0	0	0	0
Nicaragua	547	824	538	1 142	1 293	1 256	363	396	168	1 101
Puerto Rico	0	0	0	0	0	0	0	0	0	-
St Kitts Nev	0	0	0	0	0	0	0	0	0	-
St Lucia	0	0	0	0	0	0	0	0	0	0
St Pier Mq	0	0	0	0	0	0	0	0	0	0
St Vincent	0	0	0	2	1	0	0	0	0	0
Trinidad Tob	0	0	0	0	0	0	0	0	0	0
Turks Caicos	0	0	0	0	0	0	0	0	0	0
USA	1 743	1 459	2 129	283	640	301	338	650	637	579
US Virgin Is	0	0	0	0	0	0	0	0	0	0
02 *Fishing area total*	*20 266 F*	*23 829 F*	*28 965 F*	*27 826 F*	*23 737 F*	*27 091 F*	*22 970 F*	*30 844 F*	*22 692 F*	*26 553 F*
03 Argentina	1 754	1 854 F	2 908 F	3 297 F	4 135 F	4 097 F	5 138 F	6 098 F	6 517 F	6 003 F
Bolivia	4 742	4 452	4 726	4 800	4 850	4 865	4 860	4 911	4 900	4 853 F
Brazil	29 760 F	35 410 F	40 372	42 285	40 507	43 588	42 875	49 593	109 558	109 455
Colombia	17 231	19 737	9 108	16 165	14 232	6 031	9 577	7 800 F	7 800 F	7 800 F
Ecuador	372	300	300	300	400	400	400	400	400	400 F
Fr Guiana	0	0	0	0	0	0	0	0	0	0
Guyana	800	800	700 F	800	625	625	603	800	800	800
Paraguay	3 500 F	3 650 F	3 800 F	4 000 F	5 000 F	4 000 F	4 000 F	4 000 F	4 000 F	4 000 F
Peru	37 987	48 837	44 452	19 524	27 941	28 047	24 018	19 387	22 464	19 453
Suriname	187	138	140 F	150 F	200 F	200 F	200 F	200 F	200	200 F
Uruguay	150	73	45	188	40	41	342	312	254	39
Venezuela	2 185	2 532	4 524	4 149	8 321	8 141	13 204	450	5 252	13 443
03 *Fishing area total*	*98 668 F*	*117 783 F*	*111 075 F*	*95 658 F*	*106 251 F*	*100 035 F*	*105 217 F*	*93 951 F*	*162 145 F*	*166 446 F*
04 Afghanistan	1 200 F	1 300 F	1 300 F	1 300 F	1 250 F	1 200 F	1 200 F	1 000 F	800 F	900 F
Armenia	0	0	0	0	0	0	0	0	-	22
Azerbaijan	1	4	25	16	21	13	8	-	-	-
Bahrain	0	0	0	0	0	0	0	-	-	-
Bangladesh	354 459	445 676	443 319	445 377	451 055	457 055	575 609	591 300	613 860	620 185
Bhutan	320 F	310 F	310 F	300 F	300 F	300 F	300 F	300 F	300 F	300 F
Brunei Darsm	14	1	2	1	0	0	0	0	0	1
Cambodia	67 880	64 960	72 420	63 440	72 900	75 600	230 700	245 300	384 500	359 800
China	738 015	794 129	894 855	994 971	1 032 861	1 218 152	1 394 610	1 222 955	1 033 302	924 203
China, Macao	0	0	0	0	0	0	0	0	0	0
China,Taiwan	35	11	21	16	11	18	18	17	19	15
Cyprus	...	5	65	64	70	70	70	78	70 F	60 F
Georgia	0	0	0	0	0	0	5	7	2	3
India	195 021	88 484	92 964	224 990	245 595	225 100	330 448	416 490	598 798	377 060
Indonesia	110 071	125 078	122 559	115 323	103 812	96 597	110 347	112 554	107 609	107 690
Iran	7 005	851	157	1 845	1 131	2 529	1 637	1 575	5 970	4 340
Iraq	4 769	3 981	5 867	4 381	9 836	4 361	2 205	3 412	3 300 F	3 300 F
Japan	14 515	13 643	12 943	13 344	13 163	12 073	11 020	10 341	9 610	9 160
Jordan	350	350	350	350	350	350	350	400	350	350
Kazakhstan	64	8	6	269 F	358 F	364 F	10 809 F	13 715 F	1 434	532
Korea D P Rp	23 000 F	20 000	20 000	20 000	20 000	16 000 F	12 000 F	8 000 F	4 928	5 000 F
Korea Rep	4 049	2 891	3 309	3 418	2 633	2 755	1 681	6 402	5 254	4 993
Kuwait	0	0	0	0	0	0	0	0	0	0
Laos	17 000 F	20 600 F	23 370 F	19 500 F	16 057 F	16 642 F	25 541 F	24 850 F	26 350 F	28 440 F
Lebanon	10	10	10	10	10	10	10	10	10	272
Malaysia	1 971	2 064	3 939	3 683	3 949	4 626	3 366	3 549	3 446	3 185
Maldives	0	0	0	0	0	0	0	0	0	0
Mongolia	165	184	158	221	180	311	524	425	117	129
Myanmar	142 065	146 365	148 347	146 494	149 069	149 279	159 746	189 708	235 376	304 529
Nepal	7 320	7 340	11 230	11 230	11 230	12 000	12 752	16 700	16 700	17 900
Oman	0	0	0	0	0	0	0	0	0	0
Pakistan	109 185	118 703	121 405	142 092	167 530	163 524	179 865	176 468	180 100	181 000
Palest, O.T.	...	...	0	0	0	0	0	0	0	0
Philippines	9 131	7 632	9 425	7 143	10 413	6 632	8 309	9 800	9 384	11 877
Qatar	0	0	0	0	0	0	0	0	0	0
Saudi Arabia	0	0	0	0	0	0	0	0	0	0
Singapore	25	23	0	0	0	0	0	0	0	0
Sri Lanka	...	...	...	...	...	...	2 930	3 480	2 640	3 230
Syria	2 535	3 570	3 832	3 103	3 557	4 347	5 338	3 991	5 969	6 355
Tajikistan	14	7	4	40 F	75 F	100 F	32 F	19 F	-	-
Thailand	52 763	64 587	60 272	105 726	108 551	70 011	59 376	64 241	68 900	68 300
Timor-Leste	...	...	...	...	...	...	0	0	0	0
Turkey	1 882	3 139	3 424	4 621	1 800	1 700	2 434	1 697	3 290	3 478
Turkmenistan	3	9	6	27	2	2	1	2	9	7
Untd Arab Em	0	0	0	0	0	0	0	0	0	0
Uzbekistan	302	536	66	40 F	130	69	10	-	1	70
Viet Nam	145 839	79 087	94 189	163 936	176 589	137 800	168 107	169 000 F	132 280	148 200
04 *Fishing area total*	*2 010 978 F*	*2 015 538 F*	*2 150 149 F*	*2 497 271 F*	*2 604 488 F*	*2 679 590 F*	*3 311 358 F*	*3 297 786 F*	*3 454 678 F*	*3 194 886 F*
05 Albania	500 F	440 F	-	-	-	-	-	-	-	24
Andorra	0	0	0	0	0	0	0	0	0	0
Austria	420	388	404	450	465	451	432	439	362	350
Belarus	2	-	-	-	-	-	-	-	-	147
Bosnia Herzg	2 500 F	2 500 F	2 500 F	2 500 F	2 500 F	2 500 F	2 500 F	2 500 F	2 500 F	2 500 F

B-13 Miscellaneous freshwater fishes / Poissons d'eau douce divers / Peces de agua dulce diversos

Capture production by species, fishing areas and countries or areas / Captures par espèces, zones de pêche et pays ou zones / Capturas por especies, áreas de pesca y países o áreas

Species, Fishing area Espèce, Zone de pêche Especie, Area de pesca	1993 t	1994 t	1995 t	1996 t	1997 t	1998 t	1999 t	2000 t	2001 t	2002 t
Bulgaria	1 576	590 F	39	182	0	1	-	-	-	22
Channel Is	0	0	0	0	0	0	0	0	0	0
Croatia	161	184	202	217	213	5 F	5	8	12	17
Czech Rep	89	153	192	102	121	134	168	151	131	126
Denmark	0	0	0	3	-	-	0	0	1	0
Estonia	247	212	218	123	160	134	168	170	209	28
Faeroe Is	0	0	0	0	0	0	0	0	0	-
Finland	696	640	517	476	476	444	444	393	393	393
France	4 400	4 450	4 500	4 500	4 500	4 460	2 000 F	2 000 F	2 000 F	2 000 F
Germany	6 103	7 000	19 079	19 079	19 008	19 008	19 008	19 008	19 008	19 008
Greece	2 960	3 452	3 606	1 730	1 624	1 410	1 592	1 803	1 663	1 451
Hungary	595	703	733	733	776	648	714	718	89	15
Iceland	0	0	0	0	0	0	0	0	0	-
Ireland	0	0	0	0	0	0	...	...	...	9
Isle of Man	0	0	0	0	0	0	0	0	0	0
Italy	3 250	3 518	3 900	3 647	2 895	2 346	2 316	2 819	2 643	2 341
Latvia	89	9	8	9	9	9	13	41	40	38
Liechtensten	0	0	0	0	0	0	0	0	0	0
Lithuania	2	17	50	45	90	88	10	6	2	0
Luxembourg	0	0	0	0	0	0	0	0	0	0
Macedonia	70	46	146	31	68	107	113	52	7	6
Malta	0	0	0	0	0	0	0	0	0	-
Moldova Rep	10	19	11	9	11	13	-	-	-	27
Netherlands	22	228	410	350	362	176	154	150 F	150 F	134
Poland	24 769	20 898	20 871	13 345	9 500	10 000	10 200	12 120	13 035	13 035
Portugal	0	0	0	0	0	0	0	0	1	12
Romania	369	313	27	67	5	-	-	169	188	230
Russian Fed	7 835	17 991	5 926	8 006	3 584	1 753	2 731	15 849	4 440	5 211
Serbia-Monte	1 416 F	1 329 F	919	1 156	964	792	762	603	521	867
Slovenia	5	0	0	0	17	0	3	-	-	2
Spain	4 439	3 981	3 944	4 000 F	4 000 F	4 000 F	4 000 F	4 000 F	4 000 F	4 000 F
Sweden	413	243	206	126	229	134	154	197	155	179
Switzerland	0	0	8	0	0	3	0	4	1	2
Ukraine	994	485	261	213	15	1	2	...	7	-
05 Fishing area total	*63 932 F*	*69 789 F*	*68 677 F*	*61 099 F*	*51 592 F*	*48 617 F*	*47 489 F*	*63 200 F*	*51 558 F*	*52 174 F*
06 Amer Samoa	0	0	0	0	0	0	0	0	0	-
Australia	2 834	122	301	252	194	200	154	166	158	127
Christmas Is	0	0	0	0	0	0	0	0	0	0
Cocos Is	0	0	0	0	0	0	0	0	0	0
Cook Is	0	0	0	0	0	0	0	0	0	0
Fr Polynesia	0	0	0	0	50	50	50	50	50	50
Guam	0	0	0	0	-	-	-	-	-	-
Kiribati	0	0	0	0	0	0	0	0	0	-
Marshall Is	0	0	0	0	0	0	0	0	0	0
Micronesia	4 F	4 F	5 F	5 F	5 F	5 F	5 F	5 F	5 F	5 F
NewCaledonia	0	0	0	0	0	0	0	0	0	0
New Zealand	550	500	500	600	600	400	400	400	300	300
Niue	0	0	0	0	0	0	0	0	0	0
Norfolk Is	0	0	0	0	0	0	0	0	0	0
N Marianas	0	0	1	0	0	0	0	0	0	-
Palau	0	0	0	0	0	0	0	0	0	0
Papua N Guin	6 652 F	6 644 F	6 645 F	6 649 F	6 649 F	6 641 F	6 649 F	6 655 F	6 654 F	6 654 F
Pitcairn Is	0	0	0	0	0	0	0	0	0	0
Samoa	0	0	0	0	0	0	0	1	1 F	1 F
Solomon Is	0	0	0	0	0	0	0	0	0	0
Tokelau	0	0	0	0	0	0	0	0	0	0
Tonga	1	0	0	0	0	0	0	0	0	0
Tuvalu	0	0	0	0	0	0	0	0	0	0
Vanuatu	0	0	0	0	0	0	0	0	0	0
Wallis Fut I	0	0	0	0	0	0	0	0	0	0
06 Fishing area total	*10 041 F*	*7 270 F*	*7 452 F*	*7 506 F*	*7 498 F*	*7 296 F*	*7 258 F*	*7 277 F*	*7 168 F*	*7 137 F*
27 Denmark	-	-	-	0	-	-	-	-	-	-
Finland	425	412	486	481	452	229	292	179	209	254
Germany	-	-	-	-	3	1	2	3	1	3
Latvia	-	-	1	11	4	4	7	22	10	24
Lithuania	-	-	-	-	-	-	-	-	-	1
Poland	884	608	384	500	574	327	217	388	241	260
Russian Fed	1 544	2 516	1 719	1 642	2 554	2 287	1 429	-	-	2
Sweden	1	0	0	5	1	0	3	-	-	-
27 Fishing area total	*2 854*	*3 536*	*2 590*	*2 639*	*3 588*	*2 848*	*1 950*	*592*	*461*	*544*
31 Venezuela	5 101	5 404	7 705	7 303	3 656	2 870	4 130	6 772	6 937	-
31 Fishing area total	*5 101*	*5 404*	*7 705*	*7 303*	*3 656*	*2 870*	*4 130*	*6 772*	*6 937*	-
37 Russian Fed	-	-	-	-	-	-	-	-	17	-
37 Fishing area total	-	-	-	-	-	-	-	-	*17*	-
Species total	*3 073 152 F*	*3 040 902 F*	*3 207 302 F*	*3 552 457 F*	*3 649 506 F*	*3 712 400 F*	*4 373 981 F*	*4 358 824 F*	*4 549 150 F*	*4 389 297 F*
Group total	***4 233 936***	***4 174 214***	***4 471 383***	***4 692 158***	***4 828 327***	***5 061 465***	***5 596 315***	***5 578 646***	***5 700 880***	***5 503 067***

B-21

Sturgeons, paddlefishes — **Capture production by species, fishing areas and countries or areas**
Esturgeons, spatules — **Captures par espèces, zones de pêche et pays ou zones**
Esturiones, sollos — **Capturas por especies, áreas de pesca y países o áreas**

Species, Fishing area Espèce, Zone de pêche Especie, Area de pesca	1993 t	1994 t	1995 t	1996 t	1997 t	1998 t	1999 t	2000 t	2001 t	2002 t
Danube sturgeon(=Osetr)	**Esturgeon du Danube**		**Esturión del Danube**		***Acipenser gueldenstaedtii***			**1,17(01)001,02**		**APG**
05 Bulgaria	-	-	4	2	4	5	4	1	1	1
Romania	-	-	-	-	-	5	10	19	17	4
Russian Fed	2 484	1 562	1 164	482	721	646	359	250	251	219
Ukraine	...	...	...	1	2	1	...	2	-	-
05 Fishing area total	*2 484*	*1 562*	*1 168*	*485*	*727*	*657*	*373*	*272*	*269*	*224*
37 Bulgaria	-	-	1	0	2	2	2	-	-	2
Ukraine	...	...	134	129	132	112	34	20	8	8
37 Fishing area total	...	...	*135*	*129*	*134*	*114*	*36*	*20*	*8*	*10*
Species total	*2 484*	*1 562*	*1 303*	*614*	*861*	*771*	*409*	*292*	*277*	*234*
Sterlet sturgeon	**Sterlet**		**Esterlete**		***Acipenser ruthenus***			**1,17(01)001,04**		**APR**
05 Bulgaria	-	-	0	1	1	1	2	2	1	3
Hungary	14	15	36	34	14	9	35	12	11	12
Russian Fed	2	-	4	-	2	1	1	-	1	2
Slovakia	0	0	1	1	0	0	0	1	-	0
05 Fishing area total	*16*	*15*	*41*	*36*	*17*	*11*	*38*	*15*	*13*	*17*
37 Bulgaria	-	-	-	-	-	-	-	0	-	-
37 Fishing area total	-	-	-	-	-	-	-	*0*	-	-
Species total	*16*	*15*	*41*	*36*	*17*	*11*	*38*	*15*	*13*	*17*
Starry sturgeon	**Esturgeon étoilé**		**Esturión estrellado**		***Acipenser stellatus***			**1,17(01)001,05**		**APE**
05 Bulgaria	-	-	0	0	0	4	6	1	1	2
Romania	-	-	-	-	5	2	11	22	20	13
Russian Fed	1 446	1 523	980	681	448	336	234	176	172	136
05 Fishing area total	*1 446*	*1 523*	*980*	*681*	*453*	*342*	*251*	*199*	*193*	*151*
37 Ukraine	...	...	9	18	49	13	11	5	3	2
37 Fishing area total	...	...	*9*	*18*	*49*	*13*	*11*	*5*	*3*	*2*
Species total	*1 446*	*1 523*	*989*	*699*	*502*	*355*	*262*	*204*	*196*	*153*
White sturgeon	**Esturgeon blanc**		**Esturión blanco**		***Acipenser transmontanus***			**1,17(01)001,09**		**APN**
67 USA	-	-	-	-	-	-	-	206	185	200
67 Fishing area total	-	-	-	-	-	-	-	*206*	*185*	*200*
Species total	-	-	-	-	-	-	-	*206*	*185*	*200*
Green sturgeon	**Esturgeon vert**		**Esturión verde**		***Acipenser medirostris***			**1,17(01)001,15**		**AAM**
67 USA	-	-	-	-	-	-	-	36	10	7
67 Fishing area total	-	-	-	-	-	-	-	*36*	*10*	*7*
Species total	-	-	-	-	-	-	-	*36*	*10*	*7*
Beluga	**Béluga**		**Esturión beluga**		***Huso huso***			**1,17(01)005,01**		**HUH**
05 Bulgaria	...	...	21	24	31	31	27	18	7	10
Romania	-	-	-	-	2	7	7	32	20	22
Russian Fed	311	162	154	105	127	78	40	44	40	32
Slovenia	-	-	0	1	1	1	1	0	1	0
05 Fishing area total	*311*	*162*	*175*	*130*	*161*	*117*	*75*	*94*	*68*	*64*
37 Bulgaria	-	-	4	5	11	12	10	1	0	4
37 Fishing area total	-	-	*4*	*5*	*11*	*12*	*10*	*1*	*0*	*4*
Species total	*311*	*162*	*179*	*135*	*172*	*129*	*85*	*95*	*68*	*68*
Sturgeons nei	**Esturgeons nca**		**Esturiones nep**		***Acipenseridae***			**1,17(01)XXX,XX**		**STU**
02 Canada	267	281	286	212	330	349	282	281	173	172
USA	217	147	147	314	285	176	138	0	-	-
02 Fishing area total	*484*	*428*	*433*	*526*	*615*	*525*	*420*	*281*	*173*	*172*
04 Azerbaijan	84	92	76	69	63	61	69	70	76	70
Iran	1 710	1 700	1 500	1 600	1 300	1 200	1 000	1 000	870	643
Kazakhstan	1 109	635	563	510 F	370 F	270 F	240 F	215 F	282	197
Turkmenistan	...	...	...	...	...	...	11	3	3	5
04 Fishing area total	*2 903*	*2 427*	*2 139*	*2 179 F*	*1 733 F*	*1 531 F*	*1 320 F*	*1 288 F*	*1 231*	*915*
05 Bulgaria	10	5 F	-	-	-	-	-	-	-	-
Germany	-	-	-	0	0	0	0	0	0	0
Macedonia	8	0	0	0	2	6	-	-	-	-
Romania	2	2	9	5	5	1	3	-	1	1
Russian Fed	-	-	-	-	-	88	116	124	89	69
Ukraine	2	-	-	-	-	-	-	-	-	-
05 Fishing area total	*22*	*7 F*	*9*	*5*	*7*	*95*	*119*	*124*	*90*	*70*
21 Canada	181	100	157	86	91	69	9	3	4	-
USA	...	15	0	-	-	-	-	-	-	-
21 Fishing area total	*181*	*115*	*157*	*86*	*91*	*69*	*9*	*3*	*4*	-

B-21 Sturgeons, paddlefishes / Esturgeons, spatules / Esturiones, sollos

Capture production by species, fishing areas and countries or areas / Captures par espèces, zones de pêche et pays ou zones / Capturas por especies, áreas de pesca y países o áreas

Species, Fishing area Espèce, Zone de pêche Especie, Area de pesca	1993 t	1994 t	1995 t	1996 t	1997 t	1998 t	1999 t	2000 t	2001 t	2002 t
27 Denmark	-	-	-	-	-	0	-	-	-	-
France	0	0	0	0	0	0	1	-	-	-
Portugal	0	0	0	0	-	-	-	-	-	-
Spain	-	-	-	-	-	1	1	-	-	-
27 Fishing area total	*0*	*0*	*0*	*0*	*0*	*1*	*2*	-	-	-
37 Georgia	...	...	...	...	...	...	3	4	3	7
Romania	17	6	5	2	2	6	1	1	1	1
Russian Fed	904	1 012	673	430	441	284	181	54	18	15
Ukraine	284	227	...	...	...	-	-	-	-	-
37 Fishing area total	*1 205*	*1 245*	*678*	*432*	*443*	*290*	*185*	*59*	*22*	*23*
67 Canada	3	2	4	1	0	0	0	-	-	
67 Fishing area total	*3*	*2*	*4*	*1*	*0*	*0*	*0*	-	-	
Species total	*4 798*	*4 224 F*	*3 420*	*3 229 F*	*2 889 F*	*2 511 F*	*2 055 F*	*1 755 F*	*1 520*	*1 180*
Group total	***9 055***	***7 486***	***5 932***	***4 713***	***4 441***	***3 777***	***2 849***	***2 603***	***2 269***	***1 859***

B-22 River eels / Anguilles / Anguilas

Capture production by species, fishing areas and countries or areas
Captures par espèces, zones de pêche et pays ou zones
Capturas por especies, áreas de pesca y países o áreas

Species, Fishing area Espèce, Zone de pêche Especie, Area de pesca	1993 t	1994 t	1995 t	1996 t	1997 t	1998 t	1999 t	2000 t	2001 t	2002 t
European eel	**Anguille d'Europe**		**Anguila europea**		***Anguilla anguilla***				**1,43(02)002,01**	**ELE**
01 Egypt	...	...	798	537	585	501	709	2 064	1 979	1 802
Morocco	100	150	100	100	400	300	250	100	150	200
01 Fishing area total	*100*	*150*	*898*	*637*	*985*	*801*	*959*	*2 164*	*2 129*	*2 002*
04 Turkey	261	329	390	342	400	300	99	176	122	147
04 Fishing area total	*261*	*329*	*390*	*342*	*400*	*300*	*99*	*176*	*122*	*147*
05 Albania	150 F	100 F	6	10	21	58	63	70	...	23
Belarus	19	26	15	20	15	18	16	14	25	10
Belgium	30	30	30	30	30	30	30	30	30	30
Czech Rep	31	32	31	28	27	28	28	24	29	28
Denmark	57	60	62	34	39	40	30	20	23	18
Estonia	49	44	32	35	38	22	32	40	40	22
Finland	0	0	0	21	21	...	...	...	...	...
France	-	-	-	40	40	40	134	131	130 F	130 F
Germany	774	550	550	550	550	550	550	550	500	500
Greece	...	...	...	10	10	9	11	10	8	5
Hungary	263	501	411	579	124	182	179	76	27	18
Ireland	260	300	400	400	400	400	250	250	110	104
Italy	815	550	270	346	326	269	283	329	217	157
Latvia	18	38	26	25	27	25	15	13	17	9
Lithuania	10	12	10	12	11	17	18	11	12	2
Netherlands	375	310	393	300	285	322	332	330 F	340 F	353
Norway	...	...	...	...	30	22	28	...	...	...
Poland	800	800	390	373	256	223	248	257	263	235
Romania	-	-	-	-	1	1	0	26	-	-
Russian Fed	16	8	8	10	13	24	4	6	3	3
Slovakia	7	20	13	7	8	8	8	4	6	7
Slovenia	-	-	-	-	-	-	-	-	-	2
Spain	0	0	0	0	0	0	0	0	0	0
Sweden	129	171	127	97	142	112	140	113	118	102
Switzerland	4	5	5	3	2	3	3	2	2	2
Ukraine	-	-	-	-	-	-	-	-	-	5
UK	714	833	756	857	778	730	690	795	583	561
05 Fishing area total	*4 521 F*	*4 390 F*	*3 535*	*3 787*	*3 194*	*3 133*	*3 092*	*3 101 F*	*2 483 F*	*2 326 F*
27 Denmark	1 024	1 140	842	700	757	560	686	600	635	551
Estonia	10	10	6	20	18	22	28	27	27	27
Finland	0	0	0	1	1	...	...	...	...	...
France	275	346	300	237	1 435	102	78	203	180	157
Germany	253	35	35	146	196	167	197	136	138	135
Ireland	200	200	200	150	150	250	250	...	...	...
Latvia	-	1	2	1	2	2	2	2	2	2
Lithuania	...	...	...	...	...	...	...	...	...	11
Netherlands	43	49	39	36	30	23	40	21	34	20
Norway	340	472	454	352	467	341	447	281	304	310
Poland	316	290	237	266	233	231	226	172	163	127
Portugal	-	-	-	-	-	-	30	29	37	36
Russian Fed	19	25	33	36	34	25	19	40	53	52
Spain	22	30	23	31	39	23	18	23	51	21
Sweden	1 015	1 127	972	945	931	533	594	447	462	531
UK	38	40	52	38	34	11	7	1	12	10
27 Fishing area total	*3 555*	*3 765*	*3 195*	*2 959*	*4 327*	*2 290*	*2 622*	*1 982*	*2 098*	*1 990*
34 Morocco	4	0	0	0	1	4	-	-	-	-
34 Fishing area total	*4*	*0*	*0*	*0*	*1*	*4*	-	-	-	-
37 Albania	...	...	33	40	-	-	-	-	98	2
Algeria	-	-	-	-	-	10	...	...	...	...
Croatia	5	5	7	6	7	-	-	-	-	-
France	589	261	20	126	307	307	77	65	105	105
Greece	17	23	31	21	21	34	16	24	24	20 F
Italy	492	436	616	537	684	413	362	220	229	245
Serbia-Monte	...	...	5	3	3	3	...	...	...	2
Spain	55 F	50 F	45 F	37	33	-	21	47	11	72
Tunisia	373	390	130	192	202	150	206	108	135	202
37 Fishing area total	*1 531 F*	*1 165 F*	*887 F*	*962*	*1 257*	*917*	*682*	*464*	*602*	*648 F*
Species total	*9 972 F*	*9 799 F*	*8 905 F*	*8 687*	*10 164*	*7 445*	*7 454*	*7 887 F*	*7 434 F*	*7 113 F*
Japanese eel	**Anguille du Japon**		**Anguila japonesa**		***Anguilla japonica***				**1,43(02)002,04**	**ELJ**
04 Japan	970	949	899	901	860	860	817	765	677	610
Korea Rep	96	93	124	113	56	44	13	...	...	...
04 Fishing area total	*1 066*	*1 042*	*1 023*	*1 014*	*916*	*904*	*830*	*765*	*677*	*610*
Species total	*1 066*	*1 042*	*1 023*	*1 014*	*916*	*904*	*830*	*765*	*677*	*610*
American eel	**Anguille d'Amérique**		**Anguila americana**		***Anguilla rostrata***				**1,43(02)002,06**	**ELA**
02 Canada	278	272	319	289	311	278	278	286	254	241
Cuba	-	1	-	-	-	-	-	-	-	-
Dominican Rp	...	...	6	...	...	1	2	7	1	25
USA	523	724	285	441	485	460	490	649	393	3
02 Fishing area total	*801*	*997*	*610*	*730*	*796*	*739*	*770*	*942*	*648*	*269*
21 Canada	848	752	631	551	455	496	353	395	220	231

B-22 River eels / Anguilles / Anguilas

Capture production by species, fishing areas and countries or areas
Captures par espèces, zones de pêche et pays ou zones
Capturas por especies, áreas de pesca y países o áreas

Species, Fishing area Espèce, Zone de pêche Especie, Area de pesca	1993 t	1994 t	1995 t	1996 t	1997 t	1998 t	1999 t	2000 t	2001 t	2002 t
USA	-	-	-	-	-	-	-	-	-	263
21 Fishing area total	*848*	*752*	*631*	*551*	*455*	*496*	*353*	*395*	*220*	*494*
31 Mexico	-	-	43	35	19	9	2	1	0	0
USA	-	-	-	-	-	-	-	-	-	11
31 Fishing area total	-	-	*43*	*35*	*19*	*9*	*2*	*1*	*0*	*11*
Species total	*1 649*	*1 749*	*1 284*	*1 316*	*1 270*	*1 244*	*1 125*	*1 338*	*868*	*774*
Short-finned eel	**Anguille d'Australie**		**Anguila australiana**		***Anguilla australis***				**1,43(02)002,07**	**ELU**
06 Australia	39	385	480	413	320	293	296	300 F	300 F	300 F
06 Fishing area total	*39*	*385*	*480*	*413*	*320*	*293*	*296*	*300 F*	*300 F*	*300 F*
57 Australia	349	365	259	208	203	160	129	133	175	167
57 Fishing area total	*349*	*365*	*259*	*208*	*203*	*160*	*129*	*133*	*175*	*167*
81 Australia	...	...	...	...	...	...	...	37	37	30
81 Fishing area total	...	...	...	...	...	...	...	*37*	*37*	*30*
Species total	*388*	*750*	*739*	*621*	*523*	*453*	*425*	*470 F*	*512 F*	*497 F*
River eels nei	**Anguilles nca**		**Anguilas nep**		***Anguilla spp***				**1,43(02)002,XX**	**ELX**
04 Indonesia	791	3 056	926	4 339	657	787	1 212	4 553	1 907	3 790
Philippines	194	118	128	100	116	73	114	193	201	181
04 Fishing area total	*985*	*3 174*	*1 054*	*4 439*	*773*	*860*	*1 326*	*4 746*	*2 108*	*3 971*
06 Australia	138	77	40	40	98	100	178	176	182	150 F
New Zealand	610	600	614	478	392	893	735	675	762	586
06 Fishing area total	*748*	*677*	*654*	*518*	*490*	*993*	*913*	*851*	*944*	*736 F*
81 New Zealand	865	1 362	1 183	524	588	400	416	380	313	337
81 Fishing area total	*865*	*1 362*	*1 183*	*524*	*588*	*400*	*416*	*380*	*313*	*337*
Species total	*2 598*	*5 213*	*2 891*	*5 481*	*1 851*	*2 253*	*2 655*	*5 977*	*3 365*	*5 044 F*
Group total	***15 673***	***18 553***	***14 842***	***17 119***	***14 724***	***12 299***	***12 489***	***16 437***	***12 856***	***14 038***

B-23 Salmons, trouts, smelts / Saumons, truites, éperlans / Salmones, truchas, eperlanos

Capture production by species, fishing areas and countries or areas / Captures par espèces, zones de pêche et pays ou zones / Capturas por especies, áreas de pesca y países o áreas

Species, Fishing area / Espèce, Zone de pêche / Especie, Area de pesca	1993 t	1994 t	1995 t	1996 t	1997 t	1998 t	1999 t	2000 t	2001 t	2002 t
Atlantic salmon	**Saumon de l'Atlantique**		**Salmón del Atlántico**		***Salmo salar***			**1,23(01)004,01**		**SAL**
05 Denmark	0	1	0	0	0	1	0	1	0	0
Estonia	1	0	0	-	-	-	-	0	0	-
Finland	317	385	383	523	523	222	222	221	221	221
Iceland	657	448	489	358	154	166	120	85	88	88 F
Ireland	43	29	86	131	105	101	515	621	792	674
Latvia	-	-	-	-	-	-	-	1	3	1
Norway	351	351	325	267	236	331	327	423	420 F	420 F
Poland	-	-	-	-	-	-	-	20	5	10
Russian Fed	73	87	68	56	32	73	55	63	52	57
Spain	8	11	9	10 F	10 F	10 F	10 F	10 F	10 F	10 F
Sweden	86	59	51	44	47	44	34	36	56	43
UK	674	790	823	568	457	419	403	447	482	481
05 Fishing area total	*2 210*	*2 161*	*2 234*	*1 957 F*	*1 564 F*	*1 367 F*	*1 686 F*	*1 928 F*	*2 129 F*	*2 005 F*
21 Canada	164	136	103	82	77	7	1	-	-	1
Denmark	-	-	-	-	-	-	-	-	0	0
Greenland	0	0	68	82	43	...	...	24	42	-
St Pier Mq	2	3	0	1	1	1	1	1	1	1
21 Fishing area total	*166*	*139*	*171*	*165*	*121*	*8*	*2*	*25*	*43*	*2*
27 Denmark	583	745	560	528	493	486	389	412	434	321
Estonia	31	5	9	9	10	7	14	21	14	16
Faeroe Is	78	12	3	-	-	5	-	0	-	-
Finland	1 895	1 184	1 258	1 191	1 267	798	690	743	596	593
France	-	42	0	0	0	0	2	10	12	5
Germany	55	13	13	27	35	42	30	45	39	29
Greenland	-	-	2	0	1	-	-	-	-	-
Iceland	496	308	289	239	48	36	22	2	-	-
Ireland	541	804	715	685	570	624	10	-	-	-
Isle of Man	0	0	0	-	-	-	-	-	-	-
Latvia	243	130	139	151	169	125	166	150	135	110
Lithuania	15	5	2	10	4	5	6	6	4	11
Netherlands	0	1	1	2	1	1	1	-	0	0
Norway	571	649	520	526	402	422	500	631	705	602
Poland	191	184	133	125	110	114	118	125	156	188
Portugal	1	-	-	-	-	-	0	0	-	-
Russian Fed	114	120	124	117	116	91	72	91	109	91
Sweden	971	729	646	761	668	613	399	478	356	288
UK	-	-	-	-	-	-	-	33	37	31
27 Fishing area total	*5 785*	*4 931*	*4 414*	*4 371*	*3 894*	*3 369*	*2 419*	*2 747*	*2 597*	*2 285*
Species total	*8 161*	*7 231*	*6 819*	*6 493 F*	*5 579 F*	*4 744 F*	*4 107 F*	*4 700 F*	*4 769 F*	*4 292 F*
Sea trout	**Truite de mer**		**Trucha marina**		***Salmo trutta***			**1,23(01)004,02**		**TRS**
03 Falkland Is	-	-	-	1	1	1	1	1	1	1
03 Fishing area total	-	-	-	*1*	*1*	*1*	*1*	*1*	*1*	*1*
04 Turkey	479	554	594	395	200	200	263	277	364	352
04 Fishing area total	*479*	*554*	*594*	*395*	*200*	*200*	*263*	*277*	*364*	*352*
05 Belgium	100	100	100	100	100	100	150	100	100	100
Bulgaria	-	-	-	-	8	11	11	4	1	16
Denmark	3	8	6	3	3	4	3	4	0	0
Ireland	880	1 200	1 200	1 200	1 200	1 200	36	10	...	2
Lithuania	-	-	-	-	-	-	-	1	1	0
Romania	33	40	39	39	28	10	63	18	48	15
Russian Fed	31	-	-	-	-	-	-	1	3	10
Slovakia	35	32	46	34	40	42	41	37	38	33
Slovenia	15	14	14	14	13	11	10	9	7	9
Spain	2 389	2 018	2 163	2 200 F	2 200 F	2 200 F	2 200 F	2 200 F	2 200 F	2 200 F
Switzerland	-	-	10	8	13	12	11	13	13	13
UK	214	256	260	198	141	572	489	142	560	139
05 Fishing area total	*3 700*	*3 668*	*3 838*	*3 796 F*	*3 746 F*	*4 162 F*	*3 014 F*	*2 539 F*	*2 971 F*	*2 537 F*
27 Denmark	68	50	80	74	49	55	98	66	52	35
Estonia	15	8	6	15	11	8	10	13	13	16
Finland	1 256	603	615	672	661	460	441	437	416	396
France	109	890	0	0	0	0	0	-	1	1
Ireland	-	-	-	-	-	-	511	-	-	-
Isle of Man	0	0	0	-	-	-	-	-	-	-
Latvia	19	18	14	10	7	7	10	14	11	13
Lithuania	-	-	3	-	2	3	4	4	2	3
Norway	-	-	-	-	-	-	-	12	0	11
Portugal	0	0	1	0	-	-	1	1	1	8
Russian Fed	-	-	-	-	-	-	1	-	1	-
UK	0	-	-	-	-	0	0	8	11	9
27 Fishing area total	*1 467*	*1 569*	*719*	*771*	*730*	*533*	*1 076*	*555*	*508*	*492*
Species total	*5 646*	*5 791*	*5 151*	*4 963 F*	*4 677 F*	*4 896 F*	*4 354 F*	*3 372 F*	*3 844 F*	*3 382 F*
Trouts nei	**Truites nca**		**Truchas nep**		***Salmo spp***			**1,23(01)004,XX**		**TRO**
04 Armenia	5	7	0	0	6	0	163	186	180	167
Azerbaijan	-	-	-	-	-	-	-	-	5	9
Japan	1 189	1 106	1 080	1 222	1 207	1 187	1 122	1 136	1 053	1 043
04 Fishing area total	*1 194*	*1 113*	*1 080*	*1 222*	*1 213*	*1 187*	*1 285*	*1 322*	*1 238*	*1 219*

B-23

Salmons, trouts, smelts — **Capture production by species, fishing areas and countries or areas**
Saumons, truites, éperlans — **Captures par espèces, zones de pêche et pays ou zones**
Salmones, truchas, eperlanos — **Capturas por especies, áreas de pesca y países o áreas**

Species, Fishing area Espèce, Zone de pêche Especie, Area de pesca	1993 t	1994 t	1995 t	1996 t	1997 t	1998 t	1999 t	2000 t	2001 t	2002 t
05 Czech Rep	51	52	54	53	57	70	64	55	56	50
Finland	952	1 025	1 022	1 286	1 286	879	879	610	610	610
Iceland	250	250	250	250	250	250	250	91	72	72 F
Macedonia	77	137	30	37	51	18	22	131	115	120
Netherlands	0	0	0	-	-	-	-	-	-	-
Poland	-	-	-	40	-	-	-	139	48	55
Slovenia	2	2	2	3	2	2	2	2	1	0
Sweden	14	12	14	18	22	26	22	24	24	20
UK	-	-	-	-	-	-	-	-	254	426
05 Fishing area total	*1 346*	*1 478*	*1 372*	*1 687*	*1 668*	*1 245*	*1 239*	*1 052*	*1 180*	*1 353 F*
21 Canada	53	34	20	4	1	0	0	0	-	-
21 Fishing area total	*53*	*34*	*20*	*4*	*1*	*0*	*0*	*0*	-	-
27 Germany	-	0	1	7	8	6	9	12	11	13
Norway	-	-	-	-	-	-	-	4	11	0
Poland	272	222	187	150	200	329	385	579	529	811
Sweden	177	123	78	113	96	112	72	64	39	36
27 Fishing area total	*449*	*345*	*266*	*270*	*304*	*447*	*466*	*659*	*590*	*860*
Species total	*3 042*	*2 970*	*2 738*	*3 183*	*3 186*	*2 879*	*2 990*	*3 033*	*3 008*	*3 432 F*
Pink(=Humpback)salmon	**Saumon rose**		**Salmón rosado**		***Oncorhynchus gorbuscha***			**1,23(01)009,02**		**PIN**
02 USA	209	51	319	3	31	0	1	-	-	-
02 Fishing area total	*209*	*51*	*319*	*3*	*31*	*0*	*1*	-	-	-
04 Japan	729	3 598	1 511	3 134	947	2 091	927	1 947	376	926
04 Fishing area total	*729*	*3 598*	*1 511*	*3 134*	*947*	*2 091*	*927*	*1 947*	*376*	*926*
05 Russian Fed	1 190	7 420	8 862	8 804	18 597	14 057	28 582	24 277	17 961	16 927
05 Fishing area total	*1 190*	*7 420*	*8 862*	*8 804*	*18 597*	*14 057*	*28 582*	*24 277*	*17 961*	*16 927*
27 Russian Fed	-	-	-	-	-	-	39	10	184	2
27 Fishing area total	-	-	-	-	-	-	*39*	*10*	*184*	*2*
61 Japan	24 116	28 476	23 526	29 461	14 897	23 246	15 975	24 655	9 389	23 745
Russian Fed	104 855	117 533	139 369	104 377	169 070	177 382	158 560	132 851	149 421	100 655
61 Fishing area total	*128 971*	*146 009*	*162 895*	*133 838*	*183 967*	*200 628*	*174 535*	*157 506*	*158 810*	*124 400*
67 Canada	16 046	3 383	19 767	8 597	12 241	3 920	9 529	7 158	10 575	8 609
USA	155 423	165 604	201 381	140 539	102 934	150 856	173 315	94 440	173 067	116 043
67 Fishing area total	*171 469*	*168 987*	*221 148*	*149 136*	*115 175*	*154 776*	*182 844*	*101 598*	*183 642*	*124 652*
77 USA	-	-	-	-	0	-	-	-	0	-
77 Fishing area total	-	-	-	-	*0*	-	-	-	*0*	-
Species total	*302 568*	*326 065*	*394 735*	*294 915*	*318 717*	*371 552*	*386 928*	*285 338*	*360 973*	*266 907*
Chum(=Keta=Dog)salmon	**Saumon chien**		**Keta**		***Oncorhynchus keta***			**1,23(01)009,03**		**CHU**
02 USA	769	1 717	518	1 567	738	981	325	59	4	48
02 Fishing area total	*769*	*1 717*	*518*	*1 567*	*738*	*981*	*325*	*59*	*4*	*48*
04 Japan	13 608	17 460	17 736	19 026	18 346	16 069	11 684	12 326	9 599	10 000 F
04 Fishing area total	*13 608*	*17 460*	*17 736*	*19 026*	*18 346*	*16 069*	*11 684*	*12 326*	*9 599*	*10 000 F*
05 Russian Fed	6 599	10 115	10 049	6 670	6 999	7 309	8 137	14 546	14 127	13 864
05 Fishing area total	*6 599*	*10 115*	*10 049*	*6 670*	*6 999*	*7 309*	*8 137*	*14 546*	*14 127*	*13 864*
61 Japan	195 892	205 647	249 982	280 855	250 837	190 553	171 182	151 123	207 760	201 691
Russian Fed	12 610	14 603	12 632	16 413	15 899	18 737	20 025	21 944	17 940	22 698
61 Fishing area total	*208 502*	*220 250*	*262 614*	*297 268*	*266 736*	*209 290*	*191 207*	*173 067*	*225 700*	*224 389*
67 Canada	17 273	20 323	12 115	6 524	8 685	19 903	4 937	2 783	5 549	12 332
USA	40 245	58 900	121 563	80 340	46 056	58 413	64 970	73 574	52 683	50 642
67 Fishing area total	*57 518*	*79 223*	*133 678*	*86 864*	*54 741*	*78 316*	*69 907*	*76 357*	*58 232*	*62 974*
Species total	*286 996*	*328 765*	*424 595*	*411 395*	*347 560*	*311 965*	*281 260*	*276 355*	*307 662*	*311 275 F*
Masu(=Cherry) salmon	**Saumon du Japon**		**Salmón japonés**		***Oncorhynchus masou***			**1,23(01)009,05**		**CHE**
04 Japan	796	852	842	830	810	836	849	856	826	765
04 Fishing area total	*796*	*852*	*842*	*830*	*810*	*836*	*849*	*856*	*826*	*765*
05 Russian Fed	3	3	13	41	4	3	7	3	4	2
05 Fishing area total	*3*	*3*	*13*	*41*	*4*	*3*	*7*	*3*	*4*	*2*
61 Japan	1 543	1 694	1 431	1 677	990	1 734	1 130	955	779	1 114
Russian Fed	-	4	4	4	4	4	3	-	1	1
61 Fishing area total	*1 543*	*1 698*	*1 435*	*1 681*	*994*	*1 738*	*1 133*	*955*	*780*	*1 115*
Species total	*2 342*	*2 553*	*2 290*	*2 552*	*1 808*	*2 577*	*1 989*	*1 814*	*1 610*	*1 882*
Sockeye(=Red)salmon	**Saumon rouge**		**Salmón rojo**		***Oncorhynchus nerka***			**1,23(01)009,06**		**SOC**
02 USA	171	176	0	115	78	190	56	13	0	-
02 Fishing area total	*171*	*176*	*0*	*115*	*78*	*190*	*56*	*13*	*0*	-

B-23 Salmons, trouts, smelts / Saumons, truites, éperlans / Salmones, truchas, eperlanos

Capture production by species, fishing areas and countries or areas / Captures par espèces, zones de pêche et pays ou zones / Capturas por especies, áreas de pesca y países o áreas

Species, Fishing area Espèce, Zone de pêche Especie, Area de pesca	1993 t	1994 t	1995 t	1996 t	1997 t	1998 t	1999 t	2000 t	2001 t	2002 t
04 Japan	80	83	79	72	88	118	90	52	22	30 F
04 Fishing area total	*80*	*83*	*79*	*72*	*88*	*118*	*90*	*52*	*22*	*30 F*
05 Russian Fed	7 651	5 150	8 229	9 261	3 479	3 947	7 097	6 872	12 106	13 671
05 Fishing area total	*7 651*	*5 150*	*8 229*	*9 261*	*3 479*	*3 947*	*7 097*	*6 872*	*12 106*	*13 671*
61 Japan	7 725	3 704	6 155	5 651	9 158	2 650	2 660	2 095	2 718	3 192
Russian Fed	5 490	5 658	5 998	13 630	6 698	8 820	7 792	12 676	10 369	14 701
61 Fishing area total	*13 215*	*9 362*	*12 153*	*19 281*	*15 856*	*11 470*	*10 452*	*14 771*	*13 087*	*17 893*
67 Canada	42 529	30 828	10 533	15 525	25 353	5 041	1 653	8 665	6 231	10 050
USA	178 967	138 087	158 618	144 330	87 221	58 206	110 780	94 409	77 172	61 653
67 Fishing area total	*221 496*	*168 915*	*169 151*	*159 855*	*112 574*	*63 247*	*112 433*	*103 074*	*83 403*	*71 703*
Species total	*242 613*	*183 686*	*189 612*	*188 584*	*132 075*	*78 972*	*130 128*	*124 782*	*108 618*	*103 297 F*
Chinook(=Spring=King)salmon	**Saumon royal**		**Salmón real**		***Oncorhynchus tshawytscha***				**1,23(01)009,07**	**CHI**
02 USA	1 549	1 907	2 146	2 111	1 621	1 367	1 344	390	236	214
02 Fishing area total	*1 549*	*1 907*	*2 146*	*2 111*	*1 621*	*1 367*	*1 344*	*390*	*236*	*214*
05 Russian Fed	1 063	721	738	401	445	340	483	264	163	179
05 Fishing area total	*1 063*	*721*	*738*	*401*	*445*	*340*	*483*	*264*	*163*	*179*
61 Japan	615	364	195	250	825	534	270	147	111	180
Russian Fed	221	380	137	120	191	216	310	215	336	486
61 Fishing area total	*836*	*744*	*332*	*370*	*1 016*	*750*	*580*	*362*	*447*	*666*
67 Canada	4 817	3 573	1 510	455	1 662	1 386	742	507	636	1 655
USA	5 619	5 271	6 080	5 093	5 518	5 057	3 594	4 300	6 095	8 779
67 Fishing area total	*10 436*	*8 844*	*7 590*	*5 548*	*7 180*	*6 443*	*4 336*	*4 807*	*6 731*	*10 434*
77 USA	1 152	1 407	2 993	2 078	2 737	937	1 991	2 613	1 193	2 416
77 Fishing area total	*1 152*	*1 407*	*2 993*	*2 078*	*2 737*	*937*	*1 991*	*2 613*	*1 193*	*2 416*
81 New Zealand	-	-	2	1	1	3	1	1	1	0
81 Fishing area total	*-*	*-*	*2*	*1*	*1*	*3*	*1*	*1*	*1*	*0*
Species total	*15 036*	*13 623*	*13 801*	*10 509*	*13 000*	*9 840*	*8 735*	*8 437*	*8 771*	*13 909*
Coho(=Silver)salmon	**Saumon argenté**		**Salmón plateado**		***Oncorhynchus kisutch***				**1,23(01)009,08**	**COH**
02 USA	2 314	2 996	441	4 021	696	855	282	802	659	161
02 Fishing area total	*2 314*	*2 996*	*441*	*4 021*	*696*	*855*	*282*	*802*	*659*	*161*
05 Russian Fed	1 315	1 761	1 213	1 577	898	1 449	1 054	1 419	1 090	1 233
05 Fishing area total	*1 315*	*1 761*	*1 213*	*1 577*	*898*	*1 449*	*1 054*	*1 419*	*1 090*	*1 233*
61 Japan	207	-	270	701	575	746	508	376	502	562
Russian Fed	1 013	504	266	399	412	870	614	859	944	1 027
61 Fishing area total	*1 220*	*504*	*536*	*1 100*	*987*	*1 616*	*1 122*	*1 235*	*1 446*	*1 589*
67 Canada	4 316	7 713	4 866	3 871	751	16	14	31	46	453
USA	16 259	32 614	21 861	17 626	9 856	15 450	12 977	14 544	16 764	17 314
67 Fishing area total	*20 575*	*40 327*	*26 727*	*21 497*	*10 607*	*15 466*	*12 991*	*14 575*	*16 810*	*17 767*
77 USA	0	0	5	6	3	0	-	4	1	-
77 Fishing area total	*0*	*0*	*5*	*6*	*3*	*0*	*-*	*4*	*1*	*-*
87 Chile	27	14	-	-	-	-	-	-	-	-
87 Fishing area total	*27*	*14*	*-*	*-*	*-*	*-*	*-*	*-*	*-*	*-*
Species total	*25 451*	*45 602*	*28 922*	*28 201*	*13 191*	*19 386*	*15 449*	*18 035*	*20 006*	*20 750*
Rainbow trout	**Truite arc-en-ciel**		**Trucha arco iris**		***Oncorhynchus mykiss***				**1,23(01)009,09**	**TRR**
01 Morocco	15	-	-	-	-	-	-	-	-	-
01 Fishing area total	*15*	*-*	*-*	*-*	*-*	*-*	*-*	*-*	*-*	*-*
02 Mexico	297	300	1 349	1 653	102	95	91	232	223	174
USA	268	174	161	168	137	358	82	145	221	146
02 Fishing area total	*565*	*474*	*1 510*	*1 821*	*239*	*453*	*173*	*377*	*444*	*320*
03 Argentina	186	261	3	2	0	0	0	-	-	-
Bolivia	141	116	116	338	338	340	342	345	280	197
Brazil	10 F	10 F	0	0	0	0	0	0	-	-
Peru	303	...	509	63	869	392	184	220	191	380
Venezuela	50	92	57	59	...	...	...	...	186	400
03 Fishing area total	*690 F*	*479 F*	*685*	*462*	*1 207*	*732*	*526*	*565*	*657*	*977*
04 Georgia	31	5	12	3	...	...	...	...	...	...
Japan	568	567	597	728	672	618	562	536	484	447
Korea Rep	12	15	13	...	...	...	...	...	...	...
Uzbekistan	-	0	-	-	-	-	-	-	-	-
04 Fishing area total	*611*	*587*	*622*	*731*	*672*	*618*	*562*	*536*	*484*	*447*
05 Albania	-	-	-	-	-	-	-	-	-	67
Bulgaria	-	-	-	-	12	10	10	5	17	21

B-23 Salmons, trouts, smelts / Saumons, truites, éperlans / Salmones, truchas, eperlanos

Capture production by species, fishing areas and countries or areas / Captures par espèces, zones de pêche et pays ou zones / Capturas por especies, áreas de pesca y países o áreas

Species, Fishing area Espèce, Zone de pêche Especie, Area de pesca	1993 t	1994 t	1995 t	1996 t	1997 t	1998 t	1999 t	2000 t	2001 t	2002 t
Croatia	4	4	2	5	5	1 F	1	3	18	2
Czech Rep	30	24	23	28	29	30	38	39	48	44
Denmark	0	1	1	1	1	-	6	-	0	0
Finland	954	782	782	813	813	975	975	660	660	660
Ireland	53	75	75	75	99	75	...	...	...	...
Poland	...	...	...	...	...	...	...	3	1	2
Romania	...	...	...	...	3	25	25	32	65	12
Slovakia	11	10	15	16	16	17	16	19	30	31
Slovenia	33	28	36	32	33	23	19	21	21	25
Switzerland	-	-	40	0	0	0	0	0	0	0
UK	300	300	300	300	100	2 831	3 224	1 267	1 174	1 765
05 Fishing area total	*1 385*	*1 224*	*1 274*	*1 270*	*1 111*	*3 987 F*	*4 314*	*2 049*	*2 034*	*2 629*
27 Finland	307	225	252	104	105	210	171	77	128	79
Poland	-	-	-	-	35	27	14	9	11	11
Sweden	9	-	4	-	-	-	0	-	-	-
27 Fishing area total	*316*	*225*	*256*	*104*	*140*	*237*	*185*	*86*	*139*	*90*
67 Canada	8	5	4	3	1	4	1	1	-	-
67 Fishing area total	*8*	*5*	*4*	*3*	*1*	*4*	*1*	*1*	-	-
Species total	*3 590 F*	*2 994 F*	*4 351*	*4 391*	*3 370*	*6 031 F*	*5 761*	*3 614*	*3 758*	*4 463*
Pacific salmons nei	**Saumons du Pacifique nca**		**Salmones del Pacífico nep**		***Oncorhynchus spp***				**1,23(01)009,XX**	**ORC**
67 USA	-	-	-	-	-	-	-	50	204	6
67 Fishing area total	-	-	-	-	-	-	-	*50*	*204*	*6*
Species total	-	-	-	-	-	-	-	*50*	*204*	*6*
Brook trout	**Saumon de fontaine**		**Trucha de arroyo**		***Salvelinus fontinalis***				**1,23(01)010,02**	**SVF**
05 Bulgaria	-	-	-	-	1	1	1	1	3	6
Czech Rep	4	1	3	2	2	2	3	3	3	5
Slovakia	0	0	1	0	1	1	0	0	1	0
Slovenia	-	-	-	1	0	0	0	1	0	0
05 Fishing area total	*4*	*1*	*4*	*3*	*4*	*4*	*4*	*5*	*7*	*11*
Species total	*4*	*1*	*4*	*3*	*4*	*4*	*4*	*5*	*7*	*11*
Arctic char	**Omble-chevalier**		**Trucha alpina**		***Salvelinus alpinus***				**1,23(01)010,05**	**ACH**
05 Slovenia	1	1	1	0	1	0	0	1	1	2
Sweden	57	65	47	25	24	28	21	24	18	17
Switzerland	27	22	22	28	27	21	22	22	16	16
UK	-	-	-	-	-	-	-	-	1	1
05 Fishing area total	*85*	*88*	*70*	*53*	*52*	*49*	*43*	*47*	*36*	*36*
21 Greenland	75	22	55	43	78	76	24	29	20	20
21 Fishing area total	*75*	*22*	*55*	*43*	*78*	*76*	*24*	*29*	*20*	*20*
Species total	*160*	*110*	*125*	*96*	*130*	*125*	*67*	*76*	*56*	*56*
Lake trout(=Char)	**Touladi (=Omble du Canada)**		**Trucha lacustre**		***Salvelinus namaycush***				**1,23(01)010,07**	**LAT**
02 Canada	575	690	631	689	617	554	491	556	679	532
USA	324	298	406	286	492	500	494	573	451	346
02 Fishing area total	*899*	*988*	*1 037*	*975*	*1 109*	*1 054*	*985*	*1 129*	*1 130*	*878*
Species total	*899*	*988*	*1 037*	*975*	*1 109*	*1 054*	*985*	*1 129*	*1 130*	*878*
Chars nei	**Ombles nca**		**Salvelinos nep**		***Salvelinus spp***				**1,23(01)010,XX**	**CHR**
05 Norway	84	81	88	71	67	74	85	102	100 F	100 F
05 Fishing area total	*84*	*81*	*88*	*71*	*67*	*74*	*85*	*102*	*100 F*	*100 F*
21 Canada	48	60	32	11	31	35	46	46	34	19
21 Fishing area total	*48*	*60*	*32*	*11*	*31*	*35*	*46*	*46*	*34*	*19*
27 Norway	13	16	16	10	11	20	8	1	...	...
Sweden	0	1	-	-	-	-	-	-	-	-
27 Fishing area total	*13*	*17*	*16*	*10*	*11*	*20*	*8*	*1*	...	...
Species total	*145*	*158*	*136*	*92*	*109*	*129*	*139*	*149*	*134 F*	*119 F*
Huchen	**...B**		**...C**		***Hucho hucho***				**1,23(01)020,02**	**HUC**
05 Slovakia	0	0	1	1	1	1	1	1	1	1
05 Fishing area total	*0*	*0*	*1*	*1*	*1*	*1*	*1*	*1*	*1*	*1*
Species total	*0*	*0*	*1*	*1*	*1*	*1*	*1*	*1*	*1*	*1*
Grayling	**Ombre commun**		**Tímalo**		***Thymallus thymallus***				**1,23(02)005,01**	**TLV**
05 Belgium	5	5	5	5	5	5	5	5	5	5
Czech Rep	16	16	16	19	15	13	16	16	15	13
Norway	...	...	...	...	1	...	...	...	...	...
Romania	-	-	-	-	-	-	2	-	-	-
Slovakia	14	16	24	18	16	16	14	13	17	18

B-23 Salmons, trouts, smelts / Saumons, truites, éperlans / Salmones, truchas, eperlanos

Capture production by species, fishing areas and countries or areas / Captures par espèces, zones de pêche et pays ou zones / Capturas por especies, áreas de pesca y países o áreas

Species, Fishing area Espèce, Zone de pêche Especie, Area de pesca	1993 t	1994 t	1995 t	1996 t	1997 t	1998 t	1999 t	2000 t	2001 t	2002 t
Slovenia	7	6	7	7	6	4	4	3	2	2
Sweden	...	1	1	3	0	0	0	-	-	-
UK	...	1	-	-	-	-	-	0	0	-
05 Fishing area total	*42*	*45*	*53*	*52*	*43*	*38*	*41*	*37*	*39*	*38*
Species total	*42*	*45*	*53*	*52*	*43*	*38*	*41*	*37*	*39*	*38*
Ayu sweetfish	**Ayu**		**Ayu**		***Plecoglossus altivelis***			**1,23(03)016,01**		**PCA**
04 China,Taiwan	8	-	2	-	-	-	-	-	-	-
Japan	14 242	14 272	13 700	12 732	12 619	11 386	11 380	11 172	11 148	10 663
Korea Rep	-	-	-	-	5	-	-	-	-	-
04 Fishing area total	*14 250*	*14 272*	*13 702*	*12 732*	*12 624*	*11 386*	*11 380*	*11 172*	*11 148*	*10 663*
Species total	*14 250*	*14 272*	*13 702*	*12 732*	*12 624*	*11 386*	*11 380*	*11 172*	*11 148*	*10 663*
European smelt	**Eperlan européen**		**Eperlano europeo**		***Osmerus eperlanus***			**1,23(04)003,01**		**SME**
05 Denmark	-	-	-	7	0	1	-	0	-	-
Estonia	502	224	710	478	401	1 421	947	1 104	623	2 216
Finland	603	864	1 100	854	854	450	450	366	366	366
Latvia	...	3	3	2	4	18	8	-	1	1
Lithuania	136	131	105	81	190	200	147	214	178	139
Netherlands	580	1 537	2 952	856	1 033	327	1 081	1 000 F	930 F	868
Poland	-	-	-	-	38	2	33	2	1	41
Sweden	7	18	11	9	10	8	9	-	-	-
05 Fishing area total	*1 828*	*2 777*	*4 881*	*2 287*	*2 530*	*2 427*	*2 675*	*2 686 F*	*2 099 F*	*3 631*
27 Denmark	71	70	76	46	34	18	20	29	25	10
Estonia	3	3	21	6	14	10	61	90	139	104
Finland	1 427	822	1 137	1 369	1 044	759	880	413	533	663
France	60	71	84	95	106	69	75	103	96	64
Germany	46	46	12	29	87	32	46	4	6	13
Latvia	331	2	351	384	331	200	172	261	127	13
Lithuania	-	-	-	-	-	134	218	-	182	148
Netherlands	-	-	-	-	-	-	16	74	111	258
Poland	-	-	-	-	-	51	179	17	19	16
Russian Fed	1 351	1 150	805	1 022	760	835	409	844	976	730
UK	1	57	7	2 677	2 635	1 217	888	5 316	21	32
27 Fishing area total	*3 290*	*2 221*	*2 493*	*5 628*	*5 011*	*3 325*	*2 964*	*7 151*	*2 235*	*2 051*
Species total	*5 118*	*4 998*	*7 374*	*7 915*	*7 541*	*5 752*	*5 639*	*9 837 F*	*4 334 F*	*5 682*
Rainbow smelt	**Eperlan arc-en-ciel**		**Eperlano arco iris**		***Osmerus mordax***			**1,23(04)003,03**		**SMR**
02 USA	1 666	1 183	968	711	522	321	328	398	209	188
02 Fishing area total	*1 666*	*1 183*	*968*	*711*	*522*	*321*	*328*	*398*	*209*	*188*
05 Russian Fed	1 143	338	1 006	640	1 113	1 758	1 340	608	719	1 417
05 Fishing area total	*1 143*	*338*	*1 006*	*640*	*1 113*	*1 758*	*1 340*	*608*	*719*	*1 417*
21 Canada	1 036	1 396	1 132	1 078	1 227	1 255	356	970	308	314
USA	10	9	0	-	0	0	-	0	0	-
21 Fishing area total	*1 046*	*1 405*	*1 132*	*1 078*	*1 227*	*1 255*	*356*	*970*	*308*	*314*
27 Russian Fed	-	-	-	-	-	-	-	1	6	23
27 Fishing area total	-	-	-	-	-	-	-	*1*	*6*	*23*
Species total	*3 855*	*2 926*	*3 106*	*2 429*	*2 862*	*3 334*	*2 024*	*1 977*	*1 242*	*1 942*
Pond smelt	**Eperlan à petite bouche**		**Eperlano de estanque**		***Hypomesus olidus***			**1,23(04)008,01**		**PSM**
02 Canada	7 975	4 826	5 517	3 981	5 964	6 471	5 691	3 254	4 241	3 263
02 Fishing area total	*7 975*	*4 826*	*5 517*	*3 981*	*5 964*	*6 471*	*5 691*	*3 254*	*4 241*	*3 263*
Species total	*7 975*	*4 826*	*5 517*	*3 981*	*5 964*	*6 471*	*5 691*	*3 254*	*4 241*	*3 263*
Surf smelt	**Eperlan du Pacifique**		**Eperlano del Pacífico**		***Hypomesus pretiosus***			**1,23(04)008,02**		**SUS**
67 Canada	2	3	1	1	-	-	-	-	-	-
67 Fishing area total	*2*	*3*	*1*	*1*	-	-	-	-	-	-
Species total	*2*	*3*	*1*	*1*	-	-	-	-	-	-
Eulachon	**Eulakane**		**...C**		***Thaleichthys pacificus***			**1,23(04)012,01**		**EUL**
02 USA	233	20	200	4	27	3	8	-	-	-
02 Fishing area total	*233*	*20*	*200*	*4*	*27*	*3*	*8*	-	-	-
67 Canada	9	6	26	30	0	0	-	-	1	5
USA	-	-	-	-	0	3	-	13	142	324
67 Fishing area total	*9*	*6*	*26*	*30*	*0*	*3*	-	*13*	*143*	*329*
Species total	*242*	*26*	*226*	*34*	*27*	*6*	*8*	*13*	*143*	*329*
Smelts nei	**Eperlans nca**		**Eperlanos nep**		***Osmerus spp, Hypomesus spp***			**1,23(04)XXX,XX**		**SMX**
05 Russian Fed	3 750	2 701	3 118	1 927	4 095	3 102	2 740	3 825	3 568	3 965
05 Fishing area total	*3 750*	*2 701*	*3 118*	*1 927*	*4 095*	*3 102*	*2 740*	*3 825*	*3 568*	*3 965*

B-23 Salmons, trouts, smelts / Saumons, truites, éperlans / Salmones, truchas, eperlanos

Capture production by species, fishing areas and countries or areas
Captures par espèces, zones de pêche et pays ou zones
Capturas por especies, áreas de pesca y países o áreas

Species, Fishing area Espèce, Zone de pêche Especie, Area de pesca	1993 t	1994 t	1995 t	1996 t	1997 t	1998 t	1999 t	2000 t	2001 t	2002 t
61 Russian Fed	316	39	130	122	321	322	390	735	612	1 032
61 Fishing area total	*316*	*39*	*130*	*122*	*321*	*322*	*390*	*735*	*612*	*1 032*
67 USA	76	218	71	78	839	490	432	442	218	261
67 Fishing area total	*76*	*218*	*71*	*78*	*839*	*490*	*432*	*442*	*218*	*261*
77 USA	-	-	-	-	7	7	5	6	12	8
77 Fishing area total	-	-	-	-	*7*	*7*	*5*	*6*	*12*	*8*
Species total	*4 142*	*2 958*	*3 319*	*2 127*	*5 262*	*3 921*	*3 567*	*5 008*	*4 410*	*5 266*
Vendace	**Corégone blanc**		**Coregono blanco**		***Coregonus albula***				**1,23(12)001,01**	**FVE**
05 Estonia	-	-	45	127	153	159	47	1	-	-
Finland	3 609	3 555	4 174	5 845	5 845	4 990	4 990	4 837	4 837	4 837
Latvia	-	-	-	-	-	-	-	-	1	-
Lithuania	...	...	...	...	...	...	...	7	6	4
Norway	...	...	...	...	10	4	10	6	5 F	5 F
Poland	500	500	232	234	297	217	286	275	227	207
Sweden	528	593	492	604	522	257	182	261	123	124
05 Fishing area total	*4 637*	*4 648*	*4 943*	*6 810*	*6 827*	*5 627*	*5 515*	*5 387*	*5 199 F*	*5 177 F*
27 Finland	102	105	135	143	130	125	135	185	187	166
Latvia	4	5	5	5	5	6	7	4	4	3
Sweden	1 104	965	626	717	686	419	391	569	697	1 003
27 Fishing area total	*1 210*	*1 075*	*766*	*865*	*821*	*550*	*533*	*758*	*888*	*1 172*
Species total	*5 847*	*5 723*	*5 709*	*7 675*	*7 648*	*6 177*	*6 048*	*6 145*	*6 087 F*	*6 349 F*
European whitefish	**Corégone lavaret**		**Lavareto**		***Coregonus lavaretus***				**1,23(12)001,02**	**PLN**
05 Bulgaria	-	-	-	-	0	0	0	-	-	-
Czech Rep	3	-	-	2	2	1	1	1	1	1
Denmark	4	3	3	3	1	3	1	1	1	0
Finland	3 612	4 216	4 014	3 042	3 042	2 531	2 531	2 421	2 421	2 421
Lithuania	...	...	...	...	...	...	...	1	1	0
Norway	...	...	...	...	57	52	54	47	45 F	45 F
Poland	2	2	28	33	24	14	21	29	13	11
Sweden	199	247	202	180	195	185	181	171	182	194
05 Fishing area total	*3 820*	*4 468*	*4 247*	*3 260*	*3 321*	*2 786*	*2 789*	*2 671*	*2 664 F*	*2 672 F*
27 Denmark	28	36	35	27	30	26	24	6	76	51
Estonia	8	10	6	21	20	20	28	33	33	47
Finland	2 783	2 083	2 140	2 084	1 961	2 351	2 172	2 120	1 826	1 754
Germany	-	-	-	-	5	8	21	47	63	34
Lithuania	-	-	-	-	-	-	-	-	3	2
Sweden	354	571	465	350	302	304	279	249	154	222
27 Fishing area total	*3 173*	*2 700*	*2 646*	*2 482*	*2 318*	*2 709*	*2 524*	*2 455*	*2 155*	*2 110*
Species total	*6 993*	*7 168*	*6 893*	*5 742*	*5 639*	*5 495*	*5 313*	*5 126*	*4 819 F*	*4 782 F*
Houting	**Bondelle**		**Coregono picudo**		***Coregonus oxyrinchus***				**1,23(12)001,03**	**HOU**
05 Bulgaria	-	-	-	-	0	0	0	-	-	-
Denmark	-	-	-	-	-	1	-	22	-	-
Estonia	18	14	25	63	30	60	35	9	9	11
05 Fishing area total	*18*	*14*	*25*	*63*	*30*	*61*	*35*	*31*	*9*	*11*
Species total	*18*	*14*	*25*	*63*	*30*	*61*	*35*	*31*	*9*	*11*
Lake(=Common)whitefish	**Corégone de lac**		**Coregono de lago**		***Coregonus clupeaformis***				**1,23(12)001,12**	**WHL**
02 Canada	7 730	9 480	9 196	9 121	8 376	8 599	8 330	9 028	9 624	8 623
USA	5 162	5 110	5 310	5 272	5 842	5 678	5 353	5 199	4 484	4 239
02 Fishing area total	*12 892*	*14 590*	*14 506*	*14 393*	*14 218*	*14 277*	*13 683*	*14 227*	*14 108*	*12 862*
Species total	*12 892*	*14 590*	*14 506*	*14 393*	*14 218*	*14 277*	*13 683*	*14 227*	*14 108*	*12 862*
Lake cisco	**Cisco de lac**		**Coregono de artedi**		***Coregonus artedi***				**1,23(12)001,13**	**CIS**
02 Canada	608	1 301	1 101	919	1 008	764	565	581	567	503
USA	388	345	316	226	215	284	239	255	303	285
02 Fishing area total	*996*	*1 646*	*1 417*	*1 145*	*1 223*	*1 048*	*804*	*836*	*870*	*788*
Species total	*996*	*1 646*	*1 417*	*1 145*	*1 223*	*1 048*	*804*	*836*	*870*	*788*
Whitefishes nei	**Corégones nca**		**Coregonos nep**		***Coregonus spp***				**1,23(12)001,XX**	**WHF**
02 USA	1 839	2 436	2 246	1 415	1 506	2 178	1 586	1 030	768	841
02 Fishing area total	*1 839*	*2 436*	*2 246*	*1 415*	*1 506*	*2 178*	*1 586*	*1 030*	*768*	*841*
04 Armenia	1 507	970	776	428	512	605	922	881	632	202
Kazakhstan	32	27	29	27 F	20 F	15 F	20 F	30 F	35	37
Kyrgyzstan	42	62	62	53 F	40 F	26 F	15 F	16 F	12	9
04 Fishing area total	*1 581*	*1 059*	*867*	*508 F*	*572 F*	*646 F*	*957 F*	*927 F*	*679*	*248*
05 Belarus	4	1	3	4	1	2	17	18	31	46
Belgium	1	1	1	1	1	1	1	1	1	1

B-23 Salmons, trouts, smelts / Saumons, truites, éperlans / Salmones, truchas, eperlanos

Capture production by species, fishing areas and countries or areas / Captures par espèces, zones de pêche et pays ou zones / Capturas por especies, áreas de pesca y países o áreas

Species, Fishing area Espèce, Zone de pêche Especie, Area de pesca		1993 t	1994 t	1995 t	1996 t	1997 t	1998 t	1999 t	2000 t	2001 t	2002 t
	Germany	956	459	459	459	459	459	459	459	459	459
	Latvia	-	1	3	2	1	1	-	-	-	-
	Lithuania	20	13	8	18	20	10	11	-	-	-
	Russian Fed	9 705	5 408	5 781	9 307	9 350	8 783	8 867	9 132	9 053	13 170
	Switzerland	850	613	816	1 050	1 182	1 098	1 074	980	1 174	946
05	*Fishing area total*	*11 536*	*6 496*	*7 071*	*10 841*	*11 014*	*10 354*	*10 429*	*10 590*	*10 718*	*14 622*
27	Isle of Man	0	0	0	-	-	-	-	-	-	-
	Russian Fed	4	6	5	6	5	3	4	29	8	7
27	*Fishing area total*	*4*	*6*	*5*	*6*	*5*	*3*	*4*	*29*	*8*	*7*
Species total		*14 960*	*9 997*	*10 189*	*12 770 F*	*13 097 F*	*13 181 F*	*12 976 F*	*12 576 F*	*12 173*	*15 718*
Salmonoids nei		**Salmonoidés nca**		**Salmonoideos nep**		***Salmonoidei***				**1,23(XX)XXX,XX**	**SLX**
01	Kenya	112	17	196	11	16	24	29	31	67	112
01	*Fishing area total*	*112*	*17*	*196*	*11*	*16*	*24*	*29*	*31*	*67*	*112*
02	Canada	-	-	139	112	-	2	1	-	-	-
02	*Fishing area total*	*-*	*-*	*139*	*112*	*-*	*2*	*1*	*-*	*-*	*-*
04	Iran	677	242	433	8	7	8	3	5	2	9
	Korea Rep	109	90	61	18	19	102	219	...	...	...
04	*Fishing area total*	*786*	*332*	*494*	*26*	*26*	*110*	*222*	*5*	*2*	*9*
05	Albania	16 F	13 F	11	35	15	102	104	110	56	43
	Finland	364	293	293	426	426	323	323	282	282	282
	Germany	354	29	29	29	29	29	29	29	29	29
	Greece	...	...	...	1	6	38	11	50	42	40
	Italy	3 000	3 225	3 325	1 625	1 091	897	937	692	846	945
	Russian Fed	465	288	340	664	657	716	619	616	783	1 191
05	*Fishing area total*	*4 199 F*	*3 848 F*	*3 998*	*2 780*	*2 224*	*2 105*	*2 023*	*1 779*	*2 038*	*2 530*
27	Finland	...	39	39	46	46	6	6	6	0	0
	France	-	-	-	-	-	-	30	16	1	-
	Latvia	5	-	1	-	-	-	-	-	-	-
	Spain	-	-	-	0	2	-	8	33	13	1 263
	Sweden	19	9	0	0	1	0	0	-	-	-
27	*Fishing area total*	*24*	*48*	*40*	*46*	*49*	*6*	*44*	*55*	*14*	*1 263*
37	Spain	-	-	-	18	22	26	-	-	0	-
37	*Fishing area total*	*-*	*-*	*-*	*18*	*22*	*26*	*-*	*-*	*0*	*-*
61	Korea Rep	1 335	1 215	1 455	1 445	633	1 020	57	28	19	-
	Russian Fed	1 110	995	877	674	908	1 387	827	2 112	728	731
61	*Fishing area total*	*2 445*	*2 210*	*2 332*	*2 119*	*1 541*	*2 407*	*884*	*2 140*	*747*	*731*
Species total		*7 566 F*	*6 455 F*	*7 199*	*5 112*	*3 878*	*4 680*	*3 203*	*4 010*	*2 868*	*4 645*
Group total		***981 551***	***996 184***	***1 153 553***	***1 032 524***	***924 572***	***889 982***	***913 309***	***805 139***	***891 042***	***806 998***

B-24 Shads / Aloses / Sábalos

Capture production by species, fishing areas and countries or areas
Captures par espèces, zones de pêche et pays ou zones
Capturas por especies, áreas de pesca y países o áreas

Species, Fishing area Espèce, Zone de pêche Especie, Area de pesca	1993 t	1994 t	1995 t	1996 t	1997 t	1998 t	1999 t	2000 t	2001 t	2002 t
Pontic shad	**Alose de la mer Noire**		**Sábalo del Mar Negro**		***Alosa pontica***				**1,21(05)011,02**	**SHC**
05 Bulgaria	42	30 F	31	124	86	87	32	29	16	35
Romania	391	678	331	392	485	378	16	101	125	264
Russian Fed	-	2	1	1	1	-	-	-	-	2
Ukraine	220	364	223	277	329	213	21	83	146	206
05 Fishing area total	*653*	*1 074 F*	*586*	*794*	*901*	*678*	*69*	*213*	*287*	*507*
37 Bulgaria	14	15 F	112	109	79	84	41	10	16	106
Romania	-	-	-	8	2	68	4	5	3	2
Russian Fed	11	1	1	8	1	1	-	-	-	-
Ukraine	61	19	13	6	-	-	3	0	2	4
37 Fishing area total	*86*	*35 F*	*126*	*131*	*82*	*153*	*48*	*15*	*21*	*112*
Species total	*739*	*1 109 F*	*712*	*925*	*983*	*831*	*117*	*228*	*308*	*619*
American shad	**Alose savoureuse**		**Sábalo americano**		***Alosa sapidissima***				**1,21(05)011,03**	**SHA**
02 Canada	20	22	23	22	22	23	23	23	9	9
USA	673	282	308	815	523	935	557	485	768	24
02 Fishing area total	*693*	*304*	*331*	*837*	*545*	*958*	*580*	*508*	*777*	*33*
21 Canada	112	80	120	116	118	41	41	56	21	17
USA	758	457	444	558	584	707	521	323	505	340
21 Fishing area total	*870*	*537*	*564*	*674*	*702*	*748*	*562*	*379*	*526*	*357*
31 USA	265	294	273	343	246	280	117	288	171	286
31 Fishing area total	*265*	*294*	*273*	*343*	*246*	*280*	*117*	*288*	*171*	*286*
67 USA	8	34	158	15	34	62	-	105	178	66
67 Fishing area total	*8*	*34*	*158*	*15*	*34*	*62*	-	*105*	*178*	*66*
77 USA	-	-	-	-	0	-	-	-	-	-
77 Fishing area total	-	-	-	-	*0*	-	-	-	-	-
Species total	*1 836*	*1 169*	*1 326*	*1 869*	*1 527*	*2 048*	*1 259*	*1 280*	*1 652*	*742*
Allis shad	**Alose vraie (=Grande alose)**		**Sábalo común**		***Alosa alosa***				**1,21(05)011,04**	**ASD**
27 France	29	69	57	59	56	2	...	38	45	94
27 Fishing area total	*29*	*69*	*57*	*59*	*56*	*2*	...	*38*	*45*	*94*
Species total	*29*	*69*	*57*	*59*	*56*	*2*	...	*38*	*45*	*94*
Twaite shad	**Alose feinte**		**Saboga(=Alosa)**		***Alosa fallax***				**1,21(05)011,05**	**TSD**
27 France	4	3	1	2	1	...	...	11	11	9
Netherlands	-	-	-	-	-	-	-	1	5	8
Russian Fed	-	-	-	-	-	-	-	-	-	5
27 Fishing area total	*4*	*3*	*1*	*2*	*1*	...	...	*12*	*16*	*22*
Species total	*4*	*3*	*1*	*2*	*1*	...	...	*12*	*16*	*22*
Alewife	**Gaspareau**		**Pinchagua**		***Alosa pseudoharengus***				**1,21(05)011,06**	**ALE**
02 Canada	1 200	1 272	811	990	1 327	1 527	1 527	1 527	1 559	1 559
USA	67	0	4	-	5	2	23	7	21	48
02 Fishing area total	*1 267*	*1 272*	*815*	*990*	*1 332*	*1 529*	*1 550*	*1 534*	*1 580*	*1 607*
21 Canada	6 775	5 825	6 212	5 299	5 664	6 941	6 146	5 256	2 790	4 996
Cuba	26	-	-	-	-	-	-	-	-	-
USA	765	117	383	442	521	592	659	284	691	886
21 Fishing area total	*7 566*	*5 942*	*6 595*	*5 741*	*6 185*	*7 533*	*6 805*	*5 540*	*3 481*	*5 882*
27 Spain	...	...	...	...	...	...	...	61	-	-
27 Fishing area total	...	...	...	...	...	...	...	*61*	-	-
31 USA	24	321	0	0	0	0	-	0	0	-
31 Fishing area total	*24*	*321*	*0*	*0*	*0*	*0*	-	*0*	*0*	-
Species total	*8 857*	*7 535*	*7 410*	*6 731*	*7 517*	*9 062*	*8 355*	*7 135*	*5 061*	*7 489*
Hickory shad	**Alose américaine**		**...C**		***Alosa mediocris***				**1,21(05)011,08**	**SHH**
02 USA	40	31	36	88	75	47	62	51	90	42
02 Fishing area total	*40*	*31*	*36*	*88*	*75*	*47*	*62*	*51*	*90*	*42*
Species total	*40*	*31*	*36*	*88*	*75*	*47*	*62*	*51*	*90*	*42*
Shads nei	**Aloses nca**		**Sábalos nep**		***Alosa spp***				**1,21(05)011,XX**	**SHZ**
01 Morocco	2	0	0	0	0	1	-	-	-	-
01 Fishing area total	*2*	*0*	*0*	*0*	*0*	*1*	-	-	-	-
05 Albania	3 F	2 F	1	2	1	-	-	-	2	23
Greece	...	...	...	2	5	4	6	3	2	2
Romania	-	-	-	-	-	3	-	1	-	-
Switzerland	...	...	...	7	12	6	11	20	18	23
Ukraine	19	32	27	22	46	22	5	0	-	-

B-24 Shads / Aloses / Sábalos

Capture production by species, fishing areas and countries or areas
Captures par espèces, zones de pêche et pays ou zones
Capturas por especies, áreas de pesca y países o áreas

Species, Fishing area Espèce, Zone de pêche Especie, Area de pesca	1993 t	1994 t	1995 t	1996 t	1997 t	1998 t	1999 t	2000 t	2001 t	2002 t
UK	0	-	-	-	-	-	-	0	0	-
05 Fishing area total	*22 F*	*34 F*	*28*	*33*	*64*	*35*	*22*	*24*	*22*	*48*
27 France	7	3	1	2	3	4	41	6	12	15
Lithuania	-	-	-	-	-	-	-	-	-	1
Portugal	40	40	39	18	17	21	17	20	22	21
UK	0	0	0	6	0	1	1	3	8	5
27 Fishing area total	*47*	*43*	*40*	*26*	*20*	*26*	*59*	*29*	*42*	*42*
34 Greece	-	-	-	-	13	36	-	-	-	6
Morocco	4	2	-	-	-	4	-	22	62	1
34 Fishing area total	*4*	*2*	-	-	*13*	*40*	-	*22*	*62*	*7*
37 France	-	-	-	-	-	-	8	19	2	3
Greece	409	706	856	1 008	1 266	1 744	1 892	1 280	2 215	2 000 F
Morocco	1	0	7	7	10	0	-	25	0	0
Romania	189	92	106	101	43	111	60	76	22	3
Spain	-	-	-	-	-	-	-	-	-	382
Tunisia	0	-	0	0	0	-	-	-	-	-
Turkey	5 101	4 331	1 590	1 166	505	880	680	720	690	862
Ukraine	27	...	32	1	11	7	-	-	-	-
37 Fishing area total	*5 727*	*5 129*	*2 591*	*2 283*	*1 835*	*2 742*	*2 640*	*2 120*	*2 929*	*3 250 F*
Species total	*5 802 F*	*5 208 F*	*2 659*	*2 342*	*1 932*	*2 844*	*2 721*	*2 195*	*3 055*	*3 347 F*
Caspian shads	**Aloses de la mer Caspienne**		**Sábalos del Mar Caspio**		***Caspialosa spp***				**1,21(05)014,XX**	**ASP**
04 Azerbaijan	150	209	67	75	42	82	60	1	52	14
Iran	28 730	51 000	41 000	57 000	60 400	85 000	95 000	78 000	45 180	26 000
Kazakhstan	3	7	0	0	0	0	0	0	-	-
04 Fishing area total	*28 883*	*51 216*	*41 067*	*57 075*	*60 442*	*85 082*	*95 060*	*78 001*	*45 232*	*26 014*
05 Russian Fed	1 510	1 328	1 478	1 882	2 304	3 284	4 654	1 273	176	239
05 Fishing area total	*1 510*	*1 328*	*1 478*	*1 882*	*2 304*	*3 284*	*4 654*	*1 273*	*176*	*239*
Species total	*30 393*	*52 544*	*42 545*	*58 957*	*62 746*	*88 366*	*99 714*	*79 274*	*45 408*	*26 253*
American gizzard shad	**Alose noyer**		**Sábalo molleja**		***Dorosoma cepedianum***				**1,21(05)018,01**	**SHG**
02 USA	774	1 124	906	1 272	1 989	1 291	3 007	977	1 781	2 780
02 Fishing area total	*774*	*1 124*	*906*	*1 272*	*1 989*	*1 291*	*3 007*	*977*	*1 781*	*2 780*
Species total	*774*	*1 124*	*906*	*1 272*	*1 989*	*1 291*	*3 007*	*977*	*1 781*	*2 780*
Chacunda gizzard shad	**Alose chaconde**		**Sábalo chacunda**		***Anodontostoma chacunda***				**1,21(05)023,01**	**CHG**
57 Malaysia	2 633	2 595	2 338	2 167	1 916	1 474	3 009	2 645	2 769	2 350
57 Fishing area total	*2 633*	*2 595*	*2 338*	*2 167*	*1 916*	*1 474*	*3 009*	*2 645*	*2 769*	*2 350*
71 Malaysia	412	1 557	1 200	1 824	1 370	1 185	1 364	1 371	1 268	1 282
Philippines	325	288	2 955	2 761	1 260	1 191	1 115	1 140	1 346	1 413
71 Fishing area total	*737*	*1 845*	*4 155*	*4 585*	*2 630*	*2 376*	*2 479*	*2 511*	*2 614*	*2 695*
Species total	*3 370*	*4 440*	*6 493*	*6 752*	*4 546*	*3 850*	*5 488*	*5 156*	*5 383*	*5 045*
Kelee shad	**Alose palli**		**Sábalo chandano**		***Hilsa kelee***				**1,21(05)034,05**	**HIX**
04 India	47 255	36 478	49 441	48 302	44 519	53 729	44 810	174 399	64 599	38 984
04 Fishing area total	*47 255*	*36 478*	*49 441*	*48 302*	*44 519*	*53 729*	*44 810*	*174 399*	*64 599*	*38 984*
51 India	3 682	3 072	2 552	3 837	3 634	3 077	3 076	3 896	4 277	5 525
51 Fishing area total	*3 682*	*3 072*	*2 552*	*3 837*	*3 634*	*3 077*	*3 076*	*3 896*	*4 277*	*5 525*
57 India	40 857	36 311	40 714	48 452	47 524	48 475	41 468	43 604	44 502	43 674
57 Fishing area total	*40 857*	*36 311*	*40 714*	*48 452*	*47 524*	*48 475*	*41 468*	*43 604*	*44 502*	*43 674*
Species total	*91 794*	*75 861*	*92 707*	*100 591*	*95 677*	*105 281*	*89 354*	*221 899*	*113 378*	*88 183*
Hilsa shad	**Alose hilsa**		**Sábalo hilsa**		***Tenualosa ilisha***				**1,21(05)038,01**	**HIL**
04 Bangladesh	96 950	71 370	84 420	90 240	83 230	81 634	73 809	79 165	75 060	68 250
04 Fishing area total	*96 950*	*71 370*	*84 420*	*90 240*	*83 230*	*81 634*	*73 809*	*79 165*	*75 060*	*68 250*
51 Kuwait	500	928	1 198	1 148	1 034	919	970	650 F	337	340 F
Pakistan	796	658	476	562	597	611	502	190	170	174
51 Fishing area total	*1 296*	*1 586*	*1 674*	*1 710*	*1 631*	*1 530*	*1 472*	*840 F*	*507*	*514 F*
57 Bangladesh	130 295	121 161	129 115	135 358	131 204	124 105	140 710	140 367	154 654	152 343
57 Fishing area total	*130 295*	*121 161*	*129 115*	*135 358*	*131 204*	*124 105*	*140 710*	*140 367*	*154 654*	*152 343*
Species total	*228 541*	*194 117*	*215 209*	*227 308*	*216 065*	*207 269*	*215 991*	*220 372 F*	*230 221*	*221 107 F*
Toli shad	**Alose toli**		**Sábalo toli**		***Tenualosa toli***				**1,21(05)038,04**	**TOL**
57 Indonesia	519	626	456	429	476	617	630	528	681	720
57 Fishing area total	*519*	*626*	*456*	*429*	*476*	*617*	*630*	*528*	*681*	*720*
71 Indonesia	4 136	2 077	1 420	1 702	2 535	3 097	2 875	2 117	4 576	5 100
71 Fishing area total	*4 136*	*2 077*	*1 420*	*1 702*	*2 535*	*3 097*	*2 875*	*2 117*	*4 576*	*5 100*

B-24 Shads / Aloses / Sábalos

Capture production by species, fishing areas and countries or areas
Captures par espèces, zones de pêche et pays ou zones
Capturas por especies, áreas de pesca y países o áreas

Species, Fishing area Espèce, Zone de pêche Especie, Area de pesca	1993 t	1994 t	1995 t	1996 t	1997 t	1998 t	1999 t	2000 t	2001 t	2002 t
Species total	*4 655*	*2 703*	*1 876*	*2 131*	*3 011*	*3 714*	*3 505*	*2 645*	*5 257*	*5 820*
Chinese gizzard shad	**Alose à museau court**		**Alosa chata**		***Clupanodon thrissa***				**1,21(05)058,03**	**DAS**
61 China,Taiwan	4	54	30	25	10	11	27	21	74	31
Korea Rep	6 760	7 370	7 931	4 908	13 836	11 349	9 511	6 366	9 120	4 717
61 Fishing area total	*6 764*	*7 424*	*7 961*	*4 933*	*13 846*	*11 360*	*9 538*	*6 387*	*9 194*	*4 748*
Species total	*6 764*	*7 424*	*7 961*	*4 933*	*13 846*	*11 360*	*9 538*	*6 387*	*9 194*	*4 748*
Azov sea sprat	**Clupeonelle**		**Espadín del Mar d'Azov**		***Clupeonella cultriventris***				**1,21(05)059,01**	**CLA**
04 Azerbaijan	19 323	17 254	9 651	5 828	4 420	4 043	20 460	18 520	10 389	10 950
Kazakhstan	16 329	10 815	10 113	9 000 F	8 800	6 400	6 100	3 000	-	-
Turkmenistan	15 270	13 855	8 860	8 546	7 800	6 324	8 370	11 540	12 300	12 418
04 Fishing area total	*50 922*	*41 924*	*28 624*	*23 374 F*	*21 020*	*16 767*	*34 930*	*33 060*	*22 689*	*23 368*
05 Russian Fed	73 470	78 447	80 205	91 821	81 638	115 502	150 587	117 914	46 036	31 683
Ukraine	1 003	1 924	781	2 214	1 971	792	1 481	1 342	904	1 181
05 Fishing area total	*74 473*	*80 371*	*80 986*	*94 035*	*83 609*	*116 294*	*152 068*	*119 256*	*46 940*	*32 864*
37 Romania	-	47	42	4	2	52	4	5	11	4
Russian Fed	306	665	820	369	273	912	2 347	4 897	9 681	15 064
Ukraine	1 503	4 227	6 943	1 075	1 247	2 532	8 511	7 104	18 085	12 171
37 Fishing area total	*1 809*	*4 939*	*7 805*	*1 448*	*1 522*	*3 496*	*10 862*	*12 006*	*27 777*	*27 239*
Species total	*127 204*	*127 234*	*117 415*	*118 857 F*	*106 151*	*136 557*	*197 860*	*164 322*	*97 406*	*83 471*
Dotted gizzard shad	**Alose tachetée**		**Alosa manchada**		***Konosirus punctatus***				**1,21(05)060,01**	**DOD**
61 Japan	...	...	23 707	18 647	14 850	20 787	17 770	12 335	17 210	13 401
61 Fishing area total	*...*	*...*	*23 707*	*18 647*	*14 850*	*20 787*	*17 770*	*12 335*	*17 210*	*13 401*
Species total	*...*	*...*	*23 707*	*18 647*	*14 850*	*20 787*	*17 770*	*12 335*	*17 210*	*13 401*
Atlantic sabretooth anchovy	**Anchois goulard**		**Anchoa dentona**		***Lycengraulis grossidens***				**1,21(06)038,04**	**ANR**
03 Argentina	60	70 F	80 F	90 F	100 F	100 F	120 F	120 F	140 F	130 F
03 Fishing area total	*60*	*70 F*	*80 F*	*90 F*	*100 F*	*100 F*	*120 F*	*120 F*	*140 F*	*130 F*
Species total	*60*	*70 F*	*80 F*	*90 F*	*100 F*	*100 F*	*120 F*	*120 F*	*140 F*	*130 F*
Elongate ilisha	**Alose gracile**		**Sardineta grácil**		***Ilisha elongata***				**1,21(12)001,03**	**EIL**
61 China	28 624	32 540	46 635	51 339	77 532	84 290	110 359	107 689	101 342	94 509
Korea Rep	342	299	466	630	349	381	265	134	40	19
61 Fishing area total	*28 966*	*32 839*	*47 101*	*51 969*	*77 881*	*84 671*	*110 624*	*107 823*	*101 382*	*94 528*
Species total	*28 966*	*32 839*	*47 101*	*51 969*	*77 881*	*84 671*	*110 624*	*107 823*	*101 382*	*94 528*
West African ilisha	**Alose rasoir**		**Sardineta africana**		***Ilisha africana***				**1,21(12)001,12**	**ILI**
34 Benin	483 F	544 F	528 F	602 F	822 F	781 F	671	450 F	923	692
Côte dIvoire	350	802	288	491	534	452	314	152	50	417
Ghana	2 995	2 810	3 051	7 343	4 305	3 632	3 262	3 341	3 600	5 202
Liberia	44	36	6	124	26	63	242	110	198	200 F
Portugal	-	-	-	-	-	-	1	-	-	-
Senegal	...	...	...	...	...	...	...	...	...	271
Sierra Leone	3 027	3 017	3 013	3 010 F	3 009	3 110	5	2 996	4	1 363
Togo	36	152	32	15	101	35	113	418	796	667
34 Fishing area total	*6 935 F*	*7 361 F*	*6 918 F*	*11 585 F*	*8 797 F*	*8 073 F*	*4 608*	*7 467 F*	*5 571*	*8 812 F*
Species total	*6 935 F*	*7 361 F*	*6 918 F*	*11 585 F*	*8 797 F*	*8 073 F*	*4 608*	*7 467 F*	*5 571*	*8 812 F*
Indian pellona	**Alose-écaille indienne**		**Sardineta índica**		***Pellona ditchela***				**1,21(12)003,03**	**PEO**
57 Malaysia	6 225	7 260	6 091	11 227	10 919	6 113	4 448	7 069	6 884	7 236
57 Fishing area total	*6 225*	*7 260*	*6 091*	*11 227*	*10 919*	*6 113*	*4 448*	*7 069*	*6 884*	*7 236*
71 Malaysia	2 187	3 526	3 868	4 058	2 455	3 808	4 740	5 636	5 780	4 822
Philippines	1 443	1 451	1 177	792	802	842	826	839	1 221	1 458
71 Fishing area total	*3 630*	*4 977*	*5 045*	*4 850*	*3 257*	*4 650*	*5 566*	*6 475*	*7 001*	*6 280*
Species total	*9 855*	*12 237*	*11 136*	*16 077*	*14 176*	*10 763*	*10 014*	*13 544*	*13 885*	*13 516*
Diadromous clupeoids nei	**Clupéoidés diadromes nca**		**Clupeoideos diádromos nep**		***Clupeoidei***				**1,21(XX)XXX,XX**	**DCX**
01 Benin	160 F	137	260	250 F	230 F	210 F	210 F	200 F	230 F	230 F
Egypt	...	...	106	40	23	148	60	84	28	8
01 Fishing area total	*160 F*	*137*	*366*	*290 F*	*253 F*	*358 F*	*270 F*	*284 F*	*258 F*	*238 F*
03 Venezuela	155	160	345	552	355	467	403	287	200	183
03 Fishing area total	*155*	*160*	*345*	*552*	*355*	*467*	*403*	*287*	*200*	*183*
04 India	-	1 183	1 086	1 481	2 497	1 517	1 143	917	2 090	2 090
Iran	893	920	490	330	566	817	697	595	623	932
04 Fishing area total	*893*	*2 103*	*1 576*	*1 811*	*3 063*	*2 334*	*1 840*	*1 512*	*2 713*	*3 022*

B-24 Shads / Aloses / Sábalos

Capture production by species, fishing areas and countries or areas
Captures par espèces, zones de pêche et pays ou zones
Capturas por especies, áreas de pesca y países o áreas

Species, Fishing area Espèce, Zone de pêche Especie, Area de pesca	1993 t	1994 t	1995 t	1996 t	1997 t	1998 t	1999 t	2000 t	2001 t	2002 t
06 Papua N Guin	480 F	480 F	480 F	480 F	480 F	480 F	480 F	480 F	480 F	480 F
06 Fishing area total	*480 F*	*480 F*	*480 F*	*480 F*	*480 F*	*480 F*	*480 F*	*480 F*	*480 F*	*480 F*
27 Portugal	43	38	35	35	36	30	-	-	-	-
27 Fishing area total	*43*	*38*	*35*	*35*	*36*	*30*	*-*	*-*	*-*	*-*
57 Malaysia	45	17	30	107	102	22	45	13	50	305
57 Fishing area total	*45*	*17*	*30*	*107*	*102*	*22*	*45*	*13*	*50*	*305*
71 Malaysia	370	412	770	1 278	574	559	577	652	1 056	926
71 Fishing area total	*370*	*412*	*770*	*1 278*	*574*	*559*	*577*	*652*	*1 056*	*926*
Species total	*2 146 F*	*3 347 F*	*3 602 F*	*4 553 F*	*4 863 F*	*4 250 F*	*3 615 F*	*3 228 F*	*4 757 F*	*5 154 F*
Group total	***558 764***	***536 425***	***589 857***	***635 738***	***636 789***	***701 166***	***783 722***	***856 488***	***661 200***	***585 303***

B-25 Miscellaneous diadromous fishes / Poissons diadromes divers / Peces diádromos diversos

Capture production by species, fishing areas and countries or areas / Captures par espèces, zones de pêche et pays ou zones / Capturas por especies, áreas de pesca y países o áreas

Species, Fishing area Espèce, Zone de pêche Especie, Area de pesca	1993 t	1994 t	1995 t	1996 t	1997 t	1998 t	1999 t	2000 t	2001 t	2002 t
Sea lamprey	**Lamproie marine**		**Lamprea marina**		***Petromyzon marinus***				**1,02(01)001,01**	**LAU**
27 Estonia	0	-	0	18	3	4	7	8	3	2
France	42	13	13	13	15	11	27	22	23	35
Portugal	12	10	4	3	2	2	4	6	6	13
27 Fishing area total	*54*	*23*	*17*	*34*	*20*	*17*	*38*	*36*	*32*	*50*
Species total	*54*	*23*	*17*	*34*	*20*	*17*	*38*	*36*	*32*	*50*
River lamprey	**Lamproie de rivière**		**Lamprea de río**		***Lampetra fluviatilis***				**1,02(01)002,01**	**LAR**
05 Estonia	25	5	1	0	7	16	9	26	25	23
Latvia	96	113	95	140	80	79	120	135	88	79
Lithuania	1	-	-	1	-	-	-	-	3	2
UK	-	-	-	-	-	-	-	2	2	2
05 Fishing area total	*122*	*118*	*96*	*141*	*87*	*95*	*129*	*163*	*118*	*106*
Species total	*122*	*118*	*96*	*141*	*87*	*95*	*129*	*163*	*118*	*106*
Lampreys nei	**Lamproies nca**		**Lampreas nep**		***Petromyzontidae***				**1,02(01)XXX,XX**	**LAS**
02 USA	8	6	13	-	-	0	4	0	-	-
02 Fishing area total	*8*	*6*	*13*	-	-	*0*	*4*	*0*	-	-
05 Poland	6	0	0	0	-	-	-	-	-	-
Russian Fed	109	40	40	76	31	37	67	24	21	96
05 Fishing area total	*115*	*40*	*40*	*76*	*31*	*37*	*67*	*24*	*21*	*96*
27 Poland	-	-	-	-	2	2	2	6	5	3
Russian Fed	-	-	-	-	-	-	-	15	28	21
27 Fishing area total	-	-	-	-	*2*	*2*	*2*	*21*	*33*	*24*
Species total	*123*	*46*	*53*	*76*	*33*	*39*	*73*	*45*	*54*	*120*
Milkfish	**Chano**		**Chano**		***Chanos chanos***				**1,22(02)001,01**	**MIL**
04 Philippines	2 323	4 652	3 997	369	34	3 000	83	163	201	329
04 Fishing area total	*2 323*	*4 652*	*3 997*	*369*	*34*	*3 000*	*83*	*163*	*201*	*329*
31 Mexico	0	0	0	0	0	0	0	0	...	...
31 Fishing area total	*0*	*0*	*0*	*0*	*0*	*0*	*0*	*0*	...	...
51 Eritrea	...	0	2	0	2	3	1	3	1	1
Saudi Arabia	-	10	137	130	70	82	81	64	73	93
Untd Arab Em	40	35	50	51	53	54	55	58	60 F	65
51 Fishing area total	*40*	*45*	*189*	*181*	*125*	*139*	*137*	*125*	*134 F*	*159*
61 China,Taiwan	-	20	1	4	-	-	-	-	1	-
61 Fishing area total	-	*20*	*1*	*4*	-	-	-	-	*1*	-
71 Fiji Islands	19	19 F	20	34	26	21	30	30 F	32	30 F
Kiribati	280	285	285	290	290	...	80	2 175	58	57
Palau	1	2	2	1	0	1	0	1	1	0
Philippines	1 095	495	3 502	3 302	363	152	232	233	278	284
71 Fishing area total	*1 395*	*801 F*	*3 809*	*3 627*	*679*	*174*	*342*	*2 439 F*	*369*	*371 F*
77 Mexico	0	0	0	0	1	0	0	7	...	1
77 Fishing area total	*0*	*0*	*0*	*0*	*1*	*0*	*0*	*7*	...	*1*
Species total	*3 758*	*5 518 F*	*7 996*	*4 181*	*839*	*3 313*	*562*	*2 734 F*	*705 F*	*860 F*
Three-spined stickleback	**Epinoche à trois épines**		**Espinoso**		***Gasterosteus aculeatus***				**1,50(01)001,01**	**GTA**
05 Belarus	577	99	85	80	34	26	13	29	41	49
Latvia	-	3	1	1	1	-	-	-	-	-
Lithuania	16	-	-	-	-	-	13	22	18	-
Russian Fed	564	330	337	271	314	884	1 021	1 116	906	1 676
05 Fishing area total	*1 157*	*432*	*423*	*352*	*349*	*910*	*1 047*	*1 167*	*965*	*1 725*
27 Denmark	-	-	-	-	4	-	-	-	-	-
Latvia	14	7	1	33	109	82	-	-	-	-
Lithuania	-	-	-	-	-	-	-	-	-	6
Russian Fed	-	-	-	-	-	-	-	514	593	738
27 Fishing area total	*14*	*7*	*1*	*33*	*113*	*82*	-	*514*	*593*	*744*
37 Russian Fed	-	-	-	-	-	-	-	-	8	4
37 Fishing area total	-	-	-	-	-	-	-	-	*8*	*4*
Species total	*1 171*	*439*	*424*	*385*	*462*	*992*	*1 047*	*1 681*	*1 566*	*2 473*
Barramundi(=Giant seaperch)	**Perche barramundi**		**Perca gigante**		***Lates calcarifer***				**1,70(01)167,01**	**GIP**
06 Papua N Guin	350 F	350 F	350 F	350 F	350 F	350 F	350 F	350 F	350 F	350 F
06 Fishing area total	*350 F*	*350 F*	*350 F*	*350 F*	*350 F*	*350 F*	*350 F*	*350 F*	*350 F*	*350 F*
51 Pakistan	210	193	187	214	209	196	204	-	-	-
51 Fishing area total	*210*	*193*	*187*	*214*	*209*	*196*	*204*	-	-	-
57 Australia	45	43	38	46	-	-	-	-	-	-

B-25 Miscellaneous diadromous fishes / Poissons diadromes divers / Peces diádromos diversos

Capture production by species, fishing areas and countries or areas / Captures par espèces, zones de pêche et pays ou zones / Capturas por especies, áreas de pesca y países o áreas

Species, Fishing area Espèce, Zone de pêche Especie, Area de pesca	1993 t	1994 t	1995 t	1996 t	1997 t	1998 t	1999 t	2000 t	2001 t	2002 t
Indonesia	11 721	10 324	12 133	13 733	13 612	21 153	15 600	15 750	15 325	16 190
Malaysia	73	64	56	42	47	901	55	141	91	101
Thailand	-	2 688	46	168	131	1 007	30	91	249	244
57 Fishing area total	*11 839*	*13 119*	*12 273*	*13 989*	*13 790*	*23 061*	*15 685*	*15 982*	*15 665*	*16 535*
61 China,Taiwan	2	2	0	14	9	26	47	32	80	73
61 Fishing area total	*2*	*2*	*0*	*14*	*9*	*26*	*47*	*32*	*80*	*73*
71 Australia	966	984	1 002	1 226	1 178	1 306	1 690	1 747	1 915	1 783
Indonesia	25 080	28 122	35 494	34 579	42 330	44 040	49 573	53 038	48 160	52 530
Malaysia	1 529	1 604	1 263	1 280	827	979	1 342	1 560	1 427	1 339
Papua N Guin	25	19	9	6	43	45	158	73	149	130 F
Philippines	2 960	1 920	2 821	3 049	553	655	784	642	751	822
Singapore	33	64	51	39	58	39	29	41	52	51
Thailand	-	411	14	421	38	42	55	742	314	318
71 Fishing area total	*30 593*	*33 124*	*40 654*	*40 600*	*45 027*	*47 106*	*53 631*	*57 843*	*52 768*	*56 973 F*
Species total	*42 994 F*	*46 788 F*	*53 464 F*	*55 167 F*	*59 385 F*	*70 739 F*	*69 917 F*	*74 207 F*	*68 863 F*	*73 931 F*
Striped bass	**Bar d'Amérique**		**Lubina estriada**		***Morone saxatilis***				**1,70(06)006,02**	**STB**
02 USA	-	-	-	-	5	-	-	-	-	-
02 Fishing area total	-	-	-	-	*5*	-	-	-	-	-
21 Canada	3	-	18	15	0	-	-	-	-	-
USA	759	639	1 637	2 129	2 785	3 032	3 000	3 123	2 934	2 847
21 Fishing area total	*762*	*639*	*1 655*	*2 144*	*2 785*	*3 032*	*3 000*	*3 123*	*2 934*	*2 847*
31 USA	4	109	7	9	12	14	4	12	15	17
31 Fishing area total	*4*	*109*	*7*	*9*	*12*	*14*	*4*	*12*	*15*	*17*
Species total	*766*	*748*	*1 662*	*2 153*	*2 802*	*3 046*	*3 004*	*3 135*	*2 949*	*2 864*
White perch	**Bar blanc d'Amérique**		**Lubina blanca**		***Morone americana***				**1,70(06)006,03**	**PEW**
02 USA	935	747	802	995	1 164	966	1 042	1 221	1 226	1 022
02 Fishing area total	*935*	*747*	*802*	*995*	*1 164*	*966*	*1 042*	*1 221*	*1 226*	*1 022*
Species total	*935*	*747*	*802*	*995*	*1 164*	*966*	*1 042*	*1 221*	*1 226*	*1 022*
White bass	**...B**		**...C**		***Morone chrysops***				**1,70(06)006,04**	**ROY**
02 USA	191	109	169	45	50	166	109	106	145	104
02 Fishing area total	*191*	*109*	*169*	*45*	*50*	*166*	*109*	*106*	*145*	*104*
Species total	*191*	*109*	*169*	*45*	*50*	*166*	*109*	*106*	*145*	*104*
Group total	***50 114***	***54 536***	***64 683***	***63 177***	***64 842***	***79 373***	***75 921***	***83 328***	***75 658***	***81 530***

B-31 Flounders, halibuts, soles / Flets, flétans, soles / Platijas, halibuts, lenguados

Capture production by species, fishing areas and countries or areas / Captures par espèces, zones de pêche et pays ou zones / Capturas por especies, áreas de pesca y países o áreas

Species, Fishing area Espèce, Zone de pêche Especie, Area de pesca	1993 t	1994 t	1995 t	1996 t	1997 t	1998 t	1999 t	2000 t	2001 t	2002 t
Lefteye flounders nei	**Arnoglosses, rombous nca**		**Rodaballos, rombos, etc. nep**		***Bothidae***				**1,83(01)XXX,XX**	**LEF**
27 Germany	...	...	...	...	...	...	293	370	290	292
Portugal	...	...	...	...	...	...	89	117	107	106
27 Fishing area total	...	...	...	...	...	...	*382*	*487*	*397*	*398*
31 Korea Rep	-	-	-	-	-	-	-	2	-	-
31 Fishing area total	-	-	-	-	-	-	-	*2*	-	-
34 Cameroon	1	0	0	0	0	0	0	0	-	-
Korea Rep	-	7	65	2	-	-	-	15	-	-
Portugal	-	-	18	22	2	-	-	2	12	11
34 Fishing area total	*1*	*7*	*83*	*24*	*2*	*0*	*0*	*17*	*12*	*11*
41 Estonia	-	-	-	-	-	-	-	-	-	1
Korea Rep	194	-	-	-	-	-	-	-	20	-
41 Fishing area total	*194*	-	-	-	-	-	-	-	*20*	*1*
48 UK	-	-	-	-	-	-	-	0	-	-
48 Fishing area total	-	-	-	-	-	-	-	*0*	-	-
51 Eritrea	...	-	-	-	0	1	23	119	125	69
Korea Rep	-	-	-	-	-	8	4	10	-	19
51 Fishing area total	...	-	-	-	*0*	*9*	*27*	*129*	*125*	*88*
71 Korea Rep	22	115	86	38	27	30	15	8	10	1
71 Fishing area total	*22*	*115*	*86*	*38*	*27*	*30*	*15*	*8*	*10*	*1*
77 Korea Rep	-	-	-	-	-	-	9	-	-	-
77 Fishing area total	-	-	-	-	-	-	*9*	-	-	-
87 Korea Rep	-	-	-	-	-	-	-	5	-	-
87 Fishing area total	-	-	-	-	-	-	-	*5*	-	-
Species total	*217*	*122*	*169*	*62*	*29*	*39*	*433*	*648*	*564*	*499*
Atlantic halibut	**Flétan de l'Atlantique**		**Fletán del Atlántico**		***Hippoglossus hippoglossus***				**1,83(02)002,01**	**HAL**
21 Canada	1 444	1 264	895	1 097	1 362	1 300	1 188	1 219	1 647	1 649
Cuba	14	-	1	-	0	0	0	-	-	-
Estonia	651	-	-	-	-	-	-	-	0	-
Faeroe Is	-	-	-	8	-	7	-	-	-	-
Germany	0	0	-	-	-	-	0	-	-	-
Greenland	47	40	23	35	22	22	45	9	0	1
Japan	-	-	5	4	6	7	5	3	36	14
Norway	-	-	-	0	1	0	-	-	-	-
Poland	-	-	-	-	-	-	-	-	488	-
Portugal	53	45	17	11	17	31	51	29	44	46
Russian Fed	-	-	-	-	-	-	6	5	2	13
St Pier Mq	-	-	-	-	1	1	1	0	1	2
Spain	1	67	68	58	46	65	98	96	138	158
UK	-	-	-	-	-	1	-	-	-	-
USA	19	21	16	13	14	8	12	11	11	10
21 Fishing area total	*2 229*	*1 437*	*1 025*	*1 226*	*1 469*	*1 442*	*1 406*	*1 372*	*2 367*	*1 893*
27 Belgium	6	2	15	5	4	3	2	2	2	3
Denmark	59	57	61	71	70	62	51	53	53	82
Faeroe Is	722	1 090	654	451	479	384	432	318	205	360
France	26	27	27	34	16	16	21	29	44	19
Germany	138	139	30	43	23	28	42	23	20	22
Greenland	346	241	37	16	4	12	1	10	19	47
Iceland	1 486	1 427	888	837	677	501	567	493	589	683
Ireland	1	6	6	8	4	9	11	1	17	23
Netherlands	1	3	5	3	5	4	2	1	-	0
Norway	589	754	551	678	878	672	696	1 039	868	676
Portugal	0	0	0	3	-	0	0	1	1	9
Spain	-	-	-	-	-	4	5	16	14	1
Sweden	25	17	13	8	10	8	8	10	8	8
UK	169	186	405	392	367	201	254	208	159	222
27 Fishing area total	*3 568*	*3 949*	*2 692*	*2 549*	*2 537*	*1 904*	*2 092*	*2 204*	*1 999*	*2 155*
Species total	*5 797*	*5 386*	*3 717*	*3 775*	*4 006*	*3 346*	*3 498*	*3 576*	*4 366*	*4 048*
Pacific halibut	**Flétan du Pacifique**		**Fletán del Pacífico**		***Hippoglossus stenolepis***				**1,83(02)002,02**	**HAP**
67 Canada	6 379	5 862	5 745	5 430	7 448	7 714	7 103	6 095	4 766	6 487
Poland	-	-	-	-	-	-	-	-	4	-
USA	29 294	27 028	20 551	22 676	31 841	33 892	36 515	34 753	35 391	37 237
67 Fishing area total	*35 673*	*32 890*	*26 296*	*28 106*	*39 289*	*41 606*	*43 618*	*40 848*	*40 161*	*43 724*
77 USA	-	-	3	4	2	3	2	-	0	-
77 Fishing area total	-	-	*3*	*4*	*2*	*3*	*2*	-	*0*	-
Species total	*35 673*	*32 890*	*26 299*	*28 110*	*39 291*	*41 609*	*43 620*	*40 848*	*40 161*	*43 724*
European plaice	**Plie d'Europe**		**Solla europea**		***Pleuronectes platessa***				**1,83(02)004,05**	**PLE**
27 Belgium	12 775	10 398	9 290	7 684	7 469	7 369	8 186	9 148	8 487	7 043

B-31 Flounders, halibuts, soles / Flets, flétans, soles / Platijas, halibuts, lenguados

Capture production by species, fishing areas and countries or areas
Captures par espèces, zones de pêche et pays ou zones
Capturas por especies, áreas de pesca y países o áreas

Species, Fishing area Espèce, Zone de pêche Especie, Area de pesca	1993 t	1994 t	1995 t	1996 t	1997 t	1998 t	1999 t	2000 t	2001 t	2002 t
Channel Is	8 F	3	3 F	26	19	17	23	18	13	11
Denmark	27 159	27 954	24 092	22 571	24 621	18 927	23 123	23 902	26 849	22 998
Faeroe Is	319	957	425	443	506	443	325	259	250	419
France	4 582	4 667	3 982	3 843	4 326	4 667	5 642	4 703	4 247	4 701
Germany	7 038	5 832	6 533	4 935	4 304	3 050	3 462	4 501	4 842	4 148
Greenland	-	-	-	-	-	2	-	-	-	-
Iceland	12 516	11 851	10 649	11 070	10 557	7 111	7 064	5 218	4 905	5 126
Ireland	1 661	1 470	1 590	1 679	1 699	1 731	1 424	1 028	841	801
Isle of Man	13	14	20	16	11	14	5	6	1	0
Latvia	-	8	-	-	-	-	-	-	-	-
Netherlands	48 555	50 289	44 262	35 539	34 272	30 592	37 543	35 079	33 835	29 083
Norway	1 399	1 105	1 166	1 731	2 857	1 872	1 816	1 944	2 773	2 945
Portugal	218	143	147	137	89	115	95	124	145	184
Russian Fed	4 984	4 404	4 431	2 161	3 531	3 729	3 911	3 114	1 250	1 133
Spain	2	1	12	14	3	6	3	39	41	13
Sweden	475	565	511	531	558	431	405	449	416	387
UK	36 278	32 802	29 194	25 319	26 599	23 510	20 394	23 701	21 936	19 647
27 Fishing area total	*157 982 F*	*152 463*	*136 307 F*	*117 699*	*121 421*	*103 586*	*113 421*	*113 233*	*110 831*	*98 639*
37 France	-	-	-	-	-	-	0	6	7	7
37 Fishing area total	-	-	-	-	-	-	*0*	*6*	*7*	*7*
Species total	*157 982 F*	*152 463*	*136 307 F*	*117 699*	*121 421*	*103 586*	*113 421*	*113 239*	*110 838*	*98 646*
English sole	**Carlottin anglais**		**Soya inglesa**		***Pleuronectes vetulus***			**1,83(02)004,15**		**RFE**
67 USA	...	...	...	...	...	198	874	963	1 198	1 374
67 Fishing area total	...	...	...	...	...	*198*	*874*	*963*	*1 198*	*1 374*
77 USA	...	...	...	...	...	227	226	177	197	102
77 Fishing area total	...	...	...	...	...	*227*	*226*	*177*	*197*	*102*
Species total	...	...	...	...	...	*425*	*1 100*	*1 140*	*1 395*	*1 476*
Greenland halibut	**Flétan noir**		**Fletán negro**		***Reinhardtius hippoglossoides***			**1,83(02)005,01**		**GHL**
21 Canada	16 589	11 036	8 775	14 617	16 399	12 222	12 657	16 444	13 814	11 237
Cuba	1	-	-	-	-	-	-	-	-	-
Denmark	-	-	-	0	0	-	-	-	0	0
Estonia	631	-	-	-	-	-	-	181	1 100	898
Faeroe Is	1 552	850	13	870	579	272	748	1 257	643	548
Germany	46	217	-	474	445	350	415	444	537	536
Greenland	13 417	17 311	17 872	19 152	23 146	21 908	34 481	23 219	19 111	23 814
Iceland	1	2	-	-	-	-	-	507	14	-
Japan	-	-	3 145	2 532	1 876	2 053	2 420	2 512	2 814	2 779
Korea Rep	5	-	-	-	-	-	-	-	-	-
Latvia	83	-	-	-	-	-	-	215	291	11
Lithuania	-	-	-	-	-	-	-	21	392	298
Norway	2 739	3 121	2 378	1 786	1 817	1 339	1 335	1 431	1 419	2 516
Portugal	8 811	5 967	1 936	3 316	3 343	3 242	3 995	4 688	5 026	4 319
Russian Fed	4 234	3 763	677	565	-	2 433	3 669	4 128	4 687	4 581
St Pier Mq	-	-	-	-	439	1 431	1 132	2	7	0
Spain	35 640	40 772	9 135	7 311	7 945	7 238	9 022	9 537	11 571	12 833
UK	-	49	-	-	-	-	-	-	-	-
21 Fishing area total	*83 749*	*83 088*	*43 931*	*50 623*	*55 989*	*52 488*	*69 874*	*64 586*	*61 426*	*64 370*
27 Denmark	3	5	1	3	2	1	1	1	1	1
Estonia	-	-	-	-	-	-	-	-	-	227
Faeroe Is	4 469	6 361	4 481	5 371	4 545	3 515	3 363	4 327	3 485	1 814
France	177	172	542	603	647	282	268	158	230	351
Germany	438	985	848	3 450	3 401	3 520	3 117	3 298	2 865	2 093
Greenland	297	870	545	1 286	1 129	943	4 970	1 756	1 573	3 326
Iceland	34 043	27 793	27 408	22 125	18 631	10 751	11 187	14 553	16 628	19 229
Ireland	-	5	2	2	2	21	78	22	71	84
Lithuania	-	-	-	-	-	-	-	-	3	48
Norway	12 184	10 330	11 695	15 287	10 526	10 609	18 369	11 591	13 733	9 019
Poland	-	-	-	-	12	31	8	4	4	22
Portugal	43	36	84	79	50	99	49	37	40	145
Russian Fed	1 235	283	804	2 000	1 075	2 711	3 961	4 751	4 881	5 628
Spain	-	1	1 188	214	255	246	589	864	1 766	-
UK	863	682	2 890	2 358	1 515	1 777	1 611	1 746	1 690	1 346
27 Fishing area total	*53 752*	*47 523*	*50 488*	*52 778*	*41 790*	*34 506*	*47 571*	*43 108*	*46 970*	*43 333*
67 USA	8 461	7 115	5 860	4 652	...	307	14	6 186	4 391	2 937
67 Fishing area total	*8 461*	*7 115*	*5 860*	*4 652*	...	*307*	*14*	*6 186*	*4 391*	*2 937*
Species total	*145 962*	*137 726*	*100 279*	*108 053*	*97 779*	*87 301*	*117 459*	*113 880*	*112 787*	*110 640*
Arrow-tooth flounder	**Faux flétan du Pacifique**		**Halibut del Pacífico**		***Atheresthes stomias***			**1,83(02)008,01**		**ARF**
67 Canada	108	129	75	24	-	33	44	39	59	22
USA	13 589	4 335	5 091	10 656	6 319	8 246	12 037	18 735	14 342	16 588
67 Fishing area total	*13 697*	*4 464*	*5 166*	*10 680*	*6 319*	*8 279*	*12 081*	*18 774*	*14 401*	*16 610*
77 USA	-	3	1	1	1	1	2	0	0	1
77 Fishing area total	-	*3*	*1*	*1*	*1*	*1*	*2*	*0*	*0*	*1*
Species total	*13 697*	*4 467*	*5 167*	*10 681*	*6 320*	*8 280*	*12 083*	*18 774*	*14 401*	*16 611*

B-31

Flounders, halibuts, soles — **Capture production by species, fishing areas and countries or areas**
Flets, flétans, soles — **Captures par espèces, zones de pêche et pays ou zones**
Platijas, halibuts, lenguados — **Capturas por especies, áreas de pesca y países o áreas**

Species, Fishing area Espèce, Zone de pêche Especie, Area de pesca	1993 t	1994 t	1995 t	1996 t	1997 t	1998 t	1999 t	2000 t	2001 t	2002 t
Kamchatka flounder	**Faux flétan du Japon**		**Halibut japonés**		***Atheresthes evermanni***				**1,83(02)008,02**	**KAF**
61 Russian Fed	1 276	6 944	7 039	9 739	8 476	9 567	10 743	23 473	19 049	17 609
61 Fishing area total	*1 276*	*6 944*	*7 039*	*9 739*	*8 476*	*9 567*	*10 743*	*23 473*	*19 049*	*17 609*
Species total	*1 276*	*6 944*	*7 039*	*9 739*	*8 476*	*9 567*	*10 743*	*23 473*	*19 049*	*17 609*
Petrale sole	**Charlottin pétrale**		**Limanda petrale**		***Eopsetta jordani***				**1,83(02)010,01**	**EOJ**
67 USA	...	...	...	...	1 454	1 197	1 221	1 632	1 554	1 569
67 Fishing area total	...	...	...	...	*1 454*	*1 197*	*1 221*	*1 632*	*1 554*	*1 569*
77 USA	...	...	...	...	483	267	259	238	268	224
77 Fishing area total	...	...	...	...	*483*	*267*	*259*	*238*	*268*	*224*
Species total	...	...	...	...	*1 937*	*1 464*	*1 480*	*1 870*	*1 822*	*1 793*
Witch flounder	**Plie cynoglosse**		**Mendo**		***Glyptocephalus cynoglossus***				**1,83(02)011,02**	**WIT**
21 Canada	7 510	1 610	1 244	1 600	1 675	2 094	1 806	2 168	2 128	2 399
Cuba	11	-	-	-	2	-	-	-	-	-
Estonia	-	-	-	-	-	-	-	5	3	2
Faeroe Is	-	-	6	11	-	-	4	-	-	-
Japan	-	-	5	12	5	2	2	4	3	38
Lithuania	-	-	-	-	-	-	-	-	3	2
Portugal	292	573	389	240	348	381	508	228	579	433
Russian Fed	-	-	-	-	-	52	110	114	65	135
St Pier Mq	-	-	-	-	8	57	35	7	120	38
Spain	69	295	1 086	1 636	1 344	1 349	1 180	1 482	1 355	841
USA	2 595	2 693	2 197	2 082	1 775	1 855	2 123	2 439	3 020	3 189
21 Fishing area total	*10 477*	*5 171*	*4 927*	*5 581*	*5 157*	*5 790*	*5 768*	*6 447*	*7 276*	*7 077*
27 Denmark	835	864	897	999	1 563	1 906	2 116	2 338	2 049	1 907
Faeroe Is	-	2	5	2	2	3	3	...	...	...
France	570	650	734	719	816	602	484	587	582	631
Germany	5	5	10	7	9	13	8	13	8	5
Iceland	1 594	1 771	1 755	1 486	1 272	947	1 408	1 098	1 132	1 147
Ireland	396	371	601	615	605	657	713	551	915	831
Isle of Man	0	0	2	0	1	0	0	0	-	-
Netherlands	13	14	7	0	1	4	9	7	1	-
Norway	152	118	100	80	86	140	135	97	88	82
Portugal	38	25	33	30	32	22	27	26	20	21
Spain	-	2 028	2 339	1 751	4 183	4 805	4 148	2 961	2 870	3 277
Sweden	400	312	357	299	355	448	501	578	576	584
UK	2 359	2 638	3 187	3 479	3 748	3 889	3 661	4 102	4 535	3 896
27 Fishing area total	*6 362*	*8 798*	*10 027*	*9 467*	*12 673*	*13 436*	*13 213*	*12 358*	*12 776*	*12 381*
Species total	*16 839*	*13 969*	*14 954*	*15 048*	*17 830*	*19 226*	*18 981*	*18 805*	*20 052*	*19 458*
Rex sole	**Plie cynoglosse royale**		**...C**		***Glyptocephalus zachirus***				**1,83(02)011,03**	**GLZ**
67 USA	...	...	...	...	...	121	391	413	473	484
67 Fishing area total	...	...	...	...	...	*121*	*391*	*413*	*473*	*484*
77 USA	...	...	...	...	...	168	153	128	98	116
77 Fishing area total	...	...	...	...	...	*168*	*153*	*128*	*98*	*116*
Species total	...	...	...	...	...	*289*	*544*	*541*	*571*	*600*
Amer. plaice(=Long rough dab)	**Balai (=Plie canadienne)**		**Platija americana**		***Hippoglossoides platessoides***				**1,83(02)014,01**	**PLA**
21 Canada	10 900	3 093	3 095	2 525	3 583	3 546	4 359	3 904	4 466	3 714
Cuba	78	13	26	-	22	49	19	-	-	-
Estonia	-	-	-	-	-	-	-	94	54	19
Faeroe Is	244	-	-	8	-	-	3	2	-	-
Germany	-	-	-	0	-	-	-	-	-	-
Greenland	0	0	0	0	0	5	3	0	4	-
Japan	-	-	-	11	7	16	21	21	6	82
Korea Rep	16	-	-	-	-	-	-	-	-	-
Lithuania	-	-	-	-	-	-	-	-	3	26
Poland	-	-	-	-	-	-	-	-	1	-
Portugal	323	344	171	291	389	357	719	399	633	631
Russian Fed	15	-	-	-	-	-	151	368	257	418
St Pier Mq	-	0	0	0	23	27	24	41	112	115
Spain	443	394	554	660	1 002	1 141	1 461	1 438	1 158	777
USA	5 808	5 062	4 612	4 397	3 937	3 662	3 134	4 213	4 425	3 420
21 Fishing area total	*17 827*	*8 906*	*8 458*	*7 892*	*8 963*	*8 803*	*9 894*	*10 480*	*11 119*	*9 202*
27 Denmark	39	2	1	0	5	8	31	5	31	0
Estonia	-	-	-	-	-	-	-	-	-	5
France	-	17	13	13	14	11	14	10	11	-
Germany	-	-	-	-	-	-	-	0	-	3
Greenland	-	28	-	-	-	-	-	-	-	-
Iceland	1 342	2 734	5 418	7 027	6 468	3 329	3 833	3 176	3 473	3 579
Norway	-	-	-	-	119	24	15	0	15	0
Portugal	29	27	48	85	57	288	70	111	71	92
Russian Fed	-	2 920	1 157	778	1 482	6 569	7 727	2 839	3 497	2 641

B-31 Flounders, halibuts, soles / Flets, flétans, soles / Platijas, halibuts, lenguados

Capture production by species, fishing areas and countries or areas / Captures par espèces, zones de pêche et pays ou zones / Capturas por especies, áreas de pesca y países o áreas

Species, Fishing area Espèce, Zone de pêche Especie, Area de pesca	1993 t	1994 t	1995 t	1996 t	1997 t	1998 t	1999 t	2000 t	2001 t	2002 t
Spain	24	13	115	218	116	205	61	184	254	138
Sweden	0	0	0	22	6	0	-	-	-	-
UK	0	0	-	1	0	0	0	0	15	-
27 *Fishing area total*	*1 434*	*5 741*	*6 752*	*8 144*	*8 267*	*10 434*	*11 751*	*6 325*	*7 367*	*6 458*
Species total	*19 261*	*14 647*	*15 210*	*16 036*	*17 230*	*19 237*	*21 645*	*16 805*	*18 486*	*15 660*
Flathead sole	**Balai du Japon**		**Platija japonesa**		***Hippoglossoides elassodon***			**1,83(02)014,04**		**FTS**
67 USA	5 865	4 835	9 073	11 392	12 867	18 886	14 318	16 266	16 092	13 174
67 *Fishing area total*	*5 865*	*4 835*	*9 073*	*11 392*	*12 867*	*18 886*	*14 318*	*16 266*	*16 092*	*13 174*
Species total	*5 865*	*4 835*	*9 073*	*11 392*	*12 867*	*18 886*	*14 318*	*16 266*	*16 092*	*13 174*
Yellowfin sole	**Limande du Japon**		**Limanda japonesa**		***Limanda aspera***			**1,83(02)024,02**		**YES**
67 USA	105 793	107 674	96 765	101 354	149 302	80 500	56 830	69 971	54 918	63 625
67 *Fishing area total*	*105 793*	*107 674*	*96 765*	*101 354*	*149 302*	*80 500*	*56 830*	*69 971*	*54 918*	*63 625*
Species total	*105 793*	*107 674*	*96 765*	*101 354*	*149 302*	*80 500*	*56 830*	*69 971*	*54 918*	*63 625*
Yellowtail flounder	**Limande à queue jaune**		**Limanda**		***Limanda ferruginea***			**1,83(02)024,04**		**YEL**
21 Canada	8 650	2 537	1 483	1 178	1 728	5 242	8 222	12 790	15 648	13 001
Cuba	0	0	-	-	-	-	-	-	-	-
Estonia	-	-	-	-	-	-	-	53	47	14
Honduras	20	-	-	-	-	-	-	-	-	-
Korea Rep	2	-	-	-	-	-	-	-	-	-
Lithuania	-	-	-	-	-	-	-	-	1	-
Portugal	20	-	-	-	-	85	426	153	351	122
Russian Fed	-	-	-	-	-	-	96	212	148	103
St Pier Mq	-	-	-	-	18	59	33	60	152	137
Spain	14	315	65	259	656	562	752	775	622	216
USA	3 607	3 094	1 882	2 403	2 864	3 656	4 431	6 928	7 304	5 355
21 *Fishing area total*	*12 313*	*5 946*	*3 430*	*3 840*	*5 266*	*9 604*	*13 960*	*20 971*	*24 273*	*18 948*
Species total	*12 313*	*5 946*	*3 430*	*3 840*	*5 266*	*9 604*	*13 960*	*20 971*	*24 273*	*18 948*
Common dab	**Limande**		**Lenguadina**		***Limanda limanda***			**1,83(02)024,05**		**DAB**
27 Belgium	666	569	557	690	789	961	980	865	850	752
Denmark	3 957	5 067	4 484	3 952	3 211	2 646	2 514	2 113	2 300	2 648
Faeroe Is	69	7	27	37	39	39	29	...	...	...
France	1 581	1 258	1 062	1 120	1 446	1 576	1 194	1 106	950	1 182
Germany	1 102	1 944	2 074	1 880	1 384	1 129	1 104	1 124	1 074	762
Iceland	4 222	5 159	5 558	7 954	7 891	5 061	3 981	3 015	4 373	4 358
Ireland	80	90	95	76	113	109	66	39	34	32
Isle of Man	0	0	0	0	0	-	-	-	-	-
Netherlands	...	...	...	...	...	7 983	8 656	6 544	5 969	4 955
Norway	-	-	-	-	-	-	-	49	54	55
Spain	-	-	-	-	0	130	129	29	24	70
Sweden	38	38	59	37	46	33	16	10	14	10
UK	1 360	1 706	2 225	2 221	2 590	2 467	2 248	1 705	1 530	1 355
27 *Fishing area total*	*13 075*	*15 838*	*16 141*	*17 967*	*17 509*	*22 134*	*20 917*	*16 599*	*17 172*	*16 179*
Species total	*13 075*	*15 838*	*16 141*	*17 967*	*17 509*	*22 134*	*20 917*	*16 599*	*17 172*	*16 179*
Rock sole	**Fausse limande du Pacifique**		**Lenguado del Pacífico**		***Lepidopsetta bilineata***			**1,83(02)043,01**		**ROS**
67 USA	69 745	23 448	24 962	26 198	32 743	15 189	17 185	27 510	24 206	29 258
67 *Fishing area total*	*69 745*	*23 448*	*24 962*	*26 198*	*32 743*	*15 189*	*17 185*	*27 510*	*24 206*	*29 258*
77 USA	6	4	7	4	9	10	7	7	7	13
77 *Fishing area total*	*6*	*4*	*7*	*4*	*9*	*10*	*7*	*7*	*7*	*13*
Species total	*69 751*	*23 452*	*24 969*	*26 202*	*32 752*	*15 199*	*17 192*	*27 517*	*24 213*	*29 271*
Dover sole	**Limande sole du Pacifique**		**Soya escurridiza**		***Microstomus pacificus***			**1,83(02)045,03**		**MIP**
67 USA	...	...	...	...	8 496	8 174	8 384	7 546	6 038	4 659
67 *Fishing area total*	...	...	...	...	*8 496*	*8 174*	*8 384*	*7 546*	*6 038*	*4 659*
77 USA	...	...	...	...	3 846	1 818	2 174	1 866	1 403	1 994
77 *Fishing area total*	...	...	...	...	*3 846*	*1 818*	*2 174*	*1 866*	*1 403*	*1 994*
Species total	...	...	...	...	*12 342*	*9 992*	*10 558*	*9 412*	*7 441*	*6 653*
Lemon sole	**Limande sole**		**Mendo limón**		***Microstomus kitt***			**1,83(02)045,04**		**LEM**
27 Belgium	710	709	1 006	1 094	975	1 256	1 020	1 057	1 076	1 089
Channel Is	0	0	0	0	0	1	1	3	1	1
Denmark	2 008	1 291	1 208	1 108	1 172	1 591	1 812	2 037	1 820	1 447
Faeroe Is	201	276	265	236	332	464	433	389	694	1 175
France	1 898	2 005	1 944	2 396	1 782	1 522	1 349	1 308	1 366	1 444
Germany	42	29	71	67	78	151	68	74	77	121
Iceland	697	692	741	984	1 135	1 432	1 860	1 438	1 371	950
Ireland	531	390	724	581	667	527	531	468	440	482
Isle of Man	2	3	2	4	0	4	3	3	1	0

B-31 Flounders, halibuts, soles — Capture production by species, fishing areas and countries or areas
Flets, flétans, soles — Captures par espèces, zones de pêche et pays ou zones
Platijas, halibuts, lenguados — Capturas por especies, áreas de pesca y países o áreas

Species, Fishing area Espèce, Zone de pêche Especie, Area de pesca	1993 t	1994 t	1995 t	1996 t	1997 t	1998 t	1999 t	2000 t	2001 t	2002 t
Netherlands	...	...	...	...	...	839	681	492	456	402
Norway	31	33	31	47	63	59	59	60	53	60
Spain	-	-	-	-	235	1 197	1 282	2 207	4 040	408
Sweden	147	127	96	117	121	105	94	71	61	48
UK	5 443	5 117	4 974	5 444	5 723	5 230	5 010	4 367	4 005	2 719
27 Fishing area total	*11 710*	*10 672*	*11 062*	*12 078*	*12 283*	*14 378*	*14 203*	*13 974*	*15 461*	*10 346*
Species total	*11 710*	*10 672*	*11 062*	*12 078*	*12 283*	*14 378*	*14 203*	*13 974*	*15 461*	*10 346*
European flounder	**Flet d'Europe**		**Platija europea**		***Platichthys flesus***				**1,83(02)048,02**	**FLE**
27 Belgium	213	223	348	278	146	307	363	322	316	230
Denmark	4 031	3 585	4 235	5 469	5 378	4 725	3 528	4 604	6 066	5 143
Estonia	165	162	102	297	333	355	416	419	482	515
Finland	1 092	564	575	715	702	555	558	449	500	449
France	127	169	206	236	212	152	198	219	242	285
Germany	1 630	6 018	4 488	1 637	2 449	2 159	2 347	2 782	2 349	2 385
Ireland	17	19	24	13	13	13	13	12	18	19
Latvia	501	329	362	294	367	364	509	418	613	599
Lithuania	120	262	194	330	624	736	571	618	...	...
Netherlands	35	...	...	...	...	4 942	3 159	2 658	2 621	3 530
Norway	-	-	-	-	-	-	-	5	3	3
Portugal	0	0	0	0	-	-	-	-	-	-
Russian Fed	-	-	-	-	-	-	-	1 392	1 351	1 327
Spain	-	-	-	74	319	873	323	88	139	13
Sweden	210	275	459	1 262	1 073	526	274	341	467	357
UK	276	383	388	357	379	293	149	201	148	172
27 Fishing area total	*8 417*	*11 989*	*11 381*	*10 962*	*11 995*	*16 000*	*12 408*	*14 528*	*15 315*	*15 027*
37 Albania	...	...	0	3	9	42	41	45	15	3
Bulgaria	-	-	-	-	-	-	-	-	-	9
France	12	6	3	2	11	11	6	2	1	2
Slovenia	-	-	-	-	-	-	-	1	3	2
Spain	-	-	-	-	-	-	-	-	-	8
Ukraine	66	68	25	30	30	16	15	8	10	5
37 Fishing area total	*78*	*74*	*28*	*35*	*50*	*69*	*62*	*56*	*29*	*29*
Species total	*8 495*	*12 063*	*11 409*	*10 997*	*12 045*	*16 069*	*12 470*	*14 584*	*15 344*	*15 056*
Pacific sand sole	**Plie à points noirs**		**...C**		***Psettichthys melanostictus***				**1,83(02)049,01**	**YSE**
67 USA	...	...	...	...	...	15	582	44	85	152
67 Fishing area total	...	...	...	...	...	*15*	*582*	*44*	*85*	*152*
77 USA	...	...	...	...	...	19	23	35	44	35
77 Fishing area total	...	...	...	...	...	*19*	*23*	*35*	*44*	*35*
Species total	...	...	...	...	...	*34*	*605*	*79*	*129*	*187*
Yellow striped flounder	**Limande-plie du Japon**		**Acedía del Japón**		***Pseudopleuronectes herzenst.***				**1,83(02)050,01**	**YFL**
61 Korea Rep	13 505	13 343	13 683	18 066	18 079	20 135	19 569	15 423	14 503	13 816
61 Fishing area total	*13 505*	*13 343*	*13 683*	*18 066*	*18 079*	*20 135*	*19 569*	*15 423*	*14 503*	*13 816*
Species total	*13 505*	*13 343*	*13 683*	*18 066*	*18 079*	*20 135*	*19 569*	*15 423*	*14 503*	*13 816*
Winter flounder	**Limande-plie rouge**		**Solla roja**		***Pseudopleuronectes americanus***				**1,83(02)050,03**	**FLW**
21 Canada	2 667	3 448	2 804	2 468	2 548	1 804	1 763	2 342	2 257	1 775
Estonia	-	-	-	-	-	-	-	25	-	-
USA	5 267	3 592	4 002	5 687	5 765	5 100	4 654	5 818	6 930	5 890
21 Fishing area total	*7 934*	*7 040*	*6 806*	*8 155*	*8 313*	*6 904*	*6 417*	*8 185*	*9 187*	*7 665*
31 USA	-	-	-	-	-	0	-	-	-	-
31 Fishing area total	-	-	-	-	-	*0*	-	-	-	-
Species total	*7 934*	*7 040*	*6 806*	*8 155*	*8 313*	*6 904*	*6 417*	*8 185*	*9 187*	*7 665*
Sand flounders	**Rhombosoles**		**Sollas de arena**		***Rhombosolea spp***				**1,83(02)053,XX**	**FSA**
57 Australia	5	49	34	26	26	26	29	1	...	...
57 Fishing area total	*5*	*49*	*34*	*26*	*26*	*26*	*29*	*1*	...	...
81 Australia	34	32	32	1	1	2	28	...	...	...
New Zealand	...	...	103	224	251	...	...	...	37	204
81 Fishing area total	*34*	*32*	*135*	*225*	*252*	*2*	*28*	...	*37*	*204*
Species total	*39*	*81*	*169*	*251*	*278*	*28*	*57*	*1*	*37*	*204*
Curlfin sole	**...B**		**...C**		***Pleuronichthys decurrens***				**1,83(02)059,02**	**NYD**
67 USA	-	-	-	-	-	-	-	1	5	4
67 Fishing area total	-	-	-	-	-	-	-	*1*	*5*	*4*
Species total	-	-	-	-	-	-	-	*1*	*5*	*4*

B-31 Flounders, halibuts, soles / Flets, flétans, soles / Platijas, halibuts, lenguados

Capture production by species, fishing areas and countries or areas / Captures par espèces, zones de pêche et pays ou zones / Capturas por especies, áreas de pesca y países o áreas

Species, Fishing area Espèce, Zone de pêche Especie, Area de pesca	1993 t	1994 t	1995 t	1996 t	1997 t	1998 t	1999 t	2000 t	2001 t	2002 t
Common sole	**Sole commune**		**Lenguado común**		***Solea solea***				**1,83(03)007,01**	**SOL**
01 Egypt	...	...	...	...	547	381	763	2 490	2 134	1 858
01 Fishing area total	*...*	*...*	*...*	*...*	*547*	*381*	*763*	*2 490*	*2 134*	*1 858*
27 Belgium	5 231	5 703	5 457	5 150	4 514	4 102	4 492	4 479	4 975	5 089
Channel Is	2 F	2	2 F	9	26	21	21	26	25	20
Denmark	3 091	3 073	3 039	2 083	1 478	1 050	1 433	1 804	1 324	1 210
France	9 401	10 050	8 687	7 179	6 838	6 554	8 255	8 255	7 644	7 104
Germany	1 387	1 749	1 569	685	513	786	1 462	1 291	959	771
Ireland	483	573	561	463	483	526	492	376	375	334
Isle of Man	4	5	12	4	5	3	1	1	1	0
Netherlands	22 014	22 925	20 927	15 563	10 370	15 308	16 329	15 343	13 737	12 120
Norway	-	-	-	-	-	-	-	198	88	53
Portugal	118	237	235	167	151	113	121	152	189	115
Spain	103	52	39	74	178	228	248	279	224	143
Sweden	138	94	89	61	52	41	43	30	20	15
UK	3 433	3 589	3 530	3 022	2 671	2 561	2 808	2 443	2 720	2 574
27 Fishing area total	*45 405 F*	*48 052*	*44 147 F*	*34 460*	*27 279*	*31 293*	*35 705*	*34 677*	*32 281*	*29 548*
34 Congo Rep	812 F	326	230	200 F	200 F	100 F	100 F	...	50 F	50 F
France	1	1	0	0	0	-	-	-	-	-
Greece	246	92	194	148	87	125	110	66	82	185
Italy	8 384	8 429	299	193	864	731	410	291	749	384
Liberia	131	112	158	48	150	149	217	129	206	210 F
Morocco	1 152	1 342	1 451	1 340	1 633	1 520	2 594	5 925	3 775	3 304
Portugal	169	183	0	0	0	0	...	0	12	-
Russian Fed	-	-	-	-	-	-	-	-	-	190
Spain	...	...	...	-	5	5 035	...	99	4 297	...
34 Fishing area total	*10 895 F*	*10 485*	*2 332*	*1 929 F*	*2 939 F*	*7 660 F*	*3 431 F*	*6 510*	*9 171 F*	*4 323 F*
37 Albania	...	...	25	27	21	35	31	41	14	195
Algeria	350 F	400 F	313	387	333	285	234	278	271	377
Bulgaria	-	-	-	-	-	-	-	-	-	10
Egypt	814	710	473	751	762	653	965	1 012	1 041	898
France	330	187	189	159	443	519	147	192	184	199
Greece	2 068	1 742	1 259	1 087	1 082	620	509	621	480	460 F
Israel	9	100	100	...	...	...	...	...	...	...
Italy	4 002	4 097	5 766	3 404	2 221	1 907	1 842	1 874	2 217	2 482
Malta	0	0	0	0	0	0	0	0	0	-
Morocco	63	65	94	64	42	20	20	26	13	27
Romania	-	3	-	-	-	4	5	6	9	6
Slovenia	9	7	2	1	1	1	1	2	3	4
Spain	650 F	650 F	650 F	500	370	520	...	715	284	327
Tunisia	572	493	327	569	642	480	425	402	456	454
37 Fishing area total	*8 867 F*	*8 454 F*	*9 198 F*	*6 949*	*5 917*	*5 044*	*4 179*	*5 169*	*4 972*	*5 439 F*
Species total	*65 167 F*	*66 991 F*	*55 677 F*	*43 338 F*	*36 682 F*	*44 378 F*	*44 078 F*	*48 846*	*48 558 F*	*41 168 F*
Sand sole	**Sole-pole**		**Lenguado de arena**		***Solea lascaris***				**1,83(03)007,04**	**SOS**
27 Channel Is	1 F	1	1 F	2	3	3	1	1	1	1
France	140	169	160	127	150	137	118	131	153	180
Ireland	14	7	25	13	12	15	1	2	1	2
Portugal	57	51	155	77	95	118	90	116	142	97
UK	16	31	28	27	28	32	12	47	42	55
27 Fishing area total	*228 F*	*259*	*369 F*	*246*	*288*	*305*	*222*	*297*	*339*	*335*
Species total	*228 F*	*259*	*369 F*	*246*	*288*	*305*	*222*	*297*	*339*	*335*
Wedge sole	**Céteau**		**Acedía**		***Dicologlossa cuneata***				**1,83(03)017,01**	**CET**
27 France	713	847	941	798	647	488	602	686	579	606
Portugal	...	...	...	...	...	...	102	109	112	85
UK	-	-	-	-	1	0	0	-	-	-
27 Fishing area total	*713*	*847*	*941*	*798*	*648*	*488*	*704*	*795*	*691*	*691*
34 Portugal	-	39	59	87	8	0	9	10	9	12
34 Fishing area total	*-*	*39*	*59*	*87*	*8*	*0*	*9*	*10*	*9*	*12*
Species total	*713*	*886*	*1 000*	*885*	*656*	*488*	*713*	*805*	*700*	*703*
West coast sole	**Sole australe occidentale**		**Lenguado austral oeste**		***Austroglossus microlepis***				**1,83(03)057,01**	**SOW**
47 Namibia	529	661	462	339	514	393	463	571	589	644
South Africa	45	-	1	16	3	...	...	...	...	...
Spain	-	-	-	54	-	-	-	-	-	-
47 Fishing area total	*574*	*661*	*463*	*409*	*517*	*393*	*463*	*571*	*589*	*644*
Species total	*574*	*661*	*463*	*409*	*517*	*393*	*463*	*571*	*589*	*644*
Mud sole	**Sole de vase**		**Lenguado de fango**		***Austroglossus pectoralis***				**1,83(03)057,02**	**SOE**
47 South Africa	764	942	769	909	837	859	768	800 F	844	702
47 Fishing area total	*764*	*942*	*769*	*909*	*837*	*859*	*768*	*800 F*	*844*	*702*
Species total	*764*	*942*	*769*	*909*	*837*	*859*	*768*	*800 F*	*844*	*702*

B-31 Flounders, halibuts, soles / Flets, flétans, soles / Platijas, halibuts, lenguados

Capture production by species, fishing areas and countries or areas
Captures par espèces, zones de pêche et pays ou zones
Capturas por especies, áreas de pesca y países o áreas

Species, Fishing area Espèce, Zone de pêche Especie, Area de pesca	1993 t	1994 t	1995 t	1996 t	1997 t	1998 t	1999 t	2000 t	2001 t	2002 t
Southeast Atlantic soles nei	**Soles de l'Atl.sud-est nca**		**Lenguados del Atl.sudeste nep**		***Austroglossus spp***				**1,83(03)057,XX**	**SOA**
47 Korea Rep	122	362	1 365	480	1 197	773	926	965	650	492
47 Fishing area total	*122*	*362*	*1 365*	*480*	*1 197*	*773*	*926*	*965*	*650*	*492*
Species total	*122*	*362*	*1 365*	*480*	*1 197*	*773*	*926*	*965*	*650*	*492*
Thickback soles	**Soles-perdix**		**Golletas**		***Microchirus spp***				**1,83(03)081,XX**	**THS**
27 France	65	46	48	50	57	77	71	71	80	139
27 Fishing area total	*65*	*46*	*48*	*50*	*57*	*77*	*71*	*71*	*80*	*139*
Species total	*65*	*46*	*48*	*50*	*57*	*77*	*71*	*71*	*80*	*139*
Soles nei	**Soles nca**		**Lenguados nep**		***Soleidae***				**1,83(03)XXX,XX**	**SOX**
27 France	-	-	-	-	-	-	13	51	14	27
Norway	...	...	...	...	...	...	...	17	7	40
Portugal	-	-	-	-	-	-	797	1 033	831	716
27 Fishing area total	*...*	*...*	*...*	*...*	*...*	*...*	*810*	*1 101*	*852*	*783*
34 Korea Rep	263	-	29	102	-	-	256	1 016	-	-
Mauritania	600 F	1 000 F	680 F	760 F	1 120 F	830 F	449 F	444 F	...	...
Nigeria	3 711	4 511	3 807	4 640	6 084	7 633	6 583	3 301	3 171	2 507
Portugal	70	11	164	248	124	60	57	91	73	78
Spain	2 500 F	3 500 F	4 500 F	6 846	6 208	7 519	1 781	227	979	829
Togo	5	5	2	10	16	8	6	48	19	13
34 Fishing area total	*7 149 F*	*9 027 F*	*9 182 F*	*12 606 F*	*13 552 F*	*16 050 F*	*9 132 F*	*5 127 F*	*4 242*	*3 427*
37 France	-	-	-	-	-	-	18	14	12	18
37 Fishing area total	*-*	*-*	*-*	*-*	*-*	*-*	*18*	*14*	*12*	*18*
Species total	*7 149 F*	*9 027 F*	*9 182 F*	*12 606 F*	*13 552 F*	*16 050 F*	*9 960 F*	*6 242 F*	*5 106*	*4 228*
Tonguefishes	**Cynoglossidés**		**Cinoglósidos**		***Cynoglossidae***				**1,83(04)XXX,XX**	**TOX**
31 Korea Rep	-	-	-	-	-	-	59	39	20	-
31 Fishing area total	*-*	*-*	*-*	*-*	*-*	*-*	*59*	*39*	*20*	*-*
34 Cameroon	653	700 F	753	748	700 F	594	455	488	439	788
Congo Dem R	90 F	80 F	82 F	84 F	81 F	83 F	80 F	80 F	80 F	80 F
Congo Rep	...	316	300 F	270 F	240 F	200 F	180 F	158	150 F	150 F
Côte dIvoire	255	211	208	177	139	217	217	211	197	237
Gabon	242	178	209	310	411	492	386	620	538	545
Gambia	188	211	859	541	307	441	450	725	2 262	586
Ghana	247	231	407	295	339	347	284	394	227	144
Korea Rep	451	943	1 239	1 413	1 736	1 128	1 322	669	1 956	1 279
Nigeria	684	650	1 771	1 281	1 311	1 781	3 812	7 057	6 655	7 493
Sierra Leone	352	351	377	380 F	585	206	714	2 326	500	654
34 Fishing area total	*3 162 F*	*3 871 F*	*6 205 F*	*5 499 F*	*5 849 F*	*5 489 F*	*7 900 F*	*12 728 F*	*13 004 F*	*11 956 F*
41 Italy	1 118	1 124	36	23	-	-	-	-	-	-
Korea Rep	-	-	-	-	-	-	19	48	-	-
41 Fishing area total	*1 118*	*1 124*	*36*	*23*	*-*	*-*	*19*	*48*	*-*	*-*
47 Italy	-	-	4	3	-	-	-	-	-	-
South Africa	-	-	-	-	-	-	-	-	10	17
47 Fishing area total	*-*	*-*	*4*	*3*	*-*	*-*	*-*	*-*	*10*	*17*
51 Italy	1 677	1 686	61	40	-	-	-	-	-	-
Korea Rep	-	3	32	41	130	2	24	13	1	9
Pakistan	2 024	1 963	1 982	2 205	2 390	2 149	2 037	2 124	1 915	1 980
South Africa	-	-	-	-	3	7	5 F	0	7	...
51 Fishing area total	*3 701*	*3 652*	*2 075*	*2 286*	*2 523*	*2 158*	*2 066 F*	*2 137*	*1 923*	*1 989*
57 Korea Rep	-	-	-	-	-	-	7	-	-	-
Malaysia	2 403	2 644	2 430	2 123	2 602	2 390	2 298	2 120	1 551	1 696
Thailand	3 203	3 164	4 383	8 162	8 313	9 292	9 011	8 802	8 051	7 881
57 Fishing area total	*5 606*	*5 808*	*6 813*	*10 285*	*10 915*	*11 682*	*11 316*	*10 922*	*9 602*	*9 577*
61 China,H.Kong	1 008	1 583	1 326	1 160	1 261	1 177	800 F	1 000 F	1 100 F	1 050 F
Korea Rep	2 691	2 498	1 944	1 644	1 289	654	836	1 148	1 013	-
61 Fishing area total	*3 699*	*4 081*	*3 270*	*2 804*	*2 550*	*1 831*	*1 636 F*	*2 148 F*	*2 113 F*	*1 050 F*
71 Korea Rep	-	2	2	21	37	83	49	58	79	59
Malaysia	404	673	664	617	427	550	1 008	956	866	896
Thailand	4 781	7 319	8 418	6 849	6 717	6 805	6 322	7 746	7 123	7 204
71 Fishing area total	*5 185*	*7 994*	*9 084*	*7 487*	*7 181*	*7 438*	*7 379*	*8 760*	*8 068*	*8 159*
77 Korea Rep	-	-	-	-	-	-	105	95	-	-
77 Fishing area total	*-*	*-*	*-*	*-*	*-*	*-*	*105*	*95*	*-*	*-*
81 Korea Rep	-	-	-	-	-	-	9	-	-	3
81 Fishing area total	*-*	*-*	*-*	*-*	*-*	*-*	*9*	*-*	*-*	*3*
87 Korea Rep	-	-	-	-	-	-	13	38	-	-
87 Fishing area total	*-*	*-*	*-*	*-*	*-*	*-*	*13*	*38*	*-*	*-*
Species total	*22 471 F*	*26 530 F*	*27 487 F*	*28 387 F*	*29 018 F*	*28 598 F*	*30 502 F*	*36 915 F*	*34 740 F*	*32 751 F*

B-31

Flounders, halibuts, soles — **Capture production by species, fishing areas and countries or areas**
Flets, flétans, soles — **Captures par espèces, zones de pêche et pays ou zones**
Platijas, halibuts, lenguados — **Capturas por especies, áreas de pesca y países o áreas**

Species, Fishing area Espèce, Zone de pêche Especie, Area de pesca	1993 t	1994 t	1995 t	1996 t	1997 t	1998 t	1999 t	2000 t	2001 t	2002 t
Megrim	**Cardine franche**		**Gallo del Norte**		***Lepidorhombus whiffiagonis***				**1,83(05)003,01**	**MEG**
27 Belgium	63	129	225	208	188	142	143	132	87	80
Channel Is	-	1	1 F	-	-	-	-	-	1	0
Denmark	10	1	2	7	6	26	21	30	55	8
France	4 165	3 718	4 591	4 157	3 877	3 491	3 421	4 015	4 000	3 190
Germany	4	1	2	1	2	3	1	3	1	0
Iceland	213	301	405	419	281	221	124	97	96	78
Ireland	2 999	3 053	3 839	3 507	3 063	3 383	3 162	3 364	3 713	2 848
Isle of Man	-	-	-	-	3	2	...	...	-	-
Netherlands	31	29	26	11	23	31	28	...	11	8
Portugal	19	13	53	58	26	47	54	47	22	26
Sweden	0	0	1	-	-	-	-	-	-	-
UK	5 566	6 039	6 563	7 212	7 479	6 520	5 606	5 513	4 672	4 418
27 Fishing area total	*13 070*	*13 285*	*15 708 F*	*15 580*	*14 948*	*13 866*	*12 560*	*13 201*	*12 658*	*10 656*
34 Portugal	12	0	-	-	-	-	-	-	-	-
Spain	-	-	-	-	-	36	-	38	52	10
34 Fishing area total	*12*	*0*	*-*	*-*	*-*	*36*	*-*	*38*	*52*	*10*
37 Albania	...	...	1	1	0	-	-	-	1	6
France	183	103	51	110	110	110	108	139	125	121
Spain	160 F	170 F	175 F	175	185	8	-	63	79	63
37 Fishing area total	*343 F*	*273 F*	*227 F*	*286*	*295*	*118*	*108*	*202*	*205*	*190*
Species total	*13 425 F*	*13 558 F*	*15 935 F*	*15 866*	*15 243*	*14 020*	*12 668*	*13 441*	*12 915*	*10 856*
Megrims nei	**Cardines nca**		**Gallos nep**		***Lepidorhombus spp***				**1,83(05)003,XX**	**LEZ**
27 Spain	5 771	5 968	6 275	6 405	7 006	7 446	6 969	8 361	8 161	6 277
27 Fishing area total	*5 771*	*5 968*	*6 275*	*6 405*	*7 006*	*7 446*	*6 969*	*8 361*	*8 161*	*6 277*
Species total	*5 771*	*5 968*	*6 275*	*6 405*	*7 006*	*7 446*	*6 969*	*8 361*	*8 161*	*6 277*
Brill	**Barbue**		**Rémol**		***Scophthalmus rhombus***				**1,83(05)064,01**	**BLL**
27 Belgium	425	379	378	422	349	341	365	423	491	448
Channel Is	7 F	18	17 F	10	10	10	6	17	17	16
Denmark	337	363	281	220	148	189	244	237	163	148
France	509	450	452	417	357	351	373	486	464	509
Germany	73	95	72	47	48	60	54	80	65	60
Ireland	135	113	128	126	181	141	126	119	95	99
Isle of Man	1	1	1	1	0	0	1	1	0	0
Netherlands	1 647	1 235	943	736	598	811	809	1 005	1 093	908
Norway	24	23	20	21	26	26	30	27	26	23
Portugal	64	49	57	48	39	33	39	46	57	47
Spain	-	-	-	460	409	38	36	29	23	9
Sweden	16	18	15	7	11	13	18	17	13	12
UK	575	608	618	688	568	513	410	470	535	475
27 Fishing area total	*3 813 F*	*3 352*	*2 982 F*	*3 203*	*2 744*	*2 526*	*2 511*	*2 957*	*3 042*	*2 754*
37 France	...	...	...	...	...	...	24	26	26	30
Spain	-	-	-	18	20	22	-	-	-	-
37 Fishing area total	*...*	*...*	*...*	*18*	*20*	*22*	*24*	*26*	*26*	*30*
Species total	*3 813 F*	*3 352*	*2 982 F*	*3 221*	*2 764*	*2 548*	*2 535*	*2 983*	*3 068*	*2 784*
Windowpane flounder	**Turbot de sable**		**Rodaballo aranero**		***Scophthalmus aquosus***				**1,83(05)064,02**	**FLD**
21 USA	1 603	526	759	46	48	521	166	268	177	98
21 Fishing area total	*1 603*	*526*	*759*	*46*	*48*	*521*	*166*	*268*	*177*	*98*
Species total	*1 603*	*526*	*759*	*46*	*48*	*521*	*166*	*268*	*177*	*98*
Turbot	**Turbot**		**Rodaballo**		***Psetta maxima***				**1,83(05)092,01**	**TUR**
27 Belgium	480	480	499	382	337	327	368	464	506	445
Channel Is	2 F	6	6 F	5	5	3	4	6	9	6
Denmark	1 531	1 572	1 396	1 117	908	770	727	809	872	994
Faeroe Is	320	-	-	2	-	-	-	-	-	-
Finland	...	...	...	...	...	...	...	6	4	3
France	1 043	1 728	817	770	515	498	540	632	624	641
Germany	385	384	399	256	330	267	309	454	363	343
Iceland	0	0	1	0	0	0	0	0	0	0
Ireland	223	194	233	261	257	234	261	236	185	183
Isle of Man	1	0	0	1	1	0	0	-	0	-
Latvia	-	-	49	42	46	36	54	16	6	9
Lithuania	-	-	-	-	-	62	58	23	18	18
Netherlands	2 938	2 724	2 476	1 780	1 866	1 700	1 812	2 287	2 277	1 899
Norway	66	62	53	54	57	45	48	69	94	99
Portugal	65	54	57	40	28	27	34	63	83	69
Russian Fed	-	-	-	-	-	-	-	53	69	50
Spain	271	225	239	264	320	218	241	113	108	35
Sweden	114	113	195	296	294	188	159	106	64	55
UK	1 531	1 490	1 281	1 270	1 148	974	851	877	1 001	1 067
27 Fishing area total	*8 970 F*	*9 032*	*7 701 F*	*6 540*	*6 112*	*5 349*	*5 466*	*6 214*	*6 283*	*5 916*
37 Bulgaria	...	...	60	62	60	64	54	55	57	136

B-31 Flounders, halibuts, soles / Flets, flétans, soles / Platijas, halibuts, lenguados

Capture production by species, fishing areas and countries or areas
Captures par espèces, zones de pêche et pays ou zones
Capturas por especies, áreas de pesca y países o áreas

Species, Fishing area Espèce, Zone de pêche Especie, Area de pesca	1993 t	1994 t	1995 t	1996 t	1997 t	1998 t	1999 t	2000 t	2001 t	2002 t
France	89	77	5	40	131	131	13	18	15	18
Greece	182	115	102	60	60	47	65	63	77	70 F
Romania	6	6	4	6	1	-	2	2	13	17
Spain	18 F	18 F	18 F	18	19	13	11	11	14	8
Tunisia	1	2	0	0	0	-	-	-	-	-
Turkey	1 636	2 159	2 955	2 035	980	1 860	1 870	2 700	2 455	459
Ukraine	167	139	96	120	82	63	110	118	171	157
37 Fishing area total	*2 099 F*	*2 516 F*	*3 240 F*	*2 341*	*1 333*	*2 178*	*2 125*	*2 967*	*2 802*	*865 F*
Species total	*11 069 F*	*11 548 F*	*10 941 F*	*8 881*	*7 445*	*7 527*	*7 591*	*9 181*	*9 085*	*6 781 F*
Turbots nei	**Turbots nca**		**Rodaballos nep**		***Scophthalmidae***			**1,83(05)XXX,XX**		**SCF**
37 Italy	1 288	1 213	1 923	1 377	964	528	478	643	622	482
37 Fishing area total	*1 288*	*1 213*	*1 923*	*1 377*	*964*	*528*	*478*	*643*	*622*	*482*
Species total	*1 288*	*1 213*	*1 923*	*1 377*	*964*	*528*	*478*	*643*	*622*	*482*
Indian halibut	**Turbot épineux-indien**		**Lenguado espinudo-indio**		***Psettodes erumei***			**1,83(07)001,01**		**HAI**
51 Eritrea	...	9	4	8	1	0	-	-	-	-
Tanzania	40	27	43	148	50	45	50	52	75	60
Yemen	89	422	974	724	760 F	900 F	800 F	750 F	950 F	1 050 F
51 Fishing area total	*129*	*458*	*1 021*	*880*	*811 F*	*945 F*	*850 F*	*802 F*	*1 025 F*	*1 110 F*
57 Indonesia	2 819	3 881	3 352	4 937	6 102	6 167	6 954	7 081	7 012	7 970
Thailand	914	2 689	2 741	5 926	6 110	1 612	518	330	2 656	2 600
57 Fishing area total	*3 733*	*6 570*	*6 093*	*10 863*	*12 212*	*7 779*	*7 472*	*7 411*	*9 668*	*10 570*
71 Indonesia	3 131	1 979	2 317	2 460	8 973	3 772	5 117	4 062	4 361	5 540
Thailand	968	1 216	1 657	1 368	1 313	1 130	1 828	1 904	1 571	1 589
71 Fishing area total	*4 099*	*3 195*	*3 974*	*3 828*	*10 286*	*4 902*	*6 945*	*5 966*	*5 932*	*7 129*
Species total	*7 961*	*10 223*	*11 088*	*15 571*	*23 309 F*	*13 626 F*	*15 267 F*	*14 179 F*	*16 625 F*	*18 809 F*
Spottail spiny turbot	**Turbot épineux tacheté**		**Perro**		***Psettodes belcheri***			**1,83(07)001,02**		**SOT**
34 Togo	1	0	0	0	2	0	-	1	2	2
34 Fishing area total	*1*	*0*	*0*	*0*	*2*	*0*	*-*	*1*	*2*	*2*
Species total	*1*	*0*	*0*	*0*	*2*	*0*	*-*	*1*	*2*	*2*
Pacific sanddab	**...B**		**...C**		***Citharichthys sordidus***			**1,83(08)009,02**		**IYO**
67 USA	-	-	-	-	-	-	-	150	-	-
67 Fishing area total	*-*	*-*	*-*	*-*	*-*	*-*	*-*	*150*	*-*	*-*
Species total	*-*	*-*	*-*	*-*	*-*	*-*	*-*	*150*	*-*	*-*
Bastard halibut	**Cardeau hirame**		**Falso halibut del Japón**		***Paralichthys olivaceus***			**1,83(08)046,01**		**BAH**
61 Japan	6 464	6 667	7 558	8 311	8 361	7 615	7 198	7 572	6 729	6 680
Korea Rep	2 458	2 035	1 914	2 317	1 592	2 002	1 679	1 607	1 707	1 822
61 Fishing area total	*8 922*	*8 702*	*9 472*	*10 628*	*9 953*	*9 617*	*8 877*	*9 179*	*8 436*	*8 502*
Species total	*8 922*	*8 702*	*9 472*	*10 628*	*9 953*	*9 617*	*8 877*	*9 179*	*8 436*	*8 502*
California flounder	**Cardeau californien**		**Lenguado de California**		***Paralichthys californicus***			**1,83(08)046,03**		**YSF**
67 USA	...	...	...	...	...	13	3	0	4	4
67 Fishing area total	*...*	*...*	*...*	*...*	*...*	*13*	*3*	*0*	*4*	*4*
77 USA	...	...	...	...	...	510	617	390	389	424
77 Fishing area total	*...*	*...*	*...*	*...*	*...*	*510*	*617*	*390*	*389*	*424*
Species total	*...*	*...*	*...*	*...*	*...*	*523*	*620*	*390*	*393*	*428*
Summer flounder	**Cardeau d'été**		**Falso halibut del Canadá**		***Paralichthys dentatus***			**1,83(08)046,06**		**FLS**
21 Cuba	0	-	-	-	-	-	-	-	-	-
USA	5 771	4 961	6 943	5 425	4 775	5 548	5 166	5 344	5 350	6 776
21 Fishing area total	*5 771*	*4 961*	*6 943*	*5 425*	*4 775*	*5 548*	*5 166*	*5 344*	*5 350*	*6 776*
31 USA	...	...	...	...	...	800	629	719	1 204	639
31 Fishing area total	*...*	*...*	*...*	*...*	*...*	*800*	*629*	*719*	*1 204*	*639*
Species total	*5 771*	*4 961*	*6 943*	*5 425*	*4 775*	*6 348*	*5 795*	*6 063*	*6 554*	*7 415*
Bastard halibuts nei	**Cardeaux nca**		**Falsos halibuts nep**		***Paralichthys spp***			**1,83(08)046,XX**		**BAX**
31 USA	1 327	3 918	1 926	2 192	1 059	533	391	503	...	708
31 Fishing area total	*1 327*	*3 918*	*1 926*	*2 192*	*1 059*	*533*	*391*	*503*	*...*	*708*
41 Argentina	9 557	7 512	10 213	8 753	10 044	8 751	6 668	6 498	5 944	4 297
Uruguay	99	113	130	483	502	496	413	256	307	400
41 Fishing area total	*9 656*	*7 625*	*10 343*	*9 236*	*10 546*	*9 247*	*7 081*	*6 754*	*6 251*	*4 697*
Species total	*10 983*	*11 543*	*12 269*	*11 428*	*11 605*	*9 780*	*7 472*	*7 257*	*6 251*	*5 405*

B-31

Flounders, halibuts, soles — **Capture production by species, fishing areas and countries or areas**
Flets, flétans, soles — **Captures par espèces, zones de pêche et pays ou zones**
Platijas, halibuts, lenguados — **Capturas por especies, áreas de pesca y países o áreas**

Species, Fishing area Espèce, Zone de pêche Especie, Area de pesca	1993 t	1994 t	1995 t	1996 t	1997 t	1998 t	1999 t	2000 t	2001 t	2002 t
Antarctic armless flounder	**Mancoglosse antarctique**		**Mancolenguado antártico**		***Mancopsetta maculata***				**1,83(10)002,01**	**MMM**
48 Argentina	-	0	-	-	-	-	-	-	-	-
UK	-	-	-	-	-	-	-	0	-	-
48 Fishing area total	-	*0*	-	-	-	-	-	*0*	-	-
Species total	-	*0*	-	-	-	-	-	*0*	-	-
Flatfishes nei	**Poissons plats nca**		**Peces planos nep**		***Pleuronectiformes***				**1,83(XX)XXX,XX**	**FLX**
21 Canada	6 268	4 498	2 214	1 519	1 283	1 025	984	896	617	233
Cuba	19	3	13	-	-	-	1	-	-	-
Estonia	-	-	-	-	-	-	-	1	-	-
Faeroe Is	36	-	14	2	-	-	-	-	-	-
Japan	4 425	2 601	-	7	19	37	97	73	266	-
Korea Rep	13	-	-	-	-	-	-	-	-	-
Portugal	-	-	-	-	-	-	-	-	-	118
USA	58	70	49	63	70	9	54	13	5	3
21 Fishing area total	*10 819*	*7 172*	*2 290*	*1 591*	*1 372*	*1 071*	*1 136*	*983*	*888*	*354*
27 Faeroe Is	1 634	-	327	-	757	1 059	292	-	-	16
France	5	14	14	14	17	24	15	3 067	1 851	3 136
Germany	179	1 600	-	404	508	279	-	-	-	3
Iceland	30	6	10	11	13	3	5	-	2	1
Ireland	35	-	68	141	210	184	37	23	15	14
Japan	283	399	140	-	-	-	-	-	-	-
Lithuania	-	-	-	-	-	-	-	-	1 137	1 082
Norway	650	555	337	376	477	389	475	76	153	117
Poland	5 101	4 900	8 964	8 836	6 168	5 835	5 779	5 601	6 725	9 232
Portugal	1 646	1 504	1 652	1 291	1 082	964	8	10	5	12
Spain	5 800	4 070	4 251	8 307	13 955	11 569	9 972	830	653	657
Sweden	0	0	1	0	1	0	0	-	-	-
UK	-	-	132	172	172	-	-	158	171	138
27 Fishing area total	*15 363*	*13 048*	*15 896*	*19 552*	*23 360*	*20 306*	*16 583*	*9 765*	*10 712*	*14 408*
31 Korea Rep	-	-	-	-	-	-	16	7	-	-
Mexico	83	100	166	182	201	225	197	113	175	263
USA	702	728	604	153	286	333	330	371	296	312
Venezuela	...	...	...	...	...	...	51	11	...	...
31 Fishing area total	*785*	*828*	*770*	*335*	*487*	*558*	*594*	*502*	*471*	*575*
34 Benin	14 F	16 F	23 F	30 F	37 F	45 F	52	30 F	14	22
Cuba	0	0	-	-	-	-	-	-	-	-
Eq Guinea	290 F	450 F	190 F	420 F	530 F	325	380 F	40 F	40 F	40 F
Gabon	12	31	24	75	93	65	84	242	216	172
Guinea	240 F	250 F	350	254	256	179	148	1 032	914	900 F
GuineaBissau	112 F	92	87	64	70 F	60 F	50 F	50 F	50 F	50 F
Italy	1 570	1 570	-	-	261	244	96	107	117	115
Korea Rep	-	12	-	73	27	-	294	166	-	9
Lithuania	146	-	-	-	-	-	-	-	-	-
Mauritania	430 F	490 F	570 F	580 F	700 F	600 F	1 751 F	1 756 F	2 200 F	2 200 F
Morocco	3 266	3 348	3 709	3 710	2 230	4 896	5 748	5 000	3 869	5 158
Nigeria	0	0	-	-	-	-	-	-	-	-
Portugal	6	28	10	13	1	0	3	1	1	2
Senegal	7 224	11 857	10 510	8 113	8 002	7 132	7 335	8 113	9 059	7 262
Spain	800 F	1 200 F	1 600 F	1 846	3 812	3 453	5 191	9 572	6 938	9 239
34 Fishing area total	*14 110 F*	*19 344 F*	*17 073 F*	*15 178 F*	*16 019 F*	*16 999 F*	*21 132 F*	*26 109 F*	*23 418 F*	*25 169 F*
37 Croatia	217	144	124	130	133	150	65	113	111	106
France	-	-	-	-	-	-	5	8	4	7
Georgia	-	-	-	-	-	-	5	9	11	11
Lebanon	5	5	5	5	5	10	5	11	15	8
Morocco	22	17	37	40	57	19	25	38	29	72
Palest, O.T.	...	...	...	7	15	25	25 F	20 F	10 F	6
Russian Fed	72	96	44	43	44	34	36	31	25	15
Serbia-Monte	11	8	13	10	8	10	9	9	11	10
Spain	-	-	-	100	467	674	683	655	528	621
Tunisia	114	177	96	52	110	178	217	190	205	194
Turkey	1 779	1 549	1 092	1 947	2 300	2 000	2 400	1 000	1 250	585
37 Fishing area total	*2 220*	*1 996*	*1 411*	*2 334*	*3 139*	*3 100*	*3 475 F*	*2 084 F*	*2 199 F*	*1 635*
41 Brazil	2 500 F	2 500 F	1 491	1 091	1 430	1 655	1 590	1 844	1 820	3 321
Italy	209	209	-	-	-	-	-	-	-	-
Korea Rep	-	5	-	6	-	-	7	34	-	-
Spain	-	-	-	-	-	-	4	9	4	23
UK	-	-	-	-	-	-	-	-	-	5
41 Fishing area total	*2 709 F*	*2 714 F*	*1 491*	*1 097*	*1 430*	*1 655*	*1 601*	*1 887*	*1 824*	*3 349*
47 Angola	784	428	97	195	120	928	912	592	1 928	3 463
Korea Rep	1 900	2 065	132	-	-	-	-	-	-	-
Russian Fed	-	-	-	-	-	7	-	-	-	2
47 Fishing area total	*2 684*	*2 493*	*229*	*195*	*120*	*935*	*912*	*592*	*1 928*	*3 465*
51 India	20 724	27 043	18 533	17 456	21 696	17 460	12 840	24 709	11 486	15 661
Italy	314	314	-	-	-	-	-	-	-	-
Korea Rep	-	-	38	50	194	14	11	11	3	3
Saudi Arabia	30	42	51	58	75	90	84	84	85	65

B-31 Flounders, halibuts, soles / Flets, flétans, soles / Platijas, halibuts, lenguados

Capture production by species, fishing areas and countries or areas / Captures par espèces, zones de pêche et pays ou zones / Capturas por especies, áreas de pesca y países o áreas

Species, Fishing area Espèce, Zone de pêche Especie, Area de pesca	1993 t	1994 t	1995 t	1996 t	1997 t	1998 t	1999 t	2000 t	2001 t	2002 t
51 Fishing area total	*21 068*	*27 399*	*18 622*	*17 564*	*21 965*	*17 564*	*12 935*	*24 804*	*11 574*	*15 729*
57 Australia	-	1	2	2	-	-	-	-	-	-
India	4 764	4 084	6 148	6 741	7 556	6 202	5 184	5 362	3 145	3 787
Indonesia	982	1 036	1 509	2 369	2 216	2 138	2 586	2 841	3 300	3 900
Malaysia	332	281	316	538	518	377	378	342	485	394
57 Fishing area total	*6 078*	*5 402*	*7 975*	*9 650*	*10 290*	*8 717*	*8 148*	*8 545*	*6 930*	*8 081*
61 China,Taiwan	121	122	145	345	346	142	146	137	198	172
Japan	76 905	70 544	71 986	80 418	76 671	72 813	68 750	68 445	60 713	63 812
Korea D P Rp	...	2 318	2 953	1 966	6 972	4 000 F	4 000 F	3 800 F	3 800 F	3 800 F
Korea Rep	-	-	25	248	102	83	-	-	-	-
Russian Fed	66 025	38 550	46 768	57 180	63 768	79 682	97 014	103 077	95 106	79 814
61 Fishing area total	*143 051*	*111 534*	*121 877*	*140 157*	*147 859*	*156 720 F*	*169 910 F*	*175 459 F*	*159 817 F*	*147 598 F*
67 Canada	9 917	8 501	7 963	5 222	4 300	5 000	5 651	5 877	5 459	6 711
USA	44 674	23 767	26 713	29 509	18 740	18 886	13 040	7 627	4 748	5 564
67 Fishing area total	*54 591*	*32 268*	*34 676*	*34 731*	*23 040*	*23 886*	*18 691*	*13 504*	*10 207*	*12 275*
71 Indonesia	916	605	705	1 045	5 191	1 582	2 488	1 395	3 450	4 080
Korea Rep	-	-	-	-	-	-	84	-	-	-
Malaysia	1 389	1 581	1 699	1 695	1 675	2 180	2 648	1 299	1 227	1 198
Philippines	1 186	1 072	805	829	627	659	722	729	756	917
71 Fishing area total	*3 491*	*3 258*	*3 209*	*3 569*	*7 493*	*4 421*	*5 942*	*3 423*	*5 433*	*6 195*
77 Mexico	1 973	1 192	1 980	2 342	2 540	1 165	2 071	2 568	1 664	1 573
Nicaragua	...	...	...	...	...	...	...	55	67	26
USA	5 343	4 912	6 486	6 878	2 056	665	822	725	686	379
77 Fishing area total	*7 316*	*6 104*	*8 466*	*9 220*	*4 596*	*1 830*	*2 893*	*3 348*	*2 417*	*1 978*
81 Australia	5	3	3	5	6	6	12	15	12	8
New Zealand	5 751	4 873	5 247	4 284	7 747	4 270	3 505	2 939	3 220	2 810
81 Fishing area total	*5 756*	*4 876*	*5 250*	*4 289*	*7 753*	*4 276*	*3 517*	*2 954*	*3 232*	*2 818*
87 Chile	726	1 191	220	203	154	75	84	95	76	12
Colombia	15	9	11	30	45	131	48	50 F	50 F	50 F
Korea Rep	-	-	-	-	-	-	3	-	-	-
Peru	1 195	732	1 559	528	212	230	263	177	313	256
87 Fishing area total	*1 936*	*1 932*	*1 790*	*761*	*411*	*436*	*398*	*322 F*	*439 F*	*318 F*
Species total	*291 977 F*	*240 368 F*	*241 025 F*	*260 223 F*	*269 334 F*	*262 474 F*	*267 867 F*	*274 281 F*	*241 489 F*	*243 947 F*
Group total	***1 105 021***	***987 224***	***918 620***	***945 886***	***1 030 607***	***935 111***	***956 111***	***1 008 471***	***948 651***	***918 840***

B-32

Cods, hakes, haddocks
Morues, merlus, églefins
Bacalaos, merluzas, eglefinos

Capture production by species, fishing areas and countries or areas
Captures par espèces, zones de pêche et pays ou zones
Capturas por especies, áreas de pesca y países o áreas

Species, Fishing area Espèce, Zone de pêche Especie, Area de pesca	1993 t	1994 t	1995 t	1996 t	1997 t	1998 t	1999 t	2000 t	2001 t	2002 t
Smalleye moray cod	**Gadomurène petit oeil**		**Gadimorena ojichica**		***Muraenolepis microps***				**1,48(01)001,01**	**MOY**
48 Argentina	-	0	-	-	-	-	-	-	-	-
Bulgaria	-	0	-	-	-	-	-	-	-	-
UK	-	-	-	-	-	-	-	0	-	0
48 Fishing area total	-	*0*	-	-	-	-	-	*0*	-	*0*
58 South Africa	-	-	-	-	-	-	-	-	0	-
58 Fishing area total	-	-	-	-	-	-	-	-	*0*	-
88 New Zealand	-	-	-	-	-	-	4	5	-	0
South Africa	-	-	-	-	-	-	-	-	0	-
88 Fishing area total	-	-	-	-	-	-	*4*	*5*	*0*	*0*
Species total	-	*0*	-	-	-	-	*4*	*5*	*0*	*0*
Moray cods nei	**Gadomurènes nca**		**Gadimorenas nep**		***Muraenolepis spp***				**1,48(01)001,XX**	**MRL**
48 UK	-	-	-	-	-	-	0	0	-	-
48 Fishing area total	-	-	-	-	-	-	*0*	*0*	-	-
58 South Africa	-	-	-	-	-	-	-	-	0	-
58 Fishing area total	-	-	-	-	-	-	-	-	*0*	-
88 New Zealand	-	-	-	-	-	0	1	2	3	5
88 Fishing area total	-	-	-	-	-	*0*	*1*	*2*	*3*	*5*
Species total	-	-	-	-	-	*0*	*1*	*2*	*3*	*5*
Common mora	**Moro commun**		**Mollera moranella**		***Mora moro***				**1,48(02)010,01**	**RIB**
27 France	-	-	-	-	-	-	-	-	1	39
Ireland	-	-	-	-	-	-	-	-	32	44
27 Fishing area total	-	-	-	-	-	-	-	-	*33*	*83*
81 New Zealand	...	...	1 192	694	1 410	1 324	1 122	1 355	1 209	1 308
Ukraine	-	-	-	-	-	-	-	3	2	-
81 Fishing area total	...	...	*1 192*	*694*	*1 410*	*1 324*	*1 122*	*1 358*	*1 211*	*1 308*
Species total	...	...	*1 192*	*694*	*1 410*	*1 324*	*1 122*	*1 358*	*1 244*	*1 391*
Red codling	**Morue rouge**		**Brotolilla**		***Pseudophycis bachus***				**1,48(02)014,01**	**NEC**
81 Japan	467	313	105	24	14	15	27	70	26	20 F
New Zealand	12 217	10 198	15 916	10 572	11 073	16 429	12 528	5 232	4 454	4 414
Russian Fed	34	130	138	-	-	-	-	-	-	-
Ukraine	98	...	...	...	...	51	...	63	50	9
81 Fishing area total	*12 816*	*10 641*	*16 159*	*10 596*	*11 087*	*16 495*	*12 555*	*5 365*	*4 530*	*4 443 F*
Species total	*12 816*	*10 641*	*16 159*	*10 596*	*11 087*	*16 495*	*12 555*	*5 365*	*4 530*	*4 443 F*
Blue antimora	**Antimora bleu**		**Mollera azul**		***Antimora rostrata***				**1,48(02)030,01**	**ANT**
21 Spain	-	-	-	16	-	26	24	21	-	0
21 Fishing area total	-	-	-	*16*	-	*26*	*24*	*21*	-	*0*
27 Iceland	-	-	-	2	-	-	-	-	-	-
27 Fishing area total	-	-	-	*2*	-	-	-	-	-	-
48 Argentina	-	-	1	-	-	-	-	-	-	-
Chile	-	-	0	0	-	-	-	-	-	-
Korea Rep	-	-	0	0	1	-	-	-	-	-
Russian Fed	-	-	-	-	-	-	-	-	0	-
South Africa	-	-	-	-	-	1	-	-	-	-
Spain	-	-	-	-	0	-	-	-	-	-
UK	-	-	-	-	1	-	-	-	-	-
Uruguay	-	-	-	-	-	-	0	-	-	-
48 Fishing area total	-	-	*1*	*0*	*2*	*1*	*0*	-	*0*	-
58 Australia	-	-	-	-	0	-	0	-	-	-
South Africa	-	-	-	-	0	8	7	24	3	1
58 Fishing area total	-	-	-	-	*0*	*8*	*7*	*24*	*3*	*1*
88 New Zealand	-	-	-	-	-	0	0	0	3	3
South Africa	-	-	-	-	-	-	-	-	1	-
88 Fishing area total	-	-	-	-	-	*0*	*0*	*0*	*4*	*3*
Species total	-	-	*1*	*18*	*2*	*35*	*31*	*45*	*7*	*4*
Tadpole codling	**More têtard**		**Bacalao criollo**		***Salilota australis***				**1,48(02)040,01**	**SAO**
41 Argentina	...	...	2 241	1 742	2 632	3 604	6 607	8 435	1 858	2 604
Australia	-	-	-	-	-	85	60	-	-	-
Belize	-	-	-	-	-	-	28	237	42	-
Falkland Is	69	100	1 530	2 033	817	1 491	2 692	1 886	1 371	947
France	-	16	21	31	25	11	5	29	-	-
Honduras	374	107	106	189	-	-	-	-	-	-
Namibia	-	-	-	-	20	99	128	-	-	-
Panama	101	8	90	93	-	-	-	-	-	-

B-32

Cods, hakes, haddocks — **Capture production by species, fishing areas and countries or areas**
Morues, merlus, églefins — **Captures par espèces, zones de pêche et pays ou zones**
Bacalaos, merluzas, eglefinos — **Capturas por especies, áreas de pesca y países o áreas**

Species, Fishing area Espèce, Zone de pêche Especie, Area de pesca	1993 t	1994 t	1995 t	1996 t	1997 t	1998 t	1999 t	2000 t	2001 t	2002 t
Portugal	21	-	-	-	-	-	-	12	-	-
St Vincent	-	-	-	-	-	-	-	-	14	-
Seychelles	-	-	-	-	56	-	-	-	-	-
Spain	4 084	3 061	5 942	3 484	2 505	6 140	5 935	3 914	2 250	545
UK	140	6	16	18	39	24	188	30	17	15
41 Fishing area total	*4 789*	*3 298*	*9 946*	*7 590*	*6 094*	*11 454*	*15 643*	*14 543*	*5 552*	*4 111*
87 Chile	2 908	1 800	1 826	481	647	352	245	372	641	360
87 Fishing area total	*2 908*	*1 800*	*1 826*	*481*	*647*	*352*	*245*	*372*	*641*	*360*
Species total	*7 697*	*5 098*	*11 772*	*8 071*	*6 741*	*11 806*	*15 888*	*14 915*	*6 193*	*4 471*
Moras nei	**Mores nca**		**Moras nep**		***Moridae***			**1,48(02)XXX,XX**		**MOR**
27 France	-	-	-	-	-	75	67	60	73	12
Norway	-	-	-	-	-	-	-	68	277	97
UK	-	-	-	-	415	0	-	2	-	-
27 Fishing area total	-	-	-	-	*415*	*75*	*67*	*130*	*350*	*109*
61 Russian Fed	8 081	13 811	3 758	-	152	11 115	31 295	39 316	32 392	29 323
61 Fishing area total	*8 081*	*13 811*	*3 758*	-	*152*	*11 115*	*31 295*	*39 316*	*32 392*	*29 323*
Species total	*8 081*	*13 811*	*3 758*	-	*567*	*11 190*	*31 362*	*39 446*	*32 742*	*29 432*
Unicorn cod	**Bregmacère de l'océan Indien**		**Bregmacero**		***Bregmaceros mcclellandi***			**1,48(03)018,01**		**UNC**
51 India	1 271	753	770	939	971	1 113	458	1 123	2 095	2 101
Mauritius	302	309	301	312	367	336	285	347	340	271
51 Fishing area total	*1 573*	*1 062*	*1 071*	*1 251*	*1 338*	*1 449*	*743*	*1 470*	*2 435*	*2 372*
57 India	113	152	161	36	13	1 138	1 352	159	92	108
57 Fishing area total	*113*	*152*	*161*	*36*	*13*	*1 138*	*1 352*	*159*	*92*	*108*
Species total	*1 686*	*1 214*	*1 232*	*1 287*	*1 351*	*2 587*	*2 095*	*1 629*	*2 527*	*2 480*
Tusk(=Cusk)	**Brosme**		**Brosmio**		***Brosme brosme***			**1,48(04)001,01**		**USK**
21 Canada	2 957	1 692	2 010	1 405	1 801	1 641	1 065	1 097	1 498	1 288
Cuba	10	-	-	-	-	-	-	-	-	-
USA	1 428	1 081	772	468	443	354	230	188	180	150
21 Fishing area total	*4 395*	*2 773*	*2 782*	*1 873*	*2 244*	*1 995*	*1 295*	*1 285*	*1 678*	*1 438*
27 Denmark	134	92	89	130	146	105	177	225	274	221
Faeroe Is	3 313	4 678	4 490	2 562	2 593	2 389	2 714	2 585	2 922	2 177
France	525	465	433	439	439	453	428	335	282	184
Germany	39	42	29	59	26	19	16	13	10	10
Greenland	1	-	20	-	-	-	-	-	-	-
Iceland	4 747	4 612	5 245	5 226	4 847	4 118	5 796	4 741	3 425	3 935
Ireland	67	52	76	64	45	43	43	113	122	109
Norway	26 792	20 375	18 682	19 483	13 797	21 032	23 274	21 915	18 778	18 171
Portugal	0	0	0	0	-	-	-	-	-	-
Russian Fed	-	-	-	-	-	-	-	46	83	54
Spain	-	-	-	62	98	106	150	249	72	151
Sweden	12	12	5	7	3	3	3	8	6	4
UK	294	334	341	391	507	723	852	1 016	872	672
27 Fishing area total	*35 924*	*30 662*	*29 410*	*28 423*	*22 501*	*28 991*	*33 453*	*31 246*	*26 846*	*25 688*
Species total	*40 319*	*33 435*	*32 192*	*30 296*	*24 745*	*30 986*	*34 748*	*32 531*	*28 524*	*27 126*
Atlantic cod	**Morue de l'Atlantique**		**Bacalao del Atlántico**		***Gadus morhua***			**1,48(04)002,02**		**COD**
21 Canada	76 556	22 719	12 438	15 541	29 899	37 741	55 410	46 046	40 325	35 255
Cuba	15	2	1	1	0	1	5	-	-	-
Estonia	-	-	-	-	-	-	-	6	44	8
Faeroe Is	3 049	2 250	1 016	701	-	-	-	-	-	1
Germany	-	-	-	16	-	-	-	-	-	-
Greenland	1 925	2 117	1 710	944	904	326	2	764	1 680	3 813
Iceland	-	93	-	-	-	-	-	-	-	-
Korea Rep	10	-	-	-	-	-	-	-	-	-
Norway	0	1	-	-	-	-	-	-	-	-
Portugal	3 657	2 636	1 669	1 318	1 546	549	327	191	357	488
Russian Fed	298	-	-	-	-	5	26	140	254	348
St Pier Mq	103	86	60	44	1 547	3 123	3 171	4 682	2 350	2 219
Spain	6 055	3 735	563	181	2	-	3	5	-	-
UK	-	-	-	129	23	-	-	-	-	-
USA	22 908	17 533	13 440	14 253	12 982	11 119	9 727	11 367	15 064	13 128
21 Fishing area total	*114 576*	*51 172*	*30 897*	*33 128*	*46 903*	*52 864*	*68 671*	*63 201*	*60 074*	*55 260*
27 Belgium	4 347	3 611	5 938	4 491	5 677	6 893	4 540	3 693	3 207	3 506
Channel Is	2 F	3	3 F	6	14	19	21	11	6	9
Denmark	47 924	55 221	78 332	90 741	80 491	69 025	70 547	57 018	46 185	37 859
Estonia	70	905	1 049	1 392	1 174	1 070	1 059	514	755	51
Faeroe Is	24 731	33 882	44 332	59 803	57 921	39 689	33 725	32 601	38 706	38 116
Finland	230	529	1 861	3 139	1 543	1 037	1 572	1 824	1 723	1 051
France	19 037	16 607	17 669	21 500	25 123	20 640	14 431	11 886	11 336	11 660
Germany	18 602	22 055	31 892	37 613	26 491	23 075	21 990	18 414	19 222	15 412
Greenland	5 620	6 955	7 493	6 542	6 777	5 272	4 115	2 234	3 934	4 246
Iceland	260 544	214 656	202 900	204 058	208 636	242 968	260 643	238 324	240 002	213 417
Ireland	3 635	4 963	5 650	7 258	5 702	5 294	3 860	2 928	2 653	2 503

B-32 Cods, hakes, haddocks — Capture production by species, fishing areas and countries or areas
Morues, merlus, églefins — Captures par espèces, zones de pêche et pays ou zones
Bacalaos, merluzas, eglefinos — Capturas por especies, áreas de pesca y países o áreas

Species, Fishing area Espèce, Zone de pêche Especie, Area de pesca	1993 t	1994 t	1995 t	1996 t	1997 t	1998 t	1999 t	2000 t	2001 t	2002 t
Isle of Man	57	27	22	27	19	34	9	11	1	7
Japan	2	-	-	-	-	-	-	-	-	-
Latvia	1 333	2 379	6 471	8 741	6 187	7 778	6 914	6 280	6 298	4 867
Lithuania	574	1 886	3 629	5 520	4 694	3 296	4 371	4 721	3 852	2 964
Netherlands	10 220	6 512	11 189	9 307	11 838	14 724	9 075	6 000	3 656	4 714
Norway	275 238	373 577	365 333	358 395	401 277	321 428	256 555	219 192	208 856	228 672
Poland	8 909	14 426	25 001	35 968	34 295	27 705	28 056	23 340	23 310	17 209
Portugal	1 788	5 644	5 654	6 765	7 533	5 493	3 885	3 587	4 027	3 630
Russian Fed	250 567	319 679	297 770	309 391	316 147	248 714	215 590	170 878	188 630	187 865
Spain	8 800	14 929	15 580	16 175	17 226	14 392	10 156	8 765	20 283	8 409
Sweden	17 971	30 986	33 186	41 827	34 797	22 475	22 597	23 174	24 111	17 383
UK	67 289	68 855	78 650	78 235	74 614	77 182	51 695	41 750	32 840	31 548
27 Fishing area total	*1 027 490 F*	*1 198 287*	*1 239 604 F*	*1 306 894*	*1 328 176*	*1 158 203*	*1 025 406*	*877 145*	*883 593*	*835 098*
Species total	*1 142 066 F*	*1 249 459*	*1 270 501 F*	*1 340 022*	*1 375 079*	*1 211 067*	*1 094 077*	*940 346*	*943 667*	*890 358*
Pacific cod	**Morue du Pacifique**		**Bacalao del Pacífico**		***Gadus macrocephalus***			**1,48(04)002,11**		**PCO**
61 Japan	62 286	65 778	56 561	57 576	58 477	57 243	55 292	51 052	43 550	29 516
Korea Rep	11 626	3 702	2 476	2 740	3 984	6 249	6 509	10 098	13 110	9 240
Russian Fed	95 823	81 445	100 730	93 711	79 927	94 282	101 929	68 415	59 783	60 625
61 Fishing area total	*169 735*	*150 925*	*159 767*	*154 027*	*142 388*	*157 774*	*163 730*	*129 565*	*116 443*	*99 381*
67 Canada	8 123	3 550	2 172	700	1 537	1 400	836	712	474	694
Japan	200	220	-	-	-	-	-	-	-	-
Russian Fed	-	-	-	179	-	-	-	-	-	-
USA	218 996	208 785	268 257	274 569	299 970	252 197	237 679	240 635	213 967	232 617
67 Fishing area total	*227 319*	*212 555*	*270 429*	*275 448*	*301 507*	*253 597*	*238 515*	*241 347*	*214 441*	*233 311*
Species total	*397 054*	*363 480*	*430 196*	*429 475*	*443 895*	*411 371*	*402 245*	*370 912*	*330 884*	*332 692*
Greenland cod	**Morue ogac**		**Bacalao de Groenlandia**		***Gadus ogac***			**1,48(04)002,12**		**GRC**
21 Canada	0	0	0	-	-	-	-	-	-	-
Greenland	1 892	1 854	2 525	2 120	1 728	1 695	1 899	931	1 128	972
21 Fishing area total	*1 892*	*1 854*	*2 525*	*2 120*	*1 728*	*1 695*	*1 899*	*931*	*1 128*	*972*
27 Greenland	-	4	3	1	1	2	4	-	5	5
27 Fishing area total	-	*4*	*3*	*1*	*1*	*2*	*4*	-	*5*	*5*
Species total	*1 892*	*1 858*	*2 528*	*2 121*	*1 729*	*1 697*	*1 903*	*931*	*1 133*	*977*
Ling	**Lingue**		**Maruca**		***Molva molva***			**1,48(04)005,01**		**LIN**
27 Belgium	168	171	181	153	124	124	103	120	88	91
Channel Is	16 F	28	26 F	20	37	25	19	13	3	2
Denmark	1 237	950	790	868	969	823	837	741	910	829
Faeroe Is	2 047	2 799	3 686	3 132	4 056	3 547	2 998	2 358	2 558	2 139
France	5 158	5 590	5 644	5 738	5 463	5 510	5 112	3 099	2 987	2 546
Germany	973	1 041	877	1 409	965	308	247	215	110	114
Iceland	4 333	4 049	3 729	3 670	3 634	3 603	3 976	3 223	2 864	2 833
Ireland	880	1 158	1 542	1 379	1 305	1 272	1 138	1 089	1 463	1 303
Isle of Man	2	2	1	3	2	1	1	1	0	-
Netherlands	-	-	-	-	-	-	-	5	4	3
Norway	18 351	17 907	18 172	18 931	15 295	22 719	19 217	16 899	13 562	15 358
Russian Fed	-	-	-	-	-	-	-	8	2	11
Spain	3 233	4 724	5 026	5 301	6 153	9 256	8 907	6 259	4 276	6 506
Sweden	144	120	94	73	61	44	44	46	47	47
UK	10 747	11 752	14 695	13 942	12 924	13 594	11 350	9 244	8 095	8 465
27 Fishing area total	*47 289 F*	*50 291*	*54 463 F*	*54 619*	*50 988*	*60 826*	*53 949*	*43 320*	*36 969*	*40 247*
Species total	*47 289 F*	*50 291*	*54 463 F*	*54 619*	*50 988*	*60 826*	*53 949*	*43 320*	*36 969*	*40 247*
Blue ling	**Lingue bleue**		**Maruca azul**		***Molva dypterygia***			**1,48(04)005,02**		**BLI**
21 Faeroe Is	1	3	-	-	-	-	-	-	-	-
21 Fishing area total	*1*	*3*	-	-	-	-	-	-	-	-
27 Denmark	18	15	16	8	14	4	7	15	27	13
Estonia	-	-	-	-	-	-	-	-	85	22
Faeroe Is	3 116	1 781	2 398	1 624	1 172	1 274	2 136	1 756	2 454	801
France	4 772	3 278	3 590	4 127	4 736	5 955	4 794	5 571	3 666	3 223
Germany	298	176	202	119	11	16	15	110	26	10
Greenland	3	-	2	-	-	-	-	-	-	-
Iceland	5 317	1 842	1 636	1 284	1 320	1 208	2 321	1 623	765	1 274
Ireland	3	74	14	-	1	22	43	91	827	583
Lithuania	-	-	-	-	-	-	-	-	16	29
Norway	1 602	973	734	530	497	420	544	834	1 020	902
Poland	-	-	-	-	-	-	-	-	19	8
Portugal	33	42	29	25	21	14	10	14	9	13
Russian Fed	-	-	-	-	-	-	-	-	-	3
Spain	-	-	-	298	1 165	1 404	1 710	3 113	4 472	2 041
Sweden	1	0	0	0	2	0	-	-	-	-
UK	323	258	1 121	1 772	2 811	2 450	4 119	3 019	5 980	3 166
27 Fishing area total	*15 486*	*8 439*	*9 742*	*9 787*	*11 750*	*12 767*	*15 699*	*16 146*	*19 366*	*12 088*
37 Spain	-	-	-	1	1	1	-	-	-	-
37 Fishing area total	-	-	-	*1*	*1*	*1*	-	-	-	-

B-32 Cods, hakes, haddocks / Morues, merlus, églefins / Bacalaos, merluzas, eglefinos

Capture production by species, fishing areas and countries or areas
Captures par espèces, zones de pêche et pays ou zones
Capturas por especies, áreas de pesca y países o áreas

Species, Fishing area Espèce, Zone de pêche Especie, Area de pesca	1993 t	1994 t	1995 t	1996 t	1997 t	1998 t	1999 t	2000 t	2001 t	2002 t
Species total	*15 487*	*8 442*	*9 742*	*9 788*	*11 751*	*12 768*	*15 699*	*16 146*	*19 366*	*12 088*
Greater forkbeard	**Phycis de fond**		**Brótola de fango**		***Phycis blennoides***				**1,48(04)006,01**	**GFB**
27 France	637	524	499	562	605	476	526	729	743	659
Germany	-	-	-	-	-	-	1	8	12	11
Iceland	-	1	0	-	-	-	-	-	-	-
Ireland	60	111	163	154	228	318	379	399	679	720
Norway	-	-	-	-	-	-	-	709	1 340	1 334
Portugal	116	135	79	47	31	44	52	98	89	63
Russian Fed	-	-	-	-	-	-	-	2	11	-
Spain	-	-	-	1 601	5 065	4 505	3 504	2 369	1 175	1 443
UK	345	423	1 044	2 022	2 054	1 887	1 495	1 563	1 204	985
27 Fishing area total	*1 158*	*1 194*	*1 785*	*4 386*	*7 983*	*7 230*	*5 957*	*5 877*	*5 253*	*5 215*
34 Morocco	279	407	397	400	100	230	325	269	350	251
Portugal	-	6	50	66	15	1	2	0	3	-
Russian Fed	1	-	-	-	-	-	-	-	-	-
Spain	-	-	-	-	-	-	42	-	-	-
34 Fishing area total	*280*	*413*	*447*	*466*	*115*	*231*	*369*	*269*	*353*	*251*
37 Algeria	...	...	...	...	...	...	4	5	19	39
France	...	...	...	...	...	...	2	4	3	5
Malta	-	-	-	-	2	3	4	5	0	-
Morocco	48	24	31	40	45	27	20	365	26	40
Spain	300 F	250 F	200 F	140	144	166	129	-	6	314
Turkey	13	7	15	24	50	150	50	50	35	8
37 Fishing area total	*361 F*	*281 F*	*246 F*	*204*	*241*	*346*	*209*	*429*	*89*	*406*
Species total	*1 799 F*	*1 888 F*	*2 478 F*	*5 056*	*8 339*	*7 807*	*6 535*	*6 575*	*5 695*	*5 872*
Brazilian codling	**Phycis brésilien**		**Brótola brasileña**		***Urophycis brasiliensis***				**1,48(04)008,01**	**HKU**
41 Argentina	2 095	1 440	416	618	182	3 329	754	1 038	2 683	834
Brazil	1 000 F	1 000 F	2 924	2 311	2 116	2 408	1 807	2 225	6 029	8 312
Portugal	-	-	-	-	-	-	-	-	3	8
Russian Fed	-	-	-	-	-	-	-	-	-	70
Uruguay	357	280	363	223	272	344	281	235	361	382
41 Fishing area total	*3 452 F*	*2 720 F*	*3 703*	*3 152*	*2 570*	*6 081*	*2 842*	*3 498*	*9 076*	*9 606*
Species total	*3 452 F*	*2 720 F*	*3 703*	*3 152*	*2 570*	*6 081*	*2 842*	*3 498*	*9 076*	*9 606*
Red hake	**Merluche écureuil**		**Locha roja**		***Urophycis chuss***				**1,48(04)008,02**	**HKR**
21 Canada	173	100	135	372	248	120	156	81	130	115
Cuba	171	72	166	427	259	118	87	-	-	-
Portugal	366	267	230	125	56	18	77	42	273	1 969
Russian Fed	-	-	-	-	-	4	2	120	118	1 284
Spain	-	54	112	893	958	1 200	1 350	1 551	1 755	1 164
USA	1 643	1 701	1 607	1 087	1 329	1 343	1 556	1 571	1 732	911
21 Fishing area total	*2 353*	*2 194*	*2 250*	*2 904*	*2 850*	*2 803*	*3 228*	*3 365*	*4 008*	*5 443*
Species total	*2 353*	*2 194*	*2 250*	*2 904*	*2 850*	*2 803*	*3 228*	*3 365*	*4 008*	*5 443*
White hake	**Merluche blanche**		**Locha blanca**		***Urophycis tenuis***				**1,48(04)008,03**	**HKW**
21 Canada	10 209	7 348	6 492	4 827	4 309	3 102	3 318	4 246	4 140	4 585
Cuba	0	37	-	-	-	-	-	-	-	-
Estonia	-	-	-	-	-	-	-	3	2	5
Lithuania	-	-	-	-	-	-	-	-	-	2
Portugal	-	-	-	-	-	-	-	-	-	1 678
St Pier Mq	-	-	-	-	0	1	9	122	10	3
Spain	-	28	39	187	304	491	426	802	689	3 258
USA	7 460	4 695	4 279	3 289	2 217	2 365	2 624	2 984	3 482	3 268
21 Fishing area total	*17 669*	*12 108*	*10 810*	*8 303*	*6 830*	*5 959*	*6 377*	*8 157*	*8 323*	*12 799*
27 Estonia	-	-	-	-	-	-	-	-	-	1
27 Fishing area total	-	-	-	-	-	-	-	-	-	*1*
31 USA	...	...	...	...	...	...	1	1	1	1
31 Fishing area total	...	...	...	...	...	...	*1*	*1*	*1*	*1*
Species total	*17 669*	*12 108*	*10 810*	*8 303*	*6 830*	*5 959*	*6 378*	*8 158*	*8 324*	*12 801*
Haddock	**Eglefin**		**Eglefino**		***Melanogrammus aeglefinus***				**1,48(04)010,01**	**HAD**
21 Canada	12 986	6 955	7 933	10 311	9 776	11 771	10 550	12 683	15 594	14 947
Cuba	68	12	32	40	27	12	4	-	-	-
Faeroe Is	-	1	-	-	-	-	-	-	-	1
Portugal	10	10	2	-	39	6	10	13	23	78
Russian Fed	27	-	-	-	-	1	-	2	33	29
St Pier Mq	-	-	-	-	9	27	16	10	78	222
Spain	-	-	-	-	4	0	-	-	-	-
USA	879	328	398	570	1 504	2 836	3 146	4 002	5 826	7 553
21 Fishing area total	*13 970*	*7 306*	*8 365*	*10 921*	*11 359*	*14 653*	*13 726*	*16 710*	*21 554*	*22 830*
27 Belgium	835	706	648	394	746	976	569	512	840	737
Channel Is	...	...	...	44	0	0	-	-	-	0
Denmark	5 204	4 768	4 479	5 050	5 227	5 786	3 130	2 707	4 001	8 920

Species, Fishing area Espèce, Zone de pêche Especie, Area de pesca	1993 t	1994 t	1995 t	1996 t	1997 t	1998 t	1999 t	2000 t	2001 t	2002 t
Faeroe Is	5 143	7 759	8 494	13 896	21 409	22 598	19 697	16 212	16 061	23 713
France	4 597	4 187	4 167	6 027	8 060	4 983	3 582	4 379	5 970	5 883
Germany	1 522	4 259	3 978	2 718	2 441	1 712	1 039	1 225	1 368	1 786
Greenland	880	770	1 351	1 524	1 876	762	...	176	547	596
Iceland	46 932	58 426	60 125	56 223	43 256	40 712	44 729	41 698	39 825	49 951
Ireland	3 118	2 860	3 417	4 421	6 234	6 572	4 898	5 812	5 404	3 509
Isle of Man	18	24	27	38	9	13	7	19	1	0
Netherlands	192	95	146	111	494	289	115	121	295	360
Norway	43 931	73 791	79 834	97 115	106 155	79 008	53 243	45 934	51 648	55 233
Poland	-	-	-	18	35	27	24	16	96	52
Portugal	583	755	605	208	168	49	38	131	105	153
Russian Fed	35 071	51 822	54 516	73 857	41 228	20 559	30 978	24 892	34 937	38 788
Spain	76	22	62	718	501	541	780	669	2 217	156
Sweden	1 344	959	1 265	1 226	1 519	1 013	895	964	1 087	965
UK	87 625	93 698	86 315	87 420	83 388	83 436	72 001	50 644	42 865	52 869
27 Fishing area total	*237 071*	*304 901*	*309 429*	*351 008*	*322 746*	*269 036*	*235 725*	*196 111*	*207 267*	*243 671*
Species total	*251 041*	*312 207*	*317 794*	*361 929*	*334 105*	*283 689*	*249 451*	*212 821*	*228 821*	*266 501*
Navaga(=Wachna cod)	**Morue arctique**		**Bacalao navaga**		***Eleginus navaga***				**1,48(04)012,01**	**COW**
05 Russian Fed	-	112	193	98	163	177	22	11	55	127
05 Fishing area total	*-*	*112*	*193*	*98*	*163*	*177*	*22*	*11*	*55*	*127*
27 Russian Fed	624	1 106	821	561	989	1 008	458	663	1 111	835
27 Fishing area total	*624*	*1 106*	*821*	*561*	*989*	*1 008*	*458*	*663*	*1 111*	*835*
Species total	*624*	*1 218*	*1 014*	*659*	*1 152*	*1 185*	*480*	*674*	*1 166*	*962*
Saffron cod	**Morne boréale**		**Bacalao del Artico**		***Eleginus gracilis***				**1,48(04)012,02**	**SAF**
61 Russian Fed	43 079	13 213	25 566	21 110	27 803	40 426	47 032	35 763	33 753	32 591
61 Fishing area total	*43 079*	*13 213*	*25 566*	*21 110*	*27 803*	*40 426*	*47 032*	*35 763*	*33 753*	*32 591*
Species total	*43 079*	*13 213*	*25 566*	*21 110*	*27 803*	*40 426*	*47 032*	*35 763*	*33 753*	*32 591*
Pacific tomcod	**Poulamon du Pacifique**		**...C**		***Microgadus proximus***				**1,48(04)013,01**	**MGX**
67 USA	...	...	...	...	...	3	1	-	0	-
67 Fishing area total	*...*	*...*	*...*	*...*	*...*	*3*	*1*	*-*	*0*	*-*
77 USA	...	...	...	...	...	0	-	0	-	-
77 Fishing area total	*...*	*...*	*...*	*...*	*...*	*0*	*-*	*0*	*-*	*-*
Species total	*...*	*...*	*...*	*...*	*...*	*3*	*1*	*0*	*0*	*-*
Atlantic tomcod	**Poulamon atlantique**		**Microgado**		***Microgadus tomcod***				**1,48(04)013,02**	**TOM**
02 Canada	20	21	22	23	24	25	25	25	-	-
02 Fishing area total	*20*	*21*	*22*	*23*	*24*	*25*	*25*	*25*	*-*	*-*
21 Canada	74	40	109	117	66	94	17	30	57	1
21 Fishing area total	*74*	*40*	*109*	*117*	*66*	*94*	*17*	*30*	*57*	*1*
Species total	*94*	*61*	*131*	*140*	*90*	*119*	*42*	*55*	*57*	*1*
Saithe(=Pollock)	**Lieu noir**		**Carbonero(=Colín)**		***Pollachius virens***				**1,48(04)015,01**	**POK**
21 Canada	21 583	15 584	10 222	9 739	12 604	15 092	8 552	6 528	7 190	7 617
Cuba	605	13	61	125	57	8	6	2	-	-
Portugal	41	13	-	-	-	-	-	-	-	2
Russian Fed	112	-	-	-	-	1	-	-	-	-
St Pier Mq	-	-	-	-	14	13	6	38	13	135
Spain	-	-	-	-	6	-	-	-	-	-
USA	5 670	3 737	3 244	2 962	4 251	5 583	4 595	4 043	4 109	3 581
21 Fishing area total	*28 011*	*19 347*	*13 527*	*12 826*	*16 932*	*20 697*	*13 159*	*10 611*	*11 312*	*11 335*
27 Belgium	223	169	236	161	264	256	208	126	30	120
Channel Is	1 F	0	0	2	4	0	2	-	-	-
Denmark	4 314	4 331	4 395	4 708	4 517	3 973	4 501	3 536	3 592	5 686
Estonia	-	-	-	-	16	-	-	-	-	-
Faeroe Is	37 189	35 218	31 979	20 398	22 598	26 751	34 423	35 997	45 792	50 232
France	33 472	29 317	19 882	19 598	17 802	18 218	24 638	26 941	28 533	30 148
Germany	18 762	12 405	13 393	15 197	15 993	13 562	13 307	12 385	13 320	14 542
Greenland	78	15	53	165	318	437	...	601	1 526	1 580
Iceland	69 985	63 333	47 466	39 297	36 548	30 532	30 729	32 947	31 941	41 839
Ireland	2 329	2 355	2 929	2 514	1 841	1 687	1 704	1 743	2 048	1 354
Isle of Man	5	4	11	11	9	7	2	1	0	4
Japan	4	-	-	-	-	-	-	-	-	-
Netherlands	78	17	9	19	42	8	7	11	19	5
Norway	188 364	188 869	218 853	221 638	183 451	194 452	198 387	169 746	169 506	203 963
Poland	937	151	592	365	822	813	862	747	727	752
Portugal	1	0	5	24	13	49	37	64	86	130
Romania	-	31	-	-	-	-	-	-	-	-
Russian Fed	9 509	1 223	1 148	1 177	1 802	3 836	3 932	4 564	4 953	5 413
Spain	-	1	13	33	77	397	82	158	152	54
Sweden	4 955	5 366	1 998	1 773	1 649	1 857	1 929	1 468	1 628	1 868
UK	15 278	14 678	15 220	15 413	14 609	12 261	12 442	10 924	10 585	12 310
27 Fishing area total	*385 484 F*	*357 483*	*358 182*	*342 493*	*302 375*	*309 096*	*327 192*	*301 959*	*314 438*	*370 000*

B-32 Cods, hakes, haddocks / Morues, merlus, églefins / Bacalaos, merluzas, eglefinos

Capture production by species, fishing areas and countries or areas / Captures par espèces, zones de pêche et pays ou zones / Capturas por especies, áreas de pesca y países o áreas

Species, Fishing area Espèce, Zone de pêche Especie, Area de pesca	1993 t	1994 t	1995 t	1996 t	1997 t	1998 t	1999 t	2000 t	2001 t	2002 t
Species total	*413 495 F*	*376 830*	*371 709*	*355 319*	*319 307*	*329 793*	*340 351*	*312 570*	*325 750*	*381 335*
Pollack	**Lieu jaune**		**Abadejo**			***Pollachius pollachius***			**1,48(04)015,02**	**POL**
27 Belgium	109	144	158	115	119	113	108	116	137	143
Channel Is	24 F	24	22 F	27	35	52	75	97	57	40
Denmark	2 295	1 301	1 036	1 049	637	564	480	490	358	464
Faeroe Is	5	2	2	-	1	-	-	-	-	-
France	3 837	4 554	3 843	3 881	3 626	3 359	2 935	3 775	3 649	3 924
Germany	167	-	87	102	117	43	63	39	41	116
Ireland	1 149	947	1 190	1 288	1 052	946	1 049	1 131	1 382	1 334
Isle of Man	1	1	15	16	11	11	2	1	-	3
Netherlands	18	14	17	19	15	7	5	5	1	1
Norway	3 274	2 473	3 071	2 318	2 230	2 247	2 928	3 385	2 888	3 465
Portugal	1	3	2	2	2	1	1	15	41	45
Spain	189	193	216	185	213	218	175	175	436	222
Sweden	659	350	510	355	261	180	160	124	108	112
UK	3 966	3 531	4 011	3 631	3 845	3 211	2 398	2 706	2 839	2 765
27 Fishing area total	*15 694 F*	*13 537*	*14 180 F*	*12 988*	*12 164*	*10 952*	*10 379*	*12 059*	*11 937*	*12 634*
Species total	*15 694 F*	*13 537*	*14 180 F*	*12 988*	*12 164*	*10 952*	*10 379*	*12 059*	*11 937*	*12 634*
Alaska pollock(=Walleye poll.)	**Lieu de l'Alaska**		**Colín de Alaska**			***Theragra chalcogramma***			**1,48(04)016,01**	**ALK**
61 China	175 000	170 174	249 459	226 900	338 478	191 433	64 520	60 338	39 665	11 000
China,Taiwan	2	9	37	12	7	9	9	9	-	-
Japan	382 293	379 338	338 507	331 163	338 785	315 987	382 385	300 001	241 881	213 254
Korea D P Rp	...	75 065	120 219	15 369	66 578	65 000 F	62 000 F	60 000 F	60 000 F	60 000 F
Korea Rep	181 288	296 932	334 921	224 430	223 065	236 278	146 165	86 143	197 396	24 772
Poland	235 208	269 979	249 257	116 257	125 413	81 889	65 508	33 192	16 590	-
Russian Fed	2 114 456	1 746 629	2 208 410	2 439 651	2 252 742	1 930 650	1 500 450	1 215 065	1 145 016	826 706
61 Fishing area total	*3 088 247*	*2 938 126*	*3 500 810*	*3 353 782*	*3 345 068*	*2 821 246 F*	*2 221 037 F*	*1 754 748 F*	*1 700 548 F*	*1 135 732 F*
67 Canada	8 121	4 706	3 295	2 150	1 800	800	1 233	1 044	1 747	3 606
Japan	15	13	-	-	-	-	-	-	-	-
Korea Rep	44 810	14 642	10 967	2 480	-	-	-	-	-	-
Russian Fed	-	-	-	329	-	-	-	-	-	1
USA	1 477 815	1 417 278	1 293 939	1 189 844	1 139 642	1 232 177	1 055 016	1 182 438	1 442 170	1 515 515
67 Fishing area total	*1 530 761*	*1 436 639*	*1 308 201*	*1 194 803*	*1 141 442*	*1 232 977*	*1 056 249*	*1 183 482*	*1 443 917*	*1 519 122*
Species total	*4 619 008*	*4 374 765*	*4 809 011*	*4 548 585*	*4 486 510*	*4 054 223 F*	*3 277 286 F*	*2 938 230 F*	*3 144 465 F*	*2 654 854 F*
Polar cod	**Morue polaire**		**Bacalao polar**			***Boreogadus saida***			**1,48(04)019,01**	**POC**
21 Greenland	0	1	-	4	-	-	-	-	-	3
21 Fishing area total	*0*	*1*	-	*4*	-	-	-	-	-	*3*
27 Germany	-	58	-	-	-	-	-	-	-	-
Russian Fed	50 638	6 126	24 030	20 784	6 826	3 592	22 005	40 730	39 445	37 191
27 Fishing area total	*50 638*	*6 184*	*24 030*	*20 784*	*6 826*	*3 592*	*22 005*	*40 730*	*39 445*	*37 191*
61 Russian Fed	35	-	2 400	-	-	-	-	13	-	-
61 Fishing area total	*35*	-	*2 400*	-	-	-	-	*13*	-	-
Species total	*50 673*	*6 185*	*26 430*	*20 788*	*6 826*	*3 592*	*22 005*	*40 743*	*39 445*	*37 194*
Rocklings nei	**Motelles nca**		**Barbadas nep**			***Gaidropsarus spp***			**1,48(04)028,XX**	**ROL**
27 France	-	-	-	-	-	-	4	24	1	0
27 Fishing area total	-	-	-	-	-	-	*4*	*24*	*1*	*0*
37 Bulgaria	-	-	-	-	-	-	-	-	-	20
France	...	...	...	...	...	...	20	19	19	19
37 Fishing area total	...	...	...	...	...	...	*20*	*19*	*19*	*39*
Species total	...	...	...	...	...	...	*24*	*43*	*20*	*39*
Norway pout	**Tacaud norvégien**		**Faneca noruega**			***Trisopterus esmarkii***			**1,48(04)032,01**	**NOP**
27 Denmark	190 071	166 928	262 515	162 943	153 047	63 678	57 441	150 040	62 913	78 243
Faeroe Is	31 471	32 358	8 960	9 133	11 215	6 222	4 045	1 754	2 429	2 429 F
Germany	3	-	38	-	-	-	-	2	-	-
Iceland	-	-	0	0	-	-	-	-	160	253
Ireland	0	0	0	-	-	-	-	1	-	-
Netherlands	-	24	138	13	85	3	1	3	-	2
Norway	102 766	91 694	118 081	103 126	47 032	27 575	51 124	52 912	27 123	25 996
Sweden	-	-	68	237	2	-	-	133	780	-
UK	7	1	0	215	13	-	2	0	0	1
27 Fishing area total	*324 318*	*291 005*	*389 800*	*275 667*	*211 394*	*97 478*	*112 613*	*204 845*	*93 405*	*106 924 F*
Species total	*324 318*	*291 005*	*389 800*	*275 667*	*211 394*	*97 478*	*112 613*	*204 845*	*93 405*	*106 924 F*
Poor cod	**Capelan de Méditerranée**		**Capellán**			***Trisopterus minutus***			**1,48(04)032,02**	**POD**
37 France	1 084	711	586	603	428	428	637	888	754	859
Slovenia	-	-	-	-	-	-	-	2	1	1
37 Fishing area total	*1 084*	*711*	*586*	*603*	*428*	*428*	*637*	*890*	*755*	*860*

B-32 Cods, hakes, haddocks / Morues, merlus, églefins / Bacalaos, merluzas, eglefinos

Capture production by species, fishing areas and countries or areas / Captures par espèces, zones de pêche et pays ou zones / Capturas por especies, áreas de pesca y países o áreas

Species, Fishing area Espèce, Zone de pêche Especie, Area de pesca	1993 t	1994 t	1995 t	1996 t	1997 t	1998 t	1999 t	2000 t	2001 t	2002 t
Species total	*1 084*	*711*	*586*	*603*	*428*	*428*	*637*	*890*	*755*	*860*
Pouting(=Bib)	**Tacaud commun**		**Faneca**			***Trisopterus luscus***			**1,48(04)032,03**	**BIB**
27 Belgium	451	400	305	377	336	323	364	468	561	577
Channel Is	0	0	0	1	1	0	1	5	6	14
Denmark	1	2	1	1	2	2	1	1	3	2
France	6 727	6 539	5 678	6 119	6 283	6 108	6 333	7 619	7 293	7 081
Ireland	3	5	5	2	12	1	21	10	28	12
Netherlands	-	-	-	-	-	-	-	612	645	735
Portugal	3 863	2 981	3 070	2 491	2 051	2 254	2 792	3 299	4 511	3 103
Spain	3 714	4 112	4 656	2 515	3 254	1 875	2 088	2 689	2 587	3 429
UK	1 286	1 191	919	962	813	554	458	885	899	724
27 Fishing area total	*16 045*	*15 230*	*14 634*	*12 468*	*12 752*	*11 117*	*12 058*	*15 588*	*16 533*	*15 677*
34 Morocco	1 926	1 795	2 083	2 100	1 073	1 243	573	1 216	1 013	1 171
Portugal	11	2	0	0	0	-	-	-	-	-
34 Fishing area total	*1 937*	*1 797*	*2 083*	*2 100*	*1 073*	*1 243*	*573*	*1 216*	*1 013*	*1 171*
37 Morocco	1	2	5	5	5	8	21	1 001	4	0
Spain	-	-	-	308	383	394	648	946	653	660
37 Fishing area total	*1*	*2*	*5*	*313*	*388*	*402*	*669*	*1 947*	*657*	*660*
Species total	*17 983*	*17 029*	*16 722*	*14 881*	*14 213*	*12 762*	*13 300*	*18 751*	*18 203*	*17 508*
Blue whiting(=Poutassou)	**Merlan bleu**		**Bacaladilla**			***Micromesistius poutassou***			**1,48(04)033,01**	**WHB**
27 Channel Is	-	-	-	-	1	1	1	-	-	-
Denmark	69 378	22 835	46 182	52 699	33 486	69 305	79 810	62 074	65 058	51 040
Estonia	1 077	4 342	13 715	10 982	5 678	6 321	0	-	-	-
Faeroe Is	14 984	24 404	25 936	21 483	28 773	71 217	105 106	152 687	259 761	235 000 F
France	...	...	6	6 442	12 446	7 992	6 343	16 042	19 054	14 771
Germany	100	5 919	6 314	6 865	4 722	17 970	3 170	12 654	19 059	17 052
Greenland	0	-	-	-	-	-	-	-	-	-
Iceland	-	-	369	513	10 480	68 514	160 424	259 157	365 101	286 381
Ireland	0	3	222	1 709	25 987	45 538	35 880	26 067	29 910	17 825
Japan	2 002	3 248	1 127	-	-	-	-	-	-	-
Latvia	10 328	2 582	-	-	-	-	-	-	-	-
Lithuania	2 418	-	400	651	-	-	1 231	-	-	-
Netherlands	18 481	21 076	22 685	16 407	24 132	27 693	32 889	43 145	63 625	35 624
Norway	199 981	226 235	261 362	356 054	348 268	570 665	534 570	553 478	573 686	558 070
Poland	-	-	-	-	-	-	-	-	-	38
Portugal	1 222	1 987	2 346	3 565	2 448	1 900	2 676	2 169	1 763	1 698
Russian Fed	137 796	123 258	93 824	87 310	118 656	130 042	182 637	241 905	315 586	298 367
Spain	34 256	30 506	33 397	30 262	37 900	30 549	30 926	28 000	28 822	25 522
Sweden	37 265	3 705	13 000	4 038	4 568	6 034	15 511	3 362	2 058	18 483
UK	2 294	4 470	5 495	14 326	33 701	98 936	106 491	45 048	51 889	28 679
27 Fishing area total	*531 582*	*474 570*	*526 380*	*613 306*	*691 246*	*1 152 677*	*1 297 665*	*1 445 788*	*1 795 372*	*1 588 550 F*
34 Greece	-	-	1	1	-	-	-	-	-	-
Spain	-	-	-	310	68	48	-	-	-	-
34 Fishing area total	-	-	*1*	*311*	*68*	*48*	-	-	-	-
37 Albania	...	...	0	2	0	-	-	-	-	6
France	48	26	...	21	21	21	29	34	24	14
Greece	1 858	2 281	1 944	1 226	1 558	846	630	566	471	450 F
Italy	2 752	1 998	1 769	1 546	1 300	1 449	1 451	1 261	1 167	755
Spain	4 450 F	4 500 F	4 500 F	4 541	3 086	2 762	4 445	6 276	5 461	2 988
Turkey	9 734	10 598	9 716	11 518	15 000	27 200	16 975	18 180	20 810	10 500
37 Fishing area total	*18 842 F*	*19 403 F*	*17 929 F*	*18 854*	*20 965*	*32 278*	*23 530*	*26 317*	*27 933*	*14 713 F*
Species total	*550 424 F*	*493 973 F*	*544 310 F*	*632 471*	*712 279*	*1 185 003*	*1 321 195*	*1 472 105*	*1 823 305*	*1 603 263 F*
Southern blue whiting	**Merlan bleu austral**		**Polaca austral**			***Micromesistius australis***			**1,48(04)033,02**	**POS**
41 Argentina	109 829	86 084	104 208	85 040	79 945	71 643	55 097	61 313	54 311	42 453
Australia	-	-	-	-	-	23	165	-	-	-
Belize	-	-	-	-	-	-	-	257	206	-
Bulgaria	3 138	-	-	-	-	-	-	-	-	-
Chile	-	-	-	-	3 744	14 215	5 036	2 726	6 709	4 204
Falkland Is	74	279	1 616	1 083	727	1 977	2 127	2 704	4 581	2 772
France	-	23	0	67	-	-	-	-	-	-
Honduras	19	7	-	3	-	-	-	-	-	-
Japan	8 797	22 047	12 872	12 493	16 340	16 935	18 028	14 121	8 918	11 670
Latvia	298	-	-	2	-	-	-	-	-	-
Namibia	-	-	-	-	83	282	29	-	-	-
Panama	-	1	-	-	-	-	-	-	-	-
Poland	8 351	10 553	8 923	3 402	-	-	-	-	-	-
Portugal	-	124	-	-	-	-	-	1	-	-
Russian Fed	306	-	-	-	-	-	-	-	30	4
Spain	4 553	5 259	10 711	2 471	1 591	3 435	3 128	3 346	5 243	3 160
UK	2	4	30	108	20	48	85	22	30	181
Uruguay	-	-	-	-	-	-	-	9	0	5
41 Fishing area total	*135 367*	*124 381*	*138 360*	*104 669*	*102 450*	*108 558*	*83 695*	*84 499*	*80 028*	*64 449*
81 Japan	20 455	11 068	17 333	19 197	20 173	23 374	22 827	17 096	23 956	14 500 F
New Zealand	9 589	4 620	11 357	2 753	10 234	35 059	39 012	23 000	29 789	42 086
Russian Fed	2 254	1 950	22	377	610	-	-	-	-	-
Ukraine	145	934	3 610	3 713	3 607	7 730	8 306	3 502	267	-

B-32

Cods, hakes, haddocks — **Capture production by species, fishing areas and countries or areas**
Morues, merlus, églefins — **Captures par espèces, zones de pêche et pays ou zones**
Bacalaos, merluzas, eglefinos — **Capturas por especies, áreas de pesca y países o áreas**

Species, Fishing area Espèce, Zone de pêche Especie, Area de pesca	1993 t	1994 t	1995 t	1996 t	1997 t	1998 t	1999 t	2000 t	2001 t	2002 t
81 Fishing area total	*32 443*	*18 572*	*32 322*	*26 040*	*34 624*	*66 163*	*70 145*	*43 598*	*54 012*	*56 586 F*
87 Chile	27 607	4 664	20 917	25 445	29 131	26 642	31 470	24 733	22 046	25 205
87 Fishing area total	*27 607*	*4 664*	*20 917*	*25 445*	*29 131*	*26 642*	*31 470*	*24 733*	*22 046*	*25 205*
Species total	*195 417*	*147 617*	*191 599*	*156 154*	*166 205*	*201 363*	*185 310*	*152 830*	*156 086*	*146 240 F*
Whiting	**Merlan**		**Plegonero**			***Merlangius merlangus***			**1,48(04)034,01**	**WHG**
27 Belgium	1 319	1 490	1 250	1 281	989	856	1 072	826	732	500
Channel Is	0	1	1 F	1	0	3	2	3	3	1
Denmark	2 126	1 410	789	391	196	144	175	326	326	381
Faeroe Is	601	815	966	1 042	1 018	1 724	1 756	1 593	1 289	1 026
France	28 531	32 355	27 821	21 315	21 590	20 074	21 570	19 026	19 414	18 383
Germany	1 216	1 075	1 186	711	276	191	371	754	680	568
Iceland	230	315	560	430	443	531	931	1 349	1 179	1 295
Ireland	7 118	8 736	11 262	10 326	9 394	7 762	7 643	6 505	6 621	6 669
Isle of Man	55	44	41	28	24	33	5	2	1	1
Netherlands	4 799	3 863	3 640	3 411	2 554	1 981	1 806	1 899	2 619	2 448
Norway	355	287	334	210	140	116	143	145	237	171
Poland	-	-	-	-	-	1	-	-	-	-
Portugal	234	306	169	184	139	115	76	77	38	42
Spain	88	136	6	44	72	187	233	353	299	248
Sweden	871	473	670	374	101	90	128	177	153	178
UK	46 056	42 202	40 383	36 788	35 189	27 243	25 561	23 458	15 287	11 538
27 Fishing area total	*93 599*	*93 508*	*89 078 F*	*76 536*	*72 125*	*61 051*	*61 472*	*56 493*	*48 878*	*43 449*
37 Bulgaria	-	-	-	-	-	-	-	9	8	16
France	...	...	...	...	...	...	2	2	2	2
Georgia	172	187	146	223	58	53	41	...	32	37
Romania	599	432	327	372	441	640	272	275	306	85
Russian Fed	16	125	91	11	10	119	184	341	642	656
Slovenia	-	-	-	-	-	13	16	14	37	19
Turkey	20 487	16 615	18 094	21 450	15 500	13 150	14 110	18 000	10 000	8 808
Ukraine	5	64	17	3	29	55	18	20	18	9
37 Fishing area total	*21 279*	*17 423*	*18 675*	*22 059*	*16 038*	*14 030*	*14 643*	*18 661*	*11 045*	*9 632*
Species total	*114 878*	*110 931*	*107 753 F*	*98 595*	*88 163*	*75 081*	*76 115*	*75 154*	*59 923*	*53 081*
European hake	**Merlu européen**		**Merluza europea**			***Merluccius merluccius***			**1,48(05)004,01**	**HKE**
27 Belgium	116	105	76	42	54	76	92	117	124	91
Channel Is	0	0	0	0	0	0	0	-	-	-
Denmark	3 179	2 128	1 487	868	670	591	846	811	1 043	1 163
Faeroe Is	6	4	14	1	6	5	5	...	...	...
France	12 953	14 758	14 530	8 899	8 592	5 165	7 378	10 122	8 007	10 954
Germany	120	84	110	83	76	69	68	46	73	71
Ireland	2 539	2 175	2 186	1 741	2 270	1 971	2 090	2 037	1 124	698
Isle of Man	7	25	23	18	28	30	3	3	1	-
Netherlands	144	75	78	111	62	75	98	43	72	18
Norway	887	589	783	938	981	825	609	693	635	534
Portugal	3 273	2 685	3 109	2 636	2 198	2 346	3 103	3 058	3 030	2 616
Spain	21 015	20 001	27 168	18 076	20 161	20 057	22 931	24 856	12 137	13 410
Sweden	200	170	69	45	33	26	27	34	63	51
UK	7 380	6 120	5 614	5 380	5 716	5 168	5 537	4 519	2 775	2 663
27 Fishing area total	*51 819*	*48 919*	*55 247*	*38 838*	*40 847*	*36 404*	*42 787*	*46 339*	*29 084*	*32 269*
34 Greece	46	39	85	70	9	20	1	9	-	349
Italy	-	-	371	-	-	-	-	1	539	1 459
Morocco	3 735	5 090	6 038	3 086	2 473	1 161	2 313	2 331	3 812	6 492
Portugal	1 041	687	357	986	380	217	114	3	2	264
Spain	6 300 F	6 000 F	5 800 F	5 536	3 293	1 741	926	-	7	3 989
34 Fishing area total	*11 122 F*	*11 816 F*	*12 651 F*	*9 678*	*6 155*	*3 139*	*3 354*	*2 344*	*4 360*	*12 553*
37 Albania	400 F	300 F	227	293	185	340	341	330	380	200
Algeria	1 000 F	1 200 F	1 115	1 154	1 012	1 310	1 681	3	196	209
Croatia	1 589	1 230	1 129	929	828	935	650	583	557	624
Cyprus	14	13	7	3	4	2	5	6	8	3
France	4 006	1 955	1 775	1 423	1 423	1 423	1 782	1 513	2 022	2 628
Greece	5 043	6 390	5 369	4 579	4 248	3 032	3 127	2 960	2 753	2 650 F
Israel	61	120	120	131	86	134	60	62	73	68
Italy	34 001	36 334	37 680	30 707	17 971	13 166	9 754	9 219	8 765	8 459
Malta	1	1	1	2	4	5	6	6	0	-
Morocco	66	77	95	84	74	96	37	461	197	197
Serbia-Monte	1	1	13	22	20	21	19	17	18	18
Slovenia	2	1	4	4	4	2	1	0	2	2
Spain	4 000 F	3 800 F	3 600 F	3 400	2 850	5 402	5 497	6 327	7 015	4 706
Syria	134 F	129 F	128 F	250	300	125	110	87	52	63
Tunisia	818	840	744	685	922	537	1 069	974	1 284	1 538
Turkey	583	3	1	150	25	15	5	10	-	-
37 Fishing area total	*51 719 F*	*52 394 F*	*52 008 F*	*43 816*	*29 956*	*26 545*	*24 144*	*22 558*	*23 322*	*21 365 F*
Species total	*114 660 F*	*113 129 F*	*119 906 F*	*92 332*	*76 958*	*66 088*	*70 285*	*71 241*	*56 766*	*66 187 F*
Senegalese hake	**Merlu du Sénégal**		**Merluza del Senegal**			***Merluccius senegalensis***			**1,48(05)004,02**	**HKM**
34 Bulgaria	-	-	-	-	-	10	-	-	-	-
Latvia	-	43	8	68	27	16	320	280	126	105
Lithuania	-	-	...	...	...	180	307	180	43	189

B-32 Cods, hakes, haddocks / Morues, merlus, églefins / Bacalaos, merluzas, eglefinos

Capture production by species, fishing areas and countries or areas / Captures par espèces, zones de pêche et pays ou zones / Capturas por especies, áreas de pesca y países o áreas

Species, Fishing area Espèce, Zone de pêche Especie, Area de pesca	1993 t	1994 t	1995 t	1996 t	1997 t	1998 t	1999 t	2000 t	2001 t	2002 t
Poland	-	-	-	64	-	-	-	-	87	-
Portugal	28	6	38	223	102	42	17	-	-	-
Russian Fed	-	1	132	1 112	1 081	1 171	1 230	604	207	...
Senegal	33	8	1	7	162	22	335	113	98	221
Spain	13 800 F	14 200 F	14 600 F	15 001	14 478	13 594	15 678	11 211	16 716	6 978
Ukraine	11	177	...	10	580	2 361	300	1 127	648	23
34 Fishing area total	*13 872 F*	*14 435 F*	*14 779 F*	*16 485*	*16 430*	*17 396*	*18 187*	*13 515*	*17 925*	*7 516*
Species total	*13 872 F*	*14 435 F*	*14 779 F*	*16 485*	*16 430*	*17 396*	*18 187*	*13 515*	*17 925*	*7 516*
Southern hake	**Merlu austral**		**Merluza austral**		***Merluccius australis***				**1,48(05)004,03**	**HKN**
41 Argentina	3 026	1 650	3 899	4 115	3 011	3 125	3 471	7 035	4 742	5 301
Chile	5	3	-	-	-	-	-	-	1	-
41 Fishing area total	*3 031*	*1 653*	*3 899*	*4 115*	*3 011*	*3 125*	*3 471*	*7 035*	*4 743*	*5 301*
81 Korea Rep	19	229	...	454	1 178	1 976	2 512	2 142	2 224	1 645
New Zealand	6 192	2 870	9 707	8 317	9 692	15 047	15 499	12 799	12 956	12 176
Norway	542	338	57	210	117	16	-	-	-	-
Russian Fed	-	278	414	27	202	-	-	-	-	-
Ukraine	353	167	774	111	181	578	1 422	1 100	8	13
81 Fishing area total	*7 106*	*3 882*	*10 952*	*9 119*	*11 370*	*17 617*	*19 433*	*16 041*	*15 188*	*13 834*
87 Chile	20 112	23 169	24 612	23 788	24 666	22 458	24 656	29 402	28 805	23 431
87 Fishing area total	*20 112*	*23 169*	*24 612*	*23 788*	*24 666*	*22 458*	*24 656*	*29 402*	*28 805*	*23 431*
Species total	*30 249*	*28 704*	*39 463*	*37 022*	*39 047*	*43 200*	*47 560*	*52 478*	*48 736*	*42 566*
Silver hake	**Merlu argenté**		**Merluza norteamericana**		***Merluccius bilinearis***				**1,48(05)004,04**	**HKS**
21 Canada	80	283	3 485	4 209	5 333	10 490	9 676	11 870	17 929	13 725
Cuba	22 714	8 055	16 826	22 526	12 697	6 280	3 853	27	-	-
Germany	0	0	0	-	-	-	-	-	-	-
Portugal	-	-	-	-	-	-	-	-	-	29
Russian Fed	7 052	-	-	639	-	163	-	1 679	2 055	2 525
Spain	-	-	-	-	-	-	-	4	9	27
USA	16 111	16 040	15 169	16 025	15 535	14 959	14 039	12 176	13 007	7 987
21 Fishing area total	*45 957*	*24 378*	*35 480*	*43 399*	*33 565*	*31 892*	*27 568*	*25 756*	*33 000*	*24 293*
31 USA	-	2	0	-	-	-	-	-	-	-
31 Fishing area total	-	*2*	*0*	-	-	-	-	-	-	-
Species total	*45 957*	*24 380*	*35 480*	*43 399*	*33 565*	*31 892*	*27 568*	*25 756*	*33 000*	*24 293*
South Pacific hake	**Merlu du Pacifique sud**		**Merluza del Pacífico sur**		***Merluccius gayi***				**1,48(05)004,05**	**PHA**
87 Chile	64 262	68 107	75 403	88 555	87 620	80 151	103 789	110 143	121 200	116 040
Peru	88 700	135 705	181 182	234 915	177 953	82 367	37 121	83 361	125 065	46 250
87 Fishing area total	*152 962*	*203 812*	*256 585*	*323 470*	*265 573*	*162 518*	*140 910*	*193 504*	*246 265*	*162 290*
Species total	*152 962*	*203 812*	*256 585*	*323 470*	*265 573*	*162 518*	*140 910*	*193 504*	*246 265*	*162 290*
Argentine hake	**Merlu d'Argentine**		**Merluza argentina**		***Merluccius hubbsi***				**1,48(05)004,06**	**HKP**
41 Argentina	422 195	435 788	574 317	597 557	584 048	458 433	311 953	193 701	249 462	358 897
Australia	-	-	-	-	-	3	10	-	-	-
Belize	-	-	-	-	-	-	35	63	4	0
Brazil	3 000 F	1 500 F	255	0	-	-	128	226	2 436	3 739
Falkland Is	88	43	194	383	267	959	1 031	1 000	562	655
France	-	20	4	17	4	3	3	0	-	-
Honduras	223	53	78	19	-	-	-	-	-	-
Italy	-	-	44	-	-	-	-	-	-	-
Japan	68	29	72	83	53	30	27	59	3	75
Latvia	62	-	-	-	-	-	-	-	-	-
Namibia	-	-	-	-	12	15	37	-	-	-
Panama	70	18	15	14	-	-	-	-	-	-
Poland	32	-	-	-	-	-	35	-	-	-
Portugal	6	1 143	2 371	4 253	603	310	-	3	-	-
Russian Fed	44	-	-	-	-	-	-	-	197	86
St Vincent	-	-	-	-	-	-	-	-	5	-
Seychelles	-	-	-	-	27	-	-	-	-	-
Spain	1 693	967	1 161	21 697	14 816	18 298	26 810	22 196	25 302	13 572
Ukraine	-	-	-	1	-	-	-	-	-	-
UK	62	10	0	38	104	67	53	30	83	233
Uruguay	69 910	56 981	57 874	57 937	48 367	49 111	32 045	27 710	27 618	32 231
41 Fishing area total	*497 453 F*	*496 552 F*	*636 385*	*681 999*	*648 301*	*527 229*	*372 167*	*244 988*	*305 672*	*409 488*
Species total	*497 453 F*	*496 552 F*	*636 385*	*681 999*	*648 301*	*527 229*	*372 167*	*244 988*	*305 672*	*409 488*
North Pacific hake	**Merlu du Pacifique nord**		**Merluza del Pacífico norte**		***Merluccius productus***				**1,48(05)004,07**	**NHA**
67 Poland	-	-	-	-	-	-	-	977	-	-
USA	140 697	252 742	177 038	195 289	226 615	227 502	216 889	205 350	172 050	129 599
67 Fishing area total	*140 697*	*252 742*	*177 038*	*195 289*	*226 615*	*227 502*	*216 889*	*206 327*	*172 050*	*129 599*
77 Mexico	106	17	78	116	1 089	250	111	392	852	1 194
USA	1	3	1	1	1	2	-	1	1	0
77 Fishing area total	*107*	*20*	*79*	*117*	*1 090*	*252*	*111*	*393*	*853*	*1 194*

B-32 Cods, hakes, haddocks / Morues, merlus, églefins / Bacalaos, merluzas, eglefinos

Capture production by species, fishing areas and countries or areas / Captures par espèces, zones de pêche et pays ou zones / Capturas por especies, áreas de pesca y países o áreas

Species, Fishing area Espèce, Zone de pêche Especie, Area de pesca	1993 t	1994 t	1995 t	1996 t	1997 t	1998 t	1999 t	2000 t	2001 t	2002 t
Species total	*140 804*	*252 762*	*177 117*	*195 406*	*227 705*	*227 754*	*217 000*	*206 720*	*172 903*	*130 793*
Benguela hake	**Merlu d'Afrique tropicale**		**Merluza de Benguela**		***Merluccius polli***				**1,48(05)004,08**	**HKB**
34 Belize	-	-	-	-	-	-	54	...	...	...
Cameroon	3	-	-	-	-	-	-	-	-	3
Ghana	0	0	0	1	0	34	3	0	-	-
Korea Rep	-	-	-	-	-	4	-	-	-	-
Liberia	-	-	-	38	-	-	-	-	-	-
Nigeria	-	-	-	-	-	-	64	-	-	1 714
Panama	14	-	-	-	-	-	-	-	-	-
Sierra Leone	1	-	-	-	0	-	-	-	-	-
Spain	-	-	-	-	-	-	18	-	-	-
Other nei	-	-	-	-	-	-	8	44	139	504
34 Fishing area total	*18*	*0*	*0*	*39*	*0*	*38*	*147*	*44*	*139*	*2 221*
47 Russian Fed	-	-	-	-	-	-	-	-	-	1
47 Fishing area total	-	-	-	-	-	-	-	-	-	*1*
Species total	*18*	*0*	*0*	*39*	*0*	*38*	*147*	*44*	*139*	*2 222*
Offshore silver hake	**Merlu argenté du large**		**Merluza blanca de altura**		***Merluccius albidus***				**1,48(05)004,12**	**HOF**
21 USA	...	115	...	32	...	5	12	5	2	6
21 Fishing area total	...	*115*	...	*32*	...	*5*	*12*	*5*	*2*	*6*
Species total	...	*115*	...	*32*	...	*5*	*12*	*5*	*2*	*6*
Panama hake	**Merlu du Panama**		**Merluza panameña**		***Merluccius angustimanus***				**1,48(05)004,14**	**MRG**
87 Colombia	...	25	...	...	267	165	143	250 F	391	380 F
87 Fishing area total	...	*25*	...	...	*267*	*165*	*143*	*250 F*	*391*	*380 F*
Species total	...	*25*	...	...	*267*	*165*	*143*	*250 F*	*391*	*380 F*
Shallow-water Cape hake	**Merlu côtier du Cap**		**Merluza del Cabo**		***Merluccius capensis***				**1,48(05)004,19**	**HKK**
47 Angola	92	24	26	832	777	906	2 060	658	1 863	3 225
47 Fishing area total	*92*	*24*	*26*	*832*	*777*	*906*	*2 060*	*658*	*1 863*	*3 225*
Species total	*92*	*24*	*26*	*832*	*777*	*906*	*2 060*	*658*	*1 863*	*3 225*
Cape hakes	**Merlus du Cap**		**Merluzas del Cabo**		***Merluccius capensis,M.paradox.***				**1,48(05)004,XX**	**HKC**
47 Iceland	-	-	-	-	352	206	-	-	-	-
Italy	-	-	5	-	-	-	-	-	-	-
Japan	-	33	55	-	-	-	-	-	-	-
Korea Rep	44	4	-	-	-	-	-	-	-	-
Namibia	108 102	112 229	130 374	129 462	117 683	150 695	164 250	171 397	173 277	156 499
Poland	-	-	-	3	-	-	-	-	-	-
Portugal	-	-	-	-	-	-	-	-	1	-
Russian Fed	976	321	10	98	252	321	67	41	49	2
South Africa	108 336	136 917	137 742	155 155	141 076	151 317	141 165	135 000 F	146 393	149 548
Spain	18 000 F	12 500 F	9 000 F	1 724	1 592	3	2 154	2 877	3 349	338
Ukraine	11	20	-	-	-	-	6	22	4	-
Other nei	-	2	-	1	-	-	-	-	-	-
47 Fishing area total	*235 469 F*	*262 026 F*	*277 186 F*	*286 443*	*260 955*	*302 542*	*307 642*	*309 337 F*	*323 073*	*306 387*
Species total	*235 469 F*	*262 026 F*	*277 186 F*	*286 443*	*260 955*	*302 542*	*307 642*	*309 337 F*	*323 073*	*306 387*
Hakes nei	**Merlus nca**		**Merluzas nep**		***Merluccius spp***				**1,48(05)004,XX**	**HKX**
34 Belize	-	-	-	-	30	29	65	236	8	13
China	0	-	10	-	-	0	-	-	-	-
Cyprus	-	-	-	...	...	40	164	189	115	-
France	-	-	-	-	...	44	5	-	-	-
Germany	-	-	16	-	-	-	-	-	-	-
Guinea	-	-	-	-	-	-	-	-	3	...
Honduras	0	-	2	-	-	-	-	-	-	-
Korea Rep	-	-	-	-	-	-	-	-	15	-
Mauritania	90 F	100 F	40 F	150 F	110 F	818 F	940	1 558	1 324	817
Panama	-	-	-	40	29	-	-	-	-	-
Portugal	0	0	49	1 515	914	1 027	474	-	11	-
Russian Fed	-	-	-	-	-	-	-	-	-	647
St Vincent	-	-	-	-	10	106	31	28	55	130
34 Fishing area total	*90 F*	*100 F*	*117 F*	*1 705 F*	*1 093 F*	*2 064 F*	*1 679*	*2 011*	*1 531*	*1 607*
41 Korea Rep	119	50	16	370	33	-	-	-	-	-
Lithuania	-	8	-	-	-	-	-	-	-	-
Portugal	-	-	-	-	-	-	-	-	354	186
41 Fishing area total	*119*	*58*	*16*	*370*	*33*	-	-	-	*354*	*186*
Species total	*209 F*	*158 F*	*133 F*	*2 075 F*	*1 126 F*	*2 064 F*	*1 679*	*2 011*	*1 885*	*1 793*
Patagonian grenadier	**Grenadier de Patagonie**		**Merluza de cola**		***Macruronus magellanicus***				**1,48(05)017,01**	**GRM**
41 Argentina	39 373	17 251	22 796	44 065	41 835	96 157	117 571	123 684	111 885	98 723
Australia	-	-	-	-	-	31	377	-	-	-
Belize	-	-	-	-	-	-	84	1 720	374	1

B-32 Cods, hakes, haddocks / Morues, merlus, églefins / Bacalaos, merluzas, eglefinos

Capture production by species, fishing areas and countries or areas / Captures par espèces, zones de pêche et pays ou zones / Capturas por especies, áreas de pesca y países o áreas

Species, Fishing area Espèce, Zone de pêche Especie, Area de pesca	1993 t	1994 t	1995 t	1996 t	1997 t	1998 t	1999 t	2000 t	2001 t	2002 t
Brazil	...	...	30	18	20	0	0	0	-	-
Chile	-	-	-	-	-	361	181	23	1 308	1 299
Estonia	-	-	-	113	-	-	-	-	108	-
Falkland Is	58	98	864	2 569	1 829	4 246	5 109	3 404	5 452	9 768
France	-	114	15	29	-	-	2	0	-	-
Honduras	251	797	1 053	92	-	-	-	-	-	-
Japan	409	18	568	544	644	844	400	1 889	866	1 612
Korea Rep	64	-	76	33	-	31	282	1 076	1 553	1 584
Latvia	-	-	32	15	-	-	-	-	-	-
Lithuania	-	2	-	-	-	-	-	-	-	-
Namibia	-	-	-	-	98	205	308	-	-	-
Panama	214	191	223	339	-	-	-	-	-	-
Poland	258	33	-	146	-	-	86	-	73	-
Portugal	4	-	-	-	-	-	-	32	-	-
Russian Fed	128	-	-	-	-	-	-	-	173	2
Seychelles	-	-	-	-	35	5	-	-	-	-
Spain	5 608	6 736	10 601	7 733	7 422	16 104	11 132	9 794	13 599	2 280
UK	169	3	80	86	166	2	347	42	30	52
Uruguay	...	...	...	...	150	1 824	1 461	800	635	1 002
41 Fishing area total	*46 536*	*25 243*	*36 338*	*55 782*	*52 199*	*119 810*	*137 340*	*142 464*	*136 056*	*116 323*
87 Chile	82 580	81 310	206 734	379 015	71 479	353 823	309 723	91 310	160 774	132 119
87 Fishing area total	*82 580*	*81 310*	*206 734*	*379 015*	*71 479*	*353 823*	*309 723*	*91 310*	*160 774*	*132 119*
Species total	*129 116*	*106 553*	*243 072*	*434 797*	*123 678*	*473 633*	*447 063*	*233 774*	*296 830*	*248 442*
Blue grenadier	**Grenadier bleu**		**Cola de rata azul**		***Macruronus novaezelandiae***			**1,48(05)017,02**		**GRN**
57 Australia	3 039	3 048	3 239	2 680	2 851	4 752	6 209	8 964	7 313	9 013
57 Fishing area total	*3 039*	*3 048*	*3 239*	*2 680*	*2 851*	*4 752*	*6 209*	*8 964*	*7 313*	*9 013*
81 Australia	-	-	-	-	-	-	-	547	267	165
Japan	65 190	57 346	31 463	26 031	25 349	26 802	17 269	12 139	15 734	10 000 F
Korea Rep	373	376	340	1 725	6 091	4 883	5 834	8 694	9 484	9 627
New Zealand	131 086	151 718	152 161	145 308	229 890	267 616	236 652	234 029	223 703	192 482
Norway	20 236	13 321	6 100	6 614	5 576	4 633	-	-	-	-
Russian Fed	14 654	20 409	7 907	3 905	3 191	-	-	-	-	-
Ukraine	12 654	18 269	10 774	14 754	12 632	16 064	15 141	19 333	1 300	28
81 Fishing area total	*244 193*	*261 439*	*208 745*	*198 337*	*282 729*	*319 998*	*274 896*	*274 742*	*250 488*	*212 302 F*
Species total	*247 232*	*264 487*	*211 984*	*201 017*	*285 580*	*324 750*	*281 105*	*283 706*	*257 801*	*221 315 F*
Roughhead grenadier	**Grenadier berglax**		**Granadero berglax**		***Macrourus berglax***			**1,48(06)001,03**		**RHG**
21 Estonia	-	-	-	-	-	-	-	1	-	-
Lithuania	-	-	-	-	-	-	-	1	28	3
Norway	-	-	-	-	-	-	-	-	1	8
Portugal	1 993	2 223	1 377	787	762	1 090	1 299	395	610	508
Spain	-	-	-	257	3 740	6 050	5 705	8 097	1 103	905
21 Fishing area total	*1 993*	*2 223*	*1 377*	*1 044*	*4 502*	*7 140*	*7 004*	*8 494*	*1 742*	*1 424*
27 France	-	-	1	21	25	29	116	4	4	7
Iceland	-	28	6	15	4	1	-	5	3	11
Norway	-	-	-	-	-	-	-	63	147	89
UK	-	-	-	-	14	-	1 038	8	16	7
27 Fishing area total	*-*	*28*	*7*	*36*	*43*	*30*	*1 154*	*80*	*170*	*114*
Species total	*1 993*	*2 251*	*1 384*	*1 080*	*4 545*	*7 170*	*8 158*	*8 574*	*1 912*	*1 538*
Whitson's grenadier	**...B**		**...C**		***Macrourus whitsoni***			**1,48(06)001,04**		**WGR**
48 Russian Fed	-	-	-	-	-	-	-	-	0	-
UK	-	-	-	-	-	-	-	1	-	-
48 Fishing area total	*-*	*-*	*-*	*-*	*-*	*-*	*-*	*1*	*0*	*-*
58 Australia	-	-	-	-	-	-	-	0	-	-
South Africa	-	-	-	-	-	-	-	3	-	-
58 Fishing area total	*-*	*-*	*-*	*-*	*-*	*-*	*-*	*3*	*-*	*-*
88 New Zealand	-	-	-	-	-	-	1	5	48	158
88 Fishing area total	*-*	*-*	*-*	*-*	*-*	*-*	*1*	*5*	*48*	*158*
Species total	*-*	*-*	*-*	*-*	*-*	*-*	*1*	*9*	*48*	*158*
Ridge scaled rattail	**...B**		**...C**		***Macrourus carinatus***			**1,48(06)001,05**		**MCC**
88 New Zealand	-	-	-	-	-	-	20	65	-	0
88 Fishing area total	*-*	*-*	*-*	*-*	*-*	*-*	*20*	*65*	*-*	*0*
Species total	*-*	*-*	*-*	*-*	*-*	*-*	*20*	*65*	*-*	*0*
Bigeye grenadier	**Grenadier à gros yeux**		**Granadero ojisapo**		***Macrourus holotrachys***			**1,48(06)001,06**		**MCH**
58 Australia	-	-	-	-	1	-	-	-	-	-
58 Fishing area total	*-*	*-*	*-*	*-*	*1*	*-*	*-*	*-*	*-*	*-*
Species total	*-*	*-*	*-*	*-*	*1*	*-*	*-*	*-*	*-*	*-*

B-32 Cods, hakes, haddocks / Morues, merlus, églefins / Bacalaos, merluzas, eglefinos

Capture production by species, fishing areas and countries or areas / Captures par espèces, zones de pêche et pays ou zones / Capturas por especies, áreas de pesca y países o áreas

Species, Fishing area Espèce, Zone de pêche Especie, Area de pesca	1993 t	1994 t	1995 t	1996 t	1997 t	1998 t	1999 t	2000 t	2001 t	2002 t
Grenadiers nei	**Grenadiers nca**		**Granaderos nep**		***Macrourus spp***				**1,48(06)001,XX**	**GRV**
41 Argentina	16	16	664	821	1 041	2 972	2 580	10 503	3 200	5 329
Poland	3	-	-	-	-	-	13	-	-	-
Portugal	-	-	80	80	-	-	-	-	1	19
Russian Fed	-	-	-	-	-	-	-	-	-	524
Uruguay	-	-	-	-	-	-	-	-	8	180
41 Fishing area total	*19*	*16*	*744*	*901*	*1 041*	*2 972*	*2 593*	*10 503*	*3 209*	*6 052*
48 Argentina	-	-	8	8	-	-	-	-	-	-
Bulgaria	-	0	-	-	-	-	-	-	-	-
Chile	-	-	0	12	-	0	-	-	-	-
Korea Rep	-	-	0	5	10	-	-	-	-	-
Poland	-	-	-	-	-	-	-	-	3	-
Russian Fed	-	0	2	-	-	-	-	-	1	-
South Africa	-	-	-	-	-	13	0	-	-	-
Spain	-	-	-	-	2	-	-	-	-	-
UK	-	-	-	-	7	4	10	2	2	-
Uruguay	-	-	-	-	-	0	3	-	-	-
48 Fishing area total	-	*0*	*10*	*25*	*19*	*17*	*13*	*2*	*6*	-
58 Australia	0	-	-	-	0	0	1	3	-	-
France	-	-	-	-	11	14	68	143	158	371
South Africa	-	-	-	-	1	54	77	203	45	14
58 Fishing area total	*0*	-	-	-	*12*	*68*	*146*	*349*	*203*	*385*
81 Ukraine	34	121	-	-	-	-	-	-	-	-
81 Fishing area total	*34*	*121*	-	-	-	-	-	-	-	-
88 Chile	-	-	-	-	-	0	-	-	-	-
New Zealand	-	-	-	-	-	9	1	0	-	-
South Africa	-	-	-	-	-	-	-	-	6	-
88 Fishing area total	-	-	-	-	-	*9*	*1*	*0*	*6*	-
Species total	*53*	*137*	*754*	*926*	*1 072*	*3 066*	*2 753*	*10 854*	*3 424*	*6 437*
Roundnose grenadier	**Grenadier de roche**		**Granadero de roca**		***Coryphaenoides rupestris***				**1,48(06)004,01**	**RNG**
21 Canada	1 298	543	695	236	75	2	-	1	2	3
Estonia	-	-	-	-	-	-	-	20	21	24
Faeroe Is	23	-	2	1	1	-	10	23	-	1
Germany	2	12	-	5	4	5	-	-	5	-
Greenland	28	21	46	100	146	19	33	17	22	21
Japan	222	134	106	35	40	37	40	27	146	-
Lithuania	-	-	-	-	-	-	-	-	-	14
Norway	-	-	-	-	-	-	-	-	3	7
Poland	-	-	-	-	-	-	-	-	1	-
Russian Fed	17	14	191	59	-	96	53	284	186	233
St Pier Mq	-	-	-	-	4	-	-	-	-	-
Spain	2 054	1 720	2 646	3 096	-	-	-	-	5 126	4 518
21 Fishing area total	*3 644*	*2 444*	*3 686*	*3 532*	*270*	*159*	*136*	*372*	*5 512*	*4 821*
27 Denmark	1 591	1 911	2 227	1 174	2 124	4 429	2 521	1 981	2 229	3 150
Estonia	-	-	-	-	-	-	-	-	680	824
Faeroe Is	641	743	764	233	199	84	108	53	91	158
France	8 813	8 241	8 414	7 650	7 389	6 820	7 851	9 968	8 494	8 635
Germany	61	48	22	9	34	116	101	42	18	25
Greenland	18	5	14	19	26	3	1	11	5	5
Iceland	280	210	398	216	207	120	146	70	57	70
Ireland	144	6	59	1	4	-	1	45	-	-
Latvia	2 176	675	-	-	-	-	-	-	-	-
Lithuania	-	-	-	-	-	-	-	-	137	1 835
Norway	-	-	-	-	-	-	-	31	75	52
Poland	-	-	-	-	5 867	6 769	546	-	178	933
Portugal	0	0	0	0	-	-	-	-	-	-
Russian Fed	473	-	-	208	1 041	825	588	2 343	1 806	769
Spain	-	-	32	970	2 476	3 935	6 224	15 465	33 099	5 714
Sweden	-	42	1	0	42	0	-	-	258	262
UK	1	12	91	184	228	-	-	587	1 030	759
27 Fishing area total	*14 198*	*11 893*	*12 022*	*10 664*	*19 637*	*23 101*	*18 087*	*30 596*	*48 157*	*23 191*
48 Poland	-	-	-	-	-	-	-	-	-	9
48 Fishing area total	-	-	-	-	-	-	-	-	-	*9*
Species total	*17 842*	*14 337*	*15 708*	*14 196*	*19 907*	*23 260*	*18 223*	*30 968*	*53 669*	*28 021*
Grenadiers, whiptails nei	**...B**		**...C**		***Coryphaenoides spp***				**1,48(06)004,XX**	**CVY**
88 New Zealand	-	-	-	-	-	-	-	-	-	0
88 Fishing area total	-	-	-	-	-	-	-	-	-	*0*
Species total	-	-	-	-	-	-	-	-	-	*0*
Thorntooth grenadier	**...B**		**...C**		***Lepidorhynchus denticulatus***				**1,48(06)030,01**	**LDE**
81 New Zealand	...	...	745	670	2 361	4 627	3 678	3 833	4 783	5 349
81 Fishing area total	...	...	*745*	*670*	*2 361*	*4 627*	*3 678*	*3 833*	*4 783*	*5 349*
Species total	...	...	*745*	*670*	*2 361*	*4 627*	*3 678*	*3 833*	*4 783*	*5 349*

B-32 Cods, hakes, haddocks / Morues, merlus, églefins / Bacalaos, merluzas, eglefinos

Capture production by species, fishing areas and countries or areas
Captures par espèces, zones de pêche et pays ou zones
Capturas por especies, áreas de pesca y países o áreas

Species, Fishing area / Espèce, Zone de pêche / Especie, Area de pesca	1993 t	1994 t	1995 t	1996 t	1997 t	1998 t	1999 t	2000 t	2001 t	2002 t
Grenadiers, rattails nei	**...B**		**Granaderos, colas de ratón nep**		***Macrouridae***				**1,48(06)XXX,XX**	**RTX**
61 Russian Fed	764	121	222	310	1 044	62	52	579	2 232	10 817
61 Fishing area total	*764*	*121*	*222*	*310*	*1 044*	*62*	*52*	*579*	*2 232*	*10 817*
67 Russian Fed	-	-	-	-	-	-	-	-	-	36
USA	...	...	...	...	...	83	102	170	160	120
67 Fishing area total	*...*	*...*	*...*	*...*	*...*	*83*	*102*	*170*	*160*	*156*
77 Korea Rep	-	-	-	-	-	-	-	140	-	-
USA	...	...	...	...	...	417	210	145	145	156
77 Fishing area total	*...*	*...*	*...*	*...*	*...*	*417*	*210*	*285*	*145*	*156*
81 New Zealand	...	...	909	1 579	5 197	4 350	3 670	2 394	3 094	3 866
Ukraine	-	-	-	-	-	-	-	-	-	11
81 Fishing area total	*...*	*...*	*909*	*1 579*	*5 197*	*4 350*	*3 670*	*2 394*	*3 094*	*3 877*
Species total	*764*	*121*	*1 131*	*1 889*	*6 241*	*4 912*	*4 034*	*3 428*	*5 631*	*15 006*
Gadiformes nei	**Gadiformes nca**		**Gadiformes nep**		***Gadiformes***				**1,48(XX)XXX,XX**	**GAD**
21 Japan	-	1	-	2	2	-	-	13	-	-
Norway	14	-	-	-	5	7	3	7	-	-
Portugal	-	-	-	-	-	-	-	2	-	24
USA	138	98	48	1	0	18	8	16	3	2
21 Fishing area total	*152*	*99*	*48*	*3*	*7*	*25*	*11*	*38*	*3*	*26*
27 Channel Is	-	-	-	-	-	-	1	-	-	-
Faeroe Is	372	192	229	449	898	57	338	-	-	51
France	11	11	-	-	406	825	10	2 435	3 092	4 154
Germany	463	-	-	-	-	-	-	-	-	-
Ireland	90	44	55	105	190	-	279	-	55	8
Japan	4	5	12	-	-	-	-	-	-	-
Norway	307	502	263	217	257	742	435	-	-	-
Portugal	663	658	599	586	653	668	545	500	483	567
Spain	21 046	19 892	30 166	31 053	18 246	10 701	13 526	2 832	25 971	8 310
Sweden	0	1	0	0	-	-	-	57	160	-
27 Fishing area total	*22 956*	*21 305*	*31 324*	*32 410*	*20 650*	*12 993*	*15 134*	*5 824*	*29 761*	*13 090*
31 USA	-	1	0	-	4	2	1	5	2	3
31 Fishing area total	*-*	*1*	*0*	*-*	*4*	*2*	*1*	*5*	*2*	*3*
34 Eq Guinea	320 F	480 F	200 F	450 F	570 F	603	710 F	80 F	80 F	80 F
Korea Rep	-	-	-	-	59	277	89	-	-	-
Portugal	404	182	68	138	71	58	51	12	9	6
Spain	-	-	-	75	16	53	53	54	25	57
34 Fishing area total	*724 F*	*662 F*	*268 F*	*663 F*	*716 F*	*991*	*903 F*	*146 F*	*114 F*	*143 F*
37 France	31	22	27	24	36	36	0	1	-	-
Lebanon	20	20	30	30	25	30	30	32	30	28
Spain	3 500 F	4 000 F	4 500 F	4 804	3 994	1 275	759	838	1 067	289
37 Fishing area total	*3 551 F*	*4 042 F*	*4 557 F*	*4 858*	*4 055*	*1 341*	*789*	*871*	*1 097*	*317*
41 Argentina	13	13	-	-	-	-	-	-	-	-
Japan	31	42	-	148	165	132	115	108	-	-
Korea Rep	282	-	293	1 392	87	105	237	1 256	538	154
Spain	-	-	-	-	-	-	-	-	2 601	-
UK	-	-	-	-	-	-	-	-	3	5
41 Fishing area total	*326*	*55*	*293*	*1 540*	*252*	*237*	*352*	*1 364*	*3 142*	*159*
47 Lithuania	-	3	-	-	-	-	-	-	-	-
Norway	-	-	-	-	-	3	-	-	-	-
47 Fishing area total	*-*	*3*	*-*	*-*	*-*	*3*	*-*	*-*	*-*	*-*
51 Italy	-	-	76	-	-	-	-	-	-	-
South Africa	3	-	-	-	10	7	5 F	2	1	-
51 Fishing area total	*3*	*-*	*76*	*-*	*10*	*7*	*5 F*	*2*	*1*	*-*
57 Australia	-	1	-	-	-	-	5	20	5	9
Korea Rep	-	-	-	-	-	-	-	-	30	-
57 Fishing area total	*-*	*1*	*-*	*-*	*-*	*-*	*5*	*20*	*35*	*9*
61 Korea Rep	-	-	-	-	-	-	17	-	-	-
Russian Fed	-	8	3	10	-	2	5	6	1	13
61 Fishing area total	*-*	*8*	*3*	*10*	*-*	*2*	*22*	*6*	*1*	*13*
71 Australia	32	55	34	24	21	33	21	43	60	42
Korea Rep	-	-	-	-	-	-	744	7	-	-
71 Fishing area total	*32*	*55*	*34*	*24*	*21*	*33*	*765*	*50*	*60*	*42*
77 Korea Rep	-	-	-	-	-	-	26	-	-	-
Mexico	415	50	10	21	183	137	32	40	53	71
77 Fishing area total	*415*	*50*	*10*	*21*	*183*	*137*	*58*	*40*	*53*	*71*
81 Australia	-	-	-	-	-	-	-	11	3	2
Japan	7 200	3 080	5 584	3 981	3 019	3 966	3 300	2 398	2 356	1 500 F
Korea Rep	1 275	222	237	302	132	452	1 130	455	2 511	258
New Zealand	...	...	73	36	22	34	11	14	8	11
Norway	-	-	-	-	-	188	-	-	-	-

B-32 Cods, hakes, haddocks / Morues, merlus, églefins / Bacalaos, merluzas, eglefinos

Capture production by species, fishing areas and countries or areas
Captures par espèces, zones de pêche et pays ou zones
Capturas por especies, áreas de pesca y países o áreas

Species, Fishing area Espèce, Zone de pêche Especie, Area de pesca	1993 t	1994 t	1995 t	1996 t	1997 t	1998 t	1999 t	2000 t	2001 t	2002 t
Russian Fed	705	12	-	-	-	-	1	-	-	-
Ukraine	-	-	-	-	-	-	-	-	997	-
81 Fishing area total	*9 180*	*3 314*	*5 894*	*4 319*	*3 173*	*4 640*	*4 442*	*2 878*	*5 875*	*1 771 F*
Species total	*37 339 F*	*29 595 F*	*42 507 F*	*43 848 F*	*29 071 F*	*20 411*	*22 487 F*	*11 244 F*	*40 144 F*	*15 644 F*
Group total	***9 963 556***	***9 729 524***	***10 743 927***	***10 784 248***	***10 373 733***	***10 334 525***	***9 333 765***	***8 673 042***	***9 244 845***	***8 392 479***

B-33 Miscellaneous coastal fishes / Poissons côtiers divers / Peces costeros diversos

Capture production by species, fishing areas and countries or areas / Captures par espèces, zones de pêche et pays ou zones / Capturas por especies, áreas de pesca y países o áreas

Species, Fishing area Espèce, Zone de pêche Especie, Area de pesca	1993 t	1994 t	1995 t	1996 t	1997 t	1998 t	1999 t	2000 t	2001 t	2002 t
Ladyfish	**Guinée-machète**		**Malacho**		***Elops saurus***				**1,29(01)003,02**	**LAD**
21 USA	-	-	-	-	1	1	1 963	1	0	0
21 Fishing area total	*-*	*-*	*-*	*-*	*1*	*1*	*1 963*	*1*	*0*	*0*
31 Colombia	21	53	30	143	20	4	11	20 F	20 F	20 F
USA	-	-	-	-	725	975	4	120	460	655
31 Fishing area total	*21*	*53*	*30*	*143*	*745*	*979*	*15*	*140 F*	*480 F*	*675 F*
41 Brazil	80 F	80 F	147	-	-	-	-	-	-	-
41 Fishing area total	*80 F*	*80 F*	*147*	*-*	*-*	*-*	*-*	*-*	*-*	*-*
Species total	*101 F*	*133 F*	*177*	*143*	*746*	*980*	*1 978*	*141 F*	*480 F*	*675 F*
West African ladyfish	**...B**		**...C**		***Elops lacerta***				**1,29(01)003,04**	**CEC**
34 Benin	...	...	...	...	...	...	...	...	15	16
Gambia	...	...	...	...	...	...	...	...	12	26
Nigeria	579	544	1 474	379	735	1 546	1 822	755	531	1 076
34 Fishing area total	*579*	*544*	*1 474*	*379*	*735*	*1 546*	*1 822*	*755*	*558*	*1 118*
Species total	*579*	*544*	*1 474*	*379*	*735*	*1 546*	*1 822*	*755*	*558*	*1 118*
Tarpon	**Tarpon argenté**		**Tarpón**		***Megalops atlanticus***				**1,29(02)004,01**	**TAR**
31 Colombia	21	12	13	135	2	2	-	-	-	-
Dominican Rp	78	40	268	31	27	47	12	14	14	34
Mexico	-	-	2	1	14	4	4	4	1	-
Puerto Rico	...	...	...	...	...	...	-	-	1	2
31 Fishing area total	*99*	*52*	*283*	*167*	*43*	*53*	*16*	*18*	*16*	*36*
41 Brazil	1 410 F	1 410 F	1 221	348	720	551	315	331	1 169	1 201
41 Fishing area total	*1 410 F*	*1 410 F*	*1 221*	*348*	*720*	*551*	*315*	*331*	*1 169*	*1 201*
Species total	*1 509 F*	*1 462 F*	*1 504*	*515*	*763*	*604*	*331*	*349*	*1 185*	*1 237*
Indo-Pacific tarpon	**Tarpon indo-pacifique**		**Tarpón Indo-Pacífico**		***Megalops cyprinoides***				**1,29(02)004,02**	**TAI**
57 Malaysia	7	8	7	6	4	13	29	31	12	11
57 Fishing area total	*7*	*8*	*7*	*6*	*4*	*13*	*29*	*31*	*12*	*11*
71 Malaysia	92	90	126	93	510	325	197	102	92	51
Philippines	129	222	979	997	644	720	1 207	1 151	1 365	1 510
71 Fishing area total	*221*	*312*	*1 105*	*1 090*	*1 154*	*1 045*	*1 404*	*1 253*	*1 457*	*1 561*
Species total	*228*	*320*	*1 112*	*1 096*	*1 158*	*1 058*	*1 433*	*1 284*	*1 469*	*1 572*
Bonefish	**Banane de mer**		**Macabí**		***Albula vulpes***				**1,30(01)005,01**	**BOF**
31 Dominican Rp	36	65	143	151	16	34	125	10	11	6
31 Fishing area total	*36*	*65*	*143*	*151*	*16*	*34*	*125*	*10*	*11*	*6*
34 Côte dIvoire	37	20	20 F	43	1 430	1 617	29	21	13	32
Gabon	...	...	...	25	21	2	6	10	...	...
Liberia	-	-	-	-		21	104	27	6	10 F
Russian Fed	-	-	-	10	-	-	-	-	-	-
Senegal	...	...	...	...	...	...	...	...	...	9
Sierra Leone	...	...	...	...	91	...	106	181	496	153
34 Fishing area total	*37*	*20*	*20 F*	*78*	*1 542*	*1 640*	*245*	*239*	*515*	*204 F*
47 Russian Fed	-	125	-	154	-	-	-	-	-	-
47 Fishing area total	*-*	*125*	*-*	*154*	*-*	*-*	*-*	*-*	*-*	*-*
Species total	*73*	*210*	*163 F*	*383*	*1 558*	*1 674*	*370*	*249*	*526*	*210 F*
Bombay-duck	**Scopelidé**		**Bumalo**		***Harpadon nehereus***				**1,31(16)001,02**	**BUC**
04 India	-	1 380	448	-	-	-	-	-	473	230
04 Fishing area total	*-*	*1 380*	*448*	*-*	*-*	*-*	*-*	*-*	*473*	*230*
51 India	142 074	126 396	133 786	159 362	187 887	144 774	146 591	133 156	141 027	100 302
Pakistan	170	121	98	101	95	91	72	65	55	64
51 Fishing area total	*142 244*	*126 517*	*133 884*	*159 463*	*187 982*	*144 865*	*146 663*	*133 221*	*141 082*	*100 366*
57 India	6 595	11 884	25 485	25 750	25 229	35 141	35 229	35 544	34 920	35 323
Indonesia	293	168	318	1 382	2 413	2 682	2 808	539	454	670
Malaysia	...	161	-	-	-	-	-	-	-	-
57 Fishing area total	*6 888*	*12 213*	*25 803*	*27 132*	*27 642*	*37 823*	*38 037*	*36 083*	*35 374*	*35 993*
71 Indonesia	12 205	12 643	10 717	9 836	10 867	11 500	9 607	8 449	7 747	7 360
Malaysia	...	647	-	-	-	-	-	-	-	-
71 Fishing area total	*12 205*	*13 290*	*10 717*	*9 836*	*10 867*	*11 500*	*9 607*	*8 449*	*7 747*	*7 360*
Species total	*161 337*	*153 400*	*170 852*	*196 431*	*226 491*	*194 188*	*194 307*	*177 753*	*184 676*	*143 949*
Greater lizardfish	**Anoli tumbil**		**Lagarto tumbil**		***Saurida tumbil***				**1,31(16)068,01**	**LIG**
51 Pakistan	96	67	43	45	28	22	-	-	-	-
Qatar	0	1	0	1	0	0	0	0	0	0

B-33 Miscellaneous coastal fishes / Poissons côtiers divers / Peces costeros diversos

Capture production by species, fishing areas and countries or areas / Captures par espèces, zones de pêche et pays ou zones / Capturas por especies, áreas de pesca y países o áreas

Species, Fishing area Espèce, Zone de pêche Especie, Area de pesca	1993 t	1994 t	1995 t	1996 t	1997 t	1998 t	1999 t	2000 t	2001 t	2002 t
51 Fishing area total	*96*	*68*	*43*	*46*	*28*	*22*	*0*	*0*	*0*	*0*
61 China,Taiwan	6 801	4 423	6 744	6 561	3 208	3 049	3 075	3 612	2 248	1 520
Japan	9 392	9 315	9 164	8 144	7 638	7 860	7 716	7 591	6 094	5 652
61 Fishing area total	*16 193*	*13 738*	*15 908*	*14 705*	*10 846*	*10 909*	*10 791*	*11 203*	*8 342*	*7 172*
Species total	*16 289*	*13 806*	*15 951*	*14 751*	*10 874*	*10 931*	*10 791*	*11 203*	*8 342*	*7 172*
Brushtooth lizardfish	**Anoli à grandes écailles**		**Lagarto escamoso**		***Saurida undosquamis***				**1,31(16)068,04**	**LIB**
37 Israel	214	80	80	124	61	37	48	21	24	24
37 Fishing area total	*214*	*80*	*80*	*124*	*61*	*37*	*48*	*21*	*24*	*24*
51 Kuwait	23	22	31	47	53	34	20	30 F	32	32 F
51 Fishing area total	*23*	*22*	*31*	*47*	*53*	*34*	*20*	*30 F*	*32*	*32 F*
61 Korea Rep	131	296	209	162	207	215	35	795	758	33
61 Fishing area total	*131*	*296*	*209*	*162*	*207*	*215*	*35*	*795*	*758*	*33*
Species total	*368*	*398*	*320*	*333*	*321*	*286*	*103*	*846 F*	*814*	*89 F*
Lizardfishes nei	**Anolis nca**		**Lagartos nep**		***Synodontidae***				**1,31(16)XXX,XX**	**LIX**
37 Egypt	980	1 134	993	1 307	1 075	941	1 162	1 265	1 065	917
Palest, O.T.	...	...	...	16	32	32	30 F	20 F	15 F	9
Syria	...	...	...	...	...	...	...	...	...	86
Turkey	65	397	142	161	150	165	190	250	55	97
37 Fishing area total	*1 045*	*1 531*	*1 135*	*1 484*	*1 257*	*1 138*	*1 382 F*	*1 535 F*	*1 135 F*	*1 109*
51 Egypt	6 073	4 845	4 696	4 151	5 117	7 994	7 213	10 543	7 913	5 736
Eritrea	...	651	166	3	5	0	1 905	3 177	2 574	3 274
India	16 883	18 062	21 436	16 520	11 252	12 012	11 539	4 029	1 768	2 298
Saudi Arabia	76	195	195	172	188	215	214	169	199	196
51 Fishing area total	*23 032*	*23 753*	*26 493*	*20 846*	*16 562*	*20 221*	*20 871*	*17 918*	*12 454*	*11 504*
57 India	2 302	2 466	2 553	3 150	2 276	753	1 375	1 041	1 960	2 211
Indonesia	1 477	1 211	1 043	825	1 098	1 051	1 244	996	919	880
Malaysia	4 521	5 358	5 835	7 010	9 218	8 155	7 468	9 492	9 997	11 346
Thailand	11 068	13 302	11 955	10 680	8 918	41 178	12 774	17 190	16 725	16 370
57 Fishing area total	*19 368*	*22 337*	*21 386*	*21 665*	*21 510*	*51 137*	*22 861*	*28 719*	*29 601*	*30 807*
61 China,H.Kong	11 352	11 614	10 888	9 034	8 070	6 619	4 700 F	5 800 F	6 400 F	6 250 F
61 Fishing area total	*11 352*	*11 614*	*10 888*	*9 034*	*8 070*	*6 619*	*4 700 F*	*5 800 F*	*6 400 F*	*6 250 F*
71 Indonesia	5 301	5 035	6 019	6 401	14 060	10 947	11 700	12 387	11 035	12 690
Malaysia	2 437	2 849	3 018	5 039	5 272	6 269	7 582	8 075	7 627	7 858
Philippines	6 700	7 887	8 630	8 435	6 671	7 345	7 649	5 539	5 751	7 031
Singapore	286	306	186	172	125	90	51	28	6	4
Thailand	42 486	35 593	58 482	51 004	62 397	35 289	60 534	52 601	54 785	55 406
71 Fishing area total	*57 210*	*51 670*	*76 335*	*71 051*	*88 525*	*59 940*	*87 516*	*78 630*	*79 204*	*82 989*
Species total	*112 007*	*110 905*	*136 237*	*124 080*	*135 924*	*139 055*	*137 330 F*	*132 602 F*	*128 794 F*	*132 659 F*
Sea catfishes nei	**Mâchoirons nca**		**Bagres marinos nep**		***Ariidae***				**1,41(02)XXX,XX**	**CAX**
04 Philippines	3 646	2 495	2 170	2 964	2 701	3 207	3 785	4 046	3 868	4 394
04 Fishing area total	*3 646*	*2 495*	*2 170*	*2 964*	*2 701*	*3 207*	*3 785*	*4 046*	*3 868*	*4 394*
06 Papua N Guin	1 850 F	1 850 F	1 850 F	1 850 F	1 850 F	1 850 F	1 850 F	1 850 F	1 850 F	1 850 F
06 Fishing area total	*1 850 F*	*1 850 F*	*1 850 F*	*1 850 F*	*1 850 F*	*1 850 F*	*1 850 F*	*1 850 F*	*1 850 F*	*1 850 F*
31 Colombia	30	37	44	24	20	17	9	20 F	20 F	20 F
Mexico	4 129	4 365	5 038	5 085	5 902	6 270	5 333	6 008	6 308	5 829
USA	-	-	-	-	-	-	-	0	3	1
Venezuela	11 227	12 436	21 548	17 041	8 963	10 098	11 100	14 279	11 929	15 148
31 Fishing area total	*15 386*	*16 838*	*26 630*	*22 150*	*14 885*	*16 385*	*16 442*	*20 307 F*	*18 260 F*	*20 998 F*
34 Benin	21 F	23 F	22 F	20 F	17 F	15 F	12	8 F	2	2
Cameroon	334	600 F	976	886	640 F	370 F	520 F	455	410	589
Congo Rep	654 F	269	230 F	230 F	250 F	250 F	300 F	373 F	400 F	450 F
Côte dIvoire	21	18	19	16	21	25	16	20	13	62
Gabon	1 335	256	415	2 780	1 565	1 290	1 351	921	1 140	1 171
Gambia	357	302	846	158	1 234	517	540	749	950	2 238
Ghana	2	0	0	2	24	6	0	1	0	6
Guinea	3 030 F	3 110 F	4 381	3 589	2 462	2 610	2 378	4 593	6 232	6 200 F
GuineaBissau	...	175	211	195	200 F	170 F	140 F	140 F	140 F	140 F
Liberia	59	63	7	77	-	4	31	21	210	210 F
Lithuania	5	-	-	-	-	-	-	-	-	113
Mauritania	...	...	...	...	...	...	750 F	750 F	750 F	750 F
Nigeria	4	6 756	12 570	12 676	14 062	10 862	17 014	14 885	16 537	21 155
Russian Fed	25	-	-	-	-	-	2	2	-	20
Senegal	4 743	5 874	3 912	4 457	10 162	10 094	6 006	7 432	9 279	9 205
Sierra Leone	859	856	858	860 F	973	52	62	2 092	748	774
Ukraine	-	-	-	-	-	-	-	1	-	-
34 Fishing area total	*11 449 F*	*18 302 F*	*24 447 F*	*25 946 F*	*31 610 F*	*26 265 F*	*29 122 F*	*32 443 F*	*36 811 F*	*43 085 F*
41 Argentina	-	21	16	6	13	18	5	5	20	27
Brazil	13 730 F	13 760 F	14 226	16 946	17 736	20 489	25 224	29 473	34 913	34 754
Uruguay	111	169	119	18	15	24	78	51	42	96

B-33 Miscellaneous coastal fishes / Poissons côtiers divers / Peces costeros diversos

Capture production by species, fishing areas and countries or areas / Captures par espèces, zones de pêche et pays ou zones / Capturas por especies, áreas de pesca y países o áreas

Species, Fishing area Espèce, Zone de pêche Especie, Area de pesca	1993 t	1994 t	1995 t	1996 t	1997 t	1998 t	1999 t	2000 t	2001 t	2002 t
41 Fishing area total	*13 841 F*	*13 950 F*	*14 361*	*16 970*	*17 764*	*20 531*	*25 307*	*29 529*	*34 975*	*34 877*
47 Angola	35	17	8	1	43	90	69	407	2 424	3 616
Namibia	-	7	13	16	31	42	32	52	11	0
South Africa	1	1	0	2	2	0	1	-	-	-
47 Fishing area total	*36*	*25*	*21*	*19*	*76*	*132*	*102*	*459*	*2 435*	*3 616*
51 Eritrea	...	52	60	9	0	149	205	851	436	273
India	30 763	32 803	31 848	27 249	31 392	36 082	38 432	31 467	40 933	43 019
Oman	376	140	499	372	679	1 024	1 161	1 306	2 238	1 163
Pakistan	37 840	42 112	45 444	49 428	54 437	55 934	51 665	39 168	38 215	38 500
Qatar	0	0	0	0	0	0	0	0	0	0
Saudi Arabia	212	226	303	302	366	315	309	564	534	485
Tanzania	647	456	780	913	850	810	850	880	850	800
Untd Arab Em	129	139	139	140	150	150	154	763	877	723
Yemen	1 411	1 043	1 697	1 700	1 780 F	2 000 F	1 900 F	1 750 F	2 200 F	2 500 F
51 Fishing area total	*71 378*	*76 971*	*80 770*	*80 113*	*89 654 F*	*96 464 F*	*94 676 F*	*76 749 F*	*86 283 F*	*87 463 F*
57 Australia	6	14	14	218	-	-	-	-	-	-
India	67 422	53 074	42 453	45 876	46 758	36 809	47 664	49 350	51 144	44 590
Indonesia	7 082	6 847	7 154	7 698	6 235	7 994	7 992	8 653	9 247	9 630
Malaysia	4 285	4 719	4 543	5 111	4 232	4 088	4 368	4 797	5 017	4 453
Thailand	952	60	58	2 004	1 924	4 207	4 250	2 033	2 656	2 600
57 Fishing area total	*79 747*	*64 714*	*54 222*	*60 907*	*59 149*	*53 098*	*64 274*	*64 833*	*68 064*	*61 273*
61 China,Taiwan	427	302	274	122	134	208	257	725	435	529
61 Fishing area total	*427*	*302*	*274*	*122*	*134*	*208*	*257*	*725*	*435*	*529*
71 Australia	41	8	17	17	18	-	-	-	-	-
Indonesia	45 052	49 575	51 541	54 714	72 343	58 304	61 654	61 613	65 196	68 820
Malaysia	10 251	8 995	9 279	10 080	9 367	11 094	10 973	10 008	9 059	6 653
Philippines	8 194	5 284	7 990	7 533	6 324	4 590	4 527	4 350	5 176	5 392
Singapore	163	452	359	333	385	358	241	141	76	55
Thailand	6 878	839	565	3 680	3 780	3 334	8 348	9 184	5 971	6 039
71 Fishing area total	*70 579*	*65 153*	*69 751*	*76 357*	*92 217*	*77 680*	*85 743*	*85 296*	*85 478*	*86 959*
77 El Salvador	122	107	94	107	62	114	101	172	192	716
Mexico	1 023	1 079	1 234	1 101	1 338	1 755	1 603	1 377	1 772	1 445
77 Fishing area total	*1 145*	*1 186*	*1 328*	*1 208*	*1 400*	*1 869*	*1 704*	*1 549*	*1 964*	*2 161*
81 Australia	1	0	0	21	21	21	32	-	-	-
Russian Fed	1	-	-	-	-	-	-	-	-	-
81 Fishing area total	*2*	*0*	*0*	*21*	*21*	*21*	*32*	-	-	-
87 Colombia	119	223	114	425	350	121	97	100 F	100 F	100 F
87 Fishing area total	*119*	*223*	*114*	*425*	*350*	*121*	*97*	*100 F*	*100 F*	*100 F*
Species total	*269 605 F*	*262 009 F*	*275 938 F*	*289 052 F*	*311 811 F*	*297 831 F*	*323 391 F*	*317 886 F*	*340 523 F*	*347 305 F*
Eeltail catfishes	**Balibots**		**Patunas**		***Plotosus spp***				**1,41(06)064,XX**	**CAE**
57 Malaysia	916	851	1 046	807	596	634	813	1 370	1 157	1 088
Thailand	-	60	58	74	116	108	75	67	83	81
57 Fishing area total	*916*	*911*	*1 104*	*881*	*712*	*742*	*888*	*1 437*	*1 240*	*1 169*
71 Malaysia	1 217	1 649	964	996	851	1 025	936	852	830	1 030
Thailand	-	839	565	767	647	1 093	556	1 244	943	953
71 Fishing area total	*1 217*	*2 488*	*1 529*	*1 763*	*1 498*	*2 118*	*1 492*	*2 096*	*1 773*	*1 983*
Species total	*2 133*	*3 399*	*2 633*	*2 644*	*2 210*	*2 860*	*2 380*	*3 533*	*3 013*	*3 152*
Morays	**Murènes**		**Morenas**		***Muraenidae***				**1,43(06)XXX,XX**	**MUI**
27 Portugal	...	...	...	...	...	...	193	164	142	128
27 Fishing area total	...	...	...	...	...	...	*193*	*164*	*142*	*128*
34 Portugal	...	...	...	...	...	...	...	5	3	2
Senegal	166	910	837	710	268	165	142	189	209	229
34 Fishing area total	*166*	*910*	*837*	*710*	*268*	*165*	*142*	*194*	*212*	*231*
Species total	*166*	*910*	*837*	*710*	*268*	*165*	*335*	*358*	*354*	*359*
Korean sandlance	**...B**		**...C**		***Hypoptychus dybowskii***				**1,50(05)009,01**	**KLA**
61 Korea Rep	9 632	9 466	9 677	6 613	8 832	4 801	4 806	-	462	7
61 Fishing area total	*9 632*	*9 466*	*9 677*	*6 613*	*8 832*	*4 801*	*4 806*	-	*462*	*7*
Species total	*9 632*	*9 466*	*9 677*	*6 613*	*8 832*	*4 801*	*4 806*	-	*462*	*7*
Squirrelfishes nei	**Marignons nca**		**Candiles nep**		***Holocentridae***				**1,61(11)XXX,XX**	**HCZ**
31 Antigua Barb	...	...	...	...	...	...	...	...	29	45
Dominican Rp	80	144	81	70	35	56	16	24	22	130
Grenada	0	1	1	1	1	3	2	3	3	2
Puerto Rico	...	...	...	...	...	...	9	13	12	8
St Kitts Nev	...	...	3	9	5	8	9	3	3	7
31 Fishing area total	*80*	*145*	*85*	*80*	*41*	*67*	*36*	*43*	*69*	*192*
51 Saudi Arabia	...	...	7	40	50	34	53	46	40	97

B-33

Miscellaneous coastal fishes — **Capture production by species, fishing areas and countries or areas**
Poissons côtiers divers — **Captures par espèces, zones de pêche et pays ou zones**
Peces costeros diversos — **Capturas por especies, áreas de pesca y países o áreas**

Species, Fishing area / Espèce, Zone de pêche / Especie, Area de pesca	1993 t	1994 t	1995 t	1996 t	1997 t	1998 t	1999 t	2000 t	2001 t	2002 t
51 Fishing area total	...	...	*7*	*40*	*50*	*34*	*53*	*46*	*40*	*97*
77 Amer Samoa	...	0	0	1	1	0	1	1	0	1
USA	...	...	...	...	...	...	...	...	...	23
77 Fishing area total	...	*0*	*0*	*1*	*1*	*0*	*1*	*1*	*0*	*24*
Species total	*80*	*145*	*92*	*121*	*92*	*101*	*90*	*90*	*109*	*313*
Flathead grey mullet	**Mulet à grosse tête**		**Pardete**			***Mugil cephalus***			**1,65(01)001,02**	**MUF**
01 Egypt	6 567	5 441	...	...	...	...	...	...	...	...
01 Fishing area total	*6 567*	*5 441*	...	...	...	...	...	...	...	...
04 Israel	115	118	153	181	169	231	315	138	125	136
04 Fishing area total	*115*	*118*	*153*	*181*	*169*	*231*	*315*	*138*	*125*	*136*
05 Greece	-	-	-	27	54	67	58	51	46	8
05 Fishing area total	-	-	-	*27*	*54*	*67*	*58*	*51*	*46*	*8*
31 Mexico	6 295	6 016	7 636	7 882	7 128	4 299	4 872	5 190	4 838	3 545
Venezuela	9 103	10 678	9 363	8 839	6 465	6 284	5 151	5 467	5 008	4 866
31 Fishing area total	*15 398*	*16 694*	*16 999*	*16 721*	*13 593*	*10 583*	*10 023*	*10 657*	*9 846*	*8 411*
34 Greece	12	9	1	-	1	3	-	-	-	-
34 Fishing area total	*12*	*9*	*1*	-	*1*	*3*	-	-	-	-
37 Bulgaria	...	...	23	26	28	11	14	15	47	71
Croatia	134	116	123	105	105	195	120	68	14	7
Cyprus	2	2	2	5	5	2	6	7	7	8
Egypt	479	381	...	...	...	...	...	...	...	...
Greece	3 217	2 689	2 816	3 797	3 022	1 690	1 965	1 697	1 357	1 300 F
Israel	150 F	150 F	99	258	381	283	284	172	175	158
Serbia-Monte	18	19	26	25	24	26	25	26	25	27
Tunisia	2 765	2 223	2 062	1 345	1 309	1 525	238	2 048	229	952
37 Fishing area total	*6 765 F*	*5 580 F*	*5 151*	*5 561*	*4 874*	*3 732*	*2 652*	*4 033*	*1 854*	*2 523 F*
51 Kuwait	57	53	96	39	65	69	89	60 F	34	35 F
51 Fishing area total	*57*	*53*	*96*	*39*	*65*	*69*	*89*	*60 F*	*34*	*35 F*
61 China,Taiwan	1 460	932	1 638	1 302	2 446	606	863	1 890	1 559	934
Japan	4 569	4 585	4 579	4 268	3 933	4 003	3 629	3 725	3 457	2 993
Korea Rep	5 899	4 909	4 316	4 200	5 564	4 962	9 678	8 730	9 784	8 887
61 Fishing area total	*11 928*	*10 426*	*10 533*	*9 770*	*11 943*	*9 571*	*14 170*	*14 345*	*14 800*	*12 814*
Species total	*40 842 F*	*38 321 F*	*32 933*	*32 299*	*30 699*	*24 256*	*27 307*	*29 284 F*	*26 705*	*23 927 F*
Lebranche mullet	**Mulet lebranche**		**Lebranche**			***Mugil liza***			**1,65(01)001,12**	**MUB**
31 Colombia	43	16	21	39	-	2	-	-	-	-
Venezuela	4 402	3 644	3 228	3 248	1 550	2 874	2 855	3 186	2 537	2 294
31 Fishing area total	*4 445*	*3 660*	*3 249*	*3 287*	*1 550*	*2 876*	*2 855*	*3 186*	*2 537*	*2 294*
Species total	*4 445*	*3 660*	*3 249*	*3 287*	*1 550*	*2 876*	*2 855*	*3 186*	*2 537*	*2 294*
So-iuy mullet	**Mulet so-iuy**		**Lisa so-iuy**			***Mugil soiuy***			**1,65(01)001,28**	**MYZ**
05 Ukraine	-	-	-	-	-	13	190	169	382	91
05 Fishing area total	-	-	-	-	-	*13*	*190*	*169*	*382*	*91*
37 Ukraine	73	334	775	1 039	2 718	3 661	5 174	5 406	2 428	2 510
37 Fishing area total	*73*	*334*	*775*	*1 039*	*2 718*	*3 661*	*5 174*	*5 406*	*2 428*	*2 510*
Species total	*73*	*334*	*775*	*1 039*	*2 718*	*3 674*	*5 364*	*5 575*	*2 810*	*2 601*
Mountain mullet	**Mulet de fleuve**		**Lisa de río**			***Agonostomus monticola***			**1,65(01)006,02**	**AJW**
02 Dominican Rp	...	...	...	...	...	...	...	...	...	28
02 Fishing area total	...	...	...	...	...	...	...	...	...	*28*
Species total	...	...	...	...	...	...	...	...	...	*28*
Bobo mullet	**Mulet bobo**		**Bobo**			***Joturus pichardi***			**1,65(01)010,01**	**MUA**
31 Mexico	665	245	520	572	283	323	322	451	400	341
31 Fishing area total	*665*	*245*	*520*	*572*	*283*	*323*	*322*	*451*	*400*	*341*
77 Mexico	142	37	94	87	56	51	82	46	104	76
77 Fishing area total	*142*	*37*	*94*	*87*	*56*	*51*	*82*	*46*	*104*	*76*
Species total	*807*	*282*	*614*	*659*	*339*	*374*	*404*	*497*	*504*	*417*
Leaping mullet	**Mulet sauteur**		**Galúa**			***Liza saliens***			**1,65(01)012,06**	**LZS**
37 Bulgaria	6	6 F	1	3	2	2	2	0	10	25
37 Fishing area total	*6*	*6 F*	*1*	*3*	*2*	*2*	*2*	*0*	*10*	*25*
Species total	*6*	*6 F*	*1*	*3*	*2*	*2*	*2*	*0*	*10*	*25*

B-33 Miscellaneous coastal fishes / Poissons côtiers divers / Peces costeros diversos

Capture production by species, fishing areas and countries or areas / Captures par espèces, zones de pêche et pays ou zones / Capturas por especies, áreas de pesca y países o áreas

Species, Fishing area Espèce, Zone de pêche Especie, Area de pesca	1993 t	1994 t	1995 t	1996 t	1997 t	1998 t	1999 t	2000 t	2001 t	2002 t
Mullets nei	**Mulets nca**		**Lizas nep**		***Mugilidae***				**1,65(01)XXX,XX**	**MUL**
01 Benin	600 F	763	637	1 000 F	1 500 F	2 000 F	2 198	1 800 F	2 050 F	2 050 F
Egypt	5 840	6 116	15 543	11 836	14 538	15 356	17 041	16 697	20 688	21 708
01 Fishing area total	*6 440 F*	*6 879*	*16 180*	*12 836 F*	*16 038 F*	*17 356 F*	*19 239*	*18 497 F*	*22 738 F*	*23 758 F*
04 Azerbaijan	40	90	73	103	82	61	2	3	55	23
India	-	2 855	3 319	9 852	9 424	4 205	3 833	11 195	12 613	2 038
Iran	5 135	2 809	5 014	2 554	3 074	4 558	4 080	5 125	5 263	6 873
Japan	943	973	832	850	1 053	887	1 000	1 140	1 084	980
Kazakhstan	34	32	31	30 F	20 F	15 F	10 F	10 F	7	-
Korea Rep	-	-	-	-	-	-	133	-	-	-
Philippines	628	2 006	2 209	2 440	1 675	613	853	768	667	474
Turkey	13 584	13 699	13 767	14 207	22 000	21 200	20 752	16 352	16 558	15 589
Turkmenistan	...	...	...	...	10	4	3	3	1	-
04 Fishing area total	*20 364*	*22 464*	*25 245*	*30 036 F*	*37 338 F*	*31 543 F*	*30 666 F*	*34 596 F*	*36 248*	*25 977*
05 Albania	-	-	-	-	-	100	74	105	50	50
Russian Fed	2	27	31	25	55	39	129	35	11	291
05 Fishing area total	*2*	*27*	*31*	*25*	*55*	*139*	*203*	*140*	*61*	*341*
21 Portugal	-	-	-	-	-	-	-	-	-	30
USA	276	3	298	2 645	553	370	2 323	546	412	545
21 Fishing area total	*276*	*3*	*298*	*2 645*	*553*	*370*	*2 323*	*546*	*412*	*575*
27 Channel Is	5 F	9	8 F	23	14	11	13	12	9	7
Denmark	16	22	31	22	29	24	26	22	29	17
France	1 041	885	938	685	787	794	850	819	974	831
Germany	-	-	-	1	3	6	21	13	23	40
Ireland	4	5	22	40	33	15	29	11	3	8
Netherlands	...	...	4	-	0	-	17	36	184	113
Portugal	381	352	398	297	303	336	324	336	376	291
Spain	188	171	130	145	90	178	210	117	75	42
UK	89	166	203	63	126	89	115	67	65	35
27 Fishing area total	*1 724 F*	*1 610*	*1 734 F*	*1 276*	*1 385*	*1 453*	*1 605*	*1 433*	*1 738*	*1 384*
31 Colombia	5	34	10	83	34	27	16	30 F	30 F	30 F
Cuba	158	134	108	93	159	122	108	119	114	125
Dominican Rp	108	324	320	43	59	41	22	21	28	219
Mexico	5 298	6 401	6 663	6 023	8 014	7 897	8 311	9 057	7 096	6 360
Puerto Rico	...	...	...	...	...	...	39	44	41	30
USA	14 037	13 849	9 790	5 073	8 352	7 759	2 219	5 607	6 148	5 387
31 Fishing area total	*19 606*	*20 742*	*16 891*	*11 315*	*16 618*	*15 846*	*10 715*	*14 878 F*	*13 457 F*	*12 151 F*
34 Benin	...	...	...	...	...	...	3	2 F	5	6
Gabon	...	...	51	271	254	233	382	365	1 061	888
Gambia	23	18	279	475	278	66	70	123	69	208
Georgia	10 F	-	-	-	-	-	-	-	-	-
Guinea	1 240 F	1 280 F	1 800	1 901	1 244	1 534	1 600	1 894	...	1 000 F
GuineaBissau	2 800 F	2 870 F	2 930 F	3 050 F	3 200 F	2 650 F	2 200 F	2 200 F	2 200 F	2 200 F
Latvia	-	-	58	21	1	20	44	19	-	-
Liberia	4	3	-	-	22	18	63	85	...	40 F
Lithuania	-	-	-	-	-	-	-	-	-	102
Mauritania	...	...	...	...	...	...	2 000 F	2 000 F	2 000 F	2 000 F
Morocco	1 819	787	1 609	1 600	2 570	2 865	3 730	2 743	4 338	4 257
Nigeria	1 580	3 333	3 421	1 364	7 073	8 450	10 226	8 535	8 728	9 122
Portugal	4	0	-	-	-	-	-	-	-	-
Romania	10	-	-	-	-	-	-	-	-	-
Russian Fed	19	6	27	135	52	26	102	126	151	11
Senegal	3 428	3 415	1 947	2 109	2 780	1 690	2 512	4 226	2 546	2 802
Sierra Leone	590	599	598	600 F	597	150	233	595	...	...
Togo	4	4	0	0	0	1	-	3	0	-
Ukraine	...	...	...	...	...	...	78	94	27	168
34 Fishing area total	*11 531 F*	*12 315 F*	*12 720 F*	*11 526 F*	*18 071 F*	*17 703 F*	*23 243 F*	*23 010 F*	*21 125 F*	*22 804 F*
37 Albania	150 F	100 F	52	105	42	36	66	45	130	78
Egypt	1 152	2 160	2 600	2 896	2 012	2 304	1 995	1 396	2 085	5 696
France	2 104	589	206	401	505	505	608	453	511	407
Georgia	-	-	-	-	-	-	9	19	28	73
Italy	5 012	4 991	5 655	5 172	5 281	5 344	4 799	4 095	5 023	5 716
Lebanon	170	200	400	400	300	300	300	...	400	370
Malta	0	0	0	0	0	0	0	0	0	-
Morocco	7	44	27	30	58	42	20	15	9	3
Palest, O.T.	...	...	...	32	19	35	35 F	30 F	20 F	17
Romania	-	-	-	-	-	-	-	-	-	6
Russian Fed	55	224	250	575	1 163	1 730	2 380	2 499	1 421	1 518
Slovenia	34	11	3	12	2	27	13	22	49	19
Spain	-	-	-	344	366	482	-	-	-	-
Syria	...	...	...	...	...	...	...	...	...	136
Tunisia	516	594	375	932	1 472	2 272	1 383	2 306	3 016	3 010
Turkey	11 873	14 943	17 710	23 308	20 500	24 150	26 000	27 000	22 000	12 000
Ukraine	0	0	4	3	0	5	12	27	51	26
37 Fishing area total	*21 073 F*	*23 856 F*	*27 282*	*34 210*	*31 720*	*37 232*	*37 620 F*	*37 907 F*	*34 743 F*	*29 075*
41 Argentina	8	8	18	10	10	10	10	5	11	9
Brazil	14 940 F	14 970 F	13 324	7 722	10 262	10 006	10 886	11 987	11 031	11 851
Uruguay	156	206	100	346	214	272	57	194	219	282
41 Fishing area total	*15 104 F*	*15 184 F*	*13 442*	*8 078*	*10 486*	*10 288*	*10 953*	*12 186*	*11 261*	*12 142*
47 Angola	0	0	0	0	0	-	-	-	-	-

B-33 Miscellaneous coastal fishes — Capture production by species, fishing areas and countries or areas
Poissons côtiers divers — Captures par espèces, zones de pêche et pays ou zones
Peces costeros diversos — Capturas por especies, áreas de pesca y países o áreas

Species, Fishing area Espèce, Zone de pêche Especie, Area de pesca	1993 t	1994 t	1995 t	1996 t	1997 t	1998 t	1999 t	2000 t	2001 t	2002 t
Namibia	-	-	112	...	...	...	...	161	117	122
South Africa	1 253	1 106	1 147	1 086	972	908	757	400 F	8	-
47 Fishing area total	*1 253*	*1 106*	*1 259*	*1 086*	*972*	*908*	*757*	*561 F*	*125*	*122*
51 Bahrain	270	86	201	99	60	73	10	59	39	29
Djibouti	0	0	0	0	0	0	0	0	0	0
Egypt	-	-	31	67	59	352	1 811	2 365	3 760	3 326
Eritrea	...	...	...	3	4	2	4	3	3	0
India	3 761	5 616	5 255	5 747	7 623	5 760	6 104	6 799	6 083	6 635
Kenya	116	109	127	153	120	107	146	181	199	164
Kuwait	359	356	579	540	628	1 068	965	700 F	422	430 F
Mauritius	121	125	115	121	103	111	120	76	92	80
Oman	100	1 207	98	79	139	176	158	123	543	424
Pakistan	22 485	19 039	17 280	17 631	18 935	17 580	12 336	9 618	9 723	9 900
Qatar	12	8	8	8	7	9	5	4	4	4
Saudi Arabia	112	29	400	320	369	510	317	326	397	414
Tanzania	436	234	470	618	550	300	350	400	250	200
Untd Arab Em	334	460	846	1 360	1 453	1 458	1 494	86	98	93
Yemen	0	0	391	380	400 F	500 F	400 F	400 F	500 F	550 F
51 Fishing area total	*28 106*	*27 269*	*25 801*	*27 126*	*30 450 F*	*28 006 F*	*24 220 F*	*21 140 F*	*22 113 F*	*22 249 F*
57 Australia	776	773	751	607	674	679	361	704	536	599
India	11 043	9 518	6 791	8 666	9 824	9 492	9 935	10 843	12 054	9 976
Indonesia	7 305	7 876	7 365	10 158	9 556	8 720	8 791	9 046	10 937	11 610
Malaysia	1 481	1 594	3 124	3 388	2 420	3 279	2 705	3 905	2 669	1 863
Thailand	737	1 125	793	1 135	1 322	1 644	2 798	5 268	2 241	2 194
57 Fishing area total	*21 342*	*20 886*	*18 824*	*23 954*	*23 796*	*23 814*	*24 590*	*29 766*	*28 437*	*26 242*
61 China	...	...	71 426	67 423	85 967	99 023	113 454	107 243	119 246	113 879
Russian Fed	-	2	17	-	-	1	7	33	94	49
61 Fishing area total	*...*	*2*	*71 443*	*67 423*	*85 967*	*99 024*	*113 461*	*107 276*	*119 340*	*113 928*
71 Australia	1 448	1 486	1 350	1 627	1 630	1 484	2 579	1 694	2 509	1 663
Fiji Islands	2 482	2 530 F	1 795	759	860	1 195	3 067	2 800 F	2 915	2 900 F
Indonesia	21 637	23 099	24 563	25 292	25 922	26 862	26 646	27 031	22 658	22 850
Kiribati	440	440	450	450	500	149	994	611	1 300	2 534
Malaysia	1 617	2 839	2 087	1 942	2 130	2 348	2 157	1 444	1 334	1 380
NewCaledonia	90	48	33	20	61	64	63	75	88	88 F
Palau	1	1	1	1	1	2	0	1	0	0
Philippines	13 724	17 272	14 113	12 969	13 364	13 591	14 807	14 183	15 336	14 596
Singapore	34	21	173	54	55	34	32	42	27	25
Thailand	3 273	3 961	3 678	3 952	4 075	3 905	2 700	3 106	3 666	3 708
71 Fishing area total	*44 746*	*51 697 F*	*48 243*	*47 066*	*48 598*	*49 634*	*53 045*	*50 987 F*	*49 833*	*49 744 F*
77 Cook Is	5 F	5 F	5 F	5 F	5 F	5 F	5 F	5 F	5 F	5 F
Mexico	4 107	3 832	3 952	3 764	3 467	2 987	3 449	3 129	3 166	4 206
USA	5	3	3	5	2	1	-	-	-	3
77 Fishing area total	*4 117 F*	*3 840 F*	*3 960 F*	*3 774 F*	*3 474 F*	*2 993 F*	*3 454 F*	*3 134 F*	*3 171 F*	*4 214 F*
81 Australia	3 671	5 683	5 674	4 258	4 306	4 451	3 456	3 151	3 620	3 777
New Zealand	837	809	886	897	872	720	846	748	910	748
81 Fishing area total	*4 508*	*6 492*	*6 560*	*5 155*	*5 178*	*5 171*	*4 302*	*3 899*	*4 530*	*4 525*
87 Chile	496	335	278	244	78	68	134	132	93	22
Colombia	28	43	25	20	26	36	28	30 F	28	28 F
Peru	14 711	16 964	16 601	13 916	13 264	29 075	20 843	26 314	27 330	24 989
87 Fishing area total	*15 235*	*17 342*	*16 904*	*14 180*	*13 368*	*29 179*	*21 005*	*26 476 F*	*27 451*	*25 039 F*
Species total	*215 427 F*	*231 714 F*	*306 817 F*	*301 711 F*	*344 067 F*	*370 659 F*	*381 401 F*	*386 432 F*	*396 783 F*	*374 270 F*
Fusiliers	**Fusiliers**		**Fusileros**		***Caesio spp***				**1,70(00)112,XX**	**FUS**
51 Jordan	...	...	...	...	...	15	18	17	15	12
Saudi Arabia	-	-	-	-	-	-	-	-	-	19
51 Fishing area total	*...*	*...*	*...*	*...*	*...*	*15*	*18*	*17*	*15*	*31*
57 Indonesia	4 794	4 975	5 333	4 163	5 306	4 857	5 568	5 088	6 040	6 290
Malaysia	5	23	57	35	23	5	1	19	14	1
57 Fishing area total	*4 799*	*4 998*	*5 390*	*4 198*	*5 329*	*4 862*	*5 569*	*5 107*	*6 054*	*6 291*
71 Indonesia	17 204	23 325	37 360	28 551	33 052	29 285	32 376	28 624	32 272	35 830
Malaysia	1 877	1 547	1 543	1 646	1 936	1 312	1 323	961	973	846
Philippines	15 476	15 292	13 914	13 788	16 754	15 931	14 952	13 516	13 677	12 162
Singapore	22	1	1	-	9	18	13	10	1	3
71 Fishing area total	*34 579*	*40 165*	*52 818*	*43 985*	*51 751*	*46 546*	*48 664*	*43 111*	*46 923*	*48 841*
Species total	*39 378*	*45 163*	*58 208*	*48 183*	*57 080*	*51 423*	*54 251*	*48 235*	*52 992*	*55 163*
Common snook	**Crossie blanc**		**Róbalo blanco**		***Centropomus undecimalis***				**1,70(01)025,10**	**SNO**
31 Mexico	2 139	2 090	2 885	2 955	3 307	2 990	3 415	3 184	4 000	5 951
Venezuela	4 783	4 569	3 023	2 177	1 560	2 022	1 704	1 850	1 903	1 437
31 Fishing area total	*6 922*	*6 659*	*5 908*	*5 132*	*4 867*	*5 012*	*5 119*	*5 034*	*5 903*	*7 388*
Species total	*6 922*	*6 659*	*5 908*	*5 132*	*4 867*	*5 012*	*5 119*	*5 034*	*5 903*	*7 388*
Snooks(=Robalos) nei	**Crossies nca**		**Róbalos nep**		***Centropomus spp***				**1,70(01)025,XX**	**ROB**
02 Nicaragua	...	...	...	...	...	...	...	...	220	185

B-33

Miscellaneous coastal fishes — **Capture production by species, fishing areas and countries or areas**
Poissons côtiers divers — **Captures par espèces, zones de pêche et pays ou zones**
Peces costeros diversos — **Capturas por especies, áreas de pesca y países o áreas**

Species, Fishing area Espèce, Zone de pêche Especie, Area de pesca	1993 t	1994 t	1995 t	1996 t	1997 t	1998 t	1999 t	2000 t	2001 t	2002 t
02 Fishing area total	...	...	...	...	...	...	...	...	*220*	*185*
31 Colombia	86	13	22	111	5	0	-	...	...	...
Dominican Rp	52	94	35	20	24	64	12	32	17	28
Grenada	0	0	0	0	0	0	1	0	0	0
Mexico	836	852	1 124	1 313	1 140	1 017	1 125	979	1 360	913
Nicaragua	600	...	...	...	...	...	2 000	2 060	1 960	1 227
Puerto Rico	...	...	...	...	...	...	32	33	8	24
31 Fishing area total	*1 574*	*959*	*1 181*	*1 444*	*1 169*	*1 081*	*3 170*	*3 104*	*3 345*	*2 192*
41 Brazil	2 310 F	2 320 F	1 820	1 686	1 866	2 996	3 602	4 366	3 469	3 608
41 Fishing area total	*2 310 F*	*2 320 F*	*1 820*	*1 686*	*1 866*	*2 996*	*3 602*	*4 366*	*3 469*	*3 608*
77 Mexico	757	1 026	563	587	684	752	996	809	862	686
77 Fishing area total	*757*	*1 026*	*563*	*587*	*684*	*752*	*996*	*809*	*862*	*686*
87 Colombia	64	63	86	341	73	96	62	70 F	70 F	65 F
87 Fishing area total	*64*	*63*	*86*	*341*	*73*	*96*	*62*	*70 F*	*70 F*	*65 F*
Species total	*4 705 F*	*4 368 F*	*3 650*	*4 058*	*3 792*	*4 925*	*7 830*	*8 349 F*	*7 966 F*	*6 736 F*
Black grouper	**Badèche bonaci**		**Cuna bonací**			***Mycteroperca bonaci***			**1,70(02)040,01**	**MAB**
31 USA	...	...	...	...	...	...	...	9	7	6
31 Fishing area total	...	...	...	...	...	...	...	*9*	*7*	*6*
Species total	...	...	...	...	...	...	...	*9*	*7*	*6*
Gag	**Badèche baillou**		**Cuna aguají**			***Mycteroperca microlepis***			**1,70(02)040,04**	**MKM**
31 USA	...	...	...	...	...	...	...	15	10	13
31 Fishing area total	...	...	...	...	...	...	...	*15*	*10*	*13*
Species total	...	...	...	...	...	...	...	*15*	*10*	*13*
Scamp	**Badèche galopin**		**Cuna garopa**			***Mycteroperca phenax***			**1,70(02)040,06**	**MKH**
31 USA	...	...	...	...	...	1	14	27	17	26
31 Fishing area total	...	...	...	...	...	*1*	*14*	*27*	*17*	*26*
Species total	...	...	...	...	...	*1*	*14*	*27*	*17*	*26*
Yellowfin grouper	**Badèche de roche**		**Cuna de piedra**			***Mycteroperca venenosa***			**1,70(02)040,09**	**MKV**
31 USA	...	...	...	...	...	...	...	3	1	3
31 Fishing area total	...	...	...	...	...	...	...	*3*	*1*	*3*
Species total	...	...	...	...	...	...	...	*3*	*1*	*3*
Broomtail grouper	**Badèche balai**		**Garropa jaspeada**			***Mycteroperca xenarcha***			**1,70(02)040,10**	**GBS**
87 Colombia	1 447	428	207	180	125	155	82	80 F	210	200 F
87 Fishing area total	*1 447*	*428*	*207*	*180*	*125*	*155*	*82*	*80 F*	*210*	*200 F*
Species total	*1 447*	*428*	*207*	*180*	*125*	*155*	*82*	*80 F*	*210*	*200 F*
Brazilian groupers nei	**Badèches nca**		**Cunas nep**			***Mycteroperca spp***			**1,70(02)040,XX**	**GPB**
41 Brazil	1 110 F	1 110 F	1 117	423	584	518	628	536	1 040	1 347
41 Fishing area total	*1 110 F*	*1 110 F*	*1 117*	*423*	*584*	*518*	*628*	*536*	*1 040*	*1 347*
Species total	*1 110 F*	*1 110 F*	*1 117*	*423*	*584*	*518*	*628*	*536*	*1 040*	*1 347*
Dusky grouper	**Mérou noir**		**Mero moreno**			***Epinephelus marginatus***			**1,70(02)042,01**	**GPD**
27 France	-	-	-	-	-	-	1	-	-	-
Portugal	28	36	21	120	11	24	-	-	-	-
Spain	229	267	305	207	187	243	95	24	51	40
UK	-	-	-	-	12	1	-	-	-	-
27 Fishing area total	*257*	*303*	*326*	*327*	*210*	*268*	*96*	*24*	*51*	*40*
34 Greece	4	4	3	2	3	1	2	1	2	1
Italy	1 718	1 718	-	-	-	-	-	-	-	-
Portugal	0	0	2	3	2	3	8	2	2	-
34 Fishing area total	*1 722*	*1 722*	*5*	*5*	*5*	*4*	*10*	*3*	*4*	*1*
37 Greece	114	108	192	110	91	56	62	89	97	90 F
Italy	2 163	1 814	1 454	558	640	124	89	97	252	229
Spain	-	-	-	15	17	13	28	19	20	94
Turkey	500	802	620	700	463	640	124	85	79	71
37 Fishing area total	*2 777*	*2 724*	*2 266*	*1 383*	*1 211*	*833*	*303*	*290*	*448*	*484 F*
Species total	*4 756*	*4 749*	*2 597*	*1 715*	*1 426*	*1 105*	*409*	*317*	*503*	*525 F*
White grouper	**Mérou blanc**		**Cherna de ley**			***Epinephelus aeneus***			**1,70(02)042,02**	**GPW**
34 Côte dIvoire	3	1	1 F	1 F	1 F	1 F	...	2	3	121
Greece	396	343	293	306	125	123	148	137	161	109
Mauritania	250 F	330 F	420 F	340 F	450 F	390 F	450 F	450 F	450 F	450 F

B-33 Miscellaneous coastal fishes / Poissons côtiers divers / Peces costeros diversos

Capture production by species, fishing areas and countries or areas / Captures par espèces, zones de pêche et pays ou zones / Capturas por especies, áreas de pesca y países o áreas

Species, Fishing area Espèce, Zone de pêche Especie, Area de pesca	1993 t	1994 t	1995 t	1996 t	1997 t	1998 t	1999 t	2000 t	2001 t	2002 t
Morocco	94	37	21	25	23	43	29	46	33	45
Portugal	-	-	3	5	0	-	-	-	1	0
34 Fishing area total	*743 F*	*711 F*	*738 F*	*677 F*	*599 F*	*557 F*	*627 F*	*635 F*	*648 F*	*725 F*
37 Morocco	11	12	5	5	4	4	6	48	7	5
37 Fishing area total	*11*	*12*	*5*	*5*	*4*	*4*	*6*	*48*	*7*	*5*
Species total	*754 F*	*723 F*	*743 F*	*682 F*	*603 F*	*561 F*	*633 F*	*683 F*	*655 F*	*730 F*
Nassau grouper	**Mérou rayé**		**Cherna criolla**		***Epinephelus striatus***			**1,70(02)042,20**		**GPN**
31 Bahamas	...	665	358	331	514	511	381	226	281	280 F
Colombia	48	65	25	11	5	3	0	...	...	...
Cuba	70	80	81	124	79	57	57	31	44	42
31 Fishing area total	*118*	*810*	*464*	*466*	*598*	*571*	*438*	*257*	*325*	*322 F*
Species total	*118*	*810*	*464*	*466*	*598*	*571*	*438*	*257*	*325*	*322 F*
Spotted grouper	**Mérou cabrilla**		**Mero moteado**		***Epinephelus analogus***			**1,70(02)042,22**		**GPS**
87 Colombia	16	28	22	43	48	32	39	30 F	28	28 F
87 Fishing area total	*16*	*28*	*22*	*43*	*48*	*32*	*39*	*30 F*	*28*	*28 F*
Species total	*16*	*28*	*22*	*43*	*48*	*32*	*39*	*30 F*	*28*	*28 F*
Yellowedge grouper	**Mérou aile jaune**		**Mero aleta amarilla**		***Epinephelus flavolimbatus***			**1,70(02)042,24**		**EEL**
31 USA	...	...	...	...	15	18	...	73	36	67
31 Fishing area total	...	...	...	...	*15*	*18*	...	*73*	*36*	*67*
Species total	...	...	...	...	*15*	*18*	...	*73*	*36*	*67*
Red hind	**Mérou couronné**		**Mero colorado**		***Epinephelus guttatus***			**1,70(02)042,25**		**EEU**
31 Grenada	42	56	81	60	42	52	72	67	88	159
31 Fishing area total	*42*	*56*	*81*	*60*	*42*	*52*	*72*	*67*	*88*	*159*
Species total	*42*	*56*	*81*	*60*	*42*	*52*	*72*	*67*	*88*	*159*
Red grouper	**Mérou rouge**		**Mero americano**		***Epinephelus morio***			**1,70(02)042,28**		**GPR**
31 Cuba	341	298	211	269	195	69	110	171	91	86
Dominican Rp	280	504	763	29	92	11	21	33	37	0
31 Fishing area total	*621*	*802*	*974*	*298*	*287*	*80*	*131*	*204*	*128*	*86*
41 Brazil	1 810 F	1 820 F	872	320	238	212	666	1 220	...	...
41 Fishing area total	*1 810 F*	*1 820 F*	*872*	*320*	*238*	*212*	*666*	*1 220*	...	...
Species total	*2 431 F*	*2 622 F*	*1 846*	*618*	*525*	*292*	*797*	*1 424*	*128*	*86*
Snowy grouper	**Mérou neige**		**Cherna pintada**		***Epinephelus niveatus***			**1,70(02)042,30**		**EFV**
31 USA	...	...	...	...	...	...	...	6	4	9
31 Fishing area total	...	...	...	...	...	...	...	*6*	*4*	*9*
Species total	...	...	...	...	...	...	...	*6*	*4*	*9*
Dungat grouper	**Mérou de Gorée**		**Mero de Gorea**		***Epinephelus goreensis***			**1,70(02)042,34**		**EEG**
34 Sierra Leone	-	120	102	196	63	76	290	277	62	64
34 Fishing area total	-	*120*	*102*	*196*	*63*	*76*	*290*	*277*	*62*	*64*
Species total	-	*120*	*102*	*196*	*63*	*76*	*290*	*277*	*62*	*64*
Warsaw grouper	**Mérou Varsovie**		**Mero negro**		***Epinephelus nigritus***			**1,70(02)042,36**		**ELG**
31 USA	...	54	37	16	23	23	-	45	44	50
31 Fishing area total	...	*54*	*37*	*16*	*23*	*23*	-	*45*	*44*	*50*
Species total	...	*54*	*37*	*16*	*23*	*23*	-	*45*	*44*	*50*
Groupers nei	**Mérous nca**		**Meros nep**		***Epinephelus spp***			**1,70(02)042,XX**		**GPX**
31 Aruba	...	...	20	20	25	22	25	18	15	15 F
Bahamas	618	514	513	475	246	227	193	132	177	170 F
Bermuda	36	45	42	42	42	43	48	27	45	50
Br Virgin Is	...	10	12	69	1	3	4	1	1 F	1 F
Cuba	159	77	85	130	88	89	44	82	22	61
Dominican Rp	809	1 456	2 316	455	338	379	662	873	942	1 077
Mexico	13 827	13 526	15 193	12 389	11 783	11 692	13 031	13 289	10 572	10 220
Puerto Rico	...	...	...	...	...	...	89	104	111	94
St Kitts Nev	...	...	8	18	10	11	11	6	3	10
USA	6 376	9 582	4 701	4 387	4 583	4 399	1 038	5 665	5 984	5 887
US Virgin Is	...	...	...	...	...	...	21	25 F	25 F	25 F
Venezuela	1 772	1 930	2 040	2 185	2 235	1 990	1 591	1 408	811	898
31 Fishing area total	*23 597*	*27 140*	*24 930*	*20 170*	*19 351*	*18 855*	*16 757*	*21 630 F*	*18 708 F*	*18 508 F*
34 Benin	21 F	23 F	22 F	25 F	25 F	23 F	20	14 F	16	1
Cameroon	1	1 F	2	0	0	1	0	1	1	1

B-33

Miscellaneous coastal fishes — **Capture production by species, fishing areas and countries or areas**
Poissons côtiers divers — **Captures par espèces, zones de pêche et pays ou zones**
Peces costeros diversos — **Capturas por especies, áreas de pesca y países o áreas**

Species, Fishing area Espèce, Zone de pêche Especie, Area de pesca	1993 t	1994 t	1995 t	1996 t	1997 t	1998 t	1999 t	2000 t	2001 t	2002 t
Congo Rep	3 F	0	5	5 F	4 F	3 F	3 F	2	2 F	2 F
Gabon	88	114	264	250	94	190	136	105	40	113
Gambia	42	54	118	62	53	30	30	49	63	66
Ghana	385	169	306	426	478	1 361	181	94	138	233
Liberia	9	6	22	-	5	110	71	25	44	45 F
Mauritania	120 F	190 F	240 F	180 F	240 F	220 F	300 F	300 F	300 F	300 F
Portugal	0	85	116	203	36	11	9	2	5	11
Sao Tome Prn	29 F	42 F	44 F	33	31	40 F	45 F	40 F	40 F	40 F
Spain	-	-	-	119	32	21	28	42	36	5
Togo	29	33	41	14	19	215	120	16	12	5
34 Fishing area total	*727 F*	*717 F*	*1 180 F*	*1 317 F*	*1 017 F*	*2 225 F*	*943 F*	*690 F*	*697 F*	*822 F*
37 Albania	...	...	0	2	1	-	-	-	-	3
Cyprus	26	47	25	22	22	29	23	27	24	24
Egypt	614	604	721	1 010	778	905	1 943	1 621	1 123	1 453
Greece	32	51	76	54	91	65	85	63	142	130 F
Israel	336	250	298	406	384	394	381	263	188	226
Libya	3 700 F	4 000 F	4 100 F	4 000 F	4 000 F	4 000 F	4 000 F	4 000 F	4 000 F	4 000 F
Malta	1	0	15	19	27	15	0	15	15	31
Palest, O.T.	...	...	...	46	45	44	45 F	40 F	30 F	23
Syria	...	...	...	...	...	...	...	...	...	54
Tunisia	381	302	236	265	416	334	382	103	539	474
37 Fishing area total	*5 090 F*	*5 254 F*	*5 471 F*	*5 824 F*	*5 764 F*	*5 786 F*	*6 859 F*	*6 132 F*	*6 061 F*	*6 418 F*
41 Brazil	2 300 F	2 300 F	2 286	2 341	2 170	2 315	1 710	2 670	2 124	2 084
Italy	229	229	-	-	-	-	-	-	-	-
41 Fishing area total	*2 529 F*	*2 529 F*	*2 286*	*2 341*	*2 170*	*2 315*	*1 710*	*2 670*	*2 124*	*2 084*
47 Korea Rep	37	23	-	-	-	-	-	-	-	-
St Helena	20	37	32	48	41	59	33	17	18	14
47 Fishing area total	*57*	*60*	*32*	*48*	*41*	*59*	*33*	*17*	*18*	*14*
51 Bahrain	469	498	459	532	300	331	525	670	794	725
Djibouti	92 F	98 F	100 F	100 F	100 F	100 F	100 F	100 F	100 F	100 F
Egypt	334	247	387	647	689	722	1 215	3 126	3 576	3 687
Eritrea	...	...	...	1	56	50	...	48	159	109
Italy	344	344	-	-	-	-	-	-	-	-
Korea Rep	300	632	209	389	174	119	820	331	227	230
Kuwait	431	413	341	287	241	264	237	250 F	268	265 F
Oman	2 075	3 735	4 031	3 409	3 366	5 345	4 829	5 013	3 799	3 292
Pakistan	6 255	7 617	8 600	9 793	10 474	13 991	17 355	16 012	15 928	16 000
Qatar	1 145	950	728	768	736	804	913	1 215	1 820	1 567
Tanzania	278	335	290	652	400	350	500	450	500	550
51 Fishing area total	*11 723 F*	*14 869 F*	*15 145 F*	*16 578 F*	*16 536 F*	*22 076 F*	*26 494 F*	*27 215 F*	*27 171 F*	*26 525 F*
57 Indonesia	10 993	10 490	12 641	13 863	15 346	16 715	14 715	14 787	15 201	15 900
Korea Rep	-	-	-	-	-	-	234	22	32	-
Malaysia	2 166	1 671	1 513	1 533	1 654	1 487	1 347	1 491	1 156	1 281
57 Fishing area total	*13 159*	*12 161*	*14 154*	*15 396*	*17 000*	*18 202*	*16 296*	*16 300*	*16 389*	*17 181*
61 China	22 000	22 469	22 999	23 241	30 201	36 098	40 245	41 513	44 975	45 448
China,Taiwan	2 906	3 019	1 771	2 457	2 387	1 414	1 265	1 202	1 610	933
61 Fishing area total	*24 906*	*25 488*	*24 770*	*25 698*	*32 588*	*37 512*	*41 510*	*42 715*	*46 585*	*46 381*
71 Fiji Islands	1 500	1 529 F	854	1 017	1 060	1 160	1 600	1 450 F	1 544	1 530 F
Guam	...	3	1	0	1	0	1	1	1	1
Indonesia	19 022	29 431	21 363	24 423	26 818	27 051	28 757	33 635	33 315	36 420
Korea Rep	1	-	472	-	-	1	-	-	-	-
Malaysia	5 710	5 538	5 523	6 741	7 470	9 114	10 882	10 683	9 320	7 556
N Marianas	...	2	1	3	5	3	2	2	3	2
Palau	6	3	5	7	1	2	4	2	10	5
Singapore	31	129	109	120	94	72	56	43	38	33
71 Fishing area total	*26 270*	*36 635 F*	*28 328*	*32 311*	*35 449*	*37 403*	*41 302*	*45 816 F*	*44 231*	*45 547 F*
77 Amer Samoa	...	2	2	3	2	2	1	1	1	1
Cook Is	132 F	125 F	123 F	110 F	100 F	90 F	80 F	60 F	60 F	60 F
Mexico	1 928	1 414	2 588	5 254	7 126	6 158	6 800	5 306	4 037	5 115
77 Fishing area total	*2 060 F*	*1 541 F*	*2 713 F*	*5 367 F*	*7 228 F*	*6 250 F*	*6 881 F*	*5 367 F*	*4 098 F*	*5 176 F*
87 Colombia	1 447	428	207	180	125	155	82	70 F	57	55 F
Peru	16	110	191	241	60	101	151	91	87	313
87 Fishing area total	*1 463*	*538*	*398*	*421*	*185*	*256*	*233*	*161 F*	*144*	*368 F*
Species total	*111 581 F*	*126 932 F*	*119 407 F*	*125 471 F*	*137 329 F*	*150 939 F*	*159 018 F*	*168 713 F*	*166 226 F*	*169 024 F*
Orange perch	**...B**		**...C**		***Lepidoperca pulchella***				**1,70(02)072,01**	**LDP**
81 New Zealand	...	...	...	...	...	193	32	...	46	97
81 Fishing area total	...	...	...	...	...	*193*	*32*	...	*46*	*97*
Species total	...	...	...	...	...	*193*	*32*	...	*46*	*97*
Black seabass	**Fanfre noir**		**Serrano estriado**		***Centropristis striata***				**1,70(02)081,02**	**BSB**
21 USA	1 453	1 339	904	1 683	1 215	1 190	1 412	1 189	1 276	1 574
21 Fishing area total	*1 453*	*1 339*	*904*	*1 683*	*1 215*	*1 190*	*1 412*	*1 189*	*1 276*	*1 574*
31 USA	425	693	393	295	375	341	305	327	391	333
31 Fishing area total	*425*	*693*	*393*	*295*	*375*	*341*	*305*	*327*	*391*	*333*

B-33 Miscellaneous coastal fishes / Poissons côtiers divers / Peces costeros diversos

Capture production by species, fishing areas and countries or areas / Captures par espèces, zones de pêche et pays ou zones / Capturas por especies, áreas de pesca y países o áreas

Species, Fishing area Espèce, Zone de pêche Especie, Area de pesca	1993 t	1994 t	1995 t	1996 t	1997 t	1998 t	1999 t	2000 t	2001 t	2002 t
Species total	*1 878*	*2 032*	*1 297*	*1 978*	*1 590*	*1 531*	*1 717*	*1 516*	*1 667*	*1 907*
Argentine seabass	**Serran argentin**		**Mero sureño**			***Acanthistius brasilianus***			**1,70(02)259,01**	**BSZ**
41 Argentina	9 195	7 756	10 926	10 714	9 113	5 216	5 895	4 134	4 881	3 622
Uruguay	9	6	61	85	51	42	19	88	9	2
41 Fishing area total	*9 204*	*7 762*	*10 987*	*10 799*	*9 164*	*5 258*	*5 914*	*4 222*	*4 890*	*3 624*
Species total	*9 204*	*7 762*	*10 987*	*10 799*	*9 164*	*5 258*	*5 914*	*4 222*	*4 890*	*3 624*
Peruvian rock seabass	**Serran cabrilla**		**Cabrilla loca**			***Paralabrax humeralis***			**1,70(02)405,05**	**BAP**
87 Peru	3 647	3 104	5 837	4 954	2 789	2 554	3 278	4 373	2 011	1 522
87 Fishing area total	*3 647*	*3 104*	*5 837*	*4 954*	*2 789*	*2 554*	*3 278*	*4 373*	*2 011*	*1 522*
Species total	*3 647*	*3 104*	*5 837*	*4 954*	*2 789*	*2 554*	*3 278*	*4 373*	*2 011*	*1 522*
Groupers, seabasses nei	**Serranidés nca**		**Meros, chernas, nep**			***Serranidae***			**1,70(02)XXX,XX**	**BSX**
21 USA	...	...	...	...	...	23	4 702	18	18	23
21 Fishing area total	...	...	...	...	...	*23*	*4 702*	*18*	*18*	*23*
27 Germany	-	-	-	-	-	-	-	-	1	-
Netherlands	-	-	-	-	-	17	14	-	-	-
Portugal	579	488	372	437	273	329	325	283	171	159
Spain	-	-	-	142	432	442	444	593	421	194
27 Fishing area total	*579*	*488*	*372*	*579*	*705*	*788*	*783*	*876*	*593*	*353*
31 Antigua Barb	...	...	...	...	...	...	...	...	217	364
Colombia	23	39	67	86	17	37	26	30 F	30 F	30 F
Costa Rica	...	...	...	...	...	...	1	3	0	1
Grenada	7	4	4	10	11	16	20	22	38	62
Nicaragua	...	...	...	...	...	...	440	460	435	294
USA	...	...	...	...	...	...	...	55	16	...
Venezuela	133	264	279	265	322	384	328	140	123	137
31 Fishing area total	*163*	*307*	*350*	*361*	*350*	*437*	*815*	*710 F*	*859 F*	*888 F*
34 Greece	0	1	1	-	3	-	-	-	192	-
Korea Rep	-	-	-	-	177	356	413	536	351	181
Nigeria	3 113	3 468	3 102	1 285	2 001	1 741	2 486	2 117	2 487	2 354
Portugal	270	180	11	24	10	7	11	9	8	6
Senegal	4 243	5 269	4 175	4 411	4 317	3 488	3 162	3 067	3 685	2 903
Spain	-	-	-	58	12	156	176	66	82	4
34 Fishing area total	*7 626*	*8 918*	*7 289*	*5 778*	*6 520*	*5 748*	*6 248*	*5 795*	*6 805*	*5 448*
37 Algeria	...	...	...	...	...	...	6	0	0	0
Cyprus	4	6	13	16	15	11	17	3	5	10
Greece	322	756	428	410	616	330	336	286	271	250 F
Lebanon	125	150	250	250	150	250	250	230	240	250
Malta	0	0	1	1	1	1	2	2	2	1
Spain	-	-	-	14	23	30	71	4	121	400
Tunisia	49	121	75	84	130	3 161	748	98	98	97
Turkey	223	286	339	790	1 137	700	361	400	411	100
37 Fishing area total	*723*	*1 319*	*1 106*	*1 565*	*2 072*	*4 483*	*1 791*	*1 023*	*1 148*	*1 108 F*
41 Portugal	-	-	-	-	15	2	-	-	1	1
Spain	-	-	-	-	-	-	-	-	-	9
41 Fishing area total	-	-	-	-	*15*	*2*	-	-	*1*	*10*
47 Angola	5	66	235	1 002	460	1 326	347	2 341	4 214	5 774
South Africa	-	-	-	-	-	-	-	-	15	7
Spain	-	-	-	-	-	91	-	-	-	-
47 Fishing area total	*5*	*66*	*235*	*1 002*	*460*	*1 417*	*347*	*2 341*	*4 229*	*5 781*
51 Eritrea	...	14	83	140	39	67	257	378	111	41
Mauritius	1 026	1 035	1 022	905	933	931	826	863	938	885
Réunion	...	...	...	...	...	...	47	20	11	11 F
Saudi Arabia	5 752	5 707	3 514	4 207	4 403	5 053	5 430	5 402	5 273	6 838
Seychelles	249	133	71	66	124	45	143	56	107	74
South Africa	-	33	25	30	...	...	25 F	...	...	20
Spain	-	-	-	-	-	-	2	-	-	93
Untd Arab Em	5 997	6 330	6 696	6 767	7 232	7 256	7 437	24 045	27 680	22 827
Yemen	2 487	2 400	2 260	1 743	1 820 F	2 100 F	2 000 F	1 800 F	2 300 F	2 600 F
51 Fishing area total	*15 511*	*15 652*	*13 671*	*13 858*	*14 551 F*	*15 452 F*	*16 167 F*	*32 564 F*	*36 420 F*	*33 389 F*
57 Australia	95	323	377	657	1	4	60	74	84	51
Thailand	-	2 600	3 778	3 502	3 295	4 290	2 585	2 592	2 988	2 925
57 Fishing area total	*95*	*2 923*	*4 155*	*4 159*	*3 296*	*4 294*	*2 645*	*2 666*	*3 072*	*2 976*
61 China,H.Kong	2 109	1 699	1 392	1 349	1 318	1 240	900 F	1 100 F	1 200 F	1 150 F
Korea Rep	782	440	557	436	583	765	551	947	640	567
61 Fishing area total	*2 891*	*2 139*	*1 949*	*1 785*	*1 901*	*2 005*	*1 451 F*	*2 047 F*	*1 840 F*	*1 717 F*
71 Australia	32	55	34	33	22	34	0	3	0	1
Philippines	14 358	13 654	14 226	12 776	12 197	13 160	13 675	12 492	11 339	13 913
Thailand	-	5 679	5 431	5 847	5 649	5 020	5 465	5 035	5 657	5 721
71 Fishing area total	*14 390*	*19 388*	*19 691*	*18 656*	*17 868*	*18 214*	*19 140*	*17 530*	*16 996*	*19 635*

B-33 Miscellaneous coastal fishes / Poissons côtiers divers / Peces costeros diversos

Capture production by species, fishing areas and countries or areas / Captures par espèces, zones de pêche et pays ou zones / Capturas por especies, áreas de pesca y países o áreas

Species, Fishing area Espèce, Zone de pêche Especie, Area de pesca	1993 t	1994 t	1995 t	1996 t	1997 t	1998 t	1999 t	2000 t	2001 t	2002 t
77 Costa Rica	438	328	281	177	387	644	395	201	103	103
Nicaragua	...	...	...	...	...	...	600	380	335	136
USA	-	-	-	-	-	-	-	-	-	17
77 Fishing area total	*438*	*328*	*281*	*177*	*387*	*644*	*995*	*581*	*438*	*256*
81 Australia	55	58	85	170	132	141	11	22	8	61
Japan	51	25	10	3	13	21	10	11	13	10 F
81 Fishing area total	*106*	*83*	*95*	*173*	*145*	*162*	*21*	*33*	*21*	*71 F*
87 Chile	17	39	23	30	24	10	19	7	5	-
87 Fishing area total	*17*	*39*	*23*	*30*	*24*	*10*	*19*	*7*	*5*	*-*
Species total	*42 544*	*51 650*	*49 217*	*48 123*	*48 294 F*	*53 679 F*	*55 124 F*	*66 191 F*	*72 445 F*	*71 655 F*
Therapon pearch	**...B**		**...C**		***Terapon spp***				**1,70(04)089,XX**	**THO**
51 Saudi Arabia	-	-	-	-	-	1	2	0	1	2
51 Fishing area total	*-*	*-*	*-*	*-*	*-*	*1*	*2*	*0*	*1*	*2*
Species total	*-*	*-*	*-*	*-*	*-*	*1*	*2*	*0*	*1*	*2*
Giant seabass	**Bar gigantesque**		**Lubina gigante**		***Stereolepis gigas***				**1,70(05)220,02**	**TEJ**
77 USA	72	0	0	1	1	3	2	2	3	2
77 Fishing area total	*72*	*0*	*0*	*1*	*1*	*3*	*2*	*2*	*3*	*2*
Species total	*72*	*0*	*0*	*1*	*1*	*3*	*2*	*2*	*3*	*2*
Spotted seabass	**Bar tacheté**		**Baila**		***Dicentrarchus punctatus***				**1,70(06)345,01**	**SPU**
27 France	38	55	76	68	42	75	68	71	79	57
Spain	-	-	-	-	-	-	-	-	-	27
27 Fishing area total	*38*	*55*	*76*	*68*	*42*	*75*	*68*	*71*	*79*	*84*
34 Mauritania	120 F	160 F	140 F	160 F	130 F	190 F	100 F	45 F	...	...
Senegal	...	...	9	619	472	527	170	67	125	193
34 Fishing area total	*120 F*	*160 F*	*149 F*	*779 F*	*602 F*	*717 F*	*270 F*	*112 F*	*125*	*193*
37 Egypt	100	604	73	424	119	135	203	186	238	556
37 Fishing area total	*100*	*604*	*73*	*424*	*119*	*135*	*203*	*186*	*238*	*556*
51 Egypt	...	...	101	2	35	36	241	10	18	8
51 Fishing area total	*...*	*...*	*101*	*2*	*35*	*36*	*241*	*10*	*18*	*8*
Species total	*258 F*	*819 F*	*399 F*	*1 273 F*	*798 F*	*963 F*	*782 F*	*379 F*	*460*	*841*
European seabass	**Bar européen**		**Lubina**		***Dicentrarchus labrax***				**1,70(06)345,03**	**BSS**
27 Channel Is	37 F	83	77 F	56	74	79	107	129	80	73
Denmark	1	0	1	1	1	2	1	5	2	1
France	2 317	2 305	2 351	3 163	2 845	2 626	3 224	3 881	3 957	3 628
Ireland	0	0	0	-	-	-	-	-	-	-
Netherlands	-	-	-	8	1	49	32	60	79	96
Portugal	71	107	68	57	40	38	37	49	43	43
Spain	365	457	446	534	474	457	383	473	326	303
UK	249	549	722	582	572	501	687	406	457	640
27 Fishing area total	*3 040 F*	*3 501*	*3 665 F*	*4 401*	*4 007*	*3 752*	*4 471*	*5 003*	*4 944*	*4 784*
34 Greece	169	80	137	108	19	7	26	9	20	21
34 Fishing area total	*169*	*80*	*137*	*108*	*19*	*7*	*26*	*9*	*20*	*21*
37 Albania	...	...	14	32	14	30	30	50	70	64
Croatia	...	...	...	...	...	31	20	22	13	2
Egypt	419	...	429	727	453	559	662	626	800	1 336
France	984	199	179	167	167	167	279	271	251	260
Greece	412	413	392	347	361	251	263	336	280	260 F
Italy	2 127	2 528	4 633	2 481	2 030	1 889	1 881	2 195	2 735	3 428
Malta	50	0	0	0	0	0	15	0	0	-
Slovenia	-	-	2	-	-	-	1	1	5	4
37 Fishing area total	*3 992*	*3 140*	*5 649*	*3 754*	*3 025*	*2 927*	*3 151*	*3 501*	*4 154*	*5 354 F*
Species total	*7 201 F*	*6 721*	*9 451 F*	*8 263*	*7 051*	*6 686*	*7 648*	*8 513*	*9 118*	*10 159 F*
Seabasses nei	**Bars nca**		**Lubinas nep**		***Dicentrarchus spp***				**1,70(06)345,XX**	**BSE**
01 Egypt	2 413	1 272	2 706	2 397	2 233	2 738	815	4 075	3 789	3 310
01 Fishing area total	*2 413*	*1 272*	*2 706*	*2 397*	*2 233*	*2 738*	*815*	*4 075*	*3 789*	*3 310*
27 Portugal	-	-	-	-	5	6	336	405	367	357
27 Fishing area total	*-*	*-*	*-*	*-*	*5*	*6*	*336*	*405*	*367*	*357*
34 Portugal	-	-	-	-	-	-	-	-	11	-
34 Fishing area total	*-*	*-*	*-*	*-*	*-*	*-*	*-*	*-*	*11*	*-*
37 Cyprus	...	...	...	...	...	...	...	...	3	3
Serbia-Monte	2	3	3	5	4	7	6	7	7	7
Spain	-	-	-	-	-	-	158	197	259	254
Tunisia	316	316	379	430	96	529	540	573	317	949
Turkey	1 160	1 884	2 116	2 411	3 450	3 950	3 650	1 900	1 200	713

B-33 Miscellaneous coastal fishes / Poissons côtiers divers / Peces costeros diversos

Capture production by species, fishing areas and countries or areas / Captures par espèces, zones de pêche et pays ou zones / Capturas por especies, áreas de pesca y países o áreas

Species, Fishing area Espèce, Zone de pêche Especie, Area de pesca	1993 t	1994 t	1995 t	1996 t	1997 t	1998 t	1999 t	2000 t	2001 t	2002 t
37 Fishing area total	*1 478*	*2 203*	*2 498*	*2 846*	*3 550*	*4 486*	*4 354*	*2 677*	*1 786*	*1 926*
Species total	*3 891*	*3 475*	*5 204*	*5 243*	*5 788*	*7 230*	*5 505*	*7 157*	*5 953*	*5 593*
Japanese seabass	**Bar du Japon**		**Serránido japonés**		***Lateolabrax japonicus***				**1,70(08)297,01**	**BAJ**
61 Japan	6 906	7 302	7 713	8 334	9 057	9 223	9 234	9 337	10 690	10 737
Korea Rep	1 728	1 217	1 118	1 178	1 501	2 957	2 933	1 201	1 566	1 479
61 Fishing area total	*8 634*	*8 519*	*8 831*	*9 512*	*10 558*	*12 180*	*12 167*	*10 538*	*12 256*	*12 216*
Species total	*8 634*	*8 519*	*8 831*	*9 512*	*10 558*	*12 180*	*12 167*	*10 538*	*12 256*	*12 216*
Red bigeye	**Beauclaire Pacifique**		**Catalufa Pacífico**		***Priacanthus macracanthus***				**1,70(11)026,05**	**BIR**
61 China,Taiwan	5 373	3 314	5 486	7 386	4 824	3 670	2 756	3 986	3 079	2 142
61 Fishing area total	*5 373*	*3 314*	*5 486*	*7 386*	*4 824*	*3 670*	*2 756*	*3 986*	*3 079*	*2 142*
Species total	*5 373*	*3 314*	*5 486*	*7 386*	*4 824*	*3 670*	*2 756*	*3 986*	*3 079*	*2 142*
Bigeyes nei	**Beauclaires nca**		**Catalufas nep**		***Priacanthus spp***				**1,70(11)026,XX**	**BIG**
31 USA	...	...	...	...	...	...	...	0	1	1
31 Fishing area total	...	...	...	...	...	...	...	*0*	*1*	*1*
34 Côte dIvoire	12	15	15 F	20 F	20 F	25 F	29	12	1	2
34 Fishing area total	*12*	*15*	*15 F*	*20 F*	*20 F*	*25 F*	*29*	*12*	*1*	*2*
41 Brazil	...	...	70	59	55	46	60	67	45	52
41 Fishing area total	...	...	*70*	*59*	*55*	*46*	*60*	*67*	*45*	*52*
47 St Helena	3	8	3	1	1	3	7	2	1	2
47 Fishing area total	*3*	*8*	*3*	*1*	*1*	*3*	*7*	*2*	*1*	*2*
51 Eritrea	...	0	5	0	-	-	0	-	-	-
Saudi Arabia	-	-	-	-	-	2	1	0	1	0
51 Fishing area total	...	*0*	*5*	*0*	...	*2*	*1*	*0*	*1*	*0*
57 Indonesia	1 169	1 253	1 052	1 200	1 221	1 360	1 605	1 583	1 459	1 510
Thailand	10 129	7 614	11 456	14 711	15 150	15 361	11 608	10 520	12 409	12 146
57 Fishing area total	*11 298*	*8 867*	*12 508*	*15 911*	*16 371*	*16 721*	*13 213*	*12 103*	*13 868*	*13 656*
61 China,H.Kong	7 528	6 034	6 781	5 391	4 962	4 149	3 000 F	3 600 F	4 000 F	3 900 F
61 Fishing area total	*7 528*	*6 034*	*6 781*	*5 391*	*4 962*	*4 149*	*3 000 F*	*3 600 F*	*4 000 F*	*3 900 F*
71 Indonesia	2 135	2 258	2 862	2 479	3 227	3 252	3 377	4 620	3 885	4 250
Thailand	49 710	44 680	57 723	67 411	62 675	64 873	71 127	65 168	69 345	70 132
71 Fishing area total	*51 845*	*46 938*	*60 585*	*69 890*	*65 902*	*68 125*	*74 504*	*69 788*	*73 230*	*74 382*
Species total	*70 686*	*61 862*	*79 967 F*	*91 272 F*	*87 311 F*	*89 071 F*	*90 814 F*	*85 572 F*	*91 147 F*	*91 995 F*
Cardinalfishes, etc. nei	**Apogonidés nca**		**Peces cardenal, etc. nep**		***Apogonidae***				**1,70(12)XXX,XX**	**APO**
27 France	-	-	-	-	-	-	294	-	-	-
Germany	-	-	-	-	-	-	-	-	10	-
Ireland	-	-	-	-	-	-	-	-	5	55
27 Fishing area total	-	-	-	-	-	-	*294*	-	*15*	*55*
34 Portugal	-	-	11	11	0	-	-	0	-	-
34 Fishing area total	-	-	*11*	*11*	*0*	-	-	*0*	-	-
51 Yemen	...	...	...	...	...	...	...	...	262	293
51 Fishing area total	...	...	...	...	...	...	...	...	*262*	*293*
71 Fiji Islands	210	214 F	...	200	180	60	71	60 F	67	65 F
71 Fishing area total	*210*	*214 F*	...	*200*	*180*	*60*	*71*	*60 F*	*67*	*65 F*
Species total	*210*	*214 F*	*11*	*211*	*180*	*60*	*365*	*60 F*	*344*	*413 F*
Sillago-whitings	**Sillaginidés**		**Sillagínidos**		***Sillaginidae***				**1,70(15)XXX,XX**	**WHS**
51 Pakistan	321	365	423	289	266	218	201	194	204	210
51 Fishing area total	*321*	*365*	*423*	*289*	*266*	*218*	*201*	*194*	*204*	*210*
57 Australia	352	294	303	899	1 105	1 219	924	1 717	1 696	1 527
Malaysia	1 178	1 191	1 042	899	1 301	738	930	856	1 275	913
Thailand	1 444	1 593	2 781	2 532	3 144	3 345	6 240	5 659	3 860	3 778
57 Fishing area total	*2 974*	*3 078*	*4 126*	*4 330*	*5 550*	*5 302*	*8 094*	*8 232*	*6 831*	*6 218*
61 China,Taiwan	1 342	552	407	547	401	346	190	188	132	231
Korea Rep	109	157	199	248	199	48	109	100	10	15
61 Fishing area total	*1 451*	*709*	*606*	*795*	*600*	*394*	*299*	*288*	*142*	*246*
71 Australia	1 411	1 835	2 569	2 321	2 140	1 828	1 233	1 246	814	1 161
Malaysia	798	873	906	996	633	1 071	1 084	956	782	694
Philippines	8 294	8 405	7 417	6 800	8 662	8 918	9 529	9 472	10 458	11 556
Singapore	80	39	32	48	45	41	24	10	14	8
Thailand	1 956	2 490	3 785	2 901	2 594	2 156	1 746	2 012	2 409	2 437
71 Fishing area total	*12 539*	*13 642*	*14 709*	*13 066*	*14 074*	*14 014*	*13 616*	*13 696*	*14 477*	*15 856*

B-33 Miscellaneous coastal fishes / Poissons côtiers divers / Peces costeros diversos

Capture production by species, fishing areas and countries or areas / Captures par espèces, zones de pêche et pays ou zones / Capturas por especies, áreas de pesca y países o áreas

Species, Fishing area Espèce, Zone de pêche Especie, Area de pesca	1993 t	1994 t	1995 t	1996 t	1997 t	1998 t	1999 t	2000 t	2001 t	2002 t
81 Australia	1 580	2 002	1 826	1 169	700	607	1 804	1 115	1 342	1 432
81 Fishing area total	*1 580*	*2 002*	*1 826*	*1 169*	*700*	*607*	*1 804*	*1 115*	*1 342*	*1 432*
Species total	*18 865*	*19 796*	*21 690*	*19 649*	*21 190*	*20 535*	*24 014*	*23 525*	*22 996*	*23 962*
Moonfish	**Luneur**		**Lunero**		***Mene maculata***			**1,70(26)327,01**		**MOO**
61 China,Taiwan	6 147	1 823	859	725	902	1 010	1 747	1 496	1 582	1 110
61 Fishing area total	*6 147*	*1 823*	*859*	*725*	*902*	*1 010*	*1 747*	*1 496*	*1 582*	*1 110*
71 Philippines	4 656	6 175	4 201	4 810	11 933	11 516	12 372	11 703	13 268	16 945
Singapore	47	15	6	5	7	13	13	15	7	5
71 Fishing area total	*4 703*	*6 190*	*4 207*	*4 815*	*11 940*	*11 529*	*12 385*	*11 718*	*13 275*	*16 950*
Species total	*10 850*	*8 013*	*5 066*	*5 540*	*12 842*	*12 539*	*14 132*	*13 214*	*14 857*	*18 060*
Ruff	**Saumon rude**		**Salmón áspero**		***Arripis georgianus***			**1,70(29)051,01**		**RUF**
57 Australia	1 500 F	1 095	1 063	1 302	1 287	1 008	1 066	1 143	992	860
57 Fishing area total	*1 500 F*	*1 095*	*1 063*	*1 302*	*1 287*	*1 008*	*1 066*	*1 143*	*992*	*860*
Species total	*1 500 F*	*1 095*	*1 063*	*1 302*	*1 287*	*1 008*	*1 066*	*1 143*	*992*	*860*
Australian salmon	**Saumon australien**		**Salmón de Australia**		***Arripis trutta***			**1,70(29)051,02**		**ASA**
57 Australia	2 946	2 720	4 840	3 507	3 457	3 623	3 354	4 232	3 932	4 097
57 Fishing area total	*2 946*	*2 720*	*4 840*	*3 507*	*3 457*	*3 623*	*3 354*	*4 232*	*3 932*	*4 097*
81 Australia	786	489	1 086	1 162	1 316	296	172	449	802	1 065
New Zealand	6 512	4 548	4 419	3 496	2 777	2 928	4 780	3 825	2 934	3 637
81 Fishing area total	*7 298*	*5 037*	*5 505*	*4 658*	*4 093*	*3 224*	*4 952*	*4 274*	*3 736*	*4 702*
Species total	*10 244*	*7 757*	*10 345*	*8 165*	*7 550*	*6 847*	*8 306*	*8 506*	*7 668*	*8 799*
Mangrove red snapper	**Vivaneau des mangroves**		**Pargo de manglar**		***Lutjanus argentimaculatus***			**1,70(32)027,02**		**RES**
51 Pakistan	2 178	2 524	3 145	2 002	2 394	3 192	3 195	3 003	2 900	3 000
51 Fishing area total	*2 178*	*2 524*	*3 145*	*2 002*	*2 394*	*3 192*	*3 195*	*3 003*	*2 900*	*3 000*
57 Malaysia	1 110	901	659	662	658	1 227	615	789	608	627
57 Fishing area total	*1 110*	*901*	*659*	*662*	*658*	*1 227*	*615*	*789*	*608*	*627*
71 Malaysia	10 163	8 629	10 522	10 824	10 190	4 935	12 319	10 650	10 436	10 635
71 Fishing area total	*10 163*	*8 629*	*10 522*	*10 824*	*10 190*	*4 935*	*12 319*	*10 650*	*10 436*	*10 635*
Species total	*13 451*	*12 054*	*14 326*	*13 488*	*13 242*	*9 354*	*16 129*	*14 442*	*13 944*	*14 262*
Yellow snapper	**Vivaneau jaune**		**Pargo amarillo (=Huachinango)**		***Lutjanus argentiventris***			**1,70(32)027,20**		**HUS**
77 Mexico	4 357	4 081	3 810	4 917	3 123	3 390	2 994	3 392	3 388	3 733
77 Fishing area total	*4 357*	*4 081*	*3 810*	*4 917*	*3 123*	*3 390*	*2 994*	*3 392*	*3 388*	*3 733*
87 Colombia	103	63	120	43	40	100	82	80 F	80 F	80 F
87 Fishing area total	*103*	*63*	*120*	*43*	*40*	*100*	*82*	*80 F*	*80 F*	*80 F*
Species total	*4 460*	*4 144*	*3 930*	*4 960*	*3 163*	*3 490*	*3 076*	*3 472 F*	*3 468 F*	*3 813 F*
Southern red snapper	**Vivaneau rouge**		**Pargo colorado**		***Lutjanus purpureus***			**1,70(32)027,22**		**SNC**
31 Colombia	195	235	304	17	...	290	172	250 F	250 F	250 F
Cuba	476	623	809	1 002	909	807	763	689	773	885
Dominican Rp	430	774	484	316	496	157	71	126	355	337
Guyana	...	...	...	...	...	...	...	570	524	423
31 Fishing area total	*1 101*	*1 632*	*1 597*	*1 335*	*1 405*	*1 254*	*1 006*	*1 635 F*	*1 902 F*	*1 895 F*
41 Brazil	3 610 F	3 620 F	5 816	5 104	6 085	5 937	9 790	6 580	6 209	6 184
41 Fishing area total	*3 610 F*	*3 620 F*	*5 816*	*5 104*	*6 085*	*5 937*	*9 790*	*6 580*	*6 209*	*6 184*
Species total	*4 711 F*	*5 252 F*	*7 413*	*6 439*	*7 490*	*7 191*	*10 796*	*8 215 F*	*8 111 F*	*8 079 F*
Northern red snapper	**Vivaneau campèche**		**Pargo del Golfo**		***Lutjanus campechanus***			**1,70(32)027,25**		**SNR**
31 Mexico	7 198	4 903	4 714	4 555	4 219	3 392	3 445	2 726	2 717	2 566
USA	1 513	1 520	1 412	1 791	2 045	1 774	1 967	2 115	2 152	2 051
31 Fishing area total	*8 711*	*6 423*	*6 126*	*6 346*	*6 264*	*5 166*	*5 412*	*4 841*	*4 869*	*4 617*
Species total	*8 711*	*6 423*	*6 126*	*6 346*	*6 264*	*5 166*	*5 412*	*4 841*	*4 869*	*4 617*
Lane snapper	**Vivaneau gazou**		**Pargo biajaiba**		***Lutjanus synagris***			**1,70(32)027,33**		**SNL**
31 Colombia	224	58	225	236	35	12	11	20 F	20 F	20 F
Cuba	1 201	1 786	1 943	1 848	2 472	2 609	1 712	2 114	1 891	1 844
Mexico	2 745	1 875	968	950	641	591	630	430	693	280
31 Fishing area total	*4 170*	*3 719*	*3 136*	*3 034*	*3 148*	*3 212*	*2 353*	*2 564 F*	*2 604 F*	*2 144 F*
41 Brazil	...	...	506	648	659	1 038	1 031	1 014	1 437	1 392
41 Fishing area total	*...*	*...*	*506*	*648*	*659*	*1 038*	*1 031*	*1 014*	*1 437*	*1 392*
Species total	*4 170*	*3 719*	*3 642*	*3 682*	*3 807*	*4 250*	*3 384*	*3 578 F*	*4 041 F*	*3 536 F*

B-33

Miscellaneous coastal fishes — **Capture production by species, fishing areas and countries or areas**
Poissons côtiers divers — **Captures par espèces, zones de pêche et pays ou zones**
Peces costeros diversos — **Capturas por especies, áreas de pesca y países o áreas**

Species, Fishing area Espèce, Zone de pêche Especie, Area de pesca	1993 t	1994 t	1995 t	1996 t	1997 t	1998 t	1999 t	2000 t	2001 t	2002 t
Snappers nei	**Vivaneaux nca**		**Pargos tropicales nep**		***Lutjanus spp***			**1,70(32)027,XX**		**SNA**
31 Bahamas	329	258	297	341	751	781	866	721	777	780 F
Br Virgin Is	...	53	20	6	1	4	3	1	1 F	1 F
Nicaragua	100	...	...	...	...	...	540	760	910	563
St Kitts Nev	...	...	4	9	5	8	15	19	9	16
St Lucia	...	25	56	69	31	34	45	68	82	43
31 Fishing area total	*429*	*336*	*377*	*425*	*788*	*827*	*1 469*	*1 569*	*1 779 F*	*1 403 F*
34 Benin	54 F	60 F	58 F	66 F	90 F	86 F	92	60 F	32	51
Cameroon	1	300 F	510	511	400 F	320 F	337 F	337	293	297
Congo Rep	102 F	61 F	50 F	40 F	30 F	20 F	10 F	1	1 F	1 F
Gabon	192	117	463	611	480	907	941	795	765	650
Gambia	61	67	127	27	150	86	90	90	126	122
Ghana	895	716	626	328	181	294	163	491	447	774
Korea Rep	-	-	-	-	-	555	474	524	601	369
Liberia	24	13	1	9	17	27	339	132	201	210 F
Nigeria	2 222	2 902	2 999	3 779	6 473	8 685	8 345	7 235	6 619	7 537
Senegal	417	488	368	236	296	237	221	366	499	360
Sierra Leone	...	...	3	3 F	155	...	1 728	294	381	314
Togo	71	65	87	203	42	433	129	24	16	30
34 Fishing area total	*4 039 F*	*4 789 F*	*5 292 F*	*5 813 F*	*8 314 F*	*11 650 F*	*12 869 F*	*10 349 F*	*9 981 F*	*10 715 F*
51 Bahrain	97	75	117	133	207	100	294	157	103	142
Eritrea	...	43	205	281	219	294	365	149	205	158
Kuwait	95	134	117	84	78	64	51	40 F	27	30 F
Qatar	178	151	111	157	157	143	73	181	230	102
Réunion	...	...	...	...	...	...	112	38	45	45 F
Seychelles	454	515	358	381	242	424	489	213	602	490
51 Fishing area total	*824*	*918*	*908*	*1 036*	*903*	*1 025*	*1 384*	*778 F*	*1 212*	*967 F*
57 Indonesia	12 575	10 865	10 971	13 351	11 737	12 885	14 120	13 650	14 965	15 390
Malaysia	286	234	138	127	149	241	115	239	356	284
57 Fishing area total	*12 861*	*11 099*	*11 109*	*13 478*	*11 886*	*13 126*	*14 235*	*13 889*	*15 321*	*15 674*
61 Russian Fed	-	15	-	-	-	-	-	-	-	-
61 Fishing area total	*-*	*15*	*-*	*-*	*-*	*-*	*-*	*-*	*-*	*-*
71 Fiji Islands	1 650	1 682 F	1 117	1 347	1 315	1 155	1 710	1 600 F	1 728	1 700 F
Indonesia	43 278	47 473	41 856	46 987	57 848	53 395	52 372	48 656	52 808	54 460
Malaysia	2 665	2 598	2 725	3 120	3 061	4 245	4 551	4 031	3 976	3 647
NewCaledonia	48	56	38	45	36	43	22	24	23	23 F
Singapore	135	368	310	321	289	238	154	122	80	63
71 Fishing area total	*47 776*	*52 177 F*	*46 046*	*51 820*	*62 549*	*59 076*	*58 809*	*54 433 F*	*58 615*	*59 893 F*
77 Cook Is	25 F	24 F	24 F	20 F	20 F	20 F	20 F	20 F	20 F	20 F
Nicaragua	1 600	...	...	...	...	...	2 100	1 620	2 410	1 596
USA	83	191	242	215	187	157	175	191	138	181
77 Fishing area total	*1 708 F*	*215 F*	*266 F*	*235 F*	*207 F*	*177 F*	*2 295 F*	*1 831 F*	*2 568 F*	*1 797 F*
87 Colombia	387	646	278	460	43	452	515	600 F	802	780 F
Peru	87	23	300	117	75	208	573	204	55	59
87 Fishing area total	*474*	*669*	*578*	*577*	*118*	*660*	*1 088*	*804 F*	*857*	*839 F*
Species total	*68 111 F*	*70 218 F*	*64 576 F*	*73 384 F*	*84 765 F*	*86 541 F*	*92 149 F*	*83 653 F*	*90 333 F*	*91 288 F*
Yellowtail snapper	**Vivaneau queue jaune**		**Rabirrubia**		***Ocyurus chrysurus***			**1,70(32)028,01**		**SNY**
31 Br Virgin Is	...	...	...	...	5	9	9	0	0	0
Cuba	539	592	592	1 176	727	457	409	408	413	370
Dominican Rp	273	671	248	793	529	190	234	249	356	125
Mexico	910	1 184	825	858	840	1 900	1 554	1 357	1 600	1 702
Venezuela	678	684	511	338	335	272	220	291	158	213
31 Fishing area total	*2 400*	*3 131*	*2 176*	*3 165*	*2 436*	*2 828*	*2 426*	*2 305*	*2 527*	*2 410*
41 Brazil	2 800 F	2 800 F	4 766	4 167	5 000	3 317	4 541	4 165	2 002	2 158
41 Fishing area total	*2 800 F*	*2 800 F*	*4 766*	*4 167*	*5 000*	*3 317*	*4 541*	*4 165*	*2 002*	*2 158*
Species total	*5 200 F*	*5 931 F*	*6 942*	*7 332*	*7 436*	*6 145*	*6 967*	*6 470*	*4 529*	*4 568*
Vermilion snapper	**Vivaneau ti-yeux**		**Pargo cunaro**		***Rhomboplites aurorubens***			**1,70(32)225,01**		**RPU**
31 USA	...	...	...	...	678	480	731	911	1 057	980
31 Fishing area total	*...*	*...*	*...*	*...*	*678*	*480*	*731*	*911*	*1 057*	*980*
Species total	*...*	*...*	*...*	*...*	*678*	*480*	*731*	*911*	*1 057*	*980*
Snappers, jobfishes nei	**Lutianidés nca**		**Lutjánidos nep**		***Lutjanidae***			**1,70(32)XXX,XX**		**SNX**
31 Antigua Barb	...	...	...	...	...	...	...	...	284	348
Aruba	...	...	50	60	60	50	50	45	45	45 F
Barbados	19	26	41	40	26	32	24	25	20	12
Bermuda	41	39	37	38	29	29	28	23	32	33
Colombia	24	36	75	55	35	0	-	-	-	-
Cuba	301	364	416	601	403	509	388	444	641	844
Dominican Rp	84	151	84	130	235	875	49	753	231	1 848
Grenada	40	17	22	32	25	28	36	48	56	80
Mexico	850	335	920	827	1 495	3 675	3 929	3 533	1 543	1 691

B-33

Miscellaneous coastal fishes — **Capture production by species, fishing areas and countries or areas**
Poissons côtiers divers — **Captures par espèces, zones de pêche et pays ou zones**
Peces costeros diversos — **Capturas por especies, áreas de pesca y países o áreas**

Species, Fishing area Espèce, Zone de pêche Especie, Area de pesca	1993 t	1994 t	1995 t	1996 t	1997 t	1998 t	1999 t	2000 t	2001 t	2002 t
Puerto Rico	...	...	...	...	...	...	566	762	744	512
USA	3 611	18 304	2 599	2 329	1 876	1 594	1 672	1 476	1 529	1 569
US Virgin Is	...	...	...	...	...	...	57	65 F	65 F	65 F
Venezuela	8 094	6 028	7 745	8 205	9 273	8 652	3 511	7 621	4 986	2 193
31 Fishing area total	*13 064*	*25 300*	*11 989*	*12 317*	*13 457*	*15 444*	*10 310*	*14 795 F*	*10 176 F*	*9 240 F*
41 Brazil	...	...	4 550	3 690	4 488	4 390	4 549	3 859	4 176	3 911
41 Fishing area total	...	...	*4 550*	*3 690*	*4 488*	*4 390*	*4 549*	*3 859*	*4 176*	*3 911*
47 South Africa	-	-	-	-	-	-	-	-	1	1
47 Fishing area total	-	-	-	-	-	-	-	-	*1*	*1*
51 Djibouti	65 F	69 F	80 F	80 F	80 F	80 F	80 F	80 F	80 F	80 F
Egypt	...	4 875	5 053	4 044	5 165	8 784	4 265	8 830	5 014	3 756
Eritrea	...	2	0	43	59	69	95	392	103	64
Kenya	129	116	112	147	144	106	151	120	177	177
Korea Rep	-	-	-	-	-	14	-	-	-	-
Mauritius	...	...	...	...	...	...	...	...	2 184	2 113
Oman	387	254	689	426	657	474	597	669	380	337
Saudi Arabia	3 019	3 054	1 662	1 704	2 148	2 302	2 022	1 647	2 092	1 988
Seychelles	904	601	572	466	464	602	825	552	703	611
South Africa	-	-	-	2	-	-	-	-	-	-
Untd Arab Em	3 082	3 241	3 438	3 474	3 713	3 725	3 818	1 718	1 977	1 187
Yemen	3 575	2 660	2 006	1 460	1 530 F	1 700 F	1 700 F	1 500 F	1 900 F	2 150 F
51 Fishing area total	*11 161 F*	*14 872 F*	*13 612 F*	*11 846 F*	*13 960 F*	*17 856 F*	*13 553 F*	*15 508 F*	*14 610 F*	*12 463 F*
57 Australia	363	2 977	533	660	952	1 019	1 834	2 717	2 943	3 207
Malaysia	474	426	320	189	128	81	414	147	196	312
Thailand	5 798	6 476	5 532	5 473	4 359	3 188	4 280	2 717	3 694	3 615
57 Fishing area total	*6 635*	*9 879*	*6 385*	*6 322*	*5 439*	*4 288*	*6 528*	*5 581*	*6 833*	*7 134*
61 China,H.Kong	1 037	436	406	430	439	304	250 F	300 F	330 F	300 F
China,Taiwan	3 889	3 607	1 595	2 334	1 408	808	467	352	385	409
61 Fishing area total	*4 926*	*4 043*	*2 001*	*2 764*	*1 847*	*1 112*	*717 F*	*652 F*	*715 F*	*709 F*
71 Australia	16	647	496	460	734	771	722	1 401	1 742	1 738
Guam	...	20	3	1	2	3	10	6	5	3
Kiribati	1 930	1 910	1 930	1 950	1 940	1 047	1 505	2 141	2 794	947
Malaysia	2 512	1 258	3 416	5 149	4 566	6 004	7 373	5 082	4 575	3 731
N Marianas	...	2	9	10	15	7	16	5	11	10
Palau	22	9	18	15	4	11	6	3	3	4
Philippines	19 100	16 353	16 332	15 005	18 378	15 442	19 192	19 154	14 371	13 630
Singapore	68	62	52	61	66	50	31	32	16	11
Thailand	13 569	11 008	12 679	9 180	8 469	11 559	8 961	5 242	9 113	9 217
71 Fishing area total	*37 217*	*31 269*	*34 935*	*31 831*	*34 174*	*34 894*	*37 816*	*33 066*	*32 630*	*29 291*
77 Amer Samoa	...	10	7	8	7	4	5	5	9	10
Costa Rica	701	664	550	352	244	478	526	417	347	382
El Salvador	358	418	334	197	117	252	230	282	481	530
Fr Polynesia	...	...	...	...	100	81	109	127	112	80
Mexico	1 899	1 645	1 260	1 463	1 154	1 178	1 116	1 463	1 581	1 537
Panama	4 597	4 525	5 170	3 857	4 659	5 358	10 433	8 480	22 322	26 642
USA	292	53	100	78	40	54	154	148	125	53
77 Fishing area total	*7 847*	*7 315*	*7 421*	*5 955*	*6 321*	*7 405*	*12 573*	*10 922*	*24 977*	*29 234*
81 Australia	...	...	...	...	...	...	301	293	288	252
81 Fishing area total	...	...	...	...	...	...	*301*	*293*	*288*	*252*
Species total	*80 850 F*	*92 678 F*	*80 893 F*	*74 725 F*	*79 686 F*	*85 389 F*	*86 347 F*	*84 676 F*	*94 406 F*	*92 235 F*
Golden threadfin bream	**Cohana doré**		**Baga dorada**		***Nemipterus virgatus***				**1,70(33)184,05**	**THG**
61 China	...	...	224 574	238 000	258 998	262 702	246 601	291 495	287 384	306 397
China,Taiwan	9 841	4 142	3 664	4 261	4 401	3 681	3 990	4 824	3 536	2 995
61 Fishing area total	*9 841*	*4 142*	*228 238*	*242 261*	*263 399*	*266 383*	*250 591*	*296 319*	*290 920*	*309 392*
Species total	*9 841*	*4 142*	*228 238*	*242 261*	*263 399*	*266 383*	*250 591*	*296 319*	*290 920*	*309 392*
Threadfin breams nei	**Cohanas nca**		**Bagas nep**		***Nemipterus spp***				**1,70(33)184,XX**	**THB**
51 Egypt	...	...	...	...	...	...	...	2 081	4 974	3 050
Eritrea	...	928	206	4	15	0	1 674	1 757	1 115	675
Kuwait	6	4	6	10	17	49	19	20 F	24	24 F
Pakistan	526	752	952	825	884	3 192	7 166	8 940	8 466	8 600
Saudi Arabia	-	21	221	131	87	66	49	144	153	100
Tanzania	191	123	206	937	250	195	500	400	400	600
Untd Arab Em	371	348	392	396	423	424	435	432	497	302
51 Fishing area total	*1 094*	*2 176*	*1 983*	*2 303*	*1 676*	*3 926*	*9 843*	*13 774 F*	*15 629*	*13 351 F*
57 Indonesia	6 938	6 613	6 051	6 495	6 378	6 754	7 271	7 578	7 682	7 790
Malaysia	10 886	9 271	13 291	10 457	9 369	10 917	10 187	9 113	6 891	7 209
Thailand	17 424	19 260	22 148	24 843	24 773	36 912	23 088	27 738	25 315	24 779
57 Fishing area total	*35 248*	*35 144*	*41 490*	*41 795*	*40 520*	*54 583*	*40 546*	*44 429*	*39 888*	*39 778*
61 China,H.Kong	27 921	26 113	23 772	19 568	20 024	19 449	14 000 F	17 000 F	18 800 F	18 300 F
61 Fishing area total	*27 921*	*26 113*	*23 772*	*19 568*	*20 024*	*19 449*	*14 000 F*	*17 000 F*	*18 800 F*	*18 300 F*
71 Indonesia	17 582	18 665	21 409	25 098	22 962	24 183	31 926	26 640	29 497	31 750
Malaysia	19 875	19 992	18 032	19 077	20 733	29 410	29 507	23 397	22 019	23 310

B-33 Miscellaneous coastal fishes / Poissons côtiers divers / Peces costeros diversos

Capture production by species, fishing areas and countries or areas / Captures par espèces, zones de pêche et pays ou zones / Capturas por especies, áreas de pesca y países o áreas

Species, Fishing area Espèce, Zone de pêche Especie, Area de pesca	1993 t	1994 t	1995 t	1996 t	1997 t	1998 t	1999 t	2000 t	2001 t	2002 t
Philippines	40 074	34 177	35 538	32 884	29 839	30 511	29 301	29 487	27 079	49 257
Singapore	123	305	255	209	239	158	128	96	48	33
Thailand	57 903	55 850	71 637	64 749	62 994	59 683	69 949	74 544	69 450	70 238
71 Fishing area total	*135 557*	*128 989*	*146 871*	*142 017*	*136 767*	*143 945*	*160 811*	*154 164*	*148 093*	*174 588*
Species total	*199 820*	*192 422*	*214 116*	*205 683*	*198 987*	*221 903*	*225 200 F*	*229 367 F*	*222 410 F*	*246 017 F*
Monocle breams	**Mamilas**		**Besugatos**		***Scolopsis spp***			**1,70(33)230,XX**		**MOB**
57 Malaysia	3	0	2	1	1	2	96	86	88	121
Thailand	1	0	0	0	0	0	0	-	-	-
57 Fishing area total	*4*	*0*	*2*	*1*	*1*	*2*	*96*	*86*	*88*	*121*
71 Malaysia	952	1 110	1 263	1 243	949	827	940	1 615	1 554	1 321
Thailand	72	...	77	98	344	644	54	94	210	212
71 Fishing area total	*1 024*	*1 110*	*1 340*	*1 341*	*1 293*	*1 471*	*994*	*1 709*	*1 764*	*1 533*
Species total	*1 028*	*1 110*	*1 342*	*1 342*	*1 294*	*1 473*	*1 090*	*1 795*	*1 852*	*1 654*
Threadfin and dwarf breams nei	**Cohanas, mamilas nca**		**Bagas, besugatos nep**		***Nemipteridae***			**1,70(33)XXX,XX**		**THD**
51 Oman	639	415	260	233	2 068	466	1 166	317	1 206	2 865
Qatar	11	5	5	13	15	3	0	0	0	0
South Africa	-	-	3	-	-	-	-	-	-	-
Yemen	3 592	3 718	3 170	2 504	2 620 F	3 000 F	2 900 F	2 600 F	3 300 F	3 700 F
51 Fishing area total	*4 242*	*4 138*	*3 438*	*2 750*	*4 703 F*	*3 469 F*	*4 066 F*	*2 917 F*	*4 506 F*	*6 565 F*
Species total	*4 242*	*4 138*	*3 438*	*2 750*	*4 703 F*	*3 469 F*	*4 066 F*	*2 917 F*	*4 506 F*	*6 565 F*
Ponyfishes(=Slipmouths)	**Sapsap**		**Motambos**		***Leiognathus spp***			**1,70(35)169,XX**		**POY**
57 Malaysia	164	38	34	101	78	86	84	161	71	609
57 Fishing area total	*164*	*38*	*34*	*101*	*78*	*86*	*84*	*161*	*71*	*609*
71 Fiji Islands	127	129 F	37	76	75	69	84	80 F	87	85 F
Malaysia	1 754	1 895	2 250	2 438	2 284	3 004	2 965	2 300	2 214	1 731
Singapore	57	141	83	75	63	52	47	32	23	9
71 Fishing area total	*1 938*	*2 165 F*	*2 370*	*2 589*	*2 422*	*3 125*	*3 096*	*2 412 F*	*2 324*	*1 825 F*
Species total	*2 102*	*2 203 F*	*2 404*	*2 690*	*2 500*	*3 211*	*3 180*	*2 573 F*	*2 395*	*2 434 F*
Ponyfishes(=Slipmouths) nei	**Sapsap nca**		**Motambos nep**		***Leiognathidae***			**1,70(35)XXX,XX**		**PON**
51 Eritrea	...	-	-	-	-	-	19	38	137	70
India	10 919	9 348	7 192	10 343	12 341	9 115	8 934	4 551	6 952	10 758
Untd Arab Em	852	1 109	723	731	781	784	804	780	...	...
51 Fishing area total	*11 771*	*10 457*	*7 915*	*11 074*	*13 122*	*9 899*	*9 757*	*5 369*	*7 089*	*10 828*
57 India	44 490	51 638	56 202	58 239	55 028	36 953	46 528	44 729	45 610	43 082
Indonesia	12 281	12 610	12 555	15 496	17 249	20 801	22 608	17 631	18 906	20 120
57 Fishing area total	*56 771*	*64 248*	*68 757*	*73 735*	*72 277*	*57 754*	*69 136*	*62 360*	*64 516*	*63 202*
71 Indonesia	40 519	44 852	53 665	55 905	72 154	58 731	68 611	51 881	68 851	74 890
Philippines	60 046	59 547	59 134	57 867	61 254	59 862	68 371	67 255	65 007	65 816
71 Fishing area total	*100 565*	*104 399*	*112 799*	*113 772*	*133 408*	*118 593*	*136 982*	*119 136*	*133 858*	*140 706*
Species total	*169 107*	*179 104*	*189 471*	*198 581*	*218 807*	*186 246*	*215 875*	*186 865*	*205 463*	*214 736*
Pigfish	**Goret mule**		**Corocoro burro**		***Orthopristis chrysoptera***			**1,70(36)032,01**		**PIG**
21 USA	-	-	1	-	1	1	-	0	0	0
21 Fishing area total	-	-	*1*	-	*1*	*1*	-	*0*	*0*	*0*
Species total	-	-	*1*	-	*1*	*1*	-	*0*	*0*	*0*
Cabinza grunt	**Cagna cabinza**		**Cabinza**		***Isacia conceptionis***			**1,70(36)163,01**		**GRP**
87 Chile	161	240	165	93	142	54	156	42	28	149
Peru	987	505	1 342	1 955	1 892	2 079	2 791	3 251	3 293	5 506
87 Fishing area total	*1 148*	*745*	*1 507*	*2 048*	*2 034*	*2 133*	*2 947*	*3 293*	*3 321*	*5 655*
Species total	*1 148*	*745*	*1 507*	*2 048*	*2 034*	*2 133*	*2 947*	*3 293*	*3 321*	*5 655*
Rubberlip grunt	**Diagramme gris**		**Burro chiclero**		***Plectorhinchus mediterraneus***			**1,70(36)207,05**		**GBR**
27 Portugal	...	...	...	...	...	...	251	77	12	13
27 Fishing area total	...	...	...	...	...	...	*251*	*77*	*12*	*13*
34 Gambia	43	39	102	156	101	160	170	107	124	365
Mauritania	20 F	30 F	20 F	20 F	20 F	10 F	...	9 F	...	...
Portugal	1 246	707	712	1 436	544	115	16	10	-	170
Spain	225 F	235 F	245 F	253	234	170	453	94	133	61
34 Fishing area total	*1 534 F*	*1 011 F*	*1 079 F*	*1 865 F*	*899 F*	*455 F*	*639*	*220 F*	*257*	*596*
37 Spain	-	-	-	4	5	4	-	-	2	13
37 Fishing area total	-	-	-	*4*	*5*	*4*	-	-	*2*	*13*
Species total	*1 534 F*	*1 011 F*	*1 079 F*	*1 869 F*	*904 F*	*459 F*	*890*	*297 F*	*271*	*622*

B-33 Miscellaneous coastal fishes
Poissons côtiers divers
Peces costeros diversos

Capture production by species, fishing areas and countries or areas
Captures par espèces, zones de pêche et pays ou zones
Capturas por especies, áreas de pesca y países o áreas

Species, Fishing area Espèce, Zone de pêche Especie, Area de pesca	1993 t	1994 t	1995 t	1996 t	1997 t	1998 t	1999 t	2000 t	2001 t	2002 t
Silver grunt	**Grondeur argenté**		**Corocoro plateado**		***Pomadasys argenteus***			**1,70(36)209,04**		**GRL**
51 Eritrea	...	...	...	0	0	3	...	472	199	152
51 Fishing area total	*...*	*...*	*...*	*0*	*0*	*3*	*...*	*472*	*199*	*152*
57 Malaysia	501	423	323	625	539	666	548	581	507	639
57 Fishing area total	*501*	*423*	*323*	*625*	*539*	*666*	*548*	*581*	*507*	*639*
71 Malaysia	756	1 239	852	1 019	617	2 178	1 687	919	1 086	830
71 Fishing area total	*756*	*1 239*	*852*	*1 019*	*617*	*2 178*	*1 687*	*919*	*1 086*	*830*
Species total	*1 257*	*1 662*	*1 175*	*1 644*	*1 156*	*2 847*	*2 235*	*1 972*	*1 792*	*1 621*
Bastard grunt	**Grondeur métis**		**Ronco mestizo**		***Pomadasys incisus***			**1,70(36)209,18**		**BGR**
34 Portugal	-	-	1	1	0	-	-	0	0	-
Spain	-	-	-	0	5	-	-	-	-	-
34 Fishing area total	*-*	*-*	*1*	*1*	*5*	*-*	*-*	*0*	*0*	*-*
Species total	*-*	*-*	*1*	*1*	*5*	*-*	*-*	*0*	*0*	*-*
Sompat grunt	**Grondeur sompat**		**Ronco sompat**		***Pomadasys jubelini***			**1,70(36)209,19**		**BUR**
34 Cameroon	...	...	534	533	400 F	337 F	345 F	304	48	342
Congo Rep	...	6	5 F	5 F	5 F	5 F	5 F	...	4 F	4 F
Côte dIvoire	220	233	235 F	247	143	227	236	231	148	202
Gambia	142	120	307	498	350	220	230	276	423	439
Ghana	...	...	...	...	434	303	87	95	694	343
GuineaBissau	...	47	20	153	160 F	130	100 F	100 F	100 F	100 F
Togo	4	2	0	3	43	31	35	11	1	5
34 Fishing area total	*366*	*408*	*1 101 F*	*1 439 F*	*1 535 F*	*1 253 F*	*1 038 F*	*1 017 F*	*1 418 F*	*1 435 F*
Species total	*366*	*408*	*1 101 F*	*1 439 F*	*1 535 F*	*1 253 F*	*1 038 F*	*1 017 F*	*1 418 F*	*1 435 F*
Bigeye grunt	**Lippu pelon**		**Burro ojón**		***Brachydeuterus auritus***			**1,70(36)263,03**		**GRB**
34 Congo Dem R	420 F	380 F	388 F	398 F	385 F	396 F	400 F	400 F	400 F	400 F
Congo Rep	480 F	122	126	125 F	115 F	100 F	100 F	95	90 F	90 F
Côte dIvoire	4 292	4 950	4 848	4 632	4 732	1 617	3 964	5 175	1 689	2 812
Gabon	150	150	200	...	...	...	...	...	...	...
Ghana	11 940	18 216	14 807	13 552	19 816	12 059	12 724	10 032	13 845	9 267
GuineaBissau	7 F	9 F	12 F	14 F	15 F	10 F	10 F	10 F	10 F	10 F
Lithuania	2 652	461	-	-	-	-	-	-	-	78
Nigeria	...	...	...	556	2 537	2 887	2 141	3 345	275	849
Russian Fed	1 727	393	1 282	-	-	36	-	-	-	-
Senegal	1 687	1 367	1 255	610	1 445	1 481	2 414	1 949	2 701	2 294
Sierra Leone	...	...	...	...	...	...	...	...	132	129
Togo	49	44	21	0	1	5	465	742	844	1 148
34 Fishing area total	*23 404 F*	*26 092 F*	*22 939 F*	*19 887 F*	*29 046 F*	*18 591 F*	*22 218 F*	*21 748 F*	*19 986 F*	*17 077 F*
47 Angola	17	75	124	146	228	262	167	191	1 176	4 078
Russian Fed	-	-	-	-	-	-	-	-	5	561
47 Fishing area total	*17*	*75*	*124*	*146*	*228*	*262*	*167*	*191*	*1 181*	*4 639*
Species total	*23 421 F*	*26 167 F*	*23 063 F*	*20 033 F*	*29 274 F*	*18 853 F*	*22 385 F*	*21 939 F*	*21 167 F*	*21 716 F*
Barred grunt	**Cagna rayée**		**Ronco canario**		***Conodon nobilis***			**1,70(36)395,01**		**BRG**
41 Brazil	700 F	500 F	339	172	118	114	84	39	53	82
41 Fishing area total	*700 F*	*500 F*	*339*	*172*	*118*	*114*	*84*	*39*	*53*	*82*
Species total	*700 F*	*500 F*	*339*	*172*	*118*	*114*	*84*	*39*	*53*	*82*
Grunts, sweetlips nei	**Grondeurs, diagrammes nca**		**Burros, roncos nep**		***Haemulidae (=Pomadasyidae)***			**1,70(36)XXX,XX**		**GRX**
27 Ireland	-	-	-	-	-	-	-	-	5	7
Portugal	94	61	28	28	45	55	44	17	9	9
Spain	351	52	304	476	205	261	100	198	163	67
27 Fishing area total	*445*	*113*	*332*	*504*	*250*	*316*	*144*	*215*	*177*	*83*
31 Antigua Barb	...	...	...	...	...	...	...	...	167	259
Bahamas	13	9	8	14	67	90	66	62	67	67 F
Colombia	75	100	115	224	12	6	1	5 F	5 F	5 F
Cuba	1 247	1 992	2 128	1 723	1 451	1 207	1 303	1 079	986	971
Dominican Rp	398	716	278	249	325	205	423	348	385	347
Grenada	1	0	2	1	1	1	3	1	1	2
Mexico	5 224	5 512	5 032	5 901	9 645	6 076	5 039	3 594	3 715	3 011
Puerto Rico	...	...	...	...	...	...	76	97	104	77
St Kitts Nev	...	...	1	10	1	3	3	0	1	3
USA	484	536	387	293	243	292	316	403	429	316
Venezuela	5 693	5 387	6 310	5 466	6 336	7 685	4 108	4 191	9 132	3 314
31 Fishing area total	*13 135*	*14 252*	*14 261*	*13 881*	*18 081*	*15 565*	*11 338*	*9 780 F*	*14 992 F*	*8 372 F*
34 Benin	183 F	206 F	200 F	228 F	311 F	296 F	269	180 F	222	365
Cameroon	17	...	-	-	-	-	-	-	-	-
Congo Rep	19 F	6	15	12 F	9 F	6 F	3 F	-	-	-
Gabon	531	508	897	780	853	1 728	1 062	819	1 050	1 054
Ghana	659	575	523	1 191	43	108	389	55	153	197
Guinea	240 F	240 F	341	193	198	185	1 598	430	363	360 F

B-33 Miscellaneous coastal fishes / Poissons côtiers divers / Peces costeros diversos

Capture production by species, fishing areas and countries or areas / Captures par espèces, zones de pêche et pays ou zones / Capturas por especies, áreas de pesca y países o áreas

Species, Fishing area Espèce, Zone de pêche Especie, Area de pesca	1993 t	1994 t	1995 t	1996 t	1997 t	1998 t	1999 t	2000 t	2001 t	2002 t
Korea Rep	3 363	-	5 620	4 969	-	-	-	-	538	510
Latvia	-	-	-	-	2	-	-	-	-	-
Liberia	13	180	196	105	99	85	216	102	180	180 F
Lithuania	150	4	-	-	-	-	-	-	-	-
Mauritania	10 F	10 F	10 F	10 F	10 F	10 F	...	2 F	...	...
Morocco	1 588	1 538	1 665	1 670	2 590	3 049	2 228	2 715	3 158	3 079
Nigeria	4 194	5 235	4 152	2 585	2 611	5 000	4 561	5 152	4 804	3 352
Portugal	994	90	83	158	42	14	3	0	-	2
Russian Fed	23	6	9	-	-	-	9	1	-	-
Sao Tome Prn	1 F	1 F	1 F	1	0	0	0	0	0	0
Senegal	3 495	5 418	2 121	11 187	11 283	7 828	10 902	7 403	8 048	6 864
Sierra Leone	784	781	800	800 F	1 086	433	413	1 357	1 975	499
Ukraine	25	-	-	-	-	-	-	132	6	166
34 Fishing area total	*16 289 F*	*14 798 F*	*16 633 F*	*23 889 F*	*19 137 F*	*18 742 F*	*21 653 F*	*18 348 F*	*20 497*	*16 628 F*
41 Brazil	...	...	1 812	2 081	1 936	2 000	2 382	3 024	3 429	3 537
41 Fishing area total	...	...	*1 812*	*2 081*	*1 936*	*2 000*	*2 382*	*3 024*	*3 429*	*3 537*
47 Angola	67	78	149	1 031	557	438	412	799	1 688	3 513
Korea Rep	5	-	-	-	-	-	-	-	-	-
Portugal	-	-	-	-	-	-	-	18	18	-
Russian Fed	-	-	-	-	-	2	-	-	8	1
South Africa	-	-	-	-	-	-	-	-	1	0
47 Fishing area total	*72*	*78*	*149*	*1 031*	*557*	*440*	*412*	*817*	*1 715*	*3 514*
51 Bahrain	177	231	282	253	223	355	325	297	236	121
Eritrea	...	61	469	367	13	45	594	536	435	198
Kenya	65	52	67	65	72	55	78	63	85	90
Korea Rep	-	-	-	563	-	130	20	25	2	-
Kuwait	104	145	195	180	163	174	245	210 F	191	200 F
Oman	478	526	1 122	803	1 111	713	627	1 747	541	954
Pakistan	5 417	4 849	5 537	5 268	6 010	6 221	8 147	9 961	9 752	9 850
Qatar	731	622	606	653	542	583	433	789	900	673
Saudi Arabia	351	489	814	1 063	1 014	779	846	1 350	1 206	1 390
South Africa	-	5	2	-	6	11	5 F	1	3	3
Untd Arab Em	1 387	1 380	1 768	1 787	1 910	1 916	1 964	4 196	4 828	4 541
Yemen	754	1 147	1 813	1 318	1 380 F	1 600 F	1 500 F	1 350 F	1 700 F	1 900 F
51 Fishing area total	*9 464*	*9 507*	*12 675*	*12 320*	*12 444 F*	*12 582 F*	*14 784 F*	*20 525 F*	*19 879 F*	*19 920 F*
57 Indonesia	2 836	2 766	3 717	4 339	4 052	4 253	4 544	4 738	3 840	4 030
Malaysia	139	50	38	20	27	21	9	39	81	72
57 Fishing area total	*2 975*	*2 816*	*3 755*	*4 359*	*4 079*	*4 274*	*4 553*	*4 777*	*3 921*	*4 102*
61 Japan	...	...	5 555	5 303	4 801	5 703	5 282	5 136	4 995	5 606
Russian Fed	-	32	-	-	-	-	3	-	-	-
61 Fishing area total	...	*32*	*5 555*	*5 303*	*4 801*	*5 703*	*5 285*	*5 136*	*4 995*	*5 606*
71 Indonesia	9 669	9 525	9 492	10 515	10 605	10 950	10 804	12 630	13 707	14 340
Malaysia	915	730	978	1 093	1 047	1 057	1 548	1 468	1 336	1 002
Singapore	64	39	33	53	45	27	25	18	18	16
71 Fishing area total	*10 648*	*10 294*	*10 503*	*11 661*	*11 697*	*12 034*	*12 377*	*14 116*	*15 061*	*15 358*
77 Mexico	1 108	1 169	604	484	868	1 276	375	471	506	634
77 Fishing area total	*1 108*	*1 169*	*604*	*484*	*868*	*1 276*	*375*	*471*	*506*	*634*
87 Ecuador	...	...	...	...	...	1 091	500	...	...	...
Peru	416	185	890	194	379	254	318	183	419	526
87 Fishing area total	*416*	*185*	*890*	*194*	*379*	*1 345*	*818*	*183*	*419*	*526*
Species total	*54 552 F*	*53 244 F*	*67 169 F*	*75 707 F*	*74 229 F*	*74 277 F*	*74 121 F*	*77 392 F*	*85 591 F*	*78 280 F*
Corvina	**...B**		**...C**			***Sciaena gilberti***			**1,70(37)009,02**	**IAG**
87 Chile	2 150	1 868	1 239	1 179	1 350	1 069	747	1 052	1 033	733
Peru	4 098	4 275	4 353	7 920	2 215	5 085	5 695	3 692	3 295	5 318
87 Fishing area total	*6 248*	*6 143*	*5 592*	*9 099*	*3 565*	*6 154*	*6 442*	*4 744*	*4 328*	*6 051*
Species total	*6 248*	*6 143*	*5 592*	*9 099*	*3 565*	*6 154*	*6 442*	*4 744*	*4 328*	*6 051*
Brown meagre	**Corb commun**		**Corvallo**			***Sciaena umbra***			**1,70(37)009,21**	**CBM**
34 Korea Rep	-	-	234	208	...	...	...	...	...	...
34 Fishing area total	-	-	*234*	*208*	...	...	...	...	...	...
37 Albania	...	...	2	6	2	-	-	-	10	8
Cyprus	2	2	0	2	5	3	7	4	3	-
Tunisia	116	162	90	164	180	192	114	104	138	109
Turkey	233	143	43	50	35	45	65	20	20	39
37 Fishing area total	*351*	*307*	*135*	*222*	*222*	*240*	*186*	*128*	*171*	*156*
Species total	*351*	*307*	*369*	*430*	*222*	*240*	*186*	*128*	*171*	*156*
Peruvian weakfish	**Acoupa du Pérou**		**Corvinata ayanque**			***Cynoscion analis***			**1,70(37)016,04**	**WEP**
87 Chile	21	52	249	11	12	18	19	14	14	30
Colombia	193	336	255	401	284	374	352	330	317	310 F
Peru	9 676	5 248	8 902	7 475	5 501	10 795	8 558	5 995	4 107	3 147
87 Fishing area total	*9 890*	*5 636*	*9 406*	*7 887*	*5 797*	*11 187*	*8 929*	*6 339*	*4 438*	*3 487 F*

B-33 Miscellaneous coastal fishes / Poissons côtiers divers / Peces costeros diversos

Capture production by species, fishing areas and countries or areas
Captures par espèces, zones de pêche et pays ou zones
Capturas por especies, áreas de pesca y países o áreas

Species, Fishing area Espèce, Zone de pêche Especie, Area de pesca	1993 t	1994 t	1995 t	1996 t	1997 t	1998 t	1999 t	2000 t	2001 t	2002 t
Species total	*9 890*	*5 636*	*9 406*	*7 887*	*5 797*	*11 187*	*8 929*	*6 339*	*4 438*	*3 487 F*
Spotted weakfish	**Acoupa pintade**		**Corvinata pintada**		***Cynoscion nebulosus***				**1,70(37)016,09**	**SWF**
21 USA	151	19	206	86	78	79	202	119	36	39
21 Fishing area total	*151*	*19*	*206*	*86*	*78*	*79*	*202*	*119*	*36*	*39*
31 Mexico	1 716	1 253	3 107	3 213	3 448	6 600	6 549	6 227	4 799	5 957
USA	910	1 043	637	397	349	186	193	140	116	131
31 Fishing area total	*2 626*	*2 296*	*3 744*	*3 610*	*3 797*	*6 786*	*6 742*	*6 367*	*4 915*	*6 088*
Species total	*2 777*	*2 315*	*3 950*	*3 696*	*3 875*	*6 865*	*6 944*	*6 486*	*4 951*	*6 127*
Squeteague(=Gray weakfish)	**Acoupa royal**		**Corvinata real**		***Cynoscion regalis***				**1,70(37)016,14**	**STG**
21 USA	1 977	1 093	2 578	2 729	2 697	3 212	2 736	2 329	2 119	2 054
21 Fishing area total	*1 977*	*1 093*	*2 578*	*2 729*	*2 697*	*3 212*	*2 736*	*2 329*	*2 119*	*2 054*
31 USA	1 213	1 674	517	532	620	609	410	109	154	108
31 Fishing area total	*1 213*	*1 674*	*517*	*532*	*620*	*609*	*410*	*109*	*154*	*108*
Species total	*3 190*	*2 767*	*3 095*	*3 261*	*3 317*	*3 821*	*3 146*	*2 438*	*2 273*	*2 162*
Striped weakfish	**Acoupa rayé**		**Corvinata pescadilla**		***Cynoscion striatus***				**1,70(37)016,17**	**WKS**
41 Argentina	6 239	15 661	19 218	18 987	24 132	17 108	11 107	9 433	11 844	11 412
Uruguay	6 962	10 323	13 417	12 654	15 186	15 286	8 481	13 440	10 890	8 958
41 Fishing area total	*13 201*	*25 984*	*32 635*	*31 641*	*39 318*	*32 394*	*19 588*	*22 873*	*22 734*	*20 370*
Species total	*13 201*	*25 984*	*32 635*	*31 641*	*39 318*	*32 394*	*19 588*	*22 873*	*22 734*	*20 370*
Weakfishes nei	**Acoupas nca**		**Corvinatas nep**		***Cynoscion spp***				**1,70(37)016,XX**	**WKX**
21 USA	-	-	-	-	14	-	11	-	-	-
21 Fishing area total	-	-	-	-	*14*	-	*11*	-	-	-
31 Dominican Rp	26	47	83	56	14	14	7	10	10	136
Nicaragua	...	...	...	...	...	...	...	42	87	56
USA	140	190	91	76	60	56	84	74	53	66
Venezuela	18 541	19 078	20 045	13 879	10 503	13 481	6 303	12 716	10 734	10 803
31 Fishing area total	*18 707*	*19 315*	*20 219*	*14 011*	*10 577*	*13 551*	*6 394*	*12 842*	*10 884*	*11 061*
41 Brazil	30 000 F	28 000 F	27 075	26 470	25 960	26 677	39 332	43 523	2 637	1 810
41 Fishing area total	*30 000 F*	*28 000 F*	*27 075*	*26 470*	*25 960*	*26 677*	*39 332*	*43 523*	*2 637*	*1 810*
77 Mexico	2 643	2 788	2 281	3 402	4 042	4 602	5 507	2 494	5 356	5 596
Nicaragua	...	...	...	...	...	...	...	55	200	77
77 Fishing area total	*2 643*	*2 788*	*2 281*	*3 402*	*4 042*	*4 602*	*5 507*	*2 549*	*5 556*	*5 673*
Species total	*51 350 F*	*50 103 F*	*49 575*	*43 883*	*40 593*	*44 830*	*51 244*	*58 914*	*19 077*	*18 544*
Whitemouth croaker	**Tambour rayé**		**Corvinón rayado**		***Micropogonias furnieri***				**1,70(37)038,02**	**CKM**
31 Venezuela	6 697	6 058	7 043	6 065	3 694	4 871	1 900	3 262	6 848	2 892
31 Fishing area total	*6 697*	*6 058*	*7 043*	*6 065*	*3 694*	*4 871*	*1 900*	*3 262*	*6 848*	*2 892*
41 Argentina	12 709	18 261	29 989	23 514	26 108	9 451	6 641	5 296	4 479	3 652
Brazil	25 000 F	23 000 F	22 024	23 374	23 311	27 927	21 926	28 797	39 266	42 331
Uruguay	25 804	29 012	29 513	25 745	23 744	22 253	14 650	24 146	27 322	26 665
41 Fishing area total	*63 513 F*	*70 273 F*	*81 526*	*72 633*	*73 163*	*59 631*	*43 217*	*58 239*	*71 067*	*72 648*
Species total	*70 210 F*	*76 331 F*	*88 569*	*78 698*	*76 857*	*64 502*	*45 117*	*61 501*	*77 915*	*75 540*
Atlantic croaker	**Tambour brésilien**		**Corvinón brasileño**		***Micropogonias undulatus***				**1,70(37)038,04**	**CKA**
21 USA	3 270	2 787	6 462	8 250	11 038	10 438	11 451	11 632	12 410	11 460
21 Fishing area total	*3 270*	*2 787*	*6 462*	*8 250*	*11 038*	*10 438*	*11 451*	*11 632*	*12 410*	*11 460*
31 USA	687	2 205	551	799	1 396	1 084	729	506	608	302
31 Fishing area total	*687*	*2 205*	*551*	*799*	*1 396*	*1 084*	*729*	*506*	*608*	*302*
Species total	*3 957*	*4 992*	*7 013*	*9 049*	*12 434*	*11 522*	*12 180*	*12 138*	*13 018*	*11 762*
Croakers nei	**Tambours nca**		**Corvinones nep**		***Micropogonias spp***				**1,70(37)038,XX**	**CRX**
77 Mexico	3 666	3 867	3 432	5 036	4 966	5 476	4 602	3 705	3 728	2 798
77 Fishing area total	*3 666*	*3 867*	*3 432*	*5 036*	*4 966*	*5 476*	*4 602*	*3 705*	*3 728*	*2 798*
87 Chile	104	113	111	128	101	9	0	0	0	0
Colombia	168	192	105	165	179	156	173	170 F	237	230 F
Peru	1 369	602	704	733	606	1 068	616	955	825	1 184
87 Fishing area total	*1 641*	*907*	*920*	*1 026*	*886*	*1 233*	*789*	*1 125 F*	*1 062*	*1 414 F*
Species total	*5 307*	*4 774*	*4 352*	*6 062*	*5 852*	*6 709*	*5 391*	*4 830 F*	*4 790*	*4 212 F*
Northern kingfish	**Bourrugue renard**		**Lambe zorro**		***Menticirrhus saxatilis***				**1,70(37)039,07**	**KGF**
21 USA	41	34	34	49	21	17	13	28	39	8
21 Fishing area total	*41*	*34*	*34*	*49*	*21*	*17*	*13*	*28*	*39*	*8*

B-33 Miscellaneous coastal fishes / Poissons côtiers divers / Peces costeros diversos

Capture production by species, fishing areas and countries or areas
Captures par espèces, zones de pêche et pays ou zones
Capturas por especies, áreas de pesca y países o áreas

Species, Fishing area Espèce, Zone de pêche Especie, Area de pesca	1993 t	1994 t	1995 t	1996 t	1997 t	1998 t	1999 t	2000 t	2001 t	2002 t
Species total	*41*	*34*	*34*	*49*	*21*	*17*	*13*	*28*	*39*	*8*
Gulf kingcroaker	**Bourrugue du Golfe**		**Lambe verrugato**		***Menticirrhus littoralis***			**1,70(37)039,08**		**KGG**
31 Mexico	582	869	948	1 070	1 220	1 824	1 576	606	484	630
St Lucia	...	6	20	...	...	...	...	...	...	...
USA	-	-	-	-	223	237	246	355	206	156
31 Fishing area total	*582*	*875*	*968*	*1 070*	*1 443*	*2 061*	*1 822*	*961*	*690*	*786*
Species total	*582*	*875*	*968*	*1 070*	*1 443*	*2 061*	*1 822*	*961*	*690*	*786*
Kingcroakers nei	**Bourrugues nca**		**Lambes nep**		***Menticirrhus spp***			**1,70(37)039,XX**		**KIX**
41 Brazil	...	...	1 330	1 144	1 303	1 395	1 058	1 348	1 951	2 042
Uruguay	-	-	-	-	-	7	4	4	6	5
41 Fishing area total	...	...	*1 330*	*1 144*	*1 303*	*1 402*	*1 062*	*1 352*	*1 957*	*2 047*
Species total	...	...	*1 330*	*1 144*	*1 303*	*1 402*	*1 062*	*1 352*	*1 957*	*2 047*
Shi drum	**Ombrine côtière**		**Verrugato fusco**		***Umbrina cirrosa***			**1,70(37)070,01**		**COB**
34 Greece	97	29	63	86	52	43	83	93	57	54
Togo	8	0	0	0	0	0	-	-	-	-
34 Fishing area total	*105*	*29*	*63*	*86*	*52*	*43*	*83*	*93*	*57*	*54*
37 Albania	...	...	...	...	...	...	...	...	...	10
Greece	23	87	87	74	26	66	53	58	46	45 F
Italy	555	762	956	495	351	138	138	158	259	156
Slovenia	-	-	-	-	-	-	-	-	-	1
Tunisia	32	30	29	25	51	32	40	28	33	42
Turkey	748	786	91	116	65	150	155	45	55	70
37 Fishing area total	*1 358*	*1 665*	*1 163*	*710*	*493*	*386*	*386*	*289*	*393*	*324 F*
Species total	*1 463*	*1 694*	*1 226*	*796*	*545*	*429*	*469*	*382*	*450*	*378 F*
Argentine croaker	**Ombrine d'Argentine**		**Verrugato pargo**		***Umbrina canosai***			**1,70(37)070,03**		**CKY**
41 Argentina	84	884	3 561	6 825	2 753	1 956	1 410	421	2 083	1 539
Brazil	11 000 F	9 000 F	9 268	4 983	3 358	2 216	4 849	7 729	12 757	14 711
Uruguay	903	1 576	1 708	1 210	547	1 112	1 401	1 071	1 416	1 546
41 Fishing area total	*11 987 F*	*11 460 F*	*14 537*	*13 018*	*6 658*	*5 284*	*7 660*	*9 221*	*16 256*	*17 796*
Species total	*11 987 F*	*11 460 F*	*14 537*	*13 018*	*6 658*	*5 284*	*7 660*	*9 221*	*16 256*	*17 796*
Canary drum (=Baardman)	**Ombrine bronze**		**Verrugato de Canarias**		***Umbrina canariensis***			**1,70(37)070,09**		**UCA**
27 France	1	1	1	1	-	-	3	7	7	7
Spain	-	-	-	-	-	-	-	-	4	-
27 Fishing area total	*1*	*1*	*1*	*1*	-	-	*3*	*7*	*11*	*7*
34 Mauritania	5 F	5 F	5 F	10 F	10 F	10 F	6 F	2 F	...	...
34 Fishing area total	*5 F*	*5 F*	*5 F*	*10 F*	*10 F*	*10 F*	*6 F*	*2 F*	...	...
47 Angola	...	...	...	...	...	...	830	...	2 202	...
South Africa	-	-	-	-	-	-	-	-	5	13
47 Fishing area total	...	...	...	...	...	...	*830*	...	*2 207*	*13*
Species total	*6 F*	*6 F*	*6 F*	*11 F*	*10 F*	*10 F*	*839 F*	*9 F*	*2 218*	*20*
King weakfish	**Acoupa chasseur**		**Pescadilla real**		***Macrodon ancylodon***			**1,70(37)096,01**		**WKK**
41 Argentina	-	-	149	380	224	167	515	44	64	90
Brazil	...	...	3 677	5 143	6 006	6 708	4 278	5 475	2 589	4 163
Uruguay	402	1 082	1 519	1 743	1 395	2 354	802	1 123	1 486	2 008
41 Fishing area total	*402*	*1 082*	*5 345*	*7 266*	*7 625*	*9 229*	*5 595*	*6 642*	*4 139*	*6 261*
Species total	*402*	*1 082*	*5 345*	*7 266*	*7 625*	*9 229*	*5 595*	*6 642*	*4 139*	*6 261*
Meagre	**Maigre commun**		**Corvina**		***Argyrosomus regius***			**1,70(37)106,01**		**MGR**
01 Egypt	76	136	127	168	128	106	92	201	51	67
01 Fishing area total	*76*	*136*	*127*	*168*	*128*	*106*	*92*	*201*	*51*	*67*
27 France	134	100	40	250	409	457	349	189	162	153
Portugal	-	-	-	-	0	3	3	4	6	36
Russian Fed	-	-	-	-	-	-	-	5	-	-
27 Fishing area total	*134*	*100*	*40*	*250*	*409*	*460*	*352*	*198*	*168*	*189*
34 Côte dIvoire	9	15	15 F	10 F	10 F	5 F	3	1	1	1
Gambia	...	27	50	125	72	8	10	22	33	645
GuineaBissau	...	290	222	482	500 F	430 F	350 F	350 F	350 F	350 F
Mauritania	390 F	450 F	500 F	480 F	580 F	510 F	600 F	600 F	600 F	600 F
Morocco	926	1 381	904	868	1 540	1 263	2 473	1 755	1 523	2 047
Portugal	0	0	-	0	-	-	-	-	24	-
Spain	170 F	150 F	130 F	-	-	-	-	-	-	-
34 Fishing area total	*1 495 F*	*2 313 F*	*1 821 F*	*1 965 F*	*2 702 F*	*2 216 F*	*3 436 F*	*2 728 F*	*2 531 F*	*3 643 F*
37 Egypt	545	547	541	908	615	768	882	575	987	1 305

B-33

Miscellaneous coastal fishes — **Capture production by species, fishing areas and countries or areas**
Poissons côtiers divers — **Captures par espèces, zones de pêche et pays ou zones**
Peces costeros diversos — **Capturas por especies, áreas de pesca y países o áreas**

Species, Fishing area Espèce, Zone de pêche Especie, Area de pesca	1993 t	1994 t	1995 t	1996 t	1997 t	1998 t	1999 t	2000 t	2001 t	2002 t
Israel	47	80	80	33	2	2	9	288	223	273
Malta	0	-	-	-	-	-	-	-	-	-
Morocco	0	3	0	1	2	0	-	30	15	0
Palest, O.T.	...	...	...	39	7	4	5 F	5 F	5 F	3
Turkey	353	295	290	71	40	30	65	70	50	63
37 Fishing area total	*945*	*925*	*911*	*1 052*	*666*	*804*	*961 F*	*968 F*	*1 280 F*	*1 644*
51 Egypt	-	158	45	0	0	19	82	-	-	-
51 Fishing area total	-	*158*	*45*	*0*	*0*	*19*	*82*	-	-	-
Species total	*2 650 F*	*3 632 F*	*2 944 F*	*3 435 F*	*3 905 F*	*3 605 F*	*4 923 F*	*4 095 F*	*4 030 F*	*5 543 F*
Southern meagre(=Mulloway)	**Maigre du Sud**		**Corvina del Sur**		***Argyrosomus hololepidotus***				**1,70(37)106,06**	**KOB**
47 Angola	...	...	...	...	...	...	259	...	2 540	...
Namibia	664	595	496	464	184	424	273	409	596	288
Portugal	-	-	-	-	-	-	-	-	1	-
South Africa	1 052	765	772	829	748	690	671	600 F	597	323
47 Fishing area total	*1 716*	*1 360*	*1 268*	*1 293*	*932*	*1 114*	*1 203*	*1 009 F*	*3 734*	*611*
51 South Africa	1	15	22	21	17	...	...	19	20	27
51 Fishing area total	*1*	*15*	*22*	*21*	*17*	...	...	*19*	*20*	*27*
57 Australia	22	26	26	44	44	44	96	286	363	362
57 Fishing area total	*22*	*26*	*26*	*44*	*44*	*44*	*96*	*286*	*363*	*362*
71 Australia	-	54	45	44	45	48	49	93	142	132
71 Fishing area total	-	*54*	*45*	*44*	*45*	*48*	*49*	*93*	*142*	*132*
81 Australia	148	140	140	95	88	91	94	79	66	59
81 Fishing area total	*148*	*140*	*140*	*95*	*88*	*91*	*94*	*79*	*66*	*59*
Species total	*1 887*	*1 595*	*1 501*	*1 497*	*1 126*	*1 297*	*1 442*	*1 486 F*	*4 325*	*1 191*
Geelbek croaker	**Téraglin**		**Corvinata prieta**		***Atractoscion aequidens***				**1,70(37)108,01**	**AWE**
47 South Africa	579	444	353	346	474	605	415	380 F	379	294
47 Fishing area total	*579*	*444*	*353*	*346*	*474*	*605*	*415*	*380 F*	*379*	*294*
51 South Africa	-	11	17	15	18	...	15 F	...	...	12
51 Fishing area total	-	*11*	*17*	*15*	*18*	...	*15 F*	...	...	*12*
81 Australia	47	27	27	20	27	27	29	32	36	20
81 Fishing area total	*47*	*27*	*27*	*20*	*27*	*27*	*29*	*32*	*36*	*20*
Species total	*626*	*482*	*397*	*381*	*519*	*632*	*459 F*	*412 F*	*415*	*326*
White weakfish	**Acoupa blanc**		**Corvinata blanca**		***Atractoscion nobilis***				**1,70(37)108,03**	**WEW**
77 USA	44	35	33	46	28	71	112	101	124	194
77 Fishing area total	*44*	*35*	*33*	*46*	*28*	*71*	*112*	*101*	*124*	*194*
Species total	*44*	*35*	*33*	*46*	*28*	*71*	*112*	*101*	*124*	*194*
White croaker	**Courbine blanche**		**Roncador blanco**		***Genyonemus lineatus***				**1,70(37)147,01**	**KIC**
77 USA	296	205	256	242	167	64	74	105	137	97
77 Fishing area total	*296*	*205*	*256*	*242*	*167*	*64*	*74*	*105*	*137*	*97*
Species total	*296*	*205*	*256*	*242*	*167*	*64*	*74*	*105*	*137*	*97*
Boe drum	**Courbine pélin**		**Bombache boe**		***Pteroscion peli***				**1,70(37)166,07**	**DRS**
34 Benin	...	...	...	...	...	...	61	40 F	115	137
Congo Rep	942 F	566 F	637	630 F	620 F	600 F	600 F	602	600 F	600 F
Portugal	-	5	350	582	136	101	18	2	0	17
Senegal	1 227	1 802	6 833	667	527	489	581	472	444	374
Sierra Leone	...	...	...	...	8	0	...	10	-	-
34 Fishing area total	*2 169 F*	*2 373 F*	*7 820*	*1 879 F*	*1 291 F*	*1 190 F*	*1 260 F*	*1 126 F*	*1 159 F*	*1 128 F*
Species total	*2 169 F*	*2 373 F*	*7 820*	*1 879 F*	*1 291 F*	*1 190 F*	*1 260 F*	*1 126 F*	*1 159 F*	*1 128 F*
Tigertooth croaker	**Grande verrue tigre**		**Bombache tigre mayor**		***Otolithes ruber***				**1,70(37)186,03**	**LKR**
51 South Africa	-	-	-	-	-	13	10 F	...	...	...
51 Fishing area total	-	-	-	-	-	*13*	*10 F*	...	...	...
Species total	-	-	-	-	-	*13*	*10 F*	...	...	...
Peruvian banded croaker	**Bourrugue coco**		**Lambe coco**		***Paralonchurus peruanus***				**1,70(37)194,02**	**PDR**
87 Peru	7 550	3 788	5 543	4 263	2 737	4 363	6 063	5 729	4 167	1 886
87 Fishing area total	*7 550*	*3 788*	*5 543*	*4 263*	*2 737*	*4 363*	*6 063*	*5 729*	*4 167*	*1 886*
Species total	*7 550*	*3 788*	*5 543*	*4 263*	*2 737*	*4 363*	*6 063*	*5 729*	*4 167*	*1 886*
Black drum	**Grand tambour**		**Corvinón negro**		***Pogonias cromis***				**1,70(37)210,02**	**BDM**
21 USA	65	102	32	39	72	41	88	52	39	23

B-33 Miscellaneous coastal fishes / Poissons côtiers divers / Peces costeros diversos

Capture production by species, fishing areas and countries or areas / Captures par espèces, zones de pêche et pays ou zones / Capturas por especies, áreas de pesca y países o áreas

Species, Fishing area Espèce, Zone de pêche Especie, Area de pesca	1993 t	1994 t	1995 t	1996 t	1997 t	1998 t	1999 t	2000 t	2001 t	2002 t
21 Fishing area total	*65*	*102*	*32*	*39*	*72*	*41*	*88*	*52*	*39*	*23*
31 Mexico	250	204	166	134	362	392	179	207	143	165
USA	-	-	-	-	1 158	1 926	631	2 260	2 489	2 225
31 Fishing area total	*250*	*204*	*166*	*134*	*1 520*	*2 318*	*810*	*2 467*	*2 632*	*2 390*
41 Argentina	96	11	40	159	81	260	182	13	52	33
Brazil	770 F	770 F	38	1	-	-	-	-	-	-
Uruguay	273	220	574	172	82	678	100	344	336	504
41 Fishing area total	*1 139 F*	*1 001 F*	*652*	*332*	*163*	*938*	*282*	*357*	*388*	*537*
77 Mexico	6	1	0	0	0	0	1	0	0	0
77 Fishing area total	*6*	*1*	*0*	*0*	*0*	*0*	*1*	*0*	*0*	*0*
Species total	*1 460 F*	*1 308 F*	*850*	*505*	*1 755*	*3 297*	*1 181*	*2 876*	*3 059*	*2 950*
Honnibe croaker	**Tambour honnibe**		**Corvina honnibe**			***Nibea mitsukurii***			**1,70(37)298,05**	**HOC**
61 Korea Rep	1 920	2 364	2 164	1 940	1 177	1 285	1 566	1 999	2 156	1 148
61 Fishing area total	*1 920*	*2 364*	*2 164*	*1 940*	*1 177*	*1 285*	*1 566*	*1 999*	*2 156*	*1 148*
Species total	*1 920*	*2 364*	*2 164*	*1 940*	*1 177*	*1 285*	*1 566*	*1 999*	*2 156*	*1 148*
Large yellow croaker	**Tambour à gros yeux**		**Corvina japonesa**			***Larimichthys croceus***			**1,70(37)303,04**	**LYC**
61 China	34 820	69 181	67 031	80 072	69 950	70 935	65 806	122 920	76 289	83 087
Korea Rep	20 680	26 613	22 599	22 225	1 565	294	270	354	426	325
61 Fishing area total	*55 500*	*95 794*	*89 630*	*102 297*	*71 515*	*71 229*	*66 076*	*123 274*	*76 715*	*83 412*
Species total	*55 500*	*95 794*	*89 630*	*102 297*	*71 515*	*71 229*	*66 076*	*123 274*	*76 715*	*83 412*
Yellow croaker	**...B**		**Verrugato de Manchuria**			***Larimichthys polyactis***			**1,70(37)303,05**	**CRY**
61 China	78 311	102 976	153 048	253 482	142 681	191 751	243 101	281 717	245 325	260 457
China,Taiwan	1 021	2 815	2 786	2 859	2 228	1 010	1 167	780	903	742
Japan	93	225	-	-	-	-	-	-	-	-
Korea Rep	31 119	37 488	25 719	23 695	22 101	15 011	13 490	19 630	7 938	10 941
61 Fishing area total	*110 544*	*143 504*	*181 553*	*280 036*	*167 010*	*207 772*	*257 758*	*302 127*	*254 166*	*272 140*
Species total	*110 544*	*143 504*	*181 553*	*280 036*	*167 010*	*207 772*	*257 758*	*302 127*	*254 166*	*272 140*
Spot croaker	**Tambour croca**		**Verrugato croca**			***Leiostomus xanthurus***			**1,70(37)363,01**	**SPT**
21 USA	1 837	1 940	1 982	1 738	1 975	2 354	1 715	2 159	1 877	1 784
21 Fishing area total	*1 837*	*1 940*	*1 982*	*1 738*	*1 975*	*2 354*	*1 715*	*2 159*	*1 877*	*1 784*
31 USA	1 526	2 042	1 539	816	1 098	1 005	884	982	1 214	703
31 Fishing area total	*1 526*	*2 042*	*1 539*	*816*	*1 098*	*1 005*	*884*	*982*	*1 214*	*703*
Species total	*3 363*	*3 982*	*3 521*	*2 554*	*3 073*	*3 359*	*2 599*	*3 141*	*3 091*	*2 487*
Red drum	**Tambour rouge**		**Corvinón ocelado**			***Sciaenops ocellatus***			**1,70(37)411,01**	**RDM**
21 USA	4	5	2	1	2	3	6	6	3	4
21 Fishing area total	*4*	*5*	*2*	*1*	*2*	*3*	*6*	*6*	*3*	*4*
Species total	*4*	*5*	*2*	*1*	*2*	*3*	*6*	*6*	*3*	*4*
Law croaker	**Otolithe gabo**		**Corvina reina**			***Pseudotolithus brachygnathus***			**1,70(37)457,01**	**CKL**
34 Gambia	357	216	449	355	225	264	270	454	856	794
Sierra Leone	357	356	364	360 F	393	90	48	357	227	61
34 Fishing area total	*714*	*572*	*813*	*715 F*	*618*	*354*	*318*	*811*	*1 083*	*855*
Species total	*714*	*572*	*813*	*715 F*	*618*	*354*	*318*	*811*	*1 083*	*855*
Cassava croaker	**Otolithe sénégalais**		**Corvina casava**			***Pseudotolithus senegalensis***			**1,70(37)457,02**	**PSS**
34 Cameroon	...	...	2 686	2 540	2 600 F	2 900 F	3 200 F	1 972	2 679	2 100
Côte dIvoire	30	...	...	5 F	10 F	20 F	30	39	39	26
Gambia	28	37	230	182	160	109	120	58	400	620
Liberia	79	40	20	...	11	...	...	...	...	...
Senegal	...	...	...	...	...	1 630	1 487	1 880	1 682	1 159
34 Fishing area total	*137*	*77*	*2 936*	*2 727 F*	*2 781 F*	*4 659 F*	*4 837 F*	*3 949*	*4 800*	*3 905*
Species total	*137*	*77*	*2 936*	*2 727 F*	*2 781 F*	*4 659 F*	*4 837 F*	*3 949*	*4 800*	*3 905*
Bobo croaker	**Otolithe bobo**		**Corvina bobo**			***Pseudotolithus elongatus***			**1,70(37)457,05**	**PSE**
34 Cameroon	1 005	3 500 F	4 436	4 371	4 400 F	4 400 F	1 900 F	3 400	3 218	1 883
Congo Rep	5 F	3 F	3 F	3 F	3 F	3 F	3 F	...	2 F	2 F
Côte dIvoire	4	4	0	5	3	4	5	...	4	1
Gabon	176	904	986	946	769	1 584	2 079	1 784	1 555	1 238
Gambia	92	50	328	242	214	163	170	138	120	494
Guinea	2 540 F	2 610 F	3 685	3 386	2 751	2 781	2 142	4 015	5 171	5 100 F
Liberia	100	40	7	109	27	...	...	...	...	...
Senegal	...	...	...	...	...	137	146	33	114	115
Sierra Leone	3 637	3 625	3 650	3 600 F	3 330	389	410	4 035	748	3 233
34 Fishing area total	*7 559 F*	*10 736 F*	*13 095 F*	*12 662 F*	*11 497 F*	*9 461 F*	*6 855 F*	*13 405*	*10 932 F*	*12 066 F*

B-33 Miscellaneous coastal fishes / Poissons côtiers divers / Peces costeros diversos

Capture production by species, fishing areas and countries or areas / Captures par espèces, zones de pêche et pays ou zones / Capturas por especies, áreas de pesca y países o áreas

Species, Fishing area Espèce, Zone de pêche Especie, Area de pesca	1993 t	1994 t	1995 t	1996 t	1997 t	1998 t	1999 t	2000 t	2001 t	2002 t
Species total	*7 559 F*	*10 736 F*	*13 095 F*	*12 662 F*	*11 497 F*	*9 461 F*	*6 855 F*	*13 405*	*10 932 F*	*12 066 F*
West African croakers nei	**Otolithes nca**		**Corvinas africanas nep**		***Pseudotolithus spp***				**1,70(37)457,XX**	**CKW**
34 Congo Rep	2 225 F	864	581	500 F	450 F	400 F	350 F	298	300 F	300 F
Côte dIvoire	604	807	653	551	466	545	378	356	407	498
Gabon	3 365	1 570	2 095	4 020	3 068	4 653	3 367	2 344	2 197	2 554
Gambia	769	202	414	473	412	327	340	629	504	49
Georgia	20 F	10 F	-	-	-	-	-	-	-	-
Ghana	1 301	1 104	698	1 128	1 995	962	937	739	1 070	1 067
Guinea	2 840 F	2 920 F	4 115	2 856	2 206	2 570	1 796	3 423	3 644	3 600 F
Liberia	800	646	1 008	364	510	433	1 025	327	210	220 F
Mauritania	10 F	10 F	10 F	10 F	10 F	96 F	...	9 F	...	...
Nigeria	8 114	12 013	9 264	14 544	11 762	15 035	14 591	12 604	15 084	11 922
Portugal	92	254	115	157	36	15	49	635	25	31
Russian Fed	22	15	146	-	15	32	13	13	18	-
Sierra Leone	587	585	677	680 F	1 210	687	869	2 014	1 011	1 470
Togo	61	51	15	22	16	11	6	37	43	66
Ukraine	2	...	...	...	...	28	12	87	5	23
34 Fishing area total	*20 812 F*	*21 051 F*	*19 791 F*	*25 305 F*	*22 156 F*	*25 794 F*	*23 733 F*	*23 515 F*	*24 518 F*	*21 800 F*
47 Angola	1 745	2 962	1 035	4 044	1 606	7 095	8 936	8 005	8 725	25 946
Russian Fed	2	-	-	-	-	1	-	-	19	25
47 Fishing area total	*1 747*	*2 962*	*1 035*	*4 044*	*1 606*	*7 096*	*8 936*	*8 005*	*8 744*	*25 971*
Species total	*22 559 F*	*24 013 F*	*20 826 F*	*29 349 F*	*23 762 F*	*32 890 F*	*32 669 F*	*31 520 F*	*33 262 F*	*47 771 F*
Blackmouth croaker	**Maigre noire**		**Corvina negra**		***Atrobucca nibe***				**1,70(37)553,01**	**CRL**
61 China,Taiwan	2 481	495	197	297	288	237	223	450	644	535
61 Fishing area total	*2 481*	*495*	*197*	*297*	*288*	*237*	*223*	*450*	*644*	*535*
Species total	*2 481*	*495*	*197*	*297*	*288*	*237*	*223*	*450*	*644*	*535*
Silver croaker	**Maigre argenté**		**Corvina plateada**		***Pennahia argentata***				**1,70(37)561,01**	**CRV**
61 China,Taiwan	7 289	8 467	7 369	8 599	6 085	5 089	4 940	4 180	3 796	3 394
61 Fishing area total	*7 289*	*8 467*	*7 369*	*8 599*	*6 085*	*5 089*	*4 940*	*4 180*	*3 796*	*3 394*
Species total	*7 289*	*8 467*	*7 369*	*8 599*	*6 085*	*5 089*	*4 940*	*4 180*	*3 796*	*3 394*
Croakers, drums nei	**Sciaenidés nca**		**Esciénidos nep**		***Sciaenidae***				**1,70(37)XXX,XX**	**CDX**
04 India	-	201	174	3 271	748	667	686	213	864	864
04 Fishing area total	*-*	*201*	*174*	*3 271*	*748*	*667*	*686*	*213*	*864*	*864*
27 Portugal	...	...	...	...	...	...	228	132	168	340
27 Fishing area total	*...*	*...*	*...*	*...*	*...*	*...*	*228*	*132*	*168*	*340*
31 Korea Rep	-	-	-	-	-	-	818	1 037	137	-
Mexico	0	0	0	1	2	183	179	91	663	114
31 Fishing area total	*0*	*0*	*0*	*1*	*2*	*183*	*997*	*1 128*	*800*	*114*
34 Bahamas	-	-	25	-	-	-	-	-	-	-
Belize	17	3	67	-	1	65	299	580	1	...
Cameroon	2 300	3 000 F	3 679	3 539	3 500 F	3 500 F	3 500 F	3 260	2 399	1 946
China	748	113	1 065	530	490	1 568	1 027	2 301	2 061	-
Congo Rep	27 F	16 F	15 F	10 F	10 F	5 F	5 F	0	0	0
Gambia	0	-	-	-	-	-	-	-	-	-
GuineaBissau	1 010 F	900 F	1 000 F	1 020 F	1 040 F	850 F	650 F	650 F	650 F	650 F
Honduras	184	32	192	91	130	19	73	19	62	-
Italy	1 372	1 372	-	-	-	-	-	-	-	-
Korea D P Rp	276	-	-	-	-	-	-	224	-	-
Korea Rep	4 187	7 479	5 119	4 793	12 986	9 924	13 036	13 614	14 750	15 152
Liberia	7	4	87	-	-	-	-	-	-	-
Lithuania	-	-	-	-	-	-	-	-	-	61
Mauritania	200 F	200 F	200 F	200 F	200 F	200 F	300 F	300 F	300 F	300 F
Nigeria	1 453	1 543	4 605	2 165	6 335	4 056	7 846	5 826	5 900	9 384
Panama	75	-	54	4	1	0	-	-	-	-
Portugal	805	310	60	110	52	30	11	4	6	36
Russian Fed	68	15	127	58	-	-	-	-	-	-
St Vincent	14	-	33	-	-	-	-	-	-	-
Sao Tome Prn	66 F	96 F	101 F	71	30	40 F	45 F	40 F	40 F	40 F
Senegal	4 649	5 363	2 115	8 047	7 620	5 572	3 295	3 199	3 700	3 405
Spain	-	-	-	117	-	166	219	10	9	348
Other nei	5	-	-	-	-	-	-	-	12	...
34 Fishing area total	*17 463 F*	*20 446 F*	*18 544 F*	*20 755 F*	*32 395 F*	*25 995 F*	*30 306 F*	*30 027 F*	*29 890 F*	*31 322 F*
37 Korea Rep	-	-	22	-	-	-	-	-	-	-
Spain	-	-	-	-	-	18	29	164	32	56
37 Fishing area total	*-*	*-*	*22*	*-*	*-*	*18*	*29*	*164*	*32*	*56*
41 Brazil	...	...	69	150	52	57	36	34	35 590	32 426
Italy	183	183	-	-	-	-	-	-	-	-
Korea Rep	203	80	20	-	-	60	710	654	270	-
Spain	-	-	-	-	-	-	-	-	-	8
41 Fishing area total	*386*	*263*	*89*	*150*	*52*	*117*	*746*	*688*	*35 860*	*32 434*

B-33 **Miscellaneous coastal fishes** / **Poissons côtiers divers** / **Peces costeros diversos**

Capture production by species, fishing areas and countries or areas
Captures par espèces, zones de pêche et pays ou zones
Capturas por especies, áreas de pesca y países o áreas

Species, Fishing area Espèce, Zone de pêche Especie, Area de pesca	1993 t	1994 t	1995 t	1996 t	1997 t	1998 t	1999 t	2000 t	2001 t	2002 t
47 Angola	45	33	5	-	12	70	91	29	594	1 332
China	-	-	-	-	-	-	-	64	-	-
Honduras	2	-	0	-	-	-	-	-	-	-
Korea Rep	1 464	2 937	4 853	4 004	4 625	3 049	2 802	2 202	1 965	2 217
Russian Fed	-	-	-	-	-	-	-	-	59	4
Spain	-	-	-	-	-	-	27	-	17	13
47 Fishing area total	*1 511*	*2 970*	*4 858*	*4 004*	*4 637*	*3 119*	*2 920*	*2 295*	*2 635*	*3 566*
51 India	247 198	277 827	242 660	250 080	270 961	233 160	280 556	235 744	212 845	236 623
Italy	274	274	-	-	-	-	-	-	-	-
Korea Rep	633	176	381	381	2 385	751	1 127	951	361	423
Kuwait	798	1 089	1 572	974	1 127	1 211	1 385	1 100 F	853	860 F
Oman	3 444	3 859	4 070	3 709	5 468	2 218	2 121	1 926	1 873	4 152
Pakistan	19 740	22 808	25 201	19 934	20 428	19 625	24 665	21 976	21 725	22 886
Portugal	-	-	-	-	-	-	-	2	1	-
South Africa	-	1	9	8	7	...	...	1	0	-
51 Fishing area total	*272 087*	*306 034*	*273 893*	*275 086*	*300 376*	*256 965*	*309 854*	*261 700 F*	*237 658*	*264 944 F*
57 India	28 554	30 639	44 743	46 207	41 505	41 413	36 510	32 116	34 709	35 907
Indonesia	8 031	9 713	10 698	9 802	8 505	9 817	11 827	12 095	12 092	12 810
Korea Rep	-	-	92	-	-	-	273	152	202	14
Malaysia	10 631	10 089	10 313	12 221	11 763	11 484	10 956	12 216	10 507	11 593
Thailand	15 773	15 396	19 750	24 643	24 913	29 750	20 975	23 490	22 825	22 342
57 Fishing area total	*62 989*	*65 837*	*85 596*	*92 873*	*86 686*	*92 464*	*80 541*	*80 069*	*80 335*	*82 666*
61 China	...	...	107 176	-	-	-	-	-	-	-
China,H.Kong	11 367	8 725	5 547	3 776	4 881	3 992	2 800 F	3 500 F	3 900 F	3 800 F
China,Taiwan	1 483	3 488	3 269	3 070	2 497	1 422	1 580	1 708	3 841	8 428
Japan	7 518	8 310	9 008	7 062	5 998	5 430	4 850	4 791	4 362	4 112
Korea Rep	71 872	70 259	70 394	57 274	67 804	68 218	81 690	40 314	28 486	23 767
61 Fishing area total	*92 240*	*90 782*	*195 394*	*71 182*	*81 180*	*79 062*	*90 920 F*	*50 313 F*	*40 589 F*	*40 107 F*
71 Indonesia	28 329	27 687	29 100	35 431	36 332	40 297	45 164	40 159	37 555	39 080
Korea Rep	8 087	10 885	8 282	5 881	6 180	5 606	5 401	4 789	4 072	6 223
Malaysia	6 683	7 684	7 079	8 129	8 605	10 986	11 232	11 223	18 253	10 744
Singapore	280	162	136	123	180	160	114	68	45	56
Thailand	4 760	3 933	4 195	5 118	5 048	3 896	15 616	16 456	9 637	9 746
71 Fishing area total	*48 139*	*50 351*	*48 792*	*54 682*	*56 345*	*60 945*	*77 527*	*72 695*	*69 562*	*65 849*
77 El Salvador	289	407	233	313	337	252	334	279	575	826
Korea Rep	1 935	809	793	1 320	1 301	1 404	7 325	3 030	441	174
Mexico	426	450	326	710	690	763	721	511	735	345
Panama	557	548	346	662	666	1 138	1 533	4 352	4 543	2 279
77 Fishing area total	*3 207*	*2 214*	*1 698*	*3 005*	*2 994*	*3 557*	*9 913*	*8 172*	*6 294*	*3 624*
81 Korea Rep	-	-	-	-	-	9	305	-	-	-
81 Fishing area total	*-*	*-*	*-*	*-*	*-*	*9*	*305*	*-*	*-*	*-*
87 Chile	38	37	31	27	25	50	39	25	15	5
Colombia	22	8	-	24	1	0	-	-	-	-
Ecuador	2 532	10 427	30 910	2 486	316	3 300	7 320	...	1 558	1 700 F
Korea Rep	-	-	-	-	-	-	673	820	105	92
87 Fishing area total	*2 592*	*10 472*	*30 941*	*2 537*	*342*	*3 350*	*8 032*	*845*	*1 678*	*1 797 F*
Species total	*500 614 F*	*549 570 F*	*660 001 F*	*527 546 F*	*565 757 F*	*526 451 F*	*613 004 F*	*508 441 F*	*506 365 F*	*527 683 F*
Largeeye breams	**...B**		**...C**		***Gymnocranius spp***				**1,70(38)155,XX**	**LBR**
61 China,H.Kong	270	211	196	208	170	143	100 F	120 F	130 F	120 F
61 Fishing area total	*270*	*211*	*196*	*208*	*170*	*143*	*100 F*	*120 F*	*130 F*	*120 F*
71 Fiji Islands	13	13 F	19	20	22	25	28	25 F	30	30 F
71 Fishing area total	*13*	*13 F*	*19*	*20*	*22*	*25*	*28*	*25 F*	*30*	*30 F*
Species total	*283*	*224 F*	*215*	*228*	*192*	*168*	*128 F*	*145 F*	*160 F*	*150 F*
Atlantic emperor	**Empereur atlantique**		**Emperador atlántico**		***Lethrinus atlanticus***				**1,70(38)172,04**	**LTN**
34 Guinea	...	...	137	231	334	380	...	...	...	...
34 Fishing area total	*...*	*...*	*137*	*231*	*334*	*380*	*...*	*...*	*...*	*...*
Species total	*...*	*...*	*137*	*231*	*334*	*380*	*...*	*...*	*...*	*...*
Emperors(=Scavengers) nei	**Empereurs nca**		**Emperadores nep**		***Lethrinidae***				**1,70(38)XXX,XX**	**EMP**
47 South Africa	-	-	-	-	-	-	-	-	3	3
47 Fishing area total	*-*	*-*	*-*	*-*	*-*	*-*	*-*	*-*	*3*	*3*
51 Bahrain	1 603	1 194	1 486	1 506	892	944	1 227	1 403	1 377	1 675
Egypt	...	...	429	425	574	577	1 040	1 147	2 696	3 513
Eritrea	...	217	1 010	767	90	104	371	443	345	164
Jordan	...	...	...	...	...	2	2	1	2	1
Kenya	441	353	396	433	361	412	358	334	466	414
Korea Rep	-	-	-	-	-	93	160	97	459	-
Kuwait	36	17	15	15	20	38	35	50 F	78	75 F
Mauritius	6 296	6 312	6 216	5 137	5 018	4 981	4 598	4 698	4 008	4 078
Oman	4 358	3 521	6 013	4 087	5 767	6 630	6 954	7 664	6 526	7 179
Pakistan	1 466	1 660	1 643	1 549	1 911	2 334	3 323	5 173	5 044	5 100
Qatar	1 271	922	722	1 031	1 172	1 326	798	1 442	1 820	1 512

B-33

Miscellaneous coastal fishes / Poissons côtiers divers / Peces costeros diversos

Capture production by species, fishing areas and countries or areas / Captures par espèces, zones de pêche et pays ou zones / Capturas por especies, áreas de pesca y países o áreas

Species, Fishing area Espèce, Zone de pêche Especie, Area de pesca	1993 t	1994 t	1995 t	1996 t	1997 t	1998 t	1999 t	2000 t	2001 t	2002 t
Saudi Arabia	8 281	7 524	6 598	7 314	6 904	6 796	7 233	7 078	7 448	8 902
Seychelles	355	285	294	295	220	280	285	423	483	336
South Africa	-	-	-	-	-	-	-	-	-	1
Tanzania	4 308	4 566	6 490	7 304	7 350	7 025	7 500	8 000	7 850	7 800
Untd Arab Em	9 272	10 515	11 336	11 455	12 242	12 283	12 590	19 647	22 619	21 056
Yemen	5 478	4 390	3 214	2 437	2 550 F	2 900 F	2 800 F	2 500 F	3 200 F	3 600 F
51 Fishing area total	*43 165*	*41 476*	*45 862*	*43 755*	*45 071 F*	*46 725 F*	*49 274 F*	*60 100 F*	*64 421 F*	*65 406 F*
57 Indonesia	3 227	4 228	4 950	5 046	5 921	6 791	7 060	6 648	6 115	6 670
57 Fishing area total	*3 227*	*4 228*	*4 950*	*5 046*	*5 921*	*6 791*	*7 060*	*6 648*	*6 115*	*6 670*
71 Fiji Islands	3 097	2 956 F	1 359	1 230	1 780	1 731	2 990	2 700 F	2 883	2 850 F
Guam	...	8	1	1	1	1	1	1	3	3
Indonesia	17 654	19 330	26 852	29 513	25 987	24 915	25 497	24 009	23 460	24 530
Kiribati	1 950	1 930	1 950	1 970	1 960	675	1 299	2 137	3 930	1 363
Korea Rep	-	-	-	-	-	-	5	1	-	-
N Marianas	...	1	2	5	13	50	4	4	8	3
Palau	14	13	17	12	6	17	8	6	9	3
71 Fishing area total	*22 715*	*24 238 F*	*30 181*	*32 731*	*29 747*	*27 389*	*29 804*	*28 858 F*	*30 293*	*28 752 F*
77 Amer Samoa	...	4	2	5	2	0	1	4	6	6
77 Fishing area total	*...*	*4*	*2*	*5*	*2*	*0*	*1*	*4*	*6*	*6*
81 New Zealand	-	-	-	-	4	-	-	-	-	-
81 Fishing area total	-	-	-	-	*4*	-	-	-	-	-
Species total	*69 107*	*69 946 F*	*80 995*	*81 537*	*80 745 F*	*80 905 F*	*86 139 F*	*95 610 F*	*100 838 F*	*100 837 F*
Blackspot(=red) seabream	**Dorade rose**		**Besugo**			***Pagellus bogaraveo***			**1,70(39)008,01**	**SBR**
27 France	16	21	8	15	19	20	31	22	13	27
Ireland	7	0	3	8	8	6	1	...	11	7
Latvia	-	75	-	-	-	-	-	-	-	-
Netherlands	-	-	-	-	-	-	-	-	2	4
Portugal	1 030	1 159	1 364	1 305	1 242	1 370	1 373	1 019	1 116	1 306
Spain	1 567	1 754	1 206	739	3 075	1 395	947	699	94	246
UK	...	...	...	1	36	5	15	13	38	41
27 Fishing area total	*2 620*	*3 009*	*2 581*	*2 068*	*4 380*	*2 796*	*2 367*	*1 753*	*1 274*	*1 631*
34 Netherlands	-	-	-	38	-	-	28	71	-	7
Portugal	-	-	27	141	108	141	117	14	12	6
34 Fishing area total	-	-	*27*	*179*	*108*	*141*	*145*	*85*	*12*	*13*
37 France	...	42	12	12	20	20	38	38	38	50
37 Fishing area total	*...*	*42*	*12*	*12*	*20*	*20*	*38*	*38*	*38*	*50*
Species total	*2 620*	*3 051*	*2 620*	*2 259*	*4 508*	*2 957*	*2 550*	*1 876*	*1 324*	*1 694*
Common pandora	**Pageot commun**		**Breca**			***Pagellus erythrinus***			**1,70(39)008,02**	**PAC**
27 France	17	1	1	2	0	1	2	3	3	2
Portugal	201	115	87	102	152	132	104	151	128	142
27 Fishing area total	*218*	*116*	*88*	*104*	*152*	*133*	*106*	*154*	*131*	*144*
34 Portugal	69	17	7	20	6	2	1	0	-	-
Spain	-	-	-	-	34	139	204	-	-	-
34 Fishing area total	*69*	*17*	*7*	*20*	*40*	*141*	*205*	*0*	-	-
37 Algeria	...	...	1 204	1 633	2 661	1 503	1 145	1 518	1 339	1 515
Cyprus	36	32	32	32	25	19	37	42	34	31
France	281	118	12	76	77	77	165	123	94	107
Israel	34	150	150	100	65	47	67	517	...	...
Malta	2	1	3	5	6	6	6	5	2	2
Slovenia	-	-	-	-	-	1	1	3	7	8
Spain	-	-	-	-	213	183	73	305	229	514
Tunisia	2 820	1 981	2 292	2 746	2 506	2 967	3 128	3 105	3 219	2 687
37 Fishing area total	*3 173*	*2 282*	*3 693*	*4 592*	*5 553*	*4 803*	*4 622*	*5 618*	*4 924*	*4 864*
Species total	*3 460*	*2 415*	*3 788*	*4 716*	*5 745*	*5 077*	*4 933*	*5 772*	*5 055*	*5 008*
Axillary seabream	**Pageot acarne**		**Aligote**			***Pagellus acarne***			**1,70(39)008,03**	**SBA**
27 France	20	32	23	17	8	5	32	25	12	17
Portugal	1 208	977	1 044	1 180	966	878	996	1 297	1 201	1 129
27 Fishing area total	*1 228*	*1 009*	*1 067*	*1 197*	*974*	*883*	*1 028*	*1 322*	*1 213*	*1 146*
34 Portugal	554	32	8	18	4	5	1	1	1	1
Spain	-	-	-	-	7	122	-	-	-	-
34 Fishing area total	*554*	*32*	*8*	*18*	*11*	*127*	*1*	*1*	*1*	*1*
37 Algeria	...	...	332	...	...	362	...	...	...	...
Cyprus	29	32	11	22	25	23	50	25	20	19
France	104	80	95	63	31	31	286	257	186	236
Malta	...	...	7	5	4	3	3	2	-	-
Spain	-	-	-	-	417	251	-	-	-	-
37 Fishing area total	*133*	*112*	*445*	*90*	*477*	*670*	*339*	*284*	*206*	*255*
Species total	*1 915*	*1 153*	*1 520*	*1 305*	*1 462*	*1 680*	*1 368*	*1 607*	*1 420*	*1 402*

B-33 Miscellaneous coastal fishes / Poissons côtiers divers / Peces costeros diversos

Capture production by species, fishing areas and countries or areas
Captures par espèces, zones de pêche et pays ou zones
Capturas por especies, áreas de pesca y países o áreas

Species, Fishing area Espèce, Zone de pêche Especie, Area de pesca	1993 t	1994 t	1995 t	1996 t	1997 t	1998 t	1999 t	2000 t	2001 t	2002 t
Red pandora	**Pageot à tache rouge**		**Breca chata**		*Pagellus bellottii*				**1,70(39)008,07**	**PAR**
34 Congo Rep	34 F	2	7	5 F	4 F	3 F	2 F	...	2 F	2 F
Ghana	7 915	5 635	4 505	7 541	7 933	10 029	13 265	2 916	3 206	3 132
Portugal	12	-	-	1	-	-	-	-	-	-
Senegal	3 890	5 051	4 679	4 380	5 086	3 719	4 319	4 728	3 932	3 812
Spain	-	-	-	-	43	44	42	-	-	-
Togo	110	93	115	74	67	202	237	64	45	31
34 Fishing area total	*11 961 F*	*10 781*	*9 306*	*12 001 F*	*13 133 F*	*13 997 F*	*17 865 F*	*7 708*	*7 185 F*	*6 977 F*
47 Angola	...	...	...	...	...	...	188	...	744	674
47 Fishing area total	...	...	...	...	...	...	*188*	...	*744*	*674*
Species total	*11 961 F*	*10 781*	*9 306*	*12 001 F*	*13 133 F*	*13 997 F*	*18 053 F*	*7 708*	*7 929 F*	*7 651 F*
Pandoras nei	**Pageots nca**		**Brecas nep**		*Pagellus spp*				**1,70(39)008,XX**	**PAX**
34 Benin	12 F	14 F	14 F	16 F	22 F	20 F	15	10 F	...	1
Côte dIvoire	346	399	361	328	346	1 093	1 137	913	697	819
Gabon	...	68	538	200	191	301	127	376	274	444
Gambia	-	-	123	76	9	12	20	12	...	8
Greece	1 215	906	751	615	449	384	483	620	602	196
Portugal	42	42	-	-	-	-	-	-	-	-
Sao Tome Prn	79 F	115 F	121 F	88	86	100 F	110 F	100 F	100 F	100 F
Spain	600 F	1 000 F	1 500 F	2 327	47	26	764	57	101	196
Togo	48	37	208	91	65	365	1 055	120	75	17
34 Fishing area total	*2 342 F*	*2 581 F*	*3 616 F*	*3 741 F*	*1 215 F*	*2 301 F*	*3 711 F*	*2 208 F*	*1 849 F*	*1 781 F*
37 Albania	...	...	12	27	25	33	35	34	15	26
Greece	897	1 257	1 187	828	1 095	455	517	430	335	310 F
Italy	1 237	1 181	2 937	1 152	1 445	836	751	1 171	949	800
Spain	900 F	850 F	800 F	756	716	682	1 353	1 552	1 955	1 772
Syria	80 F	77 F	77 F	134	98	125	100	65	85	100
37 Fishing area total	*3 114 F*	*3 365 F*	*5 013 F*	*2 897*	*3 379*	*2 131*	*2 756*	*3 252*	*3 339*	*3 008 F*
51 Oman	419	...	...	...	...	...	...	...	...	...
Spain	-	-	-	-	-	-	5	-	-	1
51 Fishing area total	*419*	...	...	...	...	...	*5*	...	...	*1*
Species total	*5 875 F*	*5 946 F*	*8 629 F*	*6 638 F*	*4 594 F*	*4 432 F*	*6 472 F*	*5 460 F*	*5 188 F*	*4 790 F*
South American silver porgy	**...B**		**Sargo de América del Sur**		*Diplodus argenteus*				**1,70(39)033,01**	**DIG**
41 Argentina	-	6	8	39	84	15	5	2	2	1
41 Fishing area total	-	*6*	*8*	*39*	*84*	*15*	*5*	*2*	*2*	*1*
Species total	-	*6*	*8*	*39*	*84*	*15*	*5*	*2*	*2*	*1*
White seabream	**Sar commun**		**Sargo**		*Diplodus sargus*				**1,70(39)033,03**	**SWA**
27 France	54	52	45	54	41	35	55	64	51	78
27 Fishing area total	*54*	*52*	*45*	*54*	*41*	*35*	*55*	*64*	*51*	*78*
37 Greece	659	629	539	409	509	327	503	581	347	330 F
Spain	-	-	-	-	131	106	137	114	182	548
Syria	...	...	...	...	...	...	...	...	...	129
37 Fishing area total	*659*	*629*	*539*	*409*	*640*	*433*	*640*	*695*	*529*	*1 007 F*
Species total	*713*	*681*	*584*	*463*	*681*	*468*	*695*	*759*	*580*	*1 085 F*
Sargo breams nei	**Sars, sparaillons nca**		**Sargos, raspallones nep**		*Diplodus spp*				**1,70(39)033,XX**	**SRG**
27 Portugal	...	...	...	...	...	...	1 050	977	793	895
27 Fishing area total	...	...	...	...	...	...	*1 050*	*977*	*793*	*895*
34 Greece	4	12	31	19	7	22	18	117	88	50
Latvia	14	5	18	19	20	13	81	90	176	135
Mauritania	30 F	30 F	20 F	20 F	20 F	70 F	99	145	278	286
Morocco	1 063	1 194	837	840	1 002	975	1 062	1 217	1 089	1 456
Portugal	260	40	34	65	18	4	1	2	2	4
Senegal	-	-	-	-	4	221	194	364	453	313
Spain	-	-	-	151	117	111	77	74	98	7
34 Fishing area total	*1 371 F*	*1 281 F*	*940 F*	*1 114 F*	*1 188 F*	*1 416 F*	*1 532*	*2 009*	*2 184*	*2 251*
37 Bulgaria	-	-	-	-	-	-	-	-	90	-
Egypt	113	0	346	590	382	390	841	812	793	447
France	166	99	60	80	109	109	82	79	83	84
Italy	1 361	1 722	1 123	1 069	706	382	340	321	462	475
Malta	0	0	0	2	0	4	4	2	2	3
Morocco	5	15	92	100	47	7	11	240	199	23
Palest, O.T.	...	...	...	4	17	24	20 F	20 F	20 F	16
Slovenia	-	-	-	-	-	-	-	1	2	3
Tunisia	407	345	2 292	2 706	3 177	2 055	3 036	3 084	3 241	3 226
Turkey	2 420	2 887	2 398	2 345	1 750	2 210	2 110	1 420	655	751
37 Fishing area total	*4 472*	*5 068*	*6 311*	*6 896*	*6 188*	*5 181*	*6 444 F*	*5 979 F*	*5 547 F*	*5 028*
47 Portugal	-	-	-	-	-	-	-	-	4	-
47 Fishing area total	-	-	-	-	-	-	-	-	*4*	-

B-33

Miscellaneous coastal fishes
Poissons côtiers divers
Peces costeros diversos

Capture production by species, fishing areas and countries or areas
Captures par espèces, zones de pêche et pays ou zones
Capturas por especies, áreas de pesca y países o áreas

Species, Fishing area Espèce, Zone de pêche Especie, Area de pesca	1993 t	1994 t	1995 t	1996 t	1997 t	1998 t	1999 t	2000 t	2001 t	2002 t
Species total	*5 843 F*	*6 349 F*	*7 251 F*	*8 010 F*	*7 376 F*	*6 597 F*	*9 026 F*	*8 965 F*	*8 528 F*	*8 174*
Porgies	**Daubenets**		**Plumas**		***Calamus spp***			**1,70(39)034,XX**		**PRG**
31 Cuba	286	373	378	385	333	270	259	335	310	301
Dominican Rp	400	720	631	428	1 112	17	430	17	10	27
Mexico	160	168	257	239	96	266	577	235	214	121
31 Fishing area total	*846*	*1 261*	*1 266*	*1 052*	*1 541*	*553*	*1 266*	*587*	*534*	*449*
88 New Zealand	-	-	-	-	-	0	-	-	-	-
88 Fishing area total	-	-	-	-	-	*0*	-	-	-	-
Species total	*846*	*1 261*	*1 266*	*1 052*	*1 541*	*553*	*1 266*	*587*	*534*	*449*
Large-eye dentex	**Denté à gros yeux**		**Cachucho**		***Dentex macrophthalmus***			**1,70(39)060,02**		**DEL**
27 Portugal	34	25	12	8	26	2	2	0	0	1
27 Fishing area total	*34*	*25*	*12*	*8*	*26*	*2*	*2*	*0*	*0*	*1*
34 Georgia	30 F	10 F	-	-	-	-	-	-	-	-
Greece	8	15	1	2	16	9	19	20	12	14
Latvia	-	-	-	-	24	57	91	190	71	29
Liberia	12	8	...	9	39	...	...	...	...	...
Lithuania	-	80	-	-	-	-	-	-	-	121
Portugal	1 330	90	8	63	48	76	64	-	0	-
Russian Fed	142	172	329	528	445	1 506	2 217	592	374	899
Senegal	...	...	...	447	...	451	531	684	499	249
Togo	2	0	0	0	0	1	-	-	-	-
Ukraine	9	12	...	52	...	207	199	468	130	49
34 Fishing area total	*1 533 F*	*387 F*	*338*	*1 101*	*572*	*2 307*	*3 121*	*1 954*	*1 086*	*1 361*
37 Greece	221	550	710	651	1 010	535	497	378	318	300 F
37 Fishing area total	*221*	*550*	*710*	*651*	*1 010*	*535*	*497*	*378*	*318*	*300 F*
47 Estonia	-	-	181	-	-	-	-	-	-	-
Namibia	...	...	...	...	...	...	...	287	182	171
Russian Fed	113	40	229	682	1 121	736	1 786	1 793	807	66
Ukraine	...	...	...	-	4	-	73	456	55	-
47 Fishing area total	*113*	*40*	*410*	*682*	*1 125*	*736*	*1 859*	*2 536*	*1 044*	*237*
Species total	*1 901 F*	*1 002 F*	*1 470*	*2 442*	*2 733*	*3 580*	*5 479*	*4 868*	*2 448*	*1 899 F*
Canary dentex	**Denté à tache rouge**		**Chacarona de Canarias**		***Dentex canariensis***			**1,70(39)060,04**		**DEN**
34 Netherlands	-	-	-	-	-	-	-	20	-	-
34 Fishing area total	-	-	-	-	-	-	-	*20*	-	-
Species total	-	-	-	-	-	-	-	*20*	-	-
Common dentex	**Denté commun**		**Dentón**		***Dentex dentex***			**1,70(39)060,06**		**DEC**
27 Portugal	19	10	13	11	18	36	24	13	16	9
27 Fishing area total	*19*	*10*	*13*	*11*	*18*	*36*	*24*	*13*	*16*	*9*
34 Greece	780	383	623	550	228	387	254	134	317	444
Italy	3 430	3 430	-	-	-	-	-	-	-	-
Latvia	-	101	694	436	8	...	...	...	19	29
Portugal	29	8	5	9	3	2	1	0	-	1
34 Fishing area total	*4 239*	*3 922*	*1 322*	*995*	*239*	*389*	*255*	*134*	*336*	*474*
37 Albania	...	...	...	...	...	...	...	...	26	80
Croatia	49	47	53	56	76	70	50	55	10	7
Cyprus	18	47	23	38	18	18	22	18	19	16
France	8	8	6	10	9	9	5	2	1	0
Greece	227	398	291	194	220	135	220	230	163	150 F
Italy	2 250	2 279	2 270	1 253	389	190	205	309	201	207
Malta	0	1	0	0	1	0	1	1	1	2
Serbia-Monte	4	4	5	7	7	9	9	11	11	10
Spain	40 F	50 F	55 F	60	62	61	149	265	56	44
Tunisia	214	250	167	227	346	375	291	250	219	216
Turkey	229	218	191	442	330	370	220	100	60	79
37 Fishing area total	*3 039 F*	*3 302 F*	*3 061 F*	*2 287*	*1 458*	*1 237*	*1 172*	*1 241*	*767*	*811 F*
Species total	*7 297 F*	*7 234 F*	*4 396 F*	*3 293*	*1 715*	*1 662*	*1 451*	*1 388*	*1 119*	*1 294 F*
Angolan dentex	**Denté angolais**		**Dentón angoleño**		***Dentex angolensis***			**1,70(39)060,10**		**DEA**
34 Ghana	428	183	591	489	490	1 416	1 767	504	838	990
34 Fishing area total	*428*	*183*	*591*	*489*	*490*	*1 416*	*1 767*	*504*	*838*	*990*
47 Korea Rep	-	-	-	274	-	-	-	-	-	-
47 Fishing area total	-	-	-	*274*	-	-	-	-	-	-
Species total	*428*	*183*	*591*	*763*	*490*	*1 416*	*1 767*	*504*	*838*	*990*
Congo dentex	**Denté congolais**		**Dentón congolés**		***Dentex congoensis***			**1,70(39)060,11**		**DNC**
34 Ghana	364	112	43	102	119	392	1 272	350	571	445
34 Fishing area total	*364*	*112*	*43*	*102*	*119*	*392*	*1 272*	*350*	*571*	*445*

B-33

Miscellaneous coastal fishes — **Capture production by species, fishing areas and countries or areas**
Poissons côtiers divers — **Captures par espèces, zones de pêche et pays ou zones**
Peces costeros diversos — **Capturas por especies, áreas de pesca y países o áreas**

Species, Fishing area Espèce, Zone de pêche Especie, Area de pesca	1993 t	1994 t	1995 t	1996 t	1997 t	1998 t	1999 t	2000 t	2001 t	2002 t
Species total	*364*	*112*	*43*	*102*	*119*	*392*	*1 272*	*350*	*571*	*445*
Dentex nei	**Dentés nca**		**Dentones, samas, etc. nep**		***Dentex spp***			**1,70(39)060,XX**		**DEX**
27 Lithuania	-	-	8	-	-	-	-	-	-	-
Portugal	14	2	1	2	2	3	3	1	1	1
Spain	60	54	22	35	39	23	31	386	182	231
27 Fishing area total	*74*	*56*	*31*	*37*	*41*	*26*	*34*	*387*	*183*	*232*
34 Benin	76 F	86 F	83 F	95 F	100 F	90 F	83	60 F	20	19
Congo Rep	420 F	252 F	220 F	180 F	140 F	100 F	60 F	22	20 F	20 F
Estonia	91	-	-	-	-	-	-	-	-	-
Gabon	756	477	776	820	522	1 047	423	371	498	556
Ghana	1 192	813	1 372	1 584	1 084	1 278	1 874	892	675	965
Korea Rep	155	-	4	4	-	-	-	-	-	-
Liberia	72	84	313	327	346	974	936	588	671	680 F
Portugal	404	7	10	13	0	60	25	5	1	2
Senegal	2 731	1 906	4 318	1 608	2 301	1 132	788	1 033	833	884
Spain	400 F	350 F	350 F	300	299	471	618	473	687	937
34 Fishing area total	*6 297 F*	*3 975 F*	*7 446 F*	*4 931 F*	*4 792 F*	*5 152 F*	*4 807 F*	*3 444 F*	*3 405 F*	*4 063 F*
41 Italy	457	457	-	-	-	-	-	-	-	-
41 Fishing area total	*457*	*457*	-	-	-	-	-	-	-	-
47 Angola	2 490	2 883	7 283	2 624	1 567	8 683	8 758	13 765	16 080	8 948
Other nei	-	124	1 106	462	-	-	-	-	-	-
47 Fishing area total	*2 490*	*3 007*	*8 389*	*3 086*	*1 567*	*8 683*	*8 758*	*13 765*	*16 080*	*8 948*
51 Italy	686	686	-	-	-	-	-	-	-	-
Ukraine	-	-	-	-	-	2	-	-	-	-
51 Fishing area total	*686*	*686*	-	-	-	*2*	-	-	-	-
Species total	*10 004 F*	*8 181 F*	*15 866 F*	*8 054 F*	*6 400 F*	*13 863 F*	*13 599 F*	*17 596 F*	*19 668 F*	*13 243 F*
Black seabream	**Dorade grise**		**Chopa**		***Spondyliosoma cantharus***			**1,70(39)063,02**		**BRB**
27 France	2 857	2 666	2 219	2 802	2 884	3 115	3 030	3 011	2 784	3 118
Netherlands	-	-	-	-	-	-	-	-	-	1
Portugal	...	...	...	...	...	...	164	177	158	224
UK	-	-	-	-	-	5	260	240	202	294
27 Fishing area total	*2 857*	*2 666*	*2 219*	*2 802*	*2 884*	*3 120*	*3 454*	*3 428*	*3 144*	*3 637*
34 Benin	10 F	20 F	20 F	30 F	45 F	48 F	56	40 F	93	147
Greece	9	13	25	13	2	1	5	18	6	5
Liberia	-	-	-	-	-	-	-	-	10	10 F
Morocco	143	151	103	167	121	109	139	220	265	374
Portugal	33	12	13	23	5	4	1	0	0	1
Senegal	1 744	1 628	862	1 196	1 251	597	922	1 067	1 035	1 254
Spain	270 F	300 F	320 F	340	351	318	-	-	-	47
Togo	8	15	10	6	5	2	6	25	12	15
34 Fishing area total	*2 217 F*	*2 139 F*	*1 353 F*	*1 775 F*	*1 780 F*	*1 079 F*	*1 129*	*1 370 F*	*1 421*	*1 853 F*
37 Croatia	31	36	52	51	56	60	23	19	5	3
France	1	0	0	1	1	1	1	18	15	21
Greece	394	541	319	378	316	218	325	158	132	120 F
Morocco	31	9	53	62	64	17	7	215	35	41
Serbia-Monte	1	1	2	5	3	5	3	4	4	4
Spain	-	-	-	-	9	7	7	18	23	129
Turkey	346	1 208	72	73	35	40	50	45	45	40
37 Fishing area total	*804*	*1 795*	*498*	*570*	*484*	*348*	*416*	*477*	*259*	*358 F*
47 Angola	...	...	...	...	...	93	129	212	494	13
47 Fishing area total	...	...	...	...	...	*93*	*129*	*212*	*494*	*13*
Species total	*5 878 F*	*6 600 F*	*4 070 F*	*5 147 F*	*5 148 F*	*4 640 F*	*5 128*	*5 487 F*	*5 318*	*5 861 F*
Saddled seabream	**Oblade**		**Oblada**		***Oblada melanura***			**1,70(39)076,01**		**SBS**
01 Egypt	...	...	...	...	33	67	47	158	70	77
01 Fishing area total	...	...	...	...	*33*	*67*	*47*	*158*	*70*	*77*
34 Greece	-	5	29	19	-	-	-	-	-	-
Portugal	-	-	-	1	0	1	-	0	0	0
34 Fishing area total	-	*5*	*29*	*20*	*0*	*1*	-	*0*	*0*	*0*
37 Croatia	183	175	180	183	169	185	130	120	55	30
Cyprus	3	3	10	4	4	3	7	8	11	14
France	5	4	4	13	14	14	10	14	8	8
Greece	1 206	1 072	651	588	399	332	239	443	287	270 F
Malta	0	0	1	1	2	1	2	2	1	1
Serbia-Monte	1	2	3	4	4	6	7	5	6	6
Spain	-	-	-	-	30	44	78	66	81	56
Tunisia	17	22	24	43	36	29	146	84	51	53
Turkey	208	330	196	184	190	180	115	80	90	130
37 Fishing area total	*1 623*	*1 608*	*1 069*	*1 020*	*848*	*794*	*734*	*822*	*590*	*568 F*
Species total	*1 623*	*1 613*	*1 098*	*1 040*	*881*	*862*	*781*	*980*	*660*	*645 F*

B-33 Miscellaneous coastal fishes / Poissons côtiers divers / Peces costeros diversos

Capture production by species, fishing areas and countries or areas / Captures par espèces, zones de pêche et pays ou zones / Capturas por especies, áreas de pesca y países o áreas

Species, Fishing area Espèce, Zone de pêche Especie, Area de pesca	1993 t	1994 t	1995 t	1996 t	1997 t	1998 t	1999 t	2000 t	2001 t	2002 t
Sheepshead	**Rondeau mouton**		**Sargo chopa**		***Archosargus probatocephalus***				**1,70(39)102,01**	**SPH**
21 USA	13	0	29	137	31	20	87	27	19	20
21 Fishing area total	*13*	*0*	*29*	*137*	*31*	*20*	*87*	*27*	*19*	*20*
31 USA	2 167	2 130	1 806	1 519	1 712	1 346	1 096	1 501	1 195	1 091
31 Fishing area total	*2 167*	*2 130*	*1 806*	*1 519*	*1 712*	*1 346*	*1 096*	*1 501*	*1 195*	*1 091*
Species total	*2 180*	*2 130*	*1 835*	*1 656*	*1 743*	*1 366*	*1 183*	*1 528*	*1 214*	*1 111*
Western Atlantic seabream	**Rondeau brème**		**Sargo amarillo**		***Archosargus rhomboidalis***				**1,70(39)102,02**	**AVB**
31 Dominican Rp	...	...	...	...	...	...	...	...	...	53
31 Fishing area total	...	...	...	...	...	...	...	...	...	*53*
Species total	...	...	...	...	...	...	...	...	...	*53*
King soldier bream	**Spare royal**		**Sargo real**		***Argyrops spinifer***				**1,70(39)105,01**	**KBR**
51 Qatar	174	110	85	130	177	146	98	199	426	327
Untd Arab Em	2 470	2 703	2 626	2 654	2 836	2 846	2 917	3 604	4 149	4 797
51 Fishing area total	*2 644*	*2 813*	*2 711*	*2 784*	*3 013*	*2 992*	*3 015*	*3 803*	*4 575*	*5 124*
Species total	*2 644*	*2 813*	*2 711*	*2 784*	*3 013*	*2 992*	*3 015*	*3 803*	*4 575*	*5 124*
Carpenter seabream	**Denté charpentier**		**Dentón carpintero**		***Argyrozona argyrozona***				**1,70(39)107,01**	**SLF**
47 South Africa	692	16	729	883	780	505	541	500 F	287	249
47 Fishing area total	*692*	*16*	*729*	*883*	*780*	*505*	*541*	*500 F*	*287*	*249*
Species total	*692*	*16*	*729*	*883*	*780*	*505*	*541*	*500 F*	*287*	*249*
Santer seabream	**Denté nufar**		**Dentón nufar**		***Cheimerius nufar***				**1,70(39)118,02**	**SLD**
47 South Africa	81	48	41	40	34	0	29	25 F	25	15
47 Fishing area total	*81*	*48*	*41*	*40*	*34*	*0*	*29*	*25 F*	*25*	*15*
51 South Africa	-	32	25	33	28	...	...	...	...	31
51 Fishing area total	-	*32*	*25*	*33*	*28*	...	...	...	...	*31*
Species total	*81*	*80*	*66*	*73*	*62*	*0*	*29*	*25 F*	*25*	*46*
Red porgy	**Pagre rouge**		**Pargo**		***Pagrus pagrus***				**1,70(39)191,03**	**RPG**
27 France	2	2	2	3	1	1	2	1	1	0
Portugal	90	83	91	106	162	317	540	441	243	113
Spain	-	-	-	22	302	720	243	86	100	87
27 Fishing area total	*92*	*85*	*93*	*131*	*465*	*1 038*	*785*	*528*	*344*	*200*
34 Greece	1 184	840	1 011	909	527	569	587	572	820	750
Portugal	-	-	27	65	112	318	54	35	26	22
34 Fishing area total	*1 184*	*840*	*1 038*	*974*	*639*	*887*	*641*	*607*	*846*	*772*
37 Cyprus	71	103	45	35	27	25	31	23	21	15
Egypt	1 080	1 508	1 021	1 825	1 425	1 230	2 984	1 847	575	1 231
Greece	422	487	416	371	406	284	300	341	319	300 F
Malta	3	5	6	8	9	8	6	6	4	4
Palest, O.T.	...	...	...	35	47	35	35 F	25 F	10 F	4
Spain	115 F	115 F	120 F	120	123	121	114	136	81	147
Tunisia	237	230	272	254	327	362	388	460	416	400
Turkey	2 217	1 624	2 012	750	560	675	480	540	320	403
37 Fishing area total	*4 145 F*	*4 072 F*	*3 892 F*	*3 398*	*2 924*	*2 740*	*4 338 F*	*3 378 F*	*1 746 F*	*2 504 F*
41 Argentina	1 216	1 248	1 203	1 590	1 159	574	2 159	1 301	935	904
Brazil	130 F	130 F	83	884	1 572	1 448	1 362	1 497	1 440	1 587
Italy	229	229	-	-	-	-	-	-	-	-
Lithuania	-	3	-	-	-	-	-	-	-	-
Uruguay	21	19	12	1	12	4	14	27	7	8
41 Fishing area total	*1 596 F*	*1 629 F*	*1 298*	*2 475*	*2 743*	*2 026*	*3 535*	*2 825*	*2 382*	*2 499*
Species total	*7 017 F*	*6 626 F*	*6 321 F*	*6 978*	*6 771*	*6 691*	*9 299 F*	*7 338 F*	*5 318 F*	*5 975 F*
Silver seabream	**Dorade**		**Dorada del Pacífico**		***Pagrus auratus***				**1,70(39)191,15**	**GSU**
57 Australia	2 321	2 611	3 025	3 293	3 213	2 487	2 588	2 501	3 391	3 453
57 Fishing area total	*2 321*	*2 611*	*3 025*	*3 293*	*3 213*	*2 487*	*2 588*	*2 501*	*3 391*	*3 453*
61 China,Taiwan	1 585	1 270	1 580	1 809	2 445	2 073	3 333	4 726	6 722	6 737
Japan	14 159	14 442	15 007	16 468	15 611	15 375	15 731	15 041	14 633	16 527
Korea Rep	979	456	552	762	1 149	1 657	924	986	913	1 084
61 Fishing area total	*16 723*	*16 168*	*17 139*	*19 039*	*19 205*	*19 105*	*19 988*	*20 753*	*22 268*	*24 348*
71 Australia	570	1 181	783	644	649	872	395	303	388	347
71 Fishing area total	*570*	*1 181*	*783*	*644*	*649*	*872*	*395*	*303*	*388*	*347*
81 Australia	588	523	403	324	318	272	1 618	993	811	574
Japan	71	9	4	-	-	-	-	-	-	-
New Zealand	7 151	6 701	6 161	5 814	6 233	6 278	6 876	6 852	6 209	6 571
Ukraine	-	-	-	-	-	-	-	-	-	3

B-33 Miscellaneous coastal fishes / Poissons côtiers divers / Peces costeros diversos

Capture production by species, fishing areas and countries or areas / Captures par espèces, zones de pêche et pays ou zones / Capturas por especies, áreas de pesca y países o áreas

Species, Fishing area Espèce, Zone de pêche Especie, Area de pesca	1993 t	1994 t	1995 t	1996 t	1997 t	1998 t	1999 t	2000 t	2001 t	2002 t
81 Fishing area total	*7 810*	*7 233*	*6 568*	*6 138*	*6 551*	*6 550*	*8 494*	*7 845*	*7 020*	*7 148*
Species total	*27 424*	*27 193*	*27 515*	*29 114*	*29 618*	*29 014*	*31 465*	*31 402*	*33 067*	*35 296*
Pargo breams nei	**Dorades nca**		**Pargos nep**		***Pagrus spp***				**1,70(39)191,XX**	**SBP**
34 Ghana	1 862	834	1 546	1 448	1 347	1 682	...	...	2 189	2 624
Greece	1	4	-	1	5	14	-	-	-	7
Korea Rep	35	-	30	10	-	-	-	-	-	-
Morocco	67	13	16	20	19	100	92	124	156	46
Portugal	862	437	124	196	31	11	4	6	23	11
Senegal	2 261	2 394	1 775	2 570	2 763	1 881	1 931	2 206	2 488	1 419
Sierra Leone	221	220	239	240 F	933	713	752	...	...	...
Spain	10 F	40 F	70 F	100	106	223	313	177	141	5
34 Fishing area total	*5 319 F*	*3 942 F*	*3 800 F*	*4 585 F*	*5 204*	*4 624*	*3 092*	*2 513*	*4 997*	*4 112*
37 Algeria	...	...	71	81	58	255	254	153	267	394
Greece	371	720	697	530	460	357	435	386	326	310 F
Morocco	1	2	5	5	4	36	0	300	112	1
Syria	40 F	39 F	39 F	50	80	90	62	74	77	112
37 Fishing area total	*412 F*	*761 F*	*812 F*	*666*	*602*	*738*	*751*	*913*	*782*	*817 F*
47 Angola	334	246	39	95	716	1 077	183	1 840	4 484	5 773
Portugal	-	-	-	-	-	-	-	35	122	-
47 Fishing area total	*334*	*246*	*39*	*95*	*716*	*1 077*	*183*	*1 875*	*4 606*	*5 773*
Species total	*6 065 F*	*4 949 F*	*4 651 F*	*5 346 F*	*6 522*	*6 439*	*4 026*	*5 301*	*10 385*	*10 702 F*
Red steenbras	**Denté du Cap**		**Dentón del Cabo**		***Petrus rupestris***				**1,70(39)203,01**	**RER**
47 South Africa	73	37	56	28	35	...	22	10 F	7	6
47 Fishing area total	*73*	*37*	*56*	*28*	*35*	*...*	*22*	*10 F*	*7*	*6*
Species total	*73*	*37*	*56*	*28*	*35*	*...*	*22*	*10 F*	*7*	*6*
Panga seabream	**Spare panga**		**Panga**		***Pterogymnus laniarius***				**1,70(39)220,01**	**PGA**
47 Namibia	-	246	291	470	416	199	383	8 603	5 898	2 621
South Africa	1 000	690	708	730	907	843	966	900 F	1 058	1 177
47 Fishing area total	*1 000*	*936*	*999*	*1 200*	*1 323*	*1 042*	*1 349*	*9 503 F*	*6 956*	*3 798*
Species total	*1 000*	*936*	*999*	*1 200*	*1 323*	*1 042*	*1 349*	*9 503 F*	*6 956*	*3 798*
White stumpnose	**Sargue australe**		**Pargo ñato**		***Rhabdosargus globiceps***				**1,70(39)224,01**	**WSN**
47 South Africa	139	143	168	237	165	296	335	300 F	169	107
47 Fishing area total	*139*	*143*	*168*	*237*	*165*	*296*	*335*	*300 F*	*169*	*107*
Species total	*139*	*143*	*168*	*237*	*165*	*296*	*335*	*300 F*	*169*	*107*
Gilthead seabream	**Dorade royale**		**Dorada**		***Sparus aurata***				**1,70(39)235,08**	**SBG**
01 Egypt	1 197	899	908	474	554	672	557	1 727	1 711	1 506
01 Fishing area total	*1 197*	*899*	*908*	*474*	*554*	*672*	*557*	*1 727*	*1 711*	*1 506*
27 Channel Is	...	...	...	...	...	...	15	-	-	-
France	133	147	126	155	115	107	152	206	162	139
Portugal	141	164	200	209	188	173	151	183	213	268
Spain	273	377	405	511	363	258	118	100	152	203
27 Fishing area total	*547*	*688*	*731*	*875*	*666*	*538*	*436*	*489*	*527*	*610*
34 Italy	1 718	1 718	-	-	-	-	-	-	-	-
Morocco	0	0	0	0	0	4	0	6	25	9
Portugal	-	-	2	4	1	0	0	-	-	0
Spain	-	-	-	-	15	24	735	593	1 263	417
34 Fishing area total	*1 718*	*1 718*	*2*	*4*	*16*	*28*	*735*	*599*	*1 288*	*426*
37 Albania	...	...	17	27	11	20	20	23	90	181
Croatia	9	18	17	13	44	84	27	25	11	6
Cyprus	-	-	1	1	0	0	0	...	37	45
Egypt	378	475	451	754	533	553	1 398	751	601	974
France	387	79	103	132	106	106	226	170	207	316
Greece	310	445	201	199	138	125	142	248	176	170 F
Italy	1 944	3 086	2 179	1 743	1 859	1 717	1 754	1 939	2 675	3 004
Morocco	...	...	...	...	...	...	...	200	0	1
Serbia-Monte	1	1	2	4	4	4	6	6	7	7
Slovenia	-	-	2	-	-	-	1	1	4	4
Spain	230 F	210 F	190 F	170	168	226	103	536	749	554
Syria	...	...	...	...	...	...	...	...	...	63
Tunisia	330	309	125	107	265	333	409	757	399	822
Turkey	1 593	1 334	1 432	1 340	1 200	1 400	1 665	830	1 070	700
37 Fishing area total	*5 182 F*	*5 957 F*	*4 720 F*	*4 490*	*4 328*	*4 568*	*5 751*	*5 486*	*6 026*	*6 847 F*
Species total	*8 644 F*	*9 262 F*	*6 361 F*	*5 843*	*5 564*	*5 806*	*7 479*	*8 301*	*9 552*	*9 389 F*
Bogue	**Bogue**		**Boga**		***Boops boops***				**1,70(39)261,01**	**BOG**
27 France	83	85	61	17	10	26	20	76	71	42
Portugal	628	500	428	380	375	316	354	642	929	805

B-33 Miscellaneous coastal fishes / Poissons côtiers divers / Peces costeros diversos

Capture production by species, fishing areas and countries or areas
Captures par espèces, zones de pêche et pays ou zones
Capturas por especies, áreas de pesca y países o áreas

Species, Fishing area Espèce, Zone de pêche Especie, Area de pesca	1993 t	1994 t	1995 t	1996 t	1997 t	1998 t	1999 t	2000 t	2001 t	2002 t
Spain	521	399	644	406	639	1 036	2 985	1 043	933	794
27 Fishing area total	*1 232*	*984*	*1 133*	*803*	*1 024*	*1 378*	*3 359*	*1 761*	*1 933*	*1 641*
34 Greece	0	0	3	2	-	-	-	-	-	-
Lithuania	11	-	-	-	-	-	-	-	-	-
Morocco	324	393	504	500	510	250	60	2 024	785	1 102
Portugal	-	-	23	37	45	42	72	28	29	44
Sao Tome Prn	...	...	...	...	17	20 F	25 F	20 F	20 F	20 F
Senegal	66	35	139	10	20	38	7	17	14	98
Spain	-	-	-	-	6	13	98	37	30	32
Togo	27	11	3	40	18	109	198	71	99	4
34 Fishing area total	*428*	*439*	*672*	*589*	*616*	*472 F*	*460 F*	*2 197 F*	*977 F*	*1 300 F*
37 Albania	150 F	100 F	52	104	65	220	220	220	120	150
Algeria	...	...	2 282	1 425	1 631	3 234	3 452	4 132	4 135	4 697
Croatia	501	396	329	303	342	335	140	147	109	55
Cyprus	278	178	290	285	230	233	259	354	216	162
Egypt	3 118	2 638	2 173	2 609	2 499	1 956	...	1 450	1 222	1 541
France	486	197	37	172	193	193	197	223	258	258
Greece	12 398	14 592	6 842	6 742	5 973	4 228	4 658	4 096	3 674	3 500 F
Israel	123	50	102	34	34	47	73	91	37	38
Italy	5 543	4 457	5 659	5 281	4 178	4 074	3 105	3 541	3 537	3 129
Libya	2 300 F	2 500 F	2 550 F	2 500 F	2 500 F	2 500 F	2 500 F	2 500 F	2 500 F	2 500 F
Malta	43	19	19	17	16	15	12	21	27	16
Morocco	1 766	1 799	2 877	2 900	2 000	2 951	3 276	2 864	2 471	3 216
Palest, O.T.	...	...	...	9	81	162	160 F	100 F	50 F	32
Serbia-Monte	17	18	18	21	20	20	22	24	25	33
Slovenia	-	-	-	-	-	2	1	3	2	1
Spain	740 F	820 F	900 F	950	997	870	815	863	856	1 151
Syria	...	...	...	...	...	...	...	...	...	141
Tunisia	1 529	1 491	1 636	1 928	2 205	2 466	3 890	3 052	2 852	3 209
Turkey	3 154	4 236	3 196	3 736	2 450	4 100	1 620	1 500	1 000	1 126
37 Fishing area total	*32 146 F*	*33 491 F*	*28 962 F*	*29 016 F*	*25 414 F*	*27 606 F*	*24 400 F*	*25 181 F*	*23 091 F*	*24 955 F*
Species total	*33 806 F*	*34 914 F*	*30 767 F*	*30 408 F*	*27 054 F*	*29 456 F*	*28 219 F*	*29 139 F*	*26 001 F*	*27 896 F*
Daggerhead breams nei	**Spares australs nca**		**Sargos australes nep**		***Chrysoblephus spp***			**1,70(39)264,XX**		**RSX**
47 South Africa	171	231	197	176	139	154	74	70 F	78	57
47 Fishing area total	*171*	*231*	*197*	*176*	*139*	*154*	*74*	*70 F*	*78*	*57*
51 South Africa	-	109	70	92	85	...	75 F	...	...	67
51 Fishing area total	*-*	*109*	*70*	*92*	*85*	*...*	*75 F*	*...*	*...*	*67*
Species total	*171*	*340*	*267*	*268*	*224*	*154*	*149 F*	*70 F*	*78*	*124*
White steenbras	**Marbré du Cap**		**Herrera del Cabo**		***Lithognathus lithognathus***			**1,70(39)277,01**		**SNW**
47 South Africa	4	1	11	7	8	9	2	...	...	...
47 Fishing area total	*4*	*1*	*11*	*7*	*8*	*9*	*2*	*...*	*...*	*...*
Species total	*4*	*1*	*11*	*7*	*8*	*9*	*2*	*...*	*...*	*...*
Sand steenbras	**Marbré**		**Herrera**		***Lithognathus mormyrus***			**1,70(39)277,02**		**SSB**
27 France	54	47	78	36	41	30	46	59	36	34
Portugal	...	...	...	...	...	...	178	158	109	145
27 Fishing area total	*54*	*47*	*78*	*36*	*41*	*30*	*224*	*217*	*145*	*179*
34 Senegal	-	-	-	-	-	223	45	8	77	32
34 Fishing area total	*-*	*-*	*-*	*-*	*-*	*223*	*45*	*8*	*77*	*32*
37 France	45	24	2	24	80	80	66	59	61	55
Slovenia	-	-	-	-	-	-	-	1	2	2
Spain	-	-	-	-	270	323	255	290	246	203
Syria	...	...	...	...	...	...	...	...	...	25
Tunisia	...	...	917	694	661	781	683	656	781	663
37 Fishing area total	*45*	*24*	*919*	*718*	*1 011*	*1 184*	*1 004*	*1 006*	*1 090*	*948*
47 Angola	642	580	279	214	176	1 072	726	999	1 735	2 802
47 Fishing area total	*642*	*580*	*279*	*214*	*176*	*1 072*	*726*	*999*	*1 735*	*2 802*
Species total	*741*	*651*	*1 276*	*968*	*1 228*	*2 509*	*1 999*	*2 230*	*3 047*	*3 961*
Steenbrasses nei	**Marbrés nca**		**Herreras nep**		***Lithognathus spp***			**1,70(39)277,XX**		**STW**
47 Namibia	0	7	195	6	136	2	116	56	20	4
47 Fishing area total	*0*	*7*	*195*	*6*	*136*	*2*	*116*	*56*	*20*	*4*
Species total	*0*	*7*	*195*	*6*	*136*	*2*	*116*	*56*	*20*	*4*
Salema	**Saupe**		**Salema**		***Sarpa salpa***			**1,70(39)293,01**		**SLM**
27 France	2	4	3	-	5	2	4	8	10	5
Portugal	...	...	...	...	...	...	336	246	320	239
27 Fishing area total	*2*	*4*	*3*	*...*	*5*	*2*	*340*	*254*	*330*	*244*
37 Albania	...	...	...	...	...	...	...	...	1	5
Croatia	166	142	149	127	132	125	42	24	28	14

B-33 Miscellaneous coastal fishes / Poissons côtiers divers / Peces costeros diversos

Capture production by species, fishing areas and countries or areas
Captures par espèces, zones de pêche et pays ou zones
Capturas por especies, áreas de pesca y países o áreas

Species, Fishing area Espèce, Zone de pêche Especie, Area de pesca	1993 t	1994 t	1995 t	1996 t	1997 t	1998 t	1999 t	2000 t	2001 t	2002 t
Cyprus	1	0	4	5	3	4	1	4	5	6
Egypt	0	0	0	0	0	-	-	-	-	-
France	110	36	15	30	77	77	60	58	60	63
Greece	656	1 160	592	687	485	369	404	408	346	330 F
Malta	0	0	0	0	0	1	0	0	0	-
Serbia-Monte	1	4	4	7	7	9	10	10	10	10
Slovenia	-	-	-	-	-	-	-	-	-	2
Spain	-	-	-	-	242	199	167	147	164	253
Tunisia	974	972	916	750	1 127	1 151	1 124	951	1 024	1 071
Turkey	1 479	1 765	256	331	340	300	155	200	160	243
37 Fishing area total	*3 387*	*4 079*	*1 936*	*1 937*	*2 413*	*2 235*	*1 963*	*1 802*	*1 798*	*1 997 F*
47 Russian Fed	-	-	-	-	-	-	-	-	1	-
South Africa	-	1	1	4	3	0	1	...	...	...
47 Fishing area total	*-*	*1*	*1*	*4*	*3*	*0*	*1*	*...*	*1*	*...*
Species total	*3 389*	*4 084*	*1 940*	*1 941*	*2 421*	*2 237*	*2 304*	*2 056*	*2 129*	*2 241 F*
Blackhead seabream	**Pagre tête noire**		**...C**		***Acanthopagrus schlegeli***			**1,70(39)330,04**		**MLM**
61 China,Taiwan	178	153	197	207	277	404	604	717	878	784
61 Fishing area total	*178*	*153*	*197*	*207*	*277*	*404*	*604*	*717*	*878*	*784*
Species total	*178*	*153*	*197*	*207*	*277*	*404*	*604*	*717*	*878*	*784*
Yellowfin seabream	**Pagre à nageoires jaunes**		**Sargo aleta amarilla**		***Acanthopagrus latus***			**1,70(39)330,05**		**YWF**
51 Kuwait	167	210	273	234	249	280	464	350 F	271	280 F
Untd Arab Em	...	...	...	...	...	...	...	...	313	122
51 Fishing area total	*167*	*210*	*273*	*234*	*249*	*280*	*464*	*350 F*	*584*	*402 F*
Species total	*167*	*210*	*273*	*234*	*249*	*280*	*464*	*350 F*	*584*	*402 F*
Twobar seabream	**Pagre double bande**		**Sargo de dos bandas**		***Acanthopagrus bifasciatus***			**1,70(39)330,06**		**AAB**
51 Untd Arab Em	...	...	...	...	...	...	...	...	1 586	332
51 Fishing area total	*...*	*...*	*...*	*...*	*...*	*...*	*...*	*...*	*1 586*	*332*
Species total	*...*	*...*	*...*	*...*	*...*	*...*	*...*	*...*	*1 586*	*332*
Scup	**Spare doré**		**Sargo de América del Norte**		***Stenotomus chrysops***			**1,70(39)353,01**		**SCP**
21 USA	4 090	4 027	2 845	2 952	2 196	1 893	1 676	1 206	1 845	3 312
21 Fishing area total	*4 090*	*4 027*	*2 845*	*2 952*	*2 196*	*1 893*	*1 676*	*1 206*	*1 845*	*3 312*
Species total	*4 090*	*4 027*	*2 845*	*2 952*	*2 196*	*1 893*	*1 676*	*1 206*	*1 845*	*3 312*
Porgies, seabreams nei	**Dentés, spares nca**		**Dentones, sargos nep**		***Sparidae***			**1,70(39)XXX,XX**		**SBX**
27 Channel Is	2 F	14	13 F	9	48	126	132	105	108	125
France	-	-	-	-	-	-	2	-	3	-
Portugal	2 406	2 379	2 317	1 955	1 949	1 834	-	21	3	2
Spain	1 173	535	1 272	582	1 857	2 006	1 648	1 940	1 950	693
UK	475	484	766	482	567	496	134	146	111	234
27 Fishing area total	*4 056 F*	*3 412*	*4 368 F*	*3 028*	*4 421*	*4 462*	*1 916*	*2 212*	*2 175*	*1 054*
31 Antigua Barb	...	...	...	...	...	...	...	...	9	15
Cuba	76	95	104	52	53	35	7	16	10	13
Korea Rep	-	-	-	-	-	309	18	4	-	-
Puerto Rico	...	...	...	...	...	...	22	24	25	20
USA	583	644	431	176	398	336	93	168	219	209
Venezuela	...	...	...	...	...	...	...	6	...	...
31 Fishing area total	*659*	*739*	*535*	*228*	*451*	*680*	*140*	*218*	*263*	*257*
34 Belize	54	7	88	-	183	1 152	2 826	2 188	129	85
Bulgaria	-	-	-	-	-	35	-	-	-	-
Cameroon	14	0	0	2	2 F	5 F	5 F	4	34	18
Chile	-	3	-	-	-	-	-	-	-	-
China	1 028	76	1 057	904	3 011	2 442	2 783	2 544	3 638	-
Côte dIvoire	0	1	1	1 F	1 F	1 F	1	3	...	...
Cuba	-	-	-	263	-	-	-	-	-	-
Cyprus	-	-	-	...	...	38	80	93	195	-
France	-	-	-	-	...	77	53	-	-	-
Gambia	763	146	446	129	...	...	-	-	-	-
Georgia	170 F	60 F	-	-	-	-	-	-	-	-
Germany	-	-	23	-	-	-	-	-	-	21
Ghana	443	165	157	220	129	316	3 986	1 238	532	969
Greece	0	0	0	-	-	-	-	-	-	-
Guinea	3 250 F	3 340 F	4 709	3 814	3 649	3 019	3 265	1 838	4 562	4 500 F
GuineaBissau	10 F	10 F	28	14 F	15 F	10 F	10 F	10 F	10 F	10 F
Honduras	2 170	83	1 768	736	565	233	705	120	438	-
Italy	8 573	8 573	-	-	-	-	-	-	-	-
Korea D P Rp	49	-	-	-	-	-	-	273	-	-
Korea Rep	352	581	287	408	933	1 924	637	224	444	737
Latvia	-	-	-	-	-	48	80	53	-	-
Liberia	41	18	42	-	-	-	-	-	-	-
Lithuania	7	38	...	...	...	192	157	155	32	137
Mauritania	360 F	490 F	350 F	490 F	520 F	870 F	900 F	900 F	900 F	900 F
Morocco	8 447	8 195	8 396	9 588	10 326	7 415	9 320	10 169	9 568	10 518

B-33 Miscellaneous coastal fishes / Poissons côtiers divers / Peces costeros diversos

Capture production by species, fishing areas and countries or areas / Captures par espèces, zones de pêche et pays ou zones / Capturas por especies, áreas de pesca y países o áreas

Species, Fishing area Espèce, Zone de pêche Especie, Area de pesca	1993 t	1994 t	1995 t	1996 t	1997 t	1998 t	1999 t	2000 t	2001 t	2002 t
Nigeria	297	148	395	-	-	-	-	-	-	-
Norway	-	61	-	-	-	-	-	-	-	-
Panama	498	-	396	402	46	319	66	-	-	-
Poland	-	-	-	7	5	-	-	-	-	-
Portugal	40	27	298	522	95	25	6	0	0	2
Russian Fed	201	187	362	408	615	1 265	1 492	954	860	2 097
St Vincent	267	-	16	30	10	59	39	5	51	20
Sao Tome Prn	33 F	48 F	50 F	37	49	60 F	70 F	60 F	60 F	60 F
Senegal	-	-	-	-	-	-	-	-	295	79
Sierra Leone	2 836	37	885	2 565	1 967	522	500	1 106	1 189	1 007
Spain	400 F	600 F	800 F	754	100	226	145	117	163	36
Ukraine	100	197	155	153	1 245	757	1 259	1 966	363	214
Vanuatu	95	0	0	0	-	0	-	-	-	-
Other nei	-	-	-	-	-	-	63	49	341	114
34 Fishing area total	*30 498 F*	*23 091 F*	*20 709 F*	*21 447 F*	*23 466 F*	*21 010 F*	*28 448 F*	*24 069 F*	*23 804 F*	*21 524 F*
37 Albania	70 F	50 F	3	9	1	-	-	-	1	2
Algeria	5 500 F	6 000 F	-	-	-	-	7	-	-	-
Cyprus	76	130	61	65	83	64	74	63	46	55
Israel	193	100	100	100	-	-	-	-	562	494
Lebanon	250	300	450	450	350	400	450	450	400	370
Libya	3 700 F	4 000 F	4 100 F	4 000 F	4 000 F	4 000 F	4 000 F	4 000 F	4 000 F	4 000 F
Morocco	885	1 131	1 336	1 400	54	36	830	851	1 156	778
Spain	1 100 F	1 300 F	1 500 F	1 808	813	853	444	230	232	62
Tunisia	3 262	2 865	2 693	2 381	2 714	1 841	2 973	3 084	2 825	2 818
Turkey	495	792	334	412	220	180	240	110	135	137
37 Fishing area total	*15 531 F*	*16 668 F*	*10 577 F*	*10 625 F*	*8 235 F*	*7 374 F*	*9 018 F*	*8 788 F*	*9 357 F*	*8 716 F*
41 Italy	1 143	1 143	-	-	-	-	-	-	-	-
Korea Rep	...	148	-	-	-	-	2	-	-	-
Spain	-	-	-	-	-	-	-	1 054	1 221	-
41 Fishing area total	*1 143*	*1 291*	-	-	-	-	*2*	*1 054*	*1 221*	-
47 China	-	-	-	-	-	-	-	124	-	-
Honduras	27	-	5	-	-	-	-	-	-	-
Japan	-	-	-	121	-	-	-	-	-	-
Korea Rep	1 227	2 409	665	-	-	-	250	198	-	155
Namibia	717	973	2 178	2 279	2 608	1 364	4 434	...	20	...
Portugal	-	-	-	-	-	-	-	-	0	-
Russian Fed	1	-	-	-	-	76	-	407	8	5
South Africa	606	836	274	305	311	365	274	250 F	151	110
47 Fishing area total	*2 578*	*4 218*	*3 122*	*2 705*	*2 919*	*1 805*	*4 958*	*979 F*	*179*	*270*
51 Bahrain	294	283	439	496	493	757	582	591	401	488
Djibouti	34 F	36 F	40 F	40 F	40 F	40 F	40 F	40 F	40 F	40 F
Egypt	...	...	4 012	2 428	2 175	1 282	182	4 731	3 545	2 827
Eritrea	...	12	22	12	0	0	15	48	11	1
Italy	2 059	2 059	-	-	-	-	-	-	-	-
Korea Rep	3 785	2 492	2 633	2 453	1 926	1 479	3 503	1 988	1 218	4 647
Oman	4 503	5 770	5 090	4 692	4 322	4 016	6 098	4 419	3 976	8 776
Pakistan	3 939	3 866	3 358	3 097	3 058	1 255	4 220	4 510	4 411	4 430
Qatar	213	183	124	126	221	188	182	248	288	209
Saudi Arabia	1 064	1 910	1 469	1 722	2 481	2 822	2 771	2 488	2 646	3 252
South Africa	-	26	6	19	15	1	5 F	...	...	4
51 Fishing area total	*15 891 F*	*16 637 F*	*17 193 F*	*15 085 F*	*14 731 F*	*11 840 F*	*17 598 F*	*19 063 F*	*16 536 F*	*24 674 F*
57 Australia	320	643	552	278	113	159	325	794	832	663
Korea Rep	-	-	3	-	3	-	666	1 674	677	791
57 Fishing area total	*320*	*643*	*555*	*278*	*116*	*159*	*991*	*2 468*	*1 509*	*1 454*
61 China	57 565	56 785	58 576	56 174	71 043	74 905	78 149	105 785	125 692	118 576
China,H.Kong	486	932	1 215	925	883	852	600 F	700 F	780 F	750 F
China,Taiwan	6 306	6 515	5 626	5 401	6 334	6 440	7 312	6 288	9 953	7 370
Japan	11 433	10 994	11 496	11 484	11 256	11 284	10 675	9 066	9 545	10 000 F
Korea Rep	6 341	6 062	2 290	2 161	2 637	2 424	1 748	2 056	1 931	1 466
61 Fishing area total	*82 131*	*81 288*	*79 203*	*76 145*	*92 153*	*95 905*	*98 484 F*	*123 895 F*	*147 901 F*	*138 162 F*
71 Australia	136	127	133	160	169	157	156	182	199	156
Korea Rep	2 519	3 990	3 196	2 048	434	94	62	42	70	291
71 Fishing area total	*2 655*	*4 117*	*3 329*	*2 208*	*603*	*251*	*218*	*224*	*269*	*447*
77 Korea Rep	-	-	-	-	-	-	-	2	-	1
77 Fishing area total	-	-	-	-	-	-	-	*2*	-	*1*
81 Australia	575	727	722	502	498	439	378	333	309	282
Japan	2	1	6	-	-	-	-	-	-	-
Korea Rep	754	1 197	961	939	1 715	922	1 422	2 309	2 427	2 302
Russian Fed	-	9	2	494	8	-	-	-	-	-
Ukraine	-	-	-	-	-	-	5	-	-	-
81 Fishing area total	*1 331*	*1 934*	*1 691*	*1 935*	*2 221*	*1 361*	*1 805*	*2 642*	*2 736*	*2 584*
87 Korea Rep	-	-	-	-	-	-	8	8	-	-
87 Fishing area total	-	-	-	-	-	-	*8*	*8*	-	-
Species total	*156 793 F*	*154 038 F*	*141 282 F*	*133 684 F*	*149 316 F*	*144 847 F*	*163 586 F*	*185 622 F*	*205 950 F*	*199 143 F*
Blotched picarel	**Mendole**		**Chucla**			***Spicara maena***			**1,70(40)075,01**	**BPI**
37 Tunisia	135	203	218	285	505	395	419	255	381	593

B-33 Miscellaneous coastal fishes / Poissons côtiers divers / Peces costeros diversos

Capture production by species, fishing areas and countries or areas / Captures par espèces, zones de pêche et pays ou zones / Capturas por especies, áreas de pesca y países o áreas

Species, Fishing area Espèce, Zone de pêche Especie, Area de pesca	1993 t	1994 t	1995 t	1996 t	1997 t	1998 t	1999 t	2000 t	2001 t	2002 t
37 Fishing area total	*135*	*203*	*218*	*285*	*505*	*395*	*419*	*255*	*381*	*593*
Species total	*135*	*203*	*218*	*285*	*505*	*395*	*419*	*255*	*381*	*593*
Picarels nei	**Mendoles, picarels nca**		**Chuclas, carameles nep**		***Spicara spp***			**1,70(40)075,XX**		**PIC**
27 Portugal	...	...	...	...	...	...	43	22	25	36
27 Fishing area total	...	...	...	...	...	...	*43*	*22*	*25*	*36*
34 Greece	6	-	-	-	17	8	-	-	-	16
34 Fishing area total	*6*	-	-	-	*17*	*8*	-	-	-	*16*
37 Albania	100 F	50 F	...	11	0	7	7	10	5	11
Croatia	510	315	312	290	340	255	245	231	189	72
Cyprus	715	513	711	764	650	709	546	533	671	486
France	99	31	...	8	9	9	10	9	5	6
Greece	7 432	14 118	6 354	8 307	8 354	4 541	4 663	4 029	4 031	3 850 F
Italy	971	810	1 138	953	647	545	547	385	313	287
Lebanon	50	50	50	100	100	100	100	100	95	93
Malta	3	3	3	7	7	7	8	9	6	3
Serbia-Monte	7	8	8	11	12	13	16	21	27	38
Slovenia	-	-	-	-	-	3	3	4	5	6
Spain	-	-	-	1	-	-	494	712	759	346
Tunisia	406	345	382	335	485	3 258	1 005	646	572	858
Turkey	5 306	5 378	1 210	1 525	1 650	3 700	1 680	1 500	2 250	1 105
Ukraine	-	-	-	-	-	-	-	3	-	-
37 Fishing area total	*15 599 F*	*21 621 F*	*10 168*	*12 312*	*12 254*	*13 147*	*9 324*	*8 192*	*8 928*	*7 161 F*
47 Angola	407	113	9	7	0	55	...	13	-	-
47 Fishing area total	*407*	*113*	*9*	*7*	*0*	*55*	...	*13*	-	-
Species total	*16 012 F*	*21 734 F*	*10 177*	*12 319*	*12 271*	*13 210*	*9 367*	*8 227*	*8 953*	*7 213 F*
Surmullet	**Rouget de roche**		**Salmonete de roca**		***Mullus surmuletus***			**1,70(41)007,01**		**MUR**
27 Channel Is	2 F	5	5 F	4	11	17	22	23	19	15
Denmark	0	0	0	1	1	1	3	2	5	12
France	1 414	1 171	2 886	2 936	1 861	3 666	2 592	4 054	2 969	2 636
Ireland	-	-	-	40	-	38	-	-	-	-
Isle of Man	-	-	-	-	-	-	-	-	4	-
Netherlands	-	-	-	1	0	-	-	235	560	337
Portugal	69	109	64	52	69	66	180	155	191	166
Spain	-	-	-	292	300	132	136	278	352	198
UK	218	172	228	192	173	180	147	220	310	217
27 Fishing area total	*1 703 F*	*1 457*	*3 183 F*	*3 518*	*2 415*	*4 100*	*3 080*	*4 967*	*4 410*	*3 581*
37 Croatia	479	395	420	311	280	285	200	277	480	294
Cyprus	184	189	215	240	228	176	184	159	132	125
Greece	2 101	3 320	2 535	2 004	1 944	1 095	1 274	1 368	1 238	1 200 F
Libya	3 700 F	4 000 F	4 100 F	4 000 F	4 000 F	4 000 F	4 000 F	4 000 F	4 000 F	4 000 F
Serbia-Monte	3	5	10	11	10	11	9	10	11	11
Spain	...	...	...	...	...	...	414	541	473	738
Tunisia	1 448	1 506	1 140	1 161	2 406	1 459	1 504	1 527	2 163	1 803
Turkey	4 728	3 704	3 602	3 962	2 950	2 050	2 100	2 300	1 570	1 450
37 Fishing area total	*12 643 F*	*13 119 F*	*12 022 F*	*11 689 F*	*11 818 F*	*9 076 F*	*9 685 F*	*10 182 F*	*10 067 F*	*9 621 F*
Species total	*14 346 F*	*14 576 F*	*15 205 F*	*15 207 F*	*14 233 F*	*13 176 F*	*12 765 F*	*15 149 F*	*14 477 F*	*13 202 F*
Red mullet	**Rouget de vase**		**Salmonete de fango**		***Mullus barbatus***			**1,70(41)007,02**		**MUT**
37 Bulgaria	-	-	-	-	-	-	-	5	26	33
Cyprus	113	121	88	108	119	115	145	103	91	84
Spain	...	...	...	...	...	...	328	263	262	560
Tunisia	1 168	1 032	1 108	1 287	1 394	1 228	1 250	1 600	2 632	2 045
Turkey	4 774	4 447	3 906	3 936	3 000	3 500	3 865	2 450	2 455	2 395
Ukraine	12	10	13	...	18	35	30	12	201	42
37 Fishing area total	*6 067*	*5 610*	*5 115*	*5 331*	*4 531*	*4 878*	*5 618*	*4 433*	*5 667*	*5 159*
Species total	*6 067*	*5 610*	*5 115*	*5 331*	*4 531*	*4 878*	*5 618*	*4 433*	*5 667*	*5 159*
Argentine goatfish	**Rouget-barbet argentin**		**Trilla**		***Mullus argentinae***			**1,70(41)007,03**		**MWU**
41 Argentina	20	26	63	75	73	99	229	498	225	78
41 Fishing area total	*20*	*26*	*63*	*75*	*73*	*99*	*229*	*498*	*225*	*78*
Species total	*20*	*26*	*63*	*75*	*73*	*99*	*229*	*498*	*225*	*78*
Surmullets(=Red mullets) nei	**Rougets nca**		**Salmonetes nep**		***Mullus spp***			**1,70(41)007,XX**		**MUX**
27 Germany	-	-	-	-	-	-	-	13	10	9
27 Fishing area total	-	-	-	-	-	-	-	*13*	*10*	*9*
34 Mauritania	30 F	50 F	30 F	30 F	30 F	141 F	150 F	150 F	150 F	150 F
Morocco	346	319	431	266	528	385	426	884	1 062	992
Portugal	1	0	0	2	0	0	-	-	-	-
Spain	-	-	-	276	47	56	26	15	28	7
34 Fishing area total	*377 F*	*369 F*	*461 F*	*574 F*	*605 F*	*582 F*	*602 F*	*1 049 F*	*1 240 F*	*1 149 F*
37 Albania	400 F	200 F	76	64	61	143	145	140	170	97

B-33 Miscellaneous coastal fishes / Poissons côtiers divers / Peces costeros diversos

Capture production by species, fishing areas and countries or areas / Captures par espèces, zones de pêche et pays ou zones / Capturas por especies, áreas de pesca y países o áreas

Species, Fishing area Espèce, Zone de pêche Especie, Area de pesca	1993 t	1994 t	1995 t	1996 t	1997 t	1998 t	1999 t	2000 t	2001 t	2002 t
Algeria	1 400 F	1 800 F	1 576	1 950	1 320	1 480	1 420	2 170	2 081	2 140
Egypt	1 516	2 098	1 512	2 314	2 533	1 681	2 727	1 611	1 717	1 957
France	267	186	189	149	149	149	266	270	206	188
Georgia	-	-	-	-	14	11	8	3	22	67
Greece	2 808	3 016	2 550	2 495	2 426	1 744	1 735	1 810	1 541	1 500 F
Israel	300 F	250 F	162	227	185	276	350	340	401	368
Italy	10 448	9 921	9 441	11 325	7 499	7 491	8 771	9 044	7 121	6 111
Lebanon	150	200	200	200	150	200	200	250	200	200
Malta	2	3	4	7	7	8	12	7	5	4
Morocco	374	254	343	416	374	421	396	605	362	309
Palest, O.T.	...	...	...	56	57	85	85 F	70 F	50 F	35
Romania	-	5	6	1	3	3	1	2	3	2
Russian Fed	2	25	324	76	68	119	92	127	119	47
Slovenia	-	-	-	-	-	1	1	1	4	1
Spain	1 400 F	1 100 F	800 F	478	480	491	1 901	1 673	1 750	1 192
Syria	121 F	117 F	116 F	232	250	116	130	122	125	147
37 Fishing area total	*19 188 F*	*19 175 F*	*17 299 F*	*19 990*	*15 576*	*14 419*	*18 240 F*	*18 245 F*	*15 877 F*	*14 365 F*
Species total	*19 565 F*	*19 544 F*	*17 760 F*	*20 564 F*	*16 181 F*	*15 001 F*	*18 842 F*	*19 307 F*	*17 127 F*	*15 523 F*
Goatfishes	**Rougets-souris**		**Salmonetes**		***Upeneus spp***			**1,70(41)251,XX**		**GOX**
51 Bahrain	29	85	36	139	65	16	414	82	192	17
Eritrea	...	61	7	2	1	1	21	19	9	9
India	9 620	12 213	11 316	10 300	6 148	7 061	7 186	8 457	14 748	15 457
Mauritius	403	416	457	512	479	436	509	541	556	501
Saudi Arabia	10	52	65	83	73	107	123	72	122	88
51 Fishing area total	*10 062*	*12 827*	*11 881*	*11 036*	*6 766*	*7 621*	*8 253*	*9 171*	*15 627*	*16 072*
57 India	8 151	7 768	7 553	7 829	9 722	10 199	10 706	2 079	1 988	2 073
Indonesia	8 548	8 526	8 586	10 758	13 357	15 752	15 757	16 228	16 622	18 150
Malaysia	1 958	2 085	1 407	966	1 227	343	2 059	5 021	6 488	6 413
57 Fishing area total	*18 657*	*18 379*	*17 546*	*19 553*	*24 306*	*26 294*	*28 522*	*23 328*	*25 098*	*26 636*
61 China,H.Kong	1 322	706	593	718	477	393	300 F	350 F	400 F	390 F
China,Taiwan	469	602	399	508	440	335	478	477	300	344
61 Fishing area total	*1 791*	*1 308*	*992*	*1 226*	*917*	*728*	*778 F*	*827 F*	*700 F*	*734 F*
71 Fiji Islands	117	119 F	9	108	190	155	160	140 F	157	150 F
Indonesia	8 582	8 244	9 026	9 966	10 846	9 455	10 495	11 720	12 038	12 600
Kiribati	430	430	440	450	450	149	582	639	1 609	404
Malaysia	3 429	5 104	8 194	7 141	9 075	7 307	6 547	5 824	6 472	7 119
Singapore	216	35	29	24	29	29	25	8	15	1
71 Fishing area total	*12 774*	*13 932 F*	*17 698*	*17 689*	*20 590*	*17 095*	*17 809*	*18 331 F*	*20 291*	*20 274 F*
Species total	*43 284*	*46 446 F*	*48 117*	*49 504*	*52 579*	*51 738*	*55 362 F*	*51 657 F*	*61 716 F*	*63 716 F*
West African goatfish	**Rouget du Sénégal**		**Salmonete barbudo**		***Pseudupeneus prayensis***			**1,70(41)499,03**		**GOA**
34 Côte dIvoire	47	53	53 F	68	44	38	61	44	40	35
Gabon	...	177	224	100	573	76	57	448	311	392
Ghana	163	190	65	586	1 035	553	247	39	285	278
Greece	1 306	1 024	907	1 065	692	805	838	715	1 081	675
Italy	1 959	2 004	-	-	-	-	-	-	-	-
Korea Rep	12	-	18	-	-	-	-	-	-	-
Liberia	2	3	-	-	46	-	-	-	-	-
Nigeria	...	...	...	4 575	...	...	...	...	...	1 085
Portugal	12	4	-	1	-	1	0	-	-	-
Senegal	998	1 714	352	731	896	1 087	1 223	1 110	1 846	1 855
Sierra Leone	...	...	...	...	37	...	...	35	119	130
Togo	3	5	0	0	0	0	-	-	-	-
34 Fishing area total	*4 502*	*5 174*	*1 619 F*	*7 126*	*3 323*	*2 560*	*2 426*	*2 391*	*3 682*	*4 450*
47 Angola	11	26	0	-	0	-	-	-	-	-
Korea Rep	42	-	-	-	-	-	-	-	-	-
47 Fishing area total	*53*	*26*	*0*	*-*	*0*	*-*	*-*	*-*	*-*	*...*
Species total	*4 555*	*5 200*	*1 619 F*	*7 126*	*3 323*	*2 560*	*2 426*	*2 391*	*3 682*	*4 450*
Goatfishes, red mullets nei	**Rougets, etc. nca**		**Salmonetes, etc. nep**		***Mullidae***			**1,70(41)XXX,XX**		**MUM**
31 Dominican Rp	325	585	376	265	178	168	90	171	161	40
Grenada	0	1	1	1	1	0	0	0	0	0
Puerto Rico	...	...	...	...	...	...	17	17	15	10
St Kitts Nev	...	...	3	10	1	2	1	0	0	0
USA	-	-	-	-	-	-	-	-	36	6
31 Fishing area total	*325*	*586*	*380*	*276*	*180*	*170*	*108*	*188*	*212*	*56*
41 Brazil	260 F	260 F	964	530	749	840	1 117	1 756	4 167	6 040
Italy	261	267	-	-	-	-	-	-	-	-
41 Fishing area total	*521 F*	*527 F*	*964*	*530*	*749*	*840*	*1 117*	*1 756*	*4 167*	*6 040*
51 Egypt	689	777	1 077	716	744	439	876	914	2 590	1 684
Italy	392	401	-	-	-	-	-	-	-	-
Qatar	14	10	15	16	14	14	20	1	0	0
Russian Fed	-	-	-	-	-	-	-	-	-	28
Untd Arab Em	140	152	130	154	172	185	192	178	180 F	200
51 Fishing area total	*1 235*	*1 340*	*1 222*	*886*	*930*	*638*	*1 088*	*1 093*	*2 770 F*	*1 912*

B-33 Miscellaneous coastal fishes / Poissons côtiers divers / Peces costeros diversos

Capture production by species, fishing areas and countries or areas / Captures par espèces, zones de pêche et pays ou zones / Capturas por especies, áreas de pesca y países o áreas

Species, Fishing area Espèce, Zone de pêche Especie, Area de pesca	1993 t	1994 t	1995 t	1996 t	1997 t	1998 t	1999 t	2000 t	2001 t	2002 t
61 Korea Rep	58	161	-	-	-	-	-	-	-	-
61 Fishing area total	*58*	*161*	-	-	-	-	-	-	-	-
71 N Marianas	...	0	1	12	9	1	1	1	1	0
Palau	2	1	2	2	1	2	1	1	1	0
Philippines	25 919	22 091	25 510	23 662	15 884	14 182	14 527	14 718	14 291	24 440
71 Fishing area total	*25 921*	*22 092*	*25 513*	*23 676*	*15 894*	*14 185*	*14 529*	*14 720*	*14 293*	*24 440*
77 USA	-	-	-	-	-	-	-	-	-	27
77 Fishing area total	-	-	-	-	-	-	-	-	-	*27*
Species total	*28 060 F*	*24 706 F*	*28 079*	*25 368*	*17 753*	*15 833*	*16 842*	*17 757*	*21 442 F*	*32 475*
Mojarras(=Silver-biddies) nei	**Blanches nca**		**Mojarras nep**		***Gerres spp***			**1,70(46)036,XX**		**MOJ**
31 USA	-	-	-	-	-	-	-	159	180	211
31 Fishing area total	-	-	-	-	-	-	-	*159*	*180*	*211*
51 Bahrain	64	233	160	219	350	347	261	334	372	172
Qatar	115	53	53	50	56	92	77	78	82	52
Saudi Arabia	5	164	337	360	438	445	520	529	510	662
Untd Arab Em	1 150	1 383	1 222	1 235	1 320	1 324	1 351	1 651	1 900	371
51 Fishing area total	*1 334*	*1 833*	*1 772*	*1 864*	*2 164*	*2 208*	*2 209*	*2 592*	*2 864*	*1 257*
71 Fiji Islands	12	12 F	9	6	10	13	11	10 F	11	10 F
Kiribati	1 710	1 690	1 700	1 720	1 730	566	890	674	1 455	847
Palau	0	0	1	1	1	1	0	1	0	0
Philippines	7 266	7 911	6 491	5 909	4 285	4 854	5 008	5 117	5 525	6 230
71 Fishing area total	*8 988*	*9 613 F*	*8 201*	*7 636*	*6 026*	*5 434*	*5 909*	*5 802 F*	*6 991*	*7 087 F*
Species total	*10 322*	*11 446 F*	*9 973*	*9 500*	*8 190*	*7 642*	*8 118*	*8 553 F*	*10 035*	*8 555 F*
Mojarras, etc. nei	**Blanches, etc. nca**		**Mojarras, etc. nep**		***Gerreidae***			**1,70(46)XXX,XX**		**GDJ**
31 Colombia	71	90	38	125	6	23	0	5 F	5 F	5 F
Cuba	1 637	2 144	2 221	1 719	1 346	1 199	1 013	1 081	1 104	858
Dominican Rp	23	41	60	45	12	20	9	14	15	10
Mexico	7 554	7 549	7 379	7 867	4 547	4 245	3 069	2 701	2 800	2 108
Puerto Rico	...	...	...	...	...	...	14	15	13	11
Venezuela	0	0	0	0	0	0	0	0	0	-
31 Fishing area total	*9 285*	*9 824*	*9 698*	*9 756*	*5 911*	*5 487*	*4 105*	*3 816 F*	*3 937 F*	*2 992 F*
77 Mexico	4 792	4 786	7 085	6 841	3 946	1 945	2 114	1 724	1 800	1 692
77 Fishing area total	*4 792*	*4 786*	*7 085*	*6 841*	*3 946*	*1 945*	*2 114*	*1 724*	*1 800*	*1 692*
Species total	*14 077*	*14 610*	*16 783*	*16 597*	*9 857*	*7 432*	*6 219*	*5 540 F*	*5 737 F*	*4 684 F*
Parore	**...B**		**...C**		***Girella tricuspidata***			**1,70(47)003,02**		**GIY**
81 Australia	...	...	...	501	507	496	503	541	513	470
New Zealand	-	-	83	60	81	80	76	94	71	78
81 Fishing area total	...	...	*83*	*561*	*588*	*576*	*579*	*635*	*584*	*548*
Species total	...	...	*83*	*561*	*588*	*576*	*579*	*635*	*584*	*548*
Opaleye	**...B**		**...C**		***Girella nigricans***			**1,70(47)003,03**		**GIQ**
77 USA	-	-	-	-	-	-	-	3	2	1
77 Fishing area total	-	-	-	-	-	-	-	*3*	*2*	*1*
Species total	-	-	-	-	-	-	-	*3*	*2*	*1*
Blue sea chub	**Calicagère bleue**		**Chopa azul**		***Kyphosus cinerascens***			**1,70(47)035,01**		**KYC**
51 Saudi Arabia	-	-	-	-	8	-	1	-	-	17
51 Fishing area total	-	-	-	-	*8*	-	*1*	-	-	*17*
Species total	-	-	-	-	*8*	-	*1*	-	-	*17*
Sea chubs nei	**Calicagères nca**		**Chopas nep**		***Kyphosus spp***			**1,70(47)035,XX**		**KYP**
31 Antigua Barb	...	...	...	...	...	...	...	...	8	4
31 Fishing area total	...	...	...	...	...	...	...	...	*8*	*4*
41 Estonia	-	-	-	-	-	-	-	-	-	37
41 Fishing area total	-	-	-	-	-	-	-	-	-	*37*
71 Palau	1	2	1	3	2	2	1	1	1	1
71 Fishing area total	*1*	*2*	*1*	*3*	*2*	*2*	*1*	*1*	*1*	*1*
77 USA	-	-	-	-	-	-	-	-	-	12
77 Fishing area total	-	-	-	-	-	-	-	-	-	*12*
Species total	*1*	*2*	*1*	*3*	*2*	*2*	*1*	*1*	*9*	*54*
Spotted sicklefish	**Forgeron tacheté**		**Catemo manchado**		***Drepane punctata***			**1,70(50)132,01**		**SPS**
57 Malaysia	67	33	17	51	23	30	58	53	26	59
57 Fishing area total	*67*	*33*	*17*	*51*	*23*	*30*	*58*	*53*	*26*	*59*

B-33 Miscellaneous coastal fishes / Poissons côtiers divers / Peces costeros diversos

Capture production by species, fishing areas and countries or areas / Captures par espèces, zones de pêche et pays ou zones / Capturas por especies, áreas de pesca y países o áreas

Species, Fishing area Espèce, Zone de pêche Especie, Area de pesca	1993 t	1994 t	1995 t	1996 t	1997 t	1998 t	1999 t	2000 t	2001 t	2002 t
71 Malaysia	525	516	619	824	560	486	785	625	663	730
Philippines	121	369	63	64	174	96	133	114	135	138
71 Fishing area total	*646*	*885*	*682*	*888*	*734*	*582*	*918*	*739*	*798*	*868*
Species total	*713*	*918*	*699*	*939*	*757*	*612*	*976*	*792*	*824*	*927*
African sicklefish	**Forgeron ailé**		**Catemo africano**		***Drepane africana***				**1,70(50)132,02**	**SIC**
34 Benin	16 F	15 F	15 F	15 F	18 F	14 F	12	8 F	6	4
Cameroon	39	50 F	111	72	75 F	78	94	73	70	30
Congo Rep	3 F	7	15	15 F	15 F	15 F	20 F	24	20 F	20 F
Côte dIvoire	19	26	35	23	18	28	21	35	20	24
Gabon	269	189	330	270	222	281	290	86	71	107
Gambia	44	42	118	151	104	23	30	60	85	186
Ghana	20	7	34	6	46	4	24	2	8	30
Guinea	...	...	89	324	149	...	...	...	...	...
GuineaBissau	...	24	71	172	180 F	150 F	120 F	120 F	120 F	120 F
Liberia	8	12	30	15	-	70	256	94	104	110 F
Nigeria	251	440	597	954	2 047	603	...	1 316	1 429	1 511
Sao Tome Prn	...	...	...	...	15	20 F	25 F	20 F	20 F	20 F
Senegal	530	605	248	741	606	325	439	356	421	244
Sierra Leone	...	...	12	10 F	293	657	214	662	416	244
Togo	7	4	1	6	2	0	0	-	-	0
34 Fishing area total	*1 206 F*	*1 421 F*	*1 706 F*	*2 774 F*	*3 790 F*	*2 268 F*	*1 545 F*	*2 856 F*	*2 790 F*	*2 650 F*
Species total	*1 206 F*	*1 421 F*	*1 706 F*	*2 774 F*	*3 790 F*	*2 268 F*	*1 545 F*	*2 856 F*	*2 790 F*	*2 650 F*
Ballan wrasse	**Vieille commune**		**Maragota**		***Labrus bergylta***				**1,70(63)005,01**	**USB**
27 Channel Is	-	-	-	-	-	-	2	1	-	-
27 Fishing area total	-	-	-	-	-	-	*2*	*1*	-	-
Species total	-	-	-	-	-	-	*2*	*1*	-	-
Tautog	**Tautogue noir**		**...C**		***Tautoga onitis***				**1,70(63)243,01**	**TAU**
21 USA	325	206	148	119	116	116	95	111	138	160
21 Fishing area total	*325*	*206*	*148*	*119*	*116*	*116*	*95*	*111*	*138*	*160*
Species total	*325*	*206*	*148*	*119*	*116*	*116*	*95*	*111*	*138*	*160*
Cunner	**Tanche-tautogue**		**...C**		***Tautogolabrus adspersus***				**1,70(63)244,02**	**CUN**
21 USA	0	1	0	1	1	3	4	3	9	13
21 Fishing area total	*0*	*1*	*0*	*1*	*1*	*3*	*4*	*3*	*9*	*13*
Species total	*0*	*1*	*0*	*1*	*1*	*3*	*4*	*3*	*9*	*13*
California sheephead	**Labre californien**		**Vieja de California**		***Semicossyphus pulcher***				**1,70(63)331,02**	**YFH**
77 USA	...	...	...	...	...	118	58	78	68	52
77 Fishing area total	...	...	...	...	...	*118*	*58*	*78*	*68*	*52*
Species total	...	...	...	...	...	*118*	*58*	*78*	*68*	*52*
Wrasses, hogfishes, etc. nei	**Pourceaux, donzelles, etc. nca**		**Lábridos(=Tordos,maragotas)nep**		***Labridae***				**1,70(63)XXX,XX**	**WRA**
27 France	...	...	...	...	...	...	222	250	250	221
Norway	-	-	-	-	-	-	-	10	6	4
Portugal	...	...	...	...	...	...	44	36	39	54
UK	21	22	30	17	16	20	20	22	25	20
27 Fishing area total	*21*	*22*	*30*	*17*	*16*	*20*	*286*	*318*	*320*	*299*
31 Dominican Rp	741	1 334	821	529	1 456	52	52	69	22	597
Puerto Rico	...	...	...	...	...	...	30	47	46	36
31 Fishing area total	*741*	*1 334*	*821*	*529*	*1 456*	*52*	*82*	*116*	*68*	*633*
34 Portugal	-	-	-	-	-	-	-	1	1	1
34 Fishing area total	-	-	-	-	-	-	-	*1*	*1*	*1*
51 Saudi Arabia	...	...	165	108	105	66	155	91	79	153
Tanzania	2 040	2 583	2 810	3 725	3 200	3 000	3 000	3 200	3 500	3 000
51 Fishing area total	*2 040*	*2 583*	*2 975*	*3 833*	*3 305*	*3 066*	*3 155*	*3 291*	*3 579*	*3 153*
57 Australia	75	105	121	77	77	77	98	85	88	90
Malaysia	3	10	8	7	3	3	2	0	-	1
57 Fishing area total	*78*	*115*	*129*	*84*	*80*	*80*	*100*	*85*	*88*	*91*
71 Australia	0	0	0	0	0	0	-	0	1	0
Fiji Islands	383	390 F	218	204	375	180	399	360 F	400	390 F
Guam	-	-	-	-	-	1	2	0	1	0
Malaysia	78	59	192	219	2 299	2 694	2 400	2 139	1 948	292
Palau	1	0	0	0	0	0	2	5	7	11
Philippines	16 780	21 017	19 099	17 097	15 107	13 391	12 318	12 384	12 664	7 106
71 Fishing area total	*17 242*	*21 466 F*	*19 509*	*17 520*	*17 781*	*16 266*	*15 121*	*14 888 F*	*15 021*	*7 799 F*
81 Australia	-	-	-	-	-	-	1	13	12	11
New Zealand	-	-	0	0	24	2	3	4	2	3

B-33 Miscellaneous coastal fishes / Poissons côtiers divers / Peces costeros diversos

Capture production by species, fishing areas and countries or areas / Captures par espèces, zones de pêche et pays ou zones / Capturas por especies, áreas de pesca y países o áreas

Species, Fishing area Espèce, Zone de pêche Especie, Area de pesca	1993 t	1994 t	1995 t	1996 t	1997 t	1998 t	1999 t	2000 t	2001 t	2002 t
81 Fishing area total	-	-	*0*	*0*	*24*	*2*	*4*	*17*	*14*	*14*
Species total	*20 122*	*25 520 F*	*23 464*	*21 983*	*22 662*	*19 486*	*18 748*	*18 716 F*	*19 091*	*11 990 F*
Parrotfish	**Perroquet vieillard**		**Loro viejo**		***Sparisoma cretense***			**1,70(65)055,06**		**PRR**
27 Portugal	-	-	-	-	-	-	56	87	161	153
27 Fishing area total	-	-	-	-	-	-	*56*	*87*	*161*	*153*
34 Portugal	-	-	-	-	-	-	-	2	1	0
34 Fishing area total	-	-	-	-	-	-	-	*2*	*1*	*0*
Species total	-	-	-	-	-	-	*56*	*89*	*162*	*153*
Parrotfishes nei	**Perroquets nca**		**Loros nep**		***Scaridae***			**1,70(65)XXX,XX**		**PWT**
31 Antigua Barb	...	...	...	...	...	...	...	...	173	252
Dominican Rp	195	351	149	80	94	109	7	18	19	52
Grenada	0	0	0	0	1	1	42	18	41	113
Puerto Rico	...	...	...	...	...	...	52	63	66	57
St Kitts Nev	...	...	7	19	5	8	8	2	7	8
31 Fishing area total	*195*	*351*	*156*	*99*	*100*	*118*	*109*	*101*	*306*	*482*
34 Gambia	...	...	...	...	...	...	...	13	1	17
34 Fishing area total	...	...	...	...	...	...	...	*13*	*1*	*17*
37 Cyprus	44	45	29	32	37	78	58	33	42	55
37 Fishing area total	*44*	*45*	*29*	*32*	*37*	*78*	*58*	*33*	*42*	*55*
51 Bahrain	34	26	63	56	46	63	66	32	21	27
Egypt	...	...	...	...	227	283	918	...	...	...
Eritrea	...	2	15	9	0	0	3	2	2	1
Saudi Arabia	242	176	837	777	906	702	655	756	758	849
Untd Arab Em	...	...	...	...	...	...	...	333	383	712
51 Fishing area total	*276*	*204*	*915*	*842*	*1 179*	*1 048*	*1 642*	*1 123*	*1 164*	*1 589*
61 China,Taiwan	13	447	8	11	17	12	5	29	27	46
61 Fishing area total	*13*	*447*	*8*	*11*	*17*	*12*	*5*	*29*	*27*	*46*
71 Guam	...	1	0	1	1	4	1	0	0	0
N Marianas	...	5	1	3	7	1	2	4	13	2
Palau	36	25	39	25	18	12	12	12	26	12
71 Fishing area total	*36*	*31*	*40*	*29*	*26*	*17*	*15*	*16*	*39*	*14*
77 Amer Samoa	...	1	0	8	3	10	7	6	3	1
USA	-	-	-	-	-	-	-	-	-	22
77 Fishing area total	...	*1*	*0*	*8*	*3*	*10*	*7*	*6*	*3*	*23*
Species total	*564*	*1 079*	*1 148*	*1 021*	*1 362*	*1 283*	*1 836*	*1 321*	*1 582*	*2 226*
Angelfishes nei	**Demoiselles nca**		**Angeles nep**		***Pomacanthidae***			**1,70(66)XXX,XX**		**ANW**
31 Antigua Barb	...	...	...	...	...	...	...	...	...	28
31 Fishing area total	...	...	...	...	...	...	...	...	...	*28*
51 Eritrea	...	-	-	0	0	0	3	0	0	0
Saudi Arabia	-	-	-	-	-	7	14	1	3	9
51 Fishing area total	...	-	-	*0*	*0*	*7*	*17*	*1*	*3*	*9*
Species total	...	...	...	*0*	*0*	*7*	*17*	*1*	*3*	*37*
Giant African threadfin	**Gros capitaine**		**Barbudo gigante africano**		***Polydactylus quadrifilis***			**1,70(77)001,09**		**TGA**
34 Cameroon	39	50 F	76	54	40 F	12	34	43	170	75
Gambia	154	72	94	179	110	83	90	169	141	206
GuineaBissau	...	76	22	55	60 F	50 F	40 F	40 F	40 F	40 F
Nigeria	1 569	2 821	2 755	6 415	9 815	9 803	10 544	11 287	11 645	14 052
Senegal	...	...	...	...	...	...	...	...	...	650
Sierra Leone	638	636	635	640 F	679	75	61	690	445	469
Togo	1	15	16	16	1	0	1	8	6	1
34 Fishing area total	*2 401*	*3 670 F*	*3 598*	*7 359 F*	*10 705 F*	*10 023 F*	*10 770 F*	*12 237 F*	*12 447 F*	*15 493 F*
Species total	*2 401*	*3 670 F*	*3 598*	*7 359 F*	*10 705 F*	*10 023 F*	*10 770 F*	*12 237 F*	*12 447 F*	*15 493 F*
Fourfinger threadfin	**Barbure à quatre doigts**		**Barbudo de cuatro dedos**		***Eleutheronema tetradactylum***			**1,70(77)002,01**		**FOT**
51 Pakistan	753	653	812	516	1 783	969	...	63	55	60
51 Fishing area total	*753*	*653*	*812*	*516*	*1 783*	*969*	...	*63*	*55*	*60*
61 China,Taiwan	10 310	4 603	3 469	3 157	2 831	3 248	2 414	6 917	1 715	1 236
61 Fishing area total	*10 310*	*4 603*	*3 469*	*3 157*	*2 831*	*3 248*	*2 414*	*6 917*	*1 715*	*1 236*
Species total	*11 063*	*5 256*	*4 281*	*3 673*	*4 614*	*4 217*	*2 414*	*6 980*	*1 770*	*1 296*
Lesser African threadfin	**Petit capitaine**		**Barbudo enano africano**		***Galeoides decadactylus***			**1,70(77)003,02**		**GAL**
34 Cameroon	350	600 F	949	810	500 F	13	150	477	462	134
Congo Rep	379 F	418	350 F	300 F	250 F	190 F	150 F	63	60 F	60 F
Côte dIvoire	279	408	374	268	173	278	196	224	168	246

B-33 Miscellaneous coastal fishes / Poissons côtiers divers / Peces costeros diversos

Capture production by species, fishing areas and countries or areas / Captures par espèces, zones de pêche et pays ou zones / Capturas por especies, áreas de pesca y países o áreas

Species, Fishing area Espèce, Zone de pêche Especie, Area de pesca	1993 t	1994 t	1995 t	1996 t	1997 t	1998 t	1999 t	2000 t	2001 t	2002 t
Gabon	1 598	1 045	1 876	3 805	3 174	3 484	2 808	2 516	3 101	2 072
Gambia	99	41	88	140	146	57	60	87	116	120
Ghana	2 120	3 247	1 969	3 146	1 477	774	586	1 947	2 892	2 204
Guinea	300 F	300 F	431	338	204	43	88	123	62	60 F
GuineaBissau	...	45	101	108	110 F	100 F	90 F	90 F	90 F	90 F
Liberia	13	11	26	-	-	-	-	-	-	-
Nigeria	819	578	983	3 194	5 974	2 343	2 884	4 279	5 131	4 819
Senegal	915	1 456	5 357	5 255	5 453	4 849	3 016	2 844	3 849	3 037
Sierra Leone	328	327	392	390 F	921	758	716	1 149	914	753
Togo	28	24	7	14	3	1	20	149	98	88
34 Fishing area total	*7 228 F*	*8 500 F*	*12 903 F*	*17 768 F*	*18 385 F*	*12 890 F*	*10 764 F*	*13 948 F*	*16 943 F*	*13 683 F*
Species total	*7 228 F*	*8 500 F*	*12 903 F*	*17 768 F*	*18 385 F*	*12 890 F*	*10 764 F*	*13 948 F*	*16 943 F*	*13 683 F*
Royal threadfin	**Capitaine royal**		**Barbudo real**		***Pentanemus quinquarius***			**1,70(77)004,01**		**PET**
34 Cameroon	1 777	1 500 F	610	584	1 000 F	2 368	1 778	1 363	3 276	832
Congo Rep	329 F	200 F	200 F	200 F	200 F	180 F	190 F	189	180 F	180 F
Gabon	...	...	...	...	...	0	292	207	389	210
Guinea	700 F	720 F	1 020	435	233	...	...	...	...	...
Liberia	50	25	134	36	155	138	118	92	102	110 F
Sierra Leone	1 778	1 773	1 770	1 750 F	1 768	1 744	1 784	1 761	...	107
34 Fishing area total	*4 634 F*	*4 218 F*	*3 734 F*	*3 005 F*	*3 356 F*	*4 430 F*	*4 162 F*	*3 612*	*3 947 F*	*1 439 F*
Species total	*4 634 F*	*4 218 F*	*3 734 F*	*3 005 F*	*3 356 F*	*4 430 F*	*4 162 F*	*3 612*	*3 947 F*	*1 439 F*
Threadfins, tasselfishes nei	**Barbures, capitaines nca**		**Barbudos nep**		***Polynemidae***			**1,70(77)XXX,XX**		**THF**
04 India	-	-	-	467	416	243	229	66	345	345
04 Fishing area total	-	-	-	*467*	*416*	*243*	*229*	*66*	*345*	*345*
34 Benin	700 F	650 F	550 F	500 F	500 F	450 F	358	250 F	324	463
Korea Rep	195	-	-	-	-	141	-	-	-	-
Nigeria	1 638	1 845	2 156	2 930	3 184	-	-	-	-	-
Sao Tome Prn	67 F	97 F	102 F	75	75	100 F	110 F	100 F	100 F	100 F
Sierra Leone	...	...	2	2 F	-	-	-	-	-	-
34 Fishing area total	*2 600 F*	*2 592 F*	*2 810 F*	*3 507 F*	*3 759 F*	*691 F*	*468 F*	*350 F*	*424 F*	*563 F*
47 Angola	136	309	259	370	265	520	1 378	872	1 834	2 664
Korea Rep	327	272	25	-	-	-	-	-	-	-
Spain	-	-	-	-	2	-	-	-	-	-
47 Fishing area total	*463*	*581*	*284*	*370*	*267*	*520*	*1 378*	*872*	*1 834*	*2 664*
51 India	7 186	5 417	3 837	1 584	1 535	1 544	2 780	2 243	2 313	2 346
Korea Rep	-	-	-	239	-	456	-	-	-	-
Réunion	...	...	...	...	...	...	49	5	10	10 F
51 Fishing area total	*7 186*	*5 417*	*3 837*	*1 823*	*1 535*	*2 000*	*2 829*	*2 248*	*2 323*	*2 356 F*
57 Australia	193	42	255	333	-	-	-	-	-	-
India	5 610	5 707	5 810	5 001	5 144	5 977	7 049	3 467	3 094	3 736
Indonesia	5 336	4 746	6 445	6 528	6 983	7 543	7 035	7 363	7 973	8 440
Malaysia	1 272	1 037	939	1 270	1 032	1 507	928	1 624	1 951	1 341
Thailand	788	3 811	1 194	444	38	109	28	29	125	122
57 Fishing area total	*13 199*	*15 343*	*14 643*	*13 576*	*13 197*	*15 136*	*15 040*	*12 483*	*13 143*	*13 639*
71 Australia	28	475	385	573	216	272	914	1 013	1 094	1 009
Indonesia	16 683	18 050	25 834	24 183	24 726	26 677	26 300	30 919	27 390	29 440
Korea Rep	-	-	-	-	-	644	-	-	-	-
Malaysia	2 810	3 687	3 268	3 828	2 669	3 436	2 983	2 691	2 377	1 523
Philippines	1 690	2 616	2 548	2 383	2 933	3 017	2 524	2 371	2 804	2 892
Singapore	0	14	5	5	6	7	13	25	11	11
Thailand	2 365	2 132	1 010	1 106	1 041	1 196	404	358	838	848
71 Fishing area total	*23 576*	*26 974*	*33 050*	*32 078*	*31 591*	*35 249*	*33 138*	*37 377*	*34 514*	*35 723*
87 Colombia	-	-	-	16	4	3	16	10 F	10 F	10 F
87 Fishing area total	-	-	-	*16*	*4*	*3*	*16*	*10 F*	*10 F*	*10 F*
Species total	*47 024 F*	*50 907 F*	*54 624 F*	*51 837 F*	*50 769 F*	*53 842 F*	*53 098 F*	*53 406 F*	*52 593 F*	*55 300 F*
Patagonian blennie	**Guite de Patagonie**		**Róbalo patagónico**		***Eleginops maclovinus***			**1,70(92)007,02**		**BLP**
41 Argentina	156	624	66	148	57	1 194	2 023	1 745	233	50
Falkland Is	4	4	4	4	4	4	4	10	61	47
UK	-	-	-	-	-	13	-	-	-	-
41 Fishing area total	*160*	*628*	*70*	*152*	*61*	*1 211*	*2 027*	*1 755*	*294*	*97*
87 Chile	782	615	274	287	133	103	179	164	109	240
Estonia	-	-	-	-	-	-	14	-	-	-
87 Fishing area total	*782*	*615*	*274*	*287*	*133*	*103*	*193*	*164*	*109*	*240*
Species total	*942*	*1 243*	*344*	*439*	*194*	*1 314*	*2 220*	*1 919*	*403*	*337*
Marbled rockcod	**Bocasse marbrée**		**Trama jaspeada**		***Notothenia rossii***			**1,70(92)019,02**		**NOR**
48 Argentina	-	1	2	-	-	-	-	-	-	-
Japan	-	-	-	-	-	-	-	-	-	0
UK	-	1	-	-	-	-	-	0	-	5
USA	-	-	-	-	-	-	0	-	0	-
48 Fishing area total	-	*2*	*2*	-	-	-	*0*	*0*	*0*	*5*

B-33 Miscellaneous coastal fishes / Poissons côtiers divers / Peces costeros diversos

Capture production by species, fishing areas and countries or areas / Captures par espèces, zones de pêche et pays ou zones / Capturas por especies, áreas de pesca y países o áreas

Species, Fishing area Espèce, Zone de pêche Especie, Area de pesca	1993 t	1994 t	1995 t	1996 t	1997 t	1998 t	1999 t	2000 t	2001 t	2002 t
58 Australia	-	-	-	-	0	-	0	-	-	0
France	-	0	-	-	1	-	1	-	-	-
Ukraine	2	-	-	-	-	-	-	-	-	-
58 Fishing area total	*2*	*0*	-	-	*1*	-	*1*	-	-	*0*
Species total	*2*	*2*	*2*	-	*1*	-	*1*	*0*	*0*	*5*
Humped rockcod	**Bocasse bossue**		**Trama jorobada**		***Notothenia gibberifrons***				**1,70(92)019,03**	**NOG**
48 Argentina	-	1	1	-	-	-	-	-	-	-
France	-	-	-	-	-	-	-	-	0	-
UK	-	3	-	-	-	-	0	1	0	1
USA	-	-	-	-	-	-	5	-	2	-
48 Fishing area total	-	*4*	*1*	-	-	-	*5*	*1*	*2*	*1*
Species total	-	*4*	*1*	-	-	-	*5*	*1*	*2*	*1*
Yellowbelly rockcod	**Bocasse jaune**		**Trama amarilla**		***Notothenia neglecta***				**1,70(92)019,04**	**NON**
48 USA	-	-	-	-	-	-	0	-	2	-
48 Fishing area total	-	-	-	-	-	-	*0*	-	*2*	-
Species total	-	-	-	-	-	-	*0*	-	*2*	-
Grey rockcod	**Bocasse grise**		**Trama gris**		***Notothenia squamifrons***				**1,70(92)019,06**	**NOS**
48 Argentina	-	0	-	-	-	-	-	-	-	-
Bulgaria	-	0	-	-	-	-	-	-	-	-
UK	-	0	-	-	-	-	-	5	-	0
USA	-	-	-	-	-	-	5	-	0	-
48 Fishing area total	-	*0*	-	-	-	-	*5*	*5*	*0*	*0*
57 Australia	-	-	-	-	-	-	-	363	329	329
57 Fishing area total	-	-	-	-	-	-	-	*363*	*329*	*329*
58 Australia	0	-	-	-	4	3	10	0	0	1
France	-	-	-	15	-	-	-	-	-	-
Ukraine	-	-	0	-	-	-	-	-	-	-
58 Fishing area total	*0*	-	*0*	*15*	*4*	*3*	*10*	*0*	*0*	*1*
Species total	*0*	*0*	*0*	*15*	*4*	*3*	*15*	*368*	*329*	*330*
Striped-eyed rockcod	**Bocasse aux yeux rayés**		**Trama ojirayada**		***Notothenia kempi***				**1,70(92)019,10**	**NOK**
58 South Africa	-	-	-	-	-	-	-	-	-	0
58 Fishing area total	-	-	-	-	-	-	-	-	-	*0*
Species total	-	-	-	-	-	-	-	-	-	*0*
Triangular rockcod	**Bocasse triangulaire**		**Trama triangular**		***Notothenia acuta***				**1,70(92)019,11**	**NOA**
58 Australia	0	-	-	-	-	-	-	-	-	-
58 Fishing area total	*0*	-	-	-	-	-	-	-	-	-
Species total	*0*	-	-	-	-	-	-	-	-	-
Painted notie	**Bocassette écrivain**		**Doradillo escribano**		***Nototheniops larseni***				**1,70(92)430,01**	**NOL**
48 Argentina	-	0	-	-	-	-	-	-	-	-
UK	-	-	-	-	-	-	-	0	-	0
48 Fishing area total	-	*0*	-	-	-	-	-	*0*	-	*0*
Species total	-	*0*	-	-	-	-	-	*0*	-	*0*
Yellowfin notie	**Bocassette dégarnie**		**Doradillo pobre**		***Nototheniops nudifrons***				**1,70(92)430,02**	**NOD**
48 Argentina	-	0	-	-	-	-	-	-	-	-
UK	-	-	-	-	-	-	-	0	-	-
USA	-	-	-	-	-	-	-	-	0	-
48 Fishing area total	-	*0*	-	-	-	-	-	*0*	*0*	-
Species total	-	*0*	-	-	-	-	-	*0*	*0*	-
Toad notie	**Bocassette crapaud**		**Ojo de sapo**		***Nototheniops mizops***				**1,70(92)430,03**	**NOZ**
58 Australia	0	-	-	-	-	-	-	-	-	-
58 Fishing area total	*0*	-	-	-	-	-	-	-	-	-
Species total	*0*	-	-	-	-	-	-	-	-	-
Antarctic rockcods nei	**Bocassons nca**		**Austrobacalaos nep**		***Trematomus spp***				**1,70(92)448,XX**	**TRT**
48 USA	-	-	-	-	-	-	0	-	0	-
48 Fishing area total	-	-	-	-	-	-	*0*	-	*0*	-
Species total	-	-	-	-	-	-	*0*	-	*0*	-

B-33 Miscellaneous coastal fishes — Capture production by species, fishing areas and countries or areas
Poissons côtiers divers — Captures par espèces, zones de pêche et pays ou zones
Peces costeros diversos — Capturas por especies, áreas de pesca y países o áreas

Species, Fishing area Espèce, Zone de pêche Especie, Area de pesca	1993 t	1994 t	1995 t	1996 t	1997 t	1998 t	1999 t	2000 t	2001 t	2002 t
Striped rockcod	**Bocasson rayé**		**Austrobacalao rayado**		***Pagothenia hansoni***				**1,70(92)529,01**	**TRH**
48 Argentina	-	0	-	-	-	-	-	-	-	-
UK	-	-	-	-	-	-	-	0	-	-
48 Fishing area total	-	*0*	-	-	-	-	-	*0*	-	-
Species total	-	*0*	-	-	-	-	-	*0*	-	-
Antarctic silverfish	**Calandre antarctique**		**Diablillo antártico**		***Pleuragramma antarcticum***				**1,70(92)535,01**	**ANS**
48 USA	-	-	-	-	-	-	-	-	0	-
48 Fishing area total	-	-	-	-	-	-	-	-	*0*	-
Species total	-	-	-	-	-	-	-	-	*0*	-
Antarctic rockcods, noties nei	**Bocasses, bocassons nca**		**Tramas, doradillos nep**		***Nototheniidae***				**1,70(92)XXX,XX**	**NOX**
41 Argentina	-	-	2	-	-	-	-	-	182	180
Poland	-	-	-	-	-	-	207	-	-	-
41 Fishing area total	-	-	*2*	-	-	-	*207*	-	*182*	*180*
48 Argentina	-	0	2	-	-	-	-	-	-	-
Chile	-	-	-	-	-	0	-	-	-	-
South Africa	-	-	-	-	-	1	-	-	-	-
Spain	-	-	-	-	0	-	-	-	-	-
UK	-	-	-	-	-	-	-	-	0	10
USA	-	-	-	-	-	-	-	-	0	-
48 Fishing area total	-	*0*	*2*	-	*0*	*1*	-	-	*0*	*10*
58 Australia	-	-	-	-	-	3	-	-	0	-
58 Fishing area total	-	-	-	-	-	*3*	-	-	*0*	-
81 New Zealand	...	...	34	16	34	3	19	3	15	42
81 Fishing area total	...	...	*34*	*16*	*34*	*3*	*19*	*3*	*15*	*42*
88 Chile	-	-	-	-	-	0	-	-	-	-
New Zealand	-	-	-	-	-	0	0	0	1	0
South Africa	-	-	-	-	-	-	-	-	0	-
88 Fishing area total	-	-	-	-	-	*0*	*0*	*0*	*1*	*0*
Species total	...	*0*	*38*	*16*	*34*	*7*	*226*	*3*	*198*	*232*
Glassfishes	**...B**		**...C**		***Ambassidae***				**1,70(95)XXX,XX**	**AIB**
71 Philippines	4 831	3 969	3 554	3 075	2 594	2 186	1 733	1 575	1 876	1 986
71 Fishing area total	*4 831*	*3 969*	*3 554*	*3 075*	*2 594*	*2 186*	*1 733*	*1 575*	*1 876*	*1 986*
Species total	*4 831*	*3 969*	*3 554*	*3 075*	*2 594*	*2 186*	*1 733*	*1 575*	*1 876*	*1 986*
Percoids nei	**Percoides nca**		**Percoideos nep**		***Percoidei***				**1,70(XX)XXX,XX**	**PRC**
41 Brazil	1 000 F	800 F	720	664	692	1 838	947	1 465	2 018	2 358
41 Fishing area total	*1 000 F*	*800 F*	*720*	*664*	*692*	*1 838*	*947*	*1 465*	*2 018*	*2 358*
51 India	10 563	8 922	11 273	89 717	58 969	52 230	53 600	58 297	39 928	72 798
Mauritius	196	201	186	183	158	153	165	181	152	135
Tanzania	312	269	340	333	350	200	250	300	200	100
51 Fishing area total	*11 071*	*9 392*	*11 799*	*90 233*	*59 477*	*52 583*	*54 015*	*58 778*	*40 280*	*73 033*
57 Australia	0	7	6	351	-	-	-	-	-	-
India	33 640	32 090	26 710	26 162	28 741	10 268	36 350	36 264	38 142	28 885
Malaysia	-	744	-	-	-	-	-	-	-	-
57 Fishing area total	*33 640*	*32 841*	*26 716*	*26 513*	*28 741*	*10 268*	*36 350*	*36 264*	*38 142*	*28 885*
71 Australia	0	0	0	0	0	0	-	-	-	-
Kiribati	1 580	1 560	1 570	1 590	1 580	10 713	24 588	632	1 069	1 198
Malaysia	-	431	-	-	-	-	-	-	-	-
Philippines	41 228	38 425	38 843	36 825	20 212	20 036	20 717	19 912	21 207	23 813
71 Fishing area total	*42 808*	*40 416*	*40 413*	*38 415*	*21 792*	*30 749*	*45 305*	*20 544*	*22 276*	*25 011*
81 Australia	145	130	130	146	131	131	-	69	41	58
81 Fishing area total	*145*	*130*	*130*	*146*	*131*	*131*	-	*69*	*41*	*58*
Species total	*88 664 F*	*83 579 F*	*79 778*	*155 971*	*110 833*	*95 569*	*136 617*	*117 120*	*102 757*	*129 345*
Eelpout	**Loquette d'Europe**		**Viruela**		***Zoarces viviparus***				**1,71(15)004,01**	**ELP**
27 Denmark	9	9	2	3	7	2	7	3	5	4
Estonia	9	1	2	2	8	9	2	1	1	1
Germany	7	1	5	3	2	2	2	1	5	1
Latvia	79	164	143	139	80	41	32	23	26	37
27 Fishing area total	*104*	*175*	*152*	*147*	*97*	*54*	*43*	*28*	*37*	*43*
Species total	*104*	*175*	*152*	*147*	*97*	*54*	*43*	*28*	*37*	*43*
Ocean pout	**Loquette d'Amérique**		**...C**		***Macrozoarces americanus***				**1,71(15)012,02**	**OPT**
21 USA	226	196	24	41	15	17	18	19	18	12
21 Fishing area total	*226*	*196*	*24*	*41*	*15*	*17*	*18*	*19*	*18*	*12*

B-33 Miscellaneous coastal fishes / Poissons côtiers divers / Peces costeros diversos

Capture production by species, fishing areas and countries or areas / Captures par espèces, zones de pêche et pays ou zones / Capturas por especies, áreas de pesca y países o áreas

Species, Fishing area Espèce, Zone de pêche Especie, Area de pesca	1993 t	1994 t	1995 t	1996 t	1997 t	1998 t	1999 t	2000 t	2001 t	2002 t
Species total	*226*	*196*	*24*	*41*	*15*	*17*	*18*	*19*	*18*	*12*
Pacific sandlance	**Lançon du Pacifique**		**Lanzón del Pacífico**		***Ammodytes personatus***				**1,72(04)002,04**	**PAS**
61 Japan	106 568	109 344	108 124	115 766	108 666	90 688	82 918	49 819	88 164	67 564
Korea Rep	-	-	-	-	-	-	-	16 293	4 803	5 145
Russian Fed	-	-	-	-	-	-	-	17	-	192
61 Fishing area total	*106 568*	*109 344*	*108 124*	*115 766*	*108 666*	*90 688*	*82 918*	*66 129*	*92 967*	*72 901*
Species total	*106 568*	*109 344*	*108 124*	*115 766*	*108 666*	*90 688*	*82 918*	*66 129*	*92 967*	*72 901*
Sandeels(=Sandlances) nei	**Lançons nca**		**Lanzones nep**		***Ammodytes spp***				**1,72(04)002,XX**	**SAN**
21 Canada	3	-	-	2	1	0	-	2	-	-
USA	7	14	1	-	1	1	1	0	0	0
21 Fishing area total	*10*	*14*	*1*	*2*	*2*	*1*	*1*	*2*	*0*	*0*
27 Channel Is	1 F	-	-	2	63	61	41	41	41	40
Denmark	631 549	839 825	844 512	669 035	840 774	646 905	528 551	567 350	666 295	662 402
Faeroe Is	2 934	10 288	7 485	5 023	11 221	11 071	7 487	8 513	6 030	6 030 F
France	131	89	114	95	159	70	92	88	50	77
Iceland	-	-	-	-	-	-	-	-	8	-
Ireland	-	-	-	-	-	-	389	-	-	10
Norway	101 519	168 155	263 490	160 702	350 672	343 625	187 589	119 015	187 459	175 985
Portugal	9	17	64	41	18	9	13	29	40	34
Spain	-	-	-	542	492	183	390	811	584	280
Sweden	-	20	40	-	1	8 585	23 225	28 165	50 559	55 953
UK	9 121	19 123	17 528	22 936	39 271	29 080	14 102	16 530	1 264	3 691
27 Fishing area total	*745 264 F*	*1 037 517*	*1 133 233*	*858 376*	*1 242 671*	*1 039 589*	*761 879*	*740 542*	*912 330*	*904 502 F*
37 France	...	...	...	...	...	...	1	2	1	0
Spain	-	-	-	74	164	283	367	306	105	428
37 Fishing area total	*...*	*...*	*...*	*74*	*164*	*283*	*368*	*308*	*106*	*428*
Species total	*745 274 F*	*1 037 531*	*1 133 234*	*858 452*	*1 242 837*	*1 039 873*	*762 248*	*740 852*	*912 436*	*904 930 F*
Brazilian flathead	**Platête brésilien**		**Pez palo**		***Percophis brasilianus***				**1,72(08)202,01**	**FLA**
41 Argentina	5 607	6 187	8 680	8 771	11 475	9 677	6 526	7 255	7 040	4 225
Brazil	...	...	640	460	412	514	709	607	624	711
Uruguay	-	-	-	-	-	-	1	-	0	0
41 Fishing area total	*5 607*	*6 187*	*9 320*	*9 231*	*11 887*	*10 191*	*7 236*	*7 862*	*7 664*	*4 936*
Species total	*5 607*	*6 187*	*9 320*	*9 231*	*11 887*	*10 191*	*7 236*	*7 862*	*7 664*	*4 936*
Argentinian sandperch	**...B**		**Salmón de mar**		***Pseudopercis semifasciata***				**1,72(09)006,02**	**UPR**
41 Argentina	3 589	2 907	3 023	3 438	2 483	2 222	2 679	1 905	1 887	1 641
41 Fishing area total	*3 589*	*2 907*	*3 023*	*3 438*	*2 483*	*2 222*	*2 679*	*1 905*	*1 887*	*1 641*
Species total	*3 589*	*2 907*	*3 023*	*3 438*	*2 483*	*2 222*	*2 679*	*1 905*	*1 887*	*1 641*
New Zealand blue cod	**...B**		**...C**		***Parapercis colias***				**1,72(09)196,02**	**NEB**
81 New Zealand	1 986	1 910	11 262	2 115	2 227	2 313	2 286	2 130	2 441	2 376
81 Fishing area total	*1 986*	*1 910*	*11 262*	*2 115*	*2 227*	*2 313*	*2 286*	*2 130*	*2 441*	*2 376*
Species total	*1 986*	*1 910*	*11 262*	*2 115*	*2 227*	*2 313*	*2 286*	*2 130*	*2 441*	*2 376*
Greater weever	**Grande vive**		**Escorpión**		***Trachinus draco***				**1,72(12)010,02**	**WEG**
27 Denmark	120	94	76	45	50	36	52	39	48	98
France	131	149	169	127	132	147	213	218	203	210
Ireland	-	-	-	-	-	-	-	-	691	616
Netherlands	-	-	-	-	-	-	-	-	6	0
Portugal	0	0	0	0	0	2	7	12	11	22
Spain	-	-	-	-	-	-	-	2	-	-
Sweden	27	21	25	10	1	3	12	7	26	9
27 Fishing area total	*278*	*264*	*270*	*182*	*183*	*188*	*284*	*278*	*985*	*955*
37 France	4	3	2	2	37	37	5	6	3	5
Malta	-	-	1	2	2	0	3	0	0	-
Spain	-	-	-	60	69	67	-	28	57	67
Tunisia	24	8	5	19	46	103	16	19	20	13
37 Fishing area total	*28*	*11*	*8*	*83*	*154*	*207*	*24*	*53*	*80*	*85*
Species total	*306*	*275*	*278*	*265*	*337*	*395*	*308*	*331*	*1 065*	*1 040*
Black goby	**Gobie noir**		**Chaparrudo**		***Gobius niger***				**1,73(21)003,03**	**GBN**
37 Slovenia	-	-	-	-	-	-	-	1	1	1
37 Fishing area total	-	-	-	-	-	-	-	*1*	*1*	*1*
Species total	-	-	-	-	-	-	-	*1*	*1*	*1*
Gobies nei	**Gobies nca**		**Góbidos nep**		***Gobiidae***				**1,73(21)XXX,XX**	**GPA**
05 Albania	-	-	-	-	-	-	-	-	5	7
05 Fishing area total	-	-	-	-	-	-	-	-	*5*	*7*

B-33

Miscellaneous coastal fishes — **Capture production by species, fishing areas and countries or areas**
Poissons côtiers divers — **Captures par espèces, zones de pêche et pays ou zones**
Peces costeros diversos — **Capturas por especies, áreas de pesca y países o áreas**

Species, Fishing area Espèce, Zone de pêche Especie, Area de pesca	1993 t	1994 t	1995 t	1996 t	1997 t	1998 t	1999 t	2000 t	2001 t	2002 t
27 Portugal	0	0	0	0	0	-	0	-	-	-
27 Fishing area total	*0*	*0*	*0*	*0*	*0*	-	*0*	-	-	-
37 Albania	...	...	6	10	4	-	-	-	-	-
Bulgaria	10	11 F	580	477	424	381	437	145	142	142
France	3	3	0	4	2	0	4	3	1	3
Georgia	...	...	...	...	...	...	2	3	5	32
Italy	1 187	1 583	1 452	1 311	1 085	991	800	712	665	695
Malta	2	0	0	0	0	0	0	0	-	-
Romania	3	22	13	8	2	6	30	42	24	46
Russian Fed	2	-	-	2	1	3	5	12	25	44
Spain	-	-	-	-	-	-	478	311	378	294
Turkey	1 024	1 248	233	390	305	250	325	300	335	493
Ukraine	302	401	201	72	96	286	601	825	1 510	3 582
37 Fishing area total	*2 533*	*3 268 F*	*2 485*	*2 274*	*1 919*	*1 917*	*2 682*	*2 353*	*3 085*	*5 331*
47 Namibia	172	19	5	552	287	20 998	16	3	...	...
47 Fishing area total	*172*	*19*	*5*	*552*	*287*	*20 998*	*16*	*3*	...	...
61 Korea Rep	2 788	3 570	1 941	1 598	1 722	1 299	897	788	703	388
Russian Fed	27 479	12 549	14 071	14 522	16 176	20 551	30 172	32 733	27 321	22 099
61 Fishing area total	*30 267*	*16 119*	*16 012*	*16 120*	*17 898*	*21 850*	*31 069*	*33 521*	*28 024*	*22 487*
71 Philippines	4 090	7 856	8 780	8 404	7 542	6 756	8 021	7 719	9 032	9 291
71 Fishing area total	*4 090*	*7 856*	*8 780*	*8 404*	*7 542*	*6 756*	*8 021*	*7 719*	*9 032*	*9 291*
Species total	*37 062*	*27 262 F*	*27 282*	*27 350*	*27 646*	*51 521*	*41 788*	*43 596*	*40 146*	*37 116*
Surgeonfishes nei	**Chirurgiens nca**		**Navajones nep**		***Acanthuridae***			**1,74(02)XXX,XX**		**SUR**
31 Antigua Barb	...	...	...	...	...	...	...	...	158	237
Grenada	0	0	0	0	0	0	1	1	2	2
St Kitts Nev	...	...	5	11	4	9	6	1	0	4
31 Fishing area total	*0*	*0*	*5*	*11*	*4*	*9*	*7*	*2*	*160*	*243*
34 Nigeria	-	31	-	-	-	-	-	-	-	-
34 Fishing area total	-	*31*	-	-	-	-	-	-	-	-
51 Saudi Arabia	-	68	113	238	173	253	205	149	343	364
51 Fishing area total	-	*68*	*113*	*238*	*173*	*253*	*205*	*149*	*343*	*364*
71 Fiji Islands	346	353 F	278	125	105	180	206	180 F	196	195 F
Guam	...	5	0	1	...	3	0	-	0	5
N Marianas	...	...	1	3	-	0	3	3	10	2
Palau	24	26	51	42	10	7	13	13	46	36
Philippines	5 172	6 058	4 752	4 474	7 367	7 770	5 797	5 392	6 211	6 939
71 Fishing area total	*5 542*	*6 442 F*	*5 082*	*4 645*	*7 482*	*7 960*	*6 019*	*5 588 F*	*6 463*	*7 177 F*
77 Amer Samoa	...	...	...	8	-	13	16	10	3	2
USA	-	-	-	-	-	-	-	-	-	60
77 Fishing area total	...	...	...	*8*	-	*13*	*16*	*10*	*3*	*62*
Species total	*5 542*	*6 541 F*	*5 200*	*4 902*	*7 659*	*8 235*	*6 247*	*5 749 F*	*6 969*	*7 846 F*
Batfishes	**Poules d'eau**		**Dalapuganos**		***Platax spp***			**1,74(05)206,XX**		**BAT**
51 Eritrea	...	-	-	-	0	1	7	1	0	0
51 Fishing area total	...	-	-	-	*0*	*1*	*7*	*1*	*0*	*0*
71 Philippines	1 506	1 928	1 600	1 584	2 975	2 792	2 870	2 597	2 830	3 227
71 Fishing area total	*1 506*	*1 928*	*1 600*	*1 584*	*2 975*	*2 792*	*2 870*	*2 597*	*2 830*	*3 227*
Species total	*1 506*	*1 928*	*1 600*	*1 584*	*2 975*	*2 793*	*2 877*	*2 598*	*2 830*	*3 227*
Spadefishes nei	**Chèvres, disques nca**		**Pagualas nep**		***Ephippidae***			**1,74(05)XXX,XX**		**SPA**
21 USA	2	2	5	6	-	14	21	14	20	10
21 Fishing area total	*2*	*2*	*5*	*6*	-	*14*	*21*	*14*	*20*	*10*
31 USA	-	-	-	-	-	-	-	15	1	6
31 Fishing area total	-	-	-	-	-	-	-	*15*	*1*	*6*
34 Benin	1 F	1 F	2 F	3 F	5 F	4 F	4	3 F	2	1
Togo	2	0	-	-	1	1	0	1	-	-
34 Fishing area total	*3 F*	*1 F*	*2 F*	*3 F*	*6 F*	*5 F*	*4*	*4 F*	*2*	*1*
41 Brazil	...	...	40	71	55	51	283	325	381	241
41 Fishing area total	...	...	*40*	*71*	*55*	*51*	*283*	*325*	*381*	*241*
51 Saudi Arabia	-	-	-	-	-	17	1	1	2	2
51 Fishing area total	-	-	-	-	-	*17*	*1*	*1*	*2*	*2*
Species total	*5 F*	*3 F*	*47 F*	*80 F*	*61 F*	*87 F*	*309*	*359 F*	*406*	*260*
Scats	**Pavillons**		**Pingos**		***Scatophagus spp***			**1,74(06)330,XX**		**SCT**
04 Philippines	32	18	169	169	95	20	40	61	55	30
04 Fishing area total	*32*	*18*	*169*	*169*	*95*	*20*	*40*	*61*	*55*	*30*

B-33 Miscellaneous coastal fishes / Poissons côtiers divers / Peces costeros diversos

Capture production by species, fishing areas and countries or areas / Captures par espèces, zones de pêche et pays ou zones / Capturas por especies, áreas de pesca y países o áreas

Species, Fishing area Espèce, Zone de pêche Especie, Area de pesca	1993 t	1994 t	1995 t	1996 t	1997 t	1998 t	1999 t	2000 t	2001 t	2002 t
71 Philippines	2 767	3 228	4 613	4 035	2 057	2 476	2 501	2 637	3 162	3 205
71 Fishing area total	*2 767*	*3 228*	*4 613*	*4 035*	*2 057*	*2 476*	*2 501*	*2 637*	*3 162*	*3 205*
Species total	*2 799*	*3 246*	*4 782*	*4 204*	*2 152*	*2 496*	*2 541*	*2 698*	*3 217*	*3 235*
Spinefeet(=Rabbitfishes) nei	**Sigans nca**		**Síganos nep**		***Siganus spp***				**1,74(07)001,XX**	**SPI**
37 Cyprus	19	3	1	11	7	36	19	12	31	57
Egypt	...	...	246	750	372	379	481	624	904	923
Israel	85	50	50	...	...	...	...	...	...	...
Palest, O.T.	...	...	...	2	11	10	10 F	10 F	5 F	3
Syria	...	...	...	...	...	...	...	...	...	118
37 Fishing area total	*104*	*53*	*297*	*763*	*390*	*425*	*510 F*	*646 F*	*940 F*	*1 101*
51 Bahrain	1 242	1 251	1 543	2 185	1 612	1 523	1 241	2 114	1 899	2 009
Egypt	...	...	185	128	129	175	199	42	1 402	665
Eritrea	...	-	-	0	0	1	1	1	0	0
Jordan	...	...	...	...	...	10	12	10	11	8
Kenya	440	365	387	404	347	356	304	299	403	469
Mauritius	438	455	465	514	430	494	461	448	450	365
Oman	116	69	142	59	259	122	97	131	363	559
Qatar	281	194	161	225	237	285	240	387	451	400
Saudi Arabia	1 388	1 887	2 341	2 779	2 173	1 761	1 691	1 832	1 822	1 998
Seychelles	...	...	293	344	163	145	248	164	91	203
Tanzania	2 319	2 527	3 470	3 816	4 000	3 550	3 500	3 525	3 000	3 200
Untd Arab Em	503	472	535	541	578	580	595	1 825	2 100	1 234
51 Fishing area total	*6 727*	*7 220*	*9 522*	*10 995*	*9 928*	*9 002*	*8 589*	*10 778*	*11 992*	*11 110*
57 Malaysia	134	104	86	81	88	237	27	109	87	49
57 Fishing area total	*134*	*104*	*86*	*81*	*88*	*237*	*27*	*109*	*87*	*49*
71 Fiji Islands	76	77 F	62	73	80	93	100	90 F	96	95 F
Guam	...	6	0	0	...	1	2	0	0	1
Malaysia	1 354	1 090	1 058	1 071	1 247	1 227	1 181	1 284	1 407	1 683
N Marianas	...	...	0	...	...	2	3	5	4	4
Palau	14	12	13	14	5	4	7	14	11	6
Philippines	24 293	19 175	18 766	17 012	15 720	18 246	19 977	20 108	20 336	17 581
Singapore	48	20	-	14	21	24	28	5	8	12
71 Fishing area total	*25 785*	*20 380 F*	*19 899*	*18 184*	*17 073*	*19 597*	*21 298*	*21 506 F*	*21 862*	*19 382 F*
Species total	*32 750*	*27 757 F*	*29 804*	*30 023*	*27 479*	*29 261*	*30 424 F*	*33 039 F*	*34 881 F*	*31 642 F*
Lingcod	**Morue-lingue**		**Bacalao largo(=Lorcha)**		***Ophiodon elongatus***				**1,78(07)005,01**	**CLI**
67 Canada	5 234	4 568	3 793	2 500	1 700	1 900	2 523	3 042	2 511	2 673
USA	1 681	1 851	1 310	1 902	1 525	564	538	144	153	208
67 Fishing area total	*6 915*	*6 419*	*5 103*	*4 402*	*3 225*	*2 464*	*3 061*	*3 186*	*2 664*	*2 881*
77 USA	497	441	381	352	364	95	93	28	34	44
77 Fishing area total	*497*	*441*	*381*	*352*	*364*	*95*	*93*	*28*	*34*	*44*
Species total	*7 412*	*6 860*	*5 484*	*4 754*	*3 589*	*2 559*	*3 154*	*3 214*	*2 698*	*2 925*
Okhotsk atka mackerel	**Terpuga arabesque de Okhotsk**		**Lorcha de Okhotsk**		***Pleurogrammus azonus***				**1,78(07)014,02**	**ATK**
61 Japan	135 529	152 503	176 603	181 513	206 763	240 971	169 481	165 118	161 160	154 736
Korea D P Rp	...	3 477	3 480	6 487	3 535	3 500 F	3 500 F	3 000 F	3 000 F	3 000 F
Korea Rep	5 846	3 552	3 342	4 103	2 983	7 911	1 005	2 554	1 261	627
Russian Fed	16 945	18 283	19 112	21 260	36 075	40 889	40 283	52 784	49 171	55 570
61 Fishing area total	*158 320*	*177 815*	*202 537*	*213 363*	*249 356*	*293 271 F*	*214 269 F*	*223 456 F*	*214 592 F*	*213 933 F*
Species total	*158 320*	*177 815*	*202 537*	*213 363*	*249 356*	*293 271 F*	*214 269 F*	*223 456 F*	*214 592 F*	*213 933 F*
Atka mackerel	**Terpuga atka**		**Lorcha de atka**		***Pleurogrammus monopterygius***				**1,78(07)014,03**	**HUM**
67 USA	70 499	62 501	67 156	88 030	59 538	51 579	51 436	44 592	57 096	37 759
67 Fishing area total	*70 499*	*62 501*	*67 156*	*88 030*	*59 538*	*51 579*	*51 436*	*44 592*	*57 096*	*37 759*
Species total	*70 499*	*62 501*	*67 156*	*88 030*	*59 538*	*51 579*	*51 436*	*44 592*	*57 096*	*37 759*
Bartail flathead	**Platycéphale indien**		**Chato índico**		***Platycephalus indicus***				**1,78(09)018,01**	**FLI**
61 Korea Rep	3 489	3 036	2 520	2 900	4 196	2 857	2 248	2 310	1 699	1 199
61 Fishing area total	*3 489*	*3 036*	*2 520*	*2 900*	*4 196*	*2 857*	*2 248*	*2 310*	*1 699*	*1 199*
Species total	*3 489*	*3 036*	*2 520*	*2 900*	*4 196*	*2 857*	*2 248*	*2 310*	*1 699*	*1 199*
Flatheads nei	**Platycéphalidés nca**		**Platicefálidos nep**		***Platycephalidae***				**1,78(09)XXX,XX**	**FLH**
51 Eritrea	...	18	14	0	0	0	21	5	11	1
Saudi Arabia	-	-	-	-	-	18	7	5	6	329
51 Fishing area total	...	*18*	*14*	*0*	*0*	*18*	*28*	*10*	*17*	*330*
57 Australia	541	460	1 360	1 640	1 477	989	3 102	2 782	2 432	2 864
57 Fishing area total	*541*	*460*	*1 360*	*1 640*	*1 477*	*989*	*3 102*	*2 782*	*2 432*	*2 864*
71 Australia	...	...	...	...	...	...	72	71	57	46
71 Fishing area total	...	...	...	...	...	...	*72*	*71*	*57*	*46*

B-33 Miscellaneous coastal fishes / Poissons côtiers divers / Peces costeros diversos

Capture production by species, fishing areas and countries or areas / Captures par espèces, zones de pêche et pays ou zones / Capturas por especies, áreas de pesca y países o áreas

Species, Fishing area Espèce, Zone de pêche Especie, Area de pesca	1993 t	1994 t	1995 t	1996 t	1997 t	1998 t	1999 t	2000 t	2001 t	2002 t
81 Australia	2 629	2 647	2 760	2 748	3 098	2 296	1 564	1 964	1 644	1 500
New Zealand	-	-	2	1	1	2	6	7	19	51
81 Fishing area total	*2 629*	*2 647*	*2 762*	*2 749*	*3 099*	*2 298*	*1 570*	*1 971*	*1 663*	*1 551*
Species total	*3 170*	*3 125*	*4 136*	*4 389*	*4 576*	*3 305*	*4 772*	*4 834*	*4 169*	*4 791*
Cabezon	**...B**		**...C**		***Scorpaenichthys marmoratus***				**1,78(13)030,01**	**SMQ**
67 USA	...	...	...	...	...	7	33	37	50	49
67 Fishing area total	...	...	...	...	...	*7*	*33*	*37*	*50*	*49*
77 USA	...	...	...	...	...	161	140	111	69	48
77 Fishing area total	...	...	...	...	...	*161*	*140*	*111*	*69*	*48*
Species total	...	...	...	...	...	*168*	*173*	*148*	*119*	*97*
Sculpins nei	**Chabots nca**		**Cótidos nep**		***Cottidae***				**1,78(13)XXX,XX**	**SWU**
21 USA	0	-	1	-	2	5	2	1	1	0
21 Fishing area total	*0*	-	*1*	-	*2*	*5*	*2*	*1*	*1*	*0*
67 USA	-	-	-	-	-	-	1	63	71	158
67 Fishing area total	-	-	-	-	-	-	*1*	*63*	*71*	*158*
77 USA	-	-	-	-	-	-	-	-	3	3
77 Fishing area total	-	-	-	-	-	-	-	-	*3*	*3*
Species total	*0*	-	*1*	-	*2*	*5*	*3*	*64*	*75*	*161*
...A	**...B**		**Camotillo**		***Normanichthys crockeri***				**1,78(16)022,05**	**NRC**
21 Faeroe Is	-	-	2	3	-	-	-	-	-	-
21 Fishing area total	-	-	*2*	*3*	-	-	-	-	-	-
87 Chile	0	25	2 690	3 921	20 426	236	4 843	853	223	4 371
87 Fishing area total	*0*	*25*	*2 690*	*3 921*	*20 426*	*236*	*4 843*	*853*	*223*	*4 371*
Species total	*0*	*25*	*2 692*	*3 924*	*20 426*	*236*	*4 843*	*853*	*223*	*4 371*
Boxfishes nei	**Coffres nca**		**Toritos nep**		***Ostraciidae***				**1,90(01)XXX,XX**	**BXF**
31 Antigua Barb	...	...	...	...	...	...	...	...	66	38
Br Virgin Is	...	...	...	...	1	1	1	0	0	0
Puerto Rico	...	...	...	...	...	...	54	68	52	42
31 Fishing area total	...	...	...	...	*1*	*1*	*55*	*68*	*118*	*80*
Species total	...	...	...	...	*1*	*1*	*55*	*68*	*118*	*80*
Northern puffer	**Compère bigaré**		**Tamboril norteño**		***Sphoeroides maculatus***				**1,90(02)006,12**	**PUF**
21 USA	57	32	19	11	18	17	37	36	35	57
21 Fishing area total	*57*	*32*	*19*	*11*	*18*	*17*	*37*	*36*	*35*	*57*
Species total	*57*	*32*	*19*	*11*	*18*	*17*	*37*	*36*	*35*	*57*
Purple puffer	**Compère rouge**		**Tamboril rojo**		***Takifugu vermicularis***				**1,90(02)011,01**	**PUP**
61 Korea Rep	5 683	4 191	10 178	9 708	7 471	3 897	4 787	-	-	3 110
61 Fishing area total	*5 683*	*4 191*	*10 178*	*9 708*	*7 471*	*3 897*	*4 787*	-	-	*3 110*
Species total	*5 683*	*4 191*	*10 178*	*9 708*	*7 471*	*3 897*	*4 787*	-	-	*3 110*
Puffers nei	**Compères nca**		**Tamboriles nep**		***Tetraodontidae***				**1,90(02)XXX,XX**	**PUX**
34 Benin	...	...	...	...	...	...	...	...	2	5
Gambia	...	16	125	28	100	-	-	4	33	81
34 Fishing area total	...	*16*	*125*	*28*	*100*	...	...	*4*	*35*	*86*
41 Brazil	...	...	468	18	88	23	16	35	34	199
41 Fishing area total	...	...	*468*	*18*	*88*	*23*	*16*	*35*	*34*	*199*
51 Saudi Arabia	-	-	-	-	-	-	0	0	18	0
51 Fishing area total	-	-	-	-	-	-	*0*	*0*	*18*	*0*
57 Australia	160	68	62	158	122	115	3	442	432	534
57 Fishing area total	*160*	*68*	*62*	*158*	*122*	*115*	*3*	*442*	*432*	*534*
61 Japan	...	...	8 991	8 238	7 103	8 329	9 647	10 989	7 820	7 869
Korea Rep	-	-	-	-	-	-	-	2 978	3 735	-
61 Fishing area total	...	...	*8 991*	*8 238*	*7 103*	*8 329*	*9 647*	*13 967*	*11 555*	*7 869*
81 Australia	163	195	211	70	97	69	...	220	175	339
81 Fishing area total	*163*	*195*	*211*	*70*	*97*	*69*	...	*220*	*175*	*339*
Species total	*323*	*279*	*9 857*	*8 512*	*7 510*	*8 536*	*9 666*	*14 668*	*12 249*	*9 027*

B-33

Miscellaneous coastal fishes
Poissons côtiers divers
Peces costeros diversos

Capture production by species, fishing areas and countries or areas
Captures par espèces, zones de pêche et pays ou zones
Capturas por especies, áreas de pesca y países o áreas

Species, Fishing area Espèce, Zone de pêche Especie, Area de pesca	1993 t	1994 t	1995 t	1996 t	1997 t	1998 t	1999 t	2000 t	2001 t	2002 t
Filefishes nei	**Bourses nca**		**Cachúas nep**		***Cantherhines(=Navodon) spp***				**1,90(09)004,XX**	**FLF**
61 China	95 500	196 321	122 358	210 188	296 781	235 603	240 214	221 683	201 733	158 661
China,Taiwan	1 689	803	1 680	871	446	586	995	829	959	1 322
Korea Rep	-	-	-	-	-	-	-	2 891	1 578	-
61 Fishing area total	*97 189*	*197 124*	*124 038*	*211 059*	*297 227*	*236 189*	*241 209*	*225 403*	*204 270*	*159 983*
Species total	*97 189*	*197 124*	*124 038*	*211 059*	*297 227*	*236 189*	*241 209*	*225 403*	*204 270*	*159 983*
Threadsail filefish	**...B**		**...C**		***Stephanolepis cirrhifer***				**1,90(09)010,01**	**FIL**
61 Korea Rep	11 365	4 382	1 755	1 772	16 318	9 364	2 999	-	-	933
61 Fishing area total	*11 365*	*4 382*	*1 755*	*1 772*	*16 318*	*9 364*	*2 999*	-	-	*933*
Species total	*11 365*	*4 382*	*1 755*	*1 772*	*16 318*	*9 364*	*2 999*	-	-	*933*
Velvet leatherjacket	**...B**		**...C**		***Parika scaber***				**1,90(09)018,01**	**PKB**
81 New Zealand	290	351	329	442	1 095	312	738	1 279	1 142	1 012
81 Fishing area total	*290*	*351*	*329*	*442*	*1 095*	*312*	*738*	*1 279*	*1 142*	*1 012*
Species total	*290*	*351*	*329*	*442*	*1 095*	*312*	*738*	*1 279*	*1 142*	*1 012*
Filefishes, leatherjackets nei	**Poissons-bourses nca**		**Cachúas, lijas nep**		***Monacanthidae***				**1,90(09)XXX,XX**	**FFX**
31 Antigua Barb	...	...	...	...	...	...	...	...	...	15
31 Fishing area total	...	...	...	...	...	...	...	...	...	*15*
Species total	...	...	...	...	...	...	...	...	...	*15*
Grey triggerfish	**Baliste cabri**		**Pejepuerco blanco**		***Balistes carolinensis***				**1,90(10)002,01**	**TRG**
27 France	-	-	-	-	1	1	2	-	4	-
27 Fishing area total	-	-	-	-	*1*	*1*	*2*	-	*4*	-
37 Syria	...	...	...	...	...	...	...	...	...	5
Tunisia	34	23	59	38	52	58	34	50	62	96
37 Fishing area total	*34*	*23*	*59*	*38*	*52*	*58*	*34*	*50*	*62*	*101*
Species total	*34*	*23*	*59*	*38*	*53*	*59*	*36*	*50*	*66*	*101*
Triggerfishes, durgons nei	**Balistes nca**		**Peces-ballesta nep**		***Balistidae***				**1,90(10)XXX,XX**	**TRI**
21 USA	-	-	-	-	6	5	61	3	2	8
21 Fishing area total	-	-	-	-	*6*	*5*	*61*	*3*	*2*	*8*
27 Portugal	-	-	-	-	-	-	42	38	21	32
27 Fishing area total	-	-	-	-	-	-	*42*	*38*	*21*	*32*
31 Antigua Barb	...	...	...	...	...	...	...	...	18	83
Dominican Rp	450	810	607	31	26	67	7	10	21	99
Grenada	0	0	0	0	0	0	0	0	0	0
Mexico	365	385	492	446	449	401	348	73	122	53
Puerto Rico	...	...	...	...	...	...	32	34	41	28
St Kitts Nev	...	...	3	7	2	5	6	1	1	8
USA	403	374	355	326	74	244	135	165	178	191
US Virgin Is	...	...	...	...	...	...	31	35 F	35 F	35 F
31 Fishing area total	*1 218*	*1 569*	*1 457*	*810*	*551*	*717*	*559*	*318 F*	*416 F*	*497 F*
34 Gambia	-	-	3	51	9	2	2	0	3	1
Ghana	9	11	2	17	-	1	-	2	2	12
Korea Rep	2	-	10	15	-	-	16	-	-	-
Liberia	-	-	-	-	5	2	-	-	-	-
Portugal	4	3	4	9	4	1	2	0	0	-
Sao Tome Prn	...	...	...	...	34	40 F	45 F	40 F	40 F	40 F
Senegal	-	-	-	-	-	10	7	16	52	18
Sierra Leone	0	0	1	1 F	3	195	...	5	-	17
Togo	5	2	0	0	3	4	0	32	0	1
34 Fishing area total	*20*	*16*	*20*	*93 F*	*58*	*255 F*	*72 F*	*95 F*	*97 F*	*89 F*
51 Eritrea	...	9	1	0	0	0	3	1	0	0
Korea Rep	-	-	-	291	-	-	-	-	-	-
Saudi Arabia	-	-	-	-	26	8	8	5	7	24
51 Fishing area total	...	*9*	*1*	*291*	*26*	*8*	*11*	*6*	*7*	*24*
57 Malaysia	1 571	498	540	708	543	250	248	115	193	221
57 Fishing area total	*1 571*	*498*	*540*	*708*	*543*	*250*	*248*	*115*	*193*	*221*
71 Fiji Islands	35	36 F	6	4	8	10	14	10 F	...	...
Malaysia	742	956	852	898	1 398	1 543	2 578	1 313	2 002	1 705
71 Fishing area total	*777*	*992 F*	*858*	*902*	*1 406*	*1 553*	*2 592*	*1 323 F*	*2 002*	*1 705*
77 Mexico	1 279	1 349	906	1 266	5 787	3 259	2 338	1 645	833	901
77 Fishing area total	*1 279*	*1 349*	*906*	*1 266*	*5 787*	*3 259*	*2 338*	*1 645*	*833*	*901*
81 Russian Fed	-	-	-	-	154	-	-	-	-	-
81 Fishing area total	-	-	-	-	*154*	-	-	-	-	-
Species total	*4 865*	*4 433 F*	*3 782*	*4 070 F*	*8 531*	*6 047 F*	*5 923 F*	*3 543 F*	*3 571 F*	*3 477 F*

B-33 Miscellaneous coastal fishes / Poissons côtiers divers / Peces costeros diversos

Capture production by species, fishing areas and countries or areas
Captures par espèces, zones de pêche et pays ou zones
Capturas por especies, áreas de pesca y países o áreas

Species, Fishing area Espèce, Zone de pêche Especie, Area de pesca	1993 t	1994 t	1995 t	1996 t	1997 t	1998 t	1999 t	2000 t	2001 t	2002 t
Toadfishes, etc. nei	**Crapauds, etc. nca**		**Sapos, etc. nep**		***Batrachoididae***				**1,93(01)XXX,XX**	**TFD**
21 USA	...	...	...	...	1	6	4	21	27	22
21 Fishing area total	*...*	*...*	*...*	*...*	*1*	*6*	*4*	*21*	*27*	*22*
27 Portugal	-	-	-	-	-	-	70	91	75	86
27 Fishing area total	*-*	*-*	*-*	*-*	*-*	*-*	*70*	*91*	*75*	*86*
Species total	*...*	*...*	*...*	*...*	*1*	*6*	*74*	*112*	*102*	*108*
Group total	***4 810 733***	***5 349 537***	***6 008 549***	***5 938 912***	***6 452 786***	***6 151 293***	***6 060 207***	***6 039 972***	***6 199 635***	***6 171 124***

B-34 Miscellaneous demersal fishes / Poissons démersaux divers / Peces demersales diversos

Capture production by species, fishing areas and countries or areas
Captures par espèces, zones de pêche et pays ou zones
Capturas por especies, áreas de pesca y países o áreas

Species, Fishing area Espèce, Zone de pêche Especie, Area de pesca	1993 t	1994 t	1995 t	1996 t	1997 t	1998 t	1999 t	2000 t	2001 t	2002 t
Argentines	**Argentines**		**Argentinas**		***Argentina spp***				**1,23(05)015,XX**	**ARG**
21 Canada	428	-	229	259	591	51	12	8	17	20
Cuba	134	15	113	223	553	4	5	-	-	-
Japan	1	-	-	-	-	-	-	-	-	-
Russian Fed	-	-	-	-	-	-	-	5	-	-
21 Fishing area total	*563*	*15*	*342*	*482*	*1 144*	*55*	*17*	*13*	*17*	*20*
27 Denmark	2 353	13 741	1 069	1 446	1 455	748	1 420	1 039	907	614
Faeroe Is	1 069	960	7 131	9 496	8 433	17 167	8 186	6 388	9 952	7 015
France	0	0	-	-	-	-	114	55	41	1
Germany	-	43	357	1 394	1 498	633	24	483	189	150
Iceland	1 255	613	492	808	3 367	13 387	5 495	4 595	2 478	4 357
Ireland	0	-	6	295	1 089	405	396	4 709	7 505	7 592
Netherlands	1 466	6 256	4 136	3 953	4 696	4 964	8 033	3 636	3 659	4 213
Norway	8 480	6 116	6 419	6 817	5 167	8 654	7 823	6 107	14 668	7 406
Portugal	0	0	-	-	-	-	-	-	-	-
Russian Fed	-	-	-	-	-	-	-	1 214	496	293
Spain	-	-	-	-	-	-	-	34	34	3
Sweden	...	...	...	...	541	428	0	273	1 010	484
UK	465	2 009	485	-	-	-	28	-	7 955	4 862
27 Fishing area total	*15 088*	*29 738*	*20 095*	*24 209*	*26 246*	*46 386*	*31 519*	*28 533*	*48 894*	*36 990*
37 France	-	-	-	-	-	-	7	4	4	7
Spain	-	-	-	-	-	-	23	38	65	50
37 Fishing area total	-	-	-	-	-	-	*30*	*42*	*69*	*57*
81 Japan	-	-	-	1	-	-	-	-	-	-
New Zealand	...	...	26	8	41	68	63	42	56	101
Ukraine	-	-	-	-	-	-	-	-	-	1
81 Fishing area total	...	...	*26*	*9*	*41*	*68*	*63*	*42*	*56*	*102*
Species total	*15 651*	*29 753*	*20 463*	*24 700*	*27 431*	*46 509*	*31 629*	*28 630*	*49 036*	*37 169*
Deepsea smelt	**Argentina du Pacifique**		**Argentina del Pacífico**		***Glossanodon semifasciatus***				**1,23(05)029,01**	**DES**
61 Japan	7 776	8 978	7 705	8 134	7 431	7 142	6 312	5 970	5 414	4 926
61 Fishing area total	*7 776*	*8 978*	*7 705*	*8 134*	*7 431*	*7 142*	*6 312*	*5 970*	*5 414*	*4 926*
Species total	*7 776*	*8 978*	*7 705*	*8 134*	*7 431*	*7 142*	*6 312*	*5 970*	*5 414*	*4 926*
Baird's slickhead	**Alépocéphale de Baird**		**...C**		***Alepocephalus bairdii***				**1,23(14)007,01**	**ALC**
27 Estonia	-	-	-	-	-	-	-	-	153	259
Faeroe Is	2	-	-	-	-	-	-	-	-	-
Germany	-	-	-	-	-	-	-	12	1	-
Iceland	3	1	1	0	0	0	0	0	2	-
Lithuania	-	-	-	-	-	-	-	-	460	-
27 Fishing area total	*5*	*1*	*1*	*0*	*0*	*0*	*0*	*12*	*616*	*259*
Species total	*5*	*1*	*1*	*0*	*0*	*0*	*0*	*12*	*616*	*259*
Hector's lanternfish	**Lanternule de Hector**		**Linternillas de Hector**		***Lampanyctodes hectoris***				**1,32(08)017,01**	**LAN**
47 South Africa	1 177	871	0	33	243	6 553	0	...	...	...
47 Fishing area total	*1 177*	*871*	*0*	*33*	*243*	*6 553*	*0*	...	...	...
Species total	*1 177*	*871*	*0*	*33*	*243*	*6 553*	*0*	...	...	...
Lantern fish	**...B**		**Pez linterna**		***Lampanyctus achirus***				**1,32(08)017,15**	**LAC**
58 India	-	-	-	0	-	-	-	-	-	-
58 Fishing area total	-	-	-	*0*	-	-	-	-	-	-
Species total	-	-	-	*0*	-	-	-	-	-	-
Nichol's lanternfish	**...B**		**...C**		***Gymnoscopelus nicholsi***				**1,32(08)032,01**	**GYN**
48 UK	-	-	-	-	-	-	-	0	-	0
48 Fishing area total	-	-	-	-	-	-	-	*0*	-	*0*
Species total	-	-	-	-	-	-	-	*0*	-	*0*
Lanternfishes nei	**Lanternules nca**		**Peces linterna nep**		***Myctophidae***				**1,32(08)XXX,XX**	**LXX**
48 Argentina	-	0	-	-	-	-	-	-	-	-
Russian Fed	114	-	-	-	-	-	5	67	-	-
USA	-	-	-	-	-	-	0	-	-	-
48 Fishing area total	*114*	*0*	-	-	-	-	*5*	*67*	-	-
51 Iran	2 000	2 000	2 000	0	0	0	0	0	335	37
Ukraine	-	-	2	-	-	-	-	-	-	-
51 Fishing area total	*2 000*	*2 000*	*2 002*	*0*	*0*	*0*	*0*	*0*	*335*	*37*
58 Australia	0	-	-	-	-	-	-	-	-	-
58 Fishing area total	*0*	-	-	-	-	-	-	-	-	-
Species total	*2 114*	*2 000*	*2 002*	*0*	*0*	*0*	*5*	*67*	*335*	*37*

B-34 Miscellaneous demersal fishes / Poissons démersaux divers / Peces demersales diversos

Capture production by species, fishing areas and countries or areas / Captures par espèces, zones de pêche et pays ou zones / Capturas por especies, áreas de pesca y países o áreas

Species, Fishing area Espèce, Zone de pêche Especie, Area de pesca	1993 t	1994 t	1995 t	1996 t	1997 t	1998 t	1999 t	2000 t	2001 t	2002 t
Daggertooth pike conger	**Murénésoce-dague**		**Morenocio dentón**		***Muraenesox cinereus***			**1,43(09)011,02**		**DPC**
57 Malaysia	1 635	1 158	1 146	2 099	3 707	3 812	2 390	2 160	2 083	1 329
Thailand	1 034	789	1 321	694	734	465	1 178	551	664	650
57 Fishing area total	*2 669*	*1 947*	*2 467*	*2 793*	*4 441*	*4 277*	*3 568*	*2 711*	*2 747*	*1 979*
61 China	50 000	142 092	154 867	177 470	184 843	239 874	234 314	220 497	243 888	257 659
China,Taiwan	4 567	5 095	3 548	2 846	3 246	5 733	9 001	5 874	5 084	2 827
Japan	3 478	3 820	3 055	1 989	2 060	2 081	2 298	2 400	2 738	2 843
Korea Rep	3 760	2 243	1 604	1 411	2 518	1 506	190	1 862	1 080	833
61 Fishing area total	*61 805*	*153 250*	*163 074*	*183 716*	*192 667*	*249 194*	*245 803*	*230 633*	*252 790*	*264 162*
71 Malaysia	1 328	2 259	2 249	2 434	3 137	3 523	3 440	3 258	2 936	4 044
Thailand	2 155	3 112	2 732	921	866	1 178	1 167	1 141	1 152	1 165
71 Fishing area total	*3 483*	*5 371*	*4 981*	*3 355*	*4 003*	*4 701*	*4 607*	*4 399*	*4 088*	*5 209*
Species total	*67 957*	*160 568*	*170 522*	*189 864*	*201 111*	*258 172*	*253 978*	*237 743*	*259 625*	*271 350*
Pike-congers nei	**Murénésoces nca**		**Morenocios nep**		***Muraenesox spp***			**1,43(09)011,XX**		**PCX**
04 India	-	55	849	490	447	270	494	402	620	337
04 Fishing area total	*-*	*55*	*849*	*490*	*447*	*270*	*494*	*402*	*620*	*337*
51 India	14 884	6 317	6 230	6 514	7 327	5 820	5 414	5 236	5 058	4 895
Pakistan	9 484	5 725	4 692	4 904	5 637	5 627	8 377	5 937	5 834	5 940
South Africa	-	-	-	-	1	9	5 F	4	5	...
51 Fishing area total	*24 368*	*12 042*	*10 922*	*11 418*	*12 965*	*11 456*	*13 796 F*	*11 177*	*10 897*	*10 835*
57 India	4 326	4 111	2 466	4 097	4 192	5 189	8 182	5 748	6 441	3 521
57 Fishing area total	*4 326*	*4 111*	*2 466*	*4 097*	*4 192*	*5 189*	*8 182*	*5 748*	*6 441*	*3 521*
61 China,H.Kong	3 436	3 354	3 129	3 281	2 903	2 450	1 700 F	2 100 F	2 300 F	2 250 F
61 Fishing area total	*3 436*	*3 354*	*3 129*	*3 281*	*2 903*	*2 450*	*1 700 F*	*2 100 F*	*2 300 F*	*2 250 F*
Species total	*32 130*	*19 562*	*17 366*	*19 286*	*20 507*	*19 365*	*24 172 F*	*19 427 F*	*20 258 F*	*16 943 F*
European conger	**Congre d'Europe**		**Congrio común**		***Conger conger***			**1,43(13)001,01**		**COE**
27 Belgium	52	54	58	75	73	79	53	49	56	66
Channel Is	222 F	81	75 F	57	17	32	30	24	25	17
France	5 787	3 960	4 288	4 262	3 715	4 179	4 118	4 872	4 651	4 677
Ireland	77	86	144	142	202	374	295	279	253	277
Isle of Man	0	0	0	0	0	...	-	-	-	-
Netherlands	-	-	-	-	-	-	-	-	1	0
Norway	0	0	0	0	0	0	1	0	0	1
Portugal	2 772	3 628	3 247	2 865	2 541	2 577	2 254	1 986	1 888	1 522
Spain	3 341	3 544	3 542	4 130	4 424	4 299	4 029	2 993	2 537	2 425
UK	808	1 023	1 024	980	957	1 044	1 081	1 157	1 255	1 301
27 Fishing area total	*13 059 F*	*12 376*	*12 378 F*	*12 511*	*11 929*	*12 584*	*11 861*	*11 360*	*10 666*	*10 286*
34 Morocco	1 446	1 963	1 922	2 000	1 630	1 386	1 350	1 152	1 092	1 556
Portugal	807	344	192	490	326	284	219	16	8	5
Sierra Leone	...	...	...	...	...	348	...	...	...	...
Spain	0	0	0	179	75	137	71	45	66	6
Togo	-	-	-	-	0	0	1	0	0	-
34 Fishing area total	*2 253*	*2 307*	*2 114*	*2 669*	*2 031*	*2 155*	*1 641*	*1 213*	*1 166*	*1 567*
37 Albania	...	...	0	1	2	-	-	-	-	10
Croatia	197	143	145	122	114	130	49	38	25	9
France	706	497	358	395	395	395	494	598	574	524
Greece	...	1 419	1 036	1 160	1 922	1 293	1 211	1 019	1 062	1 000 F
Lebanon	5	5	5	5	5	10	5	8	10	9
Malta	5	4	2	2	4	3	2	2	3	3
Morocco	3	5	5	5	5	0	1	169	2	0
Serbia-Monte	6	6	11	20	19	19	17	16	14	12
Slovenia	-	-	-	-	-	-	-	0	1	0
Spain	810 F	850 F	880 F	910	850	757	650	630	708	762
Tunisia	-	11	6	0	14	14	5	42	7	5
Turkey	1 266	1 413	304	416	310	300	680	200	340	300
37 Fishing area total	*2 998 F*	*4 353 F*	*2 752 F*	*3 036*	*3 640*	*2 921*	*3 114*	*2 722*	*2 746*	*2 634 F*
Species total	*18 310 F*	*19 036 F*	*17 244 F*	*18 216*	*17 600*	*17 660*	*16 616*	*15 295*	*14 578*	*14 487 F*
Argentine conger	**Congre argentin**		**Congrio argentino**		***Conger orbignyanus***			**1,43(13)001,02**		**COS**
41 Argentina	13	27	21	22	28	81	18	14	95	43
Brazil	560 F	560 F	182	67	110	108	162	108	491	612
41 Fishing area total	*573 F*	*587 F*	*203*	*89*	*138*	*189*	*180*	*122*	*586*	*655*
Species total	*573 F*	*587 F*	*203*	*89*	*138*	*189*	*180*	*122*	*586*	*655*
American conger	**Congre d'Amérique**		**Congrio americano**		***Conger oceanicus***			**1,43(13)001,04**		**COA**
21 USA	80	87	32	29	17	48	43	49	40	61
21 Fishing area total	*80*	*87*	*32*	*29*	*17*	*48*	*43*	*49*	*40*	*61*
Species total	*80*	*87*	*32*	*29*	*17*	*48*	*43*	*49*	*40*	*61*

B-34 Miscellaneous demersal fishes / Poissons démersaux divers / Peces demersales diversos

Capture production by species, fishing areas and countries or areas / Captures par espèces, zones de pêche et pays ou zones / Capturas por especies, áreas de pesca y países o áreas

Species, Fishing area Espèce, Zone de pêche Especie, Area de pesca	1993 t	1994 t	1995 t	1996 t	1997 t	1998 t	1999 t	2000 t	2001 t	2002 t
Whitespotted conger	**Congre du Pacifique nord-ouest**		**Congrio del Pacífico**			***Conger myriaster***			**1,43(13)001,05**	**ELS**
61 Korea Rep	29 882	21 703	19 667	17 314	19 136	11 913	10 160	8 304	7 676	17 210
61 Fishing area total	*29 882*	*21 703*	*19 667*	*17 314*	*19 136*	*11 913*	*10 160*	*8 304*	*7 676*	*17 210*
Species total	*29 882*	*21 703*	*19 667*	*17 314*	*19 136*	*11 913*	*10 160*	*8 304*	*7 676*	*17 210*
Conger eels, etc. nei	**Congres, etc. nca**		**Congrios, etc. nep**			***Congridae***			**1,43(13)XXX,XX**	**COX**
34 Cameroon	14	10 F	5	6	6 F	5 F	5 F	8	7	11
Congo Dem R	80 F	70 F	71 F	73 F	71 F	73 F	70 F	70 F	70 F	70 F
Congo Rep	493 F	296 F	250 F	350 F	400 F	400 F	450 F	552	600 F	650 F
Côte dIvoire	29	19	20 F	22	26	32	12	12	17	12
Liberia	42	81	70	41	117	85	128	49	76	80 F
Nigeria	...	...	...	...	...	301	...	0	-	-
Senegal	11	28	127	83	226	241	227	236	112	140
34 Fishing area total	*669 F*	*504 F*	*543 F*	*575 F*	*846 F*	*1 137 F*	*892 F*	*927 F*	*882 F*	*963 F*
41 Korea Rep	-	-	-	-	-	-	-	20	-	-
Russian Fed	-	-	-	-	-	-	-	-	-	1 192
41 Fishing area total	-	-	-	-	-	-	-	*20*	-	*1 192*
47 Portugal	-	-	-	-	-	-	-	-	1	-
St Helena	7	5	5	1	3	3	3	3	5	2
47 Fishing area total	*7*	*5*	*5*	*1*	*3*	*3*	*3*	*3*	*6*	*2*
51 Korea Rep	-	-	-	-	6	64	19	53	-	-
51 Fishing area total	-	-	-	-	*6*	*64*	*19*	*53*	-	-
61 Japan	...	...	12 978	12 007	11 706	9 444	8 168	8 364	7 999	8 921
61 Fishing area total	...	...	*12 978*	*12 007*	*11 706*	*9 444*	*8 168*	*8 364*	*7 999*	*8 921*
71 Korea Rep	37	525	318	69	57	-	8	49	78	6
Philippines	2 131	2 384	3 061	2 687	2 053	2 540	2 459	2 349	2 663	2 775
71 Fishing area total	*2 168*	*2 909*	*3 379*	*2 756*	*2 110*	*2 540*	*2 467*	*2 398*	*2 741*	*2 781*
81 Korea Rep	-	-	-	-	-	-	132	-	-	-
New Zealand	...	...	93	109	113	97	88	96	106	144
81 Fishing area total	...	...	*93*	*109*	*113*	*97*	*220*	*96*	*106*	*144*
87 Korea Rep	-	-	-	-	-	23	-	-	-	-
87 Fishing area total	-	-	-	-	-	*23*	-	-	-	-
Species total	*2 844 F*	*3 418 F*	*16 998 F*	*15 448 F*	*14 784 F*	*13 308 F*	*11 769 F*	*11 861 F*	*11 734 F*	*14 003 F*
Longspine snipefish	**Bécasse de mer**		**Trompetero**			***Macroramphosus scolopax***			**1,51(03)004,01**	**SNS**
27 Portugal	2	0	0	-	-	-	-	-	-	0
27 Fishing area total	*2*	*0*	*0*	-	-	-	-	-	-	*0*
34 Lithuania	3 169	-	-	-	-	-	-	-	-	-
Russian Fed	-	-	20	-	89	443	-	-	44	-
34 Fishing area total	*3 169*	-	*20*	-	*89*	*443*	-	-	*44*	-
Species total	*3 171*	*0*	*20*	-	*89*	*443*	-	-	*44*	*0*
Banded yellowfish	**...B**		**...C**			***Centriscops humerosus***			**1,51(03)021,01**	**CUQ**
81 New Zealand	-	-	2	1	2	11	39	...	52	24
81 Fishing area total	-	-	*2*	*1*	*2*	*11*	*39*	...	*52*	*24*
Species total	-	-	*2*	*1*	*2*	*11*	*39*	...	*52*	*24*
Pink cusk-eel	**Abadèche rosé**		**Congribadejo rosado**			***Genypterus blacodes***			**1,58(02)001,01**	**CUS**
41 Argentina	23 788	20 338	23 265	21 933	21 917	25 086	21 503	15 166	19 644	17 794
Australia	-	-	-	-	-	2	10	-	-	-
Belize	-	-	-	-	-	-	15	87	8	0
Chile	-	2	-	-	-	-	-	-	-	-
Falkland Is	8	22	116	297	154	253	451	304	347	333
France	-	3	3	2	1	0	0	-	-	-
Honduras	196	39	61	59	-	-	-	-	-	-
Korea Rep	25	9	-	516	30	32	23	-	327	325
Namibia	-	-	-	-	5	24	45	-	-	-
Panama	56	5	33	46	-	-	-	-	-	-
Portugal	1	-	-	-	-	-	-	13	89	98
Seychelles	-	-	-	-	10	-	-	-	-	-
Spain	771	576	1 299	706	779	1 800	1 901	1 392	1 408	388
UK	36	1	5	6	11	7	32	7	9	7
Uruguay	1 645	436	105	43	41	86	206	368	756	569
41 Fishing area total	*26 526*	*21 431*	*24 887*	*23 608*	*22 948*	*27 290*	*24 186*	*17 337*	*22 588*	*19 514*
57 Australia	0	0	0	1 397	1 923	1 830	1 881	1 148	1 153	1 216
57 Fishing area total	*0*	*0*	*0*	*1 397*	*1 923*	*1 830*	*1 881*	*1 148*	*1 153*	*1 216*
81 Australia	78	73	85	...	...	...	...	941	561	430
Korea Rep	410	814	549	404	1 348	1 210	1 871	1 684	1 277	1 539
New Zealand	15 629	15 120	18 396	12 454	22 594	22 215	21 424	21 617	18 620	20 295
Ukraine	7	...	...	...	...	21	35	258	35	-
81 Fishing area total	*16 124*	*16 007*	*19 030*	*12 858*	*23 942*	*23 446*	*23 330*	*24 500*	*20 493*	*22 264*

B-34 Miscellaneous demersal fishes / Poissons démersaux divers / Peces demersales diversos

Capture production by species, fishing areas and countries or areas
Captures par espèces, zones de pêche et pays ou zones
Capturas por especies, áreas de pesca y países o áreas

Species, Fishing area Espèce, Zone de pêche Especie, Area de pesca	1993 t	1994 t	1995 t	1996 t	1997 t	1998 t	1999 t	2000 t	2001 t	2002 t
87 Chile	4 643	4 624	5 438	5 780	6 410	6 836	5 721	6 269	7 522	4 518
87 Fishing area total	*4 643*	*4 624*	*5 438*	*5 780*	*6 410*	*6 836*	*5 721*	*6 269*	*7 522*	*4 518*
Species total	*47 293*	*42 062*	*49 355*	*43 643*	*55 223*	*59 402*	*55 118*	*49 254*	*51 756*	*47 512*
Red cusk-eel	**Abadèche rouge**		**Congribadejo colorado**		***Genypterus chilensis***				**1,58(02)001,02**	**CUC**
87 Chile	1 411	1 712	1 082	982	745	584	415	608	730	213
87 Fishing area total	*1 411*	*1 712*	*1 082*	*982*	*745*	*584*	*415*	*608*	*730*	*213*
Species total	*1 411*	*1 712*	*1 082*	*982*	*745*	*584*	*415*	*608*	*730*	*213*
Black cusk-eel	**Abadèche noir**		**Congribadejo negro**		***Genypterus maculatus***				**1,58(02)001,03**	**CUB**
87 Chile	2 023	1 125	1 193	1 343	1 661	2 753	1 943	3 542	3 889	1 558
87 Fishing area total	*2 023*	*1 125*	*1 193*	*1 343*	*1 661*	*2 753*	*1 943*	*3 542*	*3 889*	*1 558*
Species total	*2 023*	*1 125*	*1 193*	*1 343*	*1 661*	*2 753*	*1 943*	*3 542*	*3 889*	*1 558*
Kingklip	**Abadèche du Cap**		**Congribadejo(=Rosada)del Cabo**		***Genypterus capensis***				**1,58(02)001,05**	**KCP**
47 Iceland	-	-	-	-	31	2	-	-	-	-
Korea Rep	13	69	-	-	-	-	-	-	-	-
Namibia	747	1 646	3 853	3 667	2 506	2 211	3 706	3 922	6 607	7 926
Russian Fed	-	-	-	-	-	-	-	-	5	1
South Africa	2 601	2 679	2 798	3 140	3 418	3 381	4 114	4 000 F	4 848	5 491
47 Fishing area total	*3 361*	*4 394*	*6 651*	*6 807*	*5 955*	*5 594*	*7 820*	*7 922 F*	*11 460*	*13 418*
Species total	*3 361*	*4 394*	*6 651*	*6 807*	*5 955*	*5 594*	*7 820*	*7 922 F*	*11 460*	*13 418*
Cusk-eels nei	**Abadèches nca**		**Congribadejos nep**		***Genypterus spp***				**1,58(02)001,XX**	**CEX**
41 Korea Rep	-	-	-	-	-	-	-	57	-	-
41 Fishing area total	-	-	-	-	-	-	-	*57*	-	-
87 Peru	640	639	1 631	1 121	439	425	196	557	552	1 029
87 Fishing area total	*640*	*639*	*1 631*	*1 121*	*439*	*425*	*196*	*557*	*552*	*1 029*
Species total	*640*	*639*	*1 631*	*1 121*	*439*	*425*	*196*	*614*	*552*	*1 029*
Bearded brotula	**Brotule barbée**		**Brótula de barbas**		***Brotula barbata***				**1,58(02)005,02**	**BRD**
31 USA	...	...	...	...	...	5	1	1	0	0
31 Fishing area total	...	...	...	...	...	*5*	*1*	*1*	*0*	*0*
34 Congo Rep	23 F	14 F	20 F	25 F	30 F	30 F	40 F	46	40 F	40 F
Côte dIvoire	333	154	298	192	158	342	513	280	156	186
Korea Rep	-	-	-	-	-	-	-	16	-	-
Liberia	108	216	46	4	10	66	48	52	...	40 F
Senegal	1 609	847	140	590	1 644	1 770	3 260	1 325	1 669	1 124
Spain	-	-	-	-	-	-	-	-	61	-
34 Fishing area total	*2 073 F*	*1 231 F*	*504 F*	*811 F*	*1 842 F*	*2 208 F*	*3 861 F*	*1 719*	*1 926 F*	*1 390 F*
Species total	*2 073 F*	*1 231 F*	*504 F*	*811 F*	*1 842 F*	*2 213 F*	*3 862 F*	*1 720*	*1 926 F*	*1 390 F*
Cusk-eels, brotulas nei	**Abadèches, brotules nca**		**Brótulas, congribadejos nep**		***Ophidiidae***				**1,58(02)XXX,XX**	**OPH**
31 Cuba	187	204	198	118	97	37	13	17	17	11
31 Fishing area total	*187*	*204*	*198*	*118*	*97*	*37*	*13*	*17*	*17*	*11*
41 Brazil	...	...	106	176	258	452	544	507	658	741
Russian Fed	-	-	-	-	-	-	-	-	18	-
41 Fishing area total	...	...	*106*	*176*	*258*	*452*	*544*	*507*	*676*	*741*
77 Korea Rep	-	-	-	-	-	-	-	1	-	-
77 Fishing area total	-	-	-	-	-	-	-	*1*	-	-
Species total	*187*	*204*	*304*	*294*	*355*	*489*	*557*	*525*	*693*	*752*
Alfonsinos nei	**Béryx nca**		**Alfonsinos nep**		***Beryx spp***				**1,61(02)003,XX**	**ALF**
21 Russian Fed	-	-	541	141	-	-	-	-	-	-
21 Fishing area total	-	-	*541*	*141*	-	-	-	-	-	-
27 Faeroe Is	-	1	3	-	5	-	-	-	-	-
France	3	5	0	0	3	27	75	40	52	43
Iceland	-	-	-	0	-	-	-	-	-	-
Portugal	-	-	-	-	-	-	87	87	60	79
Russian Fed	-	-	-	-	-	-	-	5	-	-
Spain	-	-	-	-	-	-	-	-	-	122
UK	-	-	-	-	4	-	-	7	16	29
27 Fishing area total	*3*	*6*	*3*	*0*	*12*	*27*	*162*	*139*	*128*	*273*
31 Iceland	-	-	-	7	-	-	-	-	-	-
Russian Fed	-	-	278	15	-	-	-	-	-	-
31 Fishing area total	-	-	*278*	*22*	-	-	-	-	-	-
34 Latvia	7	-	-	-	-	-	-	-	-	-

B-34

Miscellaneous demersal fishes — **Capture production by species, fishing areas and countries or areas**
Poissons démersaux divers — **Captures par espèces, zones de pêche et pays ou zones**
Peces demersales diversos — **Capturas por especies, áreas de pesca y países o áreas**

Species, Fishing area Espèce, Zone de pêche Especie, Area de pesca	1993 t	1994 t	1995 t	1996 t	1997 t	1998 t	1999 t	2000 t	2001 t	2002 t
Lithuania	-	-	-	-	-	-	-	-	-	98
Norway	-	-	-	-	-	-	-	71	-	-
Portugal	60	74	33	126	58	51	42	1	-	1
Russian Fed	20	-	-	-	-	10	21	12	6	52
Spain	-	-	-	24	47	18	-	-	243	310
34 Fishing area total	*87*	*74*	*33*	*150*	*105*	*79*	*63*	*84*	*249*	*461*
37 France	-	-	-	-	-	-	-	8	-	-
37 Fishing area total	-	-	-	-	-	-	-	*8*	-	-
41 Chile	-	-	-	-	-	144	-	-	-	-
Korea Rep	-	-	-	-	-	-	-	4	-	-
Russian Fed	-	-	-	-	-	-	-	-	-	749
41 Fishing area total	-	-	-	-	-	*144*	-	*4*	-	*749*
47 Iceland	-	-	-	-	466	126	-	-	-	-
Namibia	-	-	909	1 805	369	147	123	59	300	232
Norway	...	...	-	-	836	1 066	-	242	-	-
Poland	-	-	-	-	1 964	-	-	-	-	-
Portugal	-	-	-	-	-	-	3	1	7	1
Russian Fed	-	-	-	-	48	69	-	-	1	3
South Africa	-	-	-	-	-	-	-	-	10	-
Spain	-	-	-	-	186	402	-	-	-	-
Ukraine	172	-	-	747	392	-	-	-	-	-
47 Fishing area total	*172*	...	*909*	*2 552*	*4 261*	*1 810*	*126*	*302*	*318*	*236*
51 China	-	-	-	-	-	-	-	-	-	7
Norway	-	-	-	-	-	-	-	11	-	-
Russian Fed	-	-	-	-	-	-	-	-	210	-
Spain	-	-	-	-	-	-	-	79	4	3
Ukraine	462	1 534	2 249	3 079	1 031	859	1 964	1 578	371	-
51 Fishing area total	*462*	*1 534*	*2 249*	*3 079*	*1 031*	*859*	*1 964*	*1 668*	*585*	*10*
61 Russian Fed	-	17	-	6	-	4	38	18	14	12
61 Fishing area total	-	*17*	-	*6*	-	*4*	*38*	*18*	*14*	*12*
71 NewCaledonia	0	0	0	0	0	0	0	0	0	0
71 Fishing area total	*0*	*0*	*0*	*0*	*0*	*0*	*0*	*0*	*0*	*0*
81 China	-	-	-	-	-	-	-	-	-	152
Korea Rep	-	-	-	-	-	77	-	-	-	-
New Zealand	1 713	2 595	2 177	2 159	2 617	3 516	2 579	2 880	3 044	2 888
Ukraine	-	-	-	-	-	-	-	-	9	-
81 Fishing area total	*1 713*	*2 595*	*2 177*	*2 159*	*2 617*	*3 593*	*2 579*	*2 880*	*3 053*	*3 040*
87 Chile	-	-	-	-	-	-	706	4 366	5 182	8 166
87 Fishing area total	-	-	-	-	-	-	*706*	*4 366*	*5 182*	*8 166*
Species total	*2 437*	*4 226*	*6 190*	*8 109*	*8 026*	*6 516*	*5 638*	*9 469*	*9 529*	*12 947*
Redfish	**...B**		**...C**		***Centroberyx affinis***				**1,61(02)012,01**	**CXF**
57 Australia	-	-	-	-	-	-	-	337	408	237
57 Fishing area total	-	-	-	-	-	-	-	*337*	*408*	*237*
71 Australia	-	-	-	-	-	-	-	1	5	5
71 Fishing area total	-	-	-	-	-	-	-	*1*	*5*	*5*
81 Australia	...	...	...	1 284	1 357	1 812	1 555	1 069	800	916
New Zealand	...	...	89	123	194	172	134	176	202	117
81 Fishing area total	...	...	*89*	*1 407*	*1 551*	*1 984*	*1 689*	*1 245*	*1 002*	*1 033*
Species total	...	...	*89*	*1 407*	*1 551*	*1 984*	*1 689*	*1 583*	*1 415*	*1 275*
Orange roughy	**Hoplostète orange**		**Reloj anaranjado**		***Hoplostethus atlanticus***				**1,61(05)002,02**	**ORY**
27 Faeroe Is	60	259	732	950	854	747	349	155	1	29
France	2 159	1 939	998	1 067	1 012	1 110	1 330	1 048	1 254	484
Iceland	717	158	64	43	79	28	14	68	18	10
Ireland	-	-	-	-	-	-	-	3	2 759	4 647
Portugal	-	-	-	-	-	-	117	157	161	122
Russian Fed	-	-	-	-	-	-	-	14	-	-
Spain	-	-	-	22	26	26	38	20	15	55
UK	-	-	2	0	0	0	12	2	35	70
27 Fishing area total	*2 936*	*2 356*	*1 796*	*2 082*	*1 971*	*1 911*	*1 860*	*1 467*	*4 243*	*5 417*
47 Namibia	-	30	6 377	13 379	18 516	10 945	3 473	1 542	857	2 169
Norway	-	-	-	-	22	12	-	-	-	-
47 Fishing area total	-	*30*	*6 377*	*13 379*	*18 538*	*10 957*	*3 473*	*1 542*	*857*	*2 169*
51 China	-	-	-	-	-	-	-	623	710	38
Norway	-	-	-	-	-	-	-	642	-	-
Spain	-	-	-	-	-	-	-	-	1	-
51 Fishing area total	-	-	-	-	-	-	-	*1 265*	*711*	*38*
57 Australia	432	668	227	357	350	4 857	7 553	4 974	5 197	3 936
57 Fishing area total	*432*	*668*	*227*	*357*	*350*	*4 857*	*7 553*	*4 974*	*5 197*	*3 936*
71 Australia	-	-	-	-	-	-	-	717	872	656

B-34 Miscellaneous demersal fishes / Poissons démersaux divers / Peces demersales diversos

Capture production by species, fishing areas and countries or areas / Captures par espèces, zones de pêche et pays ou zones / Capturas por especies, áreas de pesca y países o áreas

Species, Fishing area Espèce, Zone de pêche Especie, Area de pesca		1993 t	1994 t	1995 t	1996 t	1997 t	1998 t	1999 t	2000 t	2001 t	2002 t
71	*Fishing area total*	-	-	-	-	-	-	-	*717*	*872*	*656*
81	Australia	12 050	9 977	7 070	4 526	3 129	3 207	28	26	17	14
	China	-	-	-	-	-	-	-	-	-	547
	Korea Rep	-	-	-	-	-	-	230	-	47	-
	New Zealand	29 681	31 718	33 077	28 639	20 545	21 485	23 780	17 879	14 044	17 954
	Norway	1 602	665	1	5	12	3	-	-	-	-
	Ukraine	-	-	-	-	-	-	-	102	195	-
81	*Fishing area total*	*43 333*	*42 360*	*40 148*	*33 170*	*23 686*	*24 695*	*24 038*	*18 007*	*14 303*	*18 515*
Species total		*46 701*	*45 414*	*48 548*	*48 988*	*44 545*	*42 420*	*36 924*	*27 972*	*26 183*	*30 731*
Slimeheads nei		**Poissons-montres nca**		**Relojes nep**		***Trachichthyidae***				**1,61(05)XXX,XX**	**TRC**
27	Portugal	-	-	-	-	-	-	-	-	235	-
	Spain	-	-	-	833	1 052	33	25	3	6	2
27	*Fishing area total*	-	-	-	*833*	*1 052*	*33*	*25*	*3*	*241*	*2*
34	Portugal	-	-	-	-	-	-	-	-	235	-
34	*Fishing area total*	-	-	-	-	-	-	-	-	*235*	-
81	New Zealand	-	-	-	-	-	7	3	4	2	12
81	*Fishing area total*	-	-	-	-	-	*7*	*3*	*4*	*2*	*12*
87	Chile	-	-	-	-	-	-	779	1 482	1 868	1 514
87	*Fishing area total*	-	-	-	-	-	-	*779*	*1 482*	*1 868*	*1 514*
Species total		-	-	-	*833*	*1 052*	*40*	*807*	*1 489*	*2 346*	*1 528*
John dory		**Saint Pierre**		**Pez de San Pedro**		***Zeus faber***				**1,62(01)001,01**	**JOD**
27	Channel Is	0	1	1 F	0	0	1	1	2	1	1
	France	658	703	715	692	670	767	900	1 268	1 355	1 223
	Ireland	95	81	147	125	112	98	145	174	169	153
	Portugal	371	300	160	154	173	288	324	431	457	417
	Spain	-	-	-	225	467	599	595	507	899	784
	UK	138	262	287	220	159	136	181	296	267	269
27	*Fishing area total*	*1 262*	*1 347*	*1 310 F*	*1 416*	*1 581*	*1 889*	*2 146*	*2 678*	*3 148*	*2 847*
34	Greece	2	34	9	40	20	15	6	11	28	7
	Lithuania	-	-	-	-	-	-	-	-	-	1
	Mauritania	5 F	10 F	10 F	5 F	10 F	20 F	...	...	...	...
	Morocco	529	478	558	459	587	564	621	940	510	733
	Portugal	91	14	6	12	6	29	42	0	0	0
	Russian Fed	-	-	-	-	-	-	-	2	-	-
	Senegal	63	794	661	1 157	142	53	139	282	161	151
	Spain	15 F	20 F	25 F	33	14	1	-	-	-	-
	Ukraine	-	-	-	-	-	99	5	59	9	-
34	*Fishing area total*	*705 F*	*1 350 F*	*1 269 F*	*1 706 F*	*779 F*	*781 F*	*813*	*1 294*	*708*	*892*
37	Albania	...	...	2	0	0	-	-	-	-	5
	France	103	18	5	20	20	20	6	7	9	7
	Greece	221	366	413	447	289	259	195	185	268	240 F
	Malta	0	0	0	0	1	1	2	1	0	-
	Morocco	4	5	3	2	3	3	1	2	2	3
	Spain	-	-	-	26	29	21	20	34	50	62
	Tunisia	20	25	19	28	32	107	59	56	43	61
	Turkey	33	44	35	73	50	120	135	100	130	55
37	*Fishing area total*	*381*	*458*	*477*	*596*	*424*	*531*	*418*	*385*	*502*	*433 F*
47	Angola	299	770	72	315	307	1 339	1 668	1 582	1 303	2 061
	Iceland	-	-	-	-	7	-	-	-	-	-
	Namibia	1	1	3	0	5	25	14	4	138	208
	South Africa	1 098	1 069	1 070	1 156	1 274	964	1 022	1 000 F	1 177	1 313
	Ukraine	...	...	...	...	...	...	2	-	-	-
47	*Fishing area total*	*1 398*	*1 840*	*1 145*	*1 471*	*1 593*	*2 328*	*2 706*	*2 586 F*	*2 618*	*3 582*
51	South Africa	3	-	-	-	3	4	5 F	6	6	...
51	*Fishing area total*	*3*	-	-	-	*3*	*4*	*5 F*	*6*	*6*	...
57	Australia	2	1	0	3	0	1	21	26	22	33
	Korea Rep	-	-	-	-	-	-	-	535	216	-
57	*Fishing area total*	*2*	*1*	*0*	*3*	*0*	*1*	*21*	*561*	*238*	*33*
81	Australia	309	403	401	119	118	102	154	173	156	132
	Japan	239	98	38	16	19	17	8	2	46	19 F
	Korea Rep	-	218	265	105	298	36	540	101	9	-
	New Zealand	777	825	841	729	800	828	882	841	914	1 141
81	*Fishing area total*	*1 325*	*1 544*	*1 545*	*969*	*1 235*	*983*	*1 584*	*1 117*	*1 125*	*1 292 F*
87	Japan	-	-	-	-	-	-	5	-	-	-
87	*Fishing area total*	-	-	-	-	-	-	*5*	-	-	-
Species total		*5 076 F*	*6 540 F*	*5 746 F*	*6 161 F*	*5 615 F*	*6 517 F*	*7 698 F*	*8 627 F*	*8 345*	*9 079 F*
Silvery John dory		**Saint Pierre argenté**		**San Pedro plateado**		***Zenopsis conchifer***				**1,62(01)004,01**	**JOS**
21	USA	2	10	34	27	6	49	19	28	62	66
21	*Fishing area total*	*2*	*10*	*34*	*27*	*6*	*49*	*19*	*28*	*62*	*66*

B-34 Miscellaneous demersal fishes / Poissons démersaux divers / Peces demersales diversos

Capture production by species, fishing areas and countries or areas / Captures par espèces, zones de pêche et pays ou zones / Capturas por especies, áreas de pesca y países o áreas

Species, Fishing area Espèce, Zone de pêche Especie, Area de pesca	1993 t	1994 t	1995 t	1996 t	1997 t	1998 t	1999 t	2000 t	2001 t	2002 t
34 Portugal	-	-	3	9	6	6	3	-	0	-
34 Fishing area total	*-*	*-*	*3*	*9*	*6*	*6*	*3*	*-*	*0*	*-*
Species total	*2*	*10*	*37*	*36*	*12*	*55*	*22*	*28*	*62*	*66*
Mirror dory	**...B**		**...C**		***Zenopsis nebulosus***				**1,62(01)004,02**	**ZNE**
57 Australia	81	0	0	4	9	37	...	70	111	279
57 Fishing area total	*81*	*0*	*0*	*4*	*9*	*37*	*...*	*70*	*111*	*279*
81 Australia	467	453	361	348	420	507	378	245	146	137
81 Fishing area total	*467*	*453*	*361*	*348*	*420*	*507*	*378*	*245*	*146*	*137*
Species total	*548*	*453*	*361*	*352*	*429*	*544*	*378*	*315*	*257*	*416*
Dories nei	**Saint Pierres nca**		**Peces de San Pedro nep**		***Zeidae***				**1,62(01)XXX,XX**	**ZEX**
81 New Zealand	-	-	338	288	641	754	763	778	685	1 012
Ukraine	-	-	-	-	-	-	-	-	2	74
81 Fishing area total	*-*	*-*	*338*	*288*	*641*	*754*	*763*	*778*	*687*	*1 086*
Species total	*-*	*-*	*338*	*288*	*641*	*754*	*763*	*778*	*687*	*1 086*
Boarfishes nei	**Sangliers nca**		**Ochavos nep**		***Caproidae***				**1,62(03)XXX,XX**	**BOR**
27 Ireland	-	-	-	-	-	-	-	-	120	91
27 Fishing area total	*-*	*-*	*-*	*-*	*-*	*-*	*-*	*-*	*120*	*91*
47 Russian Fed	-	-	-	-	-	5	-	-	-	-
47 Fishing area total	*-*	*-*	*-*	*-*	*-*	*5*	*-*	*-*	*-*	*-*
51 Spain	-	-	-	-	-	-	-	-	7	-
51 Fishing area total	*-*	*-*	*-*	*-*	*-*	*-*	*-*	*-*	*7*	*-*
Species total	*-*	*-*	*-*	*-*	*-*	*5*	*-*	*-*	*127*	*91*
Oreo dories nei	**Oréos nca**		**Oreós nep**		***Oreosomatidae***				**1,62(04)XXX,XX**	**ORD**
47 Namibia	-	-	6	17	188	...	42	10	54	335
Norway	-	-	-	-	-	6	-	-	-	-
47 Fishing area total	*-*	*-*	*6*	*17*	*188*	*6*	*42*	*10*	*54*	*335*
51 China	-	-	-	-	-	-	-	-	180	97
Norway	-	-	-	-	-	-	-	175	-	-
51 Fishing area total	*-*	*-*	*-*	*-*	*-*	*-*	*-*	*175*	*180*	*97*
61 Russian Fed	-	2	-	-	-	-	2	-	14	8
61 Fishing area total	*-*	*2*	*-*	*-*	*-*	*-*	*2*	*-*	*14*	*8*
81 China	-	-	-	-	-	-	-	-	-	10
New Zealand	23 216	22 602	21 833	18 776	21 850	21 095	22 646	22 775	24 165	17 625
Norway	1	11	-	1	-	1	-	-	-	-
Russian Fed	-	18	-	5	-	-	-	-	-	-
81 Fishing area total	*23 217*	*22 631*	*21 833*	*18 782*	*21 850*	*21 096*	*22 646*	*22 775*	*24 165*	*17 635*
Species total	*23 217*	*22 633*	*21 839*	*18 799*	*22 038*	*21 102*	*22 690*	*22 960*	*24 413*	*18 075*
Wreckfish	**Cernier commun**		**Cherna**		***Polyprion americanus***				**1,70(05)058,01**	**WRF**
27 Channel Is	-	-	-	-	2	1	0	-	-	-
France	33	4	2	4	10	13	20	30	22	15
Ireland	-	-	-	-	-	5	-	1	1	-
Portugal	647	831	619	460	347	304	275	338	306	381
Spain	21	48	42	115	265	124	152	72	84	36
UK	1	1	2	8	0	0	0	0	1	-
27 Fishing area total	*702*	*884*	*665*	*587*	*624*	*447*	*447*	*441*	*414*	*432*
31 USA	-	0	112	82	14	6	1	-	-	-
31 Fishing area total	*-*	*0*	*112*	*82*	*14*	*6*	*1*	*-*	*-*	*-*
34 Greece	0	0	0	-	-	-	-	-	-	-
Portugal	161	205	78	184	63	43	26	2	4	3
Spain	-	-	-	49	12	31	27	2	9	8
34 Fishing area total	*161*	*205*	*78*	*233*	*75*	*74*	*53*	*4*	*13*	*11*
37 Albania	...	...	0	1	0	-	-	-	-	10
France	...	...	...	...	...	...	22	22	-	-
Malta	29	16	8	9	14	8	8	8	8	16
Spain	-	-	-	2	3	7	7	5	4	37
37 Fishing area total	*29*	*16*	*8*	*12*	*17*	*15*	*37*	*35*	*12*	*63*
41 Argentina	149	88	227	378	121	75	218	129	...	...
Spain	-	-	-	-	-	-	-	-	35	-
41 Fishing area total	*149*	*88*	*227*	*378*	*121*	*75*	*218*	*129*	*35*	*...*
47 Portugal	-	-	-	-	6	42	20	8	-	2
47 Fishing area total	*-*	*-*	*-*	*-*	*6*	*42*	*20*	*8*	*-*	*2*

B-34 Miscellaneous demersal fishes / Poissons démersaux divers / Peces demersales diversos

Capture production by species, fishing areas and countries or areas / Captures par espèces, zones de pêche et pays ou zones / Capturas por especies, áreas de pesca y países o áreas

Species, Fishing area Espèce, Zone de pêche Especie, Area de pesca	1993 t	1994 t	1995 t	1996 t	1997 t	1998 t	1999 t	2000 t	2001 t	2002 t
51 Spain	-	-	-	-	-	-	-	-	1	-
51 Fishing area total	-	-	-	-	-	-	-	-	*1*	-
Species total	*1 041*	*1 193*	*1 090*	*1 292*	*857*	*659*	*776*	*617*	*475*	*508*
Hapuku wreckfish	**Cernier de Nouvelle Zélande**		**Cherna hapuku**		***Polyprion oxygeneios***				**1,70(05)058,02**	**WHA**
81 New Zealand	1 439	1 448	1 536	1 155	1 657	1 571	1 547	1 497	1 579	1 610
81 Fishing area total	*1 439*	*1 448*	*1 536*	*1 155*	*1 657*	*1 571*	*1 547*	*1 497*	*1 579*	*1 610*
87 Chile	42	37	33	23	30	26	8	7	10	2
87 Fishing area total	*42*	*37*	*33*	*23*	*30*	*26*	*8*	*7*	*10*	*2*
Species total	*1 481*	*1 485*	*1 569*	*1 178*	*1 687*	*1 597*	*1 555*	*1 504*	*1 589*	*1 612*
Ocean whitefish	**Tile fin**		**Blanquillo fino**		***Caulolatilus princeps***				**1,70(16)001,03**	**ULP**
77 Mexico	...	...	...	...	...	535	622	1 073	979	927
USA	-	-	-	-	-	-	5	4	5	2
77 Fishing area total	...	...	...	...	...	*535*	*627*	*1 077*	*984*	*929*
Species total	...	...	...	...	...	*535*	*627*	*1 077*	*984*	*929*
Atlantic goldeneye tilefish	**Tile oeil doré**		**Blanquillo ojo amarillo**		***Caulolatilus chrysops***				**1,70(16)001,04**	**CKZ**
21 USA	-	-	-	-	-	-	0	2	790	847
21 Fishing area total	-	-	-	-	-	-	*0*	*2*	*790*	*847*
31 USA	-	-	-	-	-	-	0	4	16	10
31 Fishing area total	-	-	-	-	-	-	*0*	*4*	*16*	*10*
Species total	-	-	-	-	-	-	*0*	*6*	*806*	*857*
Great Northern tilefish	**Tile chameau**		**Blanquillo camello**		***Lopholatilus chamaeleonticeps***				**1,70(16)400,02**	**TIL**
21 USA	1 873	783	673	1 349	1 493	1 339	528	515	117	98
21 Fishing area total	*1 873*	*783*	*673*	*1 349*	*1 493*	*1 339*	*528*	*515*	*117*	*98*
31 USA	800	735	611	114	397	80	88	173	168	197
31 Fishing area total	*800*	*735*	*611*	*114*	*397*	*80*	*88*	*173*	*168*	*197*
Species total	*2 673*	*1 518*	*1 284*	*1 463*	*1 890*	*1 419*	*616*	*688*	*285*	*295*
Tilefishes nei	**Tiles nca**		**Blanquillos, paletas nep**		***Branchiostegidae***				**1,70(16)XXX,XX**	**TIS**
31 Mexico	...	...	...	...	...	53	68	45	28	18
USA	-	-	-	-	28	294	318	483	314	268
31 Fishing area total	...	...	...	...	*28*	*347*	*386*	*528*	*342*	*286*
34 Côte dIvoire	22	14	14 F	20 F	30 F	40 F	45	44	15	16
34 Fishing area total	*22*	*14*	*14 F*	*20 F*	*30 F*	*40 F*	*45*	*44*	*15*	*16*
41 Brazil	340 F	340 F	812	1 098	1 000	786	524	547	1 309	1 261
41 Fishing area total	*340 F*	*340 F*	*812*	*1 098*	*1 000*	*786*	*524*	*547*	*1 309*	*1 261*
61 China,H.Kong	1 642	2 548	2 583	3 186	4 187	4 879	3 500 F	4 300 F	4 750 F	4 650 F
China,Taiwan	1 879	1 299	579	1 227	626	372	496	448	512	306
Japan	...	...	4 194	3 648	2 994	2 284	1 949	1 678	1 781	1 804
Korea Rep	-	-	-	-	-	-	1 651	1 664	1 049	1 341
61 Fishing area total	*3 521*	*3 847*	*7 356*	*8 061*	*7 807*	*7 535*	*7 596 F*	*8 090 F*	*8 092 F*	*8 101 F*
87 Chile	383	195	252	117	28	80	134	155	53	52
Peru	736	738	1 544	892	292	119	146	117	1 485	2 507
87 Fishing area total	*1 119*	*933*	*1 796*	*1 009*	*320*	*199*	*280*	*272*	*1 538*	*2 559*
Species total	*5 002 F*	*5 134 F*	*9 978 F*	*10 188 F*	*9 185 F*	*8 907 F*	*8 831 F*	*9 481 F*	*11 296 F*	*12 223 F*
Cape bonnetmouth	**Andorrève du Cap**		**Andorrero del Cabo**		***Emmelichthys nitidus***				**1,70(30)010,01**	**EMM**
47 Russian Fed	-	-	-	-	70	7	-	-	-	-
South Africa	434	121	96	216	121	117	113	50 F	37	-
47 Fishing area total	*434*	*121*	*96*	*216*	*191*	*124*	*113*	*50 F*	*37*	-
81 New Zealand	...	...	2 392	1 845	1 635	2 064	2 846	2 825	1 881	2 825
81 Fishing area total	...	...	*2 392*	*1 845*	*1 635*	*2 064*	*2 846*	*2 825*	*1 881*	*2 825*
Species total	*434*	*121*	*2 488*	*2 061*	*1 826*	*2 188*	*2 959*	*2 875 F*	*1 918*	*2 825*
Bonnetmouths, rubyfishes nei	**Andorrèves, poissons rubis nca**		**Andorreros, peces rubí nep**		***Emmelichthyidae***				**1,70(30)XXX,XX**	**EMT**
34 Lithuania	-	-	-	-	-	-	-	-	-	8
34 Fishing area total	-	-	-	-	-	-	-	-	-	*8*
47 Russian Fed	-	-	-	-	-	-	-	-	6	-
47 Fishing area total	-	-	-	-	-	-	-	-	*6*	-
51 Ukraine	551	227	144	28	7	275	181	-	86	-
51 Fishing area total	*551*	*227*	*144*	*28*	*7*	*275*	*181*	-	*86*	-

B-34 Miscellaneous demersal fishes / Poissons démersaux divers / Peces demersales diversos

Capture production by species, fishing areas and countries or areas / Captures par espèces, zones de pêche et pays ou zones / Capturas por especies, áreas de pesca y países o áreas

Species, Fishing area Espèce, Zone de pêche Especie, Area de pesca	1993 t	1994 t	1995 t	1996 t	1997 t	1998 t	1999 t	2000 t	2001 t	2002 t
81 New Zealand	...	...	616	596	431	378	271	582	434	403
Russian Fed	29	-	-	-	-	-	-	-	-	-
Ukraine	-	-	501	-	-	-	-	-	-	1 410
81 Fishing area total	*29*	*...*	*1 117*	*596*	*431*	*378*	*271*	*582*	*434*	*1 813*
Species total	*580*	*227*	*1 261*	*624*	*438*	*653*	*452*	*582*	*526*	*1 821*
Tripletail	...B		...C		***Lobotes surinamensis***				**1,70(34)029,01**	**LOB**
31 USA	-	-	-	-	-	-	-	1	1	1
31 Fishing area total	-	-	-	-	-	-	-	*1*	*1*	*1*
Species total	-	-	-	-	-	-	-	*1*	*1*	*1*
Giant boarfish	...B		...C		***Paristiopterus labiosus***				**1,70(57)004,02**	**SWH**
81 New Zealand	-	-	21	19	27	75	6	9	3	9
81 Fishing area total	-	-	*21*	*19*	*27*	*75*	*6*	*9*	*3*	*9*
Species total	-	-	*21*	*19*	*27*	*75*	*6*	*9*	*3*	*9*
Pelagic armourhead	**Tête casquée pélagique**		...C		***Pseudopentaceros richardsoni***				**1,70(57)007,01**	**EDR**
47 Ukraine	435	-	49	281	18	-	-	-	-	-
47 Fishing area total	*435*	-	*49*	*281*	*18*	-	-	-	-	-
51 China	-	-	-	-	-	-	-	44	-	-
Ukraine	...	40	54	17	33	78	108	77	12	-
51 Fishing area total	*...*	*40*	*54*	*17*	*33*	*78*	*108*	*121*	*12*	-
81 China	-	-	-	-	-	-	-	-	-	26
New Zealand	-	-	3	7	2	78	13	6	7	37
81 Fishing area total	-	-	*3*	*7*	*2*	*78*	*13*	*6*	*7*	*63*
Species total	*435*	*40*	*106*	*305*	*53*	*156*	*121*	*127*	*19*	*63*
Red bandfish	**Cépole commune**		...C		***Cepola rubescens***				**1,70(60)083,01**	**CBC**
27 Spain	-	-	-	-	-	-	-	-	-	17
27 Fishing area total	-	-	-	-	-	-	-	-	-	*17*
Species total	-	-	-	-	-	-	-	-	-	*17*
Castaneta	**Castanette pontude**		**Castañeta**		***Cheilodactylus bergi***				**1,70(70)270,02**	**CTA**
41 Argentina	1 346	20 383	10 409	204	744	1 827	155	81	98	149
Uruguay	75	4 289	3 034	2 938	4 149	9 681	3 105	1 351	1 269	349
41 Fishing area total	*1 421*	*24 672*	*13 443*	*3 142*	*4 893*	*11 508*	*3 260*	*1 432*	*1 367*	*498*
Species total	*1 421*	*24 672*	*13 443*	*3 142*	*4 893*	*11 508*	*3 260*	*1 432*	*1 367*	*498*
Peruvian morwong	**Castanette pintadille**		**Pintadilla**		***Cheilodactylus variegatus***				**1,70(70)270,03**	**HAW**
87 Chile	-	-	-	-	51	50	52	45	46	4
Peru	111	95	93	283	411	90	236	335	260	356
87 Fishing area total	*111*	*95*	*93*	*283*	*462*	*140*	*288*	*380*	*306*	*360*
Species total	*111*	*95*	*93*	*283*	*462*	*140*	*288*	*380*	*306*	*360*
Tarakihi	...B		...C		***Nemadactylus macropterus***				**1,70(70)305,04**	**TAK**
81 Japan	223	207	41	-	-	-	-	-	-	-
New Zealand	4 847	4 771	5 108	4 366	5 441	5 239	5 589	5 739	6 129	6 149
81 Fishing area total	*5 070*	*4 978*	*5 149*	*4 366*	*5 441*	*5 239*	*5 589*	*5 739*	*6 129*	*6 149*
Species total	*5 070*	*4 978*	*5 149*	*4 366*	*5 441*	*5 239*	*5 589*	*5 739*	*6 129*	*6 149*
Morwongs	...B		...C		***Nemadactylus spp***				**1,70(70)305,XX**	**MOW**
57 Australia	48	74	75	77	94	95	638	604	776	716
57 Fishing area total	*48*	*74*	*75*	*77*	*94*	*95*	*638*	*604*	*776*	*716*
81 Australia	1 146	1 199	1 225	1 108	1 419	1 083	696	512	421	339
New Zealand	127	94	120	92	107	109	78	99	92	91
81 Fishing area total	*1 273*	*1 293*	*1 345*	*1 200*	*1 526*	*1 192*	*774*	*611*	*513*	*430*
Species total	*1 321*	*1 367*	*1 420*	*1 277*	*1 620*	*1 287*	*1 412*	*1 215*	*1 289*	*1 146*
Trumpeters nei	...B		**Tromperos nep**		***Latridae***				**1,70(71)XXX,XX**	**TRU**
57 Australia	1	0	0	0	0	0	153	139	77	70
57 Fishing area total	*1*	*0*	*0*	*0*	*0*	*0*	*153*	*139*	*77*	*70*
81 Australia	7	16	16	20	15	34	8	9	13	8
New Zealand	548	512	669	727	761	567	574	476	505	493
81 Fishing area total	*555*	*528*	*685*	*747*	*776*	*601*	*582*	*485*	*518*	*501*
Species total	*556*	*528*	*685*	*747*	*776*	*601*	*735*	*624*	*595*	*571*

B-34

Miscellaneous demersal fishes — **Capture production by species, fishing areas and countries or areas**
Poissons démersaux divers — **Captures par espèces, zones de pêche et pays ou zones**
Peces demersales diversos — **Capturas por especies, áreas de pesca y países o áreas**

Species, Fishing area Espèce, Zone de pêche Especie, Area de pesca	1993 t	1994 t	1995 t	1996 t	1997 t	1998 t	1999 t	2000 t	2001 t	2002 t
Antarctic toothfish	**Légine antarctique**		**Austromerluza antártica**		***Dissostichus mawsoni***				**1,70(92)015,01**	**TOA**
48 Chile	-	-	-	-	-	1	-	-	-	-
USA	-	-	-	-	-	-	0	-	0	-
48 Fishing area total	*-*	*-*	*-*	*-*	*-*	*1*	*0*	*-*	*0*	*-*
88 Chile	-	-	-	-	-	0	-	-	-	-
New Zealand	-	-	-	-	-	41	296	751	582	1 354
South Africa	-	-	-	-	-	-	-	-	21	-
Uruguay	-	-	-	-	-	-	-	-	23	-
88 Fishing area total	*-*	*-*	*-*	*-*	*-*	*41*	*296*	*751*	*626*	*1 354*
Species total	*-*	*-*	*-*	*-*	*-*	*42*	*296*	*751*	*626*	*1 354*
Patagonian toothfish	**Légine australe**		**Austromerluza negra**		***Dissostichus eleginoides***				**1,70(92)015,02**	**TOP**
41 Argentina	3 651	10 840	19 180	14 811	8 793	9 950	7 692	7 771	6 410	8 164
Australia	-	-	-	-	-	15	24	-	-	-
Belize	-	-	-	-	-	-	16	27	11	0
Chile	-	-	-	-	-	-	-	-	831	207
Falkland Is	0	18	34	50	178	570	1 113	927	1 460	1 321
France	-	0	1	3	0	3	4	0	-	-
Honduras	30	5	26	7	-	-	-	-	-	-
Japan	...	...	...	...	...	...	...	...	...	2
Korea Rep	-	-	-	-	514	1 051	933	1 292	686	-
Namibia	-	-	-	-	2	21	28	-	-	-
Panama	7	3	9	-	-	-	-	-	-	-
Portugal	1	-	-	-	-	-	-	3	-	-
Russian Fed	-	-	-	-	-	-	-	-	1	3
Seychelles	-	-	-	-	1	-	-	-	3 800	-
Spain	259	147	191	79	109	355	572	538	277	215
UK	2	1	1	1	2	18	30	6	3	8
Uruguay	-	-	-	-	-	1 345	888	558	336	1 170
41 Fishing area total	*3 950*	*11 014*	*19 442*	*14 951*	*9 599*	*13 328*	*11 300*	*11 122*	*13 815*	*11 090*
47 Chile	-	-	-	-	-	-	-	-	5	-
Uruguay	-	-	-	-	-	-	-	320	-	906
47 Fishing area total	*-*	*-*	*-*	*-*	*-*	*-*	*-*	*320*	*5*	*906*
48 Argentina	-	0	816	101	-	-	10	-	-	-
Bulgaria	193	250	-	-	-	-	-	-	-	-
Chile	2 125	151	2 154	2 788	2 061	1 388	1 313	1 391	534	1 545
Japan	-	-	-	-	76	-	-	-	-	1
Korea Rep	-	135	383	432	526	-	308	412	787	300
Russian Fed	254	121	10	103	-	-	-	-	224	313
South Africa	-	-	-	-	-	835	128	364	359	332
Spain	-	-	-	-	487	-	184	308	643	832
Ukraine	458	-	-	-	-	-	-	164	149	-
UK	-	1	-	-	662	716	1 049	1 437	924	1 728
USA	-	-	9	178	-	-	-	-	-	-
Uruguay	-	-	-	-	-	262	644	863	428	693
48 Fishing area total	*3 030*	*658*	*3 372*	*3 602*	*3 812*	*3 201*	*3 636*	*4 939*	*4 048*	*5 744*
51 Seychelles	-	-	-	-	-	-	-	-	-	2 870
Uruguay	-	-	-	-	-	-	-	1 628	7 002	1 928
51 Fishing area total	*-*	*-*	*-*	*-*	*-*	*-*	*-*	*1 628*	*7 002*	*4 798*
57 Uruguay	-	-	-	-	-	-	-	-	-	1 847
57 Fishing area total	*-*	*-*	*-*	*-*	*-*	*-*	*-*	*-*	*-*	*1 847*
58 Australia	0	-	-	-	1 868	3 491	3 386	3 048	2 640	2 567
France	1 570	4 405	4 121	3 481	4 090	4 616	5 014	7 156	5 838	3 569
Japan	-	-	-	264	335	-	-	-	-	-
South Africa	-	-	-	942	1 246	904	576	1 094	271	165
Ukraine	2 027	1 032	1 590	969	1 048	885	593	56	8	
Uruguay	-	-	-	-	-	-	-	99	-	-
58 Fishing area total	*3 597*	*5 437*	*5 711*	*5 656*	*8 587*	*9 896*	*9 569*	*11 453*	*8 757*	*6 301*
81 New Zealand	...	...	...	1 061	5	43	1	0	14	12
81 Fishing area total	*...*	*...*	*...*	*1 061*	*5*	*43*	*1*	*0*	*14*	*12*
87 Chile	20 997	20 902	15 694	6 993	8 059	9 172	10 328	10 676	6 568	7 194
Spain	-	-	-	-	-	-	-	-	11	-
87 Fishing area total	*20 997*	*20 902*	*15 694*	*6 993*	*8 059*	*9 172*	*10 328*	*10 676*	*6 579*	*7 194*
88 Chile	-	-	-	-	-	0	-	-	-	-
New Zealand	-	-	-	-	0	1	1	0	30	12
South Africa	-	-	-	-	-	-	-	-	4	-
Uruguay	-	-	-	-	-	-	-	-	0	-
88 Fishing area total	*-*	*-*	*-*	*-*	*0*	*1*	*1*	*0*	*34*	*12*
Species total	*31 574*	*38 011*	*44 219*	*32 263*	*30 062*	*35 641*	*34 835*	*40 138*	*40 254*	*37 904*
Antarctic toothfishes nei	**Légines antarctiques nca**		**Austromerluzas nep**		***Dissostichus spp***				**1,70(92)015,XX**	**TOT**
51 Korea Rep	-	-	-	-	-	-	-	-	122	-
51 Fishing area total	*-*	*-*	*-*	*-*	*-*	*-*	*-*	*-*	*122*	*-*
57 Korea Rep	-	-	-	-	-	-	-	-	450	-

B-34 Miscellaneous demersal fishes / Poissons démersaux divers / Peces demersales diversos

Capture production by species, fishing areas and countries or areas / Captures par espèces, zones de pêche et pays ou zones / Capturas por especies, áreas de pesca y países o áreas

Species, Fishing area Espèce, Zone de pêche Especie, Area de pesca	1993 t	1994 t	1995 t	1996 t	1997 t	1998 t	1999 t	2000 t	2001 t	2002 t
57 Fishing area total	-	-	-	-	-	-	-	-	*450*	-
Species total	-	-	-	-	-	-	-	-	*572*	-
Patagonian rockcod	**Bocasse de Patagonie**		**Trama patagónica**		***Patagonotothen brevicauda***				**1,70(92)440,01**	**NOT**
41 Russian Fed	-	-	-	-	-	-	-	-	-	4
Spain	-	-	-	-	-	-	-	-	-	5
41 Fishing area total	-	-	-	-	-	-	-	-	-	*9*
48 Argentina	-	0	1	-	-	-	-	-	-	-
Russian Fed	-	-	-	-	-	-	3	0	-	-
UK	-	1	-	-	-	-	-	0	-	0
48 Fishing area total	-	*1*	*1*	-	-	-	*3*	*0*	-	*0*
Species total	-	*1*	*1*	-	-	-	*3*	*0*	-	*9*
Longtail Southern cod	**Notothénia queue longue**		**Nototenia coluda**		***Patagonotothen ramsayi***				**1,70(92)440,02**	**PAT**
41 Russian Fed	89	115	-	-	-	-	-	-	-	-
Spain	-	-	-	-	-	-	-	-	-	10
41 Fishing area total	*89*	*115*	-	-	-	-	-	-	-	*10*
Species total	*89*	*115*	-	-	-	-	-	-	-	*10*
...A	**...B**		**...C**		***Parachaenichthys georgianus***				**1,70(93)452,01**	**PGE**
48 Argentina	-	0	-	-	-	-	-	-	-	-
UK	-	-	-	-	-	-	-	0	-	-
48 Fishing area total	-	*0*	-	-	-	-	-	*0*	-	-
Species total	-	*0*	-	-	-	-	-	*0*	-	-
Blackfin icefish	**Grande-gueule antarctique**		**Draco antártico**		***Chaenocephalus aceratus***				**1,70(94)416,01**	**SSI**
48 Argentina	-	0	0	-	-	-	-	-	-	-
Korea Rep	-	-	-	-	-	-	-	-	-	1
Russian Fed	-	-	-	-	-	-	0	-	-	-
UK	-	2	-	-	-	-	-	0	-	4
USA	-	-	-	-	-	-	1	-	1	-
48 Fishing area total	-	*2*	*0*	-	-	-	*1*	*0*	*1*	*5*
88 New Zealand	-	-	-	-	-	-	-	0	-	-
88 Fishing area total	-	-	-	-	-	-	-	*0*	-	-
Species total	-	*2*	*0*	-	-	-	*1*	*0*	*1*	*5*
Mackerel icefish	**Poisson des glaces antarctique**		**Draco rayado**		***Champsocephalus gunnari***				**1,70(94)417,01**	**ANI**
48 Argentina	-	10	10	-	-	-	-	-	-	-
Chile	-	-	-	-	-	6	-	715	365	-
France	-	-	-	-	-	-	-	-	386	-
Korea Rep	-	-	-	-	-	-	-	-	-	602
Poland	-	-	-	-	-	-	-	-	-	296
Russian Fed	-	-	-	-	-	-	265	3 395	0	1 373
UK	-	3	-	-	-	-	-	4	208	396
USA	-	-	-	-	-	-	1	-	1	-
48 Fishing area total	-	*13*	*10*	-	-	*6*	*266*	*4 114*	*960*	*2 667*
58 Australia	3	-	-	-	227	128	2	87	1 073	966
France	12	-	84	5	0	-	-	-	-	-
Ukraine	-	1 228	2 624	-	-	-	-	-	-	-
58 Fishing area total	*15*	*1 228*	*2 708*	*5*	*227*	*128*	*2*	*87*	*1 073*	*966*
Species total	*15*	*1 241*	*2 718*	*5*	*227*	*134*	*268*	*4 201*	*2 033*	*3 633*
South Georgia icefish	**Crocodile de Géorgie**		**Draco cocodrilo**		***Pseudochaenichthys georgianus***				**1,70(94)418,01**	**SGI**
48 Argentina	-	0	-	-	-	-	-	-	-	-
Bulgaria	-	0	-	-	-	-	-	-	-	-
France	-	-	-	-	-	-	-	-	0	-
Korea Rep	-	-	-	-	-	-	-	-	-	1
Russian Fed	-	-	-	-	-	-	0	-	-	-
UK	-	1	-	-	-	-	-	0	6	5
USA	-	-	-	-	-	-	3	-	0	-
48 Fishing area total	-	*1*	-	-	-	-	*3*	*0*	*6*	*6*
Species total	-	*1*	-	-	-	-	*3*	*0*	*6*	*6*
Ocellated icefish	**Grande-gueule ocellée**		**Draco ocelado**		***Chionodraco rastrospinosus***				**1,70(94)466,01**	**KIF**
48 USA	-	-	-	-	-	-	1	-	1	-
48 Fishing area total	-	-	-	-	-	-	*1*	-	*1*	-
58 Australia	-	-	-	-	1	-	-	-	-	-
58 Fishing area total	-	-	-	-	*1*	-	-	-	-	-
Species total	-	-	-	-	*1*	-	*1*	-	*1*	-

B-34 Miscellaneous demersal fishes / Poissons démersaux divers / Peces demersales diversos

Capture production by species, fishing areas and countries or areas / Captures par espèces, zones de pêche et pays ou zones / Capturas por especies, áreas de pesca y países o áreas

Species, Fishing area Espèce, Zone de pêche Especie, Area de pesca	1993 t	1994 t	1995 t	1996 t	1997 t	1998 t	1999 t	2000 t	2001 t	2002 t
Unicorn icefish	**Grande-gueule à long nez**		**Draco rinoceronte**		***Channichthys rhinoceratus***				**1,70(94)470,01**	**LIC**
58 Australia	1	-	-	-	4	5	1	2	1	3
France	-	-	1	-	5	1	1	-	-	-
58 Fishing area total	*1*	-	*1*	-	*9*	*6*	*2*	*2*	*1*	*3*
Species total	*1*	-	*1*	-	*9*	*6*	*2*	*2*	*1*	*3*
Spiny icefish	**Grande-gueule épineuse**		**Draco espinudo**		***Chaenodraco wilsoni***				**1,70(94)480,01**	**WIC**
48 USA	-	-	-	-	-	-	0	-	0	-
48 Fishing area total	-	-	-	-	-	-	*0*	-	*0*	-
58 Australia	-	-	-	-	-	-	-	-	11	-
58 Fishing area total	-	-	-	-	-	-	-	-	*11*	-
Species total	-	-	-	-	-	-	*0*	-	*11*	-
Icefishes nei	**Poissons des glaces nca**		**Dracos nep**		***Channichthyidae***				**1,70(94)XXX,XX**	**ICX**
48 France	-	-	-	-	-	-	-	-	0	-
UK	-	-	-	-	-	-	-	-	-	0
USA	-	-	-	-	-	-	0	-	0	-
48 Fishing area total	-	-	-	-	-	-	*0*	-	*0*	*0*
88 New Zealand	-	-	-	-	-	0	0	0	2	2
South Africa	-	-	-	-	-	-	-	-	0	
88 Fishing area total	-	-	-	-	-	*0*	*0*	*0*	*2*	*2*
Species total	-	-	-	-	-	*0*	*0*	*0*	*2*	*2*
Black cardinal fish	**Poisson cardinal**		**Boca negra(Pez del diablo)**		***Epigonus telescopus***				**1,70(96)373,01**	**EPI**
27 Faeroe Is	41	45	38	31	129	94	4	...	...	...
France	26	231	95	52	52	232	...	197	153	63
Germany	-	-	-	-	-	-	-	50	-	-
Spain	-	-	-	-	-	-	-	-	-	70
UK	-	-	-	-	-	-	-	1	22	-
27 Fishing area total	*67*	*276*	*133*	*83*	*181*	*326*	*4*	*248*	*175*	*133*
81 New Zealand	2 049	4 291	3 650	3 002	4 334	2 568	2 869	4 095	1 957	2 741
81 Fishing area total	*2 049*	*4 291*	*3 650*	*3 002*	*4 334*	*2 568*	*2 869*	*4 095*	*1 957*	*2 741*
Species total	*2 116*	*4 567*	*3 783*	*3 085*	*4 515*	*2 894*	*2 873*	*4 343*	*2 132*	*2 874*
Cardinal fishes nei	**Poissons-cardinaux nca**		**Peces cardenal nep**		***Epigonus spp***				**1,70(96)373,XX**	**CDL**
51 China	-	-	-	-	-	-	-	-	-	3
51 Fishing area total	-	-	-	-	-	-	-	-	-	*3*
87 Chile	862	137	232	513	1 727	5 284	2 999	5 792	4 648	1 595
87 Fishing area total	*862*	*137*	*232*	*513*	*1 727*	*5 284*	*2 999*	*5 792*	*4 648*	*1 595*
Species total	*862*	*137*	*232*	*513*	*1 727*	*5 284*	*2 999*	*5 792*	*4 648*	*1 598*
Plunderfish	**...B**		**...C**		***Pogonophryne permitini***				**1,70(97)001,01**	**PGR**
88 New Zealand	-	-	-	-	-	-	-	0	0	0
88 Fishing area total	-	-	-	-	-	-	-	*0*	*0*	*0*
Species total	-	-	-	-	-	-	-	*0*	*0*	*0*
Atlantic wolffish	**Loup atlantique**		**Perro del Norte**		***Anarhichas lupus***				**1,71(02)001,01**	**CAA**
21 Spain	-	8	116	7	23	7	-	2	7	0
21 Fishing area total	-	*8*	*116*	*7*	*23*	*7*	-	*2*	*7*	*0*
27 Belgium	180	161	206	99	125	208	201	290	175	188
Denmark	569	392	248	195	220	273	298	294	223	248
Estonia	-	-	-	-	-	-	-	-	-	1
Faeroe Is	176	132	141	146	196	264	291	...	...	...
France	8	3	4	3	1	2	14	9	7	8
Germany	145	139	176	94	44	88	67	86	66	96
Greenland	4	28	6	15	6	42	7	12	16	16
Iceland	12 945	12 766	12 574	14 638	11 685	11 844	13 769	15 043	17 953	14 303
Ireland	30	45	42	39	22	39	35	66	27	50
Japan	7	5	5	-	-	-	-	-	-	-
Norway	...	...	...	...	...	...	...	917	1 111	870
Poland	-	-	-	-	19	40	6	18	8	12
Russian Fed	4 300	3 944	6 280	10 523	18 247	23 730	23 756	22 756	19 087	20 190
Sweden	262	173	146	117	174	157	163	154	95	74
UK	571	454	277	1 128	199	893	928	917	140	151
27 Fishing area total	*19 197*	*18 242*	*20 105*	*26 997*	*30 938*	*37 580*	*39 535*	*40 562*	*38 908*	*36 207*
61 Russian Fed	-	-	-	-	-	-	38	36	33	22
61 Fishing area total	-	-	-	-	-	-	*38*	*36*	*33*	*22*
Species total	*19 197*	*18 250*	*20 221*	*27 004*	*30 961*	*37 587*	*39 573*	*40 600*	*38 948*	*36 229*

B-34 Miscellaneous demersal fishes / Poissons démersaux divers / Peces demersales diversos

Capture production by species, fishing areas and countries or areas / Captures par espèces, zones de pêche et pays ou zones / Capturas por especies, áreas de pesca y países o áreas

Species, Fishing area Espèce, Zone de pêche Especie, Area de pesca	1993 t	1994 t	1995 t	1996 t	1997 t	1998 t	1999 t	2000 t	2001 t	2002 t
Spotted wolffish	**Loup tacheté**		**Perro pintado**		***Anarhichas minor***				**1,71(02)001,03**	**CAS**
27 Iceland	1 244	916	700	1 109	1 180	1 599	1 545	1 896	2 126	2 128
Norway	...	...	...	...	...	...	...	1 091	1 111	1 394
Portugal	...	...	...	...	...	...	...	...	20	-
27 Fishing area total	*1 244*	*916*	*700*	*1 109*	*1 180*	*1 599*	*1 545*	*2 987*	*3 257*	*3 522*
Species total	*1 244*	*916*	*700*	*1 109*	*1 180*	*1 599*	*1 545*	*2 987*	*3 257*	*3 522*
Wolffishes(=Catfishes) nei	**Loups nca**		**Perritos del Norte nep**		***Anarhichas spp***				**1,71(02)001,XX**	**CAT**
21 Canada	1 073	485	303	422	856	526	694	678	578	-
Cuba	0	-	-	-	-	-	-	-	-	-
Estonia	-	-	-	-	-	-	-	6	5	1
Faeroe Is	5	-	1	4	-	-	-	-	-	-
Greenland	156	101	50	47	67	30	26	47	58	106
Iceland	-	2	-	-	-	-	-	-	-	-
Japan	-	1	33	20	17	26	21	15	53	30
Norway	16	-	0	-	-	-	-	-	1	
Portugal	2 302	3 219	1 358	123	185	141	549	61	141	87
Russian Fed	14	-	57	-	-	38	-	7	23	56
St Pier Mq	-	-	-	-	3	3	2	0	1	1
Spain	-	184	195	695	535	473	435	518	785	532
UK	-	-	-	0	-	-	-	-	-	-
USA	506	479	464	363	309	296	258	200	250	155
21 Fishing area total	*4 072*	*4 471*	*2 461*	*1 674*	*1 972*	*1 533*	*1 985*	*1 532*	*1 895*	*968*
27 Faeroe Is	15	49	37	27	64	60	118	154	155	258
Netherlands	71	58	50	6	16	36	21	10	2	3
Norway	3 143	6 505	7 589	6 819	12 769	16 332	6 398	4 370	12 204	3 435
Portugal	530	594	502	169	225	476	96	168	163	627
Spain	134	140	174	-	20	37	20	66	98	109
UK	1 242	1 007	833	...	...	...	...	...	600	437
27 Fishing area total	*5 135*	*8 353*	*9 185*	*7 021*	*13 094*	*16 941*	*6 653*	*4 768*	*13 222*	*4 869*
Species total	*9 207*	*12 824*	*11 646*	*8 695*	*15 066*	*18 474*	*8 638*	*6 300*	*15 117*	*5 837*
Eelpouts	**Loquettes**		**Viruelas**		***Lycodes spp***				**1,71(15)024,XX**	**ELZ**
27 Russian Fed	-	-	-	-	-	-	-	-	1	2
27 Fishing area total	-	-	-	-	-	-	-	-	*1*	*2*
61 Russian Fed	380	5	45	18	-	2	1	28	47	60
61 Fishing area total	*380*	*5*	*45*	*18*	-	*2*	*1*	*28*	*47*	*60*
Species total	*380*	*5*	*45*	*18*	-	*2*	*1*	*28*	*48*	*62*
Stargazer	**Uranoscope**		**Rata**		***Uranoscopus scaber***				**1,72(13)352,02**	**UUC**
27 Portugal	...	...	...	...	...	...	15	50	46	104
27 Fishing area total	...	...	...	...	...	...	*15*	*50*	*46*	*104*
Species total	...	...	...	...	...	...	*15*	*50*	*46*	*104*
Giant stargazer	**Uranoscope géant**		**Miracielo gigante**		***Kathetostoma giganteum***				**1,72(13)484,01**	**STZ**
81 New Zealand	2 856	99	9 597	2 122	3 990	2 195	3 370	3 638	4 233	3 272
Norway	10	10	-	1	1	1	-	-	-	-
81 Fishing area total	*2 866*	*109*	*9 597*	*2 123*	*3 991*	*2 196*	*3 370*	*3 638*	*4 233*	*3 272*
Species total	*2 866*	*109*	*9 597*	*2 123*	*3 991*	*2 196*	*3 370*	*3 638*	*4 233*	*3 272*
Japanese sandfish	**...B**		**...C**		***Arctoscopus japonicus***				**1,72(72)103,01**	**JAS**
61 Japan	5 796	5 877	5 506	6 719	6 209	6 795	6 615	6 652	8 753	9 249
Korea Rep	3 731	1 466	2 065	2 501	2 194	1 490	2 449	1 571	1 286	3 381
61 Fishing area total	*9 527*	*7 343*	*7 571*	*9 220*	*8 403*	*8 285*	*9 064*	*8 223*	*10 039*	*12 630*
Species total	*9 527*	*7 343*	*7 571*	*9 220*	*8 403*	*8 285*	*9 064*	*8 223*	*10 039*	*12 630*
Snoek	**Escolier**		**Sierra**		***Thyrsites atun***				**1,75(05)001,01**	**SNK**
34 Russian Fed	-	26	-	-	-	-	-	-	-	-
34 Fishing area total	-	*26*	-	-	-	-	-	-	-	-
47 Iceland	-	-	-	-	1	-	-	-	-	-
Namibia	939	683	856	622	895	701	1 212	966	1 699	1 890
Russian Fed	5	103	-	15	11	-	-	-	-	-
South Africa	15 248	13 180	15 299	12 400	11 126	14 135	11 188	10 200 F	10 328	11 079
Ukraine	-	-	-	-	-	-	2	-	-	-
47 Fishing area total	*16 192*	*13 966*	*16 155*	*13 037*	*12 033*	*14 836*	*12 402*	*11 166 F*	*12 027*	*12 969*
57 Australia	370 F	370 F	400 F	400 F	300	200	88	120	154	161
57 Fishing area total	*370 F*	*370 F*	*400 F*	*400 F*	*300*	*200*	*88*	*120*	*154*	*161*
81 Australia	400 F	400 F	450 F	450 F	...	...	14	21	29	9
Japan	2 502	1 582	588	28	12	3	23	59	189	84
New Zealand	22 348	14 624	18 428	15 849	22 047	25 972	20 642	21 905	25 222	23 121
Russian Fed	1 963	91	-	-	-	-	-	-	-	-

Species, Fishing area Espèce, Zone de pêche Especie, Area de pesca	1993 t	1994 t	1995 t	1996 t	1997 t	1998 t	1999 t	2000 t	2001 t	2002 t
Ukraine	...	...	1 748	2 244	3 898	3 881	7 920	6 113	2 970	3 846
81 Fishing area total	*27 213 F*	*16 697 F*	*21 214 F*	*18 571 F*	*25 957*	*29 856*	*28 599*	*28 098*	*28 410*	*27 060*
87 Chile	427	572	687	821	1 337	1 022	604	851	830	934
87 Fishing area total	*427*	*572*	*687*	*821*	*1 337*	*1 022*	*604*	*851*	*830*	*934*
Species total	*44 202 F*	*31 631 F*	*38 456 F*	*32 829 F*	*39 627*	*45 914*	*41 693*	*40 235 F*	*41 421*	*41 124*
White snake mackerel	**Escolier blanc**		**Escolar sierra**		***Thyrsitops lepidopoides***				**1,75(05)002,01**	**WSM**
41 Argentina	32	32	66	1 232	309	285	136	-	-	-
41 Fishing area total	*32*	*32*	*66*	*1 232*	*309*	*285*	*136*	-	-	-
Species total	*32*	*32*	*66*	*1 232*	*309*	*285*	*136*	-	-	-
Escolar	**Escolier noir**		**Escolar negro**		***Lepidocybium flavobrunneum***				**1,75(05)005,01**	**LEC**
21 USA	...	...	...	...	...	1	1	2	4	1
21 Fishing area total	...	...	...	...	...	*1*	*1*	*2*	*4*	*1*
31 USA	...	...	...	...	...	51	33	70	40	57
31 Fishing area total	...	...	...	...	...	*51*	*33*	*70*	*40*	*57*
77 USA	...	...	...	...	...	2	1	6	3	3
77 Fishing area total	...	...	...	...	...	*2*	*1*	*6*	*3*	*3*
81 New Zealand	-	-	0	0	1	1	22	47	87	76
81 Fishing area total	-	-	*0*	*0*	*1*	*1*	*22*	*47*	*87*	*76*
Species total	...	...	*0*	*0*	*1*	*55*	*57*	*125*	*134*	*137*
Oilfish	**Rouvet**		**Escolar clavo**		***Ruvettus pretiosus***				**1,75(05)007,01**	**OIL**
27 Portugal	-	-	-	-	-	-	14	9	9	9
27 Fishing area total	-	-	-	-	-	-	*14*	*9*	*9*	*9*
31 USA	...	...	...	...	...	10	11	38	30	25
31 Fishing area total	...	...	...	...	...	*10*	*11*	*38*	*30*	*25*
34 Portugal	-	-	-	-	-	-	-	5	2	8
34 Fishing area total	-	-	-	-	-	-	-	*5*	*2*	*8*
47 Norway	-	-	-	-	5	-	-	-	-	-
47 Fishing area total	-	-	-	-	*5*	-	-	-	-	-
61 China,Taiwan	2 341	2 094	3 729	2 634	2 622	3 043	2 661	2 584	3 678	4 190
61 Fishing area total	*2 341*	*2 094*	*3 729*	*2 634*	*2 622*	*3 043*	*2 661*	*2 584*	*3 678*	*4 190*
77 USA	-	-	-	-	-	-	-	-	-	92
77 Fishing area total	-	-	-	-	-	-	-	-	-	*92*
81 New Zealand	-	-	38	53	34	57	53	61	86	131
81 Fishing area total	-	-	*38*	*53*	*34*	*57*	*53*	*61*	*86*	*131*
Species total	*2 341*	*2 094*	*3 767*	*2 687*	*2 661*	*3 110*	*2 739*	*2 697*	*3 805*	*4 455*
Silver gemfish	**Escolier tifiati**		**Escolar plateado**		***Rexea solandri***				**1,75(05)009,01**	**GEM**
57 Australia	3 000 F	3 000 F	3 000 F	2 000 F	...	...	4	447	482	223
57 Fishing area total	*3 000 F*	*3 000 F*	*3 000 F*	*2 000 F*	...	...	*4*	*447*	*482*	*223*
71 Australia	-	-	-	-	-	-	-	-	-	11
71 Fishing area total	-	-	-	-	-	-	-	-	-	*11*
81 Australia	3 000 F	2 000 F	2 000 F	1 000 F	339	598	454	220	87	73
New Zealand	2 861	2 365	2 026	1 344	1 914	1 301	902	1 029	827	688
81 Fishing area total	*5 861 F*	*4 365 F*	*4 026 F*	*2 344 F*	*2 253*	*1 899*	*1 356*	*1 249*	*914*	*761*
Species total	*8 861 F*	*7 365 F*	*7 026 F*	*4 344 F*	*2 253*	*1 899*	*1 360*	*1 696*	*1 396*	*995*
Roudi escolar	**Escolier clair**		**Escolar prometeo**		***Promethichthys prometheus***				**1,75(05)017,01**	**PRP**
34 Portugal	-	-	-	-	-	-	-	-	6	7
34 Fishing area total	-	-	-	-	-	-	-	-	*6*	*7*
Species total	-	-	-	-	-	-	-	-	*6*	*7*
Frostfishes	**...B**		**...C**		***Benthodesmus spp***				**1,75(06)002,XX**	**BEH**
81 New Zealand	-	-	-	-	-	-	-	-	1	80
81 Fishing area total	-	-	-	-	-	-	-	-	*1*	*80*
Species total	-	-	-	-	-	-	-	-	*1*	*80*
Largehead hairtail	**Poisson-sabre commun**		**Pez sable**		***Trichiurus lepturus***				**1,75(06)003,02**	**LHT**
21 USA	...	...	...	...	...	0	4	5	1	2
21 Fishing area total	...	...	...	...	...	*0*	*4*	*5*	*1*	*2*

B-34 Miscellaneous demersal fishes / Poissons démersaux divers / Peces demersales diversos

Capture production by species, fishing areas and countries or areas
Captures par espèces, zones de pêche et pays ou zones
Capturas por especies, áreas de pesca y países o áreas

Species, Fishing area Espèce, Zone de pêche Especie, Area de pesca		1993 t	1994 t	1995 t	1996 t	1997 t	1998 t	1999 t	2000 t	2001 t	2002 t
27	Ireland	-	-	-	-	-	-	-	-	523	831
	Portugal	-	-	-	-	-	-	3	0	0	0
27	*Fishing area total*	-	-	-	-	-	-	*3*	*0*	*523*	*831*
31	Russian Fed	-	-	-	5	-	-	-	-	-	-
	USA	4	20	32	1	10	5	2	36	6	14
	Venezuela	3 560	3 944	4 933	4 609	5 050	5 408	4 017	3 716	8 793	6 484
31	*Fishing area total*	*3 564*	*3 964*	*4 965*	*4 615*	*5 060*	*5 413*	*4 019*	*3 752*	*8 799*	*6 498*
34	Belize	-	-	-	-	300	317	1 121	248	7	56
	Benin	150 F	200 F	250 F	300 F	350 F	400 F	543	370 F	584	579
	Bulgaria	-	-	-	-	-	1 383	-	-	-	-
	Cameroon	5	2 F	1	4	10 F	16	1	6	59	6
	Côte dIvoire	221	229	231 F	180	265	492	518	321	200	171
	Cyprus	-	-	-	...	...	5 061	130	230	3 949	-
	France	-	-	-	-	...	3 400	509	-	-	-
	Georgia	40 F	10 F	-	-	-	-	-	-	-	-
	Ghana	1 445	1 140	1 823	2 543	2 866	2 047	1 267	1 664	1 849	3 154
	Guinea	...	...	324	401	409	400	400	418	...	400 F
	GuineaBissau	5 F	27	17	17	20 F	15 F	10 F	10 F	10 F	10 F
	Ireland	-	-	-	-	-	-	-	-	253	75
	Latvia	250	16	-	12	-	1 232	1 502	544	13	46
	Liberia	6	2	12	7	10	9	33	34	169	160 F
	Lithuania	-	-	...	...	...	9 708	13	32	167	8 120
	Mauritania	100 F	60 F	100 F	160 F	110 F	-	...	...	...	...
	Morocco	3 475	3 519	5 450	5 500	8 340	8 064	6 595	6 176	5 124	3 793
	Netherlands	-	-	-	33	-	103	401	115	697	547
	Portugal	0	0	0	0	0	0	-	-	-	-
	Romania	34	-	-	-	-	-	-	-	-	-
	Russian Fed	17 441	19 564	23 709	51 773	33 368	7 303	6 777	3 297	1 146	66
	St Vincent	-	-	-	-	130	1 393	35	24	716	1 441
	Senegal	195	463	345	432	1 114	948	838	781	1 083	886
	Sierra Leone	...	...	...	...	...	204	75	35	48	66
	Spain	-	-	-	-	-	-	-	-	5	36
	Togo	11	2	0	0	1	0	6	2	4	14
	Ukraine	-	-	-	-	-	2 490	1 314	1 423	2 096	59
	Other nei	-	-	-	-	-	-	147	948	660	301
34	*Fishing area total*	*23 378 F*	*25 234 F*	*32 262 F*	*61 362 F*	*47 293 F*	*44 985 F*	*22 235 F*	*16 678 F*	*18 839 F*	*19 986 F*
37	Egypt	...	...	564	914	679	774	711	809	1 107	1 096
	Morocco	-	83	1	5	19	0	-	0	0	0
37	*Fishing area total*	...	*83*	*565*	*919*	*698*	*774*	*711*	*809*	*1 107*	*1 096*
41	Brazil	1 110 F	1 110 F	1 197	736	938	1 405	1 230	1 665	2 883	3 131
41	*Fishing area total*	*1 110 F*	*1 110 F*	*1 197*	*736*	*938*	*1 405*	*1 230*	*1 665*	*2 883*	*3 131*
47	Angola	...	...	...	...	...	...	28	-	-	-
	Namibia	-	-	13	...	...	...	...	346	691	...
	Russian Fed	1	2	-	-	97	119	440	1 549	1 180	208
47	*Fishing area total*	*1*	*2*	*13*	...	*97*	*119*	*468*	*1 895*	*1 871*	*208*
51	Egypt	-	-	-	-	-	-	12	2	-	-
	Pakistan	3 474	6 320	6 093	9 073	11 583	12 337	31 623	28 754	27 355	28 440
51	*Fishing area total*	*3 474*	*6 320*	*6 093*	*9 073*	*11 583*	*12 337*	*31 635*	*28 756*	*27 355*	*28 440*
57	Malaysia	4 123	2 713	2 653	2 524	3 372	14 038	5 995	1 957	1 681	1 617
	Thailand	2 972	5 278	10 015	11 712	13 033	16 223	9 509	8 737	10 915	10 683
57	*Fishing area total*	*7 095*	*7 991*	*12 668*	*14 236*	*16 405*	*30 261*	*15 504*	*10 694*	*12 596*	*12 300*
61	China	635 315	878 144	1 039 684	1 071 914	1 014 598	1 223 360	1 222 454	1 285 469	1 282 698	1 287 798
	China,Taiwan	17 227	19 313	16 210	11 830	8 955	7 991	9 375	7 271	8 834	8 390
	Japan	31 712	31 577	28 207	26 644	20 932	22 268	26 200	22 947	16 615	14 405
	Korea Rep	58 035	101 052	94 596	74 461	67 170	74 851	64 445	81 050	79 898	60 172
61	*Fishing area total*	*742 289*	*1 030 086*	*1 178 697*	*1 184 849*	*1 111 655*	*1 328 470*	*1 322 474*	*1 396 737*	*1 388 045*	*1 370 765*
71	Malaysia	3 313	3 443	3 594	3 844	7 901	9 913	12 014	9 617	8 030	3 806
	Singapore	119	192	137	144	170	140	127	83	50	61
	Thailand	2 333	2 812	4 482	3 361	4 553	2 486	6 461	7 342	5 028	5 085
71	*Fishing area total*	*5 765*	*6 447*	*8 213*	*7 349*	*12 624*	*12 539*	*18 602*	*17 042*	*13 108*	*8 952*
Species total		*786 676 F*	*1 081 237 F*	*1 244 673 F*	*1 283 139 F*	*1 206 353 F*	*1 436 303 F*	*1 416 885 F*	*1 478 033 F*	*1 475 127 F*	*1 452 209 F*
Silver scabbardfish		**Sabre argenté**		**Pez cinto**			***Lepidopus caudatus***			**1,75(06)006,01**	**SFS**
27	France	-	-	-	-	-	-	30	8	16	24
	Germany	2	-	-	-	-	-	-	4	-	-
	Latvia	1 458	-	8	-	-	-	-	-	-	-
	Portugal	2 662	1 429	6 479	2 057	2 853	2 155	319	66	82	82
	Russian Fed	19	-	-	-	-	-	-	-	2	-
	Spain	-	-	-	9	651	1 377	1 584	14	256	1 863
	UK	-	-	-	-	-	-	-	12	5	1
27	*Fishing area total*	*4 141*	*1 429*	*6 487*	*2 066*	*3 504*	*3 532*	*1 933*	*104*	*361*	*1 970*
34	Germany	-	-	-	-	-	-	64	-	-	-
	Portugal	6 367	7 931	2 571	8 939	4 731	3 357	2 966	0	4	-
	Spain	-	-	-	636	-	-	-	-	-	-
34	*Fishing area total*	*6 367*	*7 931*	*2 571*	*9 575*	*4 731*	*3 357*	*3 030*	*0*	*4*	-
37	Albania	...	...	7	0	0	18	19	18	0	0

B-34 Miscellaneous demersal fishes / Poissons démersaux divers / Peces demersales diversos

Capture production by species, fishing areas and countries or areas / Captures par espèces, zones de pêche et pays ou zones / Capturas por especies, áreas de pesca y países o áreas

Species, Fishing area Espèce, Zone de pêche Especie, Area de pesca	1993 t	1994 t	1995 t	1996 t	1997 t	1998 t	1999 t	2000 t	2001 t	2002 t
France	-	-	-	41	11	11	225	8	15	11
Spain	-	-	-	1 552	1 347	1 627	1 163	270	3 031	779
Tunisia	263	173	-	26	7	57	423	411	375	263
37 Fishing area total	*263*	*173*	*7*	*1 619*	*1 365*	*1 713*	*1 830*	*707*	*3 421*	*1 053*
47 South Africa	12 196	4 697	4 931	3 196	2 001	4 557	2 560	2 300 F	2 316	4 523
Spain	-	-	-	-	2	-	-	-	-	-
47 Fishing area total	*12 196*	*4 697*	*4 931*	*3 196*	*2 003*	*4 557*	*2 560*	*2 300 F*	*2 316*	*4 523*
81 New Zealand	1 709	2 476	1 955	941	2 342	3 344	2 638	1 536	2 876	2 435
Russian Fed	6	-	-	-	-	-	-	-	-	-
Ukraine	-	-	-	-	-	-	-	83	11	40
81 Fishing area total	*1 715*	*2 476*	*1 955*	*941*	*2 342*	*3 344*	*2 638*	*1 619*	*2 887*	*2 475*
Species total	*24 682*	*16 706*	*15 951*	*17 397*	*13 945*	*16 503*	*11 991*	*4 730 F*	*8 989*	*10 021*
Black scabbardfish	**Sabre noir**		**Sable negro**		***Aphanopus carbo***				**1,75(06)012,01**	**BSF**
27 Estonia	-	-	-	-	-	-	-	-	224	-
Faeroe Is	1 315	893	550	256	126	89	45	116	412	1 094
France	3 421	2 512	2 448	2 868	2 118	1 710	1 833	3 707	5 070	4 628
Germany	149	94	3	2	-	-	-	-	-	-
Iceland	-	1	0	0	1	0	9	18	8	15
Ireland	8	-	-	0	1	-	1	12	299	259
Lithuania	-	-	-	-	-	-	-	-	3	9
Netherlands	-	-	-	-	-	-	11	7	-	21
Poland	-	-	-	-	-	-	-	-	-	2
Portugal	4 521	3 428	4 275	3 686	3 553	3 153	2 776	2 867	2 745	2 692
Spain	-	-	-	41	106	127	117	1 029	1 323	992
UK	-	2	20	40	2	159	201	428	742	1 065
27 Fishing area total	*9 414*	*6 930*	*7 296*	*6 893*	*5 907*	*5 238*	*4 993*	*8 184*	*10 826*	*10 777*
31 Iceland	-	-	-	17	-	-	-	-	-	-
31 Fishing area total	-	-	-	*17*	-	-	-	-	-	-
34 Portugal	3 467	3 133	3 469	6 748	4 023	4 430	4 405	4 203	4 008	3 873
Spain	-	-	-	-	-	10	-	-	-	-
34 Fishing area total	*3 467*	*3 133*	*3 469*	*6 748*	*4 023*	*4 440*	*4 405*	*4 203*	*4 008*	*3 873*
Species total	*12 881*	*10 063*	*10 765*	*13 658*	*9 930*	*9 678*	*9 398*	*12 387*	*14 834*	*14 650*
Hairtails, scabbardfishes nei	**Poissons-sabres, sabres nca**		**Peces sable, cintos nep**		***Trichiuridae***				**1,75(06)XXX,XX**	**CUT**
27 Spain	-	-	-	-	-	-	-	-	13	-
27 Fishing area total	-	-	-	-	-	-	-	-	*13*	-
31 Korea Rep	-	-	-	-	-	-	24	15	6	-
Mexico	...	...	...	...	...	6 349	6 723	7 263	5 085	3 872
31 Fishing area total	...	...	...	...	...	*6 349*	*6 747*	*7 278*	*5 091*	*3 872*
34 Congo Rep	67 F	187	150 F	120 F	90 F	60 F	30 F	-	-	-
Gambia	...	...	0	10	12	5	5	...	28	8
Korea Rep	106	646	255	349	901	706	963	904	1 001	571
Nigeria	2 557	3 806	5 175	4 826	6 399	7 750	5 366	754	1 156	851
34 Fishing area total	*2 730 F*	*4 639*	*5 580 F*	*5 305 F*	*7 402 F*	*8 521 F*	*6 364 F*	*1 658*	*2 185*	*1 430*
41 Korea Rep	103	203	-	-	-	51	1 205	1 888	155	-
41 Fishing area total	*103*	*203*	-	-	-	*51*	*1 205*	*1 888*	*155*	-
47 Korea Rep	52	224	-	-	5	-	-	-	-	-
47 Fishing area total	*52*	*224*	-	-	*5*	-	-	-	-	-
51 India	25 568	44 583	25 409	38 771	129 662	58 595	92 334	100 570	86 652	119 248
Korea Rep	3 709	3 350	4 301	5 595	7 918	7 035	7 782	5 147	1 663	2 592
Oman	2 971	3 883	8 163	8 132	10 384	4 767	1 776	4 367	2 617	6 600
South Africa	-	-	-	-	6	0	...	...	...	...
Untd Arab Em	50	58	51	52	65	72	78	80	80 F	82
51 Fishing area total	*32 298*	*51 874*	*37 924*	*52 550*	*148 035*	*70 469*	*101 970*	*110 164*	*91 012 F*	*128 522*
57 India	11 730	20 884	25 173	15 170	16 456	14 801	22 872	20 026	15 046	14 949
Indonesia	6 736	9 221	11 029	13 049	15 133	19 958	18 099	16 791	14 278	15 910
Korea Rep	-	-	17	-	29	-	1 085	1 374	90	-
57 Fishing area total	*18 466*	*30 105*	*36 219*	*28 219*	*31 618*	*34 759*	*42 056*	*38 191*	*29 414*	*30 859*
61 China,H.Kong	3 578	2 951	3 470	3 326	2 522	1 690	1 200 F	1 500 F	1 650 F	1 600 F
61 Fishing area total	*3 578*	*2 951*	*3 470*	*3 326*	*2 522*	*1 690*	*1 200 F*	*1 500 F*	*1 650 F*	*1 600 F*
71 Indonesia	12 478	12 230	15 560	15 322	17 848	17 196	18 559	21 286	24 224	26 430
Korea Rep	1 972	5 215	2 987	1 769	1 531	1 248	694	700	361	532
Philippines	15 307	16 175	12 522	12 266	9 412	9 877	10 363	8 641	8 315	9 187
71 Fishing area total	*29 757*	*33 620*	*31 069*	*29 357*	*28 791*	*28 321*	*29 616*	*30 627*	*32 900*	*36 149*
77 Korea Rep	-	-	10	1	-	-	606	538	-	9
Mexico	...	...	...	...	...	-	1	2	4	7
77 Fishing area total	...	...	*10*	*1*	...	-	*607*	*540*	*4*	*16*
81 Korea Rep	-	-	-	678	1 188	545	422	1 099	227	152
81 Fishing area total	-	-	-	*678*	*1 188*	*545*	*422*	*1 099*	*227*	*152*

B-34 Miscellaneous demersal fishes / Poissons démersaux divers / Peces demersales diversos

Capture production by species, fishing areas and countries or areas / Captures par espèces, zones de pêche et pays ou zones / Capturas por especies, áreas de pesca y países o áreas

Species, Fishing area Espèce, Zone de pêche Especie, Area de pesca	1993 t	1994 t	1995 t	1996 t	1997 t	1998 t	1999 t	2000 t	2001 t	2002 t
87 Ecuador	...	...	...	...	...	...	...	13 196	3 382	3 600 F
Korea Rep	-	-	-	-	-	-	85	470	-	1
87 Fishing area total	...	...	...	...	...	...	*85*	*13 666*	*3 382*	*3 601 F*
Species total	*86 984 F*	*123 616*	*114 272 F*	*119 436 F*	*219 561 F*	*150 705 F*	*190 272 F*	*206 611 F*	*166 033 F*	*206 201 F*
Indian driftfish	**Ariomme indienne**		**Arioma índica**		***Ariomma indica***			**1,76(04)012,01**		**DRI**
61 China,H.Kong	133	289	287	120	79	49	35 F	40 F	45 F	40 F
61 Fishing area total	*133*	*289*	*287*	*120*	*79*	*49*	*35 F*	*40 F*	*45 F*	*40 F*
Species total	*133*	*289*	*287*	*120*	*79*	*49*	*35 F*	*40 F*	*45 F*	*40 F*
Choicy ruff	**Sériolelle argentine**		**Cojinoba savorín**		***Seriolella porosa***			**1,76(08)010,01**		**SEO**
41 Argentina	787	1 182	1 536	2 042	2 593	2 462	3 322	3 542	3 990	5 368
41 Fishing area total	*787*	*1 182*	*1 536*	*2 042*	*2 593*	*2 462*	*3 322*	*3 542*	*3 990*	*5 368*
Species total	*787*	*1 182*	*1 536*	*2 042*	*2 593*	*2 462*	*3 322*	*3 542*	*3 990*	*5 368*
Common warehou	**...B**		**...C**		***Seriolella brama***			**1,76(08)010,02**		**SEM**
81 New Zealand	2 705	1 312	1 081	1 760	4 108	3 101	3 881	4 259	4 101	4 005
Ukraine	-	-	-	-	-	-	-	-	-	169
81 Fishing area total	*2 705*	*1 312*	*1 081*	*1 760*	*4 108*	*3 101*	*3 881*	*4 259*	*4 101*	*4 174*
Species total	*2 705*	*1 312*	*1 081*	*1 760*	*4 108*	*3 101*	*3 881*	*4 259*	*4 101*	*4 174*
Silver warehou	**...B**		**...C**		***Seriolella punctata***			**1,76(08)010,04**		**SEP**
81 New Zealand	4 518	5 113	13 377	4 926	11 253	10 993	9 029	11 218	11 268	8 778
Ukraine	-	-	-	-	-	-	-	-	-	397
81 Fishing area total	*4 518*	*5 113*	*13 377*	*4 926*	*11 253*	*10 993*	*9 029*	*11 218*	*11 268*	*9 175*
Species total	*4 518*	*5 113*	*13 377*	*4 926*	*11 253*	*10 993*	*9 029*	*11 218*	*11 268*	*9 175*
White warehou	**...B**		**...C**		***Seriolella caerulea***			**1,76(08)010,05**		**SEU**
81 New Zealand	819	1 072	2 348	1 467	2 432	2 296	2 366	2 407	1 962	1 975
81 Fishing area total	*819*	*1 072*	*2 348*	*1 467*	*2 432*	*2 296*	*2 366*	*2 407*	*1 962*	*1 975*
Species total	*819*	*1 072*	*2 348*	*1 467*	*2 432*	*2 296*	*2 366*	*2 407*	*1 962*	*1 975*
South Pacific breams nei	**Sériolelles nca**		**Cojinobas nep**		***Seriolella spp***			**1,76(08)010,XX**		**BSP**
57 Australia	2 242	2 603	2 558	3 406	4 020	3 356	3 327	3 449	4 192	4 009
57 Fishing area total	*2 242*	*2 603*	*2 558*	*3 406*	*4 020*	*3 356*	*3 327*	*3 449*	*4 192*	*4 009*
81 Norway	85	63	-	121	70	4	-	-	-	-
Russian Fed	922	721	150	185	352	34	28	-	-	-
Ukraine	24	4	...	484	1 183	320	328	1 558	666	34
81 Fishing area total	*1 031*	*788*	*150*	*790*	*1 605*	*358*	*356*	*1 558*	*666*	*34*
87 Chile	4 944	4 038	3 232	2 482	2 909	2 671	3 347	2 502	3 336	3 587
Colombia	...	...	...	...	...	...	...	...	8	8 F
Peru	2 795	8 892	7 698	3 704	388	505	1 589	1 473	3 192	2 192
87 Fishing area total	*7 739*	*12 930*	*10 930*	*6 186*	*3 297*	*3 176*	*4 936*	*3 975*	*6 536*	*5 787 F*
Species total	*11 012*	*16 321*	*13 638*	*10 382*	*8 922*	*6 890*	*8 619*	*8 982*	*11 394*	*9 830 F*
Bluenose warehou	**Rouffe antarctique**		**Rufo antártico**		***Hyperoglyphe antarctica***			**1,76(08)015,02**		**BWA**
81 New Zealand	2 387	42	2 719	2 432	2 974	2 630	2 755	2 793	2 954	3 001
81 Fishing area total	*2 387*	*42*	*2 719*	*2 432*	*2 974*	*2 630*	*2 755*	*2 793*	*2 954*	*3 001*
Species total	*2 387*	*42*	*2 719*	*2 432*	*2 974*	*2 630*	*2 755*	*2 793*	*2 954*	*3 001*
Black driftfish	**...B**		**...C**		***Hyperoglyphe bythites***			**1,76(08)015,03**		**HGY**
31 USA	...	...	...	...	...	...	2	7	5	6
31 Fishing area total	...	...	...	...	...	...	*2*	*7*	*5*	*6*
Species total	...	...	...	...	...	...	*2*	*7*	*5*	*6*
Pacific rudderfish	**Stromaté du Japon**		**Pámpano del Pacífico**		***Psenopsis anomala***			**1,76(08)020,01**		**BUP**
61 China,H.Kong	3 259	1 820	1 494	940	1 460	1 160	800 F	1 000 F	1 100 F	1 050 F
China,Taiwan	6 338	6 711	9 212	8 834	6 001	8 258	5 075	4 506	5 646	5 154
Japan	3 203	4 470	3 485	3 516	4 316	4 316	4 996	5 215	4 196	3 695
61 Fishing area total	*12 800*	*13 001*	*14 191*	*13 290*	*11 777*	*13 734*	*10 871 F*	*10 721 F*	*10 942 F*	*9 899 F*
Species total	*12 800*	*13 001*	*14 191*	*13 290*	*11 777*	*13 734*	*10 871 F*	*10 721 F*	*10 942 F*	*9 899 F*
Ruffs, barrelfishes nei	**Centrolophes nca**		**Rufos, romerillos nep**		***Centrolophidae***			**1,76(08)XXX,XX**		**CEN**
34 Lithuania	-	-	-	-	-	-	-	-	-	51
34 Fishing area total	-	-	-	-	-	-	-	-	-	*51*
47 Ukraine	...	...	...	36	-	-	-	-	-	-

B-34 Miscellaneous demersal fishes / Poissons démersaux divers / Peces demersales diversos

Capture production by species, fishing areas and countries or areas
Captures par espèces, zones de pêche et pays ou zones
Capturas por especies, áreas de pesca y países o áreas

Species, Fishing area Espèce, Zone de pêche Especie, Area de pesca	1993 t	1994 t	1995 t	1996 t	1997 t	1998 t	1999 t	2000 t	2001 t	2002 t
47 Fishing area total	*...*	*...*	*...*	*36*	*-*	*-*	*-*	*-*	*-*	*-*
51 Spain	-	-	-	-	-	-	-	36	-	-
Ukraine	301	732	485	254	440	395	753	360	299	-
51 Fishing area total	*301*	*732*	*485*	*254*	*440*	*395*	*753*	*396*	*299*	*-*
71 NewCaledonia	0	0	0	0	0	0	0	0	0	0
71 Fishing area total	*0*	*0*	*0*	*0*	*0*	*0*	*0*	*0*	*0*	*0*
81 Korea Rep	-	-	-	-	-	-	493	317	209	636
New Zealand	154	201	169	169	214	202	249	274	211	169
81 Fishing area total	*154*	*201*	*169*	*169*	*214*	*202*	*742*	*591*	*420*	*805*
Species total	*455*	*933*	*654*	*459*	*654*	*597*	*1 495*	*987*	*719*	*856*
Golden redfish	**Sébaste doré**		**Gallineta dorada**		***Sebastes marinus***			**1,78(01)001,01**		**REG**
27 Denmark	2	7	1	4	4	7	5	8	11	1
Faeroe Is	17 402	12 161	12 071	11 028	11 402	7 951	7 129	13 324	13 572	3 500
Greenland	...	...	...	...	...	...	...	41	13	384
Iceland	-	-	-	-	-	-	45 406	44 829	38 762	54 766
Norway	-	-	-	-	-	-	-	15 996	13 698	9 438
Portugal	0	0	1	2	7	10	2	269	2	5
Russian Fed	-	-	-	-	-	-	-	1 066	978	832
Spain	-	-	-	-	-	-	50	23	8	3
27 Fishing area total	*17 404*	*12 168*	*12 073*	*11 034*	*11 413*	*7 968*	*52 592*	*75 556*	*67 044*	*68 929*
Species total	*17 404*	*12 168*	*12 073*	*11 034*	*11 413*	*7 968*	*52 592*	*75 556*	*67 044*	*68 929*
Widow rockfish	**Sébaste rocote**		**Rocote**		***Sebastes entomelas***			**1,78(01)001,03**		**WRO**
67 USA	...	...	...	...	6 886	4 357	3 841	3 575	2 481	435
67 Fishing area total	*...*	*...*	*...*	*...*	*6 886*	*4 357*	*3 841*	*3 575*	*2 481*	*435*
77 USA	...	...	...	...	865	540	402	29	128	7
77 Fishing area total	*...*	*...*	*...*	*...*	*865*	*540*	*402*	*29*	*128*	*7*
Species total	*...*	*...*	*...*	*...*	*7 751*	*4 897*	*4 243*	*3 604*	*2 609*	*442*
Yellowtail rockfish	**Sébaste à queue jaune**		**Chancharro cola amarilla**		***Sebastes flavidus***			**1,78(01)001,04**		**YRO**
67 USA	...	...	...	...	2 512	3 349	3 427	3 138	2 073	1 204
67 Fishing area total	*...*	*...*	*...*	*...*	*2 512*	*3 349*	*3 427*	*3 138*	*2 073*	*1 204*
77 USA	...	...	...	...	232	253	41	32	4	2
77 Fishing area total	*...*	*...*	*...*	*...*	*232*	*253*	*41*	*32*	*4*	*2*
Species total	*...*	*...*	*...*	*...*	*2 744*	*3 602*	*3 468*	*3 170*	*2 077*	*1 206*
Pacific ocean perch	**Sébaste du Pacifique**		**Gallineta del Pacífico**		***Sebastes alutus***			**1,78(01)001,09**		**OPP**
61 Japan	2 760	1 520	1 449	1 730	767	896	766	458	668	967
Russian Fed	1 153	3 447	3 539	2 615	2 095	1 544	864	1 017	793	1 194
61 Fishing area total	*3 913*	*4 967*	*4 988*	*4 345*	*2 862*	*2 440*	*1 630*	*1 475*	*1 461*	*2 161*
67 Canada	4 609	5 649	5 207	6 174	5 782	6 184	5 849	6 177	5 832	5 897
Poland	-	-	-	-	-	-	-	21	-	-
Russian Fed	-	-	-	-	-	195	1 614	6	-	38
USA	20 649	14 000	15 613	21 004	19 580	18 445	20 608	17 940	17 696	20 602
67 Fishing area total	*25 258*	*19 649*	*20 820*	*27 178*	*25 362*	*24 824*	*28 071*	*24 144*	*23 528*	*26 537*
77 Estonia	-	-	-	-	-	-	8	-	-	-
USA	-	-	-	-	-	-	8	12	12	18
77 Fishing area total	*-*	*-*	*-*	*-*	*-*	*-*	*16*	*12*	*12*	*18*
Species total	*29 171*	*24 616*	*25 808*	*31 523*	*28 224*	*27 264*	*29 717*	*25 631*	*25 001*	*28 716*
Bocaccio rockfish	**Sébaste bocace**		**Chancharro bocacio**		***Sebastes paucispinis***			**1,78(01)001,10**		**SBC**
67 USA	...	...	...	...	448	491	135	3	0	1
67 Fishing area total	*...*	*...*	*...*	*...*	*448*	*491*	*135*	*3*	*0*	*1*
77 USA	...	...	...	...	276	105	62	24	33	21
77 Fishing area total	*...*	*...*	*...*	*...*	*276*	*105*	*62*	*24*	*33*	*21*
Species total	*...*	*...*	*...*	*...*	*724*	*596*	*197*	*27*	*33*	*22*
Beaked redfish	**Sébaste du Nord**		**Gallineta nórdica**		***Sebastes mentella***			**1,78(01)001,12**		**REB**
21 Denmark	-	-	-	-	-	-	-	-	0	0
Greenland	...	...	...	...	...	...	...	671	124	124
Poland	-	-	-	-	-	-	-	-	4	428
21 Fishing area total	*...*	*...*	*...*	*...*	*...*	*...*	*...*	*671*	*128*	*552*
27 Bulgaria	3 163	-	-	-	-	-	-	-	-	-
Denmark	18	35	14	17	19	20	48	35	89	42
Greenland	...	...	...	...	...	...	...	3 509	3 258	4 164
Iceland	-	-	-	-	-	-	21 463	45 231	42 440	44 504
Norway	-	-	-	-	-	-	-	9 484	14 958	6 883

B-34 Miscellaneous demersal fishes / Poissons démersaux divers / Peces demersales diversos

Capture production by species, fishing areas and countries or areas / Captures par espèces, zones de pêche et pays ou zones / Capturas por especies, áreas de pesca y países o áreas

Species, Fishing area Espèce, Zone de pêche Especie, Area de pesca	1993 t	1994 t	1995 t	1996 t	1997 t	1998 t	1999 t	2000 t	2001 t	2002 t
Poland	-	-	-	-	777	12	6	2	5	9
Russian Fed	-	-	-	-	-	-	-	14 715	32 208	35 303
Spain	-	-	4 554	4 307	4 438	4 587	3 526	2 530	5 575	1 652
Ukraine	2 782	5 561	3 185	518	-	-	-	-	-	-
27 Fishing area total	*5 963*	*5 596*	*7 753*	*4 842*	*5 234*	*4 619*	*25 043*	*75 506*	*98 533*	*92 557*
Species total	*5 963*	*5 596*	*7 753*	*4 842*	*5 234*	*4 619*	*25 043*	*76 177*	*98 661*	*93 109*
Canary rockfish	**Sébaste citron**		**Chancharro flioma**		***Sebastes pinniger***			**1,78(01)001,15**		**SPG**
67 USA	...	...	...	...	1 133	1 222	736	57	45	51
67 Fishing area total	...	...	...	...	*1 133*	*1 222*	*736*	*57*	*45*	*51*
77 USA	...	...	...	...	129	91	36	3	4	3
77 Fishing area total	...	...	...	...	*129*	*91*	*36*	*3*	*4*	*3*
Species total	...	...	...	...	*1 262*	*1 313*	*772*	*60*	*49*	*54*
Chilipepper rockfish	**Sébaste piment**		**Chancharro pimienta**		***Sebastes goodei***			**1,78(01)001,16**		**SGO**
67 USA	...	...	...	...	38	11	10	31	9	5
67 Fishing area total	...	...	...	...	*38*	*11*	*10*	*31*	*9*	*5*
77 USA	...	...	...	...	1 812	1 262	908	414	609	156
77 Fishing area total	...	...	...	...	*1 812*	*1 262*	*908*	*414*	*609*	*156*
Species total	...	...	...	...	*1 850*	*1 273*	*918*	*445*	*618*	*161*
Black rockfish	**...B**		**...C**		***Sebastes melanops***			**1,78(01)001,32**		**RMG**
67 USA	...	...	...	...	...	...	...	109	148	127
67 Fishing area total	...	...	...	...	...	...	...	*109*	*148*	*127*
Species total	...	...	...	...	...	...	...	*109*	*148*	*127*
Atlantic redfishes nei	**Sébastes de l'Atlantique nca**		**Gallinetas del Atlántico nep**		***Sebastes spp***			**1,78(01)001,XX**		**RED**
21 Canada	84 787	50 774	17 954	21 571	18 751	25 763	19 682	19 852	19 710	16 745
Cuba	2 880	23	70	42	87	12	39	0	-	-
Estonia	4 115	135	863	-	-	-	-	841	186	26
Faeroe Is	61	12	15	1	-	-	-	-	-	64
Germany	295	74	-	-	-	-	154	4 476	817	2 325
Greenland	907	1 063	922	864	969	929	779	721	305	422
Iceland	-	10	751	-	-	-	-	1	-	-
Japan	1 161	570	868	1 169	565	891	460	140	187	156
Korea Rep	3 687	-	-	-	-	-	-	-	-	-
Latvia	8 502	149	304	-	-	-	-	13	11	780
Lithuania	3 904	37	-	-	-	-	-	430	4 396	7 196
Norway	105	8	3	-	-	1	-	113	-	5
Portugal	9 831	8 609	3 291	2 153	1 125	2 368	6 081	5 675	5 625	6 344
Russian Fed	14 560	5 060	5 395	86	375	15	546	9 799	14 152	20 397
St Pier Mq	-	-	-	-	430	654	423	196	129	319
Spain	136	870	628	560	1 388	2 686	5 775	4 694	3 709	1 277
UK	-	-	-	-	-	1	-	-	-	-
USA	800	439	436	322	251	320	353	322	363	368
21 Fishing area total	*135 731*	*67 833*	*31 500*	*26 768*	*23 941*	*33 640*	*34 292*	*47 273*	*49 590*	*56 424*
27 Belgium	124	54	16	19	16	2	3	5	6	3
Estonia	6 365	17 875	16 854	7 091	3 720	3 968	2 108	7 811	599	15
France	2 449	2 293	2 520	2 244	2 567	1 604	1 254	943	1 076	569
Germany	34 462	30 404	20 472	22 512	21 096	20 342	18 263	9 802	11 973	13 229
Greenland	918	428	4 799	280	193	1 476	4 437	139	94	28
Iceland	116 325	142 101	117 999	120 751	111 652	116 132	43 475	26 243	11 325	11 606
Ireland	5	15	18	15	48	71	171	186	433	297
Japan	1 077	1 818	1 148	416	31	31	-	-	-	9
Latvia	6 803	13 205	5 003	1 084	-	-	-	-	-	1 061
Lithuania	7 899	7 404	22 893	10 649	-	1 769	3 884	6 257	15 786	14 657
Netherlands	1	8	29	41	53	20	16	19	8	15
Norway	32 974	28 934	23 279	28 679	22 687	28 559	30 856	1	1	4
Portugal	1 040	2 873	6 063	2 904	4 211	4 260	4 370	4 024	2 694	3 374
Russian Fed	18 478	30 629	51 377	47 574	41 613	31 200	29 069	13 557	-	19
Spain	65	34	67	408	4 577	2 532	2 410	1 024	935	2 827
Sweden	-	4	1	0	-	0	1	-	-	-
UK	1 384	743	1 346	1 776	1 507	1 553	2 672	2 495	2 875	2 117
27 Fishing area total	*230 369*	*278 822*	*273 884*	*246 443*	*213 971*	*213 519*	*142 989*	*72 506*	*47 805*	*49 830*
Species total	*366 100*	*346 655*	*305 384*	*273 211*	*237 912*	*247 159*	*177 281*	*119 779*	*97 395*	*106 254*
Red scorpionfish	**Rascasse rouge**		**Cabracho**		***Scorpaena scrofa***			**1,78(01)009,02**		**RSE**
27 UK	...	...	...	...	...	...	...	1	2	5
27 Fishing area total	...	...	...	...	...	...	...	*1*	*2*	*5*
Species total	...	...	...	...	...	...	...	*1*	*2*	*5*
Blackbelly rosefish	**Sébaste chèvre**		**Gallineta**		***Helicolenus dactylopterus***			**1,78(01)017,03**		**BRF**
21 USA	-	-	-	-	-	-	-	3	0	0
21 Fishing area total	-	-	-	-	-	-	-	*3*	*0*	*0*

B-34

Miscellaneous demersal fishes | **Capture production by species, fishing areas and countries or areas**
Poissons démersaux divers | **Captures par espèces, zones de pêche et pays ou zones**
Peces demersales diversos | **Capturas por especies, áreas de pesca y países o áreas**

Species, Fishing area Espèce, Zone de pêche Especie, Area de pesca	1993 t	1994 t	1995 t	1996 t	1997 t	1998 t	1999 t	2000 t	2001 t	2002 t
27 France	174	130	91	104	115	140	129	123	125	49
Portugal	...	...	...	...	...	...	334	436	313	297
UK	-	1	0	15	3	116	253	184	186	106
27 Fishing area total	*174*	*131*	*91*	*119*	*118*	*256*	*716*	*743*	*624*	*452*
41 Argentina	580	628	4 670	1 499	2 563	1 192	3 354	3 051	1 209	494
Spain	-	-	-	-	-	-	8	8	6	-
Uruguay	1 994	4 898	2 734	2 204	3 404	4 389	2 581	2 111	1 830	1 544
41 Fishing area total	*2 574*	*5 526*	*7 404*	*3 703*	*5 967*	*5 581*	*5 943*	*5 170*	*3 045*	*2 038*
47 Namibia	21	183	271	576	673	343	448	639	891	...
South Africa	1 476	723	1 082	1 183	957	752	766	750 F	782	821
47 Fishing area total	*1 497*	*906*	*1 353*	*1 759*	*1 630*	*1 095*	*1 214*	*1 389 F*	*1 673*	*821*
51 Spain	-	-	-	-	-	-	-	-	1	-
51 Fishing area total	*-*	*-*	*-*	*-*	*-*	*-*	*-*	*-*	*1*	*-*
Species total	*4 245*	*6 563*	*8 848*	*5 581*	*7 715*	*6 932*	*7 873*	*7 305 F*	*5 343*	*3 311*
Shortspine thornyhead	**...B**		**...C**			***Sebastolobus alascanus***			**1,78(01)079,01**	**SJU**
67 USA	...	...	...	...	...	...	...	331	268	295
67 Fishing area total	*...*	*...*	*...*	*...*	*...*	*...*	*...*	*331*	*268*	*295*
Species total	*...*	*...*	*...*	*...*	*...*	*...*	*...*	*331*	*268*	*295*
Scorpionfishes nei	**Rascasses, etc. nca**			**Rascacios, gallinetas nep**	***Scorpaenidae***				**1,78(01)XXX,XX**	**SCO**
27 France	50	38	22	31	63	1	8	16	18	48
Portugal	969	1 070	905	730	658	682	294	256	363	214
Spain	717	1 188	736	886	1 546	822	1 004	1 227	653	818
27 Fishing area total	*1 736*	*2 296*	*1 663*	*1 647*	*2 267*	*1 505*	*1 306*	*1 499*	*1 034*	*1 080*
34 Greece	169	24	28	64	35	36	42	53	42	21
Italy	-	-	-	-	-	-	-	-	14	-
Mauritania	5 F	10 F	5 F	5 F	5 F	0	1 F	0	...	...
Morocco	227	201	212	220	201	195	257	517	382	663
Nigeria	791	942	1 893	1 032	1 278	2 071	1 149	3 170	2 568	600
Portugal	129	127	-	-	-	-	-	15	9	5
Senegal	275	657	539	581	491	343	267	471	471	451
34 Fishing area total	*1 596 F*	*1 961 F*	*2 677 F*	*1 902 F*	*2 010 F*	*2 645*	*1 716 F*	*4 226*	*3 486*	*1 740*
37 Croatia	287	183	172	137	133	185	46	40	39	22
Cyprus	9	10	11	16	19	10	3	8	11	8
France	119	77	19	25	80	80	30	27	24	26
Greece	900	1 538	1 039	777	748	482	592	587	684	630 F
Lebanon	50	50	100	100	100	100	150	150	100	125
Malta	5	4	6	8	3	8	12	11	0	-
Morocco	10	6	7	10	25	4	6	56	10	18
Russian Fed	-	-	-	-	-	-	-	1	2	3
Serbia-Monte	1	3	6	7	7	7	7	8	7	7
Spain	-	-	-	236	287	307	239	419	332	1 272
Syria	...	...	...	...	...	...	...	...	...	38
Tunisia	420	428	311	339	472	755	442	400	444	453
Turkey	547	434	539	585	435	515	315	360	640	296
37 Fishing area total	*2 348*	*2 733*	*2 210*	*2 240*	*2 309*	*2 453*	*1 842*	*2 067*	*2 293*	*2 898 F*
41 Italy	229	229	-	-	-	-	-	-	-	-
Korea Rep	48	-	-	293	99	-	-	-	4	-
41 Fishing area total	*277*	*229*	*-*	*293*	*99*	*-*	*-*	*-*	*4*	*-*
47 Iceland	-	-	-	-	5	-	-	-	-	-
Portugal	-	-	-	-	-	3	4	2	-	6
Spain	-	-	-	-	48	18	-	-	-	-
47 Fishing area total	*-*	*-*	*-*	*-*	*53*	*21*	*4*	*2*	*-*	*6*
48 Argentina	-	0	-	-	-	-	-	-	-	-
48 Fishing area total	*-*	*0*	*-*	*-*	*-*	*-*	*-*	*-*	*-*	*-*
51 Italy	344	344	-	-	-	-	-	-	-	-
South Africa	2	-	-	-	3	1	1 F	2	2	...
51 Fishing area total	*346*	*344*	*-*	*-*	*3*	*1*	*1 F*	*2*	*2*	*...*
57 Australia	-	-	-	-	-	-	-	7	8	20
57 Fishing area total	*-*	*-*	*-*	*-*	*-*	*-*	*-*	*7*	*8*	*20*
61 Japan	3 594	3 059	2 888	2 317	2 082	1 638	1 314	1 244	1 132	1 323
Korea Rep	6 556	5 823	5 901	4 647	5 414	4 602	4 665	4 879	5 152	1 669
Russian Fed	-	-	-	-	-	-	-	45	54	57
61 Fishing area total	*10 150*	*8 882*	*8 789*	*6 964*	*7 496*	*6 240*	*5 979*	*6 168*	*6 338*	*3 049*
67 Canada	20 034	18 100	16 462	16 626	14 118	15 356	18 086	17 188	16 351	16 248
Russian Fed	-	-	-	-	-	-	-	-	-	1
USA	47 268	34 262	32 514	35 305	34 251	15 772	17 465	13 662	12 436	12 200
67 Fishing area total	*67 302*	*52 362*	*48 976*	*51 931*	*48 369*	*31 128*	*35 551*	*30 850*	*28 787*	*28 449*
71 Korea Rep	-	-	-	2	-	-	-	-	-	-
71 Fishing area total	*-*	*-*	*-*	*2*	*-*	*-*	*-*	*-*	*-*	*-*

B-34 Miscellaneous demersal fishes / Poissons démersaux divers / Peces demersales diversos

Capture production by species, fishing areas and countries or areas / Captures par espèces, zones de pêche et pays ou zones / Capturas por especies, áreas de pesca y países o áreas

Species, Fishing area Espèce, Zone de pêche Especie, Area de pesca	1993 t	1994 t	1995 t	1996 t	1997 t	1998 t	1999 t	2000 t	2001 t	2002 t
77 Korea Rep	-	-	-	-	-	4	-	-	-	-
USA	7 464	7 335	8 363	7 678	4 482	4 326	1 614	1 291	1 679	1 839
77 Fishing area total	*7 464*	*7 335*	*8 363*	*7 678*	*4 482*	*4 330*	*1 614*	*1 291*	*1 679*	*1 839*
81 Australia	...	...	...	316	349	371	356	60	28	36
Korea Rep	-	-	-	128	212	99	371	219	198	428
New Zealand	...	...	986	1 133	1 643	1 843	2 088	1 819	1 897	2 306
Russian Fed	2	-	-	5	-	-	-	-	-	-
81 Fishing area total	*2*	*...*	*986*	*1 582*	*2 204*	*2 313*	*2 815*	*2 098*	*2 123*	*2 770*
87 Chile	330	287	227	212	532	208	258	141	177	53
87 Fishing area total	*330*	*287*	*227*	*212*	*532*	*208*	*258*	*141*	*177*	*53*
Species total	*91 551 F*	*76 429 F*	*73 891 F*	*74 451 F*	*69 824 F*	*50 844*	*51 086 F*	*48 351*	*45 931*	*41 904 F*
Bluefin gurnard	**Grondin aile bleue**		**Testolín de aleta azul**		***Chelidonichthys kumu***			**1,78(02)003,01**		**KUG**
57 Australia	1	1	1	0	1	1	0	117	163	143
57 Fishing area total	*1*	*1*	*1*	*0*	*1*	*1*	*0*	*117*	*163*	*143*
61 Japan	1 038	1 132	-	-	-	-	-	-	-	-
Korea Rep	141	70	40	75	86	146	43	79	140	260
61 Fishing area total	*1 179*	*1 202*	*40*	*75*	*86*	*146*	*43*	*79*	*140*	*260*
71 Australia	-	-	-	14	0	0	0	0	-	-
71 Fishing area total	*-*	*-*	*-*	*14*	*0*	*0*	*0*	*0*	*-*	*-*
81 Australia	363	147	123	46	349	371	...	295	277	233
Japan	23	24	4	-	-	-	-	-	-	-
New Zealand	3 562	3 041	11 829	2 260	2 625	2 454	2 199	2 663	3 670	3 447
Ukraine	-	-	-	-	-	-	-	-	-	2
81 Fishing area total	*3 948*	*3 212*	*11 956*	*2 306*	*2 974*	*2 825*	*2 199*	*2 958*	*3 947*	*3 682*
Species total	*5 128*	*4 415*	*11 997*	*2 395*	*3 061*	*2 972*	*2 242*	*3 154*	*4 250*	*4 085*
Tub gurnard	**Grondin perlon**		**Begel**		***Chelidonichthys lucerna***			**1,78(02)003,02**		**GUU**
27 Denmark	-	-	-	5	4	5	19	15	12	63
France	...	...	...	...	...	...	...	1 209	1 134	1 234
Netherlands	-	-	-	-	-	-	-	1 164	-	1 438
Portugal	-	-	-	-	-	-	3	5	3	8
27 Fishing area total	*...*	*...*	*...*	*5*	*4*	*5*	*22*	*2 393*	*1 149*	*2 743*
37 France	...	...	...	...	...	...	...	79	68	27
37 Fishing area total	*...*	*...*	*...*	*...*	*...*	*...*	*...*	*79*	*68*	*27*
Species total	*...*	*...*	*...*	*5*	*4*	*5*	*22*	*2 472*	*1 217*	*2 770*
Red gurnard	**Grondin rouge**		**Arete**		***Chelidonichthys cuculus***			**1,78(02)003,03**		**GUR**
27 Belgium	113	119	116	145	128	157	262	418	493	500
Channel Is	7 F	3	3 F	10	8	6	10	7	-	15
France	...	...	...	...	...	...	...	4 201	4 707	5 005
Ireland	-	-	-	-	-	25	47	...	...	...
Isle of Man	0	-	-	-	-	-	-	-	-	-
Netherlands	-	-	-	-	-	-	-	46	1 724	53
UK	69	54	81	59	17	...	...	4	207	281
27 Fishing area total	*189 F*	*176*	*200 F*	*214*	*153*	*188*	*319*	*4 676*	*7 131*	*5 854*
37 France	...	...	...	...	...	...	...	184	162	311
37 Fishing area total	*...*	*...*	*...*	*...*	*...*	*...*	*...*	*184*	*162*	*311*
Species total	*189 F*	*176*	*200 F*	*214*	*153*	*188*	*319*	*4 860*	*7 293*	*6 165*
Streaked gurnard	**...B**		**...C**		***Chelidonichthys lastoviza***			**1,78(02)003,04**		**CTZ**
27 France	...	...	...	...	...	...	...	56	51	68
27 Fishing area total	*...*	*...*	*...*	*...*	*...*	*...*	*...*	*56*	*51*	*68*
Species total	*...*	*...*	*...*	*...*	*...*	*...*	*...*	*56*	*51*	*68*
Cape gurnard	**Grondin du Cap**		**Rubio del Cabo**		***Chelidonichthys capensis***			**1,78(02)003,07**		**GUC**
47 Namibia	20	28	57	51	214	216	71	236	144	234
South Africa	1 003	576	502	446	428	623	507	450 F	465	462
47 Fishing area total	*1 023*	*604*	*559*	*497*	*642*	*839*	*578*	*686 F*	*609*	*696*
Species total	*1 023*	*604*	*559*	*497*	*642*	*839*	*578*	*686 F*	*609*	*696*
Atlantic searobins	**Grondins**		**Rubios americanos**		***Prionotus spp***			**1,78(02)020,XX**		**SRA**
21 USA	17	49	111	20	10	31	38	24	39	66
21 Fishing area total	*17*	*49*	*111*	*20*	*10*	*31*	*38*	*24*	*39*	*66*
41 Brazil	930 F	930 F	922	792	540	858	1 149	1 839	317	352
41 Fishing area total	*930 F*	*930 F*	*922*	*792*	*540*	*858*	*1 149*	*1 839*	*317*	*352*
Species total	*947 F*	*979 F*	*1 033*	*812*	*550*	*889*	*1 187*	*1 863*	*356*	*418*

B-34 Miscellaneous demersal fishes / Poissons démersaux divers / Peces demersales diversos

Capture production by species, fishing areas and countries or areas / Captures par espèces, zones de pêche et pays ou zones / Capturas por especies, áreas de pesca y países o áreas

Species, Fishing area Espèce, Zone de pêche Especie, Area de pesca	1993 t	1994 t	1995 t	1996 t	1997 t	1998 t	1999 t	2000 t	2001 t	2002 t
Latchet(=Sharpbeak gurnard)	**Grondin pointu**		**Cabete picudo**		***Pterygotrigla polyommata***				**1,78(02)025,01**	**BEG**
57 Australia	43	26	0	0	60	25	55	66	78	88
57 Fishing area total	*43*	*26*	*0*	*0*	*60*	*25*	*55*	*66*	*78*	*88*
81 Australia	133	127	121	58	76	67	39	54	71	52
81 Fishing area total	*133*	*127*	*121*	*58*	*76*	*67*	*39*	*54*	*71*	*52*
Species total	*176*	*153*	*121*	*58*	*136*	*92*	*94*	*120*	*149*	*140*
Spotted gurnard	**...B**		**...C**		***Pterygotrigla picta***				**1,78(02)025,02**	**JGU**
81 New Zealand	-	-	107	87	71	87	74	55	65	49
81 Fishing area total	-	-	*107*	*87*	*71*	*87*	*74*	*55*	*65*	*49*
Species total	-	-	*107*	*87*	*71*	*87*	*74*	*55*	*65*	*49*
Grey gurnard	**Grondin gris**		**Borracho**		***Eutrigla gurnardus***				**1,78(02)070,01**	**GUG**
27 Belgium	125	79	59	119	61	47	49	44	33	63
Denmark	840	121	84	70	36	56	85	96	289	65
France	...	...	...	...	...	...	...	224	218	180
Iceland	-	-	-	1	0	0	0	0	-	0
Ireland	-	-	-	-	-	38	71	...	...	...
Netherlands	-	-	-	-	-	-	-	459	295	286
Sweden	9	12	7	4	5	8	133	5	5	2
UK	25	24	21	56	59	...	...	...	46	41
27 Fishing area total	*999*	*236*	*171*	*250*	*161*	*149*	*338*	*828*	*886*	*637*
37 Egypt	1 099	1 077	1 070	1 264	1 158	1 067	1 159	1 078	778	611
France	...	...	...	...	...	...	...	12	4	16
37 Fishing area total	*1 099*	*1 077*	*1 070*	*1 264*	*1 158*	*1 067*	*1 159*	*1 090*	*782*	*627*
Species total	*2 098*	*1 313*	*1 241*	*1 514*	*1 319*	*1 216*	*1 497*	*1 918*	*1 668*	*1 264*
Gurnards, searobins nei	**Grondins, cavillones nca**		**Cabetes, rubios nep**		***Triglidae***				**1,78(02)XXX,XX**	**GUX**
27 Belgium	209	270	255	235	245	200	160	161	177	183
Channel Is	...	...	...	...	10	31	22	10	41	7
Faeroe Is	-	-	-	-	-	-	1	...	...	...
France	6 148	5 999	5 902	5 845	5 633	5 394	5 644	140	87	76
Germany	140	123	133	94	136	141	187	180	150	198
Ireland	26	67	85	77	82	-	-	79	97	101
Isle of Man	0	2	5	2	1	1	1	...	1	-
Portugal	816	560	616	627	611	500	496	650	616	644
Russian Fed	-	-	-	-	-	-	2 426	26 081	3 155	60
Spain	-	-	-	1 079	1 264	620	681	755	506	409
UK	651	600	599	583	484	629	631	953	1 131	1 154
27 Fishing area total	*7 990*	*7 621*	*7 595*	*8 542*	*8 466*	*7 516*	*10 249*	*29 009*	*5 961*	*2 832*
34 Benin	...	...	...	...	...	...	11	7 F	-	-
Ghana	140	14	8	0	76	...	53	0	37	...
Greece	4	18	9	8	4	3	-	-	-	-
Italy	1 718	1 718	-	-	11	13	13	31	129	80
Mauritania	10 F	10 F	10 F	10 F	0	70 F	...	...	...	...
Morocco	1 765	1 798	2 002	2 269	2 211	2 827	3 140	4 858	4 160	3 546
Portugal	20	11	5	14	4	4	7	0	-	-
Russian Fed	175	-	-	-	-	628	-	-	-	-
Senegal	...	...	...	...	...	31	81	41	94	67
Spain	-	-	-	28	-	10	9	-	-	-
34 Fishing area total	*3 832 F*	*3 569 F*	*2 034 F*	*2 329 F*	*2 306*	*3 586 F*	*3 314*	*4 937 F*	*4 420*	*3 693*
37 Albania	...	...	2	15	0	-	-	-	34	72
France	235	148	326	50	50	50	207	113	105	14
Greece	1 829	421	321	368	376	234	229	207	169	160 F
Italy	4 269	4 304	4 116	3 908	3 462	3 280	2 655	2 137	2 135	1 861
Malta	-	-	1	2	0	2	4	2	1	1
Morocco	36	24	9	17	5	7	12	16	9	21
Slovenia	-	-	-	-	-	-	-	1	1	1
Spain	-	-	-	249	261	238	175	359	346	674
Syria	134 F	129 F	128 F	0	60	46	35	32	32	38
Tunisia	61	66	79	107	87	115	141	168	198	346
Turkey	2 253	1 921	1 500	2 319	750	700	710	260	200	271
37 Fishing area total	*8 817 F*	*7 013 F*	*6 482 F*	*7 035*	*5 051*	*4 672*	*4 168*	*3 295*	*3 230*	*3 459 F*
47 Angola	0	0	0	0	0	-	-	-	-	-
47 Fishing area total	*0*	*0*	*0*	*0*	*0*	-	-	-	-	-
51 South Africa	-	-	-	-	2	-	...	0	0	...
51 Fishing area total	-	-	-	-	*2*	-	...	*0*	*0*	...
87 Ecuador	...	...	...	...	...	...	22 634	27 129	1 428	1 500 F
87 Fishing area total	...	...	...	...	...	...	*22 634*	*27 129*	*1 428*	*1 500 F*
Species total	*20 639 F*	*18 203 F*	*16 111 F*	*17 906 F*	*15 825*	*15 774 F*	*40 365*	*64 370 F*	*15 039*	*11 484 F*

B-34 Miscellaneous demersal fishes / Poissons démersaux divers / Peces demersales diversos

Capture production by species, fishing areas and countries or areas / Captures par espèces, zones de pêche et pays ou zones / Capturas por especies, áreas de pesca y países o áreas

Species, Fishing area Espèce, Zone de pêche Especie, Area de pesca	1993 t	1994 t	1995 t	1996 t	1997 t	1998 t	1999 t	2000 t	2001 t	2002 t
Sablefish	**Morue charbonnière**		**Bacalao negro**		***Anoplopoma fimbria***			**1,78(08)004,01**		**SAB**
61 Russian Fed	-	-	8	502	-	-	-	6	6	19
61 Fishing area total	*-*	*-*	*8*	*502*	*-*	*-*	*-*	*6*	*6*	*19*
67 Canada	5 310	5 201	4 542	3 069	4 000	4 500	4 583	2 811	2 967	1 449
Russian Fed	-	-	-	-	-	-	-	-	-	31
USA	33 796	31 173	28 108	25 057	22 255	18 908	20 756	21 446	19 017	17 659
67 Fishing area total	*39 106*	*36 374*	*32 650*	*28 126*	*26 255*	*23 408*	*25 339*	*24 257*	*21 984*	*19 139*
77 USA	1 343	1 186	1 786	2 136	1 751	824	1 132	1 088	958	897
77 Fishing area total	*1 343*	*1 186*	*1 786*	*2 136*	*1 751*	*824*	*1 132*	*1 088*	*958*	*897*
Species total	*40 449*	*37 560*	*34 444*	*30 764*	*28 006*	*24 232*	*26 471*	*25 351*	*22 948*	*20 055*
Antarctic horsefish	**Cacique antarctique**		**Cacique antártico**		***Zanclorhynchus spinifer***			**1,78(11)030,01**		**ZSP**
58 Australia	-	-	-	-	0	-	-	-	-	-
58 Fishing area total	*-*	*-*	*-*	*-*	*0*	*-*	*-*	*-*	*-*	*-*
Species total	*-*	*-*	*-*	*-*	*0*	*-*	*-*	*-*	*-*	*-*
Lumpfish(=Lumpsucker)	**Lompe**		**Liebre de mar**		***Cyclopterus lumpus***			**1,78(20)003,01**		**LUM**
21 Canada	18	-	25	18	175	8	1	8	-	-
Greenland	246	579	448	426	1 157	2 142	3 056	1 211	3 216	5 795
St Pier Mq	126	158	226	218	363	249	422	536	146	3
21 Fishing area total	*390*	*737*	*699*	*662*	*1 695*	*2 399*	*3 479*	*1 755*	*3 362*	*5 798*
27 Denmark	924	1 443	1 069	1 973	1 553	188	835	425	422	592
Germany	2	2	1	2	2	0	4	1	3	2
Greenland	-	-	-	-	-	1	1	-	-	-
Iceland	3 318	5 324	4 563	4 201	6 520	3 165	3 373	2 458	412	206
Norway	4 613	5 618	4 015	4 355	5 652	1 365	2 059	2 374	5 184	5 936
Russian Fed	-	-	-	-	-	-	-	20	28	28
Sweden	75	109	70	147	129	20	72	47	83	213
UK	-	-	-	0	0	0	0	1	-	-
27 Fishing area total	*8 932*	*12 496*	*9 718*	*10 678*	*13 856*	*4 739*	*6 344*	*5 326*	*6 132*	*6 977*
Species total	*9 322*	*13 233*	*10 417*	*11 340*	*15 551*	*7 138*	*9 823*	*7 081*	*9 494*	*12 775*
Angler(=Monk)	**Baudroie commune**		**Rape**		***Lophius piscatorius***			**1,95(01)001,01**		**MON**
27 Belgium	1 024	1 638	1 772	1 369	1 113	961	818	1 047	1 354	1 312
Channel Is	1 F	0	0	2	3	3	2	3	5	2
Denmark	1 863	1 974	1 350	1 835	2 040	1 873	1 858	1 724	1 917	1 932
Faeroe Is	993	1 065	1 433	1 610	1 765	1 885	2 578	2 216	2 006	1 781
France	13 557	15 407	18 168	18 476	17 145	14 053	10 566	12 498	13 049	15 055
Germany	452	385	893	542	1 137	1 446	847	568	364	362
Iceland	685	641	552	669	787	850	977	1 570	1 353	965
Ireland	2 362	2 535	2 929	3 348	3 880	4 251	4 298	3 839	3 112	2 523
Isle of Man	16	22	27	34	27	28	9	5	2	1
Netherlands	559	567	362	227	319	259	169	170	168	86
Norway	4 454	2 721	1 731	2 071	1 447	2 646	3 239	4 357	4 974	3 172
Portugal	225	107	104	224	323	179	1 471	880	617	512
Russian Fed	-	-	-	-	-	-	-	-	1	-
Spain	3 336	3 274	3 518	3 534	5 161	6 301	6 986	5 607	5 245	4 559
Sweden	96	78	55	38	44	44	48	107	92	119
UK	18 894	19 999	24 456	31 451	29 783	21 395	16 989	15 955	16 249	14 289
27 Fishing area total	*48 517 F*	*50 413*	*57 350*	*65 430*	*64 974*	*56 174*	*50 855*	*50 546*	*50 508*	*46 670*
37 Albania	...	...	0	46	0	42	48	44	...	14
France	264	128	101	225	225	225	272	557	494	318
Greece	732	1 314	1 508	942	912	757	739	882	694	650 F
Italy	2 149	2 888	2 409	2 345	6 672	2 845	1 705	1 269	1 244	1 083
Malta	0	0	0	0	1	0	2	0	-	-
Morocco	143	92	107	110	113	83	66	785	310	185
Spain	-	-	-	-	-	-	949	1 283	1 492	1 187
Tunisia	32	21	7	6	33	30	44	39	35	43
37 Fishing area total	*3 320*	*4 443*	*4 132*	*3 674*	*7 956*	*3 982*	*3 825*	*4 859*	*4 269*	*3 480 F*
Species total	*51 837 F*	*54 856*	*61 482*	*69 104*	*72 930*	*60 156*	*54 680*	*55 405*	*54 777*	*50 150 F*
Blackbellied angler	**Baudroie rousse**		**Rape negro**		***Lophius budegassa***			**1,95(01)001,02**		**ANK**
34 Belize	-	-	-	-	-	-	-	-	-	2
Greece	-	-	-	-	4	7	4	27	-	18
Mauritania	10 F	10 F	20 F	20 F	20 F	20 F	20	45	28	36
Morocco	269	350	348	350	320	579	498	85	126	327
Portugal	41	58	82	127	55	26	16	18	2	6
Spain	250 F	350 F	450 F	524	336	524	126	208	369	491
34 Fishing area total	*570 F*	*768 F*	*900 F*	*1 021 F*	*735 F*	*1 156 F*	*664*	*383*	*525*	*880*
Species total	*570 F*	*768 F*	*900 F*	*1 021 F*	*735 F*	*1 156 F*	*664*	*383*	*525*	*880*
American angler	**Baudroie d'Amérique**		**Rape americano**		***Lophius americanus***			**1,95(01)001,03**		**ANG**
21 Canada	1 543	2 340	1 722	1 623	2 073	1 466	1 274	1 265	1 900	3 634
Cuba	52	2	5	-	-	18	-	-	-	-
Portugal	8	-	2	-	-	-	-	-	-	134
St Pier Mq	-	-	-	-	0	1	0	0	1	2

B-34 Miscellaneous demersal fishes / Poissons démersaux divers / Peces demersales diversos

Capture production by species, fishing areas and countries or areas / Captures par espèces, zones de pêche et pays ou zones / Capturas por especies, áreas de pesca y países o áreas

Species, Fishing area Espèce, Zone de pêche Especie, Area de pesca	1993 t	1994 t	1995 t	1996 t	1997 t	1998 t	1999 t	2000 t	2001 t	2002 t
Spain	-	-	-	-	1	2	0	3	9	93
USA	18 765	19 138	25 142	24 203	27 481	26 345	25 027	20 798	23 256	22 878
21 Fishing area total	*20 368*	*21 480*	*26 871*	*25 826*	*29 555*	*27 832*	*26 301*	*22 066*	*25 166*	*26 741*
31 USA	-	-	-	-	25	12	19	9	12	15
31 Fishing area total	-	-	-	-	*25*	*12*	*19*	*9*	*12*	*15*
Species total	*20 368*	*21 480*	*26 871*	*25 826*	*29 580*	*27 844*	*26 320*	*22 075*	*25 178*	*26 756*
Shortspine African angler	**Baudroie africaine**		**Rape africano**		***Lophius vaillanti***				**1,95(01)001,05**	**MVA**
34 Congo Rep	...	...	...	...	...	...	...	7	5 F	5 F
34 Fishing area total	...	...	...	...	...	...	...	*7*	*5 F*	*5 F*
Species total	...	...	...	...	...	...	...	*7*	*5 F*	*5 F*
Devil anglerfish	**Baudroie diable**		**Rape diablo**		***Lophius vomerinus***				**1,95(01)001,06**	**MVO**
47 Iceland	-	-	-	-	5	2	-	-	-	-
Namibia	9 226	12 158	10 130	9 236	10 259	16 429	14 802	14 358	12 390	15 174
South Africa	4 249	4 089	6 246	6 139	6 958	7 903	6 949	7 000 F	9 554	8 492
Spain	600 F	200 F	150 F	58	-	123	32	44	52	55
47 Fishing area total	*14 075 F*	*16 447 F*	*16 526 F*	*15 433*	*17 222*	*24 457*	*21 783*	*21 402 F*	*21 996*	*23 721*
Species total	*14 075 F*	*16 447 F*	*16 526 F*	*15 433*	*17 222*	*24 457*	*21 783*	*21 402 F*	*21 996*	*23 721*
Anglerfishes nei	**Baudroies, etc. nca**		**Rapes, etc. nep**		***Lophiidae***				**1,95(01)XXX,XX**	**ANF**
41 Brazil	...	...	366	294	399	542	794	1 937	6 824	5 258
Spain	-	-	-	-	-	-	-	-	3	-
UK	-	-	-	-	-	-	-	-	2 105	1 199
41 Fishing area total	...	...	*366*	*294*	*399*	*542*	*794*	*1 937*	*8 932*	*6 457*
61 Korea Rep	6 312	5 064	8 173	11 720	5 045	3 970	3 406	4 930	5 813	9 500
61 Fishing area total	*6 312*	*5 064*	*8 173*	*11 720*	*5 045*	*3 970*	*3 406*	*4 930*	*5 813*	*9 500*
Species total	*6 312*	*5 064*	*8 539*	*12 014*	*5 444*	*4 512*	*4 200*	*6 867*	*14 745*	*15 957*
Demersal percomorphs nei	**Percomorphes démersaux nca**		**Percomorfos demersales nep**		***Perciformes***				**1,99(XX)XXX,XX**	**DPX**
31 Colombia	14	27	11	32	16	4	-	20 F	20 F	20 F
Mexico	472	634	198	169	190	180	149	171	3 046	3 573
Puerto Rico	809	421	741	1 059	1 249	1 375	-	-	-	-
Trinidad Tob	1 462	2 011	2 400 F	2 656	2 572	2 001	1 918	1 812	2 261	2 477
Venezuela	3 071	3 703	1 210	3 532	4 760	3 616	4 118	1 322	1 687	2 084
31 Fishing area total	*5 828*	*6 796*	*4 560 F*	*7 448*	*8 787*	*7 176*	*6 185*	*3 325 F*	*7 014 F*	*8 154 F*
34 Cape Verde	...	...	...	...	1 450	1 150	1 079	1 314	1 307	1 340 F
Eq Guinea	210 F	320 F	140 F	310 F	390 F	255	300 F	30 F	30 F	30 F
Ghana	0	0	0	0	0	0	-	-	-	-
34 Fishing area total	*210 F*	*320 F*	*140 F*	*310 F*	*1 840 F*	*1 405*	*1 379 F*	*1 344 F*	*1 337 F*	*1 370 F*
37 France	205	2	1	3	3	3	19	43	26	26
Spain	-	-	-	-	-	-	-	841	758	3 954
Syria	511 F	494 F	492 F	606	450	626	530	392	449	-
37 Fishing area total	*716 F*	*496 F*	*493 F*	*609*	*453*	*629*	*549*	*1 276*	*1 233*	*3 980*
41 Brazil	7 820 F	7 830 F	5 217	5 697	9 825	9 828	12 050	7 644	14 291	15 006
Japan	17	-	62	25	4	89	6	4	-	-
Uruguay	14	18	22	3	57	-	334	51	34	71
41 Fishing area total	*7 851 F*	*7 848 F*	*5 301*	*5 725*	*9 886*	*9 917*	*12 390*	*7 699*	*14 325*	*15 077*
47 Japan	260	408	319	27	1 147	25	-	-	-	-
47 Fishing area total	*260*	*408*	*319*	*27*	*1 147*	*25*	...	...	...	...
51 Egypt	5 600	-	-	-	-	-	-	-	-	-
Japan	-	-	-	-	-	-	-	-	4 416	...
Kenya	407	418	211	1 247	1 188	1 196	1 366	1 224	2 060	1 450
Oman	4 818	6 232	5 175	7 529	4 554	6 938	8 760	5 735	2 421	11 273
Seychelles	364	208	-	-	-	-	-	208	153	181
Yemen	2 441	...	2 050	5 599	9 727 F	14 078 F	6 115 F	6 086 F	7 400	8 350 F
51 Fishing area total	*13 630*	*6 858*	*7 436*	*14 375*	*15 469 F*	*22 212 F*	*16 241 F*	*13 253 F*	*16 450*	*21 254 F*
57 Australia	2 300 F	2 000 F	3 000 F	2 500 F	-	-	-	-	-	-
Sri Lanka	15 277	10 585	7 088	8 968	9 100	9 210	10 440	14 910	14 490	16 320
57 Fishing area total	*17 577 F*	*12 585 F*	*10 088 F*	*11 468 F*	*9 100*	*9 210*	*10 440*	*14 910*	*14 490*	*16 320*
71 Australia	96	100 F	100 F	100 F	-	-	-	-	-	-
Fiji Islands	210	214 F	...	380	250	-	-	-	-	-
71 Fishing area total	*306*	*314 F*	*100 F*	*480 F*	*250*	-	-	-	-	-
77 Mexico	-	2	0	1	0	0	0	0	0	0
77 Fishing area total	-	*2*	*0*	*1*	*0*	*0*	*0*	*0*	*0*	*0*
81 Australia	1 500 F	2 000 F	1 000 F	-	-	-	-	-	-	-
Japan	2 736	4 358	5 136	4 459	3 661	3 653	3 008	2 891	2 073	1 300 F
81 Fishing area total	*4 236 F*	*6 358 F*	*6 136 F*	*4 459*	*3 661*	*3 653*	*3 008*	*2 891*	*2 073*	*1 300 F*
87 Colombia	128	148	1 058	170	220	161	211	200 F	200 F	190 F

B-34 Miscellaneous demersal fishes / Poissons démersaux divers / Peces demersales diversos

Capture production by species, fishing areas and countries or areas / Captures par espèces, zones de pêche et pays ou zones / Capturas por especies, áreas de pesca y países o áreas

Species, Fishing area Espèce, Zone de pêche Especie, Area de pesca	1993 t	1994 t	1995 t	1996 t	1997 t	1998 t	1999 t	2000 t	2001 t	2002 t
Ecuador	1 168	-	-	-	-	-	-	-	-	-
Japan	-	-	-	-	-	-	83	-	-	-
87 Fishing area total	*1 296*	*148*	*1 058*	*170*	*220*	*161*	*294*	*200 F*	*200 F*	*190 F*
Species total	*51 910 F*	*42 133 F*	*35 631 F*	*45 072 F*	*50 813 F*	*54 388 F*	*50 486 F*	*44 898 F*	*57 122 F*	*67 645 F*
Group total	***2 155 594***	***2 523 507***	***2 700 157***	***2 690 372***	***2 740 042***	***2 969 434***	***2 943 038***	***3 023 335***	***3 005 719***	***3 021 049***

B-35 Herrings, sardines, anchovies / Harengs, sardines, anchois / Arenques, sardinas, anchoas

Capture production by species, fishing areas and countries or areas / Captures par espèces, zones de pêche et pays ou zones / Capturas por especies, áreas de pesca y países o áreas

Species, Fishing area Espèce, Zone de pêche Especie, Area de pesca	1993 t	1994 t	1995 t	1996 t	1997 t	1998 t	1999 t	2000 t	2001 t	2002 t
Atlantic herring	**Hareng de l'Atlantique**		**Arenque del Atlántico**		***Clupea harengus***			**1,21(05)001,05**		**HER**
21 Canada	197 597	206 777	193 690	188 843	186 550	190 861	202 046	203 261	199 295	191 097
Cuba	50	1	6	213	253	134	121	-	-	-
Latvia	289	-	-	-	-	-	-	-	-	-
Russian Fed	2	-	-	-	-	5	-	4	1 664	-
St Pier Mq	-	-	-	-	-	-	2	0	0	0
USA	56 392	48 667	76 980	104 002	97 706	81 813	79 597	74 037	106 561	67 652
21 Fishing area total	*254 330*	*255 445*	*270 676*	*293 058*	*284 509*	*272 813*	*281 766*	*277 302*	*307 520*	*258 749*
27 Belgium	58	145	12	2	1	1	1	1	11	23
Denmark	169 477	177 529	191 415	153 009	125 300	139 710	137 578	153 899	141 263	112 476
Estonia	32 982	34 493	43 866	45 296	52 435	42 721	44 038	41 735	41 738	36 250
Faeroe Is	4 592	5 500	66 023	57 853	65 949	70 214	56 476	65 270	35 172	35 172 F
Finland	79 234	98 958	95 897	94 548	91 544	86 350	83 042	81 648	82 867	76 531
France	15 053	24 461	33 240	11 742	21 222	23 398	25 531	25 398	27 577	26 627
Germany	68 594	57 077	55 916	42 153	42 749	47 028	50 857	47 048	50 680	56 701
Iceland	116 617	151 229	284 473	265 413	291 117	277 461	298 435	287 663	178 950	223 842
Ireland	51 468	51 006	46 643	71 953	57 155	58 248	45 334	42 114	40 640	30 606
Isle of Man	775	716	615	693	821	0	1	...	35	-
Japan	1	0	-	-	-	-	-	-	-	-
Latvia	21 949	22 676	24 972	27 523	29 330	24 417	27 163	26 768	26 652	25 284
Lithuania	3 775	4 988	7 058	4 257	3 330	2 368	1 313	1 198	1 639	1 539
Netherlands	89 188	86 007	99 447	77 605	65 448	77 090	78 741	75 221	66 357	78 557
Norway	352 240	549 621	686 705	763 073	923 165	831 844	829 008	800 059	581 161	573 965
Poland	50 833	49 111	45 676	31 246	28 939	21 873	19 229	24 516	37 611	36 778
Portugal	0	0	0	2	0	0	1	0	2	0
Romania	-	5 565	6 588	1 794	-	-	-	-	-	-
Russian Fed	59 093	98 287	121 561	134 380	181 179	139 561	170 601	174 200	125 756	128 129
Spain	-	-	-	-	-	-	-	232	232	266
Sweden	165 158	153 109	157 503	132 153	166 311	201 738	157 541	174 081	125 749	97 625
UK	106 443	104 002	114 571	120 935	103 405	104 627	104 752	82 658	81 363	72 893
27 Fishing area total	*1 387 530*	*1 674 480*	*2 082 181*	*2 035 630*	*2 249 400*	*2 148 649*	*2 129 642*	*2 103 709*	*1 645 455*	*1 613 264 F*
Species total	*1 641 860*	*1 929 925*	*2 352 857*	*2 328 688*	*2 533 909*	*2 421 462*	*2 411 408*	*2 381 011*	*1 952 975*	*1 872 013 F*
Pacific herring	**Hareng du Pacifique**		**Arenque del Pacífico**		***Clupea pallasii***			**1,21(05)001,07**		**HEP**
61 China	994	1 330	2 325	1 665	15 780	21 796	17 936	15 258	51 950	48 693
Japan	1 482	2 146	3 873	2 021	1 926	2 531	2 579	2 260	2 385	1 366
Korea Rep	3 983	3 865	8 622	5 525	13 214	13 340	21 446	15 203	8 491	1 941
Russian Fed	115 148	85 218	116 787	171 810	313 397	395 595	359 194	361 241	278 511	203 411
61 Fishing area total	*121 607*	*92 559*	*131 607*	*181 021*	*344 317*	*433 262*	*401 155*	*393 962*	*341 337*	*255 411*
67 Canada	41 321	40 909	26 780	22 221	31 500	33 500	28 847	29 290	24 189	29 321
USA	44 492	48 211	48 666	49 300	52 892	39 878	39 038	31 092	38 891	32 284
67 Fishing area total	*85 813*	*89 120*	*75 446*	*71 521*	*84 392*	*73 378*	*67 885*	*60 382*	*63 080*	*61 605*
77 USA	3 849	2 972	4 623	5 328	9 188	1 987	2 267	2 853	2 521	3 282
77 Fishing area total	*3 849*	*2 972*	*4 623*	*5 328*	*9 188*	*1 987*	*2 267*	*2 853*	*2 521*	*3 282*
Species total	*211 269*	*184 651*	*211 676*	*257 870*	*437 897*	*508 627*	*471 307*	*457 197*	*406 938*	*320 298*
Goldstripe sardinella	**Sardinelle dorée**		**Sardinela dorada**		***Sardinella gibbosa***			**1,21(05)012,03**		**SAG**
57 Indonesia	28 703	32 990	26 450	29 835	26 335	31 714	34 785	37 358	40 375	42 480
57 Fishing area total	*28 703*	*32 990*	*26 450*	*29 835*	*26 335*	*31 714*	*34 785*	*37 358*	*40 375*	*42 480*
71 Indonesia	123 857	133 462	134 646	127 269	130 579	142 977	127 925	134 861	145 537	148 820
71 Fishing area total	*123 857*	*133 462*	*134 646*	*127 269*	*130 579*	*142 977*	*127 925*	*134 861*	*145 537*	*148 820*
Species total	*152 560*	*166 452*	*161 096*	*157 104*	*156 914*	*174 691*	*162 710*	*172 219*	*185 912*	*191 300*
Indian oil sardine	**Sardinelle indienne**		**Sardinela aceitera**		***Sardinella longiceps***			**1,21(05)012,04**		**IOS**
51 India	72 676	30 136	33 351	65 370	120 214	110 288	122 254	277 842	287 628	313 843
Oman	29 204	30 267	33 053	26 741	16 765	20 650	21 710	40 044	58 960	37 931
Pakistan	92 704	65 050	55 177	52 290	51 930	44 079	30 629	31 167	31 201	31 600
Untd Arab Em	2 953	6 085	3 455	3 491	4 077	4 085	4 144	...	...	...
Yemen	...	...	...	4 120	4 310 F	4 900 F	4 700 F	4 300 F	5 500 F	6 200 F
51 Fishing area total	*197 537*	*131 538*	*125 036*	*152 012*	*197 296 F*	*184 002 F*	*183 437 F*	*353 353 F*	*383 289 F*	*389 574 F*
57 India	64 765	66 179	72 565	71 343	91 846	63 063	25 461	49 583	55 448	20 320
57 Fishing area total	*64 765*	*66 179*	*72 565*	*71 343*	*91 846*	*63 063*	*25 461*	*49 583*	*55 448*	*20 320*
Species total	*262 302*	*197 717*	*197 601*	*223 355*	*289 142 F*	*247 065 F*	*208 898 F*	*402 936 F*	*438 737 F*	*409 894 F*
Round sardinella	**Allache**		**Alacha**		***Sardinella aurita***			**1,21(05)012,10**		**SAA**
21 USA	-	-	-	-	-	-	340	-	-	-
21 Fishing area total	-	-	-	-	-	-	*340*	-	-	-
31 Cuba	...	...	...	...	5	93	-	1 510	1 575	-
Mexico	782	3 119	1 643	1 073	2 028	4 253	1 384	1 424	373	332
USA	954	1 053	191	571	512	489	196	614	623	653
Venezuela	85 751	112 540	153 037	153 782	140 571	186 060	126 468	73 534	71 168	158 125
31 Fishing area total	*87 487*	*116 712*	*154 871*	*155 426*	*143 116*	*190 895*	*128 048*	*77 082*	*73 739*	*159 110*
34 Côte dIvoire	14 492	11 819	12 098	18 491	10 002	10 884	12 206	19 358	18 864	7 288

B-35 Herrings, sardines, anchovies / Harengs, sardines, anchois / Arenques, sardinas, anchoas

Capture production by species, fishing areas and countries or areas / Captures par espèces, zones de pêche et pays ou zones / Capturas por especies, áreas de pesca y países o áreas

Species, Fishing area Espèce, Zone de pêche Especie, Area de pesca	1993 t	1994 t	1995 t	1996 t	1997 t	1998 t	1999 t	2000 t	2001 t	2002 t
Georgia	2 170 F	800 F	-	-	-	-	-	-	-	-
Ghana	92 735	69 895	67 835	118 408	49 394	55 965	57 170	102 043	67 321	64 300
Lithuania	22 810	7 393	...	...	...	10 575	8 680	6 324	4 309	22 205
Netherlands	...	...	...	...	...	...	...	...	134 490	93 032
Romania	292	-	-	-	-	-	-	-	-	-
Russian Fed	31 930	17 216	53 303	116 705	101 457	106 318	109 445	42 400	22 557	18 800
Senegal	142 197	128 138	116 766	142 457	152 713	129 429	93 512	112 970	112 120	78 244
Spain	-	-	-	2 224	3 210	3 411	17	41	49	5
Togo	342	361	335	739	1 300	1 045	1 921	1 912	3 123	1 807
34 *Fishing area total*	*306 968 F*	*235 622 F*	*250 337*	*399 024*	*318 076*	*317 627*	*282 951*	*285 048*	*362 833*	*285 681*
37 Israel	470	250	250	242	435	557	306	88	117	143
37 *Fishing area total*	*470*	*250*	*250*	*242*	*435*	*557*	*306*	*88*	*117*	*143*
Species total	*394 925 F*	*352 584 F*	*405 458*	*554 692*	*461 627*	*509 079*	*411 645*	*362 218*	*436 689*	*444 934*
Madeiran sardinella	**Grande allache**		**Machuelo**		***Sardinella maderensis***				**1,21(05)012,17**	**SAE**
34 Côte dIvoire	2 827	4 492	2 273	1 500 F	1 000 F	500 F	175	426	716	2 421
Ghana	16 984	12 467	13 211	13 619	14 183	15 468	12 105	14 970	15 906	13 755
Nigeria	...	...	...	2 230	9 387	12 226	11 041	10 158	11 931	12 132
Portugal	0	0	0	0	1	5	11	2	3	1
Romania	53	-	-	-	-	-	-	-	-	-
Senegal	73 935	51 405	64 711	108 186	114 824	103 363	105 120	114 749	99 944	105 773
Togo	150	162	86	166	432	408	614	638	445	353
34 *Fishing area total*	*93 949*	*68 526*	*80 281*	*125 701 F*	*139 827 F*	*131 970 F*	*129 066*	*140 943*	*128 945*	*134 435*
Species total	*93 949*	*68 526*	*80 281*	*125 701 F*	*139 827 F*	*131 970 F*	*129 066*	*140 943*	*128 945*	*134 435*
Japanese sardinella	**Sardinelle japonaise**		**Sardinela del Japón**		***Sardinella zunasi***				**1,21(05)012,22**	**JSS**
61 Korea Rep	24 383	23 974	18 345	10 663	5 593	1 973	6 674	4 603	766	796
61 *Fishing area total*	*24 383*	*23 974*	*18 345*	*10 663*	*5 593*	*1 973*	*6 674*	*4 603*	*766*	*796*
Species total	*24 383*	*23 974*	*18 345*	*10 663*	*5 593*	*1 973*	*6 674*	*4 603*	*766*	*796*
Bali sardinella	**Sardinelle de Bali**		**Sardinela de Balí**		***Sardinella lemuru***				**1,21(05)012,23**	**SAM**
57 Indonesia	69 357	63 006	40 361	29 646	72 662	92 286	34 141	31 236	37 810	40 170
57 *Fishing area total*	*69 357*	*63 006*	*40 361*	*29 646*	*72 662*	*92 286*	*34 141*	*31 236*	*37 810*	*40 170*
71 Indonesia	52 682	65 196	58 544	58 944	65 974	61 679	55 145	57 508	65 900	68 200
71 *Fishing area total*	*52 682*	*65 196*	*58 544*	*58 944*	*65 974*	*61 679*	*55 145*	*57 508*	*65 900*	*68 200*
Species total	*122 039*	*128 202*	*98 905*	*88 590*	*138 636*	*153 965*	*89 286*	*88 744*	*103 710*	*108 370*
Brazilian sardinella	**Sardinelle de Brésil**		**Sardinela del Brasil**		***Sardinella brasiliensis***				**1,21(05)012,24**	**BSR**
31 Grenada	24	13	0	1	0	0	0	0	0	3
31 *Fishing area total*	*24*	*13*	*0*	*1*	*0*	*0*	*0*	*0*	*0*	*3*
41 Brazil	49 991	84 635	60 212	97 092	117 642	82 283	25 518	17 053	39 039	27 891
41 *Fishing area total*	*49 991*	*84 635*	*60 212*	*97 092*	*117 642*	*82 283*	*25 518*	*17 053*	*39 039*	*27 891*
Species total	*50 015*	*84 648*	*60 212*	*97 093*	*117 642*	*82 283*	*25 518*	*17 053*	*39 039*	*27 894*
Sardinellas nei	**Sardinelles nca**		**Sardinelas nep**		***Sardinella spp***				**1,21(05)012,XX**	**SIX**
27 Spain	6	1	30	-	-	3	1 003	-	1	-
27 *Fishing area total*	*6*	*1*	*30*	*-*	*-*	*3*	*1 003*	*-*	*1*	*-*
34 Belize	-	-	-	-	2 400	2 389	7 233	3 534	1 365	4 992
Benin	1 250 F	1 200 F	1 150 F	1 100 F	1 200 F	1 100 F	1 072	750 F	1 434	1 496
Bulgaria	-	-	-	-	-	3 216	-	-	-	-
Cameroon	16 006	18 600 F	24 002	24 002	23 500 F	23 000 F	23 500 F	21 611	21 645	22 540
Congo Dem R	1 680 F	1 500 F	1 551 F	1 590 F	1 538 F	1 582 F	1 595 F	1 620 F	1 700 F	1 800 F
Congo Rep	8 500 F	10 914	11 600 F	11 200 F	10 200 F	9 300 F	9 300 F	10 500 F	9 500 F	9 700 F
Cyprus	-	-	-	...	...	2 172	4 492	15 043	9 455	-
Estonia	4 786	1 076	-	252	415	3 171	2 474	-	-	-
France	-	-	-	-	...	7 228	624	5 517	1 220	-
Gabon	...	...	1 174	1 897	878	746	128	1 414	1 083	1 270
Gambia	3	0	1	11	86	64	70	10	81	1 711
Germany	-	-	-	10 115	23 866	17 448	24 150	-	1 683	1 518
Ghana	1 333	7 545	14 074	20 070	35 985	33 264	19 664	18 358	8 345	1 553
Guinea	3 650 F	3 760 F	5 281	6 107	4 176	6 292	8 909	13 288	8 027	8 000 F
Ireland	-	-	-	-	-	-	-	-	52 980	24 552
Latvia	-	7 083	17 032	24 209	6 497	6 064	15 031	7 886	7 689	6 132
Liberia	330	222	876	199	485	620	1 112	887	1 358	1 350 F
Mauritania	2 750 F	1 340 F	3 020 F	5 440 F	4 870 F	1 060 F	4 000 F	4 000 F	4 000 F	4 000 F
Morocco	251	12	19	20	18	15	81	77	45	191
Netherlands	-	-	-	41 488	86 635	110 091	115 753	122 783	-	5 131
Nigeria	26 264	25 825	76 585	102 230	97 286	95 495	83 593	80 892	61 927	59 400
Poland	-	-	-	7 166	2 553	-	-	-	3 463	4 824
St Vincent	-	-	-	-	2 000	10 038	7 962	3 369	7 716	7 053
Sierra Leone	7 705	7 681	7 669	7 670 F	7 661	7 450	7 850	7 628	9 845	13 251
Ukraine	11 192	5 862	21 578	83 333	97 009	161 821	74 729	38 120	10 613	13 714
Other nei	-	-	-	-	-	-	5 812	2 306	38 577	31 211
34 *Fishing area total*	*85 700 F*	*92 620 F*	*185 612 F*	*348 099 F*	*409 258 F*	*503 626 F*	*419 134 F*	*359 593 F*	*263 751 F*	*225 389 F*

B-35 Herrings, sardines, anchovies / Harengs, sardines, anchois / Arenques, sardinas, anchoas

Capture production by species, fishing areas and countries or areas / Captures par espèces, zones de pêche et pays ou zones / Capturas por especies, áreas de pesca y países o áreas

Species, Fishing area Espèce, Zone de pêche Especie, Area de pesca	1993 t	1994 t	1995 t	1996 t	1997 t	1998 t	1999 t	2000 t	2001 t	2002 t
37 Algeria	10 000 F	13 000 F	17 887	8 150	15 741	9 992	11 393	20 399	33 748	24 480
Croatia	435	296	243	163	344	170	25	35	37	42
Egypt	10 634	8 147	8 935	8 267	10 621	23 920	39 563	19 689	20 037	12 106
Libya	6 400 F	7 000 F	7 100 F	7 000 F	7 000 F	7 000 F	7 000 F	7 000 F	7 000 F	7 000 F
Morocco	912	1 176	3 005	2 880	2 168	1 230	661	726	3 256	1 732
Palest, O.T.	...	...	...	978	2 483	1 780	1 750 F	1 450 F	1 300 F	1 299
Serbia-Monte	7	6	10	11	11	13	17	13	16	13
Spain	4 400 F	5 200 F	6 000 F	6 879	7 156	7 993	8 782	7 579	13 555	7 368
Syria	350 F	338 F	336 F	550	300	292	338	251	197	334
Tunisia	8 856	7 569	8 591	7 738	8 084	6 327	7 491	11 800	12 942	12 465
37 Fishing area total	*41 994 F*	*42 732 F*	*52 107 F*	*42 616 F*	*53 908 F*	*58 717 F*	*77 020 F*	*68 942 F*	*92 088 F*	*66 839 F*
47 Angola	12 357	19 731	34 211	17 484	21 014	45 483	57 579	108 211	57 896	28 513
Korea Rep	42	-	-	-	-	-	-	-	-	-
Lithuania	4 324	1 088	-	-	-	-	-	-	-	-
Russian Fed	-	15 046	7 241	229	2 739	9 537	22 169	5 645	443	1 513
Ukraine	818	124	-	-	-	-	-	-	-	-
47 Fishing area total	*17 541*	*35 989*	*41 452*	*17 713*	*23 753*	*55 020*	*79 748*	*113 856*	*58 339*	*30 026*
51 Comoros	1 000	1 000	1 000 F	1 000 F	1 000 F	1 000 F	950 F	1 050 F	1 000	1 000 F
Egypt	3 763	2 973	2 822	6 833	5 639	4 973	5 383	5 705	4 343	5 431
Eritrea	...	...	...	...	...	...	...	2	0	0
Qatar	0	0	0	0	-	0	0	0	0	0
Tanzania	5 472	8 563	3 750	14 323	5 000	4 450	14 000	15 000	15 500	14 000
Untd Arab Em	7 573	10 121	8 840	8 933	9 200	9 236	9 510	6 140	4 200	3 500
51 Fishing area total	*17 808*	*22 657*	*16 412 F*	*31 089 F*	*20 839 F*	*19 659 F*	*29 843 F*	*27 897 F*	*25 043*	*23 931 F*
57 Thailand	38 440	29 455	54 849	53 086	51 084	56 813	54 321	42 276	47 476	46 471
57 Fishing area total	*38 440*	*29 455*	*54 849*	*53 086*	*51 084*	*56 813*	*54 321*	*42 276*	*47 476*	*46 471*
61 China,H.Kong	31	19	24	30	12	8	5 F	5 F	6 F	5 F
61 Fishing area total	*31*	*19*	*24*	*30*	*12*	*8*	*5 F*	*5 F*	*6 F*	*5 F*
71 Fiji Islands	120	122 F	240	120	140	30	47	35 F	...	35 F
Philippines	243 610	253 286	264 675	257 804	302 341	302 599	279 864	298 466	282 617	244 223
Thailand	113 860	125 179	141 180	161 771	151 708	129 045	128 492	121 738	145 080	146 726
71 Fishing area total	*357 590*	*378 587 F*	*406 095*	*419 695*	*454 189*	*431 674*	*408 403*	*420 239 F*	*427 697*	*390 984 F*
Species total	*559 110 F*	*602 060 F*	*756 581 F*	*912 328 F*	*1 013 043 F*	*1 125 520 F*	*1 069 477 F*	*1 032 808 F*	*914 401 F*	*783 645 F*
Japanese pilchard	**Pilchard du Japon**		**Sardina japonesa**		***Sardinops melanostictus***				**1,21(05)013,01**	**JAP**
61 China	46 846	68 453	58 434	92 918	124 844	121 120	147 125	153 944	160 825	186 281
Japan	1 713 687	1 188 848	661 391	319 354	284 054	167 073	351 207	149 616	178 423	50 313
Korea D P Rp	...	19 701	63	5	0	...	...	...	...	...
Korea Rep	31 285	36 709	13 539	18 560	9 041	7 595	17 142	2 207	129	8
Russian Fed	4 314	28	-	-	-	-	3	-	-	-
61 Fishing area total	*1 796 132*	*1 313 739*	*733 427*	*430 837*	*417 939*	*295 788*	*515 477*	*305 767*	*339 377*	*236 602*
Species total	*1 796 132*	*1 313 739*	*733 427*	*430 837*	*417 939*	*295 788*	*515 477*	*305 767*	*339 377*	*236 602*
California pilchard	**Pilchard de Californie**		**Sardina monterrey**		***Sardinops caeruleus***				**1,21(05)013,02**	**CPI**
67 USA	-	-	1	-	-	22	775	14 368	23 908	38 958
67 Fishing area total	*-*	*-*	*1*	*-*	*-*	*22*	*775*	*14 368*	*23 908*	*38 958*
77 Mexico	272 630	287 620	320 999	418 093	455 837	339 317	345 458	478 191	609 777	624 817
USA	0	13 094	42 464	32 503	42 816	42 559	59 169	53 520	51 812	58 296
77 Fishing area total	*272 630*	*300 714*	*363 463*	*450 596*	*498 653*	*381 876*	*404 627*	*531 711*	*661 589*	*683 113*
Species total	*272 630*	*300 714*	*363 464*	*450 596*	*498 653*	*381 898*	*405 402*	*546 079*	*685 497*	*722 071*
South American pilchard	**Pilchard sudaméricain**		**Sardina sudamericana**		***Sardinops sagax***				**1,21(05)013,03**	**CHP**
87 Chile	481 119	194 499	161 557	81 043	40 473	27 966	246 045	60 189	33 271	18 561
Ecuador	23 652	212	75 916	356 480	57 191	1 012	8 821	51 648	42 143	2 539
Peru	1 461 759	1 551 833	1 265 658	1 056 413	625 143	908 291	187 924	226 294	60 298	6 853
87 Fishing area total	*1 966 530*	*1 746 544*	*1 503 131*	*1 493 936*	*722 807*	*937 269*	*442 790*	*338 131*	*135 712*	*27 953*
Species total	*1 966 530*	*1 746 544*	*1 503 131*	*1 493 936*	*722 807*	*937 269*	*442 790*	*338 131*	*135 712*	*27 953*
Southern African pilchard	**Pilchard de l'Afrique australe**		**Sardina de Africa austral**		***Sardinops ocellatus***				**1,21(05)013,05**	**PIA**
47 Namibia	114 812	116 429	42 797	1 171	27 685	68 562	44 653	25 388	7 940	4 176
South Africa	50 702	93 438	115 205	105 210	116 995	128 019	131 316	136 060	192 160	260 710
Ukraine	1 261	34	-	-	-	-	-	-	-	-
47 Fishing area total	*166 775*	*209 901*	*158 002*	*106 381*	*144 680*	*196 581*	*175 969*	*161 448*	*200 100*	*264 886*
51 South Africa	-	-	-	-	1	...	...	0	0	-
51 Fishing area total	*-*	*-*	*-*	*-*	*1*	*...*	*...*	*0*	*0*	*-*
Species total	*166 775*	*209 901*	*158 002*	*106 381*	*144 681*	*196 581*	*175 969*	*161 448*	*200 100*	*264 886*
Australian pilchard	**...B**		**...C**		***Sardinops neopilchardus***				**1,21(05)013,09**	**SRP**
81 New Zealand	...	...	209	169	385	519	894	1 253	1 399	920
Ukraine	-	-	-	-	-	-	-	-	-	72
81 Fishing area total	*...*	*...*	*209*	*169*	*385*	*519*	*894*	*1 253*	*1 399*	*992*

B-35 Herrings, sardines, anchovies / Harengs, sardines, anchois / Arenques, sardinas, anchoas

Capture production by species, fishing areas and countries or areas / Captures par espèces, zones de pêche et pays ou zones / Capturas por especies, áreas de pesca y países o áreas

Species, Fishing area Espèce, Zone de pêche Especie, Area de pesca	1993 t	1994 t	1995 t	1996 t	1997 t	1998 t	1999 t	2000 t	2001 t	2002 t
Species total	*...*	*...*	*209*	*169*	*385*	*519*	*894*	*1 253*	*1 399*	*992*
Brazilian menhaden	**Menhaden du Brésil**		**Savela**		***Brevoortia aurea***				**1,21(05)024,01**	**MHS**
41 Brazil	3 400 F	3 410 F	11 133	6 294	2 396	2 936	2 202	1 123	344	490
41 Fishing area total	*3 400 F*	*3 410 F*	*11 133*	*6 294*	*2 396*	*2 936*	*2 202*	*1 123*	*344*	*490*
Species total	*3 400 F*	*3 410 F*	*11 133*	*6 294*	*2 396*	*2 936*	*2 202*	*1 123*	*344*	*490*
Argentine menhaden	**Menhaden d'Argentine**		**Lacha**		***Brevoortia pectinata***				**1,21(05)024,02**	**MHP**
41 Argentina	137	573	294	427	893	104	205	271	265	94
Uruguay	116	483	130	84	219	415	86	66	103	47
41 Fishing area total	*253*	*1 056*	*424*	*511*	*1 112*	*519*	*291*	*337*	*368*	*141*
Species total	*253*	*1 056*	*424*	*511*	*1 112*	*519*	*291*	*337*	*368*	*141*
Atlantic menhaden	**Menhaden tyran**		**Lacha tirana**		***Brevoortia tyrannus***				**1,21(05)024,03**	**MHA**
21 USA	327 241	249 047	338 422	280 496	322 239	248 451	189 185	183 310	236 246	180 506
21 Fishing area total	*327 241*	*249 047*	*338 422*	*280 496*	*322 239*	*248 451*	*189 185*	*183 310*	*236 246*	*180 506*
31 USA	20 565	37 455	27 314	24 169	...	27 779	18 815	23 812	25 155	31 068
31 Fishing area total	*20 565*	*37 455*	*27 314*	*24 169*	*...*	*27 779*	*18 815*	*23 812*	*25 155*	*31 068*
Species total	*347 806*	*286 502*	*365 736*	*304 665*	*322 239*	*276 230*	*208 000*	*207 122*	*261 401*	*211 574*
Gulf menhaden	**Menhaden écailleux**		**Lacha escamuda**		***Brevoortia patronus***				**1,21(05)024,04**	**MHG**
31 USA	551 822	767 448	472 039	491 612	597 565	497 461	694 242	591 434	528 506	582 497
31 Fishing area total	*551 822*	*767 448*	*472 039*	*491 612*	*597 565*	*497 461*	*694 242*	*591 434*	*528 506*	*582 497*
Species total	*551 822*	*767 448*	*472 039*	*491 612*	*597 565*	*497 461*	*694 242*	*591 434*	*528 506*	*582 497*
Rainbow sardine	**Sardine arc-en-ciel**		**Sardina arco iris**		***Dussumieria acuta***				**1,21(05)029,01**	**RAS**
57 Indonesia	3 021	4 659	4 604	5 013	4 909	4 729	6 072	5 081	4 951	5 360
57 Fishing area total	*3 021*	*4 659*	*4 604*	*5 013*	*4 909*	*4 729*	*6 072*	*5 081*	*4 951*	*5 360*
71 Indonesia	13 657	14 356	16 045	16 893	17 920	17 592	18 113	18 680	15 450	15 740
Philippines	9 271	21 079	19 255	19 441	15 216	20 639	24 080	12 993	16 902	8 501
71 Fishing area total	*22 928*	*35 435*	*35 300*	*36 334*	*33 136*	*38 231*	*42 193*	*31 673*	*32 352*	*24 241*
Species total	*25 949*	*40 094*	*39 904*	*41 347*	*38 045*	*42 960*	*48 265*	*36 754*	*37 303*	*29 601*
Slender rainbow sardine	**Sardine arc-en-ciel gracile**		**Sardina arco iris grácil**		***Dussumieria elopsoides***				**1,21(05)029,02**	**RAL**
61 China,H.Kong	64	64	154	48	19	17	10 F	15 F	17 F	15 F
61 Fishing area total	*64*	*64*	*154*	*48*	*19*	*17*	*10 F*	*15 F*	*17 F*	*15 F*
Species total	*64*	*64*	*154*	*48*	*19*	*17*	*10 F*	*15 F*	*17 F*	*15 F*
Bonga shad	**Ethmalose d'Afrique**		**Sábalo africano**		***Ethmalosa fimbriata***				**1,21(05)030,02**	**BOA**
01 Benin	2 100 F	2 095	1 963	1 900 F	1 850 F	1 800 F	1 798	1 500 F	1 700 F	1 700 F
01 Fishing area total	*2 100 F*	*2 095*	*1 963*	*1 900 F*	*1 850 F*	*1 800 F*	*1 798*	*1 500 F*	*1 700 F*	*1 700 F*
34 Benin	...	...	...	...	...	...	8	5 F	8	17
Cameroon	16 000	18 600 F	24 000	24 000	23 500 F	23 000 F	23 500 F	21 609	21 645	21 782
Congo Rep	1 000 F	1 903	1 684	2 000 F	2 300 F	2 300 F	2 500 F	2 700 F	3 000 F	3 300 F
Côte dIvoire	9 500	10 000	10 000 F	9 500 F	9 500 F	11 000 F	11 006 F	11 009 F	10 465 F	9 370
Gabon	10 000 F	10 000	11 787	13 046	14 695	19 284	17 408	14 788	12 733	11 428
Gambia	14 053	16 897	13 897	22 648	21 523	21 952	22 750	20 508	18 516	22 786
Ghana	1 138	573	1 073	1 197	9 762	158	766	963	282	295
Guinea	23 000 F	25 000 F	23 596	26 051	29 529	27 852	33 780	29 015	38 454	38 000 F
Liberia	14	26	9	70	17	33	123	37	123	120 F
Lithuania	224	-	-	-	-	-	-	-	-	-
Mauritania	...	...	...	...	...	...	...	2 F	...	...
Nigeria	24 501	25 645	15 072	4 643	28 000	30 216	18 529	17 570	19 049	21 987
Senegal	12 229	13 503	17 277	17 783	17 147	16 496	29 468	22 032	31 675	26 240
Sierra Leone	21 838	21 771	21 738	21 700 F	21 807	21 840	22 400	21 621	24 790	31 492
34 Fishing area total	*133 497 F*	*143 918 F*	*140 133 F*	*142 638 F*	*177 780 F*	*174 131 F*	*182 238 F*	*161 859 F*	*180 740 F*	*186 817 F*
Species total	*135 597 F*	*146 013 F*	*142 096 F*	*144 538 F*	*179 630 F*	*175 931 F*	*184 036 F*	*163 359 F*	*182 440 F*	*188 517 F*
Red-eye round herring	**Shadine ronde**		**Sardineta canalera**		***Etrumeus teres***				**1,21(05)031,01**	**RRH**
37 Egypt	-	-	-	-	-	-	1 403	...	...	...
37 Fishing area total	*-*	*-*	*-*	*-*	*-*	*-*	*1 403*	*...*	*...*	*...*
51 Egypt	-	-	-	-	-	-	2 135	...	...	...
51 Fishing area total	*-*	*-*	*-*	*-*	*-*	*-*	*2 135*	*...*	*...*	*...*
61 China,Taiwan	2 144	1 615	1 653	1 462	2 152	1 248	893	1 118	1 306	1 553
Japan	60 095	68 374	47 590	49 752	55 043	48 441	28 712	23 833	31 525	26 355
61 Fishing area total	*62 239*	*69 989*	*49 243*	*51 214*	*57 195*	*49 689*	*29 605*	*24 951*	*32 831*	*27 908*
77 USA	-	-	-	-	72	7	58	42	2	-
77 Fishing area total	*-*	*-*	*-*	*-*	*72*	*7*	*58*	*42*	*2*	*-*

B-35 Herrings, sardines, anchovies / Harengs, sardines, anchois / Arenques, sardinas, anchoas

Capture production by species, fishing areas and countries or areas / Captures par espèces, zones de pêche et pays ou zones / Capturas por especies, áreas de pesca y países o áreas

Species, Fishing area Espèce, Zone de pêche Especie, Area de pesca	1993 t	1994 t	1995 t	1996 t	1997 t	1998 t	1999 t	2000 t	2001 t	2002 t
87 Ecuador	23 484	30 748	4 946	34 349	1 095	8 873	3 636	4 414	28	-
87 Fishing area total	*23 484*	*30 748*	*4 946*	*34 349*	*1 095*	*8 873*	*3 636*	*4 414*	*28*	*-*
Species total	*85 723*	*100 737*	*54 189*	*85 563*	*58 362*	*58 569*	*36 837*	*29 407*	*32 861*	*27 908*
Whitehead's round herring	**Shadine de Angola**		**Sardina angoleña**		***Etrumeus whiteheadi***			**1,21(05)031,04**		**WRR**
47 Namibia	5 090	1 378	1 934	20 656	5 070	5 193	176	1 127	1 432	9 650
South Africa	56 329	54 147	76 858	47 117	92 209	52 476	58 856	37 750	55 330	54 800
47 Fishing area total	*61 419*	*55 525*	*78 792*	*67 773*	*97 279*	*57 669*	*59 032*	*38 877*	*56 762*	*64 450*
Species total	*61 419*	*55 525*	*78 792*	*67 773*	*97 279*	*57 669*	*59 032*	*38 877*	*56 762*	*64 450*
Scaled sardines	**Harengules**		**Sardinetas**		***Harengula spp***			**1,21(05)033,XX**		**SAS**
31 Cuba	943	1 112	1 045	707	947	649	766	1 562	1 770	506
Dominican Rp	80	44	72	111	64	55	18	27	21	0
Grenada	3	2	0	1	0	2	0	0	0	0
31 Fishing area total	*1 026*	*1 158*	*1 117*	*819*	*1 011*	*706*	*784*	*1 589*	*1 791*	*506*
41 Brazil	-	-	-	20	12	2	162	126	488	569
41 Fishing area total	*-*	*-*	*-*	*20*	*12*	*2*	*162*	*126*	*488*	*569*
Species total	*1 026*	*1 158*	*1 117*	*839*	*1 023*	*708*	*946*	*1 715*	*2 279*	*1 075*
Pacific thread herring	**Chardin du Pacifique**		**Machuelo hebra pinchagua**		***Opisthonema libertate***			**1,21(05)042,01**		**THP**
77 Panama	29 752	40 663	21 224	32 517	26 266	49 472	38 746	63 532	29 033	48 175
77 Fishing area total	*29 752*	*40 663*	*21 224*	*32 517*	*26 266*	*49 472*	*38 746*	*63 532*	*29 033*	*48 175*
87 Ecuador	20 314	69 892	40 911	41 041	43 145	40 530	22 253	20 519	19 886	10 951
87 Fishing area total	*20 314*	*69 892*	*40 911*	*41 041*	*43 145*	*40 530*	*22 253*	*20 519*	*19 886*	*10 951*
Species total	*50 066*	*110 555*	*62 135*	*73 558*	*69 411*	*90 002*	*60 999*	*84 051*	*48 919*	*59 126*
Atlantic thread herring	**Chardin fil**		**Machuelo hebra atlántico**		***Opisthonema oglinum***			**1,21(05)042,02**		**THA**
21 USA	-	-	9	1 686	9	-	1 148	0	-	-
21 Fishing area total	*-*	*-*	*9*	*1 686*	*9*	*-*	*1 148*	*0*	*-*	*-*
31 Cuba	886	1 384	2 005	2 361	1 900	1 286	1 756	1 522	1 676	1 799
Dominican Rp	121	118	369	127	188	111	99	130	106	0
Grenada	0	0	0	0	0	0	0	0	0	0
USA	5 360	2 583	5 056	2 852	7 539	2 576	422	3 056	1 256	2 607
Venezuela	206	269	307	294	5 564	10 619	14 789	9 866	3 650	106
31 Fishing area total	*6 573*	*4 354*	*7 737*	*5 634*	*15 191*	*14 592*	*17 066*	*14 574*	*6 688*	*4 512*
41 Brazil	...	...	6 661	3 762	3 712	8 082	8 204	12 455	4 397	3 191
41 Fishing area total	*...*	*...*	*6 661*	*3 762*	*3 712*	*8 082*	*8 204*	*12 455*	*4 397*	*3 191*
Species total	*6 573*	*4 354*	*14 407*	*11 082*	*18 912*	*22 674*	*26 418*	*27 029*	*11 085*	*7 703*
Silver-stripe round herring	**Hareng gracile**		**Arenquillo de banda**		***Spratelloides gracilis***			**1,21(05)049,01**		**SRH**
61 China,Taiwan	624	1 127	827	517	650	886	521	639	505	1 033
61 Fishing area total	*624*	*1 127*	*827*	*517*	*650*	*886*	*521*	*639*	*505*	*1 033*
71 Fiji Islands	80	82 F	150	140	135	10	40	30 F	...	30 F
71 Fishing area total	*80*	*82 F*	*150*	*140*	*135*	*10*	*40*	*30 F*	*...*	*30 F*
Species total	*704*	*1 209 F*	*977*	*657*	*785*	*896*	*561*	*669 F*	*505*	*1 063 F*
European pilchard(=Sardine)	**Sardine commune**		**Sardina europea**		***Sardina pilchardus***			**1,21(05)064,01**		**PIL**
27 Denmark	53 394	39 349	36 196	13 704	1 740	17 337	17 676	7 893	1 399	3 523
France	9 962	9 904	11 197	10 378	14 321	11 907	27 156	13 042	13 536	21 265
Germany	-	2	35	-	13	166	143	307	463	133
Ireland	-	-	-	-	-	-	3 195	2 592	2 475	5 824
Netherlands	109	20	116	93	518	2 709	5 698	6 825	807	2 572
Portugal	90 405	94 468	87 710	86 853	81 475	82 985	71 956	66 283	71 907	68 740
Spain	53 662	44 329	39 590	38 822	33 652	35 068	31 594	22 460	28 846	31 872
Sweden	-	-	-	-	-	-	-	-	1 031	142
UK	4 917	2 081	7 133	7 304	7 400	6 873	4 815	4 358	10 427	9 401
27 Fishing area total	*212 449*	*190 153*	*181 977*	*157 154*	*139 119*	*157 045*	*162 233*	*123 760*	*130 891*	*143 472*
34 Belize	-	-	-	-	2 126	2 229	616	414	127	1 975
Bulgaria	-	-	-	-	-	48	-		-	-
China	-	0	-	-	-	-	-	-	2	-
Cyprus	-	-	-	...	...	-	12	559	29	-
Estonia	1 152	1 231	163	274	480	-	-	-	-	-
France	-	-	-	-	...	30	24	205	-	-
Georgia	6 900 F	2 550 F	900 F	-	-	-	-	-	-	-
Germany	-	-	-	50	3 282	4 615	1 303	-	37	2 918
Honduras	-	-	1	-	-	-	-	-	-	-
Ireland	-	-	-	-	-	-	-	-	5 380	6 335
Latvia	7 878	1 387	-	474	-	7	23	633	54	362
Lithuania	7 202	...	...	...	...	20	6	292	22	206
Morocco	364 710	476 947	556 152	376 925	487 496	426 474	415 188	522 504	751 812	674 120

B-35 Herrings, sardines, anchovies / Harengs, sardines, anchois / Arenques, sardinas, anchoas

Capture production by species, fishing areas and countries or areas / Captures par espèces, zones de pêche et pays ou zones / Capturas por especies, áreas de pesca y países o áreas

Species, Fishing area Espèce, Zone de pêche Especie, Area de pesca	1993 t	1994 t	1995 t	1996 t	1997 t	1998 t	1999 t	2000 t	2001 t	2002 t
Netherlands	-	-	-	1 149	5 970	6 289	1 926	11 037	10 979	25 009
Norway	-	-	-	-	-	3 421	-	-	-	-
Poland	-	-	-	2 439	1 269	-	-	-	-	14 244
Portugal	7	3	1	2	2	7	16	36	40	23
Romania	19	-	-	-	-	-	-	-	-	-
Russian Fed	67 523	53 845	47 526	56 168	24 864	5 100	5 504	11 200	1 829	6 450
St Vincent	-	-	-	-	46	26	96	340	41	192
Sao Tome Prn	17 F	25 F	26 F	20	...	20 F	30 F	30 F	30 F	30 F
Spain	105 000 F	115 000 F	125 000 F	132 912	114 103	127 523	59 029	9 615	1 118	242
Ukraine	50 827	32 400	49 354	44 221	10 126	12 828	49 136	40 902	27 939	19 808
Other nei	-	-	-	-	-	-	226	1 876	3 949	7 677
34 Fishing area total	*611 235 F*	*683 388 F*	*779 123 F*	*614 634*	*649 764*	*588 637 F*	*533 135 F*	*599 643 F*	*803 388 F*	*759 591 F*
37 Albania	50 F	100 F	235	196	28	28	40	45	123	90
Algeria	63 000 F	87 000 F	58 989	49 906	49 142	49 295	56 724	47 677	58 966	72 547
Croatia	12 773	7 105	6 377	9 199	6 996	12 500	10 500	11 226	9 097	12 626
Cyprus	-	-	-	-	-	-	-	5	1	2
France	12 243	10 589	13 452	6 174	6 652	6 652	11 681	15 962	15 539	13 098
Greece	20 702	20 273	20 413	18 896	20 561	17 734	15 214	16 026	14 395	14 000 F
Italy	34 013	29 679	36 825	42 129	38 174	36 387	28 876	25 805	23 980	18 049
Morocco	17 478	15 371	14 762	16 437	10 325	9 325	14 544	17 281	11 411	10 862
Romania	-	-	-	2	-	-	-	-	-	-
Serbia-Monte	68	58	57	42	45	49	49	36	28	25
Slovenia	1 748	1 907	1 719	1 982	1 973	1 788	1 614	1 415	1 219	1 223
Spain	48 000 F	48 500 F	49 000 F	50 222	44 688	37 829	37 608	48 953	41 180	22 256
Tunisia	9 577	13 140	11 940	10 389	10 767	8 907	13 394	15 001	13 988	13 311
Turkey	32 911	26 399	33 812	18 972	20 500	23 600	22 000	16 500	10 000	8 684
37 Fishing area total	*252 563 F*	*260 121 F*	*247 581 F*	*224 546*	*209 851*	*204 094*	*212 244*	*215 932*	*199 927*	*186 773 F*
Species total	*1 076 247 F*	*1 133 662 F*	*1 208 681 F*	*996 334*	*998 734*	*949 776 F*	*907 612 F*	*939 335 F*	*1 134 206 F*	*1 089 836 F*
European sprat	**Sprat**		**Espadín**		***Sprattus sprattus***				**1,21(05)066,01**	**SPR**
27 Belgium	1	1	2	3	7	4	2	2	2	1
Denmark	136 871	240 233	258 179	226 135	284 489	270 439	282 299	276 878	256 517	237 466
Estonia	5 763	9 079	13 051	22 493	39 693	32 165	36 407	41 394	40 777	40 717
Faeroe Is	-	1 827	598	100	-	753	1 719	...	...	...
Finland	205	497	4 104	14 351	19 851	27 014	18 886	23 242	15 850	17 353
France	37	293	36	597	68	0	83	97	9	1
Germany	8 278	374	231	161	427	4 551	183	22	791	950
Ireland	2 352	232	799	4 214	2 085	1 578	5 826	6 032	455	1 729
Latvia	12 553	20 132	24 383	34 211	49 314	44 858	42 834	46 186	42 769	47 540
Lithuania	2 779	2 789	4 799	10 165	6 018	4 460	3 117	1 682	3 135	2 800
Netherlands	326	609	402	293	806	54	264	307	136	169
Norway	47 038	44 078	40 969	59 115	7 051	35 166	22 214	6 353	12 465	2 609
Poland	33 700	44 556	46 182	77 472	105 298	59 090	71 706	84 324	85 757	81 243
Portugal	-	-	1	0	-	-	-	-	-	-
Russian Fed	10 745	15 904	14 934	18 287	22 194	21 078	31 627	30 369	31 959	32 854
Spain	877	317	-	5	17	-	6	17	17	28
Sweden	96 920	170 900	165 604	168 582	126 361	149 664	112 452	91 164	88 562	78 365
UK	7 352	9 338	5 906	7 172	8 317	7 021	15 168	8 336	5 091	5 792
27 Fishing area total	*365 797*	*561 159*	*580 180*	*643 356*	*671 996*	*657 895*	*644 793*	*616 405*	*584 292*	*549 617*
37 Bulgaria	2 174	2 200 F	2 874	3 535	3 646	3 275	3 595	1 737	695	11 595
France	0	0	-	-	-	-	-	-	-	-
Georgia	232	308	292	185	85	24	45	...	30	43
Greece	148	169	178	262	279	216	110	266	474	400 F
Romania	2 439	2 203	1 982	2 014	3 318	3 293	1 933	1 803	1 792	1 617
Russian Fed	694	1 013	1 263	1 537	706	1 243	4 473	5 543	11 122	11 218
Slovenia	37	14	11	7	1	-	2	3	8	14
Spain	-	-	-	-	-	-	-	-	-	13
Ukraine	9 154	12 615	15 218	20 720	20 208	30 282	29 238	32 655	49 004	45 503
37 Fishing area total	*14 878*	*18 522 F*	*21 818*	*28 260*	*28 243*	*38 333*	*39 396*	*42 007*	*63 125*	*70 403 F*
Species total	*380 675*	*579 681 F*	*601 998*	*671 616*	*700 239*	*696 228*	*684 189*	*658 412*	*647 417*	*620 020 F*
Falkland sprat	**Sprat des îles Falkland**		**Espadín de las Malvinas**		***Sprattus fuegensis***				**1,21(05)066,02**	**FAS**
41 Falkland Is	0	0	0	0	0	29	17	97	4	1
41 Fishing area total	*0*	*0*	*0*	*0*	*0*	*29*	*17*	*97*	*4*	*1*
Species total	*0*	*0*	*0*	*0*	*0*	*29*	*17*	*97*	*4*	*1*
Pacific menhaden	**Menhaden du Pacifique**		**Machete**		***Ethmidium maculatum***				**1,21(05)067,02**	**MES**
87 Chile	1 553	1 992	2 347	4 023	6 106	1 534	3 588	4 977	4 931	18 399
Peru	5 860	4 348	3 140	5 769	7 135	39 311	25 848	19 014	9 085	8 938
87 Fishing area total	*7 413*	*6 340*	*5 487*	*9 792*	*13 241*	*40 845*	*29 436*	*23 991*	*14 016*	*27 337*
Species total	*7 413*	*6 340*	*5 487*	*9 792*	*13 241*	*40 845*	*29 436*	*23 991*	*14 016*	*27 337*
Bluestripe herring	**Hareng à bande bleue**		**Arenque banda azul**		***Herklotsichthys quadrimaculat.***				**1,21(05)072,01**	**HES**
71 Fiji Islands	65	66 F	185	140	120	22	20	20 F	...	20 F
71 Fishing area total	*65*	*66 F*	*185*	*140*	*120*	*22*	*20*	*20 F*	*...*	*20 F*
Species total	*65*	*66 F*	*185*	*140*	*120*	*22*	*20*	*20 F*	*...*	*20 F*

B-35 Herrings, sardines, anchovies / Harengs, sardines, anchois / Arenques, sardinas, anchoas

Capture production by species, fishing areas and countries or areas / Captures par espèces, zones de pêche et pays ou zones / Capturas por especies, áreas de pesca y países o áreas

Species, Fishing area Espèce, Zone de pêche Especie, Area de pesca	1993 t	1994 t	1995 t	1996 t	1997 t	1998 t	1999 t	2000 t	2001 t	2002 t
Araucanian herring	**Hareng araucian**		**Sardina araucana**		***Strangomera bentincki***				**1,21(05)078,01**	**CKI**
87 Chile	244 125	341 250	126 715	446 669	441 154	317 564	782 142	722 522	324 617	347 368
87 Fishing area total	*244 125*	*341 250*	*126 715*	*446 669*	*441 154*	*317 564*	*782 142*	*722 522*	*324 617*	*347 368*
Species total	*244 125*	*341 250*	*126 715*	*446 669*	*441 154*	*317 564*	*782 142*	*722 522*	*324 617*	*347 368*
European anchovy	**Anchois**		**Boquerón**		***Engraulis encrasicolus***				**1,21(06)002,01**	**ANE**
27 Denmark	-	-	759	-	-	-	-	-	0	-
France	20 120	17 247	10 296	15 506	10 995	18 945	15 301	17 233	16 339	11 387
Germany	-	-	-	0	-	16	-	-	-	-
Latvia	3	-	-	-	-	-	-	-	-	-
Netherlands	24	2	20	6	1	16	3	-	3	4
Portugal	23	244	2 530	2 775	633	1 657	1 408	310	855	934
Russian Fed	-	-	-	-	-	-	-	998	3 215	10 172
Spain	10 229	12 378	23 407	10 693	10 662	13 611	20 460	19 555	25 094	11 504
UK	...	...	...	...	...	79	3	0	274	0
27 Fishing area total	*30 399*	*29 871*	*37 012*	*28 980*	*22 291*	*34 324*	*37 175*	*38 096*	*45 780*	*34 001*
34 Belize	-	-	-	-	300	278	3 251	3 932	2 359	5 099
Benin	100 F	125 F	200 F	250 F	350 F	400 F	478	330 F	852	1 381
Bulgaria	-	-	-	-	-	848	-	-	-	-
Cyprus	-	-	-	...	...	35	11 879	14 705	13 975	-
France	-	-	-	-	...	69	1	-	-	-
Georgia	1 470 F	550 F	100 F	-	-	-	-	-	-	-
Germany	-	-	-	-	-	-	-	-	42	-
Ghana	81 350	60 519	65 497	98 341	82 724	44 644	32 107	83 501	68 175	57 639
Greece	4	6	0	-	-	2	-	-	-	-
Latvia	-	-	-	-	-	1 978	4 876	10 142	9 143	6 872
Lithuania	-	-	...	...	...	3 612	13 774	16 137	9 492	14 064
Morocco	10 353	7 516	10 489	12 042	24 272	40 442	34 280	21 708	47 130	20 757
Portugal	0	0	-	-	-	-	-	-	-	0
Russian Fed	850	-	416	549	18 298	44 893	31 139	27 850	12 344	11 593
St Vincent	-	-	-	-	1 100	7 182	3 702	4 841	7 139	9 757
Senegal	147	8 197	73	34	31	307	1 209	964	1 015	435
Sierra Leone	183	182	182	180 F	182	150	43	181	...	43
Spain	-	-	-	3 084	-	164	140	18	60	-
Togo	7 830	4 573	4 779	7 072	4 758	6 325	6 678	7 164	6 660	6 932
Other nei	-	-	-	-	-	-	3 560	4 419	4 068	9 931
34 Fishing area total	*102 287 F*	*81 668 F*	*81 736 F*	*121 552 F*	*132 015 F*	*151 329 F*	*147 117*	*195 892 F*	*182 454*	*144 503*
37 Albania	...	...	0	2	0	-	-	-	4	4
Algeria	2 700 F	3 500 F	1 913	1 330	1 855	3 511	3 141	5 721	6 085	2 182
Bulgaria	...	...	35	23	44	48	36	64	102	237
Croatia	555	298	359	220	545	990	3 000	3 735	2 850	3 187
France	6 533	5 444	5 489	3 993	5 228	6 463	9 375	9 246	7 360	9 636
Georgia	1 656	857	1 301	1 232	2 288	2 346	1 264	2 080	1 652	2 096
Greece	14 600	17 333	13 876	15 073	14 583	17 097	16 456	9 863	10 770	10 500 F
Italy	21 219	30 840	42 746	40 541	53 439	44 429	39 783	50 728	53 047	42 068
Morocco	93	379	691	405	683	512	5 933	388	318	212
Romania	374	197	189	138	45	146	155	204	186	296
Russian Fed	2 137	4 600	10 071	2 954	3 283	2 465	2 268	5 292	7 766	9 271
Serbia-Monte	0	0	1	0	1	2	4	5	10	9
Slovenia	20	18	29	24	33	51	75	96	97	72
Spain	20 600 F	18 800 F	17 000 F	16 046	15 558	11 103	10 470	8 536	11 908	16 853
Tunisia	20	106	119	19	55	114	199	2	269	258
Turkey	227 130	294 418	387 574	290 680	241 000	228 000	350 000	280 000	320 000	373 000
Ukraine	12 858	15 987	18 505	4 398	9 444	3 914	5 527	16 390	16 668	12 941
37 Fishing area total	*310 495 F*	*392 777 F*	*499 898 F*	*377 078*	*348 084*	*321 191*	*447 686*	*392 350*	*439 092*	*482 822 F*
Species total	*443 181 F*	*504 316 F*	*618 646 F*	*527 610 F*	*502 390 F*	*506 844 F*	*631 978*	*626 338 F*	*667 326*	*661 326 F*
Japanese anchovy	**Anchois japonais**		**Anchoíta japonesa**		***Engraulis japonicus***				**1,21(06)002,02**	**JAN**
61 China	557 237	438 955	489 066	671 376	1 201 964	1 373 328	1 096 916	1 142 884	1 260 712	1 173 196
China,Taiwan	415	239	305	466	515	425	650	589	695	1 267
Japan	194 511	188 034	251 958	345 517	233 113	470 616	484 230	381 020	301 168	443 158
Korea Rep	249 209	193 398	230 679	237 128	230 911	249 519	238 463	201 192	273 927	236 315
61 Fishing area total	*1 001 372*	*820 626*	*972 008*	*1 254 487*	*1 666 503*	*2 093 888*	*1 820 259*	*1 725 685*	*1 836 502*	*1 853 936*
Species total	*1 001 372*	*820 626*	*972 008*	*1 254 487*	*1 666 503*	*2 093 888*	*1 820 259*	*1 725 685*	*1 836 502*	*1 853 936*
Argentine anchovy	**Anchois d'Argentine**		**Anchoíta**		***Engraulis anchoita***				**1,21(06)002,06**	**ANA**
41 Argentina	19 149	19 458	24 457	21 001	25 198	13 350	9 832	12 158	12 815	16 192
Uruguay	29	25	41	22	13	67	3 193	6	318	11
41 Fishing area total	*19 178*	*19 483*	*24 498*	*21 023*	*25 211*	*13 417*	*13 025*	*12 164*	*13 133*	*16 203*
Species total	*19 178*	*19 483*	*24 498*	*21 023*	*25 211*	*13 417*	*13 025*	*12 164*	*13 133*	*16 203*
Californian anchovy	**Anchois de Californie**		**Anchoa de California**		***Engraulis mordax***				**1,21(06)002,07**	**NPA**
67 USA	44	78	130	86	59	103	98	0	70	232
67 Fishing area total	*44*	*78*	*130*	*86*	*59*	*103*	*98*	*0*	*70*	*232*
77 Mexico	343	195	24 071	9 598	2 147	782	5 814	7 973	418	4 145
USA	4 395	3 702	2 949	4 419	5 719	1 450	5 225	11 487	19 189	4 650
77 Fishing area total	*4 738*	*3 897*	*27 020*	*14 017*	*7 866*	*2 232*	*11 039*	*19 460*	*19 607*	*8 795*

B-35 Herrings, sardines, anchovies / Harengs, sardines, anchois / Arenques, sardinas, anchoas

Capture production by species, fishing areas and countries or areas
Captures par espèces, zones de pêche et pays ou zones
Capturas por especies, áreas de pesca y países o áreas

Species, Fishing area Espèce, Zone de pêche Especie, Area de pesca	1993 t	1994 t	1995 t	1996 t	1997 t	1998 t	1999 t	2000 t	2001 t	2002 t
Species total	*4 782*	*3 975*	*27 150*	*14 103*	*7 925*	*2 335*	*11 137*	*19 460*	*19 677*	*9 027*
Anchoveta(=Peruvian anchovy)	**Anchois du Pérou**		**Anchoveta**		***Engraulis ringens***				**1,21(06)002,08**	**VET**
87 Chile	1 472 929	2 720 388	2 086 468	1 400 567	1 757 499	522 742	1 983 040	1 700 640	852 789	1 526 872
Ecuador	-	-	-	-	-	-	-	-	2 071	71 013
Peru	7 009 534	9 800 223	6 558 108	7 463 147	5 927 599	1 206 322	6 740 225	9 575 717	6 358 217	8 104 729
87 Fishing area total	*8 482 463*	*12 520 611*	*8 644 576*	*8 863 714*	*7 685 098*	*1 729 064*	*8 723 265*	*11 276 357*	*7 213 077*	*9 702 614*
Species total	*8 482 463*	*12 520 611*	*8 644 576*	*8 863 714*	*7 685 098*	*1 729 064*	*8 723 265*	*11 276 357*	*7 213 077*	*9 702 614*
Southern African anchovy	**Anchois de l'Afrique australe**		**Anchoa de Africa austral**		***Engraulis capensis***				**1,21(06)002,12**	**ANC**
47 Namibia	63 074	25 121	48 023	1 080	2 545	2 748	412	146	2 133	41 203
South Africa	235 606	155 554	170 308	40 712	60 095	107 548	180 542	267 840	287 190	213 440
47 Fishing area total	*298 680*	*180 675*	*218 331*	*41 792*	*62 640*	*110 296*	*180 954*	*267 986*	*289 323*	*254 643*
Species total	*298 680*	*180 675*	*218 331*	*41 792*	*62 640*	*110 296*	*180 954*	*267 986*	*289 323*	*254 643*
Atlantic anchoveta	**Anchois queue jaune**		**Anchoveta rabo amarillo**		***Cetengraulis edentulus***				**1,21(06)015,01**	**AVA**
31 Venezuela	33	35	41	8	0	119	0	0	2	0
31 Fishing area total	*33*	*35*	*41*	*8*	*0*	*119*	*0*	*0*	*2*	*0*
Species total	*33*	*35*	*41*	*8*	*0*	*119*	*0*	*0*	*2*	*0*
Pacific anchoveta	**Anchois chuchueco**		**Anchoveta chuchueco**		***Cetengraulis mysticetus***				**1,21(06)015,03**	**VEP**
77 Panama	89 220	72 111	106 743	59 830	77 726	107 730	27 356	86 681	129 147	160 414
77 Fishing area total	*89 220*	*72 111*	*106 743*	*59 830*	*77 726*	*107 730*	*27 356*	*86 681*	*129 147*	*160 414*
87 Colombia	24 240	19 453	31 823	26 344	28 747	28 501	15 781	20 500 F	25 099	25 000 F
Ecuador	57 742	27 164	23 418	26 354	89 157	44 474	27 221	13 762	73 537	18 290
87 Fishing area total	*81 982*	*46 617*	*55 241*	*52 698*	*117 904*	*72 975*	*43 002*	*34 262 F*	*98 636*	*43 290 F*
Species total	*171 202*	*118 728*	*161 984*	*112 528*	*195 630*	*180 705*	*70 358*	*120 943 F*	*227 783*	*203 704 F*
Broad-striped anchovy	**Anchois rayé**		**Anchoa legítima**		***Anchoa hepsetus***				**1,21(06)020,11**	**ENP**
31 Grenada	...	...	...	...	1	16	8	0	0	0
31 Fishing area total	*...*	*...*	*...*	*...*	*1*	*16*	*8*	*0*	*0*	*0*
Species total	*...*	*...*	*...*	*...*	*1*	*16*	*8*	*0*	*0*	*0*
Stolephorus anchovies	**Anchois Stolephorus**		**Boquerones**		***Stolephorus spp***				**1,21(06)050,XX**	**STO**
51 Untd Arab Em	8 968	8 793	9 533	9 633	10 295	10 329	10 587	2 729	4 030	6 400
51 Fishing area total	*8 968*	*8 793*	*9 533*	*9 633*	*10 295*	*10 329*	*10 587*	*2 729*	*4 030*	*6 400*
57 Indonesia	50 448	55 154	60 880	60 884	67 684	70 534	66 357	63 932	70 424	73 560
Malaysia	12 802	11 936	13 703	13 764	13 481	14 908	11 422	11 152	7 881	14 963
57 Fishing area total	*63 250*	*67 090*	*74 583*	*74 648*	*81 165*	*85 442*	*77 779*	*75 084*	*78 305*	*88 523*
61 China,H.Kong	34	14	89	31	20	17	10 F	15 F	17 F	15 F
61 Fishing area total	*34*	*14*	*89*	*31*	*20*	*17*	*10 F*	*15 F*	*17 F*	*15 F*
71 Indonesia	92 338	95 414	96 336	100 897	115 907	96 274	96 760	110 012	119 758	124 250
Malaysia	11 983	10 427	8 860	10 597	10 291	10 743	11 623	11 364	9 842	8 720
Philippines	81 354	67 507	71 516	71 456	78 678	77 049	78 087	79 630	100 899	74 095
71 Fishing area total	*185 675*	*173 348*	*176 712*	*182 950*	*204 876*	*184 066*	*186 470*	*201 006*	*230 499*	*207 065*
Species total	*257 927*	*249 245*	*260 917*	*267 262*	*296 356*	*279 854*	*274 846 F*	*278 834 F*	*312 851 F*	*302 003 F*
Anchovies, etc. nei	**Anchois, etc. nca**		**Anchoas, etc. nep**		***Engraulidae***				**1,21(06)XXX,XX**	**ANX**
31 Mexico	1 027	1 083	1 564	1 464	903	1 762	1 613	1 163	1 328	1 075
31 Fishing area total	*1 027*	*1 083*	*1 564*	*1 464*	*903*	*1 762*	*1 613*	*1 163*	*1 328*	*1 075*
41 Brazil	3 400 F	3 410 F	3 806	7 672	7 102	2 654	2 682	4 039	4 242	4 511
41 Fishing area total	*3 400 F*	*3 410 F*	*3 806*	*7 672*	*7 102*	*2 654*	*2 682*	*4 039*	*4 242*	*4 511*
51 Comoros	870	870	900 F	900 F	900 F	900 F	850 F	950 F	850	850 F
India	23 637	45 989	48 104	67 492	68 198	68 487	54 308	54 987	62 023	61 992
Oman	1 000	1 272	2 073	1 017	1 189	941	485	5 126	970	2 571
Pakistan	29 260	19 098	17 564	14 091	16 113	13 165	15 154	15 191	15 001	15 400
51 Fishing area total	*54 767*	*67 229*	*68 641 F*	*83 500 F*	*86 400 F*	*83 493 F*	*70 797 F*	*76 254 F*	*78 844*	*80 813 F*
57 Australia	178	735	470	775	...	...	...	...	...	...
India	14 323	13 167	12 875	13 202	12 320	12 719	14 435	13 621	14 274	7 837
Thailand	41 584	66 630	44 892	39 547	40 112	35 771	31 295	26 080	31 831	31 157
57 Fishing area total	*56 085*	*80 532*	*58 237*	*53 524*	*52 432*	*48 490*	*45 730*	*39 701*	*46 105*	*38 994*
61 Russian Fed	-	-	19	22	5	23	34	69	114	260
61 Fishing area total	*-*	*-*	*19*	*22*	*5*	*23*	*34*	*69*	*114*	*260*
71 Thailand	123 751	102 729	123 095	122 423	117 229	121 443	103 445	117 025	121 721	123 102
71 Fishing area total	*123 751*	*102 729*	*123 095*	*122 423*	*117 229*	*121 443*	*103 445*	*117 025*	*121 721*	*123 102*
81 Australia	12	13	13	12	19	19	2	39	22	116

Species, Fishing area Espèce, Zone de pêche Especie, Area de pesca	1993 t	1994 t	1995 t	1996 t	1997 t	1998 t	1999 t	2000 t	2001 t	2002 t
81 Fishing area total	*12*	*13*	*13*	*12*	*19*	*19*	*2*	*39*	*22*	*116*
87 Peru	63 420	39 844	189 389	59 639	24 703	706 167	11 242	3 868	137 098	6 022
87 Fishing area total	*63 420*	*39 844*	*189 389*	*59 639*	*24 703*	*706 167*	*11 242*	*3 868*	*137 098*	*6 022*
Species total	*302 462 F*	*294 840 F*	*444 764 F*	*328 256 F*	*288 793 F*	*964 051 F*	*235 545 F*	*242 158 F*	*389 474*	*254 893 F*
Dorab wolf-herring	**Chirocentre dorab**		**Arencón dorab**		***Chirocentrus dorab***			**1,21(11)002,01**		**DOB**
51 Pakistan	1 070	1 204	2 289	1 580	1 931	2 051	2 266	2 775	2 604	2 720
Qatar	0	0	0	0	-	0	0	0	0	0
Saudi Arabia	-	-	-	-	-	8	7	10	9	23
51 Fishing area total	*1 070*	*1 204*	*2 289*	*1 580*	*1 931*	*2 059*	*2 273*	*2 785*	*2 613*	*2 743*
57 Thailand	3 124	3 794	6 090	7 551	7 744	7 648	8 451	6 715	7 014	6 865
57 Fishing area total	*3 124*	*3 794*	*6 090*	*7 551*	*7 744*	*7 648*	*8 451*	*6 715*	*7 014*	*6 865*
61 China,Taiwan	4	1	3	1	2	1	4	1	-	15
61 Fishing area total	*4*	*1*	*3*	*1*	*2*	*1*	*4*	*1*	*-*	*15*
71 Thailand	6 016	7 352	9 897	2 741	2 715	4 724	6 598	7 493	5 133	5 191
71 Fishing area total	*6 016*	*7 352*	*9 897*	*2 741*	*2 715*	*4 724*	*6 598*	*7 493*	*5 133*	*5 191*
Species total	*10 214*	*12 351*	*18 279*	*11 873*	*12 392*	*14 432*	*17 326*	*16 994*	*14 760*	*14 814*
Wolf-herrings nei	**Chirocentres nca**		**Arencones nep**		***Chirocentrus spp***			**1,21(11)002,XX**		**DOS**
51 India	8 325	8 942	8 351	5 802	11 921	10 201	4 663	6 813	7 646	12 306
Untd Arab Em	81	81	71	72	77	78	80	75	...	...
51 Fishing area total	*8 406*	*9 023*	*8 422*	*5 874*	*11 998*	*10 279*	*4 743*	*6 888*	*7 646*	*12 306*
57 India	14 050	13 033	12 121	13 708	13 232	4 926	12 848	9 261	9 536	8 948
Indonesia	4 374	5 342	6 615	5 701	5 027	6 487	6 610	7 089	6 851	7 320
Malaysia	2 449	1 333	1 111	1 375	1 762	1 448	1 426	1 393	1 382	1 003
57 Fishing area total	*20 873*	*19 708*	*19 847*	*20 784*	*20 021*	*12 861*	*20 884*	*17 743*	*17 769*	*17 271*
71 Indonesia	13 968	14 739	19 598	17 358	21 500	19 111	20 903	22 065	22 946	24 630
Malaysia	2 049	2 858	2 938	2 638	2 461	2 695	3 307	2 791	2 607	2 820
Philippines	1 178	193	135	114	245	317	377	354	443	460
Singapore	35	96	81	87	79	77	51	42	30	28
71 Fishing area total	*17 230*	*17 886*	*22 752*	*20 197*	*24 285*	*22 200*	*24 638*	*25 252*	*26 026*	*27 938*
Species total	*46 509*	*46 617*	*51 021*	*46 855*	*56 304*	*45 340*	*50 265*	*49 883*	*51 441*	*57 515*
Clupeoids nei	**Clupéoidés nca**		**Clupeoideos nep**		***Clupeoidei***			**1,21(XX)XXX,XX**		**CLU**
27 France	1	-	-	-	-	-	765	165	1 086	1 858
Lithuania	-	-	35	2 400	-	-	-	-	-	-
Portugal	-	-	1	-	-	-	-	-	-	-
Spain	9 839	7 593	433	7 858	6 539	15 572	19 461	8 427	20 047	8 505
UK	3	5	4	4	17	4	-	-	-	-
27 Fishing area total	*9 843*	*7 598*	*473*	*10 262*	*6 556*	*15 576*	*20 226*	*8 592*	*21 133*	*10 363*
31 Br Virgin Is	...	...	...	...	5	5	5	0	0	0
Colombia	-	-	-	179	126	196	61	150 F	150 F	150 F
Korea Rep	-	-	-	-	-	-	1	-	-	-
Martinique	60 F	60 F	50 F	50 F	100 F	500 F	3 500	3 700	4 000	4 000
Mexico	79	74	251	222	629	332	266	188	6	91
Puerto Rico	27	18	20	18	19	14	20	20	17	15
Trinidad Tob	506	56	60 F	58	196	619	859	82	172	56
31 Fishing area total	*672 F*	*208 F*	*381 F*	*527 F*	*1 075 F*	*1 666 F*	*4 712*	*4 140 F*	*4 345 F*	*4 312 F*
34 Benin	...	...	...	...	...	...	13	10 F	-	-
Eq Guinea	160 F	240 F	100 F	220 F	500 F	850	1 300 F	2 000	1 900 F	1 900 F
Gambia	...	...	53	27	8	12	10	-	-	-
Ghana	620	312	236	477	1 060	162	132	293	1 824	2 126
Korea Rep	-	-	25	-	-	-	-	-	-	-
Liberia	121	115	75	189	147	383	386	318	208	210 F
Mauritania	-	-	-	-	-	-	-	-	5	1
Portugal	0	-	-	-	-	-	-	-	-	-
Sierra Leone	705	711	713	700 F	...	315	-	6	-	-
Spain	-	-	-	-	-	-	-	2 518	-	1
Togo	528	676	453	0	0	0	0	0	0	-
34 Fishing area total	*2 134 F*	*2 054 F*	*1 655 F*	*1 613 F*	*1 715 F*	*1 722*	*1 841 F*	*5 145 F*	*3 937 F*	*4 238 F*
37 Algeria	2 900 F	3 500 F	2 012	1 182	1 261	2 279	2 634	2 608	1 074	1 541
France	-	-	-	-	-	-	72	15	17	7
Lebanon	500	500	800	800	600	600	500	700	500	650
Malta	0	0	0	0	0	0	0	-	-	-
Spain	-	-	-	15	3	-	-	-	195	975
37 Fishing area total	*3 400 F*	*4 000 F*	*2 812*	*1 997*	*1 864*	*2 879*	*3 206*	*3 323*	*1 786*	*3 173*
41 Argentina	0	0	-	-	-	-	-	-	-	-
Brazil	...	...	2 775	3 464	4 314	6 042	7 527	11 457	27 058	29 481
41 Fishing area total	*0*	*0*	*2 775*	*3 464*	*4 314*	*6 042*	*7 527*	*11 457*	*27 058*	*29 481*
51 India	28 410	40 006	38 098	51 568	66 868	70 487	63 435	36 678	34 325	33 150
Iran	20 000	5 000	8 000	10 000	10 000	9 708	13 030	14 955	17 453	12 865
Kenya	166	162	112	217	189	155	167	119	164	101

B-35 Herrings, sardines, anchovies / Harengs, sardines, anchois / Arenques, sardinas, anchoas

Capture production by species, fishing areas and countries or areas
Captures par espèces, zones de pêche et pays ou zones
Capturas por especies, áreas de pesca y países o áreas

Species, Fishing area Espèce, Zone de pêche Especie, Area de pesca	1993 t	1994 t	1995 t	1996 t	1997 t	1998 t	1999 t	2000 t	2001 t	2002 t
Mauritius	0	0	0	0	-	-	-	-	-	-
Pakistan	43 475	18 111	31 426	27 576	26 650	25 487	26 934	24 810	24 306	24 500
Réunion	3	5	5	5	6	7	10	12	4	4 F
Saudi Arabia	369	391	259	201	560	108	162	187	178	266
51 Fishing area total	*92 423*	*63 675*	*77 900*	*89 567*	*104 273*	*105 952*	*103 738*	*76 761*	*76 430*	*70 886 F*
57 Australia	11 020	11 484	12 820	12 697	13 028	7 801	4 377	5 499	8 525	14 440
India	7 686	14 483	20 943	25 072	24 625	25 906	25 727	27 205	34 618	30 892
Malaysia	5 023	3 811	3 516	4 388	4 766	3 199	4 583	3 391	4 275	4 238
Sri Lanka	37 379	38 870	49 785	48 221	47 200	50 800	51 370	53 250	49 270	52 310
57 Fishing area total	*61 108*	*68 648*	*87 064*	*90 378*	*89 619*	*87 706*	*86 057*	*89 345*	*96 688*	*101 880*
61 China,Taiwan	6 042	6 243	6 348	3 873	4 222	4 338	3 721	4 221	4 393	4 581
Japan	59 674	59 401	55 408	58 232	59 619	52 427	79 405	74 581	58 286	62 694
61 Fishing area total	*65 716*	*65 644*	*61 756*	*62 105*	*63 841*	*56 765*	*83 126*	*78 802*	*62 679*	*67 275*
71 Kiribati	3 300	3 270	3 300	3 340	3 320	613	2 638	2 725	2 079	3 619
Malaysia	21 913	30 870	35 477	40 137	33 344	43 116	40 934	30 222	36 475	36 373
Philippines	405	613	286	266	325	663	634	602	736	772
Singapore	265	353	240	379	351	329	206	78	73	63
71 Fishing area total	*25 883*	*35 106*	*39 303*	*44 122*	*37 340*	*44 721*	*44 412*	*33 627*	*39 363*	*40 827*
77 Costa Rica	323	365	311	438	1 175	906	1 788	1 628	2 207	4 170
Mexico	41	45	53	54	168	237	105	87	15	318
77 Fishing area total	*364*	*410*	*364*	*492*	*1 343*	*1 143*	*1 893*	*1 715*	*2 222*	*4 488*
81 Australia	464	432	432	333	417	417	...	65	38	34
New Zealand	230	286	...	11	8	1	12	11	11	3
Ukraine	2	-	-	-	-	-	-	-	-	-
81 Fishing area total	*696*	*718*	*432*	*344*	*425*	*418*	*12*	*76*	*49*	*37*
87 Colombia	89	2 502	120	1	429	273	0	100 F	100 F	100 F
Ghana	-	-	-	-	-	-	-	3 496	4 320	-
87 Fishing area total	*89*	*2 502*	*120*	*1*	*429*	*273*	*0*	*3 596 F*	*4 420 F*	*100 F*
Species total	*262 328 F*	*250 563 F*	*275 035 F*	*304 872 F*	*312 794 F*	*324 863 F*	*356 750 F*	*316 579 F*	*340 110 F*	*337 060 F*
Group total	***21 993 737***	***25 910 826***	***22 004 664***	***22 385 424***	***21 730 998***	***16 664 950***	***22 637 492***	***24 892 060***	***20 628 706***	***22 472 563***

B-36 Tunas, bonitos, billfishes / Thons, pélamides, marlins / Atunes, bonitos, agujas

Capture production by species, fishing areas and countries or areas / Captures par espèces, zones de pêche et pays ou zones / Capturas por especies, áreas de pesca y países o áreas

Species, Fishing area Espèce, Zone de pêche Especie, Area de pesca	1993 t	1994 t	1995 t	1996 t	1997 t	1998 t	1999 t	2000 t	2001 t	2002 t
Atlantic bonito	**Bonite à dos rayé**		**Bonito del Atlántico**		***Sarda sarda***			**1,75(01)001,01**		**BON**
21 USA	93	126	102	153	134	74	73	44	38	16
21 Fishing area total	*93*	*126*	*102*	*153*	*134*	*74*	*73*	*44*	*38*	*16*
27 France	52	-	-	-	-	-	24	32	42	18
Portugal	120	25	77	82	48	97	98	161	47	55
27 Fishing area total	*172*	*25*	*77*	*82*	*48*	*97*	*122*	*193*	*89*	*73*
31 Br Virgin Is	...	8	0	6	8	6	9	0	0	0
Grenada	-	-	-	24	6	14	16	7	10	10
Martinique	1 000	990	990 F	610	610 F	610 F	610 F	610 F	530 F	530 F
Mexico	779	674	1 143	1 279	2 040	2 194	2 314	1 721	1 506	1 401
St Lucia	4	1	1	1	0	0	0	0	0	0
Trinidad Tob	17	703	169	266	220	30	117	117	56	452
USA	294	718	110	5	27	10	10	4	10	5
Venezuela	1 541	1 646	1 503	1 348	1 294	1 647	1 597	1 376	1 815	2 948
31 Fishing area total	*3 635*	*4 740*	*3 916 F*	*3 539*	*4 205 F*	*4 511 F*	*4 673 F*	*3 835 F*	*3 927 F*	*5 346 F*
34 Germany	-	-	-	714	417	42	143	-	51	38
Greece	0	0	0	-	-	-	-	-	-	-
Latvia	-	3	19	301	887	318	510	416	396	639
Lithuania	-	-	-	-	-	-	-	-	-	793
Morocco	1 246	584	699	894	1 259	1 557	1 390	2 163	1 700	2 019
Netherlands	-	-	-	1 694	1 625	2 171	966	1 507	1 791	1 793
Poland	-	-	-	225	39	-	-	-	521	79
Portugal	25	31	1	1	1	1	-	-	-	6
Russian Fed	6	-	6	175	1 937	4 960	2 156	837	538	1 431
Senegal	402	600	354	570	564	1 723	349	179	120	344
Sierra Leone	...	...	0	0	0	4	...	11	245	44
Spain	5	3	2	2	1	-	12	12	10	5
Togo	311	254	145	197	338	294	426	423	663	846
Ukraine	...	...	...	342	2 786	1 918	1 114	399	231	656
34 Fishing area total	*1 995*	*1 475*	*1 226*	*5 115*	*9 854*	*12 988*	*7 066*	*5 947*	*6 266*	*8 693*
37 Albania	...	...	1	2	0	12	30	25	30	24
Algeria	471	418	506	277	357	511	475	405	350	597
Bulgaria	8	...	25	33	16	51	20	35	49	-
Croatia	230	70	182	159	171	158	120	120	54	28
Cyprus	-	-	-	-	-	-	-	14	13	10
Egypt	640	648	697	985	725	724	1 442	1 128	1 072	1 416
France	6	-	-	-	-	-	-	-	28	27
Greece	2 690	1 581	2 116	1 752	1 559	945	2 135	1 914	1 550	1 420
Italy	1 662	1 828	1 512	2 233	4 580	2 121	1 614	1 116	1 006	944
Libya	70	...	...	...	...	...	...	...	...	...
Malta	0	0	0	2	7	2	2	1	-	-
Morocco	25	93	37	67	45	39	120	115	5	61
Serbia-Monte	3	2	6	10	12	12	14	17	17	16
Slovenia	-	-	-	-	-	-	-	-	-	1
Spain	200	344	632	690	628	333	433	342	349	461
Tunisia	792	305	413	560	611	855	1 350	1 528	1 183	1 126
Turkey	19 548	10 093	8 944	10 284	7 810	24 000	17 900	12 000	13 460	6 286
Other nei	300	300	300	300	75	-	-	-	-	-
37 Fishing area total	*26 645*	*15 682*	*15 371*	*17 354*	*16 596*	*29 763*	*25 655*	*18 760*	*19 166*	*12 417*
41 Argentina	434	4	138	108	130	12	38	19	235	1
Brazil	142	142	137	...	...	...	...	...	...	...
Uruguay	-	-	-	-	-	-	-	1	23	15
41 Fishing area total	*576*	*146*	*275*	*108*	*130*	*12*	*38*	*20*	*258*	*16*
47 Angola	49	20	9	39	32	-	2	118	1 157	...
Russian Fed	-	-	-	-	-	-	-	41	36	10
South Africa	0	0	0	0	-	-	-	-	-	-
47 Fishing area total	*49*	*20*	*9*	*39*	*32*	*-*	*2*	*159*	*1 193*	*10*
Species total	*33 165*	*22 214*	*20 976 F*	*26 390*	*30 999 F*	*47 445 F*	*37 629 F*	*28 958 F*	*30 937 F*	*26 571 F*
Striped bonito	**Bonite oriental**		**Bonito mono**		***Sarda orientalis***			**1,75(01)001,02**		**BIP**
51 Oman	224	155	788	370	498	162	134	95	287	378
51 Fishing area total	*224*	*155*	*788*	*370*	*498*	*162*	*134*	*95*	*287*	*378*
Species total	*224*	*155*	*788*	*370*	*498*	*162*	*134*	*95*	*287*	*378*
Eastern Pacific bonito	**Bonite du Pacifique oriental**		**Bonito del Pacífico oriental**		***Sarda chiliensis***			**1,75(01)001,04**		**BEP**
77 Mexico	472	6 333	6 718	399	875	423	1 775	429	146	250
USA	390	431	71	449	290	1 094	87	44	6	33
77 Fishing area total	*862*	*6 764*	*6 789*	*848*	*1 165*	*1 517*	*1 862*	*473*	*152*	*283*
87 Chile	288	172	52	14	28	584	368	55	19	0
Colombia	...	...	3	6	5	8	9	10 F	13	10 F
Peru	36 976	31 125	28 331	23 059	17 731	5 130	948	434	1 287	865
87 Fishing area total	*37 264*	*31 297*	*28 386*	*23 079*	*17 764*	*5 722*	*1 325*	*499 F*	*1 319*	*875 F*
Species total	*38 126*	*38 061*	*35 175*	*23 927*	*18 929*	*7 239*	*3 187*	*972 F*	*1 471*	*1 158 F*

B-36 Tunas, bonitos, billfishes — Capture production by species, fishing areas and countries or areas
Thons, pélamides, marlins — Captures par espèces, zones de pêche et pays ou zones
Atunes, bonitos, agujas — Capturas por especies, áreas de pesca y países o áreas

Species, Fishing area Espèce, Zone de pêche Especie, Area de pesca	1993 t	1994 t	1995 t	1996 t	1997 t	1998 t	1999 t	2000 t	2001 t	2002 t
Plain bonito	**Palomette**		**Tasarte**			***Orcynopsis unicolor***			**1,75(01)006,01**	**BOP**
27 Portugal	-	-	-	-	-	-	-	-	-	2
27 Fishing area total	-	-	-	-	-	-	-	-	-	2
34 Benin	1 F	1 F	1 F	1 F	3	1	1	-	-	-
Morocco	348	598	524	2 003	246	28	626	1 048	830	780
Portugal	-	-	-	-	-	-	-	-	-	1
Senegal	-	-	-	-	-	65	17	6	69	126
34 Fishing area total	*349 F*	*599 F*	*525 F*	*2 004 F*	*249*	*94*	*644*	*1 054*	*899*	*907*
37 Algeria	198	153	92	119	224	128	216	135	145	128
Libya	40 F	...	...	...	...	...	...	...	...	...
Morocco	14	23	23	13	3	2	1	10	9	9
37 Fishing area total	*252 F*	*176*	*115*	*132*	*227*	*130*	*217*	*145*	*154*	*137*
Species total	*601 F*	*775 F*	*640 F*	*2 136 F*	*476*	*224*	*861*	*1 199*	*1 053*	*1 046*
Wahoo	**Thazard-bâtard**		**Peto**			***Acanthocybium solandri***			**1,75(01)010,01**	**WAH**
21 USA	...	...	...	...	1	1	2	2	1	1
21 Fishing area total	...	...	...	...	*1*	*1*	*2*	*2*	*1*	*1*
31 Aruba	80	125	40	50	65	70	60	60	60	60 F
Barbados	91	82	42	35	52	52	41	41	24	41
Bermuda	58	50	93	99	105	108	104	61	56	87
Dominica	59	59	58	58	58	58	50	46	11	20
Dominican Rp	7	...	...	...	325	112	31	42	37	0
Grenada	55	46	49	56	56	59	82	51	71	59
NethAntilles	270	250	230	230 F	230 F	230 F	230 F	230 F	230 F	230 F
Puerto Rico	...	...	...	...	...	...	...	...	6	1
St Lucia	141	98	80	221	224	223	310	243	213	243
St Vincent	41	28	16	23	10	65	52	46	311	17
Trinidad Tob	-	-	-	0	1	1	1	2	1	9
USA	...	...	...	...	1	64	29	79	56	63
Venezuela	513	538	445	479	498	349	448	150	297	275
31 Fishing area total	*1 315*	*1 276*	*1 053*	*1 251 F*	*1 625 F*	*1 391 F*	*1 438 F*	*1 051 F*	*1 373 F*	*1 105 F*
34 Cape Verde	326	361	408	503	603	429	587	487	578	552
Sao Tome Prn	-	-	-	80	52	52 F	52 F	52 F	52 F	52 F
Spain	22	20	15	25	25	29	28	32	38	46
34 Fishing area total	*348*	*381*	*423*	*608*	*680*	*510 F*	*667 F*	*571 F*	*668 F*	*650 F*
41 Brazil	33	26	1	16	58	41	...	...	...	...
41 Fishing area total	*33*	*26*	*1*	*16*	*58*	*41*	...	...	...	...
47 St Helena	35	26	25	23	19	10	15	15	22	25
47 Fishing area total	*35*	*26*	*25*	*23*	*19*	*10*	*15*	*15*	*22*	*25*
51 India	0	0	5	14	13	14	12	51	92	...
Mayotte	...	...	...	...	...	1	2	...	...	...
Réunion	67	67	67	80	66	59	57	50	45	45 F
Seychelles	-	-	10	-	-	3	4	6	1	2
Spain	-	-	-	-	-	-	4	-	-	-
51 Fishing area total	*67*	*67*	*82*	*94*	*79*	*77*	*79*	*107*	*138*	*47 F*
57 Australia	-	-	-	1	4	4	8	6	9	7
India	1	0	0	9	8	9	17	33	60	...
Indonesia	4	6	6	12	15	14	16	12	12	12
Sri Lanka	433	268	129	128	156	196	488	545	444	456
57 Fishing area total	*438*	*274*	*135*	*150*	*183*	*223*	*529*	*596*	*525*	*475*
71 Australia	-	-	-	-	-	-	-	8	9	13
Fiji Islands	92	94 F	179	130	145	148	160	150 F	167	160 F
Guam	36	48	24	19	20	30	19	20	23	22
N Marianas	1	2	3	5	4	2	4	2	2	4
Palau	...	...	1	2	1	1	1	0	2	1
71 Fishing area total	*129*	*144 F*	*207*	*156*	*170*	*181*	*184*	*180 F*	*203*	*200 F*
77 Amer Samoa	1	4	5	5	7	12	17	20	47	108
Fr Polynesia	...	...	...	...	119	188	269	229	259	291
USA	...	...	...	...	-	9	2	0	1	298
77 Fishing area total	*1*	*4*	*5*	*5*	*126*	*209*	*288*	*249*	*307*	*697*
81 Australia	-	-	-	-	-	-	-	3	1	2
New Zealand	-	-	-	-	6	-	-	-	-	-
81 Fishing area total	-	-	-	-	*6*	-	-	*3*	*1*	*2*
Species total	*2 366*	*2 198 F*	*1 931*	*2 303 F*	*2 947 F*	*2 643 F*	*3 202 F*	*2 774 F*	*3 238 F*	*3 202 F*
Dogtooth tuna	**Bonite à gros yeux**		**Casarte ojón**			***Gymnosarda unicolor***			**1,75(01)012,02**	**DOT**
51 Maldives	627	388	438	625	489	470	426	451	647	789
51 Fishing area total	*627*	*388*	*438*	*625*	*489*	*470*	*426*	*451*	*647*	*789*
Species total	*627*	*388*	*438*	*625*	*489*	*470*	*426*	*451*	*647*	*789*
Narrow-barred Spanish mackerel	**Thazard rayé indo-pacifique**		**Carite estriado Indo-Pacífico**			***Scomberomorus commerson***			**1,75(01)015,03**	**COM**
51 Bahrain	77	69	109	158	47	85	44	66	109	121
Egypt	299	270	30	203	194	227	170	340	374	418

B-36 Tunas, bonitos, billfishes / Thons, pélamides, marlins / Atunes, bonitos, agujas

Capture production by species, fishing areas and countries or areas / Captures par espèces, zones de pêche et pays ou zones / Capturas por especies, áreas de pesca y países o áreas

Species, Fishing area Espèce, Zone de pêche Especie, Area de pesca	1993 t	1994 t	1995 t	1996 t	1997 t	1998 t	1999 t	2000 t	2001 t	2002 t
Eritrea	...	...	49	191	200	210	250	217	280	177
India	16 903	20 746	24 085	15 235	14 460	19 921	17 123	22 585	17 053	32 555
Iran	2 869	3 300	11 067	3 560	4 290	4 034	4 609	7 075	6 071	8 557
Jordan	...	...	...	...	...	0	1	0	1	1
Kenya	46	103	74	93	69	139	122	94	136	150
Kuwait	36	59	68	85	73	124	127	130 F	135	135 F
Madagascar	10 000	10 000	10 000	10 000	10 000	12 000	12 000	12 000	12 000	12 000
Oman	3 143	3 764	6 185	5 243	5 944	3 145	3 390	2 559	2 785	2 088
Pakistan	12 252	7 157	8 618	10 108	12 009	12 232	11 734	9 366	8 455	7 922
Qatar	636	406	255	307	411	552	496	768	1 019	963
Saudi Arabia	10 374	10 851	6 342	5 276	5 511	6 722	6 032	6 057	5 532	6 399
South Africa	54	15	4	13	...	...	...	...	...	...
Sudan	...	...	...	...	...	19	24	19	34	34
Untd Arab Em	5 800	6 475	6 584	6 653	7 110	7 133	7 311	6 645	7 650	3 824
Yemen	3 092	3 255	3 047	3 521	3 680	3 580	3 580 F	3 580 F	3 580 F	3 580 F
Other nei	-	-	3 060	3 060	3 060	-	-	-	-	-
51 Fishing area total	*65 581*	*66 470*	*79 577*	*63 706*	*67 058*	*70 123*	*67 013 F*	*71 501 F*	*65 214 F*	*78 924 F*
57 Australia	445	513	444	471	622	540	336	307	490	468
India	3 116	3 878	4 502	9 378	8 900	12 260	23 195	13 900	10 495	31 398
Indonesia	14 731	13 481	14 090	16 350	15 164	17 002	16 215	18 986	21 970	21 970
Sri Lanka	163	203	199	817	999	1 246	856	766	1 015	589
57 Fishing area total	*18 455*	*18 075*	*19 235*	*27 016*	*25 685*	*31 048*	*40 602*	*33 959*	*33 970*	*54 425*
61 China,Taiwan	3 316	2 845	3 211	2 541	2 701	2 417	2 674	3 551	4 153	6 613
61 Fishing area total	*3 316*	*2 845*	*3 211*	*2 541*	*2 701*	*2 417*	*2 674*	*3 551*	*4 153*	*6 613*
71 Australia	779	796	785	680	830	1 054	-	251	251	-
Fiji Islands	783	828 F	1 424	1 247	1 025	1 455	2 296	2 000 F	2 120	2 100 F
Indonesia	43 375	46 833	49 342	52 105	60 106	55 064	61 496	66 444	62 543	65 650
Palau	...	...	2	1	0	1	0	0	1	0
Philippines	12 962	9 234	10 593	10 557	11 237	10 772	9 137	8 889	9 085	9 030
71 Fishing area total	*57 899*	*57 691 F*	*62 146*	*64 590*	*73 198*	*68 346*	*72 929*	*77 584 F*	*74 000*	*76 780 F*
81 Australia	47	17	17	9	23	23	-	-	-	-
81 Fishing area total	*47*	*17*	*17*	*9*	*23*	*23*	-	-	-	-
Species total	*145 298*	*145 098 F*	*164 186*	*157 862*	*168 665*	*171 957*	*183 218 F*	*186 595 F*	*177 337 F*	*216 742 F*
Indo-Pacific king mackerel	**Thazard ponctué indo-pacifique**		**Carite del Indo-Pacífico**		***Scomberomorus guttatus***				**1,75(01)015,04**	**GUT**
51 India	14 163	11 367	11 936	7 838	8 205	13 965	12 003	8 515	9 186	23 407
Iran	1 636	1 650	5 418	4 340	4 129	3 883	3 476	4 100	2 474	4 031
Kuwait	126	60	156	172	206	166	211	210 F	204	204 F
Saudi Arabia	-	-	-	-	114	300	303	303	276	320
South Africa	-	-	-	-	20	46	16	22	49	7
51 Fishing area total	*15 925*	*13 077*	*17 510*	*12 350*	*12 674*	*18 360*	*16 009*	*13 150 F*	*12 189*	*27 969 F*
57 India	7 467	4 988	5 238	4 824	5 050	8 595	16 260	5 241	5 653	638
Indonesia	8 880	5 723	10 601	13 782	11 917	11 821	11 217	11 668	13 204	13 204
Sri Lanka	6	52	1	0	-	-	-	-	-	-
57 Fishing area total	*16 353*	*10 763*	*15 840*	*18 606*	*16 967*	*20 416*	*27 477*	*16 909*	*18 857*	*13 842*
61 China,Taiwan	586	795	1 814	1 611	1 607	1 128	1 298	1 249	1 395	992
61 Fishing area total	*586*	*795*	*1 814*	*1 611*	*1 607*	*1 128*	*1 298*	*1 249*	*1 395*	*992*
71 Indonesia	9 368	8 603	9 272	9 314	9 098	10 925	10 457	12 781	13 705	14 440
71 Fishing area total	*9 368*	*8 603*	*9 272*	*9 314*	*9 098*	*10 925*	*10 457*	*12 781*	*13 705*	*14 440*
Species total	*42 232*	*33 238*	*44 436*	*41 881*	*40 346*	*50 829*	*55 241*	*44 089 F*	*46 146*	*57 243 F*
Streaked seerfish	**Thazard cirrus**		**Carite rayado**		***Scomberomorus lineolatus***				**1,75(01)015,05**	**STS**
51 India	74	31	59	59	558	67	58	34	24	...
51 Fishing area total	*74*	*31*	*59*	*59*	*558*	*67*	*58*	*34*	*24*	...
57 India	5	15	28	37	343	40	77	21	14	431
Sri Lanka	2	0	0	0	-	-	-	-	-	-
57 Fishing area total	*7*	*15*	*28*	*37*	*343*	*40*	*77*	*21*	*14*	*431*
Species total	*81*	*46*	*87*	*96*	*901*	*107*	*135*	*55*	*38*	*431*
King mackerel	**Thazard barré**		**Carite lucio**		***Scomberomorus cavalla***				**1,75(01)015,06**	**KGM**
21 USA	322	1	280	1 625	385	341	1 171	315	232	163
21 Fishing area total	*322*	*1*	*280*	*1 625*	*385*	*341*	*1 171*	*315*	*232*	*163*
31 Dominica	-	-	-	-	-	-	36	35	2	...
Dominican Rp	791	1 330	2 042	1 648	589	288	230	271	261	492
Grenada	-	-	2	2	4	28	14	9	4	6
Guyana	-	-	-	-	270	440	398	214	239	267
Mexico	3 289	3 097	3 050	4 377	5 370	4 598	5 002	4 576	5 119	5 720
Trinidad Tob	1 192	...	471	1 029	875	746	447	432	638	1 457
USA	2 136	2 012	1 769	444	2 130	2 020	1 239	1 931	1 962	1 865
Venezuela	800	2 484	2 555	2 140	3 530	340	2 424	1 498	1 861	2 324
31 Fishing area total	*8 208*	*8 923*	*9 889*	*9 640*	*12 768*	*8 460*	*9 790*	*8 966*	*10 086*	*12 131*
41 Brazil	1 380	1 365	1 328	2 890	2 398	3 595	3 595	2 344	1 251	2 316
41 Fishing area total	*1 380*	*1 365*	*1 328*	*2 890*	*2 398*	*3 595*	*3 595*	*2 344*	*1 251*	*2 316*

B-36 Tunas, bonitos, billfishes / Thons, pélamides, marlins / Atunes, bonitos, agujas

Capture production by species, fishing areas and countries or areas / Captures par espèces, zones de pêche et pays ou zones / Capturas por especies, áreas de pesca y países o áreas

Species, Fishing area Espèce, Zone de pêche Especie, Area de pesca	1993 t	1994 t	1995 t	1996 t	1997 t	1998 t	1999 t	2000 t	2001 t	2002 t
Species total	*9 910*	*10 289*	*11 497*	*14 155*	*15 551*	*12 396*	*14 556*	*11 625*	*11 569*	*14 610*
Atlantic Spanish mackerel	**Thazard atlantique**		**Carite atlántico**		***Scomberomorus maculatus***			**1,75(01)015,07**		**SSM**
21 USA	361	228	96	1 411	364	215	694	368	363	337
21 Fishing area total	*361*	*228*	*96*	*1 411*	*364*	*215*	*694*	*368*	*363*	*337*
31 Mexico	10 066	8 300	7 673	11 049	7 389	7 381	8 382	5 717	5 320	6 123
USA	1 997	2 487	1 911	134	1 331	1 244	668	1 282	1 553	1 238
31 Fishing area total	*12 063*	*10 787*	*9 584*	*11 183*	*8 720*	*8 625*	*9 050*	*6 999*	*6 873*	*7 361*
Species total	*12 424*	*11 015*	*9 680*	*12 594*	*9 084*	*8 840*	*9 744*	*7 367*	*7 236*	*7 698*
Cero	**Thazard franc**		**Carite chinigua**		***Scomberomorus regalis***			**1,75(01)015,08**		**CER**
31 Dominican Rp	50	90	29	57	231	191	125	190	147	...
Martinique	400	400 F	400 F	250	250 F	250 F	...	...	...	...
St Vincent	0	0	0	-	-	-	-	-	-	-
31 Fishing area total	*450*	*490 F*	*429 F*	*307*	*481 F*	*441 F*	*125*	*190*	*147*	*...*
Species total	*450*	*490 F*	*429 F*	*307*	*481 F*	*441 F*	*125*	*190*	*147*	*...*
Pacific sierra	**Thazard sierra (Pacifique)**		**Carite sierra**		***Scomberomorus sierra***			**1,75(01)015,09**		**SIE**
77 Mexico	5 756	5 625	5 137	5 742	5 405	3 896	5 265	6 261	5 959	4 815
Nicaragua	...	...	...	...	...	...	...	55	67	77
Panama	576	567	463	489	665	982	159	1 395	971	1 027
77 Fishing area total	*6 332*	*6 192*	*5 600*	*6 231*	*6 070*	*4 878*	*5 424*	*7 711*	*6 997*	*5 919*
87 Colombia	444	694	360	484	645	912	521	500 F	500 F	500 F
Peru	924	352	686	439	98	861	2 712	930	181	288
87 Fishing area total	*1 368*	*1 046*	*1 046*	*923*	*743*	*1 773*	*3 233*	*1 430 F*	*681 F*	*788 F*
Species total	*7 700*	*7 238*	*6 646*	*7 154*	*6 813*	*6 651*	*8 657*	*9 141 F*	*7 678 F*	*6 707 F*
West African Spanish mackerel	**Thazard blanc**		**Carite lusitánico**		***Scomberomorus tritor***			**1,75(01)015,11**		**MAW**
34 Benin	214	194	188	188 F	362	511	205	205 F	203	185
Ghana	466	-	-	-	-	-	-	-	-	-
Lithuania	-	-	-	-	-	-	-	-	-	298
Mauritania	...	...	...	...	...	...	...	12 F	...	...
Russian Fed	19	-	-	44	-	14	19	7	4	-
Sao Tome Prn	-	-	-	8	-	-	-	-	-	-
Senegal	1 060	766	1 863	1 056	1 015	931	479	589	756	661
Ukraine	-	-	-	-	-	-	-	21	...	42
34 Fishing area total	*1 759*	*960*	*2 051*	*1 296 F*	*1 377*	*1 456*	*703*	*834 F*	*963*	*1 186*
Species total	*1 759*	*960*	*2 051*	*1 296 F*	*1 377*	*1 456*	*703*	*834 F*	*963*	*1 186*
Japanese Spanish mackerel	**Thazard oriental**		**Carite oriental**		***Scomberomorus niphonius***			**1,75(01)015,12**		**NPH**
61 China	145 480	202 811	226 520	283 784	340 302	517 528	565 764	496 566	476 690	506 195
China,Taiwan	9 130	11 217	8 971	7 546	11 761	8 579	4 516	5 955	11 497	12 565
Japan	4 200	5 059	6 381	3 607	2 349	2 864	5 321	10 932	9 056	8 936
Korea Rep	13 951	8 673	17 429	6 419	11 173	22 809	19 502	25 641	25 513	25 956
61 Fishing area total	*172 761*	*227 760*	*259 301*	*301 356*	*365 585*	*551 780*	*595 103*	*539 094*	*522 756*	*553 652*
Species total	*172 761*	*227 760*	*259 301*	*301 356*	*365 585*	*551 780*	*595 103*	*539 094*	*522 756*	*553 652*
Serra Spanish mackerel	**Thazard serra**		**Serra**		***Scomberomorus brasiliensis***			**1,75(01)015,15**		**BRS**
31 Grenada	0	0	0	0	0	1	1	1	0	...
Guyana	-	-	-	211	571	625	1 143	308	329	441
Trinidad Tob	2 130	2 130	1 816	1 568	1 699	2 130	1 328	1 722	2 671	2 472
Venezuela	5 077	3 882	4 725	3 609	3 670	3 651	1 686	3 624	3 062	2 337
31 Fishing area total	*7 207*	*6 012*	*6 541*	*5 388*	*5 940*	*6 407*	*4 158*	*5 655*	*6 062*	*5 250*
41 Brazil	842	1 149	1 308	3 047	2 125	1 516	1 516	988	251	3 071
41 Fishing area total	*842*	*1 149*	*1 308*	*3 047*	*2 125*	*1 516*	*1 516*	*988*	*251*	*3 071*
Species total	*8 049*	*7 161*	*7 849*	*8 435*	*8 065*	*7 923*	*5 674*	*6 643*	*6 313*	*8 321*
Seerfishes nei	**Thazards nca**		**Carites nep**		***Scomberomorus spp***			**1,75(01)015,XX**		**KGX**
31 Barbados	55	36	42	49	...	...	...	...	...	...
Br Virgin Is	...	...	...	...	6	6	5	0	0	0
Colombia	79	217	180	539	22	30	55	50 F	50 F	50 F
Cuba	310	408	544	613	466	236	247	222	184	263
Nicaragua	...	...	...	...	...	...	240	250	260	116
Puerto Rico	84	86	134	106	119	109	123	145	124	90
St Lucia	141	98	80	51	4	0	...	60	7	29
St Vincent	-	-	-	1	1	1	1	138	0	0
31 Fishing area total	*669*	*845*	*980*	*1 359*	*618*	*382*	*671*	*865 F*	*625 F*	*548 F*
34 Gabon	-	140	145	79	-	85	-	-	-	265
Korea Rep	-	-	-	14	-	-	-	8	-	-
34 Fishing area total	*-*	*140*	*145*	*93*	*-*	*85*	*-*	*8*	*-*	*265*

B-36 Tunas, bonitos, billfishes / Thons, pélamides, marlins / Atunes, bonitos, agujas

Capture production by species, fishing areas and countries or areas / Captures par espèces, zones de pêche et pays ou zones / Capturas por especies, áreas de pesca y países o áreas

Species, Fishing area Espèce, Zone de pêche Especie, Area de pesca	1993 t	1994 t	1995 t	1996 t	1997 t	1998 t	1999 t	2000 t	2001 t	2002 t
51 Comoros	230	271	271	270	260	260	250	270 F	250	250 F
Djibouti	57	61	67	65	61	60	60 F	60 F	60 F	60 F
India	51	733	670	753	2 064	-	-	-	-	-
Korea Rep	-	-	-	38	-	-	-	-	17	8
South Africa	-	-	-	-	2	3	2	2	2	4
Tanzania	594	544	680	766	750	750	650	400	500	450
Untd Arab Em	1 400	1 566	1 267	1 594	2 500	2 886	3 117	2 876	2 870 F	...
Yemen	538	538	500	500	520	510	510 F	510 F	510 F	510 F
Other nei	-	-	431	431	431	-	-	-	-	-
51 Fishing area total	*2 870*	*3 713*	*3 886*	*4 417*	*6 588*	*4 469*	*4 589 F*	*4 118 F*	*4 209 F*	*1 282 F*
57 Australia	111	126	89	131	171	-	-	22	55	65
Bangladesh	40	50	50	40	50	60	60	60	60	60
India	18	203	181	463	1 271	-	-	-	-	-
Korea Rep	-	-	-	-	1	-	-	-	-	-
Malaysia	4 394	4 375	3 043	3 330	3 652	4 857	3 473	4 090	3 259	3 464
Sri Lanka	-	-	-	-	-	-	168	20	3	14
Thailand	3 498	6 473	7 780	5 882	5 625	5 769	4 814	4 213	4 229	4 071
57 Fishing area total	*8 061*	*11 227*	*11 143*	*9 846*	*10 770*	*10 686*	*8 515*	*8 405*	*7 606*	*7 674*
61 China,H.Kong	3 093	2 665	2 790	2 036	1 711	1 723	1 200 F	1 500 F	1 650 F	1 600 F
China,Taiwan	1 795	1 495	1 085	1 103	1 143	1 478	2 883	2 260	1 433	2 022
61 Fishing area total	*4 888*	*4 160*	*3 875*	*3 139*	*2 854*	*3 201*	*4 083 F*	*3 760 F*	*3 083 F*	*3 622 F*
71 Australia	248	398	339	495	-	-	-	-	-	-
Japan	10	40	28	7	-	-	-	-	-	-
Korea Rep	78	43	25	9	-	38	5	-	1	2
Malaysia	16 424	11 225	11 858	11 070	10 082	11 421	13 774	11 233	11 647	10 438
NewCaledonia	20	11	12	19	3	16	41	4	1	1 F
Singapore	98	90	76	76	71	70	79	78	46	38
Thailand	11 085	9 904	10 660	9 360	8 875	9 480	9 826	8 566	9 637	9 746
71 Fishing area total	*27 963*	*21 711*	*22 998*	*21 036*	*19 031*	*21 025*	*23 725*	*19 881*	*21 332*	*20 225 F*
81 Australia	493	0	0	0	-	-	-	-	-	-
Korea Rep	2 574	1 146	1 228	1 769	3 111	2 897	1 237	1 601	2 079	1 327
81 Fishing area total	*3 067*	*1 146*	*1 228*	*1 769*	*3 111*	*2 897*	*1 237*	*1 601*	*2 079*	*1 327*
Species total	*47 518*	*42 942*	*44 255*	*41 659*	*42 972*	*42 745*	*42 820 F*	*38 638 F*	*38 934 F*	*34 943 F*
Frigate tuna	**Auxide**		**Melva**			***Auxis thazard***			**1,75(01)023,01**	**FRI**
34 NethAntilles	-	-	-	590	1 157	1 030	1 159	1 122	989	710
Panama	118	341	327	240	91	-	-	-	-	-
Other nei	32	68	62	80	120	309	491	291	420	186
34 Fishing area total	*150*	*409*	*389*	*910*	*1 368*	*1 339*	*1 650*	*1 413*	*1 409*	*896*
Species total	*150*	*409*	*389*	*910*	*1 368*	*1 339*	*1 650*	*1 413*	*1 409*	*896*
Frigate and bullet tunas	**Auxide et bonitou**		**Melva y melvera**			***Auxis thazard, A.rochei***			**1,75(01)023,XX**	**FRZ**
21 USA	0	0	0	-	-	1	17	9	3	1
21 Fishing area total	*0*	*0*	*0*	*-*	*-*	*1*	*17*	*9*	*3*	*1*
27 Portugal	0	0	-	-	-	28	263	494	208	166
Spain	57	43	15	2	-	2	1	17	13	1
27 Fishing area total	*57*	*43*	*15*	*2*	*-*	*30*	*264*	*511*	*221*	*167*
31 Grenada	-	-	-	0	1	0	0	0	1	0
Spain	5	0	-	-	-	-	-	-	-	-
Trinidad Tob	17	...	56	199	368	127	138	245	...	...
Venezuela	881	2 597	2 161	2 758	2 525	1 926	1 524	1 410	1 342	1 068
31 Fishing area total	*903*	*2 597*	*2 217*	*2 957*	*2 894*	*2 053*	*1 662*	*1 655*	*1 343*	*1 068*
34 Cape Verde	115	86	13	6	22	191	154	81	171	206
France	5 237	105	126	5 430	4 605	4 648	5 772	6 859	127	91
Korea Rep	-	25	7	-	-	-	-	-	-	-
Morocco	274	122	645	543	2 614	2 137	494	582	418	441
Portugal	0	0	0	-	1	31	5	9	28	10
Russian Fed	220	505	459	46	500	2 433	460	408	1 028	760
Sao Tome Prn	-	-	-	79	323	...	...	...	...	...
Spain	300	254	371	945	581	568	22	-	709	437
Ukraine	-	-	-	-	-	-	36	48	...	43
34 Fishing area total	*6 146*	*1 097*	*1 621*	*7 049*	*8 646*	*10 008*	*6 943*	*7 987*	*2 481*	*1 988*
37 Algeria	348	306	230	237	179	299	173	225	230	481
Croatia	52	22	28	26	16	12	0	-	-	-
France	0	0	1	-	-	-	-	-	-	-
Greece	1 400	1 400	1 400	1 426	1 426	...	...	196	125	120
Italy	379	531	1 435	229	499	254	439	215	375	251
Malta	10	1	2	3	6	6	3	1	-	-
Morocco	170	1 726	621	1 673	562	1 140	682	763	256	621
Serbia-Monte	0	0	2	6	6	6	7	8	9	8
Spain	648	1 124	1 472	2 296	604	487	669	1 024	861	493
Tunisia	20	13	14	13	26	87	38	7	5	1
37 Fishing area total	*3 027*	*5 123*	*5 205*	*5 909*	*3 324*	*2 291*	*2 011*	*2 439*	*1 861*	*1 975*
41 Brazil	608	906	558	527	215	162	166	106	98	1 117
41 Fishing area total	*608*	*906*	*558*	*527*	*215*	*162*	*166*	*106*	*98*	*1 117*

B-36 Tunas, bonitos, billfishes / Thons, pélamides, marlins / Atunes, bonitos, agujas

Capture production by species, fishing areas and countries or areas
Captures par espèces, zones de pêche et pays ou zones
Capturas por especies, áreas de pesca y países o áreas

Species, Fishing area Espèce, Zone de pêche Especie, Area de pesca	1993 t	1994 t	1995 t	1996 t	1997 t	1998 t	1999 t	2000 t	2001 t	2002 t
47 Angola	4	6	21	29	12	31	2	38	206	114
Russian Fed	-	-	-	-	-	-	-	12	25	8
47 Fishing area total	*4*	*6*	*21*	*29*	*12*	*31*	*2*	*50*	*231*	*122*
51 France	-	-	-	-	-	-	-	-	15	45
India	3 968	12 440	5 906	8 906	8 462	6 152	12 722	15 148	12 933	4 864
Iran	436	200	4 438	755	544	509	590	785	562	611
Jordan	...	...	...	...	...	2	4	2	5	5
Maldives	5 455	4 018	3 938	6 484	2 489	4 218	3 401	3 991	3 981	4 187
Oman	354	391	786	613	846	611	583	488	638	170
Pakistan	2	36	36	49	54	56	59	42	52	31
Seychelles	-	-	-	-	-	-	-	42	10	7
Spain	-	-	-	1 227	208	-	-	315	201	406
Untd Arab Em	651	737	572	578	618	620	636	376	380 F	320
Yemen	25	25	20	20	20	20	20 F	20 F	20 F	20 F
Other nei	13	10	100	100	100	-	18	367	110	513
51 Fishing area total	*10 904*	*17 857*	*15 796*	*18 732*	*13 341*	*12 188*	*18 033 F*	*21 576 F*	*18 907 F*	*11 179 F*
57 India	8	23	11	2 213	2 102	1 528	4 461	3 762	3 212	1 221
Indonesia	6 756	8 187	8 041	8 525	8 739	8 405	9 124	9 644	7 934	7 934
Japan	-	-	-	-	-	1	-	-	-	-
Sri Lanka	4 011	3 226	4 006	5 334	6 521	8 133	3 515	4 583	3 157	15 197
Thailand	2 770	2 300	4 237	2 989	2 723	2 397	419	1 034	835	918
57 Fishing area total	*13 545*	*13 736*	*16 295*	*19 061*	*20 085*	*20 464*	*17 519*	*19 023*	*15 138*	*25 270*
61 China,Taiwan	2 261	2 098	2 881	2 482	4 280	4 634	4 558	3 073	2 089	2 298
Japan	26 233	24 078	27 376	20 339	32 495	21 597	29 514	27 139	37 234	37 234 F
61 Fishing area total	*28 494*	*26 176*	*30 257*	*22 821*	*36 775*	*26 231*	*34 072*	*30 212*	*39 323*	*39 532 F*
71 China,Taiwan	0	0	0	0	0	0	0	0	0	-
Japan	1 663	239	10	35	79	15	3	-	13	13 F
Philippines	110 266	109 866	88 426	88 969	108 494	106 433	111 301	112 227	111 719	163 132
Thailand	26 800	26 900	19 250	18 850	17 000	17 600	22 700	18 400	19 800	20 000
71 Fishing area total	*138 729*	*137 005*	*107 686*	*107 854*	*125 573*	*124 048*	*134 004*	*130 627*	*131 532*	*183 145 F*
77 Japan	125	-	-	-	-	-	-	-	-	-
77 Fishing area total	*125*	-	-	-	-	-	-	-	-	-
81 New Zealand	-	-	0	0	2	4	-	5	5	5
81 Fishing area total	-	-	*0*	*0*	*2*	*4*	-	*5*	*5*	*5*
87 Ecuador	...	9 051	7 396	2 537	6 857	4 201	48 913	9 648	5 738	9 806
87 Fishing area total	...	*9 051*	*7 396*	*2 537*	*6 857*	*4 201*	*48 913*	*9 648*	*5 738*	*9 806*
Species total	*202 542*	*213 597*	*187 067*	*187 478*	*217 724*	*201 712*	*263 606 F*	*223 848 F*	*216 881 F*	*275 375 F*
Little tunny(=Atl.black skipj)	**Thonine commune**		**Bacoreta**			***Euthynnus alletteratus***			**1,75(01)024,01**	**LTA**
21 USA	116	49	136	81	199	120	405	110	121	138
21 Fishing area total	*116*	*49*	*136*	*81*	*199*	*120*	*405*	*110*	*121*	*138*
27 Portugal	45	72	72	218	320	171	14	-	-	2
Spain	-	-	-	-	-	11	1	2	22	8
27 Fishing area total	*45*	*72*	*72*	*218*	*320*	*182*	*15*	*2*	*22*	*10*
31 Bermuda	5	6	6	7	6	5	4	2	4	5
Cuba	13	15	27	23	17	9	3	2	1	3
Puerto Rico	...	...	...	...	...	...	...	...	14	8
St Lucia	-	-	-	-	2	2	2	-	1	10
St Vincent	1	0	0	-	-	-	-	-	-	-
USA	241	110	14	9	252	180	109	110	236	279
Venezuela	1 889	2 115	1 627	1 840	2 064	2 815	2 389	2 040	1 948	2 023
31 Fishing area total	*2 149*	*2 246*	*1 674*	*1 879*	*2 341*	*3 011*	*2 507*	*2 154*	*2 204*	*2 328*
34 Benin	53	60	58	58 F	196	83	69	69 F	69	69
Cape Verde	17	23	72	63	86	110	776	491	178	108
Côte dIvoire	2 314	251	253	250	114	108	...	108	...	...
France	8	54	59	2 109	1 981	1 731	2 438	2 702	...	3
Gabon	-	-	-	182	-	18	159	301	213	57
Ghana	359	994	513	113	2 025	359	306	707	730	4 768
Morocco	44	43	230	588	195	189	67	101	87	308
Panama	65	-	-	-	-	-	-	-	-	-
Portugal	0	0	-	-	0	-	-	50	-	-
Russian Fed	265	189	96	49	-	88	-	-	-	74
Sao Tome Prn	-	-	-	40	159	...	...	...	...	...
Senegal	1 496	1 628	1 133	1 066	1 662	1 604	460	1 146	1 613	963
Spain	-	-	10	55	27	99	5	-	-	-
Other nei	8	20	-	-	-	-	-	-	-	33
34 Fishing area total	*4 629*	*3 262*	*2 424*	*4 573 F*	*6 445*	*4 389*	*4 280*	*5 675 F*	*2 890*	*6 383*
37 Algeria	495	459	552	554	448	384	562	494	407	148
Croatia	2	15	7	9	9	16	0	-	-	-
Cyprus	11	23	10	19	30	10	16	14	13	10
Greece	-	-	-	-	-	-	-	-	-	132
Israel	119	119	215	119	103	73	90	113	70	50
Libya	-	-	-	-	45	52	-	5	4	4
Malta	8	8	8	3	3	0	0	0	5	4
Morocco	0	0	1	0	1	14	8	-	-	3
Palest, O.T.	...	...	...	90	59	61	60 F	60 F	60 F	129

Species, Fishing area Espèce, Zone de pêche Especie, Area de pesca	1993 t	1994 t	1995 t	1996 t	1997 t	1998 t	1999 t	2000 t	2001 t	2002 t
Serbia-Monte	28	21	35	22	18	20	18	16	16	16
Spain	-	-	15	18	9	15	-	8	82	32
Syria	161	156	155	270	350	417	390	370	370	330
Tunisia	242	204	696	824	333	1 113	752	1 453	1 036	961
Other nei	200	200	200	200	200	200	200	-	-	-
37 Fishing area total	*1 266*	*1 205*	*1 894*	*2 128*	*1 608*	*2 375*	*2 096 F*	*2 533 F*	*2 063 F*	*1 819*
41 Brazil	985	1 225	1 059	834	507	920	930	615	615	615
41 Fishing area total	*985*	*1 225*	*1 059*	*834*	*507*	*920*	*930*	*615*	*615*	*615*
47 Angola	175	121	117	235	75	406	118	132	231	155
47 Fishing area total	*175*	*121*	*117*	*235*	*75*	*406*	*118*	*132*	*231*	*155*
Species total	*9 365*	*8 180*	*7 376*	*9 948 F*	*11 495*	*11 403*	*10 351 F*	*11 221 F*	*8 146 F*	*11 448*
Black skipjack	**Thonine noire**		**Barrilete negro**		***Euthynnus lineatus***				**1,75(01)024,04**	**BKJ**
77 Belize	-	-	-	40	-	-	-	-	-	-
Ecuador	0	50	-	270	-	-	-	-	2	85
Panama	-	-	1	-	-	9	-	1	-	7
USA	63	80	101	84	44	231	90	0	0	-
Venezuela	10	-	-	50	35	10	-	10	-	-
Other nei	-	-	-	-	-	-	-	-	10	152
77 Fishing area total	*73*	*130*	*102*	*444*	*79*	*250*	*90*	*11*	*12*	*244*
87 Belize	-	-	-	-	-	20	-	-	-	-
Colombia	-	-	3	70	...	...	...	79	...	...
Ecuador	80	90	160	100	50	260	10	270	1 831	1 037
Vanuatu	-	-	-	-	-	10	-	-	-	19
Venezuela	-	-	-	-	-	60	40	-	-	-
Other nei	-	-	-	-	-	-	-	-	22	4
87 Fishing area total	*80*	*90*	*163*	*170*	*50*	*350*	*50*	*349*	*1 853*	*1 060*
Species total	*153*	*220*	*265*	*614*	*129*	*600*	*140*	*360*	*1 865*	*1 304*
Kawakawa	**Thonine orientale**		**Bacoreta oriental**		***Euthynnus affinis***				**1,75(01)024,06**	**KAW**
51 Comoros	180	180	180	170	160	160	150	170 F	160	160 F
Egypt	128	43	138	318	755	841	326	344	209	313
Eritrea	...	...	...	4	6	...	...	0	36	0
India	17 989	14 683	17 559	11 837	18 763	12 376	16 757	18 917	16 646	1 410
Iran	518	2 100	3 911	5 665	8 451	7 947	10 858	13 500	12 474	16 361
Jordan	...	...	...	...	...	26	36	33	40	40
Maldives	3 569	2 656	2 694	3 789	2 089	3 624	1 692	1 898	2 150	2 242
Oman	701	1 113	2 064	2 335	2 388	1 731	1 522	1 550	1 961	1 493
Pakistan	1 933	1 716	1 449	2 351	2 571	2 684	2 715	2 340	1 817	1 210
Réunion	27	24	28	26	24	28	22	21	19	19 F
Saudi Arabia	-	-	121	162	304	332	256	264	241	353
Seychelles	163	170	125	93	97	41	158	81	52	69
South Africa	1	-	1	-	1	2	2	0	0	0
Untd Arab Em	1 400	2 725	2 394	2 418	2 500	2 512	2 605	868	999	850
UK	-	-	-	-	-	-	-	-	-	1
Yemen	504	1 164	1 226	1 183	1 240	1 210	1 210 F	1 210 F	1 210 F	1 210 F
Other nei	-	-	140	140	140	-	-	-	-	-
51 Fishing area total	*27 113*	*26 574*	*32 030*	*30 491*	*39 489*	*33 514*	*38 309 F*	*41 196 F*	*38 014 F*	*25 731 F*
57 Australia	0	0	0	0	1	-	-	-	0	-
India	1 208	1 022	1 222	2 941	4 662	3 075	4 884	4 700	4 136	6 729
Indonesia	296	359	352	373	383	368	400	423	348	348
Malaysia	3 524	1 964	2 494	4 134	4 297	6 283	5 616	7 127	6 084	10 234
Sri Lanka	412	1 546	2 086	2 262	2 765	3 449	2 167	2 167	3 434	4 011
Thailand	17 003	14 146	25 993	18 337	16 700	14 723	2 566	6 340	5 126	5 630
57 Fishing area total	*22 443*	*19 037*	*32 147*	*28 047*	*28 808*	*27 898*	*15 633*	*20 757*	*19 128*	*26 952*
61 China,Taiwan	1 000 F	800 F	500 F	0	0	0	0	0	0	0
61 Fishing area total	*1 000 F*	*800 F*	*500 F*	*0*	*0*	*0*	*0*	*0*	*0*	*0*
71 China,Taiwan	2 266 F	2 525 F	2 954 F	2 616	2 577	2 307	2 065	1 977	2 136	1 722
Malaysia	30 520	22 881	24 948	29 810	44 201	43 126	51 665	51 005	47 133	49 149
Palau	...	...	0	1	1	3	1	2	1	0
Papua N Guin	0	0	0	0	-	-	-	-	-	-
Philippines	21 714	29 669	27 308	24 345	26 573	24 424	25 406	27 963	27 280	34 681
Thailand	40 602	40 927	28 871	28 275	25 557	26 427	34 188	27 654	29 747	30 109
71 Fishing area total	*95 102 F*	*96 002 F*	*84 081 F*	*85 047*	*98 909*	*96 287*	*113 325*	*108 601*	*106 297*	*115 661*
77 USA	-	-	-	-	-	-	-	-	-	4
77 Fishing area total	*-*	*-*	*-*	*-*	*-*	*-*	*-*	*-*	*-*	*4*
Species total	*145 658 F*	*142 413 F*	*148 758 F*	*143 585*	*167 206*	*157 699*	*167 267 F*	*170 554 F*	*163 439 F*	*168 348 F*
Skipjack tuna	**Listao**		**Listado**		***Katsuwonus pelamis***				**1,75(01)025,01**	**SKJ**
21 Canada	-	-	-	-	0	-	-	-	-	-
USA	54	30	53	35	9	24	42	2	4	0
21 Fishing area total	*54*	*30*	*53*	*35*	*9*	*24*	*42*	*2*	*4*	*0*
27 Portugal	2 315	3 384	613	6 271	3 596	3 691	1 465	1 040	1 611	2 159
Spain	-	-	-	-	-	1	0	-	-	2
27 Fishing area total	*2 315*	*3 384*	*613*	*6 271*	*3 596*	*3 692*	*1 465*	*1 040*	*1 611*	*2 161*

B-36 Tunas, bonitos, billfishes — Thons, pélamides, marlins — Atunes, bonitos, agujas

Capture production by species, fishing areas and countries or areas
Captures par espèces, zones de pêche et pays ou zones
Capturas por especies, áreas de pesca y países o áreas

Species, Fishing area Espèce, Zone de pêche Especie, Area de pesca	1993 t	1994 t	1995 t	1996 t	1997 t	1998 t	1999 t	2000 t	2001 t	2002 t
31 Barbados	6	6	6	5	5	10	3	3	1	2
Bermuda	0	0	0	0	0	0	0	0	1	1
China,Taiwan	-	-	-	-	-	0	0	-	-	-
Colombia	2 074	789	1 583	...	...	...	...	...	...	...
Cuba	752	1 151	881	1 000	1 282	1 303	750	665	443	462
Dominica	24	43	33	33	33	33	85	86	45	55
Dominican Rp	143	257	146	123	...	...	...	23	32	...
Grenada	14	11	12	11	15	23	23	23	15	14
Mexico	1	0	0	2	4	6	51	25	91	108
NethAntilles	45	40	35	30	30 F	30 F	30 F	30 F	...	...
Puerto Rico	...	...	...	...	...	...	...	...	26	20
St Lucia	53	86	72	38	100	263	153	216	151	106
St Vincent	66	56	53	37	42	57	37	68	97	358
Spain	397	-	-	-	-	-	1	1	-	-
Trinidad Tob	-	-	3	...	0	-	-	-	-	-
USA	0	0	0	-	0	0	-	-	0	-
Venezuela	11 172	6 697	2 387	3 574	3 834	4 114	2 981	3 003	6 870	2 554
31 Fishing area total	*14 747*	*9 136*	*5 211*	*4 853*	*5 345 F*	*5 839 F*	*4 114 F*	*4 143 F*	*7 772*	*3 680*
34 Benin	2 F	2 F	2 F	2 F	7	3	2	2 F	...	...
Cape Verde	860	1 007	1 314	470	591	684	962	789	794	284
China	-	-	-	-	-	4	-	-	-	-
China,Taiwan	-	-	1	5	24	42	3	28	2	2
Congo Rep	10	7	7	6	6	6	6	6	...	...
Cuba	272	117	5	-	-	-	-	-	-	-
France	33 735	32 779	25 188	23 107	17 023	18 382	20 344	18 183	16 593	16 615
Gabon	1	11	51	26	...	59	76	21	101	...
Germany	-	-	-	3	-	-	-	-	-	-
Ghana	20 225	21 258	18 607	19 602	27 667	34 150	43 460	29 950	43 340	31 887
Korea Rep	-	-	-	-	-	-	7	7	-	-
Morocco	310	248	4 981	675	4 509	2 481	848	1 198	268	280
NethAntilles	-	-	-	7 096	8 444	8 553	9 932	10 008	13 370	5 427
Panama	12 770	12 761	14 561	5 755	1 294	572	1 308	1 560	281	342
Portugal	3 336	4 136	4 357	2 000	797	849	345	266	496	785
Russian Fed	540	1 471	1 466	381	1 146	2 086	1 426	374	-	-
Sao Tome Prn	15 F	...	...	...	7	-	-	-	-	-
Senegal	53	193	293	265	430	1 836	1 422	1 009	3 768	3 606
Spain	63 660	50 538	51 594	38 538	38 513	36 007	44 519	37 226	30 954	25 454
Other nei	9 482	6 521	6 146	10 220	4 901	6 749	7 701	7 128	8 121	8 544
34 Fishing area total	*145 271 F*	*131 049 F*	*128 573 F*	*108 151 F*	*105 359*	*112 463*	*132 361*	*107 755 F*	*118 088*	*93 226*
37 Algeria	-	-	-	-	-	171	43	89	77	...
France	-	-	-	-	-	-	-	-	-	22
Morocco	2	0	43	9	4	5	10	1	-	1
Spain	-	-	-	-	-	-	-	-	-	10
Tunisia	-	-	-	-	-	-	-	-	-	932
37 Fishing area total	*2*	*0*	*43*	*9*	*4*	*176*	*53*	*90*	*77*	*965*
41 Argentina	50	1	0	1	-	-	-	-	-	-
Brazil	17 771	20 588	16 560	22 528	26 564	23 789	23 188	25 164	24 146	18 338
China,Taiwan	9	6	0	2	12	1	0	16	9	6
Panama	236	206	146	70	3	-	-	-	-	-
41 Fishing area total	*18 066*	*20 801*	*16 706*	*22 601*	*26 579*	*23 790*	*23 188*	*25 180*	*24 155*	*18 344*
47 Angola	13	7	3	15	52	2	32	14	687	...
China,Taiwan	2	1	4	8	38	32	1	62	49	38
Japan	-	-	-	-	-	-	-	-	1	...
Namibia	-	5	27	1	1	...	...	...	8	0
Panama	20	11	146	30	3	-	-	-	-	-
Portugal	0	8	26	26	6	4	...	1	61	15
St Helena	65	55	115	86	294	298	13	64	205	63
South Africa	6	4	4	1	6	2	1	-	1	...
47 Fishing area total	*106*	*91*	*325*	*167*	*400*	*338*	*47*	*141*	*1 012*	*116*
51 China,Taiwan	84	10	61	48	48	5	4	17	64	48
Comoros	1 847	2 185	2 185	2 150	2 070	2 070	2 000	2 200 F	2 000	2 000 F
France	48 192	58 430	48 652	40 003	30 425	25 423	41 996	39 862	32 023	53 971
India	4 940	8 959	6 386	6 782	6 035	831	5 707	4 797	5 716	20 110
Iran	4 353	7 400	1 133	3 239	5 970	6 514	16 583	20 091	26 058	29 859
Italy	-	-	-	-	1 733	2 316	3 416	...	1 681	2 660
Japan	29 413	10 067	556	398	2 264	62	523	448	245	28
Jordan	...	...	...	...	...	35	44	45	52	52
Kenya	72	150	116	108	114	98	109	86	183	116
Korea Rep	-	-	-	7	-	-	-	-	-	-
Maldives	58 741	69 410	70 372	66 502	69 015	78 410	92 888	79 683	88 043	115 321
Mauritius	6 902	5 166	3 848	1 898	3 055	1 685	2 361	305	8	8
Mayotte	...	...	...	...	...	...	...	...	3 834	188
Oman	-	372	775	408	730	227	320	293	1	2
Pakistan	8 950	8 134	7 089	4 140	4 480	4 372	4 505	4 308	3 405	3 102
Réunion	71	64	105	91	77	92	89	84	76	76 F
Seychelles	-	-	-	-	4 695	9 108	15 846	11 567	26 219	29 891
South Africa	3	1	1	-	1	1	2	1	0	0
Spain	51 372	61 626	69 587	66 115	62 728	53 966	74 025	77 099	68 414	91 462
UK	-	-	-	-	-	-	-	-	-	1
Yemen	14	14	15	88	90	90	90 F	90 F	90 F	90 F
Other nei	25 381	32 697	42 860	34 727	35 810	41 227	52 570	61 116	41 664	47 086
51 Fishing area total	*240 335*	*264 685*	*253 741*	*226 704*	*229 340*	*226 532*	*313 078 F*	*302 092 F*	*299 776 F*	*396 071 F*

B-36 Tunas, bonitos, billfishes / Thons, pélamides, marlins / Atunes, bonitos, agujas

Capture production by species, fishing areas and countries or areas / Captures par espèces, zones de pêche et pays ou zones / Capturas por especies, áreas de pesca y países o áreas

Species, Fishing area Espèce, Zone de pêche Especie, Area de pesca	1993 t	1994 t	1995 t	1996 t	1997 t	1998 t	1999 t	2000 t	2001 t	2002 t
57 Australia	29	1 204	466	212	905	2 247	4 986	3 050	2 080	1 145
China,Taiwan	136	4	45	11	11	2	0	12	13	7
France	-	-	-	54	851	4 916	669	73	-	-
India	48	295	191	68	61	206	-	1 189	1 417	...
Indonesia	30 624	31 978	30 732	42 011	49 682	45 768	47 144	46 821	56 302	56 302
Iran	-	-	-	2	3	157	-	-	-	-
Italy	-	-	-	-	21	708	-	...	-	-
Japan	28	-	15 431	6 632	4 452	4 739	3 973	1 890	1 465	1 134
Korea Rep	-	-	-	1	-	-	-	-	-	-
Seychelles	-	-	-	-	245	1 597	-	-	-	-
Spain	-	-	-	161	185	4 680	260	88	...	...
Sri Lanka	24 832	21 548	18 288	22 754	27 815	34 691	51 940	51 940	47 241	42 968
Thailand	-	-	-	-	-	-	-	1 110	460	...
Other nei	-	-	-	103	65	1 848	-	46	-	-
57 Fishing area total	*55 697*	*55 029*	*65 153*	*72 009*	*84 296*	*101 559*	*108 972*	*106 219*	*108 978*	*101 556*
61 China	...	...	...	...	...	...	...	...	...	1 582
China,Taiwan	3 205	1 599	3 502	2 413	2 670	616	1 864	2 893	16 159	3 042
Japan	181 405	123 799	120 380	103 190	185 322	205 303	136 316	174 287	140 171	140 171 F
Russian Fed	-	-	-	-	-	-	-	-	33	-
61 Fishing area total	*184 610*	*125 398*	*123 882*	*105 603*	*187 992*	*205 919*	*138 180*	*177 180*	*156 363*	*144 795 F*
67 USA	-	-	-	-	-	-	-	0	1	-
67 Fishing area total	-	-	-	-	-	-	-	*0*	*1*	-
71 Australia	-	-	-	-	4	2	1	1	1	1
China	-	-	-	-	-	-	-	1 050	2 750	7 713
China,Taiwan	113 190	136 358	155 167	170 949	116 862	196 181	161 321	194 499	182 531	229 415
Fiji Islands	3 178	3 379	4 319	3 124	987	459	507	343	431	420 F
Guam	68	94	23	17	24	28	19	61	60	47
Indonesia	117 003	125 998	129 394	140 743	135 073	181 884	198 222	200 685	172 251	182 120
Japan	159 572	164 970	169 215	176 008	113 275	174 788	145 073	164 031	133 980	133 980 F
Kiribati	184	1 016	2 520	4 111	2 855	5 544	4 493	3 701	3 286	3 793
Korea Rep	73 989	145 541	137 848	128 434	114 042	140 685	106 068	132 966	134 766	170 988
Marshall Is	-	-	-	-	-	-	-	6 625	31 983	37 057
Micronesia	11 692	17 531	4 216	6 745	5 501	11 314	6 972	15 843	11 267	13 957
Nauru	...	...	...	...	...	...	...	...	...	2
NewCaledonia	-	1	2	0	1	1	0	0	0	0
N Marianas	35	39	60	75	64	61	48	56	61	73
Papua N Guin	-	987	9 811	9 512	11 270	37 214	29 949	52 289	64 355	89 948
Philippines	68 065	84 560	110 111	110 004	110 097	116 673	108 778	113 011	105 484	109 977
Singapore	0	6	5	5	47	12	23	2	10	6
Solomon Is	20 080	26 661	40 136	26 485	36 311	38 662	35 613	8 368	12 530	13 897
Tuvalu	292	310	259	260 F	300 F	300 F	300 F	300 F	300 F	300 F
USA	142 457	134 394	125 504	112 467	92 503	105 200	144 148	92 780	86 441	86 743
Vanuatu	-	730	5 577	8 080	16 730	28 982	36 321	33 075	...	...
71 Fishing area total	*709 805*	*842 575*	*894 167*	*897 019 F*	*755 946 F*	*1 037 990 F*	*977 856 F*	*1 019 686 F*	*942 487 F*	*1 080 437 F*
77 Amer Samoa	11	68	80	32	16	8	20	14	56	171
Belize	-	20	70	720	500	290	880	189	-	-
China	-	-	-	-	-	-	-	-	1 130	-
China,Taiwan	5	81	45	17	40	164	710	9	5	8
Colombia	...	...	...	...	...	...	...	234	73	230
Costa Rica	...	...	...	80	...	...	96	...	...	...
Cyprus	10	-	710	700	840	-	-	-	-	-
Ecuador	650	570	-	-	-	-	-	-	-	-
El Salvador	...	...	...	...	...	750	320	...	4 476	6 804
Fr Polynesia	1 125	1 004	1 400	1 400	1 126	1 560	1 386	1 189	1 557	1 470
Guatemala	-	-	-	-	-	-	3 360	4 941	6 500 F	10 138
Honduras	-	-	-	540	1 120	630	590	50	...	...
Japan	4 200	2 462	3 368	266	4 032	109	179	33	299	299 F
Korea Rep	4	-	-	1 446	1 885	2 705	-	2	138	2 225
Mexico	13 291	7 045	30 694	16 728	24 286	16 450	18 356	14 385	7 520	9 534
Nicaragua	-	-	-	-	-	-	250	400	...	...
Panama	210	230	410	1 230	820	930	2 100	2 356	1 062	4 118
Spain	90	50	520	720	1 170	14 060	23 420	13 837	9 821	12 659
Tonga	4	3	3	2	4	7	3	2	12	23
USA	15 728	18 666	30 732	13 224	17 958	17 679	7 150	2 092	2 583	1 939
Vanuatu	230	280	3 340	3 900	6 000	5 300	3 260	1 128	1 186	528
Venezuela	750	510	3 530	2 800	5 990	2 940	2 970	2 410	930	2 360
Other nei	-	-	40	-	-	10	-	68	2 398	7 453
77 Fishing area total	*36 308*	*30 989*	*74 942*	*43 805*	*65 787*	*63 592*	*65 050*	*43 339*	*39 746 F*	*59 959 F*
81 Australia	4 313	3 514	4 297	2 767	4 689	849	429	1 863	1 091	226
China,Taiwan	285	98	12	36	84	270	-	-	7	24
Japan	15	3	12	151	2 189	431	1 227	723	512	512 F
New Zealand	963	3 212	1 428	3 631	4 792	8 156	5 688	9 699	3 691	3 344
81 Fishing area total	*5 576*	*6 827*	*5 749*	*6 585*	*11 754*	*9 706*	*7 344*	*12 285*	*5 301*	*4 106 F*
87 Belize	10	2 020	5 090	4 420	5 760	4 150	3 450	5 093	-	-
Chile	-	-	-	-	53	47	9	0	57	1
Colombia	3 294	7 473	13 189	13 420	12 175	4 653	11 152	5 661	1 800	2 054
Costa Rica	...	...	120	540	...	...	...	...	...	...
Cyprus	3 290	3 850	3 490	2 510	4 360	300	-	-	-	-
Ecuador	23 210	15 431	31 599	37 468	67 400	67 453	126 992	105 146	68 217	77 791
El Salvador	...	...	...	...	...	...	2 810	-	...	-
Guatemala	-	-	-	-	-	-	3 390	7 373	...	...
Honduras	-	-	-	240	1 990	-	3 690	1 960	...	...
Japan	26	7	4	26	8	30	26	2	7	7 F

B-36

Tunas, bonitos, billfishes — **Capture production by species, fishing areas and countries or areas**
Thons, pélamides, marlins — **Captures par espèces, zones de pêche et pays ou zones**
Atunes, bonitos, agujas — **Capturas por especies, áreas de pesca y países o áreas**

Species, Fishing area Espèce, Zone de pêche Especie, Area de pesca	1993 t	1994 t	1995 t	1996 t	1997 t	1998 t	1999 t	2000 t	2001 t	2002 t
Korea Rep	-	-	-	-	-	-	3 705	4 040	2 665	480
Liberia	-	-	-	410	-	-	-	-	-	-
Mexico	152	505	1 066	1 600	881	1 241	753	440	610	517
Nicaragua	-	-	-	-	-	-	-	30	...	...
Panama	670	1 690	3 540	2 210	3 170	30	2 742	9 076	4 454	3 149
Peru	500	109	151	85	823	9 373	802	711	81	1 339
St Vincent	-	1 200	-	-	-	-	-	-	-	-
Spain	4 730	1 780	4 940	3 070	8 480	6 070	17 460	8 960	12 950	10 849
USA	-	-	-	2 822	1 684	764	-	2 555	9	1 276
Vanuatu	10 160	9 190	11 130	7 620	8 130	5 830	17 570	8 829	5 715	5 702
Venezuela	5 410	4 910	1 080	620	1 290	3 170	11 080	2 720	1 250	1 355
Other nei	850	640	730	-	2 320	650	-	958	11 314	20 712
87 Fishing area total	*52 302*	*48 805*	*76 129*	*77 061*	*118 524*	*103 761*	*205 631*	*163 554*	*109 129*	*125 232 F*
Species total	*1 465 194 F*	*1 538 799 F*	*1 645 287 F*	*1 570 873 F*	*1 594 931 F*	*1 895 381 F*	*1 977 381 F*	*1 962 706 F*	*1 814 500 F*	*2 030 648 F*
Atlantic bluefin tuna	**Thon rouge de l'Atlantique**		**Atún rojo del Atlántico**		***Thunnus thynnus***			**1,75(01)026,01**		**BFT**
21 Canada	330	412	578	599	507	595	576	550	524	604
Japan	1 170	752	657	913	594	790	1 121	939	1 056	575
St Pier Mq	-	-	-	-	-	-	1	0	0	0
USA	981	1 019	877	720	1 006	1 041	1 032	1 033	1 175	1 141
21 Fishing area total	*2 481*	*2 183*	*2 112*	*2 232*	*2 107*	*2 426*	*2 730*	*2 522*	*2 755*	*2 320*
27 China,Taiwan	-	-	6	9	9	-	146	99	100	98
Denmark	37	-	0	0	-	1	-	-	-	-
Faeroe Is	-	-	-	-	-	67	104	118	...	...
France	1 099	336	725	563	269	613	588	542	629	755
Germany	-	-	1	-	-	-	-	-	-	-
Iceland	-	-	-	-	1	2	33	29	-	1
Ireland	-	-	-	-	14	21	52	22	8	15
Japan	1 084	1 212	3 190	2 464	2 026	2 148	1 988	1 842	1 361	2 200
Libya	-	-	-	576	477	511	450	487	...	...
Norway	0	-	-	-	-	-	5	0	-	-
Portugal	23	20	23	139	372	158	408	440	402	55
Spain	5 047	3 081	3 815	6 017	5 842	3 761	3 328	3 448	3 578	4 085
Sweden	0	0	0	-	-	-	-	-	-	-
UK	-	-	1	0	1	1	12	0	-	-
Other nei	223	68	189	71	208	-	-	-	-	-
27 Fishing area total	*7 513*	*4 717*	*7 950*	*9 839*	*9 219*	*7 283*	*7 114*	*7 027*	*6 078*	*7 209*
31 Bermuda	-	-	-	1	2	2	1	1	1	1
China,Taiwan	-	-	0	0	-	456	0	-	-	-
Japan	2	-	-	34	-	-	-	1	-	-
Mexico	-	-	5	14	7	14	16	35	10	15
St Lucia	2	43	9	3	-	-	-	-	-	-
USA	64	45	29	21	18	18	10	88	57	71
Other nei	-	-	-	2	-	-	429	270	49	-
31 Fishing area total	*68*	*88*	*43*	*75*	*27*	*490*	*456*	*395*	*117*	*87*
34 China	-	-	-	-	-	85	103	80	68	39
China,Taiwan	6	11	2	91	97	-	46	277	101	123
Greece	4	-	3	4	1	1	-	-	-	-
Guinea	-	330	...	...	...	...	...	...	...	...
Japan	264	486	590	486	709	911	342	216	211	336
Korea Rep	438	642	85	76	83	-	-	-	-	-
Morocco	415	720	678	1 035	2 068	1 866	1 591	2 228	2 497	2 565
Panama	-	1	19	550	255	-	13	-	-	-
Portugal	2	220	12	60	340	165	3	1	2	124
Sierra Leone	...	...	...	...	...	...	...	93	118	...
Spain	31	56	4	157	-	39	32	26	55	5
Other nei	-	-	-	-	-	66	-	-	-	-
34 Fishing area total	*1 160*	*2 466*	*1 393*	*2 459*	*3 553*	*3 133*	*2 130*	*2 921*	*3 052*	*3 192*
37 Algeria	1 097	1 560	156	156	157	1 947	2 142	2 330	2 012	1 710
China	-	97	137	93	49	-	-	-	-	-
China,Taiwan	328	713	493	372	398	-	58	31	196	131
Croatia	1 058	1 410	1 220	1 360	1 105	906	970	930	903	977
Cyprus	14	10	10	10	10	21	31	61	90	91
France	6 995	11 843	9 604	9 127	8 201	7 100	6 153	6 780	6 119	5 810
Greece	439	886	1 004	874	1 217	286	248	622	361	438
Israel	-	-	-	14	-	-	-	-	-	-
Italy	5 328	6 882	7 062	10 006	9 548	4 059	3 279	3 845	4 377	4 628
Japan	607	521	734	664	167	410	371	142	187	390
Korea Rep	-	-	458	591	410	-	-	-	-	-
Libya	635	1 422	1 540	812	552	820	745	1 063	1 940	...
Malta	251	572	587	399	393	407	447	376	219	176
Morocco	79	1 092	1 035	586	535	564	636	695	511	421
Panama	467	1 499	1 498	2 850	236	-	-	-	-	-
Portugal	183	428	446	274	37	54	76	61	64	-
Serbia-Monte	0	0	2	4	4	6	7	4	5	6
Spain	2 018	2 741	4 607	2 588	2 205	2 000	2 003	2 772	2 234	2 215
Tunisia	2 132	2 503	1 897	2 393	2 200	1 745	2 352	2 184	2 493	2 528
Turkey	3 084	3 466	4 220	4 616	5 093	5 899	1 200	1 070	2 100	2 300
Other nei	-	1 200	850	171	1 167	1 855	2 135	126	571	508
37 Fishing area total	*24 715*	*38 845*	*37 560*	*37 960*	*33 684*	*28 079*	*22 853*	*23 092*	*24 382*	*22 329*
41 Brazil	0	0	0	0	0	0	13	0	-	-
Uruguay	1	0	2	-	-	-	-	-	1	1

B-36 Tunas, bonitos, billfishes — Capture production by species, fishing areas and countries or areas
Thons, pélamides, marlins — Captures par espèces, zones de pêche et pays ou zones
Atunes, bonitos, agujas — Capturas por especies, áreas de pesca y países o áreas

Species, Fishing area Espèce, Zone de pêche Especie, Area de pesca	1993 t	1994 t	1995 t	1996 t	1997 t	1998 t	1999 t	2000 t	2001 t	2002 t
41 Fishing area total	*1*	*0*	*2*	*0*	*0*	*0*	*13*	*0*	*1*	*1*
47 Japan	14	-	-	-	-	3	-	-	-	-
Korea Rep	-	42	120	16	120	-	-	-	-	-
Seychelles	-	-	-	-	-	-	-	-	-	2
47 Fishing area total	*14*	*42*	*120*	*16*	*120*	*3*	*-*	*-*	*-*	*2*
Species total	*35 952*	*48 341*	*49 180*	*52 581*	*48 710*	*41 414*	*35 296*	*35 957*	*36 385*	*35 140*
Pacific bluefin tuna	**Thon bleu du Pacifique**		**Atún aleta azul del Pacífico**		***Thunnus orientalis***				**1,75(01)026,02**	**PBF**
61 China,Taiwan	1 074 F	539 F	31	157	505	695	2	1 014	580	487
Japan	4 390	7 624	5 844	6 651	6 469	3 633	11 145	10 044	5 676	5 676 F
61 Fishing area total	*5 464 F*	*8 163 F*	*5 875*	*6 808*	*6 974*	*4 328*	*11 147*	*11 058*	*6 256*	*6 163 F*
67 Japan	-	1	-	-	-	-	-	-	-	-
USA	4	12	1	2	3	8	7	3	0	0
67 Fishing area total	*4*	*13*	*1*	*2*	*3*	*8*	*7*	*3*	*0*	*0*
71 China,Taiwan	30 F	20 F	283	799	1 308	1 215	3 087	1 676	1 504	1 354
Japan	494	4	344	2	6	18	17	79	150	150 F
Korea Rep	-	23	-	-	-	-	-	-	-	-
71 Fishing area total	*524 F*	*47 F*	*627*	*801*	*1 314*	*1 233*	*3 104*	*1 755*	*1 654*	*1 504 F*
77 China,Taiwan	0	0	0	0	0	-	-	-	-	-
Japan	18	14	18	12	2	9	11	2	2	2 F
Korea Rep	43	-	-	-	-	-	-	-	-	-
Mexico	20	60	83	3 700	370	34	2 370	3 030	860	1 326
USA	552	919	642	4 767	2 269	1 954	170	313	317	41
77 Fishing area total	*633*	*993*	*743*	*8 479*	*2 641*	*1 997*	*2 551*	*3 345*	*1 179*	*1 369 F*
81 Japan	4	15	3	3	8	7	14	3	4	4 F
New Zealand	-	-	2	5	12	20	21	21	50	55
81 Fishing area total	*4*	*15*	*5*	*8*	*20*	*27*	*35*	*24*	*54*	*59 F*
87 Japan	-	-	-	-	2	-	1	-	-	-
87 Fishing area total	*-*	*-*	*-*	*-*	*2*	*-*	*1*	*-*	*-*	*-*
Species total	*6 629 F*	*9 231 F*	*7 251*	*16 098*	*10 954*	*7 593*	*16 845*	*16 185*	*9 143*	*9 095 F*
Longtail tuna	**Thon mignon**		**Atún tongol**		***Thunnus tonggol***				**1,75(01)026,03**	**LOT**
51 China,Taiwan	0	0	4 544	3 437	2 611	-	-	-	-	-
Eritrea	...	...	...	...	6	22	-	-	-	-
India	4 324	4 917	6 985	3 415	4 263	3 805	2 275	2 181	704	180
Iran	8 150	12 100	27 188	17 147	20 112	19 694	23 465	41 407	34 896	29 853
Jordan	...	...	...	...	...	2	2	2	3	3
Oman	6 705	5 088	3 967	5 316	5 020	4 379	4 798	5 318	6 011	6 854
Pakistan	2 510	5 807	5 006	4 121	5 360	5 220	5 600	5 315	4 735	4 466
Saudi Arabia	-	-	234	115	101	181	136	143	131	174
Seychelles	-	-	-	-	-	-	-	1 126	-	-
Untd Arab Em	4 000	4 466	5 715	5 775	3 671	3 678	3 739	1 725	1 720 F	2 238
Yemen	1 707	2 291	2 204	1 887	1 970	1 920	1 920 F	1 920 F	1 920 F	1 920 F
Other nei	-	-	351	351	351	-	-	85	-	-
51 Fishing area total	*27 396*	*34 669*	*56 194*	*41 564*	*43 465*	*38 901*	*41 935 F*	*59 222 F*	*50 120 F*	*45 688 F*
57 Australia	0	2	1	0	0	0	31	8	58	22
China,Taiwan	0	0	1 600	1 298	986	-	-	-	-	-
India	0	36	51	848	1 059	945	293	541	175	...
Malaysia	1 930	972	1 233	2 045	2 125	3 106	2 824	3 586	2 773	4 753
Sri Lanka	12	57	0	0	-	-	-	-	-	-
Thailand	20 950	32 762	18 709	22 036	20 035	17 599	5 799	4 838	5 592	5 977
57 Fishing area total	*22 892*	*33 829*	*21 594*	*26 227*	*24 205*	*21 650*	*8 947*	*8 973*	*8 598*	*10 752*
61 China,Taiwan	19 000	15 150	595	514	391	303	6 434	10 087	14 321	859
61 Fishing area total	*19 000*	*15 150*	*595*	*514*	*391*	*303*	*6 434*	*10 087*	*14 321*	*859*
71 Australia	7	7	16	10	-	-	-	3	7	6
China,Taiwan	2 249	1 844	4 010	3 304	2 510	6 163	2 772	4 342	4 300	3 452
Papua N Guin	0	0	0	0	-	-	-	-	-	-
Thailand	39 396	32 006	38 824	32 347	29 127	34 805	45 818	53 407	40 958	41 422
71 Fishing area total	*41 652*	*33 857*	*42 850*	*35 661*	*31 637*	*40 968*	*48 590*	*57 752*	*45 265*	*44 880*
77 China,Taiwan	-	-	388	327	249	-	-	-	-	-
77 Fishing area total	*-*	*-*	*388*	*327*	*249*	*-*	*-*	*-*	*-*	*-*
81 Australia	2	3	3	3	-	-	3	2	6	5
China,Taiwan	0	0	29	20	15	-	-	-	-	-
81 Fishing area total	*2*	*3*	*32*	*23*	*15*	*-*	*3*	*2*	*6*	*5*
Species total	*110 942*	*117 508*	*121 653*	*104 316*	*99 962*	*101 822*	*105 909 F*	*136 036 F*	*118 310 F*	*102 184 F*
Blackfin tuna	**Thon à nageoires noires**		**Atún aleta negra**		***Thunnus atlanticus***				**1,75(01)026,04**	**BLF**
21 USA	1	0	2	28	5	1	20	16	2	1
21 Fishing area total	*1*	*0*	*2*	*28*	*5*	*1*	*20*	*16*	*2*	*1*
31 Bermuda	5	7	4	5	4	6	6	5	4	5
Cuba	54	223	156	287	1	226	309	237	299	209

B-36 Tunas, bonitos, billfishes / Thons, pélamides, marlins / Atunes, bonitos, agujas

Capture production by species, fishing areas and countries or areas / Captures par espèces, zones de pêche et pays ou zones / Capturas por especies, áreas de pesca y países o áreas

Species, Fishing area Espèce, Zone de pêche Especie, Area de pesca	1993 t	1994 t	1995 t	1996 t	1997 t	1998 t	1999 t	2000 t	2001 t	2002 t
Dominica	15	19	30	...	...	...	79	83	54	78
Dominican Rp	133	239	892	518	...	...	...	...	...	...
Grenada	144	189	123	164	126	233	94	164	223	255
Guadeloupe	440	440	480	500	500	500	500	500	500	500
Martinique	700 F	890 F	890 F	540 F	540 F	540 F	540 F	540 F	470 F	470 F
NethAntilles	65	60	50	45	45 F	45 F	45 F	45 F	45 F	45 F
Puerto Rico	...	...	...	...	...	...	...	...	17	14
St Lucia	16	82	47	35	40	100	41	45	108	96
St Vincent	53	19	20	18	22	17	15	23	24	24
Spain	46	-	-	-	-	-	-	-	-	-
USA	126	107	63	25	62	52	21	34	33	31
Venezuela	1 224	21	624	758	498	1 034	1 192	589	1 902	1 210
31 Fishing area total	*3 021 F*	*2 296 F*	*3 379 F*	*2 895 F*	*1 838 F*	*2 753 F*	*2 842 F*	*2 265 F*	*3 679 F*	*2 937 F*
41 Brazil	22	37	153	649	418	55	55	38	149	1 669
41 Fishing area total	*22*	*37*	*153*	*649*	*418*	*55*	*55*	*38*	*149*	*1 669*
Species total	*3 044 F*	*2 333 F*	*3 534 F*	*3 572 F*	*2 261 F*	*2 809 F*	*2 917 F*	*2 319 F*	*3 830 F*	*4 607 F*
Albacore	**Germon**		**Atún blanco**			***Thunnus alalunga***			**1,75(01)026,05**	**ALB**
21 Canada	9	32	11	24	31	24	39	122	51	113
China,Taiwan	25	358	181	70	17	8	102	290	465	451
Japan	337	365	113	185	186	148	175	222	492	470
USA	242	284	318	88	148	187	174	108	121	91
21 Fishing area total	*613*	*1 039*	*623*	*367*	*382*	*367*	*490*	*742*	*1 129*	*1 125*
27 China,Taiwan	477	1 515	1 030	3	3	0	384	941	423	395
France	6 293	5 934	5 304	4 694	4 618	3 711	7 189	6 019	6 344	4 289
Ireland	1 946	2 534	918	874	1 913	3 750	4 858	3 464	2 093	1 087
Japan	60	70	197	104	46	89	71	146	117	112
Panama	210	363	289	369	58	58	-	-	-	-
Portugal	3 161	918	6 267	834	182	49	246	264	411	30
Spain	17 775	16 837	19 538	15 580	16 245	12 964	13 355	15 726	7 668	7 838
UK	499	613	196	49	33	117	343	15	2	-
27 Fishing area total	*30 421*	*28 784*	*33 739*	*22 507*	*23 098*	*20 738*	*26 446*	*26 575*	*17 058*	*13 751*
31 Barbados	-	-	-	-	1	1	1	...	2	5
Bermuda	-	-	-	-	1	...	2	2	2	1
China,Taiwan	2 069	2 660	3 057	3 349	2 856	1 241	5 191	5 196	9 240	9 455
Cuba	-	14	54	40	5	-	-	-	-	-
Dominican Rp	...	...	...	...	323	121	73	95	...	295
Grenada	0	0	2	1	6	7	6	12	21	23
Japan	8	14	9	35	23	33	140	125	281	...
St Lucia	1	0	1	1	0	0	0	1	3	2
St Vincent	2	0	0	-	-	-	1	2 820	5 662	318
Spain	2	1	2	0	4	8	37	-	-	-
Trinidad Tob	...	...	...	...	2	1	1	2	11	9
USA	9	21	10	5	15	3	3	5	8	12
Venezuela	246	282	279	315	49	107	91	1 374	349	162
31 Fishing area total	*2 337*	*2 992*	*3 414*	*3 746*	*3 285*	*1 522*	*5 546*	*9 632*	*15 579*	*10 282*
34 Belize	-	-	2	-	-	-	8	2	...	...
China	-	14	8	20	-	-	21	16	57	196
China,Taiwan	2 028	1 891	1 741	576	492	346	76	657	849	860
Cuba	3	-	0	-	-	-	-	-	-	-
France	564	129	82	190	38	40	13	23	16	18
Ireland	-	-	-	-	-	-	-	-	-	237
Japan	93	58	73	149	193	131	51	194	275	...
Korea Rep	-	-	-	-	5	-	-	-	-	-
Liberia	...	...	...	41	-	-	-	-	-	-
Morocco	-	-	-	-	-	-	-	-	-	55
NethAntilles	-	-	-	-	9	192	-	2	...	...
Panama	168	213	12	22	-	3	14	-	-	-
Philippines	-	-	-	-	-	2	4	0	-	-
Portugal	224	56	203	800	213	42	78	14	765	1 924
St Vincent	-	-	-	-	-	-	-	-	-	27
Sierra Leone	...	...	...	...	...	...	...	...	91	...
Spain	1 252	842	918	804	1 139	494	2 128	240	1 509	1 121
Other nei	159	133	110	180	50	50	50	-	-	-
34 Fishing area total	*4 491*	*3 336*	*3 149*	*2 782*	*2 139*	*1 300*	*2 443*	*1 148*	*3 562*	*4 438*
37 Cyprus	-	-	-	-	-	-	-	6	-	12
France	64	23	3	0	5	5	-	-	-	1
Greece	1	1	0	952	741	1 152	2 005	1 786	1 840	1 352
Italy	1 275	1 107	1 109	1 769	1 414	1 414	2 561	3 630	2 826	4 032
Malta	-	-	-	-	1	1	1	4	-	-
Spain	290	218	475	404	380	126	284	152	200	209
Other nei	500	-	-	-	-	-	-	-	-	-
37 Fishing area total	*2 130*	*1 349*	*1 587*	*3 125*	*2 541*	*2 698*	*4 851*	*5 578*	*4 866*	*5 606*
41 Argentina	0	2	0	0	120	-	-	-	-	-
Brazil	3 613	1 227	923	819	652	3 418	1 872	4 414	6 862	3 228
China,Taiwan	13 345	12 429	7 408	15 088	12 866	8 087	10 221	4 040	4 644	8 172
Japan	119	233	39	36	88	41	27	86	64	...
Spain	135	149	196	123	162	12	-	-	-	-
Uruguay	28	16	49	75	56	110	69	90	40	111
41 Fishing area total	*17 240*	*14 056*	*8 615*	*16 141*	*13 944*	*11 668*	*12 189*	*8 630*	*11 610*	*11 511*

B-36 Tunas, bonitos, billfishes / Thons, pélamides, marlins / Atunes, bonitos, agujas

Capture production by species, fishing areas and countries or areas / Captures par espèces, zones de pêche et pays ou zones / Capturas por especies, áreas de pesca y países o áreas

Species, Fishing area Espèce, Zone de pêche Especie, Area de pesca	1993 t	1994 t	1995 t	1996 t	1997 t	1998 t	1999 t	2000 t	2001 t	2002 t
47 China	-	-	-	-	-	-	39	89	26	30
China,Taiwan	11 339	10 035	8 910	3 825	3 262	9 525	7 187	6 779	3 685	3 511
Japan	645	720	609	615	692	867	800	786	456	191
Namibia	3 524	3 075	1 861	1 521	1 199	1 429	1 162	2 418	3 419	2 962
Philippines	-	-	-	-	-	3	4	-	-	-
Portugal	483	1 185	655	494	256	124	232	486	41	433
St Helena	38	5	82	47	18	1	1	58	12	2
South Africa	6 881	6 931	5 214	5 634	6 708	8 412	5 101	3 610	7 236	6 507
Spain	-	-	-	-	-	-	871	282	-	829
Other nei	68	55	63	41	5	27	-	2	10	14
47 Fishing area total	*22 978*	*22 006*	*17 394*	*12 177*	*12 140*	*20 388*	*15 397*	*14 510*	*14 885*	*14 479*
51 China	-	-	-	-	-	-	88	3	2	20
China,Taiwan	13 484	11 847	11 330	12 074	10 843	18 196	18 191	8 455	4 795	7 798
France	310	292	350	391	539	460	154	350	647	194
Iran	-	-	-	10	16	9	-	-	-	-
Italy	-	-	-	-	12	58	-	...	56	39
Japan	444	840	625	1 028	1 589	2 123	1 141	1 046	1 590	1 983
Korea Rep	4	9	3	14	102	118	26	95	31	7
Mauritius	2	2	2	2	7	15	12	...	18	8
Mayotte	...	...	...	...	...	113	86	...	13	69
Philippines	-	-	-	-	-	2	180	101	68	-
Réunion	120	175	163	347	306	318	357	579	521	521 F
Seychelles	-	-	-	-	-	183	57	344	716	1 173
South Africa	2	1	2	-	-	7	1	26	21	67
Spain	904	1 773	561	826	1 031	267	275	532	570	618
Other nei	3 010	3 288	2 994	5 900	4 689	8 279	9 360	9 910	6 336	6 303
51 Fishing area total	*18 280*	*18 227*	*16 030*	*20 592*	*19 134*	*30 148*	*29 928*	*21 441*	*15 384*	*18 800 F*
57 Australia	103	127	27	4	23	26	31	28	137	76
China	-	-	-	-	-	-	101	-	20	21
China,Taiwan	2 835	2 560	2 879	4 856	4 361	3 376	4 323	1 308	290	300
Indonesia	440	604	684	1 300	1 561	1 461	1 707	2 659	2 865	2 865
Japan	509	647	1 166	1 138	1 238	690	905	1 088	1 172	1 199
Korea Rep	-	4	3	-	-	4	1	-	-	3
Philippines	-	-	-	-	-	499	10	0	-	-
Seychelles	-	-	-	-	-	-	10	79	157	65
Spain	-	-	-	-	-	6	-	-	...	35
Thailand	-	-	-	-	-	-	-	12	14	14
Other nei	754	1 983	2 191	2 645	1 078	2 598	1 624	1 677	372	372
57 Fishing area total	*4 641*	*5 925*	*6 950*	*9 943*	*8 261*	*8 660*	*8 712*	*6 851*	*5 027*	*4 950*
61 China	-	-	-	-	-	-	-	-	528	210
China,Taiwan	241	77	4 269	2 662	3 518	651	1 306	1 310	713	854
Japan	33 011	40 098	33 857	48 364	66 487	54 206	85 662	45 852	49 507	49 507 F
61 Fishing area total	*33 252*	*40 175*	*38 126*	*51 026*	*70 005*	*54 857*	*86 968*	*47 162*	*50 748*	*50 571 F*
67 Canada	551	589	792	457	80	137	308	2 535	3 061	3 495
China,Taiwan	-	-	26	0	0	0	55	1 165	310	210
USA	4 822	8 288	5 853	9 581	14 508	11 380	4 312	7 482	8 762	8 499
67 Fishing area total	*5 373*	*8 877*	*6 671*	*10 038*	*14 588*	*11 517*	*4 675*	*11 182*	*12 133*	*12 204*
71 Australia	...	...	...	...	...	...	179	147	173	345
China	1	8	5	8	2	1	3 473	2 056	2 711	2 710
China,Taiwan	4 455	11 943	8 891	5 810	6 298	5 095	3 017	5 784	6 958	9 148
Fiji Islands	463	842	702	1 446	1 842	2 121	2 279	6 065	7 971	8 026
Japan	12 173	19 144	17 490	2 124	5 027	6 481	3 527	7 926	8 678	8 678 F
Korea Rep	26	-	20	114	666	690	147	20	332	687
Micronesia	-	-	-	-	1	-	2	5	4	-
NewCaledonia	755	840	332	414	277	860	690	895	1 020	1 165
Papua N Guin	-	-	6	38	101	104	129	159	123	136
Solomon Is	...	...	24	100	109	370	136	224	54	115
USA	-	2	0	-	-	5 110	2 628	-	-	-
Vanuatu	240 F	180 F	109	192	95	10	...	...	...	225
71 Fishing area total	*18 113 F*	*32 959 F*	*27 579*	*10 246*	*14 418*	*20 842*	*16 207*	*23 281*	*28 024*	*31 235 F*
77 Amer Samoa	0	1	27	86	309	446	338	624	3 253	5 944
China	-	-	-	-	-	-	-	1 239	1 624	1 327
China,Taiwan	5 003	6 093	3 867	4 454	4 721	5 409	8 575	6 837	6 729	6 675
Cook Is	...	23	32	5	5 F	5 F	5 F	5 F	2	879
Fr Polynesia	959	913	1 100	1 750	2 717	3 235	2 642	3 580	4 432	4 678
Japan	6 450	4 839	5 098	4 401	4 717	5 187	4 506	2 372	3 285	3 285 F
Korea Rep	75	96	48	588	1 171	3 177	954	569	1 557	1 791
Mexico	11	6	5	21	53	8	32	159	40	68
Samoa	213	641	1 883	1 775	4 108	4 742	4 027	4 067	4 820	4 360
Tonga	231	343	379	431	493	616	801	862	1 268	1 199
USA	1 707	3 579	3 504	5 772	4 236	3 593	6 976	5 909	3 814	2 251
77 Fishing area total	*14 649*	*16 534*	*15 943*	*19 283*	*22 530 F*	*26 418 F*	*28 856 F*	*26 223 F*	*30 824*	*32 457 F*
81 Australia	226	351	401	468	317	418	225	212	222	332
Canada	235	235	235	136	149	167	253	351	206	144
China,Taiwan	14 624	4 006	2 875	5 479	5 797	7 849	5 619	7 452	5 491	9 710
Japan	2 759	2 167	2 307	2 267	2 876	3 287	1 565	1 052	2 415	2 415 F
Korea Rep	-	-	-	-	-	9	-	-	-	-
New Zealand	3 613	6 352	6 423	7 150	3 220	6 525	3 903	4 500	5 353	5 645
USA	1 123	605	2 170	3 223	-	-	-	-	5 942	6 735
81 Fishing area total	*22 580*	*13 716*	*14 411*	*18 723*	*12 359*	*18 255*	*11 565*	*13 567*	*19 629*	*24 981 F*
87 Chile	19	22	15	21	-	-	7	3	5	40

B-36 Tunas, bonitos, billfishes / Thons, pélamides, marlins / Atunes, bonitos, agujas

Capture production by species, fishing areas and countries or areas
Captures par espèces, zones de pêche et pays ou zones
Capturas por especies, áreas de pesca y países o áreas

Species, Fishing area Espèce, Zone de pêche Especie, Area de pesca	1993 t	1994 t	1995 t	1996 t	1997 t	1998 t	1999 t	2000 t	2001 t	2002 t
China,Taiwan	422	57	3	162	171	85	55	1 454	866	987
Japan	1 658	1 695	960	613	430	473	457	354	864	864 F
Korea Rep	-	-	-	-	-	-	51	-	-	-
87 Fishing area total	*2 099*	*1 774*	*978*	*796*	*601*	*558*	*570*	*1 811*	*1 735*	*1 891 F*
Species total	*199 197 F*	*211 749 F*	*195 209*	*201 492*	*219 425 F*	*229 936 F*	*254 843 F*	*218 333 F*	*232 193*	*238 281 F*
Southern bluefin tuna	**Thon rouge du Sud**		**Atún rojo del Sur**		***Thunnus maccoyii***			**1,75(01)026,08**		**SBF**
41 China,Taiwan	119	6	7	18	5	6	3	3	-	-
Japan	-	1	-	-	-	-	-	-	-	-
41 Fishing area total	*119*	*7*	*7*	*18*	*5*	*6*	*3*	*3*	-	-
47 China,Taiwan	119	186	161	139	42	228	158	283	76	95
Japan	3 757	1 642	2 290	2 103	409	1 840	1 908	1 986	2 421	995
Korea Rep	-	-	-	-	-	-	28	62	19	-
South Africa	-	-	-	-	-	1	-	2	0	14
47 Fishing area total	*3 876*	*1 828*	*2 451*	*2 242*	*451*	*2 069*	*2 094*	*2 333*	*2 516*	*1 104*
51 China,Taiwan	360	555	1 098	1 195	488	1 049	805	1 400	1 473	740
Japan	543	715	257	848	2 139	1 133	855	370	1 232	2 192
Korea Rep	-	98	216	314	1 056	1 415	463	784	363	513
Seychelles	-	-	-	-	-	-	11	85	169	185
South Africa	-	-	-	-	-	-	-	2	-	1
Other nei	15	38	156	223	262	361	362	37	73	73
51 Fishing area total	*918*	*1 406*	*1 727*	*2 580*	*3 945*	*3 958*	*2 496*	*2 678*	*3 310*	*3 704*
57 Australia	2 152	2 499	2 998	4 732	4 912	4 320	5 300	5 190	5 181	5 254
China,Taiwan	261	106	208	258	105	142	781	192	106	254
Indonesia	589	460	486	1 349	1 888	1 245	2 351	1 068	1 472	1 472
Japan	843	1 806	2 392	2 743	1 824	3 733	3 219	2 667	1 907	947
Korea Rep	-	-	99	597	172	147	773	112	347	136
Seychelles	-	-	-	-	-	-	4	35	69	56
Sri Lanka	-	-	-	68	83	104	121	120	101	112
Other nei	2	16	45	72	71	115	121	12	7	7
57 Fishing area total	*3 847*	*4 887*	*6 228*	*9 819*	*9 055*	*9 806*	*12 670*	*9 396*	*9 190*	*8 238*
71 Australia	...	...	...	...	...	...	0	3	-	1
71 Fishing area total	...	...	...	...	...	...	*0*	*3*	-	*1*
81 Australia	2 743	2 198	2 270	623	1 028	471	355	74	98	41
China,Taiwan	-	-	-	0	0	14	4	4	-	-
Japan	2 157	1 307	716	973	728	1 010	1 746	1 069	1 680	1 680 F
New Zealand	50	52	181	81	138	337	460	380	358	450
81 Fishing area total	*4 950*	*3 557*	*3 167*	*1 677*	*1 894*	*1 832*	*2 565*	*1 527*	*2 136*	*2 171 F*
87 Japan	-	1	17	-	-	-	-	-	-	-
87 Fishing area total	-	*1*	*17*	-	-	-	-	-	-	-
Species total	*13 710*	*11 686*	*13 597*	*16 336*	*15 350*	*17 671*	*19 828*	*15 940*	*17 152*	*15 218 F*
Yellowfin tuna	**Albacore**		**Rabil**		***Thunnus albacares***			**1,75(01)026,10**		**YFT**
21 Canada	72	52	174	155	100	57	22	105	125	70
China,Taiwan	0	17	3	1	1	0	22	40	143	95
Japan	68	277	73	19	42	190	80	20	12	...
Spain	-	0	-	-	-	23	4	46	-	-
USA	641	598	1 637	1 144	854	546	755	888	740	420
21 Fishing area total	*781*	*944*	*1 887*	*1 319*	*997*	*816*	*883*	*1 099*	*1 020*	*585*
27 China,Taiwan	0	18	1	0	0	6	62	40	11	40
Faeroe Is	-	-	-	-	-	-	-	1	...	...
Ireland	-	-	-	-	-	-	-	-	3	-
Japan	4	5	34	28	4	2	-	-	2	...
Portugal	3	6	5	3	2	3	154	125	2	2
Spain	5	5	18	19	17	22	17	14	-	-
27 Fishing area total	*12*	*34*	*58*	*50*	*23*	*33*	*233*	*180*	*18*	*42*
31 Barbados	161	156	255	160	149	150	155	155	142	115
Bermuda	58	44	44	67	55	53	59	31	37	47
Br Virgin Is	...	...	...	...	3	2	3	1	1 F	1 F
China,Taiwan	1 162 F	361	158	567	381	437	198	456	995	527
Colombia	2 404	3 418	7 172	238	46	46	46	50 F	50 F	50 F
Cuba	1	-	-	-	-	-	34	284	156	16
Dominica	30	31	9	...	...	...	80	78	120	169
Dominican Rp	-	-	-	-	-	89	220	272	263	...
Grenada	490	385	410	523	411	484	430	403	759	593
Japan	196	141	151	484	415	503	883	1 090	860	464
Korea Rep	-	-	-	11	-	-	-	-	-	-
Mexico	855	1 093	1 518	814	1 089	1 135	1 920	1 446	1 084	1 380
NethAntilles	170	155	140	130	130 F	130 F	130 F	130 F	...	...
Panama	-	-	-	-	-	-	1	-	-	-
Philippines	-	-	-	-	-	7	103	78	2	49
Puerto Rico	...	...	...	...	...	...	...	...	24	10
St Lucia	92	130	144	110	109	276	123	134	145	97
St Vincent	65	16	43	37	35	48	38	1 989	1 365	1 165
Spain	810	0	0	-	-	-	-	-	101	54
Trinidad Tob	4	120	79	183	223	213	163	112	122	125
USA	2 821	2 561	1 757	1 783	2 407	1 722	1 168	2 043	1 443	1 856

B-36 Tunas, bonitos, billfishes / Thons, pélamides, marlins / Atunes, bonitos, agujas

Capture production by species, fishing areas and countries or areas / Captures par espèces, zones de pêche et pays ou zones / Capturas por especies, áreas de pesca y países o áreas

Species, Fishing area Espèce, Zone de pêche Especie, Area de pesca	1993 t	1994 t	1995 t	1996 t	1997 t	1998 t	1999 t	2000 t	2001 t	2002 t
Venezuela	16 663	24 789	9 714	13 772	14 671	13 995	11 187	10 549	18 651	11 421
Other nei	2 671	4 404	4 202	5 962	6 100	8 339	7 409	5 269	2 883	175
31 Fishing area total	*28 653 F*	*37 804*	*25 796*	*24 841*	*26 224 F*	*27 629 F*	*24 350 F*	*24 570 F*	*29 203 F*	*18 314 F*
34 Benin	1 F	1 F	1 F	1 F	3	1	1	1 F	1	...
Cape Verde	1 536	1 727	1 781	1 448	1 721	1 418	1 663	1 851	1 684	287
China	139	156	200	124	84	71	1 535	1 652	586	262
China,Taiwan	750 F	2 338	2 040	2 835	1 903	2 161	1 667	1 512	2 072	1 671
Congo Rep	17	14	13	12	12	12	12	12	12	...
Côte dIvoire	-	-	-	...	2	...	...	...	...	...
Cuba	541	238	212	257	269	-	-	-	-	-
France	36 064	35 468	29 567	33 819	29 966	30 739	31 246	29 789	32 211	32 753
Gabon	12	88	218	225	225	295	225	162	270	245
Gambia	...	...	14	...	...	1	1	5	1	26
Georgia	10 F	-	-	-	-	-	-	-	-	-
Ghana	13 283	9 984	9 268	11 720	16 504	17 807	28 328	17 010	30 642	23 499
Japan	1 934	2 293	2 879	3 283	1 785	2 307	1 484	1 352	1 136	673
Korea Rep	176	340	266	221	60	-	-	-	-	-
Latvia	16	-	55	151	223	97	25	36	72	334
Liberia	...	...	...	...	185	310	369	227	166	170 F
Libya	-	-	-	-	-	-	-	-	208	73
Morocco	-	-	-	-	-	-	-	-	-	79
NethAntilles	-	-	-	3 183	6 082	6 110	3 962	5 441	4 793	4 035
Panama	10 972	12 066	13 442	7 713	4 293	2 111	1 315	1 103	574	1 022
Philippines	-	-	-	-	-	126	173	86	0	51
Portugal	40	10	70	26	25	51	23	8	2	4
Russian Fed	2 160	1 503	2 936	2 696	4 275	4 931	4 359	737	-	-
Sao Tome Prn	140 F	...	...	1	4	4 F	4 F	4 F	...	...
Senegal	6	83	108	68	152	222	358	218	1 118	2 095
Seychelles	-	-	-	-	-	-	-	32	-	-
Spain	40 393	40 591	38 249	34 848	24 513	31 297	19 910	24 651	30 937	31 261
Other nei	8 251	6 186	6 143	8 437	5 981	7 224	5 190	5 448	9 273	8 209
34 Fishing area total	*116 441 F*	*113 086 F*	*107 462 F*	*111 068 F*	*98 267*	*107 295 F*	*101 850 F*	*91 337 F*	*115 758*	*106 749 F*
41 Brazil	5 131	4 169	4 021	2 767	2 705	2 514	4 127	6 145	6 239	6 172
China	-	-	-	-	-	628	655	22	470	435
China,Taiwan	1 400 F	1 676	784	2 190	1 470	1 517	1 156	663	642	775
Japan	451	401	261	241	244	147	140	282	192	...
Korea Rep	-	75	105	25	1	-	-	-	-	-
Panama	-	-	-	-	-	-	4	-	-	-
Philippines	-	-	-	-	-	29	3	-	10	30
Spain	179	7	4	36	34	23	26	125	-	-
Uruguay	20	59	53	171	53	88	52	54	98	91
41 Fishing area total	*7 181 F*	*6 387*	*5 228*	*5 430*	*4 507*	*4 946*	*6 163*	*7 291*	*7 651*	*7 503*
47 Angola	211	137	216	78	70	115	170	35	...	...
China,Taiwan	404 F	2 104	1 713	1 060	711	1 206	1 306	1 174	1 140	1 058
Japan	2 976	3 643	3 753	2 738	1 624	3 999	3 221	2 708	1 117	673
Korea Rep	4	21	82	77	179	65	94	143	21	-
Namibia	-	33	19	72	69	3	147	59	165	102
Portugal	85	110	156	259	149	213	-	62	0	-
St Helena	171	150	181	151	109	181	116	136	70	90
Seychelles	-	-	-	-	-	-	-	-	-	11
South Africa	266	486	183	157	116	229	318	353	316	144
Spain	5	16	11	12	20	18	20	16	15	155
47 Fishing area total	*4 122 F*	*6 700*	*6 314*	*4 604*	*3 047*	*6 029*	*5 392*	*4 686*	*2 844*	*2 233*
51 China	-	-	-	-	-	-	129	306	484	878
China,Taiwan	69 531	24 616	19 329	25 311	16 714	23 082	15 663	18 974	18 064	21 747
Comoros	4 742	5 609	5 609	5 520	5 310	5 310	5 200	5 600 F	5 200	5 200 F
France	39 539	35 819	39 635	35 564	30 308	18 983	30 009	37 675	31 377	35 111
India	218	169	177	5 888	3 307	2 772	1 547	1 728	3 730	60
Iran	21 636	27 162	27 175	30 233	21 248	21 333	26 871	15 743	20 153	24 045
Italy	-	-	-	-	1 156	1 850	2 626	...	1 332	1 746
Japan	14 631	9 272	4 790	9 359	11 435	13 736	9 727	10 038	10 767	13 988
Jordan	...	...	...	...	...	2	5	3	5	5
Korea Rep	4 681	3 608	2 426	3 426	3 607	2 218	718	1 738	1 240	242
Maldives	9 604	12 621	12 031	11 811	12 489	13 566	13 664	11 713	15 088	20 771
Mauritius	2 537	1 858	1 725	713	1 095	1 443	742	226	125	110
Mayotte	...	...	...	...	...	194	...	...	3 120	1 313
Oman	11 366	20 707	28 477	20 718	15 905	14 897	7 377	8 377	7 945	7 114
Pakistan	30 817	4 604	5 140	5 250	3 838	3 795	8 884	4 946	3 603	2 940
Philippines	-	-	-	-	-	609	473	316	295	319
Portugal	-	-	-	-	-	-	-	9	19	29
Réunion	410	492	402	628	636	609	534	656	591	591 F
Seychelles	-	-	5	67	2 152	5 855	9 947	11 862	13 355	17 104
South Africa	24	6	26	-	28	106	149	199	96	149
Spain	47 713	43 159	65 143	59 390	60 838	32 728	51 796	52 112	47 935	53 560
Tanzania	-	-	-	-	-	-	350	700	800	700
UK	-	-	-	-	-	-	-	-	-	26
Uruguay	-	-	-	-	-	-	-	-	14	50
Yemen	804	804	800	800	840	820	820 F	820 F	820 F	820 F
Other nei	49 438	34 362	42 576	41 882	37 661	33 389	44 391	47 821	34 501	34 108
51 Fishing area total	*307 691*	*224 868*	*255 466*	*256 560*	*228 567*	*197 297*	*231 622 F*	*231 562 F*	*220 659 F*	*242 726 F*
57 Australia	91	647	263	107	305	275	478	395	1 016	362
China	-	-	138	494	750	402	2 206	2 055	1 288	447
China,Taiwan	6 287	4 655	7 690	8 708	12 391	6 441	8 240	5 014	4 636	6 113
France	-	-	-	14	920	3 398	791	19	-	-

B-36 Tunas, bonitos, billfishes / Thons, pélamides, marlins / Atunes, bonitos, agujas

Capture production by species, fishing areas and countries or areas / Captures par espèces, zones de pêche et pays ou zones / Capturas por especies, áreas de pesca y países o áreas

Species, Fishing area Espèce, Zone de pêche Especie, Area de pesca	1993 t	1994 t	1995 t	1996 t	1997 t	1998 t	1999 t	2000 t	2001 t	2002 t
India	10	16	17	1 439	799	846	526	429	1 020	...
Indonesia	12 645	17 042	18 955	34 147	40 582	38 063	44 268	32 287	30 581	30 581
Iran	-	-	-	-	2	197	-	-	-	-
Italy	-	-	-	-	184	449	-	...	-	-
Japan	399	743	6 160	5 022	3 281	3 270	4 251	3 851	2 821	1 356
Korea Rep	-	14	18	17	35	47	190	73	161	90
Malaysia	-	-	-	-	-	-	-	-	-	202
Philippines	-	-	-	-	-	14	146	24	29	27
Seychelles	-	-	-	-	726	1 597	2	18	37	37
Spain	-	-	-	41	148	5 860	123	67	...	31
Sri Lanka	11 616	11 939	8 696	12 889	15 756	19 651	27 538	22 091	15 050	17 765
Thailand	-	-	-	-	-	-	-	478	231	26
Timor-Leste	...	...	...	...	...	...	1	3	3	3 F
Other nei	1 160	2 377	877	1 502	892	5 468	2 071	2 174	345	345
57 Fishing area total	*32 208*	*37 433*	*42 814*	*64 380*	*76 771*	*85 978*	*90 831*	*68 978*	*57 218*	*57 385 F*
61 China	-	-	-	-	-	-	-	-	1 079	1 020
China,Taiwan	38 614 F	866	4 091	6 269	3 310	5 139	2 975	5 085	1 568	2 906
Japan	25 151	13 132	14 382	9 860	11 283	11 040	16 437	16 688	12 698	12 698 F
Korea Rep	-	-	-	-	-	6	30	-	-	-
61 Fishing area total	*63 765 F*	*13 998*	*18 473*	*16 129*	*14 593*	*16 185*	*19 442*	*21 773*	*15 345*	*16 624 F*
67 USA	2	-	0	-	0	0	-	1	13	-
67 Fishing area total	*2*	*-*	*0*	*-*	*0*	*0*	*-*	*1*	*13*	*-*
71 Australia	...	...	...	...	627	593	1 226	498	1 031	1 356
China	2 754	4 823	5 837	2 757	1 419	1 435	2 237	2 207	2 359	1 371
China,Taiwan	39 000 F	58 914	48 801	35 703	64 215	82 004	63 464	66 006	78 374	61 452
Fiji Islands	947	1 368	1 507	1 540	1 016	869	725	2 467	2 126	2 027
Guam	42	29	23	15	17	25	15	18	10	13
Indonesia	56 966	70 898	75 443	89 706	90 889	97 520	105 484	115 066	103 270	111 710
Japan	76 876	63 999	60 469	34 663	67 411	50 694	50 819	49 975	52 710	52 710 F
Kiribati	108	273	1 025	651	2 223	2 076	1 423	1 209	1 220	793
Korea Rep	54 224	51 985	38 545	20 163	43 885	56 876	31 954	31 127	40 428	34 728
Marshall Is	69	27	18	-	-	-	-	900	2 927	1 101
Micronesia	4 623	3 985	2 062	891	2 845	3 212	3 403	5 411	5 848	3 922
Nauru	...	...	...	...	...	...	...	...	...	10
NewCaledonia	433	437	839	554	466	185	373	250	570	572
N Marianas	5	6	9	17	11	5	11	7	7	12
Palau	...	...	3	2	0	1	1	42	34	2
Papua N Guin	8	374	2 722	971	6 968	11 924	7 875	14 659	25 779	27 912
Philippines	38 083	63 179	60 957	61 280	67 342	79 215	90 353	90 328	83 560	99 794
Solomon Is	7 193	6 671	8 114	11 003	9 588	8 114	8 843	3 334	4 692	3 913
Tuvalu	292	110	13	15 F	20 F	20 F	20 F	20 F	20 F	20 F
USA	33 793	59 821	28 334	29 849	47 384	44 300	37 975	27 154	27 217	22 528
Vanuatu	140 F	85	1 090	982	8 827	9 603	10 337	2 415	...	107
71 Fishing area total	*315 556 F*	*386 984*	*335 811*	*290 762 F*	*415 153 F*	*448 671 F*	*416 538 F*	*413 093 F*	*432 182 F*	*426 053 F*
77 Amer Samoa	1	1	2	12	22	42	64	86	183	484
Belize	-	190	10	670	590	380	90	68	-	-
China	-	-	-	-	-	-	-	580	1 994	1 457
China,Taiwan	8 090 F	1 832	380	94	108	161	192	126	1 750	685
Colombia	...	...	...	...	...	...	...	7 635	9 623	15 065
Cook Is	...	11	22	8	5 F	5 F	5 F	5 F	1	49
Costa Rica	...	...	10	50	...	...	30	...	...	...
Cyprus	940	180	100	600	920	-	-	-	-	-
Ecuador	380	400	-	-	-	-	-	-	-	-
El Salvador	...	...	...	...	...	920	260	...	2 165	3 434
Fr Polynesia	602	436	820	811	860	843	1 225	1 762	1 514	1 163
Guatemala	-	-	-	-	-	-	1 050	2 715	4 000 F	2 387
Honduras	-	-	-	40	40	-	40	-	...	...
Japan	15 771	19 035	14 501	8 974	12 714	9 089	7 003	12 752	10 392	10 392 F
Korea Rep	8 250	8 709	10 436	11 327	13 455	14 091	8 812	13 640	15 208	13 702
Mexico	100 200	113 037	93 384	108 606	127 497	107 521	109 966	95 829	127 000	144 383
Nicaragua	-	-	-	-	-	-	3 060	3 320	...	...
Panama	3 640	1 370	370	3 150	3 660	3 230	3 350	2 717	5 779	15 266
Samoa	81	73	216	573	1 327	801	681	1 120	470	388
Spain	640	110	290	690	530	3 900	6 966	5 325	3 007	2 838
Tonga	64	46	59	88	100	125	163	175	259	262
USA	10 429	19 474	12 866	12 574	12 685	14 882	3 725	3 901	5 188	7 531
Vanuatu	14 040	12 680	16 130	7 250	14 750	13 370	11 590	6 082	4 524	2 236
Venezuela	31 940	26 350	41 500	55 750	43 320	42 350	46 430	50 410	68 800	89 250
Other nei	-	50	220	-	-	-	-	1 943	10 121	26 967
77 Fishing area total	*195 068 F*	*203 984*	*191 316*	*211 267*	*232 583 F*	*211 710 F*	*204 702 F*	*210 191 F*	*271 978 F*	*337 939 F*
81 Australia	878	1 229	1 259	1 743	1 109	1 243	958	802	896	1 120
China,Taiwan	1 000 F	764	138	138	158	329	57	173	174	128
Japan	1 200	1 615	2 615	2 564	1 912	975	345	155	567	567 F
Korea Rep	7	-	-	-	-	24	202	545	419	16
New Zealand	17	46	138	181	118	127	153	107	137	25
81 Fishing area total	*3 102 F*	*3 654*	*4 150*	*4 626*	*3 297*	*2 698*	*1 715*	*1 782*	*2 193*	*1 856 F*
87 Belize	1 100	850	880	1 540	1 280	3 110	2 400	1 399	-	-
Chile	82	118	43	32	57	78	48	77	66	15
China,Taiwan	-	17	0	15	17	1	2	16	5	-
Colombia	4 942	7 961	9 011	10 141	9 564	15 672	13 402	9 758	15 899	15 721
Cyprus	2 590	520	850	2 290	1 230	100	-	-	-	-
Ecuador	17 910	23 173	15 921	19 314	19 603	31 052	50 200	36 955	57 563	36 306
El Salvador	...	...	...	...	...	...	2 990	-	-	-

B-36 Tunas, bonitos, billfishes / Thons, pélamides, marlins / Atunes, bonitos, agujas

Capture production by species, fishing areas and countries or areas / Captures par espèces, zones de pêche et pays ou zones / Capturas por especies, áreas de pesca y países o áreas

Species, Fishing area Espèce, Zone de pêche Especie, Area de pesca	1993 t	1994 t	1995 t	1996 t	1997 t	1998 t	1999 t	2000 t	2001 t	2002 t
Guatemala	-	-	-	-	-	-	610	2 054	...	...
Honduras	-	-	-	-	190	870	1 600	620	...	...
Japan	3 752	5 077	3 154	2 900	2 006	2 478	2 044	4 457	5 192	5 192 F
Korea Rep	178	73	73	-	22	27	1 255	2 195	1 480	-
Liberia	-	-	-	180	-	-	-	-	-	-
Mexico	4 911	7 244	12 256	18 395	11 700	9 106	9 998	5 065	7 410	2 820
Nicaragua	-	-	-	-	-	-	-	1 630	...	...
Panama	2 220	2 170	1 690	490	1 920	290	3 238	4 335	5 932	4 119
Peru	3 573	269	914	953	908	12 747	2 784	2 548	4 175	5 967
St Vincent	-	950	-	-	-	-	-	-	-	-
Spain	1 740	1 120	1 740	3 270	2 390	1 840	3 670	3 520	8 430	2 033
USA	-	-	-	1 837	895	461	-	289	9	496
Vanuatu	11 900	15 200	8 400	4 950	7 690	5 270	6 550	9 089	6 412	3 465
Venezuela	12 770	16 210	7 170	7 930	16 010	20 260	11 310	19 100	41 170	27 826
Other nei	990	1 040	380	-	620	220	-	3 132	20 340	19 350
87 Fishing area total	*68 658*	*81 992*	*62 482*	*74 237*	*76 102*	*103 582*	*112 101*	*106 239*	*174 083*	*123 310 F*
Species total	*1 143 240 F*	*1 117 868 F*	*1 057 257 F*	*1 065 273 F*	*1 180 131 F*	*1 212 869 F*	*1 215 822 F*	*1 182 782 F*	*1 330 165 F*	*1 341 319 F*
Bigeye tuna	**Thon obèse(=Patudo)**		**Patudo**			***Thunnus obesus***			**1,75(01)026,12**	**BET**
21 Canada	125	111	148	144	166	120	263	327	241	280
China,Taiwan	0	13	0	0	0	0	0	12	2	-
Japan	961	856	162	356	405	498	944	290	342	...
USA	683	799	846	410	440	506	659	328	453	267
21 Fishing area total	*1 769*	*1 779*	*1 156*	*910*	*1 011*	*1 124*	*1 866*	*957*	*1 038*	*547*
27 China,Taiwan	0	21	15	4	3	38	17	27	20	45
Faeroe Is	-	-	-	-	-	-	11	8	...	...
France	-	-	-	-	-	-	28	15	28	44
Iceland	-	-	-	-	-	-	-	5	-	-
Ireland	-	-	-	-	-	-	-	-	10	-
Japan	149	108	300	486	379	368	94	269	339	...
Portugal	4 180	1 902	4 964	1 771	2 590	4 246	2 078	688	425	1 467
Spain	11	12	16	77	52	304	362	562	559	379
Other nei	-	-	-	-	4	-	-	-	-	-
27 Fishing area total	*4 340*	*2 043*	*5 295*	*2 338*	*3 028*	*4 956*	*2 590*	*1 574*	*1 381*	*1 935*
31 Barbados	-	-	-	-	24	17	18	18	6	11
China,Taiwan	3 060 F	269	3	308	236	166	34	205	114	379
Dominica	...	...	...	...	...	...	...	...	5	-
Grenada	11	10	10	-	1	-	-	-	-	-
Japan	423	407	407	496	600	1 196	2 216	2 134	2 719	1 099
Korea Rep	-	-	-	33	-	-	-	-	-	-
Mexico	-	-	-	-	6	8	6	2	2	0
Panama	-	-	-	-	-	-	-	49	-	-
Philippines	-	-	-	-	-	21	442	260	34	638
Puerto Rico	-	-	-	-	54	-	-	-	-	-
St Lucia	0	0	0	0	0	0	0	-	1	2
St Vincent	3	0	0	4	2	2	1	1 216	506	14
Spain	5	2	2	5	32	35	-	-	-	-
Trinidad Tob	3	29	27	37	36	24	19	5	11	30
USA	56	143	45	19	60	65	65	72	67	63
Venezuela	809	457	457	189	274	222	140	226	708	629
31 Fishing area total	*4 370 F*	*1 317*	*951*	*1 091*	*1 325*	*1 756*	*2 941*	*4 187*	*4 173*	*2 865*
34 Benin	8	9	9 F	9 F	30	13	11	...	...	...
Cape Verde	85	209	66	16	10	1	1	2	-	-
China	70	428	476	520	427	656	2 520	393	2 897	3 044
China,Taiwan	1 769 F	7 260	7 062	11 360	8 703	6 844	5 539	9 314	8 279	6 240
Congo Rep	14	9	9	8	8	8	8	8	8	8
Cuba	36	7	7	5	-	-	-	-	-	-
France	12 719	12 263	8 363	9 171	5 980	5 624	5 501	5 934	4 948	4 249
Gabon	1	87	10	-	-	-	184	150	121	...
Ghana	3 577	4 738	5 517	5 805	7 431	13 252	11 460	5 586	14 095	5 893
Japan	15 693	13 866	17 009	16 982	10 360	14 109	10 021	13 101	8 625	6 000
Korea Rep	363	321	368	721	361	-	-	-	-	-
Liberia	65	53	57	105	340	108	112	201	175	175 F
Libya	1 085	500	400	400	400	400	400	400	31	593
Morocco	-	-	-	-	-	-	700	770	857	913
NethAntilles	-	-	-	1 893	2 890	2 919	3 428	2 359	2 803	1 879
Panama	10 410	13 087	9 927	4 777	2 098	1 252	579	468	89	63
Philippines	-	-	-	-	-	1 038	1 627	715	29	59
Portugal	1 206	944	4 423	3 723	2 767	1 956	1 235	810	1 181	1 123
Russian Fed	-	-	-	13	38	4	8	91	-	-
Sao Tome Prn	-	-	-	-	5	-	-	-	-	-
Senegal	4	126	177	135	218	791	2 007	860	2 169	1 143
Seychelles	-	-	-	-	-	-	-	58	-	-
Sierra Leone	...	...	...	...	...	...	...	6	2	...
Spain	16 632	21 943	17 673	15 160	12 245	6 469	13 261	10 091	9 364	9 813
Togo	86	23	6	33	17	6	66	32	26	66
Other nei	7 165	11 321	12 705	15 684	18 475	26 581	27 757	17 377	11 494	2 779
34 Fishing area total	*70 988 F*	*87 194*	*84 264 F*	*86 520 F*	*72 803*	*82 031*	*86 425*	*68 726*	*67 193*	*44 040 F*
41 Brazil	1 256	596	1 935	1 707	1 237	644	2 024	2 768	2 659	2 582
China	-	-	-	-	-	847	-	-	-	-
China,Taiwan	4 440 F	4 485	4 134	8 017	6 142	3 437	2 711	3 356	3 732	5 417
Japan	4 266	4 190	1 913	1 729	1 795	796	753	1 255	1 133	1 604
Korea Rep	-	65	53	109	4	-	-	-	-	-

B-36 Tunas, bonitos, billfishes — Capture production by species, fishing areas and countries or areas
Thons, pélamides, marlins — Captures par espèces, zones de pêche et pays ou zones
Atunes, bonitos, agujas — Capturas por especies, áreas de pesca y países o áreas

Species, Fishing area Espèce, Zone de pêche Especie, Area de pesca	1993 t	1994 t	1995 t	1996 t	1997 t	1998 t	1999 t	2000 t	2001 t	2002 t
Panama	-	-	-	-	-	-	-	435	-	-
Philippines	-	-	-	-	-	95	44	-	314	141
Spain	-	9	13	11	123	183	58	112	150	149
Uruguay	48	37	80	124	69	59	27	29	47	67
41 Fishing area total	*10 010 F*	*9 382*	*8 128*	*11 697*	*9 370*	*6 061*	*5 617*	*7 955*	*8 035*	*9 960*
47 China	-	-	-	-	-	-	4 827	6 170	4 313	2 795
China,Taiwan	2 612 F	5 426	6 808	5 426	4 158	5 829	8 536	9 126	8 895	9 436
Japan	20 293	20 767	16 708	13 652	10 982	9 486	8 145	6 894	4 926	6 000
Korea Rep	14	-	-	164	212	163	124	70	4	-
Namibia	-	751	352	63	46	16	423	589	640	312
Panama	28	147	-	-	-	-	-	-	-	-
Portugal	230	253	275	316	80	132	-	-	-	-
St Helena	6	6	10	10	12	17	6	8	4	5
Seychelles	-	-	-	-	-	-	-	-	-	162
South Africa	88	76	27	7	10	41	41	225	167	304
Spain	134	130	145	140	61	123	58	486	61	184
47 Fishing area total	*23 405 F*	*27 556*	*24 325*	*19 778*	*15 561*	*15 807*	*22 160*	*23 568*	*19 010*	*19 198*
51 China	-	-	-	-	-	-	69	877	889	1 917
China,Taiwan	20 340	16 119	27 210	21 358	24 371	35 002	35 156	42 712	48 908	42 737
Comoros	27	32	32	30	30	30	30	30 F	30	30 F
France	5 016	5 367	7 280	6 896	7 648	3 854	8 324	6 657	5 090	7 398
India	839	1 042	1 042	-	-	-	-	-	-	-
Iran	-	-	-	153	261	310	592	347	430	127
Italy	-	-	-	-	450	296	848	...	57	315
Japan	5 820	9 464	6 560	7 115	8 403	10 293	6 042	5 516	4 953	6 659
Korea Rep	7 146	8 179	6 106	10 737	10 129	3 154	608	3 091	1 145	178
Maldives	505	506	473	630	540	606	604	472	536	958
Mauritius	664	729	570	271	546	260	250	37	5	3
Mayotte	...	...	...	...	...	14	109	...	464	109
Philippines	-	-	-	-	-	1 465	1 177	1 354	876	587
Réunion	6	7	15	98	91	112	213	167	151	151 F
Seychelles	-	-	5	75	858	1 494	3 108	2 271	3 677	5 636
South Africa	-	-	-	-	-	8	13	29	25	203
Spain	5 414	5 950	12 233	11 339	15 863	8 447	16 024	10 803	8 028	11 250
Uruguay	-	-	-	-	-	-	-	-	16	19
Other nei	12 184	8 321	12 860	14 268	16 337	19 885	21 749	20 503	12 347	13 251
51 Fishing area total	*57 961*	*55 716*	*74 386*	*72 970*	*85 527*	*85 230*	*94 916*	*94 866 F*	*87 627*	*91 528 F*
57 Australia	475	146	84	25	57	187	476	468	553	419
China	-	-	140	466	1 652	2 165	2 113	1 822	2 105	875
China,Taiwan	5 865	7 871	9 079	13 063	17 230	8 526	6 353	10 933	4 780	6 986
France	-	-	-	12	176	2 535	192	16	-	-
India	27	34	34	-	-	4	-	-	2	-
Indonesia	7 864	10 785	12 211	23 206	27 861	26 090	30 475	20 926	21 125	21 125
Iran	-	-	-	-	1	95	-	-	-	-
Italy	-	-	-	-	7	316	-	...	-	-
Japan	4 126	7 022	12 643	9 029	7 903	7 247	8 832	7 800	8 137	7 624
Korea Rep	-	60	48	48	77	33	737	129	256	8
Malaysia	-	-	-	-	-	-	-	-	-	145
Philippines	-	-	-	-	-	86	714	107	135	292
Seychelles	-	-	-	-	78	591	4	36	72	191
Spain	-	-	-	35	46	2 833	68	13	...	13
Sri Lanka	1 009	1 012	2 108	491	600	749	462	348	336	338
Thailand	-	-	-	-	-	-	-	280	188	90
Other nei	2 845	3 597	2 936	4 781	2 895	5 348	5 177	5 465	822	822
57 Fishing area total	*22 211*	*30 527*	*39 283*	*51 156*	*58 583*	*56 805*	*55 603*	*48 343*	*38 511*	*38 928*
61 China	-	-	-	-	-	-	-	-	2 003	1 573
China,Taiwan	1 000	160	52	12	21	2 440	1 060	1 268	316	1 452
Japan	12 686	10 841	8 130	6 598	9 653	7 703	9 023	8 110	6 254	6 254 F
Korea Rep	-	-	-	-	-	2	90	-	-	-
61 Fishing area total	*13 686*	*11 001*	*8 182*	*6 610*	*9 674*	*10 145*	*10 173*	*9 378*	*8 573*	*9 279 F*
67 Japan	-	-	-	-	-	1	-	-	-	-
USA	-	-	-	-	-	-	-	-	1	-
67 Fishing area total	-	-	-	-	-	*1*	-	-	*1*	-
71 Australia	...	...	...	...	419	845	646	317	628	568
China	3 665	7 846	4 744	3 261	2 243	1 836	1 805	1 981	2 227	739
China,Taiwan	501	4 305	5 260	4 748	13 523	12 974	15 996	4 698	5 914	31 335
Fiji Islands	204	249	378	593	409	460	462	687	662	853
Japan	18 529	18 061	15 623	9 529	20 291	19 815	17 468	17 344	25 629	25 629 F
Kiribati	-	26	66	69	130	99	157	63	113	74
Korea Rep	2 235	1 980	1 483	1 120	2 624	2 313	3 131	3 147	2 370	4 316
Marshall Is	67	26	13	-	-	-	-	35	134	84
Micronesia	473	360	211	183	430	705	1 016	1 175	984	958
Nauru	...	...	...	...	...	...	...	...	...	2
NewCaledonia	106	78	103	233	234	498	553	517	128	189
Palau	...	...	...	...	...	...	...	49	19	...
Papua N Guin	...	20	161	50	1 060	1 486	1 028	1 539	5 004	3 583
Solomon Is	733	593	1 391	1 109	1 434	1 232	1 070	577	576	653
USA	3	1	2	864	504	-	3 289	1 863	-	4 818
Vanuatu	...	5	151	151	907	583	999	140	...	20
71 Fishing area total	*26 516*	*33 550*	*29 586*	*21 910*	*44 208*	*42 846*	*47 620*	*34 132*	*44 388*	*73 821 F*
77 Amer Samoa	0	0	1	4	4	10	9	21	74	196
Belize	-	-	120	-	20	10	0	43	-	-

B-36 Tunas, bonitos, billfishes / Thons, pélamides, marlins / Atunes, bonitos, agujas

Capture production by species, fishing areas and countries or areas / Captures par espèces, zones de pêche et pays ou zones / Capturas por especies, áreas de pesca y países o áreas

Species, Fishing area Espèce, Zone de pêche Especie, Area de pesca	1993 t	1994 t	1995 t	1996 t	1997 t	1998 t	1999 t	2000 t	2001 t	2002 t
China	-	-	-	-	-	-	-	750	3 074	7 614
China,Taiwan	1 540	1 042	463	93	177	465	1 032	678	169	216
Cook Is	...	8	17	3	3 F	3 F	3 F	3 F	1	66
Costa Rica	...	...	...	120	...	...	20	...	...	...
Cyprus	-	-	20	430	160	-	-	-	-	-
Ecuador	20	690	-	-	-	-	-	-	-	-
El Salvador	...	...	...	...	...	10	...	...	2 059	4 590
Fr Polynesia	163	165	184	186	310	403	278	712	746	651
Guatemala	-	-	-	-	-	-	760	2 261	2 500 F	2 624
Honduras	-	-	-	1 000	850	140	-	20	...	...
Japan	57 640	51 112	40 822	30 385	30 533	37 124	25 234	27 870	32 366	32 366 F
Korea Rep	14 633	19 557	18 599	15 486	17 597	27 275	21 236	22 704	26 868	27 438
Mexico	80	40	299	495	102	5	91	6	90	0
Nicaragua	-	-	-	-	-	-	30	10	...	...
Panama	10	-	60	200	40	8	869	225	503	794
Samoa	3	14	40	27	63	334	283	177	185	153
Spain	-	-	170	500	250	2 960	7 347	8 254	2 884	5 006
Tonga	34	19	23	60	69	86	112	120	191	219
USA	4 385	2 147	8 506	4 574	5 255	7 063	2 896	2 986	5 378	6 932
Vanuatu	-	130	1 990	2 240	1 490	1 920	1 240	412	910	309
Venezuela	-	10	-	-	75	-	-	0	-	6
Other nei	-	-	-	-	-	300	-	-	1 349	3 882
77 *Fishing area total*	*78 508*	*74 934*	*71 314*	*55 803*	*56 998 F*	*78 116 F*	*61 440 F*	*67 252 F*	*79 347 F*	*93 062 F*
81 Australia	40	120	155	293	563	1 024	257	355	367	450
China,Taiwan	200	465	31	59	112	528	325	139	107	17
Japan	586	484	587	903	668	1 203	875	468	1 108	1 108 F
Korea Rep	15	-	-	-	-	43	497	938	692	22
New Zealand	74	69	60	86	140	388	420	421	480	200
81 *Fishing area total*	*915*	*1 138*	*833*	*1 341*	*1 483*	*3 186*	*2 374*	*2 321*	*2 754*	*1 797 F*
87 Belize	-	2 360	1 150	2 600	4 700	1 150	320	555	-	-
Chile	-	8	15	16	6	29	6	20	5	7
China,Taiwan	-	42	0	50	94	1	2	16	8	-
Colombia	650	5 134	5 559	7 368	3 289	561	1 430	230	103	159
Costa Rica	-	-	150	720	-	-	-	-	-	-
Cyprus	480	2 110	1 580	750	1 290	30	-	-	-	-
Ecuador	1 806	3 276	10 193	17 892	26 148	17 909	22 278	29 398	23 440	21 265
El Salvador	...	...	...	...	...	-	230	-	-	-
Guatemala	-	-	-	-	-	-	820	10 091	...	...
Honduras	-	-	-	230	380	-	420	-	...	...
Japan	5 605	7 176	5 752	4 331	3 608	5 732	3 026	5 016	7 716	7 716 F
Korea Rep	434	442	162	-	187	59	135	-	-	-
Liberia	-	-	-	310	-	-	-	-	-	-
Panama	-	-	570	870	1 280	-	-	2 901	1 678	812
St Vincent	-	600	-	-	-	-	-	-	-	-
Spain	420	850	2 050	2 040	2 950	2 370	4 720	12 870	4 580	3 336
USA	-	-	-	656	400	343	-	466	-	443
Vanuatu	1 360	6 620	7 250	7 840	3 710	1 640	2 860	5 426	3 374	2 128
Venezuela	180	450	470	430	170	240	10	210	-	393
Other nei	70	190	100	-	420	-	-	221	5 761	7 070
87 *Fishing area total*	*11 005*	*29 258*	*35 001*	*46 103*	*48 632*	*30 064*	*36 257*	*67 420*	*46 665*	*43 329 F*
Species total	*325 684 F*	*365 395*	*382 704 F*	*378 227 F*	*408 203 F*	*418 128 F*	*429 982 F*	*430 679 F*	*408 696 F*	*430 289 F*
Indo-Pacific sailfish	**Voilier indo-pacifique**		**Pez vela del Indo-Pacífico**		***Istiophorus platypterus***			**1,75(03)004,02**		**SFA**
51 China,Taiwan	626	513	432	142	91	133	344	706	1 602	2 051
Comoros	212	250	250	250	240	240	200	250 F	200	200 F
Eritrea	...	-	-	-	1	0	0	1	-	-
India	-	-	-	13	12	-	-	8	-	-
Iran	740	1 085	3 619	2 306	1 928	1 813	3 193	3 080	3 160	4 258
Japan	7	19	29	26	125	124	148	117	98	97
Korea Rep	-	-	-	3	5	-	-	-	-	-
Mayotte	...	...	...	...	...	4	1	...	...	...
Oman	540	357	664	581	1 261	591	399	448	218	147
Pakistan	731	855	910	980	41	45	46	...	...	...
Réunion	6	8	7	10	11	17	18	30	27	27 F
Saudi Arabia	-	-	-	-	1	2	1	1	8	22
Seychelles	-	-	-	1	8	2	18	9	18	9
South Africa	1	-	-	-	0	0	0	0	0	-
Spain	-	-	-	-	-	9	7	1	1	2
51 *Fishing area total*	*2 863*	*3 087*	*5 911*	*4 312*	*3 724*	*2 980*	*4 375*	*4 651 F*	*5 332*	*6 813 F*
57 Australia	-	-	-	-	-	3	1	-	-	0
China,Taiwan	202	315	99	29	18	11	47	278	202	579
Honduras	-	-	-	-	-	-	-	-	205	-
India	-	-	-	-	-	14	9	1	-	-
Indonesia	170	233	264	502	603	564	659	366	322	322
Japan	1	3	28	7	15	23	42	57	36	9
Sri Lanka	1 872	4 638	3 335	5 360	6 552	8 172	6 979	8 878	7 574	6 827
57 *Fishing area total*	*2 245*	*5 189*	*3 726*	*5 898*	*7 188*	*8 787*	*7 737*	*9 580*	*8 339*	*7 737*
61 China,Taiwan	3 257 F	1 758	1 057	1 784	3 046	762	3 062	1 539	201	266
Japan	285	546	621	859	382	1 059	779	541	382	382 F
61 *Fishing area total*	*3 542 F*	*2 304*	*1 678*	*2 643*	*3 428*	*1 821*	*3 841*	*2 080*	*583*	*648 F*
71 Australia	...	...	...	...	...	...	...	...	...	5
China,Taiwan	320 F	200	438	302	186	3 027	153	4	63	47

B-36 Tunas, bonitos, billfishes / Thons, pélamides, marlins / Atunes, bonitos, agujas

Capture production by species, fishing areas and countries or areas / Captures par espèces, zones de pêche et pays ou zones / Capturas por especies, áreas de pesca y países o áreas

Species, Fishing area Espèce, Zone de pêche Especie, Area de pesca	1993 t	1994 t	1995 t	1996 t	1997 t	1998 t	1999 t	2000 t	2001 t	2002 t
Guam	-	-	1	0	0	1	0	1	1	1
Japan	232	305	215	11	132	76	18	56	40	40 F
Korea Rep	-	-	-	4	-	-	9	2	-	-
71 Fishing area total	*552 F*	*505*	*654*	*317*	*318*	*3 104*	*180*	*63*	*104*	*93 F*
77 Amer Samoa	...	1	3	2	3	2	3	1	1	1
China,Taiwan	-	-	7	39	7	41	4	0	112	20
Japan	207	205	226	175	140	234	358	326	295	295 F
Korea Rep	589	1 200	224	243	1 297	389	210	132	33	128
USA	-	-	-	-	-	-	-	-	-	5
77 Fishing area total	*796*	*1 406*	*460*	*459*	*1 447*	*666*	*575*	*459*	*441*	*449 F*
81 Australia	...	...	...	...	...	...	...	...	...	2
China,Taiwan	-	-	0	0	2	1	0	0	36	26
Japan	33	29	17	9	6	13	5	1	10	10 F
Korea Rep	-	-	-	-	-	-	277	51	-	-
81 Fishing area total	*33*	*29*	*17*	*9*	*8*	*14*	*282*	*52*	*46*	*38 F*
87 Japan	13	11	7	8	44	54	10	11	35	35 F
87 Fishing area total	*13*	*11*	*7*	*8*	*44*	*54*	*10*	*11*	*35*	*35 F*
Species total	*10 044 F*	*12 531*	*12 453*	*13 646*	*16 157*	*17 426*	*17 000*	*16 896 F*	*14 880*	*15 813 F*
Atlantic sailfish	**Voilier de l'Atlantique**		**Pez vela del Atlántico**		***Istiophorus albicans***			**1,75(03)004,04**		**SAI**
21 Japan	-	-	-	-	-	6	-	-	-	-
21 Fishing area total	-	-	-	-	-	*6*	-	-	-	-
27 Portugal	2	1	-	-	-	-	11	7	3	7
27 Fishing area total	*2*	*1*	-	-	-	-	*11*	*7*	*3*	*7*
31 Aruba	10	10	10	10	...	...	...	...	...	...
Barbados	50	46	74	25	71	58	44	44	...	...
China,Taiwan	223	101	38	37	17	1	8	0	-	-
Cuba	42	46	33	33	40	28	196	208	68	32
Dominican Rp	50	90	40	25	101	89	27	81	81	46
Grenada	141	151	119	56	83	151	148	164	187	151
Japan	-	-	-	1	7	2	9	6	2	3
Mexico	2	19	6	10	7	21	33	37	36	38
NethAntilles	15	15 F	15 F	15 F	15 F	15 F	15 F	15 F	...	...
St Vincent	4	4	2	1	3	-	1	-	2	161
Spain	-	-	-	-	7	-	-	-	66	96
Trinidad Tob	56	101	101	104	10	...	4	3	7	6
Venezuela	206	162	103	165	185	258	179	93	126	159
Other nei	31	30	30	30	30	-	-	-	-	-
31 Fishing area total	*830*	*775 F*	*571 F*	*512 F*	*576 F*	*623 F*	*664 F*	*651 F*	*575*	*692*
34 Benin	20	20 F	20	19	6	4	5	5 F	12	2
China	-	3	3	3	3	5	9	4	5	11
China,Taiwan	420	233	155	65	29	38	34	57	-	128
Côte dIvoire	40	54	66	91	65	35	80	45	47	65
Cuba	77	83	72	533	-	-	-	-	-	-
France	182	160	128	97	110	138	131	98	-	-
Ghana	693	450	353	303	196	351	305	275	568	529
Japan	28	39	49	48	19	47	21	37	11	12
Portugal	-	-	-	-	-	-	40	7	...	3
St Vincent	-	-	-	-	-	-	-	-	-	7
Sao Tome Prn	...	...	...	...	139	...	...	...	...	...
Senegal	41	172	204	206	509	192	79	447	266	138
Spain	2	0	1	2	6	2	-	1	75	75
Other nei	11	15	10	10	10	-	-	-	-	-
34 Fishing area total	*1 514*	*1 229 F*	*1 061*	*1 377*	*1 092*	*812*	*704*	*976 F*	*984*	*970*
41 Brazil	243	128	245	310	137	184	356	598	412	547
China	-	3	3	3	3	3	9	4	3	1
China,Taiwan	-	-	-	-	-	3	44	16	-	4
Japan	20	19	4	4	17	6	14	13	4	...
Spain	5	3	36	3	8	20	6	14	211	374
41 Fishing area total	*268*	*153*	*288*	*320*	*165*	*216*	*429*	*645*	*630*	*926*
47 China,Taiwan	-	-	-	-	-	111	13	62	-	66
Japan	28	34	31	27	17	30	27	17	16	...
Portugal	-	-	-	-	-	-	2	4	-	-
Spain	28	7	12	23	20	16	19	8	73	113
47 Fishing area total	*56*	*41*	*43*	*50*	*37*	*157*	*61*	*91*	*89*	*179*
Species total	*2 670*	*2 199 F*	*1 963 F*	*2 259 F*	*1 870 F*	*1 814 F*	*1 869 F*	*2 370 F*	*2 281*	*2 774*
Blue marlin	**Makaire bleu**		**Aguja azul**		***Makaira nigricans***			**1,75(03)005,05**		**BUM**
21 Japan	1	7	1	-	1	12	-	2	-	-
21 Fishing area total	*1*	*7*	*1*	-	*1*	*12*	-	*2*	-	-
27 China,Taiwan	-	-	-	-	-	1	-	-	-	-
Japan	-	4	3	1	-	1	-	-	-	-
Portugal	-	-	-	-	-	40	3	14	16	1
Spain	7	5	-	20	3	3	3	8	5	0
27 Fishing area total	*7*	*9*	*3*	*21*	*3*	*45*	*6*	*22*	*21*	*1*

B-36 Tunas, bonitos, billfishes / Thons, pélamides, marlins / Atunes, bonitos, agujas

Capture production by species, fishing areas and countries or areas / Captures par espèces, zones de pêche et pays ou zones / Capturas por especies, áreas de pesca y países o áreas

Species, Fishing area Espèce, Zone de pêche Especie, Area de pesca	1993 t	1994 t	1995 t	1996 t	1997 t	1998 t	1999 t	2000 t	2001 t	2002 t
31 Barbados	21	19	31	25	30	25	19	19	...	...
Bermuda	11	15	15	15	3	5	1	2	2	2
China,Taiwan	157	21	10	7	6	4	7	0	-	-
Cuba	69	12	41	1	53	12	38	55	56	11
Dominican Rp	-	-	-	-	41	71	29	23	23	204
Grenada	33	52	50	26	47	60	100	87	104	69
Japan	70	31	3	105	101	84	153	175	39	...
NethAntilles	40 F	40 F	40 F	40 F	40 F	40 F	40 F	40 F	...	...
Panama	-	-	-	-	-	-	-	3	-	-
Philippines	-	-	-	-	-	-	-	38	-	-
St Vincent	2	2	2	0	1	-	-	-	-	20
Spain	-	1	0	2	1	3	-	17	3	1
Trinidad Tob	3	27	46	21	81	70	33	55	17	16
Venezuela	74	122	106	137	130	205	220	28	72	76
31 Fishing area total	*480 F*	*342 F*	*344 F*	*379 F*	*534 F*	*579 F*	*640 F*	*542 F*	*316*	*399*
34 Benin	6 F	5	5 F	5 F	5	5	5	5 F	...	...
China	-	41	48	41	51	79	133	9	31	15
China,Taiwan	980	180	234	334	268	256	189	27	249	92
Côte dIvoire	139	212	177	157	222	182	275	206	196	78
France	146	133	126	96	82	80	83	79	-	-
Ghana	236	441	472	422	491	447	624	639	1 295	999
Japan	514	671	716	864	485	529	331	338	131	125
Korea Rep	-	-	-	-	7	-	-	-	-	-
Liberia	...	...	...	114	122	59	37	187	131	130 F
Philippines	-	-	-	-	-	5	38	-	-	-
Portugal	15	11	10	7	3	6	5	1	1	0
Sao Tome Prn	-	-	-	-	35	-	-	-	-	-
Spain	7	16	27	84	57	44	77	38	16	25
Other nei	57	100	100	100	100	-	-	-	-	-
34 Fishing area total	*2 100 F*	*1 810*	*1 915 F*	*2 224 F*	*1 928*	*1 692*	*1 797*	*1 529 F*	*2 050*	*1 464 F*
41 Brazil	147	81	180	331	193	486	509	467	780	387
China,Taiwan	584	272	124	222	178	138	107	0	106	143
Japan	194	275	118	101	127	77	53	51	46	...
Panama	-	-	-	-	-	-	-	38	-	-
Portugal	-	-	-	-	-	-	-	-	-	3
Spain	11	5	11	4	15	38	32	78	70	2
Uruguay	-	-	-	-	-	23	-	-	-	0
41 Fishing area total	*936*	*633*	*433*	*658*	*513*	*762*	*701*	*634*	*1 002*	*535*
47 China	-	21	25	21	27	41	68	15	61	73
China,Taiwan	183	127	99	97	78	179	183	62	323	210
Japan	434	556	586	614	461	395	303	293	115	155
Philippines	-	-	-	-	-	2	33	-	-	-
Portugal	-	-	-	-	-	1	-	2	1	3
South Africa	-	-	-	-	-	-	-	-	1	4
Other nei	117	100	100	100	100	-	-	-	-	-
47 Fishing area total	*734*	*804*	*810*	*832*	*666*	*618*	*587*	*372*	*501*	*445*
51 China,Taiwan	3 715	208	1 727	1 522	1 522	2 560	3 208	3 843	4 085	3 712
Japan	239	500	259	464	916	1 110	578	695	372	475
Korea Rep	-	3	7	1	75	101	10	79	16	-
Philippines	-	-	-	-	-	63	78	46	40	15
Seychelles	-	-	-	-	-	1	8	-	-	-
South Africa	-	-	-	-	0	1	1	2	-	-
Spain	-	-	-	-	-	-	16	1	3	5
Uruguay	-	-	-	-	-	-	-	-	2	32
Other nei	794	372	582	815	811	1 251	920	969	503	503
51 Fishing area total	*4 748*	*1 083*	*2 575*	*2 802*	*3 324*	*5 087*	*4 819*	*5 635*	*5 021*	*4 742*
57 Australia	1	33	4	0	2	-	-	-	-	0
China,Taiwan	1 098	54	435	421	421	240	492	1 046	282	678
Indonesia	645	884	1 001	1 903	2 285	2 140	2 499	1 384	1 220	1 220
Japan	54	97	171	112	215	213	358	304	142	112
Korea Rep	-	-	-	-	-	2	6	-	-	-
Philippines	-	-	-	-	-	6	36	22	1	1
Sri Lanka	2 031	6 589	37	0	-	-	-	-	-	-
Thailand	-	-	-	-	-	-	-	...	1	1
Other nei	204	97	149	242	229	536	453	477	35	35
57 Fishing area total	*4 033*	*7 754*	*1 797*	*2 678*	*3 152*	*3 137*	*3 844*	*3 233*	*1 681*	*2 047*
61 China,Taiwan	3 995	3 255	4 210	7 025	4 399	3 323	2 458	2 122	2 016	2 525
Japan	2 476	2 041	1 975	2 200	1 087	1 582	1 422	1 184	1 152	1 152 F
61 Fishing area total	*6 471*	*5 296*	*6 185*	*9 225*	*5 486*	*4 905*	*3 880*	*3 306*	*3 168*	*3 677 F*
71 China,Taiwan	345	2 379	4 018	4 462	4 399	6 385	7 153	7 117	6 442	5 866
Guam	32	66	31	15	24	13	14	30	15	13
Japan	3 029	3 023	2 671	1 066	1 312	1 506	1 522	1 813	1 969	1 969 F
Korea Rep	-	-	-	17	88	112	25	15	164	221
N Marianas	...	1	3	3	4	2	2	2	1	1
Philippines	679	991	1 178	1 191	1 096	1 470	2 208	3 173	3 752	3 856
71 Fishing area total	*4 085*	*6 460*	*7 901*	*6 754*	*6 923*	*9 488*	*10 924*	*12 150*	*12 343*	*11 926 F*
77 Amer Samoa	...	6	11	14	18	16	13	17	5	20
China,Taiwan	242	1 668	128	70	38	47	105	91	871	990
Japan	6 383	8 126	6 786	3 009	3 969	3 458	2 432	2 335	2 407	2 407 F
Korea Rep	84	5	1	6	223	567	256	207	170	175
USA	...	...	...	...	...	...	...	...	...	394

B-36 Tunas, bonitos, billfishes / Thons, pélamides, marlins / Atunes, bonitos, agujas

Capture production by species, fishing areas and countries or areas / Captures par espèces, zones de pêche et pays ou zones / Capturas por especies, áreas de pesca y países o áreas

Species, Fishing area Espèce, Zone de pêche Especie, Area de pesca	1993 t	1994 t	1995 t	1996 t	1997 t	1998 t	1999 t	2000 t	2001 t	2002 t
77 *Fishing area total*	*6 709*	*9 805*	*6 926*	*3 099*	*4 248*	*4 088*	*2 806*	*2 650*	*3 453*	*3 986 F*
81 China,Taiwan	15	248	1	4	2	1	2	16	26	17
Japan	62	49	66	31	54	83	26	10	15	15 F
Korea Rep	-	-	-	-	-	23	-	-	-	-
81 *Fishing area total*	*77*	*297*	*67*	*35*	*56*	*107*	*28*	*26*	*41*	*32 F*
87 China,Taiwan	37	19	0	3	1	1	0	1	1	-
Japan	301	697	493	253	271	543	120	72	163	163 F
Korea Rep	-	-	-	-	12	-	3	-	-	-
87 *Fishing area total*	*338*	*716*	*493*	*256*	*284*	*544*	*123*	*73*	*164*	*163 F*
Species total	*30 719 F*	*35 016 F*	*29 450 F*	*28 963 F*	*27 118 F*	*31 064 F*	*30 155 F*	*30 174 F*	*29 761*	*29 417 F*
Black marlin	**Makaire noir**		**Aguja negra**		***Makaira indica***			**1,75(03)005,07**		**BLM**
34 Belize	-	-	-	-	-	-	10	4	...	...
China,Taiwan	3	2	21	14	3	45	99	27	-	1
Korea Rep	-	-	-	-	3	-	-	-	-	-
34 *Fishing area total*	*3*	*2*	*21*	*14*	*6*	*45*	*109*	*31*	*...*	*1*
47 China,Taiwan	-	4	10	3	1	97	22	62	618	318
47 *Fishing area total*	-	*4*	*10*	*3*	*1*	*97*	*22*	*62*	*618*	*318*
51 China,Taiwan	977	41	376	247	247	226	171	512	94	127
Iran	-	-	-	3	3	5	1	1	-	3
Japan	38	47	35	...	...	...	...	...	...	48
Korea Rep	-	-	21	8	40	20	2	25	10	4
Mauritius	...	...	...	...	...	...	...	...	2	2
Seychelles	-	-	-	-	-	1	1	-	-	2
Spain	-	14	1	0	-	-	1	0	0	0
Other nei	56	118	164	141	78	147	88	79	54	54
51 *Fishing area total*	*1 071*	*220*	*597*	*399*	*368*	*399*	*264*	*617*	*160*	*240*
57 Australia	-	-	-	-	4	-	-	-	-	-
China,Taiwan	326	11	194	121	121	74	59	296	-	36
Indonesia	133	182	206	391	470	440	514	286	252	252
Japan	28	20	54	...	...	...	...	...	...	34
Korea Rep	-	-	-	-	-	-	11	-	13	2
Sri Lanka	2 612	46	0	39	48	59	69	68	58	64
Other nei	18	21	29	67	35	73	85	76	8	8
57 *Fishing area total*	*3 117*	*280*	*483*	*618*	*678*	*646*	*738*	*726*	*331*	*396*
61 China,Taiwan	340	1 138	483	830	659	285	681	332	193	355
61 *Fishing area total*	*340*	*1 138*	*483*	*830*	*659*	*285*	*681*	*332*	*193*	*355*
71 China,Taiwan	548	864	255	253	324	1 332	389	28	59	20
Korea Rep	-	-	76	9	10	-	7	16	115	147
71 *Fishing area total*	*548*	*864*	*331*	*262*	*334*	*1 332*	*396*	*44*	*174*	*167*
77 China,Taiwan	2	4	1	0	0	2	11	29	-	-
Korea Rep	1 375	483	395	300	182	403	793	735	1 153	997
Tonga	28	22	15	20	10	14	6	13	10	37
USA	...	...	...	...	...	...	...	...	...	4
77 *Fishing area total*	*1 405*	*509*	*411*	*320*	*192*	*419*	*810*	*777*	*1 163*	*1 038*
81 China,Taiwan	3	10	-	-	1	2	0	9	-	-
81 *Fishing area total*	*3*	*10*	-	-	*1*	*2*	*0*	*9*	-	-
87 China,Taiwan	-	-	-	1	1	0	0	1	-	-
Korea Rep	-	-	-	-	-	4	-	-	-	-
87 *Fishing area total*	-	-	-	*1*	*1*	*4*	*0*	*1*	-	-
Species total	*6 487*	*3 027*	*2 336*	*2 447*	*2 240*	*3 229*	*3 020*	*2 599*	*2 639*	*2 515*
Striped marlin	**Marlin rayé**		**Marlín rayado**		***Tetrapturus audax***			**1,75(03)009,03**		**MLS**
51 China,Taiwan	2 482	1 318	2 720	2 138	2 138	1 793	1 148	1 601	1 602	-
Japan	77	136	132	173	192	175	141	169	63	82
Korea Rep	3	2	38	-	65	43	-	20	2	-
Mauritius	...	...	...	...	...	...	...	...	1	2
Philippines	-	-	-	-	-	103	35	21	11	14
Seychelles	-	-	-	-	-	-	1	-	-	-
Spain	-	-	-	-	-	-	1	0	0	0
Other nei	1 233	771	1 019	995	462	588	510	511	287	287
51 *Fishing area total*	*3 795*	*2 227*	*3 909*	*3 306*	*2 857*	*2 702*	*1 836*	*2 322*	*1 966*	*385*
57 Australia	3	9	7	3	17	12	62	3	3	1
China,Taiwan	368	510	917	828	828	567	591	613	202	-
Indonesia	80	110	125	237	285	267	312	174	153	153
Japan	32	58	92	105	119	95	134	144	71	68
Korea Rep	-	-	-	-	-	-	1	-	1	-
Philippines	-	-	-	-	-	18	19	11	2	2
Sri Lanka	8	-	-	0	0	0	-	-	-	-
Thailand	-	-	-	-	-	-	-	...	1	1
Other nei	220	159	211	448	281	501	326	327	36	36
57 *Fishing area total*	*711*	*846*	*1 352*	*1 621*	*1 530*	*1 460*	*1 445*	*1 272*	*469*	*261*
61 China,Taiwan	227	322	173	148	160	108	505	408	1	-

B-36 Tunas, bonitos, billfishes / Thons, pélamides, marlins / Atunes, bonitos, agujas

Capture production by species, fishing areas and countries or areas / Captures par espèces, zones de pêche et pays ou zones / Capturas por especies, áreas de pesca y países o áreas

Species, Fishing area Espèce, Zone de pêche Especie, Area de pesca	1993 t	1994 t	1995 t	1996 t	1997 t	1998 t	1999 t	2000 t	2001 t	2002 t
Japan	4 484	4 347	4 297	3 712	2 308	4 220	3 588	2 973	2 875	2 875 F
61 Fishing area total	*4 711*	*4 669*	*4 470*	*3 860*	*2 468*	*4 328*	*4 093*	*3 381*	*2 876*	*2 875 F*
71 Australia	...	...	...	...	...	...	...	209	354	416
China,Taiwan	209	1 001	199	188	311	1 404	278	107	64	-
Japan	682	653	450	218	749	237	145	155	186	186 F
Korea Rep	11	-	-	51	20	11	13	17	3	9
71 Fishing area total	*902*	*1 654*	*649*	*457*	*1 080*	*1 652*	*436*	*488*	*607*	*611 F*
77 China,Taiwan	126	104	84	57	51	164	74	74	112	-
Japan	2 643	2 266	2 878	1 594	1 992	1 419	1 419	809	656	656 F
Korea Rep	651	590	374	400	1 087	739	571	600	343	266
USA	...	...	...	...	...	...	...	...	...	278
77 Fishing area total	*3 420*	*2 960*	*3 336*	*2 051*	*3 130*	*2 322*	*2 064*	*1 483*	*1 111*	*1 200 F*
81 Australia	...	...	...	...	...	...	...	307	363	353
China,Taiwan	186	171	33	56	50	80	98	189	36	-
Japan	573	637	813	728	449	914	330	133	266	266 F
New Zealand	0	0	0	0	1	-	-	-	-	-
81 Fishing area total	*759*	*808*	*846*	*784*	*500*	*994*	*428*	*629*	*665*	*619 F*
87 China,Taiwan	39	1	0	1	1	-	2	3	-	-
Japan	935	967	621	1 083	728	1 076	194	162	666	666 F
Korea Rep	-	-	-	-	-	2	4	-	-	-
87 Fishing area total	*974*	*968*	*621*	*1 084*	*729*	*1 078*	*200*	*165*	*666*	*666 F*
Species total	*15 272*	*14 132*	*15 183*	*13 163*	*12 294*	*14 536*	*10 502*	*9 740*	*8 360*	*6 617 F*
Atlantic white marlin	**Makaire blanc de l'Atlantique**		**Aguja blanca del Atlántico**		***Tetrapturus albidus***			**1,75(03)009,04**		**WHM**
21 Canada	-	4	4	8	8	8	5	5	3	2
China,Taiwan	-	4	3	0	0	-	-	-	-	-
Japan	1	1	-	-	-	2	-	-	-	-
21 Fishing area total	*1*	*9*	*7*	*8*	*8*	*10*	*5*	*5*	*3*	*2*
27 China,Taiwan	1	8	13	0	0	-	2	-	1	-
Japan	-	-	-	-	1	-	-	-	-	-
Portugal	-	-	-	-	-	-	-	-	-	1
27 Fishing area total	*1*	*8*	*13*	*0*	*1*	*-*	*2*	*-*	*1*	*1*
31 Barbados	29	26	43	15	41	33	25	25	...	...
Bermuda	1	1	1	1	1	1	1	0	1	0
China,Taiwan	75	42	133	38	29	45	22	3	11	-
Costa Rica	...	...	...	...	...	...	3	14	-	-
Cuba	0	-	-	-	-	-	-	-	-	-
Grenada	-	-	-	-	-	-	-	1	15	8
Japan	6	6	6	29	6	4	5	15	7	8
St Vincent	1	0	0	0	-	-	-	-	-	-
Spain	1	1	0	7	4	0	43	45	20	4
Venezuela	226	148	171	164	90	80	61	13	72	110
31 Fishing area total	*339*	*224*	*354*	*254*	*171*	*163*	*160*	*116*	*126*	*130*
34 Belize	-	-	-	-	1	-	1	0	...	...
China	-	6	7	6	7	10	20	1	7	4
China,Taiwan	350	293	128	141	110	124	175	321	-	-
Côte dIvoire	-	-	-	1	2	1	5	1	2	2
Cuba	0	-	-	-	-	-	-	-	-	-
France	12	11	9	7	7	9	8	7	-	-
Ghana	22	1	2	1	3	7	6	8	21	2
Japan	51	36	30	58	19	31	23	50	11	...
Philippines	-	-	-	-	-	0	4	-	-	-
Sao Tome Prn	-	-	-	-	45	-	-	-	-	-
Spain	14	24	9	68	66	64	45	72	24	-
Other nei	46	50	50	50	50	-	-	-	-	-
34 Fishing area total	*495*	*421*	*235*	*332*	*310*	*246*	*287*	*460*	*65*	*8*
37 Spain	-	-	1	-	1	1	-	1	0	0
37 Fishing area total	*-*	*-*	*1*	*-*	*1*	*1*	*-*	*1*	*0*	*0*
41 Brazil	301	91	105	75	105	217	158	106	172	407
China,Taiwan	395	366	336	311	242	115	101	118	6	-
Japan	23	24	5	5	8	3	6	6	14	...
Spain	-	0	14	5	2	4	28	61	18	-
Uruguay	0	6	1	2	50	22	-	-	0	2
41 Fishing area total	*719*	*487*	*461*	*398*	*407*	*361*	*293*	*291*	*210*	*409*
47 China	-	3	4	3	4	5	10	1	13	19
China,Taiwan	134	186	295	77	60	223	164	281	17	-
Japan	33	32	18	23	21	19	16	18	28	4
Philippines	-	-	-	-	-	1	8	-	-	-
Spain	8	-	4	27	1	-	17	7	-	2
Other nei	68	50	50	50	50	-	-	-	-	-
47 Fishing area total	*243*	*271*	*371*	*180*	*136*	*248*	*215*	*307*	*58*	*25*
Species total	*1 798*	*1 420*	*1 442*	*1 172*	*1 034*	*1 029*	*962*	*1 180*	*463*	*575*
Shortbill spearfish	**Makaire à rostre court**		**Marlín trompa corta**		***Tetrapturus angustirostris***			**1,75(03)009,05**		**SSP**
51 Japan	...	...	...	...	...	...	...	...	...	28

B-36 Tunas, bonitos, billfishes / Thons, pélamides, marlins / Atunes, bonitos, agujas

Capture production by species, fishing areas and countries or areas / Captures par espèces, zones de pêche et pays ou zones / Capturas por especies, áreas de pesca y países o áreas

Species, Fishing area Espèce, Zone de pêche Especie, Area de pesca	1993 t	1994 t	1995 t	1996 t	1997 t	1998 t	1999 t	2000 t	2001 t	2002 t
Réunion	-	-	-	2	1	3	5	5	5	5 F
Spain	-	-	-	-	1	2	2	0	0	1
51 Fishing area total	...	...	...	*2*	*2*	*5*	*7*	*5*	*5*	*34 F*
57 Australia	-	-	-	-	-	-	1	-	0	0
Japan	...	...	...	...	...	...	...	...	...	6
57 Fishing area total	...	...	...	...	...	...	*1*	...	*0*	*6*
71 Australia	...	...	...	...	...	...	...	...	...	10
71 Fishing area total	...	...	...	...	...	...	...	...	...	*10*
81 Australia	...	...	...	...	...	...	...	...	...	6
New Zealand	-	-	8	42	18	-	-	-	-	0
81 Fishing area total	...	...	*8*	*42*	*18*	...	...	...	...	*6*
Species total	...	...	*8*	*44*	*20*	*5*	*8*	*5*	*5*	*56 F*
Longbill spearfish	**Makaire bécune**		**Aguja picuda**		***Tetrapturus pfluegeri***			**1,75(03)009,06**		**SPF**
31 Japan	...	...	...	...	...	...	...	...	...	8
Spain	-	-	-	-	-	-	-	3	3	0
Trinidad Tob	62	-	-	-	-	-	-	-	-	-
Venezuela	-	0	-	1	0	1	0	-	4	0
31 Fishing area total	*62*	*0*	...	*1*	*0*	*1*	*0*	*3*	*7*	*8*
34 France	112	98	78	59	68	86	81	60	-	-
Japan	...	...	...	...	...	...	...	...	...	14
Spain	0	0	3	1	-	-	2	3	1	2
34 Fishing area total	*112*	*98*	*81*	*60*	*68*	*86*	*83*	*63*	*1*	*16*
41 Brazil	-	-	-	-	-	-	-	12	56	39
Spain	2	-	1	-	0	-	22	47	20	5
41 Fishing area total	*2*	-	*1*	-	*0*	-	*22*	*59*	*76*	*44*
47 China	-	-	-	-	-	2	-	-	-	-
Spain	8	-	-	-	1	1	28	11	7	5
47 Fishing area total	*8*	-	-	-	*1*	*3*	*28*	*11*	*7*	*5*
Species total	*184*	*98*	*82*	*61*	*69*	*90*	*133*	*136*	*91*	*73*
Marlins,sailfishes,etc. nei	**Makaires,marlins,voiliers nca**		**Agujas,marlines,peces vela nep**		***Istiophoridae***			**1,75(03)XXX,XX**		**BIL**
27 Portugal	-	-	-	-	-	-	-	68	408	626
27 Fishing area total	-	-	-	-	-	-	-	*68*	*408*	*626*
31 Barbados	-	-	-	-	-	-	-	-	85	53
Bermuda	5	-	-	-	3	1	-	-	-	-
China,Taiwan	-	-	2	15	3	-	-	-	-	-
Dominica	...	...	...	...	...	...	...	...	69	72
Korea Rep	-	-	-	17	-	-	-	-	-	-
Mexico	2	8	22	73	77	164	104	133	97	41
St Lucia	-	-	-	-	4	1	-	14	5	9
St Vincent	-	-	-	1	0	2	1	343	339	...
Spain	-	0	3	6	11	29	-	-	1	-
Trinidad Tob	-	-	-	-	-	-	-	-	2	5
Venezuela	-	-	-	-	-	-	-	-	8	-
31 Fishing area total	*7*	*8*	*27*	*112*	*98*	*197*	*105*	*490*	*606*	*180*
34 China	-	-	-	18	-	-	-	-	-	-
France	-	-	-	-	-	-	-	66	-	-
Gabon	4	5	0	523	7	-	-	-	1	-
Korea Rep	-	-	-	72	36	-	-	-	-	-
Liberia	27	112	120	145	71	781	513	683	163	165 F
Portugal	0	-	-	0	-	15	13	5	6	62
Seychelles	-	-	-	-	-	-	-	19	-	-
Spain	-	1	10	10	17	28	-	-	-	1
Togo	...	...	...	...	32	...	110	77	205	158
34 Fishing area total	*31*	*118*	*130*	*768*	*163*	*824*	*636*	*850*	*375*	*386 F*
37 Korea Rep	-	-	-	-	28	-	-	-	-	-
Malta	-	-	1	1	1	-	-	-	-	-
Portugal	-	-	-	-	-	-	-	1	25	2
37 Fishing area total	-	-	*1*	*1*	*29*	-	-	*1*	*25*	*2*
41 Brazil	-	-	-	-	-	-	-	18	2	1
China,Taiwan	3	17	78	9	2	24	0	0	8	-
Korea Rep	-	92	62	24	1	-	-	-	-	-
Spain	26	42	-	-	27	38	-	-	-	-
41 Fishing area total	*29*	*151*	*140*	*33*	*30*	*62*	*0*	*18*	*10*	*1*
47 Korea Rep	-	-	-	12	20	-	-	-	-	-
Namibia	-	-	-	-	-	-	-	-	-	4
Portugal	-	-	-	-	-	-	-	1	30	1
St Helena	...	...	2	2	1	2	4	4	3	4
South Africa	-	-	-	-	-	-	-	-	-	16
Spain	24	4	-	1	40	32	-	-	-	-
47 Fishing area total	*24*	*4*	*2*	*15*	*61*	*34*	*4*	*5*	*33*	*25*
51 China	-	-	-	-	-	-	-	142	121	218

B-36 Tunas, bonitos, billfishes / Thons, pélamides, marlins / Atunes, bonitos, agujas

Capture production by species, fishing areas and countries or areas
Captures par espèces, zones de pêche et pays ou zones
Capturas por especies, áreas de pesca y países o áreas

Species, Fishing area Espèce, Zone de pêche Especie, Area de pesca	1993 t	1994 t	1995 t	1996 t	1997 t	1998 t	1999 t	2000 t	2001 t	2002 t
China,Taiwan	-	-	-	-	-	-	-	-	57	-
Comoros	115	136	136	130	120	120	100	130 F	120	120 F
India	1 274	1 577	1 301	1 864	2 128	1 557	1 188	1 666	2 038	132
Kenya	-	-	-	73	53	38	82	80	78	55
Korea Rep	1 548	2 003	1 248	2 128	945	219	4	174	74	38
Mauritius	194	227	196	190	639	295	287	287	287	287
Mayotte	...	...	...	...	...	37	10	...	...	...
Pakistan	2 245	2 932	2 684	2 834	2 198	2 264	2 340	2 215	1 122	995
Portugal	-	-	-	-	-	3	-	-	1	-
Réunion	61	72	87	120	110	136	109	123	111	111 F
Seychelles	-	2	2	80	32	6	11	33	66	125
South Africa	1	-	-	-	2	2	4	2	4	12
Tanzania	531	760	580	656	700	800	780	800	850	700
Untd Arab Em	193	193	232	239	230	233	233	250	250 F	...
Other nei	320	176	201	126	309	279	191	202	141	141
51 Fishing area total	*6 482*	*8 078*	*6 667*	*8 440*	*7 466*	*5 989*	*5 339*	*6 104 F*	*5 320 F*	*2 934 F*
57 China	-	-	-	-	-	-	287	344	258	37
China,Taiwan	-	-	-	-	-	-	-	-	700	-
Honduras	-	-	-	-	-	-	-	-	182	-
India	585	300	242	2 076	2 374	1 824	3 014	1 904	2 355	82
Indonesia	596	723	710	753	772	742	806	851	701	701
Korea Rep	-	-	25	9	22	15	14	1	4	4
Malaysia	0	3	3	-	5	0	0	0	1	14
Seychelles	-	-	-	-	-	-	0	2	4	6
Sri Lanka	252	347	5 196	5 675	6 937	8 653	7 253	5 728	3 262	5 191
Thailand	-	-	-	-	-	-	-	16	-	-
Other nei	72	47	35	17	35	78	103	108	8	8
57 Fishing area total	*1 505*	*1 420*	*6 211*	*8 530*	*10 145*	*11 312*	*11 477*	*8 954*	*7 475*	*6 043*
61 China	-	-	-	-	-	-	-	-	146	123
Korea Rep	-	-	-	-	-	2	3	-	-	-
61 Fishing area total	*-*	*-*	*-*	*-*	*-*	*2*	*3*	*-*	*146*	*123*
71 Australia	...	...	...	199	1 084	1 929	2 185	3	4	-
China	-	-	-	-	-	-	114	469	171	61
China,Taiwan	-	-	-	-	-	-	-	-	42	-
Korea Rep	502	103	87	254	230	361	59	40	100	104
Malaysia	469	230	450	274	357	324	2 046	161	154	205
Palau	0	2	0	1	0	0	0	1	4	0
Papua N Guin	...	0	0	16	6	87	230	418	368	340 F
Philippines	4 575	2 423	4 648	4 317	4 799	5 338	5 082	4 969	6 037	6 135
Solomon Is	50 F	50 F	50 F	50 F	50 F	50 F	50 F	50 F	50 F	50 F
Vanuatu	...	...	209	98	97	99	14	...	...	...
71 Fishing area total	*5 596 F*	*2 808 F*	*5 444 F*	*5 209 F*	*6 623 F*	*8 188 F*	*9 780 F*	*6 111 F*	*6 930 F*	*6 895 F*
77 China	-	-	-	-	-	-	-	45	103	532
Cook Is	...	26	24	40 F	17 F	15 F	10 F	5 F	1 F	62 F
Costa Rica	388	1 226	1 764	1 892	1 876	1 647	2 143	2 035	2 206	2 273
Fr Polynesia	...	465	598	587	598	518	703	566	551	527
Korea Rep	2 723	2 466	3 309	2 840	1 952	3 240	1 333	1 112	1 479	1 673
Mexico	235	87	204	553	1 391	2 375	283	242	250	196
USA	1 324	1 048	1 530	1 368	100	134	214	123	250	142
77 Fishing area total	*4 670*	*5 318*	*7 429*	*7 280 F*	*5 934 F*	*7 929 F*	*4 686 F*	*4 128 F*	*4 840 F*	*5 405 F*
81 Korea Rep	-	-	-	-	-	12	-	-	-	-
81 Fishing area total	*-*	*-*	*-*	*-*	*-*	*12*	*-*	*-*	*-*	*-*
87 Ecuador	-	-	-	-	-	3 727	6 962	...	...	...
Korea Rep	22	-	-	-	-	-	-	-	-	-
Spain	-	-	-	-	-	-	-	-	-	6
87 Fishing area total	*22*	*-*	*-*	*-*	*-*	*3 727*	*6 962*	*...*	*...*	*6*
Species total	*18 366 F*	*17 905 F*	*26 051 F*	*30 388 F*	*30 549 F*	*38 276 F*	*38 992 F*	*26 729 F*	*26 168 F*	*22 626 F*
Swordfish	**Espadon**		**Pez espada**		***Xiphias gladius***				**1,75(04)003,01**	**SWO**
21 Canada	2 230	1 675	1 610	739	1 089	1 115	1 118	968	1 079	959
China,Taiwan	-	-	1	-	-	-	1	3	-	10
Japan	138	75	27	37	37	81	128	34	1	...
USA	1 531	1 248	1 215	1 661	918	1 093	1 111	1 163	948	1 075
21 Fishing area total	*3 899*	*2 998*	*2 853*	*2 437*	*2 044*	*2 289*	*2 358*	*2 168*	*2 028*	*2 044*
27 China,Taiwan	-	-	38	-	-	2	19	10	1	26
Faeroe Is	-	-	-	-	-	-	5	4	...	...
France	95	46	84	97	164	110	104	122	101	74
Iceland	-	-	-	-	-	-	2	2	-	-
Ireland	-	-	-	15	15	132	81	35	17	4
Japan	47	37	60	111	42	56	46	18	1	...
Portugal	1 931	1 542	1 462	1 158	750	602	465	667	665	485
Spain	4 834	4 561	5 001	4 654	3 827	2 517	2 076	2 422	2 274	2 041
UK	2	3	1	5	11	0	2	1	-	-
27 Fishing area total	*6 909*	*6 189*	*6 646*	*6 040*	*4 809*	*3 419*	*2 800*	*3 281*	*3 059*	*2 630*
31 Barbados	-	-	-	33	16	16	12	13	19	10
Bermuda	-	-	1	1	5	5	3	3	2	2
Br Virgin Is	...	58	19	54	6	5	5	2	2 F	2 F
China,Taiwan	90	70	19	59	41	14	15	50	10	63
Costa Rica	...	...	...	...	...	...	1	2	-	-

B-36 Tunas, bonitos, billfishes / Thons, pélamides, marlins / Atunes, bonitos, agujas

Capture production by species, fishing areas and countries or areas
Captures par espèces, zones de pêche et pays ou zones
Capturas por especies, áreas de pesca y países o áreas

Species, Fishing area Espèce, Zone de pêche Especie, Area de pesca	1993 t	1994 t	1995 t	1996 t	1997 t	1998 t	1999 t	2000 t	2001 t	2002 t
Cuba	8	10	15	5	10	10	5	11	3	1
Dominica	...	...	...	...	...	...	...	...	1	-
Grenada	7	0	1	4	47	33	42	84	73	54
Japan	28	33	27	69	50	78	104	41	8	...
Mexico	6	14	13	11	14	28	24	37	27	22
Philippines	-	-	-	-	-	-	-	-	1	4
Portugal	-	-	-	-	-	1	-	-	-	39
St Lucia	-	1	0	-	-	-	-	-	-	1
St Vincent	23	0	4	3	1	0	1	0	22	...
Spain	1 761	1 619	1 948	886	1 309	1 562	1 917	2 161	1 694	1 913
Trinidad Tob	11	180	150	158	110	130	138	41	75	92
USA	1 445	1 184	1 270	486	1 123	1 122	1 068	1 281	816	834
Venezuela	73	69	54	85	20	37	30	30	21	34
Other nei	111	-	-	-	-	-	-	-	-	-
31 Fishing area total	*3 563*	*3 238*	*3 521*	*1 854*	*2 752*	*3 041*	*3 365*	*3 756*	*2 774 F*	*3 071 F*
34 Belize	-	-	1	-	-	-	17	8	...	...
Benin	28	25	24	24 F	10	0	3	...	41	68
China	73	86	104	132	40	38	304	22	102	90
China,Taiwan	18 F	618	1 356	1 737	1 206	649	606	1 666	1 986	512
Côte dIvoire	14	20	19	26	18	25	26	20	19	19
Cuba	200	492	849	62	-	-	-	-	-	-
France	-	-	-	-	-	-	-	4	-	-
Ghana	121	51	103	140	44	106	121	117	531	372
Greece	3	6	2	1	-	-	-	-	-	3
Japan	2 724	1 870	1 986	2 116	878	1 483	955	562	262	373
Korea Rep	-	-	-	18	33	-	-	-	-	-
Liberia	14	26	28	112	543	21	39	42	34	35 F
Lithuania	-	794	-	-	-	-	-	-	-	-
Morocco	39	36	79	462	267	191	119	114	523	223
Nigeria	0	857	-	9	-	-	-	-	-	133
Panama	-	-	-	-	-	-	62	-	-	-
Portugal	30	57	382	934	153	321	662	144	69	400
Sao Tome Prn	-	-	-	-	14	14 F	14 F	...	...	...
Senegal	-	-	-	-	-	169	127	39	35	61
Seychelles	-	-	-	-	-	-	-	10	-	-
Sierra Leone	...	...	...	...	...	...	...	2	2	...
Spain	3	5	4	7	4	-	-	12	1	3
Togo	8	14	14	64	2	23	37	7	17	21
34 Fishing area total	*3 275 F*	*4 957*	*4 951*	*5 844 F*	*3 212*	*3 040 F*	*3 092 F*	*2 769*	*3 622*	*2 313 F*
37 Albania	...	...	0	13	0	-	-	-	2	15
Algeria	562	600	807	807	807	825	709	816	1 081	814
China,Taiwan	-	-	-	1	1	-	-	-	1	-
Croatia	...	...	...	...	...	10	20	...	...	...
Cyprus	116	159	89	40	51	61	92	82	135	104
France	-	-	-	-	-	-	-	-	12	27
Greece	1 568	2 520	974	1 237	750	1 650	1 520	1 960	1 730	975
Italy	6 330	7 765	6 725	5 286	6 104	6 104	6 312	7 515	6 388	3 539
Japan	6	4	6	6	3	7	4	1	1	...
Libya	-	-	-	-	-	11	...	8	6	...
Malta	91	47	72	72	100	153	187	175	102	253
Morocco	2 589	2 654	1 696	2 734	4 900	3 228	3 238	2 708	3 026	3 379
Portugal	-	-	-	-	-	-	-	13	115	8
Spain	1 358	1 503	1 379	1 186	1 264	1 443	906	1 436	1 484	1 498
Tunisia	354	298	378	352	346	414	468	483	567	1 138
Turkey	292	533	306	320	350	450	230	373	360	370
37 Fishing area total	*13 266*	*16 083*	*12 432*	*12 054*	*14 676*	*14 356*	*13 686*	*15 570*	*15 010*	*12 120*
41 Argentina	14	24	0	-	-	-	-	-	5	-
Brazil	2 013	1 571	1 975	1 892	4 100	3 847	4 721	4 697	4 082	2 910
China	-	-	-	-	-	328	-	-	-	-
China,Taiwan	541	1 354	970	1 023	710	366	212	696	430	712
Japan	587	657	174	170	149	71	93	143	166	41
Panama	-	-	-	-	-	-	48	-	-	-
Philippines	-	-	-	-	-	-	-	-	6	1
Portugal	-	-	153	-	81	61	-	90	63	47
Spain	4 545	5 186	7 339	3 671	4 475	3 119	3 150	4 496	3 770	4 247
Uruguay	260	165	499	644	760	889	661	713	636	768
41 Fishing area total	*7 960*	*8 957*	*11 110*	*7 400*	*10 275*	*8 681*	*8 885*	*10 835*	*9 158*	*8 726*
47 Angola	0	0	0	0	0	0	0	0	...	...
China	-	-	-	-	-	-	534	344	200	423
China,Taiwan	100 F	540	981	573	398	402	602	824	747	635
Japan	2 458	3 173	1 876	1 670	877	1 043	768	604	353	400
Namibia	-	-	-	-	-	-	730	469	751	744
Norway	-	-	-	-	1	-	-	-	-	-
Panama	-	-	-	-	-	-	12	-	-	-
Portugal	-	-	-	-	360	172	31	224	330	174
Seychelles	-	-	-	-	-	-	-	-	-	6
South Africa	4	1	4	1	1	169	76	230	397	500
Spain	2 429	2 751	3 951	5 951	3 986	2 713	2 608	1 892	2 019	1 494
47 Fishing area total	*4 991 F*	*6 465*	*6 812*	*8 195*	*5 623*	*4 499*	*5 361*	*4 587*	*4 797*	*4 376*
51 China	-	-	-	-	-	-	8	79	93	327
China,Taiwan	3 960 F	3 631	16 312	7 493	13 369	13 778	12 901	10 914	11 528	12 746
India	-	-	-	60	56	-	-	-	0	-
Iran	-	-	-	9	9	13	2	1	-	7
Japan	628	999	559	927	993	1 406	676	689	650	967

B-36 Tunas, bonitos, billfishes / Thons, pélamides, marlins / Atunes, bonitos, agujas

Capture production by species, fishing areas and countries or areas / Captures par espèces, zones de pêche et pays ou zones / Capturas por especies, áreas de pesca y países o áreas

	Species, Fishing area Espèce, Zone de pêche Especie, Area de pesca	1993 t	1994 t	1995 t	1996 t	1997 t	1998 t	1999 t	2000 t	2001 t	2002 t
	Korea Rep	20	17	74	51	196	147	8	63	18	9
	Mauritius	...	...	...	...	...	...	...	...	37	189
	Mayotte	...	...	...	...	...	622	314	...	...	...
	Mozambique	-	-	312	358	524	1 039	447	...	...	...
	Philippines	-	-	-	-	-	260	236	140	134	215
	Portugal	-	-	-	-	-	105	230	197	567	785
	Réunion	286	734	769	1 336	1 586	2 080	1 930	1 744	1 572	1 572 F
	Seychelles	-	-	22	141	317	216	317	435	600	532
	South Africa	-	-	-	-	-	236	48	20	229	592
	Spain	207	694	19	29	508	1 425	2 013	983	1 860	3 222
	Uruguay	-	-	-	-	-	-	-	-	80	510
	Other nei	4 370	3 366	5 410	6 526	4 388	5 835	5 100	5 640	2 736	2 736
51	*Fishing area total*	*9 471 F*	*9 441*	*23 477*	*16 930*	*21 946*	*27 162*	*24 230*	*20 905*	*20 104*	*24 409 F*
57	Australia	189	115	62	22	43	337	1 360	1 798	2 900	2 005
	China	-	-	71	238	255	117	262	294	170	70
	China,Taiwan	520	545	1 949	2 127	3 794	3 051	1 826	3 213	2 620	3 776
	Honduras	-	-	-	-	-	-	-	-	165	-
	India	-	-	-	2	2	25	15	-	30	...
	Indonesia	332	456	516	981	1 178	1 103	1 288	714	630	630
	Japan	317	360	684	758	598	538	639	576	497	412
	Korea Rep	-	-	2	-	8	2	21	-	19	3
	Malaysia	-	-	-	-	-	-	-	-	-	4
	Philippines	-	-	-	-	-	12	89	53	15	57
	Seychelles	-	-	-	-	-	-	1	11	21	46
	Spain	-	-	-	-	-	-	-	-	-	280
	Sri Lanka	4 662	2 407	2 558	2 591	3 167	3 950	2 132	5 545	4 757	2 467
	Thailand	-	-	-	-	-	-	-	19	6	6
	Other nei	343	748	762	2 023	1 676	2 269	1 984	2 193	329	329
57	*Fishing area total*	*6 363*	*4 631*	*6 604*	*8 742*	*10 721*	*11 404*	*9 617*	*14 416*	*12 159*	*10 085*
61	China	-	-	-	-	-	-	-	-	104	174
	China,Taiwan	1 180	785	508	489	763	760	577	2 620	1 710	1 504
	Japan	8 789	8 392	7 376	5 493	4 991	7 259	7 040	7 242	5 506	5 506 F
	Russian Fed	-	-	-	-	-	-	-	-	-	2
61	*Fishing area total*	*9 969*	*9 177*	*7 884*	*5 982*	*5 754*	*8 019*	*7 617*	*9 862*	*7 320*	*7 186 F*
67	Japan	-	5	-	-	-	-	-	-	-	-
	USA	313	90	22	23	34	128	27	34	14	0
67	*Fishing area total*	*313*	*95*	*22*	*23*	*34*	*128*	*27*	*34*	*14*	*0*
71	Australia	...	...	...	...	...	...	...	933	1 064	1 170
	China	-	-	-	-	-	-	396	72	95	47
	China,Taiwan	94 F	1 259	777	712	1 477	1 741	2 177	101	17	2 143
	Japan	692	789	645	406	1 000	398	369	427	589	589 F
	Korea Rep	27	-	9	5	513	19	3	24	100	185
	Palau	...	...	...	...	...	...	...	1	2	...
	Philippines	-	-	-	-	-	-	-	2 677	3 158	3 421
71	*Fishing area total*	*813 F*	*2 048*	*1 431*	*1 123*	*2 990*	*2 158*	*2 945*	*4 235*	*5 025*	*7 555 F*
77	Amer Samoa	-	-	-	-	-	2	0	1	1	1
	China	-	-	-	-	-	-	-	68	319	852
	China,Taiwan	210 F	134	93	21	22	48	162	118	592	27
	Cook Is	...	26	24	41 F	17 F	15 F	10 F	5 F	2 F	62 F
	Costa Rica	35	17	29	433	1 072	419	99	407	653	638
	Fr Polynesia	65	71	62	84	56	58	66	47	79	70
	Japan	2 709	2 101	1 979	2 114	1 678	2 620	1 608	1 952	3 029	3 029 F
	Korea Rep	82	15	13	55	198	401	439	578	894	954
	Mexico	806	567	424	428	2 351	3 575	1 112	2 179	753	443
	Tonga	5	8	10	7	6	8	5	53	8	42
	USA	6 668	4 400	3 409	3 672	4 088	4 503	5 071	5 598	2 490	2 012
77	*Fishing area total*	*10 580 F*	*7 339*	*6 043*	*6 855 F*	*9 488 F*	*11 649 F*	*8 572 F*	*11 006 F*	*8 820 F*	*8 130 F*
81	Australia	...	...	...	...	...	...	...	1 144	789	1 167
	China,Taiwan	51 F	223	26	60	61	128	117	106	17	121
	Japan	1 278	1 419	1 096	1 387	1 029	1 164	751	537	488	488 F
	New Zealand	-	89	93	152	170	564	1 004	975	1 029	929
81	*Fishing area total*	*1 329 F*	*1 731*	*1 215*	*1 599*	*1 260*	*1 856*	*1 872*	*2 762*	*2 323*	*2 705 F*
87	Chile	4 712	3 801	2 594	3 145	4 040	4 492	2 925	2 973	3 262	3 523
	China,Taiwan	-	6	-	20	21	1	1	9	2	9
	Colombia	-	-	-	-	-	6	-	-	1	...
	Ecuador	33	0	222	...	...	...	...	...	...	...
	Japan	605	824	523	388	270	694	404	355	1 267	1 267 F
	Korea Rep	-	-	-	-	19	4	3	-	-	-
	Peru	19	5	-	1	-	57	42	20	356	278
	Spain	928	576	698	772	2 039	1 346	1 121	1 807	828	6 322
87	*Fishing area total*	*6 297*	*5 212*	*4 037*	*4 326*	*6 389*	*6 600*	*4 496*	*5 164*	*5 716*	*11 399 F*
	Species total	*88 998 F*	*88 561*	*99 038*	*89 404 F*	*101 973 F*	*108 301 F*	*98 923 F*	*111 350 F*	*101 929 F*	*106 749 F*
	Tuna-like fishes nei	**Poissons type thon nca**		**Peces parec.a los atunes nep**		***Scombroidei***				**1,75(XX)XXX,XX**	**TUX**
21	Canada	8	-	5	-	-	0	-	-	-	-
	Russian Fed	-	-	-	-	-	-	-	3	-	-
	USA	117	57	45	115	18	15	29	35	13	4
21	*Fishing area total*	*125*	*57*	*50*	*115*	*18*	*15*	*29*	*38*	*13*	*4*
27	Portugal	0	25	417	252	155	263	87	101	61	43

B-36

Tunas, bonitos, billfishes — **Capture production by species, fishing areas and countries or areas**
Thons, pélamides, marlins — **Captures par espèces, zones de pêche et pays ou zones**
Atunes, bonitos, agujas — **Capturas por especies, áreas de pesca y países o áreas**

	Species, Fishing area Espèce, Zone de pêche Especie, Area de pesca	1993 t	1994 t	1995 t	1996 t	1997 t	1998 t	1999 t	2000 t	2001 t	2002 t
	Spain	-	-	-	-	-	-	-	-	206	588
27	*Fishing area total*	*0*	*25*	*417*	*252*	*155*	*263*	*87*	*101*	*267*	*631*
31	Antigua Barb	2	...	...	...	...	...	...	...	28	12
	Barbados	167	162	255	68	-	-	-	11	-	-
	Br Virgin Is	...	20	15	19	-	-	-	-	-	-
	Colombia	228	112	265	9 647 F	5 188 F	13 259 F	5 380 F	5 000 F	5 000 F	5 000 F
	Costa Rica	3	0	0	0	0	0	6	26	3	2
	Dominica	-	-	-	-	-	-	-	-	12	36
	Dominican Rp	86	155	138	85	855	354	192	249	517	668
	Jamaica	...	...	...	239	275	...	...	78	86	48
	Panama	-	-	-	-	-	-	2	1	-	-
	Puerto Rico	62	64	82	88	72	121	99	111	17	6
	St Kitts Nev	...	...	1	3	7	16	24	24	23	13
	St Lucia	71	15	10	8	1	3	3	1	-	-
	Trinidad Tob	-	-	25	134	206	92	81	138	405	482
	USA	116	29	23	6	33	17	25	33	47	20
	Venezuela	-	862	-	-	-	-	-	-	13	-
31	*Fishing area total*	*735*	*1 419*	*814*	*10 297 F*	*6 637 F*	*13 862 F*	*5 812 F*	*5 672 F*	*6 151 F*	*6 287 F*
34	Belize	-	-	-	-	-	-	107	143	125	172
	Bulgaria	-	-	-	-	-	225	-	-	-	-
	Cameroon	6	1	1	0	0	0	0	1	...	59
	Cape Verde	-	-	-	-	245	-	-	-	-	-
	China	41	68	76	80	-	-	150	6	187	80
	Cuba	3	30	3	21	-	-	-	-	-	-
	Cyprus	-	-	-	...	...	39	72	173	268	-
	Eq Guinea	390	380	340	216	300 F	392	200 F	50 F	50 F	50 F
	France	-	-	-	-	-	-	-	-	-	53
	Greece	7	4	-	-	-	-	-	-	3	-
	GuineaBissau	5 F	5 F	6 F	6 F	6 F	5 F	4 F	4 F	4 F	4 F
	Italy	-	-	-	-	1	-	-	-	-	-
	Korea Rep	24	15	39	120	8	-	-	-	-	-
	Latvia	-	-	-	-	-	147	27	-	-	-
	Liberia	-	-	-	-	-	3	2	14	8	10 F
	Lithuania	73	-	...	...	...	467	110	80	154	216
	Mauritania	746	54	263	2 479	2 170	1 304	...	...	...	...
	Morocco	-	-	-	-	-	-	82	-	-	108
	Nigeria	157	109	119	200	55	73	7	51	5	1
	Philippines	-	-	-	-	-	15	5	-	-	-
	Portugal	-	-	-	-	-	-	-	-	16	55
	St Vincent	-	-	-	-	-	-	-	-	49	...
	Sao Tome Prn	183 F	-	-	-	9	-	-	-	-	-
	Senegal	3	3	52	-	-	-	-	-	6	8
	Seychelles	-	-	-	-	-	-	-	8	-	-
	Sierra Leone	601	599	598	254	2 618	4 980	2 166	623	6 638	498
	Spain	-	-	-	-	-	-	-	789	322	16
	Ukraine	3	4	...	...	...	303	-	28	213	39
	Other nei	-	-	-	-	-	-	-	744	-	1
34	*Fishing area total*	*2 242 F*	*1 272 F*	*1 497 F*	*3 376 F*	*5 412 F*	*7 953 F*	*2 932 F*	*2 714 F*	*8 048 F*	*1 370 F*
37	Cyprus	-	-	-	-	-	-	-	7	-	8
	Egypt	-	-	530	1 071	594	576	2 004	1 676	778	1 301
	Greece	510	116	116	116	145	300	...	195	125	-
	Japan	-	-	1	-	1	-	-	-	-	-
	Korea Rep	-	-	2	-	248	-	-	-	-	-
	Lebanon	175	200	500	500	700	400	400	500	450	400
	Malta	8	0	0	0	0	0	-	-	-	-
	Morocco	-	-	-	-	-	-	71	-	-	-
	Palest, O.T.	...	...	...	50	102	92	100 F	80 F	60 F	54
	Portugal	-	-	-	-	-	-	-	-	-	1
	Spain	-	-	-	8	-	-	-	-	-	-
	Tunisia	309	105	115	215	657	6	814	905	989	104
	Other nei	-	-	-	50	-	-	-	-	-	-
37	*Fishing area total*	*1 002*	*421*	*1 264*	*2 010*	*2 447*	*1 374*	*3 389 F*	*3 363 F*	*2 402 F*	*1 868*
41	Argentina	2	0	0	0	-	-	-	-	-	-
	Brazil	287	140	58	-	446	258	570	151	5	467
	Korea Rep	-	3	9	-	-	-	-	-	-	-
	Panama	-	-	-	-	-	-	68	14	-	-
	Portugal	-	-	-	-	2	7	-	78	4	13
	Uruguay	0	0	0	2	50	94	136	95	20	51
41	*Fishing area total*	*289*	*143*	*67*	*2*	*498*	*359*	*774*	*338*	*29*	*531*
47	Angola	-	-	-	-	-	-	-	-	-	6 579
	Cambodia	-	-	-	-	-	-	56	-	-	-
	China	-	-	-	-	90	-	265	229	346	510
	Honduras	-	-	13	10	15	5	20	-	-	-
	Korea Rep	-	9	66	36	105	62	23	10	3	-
	Panama	-	-	-	-	-	-	7	-	-	-
	Philippines	-	-	-	-	-	-	7	-	-	-
	Portugal	-	-	72	0	7	19	43	31	42	23
	South Africa	-	-	-	-	-	-	-	-	1	506
	Spain	-	-	-	-	-	-	-	-	61	130
47	*Fishing area total*	*...*	*9*	*151*	*46*	*217*	*86*	*421*	*270*	*453*	*7 748*
51	China	-	-	-	-	-	-	-	179	125	89
	Comoros	434	513	513	510	490	490	450	520 F	450	450 F
	Djibouti	13	14	15	15	14	15	15 F	15 F	15 F	15 F

B-36

Tunas, bonitos, billfishes — **Capture production by species, fishing areas and countries or areas**
Thons, pélamides, marlins — **Captures par espèces, zones de pêche et pays ou zones**
Atunes, bonitos, agujas — **Capturas por especies, áreas de pesca y países o áreas**

Species, Fishing area Espèce, Zone de pêche Especie, Area de pesca		1993 t	1994 t	1995 t	1996 t	1997 t	1998 t	1999 t	2000 t	2001 t	2002 t
	Eritrea	...	7	1	30	44	111	64	42	96	136
	India	7 588	3 130	4 951	21	60	4 931	-	-	-	4 044
	Iran	-	-	-	4	5	9	-	-	-	-
	Italy	-	-	-	-	-	-	-	-	8	-
	Japan	650	-	-	-	-	-	-	-	-	-
	Korea Rep	796	584	577	1 036	1 545	705	182	529	294	22
	Mauritius	344	141	189	70	199	44	681	726	745	745
	Mayotte	...	...	333	553	601	655	590	430	520	520
	Mozambique	7 000	3 914	3 347	2 461	3 602	7 140	6 078	5 081	3 241	2 178
	Oman	477	588	240	184	366	124	99	521	188	7
	Pakistan	635	-	-	1 990	1 500	1 592	4 610	4 240	2 775	2 480
	Portugal	-	-	-	-	-	2	11	10	55	45
	Qatar	0	0	0	0	-	0	0	0	0	0
	Réunion	-	-	-	38	-	-	-	-	-	-
	Russian Fed	10 893	19 379	-	-	-	-	-	-	-	-
	Saudi Arabia	713	923	690	804	861	773	797	783	709	1 197
	Seychelles	10	29	7	1	6	12	28	84	136	32
	South Africa	44	3	11	-	8	72	25	12	7	12
	Spain	41	138	-	-	-	-	-	-	-	-
	Tanzania	538	1 002	670	757	850	650	500	250	250	250
	Uruguay	-	-	-	-	-	-	-	-	14	9
	Yemen	316	316	300	300	310	300	300 F	300 F	300 F	300 F
	Other nei	1 515	75	53	23	18	36	21	4	2	2
51	*Fishing area total*	*32 007*	*30 756*	*11 897*	*8 797*	*10 479*	*17 661*	*14 451 F*	*13 726 F*	*9 930 F*	*12 533 F*
57	Australia	19	40	24	16	9	116	34	8	9	21
	China	-	-	96	299	307	396	712	309	168	24
	India	316	147	247	1	2	1 225	-	1	-	217
	Indonesia	59 913	72 602	71 305	75 604	77 507	74 544	80 919	85 525	70 363	70 363
	Japan	-	-	-	-	-	24	-	-	236	20
	Korea Rep	-	-	-	46	14	19	62	-	29	-
	Sri Lanka	14	2	2	0	0	0	11	17	8	24
	Timor-Leste	...	...	...	...	...	...	1	3	3	3 F
	Other nei	61	18	23	7	2	5	19	3	-	-
57	*Fishing area total*	*60 323*	*72 809*	*71 697*	*75 973*	*77 841*	*76 329*	*81 758*	*85 866*	*70 816*	*70 672 F*
61	China	...	...	23 477	15 556	14 203	13 691	19 397	29 820	304	179
	China,H.Kong	0	0	18	18	1	4	0	0	0	0
	Japan	5 900	10 352	15 206	8 234	13 798	11 942	9 595	10 790	8 096	8 096 F
	Korea Rep	459	848	1 816	2 792	3 115	3 393	2 893	3 245	3 294	-
	Russian Fed	-	-	-	-	-	6	11	19	18	3
61	*Fishing area total*	*6 359*	*11 200*	*40 517*	*26 600*	*31 117*	*29 036*	*31 896*	*43 874*	*11 712*	*8 278 F*
67	USA	1	0	0	-	1 263	-	1	-	0	1
67	*Fishing area total*	*1*	*0*	*0*	*...*	*1 263*	*...*	*1*	*...*	*0*	*1*
71	Australia	0	0	0	0	-	-	-	10	13	6
	Guam	0	1	1	1	2	1	1	0	2	2
	Indonesia	103 079	115 275	116 912	136 018	142 935	150 487	148 868	168 460	158 925	168 140
	Japan	3 498	2 929	3 185	370	574	165	328	36	621	621 F
	Korea Rep	263	56	27	99	186	135	593	517	256	869
	NewCaledonia	101	300	255	244	169	285	236	294	310	265
	N Marianas	...	3	4	7	7	8	8	7	3	6
	Palau	12	11	1	3	1	1	1	1	2	51 F
	Philippines	4 627	3 641	4 202	4 002	5 554	4 789	3 902	3 621	3 809	-
	Russian Fed	18 937	5 840	4 981	-	-	-	-	-	-	-
	Singapore	0	0	0	0	-	-	-	-	-	-
	Tuvalu	219	21	15	15 F	20 F	20 F	20 F	20 F	20 F	20 F
	USA	-	-	-	168	-	-	-	-	-	-
	Viet Nam	...	...	...	...	3 200 F	7 400 F	7 000 F	6 500 F	15 800 F	15 800 F
71	*Fishing area total*	*130 736*	*128 077*	*129 583*	*140 927 F*	*152 648 F*	*163 291 F*	*160 957 F*	*179 466 F*	*179 761 F*	*185 780 F*
77	Amer Samoa	3	1	1	3	0	0	0	0	1	0
	Belize	-	30	-	-	-	-	-	-	-	-
	Costa Rica	292	613	275	2 572	990	1 213	1 057	1 110	1 160	1 583
	Cyprus	10	-	-	-	-	-	-	-	-	-
	Fr Polynesia	...	...	...	...	...	...	2	0	0	4
	Japan	137	108	60	877	12	-	-	-	-	-
	Korea Rep	1 854	1 573	2 769	2 740	2 972	4 159	2 574	2 484	3 408	4 122
	Panama	10	-	-	-	-	-	-	-	-	-
	Samoa	-	-	-	-	-	-	-	479	470 F	470 F
	Spain	10	-	-	-	-	1 040	20	-	10	-
	Tonga	28	18	8	1	8	14	34	42	85	35
	USA	10	27	7	8	15	10	14	4	5	2
	Other nei	-	-	-	-	-	-	-	-	-	3
77	*Fishing area total*	*2 354*	*2 370*	*3 120*	*6 201*	*3 997*	*6 436*	*3 701*	*4 119*	*5 139 F*	*6 219 F*
81	Australia	...	...	...	289	281	328	-	219	154	136
	Japan	-	-	-	-	-	-	1	-	-	-
	Korea Rep	4	-	-	-	-	-	78	167	228	14
	New Zealand	-	-	41	36	83	66	101	69	88	96
	Russian Fed	-	1	13	-	-	-	-	-	-	-
	Ukraine	-	-	-	-	-	-	-	-	-	15
81	*Fishing area total*	*4*	*1*	*54*	*325*	*364*	*394*	*180*	*455*	*470*	*261*
87	Colombia	35 752	2 132	2	2 851	8 111	10 000 F	1 501	141	120	100 F
	Ecuador	220	380	-	-	-	-	-	-	21 422	-
	Korea Rep	145	71	49	-	-	-	64	-	-	-
	Spain	-	-	-	1	-	1 790	-	-	-	10

B-36

Tunas, bonitos, billfishes — **Capture production by species, fishing areas and countries or areas**
Thons, pélamides, marlins — **Captures par espèces, zones de pêche et pays ou zones**
Atunes, bonitos, agujas — **Capturas por especies, áreas de pesca y países o áreas**

Species, Fishing area Espèce, Zone de pêche Especie, Area de pesca	1993 t	1994 t	1995 t	1996 t	1997 t	1998 t	1999 t	2000 t	2001 t	2002 t
Vanuatu	10	-	-	-	-	-	-	-	-	-
Other nei	-	10	-	-	-	-	-	-	-	-
87 Fishing area total	*36 127*	*2 593*	*51*	*2 852*	*8 111*	*11 790 F*	*1 565*	*141*	*21 542*	*110 F*
Species total	*272 304 F*	*251 152 F*	*261 179 F*	*277 773 F*	*301 204 F*	*328 849 F*	*307 953 F*	*340 143 F*	*316 733 F*	*302 293 F*
Group total	***4 631 593***	***4 763 798***	***4 879 077***	***4 853 169***	***5 186 555***	***5 758 303***	***5 972 769***	***5 828 375***	***5 722 174***	***6 088 337***

B-37 Miscellaneous pelagic fishes / Poissons pélagiques divers / Peces pelágicos diversos

Capture production by species, fishing areas and countries or areas / Captures par espèces, zones de pêche et pays ou zones / Capturas por especies, áreas de pesca y países o áreas

Species, Fishing area Espèce, Zone de pêche Especie, Area de pesca	1993 t	1994 t	1995 t	1996 t	1997 t	1998 t	1999 t	2000 t	2001 t	2002 t
Capelin	**Capelan**		**Capelán**		***Mallotus villosus***				**1,23(04)002,01**	**CAP**
21 Canada	48 647	2 249	293	32 628	21 945	38 249	23 495	21 352	19 747	13 559
Cuba	3	0	-	-	-	-	-	-	-	-
Greenland	110	158	68	83	41	21	34	22	3	12
St Pier Mq	1	2	1	0	1	-	2	0	1	4
21 Fishing area total	*48 761*	*2 409*	*362*	*32 711*	*21 987*	*38 270*	*23 531*	*21 374*	*19 751*	*13 575*
27 Denmark	0	294	-	60 898	48 524	40 349	3 837	20 807	17 588	23 165
Estonia	0	-	-	-	-	-	-	-	-	-
Faeroe Is	40 980	12 310	3 306	39 777	43 466	41 966	24 275	59 855	32 110	35 000 F
France	-	-	-	-	-	-	-	1	1	4
Germany	-	-	-	-	-	5 001	-	-	-	95
Greenland	11 064	1 953	1 797	7 099	12 121	16 914	24 261	24 623	18 638	30 227
Iceland	940 947	753 466	715 551	1 179 051	1 319 191	750 065	703 694	892 405	918 417	1 078 818
Ireland	-	-	-	-	0	1	-	-	-	-
Norway	530 401	113 393	27 740	207 706	157 889	88 226	91 813	370 769	482 835	522 349
Russian Fed	169 858	522	-	-	-	-	32 485	94 693	180 098	247 039
Sweden	-	-	-	-	-	-	-	-	-	7 570
UK	-	-	-	-	-	1 115	79	-	-	-
27 Fishing area total	*1 693 250*	*881 938*	*748 394*	*1 494 531*	*1 581 191*	*943 637*	*880 444*	*1 463 153*	*1 649 687*	*1 944 267 F*
61 Russian Fed	675	57	44	180	160	405	70	291	1 468	3 882
61 Fishing area total	*675*	*57*	*44*	*180*	*160*	*405*	*70*	*291*	*1 468*	*3 882*
Species total	*1 742 686*	*884 404*	*748 800*	*1 527 422*	*1 603 338*	*982 312*	*904 045*	*1 484 818*	*1 670 906*	*1 961 724 F*
Garfish	**Orphie**		**Aguja**		***Belone belone***				**1,47(01)001,01**	**GAR**
27 Channel Is	0	0	0	0	1	2	0	2	2	1
Denmark	755	641	622	708	994	648	571	714	558	580
Estonia	40	119	193	405	400	167	122	135	111	148
France	71	53	38	59	70	43	55	59	53	101
Germany	100	140	152	168	130	58	125	82	73	113
Latvia	-	-	-	-	-	-	-	-	11	-
Netherlands	-	-	-	-	-	-	-	-	2	2
Norway	1	2	3	1	2	1	1	0	0	1
Portugal	52	95	72	41	35	43	54	55	57	30
Spain	72	176	295	68	148	651	582	165	152	198
Sweden	47	35	33	14	12	27	30	7	9	8
UK	0	0	0	0	0	1	4	3	2	3
27 Fishing area total	*1 138*	*1 261*	*1 408*	*1 464*	*1 792*	*1 641*	*1 544*	*1 222*	*1 030*	*1 185*
34 Russian Fed	-	-	-	-	-	-	13	82	-	-
34 Fishing area total	-	-	-	-	-	-	*13*	*82*	-	-
37 Bulgaria	-	-	-	2	2	4	4	9	16	37
Croatia	144	88	81	63	85	60	10	5	11	4
France	-	-	-	-	-	-	-	1	2	3
Greece	371	491	323	331	253	188	83	184	126	120 F
Italy	364	304	554	243	216	238	209	134	139	139
Romania	-	2	-	-	-	-	-	-	-	-
Russian Fed	-	-	-	-	-	-	2	1	1	1
Serbia-Monte	0	0	1	1	1	2	5	5	6	6
Slovenia	-	-	-	-	-	-	-	1	1	0
Spain	-	-	-	387	361	347	18	105	228	1 015
Tunisia	209	114	48	141	273	272	382	98	307	349
Turkey	1 150	1 564	581	395	470	450	500	300	640	482
Ukraine	-	1	1	0	0	1	1	0	-	-
37 Fishing area total	*2 238*	*2 564*	*1 589*	*1 563*	*1 661*	*1 562*	*1 214*	*843*	*1 477*	*2 156 F*
Species total	*3 376*	*3 825*	*2 997*	*3 027*	*3 453*	*3 203*	*2 771*	*2 147*	*2 507*	*3 341 F*
Needlefishes nei	**Aiguilles nca**		**Maraos nep**		***Tylosurus spp***				**1,47(01)013,XX**	**NED**
51 Egypt	...	...	123	48	32	17	28	11	28	-
Oman	...	779	508	201	252	116	209	131	130	541
Qatar	54	31	19	14	25	12	20	1	2	1
Saudi Arabia	-	461	324	96	111	131	140	107	126	136
51 Fishing area total	*54*	*1 271*	*974*	*359*	*420*	*276*	*397*	*250*	*286*	*678*
57 Indonesia	6 589	5 723	6 081	6 584	8 335	8 335	7 848	9 154	8 141	8 420
57 Fishing area total	*6 589*	*5 723*	*6 081*	*6 584*	*8 335*	*8 335*	*7 848*	*9 154*	*8 141*	*8 420*
71 Indonesia	17 583	19 726	24 038	20 104	21 462	21 406	21 561	23 716	19 179	19 560
Philippines	11 730	11 386	10 171	9 454	9 967	9 532	10 323	10 084	11 819	12 139
71 Fishing area total	*29 313*	*31 112*	*34 209*	*29 558*	*31 429*	*30 938*	*31 884*	*33 800*	*30 998*	*31 699*
Species total	*35 956*	*38 106*	*41 264*	*36 501*	*40 184*	*39 549*	*40 129*	*43 204*	*39 425*	*40 797*
Needlefishes, etc. nei	**Aiguilles, orphies nca**		**Agujones, maraos nep**		***Belonidae***				**1,47(01)XXX,XX**	**BEN**
31 Dominican Rp	81	46	146	39	5	4	7	15	45	103
Grenada	2	1	1	0	2	3	0	0	0	0
St Kitts Nev	...	...	12	27	26	60	58	60	37	34
31 Fishing area total	*83*	*47*	*159*	*66*	*33*	*67*	*65*	*75*	*82*	*137*
34 Benin	55 F	50 F	45 F	40 F	35 F	30 F	27	20 F	38	32
Côte dIvoire	0	1	1 F	1 F	1 F	1 F	...	...	3	2

B-37 Miscellaneous pelagic fishes / Poissons pélagiques divers / Peces pelágicos diversos

Capture production by species, fishing areas and countries or areas
Captures par espèces, zones de pêche et pays ou zones
Capturas por especies, áreas de pesca y países o áreas

Species, Fishing area Espèce, Zone de pêche Especie, Area de pesca	1993 t	1994 t	1995 t	1996 t	1997 t	1998 t	1999 t	2000 t	2001 t	2002 t
Liberia	-	-	-	-	-	54	90	30	31	35 F
Nigeria	1 708	1 570	590	1 082	272	99	166	535	431	301
Portugal	-	-	1	1	-	19	-	-	-	-
Senegal	...	...	...	...	...	5	37	8	0	-
Spain	-	-	-	-	-	-	-	-	-	3
Togo	60	13	18	37	90	122	96	53	153	69
34 Fishing area total	*1 823 F*	*1 634 F*	*655 F*	*1 161 F*	*398 F*	*330 F*	*416*	*646 F*	*656*	*442 F*
81 Australia	2	2	2	1	0	0	-	-	-	-
81 Fishing area total	*2*	*2*	*2*	*1*	*0*	*0*	-	-	-	-
Species total	*1 908 F*	*1 683 F*	*816 F*	*1 228 F*	*431 F*	*397 F*	*481*	*721 F*	*738*	*579 F*
Atlantic saury	**Balaou atlantique**		**Paparda del Atlántico**			***Scomberesox saurus***			**1,47(02)002,01**	**SAU**
27 Portugal	0	1	0	0	0	54	0	1	1	-
Spain	1	619	526	483	639	907	574	184	992	589
27 Fishing area total	*1*	*620*	*526*	*483*	*639*	*961*	*574*	*185*	*993*	*589*
87 Chile	-	37	8	1	175	407	580	296	4 178	1 492
87 Fishing area total	-	*37*	*8*	*1*	*175*	*407*	*580*	*296*	*4 178*	*1 492*
Species total	*1*	*657*	*534*	*484*	*814*	*1 368*	*1 154*	*481*	*5 171*	*2 081*
Pacific saury	**Balaou du Japon**		**Paparda del Pacífico**			***Cololabis saira***			**1,47(02)007,01**	**SAP**
61 China,Taiwan	36 435	12 550	13 772	8 236	21 887	12 794	12 541	27 868	39 764	51 295
Japan	277 461	261 587	273 510	229 227	290 812	144 983	141 011	216 471	269 797	205 282
Korea Rep	38 435	34 810	37 865	28 368	68 853	18 531	28 784	43 280	25 782	26 953
Russian Fed	48 145	26 385	25 140	10 280	7 091	4 665	4 808	17 390	40 407	51 709
61 Fishing area total	*400 476*	*335 332*	*350 287*	*276 111*	*388 643*	*180 973*	*187 144*	*305 009*	*375 750*	*335 239*
67 Korea Rep	2 454	272	-	-	-	-	-	-	-	-
67 Fishing area total	*2 454*	*272*	-	-	-	-	-	-	-	-
77 Korea Rep	-	-	-	-	-	-	754	1 060	423	234
77 Fishing area total	-	-	-	-	-	-	*754*	*1 060*	*423*	*234*
Species total	*402 930*	*335 604*	*350 287*	*276 111*	*388 643*	*180 973*	*187 898*	*306 069*	*376 173*	*335 473*
Japanese halfbeak	**Demi-bec du Japon**		**Agujeta del Japón**			***Hyporhamphus sajori***			**1,47(03)003,11**	**HAJ**
61 Korea Rep	2 005	1 480	1 531	990	1 193	1 160	913	956	613	670
61 Fishing area total	*2 005*	*1 480*	*1 531*	*990*	*1 193*	*1 160*	*913*	*956*	*613*	*670*
Species total	*2 005*	*1 480*	*1 531*	*990*	*1 193*	*1 160*	*913*	*956*	*613*	*670*
Ballyhoo halfbeak	**Demi-bec brésilien**		**Agujeta brasileña**			***Hemiramphus brasiliensis***			**1,47(03)004,06**	**BAL**
31 Puerto Rico	...	...	...	...	...	...	32	47	41	36
31 Fishing area total	...	...	...	...	...	...	*32*	*47*	*41*	*36*
41 Brazil	500 F	500 F	631	494	442	1 500	832	989	3 342	3 652
41 Fishing area total	*500 F*	*500 F*	*631*	*494*	*442*	*1 500*	*832*	*989*	*3 342*	*3 652*
Species total	*500 F*	*500 F*	*631*	*494*	*442*	*1 500*	*864*	*1 036*	*3 383*	*3 688*
Halfbeaks nei	**Demi-becs nca**		**Agujetas nep**			***Hemiramphus spp***			**1,47(03)004,XX**	**HAX**
21 USA	-	-	-	-	-	-	258	-	-	-
21 Fishing area total	-	-	-	-	-	-	*258*	-	-	-
31 Grenada	32	2	1	1	2	2	0	1	1	1
USA	523	500	528	398	293	441	92	414	354	393
31 Fishing area total	*555*	*502*	*529*	*399*	*295*	*443*	*92*	*415*	*355*	*394*
34 Liberia	-	-	-	-	-	77	85	96	97	100 F
Sao Tome Prn	262 F	381 F	401 F	293	126	150 F	160 F	150 F	150 F	150 F
Senegal	-	-	-	-	-	70	30	14	14	48
Togo	0	3	1	2	4	102	3	4	18	18
34 Fishing area total	*262 F*	*384 F*	*402 F*	*295*	*130*	*399 F*	*278 F*	*264 F*	*279 F*	*316 F*
51 India	582	747	3 828	1 057	1 172	1 312	1 928	1 650	1 318	2 530
Tanzania	1 213	1 066	1 640	1 483	1 850	1 175	1 200	1 250	1 300	1 200
Untd Arab Em	32	35	29	36	42	58	61	55	50 F	55
51 Fishing area total	*1 827*	*1 848*	*5 497*	*2 576*	*3 064*	*2 545*	*3 189*	*2 955*	*2 668 F*	*3 785*
57 Australia	158	163	174	138	565	580	530	691	746	663
India	2 876	2 870	4 846	5 061	4 402	3 360	3 461	5 993	4 695	4 880
57 Fishing area total	*3 034*	*3 033*	*5 020*	*5 199*	*4 967*	*3 940*	*3 991*	*6 684*	*5 441*	*5 543*
71 Australia	-	-	-	-	-	-	-	103	139	156
Fiji Islands	99	101 F	49	70	62	49	61	50 F	56	55 F
Philippines	2 719	2 843	3 555	3 231	1 655	1 949	2 281	2 309	2 820	2 990
71 Fishing area total	*2 818*	*2 944 F*	*3 604*	*3 301*	*1 717*	*1 998*	*2 342*	*2 462 F*	*3 015*	*3 201 F*
81 Australia	280	193	193	99	115	115	...	44	63	46
New Zealand	14	27	18	16	17	28	24	18	13	11
81 Fishing area total	*294*	*220*	*211*	*115*	*132*	*143*	*24*	*62*	*76*	*57*

B-37 Miscellaneous pelagic fishes / Poissons pélagiques divers / Peces pelágicos diversos

Capture production by species, fishing areas and countries or areas / Captures par espèces, zones de pêche et pays ou zones / Capturas por especies, áreas de pesca y países o áreas

Species, Fishing area Espèce, Zone de pêche Especie, Area de pesca	1993 t	1994 t	1995 t	1996 t	1997 t	1998 t	1999 t	2000 t	2001 t	2002 t
Species total	*8 790 F*	*8 931 F*	*15 263 F*	*11 885*	*10 305*	*9 468 F*	*10 174 F*	*12 842 F*	*11 834 F*	*13 296 F*
Japanese flyingfish	**Poisson-volant du Japon**		**Volador japonés**		***Cypselurus agoo***				**1,47(04)010,02**	**JFL**
61 Japan	8 039	8 606	7 881	8 501	7 486	8 933	6 738	9 615	8 286	6 909
61 Fishing area total	*8 039*	*8 606*	*7 881*	*8 501*	*7 486*	*8 933*	*6 738*	*9 615*	*8 286*	*6 909*
Species total	*8 039*	*8 606*	*7 881*	*8 501*	*7 486*	*8 933*	*6 738*	*9 615*	*8 286*	*6 909*
Flyingfishes nei	**Exocets nca**		**Voladores nep**		***Exocoetidae***				**1,47(04)XXX,XX**	**FLY**
31 Barbados	1 987	1 640	1 766	2 042	1 566	2 680	2 075	1 916	1 673	1 590
Grenada	27	15	6	12	0	5	1	14	10	5
Martinique	0	0	0	0	0	0	0	0	0	0
St Kitts Nev	...	...	21	54	23	38	22	22	48	42
St Lucia	...	48	50	40	34	112	67	99	323	193
31 Fishing area total	*2 014*	*1 703*	*1 843*	*2 148*	*1 623*	*2 835*	*2 165*	*2 051*	*2 054*	*1 830*
34 Benin	30 F	33 F	32 F	28 F	25 F	20 F	15	10 F	96	92
Liberia	...	...	...	69	8	10	-	38	19	20 F
Nigeria	7	7	7	17	...	87	69	8	55	1 837
Sao Tome Prn	229 F	333 F	350 F	256	939	1 000 F	1 100 F	1 000 F	1 000 F	1 000 F
Togo	94	115	109	204	287	610	211	130	186	190
34 Fishing area total	*360 F*	*488 F*	*498 F*	*574 F*	*1 259 F*	*1 727 F*	*1 395 F*	*1 186 F*	*1 356 F*	*3 139 F*
41 Brazil	650 F	650 F	1 036	743	1 082	1 084	760	388	217	708
41 Fishing area total	*650 F*	*650 F*	*1 036*	*743*	*1 082*	*1 084*	*760*	*388*	*217*	*708*
51 Bahrain	39	91	85	92	85	145	299	52	112	28
India	42	64	82	41	99	57	102	114	47	79
51 Fishing area total	*81*	*155*	*167*	*133*	*184*	*202*	*401*	*166*	*159*	*107*
57 India	1 895	2 772	2 263	1 845	1 995	555	634	2 597	2 467	2 620
Indonesia	2 562	2 417	2 992	4 613	5 205	7 562	5 585	4 627	4 471	4 940
57 Fishing area total	*4 457*	*5 189*	*5 255*	*6 458*	*7 200*	*8 117*	*6 219*	*7 224*	*6 938*	*7 560*
61 China,Taiwan	657	699	572	662	2 497	1 077	618	287	366	502
61 Fishing area total	*657*	*699*	*572*	*662*	*2 497*	*1 077*	*618*	*287*	*366*	*502*
71 Indonesia	9 155	8 373	10 188	11 921	10 921	11 399	12 237	15 353	11 996	12 550
Japan	2	-	-	-	-	-	-	-	-	-
Kiribati	1 760	1 740	1 760	1 780	1 770	2 525	2 547	836	1 594	1 617
Philippines	13 799	14 019	16 826	17 300	27 801	36 134	38 280	36 050	33 212	34 650
71 Fishing area total	*24 716*	*24 132*	*28 774*	*31 001*	*40 492*	*50 058*	*53 064*	*52 239*	*46 802*	*48 817*
77 Cook Is	40 F	38 F	38 F	30 F	30 F	30 F	30 F	30 F	30 F	30 F
Fr Polynesia	25	26	...	...	55	92	...	88	82	79
77 Fishing area total	*65 F*	*64 F*	*38 F*	*30 F*	*85 F*	*122 F*	*30 F*	*118 F*	*112 F*	*109 F*
81 New Zealand	-	-	4	4	2	3	1	2	1	7
81 Fishing area total	*-*	*-*	*4*	*4*	*2*	*3*	*1*	*2*	*1*	*7*
87 Peru	9 210	926	35 619	639	1 056	4 506	298 373	41 059	5 035	3 198
87 Fishing area total	*9 210*	*926*	*35 619*	*639*	*1 056*	*4 506*	*298 373*	*41 059*	*5 035*	*3 198*
Species total	*42 210 F*	*34 006 F*	*73 806 F*	*42 392 F*	*55 480 F*	*69 731 F*	*363 026 F*	*104 720 F*	*63 040 F*	*65 977 F*
Opah	**Opah**		**Opa**		***Lampris guttatus***				**1,52(01)001,02**	**LAG**
21 USA	...	...	...	...	...	1	1	2	1	0
21 Fishing area total	*...*	*...*	*...*	*...*	*...*	*1*	*1*	*2*	*1*	*0*
27 Faeroe Is	-	-	-	-	-	-	1	...	...	...
27 Fishing area total	*-*	*-*	*-*	*-*	*-*	*-*	*1*	*...*	*...*	*...*
31 USA	...	...	...	...	...	-	-	-	1	0
31 Fishing area total	*...*	*...*	*...*	*...*	*...*	*-*	*-*	*-*	*1*	*0*
77 Amer Samoa	-	-	-	-	-	-	1	1	1	-
Fr Polynesia	...	77	93	96	...	...	137	124	148	140
USA	...	...	...	...	...	114	67	56	47	446
77 Fishing area total	*...*	*77*	*93*	*96*	*...*	*114*	*205*	*181*	*196*	*586*
81 New Zealand	-	-	130	76	126	254	335	283	340	329
81 Fishing area total	*-*	*-*	*130*	*76*	*126*	*254*	*335*	*283*	*340*	*329*
Species total	*...*	*77*	*223*	*172*	*126*	*369*	*542*	*466*	*538*	*915*
Southern opah	**...B**		**...C**		***Lampris immaculatus***				**1,52(01)001,03**	**LAI**
58 Australia	-	-	-	-	-	-	-	-	-	0
58 Fishing area total	*-*	*-*	*-*	*-*	*-*	*-*	*-*	*-*	*-*	*0*
Species total	*-*	*-*	*-*	*-*	*-*	*-*	*-*	*-*	*-*	*0*
Dealfishes	**...B**		**...C**		***Trachipterus spp***				**1,52(04)002,XX**	**TRP**
27 Portugal	...	...	...	...	...	...	29	20	25	16

B-37 Miscellaneous pelagic fishes / Poissons pélagiques divers / Peces pelágicos diversos

Capture production by species, fishing areas and countries or areas / Captures par espèces, zones de pêche et pays ou zones / Capturas por especies, áreas de pesca y países o áreas

Species, Fishing area Espèce, Zone de pêche Especie, Area de pesca	1993 t	1994 t	1995 t	1996 t	1997 t	1998 t	1999 t	2000 t	2001 t	2002 t
27 Fishing area total	...	...	...	...	...	...	*29*	*20*	*25*	*16*
81 New Zealand	-	-	127	49	60	74	65	67	103	63
Ukraine	-	-	-	-	-	-	-	-	-	1
81 Fishing area total	-	-	*127*	*49*	*60*	*74*	*65*	*67*	*103*	*64*
Species total	...	...	*127*	*49*	*60*	*74*	*94*	*87*	*128*	*80*
King of herrings	**Roi des harengs**		**Rey de los arenques**		***Regalecus glesne***				**1,52(05)001,01**	**REL**
81 New Zealand	-	-	5	10	64	60	34	20	1	0
81 Fishing area total	-	-	*5*	*10*	*64*	*60*	*34*	*20*	*1*	*0*
Species total	-	-	*5*	*10*	*64*	*60*	*34*	*20*	*1*	*0*
...A	**...B**		**Pejerrey de mar**		***Odontesthes regia***				**1,63(02)002,01**	**ODR**
87 Chile	341	701	558	690	494	560	3 414	1 357	833	1 396
87 Fishing area total	*341*	*701*	*558*	*690*	*494*	*560*	*3 414*	*1 357*	*833*	*1 396*
Species total	*341*	*701*	*558*	*690*	*494*	*560*	*3 414*	*1 357*	*833*	*1 396*
Big-scale sand smelt	**Joël**		**Pejerrey mediterráneo**		***Atherina boyeri***				**1,63(02)003,01**	**ATB**
05 Greece	-	-	-	8	110	350	350	350	326	390
Russian Fed	-	-	-	-	-	-	11	-	-	-
05 Fishing area total	-	-	-	*8*	*110*	*350*	*361*	*350*	*326*	*390*
37 Bulgaria	-	-	-	-	-	0	1	21	2	0
Russian Fed	9	5	15	31	13	25	19	63	95	99
37 Fishing area total	*9*	*5*	*15*	*31*	*13*	*25*	*20*	*84*	*97*	*99*
Species total	*9*	*5*	*15*	*39*	*123*	*375*	*381*	*434*	*423*	*489*
Atlantic silverside	**Capucette**		**Pejerrey del Atlántico**		***Menidia menidia***				**1,63(02)013,02**	**SSA**
21 Canada	83	60	223	151	238	231	558	304	646	135
USA	39	32	13	19	21	24	25	15	15	15
21 Fishing area total	*122*	*92*	*236*	*170*	*259*	*255*	*583*	*319*	*661*	*150*
Species total	*122*	*92*	*236*	*170*	*259*	*255*	*583*	*319*	*661*	*150*
Silversides(=Sand smelts) nei	**Athérinidés nca**		**Pejerreyes nep**		***Atherinidae***				**1,63(02)XXX,XX**	**SIL**
01 Egypt	927	1 375	1 241	1 326	1 229	1 150	1 180	3 182	5 154	5 941
01 Fishing area total	*927*	*1 375*	*1 241*	*1 326*	*1 229*	*1 150*	*1 180*	*3 182*	*5 154*	*5 941*
02 Mexico	...	...	...	...	...	372	487	392	558	379
02 Fishing area total	...	...	...	...	...	*372*	*487*	*392*	*558*	*379*
03 Bolivia	635	785	850	850	850	850	850	850	760	750 F
03 Fishing area total	*635*	*785*	*850*	*850*	*850*	*850*	*850*	*850*	*760*	*750 F*
04 Turkey	953	899	909	562	1 600	1 500	1 455	1 583	1 685	1 733
04 Fishing area total	*953*	*899*	*909*	*562*	*1 600*	*1 500*	*1 455*	*1 583*	*1 685*	*1 733*
27 France	98	78	76	68	72	65	78	54	52	27
Portugal	11	10	1 002	3	6	22	6	3	0	1
Spain	-	-	-	262	281	212	213	192	192	260
UK	-	-	-	-	-	-	-	-	1	-
27 Fishing area total	*109*	*88*	*1 078*	*333*	*359*	*299*	*297*	*249*	*245*	*288*
37 Albania	20 F	15 F	8	20	8	11	15	20	10	16
Croatia	224	176	175	148	171	260	50	44	20	8
Egypt	2 027	3 740	2 596	5 316	4 066	4 558	3 017	1 732	3 490	3 267
France	154	80	3	31	55	55	57	29	44	39
Italy	1 762	1 595	1 530	1 112	1 101	883	851	725	736	772
Lebanon	-	-	25	25	50	50	50	50	50	50
Romania	-	4	3	3	10	73	33	42	29	8
Serbia-Monte	6	3	6	6	7	9	9	13	14	22
Slovenia	-	-	-	-	-	-	-	1	2	5
Spain	-	-	-	35	32	30	31	62	20	80
Tunisia	...	254	594	141	116	338	6	57	309	-
Turkey	7 682	3 834	1 171	412	640	800	1 300	500	575	518
Ukraine	937	439	195	326	396	632	388	653	332	540
37 Fishing area total	*12 812 F*	*10 140 F*	*6 306*	*7 575*	*6 652*	*7 699*	*5 807*	*3 928*	*5 631*	*5 325*
41 Argentina	571	489	411	583	520	48	618	14	35	44
Brazil	170 F	170 F	40	52	62	52	17	39	28	5
Falkland Is	1	2	2	2	2	2	2	4	1	2
41 Fishing area total	*742 F*	*661 F*	*453*	*637*	*584*	*102*	*637*	*57*	*64*	*51*
51 Egypt	-	-	-	-	-	-	78	-	-	-
51 Fishing area total	-	-	-	-	-	-	*78*	-	-	-
71 Fiji Islands	31	32 F	30	31	25	80	91	80 F	86	85 F
Philippines	2 867	855	864	866	669	596	618	543	625	638
71 Fishing area total	*2 898*	*887 F*	*894*	*897*	*694*	*676*	*709*	*623 F*	*711*	*723 F*

B-37 Miscellaneous pelagic fishes / Poissons pélagiques divers / Peces pelágicos diversos

Capture production by species, fishing areas and countries or areas
Captures par espèces, zones de pêche et pays ou zones
Capturas por especies, áreas de pesca y países o áreas

Species, Fishing area Espèce, Zone de pêche Especie, Area de pesca	1993 t	1994 t	1995 t	1996 t	1997 t	1998 t	1999 t	2000 t	2001 t	2002 t
77 Mexico	...	...	...	...	...	1 062	906	793	704	986
USA	-	-	-	-	-	-	-	8	34	37
77 Fishing area total	...	...	...	...	...	*1 062*	*906*	*801*	*738*	*1 023*
87 Peru	1 395	2 207	2 357	3 802	5 184	45	6 692	11 215	7 528	11 220
87 Fishing area total	*1 395*	*2 207*	*2 357*	*3 802*	*5 184*	*45*	*6 692*	*11 215*	*7 528*	*11 220*
Species total	*20 471 F*	*17 042 F*	*14 088*	*15 982*	*17 152*	*13 755*	*19 098*	*22 880 F*	*23 074*	*27 433 F*
False trevally	**Péliau chanos**		**Pagapa**			***Lactarius lactarius***			**1,70(19)165,02**	**TRF**
51 India	3 169	3 934	4 517	6 401	7 056	7 995	5 503	4 623	5 992	4 268
Pakistan	3	2	3	2	4	5	-	-	-	-
51 Fishing area total	*3 172*	*3 936*	*4 520*	*6 403*	*7 060*	*8 000*	*5 503*	*4 623*	*5 992*	*4 268*
57 India	1 778	2 376	2 570	529	886	1 533	1 324	1 438	992	926
Malaysia	1 202	112	25	3	211	212	92	18	2	1
57 Fishing area total	*2 980*	*2 488*	*2 595*	*532*	*1 097*	*1 745*	*1 416*	*1 456*	*994*	*927*
71 Malaysia	580	531	393	384	350	363	589	404	437	362
Philippines	54	28	25	26	202	213	235	253	285	346
71 Fishing area total	*634*	*559*	*418*	*410*	*552*	*576*	*824*	*657*	*722*	*708*
Species total	*6 786*	*6 983*	*7 533*	*7 345*	*8 709*	*10 321*	*7 743*	*6 736*	*7 708*	*5 903*
Bluefish	**Tassergal**		**Anjova**			***Pomatomus saltatrix***			**1,70(20)213,01**	**BLU**
21 USA	3 892	3 081	3 367	4 137	3 900	3 446	3 145	3 499	3 825	3 011
21 Fishing area total	*3 892*	*3 081*	*3 367*	*4 137*	*3 900*	*3 446*	*3 145*	*3 499*	*3 825*	*3 011*
27 Portugal	123	72	57	44	40	25	21	20	15	14
Spain	601	651	674	1 833	1 874	979	787	39	38	19
27 Fishing area total	*724*	*723*	*731*	*1 877*	*1 914*	*1 004*	*808*	*59*	*53*	*33*
31 USA	905	1 376	434	107	322	318	214	162	168	152
Venezuela	987	1 130	1 024	651	825	581	542	950	1 158	813
31 Fishing area total	*1 892*	*2 506*	*1 458*	*758*	*1 147*	*899*	*756*	*1 112*	*1 326*	*965*
34 Benin	1 150 F	1 100 F	1 050 F	1 000 F	950 F	900 F	875	600 F	697	978
Gambia	...	20	23	31	8	75	80	35	70	60
Georgia	10 F	-	-	-	-	-	-	-	-	-
Greece	0	0	-	-	-	-	-	-	-	-
GuineaBissau	2 F	2 F	2 F	3 F	3 F	3 F	3 F	3 F	3 F	3 F
Latvia	17	-	7	155	116	31	116	144	17	48
Lithuania	41	32	-	-	-	-	-	-	-	742
Mauritania	...	...	...	...	...	...	...	1 F	...	...
Morocco	3	9	11	10	54	161	47	56	193	135
Portugal	21	14	13	18	8	12	1	0	0	1
Romania	13	-	-	-	-	-	-	-	-	-
Russian Fed	150	5	74	96	406	283	536	226	157	168
Senegal	2 142	1 614	461	691	2 075	3 118	327	277	1 149	5 417
Spain	-	-	-	1 801	-	-	-	-	-	-
Ukraine	-	-	-	-	209	13	238	97	29	325
34 Fishing area total	*3 549 F*	*2 796 F*	*1 641 F*	*3 805 F*	*3 829 F*	*4 596 F*	*2 223 F*	*1 439 F*	*2 315 F*	*7 877 F*
37 Bulgaria	8	8 F	12	10	12	10	8	18	2	102
Egypt	0	37	174	307	147	56	153	326	468	549
Greece	457	377	346	128	111	144	259	265	511	450 F
Morocco	17	4	1	5	20	2	-	45	5	59
Palest, O.T.	...	...	...	36	8	31	30 F	20 F	10 F	5
Romania	-	2	-	-	-	12	3	4	10	2
Spain	-	-	-	251	240	190	171	222	168	113
Tunisia	2 049	1 270	653	1 317	1 426	787	748	1 096	1 037	1 285
Turkey	16 442	8 078	5 456	4 117	3 050	3 350	2 995	4 250	13 060	25 000
Ukraine	-	1	0	-	-	-	-	-	-	-
37 Fishing area total	*18 973*	*9 777 F*	*6 642*	*6 171*	*5 014*	*4 582*	*4 367 F*	*6 246 F*	*15 271 F*	*27 565 F*
41 Argentina	1 331	202	565	342	416	15	286	416	181	701
Brazil	5 510 F	5 520 F	7 588	5 866	2 616	2 504	2 064	3 314	3 427	4 029
Uruguay	13	63	56	11	11	84	18	48	93	56
41 Fishing area total	*6 854 F*	*5 785 F*	*8 209*	*6 219*	*3 043*	*2 603*	*2 368*	*3 778*	*3 701*	*4 786*
47 Angola	-	-	-	-	-	-	-	-	167	-
South Africa	68	34	57	38	15	4	33	20 F	3	2
47 Fishing area total	*68*	*34*	*57*	*38*	*15*	*4*	*33*	*20 F*	*170*	*2*
57 Australia	45	49	54	56	-	-	-	-	-	-
57 Fishing area total	*45*	*49*	*54*	*56*	-	-	-	-	-	-
71 Australia	156	98	181	103	179	128	204	139	240	222
71 Fishing area total	*156*	*98*	*181*	*103*	*179*	*128*	*204*	*139*	*240*	*222*
81 Australia	98	96	96	101	64	64	46	70	111	87
81 Fishing area total	*98*	*96*	*96*	*101*	*64*	*64*	*46*	*70*	*111*	*87*
Species total	*36 251 F*	*24 945 F*	*22 436 F*	*23 265 F*	*19 105 F*	*17 326 F*	*13 950 F*	*16 362 F*	*27 012 F*	*44 548 F*

B-37 Miscellaneous pelagic fishes / Poissons pélagiques divers / Peces pelágicos diversos

Capture production by species, fishing areas and countries or areas / Captures par espèces, zones de pêche et pays ou zones / Capturas por especies, áreas de pesca y países o áreas

Species, Fishing area Espèce, Zone de pêche Especie, Area de pesca	1993 t	1994 t	1995 t	1996 t	1997 t	1998 t	1999 t	2000 t	2001 t	2002 t
Cobia	**Mafou**		**Cobia**			***Rachycentron canadum***			**1,70(22)221,01**	**CBA**
21 USA	4	3	18	142	6	13	72	16	13	13
21 Fishing area total	*4*	*3*	*18*	*142*	*6*	*13*	*72*	*16*	*13*	*13*
31 Mexico	363	215	347	347	630	588	565	303	298	180
USA	177	182	152	45	127	129	65	96	84	87
31 Fishing area total	*540*	*397*	*499*	*392*	*757*	*717*	*630*	*399*	*382*	*267*
34 Gambia	-	-	33	0	6	0	0	-	-	-
34 Fishing area total	-	-	*33*	*0*	*6*	*0*	*0*	-	-	-
41 Brazil	...	...	256	498	367	622	1 818	1 580	1 036	1 065
41 Fishing area total	...	...	*256*	*498*	*367*	*622*	*1 818*	*1 580*	*1 036*	*1 065*
51 Bahrain	30	17	32	38	19	6	9	9	20	11
Eritrea	...	2	10	38	2	6	8	31	31	41
Kuwait	...	...	...	...	...	...	...	...	38	40 F
Oman	202	104	234	103	180	115	124	100	54	202
Pakistan	1 459	1 541	2 306	1 574	1 449	1 254	1 136	2 896	2 797	2 808
Qatar	62	54	47	56	52	56	44	56	95	89
Saudi Arabia	48	71	124	155	155	130	137	138	167	188
Untd Arab Em	50	47	50	52	56	57	58	632	800 F	954
51 Fishing area total	*1 851*	*1 836*	*2 803*	*2 016*	*1 913*	*1 624*	*1 516*	*3 862*	*4 002 F*	*4 333 F*
57 Malaysia	131	109	62	76	111	87	111	136	150	152
57 Fishing area total	*131*	*109*	*62*	*76*	*111*	*87*	*111*	*136*	*150*	*152*
61 China,Taiwan	843	978	779	692	987	815	655	1 014	486	457
61 Fishing area total	*843*	*978*	*779*	*692*	*987*	*815*	*655*	*1 014*	*486*	*457*
71 Malaysia	476	374	279	292	274	436	645	504	407	367
Philippines	1 714	1 344	1 003	964	1 162	1 235	1 187	1 046	1 124	1 363
71 Fishing area total	*2 190*	*1 718*	*1 282*	*1 256*	*1 436*	*1 671*	*1 832*	*1 550*	*1 531*	*1 730*
Species total	*5 559*	*5 041*	*5 732*	*5 072*	*5 583*	*5 549*	*6 634*	*8 557*	*7 600 F*	*8 017 F*
Atlantic horse mackerel	**Chinchard d'Europe**		**Jurel**			***Trachurus trachurus***			**1,70(23)004,01**	**HOM**
27 Belgium	74	58	51	28	19	19	21	19	20	30
Bulgaria	226	-	-	-	-	-	-	-	-	-
Channel Is	0	0	0	5	7	11	10	8	8	9
Denmark	49 426	53 616	56 167	63 929	63 430	32 597	32 047	25 083	23 631	11 611
Estonia	-	55	-	80	203	34	0	-	-	-
Faeroe Is	10 531	505	950	863	1 005	216	3 643	2 014	180	...
France	7 986	6 152	16 000	25 396	26 862	28 103	27 087	21 628	19 852	21 241
Germany	29 381	17 277	20 407	21 815	37 584	33 750	23 803	16 778	12 464	15 926
Iceland	0	-	0	-	-	-	-	-	-	-
Ireland	68 556	85 804	178 355	127 876	75 002	74 253	58 201	55 438	54 975	33 072
Latvia	2 803	-	6	-	-	-	-	-	-	-
Lithuania	-	-	232	7 400	-	-	421	5	344	-
Netherlands	146 423	101 845	113 828	135 965	122 683	103 248	84 891	65 994	84 011	56 575
Norway	128 338	94 648	96 132	15 556	46 491	13 366	46 657	2 084	7 988	36 686
Portugal	25 307	19 040	17 701	14 054	18 720	21 364	15 533	15 471	15 305	20 683
Romania	-	-	-	360	-	-	-	-	-	-
Russian Fed	1 197	996	1 709	804	554	345	121	86	16	3
Sweden	2	207	447	166	1 761	3 418	2 004	1 162	119	575
Ukraine	264	74	-	-	-	-	-	-	-	-
UK	16 456	32 143	48 203	49 927	51 909	32 832	21 025	17 100	19 581	12 332
27 Fishing area total	*486 970*	*412 420*	*550 188*	*464 224*	*446 230*	*343 556*	*315 464*	*222 870*	*238 494*	*208 743*
34 Germany	-	-	-	213	493	626	574	-	-	-
Portugal	207	14	2	11	19	40	2	-	-	-
34 Fishing area total	*207*	*14*	*2*	*224*	*512*	*666*	*576*	-	-	-
37 Greece	582	1 628	1 860	2 484	2 956	1 360	942	774	747	700 F
Israel	200 F	300 F	340	65	226	172	178	175	90	129
Slovenia	41	12	7	9	8	4	5	4	4	2
Syria	40 F	39 F	39 F	60	77	40	30	36	56	66
Turkey	27 321	20 019	7 431	7 559	5 100	4 500	4 000	7 200	10 635	6 982
37 Fishing area total	*28 184 F*	*21 998 F*	*9 677 F*	*10 177*	*8 367*	*6 076*	*5 155*	*8 189*	*11 532*	*7 879 F*
Species total	*515 361 F*	*434 432 F*	*559 867 F*	*474 625*	*455 109*	*350 298*	*321 195*	*231 059*	*250 026*	*216 622 F*
Japanese jack mackerel	**Chinchard du Japon**		**Jurel japonés**			***Trachurus japonicus***			**1,70(23)004,03**	**JJM**
61 China,Taiwan	193	358	4 841	4 218	4 711	7 121	2 661	6 003	3 903	7 458
Japan	311 949	326 130	312 994	330 406	323 142	311 311	211 077	245 988	214 434	196 044
Korea Rep	38 095	38 433	12 269	14 542	22 766	22 132	13 552	19 510	17 537	26 037
61 Fishing area total	*350 237*	*364 921*	*330 104*	*349 166*	*350 619*	*340 564*	*227 290*	*271 501*	*235 874*	*229 539*
71 Korea Rep	3	38	-	-	-	1	-	-	-	-
71 Fishing area total	*3*	*38*	-	-	-	*1*	-	-	-	-
Species total	*350 240*	*364 959*	*330 104*	*349 166*	*350 619*	*340 565*	*227 290*	*271 501*	*235 874*	*229 539*
Chilean jack mackerel	**Chinchard du Chili**		**Jurel chileno**			***Trachurus murphyi***			**1,70(23)004,05**	**CJM**
87 Chile	3 236 244	4 041 447	4 404 193	3 883 326	2 917 064	1 612 912	1 219 689	1 234 299	1 649 933	1 518 994
China	-	-	-	-	-	-	-	-	-	76 261
Ecuador	9 946	23 723	174 393	56 781	30 302	25 900	19 072	7 144	134 011	604

B-37 Miscellaneous pelagic fishes / Poissons pélagiques divers / Peces pelágicos diversos

Capture production by species, fishing areas and countries or areas
Captures par espèces, zones de pêche et pays ou zones
Capturas por especies, áreas de pesca y países o áreas

Species, Fishing area Espèce, Zone de pêche Especie, Area de pesca	1993 t	1994 t	1995 t	1996 t	1997 t	1998 t	1999 t	2000 t	2001 t	2002 t
Ghana	-	-	-	-	-	-	-	2 472	1 157	-
Japan	-	-	-	-	-	-	7	-	-	-
Peru	130 681	196 771	376 600	438 736	649 751	386 946	184 679	296 579	723 733	154 219
87 Fishing area total	*3 376 871*	*4 261 941*	*4 955 186*	*4 378 843*	*3 597 117*	*2 025 758*	*1 423 447*	*1 540 494*	*2 508 834*	*1 750 078*
Species total	*3 376 871*	*4 261 941*	*4 955 186*	*4 378 843*	*3 597 117*	*2 025 758*	*1 423 447*	*1 540 494*	*2 508 834*	*1 750 078*
Pacific jack mackerel	**Chinchard gros yeux**		**Chicharro ojotón**			***Trachurus symmetricus***			**1,70(23)004,06**	**PJM**
67 Russian Fed	-	-	-	-	-	230	10	-	-	-
USA	277	200	147	-	2	731	166	181	215	21
67 Fishing area total	*277*	*200*	*147*	-	*2*	*961*	*176*	*181*	*215*	*21*
77 USA	1 504	2 697	1 728	2 176	1 158	832	950	1 135	3 624	1 005
77 Fishing area total	*1 504*	*2 697*	*1 728*	*2 176*	*1 158*	*832*	*950*	*1 135*	*3 624*	*1 005*
Species total	*1 781*	*2 897*	*1 875*	*2 176*	*1 160*	*1 793*	*1 126*	*1 316*	*3 839*	*1 026*
Mediterranean horse mackerel	**Chinchard à queue jaune**		**Jurel mediterráneo**			***Trachurus mediterraneus***			**1,70(23)004,08**	**HMM**
37 Bulgaria	79	80 F	70	68	36	40	30	111	130	142
Croatia	787	566	453	361	336	200	90	75	192	84
Greece	8 263	11 323	8 331	8 039	7 169	4 350	3 534	3 902	3 408	3 250 F
Malta	6	5	8	5	4	2	4	0	0	-
Romania	30	35	23	13	1	15	3	8	17	21
Serbia-Monte	7	10	15	16	14	15	17	14	15	16
Turkey	8 027	11 742	11 260	12 500	9 500	10 500	9 220	15 000	15 545	19 500
Ukraine	0	1	2	...	5	-	-	1	1	34
37 Fishing area total	*17 199*	*23 762 F*	*20 162*	*21 002*	*17 065*	*15 122*	*12 898*	*19 111*	*19 308*	*23 047 F*
Species total	*17 199*	*23 762 F*	*20 162*	*21 002*	*17 065*	*15 122*	*12 898*	*19 111*	*19 308*	*23 047 F*
Rough scad	**Chinchard frappeur**		**Chicharro garretón**			***Trachurus lathami***			**1,70(23)004,11**	**RSC**
41 Argentina	207	214	196	587	288	247	470	67	117	127
Brazil	...	...	536	389	313	81	25	40	228	328
41 Fishing area total	*207*	*214*	*732*	*976*	*601*	*328*	*495*	*107*	*345*	*455*
Species total	*207*	*214*	*732*	*976*	*601*	*328*	*495*	*107*	*345*	*455*
Cape horse mackerel	**Chinchard du Cap**		**Jurel del Cabo**			***Trachurus capensis***			**1,70(23)004,13**	**HMC**
47 Estonia	31 447	31 372	28 655	-	-	-	-	-	-	-
Georgia	2 000 F	1 000 F	-	-	-	-	-	-	-	-
Japan	-	27	18	-	-	-	-	-	-	-
Korea Rep	189	575	-	-	-	-	-	-	-	-
Lithuania	16 460	1 726	-	-	-	-	-	-	-	-
Namibia	474 611	364 801	310 836	321 322	301 847	312 422	320 394	344 314	309 381	359 183
Poland	-	-	3 058	1 700	-	-	-	-	3 098	3 557
Russian Fed	136 576	140 525	116 695	78 429	69 248	104 935	55 642	50 456	28 215	1 738
South Africa	35 429	20 031	10 262	31 995	31 206	46 384	17 970	15 000 F	9 659	21 883
Spain	-	-	-	-	2	-	-	-	-	-
Ukraine	88 706	50 719	18 060	27 121	5 179	18 345	4 592	13 837	3 693	-
Other nei	-	2 877	18 596	9 805	-	-	-	-	-	-
47 Fishing area total	*785 418 F*	*613 653 F*	*506 180*	*470 372*	*407 482*	*482 086*	*398 598*	*423 607 F*	*354 046*	*386 361*
Species total	*785 418 F*	*613 653 F*	*506 180*	*470 372*	*407 482*	*482 086*	*398 598*	*423 607 F*	*354 046*	*386 361*
Cunene horse mackerel	**Chinchard du Cunène**		**Jurel de Cunene**			***Trachurus trecae***			**1,70(23)004,14**	**HMZ**
34 Belize	1	-	-	-	-	-	-	-	-	-
China	0	-	-	-	-	-	-	-	-	-
Ghana	1 893	2 741	4 215	7 714	6 962	11 690	9 964	572	1 540	...
Honduras	3	-	2	-	-	-	-	-	-	-
Liberia	3	3	12	-	-	-	-	-	-	-
Romania	441	-	-	-	-	-	-	-	-	-
Sierra Leone	8	0	4	11	44	40	5	...	434	134
Togo	107	224	149	163	107	82	815	449	501	154
Ukraine	-	-	-	-	-	-	-	-	-	501
Vanuatu	13	-	-	-	-	-	-	-	-	-
34 Fishing area total	*2 469*	*2 968*	*4 382*	*7 888*	*7 113*	*11 812*	*10 784*	*1 021*	*2 475*	*789*
47 Angola	43 970	29 459	25 308	19 764	35 797	39 739	47 719	53 243	40 898	43 292
Russian Fed	57 980	70 408	53 760	58 466	51 583	6 265	33 973	16 387	5 934	1 934
47 Fishing area total	*101 950*	*99 867*	*79 068*	*78 230*	*87 380*	*46 004*	*81 692*	*69 630*	*46 832*	*45 226*
Species total	*104 419*	*102 835*	*83 450*	*86 118*	*94 493*	*57 816*	*92 476*	*70 651*	*49 307*	*46 015*
Greenback horse mackerel	**Chinchard dos vert**		**Jurel verde**			***Trachurus declivis***			**1,70(23)004,15**	**HMG**
57 Australia	-	-	-	-	-	-	-	323	82	594
57 Fishing area total	-	-	-	-	-	-	-	*323*	*82*	*594*
81 Australia	...	...	...	68	30	18	16	129	70	42
Russian Fed	4 260	1 804	1 602	2 280	886	52	223	-	-	-
Ukraine	7 937	4 192	8 990	13 093	9 740	9 309	15 306	12 213	7 577	5 667
81 Fishing area total	*12 197*	*5 996*	*10 592*	*15 441*	*10 656*	*9 379*	*15 545*	*12 342*	*7 647*	*5 709*
Species total	*12 197*	*5 996*	*10 592*	*15 441*	*10 656*	*9 379*	*15 545*	*12 665*	*7 729*	*6 303*

B-37 Miscellaneous pelagic fishes / Poissons pélagiques divers / Peces pelágicos diversos

Capture production by species, fishing areas and countries or areas / Captures par espèces, zones de pêche et pays ou zones / Capturas por especies, áreas de pesca y países o áreas

Species, Fishing area Espèce, Zone de pêche Especie, Area de pesca	1993 t	1994 t	1995 t	1996 t	1997 t	1998 t	1999 t	2000 t	2001 t	2002 t
Jack and horse mackerels nei	**Chinchards noirs nca**		**Jureles nep**			***Trachurus spp***			**1,70(23)004,XX**	**JAX**
27 Spain	29 828	29 895	31 863	32 283	39 846	36 894	38 160	36 989	39 824	33 429
27 Fishing area total	*29 828*	*29 895*	*31 863*	*32 283*	*39 846*	*36 894*	*38 160*	*36 989*	*39 824*	*33 429*
34 Belize	4	0	-	-	1 600	1 626	5 924	5 714	3 731	5 380
Bulgaria	-	-	-	-	-	1 669	-	-	-	-
China	4	2	-	-	-	1	-	0	3	-
Congo Dem R	1 510 F	1 370 F	1 397 F	1 432 F	1 386 F	1 426 F	1 400 F	1 430 F	1 500 F	1 600 F
Congo Rep	367 F	26	64	60 F	50 F	40 F	30 F	23	20 F	20 F
Cyprus	-	-	-	...	...	5 968	17 892	27 326	42 691	-
Estonia	32 036	12 364	1 989	1 903	2 080	1 550	-	-	-	-
France	-	-	-	-	...	4 623	1 316	162	-	-
Gambia	128	104	336	312	133	119	130	175	246	417
Georgia	1 350 F	500 F	-	-	-	-	-	-	-	-
Germany	-	-	-	-	-	-	-	-	708	278
Ghana	-	-	5 289	3 201	10 512	4 892	1 904	2 183	2 109	504
Greece	-	-	-	-	-	-	-	1	-	2
Guinea	3 470 F	3 570 F	4 781	3 576	1 546	7 109	508	6 084	5 170	5 100 F
Honduras	-	-	1	-	-	-	-	0	-	-
Ireland	-	-	-	-	-	-	-	-	8 522	3 411
Korea Rep	-	-	12	12	1	-	-	-	-	-
Latvia	34 799	34 764	38 824	14 818	4 881	8 710	14 284	22 591	18 522	7 768
Lithuania	16 580	11 716	...	...	...	11 902	20 657	25 464	17 672	46 282
Mauritania	1 360 F	970 F	1 510 F	1 220 F	1 060 F	940 F	...	...	75	20
Morocco	24 881	23 717	23 130	10 660	8 694	7 320	10 938	19 825	9 930	9 159
Netherlands	-	-	-	1 938	3 245	3 163	2 847	9 053	14 476	11 003
Panama	3	-	-	-	-	-	-	-	-	-
Poland	-	-	-	3 583	281	-	-	-	1 449	1 638
Portugal	551	299	206	599	764	660	495	562	386	358
Russian Fed	53 644	40 908	86 123	73 446	56 008	84 734	71 260	70 947	56 229	41 338
St Vincent	-	-	-	-	1 300	9 765	3 546	8 275	19 564	18 098
Senegal	1 710	1 093	1 027	521	1 078	1 061	2 108	1 020	1 877	3 886
Spain	-	-	-	345	124	52	57	133	215	432
Ukraine	43 242	41 246	56 236	42 471	38 311	64 222	53 322	66 266	37 233	12 995
Other nei	-	-	-	-	-	-	3 375	8 994	12 012	58 030
34 Fishing area total	*215 639 F*	*172 649 F*	*220 925 F*	*160 097 F*	*133 054 F*	*221 552 F*	*211 993 F*	*276 228 F*	*254 340 F*	*227 719 F*
37 Albania	...	...	50	68	18	85	92	90	21	65
Algeria	5 000 F	7 000 F	6 552	3 644	5 491	4 827	6 212	7 702	8 174	7 088
Egypt	354	...	...	...	...	...	...	...	...	...
France	519	272	311	422	2 148	2 148	571	753	1 038	705
Georgia	-	-	-	-	18	13	...	35	7	19
Italy	6 214	6 164	7 458	6 790	5 168	6 314	4 315	3 428	3 927	2 913
Libya	2 750 F	3 000 F	3 050	3 000 F	3 000 F	3 000 F	3 000 F	3 000 F	3 000 F	3 000 F
Morocco	3 916	5 073	7 344	4 811	3 818	2 496	1 849	2 270	2 237	1 853
Palest, O.T.	...	...	...	90	100	115	115 F	115 F	115 F	125
Russian Fed	-	1	1	-	-	2	2	2	6	28
Spain	-	-	-	-	-	-	5 533	7 346	6 957	4 400
Tunisia	2 382	3 704	4 116	4 755	6 326	3 899	5 635	4 917	4 204	4 746
37 Fishing area total	*21 135 F*	*25 214 F*	*28 882*	*23 580 F*	*26 087 F*	*22 899 F*	*27 324 F*	*29 658 F*	*29 686 F*	*24 942 F*
47 Korea Rep	1 000	20	-	33	45	-	-	-	-	-
Russian Fed	24 836	-	-	-	-	-	-	-	-	-
47 Fishing area total	*25 836*	*20*	*...*	*33*	*45*	*...*	*...*	*...*	*...*	*...*
51 Korea Rep	-	-	-	-	-	-	-	-	6	3
Russian Fed	-	65	-	-	-	-	-	-	-	57
Ukraine	127	...	3	...	...	15	11	-	7	-
Yemen	990	990	4 412	1 380	1 440 F	1 600 F	1 600 F	1 400 F	1 800 F	2 000 F
51 Fishing area total	*1 117*	*1 055*	*4 415*	*1 380*	*1 440 F*	*1 615 F*	*1 611 F*	*1 400 F*	*1 813 F*	*2 060 F*
81 Japan	14 044	10 034	4 111	335	29	8	11	12	655	9
Korea Rep	1 260	359	474	1 157	2 087	1 983	2 182	259	307	11
New Zealand	34 838	34 002	31 919	29 085	34 057	36 059	34 003	22 544	28 507	32 285
Norway	3	-	-	0	1	-	-	-	-	-
81 Fishing area total	*50 145*	*44 395*	*36 504*	*30 577*	*36 174*	*38 050*	*36 196*	*22 815*	*29 469*	*32 305*
Species total	*343 700 F*	*273 228 F*	*322 589 F*	*247 950 F*	*236 646 F*	*321 010 F*	*315 284 F*	*367 090 F*	*355 132 F*	*320 455 F*
White trevally	**Carangue dentue**		**Jurel dentón**			***Pseudocaranx dentex***			**1,70(23)011,27**	**TRZ**
57 Australia	-	-	-	-	-	-	-	38	26	18
57 Fishing area total	*-*	*-*	*-*	*-*	*-*	*-*	*-*	*38*	*26*	*18*
81 Australia	...	...	...	964	872	558	479	380	389	473
New Zealand	3 393	3 406	3 910	2 869	2 941	3 608	4 192	3 603	3 116	3 280
Ukraine	-	-	-	-	-	-	-	-	-	5
81 Fishing area total	*3 393*	*3 406*	*3 910*	*3 833*	*3 813*	*4 166*	*4 671*	*3 983*	*3 505*	*3 758*
Species total	*3 393*	*3 406*	*3 910*	*3 833*	*3 813*	*4 166*	*4 671*	*4 021*	*3 531*	*3 776*
Japanese scad	**Comète japonaise**		**Macarela japonesa**			***Decapterus maruadsi***			**1,70(23)043,07**	**RSA**
61 China,Taiwan	1 987	3 356	3 626	5 059	19 667	12 090	20 532	4 387	6 224	7 755
Japan	49 900	47 810	72 109	57 319	50 097	59 078	47 157	36 416	41 236	41 831
61 Fishing area total	*51 887*	*51 166*	*75 735*	*62 378*	*69 764*	*71 168*	*67 689*	*40 803*	*47 460*	*49 586*

B-37 Miscellaneous pelagic fishes / Poissons pélagiques divers / Peces pelágicos diversos

Capture production by species, fishing areas and countries or areas / Captures par espèces, zones de pêche et pays ou zones / Capturas por especies, áreas de pesca y países o áreas

Species, Fishing area Espèce, Zone de pêche Especie, Area de pesca	1993 t	1994 t	1995 t	1996 t	1997 t	1998 t	1999 t	2000 t	2001 t	2002 t
71 Japan	13	-	-	-	-	-	-	-	-	-
71 Fishing area total	*13*	-	-	-	-	-	-	-	-	-
Species total	*51 900*	*51 166*	*75 735*	*62 378*	*69 764*	*71 168*	*67 689*	*40 803*	*47 460*	*49 586*
Indian scad	**Comète indienne**		**Macarela índica**		***Decapterus russelli***				**1,70(23)043,08**	**RUS**
57 Malaysia	10 507	9 979	10 392	8 748	6 953	13 189	8 423	5 117	5 811	15 474
Thailand	8 984	35 994	22 981	32 557	30 658	28 270	28 113	25 331	26 726	26 160
57 Fishing area total	*19 491*	*45 973*	*33 373*	*41 305*	*37 611*	*41 459*	*36 536*	*30 448*	*32 537*	*41 634*
61 China,Taiwan	927	678	299	382	687	6 158	7 703	3 927	598	58
61 Fishing area total	*927*	*678*	*299*	*382*	*687*	*6 158*	*7 703*	*3 927*	*598*	*58*
71 Malaysia	54 215	51 662	45 231	50 985	64 231	40 237	61 737	79 086	71 583	74 827
Thailand	46 186	38 394	54 641	52 648	47 498	57 893	56 461	67 902	59 080	59 750
71 Fishing area total	*100 401*	*90 056*	*99 872*	*103 633*	*111 729*	*98 130*	*118 198*	*146 988*	*130 663*	*134 577*
Species total	*120 819*	*136 707*	*133 544*	*145 320*	*150 027*	*145 747*	*162 437*	*181 363*	*163 798*	*176 269*
Scads nei	**Comètes nca**		**Macarelas nep**		***Decapterus spp***				**1,70(23)043,XX**	**SDX**
31 Grenada	104	64	94	82	61	101	59	38	52	75
31 Fishing area total	*104*	*64*	*94*	*82*	*61*	*101*	*59*	*38*	*52*	*75*
34 Gabon	216	216	106	33	20	18	76	21	101	84
Ghana	1 235	963	1 760	1 462	1 654	2 989	3	263	2 466	2 944
Poland	-	-	-	54	-	-	-	-	-	-
Sao Tome Prn	...	...	...	...	43	55 F	60 F	60 F	60 F	60 F
Senegal	2 524	1 958	1 973	2 265	2 428	3 249	2 401	4 963	4 209	7 416
Sierra Leone	...	...	...	...	112	39	277	141	...	...
34 Fishing area total	*3 975*	*3 137*	*3 839*	*3 814*	*4 257*	*6 350 F*	*2 817 F*	*5 448 F*	*6 836 F*	*10 504 F*
47 Ukraine	327	-	181	...	...	-	-	-	-	-
47 Fishing area total	*327*	-	*181*	...	...	-	-	-	-	-
51 Jordan	...	...	...	...	...	20	25	25	26	25
Pakistan	1 675	1 875	1 920	1 010	1 225	3 505	4 661	4 600	4 355	4 400
Qatar	0	0	0	0	0	0	0	0	0	0
Untd Arab Em	1 600	1 783	1 790	1 809	1 933	1 939	1 987	580	565 F	550
51 Fishing area total	*3 275*	*3 658*	*3 710*	*2 819*	*3 158*	*5 464*	*6 673*	*5 205*	*4 946 F*	*4 975*
57 Indonesia	35 312	30 804	33 760	36 295	33 650	39 494	41 593	42 044	43 133	44 400
57 Fishing area total	*35 312*	*30 804*	*33 760*	*36 295*	*33 650*	*39 494*	*41 593*	*42 044*	*43 133*	*44 400*
61 China	260 758	430 860	515 298	607 686	505 991	532 986	502 590	502 289	544 728	603 015
China,H.Kong	3 874	4 600	5 254	5 368	4 143	2 962	2 100 F	2 600 F	2 900 F	2 800 F
China,Taiwan	16 863	15 041	9 168	32 586	12 743	3 172	2 836	2 491	2 670	2 486
61 Fishing area total	*281 495*	*450 501*	*529 720*	*645 640*	*522 877*	*539 120*	*507 526 F*	*507 380 F*	*550 298 F*	*608 301 F*
71 Guam	-	-	0	0	3	2	5	4	5	2
Indonesia	168 039	189 089	213 545	214 994	243 274	238 099	219 545	213 331	215 260	222 650
N Marianas	...	1	5	2	4	0	5	10	13	10
Philippines	274 029	235 973	264 472	228 757	234 849	250 809	254 178	260 999	291 774	283 594
Singapore	138	249	209	227	212	222	156	163	106	74
71 Fishing area total	*442 206*	*425 312*	*478 231*	*443 980*	*478 342*	*489 132*	*473 889*	*474 507*	*507 158*	*506 330*
77 El Salvador	138	94	156	237	130	149	226	187	270	295
USA	-	-	-	-	-	-	-	-	-	111
77 Fishing area total	*138*	*94*	*156*	*237*	*130*	*149*	*226*	*187*	*270*	*406*
81 New Zealand	...	...	33	23	11	63	29	82	87	70
81 Fishing area total	...	...	*33*	*23*	*11*	*63*	*29*	*82*	*87*	*70*
Species total	*766 832*	*913 570*	*1 049 724*	*1 132 890*	*1 042 486*	*1 079 873 F*	*1 032 812 F*	*1 034 891 F*	*1 112 780 F*	*1 175 061 F*
Blue runner	**Carangue coubali**		**Cojinúa negra**		***Caranx crysos***				**1,70(23)044,26**	**RUB**
21 USA	-	-	-	-	-	-	49	1	0	1
21 Fishing area total	-	-	-	-	-	-	*49*	*1*	*0*	*1*
31 Dominican Rp	300	540	205	194	256	132	53	74	68	130
USA	820	552	531	119	152	276	131	130	157	170
31 Fishing area total	*1 120*	*1 092*	*736*	*313*	*408*	*408*	*184*	*204*	*225*	*300*
41 Brazil	...	...	338	503	515	443	590	625	680	688
41 Fishing area total	...	...	*338*	*503*	*515*	*443*	*590*	*625*	*680*	*688*
Species total	*1 120*	*1 092*	*1 074*	*816*	*923*	*851*	*823*	*830*	*905*	*989*
Crevalle jack	**Carangue crevalle**		**Jurel común**		***Caranx hippos***				**1,70(23)044,29**	**CVJ**
21 USA	-	-	-	-	-	4	232	2	1	4
21 Fishing area total	-	-	-	-	-	*4*	*232*	*2*	*1*	*4*
31 USA	-	-	146	136	264	393	91	316	304	217
31 Fishing area total	-	-	*146*	*136*	*264*	*393*	*91*	*316*	*304*	*217*
34 Ghana	4 399	5 603	4 422	2 884	2 207	2 891	3 906	79	2 525	2 872

B-37 Miscellaneous pelagic fishes / Poissons pélagiques divers / Peces pelágicos diversos

Capture production by species, fishing areas and countries or areas
Captures par espèces, zones de pêche et pays ou zones
Capturas por especies, áreas de pesca y países o áreas

Species, Fishing area Espèce, Zone de pêche Especie, Area de pesca	1993 t	1994 t	1995 t	1996 t	1997 t	1998 t	1999 t	2000 t	2001 t	2002 t
Sao Tome Prn	...	...	...	...	42	55 F	60 F	60 F	60 F	60 F
34 Fishing area total	*4 399*	*5 603*	*4 422*	*2 884*	*2 249*	*2 946 F*	*3 966 F*	*139 F*	*2 585 F*	*2 932 F*
47 Angola	15	43	42	39	16	89	302	1 835	674	271
47 Fishing area total	*15*	*43*	*42*	*39*	*16*	*89*	*302*	*1 835*	*674*	*271*
Species total	*4 414*	*5 646*	*4 610*	*3 059*	*2 529*	*3 432 F*	*4 591 F*	*2 292 F*	*3 564 F*	*3 424 F*
Bar jack	**Carangue comade**		**Cojinúa carbonera**		***Caranx ruber***			**1,70(23)044,31**		**CXR**
31 USA	-	-	-	-	-	-	3	5	7	13
31 Fishing area total	-	-	-	-	-	-	*3*	*5*	*7*	*13*
Species total	-	-	-	-	-	-	*3*	*5*	*7*	*13*
False scad	**Comète coussut**		**Macarela real**		***Caranx rhonchus***			**1,70(23)044,42**		**HMY**
34 Ghana	2 213	4 040	3 483	3 301	3 337	2 753	2 275	3 800	147	1 605
Liberia	-	-	-	-	7	111	435	135	134	140 F
Romania	367	-	-	-	-	-	-	-	-	-
34 Fishing area total	*2 580*	*4 040*	*3 483*	*3 301*	*3 344*	*2 864*	*2 710*	*3 935*	*281*	*1 745 F*
Species total	*2 580*	*4 040*	*3 483*	*3 301*	*3 344*	*2 864*	*2 710*	*3 935*	*281*	*1 745 F*
Jacks, crevalles nei	**Chinchards, carangues nca**		**Jureles, pámpanos nep**		***Caranx spp***			**1,70(23)044,XX**		**TRE**
31 Br Virgin Is	...	...	...	...	13	15	14	1	1 F	1 F
Colombia	217	496	510	1 111	350	239	117	260 F	260 F	260 F
Cuba	268	426	344	349	234	499	167	2 559	2 138	180
Dominican Rp	270	486	144	424	453	75	23	43	43	111
Mexico	2 587	2 766	5 437	5 911	8 398	8 735	6 365	5 779	6 002	6 328
Trinidad Tob	501	328	400 F	504	562	203	189	202	201	294
USA	-	-	-	-	-	-	-	-	-	103
Venezuela	5 136	4 271	3 439	3 732	2 823	2 898	2 642	3 462	2 115	2 161
31 Fishing area total	*8 979*	*8 773*	*10 274 F*	*12 031*	*12 833*	*12 664*	*9 517*	*12 306 F*	*10 760 F*	*9 438 F*
34 Congo Rep	0	16	15 F	12 F	9 F	5 F	3 F	0	0	0
Côte dIvoire	415	377	128	69	51	282	290	144	119	690
Estonia	-	-	-	-	-	-	1 622	-	-	-
Gabon	...	31	47	594	404	91	4	29	31	39
Gambia	73	88	73	174	124	147	160	137	288	899
Ghana	31	14	57	13	11	1	-	1	-	-
GuineaBissau	46 F	50 F	100 F	100 F	100 F	80 F	70 F	70 F	70 F	70 F
Latvia	66	-	-	-	11	-	-	-	-	-
Liberia	97	96	76	62	30	178	394	235	271	280 F
Lithuania	209	-	-	-	-	-	-	-	-	-
Nigeria	3 290	2 414	636	591	2 525	414	1 099	825	317	277
Portugal	102	188	44	134	114	95	32	1	1	0
Russian Fed	126	13	53	15	24	-	32	2	-	-
Sao Tome Prn	...	...	...	...	129	150 F	160 F	150 F	150 F	150 F
Sierra Leone	469	468	467	470 F	474	18	7	587	327	726
Ukraine	47	...	...	...	...	-	-	-	-	-
34 Fishing area total	*4 971 F*	*3 755 F*	*1 696 F*	*2 234 F*	*4 006 F*	*1 461 F*	*3 873 F*	*2 181 F*	*1 574 F*	*3 131 F*
37 Egypt	-	-	433	716	441	402	326	525	1 351	1 122
Syria	...	...	...	...	...	...	...	...	...	36
37 Fishing area total	...	...	*433*	*716*	*441*	*402*	*326*	*525*	*1 351*	*1 158*
41 Brazil	4 510 F	4 520 F	3 656	3 905	5 838	5 635	4 164	6 950	5 878	6 083
41 Fishing area total	*4 510 F*	*4 520 F*	*3 656*	*3 905*	*5 838*	*5 635*	*4 164*	*6 950*	*5 878*	*6 083*
47 Russian Fed	-	-	-	-	-	-	-	-	1	-
47 Fishing area total	-	-	-	-	-	-	-	-	*1*	-
51 Egypt	14 361	11 823	...	...	...	...	...	...	-	-
India	53 501	48 378	5 263	50 436	33 042	52 315	51 741	22 808	30 894	17 624
Korea Rep	-	5	-	46	-	11	-	-	-	-
Oman	1 914	1 773	3 392	2 927	3 896	2 957	2 552	1 806	1 218	1 666
Pakistan	3 933	4 003	4 631	3 972	5 391	6 523	8 407	9 111	8 928	9 200
Qatar	361	201	175	222	247	308	289	409	388	318
Ukraine	-	-	-	-	-	-	2	-	-	-
Untd Arab Em	6 000	6 682	6 833	6 905	7 379	7 403	7 588	3 107	2 762	4 759
Yemen	174	480	431	413	430 F	500 F	500 F	400 F	500 F	550 F
51 Fishing area total	*80 244*	*73 345*	*20 725*	*64 921*	*50 385 F*	*70 017 F*	*71 079 F*	*37 641 F*	*44 690 F*	*34 117 F*
57 India	11 117	11 731	16 176	17 443	17 791	14 595	13 758	16 763	16 461	15 116
Indonesia	8 273	6 097	8 162	7 847	8 185	14 941	9 507	10 398	11 008	11 960
Malaysia	4 859	6 465	7 847	6 524	5 319	834	5 020	6 117	5 807	6 024
57 Fishing area total	*24 249*	*24 293*	*32 185*	*31 814*	*31 295*	*30 370*	*28 285*	*33 278*	*33 276*	*33 100*
61 China,H.Kong	50	72	106	66	33	20	10 F	15 F	17 F	15 F
China,Taiwan	9 643	8 270	6 595	7 065	11 433	8 228	4 882	6 355	3 273	4 614
61 Fishing area total	*9 693*	*8 342*	*6 701*	*7 131*	*11 466*	*8 248*	*4 892 F*	*6 370 F*	*3 290 F*	*4 629 F*
71 Fiji Islands	1 795	1 829 F	644	384	695	647	730	650 F	709	700 F
Indonesia	18 632	19 989	20 863	22 198	23 912	24 502	24 713	25 923	26 980	28 260
Kiribati	510	500	500	500	510	2 530	1 910	1 108	2 858	3 702
Malaysia	24 347	19 893	18 683	19 845	24 561	10 719	31 624	33 497	31 318	32 245
Singapore	331	352	297	312	313	234	175	139	66	69

B-37 Miscellaneous pelagic fishes / Poissons pélagiques divers / Peces pelágicos diversos

Capture production by species, fishing areas and countries or areas / Captures par espèces, zones de pêche et pays ou zones / Capturas por especies, áreas de pesca y países o áreas

Species, Fishing area Espèce, Zone de pêche Especie, Area de pesca	1993 t	1994 t	1995 t	1996 t	1997 t	1998 t	1999 t	2000 t	2001 t	2002 t
71 Fishing area total	*45 615*	*42 563 F*	*40 987*	*43 239*	*49 991*	*38 632*	*59 152*	*61 317 F*	*61 931*	*64 976 F*
77 Cook Is	55 F	52 F	51 F	40 F	40 F	40 F	40 F	40 F	40 F	40 F
Mexico	983	1 146	1 723	2 049	2 281	2 198	1 931	2 487	2 868	2 729
USA	-	-	-	-	-	-	-	0	-	28
77 Fishing area total	*1 038 F*	*1 198 F*	*1 774 F*	*2 089 F*	*2 321 F*	*2 238 F*	*1 971 F*	*2 527 F*	*2 908 F*	*2 797 F*
87 Colombia	194	181	129	538	504	278	273	200 F	138	140 F
Peru	...	18	-	129	16	69	607	35	11	62
87 Fishing area total	*194*	*199*	*129*	*667*	*520*	*347*	*880*	*235 F*	*149*	*202 F*
Species total	*179 493 F*	*166 988 F*	*118 560 F*	*168 747 F*	*169 096 F*	*170 014 F*	*184 139 F*	*163 330 F*	*165 808 F*	*159 631 F*
Atlantic moonfish	**Musso atlantique**		**Jorobado lamparosa**		***Selene setapinnis***				**1,70(23)046,04**	**MOA**
31 Grenada	0	0	0	0	0	0	0	0	0	0
Venezuela	2 467	2 603	2 544	1 766	2 110	2 338	1 529	976	3 443	1 345
31 Fishing area total	*2 467*	*2 603*	*2 544*	*1 766*	*2 110*	*2 338*	*1 529*	*976*	*3 443*	*1 345*
41 Brazil	...	...	2 468	1 679	1 914	1 512	1 514	1 386	1 108	1 350
41 Fishing area total	...	...	*2 468*	*1 679*	*1 914*	*1 512*	*1 514*	*1 386*	*1 108*	*1 350*
Species total	*2 467*	*2 603*	*5 012*	*3 445*	*4 024*	*3 850*	*3 043*	*2 362*	*4 551*	*2 695*
African moonfish	**Musso africain**		**Jorobado africano**		***Selene dorsalis***				**1,70(23)046,05**	**LUK**
34 Benin	22 F	25 F	24 F	27 F	37 F	35 F	57	40 F	121	144
Congo Rep	75 F	45 F	50 F	55 F	60 F	60 F	70 F	72	70 F	70 F
Georgia	10 F	-	-	-	-	-	-	-	-	-
Ghana	882	501	976	907	712	381	469	738	1 100	690
Russian Fed	194	176	20	15	12	1	-	-	-	-
Senegal	-	-	-	-	9	278	176	251	326	469
Sierra Leone	0	0	1	1 F	...	...	...	135	319	260
Togo	4	3	2	10	2	0	-	0	-	13
Ukraine	91	-	-	-	-	-	-	-	-	3
34 Fishing area total	*1 278 F*	*750 F*	*1 073 F*	*1 015 F*	*832 F*	*755 F*	*772 F*	*1 236 F*	*1 936 F*	*1 649 F*
Species total	*1 278 F*	*750 F*	*1 073 F*	*1 015 F*	*832 F*	*755 F*	*772 F*	*1 236 F*	*1 936 F*	*1 649 F*
Snubnose pompano	**Pompaneau lune**		**Pámpano lunero**		***Trachinotus blochii***				**1,70(23)047,01**	**POO**
51 Saudi Arabia	-	-	-	-	-	4	0	31	-	-
51 Fishing area total	-	-	-	-	-	*4*	*0*	*31*	-	-
Species total	-	-	-	-	-	*4*	*0*	*31*	-	-
Florida pompano	**Pompaneau sole**		**Pámpano amarillo**		***Trachinotus carolinus***				**1,70(23)047,03**	**POM**
21 USA	0	1	0	44	1	2	139	0	0	1
21 Fishing area total	*0*	*1*	*0*	*44*	*1*	*2*	*139*	*0*	*0*	*1*
31 USA	-	-	54	-	296	319	68	242	181	158
31 Fishing area total	-	-	*54*	-	*296*	*319*	*68*	*242*	*181*	*158*
Species total	*0*	*1*	*54*	*44*	*297*	*321*	*207*	*242*	*181*	*159*
Pompanos nei	**Pompaneaux nca**		**Pámpanos(=Palometas) nep**		***Trachinotus spp***				**1,70(23)047,XX**	**POX**
31 Dominican Rp	58	104	76	57	19	47	5	10	7	111
Mexico	643	602	599	735	429	556	542	410	333	605
Venezuela	140	122	244	307	285	435	146	132	117	174
31 Fishing area total	*841*	*828*	*919*	*1 099*	*733*	*1 038*	*693*	*552*	*457*	*890*
34 Estonia	10	-	-	-	-	-	-	-	-	-
Gambia	-	-	8	20	7	1	2	1	22	...
Portugal	-	-	0	0	1	2	3	-	-	0
Senegal	98	55	64	132	83	28	207	27	158	110
Sierra Leone	...	...	3	3 F	63	279	7	-	-	-
Spain	-	-	-	-	-	-	4	-	7	85
34 Fishing area total	*108*	*55*	*75*	*155 F*	*154*	*310*	*223*	*28*	*187*	*195*
41 Brazil	...	...	424	534	166	172	144	286	273	219
Estonia	-	-	-	-	-	-	-	-	-	3
41 Fishing area total	...	...	*424*	*534*	*166*	*172*	*144*	*286*	*273*	*222*
51 India	53 087	3 649	6 765	5 706	3 697	2 533	2 329	12	11	12
51 Fishing area total	*53 087*	*3 649*	*6 765*	*5 706*	*3 697*	*2 533*	*2 329*	*12*	*11*	*12*
57 India	0	1 043	2 132	1 078	-	205	-	-	-	-
57 Fishing area total	*0*	*1 043*	*2 132*	*1 078*	-	*205*	-	-	-	-
77 Mexico	303	298	286	206	200	329	309	274	222	258
77 Fishing area total	*303*	*298*	*286*	*206*	*200*	*329*	*309*	*274*	*222*	*258*
87 Peru	895	580	566	460	268	779	2 801	1 220	632	878
87 Fishing area total	*895*	*580*	*566*	*460*	*268*	*779*	*2 801*	*1 220*	*632*	*878*
Species total	*55 234*	*6 453*	*11 167*	*9 238 F*	*5 218*	*5 366*	*6 499*	*2 372*	*1 782*	*2 455*

B-37

Miscellaneous pelagic fishes — **Capture production by species, fishing areas and countries or areas**
Poissons pélagiques divers — **Captures par espèces, zones de pêche et pays ou zones**
Peces pelágicos diversos — **Capturas por especies, áreas de pesca y países o áreas**

Species, Fishing area Espèce, Zone de pêche Especie, Area de pesca	1993 t	1994 t	1995 t	1996 t	1997 t	1998 t	1999 t	2000 t	2001 t	2002 t
Greater amberjack	**Sériole couronnée**		**Pez de limón**		***Seriola dumerili***				**1,70(23)048,01**	**AMB**
21 USA	...	...	...	...	...	...	368	-	-	-
21 Fishing area total	*...*	*...*	*...*	*...*	*...*	*...*	*368*	*-*	*-*	*-*
27 Portugal	...	...	...	...	...	...	10	21	20	37
27 Fishing area total	*...*	*...*	*...*	*...*	*...*	*...*	*10*	*21*	*20*	*37*
31 USA	...	...	...	...	...	560	149	538	452	448
31 Fishing area total	*...*	*...*	*...*	*...*	*...*	*560*	*149*	*538*	*452*	*448*
37 Albania	...	...	0	2	1	-	-	-	2	19
Algeria	...	...	87	32	19	48	75	99	341	542
Croatia	30	55	78	80	64	62	30	25	19	38
Cyprus	12	38	16	28	21	8	17	25	14	15
Greece	114	199	322	697	336	252	205	176	396	350 F
Israel	34	30	79	33	80	185	98	307	223	201
Malta	0	0	4	9	6	6	6	3	2	3
Serbia-Monte	12	8	16	11	9	11	9	12	13	12
Spain	-	-	-	-	-	-	704	375	430	283
Syria	54 F	52 F	52 F	90	75	108	88	52	96	53
Tunisia	72	27	59	75	113	75	88	400	122	93
37 Fishing area total	*328 F*	*409 F*	*713 F*	*1 057*	*724*	*755*	*1 320*	*1 474*	*1 658*	*1 609 F*
Species total	*328 F*	*409 F*	*713 F*	*1 057*	*724*	*1 315*	*1 847*	*2 033*	*2 130*	*2 094 F*
Japanese amberjack	**Sériole du Japon**		**Medregal del Japón**		***Seriola quinqueradiata***				**1,70(23)048,02**	**AMJ**
61 Japan	-	-	7 564	246	...	...	...	...	...	...
61 Fishing area total	*-*	*-*	*7 564*	*246*	*...*	*...*	*...*	*...*	*...*	*...*
Species total	*-*	*-*	*7 564*	*246*	*...*	*...*	*...*	*...*	*...*	*...*
Yellowtail amberjack	**Sériole chicard**		**Medregal rabo amarillo**		***Seriola lalandi***				**1,70(23)048,06**	**YTC**
41 Argentina	24	13	9	16	17	7	9	10	33	8
Brazil	...	...	653	538	466	880	544	611	155	189
41 Fishing area total	*24*	*13*	*662*	*554*	*483*	*887*	*553*	*621*	*188*	*197*
47 Russian Fed	-	-	6	-	-	-	-	-	-	3
South Africa	858	771	769	488	477	519	302	300 F	315	213
47 Fishing area total	*858*	*771*	*775*	*488*	*477*	*519*	*302*	*300 F*	*315*	*216*
51 South Africa	-	-	-	-	1	...	...	...	...	...
51 Fishing area total	*-*	*-*	*-*	*-*	*1*	*...*	*...*	*...*	*...*	*...*
77 USA	-	-	-	-	-	111	30	50	39	34
77 Fishing area total	*-*	*-*	*-*	*-*	*-*	*111*	*30*	*50*	*39*	*34*
Species total	*882*	*784*	*1 437*	*1 042*	*961*	*1 517*	*885*	*971 F*	*542*	*447*
Amberjacks nei	**Sérioles nca**		**Medregales nep**		***Seriola spp***				**1,70(23)048,XX**	**AMX**
21 USA	...	...	...	...	...	4	5	8	3	2
21 Fishing area total	*...*	*...*	*...*	*...*	*...*	*4*	*5*	*8*	*3*	*2*
31 Dominican Rp	90	62	59	50	46	50	12	16	18	41
Mexico	981	1 035	1 148	1 109	1 590	1 574	1 615	1 805	1 405	1 293
USA	1 631	1 995	1 294	1 123	829	243	184	370	383	208
Venezuela	286	380	288	338	355	336	450	610	399	538
31 Fishing area total	*2 988*	*3 472*	*2 789*	*2 620*	*2 820*	*2 203*	*2 261*	*2 801*	*2 205*	*2 080*
34 Portugal	-	-	10	33	45	101	32	8	7	8
Russian Fed	11	-	-	82	26	-	4	-	-	-
Senegal	-	-	-	-	-	62	60	64	165	103
Spain	-	-	-	114	142	161	71	88	116	26
Ukraine	-	-	-	-	-	-	15	...	...	...
34 Fishing area total	*11*	*-*	*10*	*229*	*213*	*324*	*182*	*160*	*288*	*137*
41 Brazil	...	...	856	825	1 048	119	104	128	588	653
41 Fishing area total	*...*	*...*	*856*	*825*	*1 048*	*119*	*104*	*128*	*588*	*653*
51 Kenya	44	42	89	79	63	60	78	71	92	31
Korea Rep	-	-	-	-	21	5	1	-	-	-
Tanzania	198	143	215	51	250	275	100	75	60	50
51 Fishing area total	*242*	*185*	*304*	*130*	*334*	*340*	*179*	*146*	*152*	*81*
57 Australia	7	102	105	100	10	10	1	2	2	1
57 Fishing area total	*7*	*102*	*105*	*100*	*10*	*10*	*1*	*2*	*2*	*1*
61 Japan	43 243	53 801	54 101	50 333	47 211	45 484	54 918	77 461	66 925	51 194
Korea Rep	2 893	3 710	3 745	4 093	6 064	9 620	8 653	4 814	6 475	6 191
61 Fishing area total	*46 136*	*57 511*	*57 846*	*54 426*	*53 275*	*55 104*	*63 571*	*82 275*	*73 400*	*57 385*
71 Australia	-	-	-	-	-	-	-	-	0	-
Japan	5	1	-	-	-	-	-	-	-	-
Korea Rep	-	48	72	34	34	-	-	-	-	-
71 Fishing area total	*5*	*49*	*72*	*34*	*34*	*-*	*-*	*-*	*0*	*-*

B-37 Miscellaneous pelagic fishes / Poissons pélagiques divers / Peces pelágicos diversos

Capture production by species, fishing areas and countries or areas
Captures par espèces, zones de pêche et pays ou zones
Capturas por especies, áreas de pesca y países o áreas

Species, Fishing area Espèce, Zone de pêche Especie, Area de pesca	1993 t	1994 t	1995 t	1996 t	1997 t	1998 t	1999 t	2000 t	2001 t	2002 t
77 Mexico	5	6	11	7	12	6	13	27	28	20
Panama	864	850	341	470	333	469	159	418	510	1 433
77 Fishing area total	*869*	*856*	*352*	*477*	*345*	*475*	*172*	*445*	*538*	*1 453*
81 Australia	401	347	348	194	84	76	1	1	0	1
New Zealand	489	281	315	381	349	327	317	296	278	242
Russian Fed	-	-	-	-	206	-	209	-	-	-
Ukraine	-	-	-	-	-	-	-	-	-	17
81 Fishing area total	*890*	*628*	*663*	*575*	*639*	*403*	*527*	*297*	*278*	*260*
87 Colombia	...	17	...	...	51	109	47	70 F	91	90 F
Peru	3 084	3 325	6 598	1 558	4 648	21 104	2 084	11 159	28 025	29 787
87 Fishing area total	*3 084*	*3 342*	*6 598*	*1 558*	*4 699*	*21 213*	*2 131*	*11 229 F*	*28 116*	*29 877 F*
Species total	*54 232*	*66 145*	*69 595*	*60 974*	*63 417*	*80 195*	*69 133*	*97 491 F*	*105 570*	*91 929 F*
Leerfish	**Liche**		**Palometón**		***Lichia amia***			**1,70(23)072,02**		**LEE**
27 Lithuania	-	-	1	-	-	-	-	-	-	-
Portugal	0	0	0	0	0	-	0	0	0	-
Spain	2	0	4	-	5	2	1	1	10	6
27 Fishing area total	*2*	*0*	*5*	*0*	*5*	*2*	*1*	*1*	*10*	*6*
34 Côte dIvoire	-	-	-	-	-	-	20	16	3	4
Italy	1 029	1 029	-	-	-	-	-	-	-	-
Latvia	27	-	18	152	236	127	172	274	96	454
Liberia	-	-	-	2	0	-	-	-	-	-
Lithuania	72	11	-	-	-	-	-	-	-	386
Portugal	0	0	0	0	0	-	-	-	-	-
Russian Fed	24	40	14	84	393	736	622	200	83	333
Sao Tome Prn	22 F	32 F	34 F	25	36	45 F	50 F	50 F	50 F	50 F
Senegal	501	920	2 450	228	257	540	489	490	341	270
Spain	-	-	-	5	-	-	-	-	-	38
Ukraine	31	2	...	...	428	-	266	109	31	389
34 Fishing area total	*1 706 F*	*2 034 F*	*2 516 F*	*496*	*1 350*	*1 448 F*	*1 619 F*	*1 139 F*	*604 F*	*1 924 F*
37 Italy	776	629	948	643	400	197	249	185	183	223
Spain	-	-	-	19	27	36	27	42	35	88
Tunisia	134	100	101	93	145	160	112	90	134	138
Turkey	923	950	1 345	1 544	1 650	1 950	2 780	320	255	330
37 Fishing area total	*1 833*	*1 679*	*2 394*	*2 299*	*2 222*	*2 343*	*3 168*	*637*	*607*	*779*
47 Russian Fed	-	-	-	-	-	-	-	-	1	21
47 Fishing area total	*-*	*-*	*-*	*-*	*-*	*-*	*-*	*-*	*1*	*21*
Species total	*3 541 F*	*3 713 F*	*4 915 F*	*2 795*	*3 577*	*3 793 F*	*4 788 F*	*1 777 F*	*1 222 F*	*2 730 F*
Alexandria pompano	**Cordonnier bossu**		**Jurel de Alejandría**		***Alectis alexandrinus***			**1,70(23)090,03**		**ALA**
34 Senegal	...	324	124	160	742	1 023	563	502	862	611
34 Fishing area total	*...*	*324*	*124*	*160*	*742*	*1 023*	*563*	*502*	*862*	*611*
Species total	*...*	*324*	*124*	*160*	*742*	*1 023*	*563*	*502*	*862*	*611*
Black pomfret	**Castagnoline noire**		**Palometa negra**		***Parastromateus niger***			**1,70(23)099,01**		**POB**
51 Pakistan	1 961	2 199	3 066	2 221	2 322	2 109	2 917	2 027	1 975	2 002
51 Fishing area total	*1 961*	*2 199*	*3 066*	*2 221*	*2 322*	*2 109*	*2 917*	*2 027*	*1 975*	*2 002*
57 Indonesia	4 296	3 295	4 809	5 637	5 638	7 043	7 041	7 276	7 758	8 480
Malaysia	...	2 317	2 709	2 430	2 785	2 816	3 856	3 115	2 612	2 767
Thailand	1 181	1 957	2 080	2 900	2 981	1 013	1 219	769	1 619	1 584
57 Fishing area total	*5 477*	*7 569*	*9 598*	*10 967*	*11 404*	*10 872*	*12 116*	*11 160*	*11 989*	*12 831*
61 China,Taiwan	3 832	4 668	5 444	3 427	2 534	1 744	1 307	1 671	2 129	1 779
61 Fishing area total	*3 832*	*4 668*	*5 444*	*3 427*	*2 534*	*1 744*	*1 307*	*1 671*	*2 129*	*1 779*
71 Indonesia	19 556	19 568	21 225	22 358	27 000	25 708	24 819	26 817	35 927	39 000
Malaysia	...	2 517	2 102	2 220	1 932	2 186	1 771	1 431	1 199	1 617
Thailand	1 895	2 545	3 583	3 483	3 272	1 905	4 660	3 240	3 457	3 496
71 Fishing area total	*21 451*	*24 630*	*26 910*	*28 061*	*32 204*	*29 799*	*31 250*	*31 488*	*40 583*	*44 113*
Species total	*32 721*	*39 066*	*45 018*	*44 676*	*48 464*	*44 524*	*47 590*	*46 346*	*56 676*	*60 725*
Rainbow runner	**Comète saumon**		**Macarela salmón**		***Elagatis bipinnulata***			**1,70(23)134,01**		**RRU**
31 Grenada	19	6	32	14	20	20	36	23	15	20
31 Fishing area total	*19*	*6*	*32*	*14*	*20*	*20*	*36*	*23*	*15*	*20*
34 Togo	-	-	-	-	1	-	27	4	1	1
34 Fishing area total	*-*	*-*	*-*	*-*	*1*	*-*	*27*	*4*	*1*	*1*
51 Saudi Arabia	-	-	-	-	415	98	132	5	52	4
51 Fishing area total	*...*	*...*	*...*	*...*	*415*	*98*	*132*	*5*	*52*	*4*
57 Indonesia	1 677	2 206	1 669	1 477	1 476	6 497	1 883	1 275	2 177	3 000
Malaysia	262	479	470	262	97	68	27	446	75	28
57 Fishing area total	*1 939*	*2 685*	*2 139*	*1 739*	*1 573*	*6 565*	*1 910*	*1 721*	*2 252*	*3 028*

B-37 Miscellaneous pelagic fishes / Poissons pélagiques divers / Peces pelágicos diversos

Capture production by species, fishing areas and countries or areas
Captures par espèces, zones de pêche et pays ou zones
Capturas por especies, áreas de pesca y países o áreas

Species, Fishing area Espèce, Zone de pêche Especie, Area de pesca	1993 t	1994 t	1995 t	1996 t	1997 t	1998 t	1999 t	2000 t	2001 t	2002 t
71 Guam	-	-	-	-	-	1	2	1	2	3
Indonesia	5 156	5 071	6 046	6 016	5 579	7 636	8 431	8 708	7 842	8 340
Malaysia	369	359	267	128	1 484	322	836	119	120	32
Philippines	11 608	4 503	4 966	4 505	4 075	4 298	4 500	4 342	4 755	5 415
71 Fishing area total	*17 133*	*9 933*	*11 279*	*10 649*	*11 138*	*12 257*	*13 769*	*13 170*	*12 719*	*13 790*
77 USA	-	-	-	-	-	-	-	-	-	2
77 Fishing area total	-	-	-	-	-	-	-	-	-	*2*
Species total	*19 091*	*12 624*	*13 450*	*12 402*	*13 147*	*18 940*	*15 874*	*14 923*	*15 039*	*16 845*
Golden trevally	**Carangue royale**		**Jurel dorado**		***Gnathanodon speciosus***				**1,70(23)151,01**	**GLT**
51 Qatar	224	123	98	101	108	172	116	185	204	173
Untd Arab Em	600	673	336	340	363	364	373	940	1 082	2 471
51 Fishing area total	*824*	*796*	*434*	*441*	*471*	*536*	*489*	*1 125*	*1 286*	*2 644*
Species total	*824*	*796*	*434*	*441*	*471*	*536*	*489*	*1 125*	*1 286*	*2 644*
Torpedo scad	**Comète torpille**		**Macarela torpedo**		***Megalaspis cordyla***				**1,70(23)179,01**	**HAS**
51 Pakistan	5 863	3 369	6 511	3 000	2 100	1 100	1 450	2 017	1 825	1 950
Untd Arab Em	960	1 079	944	954	1 019	1 022	1 048	1 100	...	...
51 Fishing area total	*6 823*	*4 448*	*7 455*	*3 954*	*3 119*	*2 122*	*2 498*	*3 117*	*1 825*	*1 950*
57 Indonesia	10 498	7 395	7 438	7 182	5 720	5 628	6 764	7 250	6 555	6 260
Malaysia	4 724	9 332	12 896	8 324	7 849	10 377	5 872	3 402	2 929	4 950
Thailand	4 716	14 143	7 347	14 457	16 038	16 207	14 743	13 549	13 820	13 527
57 Fishing area total	*19 938*	*30 870*	*27 681*	*29 963*	*29 607*	*32 212*	*27 379*	*24 201*	*23 304*	*24 737*
61 China,Taiwan	454	205	65	233	6 847	52	113	327	198	229
61 Fishing area total	*454*	*205*	*65*	*233*	*6 847*	*52*	*113*	*327*	*198*	*229*
71 Indonesia	7 923	6 592	6 665	7 733	9 217	11 933	12 693	13 235	15 041	16 420
Malaysia	10 585	7 474	6 574	6 588	5 813	8 408	14 023	13 930	11 023	17 555
Philippines	8 344	7 988	6 097	5 864	11 429	14 817	16 033	16 274	19 590	21 747
Thailand	18 581	20 809	9 723	5 217	4 027	7 981	7 374	7 433	6 704	6 780
71 Fishing area total	*45 433*	*42 863*	*29 059*	*25 402*	*30 486*	*43 139*	*50 123*	*50 872*	*52 358*	*62 502*
Species total	*72 648*	*78 386*	*64 260*	*59 552*	*70 059*	*77 525*	*80 113*	*78 517*	*77 685*	*89 418*
Queenfishes	**Sauteurs**		**Jureles saltadores**		***Scomberoides spp***				**1,70(23)231,XX**	**QUE**
51 Eritrea	...	...	...	8	4	...	44	243	257	229
Oman	281	518	1 019	823	703	528	693	408	528	420
Qatar	72	45	49	57	56	80	42	57	78	54
Saudi Arabia	...	154	312	349	385	572	456	572	543	580
Untd Arab Em	1 700	1 931	1 981	2 002	2 140	2 147	2 201	609	701	453
51 Fishing area total	*2 053*	*2 648*	*3 361*	*3 239*	*3 288*	*3 327*	*3 436*	*1 889*	*2 107*	*1 736*
57 Indonesia	2 250	2 884	3 722	4 678	2 263	2 304	2 588	2 328	2 433	2 550
Malaysia	579	454	219	389	256	341	380	395	357	460
57 Fishing area total	*2 829*	*3 338*	*3 941*	*5 067*	*2 519*	*2 645*	*2 968*	*2 723*	*2 790*	*3 010*
71 Indonesia	10 743	9 662	11 185	10 945	11 310	13 568	13 391	12 224	14 852	15 560
Malaysia	1 730	2 701	3 093	2 851	2 616	2 912	3 431	3 087	3 170	2 236
Philippines	4 184	4 182	3 106	2 965	2 631	2 830	4 404	4 665	5 242	5 625
71 Fishing area total	*16 657*	*16 545*	*17 384*	*16 761*	*16 557*	*19 310*	*21 226*	*19 976*	*23 264*	*23 421*
Species total	*21 539*	*22 531*	*24 686*	*25 067*	*22 364*	*25 282*	*27 630*	*24 588*	*28 161*	*28 167*
Atlantic bumper	**Sapater**		**Casabe**		***Chloroscombrus chrysurus***				**1,70(23)268,01**	**BUA**
34 Benin	-	-	-	-	-	-	-	-	853	1 022
Côte dIvoire	1 103	1 178	1 188	1 116	1 302	1 107	1 120	1 374	2 364	2 047
Ghana	6 742	4 644	2 482	2 466	3 847	7 264	7 611	6 861	6 418	5 670
Lithuania	2 220	-	-	-	-	-	-	-	-	-
Russian Fed	3 274	702	59	-	-	49	-	-	-	-
Senegal	-	-	-	-	-	-	-	-	2 045	5 871
Sierra Leone	...	...	0	0	-	-	-	-	227	-
Ukraine	-	-	-	-	-	-	-	2	44	-
34 Fishing area total	*13 339*	*6 524*	*3 729*	*3 582*	*5 149*	*8 420*	*8 731*	*8 237*	*11 951*	*14 610*
41 Brazil	...	...	1 868	952	1 035	4 065	3 519	1 573	5 304	5 757
41 Fishing area total	...	...	*1 868*	*952*	*1 035*	*4 065*	*3 519*	*1 573*	*5 304*	*5 757*
47 Russian Fed	-	-	-	-	2	-	-	-	-	-
47 Fishing area total	-	-	-	-	*2*	-	-	-	-	-
Species total	*13 339*	*6 524*	*5 597*	*4 534*	*6 186*	*12 485*	*12 250*	*9 810*	*17 255*	*20 367*
Pacific bumper	**Sapater du Pacifique**		**Casabe orqueta**		***Chloroscombrus orqueta***				**1,70(23)268,02**	**HSO**
87 Ecuador	...	8 346	17 999	1 706	952	565	1 409	...	1 008	1 100 F
87 Fishing area total	...	*8 346*	*17 999*	*1 706*	*952*	*565*	*1 409*	...	*1 008*	*1 100 F*
Species total	...	*8 346*	*17 999*	*1 706*	*952*	*565*	*1 409*	...	*1 008*	*1 100 F*

B-37 Miscellaneous pelagic fishes / Poissons pélagiques divers / Peces pelágicos diversos

Capture production by species, fishing areas and countries or areas / Captures par espèces, zones de pêche et pays ou zones / Capturas por especies, áreas de pesca y países o áreas

Species, Fishing area Espèce, Zone de pêche Especie, Area de pesca	1993 t	1994 t	1995 t	1996 t	1997 t	1998 t	1999 t	2000 t	2001 t	2002 t
Parona leatherjacket	**Sauteur parone**		**Parona**		***Parona signata***				**1,70(23)283,01**	**PAO**
41 Argentina	648	2 128	1 478	1 925	1 805	1 744	1 710	1 483	933	868
Uruguay	491	664	391	439	431	465	360	371	618	446
41 Fishing area total	*1 139*	*2 792*	*1 869*	*2 364*	*2 236*	*2 209*	*2 070*	*1 854*	*1 551*	*1 314*
Species total	*1 139*	*2 792*	*1 869*	*2 364*	*2 236*	*2 209*	*2 070*	*1 854*	*1 551*	*1 314*
Bigeye scad	**Sélar coulisou**		**Chicharro ojón**		***Selar crumenophthalmus***				**1,70(23)291,01**	**BIS**
31 Grenada	341	191	48	100	181	53	72	137	97	67
St Kitts Nev	...	...	-	-	16	20	36	32	25	41
Venezuela	2 801	2 531	2 836	2 353	2 425	2 653	3 765	1 704	1 774	2 267
31 Fishing area total	*3 142*	*2 722*	*2 884*	*2 453*	*2 622*	*2 726*	*3 873*	*1 873*	*1 896*	*2 375*
57 Thailand	-	-	-	1 984	1 904	3 830	3 442	3 180	2 656	2 600
57 Fishing area total	-	-	-	*1 984*	*1 904*	*3 830*	*3 442*	*3 180*	*2 656*	*2 600*
71 Philippines	33 200	49 937	43 592	43 660	54 167	61 999	65 776	71 365	80 858	100 786
Thailand	-	-	-	24 533	22 188	24 931	25 966	29 075	26 502	26 803
71 Fishing area total	*33 200*	*49 937*	*43 592*	*68 193*	*76 355*	*86 930*	*91 742*	*100 440*	*107 360*	*127 589*
77 USA	-	-	-	-	-	-	-	-	-	128
77 Fishing area total	-	-	-	-	-	-	-	-	-	*128*
Species total	*36 342*	*52 659*	*46 476*	*72 630*	*80 881*	*93 486*	*99 057*	*105 493*	*111 912*	*132 692*
Yellowstripe scad	**Sélar à bande dorée**		**Chicharro banda dorada**		***Selaroides leptolepis***				**1,70(23)422,01**	**TRY**
51 Untd Arab Em	360	403	2 878	2 908	3 108	3 118	3 196	2 635	3 034	5 320
51 Fishing area total	*360*	*403*	*2 878*	*2 908*	*3 108*	*3 118*	*3 196*	*2 635*	*3 034*	*5 320*
57 Malaysia	2 946	3 684	2 903	3 656	5 505	3 643	4 871	3 437	3 693	5 861
57 Fishing area total	*2 946*	*3 684*	*2 903*	*3 656*	*5 505*	*3 643*	*4 871*	*3 437*	*3 693*	*5 861*
71 Malaysia	26 510	23 870	20 956	24 467	26 526	29 091	36 470	40 816	36 165	35 145
71 Fishing area total	*26 510*	*23 870*	*20 956*	*24 467*	*26 526*	*29 091*	*36 470*	*40 816*	*36 165*	*35 145*
Species total	*29 816*	*27 957*	*26 737*	*31 031*	*35 139*	*35 852*	*44 537*	*46 888*	*42 892*	*46 326*
Blackbanded trevally	**Sériole amourez**		**...C**		***Seriolina nigrofasciata***				**1,70(23)425,01**	**RNJ**
57 Thailand	-	1 373	2 882	3 666	3 564	2 954	2 700	2 227	2 781	2 722
57 Fishing area total	-	*1 373*	*2 882*	*3 666*	*3 564*	*2 954*	*2 700*	*2 227*	*2 781*	*2 722*
71 Thailand	-	3 934	4 048	3 603	3 532	3 088	3 345	3 188	3 457	3 496
71 Fishing area total	-	*3 934*	*4 048*	*3 603*	*3 532*	*3 088*	*3 345*	*3 188*	*3 457*	*3 496*
Species total	-	*5 307*	*6 930*	*7 269*	*7 096*	*6 042*	*6 045*	*5 415*	*6 238*	*6 218*
Carangids nei	**Carangidés nca**		**Carángidos nep**		***Carangidae***				**1,70(23)XXX,XX**	**CGX**
27 Portugal	...	...	...	...	...	...	43	28	28	19
27 Fishing area total	...	...	...	...	...	...	*43*	*28*	*28*	*19*
31 Antigua Barb	...	...	...	...	...	...	...	...	33	34
Bahamas	113	106	72	91	103	92	79	81	101	100 F
Barbados	28	15	24	28	19	14	6	28	11	10
Bermuda	60	59	86	70	50	48	51	30	41	52
Dominican Rp	18	32	80	30	10	27	6	22	28	126
Grenada	17	8	7	10	17	10	16	12	14	26
Mexico	205	217	164	201	454	364	342	111	470	157
Puerto Rico	...	...	...	...	...	...	50	70	66	53
Venezuela	41	32	25	6	23	31	11	26	5	26
31 Fishing area total	*482*	*469*	*458*	*436*	*676*	*586*	*561*	*380*	*769*	*584 F*
34 Benin	400 F	500 F	600 F	700 F	800 F	900 F	1 027	700 F	475	428
Cameroon	5	5 F	6	5	5 F	2	2	4	5	7
Côte dIvoire	1 282	626	200	169	1 400	1 267	1 025	509	-	-
Greece	-	-	-	-	-	11	-	4	1	1
Guinea	...	...	253	311	257	392	326	764	2 559	2 500 F
Latvia	-	-	-	-	36	-	-	-	-	-
Mauritania	200 F	200 F	100 F	100 F	100 F	100 F	100 F	112 F	...	...
Nigeria	9 850	5 468	5 199	2 923	1 035	700	4 451	3 521	833	630
Portugal	34	7	-	-	-	-	2	1	2	1
Senegal	3 373	4 636	4 772	3 299	4 806	7 491	4 456	5 725	2 616	2 097
Sierra Leone	...	...	...	...	52	158	-	39	-	-
Togo	326	659	332	273	260	291	1 613	2 433	2 482	1 489
34 Fishing area total	*15 470 F*	*12 101 F*	*11 462 F*	*7 780 F*	*8 751 F*	*11 312 F*	*13 002 F*	*13 812 F*	*8 973*	*7 153 F*
37 Israel	102	30	171	-	-	-	-	-	-	-
Lebanon	150	150	450	450	350	400	350	450	400	400
Malta	72	29	13	7	4	13	23	28	8	8
Spain	-	-	-	-	-	-	39	34	393	14
37 Fishing area total	*324*	*209*	*634*	*457*	*354*	*413*	*412*	*512*	*801*	*422*
41 Brazil	7 610 F	7 630 F	-	-	-	-	-	-	-	-
Italy	137	137	-	-	-	-	-	-	-	-
Latvia	61	-	-	-	-	-	-	-	-	-
41 Fishing area total	*7 808 F*	*7 767 F*	-	-	-	-	-	-	-	-

B-37 Miscellaneous pelagic fishes / Poissons pélagiques divers / Peces pelágicos diversos

Capture production by species, fishing areas and countries or areas / Captures par espèces, zones de pêche et pays ou zones / Capturas por especies, áreas de pesca y países o áreas

Species, Fishing area Espèce, Zone de pêche Especie, Area de pesca	1993 t	1994 t	1995 t	1996 t	1997 t	1998 t	1999 t	2000 t	2001 t	2002 t
47 Russian Fed	-	-	7	40	227	81	-	38	805	583
St Helena	3	2	2	2	3	2	3	1	1	1
47 Fishing area total	*3*	*2*	*9*	*42*	*230*	*83*	*3*	*39*	*806*	*584*
51 Bahrain	515	382	495	668	498	359	524	414	450	470
Comoros	500	500	500 F	490 F	470 F	470 F	450 F	500 F	500	500 F
Djibouti	15 F	16 F	20 F	20 F	20 F	20 F	20 F	20 F	20 F	20 F
Eritrea	...	48	697	818	13	184	386	2 026	1 014	730
India	6 180	9 922	13 341	28 831	28 764	21 426	22 589	5 679	5 909	20 898
Italy	206	206	-	-	-	-	-	-	-	-
Kenya	68	73	76	82	111	86	101	85	119	144
Kuwait	92	113	111	97	138	74	73	150 F	242	240 F
Mauritius	165	167	165	43	58	46	33	53	51	37
Oman	2 113	2 123	3 607	2 708	3 216	2 323	2 556	2 485	1 302	7 839
Pakistan	13 111	13 760	16 495	15 957	19 002	18 689	17 779	16 545	15 988	16 400
Qatar	452	250	291	308	308	326	287	420	552	364
Réunion	...	...	...	...	...	...	107	130	131	131 F
Saudi Arabia	4 920	5 730	4 547	5 524	5 125	4 795	5 534	6 581	5 943	6 287
Seychelles	1 492	1 066	1 284	1 568	1 562	1 008	1 474	1 764	1 285	2 042
Tanzania	1 216	1 016	1 380	3 669	1 800	2 000	2 000	2 500	2 000	1 880
Untd Arab Em	5 700	6 371	5 800	5 861	6 205	6 810	7 020	3 568	4 107	1 164
51 Fishing area total	*36 745 F*	*41 743 F*	*48 809 F*	*66 644 F*	*67 290 F*	*58 616 F*	*60 933 F*	*42 920 F*	*39 613 F*	*59 146 F*
57 India	7 249	9 301	9 148	5 394	3 988	4 845	7 327	5 450	7 061	8 744
Indonesia	24 273	24 549	28 076	27 313	27 764	29 932	29 830	30 059	30 690	31 640
Sri Lanka	10 878	8 000	6 910	6 088	6 900	8 500	8 680	10 450	9 950	10 760
Thailand	14 304	13 180	8 226	8 663	8 391	10 395	8 582	11 764	8 798	8 612
57 Fishing area total	*56 704*	*55 030*	*52 360*	*47 458*	*47 043*	*53 672*	*54 419*	*57 723*	*56 499*	*59 756*
71 Guam	...	5	1	1	1	1	2	1	1	2
Indonesia	81 673	89 381	88 693	88 781	97 740	98 527	98 965	99 854	102 308	105 310
N Marianas	...	1	0	1	2	2	2	1	3	2
Palau	7	5	10	14	4	10	4	6	3	2
Philippines	44 624	47 539	39 682	37 456	32 175	31 472	35 204	34 713	42 442	54 019
Singapore	89	9	22	37	33	36	30	21	12	16
Thailand	42 224	55 616	47 456	44 365	41 356	35 599	35 668	30 381	39 282	39 727
71 Fishing area total	*168 617*	*192 556*	*175 864*	*170 655*	*171 311*	*165 647*	*169 875*	*164 977*	*184 051*	*199 078*
77 Amer Samoa	...	2	1	1	2	1	1	2	1	1
Mexico	171	180	61	73	318	1 628	735	379	166	330
77 Fishing area total	*171*	*182*	*62*	*74*	*320*	*1 629*	*736*	*381*	*167*	*331*
87 Chile	251	143	89	96	205	348	203	307	363	769
Ecuador	497	1 328	199	65	45	839	1 632	...	68	70 F
87 Fishing area total	*748*	*1 471*	*288*	*161*	*250*	*1 187*	*1 835*	*307*	*431*	*839 F*
Species total	*287 072 F*	*311 530 F*	*289 946 F*	*293 707 F*	*296 225 F*	*293 145 F*	*301 819 F*	*281 079 F*	*292 138 F*	*327 912 F*
Atlantic pomfret	**Grande castagnole**		**Japuta**		***Brama brama***				**1,70(27)003,01**	**POA**
27 Denmark	0	0	-	-	-	-	-	-	0	0
France	40	23	10	1	0	2	2	2	-	-
Ireland	-	-	-	-	-	-	-	1	184	404
Portugal	589	37	18	34	91	9	9	5	7	76
Spain	5 191	5 478	5 592	5 313	3 334	2 821	2 233	559	244	30
27 Fishing area total	*5 820*	*5 538*	*5 620*	*5 348*	*3 425*	*2 832*	*2 244*	*567*	*435*	*510*
34 Portugal	45	14	-	-	-	-	-	3	3	4
Russian Fed	6	-	2	-	26	-	103	21	-	29
Ukraine	-	-	-	-	-	-	-	9	-	-
34 Fishing area total	*51*	*14*	*2*	*-*	*26*	*-*	*103*	*33*	*3*	*33*
37 France	...	...	...	...	...	...	5	5	5	5
Spain	-	-	-	26	157	87	113	27	104	47
37 Fishing area total	*...*	*...*	*...*	*26*	*157*	*87*	*118*	*32*	*109*	*52*
47 Namibia	62	39	20	19	9	15	99	493	684	486
Russian Fed	4	4	-	103	30	-	9	-	-	-
South Africa	1 717	2 019	1 709	1 070	381	307	324	350 F	437	404
47 Fishing area total	*1 783*	*2 062*	*1 729*	*1 192*	*420*	*322*	*432*	*843 F*	*1 121*	*890*
58 South Africa	-	-	-	-	-	-	-	-	0	0
58 Fishing area total	*-*	*-*	*-*	*-*	*-*	*-*	*-*	*-*	*0*	*0*
81 New Zealand	...	...	449	480	413	488	465	401	903	552
81 Fishing area total	*...*	*...*	*449*	*480*	*413*	*488*	*465*	*401*	*903*	*552*
87 Chile	-	1 186	3 930	5 585	5 998	6 332	6 830	8 160	15 156	4 430
87 Fishing area total	*-*	*1 186*	*3 930*	*5 585*	*5 998*	*6 332*	*6 830*	*8 160*	*15 156*	*4 430*
Species total	*7 654*	*8 800*	*11 730*	*12 631*	*10 439*	*10 061*	*10 192*	*10 036 F*	*17 727*	*6 467*
Pomfrets, ocean breams nei	**Castagnoles nca**		**Japutas nep**		***Bramidae***				**1,70(27)XXX,XX**	**BRZ**
77 Amer Samoa	-	-	-	-	-	-	-	-	1	0
USA	-	-	-	-	-	-	-	-	4	222
77 Fishing area total	*-*	*-*	*-*	*-*	*-*	*-*	*-*	*-*	*5*	*222*

B-37 Miscellaneous pelagic fishes / Poissons pélagiques divers / Peces pelágicos diversos

Capture production by species, fishing areas and countries or areas / Captures par espèces, zones de pêche et pays ou zones / Capturas por especies, áreas de pesca y países o áreas

Species, Fishing area Espèce, Zone de pêche Especie, Area de pesca	1993 t	1994 t	1995 t	1996 t	1997 t	1998 t	1999 t	2000 t	2001 t	2002 t
81 Russian Fed	2	-	-	-	-	-	-	-	-	-
Ukraine	-	-	-	-	-	-	-	2	-	3
81 Fishing area total	*2*	-	-	-	-	-	-	*2*	-	*3*
Species total	*2*	-	-	-	-	-	-	*2*	*5*	*225*
Common dolphinfish	**Coryphène commune**		**Lampuga**			**Coryphaena hippurus**			**1,70(28)071,01**	**DOL**
27 Portugal	-	-	-	-	-	-	1	1	3	4
27 Fishing area total	-	-	-	-	-	-	*1*	*1*	*3*	*4*
31 Antigua Barb	...	...	...	...	...	...	...	...	4	7
Barbados	513	499	758	849	721	482	745	728	574	553
Br Virgin Is	...	5	2	6	3	2	3	1	1 F	1 F
Cuba	-	-	-	5	362	-	-	-	-	-
Dominica	...	...	...	...	...	...	...	...	...	211
Dominican Rp	225	405	89	237	113	151	175	255	232	571
Grenada	103	124	182	130	171	153	162	167	221	178
Guadeloupe	650 F	650 F	700 F	730	800 F	670 F	670 F	700 F	700 F	700 F
Martinique	400 F	400 F	350 F	250 F	350 F	320 F	300 F	250 F	220 F	220 F
Mexico	49	175	88	244	871	1 078	376	308	315	350
Puerto Rico	...	...	...	...	...	...	83	111	74	53
St Kitts Nev	...	...	3	13	20	34	13	26	26	40
St Lucia	...	144	200	351	455	264	588	552	427	402
USA	496	620	980	739	763	322	555	498	411	389
Venezuela	244	274	447	...	...	...	290	141	...	...
31 Fishing area total	*2 680 F*	*3 296 F*	*3 799 F*	*3 554 F*	*4 629 F*	*3 476 F*	*3 960 F*	*3 737 F*	*3 205 F*	*3 675 F*
34 Benin	...	...	...	...	...	...	4	3 F	48	28
Liberia	...	...	...	...	31	20	48	45	22	25 F
Portugal	-	-	1	1	0	1	-	0	1	0
Senegal	-	-	-	-	-	103	115	150	258	198
Spain	-	-	-	-	-	-	-	-	-	14
Togo	3	2	1	3	3	0	10	36	148	147
34 Fishing area total	*3*	*2*	*2*	*4*	*34*	*124*	*177*	*234 F*	*477*	*412 F*
37 Malta	174	334	334	307	295	363	349	234	303	347
Spain	-	-	-	-	-	-	92	137	70	156
Tunisia	340	366	434	399	538	1 088	919	667	784	678
37 Fishing area total	*514*	*700*	*768*	*706*	*833*	*1 451*	*1 360*	*1 038*	*1 157*	*1 181*
41 Brazil	1 710 F	1 720 F	3 186	2 500	4 028	4 117	2 848	4 359	4 299	4 818
41 Fishing area total	*1 710 F*	*1 720 F*	*3 186*	*2 500*	*4 028*	*4 117*	*2 848*	*4 359*	*4 299*	*4 818*
47 St Helena	...	...	1	1	1	1	1	1	0	1
South Africa	-	-	-	-	-	-	-	-	1	1
47 Fishing area total	...	...	*1*	*1*	*1*	*1*	*1*	*1*	*1*	*2*
51 Oman	11	...	...	...	...	...	-	-	-	-
Pakistan	1 875	2 054	2 570	1 841	1 658	1 892	3 109	1 954	1 869	1 875
Réunion	...	...	...	...	...	...	221	194	141	141 F
South Africa	-	-	-	-	1	...	...	...	...	1
Untd Arab Em	64	46	55	56	60	61	63	55	...	...
51 Fishing area total	*1 950*	*2 100*	*2 625*	*1 897*	*1 719*	*1 953*	*3 393*	*2 203*	*2 010*	*2 017 F*
57 Japan	-	-	-	-	-	3	-	-	-	-
57 Fishing area total	-	-	-	-	-	*3*	-	-	-	-
61 China,Taiwan	10 385	6 881	12 203	3 209	8 089	17 157	8 560	5 558	7 427	7 014
Japan	9 187	7 586	9 777	8 164	10 218	14 909	9 269	8 953	9 123	9 000 F
61 Fishing area total	*19 572*	*14 467*	*21 980*	*11 373*	*18 307*	*32 066*	*17 829*	*14 511*	*16 550*	*16 014 F*
71 Guam	180	90	72	35	43	79	40	34	53	36
Japan	2 243	525	428	357	10	32	11	6	19	...
N Marianas	...	7	11	16	17	9	6	3	6	9
Palau	0	0	0	0	0	0	0	0	0	0
Philippines	1 491	1 799	4 253	3 544	359	189	223	211	258	267
71 Fishing area total	*3 914*	*2 421*	*4 764*	*3 952*	*429*	*309*	*280*	*254*	*336*	*312*
77 Amer Samoa	...	5	5	5	17	10	13	15	16	14
Costa Rica	...	...	...	...	...	...	...	8 370	11 221	7 832
Fr Polynesia	467	256	178	257	427	437	429	446	651	610
Japan	-	1	7	6	1	-	-	-	-	-
Mexico	230	9	26	70	113	115	104	201	208	345
Nicaragua	...	...	...	...	...	...	710	2 540	2 470	1 449
USA	...	...	...	...	5	3	17	43	7	621
77 Fishing area total	*697*	*271*	*216*	*338*	*563*	*565*	*1 273*	*11 615*	*14 573*	*10 871*
81 New Zealand	-	-	1	0	0	0	1	0	15	7
81 Fishing area total	-	-	*1*	*0*	*0*	*0*	*1*	*0*	*15*	*7*
87 Chile	374	206	104	179	124	80	109	131	136	134
Ecuador	30	247	104	119	34	15	75	88	158	44
Venezuela	-	-	-	-	-	10	-	-	-	-
87 Fishing area total	*404*	*453*	*208*	*298*	*158*	*105*	*184*	*219*	*294*	*178*
Species total	*31 444 F*	*25 430 F*	*37 550 F*	*24 623 F*	*30 701 F*	*44 170 F*	*31 307 F*	*38 172 F*	*42 920 F*	*39 491 F*

B-37 Miscellaneous pelagic fishes / Poissons pélagiques divers / Peces pelágicos diversos

Capture production by species, fishing areas and countries or areas / Captures par espèces, zones de pêche et pays ou zones / Capturas por especies, áreas de pesca y países o áreas

Species, Fishing area Espèce, Zone de pêche Especie, Area de pesca	1993 t	1994 t	1995 t	1996 t	1997 t	1998 t	1999 t	2000 t	2001 t	2002 t
Suckerfishes, remoras nei	**Rémoras nca**		**Remoras, pegas nep**		***Echeneidae***				**1,70(42)XXX,XX**	**ECN**
34 Liberia	-	-	-	-	-	3	1	12	...	...
34 Fishing area total	-	-	-	-	-	*3*	*1*	*12*	...	...
Species total	-	-	-	-	-	*3*	*1*	*12*	...	...
Chub mackerel	**Maquereau espagnol**		**Estornino**		***Scomber japonicus***				**1,75(01)002,01**	**MAS**
27 France	122	412	374	126	29	103	104	109	110	407
Portugal	7 606	4 810	4 351	5 385	6 116	7 218	14 061	10 703	4 492	5 516
27 Fishing area total	*7 728*	*5 222*	*4 725*	*5 511*	*6 145*	*7 321*	*14 165*	*10 812*	*4 602*	*5 923*
31 Mexico	9	9	2	2	1	0	0	3	0	0
USA	2	19	0	-	6	18	10	7	13	-
Venezuela	431	401	377	416	549	753	631	399	514	664
31 Fishing area total	*442*	*429*	*379*	*418*	*556*	*771*	*641*	*409*	*527*	*664*
34 Belize	-	-	-	-	1 100	1 051	1 732	3 194	1 041	3 698
Benin	330 F	320 F	300 F	280 F	300 F	250 F	205	140 F	314	452
Bulgaria	-	-	-	-	-	486	-	-	-	-
Cape Verde	...	...	...	...	...	182	...	...	...	...
China	-	4	-	-	2	-	8	59	36	-
Côte dIvoire	230	8	679	313	1 355	1 340	2 494	259	110	878
Cyprus	-	-	-	...	...	2 758	1 906	6 207	7 305	-
Eq Guinea	240 F	360 F	150 F	330 F	420 F	400	470 F	50 F	50 F	50 F
Estonia	3 390	692	2 991	4 486	1 980	7 215	3 416	-	-	-
France	-	-	-	-	...	1 491	213	334	-	-
Georgia	700 F	260 F	-	-	-	-	-	-	-	-
Germany	-	-	-	87	645	315	334	-	20	318
Ghana	4 286	9 769	12 473	15 590	19 883	30 160	15 482	28 465	14 338	7 018
Greece	0	1	1	-	13	-	-	-	-	-
Guinea	...	...	...	2 043	1 006	849	...	...	...	...
Korea Rep	-	-	-	-	-	-	61	37	26	281
Latvia	2 612	4 056	3 494	3 765	1 931	2 562	3 123	7 151	9 924	7 079
Liberia	37	12	134	24	8	45	218	238	34	40 F
Lithuania	1 887	2 351	...	...	...	5 420	2 105	3 871	3 074	12 119
Mauritania	180 F	230 F	-	110 F	80 F	150 F	...	...	...	...
Morocco	16 970	38 215	28 350	16 154	28 018	11 261	15 786	62 160	25 679	23 559
Netherlands	-	-	-	857	3 202	1 836	1 561	12 005	12 708	23 254
Panama	-	-	201	-	-	397	-	-	-	-
Poland	-	-	-	4 086	480	-	-	-	1 666	5 958
Portugal	1 232	1 082	859	2 273	1 665	553	897	890	446	289
Romania	67	-	-	-	-	-	-	-		-
Russian Fed	14 093	29 583	52 551	72 874	61 309	65 958	47 215	43 654	29 619	35 688
St Vincent	-	-	-	-	350	2 644	988	2 951	5 384	4 071
Senegal	1 394	1 149	1 748	947	1 670	2 117	988	2 656	2 710	6 577
Spain	-	-	-	894	99	352	302	3 697	1 138	86
Togo	129	97	158	112	252	183	454	308	204	419
Ukraine	7 361	15 072	65 926	98 944	120 272	72 959	42 283	44 961	18 988	8 015
Vanuatu	54	-	-	-	-	-	-	-	-	-
Other nei	-	-	-	-	-	-	781	4 002	4 845	15 193
34 Fishing area total	*55 192 F*	*103 261 F*	*170 015 F*	*224 169 F*	*246 040 F*	*212 934 F*	*143 022 F*	*227 289 F*	*139 659 F*	*155 042 F*
37 Cyprus	...	...	...	...	...	...	...	...	3	5
France	-	-	-	58	69	69	45	27	77	66
Greece	10 996	14 308	6 468	6 520	5 627	2 173	1 850	2 286	1 829	1 750 F
Israel	80 F	50 F	12	71	47	54	44	8	4	...
Malta	10	10	13	23	29	40	19	34	32	3
Morocco	1 341	980	1 756	834	757	360	1 597	1 221	342	722
Palest, O.T.	...	...	...	130	120	337	340 F	280 F	220 F	160
Slovenia	1	4	3	10	4	17	3	4	3	1
Tunisia	3 263	2 333	2 562	1 439	4 118	3 233	3 515	2 200	3 727	2 480
Turkey	15 908	16 748	17 410	10 444	10 850	10 120	10 200	9 000	4 500	1 500
37 Fishing area total	*31 599 F*	*34 433 F*	*28 224*	*19 529*	*21 621*	*16 403*	*17 613 F*	*15 060 F*	*10 737 F*	*6 687 F*
41 Argentina	6 689	10 297	13 442	11 195	10 468	3 224	7 012	10 122	4 602	11 616
Brazil	5 100 F	5 110 F	8 049	5 670	8 306	9 422	1 595	6 377	1 405	2 189
Estonia	24	-	-	-	-	-	-	-	-	-
Korea Rep	44	-	-	-	-	-	-	-	-	-
Uruguay	3	5	5	4	5	1	5	-	0	0
41 Fishing area total	*11 860 F*	*15 412 F*	*21 496*	*16 869*	*18 779*	*12 647*	*8 612*	*16 499*	*6 007*	*13 805*
47 Iceland	-	-	-	-	28	-	-	-	-	-
Korea Rep	69	759	9	-	31	-	-	-	-	35
Lithuania	-	6	-	-	-	-	-	-	-	-
Namibia	...	...	...	...	...	...	...	1 641	6 329	3 530
Poland	-	-	-	-	-	-	-	-	-	675
Portugal	-	-	-	-	-	-	4	-	-	-
Russian Fed	8	40	80	66	3 439	2 369	1 150	2 182	2 090	1 813
St Helena	7	8	7	7	12	2	8	4	11	7
South Africa	4 721	4 508	4 677	4 139	8 039	3 201	1 765	1 600 F	1 620	876
Spain	-	-	-	-	1	26	-	-	-	-
Ukraine	-	-	284	...	...	-	-	-	-	-
47 Fishing area total	*4 805*	*5 321*	*5 057*	*4 212*	*11 550*	*5 598*	*2 927*	*5 427 F*	*10 050*	*6 936*
51 Egypt	453	3 041	1 926	2 042	2 392	810	378	3 561	2 747	1 614
Korea Rep	-	32	-	-	-	-	4	-	-	1
South Africa	1	-	-	11	5	3	3 F	...	...	...
51 Fishing area total	*454*	*3 073*	*1 926*	*2 053*	*2 397*	*813*	*385 F*	*3 561*	*2 747*	*1 615*

B-37 Miscellaneous pelagic fishes / Poissons pélagiques divers / Peces pelágicos diversos

Capture production by species, fishing areas and countries or areas / Captures par espèces, zones de pêche et pays ou zones / Capturas por especies, áreas de pesca y países o áreas

Species, Fishing area Espèce, Zone de pêche Especie, Area de pesca	1993 t	1994 t	1995 t	1996 t	1997 t	1998 t	1999 t	2000 t	2001 t	2002 t
61 China	272 604	336 091	372 038	374 400	408 933	385 183	402 540	350 750	381 561	415 364
China,Taiwan	41 908	54 914	54 155	53 906	47 279	35 527	45 339	28 635	24 298	34 950
Japan	664 298	633 062	469 447	760 430	848 967	511 238	381 865	346 220	375 273	279 000
Korea Rep	174 684	210 442	200 481	415 003	160 448	172 925	177 540	145 908	203 717	141 751
Russian Fed	-	-	-	-	-	-	-	-	1	-
61 Fishing area total	*1 153 494*	*1 234 509*	*1 096 121*	*1 603 739*	*1 465 627*	*1 104 873*	*1 007 284*	*871 513*	*984 850*	*871 065*
67 Russian Fed	-	-	-	-	-	41	-	-	-	-
USA	30	33	1	388	204	591	75	121	322	127
67 Fishing area total	*30*	*33*	*1*	*388*	*204*	*632*	*75*	*121*	*322*	*127*
71 Philippines	4 842	6 605	3 307	2 991	2 025	1 507	1 485	1 422	1 664	1 866
71 Fishing area total	*4 842*	*6 605*	*3 307*	*2 991*	*2 025*	*1 507*	*1 485*	*1 422*	*1 664*	*1 866*
77 Korea Rep	-	-	-	-	-	-	4	-	-	-
Mexico	20 608	12 199	22 832	12 959	24 245	22 990	69 375	45 202	17 156	10 433
USA	10 873	10 007	8 606	9 589	18 186	19 814	8 639	21 221	6 925	3 368
77 Fishing area total	*31 481*	*22 206*	*31 438*	*22 548*	*42 431*	*42 804*	*78 018*	*66 423*	*24 081*	*13 801*
87 Chile	96 023	27 171	110 210	146 649	211 649	71 769	120 123	95 789	365 031	343 371
Ecuador	45 322	38 991	57 950	79 484	192 182	44 716	28 307	84 324	85 378	17 073
Ghana	-	-	-	-	-	-	-	1 148	855	-
Japan	-	-	-	-	-	-	1	-	-	-
Peru	29 504	44 115	44 259	49 221	206 183	401 903	527 729	73 263	176 202	32 698
87 Fishing area total	*170 849*	*110 277*	*212 419*	*275 354*	*610 014*	*518 388*	*676 160*	*254 524*	*627 466*	*393 142*
Species total	*1 472 776 F*	*1 540 781 F*	*1 575 108 F*	*2 177 781 F*	*2 427 389 F*	*1 924 691 F*	*1 950 387 F*	*1 473 060 F*	*1 812 712 F*	*1 470 673 F*
Atlantic mackerel	**Maquereau commun**		**Caballa del Atlántico**		***Scomber scombrus***				**1,75(01)002,05**	**MAC**
21 Canada	26 474	20 612	18 300	21 027	21 305	19 557	16 317	17 637	24 488	43 092
Cuba	666	44	61	77	115	6	13	-	-	-
Russian Fed	22	-	-	-	-	1	-	-	3	-
St Pier Mq	4	3	1	1	1	3	1	26	7	7
USA	4 691	8 959	8 495	16 022	15 395	14 429	12 041	5 649	12 339	27 364
21 Fishing area total	*31 857*	*29 618*	*26 857*	*37 127*	*36 816*	*33 996*	*28 372*	*23 312*	*36 837*	*70 463*
27 Belgium	193	353	108	64	106	125	178	151	98	23
Bulgaria	1 883	-	-	-	-	-	-	-	-	-
Channel Is	1 F	1	1 F	9	9	23	18	16	14	12
Denmark	42 056	46 839	36 758	26 238	24 054	27 415	29 705	31 642	31 370	33 046
Estonia	1 100	3 302	2 286	3 741	6 324	7 356	3 595	2 673	218	-
Faeroe Is	13 978	21 570	34 924	19 530	8 401	10 654	11 334	21 022	24 005	24 005 F
France	20 011	25 787	22 807	13 167	14 368	18 764	17 400	20 897	20 958	21 871
Germany	28 734	26 492	24 417	16 229	15 864	21 490	19 960	22 980	25 325	26 536
Iceland	-	0	0	92	927	357	144	0	1	53
Ireland	94 979	86 274	78 534	49 966	53 094	67 310	59 609	70 184	70 451	72 189
Isle of Man	0	1	1	0	0	0	4	0	8	6
Japan	4	-	-	-	-	-	-	-	-	-
Latvia	3 424	1 508	534	233	-	-	-	-	-	-
Lithuania	1 000	707	6 236	7 334	-	2 823	4 936	2 085	1 949	1 600
Netherlands	42 532	44 335	35 787	24 246	23 702	30 163	27 816	32 403	33 109	43 460
Norway	223 871	260 121	202 209	136 699	137 256	158 340	161 046	174 228	180 603	184 371
Poland	-	-	-	-	22	-	-	-	-	-
Portugal	2 015	2 149	3 073	3 009	2 083	2 898	2 035	2 254	3 121	3 090
Romania	-	2 903	30 844	7 265	-	-	-	-	-	-
Russian Fed	46 694	29 509	46 249	43 046	53 732	67 837	51 348	50 772	41 568	45 811
Spain	10 452	11 452	10 595	13 748	20 301	25 541	24 026	25 384	24 382	26 558
Sweden	3 610	7 515	6 268	5 387	4 390	5 161	5 003	4 500	5 098	5 232
Ukraine	4 514	7 193	-	-	-	-	-	-	-	-
UK	257 736	238 457	218 417	144 964	149 448	179 711	166 658	193 638	198 953	200 405
27 Fishing area total	*798 787 F*	*816 468*	*760 048 F*	*514 967*	*514 081*	*625 968*	*584 815*	*654 829*	*661 231*	*688 268 F*
34 Greece	0	0	6	6	13	-	-	-	-	-
Ireland	-	-	-	-	-	-	-	-	6 135	4 473
34 Fishing area total	*0*	*0*	*6*	*6*	*13*	*...*	*...*	*...*	*6 135*	*4 473*
37 Algeria	379 F	539 F	1 243	590	750	1 055	1 053	1 144	1 167	567
France	1 326	1 313	789	741	2 199	2 199	856	1 743	1 342	1 179
Greece	1 551	1 360	680	880	292	141	200	443	155	150 F
Slovenia	1	-	-	5	5	10	13	14	16	4
Spain	4 500 F	4 400 F	4 300 F	4 234	955	2 671	1 610	2 806	2 958	3 214
Syria	121 F	117 F	116 F	116	210	274	245	384	187	264
Turkey	2 311	1 105	296	592	300	650	850	900	550	486
37 Fishing area total	*10 189 F*	*8 834 F*	*7 424 F*	*7 158*	*4 711*	*7 000*	*4 827*	*7 434*	*6 375*	*5 864 F*
Species total	*840 833 F*	*854 920 F*	*794 335 F*	*559 258*	*555 621*	*666 964*	*618 014*	*685 575*	*710 578*	*769 068 F*
Blue mackerel	**Maquereau tacheté**		**Caballa pintoja**		***Scomber australasicus***				**1,75(01)002,07**	**MAA**
81 Japan	380	292	358	-	-	-	-	-	-	-
Korea Rep	-	-	-	1	-	5	-	-	-	-
New Zealand	9 554	5 467	7 534	2 837	8 768	7 041	12 417	10 431	9 761	13 287
Russian Fed	326	204	75	-	-	-	-	-	-	-
Ukraine	94	133	...	156	9	214	3 457	1 677	2 040	1 849
81 Fishing area total	*10 354*	*6 096*	*7 967*	*2 994*	*8 777*	*7 260*	*15 874*	*12 108*	*11 801*	*15 136*
Species total	*10 354*	*6 096*	*7 967*	*2 994*	*8 777*	*7 260*	*15 874*	*12 108*	*11 801*	*15 136*

B-37

Miscellaneous pelagic fishes — Capture production by species, fishing areas and countries or areas
Poissons pélagiques divers — Captures par espèces, zones de pêche et pays ou zones
Peces pelágicos diversos — Capturas por especies, áreas de pesca y países o áreas

Species, Fishing area Espèce, Zone de pêche Especie, Area de pesca	1993 t	1994 t	1995 t	1996 t	1997 t	1998 t	1999 t	2000 t	2001 t	2002 t
Scomber mackerels nei	**Maquereaux scomber nca**		**Caballas scomber nep**		***Scomber spp***				**1,75(01)002,XX**	**MAZ**
34 Spain	-	-	-	1 489	-	-	23	-	-	557
34 Fishing area total	-	-	-	*1 489*	-	-	*23*	-	-	*557*
37 Albania	...	...	0	10	5	4	4	4	11	11
Croatia	1 541	580	615	650	998	585	175	377	490	420
Italy	7 997	5 862	6 949	8 012	7 866	7 277	5 748	5 522	6 033	5 041
Libya	2 750 F	3 000 F	3 050 F	3 000 F	3 000 F	3 000 F	3 000 F	3 000 F	3 000 F	3 000 F
Serbia-Monte	5	5	7	8	9	13	14	13	13	13
Syria	...	...	...	...	...	...	...	...	...	41
37 Fishing area total	*12 293 F*	*9 447 F*	*10 621 F*	*11 680 F*	*11 878 F*	*10 879 F*	*8 941 F*	*8 916 F*	*9 547 F*	*8 526 F*
Species total	*12 293 F*	*9 447 F*	*10 621 F*	*13 169 F*	*11 878 F*	*10 879 F*	*8 964 F*	*8 916 F*	*9 547 F*	*9 083 F*
Short mackerel	**Maquereau trapu**		**Caballa rechoncha**		***Rastrelliger brachysoma***				**1,75(01)014,01**	**RAB**
71 Philippines	26 133	27 517	26 200	25 224	22 978	23 350	25 713	26 771	28 091	32 657
71 Fishing area total	*26 133*	*27 517*	*26 200*	*25 224*	*22 978*	*23 350*	*25 713*	*26 771*	*28 091*	*32 657*
Species total	*26 133*	*27 517*	*26 200*	*25 224*	*22 978*	*23 350*	*25 713*	*26 771*	*28 091*	*32 657*
Indian mackerel	**Maquereau des Indes**		**Caballa de la India**		***Rastrelliger kanagurta***				**1,75(01)014,03**	**RAG**
51 Egypt	...	...	1 363	1 151	1 914	652	1 004	430	1 442	2 782
Eritrea	...	40	75	58	2	0	20	197	100	13
India	113 309	180 680	168 757	267 692	157 541	142 669	138 086	62 026	32 410	38 041
Oman	1 955	1 282	2 122	1 712	2 207	1 994	2 024	2 426	3 223	3 459
Saudi Arabia	1 741	3 240	3 069	1 549	1 990	2 072	1 979	1 525	1 803	1 921
Seychelles	304	586	322	425	298	25	101	295	182	281
Tanzania	2 542	3 248	3 490	4 618	4 000	4 050	4 500	4 500	5 000	5 250
Untd Arab Em	4 750	5 382	4 309	4 354	4 653	4 669	4 786	4 775	2 000	2 104
Yemen	6 000	6 500	6 958	752	790	900 F	900 F	900 F	1 150 F	1 300 F
51 Fishing area total	*130 601*	*200 958*	*190 465*	*282 311*	*173 395*	*157 031 F*	*153 400 F*	*77 074 F*	*47 310 F*	*55 151 F*
57 India	34 418	25 642	28 476	25 834	32 831	41 750	41 292	16 795	13 748	15 033
Thailand	13 743	13 695	25 118	23 801	23 400	24 289	20 973	13 301	19 505	19 092
57 Fishing area total	*48 161*	*39 337*	*53 594*	*49 635*	*56 231*	*66 039*	*62 265*	*30 096*	*33 253*	*34 125*
71 Fiji Islands	716	730 F	452	400	312	224	721	650 F	722	700 F
Palau	...	...	2	1	0	1	1	0	0	0
Philippines	57 945	59 505	51 352	46 264	54 732	51 919	53 606	55 088	62 484	72 578
Thailand	35 986	50 898	45 338	21 328	19 276	19 393	26 912	21 902	22 836	23 095
71 Fishing area total	*94 647*	*111 133 F*	*97 144*	*67 993*	*74 320*	*71 537*	*81 240*	*77 640 F*	*86 042*	*96 373 F*
Species total	*273 409*	*351 428 F*	*341 203*	*399 939*	*303 946*	*294 607 F*	*296 905 F*	*184 810 F*	*166 605 F*	*185 649 F*
Indian mackerels nei	**Maquereaux(Indo-pacifiq.) nca**		**Caballas Indo-Pacifico nep**		***Rastrelliger spp***				**1,75(01)014,XX**	**RAX**
51 Comoros	220	220	250 F	250 F	240 F	240 F	230 F	250 F	220	220 F
Seychelles	161	260	205	159	116	111	219	174	83	62
51 Fishing area total	*381*	*480*	*455 F*	*409 F*	*356 F*	*351 F*	*449 F*	*424 F*	*303*	*282 F*
57 Indonesia	62 120	64 348	65 362	67 316	70 534	72 024	67 916	75 080	76 984	79 140
Malaysia	36 104	63 771	101 001	73 780	66 586	79 446	84 665	65 733	65 971	55 527
Thailand	66 985	65 499	46 945	48 061	46 999	43 927	38 935	32 002	38 678	37 859
57 Fishing area total	*165 209*	*193 618*	*213 308*	*189 157*	*184 119*	*195 397*	*191 516*	*172 815*	*181 633*	*172 526*
71 Indonesia	111 826	130 534	128 528	121 594	130 870	132 739	133 550	131 957	137 403	141 250
Malaysia	31 871	29 875	25 169	21 584	20 215	22 626	26 700	32 322	33 498	32 383
Singapore	101	210	151	12	51	165	129	97	68	35
Thailand	76 997	82 021	112 280	92 765	91 622	107 083	125 175	120 882	112 503	113 779
71 Fishing area total	*220 795*	*242 640*	*266 128*	*235 955*	*242 758*	*262 613*	*285 554*	*285 258*	*283 472*	*287 447*
Species total	*386 385*	*436 738*	*479 891 F*	*425 521 F*	*427 233 F*	*458 361 F*	*477 519 F*	*458 497 F*	*465 408*	*460 255 F*
Mackerels nei	**Maquereaux nca**		**Caballas nep**		***Scombridae***				**1,75(01)XXX,XX**	**MAX**
27 France	9	-	-	-	120	25	13	1	2 642	-
Portugal	47	72	106	236	340	246	-	-	-	
Spain	11 898	17 133	21 899	26 414	34 628	36 547	19 200	13 609	22 949	19 936
27 Fishing area total	*11 954*	*17 205*	*22 005*	*26 650*	*35 088*	*36 818*	*19 213*	*13 610*	*25 591*	*19 936*
31 Guadeloupe	1 280 F	1 290 F	1 400 F	1 550	1 700 F	1 430 F	1 430 F	1 600 F	1 600 F	1 600 F
31 Fishing area total	*1 280 F*	*1 290 F*	*1 400 F*	*1 550*	*1 700 F*	*1 430 F*	*1 430 F*	*1 600 F*	*1 600 F*	*1 600 F*
34 Gabon	...	81	145	64	65	114	158	304	201	267
GuineaBissau	10 F	10 F	12 F	14 F	15 F	14 F	10 F	10 F	10 F	10 F
Nigeria	2 871	2 833	2 254	3 745	3 326	3 640	2 919	2 710	3 759	3 500
Portugal	2	0	1	1	0	1	-	-	-	-
34 Fishing area total	*2 883 F*	*2 924 F*	*2 412 F*	*3 824 F*	*3 406 F*	*3 769 F*	*3 087 F*	*3 024 F*	*3 970 F*	*3 777 F*
37 France	-	-	-	-	-	-	-	3	7	-
Lebanon	150	150	500	500	300	300	300	350	350	350
37 Fishing area total	*150*	*150*	*500*	*500*	*300*	*300*	*300*	*353*	*357*	*350*
41 Portugal	-	-	-	-	-	-	21	0	-	-
41 Fishing area total	-	-	-	-	-	-	*21*	*0*	-	-

B-37 Miscellaneous pelagic fishes / Poissons pélagiques divers / Peces pelágicos diversos

Capture production by species, fishing areas and countries or areas / Captures par espèces, zones de pêche et pays ou zones / Capturas por especies, áreas de pesca y países o áreas

Species, Fishing area Espèce, Zone de pêche Especie, Area de pesca	1993 t	1994 t	1995 t	1996 t	1997 t	1998 t	1999 t	2000 t	2001 t	2002 t
47 Iceland	-	-	-	-	11	-	-	-	-	-
47 Fishing area total	*-*	*-*	*-*	*-*	*11*	*-*	*-*	*-*	*-*	*-*
51 Mauritius	12	12	11	12	10	11	11	13	13	11
Portugal	-	-	-	-	-	-	14	-	-	-
Saudi Arabia	...	...	...	...	...	12	16	11	10	3
South Africa	-	-	-	-	-	-	-	0	0	-
51 Fishing area total	*12*	*12*	*11*	*12*	*10*	*23*	*41*	*24*	*23*	*14*
57 Australia	400	462	471	502	2	...	1 478	462	445	465
Sri Lanka	10 854	16 450	17 642	17 700	20 000	20 900	21 350	22 180	16 760	17 250
57 Fishing area total	*11 254*	*16 912*	*18 113*	*18 202*	*20 002*	*20 900*	*22 828*	*22 642*	*17 205*	*17 715*
71 Australia	0	0	0	85	87	91	1 792	2 068	2 108	2 431
NewCaledonia	65	104	126	119	102	104	161	161	30	30 F
71 Fishing area total	*65*	*104*	*126*	*204*	*189*	*195*	*1 953*	*2 229*	*2 138*	*2 461 F*
81 Australia	71	414	408	400	...	...	458	781	406	990
81 Fishing area total	*71*	*414*	*408*	*400*	*...*	*...*	*458*	*781*	*406*	*990*
87 Cuba	0	-	-	-	-	-	-	-	-	-
87 Fishing area total	*0*	*-*	*-*	*-*	*-*	*-*	*-*	*-*	*-*	*-*
Species total	*27 669 F*	*39 011 F*	*44 975 F*	*51 342 F*	*60 706 F*	*63 435 F*	*49 331 F*	*44 263 F*	*51 290 F*	*46 843 F*
Blue butterfish	**Fiatole**		**Palometa fiatola**		***Stromateus fiatola***				**1,76(03)004,01**	**BLB**
34 Congo Rep	...	...	...	...	...	...	...	1	1 F	1 F
Côte dIvoire	3	6	6 F	10 F	15 F	20 F	22	45	27	21
Estonia	262	-	-	-	-	-	-	-	-	-
Korea Rep	77	-	-	-	-	-	-	-	-	-
Liberia	299	307	382	143	236	248	573	181	183	190 F
Lithuania	-	-	-	-	-	-	-	-	-	4
Nigeria	0	0	-	-	-	-	-	-	4 106	1 000
Portugal	-	-	63	87	71	3	-	1	-	-
Russian Fed	-	-	15	-	-	-	4	-	-	-
Senegal	-	-	-	-	-	-	20	43	57	190
Spain	...	...	...	...	...	...	...	20	58	50
Ukraine	-	-	-	-	-	-	-	1	-	8
34 Fishing area total	*641*	*313*	*466 F*	*240 F*	*322 F*	*271 F*	*619*	*292*	*4 432 F*	*1 464 F*
47 Angola	0	0	0	0	-	-	-	-	-	...
Portugal	-	-	-	-	1	21	17	28	-	70
Russian Fed	-	-	-	-	-	-	-	-	-	13
47 Fishing area total	*0*	*0*	*0*	*0*	*1*	*21*	*17*	*28*	*-*	*83*
Species total	*641*	*313*	*466 F*	*240 F*	*323 F*	*292 F*	*636*	*320*	*4 432 F*	*1 547 F*
Silver pomfret	**Aileron argenté**		**Palometón platero**		***Pampus argenteus***				**1,76(03)009,01**	**SIP**
51 Kuwait	1 094	1 112	1 101	862	560	501	259	200 F	133	140 F
Saudi Arabia	-	5	41	31	16	14	30	43	25	25
51 Fishing area total	*1 094*	*1 117*	*1 142*	*893*	*576*	*515*	*289*	*243 F*	*158*	*165 F*
57 Indonesia	4 547	4 446	4 985	4 849	5 100	6 304	6 236	5 919	6 330	6 620
Malaysia	...	2 501	2 924	2 623	3 006	3 040	1 487	1 201	1 400	1 067
Thailand	434	1 036	1 371	1 100	1 096	225	299	135	540	528
57 Fishing area total	*4 981*	*7 983*	*9 280*	*8 572*	*9 202*	*9 569*	*8 022*	*7 255*	*8 270*	*8 215*
61 China,Taiwan	4 258	4 049	6 978	5 113	2 703	1 621	2 477	2 852	4 245	4 182
Korea Rep	5 131	4 527	4 806	4 806	919	1 029	3 374	-	-	2 247
61 Fishing area total	*9 389*	*8 576*	*11 784*	*9 919*	*3 622*	*2 650*	*5 851*	*2 852*	*4 245*	*6 429*
71 Indonesia	12 945	12 261	13 763	15 083	15 361	15 004	15 104	19 573	23 955	26 010
Malaysia	...	2 717	2 269	2 397	2 087	2 361	2 884	2 330	2 714	2 633
Thailand	1 223	1 515	515	588	521	153	96	272	314	318
71 Fishing area total	*14 168*	*16 493*	*16 547*	*18 068*	*17 969*	*17 518*	*18 084*	*22 175*	*26 983*	*28 961*
Species total	*29 632*	*34 169*	*38 753*	*37 452*	*31 369*	*30 252*	*32 246*	*32 525 F*	*39 656*	*43 770 F*
Silver pomfrets nei	**Ailerons nca**		**Palometones nep**		***Pampus spp***				**1,76(03)009,XX**	**XPO**
61 China	116 553	138 335	209 031	220 364	242 547	303 024	337 919	338 848	352 493	386 317
61 Fishing area total	*116 553*	*138 335*	*209 031*	*220 364*	*242 547*	*303 024*	*337 919*	*338 848*	*352 493*	*386 317*
Species total	*116 553*	*138 335*	*209 031*	*220 364*	*242 547*	*303 024*	*337 919*	*338 848*	*352 493*	*386 317*
North Atlantic harvestfish	**Stromaté lune**		**Palometa mono**		***Peprilus alepidotus***				**1,76(03)011,01**	**HVF**
31 USA	36	13	25	14	15	13	16	23	18	22
31 Fishing area total	*36*	*13*	*25*	*14*	*15*	*13*	*16*	*23*	*18*	*22*
Species total	*36*	*13*	*25*	*14*	*15*	*13*	*16*	*23*	*18*	*22*
Gulf butterfishes, etc. nei	**Stromatés du Golfe, etc. nca**		**Pámpanos del Golfo, etc. nep**		***Peprilus spp***				**1,76(03)011,XX**	**BTG**
31 Venezuela	2 856	1 801	1 881	1 889	568	1 175	1 004	809	742	872
31 Fishing area total	*2 856*	*1 801*	*1 881*	*1 889*	*568*	*1 175*	*1 004*	*809*	*742*	*872*

B-37 Miscellaneous pelagic fishes / Poissons pélagiques divers / Peces pelágicos diversos

Capture production by species, fishing areas and countries or areas / Captures par espèces, zones de pêche et pays ou zones / Capturas por especies, áreas de pesca y países o áreas

Species, Fishing area Espèce, Zone de pêche Especie, Area de pesca	1993 t	1994 t	1995 t	1996 t	1997 t	1998 t	1999 t	2000 t	2001 t	2002 t
Species total	*2 856*	*1 801*	*1 881*	*1 889*	*568*	*1 175*	*1 004*	*809*	*742*	*872*
Butterfishes, pomfrets nei	**Stromatés, ailerons nca**		**Pámpanos, palometónes nep**		***Stromateidae***				**1,76(03)XXX,XX**	**BUX**
21 USA	4 607	3 641	2 127	3 611	3 357	1 944	2 116	1 438	4 416	889
21 Fishing area total	*4 607*	*3 641*	*2 127*	*3 611*	*3 357*	*1 944*	*2 116*	*1 438*	*4 416*	*889*
27 Portugal	-	-	-	-	-	-	94	35	74	61
Spain	-	-	-	-	-	-	-	-	7	
27 Fishing area total	-	-	-	-	-	-	*94*	*35*	*81*	*61*
31 USA	142	993	789	782	65	633	644	681	539	613
31 Fishing area total	*142*	*993*	*789*	*782*	*65*	*633*	*644*	*681*	*539*	*613*
41 Korea Rep	45	-	-	-	-	-	-	-	-	-
Portugal	-	-	-	-	-	5	-	-	20	-
41 Fishing area total	*45*	-	-	-	-	*5*	-	-	*20*	-
51 India	22 416	30 925	34 723	19 529	18 036	15 200	14 897	10 417	9 112	19 232
Kuwait	153	191	289	230	186	111	170	-	-	
Pakistan	3 103	2 985	4 156	2 799	3 786	4 089	4 605	3 945	3 454	3 500
Spain	-	-	-	-	-	-	-	-	-	24
51 Fishing area total	*25 672*	*34 101*	*39 168*	*22 558*	*22 008*	*19 400*	*19 672*	*14 362*	*12 566*	*22 756*
57 India	22 395	31 998	23 255	17 967	18 695	18 119	16 288	14 974	15 545	16 521
Malaysia	4 631	228	268	240	275	278	165	133	109	118
Thailand	-	-	-	-	-	-	-	16	-	-
57 Fishing area total	*27 026*	*32 226*	*23 523*	*18 207*	*18 970*	*18 397*	*16 453*	*15 123*	*15 654*	*16 639*
61 China,H.Kong	2 321	2 331	2 428	2 114	1 973	2 172	1 500 F	1 900 F	2 100 F	2 050 F
China,Taiwan	1 894	378	226	246	336	231	280	131	146	210
Korea Rep	8 101	9 849	10 896	9 484	10 770	13 210	15 235	7 838	6 819	6 191
61 Fishing area total	*12 316*	*12 558*	*13 550*	*11 844*	*13 079*	*15 613*	*17 015 F*	*9 869 F*	*9 065 F*	*8 451 F*
71 Malaysia	5 498	248	208	220	192	217	405	328	268	370
Philippines	2 434	2 044	2 382	2 281	1 001	1 366	1 547	1 522	1 759	1 871
Singapore	82	123	103	100	94	103	94	75	48	44
71 Fishing area total	*8 014*	*2 415*	*2 693*	*2 601*	*1 287*	*1 686*	*2 046*	*1 925*	*2 075*	*2 285*
77 USA	-	-	-	-	-	1	2	3	7	23
77 Fishing area total	-	-	-	-	-	*1*	*2*	*3*	*7*	*23*
81 Japan	4 232	2 769	1 314	1 518	2 126	2 194	1 837	2 077	2 211	1 300 F
Korea Rep	-	-	-	-	-	39	-	-	-	-
81 Fishing area total	*4 232*	*2 769*	*1 314*	*1 518*	*2 126*	*2 233*	*1 837*	*2 077*	*2 211*	*1 300 F*
87 Chile	4	7	7	9	6	11	12	10	11	141
Ecuador	2 748	20 545	...	...	426	...	...	...	...	...
Japan	-	-	-	-	-	-	5	-	-	
87 Fishing area total	*2 752*	*20 552*	*7*	*9*	*432*	*11*	*17*	*10*	*11*	*141*
Species total	*84 806*	*109 255*	*83 171*	*61 130*	*61 324*	*59 923*	*59 896 F*	*45 523 F*	*46 645 F*	*53 158 F*
Barracudas nei	**Bécunes nca**		**Picudas nep**			***Sphyraena spp***			**1,77(10)001,XX**	**BAR**
27 Portugal	...	...	...	...	...	...	33	41	33	37
27 Fishing area total	...	...	...	...	...	...	*33*	*41*	*33*	*37*
31 Antigua Barb	...	...	...	...	...	...	...	...	6	8
Dominican Rp	15	17	14	3	-	8	4	5	7	33
Grenada	37	28	24	40	30	41	35	57	43	52
Mexico	320	460	781	654	1 102	899	987	445	708	532
Puerto Rico	...	...	...	...	...	...	16	21	13	13
USA	...	...	...	...	0	1	-	53	55	51
Venezuela	1 039	998	923	899	998	1 123	881	731	1 165	1 311
31 Fishing area total	*1 411*	*1 503*	*1 742*	*1 596*	*2 130*	*2 072*	*1 923*	*1 312*	*1 997*	*2 000*
34 Benin	300 F	338 F	328 F	330 F	400 F	350 F	297	200 F	262	365
Cameroon	16	20 F	27	37	30 F	5	28	23	293	340
Congo Rep	0	0	0	0	0	0	0	1	1 F	1 F
Côte dIvoire	406	285	287 F	366	151	195	184	160	75	218
Gabon	215	184	420	1 055	883	1 238	975	976	1 636	1 379
Gambia	202	105	170	355	144	116	120	284	631	2 573
Ghana	1 970	1 095	1 372	1 763	1 684	872	2 591	759	1 156	1 193
Guinea	...	...	452	309	431	400	300	226	1 043	1 000 F
Korea Rep	54	-	-	9	-		-	-	-	
Liberia	42	31	202	35	-	67	343	174	196	200 F
Lithuania	11	-	-	-		-	-	-	-	40
Nigeria	4 941	5 039	3 238	3 556	5 212	7 671	10 490	10 799	10 947	11 085
Portugal	8	7	14	20	9	10	10	2	9	2
Russian Fed	-	-	-	-	-	-	-	-	-	2
Sao Tome Prn	20 F	29 F	30 F	22	56	70 F	80 F	70 F	70 F	70 F
Senegal	1 072	1 717	1 117	1 530	2 339	1 456	1 474	1 680	1 519	1 483
Sierra Leone	680	678	681	680 F	772	237	124	1 095	2 757	2 063
Spain	-	-	-	24	31	19	28	22	37	9
Togo	45	59	33	96	38	25	144	226	54	76
Ukraine	-	-	-	-	-	-	-	-	-	3
34 Fishing area total	*9 982 F*	*9 587 F*	*8 371 F*	*10 187 F*	*12 180 F*	*12 731 F*	*17 188 F*	*16 697 F*	*20 686 F*	*22 102 F*

B-37 Miscellaneous pelagic fishes / Poissons pélagiques divers / Peces pelágicos diversos

Capture production by species, fishing areas and countries or areas / Captures par espèces, zones de pêche et pays ou zones / Capturas por especies, áreas de pesca y países o áreas

Species, Fishing area Espèce, Zone de pêche Especie, Area de pesca	1993 t	1994 t	1995 t	1996 t	1997 t	1998 t	1999 t	2000 t	2001 t	2002 t
37 Algeria	100 F	100 F	...	...	...	...	6	47	392	447
Egypt	730	...	1 731	1 098	1 721	1 764	1 159	1 191	1 085	995
Israel	210	50	84	76	104	105	160	77	55	54
Lebanon	100	125	200	200	200	200	200	200	250	200
Palest, O.T.	...	...	...	49	101	90	90 F	80 F	60 F	52
Spain	-	-	-	-	-	-	29	103	130	160
Syria	54 F	52 F	52 F	82	90	88	130	76	77	60
Tunisia	52	52	36	49	69	100	103	95	96	122
Turkey	1 594	1 507	506	250	200	120	170	150	130	206
37 Fishing area total	*2 840 F*	*1 886 F*	*2 609 F*	*1 804*	*2 485*	*2 467*	*2 047 F*	*2 019 F*	*2 275 F*	*2 296*
41 Brazil	180 F	180 F	20	11	12	828	217	527	378	371
Estonia	-	-	-	-	-	-	-	-	-	1
41 Fishing area total	*180 F*	*180 F*	*20*	*11*	*12*	*828*	*217*	*527*	*378*	*372*
47 Angola	...	...	...	...	...	...	119	-	-	-
Russian Fed	-	-	-	-	-	-	79	-	1	3
47 Fishing area total	*...*	*...*	*...*	*...*	*...*	*...*	*198*	*-*	*1*	*3*
51 Bahrain	89	66	107	159	82	156	6	8	7	6
Djibouti	14 F	15 F	18 F	20 F	20 F	20 F	20 F	20 F	20 F	20 F
Egypt	-	-	12	3	1	-	3	1 199	485	260
Eritrea	...	37	109	185	21	57	150	684	2	455
India	3 036	2 893	4 692	4 202	3 588	9 903	9 926	350	148	2 347
Kenya	57	55	65	54	54	71	108	83	99	88
Oman	1 580	1 713	2 244	1 948	2 023	1 789	2 347	1 431	1 781	3 215
Pakistan	2 485	2 923	2 324	2 878	2 683	2 664	3 520	3 981	3 889	3 900
Qatar	18	52	48	54	59	21	2	62	86	0
Saudi Arabia	1 990	2 249	1 065	1 251	1 246	1 489	1 267	1 563	1 562	1 492
Seychelles	286	152	163	190	183	150	323	217	253	300
Tanzania	23	30	26	29	30	15	20	25	25	25
Untd Arab Em	2 444	2 810	2 094	2 116	2 261	2 269	2 326	543	625	487
Yemen	3 279	2 716	2 356	1 813	1 900 F	2 100 F	2 100 F	1 900 F	2 400 F	2 700 F
51 Fishing area total	*15 301 F*	*15 711 F*	*15 323 F*	*14 902 F*	*14 151 F*	*20 704 F*	*22 118 F*	*12 066 F*	*11 382 F*	*15 295 F*
57 India	3 335	4 760	4 378	4 671	4 239	5 062	5 010	3 812	4 717	4 690
Indonesia	3 886	4 439	5 135	5 624	5 732	7 004	6 480	7 442	7 740	8 470
Malaysia	1 036	815	730	844	1 012	709	872	735	554	716
Thailand	5 523	5 487	7 726	10 427	10 554	10 817	11 297	9 992	9 794	9 587
57 Fishing area total	*13 780*	*15 501*	*17 969*	*21 566*	*21 537*	*23 592*	*23 659*	*21 981*	*22 805*	*23 463*
61 China,Taiwan	927	866	429	547	761	771	519	362	439	311
61 Fishing area total	*927*	*866*	*429*	*547*	*761*	*771*	*519*	*362*	*439*	*311*
71 Fiji Islands	3 247	2 883 F	1 735	567	1 562	1 626	2 979	2 700 F	2 809	2 800 F
Guam	1	2	0	1	1	1	1	2	4	2
Indonesia	23 474	19 641	9 093	10 928	12 474	13 436	14 098	12 116	11 330	10 710
Kiribati	2 130	2 110	2 130	2 160	450	987	855	420	732	1 360
Malaysia	2 856	3 148	3 966	4 089	5 642	6 445	7 231	6 712	6 057	5 247
Palau	1	0	1	3	1	1	1	0	0	0
Philippines	11 936	11 039	10 654	10 003	6 707	7 722	8 976	7 778	9 128	10 295
Singapore	153	232	195	170	134	149	105	79	86	65
Thailand	2 389	4 011	4 335	3 962	3 691	3 208	5 563	6 397	4 819	4 873
71 Fishing area total	*46 187*	*43 066 F*	*32 109*	*31 883*	*30 662*	*33 575*	*39 809*	*36 204 F*	*34 965*	*35 352 F*
77 Amer Samoa	...	2	2	2	4	1	0	0	0	1
Mexico	195	37	33	68	485	561	164	104	224	415
USA	...	...	...	...	...	59	95	76	72	48
77 Fishing area total	*195*	*39*	*35*	*70*	*489*	*621*	*259*	*180*	*296*	*464*
81 Korea Rep	-	-	-	-	80	1	349	432	512	585
Russian Fed	911	1 597	515	1 176	399	125	765	-	-	-
81 Fishing area total	*911*	*1 597*	*515*	*1 176*	*479*	*126*	*1 114*	*432*	*512*	*585*
Species total	*91 714 F*	*89 936 F*	*79 122 F*	*83 742 F*	*84 886 F*	*97 487 F*	*109 084 F*	*91 821 F*	*95 769 F*	*102 280 F*
Ocean sunfish	**Poisson lune**		**Pez luna**		***Mola mola***			**1,90(08)002,01**		**MOX**
27 Ireland	-	-	-	-	-	-	-	-	13	-
Portugal	-	-	-	-	-	-	12	-	-	-
27 Fishing area total	*-*	*-*	*-*	*-*	*-*	*-*	*12*	*-*	*13*	*-*
47 Namibia	...	...	...	...	...	...	...	...	2	1
47 Fishing area total	*...*	*...*	*...*	*...*	*...*	*...*	*...*	*...*	*2*	*1*
Species total	*...*	*...*	*...*	*...*	*...*	*...*	*12*	*...*	*15*	*1*
Pelagic percomorphs nei	**Percomorphes pélagiques nca**		**Percomorfos pelágicos nep**		***Perciformes***			**1,99(XX)XXX,XX**		**PPX**
31 Puerto Rico	123	79	186	98	97	83	-	-	-	-
Venezuela	403	520	3 579	759	442	381	219	182	109	74
31 Fishing area total	*526*	*599*	*3 765*	*857*	*539*	*464*	*219*	*182*	*109*	*74*
34 Cape Verde	...	...	...	...	4 414	4 899	4 463	4 823	4 288	4 500 F
Eq Guinea	160 F	240 F	100 F	220 F	300 F	170	160 F	20 F	10 F	10 F
Spain	-	-	-	-	-	-	-	1	2	-
34 Fishing area total	*160 F*	*240 F*	*100 F*	*220 F*	*4 714 F*	*5 069*	*4 623 F*	*4 844 F*	*4 300 F*	*4 510 F*

B-37 Miscellaneous pelagic fishes — Capture production by species, fishing areas and countries or areas
Poissons pélagiques divers — Captures par espèces, zones de pêche et pays ou zones
Peces pelágicos diversos — Capturas por especies, áreas de pesca y países o áreas

Species, Fishing area Espèce, Zone de pêche Especie, Area de pesca	1993 t	1994 t	1995 t	1996 t	1997 t	1998 t	1999 t	2000 t	2001 t	2002 t
41 Brazil	290 F	290 F	-	-	-	-	-	-	-	-
Uruguay	15	2	3	2	243	150	10	18	-	-
41 Fishing area total	*305 F*	*292 F*	*3*	*2*	*243*	*150*	*10*	*18*	*-*	*-*
47 Cuba	-	-	-	-	-	-	2 427	-	-	-
47 Fishing area total	*-*	*-*	*-*	*-*	*-*	*-*	*2 427*	*-*	*-*	*-*
51 Kenya	290	114	286	358	351	611	528	378	1 003	982
Oman	8 447	7 788	1 269	2 045	2 949	6 055	10 790	5 353	3 391	5 468
Yemen	35 938	39 227	60 721	62 563	60 707 F	67 630 F	74 000 F	63 900 F	80 915 F	91 570 F
51 Fishing area total	*44 675*	*47 129*	*62 276*	*64 966*	*64 007 F*	*74 296 F*	*85 318 F*	*69 631 F*	*85 309 F*	*98 020 F*
57 Australia	3 700 F	3 500 F	5 000 F	4 000 F	-	-	-	-	-	-
57 Fishing area total	*3 700 F*	*3 500 F*	*5 000 F*	*4 000 F*	*-*	*-*	*-*	*-*	*-*	*-*
71 Australia	15 F	20 F	20 F	20 F	-	-	-	-	-	-
71 Fishing area total	*15 F*	*20 F*	*20 F*	*20 F*	*-*	*-*	*-*	*-*	*-*	*-*
81 Australia	2 500 F	2 000 F	1 000 F	-	-	-	-	-	-	-
Japan	12	10	2	-	-	-	-	-	-	-
81 Fishing area total	*2 512 F*	*2 010 F*	*1 002 F*	*-*	*-*	*-*	*-*	*-*	*-*	*-*
87 Peru	464	152	382	858	120	33	1 073	771	371	684
87 Fishing area total	*464*	*152*	*382*	*858*	*120*	*33*	*1 073*	*771*	*371*	*684*
Species total	*52 357 F*	*53 942 F*	*72 548 F*	*70 923 F*	*69 623 F*	*80 012 F*	*93 670 F*	*75 446 F*	*90 089 F*	*103 288 F*
Group total	***13 055 924***	***13 116 852***	***13 935 773***	***14 332 473***	***13 789 709***	***11 214 197***	***10 684 205***	***10 645 490***	***12 249 788***	***11 639 160***

B-38 Sharks, rays, chimaeras / Squales, raies, chimères / Tiburones, rayas, quimeras

Capture production by species, fishing areas and countries or areas
Captures par espèces, zones de pêche et pays ou zones
Capturas por especies, áreas de pesca y países o áreas

Species, Fishing area Espèce, Zone de pêche Especie, Area de pesca	1993 t	1994 t	1995 t	1996 t	1997 t	1998 t	1999 t	2000 t	2001 t	2002 t
Bluntnose sixgill shark	**Requin griset**		**Cañabota gris**		***Hexanchus griseus***				**1,05(02)002,01**	**SBL**
27 Portugal	-	-	-	-	-	-	-	-	1	7
27 Fishing area total	*-*	*-*	*-*	*-*	*-*	*-*	*-*	*-*	*1*	*7*
Species total	*-*	*-*	*-*	*-*	*-*	*-*	*-*	*-*	*1*	*7*
Broadnose sevengill shark	**Platnez**		**Cañabota gata**		***Notorynchus cepedianus***				**1,05(02)005,02**	**NTC**
81 New Zealand	...	...	...	...	...	2	3	4	5	4
81 Fishing area total	*...*	*...*	*...*	*...*	*...*	*2*	*3*	*4*	*5*	*4*
Species total	*...*	*...*	*...*	*...*	*...*	*2*	*3*	*4*	*5*	*4*
Basking shark	**Pèlerin**		**Peregrino**		***Cetorhinus maximus***				**1,06(01)003,01**	**BSK**
27 France	-	-	0	0	1	0	3	-	-	-
Norway	2 910	1 762	108	1 979	1 159	137	77	293	200	135
Portugal	0	1	1	1	1	-	1	1	3	1
27 Fishing area total	*2 910*	*1 763*	*109*	*1 980*	*1 161*	*137*	*81*	*294*	*203*	*136*
37 Spain	-	-	-	2	6	6	-	-	-	4
37 Fishing area total	*-*	*-*	*-*	*2*	*6*	*6*	*-*	*-*	*-*	*4*
81 New Zealand	...	...	14	2	2	49	129	95	84	40
81 Fishing area total	*...*	*...*	*14*	*2*	*2*	*49*	*129*	*95*	*84*	*40*
Species total	*2 910*	*1 763*	*123*	*1 984*	*1 169*	*192*	*210*	*389*	*287*	*180*
Sand tiger shark	**Requin taureau**		**Toro bacota**		***Carcharias taurus***				**1,06(02)005,01**	**CCT**
21 USA	-	2	-	-	-	-	-	1	-	-
21 Fishing area total	*-*	*2*	*-*	*-*	*-*	*-*	*-*	*1*	*-*	*-*
Species total	*-*	*2*	*-*	*-*	*-*	*-*	*-*	*1*	*-*	*-*
Thresher	**Renard**		**Zorro**		***Alopias vulpinus***				**1,06(06)006,01**	**ALV**
21 USA	14	23	1	-	-	-	-	8	11	11
21 Fishing area total	*14*	*23*	*1*	*-*	*-*	*-*	*-*	*8*	*11*	*11*
27 France	14	11	13	7	13	7	21	116	113	11
Portugal	...	...	...	...	...	...	13	20	27	21
27 Fishing area total	*14*	*11*	*13*	*7*	*13*	*7*	*34*	*136*	*140*	*32*
34 Liberia	...	...	...	...	...	...	151	146	...	...
Portugal	...	...	...	...	...	...	...	...	7	1
Spain	...	...	...	...	30	45	...	...	...	...
34 Fishing area total	*...*	*...*	*...*	*...*	*30*	*45*	*151*	*146*	*7*	*1*
37 France	...	...	...	...	...	...	14	12	19	13
Portugal	...	...	...	...	...	...	...	...	2	1
37 Fishing area total	*...*	*...*	*...*	*...*	*...*	*...*	*14*	*12*	*21*	*14*
41 Portugal	...	...	...	...	...	...	...	...	3	-
41 Fishing area total	*...*	*...*	*...*	*...*	*...*	*...*	*...*	*...*	*3*	*-*
67 USA	...	...	...	...	...	17	2	48	76	73
67 Fishing area total	*...*	*...*	*...*	*...*	*...*	*17*	*2*	*48*	*76*	*73*
77 USA	...	...	...	...	...	303	262	249	299	279
77 Fishing area total	*...*	*...*	*...*	*...*	*...*	*303*	*262*	*249*	*299*	*279*
81 New Zealand	...	...	15	13	24	21	32	51	57	53
Ukraine	-	-	-	-	-	-	-	-	-	1
81 Fishing area total	*...*	*...*	*15*	*13*	*24*	*21*	*32*	*51*	*57*	*54*
Species total	*28*	*34*	*29*	*20*	*67*	*393*	*495*	*650*	*614*	*464*
Bigeye thresher	**Renard à gros yeux**		**Zorro ojón**		***Alopias superciliosus***				**1,06(06)006,03**	**BTH**
31 Spain	-	-	-	-	1	...	...	...	...	...
31 Fishing area total	*-*	*-*	*-*	*-*	*1*	*...*	*...*	*...*	*...*	*...*
34 Spain	...	...	...	...	148	114	...	...	...	...
34 Fishing area total	*...*	*...*	*...*	*...*	*148*	*114*	*...*	*...*	*...*	*...*
77 USA	...	...	...	...	...	11	5	5	2	-
77 Fishing area total	*...*	*...*	*...*	*...*	*...*	*11*	*5*	*5*	*2*	*-*
Species total	*...*	*...*	*...*	*...*	*149*	*125*	*5*	*5*	*2*	
Thresher sharks nei	**Renards de mer nca**		**Zorros nep**		***Alopias spp***				**1,06(06)006,XX**	**THR**
34 Spain	...	...	...	...	34	55	66	...	...	...
34 Fishing area total	*...*	*...*	*...*	*...*	*34*	*55*	*66*	*...*	*...*	*...*
Species total	*...*	*...*	*...*	*...*	*34*	*55*	*66*	*...*	*...*	*...*

B-38

Sharks, rays, chimaeras — **Capture production by species, fishing areas and countries or areas**
Squales, raies, chimères — **Captures par espèces, zones de pêche et pays ou zones**
Tiburones, rayas, quimeras — **Capturas por especies, áreas de pesca y países o áreas**

Species, Fishing area Espèce, Zone de pêche Especie, Area de pesca	1993 t	1994 t	1995 t	1996 t	1997 t	1998 t	1999 t	2000 t	2001 t	2002 t
Shortfin mako	**Taupe bleue**		**Marrajo dientuso**		***Isurus oxyrinchus***			**1,06(08)002,01**		**SMA**
21 Portugal	...	...	...	...	...	...	...	10	-	-
USA	71	115	5	-	-	-	-	19	19	20
21 Fishing area total	*71*	*115*	*5*	...	...	...	...	*29*	*19*	*20*
27 Portugal	...	...	...	...	...	...	160	183	186	107
UK	-	-	-	-	-	0	2	3	2	1
27 Fishing area total	...	...	...	...	...	*0*	*162*	*186*	*188*	*108*
31 Portugal	-	-	-	-	-	0	-	-	-	-
Spain	-	-	-	-	73	33	...	...	...	134
USA	-	-	-	-	-	-	-	5	5	-
31 Fishing area total	-	-	-	-	*73*	*33*	...	*5*	*5*	*134*
34 Benin	...	...	...	...	...	...	4	3 F	1	-
China	-	-	-	-	...	...	...	3	...	...
Philippines	-	-	-	-	-	-	3	-	-	-
Portugal	...	...	...	...	...	...	...	42	42	68
34 Fishing area total	...	...	...	...	...	...	*7*	*48 F*	*43*	*68*
37 Portugal	-	-	-	-	-	-	-	1	6	-
Spain	-	-	-	-	-	-	-	-	-	2
37 Fishing area total	-	-	-	-	-	-	-	*1*	*6*	*2*
41 Brazil	100 F	...	...	83	190	...	100	120	...	...
Portugal	...	...	...	...	...	...	...	15	26	22
Spain	-	-	-	-	-	-	-	-	-	257
41 Fishing area total	*100 F*	...	...	*83*	*190*	...	*100*	*135*	*26*	*279*
47 China	-	-	-	-	-	-	-	150	...	11
Namibia	...	...	...	...	...	...	...	...	...	1 587
Portugal	-	-	1	...	...	4	3	350	158	0
South Africa	...	...	...	...	...	...	...	...	2	24
Spain	-	-	-	-	587	304	335	264	228	456
47 Fishing area total	...	...	*1*	...	*587*	*308*	*338*	*764*	*388*	*2 078*
51 Portugal	-	-	-	-	-	18	...	58	95	66
South Africa	...	...	...	...	...	...	...	...	...	7
Spain	-	-	-	-	-	-	-	-	-	308
51 Fishing area total	...	...	...	...	...	*18*	...	*58*	*95*	*381*
67 USA	...	...	...	...	...	3	-	0	0	0
67 Fishing area total	...	...	...	...	...	*3*	-	*0*	*0*	*0*
77 Costa Rica	91	80	51	39	32	91	56	8	10	13
Estonia	-	-	-	-	-	-	2	-	-	-
Fr Polynesia	...	...	...	...	...	...	...	27	53	41
USA	...	...	...	...	...	93	61	79	46	81
77 Fishing area total	*91*	*80*	*51*	*39*	*32*	*184*	*119*	*114*	*109*	*135*
81 New Zealand	...	...	33	52	40	74	110	208	327	238
Ukraine	-	-	-	-	-	-	-	-	-	1
81 Fishing area total	...	...	*33*	*52*	*40*	*74*	*110*	*208*	*327*	*239*
87 Chile	581	450	475	320	888	830	379	592	964	425
Spain	-	-	-	-	330	927	...	...	...	1 006
87 Fishing area total	*581*	*450*	*475*	*320*	*1 218*	*1 757*	*379*	*592*	*964*	*1 431*
Species total	*843 F*	*645*	*565*	*494*	*2 140*	*2 377*	*1 215*	*2 140 F*	*2 170*	*4 875*
Longfin mako	**Petite taupe**		**Marrajo carite**		***Isurus paucus***			**1,06(08)002,03**		**LMA**
21 USA	0	5	0	-	-	-	-	1	0	1
21 Fishing area total	*0*	*5*	*0*	-	-	-	-	*1*	*0*	*1*
31 USA	-	-	-	-	1	1	-	3	3	-
31 Fishing area total	-	-	-	-	*1*	*1*	-	*3*	*3*	-
Species total	*0*	*5*	*0*	-	*1*	*1*	-	*4*	*3*	*1*
Mako sharks	**Taupes**		**Marrajos**		***Isurus spp***			**1,06(08)002,XX**		**MAK**
21 USA	-	-	-	-	-	-	-	-	47	37
21 Fishing area total	-	-	-	-	-	-	-	-	*47*	*37*
34 Côte dIvoire	7	17	12	...	92	38	...	...	...	...
Liberia	...	...	...	...	...	...	...	116	...	...
34 Fishing area total	*7*	*17*	*12*	...	*92*	*38*	...	*116*	...	...
77 USA	-	-	-	-	-	-	-	-	-	80
77 Fishing area total	-	-	-	-	-	-	-	-	-	*80*
Species total	*7*	*17*	*12*	...	*92*	*38*	...	*116*	*47*	*117*
Porbeagle	**Requin-taupe commun**		**Marrajo sardinero**		***Lamna nasus***			**1,06(08)003,01**		**POR**
21 Canada	-	-	1 388	1 029	1 343	1 006	958	904	499	238
Faeroe Is	250	-	-	-	-	-	-	-	-	-
Portugal	-	-	-	-	-	-	-	-	-	8
St Pier Mq	-	-	7	40	13	20	0	23	2	1

B-38 Sharks, rays, chimaeras / Squales, raies, chimères / Tiburones, rayas, quimeras

Capture production by species, fishing areas and countries or areas / Captures par espèces, zones de pêche et pays ou zones / Capturas por especies, áreas de pesca y países o áreas

Species, Fishing area Espèce, Zone de pêche Especie, Area de pesca	1993 t	1994 t	1995 t	1996 t	1997 t	1998 t	1999 t	2000 t	2001 t	2002 t
USA	39	64	0	-	-	-	-	3	1	1
21 Fishing area total	*289*	*64*	*1 395*	*1 069*	*1 356*	*1 026*	*958*	*930*	*502*	*248*
27 Channel Is	...	...	...	...	...	1	0	2	2	2
Denmark	91	94	86	71	69	85	107	73	76	42
Faeroe Is	76	48	44	7	9	7	10	13	8	10
France	633	820	565	267	315	219	318	410	368	461
Germany	-	22	-	-	-	-	0	17	1	3
Iceland	3	4	6	5	3	4	2	2	3	2
Ireland	-	-	-	-	-	-	8	1	6	3
Norway	24	25	27	28	17	28	33	22	17	19
Portugal	0	0	0	0	0	0	0	6	2	-
Spain	-	-	-	31	124	679	1 001	1 184	1 007	-
Sweden	3	2	2	1	1	1	1	1	1	0
UK	-	-	-	-	-	-	6	6	10	7
27 Fishing area total	*830*	*1 015*	*730*	*410*	*538*	*1 024*	*1 486*	*1 737*	*1 501*	*549*
34 Portugal	...	...	...	...	...	...	...	10	3	8
34 Fishing area total	*...*	*...*	*...*	*...*	*...*	*...*	*...*	*10*	*3*	*8*
37 Malta	0	0	0	1	0	0	0	0	0	-
Spain	-	-	-	-	-	-	-	-	1	-
37 Fishing area total	*0*	*0*	*0*	*1*	*0*	*0*	*0*	*0*	*1*	-
51 Spain	-	-	-	-	-	-	-	-	1	-
51 Fishing area total	-	-	-	-	-	-	-	-	*1*	-
58 Australia	-	-	-	-	2	-	-	-	-	-
58 Fishing area total	-	-	-	-	*2*	-	-	-	-	-
81 New Zealand	...	...	5	16	21	164	246	188	127	130
Ukraine	-	-	-	-	-	-	-	-	-	8
81 Fishing area total	*...*	*...*	*5*	*16*	*21*	*164*	*246*	*188*	*127*	*138*
87 Spain	-	-	-	-	...	7	...	...	...	65
87 Fishing area total	-	-	-	-	*...*	*7*	*...*	*...*	*...*	*65*
Species total	*1 119*	*1 079*	*2 130*	*1 496*	*1 917*	*2 221*	*2 690*	*2 865*	*2 135*	*1 008*
Great white shark	**Grand requin blanc**		**Jaquetón blanco**		***Carcharodon carcharias***				**1,06(08)007,01**	**WSH**
21 USA	-	-	-	-	-	-	-	1	0	-
21 Fishing area total	-	-	-	-	-	-	-	*1*	*0*	-
77 USA	-	-	-	-	-	-	-	1	0	-
77 Fishing area total	-	-	-	-	-	-	-	*1*	*0*	-
Species total	-	-	-	-	-	-	-	*2*	*0*	-
Nurse shark	**Requin-nourrice**		**Gata nodriza**		***Ginglymostoma cirratum***				**1,07(03)009,01**	**GNC**
21 USA	-	0	214	-	-	-	-	0	-	0
21 Fishing area total	-	*0*	*214*	-	-	-	-	*0*	-	*0*
31 Dominican Rp	-	-	-	-	-	-	-	407	89	17
31 Fishing area total	-	-	-	-	-	-	-	*407*	*89*	*17*
Species total	-	*0*	*214*	-	-	-	-	*407*	*89*	*17*
Blackmouth catshark	**Chien espagnol**		**Pintarroja bocanegra**		***Galeus melastomus***				**1,08(01)001,04**	**SHO**
27 Spain	...	...	...	...	...	...	...	...	...	230
27 Fishing area total	*...*	*...*	*...*	*...*	*...*	*...*	*...*	*...*	*...*	*230*
37 Spain	...	...	...	...	...	...	...	...	...	58
37 Fishing area total	*...*	*...*	*...*	*...*	*...*	*...*	*...*	*...*	*...*	*58*
Species total	*...*	*...*	*...*	*...*	*...*	*...*	*...*	*...*	*...*	*288*
Small-spotted catshark	**Petite roussette**		**Pintarroja**		***Scyliorhinus canicula***				**1,08(01)003,01**	**SYC**
27 France	5 019	4 975	5 022	4 871	5 338	5 480	5 707	5 914	6 253	5 706
Ireland	...	...	...	...	...	...	...	...	633	564
Spain	...	...	...	...	...	...	...	...	...	6
UK	...	...	341	273	275	260	111	238	155	162
27 Fishing area total	*5 019*	*4 975*	*5 363*	*5 144*	*5 613*	*5 740*	*5 818*	*6 152*	*7 041*	*6 438*
37 France	...	...	...	...	...	...	...	30	31	33
37 Fishing area total	*...*	*...*	*...*	*...*	*...*	*...*	*...*	*30*	*31*	*33*
Species total	*5 019*	*4 975*	*5 363*	*5 144*	*5 613*	*5 740*	*5 818*	*6 182*	*7 072*	*6 471*
Nursehound	**Grande roussette**		**Alitán**		***Scyliorhinus stellaris***				**1,08(01)003,02**	**SYT**
27 France	370	196	174	197	292	181	135	159	180	156
UK	...	...	69	109	86	77	139	115	84	39
27 Fishing area total	*370*	*196*	*243*	*306*	*378*	*258*	*274*	*274*	*264*	*195*
Species total	*370*	*196*	*243*	*306*	*378*	*258*	*274*	*274*	*264*	*195*

B-38 Sharks, rays, chimaeras / Squales, raies, chimères / Tiburones, rayas, quimeras

Capture production by species, fishing areas and countries or areas / Captures par espèces, zones de pêche et pays ou zones / Capturas por especies, áreas de pesca y países o áreas

Species, Fishing area Espèce, Zone de pêche Especie, Area de pesca	1993 t	1994 t	1995 t	1996 t	1997 t	1998 t	1999 t	2000 t	2001 t	2002 t
Catsharks, nursehounds nei	**Roussettes nca**		**Alitanes, pintarrojas nep**		***Scyliorhinus spp***				**1,08(01)003,XX**	**SCL**
27 France	1	7	8	18	6	51	13	68	7	7
27 Fishing area total	*1*	*7*	*8*	*18*	*6*	*51*	*13*	*68*	*7*	*7*
37 Spain	-	-	-	-	-	-	213	331	379	185
Tunisia	78	84	48	36	72	...	49	126	122	139
37 Fishing area total	*78*	*84*	*48*	*36*	*72*	*...*	*262*	*457*	*501*	*324*
Species total	*79*	*91*	*56*	*54*	*78*	*51*	*275*	*525*	*508*	*331*
Dusky catshark	**Holbiche sombre**		**Pejegato oscuro**		***Halaelurus canescens***				**1,08(01)030,04**	**HAN**
48 UK	-	-	-	-	-	-	-	-	-	0
48 Fishing area total	-	-	-	-	-	-	-	-	-	*0*
Species total	-	-	-	-	-	-	-	-	-	*0*
Blue shark	**Peau bleue**		**Tiburón azul**		***Prionace glauca***				**1,08(02)004,01**	**BSH**
21 Portugal	...	...	...	...	...	...	...	169	-	-
21 Fishing area total	...	...	...	...	...	...	...	*169*	-	-
27 Channel Is	...	...	...	...	...	1	0	-	-	-
Denmark	0	-	-	3	1	1	1	2	1	13
France	322	350	266	278	213	163	230	395	205	109
Ireland	-	-	-	-	-	-	67	23	66	11
Portugal	...	...	...	...	...	...	887	21	1 006	1 209
UK	-	-	-	-	-	-	-	12	9	6
27 Fishing area total	*322*	*350*	*266*	*281*	*214*	*165*	*1 185*	*453*	*1 287*	*1 348*
31 Portugal	-	-	-	-	-	17	-	-	-	8
Spain	-	-	-	-	1 700	418	...	...	...	...
31 Fishing area total	-	-	-	-	*1 700*	*435*	...	...	...	*8*
34 Liberia	...	...	...	...	...	...	76	70	...	...
Portugal	...	...	...	...	...	...	...	351	557	668
Spain	-	-	-	-	-	-	-	-	-	4 601
34 Fishing area total	...	...	...	...	...	...	*76*	*421*	*557*	*5 269*
37 France	...	...	...	...	...	...	3	1	2	0
Portugal	-	-	-	-	-	-	-	3	40	14
Spain	-	-	-	-	-	-	-	-	-	2
37 Fishing area total	...	...	...	...	...	...	*3*	*4*	*42*	*16*
41 Brazil	800 F	...	...	743	1 103	...	500	580	...	2 000
Portugal	...	...	...	...	...	...	...	56	988	1 015
Spain	-	-	-	-	-	-	-	-	-	1 218
41 Fishing area total	*800 F*	...	...	*743*	*1 103*	...	*500*	*636*	*988*	*4 233*
47 Namibia	...	...	...	...	...	...	...	...	...	794
Portugal	-	-	21	...	...	8	18	1 908	949	513
South Africa	...	...	...	...	...	...	...	...	1	73
Spain	-	-	-	-	3 560	1 463	2 233	2 803	2 784	1 134
47 Fishing area total	...	...	*21*	...	*3 560*	*1 471*	*2 251*	*4 711*	*3 734*	*2 514*
51 Portugal	-	-	-	-	-	60	...	575	1 123	1 148
South Africa	...	...	...	...	...	...	...	...	...	21
Spain	-	-	-	-	-	-	-	-	-	2 135
51 Fishing area total	...	...	...	...	...	*60*	...	*575*	*1 123*	*3 304*
67 USA	-	-	-	-	-	-	-	1	2	0
67 Fishing area total	-	-	-	-	-	-	-	*1*	*2*	*0*
77 USA	...	...	...	...	...	1	-	0	0	1
77 Fishing area total	...	...	...	...	...	*1*	-	*0*	*0*	*1*
81 New Zealand	...	...	111	246	120	540	593	1 169	1 328	1 186
81 Fishing area total	...	...	*111*	*246*	*120*	*540*	*593*	*1 169*	*1 328*	*1 186*
87 Chile	237	33	39	11	114	10	7	262	445	605
Spain	-	-	-	-	...	814	...	...	11	1 886
87 Fishing area total	*237*	*33*	*39*	*11*	*114*	*824*	*7*	*262*	*456*	*2 491*
Species total	*1 359 F*	*383*	*437*	*1 281*	*6 811*	*3 496*	*4 615*	*8 401*	*9 517*	*20 370*
Sandbar shark	**Requin gris**		**Tiburón trozo**		***Carcharhinus plumbeus***				**1,08(02)010,01**	**CCP**
21 USA	31	24	1	-	-	-	-	41	24	28
21 Fishing area total	*31*	*24*	*1*	-	-	-	-	*41*	*24*	*28*
Species total	*31*	*24*	*1*	-	-	-	-	*41*	*24*	*28*
Blacktip shark	**Requin bordé**		**Tiburón macuira**		***Carcharhinus limbatus***				**1,08(02)010,03**	**CCL**
21 USA	-	7	-	-	-	-	-	21	1	11
21 Fishing area total	-	*7*	-	-	-	-	-	*21*	*1*	*11*
31 Trinidad Tob	...	...	...	3	8	10	11	...	...	...
USA	-	-	-	-	1	0	-	580	520	-

B-38 Sharks, rays, chimaeras / Squales, raies, chimères / Tiburones, rayas, quimeras

Capture production by species, fishing areas and countries or areas / Captures par espèces, zones de pêche et pays ou zones / Capturas por especies, áreas de pesca y países o áreas

Species, Fishing area Espèce, Zone de pêche Especie, Area de pesca	1993 t	1994 t	1995 t	1996 t	1997 t	1998 t	1999 t	2000 t	2001 t	2002 t
31 Fishing area total	...	...	...	3	9	10	11	580	520	...
Species total	...	7	...	3	9	10	11	601	521	11
Dusky shark	**Requin de sable**		**Tiburón arenero**		***Carcharhinus obscurus***				**1,08(02)010,16**	**DUS**
21 USA	23	20	0	-	-	-	-	80	0	3
21 Fishing area total	23	20	0	-	-	-	-	80	0	3
51 South Africa	-	-	-	-	7	0	...	...	...	...
51 Fishing area total	-	-	-	-	7	0	...	...	...	...
Species total	23	20	0	-	7	0	...	80	0	3
Silky shark	**Requin soyeux**		**Tiburón jaquetón**		***Carcharhinus falciformis***				**1,08(02)010,17**	**FAL**
31 Costa Rica	23	4	11	5	0	24	49	63	59	39
Spain	-	-	-	-	-	-	-	-	-	17
31 Fishing area total	23	4	11	5	0	24	49	63	59	56
34 Côte dIvoire	4	13	18	...	2	...	...	...	...	...
Liberia	...	...	...	...	...	...	110	99	...	...
Senegal	...	...	...	...	...	...	...	...	...	1 341
Spain	-	-	-	-	-	-	-	-	-	14
34 Fishing area total	4	13	18	...	2	...	110	99	...	1 355
41 Brazil	...	...	...	502	279	...	70	80	...	...
41 Fishing area total	...	...	...	502	279	...	70	80	...	...
57 Sri Lanka	21 800	25 400	21 400	21 000	15 000	20 875	20 700	14 130	15 870	18 510
57 Fishing area total	21 800	25 400	21 400	21 000	15 000	20 875	20 700	14 130	15 870	18 510
77 Costa Rica	993	1 500	1 609	1 541	1 595	2 097	2 130	1 678	1 031	1 484
77 Fishing area total	993	1 500	1 609	1 541	1 595	2 097	2 130	1 678	1 031	1 484
Species total	22 820	26 917	23 038	23 048	16 876	22 996	23 059	16 050	16 960	21 405
Copper shark	**Requin cuivre**		**Tiburón cobrizo**		***Carcharhinus brachyurus***				**1,08(02)010,20**	**BRO**
81 New Zealand	...	...	...	...	...	15	14	25	38	38
81 Fishing area total	...	...	...	...	...	15	14	25	38	38
Species total	...	...	...	...	...	15	14	25	38	38
Tiger shark	**Requin tigre commun**		**Tintorera tigre**		***Galeocerdo cuvier***				**1,08(02)017,03**	**TIG**
21 USA	-	-	-	-	-	-	-	-	1	0
21 Fishing area total	-	-	-	-	-	-	-	-	1	0
27 Netherlands	-	-	-	-	-	-	-	-	-	13
27 Fishing area total	-	-	-	-	-	-	-	-	-	13
Species total	-	-	-	-	-	-	-	-	1	13
Requiem sharks nei	**Requins nca**		**Cazones picudos,tintoreras nep**		***Carcharhinidae***				**1,08(02)XXX,XX**	**RSK**
31 Mexico	6 426	5 544	4 741	5 166	3 785	3 991	3 281	3 108	3 196	3 022
Nicaragua	...	...	...	...	...	...	40	40	175	23
Venezuela	6 101	6 656	7 468	6 979	6 000	4 597	2 973	3 343	2 536	5 663
31 Fishing area total	12 527	12 200	12 209	12 145	9 785	8 588	6 294	6 491	5 907	8 708
51 Eritrea	...	13	6	15	13	17	38	130	110	144
Mozambique	-	-	165	21	...	...	...	...	...	...
Pakistan	28 780	30 226	32 288	34 447	31 179	35 357	32 535	28 245	26 524	27 000
Qatar	0	0	0	0	-	0	0	0	0	0
Spain	-	-	-	-	43	810	-	9	8	57
51 Fishing area total	28 780	30 239	32 459	34 483	31 235	36 184	32 573	28 384	26 642	27 201
77 Mexico	6 764	5 985	6 334	5 849	3 482	2 988	2 789	3 210	2 859	2 450
Nicaragua	...	...	...	...	...	...	160	110	200	269
77 Fishing area total	6 764	5 985	6 334	5 849	3 482	2 988	2 949	3 320	3 059	2 719
Species total	48 071	48 424	51 002	52 477	44 502	47 760	41 816	38 195	35 608	38 628
Smooth hammerhead	**Requin-marteau commun**		**Cornuda cruz(=Pez martillo)**		***Sphyrna zygaena***				**1,08(03)005,01**	**SPZ**
27 Portugal	...	...	...	...	...	...	8	8	4	5
27 Fishing area total	...	...	...	...	...	...	8	8	4	5
34 Portugal	...	...	...	...	...	...	...	7	1	13
Spain	-	-	-	-	-	-	-	-	-	9
34 Fishing area total	...	...	...	...	...	...	...	7	1	22
41 Portugal	...	...	...	...	...	...	...	3	-	-
41 Fishing area total	...	...	...	...	...	...	...	3	-	-
47 Portugal	...	...	...	...	...	...	...	4	5	3
Spain	-	-	-	-	220	103	-	-	-	-
47 Fishing area total	...	...	...	...	220	103	...	4	5	3

B-38 Sharks, rays, chimaeras / Squales, raies, chimères / Tiburones, rayas, quimeras

Capture production by species, fishing areas and countries or areas / Captures par espèces, zones de pêche et pays ou zones / Capturas por especies, áreas de pesca y países o áreas

Species, Fishing area Espèce, Zone de pêche Especie, Area de pesca	1993 t	1994 t	1995 t	1996 t	1997 t	1998 t	1999 t	2000 t	2001 t	2002 t
81 New Zealand	...	...	12	10	3	6	11	13	17	10
81 Fishing area total	*...*	*...*	*12*	*10*	*3*	*6*	*11*	*13*	*17*	*10*
Species total	*...*	*...*	*12*	*10*	*223*	*109*	*19*	*35*	*27*	*40*
Scalloped hammerhead	**Requin-marteau halicorne**		**Cornuda común**		***Sphyrna lewini***				**1,08(03)005,06**	**SPL**
34 GuineaBissau	...	2	12	12	10 F	10 F	10 F	10 F	10 F	10 F
Spain	-	-	-	-	-	-	-	-	-	257
34 Fishing area total	*...*	*2*	*12*	*12*	*10 F*	*10 F*	*10 F*	*10 F*	*10 F*	*267 F*
41 Brazil	100 F	...	...	25	170	...	30	38	507	508
Spain	-	-	-	-	-	-	-	-	-	33
41 Fishing area total	*100 F*	*...*	*...*	*25*	*170*	*...*	*30*	*38*	*507*	*541*
Species total	*100 F*	*2*	*12*	*37*	*180 F*	*10 F*	*40 F*	*48 F*	*517 F*	*808 F*
Hammerhead sharks, etc. nei	**Requins marteau, etc. nca**		**Cornudas, etc. nep**		***Sphyrnidae***				**1,08(03)XXX,XX**	**SPY**
31 Spain	-	-	-	-	3	2	...	...	...	...
31 Fishing area total	*-*	*-*	*-*	*-*	*3*	*2*	*...*	*...*	*...*	*...*
34 Congo Rep	...	...	...	...	...	...	...	500 F	500 F	500 F
Côte dIvoire	55	66	69	...	190	125	...	...	...	...
Liberia	...	...	...	...	...	...	127	152	...	...
Senegal	-	-	-	-	-	156	20	1 305	1 451	1 038
Spain	...	...	...	...	805	739	...	...	...	...
34 Fishing area total	*55*	*66*	*69*	*...*	*995*	*1 020*	*147*	*1 957 F*	*1 951 F*	*1 538 F*
77 USA	...	...	...	...	...	1	-	-	-	-
77 Fishing area total	*...*	*...*	*...*	*...*	*...*	*1*	*-*	*-*	*-*	*-*
87 Spain	-	-	-	-	...	5	...	...	...	2
87 Fishing area total	*-*	*-*	*-*	*-*	*...*	*5*	*...*	*...*	*...*	*2*
Species total	*55*	*66*	*69*	*...*	*998*	*1 028*	*147*	*1 957 F*	*1 951 F*	*1 540 F*
Dusky smooth-hound	**Emissole douce**		**Boca dulce**		***Mustelus canis***				**1,08(04)007,03**	**CTI**
21 USA	...	280	0	...	...	...	...	334	321	493
21 Fishing area total	*...*	*280*	*0*	*...*	*...*	*...*	*...*	*334*	*321*	*493*
Species total	*...*	*280*	*0*	*...*	*...*	*...*	*...*	*334*	*321*	*493*
Brown smooth-hound	**Emissole brune**		**Musola parda**		***Mustelus henlei***				**1,08(04)007,07**	**CTK**
77 USA	...	...	...	...	...	3	5	3	4	3
77 Fishing area total	*...*	*...*	*...*	*...*	*...*	*3*	*5*	*3*	*4*	*3*
Species total	*...*	*...*	*...*	*...*	*...*	*3*	*5*	*3*	*4*	*3*
Spotted estuary smooth-hound	**Emissole grivelée**		**Musola manchada**		***Mustelus lenticulatus***				**1,08(04)007,09**	**MTL**
81 New Zealand	1 701	1 806	2 787	1 350	3 464	1 707	1 662	1 643	1 563	1 403
81 Fishing area total	*1 701*	*1 806*	*2 787*	*1 350*	*3 464*	*1 707*	*1 662*	*1 643*	*1 563*	*1 403*
Species total	*1 701*	*1 806*	*2 787*	*1 350*	*3 464*	*1 707*	*1 662*	*1 643*	*1 563*	*1 403*
Narrownose smooth-hound	**Emissole gatuso**		**Gatuso**		***Mustelus schmitti***				**1,08(04)007,12**	**SDP**
41 Argentina	11 070	11 450	11 057	10 252	9 956	11 266	9 062	7 120	9 613	7 019
Uruguay	329	319	286	204	174	2 156	3 212	1 037	1 153	1 121
41 Fishing area total	*11 399*	*11 769*	*11 343*	*10 456*	*10 130*	*13 422*	*12 274*	*8 157*	*10 766*	*8 140*
Species total	*11 399*	*11 769*	*11 343*	*10 456*	*10 130*	*13 422*	*12 274*	*8 157*	*10 766*	*8 140*
Smooth-hound	**Emissole lisse**		**Musola**		***Mustelus mustelus***				**1,08(04)007,13**	**SMD**
27 UK	-	-	-	-	-	-	-	15	-	56
27 Fishing area total	*-*	*-*	*-*	*-*	*-*	*-*	*-*	*15*	*-*	*56*
47 South Africa	...	...	...	...	...	...	...	...	...	2
47 Fishing area total	*...*	*...*	*...*	*...*	*...*	*...*	*...*	*...*	*...*	*2*
Species total	*...*	*...*	*...*	*...*	*...*	*...*	*...*	*15*	*...*	*58*
Smooth-hounds nei	**Emissoles nca**		**Tollos nep**		***Mustelus spp***				**1,08(04)007,XX**	**SDV**
27 France	298	352	414	578	624	749	824	1 050	1 249	1 586
Portugal	...	...	...	...	...	...	76	41	43	42
27 Fishing area total	*298*	*352*	*414*	*578*	*624*	*749*	*900*	*1 091*	*1 292*	*1 628*
31 Colombia	307	102	46	253	27	45	3	30 F	30 F	30 F
31 Fishing area total	*307*	*102*	*46*	*253*	*27*	*45*	*3*	*30 F*	*30 F*	*30 F*
34 Greece	75	64	102	87	24	43	23	39	21	29
Portugal	86	66	97	187	27	14	5	-	-	6
Senegal	...	...	...	...	...	...	...	464	1 908	2 064
Spain	-	-	-	118	-	-	-	-	-	-
34 Fishing area total	*161*	*130*	*199*	*392*	*51*	*57*	*28*	*503*	*1 929*	*2 099*

B-38 Sharks, rays, chimaeras / Squales, raies, chimères / Tiburones, rayas, quimeras

Capture production by species, fishing areas and countries or areas / Captures par espèces, zones de pêche et pays ou zones / Capturas por especies, áreas de pesca y países o áreas

Species, Fishing area Espèce, Zone de pêche Especie, Area de pesca	1993 t	1994 t	1995 t	1996 t	1997 t	1998 t	1999 t	2000 t	2001 t	2002 t
37 Albania	...	...	20	12	3	12	12	32	5	23
France	0	0	0	0	-	-	0	-	-	-
Greece	267	377	360	353	493	316	553	578	351	330 F
Italy	4 675	9 999	5 942	2 659	621	636	440	462	369	325
Palest, O.T.	...	...	...	24	11	9	10 F	10 F	5 F	2
Slovenia	-	-	-	-	-	-	-	2	4	2
Spain	-	-	-	-	-	-	21	15	19	12
Syria	40 F	39 F	39 F	50	-	-	-	-	-	-
Tunisia	187	142	128	640	132	826	997	121	192	186
Turkey	1 436	2 880	1 783	2 158	1 720	1 450	1 625	2 880	1 000	686
37 Fishing area total	*6 605 F*	*13 437 F*	*8 272 F*	*5 896*	*2 980*	*3 249*	*3 658 F*	*4 100 F*	*1 945 F*	*1 566 F*
57 Australia	...	4 626	3 911	3 878	4 169	2 858	2 462	1 796	2 485	2 415
57 Fishing area total	...	*4 626*	*3 911*	*3 878*	*4 169*	*2 858*	*2 462*	*1 796*	*2 485*	*2 415*
81 Australia	...	...	...	...	...	...	...	100	87	109
81 Fishing area total	...	...	...	...	...	...	...	*100*	*87*	*109*
87 Chile	398	588	193	225	108	56	208	143	128	57
Colombia	316	365	162	754	408	316	385	330 F	281	280 F
Peru	8 747	3 431	4 125	3 230	3 166	8 038	2 892	4 042	4 648	7 015
87 Fishing area total	*9 461*	*4 384*	*4 480*	*4 209*	*3 682*	*8 410*	*3 485*	*4 515 F*	*5 057*	*7 352 F*
Species total	*16 832 F*	*23 031 F*	*17 322 F*	*15 206*	*11 533*	*15 368*	*10 536 F*	*12 135 F*	*12 825 F*	*15 199 F*
Tope shark	**Requin-hâ**		**Cazón**		***Galeorhinus galeus***				**1,08(04)011,03**	**GAG**
27 Denmark	-	-	-	2	2	3	4	7	4	4
France	291	301	317	403	454	369	386	450	469	349
Ireland	...	...	...	...	...	...	...	...	4	1
UK	62	71	63	53	55	55	74	110	82	73
27 Fishing area total	*353*	*372*	*380*	*458*	*511*	*427*	*464*	*567*	*559*	*427*
34 Portugal	...	...	...	...	...	...	...	2	1	2
34 Fishing area total	...	...	...	...	...	...	...	*2*	*1*	*2*
41 Argentina	230	75	104	92	103	92	89	109	50	29
Spain	-	-	-	-	-	-	-	-	37	6
41 Fishing area total	*230*	*75*	*104*	*92*	*103*	*92*	*89*	*109*	*87*	*35*
47 South Africa	...	...	...	...	...	...	...	...	...	19
47 Fishing area total	...	...	...	...	...	...	...	...	...	*19*
57 Australia	...	...	...	...	...	...	...	498	325	350
57 Fishing area total	...	...	...	...	...	...	...	*498*	*325*	*350*
67 USA	...	...	...	...	...	1	-	3	1	0
67 Fishing area total	...	...	...	...	...	*1*	-	*3*	*1*	*0*
77 USA	...	...	...	...	...	51	73	45	44	32
77 Fishing area total	...	...	...	...	...	*51*	*73*	*45*	*44*	*32*
81 New Zealand	2 577	2 618	3 705	3 044	2 864	3 083	3 633	3 100	3 091	3 315
Ukraine	-	-	-	-	-	-	-	-	-	1
81 Fishing area total	*2 577*	*2 618*	*3 705*	*3 044*	*2 864*	*3 083*	*3 633*	*3 100*	*3 091*	*3 316*
Species total	*3 160*	*3 065*	*4 189*	*3 594*	*3 478*	*3 654*	*4 259*	*4 324*	*4 108*	*4 181*
Greenland shark	**Laimargue du Groenland**		**Tollo de Groenlandia**		***Somniosus microcephalus***				**1,09(01)002,01**	**GSK**
27 Iceland	41	42	44	61	73	87	51	45	57	56
Portugal	9	1	11	0	0	-	0	0	1	0
27 Fishing area total	*50*	*43*	*55*	*61*	*73*	*87*	*51*	*45*	*58*	*56*
58 South Africa	-	-	-	-	-	-	-	-	-	1
58 Fishing area total	-	-	-	-	-	-	-	-	-	*1*
Species total	*50*	*43*	*55*	*61*	*73*	*87*	*51*	*45*	*58*	*57*
Little sleeper shark	**Laimargue de la Méditerranée**		**Tollo boreal**		***Somniosus rostratus***				**1,09(01)002,02**	**SOR**
27 Spain	...	...	...	...	...	...	...	...	...	2
27 Fishing area total	...	...	...	...	...	...	...	...	...	*2*
Species total	...	...	...	...	...	...	...	...	...	*2*
Pacific sleeper shark	**Laimargue dormeur**		**Tollo negro dormilón**		***Somniosus pacificus***				**1,09(01)002,03**	**SON**
58 Australia	-	-	-	-	-	-	1	-	-	3
58 Fishing area total	-	-	-	-	-	-	*1*	-	-	*3*
Species total	-	-	-	-	-	-	*1*	-	-	*3*
Picked dogfish	**Aiguillat commun**		**Mielga**		***Squalus acanthias***				**1,09(01)007,04**	**DGS**
21 Canada	1 271	1 690	957	431	446	1 081	2 456	2 828	3 760	3 589
Cuba	-	-	-	-	6	-	-	-	-	-
St Pier Mq	0	0	0	0	0	-	-	0	0	0
Spain	-	-	-	63	0	0	-	-	1	0

B-38

Sharks, rays, chimaeras — **Capture production by species, fishing areas and countries or areas**
Squales, raies, chimères — **Captures par espèces, zones de pêche et pays ou zones**
Tiburones, rayas, quimeras — **Capturas por especies, áreas de pesca y países o áreas**

Species, Fishing area Espèce, Zone de pêche Especie, Area de pesca	1993 t	1994 t	1995 t	1996 t	1997 t	1998 t	1999 t	2000 t	2001 t	2002 t
USA	...	...	128	...	...	...	...	7 873	2 234	2 108
21 Fishing area total	*1 271*	*1 690*	*1 085*	*494*	*452*	*1 081*	*2 456*	*10 701*	*5 995*	*5 697*
27 Belgium	47	21	14	16	15	17	10	11	13	23
Denmark	486	211	146	142	196	126	131	146	156	256
Faeroe Is	...	...	308	51	212	356	484	354	613	337
France	1 911	1 661	1 349	1 719	1 708	1 410	1 192	1 097	1 333	1 126
Germany	8	-	-	-	-	-	45	188	303	119
Iceland	109	97	166	157	106	78	57	109	136	276
Ireland	3 424	3 624	2 435	2 095	1 407	1 259	962	880	1 301	1 293
Netherlands	-	-	-	-	-	-	-	28	39	27
Norway	6 945	4 546	3 939	2 749	1 567	1 293	1 461	1 644	1 424	1 126
Portugal	5	7	5	2	2	2	21	2	3	4
Spain	-	-	-	-	0	27	94	372	363	359
Sweden	188	95	104	154	197	140	114	124	238	270
UK	10 032	8 072	10 815	9 423	8 691	8 926	7 527	7 138	7 306	5 170
27 Fishing area total	*23 155*	*18 334*	*19 281*	*16 508*	*14 101*	*13 634*	*12 098*	*12 093*	*13 228*	*10 386*
37 Bulgaria	12	12 F	80	64	40	28	25	102	126	100
France	21	19	...	7	7	7	5	3	2	2
Malta	33	29	24	28	28	23	18	19	17	24
Romania	6	3	7	-	-	-	-	-	-	-
Slovenia	4	2	4	0	0	1	1	-	-	-
Spain	...	...	...	...	...	...	...	9	8	8
Ukraine	409	148	67	44	20	38	94	71	134	97
37 Fishing area total	*485*	*213 F*	*182*	*143*	*95*	*97*	*143*	*204*	*287*	*231*
67 Canada	830	1 776	2 744	4 000	2 100	2 500	5 897	4 696	4 543	4 701
USA	...	...	...	...	...	1	542	667	638	990
67 Fishing area total	*830*	*1 776*	*2 744*	*4 000*	*2 100*	*2 501*	*6 439*	*5 363*	*5 181*	*5 691*
77 USA	3	0	1	-	0	5	24	8	3	17
77 Fishing area total	*3*	*0*	*1*	*-*	*0*	*5*	*24*	*8*	*3*	*17*
81 New Zealand	5 429	3 601	2 753	2 477	7 232	3 064	4 409	3 362	4 192	6 024
Ukraine	-	-	-	-	-	-	-	-	-	162
81 Fishing area total	*5 429*	*3 601*	*2 753*	*2 477*	*7 232*	*3 064*	*4 409*	*3 362*	*4 192*	*6 186*
Species total	*31 173*	*25 614 F*	*26 046*	*23 622*	*23 980*	*20 382*	*25 569*	*31 731*	*28 886*	*28 208*
Gulper shark	**Squale-chagrin commun**		**Quelvacho**		***Centrophorus granulosus***				**1,09(01)008,01**	**GUP**
27 Portugal	...	...	...	...	...	...	73	54	93	152
Spain	-	-	-	-	-	-	-	-	-	8
27 Fishing area total	...	...	...	...	...	...	*73*	*54*	*93*	*160*
37 Spain	...	...	...	...	...	...	...	...	...	1
37 Fishing area total	...	...	...	...	...	...	...	...	...	*1*
Species total	...	...	...	...	...	...	*73*	*54*	*93*	*161*
Leafscale gulper shark	**Squale-chagrin de l'Atlantique**		**Quelvacho negro**		***Centrophorus squamosus***				**1,09(01)008,03**	**GUQ**
27 Faeroe Is	-	3	51	53	58	129	11	...	...	...
France	-	-	-	-	-	-	-	-	-	48
Norway	-	-	-	-	-	-	-	-	1	-
Portugal	...	...	...	...	...	...	440	478	510	630
Spain	-	-	-	-	-	-	-	-	-	495
27 Fishing area total	...	*3*	*51*	*53*	*58*	*129*	*451*	*478*	*511*	*1 173*
34 Portugal	...	...	...	...	...	...	...	28	27	29
34 Fishing area total	...	...	...	...	...	...	...	*28*	*27*	*29*
81 New Zealand	...	...	...	...	...	4	1	0	0	1
81 Fishing area total	...	...	...	...	...	*4*	*1*	*0*	*0*	*1*
Species total	...	*3*	*51*	*53*	*58*	*133*	*452*	*506*	*538*	*1 203*
Lanternsharks nei	**Sagres nca**		**Tollos lucero nep**		***Etmopterus spp***				**1,09(01)010,XX**	**SHL**
27 Spain	...	...	...	...	...	...	573	...	...	99
27 Fishing area total	...	...	...	...	...	...	*573*	...	...	*99*
37 Spain	...	...	...	...	...	...	...	...	...	3
37 Fishing area total	...	...	...	...	...	...	...	...	...	*3*
81 New Zealand	...	...	3	0	2	-	-	-	4	25
81 Fishing area total	...	...	*3*	*0*	*2*	*-*	*-*	*-*	*4*	*25*
Species total	...	...	*3*	*0*	*2*	...	*573*	...	*4*	*127*
Birdbeak dogfish	**Squale savate**		**Tollo pajarito**		***Deania calcea***				**1,09(01)014,01**	**DCA**
27 Portugal	...	...	...	...	...	...	...	18	50	90
Spain	-	-	-	-	-	-	-	-	-	12
UK	-	-	-	-	-	-	-	-	1	-
27 Fishing area total	...	...	...	...	...	...	...	*18*	*51*	*102*
81 New Zealand	...	...	...	...	...	36	17	28	66	86
81 Fishing area total	...	...	...	...	...	*36*	*17*	*28*	*66*	*86*

B-38 Sharks, rays, chimaeras / Squales, raies, chimères / Tiburones, rayas, quimeras

Capture production by species, fishing areas and countries or areas / Captures par espèces, zones de pêche et pays ou zones / Capturas por especies, áreas de pesca y países o áreas

Species, Fishing area Espèce, Zone de pêche Especie, Area de pesca	1993 t	1994 t	1995 t	1996 t	1997 t	1998 t	1999 t	2000 t	2001 t	2002 t
Species total	...	...	...	...	...	*36*	*17*	*46*	*117*	*188*
Portuguese dogfish	**Pailona commun**		**Pailona**		***Centroscymnus coelolepis***				**1,09(01)016,01**	**CYO**
27 Faeroe Is	22	0	60	282	228	80	35	...	...	...
France	-	-	-	-	-	-	-	-	-	456
Iceland	1	0	-	-	-	5	0	0	-	-
Ireland	...	...	...	...	...	...	...	...	216	341
Norway	-	-	-	-	-	-	-	-	13	
Portugal	...	...	...	...	...	...	607	633	620	596
Spain	-	-	-	-	-	-	-	-	-	135
UK	-	-	-	54	52	147	75	514	1 663	1 456
27 Fishing area total	*23*	*0*	*60*	*336*	*280*	*232*	*717*	*1 147*	*2 512*	*2 984*
37 Portugal	-	-	-	-	-	-	-	7	23	2
37 Fishing area total	-	-	-	-	-	-	-	*7*	*23*	*2*
Species total	*23*	*0*	*60*	*336*	*280*	*232*	*717*	*1 154*	*2 535*	*2 986*
Longnose velvet dogfish	**Pailona à long nez**		**Sapata negra**		***Centroscymnus crepidater***				**1,09(01)016,02**	**CYP**
27 France	-	-	-	-	-	-	-	-	-	12
UK	-	-	-	-	-	3	-	-	-	-
27 Fishing area total	-	-	-	-	-	*3*	-	-	-	*12*
Species total	-	-	-	-	-	*3*	-	-	-	*12*
Kitefin shark	**Squale liche**		**Carocho**		***Dalatias licha***				**1,09(01)018,01**	**SCK**
27 Portugal	...	...	...	...	...	...	45	311	189	40
27 Fishing area total	...	...	...	...	...	...	*45*	*311*	*189*	*40*
81 New Zealand	...	...	303	175	352	434	328	317	375	520
81 Fishing area total	...	...	*303*	*175*	*352*	*434*	*328*	*317*	*375*	*520*
Species total	...	...	*303*	*175*	*352*	*434*	*373*	*628*	*564*	*560*
Black dogfish	**Aiguillat noir**		**Tollo negro merga**		***Centroscyllium fabricii***				**1,09(01)019,01**	**CFB**
27 France	-	-	-	-	-	-	-	269	271	28
Iceland	0	-	1	4	0	-	-	2	-	-
27 Fishing area total	*0*	-	*1*	*4*	*0*	-	-	*271*	*271*	*28*
Species total	*0*	-	*1*	*4*	*0*	-	-	*271*	*271*	*28*
Dogfishes and hounds nei	**Squales et émissoles nca**		**Galludos, tollos y musolas nep**		***Squalidae, Scyliorhinidae***				**1,09(01)XXX,XX**	**DGH**
27 Belgium	288	379	377	415	430	365	345	390	396	448
Channel Is	14 F	3	3 F	48	30	22	15	33	51	53
Portugal	1 815	1 570	1 594	1 341	1 376	1 266	754	803	810	762
UK	638	487	132	218	247	358	999	1 806	1 443	1 795
27 Fishing area total	*2 755 F*	*2 439*	*2 106 F*	*2 022*	*2 083*	*2 011*	*2 113*	*3 032*	*2 700*	*3 058*
Species total	*2 755 F*	*2 439*	*2 106 F*	*2 022*	*2 083*	*2 011*	*2 113*	*3 032*	*2 700*	*3 058*
Dogfish sharks nei	**Squales nca**		**Galludos, tollos, nep**		***Squalidae***				**1,09(01)XXX,XX**	**DGX**
21 Cuba	29	-	-	-	-	-	-	-	-	-
Germany	-	-	-	0	-	-	-	-	-	-
Greenland	11	34	47	134	...	...	...	-	-	-
Spain	-	-	-	138	211	608	550	493	692	475
UK	-	-	-	-	-	4	-	-		
USA	20 804	14 171	22 577	27 447	20 086	21 379	15 858	1 753	287	243
21 Fishing area total	*20 844*	*14 205*	*22 624*	*27 719*	*20 297*	*21 991*	*16 408*	*2 246*	*979*	*718*
27 Faeroe Is	141	225	12	7	8	8	3	...	...	...
France	3 610	3 421	3 125	3 135	2 811	2 288	3 077	4 224	4 194	1 952
Germany	-	-	-	19	12	16	235	271	433	519
Greenland	3	0	20	2	6	...	...	-	-	-
Ireland	...	...	1 676	1 170	917	1 144	683	4	30	14
Isle of Man	60	54	24	25	25	12	19	11	3	1
Lithuania	-	-	-	-	-	-	-	-	14	40
Norway	-	-	-	-	-	-	-	118	313	10
Portugal	1 210	889	1 137	977	999	905	-	-	-	-
Spain	-	-	-	534	1 226	627	321	419	365	171
UK	-	-	11	-	-	-	-	-	477	752
27 Fishing area total	*5 024*	*4 589*	*6 005*	*5 869*	*6 004*	*5 000*	*4 338*	*5 047*	*5 829*	*3 459*
31 USA	0	4 478	26	138	310	334	222	104	8	8
31 Fishing area total	*0*	*4 478*	*26*	*138*	*310*	*334*	*222*	*104*	*8*	*8*
34 Congo Rep	202 F	260	220 F	250 F	300 F	300 F	400 F	412 F	500 F	550 F
Greece	15	3	2	5	2	1	0	0	-	14
34 Fishing area total	*217 F*	*263*	*222 F*	*255 F*	*302 F*	*301 F*	*400 F*	*412 F*	*500 F*	*564 F*
37 Albania	...	...	1	64	13	-	-	-	10	77
Croatia	535	317	315	260	239	105	53	50	74	34
France	-	-	-	-	-	-	-	12	17	14
Greece	124	205	266	285	241	289	258	270	224	210 F

B-38

Sharks, rays, chimaeras — Capture production by species, fishing areas and countries or areas
Squales, raies, chimères — Captures par espèces, zones de pêche et pays ou zones
Tiburones, rayas, quimeras — Capturas por especies, áreas de pesca y países o áreas

Species, Fishing area Espèce, Zone de pêche Especie, Area de pesca	1993 t	1994 t	1995 t	1996 t	1997 t	1998 t	1999 t	2000 t	2001 t	2002 t
Malta	7	10	5	4	5	3	1	2	3	2
Serbia-Monte	2	3	7	10	10	8	9	9	7	8
Spain	-	-	-	4	5	-	68	11	20	19
Tunisia	860	674	596	19	806	44	25	680	883	945
37 Fishing area total	*1 528*	*1 209*	*1 190*	*646*	*1 319*	*449*	*414*	*1 034*	*1 238*	*1 309 F*
67 USA	1 957	2 324	1 300	2 053	625	555	1	-	-	-
67 Fishing area total	*1 957*	*2 324*	*1 300*	*2 053*	*625*	*555*	*1*	-	-	-
81 New Zealand	...	...	413	693	1 705	701	1 010	770	705	325
81 Fishing area total	...	...	*413*	*693*	*1 705*	*701*	*1 010*	*770*	*705*	*325*
Species total	*29 570 F*	*27 068*	*31 780 F*	*37 373 F*	*30 562 F*	*29 331 F*	*22 793 F*	*9 613 F*	*9 259 F*	*6 383 F*
Sawsharks nei	**Requins scies nca**		**Tiburónes sierra nep**		***Pristiophorus spp***				**1,09(02)004,XX**	**PWS**
57 Australia	...	...	...	...	...	...	...	270	423	371
57 Fishing area total	...	...	...	...	...	...	...	*270*	*423*	*371*
Species total	...	...	...	...	...	...	...	*270*	*423*	*371*
Angelshark	**Ange de mer commun**		**Angelote**		***Squatina squatina***				**1,09(03)004,01**	**AGN**
27 Estonia	-	-	-	-	-	-	-	-	-	3
27 Fishing area total	-	-	-	-	-	-	-	-	-	*3*
37 Tunisia	53	18	35	18	34	44	25	20	22	13
37 Fishing area total	*53*	*18*	*35*	*18*	*34*	*44*	*25*	*20*	*22*	*13*
Species total	*53*	*18*	*35*	*18*	*34*	*44*	*25*	*20*	*22*	*16*
Argentine angelshark	**Ange de mer argentin**		**Angelote argentino**		***Squatina argentina***				**1,09(03)004,04**	**SUG**
41 Argentina	3 975	3 622	3 802	4 281	4 410	4 311	3 368	3 123	3 339	2 288
41 Fishing area total	*3 975*	*3 622*	*3 802*	*4 281*	*4 410*	*4 311*	*3 368*	*3 123*	*3 339*	*2 288*
Species total	*3 975*	*3 622*	*3 802*	*4 281*	*4 410*	*4 311*	*3 368*	*3 123*	*3 339*	*2 288*
Angelsharks, sand devils nei	**Anges de mer nca**		**Angelotes, peces ángel nep**		***Squatinidae***				**1,09(03)XXX,XX**	**ASK**
27 France	2	2	2	1	0	0	1	1	1	-
UK	0	0	0	0	47	-	-	-	-	-
27 Fishing area total	*2*	*2*	*2*	*1*	*47*	*0*	*1*	*1*	*1*	-
37 Albania	...	...	0	53	20	31	30	30	16	79
Malta	0	0	0	0	0	0	0	0	0	
Turkey	13	15	31	42	15	140	70	60	20	18
37 Fishing area total	*13*	*15*	*31*	*95*	*35*	*171*	*100*	*90*	*36*	*97*
41 Brazil	...	...	113	1 587	...	...	...	...	...	...
41 Fishing area total	...	...	*113*	*1 587*	...	...	...	...	...	...
57 Australia	...	47	98	102	129	120	102	98	71	118
57 Fishing area total	...	*47*	*98*	*102*	*129*	*120*	*102*	*98*	*71*	*118*
87 Peru	228	159	289	358	189	101	262	406	510	477
87 Fishing area total	*228*	*159*	*289*	*358*	*189*	*101*	*262*	*406*	*510*	*477*
Species total	*243*	*223*	*533*	*2 143*	*400*	*392*	*465*	*595*	*618*	*692*
Angular roughshark	**Centrine commune**		**Cerdo marino**		***Oxynotus centrina***				**1,09(05)006,01**	**OXY**
27 Portugal	...	...	...	...	...	...	81	33	63	86
27 Fishing area total	...	...	...	...	...	...	*81*	*33*	*63*	*86*
Species total	...	...	...	...	...	...	*81*	*33*	*63*	*86*
Sailfin roughshark	**Humantin**		**Cerdo marino velero**		***Oxynotus paradoxus***				**1,09(05)006,02**	**OXN**
27 Spain	...	...	...	...	...	...	...	...	...	1
27 Fishing area total	...	...	...	...	...	...	...	...	...	*1*
Species total	...	...	...	...	...	...	...	...	...	*1*
Dogfish sharks, etc. nei	**Squaliformes nca**		**Squaliformes nep**		***Squaliformes***				**1,09(XX)XXX,XX**	**SHX**
21 Canada	1 122	1 000	293	106	169	115	143	120	92	107
USA	1 773	1 026	1 470	3 317	1 083	772	1 530	590	784	940
21 Fishing area total	*2 895*	*2 026*	*1 763*	*3 423*	*1 252*	*887*	*1 673*	*710*	*876*	*1 047*
Species total	*2 895*	*2 026*	*1 763*	*3 423*	*1 252*	*887*	*1 673*	*710*	*876*	*1 047*
Chola guitarfish	**Poisson-guitare chola**		**Guitarra chola**		***Rhinobatos percellens***				**1,10(01)005,09**	**GUD**
41 Brazil	1 110 F	1 110 F	162	404	...	...	...	...	...	...
41 Fishing area total	*1 110 F*	*1 110 F*	*162*	*404*	...	...	...	...	...	...
Species total	*1 110 F*	*1 110 F*	*162*	*404*	...	...	...	...	...	...

B-38 Sharks, rays, chimaeras / Squales, raies, chimères / Tiburones, rayas, quimeras

Capture production by species, fishing areas and countries or areas
Captures par espèces, zones de pêche et pays ou zones
Capturas por especies, áreas de pesca y países o áreas

Species, Fishing area Espèce, Zone de pêche Especie, Area de pesca	1993 t	1994 t	1995 t	1996 t	1997 t	1998 t	1999 t	2000 t	2001 t	2002 t
Pacific guitarfish	**Poisson-guitare du Pacifique**		**Guitarra del Pacífico**		***Rhinobatos planiceps***				**1,10(01)005,10**	**GUF**
87 Peru	89	...	121	460	333	344	95	2 624	1 060	822
87 Fishing area total	*89*	...	*121*	*460*	*333*	*344*	*95*	*2 624*	*1 060*	*822*
Species total	*89*	...	*121*	*460*	*333*	*344*	*95*	*2 624*	*1 060*	*822*
Guitarfishes, etc. nei	**Guitares, etc. nca**		**Guitarras, etc. nep**		***Rhinobatidae***				**1,10(01)XXX,XX**	**GTF**
34 Liberia	...	...	...	...	...	54	175	16	...	...
Senegal	...	...	...	...	...	171	59	1 930	1 772	1 062
34 Fishing area total	...	...	...	...	...	*225*	*234*	*1 946*	*1 772*	*1 062*
37 Albania	...	...	0	1	0	-	-	-	-	2
Greece	20	117	79	112	63	87	73	94	89	86 F
Palest, O.T.	...	...	...	...	6	6	5 F	5 F	5 F	1
37 Fishing area total	*20*	*117*	*79*	*113*	*69*	*93*	*78 F*	*99 F*	*94 F*	*89 F*
51 Eritrea	...	3	1	0	0	-	-	0	1	7
Pakistan	1 500	1 442	1 208	1 422	1 481	1 564	1 643	2 185	1 944	2 004
51 Fishing area total	*1 500*	*1 445*	*1 209*	*1 422*	*1 481*	*1 564*	*1 643*	*2 185*	*1 945*	*2 011*
Species total	*1 520*	*1 562*	*1 288*	*1 535*	*1 550*	*1 882*	*1 955 F*	*4 230 F*	*3 811 F*	*3 162 F*
Sawfishes	**Poissons-scies**		**Peces sierra**		***Pristidae***				**1,10(02)XXX,XX**	**SAW**
34 Liberia	...	...	...	...	48	...	39	42	...	...
Senegal	...	...	...	...	...	...	2	-	-	-
34 Fishing area total	...	...	...	...	*48*	...	*41*	*42*	...	...
41 Brazil	690 F	690 F	...	...	...	...	...	...	...	...
41 Fishing area total	*690 F*	*690 F*	...	...	...	...	...	...	...	...
51 Pakistan	32	28	23	-	-	-	-	-	-	-
51 Fishing area total	*32*	*28*	*23*	-	-	-	-	-	-	-
Species total	*722 F*	*718 F*	*23*	...	*48*	...	*41*	*42*	...	...
Blue skate	**Pocheteau gris**		**Noriega**		***Raja batis***				**1,10(04)001,01**	**RJB**
27 Denmark	-	-	-	32	9	7	11	47	0	0
France	245	239	285	295	314	296	467	653	667	447
Iceland	274	299	245	181	118	108	80	94	85	59
Norway	...	...	...	...	...	...	...	72	65	53
27 Fishing area total	*519*	*538*	*530*	*508*	*441*	*411*	*558*	*866*	*817*	*559*
Species total	*519*	*538*	*530*	*508*	*441*	*411*	*558*	*866*	*817*	*559*
Thornback ray	**Raie bouclée**		**Raya de clavos**		***Raja clavata***				**1,10(04)001,02**	**RJC**
27 France	1 745	1 577	1 749	1 756	1 579	1 343	1 290	1 222	1 214	1 160
Norway	-	-	-	-	-	-	-	2	-	-
27 Fishing area total	*1 745*	*1 577*	*1 749*	*1 756*	*1 579*	*1 343*	*1 290*	*1 224*	*1 214*	*1 160*
37 France	...	...	...	...	...	...	45	29	17	18
Ukraine	...	...	...	17	9	24	31	24	65	79
37 Fishing area total	...	...	...	*17*	*9*	*24*	*76*	*53*	*82*	*97*
Species total	*1 745*	*1 577*	*1 749*	*1 773*	*1 588*	*1 367*	*1 366*	*1 277*	*1 296*	*1 257*
Starry ray	**Raie radiée**		**Raya radiante**		***Raja radiata***				**1,10(04)001,03**	**RJR**
27 Iceland	295	1 206	1 749	1 493	1 431	1 252	996	1 076	1 211	1 781
27 Fishing area total	*295*	*1 206*	*1 749*	*1 493*	*1 431*	*1 252*	*996*	*1 076*	*1 211*	*1 781*
Species total	*295*	*1 206*	*1 749*	*1 493*	*1 431*	*1 252*	*996*	*1 076*	*1 211*	*1 781*
Spotted ray	**Raie douce**		**Raya pintada**		***Raja montagui***				**1,10(04)001,04**	**RJM**
27 France	1 120	953	925	977	1 163	1 179	1 260	1 341	1 563	1 451
27 Fishing area total	*1 120*	*953*	*925*	*977*	*1 163*	*1 179*	*1 260*	*1 341*	*1 563*	*1 451*
Species total	*1 120*	*953*	*925*	*977*	*1 163*	*1 179*	*1 260*	*1 341*	*1 563*	*1 451*
Sandy ray	**Raie circulaire**		**Raya falsa vela**		***Raja circularis***				**1,10(04)001,06**	**RJI**
27 France	356	397	431	438	438	410	435	369	330	301
27 Fishing area total	*356*	*397*	*431*	*438*	*438*	*410*	*435*	*369*	*330*	*301*
Species total	*356*	*397*	*431*	*438*	*438*	*410*	*435*	*369*	*330*	*301*
Shagreen ray	**Raie chardon**		**Raya cardadora**		***Raja fullonica***				**1,10(04)001,07**	**RJF**
27 France	71	56	51	46	39	38	65	38	68	71
Iceland	2	12	24	19	16	12	21	27	37	32
27 Fishing area total	*73*	*68*	*75*	*65*	*55*	*50*	*86*	*65*	*105*	*103*
Species total	*73*	*68*	*75*	*65*	*55*	*50*	*86*	*65*	*105*	*103*

B-38 Sharks, rays, chimaeras / Squales, raies, chimères / Tiburones, rayas, quimeras

Capture production by species, fishing areas and countries or areas / Captures par espèces, zones de pêche et pays ou zones / Capturas por especies, áreas de pesca y países o áreas

Species, Fishing area Espèce, Zone de pêche Especie, Area de pesca	1993 t	1994 t	1995 t	1996 t	1997 t	1998 t	1999 t	2000 t	2001 t	2002 t
Small-eyed ray	**Raie mêlée**		**Raya colorada**		***Raja microocellata***				**1,10(04)001,09**	**RJE**
27 France	-	-	-	-	-	1	11	-	-	-
27 Fishing area total	*-*	*-*	*-*	*-*	*-*	*1*	*11*	*-*	*-*	*-*
Species total	*-*	*-*	*-*	*-*	*-*	*1*	*11*	*-*	*-*	*-*
Cuckoo ray	**Raie fleurie**		**Raya santiguesa**		***Raja naevus***				**1,10(04)001,10**	**RJN**
27 France	3 050	3 365	3 762	4 077	4 721	4 015	3 638	3 064	2 885	2 735
27 Fishing area total	*3 050*	*3 365*	*3 762*	*4 077*	*4 721*	*4 015*	*3 638*	*3 064*	*2 885*	*2 735*
Species total	*3 050*	*3 365*	*3 762*	*4 077*	*4 721*	*4 015*	*3 638*	*3 064*	*2 885*	*2 735*
Longnosed skate	**Pocheteau noir**		**Raya picuda**		***Raja oxyrinchus***				**1,10(04)001,11**	**RJO**
27 France	387	347	359	346	311	327	194	140	89	211
27 Fishing area total	*387*	*347*	*359*	*346*	*311*	*327*	*194*	*140*	*89*	*211*
Species total	*387*	*347*	*359*	*346*	*311*	*327*	*194*	*140*	*89*	*211*
Antarctic starry skate	**Raie étoilée antarctique**		**Raya estrellada antártica**		***Raja georgiana***				**1,10(04)001,32**	**SRR**
48 Argentina	-	0	-	-	-	-	-	-	-	-
UK	-	-	-	-	-	-	-	0	-	-
48 Fishing area total	*-*	*0*	*-*	*-*	*-*	*-*	*-*	*0*	*-*	*-*
58 South Africa	-	-	-	-	-	-	-	-	-	0
58 Fishing area total	*-*	*-*	*-*	*-*	*-*	*-*	*-*	*-*	*-*	*0*
88 New Zealand	-	-	-	-	-	-	11	36	7	24
88 Fishing area total	*-*	*-*	*-*	*-*	*-*	*-*	*11*	*36*	*7*	*24*
Species total	*-*	*0*	*-*	*-*	*-*	*-*	*11*	*36*	*7*	*24*
Raja rays nei	**Pocheteaux et raies raja nca**		**Rayas raja nep**		***Raja spp***				**1,10(04)001,XX**	**SKA**
21 Canada	319	6 362	6 263	3 900	4 373	3 044	3 119	2 066	2 625	2 969
Cuba	10	-	5	-	0	1	-	-	-	-
Estonia	-	-	-	-	-	-	-	240	1 023	328
Faeroe Is	-	-	-	-	-	-	3	12	-	-
Germany	-	2	-	0	-	-	-	-	-	-
Greenland	0	5	-	-	-	-	-	-	-	6
Korea Rep	5	-	-	-	-	-	-	-	-	-
Lithuania	-	-	-	-	-	-	-	-	4	18
Norway	7	-	-	3	-	-	-	-	-	0
Poland	-	-	-	-	-	-	-	-	2	-
Portugal	7 605	6 239	2 060	794	904	1 104	2 168	671	880	1 361
Russian Fed	14	-	6	7	-	3	160	3 578	2 575	3 168
St Pier Mq	12	4	4	3	3	9	4	21	38	238
Spain	2 126	5 485	4 511	4 578	9 329	8 106	9 390	14 075	10 547	6 172
UK	-	-	-	-	-	5	-	-	-	-
USA	8 103	8 846	6 454	13 891	10 142	13 932	12 619	13 335	13 122	13 010
21 Fishing area total	*18 201*	*26 943*	*19 303*	*23 176*	*24 751*	*26 204*	*27 463*	*33 998*	*30 816*	*27 270*
27 Belgium	1 430	1 307	1 275	1 363	1 259	1 232	1 351	1 231	1 527	1 734
Channel Is	145 F	156	144 F	180	33	223	262	181	241	235
Denmark	35	64	57	44	40	20	46	87	122	60
Estonia	-	-	-	-	-	-	-	-	56	6
Faeroe Is	208	174	230	169	187	151	180	113	90	33
France	3 438	2 967	2 752	2 902	3 196	2 870	3 408	3 111	3 195	3 126
Germany	20	57	35	65	74	81	102	130	27	26
Ireland	1 755	1 524	2 098	2 212	2 715	2 120	2 283	2 078	2 140	2 501
Isle of Man	7	6	9	10	6	6	3	5	1	0
Netherlands	-	-	-	-	-	550	480	631	748	793
Norway	1 112	1 060	951	795	591	752	791	704	725	484
Portugal	1 696	1 467	1 599	1 665	1 711	1 703	1 560	1 654	1 683	1 638
Russian Fed	502	626	-	-	476	929	815	907	339	211
Spain	1 659	3 825	1 313	3 972	10 026	11 912	14 997	11 158	9 198	3 692
Sweden	31	35	17	9	8	2	3	3	12	8
UK	7 538	7 781	8 373	9 157	8 088	7 532	6 233	6 457	6 392	6 059
27 Fishing area total	*19 576 F*	*21 049*	*18 853 F*	*22 543*	*28 410*	*30 083*	*32 514*	*28 450*	*26 496*	*20 606*
37 Greece	528	1 380	1 120	1 002	900	715	718	746	579	550 F
Syria	...	...	...	...	...	...	...	...	...	182
37 Fishing area total	*528*	*1 380*	*1 120*	*1 002*	*900*	*715*	*718*	*746*	*579*	*732 F*
41 Estonia	-	-	-	-	-	-	-	-	-	10
41 Fishing area total	*-*	*-*	*-*	*-*	*-*	*-*	*-*	*-*	*-*	*10*
47 Angola	860	568	215	21	16	750	1 399	593	1 430	2 696
Iceland	-	-	-	-	14	-	-	-	-	-
South Africa	1 110	1 130	929	1 026	1 239	1 074	1 000	1 000 F	1 151	1 300
47 Fishing area total	*1 970*	*1 698*	*1 144*	*1 047*	*1 269*	*1 824*	*2 399*	*1 593 F*	*2 581*	*3 996*
Species total	*40 275 F*	*51 070*	*40 420 F*	*47 768*	*55 330*	*58 826*	*63 094*	*64 787 F*	*60 472*	*52 614 F*
Eaton's skate	**Raie d'Eaton**		**Raya de Eaton**		***Bathyraja eatonii***				**1,10(04)002,01**	**BEA**
48 Russian Fed	-	-	-	-	-	-	-	-	0	-
48 Fishing area total	*-*	*-*	*-*	*-*	*-*	*-*	*-*	*-*	*0*	*-*

B-38 Sharks, rays, chimaeras / Squales, raies, chimères / Tiburones, rayas, quimeras

Capture production by species, fishing areas and countries or areas / Captures par espèces, zones de pêche et pays ou zones / Capturas por especies, áreas de pesca y países o áreas

Species, Fishing area Espèce, Zone de pêche Especie, Area de pesca	1993 t	1994 t	1995 t	1996 t	1997 t	1998 t	1999 t	2000 t	2001 t	2002 t
58 South Africa	-	-	-	-	-	-	-	-	-	0
58 Fishing area total	-	-	-	-	-	-	-	-	-	*0*
88 New Zealand	-	-	-	-	-	-	1	5	0	1
88 Fishing area total	-	-	-	-	-	-	*1*	*5*	*0*	*1*
Species total	-	-	-	-	-	-	*1*	*5*	*0*	*1*
Murray's skate	**Raie de Murray**		**Raya de Murray**		***Bathyraja murrayi***				**1,10(04)002,03**	**BMU**
58 South Africa	-	-	-	-	-	-	-	0	-	-
58 Fishing area total	-	-	-	-	-	-	-	*0*	-	-
Species total	-	-	-	-	-	-	-	*0*	-	-
...A	**...B**		**...C**		***Bathyraja meridionalis***				**1,10(04)002,16**	**EYE**
48 UK	-	-	-	-	-	-	-	0	-	-
48 Fishing area total	-	-	-	-	-	-	-	*0*	-	-
Species total	-	-	-	-	-	-	-	*0*	-	-
Bathyraja rays nei	**Raies bathyraja nca**		**Rayas bathyraja nep**		***Bathyraja spp***				**1,10(04)002,XX**	**BHY**
58 South Africa	-	-	-	-	-	-	-	0	-	-
58 Fishing area total	-	-	-	-	-	-	-	*0*	-	-
88 New Zealand	-	-	-	-	-	-	1	-	-	-
88 Fishing area total	-	-	-	-	-	-	*1*	-	-	-
Species total	-	-	-	-	-	-	*1*	*0*	-	-
Rays and skates nei	**Rajidés nca**		**Rayidos nep**		***Rajidae***				**1,10(04)XXX,XX**	**RAJ**
88 New Zealand	-	-	-	-	-	-	6	-	-	-
88 Fishing area total	-	-	-	-	-	-	*6*	-	-	-
Species total	-	-	-	-	-	-	*6*	-	-	-
Whip stingray	**Pastenague du Pacifique**		**Raya látigo del Pacífico**		***Dasyatis akajei***				**1,10(05)003,01**	**WST**
61 Japan	4 247	4 040	3 985	4 029	3 959	4 329	4 407	5 388	4 312	4 512
61 Fishing area total	*4 247*	*4 040*	*3 985*	*4 029*	*3 959*	*4 329*	*4 407*	*5 388*	*4 312*	*4 512*
Species total	*4 247*	*4 040*	*3 985*	*4 029*	*3 959*	*4 329*	*4 407*	*5 388*	*4 312*	*4 512*
Common stingray	**Pastenague commune**		**Raya látigo común**		***Dasyatis pastinaca***				**1,10(05)003,26**	**JDP**
37 Ukraine	...	...	...	...	...	-	-	4	11	-
37 Fishing area total	...	...	...	...	...	-	-	*4*	*11*	-
Species total	...	...	...	...	...	-	-	*4*	*11*	-
Stingrays nei	**Pastenagues nca**		**Pastinacas nep**		***Dasyatis spp***				**1,10(05)003,XX**	**STI**
27 France	1	2	-	1	2	5	6	10	7	10
27 Fishing area total	*1*	*2*	-	*1*	*2*	*5*	*6*	*10*	*7*	*10*
Species total	*1*	*2*	-	*1*	*2*	*5*	*6*	*10*	*7*	*10*
Eagle rays	**Aigles de mer**		**Aguilas de mar**		***Myliobatidae***				**1,10(07)XXX,XX**	**EAG**
27 France	3	2	2	0	0	0	2	2	2	-
Portugal	...	...	...	...	...	...	11	8	9	12
27 Fishing area total	*3*	*2*	*2*	*0*	*0*	*0*	*13*	*10*	*11*	*12*
81 New Zealand	...	...	0	0	1	1	2	2	5	9
81 Fishing area total	...	...	*0*	*0*	*1*	*1*	*2*	*2*	*5*	*9*
Species total	*3*	*2*	*2*	*0*	*1*	*1*	*15*	*12*	*16*	*21*
Mantas	**Mantes**		**Mantas**		***Mobulidae***				**1,10(08)XXX,XX**	**MAN**
34 Liberia	...	...	...	...	...	342	802	931	106	110 F
34 Fishing area total	...	...	...	...	...	*342*	*802*	*931*	*106*	*110 F*
Species total	...	...	...	...	...	*342*	*802*	*931*	*106*	*110 F*
Rays, stingrays, mantas nei	**Raies, pastenagues, mantes nca**		**Rayas, pastinacas, mantas nep**		***Rajiformes***				**1,10(XX)XXX,XX**	**SRX**
31 Cuba	12	0	5	955	1 358	1 334	1 352	1 193	1 119	1 278
Dominican Rp	10	18	90	39	96	62	134	111	123	10
Korea Rep	-	-	-	-	-	175	42	97	3	-
Martinique	5 F	5 F	5 F	3 F	5 F	5 F	5 F	5 F	5 F	5 F
Mexico	3 992	4 212	5 165	4 782	5 496	6 197	4 201	2 873	2 618	2 407
Venezuela	1 748	1 994	2 450	1 812	1 896	2 111	2 287	2 148	2 182	1 956
31 Fishing area total	*5 767 F*	*6 229 F*	*7 715 F*	*7 591 F*	*8 851 F*	*9 884 F*	*8 021 F*	*6 427 F*	*6 050 F*	*5 656 F*
34 Benin	50 F	56 F	54 F	62 F	70 F	60 F	47	30 F	38	32
Cameroon	70	100 F	160	186	170 F	161	211	150	130	133

B-38 Sharks, rays, chimaeras / Squales, raies, chimères / Tiburones, rayas, quimeras

Capture production by species, fishing areas and countries or areas / Captures par espèces, zones de pêche et pays ou zones / Capturas por especies, áreas de pesca y países o áreas

Species, Fishing area Espèce, Zone de pêche Especie, Area de pesca	1993 t	1994 t	1995 t	1996 t	1997 t	1998 t	1999 t	2000 t	2001 t	2002 t
Congo Rep	395 F	185	160 F	135 F	110 F	85 F	60 F	33	30 F	30 F
Côte dIvoire	...	...	...	218	146	202	227	241	168	240
Gabon	0	3	33	172	173	90	197	141	88	46
Ghana	272	296	338	261	185	172	869	231	814	969
Italy	1 372	1 372	26	6	-	-	-	-	12	23
Korea Rep	369	635	565	576	653	57	249	402	334	399
Liberia	50	56	33	12	38	50	119	103	29	30 F
Mauritania	20 F	20 F	10 F	10 F	20 F	180 F	350 F	350 F	350 F	350 F
Morocco	892	1 262	1 643	1 650	1 350	1 187	1 348	2 139	1 487	1 867
Nigeria	1 126	3 102	3 670	4 103	5 004	7 382	7 622	5 753	7 352	6 262
Portugal	57	24	57	126	239	155	74	0	-	-
Senegal	2 205	2 886	3 308	2 962	4 515	4 051	3 332	1 585	1 667	917
Sierra Leone	1 408	1 403	1 401	1 400 F	1 404	15	10	18	...	1
Spain	-	-	-	203	-	18	10	0	-	-
Togo	18	9	5	11	2	0	2	16	5	2
34 Fishing area total	*8 304 F*	*11 409 F*	*11 463 F*	*12 093 F*	*14 079 F*	*13 865 F*	*14 727 F*	*11 192 F*	*12 504 F*	*11 301 F*
37 Albania	10 F	15 F	67	23	24	86	78	85	14	28
Algeria	...	...	124	272	120	450	207	191	298	491
Croatia	276	224	190	141	119	120	68	57	42	34
France	135	86	15	75	75	75	46	70	64	74
Italy	3 011	2 358	4 552	2 301	5 325	2 807	1 117	507	543	498
Malta	7	5	5	7	8	5	6	7	0	-
Morocco	40	37	27	30	30	19	115	...	23	34
Palest, O.T.	...	...	...	29	16	23	20 F	25 F	25 F	28
Russian Fed	20	24	14	22	16	18	40	13	30	14
Serbia-Monte	9	8	14	12	12	12	12	11	11	10
Spain	-	-	-	248	288	289	501	536	375	835
Tunisia	614	551	460	489	803	836	922	974	1 113	1 092
Turkey	1 124	1 238	337	524	340	385	420	1 100	555	369
Ukraine	3	4	15	...	1	-	-	-	-	-
37 Fishing area total	*5 249 F*	*4 550 F*	*5 820*	*4 173*	*7 177*	*5 125*	*3 552 F*	*3 576 F*	*3 093 F*	*3 507*
41 Argentina	910	5 701	7 190	12 478	12 119	14 855	12 116	13 265	15 179	13 215
Australia	-	-	-	-	-	3	23	-	-	
Belize	-	-	-	-	-	-	519	48	201	10
Brazil	3 500 F	3 000 F	3 948	3 104	3 010	4 673	4 277	4 867	4 860	6 294
Chile	-	16	-	-	-	4	4	-	-	-
Falkland Is	98	63	117	184	204	216	314	353	417	466
France	-	20	5	9	3	1	0	0	-	
Honduras	1 948	876	615	460	-	-	-	-	-	
Italy	183	183	3	1	-	-	-	-	-	
Korea Rep	1 696	2 631	5 506	5 243	5 118	1 503	5 539	7 246	6 382	4 430
Namibia	-	-	-	-	3	14	12	-	-	-
Panama	611	372	85	170	-	-	-	-	-	
Portugal	11	28	80	135	18	34	-	0	82	54
Russian Fed	-	-	-	-	-	-	-	-	203	
Seychelles	-	-	-	-	4	1	-	-	-	-
UK	29	3	12	8	17	13	40	17	32	24
Uruguay	128	1 032	1 469	2 614	2 342	398	1 575	1 004	989	2 000
41 Fishing area total	*9 114 F*	*13 925 F*	*19 030*	*24 406*	*22 838*	*21 715*	*24 419*	*26 800*	*28 345*	*26 493*
47 Korea Rep	475	364	540	227	283	435	343	375	323	268
Namibia	1	0	62	133	191	80	377	966	1 204	...
Russian Fed	-	-	-	-	-	8	-	-	-	
47 Fishing area total	*476*	*364*	*602*	*360*	*474*	*523*	*720*	*1 341*	*1 527*	*268*
48 Argentina	-	-	28	0	-	-	-	-	-	-
Bulgaria	-	0	-	-	-	-	-	-	-	-
Chile	-	-	20	21	-	0	-	-	-	-
Korea Rep	-	11	42	24	24	-	5	-	7	-
Russian Fed	-	0	0	-	-	-	-	-	0	-
South Africa	-	-	-	-	-	8	-	-	-	-
Spain	-	-	-	-	3	-	-	-	-	-
UK	-	-	-	-	5	0	11	0	6	-
USA	-	-	-	-	-	-	0	-	0	-
Uruguay	-	-	-	-	-	0	1	-	-	-
48 Fishing area total	-	*11*	*90*	*45*	*32*	*8*	*17*	*0*	*13*	-
51 Italy	274	274	5	1	-	-	-	-	-	-
Korea Rep	315	629	27	106	114	329	162	14	23	36
Mauritius	2	2	2	2	2	2	2	2	2	2
Oman	288	188	538	372	359	189	289	240	198	199
Pakistan	16 093	18 481	16 445	15 563	15 769	17 576	20 780	20 740	20 801	20 900
Saudi Arabia	-	-	-	-	-	4	8	4	6	8
Tanzania	2 511	2 474	3 200	4 006	3 500	3 350	3 500	3 600	3 500	2 500
Yemen	-	-	156	...	100 F	100 F	100 F	100 F	130 F	150 F
51 Fishing area total	*19 483*	*22 048*	*20 373*	*20 050*	*19 844 F*	*21 550 F*	*24 841 F*	*24 700 F*	*24 660 F*	*23 795 F*
57 Australia	-	17	22	55	...	...	8	11	24	36
Indonesia	12 840	12 605	10 193	9 471	10 187	16 960	12 452	11 681	11 978	12 210
Korea Rep	-	-	7	-	-	-	18	8	-	-
Malaysia	5 608	4 645	4 211	4 541	5 766	3 915	4 833	5 389	5 176	5 058
Thailand	2 099	2 370	3 520	5 075	5 238	4 185	4 688	4 844	4 441	6 462
57 Fishing area total	*20 547*	*19 637*	*17 953*	*19 142*	*21 191*	*25 060*	*21 999*	*21 933*	*21 619*	*23 766*
58 Australia	0	-	-	-	3	3	2	-	-	-
France	-	2	-	-	4	21	30	87	119	343
South Africa	-	-	-	-	0	2	4	9	2	0
Ukraine	-	-	0	0	-	-	-	-	-	-

B-38 Sharks, rays, chimaeras / Squales, raies, chimères / Tiburones, rayas, quimeras

Capture production by species, fishing areas and countries or areas / Captures par espèces, zones de pêche et pays ou zones / Capturas por especies, áreas de pesca y países o áreas

Species, Fishing area Espèce, Zone de pêche Especie, Area de pesca	1993 t	1994 t	1995 t	1996 t	1997 t	1998 t	1999 t	2000 t	2001 t	2002 t
58 Fishing area total	*0*	*2*	*0*	*0*	*7*	*26*	*36*	*96*	*121*	*343*
61 China,Taiwan	673	533	647	2 457	1 367	246	235	351	851	1 143
Korea Rep	9 918	6 914	6 235	6 787	6 638	4 375	3 948	2 565	211	2 583
Russian Fed	-	-	6	6	9	8	304	1 427	1 602	1 358
61 Fishing area total	*10 591*	*7 447*	*6 888*	*9 250*	*8 014*	*4 629*	*4 487*	*4 343*	*2 664*	*5 084*
67 Canada	249	570	982	1 293	1 584	900	1 406	1 661	1 618	1 540
Korea Rep	-	13	-	-	-	-	-	-	-	-
USA	139	17	370	1 344	2 173	1 145	1 411	1 549	1 399	3 112
67 Fishing area total	*388*	*600*	*1 352*	*2 637*	*3 757*	*2 045*	*2 817*	*3 210*	*3 017*	*4 652*
71 Australia	-	-	-	-	-	-	-	-	0	-
Indonesia	22 846	24 001	24 624	26 986	33 137	31 331	32 875	33 579	32 473	34 030
Korea Rep	356	3 123	3 007	1 422	1 289	1 121	2 155	1 401	1 526	1 861
Malaysia	8 996	9 355	11 496	11 387	11 516	12 189	12 200	11 184	11 356	10 883
Philippines	6 230	4 906	4 980	4 756	2 125	2 174	2 299	2 248	2 733	2 848
Singapore	311	411	320	327	308	336	250	261	187	162
Thailand	3 702	7 772	6 448	4 903	5 115	4 104	7 591	8 806	6 390	4 347
71 Fishing area total	*42 441*	*49 568*	*50 875*	*49 781*	*53 490*	*51 255*	*57 370*	*57 479*	*54 665*	*54 131*
77 Korea Rep	25	92	24	167	239	162	1 069	596	106	75
Mexico	3 922	4 137	4 411	5 150	4 578	4 757	4 875	4 944	4 405	4 098
USA	29	31	60	210	315	195	205	167	101	27
77 Fishing area total	*3 976*	*4 260*	*4 495*	*5 527*	*5 132*	*5 114*	*6 149*	*5 707*	*4 612*	*4 200*
81 Australia	29	75	68	70	72	67	43	45	40	35
Korea Rep	188	340	177	135	47	21	197	82	168	121
New Zealand	2 788	2 598	2 116	1 582	2 227	2 313	2 821	2 634	2 784	2 648
81 Fishing area total	*3 005*	*3 013*	*2 361*	*1 787*	*2 346*	*2 401*	*3 061*	*2 761*	*2 992*	*2 804*
87 Chile	1 971	2 899	2 622	2 675	2 958	2 011	3 365	4 151	2 974	2 992
Colombia	-	-	-	3	2	2	1	1 F	1	1 F
Korea Rep	668	120	-	-	-	-	252	247	38	2
Peru	3 632	1 658	1 841	1 126	1 177	1 477	2 789	4 026	2 034	4 886
87 Fishing area total	*6 271*	*4 677*	*4 463*	*3 804*	*4 137*	*3 490*	*6 407*	*8 425 F*	*5 047*	*7 881 F*
88 Chile	-	-	-	-	-	0	-	-	-	-
New Zealand	-	-	-	-	-	5	-	0	-	-
88 Fishing area total	-	-	-	-	-	*5*	-	*0*	-	-
Species total	*135 612 F*	*147 740 F*	*153 480 F*	*160 646 F*	*171 369 F*	*166 695 F*	*178 623 F*	*177 990 F*	*170 929 F*	*173 881 F*
Torpedo rays	**Torpilles**		**Tremolinas**		***Torpedo spp***				**1,11(01)002,XX**	**TOE**
27 France	21	21	20	16	18	19	34	32	43	34
27 Fishing area total	*21*	*21*	*20*	*16*	*18*	*19*	*34*	*32*	*43*	*34*
Species total	*21*	*21*	*20*	*16*	*18*	*19*	*34*	*32*	*43*	*34*
Rabbit fish	**Chimère commune**		**Quimera**		***Chimaera monstrosa***				**1,12(01)003,01**	**CMO**
27 Denmark	-	-	-	-	-	-	-	-	1	0
Iceland	3	60	106	21	15	29	11	5	1	-
Ireland	...	...	...	...	...	2	-	5	15	16
Norway	-	-	-	-	-	-	-	1	70	46
UK	-	-	-	-	0	1	1	4	36	7
27 Fishing area total	*3*	*60*	*106*	*21*	*15*	*32*	*12*	*15*	*123*	*69*
Species total	*3*	*60*	*106*	*21*	*15*	*32*	*12*	*15*	*123*	*69*
Spotted ratfish	**Chimère d'Amérique**		**Quimera americana**		***Hydrolagus colliei***				**1,12(01)004,01**	**RAT**
77 USA	-	-	-	-	-	-	-	-	-	2
77 Fishing area total	-	-	-	-	-	-	-	-	-	*2*
Species total	-	-	-	-	-	-	-	-	-	*2*
Dark ghost shark	**...B**		**...C**		***Hydrolagus novaezealandiae***				**1,12(01)004,11**	**CYV**
81 New Zealand	1 089	1 455	1 593	1 614	2 064	1 956	1 975	1 819	1 572	2 055
81 Fishing area total	*1 089*	*1 455*	*1 593*	*1 614*	*2 064*	*1 956*	*1 975*	*1 819*	*1 572*	*2 055*
Species total	*1 089*	*1 455*	*1 593*	*1 614*	*2 064*	*1 956*	*1 975*	*1 819*	*1 572*	*2 055*
Ratfishes nei	**Chimères nca**		**Quimeras nep**		***Hydrolagus spp***				**1,12(01)004,XX**	**HYD**
27 France	-	-	-	-	-	-	38	573	822	627
Ireland	...	...	...	...	...	...	...	...	5	-
Norway	-	-	-	-	-	-	-	-	6	7
UK	-	-	-	-	-	-	-	-	2	-
27 Fishing area total	...	...	...	...	...	...	*38*	*573*	*835*	*634*
81 New Zealand	...	...	...	...	0	36	453	975	2 184	1 901
81 Fishing area total	...	...	...	...	*0*	*36*	*453*	*975*	*2 184*	*1 901*
Species total	...	...	...	...	*0*	*36*	*491*	*1 548*	*3 019*	*2 535*

B-38 Sharks, rays, chimaeras / Squales, raies, chimères / Tiburones, rayas, quimeras

Capture production by species, fishing areas and countries or areas / Captures par espèces, zones de pêche et pays ou zones / Capturas por especies, áreas de pesca y países o áreas

Species, Fishing area Espèce, Zone de pêche Especie, Area de pesca	1993 t	1994 t	1995 t	1996 t	1997 t	1998 t	1999 t	2000 t	2001 t	2002 t
Straightnose rabbitfish	**Chimère à nez mou**		**Narigón sierra**		***Rhinochimaera atlantica***				**1,12(02)001,01**	**RCT**
27 UK	-	-	-	-	-	-	-	-	2	1
27 Fishing area total	-	-	-	-	-	-	-	-	*2*	*1*
Species total	-	-	-	-	-	-	-	-	*2*	*1*
Ghost shark	**Masca laboureur**		**...C**		***Callorhinchus milii***				**1,12(03)001,01**	**CHB**
57 Australia	...	...	...	...	49	21	14	82	105	102
57 Fishing area total	...	...	...	...	*49*	*21*	*14*	*82*	*105*	*102*
81 New Zealand	587	639	769	595	913	951	1 260	1 228	1 189	1 085
Ukraine	-	-	-	-	-	-	-	-	-	1
81 Fishing area total	*587*	*639*	*769*	*595*	*913*	*951*	*1 260*	*1 228*	*1 189*	*1 086*
Species total	*587*	*639*	*769*	*595*	*962*	*972*	*1 274*	*1 310*	*1 294*	*1 188*
Cape elephantfish	**Masca du Cap**		**Pejegallo del Cabo**		***Callorhinchus capensis***				**1,12(03)001,03**	**CHM**
47 South Africa	983	262	386	366	484	482	356	380 F	405	422
47 Fishing area total	*983*	*262*	*386*	*366*	*484*	*482*	*356*	*380 F*	*405*	*422*
Species total	*983*	*262*	*386*	*366*	*484*	*482*	*356*	*380 F*	*405*	*422*
Elephantfishes nei	**Mascas nca**		**Pejegallos nep**		***Callorhinchus spp***				**1,12(03)001,XX**	**ELF**
41 Argentina	704	1 096	921	815	1 329	1 770	1 977	1 390	451	568
41 Fishing area total	*704*	*1 096*	*921*	*815*	*1 329*	*1 770*	*1 977*	*1 390*	*451*	*568*
87 Chile	2 516	1 570	920	1 450	822	1 416	632	603	1 125	312
87 Fishing area total	*2 516*	*1 570*	*920*	*1 450*	*822*	*1 416*	*632*	*603*	*1 125*	*312*
Species total	*3 220*	*2 666*	*1 841*	*2 265*	*2 151*	*3 186*	*2 609*	*1 993*	*1 576*	*880*
Chimaeras, etc. nei	**Chimères, etc. nca**		**Quimeras, etc. nep**		***Chimaeriformes***				**1,12(XX)XXX,XX**	**HOL**
27 Iceland	2	0	2	1	0	-	-	-	-	-
27 Fishing area total	*2*	*0*	*2*	*1*	*0*	-	-	-	-	-
81 New Zealand	...	...	5	49	5	5	21	40	76	103
81 Fishing area total	...	...	*5*	*49*	*5*	*5*	*21*	*40*	*76*	*103*
Species total	*2*	*0*	*7*	*50*	*5*	*5*	*21*	*40*	*76*	*103*
Various sharks nei	**Requins divers nca**		**Escualos diversos nep**		***Selachimorpha(Pleurotremata)***				**1,99(XX)XXX,XX**	**SKH**
27 Belgium	22	19	20	19	18	11	14	15	18	12
Channel Is	43 F	32	30 F	2	3	3	7	1	-	-
Denmark	5	3	4	-	-	-	-	-	-	-
Estonia	-	-	-	-	-	-	-	-	-	53
Germany	133	440	292	309	139	110	-	-	-	-
Ireland	17	16	40	23	32	169	90	175	455	496
Netherlands	-	-	-	-	-	-	-	-	3	-
Norway	-	0	0	0	1	0	13	1	72	1
Portugal	1 047	969	1 853	1 659	1 874	1 882	352	297	216	345
Spain	6 914	10 998	18 101	3 642	30 377	20 450	23 662	31 793	30 815	17 771
UK	1 393	1 944	2 339	2 040	3 865	2 669	2 342	954	1 640	1 198
27 Fishing area total	*9 574 F*	*14 421*	*22 679 F*	*7 694*	*36 309*	*25 294*	*26 480*	*33 236*	*33 219*	*19 876*
Species total	*9 574 F*	*14 421*	*22 679 F*	*7 694*	*36 309*	*25 294*	*26 480*	*33 236*	*33 219*	*19 876*
Sharks, rays, skates, etc. nei	**Requins, raies, etc. nca**		**Tiburones, rayas, etc. nep**		***Elasmobranchii***				**1,99(XX)XXX,XX**	**SKX**
21 China,Taiwan	-	-	18	12	12	-	2	6	8	14
Japan	553	450	397	238	99	107	123	83	116	34
Portugal	-	-	-	-	-	-	103	1	-	7
21 Fishing area total	*553*	*450*	*415*	*250*	*111*	*107*	*228*	*90*	*124*	*55*
27 China,Taiwan	-	-	22	15	15	-	10	14	8	20
France	-	-	2	-	-	-	-	-	-	-
Japan	174	168	376	132	108	211	72	35	82	4
Poland	-	-	-	-	-	-	-	-	11	8
Portugal	...	...	...	...	...	...	41	40	40	41
Sweden	0	0	0	0	0	0	0	0	-	-
Uruguay	-	-	-	-	-	-	-	-	-	314
27 Fishing area total	*174*	*168*	*400*	*147*	*123*	*211*	*123*	*89*	*141*	*387*
31 Antigua Barb	...	...	...	...	...	...	...	...	8	17
Bahamas	37	0	0	5	3	2	1	0	0	0
Barbados	18	22	24	25	14	12	10	14	10	9
Belize	...	...	...	...	1	0	2	6	...	5
Bermuda	14	10	17	13	9	12	24	10	5	5
Br Virgin Is	...	...	...	...	1	1	1	0	0	0
China,Taiwan	33 F	23 F	67	47	47	18	4	100	195	107
Costa Rica	9	7	16	6	1	68	15	81	49	27
Cuba	881	1 382	1 365	1 409	1 908	3 072	2 847	2 264	2 396	2 344
Grenada	12	4	14	4	9	18	24	29	29	12
Guyana	...	...	...	765	1 892	...	2 175	...	...	953
Japan	13	30	17	9	17	16	63	42	75	...
Korea Rep	-	-	-	-	-	-	2	-	-	-
Martinique	120 F	120 F	100 F	70 F	90 F	80 F	70 F	50 F	40 F	40 F

B-38 Sharks, rays, chimaeras / Squales, raies, chimères / Tiburones, rayas, quimeras

Capture production by species, fishing areas and countries or areas
Captures par espèces, zones de pêche et pays ou zones
Capturas por especies, áreas de pesca y países o áreas

Species, Fishing area Espèce, Zone de pêche Especie, Area de pesca	1993 t	1994 t	1995 t	1996 t	1997 t	1998 t	1999 t	2000 t	2001 t	2002 t
Mexico	6 674	6 696	6 860	7 417	4 994	4 617	4 743	4 370	3 986	3 100
Portugal	-	-	-	-	-	-	22	-	-	2
Puerto Rico	...	...	...	...	...	...	28	35	32	20
St Lucia	...	6	6	11	3	8	6	5	5	10
St Vincent	...	...	...	2	...	...	3	...	2	...
Spain	-	-	-	-	-	-	-	-	-	205
Trinidad Tob	440	488	550 F	621	545	635	701	755	715	937
USA	3 582	5 134	3 471	1 090	2 814	2 845	1 728	1 808	1 846	2 426
31 Fishing area total	*11 833 F*	*13 922 F*	*12 507 F*	*11 494 F*	*12 348 F*	*11 404 F*	*12 469 F*	*9 569 F*	*9 393 F*	*10 219 F*
34 Benin	160 F	140 F	120 F	100 F	100 F	80 F	59	40 F	87	86
Cameroon	92	80 F	59	48	50 F	55	86	67	146	85
Cape Verde	...	1	1	...	...	...	...	...	...	...
China	-	-	-	-	-	5	31	-	-	420
China,Taiwan	33 F	23 F	388	270	270	45	82	411	116	69
Congo Dem R	...	...	...	...	...	...	...	400 F	450 F	450 F
Côte dIvoire	269	177	190	405	71	42	313	521	66	132
Cuba	1 915	2 008	1 691	1 051	-	-	-	-	-	-
Eq Guinea	330 F	500 F	220 F	490 F	620 F	779	910 F	100 F	100 F	100 F
Gabon	0	2	22	1 267	626	1 933	1 338	659	375	360
Gambia	316	480	498	415	3 223	606	630	720	3 982	6 128
Ghana	1 981	1 171	1 115	1 106	709	1 764	3 998	1 670	2 092	2 451
Guinea	...	...	726	506	505	700	800	969	681	680 F
Italy	1 715	1 715	-	-	-	-	-	-	-	-
Japan	1 412	977	925	729	464	709	228	270	265	...
Korea Rep	-	34	-	2	11	1	-	2	-	-
Liberia	100	309	358	207	386	210	-	-	512	520 F
Mauritania	50 F	60 F	80 F	10 F	10 F	350 F	500 F	500 F	500 F	500 F
Morocco	1 423	1 108	1 615	1 600	1 210	2 234	1 985	3 460	2 181	2 131
Nigeria	4 723	5 951	2 801	4 285	3 817	6 587	7 751	7 485	7 274	7 187
Panama	-	-	-	-	-	-	202	...	-	-
Portugal	170	93	859	2 366	1 127	849	737	141	20	125
Russian Fed	-	-	-	-	-	101	11	-	-	-
Sao Tome Prn	221 F	321 F	337 F	247	130	175 F	190 F	180 F	180 F	180 F
Senegal	1 791	3 347	4 169	3 803	4 470	4 887	4 808	5 473	3 260	...
Sierra Leone	...	...	2	2 F	1	68	41	1 672	164	403
Spain	-	-	-	3 977	34 443	14 145	9 992	9 481	9 536	3 138
Togo	26	4	15	202	57	67	230	132	130	254
34 Fishing area total	*16 727 F*	*18 501 F*	*16 191 F*	*23 088 F*	*52 300 F*	*36 392 F*	*34 922 F*	*34 353 F*	*32 117 F*	*25 399 F*
37 Algeria	1 127	1 200 F	1 000	965	415	867	854	331	679	519
Cyprus	30	19	21	14	17	10	12	14	28	22
Egypt	1 000	1 226	1 172	1 120	1 629	1 211	1 383	1 197	2 143	2 020
France	...	...	...	28	...	...	43	-	-	-
Georgia	131	45	31	71	1	550	18	21	27	65
Israel	60 F	50 F	48	330	49	59	58	...	35	...
Japan	3	5	8	3	1	-	1	-	-	-
Lebanon	50	50	50	50	50	50	50	60	55	60
Malta	1	1	4	3	2	11	4	13	0	-
Morocco	31	44	21	25	45	9	19	...	11	30
Portugal	-	-	-	-	-	-	-	3	1	1
Russian Fed	5	11	90	19	9	6	9	12	27	19
Spain	-	-	-	1 277	1 420	2 140	663	426	377	29
37 Fishing area total	*2 438 F*	*2 651 F*	*2 445*	*3 905*	*3 638*	*4 913*	*3 114*	*2 077*	*3 383*	*2 765*
41 Argentina	2 044	1 707	2 230	2 251	1 070	1 220	905	719	798	1 103
Brazil	12 000 F	11 000 F	10 658	8 446	10 189	12 596	13 576	15 900	14 626	12 032
China,Taiwan	65 F	46 F	650	450	452	222	357	263	142	221
Estonia	-	-	-	-	-	-	-	-	-	188
Italy	229	229	-	-	-	-	-	-	-	-
Japan	185	581	65	69	81	35	34	25	46	...
Portugal	-	-	-	-	114	103	55	11	41	32
Spain	838	469	395	225	3 855	1 935	1 653	1 817	1 818	744
UK	-	-	-	-	16	31	-	-	4	26
Uruguay	803	949	1 577	1 760	2 367	444	1 901	991	868	788
41 Fishing area total	*16 164 F*	*14 981 F*	*15 575*	*13 201*	*18 144*	*16 586*	*18 481*	*19 726*	*18 343*	*15 134*
47 Angola	29	35	755	379	90	376	...	157	3 354	3 236
China,Taiwan	65 F	46 F	268	186	186	341	639	255	568	703
Honduras	...	...	...	...	10	4	...	...	...	-
Japan	1 140	1 295	676	398	369	565	603	324	205	...
Namibia	-	4	7	5	4	6	1	769	1 875	-
Poland	-	1	-	-	-	-	-	-	-	-
Portugal	-	-	12	...	...	...	488	20	239	150
St Helena	...	...	...	...	...	...	...	...	6	4
South Africa	699	630	389	317	433	500	430	350 F	314	350
Spain	35 F	50 F	60 F	-	95	28	109	20	20	81
47 Fishing area total	*1 968 F*	*2 061 F*	*2 167 F*	*1 285*	*1 187*	*1 820*	*2 270*	*1 895 F*	*6 581*	*4 524*
48 Uruguay	-	-	-	-	-	-	0	-	-	-
48 Fishing area total	-	-	-	-	-	-	*0*	-	-	-
51 China	-	-	-	-	-	-	-	4	...	...
China,Taiwan	405 F	286 F	331	360	360	261	392	802	1 826	1 979
Comoros	58	...	...	...	...	...	...	...	...	...
Egypt	89	69	137	122	180	135	182	244	263	202
Eritrea	...	...	...	...	6	7	6	-	-	-
India	32 169	32 514	36 642	93 268	37 588	33 418	34 088	37 060	28 933	33 799
Iran	...	...	...	...	...	1	...	...	...	8 071
Italy	343	343	-	-	-	-	-	-	-	-

B-38

Sharks, rays, chimaeras — **Capture production by species, fishing areas and countries or areas**
Squales, raies, chimères — **Captures par espèces, zones de pêche et pays ou zones**
Tiburones, rayas, quimeras — **Capturas por especies, áreas de pesca y países o áreas**

Species, Fishing area Espèce, Zone de pêche Especie, Area de pesca	1993 t	1994 t	1995 t	1996 t	1997 t	1998 t	1999 t	2000 t	2001 t	2002 t
Japan	196	502	282	620	430	442	276	185	158	...
Kenya	152	166	176	191	140	134	131	115	175	134
Korea Rep	-	30	82	76	28	64	33	36	49	59
Maldives	9 168	11 212	11 245	11 856	10 643	10 887	6 883	13 523	11 935	11 498
Mauritius	16	17	15	17	58	9	9	25	12	48
Mayotte	...	...	...	...	...	32	18	...	...	...
Oman	4 540	3 503	6 566	5 870	6 342	4 805	4 020	3 651	3 632	3 803
Philippines	-	-	-	-	-	11	11	7	29	66
Portugal	-	-	-	-	-	262	487	354	728	93
Réunion	36	33	37	46	89	111	81	71	64	64 F
Saudi Arabia	42	125	467	398	543	697	497	649	651	731
Seychelles	82	117	116	84	57	102	68	150	95	90
South Africa	100	110	70	10	110	130	100 F	60	35	7
Spain	-	-	39	50	410	509	3 574	5 357	8 491	12 223
Sudan	...	...	...	...	...	45	56	44	79	79
Tanzania	962	1 389	1 310	1 594	1 500	1 325	1 375	1 400	1 500	1 500
Untd Arab Em	1 600	1 802	1 553	1 902	1 832	1 881	1 945	1 530	1 762	2 541
Uruguay	-	-	-	-	-	-	-	-	22	43
Yemen	6 537	6 455	4 480	4 878	5 000 F	5 800 F	5 600 F	5 000 F	6 300 F	7 100 F
Other nei	281	196	406	453	263	271	149	178	471	471
51 Fishing area total	*56 776 F*	*58 869 F*	*63 954*	*121 795*	*65 579 F*	*61 339 F*	*59 981 F*	*70 445 F*	*67 210 F*	*84 601 F*
57 Australia	8 483	2 102	2 597	2 533	1 497	829	1 000	2 397	2 465	2 850
China	-	-	-	-	-	-	187	95	...	...
China,Taiwan	206 F	150 F	799	758	758	108	11	240	345	626
Honduras	-	-	-	-	-	-	-	-	85	-
India	44 435	51 175	40 436	38 892	34 403	41 286	42 714	38 997	39 038	33 559
Indonesia	15 982	16 949	17 905	18 048	16 811	22 786	18 694	19 572	19 927	20 670
Japan	244	185	554	437	436	227	378	323	271	...
Korea Rep	-	-	-	-	-	-	-	3	-	-
Malaysia	694	709	824	810	762	964	685	1 271	1 119	1 216
Philippines	-	-	-	-	-	22	3	2	0	0
Seychelles	-	-	-	-		-	0	2	5	5
Spain	-	-	-	-		-	-	-	-	1 279
Sri Lanka	7 311	8 475	7 077	6 954	11 920	7 625	8 660	13 884	8 240	6 830
Thailand	577	879	2 501	4 615	4 623	4 431	3 816	3 906	3 943	3 859
Other nei	36	25	51	36	96	17	45	53	89	89
57 Fishing area total	*77 968 F*	*80 649 F*	*72 744*	*73 083*	*71 306*	*78 295*	*76 193*	*80 745*	*75 527*	*70 983*
58 Australia	-	-	-	-	2	-	-	-	-	-
South Africa	-	-	-	-	-	-	1	-	-	-
58 Fishing area total	-	-	-	-	*2*	-	*1*	-	-	-
61 China,H.Kong	848	688	485	456	420	382	300 F	330 F	370 F	350 F
China,Taiwan	8 200 F	5 750 F	4 646	4 910	4 913	3 171	24 718	22 759	8 230	9 482
Japan	23 330	18 864	16 633	12 556	19 494	21 369	23 516	22 997	19 905	20 000 F
Korea Rep	5 253	1 512	871	...	565	503	449	262	389	243
Russian Fed	-	-	-	-	-	-	10	-	100	-
61 Fishing area total	*37 631 F*	*26 814 F*	*22 635*	*17 922*	*25 392*	*25 425*	*48 993 F*	*46 348 F*	*28 994 F*	*30 075 F*
67 Japan	-	14	-	-	-	-	-	-	-	-
USA	63	14	10	10	14	20	81	3	36	4
67 Fishing area total	*63*	*28*	*10*	*10*	*14*	*20*	*81*	*3*	*36*	*4*
71 Australia	529	1 006	938	1 040	1 243	1 215	2 014	1 495	1 722	1 703
China	-	-	-	-	-	-	160	-	-	-
China,Taiwan	46 400 F	32 600 F	36 224	31 687	31 703	35 559	16 431	20 650	29 961	29 952
Guam	...	5	0	0	0	-	0	0	0	0
Indonesia	35 470	39 221	45 376	39 891	35 863	39 711	44 372	48 794	45 933	47 690
Japan	780	1 124	480	246	374	227	297	203	155	...
Kiribati	1 830	1 800	1 820	1 840	1 830	2 381	3 012	1 581	1 273	2 769
Korea Rep	-	1	2	78	49	6	43	101	8	10
Malaysia	5 600	6 180	7 613	7 269	6 721	6 875	7 407	6 677	7 558	7 010
Philippines	4 698	4 175	4 079	3 839	1 690	2 086	2 174	2 071	2 542	2 616
Singapore	241	124	104	94	93	80	59	43	32	30
Solomon Is	60 F	140 F	80 F	50 F	4 000 F	600 F	310 F	300 F	300 F	300 F
Thailand	1 934	2 208	2 812	3 160	2 993	3 306	6 302	7 133	4 819	4 873
71 Fishing area total	*97 542 F*	*88 584 F*	*99 528 F*	*89 194 F*	*86 559 F*	*92 046 F*	*82 581 F*	*89 048 F*	*94 303 F*	*96 953 F*
77 Amer Samoa	...	2	0	-	4	-	-	-	-	-
China,Taiwan	-	-	4	6	6	54	52	72	105	96
Cook Is	32 F	30 F	30 F	20 F	20 F	20 F	20 F	20 F	20 F	20 F
Costa Rica	1 466	1 275	1 254	1 906	3 921	5 444	5 647	11 071	8 510	7 444
El Salvador	287	980	759	347	1 186	266	176	364	759	951
Fr Polynesia	...	420	365	387	367	347	427	582	705	1 063
Guatemala	225	225	207	81	146	237	203	151	250 F	359
Japan	3 591	3 645	5 215	2 982	2 400	3 173	1 829	1 163	956	...
Korea Rep	216	-	38	119	646	1 173	841	852	703	239
Mexico	15 825	16 348	15 959	16 841	13 330	13 982	15 350	16 755	15 654	15 811
Samoa	...	...	...	...	...	...	...	20	20 F	20 F
USA	1 443	1 183	1 466	2 543	2 861	2 892	2 930	1 547	187	35
77 Fishing area total	*23 085 F*	*24 108 F*	*25 297 F*	*25 232 F*	*24 887 F*	*27 588 F*	*27 475 F*	*32 597 F*	*27 869 F*	*26 038 F*
81 Australia	887	1 326	1 324	1 040	1 152	1 382	659	751	779	863
Japan	1 675	1 022	862	901	639	1 088	842	412	727	...
Korea Rep	858	1 396	815	636	191	380	1 051	1 105	861	1 635
New Zealand	...	...	3 129	2 375	1 580	673	1 062	6	-	13
Ukraine	...	...	...	...	...	...	...	...	1	-
81 Fishing area total	*3 420*	*3 744*	*6 130*	*4 952*	*3 562*	*3 523*	*3 614*	*2 274*	*2 368*	*2 511*

B-38

Sharks, rays, chimaeras	Capture production by species, fishing areas and countries or areas
Squales, raies, chimères	Captures par espèces, zones de pêche et pays ou zones
Tiburones, rayas, quimeras	Capturas por especies, áreas de pesca y países o áreas

Species, Fishing area Espèce, Zone de pêche Especie, Area de pesca	1993 t	1994 t	1995 t	1996 t	1997 t	1998 t	1999 t	2000 t	2001 t	2002 t
87 Japan	996	1 415	671	857	526	1 167	441	438	648	...
Korea Rep	-	-	-	-	5	-	-	-	-	-
Peru	1 212	548	694	1 506	1 915	4 335	2 951	4 307	3 618	3 433
Spain	-	-	-	-	289	-	774	1 776	2	945
87 Fishing area total	*2 208*	*1 963*	*1 365*	*2 363*	*2 735*	*5 502*	*4 166*	*6 521*	*4 268*	*4 378*
Species total	*348 550 F*	*337 493 F*	*341 363 F*	*387 921 F*	*367 887 F*	*365 171 F*	*374 692 F*	*395 780 F*	*370 657 F*	*374 026 F*
Group total	***741 816***	***757 369***	***763 310***	***815 533***	***825 970***	***817 903***	***833 397***	***857 849***	***819 327***	***818 542***

B-39 Marine fishes not identified / Poissons marins non identifiés / Peces marinos no identificados

Capture production by species, fishing areas and countries or areas
Captures par espèces, zones de pêche et pays ou zones
Capturas por especies, áreas de pesca y países o áreas

Species, Fishing area Espèce, Zone de pêche Especie, Area de pesca	1993 t	1994 t	1995 t	1996 t	1997 t	1998 t	1999 t	2000 t	2001 t	2002 t
Groundfishes nei	**Poissons de fond nca**		**Peces de fondo nep**		***Osteichthyes***				**1,99(XX)XXX,XX**	**GRO**
21 Canada	829	1 176	872	597	475	365	770	1 003	831	-
Cuba	9	4	13	12	162	-	-	144	-	-
Faeroe Is	94	-	-	-	-	-	-	-	-	-
Greenland	432	639	623	691	1 273	588	-	-	-	-
Japan	7	28	130	57	32	35	31	11	6	154
Latvia	-	-	43	-	-	-	-	-	-	-
UK	-	-	-	-	-	283	-	-	-	-
USA	0	126	0	-	-	-	-	-	0	0
21 Fishing area total	*1 371*	*1 973*	*1 681*	*1 357*	*1 942*	*1 271*	*801*	*1 158*	*837*	*154*
27 Belgium	1 081	1 158	1 027	1 290	1 389	979	774	643	501	509
Denmark	2	1	0	0	0	0	-	-	-	-
Faeroe Is	-	-	-	-	474	133	-	-	-	-
France	211	409	277	228	-	-	-	-	-	-
Iceland	244	512	337	237	179	2	82	-	45	55
Japan	11	3	-	-	-	-	-	-	-	6
Poland	7	29	165	36	39	41	-	-	-	-
Spain	12 504	16 589	13 806	12 354	2 506	5 411	3 506	7 687	828	743
UK	4 706	5 262	5 093	4 131	3 689	3 401	2 498	911	977	893
27 Fishing area total	*18 766*	*23 963*	*20 705*	*18 276*	*8 276*	*9 967*	*6 860*	*9 241*	*2 351*	*2 206*
Species total	*20 137*	*25 936*	*22 386*	*19 633*	*10 218*	*11 238*	*7 661*	*10 399*	*3 188*	*2 360*
Pelagic fishes nei	**Poissons pélagiques nca**		**Peces pelágicos nep**		***Osteichthyes***				**1,99(XX)XXX,XX**	**PEL**
21 Canada	-	-	-	36	128	25	31	29	208	-
Cuba	-	-	-	-	-	-	13	-	-	-
Japan	1	-	-	-	-	-	-	-	-	-
Russian Fed	-	-	102	8	-	-	-	-	-	-
21 Fishing area total	*1*	*-*	*102*	*44*	*128*	*25*	*44*	*29*	*208*	*-*
27 Belgium	10	20	7	4	2	1	1	2	4	30
Poland	12	50	281	60	66	61	-	-	-	-
Spain	30 034	11 375	18 450	25 609	8 169	15 685	4 426	2 947	100	4 701
UK	182	165	-	-	10	-	12	-	-	-
27 Fishing area total	*30 238*	*11 610*	*18 738*	*25 673*	*8 247*	*15 747*	*4 439*	*2 949*	*104*	*4 731*
Species total	*30 239*	*11 610*	*18 840*	*25 717*	*8 375*	*15 772*	*4 483*	*2 978*	*312*	*4 731*
Finfishes nei	**Poissons téléostéens nca**		**Peces de escama nep**		***Osteichthyes***				**1,99(XX)XXX,XX**	**FIN**
21 Canada	140	2 316	125	1	-	-	1	-	-	-
Greenland	-	-	-	-	-	-	-	769	589	584
Lithuania	-	-	-	-	-	-	-	-	1	1
Norway	-	-	-	-	-	-	-	3	-	-
Poland	-	-	-	-	-	-	-	-	1	-
Portugal	238	12	14	22	114	40	122	164	41	95
Russian Fed	11	-	17	72	-	26	57	63	162	153
Spain	601	-	72	56	20	30	35	68	74	175
USA	5 802	5 588	15 595	12 935	12 153	4 431	6 398	5 907	2 566	3 354
21 Fishing area total	*6 792*	*7 916*	*15 823*	*13 086*	*12 287*	*4 527*	*6 613*	*6 974*	*3 434*	*4 362*
27 Denmark	1 094	400	545	118	103	116	78	142	107	209
Estonia	78	60	29	26	23	28	32	41	91	125
France	5 688	5 199	5 167	3 771	3 732	2 344	2 179	8 759	12 661	5 844
Germany	-	248	316	270	291	164	150	66	37	36
Isle of Man	-	1	0	-	-	0	-	-	-	-
Lithuania	107	45	66	188	152	41	40	204	2	563
Norway	796	165	128	158	322	2 791	3 389	3 196	1 909	1 179
Poland	85	43	53	23	34	23	-	17	306	339
Portugal	15 684	10 467	12 809	9 330	9 293	10 727	4 950	2 350	1 531	636
Russian Fed	1 935	1 295	1 071	729	520	577	788	22 596	536	3 233
Spain	11 781	10 362	10 031	10 665	19 512	33 681	22 056	17 424	12 572	13 932
Sweden	755	741	12 650	3 065	3 003	338	382	540	508	300
Ukraine	316	557	167	32	-	-	-	-	-	-
27 Fishing area total	*38 319*	*29 583*	*43 032*	*28 375*	*36 985*	*50 830*	*34 044*	*55 335*	*30 260*	*26 396*
Species total	*45 111*	*37 499*	*58 855*	*41 461*	*49 272*	*55 357*	*40 657*	*62 309*	*33 694*	*30 758*
Marine fishes nei	**Poissons marins nca**		**Peces marinos nep**		***Osteichthyes***				**1,99(XX)XXX,XX**	**MZZ**
21 Japan	18	29	6	6	1	7	29	12	4	-
St Pier Mq	-	-	-	-	18	18	-	-	-	-
21 Fishing area total	*18*	*29*	*6*	*6*	*19*	*25*	*29*	*12*	*4*	*-*
27 Channel Is	16 F	-	-	22	203	27	40	-	-	-
Faeroe Is	...	...	...	1 201	1 362	858	247	3 899	2 911	2 932 F
Greenland	144	244	80	168	189	1 187	...	200	264	41
Ireland	-	-	-	-	-	575	177	1	284	575
Japan	2	-	3	3	1	1	2	5	4	-
Latvia	-	660	464	-	-	4	-	-	-	-
Netherlands	16 687	20 267	19 457	16 596	19 114	3 305	3 442	597	-	83
Romania	-	94	76	13	-	-	-	-	-	-
Svalbard Is	0	0	0	0	0	0	0	0	0	0
27 Fishing area total	*16 849 F*	*21 265*	*20 080*	*18 003*	*20 869*	*5 957*	*3 908*	*4 702*	*3 463*	*3 631 F*
31 Anguilla	232	234	105 F	140 F	180 F	180 F	180 F	180 F	180 F	180 F

B-39

Marine fishes not identified — Capture production by species, fishing areas and countries or areas
Poissons marins non identifiés — Captures par espèces, zones de pêche et pays ou zones
Peces marinos no identificados — Capturas por especies, áreas de pesca y países o áreas

	Species, Fishing area Espèce, Zone de pêche Especie, Area de pesca	1993 t	1994 t	1995 t	1996 t	1997 t	1998 t	1999 t	2000 t	2001 t	2002 t
	Antigua Barb	450	487	1 116	1 045	1 242	1 013	1 041	1 164	66	13
	Aruba	170	125	20	20	55	40	40	40	43	43 F
	Bahamas	544	428	381	385	264	156	139	110	133	150 F
	Barbados	69	83	220	113	74	62	72	60	109	89
	Belize	217	145	166	400	121	133	202	119	66	83
	Bermuda	94	101	93	96	108	113	80	55	57	72
	Br Virgin Is	291	192	389	265	39	42	41	27	34 F	34 F
	Cayman Is	125	125	125	110	125	125	125	125	125	125
	China,Taiwan	110 F	80 F	71	89	53	48	184	29	32	46
	Colombia	1 524	2 015	2 879	4 384	3 079	1 701	1 061	5 752 F	5 595 F	5 595 F
	Costa Rica	133	111	197	73	163	141	367	563	583	427
	Cuba	16 282	21 323	18 384	13 846	17 278	13 750	20 968	22 918	11 384	5 588
	Dominica	666	730	820	939	988	1 121	870 F	872 F	881 F	576
	Dominican Rp	188	7	-	2 650	1 874	628	1 896	2 401	4 391	462
	Fr Guiana	3 500	3 578	3 634	3 000 F	2 500	2 500	2 500 F	2 500 F	2 500 F	2 500 F
	Grenada	347	147	135	173	130	158	29	5	10	3
	Guadeloupe	5 590 F	5 765	6 270 F	6 220	6 860 F	5 800 F	5 800 F	6 600 F	6 500 F	6 600 F
	Guatemala	92	120	270	270	195	213	196	203	200 F	200 F
	Guyana	37 151	38 122	37 000 F	33 971	34 841	38 124	37 534	27 666	24 662	24 569
	Haiti	3 200 F	3 730 F	3 600 F	4 000 F	4 000 F	4 000 F	3 800 F	3 800 F	3 800 F	3 800 F
	Honduras	564	389	313	372	216 F	379 F	436 F	1 938 F	2 471 F	2 909 F
	Jamaica	7 300 F	7 000 F	7 300 F	8 500 F	5 305	4 161	6 283	4 586	4 616 F	4 604 F
	Japan	-	1	-	-	2	5	8	8	5	...
	Korea Rep	-	-	-	-	-	-	50	1 023	1	-
	Martinique	3 036 F	2 810 F	2 390 F	1 647 F	3 430 F	3 060 F	810 F	45 F	35 F	35 F
	Mexico	136 458	142 846	77 964	81 517	108 324	91 863	70 679	59 872	52 439	52 813
	Montserrat	58	62	48	38	45	46	50 F	50 F	50 F	46
	NethAntilles	590 F	535 F	505 F	505 F	455 F	455 F	455 F	455 F	570 F	570 F
	Nicaragua	392	1 301	1 367	2 279	2 291	4 088	741	589	518	2 220
	Puerto Rico	195	979	983	569	691	50	264	298	173	126
	Russian Fed	-	-	82	-	-	-	-	-	-	-
	St Kitts Nev	200 F	172	80	96	54	79	163	174	268	83
	St Lucia	790	442	385	308	264	243	325	352	435	324
	St Vincent	1 218	933	774	737	821	1 152	874	675	653	...
	Spain	-	-	-	-	-	-	-	-	-	111
	Suriname	9 250	14 091	11 855 F	10 400 F	11 777 F	11 845	9 800	10 500 F	11 300	11 180 F
	Trinidad Tob	1 298	6 927	4 447 F	1 609	2 818	1 424	1 928	2 074	2 397	2 668
	Turks Caicos	307	310 F	300 F	300 F	250 F	250 F	250 F	200 F	200 F	200 F
	USA	22 352	26 959	14 266	7 694	4 527	7 320	5 545	5 427	3 634	3 804
	US Virgin Is	560 F	470 F	400 F	340 F	300 F	255 F	119	139 F	139 F	139 F
	Venezuela	27 223	27 691	32 109	33 049	36 793	16 376	-	464	11 304	29 592
31	*Fishing area total*	*282 766 F*	*311 566 F*	*231 443 F*	*222 149 F*	*252 532 F*	*213 099 F*	*175 905 F*	*164 058 F*	*152 559 F*	*162 579 F*
34	Belize	80	7	27	-	372	1 611	5 211	7 551	829	1 068
	Benin	455 F	1 110 F	889 F	1 785 F	3 628 F	3 294 F	1 677	1 150 F	370	1 192
	Bulgaria	-	-	-	-	-	269	-	-	-	-
	Cameroon	3 000	3 740 F	-	-	-	-	-	954	907	830
	Cape Verde	3 980	4 774	4 780	6 612	538	333	640	719	633	700 F
	Chile	-	10	-	-	-	-	-	-	-	-
	China	2 325	653	4 389	4 295	3 691	2 713	4 536	6 604	10 803	-
	China,Taiwan	110 F	80 F	71	89	53	73	374	197	23	63
	Congo Dem R	420 F	380 F	387 F	396 F	383 F	394 F	400 F	300 F	400 F	600 F
	Congo Rep	941 F	578 F	1 403 F	2 036 F	2 645 F	2 260 F	2 829 F	7 232 F	5 455 F	5 277 F
	Côte dIvoire	22 922	20 118	23 101	17 311 F	17 602 F	22 313 F	26 088 F	21 502	25 500	26 000 F
	Cuba	0	0	0	172	-	-	-	-	-	-
	Cyprus	-	-	-	2 632	13 640	276	691	647	757	-
	Eq Guinea	277 F	649 F	96 F	789 F	725 F	700	820 F	117 F	169 F	169 F
	Estonia	15 162	2 200	31	148	-	469	-	0	-	-
	France	10 000	7 210	12 669	13 319	13 220	1 868	24	-	1	-
	Gabon	8 538 F	9 142	8 256	278	1 325	1 154	3 949	2 896	...	284
	Gambia	186	37	27	18	59	-	-	-	6	204
	Georgia	600 F	250 F	-	-	-	-	-	-	-	-
	Germany	-	-	-	40	45	53	110	-	132	77
	Ghana	23 620	19 552	19 946	25 128	27 896	43 399	115 933	22 235	40 993	26 234
	Greece	4 038	2 579	2 198	1 986	1 104	1 998	1 223	1 644	1 298	967
	Guinea	11 290 F	12 539 F	8 001	3 695	7 089	7 631	23 580	18 646	23 308	22 600 F
	GuineaBissau	-	5	17	-	-	-	-	-	-	-
	Honduras	1 710	125	1 624	1 638	1 600	481	801	752	620	-
	Ireland	-	-	-	-	-	-	-	-	425	505
	Italy	-	-	3 752	1 460	2 241	1 631	1 302	954	2 183	1 805
	Japan	60	62	50	51	22	37	31	98	46	...
	Korea D P Rp	28	-	-	-	-	-	-	154	-	-
	Korea Rep	1 377	10 584	4 335	4 856	3 501	2 393	4 403	5 398	5 303	6 399
	Latvia	17 530	18 231	22 475	19 130	3 118	1 103	3 247	1 616	996	407
	Liberia	710	598	134	362	157	539	797	281	640	650 F
	Lithuania	-	-	9 572	33 330	25 680	3 495	1 099	899	76 035	-
	Malta	-	1 236	3 465	8 197	-	-	-	-	-	-
	Marshall Is	-	-	-	2 403	-	-	-	-	-	-
	Mauritania	18 591 F	13 917 F	15 404 F	18 595 F	19 711 F	22 179 F	37 674 F	39 997 F	43 070 F	42 918 F
	Morocco	38 141	40 527	40 909	39 138	84 415	63 098	53 179	5 000	13 220	25 150
	Netherlands	-	-	-	318	292	822	455	604	1 781	1 367
	Nigeria	12 260	20 292	39 520	32 995	21 516	40 295	33 770	51 541	49 205	28 152
	Panama	257	-	34	6	49	50	166	199	-	-
	Poland	-	2 220	-	2 142	641	-	-	-	5 999	1 969
	Portugal	577	631	327	734	479	401	256	21	52	73
	Romania	9	-	-	-	-	-	-	-	-	-
	Russian Fed	3 661	840	6 275	6 815	6 140	6 304	-	6 379	2 242	857
	St Vincent	92	-	9	-	198	1 395	328	536	926	966
	Sao Tome Prn	927 F	1 837 F	1 933 F	2 579	683	1 167 F	1 284 F	1 234 F	1 238 F	1 238 F

B-39

Marine fishes not identified — **Capture production by species, fishing areas and countries or areas**
Poissons marins non identifiés — **Captures par espèces, zones de pêche et pays ou zones**
Peces marinos no identificados — **Capturas por especies, áreas de pesca y países o áreas**

Species, Fishing area / Espèce, Zone de pêche / Especie, Area de pesca	1993 t	1994 t	1995 t	1996 t	1997 t	1998 t	1999 t	2000 t	2001 t	2002 t
Senegal	38 369	10 778	18 088	13 522	13 782	7 590	6 679	12 146	14 409	8 721
Sierra Leone	0	0	0	0	0	0	4	2 732	3 609	3 547
Spain	14 701 F	19 043 F	19 522 F	10 033	15 237	18 797	11 060	7 185	9 750	2 445
Togo	341	226	-	176	823	594	2 059	1 424	1 027	807
Ukraine	5 567	4 034	20 488	15 528	7 167	26 274	57 653	53 274	88 540	34 091
Vanuatu	16	7	11	9	27	10	-	-	-	-
Westn Sahara	0	0	0	0	0	0	0	0	0	0
Other nei	3 070	6 110	124 126	13	202	-	740	675	648	1 012
34 Fishing area total	*265 938 F*	*236 911 F*	*418 341 F*	*294 759 F*	*301 696 F*	*289 463 F*	*405 072 F*	*285 493 F*	*433 548 F*	*249 344 F*
37 Albania	250 F	450 F	105	327	108	370	375	789	312	323
Algeria	1 913 F	2 737 F	3 890	3 964	3 799	2 363	3 403	9 323	6 010	6 749
Bosnia Herzg	0	0	0	0	0	0	0	0	0	0
Bulgaria	7	8 F	70	49	51	107	-	-	2	15
Croatia	1 094	771	738	619	613	590	619	1 080	863	1 625
Cyprus	405	339	322	425	363	359	253	274	238	183
Egypt	10 773	11 535	7 985	7 434	10 125	12 824	9 308	4 109	3 637	4 035
France	1 337	984	350	715	700	700	907	751	1 748	745
Georgia	-	-	-	25	-	-	1	4	5	17
Gibraltar	0	0	0	0	0	0	0	0	0	0
Greece	11 635	12 518	11 197	9 100	11 329	7 085	8 383	10 951	10 055	9 430 F
Israel	203 F	84 F	727	496	912	1 273	1 129	1 143	958	848
Italy	28 595	27 503	34 458	31 259	26 017	26 142	22 053	23 177	24 946	19 790
Korea Rep	-	-	2	110	-	-	-	-	-	-
Libya	3 955 F	4 078 F	4 410 F	3 688 F	2 903 F	3 617 F	3 755 F	3 924 F	3 550 F	5 496 F
Malta	4	3	1	2	1	12	16	29	82	59
Monaco	3 F	3 F	3 F	3 F	3 F	3 F	3 F	3 F	3 F	3 F
Morocco	998	2 279	2 885	3 135	4 266	4 679	1 649	2 600	1 241	2 419
Palest, O.T.	...	...	1 057	559	140	194	200 F	170 F	140 F	131
Portugal	-	-	-	-	-	-	-	7	12	-
Romania	244	1	12	10	2	2	1	1	5	6
Russian Fed	21	4 450	9	2	32	1	1	727	-	-
Serbia-Monte	24	31	21	19	25	20	25	27	-	22
Slovenia	35	15	12	8	7	5	4	-	-	-
Spain	28 147 F	28 704 F	27 190 F	22 864	20 406	18 105	8 741	8 889	6 200	9 893
Syria	85 F	81 F	91 F	90	150	328	342	580	462	-
Tunisia	16 251	22 469	18 828	18 813	5 112	12 965	9 617	5 125	6 546	2 228
Turkey	1 596	1 412	1 829	1 855	1 095	1 861	1 375	7 365	1 230	2 380
Ukraine	211	64	57	8	4	1	-	-	20	6
37 Fishing area total	*107 786 F*	*120 519 F*	*116 249 F*	*105 579 F*	*88 163 F*	*93 606 F*	*72 160 F*	*81 048 F*	*68 265 F*	*66 403 F*
41 Argentina	1 919	4 661	10 945	4 960	7 855	6 702	5 998	-	-	-
Australia	-	-	-	-	-	234	389	-	-	-
Belize	-	-	-	-	-	-	7	222	43	0
Brazil	195 450 F	191 029 F	145 188	143 419	146 661	138 925	169 938	176 695	151 386	161 161
Chile	-	-	-	-	-	31	5	-	-	-
Estonia	-	-	-	-	-	-	-	-	-	4
Falkland Is	78	17	50	370	181	1 033	1 217	1 250	774	604
France	-	8	15	0	0	15	-	-	-	-
Honduras	57	2	4	12	-	-	-	-	-	-
Italy	-	-	447	174	-	-	-	-	-	-
Japan	57	30	43	44	5	21	15	53	21	14
Korea Rep	16 171	12 881	11 043	6 136	5 873	4 020	1 176	5 724	10 621	10 480
Latvia	-	413	76	68	-	-	-	-	-	-
Namibia	-	-	-	-	4	16	96	-	-	-
Panama	-	-	4	43	-	-	-	-	-	-
Poland	-	72	-	-	-	-	8	-	-	-
Portugal	0	5	5	7	2	6	1	8	7	-
Russian Fed	37	-	-	-	-	-	-	-	14	199
Seychelles	-	-	-	-	6	7	-	-	-	-
Spain	550	219	1 037	-	-	-	1 379	207	370	570
UK	5	0	20	48	21	85	0	13	19	23
Uruguay	2 851	3 072	4 083	3 289	4 510	6 335	936	3 669	2 686	3 109
41 Fishing area total	*217 175 F*	*212 409 F*	*172 960*	*158 570*	*165 118*	*157 430*	*181 165*	*187 841*	*165 941*	*176 164*
47 Angola	52 634	64 541	44 922	80 633	74 578	43 038	32 185	31 674	80 740	87 406
China	-	-	-	-	-	-	-	5	-	-
Honduras	65	-	20	-	-	-	-	-	-	-
Iceland	-	-	-	-	-	4	-	-	-	-
Italy	-	-	50	20	-	-	-	-	-	-
Japan	956	527	896	922	388	825	763	676	441	...
Korea Rep	499	746	1 244	1 091	2 550	3 473	3 074	2 599	2 041	2 816
Namibia	3 411	3 143	3 846	5 658	15 551	7 371	13 009	3 205	2 029	4 125
Poland	-	-	120	33	-	-	-	-	2	86
Portugal	-	-	-	-	5	3	32	54	74	19
Russian Fed	48	98	260	301	147	3 742	8 138	3 732	146	92
St Helena	9	79	79	79	37	56	63	64	67	65
South Africa	770	2 184	1 766	1 549	13 108	1 075	3 586	1 327 F	4 505	4 573
Spain	82 F	75 F	51 F	-	56	301	79	512	444	93
Ukraine	4 690	2 459	1 185	220	108	-	-	3 781	531	-
47 Fishing area total	*63 164 F*	*73 852 F*	*54 439 F*	*90 506*	*106 528*	*59 888*	*60 929*	*47 629 F*	*91 020*	*99 275*
48 Argentina	-	-	10	-	-	-	-	-	-	-
Chile	-	-	-	-	-	7	-	-	-	-
Korea Rep	-	-	-	-	-	-	-	-	-	0
Poland	-	-	-	-	-	-	-	-	-	1
Russian Fed	-	0	-	-	-	-	-	-	-	-
South Africa	-	-	-	-	-	0	-	-	-	-
USA	-	-	-	-	-	-	0	-	0	-

B-39 Marine fishes not identified / Poissons marins non identifiés / Peces marinos no identificados

Capture production by species, fishing areas and countries or areas / Captures par espèces, zones de pêche et pays ou zones / Capturas por especies, áreas de pesca y países o áreas

Species, Fishing area Espèce, Zone de pêche Especie, Area de pesca	1993 t	1994 t	1995 t	1996 t	1997 t	1998 t	1999 t	2000 t	2001 t	2002 t
48 Fishing area total	*-*	*0*	*10*	*-*	*-*	*7*	*0*	*-*	*0*	*1*
51 Bahrain	764	785	1 043	1 185	1 029	852	938	859	1 077	803
Br Ind Oc Tr	0	0	0	0	0	0	0	0	0	0
China	-	-	-	-	-	-	-	330	169	...
China,Taiwan	3 190 F	2 240 F	2 312	1 953	1 720	267	440	2 081	905	1 421
Comoros	1 200	1 200	1 159 F	1 010 F	1 190 F	1 190 F	1 120 F	1 260 F	1 180	1 200 F
Djibouti	10 F	11 F	10 F	10 F	15 F	15 F	15 F	15 F	15 F	15 F
Egypt	18 051	18 121	13 361	24 063	30 339	27 923	48 787	22 770	23 885	28 274
Eritrea	475	449	314	202	215	211	246	8	1	100
Fr South Tr	56	75	75 F	75 F	75 F	75 F	80 F	80 F	80 F	80 F
India	413 427	404 127	492 372	256 427	307 120	264 758	288 658	293 193	418 694	377 883
Iran	175 209	148 553	149 090	157 397	143 212	139 107	130 225	122 620	124 118	121 692
Iraq	2 133	4 221	5 253	11 688	10 783	13 463	13 093	12 389	8 500 F	5 000 F
Israel	80	110	150	225	171	137	98	...	120	...
Italy	-	-	769	299	-	-	-	-	-	-
Japan	272	169	189	194	310	426	270	269	195	...
Jordan	45	60	75	90	100	6	11	12	10	24
Kenya	2 611	728	1 842	1 187	1 006	1 675	1 774	474	667	874
Korea Rep	5 536	2 350	3 048	1 979	4 057	4 878	2 269	4 952	1 426	3 353
Kuwait	1 579	753	729	893	922	1 055	231	550 F	580	590 F
Madagascar	62 029	56 512	57 944	56 365	57 996	68 688	73 417	74 107	77 601	82 731
Maldives	13 081	17 181	17 620	18 526	17 914	16 788	13 450	19 618	3 571	5 024
Mauritius	586	594	550	589	525	502	515	441	568	442
Mayotte	500 F	600 F	700 F	1 000	1 500 F	1 500 F	2 000	2 600 F	3 100	2 100
Mozambique	6 108	7 437	6 209	13 058	10 374	6 990	8 337	7 786	8 951	9 387
Norway	-	-	-	-	-	-	-	42	-	-
Oman	2 980	1 422	2	101	355	432	1	384	1 273	502
Pakistan	31 359	40 725	14 742	16 311	21 042	35 352	39 473	36 451	35 450	31 769
Portugal	-	-	-	-	-	-	10	-	4	11
Qatar	839	627	593	357	332	38	10	587	79	27
Réunion	582	845	810	775	1 279	1 000	67	140	96	161 F
Russian Fed	-	-	-	-	-	-	-	-	11	7
Saudi Arabia	4 325	3 074	1 480	1 205	820	768	586	574	804	893
Seychelles	273	295	123	221	108	122	416	245	228	253
Somalia	26 600 F	28 700 F	26 800 F	24 900 F	23 000 F	21 100 F	19 100 F	19 100 F	18 900 F	17 150 F
South Africa	77	139	66	85	228	92	125 F	176	85	55
Spain	-	-	-	-	-	-	21	661	69	359
Sudan	2 500	4 000	4 000	4 500	5 000	5 436	5 420	4 937 F	4 887	4 887
Tanzania	7 112	6 108	6 721	6 334	8 300	8 395	2 125	2 000	2 000	2 000
Ukraine	50	100	33	102	59	1	334	-	35	-
Untd Arab Em	14 874	9 908	8 548	7 441	10 015	9 007	9 019	2 046	1 213	382
51 Fishing area total	*798 513 F*	*762 219 F*	*818 732 F*	*610 747 F*	*661 111 F*	*632 249 F*	*662 681 F*	*633 757 F*	*740 547 F*	*699 449 F*
57 Australia	20 820 F	5 031 F	13 431 F	11 499 F	15 562	14 547	19 511	6 908	4 663	10 384
Bangladesh	161 860	110 314	115 122	119 484	135 860	145 260	137 285	161 977	193 746	231 041
China,Taiwan	110 F	80 F	34	29	25	11	9	96	132	29
Christmas Is	0	0	0	0	0	0	0	0	0	0
Cocos Is	0	0	0	0	0	0	0	0	0	0
India	119 554	169 430	168 837	199 275	196 114	235 105	184 787	268 175	287 900	391 993
Indonesia	69 076	65 154	59 673	57 318	46 353	68 638	56 135	54 577	43 510	41 860
Japan	209	578	1 062	1 093	1 140	1 498	1 360	1 198	921	...
Korea Rep	-	-	7	-	142	-	428	950	1 443	1 723
Malaysia	166 280	163 315	174 506	177 896	174 448	206 661	183 297	205 880	197 607	212 517
Myanmar	585 637	584 876	581 073	435 868	607 814	656 406	731 664	849 018	900 492	975 113
Spain	-	-	-	-	-	-	-	-	-	13
Sri Lanka	35 090	41 998	55 428	36 404	23 398	25 084	16 106	24 390	43 890	44 007
Thailand	425 370	300 258	321 441	312 014	302 880	339 948	329 976	308 922	324 359	321 983
Timor-Leste	...	...	...	...	...	...	400 F	350	344	338 F
57 Fishing area total	*1 584 006 F*	*1 441 034 F*	*1 490 614 F*	*1 350 880 F*	*1 503 736*	*1 693 158*	*1 660 958 F*	*1 882 441*	*1 999 007*	*2 231 001 F*
58 Australia	0	-	-	-	1	0	0	0	-	-
South Africa	-	-	-	-	-	0	-	0	-	-
Ukraine	14	0	61	-	-	-	-	-	-	-
58 Fishing area total	*14*	*0*	*61*	*-*	*1*	*0*	*0*	*0*	*-*	*-*
61 China	3 094 592	3 237 857	3 321 368	4 394 586	4 155 850	4 291 743	4 399 365	3 988 138	3 523 090	3 408 278
China,H.Kong	103 926	104 611	98 215	101 269	101 062	108 206	76 460 F	94 272 F	104 550 F	102 250 F
China, Macao	1 200	1 259	1 151	970	1 020 F	1 020 F	1 020 F	1 020 F	1 020 F	1 020 F
China,Taiwan	108 251 F	75 219 F	77 082	48 677	60 098	71 179	72 750	79 064	74 853	81 434
Japan	374 123	319 991	312 769	287 326	286 882	264 415	256 730	271 352	247 560	334 626 F
Korea D P Rp	349 900 F	220 011	155 973	166 325	109 770	114 200 F	113 000 F	112 000 F	108 991 F	107 600 F
Korea Rep	90 327	91 418	91 708	148 719	136 916	115 298	90 683	77 224	101 327	71 691
Poland	-	-	108	9	1	-	-	25	-	-
Russian Fed	3 780	5 941	-	-	1 621	1 259	104	66 076	43 018	15 174
61 Fishing area total	*4 126 099 F*	*4 056 307 F*	*4 058 374*	*5 147 881*	*4 853 220 F*	*4 967 320 F*	*5 010 112 F*	*4 689 171 F*	*4 204 409 F*	*4 122 073 F*
67 Canada	67 490	121 769	62 475	105 185	102 100	96 130	97 253	26 979	69 582	-
Japan	485	430	-	-	-	-	-	-	-	-
Russian Fed	-	-	-	32	-	-	-	-	-	2
USA	28 572	1 832	2 439	1 603	4 501	1 723	938	4 024	4 763	4 862
67 Fishing area total	*96 547*	*124 031*	*64 914*	*106 820*	*106 601*	*97 853*	*98 191*	*31 003*	*74 345*	*4 864*
71 Australia	4 866 F	3 827 F	4 500 F	5 326 F	197	422	7 335	3 900	4 748	5 129
Brunei Darsm	1 330	3 131	4 389	6 818	4 405	4 930	3 081	2 356	1 186	1 528
Cambodia	24 835	22 400	22 455	22 958	21 928	23 701	28 070	26 585	30 980	33 830
China	1 377	1 738	926	380	110	174	818	1 047	825	283
China,Taiwan	-	-	141	119	105	3 493	66	2	-	45
Fiji Islands	632	724	1 414	1 800	1 035	1 358	1 218	1 200 F	1 113	1 070 F

B-39 Marine fishes not identified / Poissons marins non identifiés / Peces marinos no identificados

Capture production by species, fishing areas and countries or areas
Captures par espèces, zones de pêche et pays ou zones
Capturas por especies, áreas de pesca y países o áreas

Species, Fishing area Espèce, Zone de pêche Especie, Area de pesca	1993 t	1994 t	1995 t	1996 t	1997 t	1998 t	1999 t	2000 t	2001 t	2002 t
Guam	72	52	4	13	18	49	85	92	89	72
Indonesia	295 620	333 283	321 392	373 113	366 727	387 908	414 441	454 389	525 084	565 460
Japan	1 743	1 606	1 237	1 273	156	396	427	355	285	...
Kiribati	2 420	2 323	2 350	2 400	2 380	3 875	4 216	2 396	3 198	3 130
Korea Rep	10 038	16 691	13 965	11 628	7 965	5 920	3 420	4 321	3 177	1 811
Malaysia	166 241	195 030	193 020	174 941	157 279	176 145	190 892	205 348	215 813	233 783
Marshall Is	335 F	340 F	350 F	369	400 F	500 F	500 F	500 F	500 F	500 F
Micronesia	1 070 F	1 010 F	1 100 F	1 200 F	1 300 F	1 300 F	1 400 F	1 400 F	1 500 F	1 500 F
Nauru	500	500	400 F	300 F	250 F	200 F	150 F	100 F	50 F	7
NewCaledonia	229	471	363	415	385	458	415	509	618	602 F
N Marianas	94	104	79	61	88	80	75	75	49	55
Palau	944	884	853	846	845	868	886	878	883	868
Papua N Guin	9 500 F	9 500 F	10 000 F	10 500 F	10 500 F	10 000 F	10 000 F	10 000 F	10 000 F	10 000 F
Philippines	9 039	14 645	19 052	37 385	10 176	13 104	12 783	12 043	14 656	16 527
Singapore	3 992	4 231	4 465	4 308	3 933	2 812	2 740	2 339	1 325	1 031
Solomon Is	14 500 F	12 500 F	12 000 F	12 000 F	12 000 F	12 000 F	12 000 F	12 000 F	12 000 F	12 000 F
Thailand	818 746	863 460	805 867	743 398	702 437	636 929	660 947	685 805	663 800	649 908
Tuvalu	657	120	112	110 F	160 F	160 F	160 F	160 F	160 F	160 F
Vanuatu	1 200 F	1 160 F	1 200 F	1 200 F	1 300 F	1 300 F	1 400 F	1 400 F	1 500 F	1 500 F
Viet Nam	669 832	719 661	722 055	808 226	832 118	849 310	922 690	927 205	1 033 623	1 025 800
Wallis Fut I	150	187	164	174	170	294	294 F	294 F	294 F	294 F
71 Fishing area total	*2 039 962 F*	*2 209 578 F*	*2 143 853 F*	*2 224 261 F*	*2 138 367 F*	*2 137 686 F*	*2 280 509 F*	*2 356 699 F*	*2 527 456 F*	*2 566 893 F*
77 Amer Samoa	11	1	3	10	9	5	6	0	0	1
China	-	-	-	-	-	-	-	-	303	712
China,Taiwan	110 F	80 F	32	27	24	124	0	0	16	-
Cook Is	384 F	268 F	396 F	333 F	343 F	337 F	347 F	347 F	343 F	300 F
Costa Rica	6 280	5 141	5 224	7 543	10 564	7 727	9 466	3 978	3 544	3 458
El Salvador	1 501	2 246	2 553	2 267	1 791	1 595	1 842	1 925	1 678	3 385
Fr Polynesia	4 665	4 339	4 200	4 332	4 820	4 585	4 538	4 274	4 362	4 526
Guatemala	458	514	375	375	305	488	537	310	400 F	418
Honduras	448	344	261	240	471 F	1 461 F	1 282 F	738 F	1 404 F	1 959 F
Japan	1 712	1 160	1 232	1 268	729	972	1 230	785	861	...
Korea Rep	651	561	327	77	65	30	573	711	744	56
Mexico	120 349	174 176	238 947	180 690	131 611	113 369	66 776	83 729	71 432	74 800
Nicaragua	494	1 891	3 894	3 580	4 595	5 442	926	592	936	4 216
Niue	120 F	150 F	150 F	200 F	200 F	200 F	200 F	200 F	200 F	200 F
Panama	4 069	6 340	6 731	6 458	17 883	15 322	15 331	17 998	38 310	27 473
Samoa	345	350	350	322	1 513	1 640	5 163	5 289	5 200 F	5 200 F
Tokelau	200 F	200 F	200 F	200 F	200 F	200 F	200 F	200 F	200 F	200 F
Tonga	1 900	1 950	2 000	2 100	2 000	2 936	2 890	2 305	2 548	2 612
USA	16 453	2 425	2 640	3 288	4 323	3 190	3 359	3 080	3 004	43
US Minor Is	0	0	0	0	0	0	0	0	0	0
77 Fishing area total	*160 150 F*	*202 136 F*	*269 515 F*	*213 310 F*	*181 446 F*	*159 623 F*	*114 666 F*	*126 461 F*	*135 485 F*	*129 559 F*
81 Australia	10 281 F	9 109 F	7 797 F	12 829 F	14 924	20 545	3 923	2 140	2 220	2 155
China	-	-	-	-	-	-	-	-	-	17
China,Taiwan	2 130 F	1 500 F	0	0	-	-	0	112	7	-
Japan	4 742	3 003	220	226	2 271	269	278	245	403	...
Korea Rep	6 472	5 964	7 952	8 420	4 361	3 962	3 478	3 297	2 880	6 721
New Zealand	9 401	8 696	5 596	26 526	26 264	54	74	122	68	125
Norfolk Is	0	0	0	0	0	0	0	0	0	0
Norway	1 494	882	208	235	155	187	-	-	-	-
Pitcairn Is	8 F	8 F	8 F	8 F	8 F	8 F	8 F	8 F	8 F	8 F
Russian Fed	2 003	3 905	175	289	210	57	754	-	-	-
Ukraine	1 412	1 557	31	805	941	3 692	1 804	3 091	34 155	33 697
81 Fishing area total	*37 943 F*	*34 624 F*	*21 987 F*	*49 338 F*	*49 134 F*	*28 774 F*	*10 319 F*	*9 015 F*	*39 741 F*	*42 723 F*
87 Chile	3 238	2 119	2 088	2 253	1 014	897	2 464	6 414	5 861	8 581
Colombia	5 862	1 947	16 604	14 529 F	19 380 F	17 293 F	34 657 F	37 251 F	28 450 F	28 000 F
Cuba	-	-	-	-	-	-	4 234	-	-	-
Ecuador	39 644	34 297	5 426	20 307	8 262	9 495	97 162	186 431	39 897	39 730
Ghana	-	-	-	-	-	-	-	13	-	-
Japan	291	420	205	211	182	600	470	386	771	...
Korea Rep	-	-	-	-	-	188	360	484	12	12
Peru	20 146	15 686	58 779	45 788	83 203	375 784	164 855	91 905	122 447	63 296
Spain	-	-	-	-	-	-	-	-	3	186
87 Fishing area total	*69 181*	*54 469*	*83 102*	*83 088 F*	*112 041 F*	*404 257 F*	*304 202 F*	*322 884 F*	*197 441 F*	*139 805 F*
88 Chile	-	-	-	-	-	0	-	-	-	-
New Zealand	-	-	-	-	-	-	-	-	-	0
88 Fishing area total	-	-	-	-	-	*0*	-	-	-	*0*
Species total	*9 866 111 F*	*9 860 949 F*	*9 964 680 F*	*10 675 897 F*	*10 540 582 F*	*10 940 395 F*	*11 040 806 F*	*10 822 214 F*	*10 833 231 F*	*10 693 764 F*
Group total	***9 961 598***	***9 935 994***	***10 064 761***	***10 762 708***	***10 608 447***	***11 022 762***	***11 093 607***	***10 897 900***	***10 870 425***	***10 731 613***

B-41 Freshwater crustaceans / Crustacés d'eau douce / Crustáceos de agua dulce

Capture production by species, fishing areas and countries or areas / Captures par espèces, zones de pêche et pays ou zones / Capturas por especies, áreas de pesca y países o áreas

Species, Fishing area Espèce, Zone de pêche Especie, Area de pesca	1993 t	1994 t	1995 t	1996 t	1997 t	1998 t	1999 t	2000 t	2001 t	2002 t
Giant river prawn	**Bouquet géant**		**Langostino de río**		***Macrobrachium rosenbergii***			**2,28(12)023,07**		**PRF**
02 Martinique	3	-	-	-	-	-	-	-	-	-
02 Fishing area total	*3*	-	-	-	-	-	-	-	-	-
04 Brunei Darsm	11	3	5	14	17	35	26	23	16	13
Indonesia	4 923	5 393	5 524	6 126	5 208	5 362	5 937	5 449	5 185	5 240
04 Fishing area total	*4 934*	*5 396*	*5 529*	*6 140*	*5 225*	*5 397*	*5 963*	*5 472*	*5 201*	*5 253*
06 Fr Polynesia	-	0	0	0	3	3	3	3	3	3
06 Fishing area total	-	*0*	*0*	*0*	*3*	*3*	*3*	*3*	*3*	*3*
Species total	*4 937*	*5 396*	*5 529*	*6 140*	*5 228*	*5 400*	*5 966*	*5 475*	*5 204*	*5 256*
River prawns nei	**Bouquets d'eau douce nca**		**Camarones de agua dulce nep**		***Macrobrachium spp***			**2,28(12)023,XX**		**PPF**
01 Benin	120 F	139	188	150 F	130 F	120 F	120 F	100 F	110 F	110 F
01 Fishing area total	*120 F*	*139*	*188*	*150 F*	*130 F*	*120 F*	*120 F*	*100 F*	*110 F*	*110 F*
02 Mexico	4 456	3 226	4 202	4 261	3 580	3 244	4 161	3 478	3 112	3 051
Nicaragua	-	-	-	-	-	-	7	0	3	3
USA	20	19	19	-	-	-	-	0	-	-
02 Fishing area total	*4 476*	*3 245*	*4 221*	*4 261*	*3 580*	*3 244*	*4 168*	*3 478*	*3 115*	*3 054*
03 Brazil	6 700 F	6 900 F	1 412	2 677	2 157	1 506	3 235	2 437	...	...
03 Fishing area total	*6 700 F*	*6 900 F*	*1 412*	*2 677*	*2 157*	*1 506*	*3 235*	*2 437*	...	...
06 Papua N Guin	5 F	5 F	5 F	5 F	5 F	5 F	5 F	5 F	5 F	5 F
06 Fishing area total	*5 F*	*5 F*	*5 F*	*5 F*	*5 F*	*5 F*	*5 F*	*5 F*	*5 F*	*5 F*
Species total	*11 301 F*	*10 289 F*	*5 826 F*	*7 093 F*	*5 872 F*	*4 875 F*	*7 528 F*	*6 020 F*	*3 230 F*	*3 169 F*
Freshwater prawns, shrimps nei	**Crevettes d'eau douce nca**		**Gambas,camaron.(agua dulce)nep**		***Palaemonidae***			**2,28(12)XXX,XX**		**PPZ**
01 Egypt	2 352	3 687	3 169	2 286	2 387	2 886	1 388	2 411	2 450	2 326
Morocco	0	0	0	0	0	2	5	3	4	5
01 Fishing area total	*2 352*	*3 687*	*3 169*	*2 286*	*2 387*	*2 888*	*1 393*	*2 414*	*2 454*	*2 331*
03 Chile	7	5	-	-	-	-	-	-	-	-
03 Fishing area total	*7*	*5*	-	-	-	-	-	-	-	-
04 Indonesia	3 336	3 738	3 944	3 356	3 242	3 871	3 900	4 210	3 898	4 000
Japan	3 013	3 140	2 717	2 222	2 413	1 872	1 942	1 676	1 158	1 002
Malaysia	-	-	-	-	-	-	-	-	-	265
Philippines	3 351	3 938	4 581	5 214	4 282	2 014	3 915	4 020	4 507	5 115
Thailand	...	676	308	1 614	1 541	36	406	200	300	300
04 Fishing area total	*9 700*	*11 492*	*11 550*	*12 406*	*11 478*	*7 793*	*10 163*	*10 106*	*9 863*	*10 682*
Species total	*12 059*	*15 184*	*14 719*	*14 692*	*13 865*	*10 681*	*11 556*	*12 520*	*12 317*	*13 013*
Noble crayfish	**Ecrevisse à pieds rouges**		**Cangrejo de río de patas rojas**		***Astacus astacus***			**2,29(03)027,02**		**AAS**
05 Norway	...	...	...	...	10	10	10	...	...	...
05 Fishing area total	...	...	...	...	*10*	*10*	*10*	...	...	...
Species total	...	...	...	...	*10*	*10*	*10*	...	...	...
Signal crayfish	**Ecrevisse signal**		**...C**		***Pacifastacus leniusculus***			**2,29(03)076,01**		**PCL**
05 UK	-	-	-	-	-	-	10	81	80	50
05 Fishing area total	-	-	-	-	-	-	*10*	*81*	*80*	*50*
Species total	-	-	-	-	-	-	*10*	*81*	*80*	*50*
White-clawed crayfish	**Ecrevisse à pattes blanches**		**Cangrejo a pinzas blancas**		***Austropotamobius pallipes***			**2,29(03)139,01**		**AUP**
05 Spain	0	0	0	0	0	0	0	0	0	0
05 Fishing area total	*0*	*0*	*0*	*0*	*0*	*0*	*0*	*0*	*0*	*0*
Species total	*0*	*0*	*0*	*0*	*0*	*0*	*0*	*0*	*0*	*0*
Red swamp crawfish	**Ecrevisse rouge de marais**		**Cangrejo de las marismas**		***Procambarus clarkii***			**2,29(04)001,01**		**RCW**
01 Kenya	41	40	20	13	24	19	21	22	3	14
01 Fishing area total	*41*	*40*	*20*	*13*	*24*	*19*	*21*	*22*	*3*	*14*
05 Spain	2 379	274	2 753	2 500 F	2 500 F	2 500 F	2 500 F	2 500 F	2 500 F	2 500 F
05 Fishing area total	*2 379*	*274*	*2 753*	*2 500 F*	*2 500 F*	*2 500 F*	*2 500 F*	*2 500 F*	*2 500 F*	*2 500 F*
Species total	*2 420*	*314*	*2 773*	*2 513 F*	*2 524 F*	*2 519 F*	*2 521 F*	*2 522 F*	*2 503 F*	*2 514 F*
Australian crayfish	**Ecrevisse d'Australie**		**Cangrejo de río de Australia**		***Euastacus armatus***			**2,29(05)158,01**		**AYA**
06 Australia	-	-	-	-	-	-	66	24	23	69
06 Fishing area total	-	-	-	-	-	-	*66*	*24*	*23*	*69*
Species total	-	-	-	-	-	-	*66*	*24*	*23*	*69*

B-41 Freshwater crustaceans / Crustacés d'eau douce / Crustáceos de agua dulce

Capture production by species, fishing areas and countries or areas / Captures par espèces, zones de pêche et pays ou zones / Capturas por especies, áreas de pesca y países o áreas

Species, Fishing area Espèce, Zone de pêche Especie, Area de pesca	1993 t	1994 t	1995 t	1996 t	1997 t	1998 t	1999 t	2000 t	2001 t	2002 t
Oceanian crayfishes nei	**...B**		**...C**			***Parastacidae***			**2,29(05)XXX,XX**	**CJF**
06 Cook Is	...	...	10		...	...	...	...	...	...
Papua N Guin	3	11	10	6	6	14	6	0	1	1 F
06 Fishing area total	*3*	*11*	*20*	*6*	*6*	*14*	*6*	*0*	*1*	*1 F*
Species total	*3*	*11*	*20*	*6*	*6*	*14*	*6*	*0*	*1*	*1 F*
Euro-American crayfishes nei	**Ecrevisses euro-américain. nca**		**Cangrejos de río nep**			***Astacidae, Cambaridae***			**2,29(XX)XXX,XX**	**AYS**
02 Mexico	...	157	170	198	101	99	163	108	17	94
USA	22 105	11 130	7 073	5 689	10 496	10 082	5 322	217	4 676	7 125
02 Fishing area total	*22 105*	*11 287*	*7 243*	*5 887*	*10 597*	*10 181*	*5 485*	*325*	*4 693*	*7 219*
04 Turkey	404	524	551	850	1 100	1 500	1 372	1 681	1 634	1 894
04 Fishing area total	*404*	*524*	*551*	*850*	*1 100*	*1 500*	*1 372*	*1 681*	*1 634*	*1 894*
05 Bulgaria	-	-	-	-	-	-	-	-	1	8
Denmark	0	0	0	0	0	0	0	0	0	0
Estonia	-	0	...	0	...	...	...	1	1	1
Finland	362	191	191	227	227	134	134	134	114	76
Greece	...	...	...	...	15	23	28	23	27	25
Lithuania	0	1	1	1	1	0	1	1	0	0
Romania	-	-	-	-	181	65	56	32	-	3
Sweden	4	5	5	6	9	6	8	17	37	49
05 Fishing area total	*366*	*197*	*197*	*234*	*433*	*228*	*227*	*208*	*180*	*162*
Species total	*22 875*	*12 008*	*7 991*	*6 971*	*12 130*	*11 909*	*7 084*	*2 214*	*6 507*	*9 275*
Freshwater crustaceans nei	**Crustacés d'eau douce nca**		**Crustáceos de agua dulce nep**		***Crustacea***				**2,99(XX)XXX,XX**	**FCX**
01 Benin	3 800 F	3 794	4 503	4 600 F	4 700 F	4 800 F	4 924	4 100 F	4 600 F	4 600 F
Côte dIvoire	1 000	1 000	500	400 F	300 F	200 F	100 F	27	130 F	1 843
Egypt	...	...	950	557	838	918	1 237	2 473	2 192	2 029
Gabon	...	...	30	36	44	50	6	4	9	4
01 Fishing area total	*4 800 F*	*4 794*	*5 983*	*5 593 F*	*5 882 F*	*5 968 F*	*6 267 F*	*6 604 F*	*6 931 F*	*8 476 F*
02 Cuba	...	...	...	...	...	2	1	22	13	11
Dominican Rp	-	-	-	21	-	-	16	1	48	4
El Salvador	177	3	5	5	0	3	11	9	16	22
02 Fishing area total	*177*	*3*	*5*	*26*	*0*	*5*	*28*	*32*	*77*	*37*
03 Brazil	10 F	10 F	0	0	0	0	0	0	-	-
03 Fishing area total	*10 F*	*10 F*	*0*	*0*	*0*	*0*	*0*	*0*	*-*	*-*
04 Cambodia	20	40	80	70	100	100	300	300	500	500
China	154 891	203 372	270 536	364 441	479 645	601 966	455 761	530 026	586 985	772 702
India	-	33	26	803	71	114	85	112	183	202
Indonesia	164	242	86	223	49	93	132	132	337	490
Korea Rep	138	181	216	248	-	2	111	93	48	61
Philippines	1 232	732	73	55	34	32	1	5	67	69
Viet Nam	1 000 F	500 F	500 F	1 000 F	1 000 F	1 000 F	1 000 F	1 000 F	1 000 F	1 000 F
04 Fishing area total	*157 445 F*	*205 100 F*	*271 517 F*	*366 840 F*	*480 899 F*	*603 307 F*	*457 390 F*	*531 668 F*	*589 120 F*	*775 024 F*
05 Albania	10 F	5 F	0	0	-	-	-	-	-	-
Germany	1	12	12	12	12	12	12	12	12	12
Russian Fed	7	5	12	28	34	33	41	957	1 733	1 754
Ukraine	...	0	...	...	...	...	...	0	2	3
05 Fishing area total	*18 F*	*22 F*	*24*	*40*	*46*	*45*	*53*	*969*	*1 747*	*1 769*
06 Fiji Islands	102	104 F	1 011	327	315	323	332	340 F	343	340 F
06 Fishing area total	*102*	*104 F*	*1 011*	*327*	*315*	*323*	*332*	*340 F*	*343*	*340 F*
Species total	*162 552 F*	*210 033 F*	*278 540 F*	*372 826 F*	*487 142 F*	*609 648 F*	*464 070 F*	*539 613 F*	*598 218 F*	*785 646 F*
Group total	***216 147***	***253 235***	***315 398***	***410 241***	***526 777***	***645 056***	***498 817***	***568 469***	***628 083***	***818 993***

B-42 Crabs, sea-spiders / Crabes, araignées de mer / Cangrejos, centollas

Capture production by species, fishing areas and countries or areas / Captures par espèces, zones de pêche et pays ou zones / Capturas por especies, áreas de pesca y países o áreas

Species, Fishing area Espèce, Zone de pêche Especie, Area de pesca	1993 t	1994 t	1995 t	1996 t	1997 t	1998 t	1999 t	2000 t	2001 t	2002 t
Atlantic rock crab	**Tourteau poïnclos**		**Jaiba de roca amarilla**		***Cancer irroratus***				**2,31(09)010,03**	**CRK**
21 Canada	2 134	3 743	5 415	4 286	6 410	6 338	5 732	7 672	7 968	7 829
USA	880	718	844	55	1 120	1 358	1 301	1 844	326	502
21 Fishing area total	*3 014*	*4 461*	*6 259*	*4 341*	*7 530*	*7 696*	*7 033*	*9 516*	*8 294*	*8 331*
Species total	*3 014*	*4 461*	*6 259*	*4 341*	*7 530*	*7 696*	*7 033*	*9 516*	*8 294*	*8 331*
Dungeness crab	**Dormeur du Pacifique**		**Buey del Pacífico**		***Cancer magister***				**2,31(09)010,04**	**DUN**
67 Canada	6 225	5 995	4 586	5 025	3 923	2 968	2 944	2 832	5 685	4 090
USA	24 594	19 202	20 116	28 581	15 735	14 034	15 346	16 337	15 685	20 177
67 Fishing area total	*30 819*	*25 197*	*24 702*	*33 606*	*19 658*	*17 002*	*18 290*	*19 169*	*21 370*	*24 267*
77 USA	559	1 666	1 580	897	1 593	1 485	734	772	840	2 008
77 Fishing area total	*559*	*1 666*	*1 580*	*897*	*1 593*	*1 485*	*734*	*772*	*840*	*2 008*
Species total	*31 378*	*26 863*	*26 282*	*34 503*	*21 251*	*18 487*	*19 024*	*19 941*	*22 210*	*26 275*
Edible crab	**Tourteau**		**Buey de mar**		***Cancer pagurus***				**2,31(09)010,06**	**CRE**
27 Belgium	274	244	268	175	222	180	81	106	108	126
Channel Is	1 633 F	1 459	1 349 F	2 043	2 268	2 214	1 655	1 440	1 560	1 423
Denmark	15	17	17	15	14	21	25	33	53	67
France	6 495	6 395	6 185	5 927	6 554	5 598	6 503	6 549	6 604	5 877
Germany	8	-	-	-	38	44	57	64	44	74
Ireland	4 218	6 374	7 049	5 649	7 572	7 392	7 772	9 598	9 738	10 099
Isle of Man	137	52	282	94	478	274	231	142	170	387
Netherlands	-	-	-	-	-	-	-	146	300	506
Norway	1 642	1 781	1 807	1 889	2 204	2 984	2 837	2 889	3 476	4 345
Portugal	11	6	5	4	15	13	21	14	13	12
Spain	127	157	149	120	51	23	49	48	35	64
Sweden	95	161	105	87	79	93	122	128	133	146
UK	11 289	14 241	16 166	13 182	18 516	21 799	19 988	21 607	21 909	20 760
27 Fishing area total	*25 944 F*	*30 887*	*33 382 F*	*29 185*	*38 011*	*40 635*	*39 341*	*42 764*	*44 143*	*43 886*
Species total	*25 944 F*	*30 887*	*33 382 F*	*29 185*	*38 011*	*40 635*	*39 341*	*42 764*	*44 143*	*43 886*
Jonah crab	**Tourteau jona**		**Jaiba de roca jonás**		***Cancer borealis***				**2,31(09)010,07**	**CRJ**
21 USA	1 099	1 120	332	334	745	1 255	1 549	1 114	1 245	1 190
21 Fishing area total	*1 099*	*1 120*	*332*	*334*	*745*	*1 255*	*1 549*	*1 114*	*1 245*	*1 190*
Species total	*1 099*	*1 120*	*332*	*334*	*745*	*1 255*	*1 549*	*1 114*	*1 245*	*1 190*
Pacific rock crab	**Tourteau du Pacifique**		**Jaiba del Pacífico**		***Cancer productus***				**2,31(09)010,08**	**ROC**
67 USA	...	...	...	...	...	3	5	4	5	6
67 Fishing area total	...	...	...	...	...	*3*	*5*	*4*	*5*	*6*
77 USA	...	...	...	...	...	571	354	490	532	546
77 Fishing area total	...	...	...	...	...	*571*	*354*	*490*	*532*	*546*
Species total	...	...	...	...	...	*574*	*359*	*494*	*537*	*552*
Black stone crab	**Crabe caillou noir**		**Cangrejo de piedra negro**		***Menippe mercenaria***				**2,31(10)050,01**	**STC**
31 Bahamas	40	47	40	25	42	39	50	46	47	50 F
Belize	14	20	28	53	210	25	29	16	8	4
Cuba	3	2	5	28	9	35	52	54	76	48
Mexico	60	58	120	353	610	681	355	120	205	378
USA	2 483	3 177	2 376	3 216	2 483	2 567	1 877	2 961	2 996	2 942
31 Fishing area total	*2 600*	*3 304*	*2 569*	*3 675*	*3 354*	*3 347*	*2 363*	*3 197*	*3 332*	*3 422 F*
Species total	*2 600*	*3 304*	*2 569*	*3 675*	*3 354*	*3 347*	*2 363*	*3 197*	*3 332*	*3 422 F*
Blue swimming crab	**Etrille bleue**		**Jaiba azul**		***Portunus pelagicus***				**2,31(11)004,01**	**SCD**
37 Egypt	318	657	...	...	...	...	...	...	...	...
37 Fishing area total	*318*	*657*	...	...	...	...	...	...	...	...
57 Australia	308	644	607	522	740	865	717	810	1 007	1 000
Indonesia	768	1 542	1 975	1 830	2 207	3 025	3 594	3 301	3 694	4 630
Thailand	6 118	4 930	5 781	5 696	5 173	9 397	7 386	6 652	7 823	8 196
57 Fishing area total	*7 194*	*7 116*	*8 363*	*8 048*	*8 120*	*13 287*	*11 697*	*10 763*	*12 524*	*13 826*
61 China	15 000	20 000	25 168	51 288	45 749	48 190	52 577	56 092	59 416	61 981
China,Taiwan	1 579	1 690	2 808	1 391	1 141	966	511	1 966	1 703	1 030
61 Fishing area total	*16 579*	*21 690*	*27 976*	*52 679*	*46 890*	*49 156*	*53 088*	*58 058*	*61 119*	*63 011*
71 Australia	2 573	4 047	4 590	4 816	4 507	4 390	3 732	4 751	5 479	4 713
Indonesia	6 470	6 787	8 909	9 728	13 181	9 345	10 682	10 752	18 346	21 500
Philippines	26 151	26 128	29 536	27 660	30 358	23 919	34 076	36 303	38 083	32 947
Thailand	33 641	35 157	35 414	36 219	34 916	37 281	33 864	37 219	37 884	38 420
71 Fishing area total	*68 835*	*72 119*	*78 449*	*78 423*	*82 962*	*74 935*	*82 354*	*89 025*	*99 792*	*97 580*
81 Australia	627	662	788	800	761	745	628	377	292	229
81 Fishing area total	*627*	*662*	*788*	*800*	*761*	*745*	*628*	*377*	*292*	*229*

B-42 Crabs, sea-spiders / Crabes, araignées de mer / Cangrejos, centollas

Capture production by species, fishing areas and countries or areas / Captures par espèces, zones de pêche et pays ou zones / Capturas por especies, áreas de pesca y países o áreas

Species, Fishing area Espèce, Zone de pêche Especie, Area de pesca	1993 t	1994 t	1995 t	1996 t	1997 t	1998 t	1999 t	2000 t	2001 t	2002 t
Species total	*93 553*	*102 244*	*115 576*	*139 950*	*138 733*	*138 123*	*147 767*	*158 223*	*173 727*	*174 646*
Gazami crab	**Crabe gazami**		**Jaiba gazami**		***Portunus trituberculatus***				**2,31(11)004,04**	**GAZ**
61 China	117 264	272 102	243 485	283 394	237 960	266 630	270 280	335 078	333 556	347 103
Japan	2 958	3 564	4 159	4 022	3 112	3 528	2 752	3 131	3 596	3 485
Korea Rep	10 419	21 483	17 651	15 754	11 430	13 813	11 819	12 842	13 016	18 659
61 Fishing area total	*130 641*	*297 149*	*265 295*	*303 170*	*252 502*	*283 971*	*284 851*	*351 051*	*350 168*	*369 247*
Species total	*130 641*	*297 149*	*265 295*	*303 170*	*252 502*	*283 971*	*284 851*	*351 051*	*350 168*	*369 247*
Portunus swimcrabs nei	**Etrilles nca**		**Jaibas, nécoras nep**		***Portunus spp***				**2,31(11)004,XX**	**CRS**
27 France	336	361	254	225	261	155	163	242	267	308
Ireland	...	...	...	...	...	314	463	...	214	...
Portugal	94	30	32	29	16	19	0	59	67	123
Spain	198	186	169	0	6	20	18	-	-	-
UK	2 515	2 886	3 835	1 459	2 929	2 513	2 282	1 672	2 418	2 397
27 Fishing area total	*3 143*	*3 463*	*4 290*	*1 713*	*3 212*	*3 021*	*2 926*	*1 973*	*2 966*	*2 828*
31 Trinidad Tob	-	-	-	-	-	0	1	1	0	0
Venezuela	650	197	1	1	289	224	4 041	4 994	4 001	2
31 Fishing area total	*650*	*197*	*1*	*1*	*289*	*224*	*4 042*	*4 995*	*4 001*	*2*
51 Bahrain	690	821	807	1 047	1 289	1 017	2 179	2 380	2 556	2 828
Eritrea	...	-	-	-	-	-	-	-	1	4
51 Fishing area total	*690*	*821*	*807*	*1 047*	*1 289*	*1 017*	*2 179*	*2 380*	*2 557*	*2 832*
Species total	*4 483*	*4 481*	*5 098*	*2 761*	*4 790*	*4 262*	*9 147*	*9 348*	*9 524*	*5 662*
Blue crab	**Crabe bleu**		**Cangrejo azul**		***Callinectes sapidus***				**2,31(11)012,02**	**CRB**
21 USA	73 171	39 760	57 740	67 337	62 784	56 834	60 392	42 408	39 511	42 997
21 Fishing area total	*73 171*	*39 760*	*57 740*	*67 337*	*62 784*	*56 834*	*60 392*	*42 408*	*39 511*	*42 997*
31 Cuba	513	840	744	897	704	1 065	566	584	528	410
Mexico	11 076	11 026	10 657	13 736	14 412	12 920	12 714	8 602	7 291	8 063
Nicaragua	-	-	-	-	-	131	106	71	69	0
USA	40 870	56 889	34 457	33 582	43 743	45 046	31 459	40 994	32 356	37 609
31 Fishing area total	*52 459*	*68 755*	*45 858*	*48 215*	*58 859*	*59 162*	*44 845*	*50 251*	*40 244*	*46 082*
Species total	*125 630*	*108 515*	*103 598*	*115 552*	*121 643*	*115 996*	*105 237*	*92 659*	*79 755*	*89 079*
Dana swimcrab	**Crabe lénée**		**Cangrejo sirí**		***Callinectes danae***				**2,31(11)012,05**	**CRZ**
41 Brazil	2 800 F	2 500 F	2 062	2 020	2 600	3 014	1 626	1 597	1 062	1 310
41 Fishing area total	*2 800 F*	*2 500 F*	*2 062*	*2 020*	*2 600*	*3 014*	*1 626*	*1 597*	*1 062*	*1 310*
Species total	*2 800 F*	*2 500 F*	*2 062*	*2 020*	*2 600*	*3 014*	*1 626*	*1 597*	*1 062*	*1 310*
Green crab	**Crabe vert**		**Cangrejo verde**		***Carcinus maenas***				**2,31(11)090,01**	**CRG**
21 USA	0	0	0	-	54	86	14	16	67	39
21 Fishing area total	*0*	*0*	*0*	*-*	*54*	*86*	*14*	*16*	*67*	*39*
27 France	327	342	292	350	359	204	465	558	541	471
Ireland	...	...	...	...	...	79	16	...	68	...
Portugal	739	382	351	200	125	156	77	111	125	335
Spain	-	-	-	1	2	8	8	10	12	11
UK	512	502	470	286	411	462	333	194	331	402
27 Fishing area total	*1 578*	*1 226*	*1 113*	*837*	*897*	*909*	*899*	*873*	*1 077*	*1 219*
Species total	*1 578*	*1 226*	*1 113*	*837*	*951*	*995*	*913*	*889*	*1 144*	*1 258*
Mediterranean shore crab	**Crabe vert de la Méditerranée**		**Cangrejo verde mediterráneo**		***Carcinus aestuarii***				**2,31(11)090,02**	**CMR**
37 France	7	28	...	16	16	16	7	14	9	9
Greece	82	59	16	49	28	50	37	16	17	17 F
Spain	-	-	-	-	-	-	-	-	-	36
Tunisia	-	-	-	-	-	-	-	-	11	12
37 Fishing area total	*89*	*87*	*16*	*65*	*44*	*66*	*44*	*30*	*37*	*74 F*
Species total	*89*	*87*	*16*	*65*	*44*	*66*	*44*	*30*	*37*	*74 F*
Indo-Pacific swamp crab	**Crabe de palétuviers**		**Cangrejo de manglares**		***Scylla serrata***				**2,31(11)140,01**	**MUD**
04 China,Taiwan	-	-	-	-	-	2	1	10	0	-
Philippines	321	145	329	363	267	42	46	67	150	219
04 Fishing area total	*321*	*145*	*329*	*363*	*267*	*44*	*47*	*77*	*150*	*219*
51 Mauritius	-	-	22	23	19	23	21	25	24	22
51 Fishing area total	*-*	*-*	*22*	*23*	*19*	*23*	*21*	*25*	*24*	*22*
57 Indonesia	2 225	3 018	2 351	2 597	2 714	3 296	3 638	3 490	3 938	4 280
Thailand	1 816	3 653	3 508	2 527	2 421	1 884	1 973	3 495	2 805	2 939
57 Fishing area total	*4 041*	*6 671*	*5 859*	*5 124*	*5 135*	*5 180*	*5 611*	*6 985*	*6 743*	*7 219*
61 China,Taiwan	84	169	1 339	935	180	213	268	289	230	337
61 Fishing area total	*84*	*169*	*1 339*	*935*	*180*	*213*	*268*	*289*	*230*	*337*

B-42 Crabs, sea-spiders / Crabes, araignées de mer / Cangrejos, centollas

Capture production by species, fishing areas and countries or areas
Captures par espèces, zones de pêche et pays ou zones
Capturas por especies, áreas de pesca y países o áreas

Species, Fishing area Espèce, Zone de pêche Especie, Area de pesca	1993 t	1994 t	1995 t	1996 t	1997 t	1998 t	1999 t	2000 t	2001 t	2002 t
71 Fiji Islands	198	202 F	234	208	290	270	281	250 F	268	265 F
Indonesia	8 284	3 423	5 629	4 745	5 584	4 865	5 069	5 284	7 814	8 310
Marshall Is	0	0	0	0	0	0	0	0	0	0
Micronesia	4 F	4 F	5 F	5 F	5 F	5 F	5 F	5 F	5 F	5 F
Palau	...	...	1	1	1	2	8	6	8	5
Papua N Guin	20 F	27 F	28 F	25 F	24 F	23 F	22 F	24 F	24 F	24 F
Philippines	6 080	5 490	4 506	3 895	866	1 082	1 165	1 180	1 454	1 473
Singapore	38	33	27	19	15	9	9	28	9	8
Thailand	2 972	2 541	2 268	1 716	1 610	1 848	3 763	3 426	2 609	2 646
71 Fishing area total	*17 596 F*	*11 720 F*	*12 698 F*	*10 614 F*	*8 395 F*	*8 104 F*	*10 322 F*	*10 203 F*	*12 191 F*	*12 736 F*
77 Fr Polynesia	-	-	-	-	2	2	2	2	2	3
77 Fishing area total	-	-	-	-	*2*	*2*	*2*	*2*	*2*	*3*
Species total	*22 042 F*	*18 705 F*	*20 247 F*	*17 059 F*	*13 998 F*	*13 566 F*	*16 271 F*	*17 581 F*	*19 340 F*	*20 536 F*
Harbour spidercrab	**Araignée portuaire**		**Araña porteña**		***Mithrax armatus***				**2,31(21)001,02**	**MXT**
31 Nicaragua	-	-	-	-	-	114	110	54	35	22
31 Fishing area total	-	-	-	-	-	*114*	*110*	*54*	*35*	*22*
Species total	-	-	-	-	-	*114*	*110*	*54*	*35*	*22*
Spinous spider crab	**Araignée européenne**		**Centolla europea**		***Maja squinado***				**2,31(21)005,01**	**SCR**
27 Channel Is	489 F	573	530 F	868	445	460	553	522	428	406
Denmark	1	1	1	0	1	1	1	1	2	2
France	3 611	3 631	3 467	3 169	3 433	3 124	3 380	4 422	5 439	4 141
Ireland	130	264	153	192	153	185	299	163	264	330
Portugal	69	50	49	40	47	58	60	59	51	49
Spain	154	137	151	181	206	193	209	198	108	114
UK	812	1 352	2 178	1 224	1 849	1 843	1 310	990	1 199	1 171
27 Fishing area total	*5 266 F*	*6 008*	*6 529 F*	*5 674*	*6 134*	*5 864*	*5 812*	*6 355*	*7 491*	*6 213*
37 Croatia	37	39	43	25	54	50	15	14	21	61
France	7	11	...	2	2	2	0	6	1	2
Serbia-Monte	0	0	0	0	0	0	0	0	0	1
Spain	-	-	-	4	3	3	1	1	5	3
37 Fishing area total	*44*	*50*	*43*	*31*	*59*	*55*	*16*	*21*	*27*	*67*
Species total	*5 310 F*	*6 058*	*6 572 F*	*5 705*	*6 193*	*5 919*	*5 828*	*6 376*	*7 518*	*6 280*
Queen crab	**Crabe des neiges**		**Cangrejo de las nieves**		***Chionoecetes opilio***				**2,31(21)145,01**	**CRQ**
21 Canada	48 099	60 400	65 372	65 825	71 416	75 216	95 148	93 505	95 299	106 766
Greenland	1	72	998	817	2 557	1 947	2 896	10 236	14 247	10 271
St Pier Mq	1	2	2	189	368	354	589	511	498	150
21 Fishing area total	*48 101*	*60 474*	*66 372*	*66 831*	*74 341*	*77 517*	*98 633*	*104 252*	*110 044*	*117 187*
Species total	*48 101*	*60 474*	*66 372*	*66 831*	*74 341*	*77 517*	*98 633*	*104 252*	*110 044*	*117 187*
Tanner crabs nei	**Crabes des neiges du Pac. nca**		**Cangrejos de las nieves (Pac.)**		***Chionoecetes spp***				**2,31(21)145,XX**	**PCR**
61 Japan	5 612	6 919	9 090	3 447	4 870	4 677	4 892	5 640	5 355	5 016
Korea Rep	-	-	-	-	-	-	-	17 037	13 974	896
Russian Fed	...	...	...	22 769	27 112	20 848	21 234	21 848	24 493	23 759
61 Fishing area total	*5 612*	*6 919*	*9 090*	*26 216*	*31 982*	*25 525*	*26 126*	*44 525*	*43 822*	*29 671*
67 USA	116 000	72 382	36 658	30 785	53 932	114 230	83 989	15 665	12 177	15 078
67 Fishing area total	*116 000*	*72 382*	*36 658*	*30 785*	*53 932*	*114 230*	*83 989*	*15 665*	*12 177*	*15 078*
Species total	*121 612*	*79 301*	*45 748*	*57 001*	*85 914*	*139 755*	*110 115*	*60 190*	*55 999*	*44 749*
Hair crab	**Crabe velu**		**Cangrejo peludo**		***Erimacrus isenbeckii***				**2,31(23)001,01**	**HBZ**
61 Russian Fed	...	...	...	789	612	409	440	198	162	117
61 Fishing area total	...	...	...	*789*	*612*	*409*	*440*	*198*	*162*	*117*
Species total	...	...	...	*789*	*612*	*409*	*440*	*198*	*162*	*117*
Red crab	**Gériocrabe rouge**		**Geriocangrejo rojo**		***Geryon quinquedens***				**2,31(43)088,01**	**CRR**
21 USA	1 440	...	521	465	96	-	-	3 130	4 004	2 169
21 Fishing area total	*1 440*	...	*521*	*465*	*96*	-	-	*3 130*	*4 004*	*2 169*
31 USA	-	-	-	-	-	-	-	2	-	-
31 Fishing area total	-	-	-	-	-	-	-	*2*	-	-
34 UK	-	-	-	-	-	-	-	-	31	-
34 Fishing area total	-	-	-	-	-	-	-	-	*31*	-
41 UK	-	-	-	-	-	61	-	-	596	417
Uruguay	254	852	783	1 513	3 432	2 682	3 382	5 259	2 089	2 004
41 Fishing area total	*254*	*852*	*783*	*1 513*	*3 432*	*2 743*	*3 382*	*5 259*	*2 685*	*2 421*
Species total	*1 694*	*852*	*1 304*	*1 978*	*3 528*	*2 743*	*3 382*	*8 391*	*6 720*	*4 590*

B-42 Crabs, sea-spiders / Crabes, araignées de mer / Cangrejos, centollas

Capture production by species, fishing areas and countries or areas
Captures par espèces, zones de pêche et pays ou zones
Capturas por especies, áreas de pesca y países o áreas

Species, Fishing area Espèce, Zone de pêche Especie, Area de pesca		1993 t	1994 t	1995 t	1996 t	1997 t	1998 t	1999 t	2000 t	2001 t	2002 t
Geryons nei		**Géryons nca**		**Geriones nep**		***Geryon spp***			**2,31(43)088,XX**		**GER**
27	UK	-	-	10	1 477	325	587	1 015	763	382	1 041
27	*Fishing area total*	-	-	*10*	*1 477*	*325*	*587*	*1 015*	*763*	*382*	*1 041*
34	Mauritania	...	...	...	...	...	...	...	6	16	5
34	*Fishing area total*	...	...	...	...	...	...	...	*6*	*16*	*5*
47	Japan	6 780	5 249	4 855	3 250	1 937	1 562	1 730	2 980	1 650	1 650 F
	Namibia	3 190	3 598	2 008	1 709	1 478	2 283	2 074	2 700	2 343	2 471
	South Africa	0	-	-	-	-	-	-	-	-	-
	Spain	300 F	300 F	300 F	370	-	300	135	156	168	-
47	*Fishing area total*	*10 270 F*	*9 147 F*	*7 163 F*	*5 329*	*3 415*	*4 145*	*3 939*	*5 836*	*4 161*	*4 121 F*
51	Mozambique	406	345	414	564	1 144	911	795	832	632	890
	South Africa	138	-	-	-	-	91	80 F	54	53	30
51	*Fishing area total*	*544*	*345*	*414*	*564*	*1 144*	*1 002*	*875 F*	*886*	*685*	*920*
Species total		*10 814 F*	*9 492 F*	*7 587 F*	*7 370*	*4 884*	*5 734*	*5 829 F*	*7 491*	*5 244*	*6 087 F*
Marine crabs nei		**Crabes de mer nca**		**Cangrejos de mar nep**		***Brachyura***			**2,31(XX)XXX,XX**		**CRA**
21	Canada	34	806	2 489	1 588	1 948	2 267	1 835	3 052	4 306	4 373
	Spain	-	-	-	11	-	-	-	-	-	-
	USA	228	295	632	1 692	3 128	1 927	2 958	2 615	6 020	6 352
21	*Fishing area total*	*262*	*1 101*	*3 121*	*3 291*	*5 076*	*4 194*	*4 793*	*5 667*	*10 326*	*10 725*
27	Channel Is	-	-	-	-	-	2	2	2	-	0
	Denmark	236	238	188	171	193	220	193	205	189	245
	Faeroe Is	-	-	1	3	1	4	1	6	1	...
	France	14	17	14	160	5	5	-	13	10	6
	Germany	4	-	-	13	6	4	-	-	-	141
	Iceland	0	0	0	0	0	-	-	-	1	0
	Ireland	518	355	487	312	272	-	1	268	-	-
	Norway	-	-	-	-	-	-	-	-	2	0
	Portugal	-	-	-	-	-	-	19	39	45	32
	Spain	1 119	178	575	1 849	9 652	1 712	738	101	74	41
	UK	1 008	1 827	1 469	639	205	57	90	17	60	132
27	*Fishing area total*	*2 899*	*2 615*	*2 734*	*3 147*	*10 334*	*2 004*	*1 044*	*651*	*382*	*597*
31	Colombia	-	65	181	21	131	32	...	100 F	100 F	100 F
	Cuba	0	216	284	348	668	763	851	907	865	257
	Dominican Rp	74	92	75	32	14	37	7	14	52	150
	Haiti	...	...	17	5	71	59	50 F	50 F	50 F	50 F
	Honduras	10	9	20	47	19	5	2	172	20 F	20 F
	Jamaica	...	...	34	11	9	9	9	8	8 F	8 F
	Korea Rep	-	-	-	5	-	-	-	-	2	-
	Puerto Rico	...	...	...	...	...	...	2	2	6	3
	Suriname	4	5	5 F	6 F	10 F	10 F	10 F	20	25	20 F
	USA	160	137	358	47	473	179	97	392	346	239
	Venezuela	3 940	6 072	5 551	4 591	5 340	4 180	196	539	155	8 459
31	*Fishing area total*	*4 188*	*6 596*	*6 525 F*	*5 113 F*	*6 735 F*	*5 274 F*	*1 224 F*	*2 204 F*	*1 629 F*	*9 306 F*
34	Benin	36 F	40 F	40 F	40 F	50 F	45 F	32	20 F	9	22
	Cameroon	40	40 F	32	18	20 F	20 F	20 F	27	26	20
	Congo Rep	1 F	0	5 F	10 F	5 F	20 F	30 F	44	10 F	10 F
	Côte dIvoire	0	0	0	0	0	0	1	2	2 F	2 F
	Gabon	26	34	65	120	145	158	283	289	142	136
	Gambia	...	...	...	...	...	...	...	6	2	...
	Ghana	195	462	218	399	576	271	145	74	155	156
	Greece	610	58	71	60	84	90	74	59	54	63
	GuineaBissau	23 F	23 F	27 F	30 F	30 F	25 F	20 F	20 F	20 F	20 F
	Italy	-	-	-	-	-	-	-	8	-	-
	Korea Rep	-	5	-	-	1	1	-	-	-	-
	Liberia	22	10	28	-	35	38	32	27	122	125 F
	Lithuania	0	-	-	-	-	-	-	-	-	-
	Morocco	28	56	46	50	286	236	400	370	364	247
	Nigeria	92	28	48	57	593	3 384	4 018	3 211	4 374	5 029
	Portugal	-	-	21	33	61	36	17	79	193	12
	Senegal	29	520	158	217	727	356	186	343	389	243
	Sierra Leone	100	100	14	15 F	140	83	107	142	269	168
	Spain	-	-	-	1 282	1 361	1 794	218	471	516	611
	Togo	0	0	0	0	0	1	3	33	23	17
34	*Fishing area total*	*1 202 F*	*1 376 F*	*773 F*	*2 331 F*	*4 114 F*	*6 558 F*	*5 586 F*	*5 225 F*	*6 670 F*	*6 881 F*
37	Egypt	848	684	1 026	981	1 063	1 249	3 524	1 725	1 160	2 222
	Morocco	1	7	0	0	0	0	-	-	1	0
	Palest, O.T.	...	...	66	72	25	65	65 F	65 F	65 F	73
	Spain	-	-	-	741	595	842	663	863	1 031	834
	Turkey	92	83	117	305	318	243	179	187	273	215
37	*Fishing area total*	*941*	*774*	*1 209*	*2 099*	*2 001*	*2 399*	*4 431 F*	*2 840 F*	*2 530 F*	*3 344*
41	Argentina	54	168	345	390	403	326	170	-	-	-
	Brazil	13 530 F	13 560 F	11 917	9 922	10 233	9 202	10 913	11 135	12 535	12 790
	Russian Fed	12	17	-	-	-	-	-	-	-	-
	Spain	-	-	-	-	-	-	-	-	-	7
	UK	-	-	-	-	-	-	-	-	36	360
41	*Fishing area total*	*13 596 F*	*13 745 F*	*12 262*	*10 312*	*10 636*	*9 528*	*11 083*	*11 135*	*12 571*	*13 157*
47	Angola	...	...	...	...	...	...	602	800	836	1 000

B-42 Crabs, sea-spiders / Crabes, araignées de mer / Cangrejos, centollas

Capture production by species, fishing areas and countries or areas / Captures par espèces, zones de pêche et pays ou zones / Capturas por especies, áreas de pesca y países o áreas

Species, Fishing area Espèce, Zone de pêche Especie, Area de pesca	1993 t	1994 t	1995 t	1996 t	1997 t	1998 t	1999 t	2000 t	2001 t	2002 t
Spain	-	-	-	-	-	1 494	314	239	254	179
47 Fishing area total	*...*	*...*	*...*	*...*	*...*	*1 494*	*916*	*1 039*	*1 090*	*1 179*
51 Egypt	-	-	10 049	0	4	149	1 177	2 141	744	678
India	5 908	25 244	13 859	521	642	369	394	6 193	1 841	2 264
Kenya	69	59	70	112	100	117	135	166	134	122
Korea Rep	-	-	-	-	10	-	-	167	5	3
Madagascar	1 085	1 300	1 300	1 000	1 000	1 500	868	1 030	1 347	1 428
Pakistan	480	650	877	3 200	3 989	5 680	5 109	5 187	5 099	6 060
Portugal	-	-	-	-	-	-	-	2	5	-
Qatar	52	30	39	47	47	69	39	26	21	17
Réunion	4	5	5	5	6	7	5	7	11	11 F
Saudi Arabia	256	1 126	1 035	1 060	1 371	1 062	959	1 021	1 002	1 209
Seychelles	14	8	9	17	20	24	11	11	26	24
South Africa	-	-	-	-	113	4	5 F	...	2	-
Spain	-	-	-	-	-	-	-	-	188	-
Untd Arab Em	50	58	47	56	60	60	62	1 710	1 969	353
Yemen	...	...	...	...	...	...	...	...	37	34
51 Fishing area total	*7 918*	*28 480*	*27 290*	*6 018*	*7 362*	*9 041*	*8 764 F*	*17 661*	*12 431*	*12 203 F*
57 India	12 157	11 627	2 846	1 026	1 595	1 420	1 299	17 457	16 164	16 905
Korea Rep	-	-	-	-	-	-	10	25	-	-
Malaysia	3 703	3 372	3 781	3 406	3 793	5 679	5 864	5 105	4 434	4 003
Thailand	2 803	2 475	3 242	5 680	5 885	5 812	6 947	6 528	7 036	7 371
Timor-Leste	...	...	...	...	...	...	1 F	1	1	1 F
57 Fishing area total	*18 663*	*17 474*	*9 869*	*10 112*	*11 273*	*12 911*	*14 121 F*	*29 116*	*27 635*	*28 280 F*
61 China,H.Kong	1 224	1 896	1 601	1 408	1 105	1 108	800 F	1 000 F	1 100 F	1 050 F
China,Taiwan	2 155	3 124	4 747	4 830	4 630	4 553	5 433	5 898	7 373	8 253
Japan	40 148	40 295	38 812	37 265	34 894	33 678	30 859	30 310	27 353	28 000 F
Korea Rep	40 620	56 547	58 804	61 308	58 741	43 879	35 375	9 372	9 001	6 753
Russian Fed	-	-	-	-	49	-	-	-	194	856
61 Fishing area total	*84 147*	*101 862*	*103 964*	*104 811*	*99 419*	*83 218*	*72 467 F*	*46 580 F*	*45 021 F*	*44 912 F*
67 USA	892	1 453	2 185	647	333	133	99	1	42	2
67 Fishing area total	*892*	*1 453*	*2 185*	*647*	*333*	*133*	*99*	*1*	*42*	*2*
71 Cambodia	2 300	2 200	2 300	2 350	2 240	2 420	2 850	3 593	4 200	4 600
Fiji Islands	57	58 F	288	200	280	255	300	270 F	290	280 F
Korea Rep	-	-	-	200	103	119	54	-	23	1
Malaysia	5 424	6 210	6 416	5 596	5 938	8 564	8 566	7 534	7 864	7 232
NewCaledonia	28	7	8	9	29	20	58	22	14	14 F
Palau	13	3	...	...	...	...	...	...	...	...
Singapore	196	235	196	176	203	264	175	189	203	173
Thailand	1 452	2 194	2 093	921	989	1 723	1 510	788	1 252	1 270
Viet Nam	10 800 F	28 000 F	32 000 F	28 300 F	26 400 F	48 000 F	35 800 F	40 000 F	36 900 F	46 500 F
Wallis Fut I	...	1	1	1	1	1	1 F	1 F	1 F	1 F
71 Fishing area total	*20 270 F*	*38 908 F*	*43 302 F*	*37 753 F*	*36 183 F*	*61 366 F*	*49 314 F*	*52 397 F*	*50 747 F*	*60 071 F*
77 Cook Is	5 F	5 F	5 F	5 F	5 F	5 F	5 F	5 F	5 F	5 F
Costa Rica	...	...	1	3	50	9	8	4	3	3
Honduras	23	47	50	70	299	30	4	60	40 F	40 F
Mexico	2 706	4 955	10 395	13 602	10 073	6 503	6 670	12 036	11 204	7 896
Panama	...	...	...	14	3	2	0	0	20	3
USA	576	507	492	554	632	58	41	33	54	52
77 Fishing area total	*3 310 F*	*5 514 F*	*10 943 F*	*14 248 F*	*11 062 F*	*6 607 F*	*6 728 F*	*12 138 F*	*11 326 F*	*7 999 F*
81 Korea Rep	-	-	-	-	-	-	-	-	-	1
New Zealand	326	492	9 231	294	359	393	407	355	360	292
Ukraine	-	-	-	-	-	-	-	-	-	6
81 Fishing area total	*326*	*492*	*9 231*	*294*	*359*	*393*	*407*	*355*	*360*	*299*
87 Chile	4 949	5 085	5 733	3 989	3 541	5 063	6 501	6 758	6 770	7 292
Colombia	-	33	57	28	35	12	11	20 F	20 F	20 F
Ecuador	683	750	750	750	750	600	600	600	600	600 F
Peru	1 027	1 383	2 553	1 605	303	752	11 397	1 974	1 568	2 838
87 Fishing area total	*6 659*	*7 251*	*9 093*	*6 372*	*4 629*	*6 427*	*18 509*	*9 352 F*	*8 958 F*	*10 750 F*
Species total	*165 273 F*	*227 641 F*	*242 501 F*	*206 548 F*	*209 516 F*	*211 547 F*	*199 486 F*	*196 361 F*	*191 718 F*	*209 705 F*
Group total	**797 655**	**985 360**	**951 913**	**999 674**	**991 140**	**1 075 725**	**1 059 348**	**1 091 717**	**1 091 958**	**1 134 205**

B-43

Lobsters, spiny-rock lobsters — **Capture production by species, fishing areas and countries or areas**
Homards, langoustes — **Captures par espèces, zones de pêche et pays ou zones**
Bogavantes, langostas — **Capturas por especies, áreas de pesca y países o áreas**

Species, Fishing area Espèce, Zone de pêche Especie, Area de pesca	1993 t	1994 t	1995 t	1996 t	1997 t	1998 t	1999 t	2000 t	2001 t	2002 t
Longlegged spiny lobster	**Langouste diablotin**		**Langosta duende**		***Panulirus longipes***			**2,29(01)001,01**		**LOJ**
61 China,Taiwan	23	21	18	14	14	14	12	11	23	13
Japan	1 238	1 118	1 136	1 092	1 068	1 084	1 154	1 244	1 486	1 378
Korea Rep	-	-	-	-	-	-	-	461	415	391
61 Fishing area total	*1 261*	*1 139*	*1 154*	*1 106*	*1 082*	*1 098*	*1 166*	*1 716*	*1 924*	*1 782*
Species total	*1 261*	*1 139*	*1 154*	*1 106*	*1 082*	*1 098*	*1 166*	*1 716*	*1 924*	*1 782*
Caribbean spiny lobster	**Langouste blanche**		**Langosta común del Caribe**		***Panulirus argus***			**2,29(01)001,08**		**SLC**
31 Anguilla	90	90	40 F	50 F	60 F	60 F	60 F	60 F	60 F	60 F
Antigua Barb	120	140	149	125	160	357	274	275	272	276
Bahamas	7 848	7 589	7 750	7 938	7 798	7 553	8 225	9 023	7 042	7 050 F
Belize	388	490	608	448	534	468	552	503	394	525
Bermuda	16	17	10	10	38	30	36	29	21	26
Br Virgin Is	52	92	32	27	5	6	4	3	3 F	3 F
Colombia	218	97	449	185	108	319	175	250 F	250 F	250 F
Costa Rica	29	32	93	196	196	40	163	271	39	57
Cuba	8 501	9 684	9 405	9 375	8 996	9 417	9 879	7 478	6 776	7 972
Dominican Rp	537	967	619	420	1 061	863	828	1 286	1 209	2 977
Grenada	26	30	57	23	14	31	72	47	32	24
Haiti	830 F	780 F	900 F	190 F	200 F	200 F	200 F	200 F	200 F	200 F
Honduras	1 045	1 096	1 257	1 450	1 176	1 237	1 541	823	866	837
Jamaica	250 F	300 F	350 F	394	271	170	330	517	500 F	500 F
Martinique	113	110 F	110 F	70 F	110 F	120 F	150 F	200	190	190
Mexico	881	945	896	756	844	613	645	747	782	1 070
Nicaragua	2 191	2 805	2 260	4 357	4 012	3 729	5 141	6 327	3 909	4 437
Puerto Rico	...	...	...	...	...	...	209	212	190	158
St Kitts Nev	...	19	12	17	8	25	29	26	33	10
Trinidad Tob	-	-	-	-	-	2	0	1	1	2
Turks Caicos	413	410 F	376	310	181	230	313	255	291	323
USA	2 548	3 420	2 934	3 373	2 783	2 343	2 749	2 571	1 527	2 047
US Virgin Is	64 F	59 F	55 F	50 F	45 F	40 F	34	35 F	35 F	35 F
Venezuela	940	763	629	648	619	260	95	105	78	88
31 Fishing area total	*27 100 F*	*29 935 F*	*28 991 F*	*30 412 F*	*29 219 F*	*28 113 F*	*31 704 F*	*31 244 F*	*24 700 F*	*29 117 F*
41 Brazil	9 100 F	9 120 F	10 817	8 026	7 502	6 002	6 334	6 469	7 135	7 016
41 Fishing area total	*9 100 F*	*9 120 F*	*10 817*	*8 026*	*7 502*	*6 002*	*6 334*	*6 469*	*7 135*	*7 016*
Species total	*36 200 F*	*39 055 F*	*39 808 F*	*38 438 F*	*36 721 F*	*34 115 F*	*38 038 F*	*37 713 F*	*31 835 F*	*36 133 F*
Australian spiny lobster	**Langouste d'Australie**		**Langosta de Australia**		***Panulirus cygnus***			**2,29(01)001,15**		**LOA**
57 Australia	12 366	11 046	10 886	9 902	9 896	10 400	13 065	14 605	11 353	9 050
57 Fishing area total	*12 366*	*11 046*	*10 886*	*9 902*	*9 896*	*10 400*	*13 065*	*14 605*	*11 353*	*9 050*
Species total	*12 366*	*11 046*	*10 886*	*9 902*	*9 896*	*10 400*	*13 065*	*14 605*	*11 353*	*9 050*
Green spiny lobster	**Langouste verte**		**Langosta barbona**		***Panulirus gracilis***			**2,29(01)001,16**		**NUG**
77 Nicaragua	9	17	14	101	152	66	131	476	652	334
77 Fishing area total	*9*	*17*	*14*	*101*	*152*	*66*	*131*	*476*	*652*	*334*
87 Colombia	9	6	3	4	7	6	1	1 F	1 F	1 F
Ecuador	33	50	50	50	50	50	50	50	50	50 F
Peru	14	52	168	52	12	669	496	278	62	20
87 Fishing area total	*56*	*108*	*221*	*106*	*69*	*725*	*547*	*329 F*	*113 F*	*71 F*
Species total	*65*	*125*	*235*	*207*	*221*	*791*	*678*	*805 F*	*765 F*	*405 F*
Tropical spiny lobsters nei	**Langoustes tropicales nca**		**Langostas tropicales nep**		***Panulirus spp***			**2,29(01)001,XX**		**SLV**
31 Bermuda	-	-	-	-	-	5	5	5	4	4
31 Fishing area total	*-*	*-*	*-*	*-*	*-*	*5*	*5*	*5*	*4*	*4*
34 Cameroon	0	1	0	0	0	1	1	0	0	1
Cape Verde	26	28	20	12	8	9	12	10	7	10 F
Ghana	369	510	230	134	203	65	...	39	342	97
Sierra Leone	10	10	2	4 F	57	29	51	56	30	43
Togo	1	1	0	0	0	0	2	2	0	0
34 Fishing area total	*406*	*550*	*252*	*150 F*	*268*	*104*	*66*	*107*	*379*	*151 F*
41 Italy	2	2	-	-	-	-	-	-	-	-
41 Fishing area total	*2*	*2*	*...*	*...*	*...*	*...*	*...*	*...*	*...*	*...*
47 St Helena	0	0	0	0	-	-	-	-	-	-
47 Fishing area total	*0*	*0*	*0*	*0*	*-*	*-*	*-*	*-*	*-*	*-*
51 Djibouti	0	0	0	0	0	0	0	0	0	0
Iran	45	35	30	49	76	65	35	25	20	20
Italy	3	3	-	-	-	-	-	-	-	-
Kenya	47	44	119	177	136	39	52	52	76	119
Madagascar	358	390	390	390	390	341	338	329	359	402
Maldives	...	...	...	...	...	...	...	7	13	...
Mauritius	13	13	12	14	17	17	17	17	19	26
Oman	701	623	608	397	263	336	180	402	379	449
Pakistan	507	669	615	724	765	782	1 077	807	756	802
Saudi Arabia	8	23	21	13	18	13	19	8	14	19

B-43 Lobsters, spiny-rock lobsters / Homards, langoustes / Bogavantes, langostas

Capture production by species, fishing areas and countries or areas / Captures par espèces, zones de pêche et pays ou zones / Capturas por especies, áreas de pesca y países o áreas

Species, Fishing area Espèce, Zone de pêche Especie, Area de pesca	1993 t	1994 t	1995 t	1996 t	1997 t	1998 t	1999 t	2000 t	2001 t	2002 t
Seychelles	0	0	2	1	-	0	7	14	5	6
Somalia	400 F	400 F	400 F	400 F	400 F	400 F	400 F	400 F	400 F	300 F
Ukraine	-	-	-	-	-	1	-	-	-	-
Yemen	1 021	475	328	323	482	828	334	178	202	227
51 Fishing area total	*3 103 F*	*2 675 F*	*2 525 F*	*2 488 F*	*2 547 F*	*2 822 F*	*2 459 F*	*2 239 F*	*2 243 F*	*2 370 F*
57 Australia	0	0	1	4	0	0	-	-	-	-
Indonesia	731	1 419	1 397	1 241	1 309	1 039	1 284	1 392	1 777	2 060
Malaysia	27	33	20	14	15	12	15	38	104	20
Timor-Leste	...	...	...	...	...	...	2 F	2	2	2 F
57 Fishing area total	*758*	*1 452*	*1 418*	*1 259*	*1 324*	*1 051*	*1 301 F*	*1 432*	*1 883*	*2 082 F*
61 China,H.Kong	60	50	20	0	0	0	0	0	0	0
61 Fishing area total	*60*	*50*	*20*	*0*	*0*	*0*	*0*	*0*	*0*	*0*
71 Australia	174	185	182	201	233	551	520	359	274	330
Fiji Islands	96	98 F	184	105	130	211	220	200 F	224	215 F
Guam	-	-	0	0	0	1	1	2	1	1
Indonesia	477	602	1 455	1 223	2 712	1 355	1 960	2 204	2 713	3 740
Malaysia	1 091	1 170	682	800	821	1 025	1 079	1 065	1 508	2 039
Marshall Is	0	0	0	0	0	0	0	-	-	-
Micronesia	10 F	15 F	15 F	20 F	20 F	20 F	20 F	20 F	20 F	20 F
NewCaledonia	6	26	16	17	14	17	17	13	9	9 F
N Marianas	1	2	1	2	0	2	1	2	2	2
Palau	5	2	1	1	1	1	1	1	1	1
Papua N Guin	104	182	206	165	205	217	203	197	131	130 F
Philippines	896	877	559	522	421	214	249	250	309	315
Singapore	54	2	1	-	5	11	8	8	7	2
Wallis Fut I	...	2	2	2	2	2	2 F	2 F	2 F	2 F
71 Fishing area total	*2 914 F*	*3 163 F*	*3 304 F*	*3 058 F*	*4 564 F*	*3 627 F*	*4 281 F*	*4 323 F*	*5 201 F*	*6 806 F*
77 Amer Samoa	0	0	-	1	1	2	1	1	1	0
Costa Rica	1	4	5	7	7	3	4	14	17	11
El Salvador	-	-	-	-	-	-	-	3	2	5
Fr Polynesia	3	3	20	20	40	51	50	51	58	54
Guatemala	2	4	3	5	7	16	18	2	2 F	1
Honduras	1	1	2	2	9	2	1	1 F	1 F	1 F
Mexico	1 136	1 294	1 421	1 799	1 708	1 599	1 328	2 052	1 727	1 923
Panama	280	279	197	288	306	415	485	612	845	687
USA	208	256	297	395	501	350	287	361	324	306
77 Fishing area total	*1 631*	*1 841*	*1 945*	*2 517*	*2 579*	*2 438*	*2 174*	*3 097 F*	*2 977 F*	*2 988 F*
Species total	*8 874 F*	*9 733 F*	*9 464 F*	*9 472 F*	*11 282 F*	*10 047 F*	*10 286 F*	*11 203 F*	*12 687 F*	*14 401 F*
Cape rock lobster	**Langouste du Cap**		**Langosta del Cabo**		***Jasus lalandii***				**2,29(01)002,01**	**LBC**
47 Namibia	136	134	224	251	199	350	304	365	363	358
South Africa	2 176	2 192	1 956	1 516	1 680	1 726	1 793	1 693	1 611	2 671
47 Fishing area total	*2 312*	*2 326*	*2 180*	*1 767*	*1 879*	*2 076*	*2 097*	*2 058*	*1 974*	*3 029*
Species total	*2 312*	*2 326*	*2 180*	*1 767*	*1 879*	*2 076*	*2 097*	*2 058*	*1 974*	*3 029*
Juan Fernandez rock lobster	**Langouste Juan Fernandez**		**Langosta de Juan Fernández**		***Jasus frontalis***				**2,29(01)002,02**	**LOF**
87 Chile	30	24	29	36	32	21	22	17	21	9
87 Fishing area total	*30*	*24*	*29*	*36*	*32*	*21*	*22*	*17*	*21*	*9*
Species total	*30*	*24*	*29*	*36*	*32*	*21*	*22*	*17*	*21*	*9*
Green rock lobster	**Langouste d'Océanie**		**Langosta de Oceanía**		***Jasus verreauxi***				**2,29(01)002,03**	**LOG**
81 Australia	100	143	151	97	104	108	110	117	104	103
New Zealand	8	5	48	64	54	16	13	35	10	8
81 Fishing area total	*108*	*148*	*199*	*161*	*158*	*124*	*123*	*152*	*114*	*111*
Species total	*108*	*148*	*199*	*161*	*158*	*124*	*123*	*152*	*114*	*111*
Tristan da Cunha rock lobster	**Langouste de Tristan da Cunha**		**Langosta de Tristán da Cunha**		***Jasus tristani***				**2,29(01)002,05**	**LBT**
47 St Helena	362	321	344	327	321	376	336	316	425	301
47 Fishing area total	*362*	*321*	*344*	*327*	*321*	*376*	*336*	*316*	*425*	*301*
Species total	*362*	*321*	*344*	*327*	*321*	*376*	*336*	*316*	*425*	*301*
Red rock lobster	**...B**		**Langosta roja**		***Jasus edwardsii***				**2,29(01)002,06**	**LOR**
81 New Zealand	3 039	2 730	3 568	3 121	5 009	2 707	2 818	2 789	2 551	2 485
81 Fishing area total	*3 039*	*2 730*	*3 568*	*3 121*	*5 009*	*2 707*	*2 818*	*2 789*	*2 551*	*2 485*
Species total	*3 039*	*2 730*	*3 568*	*3 121*	*5 009*	*2 707*	*2 818*	*2 789*	*2 551*	*2 485*
Southern rock lobster	**...B**		**...C**		***Jasus novaehollandiae***				**2,29(01)002,07**	**JSN**
57 Australia	5 164	4 635	4 507	4 856	4 888	4 615	4 655	4 756	4 677	4 386
57 Fishing area total	*5 164*	*4 635*	*4 507*	*4 856*	*4 888*	*4 615*	*4 655*	*4 756*	*4 677*	*4 386*
Species total	*5 164*	*4 635*	*4 507*	*4 856*	*4 888*	*4 615*	*4 655*	*4 756*	*4 677*	*4 386*

B-43 Lobsters, spiny-rock lobsters / Homards, langoustes / Bogavantes, langostas

Capture production by species, fishing areas and countries or areas / Captures par espèces, zones de pêche et pays ou zones / Capturas por especies, áreas de pesca y países o áreas

Species, Fishing area Espèce, Zone de pêche Especie, Area de pesca	1993 t	1994 t	1995 t	1996 t	1997 t	1998 t	1999 t	2000 t	2001 t	2002 t
St.Paul rock lobster	**Langouste de St.Paul**		**Langosta de St.Paul**		***Jasus paulensis***			**2,29(01)002,08**		**JSP**
51 Fr South Tr	403	443	439	357	295	308	345	192	183	334
51 Fishing area total	*403*	*443*	*439*	*357*	*295*	*308*	*345*	*192*	*183*	*334*
Species total	*403*	*443*	*439*	*357*	*295*	*308*	*345*	*192*	*183*	*334*
Pink spiny lobster	**Langouste rose**		**Langosta mora**		***Palinurus mauritanicus***			**2,29(01)008,02**		**PSL**
27 France	18	11	3	1	0	25	11	9	5	1
27 Fishing area total	*18*	*11*	*3*	*1*	*0*	*25*	*11*	*9*	*5*	*1*
37 France	0	0	0	0	0	0	0	-	-	-
37 Fishing area total	*0*	*0*	*0*	*0*	*0*	*0*	*0*	-	-	-
Species total	*18*	*11*	*3*	*1*	*0*	*25*	*11*	*9*	*5*	*1*
Common spiny lobster	**Langouste rouge**		**Langosta común**		***Palinurus elephas***			**2,29(01)008,04**		**SLO**
27 France	137	128	112	104	90	44	58	59	65	46
27 Fishing area total	*137*	*128*	*112*	*104*	*90*	*44*	*58*	*59*	*65*	*46*
37 Albania	...	...	...	...	...	...	...	...	1	2
France	44	33	8	9	9	9	3	3	3	3
Italy	211	198	197	312	331	174	161	123	166	152
Turkey	21	2	4	10	45	40	6	11	18	19
37 Fishing area total	*276*	*233*	*209*	*331*	*385*	*223*	*170*	*137*	*188*	*176*
Species total	*413*	*361*	*321*	*435*	*475*	*267*	*228*	*196*	*253*	*222*
Natal spiny lobster	**Langouste du Natal**		**Langosta del Natal**		***Palinurus delagoae***			**2,29(01)008,05**		**SLN**
51 South Africa	33	10	13	10	10	6	7 F	8	10	7
51 Fishing area total	*33*	*10*	*13*	*10*	*10*	*6*	*7 F*	*8*	*10*	*7*
Species total	*33*	*10*	*13*	*10*	*10*	*6*	*7 F*	*8*	*10*	*7*
Southern spiny lobster	**Langouste du Sud**		**Langosta del Sur**		***Palinurus gilchristi***			**2,29(01)008,06**		**SLS**
47 South Africa	985	1 021	966	918	892	864	429	305	1 053	651
47 Fishing area total	*985*	*1 021*	*966*	*918*	*892*	*864*	*429*	*305*	*1 053*	*651*
Species total	*985*	*1 021*	*966*	*918*	*892*	*864*	*429*	*305*	*1 053*	*651*
Palinurid spiny lobsters nei	**Langoustes palinurus nca**		**Langostas palinurus nep**		***Palinurus spp***			**2,29(01)008,XX**		**CRW**
27 Channel Is	1 F	0	0	1	0	1	1	1	3	2
France	-	-	-	-	-	-	1	-	-	-
Ireland	108	...	84	62	48	46	35	42	35	36
Portugal	27	12	13	23	15	27	20	12	7	5
Spain	4	20	7	11	22	17	26	4	21	6
UK	50	59	67	7	42	19	15	15	15	15
27 Fishing area total	*190 F*	*91*	*171*	*104*	*127*	*110*	*98*	*74*	*81*	*64*
34 Benin	4 F	4 F	3 F	3 F	5 F	4 F	3	2 F	2	7
Cape Verde	50	40	40	25	17	18	23	19	13	13 F
Congo Rep	34 F	1	3 F	4 F	3 F	7 F	9 F	10	4 F	4 F
Côte dIvoire	0	0	0	2 F	4 F	5 F	6	10	2	18
Cyprus	-	-	-	-	-	-	-	0	1	-
Gabon	23	16	42	103	33	85	50	32	57	121
Gambia	125	44	26	84	37	86	80	130	75	98
Italy	14	14	-	-	-	-	-	-	-	-
Latvia	-	-	-	-	-	-	-	-	-	10
Liberia	4	6	10	-	8	26	4	41	36	40 F
Mauritania	20 F	10 F	30 F	50 F	110 F	100 F	80 F	82 F	83 F	81 F
Morocco	3	5	13	20	58	31	38	32	394	162
Nigeria	1 049	1 751	1 496	1 133	8 315	3 071	1 245	1 939	1 699	193
Portugal	66	154	84	116	30	-	-	8	5	-
Senegal	103	787	121	130	196	144	39	37	53	50
Spain	-	-	-	9	...	...	13	3	46	54
34 Fishing area total	*1 495 F*	*2 832 F*	*1 868 F*	*1 679 F*	*8 816 F*	*3 577 F*	*1 590 F*	*2 345 F*	*2 470 F*	*851 F*
37 Algeria	25 F	30 F	22	9	44	141	70	33	34	46
Morocco	3	3	1	6	3	2	0	10	62	3
Spain	130 F	130 F	130 F	129	82	86	57	39	44	103
Tunisia	62	48	44	40	71	68	61	49	41	45
37 Fishing area total	*220 F*	*211 F*	*197 F*	*184*	*200*	*297*	*188*	*131*	*181*	*197*
Species total	*1 905 F*	*3 134 F*	*2 236 F*	*1 967 F*	*9 143 F*	*3 984 F*	*1 876 F*	*2 550 F*	*2 732 F*	*1 112 F*
Spiny lobsters nei	**Langoustes diverses nca**		**Langostas diversas nep**		***Palinuridae***			**2,29(01)XXX,XX**		**VLO**
51 Eritrea	...	...	...	1	1	2	1	0	0	0
Mozambique	313	307	248	331	232	237	203	228	213	276
51 Fishing area total	*313*	*307*	*248*	*332*	*233*	*239*	*204*	*228*	*213*	*276*
Species total	*313*	*307*	*248*	*332*	*233*	*239*	*204*	*228*	*213*	*276*

B-43 Lobsters, spiny-rock lobsters / Homards, langoustes / Bogavantes, langostas

Capture production by species, fishing areas and countries or areas / Captures par espèces, zones de pêche et pays ou zones / Capturas por especies, áreas de pesca y países o áreas

Species, Fishing area Espèce, Zone de pêche Especie, Area de pesca	1993 t	1994 t	1995 t	1996 t	1997 t	1998 t	1999 t	2000 t	2001 t	2002 t
Japanese fan lobster	**Cigale japonaise**		**Cigarra japonesa**		***Ibacus ciliatus***				**2,29(15)001,04**	**IBC**
61 China,Taiwan	456	236	1 224	1 115	642	696	676	1 600	1 607	1 005
61 Fishing area total	*456*	*236*	*1 224*	*1 115*	*642*	*696*	*676*	*1 600*	*1 607*	*1 005*
Species total	*456*	*236*	*1 224*	*1 115*	*642*	*696*	*676*	*1 600*	*1 607*	*1 005*
Flathead lobster	**Cigale raquette**		**Cigarra chata**		***Thenus orientalis***				**2,29(15)005,01**	**THQ**
57 Thailand	199	225	335	299	177	533	37	23	184	178
57 Fishing area total	*199*	*225*	*335*	*299*	*177*	*533*	*37*	*23*	*184*	*178*
71 Thailand	1 236	861	1 730	2 716	2 785	3 005	1 760	2 289	2 356	2 286
71 Fishing area total	*1 236*	*861*	*1 730*	*2 716*	*2 785*	*3 005*	*1 760*	*2 289*	*2 356*	*2 286*
Species total	*1 435*	*1 086*	*2 065*	*3 015*	*2 962*	*3 538*	*1 797*	*2 312*	*2 540*	*2 464*
Slipper lobsters nei	**Cigales nca**		**Cigarros nep**		***Scyllaridae***				**2,29(15)XXX,XX**	**LOS**
34 Senegal	-	-	-	-	-	1	2	4	6	1
34 Fishing area total	-	-	-	-	-	*1*	*2*	*4*	*6*	*1*
47 St Helena	0	0	0	0	0	1	1	1	0	-
47 Fishing area total	*0*	*0*	*0*	*0*	*0*	*1*	*1*	*1*	*0*	-
51 Bahrain	209	149	150	125	56	51	6	2	2	3
Qatar	33	23	8	8	16	15	13	5	20	17
South Africa	-	-	-	-	2	1	1 F	2	4	...
51 Fishing area total	*242*	*172*	*158*	*133*	*74*	*67*	*20 F*	*9*	*26*	*20*
57 Australia	0	18	13	21	0	0	0	0	-	-
57 Fishing area total	*0*	*18*	*13*	*21*	*0*	*0*	*0*	*0*	-	-
71 Australia	603	901	773	600	717	789	...	...	...	...
Philippines	342	938	350	334	8	65	89	90	121	123
Singapore	50	16	11	9	11	12	9	6	7	4
71 Fishing area total	*995*	*1 855*	*1 134*	*943*	*736*	*866*	*98*	*96*	*128*	*127*
81 Australia	114	28	1	1	158	156	7	0	0	-
New Zealand	-	-	0	0	0	4	0	1	4	2
81 Fishing area total	*114*	*28*	*1*	*1*	*158*	*160*	*7*	*1*	*4*	*2*
Species total	*1 351*	*2 073*	*1 306*	*1 098*	*968*	*1 095*	*128 F*	*111*	*164*	*150*
Mozambique lobster	**Langoustine du Mozambique**		**Cigala del Mozambique**		***Metanephrops mozambicus***				**2,29(42)005,02**	**NEM**
51 Mozambique	443	261	179	132	156	147	92	105	69	75
South Africa	83	-	-	-	-	45	60 F	75	72	55
51 Fishing area total	*526*	*261*	*179*	*132*	*156*	*192*	*152 F*	*180*	*141*	*130*
Species total	*526*	*261*	*179*	*132*	*156*	*192*	*152 F*	*180*	*141*	*130*
New Zealand lobster	**Langoustine de N.lle Zélande**		**Cigala de Nueva Zelandia**		***Metanephrops challengeri***				**2,29(42)005,04**	**MEC**
81 New Zealand	926	1 067	1 078	670	1 093	989	925	1 034	1 093	1 020
81 Fishing area total	*926*	*1 067*	*1 078*	*670*	*1 093*	*989*	*925*	*1 034*	*1 093*	*1 020*
Species total	*926*	*1 067*	*1 078*	*670*	*1 093*	*989*	*925*	*1 034*	*1 093*	*1 020*
Metanephrops nei	**Metanephrops nca**		**Metanephrops nep**		***Metanephrops spp***				**2,29(42)005,XX**	**MWF**
57 Australia	-	-	-	-	-	-	-	39	105	88
57 Fishing area total	-	-	-	-	-	-	-	*39*	*105*	*88*
71 Australia	-	-	-	-	-	-	-	-	-	9
71 Fishing area total	-	-	-	-	-	-	-	-	-	*9*
Species total	-	-	-	-	-	-	-	*39*	*105*	*97*
Norway lobster	**Langoustine**		**Cigala**		***Nephrops norvegicus***				**2,29(42)006,02**	**NEP**
27 Belgium	418	306	413	188	316	240	349	254	284	287
Denmark	2 942	3 178	3 608	4 176	4 282	4 982	5 455	5 083	4 813	5 437
Faeroe Is	55	105	91	66	40	57	80	73	51	29
France	9 844	9 271	9 782	8 623	7 125	6 611	5 843	6 639	6 992	6 870
Germany	16	23	17	77	70	70	110	86	141	132
Iceland	2 381	2 238	1 027	1 623	1 215	1 411	1 389	1 230	1 420	1 548
Ireland	4 765	5 312	7 241	5 171	7 020	6 950	8 492	7 709	7 074	6 983
Isle of Man	32	14	29	20	24	17	10	3	2	-
Netherlands	130	158	253	423	627	694	662	572	853	966
Norway	211	234	166	188	187	293	383	346	281	280
Portugal	237	223	282	185	162	175	216	211	282	365
Spain	1 621	1 500	1 486	959	1 256	898	994	844	907	854
Sweden	864	764	914	1 105	1 130	1 317	1 263	1 232	1 067	1 046
UK	28 803	30 462	31 893	29 218	31 713	29 212	31 312	28 422	28 536	28 502
27 Fishing area total	*52 319*	*53 788*	*57 202*	*52 022*	*55 167*	*52 927*	*56 558*	*52 704*	*52 703*	*53 299*
34 Greece	5	1	-	1	-	10	-	2	-	-
Morocco	2	1	3	6	3	4	2	2	13	14
Portugal	0	0	0	-	-	12	42	78	88	60
Spain	-	-	-	303	261	271	1 100	328	3	7

B-43
Lobsters, spiny-rock lobsters — **Capture production by species, fishing areas and countries or areas**
Homards, langoustes — **Captures par espèces, zones de pêche et pays ou zones**
Bogavantes, langostas — **Capturas por especies, áreas de pesca y países o áreas**

Species, Fishing area Espèce, Zone de pêche Especie, Area de pesca	1993 t	1994 t	1995 t	1996 t	1997 t	1998 t	1999 t	2000 t	2001 t	2002 t
34 Fishing area total	*7*	*2*	*3*	*310*	*264*	*297*	*1 144*	*410*	*104*	*81*
37 Albania	...	...	0	3	0	-	-	-	10	5
Algeria	130 F	150 F	129	100	89	138	96	92	45	56
Croatia	593	657	524	486	473	500	240	250	274	140
France	...	...	...	...	...	...	0	-	-	-
Greece	799	1 087	1 104	486	351	445	243	266	242	230 F
Italy	4 864	5 263	4 312	5 101	4 834	2 582	3 033	2 485	2 287	2 051
Morocco	0	1	0	1	1	1	1	3	2	1
Serbia-Monte	10	5	13	5	5	7	5	7	9	10
Spain	515 F	500 F	485 F	473	411	478	448	407	630	636
Tunisia	1	15	2	3	1	4	2	4	4	4
37 Fishing area total	*6 912 F*	*7 678 F*	*6 569 F*	*6 658*	*6 165*	*4 155*	*4 068*	*3 514*	*3 503*	*3 133 F*
Species total	*59 238 F*	*61 468 F*	*63 774 F*	*58 990*	*61 596*	*57 379*	*61 770*	*56 628*	*56 310*	*56 513 F*
American lobster	**Homard américain**		**Bogavante americano**		***Homarus americanus***				**2,29(42)007,01**	**LBA**
21 Canada	40 917	41 537	40 508	39 369	40 079	41 030	43 428	45 331	51 412	45 111
St Pier Mq	1	0	1	1	1	-	1	1	2	2
USA	25 634	30 126	30 122	32 496	38 066	36 125	39 676	37 730	32 389	37 309
21 Fishing area total	*66 552*	*71 663*	*70 631*	*71 866*	*78 146*	*77 155*	*83 105*	*83 062*	*83 803*	*82 422*
Species total	*66 552*	*71 663*	*70 631*	*71 866*	*78 146*	*77 155*	*83 105*	*83 062*	*83 803*	*82 422*
European lobster	**Homard européen**		**Bogavante**		***Homarus gammarus***				**2,29(42)007,18**	**LBE**
27 Belgium	-	-	3	-	-	-	1	1	6	1
Channel Is	129 F	123	114 F	260	221	214	212	153	178	200
Denmark	22	26	27	44	39	18	11	11	11	11
France	202	223	266	267	327	219	308	332	329	359
Germany	2	2	9	17	0	0	0	-	0	0
Ireland	470	824	564	567	513	611	597	606	781	740
Isle of Man	3	12	14	0	26	25	14	8	12	23
Netherlands	-	-	-	-	-	-	13	12	33	11
Norway	28	30	34	30	35	45	59	52	40	42
Portugal	4	3	2	3	2	2	1	2	2	2
Spain	22	17	18	21	14	12	11	7	8	6
Sweden	18	26	29	26	27	26	25	20	18	19
UK	1 045	1 162	1 306	1 043	1 534	1 482	1 819	1 141	1 086	1 188
27 Fishing area total	*1 945 F*	*2 448*	*2 386 F*	*2 278*	*2 738*	*2 654*	*3 071*	*2 345*	*2 504*	*2 602*
34 Greece	176	11	13	14	17	17	12	9	12	15
Morocco	1	6	7	19	12	14	7	11	11	55
Portugal	-	-	-	-	1	-	-	-	-	-
Spain	-	-	-	2	-	-	-	-	-	-
34 Fishing area total	*177*	*17*	*20*	*35*	*30*	*31*	*19*	*20*	*23*	*70*
37 Algeria	0	0	0	0	0	0	-	-	-	-
Croatia	23	43	30	31	43	40	18	18	13	3
France	3	1	0	0	0	0	0	-	-	-
Greece	118	320	510	198	357	138	158	192	221	210 F
Morocco	0	3	1	1	2	1	1	-	-	0
Serbia-Monte	0	1	1	5	5	6	5	6	7	5
Spain	-	-	-	7	3	3	2	3	3	1
Tunisia	2	0	0	1	1	-	1	1	0	0
Turkey	8	18	33	34	40	60	10	15	10	9
37 Fishing area total	*154*	*386*	*575*	*277*	*451*	*248*	*195*	*235*	*254*	*228 F*
Species total	*2 276 F*	*2 851*	*2 981 F*	*2 590*	*3 219*	*2 933*	*3 285*	*2 600*	*2 781*	*2 900 F*
Blue mud shrimp	**...B**		**...C**		***Upogebia pugettensis***				**2,29(49)001,03**	**UOP**
67 USA	...	...	...	...	...	...	...	10	4	4
67 Fishing area total	...	...	...	...	...	...	...	*10*	*4*	*4*
Species total	...	...	...	...	...	...	...	*10*	*4*	*4*
Ghost shrimps	**...B**		**...C**		***Callianassa spp***				**2,29(59)001,XX**	**CZP**
67 USA	-	-	-	-	-	-	25	17	14	14
67 Fishing area total	-	-	-	-	-	-	*25*	*17*	*14*	*14*
Species total	-	-	-	-	-	-	*25*	*17*	*14*	*14*
Lobsters nei	**Langoustes, homards nca**		**Langostas nep**		***Reptantia***				**2,29(XX)XXX,XX**	**LOX**
27 Portugal	-	-	-	-	-	-	22	11	6	5
27 Fishing area total	-	-	-	-	-	-	*22*	*11*	*6*	*5*
37 Cyprus	1	3	2	5	2	4	2	4	5	5
France	-	-	-	-	-	-	0	2	2	-
37 Fishing area total	*1*	*3*	*2*	*5*	*2*	*4*	*2*	*6*	*7*	*5*
47 Portugal	-	-	9	9	25	17	67	10	-	6
Spain	-	-	-	11	-	14	1	-	-	2
47 Fishing area total	-	-	*9*	*20*	*25*	*31*	*68*	*10*	-	*8*
51 Portugal	-	-	-	-	-	-	-	62	2	-

B-43 Lobsters, spiny-rock lobsters / Homards, langoustes / Bogavantes, langostas

Capture production by species, fishing areas and countries or areas / Captures par espèces, zones de pêche et pays ou zones / Capturas por especies, áreas de pesca y países o áreas

Species, Fishing area Espèce, Zone de pêche Especie, Area de pesca	1993 t	1994 t	1995 t	1996 t	1997 t	1998 t	1999 t	2000 t	2001 t	2002 t
51 Fishing area total	-	-	-	-	-	-	-	*62*	*2*	-
71 Viet Nam	...	...	...	...	2 000 F	1 000 F	200 F	500 F	700 F	700 F
71 Fishing area total	...	...	...	...	*2 000 F*	*1 000 F*	*200 F*	*500 F*	*700 F*	*700 F*
Species total	*1*	*3*	*11*	*25*	*2 027 F*	*1 035 F*	*292 F*	*589 F*	*715 F*	*718 F*
Group total	***206 612***	***217 277***	***219 849***	***212 914***	***233 346***	***217 075***	***228 539***	***227 800***	***222 052***	***222 132***

B-44 King crabs, squat-lobsters / Crabes royaux, galatées / Cangrejos reales, galateidos

Capture production by species, fishing areas and countries or areas / Captures par espèces, zones de pêche et pays ou zones / Capturas por especies, áreas de pesca y países o áreas

Species, Fishing area Espèce, Zone de pêche Especie, Area de pesca	1993 t	1994 t	1995 t	1996 t	1997 t	1998 t	1999 t	2000 t	2001 t	2002 t
Pelagic red crab	**Galatée pélagique**		**Langostino pelágico**		***Pleuroncodes planipes***			**2,30(19)001,01**		**LQL**
77 El Salvador	-	299	356	164	-	-	-	-	1 263	3 156
77 Fishing area total	*-*	*299*	*356*	*164*	*-*	*-*	*-*	*-*	*1 263*	*3 156*
Species total	*-*	*299*	*356*	*164*	*-*	*-*	*-*	*-*	*1 263*	*3 156*
Carrot squat lobster	**Galatée orange**		**Langostino colorado**		***Pleuroncodes monodon***			**2,30(19)001,02**		**PQG**
87 Chile	3 334	2 422	4 938	7 726	8 939	12 602	12 710	11 129	1 754	2 499
87 Fishing area total	*3 334*	*2 422*	*4 938*	*7 726*	*8 939*	*12 602*	*12 710*	*11 129*	*1 754*	*2 499*
Species total	*3 334*	*2 422*	*4 938*	*7 726*	*8 939*	*12 602*	*12 710*	*11 129*	*1 754*	*2 499*
Blue squat lobster	**Galatée bleue**		**Langostino amarillo**		***Cervimunida johni***			**2,30(19)003,01**		**CZJ**
87 Chile	2 224	4 842	5 743	6 402	10 322	9 426	7 273	5 069	2 178	925
87 Fishing area total	*2 224*	*4 842*	*5 743*	*6 402*	*10 322*	*9 426*	*7 273*	*5 069*	*2 178*	*925*
Species total	*2 224*	*4 842*	*5 743*	*6 402*	*10 322*	*9 426*	*7 273*	*5 069*	*2 178*	*925*
Craylets, squat lobsters	**Galatées**		**Camaroncillos,langostinos,etc.**		***Galatheidae***			**2,30(19)XXX,XX**		**LOQ**
27 France	104	89	85	90	94	81	102	98	84	84
UK	10	15	7	15	12	-	28	-	-	-
27 Fishing area total	*114*	*104*	*92*	*105*	*106*	*81*	*130*	*98*	*84*	*84*
87 Chile	-	-	-	-	-	-	-	254	1	724
87 Fishing area total	*-*	*-*	*-*	*-*	*-*	*-*	*-*	*254*	*1*	*724*
Species total	*114*	*104*	*92*	*105*	*106*	*81*	*130*	*352*	*85*	*808*
Red king crab	**Crabe royal du Kamtchatka**		**Cangrejo real rojo**		***Paralithodes camtschaticus***			**2,30(20)018,01**		**KCD**
27 Norway	...	32	32	70	71	124	202	211	434	414
Russian Fed	-	-	-	-	-	71	-	78	252	1 217
UK	-	-	-	-	-	37	139	89	44	87
27 Fishing area total	*...*	*32*	*32*	*70*	*71*	*232*	*341*	*378*	*730*	*1 718*
61 Russian Fed	35 330	38 068	54 986	34 300	23 262	32 486	37 072	28 554	16 064	9 691
61 Fishing area total	*35 330*	*38 068*	*54 986*	*34 300*	*23 262*	*32 486*	*37 072*	*28 554*	*16 064*	*9 691*
Species total	*35 330*	*38 100*	*55 018*	*34 370*	*23 333*	*32 718*	*37 413*	*28 932*	*16 794*	*11 409*
Blue king crab	**Crabe royal bleu**		**Cangrejo real azul**		***Paralithodes platypus***			**2,30(20)018,02**		**KCI**
61 Russian Fed	...	...	...	8 762	10 268	4 508	5 455	5 233	4 500	4 598
61 Fishing area total	*...*	*...*	*...*	*8 762*	*10 268*	*4 508*	*5 455*	*5 233*	*4 500*	*4 598*
Species total	*...*	*...*	*...*	*8 762*	*10 268*	*4 508*	*5 455*	*5 233*	*4 500*	*4 598*
Brown king crab	**Crabe royal brun**		**Cangrejo real marrón**		***Paralithodes brevipes***			**2,30(20)018,03**		**KCY**
61 Russian Fed	...	...	...	204	418	194	256	347	254	350
61 Fishing area total	*...*	*...*	*...*	*204*	*418*	*194*	*256*	*347*	*254*	*350*
Species total	*...*	*...*	*...*	*204*	*418*	*194*	*256*	*347*	*254*	*350*
King crabs	**Crabes royaux**		**Cangrejos rusos**		***Paralithodes spp***			**2,30(20)018,XX**		**KCS**
48 Bulgaria	-	0	-	-	-	-	-	-	-	-
UK	-	-	-	-	-	-	0	-	-	-
48 Fishing area total	*-*	*0*	*-*	*-*	*-*	*-*	*0*	*-*	*-*	*-*
61 Japan	313	472	260	322	154	132	117	89	181	129
Korea Rep	9	-	-	-	-	-	-	-	-	-
61 Fishing area total	*322*	*472*	*260*	*322*	*154*	*132*	*117*	*89*	*181*	*129*
67 USA	11 218	5 425	6 656	9 526	8 177	10 941	7 675	6 848	7 281	7 616
67 Fishing area total	*11 218*	*5 425*	*6 656*	*9 526*	*8 177*	*10 941*	*7 675*	*6 848*	*7 281*	*7 616*
77 USA	-	-	-	-	-	-	-	1	2	1
77 Fishing area total	*-*	*-*	*-*	*-*	*-*	*-*	*-*	*1*	*2*	*1*
Species total	*11 540*	*5 897*	*6 916*	*9 848*	*8 331*	*11 073*	*7 792*	*6 938*	*7 464*	*7 746*
Southern king crab	**Crabe royal de Patagonie**		**Centolla patagónica**		***Lithodes antarcticus***			**2,30(20)070,01**		**KCR**
41 Argentina	158	295	380	200	413	456	270	104	88	385
Uruguay	2	0	0	0	0	0	0	0	0	-
41 Fishing area total	*160*	*295*	*380*	*200*	*413*	*456*	*270*	*104*	*88*	*385*
87 Chile	1 980	1 673	1 906	1 759	2 160	2 766	2 155	2 902	2 937	2 869
87 Fishing area total	*1 980*	*1 673*	*1 906*	*1 759*	*2 160*	*2 766*	*2 155*	*2 902*	*2 937*	*2 869*
Species total	*2 140*	*1 968*	*2 286*	*1 959*	*2 573*	*3 222*	*2 425*	*3 006*	*3 025*	*3 254*
Subantarctic stone crab	**Crabe royal subantarctique**		**Centolla subantártica**		***Lithodes murrayi***			**2,30(20)070,02**		**KCM**
48 Bulgaria	-	0	-	-	-	-	-	-	-	-

B-44 King crabs, squat-lobsters / Crabes royaux, galatées / Cangrejos reales, galateidos

Capture production by species, fishing areas and countries or areas / Captures par espèces, zones de pêche et pays ou zones / Capturas por especies, áreas de pesca y países o áreas

Species, Fishing area Espèce, Zone de pêche Especie, Area de pesca	1993 t	1994 t	1995 t	1996 t	1997 t	1998 t	1999 t	2000 t	2001 t	2002 t
Russian Fed	-	0	-	-	-	-		-	-	-
South Africa	-	-	-	-	-	1	-	-	-	-
Uruguay	-	-	-	-	-	-	0	-	-	-
48 Fishing area total	-	*0*	-	-	-	*1*	*0*	-	-	-
58 South Africa	-	-	-	-	-	0	-	0	0	0
58 Fishing area total	-	-	-	-	-	*0*	-	*0*	*0*	*0*
88 New Zealand	-	-	-	-	-	-	-	-	-	0
88 Fishing area total	-	-	-	-	-	-	-	-	-	*0*
Species total	-	*0*	-	-	-	*1*	*0*	*0*	*0*	*0*
Golden king crab	**Crabe royal doré**		**Centolla dorada**		***Lithodes aequispina***				**2,30(20)070,05**	**KAQ**
61 Russian Fed	...	...	...	4 666	4 917	3 897	2 746	1 797	2 245	2 291
61 Fishing area total	...	...	...	*4 666*	*4 917*	*3 897*	*2 746*	*1 797*	*2 245*	*2 291*
Species total	...	...	...	*4 666*	*4 917*	*3 897*	*2 746*	*1 797*	*2 245*	*2 291*
King crabs nei	**Crabes royaux nca**		**Centollas nep**		***Lithodes spp***				**2,30(20)070,XX**	**KCZ**
48 Uruguay	-	-	-	-	-	-	0	-	-	-
48 Fishing area total	-	-	-	-	-	-	*0*	-	-	-
Species total	-	-	-	-	-	-	*0*	-	-	-
Softshell red crab	**Crabe royal hérisson**		**Centollón**		***Paralomis granulosa***				**2,30(20)123,01**	**PAG**
41 Argentina	...	...	...	...	...	...	...	264	213	129
Falkland Is	1	1	1	1	1	1	1	1	5	1
41 Fishing area total	*1*	*1*	*1*	*1*	*1*	*1*	*1*	*265*	*218*	*130*
87 Chile	955	2 221	1 316	1 273	1 477	1 501	1 438	4 938	6 527	4 631
87 Fishing area total	*955*	*2 221*	*1 316*	*1 273*	*1 477*	*1 501*	*1 438*	*4 938*	*6 527*	*4 631*
Species total	*956*	*2 222*	*1 317*	*1 274*	*1 478*	*1 502*	*1 439*	*5 203*	*6 745*	*4 761*
Red stone crab	**Crabe royal rouge**		**Centolla colorada**		***Paralomis aculeata***				**2,30(20)123,02**	**KCU**
48 Uruguay	-	-	-	-	-	-	0	-	-	-
48 Fishing area total	-	-	-	-	-	-	*0*	-	-	-
58 South Africa	-	-	-	-	-	-	-	0	0	-
58 Fishing area total	-	-	-	-		-	-	*0*	*0*	-
88 New Zealand	-	-	-	-	-	0	-	-	-	-
88 Fishing area total	-	-	-	-	-	*0*	-	-	-	-
Species total	-	-	-	-		*0*	*0*	*0*	*0*	
Antarctic stone crab	**Crabe royal de l'Antarctique**		**Centolla antártica**		***Paralomis spinosissima***				**2,30(20)123,03**	**KCV**
48 Japan	-	-	-	-	-	-	-	-	-	55
Korea Rep	-	-	-	0	0	-	-	-	-	-
Spain	-	-	-	-	0	-	-	-	-	-
UK	-	-	-	-	-	-	0	0	4	0
USA	-	-	283	214	-	-	-	-	-	-
48 Fishing area total	-	-	*283*	*214*	*0*	-	*0*	*0*	*4*	*55*
Species total	-	-	*283*	*214*	*0*	-	*0*	*0*	*4*	*55*
Globose king crab	**Crabe royal sphérique**		**Centolla redonda**		***Paralomis formosa***				**2,30(20)123,04**	**KCF**
48 Japan	-	-	-	-	-	-	-	-	-	56
UK	-	-	-	-	-	-	2	2	11	-
48 Fishing area total	-	-	-	-	-	-	*2*	*2*	*11*	*56*
Species total	-	-	-	-	-	-	*2*	*2*	*11*	*56*
King crabs, stone crabs nei	**Crabes royaux, etc. nca**		**Centollas, centollones nep**		***Lithodidae***				**2,30(20)XXX,XX**	**KCX**
41 UK	-	-	-	-	-	-	-	-	4	2
41 Fishing area total	-	-	-	-	-	-	-	-	*4*	*2*
48 Korea Rep	-	-	-	-	0	-	-	-	-	-
South Africa	-	-	-	-	-	0	-	-	-	-
UK	-	-	-	-	1	0	0	0	-	-
48 Fishing area total	-	-	-	-	*1*	*0*	*0*	*0*	-	-
58 Australia	-	-	-	-	0	-	-	-	-	-
South Africa	-	-	-	-	0	0	0	3	0	0
58 Fishing area total	-	-	-	-	*0*	*0*	*0*	*3*	*0*	*0*
88 New Zealand	-	-	-	-	-	0	-	-	-	-
88 Fishing area total	-	-	-	-	-	*0*	-	-	-	-
Species total	-	-	-	-	*1*	*0*	*0*	*3*	*4*	*2*
Group total	***55 638***	***55 854***	***76 949***	***75 694***	***70 686***	***79 224***	***77 641***	***68 011***	***46 326***	***41 910***

B-45 Shrimps, prawns / Crevettes / Gambas, camarones

Capture production by species, fishing areas and countries or areas
Captures par espèces, zones de pêche et pays ou zones
Capturas por especies, áreas de pesca y países o áreas

Species, Fishing area Espèce, Zone de pêche Especie, Area de pesca	1993 t	1994 t	1995 t	1996 t	1997 t	1998 t	1999 t	2000 t	2001 t	2002 t
Northern brown shrimp	**Crevette royale grise**		**Camarón café norteño**		***Penaeus aztecus***				**2,28(01)001,01**	**ABS**
21 USA	187	-	19	666	3 377	-	679	1 104	458	856
21 Fishing area total	*187*	*-*	*19*	*666*	*3 377*	*-*	*679*	*1 104*	*458*	*856*
31 Costa Rica	0	0	4	8	0	0	0	0	-	-
USA	52 851	51 574	57 107	54 695	44 459	50 722	60 527	62 713	68 411	57 335
31 Fishing area total	*52 851*	*51 574*	*57 111*	*54 703*	*44 459*	*50 722*	*60 527*	*62 713*	*68 411*	*57 335*
Species total	*53 038*	*51 574*	*57 130*	*55 369*	*47 836*	*50 722*	*61 206*	*63 817*	*68 869*	*58 191*
Banana prawn	**Crevette banane**		**Langostino banana**		***Penaeus merguiensis***				**2,28(01)001,03**	**PBA**
57 Australia	...	...	...	...	...	...	517	329	281	345
Indonesia	6 953	9 092	9 852	10 318	10 073	10 382	9 569	10 087	10 021	10 540
Thailand	4 478	3 283	6 100	6 512	5 464	4 850	4 574	4 616	4 476	4 311
57 Fishing area total	*11 431*	*12 375*	*15 952*	*16 830*	*15 537*	*15 232*	*14 660*	*15 032*	*14 778*	*15 196*
71 Australia	4 058	2 433	4 490	4 347	4 546	3 711	3 608	2 257	6 365	5 449
Indonesia	36 972	38 145	40 625	43 596	43 851	51 810	54 610	56 557	55 248	58 180
Papua N Guin	813	628	858	820	676	1 233	949	1 136	1 017	1 000 F
Solomon Is	10 F	3 F	5 F	20 F	30 F	40 F	20 F	20 F	20 F	20 F
Thailand	8 130	9 108	9 225	7 133	7 169	7 648	8 206	10 385	7 601	7 374
71 Fishing area total	*49 983 F*	*50 317 F*	*55 203 F*	*55 916 F*	*56 272 F*	*64 442 F*	*67 393 F*	*70 355 F*	*70 251 F*	*72 023 F*
Species total	*61 414 F*	*62 692 F*	*71 155 F*	*72 746 F*	*71 809 F*	*79 674 F*	*82 053 F*	*85 387 F*	*85 029 F*	*87 219 F*
Yellowleg shrimp	**Crevette pattes jaunes**		**Camarón patiamarillo**		***Penaeus californiensis***				**2,28(01)001,04**	**YPS**
77 Costa Rica	12	10	8	6	2	12	12	6	2	2
Guatemala	137	393	275	266	124	259	218	132	100 F	40
77 Fishing area total	*149*	*403*	*283*	*272*	*126*	*271*	*230*	*138*	*102 F*	*42*
87 Ecuador	-	-	73	73	60	50	20	20	40	45 F
87 Fishing area total	*-*	*-*	*73*	*73*	*60*	*50*	*20*	*20*	*40*	*45 F*
Species total	*149*	*403*	*356*	*345*	*186*	*321*	*250*	*158*	*142 F*	*87 F*
Northern pink shrimp	**Crevette rose du Nord**		**Camarón rosado norteño**		***Penaeus duorarum***				**2,28(01)001,05**	**APS**
21 USA	13	-	-	10 753	114	-	5 062	16	1	29
21 Fishing area total	*13*	*-*	*-*	*10 753*	*114*	*-*	*5 062*	*16*	*1*	*29*
31 Cuba	2 738	2 229	1 851	1 710	1 998	1 368	1 579	1 588	1 483	1 308
USA	7 477	5 757	9 270	3 047	8 760	10 718	863	5 573	7 143	8 045
31 Fishing area total	*10 215*	*7 986*	*11 121*	*4 757*	*10 758*	*12 086*	*2 442*	*7 161*	*8 626*	*9 353*
Species total	*10 228*	*7 986*	*11 121*	*15 510*	*10 872*	*12 086*	*7 504*	*7 177*	*8 627*	*9 382*
Kuruma prawn	**Crevette kuruma**		**Langostino japonés**		***Penaeus japonicus***				**2,28(01)001,09**	**KUP**
37 Israel	-	-	-	-	84	...	...	...	121	111
37 Fishing area total	*-*	*-*	*-*	*-*	*84*	*...*	*...*	*...*	*121*	*111*
61 China,Taiwan	2 379	2 178	1 960	1 561	2 665	1 906	4 071	8 168	4 179	1 590
Japan	2 263	2 685	2 668	2 262	2 144	2 069	1 523	1 447	1 271	1 123
Korea Rep	2 531	2 265	2 399	1 783	2 102	1 138	502	578	513	217
61 Fishing area total	*7 173*	*7 128*	*7 027*	*5 606*	*6 911*	*5 113*	*6 096*	*10 193*	*5 963*	*2 930*
Species total	*7 173*	*7 128*	*7 027*	*5 606*	*6 995*	*5 113*	*6 096*	*10 193*	*6 084*	*3 041*
Whiteleg shrimp	**Crevette pattes blanches**		**Camarón patiblanco**		***Penaeus vannamei***				**2,28(01)001,11**	**PNV**
77 El Salvador	830	1 175	1 258	1 238	1 046	1 277	1 082	519	460	926
77 Fishing area total	*830*	*1 175*	*1 258*	*1 238*	*1 046*	*1 277*	*1 082*	*519*	*460*	*926*
87 Ecuador	14 048	10 000	6 000	4 497	4 000	3 000	1 100	1 100	2 250	2 450 F
87 Fishing area total	*14 048*	*10 000*	*6 000*	*4 497*	*4 000*	*3 000*	*1 100*	*1 100*	*2 250*	*2 450 F*
Species total	*14 878*	*11 175*	*7 258*	*5 735*	*5 046*	*4 277*	*2 182*	*1 619*	*2 710*	*3 376 F*
Giant tiger prawn	**Crevette géante tigrée**		**Langostino jumbo**		***Penaeus monodon***				**2,28(01)001,12**	**GIT**
51 India	133 063	184 476	140 028	112 342	106 622	167 904	168 942	145 857	100 155	112 356
Pakistan	157	138	132	141	140	122	138	139	140	155
51 Fishing area total	*133 220*	*184 614*	*140 160*	*112 483*	*106 762*	*168 026*	*169 080*	*145 996*	*100 295*	*112 511*
57 Australia	-	-	-	-	-	-	-	2	14	23
India	41 611	35 159	36 261	38 251	39 489	36 661	54 761	58 731	57 868	44 543
Indonesia	4 905	6 169	8 189	4 825	8 489	8 158	8 080	8 047	9 099	10 250
Thailand	536	153	799	1 993	1 837	871	1 209	693	1 127	1 085
57 Fishing area total	*47 052*	*41 481*	*45 249*	*45 069*	*49 815*	*45 690*	*64 050*	*67 473*	*68 108*	*55 901*
61 China,Taiwan	38	28	67	49	45	38	92	57	91	54
61 Fishing area total	*38*	*28*	*67*	*49*	*45*	*38*	*92*	*57*	*91*	*54*
71 Australia	3 477	3 339	4 368	3 841	3 055	3 686	3 477	4 586	4 253	4 167
Indonesia	11 211	10 791	16 312	14 571	17 440	21 889	26 143	32 940	34 660	40 410
Papua N Guin	144	123	137	109	99	209	164	126	117	117 F

B-45 Shrimps, prawns / Crevettes / Gambas, camarones

Capture production by species, fishing areas and countries or areas
Captures par espèces, zones de pêche et pays ou zones
Capturas por especies, áreas de pesca y países o áreas

Species, Fishing area Espèce, Zone de pêche Especie, Area de pesca	1993 t	1994 t	1995 t	1996 t	1997 t	1998 t	1999 t	2000 t	2001 t	2002 t
Philippines	...	1 216	431	360	292	422	169	232	328	341
Thailand	...	488	373	572	631	341	1 043	1 580	760	737
71 Fishing area total	*14 832*	*15 957*	*21 621*	*19 453*	*21 517*	*26 547*	*30 996*	*39 464*	*40 118*	*45 772 F*
81 Australia	-	-	-	-	-	-	-	13	5	4
81 Fishing area total	-	-	-	-	-	-	-	*13*	*5*	*4*
Species total	*195 142*	*242 080*	*207 097*	*177 054*	*178 139*	*240 301*	*264 218*	*253 003*	*208 617*	*214 242 F*
Fleshy prawn	**Crevette charnue**		**Langostino carnoso**		***Penaeus chinensis***			**2,28(01)001,16**		**FLP**
61 China	16 582	45 770	43 043	55 292	69 406	78 350	69 911	84 334	97 002	99 468
Korea Rep	897	1 363	1 406	1 242	1 911	1 245	814	1 211	581	222
61 Fishing area total	*17 479*	*47 133*	*44 449*	*56 534*	*71 317*	*79 595*	*70 725*	*85 545*	*97 583*	*99 690*
Species total	*17 479*	*47 133*	*44 449*	*56 534*	*71 317*	*79 595*	*70 725*	*85 545*	*97 583*	*99 690*
Caramote prawn	**Caramote**		**Langostino**		***Penaeus kerathurus***			**2,28(01)001,17**		**TGS**
34 Greece	597	181	130	96	100	40	13	8	1	41
Italy	-	-	193	-	-	-	-	-	-	-
Spain	25 F	35 F	50 F	63	-	-	6	-	-	-
34 Fishing area total	*622 F*	*216 F*	*373 F*	*159*	*100*	*40*	*19*	*8*	*1*	*41*
37 Albania	...	...	30	8	4	18	18	20	23	84
France	...	...	...	...	...	...	...	4	6	7
Greece	473	494	737	1 165	1 554	1 459	1 103	1 440	1 502	1 400 F
Spain	150 F	150 F	150 F	150	53	196	97	63	64	250
Tunisia	1 657	2 287	3 587	3 057	4 367	3 785	4 729	6 165	3 356	2 539
37 Fishing area total	*2 280 F*	*2 931 F*	*4 504 F*	*4 380*	*5 978*	*5 458*	*5 947*	*7 692*	*4 951*	*4 280 F*
47 Italy	-	-	3	3	-	-	-	-	-	-
47 Fishing area total	-	-	*3*	*3*	-	-	-	-	-	-
Species total	*2 902 F*	*3 147 F*	*4 880 F*	*4 542*	*6 078*	*5 498*	*5 966*	*7 700*	*4 952*	*4 321 F*
Redspotted shrimp	**Crevette royale rose**		**Camarón rosado con manchas**		***Penaeus brasiliensis***			**2,28(01)001,19**		**PNB**
41 Brazil	6 150 F	6 160 F	6 565	8 743	10 758	7 796	9 092	10 728	5 010	5 903
41 Fishing area total	*6 150 F*	*6 160 F*	*6 565*	*8 743*	*10 758*	*7 796*	*9 092*	*10 728*	*5 010*	*5 903*
Species total	*6 150 F*	*6 160 F*	*6 565*	*8 743*	*10 758*	*7 796*	*9 092*	*10 728*	*5 010*	*5 903*
Green tiger prawn	**Crevette tigrée verte**		**Langostino tigre verde**		***Penaeus semisulcatus***			**2,28(01)001,20**		**TIP**
51 Qatar	0	1	0	0	-	0	0	0	0	0
51 Fishing area total	*0*	*1*	*0*	*0*	-	*0*	*0*	*0*	*0*	*0*
57 Thailand	673	760	1 909	2 427	2 449	2 180	2 212	1 584	1 867	1 798
57 Fishing area total	*673*	*760*	*1 909*	*2 427*	*2 449*	*2 180*	*2 212*	*1 584*	*1 867*	*1 798*
71 Thailand	323	642	711	1 042	1 068	839	609	714	836	811
71 Fishing area total	*323*	*642*	*711*	*1 042*	*1 068*	*839*	*609*	*714*	*836*	*811*
Species total	*996*	*1 403*	*2 620*	*3 469*	*3 517*	*3 019*	*2 821*	*2 298*	*2 703*	*2 609*
Northern white shrimp	**Crevette ligubam du Nord**		**Camarón blanco norteño**		***Penaeus setiferus***			**2,28(01)001,22**		**PST**
21 USA	96	-	-	1 347	913	-	619	313	28	171
21 Fishing area total	*96*	-	-	*1 347*	*913*	-	*619*	*313*	*28*	*171*
31 USA	32 264	36 838	39 959	27 461	31 928	39 799	44 014	52 280	40 696	43 015
31 Fishing area total	*32 264*	*36 838*	*39 959*	*27 461*	*31 928*	*39 799*	*44 014*	*52 280*	*40 696*	*43 015*
Species total	*32 360*	*36 838*	*39 959*	*28 808*	*32 841*	*39 799*	*44 633*	*52 593*	*40 724*	*43 186*
Crystal shrimp	**Crevette cristal**		**Camarón cristal**		***Penaeus brevirostris***			**2,28(01)001,23**		**CSP**
77 Costa Rica	974	1 028	526	408	392	276	548	350	456	343
El Salvador	615	955	861	397	318	483	177	80	153	93
Guatemala	72	84	58	33	18	21	19	16	8 F	...
Panama	2 150	1 990	2 232	1 298	2 048	1 310	1 718	2 158	2 286	1 636
77 Fishing area total	*3 811*	*4 057*	*3 677*	*2 136*	*2 776*	*2 090*	*2 462*	*2 604*	*2 903 F*	*2 072*
Species total	*3 811*	*4 057*	*3 677*	*2 136*	*2 776*	*2 090*	*2 462*	*2 604*	*2 903 F*	*2 072*
Western king prawn	**Crevette royale occidentale**		**Langostino marfil**		***Penaeus latisulcatus***			**2,28(01)001,28**		**WKP**
57 Thailand	739	385	1 475	1 674	1 724	1 501	1 518	407	1 175	1 132
57 Fishing area total	*739*	*385*	*1 475*	*1 674*	*1 724*	*1 501*	*1 518*	*407*	*1 175*	*1 132*
71 Australia	93	88	111	76	67	110	81	92	71	169
Thailand	1 399	1 304	1 486	1 600	1 838	1 775	2 546	2 993	2 052	1 991
71 Fishing area total	*1 492*	*1 392*	*1 597*	*1 676*	*1 905*	*1 885*	*2 627*	*3 085*	*2 123*	*2 160*
Species total	*2 231*	*1 777*	*3 072*	*3 350*	*3 629*	*3 386*	*4 145*	*3 492*	*3 298*	*3 292*

B-45 Shrimps, prawns / Crevettes / Gambas, camarones

Capture production by species, fishing areas and countries or areas
Captures par espèces, zones de pêche et pays ou zones
Capturas por especies, áreas de pesca y países o áreas

Species, Fishing area Espèce, Zone de pêche Especie, Area de pesca	1993 t	1994 t	1995 t	1996 t	1997 t	1998 t	1999 t	2000 t	2001 t	2002 t
Western white shrimp	**Crevette royale blanche**		**Camarón blanco del Pacífico**		***Penaeus occidentalis***				**2,28(01)001,29**	**WWP**
87 Colombia	549	759	619	1 091	1 445	1 211	2 686	2 000 F	1 219	1 200 F
Ecuador	-	-	67	67	60	50	20	20	40	45 F
87 Fishing area total	*549*	*759*	*686*	*1 158*	*1 505*	*1 261*	*2 706*	*2 020 F*	*1 259*	*1 245 F*
Species total	*549*	*759*	*686*	*1 158*	*1 505*	*1 261*	*2 706*	*2 020 F*	*1 259*	*1 245 F*
Redtail prawn	**Crevette queue rouge**		**Camarón rabo colorado**		***Penaeus penicillatus***				**2,28(01)001,30**	**REP**
61 China,Taiwan	2 538	2 071	3 564	5 020	2 473	647	316	308	312	509
61 Fishing area total	*2 538*	*2 071*	*3 564*	*5 020*	*2 473*	*647*	*316*	*308*	*312*	*509*
Species total	*2 538*	*2 071*	*3 564*	*5 020*	*2 473*	*647*	*316*	*308*	*312*	*509*
Southern pink shrimp	**Crevette rose du Sud**		**Camarón rosado sureño**		***Penaeus notialis***				**2,28(01)001,31**	**SOP**
34 Gabon	512	606	867	950	828	2 272	93	62	...	...
Gambia	...	559	367	339	...	399	400	301	211	324
Nigeria	13 755	8 595	14 742	12 073	14 799	12 396	27 341	18 882	18 805	13 413
Senegal	3 597	2 673	4 828	4 302	6 606	4 105	4 887	6 005	4 643	3 539
Spain	620 F	650 F	680 F	707	-	-	-	-	-	-
34 Fishing area total	*18 484 F*	*13 083 F*	*21 484 F*	*18 371*	*22 233*	*19 172*	*32 721*	*25 250*	*23 659*	*17 276*
Species total	*18 484 F*	*13 083 F*	*21 484 F*	*18 371*	*22 233*	*19 172*	*32 721*	*25 250*	*23 659*	*17 276*
Sao Paulo shrimp	**Crevette de Sao Paulo**		**Langostino de Sao Paulo**		***Penaeus paulensis***				**2,28(01)001,32**	**PPS**
41 Uruguay	0	0	0	-	177	13	12	56	23	7
41 Fishing area total	*0*	*0*	*0*	-	*177*	*13*	*12*	*56*	*23*	*7*
Species total	*0*	*0*	*0*	-	*177*	*13*	*12*	*56*	*23*	*7*
Penaeus shrimps nei	**Crevettes Penaeus nca**		**Langostinos Penaeus nep**		***Penaeus spp***				**2,28(01)001,XX**	**PEN**
01 Benin	3 800 F	3 739	3 117	3 000 F	2 850 F	2 700 F	2 574	2 100 F	2 400 F	2 400 F
01 Fishing area total	*3 800 F*	*3 739*	*3 117*	*3 000 F*	*2 850 F*	*2 700 F*	*2 574*	*2 100 F*	*2 400 F*	*2 400 F*
04 India	...	10 421	18 564	14 645	20 558	8 783	7 830	10 873	14 772	14 160
04 Fishing area total	...	*10 421*	*18 564*	*14 645*	*20 558*	*8 783*	*7 830*	*10 873*	*14 772*	*14 160*
27 Netherlands	-	-	-	-	-	-	1	1	3	5
Portugal	94	176	277	354	497	808	-	-	-	-
Spain	44	50	-	373	368	498	336	99	267	104
27 Fishing area total	*138*	*226*	*277*	*727*	*865*	*1 306*	*337*	*100*	*270*	*109*
31 Belize	28	34	49	38	43	40	35	45	69	81
Colombia	1 210	992	391	710	1 745	377	775	3 039 F	3 000 F	3 000 F
Costa Rica	2	2	0	0	27	43	61	65	83	82
Cuba	...	...	...	...	...	2	1	-	1	-
Dominican Rp	40	385	375	47	79	77	49	...	32	658
Fr Guiana	3 431	4 241	4 455	4 377	4 102	4 209	3 771	2 737	2 694	3 068
Guatemala	-	59	120	120	90	115	96	163	150 F	150 F
Guyana	558	708	400 F	84	79	1 935	1 595	1 132	1 698	1 505
Honduras	2 157	8 693	7 973	8 335	7 342	8 277	6 496 F	2 553 F	1 901 F	1 745 F
Jamaica	...	277	100 F	60	67	70	70	37	40 F	40 F
Japan	786	630	349	192	41	-	-	-	-	-
Korea Rep	907	750	653	560	143	136	502	-	-	-
Mexico	28 058	22 709	23 435	21 450	21 984	23 170	20 155	21 288	21 847	18 533
Nicaragua	1 718	2 632	2 330	3 049	3 148	3 787	3 560	3 630	4 130	4 029
Suriname	59	231	1 000 F	2 444	2 013	2 094	1 653	2 240 F	2 840	2 800 F
Trinidad Tob	1 296	946	700 F	285	751	648	658	755	856	852
USA	4 368	4 232	4 498	3 793	6 675	5 708	2 358	1 800	2 663	1 556
Venezuela	13 875	13 645	10 786	11 735	10 949	6 910	4 607	9 882	12 128	9 981
31 Fishing area total	*58 493*	*61 166*	*57 614 F*	*57 279*	*59 278*	*57 598*	*46 442 F*	*49 366 F*	*54 132 F*	*48 080 F*
34 Belize	-	-	-	-	0	549	1 283	1 191	41	...
Cameroon	462	500 F	514	442	450 F	635	326	471	168	280
Chile	-	3	-	-	-	-	-	-	-	-
China	1 758	249	2 372	2 113	2 260	2 504	1 997	1 873	2 087	-
Côte dIvoire	340	275	400	310 F	260 F	300 F	340 F	851	1	484
Gabon	-	-	-	-	556	190	987	1 682	1 892	1 720
Guinea	25	...	97	196	98	216	396	526	701	...
Honduras	162	20	106	173	193	128	86	79	-	-
Panama	118	-	28	8	39	-	-	-	-	-
St Vincent	-	-	1	-	-	-	-	-	-	-
Sierra Leone	573	169	868	973	2 147	1 690	2 155	1 663	1 280	1 237
Togo	1	1	1	12	2	2	0	2	0	-
Vanuatu	145	50	137	170	150	70	-	-	-	-
Other nei	-	-	-	-	-	-	19	31	-	-
34 Fishing area total	*3 584*	*1 267 F*	*4 524*	*4 397 F*	*6 155 F*	*6 284 F*	*7 589 F*	*8 369*	*6 170*	*3 721*
41 Brazil	25 320 F	25 380 F	27 132	15 007	14 506	13 998	12 550	16 509	10 697	10 533
Italy	286	290	23	30	-	-	-	-	-	-
41 Fishing area total	*25 606 F*	*25 670 F*	*27 155*	*15 037*	*14 506*	*13 998*	*12 550*	*16 509*	*10 697*	*10 533*
51 Bahrain	2 128	1 185	1 662	3 565	2 571	2 530	1 622	2 104	1 359	1 401
Eritrea	...	15	13	2	0	9	75	519	790	508
Mozambique	8 897	7 608	8 615	8 183	9 825	8 559	8 806	9 429	9 401	9 472
Pakistan	6 663	5 883	5 591	5 982	5 975	5 189	5 874	5 920	5 974	5 600
Saudi Arabia	2 658	4 282	5 941	7 423	7 011	7 812	3 612	5 639	4 761	3 996

B-45 Shrimps, prawns / Crevettes / Gambas, camarones

Capture production by species, fishing areas and countries or areas
Captures par espèces, zones de pêche et pays ou zones
Capturas por especies, áreas de pesca y países o áreas

Species, Fishing area Espèce, Zone de pêche Especie, Area de pesca	1993 t	1994 t	1995 t	1996 t	1997 t	1998 t	1999 t	2000 t	2001 t	2002 t
South Africa	13	-	-	-	127	178	180 F	168	249	160
Spain	-	-	-	-	-	-	-	219	166	152
Yemen	619	722	984	665	547	904	686	526	1 600	1 700
51 Fishing area total	*20 978*	*19 695*	*22 806*	*25 820*	*26 056*	*25 181*	*20 855 F*	*24 524*	*24 300*	*22 989*
57 Australia	5 221	5 689	6 080	6 226	6 051	6 884	7 516	7 332	6 322	6 263
Thailand	11 010	11 087	10 185	11 807	12 272	9 452	10 965	8 986	9 192	8 853
57 Fishing area total	*16 231*	*16 776*	*16 265*	*18 033*	*18 323*	*16 336*	*18 481*	*16 318*	*15 514*	*15 116*
71 Australia	7 207	6 368	7 440	8 986	8 172	8 303	5 539	2 838	3 681	3 775
NewCaledonia	-	-	-	-	-	-	0	0	0	0
Philippines	14 608	11 567	11 309	10 462	11 508	10 796	11 255	11 063	13 632	13 902
Thailand	59 688	59 604	62 073	62 338	59 253	37 941	34 147	37 638	43 475	42 182
71 Fishing area total	*81 503*	*77 539*	*80 822*	*81 786*	*78 933*	*57 040*	*50 941*	*51 539*	*60 788*	*59 859*
77 Costa Rica	1 666	924	1 172	1 923	1 588	1 062	1 080	706	634	813
El Salvador	133	192	167	65	96	157	24	8	19	30
Guatemala	733	376	557	284	506	1 597	1 103	880	650	443
Honduras	33	300	400	420	517	302	900 F	2 600 F	3 250 F	3 000 F
Mexico	40 280	41 134	46 599	44 114	49 083	43 416	46 336	40 309	35 662	36 100
Nicaragua	518	564	592	934	685	1 077	1 658	801	836	588
Panama	2 462	2 458	3 082	2 558	2 014	3 200	2 512	1 952	2 126	2 310
77 Fishing area total	*45 825*	*45 948*	*52 569*	*50 298*	*54 489*	*50 811*	*53 613 F*	*47 256 F*	*43 177 F*	*43 284 F*
81 Australia	2 100	2 312	2 153	1 822	1 926	1 683	2 399	2 829	2 851	2 197
81 Fishing area total	*2 100*	*2 312*	*2 153*	*1 822*	*1 926*	*1 683*	*2 399*	*2 829*	*2 851*	*2 197*
87 Ecuador	-	-	344	344	350	300	125	125	264	286 F
Peru	9 270	9 610	10 877	9 245	15 648	17 752	7 255	1 852	2 075	1 537
87 Fishing area total	*9 270*	*9 610*	*11 221*	*9 589*	*15 998*	*18 052*	*7 380*	*1 977*	*2 339*	*1 823 F*
Species total	*267 528 F*	*274 369 F*	*297 087 F*	*282 433 F*	*299 937 F*	*259 772 F*	*230 991 F*	*231 760 F*	*237 410 F*	*224 271 F*
Speckled shrimp	**Crevette mouchetée**		**Gamba moteada**		***Metapenaeus monoceros***			**2,28(01)016,01**		**MPN**
37 Tunisia	-	-	-	-	-	-	-	-	757	1 607
37 Fishing area total	-	-	-	-	-	-	-	-	*757*	*1 607*
Species total	-	-	-	-	-	-	-	-	*757*	*1 607*
Endeavour shrimp	**Crevette devo**		**Camarón devo**		***Metapenaeus endeavouri***			**2,28(01)016,06**		**ENS**
57 Australia	...	...	...	...	...	...	...	38	23	28
57 Fishing area total	...	...	...	...	...	...	...	*38*	*23*	*28*
71 Australia	1 801	1 920	1 945	2 175	2 245	2 885	2 575	3 457	3 328	2 926
Philippines	-	192	237	225	94	45	116	136	174	192
71 Fishing area total	*1 801*	*2 112*	*2 182*	*2 400*	*2 339*	*2 930*	*2 691*	*3 593*	*3 502*	*3 118*
Species total	*1 801*	*2 112*	*2 182*	*2 400*	*2 339*	*2 930*	*2 691*	*3 631*	*3 525*	*3 146*
Shiba shrimp	**Crevette siba**		**Camarón siba**		***Metapenaeus joyneri***			**2,28(01)016,07**		**SHI**
61 Korea Rep	3 834	4 993	2 168	2 211	1 976	3 651	3 633	2 621	2 385	1 736
61 Fishing area total	*3 834*	*4 993*	*2 168*	*2 211*	*1 976*	*3 651*	*3 633*	*2 621*	*2 385*	*1 736*
Species total	*3 834*	*4 993*	*2 168*	*2 211*	*1 976*	*3 651*	*3 633*	*2 621*	*2 385*	*1 736*
Metapenaeus shrimps nei	**Crevettes Metapenaeus nca**		**Camarones Metapenaeus nep**		***Metapenaeus spp***			**2,28(01)016,XX**		**MET**
51 Pakistan	9 468	7 120	6 981	7 602	6 801	6 204	6 791	7 126	7 246	6 555
51 Fishing area total	*9 468*	*7 120*	*6 981*	*7 602*	*6 801*	*6 204*	*6 791*	*7 126*	*7 246*	*6 555*
57 Indonesia	4 175	6 350	9 437	7 573	10 309	9 698	11 419	10 092	10 572	12 190
Thailand	1 386	1 634	3 978	3 587	3 613	2 726	2 559	1 162	2 348	2 261
57 Fishing area total	*5 561*	*7 984*	*13 415*	*11 160*	*13 922*	*12 424*	*13 978*	*11 254*	*12 920*	*14 451*
61 China,Taiwan	794	780	3 767	626	369	299	210	202	241	227
61 Fishing area total	*794*	*780*	*3 767*	*626*	*369*	*299*	*210*	*202*	*241*	*227*
71 Indonesia	11 639	14 014	13 426	14 715	22 279	31 019	22 428	28 833	25 786	29 230
Papua N Guin	127	138	220	219	159	187	279	328	346	320 F
Philippines	3 497	5 135	6 054	5 431	5 420	6 545	6 419	5 987	7 059	7 248
Thailand	7 209	7 888	7 673	8 569	8 357	7 153	6 806	10 110	7 677	7 448
71 Fishing area total	*22 472*	*27 175*	*27 373*	*28 934*	*36 215*	*44 904*	*35 932*	*45 258*	*40 868*	*44 246 F*
Species total	*38 295*	*43 059*	*51 536*	*48 322*	*57 307*	*63 831*	*56 911*	*63 840*	*61 275*	*65 479 F*
Deepwater rose shrimp	**Crevette rose du large**		**Gamba de altura**		***Parapenaeus longirostris***			**2,28(01)017,01**		**DPS**
27 France	1	3	4	3	...	...	...	...	...	4
Portugal	-	-	-	-	-	-	2 084	1 360	1 080	620
27 Fishing area total	*1*	*3*	*4*	*3*	...	...	*2 084*	*1 360*	*1 080*	*624*
34 Estonia	-	-	-	-	-	-	-	3	6	4
Gabon	...	...	5	6	257	56	76	356	55	48
GuineaBissau	...	0	40	124	148	120 F	100 F	100 F	100 F	100 F
Liberia	...	...	...	...	8	...	...	...	...	...
Portugal	110	70	53	534	340	646	260	396	424	104

B-45 Shrimps, prawns / Crevettes / Gambas, camarones

Capture production by species, fishing areas and countries or areas
Captures par espèces, zones de pêche et pays ou zones
Capturas por especies, áreas de pesca y países o áreas

Species, Fishing area Espèce, Zone de pêche Especie, Area de pesca	1993 t	1994 t	1995 t	1996 t	1997 t	1998 t	1999 t	2000 t	2001 t	2002 t
Russian Fed	40	61	-	-	-	-	-	-	-	-
Senegal	...	...	699	954	2 998	3 888	1 167	2 451	2 199	2 578
Spain	4 500 F	3 500 F	2 500 F	2 000	1 146	6 417	4 237	2 301	4 867	232
34 Fishing area total	*4 650 F*	*3 631 F*	*3 297 F*	*3 618*	*4 897*	*11 127 F*	*5 840 F*	*5 607 F*	*7 651 F*	*3 066 F*
37 Albania	...	...	0	20	8	-	-	-	52	57
Algeria	800 F	1 000 F	539	918	1 433	1 639	2 147	2 696	2 107	1 654
Cyprus	1	3	3	2	1	1	5	4	2	3
France	2	5	3	4	17	17	1	1	1	0
Israel	28	50	60	50	...	...	...	...	...	...
Italy	10 731	11 262	7 998	7 065	7 019	4 410	4 631	7 500	6 980	6 378
Spain	1 200 F	1 100 F	1 000 F	881	1 000	1 222	1 463	839	504	505
Tunisia	599	354	213	161	641	838	1 014	1 283	1 454	1 536
37 Fishing area total	*13 361 F*	*13 774 F*	*9 816 F*	*9 101*	*10 119*	*8 127*	*9 261*	*12 323*	*11 100*	*10 133*
47 Angola	796	904	682	1 367	1 175	1 900	1 101	1 600	1 660	2 000
Portugal	-	-	-	-	-	-	1	-	-	-
Spain	2 427	3 625	2 034	2 112	3 071	1 635	789	2 609	3 966	-
47 Fishing area total	*3 223*	*4 529*	*2 716*	*3 479*	*4 246*	*3 535*	*1 891*	*4 209*	*5 626*	*2 000*
51 Spain	-	-	-	-	-	-	-	-	9	-
51 Fishing area total	-	-	-	-	-	-	-	-	*9*	-
Species total	*21 235 F*	*21 937 F*	*15 833 F*	*16 201*	*19 262*	*22 789 F*	*19 076 F*	*23 499 F*	*25 466 F*	*15 823 F*
Parapenaeopsis shrimps nei	**Crevettes parapenaeopsis nca**		**Camarones parapenaeopsis nep**		***Parapenaeopsis spp***				**2,28(01)019,XX**	**NPP**
51 Pakistan	18 632	16 023	12 919	14 047	16 722	14 689	12 889	11 945	11 576	10 222
51 Fishing area total	*18 632*	*16 023*	*12 919*	*14 047*	*16 722*	*14 689*	*12 889*	*11 945*	*11 576*	*10 222*
Species total	*18 632*	*16 023*	*12 919*	*14 047*	*16 722*	*14 689*	*12 889*	*11 945*	*11 576*	*10 222*
Atlantic seabob	**Crevette seabob(Atlantique)**		**Camarón siete barbas**		***Xiphopenaeus kroyeri***				**2,28(01)022,01**	**BOB**
31 Guyana	5 614	6 737	9 800 F	12 752	15 720	11 091	10 396	16 733	23 771	17 659
Suriname	-	-	-	-	-	2 046	4 537	4 540 F	4 550	4 500 F
Trinidad Tob	...	...	...	...	...	69	89	94	79	88
USA	5 016	3 969	1 724	4 558	5 744	3 397	3 626	3 415	3 951	3 514
31 Fishing area total	*10 630*	*10 706*	*11 524 F*	*17 310*	*21 464*	*16 603*	*18 648*	*24 782 F*	*32 351*	*25 761 F*
41 Brazil	4 850 F	4 860 F	7 278	11 764	15 257	13 755	9 964	11 948	12 264	11 673
41 Fishing area total	*4 850 F*	*4 860 F*	*7 278*	*11 764*	*15 257*	*13 755*	*9 964*	*11 948*	*12 264*	*11 673*
Species total	*15 480 F*	*15 566 F*	*18 802 F*	*29 074*	*36 721*	*30 358*	*28 612*	*36 730 F*	*44 615*	*37 434 F*
Pacific seabob	**Crevette seabob**		**Camarón botalón**		***Xiphopenaeus riveti***				**2,28(01)022,02**	**TIT**
87 Colombia	1 932	1 835	1 037	2 683	2 830	1 970	1 752	2 000 F	2 341	2 300 F
87 Fishing area total	*1 932*	*1 835*	*1 037*	*2 683*	*2 830*	*1 970*	*1 752*	*2 000 F*	*2 341*	*2 300 F*
Species total	*1 932*	*1 835*	*1 037*	*2 683*	*2 830*	*1 970*	*1 752*	*2 000 F*	*2 341*	*2 300 F*
Southern rough shrimp	**Crevette-archer**		**Camarón fijador arquero**		***Trachypenaeus curvirostris***				**2,28(01)043,02**	**TRV**
61 China	120 000	167 165	151 746	163 060	174 967	175 618	400 786	312 436	244 202	233 736
China,Taiwan	815	701	2 877	4 663	5 288	3 926	2 241	532	593	366
Korea Rep	-	-	-	-	-	-	-	-	3 145	-
61 Fishing area total	*120 815*	*167 866*	*154 623*	*167 723*	*180 255*	*179 544*	*403 027*	*312 968*	*247 940*	*234 102*
Species total	*120 815*	*167 866*	*154 623*	*167 723*	*180 255*	*179 544*	*403 027*	*312 968*	*247 940*	*234 102*
Argentine stiletto shrimp	**Crevette stylet d'Argentine**		**Camarón estilete argentino**		***Artemesia longinaris***				**2,28(01)067,01**	**ASH**
41 Argentina	185	215	250	263	166	146	37	39	283	298
41 Fishing area total	*185*	*215*	*250*	*263*	*166*	*146*	*37*	*39*	*283*	*298*
Species total	*185*	*215*	*250*	*263*	*166*	*146*	*37*	*39*	*283*	*298*
Pacific seabobs	**Crevettes seabob(Pacifique)**		**Camaroncillos**		***Xiphopenaeus,Trachypenaeus spp***				**2,28(01)XXX,XX**	**BOS**
77 Costa Rica	538	346	544	576	314	586	310	198	228	125
El Salvador	2 487	1 827	2 812	5 038	3 079	3 015	1 656	1 472	1 525	918
Guatemala	1 621	1 638	2 352	2 203	355	1 320	1 619	365	200 F	59
Panama	3 285	4 924	4 190	4 339	3 606	4 960	2 872	2 685	3 002	1 675
77 Fishing area total	*7 931*	*8 735*	*9 898*	*12 156*	*7 354*	*9 881*	*6 457*	*4 720*	*4 955 F*	*2 777*
Species total	*7 931*	*8 735*	*9 898*	*12 156*	*7 354*	*9 881*	*6 457*	*4 720*	*4 955 F*	*2 777*
Scarlet shrimp	**Gambon écarlate**		**Gamba carabinero**		***Plesiopenaeus edwardsianus***				**2,28(02)021,01**	**SSH**
34 Spain	500 F	450 F	400 F	352	479	337	505	14	39	1
34 Fishing area total	*500 F*	*450 F*	*400 F*	*352*	*479*	*337*	*505*	*14*	*39*	*1*
37 Spain	-	-	-	-	-	-	100	40	-	33
37 Fishing area total	-	-	-	-	-	-	*100*	*40*	-	*33*
Species total	*500 F*	*450 F*	*400 F*	*352*	*479*	*337*	*605*	*54*	*39*	*34*

B-45 Shrimps, prawns / Crevettes / Gambas, camarones

Capture production by species, fishing areas and countries or areas
Captures par espèces, zones de pêche et pays ou zones
Capturas por especies, áreas de pesca y países o áreas

Species, Fishing area Espèce, Zone de pêche Especie, Area de pesca	1993 t	1994 t	1995 t	1996 t	1997 t	1998 t	1999 t	2000 t	2001 t	2002 t
Blue and red shrimp	**Crevette rouge**		**Gamba rosada**		***Aristeus antennatus***				**2,28(02)031,01**	**ARA**
27 Portugal	...	...	...	...	...	...	194	269	182	97
27 Fishing area total	...	...	...	...	...	...	*194*	*269*	*182*	*97*
37 Albania	...	...	0	3	0	-	-	-	-	34
Algeria	1 500 F	1 600 F	1 305	1 230	1 186	1 485	880	1 116	740	893
Spain	-	-	-	-	-	-	414	772	1 418	1 033
Tunisia	69	30	28	195	173	187	38	16	51	51
37 Fishing area total	*1 569 F*	*1 630 F*	*1 333*	*1 428*	*1 359*	*1 672*	*1 332*	*1 904*	*2 209*	*2 011*
Species total	*1 569 F*	*1 630 F*	*1 333*	*1 428*	*1 359*	*1 672*	*1 526*	*2 173*	*2 391*	*2 108*
Striped red shrimp	**Gambon rayé**		**Gamba listada**		***Aristeus varidens***				**2,28(02)031,02**	**ARV**
47 Angola	331	364	538	814	543	876	820	1 154	1 200	1 500
Spain	1 065	1 101	1 032	764	780	1 932	1 007	1 870	2 205	546
47 Fishing area total	*1 396*	*1 465*	*1 570*	*1 578*	*1 323*	*2 808*	*1 827*	*3 024*	*3 405*	*2 046*
Species total	*1 396*	*1 465*	*1 570*	*1 578*	*1 323*	*2 808*	*1 827*	*3 024*	*3 405*	*2 046*
Aristeid shrimps nei	**Gambons,crevet. aristeidés nca**		**Gambas aristeidos nep**		***Aristeidae***				**2,28(02)XXX,XX**	**ARI**
37 Italy	...	...	2 551	2 258	2 406	1 231	2 128	4 463	1 833	1 768
37 Fishing area total	...	...	*2 551*	*2 258*	*2 406*	*1 231*	*2 128*	*4 463*	*1 833*	*1 768*
Species total	...	...	*2 551*	*2 258*	*2 406*	*1 231*	*2 128*	*4 463*	*1 833*	*1 768*
Coonstripe shrimp	**Crevette à front rayé**		**Camarón malacho**		***Pandalus hypsinotus***				**2,28(04)002,01**	**PYX**
61 Russian Fed	...	...	...	...	467	388	288	275	359	231
61 Fishing area total	...	...	...	...	*467*	*388*	*288*	*275*	*359*	*231*
Species total	...	...	...	...	*467*	*388*	*288*	*275*	*359*	*231*
Northern prawn	**Crevette nordique**		**Camarón norteño**		***Pandalus borealis***				**2,28(04)002,03**	**PRA**
21 Canada	24 968	28 985	30 213	31 340	48 310	78 867	85 331	100 091	94 328	102 097
Cuba	-	-	-	-	-	-	120	46	-	-
Denmark	897	245	447	403	421	556	235	-	93	359
Estonia	-	1 051	2 379	1 898	3 240	5 533	10 834	12 209	9 917	13 697
Faeroe Is	9 823	6 908	5 815	8 600	7 871	9 815	9 792	7 249	12 337	8 534
Greenland	74 124	75 907	77 832	67 441	60 079	66 034	74 563	80 888	81 398	105 972
Honduras	1 265	-	-	-	-	-	-	-	-	-
Iceland	2 195	2 355	7 481	20 680	7 197	6 572	9 147	8 832	5 065	6 877
Japan	-	-	-	-	-	-	-	-	130	100
Latvia	-	324	679	1 253	997	1 191	3 080	3 169	3 028	1 951
Lithuania	-	863	980	1 585	1 785	3 107	3 370	3 595	2 768	3 388
Norway	7 074	8 625	9 391	5 648	1 886	1 339	2 975	2 742	13 291	12 000
Portugal	-	-	16	-	170	203	227	289	420	15
Russian Fed	55	350	3 397	4 444	1 090	-	1 103	7 137	5 754	1 243
Spain	-	187	279	201	423	913	1 029	1 388	885	671
Ukraine	-	-	-	-	-	-	-	-	405	-
USA	4 321	3 718	7 416	9 932	7 239	3 926	2 832	2 625	1 309	682
21 Fishing area total	*124 722*	*129 518*	*146 325*	*153 425*	*140 708*	*178 056*	*204 638*	*230 260*	*231 128*	*257 586*
27 Denmark	4 688	4 016	8 013	8 668	8 131	7 544	4 645	5 721	5 238	5 156
Estonia	-	-	-	1 322	1 751	1 673	1 614	607	1 318	539
Faeroe Is	1 815	2 351	3 475	1 992	2 997	3 170	5 051	5 362	3 838	4 000 F
Germany	-	-	-	-	-	-	1 585	-	-	-
Greenland	2 337	3 924	4 094	4 545	3 853	3 536	4 615	4 514	4 444	3 564
Iceland	53 881	72 792	76 048	68 953	75 430	56 155	33 811	24 707	25 725	29 280
Lithuania	-	-	-	-	-	233	797	2 781	2 645	3 319
Norway	41 883	29 543	29 859	35 857	40 075	55 802	60 563	63 836	53 045	58 524
Portugal	1	1	4	1	71	171	835	266	220	1
Russian Fed	22 397	7 108	3 564	5 747	1 493	4 895	10 765	19 596	5 846	3 790
Spain	-	-	-	739	1 006	453	267	906	798	267
Sweden	2 300	2 728	2 678	2 176	2 598	2 283	2 297	2 073	2 113	2 188
UK	545	77	1 541	1 996	417	595	1 815	539	996	68
27 Fishing area total	*129 847*	*122 540*	*129 276*	*131 996*	*137 822*	*136 510*	*128 660*	*130 908*	*106 226*	*110 696 F*
61 Russian Fed	-	-	-	-	1 786	3 127	5 096	8 520	8 457	7 596
61 Fishing area total	-	-	-	-	*1 786*	*3 127*	*5 096*	*8 520*	*8 457*	*7 596*
Species total	*254 569*	*252 058*	*275 601*	*285 421*	*280 316*	*317 693*	*338 394*	*369 688*	*345 811*	*375 878 F*
Aesop shrimp	**Crevette ésope**		**Camarón esópico**		***Pandalus montagui***				**2,28(04)002,05**	**AES**
21 Greenland	...	...	...	...	...	...	...	697	609	206
21 Fishing area total	...	...	...	...	...	...	...	*697*	*609*	*206*
Species total	...	...	...	...	...	...	...	*697*	*609*	*206*
Humpy shrimp	**Crevette gibbeuse**		**Camarón jiboso**		***Pandalus goniurus***				**2,28(04)002,07**	**DUJ**
61 Russian Fed	...	...	...	...	-	1 199	330	1 200	247	32
61 Fishing area total	...	...	...	...	-	*1 199*	*330*	*1 200*	*247*	*32*
Species total	...	...	...	...	-	*1 199*	*330*	*1 200*	*247*	*32*

B-45 Shrimps, prawns / Crevettes / Gambas, camarones

Capture production by species, fishing areas and countries or areas
Captures par espèces, zones de pêche et pays ou zones
Capturas por especies, áreas de pesca y países o áreas

Species, Fishing area Espèce, Zone de pêche Especie, Area de pesca	1993 t	1994 t	1995 t	1996 t	1997 t	1998 t	1999 t	2000 t	2001 t	2002 t
Hokkai shrimp	**Crevette hokkai**		**Camarón de Hokkai**		***Pandalus kessleri***				**2,28(04)002,08**	**DUK**
61 Russian Fed	...	...	...	...	123	55	97	75	94	81
61 Fishing area total	...	...	...	...	*123*	*55*	*97*	*75*	*94*	*81*
Species total	...	...	...	...	*123*	*55*	*97*	*75*	*94*	*81*
Pandalus shrimps nei	**Crevettes Pandalus nca**		**Camarones Pandalus nep**		***Pandalus spp***				**2,28(04)002,XX**	**PAN**
21 Canada	17 973	19 676	24 369	24 976	28 601	29 042	30 603	35 180	31 234	33 342
21 Fishing area total	*17 973*	*19 676*	*24 369*	*24 976*	*28 601*	*29 042*	*30 603*	*35 180*	*31 234*	*33 342*
27 UK	29	37	46	39	48	1 421	46	460	74	290
27 Fishing area total	*29*	*37*	*46*	*39*	*48*	*1 421*	*46*	*460*	*74*	*290*
61 Russian Fed	2 462	1 989	2 398	3 000	-	-	-	-	-	-
61 Fishing area total	*2 462*	*1 989*	*2 398*	*3 000*	-	-	-	-	-	-
Species total	*20 464*	*21 702*	*26 813*	*28 015*	*28 649*	*30 463*	*30 649*	*35 640*	*31 308*	*33 632*
Chilean nylon shrimp	**Crevette nylon chilienne**		**Camarón nailon**		***Heterocarpus reedi***				**2,28(04)005,01**	**CHS**
87 Chile	8 237	9 840	10 620	10 535	10 239	7 301	7 951	5 448	4 863	4 112
87 Fishing area total	*8 237*	*9 840*	*10 620*	*10 535*	*10 239*	*7 301*	*7 951*	*5 448*	*4 863*	*4 112*
Species total	*8 237*	*9 840*	*10 620*	*10 535*	*10 239*	*7 301*	*7 951*	*5 448*	*4 863*	*4 112*
Northern nylon shrimp	**Crevette nylon nordique**		**Camarón nailon norteño**		***Heterocarpus vicarius***				**2,28(04)005,08**	**HUV**
77 Nicaragua	-	-	-	-	-	40	37	20	131	0
77 Fishing area total	-	-	-	-	-	*40*	*37*	*20*	*131*	*0*
Species total	-	-	-	-	-	*40*	*37*	*20*	*131*	*0*
Morotoge shrimp	**Crevette morotoge**		**Camarón morotoje**		***Pandalopsis japonica***				**2,28(04)037,02**	**NDJ**
61 Russian Fed	...	...	...	...	12	86	35	11	36	22
61 Fishing area total	...	...	...	...	*12*	*86*	*35*	*11*	*36*	*22*
Species total	...	...	...	...	*12*	*86*	*35*	*11*	*36*	*22*
Pacific shrimps nei	**Crevettes océan Pacifique nca**		**Camarones Océano Pacífico nep**		***Pandalus spp, Pandalopsis spp***				**2,28(04)XXX,XX**	**PSH**
67 USA	24 058	16 539	13 591	15 030	19 131	6 268	14 015	16 158	18 740	26 351
67 Fishing area total	*24 058*	*16 539*	*13 591*	*15 030*	*19 131*	*6 268*	*14 015*	*16 158*	*18 740*	*26 351*
77 USA	880	1 287	1 539	1 707	1 288	622	636	372	309	556
77 Fishing area total	*880*	*1 287*	*1 539*	*1 707*	*1 288*	*622*	*636*	*372*	*309*	*556*
Species total	*24 938*	*17 826*	*15 130*	*16 737*	*20 419*	*6 890*	*14 651*	*16 530*	*19 049*	*26 907*
Akiami paste shrimp	**Chevrette akiami**		**Camaroncillo akiami**		***Acetes japonicus***				**2,28(07)009,03**	**AKS**
61 China	262 457	326 314	390 000	442 460	480 056	571 383	579 213	625 234	565 792	578 634
Korea Rep	24 324	18 510	16 495	18 411	15 624	15 993	19 389	13 985	11 705	7 013
61 Fishing area total	*286 781*	*344 824*	*406 495*	*460 871*	*495 680*	*587 376*	*598 602*	*639 219*	*577 497*	*585 647*
Species total	*286 781*	*344 824*	*406 495*	*460 871*	*495 680*	*587 376*	*598 602*	*639 219*	*577 497*	*585 647*
Sergestid shrimps nei	**Crevettes sergestid nca**		**Camarones sergéstidos nep**		***Sergestidae***				**2,28(07)XXX,XX**	**SHS**
57 Malaysia	17 310	14 654	12 799	14 057	13 241	9 684	7 893	9 094	7 026	6 362
Thailand	2 654	2 455	4 194	3 740	2 388	1 057	1 279	659	1 570	1 512
57 Fishing area total	*19 964*	*17 109*	*16 993*	*17 797*	*15 629*	*10 741*	*9 172*	*9 753*	*8 596*	*7 874*
71 Malaysia	2 968	5 012	4 855	6 770	1 999	1 652	8 587	1 354	1 341	2 198
Philippines	16 214	15 809	18 997	18 657	15 562	16 719	19 262	20 122	22 960	16 879
Thailand	19 354	21 375	19 532	18 387	15 100	14 315	6 381	3 098	10 793	10 471
71 Fishing area total	*38 536*	*42 196*	*43 384*	*43 814*	*32 661*	*32 686*	*34 230*	*24 574*	*35 094*	*29 548*
Species total	*58 500*	*59 305*	*60 377*	*61 611*	*48 290*	*43 427*	*43 402*	*34 327*	*43 690*	*37 422*
Whitebelly prawn	**Bouquet covac**		**Camarón cuac**		***Nematopalaemon schmitti***				**2,28(12)003,02**	**NLC**
31 Guyana	-	-	-	-	-	-	-	1 464	1 382	1 400
31 Fishing area total	-	-	-	-	-	-	-	*1 464*	*1 382*	*1 400*
Species total	-	-	-	-	-	-	-	*1 464*	*1 382*	*1 400*
Delta prawn	**Bouquet delta**		**Camarón delta**		***Palaemon longirostris***				**2,28(12)018,06**	**PIQ**
27 France	...	...	...	...	...	...	10	19	26	18
27 Fishing area total	...	...	...	...	...	...	*10*	*19*	*26*	*18*
Species total	...	...	...	...	...	...	*10*	*19*	*26*	*18*
Common prawn	**Bouquet commun**		**Camarón común**		***Palaemon serratus***				**2,28(12)018,10**	**CPR**
27 Denmark	248	632	195	144	182	122	141	151	141	144

B-45 Shrimps, prawns / Crevettes / Gambas, camarones

Capture production by species, fishing areas and countries or areas
Captures par espèces, zones de pêche et pays ou zones
Capturas por especies, áreas de pesca y países o áreas

Species, Fishing area Espèce, Zone de pêche Especie, Area de pesca	1993 t	1994 t	1995 t	1996 t	1997 t	1998 t	1999 t	2000 t	2001 t	2002 t
France	345	562	381	311	288	213	296	307	326	247
Spain	-	-	-	80	91	79	66	47	41	13
UK	16	14	16	8	15	6	22	15	24	13
27 Fishing area total	*609*	*1 208*	*592*	*543*	*576*	*420*	*525*	*520*	*532*	*417*
37 Algeria	...	...	37	40	35	89	9	...	...	...
France	5	0	0	0	0	0	-	-	-	-
Spain	...	...	...	...	...	...	...	...	...	97
37 Fishing area total	*5*	*0*	*37*	*40*	*35*	*89*	*9*	*...*	*...*	*97*
Species total	*614*	*1 208*	*629*	*583*	*611*	*509*	*534*	*520*	*532*	*514*
Palaemonid shrimps nei	**Crevettes palémonides nca**		**Camarones palemónidos nep**		***Palaemonidae***			**2,28(12)XXX,XX**		**PAL**
27 Ireland	273	314	312	399	358	505	551	450	268	208
Portugal	-	-	1	1	15	5	38	63	13	7
27 Fishing area total	*273*	*314*	*313*	*400*	*373*	*510*	*589*	*513*	*281*	*215*
Species total	*273*	*314*	*313*	*400*	*373*	*510*	*589*	*513*	*281*	*215*
Common shrimp	**Crevette grise**		**Quisquilla**		***Crangon crangon***			**2,28(23)003,03**		**CSH**
27 Belgium	981	1 223	1 567	903	743	378	1 053	616	988	538
Denmark	1 521	1 743	2 065	2 207	3 250	2 509	2 911	2 322	1 826	3 197
France	509	541	247	309	237	272	399	472	389	285
Germany	13 480	16 769	11 608	15 994	19 890	14 814	17 457	17 423	12 571	15 966
Netherlands	9 672	10 180	13 912	12 067	13 054	11 871	13 772	11 496	14 081	11 453
Portugal	0	0	1	0	0	-	1	3	0	-
Spain	-	-	-	-	0	-	0	-	1	4
UK	1 004	1 346	1 131	742	598	739	1 453	1 129	2 290	1 430
27 Fishing area total	*27 167*	*31 802*	*30 531*	*32 222*	*37 772*	*30 583*	*37 046*	*33 461*	*32 146*	*32 873*
37 Algeria	...	...	...	...	...	...	...	0	2	3
France	7	0	0	0	0	0	0	-	2	2
Spain	270 F	250 F	230 F	222	150	100	87	...	...	134
37 Fishing area total	*277 F*	*250 F*	*230 F*	*222*	*150*	*100*	*87*	*0*	*4*	*139*
Species total	*27 444 F*	*32 052 F*	*30 761 F*	*32 444*	*37 922*	*30 683*	*37 133*	*33 461*	*32 150*	*33 012*
Sculptured shrimps nei	**Crevettes sculptées nca**		**Camarones esculpidos nep**		***Sclerocrangon spp***			**2,28(23)006,XX**		**CVL**
61 Russian Fed	...	...	...	...	38	59	82	20	45	41
61 Fishing area total	*...*	*...*	*...*	*...*	*38*	*59*	*82*	*20*	*45*	*41*
Species total	*...*	*...*	*...*	*...*	*38*	*59*	*82*	*20*	*45*	*41*
Crangonid shrimps nei	**Crevettes crangonidés nca**		**Camarones crangónidos nep**		***Crangonidae***			**2,28(23)XXX,XX**		**CRN**
77 USA	-	-	-	-	31	41	-	42	45	38
77 Fishing area total	*-*	*-*	*-*	*-*	*31*	*41*	*-*	*42*	*45*	*38*
Species total	*-*	*-*	*-*	*-*	*31*	*41*	*-*	*42*	*45*	*38*
Rock shrimp	**Boucot ovetgernade**		**Camarón de piedra**		***Sicyonia brevirostris***			**2,28(28)028,01**		**RSH**
21 USA	-	-	-	9 470	8	-	481	-	-	-
21 Fishing area total	*-*	*-*	*-*	*9 470*	*8*	*-*	*481*	*-*	*-*	*-*
31 USA	3 075	3 993	3 848	1 079	1 781	4 409	1 345	3 254	2 909	890
31 Fishing area total	*3 075*	*3 993*	*3 848*	*1 079*	*1 781*	*4 409*	*1 345*	*3 254*	*2 909*	*890*
Species total	*3 075*	*3 993*	*3 848*	*10 549*	*1 789*	*4 409*	*1 826*	*3 254*	*2 909*	*890*
Pacific rock shrimp	**Boucot du Pacifique**		**Camarón de piedra del Pacífico**		***Sicyonia ingentis***			**2,28(28)028,06**		**YII**
77 USA	...	...	...	...	...	185	630	756	165	215
77 Fishing area total	*...*	*...*	*...*	*...*	*...*	*185*	*630*	*756*	*165*	*215*
Species total	*...*	*...*	*...*	*...*	*...*	*185*	*630*	*756*	*165*	*215*
Argentine red shrimp	**Salicoque rouge d'Argentine**		**Camarón langostín argentino**		***Pleoticus muelleri***			**2,28(29)006,01**		**LAA**
41 Argentina	17 645	15 826	6 705	9 874	6 479	23 203	15 888	37 150	78 866	51 410
Uruguay	-	-	-	-	-	-	40	0	0	2
41 Fishing area total	*17 645*	*15 826*	*6 705*	*9 874*	*6 479*	*23 203*	*15 928*	*37 150*	*78 866*	*51 412*
Species total	*17 645*	*15 826*	*6 705*	*9 874*	*6 479*	*23 203*	*15 928*	*37 150*	*78 866*	*51 412*
Royal red shrimp	**Salicoque royale rouge**		**Camarón rojo real**		***Pleoticus robustus***			**2,28(29)006,03**		**RRS**
21 USA	-	-	-	14	4	13	78	22	24	23
21 Fishing area total	*-*	*-*	*-*	*14*	*4*	*13*	*78*	*22*	*24*	*23*
31 USA	297	290	252	184	205	182	208	369	281	418
31 Fishing area total	*297*	*290*	*252*	*184*	*205*	*182*	*208*	*369*	*281*	*418*
Species total	*297*	*290*	*252*	*198*	*209*	*195*	*286*	*391*	*305*	*441*

B-45 Shrimps, prawns / Crevettes / Gambas, camarones

Capture production by species, fishing areas and countries or areas
Captures par espèces, zones de pêche et pays ou zones
Capturas por especies, áreas de pesca y países o áreas

Species, Fishing area Espèce, Zone de pêche Especie, Area de pesca	1993 t	1994 t	1995 t	1996 t	1997 t	1998 t	1999 t	2000 t	2001 t	2002 t
Kolibri shrimp	**Salicoque colibri**		**Camarón chupaflor**		***Solenocera agassizii***				**2,28(29)072,01**	**SOK**
87 Colombia	...	...	...	...	...	...	...	...	686	680 F
87 Fishing area total	...	...	...	...	...	...	...	...	*686*	*680 F*
Species total	...	...	...	...	...	...	...	...	*686*	*680 F*
Knife shrimp	**Salicoque couteau**		**Camarón navaja**		***Haliporoides triarthrus***				**2,28(29)073,01**	**KNS**
51 Mozambique	2 093	2 512	2 036	1 771	1 510	1 882	1 611	1 766	1 738	1 441
51 Fishing area total	*2 093*	*2 512*	*2 036*	*1 771*	*1 510*	*1 882*	*1 611*	*1 766*	*1 738*	*1 441*
Species total	*2 093*	*2 512*	*2 036*	*1 771*	*1 510*	*1 882*	*1 611*	*1 766*	*1 738*	*1 441*
Chilean knife shrimp	**Salicoque couteau du Chili**		**Camarón cuchilla**		***Haliporoides diomedeae***				**2,28(29)073,03**	**HJD**
87 Chile	5	20	5	15	32	29	135	169	309	482
87 Fishing area total	*5*	*20*	*5*	*15*	*32*	*29*	*135*	*169*	*309*	*482*
Species total	*5*	*20*	*5*	*15*	*32*	*29*	*135*	*169*	*309*	*482*
Knife shrimps nei	**Salicoques-couteau nca**		**Camarones navaja nep**		***Haliporoides spp***				**2,28(29)073,XX**	**KNI**
51 South Africa	218	-	-	-	-	-	-	-	-	-
51 Fishing area total	*218*	-	-	-	-	-	-	-	-	-
Species total	*218*	-	-	-	-	-	-	-	-	-
Natantian decapods nei	**Décapodes natantia nca**		**Decápodos natantia nep**		***Natantia***				**2,28(XX)XXX,XX**	**DCP**
04 China,Taiwan	76	33	38	22	12	-	-	-	-	12
India	...	6 083	7 746	13 186	17 396	7 836	7 630	15 835	14 569	80 455
Indonesia	6 323	8 965	6 768	6 565	6 330	7 281	7 905	7 478	7 721	8 030
04 Fishing area total	*6 399*	*15 081*	*14 552*	*19 773*	*23 738*	*15 117*	*15 535*	*23 313*	*22 290*	*88 497*
21 Poland	-	-	-	-	824	148	894	1 732	263	-
21 Fishing area total	-	-	-	-	*824*	*148*	*894*	*1 732*	*263*	-
27 Portugal	54	50	196	204	233	382	30	70	84	123
Spain	813	677	1 141	1 030	13 494	5 563	3 814	4 990	1 842	3 910
UK	0	3	2	2		-	-	-	-	
27 Fishing area total	*867*	*730*	*1 339*	*1 236*	*13 727*	*5 945*	*3 844*	*5 060*	*1 926*	*4 033*
31 Colombia	252	17	518	139	-	-	-	-	-	-
Haiti	120 F	110 F	150 F	150 F	150 F	150 F	150 F	150 F	150 F	150 F
Korea Rep	-	-	-	-	-	-	67	1 099	68	45
31 Fishing area total	*372 F*	*127 F*	*668 F*	*289 F*	*150 F*	*150 F*	*217 F*	*1 249 F*	*218 F*	*195 F*
34 Belize	-	-	-	-	-	2	28	6	11	...
Benin	0	0	0	0	0	0	31	20 F	3	55
Cayman Is	320	...	...	...	...	...	...	...	...	...
Congo Rep	325 F	23	322	584	320 F	420 F	374	529	400 F	400 F
Egypt	-	-	-	-	-	-	4	-	-	-
Ghana	1 148	1 507	2 228	1 554	1 602	1 448	87	1 446	1 361	973
Greece	1 534	932	884	951	746	563	573	553	511	517
GuineaBissau	1 060 F	1 040 F	1 070 F	1 100 F	1 100 F	900 F	800 F	800 F	800 F	800 F
Honduras	-	-	-	-	0	-	-	-	-	-
Italy	2 149	2 179	-	254	345	456	493	370	686	473
Korea D P Rp	-	-	-	-	-	-	-	52	-	-
Korea Rep	1	13	-	-	4	5	16	4	916	4
Latvia	-	-	-	-	-	-	-	-	19	7
Liberia	205	116	110	28	73	113	302	25	31	40 F
Lithuania	-	-	-	-	-	-	-	11	-	-
Mauritania	110 F	120 F	130 F	160 F	190 F	270 F	222	779	1 546	1 226
Morocco	6 387	8 425	8 947	8 950	7 268	9 860	8 905	11 608	7 259	3 472
Nigeria	2 612	2 207	4 474	3 417	3 456	7 364	2 690	1 564	909	17 076
Panama	-	-	-	4	2	-	-	-	-	-
Portugal	0	561	688	1 069	446	421	181	164	255	293
Senegal	51	-	-	-	-	-	-	-	19	18
Spain	2 000 F	4 000 F	6 000 F	12 842	3 790	24 723	9 003	2 160	5 661	6 671
Vanuatu	-	-	-	-	-	0	-	-	-	-
Other nei	-	-	-	-	-	-	-	17	-	-
34 Fishing area total	*17 902 F*	*21 123 F*	*24 853 F*	*30 913 F*	*19 342 F*	*46 545 F*	*23 709 F*	*20 108 F*	*20 387 F*	*32 025 F*
37 Bulgaria	0	0	1	1	3	2	2	-	-	-
Egypt	3 818	3 959	3 971	3 164	4 851	5 071	7 099	4 408	3 668	4 685
France	-	-	-	-	-	-	5	2	1	1
Greece	1 371	1 319	980	1 327	1 724	1 245	1 239	1 085	844	800 F
Israel	110	30	200	200	221	225	186	184	116	-
Malta	6	4	5	9	16	18	24	23	36	29
Morocco	83	74	196	388	8	910	935	1 044	326	767
Palest, O.T.	...	...	50	55	161	161	160 F	120 F	80 F	60
Russian Fed	-	-	-	-	2	-	-	1	3	-
Spain	900 F	1 200 F	1 500 F	1 710	1 781	1 815	2	3	8	734
Turkey	4 275	3 437	1 976	1 100	1 380	1 400	890	2 000	3 000	4 000
Ukraine	...	...	...	1	1	1	1	1	...	4
37 Fishing area total	*10 563 F*	*10 023 F*	*8 879 F*	*7 955*	*10 148*	*10 848*	*10 543 F*	*8 871 F*	*8 082 F*	*11 080 F*
41 Korea Rep	-	-	-	-	-	-	1 090	556	2 750	12 965
Spain	-	-	-	-	-	-	-	-	66	57

B-45 Shrimps, prawns / Crevettes / Gambas, camarones

Capture production by species, fishing areas and countries or areas
Captures par espèces, zones de pêche et pays ou zones
Capturas por especies, áreas de pesca y países o áreas

Species, Fishing area Espèce, Zone de pêche Especie, Area de pesca		1993 t	1994 t	1995 t	1996 t	1997 t	1998 t	1999 t	2000 t	2001 t	2002 t
41	*Fishing area total*	-	-		-	-	-	*1 090*	*556*	*2 816*	*13 022*
47	Angola	387	676	-	-	-	-	0	-	-	-
	Poland	-	-		-	-	543	-		-	-
	Spain	-	-	-	855	-	9 344	1 615	3 188	4 298	1 798
47	*Fishing area total*	*387*	*676*	-	*855*	-	*9 887*	*1 615*	*3 188*	*4 298*	*1 798*
51	Comoros	10	10	15 F	20 F	20 F	20 F	20 F	20 F	20	20 F
	Egypt	614	572	763	644	639	436	1 170	2 655	1 704	1 366
	India	86 919	95 120	85 016	113 259	90 528	97 570	90 957	90 734	119 114	102 105
	Iran	6 985	6 694	6 864	5 837	7 620	5 774	4 570	9 850	6 940	5 726
	Italy	429	436	40	52	-	-	-	-	-	-
	Kenya	208	379	207	378	491	774	513	458	690	581
	Korea Rep	44	-	-	165	49	29	-	-	279	172
	Kuwait	2 810	2 093	1 739	2 358	2 066	1 598	720	1 300 F	1 977	1 980 F
	Lithuania	-	11			-	-	-		-	-
	Madagascar	8 869	12 279	9 919	10 470	10 755	11 470	10 507	12 127	11 776	13 223
	Mauritius	22	21	1	0	1	0	1	1	1	1
	Oman	380	451	340	276	376	65	356	432	627	467
	Portugal	-	-	-	-	-	-	-	145	735	-
	Russian Fed	-	-	-	-	-	-	-	-	-	31
	Spain	-	-	-	-	-	-	164	-	-	-
	Tanzania	1 829	1 417	2 260	2 664	2 500	2 800	2 100	2 100	2 000	2 000
	Yemen	65	...	...	...	...	38	7	44	151	153
51	*Fishing area total*	*109 184*	*119 483*	*107 164 F*	*136 123 F*	*115 045 F*	*120 574 F*	*111 085 F*	*119 866 F*	*146 014*	*127 825 F*
57	India	29 167	24 174	20 984	28 736	26 428	15 638	22 077	21 830	22 463	16 117
	Indonesia	13 015	17 725	17 402	14 454	18 149	17 978	19 933	17 333	17 761	18 700
	Korea Rep	-	-	-	-	-	-	1	-	1	-
	Malaysia	55 591	44 007	43 079	45 517	47 599	20 826	41 756	50 237	39 730	37 887
	Myanmar	12 000 F	15 000 F	20 000 F	16 000 F	22 000 F	24 000 F	27 000 F	30 000 F	30 000 F	32 000 F
	Timor-Leste	...	...	...	...	...	...	1 F	1	1	1 F
57	*Fishing area total*	*109 773 F*	*100 906 F*	*101 465 F*	*104 707 F*	*114 176 F*	*78 442 F*	*110 768 F*	*119 401 F*	*109 956 F*	*104 705 F*
61	China,H.Kong	8 621	7 549	7 155	6 589	6 225	5 285	3 800 F	4 600 F	5 100 F	5 000 F
	China, Macao	406	266	181	218	230 F	230 F	230 F	230 F	230 F	230 F
	China,Taiwan	16 572	16 167	24 108	23 748	22 100	15 702	14 943	11 336	11 987	10 787
	Japan	32 756	34 275	31 764	28 641	27 114	25 283	25 630	25 898	24 276	25 000 F
	Korea Rep	34 739	29 464	18 954	16 086	17 621	24 503	16 861	15 751	8 077	7 206
	Russian Fed	-	-	-	-	-	-	16	91	80	232
61	*Fishing area total*	*93 094*	*87 721*	*82 162*	*75 282*	*73 290 F*	*71 003 F*	*61 480 F*	*57 906 F*	*49 750 F*	*48 455 F*
67	Canada	4 498	4 502	8 557	9 392	5 200	5 203	4 071	4 223	4 212	3 622
67	*Fishing area total*	*4 498*	*4 502*	*8 557*	*9 392*	*5 200*	*5 203*	*4 071*	*4 223*	*4 212*	*3 622*
71	Brunei Darsm	299	1 208	272	275	0	0	0	0	0	0
	Cambodia	4 500	4 000	4 200	4 300	4 110	4 440	5 250	5 000	5 900	6 400
	Guam	-	-	-	-	-	1	0	0	0	0
	Indonesia	66 699	73 427	63 859	74 761	77 641	69 222	83 439	81 547	95 400	100 460
	Korea Rep	36	49	2	45	94	51	124	138	380	52
	Malaysia	29 892	36 872	32 053	33 893	28 606	15 069	32 238	35 291	29 371	29 573
	Marshall Is	0	0	0	0	0	0	0	-	-	-
	Micronesia	0	0	0	0	0	0	0	0	0	0
	N Marianas	-	-	1	-	-	-	-	-	-	0
	Papua N Guin	60	94	84	104	59	50	99	135	117	110 F
	Singapore	728	908	767	857	706	621	522	422	250	222
	Viet Nam	55 170	66 981	82 758	86 166	98 401	93 541	91 500	81 700 F	90 000 F	90 000 F
71	*Fishing area total*	*157 384*	*183 539*	*183 996*	*200 401*	*209 617*	*182 995*	*213 172*	*204 233 F*	*221 418 F*	*226 817 F*
77	Costa Rica	1 015	2 356	2 118	1 362	1 022	672	1 004	1 009	508	561
	Japan	-	57	-	-	-	-	-	-	-	-
	Korea Rep	360	88	12	-	-	-	491	74	-	2
	Panama	748	1 697	1 945	3 885	4 820	833	1 674	1 661	1 147	346
77	*Fishing area total*	*2 123*	*4 198*	*4 075*	*5 247*	*5 842*	*1 505*	*3 169*	*2 744*	*1 655*	*909*
81	Japan	3	-	-	-	-	-	-	-	5	5 F
	Korea Rep	-	-	-	-	-	-	11	-	-	-
	New Zealand	-	-	1	2	23	1	-	-	0	-
81	*Fishing area total*	*3*	-	*1*	*2*	*23*	*1*	*11*	-	*5*	*5 F*
87	Colombia	113	282	369	317	559	345	555	570 F	591	580 F
	Korea Rep	-	-	-	-	-	-	37	18	-	-
87	*Fishing area total*	*113*	*282*	*369*	*317*	*559*	*345*	*592*	*588 F*	*591*	*580 F*
Species total		*512 662 F*	*548 391 F*	*538 080 F*	*592 492 F*	*591 681 F*	*548 708 F*	*561 795 F*	*573 038 F*	*593 881 F*	*663 568 F*
Group total		***2 148 470***	***2 369 882***	***2 441 954***	***2 559 141***	***2 633 493***	***2 755 710***	***3 026 265***	***3 073 108***	***2 949 714***	***2 979 336***

B-46 Krill, planktonic crustaceans / Krill, crustacés planctoniques / Krill, crustáceos planctónicos

Capture production by species, fishing areas and countries or areas
Captures par espèces, zones de pêche et pays ou zones
Capturas por especies, áreas de pesca y países o áreas

Species, Fishing area Espèce, Zone de pêche Especie, Area de pesca	1993 t	1994 t	1995 t	1996 t	1997 t	1998 t	1999 t	2000 t	2001 t	2002 t
Antarctic krill	**Krill antarctique**		**Krill antártico**			*Euphausia superba*			**2,26(01)005,01**	**KRI**
41 Japan	-	-	-	-	-	-	-	4	-	
Poland	2 506	-	-	-	-	74	-	-	-	-
41 Fishing area total	*2 506*	-	-	-	-	*74*	-	*4*	-	-
47 Poland	-	-	-	-	-	254	-	-	-	-
47 Fishing area total	-	-	-	-	-	*254*	-	-	-	-
48 Argentina	-	-	-	-	-	-	6 524	-	-	-
Chile	3 261	3 834	-	-	-	-	-	-	-	-
Japan	50 845	60 198	62 111	58 769	60 937	67 481	66 076	80 597	67 378	51 079
Korea Rep	-	-	-	-	-	2 850	27	7 233	7 525	14 353
Latvia	71	-	-	-	-	-	-	-	-	-
Panama	-	-	637	-	-	-	-	-	-	-
Poland	4 390	7 997	12 521	22 104	14 408	19 060	19 167	20 049	13 696	16 365
Russian Fed	2 216	-	-	-	-	-	-	-	-	-
South Africa	-	3	-	-	-	-	-	-	-	-
Ukraine	-	12 613	59 150	10 277	-	-	6 719	-	14 023	32 015
UK	-	-	-	-	308	634	-	-	-	-
USA	-	-	-	-	-	-	-	70	1 561	12 175
Uruguay	-	-	-	-	-	-	3 444	6 477	-	
48 Fishing area total	*60 783*	*84 645*	*134 419*	*91 150*	*75 653*	*90 025*	*101 957*	*114 426*	*104 183*	*125 987*
58 India	-	-	-	6	-	-	-	-	-	-
Japan	5 762	899	1 266	-	-	-	-	-	-	-
58 Fishing area total	*5 762*	*899*	*1 266*	*6*	-	-	-	-	-	-
Species total	*69 051*	*85 544*	*135 685*	*91 156*	*75 653*	*90 353*	*101 957*	*114 430*	*104 183*	*125 987*
Antarctic krill nei	**Krill antarctique nca**		**Krill antártico nep**			*Euphausia spp*			**2,26(01)005,XX**	**KRX**
58 India	-	-	-	0	-	-	-	-	-	-
58 Fishing area total	-	-	-	*0*	-	-	-	-	-	-
Species total	-	-	-	*0*	-	-	-	-	-	-
Norwegian krill	**Krill norvégien**		**Krill de Noruega**			*Meganyctiphanes norvegica*			**2,26(01)011,01**	**NKR**
27 Denmark	-	-	-	-	-	88	-	-	36	-
27 Fishing area total	-	-	-	-	-	*88*	-	-	*36*	-
Species total	-	-	-	-	-	*88*	-	-	*36*	-
Group total	***69 051***	***85 544***	***135 685***	***91 156***	***75 653***	***90 441***	***101 957***	***114 430***	***104 219***	***125 987***

B-47

Miscellaneous marine crustaceans — **Capture production by species, fishing areas and countries or areas**
Crustacés marins divers — **Captures par espèces, zones de pêche et pays ou zones**
Crustáceos marinos diversos — **Capturas por especies, áreas de pesca y países o áreas**

Species, Fishing area Espèce, Zone de pêche Especie, Area de pesca	1993 t	1994 t	1995 t	1996 t	1997 t	1998 t	1999 t	2000 t	2001 t	2002 t
Brine shrimp	**Crevette de salines**		**Artemia**		***Artemia salina***			**2,02(02)001,01**		**AMS**
67 USA	...	...	...	...	...	...	...	49	63	58
67 Fishing area total	...	...	...	...	...	...	...	*49*	*63*	*58*
77 USA	...	...	...	...	...	513	691	512	475	320
77 Fishing area total	...	...	...	...	...	*513*	*691*	*512*	*475*	*320*
Species total	...	...	...	...	...	*513*	*691*	*561*	*538*	*378*
Goose barnacles	**Balanes**		**Bellotas de mar**		***Lepas spp***			**2,13(02)007,XX**		**GOO**
27 France	22	12	0	2	-	-	-	-	-	-
27 Fishing area total	*22*	*12*	*0*	*2*	-	-	-	-	-	-
Species total	*22*	*12*	*0*	*2*	-	-	-	-	-	-
Giant barnacle	**Balane géante**		**Picoroco gigante**		***Megabalanus psittacus***			**2,13(03)015,01**		**MBZ**
87 Chile	1 192	809	681	879	579	683	620	620	685	199
87 Fishing area total	*1 192*	*809*	*681*	*879*	*579*	*683*	*620*	*620*	*685*	*199*
Species total	*1 192*	*809*	*681*	*879*	*579*	*683*	*620*	*620*	*685*	*199*
Spottail mantis squillid	**Squille ocellée**		**Galera ocelada**		***Squilla mantis***			**2,25(01)001,02**		**MTS**
37 France	11	11	2	15	10	10	34	44	33	43
Italy	4 037	4 317	4 611	5 431	4 497	3 670	4 767	5 244	5 570	5 374
Slovenia	-	-	-	-	-	-	-	0	3	4
Spain	-	-	-	-	-	-	1 222	1 043	1 000	810
37 Fishing area total	*4 048*	*4 328*	*4 613*	*5 446*	*4 507*	*3 680*	*6 023*	*6 331*	*6 606*	*6 231*
Species total	*4 048*	*4 328*	*4 613*	*5 446*	*4 507*	*3 680*	*6 023*	*6 331*	*6 606*	*6 231*
Squillids nei	**Squilles nca**		**Galeras nep**		***Squillidae***			**2,25(01)XXX,XX**		**SQY**
71 Philippines	1 005	835	982	695	2 133	2 219	2 068	2 141	2 612	3 344
71 Fishing area total	*1 005*	*835*	*982*	*695*	*2 133*	*2 219*	*2 068*	*2 141*	*2 612*	*3 344*
Species total	*1 005*	*835*	*982*	*695*	*2 133*	*2 219*	*2 068*	*2 141*	*2 612*	*3 344*
Stomatopods nei	**Stomatopodes nca**		**Estomatopodos nep**		***Stomatopoda***			**2,25(XX)XXX,XX**		**SVX**
21 USA	...	0	1	-	-	1	5	6	1	1
21 Fishing area total	...	*0*	*1*	-	-	*1*	*5*	*6*	*1*	*1*
57 Thailand	-	-	-	-	0	1	116	9	22	21
57 Fishing area total	-	-	-	-	*0*	*1*	*116*	*9*	*22*	*21*
71 Thailand	-	-	-	181	176	457	750	866	456	442
71 Fishing area total	-	-	-	*181*	*176*	*457*	*750*	*866*	*456*	*442*
Species total	...	*0*	*1*	*181*	*176*	*459*	*871*	*881*	*479*	*464*
Marine crustaceans nei	**Crustacés marins nca**		**Crustáceos marinos nep**		***Crustacea***			**2,99(XX)XXX,XX**		**CRU**
21 Japan	-	-	-	2	-	-	-	-	-	-
21 Fishing area total	-	-	-	*2*	-	-	-	-	-	-
27 Denmark	10	6 089	563	117	0	-	-	-	-	-
France	933	864	979	885	1 154	676	700	686	689	1 388
Ireland	-	-	-	-	-	-	-	-	401	595
Portugal	65	69	95	63	180	123	137	15	25	20
Spain	439	522	600	552	449	590	437	365	545	575
Sweden	-	-	-	-	-	-	2	1	-	-
UK	275	8	2	2	-	-	-	-	12	7
27 Fishing area total	*1 722*	*7 552*	*2 239*	*1 619*	*1 783*	*1 389*	*1 276*	*1 067*	*1 672*	*2 585*
31 Barbados	0	0	0	0	0	0	0	0	-	-
Cuba	170	104	80	25	18	13	3	6	-	3
Guadeloupe	150 F	150 F	150 F	140	150	134	134	150	150	150
Korea Rep	-	-	-	-	-	1	1	-	-	-
Puerto Rico	120	136	199	189	171	184	-	-	-	-
St Kitts Nev	50 F	...	...	...	...	...	...	...	...	...
St Lucia	15	14	12	12	12	32	30	25	36	10
31 Fishing area total	*505 F*	*404 F*	*441 F*	*366*	*351*	*364*	*168*	*181*	*186*	*163*
34 Eq Guinea	360 F	550 F	230 F	510 F	650 F	430	500 F	50 F	50 F	50 F
Italy	-	-	387	172	595	173	126	44	77	40
Korea Rep	-	40	32	6	-	-	-	6	-	-
Liberia	-	12	-	-	-	-	-	-	-	-
Mauritania	-	-	-	-	-	30 F	9	6	20	5
Morocco	4	1	1	2	1	0	-	1	1	33
Nigeria	0	0	9	0	0	0	0	0	-	-
Portugal	1 334	21	2	3	2	-	52	-	3	1
Sao Tome Prn	4 F	6 F	6 F	4	5	75	7 F	10 F	10 F	10 F
Spain	2 000 F	1 500 F	1 000 F	337	64	1 035	181	132	375	-
Togo	2	0	1	0	0	0	0	0	0	-
Westn Sahara	0	0	0	0	0	0	0	0	0	0

B-47 Miscellaneous marine crustaceans / Crustacés marins divers / Crustáceos marinos diversos

Capture production by species, fishing areas and countries or areas / Captures par espèces, zones de pêche et pays ou zones / Capturas por especies, áreas de pesca y países o áreas

Species, Fishing area Espèce, Zone de pêche Especie, Area de pesca	1993 t	1994 t	1995 t	1996 t	1997 t	1998 t	1999 t	2000 t	2001 t	2002 t
34 Fishing area total	*3 704 F*	*2 130 F*	*1 668 F*	*1 034 F*	*1 317 F*	*1 743 F*	*875 F*	*249 F*	*536 F*	*139 F*
37 Algeria	150 F	250 F	73	67	71	154	83	308	222	297
France	24	27	...	2	2	2	1	-	-	-
Italy	2 143	2 169	3 944	3 428	3 086	3 072	2 168	1 984	1 707	1 655
Lebanon	25	25	25	25	150	50	125	55	55	60
Morocco	32	22	17	27	35	17	-	33	28	58
Spain	2 000 F	1 800 F	1 600 F	711	891	751	1 456	1 656	1 349	373
Syria	94 F	91 F	90 F	90	84	75	70	60	57	66
Tunisia	1	1	1	8	1	72	1	1	0	-
37 Fishing area total	*4 469 F*	*4 385 F*	*5 750 F*	*4 358*	*4 320*	*4 193*	*3 904*	*4 097*	*3 418*	*2 509*
41 Argentina	395	4 200	80	2	288	2	6 027	-	-	-
Brazil	750 F	750 F	287	288	310	244	266	227	162	247
Italy	-	-	46	20	-	-	-	-	-	-
Korea Rep	2 161	2 292	447	-	251	1 915	-	-	1 155	-
Uruguay	0	0	0	3	-	-	-	-	-	-
41 Fishing area total	*3 306 F*	*7 242 F*	*860*	*313*	*849*	*2 161*	*6 293*	*227*	*1 317*	*247*
47 Angola	-	-	-	-	-	-	-	-	1 062	1 653
Italy	-	-	5	2	-	-	-	-	-	-
47 Fishing area total	*-*	*-*	*5*	*2*	*-*	*-*	*-*	*-*	*1 062*	*1 653*
51 India	3 839	13 441	11 940	15 333	12 130	12 658	11 272	11 047	26 503	30 769
Italy	-	-	79	35	-	-	-	102	65	-
Kenya	48	72	59	185	223	139	104	101	136	108
Korea Rep	-	-	-	47	14	-	10	-	361	412
Réunion	...	...	...	...	...	...	2	1	1	1 F
Spain	-	-	-	-	-	-	72	-	-	-
51 Fishing area total	*3 887*	*13 513*	*12 078*	*15 600*	*12 367*	*12 797*	*11 460*	*11 251*	*27 066*	*31 290 F*
57 Australia	87	277	261	1 017	1 351	1 467	858	905	1 062	1 043
Bangladesh	20 520	21 519	20 363	24 288	28 027	31 027	31 742	31 395	31 037	31 976
India	19 111	16 276	11 745	12 714	13 209	18 603	18 475	3 703	4 420	2 549
Indonesia	44	407	1 495	47	993	294	172	154	219	210
Korea Rep	-	-	-	-	-	-	2	2	155	-
Sri Lanka	7 599	4 900	1 800	2 502	2 360	880	3 080	230	4 450	7 530
57 Fishing area total	*47 361*	*43 379*	*35 664*	*40 568*	*45 940*	*52 271*	*54 329*	*36 389*	*41 343*	*43 308*
61 China	754 006	755 775	852 257	914 672	1 086 501	1 231 473	1 131 643	1 213 725	1 264 976	1 214 762
China, Macao	271	296	207	200	210 F	210 F	210 F	210 F	210 F	210 F
Japan	...	...	60 783	58 890	63 028	67 945	49 783	74 072	49 619	50 000 F
Korea D P Rp	24 000 F	26 654	21 091	33 931	15 265	15 200 F	15 400 F	15 600 F	16 181	16 000 F
Korea Rep	-	136	17	-	-	-	1 645	1 136	4 257	936
61 Fishing area total	*778 277 F*	*782 861*	*934 355*	*1 007 693*	*1 165 004 F*	*1 314 828 F*	*1 198 681 F*	*1 304 743 F*	*1 335 243 F*	*1 281 908 F*
71 Australia	0	0	0	179	3 966	3 769	964	895	614	567
Brunei Darsm	39	68	18	281	78	65	44	78	266	485
Fiji Islands	25	26 F	180	85	78	85	91	80 F	87	85 F
Indonesia	313	2 820	1 098	860	1 590	514	331	774	414	490
Kiribati	210	210	210	220	4	...	...	131	418	174
Korea Rep	864	620	225	-	-	-	-	-	30	6
NewCaledonia	0	0	0	0	0	-	0	0	0	0
Vanuatu	250 F	240 F	250 F	250 F	250 F	250 F	250 F	250 F	250 F	250 F
71 Fishing area total	*1 701 F*	*3 984 F*	*1 981 F*	*1 875 F*	*5 966 F*	*4 683 F*	*1 680 F*	*2 208 F*	*2 079 F*	*2 057 F*
77 El Salvador	405	370	564	498	392	355	294	337	409	4 370
Fr Polynesia	1	0	0	0	0	0	0	0	-	-
Guatemala	-	6	8	0	8	19	20	14	20 F	27
Mexico	21	10	0	0	2	3	1	1	-	-
Panama	4	3	...	3	3	2	1	1	2	3
Samoa	7	10	10	10	10	10	30	207	200 F	200 F
Tonga	80	90	100	120	100	177	200	175	270	337
77 Fishing area total	*518*	*489*	*682*	*631*	*515*	*566*	*546*	*735*	*901 F*	*4 937 F*
81 Australia	167	366	521	156	157	92	64	307	293	273
Japan	3	-	2	-	-	-	-	-	-	-
81 Fishing area total	*170*	*366*	*523*	*156*	*157*	*92*	*64*	*307*	*293*	*273*
87 Chile	26	51	-	1	6	15	67	7	65	70
Peru	-	-	-	-	-	-	-	-	758	1 367
Spain	-	-	-	-	-	-	-	-	2	16
87 Fishing area total	*26*	*51*	*-*	*1*	*6*	*15*	*67*	*7*	*825*	*1 453*
Species total	*845 646 F*	*866 356 F*	*996 246 F*	*1 074 218 F*	*1 238 575 F*	*1 395 102 F*	*1 279 343 F*	*1 361 461 F*	*1 415 941 F*	*1 372 522 F*
Group total	***851 913***	***872 340***	***1 002 523***	***1 081 421***	***1 245 970***	***1 402 656***	***1 289 616***	***1 371 995***	***1 426 861***	***1 383 138***

B-51 Freshwater molluscs — Capture production by species, fishing areas and countries or areas
Mollusques d'eau douce — Captures par espèces, zones de pêche et pays ou zones
Moluscos de agua dulce — Capturas por especies, áreas de pesca y países o áreas

Species, Fishing area Espèce, Zone de pêche Especie, Area de pesca	1993 t	1994 t	1995 t	1996 t	1997 t	1998 t	1999 t	2000 t	2001 t	2002 t
Japanese corbicula	**Cyrène japonaise**		**Corbicula japonesa**		***Corbicula japonica***				**3,16(21)025,02**	**CMJ**
04 China,Taiwan	21	-	-	-	-	-	-	-	-	-
Japan	27 134	23 988	26 938	26 714	21 822	19 932	20 009	19 295	17 295	17 779
Korea Rep	755	713	659	600	387	677	1 006	-	-	-
04 Fishing area total	*27 910*	*24 701*	*27 597*	*27 314*	*22 209*	*20 609*	*21 015*	*19 295*	*17 295*	*17 779*
61 Russian Fed	-	-	-	-	-	-	-	-	74	209
61 Fishing area total	-	-	-	-	-	-	-	-	*74*	*209*
Species total	*27 910*	*24 701*	*27 597*	*27 314*	*22 209*	*20 609*	*21 015*	*19 295*	*17 369*	*17 988*
Freshwater molluscs nei	**Mollusques d'eau douce nca**		**Moluscos de agua dulce nep**		***Mollusca***				**3,99(XX)XXX,XX**	**MOF**
01 Egypt	-	-	488	530	610	625	598	916	845	1 190
Morocco	-	-	-	-	-	-	8	5	6	7
01 Fishing area total	-	-	*488*	*530*	*610*	*625*	*606*	*921*	*851*	*1 197*
02 El Salvador	38	1	-	0	2	3	1	8	2	7
Mexico	120	95	198	240	120	177	239	557	1 027	1 045
USA	42	-	-	-	-	5	-	-	-	-
02 Fishing area total	*200*	*96*	*198*	*240*	*122*	*185*	*240*	*565*	*1 029*	*1 052*
04 China	289 484	330 284	441 994	403 448	374 461	460 126	434 993	480 249	529 645	551 021
China,Taiwan	-	-	-	-	-	1	2	2	-	-
Indonesia	499	2 201	932	1 342	909	806	597	734	2 200	1 180
Japan	705	787	847	1 305	1 361	1 132	900	628	583	338
Korea Rep	832	404	245	200	290	409	255	645	669	636
Philippines	134 891	147 076	120 547	122 636	101 841	90 154	87 259	85 575	68 958	52 571
Turkey	1 250	784	1 150	1 500	2 000	1 500	1 585	1 592	1 601	1 937
04 Fishing area total	*427 661*	*481 536*	*565 715*	*530 431*	*480 862*	*554 128*	*525 591*	*569 425*	*603 656*	*607 683*
05 Netherlands	-	-	-	-	-	-	-	-	-	451
05 Fishing area total	-	-	-	-	-	-	-	-	-	*451*
06 Fiji Islands	2 798	2 852 F	2 569	2 670	3 970	4 500	5 000	5 080 F	5 300	5 180 F
06 Fishing area total	*2 798*	*2 852 F*	*2 569*	*2 670*	*3 970*	*4 500*	*5 000*	*5 080 F*	*5 300*	*5 180 F*
Species total	*430 659*	*484 484 F*	*568 970*	*533 871*	*485 564*	*559 438*	*531 437*	*575 991 F*	*610 836*	*615 563 F*
Group total	***458 569***	***509 185***	***596 567***	***561 185***	***507 773***	***580 047***	***552 452***	***595 286***	***628 205***	***633 551***

B-52 Abalones, winkles, conchs — Ormeaux, bigorneaux, strombes — Orejas de mar, bígaros, estrombos

Capture production by species, fishing areas and countries or areas
Captures par espèces, zones de pêche et pays ou zones
Capturas por especies, áreas de pesca y países o áreas

Species, Fishing area Espèce, Zone de pêche Especie, Area de pesca	1993 t	1994 t	1995 t	1996 t	1997 t	1998 t	1999 t	2000 t	2001 t	2002 t
Common periwinkle	**Bigorneau**		**Bígaro**		***Littorina littorea***			**3,07(01)001,01**		**PEE**
27 Ireland	1 770	2 457	3 007	2 814	3 152	2 636	3 018	2 641	2 781	2 287
27 Fishing area total	*1 770*	*2 457*	*3 007*	*2 814*	*3 152*	*2 636*	*3 018*	*2 641*	*2 781*	*2 287*
37 Spain	-	-	-	132	147	117	126	139	...	28
37 Fishing area total	-	-	-	*132*	*147*	*117*	*126*	*139*	...	*28*
Species total	*1 770*	*2 457*	*3 007*	*2 946*	*3 299*	*2 753*	*3 144*	*2 780*	*2 781*	*2 315*
Periwinkles nei	**Bigorneaux nca**		**Bígaros nep**		***Littorina spp***			**3,07(01)001,XX**		**PER**
21 Canada	279	230	166	200	274	198	149	98	92	24
St Pier Mq	-	-	-	-	-	-	-	-	-	1
USA	384	202	32	19	-	171	223	216	649	1 061
21 Fishing area total	*663*	*432*	*198*	*219*	*274*	*369*	*372*	*314*	*741*	*1 086*
27 Denmark	0	0	0	4	1	2	-	2	0	0
France	9	3	-	2	-	-	-	-	-	-
Portugal	0	0	0	0	-	-	-	-	-	-
Spain	2	2	5	8	8	15	8	3	5	3
UK	1 910	2 263	2 331	1 756	2 870	1 925	1 336	1 059	760	271
27 Fishing area total	*1 921*	*2 268*	*2 336*	*1 770*	*2 879*	*1 942*	*1 344*	*1 064*	*765*	*274*
Species total	*2 584*	*2 700*	*2 534*	*1 989*	*3 153*	*2 311*	*1 716*	*1 378*	*1 506*	*1 360*
Murex	**Rochers**		**Murices**		***Murex spp***			**3,07(02)002,XX**		**MUE**
34 Senegal	...	...	748	1 267	1 223	2 543	1 255	1 529	2 080	1 617
34 Fishing area total	...	...	*748*	*1 267*	*1 223*	*2 543*	*1 255*	*1 529*	*2 080*	*1 617*
37 France	88	84	43	37	41	41	35	51	52	60
37 Fishing area total	*88*	*84*	*43*	*37*	*41*	*41*	*35*	*51*	*52*	*60*
Species total	*88*	*84*	*791*	*1 304*	*1 264*	*2 584*	*1 290*	*1 580*	*2 132*	*1 677*
Sea snails	**Escargots de mer**		**Caracols de mar**		***Rapana spp***			**3,07(02)018,XX**		**RPN**
37 Bulgaria	...	3 000 F	3 120	3 260	4 900	4 300	3 800	3 800	3 353	698
Georgia	-	-	700	711	118	-		-	-	60
Ukraine	3	5	303	378	476	371	619	913	400	93
37 Fishing area total	*3*	*3 005 F*	*4 123*	*4 349*	*5 494*	*4 671*	*4 419*	*4 713*	*3 753*	*851*
Species total	*3*	*3 005 F*	*4 123*	*4 349*	*5 494*	*4 671*	*4 419*	*4 713*	*3 753*	*851*
False abalone	**Rocher loco**		**Loco**		***Concholepas concholepas***			**3,07(02)023,01**		**SNE**
87 Chile	8 574	8 111	2 670	2 541	3 154	2 564	2 294	1 274	828	1 622
Peru	2 919	2 557	1 361	2 728	4 366	830	2 289	1 250	544	686
87 Fishing area total	*11 493*	*10 668*	*4 031*	*5 269*	*7 520*	*3 394*	*4 583*	*2 524*	*1 372*	*2 308*
Species total	*11 493*	*10 668*	*4 031*	*5 269*	*7 520*	*3 394*	*4 583*	*2 524*	*1 372*	*2 308*
Giant abalone	**Ormeau géant**		**Abulón gigante**		***Haliotis gigantea***			**3,07(03)001,09**		**ABG**
61 Japan	2 353	2 164	1 980	1 941	2 218	2 269	2 109	2 146	1 982	2 223
61 Fishing area total	*2 353*	*2 164*	*1 980*	*1 941*	*2 218*	*2 269*	*2 109*	*2 146*	*1 982*	*2 223*
Species total	*2 353*	*2 164*	*1 980*	*1 941*	*2 218*	*2 269*	*2 109*	*2 146*	*1 982*	*2 223*
Perlemoen abalone	**Ormeau de Mida**		**Oreja de mar**		***Haliotis midae***			**3,07(03)001,11**		**ABP**
47 South Africa	561	586	615	735	330	524	481	490	527	516
47 Fishing area total	*561*	*586*	*615*	*735*	*330*	*524*	*481*	*490*	*527*	*516*
Species total	*561*	*586*	*615*	*735*	*330*	*524*	*481*	*490*	*527*	*516*
Blacklip abalone	**Ormeau à lèvres noires**		**Oreja de mar de labios negros**		***Haliotis rubra***			**3,07(03)001,13**		**ABR**
57 Australia	4 341	4 361	4 896	5 081	4 905	4 920	5 297	5 207	5 304	5 845
57 Fishing area total	*4 341*	*4 361*	*4 896*	*5 081*	*4 905*	*4 920*	*5 297*	*5 207*	*5 304*	*5 845*
71 Solomon Is	26	0	-	-	-	-	-	-	-	-
71 Fishing area total	*26*	*0*	-	-	-	-	-	-	-	-
81 Australia	327	312	312	344	335	327	323	325	305	281
81 Fishing area total	*327*	*312*	*312*	*344*	*335*	*327*	*323*	*325*	*305*	*281*
Species total	*4 694*	*4 673*	*5 208*	*5 425*	*5 240*	*5 247*	*5 620*	*5 532*	*5 609*	*6 126*
Tuberculate abalone	**Ormeau tuberculeux**		**Oreja marina tuberculosa**		***Haliotis tuberculata***			**3,07(03)001,14**		**HLT**
27 Channel Is	...	-	-	-	-	-	-	3	2	3
France	11	3	49	62	75	36	37	61	60	31
27 Fishing area total	*11*	*3*	*49*	*62*	*75*	*36*	*37*	*64*	*62*	*34*
Species total	*11*	*3*	*49*	*62*	*75*	*36*	*37*	*64*	*62*	*34*

B-52 Abalones, winkles, conchs / Ormeaux, bigorneaux, strombes / Orejas de mar, bigaros, estrombos

Capture production by species, fishing areas and countries or areas
Captures par espèces, zones de pêche et pays ou zones
Capturas por especies, áreas de pesca y países o áreas

Species, Fishing area Espèce, Zone de pêche Especie, Area de pesca	1993 t	1994 t	1995 t	1996 t	1997 t	1998 t	1999 t	2000 t	2001 t	2002 t
Abalones nei	**Ormeaux nca**		**Orejas de mar nep**		***Haliotis spp***				**3,07(03)001,XX**	**ABX**
51 Oman	34	36	43	43	40	40	29	45	51	50
51 Fishing area total	*34*	*36*	*43*	*43*	*40*	*40*	*29*	*45*	*51*	*50*
61 China,Taiwan	14	15	14	8	6	2	2	42	1	-
Korea Rep	361	281	260	188	214	71	79	113	104	75
61 Fishing area total	*375*	*296*	*274*	*196*	*220*	*73*	*81*	*155*	*105*	*75*
67 USA	14	45	29	-	-	-	-	-	3	3
67 Fishing area total	*14*	*45*	*29*	-	-	-	-	-	*3*	*3*
71 Philippines	122	240	483	448	183	347	282	241	288	292
Solomon Is	2	-	-	-	-	1	-	-	-	-
71 Fishing area total	*124*	*240*	*483*	*448*	*183*	*348*	*282*	*241*	*288*	*292*
77 Mexico	2 180	1 536	1 227	1 075	924	709	574	535	482	494
USA	209	168	150	127	102	-	8	-	-	-
77 Fishing area total	*2 389*	*1 704*	*1 377*	*1 202*	*1 026*	*709*	*582*	*535*	*482*	*494*
81 New Zealand	1 099	1 080	1 280	1 020	1 180	1 300	1 170	1 265	1 064	1 090
81 Fishing area total	*1 099*	*1 080*	*1 280*	*1 020*	*1 180*	*1 300*	*1 170*	*1 265*	*1 064*	*1 090*
Species total	*4 035*	*3 401*	*3 486*	*2 909*	*2 649*	*2 470*	*2 144*	*2 241*	*1 993*	*2 004*
Horned turban	**Troque**		**Peonza cornuda**		***Turbo cornutus***				**3,07(05)002,02**	**TOS**
61 Japan	9 617	10 393	9 943	10 119	12 132	12 556	11 000	9 839	10 241	9 504
Korea Rep	8 764	9 128	8 921	7 361	6 878	9 192	7 397	7 281	6 274	7 193
61 Fishing area total	*18 381*	*19 521*	*18 864*	*17 480*	*19 010*	*21 748*	*18 397*	*17 120*	*16 515*	*16 697*
Species total	*18 381*	*19 521*	*18 864*	*17 480*	*19 010*	*21 748*	*18 397*	*17 120*	*16 515*	*16 697*
Stromboid conchs nei	**Strombes nca**		**Cobos nep**		***Strombus spp***				**3,07(06)002,XX**	**CON**
31 Anguilla	8	9	5 F	10 F	10 F	10 F	10 F	10 F	10 F	10 F
Antigua Barb	525	518	345	293	263	338	345	315	278	319
Bahamas	527	693	494	589	648	670	472	667	661	650 F
Belize	1 137	1 413	1 026	1 105	1 926	1 891	1 051	1 745	1 980	1 380
Br Virgin Is	...	32	43	54	8	9	8	6	6 F	6 F
Cuba	90	47	32	717	1 234	487	829	1 163	1 172	997
Dominican Rp	2 600	4 680	2 210	1 889	1 594	2 683	1 257	1 778	1 437	2 332
Grenada	11	1	2	6	1	24	6	0	2	32
Guadeloupe	480 F	500 F	500 F	430	470	550	580	550	550	550
Haiti	400 F	380 F	350 F	400 F	380 F	350 F	300 F	300 F	300 F	300 F
Honduras	485	483	675	609	868	601	721	713	742	789
Jamaica	15 000	17 250	15 998	21 375	13 658	12 750	10 245	0	-	-
Mexico	4 023	2 670	4 963	2 566	5 218	3 293	7 243	8 295	8 730	6 293
NethAntilles	5 F	5 F	5 F	5 F	5 F	5 F	5 F	5 F	5 F	5 F
Nicaragua	-	-	-	-	-	162	209	555	956	318
Puerto Rico	375 F	405	758	450	638	1 025	1 025	1 710	1 643	931
St Kitts Nev	...	21	29	49	29	81	67	73	71	36
St Lucia	10	13	15	15	25	28	25	40	41	60
St Vincent	...	32	30	25 F	10	21	7	7	37	40
Trinidad Tob	...	...	...	...	...	-	0	-	-	-
Turks Caicos	5 535	5 243	7 148	7 223	5 520	5 903	4 838	5 520	6 120	5 430
US Virgin Is	25 F	20 F	15 F	10 F	5 F	5 F	1	1 F	1 F	1 F
31 Fishing area total	*31 236 F*	*34 415 F*	*34 643 F*	*37 820 F*	*32 510 F*	*30 886 F*	*29 244 F*	*23 453 F*	*24 742 F*	*20 479 F*
77 Mexico	3 607	4 920	4 163	3 091	2 421	973	1 348	2 097	2 667	2 732
Panama	3	...	...	...	6	527	125	5	55	83
77 Fishing area total	*3 610*	*4 920*	*4 163*	*3 091*	*2 427*	*1 500*	*1 473*	*2 102*	*2 722*	*2 815*
87 Ecuador	5	5	22	10	10	10	10	10	10	10 F
87 Fishing area total	*5*	*5*	*22*	*10*	*10*	*10*	*10*	*10*	*10*	*10 F*
Species total	*34 851 F*	*39 340 F*	*38 828 F*	*40 921 F*	*34 947 F*	*32 396 F*	*30 727 F*	*25 565 F*	*27 474 F*	*23 304 F*
Whelk	**Buccin**		**Bocina**		***Buccinum undatum***				**3,07(08)001,01**	**WHE**
21 St Pier Mq	...	...	...	...	32	45	...	0	-	-
21 Fishing area total	...	...	...	...	*32*	*45*	...	*0*	-	-
27 Belgium	252	165	164	226	166	117	83	101	119	92
Channel Is	0	0	375 F	550	438	135	8	338	519	204
Denmark	-	-	-	-	-	-	1	-	1	1
Faeroe Is	-	4	-	-	-	-	-	-	-	-
France	1 074	1 363	3 679	4 406	12 887	6 266	7 691	12 724	11 030	10 532
Iceland	0	0	0	520	1 199	13	298	770	678	-
Ireland	2 562	4 489	5 952	6 575	3 852	3 667	4 561	4 942	6 364	7 901
Isle of Man	-	149	148	296	193	...	227	89	2	62
Netherlands	-	-	-	-	-	-	-	121	163	178
Spain	-	-	-	-	-	-	-	-	-	4
UK	3 514	3 315	5 534	12 150	8 892	3 525	4 925	10 733	11 270	8 669
27 Fishing area total	*7 402*	*9 485*	*15 852 F*	*24 723*	*27 627*	*13 723*	*17 794*	*29 818*	*30 146*	*27 643*
Species total	*7 402*	*9 485*	*15 852 F*	*24 723*	*27 659*	*13 768*	*17 794*	*29 818*	*30 146*	*27 643*

B-52

Abalones, winkles, conchs — **Capture production by species, fishing areas and countries or areas**
Ormeaux, bigorneaux, strombes — **Captures par espèces, zones de pêche et pays ou zones**
Orejas de mar, bígaros, estrombos — **Capturas por especies, áreas de pesca y países o áreas**

Species, Fishing area Espèce, Zone de pêche Especie, Area de pesca	1993 t	1994 t	1995 t	1996 t	1997 t	1998 t	1999 t	2000 t	2001 t	2002 t
Whelks	**Busycons**		**Busicones**		***Busycon spp***				**3,07(09)003,XX**	**WHX**
21 Canada	742	570	935	1 291	1 275	1 062	1 557	1 923	2 041	2 218
USA	3 687	2 237	1 911	3 003	2 087	1 673	3 792	2 330	3 557	2 917
21 Fishing area total	*4 429*	*2 807*	*2 846*	*4 294*	*3 362*	*2 735*	*5 349*	*4 253*	*5 598*	*5 135*
31 USA	1 181	1 489	1 491	1 193	1 017	908	832	600	480	153
31 Fishing area total	*1 181*	*1 489*	*1 491*	*1 193*	*1 017*	*908*	*832*	*600*	*480*	*153*
Species total	*5 610*	*4 296*	*4 337*	*5 487*	*4 379*	*3 643*	*6 181*	*4 853*	*6 078*	*5 288*
Volutes nei	**Volutes nca**		**Volutas nep**		***Cymbium spp***				**3,07(42)004,XX**	**CXY**
34 Senegal	4 294	4 888	7 454	6 648	5 166	4 678	5 737	4 989	5 422	4 347
34 Fishing area total	*4 294*	*4 888*	*7 454*	*6 648*	*5 166*	*4 678*	*5 737*	*4 989*	*5 422*	*4 347*
Species total	*4 294*	*4 888*	*7 454*	*6 648*	*5 166*	*4 678*	*5 737*	*4 989*	*5 422*	*4 347*
Angulate volute	**Volute angulée**		**Voluta angulosa**		***Zidona dufresnei***				**3,07(42)007,01**	**ZDF**
41 Argentina	191	223	574	558	1 322	1 010	683	631	694	414
Uruguay	...	...	...	...	...	...	...	990	825	900
41 Fishing area total	*191*	*223*	*574*	*558*	*1 322*	*1 010*	*683*	*1 621*	*1 519*	*1 314*
Species total	*191*	*223*	*574*	*558*	*1 322*	*1 010*	*683*	*1 621*	*1 519*	*1 314*
Gastropods nei	**Gastropodes nca**		**Gasterópodos nep**		***Gastropoda***				**3,07(XX)XXX,XX**	**GAS**
27 France	-	-	-	-	-	-	1	1	-	-
Portugal	...	...	...	...	...	...	141	151	146	119
27 Fishing area total	...	...	...	...	...	...	*142*	*152*	*146*	*119*
31 Colombia	121	134	329	112	169	135	121	100 F	100 F	100 F
Cuba	1	14	46	132	15	4	-	-	-	-
31 Fishing area total	*122*	*148*	*375*	*244*	*184*	*139*	*121*	*100 F*	*100 F*	*100 F*
34 Gambia	...	55	194	227	128	230	250	5	20	1
Nigeria	...	...	...	...	...	2 357	...	...	2 084	2 426
Portugal	17	23	20	48	41	41	48	49	42	61
Senegal	917	682	24	45	40	...	...	65	11	4
Spain	-	-	-	-	-	-	-	-	-	115
34 Fishing area total	*934*	*760*	*238*	*320*	*209*	*2 628*	*298*	*119*	*2 157*	*2 607*
37 France	-	-	-	-	-	-	10	14	19	26
Spain	-	-	-	-	-	-	955	799	851	774
37 Fishing area total	-	-	-	-	-	-	*965*	*813*	*870*	*800*
61 China,Taiwan	959	816	998	760	1 019	125	0	10	53	-
Korea Rep	8 352	8 131	5 224	2 975	1 617	1 325	1 670	818	730	1 555
61 Fishing area total	*9 311*	*8 947*	*6 222*	*3 735*	*2 636*	*1 450*	*1 670*	*828*	*783*	*1 555*
67 USA	...	3 565	4 228	6 369	...	...	...	-	1	4
67 Fishing area total	...	*3 565*	*4 228*	*6 369*	...	...	...	-	*1*	*4*
71 Solomon Is	0	1	-	-	-	0	0	-	-	-
71 Fishing area total	*0*	*1*	-	-	-	*0*	*0*	-	-	-
77 Panama	9	372	0	5	-	116	17	1	1	12
USA	22	22	18	11	12	49	36	35	25	47
77 Fishing area total	*31*	*394*	*18*	*16*	*12*	*165*	*53*	*36*	*26*	*59*
87 Chile	9 377	7 538	6 403	6 078	5 348	4 649	7 204	6 898	5 429	5 120
Colombia	...	9	7	17	29	20	10	20 F	20 F	20 F
Peru	2 871	2 504	3 686	2 215	7 098	3 110	4 525	2 768	4 995	2 349
87 Fishing area total	*12 248*	*10 051*	*10 096*	*8 310*	*12 475*	*7 779*	*11 739*	*9 686 F*	*10 444 F*	*7 489 F*
Species total	*22 646*	*23 866*	*21 177*	*18 994*	*15 516*	*12 161*	*14 988*	*11 734 F*	*14 527 F*	*12 733 F*
Group total	***120 967***	***131 360***	***132 910***	***141 740***	***139 241***	***115 663***	***120 050***	***119 148***	***123 398***	***110 740***

B-53 Oysters / Huîtres / Ostras

Capture production by species, fishing areas and countries or areas
Captures par espèces, zones de pêche et pays ou zones
Capturas por especies, áreas de pesca y países o áreas

Species, Fishing area Espèce, Zone de pêche Especie, Area de pesca	1993 t	1994 t	1995 t	1996 t	1997 t	1998 t	1999 t	2000 t	2001 t	2002 t
Chilean flat oyster	**Huître plate chilienne**		**Ostra chilena**		***Ostrea chilensis***				**3,16(07)002,03**	**OCH**
87 Chile	14	18	-	-	5	1	6	9	202	7
87 Fishing area total	*14*	*18*	-	-	*5*	*1*	*6*	*9*	*202*	*7*
Species total	*14*	*18*	-	-	*5*	*1*	*6*	*9*	*202*	*7*
European flat oyster	**Huître plate européenne**		**Ostra europea**		***Ostrea edulis***				**3,16(07)002,05**	**OYF**
27 Channel Is	4 F	-	-	-	-	-	-	-	-	-
Denmark	75	4	12	9	24	6	8	9	23	528
France	20	9	18	6	8	4	10	7	130	197
Ireland	350	1 208	815	415	773	...	...	...	...	...
Netherlands	123	-	-	-	-	-	-	-	-	-
Spain	38	17	11	252	388	316	443	152	37	59
Sweden	4	2	2	3	3	2	4	2	1	2
UK	432	521	527	584	553	1 047	407	439	611	586
27 Fishing area total	*1 046 F*	*1 761*	*1 385*	*1 269*	*1 749*	*1 375*	*872*	*609*	*802*	*1 372*
37 Croatia	...	...	...	...	...	35	14	12	24	5
France	80	92	22	43	10	10	2	2	2	0
Greece	1 691	1 093	1 096	1 003	344	95	47	105	120	105 F
Serbia-Monte	0	0	0	0	0	0	1	1	0	0
Slovenia	5	1	0	0	0	-	-	0	0	0
Spain	6 F	3 F	1 F	1	0	0	-	-	-	-
Tunisia	6	-	-	-	-	-	-	1	1	8
Turkey	1 222	1 803	1 836	1 140	1 495	1 050	840	150	10	70
37 Fishing area total	*3 010 F*	*2 992 F*	*2 955 F*	*2 187*	*1 849*	*1 190*	*904*	*271*	*157*	*188 F*
Species total	*4 056 F*	*4 753 F*	*4 340 F*	*3 456*	*3 598*	*2 565*	*1 776*	*880*	*959*	*1 560 F*
New Zealand dredge oyster	**Huître plate néo-zélandaise**		**Ostra de Nueva Zelandia**		***Ostrea lutaria***				**3,16(07)002,20**	**DRY**
81 New Zealand	871	584	1 077	1 931	2 172	1 000	995	766	832	816
81 Fishing area total	*871*	*584*	*1 077*	*1 931*	*2 172*	*1 000*	*995*	*766*	*832*	*816*
Species total	*871*	*584*	*1 077*	*1 931*	*2 172*	*1 000*	*995*	*766*	*832*	*816*
Olympia flat oyster	**Huître plate Olympie**		**Ostra Olimpia**		***Ostrea conchaphila***				**3,16(07)002,25**	**OYH**
67 USA	...	...	...	...	...	...	6	7	8	11
67 Fishing area total	...	...	...	...	...	...	*6*	*7*	*8*	*11*
Species total	...	...	...	...	...	...	*6*	*7*	*8*	*11*
Pacific cupped oyster	**Huître creuse du Pacifique**		**Ostión japonés**		***Crassostrea gigas***				**3,16(07)008,01**	**OYG**
27 France	-	-	-	4	0	0	2	60	63	10
Portugal	1	-	8	-	-	-	-	-	-	-
Spain	-	-	10	0	-	-	1	9	9	-
UK	99	22	82	125	98	59	6	5	20	5
27 Fishing area total	*100*	*22*	*100*	*129*	*98*	*59*	*9*	*74*	*92*	*15*
37 France	36	369	50	50	14	14	48	48	-	-
Spain	-	-	-	9	-	-	-	-	-	-
37 Fishing area total	*36*	*369*	*50*	*59*	*14*	*14*	*48*	*48*	-	-
61 China,Taiwan	8	2	7	-	-	-	-	-	4	-
Korea Rep	28 215	20 710	18 262	18 259	17 210	9 905	11 609	15 939	10 056	16 678
61 Fishing area total	*28 223*	*20 712*	*18 269*	*18 259*	*17 210*	*9 905*	*11 609*	*15 939*	*10 060*	*16 678*
67 USA	...	...	...	4 989	6 187	11	...	...	...	...
67 Fishing area total	...	...	...	*4 989*	*6 187*	*11*	...	...	...	...
77 USA	2 443	2 583	1 709	1 717	2 176	2 107	539	1 917	1 083	1 083
77 Fishing area total	*2 443*	*2 583*	*1 709*	*1 717*	*2 176*	*2 107*	*539*	*1 917*	*1 083*	*1 083*
81 New Zealand	-	-	8	2	2	0	62	0	0	-
81 Fishing area total	-	-	*8*	*2*	*2*	*0*	*62*	*0*	*0*	-
87 Ecuador	1	5	5	5	5	5	5	5	5	5 F
87 Fishing area total	*1*	*5*	*5*	*5*	*5*	*5*	*5*	*5*	*5*	*5 F*
Species total	*30 803*	*23 691*	*20 141*	*25 160*	*25 692*	*12 101*	*12 272*	*17 983*	*11 240*	*17 781 F*
Mangrove cupped oyster	**Huître creuse des Caraïbes**		**Ostión de mangle**		***Crassostrea rhizophorae***				**3,16(07)008,02**	**OYM**
31 Colombia	...	20	-	-	-	8	-	-	-	-
Cuba	744	1 647	1 885	1 888	2 011	2 282	1 865	1 631	1 482	1 300
Dominican Rp	20	36	41	6	6	12	4	7	8	135
Venezuela	2 465	3 144	3 775	2 695	1 705	2 594	2 037	1 590	3 754	800
31 Fishing area total	*3 229*	*4 847*	*5 701*	*4 589*	*3 722*	*4 896*	*3 906*	*3 228*	*5 244*	*2 235*
Species total	*3 229*	*4 847*	*5 701*	*4 589*	*3 722*	*4 896*	*3 906*	*3 228*	*5 244*	*2 235*
American cupped oyster	**Huître creuse américaine**		**Ostión virgínico**		***Crassostrea virginica***				**3,16(07)008,03**	**OYA**
21 Canada	1 590	2 636	2 312	2 132	1 726	3 124	3 225	4 133	3 009	2 410
USA	21 589	40 442	30 570	37 437	47 373	30 141	43 832	22 307	15 548	10 807

B-53 Oysters / Huîtres / Ostras

Capture production by species, fishing areas and countries or areas
Captures par espèces, zones de pêche et pays ou zones
Capturas por especies, áreas de pesca y países o áreas

Species, Fishing area Espèce, Zone de pêche Especie, Area de pesca	1993 t	1994 t	1995 t	1996 t	1997 t	1998 t	1999 t	2000 t	2001 t	2002 t
21 Fishing area total	*23 179*	*43 078*	*32 882*	*39 569*	*49 099*	*33 265*	*47 057*	*26 440*	*18 557*	*13 217*
31 Mexico	22 821	33 647	27 609	34 726	38 515	30 715	39 268	48 101	48 570	47 570
USA	62 480	51 747	98 043	74 245	57 093	71 242	45 882	147 012	107 915	98 381
31 Fishing area total	*85 301*	*85 394*	*125 652*	*108 971*	*95 608*	*101 957*	*85 150*	*195 113*	*156 485*	*145 951*
Species total	*108 480*	*128 472*	*158 534*	*148 540*	*144 707*	*135 222*	*132 207*	*221 553*	*175 042*	*159 168*
Slipper cupped oyster	**Huître creuse chausson**		**...C**		***Crassostrea iredalei***				**3,16(07)008,11**	**CSI**
71 Philippines	161	107	324	291	152	89	95	79	96	97
71 Fishing area total	*161*	*107*	*324*	*291*	*152*	*89*	*95*	*79*	*96*	*97*
Species total	*161*	*107*	*324*	*291*	*152*	*89*	*95*	*79*	*96*	*97*
Cupped oysters nei	**Huîtres creuses nca**		**Ostiones nep**		***Crassostrea spp***				**3,16(07)008,XX**	**OYC**
21 USA	6 063	...	...	...	...	64	49	35	14	19
21 Fishing area total	*6 063*	*...*	*...*	*...*	*...*	*64*	*49*	*35*	*14*	*19*
27 Norway	...	-	0	-	-	-	-	-	5	-
UK	0	3	2	2	5	-	-	-	-	-
27 Fishing area total	*0*	*3*	*2*	*2*	*5*	*-*	*-*	*-*	*5*	*-*
34 Senegal	13	86	223	100	109	89	125	101	151	120
34 Fishing area total	*13*	*86*	*223*	*100*	*109*	*89*	*125*	*101*	*151*	*120*
41 Brazil	430 F	430 F	726	873	828	744	1 547	884	895	876
41 Fishing area total	*430 F*	*430 F*	*726*	*873*	*828*	*744*	*1 547*	*884*	*895*	*876*
47 South Africa	...	...	...	...	...	...	-	-	-	1
47 Fishing area total	*...*	*...*	*...*	*...*	*...*	*...*	*-*	*-*	*-*	*1*
51 Kenya	17	8	14	32	16	9	8	2	1	1
51 Fishing area total	*17*	*8*	*14*	*32*	*16*	*9*	*8*	*2*	*1*	*1*
57 Indonesia	44	15	-	-	-	-	-	350	298	180
57 Fishing area total	*44*	*15*	*-*	*-*	*-*	*-*	*-*	*350*	*298*	*180*
61 Russian Fed	-	-	-	-	-	-	1	4	32	1
61 Fishing area total	*-*	*-*	*-*	*-*	*-*	*-*	*1*	*4*	*32*	*1*
67 USA	...	...	...	...	...	...	...	7	12	8
67 Fishing area total	*...*	*...*	*...*	*...*	*...*	*...*	*...*	*7*	*12*	*8*
71 Indonesia	102	642	1 331	1 596	1 678	2 029	2 025	1 931	896	1 600
71 Fishing area total	*102*	*642*	*1 331*	*1 596*	*1 678*	*2 029*	*2 025*	*1 931*	*896*	*1 600*
77 Mexico	2 033	615	1 801	1 354	1 903	1 276	2 489	1 828	2 234	2 218
77 Fishing area total	*2 033*	*615*	*1 801*	*1 354*	*1 903*	*1 276*	*2 489*	*1 828*	*2 234*	*2 218*
Species total	*8 702 F*	*1 799 F*	*4 097*	*3 957*	*4 539*	*4 211*	*6 244*	*5 142*	*4 538*	*5 024*
Group total	***156 316***	***164 271***	***194 214***	***187 924***	***184 587***	***160 085***	***157 507***	***249 647***	***198 161***	***186 699***

B-54 Mussels / Moules / Mejillones

Capture production by species, fishing areas and countries or areas
Captures par espèces, zones de pêche et pays ou zones
Capturas por especies, áreas de pesca y países o áreas

Species, Fishing area Espèce, Zone de pêche Especie, Area de pesca	1993 t	1994 t	1995 t	1996 t	1997 t	1998 t	1999 t	2000 t	2001 t	2002 t
Korean mussel	**Moule coréenne**		**Mejillón coreano**		***Mytilus coruscus***			**3,16(10)001,01**		**MUK**
61 Korea Rep	2 271	2 731	2 942	2 191	3 211	1 469	1 414	1 133	1 085	946
61 Fishing area total	*2 271*	*2 731*	*2 942*	*2 191*	*3 211*	*1 469*	*1 414*	*1 133*	*1 085*	*946*
Species total	*2 271*	*2 731*	*2 942*	*2 191*	*3 211*	*1 469*	*1 414*	*1 133*	*1 085*	*946*
Chilean mussel	**Moule chilienne**		**Chorito**		***Mytilus chilensis***			**3,16(10)001,03**		**MYC**
41 Falkland Is	1	1	1	1	1	1	1	1	0	0
41 Fishing area total	*1*	*1*	*1*	*1*	*1*	*1*	*1*	*1*	*0*	*0*
87 Chile	6 892	5 718	5 128	5 714	4 723	4 899	4 343	5 236	6 758	1 416
87 Fishing area total	*6 892*	*5 718*	*5 128*	*5 714*	*4 723*	*4 899*	*4 343*	*5 236*	*6 758*	*1 416*
Species total	*6 893*	*5 719*	*5 129*	*5 715*	*4 724*	*4 900*	*4 344*	*5 237*	*6 758*	*1 416*
Blue mussel	**Moule commune**		**Mejillón común**		***Mytilus edulis***			**3,16(10)001,05**		**MUS**
21 Canada	4 938	6 118	4 505	8 147	9 163	10 273	11 565	15 089	11 829	13 769
St Pier Mq	9	7	8	4	2	4	0	0	0	0
USA	14 931	10 891	11 210	9 434	8 063	4 774	4 086	6 009	6 524	9 862
21 Fishing area total	*19 878*	*17 016*	*15 723*	*17 585*	*17 228*	*15 051*	*15 651*	*21 098*	*18 353*	*23 631*
27 Channel Is	2 F	-	-	-	-	-	-	-	-	-
Denmark	136 677	129 317	107 377	86 002	90 765	108 329	96 215	110 618	122 480	110 873
France	2 050	3 908	8 910	197	6 972	1 355	9 564	8 660	8 015	4 782
Ireland	4 095	3 033	4 556	1 372	1 963	955	503	-	-	-
Norway	...	0	8	4	0	-	1	10	-	-
Portugal	59	37	45	35	46	24	87	48	74	165
Spain	38	-	-	18	176	84	27	-	33	66
Sweden	30	51	52	-	3	36	0	70	51	86
UK	7 751	10 348	9 534	12 337	18 994	11 434	7 972	7 468	14 905	16 738
27 Fishing area total	*150 702 F*	*146 694*	*130 482*	*99 965*	*118 919*	*122 217*	*114 369*	*126 874*	*145 558*	*132 710*
Species total	*170 580 F*	*163 710*	*146 205*	*117 550*	*136 147*	*137 268*	*130 020*	*147 972*	*163 911*	*156 341*
River Plata mussel	**Moule de la Plata**		**Mejillón del Plata**		***Mytilus platensis***			**3,16(10)001,08**		**MSR**
41 Argentina	648	389	388	164	180	149	332	236	217	344
Uruguay	189	183	299	206	174	226	142	176	306	96
41 Fishing area total	*837*	*572*	*687*	*370*	*354*	*375*	*474*	*412*	*523*	*440*
Species total	*837*	*572*	*687*	*370*	*354*	*375*	*474*	*412*	*523*	*440*
Mediterranean mussel	**Moule méditerranéenne**		**Mejillón mediterráneo**		***Mytilus galloprovincialis***			**3,16(10)001,12**		**MSM**
37 Albania	...	...	...	...	24	-	-	-	-	350
Bulgaria	-	-	-	-	-	-	-	-	7	55
Croatia	...	...	...	...	...	200	60	57	32	45
France	490	798	1 912	500	1 078	1 078	23	24	12	4
Greece	2 919	2 273	10 806	12 625	24 139	6 389	15 860	469	254	250 F
Italy	28 344	22 702	21 425	22 174	21 430	27 270 F	26 510 F	44 200 F	44 160 F	46 030
Morocco	1	0	0	0	0	0	14	60	32	0
Russian Fed	-	-	-	-	-	-	4	-	-	-
Slovenia	-	-	1	1	1	-	-	-	0	0
Spain	-	-	-	0	29	0	19	18	24	11
Turkey	7 086	8 033	6 042	3 500	6 450	3 880	1 800	1 200	1 500	5 000
Ukraine	210	226	578	246	159	82	163	115	71	75
37 Fishing area total	*39 050*	*34 032*	*40 764*	*39 046*	*53 310*	*38 899 F*	*44 453 F*	*46 143 F*	*46 092 F*	*51 820 F*
Species total	*39 050*	*34 032*	*40 764*	*39 046*	*53 310*	*38 899 F*	*44 453 F*	*46 143 F*	*46 092 F*	*51 820 F*
Australian mussel	**Moule d'Australie**		**Mejillón de Australia**		***Mytilus planulatus***			**3,16(10)001,17**		**MYA**
57 Australia	1 000 F	500 F	243	71	-	-	-	-	-	-
57 Fishing area total	*1 000 F*	*500 F*	*243*	*71*	-	-	-	-	-	-
81 Australia	1	4	4	4	1	1	1	1	1	0
81 Fishing area total	*1*	*4*	*4*	*4*	*1*	*1*	*1*	*1*	*1*	*0*
Species total	*1 001 F*	*504 F*	*247*	*75*	*1*	*1*	*1*	*1*	*1*	*0*
Choro mussel	**Moule choro**		**Choro**		***Choromytilus chorus***			**3,16(10)026,01**		**CHC**
87 Chile	1 201	927	307	323	266	127	155	217	166	91
87 Fishing area total	*1 201*	*927*	*307*	*323*	*266*	*127*	*155*	*217*	*166*	*91*
Species total	*1 201*	*927*	*307*	*323*	*266*	*127*	*155*	*217*	*166*	*91*
Horse mussels nei	**Modioles nca**		**Modiolos nep**		***Modiolus spp***			**3,16(10)028,XX**		**MOD**
27 Norway	14	6	7	20	30	20	7	2	2	12
27 Fishing area total	*14*	*6*	*7*	*20*	*30*	*20*	*7*	*2*	*2*	*12*
57 Thailand	-	-	-	-	-	-	-	21	20	20
57 Fishing area total	-	-	-	-	-	-	-	*21*	*20*	*20*
71 Thailand	-	-	-	-	-	-	-	73	-	-

B-54 Mussels / Moules / Mejillones

Capture production by species, fishing areas and countries or areas
Captures par espèces, zones de pêche et pays ou zones
Capturas por especies, áreas de pesca y países o áreas

Species, Fishing area Espèce, Zone de pêche Especie, Area de pesca	1993 t	1994 t	1995 t	1996 t	1997 t	1998 t	1999 t	2000 t	2001 t	2002 t
71 Fishing area total	*-*	*-*	*-*	*-*	*-*	*-*	*-*	*73*	*-*	*-*
Species total	*14*	*6*	*7*	*20*	*30*	*20*	*7*	*96*	*22*	*32*
South American rock mussel	**Moule de roche sudaméricaine**		**Mejillón de roca sudamericano**		***Perna perna***				**3,16(10)032,01**	**MSL**
31 Venezuela	325	366	155	223	295	3 802	451	316	1 081	1 272
31 Fishing area total	*325*	*366*	*155*	*223*	*295*	*3 802*	*451*	*316*	*1 081*	*1 272*
Species total	*325*	*366*	*155*	*223*	*295*	*3 802*	*451*	*316*	*1 081*	*1 272*
Green mussel	**Moule verte asiatique**		**Mejillón verde**		***Perna viridis***				**3,16(10)032,02**	**MSV**
04 Philippines	-	111	136	4	-	-	-	-	-	-
04 Fishing area total	*-*	*111*	*136*	*4*	*-*	*-*	*-*	*-*	*-*	*-*
71 Philippines	1 110	4 797	334	330	47	22	20	17	20	20
Singapore	88	-	-	188	-	-	-	-	-	-
Thailand	24 850	25 083	20 079	19 698	17 970	17 432	6 534	41 156	22 762	23 262
71 Fishing area total	*26 048*	*29 880*	*20 413*	*20 216*	*18 017*	*17 454*	*6 554*	*41 173*	*22 782*	*23 282*
Species total	*26 048*	*29 991*	*20 549*	*20 220*	*18 017*	*17 454*	*6 554*	*41 173*	*22 782*	*23 282*
Cholga mussel	**Moule cholga**		**Cholga**		***Aulacomya ater***				**3,16(10)038,01**	**MSC**
87 Chile	7 565	9 640	6 376	7 405	6 409	7 725	5 126	5 563	7 884	3 775
Peru	5 976	7 203	11 204	6 023	9 669	15 106	14 612	13 370	14 700	15 658
87 Fishing area total	*13 541*	*16 843*	*17 580*	*13 428*	*16 078*	*22 831*	*19 738*	*18 933*	*22 584*	*19 433*
Species total	*13 541*	*16 843*	*17 580*	*13 428*	*16 078*	*22 831*	*19 738*	*18 933*	*22 584*	*19 433*
Sea mussels nei	**Moules nca**		**Mejillones nep**		***Mytilidae***				**3,16(10)XXX,XX**	**MSX**
27 France	1 175	1 988	2 031	1 077	0	0	0	1 792	1 110	1 001
Iceland	0	0	0	-	-	-	-	-	-	-
Russian Fed	-	-	55	-	-	-	-	39	26	2
27 Fishing area total	*1 175*	*1 988*	*2 086*	*1 077*	*0*	*0*	*0*	*1 831*	*1 136*	*1 003*
34 Morocco	0	2	8	10	6	7	...	401	102	8
34 Fishing area total	*0*	*2*	*8*	*10*	*6*	*7*	*...*	*401*	*102*	*8*
41 Brazil	...	...	2 217	863	1 193	1 230	906	802	2 017	1 721
41 Fishing area total	*...*	*...*	*2 217*	*863*	*1 193*	*1 230*	*906*	*802*	*2 017*	*1 721*
61 Korea Rep	3 334	3 162	3 685	837	4 098	6 456	6 771	5 795	3 828	3 500
Russian Fed	107	-	28	-	-	9	80	24	42	33
61 Fishing area total	*3 441*	*3 162*	*3 713*	*837*	*4 098*	*6 465*	*6 851*	*5 819*	*3 870*	*3 533*
67 USA	709	689	1 317	455	...	...	1	675	448	251
67 Fishing area total	*709*	*689*	*1 317*	*455*	*...*	*...*	*1*	*675*	*448*	*251*
77 Mexico	414	291	118	412	2 036	2 056	737	547	733	691
USA	91	93	172	171	170	174	95	96	90	91
77 Fishing area total	*505*	*384*	*290*	*583*	*2 206*	*2 230*	*832*	*643*	*823*	*782*
81 New Zealand	...	...	191	450	0	664	2 977	4 467	2 270	1 725
81 Fishing area total	*...*	*...*	*191*	*450*	*0*	*664*	*2 977*	*4 467*	*2 270*	*1 725*
87 Ecuador	3	3	3	3	3	5	5	5	5	5 F
87 Fishing area total	*3*	*3*	*3*	*3*	*3*	*5*	*5*	*5*	*5*	*5 F*
Species total	*5 833*	*6 228*	*9 825*	*4 278*	*7 506*	*10 601*	*11 572*	*14 643*	*10 671*	*9 028 F*
Group total	***267 594***	***261 629***	***244 397***	***203 439***	***239 939***	***237 747***	***219 183***	***276 276***	***275 676***	***264 101***

B-55 Scallops, pectens / Coquilles St-Jacques / Vieiras

Capture production by species, fishing areas and countries or areas
Captures par espèces, zones de pêche et pays ou zones
Capturas por especies, áreas de pesca y países o áreas

Species, Fishing area Espèce, Zone de pêche Especie, Area de pesca	1993 t	1994 t	1995 t	1996 t	1997 t	1998 t	1999 t	2000 t	2001 t	2002 t
Queen scallop	**Vanneau**		**Volandeira**		***Aequipecten opercularis***			**3,16(08)001,05**		**QSC**
27 Channel Is	4 F	-	-	-	-	-	7	-	-	-
Faeroe Is	3 320	3 854	2 781	3 559	3 581	4 751	5 993	3 989	4 053	...
France	1 399	2 223	926	311	595	637	2 574	3 475	5 989	4 243
Ireland	55	27	11	3	7	5	29	3	13	58
Isle of Man	3 000	1 455	1 465	1 129	1 630	991	1 255	2 275	1 749	1 655
Portugal	0	0	0	0	-	-	-	-	-	1
UK	7 485	2 979	2 857	2 181	5 630	8 102	5 888	5 149	8 660	10 774
27 Fishing area total	*15 263 F*	*10 538*	*8 040*	*7 183*	*11 443*	*14 486*	*15 746*	*14 891*	*20 464*	*16 731*
Species total	*15 263 F*	*10 538*	*8 040*	*7 183*	*11 443*	*14 486*	*15 746*	*14 891*	*20 464*	*16 731*
Great Atlantic scallop	**Coquille St-Jacques atlantique**		**Vieira(=Concha de Santiago)**		***Pecten maximus***			**3,16(08)003,09**		**SCE**
27 Belgium	115	163	137	163	208	224	247	292	340	432
Channel Is	19 F	73	65 F	29	109	155	203	295	439	463
France	13 348	13 503	12 288	12 169	14 451	12 866	13 707	13 727	16 454	19 383
Ireland	543	918	423	560	633	693	1 497	1 579	1 413	1 140
Isle of Man	639	931	931	1 064	933	706	794	965	1 115	977
Netherlands	-	-	-	228	188	408	306	249	274	473
Norway	3	100	66	14	39	114	425	571	670	575
Portugal	-	4	-	1	2	0	0	0	1	-
Spain	206	282	185	675	391	299	86	508	84	223
UK	5 556	9 020	9 376	9 195	9 706	9 707	...	...	9 796	9 171
27 Fishing area total	*20 429 F*	*24 994*	*23 471 F*	*24 098*	*26 660*	*25 172*	*17 265*	*18 186*	*30 586*	*32 837*
37 France	...	...	...	...	...	...	12	18	4	6
37 Fishing area total	...	...	...	...	...	...	*12*	*18*	*4*	*6*
Species total	*20 429 F*	*24 994*	*23 471 F*	*24 098*	*26 660*	*25 172*	*17 277*	*18 204*	*30 590*	*32 843*
Great Mediterranean scallop	**Coquille St-Jacques méditerr.**		**Concha de peregrino**		***Pecten jacobaeus***			**3,16(08)003,11**		**SJA**
37 Turkey	202	308	23	52	95	50	68	570	150	470
37 Fishing area total	*202*	*308*	*23*	*52*	*95*	*50*	*68*	*570*	*150*	*470*
Species total	*202*	*308*	*23*	*52*	*95*	*50*	*68*	*570*	*150*	*470*
New Zealand scallop	**Pecten de Nouvelle Zélande**		**Vieira de Nueva Zelandia**		***Pecten novaezelandiae***			**3,16(08)003,13**		**SCZ**
81 New Zealand	8 928	9 088	14 160	5 080	18 848	4 592	6 152	2 912	6 792	4 408
81 Fishing area total	*8 928*	*9 088*	*14 160*	*5 080*	*18 848*	*4 592*	*6 152*	*2 912*	*6 792*	*4 408*
Species total	*8 928*	*9 088*	*14 160*	*5 080*	*18 848*	*4 592*	*6 152*	*2 912*	*6 792*	*4 408*
Delicate scallop	**...B**		**...C**		***Zygochlamis delicatula***			**3,16(08)012,01**		**ZYE**
81 New Zealand	...	141	135	124	201	91	128	0	222	118
81 Fishing area total	...	*141*	*135*	*124*	*201*	*91*	*128*	*0*	*222*	*118*
Species total	...	*141*	*135*	*124*	*201*	*91*	*128*	*0*	*222*	*118*
Patagonian scallop	**Peigne de Patagonie**		**Pecten patagónico**		***Zygochlamis patagonica***			**3,16(08)012,02**		**ZYP**
41 Argentina	25	0	10 592	36 952	39 817	28 441	42 700	36 514	38 960	50 966
Uruguay	...	...	...	...	...	...	...	890	3 638	215
41 Fishing area total	*25*	*0*	*10 592*	*36 952*	*39 817*	*28 441*	*42 700*	*37 404*	*42 598*	*51 181*
Species total	*25*	*0*	*10 592*	*36 952*	*39 817*	*28 441*	*42 700*	*37 404*	*42 598*	*51 181*
American sea scallop	**Pecten d'Amérique**		**Vieira americana**		***Placopecten magellanicus***			**3,16(08)014,04**		**SCA**
21 Canada	87 219	84 833	58 770	47 807	53 881	56 399	54 756	83 852	89 196	93 971
St Pier Mq	0	11	4	23	27	3	0	0	2	1
USA	56 137	60 616	61 651	61 093	47 966	42 880	77 179	113 026	164 762	186 336
21 Fishing area total	*143 356*	*145 460*	*120 425*	*108 923*	*101 874*	*99 282*	*131 935*	*196 878*	*253 960*	*280 308*
31 USA	565	572	805	459	154	150	27	115	236	264
31 Fishing area total	*565*	*572*	*805*	*459*	*154*	*150*	*27*	*115*	*236*	*264*
Species total	*143 921*	*146 032*	*121 230*	*109 382*	*102 028*	*99 432*	*131 962*	*196 993*	*254 196*	*280 572*
Calico scallop	**Peigne calicot**		**Peine percal**		***Argopecten gibbus***			**3,16(08)030,01**		**SCC**
31 USA	...	74 325	10 003	...	...	...	...	...	...	...
31 Fishing area total	...	*74 325*	*10 003*	...	...	...	...	...	...	...
Species total	...	*74 325*	*10 003*	...	...	...	...	...	...	...
Atlantic bay scallop	**Peigne baie**		**Peine caletero**		***Argopecten irradians***			**3,16(08)030,02**		**SCB**
21 USA	1 695	0	88	5	25	6	20	13	13	8
21 Fishing area total	*1 695*	*0*	*88*	*5*	*25*	*6*	*20*	*13*	*13*	*8*
31 USA	975	524	1 505	225	427	684	196	141	17	127
31 Fishing area total	*975*	*524*	*1 505*	*225*	*427*	*684*	*196*	*141*	*17*	*127*
Species total	*2 670*	*524*	*1 593*	*230*	*452*	*690*	*216*	*154*	*30*	*135*

B-55 Scallops, pectens / Coquilles St-Jacques / Vieiras

Capture production by species, fishing areas and countries or areas
Captures par espèces, zones de pêche et pays ou zones
Capturas por especies, áreas de pesca y países o áreas

Species, Fishing area Espèce, Zone de pêche Especie, Area de pesca	1993 t	1994 t	1995 t	1996 t	1997 t	1998 t	1999 t	2000 t	2001 t	2002 t
Peruvian calico scallop	**Pétoncle éventail**		**Ostión abanico**		***Argopecten purpuratus***				**3,16(08)030,03**	**SCQ**
87 Chile	-	5	-	9	4	21	0	20	272	55
Peru	2 732	837	3 113	2 086	4 009	23 525	30 141	11 810	6 272	2 031
87 Fishing area total	*2 732*	*842*	*3 113*	*2 095*	*4 013*	*23 546*	*30 141*	*11 830*	*6 544*	*2 086*
Species total	*2 732*	*842*	*3 113*	*2 095*	*4 013*	*23 546*	*30 141*	*11 830*	*6 544*	*2 086*
Pacific calico scallop	**Pétoncle volant**		**Peine volador**		***Argopecten ventricosus***				**3,16(08)030,04**	**SCH**
77 Mexico	5 883	8 570	1 256	17 290	2 320	2 726	1 864	6 287	3 100	9 875
77 Fishing area total	*5 883*	*8 570*	*1 256*	*17 290*	*2 320*	*2 726*	*1 864*	*6 287*	*3 100*	*9 875*
Species total	*5 883*	*8 570*	*1 256*	*17 290*	*2 320*	*2 726*	*1 864*	*6 287*	*3 100*	*9 875*
Iceland scallop	**Peigne islandais**		**Peine islándico**		***Chlamys islandica***				**3,16(08)036,03**	**ISC**
21 Canada	3 746	6 793	9 900	12 180	12 145	6 632	3 160	2 764	1 347	989
Greenland	1 566	2 028	5 287	1 373	1 886	2 211	...	1 630	1 593	2 457
Iceland	-	-	-	2	-	-	-	-	-	-
St Pier Mq	20	8	2	221	239	5	0	3	16	9
USA	0	0	0	-	-	-	-	-	-	-
21 Fishing area total	*5 332*	*8 829*	*15 189*	*13 776*	*14 270*	*8 848*	*3 160*	*4 397*	*2 956*	*3 455*
27 Iceland	11 466	8 401	8 381	8 976	10 403	10 098	8 858	9 074	6 499	5 192
Norway	-	-	-	-	-	-	-	14	14	16
27 Fishing area total	*11 466*	*8 401*	*8 381*	*8 976*	*10 403*	*10 098*	*8 858*	*9 088*	*6 513*	*5 208*
Species total	*16 798*	*17 230*	*23 570*	*22 752*	*24 673*	*18 946*	*12 018*	*13 485*	*9 469*	*8 663*
Weathervane scallop	**Pecten géant du Pacifique**		**Vieira gigante del Pacífico**		***Patinopecten caurinus***				**3,16(08)066,01**	**SCG**
67 USA	6 224	4 944	1 950	2 372	2	3 228	2 642	2 012	1 052	1 264
67 Fishing area total	*6 224*	*4 944*	*1 950*	*2 372*	*2*	*3 228*	*2 642*	*2 012*	*1 052*	*1 264*
Species total	*6 224*	*4 944*	*1 950*	*2 372*	*2*	*3 228*	*2 642*	*2 012*	*1 052*	*1 264*
Yesso scallop	**Pétoncle du Japon**		**Vieira japonesa**		***Patinopecten yessoensis***				**3,16(08)066,07**	**JSC**
61 China,Taiwan	59	58	22	76	63	114	112	90	47	25
Japan	223 844	270 890	274 879	271 124	261 164	287 802	299 628	304 286	290 974	306 666
Korea Rep	94	53	74	122	196	42	6	0	11	-
Russian Fed	2 783	3 289	4 407	5 084	5 534	6 253	5 764	5 728	2 236	3 612
61 Fishing area total	*226 780*	*274 290*	*279 382*	*276 406*	*266 957*	*294 211*	*305 510*	*310 104*	*293 268*	*310 303*
Species total	*226 780*	*274 290*	*279 382*	*276 406*	*266 957*	*294 211*	*305 510*	*310 104*	*293 268*	*310 303*
Scallops nei	**Peignes nca**		**Peines nep**		***Pectinidae***				**3,16(08)XXX,XX**	**SCX**
27 Channel Is	8 F	-	-	-	50	80	-	-	-	-
Denmark	0	0	0	0	0	0	0	0	0	0
France	233	258	213	268	185	149	328	313	430	262
Norway	10 252	7 967	7 310	3	16	21	12	4	13	37
Russian Fed	4 000	6 211	8 372	7 634	13 878	12 887	11 948	12 717	13 598	5 751
Spain	41	81	81	53	36	35	141	63	35	32
UK	3 771	5 000	6 297	7 421	8 995	10 371	19 108	19 507	9 722	9 553
27 Fishing area total	*18 305 F*	*19 517*	*22 273*	*15 379*	*23 160*	*23 543*	*31 537*	*32 604*	*23 798*	*15 635*
37 Spain	-	-	-	41	42	15	-	-	-	-
37 Fishing area total	-	-	-	*41*	*42*	*15*	-	-	-	-
57 Australia	27 240	17 735	7 411	6 261	6 087	5 250	7 396	8 303	4 421	2 525
Indonesia	68	3	730	606	1	290	225	-	-	-
Thailand	-	12	18	5	5	3	1	14	-	-
57 Fishing area total	*27 308*	*17 750*	*8 159*	*6 872*	*6 093*	*5 543*	*7 622*	*8 317*	*4 421*	*2 525*
71 Australia	6 384	6 738	5 869	6 054	2 501	4 623	4 177	3 711	4 776	3 133
Indonesia	195	291	373	536	423	547	614	578	419	480
Philippines	27	107	156	139	61	40	62	53	54	55
Thailand	-	534	588	547	503	1 182	550	259	652	666
71 Fishing area total	*6 606*	*7 670*	*6 986*	*7 276*	*3 488*	*6 392*	*5 403*	*4 601*	*5 901*	*4 334*
81 Australia	6	23	23	5	1	1	1	0	1	-
81 Fishing area total	*6*	*23*	*23*	*5*	*1*	*1*	*1*	*0*	*1*	-
87 Chile	1 332	1 225	1 365	1 577	2 598	3 662	1 715	332	141	373
87 Fishing area total	*1 332*	*1 225*	*1 365*	*1 577*	*2 598*	*3 662*	*1 715*	*332*	*141*	*373*
Species total	*53 557 F*	*46 185*	*38 806*	*31 150*	*35 382*	*39 156*	*46 278*	*45 854*	*34 262*	*22 867*
Group total	***503 412***	***618 011***	***537 324***	***535 166***	***532 891***	***554 767***	***612 702***	***660 700***	***702 737***	***741 516***

B-56 Clams, cockles, arkshells / Clams, coques, arches / Almejas, berberechos, arcas

Capture production by species, fishing areas and countries or areas / Captures par espèces, zones de pêche et pays ou zones / Capturas por especies, áreas de pesca y países o áreas

Species, Fishing area Espèce, Zone de pêche Especie, Area de pesca	1993 t	1994 t	1995 t	1996 t	1997 t	1998 t	1999 t	2000 t	2001 t	2002 t
Ark clams nei	**Arches nca**		**Arcas nep**		***Arca spp***			**3,16(04)001,XX**		**ARK**
31 Cuba	1 354	1 553	1 905	1 860	1 729	2 438	2 043	2 591	633	375
Venezuela	28 581	31 193	33 987	31 925	39 128	27 981	38 646	44 709	43 675	44 725
31 Fishing area total	*29 935*	*32 746*	*35 892*	*33 785*	*40 857*	*30 419*	*40 689*	*47 300*	*44 308*	*45 100*
61 Korea Rep	553	601	473	782	667	1 311	534	318	153	228
61 Fishing area total	*553*	*601*	*473*	*782*	*667*	*1 311*	*534*	*318*	*153*	*228*
77 Mexico	837	1 315	1 033	834	1 367	1 018	880	675	600	328
77 Fishing area total	*837*	*1 315*	*1 033*	*834*	*1 367*	*1 018*	*880*	*675*	*600*	*328*
Species total	*31 325*	*34 662*	*37 398*	*35 401*	*42 891*	*32 748*	*42 103*	*48 293*	*45 061*	*45 656*
Half-crenated ark	**Arche crénelée**		**Arca japonesa**		***Scapharca subcrenata***			**3,16(04)005,08**		**MCL**
61 Japan	16 689	17 373	15 426	16 328	14 133	10 120	10 413	7 308	4 899	9 828
61 Fishing area total	*16 689*	*17 373*	*15 426*	*16 328*	*14 133*	*10 120*	*10 413*	*7 308*	*4 899*	*9 828*
Species total	*16 689*	*17 373*	*15 426*	*16 328*	*14 133*	*10 120*	*10 413*	*7 308*	*4 899*	*9 828*
Blood cockle	**Arche granuleuse**		**Arca del Pacífico occidental**		***Anadara granosa***			**3,16(04)071,01**		**BLC**
57 Thailand	-	-	-	-	-	20	7	24	-	-
57 Fishing area total	-	-	-	-	-	*20*	*7*	*24*	-	-
61 Korea Rep	3 279	1 245	1 415	995	493	12 114	6 503	4 184	923	790
Russian Fed	-	-	-	-	-	-	-	-	100	650
61 Fishing area total	*3 279*	*1 245*	*1 415*	*995*	*493*	*12 114*	*6 503*	*4 184*	*1 023*	*1 440*
71 Thailand	-	-	-	-	-	421	1 750	948	948	969
71 Fishing area total	-	-	-	-	-	*421*	*1 750*	*948*	*948*	*969*
Species total	*3 279*	*1 245*	*1 415*	*995*	*493*	*12 555*	*8 260*	*5 156*	*1 971*	*2 409*
Anadara clams nei	**Arches anadara nca**		**Arcas anadara nep**		***Anadara spp***			**3,16(04)071,XX**		**BLS**
57 Indonesia	21 648	25 209	16 712	21 227	22 348	15 443	10 603	9 538	10 559	11 170
57 Fishing area total	*21 648*	*25 209*	*16 712*	*21 227*	*22 348*	*15 443*	*10 603*	*9 538*	*10 559*	*11 170*
71 Fiji Islands	446	454 F	1 513	1 044	1 950	2 800	2 990	2 750 F	2 884	2 870 F
Indonesia	20 149	21 738	26 949	24 956	19 571	16 148	23 307	25 157	53 749	56 880
Philippines	36	62	6	6	3	4	4	3	3	3
71 Fishing area total	*20 631*	*22 254 F*	*28 468*	*26 006*	*21 524*	*18 952*	*26 301*	*27 910 F*	*56 636*	*59 753 F*
Species total	*42 279*	*47 463 F*	*45 180*	*47 233*	*43 872*	*34 395*	*36 904*	*37 448 F*	*67 195*	*70 923 F*
Ocean quahog	**Cyprine d'Islande**		**Almeja de Islandia**		***Arctica islandica***			**3,16(09)045,01**		**CLQ**
21 Canada	-	-	124	114	-	-	66	-	164	92
USA	197 559	177 874	183 777	173 685	163 799	148 506	144 366	122 547	142 662	149 767
21 Fishing area total	*197 559*	*177 874*	*183 901*	*173 799*	*163 799*	*148 506*	*144 432*	*122 547*	*142 826*	*149 859*
27 Iceland	-	-	1 980	6 315	4 351	8 776	3 501	1 584	7 434	12 353
27 Fishing area total	-	-	*1 980*	*6 315*	*4 351*	*8 776*	*3 501*	*1 584*	*7 434*	*12 353*
Species total	*197 559*	*177 874*	*185 881*	*180 114*	*168 150*	*157 282*	*147 933*	*124 131*	*150 260*	*162 212*
Striped venus	**Petite praire**		**Chirla**		***Chamelea gallina***			**3,16(11)001,05**		**SVE**
27 France	573	580	651	706	810	715	790	696	627	798
Portugal	48	38	4	7	17	185	129	156	48	865
Spain	16	9	10	1 728	1 728	3 087	3 265	4 390	4 881	2 789
27 Fishing area total	*637*	*627*	*665*	*2 441*	*2 555*	*3 987*	*4 184*	*5 242*	*5 556*	*4 452*
37 Albania	50 F	20 F	0	0	-	-	-	-	-	-
Bulgaria	-	-	182	-	-	-	-	-	-	-
Italy	29 392	19 255	32 609	36 707	28 604	28 830	36 462	34 191	34 916	25 971
Slovenia	-	-	-	-	-	-	-	0	1	1
Spain	700 F	600 F	500 F	428	485	431	374	389	240	105
Turkey	30 134	31 869	11 864	10 925	7 150	3 550	3 585	10 000	7 500	10 000
37 Fishing area total	*60 276 F*	*51 744 F*	*45 155 F*	*48 060*	*36 239*	*32 811*	*40 421*	*44 580*	*42 657*	*36 077*
Species total	*60 913 F*	*52 371 F*	*45 820 F*	*50 501*	*38 794*	*36 798*	*44 605*	*49 822*	*48 213*	*40 529*
Pullet carpet shell	**Palourde bleue**		**Almeja babosa**		***Venerupis pullastra***			**3,16(11)003,01**		**CTS**
27 France	1 036	1 397	638	934	1 493	2 216	2 306	1 311	1 595	1 711
Portugal	213	573	225	153	419	240	212	87	81	73
Spain	350	432	442	1 734	2 141	2 710	2 310	963	744	1 034
27 Fishing area total	*1 599*	*2 402*	*1 305*	*2 821*	*4 053*	*5 166*	*4 828*	*2 361*	*2 420*	*2 818*
37 Spain	-	-	-	5	5	-	141	27	...	3
37 Fishing area total	-	-	-	*5*	*5*	-	*141*	*27*	...	*3*
Species total	*1 599*	*2 402*	*1 305*	*2 826*	*4 058*	*5 166*	*4 969*	*2 388*	*2 420*	*2 821*

B-56

Clams, cockles, arkshells — **Capture production by species, fishing areas and countries or areas**
Clams, coques, arches — **Captures par espèces, zones de pêche et pays ou zones**
Almejas, berberechos, arcas — **Capturas por especies, áreas de pesca y países o áreas**

Species, Fishing area Espèce, Zone de pêche Especie, Area de pesca	1993 t	1994 t	1995 t	1996 t	1997 t	1998 t	1999 t	2000 t	2001 t	2002 t
Stutchbury's venus	...B		...C		*Chione stutchburyi*				3,16(11)007,04	KNU
81 New Zealand	...	...	1 220	815	541	1 325	1 396	1 789	1 748	1 687
81 Fishing area total	...	...	*1 220*	*815*	*541*	*1 325*	*1 396*	*1 789*	*1 748*	*1 687*
Species total	...	...	*1 220*	*815*	*541*	*1 325*	*1 396*	*1 789*	*1 748*	*1 687*
Japanese hard clam	Cythérée du Japon		Mercenaria japonesa		*Meretrix lusoria*				3,16(11)017,01	HCJ
61 China,Taiwan	419	72	256	135	161	805	526	0	-	-
Japan	2 964	2 330	2 060	1 944	1 897	1 870	1 785	1 543	1 245	1 300
Korea Rep	1 407	1 561	2 715	2 315	2 075	3 472	1 799	1 430	1 044	1 704
61 Fishing area total	*4 790*	*3 963*	*5 031*	*4 394*	*4 133*	*6 147*	*4 110*	*2 973*	*2 289*	*3 004*
Species total	*4 790*	*3 963*	*5 031*	*4 394*	*4 133*	*6 147*	*4 110*	*2 973*	*2 289*	*3 004*
Hard clams nei	...B		...C		*Meretrix spp*				3,16(11)017,XX	HCX
57 Indonesia	1 392	2 126	3 074	3 693	3 722	6 639	3 420	2 606	2 383	2 730
57 Fishing area total	*1 392*	*2 126*	*3 074*	*3 693*	*3 722*	*6 639*	*3 420*	*2 606*	*2 383*	*2 730*
71 Indonesia	7 312	10 431	14 949	9 788	10 305	10 507	11 347	11 571	10 992	11 870
71 Fishing area total	*7 312*	*10 431*	*14 949*	*9 788*	*10 305*	*10 507*	*11 347*	*11 571*	*10 992*	*11 870*
Species total	*8 704*	*12 557*	*18 023*	*13 481*	*14 027*	*17 146*	*14 767*	*14 177*	*13 375*	*14 600*
Grooved carpet shell	Palourde croisée d'Europe		Almeja fina		*Ruditapes decussatus*				3,16(11)020,01	CTG
27 France	14	3	3	100	65	504	469	80	298	49
Ireland	...	...	20	21	23	3	3	3	130	-
Portugal	8 028	1 638	169	185	27	33	75	16	21	18
Spain	221	339	448	675	1 036	1 231	1 209	592	270	627
UK	-	-	-	-	2	-	-	-	29	-
27 Fishing area total	*8 263*	*1 980*	*640*	*981*	*1 153*	*1 771*	*1 756*	*691*	*748*	*694*
37 France	195	96	36	6	6	6	21	21	6	6
Spain	-	-	-	86	52	109	74	20	19	18
Tunisia	1 538	1 036	1 343	396	77	57	48	755	544	1 002
37 Fishing area total	*1 733*	*1 132*	*1 379*	*488*	*135*	*172*	*143*	*796*	*569*	*1 026*
Species total	*9 996*	*3 112*	*2 019*	*1 469*	*1 288*	*1 943*	*1 899*	*1 487*	*1 317*	*1 720*
Japanese carpet shell	Palourde japonaise		Almeja japonesa		*Ruditapes philippinarum*				3,16(11)020,02	CLJ
61 China,Taiwan	26	24	16	-	-	-	-	-	-	-
Japan	57 356	46 597	49 466	43 703	39 660	36 807	43 088	35 558	31 022	34 494
Korea Rep	31 202	14 595	15 041	12 392	16 854	14 585	13 963	20 982	20 004	14 758
61 Fishing area total	*88 584*	*61 216*	*64 523*	*56 095*	*56 514*	*51 392*	*57 051*	*56 540*	*51 026*	*49 252*
Species total	*88 584*	*61 216*	*64 523*	*56 095*	*56 514*	*51 392*	*57 051*	*56 540*	*51 026*	*49 252*
Gay's little venus	...B		Juliana		*Tawera gayi*				3,16(11)024,01	TWG
87 Chile	...	...	...	...	...	...	...	1	291	271
87 Fishing area total	...	...	...	...	...	...	...	*1*	*291*	*271*
Species total	...	...	...	...	...	...	...	*1*	*291*	*271*
Triangular tivela	...B		...C		*Tivela mactroides*				3,16(11)025,02	TVM
41 Brazil	...	...	...	126	196	664	1 684	273	280	835
41 Fishing area total	...	...	...	*126*	*196*	*664*	*1 684*	*273*	*280*	*835*
Species total	...	...	...	*126*	*196*	*664*	*1 684*	*273*	*280*	*835*
Butter clam	Coque jaune		Almeja amarilla		*Saxidomus giganteus*				3,16(11)037,02	BCL
67 Canada	1 352	1 797	1 728	1 431	1 206	1 600	945	949	1 111	1 370
USA	...	...	...	...	87	126	74	164	48	58
67 Fishing area total	*1 352*	*1 797*	*1 728*	*1 431*	*1 293*	*1 726*	*1 019*	*1 113*	*1 159*	*1 428*
Species total	*1 352*	*1 797*	*1 728*	*1 431*	*1 293*	*1 726*	*1 019*	*1 113*	*1 159*	*1 428*
Short neck clams nei	...B		...C		*Paphia spp*				3,16(11)041,XX	NCL
57 Thailand	2 738	5 070	10 800	18 329	11 625	6 204	27 366	23 130	20 364	20 060
57 Fishing area total	*2 738*	*5 070*	*10 800*	*18 329*	*11 625*	*6 204*	*27 366*	*23 130*	*20 364*	*20 060*
71 Philippines	517	1 498	30	31	2	1	1	2	2	2
Thailand	39 834	27 960	20 060	34 560	24 227	43 457	42 612	25 964	34 202	34 954
71 Fishing area total	*40 351*	*29 458*	*20 090*	*34 591*	*24 229*	*43 458*	*42 613*	*25 966*	*34 204*	*34 956*
81 New Zealand	...	...	317	211	114	204	181	131	202	212
81 Fishing area total	...	...	*317*	*211*	*114*	*204*	*181*	*131*	*202*	*212*
Species total	*43 089*	*34 528*	*31 207*	*53 131*	*35 968*	*49 866*	*70 160*	*49 227*	*54 770*	*55 228*
Pacific littleneck clam	Palourde commune		Almejuela común		*Protothaca staminea*				3,16(11)055,02	PTS
67 USA	-	-	-	-	-	-	-	-	102	146
67 Fishing area total	-	-	-	-	-	-	-	-	*102*	*146*

B-56 Clams, cockles, arkshells / Clams, coques, arches / Almejas, berberechos, arcas

Capture production by species, fishing areas and countries or areas / Captures par espèces, zones de pêche et pays ou zones / Capturas por especies, áreas de pesca y países o áreas

Species, Fishing area Espèce, Zone de pêche Especie, Area de pesca	1993 t	1994 t	1995 t	1996 t	1997 t	1998 t	1999 t	2000 t	2001 t	2002 t
Species total	-	-	-	-	-	-	-	-	*102*	*146*
Taca clam	**Palourde taca**		**Taca**		***Protothaca thaca***			**3,16(11)055,03**		**TCL**
87 Chile	23 068	16 236	17 162	20 016	12 475	24 254	16 429	16 303	26 483	5 360
87 Fishing area total	*23 068*	*16 236*	*17 162*	*20 016*	*12 475*	*24 254*	*16 429*	*16 303*	*26 483*	*5 360*
Species total	*23 068*	*16 236*	*17 162*	*20 016*	*12 475*	*24 254*	*16 429*	*16 303*	*26 483*	*5 360*
Northern quahog(=Hard clam)	**Praire**		**Chirla mercenaria**		***Mercenaria mercenaria***			**3,16(11)075,01**		**CLH**
21 Canada	1 127	1 170	830	1 109	1 656	1 234	2 536	1 252	1 657	2 519
USA	23 026	10 562	16 058	21 855	4 631	...	...	8 217	5 023	12 045
21 Fishing area total	*24 153*	*11 732*	*16 888*	*22 964*	*6 287*	*1 234*	*2 536*	*9 469*	*6 680*	*14 564*
27 UK	-	-	-	-	0	0	-	175	-	-
27 Fishing area total	-	-	-	-	*0*	*0*	-	*175*	-	-
31 USA	...	...	...	...	...	...	...	5 506	5 113	4 441
31 Fishing area total	...	...	...	...	...	...	...	*5 506*	*5 113*	*4 441*
Species total	*24 153*	*11 732*	*16 888*	*22 964*	*6 287*	*1 234*	*2 536*	*15 150*	*11 793*	*19 005*
Venus clams nei	**Petites praires nca**		**Almejas(=Veneridos) nep**		***Veneridae***			**3,16(11)XXX,XX**		**CLV**
27 France	-	-	-	1	99	-	189	234	289	355
27 Fishing area total	-	-	-	*1*	*99*	-	*189*	*234*	*289*	*355*
31 Mexico	637	805	1 000	814	746	1 001	844	1 403	1 834	646
Venezuela	354	487	517	536	533	295	378	353	388	262
31 Fishing area total	*991*	*1 292*	*1 517*	*1 350*	*1 279*	*1 296*	*1 222*	*1 756*	*2 222*	*908*
37 France	130	64	-	-	-	-	18	18	22	12
37 Fishing area total	*130*	*64*	-	-	-	-	*18*	*18*	*22*	*12*
77 Honduras	4	...	5	...	3	...	0	...	...	...
Mexico	4 882	5 134	6 239	5 038	3 912	3 672	4 486	4 835	3 582	2 930
77 Fishing area total	*4 886*	*5 134*	*6 244*	*5 038*	*3 915*	*3 672*	*4 486*	*4 835*	*3 582*	*2 930*
87 Ecuador	0	0	0	0	0	0	0	0	0	-
87 Fishing area total	*0*	*0*	*0*	*0*	*0*	*0*	*0*	*0*	*0*	-
Species total	*6 007*	*6 490*	*7 761*	*6 389*	*5 293*	*4 968*	*5 915*	*6 843*	*6 115*	*4 205*
Imperial surf clam	**Clam**		**Almeja**		***Pseudocardium sybillae***			**3,16(12)001,02**		**HCL**
61 Japan	7 768	8 003	8 018	8 738	7 541	8 227	8 808	8 883	8 914	8 861
Korea Rep	8 137	11 383	9 038	6 053	7 179	2 877	7 620	4 150	3 409	3 464
61 Fishing area total	*15 905*	*19 386*	*17 056*	*14 791*	*14 720*	*11 104*	*16 428*	*13 033*	*12 323*	*12 325*
Species total	*15 905*	*19 386*	*17 056*	*14 791*	*14 720*	*11 104*	*16 428*	*13 033*	*12 323*	*12 325*
Pacific horse clams nei	**...B**		**...C**		***Tresus spp***			**3,16(12)005,XX**		**TQZ**
67 USA	-	-	-	-	-	-	-	1	2	9
67 Fishing area total	-	-	-	-	-	-	-	*1*	*2*	*9*
Species total	-	-	-	-	-	-	-	*1*	*2*	*9*
Atlantic surf clam	**Mactre solide**		**Almeja blanca**		***Spisula solidissima***			**3,16(12)020,01**		**CLB**
21 Canada	885	921	674	1 204	800	444	303	496	602	613
USA	179 354	167 600	155 382	154 380	140 621	131 256	142 067	165 269	165 708	175 709
21 Fishing area total	*180 239*	*168 521*	*156 056*	*155 584*	*141 421*	*131 700*	*142 370*	*165 765*	*166 310*	*176 322*
Species total	*180 239*	*168 521*	*156 056*	*155 584*	*141 421*	*131 700*	*142 370*	*165 765*	*166 310*	*176 322*
Stimpson's surf clam	**Douceron de Stimpson**		**...C**		***Spisula polynyma***			**3,16(12)020,02**		**CLT**
21 Canada	19 930	20 735	24 008	25 597	27 322	25 976	26 699	22 979	20 257	19 960
USA	0	0	9	15	43	32	23	6	16	19
21 Fishing area total	*19 930*	*20 735*	*24 017*	*25 612*	*27 365*	*26 008*	*26 722*	*22 985*	*20 273*	*19 979*
Species total	*19 930*	*20 735*	*24 017*	*25 612*	*27 365*	*26 008*	*26 722*	*22 985*	*20 273*	*19 979*
Solid surf clam	**Spisule épaisse**		**...C**		***Spisula solida***			**3,16(12)020,05**		**ULO**
27 Denmark	-	-	-	-	-	-	-	55	214	1 709
France	...	...	...	...	...	...	...	127	101	101
Portugal	-	-	-	-	-	-	765	1 153	1 125	1 220
27 Fishing area total	...	...	...	...	...	...	*765*	*1 335*	*1 440*	*3 030*
Species total	...	...	...	...	...	...	*765*	*1 335*	*1 440*	*3 030*
Taquilla clams	**Mactres taquille**		**Taquillas**		***Mulinia spp***			**3,16(12)040,XX**		**MUN**
87 Chile	...	2 567	1 852	999	2 757	2 549	1 536	1 491	1 699	7 034
87 Fishing area total	...	*2 567*	*1 852*	*999*	*2 757*	*2 549*	*1 536*	*1 491*	*1 699*	*7 034*

B-56 Clams, cockles, arkshells / Clams, coques, arches / Almejas, berberechos, arcas

Capture production by species, fishing areas and countries or areas / Captures par espèces, zones de pêche et pays ou zones / Capturas por especies, áreas de pesca y países o áreas

Species, Fishing area Espèce, Zone de pêche Especie, Area de pesca	1993 t	1994 t	1995 t	1996 t	1997 t	1998 t	1999 t	2000 t	2001 t	2002 t
Species total	*...*	*2 567*	*1 852*	*999*	*2 757*	*2 549*	*1 536*	*1 491*	*1 699*	*7 034*
Donax clams	**Olives de mer**		**Coquinas**		***Donax spp***				**3,16(15)002,XX**	**DON**
27 France	124	292	197	-	-	-	34	94	88	68
Portugal	...	...	...	...	...	...	456	401	540	347
Spain	-	-	-	-	-	-	-	-	-	12
27 Fishing area total	*124*	*292*	*197*	*...*	*...*	*...*	*490*	*495*	*628*	*427*
41 Uruguay	0	0	0	0	0	0	0	0	0	-
41 Fishing area total	*0*	*0*	*0*	*0*	*0*	*0*	*0*	*0*	*0*	*-*
47 South Africa	0	0	0	0	-	-	-	-	-	36
47 Fishing area total	*0*	*0*	*0*	*0*	*-*	*-*	*-*	*-*	*-*	*36*
Species total	*124*	*292*	*197*	*0*	*0*	*0*	*490*	*495*	*628*	*463*
Razor clams nei	**Couteaux nca**		**Navajas(=Solénidos) nep**		***Solen spp***				**3,16(16)003,XX**	**RAZ**
27 Ireland	15	-	-	-	28	316	407	334	201	167
Portugal	1 106	1 225	2 603	1 729	124	15	4	12	214	195
Spain	100	80	21	86	205	166	171	50	129	-
UK	-	41	46	56	220	134	114	67	59	37
27 Fishing area total	*1 221*	*1 346*	*2 670*	*1 871*	*577*	*631*	*696*	*463*	*603*	*399*
37 France	...	...	...	...	...	...	5	5	5	5
Spain	-	-	-	2	4	1	-	-	-	-
37 Fishing area total	*...*	*...*	*...*	*2*	*4*	*1*	*5*	*5*	*5*	*5*
Species total	*1 221*	*1 346*	*2 670*	*1 873*	*581*	*632*	*701*	*468*	*608*	*404*
Atl.jackknife(=Atl.razor clam)	**Couteau de l'Atlantique**		**Navaja del Atlántico**		***Ensis directus***				**3,16(16)005,02**	**CLR**
21 USA	6	-	-	-	14	49	64	99	36	90
21 Fishing area total	*6*	*-*	*-*	*-*	*14*	*49*	*64*	*99*	*36*	*90*
Species total	*6*	*-*	*-*	*-*	*14*	*49*	*64*	*99*	*36*	*90*
Pacific razor clam	**Couteau du Pacifique**		**Navaja del Pacífico**		***Siliqua patula***				**3,16(16)007,01**	**RAP**
67 USA	-	-	-	-	-	-	-	14	61	146
67 Fishing area total	*-*	*-*	*-*	*-*	*-*	*-*	*-*	*14*	*61*	*146*
Species total	*-*	*-*	*-*	*-*	*-*	*-*	*-*	*14*	*61*	*146*
Sand gaper	**Mye des sables**		**Almeja de can**		***Mya arenaria***				**3,16(17)006,01**	**CLS**
21 Canada	2 152	3 270	3 841	1 469	2 323	2 638	2 680	3 026	3 152	3 073
USA	9 146	5 190	4 813	4 355	4 454	5 493	5 113	5 222	6 794	5 662
21 Fishing area total	*11 298*	*8 460*	*8 654*	*5 824*	*6 777*	*8 131*	*7 793*	*8 248*	*9 946*	*8 735*
67 USA	-	-	-	-	-	-	-	-	123	516
67 Fishing area total	*-*	*-*	*-*	*-*	*-*	*-*	*-*	*-*	*123*	*516*
Species total	*11 298*	*8 460*	*8 654*	*5 824*	*6 777*	*8 131*	*7 793*	*8 248*	*10 069*	*9 251*
Pacific geoduck	**Panopée du Pacifique**		**Panopea del Pacífico**		***Panopea abrupta***				**3,16(18)089,01**	**GEC**
67 Canada	2 455	2 235	2 056	1 768	1 757	1 784	1 688	1 562	1 413	1 706
USA	-	-	-	-	1 962	2 243	2 451	2 259	2 481	3 061
67 Fishing area total	*2 455*	*2 235*	*2 056*	*1 768*	*3 719*	*4 027*	*4 139*	*3 821*	*3 894*	*4 767*
Species total	*2 455*	*2 235*	*2 056*	*1 768*	*3 719*	*4 027*	*4 139*	*3 821*	*3 894*	*4 767*
Common edible cockle	**Coque commune**		**Berberecho común**		***Cerastoderma edule***				**3,16(23)002,03**	**COC**
27 Denmark	543	31	-	5	2 603	1 993	246	2 089	2 392	78
France	4 565	2 152	379	664	731	418	481	81	11	18
Ireland	17	26	20	10	64	296	1	8	6	112
Netherlands	43 635	38 350	39 594	6 300	10 923	68 133	50 888	19 633	-	-
Norway	-	-	-	-	-	-	-	38	33	19
Portugal	488	531	591	3 522	1 285	1 264	1 409	1 292	683	3 518
Spain	331	573	627	2 358	2 964	2 472	3 104	3 740	1 486	737
UK	21 360	22 330	21 796	24 176	19 493	12 035	14 123	20 306	19 048	14 268
27 Fishing area total	*70 939*	*63 993*	*63 007*	*37 035*	*38 063*	*86 611*	*70 252*	*47 187*	*23 659*	*18 750*
34 Senegal	...	...	178	54	69	139	147	117	105	152
34 Fishing area total	*...*	*...*	*178*	*54*	*69*	*139*	*147*	*117*	*105*	*152*
37 Spain	-	-	-	-	-	-	1	-	-	-
37 Fishing area total	*-*	*-*	*-*	*-*	*-*	*-*	*1*	*-*	*-*	*-*
Species total	*70 939*	*63 993*	*63 185*	*37 089*	*38 132*	*86 750*	*70 400*	*47 304*	*23 764*	*18 902*
Basket cockle	**...B**		**...C**		***Clinocardium nuttallii***				**3,16(23)004,02**	**KCL**
67 USA	-	-	-	-	22	-	18	49	67	80
67 Fishing area total	*-*	*-*	*-*	*-*	*22*	*-*	*18*	*49*	*67*	*80*

B-56

Clams, cockles, arkshells — Capture production by species, fishing areas and countries or areas
Clams, coques, arches — Captures par espèces, zones de pêche et pays ou zones
Almejas, berberechos, arcas — Capturas por especies, áreas de pesca y países o áreas

Species, Fishing area Espèce, Zone de pêche Especie, Area de pesca	1993 t	1994 t	1995 t	1996 t	1997 t	1998 t	1999 t	2000 t	2001 t	2002 t
Species total	-	-	-	-	*22*	-	*18*	*49*	*67*	*80*
Cockles nei	**Coques nca**		**Berberechos(=Cárdidos) nep**		***Cardiidae***				**3,16(23)XXX,XX**	**COZ**
61 Korea Rep	11 226	6 551	428	2 583	1 587	1 078	112	411	2 002	810
61 Fishing area total	*11 226*	*6 551*	*428*	*2 583*	*1 587*	*1 078*	*112*	*411*	*2 002*	*810*
Species total	*11 226*	*6 551*	*428*	*2 583*	*1 587*	*1 078*	*112*	*411*	*2 002*	*810*
Pipi wedge clam	**...B**		**...C**		***Paphies australis***				**3,16(24)002,01**	**AFQ**
57 Australia	304	380	351	454	830	1 041	976	1 085	1 251	1 085
57 Fishing area total	*304*	*380*	*351*	*454*	*830*	*1 041*	*976*	*1 085*	*1 251*	*1 085*
81 Australia	315	246	254	248	465	455	513	628	624	475
81 Fishing area total	*315*	*246*	*254*	*248*	*465*	*455*	*513*	*628*	*624*	*475*
Species total	*619*	*626*	*605*	*702*	*1 295*	*1 496*	*1 489*	*1 713*	*1 875*	*1 560*
Macha clam	**Mesodème chilienne**		**Macha**		***Mesodesma donacium***				**3,16(24)039,01**	**CLM**
87 Chile	8 274	6 415	6 913	6 144	6 770	6 464	1 728	1 249	1 396	1 303
Peru	1 513	1 070	1 200	1 060	1 061	578	-	10	0	85
87 Fishing area total	*9 787*	*7 485*	*8 113*	*7 204*	*7 831*	*7 042*	*1 728*	*1 259*	*1 396*	*1 388*
Species total	*9 787*	*7 485*	*8 113*	*7 204*	*7 831*	*7 042*	*1 728*	*1 259*	*1 396*	*1 388*
Chilean semele	**Sémèle chilienne**		**Almeja blanca chilena**		***Semele solida***				**3,16(39)001,03**	**TUW**
87 Chile	-	4 613	2 523	4 418	2 199	1 900	2 071	4 212	3 054	2 162
87 Fishing area total	-	*4 613*	*2 523*	*4 418*	*2 199*	*1 900*	*2 071*	*4 212*	*3 054*	*2 162*
Species total	-	*4 613*	*2 523*	*4 418*	*2 199*	*1 900*	*2 071*	*4 212*	*3 054*	*2 162*
Clams, etc. nei	**Clams, etc. nca**		**Almejas, etc. nep**		***Bivalvia***				**3,16(XX)XXX,XX**	**CLX**
04 Indonesia	1 775	763	835	844	956	914	304	303	792	850
04 Fishing area total	*1 775*	*763*	*835*	*844*	*956*	*914*	*304*	*303*	*792*	*850*
21 Canada	642	0	263	773	259	49	791	345	1 078	2 459
USA	0	789	0	-	2	17	345	4	51	66
21 Fishing area total	*642*	*789*	*263*	*773*	*261*	*66*	*1 136*	*349*	*1 129*	*2 525*
27 Channel Is	1 F	76	70 F	-	-	-	-	-	-	-
Denmark	1 651	2 716	3 136	-	-	-	-	-	-	-
France	2 370	1 465	2 637	3 746	2 893	5 822	4 273	3 482	4 005	4 595
Germany	1 301	1 034	5 410	-	-	-	-	-	-	-
Ireland	-	-	-	-	-	-	-	301	126	83
Portugal	3 367	4 656	1 489	1 482	741	1 007	493	395	398	322
Spain	1 412	3 251	3 210	1 758	2 459	2 967	2 761	2 000	1 850	2 494
UK	246	37	326	49	142	13	67	15	91	43
27 Fishing area total	*10 348 F*	*13 235*	*16 278 F*	*7 035*	*6 235*	*9 809*	*7 594*	*6 193*	*6 470*	*7 537*
31 Colombia	5	7	1	0	-	0	-	...	...	...
Martinique	...	...	...	...	...	...	...	900	700	700
USA	9 999	16 012	22 297	4 455	6 739	6 392	3 972	254	2 096	2 002
31 Fishing area total	*10 004*	*16 019*	*22 298*	*4 455*	*6 739*	*6 392*	*3 972*	*1 154*	*2 796*	*2 702*
34 Spain	-	-	-	-	-	-	1	3	7	19
34 Fishing area total	-	-	-	-	-	-	*1*	*3*	*7*	*19*
37 France	-	-	-	-	-	-	365	798	691	514
Morocco	10	0	0	0	0	0	-	-	-	-
Spain	3 000 F	3 500 F	4 000 F	4 078	2 963	3 527	2 155	1 179	985	1 283
37 Fishing area total	*3 010 F*	*3 500 F*	*4 000 F*	*4 078*	*2 963*	*3 527*	*2 520*	*1 977*	*1 676*	*1 797*
41 Spain	-	-	-	-	-	-	759	-	-	-
Uruguay	9	14	...	...	0	0	0	6	17	240
41 Fishing area total	*9*	*14*	*...*	*...*	*0*	*0*	*759*	*6*	*17*	*240*
51 Mozambique	5	11	38	57	8	19	26	30	11	19
51 Fishing area total	*5*	*11*	*38*	*57*	*8*	*19*	*26*	*30*	*11*	*19*
57 Malaysia	18 424	8 670	15 250	17 745	24 663	5 097	572	5 115	2 983	2 606
57 Fishing area total	*18 424*	*8 670*	*15 250*	*17 745*	*24 663*	*5 097*	*572*	*5 115*	*2 983*	*2 606*
61 Japan	57 728	42 515	50 015	52 361	42 987	47 584	35 319	35 529	30 133	28 198
Korea Rep	30 057	33 207	31 063	33 446	8 851	15 665	14 728	12 277	15 705	11 453
Russian Fed	-	-	-	-	-	-	-	216	317	200
61 Fishing area total	*87 785*	*75 722*	*81 078*	*85 807*	*51 838*	*63 249*	*50 047*	*48 022*	*46 155*	*39 851*
67 USA	10 550	10 244	11 215	3 912	381	2 509	1 601	1 519	1 250	1 892
67 Fishing area total	*10 550*	*10 244*	*11 215*	*3 912*	*381*	*2 509*	*1 601*	*1 519*	*1 250*	*1 892*
71 Fiji Islands	27	28 F	92	54	45	52	50	40 F	43	45 F
Malaysia	3 816	722	732	861	837	904	1 105	1 116	966	648
Philippines	1 343	4 282	921	725	375	227	277	222	223	225
Solomon Is	...	...	...	60	280 F	10 F	10 F	5 F	5 F	5 F

B-56 Clams, cockles, arkshells — Capture production by species, fishing areas and countries or areas
Clams, coques, arches — Captures par espèces, zones de pêche et pays ou zones
Almejas, berberechos, arcas — Capturas por especies, áreas de pesca y países o áreas

Species, Fishing area Espèce, Zone de pêche Especie, Area de pesca	1993 t	1994 t	1995 t	1996 t	1997 t	1998 t	1999 t	2000 t	2001 t	2002 t
71 Fishing area total	*5 186*	*5 032 F*	*1 745*	*1 700*	*1 537 F*	*1 193 F*	*1 442 F*	*1 383 F*	*1 237 F*	*923 F*
77 Panama	1 043	758	1 669	1 391	1 507	1 072	1 518	1 630	1 661	1 124
USA	16	4	4	2	3	8	29	3	5	4
77 Fishing area total	*1 059*	*762*	*1 673*	*1 393*	*1 510*	*1 080*	*1 547*	*1 633*	*1 666*	*1 128*
87 Chile	22 216	21 510	20 075	18 720	17 880	11 900	20 804	16 551	16 775	12 723
Colombia	5	1	0	4	5	6	6	10 F	10 F	10 F
Ecuador	2	2	6	10	10	10	10	10	10	10 F
Peru	668	643	569	411	236	152	338	956	949	978
87 Fishing area total	*22 891*	*22 156*	*20 650*	*19 145*	*18 131*	*12 068*	*21 158*	*17 527 F*	*17 744 F*	*13 721 F*
Species total	*171 688 F*	*156 917 F*	*175 323 F*	*146 944*	*115 222 F*	*105 923 F*	*92 679 F*	*85 214 F*	*83 933 F*	*75 810 F*
Group total	***1 058 823***	***948 748***	***960 722***	***919 100***	***814 048***	***838 118***	***841 658***	***798 339***	***824 201***	***825 651***

B-57 Squids, cuttlefishes, octopuses — Capture production by species, fishing areas and countries or areas

Encornets, seiches, poulpes — Captures par espèces, zones de pêche et pays ou zones

Calamares, jibias, pulpos — Capturas por especies, áreas de pesca y países o áreas

Species, Fishing area Espèce, Zone de pêche Especie, Area de pesca	1993 t	1994 t	1995 t	1996 t	1997 t	1998 t	1999 t	2000 t	2001 t	2002 t
Common cuttlefish	**Seiche commune**		**Sepia común**		***Sepia officinalis***			**3,21(02)002,02**		**CTC**
27 Belgium	206	288	467	386	154	252	222	463	370	741
Channel Is	4 F	2	2 F	12	10	15	22	26	8	11
Portugal	1 209	1 125	987	1 636	1 422	1 734	1 162	1 357	1 348	1 368
Spain	-	-	-	176	336	1 078	1 411	1 895	833	712
27 Fishing area total	*1 419 F*	*1 415*	*1 456 F*	*2 210*	*1 922*	*3 079*	*2 817*	*3 741*	*2 559*	*2 832*
34 Cyprus	-	-	-	-	-	-	-	1	0	-
Greece	1 349	1 125	900	654	588	472	391	157	615	1 210
Portugal	166	360	377	440	218	131	157	75	115	110
34 Fishing area total	*1 515*	*1 485*	*1 277*	*1 094*	*806*	*603*	*548*	*233*	*730*	*1 320*
37 Albania	...	...	39	33	33	51	51	50	22	52
Algeria	500 F	600 F	351	567	619	492	312	392	340	318
Cyprus	174	218	153	111	89	146	140	122	106	105
France	152	130	100	85	85	85	88	106	83	96
Greece	1 941	2 853	2 516	1 819	2 381	1 807	2 732	1 609	1 623	1 550 F
Slovenia	21	4	10	6	5	18	18	11	72	22
Tunisia	6 315	5 121	3 517	4 340	6 479	4 935	6 622	6 002	7 148	7 995
Turkey	526	717	933	644	900	750	537	550	465	909
37 Fishing area total	*9 629 F*	*9 643 F*	*7 619*	*7 605*	*10 591*	*8 284*	*10 500*	*8 842*	*9 859*	*11 047 F*
Species total	*12 563 F*	*12 543 F*	*10 352 F*	*10 909*	*13 319*	*11 966*	*13 865*	*12 816*	*13 148*	*15 199 F*
Cuttlefish,bobtail squids nei	**Seiches, sépioles nca**		**Sepias,choquitos,globitos nep**		***Sepiidae, Sepiolidae***			**3,21(02)XXX,XX**		**CTL**
27 Denmark	5	2	3	2	0	-	26	20	5	22
France	13 832	11 932	14 642	14 532	11 982	13 015	14 465	18 939	13 814	16 105
Netherlands	-	-	-	-	-	-	-	101	162	381
Spain	1 267	1 196	1 588	971	1 621	1 446	1 322	256	65	305
UK	2 164	2 089	3 631	4 607	2 202	2 760	2 260	3 076	2 705	3 535
27 Fishing area total	*17 268*	*15 219*	*19 864*	*20 112*	*15 805*	*17 221*	*18 073*	*22 392*	*16 751*	*20 348*
31 Korea Rep	-	-	-	-	-	-	32	-	-	-
31 Fishing area total	-	-	-	-	-	-	*32*	-	-	-
34 Belize	460	80	141	-	406	2 445	4 911	4 869	19	...
Benin	...	...	...	...	...	...	18	12 F	1	5
Cameroon	0	0	2	2	2 F	2 F	2 F	1	1	1
Chile	-	9	-	-	-	-	-	-	-	-
China	3 082	500	8 111	6 823	7 485	5 686	5 528	4 930	8 238	-
Congo Rep	0	0	1 F	3 F	2 F	7 F	9 F	10	3 F	3 F
Côte dIvoire	-	-	-	-	-	-	81	291	140	...
Gabon	27	56	33	157	52	186	92	281	166	512
Gambia	182	62	325	184	137	98	100	422	1 499	696
Ghana	1 673	2 396	2 891	2 967	3 355	3 288	4 095	1 805	2 866	3 649
Guinea	184	31	61	41	95	386	181	124	236	...
GuineaBissau	...	31	17	2	2 F	2 F	2 F	2 F	2 F	2 F
Honduras	2 336	291	2 873	3 108	2 076	889	1 077	687	-	-
Italy	2 017	2 017	1 017	790	2 016	1 491	1 360	791	1 239	919
Korea D P Rp	25	-	-	-	-	-	-	61	-	-
Korea Rep	184	228	288	169	112	510	215	29	487	17
Latvia	-	-	-	-	-	-	-	-	-	17
Liberia	14	8	-	-	-	12	8	14	29	30 F
Mauritania	4 150 F	4 470 F	5 710 F	4 510 F	5 240 F	5 000 F	4 129 F	4 344 F	4 981 F	2 342 F
Morocco	9 074	14 209	12 131	12 500	16 625	14 536	20 011	32 725	17 544	2 243
Nigeria	...	...	...	3	...	...	...	...	189	419
Panama	253	-	78	0	53	82	-	-	-	-
Russian Fed	21	30	-	-	-	-	-	-	-	-
St Vincent	84	-	2	-	-	-	-	-	2	-
Senegal	6 380	6 945	6 399	5 919	6 864	6 365	5 709	3 953	3 941	3 986
Sierra Leone	835	127	602	1 621	1 626	600	512	294	574	1 138
Spain	2 500 F	2 000 F	1 500 F	702	658	18	3 727	1 262	2 761	461
Togo	32	9	1	12	77	51	5	0	20	0
Vanuatu	22	13	22	33	19	11	-	-	-	-
Other nei	-	-	-	-	-	-	138	154	148	-
34 Fishing area total	*33 535 F*	*33 512 F*	*42 205 F*	*39 546 F*	*46 902 F*	*41 665 F*	*51 910 F*	*57 061 F*	*45 086 F*	*16 440 F*
37 Croatia	203	118	118	102	151	215	145	138	85	45
Egypt	1 067	1 031	1 097	1 365	1 370	1 152	1 449	1 503	1 554	1 734
Italy	7 682	13 616	11 150	7 974	7 382	6 839	5 720	5 534	6 131	4 076
Lebanon	25	25	25	25	50	25	50	25	25	25
Malta	1	1	0	5	3	3	5	4	0	-
Morocco	215	139	265	178	162	132	157	179	18	220
Palest, O.T.	...	...	41	62	98	144	145 F	120 F	80 F	55
Serbia-Monte	6	5	9	10	9	10	10	10	10	10
Spain	1 000 F	900 F	800 F	1 245	1 220	1 317	1 431	1 311	1 142	972
37 Fishing area total	*10 199 F*	*15 835 F*	*13 505 F*	*10 966*	*10 445*	*9 837*	*9 112 F*	*8 824 F*	*9 045 F*	*7 137*
47 Angola	32	-	-	-	-	-	-	-	-	-
China	-	-	-	-	-	-	-	18	-	-
Honduras	36	-	4	-	-	-	-	-	-	-
Italy	-	-	14	11	-	-	-	-	-	-
Korea Rep	44	-	-	-	-	-	-	-	-	-
Russian Fed	-	-	-	-	-	-	-	-	5	-
Spain	-	-	-	113	-	88	33	72	26	101
47 Fishing area total	*112*	-	*18*	*124*	-	*88*	*33*	*90*	*31*	*101*

B-57 Squids, cuttlefishes, octopuses / Encornets, seiches, poulpes / Calamares, jibias, pulpos

Capture production by species, fishing areas and countries or areas / Captures par espèces, zones de pêche et pays ou zones / Capturas por especies, áreas de pesca y países o áreas

Species, Fishing area Espèce, Zone de pêche Especie, Area de pesca	1993 t	1994 t	1995 t	1996 t	1997 t	1998 t	1999 t	2000 t	2001 t	2002 t
51 Bahrain	138	101	113	285	126	139	48	85	104	128
Egypt	286	486	560	399	414	237	3 525	1 043	1 365	854
Eritrea	...	...	...	51	0	3	11	31	35	9
Italy	403	403	208	162	-	-	-	-	-	-
Korea Rep	1 307	1 006	814	1 179	2 730	1 314	1 003	601	840	721
Oman	1 948	2 584	2 945	5 036	6 148	4 080	7 478	2 891	5 673	8 061
Pakistan	2 945	3 356	2 455	3 308	4 528	3 225	5 146	5 307	5 256	5 302
Saudi Arabia	35	353	308	578	598	656	760	739	796	748
South Africa	11	-	-	-	9	5	10 F	16	16	19
Spain	-	-	-	-	-	-	1	-	-	2
Untd Arab Em	20	23	22	25	26	27	28	491	465	205
Yemen	906	31	1 457	1 884	8 657	5 092	5 292	8 917	9 330	10 412
51 Fishing area total	*7 999*	*8 343*	*8 882*	*12 907*	*23 236*	*14 778*	*23 302 F*	*20 121*	*23 880*	*26 461*
57 Australia	28	42	50	56	56	56	62	32	48	46
Indonesia	2 229	1 845	2 316	4 019	3 221	5 120	5 328	4 848	7 477	9 070
Korea Rep	-	-	-	-	8	-	399	200	189	352
Malaysia	6 283	5 720	7 036	8 014	8 923	13 329	9 570	13 492	9 810	9 778
Sri Lanka	200	250	300	300	300	300	365	310	290	380
Thailand	15 911	14 333	16 561	23 286	23 244	18 493	21 238	22 367	21 419	21 296
57 Fishing area total	*24 651*	*22 190*	*26 263*	*35 675*	*35 752*	*37 298*	*36 962*	*41 249*	*39 233*	*40 922*
61 China	118 536	193 046	213 772	166 319	235 946	221 834	210 775	255 958	310 129	296 836
China,H.Kong	1 209	1 151	1 689	991	841	510	400 F	450 F	500 F	490 F
China,Taiwan	9 536	10 959	12 084	7 191	8 621	7 198	5 019	3 880	4 546	2 405
Japan	9 837	9 926	9 643	9 545	8 052	9 790	9 942	8 241	8 297	7 873
Korea Rep	5 452	3 090	2 567	1 484	2 082	2 563	6 652	1 267	1 443	1 920
61 Fishing area total	*144 570*	*218 172*	*239 755*	*185 530*	*255 542*	*241 895*	*232 788 F*	*269 796 F*	*324 915 F*	*309 524 F*
71 Australia	3	3	5	4	6	6	8	9	7	6
Indonesia	3 044	3 444	3 623	4 335	4 980	6 353	7 003	7 097	8 312	9 450
Korea Rep	262	145	22	-	14	70	8	10	7	9
Malaysia	9 034	9 382	8 440	10 762	11 668	12 853	12 498	12 822	11 657	13 823
NewCaledonia	6	1	3	3	1	2	3	1	0	0
Philippines	2 196	2 188	2 836	2 702	2 803	2 204	2 093	2 016	2 320	2 475
Singapore	237	233	196	214	235	179	142	134	56	57
Thailand	44 456	41 987	45 358	47 239	48 344	44 847	45 009	44 927	46 378	46 437
71 Fishing area total	*59 238*	*57 383*	*60 483*	*65 259*	*68 051*	*66 514*	*66 764*	*67 016*	*68 737*	*72 257*
81 Australia	328	392	392	378	374	251	381	261	237	245
81 Fishing area total	*328*	*392*	*392*	*378*	*374*	*251*	*381*	*261*	*237*	*245*
Species total	*297 900 F*	*371 046 F*	*411 367 F*	*370 497 F*	*456 107 F*	*429 547 F*	*439 357 F*	*486 810 F*	*527 915 F*	*493 435 F*
Patagonian squid	**Calmar de Patagonie**		**Calamar patagónico**		***Loligo gahi***			**3,21(04)001,02**		**SQP**
41 Australia	-	-	-	-	-	3 198	2 486	-	-	-
Belize	-	-	-	-	-	-	-	2	-	-
Falkland Is	1 494	5 266	22 310	24 366	12 710	32 029	22 502	50 270	42 909	18 598
France	-	1 636	7 245	4 394	1 512	4 146	2 309	2 024	-	-
Honduras	1	1	31	8	-	-	-	-	-	-
Namibia	-	-	-	-	74	1	0	-	-	-
Panama	-	0	1	1	-	-	-	-	-	-
Poland	3 811	593	-	-	-	19	4 875	-	-	-
St Vincent	-	-	-	-	-	-	-	-	1 795	-
Seychelles	-	-	-	-	1 114	88	-	-	-	-
Spain	39 012	46 736	53 671	35 692	3 967	8 933	8 185	9 392	9 011	3 673
UK	1	1 228	1 916	4 043	2 334	3 337	2 148	5 328	4 015	2 705
41 Fishing area total	*44 319*	*55 460*	*85 174*	*68 504*	*21 711*	*51 751*	*42 505*	*67 016*	*57 730*	*24 976*
Species total	*44 319*	*55 460*	*85 174*	*68 504*	*21 711*	*51 751*	*42 505*	*67 016*	*57 730*	*24 976*
Opalescent inshore squid	**Calmar opale**		**Calamar opalescente**		***Loligo opalescens***			**3,21(04)001,03**		**SQO**
77 USA	...	...	...	...	70 915	2 709	90 662	117 711	85 829	72 879
77 Fishing area total	...	...	...	...	*70 915*	*2 709*	*90 662*	*117 711*	*85 829*	*72 879*
Species total	...	...	...	...	*70 915*	*2 709*	*90 662*	*117 711*	*85 829*	*72 879*
Longfin squid	**Calmar totam**		**Calamar pálido**		***Loligo pealei***			**3,21(04)001,05**		**SQL**
21 USA	22 146	22 469	18 887	12 490	16 161	18 879	18 749	16 942	14 211	16 684
21 Fishing area total	*22 146*	*22 469*	*18 887*	*12 490*	*16 161*	*18 879*	*18 749*	*16 942*	*14 211*	*16 684*
31 USA	54	33	39	-	-	-	-	-	-	-
31 Fishing area total	*54*	*33*	*39*	-	-	-	-	-	-	-
Species total	*22 200*	*22 502*	*18 926*	*12 490*	*16 161*	*18 879*	*18 749*	*16 942*	*14 211*	*16 684*
Cape Hope squid	**Calmar du Cap**		**Calamar del Cabo**		***Loligo reynaudi***			**3,21(04)001,12**		**CHO**
47 South Africa	6 271	5 814	7 047	7 491	3 696	6 670	7 169	6 000 F	3 373	7 405
Spain	-	-	-	58	-	-	0	-	-	-
47 Fishing area total	*6 271*	*5 814*	*7 047*	*7 549*	*3 696*	*6 670*	*7 169*	*6 000 F*	*3 373*	*7 405*
51 Spain	-	-	-	-	-	-	-	-	-	1
51 Fishing area total	-	-	-	-	-	-	-	-	-	*1*
Species total	*6 271*	*5 814*	*7 047*	*7 549*	*3 696*	*6 670*	*7 169*	*6 000 F*	*3 373*	*7 406*

B-57 Squids, cuttlefishes, octopuses / Encornets, seiches, poulpes / Calamares, jibias, pulpos

Capture production by species, fishing areas and countries or areas
Captures par espèces, zones de pêche et pays ou zones
Capturas por especies, áreas de pesca y países o áreas

Species, Fishing area Espèce, Zone de pêche Especie, Area de pesca	1993 t	1994 t	1995 t	1996 t	1997 t	1998 t	1999 t	2000 t	2001 t	2002 t
Common squids nei	**Calmars nca**		**Calamares nep**		*Loligo spp*				**3,21(04)001,XX**	**SQC**
27 Belgium	84	57	45	21	17	25	23	59	51	137
Channel Is	1 F	2	2 F	1	6	5	11	9	1	2
Isle of Man	15	6	7	3	2	2	2	-	1	0
Portugal	879	616	1 267	787	1 514	1 499	701	1 003	1 273	1 090
Spain	1 251	1 280	1 135	1 015	2 520	2 190	1 376	1 371	1 473	505
UK	1 655	1 947	2 272	3 264	3 004	3 039	2 611	1 757	850	1 002
27 Fishing area total	*3 885 F*	*3 908*	*4 728 F*	*5 091*	*7 063*	*6 760*	*4 724*	*4 199*	*3 649*	*2 736*
31 Dominican Rp	24	3	37	38	24	84	12	56	51	23
Mexico	572	111	69	112	140	71	91	85	92	58
Venezuela	1 384	1 633	756	771	719	787	102	463	1 261	637
31 Fishing area total	*1 980*	*1 747*	*862*	*921*	*883*	*942*	*205*	*604*	*1 404*	*718*
34 Italy	2 689	2 689	188	2	11	-	-	-	151	9
Portugal	122	0	2	5	5	0	3	3	1	2
34 Fishing area total	*2 811*	*2 689*	*190*	*7*	*16*	*0*	*3*	*3*	*152*	*11*
37 Albania	...	...	7	47	34	93	93	90	65	85
Algeria	250 F	300 F	101	53	138	202	240	212	224	172
Croatia	330	268	273	233	287	290	170	127	162	53
France	220	136	150	135	465	465	241	264	177	177
Greece	1 369	1 101	945	856	623	426	397	560	405	380 F
Italy	4 372	5 282	5 483	5 366	4 141	2 237	1 909	1 890	2 201	1 776
Malta	0	0	0	2	2	2	2	3	2	2
Palest, O.T.	...	...	15	23	30	61	60 F	60 F	60 F	56
Serbia-Monte	10	7	13	12	13	13	12	13	14	13
Slovenia	4	4	2	2	4	3	2	3	8	3
Spain	800 F	700 F	700 F	505	462	405	1 002	691	806	900
Tunisia	214	277	240	230	253	288	348	310	327	510
Turkey	397	579	331	364	420	500	360	400	230	330
37 Fishing area total	*7 966 F*	*8 654 F*	*8 260 F*	*7 828*	*6 872*	*4 985*	*4 836 F*	*4 623 F*	*4 681 F*	*4 457 F*
41 Argentina	957	1 073	914	184	1 999	802	209	268	251	64
Brazil	1 310 F	1 320 F	1 317	950	1 486	758	1 696	1 187	1 186	2 137
Italy	359	359	22	0	-	-	-	-	-	-
Japan	4 118	5 683	5 561	3 187	1 562	2 618	1 847	-	-	4 106
Portugal	1 762	2 296	3 805	3 984	48	0	-	-	-	-
41 Fishing area total	*8 506 F*	*10 731 F*	*11 619*	*8 305*	*5 095*	*4 178*	*3 752*	*1 455*	*1 437*	*6 307*
51 China,Taiwan	2 500 F	2 100 F	758 F	-	-	-	-	-	-	-
Italy	538	538	38	0	-	-	-	-	-	-
Qatar	46	34	31	30	37	37	10	22	41	11
51 Fishing area total	*3 084 F*	*2 672 F*	*827 F*	*30*	*37*	*37*	*10*	*22*	*41*	*11*
57 China,Taiwan	100 F	100 F	36 F	-	-	-	-	-	-	-
Indonesia	3 777	3 846	9 656	8 000	8 359	10 498	10 204	9 748	12 100	14 860
Thailand	16 295	16 464	18 485	23 229	23 208	24 120	20 522	21 532	22 139	22 011
57 Fishing area total	*20 172 F*	*20 410 F*	*28 177 F*	*31 229*	*31 567*	*34 618*	*30 726*	*31 280*	*34 239*	*36 871*
61 China,Taiwan	20 500 F	17 000 F	6 143 F	14 853	19 221	20 074	11 083	7 325	5 791	6 232
Korea Rep	-	-	-	-	-	-	1 629	885	898	-
61 Fishing area total	*20 500 F*	*17 000 F*	*6 143 F*	*14 853*	*19 221*	*20 074*	*12 712*	*8 210*	*6 689*	*6 232*
71 China,Taiwan	45 000 F	37 400 F	13 518 F	9 868	9 181	6 853	4 815	2 248	3 308	4 445
Indonesia	17 137	22 370	17 919	19 169	33 396	21 352	26 503	30 090	48 429	57 770
Philippines	55 789	48 948	56 415	52 458	54 155	48 678	47 115	46 778	41 964	50 612
Singapore	246	1 000	679	546	470	462	376	348	186	185
Thailand	55 867	55 762	59 624	56 006	55 740	68 788	62 613	64 671	60 941	61 019
71 Fishing area total	*174 039 F*	*165 480 F*	*148 155 F*	*138 047*	*152 942*	*146 133*	*141 422*	*144 135*	*154 828*	*174 031*
87 Ecuador	85	90	90	90	100	100	100	100	100	100 F
Peru	1 316	1 215	7 766	10 250	3 806	287	1 353	24 548	18 738	6 490
87 Fishing area total	*1 401*	*1 305*	*7 856*	*10 340*	*3 906*	*387*	*1 453*	*24 648*	*18 838*	*6 590 F*
Species total	*244 344 F*	*234 596 F*	*216 817 F*	*216 651*	*227 602*	*218 114*	*199 843 F*	*219 179 F*	*225 958 F*	*237 964 F*
Neon flying squid	**Encornet volant**		**Pota saltadora**		*Ommastrephes bartrami*				**3,21(05)003,01**	**OFJ**
61 Japan	...	...	...	...	47 852	53 638	34 551	46 199	21 578	22 000 F
Russian Fed	-	-	-	-	-	-	-	405	100	483
61 Fishing area total	*...*	*...*	*...*	*...*	*47 852*	*53 638*	*34 551*	*46 604*	*21 678*	*22 483 F*
67 Japan	-	-	-	-	2 008	1 200	1 521	683	2 061	...
67 Fishing area total	*-*	*-*	*-*	*-*	*2 008*	*1 200*	*1 521*	*683*	*2 061*	*...*
77 Japan	-	-	-	-	10	113	4	81	131	...
77 Fishing area total	*-*	*-*	*-*	*-*	*10*	*113*	*4*	*81*	*131*	*...*
Species total	*...*	*...*	*...*	*...*	*49 870*	*54 951*	*36 076*	*47 368*	*23 870*	*22 483 F*
Northern shortfin squid	**Encornet rouge nordique**		**Pota norteña**		*Illex illecebrosus*				**3,21(05)010,01**	**SQI**
21 Canada	2 802	5 778	999	8 822	15 775	1 944	314	368	65	253
Cuba	2 354	4 036	959	493	2 978	1 084	280	-	-	-
Honduras	13	-	-	-	-	-	-	-	-	-
Japan	-	-	-	1	-	-	-	-	-	-

B-57

Squids, cuttlefishes, octopuses — **Capture production by species, fishing areas and countries or areas**
Encornets, seiches, poulpes — **Captures par espèces, zones de pêche et pays ou zones**
Calamares, jibias, pulpos — **Capturas por especies, áreas de pesca y países o áreas**

Species, Fishing area Espèce, Zone de pêche Especie, Area de pesca	1993 t	1994 t	1995 t	1996 t	1997 t	1998 t	1999 t	2000 t	2001 t	2002 t
Portugal	-	-	-	4	-	1	-	-	-	-
Russian Fed	92	-	-	-	-	29	-	12	-	-
St Pier Mq	3	10	0	0	6	-	-	-	-	2
Spain	-	3	-	1	1	3	-	-	-	-
USA	18 224	18 419	14 202	17 041	13 630	22 715	7 334	9 010	4 009	2 750
21 Fishing area total	*23 488*	*28 246*	*16 160*	*26 362*	*32 390*	*25 776*	*7 928*	*9 390*	*4 074*	*3 005*
27 Denmark	-	-	-	-	17	16	-	-	-	-
Faeroe Is	-	1	-	-	2	32	23	...	...	...
Germany	0	3	11	-	-	-	-	-	-	-
Ireland	...	...	...	...	...	...	...	...	121	114
Spain	2 089	2 272	2 359	2 600	2 426	762	1 932	1 831	1 503	2 406
27 Fishing area total	*2 089*	*2 276*	*2 370*	*2 600*	*2 445*	*810*	*1 955*	*1 831*	*1 624*	*2 520*
31 USA	5	148	20	9	2	2	-	1	0	0
31 Fishing area total	*5*	*148*	*20*	*9*	*2*	*2*	*-*	*1*	*0*	*0*
Species total	*25 582*	*30 670*	*18 550*	*28 971*	*34 837*	*26 588*	*9 883*	*11 222*	*5 698*	*5 525*
Broadtail shortfin squid	**Encornet rouge**		**Pota voladora**		***Illex coindetii***				**3,21(05)010,02**	**SQM**
27 France	762	702	571	395	411	216	304	360	219	450
27 Fishing area total	*762*	*702*	*571*	*395*	*411*	*216*	*304*	*360*	*219*	*450*
37 France	...	...	...	...	...	...	34	42	31	77
37 Fishing area total	*...*	*...*	*...*	*...*	*...*	*...*	*34*	*42*	*31*	*77*
Species total	*762*	*702*	*571*	*395*	*411*	*216*	*338*	*402*	*250*	*527*
Argentine shortfin squid	**Encornet rouge argentin**		**Pota argentina**		***Illex argentinus***				**3,21(05)010,03**	**SQA**
41 Argentina	193 690	196 893	199 048	292 628	411 994	291 174	342 691	279 046	230 272	177 314
Australia	-	-	-	-	-	-	167	-	-	-
Belize	-	-	-	-	-	-	3 796	4 066	1 692	124
Bulgaria	1 062	-	-	-	-	-	-	-	-	-
Cambodia	-	-	-	-	-	-	-	2 768	1 200	32
Chile	841	2	302	-	-	-	-	-	-	-
China	-	-	-	-	-	30 000	61 000	93 130	93 500	85 436
China,Taiwan	123 700 F	104 480 F	100 371	101 328	185 775	163 180	264 089	238 334	146 783	110 887
Estonia	-	-	-	-	-	-	-	-	1 833	533
Falkland Is	-	-	351	196	37	804	2 582	716	1 879	140
France	-	106	23	28	0	0	56	0	-	-
Honduras	556	1 089	679	-	-	-	-	-	-	-
Japan	131 707	92 838	75 691	73 896	126 504	77 452	154 129	119 966	70 652	7 800
Korea Rep	128 581	79 130	124 005	144 750	208 160	92 397	271 716	150 149	142 585	98 649
Namibia	-	-	-	-	3	0	63	-	-	-
Poland	6 421	2 390	282	1	-	-	-	970	683	2 754
Portugal	4	170	353	640	712	1 531	-	-	1 049	2 355
Russian Fed	47 309	23 557	11 762	20 254	884	-	-	3 404	2 578	5 453
St Vincent	-	-	-	-	-	-	-	-	4	-
Spain	776	3 021	3 889	17 091	25 374	23 829	30 694	26 140	41 318	7 750
UK	-	-	-	-	-	-	336	6	21	-
Uruguay	3 806	2 022	4 182	5 669	20 857	13 175	13 679	12 144	7 373	11 811
41 Fishing area total	*638 453 F*	*505 698 F*	*520 938*	*656 481*	*980 300*	*693 542*	*1 144 998*	*930 839*	*743 422*	*511 038*
48 Poland	-	-	-	-	-	-	-	-	18	49
48 Fishing area total	*-*	*-*	*-*	*-*	*-*	*-*	*-*	*-*	*18*	*49*
Species total	*638 453 F*	*505 698 F*	*520 938*	*656 481*	*980 300*	*693 542*	*1 144 998*	*930 839*	*743 440*	*511 087*
Jumbo flying squid	**Encornet géant**		**Jibia gigante**		***Dosidicus gigas***				**3,21(05)023,01**	**GIS**
77 Japan	-	-	-	13 096	19 988	-	348	25 904	211	...
Mexico	3 043	1 800	39 657	107 967	120 877	26 611	57 985	56 153	73 741	115 896
USA	...	...	...	...	3	107	18	1	0	4
77 Fishing area total	*3 043*	*1 800*	*39 657*	*121 063*	*140 868*	*26 718*	*58 351*	*82 058*	*73 952*	*115 900*
87 China	-	-	-	-	-	-	-	-	-	50 483
Japan	55 800	84 205	36 515	1 201	3 191	-	2 957	30 921	72 201	72 201 F
Korea Rep	57 778	66 386	34 440	11 784	2 384	201	18 813	15 625	5 797	21 382
Peru	7 769	42 838	25 676	8 138	16 061	547	54 652	53 795	71 834	146 390
87 Fishing area total	*121 347*	*193 429*	*96 631*	*21 123*	*21 636*	*748*	*76 422*	*100 341*	*149 832*	*290 456 F*
Species total	*124 390*	*195 229*	*136 288*	*142 186*	*162 504*	*27 466*	*134 773*	*182 399*	*223 784*	*406 356 F*
European flying squid	**Toutenon commun**		**Pota europea**		***Todarodes sagittatus***				**3,21(05)058,01**	**SQE**
27 Iceland	-	-	11	3	5	4	3	1	-	0
Ireland	...	...	...	...	...	...	...	...	14	40
Norway	-	-	352	0	190	2	0	0	-	-
Spain	-	-	-	973	2 508	2 137	2 481	1 866	1 588	1 735
UK	-	-	-	-	18	293	204	186	193	171
27 Fishing area total	*...*	*...*	*363*	*976*	*2 721*	*2 436*	*2 688*	*2 053*	*1 795*	*1 946*
37 France	11	21	19	87	87	87	-	-	-	-
Italy	6 451	5 737	4 789	4 672	2 614	3 995	2 056	2 516	2 346	2 034
Malta	-	-	-	-	-	2	2	2	0	-
Spain	-	-	-	140	149	119	144	338	120	1 217
37 Fishing area total	*6 462*	*5 758*	*4 808*	*4 899*	*2 850*	*4 203*	*2 202*	*2 856*	*2 466*	*3 251*

B-57 Squids, cuttlefishes, octopuses / Encornets, seiches, poulpes / Calamares, jibias, pulpos

Capture production by species, fishing areas and countries or areas / Captures par espèces, zones de pêche et pays ou zones / Capturas por especies, áreas de pesca y países o áreas

Species, Fishing area Espèce, Zone de pêche Especie, Area de pesca	1993 t	1994 t	1995 t	1996 t	1997 t	1998 t	1999 t	2000 t	2001 t	2002 t
Species total	*6 462*	*5 758*	*5 171*	*5 875*	*5 571*	*6 639*	*4 890*	*4 909*	*4 261*	*5 197*
Japanese flying squid	**Toutenon japonais**		**Pota japonesa**		***Todarodes pacificus***				**3,21(05)058,03**	**SQJ**
61 China,Taiwan	10 422 F	10 917	22 243	19 101	12 430	34 840	11 261	6 833	4 716	4 215
Japan	315 934	301 651	290 273	444 189	365 978	180 749	237 346	337 285	298 191	273 567 F
Korea Rep	222 009	191 857	200 897	252 618	224 959	163 016	249 280	226 309	225 616	226 656
61 Fishing area total	*548 365 F*	*504 425*	*513 413*	*715 908*	*603 367*	*378 605*	*497 887*	*570 427*	*528 523*	*504 438 F*
Species total	*548 365 F*	*504 425*	*513 413*	*715 908*	*603 367*	*378 605*	*497 887*	*570 427*	*528 523*	*504 438 F*
Wellington flying squid	**Encornet minami**		**Pota neozelandesa**		***Nototodarus sloani***				**3,21(05)059,01**	**TSQ**
81 China,Taiwan	6 000 F	7 000 F	8 284	14 747	6 620	3 974	761	-	-	-
Japan	8 072	10 180	19 687	11 342	5 182	3 734	1 853	1 853	1 396	988
New Zealand	25 530	51 841	59 497	23 474	44 845	42 541	27 282	20 878	35 100	50 016
Ukraine	5 546	10 428	6 630	4 136	7 955	5 321	1 462	2 872	8 623	11 230
81 Fishing area total	*45 148 F*	*79 449 F*	*94 098*	*53 699*	*64 602*	*55 570*	*31 358*	*25 603*	*45 119*	*62 234*
Species total	*45 148 F*	*79 449 F*	*94 098*	*53 699*	*64 602*	*55 570*	*31 358*	*25 603*	*45 119*	*62 234*
Sevenstar flying squid	**Encornet étoile**		**Pota festoneada**		***Martialia hyadesi***				**3,21(05)060,01**	**SQS**
41 Argentina	-	-	404	-	-	-	-	-	-	-
China,Taiwan	1 239 F	392	23 464	3 792	8 348	-	-	-	-	-
Falkland Is	-	-	0	0	0	-	0	0	-	-
Honduras	-	-	118	-	-	-	-	-	-	-
Japan	-	-	-	-	-	2	27	33	2	...
Poland	13	-	-	-	-	-	-	-	-	-
Spain	0	0	-	1	0	0	0	-	-	-
41 Fishing area total	*1 252 F*	*392*	*23 986*	*3 793*	*8 348*	*2*	*27*	*33*	*2*	*...*
48 Korea Rep	-	-	-	52	81	-	-	-	2	-
48 Fishing area total	-	-	-	*52*	*81*	-	-	-	*2*	-
Species total	*1 252 F*	*392*	*23 986*	*3 845*	*8 429*	*2*	*27*	*33*	*4*	*...*
Various squids nei	**Calmars, encornets nca**		**Calamares, jibias, potas nep**		***Loliginidae, Ommastrephidae***				**3,21(05)XXX,XX**	**SQU**
21 USA	762	1 292	531	56	182	69	90	70	124	196
21 Fishing area total	*762*	*1 292*	*531*	*56*	*182*	*69*	*90*	*70*	*124*	*196*
27 France	6 881	6 140	6 017	3 852	4 413	4 289	5 946	5 399	4 690	6 358
Germany	3	-	-	2	4	11	6	5	3	17
Ireland	364	277	1 338	481	442	610	282	135	242	356
Netherlands	-	-	-	-	-	-	-	773	171	312
Spain	-	-	-	-	-	1 459	2 267	1 022	307	234
Sweden	4	0	0	0	1	1	1	-	0	1
UK	543	273	666	13	8	8	4	3	815	1 346
27 Fishing area total	*7 795*	*6 690*	*8 021*	*4 348*	*4 868*	*6 378*	*8 506*	*7 337*	*6 228*	*8 624*
31 Belize	-	-	-	4	-	-	-	-	-	-
Colombia	45	50	27	29	20	8	-	...	...	...
Cuba	...	...	...	...	-	0	-	-	-	-
Grenada	0	0	0	0	0	0	0	0	5	3
Honduras	13	22	28	...	77	1	1	1	...	...
Korea Rep	-	-	-	-	-	-	-	1	-	-
Trinidad Tob	-	-	-	-	-	0	2	1	9	10
USA	-	-	-	-	42	110	42	62	74	51
31 Fishing area total	*58*	*72*	*55*	*33*	*139*	*119*	*45*	*65*	*88*	*64*
34 Belize	-	-	48	-	21	114	128	31	1	...
China	18	6	57	229	66	200	237	333	290	-
Cyprus	-	-	-	-	-	-	9	1	-	-
Gabon	...	29	3	5	2	0	1	8	10	10
Gambia	126	124	1	-	-	-	-	-	-	-
Georgia	10 F	-	-	-	-	-	-	-	-	-
Greece	342	269	55	31	-	1	11	16	11	13
Guinea	-	-	16	1	2	53	27	9	...	...
GuineaBissau	-	-	8	-	-	-	-	-	-	-
Honduras	1 003	160	615	212	52	29	23	-	-	-
Italy	673	686	377	307	36	54	101	95	11	105
Korea Rep	80	26	-	-	57	93	115	13	55	48
Latvia	-	-	-	-	-	-	-	-	-	12
Lithuania	-	-	...	...	...	233	2	-	-	-
Mauritania	250 F	190 F	260 F	230 F	280 F	1 472 F	2 359 F	2 354 F	2 223 F	1 206 F
Morocco	18 487	12 313	19 872	20 000	8 987	11 350	8 079	15 426	10 622	2 752
Panama	49	-	3	8	2	1	-	-	-	-
Portugal	7	0	0	0	0	2	0	9	2	-
Russian Fed	-	-	94	53	-	-	-	-	-	-
St Vincent	-	-	-	-	-	-	-	-	4	-
Senegal	...	...	7	49	92	17	140	152	107	125
Sierra Leone	401	114	166	121	90	143	...	...	...	...
Spain	5 000 F	6 500 F	8 000 F	8 614	3 770	7 062	11 707	1 637	2 548	1 863
Other nei	2	-	-	-	-	-	79	91	77	-
34 Fishing area total	*26 448 F*	*20 417 F*	*29 582 F*	*29 860 F*	*13 457 F*	*20 824 F*	*23 018 F*	*20 175 F*	*15 961 F*	*6 134 F*
37 France	86	67	...	43	43	43	-	-	-	-

B-57 Squids, cuttlefishes, octopuses / Encornets, seiches, poulpes / Calamares, jibias, pulpos

Capture production by species, fishing areas and countries or areas / Captures par espèces, zones de pêche et pays ou zones / Capturas por especies, áreas de pesca y países o áreas

Species, Fishing area Espèce, Zone de pêche Especie, Area de pesca	1993 t	1994 t	1995 t	1996 t	1997 t	1998 t	1999 t	2000 t	2001 t	2002 t
Greece	567	678	906	812	513	566	675	870	739	700 F
Morocco	25	122	32	10	2	2	3	6	11	4
Spain	-	-	-	592	547	747	673	790	302	237
37 Fishing area total	*678*	*867*	*938*	*1 457*	*1 105*	*1 358*	*1 351*	*1 666*	*1 052*	*941 F*
41 Estonia	-	-	-	2 425	-	-	-	-	-	-
Italy	90	92	45	37	-	-	-	-	-	-
Korea Rep	-	-	-	-	415	-	-	-	-	-
Latvia	4 608	6 280	1 717	3 956	-	-	-	-	-	-
Lithuania	15 171	3 262	-	3 400	-	-	-	-	-	-
Ukraine	4 719	3 202	998	339	-	-	-	-	-	-
41 Fishing area total	*24 588*	*12 836*	*2 760*	*10 157*	*415*	-	-	-	-	-
47 Angola	228	191	122	42	...	390	463	442	292	500
Honduras	6	-	1	-	-	-	-	-	-	-
Iceland	-	-	-	-	4	0	-	-	-	-
Italy	-	-	8	4	-	-	-	-	-	-
Korea Rep	142	317	79	-	176	-	158	601	-	-
Namibia	83	22	16	26	34	42	19	28	775	993
Russian Fed	-	-	-	-	-	-	-	-	-	1
South Africa	59	36	17	34	45	29	24	40 F	53	40
Spain	-	-	-	-	-	-	1	-	1	-
47 Fishing area total	*518*	*566*	*243*	*106*	*259*	*461*	*665*	*1 111 F*	*1 121*	*1 534*
51 Djibouti	0	0	0	0	0	0	0	0	0	0
Eritrea	...	13	15	0	0	0	5	38	85	19
Italy	134	137	77	63	-	-	-	-	-	-
Kenya	32	57	345	389	317	30	35	42	78	74
Korea Rep	52	114	465	210	439	757	917	441	612	749
Mozambique	188	106	142	366	602	677	616	538	395	534
Pakistan	3 317	3 126	2 832	2 600	4 460	3 300	5 062	4 070	4 024	3 200
Portugal	-	-	-	-	-	-	-	1	3	-
South Africa	6	-	-	-	2	...	...	...	...	5
Spain	-	-	-	-	-	-	76	-	-	-
Ukraine	-	-	-	-	-	-	3	-	-	-
51 Fishing area total	*3 729*	*3 553*	*3 876*	*3 628*	*5 820*	*4 764*	*6 714*	*5 130*	*5 197*	*4 581*
57 Australia	840	1 315	2 111	930	2 657	1 083	2 745	1 816	3 272	1 723
Korea Rep	-	-	-	-	1	-	44	13	235	277
Malaysia	12 041	11 011	13 990	14 568	18 028	16 873	18 164	28 039	20 128	20 779
57 Fishing area total	*12 881*	*12 326*	*16 101*	*15 498*	*20 686*	*17 956*	*20 953*	*29 868*	*23 635*	*22 779*
61 China	...	...	...	...	...	117 000	132 000	123 855	81 000	87 056
China,H.Kong	13 613	13 700	10 043	8 559	13 586	8 340	6 000 F	7 300 F	8 100 F	7 900 F
Japan	61 767	91 145	110 441	121 718	63 116	56 371	56 575	49 607	46 451	30 565
Korea D P Rp	...	4 635	3 164	8 892	14 192	10 000 F	10 000 F	9 500 F	9 500 F	9 500 F
Korea Rep	187	12 624	14 928	3 573	1 915	8 471	3 357	-	893	991
Russian Fed	24 820	15 750	33 842	33 260	54 913	50 557	54 756	69 835	44 249	72 537
61 Fishing area total	*100 387*	*137 854*	*172 418*	*176 002*	*147 722*	*250 739 F*	*262 688 F*	*260 097 F*	*190 193 F*	*208 549 F*
67 Korea Rep	-	948	-	-	-	311	1 330	-	-	-
USA	193	410	225	488	571	521	221	6	852	665
67 Fishing area total	*193*	*1 358*	*225*	*488*	*571*	*832*	*1 551*	*6*	*852*	*665*
71 Australia	211	172	133	167	508	442	512	290	594	376
Japan	6	3	-	-	-	-	-	-	-	-
Korea Rep	1 531	2 901	2 948	1 785	2 125	1 423	2 478	700	1 535	1 888
Malaysia	20 581	24 913	17 264	21 702	20 463	21 824	22 119	26 300	25 149	31 704
71 Fishing area total	*22 329*	*27 989*	*20 345*	*23 654*	*23 096*	*23 689*	*25 109*	*27 290*	*27 278*	*33 968*
77 China	-	-	-	-	-	-	-	-	6 000	-
Costa Rica	14	13	13	20	11	1	2	7	11	30
El Salvador	-	-	-	-	-	-	-	16	9	13
Guatemala	14	16	3	16	21	39	23	231	120 F	18
Japan	1 154	-	-	-	-	-	-	-	-	-
Korea Rep	-	693	461	205	203	-	316	1 429	-	377
Nicaragua	-	-	-	-	-	-	-	-	-	2
Panama	386	24	1	...	...	...	...	...	...	...
USA	32 262	55 108	70 210	78 794	...	...	...	...	...	1
77 Fishing area total	*33 830*	*55 854*	*70 688*	*79 035*	*235*	*40*	*341*	*1 683*	*6 140 F*	*441*
81 Australia	1 722	1 454	1 577	787	186	92	384	664	379	196
Korea Rep	6 652	13 110	17 436	9 836	13 068	12 278	9 951	8 801	11 380	16 991
New Zealand	-	-	10	7	17	27	48	74	45	41
Russian Fed	15 600	22 098	17 004	8 365	5 809	1 907	1 352	-	-	-
81 Fishing area total	*23 974*	*36 662*	*36 027*	*18 995*	*19 080*	*14 304*	*11 735*	*9 539*	*11 804*	*17 228*
87 Chile	7 604	462	55	26	110	179	99	64	3 594	5 661
Colombia	285	36	14	295	183	35	38	60 F	87	85 F
87 Fishing area total	*7 889*	*498*	*69*	*321*	*293*	*214*	*137*	*124 F*	*3 681*	*5 746 F*
Species total	*266 059 F*	*318 834 F*	*361 879 F*	*363 638 F*	*237 928 F*	*341 747 F*	*362 903 F*	*364 161 F*	*293 354 F*	*311 450 F*

Antarctic octopuses **Elédones antarctiques** **Pulpos antárticos** ***Pareledone spp*** **3,21(09)001,XX** **PRD**

	1993	1994	1995	1996	1997	1998	1999	2000	2001	2002
48 UK	-	-	-	-	-	-	-	0	-	-
48 Fishing area total	-	-	-	-	-	-	-	*0*	-	-
Species total	-	-	-	-	-	-	-	*0*	-	-

B-57 Squids, cuttlefishes, octopuses / Encornets, seiches, poulpes / Calamares, jibias, pulpos

Capture production by species, fishing areas and countries or areas / Captures par espèces, zones de pêche et pays ou zones / Capturas por especies, áreas de pesca y países o áreas

Species, Fishing area Espèce, Zone de pêche Especie, Area de pesca	1993 t	1994 t	1995 t	1996 t	1997 t	1998 t	1999 t	2000 t	2001 t	2002 t
Common octopus	**Pieuvre**		**Pulpo común**		***Octopus vulgaris***				**3,21(09)005,07**	**OCC**
27 Portugal	-	-	-	-	-	-	12	624	823	659
27 Fishing area total	-	-	-	-	-	-	*12*	*624*	*823*	*659*
31 Dominican Rp	21	28	33	36	33	87	39	99	57	73
Mexico	15 790	16 817	18 958	28 572	17 776	16 478	19 081	22 463	20 596	16 018
31 Fishing area total	*15 811*	*16 845*	*18 991*	*28 608*	*17 809*	*16 565*	*19 120*	*22 562*	*20 653*	*16 091*
34 GuineaBissau	...	7	34	1	1 F	1 F	1 F	1 F	1 F	1 F
Italy	1 345	1 345	213	244	612	933	1 845	1 373	1 388	784
Spain	33 000 F	32 000 F	31 000 F	26 874	22 545	38 235	21 692	11 379	14 913	8 483
34 Fishing area total	*34 345 F*	*33 352 F*	*31 247 F*	*27 119*	*23 158 F*	*39 169 F*	*23 538 F*	*12 753 F*	*16 302 F*	*9 268 F*
37 Albania	...	...	66	75	59	93	90	85	24	105
Croatia	231	187	193	180	293	435	...	...	...	...
France	1 161	873	1 146	706	439	439	1 493	1 643	1 406	2 100
Greece	2 456	2 886	2 856	2 919	2 952	1 673	1 910	2 193	2 058	1 950 F
Italy	11 565	9 661	9 757	9 057	8 907	9 478	6 999	7 800	8 648	7 373
Lebanon	...	...	...	...	...	25	25	25	25	25
Serbia-Monte	6	4	4	8	8	9	9	10	12	11
Slovenia	25	6	34	2	7	-	-	-	-	-
Tunisia	3 350	1 841	1 868	3 460	5 681	3 129	2 349	2 103	1 752	2 535
Turkey	472	659	602	802	1 000	1 450	510	680	1 400	1 502
37 Fishing area total	*19 266*	*16 117*	*16 526*	*17 209*	*19 346*	*16 731*	*13 385*	*14 539*	*15 325*	*15 601 F*
Species total	*69 422 F*	*66 314 F*	*66 764 F*	*72 936*	*60 313 F*	*72 465 F*	*56 055 F*	*50 478 F*	*53 103 F*	*41 619 F*
Horned and musky octopuses	**Elédones communes et musquées**		**Pulpos blancos y almizclados**		***Eledone spp***				**3,21(09)024,XX**	**OCM**
37 Italy	3 542	3 928	2 500	1 939	1 590	1 506	1 041	1 621	1 661	1 309
Tunisia	...	...	...	...	668	252	392	369	696	878
37 Fishing area total	*3 542*	*3 928*	*2 500*	*1 939*	*2 258*	*1 758*	*1 433*	*1 990*	*2 357*	*2 187*
Species total	*3 542*	*3 928*	*2 500*	*1 939*	*2 258*	*1 758*	*1 433*	*1 990*	*2 357*	*2 187*
Octopuses, etc. nei	**Pieuvres, poulpes nca**		**Pulpitos, pulpos nep**		***Octopodidae***				**3,21(09)XXX,XX**	**OCT**
21 USA	-	-	5	-	0	0	-	0	0	5
21 Fishing area total	-	-	*5*	-	*0*	*0*	-	*0*	*0*	*5*
27 Belgium	37	40	62	28	45	49	41	44	29	70
France	226	106	126	67	75	90	246	484	388	135
Ireland	4	6	25	13	7	3	10	8	14	13
Netherlands	-	-	-	-	-	-	-	-	7	-
Portugal	7 219	7 479	9 836	11 652	9 181	6 445	9 254	9 072	7 329	7 574
Spain	8 135	5 822	7 318	5 650	6 292	6 350	4 374	6 339	2 895	3 902
UK	182	313	333	229	148	111	63	135	164	179
27 Fishing area total	*15 803*	*13 766*	*17 700*	*17 639*	*15 748*	*13 048*	*13 988*	*16 082*	*10 826*	*11 873*
31 Korea Rep	-	-	-	-	-	-	157	3	-	-
Puerto Rico	...	...	...	...	...	...	28	39	23	15
Venezuela	901	1 196	724	1 548	1 917	6 507	526	1 110	1 064	1 226
31 Fishing area total	*901*	*1 196*	*724*	*1 548*	*1 917*	*6 507*	*711*	*1 152*	*1 087*	*1 241*
34 Bahamas	-	-	4	-	-	-	-	-	-	-
Belize	17	0	86	-	185	785	3 247	1 185	...	...
Chile	-	0	-	-	-	-	-	-	-	-
China	523	9	3 481	2 464	3 193	4 336	7 491	4 578	5 130	-
Côte dIvoire	-	-	-	-	-	-	-	-	88	...
Cyprus	-	-	-	-	-	-	-	0	3	-
Gabon	-	-	-	-	-	5	91	0	0	4
Gambia	277	141	6	...	...	...	...	...	1	-
Ghana	43	73	55	137	67	103	19	4	94	...
Greece	326	201	112	95	75	50	987	56	133	180
Guinea	1	-	114	22	12	183	1 092	96	97	...
Honduras	1 994	282	1 657	1 040	468	129	590	54	-	-
Korea D P Rp	20	-	-	-	-	-	-	86	-	-
Korea Rep	276	481	392	162	96	406	2 653	387	32	18
Latvia	-	-	-	-	-	-	-	-	2	6
Liberia	-	-	-	176	2	61	23	16	41	45 F
Mauritania	23 230 F	21 520 F	18 250 F	18 770 F	14 620 F	18 420 F	12 758 F	13 709 F	13 576 F	12 200 F
Morocco	63 813	56 239	57 765	58 600	38 089	42 465	84 464	99 279	112 589	37 872
Norway	-	26	-	-	-	-	-	-	-	-
Panama	154	-	82	89	41	102	-	-	-	-
Portugal	27	128	99	107	51	14	91	2	2	5
Russian Fed	9	-	-	-	-	-	-	-	-	-
St Vincent	9	0	-	-	-	-	-	-	-	-
Senegal	4 799	8 397	4 191	4 180	2 701	5 530	37 257	6 058	2 985	12 795
Sierra Leone	777	143	381	912	777	328	188	8	11	1
Vanuatu	-	-	-	-	-	0	-	-	-	-
Other nei	22	-	-	-	-	-	196	245	264	-
34 Fishing area total	*96 317 F*	*87 640 F*	*86 675 F*	*86 754 F*	*60 377 F*	*72 917 F*	*151 147 F*	*125 763 F*	*135 048 F*	*63 126 F*
37 Algeria	...	...	382	190	185	543	305	306	343	366
Cyprus	288	474	300	190	199	228	179	166	173	117
Greece	780	1 166	1 451	809	740	732	794	878	680	650 F
Malta	3	4	6	11	11	9	11	9	5	7

B-57 Squids, cuttlefishes, octopuses / Encornets, seiches, poulpes / Calamares, jibias, pulpos

Capture production by species, fishing areas and countries or areas
Captures par espèces, zones de pêche et pays ou zones
Capturas por especies, áreas de pesca y países o áreas

Species, Fishing area Espèce, Zone de pêche Especie, Area de pesca	1993 t	1994 t	1995 t	1996 t	1997 t	1998 t	1999 t	2000 t	2001 t	2002 t
Morocco	71	25	69	36	98	68	115	112	45	796
Spain	5 700 F	5 700 F	5 700 F	5 594	4 898	5 134	5 565	7 889	6 634	5 087
37 Fishing area total	*6 842 F*	*7 369 F*	*7 908 F*	*6 830*	*6 131*	*6 714*	*6 969*	*9 360*	*7 880*	*7 023 F*
41 Argentina	6	5	34	39	48	17	34	5	13	3
Brazil	630 F	630 F	577	459	640	554	906	1 032	1 160	962
Estonia	1 314	-	-	-	-	-	-	-	-	-
Italy	448	448	146	123	-	-	-	-	-	-
Korea Rep	-	1	-	14	-	-	516	47	2	-
41 Fishing area total	*2 398 F*	*1 084 F*	*757*	*635*	*688*	*571*	*1 456*	*1 084*	*1 175*	*965*
47 Angola	-	-	-	-	-	-	45	-	-	...
Honduras	57	-	5	-	-	-	-	-	-	-
Italy	-	-	3	3	-	-	-	-	-	-
Korea Rep	-	24	36	-	-	-	-	-	-	-
Portugal	...	...	21	26	101	56	164	144	-	133
St Helena	-	-	27	34	25	48	22	24	16	12
South Africa	149	91	33	46	64	60	87	90 F	93	68
Spain	-	-	-	282	-	418	116	47	20	78
47 Fishing area total	*206*	*115*	*125*	*391*	*190*	*582*	*434*	*305 F*	*129*	*291*
51 Fr South Tr	1	6	5 F	5 F	5 F	5 F	...	...	...	...
Italy	269	269	43	50	-	-	-	-	-	-
Kenya	78	106	460	117	393	155	169	106	154	208
Korea Rep	-	-	-	-	16	-	48	-	11	4
Mauritius	335	343	325	341	306	299	299	303	347	335
Mozambique	53	30	29	49	51	80	104	109	128	24
Portugal	-	-	-	-	-	-	-	2	1	-
Réunion	...	...	...	...	...	...	8	5	3	3 F
Saudi Arabia	-	-	-	-	1	-	-	0	-	-
Seychelles	67	32	20	32	19	40	78	29	54	32
South Africa	7	-	-	-	10	8	10 F	15	14	7
Spain	-	-	-	-	-	-	10	-	-	-
Tanzania	393	314	490	605	653	690	600	600	650	800
Yemen	...	...	...	...	...	...	...	...	21	9
51 Fishing area total	*1 203*	*1 100*	*1 372 F*	*1 199 F*	*1 454 F*	*1 277 F*	*1 326 F*	*1 169*	*1 383*	*1 422 F*
57 Australia	218	135	111	74	84	69	121	152	158	143
Indonesia	503	278	288	210	298	309	504	387	430	450
Malaysia	223	304	476	663	925	993	727	774	804	934
Thailand	7 027	4 608	5 541	10 702	10 919	14 989	12 880	13 073	12 236	12 165
57 Fishing area total	*7 971*	*5 325*	*6 416*	*11 649*	*12 226*	*16 360*	*14 232*	*14 386*	*13 628*	*13 692*
58 South Africa	-	-	-	-	-	-	-	-	-	0
58 Fishing area total	*-*	*-*	*-*	*-*	*-*	*-*	*-*	*-*	*-*	*0*
61 China,Taiwan	929	909	919	967	615	941	500	287	374	481
Japan	51 288	51 459	51 874	50 584	56 593	61 260	57 427	47 374	45 200	57 482
Korea Rep	7 618	7 916	21 700	21 384	22 952	26 432	19 262	19 148	19 873	21 591
Russian Fed	19	154	317	235	210	34	24	29	64	77
61 Fishing area total	*59 854*	*60 438*	*74 810*	*73 170*	*80 370*	*88 667*	*77 213*	*66 838*	*65 511*	*79 631*
67 Canada	145	89	108	197	208	139	54	56	61	125
USA	-	-	-	-	-	122	6	3	23	52
67 Fishing area total	*145*	*89*	*108*	*197*	*208*	*261*	*60*	*59*	*84*	*177*
71 Australia	...	...	...	...	...	...	...	15	32	26
Fiji Islands	58	59 F	91	48	35	62	61	50 F	57	55 F
Guam	-	-	0	0	-	1	2	1	2	2
Indonesia	327	416	376	435	971	3 210	1 833	1 598	2 797	2 900
Kiribati	2 200	2 180	2 200	2 230	1 874	650	687	373	69	23
Korea Rep	5	43	14	147	16	336	27	8	47	59
Malaysia	361	351	536	326	345	354	785	777	676	597
Micronesia	10 F	15 F	15 F	20 F	20 F	20 F	20 F	20 F	20 F	20 F
N Marianas	0	0	0	0	-	0	0	-	0	1
Philippines	8 915	7 109	9 729	9 025	7 991	5 235	5 813	5 502	6 945	7 179
Thailand	13 681	11 282	10 828	12 721	12 193	16 919	12 120	10 885	12 057	12 072
Wallis Fut I	...	1	1	1	1	1	1 F	1 F	1 F	1 F
71 Fishing area total	*25 557 F*	*21 456 F*	*23 790 F*	*24 953 F*	*23 446 F*	*26 788 F*	*21 349 F*	*19 230 F*	*22 703 F*	*22 935 F*
77 Amer Samoa	-	-	-	-	-	0	1	0	0	0
Cook Is	73 F	69 F	68 F	60 F	50 F	40 F	30 F	30 F	30 F	30 F
Costa Rica	15	67	26	27	58	6	76	71	66	35
Korea Rep	-	-	-	-	-	-	8	12	-	6
Mexico	1 196	966	877	1 257	944	755	1 094	883	837	676
Nicaragua	-	-	-	-	-	-	-	-	-	3
USA	-	-	-	-	-	6	2	2	3	20
77 Fishing area total	*1 284 F*	*1 102 F*	*971 F*	*1 344 F*	*1 052 F*	*807 F*	*1 211 F*	*998 F*	*936 F*	*770 F*
81 Australia	...	...	...	...	506	714	439	279	531	376
Korea Rep	-	-	-	-	4	-	3	-	-	-
New Zealand	286	183	9 064	227	267	167	148	119	140	100
81 Fishing area total	*286*	*183*	*9 064*	*227*	*777*	*881*	*590*	*398*	*671*	*476*
87 Chile	3 608	3 732	3 796	3 477	4 404	4 877	3 168	1 682	2 008	1 354
Ecuador	1	0	0	0	0	5	5	5	5	5 F
Peru	1 245	602	800	760	1 856	5 123	1 593	819	635	1 415
87 Fishing area total	*4 854*	*4 334*	*4 596*	*4 237*	*6 260*	*10 005*	*4 766*	*2 506*	*2 648*	*2 774 F*

B-57 Squids, cuttlefishes, octopuses / Encornets, seiches, poulpes / Calamares, jibias, pulpos

Capture production by species, fishing areas and countries or areas
Captures par espèces, zones de pêche et pays ou zones
Capturas por especies, áreas de pesca y países o áreas

Species, Fishing area Espèce, Zone de pêche Especie, Area de pesca	1993 t	1994 t	1995 t	1996 t	1997 t	1998 t	1999 t	2000 t	2001 t	2002 t
88 New Zealand	-	-	-	-	-	-	-	-	-	0
88 Fishing area total	-	-	-	-	-	-	-	-	-	*0*
Species total	*223 621 F*	*205 197 F*	*235 021 F*	*230 773 F*	*210 844 F*	*245 385 F*	*295 452 F*	*259 330 F*	*263 709 F*	*206 401 F*
Cephalopods nei	**Céphalopodes nca**		**Cefalópodos nep**		**Cephalopoda**			**3,21(XX)XXX,XX**		**CEP**
34 Mauritania	-	-	-	-	-	10 F	16	129	150	236
Spain	-	-	-	6 240	5 636	8 541	222	311	129	335
34 Fishing area total	-	-	-	*6 240*	*5 636*	*8 551 F*	*238*	*440*	*279*	*571*
37 Algeria	-	-	-	-	-	70	29	6	22	113
Croatia	799	455	431	483	443	265	646	873	885	403
France	-	-	-	-	-	-	55	49	48	55
Israel	64	50	50	50	98	76	120	117	146	127
Spain	-	-	-	-	-	-	20	90	61	...
37 Fishing area total	*863*	*505*	*481*	*533*	*541*	*411*	*870*	*1 135*	*1 162*	*698*
51 India	62 801	84 061	92 334	74 524	108 412	86 337	84 793	85 939	44 021	79 077
Iran	2 138	1 550	2 500	1 580	8 620	4 189	4 060	5 685	6 517	2 877
Madagascar	200	200	350	500	500	550	600	600	600	600
Somalia	500 F	500 F	500 F	500 F	500 F	500 F	500 F	500 F	500 F	400 F
51 Fishing area total	*65 639 F*	*86 311 F*	*95 684 F*	*77 104 F*	*118 032 F*	*91 576 F*	*89 953 F*	*92 724 F*	*51 638 F*	*82 954 F*
57 India	7 622	11 048	11 405	10 596	9 212	9 497	8 916	10 469	9 250	10 999
Timor-Leste	...	...	...	...	...	...	1 F	1	1	1 F
57 Fishing area total	*7 622*	*11 048*	*11 405*	*10 596*	*9 212*	*9 497*	*8 917 F*	*10 470*	*9 251*	*11 000 F*
58 Australia	-	-	-	-	1	-	-	-	-	-
58 Fishing area total	-	-	-	-	*1*	-	-	-	-	-
61 China,Taiwan	-	-	-	-	-	-	-	-	3	2
61 Fishing area total	-	-	-	-	-	-	-	-	*3*	*2*
71 Viet Nam	33 000 F	87 000 F	103 000 F	92 000 F	92 500 F	103 000 F	110 000 F	180 000 F	130 000 F	130 000 F
71 Fishing area total	*33 000 F*	*87 000 F*	*103 000 F*	*92 000 F*	*92 500 F*	*103 000 F*	*110 000 F*	*180 000 F*	*130 000 F*	*130 000 F*
Species total	*107 124 F*	*184 864 F*	*210 570 F*	*186 473 F*	*225 922 F*	*213 035 F*	*209 978 F*	*284 769 F*	*192 333 F*	*225 225 F*
Group total	***2 687 779***	***2 803 421***	***2 939 432***	***3 149 719***	***3 456 667***	***2 857 605***	***3 598 201***	***3 660 404***	***3 307 969***	***3 173 272***

B-58 Miscellaneous marine molluscs / Mollusques marins divers / Moluscos marinos diversos

Capture production by species, fishing areas and countries or areas
Captures par espèces, zones de pêche et pays ou zones
Capturas por especies, áreas de pesca y países o áreas

Species, Fishing area Espèce, Zone de pêche Especie, Area de pesca	1993 t	1994 t	1995 t	1996 t	1997 t	1998 t	1999 t	2000 t	2001 t	2002 t
Marine molluscs nei	**Mollusques marins nca**		**Moluscos marinos nep**		***Mollusca***			**3,99(XX)XXX,XX**		**MOL**
04 Indonesia	106	197	325	481	52	165	73	166	112	120
04 Fishing area total	*106*	*197*	*325*	*481*	*52*	*165*	*73*	*166*	*112*	*120*
21 Canada	52	705	1 089	663	503	168	644	1 263	1 262	3 165
St Pier Mq	-	-	-	1	0	0	17	205	115	42
USA	255	2 035	942	1 730	97	0	1 259	1	7	3
21 Fishing area total	*307*	*2 740*	*2 031*	*2 394*	*600*	*168*	*1 920*	*1 469*	*1 384*	*3 210*
27 Belgium	43	33	20	30	31	26	22	4	10	1
Norway	9	3	40	1	14	111	118	17	9	43
Portugal	522	524	640	626	687	797	28	66	118	52
Spain	195	399	319	206	268	360	382	288	286	79
Sweden	-	-	-	-	-	-	-	1	-	-
UK	114	6	37	15	23	8	113	16	1	21
27 Fishing area total	*883*	*965*	*1 056*	*878*	*1 023*	*1 302*	*663*	*392*	*424*	*196*
31 Barbados	0	0	0	0	0	0	0	0	-	-
Colombia	14	8	8	20	17	3	-	15 F	15 F	15 F
Dominican Rp	...	...	...	...	...	...	...	48	18	1 043
Honduras	-	-	-	-	2	...	3	...	...	...
Korea Rep	555	-	-	-	-	-	-	-	-	-
Puerto Rico	82	87	70	124	77	45	9	12	9	6
USA	59 613	1 740	2 660	104	1 177	1 868	886	184	815	1 721
US Virgin Is	1 F	1 F	0	0	0	0	0	0	0	0
Venezuela	993	744	1 098	425	600	1 005	5 808	2 457	1 056	186
31 Fishing area total	*61 258 F*	*2 580 F*	*3 836*	*673*	*1 873*	*2 921*	*6 706*	*2 716 F*	*1 913 F*	*2 971 F*
34 Cameroon	7	0	0	0	0	0	0	0	0	2
Cape Verde	0	0	0	0	0	0	0	0	-	-
Côte dIvoire	-	-	-	-	-	-	-	26	-	1
Eq Guinea	120 F	180 F	80 F	180 F	230 F	122	140 F	20 F	20 F	20 F
Estonia	-	-	-	-	-	-	-	2	-	-
GuineaBissau	10 F	10 F	12 F	14 F	15 F	15 F	10 F	10 F	10 F	10 F
Italy	-	-	1 244	-	187	165	99	12	64	6
Korea Rep	2 970	-	-	-	-		-	1	-	-
Mauritania	-	-	-	-	-	20 F	13	1	6	23
Morocco	277	356	172	200	273	1 800	1 750	980	114	870
Portugal	0	0	-	0	0	2	4	2	3	8
Sao Tome Prn	19 F	28 F	29 F	21	20	25 F	30 F	30 F	30 F	30 F
Senegal	3	...	268	247	748	699	212	105	75	19
Sierra Leone	-	-	-	-	-	-	-	-	63	2 434
Spain	-	-	-	84	16	9 751		-	-	-
Togo	0	0	1	0	0	0	0	0	0	-
Westn Sahara	0	0	0	0	0	0	0	0	0	0
34 Fishing area total	*3 406 F*	*574 F*	*1 806 F*	*746 F*	*1 489 F*	*12 599 F*	*2 258 F*	*1 189 F*	*385 F*	*3 423 F*
37 Croatia	33	32	16	38	313	125	40	27	69	33
Egypt	...	...	140	233	0	198	0	-	4 173	3 243
France	64	26	-	-	-	-	-	-	-	-
Greece	2 382	1 739	2 393	3 076	2 241	1 789	1 736	1 688	2 169	2 000 F
Italy	9 407	7 864	11 663	11 238	9 563	7 680	7 761	6 270	6 260	5 879
Morocco	1	3	8	53	2	120	176	68	2	150
Romania	45	-	-	-	-	-	-	-	-	-
Serbia-Monte	0	0	0	0	0	0	0	-	1	-
Slovenia	-	1	6	5	10	12	8	18	54	25
Spain	1 000 F	1 200 F	1 200 F	1 011	1 259	1 119	22	6	1	17
Syria	...	...	...	...	...	...	...	...	...	136
Tunisia	-	-	-	-	-	-	-	-	1	10
Turkey	3 689	2 632	1 224	2 466	2 092	4 077	3 646	2 168	2 671	6 274
Ukraine	-	1	0	-	-	-	-	-	-	2
37 Fishing area total	*16 621 F*	*13 498 F*	*16 650 F*	*18 120*	*15 480*	*15 120*	*13 389*	*10 245*	*15 401*	*17 769 F*
41 Argentina	72	714	13	1	50	-	1 000	-	-	-
Brazil	3 460 F	3 470 F	3 194	1 085	1 244	98	286	2 096	313	1 795
Italy	-	-	148	-	-	-	-	-	-	-
Korea Rep	40	-	-	-	-	-	-	378	769	448
Uruguay	64	170	323	89	557	1 310	3 651	-	-	91
41 Fishing area total	*3 636 F*	*4 354 F*	*3 678*	*1 175*	*1 851*	*1 408*	*4 937*	*2 474*	*1 082*	*2 334*
47 Angola	-	-	-	-	-	-	-	-	-	2 000
Italy	-	-	17	-	-	-	-	-	-	-
47 Fishing area total	-	-	*17*	-	-	-	-	-	-	*2 000*
51 Comoros	0	0	0	0	0	0	0	0	-	-
Egypt	-	-	-	-	-	-	-	1 718	361	135
India	934	1 385	2 409	15 299	4 750	2 718	1 217	815	1 675	2 078
Iran	270	115	150	150	1 992	1 583	1 640	1 235	1 144	0
Italy	-	-	254	-	-	-	-	-	-	-
Korea Rep	-	-	19	-	-	-	5	-	-	-
Madagascar	350	350	350	350	350	400	400	400	400	400
Maldives	30	110	143	140	271	314	485	866	-	-
51 Fishing area total	*1 584*	*1 960*	*3 325*	*15 939*	*7 363*	*5 015*	*3 747*	*5 034*	*3 580*	*2 613*
57 Australia	721	444	284	358	152	629	7 148	851	862	904
India	-	407	43	0	-	...	1 000	981	922	346
Indonesia	42	28	69	23	111	188	213	81	106	110

B-58 Miscellaneous marine molluscs / Mollusques marins divers / Moluscos marinos diversos

Capture production by species, fishing areas and countries or areas / Captures par espèces, zones de pêche et pays ou zones / Capturas por especies, áreas de pesca y países o áreas

Species, Fishing area Espèce, Zone de pêche Especie, Area de pesca	1993 t	1994 t	1995 t	1996 t	1997 t	1998 t	1999 t	2000 t	2001 t	2002 t
Sri Lanka	0	0	0	0	-	-	10	15	30	60
Thailand	13	-	5	-	-	27	25	12	21	20
57 Fishing area total	*776*	*879*	*401*	*381*	*263*	*844*	*8 396*	*1 940*	*1 941*	*1 440*
61 China	942 725	1 103 048	1 294 815	932 374	1 494 902	1 479 065	1 445 303	1 402 625	1 399 810	1 375 782
China,H.Kong	2 128	2 585	1 849	1 461	1 213	700	500 F	600 F	660 F	650 F
China, Macao	21	69	65	30	40 F	40 F	40 F	40 F	40 F	40 F
China,Taiwan	883	1 005	932	464	443	663	433	914	118	209
Korea Rep	14 692	11 941	633	1 011	837	2 209	564	438	501	383
Russian Fed	800	2 981	4 122	3 530	4 780	6 301	10 586	5 643	4 780	4 679
61 Fishing area total	*961 249*	*1 121 629*	*1 302 416*	*938 870*	*1 502 215 F*	*1 488 978 F*	*1 457 426 F*	*1 410 260 F*	*1 405 909 F*	*1 381 743 F*
67 Canada	1 227	1 354	1 567	1 336	2 805	1 835	989	1 272	919	1 184
USA	172	815	588	537	1 788	1 044	861	632	730	910
67 Fishing area total	*1 399*	*2 169*	*2 155*	*1 873*	*4 593*	*2 879*	*1 850*	*1 904*	*1 649*	*2 094*
71 Australia	46	47	86	134	60	187	299	342	201	118
Brunei Darsm	35	34	33	16	21	19	35	30	24	31
Cambodia	1 465	1 400	1 500	1 550	1 480	1 600	1 900	801	900	1 000
Fiji Islands	959	977 F	2 570	2 000	3 880	3 200	3 302	3 100 F	3 150	3 150 F
Indonesia	809	108	80	644	219	321	331	277	2 627	2 810
Kiribati	4 100	4 060	4 100	4 150	4 120	571	776	1 947	3 260	2 337
Korea Rep	-	48	-	20	180	99	114	-	-	-
NewCaledonia	6	7	34	130	95	150	27	6	9	9 F
Philippines	13	68	66	62	-	-	-	-	-	-
Thailand	322	-	-	-	-	99	1 664	1 391	712	728
Vanuatu	600 F	580 F	600 F	600 F	600 F	600 F	600 F	600 F	600 F	600 F
Viet Nam	16 502 F	44 680 F	50 437 F	44 016 F	44 117 F	52 903 F	50 003 F	44 685 F	50 000 F	50 000 F
71 Fishing area total	*24 857 F*	*52 009 F*	*59 506 F*	*53 322 F*	*54 772 F*	*59 749 F*	*59 051 F*	*53 179 F*	*61 483 F*	*60 783 F*
77 Cook Is	208 F	198 F	196 F	160 F	140 F	120 F	120 F	120 F	120 F	120 F
Costa Rica	19	66	74	65	109	62	55	42	-	-
El Salvador	825	670	601	598	535	620	1 119	1 115	823	750
Fr Polynesia	0	0	0	0	10	10	10	25	25	25
Honduras	...	...	146	...	7	20	24	465	...	...
Panama	84	...	45	48	71	172	16	76	19	22
Samoa	24	20	20	20	20	20	20	1 644	1 600 F	1 600 F
Tonga	...	...	...	...	1	3	7	13	22	38
USA	735	333	298	416	964	435	371	316	341	428
77 Fishing area total	*1 895 F*	*1 287 F*	*1 380 F*	*1 307 F*	*1 857 F*	*1 462 F*	*1 742 F*	*3 816 F*	*2 950 F*	*2 983 F*
81 Australia	163	256	256	326	169	230	111	52	60	75
New Zealand	419	1 422	10	16	4	-	-	2	1	15
81 Fishing area total	*582*	*1 678*	*266*	*342*	*173*	*230*	*111*	*54*	*61*	*90*
87 Chile	306	103	12	9	69	72	35	28	17	-
Colombia	568	2	77	434	840	84	19	143 F	145 F	140 F
Korea Rep	-	-	-	-	-	-	-	30	-	-
Peru	540	556	468	537	2 538	1 546	3 676	2 312	2 116	7 940
87 Fishing area total	*1 414*	*661*	*557*	*980*	*3 447*	*1 702*	*3 730*	*2 513 F*	*2 278 F*	*8 080 F*
Species total	*1 079 973 F*	*1 207 180 F*	*1 399 405 F*	*1 037 481 F*	*1 597 051 F*	*1 594 542 F*	*1 565 999 F*	*1 497 351 F*	*1 500 552 F*	*1 491 849 F*
Group total	***1 079 973***	***1 207 180***	***1 399 405***	***1 037 481***	***1 597 051***	***1 594 542***	***1 565 999***	***1 497 351***	***1 500 552***	***1 491 849***

B-61 Blue-whales, fin-whales / Baleines bleues, rorquals communs / Ballenas azules, rorcuales

Capture production by species, fishing areas and countries or areas / Captures par espèces, zones de pêche et pays ou zones / Capturas por especies, áreas de pesca y países o áreas

Species, Fishing area Espèce, Zone de pêche Especie, Area de pesca	1993 no	1994 no	1995 no	1996 no	1997 no	1998 no	1999 no	2000 no	2001 no	2002 no
Minke whale	**Petit rorqual**		**Rorcual enano**		***Balaenoptera acutorostrata***			**4,23(02)001,01**		**MIW**
98 Japan	330	330	330	440	440	438	389	439	440	440
98 Fishing area total	*330*	*330*	*330*	*440*	*440*	*438*	*389*	*439*	*440*	*440*
99 Australia	-	-	-	-	-	1	2	-	-	-
France	1	-	-	-	1	1	-	-	-	-
Greenland	110	105	156	172	157	176	179	154	151	145
Iceland	2	2	-	2	2	3	-	1	-	2
Japan	14	37	120	104	127	124	119	69	179	259
Korea Rep	-	-	-	129	78	45	56	79	149	84
Norway	226	273	218	388	503	625	589	487	543	634
Peru	2	-	-	-	-	-	-	-	-	-
South Africa	-	-	-	-	1	-	-	-	-	-
Spain	-	-	-	-	-	-	-	1	-	-
UK	-	-	-	-	1	3	-	3	1	2
USA	-	-	-	12	4	2	5	-	-	-
99 Fishing area total	*355*	*417*	*494*	*807*	*874*	*980*	*950*	*794*	*1 023*	*1 126*
Species total	*685*	*747*	*824*	*1 247*	*1 314*	*1 418*	*1 339*	*1 233*	*1 463*	*1 566*
Bryde's whale	**Rorqual de Bryde**		**Rorcual tropical**		***Balaenoptera edeni***			**4,23(02)001,02**		**BRW**
99 Australia	-	-	-	-	-	-	-	1	-	-
Japan	-	-	-	-	-	1	-	43	50	50
New Zealand	-	-	-	-	1	-	-	-	-	-
St Vincent	-	-	-	-	-	-	-	1	-	-
South Africa	-	-	-	-	1	-	-	-	-	-
Spain	-	-	-	-	-	-	-	1	-	-
99 Fishing area total	-	-	-	-	*2*	*1*	-	*46*	*50*	*50*
Species total	-	-	-	-	*2*	*1*	-	*46*	*50*	*50*
Sei whale	**Rorqual de Rudolphi**		**Rorcual del Norte**		***Balaenoptera borealis***			**4,23(02)001,03**		**SIW**
99 Japan	-	-	-	-	-	-	-	-	1	39
New Zealand	1	-	-	-	-	-	-	-	-	-
99 Fishing area total	*1*	-	-	-	-	-	-	-	*1*	*39*
Species total	*1*	...	...	...	...	...	...	...	*1*	*39*
Blue whale	**Rorqual bleu**		**Ballena azul**		***Balaenoptera musculus***			**4,23(02)001,04**		**BLW**
99 Australia	-	-	-	-	-	-	-	1	2	-
New Zealand	-	1	-	-	-	-	-	-	-	-
USA	-	-	-	-	-	1	-	-	-	-
99 Fishing area total	-	*1*	-	-	-	*1*	-	*1*	*2*	-
Species total	...	*1*	...	...	...	*1*	...	*1*	*2*	...
Fin whale	**Rorqual commun**		**Rorcual común**		***Balaenoptera physalus***			**4,23(02)001,06**		**FIW**
99 France	-	-	1	-	-	-	-	-	-	-
Greenland	13	20	12	19	11	9	7	6	7	14
Ireland	-	-	-	-	-	-	-	1	-	-
Japan	-	-	-	-	1	-	-	-	1	-
Korea Rep	-	-	-	-	-	-	-	-	-	1
USA	-	-	-	1	-	1	3	-	-	-
99 Fishing area total	*13*	*20*	*13*	*20*	*12*	*10*	*10*	*7*	*8*	*15*
Species total	*13*	*20*	*13*	*20*	*12*	*10*	*10*	*7*	*8*	*15*
Humpback whale	**Baleine à bosse**		**Rorcual jorobado**		***Megaptera novaeangliae***			**4,23(02)003,01**		**HUW**
99 Australia	2	-	1	-	-	-	-	-	5	1
Brazil	-	-	-	-	-	-	-	-	-	2
Ecuador	-	-	2	-	-	-	-	-	-	-
Greenland	-	1	-	1	-	1	1	2	2	5
Iceland	-	-	-	-	-	-	-	-	-	2
Japan	-	-	-	2	1	1	1	1	-	3
Korea Rep	-	-	-	-	-	-	1	-	-	-
Norway	-	-	-	-	-	-	1	-	-	-
St Vincent	2	-	-	1	-	2	2	2	2	-
Tanzania	-	-	-	-	-	-	-	-	-	1
USA	-	-	2	3	1	5	2	-	-	-
99 Fishing area total	*4*	*1*	*5*	*7*	*2*	*9*	*8*	*5*	*9*	*14*
Species total	*4*	*1*	*5*	*7*	*2*	*9*	*8*	*5*	*9*	*14*
Northern right whale	**Baleine de Biscaye**		**Ballena franca**		***Eubalaena glacialis***			**4,23(03)001,01**		**EUG**
21 USA	1	-	-	2	-	-	2	-	-	-
21 Fishing area total	*1*	-	-	*2*	-	-	*2*	-	-	-
Species total	*1*	-	-	*2*	-	-	*2*	-	-	-
Bowhead whale	**Baleine du Groenland**		**Ballena de cabeza arqueada**		***Balaena mysticetus***			**4,23(03)002,01**		**BMY**
61 Russian Fed	...	...	...	...	...	1	1	1	1	2

B-61 Blue-whales, fin-whales / Baleines bleues, rorquals communs / Ballenas azules, rorcuales

Capture production by species, fishing areas and countries or areas / Captures par espèces, zones de pêche et pays ou zones / Capturas por especies, áreas de pesca y países o áreas

Species, Fishing area Espèce, Zone de pêche Especie, Area de pesca	1993 no	1994 no	1995 no	1996 no	1997 no	1998 no	1999 no	2000 no	2001 no	2002 no
61 Fishing area total	...	...	...	...	...	*1*	*1*	*1*	*1*	*2*
67 USA	41	34	45	39	48	40	42	35	49	39
67 Fishing area total	*41*	*34*	*45*	*39*	*48*	*40*	*42*	*35*	*49*	*39*
Species total	*41*	*34*	*45*	*39*	*48*	*41*	*43*	*36*	*50*	*41*
Gray whale	**Baleine grise**		**Ballena gris**		***Eschrichtius robustus***			**4,23(04)001,01**		**GRW**
61 Russian Fed	-	44	89	43	79	122	121	113	112	131
61 Fishing area total	-	*44*	*89*	*43*	*79*	*122*	*121*	*113*	*112*	*131*
67 USA	...	...	3	-	-	1	1	-	-	-
67 Fishing area total	...	...	*3*	-	-	*1*	*1*	-	-	-
Species total	...	*44*	*92*	*43*	*79*	*123*	*122*	*113*	*112*	*131*
Baleen whales nei	**Baleines mysticètes nca**		**Ballenas mysticetas nep**		***Mysticeti***			**4,23(XX)XXX,XX**		**MYS**
99 Australia	2	-	3	-	-	-	-	-	3	-
Brazil	-	-	-	-	1	1	-	-	-	-
Canada	-	-	-	1	-	1	-	1	-	1
France	5	-	-	-	-	-	-	-	-	-
Japan	17	6	-	-	-	-	-	-	1	-
New Zealand	1	2	-	-	-	-	-	-	-	1
Oman	...	...	...	...	...	...	-	-	2	1
Spain	-	-	-	-	-	-	1	1	-	-
99 Fishing area total	*25*	*8*	*3*	*1*	*1*	*2*	*1*	*2*	*6*	*3*
Species total	*25*	*8*	*3*	*1*	*1*	*2*	*1*	*2*	*6*	*3*
Group total	***770***	***855***	***982***	***1 359***	***1 458***	***1 605***	***1 525***	***1 443***	***1 701***	***1 859***

Fishing area 98: Antarctic, pelagic; split-year data shown under the calendar year in which the split-year ends.

Fishing area 99: Outside the Antarctic; calendar year data.

Data are derived mainly from sources of the International Whaling Commission (IWC).

See paragraph 5 of the INTRODUCTION.

Zone de pêche 98: Antarctique, pélagique; données relatives à des années fractionnées figurant sous l'année civile durant laquelle se termine l'année fractionnée.

Zone de pêche 99: Hors de l'Antarctique; données relatives à l'année civile.

Données provenant principalement des sources de la Commission baleinière internationale.

Voir le paragraphe 5 de l'INTRODUCTION.

Area de pesca 98: Antártico, pelágico; datos correspondientes a los años emergentes que figuran en la columna del año civil en que finaliza el año emergente.

Area de pesca 99: Fuera del Antártico; datos relativos al año civil.

Datos obtenidos por la mayoría desde fuentes de la Comisión Ballenera Internacional.

Véase el párrafo 5 en la INTRODUCCIÓN.

B-62 Sperm-whales, pilot-whales / Cachalots, globicéphales / Cachalotes, calderones

Capture production by species, fishing areas and countries or areas / Captures par espèces, zones de pêche et pays ou zones / Capturas por especies, áreas de pesca y países o áreas

Species, Fishing area Espèce, Zone de pêche Especie, Area de pesca	1993 no	1994 no	1995 no	1996 no	1997 no	1998 no	1999 no	2000 no	2001 no	2002 no
Northern bottlenose whale	**Hyperoodon boréal**		**Ballena nariz de botella norte**		***Hyperoodon ampullatus***				**4,22(02)005,01**	**BOW**
99 Faeroe Is	-	-	-	-	-	-	-	-	2	-
99 Fishing area total	-	-	-	-	-	-	-	-	*2*	-
Species total	-	-	-	-	-	-	-	-	*2*	-
Baird's beaked whale	**Baleine à bec de Baird**		**Zifio de Baird**		***Berardius bairdii***				**4,22(02)019,02**	**BEW**
99 Japan	54	54	54	54	54	54	62	62	62	...
Korea Rep	-	-	-	-	1	-	1	-	1	-
USA	-	11	-	-	-	-	-	-	-	-
99 Fishing area total	*54*	*65*	*54*	*54*	*55*	*54*	*63*	*62*	*63*	...
Species total	*54*	*65*	*54*	*54*	*55*	*54*	*63*	*62*	*63*	...
Sperm whale	**Cachalot**		**Cachalote**		***Physeter catodon***				**4,22(03)001,01**	**SPW**
99 Brazil	-	-	1	-	-	-	-	-	-	-
Ecuador	-	-	4	-	-	-	-	-	-	-
France	6	1	-	-	-	-	-	-	-	-
Japan	-	-	1	-	-	-	-	5	8	5
Oman	-	-	-	-	-	-	-	-	1	1
St Vincent	-	-	-	-	-	-	-	-	-	2
Spain	-	-	-	-	-	-	3	5	3	-
USA	22	-	-	1	-	5	1	-	-	-
99 Fishing area total	*28*	*1*	*6*	*1*	-	*5*	*4*	*10*	*12*	*8*
Species total	*28*	*1*	*6*	*1*	...	*5*	*4*	*10*	*12*	*8*
Long-finned pilot whale	**Globicéphale commun**		**Calderón común**		***Globicephala melas***				**4,22(04)004,02**	**PIW**
99 Brazil	6	-	-	-	-	-	-	-	-	-
Canada	14	3	9	6	15	-	-	-	-	-
Chile	-	-	-	-	-	-	1	-	-	-
Faeroe Is	-	1 201	228	1 524	1 162	815	608	588	918	-
France	19	2	2	2	3	1	5	1	2	1
Germany	-	-	-	8	-	-	-	-	-	-
Greenland	20	-	-	67	208	365	115	5	43	-
Ireland	-	-	-	4	-	-	8	-	-	-
Netherlands	2	15	-	16	1	-	-	-	-	-
New Zealand	1	-	7	-	-	1	3	-	-	1
Spain	-	2	-	-	-	-	1	-	-	4
USA	31	22	31	12	93	104	371	58	-	-
99 Fishing area total	*93*	*1 245*	*277*	*1 639*	*1 482*	*1 286*	*1 112*	*652*	*963*	*6*
Species total	*93*	*1 245*	*277*	*1 639*	*1 482*	*1 286*	*1 112*	*652*	*963*	*6*
Short-finned pilot whale	**Globicéphale tropical**		**Calderón de aletas cortas**		***Globicephala macrorhynchus***				**4,22(04)004,03**	**SHW**
99 Ecuador	20	-	-	-	-	-	-	-	-	-
Japan	337	196	239	482	347	229	394	304	389	...
Korea Rep	-	-	-	-	2	-	-	-	-	-
Peru	1	-	-	-	-	-	-	-	-	-
St Lucia	-	-	-	-	-	-	35	-	-	-
Spain	-	-	-	-	-	-	2	-	-	1
USA	81	-	-	-	6	-	-	-	-	-
Other nei	-	-	-	-	5	-	-	1	-	-
99 Fishing area total	*439*	*196*	*239*	*482*	*360*	*229*	*431*	*305*	*389*	*1*
Species total	*439*	*196*	*239*	*482*	*360*	*229*	*431*	*305*	*389*	*1*
Killer whale	**Orque**		**Orca**		***Orcinus orca***				**4,22(04)022,01**	**KIW**
99 Brazil	1	1	-	-	-	-	-	-	-	-
Japan	-	-	-	-	1	-	-	-	-	-
Korea Rep	-	-	-	1	-	-	-	1	-	3
New Zealand	1	-	-	-	-	-	-	-	-	1
Spain	-	-	-	-	-	-	-	-	-	1
USA	1	-	6	-	4	1	4	-	-	-
99 Fishing area total	*3*	*1*	*6*	*1*	*5*	*1*	*4*	*1*	-	*5*
Species total	*3*	*1*	*6*	*1*	*5*	*1*	*4*	*1*	-	*5*
Harbour porpoise	**Marsouin commun**		**Marsopa común**		***Phocoena phocoena***				**4,22(05)002,01**	**PHR**
99 Canada	257	98	-	-	-	-	-	-	-	-
Denmark	4 449	4 449	7 000	-	-	2	4 427	4 149	3 887	-
Faeroe Is	-	-	-	3	-	-	-	-	-	-
France	-	-	-	-	9	1	8	11	12	3
Germany	12	18	8	6	4	5	3	5	8	8
Greenland	-	1 716	1 135	1 824	1 592	2 131	1 830	1 607	1 628	-
Ireland	1 497	-	1	2	3	2	4	-	1	5
Japan	-	-	-	-	-	-	-	-	1	-
Korea Rep	-	-	-	1	-	-	1	-	87	34
Netherlands	4	-	1	-	4	4	-	2	-	-
Spain	-	-	1	-	-	-	1	2	1	-
Sweden	-	25	53	124	8	14	2	3	-	3
Turkey	-	-	-	-	-	-	-	-	-	20

B-62 Sperm-whales, pilot-whales / Cachalots, globicéphales / Cachalotes, calderones

Capture production by species, fishing areas and countries or areas / Captures par espèces, zones de pêche et pays ou zones / Capturas por especies, áreas de pesca y países o áreas

Species, Fishing area Espèce, Zone de pêche Especie, Area de pesca	1993 no	1994 no	1995 no	1996 no	1997 no	1998 no	1999 no	2000 no	2001 no	2002 no
UK	740	761	933	752	791	33	19	34	11	29
USA	1 414	2 129	1 537	1 540	1 430	842	475	554	-	16
99 Fishing area total	*8 373*	*9 196*	*10 669*	*4 252*	*3 841*	*3 034*	*6 770*	*6 367*	*5 636*	*118*
Species total	*8 373*	*9 196*	*10 669*	*4 252*	*3 841*	*3 034*	*6 770*	*6 367*	*5 636*	*118*
White whale	**Bélouga**		**Beluga**		***Delphinapterus leucas***				**4,22(06)014,01**	**BEL**
27 Russian Fed	-	-	-	-	-	-	-	-	-	5
27 Fishing area total	-	-	-	-	-	-	-	-	-	*5*
61 Russian Fed	26	71	20	3	3	27	70	22	7	15
61 Fishing area total	*26*	*71*	*20*	*3*	*3*	*27*	*70*	*22*	*7*	*15*
67 Russian Fed	-	-	-	-	-	-	6	-	-	-
67 Fishing area total	-	-	-	-	-	-	*6*	-	-	-
99 Canada	-	-	-	-	-	-	-	-	375	-
Greenland	475	488	606	542	577	746	493	610	260	-
USA	312	276	229	304	232	327	238	212	412	364
99 Fishing area total	*787*	*764*	*835*	*846*	*809*	*1 073*	*731*	*822*	*1 047*	*364*
Species total	*813*	*835*	*855*	*849*	*812*	*1 100*	*807*	*844*	*1 054*	*384*
Narwhal	**Narval**		**Narval**		***Monodon monoceros***				**4,22(06)018,01**	**NAR**
99 Canada	-	-	-	-	-	-	-	-	559	-
Greenland	741	847	461	738	797	822	912	600	449	-
99 Fishing area total	*741*	*847*	*461*	*738*	*797*	*822*	*912*	*600*	*1 008*	-
Species total	*741*	*847*	*461*	*738*	*797*	*822*	*912*	*600*	*1 008*	-
Toothed whales nei	**Baleines odontocètes nca**		**Ballenas odontocetas nep**		***Odontoceti***				**4,22(XX)XXX,XX**	**ODN**
99 Argentina	364	117	162	-	-	15	72	490	197	217
Australia	30	30	30	22	31	50	40	25	24	30
Brazil	255	335	223	495	163	132	1 222	1 507	23	131
Canada	-	3	1	3	-	-	-	-	-	-
Chile	-	-	-	-	-	1	1	-	-	-
Ecuador	3 741	227	-	-	-	-	-	-	-	-
Faeroe Is	-	266	151	173	350	438	-	255	552	-
France	1 735	56	16	30	312	47	179	212	140	234
Ireland	321	-	-	493	2	29	146	4	1	1
Italy	-	-	-	-	-	-	19	24	-	-
Japan	16 126	17 577	14 474	17 173	19 739	13 037	16 764	18 404	17 902	...
Korea Rep	-	-	-	90	75	33	46	94	140	159
Mexico	14	-	-	-	1	-	-	6	3	3
Netherlands	14	103	10	32	43	29	-	-	-	-
New Zealand	34	19	21	3	8	15	6	13	19	10
Oman	...	...	...	...	...	...	...	13	20	3
Peru	4 154	1 877	252	15	8	48	209	77	24	304
St Lucia	-	-	-	-	-	-	126	-	-	-
South Africa	87	165	95	100	149	45	60	-	37	78
Spain	-	33	3	3	7	6	9	20	11	44
Tanzania	-	-	-	-	-	-	-	-		13
Turkey	-	-	-	-	-	-	-	-	-	80
UK	9	64	168	2	9	6	6	12	75	37
USA	1 170	1 041	947	1 723	895	467	624	839	-	87
Other nei	3 486	4 096	3 274	2 547	3 000	1 877	1 348	1 635	2 133	1 513
99 Fishing area total	*31 540*	*26 009*	*19 827*	*22 904*	*24 792*	*16 275*	*20 877*	*23 630*	*21 301*	*2 944*
Species total	*31 540*	*26 009*	*19 827*	*22 904*	*24 792*	*16 275*	*20 877*	*23 630*	*21 301*	*2 944*
Group total	***42 084***	***38 395***	***32 394***	***30 920***	***32 144***	***22 806***	***30 980***	***32 471***	***30 428***	***3 466***

Fishing area 99: Outside the Antarctic; calendar year data.

Data are derived mainly from sources of the International Whaling Commission (IWC).

See paragraph 5 of the INTRODUCTION.

Zone de pêche 99: Hors de l'Antarctique; données relatives à l'année civile.

Données provenant principalement des sources de la Commission baleinière internationale.

Voir le paragraphe 5 de l'INTRODUCTION.

Area de pesca 99: Fuera del Antártico; datos relativos al año civil.

Datos obtenidos por la mayoría desde fuentes de la Comisión Ballenera Internacional.

Véase el párrafo 5 en la INTRODUCCIÓN.

B-63 Eared seals, hair seals, walruses / Otaries, phoques, morses / Lobos marinos, focas, morsas

Capture production by species, fishing areas and countries or areas / Captures par espèces, zones de pêche et pays ou zones / Capturas por especies, áreas de pesca y países o áreas

Species, Fishing area Espèce, Zone de pêche Especie, Area de pesca	1993 no	1994 no	1995 no	1996 no	1997 no	1998 no	1999 no	2000 no	2001 no	2002 no
Steller sea lion	**Lion de mer de Steller**		**Lobo marino de Steller**		***Eumetopias jubatus***				**4,06(01)001,01**	**SSL**
67 USA	487	416	339	186	164	178	...	171	198	185
67 Fishing area total	*487*	*416*	*339*	*186*	*164*	*178*	...	*171*	*198*	*185*
Species total	*487*	*416*	*339*	*186*	*164*	*178*	...	*171*	*198*	*185*
Northern fur seal	**Otarie des Pribilofs**		**Lobo fino del Norte**		***Callorhinus ursinus***				**4,06(01)002,01**	**SEN**
61 Russian Fed	8 738	10 095	8 103	6 487	6 537	7 742	4 500	2 919	1 602	2 895
61 Fishing area total	*8 738*	*10 095*	*8 103*	*6 487*	*6 537*	*7 742*	*4 500*	*2 919*	*1 602*	*2 895*
67 USA	1 837	1 777	1 785	1 823	1 380	1 553	1 193	...	...	...
67 Fishing area total	*1 837*	*1 777*	*1 785*	*1 823*	*1 380*	*1 553*	*1 193*	...	...	...
Species total	*10 575*	*11 872*	*9 888*	*8 310*	*7 917*	*9 295*	*5 693*	*2 919*	*1 602*	*2 895*
South American fur seal	**Otarie d'Amérique du Sud**		**Lobo fino austral**		***Arctocephalus australis***				**4,06(01)006,01**	**SEF**
41 Uruguay	112	-	-	-	-	-	-	-	-	-
41 Fishing area total	*112*	-	-	-	-	-	-	-	-	-
Species total	*112*	-	-	-	-	-	-	-	-	-
South African fur seal	**Otarie du Cap**		**Lobo marino de dos pelos**		***Arctocephalus pusillus***				**4,06(01)006,03**	**SEK**
47 Namibia	35 730	37 853	20 450	20 814	25 783	29 475	25 161	41 753	44 223	37 670
47 Fishing area total	*35 730*	*37 853*	*20 450*	*20 814*	*25 783*	*29 475*	*25 161*	*41 753*	*44 223*	*37 670*
Species total	*35 730*	*37 853*	*20 450*	*20 814*	*25 783*	*29 475*	*25 161*	*41 753*	*44 223*	*37 670*
South American sea lion	**Lion de mer d'Amérique du Sud**		**Lobo común**		***Otaria byronia***				**4,06(01)016,01**	**SEL**
87 Chile	-	-	389	96	16	-	-	-	-	-
87 Fishing area total	-	-	*389*	*96*	*16*	-	-	-	-	-
Species total	-	-	*389*	*96*	*16*	-	-	-	-	-
Walrus	**Morse**		**Morsa**		***Odobenus rosmarus***				**4,06(02)004,01**	**WAL**
21 Greenland	-	-	-	305	317	610	311	329	335	318
21 Fishing area total	-	-	-	*305*	*317*	*610*	*311*	*329*	*335*	*318*
61 Russian Fed	856	1 013	1 091	955	731	950	657	534	738	1 369
61 Fishing area total	*856*	*1 013*	*1 091*	*955*	*731*	*950*	*657*	*534*	*738*	*1 369*
67 Russian Fed	-	-	-	-	-	-	688	210	-	-
67 Fishing area total	-	-	-	-	-	-	*688*	*210*	-	-
Species total	*856*	*1 013*	*1 091*	*1 260*	*1 048*	*1 560*	*1 656*	*1 073*	*1 073*	*1 687*
Harp seal	**Phoque du Groenland**		**Foca de Groenlandia**		***Phoca groenlandica***				**4,06(03)005,01**	**SEH**
21 Canada	25 175	61 176	65 391	242 717	264 204	249 053	228 828	87 084	193 366	285 510
Greenland	53 500	54 500	59 500	70 000	65 000	77 000	89 000	93 500	74 000	45 500
21 Fishing area total	*78 675*	*115 676*	*124 891*	*312 717*	*329 204*	*326 053*	*317 828*	*180 584*	*267 366*	*331 010*
27 Greenland	3 247	3 408	3 905	4 945	4 663	5 491	6 097	6 347	5 276	2 998
Norway	12 278	17 621	15 048	15 948	7 163	2 716	1 953	18 700	8 192	3 580
Russian Fed	31 500	35 272	29 644	31 528	31 380	13 370	34 850	38 413	39 116	34 187
27 Fishing area total	*47 025*	*56 301*	*48 597*	*52 421*	*43 206*	*21 577*	*42 900*	*63 460*	*52 584*	*40 765*
Species total	*125 700*	*171 977*	*173 488*	*365 138*	*372 410*	*347 630*	*360 728*	*244 044*	*319 950*	*371 775*
Harbour seal	**Phoque veau marin**		**Foca común**		***Phoca vitulina***				**4,06(03)005,02**	**SEC**
21 Greenland	230	235	230	220	250	190	130	110	65	110
21 Fishing area total	*230*	*235*	*230*	*220*	*250*	*190*	*130*	*110*	*65*	*110*
27 Greenland	44	50	46	36	45	27	18	14	8	13
UK	2	8	-	-	-	-	-	-	-	-
27 Fishing area total	*46*	*58*	*46*	*36*	*45*	*27*	*18*	*14*	*8*	*13*
67 USA	2 736	2 621	2 742	2 741	2 546	2 597	...	2 224	2 031	1 834
67 Fishing area total	*2 736*	*2 621*	*2 742*	*2 741*	*2 546*	*2 597*	...	*2 224*	*2 031*	*1 834*
Species total	*3 012*	*2 914*	*3 018*	*2 997*	*2 841*	*2 814*	*148*	*2 348*	*2 104*	*1 957*
Ringed seal	**Phoque annelé ou marbré**		**Foca marbreada**		***Phoca hispida***				**4,06(03)005,03**	**SER**
21 Canada	-	-	-	670	-	1 234	742	1 747	2 035	1 340
Greenland	54 000	50 000	55 000	63 000	56 000	57 000	58 000	56 000	50 000	41 000
21 Fishing area total	*54 000*	*50 000*	*55 000*	*63 670*	*56 000*	*58 234*	*58 742*	*57 747*	*52 035*	*42 340*
27 Finland	16	-	-	-	-	-	-	-	-	-
Greenland	23 154	22 863	24 290	27 309	24 387	25 108	25 453	24 365	22 017	18 032
Russian Fed	530	628	399	155	705	532	599	672	687	581
27 Fishing area total	*23 700*	*23 491*	*24 689*	*27 464*	*25 092*	*25 640*	*26 052*	*25 037*	*22 704*	*18 613*

B-63 Eared seals, hair seals, walruses / Otaries, phoques, morses / Lobos marinos, focas, morsas

Capture production by species, fishing areas and countries or areas / Captures par espèces, zones de pêche et pays ou zones / Capturas por especies, áreas de pesca y países o áreas

Species, Fishing area Espèce, Zone de pêche Especie, Area de pesca	1993 no	1994 no	1995 no	1996 no	1997 no	1998 no	1999 no	2000 no	2001 no	2002 no
61 Russian Fed	16 148	9 198	3 142	1 973	1 758	2 443	1 673	2 645	2 030	1 580
61 Fishing area total	*16 148*	*9 198*	*3 142*	*1 973*	*1 758*	*2 443*	*1 673*	*2 645*	*2 030*	*1 580*
67 Russian Fed	-	-	-	-	-	-	1 641	1 310	-	2 453
67 Fishing area total	-	-	-	-	-	-	*1 641*	*1 310*	-	*2 453*
Species total	*93 848*	*82 689*	*82 831*	*93 107*	*82 850*	*86 317*	*88 108*	*86 739*	*76 769*	*64 986*
Ribbon seal	**Phoque à rubans**		**Foca fajada**		***Phoca fasciata***			**4,06(03)005,04**		**SLR**
61 Russian Fed	13 543	3 585	18	24	20	-	8	8	-	1
61 Fishing area total	*13 543*	*3 585*	*18*	*24*	*20*	-	*8*	*8*	-	*1*
Species total	*13 543*	*3 585*	*18*	*24*	*20*	-	*8*	*8*	-	*1*
Caspian seal	**Phoque de la Mer Caspienne**		**Foca del Caspio**		***Phoca caspica***			**4,06(03)005,05**		**SAC**
05 Russian Fed	23 000	11 700	8 110	4 930	4 320	4 500	4 300	-	9	195
05 Fishing area total	*23 000*	*11 700*	*8 110*	*4 930*	*4 320*	*4 500*	*4 300*	-	*9*	*195*
Species total	*23 000*	*11 700*	*8 110*	*4 930*	*4 320*	*4 500*	*4 300*	-	*9*	*195*
Baikal seal	**Phoque du lac Baikal**		**Foca de Baikal**		***Phoca sibirica***			**4,06(03)005,06**		**SBK**
05 Russian Fed	4 902	2 951	2 272	2 450	2 991	-	193	2 381	2 817	729
05 Fishing area total	*4 902*	*2 951*	*2 272*	*2 450*	*2 991*	-	*193*	*2 381*	*2 817*	*729*
Species total	*4 902*	*2 951*	*2 272*	*2 450*	*2 991*	-	*193*	*2 381*	*2 817*	*729*
Larga seal	**Veau marin du Pacifique**		**Foca largha**		***Phoca largha***			**4,06(03)005,07**		**SST**
61 Russian Fed	4 839	1 778	350	317	167	181	231	331	179	1 824
61 Fishing area total	*4 839*	*1 778*	*350*	*317*	*167*	*181*	*231*	*331*	*179*	*1 824*
67 Russian Fed	-	-	-	-	-	-	222	-	-	159
67 Fishing area total	-	-	-	-	-	-	*222*	-	-	*159*
Species total	*4 839*	*1 778*	*350*	*317*	*167*	*181*	*453*	*331*	*179*	*1 983*
Bearded seal	**Phoque barbu**		**Foca barbuda**		***Erignathus barbatus***			**4,06(03)008,01**		**SEB**
21 Canada	-	-	-	45	-	59	50	75	170	70
Greenland	1 400	1 550	1 550	1 630	1 800	1 800	1 790	2 060	1 700	1 000
21 Fishing area total	*1 400*	*1 550*	*1 550*	*1 675*	*1 800*	*1 859*	*1 840*	*2 135*	*1 870*	*1 070*
27 Greenland	407	424	482	504	549	554	546	635	505	329
Russian Fed	21	31	15	6	24	30	29	24	26	19
27 Fishing area total	*428*	*455*	*497*	*510*	*573*	*584*	*575*	*659*	*531*	*348*
61 Russian Fed	3 255	1 719	1 013	1 114	1 014	890	802	528	340	501
61 Fishing area total	*3 255*	*1 719*	*1 013*	*1 114*	*1 014*	*890*	*802*	*528*	*340*	*501*
67 Russian Fed	-	-	-	-	-	-	338	119	-	326
67 Fishing area total	-	-	-	-	-	-	*338*	*119*	-	*326*
Species total	*5 083*	*3 724*	*3 060*	*3 299*	*3 387*	*3 333*	*3 555*	*3 441*	*2 741*	*2 245*
Hooded seal	**Phoque à crete**		**Foca capuchina**		***Cystophora cristata***			**4,06(03)010,01**		**SEZ**
21 Canada	19	149	857	25 754	7 058	10 020	49	10	257	11
Greenland	4 150	4 800	4 200	5 800	4 400	3 700	4 400	3 400	3 700	2 000
21 Fishing area total	*4 169*	*4 949*	*5 057*	*31 554*	*11 458*	*13 720*	*4 449*	*3 410*	*3 957*	*2 011*
27 Greenland	2 832	3 341	2 997	4 106	3 100	2 628	3 058	2 434	2 567	1 419
Norway	384	492	933	811	2 934	6 351	4 446	1 936	3 820	7 116
Russian Fed	-	4 252	-	-	-	-	-	-	-	-
27 Fishing area total	*3 216*	*8 085*	*3 930*	*4 917*	*6 034*	*8 979*	*7 504*	*4 370*	*6 387*	*8 535*
Species total	*7 385*	*13 034*	*8 987*	*36 471*	*17 492*	*22 699*	*11 953*	*7 780*	*10 344*	*10 546*
Grey seal	**Phoque gris**		**Foca de gris**		***Halichoerus grypus***			**4,06(03)011,01**		**SEG**
27 Finland	23	-	-	-	-	4	30	62	73	86
UK	35	23	-	-	-	-	-	-	-	-
27 Fishing area total	*58*	*23*	-	-	-	*4*	*30*	*62*	*73*	*86*
Species total	*58*	*23*	-	-	-	*4*	*30*	*62*	*73*	*86*
Seals nei	**Phoques nca**		**Focas nep**		***Otariidae, Phocidae***			**4,06(XX)XXX,XX**		**SXX**
21 Canada	1 125	1 798	1 799	190	1 838	55	28	7	39	1 005
21 Fishing area total	*1 125*	*1 798*	*1 799*	*190*	*1 838*	*55*	*28*	*7*	*39*	*1 005*
Species total	*1 125*	*1 798*	*1 799*	*190*	*1 838*	*55*	*28*	*7*	*39*	*1 005*
Group total	***330 255***	***347 327***	***316 090***	***539 589***	***523 244***	***508 041***	***502 014***	***393 057***	***462 121***	***497 945***

See paragraph 5 of the INTRODUCTION. Voir le paragraphe 5 de l'INTRODUCTION. Véase el párrafo 5 en la INTRODUCCIÓN.

B-64 Miscellaneous aquatic mammals / Mammifères aquatiques divers / Mamíferos acuáticos diversos

Capture production by species, fishing areas and countries or areas / Captures par espèces, zones de pêche et pays ou zones / Capturas por especies, áreas de pesca y países o áreas

Species, Fishing area Espèce, Zone de pêche Especie, Area de pesca	1993 t	1994 t	1995 t	1996 t	1997 t	1998 t	1999 t	2000 t	2001 t	2002 t
Aquatic mammals nei	**Mammifères aquatiques nca**		**Mamíferos acuáticos nep**		***Mammalia***				**4,99(XX)XXX,XX**	**MAM**
61 Japan	1 522	1 605	1 259	1 748	1 883	1 242	1 705	1 767	1 874	1 997
61 Fishing area total	*1 522*	*1 605*	*1 259*	*1 748*	*1 883*	*1 242*	*1 705*	*1 767*	*1 874*	*1 997*
Species total	*1 522*	*1 605*	*1 259*	*1 748*	*1 883*	*1 242*	*1 705*	*1 767*	*1 874*	*1 997*
Group total	***1 522***	***1 605***	***1 259***	***1 748***	***1 883***	***1 242***	***1 705***	***1 767***	***1 874***	***1 997***

See paragraph 5 of the INTRODUCTION. Voir le paragraphe 5 de l'INTRODUCTION. Véase el párrafo 5 en la INTRODUCCIÓN.

B-71 Frogs and other amphibians — Capture production by species, fishing areas and countries or areas
Grenouilles et autres amphibies — Captures par espèces, zones de pêche et pays ou zones
Ranas y otros anfibios — Capturas por especies, áreas de pesca y países o áreas

Species, Fishing area Espèce, Zone de pêche Especie, Area de pesca	1993 t	1994 t	1995 t	1996 t	1997 t	1998 t	1999 t	2000 t	2001 t	2002 t
Frogs	**Grenouilles**		**Ranas**		***Rana spp***				**5,12(01)001,XX**	**FRG**
02 Cuba	52	52	62	69	46	28	26	43	45	21
Mexico	312	335	497	351	1 979	1 167	311	295	315	125
USA	20	17	5	2	9	6	-	0	1	-
02 Fishing area total	*384*	*404*	*564*	*422*	*2 034*	*1 201*	*337*	*338*	*361*	*146*
03 Uruguay	-	-	5	0	0	0	0	7	9	5
03 Fishing area total	*-*	*-*	*5*	*0*	*0*	*0*	*0*	*7*	*9*	*5*
04 Bangladesh	700	700	-	-	-	-	-	-	-	-
Indonesia	2 411	2 111	2 194	1 793	1 390	1 667	1 317	1 880	1 243	1 180
Turkey	750	851	864	740	160	100	118	77	873	898
04 Fishing area total	*3 861*	*3 662*	*3 058*	*2 533*	*1 550*	*1 767*	*1 435*	*1 957*	*2 116*	*2 078*
05 Romania	...	...	...	...	38	41	35	26	-	29
05 Fishing area total	*...*	*...*	*...*	*...*	*38*	*41*	*35*	*26*	*-*	*29*
Species total	*4 245*	*4 066*	*3 627*	*2 955*	*3 622*	*3 009*	*1 807*	*2 328*	*2 486*	*2 258*
Group total	***4 245***	***4 066***	***3 627***	***2 955***	***3 622***	***3 009***	***1 807***	***2 328***	***2 486***	***2 258***

B-72 Turtles / Tortues / Tortugas

Capture production by species, fishing areas and countries or areas
Captures par espèces, zones de pêche et pays ou zones
Capturas por especies, áreas de pesca y países o áreas

Species, Fishing area Espèce, Zone de pêche Especie, Area de pesca	1993 t	1994 t	1995 t	1996 t	1997 t	1998 t	1999 t	2000 t	2001 t	2002 t
Diamond back terrapins	**Tortues diamantées**		**Tortugas comestibles**		***Malaclemys spp***				**5,31(06)021,XX**	**TTG**
02 USA	0	0	0	-	-	0	-	0	-	-
02 Fishing area total	*0*	*0*	*0*	-	-	*0*	-	*0*	-	-
Species total	*0*	*0*	*0*	-	-	*0*	-	*0*	-	-
Green turtle	**Tortue verte**		**Tortuga verde**		***Chelonia mydas***				**5,31(07)005,02**	**TUG**
31 Cuba	122	115	46	33	18	14	8	10	5	6
Grenada	8	4	7	8	6	6	5	5	7	8
Venezuela	0	0	0	0	0	0	-	-	-	-
31 Fishing area total	*130*	*119*	*53*	*41*	*24*	*20*	*13*	*15*	*12*	*14*
71 Fiji Islands	49	25 F	6	22	4	2	2	2 F	3	2 F
71 Fishing area total	*49*	*25 F*	*6*	*22*	*4*	*2*	*2*	*2 F*	*3*	*2 F*
Species total	*179*	*144 F*	*59*	*63*	*28*	*22*	*15*	*17 F*	*15*	*16 F*
Hawksbill turtle	**Tortue caret**		**Tortuga carey**		***Eretmochelys imbricata***				**5,31(07)017,01**	**TTH**
31 Cuba	117	45	20	23	19	12	12	18	11	17
Venezuela	0	0	0	0	0	0	-	-	-	-
31 Fishing area total	*117*	*45*	*20*	*23*	*19*	*12*	*12*	*18*	*11*	*17*
Species total	*117*	*45*	*20*	*23*	*19*	*12*	*12*	*18*	*11*	*17*
Loggerhead turtle	**Caouane**		**Caguama**		***Caretta caretta***				**5,31(07)018,01**	**TTL**
31 Cuba	48	23	11	10	7	5	5	9	7	8
Venezuela	0	0	0	0	0	0	-	-	-	-
31 Fishing area total	*48*	*23*	*11*	*10*	*7*	*5*	*5*	*9*	*7*	*8*
Species total	*48*	*23*	*11*	*10*	*7*	*5*	*5*	*9*	*7*	*8*
River and lake turtles nei	**Tortues d'eau douce nca**		**Galápagos nep**		***Testudinata***				**5,31(XX)XXX,XX**	**TUL**
02 Cuba	0	1	9	3	4	2	-	-	1	-
USA	28	13	6	15	13	24	15	53	51	27
02 Fishing area total	*28*	*14*	*15*	*18*	*17*	*26*	*15*	*53*	*52*	*27*
03 Brazil	10 F	5 F	0	0	0	0	0	0	-	-
Venezuela	0	0	0	0	0	0	-	-	-	-
03 Fishing area total	*10 F*	*5 F*	*0*	*0*	*0*	*0*	*0*	*0*	-	-
04 Indonesia	-	56	-	36	48	27	1	3	42	40
04 Fishing area total	-	*56*	-	*36*	*48*	*27*	*1*	*3*	*42*	*40*
Species total	*38 F*	*75 F*	*15*	*54*	*65*	*53*	*16*	*56*	*94*	*67*
Marine turtles nei	**Tortues de mer nca**		**Tortugas de mar nep**		***Testudinata***				**5,31(XX)XXX,XX**	**TTX**
31 Bahamas	4	2	2	3	3	3	1	2	4	3 F
Costa Rica	...	113	101	149	33	86	0	0	-	-
Cuba	8	3	13	1	0	10	-	-	-	-
Guadeloupe	10 F	5 F	0	0	0	0	-	-	-	-
Martinique	0	0	0	0	0	0	-	-	-	-
Puerto Rico	0	0	0	0	0	0	-	-	-	-
Venezuela	0	0	0	0	0	0	-	-	-	-
31 Fishing area total	*22 F*	*123 F*	*116*	*153*	*36*	*99*	*1*	*2*	*4*	*3 F*
34 Benin	...	...	...	...	...	...	29	20 F	5	1
Cape Verde	5	0	0	0	0	0	-	-	-	-
Côte dIvoire	...	...	...	...	...	...	...	50	71	...
Eq Guinea	50 F	20 F	10 F	5 F	5 F	9	10 F	1 F	1 F	1 F
Gabon	...	...	12	37	159	180	424	51	238	843
Liberia	4	-	-	-	-	-	-	-	-	-
34 Fishing area total	*59 F*	*20 F*	*22 F*	*42 F*	*164 F*	*189*	*463 F*	*122 F*	*315 F*	*845 F*
41 Brazil	5 F	0	0	0	0	0	0	0	-	-
41 Fishing area total	*5 F*	*0*	*0*	*0*	*0*	*0*	*0*	*0*	-	-
51 Madagascar	20	-	-	-	-	-	-	-	-	-
Seychelles	...	10	-	-	-	-	-	-	-	-
51 Fishing area total	*20*	*10*	-	-	-	-	-	-	-	-
57 Indonesia	52	337	291	353	248	337	272	371	44	70
Timor-Leste	...	...	...	...	...	...	1 F	1	1	1 F
57 Fishing area total	*52*	*337*	*291*	*353*	*248*	*337*	*273 F*	*372*	*45*	*71 F*
71 Fiji Islands	20	15 F	7	24	7	2	8	6 F	7	7 F
Indonesia	220	207	284	368	390	444	434	373	306	320
Micronesia	2 F	2 F	0	0	0	0	0	0	0	0
N Marianas	0	0	0	0	-	-	-	-	-	-
Philippines	-	-	1	1	-	2	2	1	1	1
Wallis Fut I	...	2	2	2	2	2	2 F	2 F	2 F	2 F
71 Fishing area total	*242 F*	*226 F*	*294*	*395*	*399*	*450*	*446 F*	*382 F*	*316 F*	*330 F*
77 Honduras	-	-	-	-	-	3 F	1	21	...	...

B-72 Turtles / Tortues / Tortugas

Capture production by species, fishing areas and countries or areas
Captures par espèces, zones de pêche et pays ou zones
Capturas por especies, áreas de pesca y países o áreas

Species, Fishing area Espèce, Zone de pêche Especie, Area de pesca	1993 t	1994 t	1995 t	1996 t	1997 t	1998 t	1999 t	2000 t	2001 t	2002 t
Panama	0	0	0	0	0	0	0	0	...	...
77 Fishing area total	*0*	*0*	*0*	*0*	*0*	*3 F*	*1*	*21*	*...*	*...*
87 Ecuador	-	-	10	10	10	10	10	10	10	10 F
Peru	28	6	4	0	1	2	1	1	2	2
87 Fishing area total	*28*	*6*	*14*	*10*	*11*	*12*	*11*	*11*	*12*	*12 F*
Species total	*428 F*	*722 F*	*737 F*	*953 F*	*858 F*	*1 090 F*	*1 195 F*	*910 F*	*692 F*	*1 261 F*
Group total	***810***	***1 009***	***842***	***1 103***	***977***	***1 182***	***1 243***	***1 010***	***819***	***1 369***

B-73 Crocodiles and alligators — Capture production by species, fishing areas and countries or areas
Crocodiles et alligators — Captures par espèces, zones de pêche et pays ou zones
Cocodrilos y aligatores — Capturas por especies, áreas de pesca y países o áreas

Species, Fishing area Espèce, Zone de pêche Especie, Area de pesca	1993 no	1994 no	1995 no	1996 no	1997 no	1998 no	1999 no	2000 no	2001 no	2002 no
Spectacled caiman	**Caïman à lunettes**		**Caimán de anteojos**		***Caiman crocodilus***				**5,36(01)001,03**	**CAI**
02 Costa Rica	-	-	-	-	-	40	-	-	-	-
Cuba	-	-	-	302	506	5	3	-	-	-
Honduras	4 000 F	3 000 F	2 000	6 000	...	...	-	-	-	-
Nicaragua	9 963	8 919	4 238	10 795	1 590	3 927	250	6 440	...	...
Panama	7 869	2 840	2 005	46	500	3 022	10	10 250	11 700	13 298
02 Fishing area total	*21 832 F*	*14 759 F*	*8 243*	*17 143*	*2 596*	*6 994*	*263*	*16 690*	*11 700*	*13 298*
03 Bolivia	...	...	...	...	15 961	1 757	17 500	...	28 170	31 021
Brazil	7 034	43 633	369	659	7 307	2 092	4 619	8 286	1 253	6 048
Colombia	457 749	536 501	828 533	656 522	452 707	670 389	771 456	832 203	704 313	585 000
Guyana	1 558	685	1 556	2 650	910	...	9 880	9 880	5 917	8 222
Paraguay	...	5 466	19 793	725	503	4 445	-	9 750	3 793	8 373
Venezuela	78 972	54 038	55 195	29 996	33 528	35 579	24 640	23 655	19 215	20 349
03 Fishing area total	*545 313*	*640 323*	*905 446*	*690 552*	*510 916*	*714 262*	*828 095*	*883 774*	*762 661*	*659 013*
04 China,Taiwan	21	-	-	-	-	-	-	-	-	-
04 Fishing area total	*21*	-	-	-	-	-	-	-	-	-
Species total	*567 166 F*	*655 082 F*	*913 689*	*707 695*	*513 512*	*721 256*	*828 358*	*900 464*	*774 361*	*672 311*
American alligator	**Alligator américain**		**Caimán americano**		***Alligator mississippiensis***				**5,36(01)002,01**	**AGM**
02 USA	201 431	185 853	197 368	190 841	223 093	206 620	239 519	248 922	265 470	266 628
02 Fishing area total	*201 431*	*185 853*	*197 368*	*190 841*	*223 093*	*206 620*	*239 519*	*248 922*	*265 470*	*266 628*
04 Israel	1 055	1 815	348	944	210	401	425	233	6	-
04 Fishing area total	*1 055*	*1 815*	*348*	*944*	*210*	*401*	*425*	*233*	*6*	-
Species total	*202 486*	*187 668*	*197 716*	*191 785*	*223 303*	*207 021*	*239 944*	*249 155*	*265 476*	*266 628*
Estuarine crocodile	**Crocodile d'estuaires**		**Cocodrilo estuarino**		***Crocodylus porosus***				**5,36(01)003,01**	**CDP**
04 Cambodia	4 816	6 200	14 691	20 200	17 000	40 700	25 380	26 300	35 970	50 850
Indonesia	1 064	3 346	...	...	150	3 141	1 087	3 172	3 456	...
Malaysia	2 090	2 522	398	120	120	320	120	559	375	122
Singapore	286	301	1 004	411	296	416	350	481	1 041	1 058
Thailand	-	1	419	160	440	...	...	...	...	...
04 Fishing area total	*8 256*	*12 370*	*16 512*	*20 891*	*18 006*	*44 577*	*26 937*	*30 512*	*40 842*	*52 030*
06 Australia	6 886	5 356	7 251	9 054	8 777	9 896	5 048	13 296	11 849	7 205
Papua N Guin	8 415	7 551	12 908	10 597	8 578	12 138	9 388	8 336	10 677	9 332
06 Fishing area total	*15 301*	*12 907*	*20 159*	*19 651*	*17 355*	*22 034*	*14 436*	*21 632*	*22 526*	*16 537*
Species total	*23 557*	*25 277*	*36 671*	*40 542*	*35 361*	*66 611*	*41 373*	*52 144*	*63 368*	*68 567*
Siamese crocodile	**Crocodile siamois**		**Cocodrilo del Siam**		***Crocodylus siamensis***				**5,36(01)003,02**	**CDS**
04 Cambodia	-	-	-	-	-	-	-	-	30	...
Thailand	19	2 067	4 372	3 186	5 452	...	...	4 945	2 399	...
04 Fishing area total	*19*	*2 067*	*4 372*	*3 186*	*5 452*	...	...	*4 945*	*2 429*	...
Species total	*19*	*2 067*	*4 372*	*3 186*	*5 452*	...	...	*4 945*	*2 429*	...
Australian crocodile	**Crocodile australien**		**Cocodrilo de Australia**		***Crocodylus johnstoni***				**5,36(01)003,03**	**CRH**
06 Australia	4 290	2 381	3 132	1 641	194	309	44	10	-	2
06 Fishing area total	*4 290*	*2 381*	*3 132*	*1 641*	*194*	*309*	*44*	*10*	-	*2*
Species total	*4 290*	*2 381*	*3 132*	*1 641*	*194*	*309*	*44*	*10*	-	*2*
Nile crocodile	**Crocodile du Nil**		**Cocodrilo del Nilo**		***Crocodylus niloticus***				**5,36(01)003,04**	**CRI**
01 Botswana	7 414	587	699	347	338	2	9	11	152	-
Ethiopia	594	2	2 005	...	...	...	991	930	42	59
Guinea	-	-	100	...	...	...	...	...	-	-
Kenya	3 721	4 258	2 250	300	1 445	714	3 350	3 460	4 250	3 967
Madagascar	1 909	2 800	2 411	4 589	5 814	6 520	4 302	6 606	9 408	6 936
Malawi	2 036	1 732	950	636	400	200	199	200	1 256	60
Mozambique	3 164	1 042	3 021	523	1 430	810	585	718	477	7 322
Namibia	...	277	515	210	120	53	115	105	-	-
South Africa	18 451	25 416	14 805	2 280	13 322	8 863	27 631	29 733	33 174	75 882
Tanzania	144	342	915	1 185	630	777	827	1 302	1 589	1 369
Uganda	4 019	4 817	...	...	...	...	...	508	900	...
Zambia	8 645	6 059	11 644	2 415	12 228	9 250	19 702	19 740	20 900	22 259
Zimbabwe	54 111	39 271	39 590	38 295	52 829	40 720	63 064	81 898	76 157	85 339
01 Fishing area total	*104 208*	*86 603*	*78 905*	*50 780*	*88 556*	*67 909*	*120 775*	*145 211*	*148 305*	*203 193*
03 Brazil	-	-	-	-	-	-	720	1 477	50	-
03 Fishing area total	-	-	-	-	-	-	*720*	*1 477*	*50*	-
04 Israel	-	-	-	-	-	...	552	1 661	2 289	699
04 Fishing area total	-	-	-	-	-	...	*552*	*1 661*	*2 289*	*699*
Species total	*104 208*	*86 603*	*78 905*	*50 780*	*88 556*	*67 909*	*122 047*	*148 349*	*150 644*	*203 892*

B-73

Crocodiles and alligators — **Capture production by species, fishing areas and countries or areas**
Crocodiles et alligators — **Captures par espèces, zones de pêche et pays ou zones**
Cocodrilos y aligatores — **Capturas por especies, áreas de pesca y países o áreas**

Species, Fishing area Espèce, Zone de pêche Especie, Area de pesca	1993 no	1994 no	1995 no	1996 no	1997 no	1998 no	1999 no	2000 no	2001 no	2002 no
New Guinea crocodile	**Crocodile de Nouvelle-Guinée**		**Cocodrilo de Nueva Guinea**		***Crocodylus novaeguineae***				**5,36(01)003,05**	**CNG**
04 Indonesia	2 263	9 016	...	...	100	8 506	6 574	7 215	9 946	...
04 Fishing area total	*2 263*	*9 016*	...	...	*100*	*8 506*	*6 574*	*7 215*	*9 946*	...
06 Australia	-	-	-	-	-	-	139	-	-	-
Papua N Guin	19 133	22 596	19 556	14 234	32 911	15 078	15 617	16 018	20 668	18 798
06 Fishing area total	*19 133*	*22 596*	*19 556*	*14 234*	*32 911*	*15 078*	*15 756*	*16 018*	*20 668*	*18 798*
Species total	*21 396*	*31 612*	*19 556*	*14 234*	*33 011*	*23 584*	*22 330*	*23 233*	*30 614*	*18 798*
Cuban crocodile	**Crocodile cubain**		**Cocodrilo de Cuba**		***Crocodylus rhombifer***				**5,36(01)003,06**	**CMB**
02 Cuba	-	-	99	40	-	3	-	-	-	-
02 Fishing area total	-	-	*99*	*40*	-	*3*	-	-	-	-
Species total	-	-	*99*	*40*	-	*3*	-	-	-	-
Morelet's crocodile	**Crocodile de Morelet**		**...C**		***Crocodylus moreletii***				**5,36(01)003,07**	**CME**
02 Mexico	-	-	2	20	146	193	2	1 228	3 643	1 588
02 Fishing area total	-	-	*2*	*20*	*146*	*193*	*2*	*1 228*	*3 643*	*1 588*
Species total	-	-	*2*	*20*	*146*	*193*	*2*	*1 228*	*3 643*	*1 588*
American crocodile	**...B**		**...C**		***Crocodylus acutus***				**5,36(01)003,08**	**YUU**
03 Colombia	-	-	-	-	-	-	-	-	100	...
03 Fishing area total	-	-	-	-	-	-	-	-	*100*	...
Species total	-	-	-	-	-	-	-	-	*100*	...
Cuvier's Dwarf caiman	**...B**		**...C**		***Paleosuchus palpebrosus***				**5,36(01)008,01**	**UCB**
03 Guyana	-	-	-	-	-	-	409	409	476	486
03 Fishing area total	-	-	-	-	-	-	*409*	*409*	*476*	*486*
Species total	-	-	-	-	-	-	*409*	*409*	*476*	*486*
Smooth-fronted caiman	**...B**		**...C**		***Paleosuchus trigonatus***				**5,36(01)008,02**	**UCI**
03 Guyana	-	-	-	-	-	-	270	270	423	500
03 Fishing area total	-	-	-	-	-	-	*270*	*270*	*423*	*500*
Species total	-	-	-	-	-	-	*270*	*270*	*423*	*500*
Group total	***923 122***	***990 690***	***1 254 142***	***1 009 923***	***899 535***	***1 086 886***	***1 254 777***	***1 380 207***	***1 291 534***	***1 232 772***

The data, derived mainly from sources of the UNEP World Conservation Monitoring Centre, refer to captive breeding, ranching and wild harvest.

See paragraph 5 of the INTRODUCTION.

Les données, provenant principalement des sources du Centre mondial de surveillance de la conservation du PNUE, ont trait à l'élevage en captivité, au 'ranching' et à la récolte sauvage.

Voir le paragraphe 5 de l'INTRODUCTION.

Los datos, obtenidos por la mayoría desde fuentes del Centro Mundial de Vigilancia de la Conservación del PNUMA, se refieren a la cría en cautividad, las grandes explotaciones y las capturas en libertad.

Véase el párrafo 5 en la INTRODUCCIÓN.

B-74 Sea-squirts and other tunicates / Ascidiens et autres tuniciers / Ascidias y otros tunicados

Capture production by species, fishing areas and countries or areas / Captures par espèces, zones de pêche et pays ou zones / Capturas por especies, áreas de pesca y países o áreas

Species, Fishing area Espèce, Zone de pêche Especie, Area de pesca	1993 t	1994 t	1995 t	1996 t	1997 t	1998 t	1999 t	2000 t	2001 t	2002 t
Red sea squirt	**Violet chilien**		**Piure chileno**		***Pyura chilensis***				**6,96(09)005,02**	**SSE**
87 Chile	3 992	3 009	3 297	4 549	3 174	2 530	2 704	2 290	1 298	1 217
87 Fishing area total	*3 992*	*3 009*	*3 297*	*4 549*	*3 174*	*2 530*	*2 704*	*2 290*	*1 298*	*1 217*
Species total	*3 992*	*3 009*	*3 297*	*4 549*	*3 174*	*2 530*	*2 704*	*2 290*	*1 298*	*1 217*
Grooved sea squirt	**Violet**		**Provecho**		***Microcosmus sulcatus***				**6,96(09)037,01**	**SSG**
37 France	147	127	...	28	22	22	30	30	76	78
37 Fishing area total	*147*	*127*	*...*	*28*	*22*	*22*	*30*	*30*	*76*	*78*
Species total	*147*	*127*	*...*	*28*	*22*	*22*	*30*	*30*	*76*	*78*
Sea squirts nei	**Ascidiens nca**		**Ascidias nep**		***Ascidiacea***				**6,96(XX)XXX,XX**	**SSX**
51 India	-	-	-	-	-	-	-	95	-	-
51 Fishing area total	-	-	-	-	-	-	-	*95*	-	-
61 Korea Rep	402	2 178	5 767	16 747	2 780	891	1 171	1 443	1 053	1 025
61 Fishing area total	*402*	*2 178*	*5 767*	*16 747*	*2 780*	*891*	*1 171*	*1 443*	*1 053*	*1 025*
Species total	*402*	*2 178*	*5 767*	*16 747*	*2 780*	*891*	*1 171*	*1 538*	*1 053*	*1 025*
Salps	**Salpes**		**...C**		***Salpidae***				**6,97(01)XXX,XX**	**SPX**
58 India	-	-	-	7	-	-	-	-	-	-
58 Fishing area total	-	-	-	*7*	-	-	-	-	-	-
Species total	-	-	-	*7*	-	-	-	-	-	-
Group total	***4 541***	***5 314***	***9 064***	***21 331***	***5 976***	***3 443***	***3 905***	***3 858***	***2 427***	***2 320***

B-75 Horseshoe crabs and other arachnoids / Limules et autres arachnoïdés / Límulos y otros arácnidos

Capture production by species, fishing areas and countries or areas / Captures par espèces, zones de pêche et pays ou zones / Capturas por especies, áreas de pesca y países o áreas

Species, Fishing area Espèce, Zone de pêche Especie, Area de pesca	1993 t	1994 t	1995 t	1996 t	1997 t	1998 t	1999 t	2000 t	2001 t	2002 t
Horseshoe crab	**Limule**		**Límulo(=Cangrejo cacerola)**			***Limulus polyphemus***			**6,56(01)001,02**	**HSC**
21 USA	810	634	926	1 598	2 606	3 182	2 238	1 557	1 071	1 257
21 Fishing area total	*810*	*634*	*926*	*1 598*	*2 606*	*3 182*	*2 238*	*1 557*	*1 071*	*1 257*
31 USA	-	-	-	-	1	70	159	139	228	130
31 Fishing area total	-	-	-	-	*1*	*70*	*159*	*139*	*228*	*130*
Species total	*810*	*634*	*926*	*1 598*	*2 607*	*3 252*	*2 397*	*1 696*	*1 299*	*1 387*
Group total	***810***	***634***	***926***	***1 598***	***2 607***	***3 252***	***2 397***	***1 696***	***1 299***	***1 387***

B-76 Sea-urchins and other echinoderms / Oursins et autres échinodermes / Erizos de mar y otros equinodermos

Capture production by species, fishing areas and countries or areas / Captures par espèces, zones de pêche et pays ou zones / Capturas por especies, áreas de pesca y países o áreas

Species, Fishing area Espèce, Zone de pêche Especie, Area de pesca	1993 t	1994 t	1995 t	1996 t	1997 t	1998 t	1999 t	2000 t	2001 t	2002 t
Echinoderms	**Oursins, bèches-de-mer**		**Erizos, cohombros de mar**		***Echinodermata***			**6,89(XX)XXX,XX**		**ECH**
27 Portugal	-	-	-	-	-	-	-	-	15	1
Spain	-	-	-	485	586	551	616	306	304	603
27 Fishing area total	-	-	-	*485*	*586*	*551*	*616*	*306*	*319*	*604*
31 Mexico	8	2	0	0	0	0	0	0	0	-
31 Fishing area total	*8*	*2*	*0*	*0*	*0*	*0*	*0*	*0*	*0*	-
37 Spain	-	-	-	2	4	9	5	3	2	-
37 Fishing area total	-	-	-	*2*	*4*	*9*	*5*	*3*	*2*	-
61 Korea Rep	3 944	3 714	3 707	2 802	2 771	1 410	1 182	1 461	1 454	1 459
61 Fishing area total	*3 944*	*3 714*	*3 707*	*2 802*	*2 771*	*1 410*	*1 182*	*1 461*	*1 454*	*1 459*
77 Fr Polynesia	0	0	0	0	10	10	10	15	15	15
Mexico	2 809	3 419	2 791	3 027	2 099	1 138	2 042	2 813	2 278	2 068
77 Fishing area total	*2 809*	*3 419*	*2 791*	*3 027*	*2 109*	*1 148*	*2 052*	*2 828*	*2 293*	*2 083*
81 New Zealand	848	944	804	277	627	832	643	712	853	738
81 Fishing area total	*848*	*944*	*804*	*277*	*627*	*832*	*643*	*712*	*853*	*738*
87 Peru	13	15	131	461	424	90	1 204	1 626	2 114	2 245
87 Fishing area total	*13*	*15*	*131*	*461*	*424*	*90*	*1 204*	*1 626*	*2 114*	*2 245*
Species total	*7 622*	*8 094*	*7 433*	*7 054*	*6 521*	*4 040*	*5 702*	*6 936*	*7 035*	*7 129*
Starfishes nei	**Astéridés nca**		**Estrellas nep**		***Asteroidea***			**6,91(XX)XXX,XX**		**STF**
21 USA	1	9	0	-	-	-	-	0	-	-
21 Fishing area total	*1*	*9*	*0*	-	-	-	-	*0*	-	-
27 Denmark	0	0	0	-	-	-	0	0	0	0
27 Fishing area total	*0*	*0*	*0*	-	-	-	*0*	*0*	*0*	*0*
77 USA	-	-	-	-	-	-	1	-	0	1
77 Fishing area total	-	-	-	-	-	-	*1*	-	*0*	*1*
81 New Zealand	-	-	9	4	0	13	2	6	9	39
81 Fishing area total	-	-	*9*	*4*	*0*	*13*	*2*	*6*	*9*	*39*
88 New Zealand	-	-	-	-	-	-	-	-	2	0
88 Fishing area total	-	-	-	-	-	-	-	-	*2*	*0*
Species total	*1*	*9*	*9*	*4*	*0*	*13*	*3*	*6*	*11*	*40*
Sea urchins nei	**Oursins nca**		**Erizos nep**		***Strongylocentrotus spp***			**6,93(02)004,XX**		**URC**
21 Canada	1 153	2 318	3 167	3 798	3 679	3 707	3 662	3 125	2 775	2 295
St Pier Mq	-	-	1	1	-	-	-	0	0	-
USA	19 237	16 445	14 558	10 052	8 522	7 013	7 132	6 006	4 499	2 412
21 Fishing area total	*20 390*	*18 763*	*17 726*	*13 851*	*12 201*	*10 720*	*10 794*	*9 131*	*7 274*	*4 707*
27 Faeroe Is	-	14	-	-	-	-	-	-	-	-
Norway	-	-	-	-	-	-	-	1	0	0
Russian Fed	-	-	-	-	-	-	-	149	151	38
UK	-	-	-	1	0	-	-	-	-	-
27 Fishing area total	-	*14*	-	*1*	*0*	-	-	*150*	*151*	*38*
31 Grenada	5	36	0	0	0	0	0	0	-	-
Martinique	16	15 F	15 F	10 F	15 F	15 F	15 F	10	10	10
31 Fishing area total	*21*	*51 F*	*15 F*	*10 F*	*15 F*	*15 F*	*15 F*	*10*	*10*	*10*
61 China	100	150	150	200	200	200	200	200	200	200
China,Taiwan	31	51	63	59	61	39	33	41	50	63
Japan	13 713	15 525	13 735	12 996	14 297	13 653	13 530	12 455	11 208	12 733
Korea D P Rp	100 F	100	140	150	150	100 F	100 F	100 F	100 F	100 F
Russian Fed	2 460	2 069	2 344	1 608	1 153	1 560	1 245	1 528	1 612	2 582
61 Fishing area total	*16 404 F*	*17 895*	*16 432*	*15 013*	*15 861*	*15 552 F*	*15 108 F*	*14 324 F*	*13 170 F*	*15 678 F*
67 Canada	7 102	6 161	6 666	5 867	5 542	6 160	5 390	4 887	4 288	4 146
USA	4 048	3 321	1 974	1 218	3 502	1 921	1 711	1 954	2 090	2 053
67 Fishing area total	*11 150*	*9 482*	*8 640*	*7 085*	*9 044*	*8 081*	*7 101*	*6 841*	*6 378*	*6 199*
71 Fiji Islands	55	56 F	59	40	95	103	100	90 F	96	95 F
Philippines	74	151	466	452	296	161	143	125	127	129
71 Fishing area total	*129*	*207 F*	*525*	*492*	*391*	*264*	*243*	*215 F*	*223*	*224 F*
77 Cook Is	26 F	25 F	25 F	20 F	20 F	20 F	20 F	20 F	20 F	20 F
USA	9 084	9 401	9 991	9 111	8 192	4 692	6 375	6 054	5 871	6 162
77 Fishing area total	*9 110 F*	*9 426 F*	*10 016 F*	*9 131 F*	*8 212 F*	*4 712 F*	*6 395 F*	*6 074 F*	*5 891 F*	*6 182 F*
87 Ecuador	0	0	0	0	0	0	0	0	0	-
87 Fishing area total	*0*	*0*	*0*	*0*	*0*	*0*	*0*	*0*	*0*	-
Species total	*57 204 F*	*55 838 F*	*53 354 F*	*45 583 F*	*45 724 F*	*39 344 F*	*39 656 F*	*36 745 F*	*33 097 F*	*33 038 F*

B-76

Sea-urchins and other echinoderms — **Capture production by species, fishing areas and countries or areas**
Oursins et autres échinodermes — **Captures par espèces, zones de pêche et pays ou zones**
Erizos de mar y otros equinodermos — **Capturas por especies, áreas de pesca y países o áreas**

Species, Fishing area Espèce, Zone de pêche Especie, Area de pesca	1993 t	1994 t	1995 t	1996 t	1997 t	1998 t	1999 t	2000 t	2001 t	2002 t
Stony sea urchin	**Oursin-pierre**		**Erizo de mar**		***Paracentrotus lividus***				**6,93(04)007,01**	**URM**
27 France	17	17	7	0	1	12	14	6	3	-
27 Fishing area total	*17*	*17*	*7*	*0*	*1*	*12*	*14*	*6*	*3*	*-*
37 France	240	142	71	63	47	47	70	192	98	164
37 Fishing area total	*240*	*142*	*71*	*63*	*47*	*47*	*70*	*192*	*98*	*164*
Species total	*257*	*159*	*78*	*63*	*48*	*59*	*84*	*198*	*101*	*164*
European edible sea urchin	**Oursin d'Europe**		**Erizo europeo**		***Echinus esculentus***				**6,93(04)014,01**	**URS**
27 Denmark	0	0	0	0	0	0	1	0	0	0
Iceland	713	1 409	923	423	20	-	10	-	-	-
Ireland	26	34	10	2	5	1	2	1	5	-
27 Fishing area total	*739*	*1 443*	*933*	*425*	*25*	*1*	*13*	*1*	*5*	*0*
Species total	*739*	*1 443*	*933*	*425*	*25*	*1*	*13*	*1*	*5*	*0*
Chilean sea urchin	**Oursin chilien**		**Erizo blanco**		***Loxechinus albus***				**6,93(04)017,01**	**UCH**
87 Chile	31 300	39 705	54 609	51 437	45 560	44 843	55 654	54 096	46 794	60 166
87 Fishing area total	*31 300*	*39 705*	*54 609*	*51 437*	*45 560*	*44 843*	*55 654*	*54 096*	*46 794*	*60 166*
Species total	*31 300*	*39 705*	*54 609*	*51 437*	*45 560*	*44 843*	*55 654*	*54 096*	*46 794*	*60 166*
Japanese sea cucumber	**Bèche-de-mer japonaise**		**Cohombro de mar japonés**		***Stichopus japonicus***				**6,94(14)004,01**	**CUJ**
61 Japan	5 996	6 106	6 602	7 226	7 160	6 952	6 662	6 957	7 229	7 259
Korea Rep	2 068	2 117	1 892	1 979	2 217	1 439	1 204	1 419	900	833
61 Fishing area total	*8 064*	*8 223*	*8 494*	*9 205*	*9 377*	*8 391*	*7 866*	*8 376*	*8 129*	*8 092*
Species total	*8 064*	*8 223*	*8 494*	*9 205*	*9 377*	*8 391*	*7 866*	*8 376*	*8 129*	*8 092*
Sea cucumbers nei	**Bèches-de-mer nca**		**Cohombros de mar nep**		***Holothurioidea***				**6,94(XX)XXX,XX**	**CUX**
21 USA	0	1 505	0	1 288	-	2 406	3 504	4 309	1 504	4 656
21 Fishing area total	*0*	*1 505*	*0*	*1 288*	*-*	*2 406*	*3 504*	*4 309*	*1 504*	*4 656*
27 Iceland	-	-	2	-	-	-	-	-	-	-
Spain	-	-	-	-	-	-	-	-	-	8
27 Fishing area total	*-*	*-*	*2*	*-*	*-*	*-*	*-*	*-*	*-*	*8*
31 Mexico	-	-	-	-	-	-	-	-	39	-
31 Fishing area total	*-*	*-*	*-*	*-*	*-*	*-*	*-*	*-*	*39*	*-*
37 Spain	-	-	-	4	4	4	1	9	4	2
37 Fishing area total	*-*	*-*	*-*	*4*	*4*	*4*	*1*	*9*	*4*	*2*
51 Egypt	-	-	-	-	-	-	-	20	139	2 310
Kenya	14	41	55	15	41	38	15	30	13	68
Madagascar	1 350	5 400	5 400	5 400	5 400	1 446	1 500	1 500	1 500	500
Maldives	72	66	94	145	318	85	54	205	226	191
Mozambique	0	0	6	54	7	2	8	12	11	10
Tanzania	980	1 591	1 460	1 644	1 527	1 800	189	372	340	65
Yemen	-	102	-	-	-	-	1	-	-	14
51 Fishing area total	*2 416*	*7 200*	*7 015*	*7 258*	*7 293*	*3 371*	*1 767*	*2 139*	*2 229*	*3 158*
57 Indonesia	608	548	227	269	338	630	689	903	697	770
Sri Lanka	65	92	100	150	272	203	170	145	90	150
57 Fishing area total	*673*	*640*	*327*	*419*	*610*	*833*	*859*	*1 048*	*787*	*920*
67 USA	472	636	729	491	-	-	228	274	300	249
67 Fishing area total	*472*	*636*	*729*	*491*	*-*	*-*	*228*	*274*	*300*	*249*
71 Cambodia	23	18	45	42	42	39	30	21	20	20
Fiji Islands	191	400 F	835	850	790	400	880	800 F	824	820 F
Indonesia	1 756	2 584	2 335	2 174	2 800	2 428	1 928	2 138	2 820	3 060
Kiribati	...	...	...	...	136	154	89	64	60	260
NewCaledonia	777	798	480	776	565	402	493	615	489	450 F
Palau	0	0	0	0	0	0	0	0	0	0
Papua N Guin	1 440	627	1 335	1 788	1 515	2 037	1 185	1 824	1 453	1 450 F
Philippines	3 109	1 497	2 062	2 123	1 191	830	849	730	791	801
Solomon Is	316	285	219	113	203	253	376	48	50 F	50 F
Vanuatu	40 F	40 F	50 F	50 F	50 F	50 F	50 F	50 F	50 F	50 F
71 Fishing area total	*7 652 F*	*6 249 F*	*7 361 F*	*7 916 F*	*7 292 F*	*6 593 F*	*5 880 F*	*6 290 F*	*6 557 F*	*6 961 F*
77 Mexico	...	...	...	...	...	271	234	426	442	290
Tonga	...	...	...	86	80	90	0	0	0	0
77 Fishing area total	*...*	*...*	*...*	*86*	*80*	*361*	*234*	*426*	*442*	*290*
81 New Zealand	-	-	4	1	0	-	-	-	0	1
81 Fishing area total	*-*	*-*	*4*	*1*	*0*	*-*	*-*	*-*	*0*	*1*
87 Chile	13	4	106	115	1	30	108	1 510	107	106
Ecuador	12	12	12	12	15	15	15	15	15	15 F
87 Fishing area total	*25*	*16*	*118*	*127*	*16*	*45*	*123*	*1 525*	*122*	*121 F*
Species total	*11 238 F*	*16 246 F*	*15 556 F*	*17 590 F*	*15 295 F*	*13 613 F*	*12 596 F*	*16 020 F*	*11 984 F*	*16 366 F*
Group total	***116 425***	***129 717***	***140 466***	***131 361***	***122 550***	***110 304***	***121 574***	***122 378***	***107 156***	***124 995***

B-77 Miscellaneous aquatic invertebrates / Invertébrés aquatiques divers / Invertebrados acuáticos diversos

Capture production by species, fishing areas and countries or areas / Captures par espèces, zones de pêche et pays ou zones / Capturas por especies, áreas de pesca y países o áreas

Species, Fishing area Espèce, Zone de pêche Especie, Area de pesca	1993 t	1994 t	1995 t	1996 t	1997 t	1998 t	1999 t	2000 t	2001 t	2002 t
Jellyfishes	**Méduses**		**Medusas**		***Rhopilema spp***				**6,18(41)007,XX**	**JEL**
21 USA	-	-	-	-	-	-	578	-	-	-
21 Fishing area total	-	-	-	-	-	-	*578*	-	-	-
31 USA	...	...	...	...	...	...	...	137	-	-
31 Fishing area total	...	...	...	...	...	...	...	*137*	-	-
37 Turkey	781	814	487	904	900	1 750	1 203	900	2 000	500
37 Fishing area total	*781*	*814*	*487*	*904*	*900*	*1 750*	*1 203*	*900*	*2 000*	*500*
47 Namibia	-	-	-	-	-	-	-	106	44	2 193
47 Fishing area total	-	-	-	-	-	-	-	*106*	*44*	*2 193*
48 UK	-	-	-	-	-	-	-	5	-	0
48 Fishing area total	-	-	-	-	-	-	-	*5*	-	*0*
57 Indonesia	675	1 744	1 309	4 018	5 354	-	29 377	28 181	16 159	17 540
Malaysia	2 001	10	205	2 823	2 813	633	4 555	5 926	6 367	2 607
Myanmar	...	...	1 812	3 426	1 412	432	1 000 F	1 000 F	1 000 F	1 000 F
Thailand	5 778	7 804	17 060	12 540	19 968	5 595	48	986	945	898
57 Fishing area total	*8 454*	*9 558*	*20 386*	*22 807*	*29 547*	*6 660*	*34 980 F*	*36 093 F*	*24 471 F*	*22 045 F*
58 Australia	-	-	-	-	11	2	-	-	-	-
India	-	-	-	0	-	-	-	-	-	-
58 Fishing area total	-	-	-	*0*	*11*	*2*	-	-	-	-
61 China	132 572	113 354	171 905	265 325	400 483	430 784	402 206	334 869	331 297	288 113
Russian Fed	-	-	-	-	-	-	-	-	142	-
61 Fishing area total	*132 572*	*113 354*	*171 905*	*265 325*	*400 483*	*430 784*	*402 206*	*334 869*	*331 439*	*288 113*
71 Indonesia	25 768	1 059	121 767	2 722	12 365	3 861	3 275	1 335	5 306	5 760
Malaysia	14 292	10 128	7 487	17 079	50 998	11 169	2 627	3 110	3 932	4 041
Philippines	73	...	61	57	20	10	12	12	12	12
Thailand	9 857	78 308	16 668	17 956	22 425	59 165	84 520	138 636	65 682	67 033
71 Fishing area total	*49 990*	*89 495*	*145 983*	*37 814*	*85 808*	*74 205*	*90 434*	*143 093*	*74 932*	*76 846*
Species total	*191 797*	*213 221*	*338 761*	*326 850*	*516 749*	*513 401*	*529 401 F*	*515 203 F*	*432 886 F*	*389 697 F*
Marine worms	**Vers marins**		**Poliquetos**		***Polychaeta***				**6,49(XX)XXX,XX**	**WOR**
02 Mexico	98	49	59	58	57	58	48	27	26	27
02 Fishing area total	*98*	*49*	*59*	*58*	*57*	*58*	*48*	*27*	*26*	*27*
21 USA	400	379	0	-	223	299	343	202	481	507
21 Fishing area total	*400*	*379*	*0*	-	*223*	*299*	*343*	*202*	*481*	*507*
77 Panama	...	...	...	42	67	51	31	28	36	25
77 Fishing area total	...	...	...	*42*	*67*	*51*	*31*	*28*	*36*	*25*
Species total	*498*	*428*	*59*	*100*	*347*	*408*	*422*	*257*	*543*	*559*
Aquatic invertebrates nei	**Invertébrés aquatiques nca**		**Invertebrados acuáticos nep**		***Invertebrata***				**6,99(XX)XXX,XX**	**INV**
02 Mexico	138	170	124	279	240	425	569	204	93	126
02 Fishing area total	*138*	*170*	*124*	*279*	*240*	*425*	*569*	*204*	*93*	*126*
04 Indonesia	432	198	1 067	49	714	72	830	790	645	700
Japan	2 281	1 431	1 552	1 445	1 636	1 403	1 216	1 240	933	1 035
Korea Rep	10	10	24	-	110	5	2	1	-	-
04 Fishing area total	*2 723*	*1 639*	*2 643*	*1 494*	*2 460*	*1 480*	*2 048*	*2 031*	*1 578*	*1 735*
27 Denmark	0	1	1	0	-	-	-	-	-	-
UK	-	-	-	-	-	-	-	-	24	4
27 Fishing area total	*0*	*1*	*1*	*0*	-	-	-	-	*24*	*4*
31 Mexico	5	7	11	6	2	4	0	2	-	-
31 Fishing area total	*5*	*7*	*11*	*6*	*2*	*4*	*0*	*2*	-	-
37 Croatia	-	-	-	-	-	4	5	8	592	59
Spain	-	-	-	5	4	10	529	-	-	11
Ukraine	-	-	-	-	-	-	-	-	3	9
37 Fishing area total	-	-	-	*5*	*4*	*14*	*534*	*8*	*595*	*79*
47 Spain	-	-	-	-	-	-	-	-	-	19
47 Fishing area total	-	-	-	-	-	-	-	-	-	*19*
57 Australia	-	-	-	-	-	-	-	63	96	72
Indonesia	167	139	211	720	200	417	544	530	661	720
57 Fishing area total	*167*	*139*	*211*	*720*	*200*	*417*	*544*	*593*	*757*	*792*
58 Australia	-	-	-	-	1	-	-	-	-	-
58 Fishing area total	-	-	-	-	*1*	-	-	-	-	-
61 Japan	78 567	85 215	37 360	12 537	9 134	9 618	7 130	9 692	7 406	7 500 F
Korea Rep	913	2 161	1 016	873	984	1 042	1 001	1 882	1 302	1 936
61 Fishing area total	*79 480*	*87 376*	*38 376*	*13 410*	*10 118*	*10 660*	*8 131*	*11 574*	*8 708*	*9 436 F*
71 Australia	-	-	-	-	-	-	-	25	44	14

B-77 Miscellaneous aquatic invertebrates / Invertébrés aquatiques divers / Invertebrados acuáticos diversos

Capture production by species, fishing areas and countries or areas / Captures par espèces, zones de pêche et pays ou zones / Capturas por especies, áreas de pesca y países o áreas

Species, Fishing area Espèce, Zone de pêche Especie, Area de pesca	1993 t	1994 t	1995 t	1996 t	1997 t	1998 t	1999 t	2000 t	2001 t	2002 t
Indonesia	262	271	237	471	771	220	1 275	1 529	612	660
Palau	...	...	0	0	0	0	1	0	0	0
Thailand	-	-	-	-	-	-	34	219	1 749	-
71 Fishing area total	*262*	*271*	*237*	*471*	*771*	*220*	*1 310*	*1 773*	*2 405*	*674*
77 Mexico	973	3 326	1 346	830	284	8	0	220	-	-
77 Fishing area total	*973*	*3 326*	*1 346*	*830*	*284*	*8*	*0*	*220*	-	-
81 Australia	-	-	-	-	-	-	-	22	29	20
New Zealand	-	-	-	-	-	-	-	-	-	1
81 Fishing area total	-	-	-	-	-	-	-	*22*	*29*	*21*
Species total	*83 748*	*92 929*	*42 949*	*17 215*	*14 080*	*13 228*	*13 136*	*16 427*	*14 189*	*12 886 F*
Group total	***276 043***	***306 578***	***381 769***	***344 165***	***531 176***	***527 037***	***542 959***	***531 887***	***447 618***	***403 142***

B-81

Pearls, mother-of-pearl, shells — **Capture production by species, fishing areas and countries or areas**
Perles, nacres, coquilles — **Captures par espèces, zones de pêche et pays ou zones**
Perlas, madreperlas, conchas — **Capturas por especies, áreas de pesca y países o áreas**

Species, Fishing area Espèce, Zone de pêche Especie, Area de pesca	1993 kg	1994 kg	1995 kg	1996 kg	1997 kg	1998 kg	1999 kg	2000 kg	2001 kg	2002 kg
Trochus shells	**Troques**		**Tróquidos**		***Ex Trochus spp***				**3,07(04)006,XX**	**TSH**
51 Eritrea	-	-	-	-	-	223 900	106 050	100 000 F	100 000 F	100 000 F
51 Fishing area total	-	-	-	-	-	*223 900*	*106 050*	*100 000 F*	*100 000 F*	*100 000 F*
71 Australia	0	0	0	0	0	0	0	0	0	-
Fiji Islands	65 899	100 000 F	160 000	160 000 F	150 000 F	115 000	165 000	158 000 F	150 000 F	150 000 F
Marshall Is	100 000 F	120 000 F	120 000 F	120 000 F	110 000 F	100 000 F	100 000 F	100 000 F	100 000 F	100 000 F
Micronesia	132 000	266 000	150 000 F	150 000 F	150 000 F	150 000 F	150 000 F	150 000 F	150 000 F	150 000 F
NewCaledonia	223 000	274 000	250 000	197 400	124 700	151 300	98 100	96 400	342 700	77 900 F
Palau	26 636	...	389 090	69 450	...	...	...	175 000	...	...
Papua N Guin	180 946	221 635	320 160	177 987	209 870	281 590	233 319	224 400	345 062	300 000
Solomon Is	394 000	306 000	80 405	30 625	181 670	60 910	261 293	54 431	54 000 F	54 000 F
Vanuatu	160 000	107 000	100 000 F	100 000 F	100 000 F	100 000 F	100 000 F	100 000 F	100 000 F	100 000 F
Wallis Fut I	16 000	34 000	30 000 F	30 000 F	30 000 F	25 000 F	25 000 F	25 000 F	25 000 F	25 000 F
71 Fishing area total	*1 298 481 F*	*1 428 635 F*	*1 599 655 F*	*1 035 462 F*	*1 056 240 F*	*983 800 F*	*1 132 712 F*	*1 083 231 F*	*1 266 762 F*	*956 900 F*
77 Cook Is	25 000 F	25 000 F	26 000 F	26 000 F	26 000 F	25 000 F	25 000 F	25 000 F	15 000 F	15 000 F
77 Fishing area total	*25 000 F*	*25 000 F*	*26 000 F*	*26 000 F*	*26 000 F*	*25 000 F*	*25 000 F*	*25 000 F*	*15 000 F*	*15 000 F*
Species total	*1 323 481 F*	*1 453 635 F*	*1 625 655 F*	*1 061 462 F*	*1 082 240 F*	*1 232 700 F*	*1 263 762 F*	*1 208 231 F*	*1 381 762 F*	*1 071 900 F*
Turban shells nei	**Coquilles turbo nca**		**Turbinas nep**		***Ex Turbo spp***				**3,07(05)002,XX**	**GSH**
71 Malaysia	20 000	17 800	97 000	90 000 F	90 000 F	80 000 F	80 000 F	80 000 F	80 000 F	80 000 F
71 Fishing area total	*20 000*	*17 800*	*97 000*	*90 000 F*	*90 000 F*	*80 000 F*	*80 000 F*	*80 000 F*	*80 000 F*	*80 000 F*
Species total	*20 000*	*17 800*	*97 000*	*90 000 F*	*90 000 F*	*80 000 F*	*80 000 F*	*80 000 F*	*80 000 F*	*80 000 F*
Freshwater mussel shells	**Coquilles des moules eau douce**		**Conchas de mejillón agua dulce**		***Ex Unionidae***				**3,16(05)XXX,XX**	**FSH**
02 USA	499	1 814	-	19 000	1 183	...	2 716	...	...	...
02 Fishing area total	*499*	*1 814*	-	*19 000*	*1 183*	...	*2 716*	...	...	...
Species total	*499*	*1 814*	-	*19 000*	*1 183*	...	*2 716*	...	...	...
Pearl oyster shells nei	**Coquilles d'huîtres perl. nca**		**Conchas de ostras perleras nep**		***Ex Pinctada spp***				**3,16(06)006,XX**	**OSH**
04 Japan	135	153	165	190	204	214	187	181	...	...
04 Fishing area total	*135*	*153*	*165*	*190*	*204*	*214*	*187*	*181*	...	...
51 Sudan	14 000	13 000	3 000	9 600	10 000	13 000	13 500	10 000 F	7 500	7 500
51 Fishing area total	*14 000*	*13 000*	*3 000*	*9 600*	*10 000*	*13 000*	*13 500*	*10 000 F*	*7 500*	*7 500*
57 Australia	220 000 F	230 000 F	240 000 F	240 000 F	250 000 F	250 000 F	250 000 F	240 000 F	190 000 F	220 000 F
57 Fishing area total	*220 000 F*	*230 000 F*	*240 000 F*	*240 000 F*	*250 000 F*	*250 000 F*	*250 000 F*	*240 000 F*	*190 000 F*	*220 000 F*
61 China,Taiwan	-	-	-	-	-	-	-	-	1 251	1 518
Japan	72 711	64 892	63 330	56 565	48 307	28 893	24 576	29 905	34 516	31 910
61 Fishing area total	*72 711*	*64 892*	*63 330*	*56 565*	*48 307*	*28 893*	*24 576*	*29 905*	*35 767*	*33 428*
71 Fiji Islands	17 034	17 000 F	20 000 F	20 000 F	20 000 F	10 000	12 020	12 000 F	10 000 F	10 000 F
Indonesia	...	...	...	...	...	...	500	68 600	1 000	1 000 F
Papua N Guin	14 739	7 681	10 498	16 460	20 561	35 960	17 051	21 521	1 484	1 500 F
71 Fishing area total	*31 773*	*24 681 F*	*30 498 F*	*36 460 F*	*40 561 F*	*45 960*	*29 571*	*102 121 F*	*12 484 F*	*12 500 F*
77 Fr Polynesia	...	...	...	...	752 349	562 213	650 000 F	752 364	820 724	1 289 488
77 Fishing area total	...	...	...	...	*752 349*	*562 213*	*650 000 F*	*752 364*	*820 724*	*1 289 488*
Species total	*338 619 F*	*332 726 F*	*336 993 F*	*342 815 F*	*1 101 421 F*	*900 280 F*	*967 834 F*	*1 134 571 F*	*1 066 475 F*	*1 562 916 F*
Marine shells nei	**Coquilles marines nca**		**Conchas marinas nep**		***Ex Mollusca***				**3,99(XX)XXX,XX**	**MSH**
31 Bahamas	150 294	60 490	35 671	10 807	8 430	39 168	40 000 F	40 000 F	13 000	13 000 F
Mexico	...	...	15 055	60 067	80 838	50 977	36 150	43 570	33 000 F	21 580
31 Fishing area total	*150 294*	*60 490*	*50 726*	*70 874*	*89 268*	*90 145*	*76 150 F*	*83 570 F*	*46 000 F*	*34 580 F*
37 Croatia	274	308	254	296	492	1 078	1 192	1 244	3 125	2 539
Russian Fed	29 000	2 000	...	1 000	440 000	46 000	45 000	182 000	224 000	56 000
Serbia-Monte	5 000	3 585	7 520	1 990	2 375	2 890	3 735	3 200	2 285	3 075
37 Fishing area total	*34 274*	*5 893*	*7 774*	*3 286*	*442 867*	*49 968*	*49 927*	*186 444*	*229 410*	*61 614*
51 Kenya	329 000	329 000	178 743	166 351	148 658	170 584	262 386	253 911	260 324	271 334
Madagascar	76 000	187 000	187 000	169 000	156 000	10 000	8 000	74 000	32 000	26 000
Tanzania	183 358	122 077	54 000	95 200	183 172	154 893	250 006	250 000 F	436 743	325 000
51 Fishing area total	*588 358*	*638 077*	*419 743*	*430 551*	*487 830*	*335 477*	*520 392*	*577 911 F*	*729 067*	*622 334*
57 Sri Lanka	121 890	235 670	745 920	723 310	430 840	602 860	652 700	697 600	498 000	324 000
57 Fishing area total	*121 890*	*235 670*	*745 920*	*723 310*	*430 840*	*602 860*	*652 700*	*697 600*	*498 000*	*324 000*
71 Philippines	2 855 000	5 512 000	3 842 000	3 824 000	2 801 000	2 389 000	2 358 000	2 373 000	2 433 000	2 465 000
Solomon Is	3 412	3 286	219 997	54 825	75	15 575	470	35	50 F	50 F
71 Fishing area total	*2 858 412*	*5 515 286*	*4 061 997*	*3 878 825*	*2 801 075*	*2 404 575*	*2 358 470*	*2 373 035*	*2 433 050 F*	*2 465 050 F*
77 Fr Polynesia	87 000	27 000	40 000 F	55 000 F	68 134	...	90 000 F	105 204	...	...
Mexico	...	...	833 232	481 813	1 321 742	918 563	928 808	420 242	330 000 F	243 930
77 Fishing area total	*87 000*	*27 000*	*873 232 F*	*536 813 F*	*1 389 876*	*918 563*	*1 018 808 F*	*525 446*	*330 000 F*	*243 930*
Species total	*3 840 228*	*6 482 416*	*6 159 392 F*	*5 643 659 F*	*5 641 756*	*4 401 588*	*4 676 447 F*	*4 444 006 F*	*4 265 527 F*	*3 751 508 F*
Group total	***5 522 827***	***8 288 391***	***8 219 040***	***7 156 936***	***7 916 600***	***6 614 568***	***6 990 759***	***6 866 808***	***6 793 764***	***6 466 324***

B-81

Pearls, mother-of-pearl, shells — **Capture production by species, fishing areas and countries or areas**
Perles, nacres, coquilles — **Captures par espèces, zones de pêche et pays ou zones**
Perlas, madreperlas, conchas — **Capturas por especies, áreas de pesca y países o áreas**

Species, Fishing area Espèce, Zone de pêche Especie, Area de pesca	1993 kg	1994 kg	1995 kg	1996 kg	1997 kg	1998 kg	1999 kg	2000 kg	2001 kg	2002 kg

Data of "Pearl oyster shells nei" include aquaculture production.

Data by Japan for "Pearl oyster shells nei" refer only to production of pearls.

See paragraph 5 of the INTRODUCTION.

Les données de "Coquilles d'huîtres perlières nca" comprennent les quantités provenant de l'aquaculture.

Les données du Japon concernant les "Coquilles d'huîtres perlières nca" se rapportent à la production des perles.

Voir le paragraphe 5 de l'INTRODUCTION.

Los datos correspondientes a "Conchas de ostras perleras nep" incluyen las cantidades procedentes de la acuicultura.

Los datos de Japón referidos a "Conchas de ostras perleras nep" se refieren solamente a la producción de perlas.

Véase el párrafo 5 en la INTRODUCCIÓN.

B-82 Corals / Coraux / Corales

Capture production by species, fishing areas and countries or areas
Captures par espèces, zones de pêche et pays ou zones
Capturas por especies, áreas de pesca y países o áreas

Species, Fishing area Espèce, Zone de pêche Especie, Area de pesca	1993 kg	1994 kg	1995 kg	1996 kg	1997 kg	1998 kg	1999 kg	2000 kg	2001 kg	2002 kg
Sardinia coral	**Corail Sardaigne**		**Coral Cerdeña**		***Corallium rubrum***				**6,19(01)003,01**	**COL**
27 Spain	-	-	-	3 200	3 200	1 000	2 500	3 000	1 600	2 300
27 Fishing area total	-	-	-	*3 200*	*3 200*	*1 000*	*2 500*	*3 000*	*1 600*	*2 300*
34 Morocco	3 800	2 900	2 600	1 800	1 500	600	...	1 000	600	800
34 Fishing area total	*3 800*	*2 900*	*2 600*	*1 800*	*1 500*	*600*	...	*1 000*	*600*	*800*
37 Albania	400 F	800 F	1 300	1 200	800	500	900	1 000	1 200	1 800
Algeria	5 971	7 857	6 666	6 800	5 400	3 000	3 000	2 900	5 000	4 800
Croatia	3 500	3 200	1 800	1 700	1 600	1 600	1 500	2 000	1 900	2 000
France	2 500	2 600	2 300	1 900	2 500	4 500	3 800	3 600	3 700	5 600
Greece	500	900	1 500	1 200	1 100	1 400	1 750	1 700	1 400	1 900
Italy	4 400	4 800	3 100	2 500	4 100	4 700	3 850	4 350	4 700	5 100
Morocco	600	300	200	200	100	500	3 100	3 200	3 300	3 800
Spain	4 700	4 400	6 300	5 000	5 200	1 500	4 350	4 300	5 000	5 500
Tunisia	1 200	1 071	1 100	975	1 975	1 196	1 783	2 399	2 005	2 371
37 Fishing area total	*23 771 F*	*25 928 F*	*24 266*	*21 475*	*22 775*	*18 896*	*24 033*	*25 449*	*28 205*	*32 871*
Species total	*27 571 F*	*28 828 F*	*26 866*	*26 475*	*27 475*	*20 496*	*26 533*	*29 449*	*30 405*	*35 971*
Aka coral	**Corail aka**		**Coral aka**		***Corallium japonicum***				**6,19(01)003,02**	**COJ**
61 China,Taiwan	-	-	-	600	5	-	650	400	350	250
Japan	347	358	565	475	49	492	990	1 100	1 200	680
61 Fishing area total	*347*	*358*	*565*	*1 075*	*54*	*492*	*1 640*	*1 500*	*1 550*	*930*
Species total	*347*	*358*	*565*	*1 075*	*54*	*492*	*1 640*	*1 500*	*1 550*	*930*
Momo, boke magai, misu coral	**Corail momo**		**Coral momo**		***Corallium elatius***				**6,19(01)003,03**	**CEL**
61 China,Taiwan	6 000	5 550	6 000	12 000	240	5 500	4 650	4 100	4 500	8 500
Japan	62	1 512	2 220	815	12	1 169	248	2 700	2 300	3 240
61 Fishing area total	*6 062*	*7 062*	*8 220*	*12 815*	*252*	*6 669*	*4 898*	*6 800*	*6 800*	*11 740*
Species total	*6 062*	*7 062*	*8 220*	*12 815*	*252*	*6 669*	*4 898*	*6 800*	*6 800*	*11 740*
Shiro, white coral	**Corail blanc**		**Coral blanco**		***Corallium konojoi***				**6,19(01)003,04**	**COK**
61 China,Taiwan	-	-	-	-	-	-	-	-	-	100
Japan	-	-	145	76	53	111	-	150	-	260
61 Fishing area total	-	-	*145*	*76*	*53*	*111*	-	*150*	-	*360*
Species total	-	-	*145*	*76*	*53*	*111*	-	*150*	-	*360*
Midway deep sea coral	**Corail de profondeur de Midway**		**Coral de profundidad de Midway**		***Corallium sp. nov.***				**6,19(01)003,07**	**CDE**
61 Japan	-	-	-	-	-	623	-	100	-	60
61 Fishing area total	-	-	-	-	-	*623*	-	*100*	-	*60*
Species total	-	-	-	-	-	*623*	-	*100*	-	*60*
Soft corals nei	**Corails mous nca**		**Corales muelles nep**		***Non-Scleractinia***				**6,19(XX)XXX,XX**	**CBL**
31 Mexico	350	215	100	582	775	694	574	593	500 F	370
31 Fishing area total	*350*	*215*	*100*	*582*	*775*	*694*	*574*	*593*	*500 F*	*370*
77 Mexico	1 230	556	-	-	-	-	-	-	-	-
USA	1 500	2 000	3 000	-	907	14	1 292	...	8 533	...
77 Fishing area total	*2 730*	*2 556*	*3 000*	-	*907*	*14*	*1 292*	...	*8 533*	...
Species total	*3 080*	*2 771*	*3 100*	*582*	*1 682*	*708*	*1 866*	*593*	*9 033 F*	*370*
Hard corals, madrepores nei	**Madrépores nca**		**Madréporas nep**		***Scleractinia***				**6,19(XX)XXX,XX**	**CSS**
31 Haiti	15 000 F	15 000 F	14 000 F	14 000 F	13 000 F	10 000 F	10 000 F	10 000 F	10 000 F	10 000 F
31 Fishing area total	*15 000 F*	*15 000 F*	*14 000 F*	*14 000 F*	*13 000 F*	*10 000 F*	*10 000 F*	*10 000 F*	*10 000 F*	*10 000 F*
57 India	...	...	...	...	...	...	...	...	...	8 750 000
57 Fishing area total	...	...	...	...	...	...	...	...	...	*8 750 000*
61 Japan	6 000 F	5 000 F	5 000 F	5 000 F	5 000 F	5 000 F	5 000 F	5 000 F	...	...
61 Fishing area total	*6 000 F*	*5 000 F*	*5 000 F*	*5 000 F*	*5 000 F*	*5 000 F*	*5 000 F*	*5 000 F*	...	...
71 Fiji Islands	120 000	150 000 F	300 000 F	600 000 F	900 000 F	1 279 609	1 000 000	1 000 000 F	1 000 000 F	1 000 000 F
Indonesia	2 000 000 F	1 800 000 F	1 700 000 F	1 700 000 F	1 500 000 F	1 500 000 F	1 500 000 F	1 200 000 F	1 000 000 F	1 000 000 F
Malaysia	271 140	872 690	4 743 000	4 500 000 F	4 000 000 F	4 000 000 F	4 000 000 F	4 000 000 F	4 000 000 F	4 000 000 F
Philippines	600 000 F	550 000 F	550 000 F	550 000 F	500 000 F	-	-	-	-	-
71 Fishing area total	*2 991 140 F*	*3 372 690 F*	*7 293 000 F*	*7 350 000 F*	*6 900 000 F*	*6 779 609 F*	*6 500 000 F*	*6 200 000 F*	*6 000 000 F*	*6 000 000 F*
Species total	*3 012 140 F*	*3 392 690 F*	*7 312 000 F*	*7 369 000 F*	*6 918 000 F*	*6 794 609 F*	*6 515 000 F*	*6 215 000 F*	*6 010 000 F*	*14 760 000 F*
Group total	***3 049 200***	***3 431 709***	***7 350 896***	***7 410 023***	***6 947 516***	***6 823 708***	***6 549 937***	***6 253 592***	***6 057 788***	***14 809 431***

See paragraph 5 of the INTRODUCTION. Voir le paragraphe 5 de l'INTRODUCTION. Véase el párrafo 5 en la INTRODUCCIÓN.

B-83 Sponges / Eponges / Esponjas

Capture production by species, fishing areas and countries or areas
Captures par espèces, zones de pêche et pays ou zones
Capturas por especies, áreas de pesca y países o áreas

Species, Fishing area Espèce, Zone de pêche Especie, Area de pesca	1993 kg	1994 kg	1995 kg	1996 kg	1997 kg	1998 kg	1999 kg	2000 kg	2001 kg	2002 kg
Sponges	**Eponges**		**Esponjas**		***Spongidae***				**6,15(01)XXX,XX**	**SPO**
31 Bahamas	46 650	50 526	54 812	62 337	60 162	59 517	86 000	81 000	71 000	71 000 F
Colombia	2 500 F	2 500 F	2 000 F	2 000 F	2 000 F	2 000 F	2 000 F	2 000 F	2 000 F	2 000 F
Cuba	24 300	40 400	55 200	56 800	81 500	72 100	50 100	57 000	51 000	54 000
USA	130 740	378 747	338 609	196 157	700 911	1 199 131	2 399 327	303 679	291 326	260 688
31 Fishing area total	*204 190 F*	*472 173 F*	*450 621 F*	*317 294 F*	*844 573 F*	*1 332 748 F*	*2 537 427 F*	*443 679 F*	*415 326 F*	*387 688 F*
37 Croatia	500 F	500 F	1 000 F	2 000 F	3 900	5 400	6 500	3 000	4 000	1 000
Egypt	1 700 F	1 600 F	1 500 F	1 400 F	1 200 F	1 000 F	1 000 F	1 000 F	1 000 F	1 000 F
France	12 000 F	10 000 F	8 000 F	7 000 F	5 000 F	5 000 F	4 000 F	3 000 F	2 000 F	1 400 F
Greece	2 000	2 000	10 000	11 000	10 000	10 000	10 000	10 000	5 000	2 000
Italy	4 000 F	4 000 F	3 000 F	3 000 F	2 500 F	2 000 F	1 500 F	1 500 F	1 000 F	1 000 F
Lebanon	12 000	12 000	0	0	0	0	0	0	0	-
Libya	14 000 F	13 000 F	10 000 F	10 000 F	10 000 F	10 000 F	10 000 F	10 000 F	10 000 F	10 000 F
Spain	900 F	900 F	800 F	800 F	800 F	800 F	750 F	700 F	700 F	700 F
Tunisia	16 046	33 743	14 803	14 035	27 330	21 400	14 450	15 431	23 092	44 200
Turkey	3 000	2 000	1 000	1 500 F	2 000	1 000	3 000	7 000	3 000	-
37 Fishing area total	*66 146 F*	*79 743 F*	*50 103 F*	*50 735 F*	*62 730 F*	*56 600 F*	*51 200 F*	*51 631 F*	*49 792 F*	*61 300 F*
61 Japan	7 000 F	6 000 F	5 000 F	5 000 F	5 000 F	4 000 F	4 000 F	4 000 F	...	...
61 Fishing area total	*7 000 F*	*6 000 F*	*5 000 F*	*5 000 F*	*5 000 F*	*4 000 F*	*4 000 F*	*4 000 F*	...	...
71 Philippines	3 000	1 000	23 000	2 000	16 000	7 000	8 000	8 000	8 000	8 000
71 Fishing area total	*3 000*	*1 000*	*23 000*	*2 000*	*16 000*	*7 000*	*8 000*	*8 000*	*8 000*	*8 000*
77 Fr Polynesia	...	...	...	...	...	...	...	116	-	-
77 Fishing area total	...	...	...	...	...	...	...	*116*	-	-
81 New Zealand	-	-	-	-	-	4 100	14 000	18 000 F	23 000	-
81 Fishing area total	-	-	-	-	-	*4 100*	*14 000*	*18 000 F*	*23 000*	-
Species total	*280 336 F*	*558 916 F*	*528 724 F*	*375 029 F*	*928 303 F*	*1 404 448 F*	*2 614 627 F*	*525 426 F*	*496 118 F*	*456 988 F*
Group total	***280 336***	***558 916***	***528 724***	***375 029***	***928 303***	***1 404 448***	***2 614 627***	***525 426***	***496 118***	***456 988***

See paragraph 5 of the INTRODUCTION.

Voir le paragraphe 5 de l'INTRODUCTION.

Véase el párrafo 5 en la INTRODUCCIÓN.

B-91 Brown seaweeds / Algues brunes / Algas pardas

Capture production by species, fishing areas and countries or areas / Captures par espèces, zones de pêche et pays ou zones / Capturas por especies, áreas de pesca y países o áreas

Species, Fishing area Espèce, Zone de pêche Especie, Area de pesca	1993 t	1994 t	1995 t	1996 t	1997 t	1998 t	1999 t	2000 t	2001 t	2002 t
Tangle	**Laminaire digitée**		**...C**		***Laminaria digitata***				**7,71(02)002,01**	**LQD**
27 Iceland	...	...	...	...	...	4 500	5 239	2 764	4 679	3
27 Fishing area total	...	...	...	...	...	*4 500*	*5 239*	*2 764*	*4 679*	*3*
Species total	...	...	...	...	...	*4 500*	*5 239*	*2 764*	*4 679*	*3*
Japanese kelp	**Laminaire du Japon**		**Laminaria del Japón**		***Laminaria japonica***				**7,71(02)002,02**	**LNJ**
61 Japan	134 147	103 603	120 957	120 194	122 976	91 752	94 371	93 611	97 261	104 402
61 Fishing area total	*134 147*	*103 603*	*120 957*	*120 194*	*122 976*	*91 752*	*94 371*	*93 611*	*97 261*	*104 402*
Species total	*134 147*	*103 603*	*120 957*	*120 194*	*122 976*	*91 752*	*94 371*	*93 611*	*97 261*	*104 402*
North European kelp	**...B**		**...C**		***Laminaria hyperborea***				**7,71(02)002,04**	**LAH**
27 Ireland	2 700	2 600	2 400	2 300	1 900	2 000 F	2 000 F	2 000 F	2 000 F	2 000 F
27 Fishing area total	*2 700*	*2 600*	*2 400*	*2 300*	*1 900*	*2 000 F*	*2 000 F*	*2 000 F*	*2 000 F*	*2 000 F*
Species total	*2 700*	*2 600*	*2 400*	*2 300*	*1 900*	*2 000 F*	*2 000 F*	*2 000 F*	*2 000 F*	*2 000 F*
Wakame	**Wakamé**		**Abeto marino**		***Undaria pinnatifida***				**7,71(04)003,01**	**UDP**
61 Japan	3 034	3 265	3 148	4 044	2 936	2 839	3 431	3 396	3 034	3 179
61 Fishing area total	*3 034*	*3 265*	*3 148*	*4 044*	*2 936*	*2 839*	*3 431*	*3 396*	*3 034*	*3 179*
Species total	*3 034*	*3 265*	*3 148*	*4 044*	*2 936*	*2 839*	*3 431*	*3 396*	*3 034*	*3 179*
Chilean kelp	**...B**		**Chascón**		***Lessonia nigrescens***				**7,71(05)001,01**	**LJX**
87 Chile	70 565	72 029	123 772	140 770	125 535	136 313	111 766	61 954	87 508	96 428
87 Fishing area total	*70 565*	*72 029*	*123 772*	*140 770*	*125 535*	*136 313*	*111 766*	*61 954*	*87 508*	*96 428*
Species total	*70 565*	*72 029*	*123 772*	*140 770*	*125 535*	*136 313*	*111 766*	*61 954*	*87 508*	*96 428*
...A	**...B**		**Huiro palo**		***Lessonia trabeculata***				**7,71(05)001,02**	**LJZ**
87 Chile	...	...	...	...	...	...	...	18 107	18 457	25 956
87 Fishing area total	...	...	...	...	...	...	...	*18 107*	*18 457*	*25 956*
Species total	...	...	...	...	...	...	...	*18 107*	*18 457*	*25 956*
...A	**...B**		**Chascón nep**		***Lessonia spp***				**7,71(05)001,XX**	**EOZ**
81 New Zealand	...	...	0	2	3	2	0	-	0	3
81 Fishing area total	...	...	*0*	*2*	*3*	*2*	*0*	-	*0*	*3*
Species total	...	...	*0*	*2*	*3*	*2*	*0*	-	*0*	*3*
Kelp nei	**Kelps nca**		**Huiros nep**		***Macrocystis spp***				**7,71(05)002,XX**	**GQO**
41 Argentina	57	52	102	110	40	40 F	40 F	...	...	...
41 Fishing area total	*57*	*52*	*102*	*110*	*40*	*40 F*	*40 F*	...	...	...
67 USA	435	185	191	44	149	107	202	1	2	-
67 Fishing area total	*435*	*185*	*191*	*44*	*149*	*107*	*202*	*1*	*2*	-
77 USA	84 673	116 656	73 604	58 989	70 017	24 879	78 953	41 998	34 544	47 104
77 Fishing area total	*84 673*	*116 656*	*73 604*	*58 989*	*70 017*	*24 879*	*78 953*	*41 998*	*34 544*	*47 104*
81 New Zealand	...	...	5	139	32 675	7 170	671	0	0	1
81 Fishing area total	...	...	*5*	*139*	*32 675*	*7 170*	*671*	*0*	*0*	*1*
87 Chile	4 580	4 325	12 451	18 333	11 903	10 104	11 928	6 084	9 672	9 774
87 Fishing area total	*4 580*	*4 325*	*12 451*	*18 333*	*11 903*	*10 104*	*11 928*	*6 084*	*9 672*	*9 774*
Species total	*89 745*	*121 218*	*86 353*	*77 615*	*114 784*	*42 300 F*	*91 794 F*	*48 083*	*44 218*	*56 879*
North Atlantic rockweed	**...B**		**...C**		***Ascophyllum nodosum***				**7,71(06)006,01**	**ASN**
21 Canada	14 110	18 713	15 438	18 958	25 203	21 133	24 157	26 521	9 911	11 262
USA	...	...	...	...	...	...	...	...	2 394	59
21 Fishing area total	*14 110*	*18 713*	*15 438*	*18 958*	*25 203*	*21 133*	*24 157*	*26 521*	*12 305*	*11 321*
27 Iceland	...	...	...	...	...	13 270	12 908	14 737	15 688	14 959
Ireland	27 152	30 748	32 976	30 980	33 400	34 000 F	34 000 F	34 000 F	34 000 F	34 000 F
27 Fishing area total	*27 152*	*30 748*	*32 976*	*30 980*	*33 400*	*47 270 F*	*46 908 F*	*48 737 F*	*49 688 F*	*48 959 F*
Species total	*41 262*	*49 461*	*48 414*	*49 938*	*58 603*	*68 403 F*	*71 065 F*	*75 258 F*	*61 993 F*	*60 280 F*
Bull kelp	**Durvillée antarctique**		**Cochayuyo**		***Durvillaea antartica***				**7,71(08)001,01**	**DVA**
81 New Zealand	-	-	0	3	1 599	658	60	-	0	3
81 Fishing area total	-	-	*0*	*3*	*1 599*	*658*	*60*	-	*0*	*3*
87 Chile	1 749	2 283	464	1 924	1 691	3 934	4 567	2 122	2 098	2 312
87 Fishing area total	*1 749*	*2 283*	*464*	*1 924*	*1 691*	*3 934*	*4 567*	*2 122*	*2 098*	*2 312*
Species total	*1 749*	*2 283*	*464*	*1 927*	*3 290*	*4 592*	*4 627*	*2 122*	*2 098*	*2 315*

B-91 Brown seaweeds / Algues brunes / Algas pardas

Capture production by species, fishing areas and countries or areas
Captures par espèces, zones de pêche et pays ou zones
Capturas por especies, áreas de pesca y países o áreas

Species, Fishing area Espèce, Zone de pêche Especie, Area de pesca	1993 t	1994 t	1995 t	1996 t	1997 t	1998 t	1999 t	2000 t	2001 t	2002 t
Golden Cystoseira	**Cystoseire dorée**		**Cistosira barbuda**		***Cystoseira barbata***				**7,71(09)001,01**	**YQT**
37 Ukraine	-	-	-	-	-	4	3	...	...	36
37 Fishing area total	-	-	-	-	-	*4*	*3*	...	...	*36*
Species total	-	-	-	-	-	*4*	*3*	...	...	*36*
Brown seaweeds	**Algues brunes**		**Algas pardas**		***Phaeophyceae***				**7,71(XX)XXX,XX**	**SWB**
21 USA	...	...	...	...	...	...	...	...	225	17
21 Fishing area total	...	...	...	...	...	...	...	...	*225*	*17*
27 France	54 728	74 807	71 238	81 293	72 785	64 756	68 397	66 172	64 516	79 094
Iceland	11 303	13 982	11 841	14 411	19 505	-	-	-	-	-
Norway	169 606	185 065	185 033	173 160	191 681	179 762	178 542	192 426	175 210	182 641
Russian Fed	3 170	2 732	1 085	4 632	1 627	2 346	4 893	11 583	6 572	4 046
Spain	650 F	500 F	350 F	200 F	100 F	5	28	239	239	233
UK	4 769	5 683	7 400	8 400	8 100	...	-	-	-	-
27 Fishing area total	*244 226 F*	*282 769 F*	*276 947 F*	*282 096 F*	*293 798 F*	*246 869*	*251 860*	*270 420*	*246 537*	*266 014*
37 Spain	-	-	-	-	-	-	-	-	-	1
37 Fishing area total	-	-	-	-	-	-	-	-	-	*1*
47 South Africa	3 768	2 579	3 689	6 000 F	9 000 F	12 000 F	15 000 F	18 216	30 518	30 000 F
47 Fishing area total	*3 768*	*2 579*	*3 689*	*6 000 F*	*9 000 F*	*12 000 F*	*15 000 F*	*18 216*	*30 518*	*30 000 F*
51 India	16 000 F	16 000 F	16 000 F	16 000 F	16 000 F	16 000 F	16 000 F	16 000 F	16 000 F	15 000 F
51 Fishing area total	*16 000 F*	*16 000 F*	*16 000 F*	*16 000 F*	*16 000 F*	*16 000 F*	*16 000 F*	*16 000 F*	*16 000 F*	*15 000 F*
57 Australia	18 275	17 519	22 316	18 570	21 152	20 811	20 774	13 650	13 547	9 916
57 Fishing area total	*18 275*	*17 519*	*22 316*	*18 570*	*21 152*	*20 811*	*20 774*	*13 650*	*13 547*	*9 916*
61 Japan	...	...	8 936	10 834	7 933	7 553	8 326	7 247	7 238	6 973
Korea Rep	13 095	11 366	11 026	16 523	13 131	5 159	8 624	8 158	10 067	10 000 F
Russian Fed	2 504	2 561	7 526	11 490	20 345	23 639	16 247	32 557	16 406	44 904
61 Fishing area total	*15 599*	*13 927*	*27 488*	*38 847*	*41 409*	*36 351*	*33 197*	*47 962*	*33 711*	*61 877 F*
77 Mexico	52 343	32 456	44 230	27 663	34 516	6 119	26 470	28 251	38 233	24 923
USA	...	...	...	...	...	...	...	...	...	3
77 Fishing area total	*52 343*	*32 456*	*44 230*	*27 663*	*34 516*	*6 119*	*26 470*	*28 251*	*38 233*	*24 926*
Species total	*350 211 F*	*365 250 F*	*390 670 F*	*389 176 F*	*415 875 F*	*338 150 F*	*363 301 F*	*394 499 F*	*378 771 F*	*407 751 F*
Group total	***693 413***	***719 709***	***776 178***	***785 966***	***845 902***	***690 855***	***747 597***	***701 794***	***700 019***	***759 232***

See paragraph 5 of the INTRODUCTION. Voir le paragraphe 5 de l'INTRODUCTION. Véase el párrafo 5 en la INTRODUCCIÓN.

B-92 Red seaweeds / Algues rouges / Algas rojas

Capture production by species, fishing areas and countries or areas
Captures par espèces, zones de pêche et pays ou zones
Capturas por especies, áreas de pesca y países o áreas

Species, Fishing area Espèce, Zone de pêche Especie, Area de pesca	1993 t	1994 t	1995 t	1996 t	1997 t	1998 t	1999 t	2000 t	2001 t	2002 t
Phyllophora nei	**Phyllophora nca**		**Phyllophora nep**		***Phyllophora spp***			**7,87(02)001,XX**		**YFQ**
37 Ukraine	...	3 974	2 929	820	...	...	...	20	...	...
37 Fishing area total	*...*	*3 974*	*2 929*	*820*	*...*	*...*	*...*	*20*	*...*	*...*
Species total	*...*	*3 974*	*2 929*	*820*	*...*	*...*	*...*	*20*	*...*	*...*
Gracilaria seaweeds	**Algues gracilaires**		**Gracilarias**		***Gracilaria spp***			**7,87(12)004,XX**		**GLS**
41 Argentina	2 350	3 060	2 276	2 360	850	850 F	850 F	3	...	...
41 Fishing area total	*2 350*	*3 060*	*2 276*	*2 360*	*850*	*850 F*	*850 F*	*3*	*...*	*...*
47 Namibia	226	175	799	936	851	897	660	829	800	500
47 Fishing area total	*226*	*175*	*799*	*936*	*851*	*897*	*660*	*829*	*800*	*500*
61 China,Taiwan	8	1	-	-	-	-	-	-	-	-
61 Fishing area total	*8*	*1*	*-*	*-*	*-*	*-*	*-*	*-*	*-*	*-*
87 Chile	4 789	3 596	68 436	15 738	4 809	4 601	54 800	103 629	52 431	54 536
87 Fishing area total	*4 789*	*3 596*	*68 436*	*15 738*	*4 809*	*4 601*	*54 800*	*103 629*	*52 431*	*54 536*
Species total	*7 373*	*6 832*	*71 511*	*19 034*	*6 510*	*6 348 F*	*56 310 F*	*104 461*	*53 231*	*55 036*
Carragheen (Irish) moss	**Mousse perle**		**...C**		***Chondrus crispus***			**7,87(16)001,04**		**IMS**
21 USA	-	-	204	262	563	703	81	59	-	-
21 Fishing area total	*-*	*-*	*204*	*262*	*563*	*703*	*81*	*59*	*-*	*-*
Species total	*-*	*-*	*204*	*262*	*563*	*703*	*81*	*59*	*-*	*-*
Skottsberg's gigartina	**Gigartine de Skottsberg**		**Gigartina de Skottsberg**		***Gigartina skottsbergii***			**7,87(16)002,03**		**GJK**
87 Chile	...	...	...	...	...	...	...	...	22 717	21 301
87 Fishing area total	*...*	*...*	*...*	*...*	*...*	*...*	*...*	*...*	*22 717*	*21 301*
Species total	*...*	*...*	*...*	*...*	*...*	*...*	*...*	*...*	*22 717*	*21 301*
Gigartina seaweeds	**...B**		**Chicorea de mar**		***Gigartina spp***			**7,87(16)002,XX**		**GIG**
41 Argentina	8	24	22	25	8	8 F	8 F	...	...	...
41 Fishing area total	*8*	*24*	*22*	*25*	*8*	*8 F*	*8 F*	*...*	*...*	*...*
87 Chile	4 163	4 572	6 389	6 690	11 745	15 453	23 458	24 778	3 325	5 677
87 Fishing area total	*4 163*	*4 572*	*6 389*	*6 690*	*11 745*	*15 453*	*23 458*	*24 778*	*3 325*	*5 677*
Species total	*4 171*	*4 596*	*6 411*	*6 715*	*11 753*	*15 461 F*	*23 466 F*	*24 778*	*3 325*	*5 677*
Iridea nei	**Iridea nca**		**Luga-luga**		***Iridaea spp***			**7,87(16)003,XX**		**LGY**
87 Chile	19 513	28 331	37 376	32 438	22 679	26 181	23 099	30 118	37 606	27 475
87 Fishing area total	*19 513*	*28 331*	*37 376*	*32 438*	*22 679*	*26 181*	*23 099*	*30 118*	*37 606*	*27 475*
Species total	*19 513*	*28 331*	*37 376*	*32 438*	*22 679*	*26 181*	*23 099*	*30 118*	*37 606*	*27 475*
Laver (Nori)	**Algue nori**		**Lechuga nori**		***Porphyra tenera***			**7,87(20)002,08**		**PRT**
61 China,Taiwan	24	6	3	23	2	3	3	-	-	8
61 Fishing area total	*24*	*6*	*3*	*23*	*2*	*3*	*3*	*-*	*-*	*8*
Species total	*24*	*6*	*3*	*23*	*2*	*3*	*3*	*-*	*-*	*8*
Nori nei	**Nori nca**		**Luche**		***Porphyra spp***			**7,87(20)002,XX**		**FYS**
87 Chile	181	119	6	3	53	45	9	-	-	8
87 Fishing area total	*181*	*119*	*6*	*3*	*53*	*45*	*9*	*-*	*-*	*8*
Species total	*181*	*119*	*6*	*3*	*53*	*45*	*9*	*-*	*-*	*8*
Gelidium seaweeds	**Algues gélidium**		**Gelidios**		***Gelidium spp***			**7,87(26)002,XX**		**GEL**
61 China,Taiwan	19	38	3	4	9	10	33	29	26	25
61 Fishing area total	*19*	*38*	*3*	*4*	*9*	*10*	*33*	*29*	*26*	*25*
87 Chile	1 569	1 500	1 144	867	405	762	491	525	402	533
87 Fishing area total	*1 569*	*1 500*	*1 144*	*867*	*405*	*762*	*491*	*525*	*402*	*533*
Species total	*1 588*	*1 538*	*1 147*	*871*	*414*	*772*	*524*	*554*	*428*	*558*
Manifold callophyllis	**Callophyllis variable**		**Calofila variable**		***Callophyllis variegata***			**7,87(27)001,01**		**KFV**
87 Chile	-	-	-	49	11	73	84	56	5	12
87 Fishing area total	*-*	*-*	*-*	*49*	*11*	*73*	*84*	*56*	*5*	*12*
Species total	*-*	*-*	*-*	*49*	*11*	*73*	*84*	*56*	*5*	*12*
Red seaweeds	**Algues rouges**		**Algas rojas**		***Rhodophyceae***			**7,87(XX)XXX,XX**		**SWR**
21 Canada	7 450	11 501	11 233	5 374	9 493	6 339	6 503	7 986	5 021	5 489
21 Fishing area total	*7 450*	*11 501*	*11 233*	*5 374*	*9 493*	*6 339*	*6 503*	*7 986*	*5 021*	*5 489*

B-92 Red seaweeds / Algues rouges / Algas rojas

Capture production by species, fishing areas and countries or areas
Captures par espèces, zones de pêche et pays ou zones
Capturas por especies, áreas de pesca y países o áreas

Species, Fishing area Espèce, Zone de pêche Especie, Area de pesca	1993 t	1994 t	1995 t	1996 t	1997 t	1998 t	1999 t	2000 t	2001 t	2002 t
27 Estonia	0	411	548	163	2 444	2 880	1 319	201	325	912
France	5 546	4 633	4 239	2 995	2 637	2 164	2 481	1 977	1 653	1 179
Ireland	71	69	80	80	86	100 F	100 F	100 F	100 F	100 F
Latvia	1	0	0	0	0	0	0	0	0	0
Portugal	3 000 F	2 686	2 816	1 703	1 743	968	1 949	1 224	1 198	...
Russian Fed	192	95	40	95	75	85	66	6 525	135	33
Spain	13 289	13 791	14 042	14 671	14 785	13 935	13 790	13 975	13 772	...
27 Fishing area total	*22 099 F*	*21 685*	*21 765*	*19 707*	*21 770*	*20 132 F*	*19 705 F*	*24 002 F*	*17 183 F*	*2 224 F*
34 Morocco	7 108	5 357	7 858	7 625	8 094	7 049	5 920	6 080	10 015	7 919
34 Fishing area total	*7 108*	*5 357*	*7 858*	*7 625*	*8 094*	*7 049*	*5 920*	*6 080*	*10 015*	*7 919*
37 Italy	...	600	500	600	700	750	750 F	750 F	750 F	750 F
37 Fishing area total	*...*	*600*	*500*	*600*	*700*	*750*	*750 F*	*750 F*	*750 F*	*750 F*
41 Argentina	3	3	3	5	2	2 F	2 F	...	...	...
41 Fishing area total	*3*	*3*	*3*	*5*	*2*	*2 F*	*2 F*	*...*	*...*	*...*
47 South Africa	4 192	1 797	2 563	2 500 F	2 500 F	2 400 F	2 300 F	2 295	1 620	2 000 F
47 Fishing area total	*4 192*	*1 797*	*2 563*	*2 500 F*	*2 500 F*	*2 400 F*	*2 300 F*	*2 295*	*1 620*	*2 000 F*
51 India	19 000 F	20 000 F	21 000 F	22 000 F	23 000 F	24 000 F	24 000 F	24 000 F	24 000 F	20 000 F
Madagascar	423	702	787	787	1 000	2 510	1 933	5 792	5 045	2 909
Tanzania	3 600	3 500 F	3 500 F	3 500 F	3 610	4 000	4 500	5 000	5 000 F	4 500
51 Fishing area total	*23 023 F*	*24 202 F*	*25 287 F*	*26 287 F*	*27 610 F*	*30 510 F*	*30 433 F*	*34 792 F*	*34 045 F*	*27 409 F*
57 Indonesia	109 986	96 437	97 483	130 199	96 831	23 641	2 632	3 921	4 990	3 890
Timor-Leste	...	...	...	...	...	...	1	1	1	-
57 Fishing area total	*109 986*	*96 437*	*97 483*	*130 199*	*96 831*	*23 641*	*2 633*	*3 922*	*4 991*	*3 890*
61 Japan	6 795	5 714	4 204	4 136	3 722	3 489	3 207	2 824	3 218	2 614
Korea Rep	4 730	9 089	8 185	3 971	6 272	3 903	2 904	2 873	3 779	3 700 F
Russian Fed	887	2 273	1 895	2 250	4 244	4 109	4 509	2 988	4 428	6 404
61 Fishing area total	*12 412*	*17 076*	*14 284*	*10 357*	*14 238*	*11 501*	*10 620*	*8 685*	*11 425*	*12 718 F*
71 Fiji Islands	-	24 F	-	0	0	0	16	16 F	16 F	16 F
Indonesia	8 409	14 001	14 092	31 345	29 148	23 874	20 520	38 791	29 460	31 880
Philippines	1 144	1 062	919	884	494	417	433	413	447	454
71 Fishing area total	*9 553*	*15 087 F*	*15 011*	*32 229*	*29 642*	*24 291*	*20 969*	*39 220 F*	*29 923 F*	*32 350 F*
77 Mexico	4 588	4 250	4 977	6 833	7 459	6 324	5 620	5 304	8 694	5 201
77 Fishing area total	*4 588*	*4 250*	*4 977*	*6 833*	*7 459*	*6 324*	*5 620*	*5 304*	*8 694*	*5 201*
81 New Zealand	0	0	11	4	6	1	2	0	1	7
81 Fishing area total	*0*	*0*	*11*	*4*	*6*	*1*	*2*	*0*	*1*	*7*
87 Peru	243	170	413	307	155	1 650	1 733	1 312	5 505	6 176
87 Fishing area total	*243*	*170*	*413*	*307*	*155*	*1 650*	*1 733*	*1 312*	*5 505*	*6 176*
Species total	*200 657 F*	*198 165 F*	*201 388 F*	*242 027 F*	*218 500 F*	*134 590 F*	*107 190 F*	*134 348 F*	*129 173 F*	*106 133 F*
Group total	***233 507***	***243 561***	***320 975***	***302 242***	***260 485***	***184 176***	***210 766***	***294 394***	***246 485***	***216 208***

See paragraph 5 of the INTRODUCTION. Voir le paragraphe 5 de l'INTRODUCTION. Véase el párrafo 5 en la INTRODUCCIÓN.

B-93 Green seaweeds / Algues vertes / Algas verdes

Capture production by species, fishing areas and countries or areas
Captures par espèces, zones de pêche et pays ou zones
Capturas por especies, áreas de pesca y países o áreas

Species, Fishing area Espèce, Zone de pêche Especie, Area de pesca	1993 t	1994 t	1995 t	1996 t	1997 t	1998 t	1999 t	2000 t	2001 t	2002 t
Lacy sea lettuce	**...B**		**...C**		***Ulva pertusa***				**7,41(08)002,04**	**UVP**
61 China,Taiwan	86	74	66	61	61	58	47	79	46	99
61 Fishing area total	*86*	*74*	*66*	*61*	*61*	*58*	*47*	*79*	*46*	*99*
Species total	*86*	*74*	*66*	*61*	*61*	*58*	*47*	*79*	*46*	*99*
Green seaweeds	**Algues vertes**		**Algas verdes**		***Chlorophyceae***				**7,41(XX)XXX,XX**	**SWG**
04 Japan	160	286	301	252	296	98	83	144	145	119
04 Fishing area total	*160*	*286*	*301*	*252*	*296*	*98*	*83*	*144*	*145*	*119*
27 France	23	45	32	60	80	112	118	83	114	93
27 Fishing area total	*23*	*45*	*32*	*60*	*80*	*112*	*118*	*83*	*114*	*93*
37 Italy	...	...	...	1 250	1 250	1 250	1 250 F	1 250 F	1 250 F	1 250 F
37 Fishing area total	*...*	*...*	*...*	*1 250*	*1 250*	*1 250*	*1 250 F*	*1 250 F*	*1 250 F*	*1 250 F*
51 India	52 500 F	54 000 F	55 500 F	57 000 F	58 500 F	60 000 F	60 000 F	60 000 F	60 000 F	55 000 F
51 Fishing area total	*52 500 F*	*54 000 F*	*55 500 F*	*57 000 F*	*58 500 F*	*60 000 F*	*60 000 F*	*60 000 F*	*60 000 F*	*55 000 F*
61 Korea Rep	962	1 042	1 029	757	582	341	574	284	444	450 F
61 Fishing area total	*962*	*1 042*	*1 029*	*757*	*582*	*341*	*574*	*284*	*444*	*450 F*
71 Fiji Islands	376	376 F	400 F	320 F	400 F	480	520	504 F	464 F	464 F
71 Fishing area total	*376*	*376 F*	*400 F*	*320 F*	*400 F*	*480*	*520*	*504 F*	*464 F*	*464 F*
Species total	*54 021 F*	*55 749 F*	*57 262 F*	*59 639 F*	*61 108 F*	*62 281 F*	*62 545 F*	*62 265 F*	*62 417 F*	*57 376 F*
Group total	***54 107***	***55 823***	***57 328***	***59 700***	***61 169***	***62 339***	***62 592***	***62 344***	***62 463***	***57 475***

See paragraph 5 of the INTRODUCTION. Voir le paragraphe 5 de l'INTRODUCTION. Véase el párrafo 5 en la INTRODUCCIÓN.

B-94 Miscellaneous aquatic plants / Plantes aquatiques diverses / Diversas plantas acuáticas

Capture production by species, fishing areas and countries or areas / Captures par espèces, zones de pêche et pays ou zones / Capturas por especies, áreas de pesca y países o áreas

Species, Fishing area Espèce, Zone de pêche Especie, Area de pesca	1993 t	1994 t	1995 t	1996 t	1997 t	1998 t	1999 t	2000 t	2001 t	2002 t
Spirulina nei	**Spirulina nca**		**Spirulina nep**		***Spirulina spp***				**7,11(01)001,XX**	**SIZ**
02 Mexico	3 435	-	-	-	-	-	-	-	-	-
02 Fishing area total	*3 435*	-	-	-	-	-	-	-	-	-
Species total	*3 435*	-	-	-	-	-	-	-	-	-
Eel-grass	**Grande zostère**		**Gran zostera**		***Zostera marina***				**7,92(02)001,01**	**ZOM**
37 Ukraine	424	264	243	370	331	310	157	6	...	111
37 Fishing area total	*424*	*264*	*243*	*370*	*331*	*310*	*157*	*6*	...	*111*
Species total	*424*	*264*	*243*	*370*	*331*	*310*	*157*	*6*	...	*111*
Tule nei	**Tule nca**		**Tule nep**		***Scirpus spp***				**7,92(04)001,XX**	**CJW**
02 Mexico	234	71	127	102	92	13	-	-	-	-
02 Fishing area total	*234*	*71*	*127*	*102*	*92*	*13*	-	-	-	-
Species total	*234*	*71*	*127*	*102*	*92*	*13*	-	-	-	-
Aquatic plants nei	**Plantes aquatiques nca**		**Plantas acuáticas nep**		***Plantae aquaticae***				**7,99(XX)XXX,XX**	**APL**
04 Korea Rep	-	-	5	11	1	5	-	-	-	-
Philippines	-	-	-	-	-	-	-	-	-	230
04 Fishing area total	-	-	*5*	*11*	*1*	*5*	-	-	-	*230*
27 France	36	42	41	24	-	-	-	-	-	-
27 Fishing area total	*36*	*42*	*41*	*24*	...	...	...	...	...	...
34 Senegal	50 F	50 F	50 F	50 F	50 F	100 F	150 F	200 F	223	232
34 Fishing area total	*50 F*	*50 F*	*50 F*	*50 F*	*50 F*	*100 F*	*150 F*	*200 F*	*223*	*232*
61 China	124 250	151 000	150 000	151 630	184 270	170 520	215 550	204 290	266 870	297 160
China,H.Kong	1	0	0	0	0	0	0	0	0	0
China,Taiwan	69	93	24	18	52	16	40	17	19	16
Japan	23 468	24 854	13 496	14 618	12 047	11 161	11 459	11 808	11 249	10 610
Korea Rep	3 309	5 402	2 128	1 654	3 248	3 191	793	1 708	643	650 F
Russian Fed	-	-	-	200	150	100	-	-	-	-
61 Fishing area total	*151 097*	*181 349*	*165 648*	*168 120*	*199 767*	*184 988*	*227 842*	*217 823*	*278 781*	*308 436 F*
77 Cook Is	40 F	40 F	40 F	45 F	50 F	50 F	50 F	50 F	50 F	50 F
77 Fishing area total	*40 F*	*40 F*	*40 F*	*45 F*	*50 F*	*50 F*	*50 F*	*50 F*	*50 F*	*50 F*
87 Chile	-	-	-	3	8	29	1	3	32	8
87 Fishing area total	-	-	-	*3*	*8*	*29*	*1*	*3*	*32*	*8*
Species total	*151 223 F*	*181 481 F*	*165 784 F*	*168 253 F*	*199 876 F*	*185 172 F*	*228 043 F*	*218 076 F*	*279 086 F*	*308 956 F*
Group total	***155 316***	***181 816***	***166 154***	***168 725***	***200 299***	***185 495***	***228 200***	***218 082***	***279 086***	***309 067***

See paragraph 5 of the INTRODUCTION. Voir le paragraphe 5 de l'INTRODUCTION. Véase el párrafo 5 en la INTRODUCCIÓN.

C - Capture production: by fishing areas

C - Captures: par zones de pêche

C - Capturas: por áreas de pesca

C-00 Fish, crustaceans, molluscs, etc — Capture production by fishing areas and species groups
Poissons, crustacés, mollusques, etc — Captures par zones de pêche et groupes d'espèces
Peces, crustáceos, moluscos, etc — Capturas por áreas de pesca y grupos de especies

Fishing area Zone de pêche Area de pesca	Species group Groupe d'espèces Grupo de especies	1996 t	1997 t	1998 t	1999 t	2000 t	2001 t	2002 t
01 Africa - Inland waters	11	81 926	86 704	78 960	96 396	97 962	98 307	92 378
Afrique - Eaux continentales	12	366 526	361 520	394 271	423 730	467 477	475 290	472 135
Africa - Aguas continentales	13	1 376 824	1 443 241	1 492 828	1 462 412	1 441 072	1 425 036	1 473 314
	22	637	985	801	959	2 164	2 129	2 002
	23	11	16	24	29	31	67	112
	24	290	253	359	270	284	258	238
	31	...	547	381	763	2 490	2 134	1 858
	33	15 875	18 986	20 939	20 750	24 658	28 359	28 718
	35	1 900	1 850	1 800	1 798	1 500	1 700	1 700
	37	1 326	1 229	1 150	1 180	3 182	5 154	5 941
	41	8 042	8 423	8 995	7 801	9 140	9 498	10 931
	45	3 000	2 850	2 700	2 574	2 100	2 400	2 400
	51	530	610	625	606	921	851	1 197
	Area total	1 856 887	1 927 214	2 003 833	2 019 268	2 052 981	2 051 183	2 092 924
02 America, North - Inland waters	11	28 105	24 924	21 241	20 971	22 094	21 943	19 457
Amérique du Nord - Eaux continentales	12	86 191	84 500	72 068	69 657	75 289	66 090	61 990
América del Norte - Aguas continentales	13	51 261	49 959	51 319	51 767	58 297	47 613	51 295
	21	526	615	525	420	281	173	172
	22	730	796	739	770	942	648	269
	23	32 374	27 972	29 200	25 267	22 515	22 669	19 563
	24	3 187	3 941	3 825	5 199	3 070	4 228	4 462
	25	1 040	1 219	1 132	1 155	1 327	1 371	1 126
	32	23	24	25	25	25	-	-
	33	...	...	...	...	...	220	213
	37	...	...	372	487	392	558	379
	41	10 174	14 177	13 430	9 681	3 835	7 885	10 310
	51	240	122	185	240	565	1 029	1 052
	71	422	2 034	1 201	337	338	361	146
	72	18	17	26	15	53	52	27
	77	337	297	483	617	231	119	153
	Area total	214 628	210 597	195 771	186 608	189 254	174 959	170 614
03 America, South - Inland waters	11	502	213	265	307	1 745	390	392
Amérique du Sud - Eaux continentales	12	16 173	15 460	15 370	14 242	16 226	22 123	22 194
América del Sur - Aguas continentales	13	322 890	312 455	315 347	328 254	323 038	344 523	352 681
	23	463	1 208	733	527	566	658	978
	24	642	455	567	523	407	340	313
	37	850	850	850	850	850	760	750
	41	2 677	2 157	1 506	3 235	2 437	...	...
	71	0	0	0	0	7	9	5
	72	0	0	0	0	0	-	-
	Area total	344 197	332 798	334 638	347 938	345 276	368 803	377 313
04 Asia - Inland waters	11	429 436	376 916	377 296	382 408	336 298	310 843	355 909
Asie - Eaux continentales	12	86 676	95 384	112 104	125 424	122 422	119 686	121 397
Asia - Aguas continentales	13	2 791 199	2 883 899	3 067 956	3 625 592	3 611 382	3 751 883	3 494 506
	21	2 179	1 733	1 531	1 320	1 288	1 231	915
	22	5 795	2 089	2 064	2 255	5 687	2 907	4 728
	23	38 676	35 498	33 261	28 219	29 420	24 738	24 659
	24	220 802	212 274	239 546	250 449	366 137	210 293	159 638
	25	369	34	3 000	83	163	201	329
	33	37 088	41 467	35 911	35 721	39 120	41 978	31 976
	34	490	447	270	494	402	620	337
	37	562	1 600	1 500	1 455	1 583	1 685	1 733
	41	386 236	498 702	617 997	474 888	548 927	605 818	792 853
	42	363	267	44	47	77	150	219
	45	34 418	44 296	23 900	23 365	34 186	37 062	102 657
	51	557 745	503 071	574 737	546 606	588 720	620 951	625 462
	54	4	-	-	-	-	-	-
	56	844	956	914	304	303	792	850
	58	481	52	165	73	166	112	120
	71	2 533	1 550	1 767	1 435	1 957	2 116	2 078
	72	36	48	27	1	3	42	40
	77	1 494	2 460	1 480	2 048	2 031	1 578	1 735
	Area total	4 597 426	4 702 743	5 095 470	5 502 187	5 690 272	5 734 686	5 722 141
05 Europe - Inland waters	11	117 076	111 636	120 232	106 265	103 828	108 379	114 431
Europe - Eaux continentales	13	115 531	105 834	101 028	95 408	114 859	100 874	106 866
Europa - Aguas continentales	21	1 337	1 365	1 222	856	704	633	526
	22	3 787	3 194	3 133	3 092	3 101	2 483	2 326
	23	64 252	69 832	66 252	83 333	82 718	80 962	88 611
	24	96 744	86 878	120 291	156 813	120 766	47 425	33 658
	25	569	467	1 042	1 243	1 354	1 104	1 927
	32	98	163	177	22	11	55	127
	33	52	109	219	451	360	494	447
	37	8	110	350	361	350	326	390
	41	2 774	2 989	2 783	2 800	3 758	4 507	4 481
	51	-	-	-	-	-	-	451
	71	...	38	41	35	26	-	29

C-00 Fish, crustaceans, molluscs, etc — Capture production by fishing areas and species groups
Poissons, crustacés, mollusques, etc — Captures par zones de pêche et groupes d'espèces
Peces, crustáceos, moluscos, etc — Capturas por áreas de pesca y grupos de especies

Fishing area Zone de pêche Area de pesca	Species group Groupe d'espèces Grupo de especies	1996 t	1997 t	1998 t	1999 t	2000 t	2001 t	2002 t
	Area total	402 228	382 615	416 770	450 679	431 835	347 242	354 270
06 Oceania - Inland waters	11	0	28	2	3	4	6	7
Océanie - Eaux continentales	12	2 347	2 350	2 604	2 600	2 590	2 588	2 590
Oceanía - Aguas continentales	13	9 357	9 353	9 158	9 122	9 137	9 026	8 994
	22	931	810	1 286	1 209	1 151	1 244	1 036
	24	480	480	480	480	480	480	480
	25	350	350	350	350	350	350	350
	33	1 850	1 850	1 850	1 850	1 850	1 850	1 850
	41	338	329	345	412	372	375	418
	51	2 670	3 970	4 500	5 000	5 080	5 300	5 180
	Area total	18 323	19 520	20 575	21 026	21 014	21 219	20 905
21 Atlantic, Northwest	21	86	91	69	9	3	4	-
Atlantique, nord-ouest	22	551	455	496	353	395	220	494
Atlántico, noroeste	23	1 301	1 458	1 374	428	1 070	405	355
	24	6 415	6 887	8 281	7 367	5 919	4 007	6 239
	25	2 144	2 785	3 032	3 000	3 123	2 934	2 847
	31	84 379	91 352	92 171	113 787	118 636	122 063	116 383
	32	120 222	127 256	140 007	143 127	138 976	148 393	140 651
	33	20 492	20 055	19 826	28 628	19 522	20 343	21 158
	34	56 985	59 856	66 934	66 707	73 940	81 218	91 644
	35	575 240	606 757	521 264	472 439	460 612	543 766	439 255
	36	10 721	7 665	7 842	10 785	8 399	8 750	7 284
	37	77 942	66 326	77 935	58 871	49 971	65 508	88 109
	38	56 131	48 219	51 296	49 186	49 360	39 716	35 639
	39	14 493	14 376	5 848	7 487	8 173	4 483	4 516
	42	142 599	150 626	147 582	172 414	166 103	173 491	182 638
	43	71 866	78 146	77 155	83 105	83 062	83 803	82 422
	45	200 651	174 549	207 259	243 054	269 324	263 745	292 213
	47	2	-	1	5	6	1	1
	52	4 513	3 668	3 149	5 721	4 567	6 339	6 221
	53	39 569	49 099	33 329	47 106	26 475	18 571	13 236
	54	17 585	17 228	15 051	15 651	21 098	18 353	23 631
	55	122 704	116 169	108 136	135 115	201 288	256 929	283 771
	56	384 556	345 924	315 694	325 053	329 462	347 200	372 074
	57	38 908	48 733	44 724	26 767	26 402	18 409	19 890
	58	2 394	600	168	1 920	1 469	1 384	3 210
	75	1 598	2 606	3 182	2 238	1 557	1 071	1 257
	76	15 139	12 201	13 126	14 298	13 440	8 778	9 363
	77	-	223	299	921	202	481	507
	Area total	2 069 186	2 053 310	1 965 230	2 035 542	2 082 554	2 240 365	2 245 008
27 Atlantic, Northeast	11	5 834	6 674	6 311	6 574	8 172	8 399	9 975
Atlantique, nord-est	13	14 971	17 105	17 928	17 062	11 133	11 466	12 099
Atlántico, nordeste	21	0	0	1	2	-	-	-
	22	2 959	4 327	2 290	2 622	1 982	2 098	1 990
	23	14 553	13 283	11 199	10 262	14 507	9 324	10 355
	24	122	113	58	59	140	103	158
	25	67	135	101	40	571	658	818
	31	318 478	310 917	298 034	317 559	300 255	303 387	276 423
	32	3 191 871	3 135 608	3 256 629	3 291 268	3 330 963	3 607 974	3 406 129
	33	880 800	1 267 534	1 065 549	789 791	768 549	939 102	929 214
	34	435 011	418 836	425 131	393 488	420 190	422 671	393 672
	35	2 875 382	3 089 362	3 013 492	2 995 072	2 890 562	2 427 552	2 350 717
	36	47 620	44 300	40 738	41 155	40 581	30 237	29 246
	37	2 547 671	2 630 715	2 000 933	1 857 952	2 404 692	2 622 398	2 903 351
	38	74 119	106 720	94 286	98 654	104 071	107 084	82 758
	39	90 327	74 377	82 501	49 251	72 227	36 178	36 964
	42	42 033	58 913	53 020	51 037	53 379	56 441	55 784
	43	54 509	58 122	55 760	59 818	55 202	55 364	56 017
	44	175	177	313	471	476	814	1 802
	45	167 166	191 183	176 695	173 335	172 670	142 743	149 372
	46	-	-	88	-	-	36	-
	47	1 621	1 783	1 389	1 276	1 067	1 672	2 585
	52	29 369	33 733	18 337	22 335	33 739	33 900	30 357
	53	1 400	1 852	1 434	881	683	899	1 387
	54	101 062	118 949	122 237	114 376	128 707	146 696	133 725
	55	55 636	71 666	73 299	73 406	74 769	81 361	70 411
	56	58 500	57 086	116 751	94 255	65 960	49 247	50 815
	57	53 371	50 983	49 948	53 067	58 619	44 474	51 988
	58	878	1 023	1 302	663	392	424	196
	76	911	612	564	643	463	478	650
	77	0	-	-	-	-	24	4
	Area total	11 066 416	11 766 088	10 986 318	10 516 374	11 014 721	11 143 204	11 048 962
31 Atlantic, Western Central	13	7 303	3 656	2 870	4 130	6 772	6 937	-
Atlantique, centre-ouest	22	35	19	9	2	1	0	11
Atlántico, centro-occidental	24	343	246	280	117	288	171	286
	25	9	12	14	4	12	15	17

C-00

Fish, crustaceans, molluscs, etc — Capture production by fishing areas and species groups
Poissons, crustacés, mollusques, etc — Captures par zones de pêche et groupes d'espèces
Peces, crustáceos, moluscos, etc — Capturas por áreas de pesca y grupos de especies

Fishing area Zone de pêche Area de pesca	Species group Groupe d'espèces Grupo de especies	1996 t	1997 t	1998 t	1999 t	2000 t	2001 t	2002 t
	31	2 527	1 546	1 891	1 673	1 765	1 695	1 922
	32	...	4	2	2	6	3	4
	33	164 649	157 162	159 994	130 630	155 314	150 992	137 757
	34	12 416	14 408	19 486	17 505	15 203	21 535	19 132
	35	679 660	758 862	734 996	865 288	713 794	641 554	783 083
	36	88 413	88 404	93 736	85 069	87 792	103 718	84 069
	37	35 373	36 900	38 351	32 601	33 106	33 744	31 092
	38	31 629	33 108	30 760	27 069	23 679	22 064	24 836
	39	222 149	252 532	213 099	175 905	164 058	152 559	162 579
	42	57 004	69 237	68 121	52 584	60 703	49 241	58 834
	43	30 412	29 219	28 118	31 709	31 249	24 704	29 121
	45	163 062	170 023	181 549	173 843	202 638	209 006	186 447
	47	366	351	364	168	181	186	163
	52	39 257	33 711	31 933	30 197	24 153	25 322	20 732
	53	113 560	99 330	106 853	89 056	198 341	161 729	148 186
	54	223	295	3 802	451	316	1 081	1 272
	55	684	581	834	223	256	253	391
	56	39 590	48 875	38 107	45 883	55 716	54 439	53 151
	57	31 119	20 750	24 135	20 113	24 384	23 232	18 114
	58	673	1 873	2 921	6 706	2 716	1 913	2 971
	72	227	86	136	31	44	34	42
	75	-	1	70	159	139	228	130
	76	10	15	15	15	10	49	10
	77	6	2	4	0	139	-	-
	Area total	1 720 699	1 821 208	1 782 450	1 791 133	1 802 775	1 686 404	1 764 352
34 Atlantic, Eastern Central Atlantique, centre-est Atlántico, centro-oriental	12	2 603	2 582	1 365	1 760	2 062	2 324	2 333
	22	0	1	4	-	-	-	-
	24	11 585	8 810	8 113	4 608	7 489	5 633	8 819
	31	35 323	38 371	46 234	41 604	50 540	49 910	44 910
	32	31 447	25 650	25 150	25 212	19 545	25 435	25 462
	33	263 289	291 576	267 289	292 312	277 854	287 190	283 954
	34	94 725	76 143	77 018	50 478	38 726	40 055	38 351
	35	1 753 261	1 828 435	1 869 042	1 695 482	1 748 123	1 926 048	1 740 654
	36	346 623	322 931	351 789	356 802	304 768	338 374	279 141
	37	439 833	444 084	513 196	435 006	570 112	477 161	477 244
	38	35 840	68 091	52 464	51 721	52 223	51 528	49 094
	39	294 759	301 696	289 463	405 072	285 493	433 548	249 344
	42	2 331	4 114	6 558	5 586	5 231	6 717	6 886
	43	2 174	9 378	4 010	2 821	2 886	2 982	1 154
	45	57 810	53 206	83 505	70 383	59 356	57 907	56 130
	47	1 034	1 317	1 743	875	249	536	139
	52	8 235	6 598	9 849	7 290	6 637	9 659	8 571
	53	100	109	89	125	101	151	120
	54	10	6	7	...	401	102	8
	56	54	69	139	148	120	112	171
	57	190 620	150 352	183 729	250 402	216 428	213 558	96 870
	58	746	1 489	12 599	2 258	1 189	385	3 423
	72	42	164	189	463	122	315	845
	Area total	3 572 444	3 635 172	3 803 544	3 700 408	3 649 655	3 929 630	3 373 623
37 Mediterranean and Black Sea Méditerranée et mer Noire Mediterráneo y Mar Negro	11	476	882	617	643	854	575	375
	13	2 822	2 825	3 031	2 568	2 956	3 522	3 312
	21	584	637	429	242	85	33	39
	22	962	1 257	917	682	464	602	648
	23	18	22	26	-	-	0	-
	24	3 862	3 439	6 391	13 550	14 141	30 727	30 601
	25	-	-	-	-	-	8	4
	31	13 340	11 718	11 059	10 469	11 167	10 874	8 695
	32	90 708	72 072	75 371	64 641	71 692	64 917	47 992
	33	183 355	170 147	173 388	176 952	174 397	161 557	160 344
	34	21 004	23 071	18 757	17 683	17 558	19 894	20 118
	35	674 739	642 385	625 771	781 261	722 642	796 135	810 153
	36	80 682	75 137	81 243	74 811	71 572	70 006	59 238
	37	116 031	110 585	100 465	97 217	106 059	117 976	119 937
	38	16 047	16 334	14 886	12 157	12 514	11 395	10 863
	39	105 579	88 163	93 606	72 160	81 048	68 265	66 403
	42	2 195	2 104	2 520	4 491	2 891	2 594	3 485
	43	7 455	7 203	4 927	4 623	4 023	4 133	3 739
	45	25 384	30 279	27 525	29 407	35 293	29 057	31 259
	47	9 804	8 827	7 873	9 927	10 428	10 024	8 740
	52	4 518	5 682	4 829	5 545	5 716	4 675	1 739
	53	2 246	1 863	1 204	952	319	157	188
	54	39 046	53 310	38 899	44 453	46 143	46 092	51 820
	55	93	137	65	80	588	154	476
	56	52 633	39 346	36 511	43 249	47 403	44 929	38 920
	57	59 266	60 139	54 281	50 692	53 877	53 858	52 419
	58	18 120	15 480	15 120	13 389	10 245	15 401	17 769
	74	28	22	22	30	30	76	78
	76	69	55	60	76	204	104	166
	77	909	904	1 764	1 737	908	2 595	579
	Area total	1 531 975	1 444 025	1 401 557	1 533 687	1 505 217	1 570 335	1 550 099

C-00 Fish, crustaceans, molluscs, etc / Poissons, crustacés, mollusques, etc / Peces, crustáceos, moluscos, etc

Capture production by fishing areas and species groups / Captures par zones de pêche et groupes d'espèces / Capturas por áreas de pesca y grupos de especies

Fishing area Zone de pêche Area de pesca	Species group Groupe d'espèces Grupo de especies	1996 t	1997 t	1998 t	1999 t	2000 t	2001 t	2002 t
41 Atlantic, Southwest Atlantique, sud-ouest Atlántico, sudoccidental	31	10 356	11 976	10 902	8 701	8 689	8 095	8 047
	32	860 118	815 951	779 466	618 103	508 894	547 832	615 675
	33	226 233	234 505	213 701	207 297	236 185	251 775	248 419
	34	58 259	59 688	74 873	66 381	55 017	74 027	68 142
	35	139 838	161 501	115 964	59 628	58 851	89 073	82 478
	36	72 769	72 144	63 213	64 577	65 990	65 189	67 295
	37	40 265	42 416	39 028	31 276	41 725	34 979	45 976
	38	56 595	58 696	57 896	61 308	60 197	62 855	57 721
	39	158 570	165 118	157 430	181 165	187 841	165 941	176 164
	42	13 845	16 668	15 285	16 091	17 991	16 318	16 888
	43	8 026	7 502	6 002	6 334	6 469	7 135	7 016
	44	201	414	457	271	369	310	517
	45	45 681	47 343	58 911	48 673	76 986	109 959	92 848
	46	-	-	74	-	4	-	-
	47	313	849	2 161	6 293	227	1 317	247
	52	558	1 322	1 010	683	1 621	1 519	1 314
	53	873	828	744	1 547	884	895	876
	54	1 234	1 548	1 606	1 381	1 215	2 540	2 161
	55	36 952	39 817	28 441	42 700	37 404	42 598	51 181
	56	126	196	664	2 443	279	297	1 075
	57	747 875	1 016 557	750 044	1 192 738	1 000 427	803 766	543 286
	58	1 175	1 851	1 408	4 937	2 474	1 082	2 334
	72	0	0	0	0	0	...	...
	Area total	2 479 862	2 756 890	2 379 280	2 622 527	2 369 739	2 287 502	2 089 660
47 Atlantic, Southeast Atlantique, sud-est Atlántico, sudoriental	31	1 996	2 671	2 960	3 069	2 928	4 021	5 320
	32	287 275	261 732	303 451	309 702	309 995	324 936	309 613
	33	23 740	19 664	52 202	36 793	47 795	61 670	73 764
	34	58 742	65 833	73 371	53 312	51 583	55 853	63 594
	35	233 659	328 352	419 566	495 703	582 167	604 524	614 005
	36	48 631	38 599	50 823	51 926	51 299	48 500	50 565
	37	554 647	507 630	534 727	486 932	501 730	414 020	440 596
	38	3 058	7 781	6 531	8 334	10 688	15 221	13 826
	39	90 506	106 528	59 888	60 929	47 629	91 020	99 275
	42	5 329	3 415	5 639	4 855	6 875	5 251	5 300
	43	3 032	3 117	3 348	2 931	2 690	3 452	3 989
	45	5 915	5 569	16 230	5 333	10 421	13 329	5 844
	46	-	-	254	-	-	-	-
	47	2	-	-	-	-	1 062	1 653
	52	735	330	524	481	490	527	516
	53	...	...	...	-	-	-	1
	56	0	-	-	-	-	-	36
	57	8 170	4 145	7 801	8 301	7 506	4 654	9 331
	58	-	-	-	-	-	-	2 000
	77	-	-	-	-	106	44	2 212
	Area total	1 325 437	1 355 366	1 537 315	1 528 601	1 633 902	1 648 084	1 701 440
48 Atlantic, Antarctic Atlantique, Antarctique Atlántico, Antártico	31	-	-	-	-	0	-	-
	32	25	21	18	13	3	6	9
	33	-	0	1	10	6	4	16
	34	3 602	3 812	3 208	3 915	9 120	5 016	8 422
	38	45	32	8	17	0	13	0
	39	-	-	7	0	-	0	1
	44	214	1	1	2	2	15	111
	46	91 150	75 653	90 025	101 957	114 426	104 183	125 987
	57	52	81	-	-	0	20	49
	77	-	-	-	-	5	-	0
	Area total	95 088	79 600	93 268	105 914	123 562	109 257	134 595
51 Indian Ocean, Western Océan Indien, ouest Océano Indico, occidental	24	5 547	5 265	4 607	4 548	4 736	4 784	6 039
	25	395	334	335	341	125	134	159
	31	20 730	25 299	20 676	15 878	27 872	14 647	18 916
	32	1 251	1 348	1 456	748	1 472	2 436	2 372
	33	821 440	856 296	777 742	848 740	805 850	798 578	821 758
	34	90 794	189 577	118 150	166 673	168 664	155 063	194 034
	35	373 255	433 033	415 773	407 553	546 667	577 895	586 653
	36	796 802	800 918	783 481	913 946	918 054	864 343	996 906
	37	555 851	430 288	437 632	451 621	289 597	276 419	322 533
	38	177 750	118 146	120 715	119 038	126 347	121 676	141 293
	39	610 747	661 111	632 249	662 681	633 757	740 547	699 449
	42	7 652	9 814	11 083	11 839	20 952	15 697	15 977
	43	3 452	3 315	3 634	3 187	2 918	2 818	3 137
	45	297 846	272 896	336 556	322 311	311 223	291 178	281 543
	47	15 600	12 367	12 797	11 460	11 251	27 066	31 290
	52	43	40	40	29	45	51	50
	53	32	16	9	8	2	1	1
	56	57	8	19	26	30	11	19
	57	94 868	148 579	112 432	121 305	119 166	82 139	115 430
	58	15 939	7 363	5 015	3 747	5 034	3 580	2 613
	74	-	-	-	-	95	-	-
	76	7 258	7 293	3 371	1 767	2 139	2 229	3 158

C-00 Fish, crustaceans, molluscs, etc / Poissons, crustacés, mollusques, etc / Peces, crustáceos, moluscos, etc

Capture production by fishing areas and species groups / Captures par zones de pêche et groupes d'espèces / Capturas por áreas de pesca y grupos de especies

Fishing area Zone de pêche Area de pesca	Species group Groupe d'espèces Grupo de especies	1996 t	1997 t	1998 t	1999 t	2000 t	2001 t	2002 t
	Area total	3 897 309	3 983 306	3 797 772	4 067 446	3 995 996	3 981 292	4 243 330
57 Indian Ocean, Eastern	22	208	203	160	129	133	175	167
Océan Indien, est	24	197 740	192 141	180 806	190 310	194 226	209 540	206 628
Océano Indico, oriental	25	13 989	13 790	23 061	15 685	15 982	15 665	16 535
	31	30 824	33 443	28 204	26 965	26 879	26 200	28 228
	32	2 716	2 864	5 890	7 566	9 143	7 440	9 130
	33	488 324	485 137	511 989	511 983	511 301	517 302	506 572
	34	68 457	72 513	94 098	93 470	84 293	79 175	77 957
	35	435 808	497 817	490 752	393 681	394 122	431 941	408 334
	36	440 357	475 267	508 308	512 704	472 477	424 030	448 125
	37	543 336	537 456	584 052	560 548	507 071	515 504	529 577
	38	117 205	111 844	127 229	121 470	119 552	116 425	116 615
	39	1 350 880	1 503 736	1 693 158	1 660 958	1 882 441	1 999 007	2 231 001
	42	23 284	24 528	31 378	31 429	46 864	46 902	49 325
	43	16 337	16 285	16 599	19 058	20 855	18 202	15 784
	45	217 697	231 575	182 546	234 839	241 260	232 937	216 201
	47	40 568	45 940	52 272	54 445	36 398	41 365	43 329
	52	5 081	4 905	4 920	5 297	5 207	5 304	5 845
	53	-	-	-	-	350	298	180
	54	71	-	-	-	21	20	20
	55	6 872	6 093	5 543	7 622	8 317	4 421	2 525
	56	61 448	63 188	34 444	42 944	41 498	37 540	37 651
	57	104 647	109 443	115 729	111 790	127 253	119 986	125 264
	58	381	263	844	8 396	1 940	1 941	1 440
	72	353	248	337	273	372	45	71
	76	419	610	833	859	1 048	787	920
	77	23 527	29 747	7 077	35 524	36 686	25 228	22 837
	Area total	4 190 529	4 459 036	4 700 229	4 647 945	4 785 689	4 877 380	5 100 261
58 Indian Ocean, Antarctic	32	-	13	76	153	376	206	386
Océan Indien, Antarctique	33	15	5	6	11	0	0	1
Océano Indico, Antártico	34	5 661	8 824	10 030	9 573	11 542	9 842	7 270
	37	-	-	-	-	-	0	0
	38	0	11	26	38	96	121	347
	39	-	1	0	0	0	-	-
	44	-	0	0	0	3	0	0
	46	6	-	-	-	-	-	-
	57	-	1	-	-	-	-	0
	74	7	-	-	-	-	-	-
	77	0	12	2	-	-	-	-
	Area total	5 689	8 867	10 140	9 775	12 017	10 169	8 004
61 Pacific, Northwest	11	110	101	100	56	8	52	38
Pacifique, nord-ouest	23	455 779	471 418	428 221	380 303	350 771	401 629	371 815
Pacífico, noroeste	24	75 549	106 577	116 818	137 932	126 545	127 786	112 677
	25	18	9	26	47	32	81	73
	31	181 394	186 917	197 870	210 735	225 682	203 918	188 575
	32	3 529 239	3 516 455	3 030 625	2 463 168	1 959 990	1 885 369	1 307 857
	33	1 572 287	1 623 465	1 635 578	1 614 745	1 724 308	1 640 243	1 591 033
	34	1 469 582	1 394 197	1 655 751	1 637 181	1 696 006	1 712 536	1 719 785
	35	1 990 976	2 556 096	2 932 317	2 856 880	2 534 514	2 614 151	2 443 271
	36	567 298	748 063	924 890	961 585	917 339	848 314	855 844
	37	3 267 951	3 162 978	2 674 518	2 462 646	2 469 381	2 667 858	2 587 752
	38	31 201	37 365	34 383	57 887	56 079	35 970	39 671
	39	5 147 881	4 853 220	4 967 320	5 010 112	4 689 171	4 204 409	4 122 073
	42	488 600	431 585	442 492	437 240	500 701	500 522	507 295
	43	2 221	1 724	1 794	1 842	3 316	3 531	2 787
	44	48 254	39 019	41 217	45 646	36 020	23 244	17 059
	45	776 922	834 742	932 180	1 150 109	1 119 120	991 000	981 353
	47	1 007 693	1 165 004	1 314 828	1 198 681	1 304 743	1 335 243	1 281 908
	51	-	-	-	-	-	74	209
	52	23 352	24 084	25 540	22 257	20 249	19 385	20 550
	53	18 259	17 210	9 905	11 610	15 943	10 092	16 679
	54	3 028	7 309	7 934	8 265	6 952	4 955	4 479
	55	276 406	266 957	294 211	305 510	310 104	293 268	310 303
	56	181 775	144 085	156 515	145 198	132 789	119 870	116 738
	57	1 165 463	1 154 074	1 033 618	1 117 839	1 221 972	1 137 512	1 130 859
	58	938 870	1 502 215	1 488 978	1 457 426	1 410 260	1 405 909	1 381 743
	74	16 747	2 780	891	1 171	1 443	1 053	1 025
	76	27 020	28 009	25 353	24 156	24 161	22 753	25 229
	77	278 735	410 601	441 444	410 337	346 443	340 147	297 549
	Area total	23 542 610	24 686 259	24 815 317	24 130 564	23 204 042	22 550 874	21 436 229
67 Pacific, Northeast	21	1	0	0	0	242	195	207
Pacifique, nord-est	23	423 012	301 117	318 745	382 944	300 917	349 383	288 126
Pacífico, nordeste	24	15	34	62	-	105	178	66
	31	217 113	273 510	198 371	174 192	203 808	173 733	189 849
	32	1 665 540	1 669 564	1 714 162	1 511 756	1 631 326	1 830 568	1 882 188
	33	92 432	62 763	54 050	54 531	47 878	59 881	40 847

C-00 Fish, crustaceans, molluscs, etc — Capture production by fishing areas and species groups
Poissons, crustacés, mollusques, etc — Captures par zones de pêche et groupes d'espèces
Peces, crustáceos, moluscos, etc — Capturas por áreas de pesca y grupos de especies

Fishing area Zone de pêche Area de pesca	Species group Groupe d'espèces Grupo de especies	1996 t	1997 t	1998 t	1999 t	2000 t	2001 t	2002 t
	34	107 235	111 003	88 790	97 110	86 495	79 323	76 243
	35	71 607	84 451	73 503	68 758	74 750	87 058	100 795
	36	10 063	15 888	11 654	4 710	11 220	12 162	12 205
	37	388	206	1 593	251	302	537	148
	38	8 700	6 496	5 142	9 340	8 628	8 313	10 420
	39	106 820	106 601	97 853	98 191	31 003	74 345	4 864
	42	65 038	73 923	131 368	102 383	34 839	33 594	39 353
	43	...	...	...	25	27	18	18
	44	9 526	8 177	10 941	7 675	6 848	7 281	7 616
	45	24 422	24 331	11 471	18 086	20 381	22 952	29 973
	47	...	...	...	...	49	63	58
	52	6 369	...	...	...	-	4	7
	53	4 989	6 187	11	6	14	20	19
	54	455	...	...	1	675	448	251
	55	2 372	2	3 228	2 642	2 012	1 052	1 264
	56	7 111	5 415	8 262	6 777	6 517	6 658	8 984
	57	685	2 787	2 293	3 132	748	2 997	842
	58	1 873	4 593	2 879	1 850	1 904	1 649	2 094
	76	7 576	9 044	8 081	7 329	7 115	6 678	6 448
	Area total	2 833 342	2 766 092	2 742 459	2 551 689	2 477 803	2 759 090	2 702 885
71 Pacific, Western Central Pacifique, centre-ouest Pacífico, centro-occidental	24	12 415	8 996	10 682	11 497	11 755	15 247	15 001
	25	44 227	45 706	47 280	53 973	60 282	53 137	57 344
	31	14 922	24 987	16 791	20 281	18 157	19 443	21 484
	32	24	21	33	765	50	60	42
	33	1 026 052	1 076 116	1 030 488	1 159 252	1 076 667	1 100 750	1 149 324
	34	43 313	47 778	48 101	55 292	55 184	53 714	53 763
	35	1 014 955	1 070 578	1 051 747	999 289	1 028 734	1 094 228	1 036 418
	36	1 699 445	1 759 571	2 102 575	2 050 157	2 101 913	2 046 013	2 281 419
	37	1 394 094	1 477 321	1 513 491	1 616 062	1 653 087	1 709 494	1 789 831
	38	138 975	140 049	143 301	139 951	146 527	148 968	151 084
	39	2 224 261	2 138 367	2 137 686	2 280 509	2 356 699	2 527 456	2 566 893
	42	126 790	127 540	144 405	141 990	151 625	162 730	170 387
	43	6 717	10 085	8 498	6 339	7 208	8 385	9 928
	45	435 422	440 527	414 268	438 591	442 815	474 998	484 354
	47	2 751	8 275	7 359	4 498	5 215	5 147	5 843
	52	448	183	348	282	241	288	292
	53	1 887	1 830	2 118	2 120	2 010	992	1 697
	54	20 216	18 017	17 454	6 554	41 246	22 782	23 282
	55	7 276	3 488	6 392	5 403	4 601	5 901	4 334
	56	72 085	57 595	74 531	83 453	67 778	104 017	108 471
	57	343 913	360 035	366 124	364 644	437 671	403 546	433 191
	58	53 322	54 772	59 749	59 051	53 179	61 483	60 783
	72	417	403	452	448	384	319	332
	76	8 408	7 683	6 857	6 123	6 505	6 780	7 185
	77	38 285	86 579	74 425	91 744	144 866	77 337	77 520
	Area total	8 730 620	8 966 502	9 285 155	9 598 268	9 874 399	10 103 215	10 510 202
77 Pacific, Eastern Central Pacifique, centre-est Pacífico, centro-oriental	23	2 084	2 747	944	1 996	2 623	1 206	2 424
	24	-	0	-	-	-	-	-
	25	0	1	0	0	7	...	1
	31	9 229	8 937	4 853	6 470	6 284	4 823	4 887
	32	138	1 273	806	379	718	1 051	1 421
	33	43 004	46 049	44 181	57 328	46 348	60 739	63 251
	34	9 816	9 547	7 942	5 446	4 517	4 418	3 983
	35	562 780	621 114	544 447	485 986	705 994	844 121	908 267
	36	372 757	416 664	422 196	393 477	383 416	455 509	558 360
	37	28 341	48 042	51 052	85 841	85 380	48 199	33 747
	38	38 188	35 128	38 346	39 191	43 727	37 032	34 990
	39	213 310	181 446	159 623	114 666	126 461	135 485	129 559
	42	15 145	12 657	8 665	7 818	13 402	12 700	10 556
	43	2 618	2 731	2 504	2 305	3 573	3 629	3 322
	44	164	-	-	-	1	1 265	3 157
	45	73 054	72 952	66 723	68 316	59 171	53 902	50 819
	47	631	515	1 079	1 237	1 247	1 376	5 257
	52	4 309	3 465	2 374	2 108	2 673	3 230	3 368
	53	3 071	4 079	3 383	3 028	3 745	3 317	3 301
	54	583	2 206	2 230	832	643	823	782
	55	17 290	2 320	2 726	1 864	6 287	3 100	9 875
	56	7 265	6 792	5 770	6 913	7 143	5 848	4 386
	57	201 442	213 080	30 387	150 569	202 531	166 988	189 990
	58	1 307	1 857	1 462	1 742	3 816	2 950	2 983
	72	0	0	3	1	21	...	...
	76	12 244	10 401	6 221	8 682	9 328	8 626	8 556
	77	872	351	59	31	248	36	25
	Area total	1 619 642	1 704 354	1 407 976	1 446 226	1 719 304	1 860 373	2 037 267
81 Pacific, Southwest Pacifique, sud-ouest Pacífico, sudoccidental	22	524	588	400	416	417	350	367
	23	1	1	3	1	1	1	0
	31	4 514	8 005	4 278	3 554	2 954	3 269	3 025
	32	251 354	351 951	435 214	389 941	350 209	339 181	299 470
	33	25 463	26 477	23 120	27 367	26 536	25 896	26 830

C-00 Fish, crustaceans, molluscs, etc — Capture production by fishing areas and species groups
Poissons, crustacés, mollusques, etc — Captures par zones de pêche et groupes d'espèces
Peces, crustáceos, moluscos, etc — Capturas por áreas de pesca y grupos de especies

Fishing area Zone de pêche Area de pesca	Species group Groupe d'espèces Grupo de especies	1996 t	1997 t	1998 t	1999 t	2000 t	2001 t	2002 t
	34	128 844	159 270	157 473	155 554	154 231	144 720	142 461
	35	525	829	956	908	1 368	1 470	1 145
	36	37 555	36 171	42 007	29 628	37 050	37 703	39 970
	37	57 373	63 536	62 766	77 222	55 824	57 466	61 219
	38	17 075	24 720	18 773	22 584	20 172	22 452	24 149
	39	49 338	49 134	28 774	10 319	9 015	39 741	42 723
	42	1 094	1 120	1 138	1 035	732	652	528
	43	3 953	6 418	3 980	3 873	3 976	3 762	3 618
	45	1 824	1 949	1 684	2 410	2 842	2 861	2 206
	47	156	157	92	64	307	293	273
	52	1 364	1 515	1 627	1 493	1 590	1 369	1 371
	53	1 933	2 174	1 000	1 057	766	832	816
	54	454	1	665	2 978	4 468	2 271	1 725
	55	5 209	19 050	4 684	6 281	2 912	7 015	4 526
	56	1 274	1 120	1 984	2 090	2 548	2 574	2 374
	57	73 299	84 833	71 006	44 064	35 801	57 831	80 183
	58	342	173	230	111	54	61	90
	76	282	627	845	645	718	862	778
	77	-	-	-	-	22	29	21
	Area total	663 750	839 819	862 699	783 595	714 513	752 661	739 868
87 Pacific, Southeast Pacifique, sud-est Pacífico, sudoriental	31	761	411	436	411	365	439	318
	32	752 199	391 763	565 958	507 147	339 571	458 922	343 785
	33	52 472	53 423	63 270	65 065	55 474	50 711	53 678
	34	25 436	25 239	30 009	52 479	79 913	45 377	40 773
	35	11 001 839	9 049 576	3 853 560	10 057 766	12 427 660	7 947 490	10 165 635
	36	233 433	284 833	273 808	421 437	356 505	369 326	318 670
	37	4 670 631	4 227 437	2 580 236	2 425 826	1 871 096	3 200 042	2 198 855
	38	12 975	13 230	21 856	15 433	23 948	18 487	25 211
	39	83 088	112 041	404 257	304 202	322 884	197 441	139 805
	42	6 372	4 629	6 427	18 509	9 352	8 958	10 750
	43	142	101	746	569	346	134	80
	44	17 160	22 898	26 295	23 576	24 292	13 397	11 648
	45	28 867	35 223	32 008	21 636	13 322	14 678	13 717
	47	880	585	698	687	627	1 510	1 652
	52	13 589	20 005	11 183	16 332	12 220	11 826	9 807
	53	5	10	6	11	14	207	12
	54	19 468	21 070	27 862	24 241	24 391	29 513	20 945
	55	3 672	6 611	27 208	31 856	12 162	6 685	2 459
	56	51 782	43 393	47 813	42 922	40 793	50 667	29 936
	57	36 021	32 095	11 354	82 778	127 619	174 999	305 566
	58	980	3 447	1 702	3 730	2 513	2 278	8 080
	72	10	11	12	11	11	12	12
	74	4 549	3 174	2 530	2 704	2 290	1 298	1 217
	76	52 025	46 000	44 978	56 981	57 247	49 030	62 532
	Area total	17 068 356	14 397 205	8 034 212	14 176 309	15 804 615	12 653 427	13 765 143
88 Pacific, Antarctic Pacifique, Antarctique Pacífico, Antártico	32	-	-	9	27	77	61	166
	33	-	-	0	0	0	1	0
	34	-	0	42	297	751	662	1 368
	38	-	-	5	19	41	7	25
	39	-	-	0	-	-	-	0
	44	-	-	0	-	-	-	0
	57	-	-	-	-	-	-	0
	76	-	-	-	-	-	2	0
	Area total	-	0	56	343	869	733	1 559
World total Total mondial Total mundial		***93 846 643***	***94 298 586***	***87 672 034***	***93 774 052***	***95 502 004***	***92 862 087***	***93 190 654***

C-01 (a)

Fish, crustaceans, molluscs, etc — **Capture production by species items** — **Africa - Inland waters**
Poissons, crustacés, mollusques, etc — **Captures par catégories d'espèces** — **Afrique - Eaux continentales**
Peces, crustáceos, moluscos, etc — **Capturas por categorías de especies** — **Africa - Aguas continentales**

English name Nom anglais Nombre inglés	Scientific name Nom scientifique Nombre científico	Species group Groupe d'espèces Grupo de especies	1996 t	1997 t	1998 t	1999 t	2000 t	2001 t	2002 t
Common carp	*Cyprinus carpio*	11	444	261	128	141	140	131	137
Crucian carp	*Carassius carassius*	11	61	191	88	101	110	108	103
Rhinofishes nei	*Labeo spp*	11	3 456	2 362	3 375	3 840	3 672	3 609	2 506
Grass carp(=White amur)	*Ctenopharyngodon idellus*	11	15 343	16 553	4 233	707	12 826	17 918	19 528
Silver cyprinid	*Rastrineobola argentea*	11	49 670	40 315	42 336	48 816	49 618	41 384	35 455
Cyprinids nei	*Cyprinidae*	11	12 952	27 022	28 800	42 791	31 596	35 157	34 649
Nile tilapia	*Oreochromis niloticus*	12	177 072	157 714	188 056	175 846	204 350	205 353	207 572
Tilapias nei	*Oreochromis (=Tilapia) spp*	12	157 482	169 253	171 638	196 812	210 928	218 209	213 412
Mouthbrooding cichlids	*Haplochromis spp*	12	10 472	13 053	13 077	10 731	11 699	9 695	8 759
Cichlids nei	*Cichlidae*	12	21 500	21 500	21 500	40 341	40 500	42 033	42 392
African lungfishes	*Protopterus spp*	13	7 653	12 313	11 091	11 428	8 827	9 911	7 337
Dagaas	*Stolothrissa, Limnothrissa spp*	13	70 277	100 636	87 869	78 245	81 992	73 322	80 216
Northern pike	*Esox lucius*	13	-	-	2	...	...	...	...
Bonytongues nei	*Heterotis spp*	13	4 451	5 716	8 385	9 872	9 514	10 127	8 382
Knifefishes	*Notopterus spp*	13	10	1	144	33	39	22	19
Elephantsnout fishes nei	*Mormyridae*	13	...	4 481	18 781	16 030	7 691	24 509	12 567
Aba	*Gymnarchus niloticus*	13	1 108	3 952	7 685	5 887	...	...	9 097
Characins nei	*Characidae*	13	28 805	26 333	39 895	33 410	28 324	25 605	19 089
Bagrid catfish	*Chrysichthys nigrodigitatus*	13	8 505	5 814	15 538	10 679	7 311	12 413	5 678
Black catfishes nei	*Chrysichthys spp*	13	7 596	6 513	7 100	5 592	7 800	7 397	7 397
Naked catfishes	*Bagrus spp*	13	9 865	12 458	13 386	16 239	11 495	11 574	8 001
North African catfish	*Clarias gariepinus*	13	20 601	24 703	21 956	27 004	35 944	39 347	36 695
Mudfish	*Clarias anguillaris*	13	20 370	22 433	21 618	21 511	31 506	39 535	39 208
Torpedo-shaped catfishes nei	*Clarias spp*	13	19 498	23 706	34 057	27 874	29 491	25 890	21 699
Upsidedown catfishes	*Synodontis spp*	13	9 267	11 560	13 384	12 856	12 571	13 737	12 827
Freshwater siluroids nei	*Siluroidei*	13	620	1 368	1 387	1 500	500	500	500
Nile perch	*Lates niloticus*	13	312 017	329 244	339 183	306 282	302 905	282 245	257 272
Freshwater perches nei	*Lates spp*	13	1 205	2 378	3 925	1 523	3 070	1 657	1 660
Freshwater gobies nei	*Gobiidae*	13	900	850	800	800	700	800	800
Snakeheads(=Murrels) nei	*Channa spp*	13	921	86	2 589	2 038	2 990	2 951	3 313
Freshwater fishes nei	*Osteichthyes*	13	853 155	848 696	844 053	873 609	858 402	843 494	941 557
European eel	*Anguilla anguilla*	22	637	985	801	959	2 164	2 129	2 002
Salmonoids nei	*Salmonoidei*	23	11	16	24	29	31	67	112
Shads nei	*Alosa spp*	24	0	0	1	-	-	-	-
Diadromous clupeoids nei	*Clupeoidei*	24	290	253	358	270	284	258	238
Common sole	*Solea solea*	31	...	547	381	763	2 490	2 134	1 858
Mullets nei	*Mugilidae*	33	12 836	16 038	17 356	19 239	18 497	22 738	23 758
Seabasses nei	*Dicentrarchus spp*	33	2 397	2 233	2 738	815	4 075	3 789	3 310
Meagre	*Argyrosomus regius*	33	168	128	106	92	201	51	67
Saddled seabream	*Oblada melanura*	33	...	33	67	47	158	70	77
Gilthead seabream	*Sparus aurata*	33	474	554	672	557	1 727	1 711	1 506
Bonga shad	*Ethmalosa fimbriata*	35	1 900	1 850	1 800	1 798	1 500	1 700	1 700
Silversides(=Sand smelts) nei	*Atherinidae*	37	1 326	1 229	1 150	1 180	3 182	5 154	5 941
River prawns nei	*Macrobrachium spp*	41	150	130	120	120	100	110	110
Freshwater prawns, shrimps nei	*Palaemonidae*	41	2 286	2 387	2 888	1 393	2 414	2 454	2 331
Red swamp crawfish	*Procambarus clarkii*	41	13	24	19	21	22	3	14
Freshwater crustaceans nei	*Crustacea*	41	5 593	5 882	5 968	6 267	6 604	6 931	8 476
Penaeus shrimps nei	*Penaeus spp*	45	3 000	2 850	2 700	2 574	2 100	2 400	2 400
Freshwater molluscs nei	*Mollusca*	51	530	610	625	606	921	851	1 197
Total			***1 856 887***	***1 927 214***	***2 003 833***	***2 019 268***	***2 052 981***	***2 051 183***	***2 092 924***

C-01 (b)

Fish, crustaceans, molluscs, etc — **Capture production by countries or areas** — **Africa - Inland waters**
Poissons, crustacés, mollusques, etc — **Captures par pays ou zones** — **Afrique - Eaux continentales**
Peces, crustáceos, moluscos, etc — **Capturas por países o áreas** — **Africa - Aguas continentales**

Country or area Pays ou zone País o área	1993 t	1994 t	1995 t	1996 t	1997 t	1998 t	1999 t	2000 t	2001 t	2002 t
Algeria	0	0	0	0	0	2	0	0	0	0
Angola	7 000	7 000	6 000	6 000	6 000	6 000	6 000 F	6 000 F	6 000 F	6 000 F
Benin	32 805	32 707	37 449	34 193	32 871	31 778	31 894	26 400	30 000	29 993
Botswana	600 F	400 F	200 F	81	160	191	157	166	118	139
Br Ind Oc Tr	0	0	0	0	0	0	0	0	0	0
Burkina Faso	7 000	8 000	8 000	8 000	8 000	8 335	7 600	8 500	8 500 F	8 500 F
Burundi	17 000 F	22 000 F	21 101	3 041	20 296	13 426	9 199	17 315	8 964	9 000 F
Cameroon	23 000	27 000 F	30 000 F	35 000 F	40 000 F	45 000 F	50 000 F	55 000	52 500 F	65 000
Cape Verde	0	0	0	0	0	0	0	0	0	0
Cent Afr Rep	13 250 F	13 500 F	13 750 F	14 000 F	14 250 F	14 500 F	15 000	15 000 F	15 000 F	15 000 F
Chad	87 300	80 000	90 000	100 000	85 000	84 000	84 000 F	84 000 F	84 000 F	84 000 F
Comoros	0	0	0	0	0	0	0	0	0	0
Congo Dem R	192 589	152 117	154 751	159 037	158 367	174 087	204 503	205 000 F	210 000 F	215 000 F
Congo Rep	27 850	24 752	26 811	25 873	18 987	25 455	25 268	20 204 F	20 200 F	20 500 F
Côte dIvoire	13 477	15 604	11 335	11 562	12 032	12 501	10 656	10 502	10 630	23 843
Djibouti	0	0	0	0	0	0	0	0	0	0
Egypt	207 473	217 700	244 300	240 900	261 167	281 141	250 318	253 470	295 422	292 645
Eq Guinea	600	700	450	900	850	970	1 101 F	1 076	1 000 F	1 000 F
Eritrea	0	0	0	0	0	0	0	0	0	-
Ethiopia	4 175	5 285	6 325	8 770	10 370	14 000	15 858	15 681	15 390	12 300
Fr South Tr	0	0	0	0	0	0	0	0	0	0
Gabon	3 500 F	4 500 F	7 648 F	9 433	9 441	9 000	10 000	10 417	9 850	9 400
Gambia	2 400	2 400	2 500	2 500	2 500	2 500	2 500 F	2 500 F	2 500 F	2 500 F
Ghana	52 000	54 700	60 000	73 580	70 000	74 500	74 500	74 500	74 500	74 500
Guinea	4 600	3 800	3 100	2 780	3 600	4 000	4 000	4 000	4 000	4 000 F
GuineaBissau	250 F	250 F	250 F	250 F	250 F	200 F	200 F	200 F	200 F	200 F
Kenya	176 435	198 805	187 241	174 692	154 955	165 992	198 653	210 343	156 763	137 792
Lesotho	22 F	22 F	26 F	28 F	30 F	30 F	30	32	24	24 F
Liberia	4 000	4 000	4 000	4 000	4 000	4 000	4 000	4 000	4 000	4 000 F
Libya	0	0	0	0	0	0	0	0	0	0
Madagascar	30 000	30 000	30 000	30 000	30 000	30 000	30 000	30 000	30 000	30 000
Malawi	67 951	58 579	53 664	63 569	56 340	41 111	45 392	43 000 F	40 619	41 329
Mali	64 300	62 850	132 900	111 910	99 550	98 000	98 536	109 870	100 000 F	100 000 F
Mauritania	5 000 F	5 000 F	5 000 F	5 000 F	5 000 F	5 000 F	5 000 F	5 000 F	5 000 F	5 000 F
Mauritius	3	0	0	0	0	0	0	0	0	0
Morocco	1 617	1 750	1 500	1 500	2 100	1 703	2 163	1 608	1 660	2 112
Mozambique	4 689	4 925	5 173	7 510	11 668	8 994	9 052	11 813	5 284	12 156
Namibia	1 200	1 200	1 200	1 200	1 500	1 500	1 500	1 500	1 500	1 500
Niger	2 162	2 516	3 616	4 156	6 328	7 013	11 000	16 250	20 800	23 520
Nigeria	95 627	103 800	117 903	89 521	93 644	139 020	139 393	132 315	154 175	187 233
Réunion	0	0	0	0	0	0	0	0	0	0
Rwanda	3 500 F	3 400 F	3 300 F	2 952	4 428	6 641	6 433	6 726	6 828	7 000 F
St Helena	0	0	0	0	-	-	-	-	-	
Sao Tome Prn	0	0	0	0	0	0	0	0	0	0
Senegal	27 650	30 000 F	31 000 F	23 000 F	31 000 F	21 000 F	34 000 F	22 450	20 000 F	20 000 F
Seychelles	0	0	0	0	0	0	0	0	0	0
Sierra Leone	14 000	15 000	15 000	14 500 F	14 500	14 190	14 480	14 000	14 000	14 000
Somalia	250 F	250 F	250 F	250 F	250 F	250 F	250 F	200 F	200 F	150 F
South Africa	832	800	800	850	850	900	900 F	900 F	900 F	900 F
Sudan	37 500	40 000	40 000	40 500	42 000	44 000	44 000	48 000 F	53 000	53 000
Swaziland	68 F	65 F	60 F	60 F	65 F	70 F	70 F	70 F	70 F	70 F
Tanzania	294 582	247 614	317 029	262 276	306 750	300 000	260 020	280 000	283 000	273 850
Togo	6 000	5 000	4 998	5 000	5 000	5 000	5 000	5 000	5 000	5 000
Tunisia	400	243	440	706	1 010	896	808	832	860	870
Uganda	219 814	213 129	208 789	195 088	218 026	220 628	226 097	219 356	220 726	221 898
Zambia	65 768	70 057	70 546	66 332	65 923	69 938	67 327	66 671	65 000 F	65 000 F
Zimbabwe	21 230	20 219	16 463	16 387	18 156	16 371	12 410	13 114	13 000 F	13 000 F
Total	***1 841 469***	***1 791 639***	***1 974 868***	***1 856 887***	***1 927 214***	***2 003 833***	***2 019 268***	***2 052 981***	***2 051 183***	***2 092 924***

C-02 (a)

Fish, crustaceans, molluscs, etc — **Capture production by species items** — **America, North - Inland waters**

Poissons, crustacés, mollusques, etc — **Captures par catégories d'espèces** — **Amérique du Nord - Eaux continentales**

Peces, crustáceos, moluscos, etc — **Capturas por categorías de especies** — **América del Norte - Aguas continentales**

English name Nom anglais Nombre inglés	Scientific name Nom scientifique Nombre científico	Species group Groupe d'espèces Grupo de especies	1996 t	1997 t	1998 t	1999 t	2000 t	2001 t	2002 t
Buffalofishes nei	*Ictiobus spp*	11	793	991	959	697	1 280	1 569	1 558
Suckers nei	*Catostomidae*	11	542	550	464	465	231	187	173
Common carp	*Cyprinus carpio*	11	23 020	18 515	15 798	12 969	16 540	16 307	15 945
Goldfish	*Carassius auratus*	11	5	10	10	7	9	10	12
Grass carp(=White amur)	*Ctenopharyngodon idellus*	11	10	15	13	11	17	31	31
Cyprinids nei	*Cyprinidae*	11	3 735	4 843	3 997	6 822	4 017	3 839	1 738
Nile tilapia	*Oreochromis niloticus*	12	1 415	1 447	1 167	1 366	1 321	710	1 319
Blue tilapia	*Oreochromis aureus*	12	9 672	7 553	5 122	4 564	2 940	1 948	2 002
Tilapias nei	*Oreochromis (=Tilapia) spp*	12	74 473	75 077	65 266	62 446	69 828	62 158	57 844
Jaguar guapote	*Cichlasoma managuense*	12	566	367	344	351	324	410	318
Peacock cichlid	*Cichla ocellaris*	12	65	35	12	7	4	4	4
Cichlids nei	*Cichlidae*	12	0	21	157	923	872	860	503
Northern pike	*Esox lucius*	13	2 406	2 334	2 568	2 615	2 808	2 464	2 684
Catfishes nei	*Ictalurus spp*	13	9 745	11 822	9 524	13 710	11 567	11 126	9 027
Burbot	*Lota lota*	13	14	9	28	13	20	9	10
Largemouth black bass	*Micropterus salmoides*	13	1 014	1 058	684	784	903	693	1 191
American yellow perch	*Perca flavescens*	13	2 151	2 695	2 639	2 421	2 414	3 164	4 307
Walleye	*Stizostedion vitreum*	13	7 736	7 755	8 212	8 784	9 164	6 959	7 021
Sauger	*Stizostedion canadense*	13	1	...	0	...	...	...	...
Freshwater drum	*Aplodinotus grunniens*	13	368	549	573	470	577	506	502
Freshwater fishes nei	*Osteichthyes*	13	27 826	23 737	27 091	22 970	30 844	22 692	26 553
Sturgeons nei	*Acipenseridae*	21	526	615	525	420	281	173	172
American eel	*Anguilla rostrata*	22	730	796	739	770	942	648	269
Pink(=Humpback)salmon	*Oncorhynchus gorbuscha*	23	3	31	0	1	-	-	-
Chum(=Keta=Dog)salmon	*Oncorhynchus keta*	23	1 567	738	981	325	59	4	48
Sockeye(=Red)salmon	*Oncorhynchus nerka*	23	115	78	190	56	13	0	-
Chinook(=Spring=King)salmon	*Oncorhynchus tshawytscha*	23	2 111	1 621	1 367	1 344	390	236	214
Coho(=Silver)salmon	*Oncorhynchus kisutch*	23	4 021	696	855	282	802	659	161
Rainbow trout	*Oncorhynchus mykiss*	23	1 821	239	453	173	377	444	320
Lake trout(=Char)	*Salvelinus namaycush*	23	975	1 109	1 054	985	1 129	1 130	878
Rainbow smelt	*Osmerus mordax*	23	711	522	321	328	398	209	188
Pond smelt	*Hypomesus olidus*	23	3 981	5 964	6 471	5 691	3 254	4 241	3 263
Eulachon	*Thaleichthys pacificus*	23	4	27	3	8	-	-	-
Lake(=Common)whitefish	*Coregonus clupeaformis*	23	14 393	14 218	14 277	13 683	14 227	14 108	12 862
Lake cisco	*Coregonus artedi*	23	1 145	1 223	1 048	804	836	870	788
Whitefishes nei	*Coregonus spp*	23	1 415	1 506	2 178	1 586	1 030	768	841
Salmonoids nei	*Salmonoidei*	23	112	-	2	1	-	-	-
American shad	*Alosa sapidissima*	24	837	545	958	580	508	777	33
Alewife	*Alosa pseudoharengus*	24	990	1 332	1 529	1 550	1 534	1 580	1 607
Hickory shad	*Alosa mediocris*	24	88	75	47	62	51	90	42
American gizzard shad	*Dorosoma cepedianum*	24	1 272	1 989	1 291	3 007	977	1 781	2 780
Lampreys nei	*Petromyzontidae*	25	-	-	0	4	0	-	-
Striped bass	*Morone saxatilis*	25	-	5	-	-	-	-	-
White perch	*Morone americana*	25	995	1 164	966	1 042	1 221	1 226	1 022
White bass	*Morone chrysops*	25	45	50	166	109	106	145	104
Atlantic tomcod	*Microgadus tomcod*	32	23	24	25	25	25	-	-
Mountain mullet	*Agonostomus monticola*	33	...	...	...	...	...	...	28
Snooks(=Robalos) nei	*Centropomus spp*	33	...	...	...	...	...	220	185
Silversides(=Sand smelts) nei	*Atherinidae*	37	...	...	372	487	392	558	379
River prawns nei	*Macrobrachium spp*	41	4 261	3 580	3 244	4 168	3 478	3 115	3 054
Euro-American crayfishes nei	*Astacidae, Cambaridae*	41	5 887	10 597	10 181	5 485	325	4 693	7 219
Freshwater crustaceans nei	*Crustacea*	41	26	0	5	28	32	77	37
Freshwater molluscs nei	*Mollusca*	51	240	122	185	240	565	1 029	1 052
Frogs	*Rana spp*	71	422	2 034	1 201	337	338	361	146
Diamond back terrapins	*Malaclemys spp*	72	-	-	0	-	0	-	-
River and lake turtles nei	*Testudinata*	72	18	17	26	15	53	52	27
Marine worms	*Polychaeta*	77	58	57	58	48	27	26	27
Aquatic invertebrates nei	*Invertebrata*	77	279	240	425	569	204	93	126
Total			***214 628***	***210 597***	***195 771***	***186 608***	***189 254***	***174 959***	***170 614***

C-02 (b)

Fish, crustaceans, molluscs, etc	**Capture production by countries or areas**	**America, North - Inland waters**
Poissons, crustacés, mollusques, etc	**Captures par pays ou zones**	**Amérique du Nord - Eaux continentales**
Peces, crustáceos, moluscos, etc	**Capturas por países o áreas**	**América del Norte - Aguas continentales**

Country or area Pays ou zone País o área	1993 t	1994 t	1995 t	1996 t	1997 t	1998 t	1999 t	2000 t	2001 t	2002 t
Anguilla	0	0	0	0	0	0	0	0	0	0
Antigua Barb	0	0	0	0	0	0	0	0	0	0
Aruba	0	0	0	0	0	0	-	-	-	-
Bahamas	0	0	0	0	0	0	0	0	0	0
Barbados	0	0	0	0	0	0	0	0	-	-
Belize	1	1	0	0	0	0	0	0	0	-
Bermuda	0	0	0	0	0	0	0	0	0	-
Br Virgin Is	0	0	0	0	0	0	0	0	0	0
Canada	36 327	36 333	38 756	38 295	38 798	40 744	40 587	40 667	38 140	39 999
Cayman Is	0	0	0	0	0	0	0	0	0	0
Costa Rica	710	840	900	1 090	840	1 000 F	1 000 F	1 000 F	1 000 F	1 000 F
Cuba	9 886	9 537	8 270	9 744	7 603	5 154	4 591	3 005	2 007	2 034
Dominica	0	0	0	0	0	0	0	0	0	0
Dominican Rp	2 037	3 774	2 106	288	1 067	1 095	733	187	1 158	2 363
El Salvador	4 461	3 819	4 324	2 967	2 808	2 443	2 654	2 831	1 692	2 663
Greenland	0	0	0	0	0	0	0	0	0	0
Grenada	0	0	0	0	0	0	0	0	0	0
Guadeloupe	0	0	0	0	0	0	0	0	0	0
Guatemala	4 228	3 776	4 025	4 000	5 121	6 523	6 976	7 301	7 300 F	7 300 F
Haiti	600 F	500 F	500 F	500 F	500 F	500 F	500 F	500 F	500 F	500 F
Honduras	86	92	127	98	126	119	102	61	111	102 F
Jamaica	450 F	450 F	450 F	450 F	450 F	450 F	450 F	450 F	450 F	450 F
Martinique	3	0	0	0	0	0	0	0	0	0
Mexico	111 178	111 125	122 020	122 501	113 552	100 335	91 462	106 817	91 952	82 648
Montserrat	0	0	0	0	0	0	0	0	0	0
NethAntilles	0	0	0	0	0	0	0	0	0	0
Nicaragua	547	824	538	1 142	1 293	1 256	1 120	1 076	1 051	1 592
Panama	28	285	130	80	91	23	20	20	20 F	20 F
Puerto Rico	0	0	0	0	0	0	0	0	0	0
St Kitts Nev	0	0	0	0	0	0	0	0	0	-
St Lucia	0	0	0	0	0	0	0	0	0	-
St Pier Mq	0	0	0	0	0	0	0	0	0	0
St Vincent	0	0	0	2	1	0	0	0	0	0
Trinidad Tob	0	0	0	0	0	0	0	0	0	0
Turks Caicos	0	0	0	0	0	0	0	0	0	0
USA	54 377	42 648	36 688	33 471	38 347	36 129	36 413	25 339	29 578	29 943
US Virgin Is	0	0	0	0	0	0	0	0	0	0
Total	***224 919***	***214 004***	***218 834***	***214 628***	***210 597***	***195 771***	***186 608***	***189 254***	***174 959***	***170 614***

C-03 (a)

Fish, crustaceans, molluscs, etc
Poissons, crustacés, mollusques, etc
Peces, crustáceos, moluscos, etc

Capture production by species items
Captures par catégories d'espèces
Capturas por categorías de especies

America, South - Inland waters
Amérique du Sud - Eaux continentales
América del Sur - Aguas continentales

English name Nom anglais Nombre inglés	Scientific name Nom scientifique Nombre científico	Species group Groupe d'espèces Grupo de especies	1996 t	1997 t	1998 t	1999 t	2000 t	2001 t	2002 t
Common carp	*Cyprinus carpio*	11	218	1	7	5	1 390	0	0
Cyprinids nei	*Cyprinidae*	11	284	212	258	302	355	390	392
Tilapias nei	*Oreochromis (=Tilapia) spp*	12	7 300	6 927	7 444	6 616	7 893	8 329	8 248
Velvety cichlids	*Astronotus spp*	12	308	177	141	251	188	183	...
Cichlids nei	*Cichlidae*	12	8 565	8 356	7 785	7 375	8 145	13 611	13 946
Arapaima	*Arapaima gigas*	13	457	465	210	338	273	204	...
Cachama	*Colossoma macropomum*	13	4 991	3 600	3 014	3 921	5 928	4 693	4 068
Pirapatinga	*Piaractus brachypomus*	13	397	...	166	648	324	487	...
Characins nei	*Characidae*	13	115 490	102 456	101 018	92 195	93 360	44 935	44 974
Netted prochilod	*Prochilodus reticulatus*	13	7 448	2 007	6 117	9 768	10 942	11 220	10 133
Prochilods nei	*Prochilodus spp*	13	18 740	10 927	15 670	25 320	27 313	26 343	32 261
Freshwater siluroids nei	*Siluroidei*	13	79 709	86 749	89 117	90 847	90 947	94 496	94 799
Freshwater fishes nei	*Osteichthyes*	13	95 658	106 251	100 035	105 217	93 951	162 145	166 446
Sea trout	*Salmo trutta*	23	1	1	1	1	1	1	1
Rainbow trout	*Oncorhynchus mykiss*	23	462	1 207	732	526	565	657	977
Atlantic sabretooth anchovy	*Lycengraulis grossidens*	24	90	100	100	120	120	140	130
Diadromous clupeoids nei	*Clupeoidei*	24	552	355	467	403	287	200	183
Silversides(=Sand smelts) nei	*Atherinidae*	37	850	850	850	850	850	760	750
River prawns nei	*Macrobrachium spp*	41	2 677	2 157	1 506	3 235	2 437	...	...
Freshwater crustaceans nei	*Crustacea*	41	0	0	0	0	0	-	-
Frogs	*Rana spp*	71	0	0	0	0	7	9	5
River and lake turtles nei	*Testudinata*	72	0	0	0	0	0	-	-
Total			***344 197***	***332 798***	***334 638***	***347 938***	***345 276***	***368 803***	***377 313***

C-03 (b)

Fish, crustaceans, molluscs, etc
Poissons, crustacés, mollusques, etc
Peces, crustáceos, moluscos, etc

Capture production by countries or areas
Captures par pays ou zones
Capturas por países o áreas

America, South - Inland waters
Amérique du Sud - Eaux continentales
América del Sur - Aguas continentales

Country or area Pays ou zone País o área	1993 t	1994 t	1995 t	1996 t	1997 t	1998 t	1999 t	2000 t	2001 t	2002 t
Argentina	11 800	12 785	17 191	19 189	22 735	23 197	27 558	30 418	33 757	31 233
Bolivia	5 518	5 353	5 692	5 988	6 038	6 055	6 052	6 106	5 940	5 800 F
Brazil	186 990 F	191 485 F	193 042	193 309	178 871	174 190	185 471	199 159	217 275	217 750
Chile	7	5	...	...	...	4	...	...	...	...
Colombia	30 538	34 983	23 524	23 061	20 610	21 673	28 788	24 854	25 000 F	25 000 F
Ecuador	372	300	300	300	400	400	400	400	400	400 F
Falkland Is	-	-	-	1	1	1	1	1	1	1
Fr Guiana	0	0	0	0	0	0	0	0	0	0
Guyana	800	800	700 F	800	625	625	603	800	800	800
Paraguay	19 000 F	20 000 F	21 000 F	22 000	28 000	25 000 F	25 000 F	25 000 F	25 000 F	25 000 F
Peru	38 290	48 837	50 789	28 890	32 221	35 327	36 223	32 297	35 653	29 966
Suriname	187	138	140 F	150 F	200 F	200 F	200 F	200 F	200	200 F
Uruguay	621	966	849	598	2 216	1 931	2 423	2 302	451	387
Venezuela	28 251	35 412	54 175	49 911	40 881	46 035	35 219	23 739	24 326	40 776
Total	***322 374***	***351 064***	***367 402***	***344 197***	***332 798***	***334 638***	***347 938***	***345 276***	***368 803***	***377 313***

C-04 (a)

Fish, crustaceans, molluscs, etc — Capture production by species items — Asia - Inland waters
Poissons, crustacés, mollusques, etc — Captures par catégories d'espèces — Asie - Eaux continentales
Peces, crustáceos, moluscos, etc — Capturas por categorías de especies — Asia - Aguas continentales

English name Nom anglais Nombre inglés	Scientific name Nom scientifique Nombre científico	Species group Groupe d'espèces Grupo de especies	1996 t	1997 t	1998 t	1999 t	2000 t	2001 t	2002 t
Freshwater bream	*Abramis brama*	11	19 479	11 743	10 650	11 815	12 749	13 616	14 629
Freshwater breams nei	*Abramis spp*	11	3	7	20	9	20	10	38
Common carp	*Cyprinus carpio*	11	43 435	41 152	47 741	45 168	34 044	32 805	34 014
Tench	*Tinca tinca*	11	2	0	0	0	690	778	983
Crucian carp	*Carassius carassius*	11	8 273	6 877	6 213	5 686	5 528	5 226	4 602
Goldfish	*Carassius auratus*	11	172	1	1	1	1	2	1
Roach	*Rutilus rutilus*	11	235	537	502	733	1 084	1 365	1 498
Kutum	*Rutilus frisii*	11	9 210	2 320	6 624	6 905	10 120	7 199	6 433
Roaches nei	*Rutilus spp*	11	3 298	1 943	1 867	1 626	2 365	1 895	2 469
Rudd	*Scardinius erythrophthalmus*	11	...	...	...	...	...	...	485
Common dace	*Leuciscus leuciscus*	11	215	250	300	176	104	91	73
Mud carp	*Cirrhinus molitorella*	11	3	4	7	7	6	6	9
Grass carp(=White amur)	*Ctenopharyngodon idellus*	11	4 296	4 631	5 213	4 703	2 621	725	1 462
Silver carp	*Hypophthalmichthys molitrix*	11	18 324	16 958	12 055	14 819	9 415	4 270	4 451
Bighead carp	*Hypophthalmichthys nobilis*	11	1 818	1 637	2 621	3 841	4 454	447	703
Kinneret bleak	*Acanthobrama terraesanctae*	11	1 155	626	1 171	1 048	1 052	811	1 170
Sichel	*Pelecus cultratus*	11	38	21	15	10	10	16	4
Asp	*Aspius aspius*	11	419	282	211	260	362	503	1 056
Hoven's carp	*Leptobarbus hoeveni*	11	6 892	5 836	3 241	4 608	3 382	2 362	2 270
Black carp	*Mylopharyngodon piceus*	11	15	16	22	27	34	36	25
Java barb	*Puntius javanicus*	11	19 622	19 469	20 189	17 939	17 791	17 425	17 840
Asian barbs nei	*Puntius spp*	11	39 003	37 272	56 480	56 774	53 430	55 049	55 940
Cyprinids nei	*Cyprinidae*	11	253 529	225 334	202 153	206 253	177 036	166 206	205 754
Mozambique tilapia	*Oreochromis mossambicus*	12	16 943	17 715	17 865	21 172	19 831	20 210	20 980
Nile tilapia	*Oreochromis niloticus*	12	29 253	28 727	40 173	49 840	40 037	43 100	44 700
Blue tilapia	*Oreochromis aureus*	12	32	101	88	66	67	33	18
Tilapias nei	*Oreochromis (=Tilapia) spp*	12	40 058	48 331	53 529	53 884	62 173	56 209	55 583
Mango tilapia	*Sarotherodon galilaeus*	12	350	462	391	405	262	110	93
Cichlids nei	*Cichlidae*	12	40	48	58	57	52	24	23
Northern pike	*Esox lucius*	13	832	800	528	907	975	1 060	1 149
Knifefishes	*Notopterus spp*	13	3 880	3 455	3 599	3 845	4 255	3 820	3 800
Wels(=Som)catfish	*Silurus glanis*	13	2 148	2 015	1 730	1 741	1 793	1 630	1 613
Glass catfishes	*Kryptopterus spp*	13	17 560	15 938	13 943	15 927	13 110	13 522	14 980
North African catfish	*Clarias gariepinus*	13	...	...	...	216	576	520	495
Torpedo-shaped catfishes nei	*Clarias spp*	13	16 881	13 063	20 103	20 582	27 296	24 075	24 011
Pangas catfishes nei	*Pangasius spp*	13	541	522	917	1 061	1 274	1 100	1 200
Freshwater siluroids nei	*Siluroidei*	13	15 245	12 112	76 933	86 568	78 787	72 103	82 032
Lai	*Monopterus albus*	13	-	-	-	-	162	100	100
European perch	*Perca fluviatilis*	13	670	480	350	450	500	547	422
Pike-perch	*Stizostedion lucioperca*	13	13 806	5 665	6 299	4 637	3 883	3 517	3 725
Gudgeons, sleepers nei	*Eleotridae*	13	1 109	1 290	1 078	1 243	1 125	2 604	3 080
Freshwater gobies nei	*Gobiidae*	13	3 770	4 500	4 003	4 145	4 670	4 396	6 005
Climbing perch	*Anabas testudineus*	13	3 905	3 754	4 637	6 340	6 730	5 800	6 300
Snakeskin gourami	*Trichogaster pectoralis*	13	30 792	21 728	22 422	23 776	22 456	21 357	20 940
Kissing gourami	*Helostoma temminckii*	13	12 611	18 376	16 598	23 320	18 771	19 429	20 140
Snakehead	*Channa argus*	13	169	120	...	28	...	...	...
Striped snakehead	*Channa striata*	13	30 966	28 646	21 520	23 784	26 839	24 998	25 988
Indonesian snakehead	*Channa micropeltes*	13	11 614	10 117	8 253	8 787	8 149	6 357	6 440
Snakeheads(=Murrels) nei	*Channa spp*	13	127 429	136 830	185 453	86 877	92 245	90 270	77 200
Freshwater fishes nei	*Osteichthyes*	13	2 497 271	2 604 488	2 679 590	3 311 358	3 297 786	3 454 678	3 194 886
Sturgeons nei	*Acipenseridae*	21	2 179	1 733	1 531	1 320	1 288	1 231	915
European eel	*Anguilla anguilla*	22	342	400	300	99	176	122	147
Japanese eel	*Anguilla japonica*	22	1 014	916	904	830	765	677	610
River eels nei	*Anguilla spp*	22	4 439	773	860	1 326	4 746	2 108	3 971
Sea trout	*Salmo trutta*	23	395	200	200	263	277	364	352
Trouts nei	*Salmo spp*	23	1 222	1 213	1 187	1 285	1 322	1 238	1 219
Pink(=Humpback)salmon	*Oncorhynchus gorbuscha*	23	3 134	947	2 091	927	1 947	376	926
Chum(=Keta=Dog)salmon	*Oncorhynchus keta*	23	19 026	18 346	16 069	11 684	12 326	9 599	10 000
Masu(=Cherry) salmon	*Oncorhynchus masou*	23	830	810	836	849	856	826	765
Sockeye(=Red)salmon	*Oncorhynchus nerka*	23	72	88	118	90	52	22	30
Rainbow trout	*Oncorhynchus mykiss*	23	731	672	618	562	536	484	447
Ayu sweetfish	*Plecoglossus altivelis*	23	12 732	12 624	11 386	11 380	11 172	11 148	10 663
Whitefishes nei	*Coregonus spp*	23	508	572	646	957	927	679	248
Salmonoids nei	*Salmonoidei*	23	26	26	110	222	5	2	9
Caspian shads	*Caspialosa spp*	24	57 075	60 442	85 082	95 060	78 001	45 232	26 014
Kelee shad	*Hilsa kelee*	24	48 302	44 519	53 729	44 810	174 399	64 599	38 984
Hilsa shad	*Tenualosa ilisha*	24	90 240	83 230	81 634	73 809	79 165	75 060	68 250
Azov sea sprat	*Clupeonella cultriventris*	24	23 374	21 020	16 767	34 930	33 060	22 689	23 368
Diadromous clupeoids nei	*Clupeoidei*	24	1 811	3 063	2 334	1 840	1 512	2 713	3 022
Milkfish	*Chanos chanos*	25	369	34	3 000	83	163	201	329
Bombay-duck	*Harpadon nehereus*	33	-	-	-	-	-	473	230
Sea catfishes nei	*Ariidae*	33	2 964	2 701	3 207	3 785	4 046	3 868	4 394
Flathead grey mullet	*Mugil cephalus*	33	181	169	231	315	138	125	136
Mullets nei	*Mugilidae*	33	30 036	37 338	31 543	30 666	34 596	36 248	25 977
Croakers, drums nei	*Sciaenidae*	33	3 271	748	667	686	213	864	864
Threadfins, tasselfishes nei	*Polynemidae*	33	467	416	243	229	66	345	345
Scats	*Scatophagus spp*	33	169	95	20	40	61	55	30
Pike-congers nei	*Muraenesox spp*	34	490	447	270	494	402	620	337
Silversides(=Sand smelts) nei	*Atherinidae*	37	562	1 600	1 500	1 455	1 583	1 685	1 733
Giant river prawn	*Macrobrachium rosenbergii*	41	6 140	5 225	5 397	5 963	5 472	5 201	5 253
Freshwater prawns, shrimps nei	*Palaemonidae*	41	12 406	11 478	7 793	10 163	10 106	9 863	10 682
Euro-American crayfishes nei	*Astacidae, Cambaridae*	41	850	1 100	1 500	1 372	1 681	1 634	1 894

C-04 (a)

Fish, crustaceans, molluscs, etc — Capture production by species items — Asia - Inland waters
Poissons, crustacés, mollusques, etc — Captures par catégories d'espèces — Asie - Eaux continentales
Peces, crustáceos, moluscos, etc — Capturas por categorías de especies — Asia - Aguas continentales

English name Nom anglais Nombre inglés	Scientific name Nom scientifique Nombre científico	Species group Groupe d'espèces Grupo de especies	1996 t	1997 t	1998 t	1999 t	2000 t	2001 t	2002 t
Freshwater crustaceans nei	*Crustacea*	41	366 840	480 899	603 307	457 390	531 668	589 120	775 024
Indo-Pacific swamp crab	*Scylla serrata*	42	363	267	44	47	77	150	219
Penaeus shrimps nei	*Penaeus spp*	45	14 645	20 558	8 783	7 830	10 873	14 772	14 160
Natantian decapods nei	*Natantia*	45	19 773	23 738	15 117	15 535	23 313	22 290	88 497
Japanese corbicula	*Corbicula japonica*	51	27 314	22 209	20 609	21 015	19 295	17 295	17 779
Freshwater molluscs nei	*Mollusca*	51	530 431	480 862	554 128	525 591	569 425	603 656	607 683
Green mussel	*Perna viridis*	54	4	-	-	-	-	-	-
Clams, etc. nei	*Bivalvia*	56	844	956	914	304	303	792	850
Marine molluscs nei	*Mollusca*	58	481	52	165	73	166	112	120
Frogs	*Rana spp*	71	2 533	1 550	1 767	1 435	1 957	2 116	2 078
River and lake turtles nei	*Testudinata*	72	36	48	27	1	3	42	40
Aquatic invertebrates nei	*Invertebrata*	77	1 494	2 460	1 480	2 048	2 031	1 578	1 735
Total			***4 597 426***	***4 702 743***	***5 095 470***	***5 502 187***	***5 690 272***	***5 734 686***	***5 722 141***

C-04 (b)

Fish, crustaceans, molluscs, etc — Capture production by countries or areas — Asia - Inland waters
Poissons, crustacés, mollusques, etc — Captures par pays ou zones — Asie - Eaux continentales
Peces, crustáceos, moluscos, etc — Capturas por países o áreas — Asia - Aguas continentales

Country or area Pays ou zone País o área	1993 t	1994 t	1995 t	1996 t	1997 t	1998 t	1999 t	2000 t	2001 t	2002 t
Afghanistan	1 200 F	1 300 F	1 300 F	1 300 F	1 250 F	1 200 F	1 200 F	1 000 F	800 F	900 F
Armenia	1 850	1 033	821	580	580	698	1 144	1 133	866	465
Azerbaijan	21 733	18 901	10 545	6 702	5 161	4 760	20 866	18 797	10 893	11 334
Bahrain	0	0	0	0	0	0	0	-	-	-
Bangladesh	452 109	517 746	527 739	535 617	534 285	538 689	649 418	670 465	688 920	688 435
Bhutan	320 F	310 F	310 F	300 F	300 F	300 F	300 F	300 F	300 F	300 F
Brunei Darsm	25	4	7	15	17	35	26	23	16	14
Cambodia	67 900	65 000	72 500	63 510	73 000	75 700	231 000	245 600	385 000	360 300
China	1 182 390	1 327 785	1 607 385	1 762 860	1 886 967	2 280 244	2 285 364	2 233 230	2 149 932	2 247 926
China,H.Kong	0	0	0	0	0	0	0	0	0	0
China, Macao	0	0	0	0	0	0	0	0	0	0
China,Taiwan	1 216	1 059	874	407	403	449	561	549	591	599
Cyprus	...	5	65	64	70	70	70	78	70 F	60 F
Georgia	549	16	90	6	1	4	17	22	8	10
India	575 905	552 874	608 378	633 425	641 775	692 439	696 083	905 700	975 403	813 755
Indonesia	308 649	336 141	329 710	335 707	304 258	288 666	327 627	318 334	310 240	316 030
Iran	75 021	89 157	88 800	109 286	103 795	140 263	143 400	123 500	73 645	55 853
Iraq	17 808	20 926	22 955	19 049	20 519	9 111	9 330	8 378	8 000 F	8 000 F
Israel	1 813	1 478	1 214	1 845	1 476	2 164	2 145	1 852	1 286	1 568
Japan	91 032	92 219	91 455	93 501	85 612	78 822	71 270	70 612	61 354	60 843 F
Jordan	350	350	350	350	350	350	350	400	350	350
Kazakhstan	56 700	46 433	48 402	44 273	31 826	25 000 F	36 170	36 620	21 654	24 910
Korea D P Rp	23 000 F	20 000	20 000	20 000	20 000	16 000 F	12 000 F	8 000 F	4 928	5 000 F
Korea Rep	12 263	10 492	9 646	8 034	6 934	6 845	6 316	7 141	5 971	5 690
Kuwait	0	0	0	0	0	0	0	0	0	0
Kyrgyzstan	127	131	185	160 F	120 F	80 F	48	52	57	48
Laos	19 500 F	23 800	27 370	23 000 F	18 857	19 642	30 041	29 250	31 000 F	33 440
Lebanon	20	20	20	20	20	20	20	20	20	297
Malaysia	1 971	2 064	3 939	3 683	3 949	4 626	3 366	3 549	3 446	3 450
Maldives	0	0	0	0	0	0	0	0	0	0
Mongolia	165	184	158	221	180	311	524	425	117	129
Myanmar	142 065	146 365	148 347	146 494	149 069	149 279	159 746	189 708	235 376	304 529
Nepal	7 418	7 340	11 230	11 230	11 230	12 000	12 752	16 700	16 700	17 900
Oman	0	0	0	0	0	0	0	0	0	0
Pakistan	109 185	118 703	121 405	142 092	167 530	163 524	179 865	176 468	180 100	181 000
Palest, O.T.	...	...	0	0	0	0	0	0	0	0
Philippines	210 775	204 325	186 006	177 355	159 353	146 004	146 234	151 753	135 845	130 881
Qatar	0	0	0	0	0	0	0	0	0	0
Saudi Arabia	0	0	0	0	0	0	0	0	0	0
Singapore	25	23	0	0	0	0	0	0	0	0
Sri Lanka	15 000	9 500	15 000	22 250	27 250	29 900	31 450	36 700	29 870	28 130
Syria	2 535	3 570	3 832	3 103	3 557	4 347	5 338	3 991	5 969	6 355
Tajikistan	253	127	100	40 F	75 F	100 F	80 F	78 F	137	181
Thailand	175 140	196 397	186 665	205 903	203 671	200 715	206 840	201 405	202 400	205 500
Timor-Leste	...	...	...	...	...	...	0	0	0	0
Turkey	44 801	45 067	47 976	49 600	50 460	54 500	50 190	42 824	43 323	43 938
Turkmenistan	16 080	15 140	9 740	9 014	8 179	7 014	9 058	12 228	12 749	12 812
Untd Arab Em	0	0	0	0	0	0	0	0	0	0
Uzbekistan	4 358	3 095	3 611	1 494	3 075	2 799	2 871	3 387	4 070	2 009
Viet Nam	146 839	79 587	94 689	164 936	177 589	138 800	169 107	170 000 F	133 280 F	149 200 F
Total	***3 788 090***	***3 958 667***	***4 302 819***	***4 597 426***	***4 702 743***	***5 095 470***	***5 502 187***	***5 690 272***	***5 734 686***	***5 722 141***

English name Nom anglais Nombre inglés	Scientific name Nom scientifique Nombre científico	Species group Groupe d'espèces Grupo de especies	1996 t	1997 t	1998 t	1999 t	2000 t	2001 t	2002 t
Freshwater bream	*Abramis brama*	11	37 805	36 097	37 939	31 936	30 758	34 321	34 193
Freshwater breams nei	*Abramis spp*	11	2 261	1 225	2 263	2 422	2 270	2 408	2 524
Common carp	*Cyprinus carpio*	11	12 473	9 905	12 100	12 463	13 515	12 475	16 767
Tench	*Tinca tinca*	11	1 155	1 879	1 211	1 418	1 646	1 696	1 825
Bleak	*Alburnus alburnus*	11	168	130	240	268	247	545	291
Barbel	*Barbus barbus*	11	186	239	215	176	118	119	83
Common nase	*Chondrostoma nasus*	11	24	26	28	28	32	27	32
Crucian carp	*Carassius carassius*	11	250	622	795	749	765	984	1 165
Goldfish	*Carassius auratus*	11	3 277	1 697	1 932	2 523	2 369	2 460	2 349
Roach	*Rutilus rutilus*	11	3 997	2 739	2 960	2 565	2 659	2 599	2 203
Roaches nei	*Rutilus spp*	11	29 329	23 019	21 996	16 389	18 601	19 266	21 040
Rudd	*Scardinius erythrophthalmus*	11	39	113	192	175	147	102	150
Orfe(=Ide)	*Leuciscus idus*	11	3 216	3 309	3 487	2 805	2 593	2 483	2 728
Common dace	*Leuciscus leuciscus*	11	5	5	5	4	0	0	0
Chub	*Leuciscus cephalus*	11	...	92	166	167	70	35	36
Chubs nei	*Leuciscus spp*	11	47	39	32	38	33	31	24
Grass carp(=White amur)	*Ctenopharyngodon idellus*	11	557	416	452	460	583	490	546
Silver carp	*Hypophthalmichthys molitrix*	11	6 155	5 569	3 045	3 239	1 864	2 590	1 539
Bighead carp	*Hypophthalmichthys nobilis*	11	538	399	387	410	372	331	517
Vimba bream	*Vimba vimba*	11	56	162	87	133	105	97	47
Sichel	*Pelecus cultratus*	11	414	269	1 660	1 588	565	1 041	642
Asp	*Aspius aspius*	11	161	134	142	156	174	130	203
Wuchang bream	*Megalobrama amblycephala*	11	0	0	-	-	-	-	3
Cyprinids nei	*Cyprinidae*	11	14 963	23 551	28 898	26 153	24 342	24 149	25 524
Northern pike	*Esox lucius*	13	19 668	17 946	16 635	15 423	18 688	18 160	19 707
Wels(=Som)catfish	*Silurus glanis*	13	6 286	7 377	9 681	7 525	7 603	6 943	6 457
Burbot	*Lota lota*	13	2 888	2 885	2 920	2 599	2 667	2 550	2 617
European perch	*Perca fluviatilis*	13	18 228	18 900	16 697	15 398	15 616	15 196	18 004
Pike-perch	*Stizostedion lucioperca*	13	7 344	7 124	6 409	6 902	6 986	6 402	7 836
Ruffe	*Gymnocephalus cernuus*	13	18	10	9	56	99	65	1
Freshwater gobies nei	*Gobiidae*	13	...	...	60	16	...	...	70
Freshwater fishes nei	*Osteichthyes*	13	61 099	51 592	48 617	47 489	63 200	51 558	52 174
Danube sturgeon(=Osetr)	*Acipenser gueldenstaedtii*	21	485	727	657	373	272	269	224
Sterlet sturgeon	*Acipenser ruthenus*	21	36	17	11	38	15	13	17
Starry sturgeon	*Acipenser stellatus*	21	681	453	342	251	199	193	151
Beluga	*Huso huso*	21	130	161	117	75	94	68	64
Sturgeons nei	*Acipenseridae*	21	5	7	95	119	124	90	70
European eel	*Anguilla anguilla*	22	3 787	3 194	3 133	3 092	3 101	2 483	2 326
Atlantic salmon	*Salmo salar*	23	1 957	1 564	1 367	1 686	1 928	2 129	2 005
Sea trout	*Salmo trutta*	23	3 796	3 746	4 162	3 014	2 539	2 971	2 537
Trouts nei	*Salmo spp*	23	1 687	1 668	1 245	1 239	1 052	1 180	1 353
Pink(=Humpback)salmon	*Oncorhynchus gorbuscha*	23	8 804	18 597	14 057	28 582	24 277	17 961	16 927
Chum(=Keta=Dog)salmon	*Oncorhynchus keta*	23	6 670	6 999	7 309	8 137	14 546	14 127	13 864
Masu(=Cherry) salmon	*Oncorhynchus masou*	23	41	4	3	7	3	4	2
Sockeye(=Red)salmon	*Oncorhynchus nerka*	23	9 261	3 479	3 947	7 097	6 872	12 106	13 671
Chinook(=Spring=King)salmon	*Oncorhynchus tshawytscha*	23	401	445	340	483	264	163	179
Coho(=Silver)salmon	*Oncorhynchus kisutch*	23	1 577	898	1 449	1 054	1 419	1 090	1 233
Rainbow trout	*Oncorhynchus mykiss*	23	1 270	1 111	3 987	4 314	2 049	2 034	2 629
Brook trout	*Salvelinus fontinalis*	23	3	4	4	4	5	7	11
Arctic char	*Salvelinus alpinus*	23	53	52	49	43	47	36	36
Chars nei	*Salvelinus spp*	23	71	67	74	85	102	100	100
Huchen	*Hucho hucho*	23	1	1	1	1	1	1	1
Grayling	*Thymallus thymallus*	23	52	43	38	41	37	39	38
European smelt	*Osmerus eperlanus*	23	2 287	2 530	2 427	2 675	2 686	2 099	3 631
Rainbow smelt	*Osmerus mordax*	23	640	1 113	1 758	1 340	608	719	1 417
Smelts nei	*Osmerus spp, Hypomesus spp*	23	1 927	4 095	3 102	2 740	3 825	3 568	3 965
Vendace	*Coregonus albula*	23	6 810	6 827	5 627	5 515	5 387	5 199	5 177
European whitefish	*Coregonus lavaretus*	23	3 260	3 321	2 786	2 789	2 671	2 664	2 672
Houting	*Coregonus oxyrinchus*	23	63	30	61	35	31	9	11
Whitefishes nei	*Coregonus spp*	23	10 841	11 014	10 354	10 429	10 590	10 718	14 622
Salmonoids nei	*Salmonoidei*	23	2 780	2 224	2 105	2 023	1 779	2 038	2 530
Pontic shad	*Alosa pontica*	24	794	901	678	69	213	287	507
Shads nei	*Alosa spp*	24	33	64	35	22	24	22	48
Caspian shads	*Caspialosa spp*	24	1 882	2 304	3 284	4 654	1 273	176	239
Azov sea sprat	*Clupeonella cultriventris*	24	94 035	83 609	116 294	152 068	119 256	46 940	32 864
River lamprey	*Lampetra fluviatilis*	25	141	87	95	129	163	118	106
Lampreys nei	*Petromyzontidae*	25	76	31	37	67	24	21	96
Three-spined stickleback	*Gasterosteus aculeatus*	25	352	349	910	1 047	1 167	965	1 725
Navaga(=Wachna cod)	*Eleginus navaga*	32	98	163	177	22	11	55	127
Flathead grey mullet	*Mugil cephalus*	33	27	54	67	58	51	46	8
So-iuy mullet	*Mugil soiuy*	33	-	-	13	190	169	382	91
Mullets nei	*Mugilidae*	33	25	55	139	203	140	61	341
Gobies nei	*Gobiidae*	33	-	-	-	-	-	5	7
Big-scale sand smelt	*Atherina boyeri*	37	8	110	350	361	350	326	390
Noble crayfish	*Astacus astacus*	41	...	10	10	10	...	...	...
Signal crayfish	*Pacifastacus leniusculus*	41	-	-	-	10	81	80	50
White-clawed crayfish	*Austropotamobius pallipes*	41	0	0	0	0	0	0	0
Red swamp crawfish	*Procambarus clarkii*	41	2 500	2 500	2 500	2 500	2 500	2 500	2 500
Euro-American crayfishes nei	*Astacidae, Cambaridae*	41	234	433	228	227	208	180	162
Freshwater crustaceans nei	*Crustacea*	41	40	46	45	53	969	1 747	1 769
Freshwater molluscs nei	*Mollusca*	51	-	-	-	-	-	-	451

C-05 (a)

Fish, crustaceans, molluscs, etc — Capture production by species items — Europe - Inland waters
Poissons, crustacés, mollusques, etc — Captures par catégories d'espèces — Europe - Eaux continentales
Peces, crustáceos, moluscos, etc — Capturas por categorías de especies — Europa - Aguas continentales

English name Nom anglais Nombre inglés	Scientific name Nom scientifique Nombre científico	Species group Groupe d'espèces Grupo de especies	1996 t	1997 t	1998 t	1999 t	2000 t	2001 t	2002 t
Frogs	*Rana spp*	71	...	38	41	35	26	-	29
Total			***402 228***	***382 615***	***416 770***	***450 679***	***431 835***	***347 242***	***354 270***

C-05 (b)

Fish, crustaceans, molluscs, etc — Capture production by countries or areas — Europe - Inland waters
Poissons, crustacés, mollusques, etc — Captures par pays ou zones — Europe - Eaux continentales
Peces, crustáceos, moluscos, etc — Capturas por países o áreas — Europa - Aguas continentales

Country or area Pays ou zone País o área	1993 t	1994 t	1995 t	1996 t	1997 t	1998 t	1999 t	2000 t	2001 t	2002 t
Albania	850 F	700 F	219	317	180	823	814	955	1 373	1 167
Andorra	0	0	0	0	0	0	0	0	0	0
Austria	420	388	404	450	465	451	432	439	362	350
Belarus	2 993	786	715	821	499	457	514	553	943	5 575
Belgium	511	511	511	511	511	511	536	511	511	511
Bosnia Herzg	2 500 F	2 500 F	2 500 F	2 500 F	2 500 F	2 500 F	2 500 F	2 500 F	2 500 F	2 500 F
Bulgaria	1 675	995 F	762	1 127	1 881	2 336	2 475	861	1 540	1 452
Channel Is	0	0	0	0	0	0	0	0	0	0
Croatia	284	340	364	434	408	10 F	10	17	34	25
Czech Rep	3 185	3 955	3 929	3 524	3 321	3 952	4 190	4 654	4 646	4 983
Denmark	337	243	264	196	232	349	206	183	99	77
Estonia	2 411	1 909	2 366	2 361	2 439	3 878	3 108	3 190	2 461	4 578
Faeroe Is	0	0	0	0	0	0	0	0	0	-
Finland	51 522	47 895	48 436	47 618	47 618	36 813	36 813	34 840	34 820	34 782
France	4 400	4 450	4 500	4 540	4 540	4 500	2 134 F	2 131 F	2 130 F	2 130 F
Germany	10 837	10 908	22 987	22 987	22 916	22 916	22 868	22 868	22 818	22 798
Greece	2 960	3 452	3 606	2 903	2 601	2 818	3 280	3 433	3 181	3 000
Hungary	7 886	8 307	7 314	7 606	7 406	7 265	7 514	7 101	6 638	6 750
Iceland	907	698	739	608	404	416	370	176	160	160 F
Ireland	2 929	3 604	3 761	3 806	3 804	3 976	801	881	902	789
Isle of Man	0	0	0	0	0	0	0	0	0	0
Italy	9 515	9 921	10 035	6 764	6 690	4 667	5 436	4 565	5 527	4 242
Latvia	553	495	514	536	544	501	610	612	581	581
Liechtensten	0	0	0	0	0	0	0	0	0	0
Lithuania	1 146	1 187	1 260	1 295	1 713	1 737	1 715	1 912	1 854	758
Luxembourg	0	0	0	0	0	0	0	0	0	0
Macedonia	164	196	208	78	130	131	135	208	128	148
Malta	0	0	0	0	0	0	0	0	0	
Moldova Rep	630	708	709	603	569	491	309	344	387	565
Netherlands	1 601	2 446	4 107	2 157	2 293	1 547	2 303	2 250 F	2 200 F	2 578
Norway	435	432	413	338	439	507	514	578	570 F	570 F
Poland	31 391	27 500	24 889	22 037	13 832	13 236	13 875	17 543	17 789	17 947
Portugal	3	4	2	0	0	0	0	0	1	12
Romania	8 562	10 598	9 048	6 145	4 574	4 630	5 336	4 896	5 206	4 867
Russian Fed	216 866	217 950	212 874	233 272	227 091	271 311	307 823	292 368	206 430	208 522
Serbia-Monte	1 416 F	1 329 F	1 069	1 232	1 063	866	828	672	570	937
Slovakia	1 185	1 627	1 950	1 414	1 364	1 361	1 396	1 368	1 531	1 746
Slovenia	297	339	316	289	302	269	243	226	206	226
Spain	9 215	6 284	8 869	8 710 F	8 710 F	8 710 F	8 710 F	8 710 F	8 710 F	8 710 F
Sweden	2 273	2 254	1 934	1 810	2 011	1 559	1 478	1 459	1 234	1 436
Switzerland	1 822	1 481	1 588	1 841	1 859	1 809	1 840	1 659	1 715	1 544
Ukraine	13 189	14 786	6 847	9 468	6 215	4 898	4 728	4 429	4 343	3 823
UK	1 909	2 191	2 146	1 930	1 491	4 569	4 835	2 743	3 142	3 431
Total	***398 779***	***393 369***	***392 155***	***402 228***	***382 615***	***416 770***	***450 679***	***431 835***	***347 242***	***354 270***

C-06 (a)

Fish, crustaceans, molluscs, etc — Capture production by species items — Oceania - Inland waters
Poissons, crustacés, mollusques, etc — Captures par catégories d'espèces — Océanie - Eaux continentales
Peces, crustáceos, moluscos, etc — Capturas por categorías de especies — Oceanía - Aguas continentales

English name Nom anglais Nombre inglés	Scientific name Nom scientifique Nombre científico	Species group Groupe d'espèces Grupo de especies	1996 t	1997 t	1998 t	1999 t	2000 t	2001 t	2002 t
Common carp	*Cyprinus carpio*	11	0	28	2	3	4	6	7
Mozambique tilapia	*Oreochromis mossambicus*	12	2 310	2 310	2 310	2 310	2 310	2 310	2 310
Nile tilapia	*Oreochromis niloticus*	12	37	40	288	290	280	278	280
Tilapias nei	*Oreochromis (=Tilapia) spp*	12	-	-	6	-	-	-	-
Brown bullhead	*Ameiurus nebulosus*	13	1	5	12	14	10	8	7
Gudgeons, sleepers nei	*Eleotridae*	13	1 850	1 850	1 850	1 850	1 850	1 850	1 850
Freshwater fishes nei	*Osteichthyes*	13	7 506	7 498	7 296	7 258	7 277	7 168	7 137
Short-finned eel	*Anguilla australis*	22	413	320	293	296	300	300	300
River eels nei	*Anguilla spp*	22	518	490	993	913	851	944	736
Diadromous clupeoids nei	*Clupeoidei*	24	480	480	480	480	480	480	480
Barramundi(=Giant seaperch)	*Lates calcarifer*	25	350	350	350	350	350	350	350
Sea catfishes nei	*Ariidae*	33	1 850	1 850	1 850	1 850	1 850	1 850	1 850
Giant river prawn	*Macrobrachium rosenbergii*	41	0	3	3	3	3	3	3
River prawns nei	*Macrobrachium spp*	41	5	5	5	5	5	5	5
Australian crayfish	*Euastacus armatus*	41	-	-	-	66	24	23	69
Oceanian crayfishes nei	*Parastacidae*	41	6	6	14	6	0	1	1
Freshwater crustaceans nei	*Crustacea*	41	327	315	323	332	340	343	340
Freshwater molluscs nei	*Mollusca*	51	2 670	3 970	4 500	5 000	5 080	5 300	5 180
Total			***18 323***	***19 520***	***20 575***	***21 026***	***21 014***	***21 219***	***20 905***

C-06 (b)

Fish, crustaceans, molluscs, etc — Capture production by countries or areas — Oceania - Inland waters
Poissons, crustacés, mollusques, etc — Captures par pays ou zones — Océanie - Eaux continentales
Peces, crustáceos, moluscos, etc — Capturas por países o áreas — Oceanía - Aguas continentales

Country or area Pays ou zone País o área	1993 t	1994 t	1995 t	1996 t	1997 t	1998 t	1999 t	2000 t	2001 t	2002 t
Amer Samoa	0	0	0	0	0	0	0	0	0	-
Australia	3 011	584	821	705	612	593	694	666 F	663 F	646 F
Christmas Is	0	0	0	0	0	0	0	0	0	0
Cocos Is	0	0	0	0	0	0	0	0	0	0
Cook Is	0	0	10	0	0	0	0	0	0	0
Fiji Islands	2 907	2 964 F	3 586	3 034	4 325	5 111	5 622	5 700 F	5 921	5 800 F
Fr Polynesia	0	0	0	0	53	53	53	53	53	53
Guam	0	0	0	0	-	6	-	-	-	-
Kiribati	0	0	0	0	0	0	0	0	0	-
Marshall Is	0	0	0	0	0	0	0	0	0	0
Micronesia	4 F	4 F	5 F	5 F	5 F	5 F	5 F	5 F	5 F	5 F
NewCaledonia	0	0	0	0	0	0	0	0	0	0
New Zealand	1 160	1 100	1 115	1 079	1 025	1 307	1 152	1 089	1 076	900
Niue	0	0	0	0	0	0	0	0	0	0
Norfolk Is	0	0	0	0	0	0	0	0	0	0
N Marianas	0	0	1	0	0	0	0	0	0	-
Palau	0	0	0	0	0	0	0	0	0	0
Papua N Guin	13 500 F	13 500 F	13 500 F	13 500 F	13 500 F	13 500 F	13 500 F	13 500 F	13 500 F	13 500 F
Pitcairn Is	0	0	0	0	0	0	0	0	0	0
Samoa	0	0	0	0	0	0	0	1	1 F	1 F
Solomon Is	0	0	0	0	0	0	0	0	0	0
Tokelau	0	0	0	0	0	0	0	0	0	0
Tonga	1	0	0	0	0	0	0	0	0	0
Tuvalu	0	0	0	0	0	0	0	0	0	0
Vanuatu	0	0	0	0	0	0	0	0	0	0
Wallis Fut I	0	0	0	0	0	0	0	0	0	0
Total	***20 583***	***18 152***	***19 038***	***18 323***	***19 520***	***20 575***	***21 026***	***21 014***	***21 219***	***20 905***

C-21 (a)
Fish, crustaceans, molluscs, etc — Capture production by species items — Atlantic, Northwest
Poissons, crustacés, mollusques, etc — Captures par catégories d'espèces — Atlantique, nord-ouest
Peces, crustáceos, moluscos, etc — Capturas por categorías de especies — Atlántico, noroeste

English name Nom anglais Nombre inglés	Scientific name Nom scientifique Nombre científico	Species group Groupe d'espèces Grupo de especies	1996 t	1997 t	1998 t	1999 t	2000 t	2001 t	2002 t
Sturgeons nei	*Acipenseridae*	21	86	91	69	9	3	4	-
American eel	*Anguilla rostrata*	22	551	455	496	353	395	220	494
Atlantic salmon	*Salmo salar*	23	165	121	8	2	25	43	2
Trouts nei	*Salmo spp*	23	4	1	0	0	0	-	-
Arctic char	*Salvelinus alpinus*	23	43	78	76	24	29	20	20
Chars nei	*Salvelinus spp*	23	11	31	35	46	46	34	19
Rainbow smelt	*Osmerus mordax*	23	1 078	1 227	1 255	356	970	308	314
American shad	*Alosa sapidissima*	24	674	702	748	562	379	526	357
Alewife	*Alosa pseudoharengus*	24	5 741	6 185	7 533	6 805	5 540	3 481	5 882
Striped bass	*Morone saxatilis*	25	2 144	2 785	3 032	3 000	3 123	2 934	2 847
Atlantic halibut	*Hippoglossus hippoglossus*	31	1 226	1 469	1 442	1 406	1 372	2 367	1 893
Greenland halibut	*Reinhardtius hippoglossoides*	31	50 623	55 989	52 488	69 874	64 586	61 426	64 370
Witch flounder	*Glyptocephalus cynoglossus*	31	5 581	5 157	5 790	5 768	6 447	7 276	7 077
Amer. plaice(=Long rough dab)	*Hippoglossoides platessoides*	31	7 892	8 963	8 803	9 894	10 480	11 119	9 202
Yellowtail flounder	*Limanda ferruginea*	31	3 840	5 266	9 604	13 960	20 971	24 273	18 948
Winter flounder	*Pseudopleuronectes americanus*	31	8 155	8 313	6 904	6 417	8 185	9 187	7 665
Windowpane flounder	*Scophthalmus aquosus*	31	46	48	521	166	268	177	98
Summer flounder	*Paralichthys dentatus*	31	5 425	4 775	5 548	5 166	5 344	5 350	6 776
Flatfishes nei	*Pleuronectiformes*	31	1 591	1 372	1 071	1 136	983	888	354
Blue antimora	*Antimora rostrata*	32	16	-	26	24	21	-	0
Tusk(=Cusk)	*Brosme brosme*	32	1 873	2 244	1 995	1 295	1 285	1 678	1 438
Atlantic cod	*Gadus morhua*	32	33 128	46 903	52 864	68 671	63 201	60 074	55 260
Greenland cod	*Gadus ogac*	32	2 120	1 728	1 695	1 899	931	1 128	972
Red hake	*Urophycis chuss*	32	2 904	2 850	2 803	3 228	3 365	4 008	5 443
White hake	*Urophycis tenuis*	32	8 303	6 830	5 959	6 377	8 157	8 323	12 799
Haddock	*Melanogrammus aeglefinus*	32	10 921	11 359	14 653	13 726	16 710	21 554	22 830
Atlantic tomcod	*Microgadus tomcod*	32	117	66	94	17	30	57	1
Saithe(=Pollock)	*Pollachius virens*	32	12 826	16 932	20 697	13 159	10 611	11 312	11 335
Polar cod	*Boreogadus saida*	32	4	-	-	-	-	-	3
Silver hake	*Merluccius bilinearis*	32	43 399	33 565	31 892	27 568	25 756	33 000	24 293
Offshore silver hake	*Merluccius albidus*	32	32	...	5	12	5	2	6
Roughhead grenadier	*Macrourus berglax*	32	1 044	4 502	7 140	7 004	8 494	1 742	1 424
Roundnose grenadier	*Coryphaenoides rupestris*	32	3 532	270	159	136	372	5 512	4 821
Gadiformes nei	*Gadiformes*	32	3	7	25	11	38	3	26
Ladyfish	*Elops saurus*	33	-	1	1	1 963	1	0	0
Mullets nei	*Mugilidae*	33	2 645	553	370	2 323	546	412	575
Black seabass	*Centropristis striata*	33	1 683	1 215	1 190	1 412	1 189	1 276	1 574
Groupers, seabasses nei	*Serranidae*	33	...	...	23	4 702	18	18	23
Pigfish	*Orthopristis chrysoptera*	33	-	1	1	-	0	0	0
Spotted weakfish	*Cynoscion nebulosus*	33	86	78	79	202	119	36	39
Squeteague(=Gray weakfish)	*Cynoscion regalis*	33	2 729	2 697	3 212	2 736	2 329	2 119	2 054
Weakfishes nei	*Cynoscion spp*	33	-	14	-	11	-	-	-
Atlantic croaker	*Micropogonias undulatus*	33	8 250	11 038	10 438	11 451	11 632	12 410	11 460
Northern kingfish	*Menticirrhus saxatilis*	33	49	21	17	13	28	39	8
Black drum	*Pogonias cromis*	33	39	72	41	88	52	39	23
Spot croaker	*Leiostomus xanthurus*	33	1 738	1 975	2 354	1 715	2 159	1 877	1 784
Red drum	*Sciaenops ocellatus*	33	1	2	3	6	6	3	4
Sheepshead	*Archosargus probatocephalus*	33	137	31	20	87	27	19	20
Scup	*Stenotomus chrysops*	33	2 952	2 196	1 893	1 676	1 206	1 845	3 312
Tautog	*Tautoga onitis*	33	119	116	116	95	111	138	160
Cunner	*Tautogolabrus adspersus*	33	1	1	3	4	3	9	13
Ocean pout	*Macrozoarces americanus*	33	41	15	17	18	19	18	12
Sandeels(=Sandlances) nei	*Ammodytes spp*	33	2	2	1	1	2	0	0
Spadefishes nei	*Ephippidae*	33	6	-	14	21	14	20	10
Sculpins nei	*Cottidae*	33	-	2	5	2	1	1	0
...A	*Normanichthys crockeri*	33	3	-	-	-	-	-	-
Northern puffer	*Sphoeroides maculatus*	33	11	18	17	37	36	35	57
Triggerfishes, durgons nei	*Balistidae*	33	-	6	5	61	3	2	8
Toadfishes, etc. nei	*Batrachoididae*	33	...	1	6	4	21	27	22
Argentines	*Argentina spp*	34	482	1 144	55	17	13	17	20
American conger	*Conger oceanicus*	34	29	17	48	43	49	40	61
Alfonsinos nei	*Beryx spp*	34	141	-	-	-	-	-	-
Silvery John dory	*Zenopsis conchifer*	34	27	6	49	19	28	62	66
Atlantic goldeneye tilefish	*Caulolatilus chrysops*	34	-	-	-	0	2	790	847
Great Northern tilefish	*Lopholatilus chamaeleonticeps*	34	1 349	1 493	1 339	528	515	117	98
Atlantic wolffish	*Anarhichas lupus*	34	7	23	7	-	2	7	0
Wolffishes(=Catfishes) nei	*Anarhichas spp*	34	1 674	1 972	1 533	1 985	1 532	1 895	968
Escolar	*Lepidocybium flavobrunneum*	34	...	...	1	1	2	4	1
Largehead hairtail	*Trichiurus lepturus*	34	...	...	0	4	5	1	2
Beaked redfish	*Sebastes mentella*	34	...	...	...	...	671	128	552
Atlantic redfishes nei	*Sebastes spp*	34	26 768	23 941	33 640	34 292	47 273	49 590	56 424
Blackbelly rosefish	*Helicolenus dactylopterus*	34	-	-	-	-	3	0	0
Atlantic searobins	*Prionotus spp*	34	20	10	31	38	24	39	66
Lumpfish(=Lumpsucker)	*Cyclopterus lumpus*	34	662	1 695	2 399	3 479	1 755	3 362	5 798
American angler	*Lophius americanus*	34	25 826	29 555	27 832	26 301	22 066	25 166	26 741
Atlantic herring	*Clupea harengus*	35	293 058	284 509	272 813	281 766	277 302	307 520	258 749
Round sardinella	*Sardinella aurita*	35	-	-	-	340	-	-	-
Atlantic menhaden	*Brevoortia tyrannus*	35	280 496	322 239	248 451	189 185	183 310	236 246	180 506
Atlantic thread herring	*Opisthonema oglinum*	35	1 686	9	-	1 148	0	-	-
Atlantic bonito	*Sarda sarda*	36	153	134	74	73	44	38	16
Wahoo	*Acanthocybium solandri*	36	...	1	1	2	2	1	1

C-21 (a)

Fish, crustaceans, molluscs, etc — Capture production by species items — Atlantic, Northwest
Poissons, crustacés, mollusques, etc — Captures par catégories d'espèces — Atlantique, nord-ouest
Peces, crustáceos, moluscos, etc — Capturas por categorías de especies — Atlántico, noroeste

English name Nom anglais Nombre inglés	Scientific name Nom scientifique Nombre científico	Species group Groupe d'espèces Grupo de especies	1996 t	1997 t	1998 t	1999 t	2000 t	2001 t	2002 t
King mackerel	*Scomberomorus cavalla*	36	1 625	385	341	1 171	315	232	163
Atlantic Spanish mackerel	*Scomberomorus maculatus*	36	1 411	364	215	694	368	363	337
Frigate and bullet tunas	*Auxis thazard, A.rochei*	36	-	-	1	17	9	3	1
Little tunny(=Atl.black skipj)	*Euthynnus alletteratus*	36	81	199	120	405	110	121	138
Skipjack tuna	*Katsuwonus pelamis*	36	35	9	24	42	2	4	0
Atlantic bluefin tuna	*Thunnus thynnus*	36	2 232	2 107	2 426	2 730	2 522	2 755	2 320
Blackfin tuna	*Thunnus atlanticus*	36	28	5	1	20	16	2	1
Albacore	*Thunnus alalunga*	36	367	382	367	490	742	1 129	1 125
Yellowfin tuna	*Thunnus albacares*	36	1 319	997	816	883	1 099	1 020	585
Bigeye tuna	*Thunnus obesus*	36	910	1 011	1 124	1 866	957	1 038	547
Atlantic sailfish	*Istiophorus albicans*	36	-	-	6	-	-	-	-
Blue marlin	*Makaira nigricans*	36	-	1	12	-	2	-	-
Atlantic white marlin	*Tetrapturus albidus*	36	8	8	10	5	5	3	2
Swordfish	*Xiphias gladius*	36	2 437	2 044	2 289	2 358	2 168	2 028	2 044
Tuna-like fishes nei	*Scombroidei*	36	115	18	15	29	38	13	4
Capelin	*Mallotus villosus*	37	32 711	21 987	38 270	23 531	21 374	19 751	13 575
Halfbeaks nei	*Hemiramphus spp*	37	-	-	-	258	-	-	-
Opah	*Lampris guttatus*	37	...	...	1	1	2	1	0
Atlantic silverside	*Menidia menidia*	37	170	259	255	583	319	661	150
Bluefish	*Pomatomus saltatrix*	37	4 137	3 900	3 446	3 145	3 499	3 825	3 011
Cobia	*Rachycentron canadum*	37	142	6	13	72	16	13	13
Blue runner	*Caranx crysos*	37	-	-	-	49	1	0	1
Crevalle jack	*Caranx hippos*	37	-	-	4	232	2	1	4
Florida pompano	*Trachinotus carolinus*	37	44	1	2	139	0	0	1
Greater amberjack	*Seriola dumerili*	37	...	...	...	368	-	-	-
Amberjacks nei	*Seriola spp*	37	...	...	4	5	8	3	2
Atlantic mackerel	*Scomber scombrus*	37	37 127	36 816	33 996	28 372	23 312	36 837	70 463
Butterfishes, pomfrets nei	*Stromateidae*	37	3 611	3 357	1 944	2 116	1 438	4 416	889
Sand tiger shark	*Carcharias taurus*	38	-	-	-	-	1	-	-
Thresher	*Alopias vulpinus*	38	-	-	-	-	8	11	11
Shortfin mako	*Isurus oxyrinchus*	38	...	...	...	...	29	19	20
Longfin mako	*Isurus paucus*	38	-	-	-	-	1	0	1
Mako sharks	*Isurus spp*	38	-	-	-	-	-	47	37
Porbeagle	*Lamna nasus*	38	1 069	1 356	1 026	958	930	502	248
Great white shark	*Carcharodon carcharias*	38	-	-	-	-	1	0	
Nurse shark	*Ginglymostoma cirratum*	38	-	-	-	-	0	-	0
Blue shark	*Prionace glauca*	38	...	...	...	...	169	-	-
Sandbar shark	*Carcharhinus plumbeus*	38	-	-	-	-	41	24	28
Blacktip shark	*Carcharhinus limbatus*	38	-	-	-	-	21	1	11
Dusky shark	*Carcharhinus obscurus*	38	-	-	-	-	80	0	3
Tiger shark	*Galeocerdo cuvier*	38	-	-	-	-	-	1	0
Dusky smooth-hound	*Mustelus canis*	38	...	...	...	...	334	321	493
Picked dogfish	*Squalus acanthias*	38	494	452	1 081	2 456	10 701	5 995	5 697
Dogfish sharks nei	*Squalidae*	38	27 719	20 297	21 991	16 408	2 246	979	718
Dogfish sharks, etc. nei	*Squaliformes*	38	3 423	1 252	887	1 673	710	876	1 047
Raja rays nei	*Raja spp*	38	23 176	24 751	26 204	27 463	33 998	30 816	27 270
Sharks, rays, skates, etc. nei	*Elasmobranchii*	38	250	111	107	228	90	124	55
Groundfishes nei	*Osteichthyes*	39	1 357	1 942	1 271	801	1 158	837	154
Pelagic fishes nei	*Osteichthyes*	39	44	128	25	44	29	208	-
Finfishes nei	*Osteichthyes*	39	13 086	12 287	4 527	6 613	6 974	3 434	4 362
Marine fishes nei	*Osteichthyes*	39	6	19	25	29	12	4	-
Atlantic rock crab	*Cancer irroratus*	42	4 341	7 530	7 696	7 033	9 516	8 294	8 331
Jonah crab	*Cancer borealis*	42	334	745	1 255	1 549	1 114	1 245	1 190
Blue crab	*Callinectes sapidus*	42	67 337	62 784	56 834	60 392	42 408	39 511	42 997
Green crab	*Carcinus maenas*	42	-	54	86	14	16	67	39
Queen crab	*Chionoecetes opilio*	42	66 831	74 341	77 517	98 633	104 252	110 044	117 187
Red crab	*Geryon quinquedens*	42	465	96	-	-	3 130	4 004	2 169
Marine crabs nei	*Brachyura*	42	3 291	5 076	4 194	4 793	5 667	10 326	10 725
American lobster	*Homarus americanus*	43	71 866	78 146	77 155	83 105	83 062	83 803	82 422
Northern brown shrimp	*Penaeus aztecus*	45	666	3 377	-	679	1 104	458	856
Northern pink shrimp	*Penaeus duorarum*	45	10 753	114	-	5 062	16	1	29
Northern white shrimp	*Penaeus setiferus*	45	1 347	913	-	619	313	28	171
Northern prawn	*Pandalus borealis*	45	153 425	140 708	178 056	204 638	230 260	231 128	257 586
Aesop shrimp	*Pandalus montagui*	45	...	...	...	...	697	609	206
Pandalus shrimps nei	*Pandalus spp*	45	24 976	28 601	29 042	30 603	35 180	31 234	33 342
Rock shrimp	*Sicyonia brevirostris*	45	9 470	8	-	481	-	-	-
Royal red shrimp	*Pleoticus robustus*	45	14	4	13	78	22	24	23
Natantian decapods nei	*Natantia*	45	-	824	148	894	1 732	263	-
Stomatopods nei	*Stomatopoda*	47	-	-	1	5	6	1	1
Marine crustaceans nei	*Crustacea*	47	2	-	-	-	-	-	-
Periwinkles nei	*Littorina spp*	52	219	274	369	372	314	741	1 086
Whelk	*Buccinum undatum*	52	...	32	45	...	0	-	-
Whelks	*Busycon spp*	52	4 294	3 362	2 735	5 349	4 253	5 598	5 135
American cupped oyster	*Crassostrea virginica*	53	39 569	49 099	33 265	47 057	26 440	18 557	13 217
Cupped oysters nei	*Crassostrea spp*	53	...	...	64	49	35	14	19
Blue mussel	*Mytilus edulis*	54	17 585	17 228	15 051	15 651	21 098	18 353	23 631
American sea scallop	*Placopecten magellanicus*	55	108 923	101 874	99 282	131 935	196 878	253 960	280 308
Atlantic bay scallop	*Argopecten irradians*	55	5	25	6	20	13	13	8
Iceland scallop	*Chlamys islandica*	55	13 776	14 270	8 848	3 160	4 397	2 956	3 455
Ocean quahog	*Arctica islandica*	56	173 799	163 799	148 506	144 432	122 547	142 826	149 859

C-21 (a)

Fish, crustaceans, molluscs, etc — Capture production by species items — Atlantic, Northwest
Poissons, crustacés, mollusques, etc — Captures par catégories d'espèces — Atlantique, nord-ouest
Peces, crustáceos, moluscos, etc — Capturas por categorías de especies — Atlántico, noroeste

English name Nom anglais Nombre inglés	Scientific name Nom scientifique Nombre científico	Species group Groupe d'espèces Grupo de especies	1996 t	1997 t	1998 t	1999 t	2000 t	2001 t	2002 t
Northern quahog(=Hard clam)	*Mercenaria mercenaria*	56	22 964	6 287	1 234	2 536	9 469	6 680	14 564
Atlantic surf clam	*Spisula solidissima*	56	155 584	141 421	131 700	142 370	165 765	166 310	176 322
Stimpson's surf clam	*Spisula polynyma*	56	25 612	27 365	26 008	26 722	22 985	20 273	19 979
Atl.jackknife(=Atl.razor clam)	*Ensis directus*	56	-	14	49	64	99	36	90
Sand gaper	*Mya arenaria*	56	5 824	6 777	8 131	7 793	8 248	9 946	8 735
Clams, etc. nei	*Bivalvia*	56	773	261	66	1 136	349	1 129	2 525
Longfin squid	*Loligo pealei*	57	12 490	16 161	18 879	18 749	16 942	14 211	16 684
Northern shortfin squid	*Illex illecebrosus*	57	26 362	32 390	25 776	7 928	9 390	4 074	3 005
Various squids nei	*Loliginidae, Ommastrephidae*	57	56	182	69	90	70	124	196
Octopuses, etc. nei	*Octopodidae*	57	-	0	0	-	0	0	5
Marine molluscs nei	*Mollusca*	58	2 394	600	168	1 920	1 469	1 384	3 210
Horseshoe crab	*Limulus polyphemus*	75	1 598	2 606	3 182	2 238	1 557	1 071	1 257
Starfishes nei	*Asteroidea*	76	-	-	-	-	0	-	-
Sea urchins nei	*Strongylocentrotus spp*	76	13 851	12 201	10 720	10 794	9 131	7 274	4 707
Sea cucumbers nei	*Holothurioidea*	76	1 288	-	2 406	3 504	4 309	1 504	4 656
Jellyfishes	*Rhopilema spp*	77	-	-	-	578		-	-
Marine worms	*Polychaeta*	77	-	223	299	343	202	481	507
Total			***2 069 186***	***2 053 310***	***1 965 230***	***2 035 542***	***2 082 554***	***2 240 365***	***2 245 008***

C-21 (b)

Fish, crustaceans, molluscs, etc — Capture production by countries or areas — Atlantic, Northwest
Poissons, crustacés, mollusques, etc — Captures par pays ou zones — Atlantique, nord-ouest
Peces, crustáceos, moluscos, etc — Capturas por países o áreas — Atlántico, noroeste

Country or area Pays ou zone País o área	1993 t	1994 t	1995 t	1996 t	1997 t	1998 t	1999 t	2000 t	2001 t	2002 t
Canada	817 654	680 484	597 570	636 006	689 218	747 005	774 439	828 009	830 311	841 663
China,Taiwan	25	392	206	83	30	8	127	351	618	570
Cuba	29 919	12 332	18 358	24 179	17 218	7 727	4 566	219	-	-
Denmark	897	245	447	403	421	556	235		93	359
Estonia	5 397	1 186	3 242	1 898	3 240	5 533	10 834	13 685	12 402	15 022
Faeroe Is	15 138	10 024	6 884	10 209	8 451	10 094	10 560	8 543	12 980	9 149
Germany	343	305	0	495	449	355	569	4 920	1 359	2 861
Greenland	94 937	101 952	108 574	94 356	94 096	97 953	117 841	121 885	124 149	154 599
Honduras	1 298	-	-	-	-	-	-	-	-	-
Iceland	2 196	2 462	8 232	20 682	7 197	6 572	9 147	9 340	5 079	6 877
Japan	9 064	6 147	5 728	5 606	3 934	4 945	5 697	4 421	5 670	4 432
Korea Rep	3 738	-	-	-	-	-	-	-	-	-
Latvia	8 874	473	1 026	1 253	997	1 191	3 080	3 397	3 330	2 742
Lithuania	3 904	900	980	1 585	1 785	3 107	3 370	4 047	7 596	10 948
Norway	9 955	11 755	11 772	7 437	3 709	2 686	4 313	4 296	14 715	14 536
Poland	-	-	-	-	824	148	894	1 732	760	428
Portugal	35 550	30 157	12 532	9 184	8 998	9 616	16 662	13 180	15 003	18 526
Russian Fed	26 525	9 187	10 383	6 021	1 465	2 872	5 979	27 660	32 138	34 686
St Pier Mq	282	294	317	747	3 571	6 108	5 892	6 485	3 802	3 654
Spain	47 139	54 117	20 069	20 864	27 938	30 973	37 239	45 095	40 235	34 092
Ukraine	-	-	-	-	-	-	-	-	405	-
UK	-	49	-	129	23	294	-	-	-	-
USA	1 240 243	1 062 863	1 199 006	1 228 049	1 179 746	1 027 487	1 024 098	985 289	1 129 720	1 089 864
Total	***2 353 078***	***1 985 324***	***2 005 326***	***2 069 186***	***2 053 310***	***1 965 230***	***2 035 542***	***2 082 554***	***2 240 365***	***2 245 008***

English name Nom anglais Nombre inglés	Scientific name Nom scientifique Nombre científico	Species group Groupe d'espèces Grupo de especies	1996 t	1997 t	1998 t	1999 t	2000 t	2001 t	2002 t
Freshwater bream	*Abramis brama*	11	1 480	1 890	2 376	2 384	3 379	3 194	3 763
Freshwater breams nei	*Abramis spp*	11	162	275	144	249	124	176	116
Common carp	*Cyprinus carpio*	11	-	0	0	1	-	1	11
Tench	*Tinca tinca*	11	0	-	0	1	-	1	1
Crucian carp	*Carassius carassius*	11	-	1	0	3	3	3	19
Roach	*Rutilus rutilus*	11	3 574	3 921	3 235	3 418	3 184	3 407	4 231
Roaches nei	*Rutilus spp*	11	-	-	-	-	501	653	585
Rudd	*Scardinius erythrophthalmus*	11	...	...	...	...	...	...	2
Orfe(=Ide)	*Leuciscus idus*	11	396	351	312	290	350	328	311
Vimba bream	*Vimba vimba*	11	214	228	244	228	174	172	277
Sichel	*Pelecus cultratus*	11	...	...	...	...	384	348	523
Asp	*Aspius aspius*	11	...	...	...	...	6	9	23
Cyprinids nei	*Cyprinidae*	11	8	8	-	-	67	107	113
Northern pike	*Esox lucius*	13	2 917	2 934	4 312	4 340	2 840	2 790	2 799
Burbot	*Lota lota*	13	432	460	489	495	383	385	399
European perch	*Perca fluviatilis*	13	6 696	7 715	7 310	7 218	5 147	5 719	5 837
Pike-perch	*Stizostedion lucioperca*	13	2 287	2 408	2 969	3 059	2 171	2 111	2 489
Ruffe	*Gymnocephalus cernuus*	13	...	...	...	...	...	...	31
Freshwater fishes nei	*Osteichthyes*	13	2 639	3 588	2 848	1 950	592	461	544
Sturgeons nei	*Acipenseridae*	21	0	0	1	2	-	-	-
European eel	*Anguilla anguilla*	22	2 959	4 327	2 290	2 622	1 982	2 098	1 990
Atlantic salmon	*Salmo salar*	23	4 371	3 894	3 369	2 419	2 747	2 597	2 285
Sea trout	*Salmo trutta*	23	771	730	533	1 076	555	508	492
Trouts nei	*Salmo spp*	23	270	304	447	466	659	590	860
Pink(=Humpback)salmon	*Oncorhynchus gorbuscha*	23	-	-	-	39	10	184	2
Rainbow trout	*Oncorhynchus mykiss*	23	104	140	237	185	86	139	90
Chars nei	*Salvelinus spp*	23	10	11	20	8	1	...	...
European smelt	*Osmerus eperlanus*	23	5 628	5 011	3 325	2 964	7 151	2 235	2 051
Rainbow smelt	*Osmerus mordax*	23	-	-	-	-	1	6	23
Vendace	*Coregonus albula*	23	865	821	550	533	758	888	1 172
European whitefish	*Coregonus lavaretus*	23	2 482	2 318	2 709	2 524	2 455	2 155	2 110
Whitefishes nei	*Coregonus spp*	23	6	5	3	4	29	8	7
Salmonoids nei	*Salmonoidei*	23	46	49	6	44	55	14	1 263
Allis shad	*Alosa alosa*	24	59	56	2	...	38	45	94
Twaite shad	*Alosa fallax*	24	2	1	...	...	12	16	22
Alewife	*Alosa pseudoharengus*	24	...	...	...	...	61	-	-
Shads nei	*Alosa spp*	24	26	20	26	59	29	42	42
Diadromous clupeoids nei	*Clupeoidei*	24	35	36	30	-	-	-	-
Sea lamprey	*Petromyzon marinus*	25	34	20	17	38	36	32	50
Lampreys nei	*Petromyzontidae*	25	-	2	2	2	21	33	24
Three-spined stickleback	*Gasterosteus aculeatus*	25	33	113	82	-	514	593	744
Lefteye flounders nei	*Bothidae*	31	...	...	...	382	487	397	398
Atlantic halibut	*Hippoglossus hippoglossus*	31	2 549	2 537	1 904	2 092	2 204	1 999	2 155
European plaice	*Pleuronectes platessa*	31	117 699	121 421	103 586	113 421	113 233	110 831	98 639
Greenland halibut	*Reinhardtius hippoglossoides*	31	52 778	41 790	34 506	47 571	43 108	46 970	43 333
Witch flounder	*Glyptocephalus cynoglossus*	31	9 467	12 673	13 436	13 213	12 358	12 776	12 381
Amer. plaice(=Long rough dab)	*Hippoglossoides platessoides*	31	8 144	8 267	10 434	11 751	6 325	7 367	6 458
Common dab	*Limanda limanda*	31	17 967	17 509	22 134	20 917	16 599	17 172	16 179
Lemon sole	*Microstomus kitt*	31	12 078	12 283	14 378	14 203	13 974	15 461	10 346
European flounder	*Platichthys flesus*	31	10 962	11 995	16 000	12 408	14 528	15 315	15 027
Common sole	*Solea solea*	31	34 460	27 279	31 293	35 705	34 677	32 281	29 548
Sand sole	*Solea lascaris*	31	246	288	305	222	297	339	335
Wedge sole	*Dicologlossa cuneata*	31	798	648	488	704	795	691	691
Thickback soles	*Microchirus spp*	31	50	57	77	71	71	80	139
Soles nei	*Soleidae*	31	...	...	...	810	1 101	852	783
Megrim	*Lepidorhombus whiffiagonis*	31	15 580	14 948	13 866	12 560	13 201	12 658	10 656
Megrims nei	*Lepidorhombus spp*	31	6 405	7 006	7 446	6 969	8 361	8 161	6 277
Brill	*Scophthalmus rhombus*	31	3 203	2 744	2 526	2 511	2 957	3 042	2 754
Turbot	*Psetta maxima*	31	6 540	6 112	5 349	5 466	6 214	6 283	5 916
Flatfishes nei	*Pleuronectiformes*	31	19 552	23 360	20 306	16 583	9 765	10 712	14 408
Common mora	*Mora moro*	32	-	-	-	-	-	33	83
Blue antimora	*Antimora rostrata*	32	2	-	-	-	-	-	-
Moras nei	*Moridae*	32	-	415	75	67	130	350	109
Tusk(=Cusk)	*Brosme brosme*	32	28 423	22 501	28 991	33 453	31 246	26 846	25 688
Atlantic cod	*Gadus morhua*	32	1 306 894	1 328 176	1 158 203	1 025 406	877 145	883 593	835 098
Greenland cod	*Gadus ogac*	32	1	1	2	4	-	5	5
Ling	*Molva molva*	32	54 619	50 988	60 826	53 949	43 320	36 969	40 247
Blue ling	*Molva dypterygia*	32	9 787	11 750	12 767	15 699	16 146	19 366	12 088
Greater forkbeard	*Phycis blennoides*	32	4 386	7 983	7 230	5 957	5 877	5 253	5 215
White hake	*Urophycis tenuis*	32	-	-	-	-	-	-	1
Haddock	*Melanogrammus aeglefinus*	32	351 008	322 746	269 036	235 725	196 111	207 267	243 671
Navaga(=Wachna cod)	*Eleginus navaga*	32	561	989	1 008	458	663	1 111	835
Saithe(=Pollock)	*Pollachius virens*	32	342 493	302 375	309 096	327 192	301 959	314 438	370 000
Pollack	*Pollachius pollachius*	32	12 988	12 164	10 952	10 379	12 059	11 937	12 634
Polar cod	*Boreogadus saida*	32	20 784	6 826	3 592	22 005	40 730	39 445	37 191
Rocklings nei	*Gaidropsarus spp*	32	-	-	-	4	24	1	0
Norway pout	*Trisopterus esmarkii*	32	275 667	211 394	97 478	112 613	204 845	93 405	106 924
Pouting(=Bib)	*Trisopterus luscus*	32	12 468	12 752	11 117	12 058	15 588	16 533	15 677
Blue whiting(=Poutassou)	*Micromesistius poutassou*	32	613 306	691 246	1 152 677	1 297 665	1 445 788	1 795 372	1 588 550
Whiting	*Merlangius merlangus*	32	76 536	72 125	61 051	61 472	56 493	48 878	43 449
European hake	*Merluccius merluccius*	32	38 838	40 847	36 404	42 787	46 339	29 084	32 269
Roughhead grenadier	*Macrourus berglax*	32	36	43	30	1 154	80	170	114
Roundnose grenadier	*Coryphaenoides rupestris*	32	10 664	19 637	23 101	18 087	30 596	48 157	23 191

English name Nom anglais Nombre inglés	Scientific name Nom scientifique Nombre científico	Species group Groupe d'espèces Grupo de especies	1996 t	1997 t	1998 t	1999 t	2000 t	2001 t	2002 t
Gadiformes nei	*Gadiformes*	32	32 410	20 650	12 993	15 134	5 824	29 761	13 090
Morays	*Muraenidae*	33	...	...	...	193	164	142	128
Mullets nei	*Mugilidae*	33	1 276	1 385	1 453	1 605	1 433	1 738	1 384
Dusky grouper	*Epinephelus marginatus*	33	327	210	268	96	24	51	40
Groupers, seabasses nei	*Serranidae*	33	579	705	788	783	876	593	353
Spotted seabass	*Dicentrarchus punctatus*	33	68	42	75	68	71	79	84
European seabass	*Dicentrarchus labrax*	33	4 401	4 007	3 752	4 471	5 003	4 944	4 784
Seabasses nei	*Dicentrarchus spp*	33	-	5	6	336	405	367	357
Cardinalfishes, etc. nei	*Apogonidae*	33	-	-	-	294	-	15	55
Rubberlip grunt	*Plectorhinchus mediterraneus*	33	...	...	...	251	77	12	13
Grunts, sweetlips nei	*Haemulidae (=Pomadasyidae)*	33	504	250	316	144	215	177	83
Canary drum (=Baardman)	*Umbrina canariensis*	33	1	-	-	3	7	11	7
Meagre	*Argyrosomus regius*	33	250	409	460	352	198	168	189
Croakers, drums nei	*Sciaenidae*	33	...	...	...	228	132	168	340
Blackspot(=red) seabream	*Pagellus bogaraveo*	33	2 068	4 380	2 796	2 367	1 753	1 274	1 631
Common pandora	*Pagellus erythrinus*	33	104	152	133	106	154	131	144
Axillary seabream	*Pagellus acarne*	33	1 197	974	883	1 028	1 322	1 213	1 146
White seabream	*Diplodus sargus*	33	54	41	35	55	64	51	78
Sargo breams nei	*Diplodus spp*	33	...	...	...	1 050	977	793	895
Large-eye dentex	*Dentex macrophthalmus*	33	8	26	2	2	0	0	1
Common dentex	*Dentex dentex*	33	11	18	36	24	13	16	9
Dentex nei	*Dentex spp*	33	37	41	26	34	387	183	232
Black seabream	*Spondyliosoma cantharus*	33	2 802	2 884	3 120	3 454	3 428	3 144	3 637
Red porgy	*Pagrus pagrus*	33	131	465	1 038	785	528	344	200
Gilthead seabream	*Sparus aurata*	33	875	666	538	436	489	527	610
Bogue	*Boops boops*	33	803	1 024	1 378	3 359	1 761	1 933	1 641
Sand steenbras	*Lithognathus mormyrus*	33	36	41	30	224	217	145	179
Salema	*Sarpa salpa*	33	...	5	2	340	254	330	244
Porgies, seabreams nei	*Sparidae*	33	3 028	4 421	4 462	1 916	2 212	2 175	1 054
Picarels nei	*Spicara spp*	33	...	...	...	43	22	25	36
Surmullet	*Mullus surmuletus*	33	3 518	2 415	4 100	3 080	4 967	4 410	3 581
Surmullets(=Red mullets) nei	*Mullus spp*	33	-	-	-	-	13	10	9
Ballan wrasse	*Labrus bergylta*	33	-	-	-	2	1	-	-
Wrasses, hogfishes, etc. nei	*Labridae*	33	17	16	20	286	318	320	299
Parrotfish	*Sparisoma cretense*	33	-	-	-	56	87	161	153
Eelpout	*Zoarces viviparus*	33	147	97	54	43	28	37	43
Sandeels(=Sandlances) nei	*Ammodytes spp*	33	858 376	1 242 671	1 039 589	761 879	740 542	912 330	904 502
Greater weever	*Trachinus draco*	33	182	183	188	284	278	985	955
Gobies nei	*Gobiidae*	33	0	0	-	0	-	-	-
Grey triggerfish	*Balistes carolinensis*	33	-	1	1	2	-	4	-
Triggerfishes, durgons nei	*Balistidae*	33	-	-	-	42	38	21	32
Toadfishes, etc. nei	*Batrachoididae*	33	-	-	-	70	91	75	86
Argentines	*Argentina spp*	34	24 209	26 246	46 386	31 519	28 533	48 894	36 990
Baird's slickhead	*Alepocephalus bairdii*	34	0	0	0	0	12	616	259
European conger	*Conger conger*	34	12 511	11 929	12 584	11 861	11 360	10 666	10 286
Longspine snipefish	*Macroramphosus scolopax*	34	-	-	-	-	-	-	0
Alfonsinos nei	*Beryx spp*	34	0	12	27	162	139	128	273
Orange roughy	*Hoplostethus atlanticus*	34	2 082	1 971	1 911	1 860	1 467	4 243	5 417
Slimeheads nei	*Trachichthyidae*	34	833	1 052	33	25	3	241	2
John dory	*Zeus faber*	34	1 416	1 581	1 889	2 146	2 678	3 148	2 847
Boarfishes nei	*Caproidae*	34	-	-	-	-	-	120	91
Wreckfish	*Polyprion americanus*	34	587	624	447	447	441	414	432
Red bandfish	*Cepola rubescens*	34	-	-	-	-	-	-	17
Black cardinal fish	*Epigonus telescopus*	34	83	181	326	4	248	175	133
Atlantic wolffish	*Anarhichas lupus*	34	26 997	30 938	37 580	39 535	40 562	38 908	36 207
Spotted wolffish	*Anarhichas minor*	34	1 109	1 180	1 599	1 545	2 987	3 257	3 522
Wolffishes(=Catfishes) nei	*Anarhichas spp*	34	7 021	13 094	16 941	6 653	4 768	13 222	4 869
Eelpouts	*Lycodes spp*	34	-	-	-	-	-	1	2
Stargazer	*Uranoscopus scaber*	34	...	...	...	15	50	46	104
Oilfish	*Ruvettus pretiosus*	34	-	-	-	14	9	9	9
Largehead hairtail	*Trichiurus lepturus*	34	-	-	-	3	0	523	831
Silver scabbardfish	*Lepidopus caudatus*	34	2 066	3 504	3 532	1 933	104	361	1 970
Black scabbardfish	*Aphanopus carbo*	34	6 893	5 907	5 238	4 993	8 184	10 826	10 777
Hairtails, scabbardfishes nei	*Trichiuridae*	34	-	-	-	-	-	13	-
Golden redfish	*Sebastes marinus*	34	11 034	11 413	7 968	52 592	75 556	67 044	68 929
Beaked redfish	*Sebastes mentella*	34	4 842	5 234	4 619	25 043	75 506	98 533	92 557
Atlantic redfishes nei	*Sebastes spp*	34	246 443	213 971	213 519	142 989	72 506	47 805	49 830
Red scorpionfish	*Scorpaena scrofa*	34	...	...	...	...	1	2	5
Blackbelly rosefish	*Helicolenus dactylopterus*	34	119	118	256	716	743	624	452
Scorpionfishes nei	*Scorpaenidae*	34	1 647	2 267	1 505	1 306	1 499	1 034	1 080
Tub gurnard	*Chelidonichthys lucerna*	34	5	4	5	22	2 393	1 149	2 743
Red gurnard	*Chelidonichthys cuculus*	34	214	153	188	319	4 676	7 131	5 854
Streaked gurnard	*Chelidonichthys lastoviza*	34	...	...	...	...	56	51	68
Grey gurnard	*Eutrigla gurnardus*	34	250	161	149	338	828	886	637
Gurnards, searobins nei	*Triglidae*	34	8 542	8 466	7 516	10 249	29 009	5 961	2 832
Lumpfish(=Lumpsucker)	*Cyclopterus lumpus*	34	10 678	13 856	4 739	6 344	5 326	6 132	6 977
Angler(=Monk)	*Lophius piscatorius*	34	65 430	64 974	56 174	50 855	50 546	50 508	46 670
Atlantic herring	*Clupea harengus*	35	2 035 630	2 249 400	2 148 649	2 129 642	2 103 709	1 645 455	1 613 264
Sardinellas nei	*Sardinella spp*	35	-	-	3	1 003	-	1	-
European pilchard(=Sardine)	*Sardina pilchardus*	35	157 154	139 119	157 045	162 233	123 760	130 891	143 472
European sprat	*Sprattus sprattus*	35	643 356	671 996	657 895	644 793	616 405	584 292	549 617
European anchovy	*Engraulis encrasicolus*	35	28 980	22 291	34 324	37 175	38 096	45 780	34 001
Clupeoids nei	*Clupeoidei*	35	10 262	6 556	15 576	20 226	8 592	21 133	10 363
Atlantic bonito	*Sarda sarda*	36	82	48	97	122	193	89	73
Plain bonito	*Orcynopsis unicolor*	36	-	-	-	-	-	-	2

C-27 (a)

Fish, crustaceans, molluscs, etc	Capture production by species items	Atlantic, Northeast
Poissons, crustacés, mollusques, etc	Captures par catégories d'espèces	Atlantique, nord-est
Peces, crustáceos, moluscos, etc	Capturas por categorías de especies	Atlántico, nordeste

English name Nom anglais Nombre inglés	Scientific name Nom scientifique Nombre científico	Species group Groupe d'espèces Grupo de especies	1996 t	1997 t	1998 t	1999 t	2000 t	2001 t	2002 t
Frigate and bullet tunas	*Auxis thazard, A.rochei*	36	2	-	30	264	511	221	167
Little tunny(=Atl.black skipj)	*Euthynnus alletteratus*	36	218	320	182	15	2	22	10
Skipjack tuna	*Katsuwonus pelamis*	36	6 271	3 596	3 692	1 465	1 040	1 611	2 161
Atlantic bluefin tuna	*Thunnus thynnus*	36	9 839	9 219	7 283	7 114	7 027	6 078	7 209
Albacore	*Thunnus alalunga*	36	22 507	23 098	20 738	26 446	26 575	17 058	13 751
Yellowfin tuna	*Thunnus albacares*	36	50	23	33	233	180	18	42
Bigeye tuna	*Thunnus obesus*	36	2 338	3 028	4 956	2 590	1 574	1 381	1 935
Atlantic sailfish	*Istiophorus albicans*	36	-	-	-	11	7	3	7
Blue marlin	*Makaira nigricans*	36	21	3	45	6	22	21	1
Atlantic white marlin	*Tetrapturus albidus*	36	0	1	-	2	-	1	1
Marlins,sailfishes,etc. nei	*Istiophoridae*	36	-	-	-	-	68	408	626
Swordfish	*Xiphias gladius*	36	6 040	4 809	3 419	2 800	3 281	3 059	2 630
Tuna-like fishes nei	*Scombroidei*	36	252	155	263	87	101	267	631
Capelin	*Mallotus villosus*	37	1 494 531	1 581 191	943 637	880 444	1 463 153	1 649 687	1 944 267
Garfish	*Belone belone*	37	1 464	1 792	1 641	1 544	1 222	1 030	1 185
Atlantic saury	*Scomberesox saurus*	37	483	639	961	574	185	993	589
Opah	*Lampris guttatus*	37	-	-	-	1	...	...	...
Dealfishes	*Trachipterus spp*	37	...	...	...	29	20	25	16
Silversides(=Sand smelts) nei	*Atherinidae*	37	333	359	299	297	249	245	288
Bluefish	*Pomatomus saltatrix*	37	1 877	1 914	1 004	808	59	53	33
Atlantic horse mackerel	*Trachurus trachurus*	37	464 224	446 230	343 556	315 464	222 870	238 494	208 743
Jack and horse mackerels nei	*Trachurus spp*	37	32 283	39 846	36 894	38 160	36 989	39 824	33 429
Greater amberjack	*Seriola dumerili*	37	...	...	...	10	21	20	37
Leerfish	*Lichia amia*	37	0	5	2	1	1	10	6
Carangids nei	*Carangidae*	37	...	...	...	43	28	28	19
Atlantic pomfret	*Brama brama*	37	5 348	3 425	2 832	2 244	567	435	510
Common dolphinfish	*Coryphaena hippurus*	37	-	-	-	1	1	3	4
Chub mackerel	*Scomber japonicus*	37	5 511	6 145	7 321	14 165	10 812	4 602	5 923
Atlantic mackerel	*Scomber scombrus*	37	514 967	514 081	625 968	584 815	654 829	661 231	688 268
Mackerels nei	*Scombridae*	37	26 650	35 088	36 818	19 213	13 610	25 591	19 936
Butterfishes, pomfrets nei	*Stromateidae*	37	-	-	-	94	35	81	61
Barracudas nei	*Sphyraena spp*	37	...	...	...	33	41	33	37
Ocean sunfish	*Mola mola*	37	-	-	-	12	-	13	-
Bluntnose sixgill shark	*Hexanchus griseus*	38	-	-	-	-	-	1	7
Basking shark	*Cetorhinus maximus*	38	1 980	1 161	137	81	294	203	136
Thresher	*Alopias vulpinus*	38	7	13	7	34	136	140	32
Shortfin mako	*Isurus oxyrinchus*	38	...	...	0	162	186	188	108
Porbeagle	*Lamna nasus*	38	410	538	1 024	1 486	1 737	1 501	549
Blackmouth catshark	*Galeus melastomus*	38	...	...	...	...	...	...	230
Small-spotted catshark	*Scyliorhinus canicula*	38	5 144	5 613	5 740	5 818	6 152	7 041	6 438
Nursehound	*Scyliorhinus stellaris*	38	306	378	258	274	274	264	195
Catsharks, nursehounds nei	*Scyliorhinus spp*	38	18	6	51	13	68	7	7
Blue shark	*Prionace glauca*	38	281	214	165	1 185	453	1 287	1 348
Tiger shark	*Galeocerdo cuvier*	38	-	-	-	-	-	-	13
Smooth hammerhead	*Sphyrna zygaena*	38	...	...	...	8	8	4	5
Smooth-hound	*Mustelus mustelus*	38	-	-	-	-	15	-	56
Smooth-hounds nei	*Mustelus spp*	38	578	624	749	900	1 091	1 292	1 628
Tope shark	*Galeorhinus galeus*	38	458	511	427	464	567	559	427
Greenland shark	*Somniosus microcephalus*	38	61	73	87	51	45	58	56
Little sleeper shark	*Somniosus rostratus*	38	...	...	...	...	...	...	2
Picked dogfish	*Squalus acanthias*	38	16 508	14 101	13 634	12 098	12 093	13 228	10 386
Gulper shark	*Centrophorus granulosus*	38	...	...	...	73	54	93	160
Leafscale gulper shark	*Centrophorus squamosus*	38	53	58	129	451	478	511	1 173
Lanternsharks nei	*Etmopterus spp*	38	...	...	...	573	...	...	99
Birdbeak dogfish	*Deania calcea*	38	...	...	...	...	18	51	102
Portuguese dogfish	*Centroscymnus coelolepis*	38	336	280	232	717	1 147	2 512	2 984
Longnose velvet dogfish	*Centroscymnus crepidater*	38	-	-	3	-	-	-	12
Kitefin shark	*Dalatias licha*	38	...	...	...	45	311	189	40
Black dogfish	*Centroscyllium fabricii*	38	4	0	-	-	271	271	28
Dogfish sharks nei	*Squalidae*	38	5 869	6 004	5 000	4 338	5 047	5 829	3 459
Dogfishes and hounds nei	*Squalidae, Scyliorhinidae*	38	2 022	2 083	2 011	2 113	3 032	2 700	3 058
Angelshark	*Squatina squatina*	38	-	-	-	-	-	-	3
Angelsharks, sand devils nei	*Squatinidae*	38	1	47	0	1	1	1	-
Angular roughshark	*Oxynotus centrina*	38	...	...	...	81	33	63	86
Sailfin roughshark	*Oxynotus paradoxus*	38	...	...	...	...	...	...	1
Blue skate	*Raja batis*	38	508	441	411	558	866	817	559
Thornback ray	*Raja clavata*	38	1 756	1 579	1 343	1 290	1 224	1 214	1 160
Starry ray	*Raja radiata*	38	1 493	1 431	1 252	996	1 076	1 211	1 781
Spotted ray	*Raja montagui*	38	977	1 163	1 179	1 260	1 341	1 563	1 451
Sandy ray	*Raja circularis*	38	438	438	410	435	369	330	301
Shagreen ray	*Raja fullonica*	38	65	55	50	86	65	105	103
Small-eyed ray	*Raja microocellata*	38	-	-	1	11	-	-	-
Cuckoo ray	*Raja naevus*	38	4 077	4 721	4 015	3 638	3 064	2 885	2 735
Longnosed skate	*Raja oxyrinchus*	38	346	311	327	194	140	89	211
Raja rays nei	*Raja spp*	38	22 543	28 410	30 083	32 514	28 450	26 496	20 606
Stingrays nei	*Dasyatis spp*	38	1	2	5	6	10	7	10
Eagle rays	*Myliobatidae*	38	0	0	0	13	10	11	12
Torpedo rays	*Torpedo spp*	38	16	18	19	34	32	43	34
Rabbit fish	*Chimaera monstrosa*	38	21	15	32	12	15	123	69
Ratfishes nei	*Hydrolagus spp*	38	...	...	...	38	573	835	634
Straightnose rabbitfish	*Rhinochimaera atlantica*	38	-	-	-	-	-	2	1
Chimaeras, etc. nei	*Chimaeriformes*	38	1	0	-	-	-	-	-
Various sharks nei	*Selachimorpha(Pleurotremata)*	38	7 694	36 309	25 294	26 480	33 236	33 219	19 876
Sharks, rays, skates, etc. nei	*Elasmobranchii*	38	147	123	211	123	89	141	387
Groundfishes nei	*Osteichthyes*	39	18 276	8 276	9 967	6 860	9 241	2 351	2 206
Pelagic fishes nei	*Osteichthyes*	39	25 673	8 247	15 747	4 439	2 949	104	4 731

C-27 (a)

Fish, crustaceans, molluscs, etc	Capture production by species items	Atlantic, Northeast
Poissons, crustacés, mollusques, etc	Captures par catégories d'espèces	Atlantique, nord-est
Peces, crustáceos, moluscos, etc	Capturas por categorías de especies	Atlántico, nordeste

English name Nom anglais Nombre inglés	Scientific name Nom scientifique Nombre científico	Species group Groupe d'espèces Grupo de especies	1996 t	1997 t	1998 t	1999 t	2000 t	2001 t	2002 t
Finfishes nei	*Osteichthyes*	39	28 375	36 985	50 830	34 044	55 335	30 260	26 396
Marine fishes nei	*Osteichthyes*	39	18 003	20 869	5 957	3 908	4 702	3 463	3 631
Edible crab	*Cancer pagurus*	42	29 185	38 011	40 635	39 341	42 764	44 143	43 886
Portunus swimcrabs nei	*Portunus spp*	42	1 713	3 212	3 021	2 926	1 973	2 966	2 828
Green crab	*Carcinus maenas*	42	837	897	909	899	873	1 077	1 219
Spinous spider crab	*Maja squinado*	42	5 674	6 134	5 864	5 812	6 355	7 491	6 213
Geryons nei	*Geryon spp*	42	1 477	325	587	1 015	763	382	1 041
Marine crabs nei	*Brachyura*	42	3 147	10 334	2 004	1 044	651	382	597
Pink spiny lobster	*Palinurus mauritanicus*	43	1	0	25	11	9	5	1
Common spiny lobster	*Palinurus elephas*	43	104	90	44	58	59	65	46
Palinurid spiny lobsters nei	*Palinurus spp*	43	104	127	110	98	74	81	64
Norway lobster	*Nephrops norvegicus*	43	52 022	55 167	52 927	56 558	52 704	52 703	53 299
European lobster	*Homarus gammarus*	43	2 278	2 738	2 654	3 071	2 345	2 504	2 602
Lobsters nei	*Reptantia*	43	-	-	-	22	11	6	5
Craylets, squat lobsters	*Galatheidae*	44	105	106	81	130	98	84	84
Red king crab	*Paralithodes camtschaticus*	44	70	71	232	341	378	730	1 718
Penaeus shrimps nei	*Penaeus spp*	45	727	865	1 306	337	100	270	109
Deepwater rose shrimp	*Parapenaeus longirostris*	45	3	...	...	2 084	1 360	1 080	624
Blue and red shrimp	*Aristeus antennatus*	45	...	...	...	194	269	182	97
Northern prawn	*Pandalus borealis*	45	131 996	137 822	136 510	128 660	130 908	106 226	110 696
Pandalus shrimps nei	*Pandalus spp*	45	39	48	1 421	46	460	74	290
Delta prawn	*Palaemon longirostris*	45	...	...	...	10	19	26	18
Common prawn	*Palaemon serratus*	45	543	576	420	525	520	532	417
Palaemonid shrimps nei	*Palaemonidae*	45	400	373	510	589	513	281	215
Common shrimp	*Crangon crangon*	45	32 222	37 772	30 583	37 046	33 461	32 146	32 873
Natantian decapods nei	*Natantia*	45	1 236	13 727	5 945	3 844	5 060	1 926	4 033
Norwegian krill	*Meganyctiphanes norvegica*	46	-	-	88	-	-	36	-
Goose barnacles	*Lepas spp*	47	2	-	-	-	-	-	-
Marine crustaceans nei	*Crustacea*	47	1 619	1 783	1 389	1 276	1 067	1 672	2 585
Common periwinkle	*Littorina littorea*	52	2 814	3 152	2 636	3 018	2 641	2 781	2 287
Periwinkles nei	*Littorina spp*	52	1 770	2 879	1 942	1 344	1 064	765	274
Tuberculate abalone	*Haliotis tuberculata*	52	62	75	36	37	64	62	34
Whelk	*Buccinum undatum*	52	24 723	27 627	13 723	17 794	29 818	30 146	27 643
Gastropods nei	*Gastropoda*	52	...	...	...	142	152	146	119
European flat oyster	*Ostrea edulis*	53	1 269	1 749	1 375	872	609	802	1 372
Pacific cupped oyster	*Crassostrea gigas*	53	129	98	59	9	74	92	15
Cupped oysters nei	*Crassostrea spp*	53	2	5	-	-	-	5	-
Blue mussel	*Mytilus edulis*	54	99 965	118 919	122 217	114 369	126 874	145 558	132 710
Horse mussels nei	*Modiolus spp*	54	20	30	20	7	2	2	12
Sea mussels nei	*Mytilidae*	54	1 077	0	0	0	1 831	1 136	1 003
Queen scallop	*Aequipecten opercularis*	55	7 183	11 443	14 486	15 746	14 891	20 464	16 731
Great Atlantic scallop	*Pecten maximus*	55	24 098	26 660	25 172	17 265	18 186	30 586	32 837
Iceland scallop	*Chlamys islandica*	55	8 976	10 403	10 098	8 858	9 088	6 513	5 208
Scallops nei	*Pectinidae*	55	15 379	23 160	23 543	31 537	32 604	23 798	15 635
Ocean quahog	*Arctica islandica*	56	6 315	4 351	8 776	3 501	1 584	7 434	12 353
Striped venus	*Chamelea gallina*	56	2 441	2 555	3 987	4 184	5 242	5 556	4 452
Pullet carpet shell	*Venerupis pullastra*	56	2 821	4 053	5 166	4 828	2 361	2 420	2 818
Grooved carpet shell	*Ruditapes decussatus*	56	981	1 153	1 771	1 756	691	748	694
Northern quahog(=Hard clam)	*Mercenaria mercenaria*	56	-	0	0	-	175	-	-
Venus clams nei	*Veneridae*	56	1	99	-	189	234	289	355
Solid surf clam	*Spisula solida*	56	...	...	...	765	1 335	1 440	3 030
Donax clams	*Donax spp*	56	...	...	...	490	495	628	427
Razor clams nei	*Solen spp*	56	1 871	577	631	696	463	603	399
Common edible cockle	*Cerastoderma edule*	56	37 035	38 063	86 611	70 252	47 187	23 659	18 750
Clams, etc. nei	*Bivalvia*	56	7 035	6 235	9 809	7 594	6 193	6 470	7 537
Common cuttlefish	*Sepia officinalis*	57	2 210	1 922	3 079	2 817	3 741	2 559	2 832
Cuttlefish,bobtail squids nei	*Sepiidae, Sepiolidae*	57	20 112	15 805	17 221	18 073	22 392	16 751	20 348
Common squids nei	*Loligo spp*	57	5 091	7 063	6 760	4 724	4 199	3 649	2 736
Northern shortfin squid	*Illex illecebrosus*	57	2 600	2 445	810	1 955	1 831	1 624	2 520
Broadtail shortfin squid	*Illex coindetii*	57	395	411	216	304	360	219	450
European flying squid	*Todarodes sagittatus*	57	976	2 721	2 436	2 688	2 053	1 795	1 946
Various squids nei	*Loliginidae, Ommastrephidae*	57	4 348	4 868	6 378	8 506	7 337	6 228	8 624
Common octopus	*Octopus vulgaris*	57	-	-	-	12	624	823	659
Octopuses, etc. nei	*Octopodidae*	57	17 639	15 748	13 048	13 988	16 082	10 826	11 873
Marine molluscs nei	*Mollusca*	58	878	1 023	1 302	663	392	424	196
Echinoderms	*Echinodermata*	76	485	586	551	616	306	319	604
Starfishes nei	*Asteroidea*	76	-	-	-	0	0	0	0
Sea urchins nei	*Strongylocentrotus spp*	76	1	0	-	-	150	151	38
Stony sea urchin	*Paracentrotus lividus*	76	0	1	12	14	6	3	-
European edible sea urchin	*Echinus esculentus*	76	425	25	1	13	1	5	0
Sea cucumbers nei	*Holothurioidea*	76	-	-	-	-	-	-	8
Aquatic invertebrates nei	*Invertebrata*	77	0	-	-	-	-	24	4
Total			***11 066 416***	***11 766 088***	***10 986 318***	***10 516 374***	***11 014 721***	***11 143 204***	***11 048 962***

C-27 (b)

Fish, crustaceans, molluscs, etc — Capture production by countries or areas — Atlantic, Northeast
Poissons, crustacés, mollusques, etc — Captures par pays ou zones — Atlantique, nord-est
Peces, crustáceos, moluscos, etc — Capturas por países o áreas — Atlántico, nordeste

Country or area Pays ou zone País o área	1993 t	1994 t	1995 t	1996 t	1997 t	1998 t	1999 t	2000 t	2001 t	2002 t
Belgium	35 578	33 743	35 088	30 312	29 989	30 324	29 340	29 289	29 698	28 517
Bulgaria	5 272	-	-	-	-	-	-	-	-	-
Channel Is	2 854 F	2 783	2 949 F	4 346	4 238	4 117	3 601	3 589	3 927	3 449
China,Taiwan	478	1 562	1 125	31	30	47	640	1 131	564	624
Denmark	1 613 055	1 872 828	1 998 322	1 680 918	1 826 199	1 556 430	1 404 564	1 533 906	1 510 269	1 441 632
Estonia	49 703	71 650	92 410	94 586	112 979	96 898	90 315	96 266	88 357	81 071
Faeroe Is	231 156	227 684	280 912	294 378	320 694	354 403	347 802	445 856	511 857	480 851 F
Finland	104 772	116 374	119 048	131 459	132 567	134 868	123 747	121 661	115 276	110 026
France	379 989	383 736	384 524	353 922	386 921	363 118	394 383	423 035	429 671	420 802
Germany	241 830	218 948	216 864	201 707	207 239	220 252	188 810	177 901	184 432	193 624
Greenland	21 713	15 465	20 316	21 662	26 500	30 589	42 412	37 826	34 336	48 229
Iceland	1 712 478	1 553 802	1 603 577	2 038 854	2 197 419	1 674 623	1 726 750	1 973 008	1 975 476	2 122 618
Ireland	275 845	290 550	384 542	328 427	289 005	321 044	282 557	275 411	281 833	241 954
Isle of Man	4 850	3 571	3 734	3 537	4 289	2 214	2 609	3 552	3 112	3 127
Japan	4 915	7 082	6 595	3 745	2 638	2 907	2 273	2 315	1 906	2 331
Latvia	64 230	64 755	63 127	73 103	86 123	78 109	78 147	80 329	76 930	79 863
Libya	-	-	-	576	477	511	450	487	...	...
Lithuania	18 697	18 086	45 556	48 904	14 824	15 930	20 967	19 584	31 381	32 006
Netherlands	460 155	417 607	433 985	361 126	348 537	410 604	388 371	336 309	339 040	300 314
Norway	2 380 768	2 338 555	2 505 560	2 633 495	2 852 115	2 848 489	2 622 707	2 693 364	2 672 018	2 728 078
Panama	210	363	289	369	58	58	-	-	-	-
Poland	103 022	116 694	130 214	157 335	187 203	126 787	131 056	143 173	159 164	151 562
Portugal	219 685	202 539	216 513	199 833	187 960	192 861	174 924	159 680	157 673	162 355
Romania	-	8 593	37 508	9 432	-	-	-	-	-	-
Russian Fed	833 147	709 669	736 359	769 774	830 132	727 936	839 509	989 122	1 061 969	1 119 911
Spain	360 270	346 392	395 247	393 622	476 637	470 576	439 111	392 982	434 029	323 128
Svalbard Is	0	0	0	0	0	0	0	0	0	0
Sweden	339 624	384 560	402 638	369 071	355 395	409 327	349 776	337 075	310 582	293 527
Ukraine	7 876	13 385	3 352	550	-	-	-	-	-	-
UK	857 874	874 510	905 737	861 271	885 708	913 296	831 553	737 870	729 704	679 049
Uruguay	-	-	-	-	-	-	-	-	-	314
Other nei	223	68	189	71	212	-	-	-	-	-
Total	***10 330 269***	***10 295 554***	***11 026 280***	***11 066 416***	***11 766 088***	***10 986 318***	***10 516 374***	***11 014 721***	***11 143 204***	***11 048 962***

C-31 (a)

Fish, crustaceans, molluscs, etc — Capture production by species items — Atlantic, Western Central
Poissons, crustacés, mollusques, etc — Captures par catégories d'espèces — Atlantique, centre-ouest
Peces, crustáceos, moluscos, etc — Capturas por categorías de especies — Atlántico, centro-occidental

English name Nom anglais Nombre inglés	Scientific name Nom scientifique Nombre científico	Species group Groupe d'espèces Grupo de especies	1996 t	1997 t	1998 t	1999 t	2000 t	2001 t	2002 t
Freshwater fishes nei	*Osteichthyes*	13	7 303	3 656	2 870	4 130	6 772	6 937	-
American eel	*Anguilla rostrata*	22	35	19	9	2	1	0	11
American shad	*Alosa sapidissima*	24	343	246	280	117	288	171	286
Alewife	*Alosa pseudoharengus*	24	0	0	0	-	0	0	-
Milkfish	*Chanos chanos*	25	0	0	0	0	0	...	...
Striped bass	*Morone saxatilis*	25	9	12	14	4	12	15	17
Lefteye flounders nei	*Bothidae*	31	-	-	-	-	2	-	-
Winter flounder	*Pseudopleuronectes americanus*	31	-	-	0	-	-	-	-
Tonguefishes	*Cynoglossidae*	31	-	-	-	59	39	20	-
Summer flounder	*Paralichthys dentatus*	31	...	...	800	629	719	1 204	639
Bastard halibuts nei	*Paralichthys spp*	31	2 192	1 059	533	391	503	...	708
Flatfishes nei	*Pleuronectiformes*	31	335	487	558	594	502	471	575
White hake	*Urophycis tenuis*	32	...	...	...	1	1	1	1
Gadiformes nei	*Gadiformes*	32	-	4	2	1	5	2	3
Ladyfish	*Elops saurus*	33	143	745	979	15	140	480	675
Tarpon	*Megalops atlanticus*	33	167	43	53	16	18	16	36
Bonefish	*Albula vulpes*	33	151	16	34	125	10	11	6
Sea catfishes nei	*Ariidae*	33	22 150	14 885	16 385	16 442	20 307	18 260	20 998
Squirrelfishes nei	*Holocentridae*	33	80	41	67	36	43	69	192
Flathead grey mullet	*Mugil cephalus*	33	16 721	13 593	10 583	10 023	10 657	9 846	8 411
Lebranche mullet	*Mugil liza*	33	3 287	1 550	2 876	2 855	3 186	2 537	2 294
Bobo mullet	*Joturus pichardi*	33	572	283	323	322	451	400	341
Mullets nei	*Mugilidae*	33	11 315	16 618	15 846	10 715	14 878	13 457	12 151
Common snook	*Centropomus undecimalis*	33	5 132	4 867	5 012	5 119	5 034	5 903	7 388
Snooks(=Robalos) nei	*Centropomus spp*	33	1 444	1 169	1 081	3 170	3 104	3 345	2 192
Black grouper	*Mycteroperca bonaci*	33	...	...	...	...	9	7	6
Gag	*Mycteroperca microlepis*	33	...	...	...	...	15	10	13
Scamp	*Mycteroperca phenax*	33	...	...	1	14	27	17	26
Yellowfin grouper	*Mycteroperca venenosa*	33	...	...	...	...	3	1	3
Nassau grouper	*Epinephelus striatus*	33	466	598	571	438	257	325	322
Yellowedge grouper	*Epinephelus flavolimbatus*	33	...	15	18	...	73	36	67
Red hind	*Epinephelus guttatus*	33	60	42	52	72	67	88	159
Red grouper	*Epinephelus morio*	33	298	287	80	131	204	128	86
Snowy grouper	*Epinephelus niveatus*	33	...	...	...	...	6	4	9
Warsaw grouper	*Epinephelus nigritus*	33	16	23	23	-	45	44	50
Groupers nei	*Epinephelus spp*	33	20 170	19 351	18 855	16 757	21 630	18 708	18 508
Black seabass	*Centropristis striata*	33	295	375	341	305	327	391	333
Groupers, seabasses nei	*Serranidae*	33	361	350	437	815	710	859	888
Bigeyes nei	*Priacanthus spp*	33	...	...	...	...	0	1	1
Southern red snapper	*Lutjanus purpureus*	33	1 335	1 405	1 254	1 006	1 635	1 902	1 895
Northern red snapper	*Lutjanus campechanus*	33	6 346	6 264	5 166	5 412	4 841	4 869	4 617
Lane snapper	*Lutjanus synagris*	33	3 034	3 148	3 212	2 353	2 564	2 604	2 144
Snappers nei	*Lutjanus spp*	33	425	788	827	1 469	1 569	1 779	1 403
Yellowtail snapper	*Ocyurus chrysurus*	33	3 165	2 436	2 828	2 426	2 305	2 527	2 410
Vermilion snapper	*Rhomboplites aurorubens*	33	...	678	480	731	911	1 057	980
Snappers, jobfishes nei	*Lutjanidae*	33	12 317	13 457	15 444	10 310	14 795	10 176	9 240
Grunts, sweetlips nei	*Haemulidae (=Pomadasyidae)*	33	13 881	18 081	15 565	11 338	9 780	14 992	8 372
Spotted weakfish	*Cynoscion nebulosus*	33	3 610	3 797	6 786	6 742	6 367	4 915	6 088
Squeteague(=Gray weakfish)	*Cynoscion regalis*	33	532	620	609	410	109	154	108
Weakfishes nei	*Cynoscion spp*	33	14 011	10 577	13 551	6 394	12 842	10 884	11 061
Whitemouth croaker	*Micropogonias furnieri*	33	6 065	3 694	4 871	1 900	3 262	6 848	2 892
Atlantic croaker	*Micropogonias undulatus*	33	799	1 396	1 084	729	506	608	302
Gulf kingcroaker	*Menticirrhus littoralis*	33	1 070	1 443	2 061	1 822	961	690	786
Black drum	*Pogonias cromis*	33	134	1 520	2 318	810	2 467	2 632	2 390
Spot croaker	*Leiostomus xanthurus*	33	816	1 098	1 005	884	982	1 214	703
Croakers, drums nei	*Sciaenidae*	33	1	2	183	997	1 128	800	114
Porgies	*Calamus spp*	33	1 052	1 541	553	1 266	587	534	449
Sheepshead	*Archosargus probatocephalus*	33	1 519	1 712	1 346	1 096	1 501	1 195	1 091
Western Atlantic seabream	*Archosargus rhomboidalis*	33	...	...	...	...	...	...	53
Porgies, seabreams nei	*Sparidae*	33	228	451	680	140	218	263	257
Goatfishes, red mullets nei	*Mullidae*	33	276	180	170	108	188	212	56
Mojarras(=Silver-biddies) nei	*Gerres spp*	33	-	-	-	-	159	180	211
Mojarras, etc. nei	*Gerreidae*	33	9 756	5 911	5 487	4 105	3 816	3 937	2 992
Sea chubs nei	*Kyphosus spp*	33	...	...	...	...	...	8	4
Wrasses, hogfishes, etc. nei	*Labridae*	33	529	1 456	52	82	116	68	633
Parrotfishes nei	*Scaridae*	33	99	100	118	109	101	306	482
Angelfishes nei	*Pomacanthidae*	33	...	...	...	...	...	...	28
Surgeonfishes nei	*Acanthuridae*	33	11	4	9	7	2	160	243
Spadefishes nei	*Ephippidae*	33	-	-	-	-	15	1	6
Boxfishes nei	*Ostraciidae*	33	...	1	1	55	68	118	80
Filefishes, leatherjackets nei	*Monacanthidae*	33	...	...	...	...	...	...	15
Triggerfishes, durgons nei	*Balistidae*	33	810	551	717	559	318	416	497
Bearded brotula	*Brotula barbata*	34	...	...	5	1	1	0	0
Cusk-eels, brotulas nei	*Ophidiidae*	34	118	97	37	13	17	17	11
Alfonsinos nei	*Beryx spp*	34	22	-	-	-	-	-	-
Wreckfish	*Polyprion americanus*	34	82	14	6	1	-	-	-
Atlantic goldeneye tilefish	*Caulolatilus chrysops*	34	-	-	-	0	4	16	10
Great Northern tilefish	*Lopholatilus chamaeleonticeps*	34	114	397	80	88	173	168	197
Tilefishes nei	*Branchiostegidae*	34	...	28	347	386	528	342	286
Tripletail	*Lobotes surinamensis*	34	-	-	-	-	1	1	1
Escolar	*Lepidocybium flavobrunneum*	34	...	...	51	33	70	40	57
Oilfish	*Ruvettus pretiosus*	34	...	...	10	11	38	30	25
Largehead hairtail	*Trichiurus lepturus*	34	4 615	5 060	5 413	4 019	3 752	8 799	6 498
Black scabbardfish	*Aphanopus carbo*	34	17	-	-	-	-	-	-

C-31 (a)

Fish, crustaceans, molluscs, etc / **Poissons, crustacés, mollusques, etc** / **Peces, crustáceos, moluscos, etc**

Capture production by species items / **Captures par catégories d'espèces** / **Capturas por categorías de especies**

Atlantic, Western Central / **Atlantique, centre-ouest** / **Atlántico, centro-occidental**

English name Nom anglais Nombre inglés	Scientific name Nom scientifique Nombre científico	Species group Groupe d'espèces Grupo de especies	1996 t	1997 t	1998 t	1999 t	2000 t	2001 t	2002 t
Hairtails, scabbardfishes nei	*Trichiuridae*	34	...	...	6 349	6 747	7 278	5 091	3 872
Black driftfish	*Hyperoglyphe bythites*	34	...	...	...	2	7	5	6
American angler	*Lophius americanus*	34	-	25	12	19	9	12	15
Demersal percomorphs nei	*Perciformes*	34	7 448	8 787	7 176	6 185	3 325	7 014	8 154
Round sardinella	*Sardinella aurita*	35	155 426	143 116	190 895	128 048	77 082	73 739	159 110
Brazilian sardinella	*Sardinella brasiliensis*	35	1	0	0	0	0	0	3
Atlantic menhaden	*Brevoortia tyrannus*	35	24 169	...	27 779	18 815	23 812	25 155	31 068
Gulf menhaden	*Brevoortia patronus*	35	491 612	597 565	497 461	694 242	591 434	528 506	582 497
Scaled sardines	*Harengula spp*	35	819	1 011	706	784	1 589	1 791	506
Atlantic thread herring	*Opisthonema oglinum*	35	5 634	15 191	14 592	17 066	14 574	6 688	4 512
Atlantic anchoveta	*Cetengraulis edentulus*	35	8	0	119	0	0	2	0
Broad-striped anchovy	*Anchoa hepsetus*	35	...	1	16	8	0	0	0
Anchovies, etc. nei	*Engraulidae*	35	1 464	903	1 762	1 613	1 163	1 328	1 075
Clupeoids nei	*Clupeoidei*	35	527	1 075	1 666	4 712	4 140	4 345	4 312
Atlantic bonito	*Sarda sarda*	36	3 539	4 205	4 511	4 673	3 835	3 927	5 346
Wahoo	*Acanthocybium solandri*	36	1 251	1 625	1 391	1 438	1 051	1 373	1 105
King mackerel	*Scomberomorus cavalla*	36	9 640	12 768	8 460	9 790	8 966	10 086	12 131
Atlantic Spanish mackerel	*Scomberomorus maculatus*	36	11 183	8 720	8 625	9 050	6 999	6 873	7 361
Cero	*Scomberomorus regalis*	36	307	481	441	125	190	147	...
Serra Spanish mackerel	*Scomberomorus brasiliensis*	36	5 388	5 940	6 407	4 158	5 655	6 062	5 250
Seerfishes nei	*Scomberomorus spp*	36	1 359	618	382	671	865	625	548
Frigate and bullet tunas	*Auxis thazard, A.rochei*	36	2 957	2 894	2 053	1 662	1 655	1 343	1 068
Little tunny(=Atl.black skipj)	*Euthynnus alletteratus*	36	1 879	2 341	3 011	2 507	2 154	2 204	2 328
Skipjack tuna	*Katsuwonus pelamis*	36	4 853	5 345	5 839	4 114	4 143	7 772	3 680
Atlantic bluefin tuna	*Thunnus thynnus*	36	75	27	490	456	395	117	87
Blackfin tuna	*Thunnus atlanticus*	36	2 895	1 838	2 753	2 842	2 265	3 679	2 937
Albacore	*Thunnus alalunga*	36	3 746	3 285	1 522	5 546	9 632	15 579	10 282
Yellowfin tuna	*Thunnus albacares*	36	24 841	26 224	27 629	24 350	24 570	29 203	18 314
Bigeye tuna	*Thunnus obesus*	36	1 091	1 325	1 756	2 941	4 187	4 173	2 865
Atlantic sailfish	*Istiophorus albicans*	36	512	576	623	664	651	575	692
Blue marlin	*Makaira nigricans*	36	379	534	579	640	542	316	399
Atlantic white marlin	*Tetrapturus albidus*	36	254	171	163	160	116	126	130
Longbill spearfish	*Tetrapturus pfluegeri*	36	1	0	1	0	3	7	8
Marlins,sailfishes,etc. nei	*Istiophoridae*	36	112	98	197	105	490	606	180
Swordfish	*Xiphias gladius*	36	1 854	2 752	3 041	3 365	3 756	2 774	3 071
Tuna-like fishes nei	*Scombroidei*	36	10 297	6 637	13 862	5 812	5 672	6 151	6 287
Needlefishes, etc. nei	*Belonidae*	37	66	33	67	65	75	82	137
Ballyhoo halfbeak	*Hemiramphus brasiliensis*	37	...	...	...	32	47	41	36
Halfbeaks nei	*Hemiramphus spp*	37	399	295	443	92	415	355	394
Flyingfishes nei	*Exocoetidae*	37	2 148	1 623	2 835	2 165	2 051	2 054	1 830
Opah	*Lampris guttatus*	37	...	...	-	-	-	1	0
Bluefish	*Pomatomus saltatrix*	37	758	1 147	899	756	1 112	1 326	965
Cobia	*Rachycentron canadum*	37	392	757	717	630	399	382	267
Scads nei	*Decapterus spp*	37	82	61	101	59	38	52	75
Blue runner	*Caranx crysos*	37	313	408	408	184	204	225	300
Crevalle jack	*Caranx hippos*	37	136	264	393	91	316	304	217
Bar jack	*Caranx ruber*	37	-	-	-	3	5	7	13
Jacks, crevalles nei	*Caranx spp*	37	12 031	12 833	12 664	9 517	12 306	10 760	9 438
Atlantic moonfish	*Selene setapinnis*	37	1 766	2 110	2 338	1 529	976	3 443	1 345
Florida pompano	*Trachinotus carolinus*	37	-	296	319	68	242	181	158
Pompanos nei	*Trachinotus spp*	37	1 099	733	1 038	693	552	457	890
Greater amberjack	*Seriola dumerili*	37	...	...	560	149	538	452	448
Amberjacks nei	*Seriola spp*	37	2 620	2 820	2 203	2 261	2 801	2 205	2 080
Rainbow runner	*Elagatis bipinnulata*	37	14	20	20	36	23	15	20
Bigeye scad	*Selar crumenophthalmus*	37	2 453	2 622	2 726	3 873	1 873	1 896	2 375
Carangids nei	*Carangidae*	37	436	676	586	561	380	769	584
Common dolphinfish	*Coryphaena hippurus*	37	3 554	4 629	3 476	3 960	3 737	3 205	3 675
Chub mackerel	*Scomber japonicus*	37	418	556	771	641	409	527	664
Mackerels nei	*Scombridae*	37	1 550	1 700	1 430	1 430	1 600	1 600	1 600
North Atlantic harvestfish	*Peprilus alepidotus*	37	14	15	13	16	23	18	22
Gulf butterfishes, etc. nei	*Peprilus spp*	37	1 889	568	1 175	1 004	809	742	872
Butterfishes, pomfrets nei	*Stromateidae*	37	782	65	633	644	681	539	613
Barracudas nei	*Sphyraena spp*	37	1 596	2 130	2 072	1 923	1 312	1 997	2 000
Pelagic percomorphs nei	*Perciformes*	37	857	539	464	219	182	109	74
Bigeye thresher	*Alopias superciliosus*	38	-	1	...	...	...	...	...
Shortfin mako	*Isurus oxyrinchus*	38	-	73	33	...	5	5	134
Longfin mako	*Isurus paucus*	38	-	1	1	-	3	3	-
Nurse shark	*Ginglymostoma cirratum*	38	-	-	-	-	407	89	17
Blue shark	*Prionace glauca*	38	-	1 700	435	...	...	...	8
Blacktip shark	*Carcharhinus limbatus*	38	3	9	10	11	580	520	...
Silky shark	*Carcharhinus falciformis*	38	5	0	24	49	63	59	56
Requiem sharks nei	*Carcharhinidae*	38	12 145	9 785	8 588	6 294	6 491	5 907	8 708
Hammerhead sharks, etc. nei	*Sphyrnidae*	38	-	3	2	...	...	...	...
Smooth-hounds nei	*Mustelus spp*	38	253	27	45	3	30	30	30
Dogfish sharks nei	*Squalidae*	38	138	310	334	222	104	8	8
Rays, stingrays, mantas nei	*Rajiformes*	38	7 591	8 851	9 884	8 021	6 427	6 050	5 656
Sharks, rays, skates, etc. nei	*Elasmobranchii*	38	11 494	12 348	11 404	12 469	9 569	9 393	10 219
Marine fishes nei	*Osteichthyes*	39	222 149	252 532	213 099	175 905	164 058	152 559	162 579
Black stone crab	*Menippe mercenaria*	42	3 675	3 354	3 347	2 363	3 197	3 332	3 422
Portunus swimcrabs nei	*Portunus spp*	42	1	289	224	4 042	4 995	4 001	2
Blue crab	*Callinectes sapidus*	42	48 215	58 859	59 162	44 845	50 251	40 244	46 082
Harbour spidercrab	*Mithrax armatus*	42	-	-	114	110	54	35	22
Red crab	*Geryon quinquedens*	42	-	-	-	-	2	-	-

C-31 (a)

Fish, crustaceans, molluscs, etc — Capture production by species items — Atlantic, Western Central
Poissons, crustacés, mollusques, etc — Captures par catégories d'espèces — Atlantique, centre-ouest
Peces, crustáceos, moluscos, etc — Capturas por categorías de especies — Atlántico, centro-occidental

English name Nom anglais Nombre inglés	Scientific name Nom scientifique Nombre científico	Species group Groupe d'espèces Grupo de especies	1996 t	1997 t	1998 t	1999 t	2000 t	2001 t	2002 t
Marine crabs nei	*Brachyura*	42	5 113	6 735	5 274	1 224	2 204	1 629	9 306
Caribbean spiny lobster	*Panulirus argus*	43	30 412	29 219	28 113	31 704	31 244	24 700	29 117
Tropical spiny lobsters nei	*Panulirus spp*	43	-	-	5	5	5	4	4
Northern brown shrimp	*Penaeus aztecus*	45	54 703	44 459	50 722	60 527	62 713	68 411	57 335
Northern pink shrimp	*Penaeus duorarum*	45	4 757	10 758	12 086	2 442	7 161	8 626	9 353
Northern white shrimp	*Penaeus setiferus*	45	27 461	31 928	39 799	44 014	52 280	40 696	43 015
Penaeus shrimps nei	*Penaeus spp*	45	57 279	59 278	57 598	46 442	49 366	54 132	48 080
Atlantic seabob	*Xiphopenaeus kroyeri*	45	17 310	21 464	16 603	18 648	24 782	32 351	25 761
Whitebelly prawn	*Nematopalaemon schmitti*	45	-	-	-	-	1 464	1 382	1 400
Rock shrimp	*Sicyonia brevirostris*	45	1 079	1 781	4 409	1 345	3 254	2 909	890
Royal red shrimp	*Pleoticus robustus*	45	184	205	182	208	369	281	418
Natantian decapods nei	*Natantia*	45	289	150	150	217	1 249	218	195
Marine crustaceans nei	*Crustacea*	47	366	351	364	168	181	186	163
Stromboid conchs nei	*Strombus spp*	52	37 820	32 510	30 886	29 244	23 453	24 742	20 479
Whelks	*Busycon spp*	52	1 193	1 017	908	832	600	480	153
Gastropods nei	*Gastropoda*	52	244	184	139	121	100	100	100
Mangrove cupped oyster	*Crassostrea rhizophorae*	53	4 589	3 722	4 896	3 906	3 228	5 244	2 235
American cupped oyster	*Crassostrea virginica*	53	108 971	95 608	101 957	85 150	195 113	156 485	145 951
South American rock mussel	*Perna perna*	54	223	295	3 802	451	316	1 081	1 272
American sea scallop	*Placopecten magellanicus*	55	459	154	150	27	115	236	264
Atlantic bay scallop	*Argopecten irradians*	55	225	427	684	196	141	17	127
Ark clams nei	*Arca spp*	56	33 785	40 857	30 419	40 689	47 300	44 308	45 100
Northern quahog(=Hard clam)	*Mercenaria mercenaria*	56	...	...	...	...	5 506	5 113	4 441
Venus clams nei	*Veneridae*	56	1 350	1 279	1 296	1 222	1 756	2 222	908
Clams, etc. nei	*Bivalvia*	56	4 455	6 739	6 392	3 972	1 154	2 796	2 702
Cuttlefish,bobtail squids nei	*Sepiidae, Sepiolidae*	57	-	-	-	32	-	-	-
Common squids nei	*Loligo spp*	57	921	883	942	205	604	1 404	718
Northern shortfin squid	*Illex illecebrosus*	57	9	2	2	-	1	0	0
Various squids nei	*Loliginidae, Ommastrephidae*	57	33	139	119	45	65	88	64
Common octopus	*Octopus vulgaris*	57	28 608	17 809	16 565	19 120	22 562	20 653	16 091
Octopuses, etc. nei	*Octopodidae*	57	1 548	1 917	6 507	711	1 152	1 087	1 241
Marine molluscs nei	*Mollusca*	58	673	1 873	2 921	6 706	2 716	1 913	2 971
Green turtle	*Chelonia mydas*	72	41	24	20	13	15	12	14
Hawksbill turtle	*Eretmochelys imbricata*	72	23	19	12	12	18	11	17
Loggerhead turtle	*Caretta caretta*	72	10	7	5	5	9	7	8
Marine turtles nei	*Testudinata*	72	153	36	99	1	2	4	3
Horseshoe crab	*Limulus polyphemus*	75	-	1	70	159	139	228	130
Echinoderms	*Echinodermata*	76	0	0	0	0	0	0	-
Sea urchins nei	*Strongylocentrotus spp*	76	10	15	15	15	10	10	10
Sea cucumbers nei	*Holothurioidea*	76	-	-	-	-	-	39	-
Jellyfishes	*Rhopilema spp*	77	...	...	...	...	137	-	-
Aquatic invertebrates nei	*Invertebrata*	77	6	2	4	0	2	-	-
Total			***1 720 699***	***1 821 208***	***1 782 450***	***1 791 133***	***1 802 775***	***1 686 404***	***1 764 352***

C-31 (b)

Fish, crustaceans, molluscs, etc — Capture production by countries or areas — Atlantic, Western Central
Poissons, crustacés, mollusques, etc — Captures par pays ou zones — Atlantique, centre-ouest
Peces, crustáceos, moluscos, etc — Capturas por países o áreas — Atlántico, centro-occidental

Country or area Pays ou zone País o área	1993 t	1994 t	1995 t	1996 t	1997 t	1998 t	1999 t	2000 t	2001 t	2002 t
Anguilla	330	333	150 F	200 F	250 F	250 F	250 F	250 F	250 F	250 F
Antigua Barb	1 097	1 145	1 610	1 463	1 665	1 708	1 660	1 754	1 824	2 374
Aruba	260	260	140	160	205	182	175	163	163	163 F
Bahamas	10 073	10 311	9 915	10 197	10 439	10 124	10 473	11 070	9 290	9 300 F
Barbados	3 214	2 818	3 581	3 512	2 809	3 644	3 250	3 100	2 676	2 500
Belize	1 784	2 102	1 877	2 048	2 835	2 557	1 871	2 434	2 517	2 078
Bermuda	404	394	449	465	461	466	453	286	315	393
Br Virgin Is	343	470	532	506	105	116	115	43	50 F	50 F
Cayman Is	125	125	125	110	125	125	125	125	125	125
China,Taiwan	6 979 F	3 627 F	3 558	4 516	3 669	2 430	5 663	6 039	10 597	10 577
Colombia	9 578	9 354	15 539	18 888 F	11 235 F	16 825 F	8 040 F	15 196	15 000 F	15 000 F
Costa Rica	199	269	422	437	420	402	666	1 088	816	635
Cuba	41 608	52 790	50 500	47 921	51 604	47 889	53 317	57 837	42 723	31 804
Dominica	794	882	950	1 030	1 079	1 212	1 200 F	1 200 F	1 200 F	1 217
Dominican Rp	10 820	19 058	15 768	12 606	13 468	9 014	7 804	10 842	12 059	15 976
Fr Guiana	6 931	7 819	8 089	7 377 F	6 602	6 709	6 271 F	5 237 F	5 194 F	5 568 F
Grenada	2 103	1 599	1 499	1 577	1 530	1 837	1 658	1 701	2 250	2 171
Guadeloupe	8 600	8 800	9 500	9 570	10 480	9 084	9 114	10 100	10 000 F	10 100 F
Guatemala	92	179	390	390	285	328	292	366	350 F	350 F
Guyana	43 323	45 567	47 200 F	47 783	53 373	52 215	53 241	48 087	52 605	47 217
Haiti	4 550 F	5 000 F	5 017 F	4 745 F	4 801 F	4 759 F	4 500 F	4 500 F	4 500 F	4 500 F
Honduras	4 274	10 692	10 266	10 813	9 700 F	10 500 F	9 200 F	6 200 F	6 000 F	6 300 F
Iceland	-	-	-	24	-	-	-	-	-	-
Jamaica	22 550 F	24 827 F	23 782 F	30 579 F	19 585	17 160	16 937	5 226	5 250 F	5 200 F
Japan	1 532	1 293	969	1 454	1 262	1 921	3 581	3 637	3 996	1 582
Korea Rep	1 462	750	653	626	143	621	1 789	3 327	237	45
Martinique	5 850	5 800 F	5 300	3 500 F	5 500	5 500	6 000	6 310	6 200	6 200 F
Mexico	320 824	329 521	272 178	294 231	320 829	302 157	285 833	274 532	259 157	245 045
Montserrat	58	62	48	38	45	46	50 F	50 F	50 F	46
NethAntilles	1 200 F	1 100 F	1 020 F	1 000 F	950 F	950 F	950 F	950 F	850 F	850 F
Nicaragua	5 001	6 738	5 957	9 685	9 451	12 011	13 127	14 838	13 444	13 305
Panama	-	-	-	-	-	-	3	53	-	-
Philippines	-	-	-	-	-	28	545	376	37	691
Portugal	-	-	-	-	-	18	22	-	-	49
Puerto Rico	1 877	2 275	3 173	2 701	3 187	3 006	3 020	4 154	3 794	2 529
Russian Fed	-	-	360	20	-	-	-	-	-	-
St Kitts Nev	250 F	212	192	352	216	407	471	469	555	355
St Lucia	1 336	1 252	1 188	1 274	1 308	1 589	1 718	1 855	1 983	1 637
St Vincent	1 479	1 090	944	889 F	948	1 365	1 032	7 325	9 020	2 117
Spain	3 027	1 624	1 955	906	3 145	2 090	1 998	2 227	1 888	2 535
Suriname	9 313	14 327	12 860 F	12 850 F	13 800 F	15 995	16 000	17 300 F	18 715	18 500 F
Trinidad Tob	8 998	14 046	11 500 F	9 435	11 283	9 175	8 826	8 651	10 707	12 539
Turks Caicos	6 255	5 963 F	7 824	7 833	5 951	6 383	5 401	5 975	6 611	5 953
USA	942 154	1 238 611	856 754	771 970	867 630	822 594	943 641	993 827	878 443	922 633
US Virgin Is	650 F	550 F	470 F	400 F	350 F	300 F	263	300 F	300 F	300 F
Venezuela	316 550	354 366	392 892	378 624	362 355	388 419	292 750	258 236	281 731	353 418
Other nei	2 813	4 434	4 232	5 994	6 130	8 339	7 838	5 539	2 932	175
Total	***1 810 660***	***2 192 435***	***1 791 328***	***1 720 699***	***1 821 208***	***1 782 450***	***1 791 133***	***1 802 775***	***1 686 404***	***1 764 352***

C-34 (a)

Fish, crustaceans, molluscs, etc — **Capture production by species items** — **Atlantic, Eastern Central**
Poissons, crustacés, mollusques, etc — **Captures par catégories d'espèces** — **Atlantique, centre-est**
Peces, crustáceos, moluscos, etc — **Capturas por categorías de especies** — **Atlántico, centro-oriental**

English name Nom anglais Nombre inglés	Scientific name Nom scientifique Nombre científico	Species group Groupe d'espèces Grupo de especies	1996 t	1997 t	1998 t	1999 t	2000 t	2001 t	2002 t
Tilapias nei	*Oreochromis (=Tilapia) spp*	12	2 603	2 582	1 365	1 760	2 062	2 324	2 333
European eel	*Anguilla anguilla*	22	0	1	4	-	-	-	-
Shads nei	*Alosa spp*	24	-	13	40	-	22	62	7
West African ilisha	*Ilisha africana*	24	11 585	8 797	8 073	4 608	7 467	5 571	8 812
Lefteye flounders nei	*Bothidae*	31	24	2	0	0	17	12	11
Common sole	*Solea solea*	31	1 929	2 939	7 660	3 431	6 510	9 171	4 323
Wedge sole	*Dicologlossa cuneata*	31	87	8	0	9	10	9	12
Soles nei	*Soleidae*	31	12 606	13 552	16 050	9 132	5 127	4 242	3 427
Tonguefishes	*Cynoglossidae*	31	5 499	5 849	5 489	7 900	12 728	13 004	11 956
Megrim	*Lepidorhombus whiffiagonis*	31	-	-	36	-	38	52	10
Spottail spiny turbot	*Psettodes belcheri*	31	0	2	0	-	1	2	2
Flatfishes nei	*Pleuronectiformes*	31	15 178	16 019	16 999	21 132	26 109	23 418	25 169
Greater forkbeard	*Phycis blennoides*	32	466	115	231	369	269	353	251
Pouting(=Bib)	*Trisopterus luscus*	32	2 100	1 073	1 243	573	1 216	1 013	1 171
Blue whiting(=Poutassou)	*Micromesistius poutassou*	32	311	68	48	-	-	-	-
European hake	*Merluccius merluccius*	32	9 678	6 155	3 139	3 354	2 344	4 360	12 553
Senegalese hake	*Merluccius senegalensis*	32	16 485	16 430	17 396	18 187	13 515	17 925	7 516
Benguela hake	*Merluccius polli*	32	39	0	38	147	44	139	2 221
Hakes nei	*Merluccius spp*	32	1 705	1 093	2 064	1 679	2 011	1 531	1 607
Gadiformes nei	*Gadiformes*	32	663	716	991	903	146	114	143
West African ladyfish	*Elops lacerta*	33	379	735	1 546	1 822	755	558	1 118
Bonefish	*Albula vulpes*	33	78	1 542	1 640	245	239	515	204
Sea catfishes nei	*Ariidae*	33	25 946	31 610	26 265	29 122	32 443	36 811	43 085
Morays	*Muraenidae*	33	710	268	165	142	194	212	231
Flathead grey mullet	*Mugil cephalus*	33	-	1	3	-	-	-	-
Mullets nei	*Mugilidae*	33	11 526	18 071	17 703	23 243	23 010	21 125	22 804
Dusky grouper	*Epinephelus marginatus*	33	5	5	4	10	3	4	1
White grouper	*Epinephelus aeneus*	33	677	599	557	627	635	648	725
Dungat grouper	*Epinephelus goreensis*	33	196	63	76	290	277	62	64
Groupers nei	*Epinephelus spp*	33	1 317	1 017	2 225	943	690	697	822
Groupers, seabasses nei	*Serranidae*	33	5 778	6 520	5 748	6 248	5 795	6 805	5 448
Spotted seabass	*Dicentrarchus punctatus*	33	779	602	717	270	112	125	193
European seabass	*Dicentrarchus labrax*	33	108	19	7	26	9	20	21
Seabasses nei	*Dicentrarchus spp*	33	-	-	-	-	-	11	-
Bigeyes nei	*Priacanthus spp*	33	20	20	25	29	12	1	2
Cardinalfishes, etc. nei	*Apogonidae*	33	11	0	-	-	0	-	-
Snappers nei	*Lutjanus spp*	33	5 813	8 314	11 650	12 869	10 349	9 981	10 715
Rubberlip grunt	*Plectorhinchus mediterraneus*	33	1 865	899	455	639	220	257	596
Bastard grunt	*Pomadasys incisus*	33	1	5	-	-	0	0	-
Sompat grunt	*Pomadasys jubelini*	33	1 439	1 535	1 253	1 038	1 017	1 418	1 435
Bigeye grunt	*Brachydeuterus auritus*	33	19 887	29 046	18 591	22 218	21 748	19 986	17 077
Grunts, sweetlips nei	*Haemulidae (=Pomadasyidae)*	33	23 889	19 137	18 742	21 653	18 348	20 497	16 628
Brown meagre	*Sciaena umbra*	33	208	...	...	...	...	...	...
Shi drum	*Umbrina cirrosa*	33	86	52	43	83	93	57	54
Canary drum (=Baardman)	*Umbrina canariensis*	33	10	10	10	6	2	...	...
Meagre	*Argyrosomus regius*	33	1 965	2 702	2 216	3 436	2 728	2 531	3 643
Boe drum	*Pteroscion peli*	33	1 879	1 291	1 190	1 260	1 126	1 159	1 128
Law croaker	*Pseudotolithus brachygnathus*	33	715	618	354	318	811	1 083	855
Cassava croaker	*Pseudotolithus senegalensis*	33	2 727	2 781	4 659	4 837	3 949	4 800	3 905
Bobo croaker	*Pseudotolithus elongatus*	33	12 662	11 497	9 461	6 855	13 405	10 932	12 066
West African croakers nei	*Pseudotolithus spp*	33	25 305	22 156	25 794	23 733	23 515	24 518	21 800
Croakers, drums nei	*Sciaenidae*	33	20 755	32 395	25 995	30 306	30 027	29 890	31 322
Atlantic emperor	*Lethrinus atlanticus*	33	231	334	380	...	...	...	...
Blackspot(=red) seabream	*Pagellus bogaraveo*	33	179	108	141	145	85	12	13
Common pandora	*Pagellus erythrinus*	33	20	40	141	205	0	-	-
Axillary seabream	*Pagellus acarne*	33	18	11	127	1	1	1	1
Red pandora	*Pagellus bellottii*	33	12 001	13 133	13 997	17 865	7 708	7 185	6 977
Pandoras nei	*Pagellus spp*	33	3 741	1 215	2 301	3 711	2 208	1 849	1 781
Sargo breams nei	*Diplodus spp*	33	1 114	1 188	1 416	1 532	2 009	2 184	2 251
Large-eye dentex	*Dentex macrophthalmus*	33	1 101	572	2 307	3 121	1 954	1 086	1 361
Canary dentex	*Dentex canariensis*	33	-	-	-	-	20	-	-
Common dentex	*Dentex dentex*	33	995	239	389	255	134	336	474
Angolan dentex	*Dentex angolensis*	33	489	490	1 416	1 767	504	838	990
Congo dentex	*Dentex congoensis*	33	102	119	392	1 272	350	571	445
Dentex nei	*Dentex spp*	33	4 931	4 792	5 152	4 807	3 444	3 405	4 063
Black seabream	*Spondyliosoma cantharus*	33	1 775	1 780	1 079	1 129	1 370	1 421	1 853
Saddled seabream	*Oblada melanura*	33	20	0	1	-	0	0	0
Red porgy	*Pagrus pagrus*	33	974	639	887	641	607	846	772
Pargo breams nei	*Pagrus spp*	33	4 585	5 204	4 624	3 092	2 513	4 997	4 112
Gilthead seabream	*Sparus aurata*	33	4	16	28	735	599	1 288	426
Bogue	*Boops boops*	33	589	616	472	460	2 197	977	1 300
Sand steenbras	*Lithognathus mormyrus*	33	-	-	223	45	8	77	32
Porgies, seabreams nei	*Sparidae*	33	21 447	23 466	21 010	28 448	24 069	23 804	21 524
Picarels nei	*Spicara spp*	33	-	17	8	-	-	-	16
Surmullets(=Red mullets) nei	*Mullus spp*	33	574	605	582	602	1 049	1 240	1 149
West African goatfish	*Pseudupeneus prayensis*	33	7 126	3 323	2 560	2 426	2 391	3 682	4 450
African sicklefish	*Drepane africana*	33	2 774	3 790	2 268	1 545	2 856	2 790	2 650
Wrasses, hogfishes, etc. nei	*Labridae*	33	-	-	-	-	1	1	1
Parrotfish	*Sparisoma cretense*	33	-	-	-	-	2	1	0
Parrotfishes nei	*Scaridae*	33	...	...	...	...	13	1	17
Giant African threadfin	*Polydactylus quadrifilis*	33	7 359	10 705	10 023	10 770	12 237	12 447	15 493
Lesser African threadfin	*Galeoides decadactylus*	33	17 768	18 385	12 890	10 764	13 948	16 943	13 683
Royal threadfin	*Pentanemus quinquarius*	33	3 005	3 356	4 430	4 162	3 612	3 947	1 439
Threadfins, tasselfishes nei	*Polynemidae*	33	3 507	3 759	691	468	350	424	563
Spadefishes nei	*Ephippidae*	33	3	6	5	4	4	2	1

C-34 (a)

Fish, crustaceans, molluscs, etc	Capture production by species items	Atlantic, Eastern Central
Poissons, crustacés, mollusques, etc	Captures par catégories d'espèces	Atlantique, centre-est
Peces, crustáceos, moluscos, etc	Capturas por categorías de especies	Atlántico, centro-oriental

English name Nom anglais Nombre inglés	Scientific name Nom scientifique Nombre científico	Species group Groupe d'espèces Grupo de especies	1996 t	1997 t	1998 t	1999 t	2000 t	2001 t	2002 t
Puffers nei	*Tetraodontidae*	33	28	100	...	...	4	35	86
Triggerfishes, durgons nei	*Balistidae*	33	93	58	255	72	95	97	89
European conger	*Conger conger*	34	2 669	2 031	2 155	1 641	1 213	1 166	1 567
Conger eels, etc. nei	*Congridae*	34	575	846	1 137	892	927	882	963
Longspine snipefish	*Macroramphosus scolopax*	34	-	89	443	-	-	44	-
Bearded brotula	*Brotula barbata*	34	811	1 842	2 208	3 861	1 719	1 926	1 390
Alfonsinos nei	*Beryx spp*	34	150	105	79	63	84	249	461
Slimeheads nei	*Trachichthyidae*	34	-	-	-	-	-	235	-
John dory	*Zeus faber*	34	1 706	779	781	813	1 294	708	892
Silvery John dory	*Zenopsis conchifer*	34	9	6	6	3	-	0	-
Wreckfish	*Polyprion americanus*	34	233	75	74	53	4	13	11
Tilefishes nei	*Branchiostegidae*	34	20	30	40	45	44	15	16
Bonnetmouths, rubyfishes nei	*Emmelichthyidae*	34	-	-	-	-	-	-	8
Oilfish	*Ruvettus pretiosus*	34	-	-	-	-	5	2	8
Roudi escolar	*Promethichthys prometheus*	34	-	-	-	-	-	6	7
Largehead hairtail	*Trichiurus lepturus*	34	61 362	47 293	44 985	22 235	16 678	18 839	19 986
Silver scabbardfish	*Lepidopus caudatus*	34	9 575	4 731	3 357	3 030	0	4	-
Black scabbardfish	*Aphanopus carbo*	34	6 748	4 023	4 440	4 405	4 203	4 008	3 873
Hairtails, scabbardfishes nei	*Trichiuridae*	34	5 305	7 402	8 521	6 364	1 658	2 185	1 430
Ruffs, barrelfishes nei	*Centrolophidae*	34	-	-	-	-	-	-	51
Scorpionfishes nei	*Scorpaenidae*	34	1 902	2 010	2 645	1 716	4 226	3 486	1 740
Gurnards, searobins nei	*Triglidae*	34	2 329	2 306	3 586	3 314	4 937	4 420	3 693
Blackbellied angler	*Lophius budegassa*	34	1 021	735	1 156	664	383	525	880
Shortspine African angler	*Lophius vaillanti*	34	...	...	...	...	7	5	5
Demersal percomorphs nei	*Perciformes*	34	310	1 840	1 405	1 379	1 344	1 337	1 370
Round sardinella	*Sardinella aurita*	35	399 024	318 076	317 627	282 951	285 048	362 833	285 681
Madeiran sardinella	*Sardinella maderensis*	35	125 701	139 827	131 970	129 066	140 943	128 945	134 435
Sardinellas nei	*Sardinella spp*	35	348 099	409 258	503 626	419 134	359 593	263 751	225 389
Bonga shad	*Ethmalosa fimbriata*	35	142 638	177 780	174 131	182 238	161 859	180 740	186 817
European pilchard(=Sardine)	*Sardina pilchardus*	35	614 634	649 764	588 637	533 135	599 643	803 388	759 591
European anchovy	*Engraulis encrasicolus*	35	121 552	132 015	151 329	147 117	195 892	182 454	144 503
Clupeoids nei	*Clupeoidei*	35	1 613	1 715	1 722	1 841	5 145	3 937	4 238
Atlantic bonito	*Sarda sarda*	36	5 115	9 854	12 988	7 066	5 947	6 266	8 693
Plain bonito	*Orcynopsis unicolor*	36	2 004	249	94	644	1 054	899	907
Wahoo	*Acanthocybium solandri*	36	608	680	510	667	571	668	650
West African Spanish mackerel	*Scomberomorus tritor*	36	1 296	1 377	1 456	703	834	963	1 186
Seerfishes nei	*Scomberomorus spp*	36	93	-	85	-	8	-	265
Frigate tuna	*Auxis thazard*	36	910	1 368	1 339	1 650	1 413	1 409	896
Frigate and bullet tunas	*Auxis thazard, A.rochei*	36	7 049	8 646	10 008	6 943	7 987	2 481	1 988
Little tunny(=Atl.black skipj)	*Euthynnus alletteratus*	36	4 573	6 445	4 389	4 280	5 675	2 890	6 383
Skipjack tuna	*Katsuwonus pelamis*	36	108 151	105 359	112 463	132 361	107 755	118 088	93 226
Atlantic bluefin tuna	*Thunnus thynnus*	36	2 459	3 553	3 133	2 130	2 921	3 052	3 192
Albacore	*Thunnus alalunga*	36	2 782	2 139	1 300	2 443	1 148	3 562	4 438
Yellowfin tuna	*Thunnus albacares*	36	111 068	98 267	107 295	101 850	91 337	115 758	106 749
Bigeye tuna	*Thunnus obesus*	36	86 520	72 803	82 031	86 425	68 726	67 193	44 040
Atlantic sailfish	*Istiophorus albicans*	36	1 377	1 092	812	704	976	984	970
Blue marlin	*Makaira nigricans*	36	2 224	1 928	1 692	1 797	1 529	2 050	1 464
Black marlin	*Makaira indica*	36	14	6	45	109	31	...	1
Atlantic white marlin	*Tetrapturus albidus*	36	332	310	246	287	460	65	8
Longbill spearfish	*Tetrapturus pfluegeri*	36	60	68	86	83	63	1	16
Marlins,sailfishes,etc. nei	*Istiophoridae*	36	768	163	824	636	850	375	386
Swordfish	*Xiphias gladius*	36	5 844	3 212	3 040	3 092	2 769	3 622	2 313
Tuna-like fishes nei	*Scombroidei*	36	3 376	5 412	7 953	2 932	2 714	8 048	1 370
Garfish	*Belone belone*	37	-	-	-	13	82	-	-
Needlefishes, etc. nei	*Belonidae*	37	1 161	398	330	416	646	656	442
Halfbeaks nei	*Hemiramphus spp*	37	295	130	399	278	264	279	316
Flyingfishes nei	*Exocoetidae*	37	574	1 259	1 727	1 395	1 186	1 356	3 139
Bluefish	*Pomatomus saltatrix*	37	3 805	3 829	4 596	2 223	1 439	2 315	7 877
Cobia	*Rachycentron canadum*	37	0	6	0	0	-	-	-
Atlantic horse mackerel	*Trachurus trachurus*	37	224	512	666	576	-	-	-
Cunene horse mackerel	*Trachurus trecae*	37	7 888	7 113	11 812	10 784	1 021	2 475	789
Jack and horse mackerels nei	*Trachurus spp*	37	160 097	133 054	221 552	211 993	276 228	254 340	227 719
Scads nei	*Decapterus spp*	37	3 814	4 257	6 350	2 817	5 448	6 836	10 504
Crevalle jack	*Caranx hippos*	37	2 884	2 249	2 946	3 966	139	2 585	2 932
False scad	*Caranx rhonchus*	37	3 301	3 344	2 864	2 710	3 935	281	1 745
Jacks, crevalles nei	*Caranx spp*	37	2 234	4 006	1 461	3 873	2 181	1 574	3 131
African moonfish	*Selene dorsalis*	37	1 015	832	755	772	1 236	1 936	1 649
Pompanos nei	*Trachinotus spp*	37	155	154	310	223	28	187	195
Amberjacks nei	*Seriola spp*	37	229	213	324	182	160	288	137
Leerfish	*Lichia amia*	37	496	1 350	1 448	1 619	1 139	604	1 924
Alexandria pompano	*Alectis alexandrinus*	37	160	742	1 023	563	502	862	611
Rainbow runner	*Elagatis bipinnulata*	37	-	1	-	27	4	1	1
Atlantic bumper	*Chloroscombrus chrysurus*	37	3 582	5 149	8 420	8 731	8 237	11 951	14 610
Carangids nei	*Carangidae*	37	7 780	8 751	11 312	13 002	13 812	8 973	7 153
Atlantic pomfret	*Brama brama*	37	-	26	-	103	33	3	33
Common dolphinfish	*Coryphaena hippurus*	37	4	34	124	177	234	477	412
Suckerfishes, remoras nei	*Echeneidae*	37	-	-	3	1	12	...	...
Chub mackerel	*Scomber japonicus*	37	224 169	246 040	212 934	143 022	227 289	139 659	155 042
Atlantic mackerel	*Scomber scombrus*	37	6	13	...	...	...	6 135	4 473
Scomber mackerels nei	*Scomber spp*	37	1 489	-	-	23	-	-	557
Mackerels nei	*Scombridae*	37	3 824	3 406	3 769	3 087	3 024	3 970	3 777
Blue butterfish	*Stromateus fiatola*	37	240	322	271	619	292	4 432	1 464
Barracudas nei	*Sphyraena spp*	37	10 187	12 180	12 731	17 188	16 697	20 686	22 102
Pelagic percomorphs nei	*Perciformes*	37	220	4 714	5 069	4 623	4 844	4 300	4 510
Thresher	*Alopias vulpinus*	38	...	30	45	151	146	7	1

C-34 (a)

Fish, crustaceans, molluscs, etc	Capture production by species items	Atlantic, Eastern Central
Poissons, crustacés, mollusques, etc	Captures par catégories d'espèces	Atlantique, centre-est
Peces, crustáceos, moluscos, etc	Capturas por categorías de especies	Atlántico, centro-oriental

English name Nom anglais Nombre inglés	Scientific name Nom scientifique Nombre científico	Species group Groupe d'espèces Grupo de especies	1996 t	1997 t	1998 t	1999 t	2000 t	2001 t	2002 t
Bigeye thresher	*Alopias superciliosus*	38	...	148	114	...	...	...	...
Thresher sharks nei	*Alopias spp*	38	...	34	55	66	...	...	...
Shortfin mako	*Isurus oxyrinchus*	38	...	...	...	7	48	43	68
Mako sharks	*Isurus spp*	38	...	92	38	...	116	...	...
Porbeagle	*Lamna nasus*	38	...	...	...	...	10	3	8
Blue shark	*Prionace glauca*	38	...	...	...	76	421	557	5 269
Silky shark	*Carcharhinus falciformis*	38	...	2	...	110	99	...	1 355
Smooth hammerhead	*Sphyrna zygaena*	38	...	...	...	...	7	1	22
Scalloped hammerhead	*Sphyrna lewini*	38	12	10	10	10	10	10	267
Hammerhead sharks, etc. nei	*Sphyrnidae*	38	...	995	1 020	147	1 957	1 951	1 538
Smooth-hounds nei	*Mustelus spp*	38	392	51	57	28	503	1 929	2 099
Tope shark	*Galeorhinus galeus*	38	...	...	...	...	2	1	2
Leafscale gulper shark	*Centrophorus squamosus*	38	...	...	...	...	28	27	29
Dogfish sharks nei	*Squalidae*	38	255	302	301	400	412	500	564
Guitarfishes, etc. nei	*Rhinobatidae*	38	...	...	225	234	1 946	1 772	1 062
Sawfishes	*Pristidae*	38	...	48	...	41	42	...	...
Mantas	*Mobulidae*	38	...	...	342	802	931	106	110
Rays, stingrays, mantas nei	*Rajiformes*	38	12 093	14 079	13 865	14 727	11 192	12 504	11 301
Sharks, rays, skates, etc. nei	*Elasmobranchii*	38	23 088	52 300	36 392	34 922	34 353	32 117	25 399
Marine fishes nei	*Osteichthyes*	39	294 759	301 696	289 463	405 072	285 493	433 548	249 344
Red crab	*Geryon quinquedens*	42	-	-	-	-	-	31	-
Geryons nei	*Geryon spp*	42	...	...	...	...	6	16	5
Marine crabs nei	*Brachyura*	42	2 331	4 114	6 558	5 586	5 225	6 670	6 881
Tropical spiny lobsters nei	*Panulirus spp*	43	150	268	104	66	107	379	151
Palinurid spiny lobsters nei	*Palinurus spp*	43	1 679	8 816	3 577	1 590	2 345	2 470	851
Slipper lobsters nei	*Scyllaridae*	43	-	-	1	2	4	6	1
Norway lobster	*Nephrops norvegicus*	43	310	264	297	1 144	410	104	81
European lobster	*Homarus gammarus*	43	35	30	31	19	20	23	70
Caramote prawn	*Penaeus kerathurus*	45	159	100	40	19	8	1	41
Southern pink shrimp	*Penaeus notialis*	45	18 371	22 233	19 172	32 721	25 250	23 659	17 276
Penaeus shrimps nei	*Penaeus spp*	45	4 397	6 155	6 284	7 589	8 369	6 170	3 721
Deepwater rose shrimp	*Parapenaeus longirostris*	45	3 618	4 897	11 127	5 840	5 607	7 651	3 066
Scarlet shrimp	*Plesiopenaeus edwardsianus*	45	352	479	337	505	14	39	1
Natantian decapods nei	*Natantia*	45	30 913	19 342	46 545	23 709	20 108	20 387	32 025
Marine crustaceans nei	*Crustacea*	47	1 034	1 317	1 743	875	249	536	139
Murex	*Murex spp*	52	1 267	1 223	2 543	1 255	1 529	2 080	1 617
Volutes nei	*Cymbium spp*	52	6 648	5 166	4 678	5 737	4 989	5 422	4 347
Gastropods nei	*Gastropoda*	52	320	209	2 628	298	119	2 157	2 607
Cupped oysters nei	*Crassostrea spp*	53	100	109	89	125	101	151	120
Sea mussels nei	*Mytilidae*	54	10	6	7	...	401	102	8
Common edible cockle	*Cerastoderma edule*	56	54	69	139	147	117	105	152
Clams, etc. nei	*Bivalvia*	56	-	-	-	1	3	7	19
Common cuttlefish	*Sepia officinalis*	57	1 094	806	603	548	233	730	1 320
Cuttlefish,bobtail squids nei	*Sepiidae, Sepiolidae*	57	39 546	46 902	41 665	51 910	57 061	45 086	16 440
Common squids nei	*Loligo spp*	57	7	16	0	3	3	152	11
Various squids nei	*Loliginidae, Ommastrephidae*	57	29 860	13 457	20 824	23 018	20 175	15 961	6 134
Common octopus	*Octopus vulgaris*	57	27 119	23 158	39 169	23 538	12 753	16 302	9 268
Octopuses, etc. nei	*Octopodidae*	57	86 754	60 377	72 917	151 147	125 763	135 048	63 126
Cephalopods nei	*Cephalopoda*	57	6 240	5 636	8 551	238	440	279	571
Marine molluscs nei	*Mollusca*	58	746	1 489	12 599	2 258	1 189	385	3 423
Marine turtles nei	*Testudinata*	72	42	164	189	463	122	315	845
Total			***3 572 444***	***3 635 172***	***3 803 544***	***3 700 408***	***3 649 655***	***3 929 630***	***3 373 623***

C-34 (b)

Fish, crustaceans, molluscs, etc — Capture production by countries or areas — Atlantic, Eastern Central
Poissons, crustacés, mollusques, etc — Captures par pays ou zones — Atlantique, centre-est
Peces, crustáceos, moluscos, etc — Capturas por países o áreas — Atlántico, centro-oriental

Country or area Pays ou zone País o área	1993 t	1994 t	1995 t	1996 t	1997 t	1998 t	1999 t	2000 t	2001 t	2002 t
Bahamas	-	-	29	-	-	-	-	-	-	-
Belize	633	97	460	-	9 025	14 642	38 072	35 030	9 794	22 540
Benin	6 416	7 216	6 930	7 982	10 914	10 361	8 542	5 924	8 415	10 670
Bulgaria	-	-	-	-	-	8 189	-	-	-	-
Cameroon	42 257	52 000 F	64 131	63 400	62 000 F	61 800 F	60 000 F	57 109	58 531	55 135
Cape Verde	7 000	8 256	8 495	9 155	9 705	9 424	10 360	10 586	9 653	8 000 F
Cayman Is	320	...	...	...	...	...	...	...	...	...
Chile	-	25	-	-	-	-	-	-	-	-
China	9 809	2 414	21 464	18 302	20 810	20 403	28 433	25 408	36 228	4 161
China,Taiwan	6 467 F	12 929 F	13 199	17 517	13 158	10 623	8 890	14 494	13 677	9 761
Congo Dem R	4 200	3 780	3 876	3 973	3 844	3 954	3 945	4 300 F	4 600 F	5 000 F
Congo Rep	18 898	17 912	18 965	19 600	19 095	17 500 F	18 241	24 525	22 040 F	22 500 F
Côte dIvoire	63 495	58 374	58 854	57 606	52 137	57 071	63 709	65 270	62 926	55 900 F
Cuba	3 047	2 975	2 839	2 364	269	-	-	-	-	-
Cyprus	-	-	-	2 632	13 640	16 387	37 327	65 174	78 743	-
Egypt	-	-	-	-	-	-	4	-	-	-
Eq Guinea	2 907	4 369	1 856	4 140	5 240	5 035	5 900 F	2 558	2 500 F	2 500 F
Estonia	56 889	17 563	5 174	7 063	4 955	12 405	7 512	5	6	4
France	98 780	88 411	76 395	87 404	73 080	80 307	68 386	70 022	55 116	53 782
Gabon	28 289	26 515	32 789	36 680	34 143	44 609	41 143	37 053	33 021	31 475
Gambia	18 908	20 381	21 252	29 101	29 754	26 502	27 500	26 516	32 027	43 269
Georgia	13 500 F	5 000 F	1 000 F	-	-	-	-	-	-	-
Germany	-	-	39	11 222	28 748	23 099	26 678	-	2 673	5 168
Ghana	320 619	280 737	292 844	403 153	377 088	368 141	418 276	370 441	366 849	296 678
Greece	14 743	9 316	8 684	8 019	5 085	5 917	5 933	5 150	6 170	5 983
Guinea	56 000 F	60 000 F	64 760	60 580	58 841	65 764	83 314	87 513	101 227	100 000 F
GuineaBissau	5 100 F	5 750 F	6 078 F	6 750 F	7 000 F	5 800 F	4 800 F	4 800 F	4 800 F	4 800 F
Honduras	9 562	993	8 841	6 998	5 084	1 908	3 355	1 711	1 120	-
Ireland	-	-	-	-	-	-	-	-	73 695	39 588
Italy	43 445	43 578	8 067	3 428	7 180	5 891	5 845	4 077	7 359	6 202
Japan	22 773	20 358	24 307	24 766	14 934	20 294	13 487	16 218	10 973	7 533
Korea D P Rp	398	-	-	-	-	-	-	850	-	-
Korea Rep	15 510	23 057	19 324	19 380	21 851	18 481	25 219	23 978	26 809	25 974
Latvia	63 216	65 689	82 702	63 711	18 018	22 530	43 552	52 065	47 335	30 491
Liberia	3 782	3 721	4 829	3 408	4 580	6 830	11 472	7 726	7 286	7 500 F
Libya	1 085	500	400	400	400	400	400	400	239	666
Lithuania	57 469	22 880	9 572	33 330	25 680	45 804	46 910	53 445	111 000	106 434
Malta	-	1 236	3 465	8 197	-	-	-	-	-	-
Marshall Is	-	-	-	2 403	-	-	-	-	-	-
Mauritania	54 452 F	46 746 F	48 147 F	55 324 F	52 756 F	56 660 F	71 026 F	75 849 F	79 881 F	73 902 F
Morocco	591 270	717 578	807 775	601 734	758 321	680 075	705 978	856 362	1 054 147	863 488
Netherlands	...	...	...	47 515	100 969	124 475	123 937	157 195	176 922	161 143
NethAntilles	-	-	-	12 762	18 582	18 804	18 481	18 932	21 955	12 051
Nigeria	142 782	163 259	231 579	248 472	294 279	324 004	316 235	309 062	297 971	293 823
Norway	-	87	-	-	-	3 421	-	71	-	-
Panama	35 924	38 469	39 164	19 618	8 293	4 889	3 725	3 330	944	1 427
Philippines	-	-	-	-	-	1 186	1 854	801	29	110
Poland	-	2 220	-	19 766	5 268	-	-	-	13 185	28 712
Portugal	29 773	24 615	22 665	40 326	22 293	18 294	14 765	9 430	9 898	11 484
Romania	1 305	...	...	...	...	...	...	...	...	...
Russian Fed	198 630	167 472	277 631	384 330	312 524	341 413	286 179	211 018	129 436	121 505
St Vincent	466	0	61	30	5 144	32 608	16 727	20 369	41 647	41 762
Sao Tome Prn	2 334	3 391	3 565	3 980	3 338	3 477	3 756	3 500 F	3 500 F	3 500 F
Senegal	354 562	322 421	323 617	388 759	426 366	382 872	378 125	379 597	383 202	355 824
Seychelles	-	-	-	-	-	-	-	127	-	-
Sierra Leone	52 288	47 439	49 870	52 804 F	58 128	48 875	44 927	60 730	61 210	68 990
Spain	317 907 F	326 916 F	334 932 F	344 222	315 145	373 415	241 272	149 224	160 504	123 183
Togo	10 964	8 052	7 203	10 098	9 290	11 655	17 924	17 277	18 163	15 946
Ukraine	118 508	99 006	213 737	285 054	278 133	346 280	281 959	249 594	187 145	91 334
UK	-	-	-	-	-	-	-	-	31	-
Vanuatu	345	70	170	212	196	91	-	-	-	-
Westn Sahara	0	0	0	0	0	0	0	0	0	0
Other nei	28 310	30 524	149 452	34 774	29 889	40 979	56 333	54 839	95 048	143 725
Total	***2 935 337***	***2 864 297***	***3 381 188***	***3 572 444***	***3 635 172***	***3 803 544***	***3 700 408***	***3 649 655***	***3 929 630***	***3 373 623***

C-37 (a)

Fish, crustaceans, molluscs, etc — Capture production by species items — Mediterranean and Black Sea
Poissons, crustacés, mollusques, etc — Captures par catégories d'espèces — Méditerranée et mer Noire
Peces, crustáceos, moluscos, etc — Capturas por categorías de especies — Mediterráneo y Mar Negro

English name Nom anglais Nombre inglés	Scientific name Nom scientifique Nombre científico	Species group Groupe d'espèces Grupo de especies	1996 t	1997 t	1998 t	1999 t	2000 t	2001 t	2002 t
Freshwater bream	*Abramis brama*	11	346	658	495	335	336	108	47
Common carp	*Cyprinus carpio*	11	2	3	3	1	-	2	3
Roach	*Rutilus rutilus*	11	1	0	1	1	2	7	11
Roaches nei	*Rutilus spp*	11	80	107	13	78	73	114	72
Silver carp	*Hypophthalmichthys molitrix*	11	1	-	-	-	-	-	-
Sichel	*Pelecus cultratus*	11	46	113	105	228	276	185	147
Cyprinids nei	*Cyprinidae*	11	-	1	-	-	167	159	95
European perch	*Perca fluviatilis*	13	-	-	-	-	-	1	1
Percarina	*Percarina demidoffi*	13	-	-	-	-	-	-	18
Pike-perch	*Stizostedion lucioperca*	13	2 822	2 825	3 031	2 568	2 956	3 504	3 293
Freshwater fishes nei	*Osteichthyes*	13	-	-	-	-	-	17	-
Danube sturgeon(=Osetr)	*Acipenser gueldenstaedtii*	21	129	134	114	36	20	8	10
Sterlet sturgeon	*Acipenser ruthenus*	21	-	-	-	-	0	-	-
Starry sturgeon	*Acipenser stellatus*	21	18	49	13	11	5	3	2
Beluga	*Huso huso*	21	5	11	12	10	1	0	4
Sturgeons nei	*Acipenseridae*	21	432	443	290	185	59	22	23
European eel	*Anguilla anguilla*	22	962	1 257	917	682	464	602	648
Salmonoids nei	*Salmonoidei*	23	18	22	26	-	-	0	-
Pontic shad	*Alosa pontica*	24	131	82	153	48	15	21	112
Shads nei	*Alosa spp*	24	2 283	1 835	2 742	2 640	2 120	2 929	3 250
Azov sea sprat	*Clupeonella cultriventris*	24	1 448	1 522	3 496	10 862	12 006	27 777	27 239
Three-spined stickleback	*Gasterosteus aculeatus*	25	-	-	-	-	-	8	4
European plaice	*Pleuronectes platessa*	31	-	-	-	0	6	7	7
European flounder	*Platichthys flesus*	31	35	50	69	62	56	29	29
Common sole	*Solea solea*	31	6 949	5 917	5 044	4 179	5 169	4 972	5 439
Soles nei	*Soleidae*	31	-	-	-	18	14	12	18
Megrim	*Lepidorhombus whiffiagonis*	31	286	295	118	108	202	205	190
Brill	*Scophthalmus rhombus*	31	18	20	22	24	26	26	30
Turbot	*Psetta maxima*	31	2 341	1 333	2 178	2 125	2 967	2 802	865
Turbots nei	*Scophthalmidae*	31	1 377	964	528	478	643	622	482
Flatfishes nei	*Pleuronectiformes*	31	2 334	3 139	3 100	3 475	2 084	2 199	1 635
Blue ling	*Molva dypterygia*	32	1	1	1	-	-	-	-
Greater forkbeard	*Phycis blennoides*	32	204	241	346	209	429	89	406
Rocklings nei	*Gaidropsarus spp*	32	...	...	...	20	19	19	39
Poor cod	*Trisopterus minutus*	32	603	428	428	637	890	755	860
Pouting(=Bib)	*Trisopterus luscus*	32	313	388	402	669	1 947	657	660
Blue whiting(=Poutassou)	*Micromesistius poutassou*	32	18 854	20 965	32 278	23 530	26 317	27 933	14 713
Whiting	*Merlangius merlangus*	32	22 059	16 038	14 030	14 643	18 661	11 045	9 632
European hake	*Merluccius merluccius*	32	43 816	29 956	26 545	24 144	22 558	23 322	21 365
Gadiformes nei	*Gadiformes*	32	4 858	4 055	1 341	789	871	1 097	317
Brushtooth lizardfish	*Saurida undosquamis*	33	124	61	37	48	21	24	24
Lizardfishes nei	*Synodontidae*	33	1 484	1 257	1 138	1 382	1 535	1 135	1 109
Flathead grey mullet	*Mugil cephalus*	33	5 561	4 874	3 732	2 652	4 033	1 854	2 523
So-iuy mullet	*Mugil soiuy*	33	1 039	2 718	3 661	5 174	5 406	2 428	2 510
Leaping mullet	*Liza saliens*	33	3	2	2	2	0	10	25
Mullets nei	*Mugilidae*	33	34 210	31 720	37 232	37 620	37 907	34 743	29 075
Dusky grouper	*Epinephelus marginatus*	33	1 383	1 211	833	303	290	448	484
White grouper	*Epinephelus aeneus*	33	5	4	4	6	48	7	5
Groupers nei	*Epinephelus spp*	33	5 824	5 764	5 786	6 859	6 132	6 061	6 418
Groupers, seabasses nei	*Serranidae*	33	1 565	2 072	4 483	1 791	1 023	1 148	1 108
Spotted seabass	*Dicentrarchus punctatus*	33	424	119	135	203	186	238	556
European seabass	*Dicentrarchus labrax*	33	3 754	3 025	2 927	3 151	3 501	4 154	5 354
Seabasses nei	*Dicentrarchus spp*	33	2 846	3 550	4 486	4 354	2 677	1 786	1 926
Rubberlip grunt	*Plectorhinchus mediterraneus*	33	4	5	4	-	-	2	13
Brown meagre	*Sciaena umbra*	33	222	222	240	186	128	171	156
Shi drum	*Umbrina cirrosa*	33	710	493	386	386	289	393	324
Meagre	*Argyrosomus regius*	33	1 052	666	804	961	968	1 280	1 644
Croakers, drums nei	*Sciaenidae*	33	-	-	18	29	164	32	56
Blackspot(=red) seabream	*Pagellus bogaraveo*	33	12	20	20	38	38	38	50
Common pandora	*Pagellus erythrinus*	33	4 592	5 553	4 803	4 622	5 618	4 924	4 864
Axillary seabream	*Pagellus acarne*	33	90	477	670	339	284	206	255
Pandoras nei	*Pagellus spp*	33	2 897	3 379	2 131	2 756	3 252	3 339	3 008
White seabream	*Diplodus sargus*	33	409	640	433	640	695	529	1 007
Sargo breams nei	*Diplodus spp*	33	6 896	6 188	5 181	6 444	5 979	5 547	5 028
Large-eye dentex	*Dentex macrophthalmus*	33	651	1 010	535	497	378	318	300
Common dentex	*Dentex dentex*	33	2 287	1 458	1 237	1 172	1 241	767	811
Black seabream	*Spondyliosoma cantharus*	33	570	484	348	416	477	259	358
Saddled seabream	*Oblada melanura*	33	1 020	848	794	734	822	590	568
Red porgy	*Pagrus pagrus*	33	3 398	2 924	2 740	4 338	3 378	1 746	2 504
Pargo breams nei	*Pagrus spp*	33	666	602	738	751	913	782	817
Gilthead seabream	*Sparus aurata*	33	4 490	4 328	4 568	5 751	5 486	6 026	6 847
Bogue	*Boops boops*	33	29 016	25 414	27 606	24 400	25 181	23 091	24 955
Sand steenbras	*Lithognathus mormyrus*	33	718	1 011	1 184	1 004	1 006	1 090	948
Salema	*Sarpa salpa*	33	1 937	2 413	2 235	1 963	1 802	1 798	1 997
Porgies, seabreams nei	*Sparidae*	33	10 625	8 235	7 374	9 018	8 788	9 357	8 716
Blotched picarel	*Spicara maena*	33	285	505	395	419	255	381	593
Picarels nei	*Spicara spp*	33	12 312	12 254	13 147	9 324	8 192	8 928	7 161
Surmullet	*Mullus surmuletus*	33	11 689	11 818	9 076	9 685	10 182	10 067	9 621
Red mullet	*Mullus barbatus*	33	5 331	4 531	4 878	5 618	4 433	5 667	5 159
Surmullets(=Red mullets) nei	*Mullus spp*	33	19 990	15 576	14 419	18 240	18 245	15 877	14 365
Parrotfishes nei	*Scaridae*	33	32	37	78	58	33	42	55
Sandeels(=Sandlances) nei	*Ammodytes spp*	33	74	164	283	368	308	106	428

English name Nom anglais Nombre inglés	Scientific name Nom scientifique Nombre científico	Species group Groupe d'espèces Grupo de especies	1996 t	1997 t	1998 t	1999 t	2000 t	2001 t	2002 t
Greater weever	*Trachinus draco*	33	83	154	207	24	53	80	85
Black goby	*Gobius niger*	33	-	-	-	-	1	1	1
Gobies nei	*Gobiidae*	33	2 274	1 919	1 917	2 682	2 353	3 085	5 331
Spinefeet(=Rabbitfishes) nei	*Siganus spp*	33	763	390	425	510	646	940	1 101
Grey triggerfish	*Balistes carolinensis*	33	38	52	58	34	50	62	101
Argentines	*Argentina spp*	34	-	-	-	30	42	69	57
European conger	*Conger conger*	34	3 036	3 640	2 921	3 114	2 722	2 746	2 634
Alfonsinos nei	*Beryx spp*	34	-	-	-	-	8	-	-
John dory	*Zeus faber*	34	596	424	531	418	385	502	433
Wreckfish	*Polyprion americanus*	34	12	17	15	37	35	12	63
Largehead hairtail	*Trichiurus lepturus*	34	919	698	774	711	809	1 107	1 096
Silver scabbardfish	*Lepidopus caudatus*	34	1 619	1 365	1 713	1 830	707	3 421	1 053
Scorpionfishes nei	*Scorpaenidae*	34	2 240	2 309	2 453	1 842	2 067	2 293	2 898
Tub gurnard	*Chelidonichthys lucerna*	34	...	...	...	...	79	68	27
Red gurnard	*Chelidonichthys cuculus*	34	...	...	...	...	184	162	311
Grey gurnard	*Eutrigla gurnardus*	34	1 264	1 158	1 067	1 159	1 090	782	627
Gurnards, searobins nei	*Triglidae*	34	7 035	5 051	4 672	4 168	3 295	3 230	3 459
Angler(=Monk)	*Lophius piscatorius*	34	3 674	7 956	3 982	3 825	4 859	4 269	3 480
Demersal percomorphs nei	*Perciformes*	34	609	453	629	549	1 276	1 233	3 980
Round sardinella	*Sardinella aurita*	35	242	435	557	306	88	117	143
Sardinellas nei	*Sardinella spp*	35	42 616	53 908	58 717	77 020	68 942	92 088	66 839
Red-eye round herring	*Etrumeus teres*	35	-	-	-	1 403	...	...	...
European pilchard(=Sardine)	*Sardina pilchardus*	35	224 546	209 851	204 094	212 244	215 932	199 927	186 773
European sprat	*Sprattus sprattus*	35	28 260	28 243	38 333	39 396	42 007	63 125	70 403
European anchovy	*Engraulis encrasicolus*	35	377 078	348 084	321 191	447 686	392 350	439 092	482 822
Clupeoids nei	*Clupeoidei*	35	1 997	1 864	2 879	3 206	3 323	1 786	3 173
Atlantic bonito	*Sarda sarda*	36	17 354	16 596	29 763	25 655	18 760	19 166	12 417
Plain bonito	*Orcynopsis unicolor*	36	132	227	130	217	145	154	137
Frigate and bullet tunas	*Auxis thazard, A.rochei*	36	5 909	3 324	2 291	2 011	2 439	1 861	1 975
Little tunny(=Atl.black skipj)	*Euthynnus alletteratus*	36	2 128	1 608	2 375	2 096	2 533	2 063	1 819
Skipjack tuna	*Katsuwonus pelamis*	36	9	4	176	53	90	77	965
Atlantic bluefin tuna	*Thunnus thynnus*	36	37 960	33 684	28 079	22 853	23 092	24 382	22 329
Albacore	*Thunnus alalunga*	36	3 125	2 541	2 698	4 851	5 578	4 866	5 606
Atlantic white marlin	*Tetrapturus albidus*	36	-	1	1	-	1	0	0
Marlins,sailfishes,etc. nei	*Istiophoridae*	36	1	29	-	-	1	25	2
Swordfish	*Xiphias gladius*	36	12 054	14 676	14 356	13 686	15 570	15 010	12 120
Tuna-like fishes nei	*Scombroidei*	36	2 010	2 447	1 374	3 389	3 363	2 402	1 868
Garfish	*Belone belone*	37	1 563	1 661	1 562	1 214	843	1 477	2 156
Big-scale sand smelt	*Atherina boyeri*	37	31	13	25	20	84	97	99
Silversides(=Sand smelts) nei	*Atherinidae*	37	7 575	6 652	7 699	5 807	3 928	5 631	5 325
Bluefish	*Pomatomus saltatrix*	37	6 171	5 014	4 582	4 367	6 246	15 271	27 565
Atlantic horse mackerel	*Trachurus trachurus*	37	10 177	8 367	6 076	5 155	8 189	11 532	7 879
Mediterranean horse mackerel	*Trachurus mediterraneus*	37	21 002	17 065	15 122	12 898	19 111	19 308	23 047
Jack and horse mackerels nei	*Trachurus spp*	37	23 580	26 087	22 899	27 324	29 658	29 686	24 942
Jacks, crevalles nei	*Caranx spp*	37	716	441	402	326	525	1 351	1 158
Greater amberjack	*Seriola dumerili*	37	1 057	724	755	1 320	1 474	1 658	1 609
Leerfish	*Lichia amia*	37	2 299	2 222	2 343	3 168	637	607	779
Carangids nei	*Carangidae*	37	457	354	413	412	512	801	422
Atlantic pomfret	*Brama brama*	37	26	157	87	118	32	109	52
Common dolphinfish	*Coryphaena hippurus*	37	706	833	1 451	1 360	1 038	1 157	1 181
Chub mackerel	*Scomber japonicus*	37	19 529	21 621	16 403	17 613	15 060	10 737	6 687
Atlantic mackerel	*Scomber scombrus*	37	7 158	4 711	7 000	4 827	7 434	6 375	5 864
Scomber mackerels nei	*Scomber spp*	37	11 680	11 878	10 879	8 941	8 916	9 547	8 526
Mackerels nei	*Scombridae*	37	500	300	300	300	353	357	350
Barracudas nei	*Sphyraena spp*	37	1 804	2 485	2 467	2 047	2 019	2 275	2 296
Basking shark	*Cetorhinus maximus*	38	2	6	6	-	-	-	4
Thresher	*Alopias vulpinus*	38	...	...	...	14	12	21	14
Shortfin mako	*Isurus oxyrinchus*	38	-	-	-	-	1	6	2
Porbeagle	*Lamna nasus*	38	1	0	0	0	0	1	-
Blackmouth catshark	*Galeus melastomus*	38	...	...	...	...	...	...	58
Small-spotted catshark	*Scyliorhinus canicula*	38	...	...	...	...	30	31	33
Catsharks, nursehounds nei	*Scyliorhinus spp*	38	36	72	...	262	457	501	324
Blue shark	*Prionace glauca*	38	...	...	...	3	4	42	16
Smooth-hounds nei	*Mustelus spp*	38	5 896	2 980	3 249	3 658	4 100	1 945	1 566
Picked dogfish	*Squalus acanthias*	38	143	95	97	143	204	287	231
Gulper shark	*Centrophorus granulosus*	38	...	...	...	...	...	...	1
Lanternsharks nei	*Etmopterus spp*	38	...	...	...	...	...	...	3
Portuguese dogfish	*Centroscymnus coelolepis*	38	-	-	-	-	7	23	2
Dogfish sharks nei	*Squalidae*	38	646	1 319	449	414	1 034	1 238	1 309
Angelshark	*Squatina squatina*	38	18	34	44	25	20	22	13
Angelsharks, sand devils nei	*Squatinidae*	38	95	35	171	100	90	36	97
Guitarfishes, etc. nei	*Rhinobatidae*	38	113	69	93	78	99	94	89
Thornback ray	*Raja clavata*	38	17	9	24	76	53	82	97
Raja rays nei	*Raja spp*	38	1 002	900	715	718	746	579	732
Common stingray	*Dasyatis pastinaca*	38	...	...	-	-	4	11	-
Rays, stingrays, mantas nei	*Rajiformes*	38	4 173	7 177	5 125	3 552	3 576	3 093	3 507
Sharks, rays, skates, etc. nei	*Elasmobranchii*	38	3 905	3 638	4 913	3 114	2 077	3 383	2 765
Marine fishes nei	*Osteichthyes*	39	105 579	88 163	93 606	72 160	81 048	68 265	66 403
Mediterranean shore crab	*Carcinus aestuarii*	42	65	44	66	44	30	37	74
Spinous spider crab	*Maja squinado*	42	31	59	55	16	21	27	67
Marine crabs nei	*Brachyura*	42	2 099	2 001	2 399	4 431	2 840	2 530	3 344
Pink spiny lobster	*Palinurus mauritanicus*	43	0	0	0	0	-	-	-

C-37 (a)

Fish, crustaceans, molluscs, etc — Capture production by species items — Mediterranean and Black Sea
Poissons, crustacés, mollusques, etc — Captures par catégories d'espèces — Méditerranée et mer Noire
Peces, crustáceos, moluscos, etc — Capturas por categorías de especies — Mediterráneo y Mar Negro

English name Nom anglais Nombre inglés	Scientific name Nom scientifique Nombre científico	Species group Groupe d'espèces Grupo de especies	1996 t	1997 t	1998 t	1999 t	2000 t	2001 t	2002 t
Common spiny lobster	*Palinurus elephas*	43	331	385	223	170	137	188	176
Palinurid spiny lobsters nei	*Palinurus spp*	43	184	200	297	188	131	181	197
Norway lobster	*Nephrops norvegicus*	43	6 658	6 165	4 155	4 068	3 514	3 503	3 133
European lobster	*Homarus gammarus*	43	277	451	248	195	235	254	228
Lobsters nei	*Reptantia*	43	5	2	4	2	6	7	5
Kuruma prawn	*Penaeus japonicus*	45	-	84	...	...	...	121	111
Caramote prawn	*Penaeus kerathurus*	45	4 380	5 978	5 458	5 947	7 692	4 951	4 280
Speckled shrimp	*Metapenaeus monoceros*	45	-	-	-	-	-	757	1 607
Deepwater rose shrimp	*Parapenaeus longirostris*	45	9 101	10 119	8 127	9 261	12 323	11 100	10 133
Scarlet shrimp	*Plesiopenaeus edwardsianus*	45	-	-	-	100	40	-	33
Blue and red shrimp	*Aristeus antennatus*	45	1 428	1 359	1 672	1 332	1 904	2 209	2 011
Aristeid shrimps nei	*Aristeidae*	45	2 258	2 406	1 231	2 128	4 463	1 833	1 768
Common prawn	*Palaemon serratus*	45	40	35	89	9	...	...	97
Common shrimp	*Crangon crangon*	45	222	150	100	87	0	4	139
Natantian decapods nei	*Natantia*	45	7 955	10 148	10 848	10 543	8 871	8 082	11 080
Spottail mantis squillid	*Squilla mantis*	47	5 446	4 507	3 680	6 023	6 331	6 606	6 231
Marine crustaceans nei	*Crustacea*	47	4 358	4 320	4 193	3 904	4 097	3 418	2 509
Common periwinkle	*Littorina littorea*	52	132	147	117	126	139	...	28
Murex	*Murex spp*	52	37	41	41	35	51	52	60
Sea snails	*Rapana spp*	52	4 349	5 494	4 671	4 419	4 713	3 753	851
Gastropods nei	*Gastropoda*	52	-	-	-	965	813	870	800
European flat oyster	*Ostrea edulis*	53	2 187	1 849	1 190	904	271	157	188
Pacific cupped oyster	*Crassostrea gigas*	53	59	14	14	48	48	-	-
Mediterranean mussel	*Mytilus galloprovincialis*	54	39 046	53 310	38 899	44 453	46 143	46 092	51 820
Great Atlantic scallop	*Pecten maximus*	55	...	...	...	12	18	4	6
Great Mediterranean scallop	*Pecten jacobaeus*	55	52	95	50	68	570	150	470
Scallops nei	*Pectinidae*	55	41	42	15	-	-	-	-
Striped venus	*Chamelea gallina*	56	48 060	36 239	32 811	40 421	44 580	42 657	36 077
Pullet carpet shell	*Venerupis pullastra*	56	5	5	-	141	27	...	3
Grooved carpet shell	*Ruditapes decussatus*	56	488	135	172	143	796	569	1 026
Venus clams nei	*Veneridae*	56	-	-	-	18	18	22	12
Razor clams nei	*Solen spp*	56	2	4	1	5	5	5	5
Common edible cockle	*Cerastoderma edule*	56	-	-	-	1	-	-	-
Clams, etc. nei	*Bivalvia*	56	4 078	2 963	3 527	2 520	1 977	1 676	1 797
Common cuttlefish	*Sepia officinalis*	57	7 605	10 591	8 284	10 500	8 842	9 859	11 047
Cuttlefish,bobtail squids nei	*Sepiidae, Sepiolidae*	57	10 966	10 445	9 837	9 112	8 824	9 045	7 137
Common squids nei	*Loligo spp*	57	7 828	6 872	4 985	4 836	4 623	4 681	4 457
Broadtail shortfin squid	*Illex coindetii*	57	...	...	...	34	42	31	77
European flying squid	*Todarodes sagittatus*	57	4 899	2 850	4 203	2 202	2 856	2 466	3 251
Various squids nei	*Loliginidae, Ommastrephidae*	57	1 457	1 105	1 358	1 351	1 666	1 052	941
Common octopus	*Octopus vulgaris*	57	17 209	19 346	16 731	13 385	14 539	15 325	15 601
Horned and musky octopuses	*Eledone spp*	57	1 939	2 258	1 758	1 433	1 990	2 357	2 187
Octopuses, etc. nei	*Octopodidae*	57	6 830	6 131	6 714	6 969	9 360	7 880	7 023
Cephalopods nei	*Cephalopoda*	57	533	541	411	870	1 135	1 162	698
Marine molluscs nei	*Mollusca*	58	18 120	15 480	15 120	13 389	10 245	15 401	17 769
Grooved sea squirt	*Microcosmus sulcatus*	74	28	22	22	30	30	76	78
Echinoderms	*Echinodermata*	76	2	4	9	5	3	2	-
Stony sea urchin	*Paracentrotus lividus*	76	63	47	47	70	192	98	164
Sea cucumbers nei	*Holothurioidea*	76	4	4	4	1	9	4	2
Jellyfishes	*Rhopilema spp*	77	904	900	1 750	1 203	900	2 000	500
Aquatic invertebrates nei	*Invertebrata*	77	5	4	14	534	8	595	79
Total			***1 531 975***	***1 444 025***	***1 401 557***	***1 533 687***	***1 505 217***	***1 570 335***	***1 550 099***

C-37 (b)

Fish, crustaceans, molluscs, etc — **Capture production by countries or areas** — **Mediterranean and Black Sea**
Poissons, crustacés, mollusques, etc — **Captures par pays ou zones** — **Méditerranée et mer Noire**
Peces, crustáceos, moluscos, etc — **Capturas por países o áreas** — **Mediterráneo y Mar Negro**

Country or area Pays ou zone País o área	1993 t	1994 t	1995 t	1996 t	1997 t	1998 t	1999 t	2000 t	2001 t	2002 t
Albania	1 650 F	1 400 F	1 160	1 808	833	1 860	1 931	2 365	1 937	2 788
Algeria	101 895	135 402	105 872	81 989	91 580	92 344	102 396	113 156	133 623	134 320
Bosnia Herzg	0	0	0	0	0	0	0	0	0	0
Bulgaria	2 318	5 340 F	7 250	7 727	9 356	8 421	8 081	6 137	4 880	13 555
China	-	97	137	93	49	-	-	-	-	-
China,Taiwan	328	713	493	373	399	-	58	31	197	131
Croatia	26 342	17 137	15 901	17 799	16 627	21 928	18 890	21 045	18 455	21 205
Cyprus	2 696	2 762	2 505	2 550	2 309	2 408	2 241	2 230	2 258	1 918
Egypt	44 616	45 600	43 700	51 100	52 748	68 000	89 943	54 872	59 652	59 636
France	45 601	39 710	37 967	27 813	33 012	33 220	38 948	45 540	43 059	42 362
Georgia	2 191	1 397	2 470	2 447	2 582	2 997	1 396	2 178	1 822	2 527
Gibraltar	0	0	0	0	0	0	0	0	0	0
Greece	141 395	168 354	139 498	138 513	149 402	99 845	109 558	90 697	85 037	80 000 F
Israel	3 232	2 473	3 577	3 159	3 557	3 999	3 641	3 966	3 618	3 312
Italy	330 092	330 704	375 970	354 551	326 260	289 545	264 619	293 405	294 312	254 642
Japan	616	530	749	673	172	417	376	143	188	390
Korea Rep	-	-	484	701	686	-	-	-	-	-
Lebanon	2 000	2 205	4 065	4 115	3 635	3 500	3 540	3 646	3 650	3 673
Libya	30 000 F	33 000 F	34 000 F	32 000 F	31 000 F	32 000 F	32 000 F	32 500 F	33 000 F	33 000 F
Malta	838	1 120	1 170	1 000	1 036	1 180	1 244	1 074	893	1 004
Monaco	3 F	3 F	3 F	3 F	3 F	3 F	3 F	3 F	3 F	3 F
Morocco	31 624	34 999	39 676	39 652	31 485	28 658	37 290	38 650	28 146	29 357
Palest, O.T.	...	...	1 229	2 493	3 791	3 625	3 600 F	3 000 F	2 500 F	2 378
Panama	467	1 499	1 498	2 850	236	-	-	-	-	-
Portugal	183	428	446	274	37	54	76	96	288	29
Romania	3 952	3 060	2 719	2 682	3 872	4 431	2 507	2 476	2 431	2 122
Russian Fed	6 384	13 888	15 540	8 745	8 917	9 760	14 340	22 294	33 444	39 948
Serbia-Monte	277	260	369	380	376	413	423	424	416	454
Slovenia	1 987	2 007	1 851	2 078	2 065	1 959	1 784	1 630	1 621	1 460
Spain	146 000 F	148 000 F	149 000 F	150 496	133 302	123 325	122 360	140 203	139 143	118 341
Syria	2 019	1 950	1 950	2 670	2 574	2 750	2 600	2 581	2 322	2 823
Tunisia	82 514	85 367	82 915	83 028	86 002	87 179	92 378	94 718	97 622	95 815
Turkey	503 957	544 736	585 994	478 226	408 693	432 700	523 634	460 524	484 407	522 744
Ukraine	26 394	35 213	43 570	29 266	35 987	42 981	51 495	65 507	90 840	79 654
Other nei	1 000	1 700	1 350	721	1 442	2 055	2 335	126	571	508
Total	***1 542 571***	***1 661 054***	***1 705 078***	***1 531 975***	***1 444 025***	***1 401 557***	***1 533 687***	***1 505 217***	***1 570 335***	***1 550 099***

English name Nom anglais Nombre inglés	Scientific name Nom scientifique Nombre científico	Species group Groupe d'espèces Grupo de especies	1996 t	1997 t	1998 t	1999 t	2000 t	2001 t	2002 t
Lefteye flounders nei	*Bothidae*	31	-	-	-	-	-	20	1
Tonguefishes	*Cynoglossidae*	31	23	-	-	19	48	-	-
Bastard halibuts nei	*Paralichthys spp*	31	9 236	10 546	9 247	7 081	6 754	6 251	4 697
Flatfishes nei	*Pleuronectiformes*	31	1 097	1 430	1 655	1 601	1 887	1 824	3 349
Tadpole codling	*Salilota australis*	32	7 590	6 094	11 454	15 643	14 543	5 552	4 111
Brazilian codling	*Urophycis brasiliensis*	32	3 152	2 570	6 081	2 842	3 498	9 076	9 606
Southern blue whiting	*Micromesistius australis*	32	104 669	102 450	108 558	83 695	84 499	80 028	64 449
Southern hake	*Merluccius australis*	32	4 115	3 011	3 125	3 471	7 035	4 743	5 301
Argentine hake	*Merluccius hubbsi*	32	681 999	648 301	527 229	372 167	244 988	305 672	409 488
Hakes nei	*Merluccius spp*	32	370	33	-	-	-	354	186
Patagonian grenadier	*Macruronus magellanicus*	32	55 782	52 199	119 810	137 340	142 464	136 056	116 323
Grenadiers nei	*Macrourus spp*	32	901	1 041	2 972	2 593	10 503	3 209	6 052
Gadiformes nei	*Gadiformes*	32	1 540	252	237	352	1 364	3 142	159
Tarpon	*Megalops atlanticus*	33	348	720	551	315	331	1 169	1 201
Sea catfishes nei	*Ariidae*	33	16 970	17 764	20 531	25 307	29 529	34 975	34 877
Mullets nei	*Mugilidae*	33	8 078	10 486	10 288	10 953	12 186	11 261	12 142
Snooks(=Robalos) nei	*Centropomus spp*	33	1 686	1 866	2 996	3 602	4 366	3 469	3 608
Brazilian groupers nei	*Mycteroperca spp*	33	423	584	518	628	536	1 040	1 347
Red grouper	*Epinephelus morio*	33	320	238	212	666	1 220	...	...
Groupers nei	*Epinephelus spp*	33	2 341	2 170	2 315	1 710	2 670	2 124	2 084
Argentine seabass	*Acanthistius brasilianus*	33	10 799	9 164	5 258	5 914	4 222	4 890	3 624
Groupers, seabasses nei	*Serranidae*	33	-	15	2	-	-	1	10
Bigeyes nei	*Priacanthus spp*	33	59	55	46	60	67	45	52
Southern red snapper	*Lutjanus purpureus*	33	5 104	6 085	5 937	9 790	6 580	6 209	6 184
Lane snapper	*Lutjanus synagris*	33	648	659	1 038	1 031	1 014	1 437	1 392
Yellowtail snapper	*Ocyurus chrysurus*	33	4 167	5 000	3 317	4 541	4 165	2 002	2 158
Snappers, jobfishes nei	*Lutjanidae*	33	3 690	4 488	4 390	4 549	3 859	4 176	3 911
Barred grunt	*Conodon nobilis*	33	172	118	114	84	39	53	82
Grunts, sweetlips nei	*Haemulidae (=Pomadasyidae)*	33	2 081	1 936	2 000	2 382	3 024	3 429	3 537
Striped weakfish	*Cynoscion striatus*	33	31 641	39 318	32 394	19 588	22 873	22 734	20 370
Weakfishes nei	*Cynoscion spp*	33	26 470	25 960	26 677	39 332	43 523	2 637	1 810
Whitemouth croaker	*Micropogonias furnieri*	33	72 633	73 163	59 631	43 217	58 239	71 067	72 648
Kingcroakers nei	*Menticirrhus spp*	33	1 144	1 303	1 402	1 062	1 352	1 957	2 047
Argentine croaker	*Umbrina canosai*	33	13 018	6 658	5 284	7 660	9 221	16 256	17 796
King weakfish	*Macrodon ancylodon*	33	7 266	7 625	9 229	5 595	6 642	4 139	6 261
Black drum	*Pogonias cromis*	33	332	163	938	282	357	388	537
Croakers, drums nei	*Sciaenidae*	33	150	52	117	746	688	35 860	32 434
South American silver porgy	*Diplodus argenteus*	33	39	84	15	5	2	2	1
Red porgy	*Pagrus pagrus*	33	2 475	2 743	2 026	3 535	2 825	2 382	2 499
Porgies, seabreams nei	*Sparidae*	33	-	-	-	2	1 054	1 221	-
Argentine goatfish	*Mullus argentinae*	33	75	73	99	229	498	225	78
Goatfishes, red mullets nei	*Mullidae*	33	530	749	840	1 117	1 756	4 167	6 040
Sea chubs nei	*Kyphosus spp*	33	-	-	-	-	-	-	37
Patagonian blennie	*Eleginops maclovinus*	33	152	61	1 211	2 027	1 755	294	97
Antarctic rockcods, noties nei	*Nototheniidae*	33	-	-	-	207	-	182	180
Percoids nei	*Percoidei*	33	664	692	1 838	947	1 465	2 018	2 358
Brazilian flathead	*Percophis brasilianus*	33	9 231	11 887	10 191	7 236	7 862	7 664	4 936
Argentinian sandperch	*Pseudopercis semifasciata*	33	3 438	2 483	2 222	2 679	1 905	1 887	1 641
Spadefishes nei	*Ephippidae*	33	71	55	51	283	325	381	241
Puffers nei	*Tetraodontidae*	33	18	88	23	16	35	34	199
Argentine conger	*Conger orbignyanus*	34	89	138	189	180	122	586	655
Conger eels, etc. nei	*Congridae*	34	-	-	-	-	20	-	1 192
Pink cusk-eel	*Genypterus blacodes*	34	23 608	22 948	27 290	24 186	17 337	22 588	19 514
Cusk-eels nei	*Genypterus spp*	34	-	-	-	-	57	-	-
Cusk-eels, brotulas nei	*Ophidiidae*	34	176	258	452	544	507	676	741
Alfonsinos nei	*Beryx spp*	34	-	-	144	-	4	-	749
Wreckfish	*Polyprion americanus*	34	378	121	75	218	129	35	...
Tilefishes nei	*Branchiostegidae*	34	1 098	1 000	786	524	547	1 309	1 261
Castaneta	*Cheilodactylus bergi*	34	3 142	4 893	11 508	3 260	1 432	1 367	498
Patagonian toothfish	*Dissostichus eleginoides*	34	14 951	9 599	13 328	11 300	11 122	13 815	11 090
Patagonian rockcod	*Patagonotothen brevicauda*	34	-	-	-	-	-	-	9
Longtail Southern cod	*Patagonotothen ramsayi*	34	-	-	-	-	-	-	10
White snake mackerel	*Thyrsitops lepidopoides*	34	1 232	309	285	136	-	-	-
Largehead hairtail	*Trichiurus lepturus*	34	736	938	1 405	1 230	1 665	2 883	3 131
Hairtails, scabbardfishes nei	*Trichiuridae*	34	-	-	51	1 205	1 888	155	-
Choicy ruff	*Seriolella porosa*	34	2 042	2 593	2 462	3 322	3 542	3 990	5 368
Blackbelly rosefish	*Helicolenus dactylopterus*	34	3 703	5 967	5 581	5 943	5 170	3 045	2 038
Scorpionfishes nei	*Scorpaenidae*	34	293	99	-	-	-	4	-
Atlantic searobins	*Prionotus spp*	34	792	540	858	1 149	1 839	317	352
Anglerfishes nei	*Lophiidae*	34	294	399	542	794	1 937	8 932	6 457
Demersal percomorphs nei	*Perciformes*	34	5 725	9 886	9 917	12 390	7 699	14 325	15 077
Brazilian sardinella	*Sardinella brasiliensis*	35	97 092	117 642	82 283	25 518	17 053	39 039	27 891
Brazilian menhaden	*Brevoortia aurea*	35	6 294	2 396	2 936	2 202	1 123	344	490
Argentine menhaden	*Brevoortia pectinata*	35	511	1 112	519	291	337	368	141
Scaled sardines	*Harengula spp*	35	20	12	2	162	126	488	569
Atlantic thread herring	*Opisthonema oglinum*	35	3 762	3 712	8 082	8 204	12 455	4 397	3 191
Falkland sprat	*Sprattus fuegensis*	35	0	0	29	17	97	4	1
Argentine anchovy	*Engraulis anchoita*	35	21 023	25 211	13 417	13 025	12 164	13 133	16 203
Anchovies, etc. nei	*Engraulidae*	35	7 672	7 102	2 654	2 682	4 039	4 242	4 511
Clupeoids nei	*Clupeoidei*	35	3 464	4 314	6 042	7 527	11 457	27 058	29 481
Atlantic bonito	*Sarda sarda*	36	108	130	12	38	20	258	16
Wahoo	*Acanthocybium solandri*	36	16	58	41	...	...	...	...
King mackerel	*Scomberomorus cavalla*	36	2 890	2 398	3 595	3 595	2 344	1 251	2 316
Serra Spanish mackerel	*Scomberomorus brasiliensis*	36	3 047	2 125	1 516	1 516	988	251	3 071
Frigate and bullet tunas	*Auxis thazard, A.rochei*	36	527	215	162	166	106	98	1 117

C-41 (a)

Fish, crustaceans, molluscs, etc — Capture production by species items — Atlantic, Southwest
Poissons, crustacés, mollusques, etc — Captures par catégories d'espèces — Atlantique, sud-ouest
Peces, crustáceos, moluscos, etc — Capturas por categorías de especies — Atlántico, sudoccidental

English name Nom anglais Nombre inglés	Scientific name Nom scientifique Nombre científico	Species group Groupe d'espèces Grupo de especies	1996 t	1997 t	1998 t	1999 t	2000 t	2001 t	2002 t
Little tunny(=Atl.black skipj)	*Euthynnus alletteratus*	36	834	507	920	930	615	615	615
Skipjack tuna	*Katsuwonus pelamis*	36	22 601	26 579	23 790	23 188	25 180	24 155	18 344
Atlantic bluefin tuna	*Thunnus thynnus*	36	0	0	0	13	0	1	1
Blackfin tuna	*Thunnus atlanticus*	36	649	418	55	55	38	149	1 669
Albacore	*Thunnus alalunga*	36	16 141	13 944	11 668	12 189	8 630	11 610	11 511
Southern bluefin tuna	*Thunnus maccoyii*	36	18	5	6	3	3	-	-
Yellowfin tuna	*Thunnus albacares*	36	5 430	4 507	4 946	6 163	7 291	7 651	7 503
Bigeye tuna	*Thunnus obesus*	36	11 697	9 370	6 061	5 617	7 955	8 035	9 960
Atlantic sailfish	*Istiophorus albicans*	36	320	165	216	429	645	630	926
Blue marlin	*Makaira nigricans*	36	658	513	762	701	634	1 002	535
Atlantic white marlin	*Tetrapturus albidus*	36	398	407	361	293	291	210	409
Longbill spearfish	*Tetrapturus pfluegeri*	36	-	0	-	22	59	76	44
Marlins,sailfishes,etc. nei	*Istiophoridae*	36	33	30	62	0	18	10	1
Swordfish	*Xiphias gladius*	36	7 400	10 275	8 681	8 885	10 835	9 158	8 726
Tuna-like fishes nei	*Scombroidei*	36	2	498	359	774	338	29	531
Ballyhoo halfbeak	*Hemiramphus brasiliensis*	37	494	442	1 500	832	989	3 342	3 652
Flyingfishes nei	*Exocoetidae*	37	743	1 082	1 084	760	388	217	708
Silversides(=Sand smelts) nei	*Atherinidae*	37	637	584	102	637	57	64	51
Bluefish	*Pomatomus saltatrix*	37	6 219	3 043	2 603	2 368	3 778	3 701	4 786
Cobia	*Rachycentron canadum*	37	498	367	622	1 818	1 580	1 036	1 065
Rough scad	*Trachurus lathami*	37	976	601	328	495	107	345	455
Blue runner	*Caranx crysos*	37	503	515	443	590	625	680	688
Jacks, crevalles nei	*Caranx spp*	37	3 905	5 838	5 635	4 164	6 950	5 878	6 083
Atlantic moonfish	*Selene setapinnis*	37	1 679	1 914	1 512	1 514	1 386	1 108	1 350
Pompanos nei	*Trachinotus spp*	37	534	166	172	144	286	273	222
Yellowtail amberjack	*Seriola lalandi*	37	554	483	887	553	621	188	197
Amberjacks nei	*Seriola spp*	37	825	1 048	119	104	128	588	653
Atlantic bumper	*Chloroscombrus chrysurus*	37	952	1 035	4 065	3 519	1 573	5 304	5 757
Parona leatherjacket	*Parona signata*	37	2 364	2 236	2 209	2 070	1 854	1 551	1 314
Common dolphinfish	*Coryphaena hippurus*	37	2 500	4 028	4 117	2 848	4 359	4 299	4 818
Chub mackerel	*Scomber japonicus*	37	16 869	18 779	12 647	8 612	16 499	6 007	13 805
Mackerels nei	*Scombridae*	37	-	-	-	21	0	-	-
Butterfishes, pomfrets nei	*Stromateidae*	37	-	-	5	-	-	20	-
Barracudas nei	*Sphyraena spp*	37	11	12	828	217	527	378	372
Pelagic percomorphs nei	*Perciformes*	37	2	243	150	10	18	-	-
Thresher	*Alopias vulpinus*	38	...	...	...	...	...	3	-
Shortfin mako	*Isurus oxyrinchus*	38	83	190	...	100	135	26	279
Blue shark	*Prionace glauca*	38	743	1 103	...	500	636	988	4 233
Silky shark	*Carcharhinus falciformis*	38	502	279	...	70	80	...	...
Smooth hammerhead	*Sphyrna zygaena*	38	...	...	...	...	3	-	-
Scalloped hammerhead	*Sphyrna lewini*	38	25	170	...	30	38	507	541
Narrownose smooth-hound	*Mustelus schmitti*	38	10 456	10 130	13 422	12 274	8 157	10 766	8 140
Tope shark	*Galeorhinus galeus*	38	92	103	92	89	109	87	35
Argentine angelshark	*Squatina argentina*	38	4 281	4 410	4 311	3 368	3 123	3 339	2 288
Angelsharks, sand devils nei	*Squatinidae*	38	1 587	...	...	...	...	...	...
Chola guitarfish	*Rhinobatos percellens*	38	404	...	...	...	...	...	...
Raja rays nei	*Raja spp*	38	-	-	-	-	-	-	10
Rays, stingrays, mantas nei	*Rajiformes*	38	24 406	22 838	21 715	24 419	26 800	28 345	26 493
Elephantfishes nei	*Callorhinchus spp*	38	815	1 329	1 770	1 977	1 390	451	568
Sharks, rays, skates, etc. nei	*Elasmobranchii*	38	13 201	18 144	16 586	18 481	19 726	18 343	15 134
Marine fishes nei	*Osteichthyes*	39	158 570	165 118	157 430	181 165	187 841	165 941	176 164
Dana swimcrab	*Callinectes danae*	42	2 020	2 600	3 014	1 626	1 597	1 062	1 310
Red crab	*Geryon quinquedens*	42	1 513	3 432	2 743	3 382	5 259	2 685	2 421
Marine crabs nei	*Brachyura*	42	10 312	10 636	9 528	11 083	11 135	12 571	13 157
Caribbean spiny lobster	*Panulirus argus*	43	8 026	7 502	6 002	6 334	6 469	7 135	7 016
Southern king crab	*Lithodes antarcticus*	44	200	413	456	270	104	88	385
Softshell red crab	*Paralomis granulosa*	44	1	1	1	1	265	218	130
King crabs, stone crabs nei	*Lithodidae*	44	-	-	-	-	-	4	2
Redspotted shrimp	*Penaeus brasiliensis*	45	8 743	10 758	7 796	9 092	10 728	5 010	5 903
Sao Paulo shrimp	*Penaeus paulensis*	45	-	177	13	12	56	23	7
Penaeus shrimps nei	*Penaeus spp*	45	15 037	14 506	13 998	12 550	16 509	10 697	10 533
Atlantic seabob	*Xiphopenaeus kroyeri*	45	11 764	15 257	13 755	9 964	11 948	12 264	11 673
Argentine stiletto shrimp	*Artemesia longinaris*	45	263	166	146	37	39	283	298
Argentine red shrimp	*Pleoticus muelleri*	45	9 874	6 479	23 203	15 928	37 150	78 866	51 412
Natantian decapods nei	*Natantia*	45	-	-	-	1 090	556	2 816	13 022
Antarctic krill	*Euphausia superba*	46	-	-	74	-	4	-	-
Marine crustaceans nei	*Crustacea*	47	313	849	2 161	6 293	227	1 317	247
Angulate volute	*Zidona dufresnei*	52	558	1 322	1 010	683	1 621	1 519	1 314
Cupped oysters nei	*Crassostrea spp*	53	873	828	744	1 547	884	895	876
Chilean mussel	*Mytilus chilensis*	54	1	1	1	1	1	0	0
River Plata mussel	*Mytilus platensis*	54	370	354	375	474	412	523	440
Sea mussels nei	*Mytilidae*	54	863	1 193	1 230	906	802	2 017	1 721
Patagonian scallop	*Zygochlamis patagonica*	55	36 952	39 817	28 441	42 700	37 404	42 598	51 181
Triangular tivela	*Tivela mactroides*	56	126	196	664	1 684	273	280	835
Donax clams	*Donax spp*	56	0	0	0	0	0	0	-
Clams, etc. nei	*Bivalvia*	56	...	0	0	759	6	17	240
Patagonian squid	*Loligo gahi*	57	68 504	21 711	51 751	42 505	67 016	57 730	24 976

C-41 (a)

Fish, crustaceans, molluscs, etc	Capture production by species items	Atlantic, Southwest
Poissons, crustacés, mollusques, etc	Captures par catégories d'espèces	Atlantique, sud-ouest
Peces, crustáceos, moluscos, etc	Capturas por categorías de especies	Atlántico, sudoccidental

English name Nom anglais Nombre inglés	Scientific name Nom scientifique Nombre científico	Species group Groupe d'espèces Grupo de especies	1996 t	1997 t	1998 t	1999 t	2000 t	2001 t	2002 t
Common squids nei	*Loligo spp*	57	8 305	5 095	4 178	3 752	1 455	1 437	6 307
Argentine shortfin squid	*Illex argentinus*	57	656 481	980 300	693 542	1 144 998	930 839	743 422	511 038
Sevenstar flying squid	*Martialia hyadesi*	57	3 793	8 348	2	27	33	2	...
Various squids nei	*Loliginidae, Ommastrephidae*	57	10 157	415	-	-	-	-	-
Octopuses, etc. nei	*Octopodidae*	57	635	688	571	1 456	1 084	1 175	965
Marine molluscs nei	*Mollusca*	58	1 175	1 851	1 408	4 937	2 474	1 082	2 334
Marine turtles nei	*Testudinata*	72	0	0	0	0	0	-	-
Total			***2 479 862***	***2 756 890***	***2 379 280***	***2 622 527***	***2 369 739***	***2 287 502***	***2 089 660***

C-41 (b)

Fish, crustaceans, molluscs, etc	Capture production by countries or areas	Atlantic, Southwest
Poissons, crustacés, mollusques, etc	Captures par pays ou zones	Atlantique, sud-ouest
Peces, crustáceos, moluscos, etc	Capturas por países o áreas	Atlántico, sudoccidental

Country or area Pays ou zone País o área	1993 t	1994 t	1995 t	1996 t	1997 t	1998 t	1999 t	2000 t	2001 t	2002 t
Argentina	919 503	938 590	1 152 056	1 272 057	1 377 427	1 141 632	1 044 221	878 561	897 977	913 113
Australia	-	-	-	-	-	3 594	3 711	-	-	-
Belize	-	-	-	-	-	-	4 500	6 729	2 581	135
Brazil	530 100 F	548 615 F	513 666	522 173	565 714	532 599	518 470	567 687	589 397	604 409
Bulgaria	4 200	-	-	-	-	-	-	-	-	-
Cambodia	-	-	-	-	-	-	-	2 768	1 200	32
Chile	846	23	302	-	3 744	14 755	5 226	2 749	8 849	5 710
China	-	3	3	3	3	31 806	61 664	93 156	93 973	85 872
China,Taiwan	145 840 F	125 529 F	138 326	132 450	216 202	177 096	279 001	247 505	156 502	126 337
Estonia	1 338	-	-	2 538	-	-	-	-	1 941	777
Falkland Is	1 974	5 914	27 190	31 539	17 112	43 615	39 163	62 927	59 823	35 655
France	-	1 946	7 332	4 580	1 545	4 179	2 379	2 053	-	-
Honduras	3 655	2 976	2 771	849	-	-	-	-	-	-
Italy	5 792	5 810	960	408	-	-	-	-	-	-
Japan	151 049	127 068	97 448	92 775	147 786	99 299	175 714	138 098	82 127	26 924
Korea Rep	149 776	97 665	141 635	158 911	220 586	101 165	283 455	170 429	167 817	129 035
Latvia	5 029	6 693	1 825	4 041	-	-	-	-	-	-
Lithuania	15 171	3 275	-	3 400	-	-	-	-	-	-
Namibia	-	-	-	-	304	677	746	-	-	-
Panama	1 295	804	606	776	3	-	120	487	-	-
Philippines	-	-	-	-	-	124	47	-	330	172
Poland	21 395	13 641	9 205	3 549	-	93	5 224	970	756	2 754
Portugal	1 810	3 766	6 847	9 099	1 595	2 059	77	325	2 731	3 853
Russian Fed	47 925	23 689	11 762	20 254	884	-	-	3 404	3 214	8 286
St Vincent	-	-	-	-	-	-	-	-	1 818	-
Seychelles	-	-	-	-	1 253	101	-	-	3 800	-
Spain	63 047	72 592	96 511	93 032	65 264	84 266	95 482	84 740	108 808	39 307
Ukraine	4 719	3 202	998	340	-	-	-	-	-	-
UK	446	1 256	2 080	4 356	2 730	3 706	3 259	5 501	7 007	5 262
Uruguay	118 194	119 766	125 597	122 732	134 738	138 514	100 068	101 650	96 851	102 027
Total	***2 193 104***	***2 102 823***	***2 337 120***	***2 479 862***	***2 756 890***	***2 379 280***	***2 622 527***	***2 369 739***	***2 287 502***	***2 089 660***

C-47 (a)

Fish, crustaceans, molluscs, etc — **Capture production by species items** — **Atlantic, Southeast**
Poissons, crustacés, mollusques, etc — **Captures par catégories d'espèces** — **Atlantique, sud-est**
Peces, crustáceos, moluscos, etc — **Capturas por categorías de especies** — **Atlántico, sudoriental**

English name Nom anglais Nombre inglés	Scientific name Nom scientifique Nombre científico	Species group Groupe d'espèces Grupo de especies	1996 t	1997 t	1998 t	1999 t	2000 t	2001 t	2002 t
West coast sole	*Austroglossus microlepis*	31	409	517	393	463	571	589	644
Mud sole	*Austroglossus pectoralis*	31	909	837	859	768	800	844	702
Southeast Atlantic soles nei	*Austroglossus spp*	31	480	1 197	773	926	965	650	492
Tonguefishes	*Cynoglossidae*	31	3	-	-	-	-	10	17
Flatfishes nei	*Pleuronectiformes*	31	195	120	935	912	592	1 928	3 465
Benguela hake	*Merluccius polli*	32	-	-	-	-	-	-	1
Shallow-water Cape hake	*Merluccius capensis*	32	832	777	906	2 060	658	1 863	3 225
Cape hakes	*Merluccius capensis,M.paradox.*	32	286 443	260 955	302 542	307 642	309 337	323 073	306 387
Gadiformes nei	*Gadiformes*	32	-	-	3	-	-	-	-
Bonefish	*Albula vulpes*	33	154	-	-	-	-	-	-
Sea catfishes nei	*Ariidae*	33	19	76	132	102	459	2 435	3 616
Mullets nei	*Mugilidae*	33	1 086	972	908	757	561	125	122
Groupers nei	*Epinephelus spp*	33	48	41	59	33	17	18	14
Groupers, seabasses nei	*Serranidae*	33	1 002	460	1 417	347	2 341	4 229	5 781
Bigeyes nei	*Priacanthus spp*	33	1	1	3	7	2	1	2
Snappers, jobfishes nei	*Lutjanidae*	33	-	-	-	-	-	1	1
Bigeye grunt	*Brachydeuterus auritus*	33	146	228	262	167	191	1 181	4 639
Grunts, sweetlips nei	*Haemulidae (=Pomadasyidae)*	33	1 031	557	440	412	817	1 715	3 514
Canary drum (=Baardman)	*Umbrina canariensis*	33	...	...	...	830	...	2 207	13
Southern meagre(=Mulloway)	*Argyrosomus hololepidotus*	33	1 293	932	1 114	1 203	1 009	3 734	611
Geelbek croaker	*Atractoscion aequidens*	33	346	474	605	415	380	379	294
West African croakers nei	*Pseudotolithus spp*	33	4 044	1 606	7 096	8 936	8 005	8 744	25 971
Croakers, drums nei	*Sciaenidae*	33	4 004	4 637	3 119	2 920	2 295	2 635	3 566
Emperors(=Scavengers) nei	*Lethrinidae*	33	-	-	-	-	-	3	3
Red pandora	*Pagellus bellottii*	33	...	...	...	188	...	744	674
Sargo breams nei	*Diplodus spp*	33	-	-	-	-	-	4	-
Large-eye dentex	*Dentex macrophthalmus*	33	682	1 125	736	1 859	2 536	1 044	237
Angolan dentex	*Dentex angolensis*	33	274	-	-	-	-	-	-
Dentex nei	*Dentex spp*	33	3 086	1 567	8 683	8 758	13 765	16 080	8 948
Black seabream	*Spondyliosoma cantharus*	33	...	...	93	129	212	494	13
Carpenter seabream	*Argyrozona argyrozona*	33	883	780	505	541	500	287	249
Santer seabream	*Cheimerius nufar*	33	40	34	0	29	25	25	15
Pargo breams nei	*Pagrus spp*	33	95	716	1 077	183	1 875	4 606	5 773
Red steenbras	*Petrus rupestris*	33	28	35	...	22	10	7	6
Panga seabream	*Pterogymnus laniarius*	33	1 200	1 323	1 042	1 349	9 503	6 956	3 798
White stumpnose	*Rhabdosargus globiceps*	33	237	165	296	335	300	169	107
Daggerhead breams nei	*Chrysoblephus spp*	33	176	139	154	74	70	78	57
White steenbras	*Lithognathus lithognathus*	33	7	8	9	2	...	...	...
Sand steenbras	*Lithognathus mormyrus*	33	214	176	1 072	726	999	1 735	2 802
Steenbrasses nei	*Lithognathus spp*	33	6	136	2	116	56	20	4
Salema	*Sarpa salpa*	33	4	3	0	1	...	1	...
Porgies, seabreams nei	*Sparidae*	33	2 705	2 919	1 805	4 958	979	179	270
Picarels nei	*Spicara spp*	33	7	0	55	...	13	-	-
West African goatfish	*Pseudupeneus prayensis*	33	-	0	-	-	-	-	...
Threadfins, tasselfishes nei	*Polynemidae*	33	370	267	520	1 378	872	1 834	2 664
Gobies nei	*Gobiidae*	33	552	287	20 998	16	3	...	...
Hector's lanternfish	*Lampanyctodes hectoris*	34	33	243	6 553	0	...	...	...
Conger eels, etc. nei	*Congridae*	34	1	3	3	3	3	6	2
Kingklip	*Genypterus capensis*	34	6 807	5 955	5 594	7 820	7 922	11 460	13 418
Alfonsinos nei	*Beryx spp*	34	2 552	4 261	1 810	126	302	318	236
Orange roughy	*Hoplostethus atlanticus*	34	13 379	18 538	10 957	3 473	1 542	857	2 169
John dory	*Zeus faber*	34	1 471	1 593	2 328	2 706	2 586	2 618	3 582
Boarfishes nei	*Caproidae*	34	-	-	5	-	-	-	-
Oreo dories nei	*Oreosomatidae*	34	17	188	6	42	10	54	335
Wreckfish	*Polyprion americanus*	34	-	6	42	20	8	-	2
Cape bonnetmouth	*Emmelichthys nitidus*	34	216	191	124	113	50	37	-
Bonnetmouths, rubyfishes nei	*Emmelichthyidae*	34	-	-	-	-	-	6	-
Pelagic armourhead	*Pseudopentaceros richardsoni*	34	281	18	-	-	-	-	-
Patagonian toothfish	*Dissostichus eleginoides*	34	-	-	-	-	320	5	906
Snoek	*Thyrsites atun*	34	13 037	12 033	14 836	12 402	11 166	12 027	12 969
Oilfish	*Ruvettus pretiosus*	34	-	5	-	-	-	-	-
Largehead hairtail	*Trichiurus lepturus*	34	...	97	119	468	1 895	1 871	208
Silver scabbardfish	*Lepidopus caudatus*	34	3 196	2 003	4 557	2 560	2 300	2 316	4 523
Hairtails, scabbardfishes nei	*Trichiuridae*	34	-	5	-	-	-	-	-
Ruffs, barrelfishes nei	*Centrolophidae*	34	36	-	-	-	-	-	-
Blackbelly rosefish	*Helicolenus dactylopterus*	34	1 759	1 630	1 095	1 214	1 389	1 673	821
Scorpionfishes nei	*Scorpaenidae*	34	-	53	21	4	2	-	6
Cape gurnard	*Chelidonichthys capensis*	34	497	642	839	578	686	609	696
Gurnards, searobins nei	*Triglidae*	34	0	0	-	-	-	-	-
Devil anglerfish	*Lophius vomerinus*	34	15 433	17 222	24 457	21 783	21 402	21 996	23 721
Demersal percomorphs nei	*Perciformes*	34	27	1 147	25	...	...	...	...
Sardinellas nei	*Sardinella spp*	35	17 713	23 753	55 020	79 748	113 856	58 339	30 026
Southern African pilchard	*Sardinops ocellatus*	35	106 381	144 680	196 581	175 969	161 448	200 100	264 886
Whitehead's round herring	*Etrumeus whiteheadi*	35	67 773	97 279	57 669	59 032	38 877	56 762	64 450
Southern African anchovy	*Engraulis capensis*	35	41 792	62 640	110 296	180 954	267 986	289 323	254 643
Atlantic bonito	*Sarda sarda*	36	39	32	-	2	159	1 193	10
Wahoo	*Acanthocybium solandri*	36	23	19	10	15	15	22	25
Frigate and bullet tunas	*Auxis thazard, A.rochei*	36	29	12	31	2	50	231	122
Little tunny(=Atl.black skipj)	*Euthynnus alletteratus*	36	235	75	406	118	132	231	155
Skipjack tuna	*Katsuwonus pelamis*	36	167	400	338	47	141	1 012	116
Atlantic bluefin tuna	*Thunnus thynnus*	36	16	120	3	-	-	-	2
Albacore	*Thunnus alalunga*	36	12 177	12 140	20 388	15 397	14 510	14 885	14 479
Southern bluefin tuna	*Thunnus maccoyii*	36	2 242	451	2 069	2 094	2 333	2 516	1 104
Yellowfin tuna	*Thunnus albacares*	36	4 604	3 047	6 029	5 392	4 686	2 844	2 233
Bigeye tuna	*Thunnus obesus*	36	19 778	15 561	15 807	22 160	23 568	19 010	19 198

C-47 (a)

Fish, crustaceans, molluscs, etc | **Capture production by species items** | **Atlantic, Southeast**
Poissons, crustacés, mollusques, etc | **Captures par catégories d'espèces** | **Atlantique, sud-est**
Peces, crustáceos, moluscos, etc | **Capturas por categorías de especies** | **Atlántico, sudoriental**

English name Nom anglais Nombre inglés	Scientific name Nom scientifique Nombre científico	Species group Groupe d'espèces Grupo de especies	1996 t	1997 t	1998 t	1999 t	2000 t	2001 t	2002 t
Atlantic sailfish	*Istiophorus albicans*	36	50	37	157	61	91	89	179
Blue marlin	*Makaira nigricans*	36	832	666	618	587	372	501	445
Black marlin	*Makaira indica*	36	3	1	97	22	62	618	318
Atlantic white marlin	*Tetrapturus albidus*	36	180	136	248	215	307	58	25
Longbill spearfish	*Tetrapturus pfluegeri*	36	-	1	3	28	11	7	5
Marlins,sailfishes,etc. nei	*Istiophoridae*	36	15	61	34	4	5	33	25
Swordfish	*Xiphias gladius*	36	8 195	5 623	4 499	5 361	4 587	4 797	4 376
Tuna-like fishes nei	*Scombroidei*	36	46	217	86	421	270	453	7 748
Bluefish	*Pomatomus saltatrix*	37	38	15	4	33	20	170	2
Cape horse mackerel	*Trachurus capensis*	37	470 372	407 482	482 086	398 598	423 607	354 046	386 361
Cunene horse mackerel	*Trachurus trecae*	37	78 230	87 380	46 004	81 692	69 630	46 832	45 226
Jack and horse mackerels nei	*Trachurus spp*	37	33	45	...	...	...	...	...
Crevalle jack	*Caranx hippos*	37	39	16	89	302	1 835	674	271
Jacks, crevalles nei	*Caranx spp*	37	-	-	-	-	-	1	-
Yellowtail amberjack	*Seriola lalandi*	37	488	477	519	302	300	315	216
Leerfish	*Lichia amia*	37	-	-	-	-	-	1	21
Atlantic bumper	*Chloroscombrus chrysurus*	37	-	2	-	-	-	-	-
Carangids nei	*Carangidae*	37	42	230	83	3	39	806	584
Atlantic pomfret	*Brama brama*	37	1 192	420	322	432	843	1 121	890
Common dolphinfish	*Coryphaena hippurus*	37	1	1	1	1	1	1	2
Chub mackerel	*Scomber japonicus*	37	4 212	11 550	5 598	2 927	5 427	10 050	6 936
Mackerels nei	*Scombridae*	37	-	11	-	-	-	-	-
Blue butterfish	*Stromateus fiatola*	37	0	1	21	17	28	-	83
Barracudas nei	*Sphyraena spp*	37	...	...	...	198	-	1	3
Ocean sunfish	*Mola mola*	37	...	...	...	...	...	2	1
Pelagic percomorphs nei	*Perciformes*	37	-	-	-	2 427	-	-	-
Shortfin mako	*Isurus oxyrinchus*	38	...	587	308	338	764	388	2 078
Blue shark	*Prionace glauca*	38	...	3 560	1 471	2 251	4 711	3 734	2 514
Smooth hammerhead	*Sphyrna zygaena*	38	...	220	103	...	4	5	3
Smooth-hound	*Mustelus mustelus*	38	...	...	...	...	...	...	2
Tope shark	*Galeorhinus galeus*	38	...	...	...	...	...	...	19
Raja rays nei	*Raja spp*	38	1 047	1 269	1 824	2 399	1 593	2 581	3 996
Rays, stingrays, mantas nei	*Rajiformes*	38	360	474	523	720	1 341	1 527	268
Cape elephantfish	*Callorhinchus capensis*	38	366	484	482	356	380	405	422
Sharks, rays, skates, etc. nei	*Elasmobranchii*	38	1 285	1 187	1 820	2 270	1 895	6 581	4 524
Marine fishes nei	*Osteichthyes*	39	90 506	106 528	59 888	60 929	47 629	91 020	99 275
Geryons nei	*Geryon spp*	42	5 329	3 415	4 145	3 939	5 836	4 161	4 121
Marine crabs nei	*Brachyura*	42	...	...	1 494	916	1 039	1 090	1 179
Tropical spiny lobsters nei	*Panulirus spp*	43	0	-	-	-	-	-	-
Cape rock lobster	*Jasus lalandii*	43	1 767	1 879	2 076	2 097	2 058	1 974	3 029
Tristan da Cunha rock lobster	*Jasus tristani*	43	327	321	376	336	316	425	301
Southern spiny lobster	*Palinurus gilchristi*	43	918	892	864	429	305	1 053	651
Slipper lobsters nei	*Scyllaridae*	43	0	0	1	1	1	0	-
Lobsters nei	*Reptantia*	43	20	25	31	68	10	-	8
Caramote prawn	*Penaeus kerathurus*	45	3	-	-	-	-	-	-
Deepwater rose shrimp	*Parapenaeus longirostris*	45	3 479	4 246	3 535	1 891	4 209	5 626	2 000
Striped red shrimp	*Aristeus varidens*	45	1 578	1 323	2 808	1 827	3 024	3 405	2 046
Natantian decapods nei	*Natantia*	45	855	-	9 887	1 615	3 188	4 298	1 798
Antarctic krill	*Euphausia superba*	46	-	-	254	-	-	-	-
Marine crustaceans nei	*Crustacea*	47	2	-	-	-	-	1 062	1 653
Perlemoen abalone	*Haliotis midae*	52	735	330	524	481	490	527	516
Cupped oysters nei	*Crassostrea spp*	53	...	...	...	-	-	-	1
Donax clams	*Donax spp*	56	0	-	-	-	-	-	36
Cuttlefish,bobtail squids nei	*Sepiidae, Sepiolidae*	57	124	-	88	33	90	31	101
Cape Hope squid	*Loligo reynaudi*	57	7 549	3 696	6 670	7 169	6 000	3 373	7 405
Various squids nei	*Loliginidae, Ommastrephidae*	57	106	259	461	665	1 111	1 121	1 534
Octopuses, etc. nei	*Octopodidae*	57	391	190	582	434	305	129	291
Marine molluscs nei	*Mollusca*	58	-	-	-	-	-	-	2 000
Jellyfishes	*Rhopilema spp*	77	-	-	-	-	106	44	2 193
Aquatic invertebrates nei	*Invertebrata*	77	-	-	-	-	-	-	19
Total			***1 325 437***	***1 355 366***	***1 537 315***	***1 528 601***	***1 633 902***	***1 648 084***	***1 701 440***

C-47 (b)

Fish, crustaceans, molluscs, etc — Capture production by countries or areas — Atlantic, Southeast
Poissons, crustacés, mollusques, etc — Captures par pays ou zones — Atlantique, sud-est
Peces, crustáceos, moluscos, etc — Capturas por países o áreas — Atlántico, sudoriental

Country or area Pays ou zone País o área	1993 t	1994 t	1995 t	1996 t	1997 t	1998 t	1999 t	2000 t	2001 t	2002 t
Angola	119 200	125 413	116 781	131 815	140 304	157 149	169 799	232 351	246 518	254 797
Cambodia	-	-	-	-	-	-	56	-	-	-
Chile	-	-	-	-	-	-	-	-	5	-
China	-	24	29	24	121	48	5 743	7 209	4 959	3 861
China,Taiwan	14 958 F	18 655 F	19 249	11 394	8 934	18 173	18 811	18 970	16 118	16 070
Cuba	-	-	-	-	-	-	2 427	-	-	-
Estonia	31 447	31 372	28 836	-	-	-	-	-	-	-
Georgia	2 000 F	1 000 F	-	-	-	-	-	-	-	-
Honduras	193	...	48	10	25	9	20	...	...	-
Iceland	-	-	-	-	924	340	-	-	-	-
Italy	-	-	109	46	-	-	-	-	-	-
Japan	39 774	38 106	32 690	26 160	18 924	20 659	18 284	17 286	11 729	10 068 F
Korea Rep	7 711	11 242	9 216	6 414	9 548	8 020	7 822	7 225	5 026	5 983
Lithuania	20 784	2 823	-	-	-	-	-	-	-	-
Namibia	789 132	647 999	568 633	516 628	511 412	605 654	577 838	588 404	545 998	623 391
Norway	...	...	-	-	864	1 087	-	242	-	-
Panama	48	158	146	30	3	-	19	-	-	-
Philippines	-	-	-	-	-	6	52	-	-	-
Poland	-	1	3 178	1 736	1 964	797	-	-	3 100	4 318
Portugal	798	1 556	1 248	1 130	996	819	1 129	3 393	2 084	1 552
Russian Fed	220 550	226 712	178 288	138 583	129 014	128 280	123 453	82 283	39 850	8 611
St Helena	726	702	915	819	897	1 060	632	718	866	598
Seychelles	-	-	-	-	-	-	-	-	-	181
South Africa	560 716	522 129	574 374	439 229	511 249	555 852	585 240	640 000 F	747 470	763 285
Spain	25 145 F	20 759 F	16 750 F	12 555	14 331	20 990	12 601	17 403	20 068	7 805
Ukraine	96 420	53 356	19 759	28 405	5 701	18 345	4 675	18 096	4 283	-
Uruguay	-	-	-	-	-	-	-	320	-	906
Other nei	253	3 208	19 915	10 459	155	27	-	2	10	14
Total	***1 929 855***	***1 705 215***	***1 590 164***	***1 325 437***	***1 355 366***	***1 537 315***	***1 528 601***	***1 633 902***	***1 648 084***	***1 701 440***

C-48 (a)

Fish, crustaceans, molluscs, etc — Capture production by species items — Atlantic, Antarctic
Poissons, crustacés, mollusques, etc — Captures par catégories d'espèces — Atlantique, Antarctique
Peces, crustáceos, moluscos, etc — Capturas por categorías de especies — Atlántico, Antártico

English name Nom anglais Nombre inglés	Scientific name Nom scientifique Nombre científico	Species group Groupe d'espèces Grupo de especies	1996 t	1997 t	1998 t	1999 t	2000 t	2001 t	2002 t
Lefteye flounders nei	*Bothidae*	31	-	-	-	-	0	-	-
Antarctic armless flounder	*Mancopsetta maculata*	31	-	-	-	-	0	-	-
Smalleye moray cod	*Muraenolepis microps*	32	-	-	-	-	0	-	0
Moray cods nei	*Muraenolepis spp*	32	-	-	-	0	0	-	-
Blue antimora	*Antimora rostrata*	32	0	2	1	0	-	0	-
Whitson's grenadier	*Macrourus whitsoni*	32	-	-	-	-	1	0	-
Grenadiers nei	*Macrourus spp*	32	25	19	17	13	2	6	-
Roundnose grenadier	*Coryphaenoides rupestris*	32	-	-	-	-	-	-	9
Marbled rockcod	*Notothenia rossii*	33	-	-	-	0	0	0	5
Humped rockcod	*Notothenia gibberifrons*	33	-	-	-	5	1	2	1
Yellowbelly rockcod	*Notothenia neglecta*	33	-	-	-	0	-	2	-
Grey rockcod	*Notothenia squamifrons*	33	-	-	-	5	5	0	0
Painted notie	*Nototheniops larseni*	33	-	-	-	-	0	-	0
Yellowfin notie	*Nototheniops nudifrons*	33	-	-	-	-	0	0	-
Antarctic rockcods nei	*Trematomus spp*	33	-	-	-	0	-	0	-
Striped rockcod	*Pagothenia hansoni*	33	-	-	-	-	0	-	-
Antarctic silverfish	*Pleuragramma antarcticum*	33	-	-	-	-	-	0	-
Antarctic rockcods, noties nei	*Nototheniidae*	33	-	0	1	-	-	0	10
Nichol's lanternfish	*Gymnoscopelus nicholsi*	34	-	-	-	-	0	-	0
Lanternfishes nei	*Myctophidae*	34	-	-	-	5	67	-	-
Antarctic toothfish	*Dissostichus mawsoni*	34	-	-	1	0	-	0	-
Patagonian toothfish	*Dissostichus eleginoides*	34	3 602	3 812	3 201	3 636	4 939	4 048	5 744
Patagonian rockcod	*Patagonotothen brevicauda*	34	-	-	-	3	0	-	0
...A	*Parachaenichthys georgianus*	34	-	-	-	-	0	-	-
Blackfin icefish	*Chaenocephalus aceratus*	34	-	-	-	1	0	1	5
Mackerel icefish	*Champsocephalus gunnari*	34	-	-	6	266	4 114	960	2 667
South Georgia icefish	*Pseudochaenichthys georgianus*	34	-	-	-	3	0	6	6
Ocellated icefish	*Chionodraco rastrospinosus*	34	-	-	-	1	-	1	-
Spiny icefish	*Chaenodraco wilsoni*	34	-	-	-	0	-	0	-
Icefishes nei	*Channichthyidae*	34	-	-	-	0	-	0	0
Dusky catshark	*Halaelurus canescens*	38	-	-	-	-	-	-	0
Antarctic starry skate	*Raja georgiana*	38	-	-	-	-	0	-	-
Eaton's skate	*Bathyraja eatonii*	38	-	-	-	-	-	0	-
...A	*Bathyraja meridionalis*	38	-	-	-	-	0	-	-
Rays, stingrays, mantas nei	*Rajiformes*	38	45	32	8	17	0	13	-
Sharks, rays, skates, etc. nei	*Elasmobranchii*	38	-	-	-	0	-	-	-
Marine fishes nei	*Osteichthyes*	39	-	-	7	0	-	0	1
King crabs	*Paralithodes spp*	44	-	-	-	0	-	-	-
Subantarctic stone crab	*Lithodes murrayi*	44	-	-	1	0	-	-	-
King crabs nei	*Lithodes spp*	44	-	-	-	0	-	-	-
Red stone crab	*Paralomis aculeata*	44	-	-	-	0	-	-	-
Antarctic stone crab	*Paralomis spinosissima*	44	214	0	-	0	0	4	55
Globose king crab	*Paralomis formosa*	44	-	-	-	2	2	11	56
King crabs, stone crabs nei	*Lithodidae*	44	-	1	0	0	0	-	-
Antarctic krill	*Euphausia superba*	46	91 150	75 653	90 025	101 957	114 426	104 183	125 987
Argentine shortfin squid	*Illex argentinus*	57	-	-	-	-	-	18	49
Sevenstar flying squid	*Martialia hyadesi*	57	52	81	-	-	-	2	-
Antarctic octopuses	*Pareledone spp*	57	-	-	-	-	0	-	-
Jellyfishes	*Rhopilema spp*	77	-	-	-	-	5	-	0
Total			***95 088***	***79 600***	***93 268***	***105 914***	***123 562***	***109 257***	***134 595***

Data in this table refer to catches for which the new fishing season (1 December - 30 November) of the Commission for the Conservation of Antarctic Marine Living Resources (CCAMLR) has been adopted. Data are shown under the calendar year in which the split-year ends.

Les données dans ce tableau se réfèrent aux captures pour lesquelles on utilise la nouvelle période de pêche (1er décembre - 30 novembre) de la Commission pour la conservation de la faune et la flore marines de l'Antarctique (CCAMLR). Les données figurent sous l'année civile durant laquelle se termine l'année fractionnée.

Los datos en esto cuadro se refieren a las capturas para las que se utiliza el nuevo período de pesca (1 de diciembre - 30 de noviembre) de la Comisión para la Conservación de los Recursos Vivos Marinos Antárticos (CCAMLR). Los datos se incluyen en el año civil en que termina el año emergente.

C-48 (b)

Fish, crustaceans, molluscs, etc — Capture production by countries or areas — Atlantic, Antarctic
Poissons, crustacés, mollusques, etc — Captures par pays ou zones — Atlantique, Antarctique
Peces, crustáceos, moluscos, etc — Capturas por países o áreas — Atlántico, Antártico

Country or area Pays ou zone País o área	1993 t	1994 t	1995 t	1996 t	1997 t	1998 t	1999 t	2000 t	2001 t	2002 t
Argentina	-	12	879	109	-	-	6 534	-	-	-
Bulgaria	193	250	-	-	-	-	-	-	-	-
Chile	5 386	3 985	2 174	2 821	2 061	1 402	1 313	2 106	899	1 545
France	-	-	-	-	-	-	-	-	386	-
Japan	50 845	60 198	62 111	58 769	61 013	67 481	66 076	80 597	67 378	51 191
Korea Rep	-	146	425	513	642	2 850	340	7 645	8 321	15 257
Latvia	71	-	-	-	-	-	-	-	-	-
Panama	-	-	637	-	-	-	-	-	-	-
Poland	4 390	7 997	12 521	22 104	14 408	19 060	19 167	20 049	13 717	16 720
Russian Fed	2 584	121	12	103	-	-	273	3 462	225	1 686
South Africa	-	3	-	-	-	859	128	364	359	332
Spain	-	-	-	-	492	-	184	308	643	832
Ukraine	458	12 613	59 150	10 277	-	-	6 719	164	14 172	32 015
UK	-	12	-	-	984	1 354	1 072	1 457	1 161	2 149
USA	-	-	292	392	-	-	16	70	1 568	12 175
Uruguay	-	-	-	-	-	262	4 092	7 340	428	693
Total	***63 927***	***85 337***	***138 201***	***95 088***	***79 600***	***93 268***	***105 914***	***123 562***	***109 257***	***134 595***

Data in this table refer to catches for which the new fishing season (1 December - 30 November) of the Commission for the Conservation of Antarctic Marine Living Resources (CCAMLR) has been adopted. Data are shown under the calendar year in which the split-year ends.

Les données dans ce tableau se réfèrent aux captures pour lesquelles on utilise la nouvelle période de pêche (1er décembre - 30 novembre) de la Commission pour la conservation de la faune et la flore marines de l'Antarctique (CCAMLR). Les données figurent sous l'année civile durant laquelle se termine l'année fractionnée.

Los datos en esto cuadro se refieren a las capturas para las que se utiliza el nuevo período de pesca (1 de diciembre - 30 de noviembre) de la Comisión para la Conservación de los Recursos Vivos Marinos Antárticos (CCAMLR). Los datos se incluyen en el año civil en que termina el año emergente.

C-51 (a)

Fish, crustaceans, molluscs, etc — Capture production by species items — Indian Ocean, Western
Poissons, crustacés, mollusques, etc — Captures par catégories d'espèces — Océan Indien, ouest
Peces, crustáceos, moluscos, etc — Capturas por categorías de especies — Océano Indico, occidental

English name Nom anglais Nombre inglés	Scientific name Nom scientifique Nombre científico	Species group Groupe d'espèces Grupo de especies	1996 t	1997 t	1998 t	1999 t	2000 t	2001 t	2002 t
Kelee shad	*Hilsa kelee*	24	3 837	3 634	3 077	3 076	3 896	4 277	5 525
Hilsa shad	*Tenualosa ilisha*	24	1 710	1 631	1 530	1 472	840	507	514
Milkfish	*Chanos chanos*	25	181	125	139	137	125	134	159
Barramundi(=Giant seaperch)	*Lates calcarifer*	25	214	209	196	204	-	-	-
Lefteye flounders nei	*Bothidae*	31	-	0	9	27	129	125	88
Tonguefishes	*Cynoglossidae*	31	2 286	2 523	2 158	2 066	2 137	1 923	1 989
Indian halibut	*Psettodes erumei*	31	880	811	945	850	802	1 025	1 110
Flatfishes nei	*Pleuronectiformes*	31	17 564	21 965	17 564	12 935	24 804	11 574	15 729
Unicorn cod	*Bregmaceros mcclellandi*	32	1 251	1 338	1 449	743	1 470	2 435	2 372
Gadiformes nei	*Gadiformes*	32	-	10	7	5	2	1	-
Bombay-duck	*Harpadon nehereus*	33	159 463	187 982	144 865	146 663	133 221	141 082	100 366
Greater lizardfish	*Saurida tumbil*	33	46	28	22	0	0	0	0
Brushtooth lizardfish	*Saurida undosquamis*	33	47	53	34	20	30	32	32
Lizardfishes nei	*Synodontidae*	33	20 846	16 562	20 221	20 871	17 918	12 454	11 504
Sea catfishes nei	*Ariidae*	33	80 113	89 654	96 464	94 676	76 749	86 283	87 463
Squirrelfishes nei	*Holocentridae*	33	40	50	34	53	46	40	97
Flathead grey mullet	*Mugil cephalus*	33	39	65	69	89	60	34	35
Mullets nei	*Mugilidae*	33	27 126	30 450	28 006	24 220	21 140	22 113	22 249
Fusiliers	*Caesio spp*	33	...	...	15	18	17	15	31
Groupers nei	*Epinephelus spp*	33	16 578	16 536	22 076	26 494	27 215	27 171	26 525
Groupers, seabasses nei	*Serranidae*	33	13 858	14 551	15 452	16 167	32 564	36 420	33 389
Therapon pearch	*Terapon spp*	33	-	-	1	2	0	1	2
Spotted seabass	*Dicentrarchus punctatus*	33	2	35	36	241	10	18	8
Bigeyes nei	*Priacanthus spp*	33	0	...	2	1	0	1	0
Cardinalfishes, etc. nei	*Apogonidae*	33	...	...	...	...	...	262	293
Sillago-whitings	*Sillaginidae*	33	289	266	218	201	194	204	210
Mangrove red snapper	*Lutjanus argentimaculatus*	33	2 002	2 394	3 192	3 195	3 003	2 900	3 000
Snappers nei	*Lutjanus spp*	33	1 036	903	1 025	1 384	778	1 212	967
Snappers, jobfishes nei	*Lutjanidae*	33	11 846	13 960	17 856	13 553	15 508	14 610	12 463
Threadfin breams nei	*Nemipterus spp*	33	2 303	1 676	3 926	9 843	13 774	15 629	13 351
Threadfin and dwarf breams nei	*Nemipteridae*	33	2 750	4 703	3 469	4 066	2 917	4 506	6 565
Ponyfishes(=Slipmouths) nei	*Leiognathidae*	33	11 074	13 122	9 899	9 757	5 369	7 089	10 828
Silver grunt	*Pomadasys argenteus*	33	0	0	3	...	472	199	152
Grunts, sweetlips nei	*Haemulidae (=Pomadasyidae)*	33	12 320	12 444	12 582	14 784	20 525	19 879	19 920
Meagre	*Argyrosomus regius*	33	0	0	19	82	-	-	-
Southern meagre(=Mulloway)	*Argyrosomus hololepidotus*	33	21	17	...	...	19	20	27
Geelbek croaker	*Atractoscion aequidens*	33	15	18	...	15	...	...	12
Tigertooth croaker	*Otolithes ruber*	33	-	-	13	10	...	...	...
Croakers, drums nei	*Sciaenidae*	33	275 086	300 376	256 965	309 854	261 700	237 658	264 944
Emperors(=Scavengers) nei	*Lethrinidae*	33	43 755	45 071	46 725	49 274	60 100	64 421	65 406
Pandoras nei	*Pagellus spp*	33	...	...	...	5	...	...	1
Dentex nei	*Dentex spp*	33	-	-	2	-	-	-	-
King soldier bream	*Argyrops spinifer*	33	2 784	3 013	2 992	3 015	3 803	4 575	5 124
Santer seabream	*Cheimerius nufar*	33	33	28	...	...	...	...	31
Daggerhead breams nei	*Chrysoblephus spp*	33	92	85	...	75	...	...	67
Yellowfin seabream	*Acanthopagrus latus*	33	234	249	280	464	350	584	402
Twobar seabream	*Acanthopagrus bifasciatus*	33	...	...	...	...	...	1 586	332
Porgies, seabreams nei	*Sparidae*	33	15 085	14 731	11 840	17 598	19 063	16 536	24 674
Goatfishes	*Upeneus spp*	33	11 036	6 766	7 621	8 253	9 171	15 627	16 072
Goatfishes, red mullets nei	*Mullidae*	33	886	930	638	1 088	1 093	2 770	1 912
Mojarras(=Silver-biddies) nei	*Gerres spp*	33	1 864	2 164	2 208	2 209	2 592	2 864	1 257
Blue sea chub	*Kyphosus cinerascens*	33	-	8	-	1	-	-	17
Wrasses, hogfishes, etc. nei	*Labridae*	33	3 833	3 305	3 066	3 155	3 291	3 579	3 153
Parrotfishes nei	*Scaridae*	33	842	1 179	1 048	1 642	1 123	1 164	1 589
Angelfishes nei	*Pomacanthidae*	33	0	0	7	17	1	3	9
Fourfinger threadfin	*Eleutheronema tetradactylum*	33	516	1 783	969	...	63	55	60
Threadfins, tasselfishes nei	*Polynemidae*	33	1 823	1 535	2 000	2 829	2 248	2 323	2 356
Percoids nei	*Percoidei*	33	90 233	59 477	52 583	54 015	58 778	40 280	73 033
Surgeonfishes nei	*Acanthuridae*	33	238	173	253	205	149	343	364
Batfishes	*Platax spp*	33	-	0	1	7	1	0	0
Spadefishes nei	*Ephippidae*	33	-	-	17	1	1	2	2
Spinefeet(=Rabbitfishes) nei	*Siganus spp*	33	10 995	9 928	9 002	8 589	10 778	11 992	11 110
Flatheads nei	*Platycephalidae*	33	0	0	18	28	10	17	330
Puffers nei	*Tetraodontidae*	33	-	-	-	0	0	18	0
Triggerfishes, durgons nei	*Balistidae*	33	291	26	8	11	6	7	24
Lanternfishes nei	*Myctophidae*	34	0	0	0	0	0	335	37
Pike-congers nei	*Muraenesox spp*	34	11 418	12 965	11 456	13 796	11 177	10 897	10 835
Conger eels, etc. nei	*Congridae*	34	-	6	64	19	53	-	-
Alfonsinos nei	*Beryx spp*	34	3 079	1 031	859	1 964	1 668	585	10
Orange roughy	*Hoplostethus atlanticus*	34	-	-	-	-	1 265	711	38
John dory	*Zeus faber*	34	-	3	4	5	6	6	...
Boarfishes nei	*Caproidae*	34	-	-	-	-	-	7	-
Oreo dories nei	*Oreosomatidae*	34	-	-	-	-	175	180	97
Wreckfish	*Polyprion americanus*	34	-	-	-	-	-	1	-
Bonnetmouths, rubyfishes nei	*Emmelichthyidae*	34	28	7	275	181	-	86	-
Pelagic armourhead	*Pseudopentaceros richardsoni*	34	17	33	78	108	121	12	-
Patagonian toothfish	*Dissostichus eleginoides*	34	-	-	-	-	1 628	7 002	4 798
Antarctic toothfishes nei	*Dissostichus spp*	34	-	-	-	-	-	122	-
Cardinal fishes nei	*Epigonus spp*	34	-	-	-	-	-	-	3
Largehead hairtail	*Trichiurus lepturus*	34	9 073	11 583	12 337	31 635	28 756	27 355	28 440
Hairtails, scabbardfishes nei	*Trichiuridae*	34	52 550	148 035	70 469	101 970	110 164	91 012	128 522
Ruffs, barrelfishes nei	*Centrolophidae*	34	254	440	395	753	396	299	-
Blackbelly rosefish	*Helicolenus dactylopterus*	34	-	-	-	-	-	1	-
Scorpionfishes nei	*Scorpaenidae*	34	-	3	1	1	2	2	...
Gurnards, searobins nei	*Triglidae*	34	-	2	-	...	0	0	...

English name Nom anglais Nombre inglés	Scientific name Nom scientifique Nombre científico	Species group Groupe d'espèces Grupo de especies	1996 t	1997 t	1998 t	1999 t	2000 t	2001 t	2002 t
Demersal percomorphs nei	*Perciformes*	34	14 375	15 469	22 212	16 241	13 253	16 450	21 254
Indian oil sardine	*Sardinella longiceps*	35	152 012	197 296	184 002	183 437	353 353	383 289	389 574
Sardinellas nei	*Sardinella spp*	35	31 089	20 839	19 659	29 843	27 897	25 043	23 931
Southern African pilchard	*Sardinops ocellatus*	35	-	1	...	...	0	0	-
Red-eye round herring	*Etrumeus teres*	35	-	-	-	2 135	...	...	...
Stolephorus anchovies	*Stolephorus spp*	35	9 633	10 295	10 329	10 587	2 729	4 030	6 400
Anchovies, etc. nei	*Engraulidae*	35	83 500	86 400	83 493	70 797	76 254	78 844	80 813
Dorab wolf-herring	*Chirocentrus dorab*	35	1 580	1 931	2 059	2 273	2 785	2 613	2 743
Wolf-herrings nei	*Chirocentrus spp*	35	5 874	11 998	10 279	4 743	6 888	7 646	12 306
Clupeoids nei	*Clupeoidei*	35	89 567	104 273	105 952	103 738	76 761	76 430	70 886
Striped bonito	*Sarda orientalis*	36	370	498	162	134	95	287	378
Wahoo	*Acanthocybium solandri*	36	94	79	77	79	107	138	47
Dogtooth tuna	*Gymnosarda unicolor*	36	625	489	470	426	451	647	789
Narrow-barred Spanish mackerel	*Scomberomorus commerson*	36	63 706	67 058	70 123	67 013	71 501	65 214	78 924
Indo-Pacific king mackerel	*Scomberomorus guttatus*	36	12 350	12 674	18 360	16 009	13 150	12 189	27 969
Streaked seerfish	*Scomberomorus lineolatus*	36	59	558	67	58	34	24	...
Seerfishes nei	*Scomberomorus spp*	36	4 417	6 588	4 469	4 589	4 118	4 209	1 282
Frigate and bullet tunas	*Auxis thazard, A.rochei*	36	18 732	13 341	12 188	18 033	21 576	18 907	11 179
Kawakawa	*Euthynnus affinis*	36	30 491	39 489	33 514	38 309	41 196	38 014	25 731
Skipjack tuna	*Katsuwonus pelamis*	36	226 704	229 340	226 532	313 078	302 092	299 776	396 071
Longtail tuna	*Thunnus tonggol*	36	41 564	43 465	38 901	41 935	59 222	50 120	45 688
Albacore	*Thunnus alalunga*	36	20 592	19 134	30 148	29 928	21 441	15 384	18 800
Southern bluefin tuna	*Thunnus maccoyii*	36	2 580	3 945	3 958	2 496	2 678	3 310	3 704
Yellowfin tuna	*Thunnus albacares*	36	256 560	228 567	197 297	231 622	231 562	220 659	242 726
Bigeye tuna	*Thunnus obesus*	36	72 970	85 527	85 230	94 916	94 866	87 627	91 528
Indo-Pacific sailfish	*Istiophorus platypterus*	36	4 312	3 724	2 980	4 375	4 651	5 332	6 813
Blue marlin	*Makaira nigricans*	36	2 802	3 324	5 087	4 819	5 635	5 021	4 742
Black marlin	*Makaira indica*	36	399	368	399	264	617	160	240
Striped marlin	*Tetrapturus audax*	36	3 306	2 857	2 702	1 836	2 322	1 966	385
Shortbill spearfish	*Tetrapturus angustirostris*	36	2	2	5	7	5	5	34
Marlins,sailfishes,etc. nei	*Istiophoridae*	36	8 440	7 466	5 989	5 339	6 104	5 320	2 934
Swordfish	*Xiphias gladius*	36	16 930	21 946	27 162	24 230	20 905	20 104	24 409
Tuna-like fishes nei	*Scombroidei*	36	8 797	10 479	17 661	14 451	13 726	9 930	12 533
Needlefishes nei	*Tylosurus spp*	37	359	420	276	397	250	286	678
Halfbeaks nei	*Hemiramphus spp*	37	2 576	3 064	2 545	3 189	2 955	2 668	3 785
Flyingfishes nei	*Exocoetidae*	37	133	184	202	401	166	159	107
Silversides(=Sand smelts) nei	*Atherinidae*	37	-	-	-	78	-	-	-
False trevally	*Lactarius lactarius*	37	6 403	7 060	8 000	5 503	4 623	5 992	4 268
Cobia	*Rachycentron canadum*	37	2 016	1 913	1 624	1 516	3 862	4 002	4 333
Jack and horse mackerels nei	*Trachurus spp*	37	1 380	1 440	1 615	1 611	1 400	1 813	2 060
Scads nei	*Decapterus spp*	37	2 819	3 158	5 464	6 673	5 205	4 946	4 975
Jacks, crevalles nei	*Caranx spp*	37	64 921	50 385	70 017	71 079	37 641	44 690	34 117
Snubnose pompano	*Trachinotus blochii*	37	-	-	4	0	31	-	-
Pompanos nei	*Trachinotus spp*	37	5 706	3 697	2 533	2 329	12	11	12
Yellowtail amberjack	*Seriola lalandi*	37	-	1	...	...	...	...	...
Amberjacks nei	*Seriola spp*	37	130	334	340	179	146	152	81
Black pomfret	*Parastromateus niger*	37	2 221	2 322	2 109	2 917	2 027	1 975	2 002
Rainbow runner	*Elagatis bipinnulata*	37	...	415	98	132	5	52	4
Golden trevally	*Gnathanodon speciosus*	37	441	471	536	489	1 125	1 286	2 644
Torpedo scad	*Megalaspis cordyla*	37	3 954	3 119	2 122	2 498	3 117	1 825	1 950
Queenfishes	*Scomberoides spp*	37	3 239	3 288	3 327	3 436	1 889	2 107	1 736
Yellowstripe scad	*Selaroides leptolepis*	37	2 908	3 108	3 118	3 196	2 635	3 034	5 320
Carangids nei	*Carangidae*	37	66 644	67 290	58 616	60 933	42 920	39 613	59 146
Common dolphinfish	*Coryphaena hippurus*	37	1 897	1 719	1 953	3 393	2 203	2 010	2 017
Chub mackerel	*Scomber japonicus*	37	2 053	2 397	813	385	3 561	2 747	1 615
Indian mackerel	*Rastrelliger kanagurta*	37	282 311	173 395	157 031	153 400	77 074	47 310	55 151
Indian mackerels nei	*Rastrelliger spp*	37	409	356	351	449	424	303	282
Mackerels nei	*Scombridae*	37	12	10	23	41	24	23	14
Silver pomfret	*Pampus argenteus*	37	893	576	515	289	243	158	165
Butterfishes, pomfrets nei	*Stromateidae*	37	22 558	22 008	19 400	19 672	14 362	12 566	22 756
Barracudas nei	*Sphyraena spp*	37	14 902	14 151	20 704	22 118	12 066	11 382	15 295
Pelagic percomorphs nei	*Perciformes*	37	64 966	64 007	74 296	85 318	69 631	85 309	98 020
Shortfin mako	*Isurus oxyrinchus*	38	...	...	18	...	58	95	381
Porbeagle	*Lamna nasus*	38	-	-	-	-	-	1	-
Blue shark	*Prionace glauca*	38	...	...	60	...	575	1 123	3 304
Dusky shark	*Carcharhinus obscurus*	38	-	7	0	...	...	...	...
Requiem sharks nei	*Carcharhinidae*	38	34 483	31 235	36 184	32 573	28 384	26 642	27 201
Guitarfishes, etc. nei	*Rhinobatidae*	38	1 422	1 481	1 564	1 643	2 185	1 945	2 011
Rays, stingrays, mantas nei	*Rajiformes*	38	20 050	19 844	21 550	24 841	24 700	24 660	23 795
Sharks, rays, skates, etc. nei	*Elasmobranchii*	38	121 795	65 579	61 339	59 981	70 445	67 210	84 601
Marine fishes nei	*Osteichthyes*	39	610 747	661 111	632 249	662 681	633 757	740 547	699 449
Portunus swimcrabs nei	*Portunus spp*	42	1 047	1 289	1 017	2 179	2 380	2 557	2 832
Indo-Pacific swamp crab	*Scylla serrata*	42	23	19	23	21	25	24	22
Geryons nei	*Geryon spp*	42	564	1 144	1 002	875	886	685	920
Marine crabs nei	*Brachyura*	42	6 018	7 362	9 041	8 764	17 661	12 431	12 203
Tropical spiny lobsters nei	*Panulirus spp*	43	2 488	2 547	2 822	2 459	2 239	2 243	2 370
St.Paul rock lobster	*Jasus paulensis*	43	357	295	308	345	192	183	334
Natal spiny lobster	*Palinurus delagoae*	43	10	10	6	7	8	10	7
Spiny lobsters nei	*Palinuridae*	43	332	233	239	204	228	213	276
Slipper lobsters nei	*Scyllaridae*	43	133	74	67	20	9	26	20
Mozambique lobster	*Metanephrops mozambicus*	43	132	156	192	152	180	141	130
Lobsters nei	*Reptantia*	43	-	-	-	-	62	2	-
Giant tiger prawn	*Penaeus monodon*	45	112 483	106 762	168 026	169 080	145 996	100 295	112 511

C-51 (a)

Fish, crustaceans, molluscs, etc — Capture production by species items — Indian Ocean, Western
Poissons, crustacés, mollusques, etc — Captures par catégories d'espèces — Océan Indien, ouest
Peces, crustáceos, moluscos, etc — Capturas por categorías de especies — Océano Indico, occidental

English name Nom anglais Nombre inglés	Scientific name Nom scientifique Nombre científico	Species group Groupe d'espèces Grupo de especies	1996 t	1997 t	1998 t	1999 t	2000 t	2001 t	2002 t
Green tiger prawn	*Penaeus semisulcatus*	45	0	-	0	0	0	0	0
Penaeus shrimps nei	*Penaeus spp*	45	25 820	26 056	25 181	20 855	24 524	24 300	22 989
Metapenaeus shrimps nei	*Metapenaeus spp*	45	7 602	6 801	6 204	6 791	7 126	7 246	6 555
Deepwater rose shrimp	*Parapenaeus longirostris*	45	-	-	-	-	-	9	-
Parapenaeopsis shrimps nei	*Parapenaeopsis spp*	45	14 047	16 722	14 689	12 889	11 945	11 576	10 222
Knife shrimp	*Haliporoides triarthrus*	45	1 771	1 510	1 882	1 611	1 766	1 738	1 441
Natantian decapods nei	*Natantia*	45	136 123	115 045	120 574	111 085	119 866	146 014	127 825
Marine crustaceans nei	*Crustacea*	47	15 600	12 367	12 797	11 460	11 251	27 066	31 290
Abalones nei	*Haliotis spp*	52	43	40	40	29	45	51	50
Cupped oysters nei	*Crassostrea spp*	53	32	16	9	8	2	1	1
Clams, etc. nei	*Bivalvia*	56	57	8	19	26	30	11	19
Cuttlefish,bobtail squids nei	*Sepiidae, Sepiolidae*	57	12 907	23 236	14 778	23 302	20 121	23 880	26 461
Cape Hope squid	*Loligo reynaudi*	57	-	-	-	-	-	-	1
Common squids nei	*Loligo spp*	57	30	37	37	10	22	41	11
Various squids nei	*Loliginidae, Ommastrephidae*	57	3 628	5 820	4 764	6 714	5 130	5 197	4 581
Octopuses, etc. nei	*Octopodidae*	57	1 199	1 454	1 277	1 326	1 169	1 383	1 422
Cephalopods nei	*Cephalopoda*	57	77 104	118 032	91 576	89 953	92 724	51 638	82 954
Marine molluscs nei	*Mollusca*	58	15 939	7 363	5 015	3 747	5 034	3 580	2 613
Sea squirts nei	*Ascidiacea*	74	-	-	-	-	95	-	-
Sea cucumbers nei	*Holothurioidea*	76	7 258	7 293	3 371	1 767	2 139	2 229	3 158
Total			***3 897 309***	***3 983 306***	***3 797 772***	***4 067 446***	***3 995 996***	***3 981 292***	***4 243 330***

C-51 (b)

Fish, crustaceans, molluscs, etc | **Capture production by countries or areas** | **Indian Ocean, Western**
Poissons, crustacés, mollusques, etc | **Captures par pays ou zones** | **Océan Indien, ouest**
Peces, crustáceos, moluscos, etc | **Capturas por países o áreas** | **Océano Indico, occidental**

Country or area Pays ou zone País o área	1993 t	1994 t	1995 t	1996 t	1997 t	1998 t	1999 t	2000 t	2001 t	2002 t
Bahrain	8 958	7 628	9 389	12 940	10 050	9 849	10 620	11 718	11 230	11 204
Br Ind Oc Tr	0	0	0	0	0	0	0	0	0	0
China	-	-	-	-	-	-	294	2 587	2 773	3 594
China,Taiwan	121 654 F	63 484 F	88 540 F	77 278	74 522	96 352	88 423	92 017	95 003	95 106
Comoros	11 645	12 976	13 000 F	12 700 F	12 500 F	12 500 F	12 000	13 200	12 180	12 200 F
Djibouti	300 F	320 F	350 F	350 F	350 F	350 F	350 F	350 F	350 F	350 F
Egypt	50 740	48 300	47 300	48 434	57 417	57 063	82 400	75 972	73 577	72 889
Eritrea	475	2 706	3 559	3 252	1 038	1 629	6 891	12 612	8 820	7 832
France	93 057	99 908	95 917	82 854	68 920	48 720	80 483	84 544	69 152	96 719
Fr South Tr	460	524	519 F	437 F	375 F	388 F	425 F	272 F	263 F	414 F
India	1 744 857	1 904 799	1 847 631	1 956 204	2 007 772	1 832 799	1 911 318	1 850 839	1 862 686	1 991 081
Iran	246 985	218 944	252 583	242 437	238 486	226 500	243 800	260 500	262 805	269 000
Iraq	2 133	4 221	5 253	11 688	10 783	13 463	13 093	12 389	8 500 F	5 000 F
Israel	80	110	150	225	171	137	98	...	120	...
Italy	8 689	8 717	1 650	702	3 351	4 520	6 890	102	3 199	4 760
Japan	52 958	32 730	14 273	21 152	28 796	31 030	20 377	19 542	24 739	26 547
Jordan	45	60	75	90	100	120	160	150	170	176
Kenya	5 617	3 772	5 465	6 296	6 099	6 600	6 634	4 763	7 388	6 720
Korea Rep	29 879	25 322	22 765	31 573	37 976	25 692	19 943	21 435	10 878	14 457
Kuwait	8 466	7 752	8 616	8 255	7 826	7 799	6 271	6 000 F	5 846	5 900 F
Lithuania	-	11	-	-	-	-	-	-	-	-
Madagascar	84 261	86 431	85 653	84 475	86 391	96 395	99 630	102 093	105 583	111 284
Maldives	100 852	118 168	119 048	120 508	116 257	128 968	133 547	132 427	126 190	160 981
Mauritius	20 576	18 145	16 395	11 869	14 025	12 093	12 205	9 615	10 985	10 706
Mayotte	500 F	600 F	1 033 F	1 553	2 101 F	3 172 F	3 130	3 030 F	11 051	4 299
Mozambique	25 506	22 531	21 740	27 405	28 035	27 683	27 123	25 916	24 790	24 306
Norway	-	-	-	-	-	-	-	870	-	-
Oman	105 772	118 572	139 861	121 618	118 995	106 171	108 809	120 421	129 907	142 670
Pakistan	499 159	418 574	404 444	395 397	422 265	433 456	474 665	437 601	420 698	418 104
Philippines	-	-	-	-	-	2 513	2 190	1 985	1 453	1 216
Portugal	-	-	-	-	-	450	752	1 417	3 339	2 177
Qatar	6 994	5 086	4 271	4 739	5 032	5 279	4 207	7 142	8 606	6 880
Réunion	1 679	2 531	2 500	3 607	4 288	4 579	4 043	4 082	3 635	3 700
Russian Fed	10 893	19 444	-	-	-	-	-	-	221	123
Saudi Arabia	48 021	54 612	45 609	47 698	49 314	51 206	46 618	49 761	49 167	55 330
Seychelles	5 178	4 469	4 008	4 707	11 741	20 000	34 202	32 460	49 369	62 622
Somalia	27 500 F	29 600	27 700 F	25 800 F	23 900 F	22 000 F	20 000 F	20 000 F	19 800 F	17 850 F
South Africa	826	517	373	349	892	1 100	1 000 F	926	1 017	1 586
Spain	105 651	113 354	147 583	138 976	141 630	98 163	148 089	148 207	135 958	175 884
Sudan	2 500	4 000	4 000	4 500	5 000	5 500	5 500	5 000 F	5 000	5 000
Tanzania	36 685	40 785	42 771	61 645	50 210	48 000	50 489	52 779	52 900	49 680
Ukraine	1 491	2 633	2 970	3 480	1 570	1 626	3 356	2 015	810	-
Untd Arab Em	99 600	108 600	105 884	107 000	114 358	114 739	117 607	105 456	112 561	97 574
UK	-	-	-	-	-	-	-	-	-	28
Uruguay	-	-	-	-	-	-	-	1 628	7 150	2 591
Yemen	82 356	81 885	107 970	104 955	115 600	127 620	124 385	114 751	142 198	159 262
Other nei	98 610	83 790	113 363	110 161	105 170	111 548	135 429	147 422	99 225	105 528
Total	***3 751 608***	***3 776 611***	***3 814 211***	***3 897 309***	***3 983 306***	***3 797 772***	***4 067 446***	***3 995 996***	***3 981 292***	***4 243 330***

C-57 (a)

Fish, crustaceans, molluscs, etc	Capture production by species items	Indian Ocean, Eastern
Poissons, crustacés, mollusques, etc	Captures par catégories d'espèces	Océan Indien, est
Peces, crustáceos, moluscos, etc	Capturas por categorías de especies	Océano Indico, oriental

English name Nom anglais Nombre inglés	Scientific name Nom scientifique Nombre científico	Species group Groupe d'espèces Grupo de especies	1996 t	1997 t	1998 t	1999 t	2000 t	2001 t	2002 t
Short-finned eel	*Anguilla australis*	22	208	203	160	129	133	175	167
Chacunda gizzard shad	*Anodontostoma chacunda*	24	2 167	1 916	1 474	3 009	2 645	2 769	2 350
Kelee shad	*Hilsa kelee*	24	48 452	47 524	48 475	41 468	43 604	44 502	43 674
Hilsa shad	*Tenualosa ilisha*	24	135 358	131 204	124 105	140 710	140 367	154 654	152 343
Toli shad	*Tenualosa toli*	24	429	476	617	630	528	681	720
Indian pellona	*Pellona ditchela*	24	11 227	10 919	6 113	4 448	7 069	6 884	7 236
Diadromous clupeoids nei	*Clupeoidei*	24	107	102	22	45	13	50	305
Barramundi(=Giant seaperch)	*Lates calcarifer*	25	13 989	13 790	23 061	15 685	15 982	15 665	16 535
Sand flounders	*Rhombosolea spp*	31	26	26	26	29	1	...	...
Tonguefishes	*Cynoglossidae*	31	10 285	10 915	11 682	11 316	10 922	9 602	9 577
Indian halibut	*Psettodes erumei*	31	10 863	12 212	7 779	7 472	7 411	9 668	10 570
Flatfishes nei	*Pleuronectiformes*	31	9 650	10 290	8 717	8 148	8 545	6 930	8 081
Unicorn cod	*Bregmaceros mcclellandi*	32	36	13	1 138	1 352	159	92	108
Blue grenadier	*Macruronus novaezelandiae*	32	2 680	2 851	4 752	6 209	8 964	7 313	9 013
Gadiformes nei	*Gadiformes*	32	-	-	-	5	20	35	9
Indo-Pacific tarpon	*Megalops cyprinoides*	33	6	4	13	29	31	12	11
Bombay-duck	*Harpadon nehereus*	33	27 132	27 642	37 823	38 037	36 083	35 374	35 993
Lizardfishes nei	*Synodontidae*	33	21 665	21 510	51 137	22 861	28 719	29 601	30 807
Sea catfishes nei	*Ariidae*	33	60 907	59 149	53 098	64 274	64 833	68 064	61 273
Eeltail catfishes	*Plotosus spp*	33	881	712	742	888	1 437	1 240	1 169
Mullets nei	*Mugilidae*	33	23 954	23 796	23 814	24 590	29 766	28 437	26 242
Fusiliers	*Caesio spp*	33	4 198	5 329	4 862	5 569	5 107	6 054	6 291
Groupers nei	*Epinephelus spp*	33	15 396	17 000	18 202	16 296	16 300	16 389	17 181
Groupers, seabasses nei	*Serranidae*	33	4 159	3 296	4 294	2 645	2 666	3 072	2 976
Bigeyes nei	*Priacanthus spp*	33	15 911	16 371	16 721	13 213	12 103	13 868	13 656
Sillago-whitings	*Sillaginidae*	33	4 330	5 550	5 302	8 094	8 232	6 831	6 218
Ruff	*Arripis georgianus*	33	1 302	1 287	1 008	1 066	1 143	992	860
Australian salmon	*Arripis trutta*	33	3 507	3 457	3 623	3 354	4 232	3 932	4 097
Mangrove red snapper	*Lutjanus argentimaculatus*	33	662	658	1 227	615	789	608	627
Snappers nei	*Lutjanus spp*	33	13 478	11 886	13 126	14 235	13 889	15 321	15 674
Snappers, jobfishes nei	*Lutjanidae*	33	6 322	5 439	4 288	6 528	5 581	6 833	7 134
Threadfin breams nei	*Nemipterus spp*	33	41 795	40 520	54 583	40 546	44 429	39 888	39 778
Monocle breams	*Scolopsis spp*	33	1	1	2	96	86	88	121
Ponyfishes(=Slipmouths)	*Leiognathus spp*	33	101	78	86	84	161	71	609
Ponyfishes(=Slipmouths) nei	*Leiognathidae*	33	73 735	72 277	57 754	69 136	62 360	64 516	63 202
Silver grunt	*Pomadasys argenteus*	33	625	539	666	548	581	507	639
Grunts, sweetlips nei	*Haemulidae (=Pomadasyidae)*	33	4 359	4 079	4 274	4 553	4 777	3 921	4 102
Southern meagre(=Mulloway)	*Argyrosomus hololepidotus*	33	44	44	44	96	286	363	362
Croakers, drums nei	*Sciaenidae*	33	92 873	86 686	92 464	80 541	80 069	80 335	82 666
Emperors(=Scavengers) nei	*Lethrinidae*	33	5 046	5 921	6 791	7 060	6 648	6 115	6 670
Silver seabream	*Pagrus auratus*	33	3 293	3 213	2 487	2 588	2 501	3 391	3 453
Porgies, seabreams nei	*Sparidae*	33	278	116	159	991	2 468	1 509	1 454
Goatfishes	*Upeneus spp*	33	19 553	24 306	26 294	28 522	23 328	25 098	26 636
Spotted sicklefish	*Drepane punctata*	33	51	23	30	58	53	26	59
Wrasses, hogfishes, etc. nei	*Labridae*	33	84	80	80	100	85	88	91
Threadfins, tasselfishes nei	*Polynemidae*	33	13 576	13 197	15 136	15 040	12 483	13 143	13 639
Grey rockcod	*Notothenia squamifrons*	33	-	-	-	-	363	329	329
Percoids nei	*Percoidei*	33	26 513	28 741	10 268	36 350	36 264	38 142	28 885
Spinefeet(=Rabbitfishes) nei	*Siganus spp*	33	81	88	237	27	109	87	49
Flatheads nei	*Platycephalidae*	33	1 640	1 477	989	3 102	2 782	2 432	2 864
Puffers nei	*Tetraodontidae*	33	158	122	115	3	442	432	534
Triggerfishes, durgons nei	*Balistidae*	33	708	543	250	248	115	193	221
Daggertooth pike conger	*Muraenesox cinereus*	34	2 793	4 441	4 277	3 568	2 711	2 747	1 979
Pike-congers nei	*Muraenesox spp*	34	4 097	4 192	5 189	8 182	5 748	6 441	3 521
Pink cusk-eel	*Genypterus blacodes*	34	1 397	1 923	1 830	1 881	1 148	1 153	1 216
Redfish	*Centroberyx affinis*	34	-	-	-	-	337	408	237
Orange roughy	*Hoplostethus atlanticus*	34	357	350	4 857	7 553	4 974	5 197	3 936
John dory	*Zeus faber*	34	3	0	1	21	561	238	33
Mirror dory	*Zenopsis nebulosus*	34	4	9	37	...	70	111	279
Morwongs	*Nemadactylus spp*	34	77	94	95	638	604	776	716
Trumpeters nei	*Latridae*	34	0	0	0	153	139	77	70
Patagonian toothfish	*Dissostichus eleginoides*	34	-	-	-	-	-	-	1 847
Antarctic toothfishes nei	*Dissostichus spp*	34	-	-	-	-	-	450	-
Snoek	*Thyrsites atun*	34	400	300	200	88	120	154	161
Silver gemfish	*Rexea solandri*	34	2 000	...	...	4	447	482	223
Largehead hairtail	*Trichiurus lepturus*	34	14 236	16 405	30 261	15 504	10 694	12 596	12 300
Hairtails, scabbardfishes nei	*Trichiuridae*	34	28 219	31 618	34 759	42 056	38 191	29 414	30 859
South Pacific breams nei	*Seriolella spp*	34	3 406	4 020	3 356	3 327	3 449	4 192	4 009
Scorpionfishes nei	*Scorpaenidae*	34	-	-	-	-	7	8	20
Bluefin gurnard	*Chelidonichthys kumu*	34	0	1	1	0	117	163	143
Latchet(=Sharpbeak gurnard)	*Pterygotrigla polyommata*	34	0	60	25	55	66	78	88
Demersal percomorphs nei	*Perciformes*	34	11 468	9 100	9 210	10 440	14 910	14 490	16 320
Goldstripe sardinella	*Sardinella gibbosa*	35	29 835	26 335	31 714	34 785	37 358	40 375	42 480
Indian oil sardine	*Sardinella longiceps*	35	71 343	91 846	63 063	25 461	49 583	55 448	20 320
Bali sardinella	*Sardinella lemuru*	35	29 646	72 662	92 286	34 141	31 236	37 810	40 170
Sardinellas nei	*Sardinella spp*	35	53 086	51 084	56 813	54 321	42 276	47 476	46 471
Rainbow sardine	*Dussumieria acuta*	35	5 013	4 909	4 729	6 072	5 081	4 951	5 360
Stolephorus anchovies	*Stolephorus spp*	35	74 648	81 165	85 442	77 779	75 084	78 305	88 523
Anchovies, etc. nei	*Engraulidae*	35	53 524	52 432	48 490	45 730	39 701	46 105	38 994
Dorab wolf-herring	*Chirocentrus dorab*	35	7 551	7 744	7 648	8 451	6 715	7 014	6 865
Wolf-herrings nei	*Chirocentrus spp*	35	20 784	20 021	12 861	20 884	17 743	17 769	17 271
Clupeoids nei	*Clupeoidei*	35	90 378	89 619	87 706	86 057	89 345	96 688	101 880
Wahoo	*Acanthocybium solandri*	36	150	183	223	529	596	525	475

C-57 (a) Fish, crustaceans, molluscs, etc / Poissons, crustacés, mollusques, etc / Peces, crustáceos, moluscos, etc — Capture production by species items / Captures par catégories d'espèces / Capturas por categorías de especies — Indian Ocean, Eastern / Océan Indien, est / Océano Indico, oriental

English name Nom anglais Nombre inglés	Scientific name Nom scientifique Nombre científico	Species group Groupe d'espèces Grupo de especies	1996 t	1997 t	1998 t	1999 t	2000 t	2001 t	2002 t
Narrow-barred Spanish mackerel	*Scomberomorus commerson*	36	27 016	25 685	31 048	40 602	33 959	33 970	54 425
Indo-Pacific king mackerel	*Scomberomorus guttatus*	36	18 606	16 967	20 416	27 477	16 909	18 857	13 842
Streaked seerfish	*Scomberomorus lineolatus*	36	37	343	40	77	21	14	431
Seerfishes nei	*Scomberomorus spp*	36	9 846	10 770	10 686	8 515	8 405	7 606	7 674
Frigate and bullet tunas	*Auxis thazard, A.rochei*	36	19 061	20 085	20 464	17 519	19 023	15 138	25 270
Kawakawa	*Euthynnus affinis*	36	28 047	28 808	27 898	15 633	20 757	19 128	26 952
Skipjack tuna	*Katsuwonus pelamis*	36	72 009	84 296	101 559	108 972	106 219	108 978	101 556
Longtail tuna	*Thunnus tonggol*	36	26 227	24 205	21 650	8 947	8 973	8 598	10 752
Albacore	*Thunnus alalunga*	36	9 943	8 261	8 660	8 712	6 851	5 027	4 950
Southern bluefin tuna	*Thunnus maccoyii*	36	9 819	9 055	9 806	12 670	9 396	9 190	8 238
Yellowfin tuna	*Thunnus albacares*	36	64 380	76 771	85 978	90 831	68 978	57 218	57 385
Bigeye tuna	*Thunnus obesus*	36	51 156	58 583	56 805	55 603	48 343	38 511	38 928
Indo-Pacific sailfish	*Istiophorus platypterus*	36	5 898	7 188	8 787	7 737	9 580	8 339	7 737
Blue marlin	*Makaira nigricans*	36	2 678	3 152	3 137	3 844	3 233	1 681	2 047
Black marlin	*Makaira indica*	36	618	678	646	738	726	331	396
Striped marlin	*Tetrapturus audax*	36	1 621	1 530	1 460	1 445	1 272	469	261
Shortbill spearfish	*Tetrapturus angustirostris*	36	...	...	...	1	...	0	6
Marlins,sailfishes,etc. nei	*Istiophoridae*	36	8 530	10 145	11 312	11 477	8 954	7 475	6 043
Swordfish	*Xiphias gladius*	36	8 742	10 721	11 404	9 617	14 416	12 159	10 085
Tuna-like fishes nei	*Scombroidei*	36	75 973	77 841	76 329	81 758	85 866	70 816	70 672
Needlefishes nei	*Tylosurus spp*	37	6 584	8 335	8 335	7 848	9 154	8 141	8 420
Halfbeaks nei	*Hemiramphus spp*	37	5 199	4 967	3 940	3 991	6 684	5 441	5 543
Flyingfishes nei	*Exocoetidae*	37	6 458	7 200	8 117	6 219	7 224	6 938	7 560
False trevally	*Lactarius lactarius*	37	532	1 097	1 745	1 416	1 456	994	927
Bluefish	*Pomatomus saltatrix*	37	56	-	-	-	-	-	-
Cobia	*Rachycentron canadum*	37	76	111	87	111	136	150	152
Greenback horse mackerel	*Trachurus declivis*	37	-	-	-	-	323	82	594
White trevally	*Pseudocaranx dentex*	37	-	-	-	-	38	26	18
Indian scad	*Decapterus russelli*	37	41 305	37 611	41 459	36 536	30 448	32 537	41 634
Scads nei	*Decapterus spp*	37	36 295	33 650	39 494	41 593	42 044	43 133	44 400
Jacks, crevalles nei	*Caranx spp*	37	31 814	31 295	30 370	28 285	33 278	33 276	33 100
Pompanos nei	*Trachinotus spp*	37	1 078	-	205	-	-	-	-
Amberjacks nei	*Seriola spp*	37	100	10	10	1	2	2	1
Black pomfret	*Parastromateus niger*	37	10 967	11 404	10 872	12 116	11 160	11 989	12 831
Rainbow runner	*Elagatis bipinnulata*	37	1 739	1 573	6 565	1 910	1 721	2 252	3 028
Torpedo scad	*Megalaspis cordyla*	37	29 963	29 607	32 212	27 379	24 201	23 304	24 737
Queenfishes	*Scomberoides spp*	37	5 067	2 519	2 645	2 968	2 723	2 790	3 010
Bigeye scad	*Selar crumenophthalmus*	37	1 984	1 904	3 830	3 442	3 180	2 656	2 600
Yellowstripe scad	*Selaroides leptolepis*	37	3 656	5 505	3 643	4 871	3 437	3 693	5 861
Blackbanded trevally	*Seriolina nigrofasciata*	37	3 666	3 564	2 954	2 700	2 227	2 781	2 722
Carangids nei	*Carangidae*	37	47 458	47 043	53 672	54 419	57 723	56 499	59 756
Common dolphinfish	*Coryphaena hippurus*	37	-	-	3	-	-	-	-
Indian mackerel	*Rastrelliger kanagurta*	37	49 635	56 231	66 039	62 265	30 096	33 253	34 125
Indian mackerels nei	*Rastrelliger spp*	37	189 157	184 119	195 397	191 516	172 815	181 633	172 526
Mackerels nei	*Scombridae*	37	18 202	20 002	20 900	22 828	22 642	17 205	17 715
Silver pomfret	*Pampus argenteus*	37	8 572	9 202	9 569	8 022	7 255	8 270	8 215
Butterfishes, pomfrets nei	*Stromateidae*	37	18 207	18 970	18 397	16 453	15 123	15 654	16 639
Barracudas nei	*Sphyraena spp*	37	21 566	21 537	23 592	23 659	21 981	22 805	23 463
Pelagic percomorphs nei	*Perciformes*	37	4 000	-	-	-	-	-	-
Silky shark	*Carcharhinus falciformis*	38	21 000	15 000	20 875	20 700	14 130	15 870	18 510
Smooth-hounds nei	*Mustelus spp*	38	3 878	4 169	2 858	2 462	1 796	2 485	2 415
Tope shark	*Galeorhinus galeus*	38	...	...	...	...	498	325	350
Sawsharks nei	*Pristiophorus spp*	38	...	...	...	...	270	423	371
Angelsharks, sand devils nei	*Squatinidae*	38	102	129	120	102	98	71	118
Rays, stingrays, mantas nei	*Rajiformes*	38	19 142	21 191	25 060	21 999	21 933	21 619	23 766
Ghost shark	*Callorhinchus milii*	38	...	49	21	14	82	105	102
Sharks, rays, skates, etc. nei	*Elasmobranchii*	38	73 083	71 306	78 295	76 193	80 745	75 527	70 983
Marine fishes nei	*Osteichthyes*	39	1 350 880	1 503 736	1 693 158	1 660 958	1 882 441	1 999 007	2 231 001
Blue swimming crab	*Portunus pelagicus*	42	8 048	8 120	13 287	11 697	10 763	12 524	13 826
Indo-Pacific swamp crab	*Scylla serrata*	42	5 124	5 135	5 180	5 611	6 985	6 743	7 219
Marine crabs nei	*Brachyura*	42	10 112	11 273	12 911	14 121	29 116	27 635	28 280
Australian spiny lobster	*Panulirus cygnus*	43	9 902	9 896	10 400	13 065	14 605	11 353	9 050
Tropical spiny lobsters nei	*Panulirus spp*	43	1 259	1 324	1 051	1 301	1 432	1 883	2 082
Southern rock lobster	*Jasus novaehollandiae*	43	4 856	4 888	4 615	4 655	4 756	4 677	4 386
Flathead lobster	*Thenus orientalis*	43	299	177	533	37	23	184	178
Slipper lobsters nei	*Scyllaridae*	43	21	0	0	0	0	-	-
Metanephrops nei	*Metanephrops spp*	43	-	-	-	-	39	105	88
Banana prawn	*Penaeus merguiensis*	45	16 830	15 537	15 232	14 660	15 032	14 778	15 196
Giant tiger prawn	*Penaeus monodon*	45	45 069	49 815	45 690	64 050	67 473	68 108	55 901
Green tiger prawn	*Penaeus semisulcatus*	45	2 427	2 449	2 180	2 212	1 584	1 867	1 798
Western king prawn	*Penaeus latisulcatus*	45	1 674	1 724	1 501	1 518	407	1 175	1 132
Penaeus shrimps nei	*Penaeus spp*	45	18 033	18 323	16 336	18 481	16 318	15 514	15 116
Endeavour shrimp	*Metapenaeus endeavouri*	45	...	...	...	...	38	23	28
Metapenaeus shrimps nei	*Metapenaeus spp*	45	11 160	13 922	12 424	13 978	11 254	12 920	14 451
Sergestid shrimps nei	*Sergestidae*	45	17 797	15 629	10 741	9 172	9 753	8 596	7 874
Natantian decapods nei	*Natantia*	45	104 707	114 176	78 442	110 768	119 401	109 956	104 705
Stomatopods nei	*Stomatopoda*	47	-	0	1	116	9	22	21
Marine crustaceans nei	*Crustacea*	47	40 568	45 940	52 271	54 329	36 389	41 343	43 308
Blacklip abalone	*Haliotis rubra*	52	5 081	4 905	4 920	5 297	5 207	5 304	5 845
Cupped oysters nei	*Crassostrea spp*	53	-	-	-	-	350	298	180
Australian mussel	*Mytilus planulatus*	54	71	-	-	-	-	-	-

C-57 (a)

Fish, crustaceans, molluscs, etc — **Capture production by species items** — **Indian Ocean, Eastern**
Poissons, crustacés, mollusques, etc — **Captures par catégories d'espèces** — **Océan Indien, est**
Peces, crustáceos, moluscos, etc — **Capturas por categorías de especies** — **Océano Indico, oriental**

English name Nom anglais Nombre inglés	Scientific name Nom scientifique Nombre científico	Species group Groupe d'espèces Grupo de especies	1996 t	1997 t	1998 t	1999 t	2000 t	2001 t	2002 t
Horse mussels nei	*Modiolus spp*	54	-	-	-	-	21	20	20
Scallops nei	*Pectinidae*	55	6 872	6 093	5 543	7 622	8 317	4 421	2 525
Blood cockle	*Anadara granosa*	56	-	-	20	7	24	-	-
Anadara clams nei	*Anadara spp*	56	21 227	22 348	15 443	10 603	9 538	10 559	11 170
Hard clams nei	*Meretrix spp*	56	3 693	3 722	6 639	3 420	2 606	2 383	2 730
Short neck clams nei	*Paphia spp*	56	18 329	11 625	6 204	27 366	23 130	20 364	20 060
Pipi wedge clam	*Paphies australis*	56	454	830	1 041	976	1 085	1 251	1 085
Clams, etc. nei	*Bivalvia*	56	17 745	24 663	5 097	572	5 115	2 983	2 606
Cuttlefish,bobtail squids nei	*Sepiidae, Sepiolidae*	57	35 675	35 752	37 298	36 962	41 249	39 233	40 922
Common squids nei	*Loligo spp*	57	31 229	31 567	34 618	30 726	31 280	34 239	36 871
Various squids nei	*Loliginidae, Ommastrephidae*	57	15 498	20 686	17 956	20 953	29 868	23 635	22 779
Octopuses, etc. nei	*Octopodidae*	57	11 649	12 226	16 360	14 232	14 386	13 628	13 692
Cephalopods nei	*Cephalopoda*	57	10 596	9 212	9 497	8 917	10 470	9 251	11 000
Marine molluscs nei	*Mollusca*	58	381	263	844	8 396	1 940	1 941	1 440
Marine turtles nei	*Testudinata*	72	353	248	337	273	372	45	71
Sea cucumbers nei	*Holothurioidea*	76	419	610	833	859	1 048	787	920
Jellyfishes	*Rhopilema spp*	77	22 807	29 547	6 660	34 980	36 093	24 471	22 045
Aquatic invertebrates nei	*Invertebrata*	77	720	200	417	544	593	757	792
Total			***4 190 529***	***4 459 036***	***4 700 229***	***4 647 945***	***4 785 689***	***4 877 380***	***5 100 261***

C-57 (b)

Fish, crustaceans, molluscs, etc — **Capture production by countries or areas** — **Indian Ocean, Eastern**
Poissons, crustacés, mollusques, etc — **Captures par pays ou zones** — **Océan Indien, est**
Peces, crustáceos, moluscos, etc — **Capturas por países o áreas** — **Océano Indico, oriental**

Country or area Pays ou zone País o área	1993 t	1994 t	1995 t	1996 t	1997 t	1998 t	1999 t	2000 t	2001 t	2002 t
Australia	127 895 F	105 408 F	107 555 F	105 022 F	102 142	98 875	127 563	115 145	111 498	116 177
Bangladesh	312 715	253 044	264 650	279 170	295 141	300 452	309 797	333 799	379 497	415 420
China	-	-	445	1 497	2 964	3 080	5 868	4 919	4 009	1 474
China,Taiwan	18 314 F	16 961 F	25 964 F	32 507	41 049	22 549	22 732	23 241	14 308	19 384
Christmas Is	0	0	0	0	0	0	0	0	0	0
Cocos Is	0	0	0	0	0	0	0	0	0	0
France	-	-	-	80	1 947	10 849	1 652	108	-	-
Honduras	-	-	-	-	-	-	-	-	637	-
India	741 950	799 934	809 231	858 312	873 901	848 254	864 749	909 889	939 003	966 076
Indonesia	744 465	778 092	783 496	848 401	921 801	1 022 988	974 313	948 225	953 743	994 074
Iran	-	-	-	2	6	449	-	-	-	-
Italy	-	-	-	-	212	1 473	-	...	-	-
Japan	6 790	11 519	40 437	27 076	21 221	22 301	24 091	19 898	17 676	12 921
Korea Rep	-	78	321	718	512	269	4 983	5 273	4 550	3 403
Malaysia	446 515	439 341	504 068	494 091	499 288	508 128	485 741	519 785	475 354	501 466
Myanmar	597 637	599 876	602 885	455 294	631 226	680 838	759 664	880 018	931 492	1 008 113
Philippines	-	-	-	-	-	657	1 017	219	182	379
Seychelles	-	-	-	-	1 049	3 785	21	183	365	406
Spain	-	-	-	237	379	13 379	451	168	...	1 651
Sri Lanka	200 400	208 900	214 171	206 695	207 849	233 430	244 630	256 710	249 770	270 130
Thailand	823 696	776 667	822 673	869 484	850 994	909 617	808 257	755 136	792 889	784 939
Timor-Leste	...	...	...	...	...	...	408 F	362	356	350 F
Uruguay	-	-	-	-	-	-	-	-	-	1 847
Other nei	5 715	9 088	7 309	11 943	7 355	18 856	12 008	12 611	2 051	2 051
Total	***4 026 092***	***3 998 908***	***4 183 205***	***4 190 529***	***4 459 036***	***4 700 229***	***4 647 945***	***4 785 689***	***4 877 380***	***5 100 261***

C-58 (a)

Fish, crustaceans, molluscs, etc — **Capture production by species items** — **Indian Ocean, Antarctic**
Poissons, crustacés, mollusques, etc — **Captures par catégories d'espèces** — **Océan Indien, Antarctique**
Peces, crustáceos, moluscos, etc — **Capturas por categorías de especies** — **Océano Indico, Antártico**

English name Nom anglais Nombre inglés	Scientific name Nom scientifique Nombre científico	Species group Groupe d'espèces Grupo de especies	1996 t	1997 t	1998 t	1999 t	2000 t	2001 t	2002 t
Smalleye moray cod	*Muraenolepis microps*	32	-	-	-	-	-	0	-
Moray cods nei	*Muraenolepis spp*	32	-	-	-	-	-	0	-
Blue antimora	*Antimora rostrata*	32	-	0	8	7	24	3	1
Whitson's grenadier	*Macrourus whitsoni*	32	-	-	-	-	3	-	-
Bigeye grenadier	*Macrourus holotrachys*	32	-	1	-	-	-	-	-
Grenadiers nei	*Macrourus spp*	32	-	12	68	146	349	203	385
Marbled rockcod	*Notothenia rossii*	33	-	1	-	1	-	-	0
Grey rockcod	*Notothenia squamifrons*	33	15	4	3	10	0	0	1
Striped-eyed rockcod	*Notothenia kempi*	33	-	-	-	-	-	-	0
Antarctic rockcods, noties nei	*Nototheniidae*	33	-	-	3	-	-	0	-
Lantern fish	*Lampanyctus achirus*	34	0	-	-	-	-	-	-
Patagonian toothfish	*Dissostichus eleginoides*	34	5 656	8 587	9 896	9 569	11 453	8 757	6 301
Mackerel icefish	*Champsocephalus gunnari*	34	5	227	128	2	87	1 073	966
Ocellated icefish	*Chionodraco rastrospinosus*	34	-	1	-	-	-	-	-
Unicorn icefish	*Channichthys rhinoceratus*	34	-	9	6	2	2	1	3
Spiny icefish	*Chaenodraco wilsoni*	34	-	-	-	-	-	11	-
Antarctic horsefish	*Zanclorhynchus spinifer*	34	-	0	-	-	-	-	-
Southern opah	*Lampris immaculatus*	37	-	-	-	-	-	-	0
Atlantic pomfret	*Brama brama*	37	-	-	-	-	-	0	0
Porbeagle	*Lamna nasus*	38	-	2	-	-	-	-	-
Greenland shark	*Somniosus microcephalus*	38	-	-	-	-	-	-	1
Pacific sleeper shark	*Somniosus pacificus*	38	-	-	-	1	-	-	3
Antarctic starry skate	*Raja georgiana*	38	-	-	-	-	-	-	0
Eaton's skate	*Bathyraja eatonii*	38	-	-	-	-	-	-	0
Murray's skate	*Bathyraja murrayi*	38	-	-	-	-	0	-	-
Bathyraja rays nei	*Bathyraja spp*	38	-	-	-	-	0	-	-
Rays, stingrays, mantas nei	*Rajiformes*	38	0	7	26	36	96	121	343
Sharks, rays, skates, etc. nei	*Elasmobranchii*	38	-	2	-	1	-	-	-
Marine fishes nei	*Osteichthyes*	39	-	1	0	0	0	-	-
Subantarctic stone crab	*Lithodes murrayi*	44	-	-	0	-	0	0	0
Red stone crab	*Paralomis aculeata*	44	-	-	-	-	0	0	-
King crabs, stone crabs nei	*Lithodidae*	44	-	0	0	0	3	0	0
Antarctic krill	*Euphausia superba*	46	6	-	-	-	-	-	-
Antarctic krill nei	*Euphausia spp*	46	0	-	-	-	-	-	-
Octopuses, etc. nei	*Octopodidae*	57	-	-	-	-	-	-	0
Cephalopods nei	*Cephalopoda*	57	-	1	-	-	-	-	-
Salps	*Salpidae*	74	7	-	-	-	-	-	-
Jellyfishes	*Rhopilema spp*	77	0	11	2	-	-	-	-
Aquatic invertebrates nei	*Invertebrata*	77	-	1	-	-	-	-	-
Total			***5 689***	***8 867***	***10 140***	***9 775***	***12 017***	***10 169***	***8 004***

C-58 (b)

Fish, crustaceans, molluscs, etc — **Capture production by countries or areas** — **Indian Ocean, Antarctic**
Poissons, crustacés, mollusques, etc — **Captures par pays ou zones** — **Océan Indien, Antarctique**
Peces, crustáceos, moluscos, etc — **Capturas por países o áreas** — **Océano Indico, Antártico**

Country or area Pays ou zone País o área	1993 t	1994 t	1995 t	1996 t	1997 t	1998 t	1999 t	2000 t	2001 t	2002 t
Australia	4	-	-	-	2 126	3 635	3 403	3 140	3 725	3 540
France	1 582	4 407	4 206	3 501	4 111	4 652	5 114	7 386	6 115	4 283
India	-	-	-	13	-	-	-	-	-	-
Japan	5 762	899	1 266	264	335	-	-	-	-	-
South Africa	-	-	-	942	1 247	968	665	1 336	321	181
Ukraine	2 043	2 260	4 275	969	1 048	885	593	56	8	-
Uruguay	-	-	-	-	-	-	-	99	-	-
Total	***9 391***	***7 566***	***9 747***	***5 689***	***8 867***	***10 140***	***9 775***	***12 017***	***10 169***	***8 004***

Data in these tables refer to catches for which the new fishing season (1 December - 30 November) of the Commission for the Conservation of Antarctic Marine Living Resources (CCAMLR) has been adopted. Data are shown under the calendar year in which the split-year ends.

Les données dans ces tableaux se réfèrent aux captures pour lesquelles on utilise la nouvelle période de pêche (1er décembre - 30 novembre) de la Commission pour la conservation de la faune et la flore marines de l'Antarctique (CCAMLR). Les données figurent sous l'année civile durant laquelle se termine l'année fractionnée.

Los datos en estos cuadros se refieren a las capturas para las que se utiliza el nuevo período de pesca (1 de diciembre - 30 de noviembre) de la Comisión para la Conservación de los Recursos Vivos Marinos Antárticos (CCAMLR). Los datos se incluyen en el año civil en que termina el año emergente.

C-61 (a)

Fish, crustaceans, molluscs, etc — **Capture production by species items** — **Pacific, Northwest**
Poissons, crustacés, mollusques, etc — **Captures par catégories d'espèces** — **Pacifique, nord-ouest**
Peces, crustáceos, moluscos, etc — **Capturas por categorías de especies** — **Pacífico, noroeste**

English name Nom anglais Nombre inglés	Scientific name Nom scientifique Nombre científico	Species group Groupe d'espèces Grupo de especies	1996 t	1997 t	1998 t	1999 t	2000 t	2001 t	2002 t
Cyprinids nei	*Cyprinidae*	11	110	101	100	56	8	52	38
Pink(=Humpback)salmon	*Oncorhynchus gorbuscha*	23	133 838	183 967	200 628	174 535	157 506	158 810	124 400
Chum(=Keta=Dog)salmon	*Oncorhynchus keta*	23	297 268	266 736	209 290	191 207	173 067	225 700	224 389
Masu(=Cherry) salmon	*Oncorhynchus masou*	23	1 681	994	1 738	1 133	955	780	1 115
Sockeye(=Red)salmon	*Oncorhynchus nerka*	23	19 281	15 856	11 470	10 452	14 771	13 087	17 893
Chinook(=Spring=King)salmon	*Oncorhynchus tshawytscha*	23	370	1 016	750	580	362	447	666
Coho(=Silver)salmon	*Oncorhynchus kisutch*	23	1 100	987	1 616	1 122	1 235	1 446	1 589
Smelts nei	*Osmerus spp, Hypomesus spp*	23	122	321	322	390	735	612	1 032
Salmonoids nei	*Salmonoidei*	23	2 119	1 541	2 407	884	2 140	747	731
Chinese gizzard shad	*Clupanodon thrissa*	24	4 933	13 846	11 360	9 538	6 387	9 194	4 748
Dotted gizzard shad	*Konosirus punctatus*	24	18 647	14 850	20 787	17 770	12 335	17 210	13 401
Elongate ilisha	*Ilisha elongata*	24	51 969	77 881	84 671	110 624	107 823	101 382	94 528
Milkfish	*Chanos chanos*	25	4	-	-	-	-	1	-
Barramundi(=Giant seaperch)	*Lates calcarifer*	25	14	9	26	47	32	80	73
Kamchatka flounder	*Atheresthes evermanni*	31	9 739	8 476	9 567	10 743	23 473	19 049	17 609
Yellow striped flounder	*Pseudopleuronectes herzenst.*	31	18 066	18 079	20 135	19 569	15 423	14 503	13 816
Tonguefishes	*Cynoglossidae*	31	2 804	2 550	1 831	1 636	2 148	2 113	1 050
Bastard halibut	*Paralichthys olivaceus*	31	10 628	9 953	9 617	8 877	9 179	8 436	8 502
Flatfishes nei	*Pleuronectiformes*	31	140 157	147 859	156 720	169 910	175 459	159 817	147 598
Moras nei	*Moridae*	32	-	152	11 115	31 295	39 316	32 392	29 323
Pacific cod	*Gadus macrocephalus*	32	154 027	142 388	157 774	163 730	129 565	116 443	99 381
Saffron cod	*Eleginus gracilis*	32	21 110	27 803	40 426	47 032	35 763	33 753	32 591
Alaska pollock(=Walleye poll.)	*Theragra chalcogramma*	32	3 353 782	3 345 068	2 821 246	2 221 037	1 754 748	1 700 548	1 135 732
Polar cod	*Boreogadus saida*	32	-	-	-	-	13	-	-
Grenadiers, rattails nei	*Macrouridae*	32	310	1 044	62	52	579	2 232	10 817
Gadiformes nei	*Gadiformes*	32	10	-	2	22	6	1	13
Greater lizardfish	*Saurida tumbil*	33	14 705	10 846	10 909	10 791	11 203	8 342	7 172
Brushtooth lizardfish	*Saurida undosquamis*	33	162	207	215	35	795	758	33
Lizardfishes nei	*Synodontidae*	33	9 034	8 070	6 619	4 700	5 800	6 400	6 250
Sea catfishes nei	*Ariidae*	33	122	134	208	257	725	435	529
Korean sandlance	*Hypoptychus dybowskii*	33	6 613	8 832	4 801	4 806	-	462	7
Flathead grey mullet	*Mugil cephalus*	33	9 770	11 943	9 571	14 170	14 345	14 800	12 814
Mullets nei	*Mugilidae*	33	67 423	85 967	99 024	113 461	107 276	119 340	113 928
Groupers nei	*Epinephelus spp*	33	25 698	32 588	37 512	41 510	42 715	46 585	46 381
Groupers, seabasses nei	*Serranidae*	33	1 785	1 901	2 005	1 451	2 047	1 840	1 717
Japanese seabass	*Lateolabrax japonicus*	33	9 512	10 558	12 180	12 167	10 538	12 256	12 216
Red bigeye	*Priacanthus macracanthus*	33	7 386	4 824	3 670	2 756	3 986	3 079	2 142
Bigeyes nei	*Priacanthus spp*	33	5 391	4 962	4 149	3 000	3 600	4 000	3 900
Sillago-whitings	*Sillaginidae*	33	795	600	394	299	288	142	246
Moonfish	*Mene maculata*	33	725	902	1 010	1 747	1 496	1 582	1 110
Snappers, jobfishes nei	*Lutjanidae*	33	2 764	1 847	1 112	717	652	715	709
Golden threadfin bream	*Nemipterus virgatus*	33	242 261	263 399	266 383	250 591	296 319	290 920	309 392
Threadfin breams nei	*Nemipterus spp*	33	19 568	20 024	19 449	14 000	17 000	18 800	18 300
Grunts, sweetlips nei	*Haemulidae (=Pomadasyidae)*	33	5 303	4 801	5 703	5 285	5 136	4 995	5 606
Honnibe croaker	*Nibea mitsukurii*	33	1 940	1 177	1 285	1 566	1 999	2 156	1 148
Large yellow croaker	*Larimichthys croceus*	33	102 297	71 515	71 229	66 076	123 274	76 715	83 412
Yellow croaker	*Larimichthys polyactis*	33	280 036	167 010	207 772	257 758	302 127	254 166	272 140
Blackmouth croaker	*Atrobucca nibe*	33	297	288	237	223	450	644	535
Silver croaker	*Pennahia argentata*	33	8 599	6 085	5 089	4 940	4 180	3 796	3 394
Croakers, drums nei	*Sciaenidae*	33	71 182	81 180	79 062	90 920	50 313	40 589	40 107
Largeeye breams	*Gymnocranius spp*	33	208	170	143	100	120	130	120
Silver seabream	*Pagrus auratus*	33	19 039	19 205	19 105	19 988	20 753	22 268	24 348
Blackhead seabream	*Acanthopagrus schlegeli*	33	207	277	404	604	717	878	784
Porgies, seabreams nei	*Sparidae*	33	76 145	92 153	95 905	98 484	123 895	147 901	138 162
Goatfishes	*Upeneus spp*	33	1 226	917	728	778	827	700	734
Parrotfishes nei	*Scaridae*	33	11	17	12	5	29	27	46
Fourfinger threadfin	*Eleutheronema tetradactylum*	33	3 157	2 831	3 248	2 414	6 917	1 715	1 236
Pacific sandlance	*Ammodytes personatus*	33	115 766	108 666	90 688	82 918	66 129	92 967	72 901
Gobies nei	*Gobiidae*	33	16 120	17 898	21 850	31 069	33 521	28 024	22 487
Okhotsk atka mackerel	*Pleurogrammus azonus*	33	213 363	249 356	293 271	214 269	223 456	214 592	213 933
Bartail flathead	*Platycephalus indicus*	33	2 900	4 196	2 857	2 248	2 310	1 699	1 199
Purple puffer	*Takifugu vermicularis*	33	9 708	7 471	3 897	4 787	-	-	3 110
Puffers nei	*Tetraodontidae*	33	8 238	7 103	8 329	9 647	13 967	11 555	7 869
Filefishes nei	*Cantherhines(=Navodon) spp*	33	211 059	297 227	236 189	241 209	225 403	204 270	159 983
Threadsail filefish	*Stephanolepis cirrhifer*	33	1 772	16 318	9 364	2 999	-	-	933
Deepsea smelt	*Glossanodon semifasciatus*	34	8 134	7 431	7 142	6 312	5 970	5 414	4 926
Daggertooth pike conger	*Muraenesox cinereus*	34	183 716	192 667	249 194	245 803	230 633	252 790	264 162
Pike-congers nei	*Muraenesox spp*	34	3 281	2 903	2 450	1 700	2 100	2 300	2 250
Whitespotted conger	*Conger myriaster*	34	17 314	19 136	11 913	10 160	8 304	7 676	17 210
Conger eels, etc. nei	*Congridae*	34	12 007	11 706	9 444	8 168	8 364	7 999	8 921
Alfonsinos nei	*Beryx spp*	34	6	-	4	38	18	14	12
Oreo dories nei	*Oreosomatidae*	34	-	-	-	2	-	14	8
Tilefishes nei	*Branchiostegidae*	34	8 061	7 807	7 535	7 596	8 090	8 092	8 101
Atlantic wolffish	*Anarhichas lupus*	34	-	-	-	38	36	33	22
Eelpouts	*Lycodes spp*	34	18	-	2	1	28	47	60
Japanese sandfish	*Arctoscopus japonicus*	34	9 220	8 403	8 285	9 064	8 223	10 039	12 630
Oilfish	*Ruvettus pretiosus*	34	2 634	2 622	3 043	2 661	2 584	3 678	4 190
Largehead hairtail	*Trichiurus lepturus*	34	1 184 849	1 111 655	1 328 470	1 322 474	1 396 737	1 388 045	1 370 765
Hairtails, scabbardfishes nei	*Trichiuridae*	34	3 326	2 522	1 690	1 200	1 500	1 650	1 600
Indian driftfish	*Ariomma indica*	34	120	79	49	35	40	45	40
Pacific rudderfish	*Psenopsis anomala*	34	13 290	11 777	13 734	10 871	10 721	10 942	9 899
Pacific ocean perch	*Sebastes alutus*	34	4 345	2 862	2 440	1 630	1 475	1 461	2 161
Scorpionfishes nei	*Scorpaenidae*	34	6 964	7 496	6 240	5 979	6 168	6 338	3 049
Bluefin gurnard	*Chelidonichthys kumu*	34	75	86	146	43	79	140	260

C-61 (a)

Fish, crustaceans, molluscs, etc — Capture production by species items — Pacific, Northwest
Poissons, crustacés, mollusques, etc — Captures par catégories d'espèces — Pacifique, nord-ouest
Peces, crustáceos, moluscos, etc — Capturas por categorías de especies — Pacífico, noroeste

English name Nom anglais Nombre inglés	Scientific name Nom scientifique Nombre científico	Species group Groupe d'espèces Grupo de especies	1996 t	1997 t	1998 t	1999 t	2000 t	2001 t	2002 t
Sablefish	*Anoplopoma fimbria*	34	502	-	-	-	6	6	19
Anglerfishes nei	*Lophiidae*	34	11 720	5 045	3 970	3 406	4 930	5 813	9 500
Pacific herring	*Clupea pallasii*	35	181 021	344 317	433 262	401 155	393 962	341 337	255 411
Japanese sardinella	*Sardinella zunasi*	35	10 663	5 593	1 973	6 674	4 603	766	796
Sardinellas nei	*Sardinella spp*	35	30	12	8	5	5	6	5
Japanese pilchard	*Sardinops melanostictus*	35	430 837	417 939	295 788	515 477	305 767	339 377	236 602
Slender rainbow sardine	*Dussumieria elopsoides*	35	48	19	17	10	15	17	15
Red-eye round herring	*Etrumeus teres*	35	51 214	57 195	49 689	29 605	24 951	32 831	27 908
Silver-stripe round herring	*Spratelloides gracilis*	35	517	650	886	521	639	505	1 033
Japanese anchovy	*Engraulis japonicus*	35	1 254 487	1 666 503	2 093 888	1 820 259	1 725 685	1 836 502	1 853 936
Stolephorus anchovies	*Stolephorus spp*	35	31	20	17	10	15	17	15
Anchovies, etc. nei	*Engraulidae*	35	22	5	23	34	69	114	260
Dorab wolf-herring	*Chirocentrus dorab*	35	1	2	1	4	1	-	15
Clupeoids nei	*Clupeoidei*	35	62 105	63 841	56 765	83 126	78 802	62 679	67 275
Narrow-barred Spanish mackerel	*Scomberomorus commerson*	36	2 541	2 701	2 417	2 674	3 551	4 153	6 613
Indo-Pacific king mackerel	*Scomberomorus guttatus*	36	1 611	1 607	1 128	1 298	1 249	1 395	992
Japanese Spanish mackerel	*Scomberomorus niphonius*	36	301 356	365 585	551 780	595 103	539 094	522 756	553 652
Seerfishes nei	*Scomberomorus spp*	36	3 139	2 854	3 201	4 083	3 760	3 083	3 622
Frigate and bullet tunas	*Auxis thazard, A.rochei*	36	22 821	36 775	26 231	34 072	30 212	39 323	39 532
Kawakawa	*Euthynnus affinis*	36	0	0	0	0	0	0	0
Skipjack tuna	*Katsuwonus pelamis*	36	105 603	187 992	205 919	138 180	177 180	156 363	144 795
Pacific bluefin tuna	*Thunnus orientalis*	36	6 808	6 974	4 328	11 147	11 058	6 256	6 163
Longtail tuna	*Thunnus tonggol*	36	514	391	303	6 434	10 087	14 321	859
Albacore	*Thunnus alalunga*	36	51 026	70 005	54 857	86 968	47 162	50 748	50 571
Yellowfin tuna	*Thunnus albacares*	36	16 129	14 593	16 185	19 442	21 773	15 345	16 624
Bigeye tuna	*Thunnus obesus*	36	6 610	9 674	10 145	10 173	9 378	8 573	9 279
Indo-Pacific sailfish	*Istiophorus platypterus*	36	2 643	3 428	1 821	3 841	2 080	583	648
Blue marlin	*Makaira nigricans*	36	9 225	5 486	4 905	3 880	3 306	3 168	3 677
Black marlin	*Makaira indica*	36	830	659	285	681	332	193	355
Striped marlin	*Tetrapturus audax*	36	3 860	2 468	4 328	4 093	3 381	2 876	2 875
Marlins,sailfishes,etc. nei	*Istiophoridae*	36	-	-	2	3	-	146	123
Swordfish	*Xiphias gladius*	36	5 982	5 754	8 019	7 617	9 862	7 320	7 186
Tuna-like fishes nei	*Scombroidei*	36	26 600	31 117	29 036	31 896	43 874	11 712	8 278
Capelin	*Mallotus villosus*	37	180	160	405	70	291	1 468	3 882
Pacific saury	*Cololabis saira*	37	276 111	388 643	180 973	187 144	305 009	375 750	335 239
Japanese halfbeak	*Hyporhamphus sajori*	37	990	1 193	1 160	913	956	613	670
Japanese flyingfish	*Cypselurus agoo*	37	8 501	7 486	8 933	6 738	9 615	8 286	6 909
Flyingfishes nei	*Exocoetidae*	37	662	2 497	1 077	618	287	366	502
Cobia	*Rachycentron canadum*	37	692	987	815	655	1 014	486	457
Japanese jack mackerel	*Trachurus japonicus*	37	349 166	350 619	340 564	227 290	271 501	235 874	229 539
Japanese scad	*Decapterus maruadsi*	37	62 378	69 764	71 168	67 689	40 803	47 460	49 586
Indian scad	*Decapterus russelli*	37	382	687	6 158	7 703	3 927	598	58
Scads nei	*Decapterus spp*	37	645 640	522 877	539 120	507 526	507 380	550 298	608 301
Jacks, crevalles nei	*Caranx spp*	37	7 131	11 466	8 248	4 892	6 370	3 290	4 629
Japanese amberjack	*Seriola quinqueradiata*	37	246	...	...	...	...	...	...
Amberjacks nei	*Seriola spp*	37	54 426	53 275	55 104	63 571	82 275	73 400	57 385
Black pomfret	*Parastromateus niger*	37	3 427	2 534	1 744	1 307	1 671	2 129	1 779
Torpedo scad	*Megalaspis cordyla*	37	233	6 847	52	113	327	198	229
Common dolphinfish	*Coryphaena hippurus*	37	11 373	18 307	32 066	17 829	14 511	16 550	16 014
Chub mackerel	*Scomber japonicus*	37	1 603 739	1 465 627	1 104 873	1 007 284	871 513	984 850	871 065
Silver pomfret	*Pampus argenteus*	37	9 919	3 622	2 650	5 851	2 852	4 245	6 429
Silver pomfrets nei	*Pampus spp*	37	220 364	242 547	303 024	337 919	338 848	352 493	386 317
Butterfishes, pomfrets nei	*Stromateidae*	37	11 844	13 079	15 613	17 015	9 869	9 065	8 451
Barracudas nei	*Sphyraena spp*	37	547	761	771	519	362	439	311
Whip stingray	*Dasyatis akajei*	38	4 029	3 959	4 329	4 407	5 388	4 312	4 512
Rays, stingrays, mantas nei	*Rajiformes*	38	9 250	8 014	4 629	4 487	4 343	2 664	5 084
Sharks, rays, skates, etc. nei	*Elasmobranchii*	38	17 922	25 392	25 425	48 993	46 348	28 994	30 075
Marine fishes nei	*Osteichthyes*	39	5 147 881	4 853 220	4 967 320	5 010 112	4 689 171	4 204 409	4 122 073
Blue swimming crab	*Portunus pelagicus*	42	52 679	46 890	49 156	53 088	58 058	61 119	63 011
Gazami crab	*Portunus trituberculatus*	42	303 170	252 502	283 971	284 851	351 051	350 168	369 247
Indo-Pacific swamp crab	*Scylla serrata*	42	935	180	213	268	289	230	337
Tanner crabs nei	*Chionoecetes spp*	42	26 216	31 982	25 525	26 126	44 525	43 822	29 671
Hair crab	*Erimacrus isenbeckii*	42	789	612	409	440	198	162	117
Marine crabs nei	*Brachyura*	42	104 811	99 419	83 218	72 467	46 580	45 021	44 912
Longlegged spiny lobster	*Panulirus longipes*	43	1 106	1 082	1 098	1 166	1 716	1 924	1 782
Tropical spiny lobsters nei	*Panulirus spp*	43	0	0	0	0	0	0	0
Japanese fan lobster	*Ibacus ciliatus*	43	1 115	642	696	676	1 600	1 607	1 005
Red king crab	*Paralithodes camtschaticus*	44	34 300	23 262	32 486	37 072	28 554	16 064	9 691
Blue king crab	*Paralithodes platypus*	44	8 762	10 268	4 508	5 455	5 233	4 500	4 598
Brown king crab	*Paralithodes brevipes*	44	204	418	194	256	347	254	350
King crabs	*Paralithodes spp*	44	322	154	132	117	89	181	129
Golden king crab	*Lithodes aequispina*	44	4 666	4 917	3 897	2 746	1 797	2 245	2 291
Kuruma prawn	*Penaeus japonicus*	45	5 606	6 911	5 113	6 096	10 193	5 963	2 930
Giant tiger prawn	*Penaeus monodon*	45	49	45	38	92	57	91	54
Fleshy prawn	*Penaeus chinensis*	45	56 534	71 317	79 595	70 725	85 545	97 583	99 690
Redtail prawn	*Penaeus penicillatus*	45	5 020	2 473	647	316	308	312	509
Shiba shrimp	*Metapenaeus joyneri*	45	2 211	1 976	3 651	3 633	2 621	2 385	1 736
Metapenaeus shrimps nei	*Metapenaeus spp*	45	626	369	299	210	202	241	227
Southern rough shrimp	*Trachypenaeus curvirostris*	45	167 723	180 255	179 544	403 027	312 968	247 940	234 102
Coonstripe shrimp	*Pandalus hypsinotus*	45	...	467	388	288	275	359	231
Northern prawn	*Pandalus borealis*	45	-	1 786	3 127	5 096	8 520	8 457	7 596

C-61 (a)

Fish, crustaceans, molluscs, etc — Capture production by species items — Pacific, Northwest
Poissons, crustacés, mollusques, etc — Captures par catégories d'espèces — Pacifique, nord-ouest
Peces, crustáceos, moluscos, etc — Capturas por categorías de especies — Pacífico, noroeste

English name Nom anglais Nombre inglés	Scientific name Nom scientifique Nombre científico	Species group Groupe d'espèces Grupo de especies	1996 t	1997 t	1998 t	1999 t	2000 t	2001 t	2002 t
Humpy shrimp	*Pandalus goniurus*	45	...	-	1 199	330	1 200	247	32
Hokkai shrimp	*Pandalus kessleri*	45	...	123	55	97	75	94	81
Pandalus shrimps nei	*Pandalus spp*	45	3 000	-	-	-	-	-	-
Morotoge shrimp	*Pandalopsis japonica*	45	...	12	86	35	11	36	22
Akiami paste shrimp	*Acetes japonicus*	45	460 871	495 680	587 376	598 602	639 219	577 497	585 647
Sculptured shrimps nei	*Sclerocrangon spp*	45	...	38	59	82	20	45	41
Natantian decapods nei	*Natantia*	45	75 282	73 290	71 003	61 480	57 906	49 750	48 455
Marine crustaceans nei	*Crustacea*	47	1 007 693	1 165 004	1 314 828	1 198 681	1 304 743	1 335 243	1 281 908
Japanese corbicula	*Corbicula japonica*	51	-	-	-	-	-	74	209
Giant abalone	*Haliotis gigantea*	52	1 941	2 218	2 269	2 109	2 146	1 982	2 223
Abalones nei	*Haliotis spp*	52	196	220	73	81	155	105	75
Horned turban	*Turbo cornutus*	52	17 480	19 010	21 748	18 397	17 120	16 515	16 697
Gastropods nei	*Gastropoda*	52	3 735	2 636	1 450	1 670	828	783	1 555
Pacific cupped oyster	*Crassostrea gigas*	53	18 259	17 210	9 905	11 609	15 939	10 060	16 678
Cupped oysters nei	*Crassostrea spp*	53	-	-	-	1	4	32	1
Korean mussel	*Mytilus coruscus*	54	2 191	3 211	1 469	1 414	1 133	1 085	946
Sea mussels nei	*Mytilidae*	54	837	4 098	6 465	6 851	5 819	3 870	3 533
Yesso scallop	*Patinopecten yessoensis*	55	276 406	266 957	294 211	305 510	310 104	293 268	310 303
Ark clams nei	*Arca spp*	56	782	667	1 311	534	318	153	228
Half-crenated ark	*Scapharca subcrenata*	56	16 328	14 133	10 120	10 413	7 308	4 899	9 828
Blood cockle	*Anadara granosa*	56	995	493	12 114	6 503	4 184	1 023	1 440
Japanese hard clam	*Meretrix lusoria*	56	4 394	4 133	6 147	4 110	2 973	2 289	3 004
Japanese carpet shell	*Ruditapes philippinarum*	56	56 095	56 514	51 392	57 051	56 540	51 026	49 252
Imperial surf clam	*Pseudocardium sybillae*	56	14 791	14 720	11 104	16 428	13 033	12 323	12 325
Cockles nei	*Cardiidae*	56	2 583	1 587	1 078	112	411	2 002	810
Clams, etc. nei	*Bivalvia*	56	85 807	51 838	63 249	50 047	48 022	46 155	39 851
Cuttlefish,bobtail squids nei	*Sepiidae, Sepiolidae*	57	185 530	255 542	241 895	232 788	269 796	324 915	309 524
Common squids nei	*Loligo spp*	57	14 853	19 221	20 074	12 712	8 210	6 689	6 232
Neon flying squid	*Ommastrephes bartrami*	57	...	47 852	53 638	34 551	46 604	21 678	22 483
Japanese flying squid	*Todarodes pacificus*	57	715 908	603 367	378 605	497 887	570 427	528 523	504 438
Various squids nei	*Loliginidae, Ommastrephidae*	57	176 002	147 722	250 739	262 688	260 097	190 193	208 549
Octopuses, etc. nei	*Octopodidae*	57	73 170	80 370	88 667	77 213	66 838	65 511	79 631
Cephalopods nei	*Cephalopoda*	57	-	-	-	-	-	3	2
Marine molluscs nei	*Mollusca*	58	938 870	1 502 215	1 488 978	1 457 426	1 410 260	1 405 909	1 381 743
Sea squirts nei	*Ascidiacea*	74	16 747	2 780	891	1 171	1 443	1 053	1 025
Echinoderms	*Echinodermata*	76	2 802	2 771	1 410	1 182	1 461	1 454	1 459
Sea urchins nei	*Strongylocentrotus spp*	76	15 013	15 861	15 552	15 108	14 324	13 170	15 678
Japanese sea cucumber	*Stichopus japonicus*	76	9 205	9 377	8 391	7 866	8 376	8 129	8 092
Jellyfishes	*Rhopilema spp*	77	265 325	400 483	430 784	402 206	334 869	331 439	288 113
Aquatic invertebrates nei	*Invertebrata*	77	13 410	10 118	10 660	8 131	11 574	8 708	9 436
Total			***23 542 610***	***24 686 259***	***24 815 317***	***24 130 564***	***23 204 042***	***22 550 874***	***21 436 229***

C-61 (b)

Fish, crustaceans, molluscs, etc — Capture production by countries or areas — Pacific, Northwest
Poissons, crustacés, mollusques, etc — Captures par pays ou zones — Pacifique, nord-ouest
Peces, crustáceos, moluscos, etc — Capturas por países o áreas — Pacífico, noroeste

Country or area Pays ou zone País o área	1993 t	1994 t	1995 t	1996 t	1997 t	1998 t	1999 t	2000 t	2001 t	2002 t
China	8 151 441	9 522 098	10 921 731	12 392 922	13 807 656	14 890 900	14 843 663	14 609 252	14 211 830	14 053 342
China,H.Kong	217 544	211 010	194 999	183 856	186 000	180 000	127 780	157 012	173 972	169 790
China, Macao	1 898	1 890	1 604	1 418	1 500 F	1 500 F	1 500 F	1 500 F	1 500 F	1 500 F
China,Taiwan	522 490 F	405 001 F	420 781 F	393 351	406 022	378 364	373 608	361 423	357 797	363 235
Japan	6 218 515	5 628 874	5 092 787	5 191 508	5 162 694	4 546 845	4 451 587	4 206 983	3 996 910	3 836 031 F
Korea D P Rp	374 000 F	351 961	307 083	233 125	216 462	212 000 F	208 000 F	204 000 F	201 572	200 000 F
Korea Rep	1 718 865	1 799 694	1 787 770	1 933 176	1 640 712	1 553 273	1 500 272	1 288 737	1 469 171	1 123 423
Poland	235 208	269 979	249 365	116 266	125 414	81 889	65 508	33 217	16 590	-
Russian Fed	2 741 954	2 256 773	2 835 602	3 096 988	3 139 799	2 970 546	2 558 646	2 341 918	2 121 532	1 688 908
Total	***20 181 915***	***20 447 280***	***21 811 722***	***23 542 610***	***24 686 259***	***24 815 317***	***24 130 564***	***23 204 042***	***22 550 874***	***21 436 229***

English name Nom anglais Nombre inglés	Scientific name Nom scientifique Nombre científico	Species group Groupe d'espèces Grupo de especies	1996 t	1997 t	1998 t	1999 t	2000 t	2001 t	2002 t
White sturgeon	*Acipenser transmontanus*	21	-	-	-	-	206	185	200
Green sturgeon	*Acipenser medirostris*	21	-	-	-	-	36	10	7
Sturgeons nei	*Acipenseridae*	21	1	0	0	0	-	-	-
Pink(=Humpback)salmon	*Oncorhynchus gorbuscha*	23	149 136	115 175	154 776	182 844	101 598	183 642	124 652
Chum(=Keta=Dog)salmon	*Oncorhynchus keta*	23	86 864	54 741	78 316	69 907	76 357	58 232	62 974
Sockeye(=Red)salmon	*Oncorhynchus nerka*	23	159 855	112 574	63 247	112 433	103 074	83 403	71 703
Chinook(=Spring=King)salmon	*Oncorhynchus tshawytscha*	23	5 548	7 180	6 443	4 336	4 807	6 731	10 434
Coho(=Silver)salmon	*Oncorhynchus kisutch*	23	21 497	10 607	15 466	12 991	14 575	16 810	17 767
Rainbow trout	*Oncorhynchus mykiss*	23	3	1	4	1	1	-	-
Pacific salmons nei	*Oncorhynchus spp*	23	-	-	-	-	50	204	6
Surf smelt	*Hypomesus pretiosus*	23	1	-	-	-	-	-	-
Eulachon	*Thaleichthys pacificus*	23	30	0	3	-	13	143	329
Smelts nei	*Osmerus spp, Hypomesus spp*	23	78	839	490	432	442	218	261
American shad	*Alosa sapidissima*	24	15	34	62	-	105	178	66
Pacific halibut	*Hippoglossus stenolepis*	31	28 106	39 289	41 606	43 618	40 848	40 161	43 724
English sole	*Pleuronectes vetulus*	31	...	...	198	874	963	1 198	1 374
Greenland halibut	*Reinhardtius hippoglossoides*	31	4 652	...	307	14	6 186	4 391	2 937
Arrow-tooth flounder	*Atheresthes stomias*	31	10 680	6 319	8 279	12 081	18 774	14 401	16 610
Petrale sole	*Eopsetta jordani*	31	...	1 454	1 197	1 221	1 632	1 554	1 569
Rex sole	*Glyptocephalus zachirus*	31	...	...	121	391	413	473	484
Flathead sole	*Hippoglossoides elassodon*	31	11 392	12 867	18 886	14 318	16 266	16 092	13 174
Yellowfin sole	*Limanda aspera*	31	101 354	149 302	80 500	56 830	69 971	54 918	63 625
Rock sole	*Lepidopsetta bilineata*	31	26 198	32 743	15 189	17 185	27 510	24 206	29 258
Dover sole	*Microstomus pacificus*	31	...	8 496	8 174	8 384	7 546	6 038	4 659
Pacific sand sole	*Psettichthys melanostictus*	31	...	...	15	582	44	85	152
Curlfin sole	*Pleuronichthys decurrens*	31	-	-	-	-	1	5	4
Pacific sanddab	*Citharichthys sordidus*	31	-	-	-	-	150	-	-
California flounder	*Paralichthys californicus*	31	...	...	13	3	0	4	4
Flatfishes nei	*Pleuronectiformes*	31	34 731	23 040	23 886	18 691	13 504	10 207	12 275
Pacific cod	*Gadus macrocephalus*	32	275 448	301 507	253 597	238 515	241 347	214 441	233 311
Pacific tomcod	*Microgadus proximus*	32	...	...	3	1	-	0	-
Alaska pollock(=Walleye poll.)	*Theragra chalcogramma*	32	1 194 803	1 141 442	1 232 977	1 056 249	1 183 482	1 443 917	1 519 122
North Pacific hake	*Merluccius productus*	32	195 289	226 615	227 502	216 889	206 327	172 050	129 599
Grenadiers, rattails nei	*Macrouridae*	32	...	...	83	102	170	160	156
Lingcod	*Ophiodon elongatus*	33	4 402	3 225	2 464	3 061	3 186	2 664	2 881
Atka mackerel	*Pleurogrammus monopterygius*	33	88 030	59 538	51 579	51 436	44 592	57 096	37 759
Cabezon	*Scorpaenichthys marmoratus*	33	...	...	7	33	37	50	49
Sculpins nei	*Cottidae*	33	-	-	-	1	63	71	158
Widow rockfish	*Sebastes entomelas*	34	...	6 886	4 357	3 841	3 575	2 481	435
Yellowtail rockfish	*Sebastes flavidus*	34	...	2 512	3 349	3 427	3 138	2 073	1 204
Pacific ocean perch	*Sebastes alutus*	34	27 178	25 362	24 824	28 071	24 144	23 528	26 537
Bocaccio rockfish	*Sebastes paucispinis*	34	...	448	491	135	3	0	1
Canary rockfish	*Sebastes pinniger*	34	...	1 133	1 222	736	57	45	51
Chilipepper rockfish	*Sebastes goodei*	34	...	38	11	10	31	9	5
Black rockfish	*Sebastes melanops*	34	...	...	...	...	109	148	127
Shortspine thornyhead	*Sebastolobus alascanus*	34	...	...	...	...	331	268	295
Scorpionfishes nei	*Scorpaenidae*	34	51 931	48 369	31 128	35 551	30 850	28 787	28 449
Sablefish	*Anoplopoma fimbria*	34	28 126	26 255	23 408	25 339	24 257	21 984	19 139
Pacific herring	*Clupea pallasii*	35	71 521	84 392	73 378	67 885	60 382	63 080	61 605
California pilchard	*Sardinops caeruleus*	35	-	-	22	775	14 368	23 908	38 958
Californian anchovy	*Engraulis mordax*	35	86	59	103	98	0	70	232
Skipjack tuna	*Katsuwonus pelamis*	36	-	-	-	-	0	1	-
Pacific bluefin tuna	*Thunnus orientalis*	36	2	3	8	7	3	0	0
Albacore	*Thunnus alalunga*	36	10 038	14 588	11 517	4 675	11 182	12 133	12 204
Yellowfin tuna	*Thunnus albacares*	36	-	0	0	-	1	13	-
Bigeye tuna	*Thunnus obesus*	36	-	-	1	-	-	1	-
Swordfish	*Xiphias gladius*	36	23	34	128	27	34	14	0
Tuna-like fishes nei	*Scombroidei*	36	...	1 263	...	1	...	0	1
Pacific jack mackerel	*Trachurus symmetricus*	37	-	2	961	176	181	215	21
Chub mackerel	*Scomber japonicus*	37	388	204	632	75	121	322	127
Thresher	*Alopias vulpinus*	38	...	...	17	2	48	76	73
Shortfin mako	*Isurus oxyrinchus*	38	...	...	3	-	0	0	0
Blue shark	*Prionace glauca*	38	-	-	-	-	1	2	0
Tope shark	*Galeorhinus galeus*	38	...	...	1	-	3	1	0
Picked dogfish	*Squalus acanthias*	38	4 000	2 100	2 501	6 439	5 363	5 181	5 691
Dogfish sharks nei	*Squalidae*	38	2 053	625	555	1	-	-	-
Rays, stingrays, mantas nei	*Rajiformes*	38	2 637	3 757	2 045	2 817	3 210	3 017	4 652
Sharks, rays, skates, etc. nei	*Elasmobranchii*	38	10	14	20	81	3	36	4
Marine fishes nei	*Osteichthyes*	39	106 820	106 601	97 853	98 191	31 003	74 345	4 864
Dungeness crab	*Cancer magister*	42	33 606	19 658	17 002	18 290	19 169	21 370	24 267
Pacific rock crab	*Cancer productus*	42	...	...	3	5	4	5	6
Tanner crabs nei	*Chionoecetes spp*	42	30 785	53 932	114 230	83 989	15 665	12 177	15 078
Marine crabs nei	*Brachyura*	42	647	333	133	99	1	42	2
Blue mud shrimp	*Upogebia pugettensis*	43	...	...	...	...	10	4	4
Ghost shrimps	*Callianassa spp*	43	-	-	-	25	17	14	14
King crabs	*Paralithodes spp*	44	9 526	8 177	10 941	7 675	6 848	7 281	7 616
Pacific shrimps nei	*Pandalus spp, Pandalopsis spp*	45	15 030	19 131	6 268	14 015	16 158	18 740	26 351
Natantian decapods nei	*Natantia*	45	9 392	5 200	5 203	4 071	4 223	4 212	3 622

C-67 (a)

Fish, crustaceans, molluscs, etc — **Capture production by species items** — **Pacific, Northeast**
Poissons, crustacés, mollusques, etc — **Captures par catégories d'espèces** — **Pacifique, nord-est**
Peces, crustáceos, moluscos, etc — **Capturas por categorías de especies** — **Pacífico, nordeste**

English name Nom anglais Nombre inglés	Scientific name Nom scientifique Nombre científico	Species group Groupe d'espèces Grupo de especies	1996 t	1997 t	1998 t	1999 t	2000 t	2001 t	2002 t
Brine shrimp	*Artemia salina*	47	...	...	...	...	49	63	58
Abalones nei	*Haliotis spp*	52	-	-	-	-	-	3	3
Gastropods nei	*Gastropoda*	52	6 369	...	...	...	-	1	4
Olympia flat oyster	*Ostrea conchaphila*	53	...	...	...	6	7	8	11
Pacific cupped oyster	*Crassostrea gigas*	53	4 989	6 187	11	...	...	...	...
Cupped oysters nei	*Crassostrea spp*	53	...	...	...	...	7	12	8
Sea mussels nei	*Mytilidae*	54	455	...	...	1	675	448	251
Weathervane scallop	*Patinopecten caurinus*	55	2 372	2	3 228	2 642	2 012	1 052	1 264
Butter clam	*Saxidomus giganteus*	56	1 431	1 293	1 726	1 019	1 113	1 159	1 428
Pacific littleneck clam	*Protothaca staminea*	56	-	-	-	-	-	102	146
Pacific horse clams nei	*Tresus spp*	56	-	-	-	-	1	2	9
Pacific razor clam	*Siliqua patula*	56	-	-	-	-	14	61	146
Sand gaper	*Mya arenaria*	56	-	-	-	-	-	123	516
Pacific geoduck	*Panopea abrupta*	56	1 768	3 719	4 027	4 139	3 821	3 894	4 767
Basket cockle	*Clinocardium nuttallii*	56	-	22	-	18	49	67	80
Clams, etc. nei	*Bivalvia*	56	3 912	381	2 509	1 601	1 519	1 250	1 892
Neon flying squid	*Ommastrephes bartrami*	57	-	2 008	1 200	1 521	683	2 061	...
Various squids nei	*Loliginidae, Ommastrephidae*	57	488	571	832	1 551	6	852	665
Octopuses, etc. nei	*Octopodidae*	57	197	208	261	60	59	84	177
Marine molluscs nei	*Mollusca*	58	1 873	4 593	2 879	1 850	1 904	1 649	2 094
Sea urchins nei	*Strongylocentrotus spp*	76	7 085	9 044	8 081	7 101	6 841	6 378	6 199
Sea cucumbers nei	*Holothurioidea*	76	491	-	-	228	274	300	249
Total			***2 833 342***	***2 766 092***	***2 742 459***	***2 551 689***	***2 477 803***	***2 759 090***	***2 702 885***

C-67 (b)

Fish, crustaceans, molluscs, etc — **Capture production by countries or areas** — **Pacific, Northeast**
Poissons, crustacés, mollusques, etc — **Captures par pays ou zones** — **Pacifique, nord-est**
Peces, crustáceos, moluscos, etc — **Capturas por países o áreas** — **Pacífico, nordeste**

Country or area Pays ou zone País o área	1993 t	1994 t	1995 t	1996 t	1997 t	1998 t	1999 t	2000 t	2001 t	2002 t
Canada	286 283	309 848	217 121	235 074	247 383	226 013	212 576	143 072	183 886	132 191
China,Taiwan	-	-	26	0	0	0	55	1 165	310	210
Japan	700	683	...	...	2 008	1 201	1 521	683	2 061	...
Korea Rep	47 264	15 875	10 967	2 480	-	311	1 330	-	-	-
Poland	-	-	-	-	-	-	-	998	4	-
Russian Fed	-	-	-	540	-	466	1 624	6	-	109
USA	2 965 279	2 822 291	2 743 083	2 595 248	2 516 701	2 514 468	2 334 583	2 331 879	2 572 829	2 570 375
Total	***3 299 526***	***3 148 697***	***2 971 197***	***2 833 342***	***2 766 092***	***2 742 459***	***2 551 689***	***2 477 803***	***2 759 090***	***2 702 885***

C-71 (a)

Fish, crustaceans, molluscs, etc — **Capture production by species items** — **Pacific, Western Central**
Poissons, crustacés, mollusques, etc — **Captures par catégories d'espèces** — **Pacifique, centre-ouest**
Peces, crustáceos, moluscos, etc — **Capturas por categorías de especies** — **Pacífico, centro-occidental**

English name Nom anglais Nombre inglés	Scientific name Nom scientifique Nombre científico	Species group Groupe d'espèces Grupo de especies	1996 t	1997 t	1998 t	1999 t	2000 t	2001 t	2002 t
Chacunda gizzard shad	*Anodontostoma chacunda*	24	4 585	2 630	2 376	2 479	2 511	2 614	2 695
Toli shad	*Tenualosa toli*	24	1 702	2 535	3 097	2 875	2 117	4 576	5 100
Indian pellona	*Pellona ditchela*	24	4 850	3 257	4 650	5 566	6 475	7 001	6 280
Diadromous clupeoids nei	*Clupeoidei*	24	1 278	574	559	577	652	1 056	926
Milkfish	*Chanos chanos*	25	3 627	679	174	342	2 439	369	371
Barramundi(=Giant seaperch)	*Lates calcarifer*	25	40 600	45 027	47 106	53 631	57 843	52 768	56 973
Lefteye flounders nei	*Bothidae*	31	38	27	30	15	8	10	1
Tonguefishes	*Cynoglossidae*	31	7 487	7 181	7 438	7 379	8 760	8 068	8 159
Indian halibut	*Psettodes erumei*	31	3 828	10 286	4 902	6 945	5 966	5 932	7 129
Flatfishes nei	*Pleuronectiformes*	31	3 569	7 493	4 421	5 942	3 423	5 433	6 195
Gadiformes nei	*Gadiformes*	32	24	21	33	765	50	60	42
Indo-Pacific tarpon	*Megalops cyprinoides*	33	1 090	1 154	1 045	1 404	1 253	1 457	1 561
Bombay-duck	*Harpadon nehereus*	33	9 836	10 867	11 500	9 607	8 449	7 747	7 360
Lizardfishes nei	*Synodontidae*	33	71 051	88 525	59 940	87 516	78 630	79 204	82 989
Sea catfishes nei	*Ariidae*	33	76 357	92 217	77 680	85 743	85 296	85 478	86 959
Eeltail catfishes	*Plotosus spp*	33	1 763	1 498	2 118	1 492	2 096	1 773	1 983
Mullets nei	*Mugilidae*	33	47 066	48 598	49 634	53 045	50 987	49 833	49 744
Fusiliers	*Caesio spp*	33	43 985	51 751	46 546	48 664	43 111	46 923	48 841
Groupers nei	*Epinephelus spp*	33	32 311	35 449	37 403	41 302	45 816	44 231	45 547
Groupers, seabasses nei	*Serranidae*	33	18 656	17 868	18 214	19 140	17 530	16 996	19 635
Bigeyes nei	*Priacanthus spp*	33	69 890	65 902	68 125	74 504	69 788	73 230	74 382
Cardinalfishes, etc. nei	*Apogonidae*	33	200	180	60	71	60	67	65
Sillago-whitings	*Sillaginidae*	33	13 066	14 074	14 014	13 616	13 696	14 477	15 856
Moonfish	*Mene maculata*	33	4 815	11 940	11 529	12 385	11 718	13 275	16 950
Mangrove red snapper	*Lutjanus argentimaculatus*	33	10 824	10 190	4 935	12 319	10 650	10 436	10 635
Snappers nei	*Lutjanus spp*	33	51 820	62 549	59 076	58 809	54 433	58 615	59 893
Snappers, jobfishes nei	*Lutjanidae*	33	31 831	34 174	34 894	37 816	33 066	32 630	29 291
Threadfin breams nei	*Nemipterus spp*	33	142 017	136 767	143 945	160 811	154 164	148 093	174 588
Monocle breams	*Scolopsis spp*	33	1 341	1 293	1 471	994	1 709	1 764	1 533
Ponyfishes(=Slipmouths)	*Leiognathus spp*	33	2 589	2 422	3 125	3 096	2 412	2 324	1 825
Ponyfishes(=Slipmouths) nei	*Leiognathidae*	33	113 772	133 408	118 593	136 982	119 136	133 858	140 706
Silver grunt	*Pomadasys argenteus*	33	1 019	617	2 178	1 687	919	1 086	830
Grunts, sweetlips nei	*Haemulidae (=Pomadasyidae)*	33	11 661	11 697	12 034	12 377	14 116	15 061	15 358
Southern meagre(=Mulloway)	*Argyrosomus hololepidotus*	33	44	45	48	49	93	142	132
Croakers, drums nei	*Sciaenidae*	33	54 682	56 345	60 945	77 527	72 695	69 562	65 849
Largeeye breams	*Gymnocranius spp*	33	20	22	25	28	25	30	30
Emperors(=Scavengers) nei	*Lethrinidae*	33	32 731	29 747	27 389	29 804	28 858	30 293	28 752
Silver seabream	*Pagrus auratus*	33	644	649	872	395	303	388	347
Porgies, seabreams nei	*Sparidae*	33	2 208	603	251	218	224	269	447
Goatfishes	*Upeneus spp*	33	17 689	20 590	17 095	17 809	18 331	20 291	20 274
Goatfishes, red mullets nei	*Mullidae*	33	23 676	15 894	14 185	14 529	14 720	14 293	24 440
Mojarras(=Silver-biddies) nei	*Gerres spp*	33	7 636	6 026	5 434	5 909	5 802	6 991	7 087
Sea chubs nei	*Kyphosus spp*	33	3	2	2	1	1	1	1
Spotted sicklefish	*Drepane punctata*	33	888	734	582	918	739	798	868
Wrasses, hogfishes, etc. nei	*Labridae*	33	17 520	17 781	16 266	15 121	14 888	15 021	7 799
Parrotfishes nei	*Scaridae*	33	29	26	17	15	16	39	14
Threadfins, tasselfishes nei	*Polynemidae*	33	32 078	31 591	35 249	33 138	37 377	34 514	35 723
Glassfishes	*Ambassidae*	33	3 075	2 594	2 186	1 733	1 575	1 876	1 986
Percoids nei	*Percoidei*	33	38 415	21 792	30 749	45 305	20 544	22 276	25 011
Gobies nei	*Gobiidae*	33	8 404	7 542	6 756	8 021	7 719	9 032	9 291
Surgeonfishes nei	*Acanthuridae*	33	4 645	7 482	7 960	6 019	5 588	6 463	7 177
Batfishes	*Platax spp*	33	1 584	2 975	2 792	2 870	2 597	2 830	3 227
Scats	*Scatophagus spp*	33	4 035	2 057	2 476	2 501	2 637	3 162	3 205
Spinefeet(=Rabbitfishes) nei	*Siganus spp*	33	18 184	17 073	19 597	21 298	21 506	21 862	19 382
Flatheads nei	*Platycephalidae*	33	...	...	...	72	71	57	46
Triggerfishes, durgons nei	*Balistidae*	33	902	1 406	1 553	2 592	1 323	2 002	1 705
Daggertooth pike conger	*Muraenesox cinereus*	34	3 355	4 003	4 701	4 607	4 399	4 088	5 209
Conger eels, etc. nei	*Congridae*	34	2 756	2 110	2 540	2 467	2 398	2 741	2 781
Alfonsinos nei	*Beryx spp*	34	0	0	0	0	0	0	0
Redfish	*Centroberyx affinis*	34	-	-	-	-	1	5	5
Orange roughy	*Hoplostethus atlanticus*	34	-	-	-	-	717	872	656
Silver gemfish	*Rexea solandri*	34	-	-	-	-	-	-	11
Largehead hairtail	*Trichiurus lepturus*	34	7 349	12 624	12 539	18 602	17 042	13 108	8 952
Hairtails, scabbardfishes nei	*Trichiuridae*	34	29 357	28 791	28 321	29 616	30 627	32 900	36 149
Ruffs, barrelfishes nei	*Centrolophidae*	34	0	0	0	0	0	0	0
Scorpionfishes nei	*Scorpaenidae*	34	2	-	-	-	-	-	-
Bluefin gurnard	*Chelidonichthys kumu*	34	14	0	0	0	0	-	-
Demersal percomorphs nei	*Perciformes*	34	480	250	-	-	-	-	-
Goldstripe sardinella	*Sardinella gibbosa*	35	127 269	130 579	142 977	127 925	134 861	145 537	148 820
Bali sardinella	*Sardinella lemuru*	35	58 944	65 974	61 679	55 145	57 508	65 900	68 200
Sardinellas nei	*Sardinella spp*	35	419 695	454 189	431 674	408 403	420 239	427 697	390 984
Rainbow sardine	*Dussumieria acuta*	35	36 334	33 136	38 231	42 193	31 673	32 352	24 241
Silver-stripe round herring	*Spratelloides gracilis*	35	140	135	10	40	30	...	30
Bluestripe herring	*Herklotsichthys quadrimaculat.*	35	140	120	22	20	20	...	20
Stolephorus anchovies	*Stolephorus spp*	35	182 950	204 876	184 066	186 470	201 006	230 499	207 065
Anchovies, etc. nei	*Engraulidae*	35	122 423	117 229	121 443	103 445	117 025	121 721	123 102
Dorab wolf-herring	*Chirocentrus dorab*	35	2 741	2 715	4 724	6 598	7 493	5 133	5 191
Wolf-herrings nei	*Chirocentrus spp*	35	20 197	24 285	22 200	24 638	25 252	26 026	27 938
Clupeoids nei	*Clupeoidei*	35	44 122	37 340	44 721	44 412	33 627	39 363	40 827
Wahoo	*Acanthocybium solandri*	36	156	170	181	184	180	203	200
Narrow-barred Spanish mackerel	*Scomberomorus commerson*	36	64 590	73 198	68 346	72 929	77 584	74 000	76 780
Indo-Pacific king mackerel	*Scomberomorus guttatus*	36	9 314	9 098	10 925	10 457	12 781	13 705	14 440
Seerfishes nei	*Scomberomorus spp*	36	21 036	19 031	21 025	23 725	19 881	21 332	20 225
Frigate and bullet tunas	*Auxis thazard, A.rochei*	36	107 854	125 573	124 048	134 004	130 627	131 532	183 145

C-71 (a)

Fish, crustaceans, molluscs, etc — **Capture production by species items** — **Pacific, Western Central**
Poissons, crustacés, mollusques, etc — **Captures par catégories d'espèces** — **Pacifique, centre-ouest**
Peces, crustáceos, moluscos, etc — **Capturas por categorías de especies** — **Pacífico, centro-occidental**

English name Nom anglais Nombre inglés	Scientific name Nom scientifique Nombre científico	Species group Groupe d'espèces Grupo de especies	1996 t	1997 t	1998 t	1999 t	2000 t	2001 t	2002 t
Kawakawa	*Euthynnus affinis*	36	85 047	98 909	96 287	113 325	108 601	106 297	115 661
Skipjack tuna	*Katsuwonus pelamis*	36	897 019	755 946	1 037 990	977 856	1 019 686	942 487	1 080 437
Pacific bluefin tuna	*Thunnus orientalis*	36	801	1 314	1 233	3 104	1 755	1 654	1 504
Longtail tuna	*Thunnus tonggol*	36	35 661	31 637	40 968	48 590	57 752	45 265	44 880
Albacore	*Thunnus alalunga*	36	10 246	14 418	20 842	16 207	23 281	28 024	31 235
Southern bluefin tuna	*Thunnus maccoyii*	36	...	...	...	0	3	-	1
Yellowfin tuna	*Thunnus albacares*	36	290 762	415 153	448 671	416 538	413 093	432 182	426 053
Bigeye tuna	*Thunnus obesus*	36	21 910	44 208	42 846	47 620	34 132	44 388	73 821
Indo-Pacific sailfish	*Istiophorus platypterus*	36	317	318	3 104	180	63	104	93
Blue marlin	*Makaira nigricans*	36	6 754	6 923	9 488	10 924	12 150	12 343	11 926
Black marlin	*Makaira indica*	36	262	334	1 332	396	44	174	167
Striped marlin	*Tetrapturus audax*	36	457	1 080	1 652	436	488	607	611
Shortbill spearfish	*Tetrapturus angustirostris*	36	...	...	...	...	...	...	10
Marlins,sailfishes,etc. nei	*Istiophoridae*	36	5 209	6 623	8 188	9 780	6 111	6 930	6 895
Swordfish	*Xiphias gladius*	36	1 123	2 990	2 158	2 945	4 235	5 025	7 555
Tuna-like fishes nei	*Scombroidei*	36	140 927	152 648	163 291	160 957	179 466	179 761	185 780
Needlefishes nei	*Tylosurus spp*	37	29 558	31 429	30 938	31 884	33 800	30 998	31 699
Halfbeaks nei	*Hemiramphus spp*	37	3 301	1 717	1 998	2 342	2 462	3 015	3 201
Flyingfishes nei	*Exocoetidae*	37	31 001	40 492	50 058	53 064	52 239	46 802	48 817
Silversides(=Sand smelts) nei	*Atherinidae*	37	897	694	676	709	623	711	723
False trevally	*Lactarius lactarius*	37	410	552	576	824	657	722	708
Bluefish	*Pomatomus saltatrix*	37	103	179	128	204	139	240	222
Cobia	*Rachycentron canadum*	37	1 256	1 436	1 671	1 832	1 550	1 531	1 730
Japanese jack mackerel	*Trachurus japonicus*	37	-	-	1	-	-	-	-
Indian scad	*Decapterus russelli*	37	103 633	111 729	98 130	118 198	146 988	130 663	134 577
Scads nei	*Decapterus spp*	37	443 980	478 342	489 132	473 889	474 507	507 158	506 330
Jacks, crevalles nei	*Caranx spp*	37	43 239	49 991	38 632	59 152	61 317	61 931	64 976
Amberjacks nei	*Seriola spp*	37	34	34	-	-	-	0	-
Black pomfret	*Parastromateus niger*	37	28 061	32 204	29 799	31 250	31 488	40 583	44 113
Rainbow runner	*Elagatis bipinnulata*	37	10 649	11 138	12 257	13 769	13 170	12 719	13 790
Torpedo scad	*Megalaspis cordyla*	37	25 402	30 486	43 139	50 123	50 872	52 358	62 502
Queenfishes	*Scomberoides spp*	37	16 761	16 557	19 310	21 226	19 976	23 264	23 421
Bigeye scad	*Selar crumenophthalmus*	37	68 193	76 355	86 930	91 742	100 440	107 360	127 589
Yellowstripe scad	*Selaroides leptolepis*	37	24 467	26 526	29 091	36 470	40 816	36 165	35 145
Blackbanded trevally	*Seriolina nigrofasciata*	37	3 603	3 532	3 088	3 345	3 188	3 457	3 496
Carangids nei	*Carangidae*	37	170 655	171 311	165 647	169 875	164 977	184 051	199 078
Common dolphinfish	*Coryphaena hippurus*	37	3 952	429	309	280	254	336	312
Chub mackerel	*Scomber japonicus*	37	2 991	2 025	1 507	1 485	1 422	1 664	1 866
Short mackerel	*Rastrelliger brachysoma*	37	25 224	22 978	23 350	25 713	26 771	28 091	32 657
Indian mackerel	*Rastrelliger kanagurta*	37	67 993	74 320	71 537	81 240	77 640	86 042	96 373
Indian mackerels nei	*Rastrelliger spp*	37	235 955	242 758	262 613	285 554	285 258	283 472	287 447
Mackerels nei	*Scombridae*	37	204	189	195	1 953	2 229	2 138	2 461
Silver pomfret	*Pampus argenteus*	37	18 068	17 969	17 518	18 084	22 175	26 983	28 961
Butterfishes, pomfrets nei	*Stromateidae*	37	2 601	1 287	1 686	2 046	1 925	2 075	2 285
Barracudas nei	*Sphyraena spp*	37	31 883	30 662	33 575	39 809	36 204	34 965	35 352
Pelagic percomorphs nei	*Perciformes*	37	20	-	-	-	-	-	-
Rays, stingrays, mantas nei	*Rajiformes*	38	49 781	53 490	51 255	57 370	57 479	54 665	54 131
Sharks, rays, skates, etc. nei	*Elasmobranchii*	38	89 194	86 559	92 046	82 581	89 048	94 303	96 953
Marine fishes nei	*Osteichthyes*	39	2 224 261	2 138 367	2 137 686	2 280 509	2 356 699	2 527 456	2 566 893
Blue swimming crab	*Portunus pelagicus*	42	78 423	82 962	74 935	82 354	89 025	99 792	97 580
Indo-Pacific swamp crab	*Scylla serrata*	42	10 614	8 395	8 104	10 322	10 203	12 191	12 736
Marine crabs nei	*Brachyura*	42	37 753	36 183	61 366	49 314	52 397	50 747	60 071
Tropical spiny lobsters nei	*Panulirus spp*	43	3 058	4 564	3 627	4 281	4 323	5 201	6 806
Flathead lobster	*Thenus orientalis*	43	2 716	2 785	3 005	1 760	2 289	2 356	2 286
Slipper lobsters nei	*Scyllaridae*	43	943	736	866	98	96	128	127
Metanephrops nei	*Metanephrops spp*	43	-	-	-	-	-	-	9
Lobsters nei	*Reptantia*	43	...	2 000	1 000	200	500	700	700
Banana prawn	*Penaeus merguiensis*	45	55 916	56 272	64 442	67 393	70 355	70 251	72 023
Giant tiger prawn	*Penaeus monodon*	45	19 453	21 517	26 547	30 996	39 464	40 118	45 772
Green tiger prawn	*Penaeus semisulcatus*	45	1 042	1 068	839	609	714	836	811
Western king prawn	*Penaeus latisulcatus*	45	1 676	1 905	1 885	2 627	3 085	2 123	2 160
Penaeus shrimps nei	*Penaeus spp*	45	81 786	78 933	57 040	50 941	51 539	60 788	59 859
Endeavour shrimp	*Metapenaeus endeavouri*	45	2 400	2 339	2 930	2 691	3 593	3 502	3 118
Metapenaeus shrimps nei	*Metapenaeus spp*	45	28 934	36 215	44 904	35 932	45 258	40 868	44 246
Sergestid shrimps nei	*Sergestidae*	45	43 814	32 661	32 686	34 230	24 574	35 094	29 548
Natantian decapods nei	*Natantia*	45	200 401	209 617	182 995	213 172	204 233	221 418	226 817
Squillids nei	*Squillidae*	47	695	2 133	2 219	2 068	2 141	2 612	3 344
Stomatopods nei	*Stomatopoda*	47	181	176	457	750	866	456	442
Marine crustaceans nei	*Crustacea*	47	1 875	5 966	4 683	1 680	2 208	2 079	2 057
Abalones nei	*Haliotis spp*	52	448	183	348	282	241	288	292
Gastropods nei	*Gastropoda*	52	-	-	0	0	-	-	-
Slipper cupped oyster	*Crassostrea iredalei*	53	291	152	89	95	79	96	97
Cupped oysters nei	*Crassostrea spp*	53	1 596	1 678	2 029	2 025	1 931	896	1 600
Horse mussels nei	*Modiolus spp*	54	-	-	-	-	73	-	-
Green mussel	*Perna viridis*	54	20 216	18 017	17 454	6 554	41 173	22 782	23 282
Scallops nei	*Pectinidae*	55	7 276	3 488	6 392	5 403	4 601	5 901	4 334
Blood cockle	*Anadara granosa*	56	-	-	421	1 750	948	948	969
Anadara clams nei	*Anadara spp*	56	26 006	21 524	18 952	26 301	27 910	56 636	59 753
Hard clams nei	*Meretrix spp*	56	9 788	10 305	10 507	11 347	11 571	10 992	11 870

C-71 (a)

Fish, crustaceans, molluscs, etc — Capture production by species items — Pacific, Western Central
Poissons, crustacés, mollusques, etc — Captures par catégories d'espèces — Pacifique, centre-ouest
Peces, crustáceos, moluscos, etc — Capturas por categorías de especies — Pacífico, centro-occidental

English name Nom anglais Nombre inglés	Scientific name Nom scientifique Nombre científico	Species group Groupe d'espèces Grupo de especies	1996 t	1997 t	1998 t	1999 t	2000 t	2001 t	2002 t
Short neck clams nei	*Paphia spp*	56	34 591	24 229	43 458	42 613	25 966	34 204	34 956
Clams, etc. nei	*Bivalvia*	56	1 700	1 537	1 193	1 442	1 383	1 237	923
Cuttlefish,bobtail squids nei	*Sepiidae, Sepiolidae*	57	65 259	68 051	66 514	66 764	67 016	68 737	72 257
Common squids nei	*Loligo spp*	57	138 047	152 942	146 133	141 422	144 135	154 828	174 031
Various squids nei	*Loliginidae, Ommastrephidae*	57	23 654	23 096	23 689	25 109	27 290	27 278	33 968
Octopuses, etc. nei	*Octopodidae*	57	24 953	23 446	26 788	21 349	19 230	22 703	22 935
Cephalopods nei	*Cephalopoda*	57	92 000	92 500	103 000	110 000	180 000	130 000	130 000
Marine molluscs nei	*Mollusca*	58	53 322	54 772	59 749	59 051	53 179	61 483	60 783
Green turtle	*Chelonia mydas*	72	22	4	2	2	2	3	2
Marine turtles nei	*Testudinata*	72	395	399	450	446	382	316	330
Sea urchins nei	*Strongylocentrotus spp*	76	492	391	264	243	215	223	224
Sea cucumbers nei	*Holothurioidea*	76	7 916	7 292	6 593	5 880	6 290	6 557	6 961
Jellyfishes	*Rhopilema spp*	77	37 814	85 808	74 205	90 434	143 093	74 932	76 846
Aquatic invertebrates nei	*Invertebrata*	77	471	771	220	1 310	1 773	2 405	674
Total			***8 730 620***	***8 966 502***	***9 285 155***	***9 598 268***	***9 874 399***	***10 103 215***	***10 510 202***

C-71 (b)

Fish, crustaceans, molluscs, etc — Capture production by countries or areas — Pacific, Western Central
Poissons, crustacés, mollusques, etc — Captures par pays ou zones — Pacifique, centre-ouest
Peces, crustáceos, moluscos, etc — Capturas por países o áreas — Pacífico, centro-occidental

Country or area Pays ou zone País o área	1993 t	1994 t	1995 t	1996 t	1997 t	1998 t	1999 t	2000 t	2001 t	2002 t
Australia	38 006 F	39 400 F	43 719 F	46 781 F	42 075	46 536	48 905	42 229	52 009	48 066
Brunei Darsm	1 703	4 441	4 712	7 390	4 504	5 014	3 160	2 464	1 476	2 044
Cambodia	33 123	30 018	30 500	31 200	29 800	32 200	38 100	36 000	42 000	45 850
China	7 797	14 415	11 512	6 406	3 774	3 446	9 003	8 882	11 138	12 924
China,Taiwan	254 607 F	291 612 F	280 936 F	271 520	254 979	365 733	283 184	309 239	321 673	380 396
Fiji Islands	24 910	26 000 F	25 250	21 995	23 430	23 047	31 091	34 300 F	37 051	37 000 F
Guam	431	435	185	121	158	247	223	275	278	231
Indonesia	2 036 673	2 214 941	2 419 447	2 420 687	2 631 601	2 649 350	2 742 603	2 853 567	3 009 679	3 195 370
Japan	282 242	277 415	272 490	226 315	210 396	254 848	220 024	242 406	225 024	224 565 F
Kiribati	27 072	27 753	30 306	31 871	30 052	35 304	52 741	25 563	32 375	31 001
Korea Rep	157 088	244 170	213 720	175 466	182 399	218 092	157 499	180 224	190 038	225 065
Malaysia	600 835	626 245	604 368	632 598	669 685	640 965	762 661	765 911	755 933	770 639
Marshall Is	471 F	393 F	381 F	369	400 F	500 F	500 F	8 060	35 544	38 742
Micronesia	17 884 F	22 922 F	7 624 F	9 064 F	10 122 F	16 576 F	12 838 F	23 879 F	19 648 F	20 382 F
Nauru	500	500	400 F	300 F	250 F	200 F	150 F	100 F	50 F	21
NewCaledonia	2 670	3 185	2 644	2 998	2 438	3 105	3 152	3 386	3 309	3 417 F
N Marianas	136	176	191	225	250	235	193	189	197	198
Palau	1 104	1 001	1 025	999	904	951	960	1 047	1 075	1 008 F
Papua N Guin	12 241 F	12 719 F	25 577 F	24 323 F	32 685 F	64 816 F	52 270 F	82 907 F	108 983 F	135 200 F
Philippines	1 623 548	1 641 010	1 674 695	1 606 246	1 646 453	1 682 862	1 720 879	1 741 504	1 811 150	1 897 093
Russian Fed	18 937	5 840	4 981	-	-	-	-	-	-	-
Singapore	9 279	11 278	10 102	9 943	9 250	7 733	6 489	5 371	3 342	2 769
Solomon Is	42 970 F	46 904 F	62 019 F	50 990 F	64 005 F	61 332 F	58 428 F	24 926 F	30 277 F	31 003 F
Thailand	1 928 853	2 042 132	2 021 736	1 938 574	1 848 233	1 820 022	1 937 211	2 040 853	1 937 085	1 930 777
Tuvalu	1 460	561	399	400 F	500 F	500 F	500 F	500 F	500 F	500 F
USA	176 253	194 218	153 840	143 348	140 391	154 610	188 040	121 797	113 658	114 089
Vanuatu	2 470 F	3 020 F	9 236 F	11 603 F	28 856 F	41 477 F	49 971 F	37 930 F	2 400 F	2 752 F
Viet Nam	785 304	946 322	990 250	1 058 708	1 098 736	1 155 154	1 217 193	1 280 590 F	1 357 023 F	1 358 800 F
Wallis Fut I	150	193	170	180	176	300	300 F	300 F	300 F	300 F
Total	***8 088 717***	***8 729 219***	***8 902 415***	***8 730 620***	***8 966 502***	***9 285 155***	***9 598 268***	***9 874 399***	***10 103 215***	***10 510 202***

C-77 (a)

Fish, crustaceans, molluscs, etc — **Capture production by species items** — **Pacific, Eastern Central**
Poissons, crustacés, mollusques, etc — **Captures par catégories d'espèces** — **Pacifique, centre-est**
Peces, crustáceos, moluscos, etc — **Capturas por categorías de especies** — **Pacífico, centro-oriental**

English name Nom anglais Nombre inglés	Scientific name Nom scientifique Nombre científico	Species group Groupe d'espèces Grupo de especies	1996 t	1997 t	1998 t	1999 t	2000 t	2001 t	2002 t
Pink(=Humpback)salmon	*Oncorhynchus gorbuscha*	23	-	0	-	-	-	0	-
Chinook(=Spring=King)salmon	*Oncorhynchus tshawytscha*	23	2 078	2 737	937	1 991	2 613	1 193	2 416
Coho(=Silver)salmon	*Oncorhynchus kisutch*	23	6	3	0	-	4	1	-
Smelts nei	*Osmerus spp, Hypomesus spp*	23	-	7	7	5	6	12	8
American shad	*Alosa sapidissima*	24	-	0	-	-	-	-	-
Milkfish	*Chanos chanos*	25	0	1	0	0	7	...	1
Lefteye flounders nei	*Bothidae*	31	-	-	-	9	-	-	-
Pacific halibut	*Hippoglossus stenolepis*	31	4	2	3	2	-	0	-
English sole	*Pleuronectes vetulus*	31	...	...	227	226	177	197	102
Arrow-tooth flounder	*Atheresthes stomias*	31	1	1	1	2	0	0	1
Petrale sole	*Eopsetta jordani*	31	...	483	267	259	238	268	224
Rex sole	*Glyptocephalus zachirus*	31	...	...	168	153	128	98	116
Rock sole	*Lepidopsetta bilineata*	31	4	9	10	7	7	7	13
Dover sole	*Microstomus pacificus*	31	...	3 846	1 818	2 174	1 866	1 403	1 994
Pacific sand sole	*Psettichthys melanostictus*	31	...	...	19	23	35	44	35
Tonguefishes	*Cynoglossidae*	31	-	-	-	105	95	-	-
California flounder	*Paralichthys californicus*	31	...	...	510	617	390	389	424
Flatfishes nei	*Pleuronectiformes*	31	9 220	4 596	1 830	2 893	3 348	2 417	1 978
Pacific tomcod	*Microgadus proximus*	32	...	...	0	-	0	-	-
North Pacific hake	*Merluccius productus*	32	117	1 090	252	111	393	853	1 194
Grenadiers, rattails nei	*Macrouridae*	32	...	...	417	210	285	145	156
Gadiformes nei	*Gadiformes*	32	21	183	137	58	40	53	71
Sea catfishes nei	*Ariidae*	33	1 208	1 400	1 869	1 704	1 549	1 964	2 161
Squirrelfishes nei	*Holocentridae*	33	1	1	0	1	1	0	24
Bobo mullet	*Joturus pichardi*	33	87	56	51	82	46	104	76
Mullets nei	*Mugilidae*	33	3 774	3 474	2 993	3 454	3 134	3 171	4 214
Snooks(=Robalos) nei	*Centropomus spp*	33	587	684	752	996	809	862	686
Groupers nei	*Epinephelus spp*	33	5 367	7 228	6 250	6 881	5 367	4 098	5 176
Groupers, seabasses nei	*Serranidae*	33	177	387	644	995	581	438	256
Giant seabass	*Stereolepis gigas*	33	1	1	3	2	2	3	2
Yellow snapper	*Lutjanus argentiventris*	33	4 917	3 123	3 390	2 994	3 392	3 388	3 733
Snappers nei	*Lutjanus spp*	33	235	207	177	2 295	1 831	2 568	1 797
Snappers, jobfishes nei	*Lutjanidae*	33	5 955	6 321	7 405	12 573	10 922	24 977	29 234
Grunts, sweetlips nei	*Haemulidae (=Pomadasyidae)*	33	484	868	1 276	375	471	506	634
Weakfishes nei	*Cynoscion spp*	33	3 402	4 042	4 602	5 507	2 549	5 556	5 673
Croakers nei	*Micropogonias spp*	33	5 036	4 966	5 476	4 602	3 705	3 728	2 798
White weakfish	*Atractoscion nobilis*	33	46	28	71	112	101	124	194
White croaker	*Genyonemus lineatus*	33	242	167	64	74	105	137	97
Black drum	*Pogonias cromis*	33	0	0	0	1	0	0	0
Croakers, drums nei	*Sciaenidae*	33	3 005	2 994	3 557	9 913	8 172	6 294	3 624
Emperors(=Scavengers) nei	*Lethrinidae*	33	5	2	0	1	4	6	6
Porgies, seabreams nei	*Sparidae*	33	-	-	-	-	2	-	1
Goatfishes, red mullets nei	*Mullidae*	33	-	-	-	-	-	-	27
Mojarras, etc. nei	*Gerreidae*	33	6 841	3 946	1 945	2 114	1 724	1 800	1 692
Opaleye	*Girella nigricans*	33	-	-	-	-	3	2	1
Sea chubs nei	*Kyphosus spp*	33	-	-	-	-	-	-	12
California sheephead	*Semicossyphus pulcher*	33	...	...	118	58	78	68	52
Parrotfishes nei	*Scaridae*	33	8	3	10	7	6	3	23
Surgeonfishes nei	*Acanthuridae*	33	8	-	13	16	10	3	62
Lingcod	*Ophiodon elongatus*	33	352	364	95	93	28	34	44
Cabezon	*Scorpaenichthys marmoratus*	33	...	...	161	140	111	69	48
Sculpins nei	*Cottidae*	33	-	-	-	-	-	3	3
Triggerfishes, durgons nei	*Balistidae*	33	1 266	5 787	3 259	2 338	1 645	833	901
Cusk-eels, brotulas nei	*Ophidiidae*	34	-	-	-	-	1	-	-
Ocean whitefish	*Caulolatilus princeps*	34	...	...	535	627	1 077	984	929
Escolar	*Lepidocybium flavobrunneum*	34	...	...	2	1	6	3	3
Oilfish	*Ruvettus pretiosus*	34	-	-	-	-	-	-	92
Hairtails, scabbardfishes nei	*Trichiuridae*	34	1	...	-	607	540	4	16
Widow rockfish	*Sebastes entomelas*	34	...	865	540	402	29	128	7
Yellowtail rockfish	*Sebastes flavidus*	34	...	232	253	41	32	4	2
Pacific ocean perch	*Sebastes alutus*	34	-	-	-	16	12	12	18
Bocaccio rockfish	*Sebastes paucispinis*	34	...	276	105	62	24	33	21
Canary rockfish	*Sebastes pinniger*	34	...	129	91	36	3	4	3
Chilipepper rockfish	*Sebastes goodei*	34	...	1 812	1 262	908	414	609	156
Scorpionfishes nei	*Scorpaenidae*	34	7 678	4 482	4 330	1 614	1 291	1 679	1 839
Sablefish	*Anoplopoma fimbria*	34	2 136	1 751	824	1 132	1 088	958	897
Demersal percomorphs nei	*Perciformes*	34	1	0	0	0	0	0	0
Pacific herring	*Clupea pallasii*	35	5 328	9 188	1 987	2 267	2 853	2 521	3 282
California pilchard	*Sardinops caeruleus*	35	450 596	498 653	381 876	404 627	531 711	661 589	683 113
Red-eye round herring	*Etrumeus teres*	35	-	72	7	58	42	2	-
Pacific thread herring	*Opisthonema libertate*	35	32 517	26 266	49 472	38 746	63 532	29 033	48 175
Californian anchovy	*Engraulis mordax*	35	14 017	7 866	2 232	11 039	19 460	19 607	8 795
Pacific anchoveta	*Cetengraulis mysticetus*	35	59 830	77 726	107 730	27 356	86 681	129 147	160 414
Clupeoids nei	*Clupeoidei*	35	492	1 343	1 143	1 893	1 715	2 222	4 488
Eastern Pacific bonito	*Sarda chiliensis*	36	848	1 165	1 517	1 862	473	152	283
Wahoo	*Acanthocybium solandri*	36	5	126	209	288	249	307	697
Pacific sierra	*Scomberomorus sierra*	36	6 231	6 070	4 878	5 424	7 711	6 997	5 919
Black skipjack	*Euthynnus lineatus*	36	444	79	250	90	11	12	244
Kawakawa	*Euthynnus affinis*	36	-	-	-	-	-	-	4
Skipjack tuna	*Katsuwonus pelamis*	36	43 805	65 787	63 592	65 050	43 339	39 746	59 959
Pacific bluefin tuna	*Thunnus orientalis*	36	8 479	2 641	1 997	2 551	3 345	1 179	1 369
Longtail tuna	*Thunnus tonggol*	36	327	249	-	-	-	-	-
Albacore	*Thunnus alalunga*	36	19 283	22 530	26 418	28 856	26 223	30 824	32 457

C-77 (a)

Fish, crustaceans, molluscs, etc | Capture production by species items | Pacific, Eastern Central
Poissons, crustacés, mollusques, etc | Captures par catégories d'espèces | Pacifique, centre-est
Peces, crustáceos, moluscos, etc | Capturas por categorías de especies | Pacífico, centro-oriental

English name Nom anglais Nombre inglés	Scientific name Nom scientifique Nombre científico	Species group Groupe d'espèces Grupo de especies	1996 t	1997 t	1998 t	1999 t	2000 t	2001 t	2002 t
Yellowfin tuna	*Thunnus albacares*	36	211 267	232 583	211 710	204 702	210 191	271 978	337 939
Bigeye tuna	*Thunnus obesus*	36	55 803	56 998	78 116	61 440	67 252	79 347	93 062
Indo-Pacific sailfish	*Istiophorus platypterus*	36	459	1 447	666	575	459	441	449
Blue marlin	*Makaira nigricans*	36	3 099	4 248	4 088	2 806	2 650	3 453	3 986
Black marlin	*Makaira indica*	36	320	192	419	810	777	1 163	1 038
Striped marlin	*Tetrapturus audax*	36	2 051	3 130	2 322	2 064	1 483	1 111	1 200
Marlins,sailfishes,etc. nei	*Istiophoridae*	36	7 280	5 934	7 929	4 686	4 128	4 840	5 405
Swordfish	*Xiphias gladius*	36	6 855	9 488	11 649	8 572	11 006	8 820	8 130
Tuna-like fishes nei	*Scombroidei*	36	6 201	3 997	6 436	3 701	4 119	5 139	6 219
Pacific saury	*Cololabis saira*	37	-	-	-	754	1 060	423	234
Flyingfishes nei	*Exocoetidae*	37	30	85	122	30	118	112	109
Opah	*Lampris guttatus*	37	96	...	114	205	181	196	586
Silversides(=Sand smelts) nei	*Atherinidae*	37	...	...	1 062	906	801	738	1 023
Pacific jack mackerel	*Trachurus symmetricus*	37	2 176	1 158	832	950	1 135	3 624	1 005
Scads nei	*Decapterus spp*	37	237	130	149	226	187	270	406
Jacks, crevalles nei	*Caranx spp*	37	2 089	2 321	2 238	1 971	2 527	2 908	2 797
Pompanos nei	*Trachinotus spp*	37	206	200	329	309	274	222	258
Yellowtail amberjack	*Seriola lalandi*	37	-	-	111	30	50	39	34
Amberjacks nei	*Seriola spp*	37	477	345	475	172	445	538	1 453
Rainbow runner	*Elagatis bipinnulata*	37	-	-	-	-	-	-	2
Bigeye scad	*Selar crumenophthalmus*	37	-	-	-	-	-	-	128
Carangids nei	*Carangidae*	37	74	320	1 629	736	381	167	331
Pomfrets, ocean breams nei	*Bramidae*	37	-	-	-	-	-	5	222
Common dolphinfish	*Coryphaena hippurus*	37	338	563	565	1 273	11 615	14 573	10 871
Chub mackerel	*Scomber japonicus*	37	22 548	42 431	42 804	78 018	66 423	24 081	13 801
Butterfishes, pomfrets nei	*Stromateidae*	37	-	-	1	2	3	7	23
Barracudas nei	*Sphyraena spp*	37	70	489	621	259	180	296	464
Thresher	*Alopias vulpinus*	38	...	...	303	262	249	299	279
Bigeye thresher	*Alopias superciliosus*	38	...	...	11	5	5	2	-
Shortfin mako	*Isurus oxyrinchus*	38	39	32	184	119	114	109	135
Mako sharks	*Isurus spp*	38	-	-	-	-	-	-	80
Great white shark	*Carcharodon carcharias*	38	-	-	-	-	1	0	-
Blue shark	*Prionace glauca*	38	...	...	1	-	0	0	1
Silky shark	*Carcharhinus falciformis*	38	1 541	1 595	2 097	2 130	1 678	1 031	1 484
Requiem sharks nei	*Carcharhinidae*	38	5 849	3 482	2 988	2 949	3 320	3 059	2 719
Hammerhead sharks, etc. nei	*Sphyrnidae*	38	...	...	1	-	-	-	-
Brown smooth-hound	*Mustelus henlei*	38	...	...	3	5	3	4	3
Tope shark	*Galeorhinus galeus*	38	...	...	51	73	45	44	32
Picked dogfish	*Squalus acanthias*	38	-	0	5	24	8	3	17
Rays, stingrays, mantas nei	*Rajiformes*	38	5 527	5 132	5 114	6 149	5 707	4 612	4 200
Spotted ratfish	*Hydrolagus colliei*	38	-	-	-	-	-	-	2
Sharks, rays, skates, etc. nei	*Elasmobranchii*	38	25 232	24 887	27 588	27 475	32 597	27 869	26 038
Marine fishes nei	*Osteichthyes*	39	213 310	181 446	159 623	114 666	126 461	135 485	129 559
Dungeness crab	*Cancer magister*	42	897	1 593	1 485	734	772	840	2 008
Pacific rock crab	*Cancer productus*	42	...	...	571	354	490	532	546
Indo-Pacific swamp crab	*Scylla serrata*	42	-	2	2	2	2	2	3
Marine crabs nei	*Brachyura*	42	14 248	11 062	6 607	6 728	12 138	11 326	7 999
Green spiny lobster	*Panulirus gracilis*	43	101	152	66	131	476	652	334
Tropical spiny lobsters nei	*Panulirus spp*	43	2 517	2 579	2 438	2 174	3 097	2 977	2 988
Pelagic red crab	*Pleuroncodes planipes*	44	164	-	-	-	-	1 263	3 156
King crabs	*Paralithodes spp*	44	-	-	-	-	1	2	1
Yellowleg shrimp	*Penaeus californiensis*	45	272	126	271	230	138	102	42
Whiteleg shrimp	*Penaeus vannamei*	45	1 238	1 046	1 277	1 082	519	460	926
Crystal shrimp	*Penaeus brevirostris*	45	2 136	2 776	2 090	2 462	2 604	2 903	2 072
Penaeus shrimps nei	*Penaeus spp*	45	50 298	54 489	50 811	53 613	47 256	43 177	43 284
Pacific seabobs	*Xiphopenaeus,Trachypenaeus spp*	45	12 156	7 354	9 881	6 457	4 720	4 955	2 777
Northern nylon shrimp	*Heterocarpus vicarius*	45	-	-	40	37	20	131	0
Pacific shrimps nei	*Pandalus spp, Pandalopsis spp*	45	1 707	1 288	622	636	372	309	556
Crangonid shrimps nei	*Crangonidae*	45	-	31	41	-	42	45	38
Pacific rock shrimp	*Sicyonia ingentis*	45	...	...	185	630	756	165	215
Natantian decapods nei	*Natantia*	45	5 247	5 842	1 505	3 169	2 744	1 655	909
Brine shrimp	*Artemia salina*	47	...	...	513	691	512	475	320
Marine crustaceans nei	*Crustacea*	47	631	515	566	546	735	901	4 937
Abalones nei	*Haliotis spp*	52	1 202	1 026	709	582	535	482	494
Stromboid conchs nei	*Strombus spp*	52	3 091	2 427	1 500	1 473	2 102	2 722	2 815
Gastropods nei	*Gastropoda*	52	16	12	165	53	36	26	59
Pacific cupped oyster	*Crassostrea gigas*	53	1 717	2 176	2 107	539	1 917	1 083	1 083
Cupped oysters nei	*Crassostrea spp*	53	1 354	1 903	1 276	2 489	1 828	2 234	2 218
Sea mussels nei	*Mytilidae*	54	583	2 206	2 230	832	643	823	782
Pacific calico scallop	*Argopecten ventricosus*	55	17 290	2 320	2 726	1 864	6 287	3 100	9 875
Ark clams nei	*Arca spp*	56	834	1 367	1 018	880	675	600	328
Venus clams nei	*Veneridae*	56	5 038	3 915	3 672	4 486	4 835	3 582	2 930
Clams, etc. nei	*Bivalvia*	56	1 393	1 510	1 080	1 547	1 633	1 666	1 128
Opalescent inshore squid	*Loligo opalescens*	57	...	70 915	2 709	90 662	117 711	85 829	72 879
Neon flying squid	*Ommastrephes bartrami*	57	-	10	113	4	81	131	...
Jumbo flying squid	*Dosidicus gigas*	57	121 063	140 868	26 718	58 351	82 058	73 952	115 900
Various squids nei	*Loliginidae, Ommastrephidae*	57	79 035	235	40	341	1 683	6 140	441

C-77 (a)

Fish, crustaceans, molluscs, etc — **Capture production by species items** — **Pacific, Eastern Central**
Poissons, crustacés, mollusques, etc — **Captures par catégories d'espèces** — **Pacifique, centre-est**
Peces, crustáceos, moluscos, etc — **Capturas por categorías de especies** — **Pacífico, centro-oriental**

English name Nom anglais Nombre inglés	Scientific name Nom scientifique Nombre científico	Species group Groupe d'espèces Grupo de especies	1996 t	1997 t	1998 t	1999 t	2000 t	2001 t	2002 t
Octopuses, etc. nei	*Octopodidae*	57	1 344	1 052	807	1 211	998	936	770
Marine molluscs nei	*Mollusca*	58	1 307	1 857	1 462	1 742	3 816	2 950	2 983
Marine turtles nei	*Testudinata*	72	0	0	3	1	21	...	...
Echinoderms	*Echinodermata*	76	3 027	2 109	1 148	2 052	2 828	2 293	2 083
Starfishes nei	*Asteroidea*	76	-	-	-	1	-	0	1
Sea urchins nei	*Strongylocentrotus spp*	76	9 131	8 212	4 712	6 395	6 074	5 891	6 182
Sea cucumbers nei	*Holothurioidea*	76	86	80	361	234	426	442	290
Marine worms	*Polychaeta*	77	42	67	51	31	28	36	25
Aquatic invertebrates nei	*Invertebrata*	77	830	284	8	0	220	-	
Total			***1 619 642***	***1 704 354***	***1 407 976***	***1 446 226***	***1 719 304***	***1 860 373***	***2 037 267***

C-77 (b)

Fish, crustaceans, molluscs, etc — **Capture production by countries or areas** — **Pacific, Eastern Central**
Poissons, crustacés, mollusques, etc — **Captures par pays ou zones** — **Pacifique, centre-est**
Peces, crustáceos, moluscos, etc — **Capturas por países o áreas** — **Pacífico, centro-oriental**

Country or area Pays ou zone País o área	1993 t	1994 t	1995 t	1996 t	1997 t	1998 t	1999 t	2000 t	2001 t	2002 t
Amer Samoa	27	111	152	210	431	586	518	830	3 663	6 963
Belize	-	240	200	1 430	1 110	680	970	300	-	-
China	-	-	-	-	-	-	-	2 682	14 547	12 494
China,Taiwan	15 328 F	11 038 F	5 492	5 205	5 443	6 679	10 917	8 034	10 461	8 717
Colombia	...	...	...	...	...	...	...	7 869	9 696	15 295
Cook Is	985 F	933 F	1 080	900 F	820 F	770 F	750 F	720 F	700 F	1 768 F
Costa Rica	15 261	16 023	15 845	21 540	25 409	23 355	26 552	33 310	32 917	31 303
Cyprus	960	180	830	1 730	1 920	-	-	-	-	-
Ecuador	1 050	1 710	-	270	-	-	-	-	2	85
El Salvador	7 990	9 740	10 748	11 466	9 089	10 215	7 841	6 759	17 318	31 792
Estonia	-	-	-	-	-	-	10	-	-	-
Fr Polynesia	8 075	8 175	9 020	9 910	11 617	12 420	12 283	13 846	15 351	15 490
Guatemala	3 262	3 256	3 838	3 263	1 490	3 996	8 930	12 018	14 750 F	16 514
Honduras	509	692	864	2 312	3 316 F	2 588 F	2 842 F	3 955 F	4 695 F	5 000 F
Japan	102 740	95 131	82 190	69 159	82 917	63 507	46 161	76 384	54 890	52 731 F
Korea Rep	33 550	36 937	37 833	37 320	44 473	59 919	49 305	51 303	53 668	54 644
Mexico	665 867	743 251	921 820	1 027 461	1 042 150	767 021	817 557	928 727	1 039 463	1 119 624
Nicaragua	2 621	2 472	4 500	4 615	5 432	6 625	9 662	10 434	8 304	8 773
Niue	120 F	150 F	150 F	200 F	200 F	200 F	200 F	200 F	200 F	200 F
Panama	143 959	139 709	155 220	122 734	147 169	197 308	111 005	198 964	243 380	295 554
Samoa	673	1 108	2 519	2 727	7 041	7 547	10 204	13 003	12 965 F	12 391 F
Spain	740	160	980	1 910	1 950	21 960	37 753	27 416	15 722	20 503
Tokelau	200 F	200 F	200 F	200 F	200 F	200 F	200 F	200 F	200 F	200 F
Tonga	2 374	2 499	2 597	2 915	2 871	4 076	4 221	3 760	4 673	4 804
USA	143 488	174 088	233 025	220 175	237 646	152 124	222 855	256 127	212 580	189 276
US Minor Is	0	0	0	0	0	0	0	0	0	0
Vanuatu	14 270	13 090	21 460	13 390	22 240	20 590	16 090	7 622	6 620	3 073
Venezuela	32 700	26 870	45 030	58 600	49 420	45 300	49 400	52 830	69 730	91 616
Other nei	-	50	260	-	-	310	-	2 011	13 878	38 457
Total	***1 196 749***	***1 287 813***	***1 555 853***	***1 619 642***	***1 704 354***	***1 407 976***	***1 446 226***	***1 719 304***	***1 860 373***	***2 037 267***

C-81 (a)

Fish, crustaceans, molluscs, etc — Capture production by species items — Pacific, Southwest
Poissons, crustacés, mollusques, etc — Captures par catégories d'espèces — Pacifique, sud-ouest
Peces, crustáceos, moluscos, etc — Capturas por categorías de especies — Pacífico, sudoccidental

English name Nom anglais Nombre inglés	Scientific name Nom scientifique Nombre científico	Species group Groupe d'espèces Grupo de especies	1996 t	1997 t	1998 t	1999 t	2000 t	2001 t	2002 t
Short-finned eel	*Anguilla australis*	22	...	...	...	...	37	37	30
River eels nei	*Anguilla spp*	22	524	588	400	416	380	313	337
Chinook(=Spring=King)salmon	*Oncorhynchus tshawytscha*	23	1	1	3	1	1	1	0
Sand flounders	*Rhombosolea spp*	31	225	252	2	28	...	37	204
Tonguefishes	*Cynoglossidae*	31	-	-	-	9	-	-	3
Flatfishes nei	*Pleuronectiformes*	31	4 289	7 753	4 276	3 517	2 954	3 232	2 818
Common mora	*Mora moro*	32	694	1 410	1 324	1 122	1 358	1 211	1 308
Red codling	*Pseudophycis bachus*	32	10 596	11 087	16 495	12 555	5 365	4 530	4 443
Southern blue whiting	*Micromesistius australis*	32	26 040	34 624	66 163	70 145	43 598	54 012	56 586
Southern hake	*Merluccius australis*	32	9 119	11 370	17 617	19 433	16 041	15 188	13 834
Blue grenadier	*Macruronus novaezelandiae*	32	198 337	282 729	319 998	274 896	274 742	250 488	212 302
Thorntooth grenadier	*Lepidorhynchus denticulatus*	32	670	2 361	4 627	3 678	3 833	4 783	5 349
Grenadiers, rattails nei	*Macrouridae*	32	1 579	5 197	4 350	3 670	2 394	3 094	3 877
Gadiformes nei	*Gadiformes*	32	4 319	3 173	4 640	4 442	2 878	5 875	1 771
Sea catfishes nei	*Ariidae*	33	21	21	21	32	-	-	-
Mullets nei	*Mugilidae*	33	5 155	5 178	5 171	4 302	3 899	4 530	4 525
Orange perch	*Lepidoperca pulchella*	33	...	...	193	32	...	46	97
Groupers, seabasses nei	*Serranidae*	33	173	145	162	21	33	21	71
Sillago-whitings	*Sillaginidae*	33	1 169	700	607	1 804	1 115	1 342	1 432
Australian salmon	*Arripis trutta*	33	4 658	4 093	3 224	4 952	4 274	3 736	4 702
Snappers, jobfishes nei	*Lutjanidae*	33	...	...	...	301	293	288	252
Southern meagre(=Mulloway)	*Argyrosomus hololepidotus*	33	95	88	91	94	79	66	59
Geelbek croaker	*Atractoscion aequidens*	33	20	27	27	29	32	36	20
Croakers, drums nei	*Sciaenidae*	33	-	-	9	305	-	-	-
Emperors(=Scavengers) nei	*Lethrinidae*	33	-	4	-	-	-	-	-
Silver seabream	*Pagrus auratus*	33	6 138	6 551	6 550	8 494	7 845	7 020	7 148
Porgies, seabreams nei	*Sparidae*	33	1 935	2 221	1 361	1 805	2 642	2 736	2 584
Parore	*Girella tricuspidata*	33	561	588	576	579	635	584	548
Wrasses, hogfishes, etc. nei	*Labridae*	33	0	24	2	4	17	14	14
Antarctic rockcods, noties nei	*Nototheniidae*	33	16	34	3	19	3	15	42
Percoids nei	*Percoidei*	33	146	131	131	-	69	41	58
New Zealand blue cod	*Parapercis colias*	33	2 115	2 227	2 313	2 286	2 130	2 441	2 376
Flatheads nei	*Platycephalidae*	33	2 749	3 099	2 298	1 570	1 971	1 663	1 551
Puffers nei	*Tetraodontidae*	33	70	97	69	...	220	175	339
Velvet leatherjacket	*Parika scaber*	33	442	1 095	312	738	1 279	1 142	1 012
Triggerfishes, durgons nei	*Balistidae*	33	-	154	-	-	-	-	-
Argentines	*Argentina spp*	34	9	41	68	63	42	56	102
Conger eels, etc. nei	*Congridae*	34	109	113	97	220	96	106	144
Banded yellowfish	*Centriscops humerosus*	34	1	2	11	39	...	52	24
Pink cusk-eel	*Genypterus blacodes*	34	12 858	23 942	23 446	23 330	24 500	20 493	22 264
Alfonsinos nei	*Beryx spp*	34	2 159	2 617	3 593	2 579	2 880	3 053	3 040
Redfish	*Centroberyx affinis*	34	1 407	1 551	1 984	1 689	1 245	1 002	1 033
Orange roughy	*Hoplostethus atlanticus*	34	33 170	23 686	24 695	24 038	18 007	14 303	18 515
Slimeheads nei	*Trachichthyidae*	34	-	-	7	3	4	2	12
John dory	*Zeus faber*	34	969	1 235	983	1 584	1 117	1 125	1 292
Mirror dory	*Zenopsis nebulosus*	34	348	420	507	378	245	146	137
Dories nei	*Zeidae*	34	288	641	754	763	778	687	1 086
Oreo dories nei	*Oreosomatidae*	34	18 782	21 850	21 096	22 646	22 775	24 165	17 635
Hapuku wreckfish	*Polyprion oxygeneios*	34	1 155	1 657	1 571	1 547	1 497	1 579	1 610
Cape bonnetmouth	*Emmelichthys nitidus*	34	1 845	1 635	2 064	2 846	2 825	1 881	2 825
Bonnetmouths, rubyfishes nei	*Emmelichthyidae*	34	596	431	378	271	582	434	1 813
Giant boarfish	*Paristiopterus labiosus*	34	19	27	75	6	9	3	9
Pelagic armourhead	*Pseudopentaceros richardsoni*	34	7	2	78	13	6	7	63
Tarakihi	*Nemadactylus macropterus*	34	4 366	5 441	5 239	5 589	5 739	6 129	6 149
Morwongs	*Nemadactylus spp*	34	1 200	1 526	1 192	774	611	513	430
Trumpeters nei	*Latridae*	34	747	776	601	582	485	518	501
Patagonian toothfish	*Dissostichus eleginoides*	34	1 061	5	43	1	0	14	12
Black cardinal fish	*Epigonus telescopus*	34	3 002	4 334	2 568	2 869	4 095	1 957	2 741
Giant stargazer	*Kathetostoma giganteum*	34	2 123	3 991	2 196	3 370	3 638	4 233	3 272
Snoek	*Thyrsites atun*	34	18 571	25 957	29 856	28 599	28 098	28 410	27 060
Escolar	*Lepidocybium flavobrunneum*	34	0	1	1	22	47	87	76
Oilfish	*Ruvettus pretiosus*	34	53	34	57	53	61	86	131
Silver gemfish	*Rexea solandri*	34	2 344	2 253	1 899	1 356	1 249	914	761
Frostfishes	*Benthodesmus spp*	34	-	-	-	-	-	1	80
Silver scabbardfish	*Lepidopus caudatus*	34	941	2 342	3 344	2 638	1 619	2 887	2 475
Hairtails, scabbardfishes nei	*Trichiuridae*	34	678	1 188	545	422	1 099	227	152
Common warehou	*Seriolella brama*	34	1 760	4 108	3 101	3 881	4 259	4 101	4 174
Silver warehou	*Seriolella punctata*	34	4 926	11 253	10 993	9 029	11 218	11 268	9 175
White warehou	*Seriolella caerulea*	34	1 467	2 432	2 296	2 366	2 407	1 962	1 975
South Pacific breams nei	*Seriolella spp*	34	790	1 605	358	356	1 558	666	34
Bluenose warehou	*Hyperoglyphe antarctica*	34	2 432	2 974	2 630	2 755	2 793	2 954	3 001
Ruffs, barrelfishes nei	*Centrolophidae*	34	169	214	202	742	591	420	805
Scorpionfishes nei	*Scorpaenidae*	34	1 582	2 204	2 313	2 815	2 098	2 123	2 770
Bluefin gurnard	*Chelidonichthys kumu*	34	2 306	2 974	2 825	2 199	2 958	3 947	3 682
Latchet(=Sharpbeak gurnard)	*Pterygotrigla polyommata*	34	58	76	67	39	54	71	52
Spotted gurnard	*Pterygotrigla picta*	34	87	71	87	74	55	65	49
Demersal percomorphs nei	*Perciformes*	34	4 459	3 661	3 653	3 008	2 891	2 073	1 300
Australian pilchard	*Sardinops neopilchardus*	35	169	385	519	894	1 253	1 399	992
Anchovies, etc. nei	*Engraulidae*	35	12	19	19	2	39	22	116
Clupeoids nei	*Clupeoidei*	35	344	425	418	12	76	49	37
Wahoo	*Acanthocybium solandri*	36	-	6	-	-	3	1	2
Narrow-barred Spanish mackerel	*Scomberomorus commerson*	36	9	23	23	-	-	-	-
Seerfishes nei	*Scomberomorus spp*	36	1 769	3 111	2 897	1 237	1 601	2 079	1 327
Frigate and bullet tunas	*Auxis thazard, A.rochei*	36	0	2	4	-	5	5	5

C-81 (a)

Fish, crustaceans, molluscs, etc — **Capture production by species items** — **Pacific, Southwest**
Poissons, crustacés, mollusques, etc — **Captures par catégories d'espèces** — **Pacifique, sud-ouest**
Peces, crustáceos, moluscos, etc — **Capturas por categorías de especies** — **Pacífico, sudoccidental**

English name Nom anglais Nombre inglés	Scientific name Nom scientifique Nombre científico	Species group Groupe d'espèces Grupo de especies	1996 t	1997 t	1998 t	1999 t	2000 t	2001 t	2002 t
Skipjack tuna	*Katsuwonus pelamis*	36	6 585	11 754	9 706	7 344	12 285	5 301	4 106
Pacific bluefin tuna	*Thunnus orientalis*	36	8	20	27	35	24	54	59
Longtail tuna	*Thunnus tonggol*	36	23	15	-	3	2	6	5
Albacore	*Thunnus alalunga*	36	18 723	12 359	18 255	11 565	13 567	19 629	24 981
Southern bluefin tuna	*Thunnus maccoyii*	36	1 677	1 894	1 832	2 565	1 527	2 136	2 171
Yellowfin tuna	*Thunnus albacares*	36	4 626	3 297	2 698	1 715	1 782	2 193	1 856
Bigeye tuna	*Thunnus obesus*	36	1 341	1 483	3 186	2 374	2 321	2 754	1 797
Indo-Pacific sailfish	*Istiophorus platypterus*	36	9	8	14	282	52	46	38
Blue marlin	*Makaira nigricans*	36	35	56	107	28	26	41	32
Black marlin	*Makaira indica*	36	-	1	2	0	9	-	-
Striped marlin	*Tetrapturus audax*	36	784	500	994	428	629	665	619
Shortbill spearfish	*Tetrapturus angustirostris*	36	42	18	...	...	...	...	6
Marlins,sailfishes,etc. nei	*Istiophoridae*	36	-	-	12	-	-	-	-
Swordfish	*Xiphias gladius*	36	1 599	1 260	1 856	1 872	2 762	2 323	2 705
Tuna-like fishes nei	*Scombroidei*	36	325	364	394	180	455	470	261
Needlefishes, etc. nei	*Belonidae*	37	1	0	0	-	-	-	-
Halfbeaks nei	*Hemiramphus spp*	37	115	132	143	24	62	76	57
Flyingfishes nei	*Exocoetidae*	37	4	2	3	1	2	1	7
Opah	*Lampris guttatus*	37	76	126	254	335	283	340	329
Dealfishes	*Trachipterus spp*	37	49	60	74	65	67	103	64
King of herrings	*Regalecus glesne*	37	10	64	60	34	20	1	0
Bluefish	*Pomatomus saltatrix*	37	101	64	64	46	70	111	87
Greenback horse mackerel	*Trachurus declivis*	37	15 441	10 656	9 379	15 545	12 342	7 647	5 709
Jack and horse mackerels nei	*Trachurus spp*	37	30 577	36 174	38 050	36 196	22 815	29 469	32 305
White trevally	*Pseudocaranx dentex*	37	3 833	3 813	4 166	4 671	3 983	3 505	3 758
Scads nei	*Decapterus spp*	37	23	11	63	29	82	87	70
Amberjacks nei	*Seriola spp*	37	575	639	403	527	297	278	260
Atlantic pomfret	*Brama brama*	37	480	413	488	465	401	903	552
Pomfrets, ocean breams nei	*Bramidae*	37	-	-	-	-	2	-	3
Common dolphinfish	*Coryphaena hippurus*	37	0	0	0	1	0	15	7
Blue mackerel	*Scomber australasicus*	37	2 994	8 777	7 260	15 874	12 108	11 801	15 136
Mackerels nei	*Scombridae*	37	400	...	...	458	781	406	990
Butterfishes, pomfrets nei	*Stromateidae*	37	1 518	2 126	2 233	1 837	2 077	2 211	1 300
Barracudas nei	*Sphyraena spp*	37	1 176	479	126	1 114	432	512	585
Broadnose sevengill shark	*Notorynchus cepedianus*	38	...	...	2	3	4	5	4
Basking shark	*Cetorhinus maximus*	38	2	2	49	129	95	84	40
Thresher	*Alopias vulpinus*	38	13	24	21	32	51	57	54
Shortfin mako	*Isurus oxyrinchus*	38	52	40	74	110	208	327	239
Porbeagle	*Lamna nasus*	38	16	21	164	246	188	127	138
Blue shark	*Prionace glauca*	38	246	120	540	593	1 169	1 328	1 186
Copper shark	*Carcharhinus brachyurus*	38	...	...	15	14	25	38	38
Smooth hammerhead	*Sphyrna zygaena*	38	10	3	6	11	13	17	10
Spotted estuary smooth-hound	*Mustelus lenticulatus*	38	1 350	3 464	1 707	1 662	1 643	1 563	1 403
Smooth-hounds nei	*Mustelus spp*	38	...	...	...	...	100	87	109
Tope shark	*Galeorhinus galeus*	38	3 044	2 864	3 083	3 633	3 100	3 091	3 316
Picked dogfish	*Squalus acanthias*	38	2 477	7 232	3 064	4 409	3 362	4 192	6 186
Leafscale gulper shark	*Centrophorus squamosus*	38	...	...	4	1	0	0	1
Lanternsharks nei	*Etmopterus spp*	38	0	2	-	-	-	4	25
Birdbeak dogfish	*Deania calcea*	38	...	...	36	17	28	66	86
Kitefin shark	*Dalatias licha*	38	175	352	434	328	317	375	520
Dogfish sharks nei	*Squalidae*	38	693	1 705	701	1 010	770	705	325
Eagle rays	*Myliobatidae*	38	0	1	1	2	2	5	9
Rays, stingrays, mantas nei	*Rajiformes*	38	1 787	2 346	2 401	3 061	2 761	2 992	2 804
Dark ghost shark	*Hydrolagus novaezealandiae*	38	1 614	2 064	1 956	1 975	1 819	1 572	2 055
Ratfishes nei	*Hydrolagus spp*	38	...	0	36	453	975	2 184	1 901
Ghost shark	*Callorhinchus milii*	38	595	913	951	1 260	1 228	1 189	1 086
Chimaeras, etc. nei	*Chimaeriformes*	38	49	5	5	21	40	76	103
Sharks, rays, skates, etc. nei	*Elasmobranchii*	38	4 952	3 562	3 523	3 614	2 274	2 368	2 511
Marine fishes nei	*Osteichthyes*	39	49 338	49 134	28 774	10 319	9 015	39 741	42 723
Blue swimming crab	*Portunus pelagicus*	42	800	761	745	628	377	292	229
Marine crabs nei	*Brachyura*	42	294	359	393	407	355	360	299
Green rock lobster	*Jasus verreauxi*	43	161	158	124	123	152	114	111
Red rock lobster	*Jasus edwardsii*	43	3 121	5 009	2 707	2 818	2 789	2 551	2 485
Slipper lobsters nei	*Scyllaridae*	43	1	158	160	7	1	4	2
New Zealand lobster	*Metanephrops challengeri*	43	670	1 093	989	925	1 034	1 093	1 020
Giant tiger prawn	*Penaeus monodon*	45	-	-	-	-	13	5	4
Penaeus shrimps nei	*Penaeus spp*	45	1 822	1 926	1 683	2 399	2 829	2 851	2 197
Natantian decapods nei	*Natantia*	45	2	23	1	11	-	5	5
Marine crustaceans nei	*Crustacea*	47	156	157	92	64	307	293	273
Blacklip abalone	*Haliotis rubra*	52	344	335	327	323	325	305	281
Abalones nei	*Haliotis spp*	52	1 020	1 180	1 300	1 170	1 265	1 064	1 090
New Zealand dredge oyster	*Ostrea lutaria*	53	1 931	2 172	1 000	995	766	832	816
Pacific cupped oyster	*Crassostrea gigas*	53	2	2	0	62	0	0	-
Australian mussel	*Mytilus planulatus*	54	4	1	1	1	1	1	0
Sea mussels nei	*Mytilidae*	54	450	0	664	2 977	4 467	2 270	1 725
New Zealand scallop	*Pecten novaezelandiae*	55	5 080	18 848	4 592	6 152	2 912	6 792	4 408
Delicate scallop	*Zygochlamis delicatula*	55	124	201	91	128	0	222	118
Scallops nei	*Pectinidae*	55	5	1	1	1	0	1	-
Stutchbury's venus	*Chione stutchburyi*	56	815	541	1 325	1 396	1 789	1 748	1 687

C-81 (a)

Fish, crustaceans, molluscs, etc — Capture production by species items — Pacific, Southwest
Poissons, crustacés, mollusques, etc — Captures par catégories d'espèces — Pacifique, sud-ouest
Peces, crustáceos, moluscos, etc — Capturas por categorías de especies — Pacífico, sudoccidental

English name Nom anglais Nombre inglés	Scientific name Nom scientifique Nombre científico	Species group Groupe d'espèces Grupo de especies	1996 t	1997 t	1998 t	1999 t	2000 t	2001 t	2002 t
Short neck clams nei	*Paphia spp*	56	211	114	204	181	131	202	212
Pipi wedge clam	*Paphies australis*	56	248	465	455	513	628	624	475
Cuttlefish,bobtail squids nei	*Sepiidae, Sepiolidae*	57	378	374	251	381	261	237	245
Wellington flying squid	*Nototodarus sloani*	57	53 699	64 602	55 570	31 358	25 603	45 119	62 234
Various squids nei	*Loliginidae, Ommastrephidae*	57	18 995	19 080	14 304	11 735	9 539	11 804	17 228
Octopuses, etc. nei	*Octopodidae*	57	227	777	881	590	398	671	476
Marine molluscs nei	*Mollusca*	58	342	173	230	111	54	61	90
Echinoderms	*Echinodermata*	76	277	627	832	643	712	853	738
Starfishes nei	*Asteroidea*	76	4	0	13	2	6	9	39
Sea cucumbers nei	*Holothurioidea*	76	1	0	-	-	-	0	1
Aquatic invertebrates nei	*Invertebrata*	77	-	-	-	-	22	29	21
Total			***663 750***	***839 819***	***862 699***	***783 595***	***714 513***	***752 661***	***739 868***

C-81 (b)

Fish, crustaceans, molluscs, etc — Capture production by countries or areas — Pacific, Southwest
Poissons, crustacés, mollusques, etc — Captures par pays ou zones — Pacifique, sud-ouest
Peces, crustáceos, moluscos, etc — Capturas por países o áreas — Pacífico, sudoccidental

Country or area Pays ou zone País o área	1993 t	1994 t	1995 t	1996 t	1997 t	1998 t	1999 t	2000 t	2001 t	2002 t
Australia	59 617 F	57 088 F	52 466 F	47 741 F	49 719	49 969	26 900	28 910	26 516	25 749
Canada	235	235	235	136	149	167	253	351	206	144
China	-	-	-	-	-	-	-	-	-	752
China,Taiwan	24 494 F	14 485 F	11 429	20 599	12 902	13 176	6 983	8 200	5 901	10 043
Japan	140 989	113 146	95 100	77 078	72 426	74 231	58 178	43 416	56 855	36 800 F
Korea Rep	20 861	25 371	30 434	26 689	35 111	31 886	34 784	34 298	38 139	44 034
New Zealand	425 508	448 560	555 520	424 485	611 399	638 873	597 971	547 360	560 186	556 830
Norfolk Is	0	0	0	0	0	0	0	0	0	0
Norway	23 973	15 290	6 366	7 187	5 932	5 033	-	-	-	-
Pitcairn Is	8 F	8 F	8 F	8 F	8 F	8 F	8 F	8 F	8 F	8 F
Russian Fed	43 672	53 227	28 017	17 108	12 027	2 175	3 332	-	-	-
Ukraine	28 306	35 805	33 058	39 496	40 146	47 181	55 186	51 970	58 908	58 773
USA	1 123	605	2 170	3 223	-	-	-	-	5 942	6 735
Total	***768 786***	***763 820***	***814 803***	***663 750***	***839 819***	***862 699***	***783 595***	***714 513***	***752 661***	***739 868***

C-87 (a)

Fish, crustaceans, molluscs, etc — Capture production by species items — Pacific, Southeast
Poissons, crustacés, mollusques, etc — Captures par catégories d'espèces — Pacifique, sud-est
Peces, crustáceos, moluscos, etc — Capturas por categorías de especies — Pacífico, sudoriental

English name Nom anglais Nombre inglés	Scientific name Nom scientifique Nombre científico	Species group Groupe d'espèces Grupo de especies	1996 t	1997 t	1998 t	1999 t	2000 t	2001 t	2002 t
Lefteye flounders nei	*Bothidae*	31	-	-	-	-	5	-	-
Tonguefishes	*Cynoglossidae*	31	-	-	-	13	38	-	-
Flatfishes nei	*Pleuronectiformes*	31	761	411	436	398	322	439	318
Tadpole codling	*Salilota australis*	32	481	647	352	245	372	641	360
Southern blue whiting	*Micromesistius australis*	32	25 445	29 131	26 642	31 470	24 733	22 046	25 205
Southern hake	*Merluccius australis*	32	23 788	24 666	22 458	24 656	29 402	28 805	23 431
South Pacific hake	*Merluccius gayi*	32	323 470	265 573	162 518	140 910	193 504	246 265	162 290
Panama hake	*Merluccius angustimanus*	32	...	267	165	143	250	391	380
Patagonian grenadier	*Macruronus magellanicus*	32	379 015	71 479	353 823	309 723	91 310	160 774	132 119
Sea catfishes nei	*Ariidae*	33	425	350	121	97	100	100	100
Mullets nei	*Mugilidae*	33	14 180	13 368	29 179	21 005	26 476	27 451	25 039
Snooks(=Robalos) nei	*Centropomus spp*	33	341	73	96	62	70	70	65
Broomtail grouper	*Mycteroperca xenarcha*	33	180	125	155	82	80	210	200
Spotted grouper	*Epinephelus analogus*	33	43	48	32	39	30	28	28
Groupers nei	*Epinephelus spp*	33	421	185	256	233	161	144	368
Peruvian rock seabass	*Paralabrax humeralis*	33	4 954	2 789	2 554	3 278	4 373	2 011	1 522
Groupers, seabasses nei	*Serranidae*	33	30	24	10	19	7	5	-
Yellow snapper	*Lutjanus argentiventris*	33	43	40	100	82	80	80	80
Snappers nei	*Lutjanus spp*	33	577	118	660	1 088	804	857	839
Cabinza grunt	*Isacia conceptionis*	33	2 048	2 034	2 133	2 947	3 293	3 321	5 655
Grunts, sweetlips nei	*Haemulidae (=Pomadasyidae)*	33	194	379	1 345	818	183	419	526
Corvina	*Sciaena gilberti*	33	9 099	3 565	6 154	6 442	4 744	4 328	6 051
Peruvian weakfish	*Cynoscion analis*	33	7 887	5 797	11 187	8 929	6 339	4 438	3 487
Croakers nei	*Micropogonias spp*	33	1 026	886	1 233	789	1 125	1 062	1 414
Peruvian banded croaker	*Paralonchurus peruanus*	33	4 263	2 737	4 363	6 063	5 729	4 167	1 886
Croakers, drums nei	*Sciaenidae*	33	2 537	342	3 350	8 032	845	1 678	1 797
Porgies, seabreams nei	*Sparidae*	33	-	-	-	8	8	-	-
Threadfins, tasselfishes nei	*Polynemidae*	33	16	4	3	16	10	10	10
Patagonian blennie	*Eleginops maclovinus*	33	287	133	103	193	164	109	240
...A	*Normanichthys crockeri*	33	3 921	20 426	236	4 843	853	223	4 371
Conger eels, etc. nei	*Congridae*	34	-	-	23	-	-	-	-
Pink cusk-eel	*Genypterus blacodes*	34	5 780	6 410	6 836	5 721	6 269	7 522	4 518
Red cusk-eel	*Genypterus chilensis*	34	982	745	584	415	608	730	213
Black cusk-eel	*Genypterus maculatus*	34	1 343	1 661	2 753	1 943	3 542	3 889	1 558
Cusk-eels nei	*Genypterus spp*	34	1 121	439	425	196	557	552	1 029
Alfonsinos nei	*Beryx spp*	34	-	-	-	706	4 366	5 182	8 166
Slimeheads nei	*Trachichthyidae*	34	-	-	-	779	1 482	1 868	1 514
John dory	*Zeus faber*	34	-	-	-	5	-	-	-
Hapuku wreckfish	*Polyprion oxygeneios*	34	23	30	26	8	7	10	2
Tilefishes nei	*Branchiostegidae*	34	1 009	320	199	280	272	1 538	2 559
Peruvian morwong	*Cheilodactylus variegatus*	34	283	462	140	288	380	306	360
Patagonian toothfish	*Dissostichus eleginoides*	34	6 993	8 059	9 172	10 328	10 676	6 579	7 194
Cardinal fishes nei	*Epigonus spp*	34	513	1 727	5 284	2 999	5 792	4 648	1 595
Snoek	*Thyrsites atun*	34	821	1 337	1 022	604	851	830	934
Hairtails, scabbardfishes nei	*Trichiuridae*	34	...	...	...	85	13 666	3 382	3 601
South Pacific breams nei	*Seriolella spp*	34	6 186	3 297	3 176	4 936	3 975	6 536	5 787
Scorpionfishes nei	*Scorpaenidae*	34	212	532	208	258	141	177	53
Gurnards, searobins nei	*Triglidae*	34	...	...	...	22 634	27 129	1 428	1 500
Demersal percomorphs nei	*Perciformes*	34	170	220	161	294	200	200	190
South American pilchard	*Sardinops sagax*	35	1 493 936	722 807	937 269	442 790	338 131	135 712	27 953
Red-eye round herring	*Etrumeus teres*	35	34 349	1 095	8 873	3 636	4 414	28	-
Pacific thread herring	*Opisthonema libertate*	35	41 041	43 145	40 530	22 253	20 519	19 886	10 951
Pacific menhaden	*Ethmidium maculatum*	35	9 792	13 241	40 845	29 436	23 991	14 016	27 337
Araucanian herring	*Strangomera bentincki*	35	446 669	441 154	317 564	782 142	722 522	324 617	347 368
Anchoveta(=Peruvian anchovy)	*Engraulis ringens*	35	8 863 714	7 685 098	1 729 064	8 723 265	11 276 357	7 213 077	9 702 614
Pacific anchoveta	*Cetengraulis mysticetus*	35	52 698	117 904	72 975	43 002	34 262	98 636	43 290
Anchovies, etc. nei	*Engraulidae*	35	59 639	24 703	706 167	11 242	3 868	137 098	6 022
Clupeoids nei	*Clupeoidei*	35	1	429	273	0	3 596	4 420	100
Eastern Pacific bonito	*Sarda chiliensis*	36	23 079	17 764	5 722	1 325	499	1 319	875
Pacific sierra	*Scomberomorus sierra*	36	923	743	1 773	3 233	1 430	681	788
Frigate and bullet tunas	*Auxis thazard, A.rochei*	36	2 537	6 857	4 201	48 913	9 648	5 738	9 806
Black skipjack	*Euthynnus lineatus*	36	170	50	350	50	349	1 853	1 060
Skipjack tuna	*Katsuwonus pelamis*	36	77 061	118 524	103 761	205 631	163 554	109 129	125 232
Pacific bluefin tuna	*Thunnus orientalis*	36	-	2	-	1	-	-	-
Albacore	*Thunnus alalunga*	36	796	601	558	570	1 811	1 735	1 891
Yellowfin tuna	*Thunnus albacares*	36	74 237	76 102	103 582	112 101	106 239	174 083	123 310
Bigeye tuna	*Thunnus obesus*	36	46 103	48 632	30 064	36 257	67 420	46 665	43 329
Indo-Pacific sailfish	*Istiophorus platypterus*	36	8	44	54	10	11	35	35
Blue marlin	*Makaira nigricans*	36	256	284	544	123	73	164	163
Black marlin	*Makaira indica*	36	1	1	4	0	1	-	-
Striped marlin	*Tetrapturus audax*	36	1 084	729	1 078	200	165	666	666
Marlins,sailfishes,etc. nei	*Istiophoridae*	36	-	-	3 727	6 962	...	...	6
Swordfish	*Xiphias gladius*	36	4 326	6 389	6 600	4 496	5 164	5 716	11 399
Tuna-like fishes nei	*Scombroidei*	36	2 852	8 111	11 790	1 565	141	21 542	110
Atlantic saury	*Scomberesox saurus*	37	1	175	407	580	296	4 178	1 492
Flyingfishes nei	*Exocoetidae*	37	639	1 056	4 506	298 373	41 059	5 035	3 198
...A	*Odontesthes regia*	37	690	494	560	3 414	1 357	833	1 396
Silversides(=Sand smelts) nei	*Atherinidae*	37	3 802	5 184	45	6 692	11 215	7 528	11 220
Chilean jack mackerel	*Trachurus murphyi*	37	4 378 843	3 597 117	2 025 758	1 423 447	1 540 494	2 508 834	1 750 078
Jacks, crevalles nei	*Caranx spp*	37	667	520	347	880	235	149	202
Pompanos nei	*Trachinotus spp*	37	460	268	779	2 801	1 220	632	878
Amberjacks nei	*Seriola spp*	37	1 558	4 699	21 213	2 131	11 229	28 116	29 877
Pacific bumper	*Chloroscombrus orqueta*	37	1 706	952	565	1 409	...	1 008	1 100
Carangids nei	*Carangidae*	37	161	250	1 187	1 835	307	431	839

C-87 (a)

Fish, crustaceans, molluscs, etc — **Capture production by species items** — **Pacific, Southeast**
Poissons, crustacés, mollusques, etc — **Captures par catégories d'espèces** — **Pacifique, sud-est**
Peces, crustáceos, moluscos, etc — **Capturas por categorías de especies** — **Pacífico, sudoriental**

English name Nom anglais Nombre inglés	Scientific name Nom scientifique Nombre científico	Species group Groupe d'espèces Grupo de especies	1996 t	1997 t	1998 t	1999 t	2000 t	2001 t	2002 t
Atlantic pomfret	*Brama brama*	37	5 585	5 998	6 332	6 830	8 160	15 156	4 430
Common dolphinfish	*Coryphaena hippurus*	37	298	158	105	184	219	294	178
Chub mackerel	*Scomber japonicus*	37	275 354	610 014	518 388	676 160	254 524	627 466	393 142
Butterfishes, pomfrets nei	*Stromateidae*	37	9	432	11	17	10	11	141
Pelagic percomorphs nei	*Perciformes*	37	858	120	33	1 073	771	371	684
Shortfin mako	*Isurus oxyrinchus*	38	320	1 218	1 757	379	592	964	1 431
Porbeagle	*Lamna nasus*	38	-	...	7	...	...	...	65
Blue shark	*Prionace glauca*	38	11	114	824	7	262	456	2 491
Hammerhead sharks, etc. nei	*Sphyrnidae*	38	-	...	5	...	...	...	2
Smooth-hounds nei	*Mustelus spp*	38	4 209	3 682	8 410	3 485	4 515	5 057	7 352
Angelsharks, sand devils nei	*Squatinidae*	38	358	189	101	262	406	510	477
Pacific guitarfish	*Rhinobatos planiceps*	38	460	333	344	95	2 624	1 060	822
Rays, stingrays, mantas nei	*Rajiformes*	38	3 804	4 137	3 490	6 407	8 425	5 047	7 881
Elephantfishes nei	*Callorhinchus spp*	38	1 450	822	1 416	632	603	1 125	312
Sharks, rays, skates, etc. nei	*Elasmobranchii*	38	2 363	2 735	5 502	4 166	6 521	4 268	4 378
Marine fishes nei	*Osteichthyes*	39	83 088	112 041	404 257	304 202	322 884	197 441	139 805
Marine crabs nei	*Brachyura*	42	6 372	4 629	6 427	18 509	9 352	8 958	10 750
Green spiny lobster	*Panulirus gracilis*	43	106	69	725	547	329	113	71
Juan Fernandez rock lobster	*Jasus frontalis*	43	36	32	21	22	17	21	9
Carrot squat lobster	*Pleuroncodes monodon*	44	7 726	8 939	12 602	12 710	11 129	1 754	2 499
Blue squat lobster	*Cervimunida johni*	44	6 402	10 322	9 426	7 273	5 069	2 178	925
Craylets, squat lobsters	*Galatheidae*	44	-	-	-	-	254	1	724
Southern king crab	*Lithodes antarcticus*	44	1 759	2 160	2 766	2 155	2 902	2 937	2 869
Softshell red crab	*Paralomis granulosa*	44	1 273	1 477	1 501	1 438	4 938	6 527	4 631
Yellowleg shrimp	*Penaeus californiensis*	45	73	60	50	20	20	40	45
Whiteleg shrimp	*Penaeus vannamei*	45	4 497	4 000	3 000	1 100	1 100	2 250	2 450
Western white shrimp	*Penaeus occidentalis*	45	1 158	1 505	1 261	2 706	2 020	1 259	1 245
Penaeus shrimps nei	*Penaeus spp*	45	9 589	15 998	18 052	7 380	1 977	2 339	1 823
Pacific seabob	*Xiphopenaeus riveti*	45	2 683	2 830	1 970	1 752	2 000	2 341	2 300
Chilean nylon shrimp	*Heterocarpus reedi*	45	10 535	10 239	7 301	7 951	5 448	4 863	4 112
Kolibri shrimp	*Solenocera agassizii*	45	...	...	...	...	...	686	680
Chilean knife shrimp	*Haliporoides diomedeae*	45	15	32	29	135	169	309	482
Natantian decapods nei	*Natantia*	45	317	559	345	592	588	591	580
Giant barnacle	*Megabalanus psittacus*	47	879	579	683	620	620	685	199
Marine crustaceans nei	*Crustacea*	47	1	6	15	67	7	825	1 453
False abalone	*Concholepas concholepas*	52	5 269	7 520	3 394	4 583	2 524	1 372	2 308
Stromboid conchs nei	*Strombus spp*	52	10	10	10	10	10	10	10
Gastropods nei	*Gastropoda*	52	8 310	12 475	7 779	11 739	9 686	10 444	7 489
Chilean flat oyster	*Ostrea chilensis*	53	-	5	1	6	9	202	7
Pacific cupped oyster	*Crassostrea gigas*	53	5	5	5	5	5	5	5
Chilean mussel	*Mytilus chilensis*	54	5 714	4 723	4 899	4 343	5 236	6 758	1 416
Choro mussel	*Choromytilus chorus*	54	323	266	127	155	217	166	91
Cholga mussel	*Aulacomya ater*	54	13 428	16 078	22 831	19 738	18 933	22 584	19 433
Sea mussels nei	*Mytilidae*	54	3	3	5	5	5	5	5
Peruvian calico scallop	*Argopecten purpuratus*	55	2 095	4 013	23 546	30 141	11 830	6 544	2 086
Scallops nei	*Pectinidae*	55	1 577	2 598	3 662	1 715	332	141	373
Gay's little venus	*Tawera gayi*	56	...	...	...	...	1	291	271
Taca clam	*Protothaca thaca*	56	20 016	12 475	24 254	16 429	16 303	26 483	5 360
Venus clams nei	*Veneridae*	56	0	0	0	0	0	0	-
Taquilla clams	*Mulinia spp*	56	999	2 757	2 549	1 536	1 491	1 699	7 034
Macha clam	*Mesodesma donacium*	56	7 204	7 831	7 042	1 728	1 259	1 396	1 388
Chilean semele	*Semele solida*	56	4 418	2 199	1 900	2 071	4 212	3 054	2 162
Clams, etc. nei	*Bivalvia*	56	19 145	18 131	12 068	21 158	17 527	17 744	13 721
Common squids nei	*Loligo spp*	57	10 340	3 906	387	1 453	24 648	18 838	6 590
Jumbo flying squid	*Dosidicus gigas*	57	21 123	21 636	748	76 422	100 341	149 832	290 456
Various squids nei	*Loliginidae, Ommastrephidae*	57	321	293	214	137	124	3 681	5 746
Octopuses, etc. nei	*Octopodidae*	57	4 237	6 260	10 005	4 766	2 506	2 648	2 774
Marine molluscs nei	*Mollusca*	58	980	3 447	1 702	3 730	2 513	2 278	8 080
Marine turtles nei	*Testudinata*	72	10	11	12	11	11	12	12
Red sea squirt	*Pyura chilensis*	74	4 549	3 174	2 530	2 704	2 290	1 298	1 217
Echinoderms	*Echinodermata*	76	461	424	90	1 204	1 626	2 114	2 245
Sea urchins nei	*Strongylocentrotus spp*	76	0	0	0	0	0	0	-
Chilean sea urchin	*Loxechinus albus*	76	51 437	45 560	44 843	55 654	54 096	46 794	60 166
Sea cucumbers nei	*Holothurioidea*	76	127	16	45	123	1 525	122	121
Total			***17 068 356***	***14 397 205***	***8 034 212***	***14 176 309***	***15 804 615***	***12 653 427***	***13 765 143***

C-87 (b)

Fish, crustaceans, molluscs, etc	Capture production by countries or areas	Pacific, Southeast
Poissons, crustacés, mollusques, etc	Captures par pays ou zones	Pacifique, sud-est
Peces, crustáceos, moluscos, etc	Capturas por países o áreas	Pacífico, sudoriental

Country or area Pays ou zone País o área	1993 t	1994 t	1995 t	1996 t	1997 t	1998 t	1999 t	2000 t	2001 t	2002 t
Belize	1 110	5 230	7 120	8 560	11 740	8 430	6 170	7 047	-	-
Chile	5 943 326	7 716 540	7 431 704	6 687 844	5 805 745	3 249 132	5 043 641	4 295 087	3 787 387	4 264 220
China	-	-	-	-	-	-	-	-	-	126 744
China,Taiwan	498	142	3	252	306	89	62	1 500	882	996
Colombia	83 381	53 462	81 636	83 880 F	91 073 F	84 410 F	86 167 F	81 725 F	80 304 F	79 705 F
Costa Rica	...	...	270	1 260	...	...	...	...	...	...
Cuba	0	-	-	-	-	-	4 234	-	-	-
Cyprus	6 360	6 480	5 920	5 550	6 880	430	-	-	-	-
Ecuador	285 211	328 238	505 095	702 404	548 588	309 622	497 472	592 147	586 168	318 055
El Salvador	...	...	...	...	...	...	6 030	-	-	-
Estonia	-	-	-	-	-	-	14	-	-	-
Ghana	-	-	-	-	-	-	-	7 129	6 332	-
Guatemala	-	-	-	-	-	-	4 820	19 518	...	...
Honduras	-	-	-	470	2 560	870	5 710	2 580	...	...
Japan	69 982	102 495	48 922	11 871	11 266	12 847	10 251	42 174	89 530	88 111 F
Korea Rep	59 225	67 092	34 724	11 784	2 629	508	25 464	23 980	10 097	21 969
Liberia	-	-	-	900	-	-	-	-	-	-
Mexico	5 063	7 749	13 322	19 995	12 581	10 347	10 751	5 505	8 020	3 337
Nicaragua	-	-	-	-	-	-	-	1 660	...	...
Panama	2 890	3 860	5 800	3 570	6 370	320	5 980	16 312	12 064	8 080
Peru	8 966 487	11 950 380	8 886 553	9 486 158	7 837 650	4 303 110	8 392 378	10 626 323	7 950 450	8 737 025
St Vincent	-	2 750	-	-	-	-	-	-	-	-
Spain	7 818	4 326	9 428	9 153	16 478	15 169	27 745	28 933	26 817	26 662
USA	-	-	-	5 315	2 979	1 568	-	3 310	18	2 215
Vanuatu	23 430	31 010	26 780	20 410	19 530	12 750	26 980	23 344	15 501	11 314
Venezuela	18 360	21 570	8 720	8 980	17 470	23 740	22 440	22 030	42 420	29 574
Other nei	1 910	1 880	1 210	-	3 360	870	-	4 311	37 437	47 136
Total	***15 475 051***	***20 303 204***	***17 067 207***	***17 068 356***	***14 397 205***	***8 034 212***	***14 176 309***	***15 804 615***	***12 653 427***	***13 765 143***

C-88 (a)

Fish, crustaceans, molluscs, etc — **Capture production by species items** — **Pacific, Antarctic**
Poissons, crustacés, mollusques, etc — **Captures par catégories d'espèces** — **Pacifique, Antarctique**
Peces, crustáceos, moluscos, etc — **Capturas por categorías de especies** — **Pacífico, Antártico**

English name Nom anglais Nombre inglés	Scientific name Nom scientifique Nombre científico	Species group Groupe d'espèces Grupo de especies	1996 t	1997 t	1998 t	1999 t	2000 t	2001 t	2002 t
Smalleye moray cod	*Muraenolepis microps*	32	-	-	-	4	5	0	0
Moray cods nei	*Muraenolepis spp*	32	-	-	0	1	2	3	5
Blue antimora	*Antimora rostrata*	32	-	-	0	0	0	4	3
Whitson's grenadier	*Macrourus whitsoni*	32	-	-	-	1	5	48	158
Ridge scaled rattail	*Macrourus carinatus*	32	-	-	-	20	65	-	0
Grenadiers nei	*Macrourus spp*	32	-	-	9	1	0	6	-
Grenadiers, whiptails nei	*Coryphaenoides spp*	32	-	-	-	-	-	-	0
Porgies	*Calamus spp*	33	-	-	0	-	-	-	-
Antarctic rockcods, noties nei	*Nototheniidae*	33	-	-	0	0	0	1	0
Antarctic toothfish	*Dissostichus mawsoni*	34	-	-	41	296	751	626	1 354
Patagonian toothfish	*Dissostichus eleginoides*	34	-	0	1	1	0	34	12
Blackfin icefish	*Chaenocephalus aceratus*	34	-	-	-	-	0	-	-
Icefishes nei	*Channichthyidae*	34	-	-	0	0	0	2	2
Plunderfish	*Pogonophryne permitini*	34	-	-	-	-	0	0	0
Antarctic starry skate	*Raja georgiana*	38	-	-	-	11	36	7	24
Eaton's skate	*Bathyraja eatonii*	38	-	-	-	1	5	0	1
Bathyraja rays nei	*Bathyraja spp*	38	-	-	-	1	-	-	-
Rays and skates nei	*Rajidae*	38	-	-	-	6	-	-	-
Rays, stingrays, mantas nei	*Rajiformes*	38	-	-	5	-	0	-	-
Marine fishes nei	*Osteichthyes*	39	-	-	0	-	-	-	0
Subantarctic stone crab	*Lithodes murrayi*	44	-	-	-	-	-	-	0
Red stone crab	*Paralomis aculeata*	44	-	-	0	-	-	-	-
King crabs, stone crabs nei	*Lithodidae*	44	-	-	0	-	-	-	-
Octopuses, etc. nei	*Octopodidae*	57	-	-	-	-	-	-	0
Starfishes nei	*Asteroidea*	76	-	-	-	-	-	2	0
Total			***-***	***0***	***56***	***343***	***869***	***733***	***1 559***

C-88 (b)

Fish, crustaceans, molluscs, etc — **Capture production by countries or areas** — **Pacific, Antarctic**
Poissons, crustacés, mollusques, etc — **Captures par pays ou zones** — **Pacifique, Antarctique**
Peces, crustáceos, moluscos, etc — **Capturas por países o áreas** — **Pacífico, Antártico**

Country or area Pays ou zone País o área	1993 t	1994 t	1995 t	1996 t	1997 t	1998 t	1999 t	2000 t	2001 t	2002 t
Chile	-	-	-	-	-	0	-	-	-	-
New Zealand	-	-	-	-	0	56	343	869	678	1 559
South Africa	-	-	-	-	-	-	-	-	32	-
Uruguay	-	-	-	-	-	-	-	-	23	-
Total	***-***	***-***	***-***	***-***	***0***	***56***	***343***	***869***	***733***	***1 559***

Data in these tables refer to catches for which the new fishing season (1 December - 30 November) of the Commission for the Conservation of Antarctic Marine Living Resources (CCAMLR) has been adopted. Data are shown under the calendar year in which the split-year ends.

Les données dans ces tableaux se réfèrent aux captures pour lesquelles on utilise la nouvelle période de pêche (1er décembre - 30 novembre) de la Commission pour la conservation de la faune et la flore marines de l'Antarctique (CCAMLR). Les données figurent sous l'année civile durant laquelle se termine l'année fractionnée.

Los datos en estos cuadros se refieren a las capturas para las que se utiliza el nuevo período de pesca (1 de diciembre - 30 de noviembre) de la Comisión para la Conservación de los Recursos Vivos Marinos Antárticos (CCAMLR). Los datos se incluyen en el año civil en que termina el año emergente.

D - Capture production: by continents

D - Captures: par continents

D - Capturas: por continentes

D-1

Fish crustaceans, molluscs, etc — Capture production by countries or areas and fishing areas — Africa
Poissons, crustacés, mollusques, etc — Captures par pays ou zones et zones de pêche — Afrique
Peces, crustáceos, moluscos, etc — Capturas por países o áreas y áreas de pesca — Africa

	Fishing area Zone de pêche Area de pesca	1993 t	1994 t	1995 t	1996 t	1997 t	1998 t	1999 t	2000 t	2001 t	2002 t
Algeria	01	0	0	0	0	0	2	0	0	0	0
	37	101 895	135 402	105 872	81 989	91 580	92 344	102 396	113 156	133 623	134 320
	Country total	*101 895*	*135 402*	*105 872*	*81 989*	*91 580*	*92 346*	*102 396*	*113 156*	*133 623*	*134 320*
Angola	01	7 000	7 000	6 000	6 000	6 000	6 000	6 000 F	6 000 F	6 000 F	6 000 F
	47	119 200	125 413	116 781	131 815	140 304	157 149	169 799	232 351	246 518	254 797
	Country total	*126 200*	*132 413*	*122 781*	*137 815*	*146 304*	*163 149*	*175 799*	*238 351*	*252 518*	*260 797*
Benin	01	32 805	32 707	37 449	34 193	32 871	31 778	31 894	26 400	30 000	29 993
	34	6 416	7 216	6 930	7 982	10 914	10 361	8 542	5 924	8 415	10 670
	Country total	*39 221*	*39 923*	*44 379*	*42 175*	*43 785*	*42 139*	*40 436*	*32 324*	*38 415*	*40 663*
Botswana	01	600 F	400 F	200 F	81	160	191	157	166	118	139
	Country total	*600 F*	*400 F*	*200 F*	*81*	*160*	*191*	*157*	*166*	*118*	*139*
Br Ind Oc Tr	01	0	0	0	0	0	0	0	0	0	0
	51	0	0	0	0	0	0	0	0	0	0
	Country total	*0*	*0*	*0*	*0*	*0*	*0*	*0*	*0*	*0*	*0*
Burkina Faso	01	7 000	8 000	8 000	8 000	8 000	8 335	7 600	8 500	8 500 F	8 500 F
	Country total	*7 000*	*8 000*	*8 000*	*8 000*	*8 000*	*8 335*	*7 600*	*8 500*	*8 500 F*	*8 500 F*
Burundi	01	17 000 F	22 000 F	21 101	3 041	20 296	13 426	9 199	17 315	8 964	9 000 F
	Country total	*17 000 F*	*22 000 F*	*21 101*	*3 041*	*20 296*	*13 426*	*9 199*	*17 315*	*8 964*	*9 000 F*
Cameroon	01	23 000	27 000 F	30 000 F	35 000 F	40 000 F	45 000 F	50 000 F	55 000	52 500 F	65 000
	34	42 257	52 000 F	64 131	63 400	62 000 F	61 800 F	60 000 F	57 109	58 531	55 135
	Country total	*65 257*	*79 000 F*	*94 131 F*	*98 400 F*	*102 000 F*	*106 800 F*	*110 000 F*	*112 109*	*111 031*	*120 135*
Cape Verde	01	0	0	0	0	0	0	0	0	0	0
	34	7 000	8 256	8 495	9 155	9 705	9 424	10 360	10 586	9 653	8 000 F
	Country total	*7 000*	*8 256*	*8 495*	*9 155*	*9 705*	*9 424*	*10 360*	*10 586*	*9 653*	*8 000 F*
Cent Afr Rep	01	13 250 F	13 500 F	13 750 F	14 000 F	14 250 F	14 500 F	15 000	15 000 F	15 000 F	15 000 F
	Country total	*13 250 F*	*13 500 F*	*13 750 F*	*14 000 F*	*14 250 F*	*14 500 F*	*15 000*	*15 000 F*	*15 000 F*	*15 000 F*
Chad	01	87 300	80 000	90 000	100 000	85 000	84 000	84 000 F	84 000 F	84 000 F	84 000 F
	Country total	*87 300*	*80 000*	*90 000*	*100 000*	*85 000*	*84 000*	*84 000 F*	*84 000 F*	*84 000 F*	*84 000 F*
Comoros	01	0	0	0	0	0	0	0	0	0	0
	51	11 645	12 976	13 000 F	12 700 F	12 500 F	12 500 F	12 000	13 200	12 180	12 200 F
	Country total	*11 645*	*12 976*	*13 000 F*	*12 700 F*	*12 500 F*	*12 500 F*	*12 000*	*13 200*	*12 180*	*12 200 F*
Congo Dem R	01	192 589	152 117	154 751	159 037	158 367	174 087	204 503	205 000 F	210 000 F	215 000 F
	34	4 200	3 780	3 876	3 973	3 844	3 954	3 945	4 300 F	4 600 F	5 000 F
	Country total	*196 789*	*155 897*	*158 627*	*163 010*	*162 211*	*178 041*	*208 448*	*209 300 F*	*214 600 F*	*220 000 F*
Congo Rep	01	27 850	24 752	26 811	25 873	18 987	25 455	25 268	20 204 F	20 200 F	20 500 F
	34	18 898	17 912	18 965	19 600	19 095	17 500 F	18 241	24 525	22 040 F	22 500 F
	Country total	*46 748*	*42 664*	*45 776*	*45 473*	*38 082*	*42 955 F*	*43 509*	*44 729*	*42 240 F*	*43 000 F*
Côte dIvoire	01	13 477	15 604	11 335	11 562	12 032	12 501	10 656	10 502	10 630	23 843
	34	63 495	58 374	58 854	57 606	52 137	57 071	63 709	65 270	62 926	55 900 F
	Country total	*76 972*	*73 978*	*70 189*	*69 168*	*64 169*	*69 572*	*74 365*	*75 772*	*73 556*	*79 743 F*
Djibouti	01	0	0	0	0	0	0	0	0	0	0
	51	300 F	320 F	350 F	350 F	350 F	350 F	350 F	350 F	350 F	350 F
	Country total	*300 F*	*320 F*	*350 F*	*350 F*	*350 F*	*350 F*	*350 F*	*350 F*	*350 F*	*350 F*
Egypt	01	207 473	217 700	244 300	240 900	261 167	281 141	250 318	253 470	295 422	292 645
	34	-	-	-	-	-	-	4	-	-	-
	37	44 616	45 600	43 700	51 100	52 748	68 000	89 943	54 872	59 652	59 636
	51	50 740	48 300	47 300	48 434	57 417	57 063	82 400	75 972	73 577	72 889
	Country total	*302 829*	*311 600*	*335 300*	*340 434*	*371 332*	*406 204*	*422 665*	*384 314*	*428 651*	*425 170*
Eq Guinea	01	600	700	450	900	850	970	1 101 F	1 076	1 000 F	1 000 F
	34	2 907	4 369	1 856	4 140	5 240	5 035	5 900 F	2 558	2 500 F	2 500 F
	Country total	*3 507*	*5 069*	*2 306*	*5 040*	*6 090*	*6 005*	*7 001*	*3 634*	*3 500 F*	*3 500 F*
Eritrea	01	0	0	0	0	0	0	0	0	0	-
	51	475	2 706	3 559	3 252	1 038	1 629	6 891	12 612	8 820	7 832
	Country total	*475*	*2 706*	*3 559*	*3 252*	*1 038*	*1 629*	*6 891*	*12 612*	*8 820*	*7 832*
Ethiopia	01	4 175	5 285	6 325	8 770	10 370	14 000	15 858	15 681	15 390	12 300
	Country total	*4 175*	*5 285*	*6 325*	*8 770*	*10 370*	*14 000*	*15 858*	*15 681*	*15 390*	*12 300*
Fr South Tr	01	0	0	0	0	0	0	0	0	0	0
	51	460	524	519 F	437 F	375 F	388 F	425 F	272 F	263 F	414 F
	Country total	*460*	*524*	*519 F*	*437 F*	*375 F*	*388 F*	*425 F*	*272 F*	*263 F*	*414 F*
Gabon	01	3 500 F	4 500 F	7 648 F	9 433	9 441	9 000	10 000	10 417	9 850	9 400
	34	28 289	26 515	32 789	36 680	34 143	44 609	41 143	37 053	33 021	31 475
	Country total	*31 789 F*	*31 015 F*	*40 437 F*	*46 113*	*43 584*	*53 609*	*51 143*	*47 470*	*42 871*	*40 875*
Gambia	01	2 400	2 400	2 500	2 500	2 500	2 500	2 500 F	2 500 F	2 500 F	2 500 F
	34	18 908	20 381	21 252	29 101	29 754	26 502	27 500	26 516	32 027	43 269
	Country total	*21 308*	*22 781*	*23 752*	*31 601*	*32 254*	*29 002*	*30 000*	*29 016*	*34 527*	*45 769*

D-1

Fish crustaceans, molluscs, etc — **Capture production by countries or areas and fishing areas** — **Africa**
Poissons, crustacés, mollusques, etc — **Captures par pays ou zones et zones de pêche** — **Afrique**
Peces, crustáceos, moluscos, etc — **Capturas por países o áreas y áreas de pesca** — **Africa**

	Fishing area Zone de pêche Area de pesca	1993 t	1994 t	1995 t	1996 t	1997 t	1998 t	1999 t	2000 t	2001 t	2002 t
Ghana	01	52 000	54 700	60 000	73 580	70 000	74 500	74 500	74 500	74 500	74 500
	34	320 619	280 737	292 844	403 153	377 088	368 141	418 276	370 441	366 849	296 678
	87	-	-	-	-	-	-	-	7 129	6 332	-
	Country total	*372 619*	*335 437*	*352 844*	*476 733*	*447 088*	*442 641*	*492 776*	*452 070*	*447 681*	*371 178*
Guinea	01	4 600	3 800	3 100	2 780	3 600	4 000	4 000	4 000	4 000	4 000 F
	34	56 000 F	60 000 F	64 760	60 580	58 841	65 764	83 314	87 513	101 227	100 000 F
	Country total	*60 600 F*	*63 800 F*	*67 860*	*63 360*	*62 441*	*69 764*	*87 314*	*91 513*	*105 227*	*104 000 F*
GuineaBissau	01	250 F	250 F	250 F	250 F	250 F	200 F	200 F	200 F	200 F	200 F
	34	5 100 F	5 750 F	6 078 F	6 750 F	7 000 F	5 800 F	4 800 F	4 800 F	4 800 F	4 800 F
	Country total	*5 350 F*	*6 000 F*	*6 328*	*7 000 F*	*7 250 F*	*6 000 F*	*5 000 F*	*5 000 F*	*5 000 F*	*5 000 F*
Kenya	01	176 435	198 805	187 241	174 692	154 955	165 992	198 653	210 343	156 763	137 792
	51	5 617	3 772	5 465	6 296	6 099	6 600	6 634	4 763	7 388	6 720
	Country total	*182 052*	*202 577*	*192 706*	*180 988*	*161 054*	*172 592*	*205 287*	*215 106*	*164 151*	*144 512*
Lesotho	01	22 F	22 F	26 F	28 F	30 F	30 F	30	32	24	24 F
	Country total	*22 F*	*22 F*	*26 F*	*28 F*	*30 F*	*30 F*	*30*	*32*	*24*	*24 F*
Liberia	01	4 000	4 000	4 000	4 000	4 000	4 000	4 000	4 000	4 000	4 000 F
	34	3 782	3 721	4 829	3 408	4 580	6 830	11 472	7 726	7 286	7 500 F
	87	-	-	-	900	-	-	-	-	-	-
	Country total	*7 782*	*7 721*	*8 829*	*8 308*	*8 580*	*10 830*	*15 472*	*11 726*	*11 286*	*11 500 F*
Libya	01	0	0	0	0	0	0	0	0	0	0
	27	-	-	-	576	477	511	450	487	...	...
	34	1 085	500	400	400	400	400	400	400	239	666
	37	30 000 F	33 000 F	34 000 F	32 000 F	31 000 F	32 000 F	32 000 F	32 500 F	33 000 F	33 000 F
	Country total	*31 085 F*	*33 500 F*	*34 400 F*	*32 976 F*	*31 877 F*	*32 911 F*	*32 850 F*	*33 387 F*	*33 239 F*	*33 666 F*
Madagascar	01	30 000	30 000	30 000	30 000	30 000	30 000	30 000	30 000	30 000	30 000
	51	84 261	86 431	85 653	84 475	86 391	96 395	99 630	102 093	105 583	111 284
	Country total	*114 261*	*116 431*	*115 653*	*114 475*	*116 391*	*126 395*	*129 630*	*132 093*	*135 583*	*141 284*
Malawi	01	67 951	58 579	53 664	63 569	56 340	41 111	45 392	43 000 F	40 619	41 329
	Country total	*67 951*	*58 579*	*53 664*	*63 569*	*56 340*	*41 111*	*45 392*	*43 000 F*	*40 619*	*41 329*
Mali	01	64 300	62 850	132 900	111 910	99 550	98 000	98 536	109 870	100 000 F	100 000 F
	Country total	*64 300*	*62 850*	*132 900*	*111 910*	*99 550*	*98 000*	*98 536*	*109 870*	*100 000 F*	*100 000 F*
Mauritania	01	5 000 F	5 000 F	5 000 F	5 000 F	5 000 F	5 000 F	5 000 F	5 000 F	5 000 F	5 000 F
	34	54 452 F	46 746 F	48 147 F	55 324 F	52 756 F	56 660 F	71 026 F	75 849 F	79 881 F	73 902 F
	Country total	*59 452 F*	*51 746 F*	*53 147 F*	*60 324 F*	*57 756 F*	*61 660 F*	*76 026 F*	*80 849 F*	*84 881 F*	*78 902 F*
Mauritius	01	3	0	0	0	0	0	0	0	0	0
	51	20 576	18 145	16 395	11 869	14 025	12 093	12 205	9 615	10 985	10 706
	Country total	*20 579*	*18 145*	*16 395*	*11 869*	*14 025*	*12 093*	*12 205*	*9 615*	*10 985*	*10 706*
Mayotte	51	500 F	600 F	1 033 F	1 553	2 101 F	3 172 F	3 130	3 030 F	11 051	4 299
	Country total	*500 F*	*600 F*	*1 033 F*	*1 553*	*2 101 F*	*3 172 F*	*3 130*	*3 030 F*	*11 051*	*4 299*
Morocco	01	1 617	1 750	1 500	1 500	2 100	1 703	2 163	1 608	1 660	2 112
	34	591 270	717 578	807 775	601 734	758 321	680 075	705 978	856 362	1 054 147	863 488
	37	31 624	34 999	39 676	39 652	31 485	28 658	37 290	38 650	28 146	29 357
	Country total	*624 511*	*754 327*	*848 951*	*642 886*	*791 906*	*710 436*	*745 431*	*896 620*	*1 083 953*	*894 957*
Mozambique	01	4 689	4 925	5 173	7 510	11 668	8 994	9 052	11 813	5 284	12 156
	51	25 506	22 531	21 740	27 405	28 035	27 683	27 123	25 916	24 790	24 306
	Country total	*30 195*	*27 456*	*26 913*	*34 915*	*39 703*	*36 677*	*36 175*	*37 729*	*30 074*	*36 462*
Namibia	01	1 200	1 200	1 200	1 200	1 500	1 500	1 500	1 500	1 500	1 500
	41	-	-	-	-	304	677	746	-	-	-
	47	789 132	647 999	568 633	516 628	511 412	605 654	577 838	588 404	545 998	623 391
	Country total	*790 332*	*649 199*	*569 833*	*517 828*	*513 216*	*607 831*	*580 084*	*589 904*	*547 498*	*624 891*
Niger	01	2 162	2 516	3 616	4 156	6 328	7 013	11 000	16 250	20 800	23 520
	Country total	*2 162*	*2 516*	*3 616*	*4 156*	*6 328*	*7 013*	*11 000*	*16 250*	*20 800*	*23 520*
Nigeria	01	95 627	103 800	117 903	89 521	93 644	139 020	139 393	132 315	154 175	187 233
	34	142 782	163 259	231 579	248 472	294 279	324 004	316 235	309 062	297 971	293 823
	Country total	*238 409*	*267 059*	*349 482*	*337 993*	*387 923*	*463 024*	*455 628*	*441 377*	*452 146*	*481 056*
Réunion	01	0	0	0	0	0	0	0	0	0	0
	51	1 679	2 531	2 500	3 607	4 288	4 579	4 043	4 082	3 635	3 700
	Country total	*1 679*	*2 531*	*2 500*	*3 607*	*4 288*	*4 579*	*4 043*	*4 082*	*3 635*	*3 700*
Rwanda	01	3 500 F	3 400 F	3 300 F	2 952	4 428	6 641	6 433	6 726	6 828	7 000 F
	Country total	*3 500 F*	*3 400 F*	*3 300 F*	*2 952*	*4 428*	*6 641*	*6 433*	*6 726*	*6 828*	*7 000 F*
St Helena	01	0	0	0	0	-	-	-	-	-	-
	47	726	702	915	819	897	1 060	632	718	866	598
	Country total	*726*	*702*	*915*	*819*	*897*	*1 060*	*632*	*718*	*866*	*598*
Sao Tome Prn	01	0	0	0	0	0	0	0	0	0	0
	34	2 334	3 391	3 565	3 980	3 338	3 477	3 756	3 500 F	3 500 F	3 500 F
	Country total	*2 334*	*3 391*	*3 565*	*3 980*	*3 338*	*3 477*	*3 756*	*3 500 F*	*3 500 F*	*3 500 F*

D-1 Fish crustaceans, molluscs, etc — Capture production by countries or areas and fishing areas — Africa
Poissons, crustacés, mollusques, etc — Captures par pays ou zones et zones de pêche — Afrique
Peces, crustáceos, moluscos, etc — Capturas por países o áreas y áreas de pesca — Africa

	Fishing area Zone de pêche Area de pesca	1993 t	1994 t	1995 t	1996 t	1997 t	1998 t	1999 t	2000 t	2001 t	2002 t
Senegal	01	27 650	30 000 F	31 000 F	23 000 F	31 000 F	21 000 F	34 000 F	22 450	20 000 F	20 000 F
	34	354 562	322 421	323 617	388 759	426 366	382 872	378 125	379 597	383 202	355 824
	Country total	*382 212*	*352 421*	*354 617*	*411 759*	*457 366*	*403 872*	*412 125*	*402 047*	*403 202*	*375 824*
Seychelles	01	0	0	0	0	0	0	0	0	0	0
	34	-	-	-	-	-	-	-	127	-	-
	41	-	-	-	-	1 253	101	-	-	3 800	-
	47	-	-	-	-	-	-	-	-	-	181
	51	5 178	4 469	4 008	4 707	11 741	20 000	34 202	32 460	49 369	62 622
	57	-	-	-	-	1 049	3 785	21	183	365	406
	Country total	*5 178*	*4 469*	*4 008*	*4 707*	*14 043*	*23 886*	*34 223*	*32 770*	*53 534*	*63 209*
Sierra Leone	01	14 000	15 000	15 000	14 500 F	14 500	14 190	14 480	14 000	14 000	14 000
	34	52 288	47 439	49 870	52 804 F	58 128	48 875	44 927	60 730	61 210	68 990
	Country total	*66 288*	*62 439*	*64 870*	*67 304 F*	*72 628*	*63 065*	*59 407*	*74 730*	*75 210*	*82 990*
Somalia	01	250 F	250 F	250 F	250 F	250 F	250 F	250 F	200 F	200 F	150 F
	51	27 500 F	29 600	27 700 F	25 800 F	23 900 F	22 000 F	20 000 F	20 000 F	19 800 F	17 850 F
	Country total	*27 750 F*	*29 850 F*	*27 950 F*	*26 050 F*	*24 150 F*	*22 250 F*	*20 250 F*	*20 200 F*	*20 000 F*	*18 000 F*
South Africa	01	832	800	800	850	850	900	900 F	900 F	900 F	900 F
	47	560 716	522 129	574 374	439 229	511 249	555 852	585 240	640 000 F	747 470	763 285
	48	-	3	-	-	-	859	128	364	359	332
	51	826	517	373	349	892	1 100	1 000 F	926	1 017	1 586
	58	-	-	-	942	1 247	968	665	1 336	321	181
	88	-	-	-	-	-	-	-	-	32	-
	Country total	*562 374*	*523 449*	*575 547*	*441 370*	*514 238*	*559 679*	*587 933*	*643 526*	*750 099*	*766 284*
Sudan	01	37 500	40 000	40 000	40 500	42 000	44 000	44 000	48 000 F	53 000	53 000
	51	2 500	4 000	4 000	4 500	5 000	5 500	5 500	5 000 F	5 000	5 000
	Country total	*40 000*	*44 000*	*44 000*	*45 000*	*47 000*	*49 500*	*49 500*	*53 000 F*	*58 000*	*58 000*
Swaziland	01	68 F	65 F	60 F	60 F	65 F	70 F	70 F	70 F	70 F	70 F
	Country total	*68 F*	*65 F*	*60 F*	*60 F*	*65 F*	*70 F*	*70 F*	*70 F*	*70 F*	*70 F*
Tanzania	01	294 582	247 614	317 029	262 276	306 750	300 000	260 020	280 000	283 000	273 850
	51	36 685	40 785	42 771	61 645	50 210	48 000	50 489	52 779	52 900	49 680
	Country total	*331 267*	*288 399*	*359 800*	*323 921*	*356 960*	*348 000*	*310 509*	*332 779*	*335 900*	*323 530*
Togo	01	6 000	5 000	4 998	5 000	5 000	5 000	5 000	5 000	5 000	5 000
	34	10 964	8 052	7 203	10 098	9 290	11 655	17 924	17 277	18 163	15 946
	Country total	*16 964*	*13 052*	*12 201*	*15 098*	*14 290*	*16 655*	*22 924*	*22 277*	*23 163*	*20 946*
Tunisia	01	400	243	440	706	1 010	896	808	832	860	870
	37	82 514	85 367	82 915	83 028	86 002	87 179	92 378	94 718	97 622	95 815
	Country total	*82 914*	*85 610*	*83 355*	*83 734*	*87 012*	*88 075*	*93 186*	*95 550*	*98 482*	*96 685*
Uganda	01	219 814	213 129	208 789	195 088	218 026	220 628	226 097	219 356	220 726	221 898
	Country total	*219 814*	*213 129*	*208 789*	*195 088*	*218 026*	*220 628*	*226 097*	*219 356*	*220 726*	*221 898*
Westn Sahara	34	0	0	0	0	0	0	0	0	0	0
	Country total	*0*	*0*	*0*	*0*	*0*	*0*	*0*	*0*	*0*	*0*
Zambia	01	65 768	70 057	70 546	66 332	65 923	69 938	67 327	66 671	65 000 F	65 000 F
	Country total	*65 768*	*70 057*	*70 546*	*66 332*	*65 923*	*69 938*	*67 327*	*66 671*	*65 000 F*	*65 000 F*
Zimbabwe	01	21 230	20 219	16 463	16 387	18 156	16 371	12 410	13 114	13 000 F	13 000 F
	Country total	*21 230*	*20 219*	*16 463*	*16 387*	*18 156*	*16 371*	*12 410*	*13 114*	*13 000 F*	*13 000 F*
Total		***5 663 948***	***5 558 857***	***5 875 915***	***5 600 043***	***5 969 802***	***6 148 491***	***6 370 393***	***6 628 144***	***6 954 183***	***6 799 227***

	Fishing area / Zone de pêche / Area de pesca	1993 t	1994 t	1995 t	1996 t	1997 t	1998 t	1999 t	2000 t	2001 t	2002 t
Anguilla	02	0	0	0	0	0	0	0	0	0	0
	31	330	333	150 F	200 F	250 F	250 F	250 F	250 F	250 F	250 F
	Country total	*330*	*333*	*150 F*	*200 F*	*250 F*	*250 F*	*250 F*	*250 F*	*250 F*	*250 F*
Antigua Barb	02	0	0	0	0	0	0	0	0	0	0
	31	1 097	1 145	1 610	1 463	1 665	1 708	1 660	1 754	1 824	2 374
	Country total	*1 097*	*1 145*	*1 610*	*1 463*	*1 665*	*1 708*	*1 660*	*1 754*	*1 824*	*2 374*
Aruba	02	0	0	0	0	0	0	-	-	-	-
	31	260	260	140	160	205	182	175	163	163	163 F
	Country total	*260*	*260*	*140*	*160*	*205*	*182*	*175*	*163*	*163*	*163 F*
Bahamas	02	0	0	0	0	0	0	0	0	0	0
	31	10 073	10 311	9 915	10 197	10 439	10 124	10 473	11 070	9 290	9 300 F
	34	-	-	29	-	-	-	-	-	-	-
	Country total	*10 073*	*10 311*	*9 944*	*10 197*	*10 439*	*10 124*	*10 473*	*11 070*	*9 290*	*9 300 F*
Barbados	02	0	0	0	0	0	0	0	0	-	-
	31	3 214	2 818	3 581	3 512	2 809	3 644	3 250	3 100	2 676	2 500
	Country total	*3 214*	*2 818*	*3 581*	*3 512*	*2 809*	*3 644*	*3 250*	*3 100*	*2 676*	*2 500*
Belize	02	1	1	0	0	0	0	0	0	0	-
	31	1 784	2 102	1 877	2 048	2 835	2 557	1 871	2 434	2 517	2 078
	34	633	97	460	-	9 025	14 642	38 072	35 030	9 794	22 540
	41	-	-	-	-	-	-	4 500	6 729	2 581	135
	77	-	240	200	1 430	1 110	680	970	300	-	-
	87	1 110	5 230	7 120	8 560	11 740	8 430	6 170	7 047	-	-
	Country total	*3 528*	*7 670*	*9 657*	*12 038*	*24 710*	*26 309*	*51 583*	*51 540*	*14 892*	*24 753*
Bermuda	02	0	0	0	0	0	0	0	0	0	-
	31	404	394	449	465	461	466	453	286	315	393
	Country total	*404*	*394*	*449*	*465*	*461*	*466*	*453*	*286*	*315*	*393*
Br Virgin Is	02	0	0	0	0	0	0	0	0	0	0
	31	343	470	532	506	105	116	115	43	50 F	50 F
	Country total	*343*	*470*	*532*	*506*	*105*	*116*	*115*	*43*	*50 F*	*50 F*
Canada	02	36 327	36 333	38 756	38 295	38 798	40 744	40 587	40 667	38 140	39 999
	21	817 654	680 484	597 570	636 006	689 218	747 005	774 439	828 009	830 311	841 663
	67	286 283	309 848	217 121	235 074	247 383	226 013	212 576	143 072	183 886	132 191
	81	235	235	235	136	149	167	253	351	206	144
	Country total	*1 140 499*	*1 026 900*	*853 682*	*909 511*	*975 548*	*1 013 929*	*1 027 855*	*1 012 099*	*1 052 543*	*1 013 997*
Cayman Is	02	0	0	0	0	0	0	0	0	0	0
	31	125	125	125	110	125	125	125	125	125	125
	34	320	...	...	...	...	...	...	...	...	...
	Country total	*445*	*125*	*125*	*110*	*125*	*125*	*125*	*125*	*125*	*125*
Costa Rica	02	710	840	900	1 090	840	1 000 F	1 000 F	1 000 F	1 000 F	1 000 F
	31	199	269	422	437	420	402	666	1 088	816	635
	77	15 261	16 023	15 845	21 540	25 409	23 355	26 552	33 310	32 917	31 303
	87	...	...	270	1 260	...	...	...	...	...	...
	Country total	*16 170*	*17 132*	*17 437*	*24 327*	*26 669*	*24 757*	*28 218*	*35 398*	*34 733*	*32 938*
Cuba	02	9 886	9 537	8 270	9 744	7 603	5 154	4 591	3 005	2 007	2 034
	21	29 919	12 332	18 358	24 179	17 218	7 727	4 566	219	-	-
	31	41 608	52 790	50 500	47 921	51 604	47 889	53 317	57 837	42 723	31 804
	34	3 047	2 975	2 839	2 364	269	-	-	-	-	-
	47	-	-	-	-	-	-	2 427	-	-	-
	87	0	-	-	-	-	-	4 234	-	-	-
	Country total	*84 460*	*77 634*	*79 967*	*84 208*	*76 694*	*60 770*	*69 135*	*61 061*	*44 730*	*33 838*
Dominica	02	0	0	0	0	0	0	0	0	0	0
	31	794	882	950	1 030	1 079	1 212	1 200 F	1 200 F	1 200 F	1 217
	Country total	*794*	*882*	*950*	*1 030*	*1 079*	*1 212*	*1 200 F*	*1 200 F*	*1 200 F*	*1 217*
Dominican Rp	02	2 037	3 774	2 106	288	1 067	1 095	733	187	1 158	2 363
	31	10 820	19 058	15 768	12 606	13 468	9 014	7 804	10 842	12 059	15 976
	Country total	*12 857*	*22 832*	*17 874*	*12 894*	*14 535*	*10 109*	*8 537*	*11 029*	*13 217*	*18 339*
El Salvador	02	4 461	3 819	4 324	2 967	2 808	2 443	2 654	2 831	1 692	2 663
	77	7 990	9 740	10 748	11 466	9 089	10 215	7 841	6 759	17 318	31 792
	87	...	...	...	...	...	...	6 030	-	-	-
	Country total	*12 451*	*13 559*	*15 072*	*14 433*	*11 897*	*12 658*	*16 525*	*9 590*	*19 010*	*34 455*
Greenland	02	0	0	0	0	0	0	0	0	0	0
	21	94 937	101 952	108 574	94 356	94 096	97 953	117 841	121 885	124 149	154 599
	27	21 713	15 465	20 316	21 662	26 500	30 589	42 412	37 826	34 336	48 229
	Country total	*116 650*	*117 417*	*128 890*	*116 018*	*120 596*	*128 542*	*160 253*	*159 711*	*158 485*	*202 828*
Grenada	02	0	0	0	0	0	0	0	0	0	0
	31	2 103	1 599	1 499	1 577	1 530	1 837	1 658	1 701	2 250	2 171
	Country total	*2 103*	*1 599*	*1 499*	*1 577*	*1 530*	*1 837*	*1 658*	*1 701*	*2 250*	*2 171*
Guadeloupe	02	0	0	0	0	0	0	0	0	0	0
	31	8 600	8 800	9 500	9 570	10 480	9 084	9 114	10 100	10 000 F	10 100 F
	Country total	*8 600*	*8 800*	*9 500*	*9 570*	*10 480*	*9 084*	*9 114*	*10 100*	*10 000 F*	*10 100 F*

	Fishing area Zone de pêche Area de pesca	1993 t	1994 t	1995 t	1996 t	1997 t	1998 t	1999 t	2000 t	2001 t	2002 t
Guatemala	02	4 228	3 776	4 025	4 000	5 121	6 523	6 976	7 301	7 300 F	7 300 F
	31	92	179	390	390	285	328	292	366	350 F	350 F
	77	3 262	3 256	3 838	3 263	1 490	3 996	8 930	12 018	14 750 F	16 514
	87	-	-	-	-	-	-	4 820	19 518	...	...
	Country total	*7 582*	*7 211*	*8 253*	*7 653*	*6 896*	*10 847*	*21 018*	*39 203*	*22 400 F*	*24 164*
Haiti	02	600 F	500 F	500 F	500 F	500 F	500 F	500 F	500 F	500 F	500 F
	31	4 550 F	5 000 F	5 017 F	4 745 F	4 801 F	4 759 F	4 500 F	4 500 F	4 500 F	4 500 F
	Country total	*5 150 F*	*5 500 F*	*5 517 F*	*5 245 F*	*5 301 F*	*5 259 F*	*5 000 F*	*5 000 F*	*5 000 F*	*5 000 F*
Honduras	02	86	92	127	98	126	119	102	61	111	102 F
	21	1 298	-	-	-	-	-	-	-	-	-
	31	4 274	10 692	10 266	10 813	9 700 F	10 500 F	9 200 F	6 200 F	6 000 F	6 300 F
	34	9 562	993	8 841	6 998	5 084	1 908	3 355	1 711	1 120	-
	41	3 655	2 976	2 771	849	-	-	-	-	-	-
	47	193	...	48	10	25	9	20	...	...	-
	57	-	-	-	-	-	-	-	-	637	-
	77	509	692	864	2 312	3 316 F	2 588 F	2 842 F	3 955 F	4 695 F	5 000 F
	87	-	-	-	470	2 560	870	5 710	2 580	...	...
	Country total	*19 577*	*15 445*	*22 917*	*21 550*	*20 811*	*15 994*	*21 229*	*14 507*	*12 563*	*11 402*
Jamaica	02	450 F	450 F	450 F	450 F	450 F	450 F	450 F	450 F	450 F	450 F
	31	22 550 F	24 827 F	23 782 F	30 579 F	19 585	17 160	16 937	5 226	5 250 F	5 200 F
	Country total	*23 000 F*	*25 277 F*	*24 232 F*	*31 029 F*	*20 035*	*17 610*	*17 387*	*5 676*	*5 700 F*	*5 650 F*
Martinique	02	3	0	0	0	0	0	0	0	0	0
	31	5 850	5 800 F	5 300	3 500 F	5 500	5 500	6 000	6 310	6 200	6 200 F
	Country total	*5 853*	*5 800 F*	*5 300*	*3 500 F*	*5 500*	*5 500*	*6 000*	*6 310*	*6 200*	*6 200 F*
Mexico	02	111 178	111 125	122 020	122 501	113 552	100 335	91 462	106 817	91 952	82 648
	31	320 824	329 521	272 178	294 231	320 829	302 157	285 833	274 532	259 157	245 045
	77	665 867	743 251	921 820	1 027 461	1 042 150	767 021	817 557	928 727	1 039 463	1 119 624
	87	5 063	7 749	13 322	19 995	12 581	10 347	10 751	5 505	8 020	3 337
	Country total	*1 102 932*	*1 191 646*	*1 329 340*	*1 464 188*	*1 489 112*	*1 179 860*	*1 205 603*	*1 315 581*	*1 398 592*	*1 450 654*
Montserrat	02	0	0	0	0	0	0	0	0	0	0
	31	58	62	48	38	45	46	50 F	50 F	50 F	46
	Country total	*58*	*62*	*48*	*38*	*45*	*46*	*50 F*	*50 F*	*50 F*	*46*
NethAntilles	02	0	0	0	0	0	0	0	0	0	0
	31	1 200 F	1 100 F	1 020 F	1 000 F	950 F	950 F	950 F	950 F	850 F	850 F
	34	-	-	-	12 762	18 582	18 804	18 481	18 932	21 955	12 051
	Country total	*1 200 F*	*1 100 F*	*1 020 F*	*13 762 F*	*19 532 F*	*19 754 F*	*19 431 F*	*19 882 F*	*22 805 F*	*12 901 F*
Nicaragua	02	547	824	538	1 142	1 293	1 256	1 120	1 076	1 051	1 592
	31	5 001	6 738	5 957	9 685	9 451	12 011	13 127	14 838	13 444	13 305
	77	2 621	2 472	4 500	4 615	5 432	6 625	9 662	10 434	8 304	8 773
	87	-	-	-	-	-	-	-	1 660	...	...
	Country total	*8 169*	*10 034*	*10 995*	*15 442*	*16 176*	*19 892*	*23 909*	*28 008*	*22 799*	*23 670*
Panama	02	28	285	130	80	91	23	20	20	20 F	20 F
	27	210	363	289	369	58	58	-	-	-	-
	31	-	-	-	-	-	-	3	53	-	-
	34	35 924	38 469	39 164	19 618	8 293	4 889	3 725	3 330	944	1 427
	37	467	1 499	1 498	2 850	236	-	-	-	-	-
	41	1 295	804	606	776	3	-	120	487	-	-
	47	48	158	146	30	3	-	19	-	-	-
	48	-	-	637	-	-	-	-	-	-	-
	77	143 959	139 709	155 220	122 734	147 169	197 308	111 005	198 964	243 380	295 554
	87	2 890	3 860	5 800	3 570	6 370	320	5 980	16 312	12 064	8 080
	Country total	*184 821*	*185 147*	*203 490*	*150 027*	*162 223*	*202 598*	*120 872*	*219 166*	*256 408 F*	*305 081*
Puerto Rico	02	0	0	0	0	0	0	0	0	0	-
	31	1 877	2 275	3 173	2 701	3 187	3 006	3 020	4 154	3 794	2 529
	Country total	*1 877*	*2 275*	*3 173*	*2 701*	*3 187*	*3 006*	*3 020*	*4 154*	*3 794*	*2 529*
St Kitts Nev	02	0	0	0	0	0	0	0	0	0	-
	31	250 F	212	192	352	216	407	471	469	555	355
	Country total	*250 F*	*212*	*192*	*352*	*216*	*407*	*471*	*469*	*555*	*355*
St Lucia	02	0	0	0	0	0	0	0	0	0	0
	31	1 336	1 252	1 188	1 274	1 308	1 589	1 718	1 855	1 983	1 637
	Country total	*1 336*	*1 252*	*1 188*	*1 274*	*1 308*	*1 589*	*1 718*	*1 855*	*1 983*	*1 637*
St Pier Mq	02	0	0	0	0	0	0	0	0	0	0
	21	282	294	317	747	3 571	6 108	5 892	6 485	3 802	3 654
	Country total	*282*	*294*	*317*	*747*	*3 571*	*6 108*	*5 892*	*6 485*	*3 802*	*3 654*
St Vincent	02	0	0	0	2	1	0	0	0	0	0
	31	1 479	1 090	944	889 F	948	1 365	1 032	7 325	9 020	2 117
	34	466	0	61	30	5 144	32 608	16 727	20 369	41 647	41 762
	41	-	-	-	-	-	-	-	-	1 818	-
	87	-	2 750	-	-	-	-	-	-	-	-
	Country total	*1 945*	*3 840*	*1 005*	*921 F*	*6 093*	*33 973*	*17 759*	*27 694*	*52 485*	*43 879*
Trinidad Tob	02	0	0	0	0	0	0	0	0	0	0
	31	8 998	14 046	11 500 F	9 435	11 283	9 175	8 826	8 651	10 707	12 539
	Country total	*8 998*	*14 046*	*11 500 F*	*9 435*	*11 283*	*9 175*	*8 826*	*8 651*	*10 707*	*12 539*

D-2 Fish crustaceans, molluscs, etc — Capture production by countries or areas and fishing areas — America, North
Poissons, crustacés, mollusques, etc — Captures par pays ou zones et zones de pêche — Amérique du Nord
Peces, crustáceos, moluscos, etc — Capturas por países o áreas y áreas de pesca — América del Norte

	Fishing area Zone de pêche Area de pesca	1993 t	1994 t	1995 t	1996 t	1997 t	1998 t	1999 t	2000 t	2001 t	2002 t
Turks Caicos	02	0	0	0	0	0	0	0	0	0	0
	31	6 255	5 963 F	7 824	7 833	5 951	6 383	5 401	5 975	6 611	5 953
	Country total	*6 255*	*5 963 F*	*7 824*	*7 833*	*5 951*	*6 383*	*5 401*	*5 975*	*6 611*	*5 953*
USA	02	54 377	42 648	36 688	33 471	38 347	36 129	36 413	25 339	29 578	29 943
	21	1 240 243	1 062 863	1 199 006	1 228 049	1 179 746	1 027 487	1 024 098	985 289	1 129 720	1 089 864
	31	942 154	1 238 611	856 754	771 970	867 630	822 594	943 641	993 827	878 443	922 633
	48	-	-	292	392	-	-	16	70	1 568	12 175
	67	2 965 279	2 822 291	2 743 083	2 595 248	2 516 701	2 514 468	2 334 583	2 331 879	2 572 829	2 570 375
	71	176 253	194 218	153 840	143 348	140 391	154 610	188 040	121 797	113 658	114 089
	77	143 488	174 088	233 025	220 175	237 646	152 124	222 855	256 127	212 580	189 276
	81	1 123	605	2 170	3 223	-	-	-	-	5 942	6 735
	87	-	-	-	5 315	2 979	1 568	-	3 310	18	2 215
	Country total	*5 522 917*	*5 535 324*	*5 224 858*	*5 001 191*	*4 983 440*	*4 708 980*	*4 749 646*	*4 717 638*	*4 944 336*	*4 937 305*
US Virgin Is	02	0	0	0	0	0	0	0	0	0	0
	31	650 F	550 F	470 F	400 F	350 F	300 F	263	300 F	300 F	300 F
	Country total	*650 F*	*550 F*	*470 F*	*400 F*	*350 F*	*300 F*	*263*	*300 F*	*300 F*	*300 F*
Total		***8 317 130***	***8 321 259***	***8 012 698***	***7 939 507***	***8 040 827***	***7 553 103***	***7 624 074***	***7 796 824***	***8 142 843***	***8 242 710***

	Fishing area Zone de pêche Area de pesca	1993 t	1994 t	1995 t	1996 t	1997 t	1998 t	1999 t	2000 t	2001 t	2002 t
Argentina	03	11 800	12 785	17 191	19 189	22 735	23 197	27 558	30 418	33 757	31 233
	41	919 503	938 590	1 152 056	1 272 057	1 377 427	1 141 632	1 044 221	878 561	897 977	913 113
	48	-	12	879	109	-	-	6 534	-	-	-
	Country total	*931 303*	*951 387*	*1 170 126*	*1 291 355*	*1 400 162*	*1 164 829*	*1 078 313*	*908 979*	*931 734*	*944 346*
Bolivia	03	5 518	5 353	5 692	5 988	6 038	6 055	6 052	6 106	5 940	5 800 F
	Country total	*5 518*	*5 353*	*5 692*	*5 988*	*6 038*	*6 055*	*6 052*	*6 106*	*5 940*	*5 800 F*
Brazil	03	186 990 F	191 485 F	193 042	193 309	178 871	174 190	185 471	199 159	217 275	217 750
	41	530 100 F	548 615 F	513 666	522 173	565 714	532 599	518 470	567 687	589 397	604 409
	Country total	*717 090 F*	*740 100 F*	*706 708*	*715 482*	*744 585*	*706 789*	*703 941*	*766 846*	*806 672*	*822 159*
Chile	03	7	5	...	...	...	4	...	...	...	...
	34	-	25	-	-	-	-	-	-	-	-
	41	846	23	302	-	3 744	14 755	5 226	2 749	8 849	5 710
	47	-	-	-	-	-	-	-	-	5	-
	48	5 386	3 985	2 174	2 821	2 061	1 402	1 313	2 106	899	1 545
	87	5 943 326	7 716 540	7 431 704	6 687 844	5 805 745	3 249 132	5 043 641	4 295 087	3 787 387	4 264 220
	88	-	-	-	-	-	0	-	-	-	-
	Country total	*5 949 565*	*7 720 578*	*7 434 180*	*6 690 665*	*5 811 550*	*3 265 293*	*5 050 180*	*4 299 942*	*3 797 140*	*4 271 475*
Colombia	03	30 538	34 983	23 524	23 061	20 610	21 673	28 788	24 854	25 000 F	25 000 F
	31	9 578	9 354	15 539	18 888 F	11 235 F	16 825 F	8 040 F	15 196	15 000 F	15 000 F
	77	...	...	...	...	...	...	...	7 869	9 696	15 295
	87	83 381	53 462	81 636	83 880 F	91 073 F	84 410 F	86 167 F	81 725 F	80 304 F	79 705 F
	Country total	*123 497*	*97 799*	*120 699*	*125 829 F*	*122 918 F*	*122 908 F*	*122 995 F*	*129 644*	*130 000 F*	*135 000 F*
Ecuador	03	372	300	300	300	400	400	400	400	400	400 F
	77	1 050	1 710	-	270	-	-	-	-	2	85
	87	285 211	328 238	505 095	702 404	548 588	309 622	497 472	592 147	586 168	318 055
	Country total	*286 633*	*330 248*	*505 395*	*702 974*	*548 988*	*310 022*	*497 872*	*592 547*	*586 570*	*318 540*
Falkland Is	03	-	-	-	1	1	1	1	1	1	1
	41	1 974	5 914	27 190	31 539	17 112	43 615	39 163	62 927	59 823	35 655
	Country total	*1 974*	*5 914*	*27 190*	*31 540*	*17 113*	*43 616*	*39 164*	*62 928*	*59 824*	*35 656*
Fr Guiana	03	0	0	0	0	0	0	0	0	0	0
	31	6 931	7 819	8 089	7 377 F	6 602	6 709	6 271 F	5 237 F	5 194 F	5 568 F
	Country total	*6 931*	*7 819*	*8 089*	*7 377 F*	*6 602*	*6 709*	*6 271 F*	*5 237 F*	*5 194 F*	*5 568 F*
Guyana	03	800	800	700 F	800	625	625	603	800	800	800
	31	43 323	45 567	47 200 F	47 783	53 373	52 215	53 241	48 087	52 605	47 217
	Country total	*44 123*	*46 367*	*47 900 F*	*48 583*	*53 998*	*52 840*	*53 844*	*48 887*	*53 405*	*48 017*
Paraguay	03	19 000 F	20 000 F	21 000 F	22 000	28 000	25 000 F	25 000 F	25 000 F	25 000 F	25 000 F
	Country total	*19 000 F*	*20 000 F*	*21 000 F*	*22 000*	*28 000*	*25 000 F*	*25 000 F*	*25 000 F*	*25 000 F*	*25 000 F*
Peru	03	38 290	48 837	50 789	28 890	32 221	35 327	36 223	32 297	35 653	29 966
	87	8 966 487	11 950 380	8 886 553	9 486 158	7 837 650	4 303 110	8 392 378	10 626 323	7 950 450	8 737 025
	Country total	*9 004 777*	*11 999 217*	*8 937 342*	*9 515 048*	*7 869 871*	*4 338 437*	*8 428 601*	*10 658 620*	*7 986 103*	*8 766 991*
Suriname	03	187	138	140 F	150 F	200 F	200 F	200 F	200 F	200	200 F
	31	9 313	14 327	12 860 F	12 850 F	13 800 F	15 995	16 000	17 300 F	18 715	18 500 F
	Country total	*9 500*	*14 465*	*13 000 F*	*13 000 F*	*14 000 F*	*16 195*	*16 200*	*17 500 F*	*18 915*	*18 700 F*
Uruguay	03	621	966	849	598	2 216	1 931	2 423	2 302	451	387
	27	-	-	-	-	-	-	-	-	-	314
	41	118 194	119 766	125 597	122 732	134 738	138 514	100 068	101 650	96 851	102 027
	47	-	-	-	-	-	-	-	320	-	906
	48	-	-	-	-	-	262	4 092	7 340	428	693
	51	-	-	-	-	-	-	-	1 628	7 150	2 591
	57	-	-	-	-	-	-	-	-	-	1 847
	58	-	-	-	-	-	-	-	99	-	-
	88	-	-	-	-	-	-	-	-	23	-
	Country total	*118 815*	*120 732*	*126 446*	*123 330*	*136 954*	*140 707*	*106 583*	*113 339*	*104 903*	*108 765*
Venezuela	03	28 251	35 412	54 175	49 911	40 881	46 035	35 219	23 739	24 326	40 776
	31	316 550	354 366	392 892	378 624	362 355	388 419	292 750	258 236	281 731	353 418
	77	32 700	26 870	45 030	58 600	49 420	45 300	49 400	52 830	69 730	91 616
	87	18 360	21 570	8 720	8 980	17 470	23 740	22 440	22 030	42 420	29 574
	Country total	*395 861*	*438 218*	*500 817*	*496 115*	*470 126*	*503 494*	*399 809*	*356 835*	*418 207*	*515 384*
Total		***17 614 587***	***22 498 197***	***19 624 584***	***19 789 286***	***17 230 905***	***10 702 894***	***16 534 825***	***17 992 410***	***14 929 607***	***16 021 401***

D-4

Fish crustaceans, molluscs, etc — Capture production by countries or areas and fishing areas — Asia
Poissons, crustacés, mollusques, etc — Captures par pays ou zones et zones de pêche — Asie
Peces, crustáceos, moluscos, etc — Capturas por países o áreas y áreas de pesca — Asia

	Fishing area Zone de pêche Area de pesca	1993 t	1994 t	1995 t	1996 t	1997 t	1998 t	1999 t	2000 t	2001 t	2002 t
Afghanistan	04	1 200 F	1 300 F	1 300 F	1 300 F	1 250 F	1 200 F	1 200 F	1 000 F	800 F	900 F
	Country total	*1 200 F*	*1 300 F*	*1 300 F*	*1 300 F*	*1 250 F*	*1 200 F*	*1 200 F*	*1 000 F*	*800 F*	*900 F*
Armenia	04	1 850	1 033	821	580	580	698	1 144	1 133	866	465
	Country total	*1 850*	*1 033*	*821*	*580*	*580*	*698*	*1 144*	*1 133*	*866*	*465*
Azerbaijan	04	21 733	18 901	10 545	6 702	5 161	4 760	20 866	18 797	10 893	11 334
	Country total	*21 733*	*18 901*	*10 545*	*6 702*	*5 161*	*4 760*	*20 866*	*18 797*	*10 893*	*11 334*
Bahrain	04	0	0	0	0	0	0	0	-	-	-
	51	8 958	7 628	9 389	12 940	10 050	9 849	10 620	11 718	11 230	11 204
	Country total	*8 958*	*7 628*	*9 389*	*12 940*	*10 050*	*9 849*	*10 620*	*11 718*	*11 230*	*11 204*
Bangladesh	04	452 109	517 746	527 739	535 617	534 285	538 689	649 418	670 465	688 920	688 435
	57	312 715	253 044	264 650	279 170	295 141	300 452	309 797	333 799	379 497	415 420
	Country total	*764 824*	*770 790*	*792 389*	*814 787*	*829 426*	*839 141*	*959 215*	*1 004 264*	*1 068 417*	*1 103 855*
Bhutan	04	320 F	310 F	310 F	300 F	300 F	300 F	300 F	300 F	300 F	300 F
	Country total	*320 F*	*310 F*	*310 F*	*300 F*	*300 F*	*300 F*	*300 F*	*300 F*	*300 F*	*300 F*
Brunei Darsm	04	25	4	7	15	17	35	26	23	16	14
	71	1 703	4 441	4 712	7 390	4 504	5 014	3 160	2 464	1 476	2 044
	Country total	*1 728*	*4 445*	*4 719*	*7 405*	*4 521*	*5 049*	*3 186*	*2 487*	*1 492*	*2 058*
Cambodia	04	67 900	65 000	72 500	63 510	73 000	75 700	231 000	245 600	385 000	360 300
	41	-	-	-	-	-	-	-	2 768	1 200	32
	47	-	-	-	-	-	-	56	-	-	-
	71	33 123	30 018	30 500	31 200	29 800	32 200	38 100	36 000	42 000	45 850
	Country total	*101 023*	*95 018*	*103 000*	*94 710*	*102 800*	*107 900*	*269 156*	*284 368*	*428 200*	*406 182*
China	04	1 182 390	1 327 785	1 607 385	1 762 860	1 886 967	2 280 244	2 285 364	2 233 230	2 149 932	2 247 926
	34	9 809	2 414	21 464	18 302	20 810	20 403	28 433	25 408	36 228	4 161
	37	-	97	137	93	49	-	-	-	-	-
	41	-	3	3	3	3	31 806	61 664	93 156	93 973	85 872
	47	-	24	29	24	121	48	5 743	7 209	4 959	3 861
	51	-	-	-	-	-	-	294	2 587	2 773	3 594
	57	-	-	445	1 497	2 964	3 080	5 868	4 919	4 009	1 474
	61	8 151 441	9 522 098	10 921 731	12 392 922	13 807 656	14 890 900	14 843 663	14 609 252	14 211 830	14 053 342
	71	7 797	14 415	11 512	6 406	3 774	3 446	9 003	8 882	11 138	12 924
	77	-	-	-	-	-	-	-	2 682	14 547	12 494
	81	-	-	-	-	-	-	-	-	-	752
	87	-	-	-	-	-	-	-	-	-	126 744
	Country total	*9 351 437*	*10 866 836*	*12 562 706*	*14 182 107*	*15 722 344*	*17 229 927*	*17 240 032*	*16 987 325*	*16 529 389*	*16 553 144*
China,H.Kong	04	0	0	0	0	0	0	0	0	0	0
	61	217 544	211 010	194 999	183 856	186 000	180 000	127 780	157 012	173 972	169 790
	Country total	*217 544*	*211 010*	*194 999*	*183 856*	*186 000*	*180 000*	*127 780*	*157 012*	*173 972*	*169 790*
China, Macao	04	0	0	0	0	0	0	0	0	0	0
	61	1 898	1 890	1 604	1 418	1 500 F	1 500 F	1 500 F	1 500 F	1 500 F	1 500 F
	Country total	*1 898*	*1 890*	*1 604*	*1 418*	*1 500 F*	*1 500 F*	*1 500 F*	*1 500 F*	*1 500 F*	*1 500 F*
China,Taiwan	04	1 216	1 059	874	407	403	449	561	549	591	599
	21	25	392	206	83	30	8	127	351	618	570
	27	478	1 562	1 125	31	30	47	640	1 131	564	624
	31	6 979 F	3 627 F	3 558	4 516	3 669	2 430	5 663	6 039	10 597	10 577
	34	6 467 F	12 929 F	13 199	17 517	13 158	10 623	8 890	14 494	13 677	9 761
	37	328	713	493	373	399	-	58	31	197	131
	41	145 840 F	125 529 F	138 326	132 450	216 202	177 096	279 001	247 505	156 502	126 337
	47	14 958 F	18 655 F	19 249	11 394	8 934	18 173	18 811	18 970	16 118	16 070
	51	121 654 F	63 484 F	88 540 F	77 278	74 522	96 352	88 423	92 017	95 003	95 106
	57	18 314 F	16 961 F	25 964 F	32 507	41 049	22 549	22 732	23 241	14 308	19 384
	61	522 490 F	405 001 F	420 781 F	393 351	406 022	378 364	373 608	361 423	357 797	363 235
	67	-	-	26	0	0	0	55	1 165	310	210
	71	254 607 F	291 612 F	280 936 F	271 520	254 979	365 733	283 184	309 239	321 673	380 396
	77	15 328 F	11 038 F	5 492	5 205	5 443	6 679	10 917	8 034	10 461	8 717
	81	24 494 F	14 485 F	11 429	20 599	12 902	13 176	6 983	8 200	5 901	10 043
	87	498	142	3	252	306	89	62	1 500	882	996
	Country total	*1 133 676*	*967 189*	*1 010 201*	*967 483*	*1 038 048*	*1 091 768*	*1 099 715*	*1 093 889*	*1 005 199*	*1 042 756*
Cyprus	04	...	5	65	64	70	70	70	78	70 F	60 F
	34	-	-	-	2 632	13 640	16 387	37 327	65 174	78 743	-
	37	2 696	2 762	2 505	2 550	2 309	2 408	2 241	2 230	2 258	1 918
	77	960	180	830	1 730	1 920	-	-	-	-	-
	87	6 360	6 480	5 920	5 550	6 880	430	-	-	-	-
	Country total	*10 016*	*9 427*	*9 320*	*12 526*	*24 819*	*19 295*	*39 638*	*67 482*	*81 071*	*1 978 F*
Georgia	04	549	16	90	6	1	4	17	22	8	10
	34	13 500 F	5 000 F	1 000 F	-	-	-	-	-	-	-
	37	2 191	1 397	2 470	2 447	2 582	2 997	1 396	2 178	1 822	2 527
	47	2 000 F	1 000 F	-	-	-	-	-	-	-	-
	Country total	*18 240 F*	*7 413 F*	*3 560 F*	*2 453*	*2 583*	*3 001*	*1 413*	*2 200*	*1 830*	*2 537*
India	04	575 905	552 874	608 378	633 425	641 775	692 439	696 083	905 700	975 403	813 755
	51	1 744 857	1 904 799	1 847 631	1 956 204	2 007 772	1 832 799	1 911 318	1 850 839	1 862 686	1 991 081
	57	741 950	799 934	809 231	858 312	873 901	848 254	864 749	909 889	939 003	966 076
	58	-	-	-	13	-	-	-	-	-	-
	Country total	*3 062 712*	*3 257 607*	*3 265 240*	*3 447 954*	*3 523 448*	*3 373 492*	*3 472 150*	*3 666 428*	*3 777 092*	*3 770 912*

D-4

Fish crustaceans, molluscs, etc — Capture production by countries or areas and fishing areas — Asia
Poissons, crustacés, mollusques, etc — Captures par pays ou zones et zones de pêche — Asie
Peces, crustáceos, moluscos, etc — Capturas por países o áreas y áreas de pesca — Asia

	Fishing area Zone de pêche Area de pesca	1993 t	1994 t	1995 t	1996 t	1997 t	1998 t	1999 t	2000 t	2001 t	2002 t
Indonesia	04	308 649	336 141	329 710	335 707	304 258	288 666	327 627	318 334	310 240	316 030
	57	744 465	778 092	783 496	848 401	921 801	1 022 988	974 313	948 225	953 743	994 074
	71	2 036 673	2 214 941	2 419 447	2 420 687	2 631 601	2 649 350	2 742 603	2 853 567	3 009 679	3 195 370
	Country total	*3 089 787*	*3 329 174*	*3 532 653*	*3 604 795*	*3 857 660*	*3 961 004*	*4 044 543*	*4 120 126*	*4 273 662*	*4 505 474*
Iran	04	75 021	89 157	88 800	109 286	103 795	140 263	143 400	123 500	73 645	55 853
	51	246 985	218 944	252 583	242 437	238 486	226 500	243 800	260 500	262 805	269 000
	57	-	-	-	2	6	449	-	-	-	-
	Country total	*322 006*	*308 101*	*341 383*	*351 725*	*342 287*	*367 212*	*387 200*	*384 000*	*336 450*	*324 853*
Iraq	04	17 808	20 926	22 955	19 049	20 519	9 111	9 330	8 378	8 000 F	8 000 F
	51	2 133	4 221	5 253	11 688	10 783	13 463	13 093	12 389	8 500 F	5 000 F
	Country total	*19 941*	*25 147*	*28 208*	*30 737*	*31 302*	*22 574*	*22 423*	*20 767*	*16 500 F*	*13 000 F*
Israel	04	1 813	1 478	1 214	1 845	1 476	2 164	2 145	1 852	1 286	1 568
	37	3 232	2 473	3 577	3 159	3 557	3 999	3 641	3 966	3 618	3 312
	51	80	110	150	225	171	137	98	...	120	...
	Country total	*5 125*	*4 061*	*4 941*	*5 229*	*5 204*	*6 300*	*5 884*	*5 818*	*5 024*	*4 880*
Japan	04	91 032	92 219	91 455	93 501	85 612	78 822	71 270	70 612	61 354	60 843 F
	21	9 064	6 147	5 728	5 606	3 934	4 945	5 697	4 421	5 670	4 432
	27	4 915	7 082	6 595	3 745	2 638	2 907	2 273	2 315	1 906	2 331
	31	1 532	1 293	969	1 454	1 262	1 921	3 581	3 637	3 996	1 582
	34	22 773	20 358	24 307	24 766	14 934	20 294	13 487	16 218	10 973	7 533
	37	616	530	749	673	172	417	376	143	188	390
	41	151 049	127 068	97 448	92 775	147 786	99 299	175 714	138 098	82 127	26 924
	47	39 774	38 106	32 690	26 160	18 924	20 659	18 284	17 286	11 729	10 068 F
	48	50 845	60 198	62 111	58 769	61 013	67 481	66 076	80 597	67 378	51 191
	51	52 958	32 730	14 273	21 152	28 796	31 030	20 377	19 542	24 739	26 547
	57	6 790	11 519	40 437	27 076	21 221	22 301	24 091	19 898	17 676	12 921
	58	5 762	899	1 266	264	335	-	-	-	-	-
	61	6 218 515	5 628 874	5 092 787	5 191 508	5 162 694	4 546 845	4 451 587	4 206 983	3 996 910	3 836 031 F
	67	700	683	...	...	2 008	1 201	1 521	683	2 061	...
	71	282 242	277 415	272 490	226 315	210 396	254 848	220 024	242 406	225 024	224 565 F
	77	102 740	95 131	82 190	69 159	82 917	63 507	46 161	76 384	54 890	52 731 F
	81	140 989	113 146	95 100	77 078	72 426	74 231	58 178	43 416	56 855	36 800 F
	87	69 982	102 495	48 922	11 871	11 266	12 847	10 251	42 174	89 530	88 111 F
	Country total	*7 252 278*	*6 615 893*	*5 969 517*	*5 931 872*	*5 928 334*	*5 303 555*	*5 188 948*	*4 984 813*	*4 713 006*	*4 443 000*
Jordan	04	350	350	350	350	350	350	350	400	350	350
	51	45	60	75	90	100	120	160	150	170	176
	Country total	*395*	*410*	*425*	*440*	*450*	*470*	*510*	*550*	*520*	*526*
Kazakhstan	04	56 700	46 433	48 402	44 273	31 826	25 000 F	36 170	36 620	21 654	24 910
	Country total	*56 700*	*46 433*	*48 402*	*44 273*	*31 826*	*25 000 F*	*36 170*	*36 620*	*21 654*	*24 910*
Korea D P Rp	04	23 000 F	20 000	20 000	20 000	20 000	16 000 F	12 000 F	8 000 F	4 928	5 000 F
	34	398	-	-	-	-	-	-	850	-	-
	61	374 000 F	351 961	307 083	233 125	216 462	212 000 F	208 000 F	204 000 F	201 572	200 000 F
	Country total	*397 398 F*	*371 961*	*327 083*	*253 125*	*236 462*	*228 000 F*	*220 000 F*	*212 850 F*	*206 500*	*205 000 F*
Korea Rep	04	12 263	10 492	9 646	8 034	6 934	6 845	6 316	7 141	5 971	5 690
	21	3 738	-	-	-	-	-	-	-	-	-
	31	1 462	750	653	626	143	621	1 789	3 327	237	45
	34	15 510	23 057	19 324	19 380	21 851	18 481	25 219	23 978	26 809	25 974
	37	-	-	484	701	686	-	-	-	-	-
	41	149 776	97 665	141 635	158 911	220 586	101 165	283 455	170 429	167 817	129 035
	47	7 711	11 242	9 216	6 414	9 548	8 020	7 822	7 225	5 026	5 983
	48	-	146	425	513	642	2 850	340	7 645	8 321	15 257
	51	29 879	25 322	22 765	31 573	37 976	25 692	19 943	21 435	10 878	14 457
	57	-	78	321	718	512	269	4 983	5 273	4 550	3 403
	61	1 718 865	1 799 694	1 787 770	1 933 176	1 640 712	1 553 273	1 500 272	1 288 737	1 469 171	1 123 423
	67	47 264	15 875	10 967	2 480	-	311	1 330	-	-	-
	71	157 088	244 170	213 720	175 466	182 399	218 092	157 499	180 224	190 038	225 065
	77	33 550	36 937	37 833	37 320	44 473	59 919	49 305	51 303	53 668	54 644
	81	20 861	25 371	30 434	26 689	35 111	31 886	34 784	34 298	38 139	44 034
	87	59 225	67 092	34 724	11 784	2 629	508	25 464	23 980	10 097	21 969
	Country total	*2 257 192*	*2 357 891*	*2 319 917*	*2 413 785*	*2 204 202*	*2 027 932*	*2 118 521*	*1 824 995*	*1 990 722*	*1 668 979*
Kuwait	04	0	0	0	0	0	0	0	0	0	0
	51	8 466	7 752	8 616	8 255	7 826	7 799	6 271	6 000 F	5 846	5 900 F
	Country total	*8 466*	*7 752*	*8 616*	*8 255*	*7 826*	*7 799*	*6 271*	*6 000 F*	*5 846*	*5 900 F*
Kyrgyzstan	04	127	131	185	160 F	120 F	80 F	48	52	57	48
	Country total	*127*	*131*	*185*	*160 F*	*120 F*	*80 F*	*48*	*52*	*57*	*48*
Laos	04	19 500 F	23 800	27 370	23 000 F	18 857	19 642	30 041	29 250	31 000 F	33 440
	Country total	*19 500 F*	*23 800*	*27 370*	*23 000 F*	*18 857*	*19 642*	*30 041*	*29 250*	*31 000 F*	*33 440*
Lebanon	04	20	20	20	20	20	20	20	20	20	297
	37	2 000	2 205	4 065	4 115	3 635	3 500	3 540	3 646	3 650	3 673
	Country total	*2 020*	*2 225*	*4 085*	*4 135*	*3 655*	*3 520*	*3 560*	*3 666*	*3 670*	*3 970*
Malaysia	04	1 971	2 064	3 939	3 683	3 949	4 626	3 366	3 549	3 446	3 450
	57	446 515	439 341	504 068	494 091	499 288	508 128	485 741	519 785	475 354	501 466
	71	600 835	626 245	604 368	632 598	669 685	640 965	762 661	765 911	755 933	770 639
	Country total	*1 049 321*	*1 067 650*	*1 112 375*	*1 130 372*	*1 172 922*	*1 153 719*	*1 251 768*	*1 289 245*	*1 234 733*	*1 275 555*

D-4

Fish crustaceans, molluscs, etc	Capture production by countries or areas and fishing areas	Asia
Poissons, crustacés, mollusques, etc	Captures par pays ou zones et zones de pêche	Asie
Peces, crustáceos, moluscos, etc	Capturas por países o áreas y áreas de pesca	Asia

	Fishing area Zone de pêche Area de pesca	1993 t	1994 t	1995 t	1996 t	1997 t	1998 t	1999 t	2000 t	2001 t	2002 t
Maldives	04	0	0	0	0	0	0	0	0	0	0
	51	100 852	118 168	119 048	120 508	116 257	128 968	133 547	132 427	126 190	160 981
	Country total	*100 852*	*118 168*	*119 048*	*120 508*	*116 257*	*128 968*	*133 547*	*132 427*	*126 190*	*160 981*
Mongolia	04	165	184	158	221	180	311	524	425	117	129
	Country total	*165*	*184*	*158*	*221*	*180*	*311*	*524*	*425*	*117*	*129*
Myanmar	04	142 065	146 365	148 347	146 494	149 069	149 279	159 746	189 708	235 376	304 529
	57	597 637	599 876	602 885	455 294	631 226	680 838	759 664	880 018	931 492	1 008 113
	Country total	*739 702*	*746 241*	*751 232*	*601 788*	*780 295*	*830 117*	*919 410*	*1 069 726*	*1 166 868*	*1 312 642*
Nepal	04	7 418	7 340	11 230	11 230	11 230	12 000	12 752	16 700	16 700	17 900
	Country total	*7 418*	*7 340*	*11 230*	*11 230*	*11 230*	*12 000*	*12 752*	*16 700*	*16 700*	*17 900*
Oman	04	0	0	0	0	0	0	0	0	0	0
	51	105 772	118 572	139 861	121 618	118 995	106 171	108 809	120 421	129 907	142 670
	Country total	*105 772*	*118 572*	*139 861*	*121 618*	*118 995*	*106 171*	*108 809*	*120 421*	*129 907*	*142 670*
Pakistan	04	109 185	118 703	121 405	142 092	167 530	163 524	179 865	176 468	180 100	181 000
	51	499 159	418 574	404 444	395 397	422 265	433 456	474 665	437 601	420 698	418 104
	Country total	*608 344*	*537 277*	*525 849*	*537 489*	*589 795*	*596 980*	*654 530*	*614 069*	*600 798*	*599 104*
Palest, O.T.	04	...	...	0	0	0	0	0	0	0	0
	37	...	...	1 229	2 493	3 791	3 625	3 600 F	3 000 F	2 500 F	2 378
	Country total	*...*	*...*	*1 229*	*2 493*	*3 791*	*3 625*	*3 600 F*	*3 000 F*	*2 500 F*	*2 378*
Philippines	04	210 775	204 325	186 006	177 355	159 353	146 004	146 234	151 753	135 845	130 881
	31	-	-	-	-	-	28	545	376	37	691
	34	-	-	-	-	-	1 186	1 854	801	29	110
	41	-	-	-	-	-	124	47	-	330	172
	47	-	-	-	-	-	6	52	-	-	-
	51	-	-	-	-	-	2 513	2 190	1 985	1 453	1 216
	57	-	-	-	-	-	657	1 017	219	182	379
	71	1 623 548	1 641 010	1 674 695	1 606 246	1 646 453	1 682 862	1 720 879	1 741 504	1 811 150	1 897 093
	Country total	*1 834 323*	*1 845 335*	*1 860 701*	*1 783 601*	*1 805 806*	*1 833 380*	*1 872 818*	*1 896 638*	*1 949 026*	*2 030 542*
Qatar	04	0	0	0	0	0	0	0	0	0	0
	51	6 994	5 086	4 271	4 739	5 032	5 279	4 207	7 142	8 606	6 880
	Country total	*6 994*	*5 086*	*4 271*	*4 739*	*5 032*	*5 279*	*4 207*	*7 142*	*8 606*	*6 880*
Saudi Arabia	04	0	0	0	0	0	0	0	0	0	0
	51	48 021	54 612	45 609	47 698	49 314	51 206	46 618	49 761	49 167	55 330
	Country total	*48 021*	*54 612*	*45 609*	*47 698*	*49 314*	*51 206*	*46 618*	*49 761*	*49 167*	*55 330*
Singapore	04	25	23	0	0	0	0	0	0	0	0
	71	9 279	11 278	10 102	9 943	9 250	7 733	6 489	5 371	3 342	2 769
	Country total	*9 304*	*11 301*	*10 102*	*9 943*	*9 250*	*7 733*	*6 489*	*5 371*	*3 342*	*2 769*
Sri Lanka	04	15 000	9 500	15 000	22 250	27 250	29 900	31 450	36 700	29 870	28 130
	57	200 400	208 900	214 171	206 695	207 849	233 430	244 630	256 710	249 770	270 130
	Country total	*215 400*	*218 400*	*229 171*	*228 945*	*235 099*	*263 330*	*276 080*	*293 410*	*279 640*	*298 260*
Syria	04	2 535	3 570	3 832	3 103	3 557	4 347	5 338	3 991	5 969	6 355
	37	2 019	1 950	1 950	2 670	2 574	2 750	2 600	2 581	2 322	2 823
	Country total	*4 554*	*5 520*	*5 782*	*5 773*	*6 131*	*7 097*	*7 938*	*6 572*	*8 291*	*9 178*
Tajikistan	04	253	127	100	40 F	75 F	100 F	80 F	78 F	137	181
	Country total	*253*	*127*	*100*	*40 F*	*75 F*	*100 F*	*80 F*	*78 F*	*137*	*181*
Thailand	04	175 140	196 397	186 665	205 903	203 671	200 715	206 840	201 405	202 400	205 500
	57	823 696	776 667	822 673	869 484	850 994	909 617	808 257	755 136	792 889	784 939
	71	1 928 853	2 042 132	2 021 736	1 938 574	1 848 233	1 820 022	1 937 211	2 040 853	1 937 085	1 930 777
	Country total	*2 927 689*	*3 015 196*	*3 031 074*	*3 013 961*	*2 902 898*	*2 930 354*	*2 952 308*	*2 997 394*	*2 932 374*	*2 921 216*
Timor-Leste	04	...	...	...	...	...	...	0	0	0	0
	57	...	...	...	...	...	...	408 F	362	356	350 F
	Country total	*...*	*...*	*...*	*...*	*...*	*...*	*408 F*	*362*	*356*	*350 F*
Turkey	04	44 801	45 067	47 976	49 600	50 460	54 500	50 190	42 824	43 323	43 938
	37	503 957	544 736	585 994	478 226	408 693	432 700	523 634	460 524	484 407	522 744
	Country total	*548 758*	*589 803*	*633 970*	*527 826*	*459 153*	*487 200*	*573 824*	*503 348*	*527 730*	*566 682*
Turkmenistan	04	16 080	15 140	9 740	9 014	8 179	7 014	9 058	12 228	12 749	12 812
	Country total	*16 080*	*15 140*	*9 740*	*9 014*	*8 179*	*7 014*	*9 058*	*12 228*	*12 749*	*12 812*
Untd Arab Em	04	0	0	0	0	0	0	0	0	0	0
	51	99 600	108 600	105 884	107 000	114 358	114 739	117 607	105 456	112 561	97 574
	Country total	*99 600*	*108 600*	*105 884*	*107 000*	*114 358*	*114 739*	*117 607*	*105 456*	*112 561*	*97 574*
Uzbekistan	04	4 358	3 095	3 611	1 494	3 075	2 799	2 871	3 387	4 070	2 009
	Country total	*4 358*	*3 095*	*3 611*	*1 494*	*3 075*	*2 799*	*2 871*	*3 387*	*4 070*	*2 009*
Viet Nam	04	146 839	79 587	94 689	164 936	177 589	138 800	169 107	170 000 F	133 280 F	149 200 F
	71	785 304	946 322	990 250	1 058 708	1 098 736	1 155 154	1 217 193	1 280 590 F	1 357 023 F	1 358 800 F
	Country total	*932 143*	*1 025 909*	*1 084 939*	*1 223 644*	*1 276 325*	*1 293 954*	*1 386 300*	*1 450 590*	*1 490 303*	*1 508 000 F*
Yemen	51	82 356	81 885	107 970	104 955	115 600	127 620	124 385	114 751	142 198	159 262

D-4 Fish crustaceans, molluscs, etc — Capture production by countries or areas and fishing areas — Asia
Poissons, crustacés, mollusques, etc — Captures par pays ou zones et zones de pêche — Asie
Peces, crustáceos, moluscos, etc — Capturas por países o áreas y áreas de pesca — Asia

	Fishing area Zone de pêche Area de pesca	1993 t	1994 t	1995 t	1996 t	1997 t	1998 t	1999 t	2000 t	2001 t	2002 t
	Country total	*82 356*	*81 885*	*107 970*	*104 955*	*115 600*	*127 620*	*124 385*	*114 751*	*142 198*	*159 262*
Total		***37 469 501***	***38 889 217***	***40 376 794***	***42 002 404***	***43 904 775***	***44 804 664***	***45 777 965***	***45 621 911***	***45 495 225***	***45 495 299***

D-5 Fish crustaceans, molluscs, etc — Capture production by countries or areas and fishing areas — Europe
Poissons, crustacés, mollusques, etc — Captures par pays ou zones et zones de pêche — Europe
Peces, crustáceos, moluscos, etc — Capturas por países o áreas y áreas de pesca — Europa

	Fishing area Zone de pêche Area de pesca	1993 t	1994 t	1995 t	1996 t	1997 t	1998 t	1999 t	2000 t	2001 t	2002 t
Albania	05	850 F	700 F	219	317	180	823	814	955	1 373	1 167
	37	1 650 F	1 400 F	1 160	1 808	833	1 860	1 931	2 365	1 937	2 788
	Country total	*2 500 F*	*2 100 F*	*1 379*	*2 125*	*1 013*	*2 683*	*2 745*	*3 320*	*3 310*	*3 955*
Andorra	05	0	0	0	0	0	0	0	0	0	0
	Country total	*0*	*0*	*0*	*0*	*0*	*0*	*0*	*0*	*0*	*0*
Austria	05	420	388	404	450	465	451	432	439	362	350
	Country total	*420*	*388*	*404*	*450*	*465*	*451*	*432*	*439*	*362*	*350*
Belarus	05	2 993	786	715	821	499	457	514	553	943	5 575
	Country total	*2 993*	*786*	*715*	*821*	*499*	*457*	*514*	*553*	*943*	*5 575*
Belgium	05	511	511	511	511	511	511	536	511	511	511
	27	35 578	33 743	35 088	30 312	29 989	30 324	29 340	29 289	29 698	28 517
	Country total	*36 089*	*34 254*	*35 599*	*30 823*	*30 500*	*30 835*	*29 876*	*29 800*	*30 209*	*29 028*
Bosnia Herzg	05	2 500 F	2 500 F	2 500 F	2 500 F	2 500 F	2 500 F	2 500 F	2 500 F	2 500 F	2 500 F
	37	0	0	0	0	0	0	0	0	0	0
	Country total	*2 500 F*	*2 500 F*	*2 500 F*	*2 500 F*	*2 500 F*	*2 500 F*	*2 500 F*	*2 500 F*	*2 500 F*	*2 500 F*
Bulgaria	05	1 675	995 F	762	1 127	1 881	2 336	2 475	861	1 540	1 452
	27	5 272	-	-	-	-	-	-	-	-	-
	34	-	-	-	-	-	8 189	-	-	-	-
	37	2 318	5 340 F	7 250	7 727	9 356	8 421	8 081	6 137	4 880	13 555
	41	4 200	-	-	-	-	-	-	-	-	-
	48	193	250	-	-	-	-	-	-	-	-
	Country total	*13 658*	*6 585 F*	*8 012*	*8 854*	*11 237*	*18 946*	*10 556*	*6 998*	*6 420*	*15 007*
Channel Is	05	0	0	0	0	0	0	0	0	0	0
	27	2 854 F	2 783	2 949 F	4 346	4 238	4 117	3 601	3 589	3 927	3 449
	Country total	*2 854 F*	*2 783*	*2 949 F*	*4 346*	*4 238*	*4 117*	*3 601*	*3 589*	*3 927*	*3 449*
Croatia	05	284	340	364	434	408	10 F	10	17	34	25
	37	26 342	17 137	15 901	17 799	16 627	21 928	18 890	21 045	18 455	21 205
	Country total	*26 626*	*17 477*	*16 265*	*18 233*	*17 035*	*21 938*	*18 900*	*21 062*	*18 489*	*21 230*
Czech Rep	05	3 185	3 955	3 929	3 524	3 321	3 952	4 190	4 654	4 646	4 983
	Country total	*3 185*	*3 955*	*3 929*	*3 524*	*3 321*	*3 952*	*4 190*	*4 654*	*4 646*	*4 983*
Denmark	05	337	243	264	196	232	349	206	183	99	77
	21	897	245	447	403	421	556	235	-	93	359
	27	1 613 055	1 872 828	1 998 322	1 680 918	1 826 199	1 556 430	1 404 564	1 533 906	1 510 269	1 441 632
	Country total	*1 614 289*	*1 873 316*	*1 999 033*	*1 681 517*	*1 826 852*	*1 557 335*	*1 405 005*	*1 534 089*	*1 510 461*	*1 442 068*
Estonia	05	2 411	1 909	2 366	2 361	2 439	3 878	3 108	3 190	2 461	4 578
	21	5 397	1 186	3 242	1 898	3 240	5 533	10 834	13 685	12 402	15 022
	27	49 703	71 650	92 410	94 586	112 979	96 898	90 315	96 266	88 357	81 071
	34	56 889	17 563	5 174	7 063	4 955	12 405	7 512	5	6	4
	41	1 338	-	-	2 538	-	-	-	-	1 941	777
	47	31 447	31 372	28 836	-	-	-	-	-	-	-
	77	-	-	-	-	-	-	10	-	-	-
	87	-	-	-	-	-	-	14	-	-	-
	Country total	*147 185*	*123 680*	*132 028*	*108 446*	*123 613*	*118 714*	*111 793*	*113 146*	*105 167*	*101 452*
Faeroe Is	05	0	0	0	0	0	0	0	0	0	-
	21	15 138	10 024	6 884	10 209	8 451	10 094	10 560	8 543	12 980	9 149
	27	231 156	227 684	280 912	294 378	320 694	354 403	347 802	445 856	511 857	480 851 F
	Country total	*246 294*	*237 708*	*287 796*	*304 587*	*329 145*	*364 497*	*358 362*	*454 399*	*524 837*	*490 000 F*
Finland	05	51 522	47 895	48 436	47 618	47 618	36 813	36 813	34 840	34 820	34 782
	27	104 772	116 374	119 048	131 459	132 567	134 868	123 747	121 661	115 276	110 026
	Country total	*156 294*	*164 269*	*167 484*	*179 077*	*180 185*	*171 681*	*160 560*	*156 501*	*150 096*	*144 808*
France	05	4 400	4 450	4 500	4 540	4 540	4 500	2 134 F	2 131 F	2 130 F	2 130 F
	27	379 989	383 736	384 524	353 922	386 921	363 118	394 383	423 035	429 671	420 802
	34	98 780	88 411	76 395	87 404	73 080	80 307	68 386	70 022	55 116	53 782
	37	45 601	39 710	37 967	27 813	33 012	33 220	38 948	45 540	43 059	42 362
	41	-	1 946	7 332	4 580	1 545	4 179	2 379	2 053	-	-
	48	-	-	-	-	-	-	-	-	386	-
	51	93 057	99 908	95 917	82 854	68 920	48 720	80 483	84 544	69 152	96 719
	57	-	-	-	80	1 947	10 849	1 652	108	-	-
	58	1 582	4 407	4 206	3 501	4 111	4 652	5 114	7 386	6 115	4 283
	Country total	*623 409*	*622 568*	*610 841*	*564 694*	*574 076*	*549 545*	*593 479*	*634 819*	*605 629*	*620 078*
Germany	05	10 837	10 908	22 987	22 987	22 916	22 916	22 868	22 868	22 818	22 798
	21	343	305	0	495	449	355	569	4 920	1 359	2 861
	27	241 830	218 948	216 864	201 707	207 239	220 252	188 810	177 901	184 432	193 624
	34	-	-	39	11 222	28 748	23 099	26 678	-	2 673	5 168
	Country total	*253 010*	*230 161*	*239 890*	*236 411*	*259 352*	*266 622*	*238 925*	*205 689*	*211 282*	*224 451*
Gibraltar	37	0	0	0	0	0	0	0	0	0	0
	Country total	*0*	*0*	*0*	*0*	*0*	*0*	*0*	*0*	*0*	*0*
Greece	05	2 960	3 452	3 606	2 903	2 601	2 818	3 280	3 433	3 181	3 000
	34	14 743	9 316	8 684	8 019	5 085	5 917	5 933	5 150	6 170	5 983
	37	141 395	168 354	139 498	138 513	149 402	99 845	109 558	90 697	85 037	80 000 F
	Country total	*159 098*	*181 122*	*151 788*	*149 435*	*157 088*	*108 580*	*118 771*	*99 280*	*94 388*	*88 983 F*

	Fishing area Zone de pêche Area de pesca	1993 t	1994 t	1995 t	1996 t	1997 t	1998 t	1999 t	2000 t	2001 t	2002 t
Hungary	05	7 886	8 307	7 314	7 606	7 406	7 265	7 514	7 101	6 638	6 750
	Country total	*7 886*	*8 307*	*7 314*	*7 606*	*7 406*	*7 265*	*7 514*	*7 101*	*6 638*	*6 750*
Iceland	05	907	698	739	608	404	416	370	176	160	160 F
	21	2 196	2 462	8 232	20 682	7 197	6 572	9 147	9 340	5 079	6 877
	27	1 712 478	1 553 802	1 603 577	2 038 854	2 197 419	1 674 623	1 726 750	1 973 008	1 975 476	2 122 618
	31	-	-	-	24	-	-	-	-	-	-
	47	-	-	-	-	924	340	-	-	-	-
	Country total	*1 715 581*	*1 556 962*	*1 612 548*	*2 060 168*	*2 205 944*	*1 681 951*	*1 736 267*	*1 982 524*	*1 980 715*	*2 129 655*
Ireland	05	2 929	3 604	3 761	3 806	3 804	3 976	801	881	902	789
	27	275 845	290 550	384 542	328 427	289 005	321 044	282 557	275 411	281 833	241 954
	34	-	-	-	-	-	-	-	-	73 695	39 588
	Country total	*278 774*	*294 154*	*388 303*	*332 233*	*292 809*	*325 020*	*283 358*	*276 292*	*356 430*	*282 331*
Isle of Man	05	0	0	0	0	0	0	0	0	0	0
	27	4 850	3 571	3 734	3 537	4 289	2 214	2 609	3 552	3 112	3 127
	Country total	*4 850*	*3 571*	*3 734*	*3 537*	*4 289*	*2 214*	*2 609*	*3 552*	*3 112*	*3 127*
Italy	05	9 515	9 921	10 035	6 764	6 690	4 667	5 436	4 565	5 527	4 242
	34	43 445	43 578	8 067	3 428	7 180	5 891	5 845	4 077	7 359	6 202
	37	330 092	330 704	375 970	354 551	326 260	289 545	264 619	293 405	294 312	254 642
	41	5 792	5 810	960	408	-	-	-	-	-	-
	47	-	-	109	46	-	-	-	-	-	-
	51	8 689	8 717	1 650	702	3 351	4 520	6 890	102	3 199	4 760
	57	-	-	-	-	212	1 473	-	...	-	-
	Country total	*397 533*	*398 730*	*396 791*	*365 899*	*343 693*	*306 096*	*282 790*	*302 149*	*310 397*	*269 846*
Latvia	05	553	495	514	536	544	501	610	612	581	581
	21	8 874	473	1 026	1 253	997	1 191	3 080	3 397	3 330	2 742
	27	64 230	64 755	63 127	73 103	86 123	78 109	78 147	80 329	76 930	79 863
	34	63 216	65 689	82 702	63 711	18 018	22 530	43 552	52 065	47 335	30 491
	41	5 029	6 693	1 825	4 041	-	-	-	-	-	-
	48	71	-	-	-	-	-	-	-	-	-
	Country total	*141 973*	*138 105*	*149 194*	*142 644*	*105 682*	*102 331*	*125 389*	*136 403*	*128 176*	*113 677*
Liechtensten	05	0	0	0	0	0	0	0	0	0	0
	Country total	*0*	*0*	*0*	*0*	*0*	*0*	*0*	*0*	*0*	*0*
Lithuania	05	1 146	1 187	1 260	1 295	1 713	1 737	1 715	1 912	1 854	758
	21	3 904	900	980	1 585	1 785	3 107	3 370	4 047	7 596	10 948
	27	18 697	18 086	45 556	48 904	14 824	15 930	20 967	19 584	31 381	32 006
	34	57 469	22 880	9 572	33 330	25 680	45 804	46 910	53 445	111 000	106 434
	41	15 171	3 275	-	3 400	-	-	-	-	-	-
	47	20 784	2 823	-	-	-	-	-	-	-	-
	51	-	11	-	-	-	-	-	-	-	-
	Country total	*117 171*	*49 162*	*57 368*	*88 514*	*44 002*	*66 578*	*72 962*	*78 988*	*151 831*	*150 146*
Luxembourg	05	0	0	0	0	0	0	0	0	0	0
	Country total	*0*	*0*	*0*	*0*	*0*	*0*	*0*	*0*	*0*	*0*
Macedonia	05	164	196	208	78	130	131	135	208	128	148
	Country total	*164*	*196*	*208*	*78*	*130*	*131*	*135*	*208*	*128*	*148*
Malta	05	0	0	0	0	0	0	0	0	0	-
	34	-	1 236	3 465	8 197	-	-	-	-	-	-
	37	838	1 120	1 170	1 000	1 036	1 180	1 244	1 074	893	1 004
	Country total	*838*	*2 356*	*4 635*	*9 197*	*1 036*	*1 180*	*1 244*	*1 074*	*893*	*1 004*
Moldova Rep	05	630	708	709	603	569	491	309	344	387	565
	Country total	*630*	*708*	*709*	*603*	*569*	*491*	*309*	*344*	*387*	*565*
Monaco	37	3 F	3 F	3 F	3 F	3 F	3 F	3 F	3 F	3 F	3 F
	Country total	*3 F*	*3 F*	*3 F*	*3 F*	*3 F*	*3 F*	*3 F*	*3 F*	*3 F*	*3 F*
Netherlands	05	1 601	2 446	4 107	2 157	2 293	1 547	2 303	2 250 F	2 200 F	2 578
	27	460 155	417 607	433 985	361 126	348 537	410 604	388 371	336 309	339 040	300 314
	34	...	...	...	47 515	100 969	124 475	123 937	157 195	176 922	161 143
	Country total	*461 756*	*420 053*	*438 092*	*410 798*	*451 799*	*536 626*	*514 611*	*495 754*	*518 162*	*464 035*
Norway	05	435	432	413	338	439	507	514	578	570 F	570 F
	21	9 955	11 755	11 772	7 437	3 709	2 686	4 313	4 296	14 715	14 536
	27	2 380 768	2 338 555	2 505 560	2 633 495	2 852 115	2 848 489	2 622 707	2 693 364	2 672 018	2 728 078
	34	-	87	-	-	-	3 421	-	71	-	-
	47	...	...	-	-	864	1 087	-	242	-	-
	51	-	-	-	-	-	-	-	870	-	-
	81	23 973	15 290	6 366	7 187	5 932	5 033	-	-	-	-
	Country total	*2 415 131*	*2 366 119*	*2 524 111*	*2 648 457*	*2 863 059*	*2 861 223*	*2 627 534*	*2 699 421*	*2 687 303*	*2 743 184*
Poland	05	31 391	27 500	24 889	22 037	13 832	13 236	13 875	17 543	17 789	17 947
	21	-	-	-	-	824	148	894	1 732	760	428
	27	103 022	116 694	130 214	157 335	187 203	126 787	131 056	143 173	159 164	151 562
	34	-	2 220	-	19 766	5 268	-	-	-	13 185	28 712
	41	21 395	13 641	9 205	3 549	-	93	5 224	970	756	2 754
	47	-	1	3 178	1 736	1 964	797	-	-	3 100	4 318
	48	4 390	7 997	12 521	22 104	14 408	19 060	19 167	20 049	13 717	16 720
	61	235 208	269 979	249 365	116 266	125 414	81 889	65 508	33 217	16 590	-

D-5

Fish crustaceans, molluscs, etc — Capture production by countries or areas and fishing areas — Europe
Poissons, crustacés, mollusques, etc — Captures par pays ou zones et zones de pêche — Europe
Peces, crustáceos, moluscos, etc — Capturas por países o áreas y áreas de pesca — Europa

	Fishing area Zone de pêche Area de pesca	1993 t	1994 t	1995 t	1996 t	1997 t	1998 t	1999 t	2000 t	2001 t	2002 t
	67	-	-	-	-	-	-	-	998	4	-
	Country total	*395 406*	*438 032*	*429 372*	*342 793*	*348 913*	*242 010*	*235 724*	*217 682*	*225 065*	*222 441*
Portugal	05	3	4	2	0	0	0	0	0	1	12
	21	35 550	30 157	12 532	9 184	8 998	9 616	16 662	13 180	15 003	18 526
	27	219 685	202 539	216 513	199 833	187 960	192 861	174 924	159 680	157 673	162 355
	31	-	-	-	-	-	18	22	-	-	49
	34	29 773	24 615	22 665	40 326	22 293	18 294	14 765	9 430	9 898	11 484
	37	183	428	446	274	37	54	76	96	288	29
	41	1 810	3 766	6 847	9 099	1 595	2 059	77	325	2 731	3 853
	47	798	1 556	1 248	1 130	996	819	1 129	3 393	2 084	1 552
	51	-	-	-	-	-	450	752	1 417	3 339	2 177
	Country total	*287 802*	*263 065*	*260 253*	*259 846*	*221 879*	*224 171*	*208 407*	*187 521*	*191 017*	*200 037*
Romania	05	8 562	10 598	9 048	6 145	4 574	4 630	5 336	4 896	5 206	4 867
	27	-	8 593	37 508	9 432	-	-	-	-	-	-
	34	1 305	...	...	...	...	...	...	...	...	...
	37	3 952	3 060	2 719	2 682	3 872	4 431	2 507	2 476	2 431	2 122
	Country total	*13 819*	*22 251*	*49 275*	*18 259*	*8 446*	*9 061*	*7 843*	*7 372*	*7 637*	*6 989*
Russian Fed	05	216 866	217 950	212 874	233 272	227 091	271 311	307 823	292 368	206 430	208 522
	21	26 525	9 187	10 383	6 021	1 465	2 872	5 979	27 660	32 138	34 686
	27	833 147	709 669	736 359	769 774	830 132	727 936	839 509	989 122	1 061 969	1 119 911
	31	-	-	360	20	-	-	-	-	-	-
	34	198 630	167 472	277 631	384 330	312 524	341 413	286 179	211 018	129 436	121 505
	37	6 384	13 888	15 540	8 745	8 917	9 760	14 340	22 294	33 444	39 948
	41	47 925	23 689	11 762	20 254	884	-	-	3 404	3 214	8 286
	47	220 550	226 712	178 288	138 583	129 014	128 280	123 453	82 283	39 850	8 611
	48	2 584	121	12	103	-	-	273	3 462	225	1 686
	51	10 893	19 444	-	-	-	-	-	-	221	123
	61	2 741 954	2 256 773	2 835 602	3 096 988	3 139 799	2 970 546	2 558 646	2 341 918	2 121 532	1 688 908
	67	-	-	-	540	-	466	1 624	6	-	109
	71	18 937	5 840	4 981	-	-	-	-	-	-	-
	81	43 672	53 227	28 017	17 108	12 027	2 175	3 332	-	-	-
	Country total	*4 368 067*	*3 703 972*	*4 311 809*	*4 675 738*	*4 661 853*	*4 454 759*	*4 141 158*	*3 973 535*	*3 628 459*	*3 232 295*
Serbia-Monte	05	1 416 F	1 329 F	1 069	1 232	1 063	866	828	672	570	937
	37	277	260	369	380	376	413	423	424	416	454
	Country total	*1 693*	*1 589*	*1 438*	*1 612*	*1 439*	*1 279*	*1 251*	*1 096*	*986*	*1 391*
Slovakia	05	1 185	1 627	1 950	1 414	1 364	1 361	1 396	1 368	1 531	1 746
	Country total	*1 185*	*1 627*	*1 950*	*1 414*	*1 364*	*1 361*	*1 396*	*1 368*	*1 531*	*1 746*
Slovenia	05	297	339	316	289	302	269	243	226	206	226
	37	1 987	2 007	1 851	2 078	2 065	1 959	1 784	1 630	1 621	1 460
	Country total	*2 284*	*2 346*	*2 167*	*2 367*	*2 367*	*2 228*	*2 027*	*1 856*	*1 827*	*1 686*
Spain	05	9 215	6 284	8 869	8 710 F	8 710 F	8 710 F	8 710 F	8 710 F	8 710 F	8 710 F
	21	47 139	54 117	20 069	20 864	27 938	30 973	37 239	45 095	40 235	34 092
	27	360 270	346 392	395 247	393 622	476 637	470 576	439 111	392 982	434 029	323 128
	31	3 027	1 624	1 955	906	3 145	2 090	1 998	2 227	1 888	2 535
	34	317 907 F	326 916 F	334 932 F	344 222	315 145	373 415	241 272	149 224	160 504	123 183
	37	146 000 F	148 000 F	149 000 F	150 496	133 302	123 325	122 360	140 203	139 143	118 341
	41	63 047	72 592	96 511	93 032	65 264	84 266	95 482	84 740	108 808	39 307
	47	25 145 F	20 759 F	16 750 F	12 555	14 331	20 990	12 601	17 403	20 068	7 805
	48	-	-	-	-	492	-	184	308	643	832
	51	105 651	113 354	147 583	138 976	141 630	98 163	148 089	148 207	135 958	175 884
	57	-	-	-	237	379	13 379	451	168	...	1 651
	77	740	160	980	1 910	1 950	21 960	37 753	27 416	15 722	20 503
	87	7 818	4 326	9 428	9 153	16 478	15 169	27 745	28 933	26 817	26 662
	Country total	*1 085 959 F*	*1 094 524 F*	*1 181 324 F*	*1 174 683*	*1 205 401*	*1 263 016*	*1 172 995*	*1 045 616*	*1 092 525*	*882 633*
Svalbard Is	27	0	0	0	0	0	0	0	0	0	0
	Country total	*0*	*0*	*0*	*0*	*0*	*0*	*0*	*0*	*0*	*0*
Sweden	05	2 273	2 254	1 934	1 810	2 011	1 559	1 478	1 459	1 234	1 436
	27	339 624	384 560	402 638	369 071	355 395	409 327	349 776	337 075	310 582	293 527
	Country total	*341 897*	*386 814*	*404 572*	*370 881*	*357 406*	*410 886*	*351 254*	*338 534*	*311 816*	*294 963*
Switzerland	05	1 822	1 481	1 588	1 841	1 859	1 809	1 840	1 659	1 715	1 544
	Country total	*1 822*	*1 481*	*1 588*	*1 841*	*1 859*	*1 809*	*1 840*	*1 659*	*1 715*	*1 544*
Ukraine	05	13 189	14 786	6 847	9 468	6 215	4 898	4 728	4 429	4 343	3 823
	21	-	-	-	-	-	-	-	-	405	-
	27	7 876	13 385	3 352	550	-	-	-	-	-	-
	34	118 508	99 006	213 737	285 054	278 133	346 280	281 959	249 594	187 145	91 334
	37	26 394	35 213	43 570	29 266	35 987	42 981	51 495	65 507	90 840	79 654
	41	4 719	3 202	998	340	-	-	-	-	-	-
	47	96 420	53 356	19 759	28 405	5 701	18 345	4 675	18 096	4 283	-
	48	458	12 613	59 150	10 277	-	-	6 719	164	14 172	32 015
	51	1 491	2 633	2 970	3 480	1 570	1 626	3 356	2 015	810	-
	58	2 043	2 260	4 275	969	1 048	885	593	56	8	-
	81	28 306	35 805	33 058	39 496	40 146	47 181	55 186	51 970	58 908	58 773
	Country total	*299 404*	*272 259*	*387 716*	*407 305*	*368 800*	*462 196*	*408 711*	*391 831*	*360 914*	*265 599*
UK	05	1 909	2 191	2 146	1 930	1 491	4 569	4 835	2 743	3 142	3 431
	21	-	49	-	129	23	294	-	-	-	-
	27	857 874	874 510	905 737	861 271	885 708	913 296	831 553	737 870	729 704	679 049

D-5 Fish crustaceans, molluscs, etc — Capture production by countries or areas and fishing areas — Europe
Poissons, crustacés, mollusques, etc — Captures par pays ou zones et zones de pêche — Europe
Peces, crustáceos, moluscos, etc — Capturas por países o áreas y áreas de pesca — Europa

	Fishing area Zone de pêche Area de pesca	1993 t	1994 t	1995 t	1996 t	1997 t	1998 t	1999 t	2000 t	2001 t	2002 t
	34	-	-	-	-	-	-	-	-	31	-
	41	446	1 256	2 080	4 356	2 730	3 706	3 259	5 501	7 007	5 262
	48	-	12	-	-	984	1 354	1 072	1 457	1 161	2 149
	51	-	-	-	-	-	-	-	-	-	28
	Country total	*860 229*	*878 018*	*909 963*	*867 686*	*890 936*	*923 219*	*840 719*	*747 571*	*741 045*	*689 919*
Total		***16 492 261***	***15 808 056***	***17 185 049***	***17 490 005***	***17 912 203***	***17 109 957***	***16 088 259***	***16 170 292***	***15 981 378***	***15 163 631***

D-6

Fish crustaceans, molluscs, etc	Capture production by countries or areas and fishing areas	Oceania
Poissons, crustacés, mollusques, etc	Captures par pays ou zones et zones de pêche	Océanie
Peces, crustáceos, moluscos, etc	Capturas por países o áreas y áreas de pesca	Oceanía

	Fishing area Zone de pêche Area de pesca	1993 t	1994 t	1995 t	1996 t	1997 t	1998 t	1999 t	2000 t	2001 t	2002 t
Amer Samoa	06	0	0	0	0	0	0	0	0	0	-
	77	27	111	152	210	431	586	518	830	3 663	6 963
	Country total	*27*	*111*	*152*	*210*	*431*	*586*	*518*	*830*	*3 663*	*6 963*
Australia	06	3 011	584	821	705	612	593	694	666 F	663 F	646 F
	41	-	-	-	-	-	3 594	3 711	-	-	-
	57	127 895 F	105 408 F	107 555 F	105 022 F	102 142	98 875	127 563	115 145	111 498	116 177
	58	4	-	-	-	2 126	3 635	3 403	3 140	3 725	3 540
	71	38 006 F	39 400 F	43 719 F	46 781 F	42 075	46 536	48 905	42 229	52 009	48 066
	81	59 617 F	57 088 F	52 466 F	47 741 F	49 719	49 969	26 900	28 910	26 516	25 749
	Country total	*228 533*	*202 480*	*204 561*	*200 249*	*196 674*	*203 202*	*211 176*	*190 090*	*194 411*	*194 178*
Christmas Is	06	0	0	0	0	0	0	0	0	0	0
	57	0	0	0	0	0	0	0	0	0	0
	Country total	*0*	*0*	*0*	*0*	*0*	*0*	*0*	*0*	*0*	*0*
Cocos Is	06	0	0	0	0	0	0	0	0	0	0
	57	0	0	0	0	0	0	0	0	0	0
	Country total	*0*	*0*	*0*	*0*	*0*	*0*	*0*	*0*	*0*	*0*
Cook Is	06	0	0	10	0	0	0	0	0	0	0
	77	985 F	933 F	1 080	900 F	820 F	770 F	750 F	720 F	700 F	1 768 F
	Country total	*985 F*	*933 F*	*1 090*	*900 F*	*820 F*	*770 F*	*750 F*	*720 F*	*700 F*	*1 768 F*
Fiji Islands	06	2 907	2 964 F	3 586	3 034	4 325	5 111	5 622	5 700 F	5 921	5 800 F
	71	24 910	26 000 F	25 250	21 995	23 430	23 047	31 091	34 300 F	37 051	37 000 F
	Country total	*27 817*	*28 964 F*	*28 836*	*25 029*	*27 755*	*28 158*	*36 713*	*40 000 F*	*42 972*	*42 800 F*
Fr Polynesia	06	0	0	0	0	53	53	53	53	53	53
	77	8 075	8 175	9 020	9 910	11 617	12 420	12 283	13 846	15 351	15 490
	Country total	*8 075*	*8 175*	*9 020*	*9 910*	*11 670*	*12 473*	*12 336*	*13 899*	*15 404*	*15 543*
Guam	06	0	0	0	0	-	6	-	-	-	-
	71	431	435	185	121	158	247	223	275	278	231
	Country total	*431*	*435*	*185*	*121*	*158*	*253*	*223*	*275*	*278*	*231*
Kiribati	06	0	0	0	0	0	0	0	0	0	-
	71	27 072	27 753	30 306	31 871	30 052	35 304	52 741	25 563	32 375	31 001
	Country total	*27 072*	*27 753*	*30 306*	*31 871*	*30 052*	*35 304*	*52 741*	*25 563*	*32 375*	*31 001*
Marshall Is	06	0	0	0	0	0	0	0	0	0	0
	34	-	-	-	2 403	-	-	-	-	-	-
	71	471 F	393 F	381 F	369	400 F	500 F	500 F	8 060	35 544	38 742
	Country total	*471 F*	*393 F*	*381 F*	*2 772*	*400 F*	*500 F*	*500 F*	*8 060*	*35 544*	*38 742*
Micronesia	06	4 F	4 F	5 F	5 F	5 F	5 F	5 F	5 F	5 F	5 F
	71	17 884 F	22 922 F	7 624 F	9 064 F	10 122 F	16 576 F	12 838 F	23 879 F	19 648 F	20 382 F
	Country total	*17 888 F*	*22 926 F*	*7 629 F*	*9 069 F*	*10 127 F*	*16 581 F*	*12 843 F*	*23 884 F*	*19 653 F*	*20 387 F*
Nauru	71	500	500	400 F	300 F	250 F	200 F	150 F	100 F	50 F	21
	Country total	*500*	*500*	*400 F*	*300 F*	*250 F*	*200 F*	*150 F*	*100 F*	*50 F*	*21*
NewCaledonia	06	0	0	0	0	0	0	0	0	0	0
	71	2 670	3 185	2 644	2 998	2 438	3 105	3 152	3 386	3 309	3 417 F
	Country total	*2 670*	*3 185*	*2 644*	*2 998*	*2 438*	*3 105*	*3 152*	*3 386*	*3 309*	*3 417 F*
New Zealand	06	1 160	1 100	1 115	1 079	1 025	1 307	1 152	1 089	1 076	900
	81	425 508	448 560	555 520	424 485	611 399	638 873	597 971	547 360	560 186	556 830
	88	-	-	-	-	0	56	343	869	678	1 559
	Country total	*426 668*	*449 660*	*556 635*	*425 564*	*612 424*	*640 236*	*599 466*	*549 318*	*561 940*	*559 289*
Niue	06	0	0	0	0	0	0	0	0	0	0
	77	120 F	150 F	150 F	200 F	200 F	200 F	200 F	200 F	200 F	200 F
	Country total	*120 F*	*150 F*	*150 F*	*200 F*	*200 F*	*200 F*	*200 F*	*200 F*	*200 F*	*200 F*
Norfolk Is	06	0	0	0	0	0	0	0	0	0	0
	81	0	0	0	0	0	0	0	0	0	0
	Country total	*0*	*0*	*0*	*0*	*0*	*0*	*0*	*0*	*0*	*0*
N Marianas	06	0	0	1	0	0	0	0	0	0	-
	71	136	176	191	225	250	235	193	189	197	198
	Country total	*136*	*176*	*192*	*225*	*250*	*235*	*193*	*189*	*197*	*198*
Palau	06	0	0	0	0	0	0	0	0	0	0
	71	1 104	1 001	1 025	999	904	951	960	1 047	1 075	1 008 F
	Country total	*1 104*	*1 001*	*1 025*	*999*	*904*	*951*	*960*	*1 047*	*1 075*	*1 008 F*
Papua N Guin	06	13 500 F	13 500 F	13 500 F	13 500 F	13 500 F	13 500 F	13 500 F	13 500 F	13 500 F	13 500 F
	71	12 241 F	12 719 F	25 577 F	24 323 F	32 685 F	64 816 F	52 270 F	82 907 F	108 983 F	135 200 F
	Country total	*25 741 F*	*26 219 F*	*39 077 F*	*37 823 F*	*46 185 F*	*78 316 F*	*65 770 F*	*96 407 F*	*122 483 F*	*148 700 F*
Pitcairn Is	06	0	0	0	0	0	0	0	0	0	0
	81	8 F	8 F	8 F	8 F	8 F	8 F	8 F	8 F	8 F	8 F
	Country total	*8 F*	*8 F*	*8 F*	*8 F*	*8 F*	*8 F*	*8 F*	*8 F*	*8 F*	*8 F*
Samoa	06	0	0	0	0	0	0	0	1	1 F	1 F
	77	673	1 108	2 519	2 727	7 041	7 547	10 204	13 003	12 965 F	12 391 F
	Country total	*673*	*1 108*	*2 519*	*2 727*	*7 041*	*7 547*	*10 204*	*13 004*	*12 966 F*	*12 392 F*

	Fishing area Zone de pêche Area de pesca	1993 t	1994 t	1995 t	1996 t	1997 t	1998 t	1999 t	2000 t	2001 t	2002 t
Solomon Is	06	0	0	0	0	0	0	0	0	0	0
	71	42 970 F	46 904 F	62 019 F	50 990 F	64 005 F	61 332 F	58 428 F	24 926 F	30 277 F	31 003 F
	Country total	*42 970 F*	*46 904 F*	*62 019 F*	*50 990 F*	*64 005 F*	*61 332 F*	*58 428 F*	*24 926 F*	*30 277 F*	*31 003 F*
Tokelau	06	0	0	0	0	0	0	0	0	0	0
	77	200 F	200 F	200 F	200 F	200 F	200 F	200 F	200 F	200 F	200 F
	Country total	*200 F*	*200 F*	*200 F*	*200 F*	*200 F*	*200 F*	*200 F*	*200 F*	*200 F*	*200 F*
Tonga	06	1	0	0	0	0	0	0	0	0	0
	77	2 374	2 499	2 597	2 915	2 871	4 076	4 221	3 760	4 673	4 804
	Country total	*2 375*	*2 499*	*2 597*	*2 915*	*2 871*	*4 076*	*4 221*	*3 760*	*4 673*	*4 804*
Tuvalu	06	0	0	0	0	0	0	0	0	0	0
	71	1 460	561	399	400 F	500 F	500 F	500 F	500 F	500 F	500 F
	Country total	*1 460*	*561*	*399*	*400 F*	*500 F*	*500 F*	*500 F*	*500 F*	*500 F*	*500 F*
US Minor Is	77	0	0	0	0	0	0	0	0	0	0
	Country total	*0*	*0*	*0*	*0*	*0*	*0*	*0*	*0*	*0*	*0*
Vanuatu	06	0	0	0	0	0	0	0	0	0	0
	34	345	70	170	212	196	91	-	-	-	-
	71	2 470 F	3 020 F	9 236 F	11 603 F	28 856 F	41 477 F	49 971 F	37 930 F	2 400 F	2 752 F
	77	14 270	13 090	21 460	13 390	22 240	20 590	16 090	7 622	6 620	3 073
	87	23 430	31 010	26 780	20 410	19 530	12 750	26 980	23 344	15 501	11 314
	Country total	*40 515 F*	*47 190 F*	*57 646 F*	*45 615 F*	*70 822 F*	*74 908 F*	*93 041 F*	*68 896 F*	*24 521 F*	*17 139 F*
Wallis Fut I	06	0	0	0	0	0	0	0	0	0	0
	71	150	193	170	180	176	300	300 F	300 F	300 F	300 F
	Country total	*150*	*193*	*170*	*180*	*176*	*300*	*300 F*	*300 F*	*300 F*	*300 F*
Total		***856 589***	***871 724***	***1 007 841***	***851 275***	***1 086 361***	***1 169 941***	***1 164 593***	***1 065 562***	***1 107 699***	***1 130 792***

D-9

Fish crustaceans, molluscs, etc — Capture production by countries or areas and fishing areas — Other nei
Poissons, crustacés, mollusques, etc — Captures par pays ou zones et zones de pêche — Autres nca
Peces, crustáceos, moluscos, etc — Capturas por países o áreas y áreas de pesca — Otros nep

	Fishing area Zone de pêche Area de pesca	1993 t	1994 t	1995 t	1996 t	1997 t	1998 t	1999 t	2000 t	2001 t	2002 t
Other nei	27	223	68	189	71	212	-	-	-	-	-
	31	2 813	4 434	4 232	5 994	6 130	8 339	7 838	5 539	2 932	175
	34	28 310	30 524	149 452	34 774	29 889	40 979	56 333	54 839	95 048	143 725
	37	1 000	1 700	1 350	721	1 442	2 055	2 335	126	571	508
	47	253	3 208	19 915	10 459	155	27	-	2	10	14
	51	98 610	83 790	113 363	110 161	105 170	111 548	135 429	147 422	99 225	105 528
	57	5 715	9 088	7 309	11 943	7 355	18 856	12 008	12 611	2 051	2 051
	77	-	50	260	-	-	310	-	2 011	13 878	38 457
	87	1 910	1 880	1 210	-	3 360	870	-	4 311	37 437	47 136
	Country total	*138 834*	*134 742*	*297 280*	*174 123*	*153 713*	*182 984*	*213 943*	*226 861*	*251 152*	*337 594*
Total		***138 834***	***134 742***	***297 280***	***174 123***	***153 713***	***182 984***	***213 943***	***226 861***	***251 152***	***337 594***

E - Capture production: by countries or areas

E - Captures: par pays ou zones

E - Capturas: por países o áreas

English name Nom anglais Nombre inglés	Scientific name Nom scientifique Nombre científico	1996 t	1997 t	1998 t	1999 t	2000 t	2001 t	2002 t
Algeria								
Northern pike	*Esox lucius*	-	-	2	...	...	...	...
Freshwater fishes nei	*Osteichthyes*	0	0	0	0	0	0	0
European eel	*Anguilla anguilla*	...	...	10	...	...	...	...
Common sole	*Solea solea*	387	333	285	234	278	271	377
Greater forkbeard	*Phycis blennoides*	...	...	...	4	5	19	39
European hake	*Merluccius merluccius*	1 154	1 012	1 310	1 681	3	196	209
Groupers, seabasses nei	*Serranidae*	...	...	...	6	0	0	0
Common pandora	*Pagellus erythrinus*	1 633	2 661	1 503	1 145	1 518	1 339	1 515
Axillary seabream	*Pagellus acarne*	...	...	362	...	...	...	...
Pargo breams nei	*Pagrus spp*	81	58	255	254	153	267	394
Bogue	*Boops boops*	1 425	1 631	3 234	3 452	4 132	4 135	4 697
Porgies, seabreams nei	*Sparidae*	-	-	-	7	-	-	-
Surmullets(=Red mullets) nei	*Mullus spp*	1 950	1 320	1 480	1 420	2 170	2 081	2 140
Sardinellas nei	*Sardinella spp*	8 150	15 741	9 992	11 393	20 399	33 748	24 480
European pilchard(=Sardine)	*Sardina pilchardus*	49 906	49 142	49 295	56 724	47 677	58 966	72 547
European anchovy	*Engraulis encrasicolus*	1 330	1 855	3 511	3 141	5 721	6 085	2 182
Clupeoids nei	*Clupeoidei*	1 182	1 261	2 279	2 634	2 608	1 074	1 541
Atlantic bonito	*Sarda sarda*	277	357	511	475	405	350	597
Plain bonito	*Orcynopsis unicolor*	119	224	128	216	135	145	128
Frigate and bullet tunas	*Auxis thazard, A.rochei*	237	179	299	173	225	230	481
Little tunny(=Atl.black skipj)	*Euthynnus alletteratus*	554	448	384	562	494	407	148
Skipjack tuna	*Katsuwonus pelamis*	-	-	171	43	89	77	...
Atlantic bluefin tuna	*Thunnus thynnus*	156	157	1 947	2 142	2 330	2 012	1 710
Swordfish	*Xiphias gladius*	807	807	825	709	816	1 081	814
Jack and horse mackerels nei	*Trachurus spp*	3 644	5 491	4 827	6 212	7 702	8 174	7 088
Greater amberjack	*Seriola dumerili*	32	19	48	75	99	341	542
Atlantic mackerel	*Scomber scombrus*	590	750	1 055	1 053	1 144	1 167	567
Barracudas nei	*Sphyraena spp*	...	...	...	6	47	392	447
Rays, stingrays, mantas nei	*Rajiformes*	272	120	450	207	191	298	491
Sharks, rays, skates, etc. nei	*Elasmobranchii*	965	415	867	854	331	679	519
Marine fishes nei	*Osteichthyes*	3 964	3 799	2 363	3 403	9 323	6 010	6 749
Palinurid spiny lobsters nei	*Palinurus spp*	9	44	141	70	33	34	46
Norway lobster	*Nephrops norvegicus*	100	89	138	96	92	45	56
European lobster	*Homarus gammarus*	0	0	0	-	-	-	-
Deepwater rose shrimp	*Parapenaeus longirostris*	918	1 433	1 639	2 147	2 696	2 107	1 654
Blue and red shrimp	*Aristeus antennatus*	1 230	1 186	1 485	880	1 116	740	893
Common prawn	*Palaemon serratus*	40	35	89	9	...	...	...
Common shrimp	*Crangon crangon*	...	...	...	...	0	2	3
Marine crustaceans nei	*Crustacea*	67	71	154	83	308	222	297
Common cuttlefish	*Sepia officinalis*	567	619	492	312	392	340	318
Common squids nei	*Loligo spp*	53	138	202	240	212	224	172
Octopuses, etc. nei	*Octopodidae*	190	185	543	305	306	343	366
Cephalopods nei	*Cephalopoda*	-	-	70	29	6	22	113
	Country total	*81 989*	*91 580*	*92 346*	*102 396*	*113 156*	*133 623*	*134 320*
Angola								
Freshwater fishes nei	*Osteichthyes*	6 000	6 000	6 000	6 000 F	6 000 F	6 000 F	6 000 F
Flatfishes nei	*Pleuronectiformes*	195	120	928	912	592	1 928	3 463
Shallow-water Cape hake	*Merluccius capensis*	832	777	906	2 060	658	1 863	3 225
Sea catfishes nei	*Ariidae*	1	43	90	69	407	2 424	3 616
Mullets nei	*Mugilidae*	0	0	-	-	-	-	-
Groupers, seabasses nei	*Serranidae*	1 002	460	1 326	347	2 341	4 214	5 774
Bigeye grunt	*Brachydeuterus auritus*	146	228	262	167	191	1 176	4 078
Grunts, sweetlips nei	*Haemulidae (=Pomadasyidae)*	1 031	557	438	412	799	1 688	3 513
Canary drum (=Baardman)	*Umbrina canariensis*	...	...	...	830	...	2 202	...
Southern meagre(=Mulloway)	*Argyrosomus hololepidotus*	...	...	...	259	...	2 540	...
West African croakers nei	*Pseudotolithus spp*	4 044	1 606	7 095	8 936	8 005	8 725	25 946
Croakers, drums nei	*Sciaenidae*	-	12	70	91	29	594	1 332
Red pandora	*Pagellus bellottii*	...	...	...	188	...	744	674
Dentex nei	*Dentex spp*	2 624	1 567	8 683	8 758	13 765	16 080	8 948
Black seabream	*Spondyliosoma cantharus*	...	...	93	129	212	494	13
Pargo breams nei	*Pagrus spp*	95	716	1 077	183	1 840	4 484	5 773
Sand steenbras	*Lithognathus mormyrus*	214	176	1 072	726	999	1 735	2 802
Picarels nei	*Spicara spp*	7	0	55	...	13	-	-
West African goatfish	*Pseudupeneus prayensis*	-	0	-	-	-	-	-
Threadfins, tasselfishes nei	*Polynemidae*	370	265	520	1 378	872	1 834	2 664
John dory	*Zeus faber*	315	307	1 339	1 668	1 582	1 303	2 061
Largehead hairtail	*Trichiurus lepturus*	...	...	...	28	-	-	-
Gurnards, searobins nei	*Triglidae*	0	0	-	-	-	-	-
Sardinellas nei	*Sardinella spp*	17 484	21 014	45 483	57 579	108 211	57 896	28 513
Atlantic bonito	*Sarda sarda*	39	32	-	2	118	1 157	...
Frigate and bullet tunas	*Auxis thazard, A.rochei*	29	12	31	2	38	206	114
Little tunny(=Atl.black skipj)	*Euthynnus alletteratus*	235	75	406	118	132	231	155
Skipjack tuna	*Katsuwonus pelamis*	15	52	2	32	14	687	...
Yellowfin tuna	*Thunnus albacares*	78	70	115	170	35	...	...
Swordfish	*Xiphias gladius*	0	0	0	0	0	...	...
Tuna-like fishes nei	*Scombroidei*	-	-	-	-	-	-	6 579
Bluefish	*Pomatomus saltatrix*	-	-	-	-	-	167	-
Cunene horse mackerel	*Trachurus trecae*	19 764	35 797	39 739	47 719	53 243	40 898	43 292
Crevalle jack	*Caranx hippos*	39	16	89	302	1 835	674	271
Blue butterfish	*Stromateus fiatola*	0	-	-	-	-	-	...
Barracudas nei	*Sphyraena spp*	...	...	...	119	-	-	-
Raja rays nei	*Raja spp*	21	16	750	1 399	593	1 430	2 696
Sharks, rays, skates, etc. nei	*Elasmobranchii*	379	90	376	...	157	3 354	3 236
Marine fishes nei	*Osteichthyes*	80 633	74 578	43 038	32 185	31 674	80 740	87 406
Marine crabs nei	*Brachyura*	...	...	...	602	800	836	1 000

English name Nom anglais Nombre inglés	Scientific name Nom scientifique Nombre científico	1996 t	1997 t	1998 t	1999 t	2000 t	2001 t	2002 t
Deepwater rose shrimp	*Parapenaeus longirostris*	1 367	1 175	1 900	1 101	1 600	1 660	2 000
Striped red shrimp	*Aristeus varidens*	814	543	876	820	1 154	1 200	1 500
Natantian decapods nei	*Natantia*	-	-	-	0	-	-	-
Marine crustaceans nei	*Crustacea*	-	-	-	-	-	1 062	1 653
Various squids nei	*Loliginidae, Ommastrephidae*	42	...	390	463	442	292	500
Octopuses, etc. nei	*Octopodidae*	-	-	-	45	-	-	...
Marine molluscs nei	*Mollusca*	-	-	-	-	-	-	2 000
	Country total	*137 815*	*146 304*	*163 149*	*175 799*	*238 351*	*252 518*	*260 797*
Benin								
Tilapias nei	*Oreochromis (=Tilapia) spp*	11 500 F	11 500 F	11 500 F	11 648	9 600 F	10 900 F	10 900 F
African lungfishes	*Protopterus spp*	110 F	105 F	100 F	100 F	100 F	110 F	110 F
Bonytongues nei	*Heterotis spp*	700 F	650 F	600 F	600 F	500 F	550 F	550 F
Black catfishes nei	*Chrysichthys spp*	1 300 F	1 400 F	1 500 F	1 542	1 300 F	1 500 F	1 500 F
Torpedo-shaped catfishes nei	*Clarias spp*	2 100 F	1 900 F	1 700 F	1 537	1 300 F	1 500 F	1 500 F
Upsidedown catfishes	*Synodontis spp*	120 F	110 F	110 F	110 F	100 F	110 F	110 F
Freshwater gobies nei	*Gobiidae*	900 F	850 F	800 F	800 F	700 F	800 F	800 F
Freshwater fishes nei	*Osteichthyes*	6 563 F	5 096 F	3 838 F	3 733 F	3 000 F	3 440 F	3 433 F
West African ilisha	*Ilisha africana*	602 F	822 F	781 F	671	450 F	923	692
Diadromous clupeoids nei	*Clupeoidei*	250 F	230 F	210 F	210 F	200 F	230 F	230 F
Flatfishes nei	*Pleuronectiformes*	30 F	37 F	45 F	52	30 F	14	22
West African ladyfish	*Elops lacerta*	...	...	...	...	...	15	16
Sea catfishes nei	*Ariidae*	20 F	17 F	15 F	12	8 F	2	2
Mullets nei	*Mugilidae*	1 000 F	1 500 F	2 000 F	2 201	1 802 F	2 055 F	2 056 F
Groupers nei	*Epinephelus spp*	25 F	25 F	23 F	20	14 F	16	1
Snappers nei	*Lutjanus spp*	66 F	90 F	86 F	92	60 F	32	51
Grunts, sweetlips nei	*Haemulidae (=Pomadasyidae)*	228 F	311 F	296 F	269	180 F	222	365
Boe drum	*Pteroscion peli*	...	...	...	61	40 F	115	137
Pandoras nei	*Pagellus spp*	16 F	22 F	20 F	15	10 F	...	1
Dentex nei	*Dentex spp*	95 F	100 F	90 F	83	60 F	20	19
Black seabream	*Spondyliosoma cantharus*	30 F	45 F	48 F	56	40 F	93	147
African sicklefish	*Drepane africana*	15 F	18 F	14 F	12	8 F	6	4
Threadfins, tasselfishes nei	*Polynemidae*	500 F	500 F	450 F	358	250 F	324	463
Spadefishes nei	*Ephippidae*	3 F	5 F	4 F	4	3 F	2	1
Puffers nei	*Tetraodontidae*	...	...	...	...	...	2	5
Largehead hairtail	*Trichiurus lepturus*	300 F	350 F	400 F	543	370 F	584	579
Gurnards, searobins nei	*Triglidae*	...	...	...	11	7 F	-	-
Sardinellas nei	*Sardinella spp*	1 100 F	1 200 F	1 100 F	1 072	750 F	1 434	1 496
Bonga shad	*Ethmalosa fimbriata*	1 900 F	1 850 F	1 800 F	1 806	1 505 F	1 708 F	1 717 F
European anchovy	*Engraulis encrasicolus*	250 F	350 F	400 F	478	330 F	852	1 381
Clupeoids nei	*Clupeoidei*	...	...	...	13	10 F	-	-
Plain bonito	*Orcynopsis unicolor*	1 F	3	1	1	-	-	-
West African Spanish mackerel	*Scomberomorus tritor*	188 F	362	511	205	205 F	203	185
Little tunny(=Atl.black skipj)	*Euthynnus alletteratus*	58 F	196	83	69	69 F	69	69
Skipjack tuna	*Katsuwonus pelamis*	2 F	7	3	2	2 F	...	...
Yellowfin tuna	*Thunnus albacares*	1 F	3	1	1	1 F	1	...
Bigeye tuna	*Thunnus obesus*	9 F	30	13	11	...	...	...
Atlantic sailfish	*Istiophorus albicans*	19	6	4	5	5 F	12	2
Blue marlin	*Makaira nigricans*	5 F	5	5	5	5 F	...	...
Swordfish	*Xiphias gladius*	24 F	10	0	3	...	41	68
Needlefishes, etc. nei	*Belonidae*	40 F	35 F	30 F	27	20 F	38	32
Flyingfishes nei	*Exocoetidae*	28 F	25 F	20 F	15	10 F	96	92
Bluefish	*Pomatomus saltatrix*	1 000 F	950 F	900 F	875	600 F	697	978
African moonfish	*Selene dorsalis*	27 F	37 F	35 F	57	40 F	121	144
Atlantic bumper	*Chloroscombrus chrysurus*	-	-	-	-	-	853	1 022
Carangids nei	*Carangidae*	700 F	800 F	900 F	1 027	700 F	475	428
Common dolphinfish	*Coryphaena hippurus*	...	...	...	4	3 F	48	28
Chub mackerel	*Scomber japonicus*	280 F	300 F	250 F	205	140 F	314	452
Barracudas nei	*Sphyraena spp*	330 F	400 F	350 F	297	200 F	262	365
Shortfin mako	*Isurus oxyrinchus*	...	...	...	4	3 F	1	-
Rays, stingrays, mantas nei	*Rajiformes*	62 F	70 F	60 F	47	30 F	38	32
Sharks, rays, skates, etc. nei	*Elasmobranchii*	100 F	100 F	80 F	59	40 F	87	86
Marine fishes nei	*Osteichthyes*	1 785 F	3 628 F	3 294 F	1 677	1 150 F	370	1 192
River prawns nei	*Macrobrachium spp*	150 F	130 F	120 F	120 F	100 F	110 F	110 F
Freshwater crustaceans nei	*Crustacea*	4 600 F	4 700 F	4 800 F	4 924	4 100 F	4 600 F	4 600 F
Marine crabs nei	*Brachyura*	40 F	50 F	45 F	32	20 F	9	22
Palinurid spiny lobsters nei	*Palinurus spp*	3 F	5 F	4 F	3	2 F	2	7
Penaeus shrimps nei	*Penaeus spp*	3 000 F	2 850 F	2 700 F	2 574	2 100 F	2 400 F	2 400 F
Natantian decapods nei	*Natantia*	0	0	0	31	20 F	3	55
Cuttlefish,bobtail squids nei	*Sepiidae, Sepiolidae*	...	...	...	18	12 F	1	5
Marine turtles nei	*Testudinata*	...	...	...	29	20 F	5	1
	Country total	*42 175*	*43 785*	*42 139*	*40 436*	*32 324*	*38 415*	*40 663*
Botswana								
Tilapias nei	*Oreochromis (=Tilapia) spp*	48	80	88	93	92	88	101
Freshwater fishes nei	*Osteichthyes*	33	80	103	64	74	30	38
	Country total	*81*	*160*	*191*	*157*	*166*	*118*	*139*
Br Ind Oc Tr								
Freshwater fishes nei	*Osteichthyes*	0	0	0	0	0	0	0
Marine fishes nei	*Osteichthyes*	0	0	0	0	0	0	0
	Country total	*0*	*0*	*0*	*0*	*0*	*0*	*0*
Burkina Faso								
Freshwater fishes nei	*Osteichthyes*	8 000	8 000	8 335	7 600	8 500	8 500 F	8 500 F

English name Nom anglais Nombre inglés	Scientific name Nom scientifique Nombre científico	1996 t	1997 t	1998 t	1999 t	2000 t	2001 t	2002 t
	Country total	*8 000*	*8 000*	*8 335*	*7 600*	*8 500*	*8 500 F*	*8 500 F*
Burundi								
Nile tilapia	*Oreochromis niloticus*	50	50	50	50	120	120	120 F
Dagaas	*Stolothrissa, Limnothrissa spp*	1 508	17 868	8 646	7 030	10 839	6 138	6 160 F
Freshwater perches nei	*Lates spp*	1 205	2 378	3 925	1 523	3 070	1 657	1 660 F
Freshwater fishes nei	*Osteichthyes*	278	...	805	596	3 286	1 049	1 060 F
	Country total	*3 041*	*20 296*	*13 426*	*9 199*	*17 315*	*8 964*	*9 000 F*
Cameroon								
Freshwater fishes nei	*Osteichthyes*	35 000 F	40 000 F	45 000 F	50 000 F	55 000	52 500 F	65 000
Lefteye flounders nei	*Bothidae*	0	0	0	0	0	-	-
Tonguefishes	*Cynoglossidae*	748	700 F	594	455	488	439	788
Benguela hake	*Merluccius polli*	-	-	-	-	-	-	3
Sea catfishes nei	*Ariidae*	886	640 F	370 F	520 F	455	410	589
Groupers nei	*Epinephelus spp*	0	0	1	0	1	1	1
Snappers nei	*Lutjanus spp*	511	400 F	320 F	337 F	337	293	297
Sompat grunt	*Pomadasys jubelini*	533	400 F	337 F	345 F	304	48	342
Cassava croaker	*Pseudotolithus senegalensis*	2 540	2 600 F	2 900 F	3 200 F	1 972	2 679	2 100
Bobo croaker	*Pseudotolithus elongatus*	4 371	4 400 F	4 400 F	1 900 F	3 400	3 218	1 883
Croakers, drums nei	*Sciaenidae*	3 539	3 500 F	3 500 F	3 500 F	3 260	2 399	1 946
Porgies, seabreams nei	*Sparidae*	2	2 F	5 F	5 F	4	34	18
African sicklefish	*Drepane africana*	72	75 F	78	94	73	70	30
Giant African threadfin	*Polydactylus quadrifilis*	54	40 F	12	34	43	170	75
Lesser African threadfin	*Galeoides decadactylus*	810	500 F	13	150	477	462	134
Royal threadfin	*Pentanemus quinquarius*	584	1 000 F	2 368	1 778	1 363	3 276	832
Conger eels, etc. nei	*Congridae*	6	6 F	5 F	5 F	8	7	11
Largehead hairtail	*Trichiurus lepturus*	4	10 F	16	1	6	59	6
Sardinellas nei	*Sardinella spp*	24 002	23 500 F	23 000 F	23 500 F	21 611	21 645	22 540
Bonga shad	*Ethmalosa fimbriata*	24 000	23 500 F	23 000 F	23 500 F	21 609	21 645	21 782
Tuna-like fishes nei	*Scombroidei*	0	0	0	0	1	...	59
Carangids nei	*Carangidae*	5	5 F	2	2	4	5	7
Barracudas nei	*Sphyraena spp*	37	30 F	5	28	23	293	340
Rays, stingrays, mantas nei	*Rajiformes*	186	170 F	161	211	150	130	133
Sharks, rays, skates, etc. nei	*Elasmobranchii*	48	50 F	55	86	67	146	85
Marine fishes nei	*Osteichthyes*	-	-	-	-	954	907	830
Marine crabs nei	*Brachyura*	18	20 F	20 F	20 F	27	26	20
Tropical spiny lobsters nei	*Panulirus spp*	0	0	1	1	0	0	1
Penaeus shrimps nei	*Penaeus spp*	442	450 F	635	326	471	168	280
Cuttlefish,bobtail squids nei	*Sepiidae, Sepiolidae*	2	2 F	2 F	2 F	1	1	1
Marine molluscs nei	*Mollusca*	0	0	0	0	0	0	2
	Country total	*98 400 F*	*102 000 F*	*106 800 F*	*110 000 F*	*112 109*	*111 031*	*120 135*
Cape Verde								
Freshwater fishes nei	*Osteichthyes*	0	0	0	0	0	0	0
Demersal percomorphs nei	*Perciformes*	...	1 450	1 150	1 079	1 314	1 307	1 340 F
Wahoo	*Acanthocybium solandri*	503	603	429	587	487	578	552
Frigate and bullet tunas	*Auxis thazard, A.rochei*	6	22	191	154	81	171	206
Little tunny(=Atl.black skipj)	*Euthynnus alletteratus*	63	86	110	776	491	178	108
Skipjack tuna	*Katsuwonus pelamis*	470	591	684	962	789	794	284
Yellowfin tuna	*Thunnus albacares*	1 448	1 721	1 418	1 663	1 851	1 684	287
Bigeye tuna	*Thunnus obesus*	16	10	1	1	2	-	-
Tuna-like fishes nei	*Scombroidei*	-	245	-	-	-	-	-
Chub mackerel	*Scomber japonicus*	...	...	182	...	...	...	...
Pelagic percomorphs nei	*Perciformes*	...	4 414	4 899	4 463	4 823	4 288	4 500 F
Marine fishes nei	*Osteichthyes*	6 612	538	333	640	719	633	700 F
Tropical spiny lobsters nei	*Panulirus spp*	12	8	9	12	10	7	10 F
Palinurid spiny lobsters nei	*Palinurus spp*	25	17	18	23	19	13	13 F
Marine molluscs nei	*Mollusca*	0	0	0	0	0	-	-
Marine turtles nei	*Testudinata*	0	0	0	-	-	-	-
	Country total	*9 155*	*9 705*	*9 424*	*10 360*	*10 586*	*9 653*	*8 000 F*
Cent Afr Rep								
Freshwater fishes nei	*Osteichthyes*	14 000 F	14 250 F	14 500 F	15 000 F	15 000 F	15 000 F	15 000 F
	Country total	*14 000 F*	*14 250 F*	*14 500 F*	*15 000*	*15 000 F*	*15 000 F*	*15 000 F*
Chad								
Freshwater fishes nei	*Osteichthyes*	100 000	85 000	84 000	84 000 F	84 000 F	84 000 F	84 000 F
	Country total	*100 000*	*85 000*	*84 000*	*84 000 F*	*84 000 F*	*84 000 F*	*84 000 F*
Comoros								
Freshwater fishes nei	*Osteichthyes*	0	0	0	0	0	0	0
Sardinellas nei	*Sardinella spp*	1 000 F	1 000 F	1 000 F	950 F	1 050 F	1 000	1 000 F
Anchovies, etc. nei	*Engraulidae*	900 F	900 F	900 F	850 F	950 F	850	850 F
Seerfishes nei	*Scomberomorus spp*	270	260	260	250	270 F	250	250 F
Kawakawa	*Euthynnus affinis*	170	160	160	150	170 F	160	160 F
Skipjack tuna	*Katsuwonus pelamis*	2 150	2 070	2 070	2 000	2 200 F	2 000	2 000 F
Yellowfin tuna	*Thunnus albacares*	5 520	5 310	5 310	5 200	5 600 F	5 200	5 200 F
Bigeye tuna	*Thunnus obesus*	30	30	30	30	30 F	30	30 F
Indo-Pacific sailfish	*Istiophorus platypterus*	250	240	240	200	250 F	200	200 F
Marlins,sailfishes,etc. nei	*Istiophoridae*	130	120	120	100	130 F	120	120 F
Tuna-like fishes nei	*Scombroidei*	510	490	490	450	520 F	450	450 F
Carangids nei	*Carangidae*	490 F	470 F	470 F	450 F	500 F	500	500 F

E-1

Fish crustaceans, molluscs, etc	Capture production by countries or areas and species	Africa
Poissons, crustacés, mollusques, etc	Captures par pays ou zones et espèces	Afrique
Peces, crustáceos, moluscos, etc	Capturas por países o áreas y especies	Africa

English name Nom anglais Nombre inglés	Scientific name Nom scientifique Nombre científico	1996 t	1997 t	1998 t	1999 t	2000 t	2001 t	2002 t
Indian mackerels nei	*Rastrelliger spp*	250 F	240 F	240 F	230 F	250 F	220	220 F
Marine fishes nei	*Osteichthyes*	1 010 F	1 190 F	1 190 F	1 120 F	1 260 F	1 180	1 200 F
Natantian decapods nei	*Natantia*	20 F	20 F	20 F	20 F	20 F	20	20 F
Marine molluscs nei	*Mollusca*	0	0	0	0	0	-	-
	Country total	*12 700 F*	*12 500 F*	*12 500 F*	*12 000*	*13 200*	*12 180*	*12 200 F*
Congo Dem R								
Freshwater fishes nei	*Osteichthyes*	159 037	158 367	174 087	204 503	205 000 F	210 000 F	215 000 F
Tonguefishes	*Cynoglossidae*	84 F	81 F	83 F	80 F	80 F	80 F	80 F
Bigeye grunt	*Brachydeuterus auritus*	398 F	385 F	396 F	400 F	400 F	400 F	400 F
Conger eels, etc. nei	*Congridae*	73 F	71 F	73 F	70 F	70 F	70 F	70 F
Sardinellas nei	*Sardinella spp*	1 590 F	1 538 F	1 582 F	1 595 F	1 620 F	1 700 F	1 800 F
Jack and horse mackerels nei	*Trachurus spp*	1 432 F	1 386 F	1 426 F	1 400 F	1 430 F	1 500 F	1 600 F
Sharks, rays, skates, etc. nei	*Elasmobranchii*	...	...	...	...	400 F	450 F	450 F
Marine fishes nei	*Osteichthyes*	396 F	383 F	394 F	400 F	300 F	400 F	600 F
	Country total	*163 010*	*162 211*	*178 041*	*208 448*	*209 300 F*	*214 600 F*	*220 000 F*
Congo Rep								
Freshwater fishes nei	*Osteichthyes*	25 873	18 987	25 455	25 268	20 204 F	20 200 F	20 500 F
Common sole	*Solea solea*	200 F	200 F	100 F	100 F	...	50 F	50 F
Tonguefishes	*Cynoglossidae*	270 F	240 F	200 F	180 F	158	150 F	150 F
Sea catfishes nei	*Ariidae*	230 F	250 F	250 F	300 F	373 F	400 F	450 F
Groupers nei	*Epinephelus spp*	5 F	4 F	3 F	3 F	2	2 F	2 F
Snappers nei	*Lutjanus spp*	40 F	30 F	20 F	10 F	1	1 F	1 F
Sompat grunt	*Pomadasys jubelini*	5 F	5 F	5 F	5 F	...	4 F	4 F
Bigeye grunt	*Brachydeuterus auritus*	125 F	115 F	100 F	100 F	95	90 F	90 F
Grunts, sweetlips nei	*Haemulidae (=Pomadasyidae)*	12 F	9 F	6 F	3 F	-	-	-
Boe drum	*Pteroscion peli*	630 F	620 F	600 F	600 F	602	600 F	600 F
Bobo croaker	*Pseudotolithus elongatus*	3 F	3 F	3 F	3 F	...	2 F	2 F
West African croakers nei	*Pseudotolithus spp*	500 F	450 F	400 F	350 F	298	300 F	300 F
Croakers, drums nei	*Sciaenidae*	10 F	10 F	5 F	5 F	0	0	0
Red pandora	*Pagellus bellottii*	5 F	4 F	3 F	2 F	...	2 F	2 F
Dentex nei	*Dentex spp*	180 F	140 F	100 F	60 F	22	20 F	20 F
African sicklefish	*Drepane africana*	15 F	15 F	15 F	20 F	24	20 F	20 F
Lesser African threadfin	*Galeoides decadactylus*	300 F	250 F	190 F	150 F	63	60 F	60 F
Royal threadfin	*Pentanemus quinquarius*	200 F	200 F	180 F	190 F	189	180 F	180 F
Conger eels, etc. nei	*Congridae*	350 F	400 F	400 F	450 F	552	600 F	650 F
Bearded brotula	*Brotula barbata*	25 F	30 F	30 F	40 F	46	40 F	40 F
Hairtails, scabbardfishes nei	*Trichiuridae*	120 F	90 F	60 F	30 F	-	-	-
Shortspine African angler	*Lophius vaillanti*	...	...	...	...	7	5 F	5 F
Sardinellas nei	*Sardinella spp*	11 200 F	10 200 F	9 300 F	9 300 F	10 500 F	9 500 F	9 700 F
Bonga shad	*Ethmalosa fimbriata*	2 000 F	2 300 F	2 300 F	2 500 F	2 700 F	3 000 F	3 300 F
Skipjack tuna	*Katsuwonus pelamis*	6	6	6	6	6	...	...
Yellowfin tuna	*Thunnus albacares*	12	12	12	12	12	12	...
Bigeye tuna	*Thunnus obesus*	8	8	8	8	8	8	8
Jack and horse mackerels nei	*Trachurus spp*	60 F	50 F	40 F	30 F	23	20 F	20 F
Jacks, crevalles nei	*Caranx spp*	12 F	9 F	5 F	3 F	0	0	0
African moonfish	*Selene dorsalis*	55 F	60 F	60 F	70 F	72	70 F	70 F
Blue butterfish	*Stromateus fiatola*	...	...	...	...	1	1 F	1 F
Barracudas nei	*Sphyraena spp*	0	0	0	0	1	1 F	1 F
Hammerhead sharks, etc. nei	*Sphyrnidae*	...	...	...	...	500 F	500 F	500 F
Dogfish sharks nei	*Squalidae*	250 F	300 F	300 F	400 F	412 F	500 F	550 F
Rays, stingrays, mantas nei	*Rajiformes*	135 F	110 F	85 F	60 F	33	30 F	30 F
Marine fishes nei	*Osteichthyes*	2 036 F	2 645 F	2 260 F	2 829 F	7 232 F	5 455 F	5 277 F
Marine crabs nei	*Brachyura*	10 F	5 F	20 F	30 F	44	10 F	10 F
Palinurid spiny lobsters nei	*Palinurus spp*	4 F	3 F	7 F	9 F	10	4 F	4 F
Natantian decapods nei	*Natantia*	584	320 F	420 F	374	529	400 F	400 F
Cuttlefish,bobtail squids nei	*Sepiidae, Sepiolidae*	3 F	2 F	7 F	9 F	10	3 F	3 F
	Country total	*45 473*	*38 082*	*42 955 F*	*43 509*	*44 729*	*42 240 F*	*43 000 F*
Côte dIvoire								
Freshwater fishes nei	*Osteichthyes*	11 162 F	11 732 F	12 301 F	10 556 F	10 475	10 500	22 000
West African ilisha	*Ilisha africana*	491	534	452	314	152	50	417
Tonguefishes	*Cynoglossidae*	177	139	217	217	211	197	237
Bonefish	*Albula vulpes*	43	1 430	1 617	29	21	13	32
Sea catfishes nei	*Ariidae*	16	21	25	16	20	13	62
White grouper	*Epinephelus aeneus*	1 F	1 F	1 F	...	2	3	121
Bigeyes nei	*Priacanthus spp*	20 F	20 F	25 F	29	12	1	2
Sompat grunt	*Pomadasys jubelini*	247	143	227	236	231	148	202
Bigeye grunt	*Brachydeuterus auritus*	4 632	4 732	1 617	3 964	5 175	1 689	2 812
Meagre	*Argyrosomus regius*	10 F	10 F	5 F	3	1	1	1
Cassava croaker	*Pseudotolithus senegalensis*	5 F	10 F	20 F	30	39	39	26
Bobo croaker	*Pseudotolithus elongatus*	5	3	4	5	...	4	1
West African croakers nei	*Pseudotolithus spp*	551	466	545	378	356	407	498
Pandoras nei	*Pagellus spp*	328	346	1 093	1 137	913	697	819
Porgies, seabreams nei	*Sparidae*	1 F	1 F	1 F	1	3	...	...
West African goatfish	*Pseudupeneus prayensis*	68	44	38	61	44	40	35
African sicklefish	*Drepane africana*	23	18	28	21	35	20	24
Lesser African threadfin	*Galeoides decadactylus*	268	173	278	196	224	168	246
Conger eels, etc. nei	*Congridae*	22	26	.32	12	12	17	12
Bearded brotula	*Brotula barbata*	192	158	342	513	280	156	186
Tilefishes nei	*Branchiostegidae*	20 F	30 F	40 F	45	44	15	16
Largehead hairtail	*Trichiurus lepturus*	180	265	492	518	321	200	171
Round sardinella	*Sardinella aurita*	18 491	10 002	10 884	12 206	19 358	18 864	7 288
Madeiran sardinella	*Sardinella maderensis*	1 500 F	1 000 F	500 F	175	426	716	2 421
Bonga shad	*Ethmalosa fimbriata*	9 500 F	9 500 F	11 000 F	11 006 F	11 009 F	10 465 F	9 370

English name Nom anglais Nombre inglés	Scientific name Nom scientifique Nombre científico	1996 t	1997 t	1998 t	1999 t	2000 t	2001 t	2002 t
Little tunny(=Atl.black skipj)	*Euthynnus alletteratus*	250	114	108	...	108	...	...
Yellowfin tuna	*Thunnus albacares*	...	2	...	...	...	...	...
Atlantic sailfish	*Istiophorus albicans*	91	65	35	80	45	47	65
Blue marlin	*Makaira nigricans*	157	222	182	275	206	196	78
Atlantic white marlin	*Tetrapturus albidus*	1	2	1	5	1	2	2
Swordfish	*Xiphias gladius*	26	18	25	26	20	19	19
Needlefishes, etc. nei	*Belonidae*	1 F	1 F	1 F	...	...	3	2
Jacks, crevalles nei	*Caranx spp*	69	51	282	290	144	119	690
Leerfish	*Lichia amia*	-	-	-	20	16	3	4
Atlantic bumper	*Chloroscombrus chrysurus*	1 116	1 302	1 107	1 120	1 374	2 364	2 047
Carangids nei	*Carangidae*	169	1 400	1 267	1 025	509	-	-
Chub mackerel	*Scomber japonicus*	313	1 355	1 340	2 494	259	110	878
Blue butterfish	*Stromateus fiatola*	10 F	15 F	20 F	22	45	27	21
Barracudas nei	*Sphyraena spp*	366	151	195	184	160	75	218
Mako sharks	*Isurus spp*	...	92	38	...	...	...	...
Silky shark	*Carcharhinus falciformis*	...	2	...	...	...	...	...
Hammerhead sharks, etc. nei	*Sphyrnidae*	...	190	125	...	...	...	...
Rays, stingrays, mantas nei	*Rajiformes*	218	146	202	227	241	168	240
Sharks, rays, skates, etc. nei	*Elasmobranchii*	405	71	42	313	521	66	132
Marine fishes nei	*Osteichthyes*	17 311 F	17 602 F	22 313 F	26 088 F	21 502	25 500	26 000 F
Freshwater crustaceans nei	*Crustacea*	400 F	300 F	200 F	100 F	27	130 F	1 843
Marine crabs nei	*Brachyura*	0	0	0	1	2	2 F	2 F
Palinurid spiny lobsters nei	*Palinurus spp*	2 F	4 F	5 F	6	10	2	18
Penaeus shrimps nei	*Penaeus spp*	310 F	260 F	300 F	340 F	851	1	484
Cuttlefish,bobtail squids nei	*Sepiidae, Sepiolidae*	-	-	-	81	291	140	...
Octopuses, etc. nei	*Octopodidae*	-	-	-	-	-	88	...
Marine molluscs nei	*Mollusca*	-	-	-	-	26	-	1
Marine turtles nei	*Testudinata*	...	...	...	...	50	71	...
	Country total	*69 168*	*64 169*	*69 572*	*74 365*	*75 772*	*73 556*	*79 743 F*
Djibouti								
Freshwater fishes nei	*Osteichthyes*	0	0	0	0	0	0	0
Mullets nei	*Mugilidae*	0	0	0	0	0	0	0
Groupers nei	*Epinephelus spp*	100 F	100 F	100 F	100 F	100 F	100 F	100 F
Snappers, jobfishes nei	*Lutjanidae*	80 F	80 F	80 F	80 F	80 F	80 F	80 F
Porgies, seabreams nei	*Sparidae*	40 F	40 F	40 F	40 F	40 F	40 F	40 F
Seerfishes nei	*Scomberomorus spp*	65	61	60	60 F	60 F	60 F	60 F
Tuna-like fishes nei	*Scombroidei*	15	14	15	15 F	15 F	15 F	15 F
Carangids nei	*Carangidae*	20 F	20 F	20 F	20 F	20 F	20 F	20 F
Barracudas nei	*Sphyraena spp*	20 F	20 F	20 F	20 F	20 F	20 F	20 F
Marine fishes nei	*Osteichthyes*	10 F	15 F	15 F	15 F	15 F	15 F	15 F
Tropical spiny lobsters nei	*Panulirus spp*	0	0	0	0	0	0	0
Various squids nei	*Loliginidae, Ommastrephidae*	0	0	0	0	0	0	0
	Country total	*350 F*	*350 F*	*350 F*	*350 F*	*350 F*	*350 F*	*350 F*
Egypt								
Grass carp(=White amur)	*Ctenopharyngodon idellus*	15 343	16 553	4 233	707	12 826	17 918	19 528
Cyprinids nei	*Cyprinidae*	1 261	1 386	2 117	4 515	1 232	5 040	4 385
Nile tilapia	*Oreochromis niloticus*	125 307	130 992	128 446	112 811	131 276	145 291	138 450
Mudfish	*Clarias anguillaris*	20 370	22 433	21 618	21 511	31 506	39 535	39 208
Upsidedown catfishes	*Synodontis spp*	...	...	...	...	2 678	3 937	5 520
Nile perch	*Lates niloticus*	2 730	2 551	3 255	2 572	3 278	5 328	5 775
Freshwater fishes nei	*Osteichthyes*	55 738	63 547	95 924	83 715	34 196	37 282	37 957
European eel	*Anguilla anguilla*	537	585	501	709	2 064	1 979	1 802
Diadromous clupeoids nei	*Clupeoidei*	40	23	148	60	84	28	8
Common sole	*Solea solea*	751	1 309	1 034	1 728	3 502	3 175	2 756
Lizardfishes nei	*Synodontidae*	5 458	6 192	8 935	8 375	11 808	8 978	6 653
Mullets nei	*Mugilidae*	14 799	16 609	18 012	20 847	20 458	26 533	30 730
Groupers nei	*Epinephelus spp*	1 657	1 467	1 627	3 158	4 747	4 699	5 140
Spotted seabass	*Dicentrarchus punctatus*	426	154	171	444	196	256	564
European seabass	*Dicentrarchus labrax*	727	453	559	662	626	800	1 336
Seabasses nei	*Dicentrarchus spp*	2 397	2 233	2 738	815	4 075	3 789	3 310
Snappers, jobfishes nei	*Lutjanidae*	4 044	5 165	8 784	4 265	8 830	5 014	3 756
Threadfin breams nei	*Nemipterus spp*	...	...	...	...	2 081	4 974	3 050
Meagre	*Argyrosomus regius*	1 076	743	893	1 056	776	1 038	1 372
Emperors(=Scavengers) nei	*Lethrinidae*	425	574	577	1 040	1 147	2 696	3 513
Sargo breams nei	*Diplodus spp*	590	382	390	841	812	793	447
Saddled seabream	*Oblada melanura*	...	33	67	47	158	70	77
Red porgy	*Pagrus pagrus*	1 825	1 425	1 230	2 984	1 847	575	1 231
Gilthead seabream	*Sparus aurata*	1 228	1 087	1 225	1 955	2 478	2 312	2 480
Bogue	*Boops boops*	2 609	2 499	1 956	...	1 450	1 222	1 541
Salema	*Sarpa salpa*	0	0	-	-	-	-	-
Porgies, seabreams nei	*Sparidae*	2 428	2 175	1 282	182	4 731	3 545	2 827
Surmullets(=Red mullets) nei	*Mullus spp*	2 314	2 533	1 681	2 727	1 611	1 717	1 957
Goatfishes, red mullets nei	*Mullidae*	716	744	439	876	914	2 590	1 684
Parrotfishes nei	*Scaridae*	...	227	283	918	...	...	...
Spinefeet(=Rabbitfishes) nei	*Siganus spp*	878	501	554	680	666	2 306	1 588
Largehead hairtail	*Trichiurus lepturus*	914	679	774	723	811	1 107	1 096
Grey gurnard	*Eutrigla gurnardus*	1 264	1 158	1 067	1 159	1 078	778	611
Sardinellas nei	*Sardinella spp*	15 100	16 260	28 893	44 946	25 394	24 380	17 537
Red-eye round herring	*Etrumeus teres*	-	-	-	3 538	...	...	...
Atlantic bonito	*Sarda sarda*	985	725	724	1 442	1 128	1 072	1 416
Narrow-barred Spanish mackerel	*Scomberomorus commerson*	203	194	227	170	340	374	418
Kawakawa	*Euthynnus affinis*	318	755	841	326	344	209	313
Tuna-like fishes nei	*Scombroidei*	1 071	594	576	2 004	1 676	778	1 301
Needlefishes nei	*Tylosurus spp*	48	32	17	28	11	28	-
Silversides(=Sand smelts) nei	*Atherinidae*	6 642	5 295	5 708	4 275	4 914	8 644	9 208

E-1

Fish crustaceans, molluscs, etc — Capture production by countries or areas and species — Africa
Poissons, crustacés, mollusques, etc — Captures par pays ou zones et espèces — Afrique
Peces, crustáceos, moluscos, etc — Capturas por países o áreas y especies — Africa

English name / Nom anglais / Nombre inglés	Scientific name / Nom scientifique / Nombre científico	1996 t	1997 t	1998 t	1999 t	2000 t	2001 t	2002 t
Bluefish	*Pomatomus saltatrix*	307	147	56	153	326	468	549
Jacks, crevalles nei	*Caranx spp*	716	441	402	326	525	1 351	1 122
Chub mackerel	*Scomber japonicus*	2 042	2 392	810	378	3 561	2 747	1 614
Indian mackerel	*Rastrelliger kanagurta*	1 151	1 914	652	1 004	430	1 442	2 782
Barracudas nei	*Sphyraena spp*	1 101	1 722	1 764	1 162	2 390	1 570	1 255
Sharks, rays, skates, etc. nei	*Elasmobranchii*	1 242	1 809	1 346	1 565	1 441	2 406	2 222
Marine fishes nei	*Osteichthyes*	31 497	40 464	40 747	58 095	26 879	27 522	32 309
Freshwater prawns, shrimps nei	*Palaemonidae*	2 286	2 387	2 886	1 388	2 411	2 450	2 326
Freshwater crustaceans nei	*Crustacea*	557	838	918	1 237	2 473	2 192	2 029
Marine crabs nei	*Brachyura*	981	1 067	1 398	4 701	3 866	1 904	2 900
Natantian decapods nei	*Natantia*	3 808	5 490	5 507	8 273	7 063	5 372	6 051
Freshwater molluscs nei	*Mollusca*	530	610	625	598	916	845	1 190
Cuttlefish,bobtail squids nei	*Sepiidae, Sepiolidae*	1 764	1 784	1 389	4 974	2 546	2 919	2 588
Marine molluscs nei	*Mollusca*	233	0	198	0	1 718	4 534	3 378
Sea cucumbers nei	*Holothurioidea*	-	-	-	-	20	139	2 310
	Country total	*340 434*	*371 332*	*406 204*	*422 665*	*384 314*	*428 651*	*425 170*
Eq Guinea								
Freshwater fishes nei	*Osteichthyes*	900	850	970	1 101 F	1 076	1 000 F	1 000 F
Flatfishes nei	*Pleuronectiformes*	420 F	530 F	325	380 F	40 F	40 F	40 F
Gadiformes nei	*Gadiformes*	450 F	570 F	603	710 F	80 F	80 F	80 F
Demersal percomorphs nei	*Perciformes*	310 F	390 F	255	300 F	30 F	30 F	30 F
Clupeoids nei	*Clupeoidei*	220 F	500 F	850	1 300 F	2 000	1 900 F	1 900 F
Tuna-like fishes nei	*Scombroidei*	216	300 F	392	200 F	50 F	50 F	50 F
Chub mackerel	*Scomber japonicus*	330 F	420 F	400	470 F	50 F	50 F	50 F
Pelagic percomorphs nei	*Perciformes*	220 F	300 F	170	160 F	20 F	10 F	10 F
Sharks, rays, skates, etc. nei	*Elasmobranchii*	490 F	620 F	779	910 F	100 F	100 F	100 F
Marine fishes nei	*Osteichthyes*	789 F	725 F	700	820 F	117 F	169 F	169 F
Marine crustaceans nei	*Crustacea*	510 F	650 F	430	500 F	50 F	50 F	50 F
Marine molluscs nei	*Mollusca*	180 F	230 F	122	140 F	20 F	20 F	20 F
Marine turtles nei	*Testudinata*	5 F	5 F	9	10 F	1 F	1 F	1 F
	Country total	*5 040*	*6 090*	*6 005*	*7 001*	*3 634*	*3 500 F*	*3 500 F*
Eritrea								
Freshwater fishes nei	*Osteichthyes*	0	0	0	0	0	0	-
Milkfish	*Chanos chanos*	0	2	3	1	3	1	1
Lefteye flounders nei	*Bothidae*	-	0	1	23	119	125	69
Indian halibut	*Psettodes erumei*	8	1	0	-	-	-	-
Lizardfishes nei	*Synodontidae*	3	5	0	1 905	3 177	2 574	3 274
Sea catfishes nei	*Ariidae*	9	0	149	205	851	436	273
Mullets nei	*Mugilidae*	3	4	2	4	3	3	0
Groupers nei	*Epinephelus spp*	1	56	50	...	48	159	109
Groupers, seabasses nei	*Serranidae*	140	39	67	257	378	111	41
Bigeyes nei	*Priacanthus spp*	0	-	-	0	-	-	-
Snappers nei	*Lutjanus spp*	281	219	294	365	149	205	158
Snappers, jobfishes nei	*Lutjanidae*	43	59	69	95	392	103	64
Threadfin breams nei	*Nemipterus spp*	4	15	0	1 674	1 757	1 115	675
Ponyfishes(=Slipmouths) nei	*Leiognathidae*	-	-	-	19	38	137	70
Silver grunt	*Pomadasys argenteus*	0	0	3	...	472	199	152
Grunts, sweetlips nei	*Haemulidae (=Pomadasyidae)*	367	13	45	594	536	435	198
Emperors(=Scavengers) nei	*Lethrinidae*	767	90	104	371	443	345	164
Porgies, seabreams nei	*Sparidae*	12	0	0	15	48	11	1
Goatfishes	*Upeneus spp*	2	1	1	21	19	9	9
Parrotfishes nei	*Scaridae*	9	0	0	3	2	2	1
Angelfishes nei	*Pomacanthidae*	0	0	0	3	0	0	0
Batfishes	*Platax spp*	-	0	1	7	1	0	0
Spinefeet(=Rabbitfishes) nei	*Siganus spp*	0	0	1	1	1	0	0
Flatheads nei	*Platycephalidae*	0	0	0	21	5	11	1
Triggerfishes, durgons nei	*Balistidae*	0	0	0	3	1	0	0
Sardinellas nei	*Sardinella spp*	...	...	...	...	2	0	0
Narrow-barred Spanish mackerel	*Scomberomorus commerson*	191	200	210	250	217	280	177
Kawakawa	*Euthynnus affinis*	4	6	...	...	0	36	0
Longtail tuna	*Thunnus tonggol*	...	6	22	-	-	-	-
Indo-Pacific sailfish	*Istiophorus platypterus*	-	1	0	0	1	-	-
Tuna-like fishes nei	*Scombroidei*	30	44	111	64	42	96	136
Cobia	*Rachycentron canadum*	38	2	6	8	31	31	41
Queenfishes	*Scomberoides spp*	8	4	...	44	243	257	229
Carangids nei	*Carangidae*	818	13	184	386	2 026	1 014	730
Indian mackerel	*Rastrelliger kanagurta*	58	2	0	20	197	100	13
Barracudas nei	*Sphyraena spp*	185	21	57	150	684	2	455
Requiem sharks nei	*Carcharhinidae*	15	13	17	38	130	110	144
Guitarfishes, etc. nei	*Rhinobatidae*	0	0	-	-	0	1	7
Sharks, rays, skates, etc. nei	*Elasmobranchii*	...	6	7	6	-	-	-
Marine fishes nei	*Osteichthyes*	202	215	211	246	8	1	100
Portunus swimcrabs nei	*Portunus spp*	-	-	-	-	-	1	4
Spiny lobsters nei	*Palinuridae*	1	1	2	1	0	0	0
Penaeus shrimps nei	*Penaeus spp*	2	0	9	75	519	790	508
Cuttlefish,bobtail squids nei	*Sepiidae, Sepiolidae*	51	0	3	11	31	35	9
Various squids nei	*Loliginidae, Ommastrephidae*	0	0	0	5	38	85	19
	Country total	*3 252*	*1 038*	*1 629*	*6 891*	*12 612*	*8 820*	*7 832*
Ethiopia								
Common carp	*Cyprinus carpio*	94	27	62	71	75	74	79
Crucian carp	*Carassius carassius*	61	191	88	101	110	108	103
Rhinofishes nei	*Labeo spp*	1 994	2 007	3 168	3 621	3 451	3 387	2 303
Cyprinids nei	*Cyprinidae*	387	639	768	878	860	843	759

English name Nom anglais Nombre inglés	Scientific name Nom scientifique Nombre científico	1996 t	1997 t	1998 t	1999 t	2000 t	2001 t	2002 t
Tilapias nei	*Oreochromis (=Tilapia) spp*	5 066	6 076	7 952	9 088	7 000	6 870	4 603
Naked catfishes	*Bagrus spp*	...	...	94	107	120	119	98
North African catfish	*Clarias gariepinus*	812	1 200	1 677	1 917	4 000	3 926	4 318
Nile perch	*Lates niloticus*	270	230	191	75	65	63	37
Freshwater fishes nei	*Osteichthyes*	86	-	-	-	-	-	-
	Country total	*8 770*	*10 370*	*14 000*	*15 858*	*15 681*	*15 390*	*12 300*
Fr South Tr								
Freshwater fishes nei	*Osteichthyes*	0	0	0	0	0	0	0
Marine fishes nei	*Osteichthyes*	75 F	75 F	75 F	80 F	80 F	80 F	80 F
St.Paul rock lobster	*Jasus paulensis*	357	295	308	345	192	183	334
Octopuses, etc. nei	*Octopodidae*	5 F	5 F	5 F	...	...	...	...
	Country total	*437 F*	*375 F*	*388 F*	*425 F*	*272 F*	*263 F*	*414 F*
Gabon								
Tilapias nei	*Oreochromis (=Tilapia) spp*	3 620	4 150	3 519	4 178	4 221	3 423	3 669
Black catfishes nei	*Chrysichthys spp*	2 124	1 131	1 950	380	887	797	797
Freshwater fishes nei	*Osteichthyes*	4 251	4 636	3 500	5 814	5 726	5 922	5 121
Tonguefishes	*Cynoglossidae*	310	411	492	386	620	538	545
Flatfishes nei	*Pleuronectiformes*	75	93	65	84	242	216	172
Bonefish	*Albula vulpes*	25	21	2	6	10	...	...
Sea catfishes nei	*Ariidae*	2 780	1 565	1 290	1 351	921	1 140	1 171
Mullets nei	*Mugilidae*	271	254	233	382	365	1 061	888
Groupers nei	*Epinephelus spp*	250	94	190	136	105	40	113
Snappers nei	*Lutjanus spp*	611	480	907	941	795	765	650
Grunts, sweetlips nei	*Haemulidae (=Pomadasyidae)*	780	853	1 728	1 062	819	1 050	1 054
Bobo croaker	*Pseudotolithus elongatus*	946	769	1 584	2 079	1 784	1 555	1 238
West African croakers nei	*Pseudotolithus spp*	4 020	3 068	4 653	3 367	2 344	2 197	2 554
Pandoras nei	*Pagellus spp*	200	191	301	127	376	274	444
Dentex nei	*Dentex spp*	820	522	1 047	423	371	498	556
West African goatfish	*Pseudupeneus prayensis*	100	573	76	57	448	311	392
African sicklefish	*Drepane africana*	270	222	281	290	86	71	107
Lesser African threadfin	*Galeoides decadactylus*	3 805	3 174	3 484	2 808	2 516	3 101	2 072
Royal threadfin	*Pentanemus quinquarius*	...	...	0	292	207	389	210
Sardinellas nei	*Sardinella spp*	1 897	878	746	128	1 414	1 083	1 270
Bonga shad	*Ethmalosa fimbriata*	13 046	14 695	19 284	17 408	14 788	12 733	11 428
Seerfishes nei	*Scomberomorus spp*	79	-	85	-	-	-	265
Little tunny(=Atl.black skipj)	*Euthynnus alletteratus*	182	-	18	159	301	213	57
Skipjack tuna	*Katsuwonus pelamis*	26	...	59	76	21	101	...
Yellowfin tuna	*Thunnus albacares*	225	225	295	225	162	270	245
Bigeye tuna	*Thunnus obesus*	-	-	-	184	150	121	...
Marlins,sailfishes,etc. nei	*Istiophoridae*	523	7	-	-	-	1	-
Scads nei	*Decapterus spp*	33	20	18	76	21	101	84
Jacks, crevalles nei	*Caranx spp*	594	404	91	4	29	31	39
Mackerels nei	*Scombridae*	64	65	114	158	304	201	267
Barracudas nei	*Sphyraena spp*	1 055	883	1 238	975	976	1 636	1 379
Rays, stingrays, mantas nei	*Rajiformes*	172	173	90	197	141	88	46
Sharks, rays, skates, etc. nei	*Elasmobranchii*	1 267	626	1 933	1 338	659	375	360
Marine fishes nei	*Osteichthyes*	278	1 325	1 154	3 949	2 896	...	284
Freshwater crustaceans nei	*Crustacea*	36	44	50	6	4	9	4
Marine crabs nei	*Brachyura*	120	145	158	283	289	142	136
Palinurid spiny lobsters nei	*Palinurus spp*	103	33	85	50	32	57	121
Southern pink shrimp	*Penaeus notialis*	950	828	2 272	93	62	...	...
Penaeus shrimps nei	*Penaeus spp*	-	556	190	987	1 682	1 892	1 720
Deepwater rose shrimp	*Parapenaeus longirostris*	6	257	56	76	356	55	48
Cuttlefish,bobtail squids nei	*Sepiidae, Sepiolidae*	157	52	186	92	281	166	512
Various squids nei	*Loliginidae, Ommastrephidae*	5	2	0	1	8	10	10
Octopuses, etc. nei	*Octopodidae*	-	-	5	91	0	0	4
Marine turtles nei	*Testudinata*	37	159	180	424	51	238	843
	Country total	*46 113*	*43 584*	*53 609*	*51 143*	*47 470*	*42 871*	*40 875*
Gambia								
Tilapias nei	*Oreochromis (=Tilapia) spp*	1 248	1 135	1 073	1 070 F	1 062 F	1 077 F	1 117 F
Freshwater fishes nei	*Osteichthyes*	1 450	1 450	1 450	1 450 F	1 450 F	1 450 F	1 450 F
Tonguefishes	*Cynoglossidae*	541	307	441	450	725	2 262	586
West African ladyfish	*Elops lacerta*	...	...	...	...	...	12	26
Sea catfishes nei	*Ariidae*	158	1 234	517	540	749	950	2 238
Mullets nei	*Mugilidae*	475	278	66	70	123	69	208
Groupers nei	*Epinephelus spp*	62	53	30	30	49	63	66
Snappers nei	*Lutjanus spp*	27	150	86	90	90	126	122
Rubberlip grunt	*Plectorhinchus mediterraneus*	156	101	160	170	107	124	365
Sompat grunt	*Pomadasys jubelini*	498	350	220	230	276	423	439
Meagre	*Argyrosomus regius*	125	72	8	10	22	33	645
Law croaker	*Pseudotolithus brachygnathus*	355	225	264	270	454	856	794
Cassava croaker	*Pseudotolithus senegalensis*	182	160	109	120	58	400	620
Bobo croaker	*Pseudotolithus elongatus*	242	214	163	170	138	120	494
West African croakers nei	*Pseudotolithus spp*	473	412	327	340	629	504	49
Pandoras nei	*Pagellus spp*	76	9	12	20	12	...	8
Porgies, seabreams nei	*Sparidae*	129	...	...	-	-	-	-
African sicklefish	*Drepane africana*	151	104	23	30	60	85	186
Parrotfishes nei	*Scaridae*	...	...	...	...	13	1	17
Giant African threadfin	*Polydactylus quadrifilis*	179	110	83	90	169	141	206
Lesser African threadfin	*Galeoides decadactylus*	140	146	57	60	87	116	120
Puffers nei	*Tetraodontidae*	28	100	-	-	4	33	81
Triggerfishes, durgons nei	*Balistidae*	51	9	2	2	0	3	1
Hairtails, scabbardfishes nei	*Trichiuridae*	10	12	5	5	...	28	8

English name Nom anglais Nombre inglés	Scientific name Nom scientifique Nombre científico	1996 t	1997 t	1998 t	1999 t	2000 t	2001 t	2002 t
Sardinellas nei	*Sardinella spp*	11	86	64	70	10	81	1 711
Bonga shad	*Ethmalosa fimbriata*	22 648	21 523	21 952	22 750	20 508	18 516	22 786
Clupeoids nei	*Clupeoidei*	27	8	12	10	-	-	-
Yellowfin tuna	*Thunnus albacares*	...	...	1	1	5	1	26
Bluefish	*Pomatomus saltatrix*	31	8	75	80	35	70	60
Cobia	*Rachycentron canadum*	0	6	0	0	-	-	-
Jack and horse mackerels nei	*Trachurus spp*	312	133	119	130	175	246	417
Jacks, crevalles nei	*Caranx spp*	174	124	147	160	137	288	899
Pompanos nei	*Trachinotus spp*	20	7	1	2	1	22	...
Barracudas nei	*Sphyraena spp*	355	144	116	120	284	631	2 573
Sharks, rays, skates, etc. nei	*Elasmobranchii*	415	3 223	606	630	720	3 982	6 128
Marine fishes nei	*Osteichthyes*	18	59	-	-	-	6	204
Marine crabs nei	*Brachyura*	...	...	...	...	6	2	...
Palinurid spiny lobsters nei	*Palinurus spp*	84	37	86	80	130	75	98
Southern pink shrimp	*Penaeus notialis*	339	...	399	400	301	211	324
Gastropods nei	*Gastropoda*	227	128	230	250	5	20	1
Cuttlefish,bobtail squids nei	*Sepiidae, Sepiolidae*	184	137	98	100	422	1 499	696
Octopuses, etc. nei	*Octopodidae*	...	...	...	...	...	1	-
	Country total	*31 601*	*32 254*	*29 002*	*30 000*	*29 016*	*34 527*	*45 769*
Ghana								
Freshwater fishes nei	*Osteichthyes*	73 580	70 000	74 500	74 500	74 500	74 500	74 500
West African ilisha	*Ilisha africana*	7 343	4 305	3 632	3 262	3 341	3 600	5 202
Tonguefishes	*Cynoglossidae*	295	339	347	284	394	227	144
Benguela hake	*Merluccius polli*	1	0	34	3	0	-	-
Sea catfishes nei	*Ariidae*	2	24	6	0	1	0	6
Groupers nei	*Epinephelus spp*	426	478	1 361	181	94	138	233
Snappers nei	*Lutjanus spp*	328	181	294	163	491	447	774
Sompat grunt	*Pomadasys jubelini*	...	434	303	87	95	694	343
Bigeye grunt	*Brachydeuterus auritus*	13 552	19 816	12 059	12 724	10 032	13 845	9 267
Grunts, sweetlips nei	*Haemulidae (=Pomadasyidae)*	1 191	43	108	389	55	153	197
West African croakers nei	*Pseudotolithus spp*	1 128	1 995	962	937	739	1 070	1 067
Red pandora	*Pagellus bellottii*	7 541	7 933	10 029	13 265	2 916	3 206	3 132
Angolan dentex	*Dentex angolensis*	489	490	1 416	1 767	504	838	990
Congo dentex	*Dentex congoensis*	102	119	392	1 272	350	571	445
Dentex nei	*Dentex spp*	1 584	1 084	1 278	1 874	892	675	965
Pargo breams nei	*Pagrus spp*	1 448	1 347	1 682	...	...	2 189	2 624
Porgies, seabreams nei	*Sparidae*	220	129	316	3 986	1 238	532	969
West African goatfish	*Pseudupeneus prayensis*	586	1 035	553	247	39	285	278
African sicklefish	*Drepane africana*	6	46	4	24	2	8	30
Lesser African threadfin	*Galeoides decadactylus*	3 146	1 477	774	586	1 947	2 892	2 204
Triggerfishes, durgons nei	*Balistidae*	17	-	1	-	2	2	12
Largehead hairtail	*Trichiurus lepturus*	2 543	2 866	2 047	1 267	1 664	1 849	3 154
Gurnards, searobins nei	*Triglidae*	0	76	...	53	0	37	...
Demersal percomorphs nei	*Perciformes*	0	0	0	-	-	-	-
Round sardinella	*Sardinella aurita*	118 408	49 394	55 965	57 170	102 043	67 321	64 300
Madeiran sardinella	*Sardinella maderensis*	13 619	14 183	15 468	12 105	14 970	15 906	13 755
Sardinellas nei	*Sardinella spp*	20 070	35 985	33 264	19 664	18 358	8 345	1 553
Bonga shad	*Ethmalosa fimbriata*	1 197	9 762	158	766	963	282	295
European anchovy	*Engraulis encrasicolus*	98 341	82 724	44 644	32 107	83 501	68 175	57 639
Clupeoids nei	*Clupeoidei*	477	1 060	162	132	3 789	6 144	2 126
Little tunny(=Atl.black skipj)	*Euthynnus alletteratus*	113	2 025	359	306	707	730	4 768
Skipjack tuna	*Katsuwonus pelamis*	19 602	27 667	34 150	43 460	29 950	43 340	31 887
Yellowfin tuna	*Thunnus albacares*	11 720	16 504	17 807	28 328	17 010	30 642	23 499
Bigeye tuna	*Thunnus obesus*	5 805	7 431	13 252	11 460	5 586	14 095	5 893
Atlantic sailfish	*Istiophorus albicans*	303	196	351	305	275	568	529
Blue marlin	*Makaira nigricans*	422	491	447	624	639	1 295	999
Atlantic white marlin	*Tetrapturus albidus*	1	3	7	6	8	21	2
Swordfish	*Xiphias gladius*	140	44	106	121	117	531	372
Chilean jack mackerel	*Trachurus murphyi*	-	-	-	-	2 472	1 157	-
Cunene horse mackerel	*Trachurus trecae*	7 714	6 962	11 690	9 964	572	1 540	...
Jack and horse mackerels nei	*Trachurus spp*	3 201	10 512	4 892	1 904	2 183	2 109	504
Scads nei	*Decapterus spp*	1 462	1 654	2 989	3	263	2 466	2 944
Crevalle jack	*Caranx hippos*	2 884	2 207	2 891	3 906	79	2 525	2 872
False scad	*Caranx rhonchus*	3 301	3 337	2 753	2 275	3 800	147	1 605
Jacks, crevalles nei	*Caranx spp*	13	11	1	-	1	-	-
African moonfish	*Selene dorsalis*	907	712	381	469	738	1 100	690
Atlantic bumper	*Chloroscombrus chrysurus*	2 466	3 847	7 264	7 611	6 861	6 418	5 670
Chub mackerel	*Scomber japonicus*	15 590	19 883	30 160	15 482	29 613	15 193	7 018
Barracudas nei	*Sphyraena spp*	1 763	1 684	872	2 591	759	1 156	1 193
Rays, stingrays, mantas nei	*Rajiformes*	261	185	172	869	231	814	969
Sharks, rays, skates, etc. nei	*Elasmobranchii*	1 106	709	1 764	3 998	1 670	2 092	2 451
Marine fishes nei	*Osteichthyes*	25 128	27 896	43 399	115 933	22 248	40 993	26 234
Marine crabs nei	*Brachyura*	399	576	271	145	74	155	156
Tropical spiny lobsters nei	*Panulirus spp*	134	203	65	...	39	342	97
Natantian decapods nei	*Natantia*	1 554	1 602	1 448	87	1 446	1 361	973
Cuttlefish,bobtail squids nei	*Sepiidae, Sepiolidae*	2 967	3 355	3 288	4 095	1 805	2 866	3 649
Octopuses, etc. nei	*Octopodidae*	137	67	103	19	4	94	...
	Country total	*476 733*	*447 088*	*442 641*	*492 776*	*452 070*	*447 681*	*371 178*
Guinea								
Freshwater fishes nei	*Osteichthyes*	2 780	3 600	4 000	4 000	4 000	4 000	4 000 F
Flatfishes nei	*Pleuronectiformes*	254	256	179	148	1 032	914	900 F
Hakes nei	*Merluccius spp*	-	-	-	-	-	3	...
Sea catfishes nei	*Ariidae*	3 589	2 462	2 610	2 378	4 593	6 232	6 200 F
Mullets nei	*Mugilidae*	1 901	1 244	1 534	1 600	1 894	...	1 000 F
Grunts, sweetlips nei	*Haemulidae (=Pomadasyidae)*	193	198	185	1 598	430	363	360 F

English name Nom anglais Nombre inglés	Scientific name Nom scientifique Nombre científico	1996 t	1997 t	1998 t	1999 t	2000 t	2001 t	2002 t
Bobo croaker	*Pseudotolithus elongatus*	3 386	2 751	2 781	2 142	4 015	5 171	5 100 F
West African croakers nei	*Pseudotolithus spp*	2 856	2 206	2 570	1 796	3 423	3 644	3 600 F
Atlantic emperor	*Lethrinus atlanticus*	231	334	380	...	...	...	...
Porgies, seabreams nei	*Sparidae*	3 814	3 649	3 019	3 265	1 838	4 562	4 500 F
African sicklefish	*Drepane africana*	324	149	...	...	...	...	...
Lesser African threadfin	*Galeoides decadactylus*	338	204	43	88	123	62	60 F
Royal threadfin	*Pentanemus quinquarius*	435	233	...	...	...	...	...
Largehead hairtail	*Trichiurus lepturus*	401	409	400	400	418	...	400 F
Sardinellas nei	*Sardinella spp*	6 107	4 176	6 292	8 909	13 288	8 027	8 000 F
Bonga shad	*Ethmalosa fimbriata*	26 051	29 529	27 852	33 780	29 015	38 454	38 000 F
Jack and horse mackerels nei	*Trachurus spp*	3 576	1 546	7 109	508	6 084	5 170	5 100 F
Carangids nei	*Carangidae*	311	257	392	326	764	2 559	2 500 F
Chub mackerel	*Scomber japonicus*	2 043	1 006	849	...	...	...	...
Barracudas nei	*Sphyraena spp*	309	431	400	300	226	1 043	1 000 F
Sharks, rays, skates, etc. nei	*Elasmobranchii*	506	505	700	800	969	681	680 F
Marine fishes nei	*Osteichthyes*	3 695	7 089	7 631	23 580	18 646	23 308	22 600 F
Penaeus shrimps nei	*Penaeus spp*	196	98	216	396	526	701	...
Cuttlefish,bobtail squids nei	*Sepiidae, Sepiolidae*	41	95	386	181	124	236	...
Various squids nei	*Loliginidae, Ommastrephidae*	1	2	53	27	9	...	...
Octopuses, etc. nei	*Octopodidae*	22	12	183	1 092	96	97	...
	Country total	*63 360*	*62 441*	*69 764*	*87 314*	*91 513*	*105 227*	*104 000 F*
GuineaBissau								
Freshwater fishes nei	*Osteichthyes*	250 F	250 F	200 F	200 F	200 F	200 F	200 F
Flatfishes nei	*Pleuronectiformes*	64	70 F	60 F	50 F	50 F	50 F	50 F
Sea catfishes nei	*Ariidae*	195	200 F	170 F	140 F	140 F	140 F	140 F
Mullets nei	*Mugilidae*	3 050 F	3 200 F	2 650 F	2 200 F	2 200 F	2 200 F	2 200 F
Sompat grunt	*Pomadasys jubelini*	153	160 F	130	100 F	100 F	100 F	100 F
Bigeye grunt	*Brachydeuterus auritus*	14 F	15 F	10 F	10 F	10 F	10 F	10 F
Meagre	*Argyrosomus regius*	482	500 F	430 F	350 F	350 F	350 F	350 F
Croakers, drums nei	*Sciaenidae*	1 020 F	1 040 F	850 F	650 F	650 F	650 F	650 F
Porgies, seabreams nei	*Sparidae*	14 F	15 F	10 F	10 F	10 F	10 F	10 F
African sicklefish	*Drepane africana*	172	180 F	150 F	120 F	120 F	120 F	120 F
Giant African threadfin	*Polydactylus quadrifilis*	55	60 F	50 F	40 F	40 F	40 F	40 F
Lesser African threadfin	*Galeoides decadactylus*	108	110 F	100 F	90 F	90 F	90 F	90 F
Largehead hairtail	*Trichiurus lepturus*	17	20 F	15 F	10 F	10 F	10 F	10 F
Tuna-like fishes nei	*Scombroidei*	6 F	6 F	5 F	4 F	4 F	4 F	4 F
Bluefish	*Pomatomus saltatrix*	3 F	3 F	3 F	3 F	3 F	3 F	3 F
Jacks, crevalles nei	*Caranx spp*	100 F	100 F	80 F	70 F	70 F	70 F	70 F
Mackerels nei	*Scombridae*	14 F	15 F	14 F	10 F	10 F	10 F	10 F
Scalloped hammerhead	*Sphyrna lewini*	12	10 F	10 F	10 F	10 F	10 F	10 F
Marine crabs nei	*Brachyura*	30 F	30 F	25 F	20 F	20 F	20 F	20 F
Deepwater rose shrimp	*Parapenaeus longirostris*	124	148	120 F	100 F	100 F	100 F	100 F
Natantian decapods nei	*Natantia*	1 100 F	1 100 F	900 F	800 F	800 F	800 F	800 F
Cuttlefish,bobtail squids nei	*Sepiidae, Sepiolidae*	2	2 F	2 F	2 F	2 F	2 F	2 F
Common octopus	*Octopus vulgaris*	1	1 F	1 F	1 F	1 F	1 F	1 F
Marine molluscs nei	*Mollusca*	14 F	15 F	15 F	10 F	10 F	10 F	10 F
	Country total	*7 000 F*	*7 250 F*	*6 000 F*	*5 000 F*	*5 000 F*	*5 000 F*	*5 000 F*
Kenya								
Common carp	*Cyprinus carpio*	334	216	48	52	47	49	50
Rhinofishes nei	*Labeo spp*	1 462	355	207	219	221	222	203
Silver cyprinid	*Rastrineobola argentea*	49 670	40 315	42 336	48 816	49 618	41 384	35 455
Cyprinids nei	*Cyprinidae*	0	0	0	0	0	0	0
Nile tilapia	*Oreochromis niloticus*	10 765	13 953	14 652	17 524	19 347	7 292	16 252
Tilapias nei	*Oreochromis (=Tilapia) spp*	7 635	19 329	19 732	18 203	19 853	19 121	17 418
Mouthbrooding cichlids	*Haplochromis spp*	3 914	2 453	2 577	2 731	2 699	1 195	1 259
African lungfishes	*Protopterus spp*	164	1 704	1 717	1 600	1 608	1 822	1 707
Naked catfishes	*Bagrus spp*	157	158	107	147	161	178	183
Torpedo-shaped catfishes nei	*Clarias spp*	339	1 724	1 532	1 570	1 592	1 611	1 581
Upsidedown catfishes	*Synodontis spp*	24	408	418	506	443	48	142
Nile perch	*Lates niloticus*	97 145	73 555	76 663	103 014	109 815	78 534	58 432
Freshwater fishes nei	*Osteichthyes*	3 059	745	5 960	4 221	4 886	5 237	4 984
Salmonoids nei	*Salmonoidei*	11	16	24	29	31	67	112
Mullets nei	*Mugilidae*	153	120	107	146	181	199	164
Snappers, jobfishes nei	*Lutjanidae*	147	144	106	151	120	177	177
Grunts, sweetlips nei	*Haemulidae (=Pomadasyidae)*	65	72	55	78	63	85	90
Emperors(=Scavengers) nei	*Lethrinidae*	433	361	412	358	334	466	414
Spinefeet(=Rabbitfishes) nei	*Siganus spp*	404	347	356	304	299	403	469
Demersal percomorphs nei	*Perciformes*	1 247	1 188	1 196	1 366	1 224	2 060	1 450
Clupeoids nei	*Clupeoidei*	217	189	155	167	119	164	101
Narrow-barred Spanish mackerel	*Scomberomorus commerson*	93	69	139	122	94	136	150
Skipjack tuna	*Katsuwonus pelamis*	108	114	98	109	86	183	116
Marlins,sailfishes,etc. nei	*Istiophoridae*	73	53	38	82	80	78	55
Amberjacks nei	*Seriola spp*	79	63	60	78	71	92	31
Carangids nei	*Carangidae*	82	111	86	101	85	119	144
Barracudas nei	*Sphyraena spp*	54	54	71	108	83	99	88
Pelagic percomorphs nei	*Perciformes*	358	351	611	528	378	1 003	982
Sharks, rays, skates, etc. nei	*Elasmobranchii*	191	140	134	131	115	175	134
Marine fishes nei	*Osteichthyes*	1 187	1 006	1 675	1 774	474	667	874
Red swamp crawfish	*Procambarus clarkii*	13	24	19	21	22	3	14
Marine crabs nei	*Brachyura*	112	100	117	135	166	134	122
Tropical spiny lobsters nei	*Panulirus spp*	177	136	39	52	52	76	119
Natantian decapods nei	*Natantia*	378	491	774	513	458	690	581
Marine crustaceans nei	*Crustacea*	185	223	139	104	101	136	108
Cupped oysters nei	*Crassostrea spp*	32	16	9	8	2	1	1
Various squids nei	*Loliginidae, Ommastrephidae*	389	317	30	35	42	78	74

English name Nom anglais Nombre inglés	Scientific name Nom scientifique Nombre científico	1996 t	1997 t	1998 t	1999 t	2000 t	2001 t	2002 t
Octopuses, etc. nei	*Octopodidae*	117	393	155	169	106	154	208
Sea cucumbers nei	*Holothurioidea*	15	41	38	15	30	13	68
	Country total	*180 988*	*161 054*	*172 592*	*205 287*	*215 106*	*164 151*	*144 512*
Lesotho								
Common carp	*Cyprinus carpio*	16 F	18 F	18 F	18 F	18 F	8	8 F
North African catfish	*Clarias gariepinus*	2 F	2 F	2 F	2 F	2 F	2	2 F
Freshwater fishes nei	*Osteichthyes*	10 F	10 F	10 F	10 F	12 F	14	14 F
	Country total	*28 F*	*30 F*	*30 F*	*30*	*32*	*24*	*24 F*
Liberia								
Freshwater fishes nei	*Osteichthyes*	4 000	4 000	4 000	4 000	4 000	4 000	4 000 F
West African ilisha	*Ilisha africana*	124	26	63	242	110	198	200 F
Common sole	*Solea solea*	48	150	149	217	129	206	210 F
Benguela hake	*Merluccius polli*	38	-	-	-	-	-	-
Bonefish	*Albula vulpes*	-	-	21	104	27	6	10 F
Sea catfishes nei	*Ariidae*	77	-	4	31	21	210	210 F
Mullets nei	*Mugilidae*	-	22	18	63	85	...	40 F
Groupers nei	*Epinephelus spp*	-	5	110	71	25	44	45 F
Snappers nei	*Lutjanus spp*	9	17	27	339	132	201	210 F
Grunts, sweetlips nei	*Haemulidae (=Pomadasyidae)*	105	99	85	216	102	180	180 F
Cassava croaker	*Pseudotolithus senegalensis*	...	11	...	...	...	...	...
Bobo croaker	*Pseudotolithus elongatus*	109	27	...	...	...	...	...
West African croakers nei	*Pseudotolithus spp*	364	510	433	1 025	327	210	220 F
Large-eye dentex	*Dentex macrophthalmus*	9	39	...	...	...	...	...
Dentex nei	*Dentex spp*	327	346	974	936	588	671	680 F
Black seabream	*Spondyliosoma cantharus*	-	-	-	-	-	10	10 F
West African goatfish	*Pseudupeneus prayensis*	-	46	-	-	-	-	-
African sicklefish	*Drepane africana*	15	-	70	256	94	104	110 F
Royal threadfin	*Pentanemus quinquarius*	36	155	138	118	92	102	110 F
Triggerfishes, durgons nei	*Balistidae*	-	5	2	-	-	-	-
Conger eels, etc. nei	*Congridae*	41	117	85	128	49	76	80 F
Bearded brotula	*Brotula barbata*	4	10	66	48	52	...	40 F
Largehead hairtail	*Trichiurus lepturus*	7	10	9	33	34	169	160 F
Sardinellas nei	*Sardinella spp*	199	485	620	1 112	887	1 358	1 350 F
Bonga shad	*Ethmalosa fimbriata*	70	17	33	123	37	123	120 F
Clupeoids nei	*Clupeoidei*	189	147	383	386	318	208	210 F
Skipjack tuna	*Katsuwonus pelamis*	410	-	-	-	-	-	-
Albacore	*Thunnus alalunga*	41	-	-	-	-	-	-
Yellowfin tuna	*Thunnus albacares*	180	185	310	369	227	166	170 F
Bigeye tuna	*Thunnus obesus*	415	340	108	112	201	175	175 F
Blue marlin	*Makaira nigricans*	114	122	59	37	187	131	130 F
Marlins,sailfishes,etc. nei	*Istiophoridae*	145	71	781	513	683	163	165 F
Swordfish	*Xiphias gladius*	112	543	21	39	42	34	35 F
Tuna-like fishes nei	*Scombroidei*	-	-	3	2	14	8	10 F
Needlefishes, etc. nei	*Belonidae*	-	-	54	90	30	31	35 F
Halfbeaks nei	*Hemiramphus spp*	-	-	77	85	96	97	100 F
Flyingfishes nei	*Exocoetidae*	69	8	10	-	38	19	20 F
False scad	*Caranx rhonchus*	-	7	111	435	135	134	140 F
Jacks, crevalles nei	*Caranx spp*	62	30	178	394	235	271	280 F
Leerfish	*Lichia amia*	2	0	-	-	-	-	-
Common dolphinfish	*Coryphaena hippurus*	...	31	20	48	45	22	25 F
Suckerfishes, remoras nei	*Echeneidae*	-	-	3	1	12	...	...
Chub mackerel	*Scomber japonicus*	24	8	45	218	238	34	40 F
Blue butterfish	*Stromateus fiatola*	143	236	248	573	181	183	190 F
Barracudas nei	*Sphyraena spp*	35	-	67	343	174	196	200 F
Thresher	*Alopias vulpinus*	...	...	...	151	146	...	...
Mako sharks	*Isurus spp*	...	...	...	...	116	...	...
Blue shark	*Prionace glauca*	...	...	...	76	70	...	...
Silky shark	*Carcharhinus falciformis*	...	...	...	110	99	...	...
Hammerhead sharks, etc. nei	*Sphyrnidae*	...	...	...	127	152	...	...
Guitarfishes, etc. nei	*Rhinobatidae*	...	...	54	175	16	...	...
Sawfishes	*Pristidae*	...	48	...	39	42	...	...
Mantas	*Mobulidae*	...	...	342	802	931	106	110 F
Rays, stingrays, mantas nei	*Rajiformes*	12	38	50	119	103	29	30 F
Sharks, rays, skates, etc. nei	*Elasmobranchii*	207	386	210	-	-	512	520 F
Marine fishes nei	*Osteichthyes*	362	157	539	797	281	640	650 F
Marine crabs nei	*Brachyura*	-	35	38	32	27	122	125 F
Palinurid spiny lobsters nei	*Palinurus spp*	-	8	26	4	41	36	40 F
Deepwater rose shrimp	*Parapenaeus longirostris*	...	8	...	...	...	...	...
Natantian decapods nei	*Natantia*	28	73	113	302	25	31	40 F
Cuttlefish,bobtail squids nei	*Sepiidae, Sepiolidae*	-	-	12	8	14	29	30 F
Octopuses, etc. nei	*Octopodidae*	176	2	61	23	16	41	45 F
	Country total	*8 308*	*8 580*	*10 830*	*15 472*	*11 726*	*11 286*	*11 500 F*
Libya								
Freshwater fishes nei	*Osteichthyes*	0	0	0	0	0	0	0
Groupers nei	*Epinephelus spp*	4 000 F	4 000 F	4 000 F	4 000 F	4 000 F	4 000 F	4 000 F
Bogue	*Boops boops*	2 500 F	2 500 F	2 500 F	2 500 F	2 500 F	2 500 F	2 500 F
Porgies, seabreams nei	*Sparidae*	4 000 F	4 000 F	4 000 F	4 000 F	4 000 F	4 000 F	4 000 F
Surmullet	*Mullus surmuletus*	4 000 F	4 000 F	4 000 F	4 000 F	4 000 F	4 000 F	4 000 F
Sardinellas nei	*Sardinella spp*	7 000 F	7 000 F	7 000 F	7 000 F	7 000 F	7 000 F	7 000 F
Little tunny(=Atl.black skipj)	*Euthynnus alletteratus*	-	45	52	-	5	4	4
Atlantic bluefin tuna	*Thunnus thynnus*	1 388	1 029	1 331	1 195	1 550	1 940	...
Yellowfin tuna	*Thunnus albacares*	-	-	-	-	-	208	73
Bigeye tuna	*Thunnus obesus*	400	400	400	400	400	31	593

English name Nom anglais Nombre inglés	Scientific name Nom scientifique Nombre científico	1996 t	1997 t	1998 t	1999 t	2000 t	2001 t	2002 t
Swordfish	*Xiphias gladius*	-	-	11	...	8	6	...
Jack and horse mackerels nei	*Trachurus spp*	3 000 F	3 000 F	3 000 F	3 000 F	3 000 F	3 000 F	3 000 F
Scomber mackerels nei	*Scomber spp*	3 000 F	3 000 F	3 000 F	3 000 F	3 000 F	3 000 F	3 000 F
Marine fishes nei	*Osteichthyes*	3 688 F	2 903 F	3 617 F	3 755 F	3 924 F	3 550 F	5 496 F
	Country total	*32 976 F*	*31 877 F*	*32 911 F*	*32 850 F*	*33 387 F*	*33 239 F*	*33 666 F*
Madagascar								
Cyprinids nei	*Cyprinidae*	4 000	4 000	4 000	4 000	4 000	4 000	4 000
Cichlids nei	*Cichlidae*	21 500	21 500	21 500	21 500	21 500	21 500	21 500
Freshwater fishes nei	*Osteichthyes*	4 500	4 500	4 500	4 500	4 500	4 500	4 500
Narrow-barred Spanish mackerel	*Scomberomorus commerson*	10 000	10 000	12 000	12 000	12 000	12 000	12 000
Marine fishes nei	*Osteichthyes*	56 365	57 996	68 688	73 417	74 107	77 601	82 731
Marine crabs nei	*Brachyura*	1 000	1 000	1 500	868	1 030	1 347	1 428
Tropical spiny lobsters nei	*Panulirus spp*	390	390	341	338	329	359	402
Natantian decapods nei	*Natantia*	10 470	10 755	11 470	10 507	12 127	11 776	13 223
Cephalopods nei	*Cephalopoda*	500	500	550	600	600	600	600
Marine molluscs nei	*Mollusca*	350	350	400	400	400	400	400
Sea cucumbers nei	*Holothurioidea*	5 400	5 400	1 446	1 500	1 500	1 500	500
	Country total	*114 475*	*116 391*	*126 395*	*129 630*	*132 093*	*135 583*	*141 284*
Malawi								
Cyprinids nei	*Cyprinidae*	505	522	299	8 302	7 500 F	6 291	6 401
Tilapias nei	*Oreochromis (=Tilapia) spp*	5 080	4 472	5 104	6 808	6 200 F	5 154	5 244
Cichlids nei	*Cichlidae*	...	...	...	18 841	19 000 F	20 533	20 892
Torpedo-shaped catfishes nei	*Clarias spp*	4 556	...	3 935	8 471	7 600 F	6 367	6 478
Freshwater fishes nei	*Osteichthyes*	53 428	51 346	31 773	2 970	2 700 F	2 274	2 314
	Country total	*63 569*	*56 340*	*41 111*	*45 392*	*43 000 F*	*40 619*	*41 329*
Mali								
Nile tilapia	*Oreochromis niloticus*	30 450	1 719	26 675	26 821	32 961	30 000 F	30 000 F
Bonytongues nei	*Heterotis spp*	453	329	397	399	1 099	1 000 F	1 000 F
Elephantsnout fishes nei	*Mormyridae*	...	...	...	...	7 691	7 000 F	7 000 F
Characins nei	*Characidae*	9 504	6 969	8 316	8 362	5 493	5 000 F	5 000 F
Black catfishes nei	*Chrysichthys spp*	4 172	3 982	3 650	3 670	4 395	4 000 F	4 000 F
North African catfish	*Clarias gariepinus*	17 129	14 933	15 009	15 091	27 468	25 000 F	25 000 F
Upsidedown catfishes	*Synodontis spp*	4 657	5 973	4 075	4 097	3 296	3 000 F	3 000 F
Nile perch	*Lates niloticus*	5 712	4 978	5 024	5 052	6 592	6 000 F	6 000 F
Freshwater fishes nei	*Osteichthyes*	39 833	60 667	34 854	35 044	20 875	19 000 F	19 000 F
	Country total	*111 910*	*99 550*	*98 000*	*98 536*	*109 870*	*100 000 F*	*100 000 F*
Mauritania								
Freshwater fishes nei	*Osteichthyes*	5 000 F	5 000 F	5 000 F	5 000 F	5 000 F	5 000 F	5 000 F
Soles nei	*Soleidae*	760 F	1 120 F	830 F	449 F	444 F	...	...
Flatfishes nei	*Pleuronectiformes*	580 F	700 F	600 F	1 751 F	1 756 F	2 200 F	2 200 F
Hakes nei	*Merluccius spp*	150 F	110 F	818 F	940	1 558	1 324	817
Sea catfishes nei	*Ariidae*	...	...	...	750 F	750 F	750 F	750 F
Mullets nei	*Mugilidae*	...	...	...	2 000 F	2 000 F	2 000 F	2 000 F
White grouper	*Epinephelus aeneus*	340 F	450 F	390 F	450 F	450 F	450 F	450 F
Groupers nei	*Epinephelus spp*	180 F	240 F	220 F	300 F	300 F	300 F	300 F
Spotted seabass	*Dicentrarchus punctatus*	160 F	130 F	190 F	100 F	45 F	...	...
Rubberlip grunt	*Plectorhinchus mediterraneus*	20 F	20 F	10 F	...	9 F	...	...
Grunts, sweetlips nei	*Haemulidae (=Pomadasyidae)*	10 F	10 F	10 F	...	2 F	...	...
Canary drum (=Baardman)	*Umbrina canariensis*	10 F	10 F	10 F	6 F	2 F	...	...
Meagre	*Argyrosomus regius*	480 F	580 F	510 F	600 F	600 F	600 F	600 F
West African croakers nei	*Pseudotolithus spp*	10 F	10 F	96 F	...	9 F	...	...
Croakers, drums nei	*Sciaenidae*	200 F	200 F	200 F	300 F	300 F	300 F	300 F
Sargo breams nei	*Diplodus spp*	20 F	20 F	70 F	99	145	278	286
Porgies, seabreams nei	*Sparidae*	490 F	520 F	870 F	900 F	900 F	900 F	900 F
Surmullets(=Red mullets) nei	*Mullus spp*	30 F	30 F	141 F	150 F	150 F	150 F	150 F
John dory	*Zeus faber*	5 F	10 F	20 F	...	...	...	...
Largehead hairtail	*Trichiurus lepturus*	160 F	110 F	-	...	...	...	...
Scorpionfishes nei	*Scorpaenidae*	5 F	5 F	0	1 F	0	...	...
Gurnards, searobins nei	*Triglidae*	10 F	0	70 F	...	...	...	...
Blackbellied angler	*Lophius budegassa*	20 F	20 F	20 F	20	45	28	36
Sardinellas nei	*Sardinella spp*	5 440 F	4 870 F	1 060 F	4 000 F	4 000 F	4 000 F	4 000 F
Bonga shad	*Ethmalosa fimbriata*	...	...	...	...	2 F	...	...
Clupeoids nei	*Clupeoidei*	-	-	-	-	-	5	1
West African Spanish mackerel	*Scomberomorus tritor*	...	...	...	...	12 F	...	...
Tuna-like fishes nei	*Scombroidei*	2 479	2 170	1 304	...	...	...	...
Bluefish	*Pomatomus saltatrix*	...	...	...	...	1 F	...	...
Jack and horse mackerels nei	*Trachurus spp*	1 220 F	1 060 F	940 F	...	...	75	20
Carangids nei	*Carangidae*	100 F	100 F	100 F	100 F	112 F	...	...
Chub mackerel	*Scomber japonicus*	110 F	80 F	150 F	...	...	...	...
Rays, stingrays, mantas nei	*Rajiformes*	10 F	20 F	180 F	350 F	350 F	350 F	350 F
Sharks, rays, skates, etc. nei	*Elasmobranchii*	10 F	10 F	350 F	500 F	500 F	500 F	500 F
Marine fishes nei	*Osteichthyes*	18 595 F	19 711 F	22 179 F	37 674 F	39 997 F	43 070 F	42 918 F
Geryons nei	*Geryon spp*	...	...	...	...	6	16	5
Palinurid spiny lobsters nei	*Palinurus spp*	50 F	110 F	100 F	80 F	82 F	83 F	81 F
Natantian decapods nei	*Natantia*	160 F	190 F	270 F	222	779	1 546	1 226
Marine crustaceans nei	*Crustacea*	-	-	30 F	9	6	20	5
Cuttlefish,bobtail squids nei	*Sepiidae, Sepiolidae*	4 510 F	5 240 F	5 000 F	4 129 F	4 344 F	4 981 F	2 342 F
Various squids nei	*Loliginidae, Ommastrephidae*	230 F	280 F	1 472 F	2 359 F	2 354 F	2 223 F	1 206 F
Octopuses, etc. nei	*Octopodidae*	18 770 F	14 620 F	18 420 F	12 758 F	13 709 F	13 576 F	12 200 F
Cephalopods nei	*Cephalopoda*	-	-	10 F	16	129	150	236
Marine molluscs nei	*Mollusca*	-	-	20 F	13	1	6	23

English name Nom anglais Nombre inglés	Scientific name Nom scientifique Nombre científico	1996 t	1997 t	1998 t	1999 t	2000 t	2001 t	2002 t
	Country total	*60 324 F*	*57 756 F*	*61 660 F*	*76 026 F*	*80 849 F*	*84 881 F*	*78 902 F*
Mauritius								
Tilapias nei	*Oreochromis (=Tilapia) spp*	0	0	0	0	0	0	0
Unicorn cod	*Bregmaceros mcclellandi*	312	367	336	285	347	340	271
Mullets nei	*Mugilidae*	121	103	111	120	76	92	80
Groupers, seabasses nei	*Serranidae*	905	933	931	826	863	938	885
Snappers, jobfishes nei	*Lutjanidae*	...	...	...	...	...	2 184	2 113
Emperors(=Scavengers) nei	*Lethrinidae*	5 137	5 018	4 981	4 598	4 698	4 008	4 078
Goatfishes	*Upeneus spp*	512	479	436	509	541	556	501
Percoids nei	*Percoidei*	183	158	153	165	181	152	135
Spinefeet(=Rabbitfishes) nei	*Siganus spp*	514	430	494	461	448	450	365
Clupeoids nei	*Clupeoidei*	0	-	-	-	-	-	-
Skipjack tuna	*Katsuwonus pelamis*	1 898	3 055	1 685	2 361	305	8	8
Albacore	*Thunnus alalunga*	2	7	15	12	...	18	8
Yellowfin tuna	*Thunnus albacares*	713	1 095	1 443	742	226	125	110
Bigeye tuna	*Thunnus obesus*	271	546	260	250	37	5	3
Black marlin	*Makaira indica*	...	...	...	...	...	2	2
Striped marlin	*Tetrapturus audax*	...	...	...	...	...	1	2
Marlins,sailfishes,etc. nei	*Istiophoridae*	190	639	295	287	287	287	287
Swordfish	*Xiphias gladius*	...	...	...	...	...	37	189
Tuna-like fishes nei	*Scombroidei*	70	199	44	681	726	745	745
Carangids nei	*Carangidae*	43	58	46	33	53	51	37
Mackerels nei	*Scombridae*	12	10	11	11	13	13	11
Rays, stingrays, mantas nei	*Rajiformes*	2	2	2	2	2	2	2
Sharks, rays, skates, etc. nei	*Elasmobranchii*	17	58	9	9	25	12	48
Marine fishes nei	*Osteichthyes*	589	525	502	515	441	568	442
Indo-Pacific swamp crab	*Scylla serrata*	23	19	23	21	25	24	22
Tropical spiny lobsters nei	*Panulirus spp*	14	17	17	17	17	19	26
Natantian decapods nei	*Natantia*	0	1	0	1	1	1	1
Octopuses, etc. nei	*Octopodidae*	341	306	299	299	303	347	335
	Country total	*11 869*	*14 025*	*12 093*	*12 205*	*9 615*	*10 985*	*10 706*
Mayotte								
Wahoo	*Acanthocybium solandri*	...	...	1	2	...	...	...
Skipjack tuna	*Katsuwonus pelamis*	...	...	...	...	...	3 834	188
Albacore	*Thunnus alalunga*	...	...	113	86	...	13	69
Yellowfin tuna	*Thunnus albacares*	...	...	194	...	...	3 120	1 313
Bigeye tuna	*Thunnus obesus*	...	...	14	109	...	464	109
Indo-Pacific sailfish	*Istiophorus platypterus*	...	...	4	1	...	...	...
Marlins,sailfishes,etc. nei	*Istiophoridae*	...	...	37	10	...	...	...
Swordfish	*Xiphias gladius*	...	...	622	314	...	...	...
Tuna-like fishes nei	*Scombroidei*	553	601	655	590	430	520	520
Sharks, rays, skates, etc. nei	*Elasmobranchii*	...	...	32	18	...	...	...
Marine fishes nei	*Osteichthyes*	1 000	1 500 F	1 500 F	2 000	2 600 F	3 100	2 100
	Country total	*1 553*	*2 101 F*	*3 172 F*	*3 130*	*3 030 F*	*11 051*	*4 299*
Morocco								
Cyprinids nei	*Cyprinidae*	800	800	900	1 200	1 000	900	1 100
Freshwater fishes nei	*Osteichthyes*	600	900	500	700	500	600	800
European eel	*Anguilla anguilla*	100	401	304	250	100	150	200
Shads nei	*Alosa spp*	7	10	5	-	47	62	1
Common sole	*Solea solea*	1 404	1 675	1 540	2 614	5 951	3 788	3 331
Flatfishes nei	*Pleuronectiformes*	3 750	2 287	4 915	5 773	5 038	3 898	5 230
Greater forkbeard	*Phycis blennoides*	440	145	257	345	634	376	291
Pouting(=Bib)	*Trisopterus luscus*	2 105	1 078	1 251	594	2 217	1 017	1 171
European hake	*Merluccius merluccius*	3 170	2 547	1 257	2 350	2 792	4 009	6 689
Mullets nei	*Mugilidae*	1 630	2 628	2 907	3 750	2 758	4 347	4 260
White grouper	*Epinephelus aeneus*	30	27	47	35	94	40	50
Grunts, sweetlips nei	*Haemulidae (=Pomadasyidae)*	1 670	2 590	3 049	2 228	2 715	3 158	3 079
Meagre	*Argyrosomus regius*	869	1 542	1 263	2 473	1 785	1 538	2 047
Sargo breams nei	*Diplodus spp*	940	1 049	982	1 073	1 457	1 288	1 479
Black seabream	*Spondyliosoma cantharus*	229	185	126	146	435	300	415
Pargo breams nei	*Pagrus spp*	25	23	136	92	424	268	47
Gilthead seabream	*Sparus aurata*	0	0	4	0	206	25	10
Bogue	*Boops boops*	3 400	2 510	3 201	3 336	4 888	3 256	4 318
Porgies, seabreams nei	*Sparidae*	10 988	10 380	7 451	10 150	11 020	10 724	11 296
Surmullets(=Red mullets) nei	*Mullus spp*	682	902	806	822	1 489	1 424	1 301
European conger	*Conger conger*	2 005	1 635	1 386	1 351	1 321	1 094	1 556
John dory	*Zeus faber*	461	590	567	622	942	512	736
Largehead hairtail	*Trichiurus lepturus*	5 505	8 359	8 064	6 595	6 176	5 124	3 793
Scorpionfishes nei	*Scorpaenidae*	230	226	199	263	573	392	681
Gurnards, searobins nei	*Triglidae*	2 286	2 216	2 834	3 152	4 874	4 169	3 567
Angler(=Monk)	*Lophius piscatorius*	110	113	83	66	785	310	185
Blackbellied angler	*Lophius budegassa*	350	320	579	498	85	126	327
Sardinellas nei	*Sardinella spp*	2 900	2 186	1 245	742	803	3 301	1 923
European pilchard(=Sardine)	*Sardina pilchardus*	393 362	497 821	435 799	429 732	539 785	763 223	684 982
European anchovy	*Engraulis encrasicolus*	12 447	24 955	40 954	40 213	22 096	47 448	20 969
Atlantic bonito	*Sarda sarda*	961	1 304	1 596	1 510	2 278	1 705	2 080
Plain bonito	*Orcynopsis unicolor*	2 016	249	30	627	1 058	839	789
Frigate and bullet tunas	*Auxis thazard, A.rochei*	2 216	3 176	3 277	1 176	1 345	674	1 062
Little tunny(=Atl.black skipj)	*Euthynnus alletteratus*	588	196	203	75	101	87	311
Skipjack tuna	*Katsuwonus pelamis*	684	4 513	2 486	858	1 199	268	281
Atlantic bluefin tuna	*Thunnus thynnus*	1 621	2 603	2 430	2 227	2 923	3 008	2 986
Albacore	*Thunnus alalunga*	-	-	-	-	-	-	55
Yellowfin tuna	*Thunnus albacares*	-	-	-	-	-	-	79

English name Nom anglais Nombre inglés	Scientific name Nom scientifique Nombre científico	1996 t	1997 t	1998 t	1999 t	2000 t	2001 t	2002 t
Bigeye tuna	*Thunnus obesus*	-	-	-	700	770	857	913
Swordfish	*Xiphias gladius*	3 196	5 167	3 419	3 357	2 822	3 549	3 602
Tuna-like fishes nei	*Scombroidei*	-	-	-	153	-	-	108
Bluefish	*Pomatomus saltatrix*	15	74	163	47	101	198	194
Jack and horse mackerels nei	*Trachurus spp*	15 471	12 512	9 816	12 787	22 095	12 167	11 012
Chub mackerel	*Scomber japonicus*	16 988	28 775	11 621	17 383	63 381	26 021	24 281
Rays, stingrays, mantas nei	*Rajiformes*	1 680	1 380	1 206	1 463	2 139	1 510	1 901
Sharks, rays, skates, etc. nei	*Elasmobranchii*	1 625	1 255	2 243	2 004	3 460	2 192	2 161
Marine fishes nei	*Osteichthyes*	42 273	88 681	67 777	54 828	7 600	14 461	27 569
Freshwater prawns, shrimps nei	*Palaemonidae*	0	0	2	5	3	4	5
Marine crabs nei	*Brachyura*	50	286	236	400	370	365	247
Palinurid spiny lobsters nei	*Palinurus spp*	26	61	33	38	42	456	165
Norway lobster	*Nephrops norvegicus*	7	4	5	3	5	15	15
European lobster	*Homarus gammarus*	20	14	15	8	11	11	55
Natantian decapods nei	*Natantia*	9 338	7 276	10 770	9 840	12 652	7 585	4 239
Marine crustaceans nei	*Crustacea*	29	36	17	-	34	29	91
Freshwater molluscs nei	*Mollusca*	-	-	-	8	5	6	7
Mediterranean mussel	*Mytilus galloprovincialis*	0	0	0	14	60	32	0
Sea mussels nei	*Mytilidae*	10	6	7	...	401	102	8
Clams, etc. nei	*Bivalvia*	0	0	0	-	-	-	-
Cuttlefish,bobtail squids nei	*Sepiidae, Sepiolidae*	12 678	16 787	14 668	20 168	32 904	17 562	2 463
Various squids nei	*Loliginidae, Ommastrephidae*	20 010	8 989	11 352	8 082	15 432	10 633	2 756
Octopuses, etc. nei	*Octopodidae*	58 636	38 187	42 533	84 579	99 391	112 634	38 668
Marine molluscs nei	*Mollusca*	253	275	1 920	1 926	1 048	116	1 020
	Country total	*642 886*	*791 906*	*710 436*	*745 431*	*896 620*	*1 083 953*	*894 957*
Mozambique								
Dagaas	*Stolothrissa, Limnothrissa spp*	5 574	9 921	7 313	9 052	11 813	5 284	12 156
Freshwater fishes nei	*Osteichthyes*	1 936	1 747	1 681	...	...	...	...
Swordfish	*Xiphias gladius*	358	524	1 039	447	...	...	...
Tuna-like fishes nei	*Scombroidei*	2 461	3 602	7 140	6 078	5 081	3 241	2 178
Requiem sharks nei	*Carcharhinidae*	21	...	...	...	...	...	...
Marine fishes nei	*Osteichthyes*	13 058	10 374	6 990	8 337	7 786	8 951	9 387
Geryons nei	*Geryon spp*	564	1 144	911	795	832	632	890
Spiny lobsters nei	*Palinuridae*	331	232	237	203	228	213	276
Mozambique lobster	*Metanephrops mozambicus*	132	156	147	92	105	69	75
Penaeus shrimps nei	*Penaeus spp*	8 183	9 825	8 559	8 806	9 429	9 401	9 472
Knife shrimp	*Haliporoides triarthrus*	1 771	1 510	1 882	1 611	1 766	1 738	1 441
Clams, etc. nei	*Bivalvia*	57	8	19	26	30	11	19
Various squids nei	*Loliginidae, Ommastrephidae*	366	602	677	616	538	395	534
Octopuses, etc. nei	*Octopodidae*	49	51	80	104	109	128	24
Sea cucumbers nei	*Holothurioidea*	54	7	2	8	12	11	10
	Country total	*34 915*	*39 703*	*36 677*	*36 175*	*37 729*	*30 074*	*36 462*
Namibia								
Freshwater fishes nei	*Osteichthyes*	1 200	1 500	1 500	1 500	1 500	1 500	1 500
West coast sole	*Austroglossus microlepis*	339	514	393	463	571	589	644
Tadpole codling	*Salilota australis*	-	20	99	128	-	-	-
Southern blue whiting	*Micromesistius australis*	-	83	282	29	-	-	-
Argentine hake	*Merluccius hubbsi*	-	12	15	37	-	-	-
Cape hakes	*Merluccius capensis,M.paradox.*	129 462	117 683	150 695	164 250	171 397	173 277	156 499
Patagonian grenadier	*Macruronus magellanicus*	-	98	205	308	-	-	-
Sea catfishes nei	*Ariidae*	16	31	42	32	52	11	0
Mullets nei	*Mugilidae*	...	...	...	...	161	117	122
Southern meagre(=Mulloway)	*Argyrosomus hololepidotus*	464	184	424	273	409	596	288
Large-eye dentex	*Dentex macrophthalmus*	...	...	...	...	287	182	171
Panga seabream	*Pterogymnus laniarius*	470	416	199	383	8 603	5 898	2 621
Steenbrasses nei	*Lithognathus spp*	6	136	2	116	56	20	4
Porgies, seabreams nei	*Sparidae*	2 279	2 608	1 364	4 434	...	20	...
Gobies nei	*Gobiidae*	552	287	20 998	16	3	...	...
Pink cusk-eel	*Genypterus blacodes*	-	5	24	45	-	-	-
Kingklip	*Genypterus capensis*	3 667	2 506	2 211	3 706	3 922	6 607	7 926
Alfonsinos nei	*Beryx spp*	1 805	369	147	123	59	300	232
Orange roughy	*Hoplostethus atlanticus*	13 379	18 516	10 945	3 473	1 542	857	2 169
John dory	*Zeus faber*	0	5	25	14	4	138	208
Oreo dories nei	*Oreosomatidae*	17	188	...	42	10	54	335
Patagonian toothfish	*Dissostichus eleginoides*	-	2	21	28	-	-	-
Snoek	*Thyrsites atun*	622	895	701	1 212	966	1 699	1 890
Largehead hairtail	*Trichiurus lepturus*	...	...	...	...	346	691	...
Blackbelly rosefish	*Helicolenus dactylopterus*	576	673	343	448	639	891	...
Cape gurnard	*Chelidonichthys capensis*	51	214	216	71	236	144	234
Devil anglerfish	*Lophius vomerinus*	9 236	10 259	16 429	14 802	14 358	12 390	15 174
Southern African pilchard	*Sardinops ocellatus*	1 171	27 685	68 562	44 653	25 388	7 940	4 176
Whitehead's round herring	*Etrumeus whiteheadi*	20 656	5 070	5 193	176	1 127	1 432	9 650
Southern African anchovy	*Engraulis capensis*	1 080	2 545	2 748	412	146	2 133	41 203
Skipjack tuna	*Katsuwonus pelamis*	1	1	...	...	...	8	0
Albacore	*Thunnus alalunga*	1 521	1 199	1 429	1 162	2 418	3 419	2 962
Yellowfin tuna	*Thunnus albacares*	72	69	3	147	59	165	102
Bigeye tuna	*Thunnus obesus*	63	46	16	423	589	640	312
Marlins,sailfishes,etc. nei	*Istiophoridae*	-	-	-	-	-	-	4
Swordfish	*Xiphias gladius*	-	-	-	730	469	751	744
Cape horse mackerel	*Trachurus capensis*	321 322	301 847	312 422	320 394	344 314	309 381	359 183
Atlantic pomfret	*Brama brama*	19	9	15	99	493	684	486
Chub mackerel	*Scomber japonicus*	...	...	...	...	1 641	6 329	3 530
Ocean sunfish	*Mola mola*	...	...	...	...	...	2	1
Shortfin mako	*Isurus oxyrinchus*	...	...	...	...	...	...	1 587
Blue shark	*Prionace glauca*	...	...	...	...	...	...	794

English name Nom anglais Nombre inglés	Scientific name Nom scientifique Nombre científico	1996 t	1997 t	1998 t	1999 t	2000 t	2001 t	2002 t
Rays, stingrays, mantas nei	*Rajiformes*	133	194	94	389	966	1 204	...
Sharks, rays, skates, etc. nei	*Elasmobranchii*	5	4	6	1	769	1 875	-
Marine fishes nei	*Osteichthyes*	5 658	15 555	7 387	13 105	3 205	2 029	4 125
Geryons nei	*Geryon spp*	1 709	1 478	2 283	2 074	2 700	2 343	2 471
Cape rock lobster	*Jasus lalandii*	251	199	350	304	365	363	358
Patagonian squid	*Loligo gahi*	-	74	1	0	-	-	-
Argentine shortfin squid	*Illex argentinus*	-	3	0	63	-	-	-
Various squids nei	*Loliginidae, Ommastrephidae*	26	34	42	19	28	775	993
Jellyfishes	*Rhopilema spp*	-	-	-	-	106	44	2 193
	Country total	*517 828*	*513 216*	*607 831*	*580 084*	*589 904*	*547 498*	*624 891*
Niger								
Freshwater fishes nei	*Osteichthyes*	4 156	6 328	7 013	11 000	16 250	20 800	23 520
	Country total	*4 156*	*6 328*	*7 013*	*11 000*	*16 250*	*20 800*	*23 520*
Nigeria								
Cyprinids nei	*Cyprinidae*	4 460	6 199	5 006	6 296	4 823	5 901	6 004
Tilapias nei	*Oreochromis (=Tilapia) spp*	10 074	12 614	16 300	19 662	13 402	18 332	17 218
African lungfishes	*Protopterus spp*	1 284	3 950	1 679	536	1 364	2 183	520
Bonytongues nei	*Heterotis spp*	3 298	4 737	7 388	8 873	7 915	8 577	6 832
Knifefishes	*Notopterus spp*	10	1	144	33	39	22	19
Elephantsnout fishes nei	*Mormyridae*	...	4 481	18 781	16 030	...	17 509	5 567
Aba	*Gymnarchus niloticus*	1 108	3 952	7 685	5 887	...	...	9 097
Characins nei	*Characidae*	7 141	7 646	19 835	11 883	12 500	10 274	6 989
Bagrid catfish	*Chrysichthys nigrodigitatus*	8 505	5 814	15 538	10 679	7 311	12 413	5 678
Naked catfishes	*Bagrus spp*	3 594	4 726	5 463	5 421	4 539	4 902	920
North African catfish	*Clarias gariepinus*	2 658	8 568	5 268	9 994	4 474	10 419	7 375
Torpedo-shaped catfishes nei	*Clarias spp*	8 480	10 202	17 220	10 963	14 012	10 925	7 640
Upsidedown catfishes	*Synodontis spp*	4 466	5 069	8 781	8 143	6 054	6 642	4 055
Nile perch	*Lates niloticus*	3 746	4 224	5 250	6 366	4 447	6 139	3 030
Snakeheads(=Murrels) nei	*Channa spp*	921	86	2 589	2 038	2 990	2 951	3 313
Freshwater fishes nei	*Osteichthyes*	29 776	11 375	2 093	16 589	48 445	36 986	102 976
Soles nei	*Soleidae*	4 640	6 084	7 633	6 583	3 301	3 171	2 507
Tonguefishes	*Cynoglossidae*	1 281	1 311	1 781	3 812	7 057	6 655	7 493
Benguela hake	*Merluccius polli*	-	-	-	64	-	-	1 714
West African ladyfish	*Elops lacerta*	379	735	1 546	1 822	755	531	1 076
Sea catfishes nei	*Ariidae*	12 676	14 062	10 862	17 014	14 885	16 537	21 155
Mullets nei	*Mugilidae*	1 364	7 073	8 450	10 226	8 535	8 728	9 122
Groupers, seabasses nei	*Serranidae*	1 285	2 001	1 741	2 486	2 117	2 487	2 354
Snappers nei	*Lutjanus spp*	3 779	6 473	8 685	8 345	7 235	6 619	7 537
Bigeye grunt	*Brachydeuterus auritus*	556	2 537	2 887	2 141	3 345	275	849
Grunts, sweetlips nei	*Haemulidae (=Pomadasyidae)*	2 585	2 611	5 000	4 561	5 152	4 804	3 352
West African croakers nei	*Pseudotolithus spp*	14 544	11 762	15 035	14 591	12 604	15 084	11 922
Croakers, drums nei	*Sciaenidae*	2 165	6 335	4 056	7 846	5 826	5 900	9 384
West African goatfish	*Pseudupeneus prayensis*	4 575	...	...	...	...	...	1 085
African sicklefish	*Drepane africana*	954	2 047	603	...	1 316	1 429	1 511
Giant African threadfin	*Polydactylus quadrifilis*	6 415	9 815	9 803	10 544	11 287	11 645	14 052
Lesser African threadfin	*Galeoides decadactylus*	3 194	5 974	2 343	2 884	4 279	5 131	4 819
Threadfins, tasselfishes nei	*Polynemidae*	2 930	3 184	-	-	-	-	-
Conger eels, etc. nei	*Congridae*	...	...	301	...	0	-	-
Hairtails, scabbardfishes nei	*Trichiuridae*	4 826	6 399	7 750	5 366	754	1 156	851
Scorpionfishes nei	*Scorpaenidae*	1 032	1 278	2 071	1 149	3 170	2 568	600
Madeiran sardinella	*Sardinella maderensis*	2 230	9 387	12 226	11 041	10 158	11 931	12 132
Sardinellas nei	*Sardinella spp*	102 230	97 286	95 495	83 593	80 892	61 927	59 400
Bonga shad	*Ethmalosa fimbriata*	4 643	28 000	30 216	18 529	17 570	19 049	21 987
Swordfish	*Xiphias gladius*	9	-	-	-	-	-	133
Tuna-like fishes nei	*Scombroidei*	200	55	73	7	51	5	1
Needlefishes, etc. nei	*Belonidae*	1 082	272	99	166	535	431	301
Flyingfishes nei	*Exocoetidae*	17	...	87	69	8	55	1 837
Jacks, crevalles nei	*Caranx spp*	591	2 525	414	1 099	825	317	277
Carangids nei	*Carangidae*	2 923	1 035	700	4 451	3 521	833	630
Mackerels nei	*Scombridae*	3 745	3 326	3 640	2 919	2 710	3 759	3 500
Blue butterfish	*Stromateus fiatola*	-	-	-	-	-	4 106	1 000
Barracudas nei	*Sphyraena spp*	3 556	5 212	7 671	10 490	10 799	10 947	11 085
Rays, stingrays, mantas nei	*Rajiformes*	4 103	5 004	7 382	7 622	5 753	7 352	6 262
Sharks, rays, skates, etc. nei	*Elasmobranchii*	4 285	3 817	6 587	7 751	7 485	7 274	7 187
Marine fishes nei	*Osteichthyes*	32 995	21 516	40 295	33 770	51 541	49 205	28 152
Marine crabs nei	*Brachyura*	57	593	3 384	4 018	3 211	4 374	5 029
Palinurid spiny lobsters nei	*Palinurus spp*	1 133	8 315	3 071	1 245	1 939	1 699	193
Southern pink shrimp	*Penaeus notialis*	12 073	14 799	12 396	27 341	18 882	18 805	13 413
Natantian decapods nei	*Natantia*	3 417	3 456	7 364	2 690	1 564	909	17 076
Marine crustaceans nei	*Crustacea*	0	0	0	0	0	-	-
Gastropods nei	*Gastropoda*	...	...	2 357	...	...	2 084	2 426
Cuttlefish,bobtail squids nei	*Sepiidae, Sepiolidae*	3	...	...	...	...	189	419
	Country total	*337 993*	*387 923*	*463 024*	*455 628*	*441 377*	*452 146*	*481 056*
Réunion								
Freshwater fishes nei	*Osteichthyes*	0	0	0	0	0	0	0
Groupers, seabasses nei	*Serranidae*	...	...	...	47	20	11	11 F
Snappers nei	*Lutjanus spp*	...	...	...	112	38	45	45 F
Threadfins, tasselfishes nei	*Polynemidae*	...	...	...	49	5	10	10 F
Clupeoids nei	*Clupeoidei*	5	6	7	10	12	4	4 F
Wahoo	*Acanthocybium solandri*	80	66	59	57	50	45	45 F
Kawakawa	*Euthynnus affinis*	26	24	28	22	21	19	19 F
Skipjack tuna	*Katsuwonus pelamis*	91	77	92	89	84	76	76 F
Albacore	*Thunnus alalunga*	347	306	318	357	579	521	521 F

English name Nom anglais Nombre inglés	Scientific name Nom scientifique Nombre científico	1996 t	1997 t	1998 t	1999 t	2000 t	2001 t	2002 t
Yellowfin tuna	*Thunnus albacares*	628	636	609	534	656	591	591 F
Bigeye tuna	*Thunnus obesus*	98	91	112	213	167	151	151 F
Indo-Pacific sailfish	*Istiophorus platypterus*	10	11	17	18	30	27	27 F
Shortbill spearfish	*Tetrapturus angustirostris*	2	1	3	5	5	5	5 F
Marlins,sailfishes,etc. nei	*Istiophoridae*	120	110	136	109	123	111	111 F
Swordfish	*Xiphias gladius*	1 336	1 586	2 080	1 930	1 744	1 572	1 572 F
Tuna-like fishes nei	*Scombroidei*	38	-	-	-	-	-	-
Carangids nei	*Carangidae*	...	...	...	107	130	131	131 F
Common dolphinfish	*Coryphaena hippurus*	...	...	...	221	194	141	141 F
Sharks, rays, skates, etc. nei	*Elasmobranchii*	46	89	111	81	71	64	64 F
Marine fishes nei	*Osteichthyes*	775	1 279	1 000	67	140	96	161 F
Marine crabs nei	*Brachyura*	5	6	7	5	7	11	11 F
Marine crustaceans nei	*Crustacea*	...	...	...	2	1	1	1 F
Octopuses, etc. nei	*Octopodidae*	...	...	...	8	5	3	3 F
	Country total	*3 607*	*4 288*	*4 579*	*4 043*	*4 082*	*3 635*	*3 700*
Rwanda								
Nile tilapia	*Oreochromis niloticus*	...	...	2 233	2 640	2 646	2 650	2 750 F
Freshwater fishes nei	*Osteichthyes*	2 952	4 428	4 408	3 793	4 080	4 178	4 250 F
	Country total	*2 952*	*4 428*	*6 641*	*6 433*	*6 726*	*6 828*	*7 000 F*
St Helena								
Freshwater fishes nei	*Osteichthyes*	0	-	-	-	-	-	-
Groupers nei	*Epinephelus spp*	48	41	59	33	17	18	14
Bigeyes nei	*Priacanthus spp*	1	1	3	7	2	1	2
Conger eels, etc. nei	*Congridae*	1	3	3	3	3	5	2
Wahoo	*Acanthocybium solandri*	23	19	10	15	15	22	25
Skipjack tuna	*Katsuwonus pelamis*	86	294	298	13	64	205	63
Albacore	*Thunnus alalunga*	47	18	1	1	58	12	2
Yellowfin tuna	*Thunnus albacares*	151	109	181	116	136	70	90
Bigeye tuna	*Thunnus obesus*	10	12	17	6	8	4	5
Marlins,sailfishes,etc. nei	*Istiophoridae*	2	1	2	4	4	3	4
Carangids nei	*Carangidae*	2	3	2	3	1	1	1
Common dolphinfish	*Coryphaena hippurus*	1	1	1	1	1	0	1
Chub mackerel	*Scomber japonicus*	7	12	2	8	4	11	7
Sharks, rays, skates, etc. nei	*Elasmobranchii*	...	...	...	...	...	6	4
Marine fishes nei	*Osteichthyes*	79	37	56	63	64	67	65
Tropical spiny lobsters nei	*Panulirus spp*	0	-	-	-	-	-	-
Tristan da Cunha rock lobster	*Jasus tristani*	327	321	376	336	316	425	301
Slipper lobsters nei	*Scyllaridae*	0	0	1	1	1	0	-
Octopuses, etc. nei	*Octopodidae*	34	25	48	22	24	16	12
	Country total	*819*	*897*	*1 060*	*632*	*718*	*866*	*598*
Sao Tome Prn								
Freshwater fishes nei	*Osteichthyes*	0	0	0	0	0	0	0
Groupers nei	*Epinephelus spp*	33	31	40 F	45 F	40 F	40 F	40 F
Grunts, sweetlips nei	*Haemulidae (=Pomadasyidae)*	1	0	0	0	0	0	0
Croakers, drums nei	*Sciaenidae*	71	30	40 F	45 F	40 F	40 F	40 F
Pandoras nei	*Pagellus spp*	88	86	100 F	110 F	100 F	100 F	100 F
Bogue	*Boops boops*	...	17	20 F	25 F	20 F	20 F	20 F
Porgies, seabreams nei	*Sparidae*	37	49	60 F	70 F	60 F	60 F	60 F
African sicklefish	*Drepane africana*	...	15	20 F	25 F	20 F	20 F	20 F
Threadfins, tasselfishes nei	*Polynemidae*	75	75	100 F	110 F	100 F	100 F	100 F
Triggerfishes, durgons nei	*Balistidae*	...	34	40 F	45 F	40 F	40 F	40 F
European pilchard(=Sardine)	*Sardina pilchardus*	20	...	20 F	30 F	30 F	30 F	30 F
Wahoo	*Acanthocybium solandri*	80	52	52 F	52 F	52 F	52 F	52 F
West African Spanish mackerel	*Scomberomorus tritor*	8	-	-	-	-	-	-
Frigate and bullet tunas	*Auxis thazard, A.rochei*	79	323	...	...	...	...	...
Little tunny(=Atl.black skipj)	*Euthynnus alletteratus*	40	159	...	...	...	...	...
Skipjack tuna	*Katsuwonus pelamis*	...	7	-	-	-	-	-
Yellowfin tuna	*Thunnus albacares*	1	4	4 F	4 F	4 F	...	...
Bigeye tuna	*Thunnus obesus*	-	5	-	-	-	-	-
Atlantic sailfish	*Istiophorus albicans*	...	139	...	...	...	...	...
Blue marlin	*Makaira nigricans*	-	35	-	-	-	-	-
Atlantic white marlin	*Tetrapturus albidus*	-	45	-	-	-	-	-
Swordfish	*Xiphias gladius*	-	14	14 F	14 F	...	...	...
Tuna-like fishes nei	*Scombroidei*	-	9	-	-	-	-	-
Halfbeaks nei	*Hemiramphus spp*	293	126	150 F	160 F	150 F	150 F	150 F
Flyingfishes nei	*Exocoetidae*	256	939	1 000 F	1 100 F	1 000 F	1 000 F	1 000 F
Scads nei	*Decapterus spp*	...	43	55 F	60 F	60 F	60 F	60 F
Crevalle jack	*Caranx hippos*	...	42	55 F	60 F	60 F	60 F	60 F
Jacks, crevalles nei	*Caranx spp*	...	129	150 F	160 F	150 F	150 F	150 F
Leerfish	*Lichia amia*	25	36	45 F	50 F	50 F	50 F	50 F
Barracudas nei	*Sphyraena spp*	22	56	70 F	80 F	70 F	70 F	70 F
Sharks, rays, skates, etc. nei	*Elasmobranchii*	247	130	175 F	190 F	180 F	180 F	180 F
Marine fishes nei	*Osteichthyes*	2 579	683	1 167 F	1 284 F	1 234 F	1 238 F	1 238 F
Marine crustaceans nei	*Crustacea*	4	5	75	7 F	10 F	10 F	10 F
Marine molluscs nei	*Mollusca*	21	20	25 F	30 F	30 F	30 F	30 F
	Country total	*3 980*	*3 338*	*3 477*	*3 756*	*3 500 F*	*3 500 F*	*3 500 F*
Senegal								
Tilapias nei	*Oreochromis (=Tilapia) spp*	1 807	1 977	1 323	1 362	10 762	10 096 F	10 175 F
Black catfishes nei	*Chrysichthys spp*	...	...	...	...	1 218	1 100 F	1 100 F
Nile perch	*Lates niloticus*	...	...	...	...	1 451	1 300 F	1 300 F
Freshwater fishes nei	*Osteichthyes*	23 000 F	31 000 F	21 000 F	34 000 F	10 648	9 500 F	9 500 F

E-1

Fish crustaceans, molluscs, etc	Capture production by countries or areas and species	Africa
Poissons, crustacés, mollusques, etc	Captures par pays ou zones et espèces	Afrique
Peces, crustáceos, moluscos, etc	Capturas por países o áreas y especies	Africa

English name Nom anglais Nombre inglés	Scientific name Nom scientifique Nombre científico	1996 t	1997 t	1998 t	1999 t	2000 t	2001 t	2002 t
West African ilisha	*Ilisha africana*	...	...	...	...	...	...	271
Flatfishes nei	*Pleuronectiformes*	8 113	8 002	7 132	7 335	8 113	9 059	7 262
Senegalese hake	*Merluccius senegalensis*	7	162	22	335	113	98	221
Bonefish	*Albula vulpes*	...	...	...	...	...	...	9
Sea catfishes nei	*Ariidae*	4 457	10 162	10 094	6 006	7 432	9 279	9 205
Morays	*Muraenidae*	710	268	165	142	189	209	229
Mullets nei	*Mugilidae*	2 109	2 780	1 690	2 512	4 226	2 546	2 802
Groupers, seabasses nei	*Serranidae*	4 411	4 317	3 488	3 162	3 067	3 685	2 903
Spotted seabass	*Dicentrarchus punctatus*	619	472	527	170	67	125	193
Snappers nei	*Lutjanus spp*	236	296	237	221	366	499	360
Bigeye grunt	*Brachydeuterus auritus*	610	1 445	1 481	2 414	1 949	2 701	2 294
Grunts, sweetlips nei	*Haemulidae (=Pomadasyidae)*	11 187	11 283	7 828	10 902	7 403	8 048	6 864
Boe drum	*Pteroscion peli*	667	527	489	581	472	444	374
Cassava croaker	*Pseudotolithus senegalensis*	...	...	1 630	1 487	1 880	1 682	1 159
Bobo croaker	*Pseudotolithus elongatus*	...	...	137	146	33	114	115
Croakers, drums nei	*Sciaenidae*	8 047	7 620	5 572	3 295	3 199	3 700	3 405
Red pandora	*Pagellus bellottii*	4 380	5 086	3 719	4 319	4 728	3 932	3 812
Sargo breams nei	*Diplodus spp*	-	4	221	194	364	453	313
Large-eye dentex	*Dentex macrophthalmus*	447	...	451	531	684	499	249
Dentex nei	*Dentex spp*	1 608	2 301	1 132	788	1 033	833	884
Black seabream	*Spondyliosoma cantharus*	1 196	1 251	597	922	1 067	1 035	1 254
Pargo breams nei	*Pagrus spp*	2 570	2 763	1 881	1 931	2 206	2 488	1 419
Bogue	*Boops boops*	10	20	38	7	17	14	98
Sand steenbras	*Lithognathus mormyrus*	-	-	223	45	8	77	32
Porgies, seabreams nei	*Sparidae*	-	-	-	-	-	295	79
West African goatfish	*Pseudupeneus prayensis*	731	896	1 087	1 223	1 110	1 846	1 855
African sicklefish	*Drepane africana*	741	606	325	439	356	421	244
Giant African threadfin	*Polydactylus quadrifilis*	...	...	...	...	...	...	650
Lesser African threadfin	*Galeoides decadactylus*	5 255	5 453	4 849	3 016	2 844	3 849	3 037
Triggerfishes, durgons nei	*Balistidae*	-	-	10	7	16	52	18
Conger eels, etc. nei	*Congridae*	83	226	241	227	236	112	140
Bearded brotula	*Brotula barbata*	590	1 644	1 770	3 260	1 325	1 669	1 124
John dory	*Zeus faber*	1 157	142	53	139	282	161	151
Largehead hairtail	*Trichiurus lepturus*	432	1 114	948	838	781	1 083	886
Scorpionfishes nei	*Scorpaenidae*	581	491	343	267	471	471	451
Gurnards, searobins nei	*Triglidae*	...	...	31	81	41	94	67
Round sardinella	*Sardinella aurita*	142 457	152 713	129 429	93 512	112 970	112 120	78 244
Madeiran sardinella	*Sardinella maderensis*	108 186	114 824	103 363	105 120	114 749	99 944	105 773
Bonga shad	*Ethmalosa fimbriata*	17 783	17 147	16 496	29 468	22 032	31 675	26 240
European anchovy	*Engraulis encrasicolus*	34	31	307	1 209	964	1 015	435
Atlantic bonito	*Sarda sarda*	570	564	1 723	349	179	120	344
Plain bonito	*Orcynopsis unicolor*	-	-	65	17	6	69	126
West African Spanish mackerel	*Scomberomorus tritor*	1 056	1 015	931	479	589	756	661
Little tunny(=Atl.black skipj)	*Euthynnus alletteratus*	1 066	1 662	1 604	460	1 146	1 613	963
Skipjack tuna	*Katsuwonus pelamis*	265	430	1 836	1 422	1 009	3 768	3 606
Yellowfin tuna	*Thunnus albacares*	68	152	222	358	218	1 118	2 095
Bigeye tuna	*Thunnus obesus*	135	218	791	2 007	860	2 169	1 143
Atlantic sailfish	*Istiophorus albicans*	206	509	192	79	447	266	138
Swordfish	*Xiphias gladius*	-	-	169	127	39	35	61
Tuna-like fishes nei	*Scombroidei*	-	-	-	-	-	6	8
Needlefishes, etc. nei	*Belonidae*	...	...	5	37	8	0	-
Halfbeaks nei	*Hemiramphus spp*	-	-	70	30	14	14	48
Bluefish	*Pomatomus saltatrix*	691	2 075	3 118	327	277	1 149	5 417
Jack and horse mackerels nei	*Trachurus spp*	521	1 078	1 061	2 108	1 020	1 877	3 886
Scads nei	*Decapterus spp*	2 265	2 428	3 249	2 401	4 963	4 209	7 416
African moonfish	*Selene dorsalis*	-	9	278	176	251	326	469
Pompanos nei	*Trachinotus spp*	132	83	28	207	27	158	110
Amberjacks nei	*Seriola spp*	-	-	62	60	64	165	103
Leerfish	*Lichia amia*	228	257	540	489	490	341	270
Alexandria pompano	*Alectis alexandrinus*	160	742	1 023	563	502	862	611
Atlantic bumper	*Chloroscombrus chrysurus*	-	-	-	-	-	2 045	5 871
Carangids nei	*Carangidae*	3 299	4 806	7 491	4 456	5 725	2 616	2 097
Common dolphinfish	*Coryphaena hippurus*	-	-	103	115	150	258	198
Chub mackerel	*Scomber japonicus*	947	1 670	2 117	988	2 656	2 710	6 577
Blue butterfish	*Stromateus fiatola*	-	-	-	20	43	57	190
Barracudas nei	*Sphyraena spp*	1 530	2 339	1 456	1 474	1 680	1 519	1 483
Silky shark	*Carcharhinus falciformis*	...	...	...	...	...	...	1 341
Hammerhead sharks, etc. nei	*Sphyrnidae*	-	-	156	20	1 305	1 451	1 038
Smooth-hounds nei	*Mustelus spp*	...	...	...	...	464	1 908	2 064
Guitarfishes, etc. nei	*Rhinobatidae*	...	...	171	59	1 930	1 772	1 062
Sawfishes	*Pristidae*	...	...	...	2	-	-	-
Rays, stingrays, mantas nei	*Rajiformes*	2 962	4 515	4 051	3 332	1 585	1 667	917
Sharks, rays, skates, etc. nei	*Elasmobranchii*	3 803	4 470	4 887	4 808	5 473	3 260	...
Marine fishes nei	*Osteichthyes*	13 522	13 782	7 590	6 679	12 146	14 409	8 721
Marine crabs nei	*Brachyura*	217	727	356	186	343	389	243
Palinurid spiny lobsters nei	*Palinurus spp*	130	196	144	39	37	53	50
Slipper lobsters nei	*Scyllaridae*	-	-	1	2	4	6	1
Southern pink shrimp	*Penaeus notialis*	4 302	6 606	4 105	4 887	6 005	4 643	3 539
Deepwater rose shrimp	*Parapenaeus longirostris*	954	2 998	3 888	1 167	2 451	2 199	2 578
Natantian decapods nei	*Natantia*	-	-	-	-	-	19	18
Murex	*Murex spp*	1 267	1 223	2 543	1 255	1 529	2 080	1 617
Volutes nei	*Cymbium spp*	6 648	5 166	4 678	5 737	4 989	5 422	4 347
Gastropods nei	*Gastropoda*	45	40	...	...	65	11	4
Cupped oysters nei	*Crassostrea spp*	100	109	89	125	101	151	120
Common edible cockle	*Cerastoderma edule*	54	69	139	147	117	105	152
Cuttlefish,bobtail squids nei	*Sepiidae, Sepiolidae*	5 919	6 864	6 365	5 709	3 953	3 941	3 986
Various squids nei	*Loliginidae, Ommastrephidae*	49	92	17	140	152	107	125
Octopuses, etc. nei	*Octopodidae*	4 180	2 701	5 530	37 257	6 058	2 985	12 795
Marine molluscs nei	*Mollusca*	247	748	699	212	105	75	19

English name Nom anglais Nombre inglés	Scientific name Nom scientifique Nombre científico	1996 t	1997 t	1998 t	1999 t	2000 t	2001 t	2002 t
	Country total	*411 759*	*457 366*	*403 872*	*412 125*	*402 047*	*403 202*	*375 824*
Seychelles								
Freshwater fishes nei	*Osteichthyes*	0	0	0	0	0	0	0
Tadpole codling	*Salilota australis*	-	56	-	-	-	-	-
Argentine hake	*Merluccius hubbsi*	-	27	-	-	-	-	-
Patagonian grenadier	*Macruronus magellanicus*	-	35	5	-	-	-	-
Groupers, seabasses nei	*Serranidae*	66	124	45	143	56	107	74
Snappers nei	*Lutjanus spp*	381	242	424	489	213	602	490
Snappers, jobfishes nei	*Lutjanidae*	466	464	602	825	552	703	611
Emperors(=Scavengers) nei	*Lethrinidae*	295	220	280	285	423	483	336
Spinefeet(=Rabbitfishes) nei	*Siganus spp*	344	163	145	248	164	91	203
Pink cusk-eel	*Genypterus blacodes*	-	10	-	-	-	-	-
Patagonian toothfish	*Dissostichus eleginoides*	-	1	-	-	-	3 800	2 870
Demersal percomorphs nei	*Perciformes*	-	-	-	-	208	153	181
Wahoo	*Acanthocybium solandri*	-	-	3	4	6	1	2
Frigate and bullet tunas	*Auxis thazard, A.rochei*	-	-	-	-	42	10	7
Kawakawa	*Euthynnus affinis*	93	97	41	158	81	52	69
Skipjack tuna	*Katsuwonus pelamis*	-	4 940	10 705	15 846	11 567	26 219	29 891
Atlantic bluefin tuna	*Thunnus thynnus*	-	-	-	-	-	-	2
Longtail tuna	*Thunnus tonggol*	-	-	-	-	1 126	-	-
Albacore	*Thunnus alalunga*	-	-	183	67	423	873	1 238
Southern bluefin tuna	*Thunnus maccoyii*	-	-	-	15	120	238	241
Yellowfin tuna	*Thunnus albacares*	67	2 878	7 452	9 949	11 912	13 392	17 152
Bigeye tuna	*Thunnus obesus*	75	936	2 085	3 112	2 365	3 749	5 989
Indo-Pacific sailfish	*Istiophorus platypterus*	1	8	2	18	9	18	9
Blue marlin	*Makaira nigricans*	-	-	1	8	-	-	-
Black marlin	*Makaira indica*	-	-	1	1	-	-	2
Striped marlin	*Tetrapturus audax*	-	-	-	1	-	-	-
Marlins,sailfishes,etc. nei	*Istiophoridae*	80	32	6	11	54	70	131
Swordfish	*Xiphias gladius*	141	317	216	318	456	621	584
Tuna-like fishes nei	*Scombroidei*	1	6	12	28	92	136	32
Carangids nei	*Carangidae*	1 568	1 562	1 008	1 474	1 764	1 285	2 042
Indian mackerel	*Rastrelliger kanagurta*	425	298	25	101	295	182	281
Indian mackerels nei	*Rastrelliger spp*	159	116	111	219	174	83	62
Barracudas nei	*Sphyraena spp*	190	183	150	323	217	253	300
Rays, stingrays, mantas nei	*Rajiformes*	-	4	1	-	-	-	-
Sharks, rays, skates, etc. nei	*Elasmobranchii*	84	57	102	68	152	100	95
Marine fishes nei	*Osteichthyes*	221	114	129	416	245	228	253
Marine crabs nei	*Brachyura*	17	20	24	11	11	26	24
Tropical spiny lobsters nei	*Panulirus spp*	1	-	0	7	14	5	6
Patagonian squid	*Loligo gahi*	-	1 114	88	-	-	-	-
Octopuses, etc. nei	*Octopodidae*	32	19	40	78	29	54	32
	Country total	*4 707*	*14 043*	*23 886*	*34 223*	*32 770*	*53 534*	*63 209*
Sierra Leone								
Freshwater fishes nei	*Osteichthyes*	14 500 F	14 500	14 190	14 480	14 000	14 000	14 000
West African ilisha	*Ilisha africana*	3 010 F	3 009	3 110	5	2 996	4	1 363
Tonguefishes	*Cynoglossidae*	380 F	585	206	714	2 326	500	654
Benguela hake	*Merluccius polli*	-	0	-	-	-	-	-
Bonefish	*Albula vulpes*	...	91	...	106	181	496	153
Sea catfishes nei	*Ariidae*	860 F	973	52	62	2 092	748	774
Mullets nei	*Mugilidae*	600 F	597	150	233	595	...	...
Dungat grouper	*Epinephelus goreensis*	196	63	76	290	277	62	64
Snappers nei	*Lutjanus spp*	3 F	155	...	1 728	294	381	314
Bigeye grunt	*Brachydeuterus auritus*	...	...	...	...	...	132	129
Grunts, sweetlips nei	*Haemulidae (=Pomadasyidae)*	800 F	1 086	433	413	1 357	1 975	499
Boe drum	*Pteroscion peli*	...	8	0	...	10	-	-
Law croaker	*Pseudotolithus brachygnathus*	360 F	393	90	48	357	227	61
Bobo croaker	*Pseudotolithus elongatus*	3 600 F	3 330	389	410	4 035	748	3 233
West African croakers nei	*Pseudotolithus spp*	680 F	1 210	687	869	2 014	1 011	1 470
Pargo breams nei	*Pagrus spp*	240 F	933	713	752	...	...	...
Porgies, seabreams nei	*Sparidae*	2 565	1 967	522	500	1 106	1 189	1 007
West African goatfish	*Pseudupeneus prayensis*	...	37	...	...	35	119	130
African sicklefish	*Drepane africana*	10 F	293	657	214	662	416	244
Giant African threadfin	*Polydactylus quadrifilis*	640 F	679	75	61	690	445	469
Lesser African threadfin	*Galeoides decadactylus*	390 F	921	758	716	1 149	914	753
Royal threadfin	*Pentanemus quinquarius*	1 750 F	1 768	1 744	1 784	1 761	...	107
Threadfins, tasselfishes nei	*Polynemidae*	2 F	-	-	-	-	-	-
Triggerfishes, durgons nei	*Balistidae*	1 F	3	195	...	5	-	17
European conger	*Conger conger*	...	...	348	...	...	...	...
Largehead hairtail	*Trichiurus lepturus*	...	...	204	75	35	48	66
Sardinellas nei	*Sardinella spp*	7 670 F	7 661	7 450	7 850	7 628	9 845	13 251
Bonga shad	*Ethmalosa fimbriata*	21 700 F	21 807	21 840	22 400	21 621	24 790	31 492
European anchovy	*Engraulis encrasicolus*	180 F	182	150	43	181	...	43
Clupeoids nei	*Clupeoidei*	700 F	...	315	-	6	-	-
Atlantic bonito	*Sarda sarda*	0	0	4	...	11	245	44
Atlantic bluefin tuna	*Thunnus thynnus*	...	...	...	...	93	118	...
Albacore	*Thunnus alalunga*	...	...	...	...	...	91	...
Bigeye tuna	*Thunnus obesus*	...	...	...	...	6	2	...
Swordfish	*Xiphias gladius*	...	...	...	...	2	2	...
Tuna-like fishes nei	*Scombroidei*	254	2 618	4 980	2 166	623	6 638	498
Cunene horse mackerel	*Trachurus trecae*	11	44	40	5	...	434	134
Scads nei	*Decapterus spp*	...	112	39	277	141	...	...
Jacks, crevalles nei	*Caranx spp*	470 F	474	18	7	587	327	726
African moonfish	*Selene dorsalis*	1 F	...	...	...	135	319	260
Pompanos nei	*Trachinotus spp*	3 F	63	279	7	-	-	-

E-1 Fish crustaceans, molluscs, etc — Capture production by countries or areas and species — Africa
Poissons, crustacés, mollusques, etc — Captures par pays ou zones et espèces — Afrique
Peces, crustáceos, moluscos, etc — Capturas por países o áreas y especies — Africa

English name Nom anglais Nombre inglés	Scientific name Nom scientifique Nombre científico	1996 t	1997 t	1998 t	1999 t	2000 t	2001 t	2002 t
Atlantic bumper	*Chloroscombrus chrysurus*	0	-	-	-	-	227	-
Carangids nei	*Carangidae*	...	52	158	-	39	-	-
Barracudas nei	*Sphyraena spp*	680 F	772	237	124	1 095	2 757	2 063
Rays, stingrays, mantas nei	*Rajiformes*	1 400 F	1 404	15	10	18	...	1
Sharks, rays, skates, etc. nei	*Elasmobranchii*	2 F	1	68	41	1 672	164	403
Marine fishes nei	*Osteichthyes*	0	0	0	4	2 732	3 609	3 547
Marine crabs nei	*Brachyura*	15 F	140	83	107	142	269	168
Tropical spiny lobsters nei	*Panulirus spp*	4 F	57	29	51	56	30	43
Penaeus shrimps nei	*Penaeus spp*	973	2 147	1 690	2 155	1 663	1 280	1 237
Cuttlefish,bobtail squids nei	*Sepiidae, Sepiolidae*	1 621	1 626	600	512	294	574	1 138
Various squids nei	*Loliginidae, Ommastrephidae*	121	90	143	...	...	...	...
Octopuses, etc. nei	*Octopodidae*	912	777	328	188	8	11	1
Marine molluscs nei	*Mollusca*	-	-	-	-	-	63	2 434
	Country total	*67 304 F*	*72 628*	*63 065*	*59 407*	*74 730*	*75 210*	*82 990*
Somalia								
Freshwater fishes nei	*Osteichthyes*	250 F	250 F	250 F	250 F	200 F	200 F	150 F
Marine fishes nei	*Osteichthyes*	24 900 F	23 000 F	21 100 F	19 100 F	19 100 F	18 900 F	17 150 F
Tropical spiny lobsters nei	*Panulirus spp*	400 F	400 F	400 F	400 F	400 F	400 F	300 F
Cephalopods nei	*Cephalopoda*	500 F	500 F	500 F	500 F	500 F	500 F	400 F
	Country total	*26 050 F*	*24 150 F*	*22 250 F*	*20 250 F*	*20 200 F*	*20 000 F*	*18 000 F*
South Africa								
Freshwater fishes nei	*Osteichthyes*	850	850	900	900 F	900 F	900 F	900 F
West coast sole	*Austroglossus microlepis*	16	3	...	...	...	...	...
Mud sole	*Austroglossus pectoralis*	909	837	859	768	800 F	844	702
Tonguefishes	*Cynoglossidae*	-	3	7	5 F	0	17	17
Smalleye moray cod	*Muraenolepis microps*	-	-	-	-	-	0	-
Moray cods nei	*Muraenolepis spp*	-	-	-	-	-	0	-
Blue antimora	*Antimora rostrata*	-	0	9	7	24	4	1
Cape hakes	*Merluccius capensis,M.paradox.*	155 155	141 076	151 317	141 165	135 000 F	146 393	149 548
Whitson's grenadier	*Macrourus whitsoni*	-	-	-	-	3	-	-
Grenadiers nei	*Macrourus spp*	-	1	67	77	203	51	14
Gadiformes nei	*Gadiformes*	-	10	7	5 F	2	1	-
Sea catfishes nei	*Ariidae*	2	2	0	1	-	-	-
Mullets nei	*Mugilidae*	1 086	972	908	757	400 F	8	-
Groupers, seabasses nei	*Serranidae*	30	...	...	25 F	...	15	27
Snappers, jobfishes nei	*Lutjanidae*	2	-	-	-	-	1	1
Grunts, sweetlips nei	*Haemulidae (=Pomadasyidae)*	-	6	11	5 F	1	4	3
Canary drum (=Baardman)	*Umbrina canariensis*	-	-	-	-	-	5	13
Southern meagre(=Mulloway)	*Argyrosomus hololepidotus*	850	765	690	671	619 F	617	350
Geelbek croaker	*Atractoscion aequidens*	361	492	605	430 F	380 F	379	306
Tigertooth croaker	*Otolithes ruber*	-	-	13	10 F	...	...	...
Croakers, drums nei	*Sciaenidae*	8	7	...	...	1	0	-
Emperors(=Scavengers) nei	*Lethrinidae*	-	-	-	-	-	3	4
Carpenter seabream	*Argyrozona argyrozona*	883	780	505	541	500 F	287	249
Santer seabream	*Cheimerius nufar*	73	62	0	29	25 F	25	46
Red steenbras	*Petrus rupestris*	28	35	...	22	10 F	7	6
Panga seabream	*Pterogymnus laniarius*	730	907	843	966	900 F	1 058	1 177
White stumpnose	*Rhabdosargus globiceps*	237	165	296	335	300 F	169	107
Daggerhead breams nei	*Chrysoblephus spp*	268	224	154	149 F	70 F	78	124
White steenbras	*Lithognathus lithognathus*	7	8	9	2	...	...	...
Salema	*Sarpa salpa*	4	3	0	1	...	...	...
Porgies, seabreams nei	*Sparidae*	324	326	366	279 F	250 F	151	114
Striped-eyed rockcod	*Notothenia kempi*	-	-	-	-	-	-	0
Antarctic rockcods, noties nei	*Nototheniidae*	-	-	1	-	-	0	-
Hector's lanternfish	*Lampanyctodes hectoris*	33	243	6 553	0	...	...	...
Pike-congers nei	*Muraenesox spp*	-	1	9	5 F	4	5	...
Kingklip	*Genypterus capensis*	3 140	3 418	3 381	4 114	4 000 F	4 848	5 491
Alfonsinos nei	*Beryx spp*	-	-	-	-	-	10	-
John dory	*Zeus faber*	1 156	1 277	968	1 027 F	1 006 F	1 183	1 313
Cape bonnetmouth	*Emmelichthys nitidus*	216	121	117	113	50 F	37	-
Antarctic toothfish	*Dissostichus mawsoni*	-	-	-	-	-	21	-
Patagonian toothfish	*Dissostichus eleginoides*	942	1 246	1 739	704	1 458	634	497
Icefishes nei	*Channichthyidae*	-	-	-	-	-	0	-
Snoek	*Thyrsites atun*	12 400	11 126	14 135	11 188	10 200 F	10 328	11 079
Silver scabbardfish	*Lepidopus caudatus*	3 196	2 001	4 557	2 560	2 300 F	2 316	4 523
Hairtails, scabbardfishes nei	*Trichiuridae*	-	6	0	...	...	...	...
Blackbelly rosefish	*Helicolenus dactylopterus*	1 183	957	752	766	750 F	782	821
Scorpionfishes nei	*Scorpaenidae*	-	3	1	1 F	2	2	...
Cape gurnard	*Chelidonichthys capensis*	446	428	623	507	450 F	465	462
Gurnards, searobins nei	*Triglidae*	-	2	-	...	0	0	...
Devil anglerfish	*Lophius vomerinus*	6 139	6 958	7 903	6 949	7 000 F	9 554	8 492
Southern African pilchard	*Sardinops ocellatus*	105 210	116 996	128 019	131 316	136 060	192 160	260 710
Whitehead's round herring	*Etrumeus whiteheadi*	47 117	92 209	52 476	58 856	37 750	55 330	54 800
Southern African anchovy	*Engraulis capensis*	40 712	60 095	107 548	180 542	267 840	287 190	213 440
Atlantic bonito	*Sarda sarda*	0	-	-	-	-	-	-
Narrow-barred Spanish mackerel	*Scomberomorus commerson*	13	...	...	...	...	...	...
Indo-Pacific king mackerel	*Scomberomorus guttatus*	-	20	46	16	22	49	7
Seerfishes nei	*Scomberomorus spp*	-	2	3	2	2	2	4
Kawakawa	*Euthynnus affinis*	-	1	2	2	0	0	0
Skipjack tuna	*Katsuwonus pelamis*	1	7	3	3	1	1	0
Albacore	*Thunnus alalunga*	5 634	6 708	8 419	5 102	3 636	7 257	6 574
Southern bluefin tuna	*Thunnus maccoyii*	-	-	1	-	4	0	15
Yellowfin tuna	*Thunnus albacares*	157	144	335	467	552	412	293
Bigeye tuna	*Thunnus obesus*	7	10	49	54	254	192	507
Indo-Pacific sailfish	*Istiophorus platypterus*	-	0	0	0	0	0	-

English name Nom anglais Nombre inglés	Scientific name Nom scientifique Nombre científico	1996 t	1997 t	1998 t	1999 t	2000 t	2001 t	2002 t
Blue marlin	*Makaira nigricans*	-	0	1	1	2	1	4
Marlins,sailfishes,etc. nei	*Istiophoridae*	-	2	2	4	2	4	28
Swordfish	*Xiphias gladius*	1	1	405	124	250	626	1 092
Tuna-like fishes nei	*Scombroidei*	-	8	72	25	12	8	518
Bluefish	*Pomatomus saltatrix*	38	15	4	33	20 F	3	2
Cape horse mackerel	*Trachurus capensis*	31 995	31 206	46 384	17 970	15 000 F	9 659	21 883
Yellowtail amberjack	*Seriola lalandi*	488	478	519	302	300 F	315	213
Atlantic pomfret	*Brama brama*	1 070	381	307	324	350 F	437	404
Common dolphinfish	*Coryphaena hippurus*	-	1	...	...	...	1	2
Chub mackerel	*Scomber japonicus*	4 150	8 044	3 204	1 768 F	1 600 F	1 620	876
Mackerels nei	*Scombridae*	-	-	-	-	0	0	-
Shortfin mako	*Isurus oxyrinchus*	...	...	...	...	...	2	31
Blue shark	*Prionace glauca*	...	...	...	...	...	1	94
Dusky shark	*Carcharhinus obscurus*	-	7	0	...	...	...	...
Smooth-hound	*Mustelus mustelus*	...	...	...	...	...	...	2
Tope shark	*Galeorhinus galeus*	...	...	...	...	...	...	19
Greenland shark	*Somniosus microcephalus*	-	-	-	-	-	-	1
Antarctic starry skate	*Raja georgiana*	-	-	-	-	-	-	0
Raja rays nei	*Raja spp*	1 026	1 239	1 074	1 000	1 000 F	1 151	1 300
Eaton's skate	*Bathyraja eatonii*	-	-	-	-	-	-	0
Murray's skate	*Bathyraja murrayi*	-	-	-	-	0	-	-
Bathyraja rays nei	*Bathyraja spp*	-	-	-	-	0	-	-
Rays, stingrays, mantas nei	*Rajiformes*	-	0	10	4	9	2	0
Cape elephantfish	*Callorhinchus capensis*	366	484	482	356	380 F	405	422
Sharks, rays, skates, etc. nei	*Elasmobranchii*	327	543	630	531 F	410 F	349	357
Marine fishes nei	*Osteichthyes*	1 634	13 336	1 167	3 711 F	1 503 F	4 590	4 628
Geryons nei	*Geryon spp*	-	-	91	80 F	54	53	30
Marine crabs nei	*Brachyura*	-	113	4	5 F	...	2	-
Cape rock lobster	*Jasus lalandii*	1 516	1 680	1 726	1 793	1 693	1 611	2 671
Natal spiny lobster	*Palinurus delagoae*	10	10	6	7 F	8	10	7
Southern spiny lobster	*Palinurus gilchristi*	918	892	864	429	305	1 053	651
Slipper lobsters nei	*Scyllaridae*	-	2	1	1 F	2	4	...
Mozambique lobster	*Metanephrops mozambicus*	-	-	45	60 F	75	72	55
Subantarctic stone crab	*Lithodes murrayi*	-	-	1	-	0	0	0
Red stone crab	*Paralomis aculeata*	-	-	-	-	0	0	-
King crabs, stone crabs nei	*Lithodidae*	-	0	0	0	3	0	0
Penaeus shrimps nei	*Penaeus spp*	-	127	178	180 F	168	249	160
Perlemoen abalone	*Haliotis midae*	735	330	524	481	490	527	516
Cupped oysters nei	*Crassostrea spp*	...	...	...	-	-	-	1
Donax clams	*Donax spp*	0	-	-	-	-	-	36
Cuttlefish,bobtail squids nei	*Sepiidae, Sepiolidae*	-	9	5	10 F	16	16	19
Cape Hope squid	*Loligo reynaudi*	7 491	3 696	6 670	7 169	6 000 F	3 373	7 405
Various squids nei	*Loliginidae, Ommastrephidae*	34	47	29	24	40 F	53	45
Octopuses, etc. nei	*Octopodidae*	46	74	68	97 F	105 F	107	75
	Country total	*441 370*	*514 238*	*559 679*	*587 933*	*643 526*	*750 099*	*766 284*
Sudan								
Nile tilapia	*Oreochromis niloticus*	10 500	11 000	16 000	16 000	18 000 F	20 000	20 000
Freshwater fishes nei	*Osteichthyes*	30 000	31 000	28 000	28 000	30 000 F	33 000	33 000
Narrow-barred Spanish mackerel	*Scomberomorus commerson*	...	...	19	24	19	34	34
Sharks, rays, skates, etc. nei	*Elasmobranchii*	...	...	45	56	44	79	79
Marine fishes nei	*Osteichthyes*	4 500	5 000	5 436	5 420	4 937 F	4 887	4 887
	Country total	*45 000*	*47 000*	*49 500*	*49 500*	*53 000 F*	*58 000*	*58 000*
Swaziland								
Freshwater fishes nei	*Osteichthyes*	60 F	65 F	70 F	70 F	70 F	70 F	70 F
	Country total	*60 F*	*65 F*	*70 F*	*70 F*	*70 F*	*70 F*	*70 F*
Tanzania								
Tilapias nei	*Oreochromis (=Tilapia) spp*	35 160	25 100	24 000	38 000	40 000	45 000	43 000
Mouthbrooding cichlids	*Haplochromis spp*	6 558	10 600	10 500	8 000	9 000	8 500	7 500
Dagaas	*Stolothrissa, Limnothrissa spp*	40 179	48 000	46 800	42 000	40 000	43 000	43 000
Naked catfishes	*Bagrus spp*	1 136	2 800	2 500	2 000	2 300	2 000	2 000
Torpedo-shaped catfishes nei	*Clarias spp*	2 107	7 750	7 500	2 500	2 000	2 500	2 000
Nile perch	*Lates niloticus*	121 161	152 000	150 000	100 000	90 000	96 000	92 000
Freshwater fishes nei	*Osteichthyes*	55 975	60 500	58 700	67 520	96 700	86 000	84 350
Indian halibut	*Psettodes erumei*	148	50	45	50	52	75	60
Sea catfishes nei	*Ariidae*	913	850	810	850	880	850	800
Mullets nei	*Mugilidae*	618	550	300	350	400	250	200
Groupers nei	*Epinephelus spp*	652	400	350	500	450	500	550
Threadfin breams nei	*Nemipterus spp*	937	250	195	500	400	400	600
Emperors(=Scavengers) nei	*Lethrinidae*	7 304	7 350	7 025	7 500	8 000	7 850	7 800
Wrasses, hogfishes, etc. nei	*Labridae*	3 725	3 200	3 000	3 000	3 200	3 500	3 000
Percoids nei	*Percoidei*	333	350	200	250	300	200	100
Spinefeet(=Rabbitfishes) nei	*Siganus spp*	3 816	4 000	3 550	3 500	3 525	3 000	3 200
Sardinellas nei	*Sardinella spp*	14 323	5 000	4 450	14 000	15 000	15 500	14 000
Seerfishes nei	*Scomberomorus spp*	766	750	750	650	400	500	450
Yellowfin tuna	*Thunnus albacares*	-	-	-	350	700	800	700
Marlins,sailfishes,etc. nei	*Istiophoridae*	656	700	800	780	800	850	700
Tuna-like fishes nei	*Scombroidei*	757	850	650	500	250	250	250
Halfbeaks nei	*Hemiramphus spp*	1 483	1 850	1 175	1 200	1 250	1 300	1 200
Amberjacks nei	*Seriola spp*	51	250	275	100	75	60	50
Carangids nei	*Carangidae*	3 669	1 800	2 000	2 000	2 500	2 000	1 880
Indian mackerel	*Rastrelliger kanagurta*	4 618	4 000	4 050	4 500	4 500	5 000	5 250
Barracudas nei	*Sphyraena spp*	29	30	15	20	25	25	25
Rays, stingrays, mantas nei	*Rajiformes*	4 006	3 500	3 350	3 500	3 600	3 500	2 500

English name Nom anglais Nombre inglés	Scientific name Nom scientifique Nombre científico	1996 t	1997 t	1998 t	1999 t	2000 t	2001 t	2002 t
Sharks, rays, skates, etc. nei	*Elasmobranchii*	1 594	1 500	1 325	1 375	1 400	1 500	1 500
Marine fishes nei	*Osteichthyes*	6 334	8 300	8 395	2 125	2 000	2 000	2 000
Natantian decapods nei	*Natantia*	2 664	2 500	2 800	2 100	2 100	2 000	2 000
Octopuses, etc. nei	*Octopodidae*	605	653	690	600	600	650	800
Sea cucumbers nei	*Holothurioidea*	1 644	1 527	1 800	189	372	340	65
	Country total	*323 921*	*356 960*	*348 000*	*310 509*	*332 779*	*335 900*	*323 530*
Togo								
Tilapias nei	*Oreochromis (=Tilapia) spp*	3 500	3 500	3 500	3 500	3 500	3 500	3 500
Freshwater siluroids nei	*Siluroidei*	500	500	500	500	500	500	500
Freshwater fishes nei	*Osteichthyes*	1 000	1 000	1 000	1 000	1 000	1 000	1 000
West African ilisha	*Ilisha africana*	15	101	35	113	418	796	667
Soles nei	*Soleidae*	10	16	8	6	48	19	13
Spottail spiny turbot	*Psettodes belcheri*	0	2	0	-	1	2	2
Mullets nei	*Mugilidae*	0	0	1	-	3	0	-
Groupers nei	*Epinephelus spp*	14	19	215	120	16	12	5
Snappers nei	*Lutjanus spp*	203	42	433	129	24	16	30
Sompat grunt	*Pomadasys jubelini*	3	43	31	35	11	1	5
Bigeye grunt	*Brachydeuterus auritus*	0	1	5	465	742	844	1 148
Shi drum	*Umbrina cirrosa*	0	0	0	-	-	-	-
West African croakers nei	*Pseudotolithus spp*	22	16	11	6	37	43	66
Red pandora	*Pagellus bellottii*	74	67	202	237	64	45	31
Pandoras nei	*Pagellus spp*	91	65	365	1 055	120	75	17
Large-eye dentex	*Dentex macrophthalmus*	0	0	1	-	-	-	-
Black seabream	*Spondyliosoma cantharus*	6	5	2	6	25	12	15
Bogue	*Boops boops*	40	18	109	198	71	99	4
West African goatfish	*Pseudupeneus prayensis*	0	0	0	-	-	-	-
African sicklefish	*Drepane africana*	6	2	0	0	-	-	0
Giant African threadfin	*Polydactylus quadrifilis*	16	1	0	1	8	6	1
Lesser African threadfin	*Galeoides decadactylus*	14	3	1	20	149	98	88
Spadefishes nei	*Ephippidae*	-	1	1	0	1	-	-
Triggerfishes, durgons nei	*Balistidae*	0	3	4	0	32	0	1
European conger	*Conger conger*	-	0	0	1	0	0	-
Largehead hairtail	*Trichiurus lepturus*	0	1	0	6	2	4	14
Round sardinella	*Sardinella aurita*	739	1 300	1 045	1 921	1 912	3 123	1 807
Madeiran sardinella	*Sardinella maderensis*	166	432	408	614	638	445	353
European anchovy	*Engraulis encrasicolus*	7 072	4 758	6 325	6 678	7 164	6 660	6 932
Clupeoids nei	*Clupeoidei*	0	0	0	0	0	0	-
Atlantic bonito	*Sarda sarda*	197	338	294	426	423	663	846
Bigeye tuna	*Thunnus obesus*	33	17	6	66	32	26	66
Marlins,sailfishes,etc. nei	*Istiophoridae*	...	32	...	110	77	205	158
Swordfish	*Xiphias gladius*	64	2	23	37	7	17	21
Needlefishes, etc. nei	*Belonidae*	37	90	122	96	53	153	69
Halfbeaks nei	*Hemiramphus spp*	2	4	102	3	4	18	18
Flyingfishes nei	*Exocoetidae*	204	287	610	211	130	186	190
Cunene horse mackerel	*Trachurus trecae*	163	107	82	815	449	501	154
African moonfish	*Selene dorsalis*	10	2	0	-	0	-	13
Rainbow runner	*Elagatis bipinnulata*	-	1	-	27	4	1	1
Carangids nei	*Carangidae*	273	260	291	1 613	2 433	2 482	1 489
Common dolphinfish	*Coryphaena hippurus*	3	3	0	10	36	148	147
Chub mackerel	*Scomber japonicus*	112	252	183	454	308	204	419
Barracudas nei	*Sphyraena spp*	96	38	25	144	226	54	76
Rays, stingrays, mantas nei	*Rajiformes*	11	2	0	2	16	5	2
Sharks, rays, skates, etc. nei	*Elasmobranchii*	202	57	67	230	132	130	254
Marine fishes nei	*Osteichthyes*	176	823	594	2 059	1 424	1 027	807
Marine crabs nei	*Brachyura*	0	0	1	3	33	23	17
Tropical spiny lobsters nei	*Panulirus spp*	0	0	0	2	2	0	0
Penaeus shrimps nei	*Penaeus spp*	12	2	2	0	2	0	-
Marine crustaceans nei	*Crustacea*	0	0	0	0	0	0	-
Cuttlefish,bobtail squids nei	*Sepiidae, Sepiolidae*	12	77	51	5	0	20	0
Marine molluscs nei	*Mollusca*	0	0	0	0	0	0	-
	Country total	*15 098*	*14 290*	*16 655*	*22 924*	*22 277*	*23 163*	*20 946*
Tunisia								
Freshwater fishes nei	*Osteichthyes*	706	1 010	896	808	832	860	870
European eel	*Anguilla anguilla*	192	202	150	206	108	135	202
Shads nei	*Alosa spp*	0	0	-	-	-	-	-
Common sole	*Solea solea*	569	642	480	425	402	456	454
Turbot	*Psetta maxima*	0	0	-	-	-	-	-
Flatfishes nei	*Pleuronectiformes*	52	110	178	217	190	205	194
European hake	*Merluccius merluccius*	685	922	537	1 069	974	1 284	1 538
Flathead grey mullet	*Mugil cephalus*	1 345	1 309	1 525	238	2 048	229	952
Mullets nei	*Mugilidae*	932	1 472	2 272	1 383	2 306	3 016	3 010
Groupers nei	*Epinephelus spp*	265	416	334	382	103	539	474
Groupers, seabasses nei	*Serranidae*	84	130	3 161	748	98	98	97
Seabasses nei	*Dicentrarchus spp*	430	96	529	540	573	317	949
Brown meagre	*Sciaena umbra*	164	180	192	114	104	138	109
Shi drum	*Umbrina cirrosa*	25	51	32	40	28	33	42
Common pandora	*Pagellus erythrinus*	2 746	2 506	2 967	3 128	3 105	3 219	2 687
Sargo breams nei	*Diplodus spp*	2 706	3 177	2 055	3 036	3 084	3 241	3 226
Common dentex	*Dentex dentex*	227	346	375	291	250	219	216
Saddled seabream	*Oblada melanura*	43	36	29	146	84	51	53
Red porgy	*Pagrus pagrus*	254	327	362	388	460	416	400
Gilthead seabream	*Sparus aurata*	107	265	333	409	757	399	822
Bogue	*Boops boops*	1 928	2 205	2 466	3 890	3 052	2 852	3 209
Sand steenbras	*Lithognathus mormyrus*	694	661	781	683	656	781	663
Salema	*Sarpa salpa*	750	1 127	1 151	1 124	951	1 024	1 071

English name Nom anglais Nombre inglés	Scientific name Nom scientifique Nombre científico	1996 t	1997 t	1998 t	1999 t	2000 t	2001 t	2002 t
Porgies, seabreams nei	*Sparidae*	2 381	2 714	1 841	2 973	3 084	2 825	2 818
Blotched picarel	*Spicara maena*	285	505	395	419	255	381	593
Picarels nei	*Spicara spp*	335	485	3 258	1 005	646	572	858
Surmullet	*Mullus surmuletus*	1 161	2 406	1 459	1 504	1 527	2 163	1 803
Red mullet	*Mullus barbatus*	1 287	1 394	1 228	1 250	1 600	2 632	2 045
Greater weever	*Trachinus draco*	19	46	103	16	19	20	13
Grey triggerfish	*Balistes carolinensis*	38	52	58	34	50	62	96
European conger	*Conger conger*	0	14	14	5	42	7	5
John dory	*Zeus faber*	28	32	107	59	56	43	61
Silver scabbardfish	*Lepidopus caudatus*	26	7	57	423	411	375	263
Scorpionfishes nei	*Scorpaenidae*	339	472	755	442	400	444	453
Gurnards, searobins nei	*Triglidae*	107	87	115	141	168	198	346
Angler(=Monk)	*Lophius piscatorius*	6	33	30	44	39	35	43
Sardinellas nei	*Sardinella spp*	7 738	8 084	6 327	7 491	11 800	12 942	12 465
European pilchard(=Sardine)	*Sardina pilchardus*	10 389	10 767	8 907	13 394	15 001	13 988	13 311
European anchovy	*Engraulis encrasicolus*	19	55	114	199	2	269	258
Atlantic bonito	*Sarda sarda*	560	611	855	1 350	1 528	1 183	1 126
Frigate and bullet tunas	*Auxis thazard, A.rochei*	13	26	87	38	7	5	1
Little tunny(=Atl.black skipj)	*Euthynnus alletteratus*	824	333	1 113	752	1 453	1 036	961
Skipjack tuna	*Katsuwonus pelamis*	-	-	-	-	-	-	932
Atlantic bluefin tuna	*Thunnus thynnus*	2 393	2 200	1 745	2 352	2 184	2 493	2 528
Swordfish	*Xiphias gladius*	352	346	414	468	483	567	1 138
Tuna-like fishes nei	*Scombroidei*	215	657	6	814	905	989	104
Garfish	*Belone belone*	141	273	272	382	98	307	349
Silversides(=Sand smelts) nei	*Atherinidae*	141	116	338	6	57	309	-
Bluefish	*Pomatomus saltatrix*	1 317	1 426	787	748	1 096	1 037	1 285
Jack and horse mackerels nei	*Trachurus spp*	4 755	6 326	3 899	5 635	4 917	4 204	4 746
Greater amberjack	*Seriola dumerili*	75	113	75	88	400	122	93
Leerfish	*Lichia amia*	93	145	160	112	90	134	138
Common dolphinfish	*Coryphaena hippurus*	399	538	1 088	919	667	784	678
Chub mackerel	*Scomber japonicus*	1 439	4 118	3 233	3 515	2 200	3 727	2 480
Barracudas nei	*Sphyraena spp*	49	69	100	103	95	96	122
Catsharks, nursehounds nei	*Scyliorhinus spp*	36	72	...	49	126	122	139
Smooth-hounds nei	*Mustelus spp*	640	132	826	997	121	192	186
Dogfish sharks nei	*Squalidae*	19	806	44	25	680	883	945
Angelshark	*Squatina squatina*	18	34	44	25	20	22	13
Rays, stingrays, mantas nei	*Rajiformes*	489	803	836	922	974	1 113	1 092
Marine fishes nei	*Osteichthyes*	18 813	5 112	12 965	9 617	5 125	6 546	2 228
Mediterranean shore crab	*Carcinus aestuarii*	-	-	-	-	-	11	12
Palinurid spiny lobsters nei	*Palinurus spp*	40	71	68	61	49	41	45
Norway lobster	*Nephrops norvegicus*	3	1	4	2	4	4	4
European lobster	*Homarus gammarus*	1	1	-	1	1	0	0
Caramote prawn	*Penaeus kerathurus*	3 057	4 367	3 785	4 729	6 165	3 356	2 539
Speckled shrimp	*Metapenaeus monoceros*	-	-	-	-	-	757	1 607
Deepwater rose shrimp	*Parapenaeus longirostris*	161	641	838	1 014	1 283	1 454	1 536
Blue and red shrimp	*Aristeus antennatus*	195	173	187	38	16	51	51
Marine crustaceans nei	*Crustacea*	8	1	72	1	1	0	-
European flat oyster	*Ostrea edulis*	-	-	-	-	1	1	8
Grooved carpet shell	*Ruditapes decussatus*	396	77	57	48	755	544	1 002
Common cuttlefish	*Sepia officinalis*	4 340	6 479	4 935	6 622	6 002	7 148	7 995
Common squids nei	*Loligo spp*	230	253	288	348	310	327	510
Common octopus	*Octopus vulgaris*	3 460	5 681	3 129	2 349	2 103	1 752	2 535
Horned and musky octopuses	*Eledone spp*	...	668	252	392	369	696	878
Marine molluscs nei	*Mollusca*	-	-	-	-	-	1	10
	Country total	*83 734*	*87 012*	*88 075*	*93 186*	*95 550*	*98 482*	*96 685*
Uganda								
Cyprinids nei	*Cyprinidae*	1 539	13 476	15 710	17 600	12 181	12 182	12 000
Tilapias nei	*Oreochromis (=Tilapia) spp*	75 027	81 379	78 500	84 540	96 468	96 172	98 000
African lungfishes	*Protopterus spp*	6 095	6 554	7 595	9 192	5 755	5 796	5 000
Characins nei	*Characidae*	12 160	11 718	11 744	13 165	10 331	10 331	7 100
Naked catfishes	*Bagrus spp*	4 978	4 774	5 222	8 564	4 375	4 375	4 800
Torpedo-shaped catfishes nei	*Clarias spp*	1 916	2 130	2 170	2 833	2 987	2 987	2 500
Freshwater siluroids nei	*Siluroidei*	120	868	887	1 000	...	...	...
Nile perch	*Lates niloticus*	81 253	91 706	98 800	89 203	87 257	88 881	90 698
Freshwater fishes nei	*Osteichthyes*	12 000	5 421	-	-	2	2	1 800
	Country total	*195 088*	*218 026*	*220 628*	*226 097*	*219 356*	*220 726*	*221 898*
Westn Sahara								
Marine fishes nei	*Osteichthyes*	0	0	0	0	0	0	0
Marine crustaceans nei	*Crustacea*	0	0	0	0	0	0	0
Marine molluscs nei	*Mollusca*	0	0	0	0	0	0	0
	Country total	*0*	*0*	*0*	*0*	*0*	*0*	*0*
Zambia								
Dagaas	*Stolothrissa, Limnothrissa spp*	7 593	7 813	9 822	8 955	8 863	8 500 F	8 500 F
Freshwater fishes nei	*Osteichthyes*	58 739	58 110	60 116	58 372	57 808	56 500 F	56 500 F
	Country total	*66 332*	*65 923*	*69 938*	*67 327*	*66 671*	*65 000 F*	*65 000 F*
Zimbabwe								
Tilapias nei	*Oreochromis (=Tilapia) spp*	320	523	412	420	830	800 F	800 F
Dagaas	*Stolothrissa, Limnothrissa spp*	15 423	17 034	15 288	11 208	10 477	10 400 F	10 400 F
Freshwater fishes nei	*Osteichthyes*	644	599	671	782	1 807	1 800 F	1 800 F
	Country total	*16 387*	*18 156*	*16 371*	*12 410*	*13 114*	*13 000 F*	*13 000 F*

E-1

Fish crustaceans, molluscs, etc	Capture production by countries or areas and species	Africa
Poissons, crustacés, mollusques, etc	Captures par pays ou zones et espèces	Afrique
Peces, crustáceos, moluscos, etc	Capturas por países o áreas y especies	Africa

English name Nom anglais Nombre inglés	Scientific name Nom scientifique Nombre científico	1996 t	1997 t	1998 t	1999 t	2000 t	2001 t	2002 t
Total		*5 600 043*	*5 969 802*	*6 148 491*	*6 370 393*	*6 628 144*	*6 954 183*	*6 799 227*

English name Nom anglais Nombre inglés	Scientific name Nom scientifique Nombre científico	1996 t	1997 t	1998 t	1999 t	2000 t	2001 t	2002 t
Anguilla								
Freshwater fishes nei	*Osteichthyes*	0	0	0	0	0	0	0
Marine fishes nei	*Osteichthyes*	140 F	180 F	180 F	180 F	180 F	180 F	180 F
Caribbean spiny lobster	*Panulirus argus*	50 F	60 F	60 F	60 F	60 F	60 F	60 F
Stromboid conchs nei	*Strombus spp*	10 F	10 F	10 F	10 F	10 F	10 F	10 F
	Country total	*200 F*	*250 F*	*250 F*	*250 F*	*250 F*	*250 F*	*250 F*
Antigua Barb								
Freshwater fishes nei	*Osteichthyes*	0	0	0	0	0	0	0
Squirrelfishes nei	*Holocentridae*	...	...	...	...	...	29	45
Groupers, seabasses nei	*Serranidae*	...	...	...	...	...	217	364
Snappers, jobfishes nei	*Lutjanidae*	...	...	...	...	...	284	348
Grunts, sweetlips nei	*Haemulidae (=Pomadasyidae)*	...	...	...	...	...	167	259
Porgies, seabreams nei	*Sparidae*	...	...	...	...	...	9	15
Sea chubs nei	*Kyphosus spp*	...	...	...	...	...	8	4
Parrotfishes nei	*Scaridae*	...	...	...	...	...	173	252
Angelfishes nei	*Pomacanthidae*	...	...	...	...	...	...	28
Surgeonfishes nei	*Acanthuridae*	...	...	...	...	...	158	237
Boxfishes nei	*Ostraciidae*	...	...	...	...	...	66	38
Filefishes, leatherjackets nei	*Monacanthidae*	...	...	...	...	...	...	15
Triggerfishes, durgons nei	*Balistidae*	...	...	...	...	...	18	83
Tuna-like fishes nei	*Scombroidei*	...	...	...	...	...	28	12
Carangids nei	*Carangidae*	...	...	...	...	...	33	34
Common dolphinfish	*Coryphaena hippurus*	...	...	...	...	...	4	7
Barracudas nei	*Sphyraena spp*	...	...	...	...	...	6	8
Sharks, rays, skates, etc. nei	*Elasmobranchii*	...	...	...	...	...	8	17
Marine fishes nei	*Osteichthyes*	1 045	1 242	1 013	1 041	1 164	66	13
Caribbean spiny lobster	*Panulirus argus*	125	160	357	274	275	272	276
Stromboid conchs nei	*Strombus spp*	293	263	338	345	315	278	319
	Country total	*1 463*	*1 665*	*1 708*	*1 660*	*1 754*	*1 824*	*2 374*
Aruba								
Freshwater fishes nei	*Osteichthyes*	0	0	0	-	-	-	-
Groupers nei	*Epinephelus spp*	20	25	22	25	18	15	15 F
Snappers, jobfishes nei	*Lutjanidae*	60	60	50	50	45	45	45 F
Wahoo	*Acanthocybium solandri*	50	65	70	60	60	60	60 F
Atlantic sailfish	*Istiophorus albicans*	10	...	...	...	...	...	...
Marine fishes nei	*Osteichthyes*	20	55	40	40	40	43	43 F
	Country total	*160*	*205*	*182*	*175*	*163*	*163*	*163 F*
Bahamas								
Freshwater fishes nei	*Osteichthyes*	0	0	0	0	0	0	0
Nassau grouper	*Epinephelus striatus*	331	514	511	381	226	281	280 F
Groupers nei	*Epinephelus spp*	475	246	227	193	132	177	170 F
Snappers nei	*Lutjanus spp*	341	751	781	866	721	777	780 F
Grunts, sweetlips nei	*Haemulidae (=Pomadasyidae)*	14	67	90	66	62	67	67 F
Carangids nei	*Carangidae*	91	103	92	79	81	101	100 F
Sharks, rays, skates, etc. nei	*Elasmobranchii*	5	3	2	1	0	0	0
Marine fishes nei	*Osteichthyes*	385	264	156	139	110	133	150 F
Black stone crab	*Menippe mercenaria*	25	42	39	50	46	47	50 F
Caribbean spiny lobster	*Panulirus argus*	7 938	7 798	7 553	8 225	9 023	7 042	7 050 F
Stromboid conchs nei	*Strombus spp*	589	648	670	472	667	661	650 F
Marine turtles nei	*Testudinata*	3	3	3	1	2	4	3 F
	Country total	*10 197*	*10 439*	*10 124*	*10 473*	*11 070*	*9 290*	*9 300 F*
Barbados								
Freshwater fishes nei	*Osteichthyes*	0	0	0	0	0	-	-
Snappers, jobfishes nei	*Lutjanidae*	40	26	32	24	25	20	12
Wahoo	*Acanthocybium solandri*	35	52	52	41	41	24	41
Seerfishes nei	*Scomberomorus spp*	49	...	...	...	...	...	...
Skipjack tuna	*Katsuwonus pelamis*	5	5	10	3	3	1	2
Albacore	*Thunnus alalunga*	-	1	1	1	...	2	5
Yellowfin tuna	*Thunnus albacares*	160	149	150	155	155	142	115
Bigeye tuna	*Thunnus obesus*	-	24	17	18	18	6	11
Atlantic sailfish	*Istiophorus albicans*	25	71	58	44	44	...	...
Blue marlin	*Makaira nigricans*	25	30	25	19	19	...	...
Atlantic white marlin	*Tetrapturus albidus*	15	41	33	25	25	...	...
Marlins,sailfishes,etc. nei	*Istiophoridae*	-	-	-	-	-	85	53
Swordfish	*Xiphias gladius*	33	16	16	12	13	19	10
Tuna-like fishes nei	*Scombroidei*	68	-	-	-	11	-	-
Flyingfishes nei	*Exocoetidae*	2 042	1 566	2 680	2 075	1 916	1 673	1 590
Carangids nei	*Carangidae*	28	19	14	6	28	11	10
Common dolphinfish	*Coryphaena hippurus*	849	721	482	745	728	574	553
Sharks, rays, skates, etc. nei	*Elasmobranchii*	25	14	12	10	14	10	9
Marine fishes nei	*Osteichthyes*	113	74	62	72	60	109	89
Marine crustaceans nei	*Crustacea*	0	0	0	0	0	-	-
Marine molluscs nei	*Mollusca*	0	0	0	0	0	-	-
	Country total	*3 512*	*2 809*	*3 644*	*3 250*	*3 100*	*2 676*	*2 500*
Belize								
Freshwater fishes nei	*Osteichthyes*	0	0	0	0	0	0	-
Tadpole codling	*Salilota australis*	-	-	-	28	237	42	-
Southern blue whiting	*Micromesistius australis*	-	-	-	-	257	206	-

English name Nom anglais Nombre inglés	Scientific name Nom scientifique Nombre científico	1996 t	1997 t	1998 t	1999 t	2000 t	2001 t	2002 t
Argentine hake	*Merluccius hubbsi*	-	-	-	35	63	4	0
Benguela hake	*Merluccius polli*	-	-	-	54	...	...	...
Hakes nei	*Merluccius spp*	-	30	29	65	236	8	13
Patagonian grenadier	*Macruronus magellanicus*	-	-	-	84	1 720	374	1
Croakers, drums nei	*Sciaenidae*	-	1	65	299	580	1	...
Porgies, seabreams nei	*Sparidae*	-	183	1 152	2 826	2 188	129	85
Pink cusk-eel	*Genypterus blacodes*	-	-	-	15	87	8	0
Patagonian toothfish	*Dissostichus eleginoides*	-	-	-	16	27	11	0
Largehead hairtail	*Trichiurus lepturus*	-	300	317	1 121	248	7	56
Blackbellied angler	*Lophius budegassa*	-	-	-	-	-	-	2
Sardinellas nei	*Sardinella spp*	-	2 400	2 389	7 233	3 534	1 365	4 992
European pilchard(=Sardine)	*Sardina pilchardus*	-	2 126	2 229	616	414	127	1 975
European anchovy	*Engraulis encrasicolus*	-	300	278	3 251	3 932	2 359	5 099
Black skipjack	*Euthynnus lineatus*	40	-	20	-	-	-	-
Skipjack tuna	*Katsuwonus pelamis*	5 140	6 260	4 440	4 330	5 282	-	-
Albacore	*Thunnus alalunga*	-	-	-	8	2	...	...
Yellowfin tuna	*Thunnus albacares*	2 210	1 870	3 490	2 490	1 467	-	-
Bigeye tuna	*Thunnus obesus*	2 600	4 720	1 160	320	598	-	-
Black marlin	*Makaira indica*	-	-	-	10	4	...	...
Atlantic white marlin	*Tetrapturus albidus*	-	1	-	1	0	...	...
Swordfish	*Xiphias gladius*	-	-	-	17	8	...	...
Tuna-like fishes nei	*Scombroidei*	-	-	-	107	143	125	172
Jack and horse mackerels nei	*Trachurus spp*	-	1 600	1 626	5 924	5 714	3 731	5 380
Chub mackerel	*Scomber japonicus*	-	1 100	1 051	1 732	3 194	1 041	3 698
Rays, stingrays, mantas nei	*Rajiformes*	-	-	-	519	48	201	10
Sharks, rays, skates, etc. nei	*Elasmobranchii*	...	1	0	2	6	...	5
Marine fishes nei	*Osteichthyes*	400	493	1 744	5 420	7 892	938	1 151
Black stone crab	*Menippe mercenaria*	53	210	25	29	16	8	4
Caribbean spiny lobster	*Panulirus argus*	448	534	468	552	503	394	525
Penaeus shrimps nei	*Penaeus spp*	38	43	589	1 318	1 236	110	81
Natantian decapods nei	*Natantia*	-	-	2	28	6	11	...
Stromboid conchs nei	*Strombus spp*	1 105	1 926	1 891	1 051	1 745	1 980	1 380
Cuttlefish,bobtail squids nei	*Sepiidae, Sepiolidae*	-	406	2 445	4 911	4 869	19	...
Patagonian squid	*Loligo gahi*	-	-	-	-	2	-	-
Argentine shortfin squid	*Illex argentinus*	-	-	-	3 796	4 066	1 692	124
Various squids nei	*Loliginidae, Ommastrephidae*	4	21	114	128	31	1	...
Octopuses, etc. nei	*Octopodidae*	-	185	785	3 247	1 185	...	...
	Country total	*12 038*	*24 710*	*26 309*	*51 583*	*51 540*	*14 892*	*24 753*
Bermuda								
Freshwater fishes nei	*Osteichthyes*	0	0	0	0	0	0	-
Groupers nei	*Epinephelus spp*	42	42	43	48	27	45	50
Snappers, jobfishes nei	*Lutjanidae*	38	29	29	28	23	32	33
Wahoo	*Acanthocybium solandri*	99	105	108	104	61	56	87
Little tunny(=Atl.black skipj)	*Euthynnus alletteratus*	7	6	5	4	2	4	5
Skipjack tuna	*Katsuwonus pelamis*	0	0	0	0	0	1	1
Atlantic bluefin tuna	*Thunnus thynnus*	1	2	2	1	1	1	1
Blackfin tuna	*Thunnus atlanticus*	5	4	6	6	5	4	5
Albacore	*Thunnus alalunga*	-	1	...	2	2	2	1
Yellowfin tuna	*Thunnus albacares*	67	55	53	59	31	37	47
Blue marlin	*Makaira nigricans*	15	3	5	1	2	2	2
Atlantic white marlin	*Tetrapturus albidus*	1	1	1	1	0	1	0
Marlins,sailfishes,etc. nei	*Istiophoridae*	-	3	1	-	-	-	-
Swordfish	*Xiphias gladius*	1	5	5	3	3	2	2
Carangids nei	*Carangidae*	70	50	48	51	30	41	52
Sharks, rays, skates, etc. nei	*Elasmobranchii*	13	9	12	24	10	5	5
Marine fishes nei	*Osteichthyes*	96	108	113	80	55	57	72
Caribbean spiny lobster	*Panulirus argus*	10	38	30	36	29	21	26
Tropical spiny lobsters nei	*Panulirus spp*	-	-	5	5	5	4	4
	Country total	*465*	*461*	*466*	*453*	*286*	*315*	*393*
Br Virgin Is								
Freshwater fishes nei	*Osteichthyes*	0	0	0	0	0	0	0
Groupers nei	*Epinephelus spp*	69	1	3	4	1	1 F	1 F
Snappers nei	*Lutjanus spp*	6	1	4	3	1	1 F	1 F
Yellowtail snapper	*Ocyurus chrysurus*	...	5	9	9	0	0	0
Boxfishes nei	*Ostraciidae*	...	1	1	1	0	0	0
Clupeoids nei	*Clupeoidei*	...	5	5	5	0	0	0
Atlantic bonito	*Sarda sarda*	6	8	6	9	0	0	0
Seerfishes nei	*Scomberomorus spp*	...	6	6	5	0	0	0
Yellowfin tuna	*Thunnus albacares*	...	3	2	3	1	1 F	1 F
Swordfish	*Xiphias gladius*	54	6	5	5	2	2 F	2 F
Tuna-like fishes nei	*Scombroidei*	19	-	-	-	-	-	-
Jacks, crevalles nei	*Caranx spp*	...	13	15	14	1	1 F	1 F
Common dolphinfish	*Coryphaena hippurus*	6	3	2	3	1	1 F	1 F
Sharks, rays, skates, etc. nei	*Elasmobranchii*	...	1	1	1	0	0	0
Marine fishes nei	*Osteichthyes*	265	39	42	41	27	34 F	34 F
Caribbean spiny lobster	*Panulirus argus*	27	5	6	4	3	3 F	3 F
Stromboid conchs nei	*Strombus spp*	54	8	9	8	6	6 F	6 F
	Country total	*506*	*105*	*116*	*115*	*43*	*50 F*	*50 F*
Canada								
Common carp	*Cyprinus carpio*	687	543	354	741	516	506	878
Northern pike	*Esox lucius*	2 406	2 334	2 568	2 615	2 808	2 464	2 684
Catfishes nei	*Ictalurus spp*	37	29	27	35	19	344	338
American yellow perch	*Perca flavescens*	1 440	2 073	2 083	1 884	1 847	2 524	3 622

English name Nom anglais Nombre inglés	Scientific name Nom scientifique Nombre científico	1996 t	1997 t	1998 t	1999 t	2000 t	2001 t	2002 t
Walleye	*Stizostedion vitreum*	7 718	7 751	8 206	8 782	9 160	6 949	7 007
Freshwater fishes nei	*Osteichthyes*	9 649	8 089	8 914	9 317	10 756	8 247	10 568
Sturgeons nei	*Acipenseridae*	299	421	418	291	284	177	172
American eel	*Anguilla rostrata*	840	766	774	631	681	474	472
Atlantic salmon	*Salmo salar*	82	77	7	1	-	-	1
Trouts nei	*Salmo spp*	4	1	0	0	0	-	-
Pink(=Humpback)salmon	*Oncorhynchus gorbuscha*	8 597	12 241	3 920	9 529	7 158	10 575	8 609
Chum(=Keta=Dog)salmon	*Oncorhynchus keta*	6 524	8 685	19 903	4 937	2 783	5 549	12 332
Sockeye(=Red)salmon	*Oncorhynchus nerka*	15 525	25 353	5 041	1 653	8 665	6 231	10 050
Chinook(=Spring=King)salmon	*Oncorhynchus tshawytscha*	455	1 662	1 386	742	507	636	1 655
Coho(=Silver)salmon	*Oncorhynchus kisutch*	3 871	751	16	14	31	46	453
Rainbow trout	*Oncorhynchus mykiss*	3	1	4	1	1	-	-
Lake trout(=Char)	*Salvelinus namaycush*	689	617	554	491	556	679	532
Chars nei	*Salvelinus spp*	11	31	35	46	46	34	19
Rainbow smelt	*Osmerus mordax*	1 078	1 227	1 255	356	970	308	314
Pond smelt	*Hypomesus olidus*	3 981	5 964	6 471	5 691	3 254	4 241	3 263
Surf smelt	*Hypomesus pretiosus*	1	-	-	-	-	-	-
Eulachon	*Thaleichthys pacificus*	30	0	0	-	-	1	5
Lake(=Common)whitefish	*Coregonus clupeaformis*	9 121	8 376	8 599	8 330	9 028	9 624	8 623
Lake cisco	*Coregonus artedi*	919	1 008	764	565	581	567	503
Salmonoids nei	*Salmonoidei*	112	-	2	1	-	-	-
American shad	*Alosa sapidissima*	138	140	64	64	79	30	26
Alewife	*Alosa pseudoharengus*	6 289	6 991	8 468	7 673	6 783	4 349	6 555
Striped bass	*Morone saxatilis*	15	0	-	-	-	-	-
Atlantic halibut	*Hippoglossus hippoglossus*	1 097	1 362	1 300	1 188	1 219	1 647	1 649
Pacific halibut	*Hippoglossus stenolepis*	5 430	7 448	7 714	7 103	6 095	4 766	6 487
Greenland halibut	*Reinhardtius hippoglossoides*	14 617	16 399	12 222	12 657	16 444	13 814	11 237
Arrow-tooth flounder	*Atheresthes stomias*	24	-	33	44	39	59	22
Witch flounder	*Glyptocephalus cynoglossus*	1 600	1 675	2 094	1 806	2 168	2 128	2 399
Amer. plaice(=Long rough dab)	*Hippoglossoides platessoides*	2 525	3 583	3 546	4 359	3 904	4 466	3 714
Yellowtail flounder	*Limanda ferruginea*	1 178	1 728	5 242	8 222	12 790	15 648	13 001
Winter flounder	*Pseudopleuronectes americanus*	2 468	2 548	1 804	1 763	2 342	2 257	1 775
Flatfishes nei	*Pleuronectiformes*	6 741	5 583	6 025	6 635	6 773	6 076	6 944
Tusk(=Cusk)	*Brosme brosme*	1 405	1 801	1 641	1 065	1 097	1 498	1 288
Atlantic cod	*Gadus morhua*	15 541	29 899	37 741	55 410	46 046	40 325	35 255
Pacific cod	*Gadus macrocephalus*	700	1 537	1 400	836	712	474	694
Red hake	*Urophycis chuss*	372	248	120	156	81	130	115
White hake	*Urophycis tenuis*	4 827	4 309	3 102	3 318	4 246	4 140	4 585
Haddock	*Melanogrammus aeglefinus*	10 311	9 776	11 771	10 550	12 683	15 594	14 947
Atlantic tomcod	*Microgadus tomcod*	140	90	119	42	55	57	1
Saithe(=Pollock)	*Pollachius virens*	9 739	12 604	15 092	8 552	6 528	7 190	7 617
Alaska pollock(=Walleye poll.)	*Theragra chalcogramma*	2 150	1 800	800	1 233	1 044	1 747	3 606
Silver hake	*Merluccius bilinearis*	4 209	5 333	10 490	9 676	11 870	17 929	13 725
Roundnose grenadier	*Coryphaenoides rupestris*	236	75	2	-	1	2	3
Sandeels(=Sandlances) nei	*Ammodytes spp*	2	1	0	-	2	-	-
Lingcod	*Ophiodon elongatus*	2 500	1 700	1 900	2 523	3 042	2 511	2 673
Argentines	*Argentina spp*	259	591	51	12	8	17	20
Wolffishes(=Catfishes) nei	*Anarhichas spp*	422	856	526	694	678	578	-
Pacific ocean perch	*Sebastes alutus*	6 174	5 782	6 184	5 849	6 177	5 832	5 897
Atlantic redfishes nei	*Sebastes spp*	21 571	18 751	25 763	19 682	19 852	19 710	16 745
Scorpionfishes nei	*Scorpaenidae*	16 626	14 118	15 356	18 086	17 188	16 351	16 248
Sablefish	*Anoplopoma fimbria*	3 069	4 000	4 500	4 583	2 811	2 967	1 449
Lumpfish(=Lumpsucker)	*Cyclopterus lumpus*	18	175	8	1	8	-	-
American angler	*Lophius americanus*	1 623	2 073	1 466	1 274	1 265	1 900	3 634
Atlantic herring	*Clupea harengus*	188 843	186 550	190 861	202 046	203 261	199 295	191 097
Pacific herring	*Clupea pallasii*	22 221	31 500	33 500	28 847	29 290	24 189	29 321
Skipjack tuna	*Katsuwonus pelamis*	-	0	-	-	-	-	-
Atlantic bluefin tuna	*Thunnus thynnus*	599	507	595	576	550	524	604
Albacore	*Thunnus alalunga*	617	260	328	600	3 008	3 318	3 752
Yellowfin tuna	*Thunnus albacares*	155	100	57	22	105	125	70
Bigeye tuna	*Thunnus obesus*	144	166	120	263	327	241	280
Atlantic white marlin	*Tetrapturus albidus*	8	8	8	5	5	3	2
Swordfish	*Xiphias gladius*	739	1 089	1 115	1 118	968	1 079	959
Tuna-like fishes nei	*Scombroidei*	-	-	0	-	-	-	-
Capelin	*Mallotus villosus*	32 628	21 945	38 249	23 495	21 352	19 747	13 559
Atlantic silverside	*Menidia menidia*	151	238	231	558	304	646	135
Atlantic mackerel	*Scomber scombrus*	21 027	21 305	19 557	16 317	17 637	24 488	43 092
Porbeagle	*Lamna nasus*	1 029	1 343	1 006	958	904	499	238
Picked dogfish	*Squalus acanthias*	4 431	2 546	3 581	8 353	7 524	8 303	8 290
Dogfish sharks, etc. nei	*Squaliformes*	106	169	115	143	120	92	107
Raja rays nei	*Raja spp*	3 900	4 373	3 044	3 119	2 066	2 625	2 969
Rays, stingrays, mantas nei	*Rajiformes*	1 293	1 584	900	1 406	1 661	1 618	1 540
Groundfishes nei	*Osteichthyes*	597	475	365	770	1 003	831	-
Pelagic fishes nei	*Osteichthyes*	36	128	25	31	29	208	-
Finfishes nei	*Osteichthyes*	1	-	-	1	-	-	-
Marine fishes nei	*Osteichthyes*	105 185	102 100	96 130	97 253	26 979	69 582	-
Atlantic rock crab	*Cancer irroratus*	4 286	6 410	6 338	5 732	7 672	7 968	7 829
Dungeness crab	*Cancer magister*	5 025	3 923	2 968	2 944	2 832	5 685	4 090
Queen crab	*Chionoecetes opilio*	65 825	71 416	75 216	95 148	93 505	95 299	106 766
Marine crabs nei	*Brachyura*	1 588	1 948	2 267	1 835	3 052	4 306	4 373
American lobster	*Homarus americanus*	39 369	40 079	41 030	43 428	45 331	51 412	45 111
Northern prawn	*Pandalus borealis*	31 340	48 310	78 867	85 331	100 091	94 328	102 097
Pandalus shrimps nei	*Pandalus spp*	24 976	28 601	29 042	30 603	35 180	31 234	33 342
Natantian decapods nei	*Natantia*	9 392	5 200	5 203	4 071	4 223	4 212	3 622
Periwinkles nei	*Littorina spp*	200	274	198	149	98	92	24
Whelks	*Busycon spp*	1 291	1 275	1 062	1 557	1 923	2 041	2 218
American cupped oyster	*Crassostrea virginica*	2 132	1 726	3 124	3 225	4 133	3 009	2 410
Blue mussel	*Mytilus edulis*	8 147	9 163	10 273	11 565	15 089	11 829	13 769
American sea scallop	*Placopecten magellanicus*	47 807	53 881	56 399	54 756	83 852	89 196	93 971

English name Nom anglais Nombre inglés	Scientific name Nom scientifique Nombre científico	1996 t	1997 t	1998 t	1999 t	2000 t	2001 t	2002 t
Iceland scallop	*Chlamys islandica*	12 180	12 145	6 632	3 160	2 764	1 347	989
Ocean quahog	*Arctica islandica*	114	-	-	66	-	164	92
Butter clam	*Saxidomus giganteus*	1 431	1 206	1 600	945	949	1 111	1 370
Northern quahog(=Hard clam)	*Mercenaria mercenaria*	1 109	1 656	1 234	2 536	1 252	1 657	2 519
Atlantic surf clam	*Spisula solidissima*	1 204	800	444	303	496	602	613
Stimpson's surf clam	*Spisula polynyma*	25 597	27 322	25 976	26 699	22 979	20 257	19 960
Sand gaper	*Mya arenaria*	1 469	2 323	2 638	2 680	3 026	3 152	3 073
Pacific geoduck	*Panopea abrupta*	1 768	1 757	1 784	1 688	1 562	1 413	1 706
Clams, etc. nei	*Bivalvia*	773	259	49	791	345	1 078	2 459
Northern shortfin squid	*Illex illecebrosus*	8 822	15 775	1 944	314	368	65	253
Octopuses, etc. nei	*Octopodidae*	197	208	139	54	56	61	125
Marine molluscs nei	*Mollusca*	1 999	3 308	2 003	1 633	2 535	2 181	4 349
Sea urchins nei	*Strongylocentrotus spp*	9 665	9 221	9 867	9 052	8 012	7 063	6 441
	Country total	*909 511*	*975 548*	*1 013 929*	*1 027 855*	*1 012 099*	*1 052 543*	*1 013 997*
Cayman Is								
Freshwater fishes nei	*Osteichthyes*	0	0	0	0	0	0	0
Marine fishes nei	*Osteichthyes*	110	125	125	125	125	125	125
	Country total	*110*	*125*	*125*	*125*	*125*	*125*	*125*
Costa Rica								
Freshwater fishes nei	*Osteichthyes*	1 090	840	1 000 F	1 000 F	1 000 F	1 000 F	1 000 F
Groupers, seabasses nei	*Serranidae*	177	387	644	396	204	103	104
Snappers, jobfishes nei	*Lutjanidae*	352	244	478	526	417	347	382
Clupeoids nei	*Clupeoidei*	438	1 175	906	1 788	1 628	2 207	4 170
Skipjack tuna	*Katsuwonus pelamis*	620	...	...	96	...	...	...
Yellowfin tuna	*Thunnus albacares*	50	...	...	30	...	...	...
Bigeye tuna	*Thunnus obesus*	840	...	...	20	...	...	...
Atlantic white marlin	*Tetrapturus albidus*	...	...	...	3	14	-	-
Marlins,sailfishes,etc. nei	*Istiophoridae*	1 892	1 876	1 647	2 143	2 035	2 206	2 273
Swordfish	*Xiphias gladius*	433	1 072	419	100	409	653	638
Tuna-like fishes nei	*Scombroidei*	2 572	990	1 213	1 063	1 136	1 163	1 585
Common dolphinfish	*Coryphaena hippurus*	...	...	...	...	8 370	11 221	7 832
Shortfin mako	*Isurus oxyrinchus*	39	32	91	56	8	10	13
Silky shark	*Carcharhinus falciformis*	1 546	1 595	2 121	2 179	1 741	1 090	1 523
Sharks, rays, skates, etc. nei	*Elasmobranchii*	1 912	3 922	5 512	5 662	11 152	8 559	7 471
Marine fishes nei	*Osteichthyes*	7 616	10 727	7 868	9 833	4 541	4 127	3 885
Marine crabs nei	*Brachyura*	3	50	9	8	4	3	3
Caribbean spiny lobster	*Panulirus argus*	196	196	40	163	271	39	57
Tropical spiny lobsters nei	*Panulirus spp*	7	7	3	4	14	17	11
Northern brown shrimp	*Penaeus aztecus*	8	0	0	0	0	-	-
Yellowleg shrimp	*Penaeus californiensis*	6	2	12	12	6	2	2
Crystal shrimp	*Penaeus brevirostris*	408	392	276	548	350	456	343
Penaeus shrimps nei	*Penaeus spp*	1 923	1 615	1 105	1 141	771	717	895
Pacific seabobs	*Xiphopenaeus,Trachypenaeus spp*	576	314	586	310	198	228	125
Natantian decapods nei	*Natantia*	1 362	1 022	672	1 004	1 009	508	561
Various squids nei	*Loliginidae, Ommastrephidae*	20	11	1	2	7	11	30
Octopuses, etc. nei	*Octopodidae*	27	58	6	76	71	66	35
Marine molluscs nei	*Mollusca*	65	109	62	55	42	-	-
Marine turtles nei	*Testudinata*	149	33	86	0	0		
	Country total	*24 327*	*26 669*	*24 757*	*28 218*	*35 398*	*34 733*	*32 938*
Cuba								
Blue tilapia	*Oreochromis aureus*	9 672	7 553	5 122	4 564	2 940	1 948	2 002
Atlantic halibut	*Hippoglossus hippoglossus*	-	0	0	0	-	-	-
Witch flounder	*Glyptocephalus cynoglossus*	-	2	-	-	-	-	-
Amer. plaice(=Long rough dab)	*Hippoglossoides platessoides*	-	22	49	19	-	-	-
Flatfishes nei	*Pleuronectiformes*	-	-	-	1	-	-	-
Atlantic cod	*Gadus morhua*	1	0	1	5	-	-	-
Red hake	*Urophycis chuss*	427	259	118	87	-	-	-
Haddock	*Melanogrammus aeglefinus*	40	27	12	4	-	-	-
Saithe(=Pollock)	*Pollachius virens*	125	57	8	6	2	-	-
Silver hake	*Merluccius bilinearis*	22 526	12 697	6 280	3 853	27	-	-
Mullets nei	*Mugilidae*	93	159	122	108	119	114	125
Nassau grouper	*Epinephelus striatus*	124	79	57	57	31	44	42
Red grouper	*Epinephelus morio*	269	195	69	110	171	91	86
Groupers nei	*Epinephelus spp*	130	88	89	44	82	22	61
Southern red snapper	*Lutjanus purpureus*	1 002	909	807	763	689	773	885
Lane snapper	*Lutjanus synagris*	1 848	2 472	2 609	1 712	2 114	1 891	1 844
Yellowtail snapper	*Ocyurus chrysurus*	1 176	727	457	409	408	413	370
Snappers, jobfishes nei	*Lutjanidae*	601	403	509	388	444	641	844
Grunts, sweetlips nei	*Haemulidae (=Pomadasyidae)*	1 723	1 451	1 207	1 303	1 079	986	971
Porgies	*Calamus spp*	385	333	270	259	335	310	301
Porgies, seabreams nei	*Sparidae*	315	53	35	7	16	10	13
Mojarras, etc. nei	*Gerreidae*	1 719	1 346	1 199	1 013	1 081	1 104	858
Argentines	*Argentina spp*	223	553	4	5	-	-	-
Cusk-eels, brotulas nei	*Ophidiidae*	118	97	37	13	17	17	11
Atlantic redfishes nei	*Sebastes spp*	42	87	12	39	0	-	-
American angler	*Lophius americanus*	-	-	18	-	-	-	-
Atlantic herring	*Clupea harengus*	213	253	134	121	-	-	-
Round sardinella	*Sardinella aurita*	...	5	93	-	1 510	1 575	-
Scaled sardines	*Harengula spp*	707	947	649	766	1 562	1 770	506
Atlantic thread herring	*Opisthonema oglinum*	2 361	1 900	1 286	1 756	1 522	1 676	1 799
Seerfishes nei	*Scomberomorus spp*	613	466	236	247	222	184	263
Little tunny(=Atl.black skipj)	*Euthynnus alletteratus*	23	17	9	3	2	1	3
Skipjack tuna	*Katsuwonus pelamis*	1 000	1 282	1 303	750	665	443	462

English name Nom anglais Nombre inglés	Scientific name Nom scientifique Nombre científico	1996 t	1997 t	1998 t	1999 t	2000 t	2001 t	2002 t
Blackfin tuna	*Thunnus atlanticus*	287	1	226	309	237	299	209
Albacore	*Thunnus alalunga*	40	5	-	-	-	-	-
Yellowfin tuna	*Thunnus albacares*	257	269	-	34	284	156	16
Bigeye tuna	*Thunnus obesus*	5	-	-	-	-	-	-
Atlantic sailfish	*Istiophorus albicans*	566	40	28	196	208	68	32
Blue marlin	*Makaira nigricans*	1	53	12	38	55	56	11
Swordfish	*Xiphias gladius*	67	10	10	5	11	3	1
Tuna-like fishes nei	*Scombroidei*	21	-	-	-	-	-	-
Jacks, crevalles nei	*Caranx spp*	349	234	499	167	2 559	2 138	180
Common dolphinfish	*Coryphaena hippurus*	5	362	-	-	-	-	-
Atlantic mackerel	*Scomber scombrus*	77	115	6	13	-	-	-
Pelagic percomorphs nei	*Perciformes*	-	-	-	2 427	-	-	-
Picked dogfish	*Squalus acanthias*	-	6	-	-	-	-	-
Raja rays nei	*Raja spp*	-	0	1	-	-	-	-
Rays, stingrays, mantas nei	*Rajiformes*	955	1 358	1 334	1 352	1 193	1 119	1 278
Sharks, rays, skates, etc. nei	*Elasmobranchii*	2 460	1 908	3 072	2 847	2 264	2 396	2 344
Groundfishes nei	*Osteichthyes*	12	162	-	-	144	-	-
Pelagic fishes nei	*Osteichthyes*	-	-	-	13	-	-	-
Marine fishes nei	*Osteichthyes*	14 018	17 278	13 750	25 202	22 918	11 384	5 588
Freshwater crustaceans nei	*Crustacea*	...	...	2	1	22	13	11
Black stone crab	*Menippe mercenaria*	28	9	35	52	54	76	48
Blue crab	*Callinectes sapidus*	897	704	1 065	566	584	528	410
Marine crabs nei	*Brachyura*	348	668	763	851	907	865	257
Caribbean spiny lobster	*Panulirus argus*	9 375	8 996	9 417	9 879	7 478	6 776	7 972
Northern pink shrimp	*Penaeus duorarum*	1 710	1 998	1 368	1 579	1 588	1 483	1 308
Penaeus shrimps nei	*Penaeus spp*	...	...	2	1	-	1	-
Northern prawn	*Pandalus borealis*	-	-	-	120	46	-	-
Marine crustaceans nei	*Crustacea*	25	18	13	3	6	-	3
Stromboid conchs nei	*Strombus spp*	717	1 234	487	829	1 163	1 172	997
Gastropods nei	*Gastropoda*	132	15	4	-	-	-	-
Mangrove cupped oyster	*Crassostrea rhizophorae*	1 888	2 011	2 282	1 865	1 631	1 482	1 300
Ark clams nei	*Arca spp*	1 860	1 729	2 438	2 043	2 591	633	375
Northern shortfin squid	*Illex illecebrosus*	493	2 978	1 084	280	-	-	-
Various squids nei	*Loliginidae, Ommastrephidae*	...	-	0	-	-	-	-
Frogs	*Rana spp*	69	46	28	26	43	45	21
Green turtle	*Chelonia mydas*	33	18	14	8	10	5	6
Hawksbill turtle	*Eretmochelys imbricata*	23	19	12	12	18	11	17
Loggerhead turtle	*Caretta caretta*	10	7	5	5	9	7	8
River and lake turtles nei	*Testudinata*	3	4	2	-	-	1	-
Marine turtles nei	*Testudinata*	1	0	10	-	-	-	-
	Country total	*84 208*	*76 694*	*60 770*	*69 135*	*61 061*	*44 730*	*33 838*
Dominica								
Freshwater fishes nei	*Osteichthyes*	0	0	0	0	0	0	0
Wahoo	*Acanthocybium solandri*	58	58	58	50	46	11	20
King mackerel	*Scomberomorus cavalla*	-	-	-	36	35	2	...
Skipjack tuna	*Katsuwonus pelamis*	33	33	33	85	86	45	55
Blackfin tuna	*Thunnus atlanticus*	...	...	...	79	83	54	78
Yellowfin tuna	*Thunnus albacares*	...	...	...	80	78	120	169
Bigeye tuna	*Thunnus obesus*	...	...	...	...	...	5	-
Marlins,sailfishes,etc. nei	*Istiophoridae*	...	...	...	...	...	69	72
Swordfish	*Xiphias gladius*	...	...	...	...	...	1	-
Tuna-like fishes nei	*Scombroidei*	-	-	-	-	-	12	36
Common dolphinfish	*Coryphaena hippurus*	...	...	...	...	...	...	211
Marine fishes nei	*Osteichthyes*	939	988	1 121	870 F	872 F	881 F	576
	Country total	*1 030*	*1 079*	*1 212*	*1 200 F*	*1 200 F*	*1 200 F*	*1 217*
Dominican Rp								
Common carp	*Cyprinus carpio*	62	109	27	180	...	397	705
Tilapias nei	*Oreochromis (=Tilapia) spp*	100	207	77	439	...	529	1 175
Largemouth black bass	*Micropterus salmoides*	...	...	...	...	...	...	425
Freshwater fishes nei	*Osteichthyes*	105	751	990	96	179	183	1
American eel	*Anguilla rostrata*	...	...	1	2	7	1	25
Tarpon	*Megalops atlanticus*	31	27	47	12	14	14	34
Bonefish	*Albula vulpes*	151	16	34	125	10	11	6
Squirrelfishes nei	*Holocentridae*	70	35	56	16	24	22	130
Mountain mullet	*Agonostomus monticola*	...	...	...	...	...	...	28
Mullets nei	*Mugilidae*	43	59	41	22	21	28	219
Snooks(=Robalos) nei	*Centropomus spp*	20	24	64	12	32	17	28
Red grouper	*Epinephelus morio*	29	92	11	21	33	37	0
Groupers nei	*Epinephelus spp*	455	338	379	662	873	942	1 077
Southern red snapper	*Lutjanus purpureus*	316	496	157	71	126	355	337
Yellowtail snapper	*Ocyurus chrysurus*	793	529	190	234	249	356	125
Snappers, jobfishes nei	*Lutjanidae*	130	235	875	49	753	231	1 848
Grunts, sweetlips nei	*Haemulidae (=Pomadasyidae)*	249	325	205	423	348	385	347
Weakfishes nei	*Cynoscion spp*	56	14	14	7	10	10	136
Porgies	*Calamus spp*	428	1 112	17	430	17	10	27
Western Atlantic seabream	*Archosargus rhomboidalis*	...	...	...	...	...	...	53
Goatfishes, red mullets nei	*Mullidae*	265	178	168	90	171	161	40
Mojarras, etc. nei	*Gerreidae*	45	12	20	9	14	15	10
Wrasses, hogfishes, etc. nei	*Labridae*	529	1 456	52	52	69	22	597
Parrotfishes nei	*Scaridae*	80	94	109	7	18	19	52
Triggerfishes, durgons nei	*Balistidae*	31	26	67	7	10	21	99
Scaled sardines	*Harengula spp*	111	64	55	18	27	21	0
Atlantic thread herring	*Opisthonema oglinum*	127	188	111	99	130	106	0
Wahoo	*Acanthocybium solandri*	...	325	112	31	42	37	0
King mackerel	*Scomberomorus cavalla*	1 648	589	288	230	271	261	492

English name Nom anglais Nombre inglés	Scientific name Nom scientifique Nombre científico	1996 t	1997 t	1998 t	1999 t	2000 t	2001 t	2002 t
Cero	*Scomberomorus regalis*	57	231	191	125	190	147	...
Skipjack tuna	*Katsuwonus pelamis*	123	...	...	...	23	32	...
Blackfin tuna	*Thunnus atlanticus*	518	...	...	...	...	...	...
Albacore	*Thunnus alalunga*	...	323	121	73	95	...	295
Yellowfin tuna	*Thunnus albacares*	-	-	89	220	272	263	...
Atlantic sailfish	*Istiophorus albicans*	25	101	89	27	81	81	46
Blue marlin	*Makaira nigricans*	-	41	71	29	23	23	204
Tuna-like fishes nei	*Scombroidei*	85	855	354	192	249	517	668
Needlefishes, etc. nei	*Belonidae*	39	5	4	7	15	45	103
Blue runner	*Caranx crysos*	194	256	132	53	74	68	130
Jacks, crevalles nei	*Caranx spp*	424	453	75	23	43	43	111
Pompanos nei	*Trachinotus spp*	57	19	47	5	10	7	111
Amberjacks nei	*Seriola spp*	50	46	50	12	16	18	41
Carangids nei	*Carangidae*	30	10	27	6	22	28	126
Common dolphinfish	*Coryphaena hippurus*	237	113	151	175	255	232	571
Barracudas nei	*Sphyraena spp*	3	-	8	4	5	7	33
Nurse shark	*Ginglymostoma cirratum*	-	-	-	-	407	89	17
Rays, stingrays, mantas nei	*Rajiformes*	39	96	62	134	111	123	10
Marine fishes nei	*Osteichthyes*	2 650	1 874	628	1 896	2 401	4 391	462
Freshwater crustaceans nei	*Crustacea*	21	-	-	16	1	48	4
Marine crabs nei	*Brachyura*	32	14	37	7	14	52	150
Caribbean spiny lobster	*Panulirus argus*	420	1 061	863	828	1 286	1 209	2 977
Penaeus shrimps nei	*Penaeus spp*	47	79	77	49	...	32	658
Stromboid conchs nei	*Strombus spp*	1 889	1 594	2 683	1 257	1 778	1 437	2 332
Mangrove cupped oyster	*Crassostrea rhizophorae*	6	6	12	4	7	8	135
Common squids nei	*Loligo spp*	38	24	84	12	56	51	23
Common octopus	*Octopus vulgaris*	36	33	87	39	99	57	73
Marine molluscs nei	*Mollusca*	...	...	...	...	48	18	1 043
	Country total	*12 894*	*14 535*	*10 109*	*8 537*	*11 029*	*13 217*	*18 339*
El Salvador								
Nile tilapia	*Oreochromis niloticus*	1 265	1 297	1 017	1 216	1 171	560	1 169
Jaguar guapote	*Cichlasoma managuense*	566	367	344	351	324	410	318
Catfishes nei	*Ictalurus spp*	240	230	213	208	142	169	125
Freshwater fishes nei	*Osteichthyes*	891	912	863	867	1 177	535	1 022
Sea catfishes nei	*Ariidae*	107	62	114	101	172	192	716
Snappers, jobfishes nei	*Lutjanidae*	197	117	252	230	282	481	530
Croakers, drums nei	*Sciaenidae*	313	337	252	334	279	575	826
Skipjack tuna	*Katsuwonus pelamis*	...	...	750	3 130	...	4 476	6 804
Yellowfin tuna	*Thunnus albacares*	...	...	920	3 250	...	2 165	3 434
Bigeye tuna	*Thunnus obesus*	...	...	10	230	...	2 059	4 590
Scads nei	*Decapterus spp*	237	130	149	226	187	270	295
Sharks, rays, skates, etc. nei	*Elasmobranchii*	347	1 186	266	176	364	759	951
Marine fishes nei	*Osteichthyes*	2 267	1 791	1 595	1 842	1 925	1 678	3 385
Freshwater crustaceans nei	*Crustacea*	5	0	3	11	9	16	22
Tropical spiny lobsters nei	*Panulirus spp*	-	-	-	-	3	2	5
Pelagic red crab	*Pleuroncodes planipes*	164	-	-	-	-	1 263	3 156
Whiteleg shrimp	*Penaeus vannamei*	1 238	1 046	1 277	1 082	519	460	926
Crystal shrimp	*Penaeus brevirostris*	397	318	483	177	80	153	93
Penaeus shrimps nei	*Penaeus spp*	65	96	157	24	8	19	30
Pacific seabobs	*Xiphopenaeus,Trachypenaeus spp*	5 038	3 079	3 015	1 656	1 472	1 525	918
Marine crustaceans nei	*Crustacea*	498	392	355	294	337	409	4 370
Freshwater molluscs nei	*Mollusca*	0	2	3	1	8	2	7
Various squids nei	*Loliginidae, Ommastrephidae*	-	-	-	-	16	9	13
Marine molluscs nei	*Mollusca*	598	535	620	1 119	1 115	823	750
	Country total	*14 433*	*11 897*	*12 658*	*16 525*	*9 590*	*19 010*	*34 455*
Greenland								
Freshwater fishes nei	*Osteichthyes*	0	0	0	0	0	0	0
Atlantic salmon	*Salmo salar*	82	44	...	...	24	42	-
Arctic char	*Salvelinus alpinus*	43	78	76	24	29	20	20
Atlantic halibut	*Hippoglossus hippoglossus*	51	26	34	46	19	19	48
European plaice	*Pleuronectes platessa*	-	-	2	-	-	-	-
Greenland halibut	*Reinhardtius hippoglossoides*	20 438	24 275	22 851	39 451	24 975	20 684	27 140
Amer. plaice(=Long rough dab)	*Hippoglossoides platessoides*	0	0	5	3	0	4	-
Atlantic cod	*Gadus morhua*	7 486	7 681	5 598	4 117	2 998	5 614	8 059
Greenland cod	*Gadus ogac*	2 121	1 729	1 697	1 903	931	1 133	977
Haddock	*Melanogrammus aeglefinus*	1 524	1 876	762	...	176	547	596
Saithe(=Pollock)	*Pollachius virens*	165	318	437	...	601	1 526	1 580
Polar cod	*Boreogadus saida*	4	-	-	-	-	-	3
Roundnose grenadier	*Coryphaenoides rupestris*	119	172	22	34	28	27	26
Atlantic wolffish	*Anarhichas lupus*	15	6	42	7	12	16	16
Wolffishes(=Catfishes) nei	*Anarhichas spp*	47	67	30	26	47	58	106
Golden redfish	*Sebastes marinus*	...	...	...	...	41	13	384
Beaked redfish	*Sebastes mentella*	...	...	...	...	4 180	3 382	4 288
Atlantic redfishes nei	*Sebastes spp*	1 144	1 162	2 405	5 216	860	399	450
Lumpfish(=Lumpsucker)	*Cyclopterus lumpus*	426	1 157	2 143	3 057	1 211	3 216	5 795
Capelin	*Mallotus villosus*	7 182	12 162	16 935	24 295	24 645	18 641	30 239
Dogfish sharks nei	*Squalidae*	136	6	...	...	-	-	-
Raja rays nei	*Raja spp*	-	-	-	-	-	-	6
Groundfishes nei	*Osteichthyes*	691	1 273	588	-	-	-	-
Finfishes nei	*Osteichthyes*	-	-	-	-	769	589	584
Marine fishes nei	*Osteichthyes*	168	189	1 187	...	200	264	41
Queen crab	*Chionoecetes opilio*	817	2 557	1 947	2 896	10 236	14 247	10 271
Northern prawn	*Pandalus borealis*	71 986	63 932	69 570	79 178	85 402	85 842	109 536
Aesop shrimp	*Pandalus montagui*	...	...	...	...	697	609	206
Iceland scallop	*Chlamys islandica*	1 373	1 886	2 211	...	1 630	1 593	2 457

English name Nom anglais Nombre inglés	Scientific name Nom scientifique Nombre científico	1996 t	1997 t	1998 t	1999 t	2000 t	2001 t	2002 t
	Country total	*116 018*	*120 596*	*128 542*	*160 253*	*159 711*	*158 485*	*202 828*
Grenada								
Freshwater fishes nei	*Osteichthyes*	0	0	0	0	0	0	0
Squirrelfishes nei	*Holocentridae*	1	1	3	2	3	3	2
Snooks(=Robalos) nei	*Centropomus spp*	0	0	0	1	0	0	0
Red hind	*Epinephelus guttatus*	60	42	52	72	67	88	159
Groupers, seabasses nei	*Serranidae*	10	11	16	20	22	38	62
Snappers, jobfishes nei	*Lutjanidae*	32	25	28	36	48	56	80
Grunts, sweetlips nei	*Haemulidae (=Pomadasyidae)*	1	1	1	3	1	1	2
Goatfishes, red mullets nei	*Mullidae*	1	1	0	0	0	0	0
Parrotfishes nei	*Scaridae*	0	1	1	42	18	41	113
Surgeonfishes nei	*Acanthuridae*	0	0	0	1	1	2	2
Triggerfishes, durgons nei	*Balistidae*	0	0	0	0	0	0	0
Brazilian sardinella	*Sardinella brasiliensis*	1	0	0	0	0	0	0
Scaled sardines	*Harengula spp*	1	0	2	0	0	0	3
Atlantic thread herring	*Opisthonema oglinum*	0	0	0	0	0	0	0
Broad-striped anchovy	*Anchoa hepsetus*	...	1	16	8	0	0	0
Atlantic bonito	*Sarda sarda*	24	6	14	16	7	10	10
Wahoo	*Acanthocybium solandri*	56	56	59	82	51	71	59
King mackerel	*Scomberomorus cavalla*	2	4	28	14	9	4	6
Serra Spanish mackerel	*Scomberomorus brasiliensis*	0	0	1	1	1	0	...
Frigate and bullet tunas	*Auxis thazard, A.rochei*	0	1	0	0	0	1	0
Skipjack tuna	*Katsuwonus pelamis*	11	15	23	23	23	15	14
Blackfin tuna	*Thunnus atlanticus*	164	126	233	94	164	223	255
Albacore	*Thunnus alalunga*	1	6	7	6	12	21	23
Yellowfin tuna	*Thunnus albacares*	523	411	484	430	403	759	593
Bigeye tuna	*Thunnus obesus*	-	1	-	-	-	-	-
Atlantic sailfish	*Istiophorus albicans*	56	83	151	148	164	187	151
Blue marlin	*Makaira nigricans*	26	47	60	100	87	104	69
Atlantic white marlin	*Tetrapturus albidus*	-	-	-	-	1	15	8
Swordfish	*Xiphias gladius*	4	47	33	42	84	73	54
Needlefishes, etc. nei	*Belonidae*	0	2	3	0	0	0	0
Halfbeaks nei	*Hemiramphus spp*	1	2	2	0	1	1	1
Flyingfishes nei	*Exocoetidae*	12	0	5	1	14	10	5
Scads nei	*Decapterus spp*	82	61	101	59	38	52	75
Atlantic moonfish	*Selene setapinnis*	0	0	0	0	0	0	0
Rainbow runner	*Elagatis bipinnulata*	14	20	20	36	23	15	20
Bigeye scad	*Selar crumenophthalmus*	100	181	53	72	137	97	67
Carangids nei	*Carangidae*	10	17	10	16	12	14	26
Common dolphinfish	*Coryphaena hippurus*	130	171	153	162	167	221	178
Barracudas nei	*Sphyraena spp*	40	30	41	35	57	43	52
Sharks, rays, skates, etc. nei	*Elasmobranchii*	4	9	18	24	29	29	12
Marine fishes nei	*Osteichthyes*	173	130	158	29	5	10	3
Caribbean spiny lobster	*Panulirus argus*	23	14	31	72	47	32	24
Stromboid conchs nei	*Strombus spp*	6	1	24	6	0	2	32
Various squids nei	*Loliginidae, Ommastrephidae*	0	0	0	0	0	5	3
Green turtle	*Chelonia mydas*	8	6	6	5	5	7	8
Sea urchins nei	*Strongylocentrotus spp*	0	0	0	0	0	-	-
	Country total	*1 577*	*1 530*	*1 837*	*1 658*	*1 701*	*2 250*	*2 171*
Guadeloupe								
Freshwater fishes nei	*Osteichthyes*	0	0	0	0	0	0	0
Blackfin tuna	*Thunnus atlanticus*	500	500	500	500	500	500	500
Common dolphinfish	*Coryphaena hippurus*	730	800 F	670 F	670 F	700 F	700 F	700 F
Mackerels nei	*Scombridae*	1 550	1 700 F	1 430 F	1 430 F	1 600 F	1 600 F	1 600 F
Marine fishes nei	*Osteichthyes*	6 220	6 860 F	5 800 F	5 800 F	6 600 F	6 500 F	6 600 F
Marine crustaceans nei	*Crustacea*	140	150	134	134	150	150	150
Stromboid conchs nei	*Strombus spp*	430	470	550	580	550	550	550
Marine turtles nei	*Testudinata*	0	0	0	-	-	-	-
	Country total	*9 570*	*10 480*	*9 084*	*9 114*	*10 100*	*10 000 F*	*10 100 F*
Guatemala								
Cichlids nei	*Cichlidae*	0	21	157	173	192	200 F	200 F
Freshwater fishes nei	*Osteichthyes*	4 000	5 100	6 366	6 803	7 109	7 100 F	7 100 F
Skipjack tuna	*Katsuwonus pelamis*	-	-	-	6 750	12 314	6 500 F	10 138
Yellowfin tuna	*Thunnus albacares*	-	-	-	1 660	4 769	4 000 F	2 387
Bigeye tuna	*Thunnus obesus*	-	-	-	1 580	12 352	2 500 F	2 624
Sharks, rays, skates, etc. nei	*Elasmobranchii*	81	146	237	203	151	250 F	359
Marine fishes nei	*Osteichthyes*	645	500	701	733	513	600 F	618 F
Tropical spiny lobsters nei	*Panulirus spp*	5	7	16	18	2	2 F	1
Yellowleg shrimp	*Penaeus californiensis*	266	124	259	218	132	100 F	40
Crystal shrimp	*Penaeus brevirostris*	33	18	21	19	16	8 F	...
Penaeus shrimps nei	*Penaeus spp*	404	596	1 712	1 199	1 043	800 F	593 F
Pacific seabobs	*Xiphopenaeus,Trachypenaeus spp*	2 203	355	1 320	1 619	365	200 F	59
Marine crustaceans nei	*Crustacea*	0	8	19	20	14	20 F	27
Various squids nei	*Loliginidae, Ommastrephidae*	16	21	39	23	231	120 F	18
	Country total	*7 653*	*6 896*	*10 847*	*21 018*	*39 203*	*22 400 F*	*24 164*
Haiti								
Freshwater fishes nei	*Osteichthyes*	500 F	500 F	500 F	500 F	500 F	500 F	500 F
Marine fishes nei	*Osteichthyes*	4 000 F	4 000 F	4 000 F	3 800 F	3 800 F	3 800 F	3 800 F
Marine crabs nei	*Brachyura*	5	71	59	50 F	50 F	50 F	50 F
Caribbean spiny lobster	*Panulirus argus*	190 F	200 F	200 F	200 F	200 F	200 F	200 F
Natantian decapods nei	*Natantia*	150 F	150 F	150 F	150 F	150 F	150 F	150 F

English name / Nom anglais / Nombre inglés	Scientific name / Nom scientifique / Nombre científico	1996 t	1997 t	1998 t	1999 t	2000 t	2001 t	2002 t
Stromboid conchs nei	*Strombus spp*	400 F	380 F	350 F	300 F	300 F	300 F	300 F
	Country total	*5 245 F*	*5 301 F*	*5 259 F*	*5 000 F*	*5 000 F*	*5 000 F*	*5 000 F*
Honduras								
Freshwater fishes nei	*Osteichthyes*	98	126	119	102	61	111	102 F
Tadpole codling	*Salilota australis*	189	-	-	-	-	-	-
Southern blue whiting	*Micromesistius australis*	3	-	-	-	-	-	-
Argentine hake	*Merluccius hubbsi*	19	-	-	-	-	-	-
Patagonian grenadier	*Macruronus magellanicus*	92	-	-	-	-	-	-
Croakers, drums nei	*Sciaenidae*	91	130	19	73	19	62	-
Porgies, seabreams nei	*Sparidae*	736	565	233	705	120	438	-
Pink cusk-eel	*Genypterus blacodes*	59	-	-	-	-	-	-
Patagonian toothfish	*Dissostichus eleginoides*	7	-	-	-	-	-	-
Skipjack tuna	*Katsuwonus pelamis*	780	3 110	630	4 280	2 010	...	...
Yellowfin tuna	*Thunnus albacares*	40	230	870	1 640	620	...	...
Bigeye tuna	*Thunnus obesus*	1 230	1 230	140	420	20	...	...
Indo-Pacific sailfish	*Istiophorus platypterus*	-	-	-	-	-	205	-
Marlins,sailfishes,etc. nei	*Istiophoridae*	-	-	-	-	-	182	-
Swordfish	*Xiphias gladius*	-	-	-	-	-	165	-
Tuna-like fishes nei	*Scombroidei*	10	15	5	20	-	-	-
Jack and horse mackerels nei	*Trachurus spp*	-	-	-	-	0	-	-
Rays, stingrays, mantas nei	*Rajiformes*	460	-	-	-	-	-	-
Sharks, rays, skates, etc. nei	*Elasmobranchii*	...	10	4	...	...	85	-
Marine fishes nei	*Osteichthyes*	2 262	2 287 F	2 321 F	2 519 F	3 428 F	4 495 F	4 868 F
Marine crabs nei	*Brachyura*	117	318	35	6	232	60 F	60 F
Caribbean spiny lobster	*Panulirus argus*	1 450	1 176	1 237	1 541	823	866	837
Tropical spiny lobsters nei	*Panulirus spp*	2	9	2	1	1 F	1 F	1 F
Penaeus shrimps nei	*Penaeus spp*	8 928	8 052	8 707	7 482 F	5 232 F	5 151 F	4 745 F
Natantian decapods nei	*Natantia*	-	0	-	-	-	-	-
Stromboid conchs nei	*Strombus spp*	609	868	601	721	713	742	789
Venus clams nei	*Veneridae*	...	3	...	0	...	...	...
Cuttlefish,bobtail squids nei	*Sepiidae, Sepiolidae*	3 108	2 076	889	1 077	687	-	-
Patagonian squid	*Loligo gahi*	8	-	-	-	-	-	-
Various squids nei	*Loliginidae, Ommastrephidae*	212	129	30	24	1	...	...
Octopuses, etc. nei	*Octopodidae*	1 040	468	129	590	54	-	-
Marine molluscs nei	*Mollusca*	...	9	20	27	465	...	...
Marine turtles nei	*Testudinata*	-	-	3 F	1	21	...	...
	Country total	*21 550*	*20 811*	*15 994*	*21 229*	*14 507*	*12 563*	*11 402*
Jamaica								
Nile tilapia	*Oreochromis niloticus*	150 F	150 F	150 F	150 F	150 F	150 F	150 F
Freshwater fishes nei	*Osteichthyes*	300 F	300 F	300 F	300 F	300 F	300 F	300 F
Tuna-like fishes nei	*Scombroidei*	239	275	...	...	78	86	48
Marine fishes nei	*Osteichthyes*	8 500 F	5 305	4 161	6 283	4 586	4 616 F	4 604 F
Marine crabs nei	*Brachyura*	11	9	9	9	8	8 F	8 F
Caribbean spiny lobster	*Panulirus argus*	394	271	170	330	517	500 F	500 F
Penaeus shrimps nei	*Penaeus spp*	60	67	70	70	37	40 F	40 F
Stromboid conchs nei	*Strombus spp*	21 375	13 658	12 750	10 245	0	-	-
	Country total	*31 029 F*	*20 035*	*17 610*	*17 387*	*5 676*	*5 700 F*	*5 650 F*
Martinique								
Freshwater fishes nei	*Osteichthyes*	0	0	0	0	0	0	0
Clupeoids nei	*Clupeoidei*	50 F	100 F	500 F	3 500	3 700	4 000	4 000
Atlantic bonito	*Sarda sarda*	610	610 F	610 F	610 F	610 F	530 F	530 F
Cero	*Scomberomorus regalis*	250	250 F	250 F	...	...	...	...
Blackfin tuna	*Thunnus atlanticus*	540 F	540 F	540 F	540 F	540 F	470 F	470 F
Flyingfishes nei	*Exocoetidae*	0	0	0	0	0	0	0
Common dolphinfish	*Coryphaena hippurus*	250 F	350 F	320 F	300 F	250 F	220 F	220 F
Rays, stingrays, mantas nei	*Rajiformes*	3 F	5 F	5 F	5 F	5 F	5 F	5 F
Sharks, rays, skates, etc. nei	*Elasmobranchii*	70 F	90 F	80 F	70 F	50 F	40 F	40 F
Marine fishes nei	*Osteichthyes*	1 647 F	3 430 F	3 060 F	810 F	45 F	35 F	35 F
Caribbean spiny lobster	*Panulirus argus*	70 F	110 F	120 F	150 F	200	190	190
Clams, etc. nei	*Bivalvia*	...	...	...	...	900	700	700
Marine turtles nei	*Testudinata*	0	0	0	-	-	-	-
Sea urchins nei	*Strongylocentrotus spp*	10 F	15 F	15 F	15 F	10	10	10
	Country total	*3 500 F*	*5 500*	*5 500*	*6 000*	*6 310*	*6 200*	*6 200 F*
Mexico								
Common carp	*Cyprinus carpio*	21 237	16 787	14 345	10 945	15 300	14 700	13 706
Cyprinids nei	*Cyprinidae*	3 732	4 838	3 982	6 805	3 988	3 806	1 711
Tilapias nei	*Oreochromis (=Tilapia) spp*	74 354	74 814	65 178	59 343	68 772	60 336	54 901
Catfishes nei	*Ictalurus spp*	5 358	4 691	4 027	4 232	3 845	3 135	2 263
Largemouth black bass	*Micropterus salmoides*	1 014	1 058	684	784	903	693	766
Freshwater fishes nei	*Osteichthyes*	9 766	5 185	6 482	3 284	8 716	3 911	4 280
American eel	*Anguilla rostrata*	35	19	9	2	1	0	0
Rainbow trout	*Oncorhynchus mykiss*	1 653	102	95	91	232	223	174
Milkfish	*Chanos chanos*	0	1	0	0	7	...	1
Flatfishes nei	*Pleuronectiformes*	2 524	2 741	1 390	2 268	2 681	1 839	1 836
North Pacific hake	*Merluccius productus*	116	1 089	250	111	392	852	1 194
Gadiformes nei	*Gadiformes*	21	183	137	32	40	53	71
Tarpon	*Megalops atlanticus*	1	14	4	4	4	1	-
Sea catfishes nei	*Ariidae*	6 186	7 240	8 025	6 936	7 385	8 080	7 274
Flathead grey mullet	*Mugil cephalus*	7 882	7 128	4 299	4 872	5 190	4 838	3 545
Bobo mullet	*Joturus pichardi*	659	339	374	404	497	504	417
Mullets nei	*Mugilidae*	9 787	11 481	10 884	11 760	12 186	10 262	10 566

English name Nom anglais Nombre inglés	Scientific name Nom scientifique Nombre científico	1996 t	1997 t	1998 t	1999 t	2000 t	2001 t	2002 t
Common snook	*Centropomus undecimalis*	2 955	3 307	2 990	3 415	3 184	4 000	5 951
Snooks(=Robalos) nei	*Centropomus spp*	1 900	1 824	1 769	2 121	1 788	2 222	1 599
Groupers nei	*Epinephelus spp*	17 643	18 909	17 850	19 831	18 595	14 609	15 335
Yellow snapper	*Lutjanus argentiventris*	4 917	3 123	3 390	2 994	3 392	3 388	3 733
Northern red snapper	*Lutjanus campechanus*	4 555	4 219	3 392	3 445	2 726	2 717	2 566
Lane snapper	*Lutjanus synagris*	950	641	591	630	430	693	280
Yellowtail snapper	*Ocyurus chrysurus*	858	840	1 900	1 554	1 357	1 600	1 702
Snappers, jobfishes nei	*Lutjanidae*	2 290	2 649	4 853	5 045	4 996	3 124	3 228
Grunts, sweetlips nei	*Haemulidae (=Pomadasyidae)*	6 385	10 513	7 352	5 414	4 065	4 221	3 645
Spotted weakfish	*Cynoscion nebulosus*	3 213	3 448	6 600	6 549	6 227	4 799	5 957
Weakfishes nei	*Cynoscion spp*	3 402	4 042	4 602	5 507	2 494	5 356	5 596
Croakers nei	*Micropogonias spp*	5 036	4 966	5 476	4 602	3 705	3 728	2 798
Gulf kingcroaker	*Menticirrhus littoralis*	1 070	1 220	1 824	1 576	606	484	630
Black drum	*Pogonias cromis*	134	362	392	180	207	143	165
Croakers, drums nei	*Sciaenidae*	711	692	946	900	602	1 398	459
Porgies	*Calamus spp*	239	96	266	577	235	214	121
Mojarras, etc. nei	*Gerreidae*	14 708	8 493	6 190	5 183	4 425	4 600	3 800
Triggerfishes, durgons nei	*Balistidae*	1 712	6 236	3 660	2 686	1 718	955	954
Ocean whitefish	*Caulolatilus princeps*	...	...	535	622	1 073	979	927
Tilefishes nei	*Branchiostegidae*	...	...	53	68	45	28	18
Hairtails, scabbardfishes nei	*Trichiuridae*	...	...	6 349	6 724	7 265	5 089	3 879
Demersal percomorphs nei	*Perciformes*	170	190	180	149	171	3 046	3 573
Round sardinella	*Sardinella aurita*	1 073	2 028	4 253	1 384	1 424	373	332
California pilchard	*Sardinops caeruleus*	418 093	455 837	339 317	345 458	478 191	609 777	624 817
Californian anchovy	*Engraulis mordax*	9 598	2 147	782	5 814	7 973	418	4 145
Anchovies, etc. nei	*Engraulidae*	1 464	903	1 762	1 613	1 163	1 328	1 075
Clupeoids nei	*Clupeoidei*	276	797	569	371	275	21	409
Atlantic bonito	*Sarda sarda*	1 279	2 040	2 194	2 314	1 721	1 506	1 401
Eastern Pacific bonito	*Sarda chiliensis*	399	875	423	1 775	429	146	250
King mackerel	*Scomberomorus cavalla*	4 377	5 370	4 598	5 002	4 576	5 119	5 720
Atlantic Spanish mackerel	*Scomberomorus maculatus*	11 049	7 389	7 381	8 382	5 717	5 320	6 123
Pacific sierra	*Scomberomorus sierra*	5 742	5 405	3 896	5 265	6 261	5 959	4 815
Skipjack tuna	*Katsuwonus pelamis*	18 330	25 171	17 697	19 160	14 850	8 221	10 159
Atlantic bluefin tuna	*Thunnus thynnus*	14	7	14	16	35	10	15
Pacific bluefin tuna	*Thunnus orientalis*	3 700	370	34	2 370	3 030	860	1 326
Albacore	*Thunnus alalunga*	21	53	8	32	159	40	68
Yellowfin tuna	*Thunnus albacares*	127 815	140 286	117 762	121 884	102 340	135 494	148 583
Bigeye tuna	*Thunnus obesus*	495	108	13	97	8	92	0
Atlantic sailfish	*Istiophorus albicans*	10	7	21	33	37	36	38
Marlins,sailfishes,etc. nei	*Istiophoridae*	626	1 468	2 539	387	375	347	237
Swordfish	*Xiphias gladius*	439	2 365	3 603	1 136	2 216	780	465
Silversides(=Sand smelts) nei	*Atherinidae*	...	...	1 434	1 393	1 185	1 262	1 365
Cobia	*Rachycentron canadum*	347	630	588	565	303	298	180
Jacks, crevalles nei	*Caranx spp*	7 960	10 679	10 933	8 296	8 266	8 870	9 057
Pompanos nei	*Trachinotus spp*	941	629	885	851	684	555	863
Amberjacks nei	*Seriola spp*	1 116	1 602	1 580	1 628	1 832	1 433	1 313
Carangids nei	*Carangidae*	274	772	1 992	1 077	490	636	487
Common dolphinfish	*Coryphaena hippurus*	314	984	1 193	480	509	523	695
Chub mackerel	*Scomber japonicus*	12 961	24 246	22 990	69 375	45 205	17 156	10 433
Barracudas nei	*Sphyraena spp*	722	1 587	1 460	1 151	549	932	947
Requiem sharks nei	*Carcharhinidae*	11 015	7 267	6 979	6 070	6 318	6 055	5 472
Rays, stingrays, mantas nei	*Rajiformes*	9 932	10 074	10 954	9 076	7 817	7 023	6 505
Sharks, rays, skates, etc. nei	*Elasmobranchii*	24 258	18 324	18 599	20 093	21 125	19 640	18 911
Marine fishes nei	*Osteichthyes*	262 207	239 935	205 232	137 455	143 601	123 871	127 613
River prawns nei	*Macrobrachium spp*	4 261	3 580	3 244	4 161	3 478	3 112	3 051
Euro-American crayfishes nei	*Astacidae, Cambaridae*	198	101	99	163	108	17	94
Black stone crab	*Menippe mercenaria*	353	610	681	355	120	205	378
Blue crab	*Callinectes sapidus*	13 736	14 412	12 920	12 714	8 602	7 291	8 063
Marine crabs nei	*Brachyura*	13 602	10 073	6 503	6 670	12 036	11 204	7 896
Caribbean spiny lobster	*Panulirus argus*	756	844	613	645	747	782	1 070
Tropical spiny lobsters nei	*Panulirus spp*	1 799	1 708	1 599	1 328	2 052	1 727	1 923
Penaeus shrimps nei	*Penaeus spp*	65 564	71 067	66 586	66 491	61 597	57 509	54 633
Marine crustaceans nei	*Crustacea*	0	2	3	1	1	-	-
Freshwater molluscs nei	*Mollusca*	240	120	177	239	557	1 027	1 045
Abalones nei	*Haliotis spp*	1 075	924	709	574	535	482	494
Stromboid conchs nei	*Strombus spp*	5 657	7 639	4 266	8 591	10 392	11 397	9 025
American cupped oyster	*Crassostrea virginica*	34 726	38 515	30 715	39 268	48 101	48 570	47 570
Cupped oysters nei	*Crassostrea spp*	1 354	1 903	1 276	2 489	1 828	2 234	2 218
Sea mussels nei	*Mytilidae*	412	2 036	2 056	737	547	733	691
Pacific calico scallop	*Argopecten ventricosus*	17 290	2 320	2 726	1 864	6 287	3 100	9 875
Ark clams nei	*Arca spp*	834	1 367	1 018	880	675	600	328
Venus clams nei	*Veneridae*	5 852	4 658	4 673	5 330	6 238	5 416	3 576
Common squids nei	*Loligo spp*	112	140	71	91	85	92	58
Jumbo flying squid	*Dosidicus gigas*	107 967	120 877	26 611	57 985	56 153	73 741	115 896
Common octopus	*Octopus vulgaris*	28 572	17 776	16 478	19 081	22 463	20 596	16 018
Octopuses, etc. nei	*Octopodidae*	1 257	944	755	1 094	883	837	676
Frogs	*Rana spp*	351	1 979	1 167	311	295	315	125
Echinoderms	*Echinodermata*	3 027	2 099	1 138	2 042	2 813	2 278	2 068
Sea cucumbers nei	*Holothurioidea*	...	...	271	234	426	481	290
Marine worms	*Polychaeta*	58	57	58	48	27	26	27
Aquatic invertebrates nei	*Invertebrata*	1 115	526	437	569	426	93	126
	Country total	*1 464 188*	*1 489 112*	*1 179 860*	*1 205 603*	*1 315 581*	*1 398 592*	*1 450 654*

Montserrat

English name	Scientific name	1996	1997	1998	1999	2000	2001	2002
Freshwater fishes nei	*Osteichthyes*	0	0	0	0	0	0	0
Marine fishes nei	*Osteichthyes*	38	45	46	50 F	50 F	50 F	46
	Country total	*38*	*45*	*46*	*50 F*	*50 F*	*50 F*	*46*

English name Nom anglais Nombre inglés	Scientific name Nom scientifique Nombre científico	1996 t	1997 t	1998 t	1999 t	2000 t	2001 t	2002 t
NethAntilles								
Freshwater fishes nei	*Osteichthyes*	0	0	0	0	0	0	0
Wahoo	*Acanthocybium solandri*	230 F	230 F	230 F	230 F	230 F	230 F	230 F
Frigate tuna	*Auxis thazard*	590	1 157	1 030	1 159	1 122	989	710
Skipjack tuna	*Katsuwonus pelamis*	7 126	8 474 F	8 583 F	9 962 F	10 038 F	13 370	5 427
Blackfin tuna	*Thunnus atlanticus*	45	45 F	45 F	45 F	45 F	45 F	45 F
Albacore	*Thunnus alalunga*	-	9	192	-	2	...	...
Yellowfin tuna	*Thunnus albacares*	3 313	6 212 F	6 240 F	4 092 F	5 571 F	4 793	4 035
Bigeye tuna	*Thunnus obesus*	1 893	2 890	2 919	3 428	2 359	2 803	1 879
Atlantic sailfish	*Istiophorus albicans*	15 F	15 F	15 F	15 F	15 F	...	...
Blue marlin	*Makaira nigricans*	40 F	40 F	40 F	40 F	40 F	...	...
Marine fishes nei	*Osteichthyes*	505 F	455 F	455 F	455 F	455 F	570 F	570 F
Stromboid conchs nei	*Strombus spp*	5 F	5 F	5 F	5 F	5 F	5 F	5 F
	Country total	*13 762 F*	*19 532 F*	*19 754 F*	*19 431 F*	*19 882 F*	*22 805 F*	*12 901 F*
Nicaragua								
Cichlids nei	*Cichlidae*	...	...	...	750	680	660	303
Freshwater fishes nei	*Osteichthyes*	1 142	1 293	1 256	363	396	168	1 101
Flatfishes nei	*Pleuronectiformes*	...	...	...	...	55	67	26
Snooks(=Robalos) nei	*Centropomus spp*	...	...	...	2 000	2 060	2 180	1 412
Groupers, seabasses nei	*Serranidae*	...	...	...	1 040	840	770	430
Snappers nei	*Lutjanus spp*	...	...	...	2 640	2 380	3 320	2 159
Weakfishes nei	*Cynoscion spp*	...	...	...	...	97	287	133
Pacific sierra	*Scomberomorus sierra*	...	...	...	...	55	67	77
Seerfishes nei	*Scomberomorus spp*	...	...	...	240	250	260	116
Skipjack tuna	*Katsuwonus pelamis*	-	-	-	250	430	...	...
Yellowfin tuna	*Thunnus albacares*	-	-	-	3 060	4 950	...	...
Bigeye tuna	*Thunnus obesus*	-	-	-	30	10	...	...
Common dolphinfish	*Coryphaena hippurus*	...	...	...	710	2 540	2 470	1 449
Requiem sharks nei	*Carcharhinidae*	...	...	...	200	150	375	292
Marine fishes nei	*Osteichthyes*	5 859	6 886	9 530	1 667	1 181	1 454	6 436
River prawns nei	*Macrobrachium spp*	-	-	-	7	0	3	3
Blue crab	*Callinectes sapidus*	-	-	131	106	71	69	0
Harbour spidercrab	*Mithrax armatus*	-	-	114	110	54	35	22
Caribbean spiny lobster	*Panulirus argus*	4 357	4 012	3 729	5 141	6 327	3 909	4 437
Green spiny lobster	*Panulirus gracilis*	101	152	66	131	476	652	334
Penaeus shrimps nei	*Penaeus spp*	3 983	3 833	4 864	5 218	4 431	4 966	4 617
Northern nylon shrimp	*Heterocarpus vicarius*	-	-	40	37	20	131	0
Stromboid conchs nei	*Strombus spp*	-	-	162	209	555	956	318
Various squids nei	*Loliginidae, Ommastrephidae*	-	-	-	-	-	-	2
Octopuses, etc. nei	*Octopodidae*	-	-	-	-	-	-	3
	Country total	*15 442*	*16 176*	*19 892*	*23 909*	*28 008*	*22 799*	*23 670*
Panama								
Tilapias nei	*Oreochromis (=Tilapia) spp*	15	56	11	13	16	16 F	16 F
Peacock cichlid	*Cichla ocellaris*	65	35	12	7	4	4 F	4 F
Tadpole codling	*Salilota australis*	93	-	-	-	-	-	-
Argentine hake	*Merluccius hubbsi*	14	-	-	-	-	-	-
Hakes nei	*Merluccius spp*	40	29	-	-	-	-	-
Patagonian grenadier	*Macruronus magellanicus*	339	-	-	-	-	-	-
Snappers, jobfishes nei	*Lutjanidae*	3 857	4 659	5 358	10 433	8 480	22 322	26 642
Croakers, drums nei	*Sciaenidae*	666	667	1 138	1 533	4 352	4 543	2 279
Porgies, seabreams nei	*Sparidae*	402	46	319	66	-	-	-
Pink cusk-eel	*Genypterus blacodes*	46	-	-	-	-	-	-
Pacific thread herring	*Opisthonema libertate*	32 517	26 266	49 472	38 746	63 532	29 033	48 175
Pacific anchoveta	*Cetengraulis mysticetus*	59 830	77 726	107 730	27 356	86 681	129 147	160 414
Pacific sierra	*Scomberomorus sierra*	489	665	982	159	1 395	971	1 027
Frigate tuna	*Auxis thazard*	240	91	-	-	-	-	-
Black skipjack	*Euthynnus lineatus*	-	-	9	-	1	-	7
Skipjack tuna	*Katsuwonus pelamis*	9 295	5 290	1 532	6 150	12 992	5 797	7 609
Atlantic bluefin tuna	*Thunnus thynnus*	3 400	491	-	13	-	-	-
Albacore	*Thunnus alalunga*	391	58	61	14	-	-	-
Yellowfin tuna	*Thunnus albacares*	11 353	9 873	5 631	7 908	8 155	12 285	20 407
Bigeye tuna	*Thunnus obesus*	5 847	3 418	1 260	1 448	4 078	2 270	1 669
Blue marlin	*Makaira nigricans*	-	-	-	-	41	-	-
Swordfish	*Xiphias gladius*	-	-	-	122	-	-	-
Tuna-like fishes nei	*Scombroidei*	-	-	-	77	15	-	-
Amberjacks nei	*Seriola spp*	470	333	469	159	418	510	1 433
Chub mackerel	*Scomber japonicus*	-	-	397	-	-	-	-
Rays, stingrays, mantas nei	*Rajiformes*	170	-	-	-	-	-	-
Sharks, rays, skates, etc. nei	*Elasmobranchii*	-	-	-	202	...	-	-
Marine fishes nei	*Osteichthyes*	6 507	17 932	15 372	15 497	18 197	38 310	27 473
Marine crabs nei	*Brachyura*	14	3	2	0	0	20	3
Tropical spiny lobsters nei	*Panulirus spp*	288	306	415	485	612	845	687
Crystal shrimp	*Penaeus brevirostris*	1 298	2 048	1 310	1 718	2 158	2 286	1 636
Penaeus shrimps nei	*Penaeus spp*	2 566	2 053	3 200	2 512	1 952	2 126	2 310
Pacific seabobs	*Xiphopenaeus,Trachypenaeus spp*	4 339	3 606	4 960	2 872	2 685	3 002	1 675
Natantian decapods nei	*Natantia*	3 889	4 822	833	1 674	1 661	1 147	346
Marine crustaceans nei	*Crustacea*	3	3	2	1	1	2	3
Stromboid conchs nei	*Strombus spp*	...	6	527	125	5	55	83
Gastropods nei	*Gastropoda*	5	-	116	17	1	1	12
Clams, etc. nei	*Bivalvia*	1 391	1 507	1 072	1 518	1 630	1 661	1 124
Cuttlefish,bobtail squids nei	*Sepiidae, Sepiolidae*	0	53	82	-	-	-	-
Patagonian squid	*Loligo gahi*	1	-	-	-	-	-	-
Various squids nei	*Loliginidae, Ommastrephidae*	8	2	1	...	...	...	...
Octopuses, etc. nei	*Octopodidae*	89	41	102	-	-	-	-

English name Nom anglais Nombre inglés	Scientific name Nom scientifique Nombre científico	1996 t	1997 t	1998 t	1999 t	2000 t	2001 t	2002 t
Marine molluscs nei	*Mollusca*	48	71	172	16	76	19	22
Marine turtles nei	*Testudinata*	0	0	0	0	0	...	...
Marine worms	*Polychaeta*	42	67	51	31	28	36	25
	Country total	*150 027*	*162 223*	*202 598*	*120 872*	*219 166*	*256 408 F*	*305 081*
Puerto Rico								
Freshwater fishes nei	*Osteichthyes*	0	0	0	0	0	0	-
Tarpon	*Megalops atlanticus*	...	...	...	-	-	1	2
Squirrelfishes nei	*Holocentridae*	...	...	...	9	13	12	8
Mullets nei	*Mugilidae*	...	...	...	39	44	41	30
Snooks(=Robalos) nei	*Centropomus spp*	...	...	...	32	33	8	24
Groupers nei	*Epinephelus spp*	...	...	...	89	104	111	94
Snappers, jobfishes nei	*Lutjanidae*	...	...	...	566	762	744	512
Grunts, sweetlips nei	*Haemulidae (=Pomadasyidae)*	...	...	...	76	97	104	77
Porgies, seabreams nei	*Sparidae*	...	...	...	22	24	25	20
Goatfishes, red mullets nei	*Mullidae*	...	...	...	17	17	15	10
Mojarras, etc. nei	*Gerreidae*	...	...	...	14	15	13	11
Wrasses, hogfishes, etc. nei	*Labridae*	...	...	...	30	47	46	36
Parrotfishes nei	*Scaridae*	...	...	...	52	63	66	57
Boxfishes nei	*Ostraciidae*	...	...	...	54	68	52	42
Triggerfishes, durgons nei	*Balistidae*	...	...	...	32	34	41	28
Demersal percomorphs nei	*Perciformes*	1 059	1 249	1 375	-	-	-	-
Clupeoids nei	*Clupeoidei*	18	19	14	20	20	17	15
Wahoo	*Acanthocybium solandri*	...	...	...	...	...	6	1
Seerfishes nei	*Scomberomorus spp*	106	119	109	123	145	124	90
Little tunny(=Atl.black skipj)	*Euthynnus alletteratus*	...	...	...	...	...	14	8
Skipjack tuna	*Katsuwonus pelamis*	...	...	...	...	...	26	20
Blackfin tuna	*Thunnus atlanticus*	...	...	...	...	...	17	14
Yellowfin tuna	*Thunnus albacares*	...	...	...	...	...	24	10
Bigeye tuna	*Thunnus obesus*	-	54	-	-	-	-	-
Tuna-like fishes nei	*Scombroidei*	88	72	121	99	111	17	6
Ballyhoo halfbeak	*Hemiramphus brasiliensis*	...	...	...	32	47	41	36
Carangids nei	*Carangidae*	...	...	...	50	70	66	53
Common dolphinfish	*Coryphaena hippurus*	...	...	...	83	111	74	53
Barracudas nei	*Sphyraena spp*	...	...	...	16	21	13	13
Pelagic percomorphs nei	*Perciformes*	98	97	83	-	-	-	-
Sharks, rays, skates, etc. nei	*Elasmobranchii*	...	...	...	28	35	32	20
Marine fishes nei	*Osteichthyes*	569	691	50	264	298	173	126
Marine crabs nei	*Brachyura*	...	...	...	2	2	6	3
Caribbean spiny lobster	*Panulirus argus*	...	...	...	209	212	190	158
Marine crustaceans nei	*Crustacea*	189	171	184	-	-	-	-
Stromboid conchs nei	*Strombus spp*	450	638	1 025	1 025	1 710	1 643	931
Octopuses, etc. nei	*Octopodidae*	...	...	...	28	39	23	15
Marine molluscs nei	*Mollusca*	124	77	45	9	12	9	6
Marine turtles nei	*Testudinata*	0	0	0	-	-	-	-
	Country total	*2 701*	*3 187*	*3 006*	*3 020*	*4 154*	*3 794*	*2 529*
St Kitts Nev								
Freshwater fishes nei	*Osteichthyes*	0	0	0	0	0	0	-
Squirrelfishes nei	*Holocentridae*	9	5	8	9	3	3	7
Groupers nei	*Epinephelus spp*	18	10	11	11	6	3	10
Snappers nei	*Lutjanus spp*	9	5	8	15	19	9	16
Grunts, sweetlips nei	*Haemulidae (=Pomadasyidae)*	10	1	3	3	0	1	3
Goatfishes, red mullets nei	*Mullidae*	10	1	2	1	0	0	0
Parrotfishes nei	*Scaridae*	19	5	8	8	2	7	8
Surgeonfishes nei	*Acanthuridae*	11	4	9	6	1	0	4
Triggerfishes, durgons nei	*Balistidae*	7	2	5	6	1	1	8
Tuna-like fishes nei	*Scombroidei*	3	7	16	24	24	23	13
Needlefishes, etc. nei	*Belonidae*	27	26	60	58	60	37	34
Flyingfishes nei	*Exocoetidae*	54	23	38	22	22	48	42
Bigeye scad	*Selar crumenophthalmus*	-	16	20	36	32	25	41
Common dolphinfish	*Coryphaena hippurus*	13	20	34	13	26	26	40
Marine fishes nei	*Osteichthyes*	96	54	79	163	174	268	83
Caribbean spiny lobster	*Panulirus argus*	17	8	25	29	26	33	10
Stromboid conchs nei	*Strombus spp*	49	29	81	67	73	71	36
	Country total	*352*	*216*	*407*	*471*	*469*	*555*	*355*
St Lucia								
Freshwater fishes nei	*Osteichthyes*	0	0	0	0	0	0	0
Snappers nei	*Lutjanus spp*	69	31	34	45	68	82	43
Atlantic bonito	*Sarda sarda*	1	0	0	0	0	0	0
Wahoo	*Acanthocybium solandri*	221	224	223	310	243	213	243
Seerfishes nei	*Scomberomorus spp*	51	4	0	...	60	7	29
Little tunny(=Atl.black skipj)	*Euthynnus alletteratus*	-	2	2	2	-	1	10
Skipjack tuna	*Katsuwonus pelamis*	38	100	263	153	216	151	106
Atlantic bluefin tuna	*Thunnus thynnus*	3	-	-	-	-	-	-
Blackfin tuna	*Thunnus atlanticus*	35	40	100	41	45	108	96
Albacore	*Thunnus alalunga*	1	0	0	0	1	3	2
Yellowfin tuna	*Thunnus albacares*	110	109	276	123	134	145	97
Bigeye tuna	*Thunnus obesus*	0	0	0	0	-	1	2
Marlins,sailfishes,etc. nei	*Istiophoridae*	-	4	1	-	14	5	9
Swordfish	*Xiphias gladius*	-	-	-	-	-	-	1
Tuna-like fishes nei	*Scombroidei*	8	1	3	3	1	-	-
Flyingfishes nei	*Exocoetidae*	40	34	112	67	99	323	193
Common dolphinfish	*Coryphaena hippurus*	351	455	264	588	552	427	402
Sharks, rays, skates, etc. nei	*Elasmobranchii*	11	3	8	6	5	5	10

English name Nom anglais Nombre inglés	Scientific name Nom scientifique Nombre científico	1996 t	1997 t	1998 t	1999 t	2000 t	2001 t	2002 t
Marine fishes nei	*Osteichthyes*	308	264	243	325	352	435	324
Marine crustaceans nei	*Crustacea*	12	12	32	30	25	36	10
Stromboid conchs nei	*Strombus spp*	15	25	28	25	40	41	60
	Country total	*1 274*	*1 308*	*1 589*	*1 718*	*1 855*	*1 983*	*1 637*
St Pier Mq								
Freshwater fishes nei	*Osteichthyes*	0	0	0	0	0	0	0
Atlantic salmon	*Salmo salar*	1	1	1	1	1	1	1
Atlantic halibut	*Hippoglossus hippoglossus*	-	1	1	1	0	1	2
Greenland halibut	*Reinhardtius hippoglossoides*	-	439	1 431	1 132	2	7	0
Witch flounder	*Glyptocephalus cynoglossus*	-	8	57	35	7	120	38
Amer. plaice(=Long rough dab)	*Hippoglossoides platessoides*	0	23	27	24	41	112	115
Yellowtail flounder	*Limanda ferruginea*	-	18	59	33	60	152	137
Atlantic cod	*Gadus morhua*	44	1 547	3 123	3 171	4 682	2 350	2 219
White hake	*Urophycis tenuis*	-	0	1	9	122	10	3
Haddock	*Melanogrammus aeglefinus*	-	9	27	16	10	78	222
Saithe(=Pollock)	*Pollachius virens*	-	14	13	6	38	13	135
Roundnose grenadier	*Coryphaenoides rupestris*	-	4	-	-	-	-	-
Wolffishes(=Catfishes) nei	*Anarhichas spp*	-	3	3	2	0	1	1
Atlantic redfishes nei	*Sebastes spp*	-	430	654	423	196	129	319
Lumpfish(=Lumpsucker)	*Cyclopterus lumpus*	218	363	249	422	536	146	3
American angler	*Lophius americanus*	-	0	1	0	0	1	2
Atlantic herring	*Clupea harengus*	-	-	-	2	0	0	0
Atlantic bluefin tuna	*Thunnus thynnus*	-	-	-	1	0	0	0
Capelin	*Mallotus villosus*	0	1	-	2	0	1	4
Atlantic mackerel	*Scomber scombrus*	1	1	3	1	26	7	7
Porbeagle	*Lamna nasus*	40	13	20	0	23	2	1
Picked dogfish	*Squalus acanthias*	0	0	-	-	0	0	0
Raja rays nei	*Raja spp*	3	3	9	4	21	38	238
Marine fishes nei	*Osteichthyes*	-	18	18	-	-	-	-
Queen crab	*Chionoecetes opilio*	189	368	354	589	511	498	150
American lobster	*Homarus americanus*	1	1	-	1	1	2	2
Periwinkles nei	*Littorina spp*	-	-	-	-	-	-	1
Whelk	*Buccinum undatum*	...	32	45	...	0	-	-
Blue mussel	*Mytilus edulis*	4	2	4	0	0	0	0
American sea scallop	*Placopecten magellanicus*	23	27	3	0	0	2	1
Iceland scallop	*Chlamys islandica*	221	239	5	0	3	16	9
Northern shortfin squid	*Illex illecebrosus*	0	6	-	-	-	-	2
Marine molluscs nei	*Mollusca*	1	0	0	17	205	115	42
Sea urchins nei	*Strongylocentrotus spp*	1	-	-	-	0	0	-
	Country total	*747*	*3 571*	*6 108*	*5 892*	*6 485*	*3 802*	*3 654*
St Vincent								
Freshwater fishes nei	*Osteichthyes*	2	1	0	0	0	0	0
Tadpole codling	*Salilota australis*	-	-	-	-	-	14	-
Argentine hake	*Merluccius hubbsi*	-	-	-	-	-	5	-
Hakes nei	*Merluccius spp*	-	10	106	31	28	55	130
Porgies, seabreams nei	*Sparidae*	30	10	59	39	5	51	20
Largehead hairtail	*Trichiurus lepturus*	-	130	1 393	35	24	716	1 441
Sardinellas nei	*Sardinella spp*	-	2 000	10 038	7 962	3 369	7 716	7 053
European pilchard(=Sardine)	*Sardina pilchardus*	-	46	26	96	340	41	192
European anchovy	*Engraulis encrasicolus*	-	1 100	7 182	3 702	4 841	7 139	9 757
Wahoo	*Acanthocybium solandri*	23	10	65	52	46	311	17
Seerfishes nei	*Scomberomorus spp*	1	1	1	1	138	0	0
Skipjack tuna	*Katsuwonus pelamis*	37	42	57	37	68	97	358
Blackfin tuna	*Thunnus atlanticus*	18	22	17	15	23	24	24
Albacore	*Thunnus alalunga*	-	-	-	1	2 820	5 662	345
Yellowfin tuna	*Thunnus albacares*	37	35	48	38	1 989	1 365	1 165
Bigeye tuna	*Thunnus obesus*	4	2	2	1	1 216	506	14
Atlantic sailfish	*Istiophorus albicans*	1	3	-	1	-	2	168
Blue marlin	*Makaira nigricans*	0	1	-	-	-	-	20
Atlantic white marlin	*Tetrapturus albidus*	0	-	-	-	-	-	-
Marlins,sailfishes,etc. nei	*Istiophoridae*	1	0	2	1	343	339	...
Swordfish	*Xiphias gladius*	3	1	0	1	0	22	...
Tuna-like fishes nei	*Scombroidei*	-	-	-	-	-	49	...
Jack and horse mackerels nei	*Trachurus spp*	-	1 300	9 765	3 546	8 275	19 564	18 098
Chub mackerel	*Scomber japonicus*	-	350	2 644	988	2 951	5 384	4 071
Sharks, rays, skates, etc. nei	*Elasmobranchii*	2	...	...	3	...	2	...
Marine fishes nei	*Osteichthyes*	737	1 019	2 547	1 202	1 211	1 579	966
Stromboid conchs nei	*Strombus spp*	25 F	10	21	7	7	37	40
Cuttlefish,bobtail squids nei	*Sepiidae, Sepiolidae*	-	-	-	-	-	2	-
Patagonian squid	*Loligo gahi*	-	-	-	-	-	1 795	-
Argentine shortfin squid	*Illex argentinus*	-	-	-	-	-	4	-
Various squids nei	*Loliginidae, Ommastrephidae*	-	-	-	-	-	4	-
	Country total	*921 F*	*6 093*	*33 973*	*17 759*	*27 694*	*52 485*	*43 879*
Trinidad Tob								
Freshwater fishes nei	*Osteichthyes*	0	0	0	0	0	0	0
Demersal percomorphs nei	*Perciformes*	2 656	2 572	2 001	1 918	1 812	2 261	2 477
Clupeoids nei	*Clupeoidei*	58	196	619	859	82	172	56
Atlantic bonito	*Sarda sarda*	266	220	30	117	117	56	452
Wahoo	*Acanthocybium solandri*	0	1	1	1	2	1	9
King mackerel	*Scomberomorus cavalla*	1 029	875	746	447	432	638	1 457
Serra Spanish mackerel	*Scomberomorus brasiliensis*	1 568	1 699	2 130	1 328	1 722	2 671	2 472
Frigate and bullet tunas	*Auxis thazard, A.rochei*	199	368	127	138	245	...	...
Skipjack tuna	*Katsuwonus pelamis*	...	0	-	-	-	-	-

English name Nom anglais Nombre inglés	Scientific name Nom scientifique Nombre científico	1996 t	1997 t	1998 t	1999 t	2000 t	2001 t	2002 t
Albacore	*Thunnus alalunga*	...	2	1	1	2	11	9
Yellowfin tuna	*Thunnus albacares*	183	223	213	163	112	122	125
Bigeye tuna	*Thunnus obesus*	37	36	24	19	5	11	30
Atlantic sailfish	*Istiophorus albicans*	104	10	...	4	3	7	6
Blue marlin	*Makaira nigricans*	21	81	70	33	55	17	16
Marlins,sailfishes,etc. nei	*Istiophoridae*	-	-	-	-	-	2	5
Swordfish	*Xiphias gladius*	158	110	130	138	41	75	92
Tuna-like fishes nei	*Scombroidei*	134	206	92	81	138	405	482
Jacks, crevalles nei	*Caranx spp*	504	562	203	189	202	201	294
Blacktip shark	*Carcharhinus limbatus*	3	8	10	11	...	...	...
Sharks, rays, skates, etc. nei	*Elasmobranchii*	621	545	635	701	755	715	937
Marine fishes nei	*Osteichthyes*	1 609	2 818	1 424	1 928	2 074	2 397	2 668
Portunus swimcrabs nei	*Portunus spp*	-	-	0	1	1	0	0
Caribbean spiny lobster	*Panulirus argus*	-	-	2	0	1	1	2
Penaeus shrimps nei	*Penaeus spp*	285	751	648	658	755	856	852
Atlantic seabob	*Xiphopenaeus kroyeri*	...	...	69	89	94	79	88
Stromboid conchs nei	*Strombus spp*	...	...	-	0	-	-	-
Various squids nei	*Loliginidae, Ommastrephidae*	-	-	0	2	1	9	10
	Country total	*9 435*	*11 283*	*9 175*	*8 826*	*8 651*	*10 707*	*12 539*
Turks Caicos								
Freshwater fishes nei	*Osteichthyes*	0	0	0	0	0	0	0
Marine fishes nei	*Osteichthyes*	300 F	250 F	250 F	250 F	200 F	200 F	200 F
Caribbean spiny lobster	*Panulirus argus*	310	181	230	313	255	291	323
Stromboid conchs nei	*Strombus spp*	7 223	5 520	5 903	4 838	5 520	6 120	5 430
	Country total	*7 833*	*5 951*	*6 383*	*5 401*	*5 975*	*6 611*	*5 953*
USA								
Buffalofishes nei	*Ictiobus spp*	793	991	959	697	1 280	1 569	1 558
Suckers nei	*Catostomidae*	542	550	464	465	231	187	173
Common carp	*Cyprinus carpio*	1 034	1 076	1 072	1 103	724	704	656
Goldfish	*Carassius auratus*	5	10	10	7	9	10	12
Grass carp(=White amur)	*Ctenopharyngodon idellus*	10	15	13	11	17	31	31
Cyprinids nei	*Cyprinidae*	3	5	15	17	29	33	27
Tilapias nei	*Oreochromis (=Tilapia) spp*	4	0	0	2 651	1 040	1 277	1 752
Catfishes nei	*Ictalurus spp*	4 110	6 872	5 257	9 235	7 561	7 478	6 301
Burbot	*Lota lota*	14	9	28	13	20	9	10
American yellow perch	*Perca flavescens*	711	622	556	537	567	640	685
Walleye	*Stizostedion vitreum*	18	4	6	2	4	10	14
Sauger	*Stizostedion canadense*	1	-	0	-	-	-	-
Freshwater drum	*Aplodinotus grunniens*	368	549	573	470	577	506	502
Freshwater fishes nei	*Osteichthyes*	283	640	301	338	650	637	579
White sturgeon	*Acipenser transmontanus*	-	-	-	-	206	185	200
Green sturgeon	*Acipenser medirostris*	-	-	-	-	36	10	7
Sturgeons nei	*Acipenseridae*	314	285	176	138	0	-	-
American eel	*Anguilla rostrata*	441	485	460	490	649	393	277
Pink(=Humpback)salmon	*Oncorhynchus gorbuscha*	140 542	102 965	150 856	173 316	94 440	173 067	116 043
Chum(=Keta=Dog)salmon	*Oncorhynchus keta*	81 907	46 794	59 394	65 295	73 633	52 687	50 690
Sockeye(=Red)salmon	*Oncorhynchus nerka*	144 445	87 299	58 396	110 836	94 422	77 172	61 653
Chinook(=Spring=King)salmon	*Oncorhynchus tshawytscha*	9 282	9 876	7 361	6 929	7 303	7 524	11 409
Coho(=Silver)salmon	*Oncorhynchus kisutch*	21 653	10 555	16 305	13 259	15 350	17 424	17 475
Rainbow trout	*Oncorhynchus mykiss*	168	137	358	82	145	221	146
Pacific salmons nei	*Oncorhynchus spp*	-	-	-	-	50	204	6
Lake trout(=Char)	*Salvelinus namaycush*	286	492	500	494	573	451	346
Rainbow smelt	*Osmerus mordax*	711	522	321	328	398	209	188
Eulachon	*Thaleichthys pacificus*	4	27	6	8	13	142	324
Smelts nei	*Osmerus spp, Hypomesus spp*	78	846	497	437	448	230	269
Lake(=Common)whitefish	*Coregonus clupeaformis*	5 272	5 842	5 678	5 353	5 199	4 484	4 239
Lake cisco	*Coregonus artedi*	226	215	284	239	255	303	285
Whitefishes nei	*Coregonus spp*	1 415	1 506	2 178	1 586	1 030	768	841
American shad	*Alosa sapidissima*	1 731	1 387	1 984	1 195	1 201	1 622	716
Alewife	*Alosa pseudoharengus*	442	526	594	682	291	712	934
Hickory shad	*Alosa mediocris*	88	75	47	62	51	90	42
American gizzard shad	*Dorosoma cepedianum*	1 272	1 989	1 291	3 007	977	1 781	2 780
Lampreys nei	*Petromyzontidae*	-	-	0	4	0	-	-
Striped bass	*Morone saxatilis*	2 138	2 802	3 046	3 004	3 135	2 949	2 864
White perch	*Morone americana*	995	1 164	966	1 042	1 221	1 226	1 022
White bass	*Morone chrysops*	45	50	166	109	106	145	104
Atlantic halibut	*Hippoglossus hippoglossus*	13	14	8	12	11	11	10
Pacific halibut	*Hippoglossus stenolepis*	22 680	31 843	33 895	36 517	34 753	35 391	37 237
English sole	*Pleuronectes vetulus*	...	...	425	1 100	1 140	1 395	1 476
Greenland halibut	*Reinhardtius hippoglossoides*	4 652	...	307	14	6 186	4 391	2 937
Arrow-tooth flounder	*Atheresthes stomias*	10 657	6 320	8 247	12 039	18 735	14 342	16 589
Petrale sole	*Eopsetta jordani*	...	1 937	1 464	1 480	1 870	1 822	1 793
Witch flounder	*Glyptocephalus cynoglossus*	2 082	1 775	1 855	2 123	2 439	3 020	3 189
Rex sole	*Glyptocephalus zachirus*	...	...	289	544	541	571	600
Amer. plaice(=Long rough dab)	*Hippoglossoides platessoides*	4 397	3 937	3 662	3 134	4 213	4 425	3 420
Flathead sole	*Hippoglossoides elassodon*	11 392	12 867	18 886	14 318	16 266	16 092	13 174
Yellowfin sole	*Limanda aspera*	101 354	149 302	80 500	56 830	69 971	54 918	63 625
Yellowtail flounder	*Limanda ferruginea*	2 403	2 864	3 656	4 431	6 928	7 304	5 355
Rock sole	*Lepidopsetta bilineata*	26 202	32 752	15 199	17 192	27 517	24 213	29 271
Dover sole	*Microstomus pacificus*	...	12 342	9 992	10 558	9 412	7 441	6 653
Pacific sand sole	*Psettichthys melanostictus*	...	...	34	605	79	129	187
Winter flounder	*Pseudopleuronectes americanus*	5 687	5 765	5 100	4 654	5 818	6 930	5 890
Curlfin sole	*Pleuronichthys decurrens*	-	-	-	-	1	5	4
Windowpane flounder	*Scophthalmus aquosus*	46	48	521	166	268	177	98
Pacific sanddab	*Citharichthys sordidus*	-	-	-	-	150	-	-

E-2

Fish crustaceans, molluscs, etc — Capture production by countries or areas and species — America, North
Poissons, crustacés, mollusques, etc — Captures par pays ou zones et espèces — Amérique du Nord
Peces, crustáceos, moluscos, etc — Capturas por países o áreas y especies — América del Norte

English name Nom anglais Nombre inglés	Scientific name Nom scientifique Nombre científico	1996 t	1997 t	1998 t	1999 t	2000 t	2001 t	2002 t
California flounder	*Paralichthys californicus*	...	...	523	620	390	393	428
Summer flounder	*Paralichthys dentatus*	5 425	4 775	6 348	5 795	6 063	6 554	7 415
Bastard halibuts nei	*Paralichthys spp*	2 192	1 059	533	391	503	...	708
Flatfishes nei	*Pleuronectiformes*	36 603	21 152	19 893	14 246	8 736	5 735	6 258
Tusk(=Cusk)	*Brosme brosme*	468	443	354	230	188	180	150
Atlantic cod	*Gadus morhua*	14 253	12 982	11 119	9 727	11 367	15 064	13 128
Pacific cod	*Gadus macrocephalus*	274 569	299 970	252 197	237 679	240 635	213 967	232 617
Red hake	*Urophycis chuss*	1 087	1 329	1 343	1 556	1 571	1 732	911
White hake	*Urophycis tenuis*	3 289	2 217	2 365	2 625	2 985	3 483	3 269
Haddock	*Melanogrammus aeglefinus*	570	1 504	2 836	3 146	4 002	5 826	7 553
Pacific tomcod	*Microgadus proximus*	...	...	3	1	0	0	-
Saithe(=Pollock)	*Pollachius virens*	2 962	4 251	5 583	4 595	4 043	4 109	3 581
Alaska pollock(=Walleye poll.)	*Theragra chalcogramma*	1 189 844	1 139 642	1 232 177	1 055 016	1 182 438	1 442 170	1 515 515
Silver hake	*Merluccius bilinearis*	16 025	15 535	14 959	14 039	12 176	13 007	7 987
North Pacific hake	*Merluccius productus*	195 290	226 616	227 504	216 889	205 351	172 051	129 599
Offshore silver hake	*Merluccius albidus*	32	...	5	12	5	2	6
Grenadiers, rattails nei	*Macrouridae*	...	...	500	312	315	305	276
Gadiformes nei	*Gadiformes*	1	4	20	9	21	5	5
Ladyfish	*Elops saurus*	-	726	976	1 967	121	460	655
Sea catfishes nei	*Ariidae*	-	-	-	-	0	3	1
Squirrelfishes nei	*Holocentridae*	...	...	...	...	...	...	23
Mullets nei	*Mugilidae*	7 723	8 907	8 130	4 542	6 153	6 560	5 935
Black grouper	*Mycteroperca bonaci*	...	...	...	...	9	7	6
Gag	*Mycteroperca microlepis*	...	...	...	...	15	10	13
Scamp	*Mycteroperca phenax*	...	...	1	14	27	17	26
Yellowfin grouper	*Mycteroperca venenosa*	...	...	...	...	3	1	3
Yellowedge grouper	*Epinephelus flavolimbatus*	...	15	18	...	73	36	67
Snowy grouper	*Epinephelus niveatus*	...	...	...	...	6	4	9
Warsaw grouper	*Epinephelus nigritus*	16	23	23	-	45	44	50
Groupers nei	*Epinephelus spp*	4 387	4 583	4 399	1 038	5 665	5 984	5 887
Black seabass	*Centropristis striata*	1 978	1 590	1 531	1 717	1 516	1 667	1 907
Groupers, seabasses nei	*Serranidae*	...	...	23	4 702	73	34	40
Giant seabass	*Stereolepis gigas*	1	1	3	2	2	3	2
Bigeyes nei	*Priacanthus spp*	...	...	...	...	0	1	1
Northern red snapper	*Lutjanus campechanus*	1 791	2 045	1 774	1 967	2 115	2 152	2 051
Snappers nei	*Lutjanus spp*	215	187	157	175	191	138	181
Vermilion snapper	*Rhomboplites aurorubens*	...	678	480	731	911	1 057	980
Snappers, jobfishes nei	*Lutjanidae*	2 407	1 916	1 648	1 826	1 624	1 654	1 622
Pigfish	*Orthopristis chrysoptera*	-	1	1	-	0	0	0
Grunts, sweetlips nei	*Haemulidae (=Pomadasyidae)*	293	243	292	316	403	429	316
Spotted weakfish	*Cynoscion nebulosus*	483	427	265	395	259	152	170
Squeteague(=Gray weakfish)	*Cynoscion regalis*	3 261	3 317	3 821	3 146	2 438	2 273	2 162
Weakfishes nei	*Cynoscion spp*	76	74	56	95	74	53	66
Atlantic croaker	*Micropogonias undulatus*	9 049	12 434	11 522	12 180	12 138	13 018	11 762
Northern kingfish	*Menticirrhus saxatilis*	49	21	17	13	28	39	8
Gulf kingcroaker	*Menticirrhus littoralis*	-	223	237	246	355	206	156
White weakfish	*Atractoscion nobilis*	46	28	71	112	101	124	194
White croaker	*Genyonemus lineatus*	242	167	64	74	105	137	97
Black drum	*Pogonias cromis*	39	1 230	1 967	719	2 312	2 528	2 248
Spot croaker	*Leiostomus xanthurus*	2 554	3 073	3 359	2 599	3 141	3 091	2 487
Red drum	*Sciaenops ocellatus*	1	2	3	6	6	3	4
Sheepshead	*Archosargus probatocephalus*	1 656	1 743	1 366	1 183	1 528	1 214	1 111
Scup	*Stenotomus chrysops*	2 952	2 196	1 893	1 676	1 206	1 845	3 312
Porgies, seabreams nei	*Sparidae*	176	398	336	93	168	219	209
Goatfishes, red mullets nei	*Mullidae*	-	-	-	-	-	36	33
Mojarras(=Silver-biddies) nei	*Gerres spp*	-	-	-	-	159	180	211
Opaleye	*Girella nigricans*	-	-	-	-	3	2	1
Sea chubs nei	*Kyphosus spp*	-	-	-	-	-	-	12
Tautog	*Tautoga onitis*	119	116	116	95	111	138	160
Cunner	*Tautogolabrus adspersus*	1	1	3	4	3	9	13
California sheephead	*Semicossyphus pulcher*	...	...	118	58	78	68	52
Parrotfishes nei	*Scaridae*	-	-	-	-	-	-	22
Marbled rockcod	*Notothenia rossii*	-	-	-	0	-	0	-
Humped rockcod	*Notothenia gibberifrons*	-	-	-	5	-	2	-
Yellowbelly rockcod	*Notothenia neglecta*	-	-	-	0	-	2	-
Grey rockcod	*Notothenia squamifrons*	-	-	-	5	-	0	-
Yellowfin notie	*Nototheniops nudifrons*	-	-	-	-	-	0	-
Antarctic rockcods nei	*Trematomus spp*	-	-	-	0	-	0	-
Antarctic silverfish	*Pleuragramma antarcticum*	-	-	-	-	-	0	-
Antarctic rockcods, noties nei	*Nototheniidae*	-	-	-	-	-	0	-
Ocean pout	*Macrozoarces americanus*	41	15	17	18	19	18	12
Sandeels(=Sandlances) nei	*Ammodytes spp*	-	1	1	1	0	0	0
Surgeonfishes nei	*Acanthuridae*	-	-	-	-	-	-	60
Spadefishes nei	*Ephippidae*	6	-	14	21	29	21	16
Lingcod	*Ophiodon elongatus*	2 254	1 889	659	631	172	187	252
Atka mackerel	*Pleurogrammus monopterygius*	88 030	59 538	51 579	51 436	44 592	57 096	37 759
Cabezon	*Scorpaenichthys marmoratus*	...	...	168	173	148	119	97
Sculpins nei	*Cottidae*	-	2	5	3	64	75	161
Northern puffer	*Sphoeroides maculatus*	11	18	17	37	36	35	57
Triggerfishes, durgons nei	*Balistidae*	326	80	249	196	168	180	199
Toadfishes, etc. nei	*Batrachoididae*	...	1	6	4	21	27	22
Lanternfishes nei	*Myctophidae*	-	-	-	0	-	-	-
American conger	*Conger oceanicus*	29	17	48	43	49	40	61
Bearded brotula	*Brotula barbata*	...	...	5	1	1	0	0
Silvery John dory	*Zenopsis conchifer*	27	6	49	19	28	62	66
Wreckfish	*Polyprion americanus*	82	14	6	1	-	-	-
Ocean whitefish	*Caulolatilus princeps*	-	-	-	5	4	5	2
Atlantic goldeneye tilefish	*Caulolatilus chrysops*	-	-	-	0	6	806	857
Great Northern tilefish	*Lopholatilus chamaeleonticeps*	1 463	1 890	1 419	616	688	285	295

English name / Nom anglais / Nombre inglés	Scientific name / Nom scientifique / Nombre científico	1996 t	1997 t	1998 t	1999 t	2000 t	2001 t	2002 t
Tilefishes nei	*Branchiostegidae*	-	28	294	318	483	314	268
Tripletail	*Lobotes surinamensis*	-	-	-	-	1	1	1
Antarctic toothfish	*Dissostichus mawsoni*	-	-	-	0	-	0	-
Patagonian toothfish	*Dissostichus eleginoides*	178	-	-	-	-	-	-
Blackfin icefish	*Chaenocephalus aceratus*	-	-	-	1	-	1	-
Mackerel icefish	*Champsocephalus gunnari*	-	-	-	1	-	1	-
South Georgia icefish	*Pseudochaenichthys georgianus*	-	-	-	3	-	0	-
Ocellated icefish	*Chionodraco rastrospinosus*	-	-	-	1	-	1	-
Spiny icefish	*Chaenodraco wilsoni*	-	-	-	0	-	0	-
Icefishes nei	*Channichthyidae*	-	-	-	0	-	0	-
Wolffishes(=Catfishes) nei	*Anarhichas spp*	363	309	296	258	200	250	155
Escolar	*Lepidocybium flavobrunneum*	...	...	54	35	78	47	61
Oilfish	*Ruvettus pretiosus*	...	...	10	11	38	30	117
Largehead hairtail	*Trichiurus lepturus*	1	10	5	6	41	7	16
Black driftfish	*Hyperoglyphe bythites*	...	...	...	2	7	5	6
Widow rockfish	*Sebastes entomelas*	...	7 751	4 897	4 243	3 604	2 609	442
Yellowtail rockfish	*Sebastes flavidus*	...	2 744	3 602	3 468	3 170	2 077	1 206
Pacific ocean perch	*Sebastes alutus*	21 004	19 580	18 445	20 616	17 952	17 708	20 620
Bocaccio rockfish	*Sebastes paucispinis*	...	724	596	197	27	33	22
Canary rockfish	*Sebastes pinniger*	...	1 262	1 313	772	60	49	54
Chilipepper rockfish	*Sebastes goodei*	...	1 850	1 273	918	445	618	161
Black rockfish	*Sebastes melanops*	...	...	...	...	109	148	127
Atlantic redfishes nei	*Sebastes spp*	322	251	320	353	322	363	368
Blackbelly rosefish	*Helicolenus dactylopterus*	-	-	-	-	3	0	0
Shortspine thornyhead	*Sebastolobus alascanus*	...	...	...	...	331	268	295
Scorpionfishes nei	*Scorpaenidae*	42 983	38 733	20 098	19 079	14 953	14 115	14 039
Atlantic searobins	*Prionotus spp*	20	10	31	38	24	39	66
Sablefish	*Anoplopoma fimbria*	27 193	24 006	19 732	21 888	22 534	19 975	18 556
American angler	*Lophius americanus*	24 203	27 506	26 357	25 046	20 807	23 268	22 893
Atlantic herring	*Clupea harengus*	104 002	97 706	81 813	79 597	74 037	106 561	67 652
Pacific herring	*Clupea pallasii*	54 628	62 080	41 865	41 305	33 945	41 412	35 566
Round sardinella	*Sardinella aurita*	571	512	489	536	614	623	653
California pilchard	*Sardinops caeruleus*	32 503	42 816	42 581	59 944	67 888	75 720	97 254
Atlantic menhaden	*Brevoortia tyrannus*	304 665	322 239	276 230	208 000	207 122	261 401	211 574
Gulf menhaden	*Brevoortia patronus*	491 612	597 565	497 461	694 242	591 434	528 506	582 497
Red-eye round herring	*Etrumeus teres*	-	72	7	58	42	2	-
Atlantic thread herring	*Opisthonema oglinum*	4 538	7 548	2 576	1 570	3 056	1 256	2 607
Californian anchovy	*Engraulis mordax*	4 505	5 778	1 553	5 323	11 487	19 259	4 882
Atlantic bonito	*Sarda sarda*	158	161	84	83	48	48	21
Eastern Pacific bonito	*Sarda chiliensis*	449	290	1 094	87	44	6	33
Wahoo	*Acanthocybium solandri*	...	2	74	33	81	58	362
King mackerel	*Scomberomorus cavalla*	2 069	2 515	2 361	2 410	2 246	2 194	2 028
Atlantic Spanish mackerel	*Scomberomorus maculatus*	1 545	1 695	1 459	1 362	1 650	1 916	1 575
Frigate and bullet tunas	*Auxis thazard, A.rochei*	-	-	1	17	9	3	1
Little tunny(=Atl.black skipj)	*Euthynnus alletteratus*	90	451	300	514	220	357	417
Black skipjack	*Euthynnus lineatus*	84	44	231	90	0	0	-
Kawakawa	*Euthynnus affinis*	-	-	-	-	-	-	4
Skipjack tuna	*Katsuwonus pelamis*	128 548	112 154	123 667	151 340	97 429	89 038	89 958
Atlantic bluefin tuna	*Thunnus thynnus*	741	1 024	1 059	1 042	1 121	1 232	1 212
Pacific bluefin tuna	*Thunnus orientalis*	4 769	2 272	1 962	177	316	317	41
Blackfin tuna	*Thunnus atlanticus*	53	67	53	41	50	35	32
Albacore	*Thunnus alalunga*	18 669	18 907	20 273	14 093	13 504	18 647	17 588
Yellowfin tuna	*Thunnus albacares*	47 187	64 225	61 911	43 623	34 276	34 610	32 831
Bigeye tuna	*Thunnus obesus*	6 523	6 659	7 977	6 909	5 715	5 899	12 523
Indo-Pacific sailfish	*Istiophorus platypterus*	-	-	-	-	-	-	5
Blue marlin	*Makaira nigricans*	...	...	...	...	...	...	394
Black marlin	*Makaira indica*	...	...	...	...	...	...	4
Striped marlin	*Tetrapturus audax*	...	...	...	...	...	...	278
Marlins,sailfishes,etc. nei	*Istiophoridae*	1 368	100	134	214	123	250	142
Swordfish	*Xiphias gladius*	5 842	6 163	6 846	7 277	8 076	4 268	3 921
Tuna-like fishes nei	*Scombroidei*	297	1 329	42	69	72	65	27
Halfbeaks nei	*Hemiramphus spp*	398	293	441	350	414	354	393
Opah	*Lampris guttatus*	...	...	115	68	58	49	446
Atlantic silverside	*Menidia menidia*	19	21	24	25	15	15	15
Silversides(=Sand smelts) nei	*Atherinidae*	-	-	-	-	8	34	37
Bluefish	*Pomatomus saltatrix*	4 244	4 222	3 764	3 359	3 661	3 993	3 163
Cobia	*Rachycentron canadum*	187	133	142	137	112	97	100
Pacific jack mackerel	*Trachurus symmetricus*	2 176	1 160	1 563	1 116	1 316	3 839	1 026
Scads nei	*Decapterus spp*	-	-	-	-	-	-	111
Blue runner	*Caranx crysos*	119	152	276	180	131	157	171
Crevalle jack	*Caranx hippos*	136	264	397	323	318	305	221
Bar jack	*Caranx ruber*	-	-	-	3	5	7	13
Jacks, crevalles nei	*Caranx spp*	-	-	-	-	0	-	131
Florida pompano	*Trachinotus carolinus*	44	297	321	207	242	181	159
Greater amberjack	*Seriola dumerili*	...	...	560	517	538	452	448
Yellowtail amberjack	*Seriola lalandi*	-	-	111	30	50	39	34
Amberjacks nei	*Seriola spp*	1 123	829	247	189	378	386	210
Rainbow runner	*Elagatis bipinnulata*	-	-	-	-	-	-	2
Bigeye scad	*Selar crumenophthalmus*	-	-	-	-	-	-	128
Pomfrets, ocean breams nei	*Bramidae*	-	-	-	-	-	4	222
Common dolphinfish	*Coryphaena hippurus*	739	768	325	572	541	418	1 010
Chub mackerel	*Scomber japonicus*	9 977	18 396	20 423	8 724	21 349	7 260	3 495
Atlantic mackerel	*Scomber scombrus*	16 022	15 395	14 429	12 041	5 649	12 339	27 364
North Atlantic harvestfish	*Peprilus alepidotus*	14	15	13	16	23	18	22
Butterfishes, pomfrets nei	*Stromateidae*	4 393	3 422	2 578	2 762	2 122	4 962	1 525
Barracudas nei	*Sphyraena spp*	...	0	60	95	129	127	99
Sand tiger shark	*Carcharias taurus*	-	-	-	-	1	-	-
Thresher	*Alopias vulpinus*	...	...	320	264	305	386	363
Bigeye thresher	*Alopias superciliosus*	...	...	11	5	5	2	-

E-2 Fish crustaceans, molluscs, etc / Poissons, crustacés, mollusques, etc / Peces, crustáceos, moluscos, etc

Capture production by countries or areas and species / Captures par pays ou zones et espèces / Capturas por países o áreas y especies

America, North / Amérique du Nord / América del Norte

English name Nom anglais Nombre inglés	Scientific name Nom scientifique Nombre científico	1996 t	1997 t	1998 t	1999 t	2000 t	2001 t	2002 t
Shortfin mako	*Isurus oxyrinchus*	...	...	96	61	103	70	101
Longfin mako	*Isurus paucus*	-	1	1	-	4	3	1
Mako sharks	*Isurus spp*	-	-	-	-	-	47	117
Porbeagle	*Lamna nasus*	-	-	-	-	3	1	1
Great white shark	*Carcharodon carcharias*	-	-	-	-	2	0	-
Nurse shark	*Ginglymostoma cirratum*	-	-	-	-	0	-	0
Blue shark	*Prionace glauca*	...	...	1	-	1	2	1
Sandbar shark	*Carcharhinus plumbeus*	-	-	-	-	41	24	28
Blacktip shark	*Carcharhinus limbatus*	-	1	0	-	601	521	11
Dusky shark	*Carcharhinus obscurus*	-	-	-	-	80	0	3
Tiger shark	*Galeocerdo cuvier*	-	-	-	-	-	1	0
Hammerhead sharks, etc. nei	*Sphyrnidae*	...	...	1	-	-	-	-
Dusky smooth-hound	*Mustelus canis*	...	...	...	...	334	321	493
Brown smooth-hound	*Mustelus henlei*	...	...	3	5	3	4	3
Tope shark	*Galeorhinus galeus*	...	...	52	73	48	45	32
Picked dogfish	*Squalus acanthias*	...	0	6	566	8 548	2 875	3 115
Dogfish sharks nei	*Squalidae*	29 638	21 021	22 268	16 081	1 857	295	251
Dogfish sharks, etc. nei	*Squaliformes*	3 317	1 083	772	1 530	590	784	940
Raja rays nei	*Raja spp*	13 891	10 142	13 932	12 619	13 335	13 122	13 010
Rays, stingrays, mantas nei	*Rajiformes*	1 554	2 488	1 340	1 616	1 716	1 500	3 139
Spotted ratfish	*Hydrolagus colliei*	-	-	-	-	-	-	2
Sharks, rays, skates, etc. nei	*Elasmobranchii*	3 643	5 689	5 757	4 739	3 358	2 069	2 465
Groundfishes nei	*Osteichthyes*	-	-	-	-	-	0	0
Finfishes nei	*Osteichthyes*	12 935	12 153	4 431	6 398	5 907	2 566	3 354
Marine fishes nei	*Osteichthyes*	12 585	13 351	12 233	9 842	12 531	11 401	8 709
River prawns nei	*Macrobrachium spp*	-	-	-	-	0	-	-
Euro-American crayfishes nei	*Astacidae, Cambaridae*	5 689	10 496	10 082	5 322	217	4 676	7 125
Atlantic rock crab	*Cancer irroratus*	55	1 120	1 358	1 301	1 844	326	502
Dungeness crab	*Cancer magister*	29 478	17 328	15 519	16 080	17 109	16 525	22 185
Jonah crab	*Cancer borealis*	334	745	1 255	1 549	1 114	1 245	1 190
Pacific rock crab	*Cancer productus*	...	...	574	359	494	537	552
Black stone crab	*Menippe mercenaria*	3 216	2 483	2 567	1 877	2 961	2 996	2 942
Blue crab	*Callinectes sapidus*	100 919	106 527	101 880	91 851	83 402	71 867	80 606
Green crab	*Carcinus maenas*	-	54	86	14	16	67	39
Tanner crabs nei	*Chionoecetes spp*	30 785	53 932	114 230	83 989	15 665	12 177	15 078
Red crab	*Geryon quinquedens*	465	96	-	-	3 132	4 004	2 169
Marine crabs nei	*Brachyura*	2 940	4 566	2 297	3 195	3 041	6 462	6 645
Caribbean spiny lobster	*Panulirus argus*	3 373	2 783	2 343	2 749	2 571	1 527	2 047
Tropical spiny lobsters nei	*Panulirus spp*	395	501	350	287	361	324	306
American lobster	*Homarus americanus*	32 496	38 066	36 125	39 676	37 730	32 389	37 309
Blue mud shrimp	*Upogebia pugettensis*	...	...	...	...	10	4	4
Ghost shrimps	*Callianassa spp*	-	-	-	25	17	14	14
King crabs	*Paralithodes spp*	9 526	8 177	10 941	7 675	6 849	7 283	7 617
Antarctic stone crab	*Paralomis spinosissima*	214	-	-	-	-	-	-
Northern brown shrimp	*Penaeus aztecus*	55 361	47 836	50 722	61 206	63 817	68 869	58 191
Northern pink shrimp	*Penaeus duorarum*	13 800	8 874	10 718	5 925	5 589	7 144	8 074
Northern white shrimp	*Penaeus setiferus*	28 808	32 841	39 799	44 633	52 593	40 724	43 186
Penaeus shrimps nei	*Penaeus spp*	3 793	6 675	5 708	2 358	1 800	2 663	1 556
Atlantic seabob	*Xiphopenaeus kroyeri*	4 558	5 744	3 397	3 626	3 415	3 951	3 514
Northern prawn	*Pandalus borealis*	9 932	7 239	3 926	2 832	2 625	1 309	682
Pacific shrimps nei	*Pandalus spp, Pandalopsis spp*	16 737	20 419	6 890	14 651	16 530	19 049	26 907
Crangonid shrimps nei	*Crangonidae*	-	31	41	-	42	45	38
Rock shrimp	*Sicyonia brevirostris*	10 549	1 789	4 409	1 826	3 254	2 909	890
Pacific rock shrimp	*Sicyonia ingentis*	...	...	185	630	756	165	215
Royal red shrimp	*Pleoticus robustus*	198	209	195	286	391	305	441
Antarctic krill	*Euphausia superba*	-	-	-	-	70	1 561	12 175
Brine shrimp	*Artemia salina*	...	...	513	691	561	538	378
Stomatopods nei	*Stomatopoda*	-	-	1	5	6	1	1
Freshwater molluscs nei	*Mollusca*	-	-	5	-	-	-	-
Periwinkles nei	*Littorina spp*	19	-	171	223	216	649	1 061
Abalones nei	*Haliotis spp*	127	102	-	8	-	3	3
Whelks	*Busycon spp*	4 196	3 104	2 581	4 624	2 930	4 037	3 070
Gastropods nei	*Gastropoda*	6 380	12	49	36	35	26	51
Olympia flat oyster	*Ostrea conchaphila*	...	...	...	6	7	8	11
Pacific cupped oyster	*Crassostrea gigas*	6 706	8 363	2 118	539	1 917	1 083	1 083
American cupped oyster	*Crassostrea virginica*	111 682	104 466	101 383	89 714	169 319	123 463	109 188
Cupped oysters nei	*Crassostrea spp*	...	...	64	49	42	26	27
Blue mussel	*Mytilus edulis*	9 434	8 063	4 774	4 086	6 009	6 524	9 862
Sea mussels nei	*Mytilidae*	626	170	174	96	771	538	342
American sea scallop	*Placopecten magellanicus*	61 552	48 120	43 030	77 206	113 141	164 998	186 600
Atlantic bay scallop	*Argopecten irradians*	230	452	690	216	154	30	135
Weathervane scallop	*Patinopecten caurinus*	2 372	2	3 228	2 642	2 012	1 052	1 264
Ocean quahog	*Arctica islandica*	173 685	163 799	148 506	144 366	122 547	142 662	149 767
Butter clam	*Saxidomus giganteus*	...	87	126	74	164	48	58
Pacific littleneck clam	*Protothaca staminea*	-	-	-	-	-	102	146
Northern quahog(=Hard clam)	*Mercenaria mercenaria*	21 855	4 631	...	...	13 723	10 136	16 486
Pacific horse clams nei	*Tresus spp*	-	-	-	-	1	2	9
Atlantic surf clam	*Spisula solidissima*	154 380	140 621	131 256	142 067	165 269	165 708	175 709
Stimpson's surf clam	*Spisula polynyma*	15	43	32	23	6	16	19
Atl.jackknife(=Atl.razor clam)	*Ensis directus*	-	14	49	64	99	36	90
Pacific razor clam	*Siliqua patula*	-	-	-	-	14	61	146
Sand gaper	*Mya arenaria*	4 355	4 454	5 493	5 113	5 222	6 917	6 178
Pacific geoduck	*Panopea abrupta*	-	1 962	2 243	2 451	2 259	2 481	3 061
Basket cockle	*Clinocardium nuttallii*	-	22	-	18	49	67	80
Clams, etc. nei	*Bivalvia*	8 369	7 125	8 926	5 947	1 780	3 402	3 964
Opalescent inshore squid	*Loligo opalescens*	...	70 915	2 709	90 662	117 711	85 829	72 879
Longfin squid	*Loligo pealei*	12 490	16 161	18 879	18 749	16 942	14 211	16 684
Northern shortfin squid	*Illex illecebrosus*	17 050	13 632	22 717	7 334	9 011	4 009	2 750
Jumbo flying squid	*Dosidicus gigas*	...	3	107	18	1	0	4

English name Nom anglais Nombre inglés	Scientific name Nom scientifique Nombre científico	1996 t	1997 t	1998 t	1999 t	2000 t	2001 t	2002 t
Various squids nei	*Loliginidae, Ommastrephidae*	79 338	795	700	353	138	1 050	913
Octopuses, etc. nei	*Octopodidae*	-	0	128	8	5	26	77
Marine molluscs nei	*Mollusca*	2 787	4 026	3 347	3 377	1 133	1 893	3 062
Frogs	*Rana spp*	2	9	6	-	0	1	-
Diamond back terrapins	*Malaclemys spp*	-	-	0	-	0	-	-
River and lake turtles nei	*Testudinata*	15	13	24	15	53	51	27
Horseshoe crab	*Limulus polyphemus*	1 598	2 607	3 252	2 397	1 696	1 299	1 387
Starfishes nei	*Asteroidea*	-	-	-	1	0	0	1
Sea urchins nei	*Strongylocentrotus spp*	20 381	20 216	13 626	15 218	14 014	12 460	10 627
Sea cucumbers nei	*Holothurioidea*	1 779	-	2 406	3 732	4 583	1 804	4 905
Jellyfishes	*Rhopilema spp*	...	...	...	578	137	-	-
Marine worms	*Polychaeta*	-	223	299	343	202	481	507
	Country total	*5 001 191*	*4 983 440*	*4 708 980*	*4 749 646*	*4 717 638*	*4 944 336*	*4 937 305*
US Virgin Is								
Freshwater fishes nei	*Osteichthyes*	0	0	0	0	0	0	0
Groupers nei	*Epinephelus spp*	...	...	...	21	25 F	25 F	25 F
Snappers, jobfishes nei	*Lutjanidae*	...	...	...	57	65 F	65 F	65 F
Triggerfishes, durgons nei	*Balistidae*	...	...	...	31	35 F	35 F	35 F
Marine fishes nei	*Osteichthyes*	340 F	300 F	255 F	119	139 F	139 F	139 F
Caribbean spiny lobster	*Panulirus argus*	50 F	45 F	40 F	34	35 F	35 F	35 F
Stromboid conchs nei	*Strombus spp*	10 F	5 F	5 F	1	1 F	1 F	1 F
Marine molluscs nei	*Mollusca*	0	0	0	0	0	0	0
	Country total	*400 F*	*350 F*	*300 F*	*263*	*300 F*	*300 F*	*300 F*
Total		***7 939 507***	***8 040 827***	***7 553 103***	***7 624 074***	***7 796 824***	***8 142 843***	***8 242 710***

E-3 Fish crustaceans, molluscs, etc / Poissons, crustacés, mollusques, etc / Peces, crustáceos, moluscos, etc

Capture production by countries or areas and species / Captures par pays ou zones et espèces / Capturas por países o áreas y especies

America, South / Amérique du Sud / América del Sur

English name Nom anglais Nombre inglés	Scientific name Nom scientifique Nombre científico	1996 t	1997 t	1998 t	1999 t	2000 t	2001 t	2002 t
Argentina								
Characins nei	*Characidae*	13 000 F	15 500 F	16 000 F	2 800 F	3 000 F	3 400 F	3 100 F
Prochilods nei	*Prochilodus spp*	...	...	...	16 000 F	17 700 F	19 600 F	18 200 F
Freshwater siluroids nei	*Siluroidei*	2 800 F	3 000 F	3 000 F	3 500 F	3 500 F	4 100 F	3 800 F
Freshwater fishes nei	*Osteichthyes*	3 297 F	4 135 F	4 097 F	5 138 F	6 098 F	6 517 F	6 003 F
Rainbow trout	*Oncorhynchus mykiss*	2	0	0	0	-	-	-
Atlantic sabretooth anchovy	*Lycengraulis grossidens*	90 F	100 F	100 F	120 F	120 F	140 F	130 F
Bastard halibuts nei	*Paralichthys spp*	8 753	10 044	8 751	6 668	6 498	5 944	4 297
Tadpole codling	*Salilota australis*	1 742	2 632	3 604	6 607	8 435	1 858	2 604
Brazilian codling	*Urophycis brasiliensis*	618	182	3 329	754	1 038	2 683	834
Southern blue whiting	*Micromesistius australis*	85 040	79 945	71 643	55 097	61 313	54 311	42 453
Southern hake	*Merluccius australis*	4 115	3 011	3 125	3 471	7 035	4 742	5 301
Argentine hake	*Merluccius hubbsi*	597 557	584 048	458 433	311 953	193 701	249 462	358 897
Patagonian grenadier	*Macruronus magellanicus*	44 065	41 835	96 157	117 571	123 684	111 885	98 723
Grenadiers nei	*Macrourus spp*	829	1 041	2 972	2 580	10 503	3 200	5 329
Sea catfishes nei	*Ariidae*	6	13	18	5	5	20	27
Mullets nei	*Mugilidae*	10	10	10	10	5	11	9
Argentine seabass	*Acanthistius brasilianus*	10 714	9 113	5 216	5 895	4 134	4 881	3 622
Striped weakfish	*Cynoscion striatus*	18 987	24 132	17 108	11 107	9 433	11 844	11 412
Whitemouth croaker	*Micropogonias furnieri*	23 514	26 108	9 451	6 641	5 296	4 479	3 652
Argentine croaker	*Umbrina canosai*	6 825	2 753	1 956	1 410	421	2 083	1 539
King weakfish	*Macrodon ancylodon*	380	224	167	515	44	64	90
Black drum	*Pogonias cromis*	159	81	260	182	13	52	33
South American silver porgy	*Diplodus argenteus*	39	84	15	5	2	2	1
Red porgy	*Pagrus pagrus*	1 590	1 159	574	2 159	1 301	935	904
Argentine goatfish	*Mullus argentinae*	75	73	99	229	498	225	78
Patagonian blennie	*Eleginops maclovinus*	148	57	1 194	2 023	1 745	233	50
Antarctic rockcods, noties nei	*Nototheniidae*	-	-	-	-	-	182	180
Brazilian flathead	*Percophis brasilianus*	8 771	11 475	9 677	6 526	7 255	7 040	4 225
Argentinian sandperch	*Pseudopercis semifasciata*	3 438	2 483	2 222	2 679	1 905	1 887	1 641
Argentine conger	*Conger orbignyanus*	22	28	81	18	14	95	43
Pink cusk-eel	*Genypterus blacodes*	21 933	21 917	25 086	21 503	15 166	19 644	17 794
Wreckfish	*Polyprion americanus*	378	121	75	218	129	...	...
Castaneta	*Cheilodactylus bergi*	204	744	1 827	155	81	98	149
Patagonian toothfish	*Dissostichus eleginoides*	14 912	8 793	9 950	7 702	7 771	6 410	8 164
White snake mackerel	*Thyrsitops lepidopoides*	1 232	309	285	136	-	-	-
Choicy ruff	*Seriolella porosa*	2 042	2 593	2 462	3 322	3 542	3 990	5 368
Blackbelly rosefish	*Helicolenus dactylopterus*	1 499	2 563	1 192	3 354	3 051	1 209	494
Argentine menhaden	*Brevoortia pectinata*	427	893	104	205	271	265	94
Argentine anchovy	*Engraulis anchoita*	21 001	25 198	13 350	9 832	12 158	12 815	16 192
Atlantic bonito	*Sarda sarda*	108	130	12	38	19	235	1
Skipjack tuna	*Katsuwonus pelamis*	1	-	-	-	-	-	-
Albacore	*Thunnus alalunga*	0	120	-	-	-	-	-
Swordfish	*Xiphias gladius*	-	-	-	-	-	5	-
Tuna-like fishes nei	*Scombroidei*	0	-	-	-	-	-	-
Silversides(=Sand smelts) nei	*Atherinidae*	583	520	48	618	14	35	44
Bluefish	*Pomatomus saltatrix*	342	416	15	286	416	181	701
Rough scad	*Trachurus lathami*	587	288	247	470	67	117	127
Yellowtail amberjack	*Seriola lalandi*	16	17	7	9	10	33	8
Parona leatherjacket	*Parona signata*	1 925	1 805	1 744	1 710	1 483	933	868
Chub mackerel	*Scomber japonicus*	11 195	10 468	3 224	7 012	10 122	4 602	11 616
Narrownose smooth-hound	*Mustelus schmitti*	10 252	9 956	11 266	9 062	7 120	9 613	7 019
Tope shark	*Galeorhinus galeus*	92	103	92	89	109	50	29
Argentine angelshark	*Squatina argentina*	4 281	4 410	4 311	3 368	3 123	3 339	2 288
Rays, stingrays, mantas nei	*Rajiformes*	12 478	12 119	14 855	12 116	13 265	15 179	13 215
Elephantfishes nei	*Callorhinchus spp*	815	1 329	1 770	1 977	1 390	451	568
Sharks, rays, skates, etc. nei	*Elasmobranchii*	2 251	1 070	1 220	905	719	798	1 103
Marine fishes nei	*Osteichthyes*	4 960	7 855	6 702	5 998	-	-	-
Marine crabs nei	*Brachyura*	390	403	326	170	-	-	-
Southern king crab	*Lithodes antarcticus*	200	413	456	270	104	88	385
Softshell red crab	*Paralomis granulosa*	...	...	...	...	264	213	129
Argentine stiletto shrimp	*Artemesia longinaris*	263	166	146	37	39	283	298
Argentine red shrimp	*Pleoticus muelleri*	9 874	6 479	23 203	15 888	37 150	78 866	51 410
Antarctic krill	*Euphausia superba*	-	-	-	6 524	-	-	-
Marine crustaceans nei	*Crustacea*	2	288	2	6 027	-	-	-
Angulate volute	*Zidona dufresnei*	558	1 322	1 010	683	631	694	414
River Plata mussel	*Mytilus platensis*	164	180	149	332	236	217	344
Patagonian scallop	*Zygochlamis patagonica*	36 952	39 817	28 441	42 700	36 514	38 960	50 966
Common squids nei	*Loligo spp*	184	1 999	802	209	268	251	64
Argentine shortfin squid	*Illex argentinus*	292 628	411 994	291 174	342 691	279 046	230 272	177 314
Octopuses, etc. nei	*Octopodidae*	39	48	17	34	5	13	3
Marine molluscs nei	*Mollusca*	1	50	-	1 000	-	-	-
	Country total	*1 291 355*	*1 400 162*	*1 164 829*	*1 078 313*	*908 979*	*931 734*	*944 346*
Bolivia								
Freshwater fishes nei	*Osteichthyes*	4 800	4 850	4 865	4 860	4 911	4 900	4 853 F
Rainbow trout	*Oncorhynchus mykiss*	338	338	340	342	345	280	197
Silversides(=Sand smelts) nei	*Atherinidae*	850	850	850	850	850	760	750 F
	Country total	*5 988*	*6 038*	*6 055*	*6 052*	*6 106*	*5 940*	*5 800 F*
Brazil								
Cyprinids nei	*Cyprinidae*	284	212	258	302	355	390	392
Tilapias nei	*Oreochromis (=Tilapia) spp*	7 300	6 927	7 444	6 616	7 893	8 329	8 248
Cichlids nei	*Cichlidae*	8 565	8 356	7 785	7 375	8 145	13 611	13 946
Cachama	*Colossoma macropomum*	4 298	2 838	2 760	2 905	4 965	3 789	4 068
Characins nei	*Characidae*	80 598	65 273	60 983	65 149	67 297	22 850	22 000

English name Nom anglais Nombre inglés	Scientific name Nom scientifique Nombre científico	1996 t	1997 t	1998 t	1999 t	2000 t	2001 t	2002 t
Freshwater siluroids nei	*Siluroidei*	47 302	52 601	49 866	57 014	58 474	58 748	59 641
Freshwater fishes nei	*Osteichthyes*	42 285	40 507	43 588	42 875	49 593	109 558	109 455
Rainbow trout	*Oncorhynchus mykiss*	0	0	0	0	0	-	-
Flatfishes nei	*Pleuronectiformes*	1 091	1 430	1 655	1 590	1 844	1 820	3 321
Brazilian codling	*Urophycis brasiliensis*	2 311	2 116	2 408	1 807	2 225	6 029	8 312
Argentine hake	*Merluccius hubbsi*	0	-	-	128	226	2 436	3 739
Patagonian grenadier	*Macruronus magellanicus*	18	20	0	0	0	-	-
Tarpon	*Megalops atlanticus*	348	720	551	315	331	1 169	1 201
Sea catfishes nei	*Ariidae*	16 946	17 736	20 489	25 224	29 473	34 913	34 754
Mullets nei	*Mugilidae*	7 722	10 262	10 006	10 886	11 987	11 031	11 851
Snooks(=Robalos) nei	*Centropomus spp*	1 686	1 866	2 996	3 602	4 366	3 469	3 608
Brazilian groupers nei	*Mycteroperca spp*	423	584	518	628	536	1 040	1 347
Red grouper	*Epinephelus morio*	320	238	212	666	1 220	...	...
Groupers nei	*Epinephelus spp*	2 341	2 170	2 315	1 710	2 670	2 124	2 084
Bigeyes nei	*Priacanthus spp*	59	55	46	60	67	45	52
Southern red snapper	*Lutjanus purpureus*	5 104	6 085	5 937	9 790	6 580	6 209	6 184
Lane snapper	*Lutjanus synagris*	648	659	1 038	1 031	1 014	1 437	1 392
Yellowtail snapper	*Ocyurus chrysurus*	4 167	5 000	3 317	4 541	4 165	2 002	2 158
Snappers, jobfishes nei	*Lutjanidae*	3 690	4 488	4 390	4 549	3 859	4 176	3 911
Barred grunt	*Conodon nobilis*	172	118	114	84	39	53	82
Grunts, sweetlips nei	*Haemulidae (=Pomadasyidae)*	2 081	1 936	2 000	2 382	3 024	3 429	3 537
Weakfishes nei	*Cynoscion spp*	26 470	25 960	26 677	39 332	43 523	2 637	1 810
Whitemouth croaker	*Micropogonias furnieri*	23 374	23 311	27 927	21 926	28 797	39 266	42 331
Kingcroakers nei	*Menticirrhus spp*	1 144	1 303	1 395	1 058	1 348	1 951	2 042
Argentine croaker	*Umbrina canosai*	4 983	3 358	2 216	4 849	7 729	12 757	14 711
King weakfish	*Macrodon ancylodon*	5 143	6 006	6 708	4 278	5 475	2 589	4 163
Black drum	*Pogonias cromis*	1	-	-	-	-	-	-
Croakers, drums nei	*Sciaenidae*	150	52	57	36	34	35 590	32 426
Red porgy	*Pagrus pagrus*	884	1 572	1 448	1 362	1 497	1 440	1 587
Goatfishes, red mullets nei	*Mullidae*	530	749	840	1 117	1 756	4 167	6 040
Percoids nei	*Percoidei*	664	692	1 838	947	1 465	2 018	2 358
Brazilian flathead	*Percophis brasilianus*	460	412	514	709	607	624	711
Spadefishes nei	*Ephippidae*	71	55	51	283	325	381	241
Puffers nei	*Tetraodontidae*	18	88	23	16	35	34	199
Argentine conger	*Conger orbignyanus*	67	110	108	162	108	491	612
Cusk-eels, brotulas nei	*Ophidiidae*	176	258	452	544	507	658	741
Tilefishes nei	*Branchiostegidae*	1 098	1 000	786	524	547	1 309	1 261
Largehead hairtail	*Trichiurus lepturus*	736	938	1 405	1 230	1 665	2 883	3 131
Atlantic searobins	*Prionotus spp*	792	540	858	1 149	1 839	317	352
Anglerfishes nei	*Lophiidae*	294	399	542	794	1 937	6 824	5 258
Demersal percomorphs nei	*Perciformes*	5 697	9 825	9 828	12 050	7 644	14 291	15 006
Brazilian sardinella	*Sardinella brasiliensis*	97 092	117 642	82 283	25 518	17 053	39 039	27 891
Brazilian menhaden	*Brevoortia aurea*	6 294	2 396	2 936	2 202	1 123	344	490
Scaled sardines	*Harengula spp*	20	12	2	162	126	488	569
Atlantic thread herring	*Opisthonema oglinum*	3 762	3 712	8 082	8 204	12 455	4 397	3 191
Anchovies, etc. nei	*Engraulidae*	7 672	7 102	2 654	2 682	4 039	4 242	4 511
Clupeoids nei	*Clupeoidei*	3 464	4 314	6 042	7 527	11 457	27 058	29 481
Wahoo	*Acanthocybium solandri*	16	58	41	...	...	...	...
King mackerel	*Scomberomorus cavalla*	2 890	2 398	3 595	3 595	2 344	1 251	2 316
Serra Spanish mackerel	*Scomberomorus brasiliensis*	3 047	2 125	1 516	1 516	988	251	3 071
Frigate and bullet tunas	*Auxis thazard, A.rochei*	527	215	162	166	106	98	1 117
Little tunny(=Atl.black skipj)	*Euthynnus alletteratus*	834	507	920	930	615	615	615
Skipjack tuna	*Katsuwonus pelamis*	22 528	26 564	23 789	23 188	25 164	24 146	18 338
Atlantic bluefin tuna	*Thunnus thynnus*	0	0	0	13	0	-	-
Blackfin tuna	*Thunnus atlanticus*	649	418	55	55	38	149	1 669
Albacore	*Thunnus alalunga*	819	652	3 418	1 872	4 414	6 862	3 228
Yellowfin tuna	*Thunnus albacares*	2 767	2 705	2 514	4 127	6 145	6 239	6 172
Bigeye tuna	*Thunnus obesus*	1 707	1 237	644	2 024	2 768	2 659	2 582
Atlantic sailfish	*Istiophorus albicans*	310	137	184	356	598	412	547
Blue marlin	*Makaira nigricans*	331	193	486	509	467	780	387
Atlantic white marlin	*Tetrapturus albidus*	75	105	217	158	106	172	407
Longbill spearfish	*Tetrapturus pfluegeri*	-	-	-	-	12	56	39
Marlins,sailfishes,etc. nei	*Istiophoridae*	-	-	-	-	18	2	1
Swordfish	*Xiphias gladius*	1 892	4 100	3 847	4 721	4 697	4 082	2 910
Tuna-like fishes nei	*Scombroidei*	-	446	258	570	151	5	467
Ballyhoo halfbeak	*Hemiramphus brasiliensis*	494	442	1 500	832	989	3 342	3 652
Flyingfishes nei	*Exocoetidae*	743	1 082	1 084	760	388	217	708
Silversides(=Sand smelts) nei	*Atherinidae*	52	62	52	17	39	28	5
Bluefish	*Pomatomus saltatrix*	5 866	2 616	2 504	2 064	3 314	3 427	4 029
Cobia	*Rachycentron canadum*	498	367	622	1 818	1 580	1 036	1 065
Rough scad	*Trachurus lathami*	389	313	81	25	40	228	328
Blue runner	*Caranx crysos*	503	515	443	590	625	680	688
Jacks, crevalles nei	*Caranx spp*	3 905	5 838	5 635	4 164	6 950	5 878	6 083
Atlantic moonfish	*Selene setapinnis*	1 679	1 914	1 512	1 514	1 386	1 108	1 350
Pompanos nei	*Trachinotus spp*	534	166	172	144	286	273	219
Yellowtail amberjack	*Seriola lalandi*	538	466	880	544	611	155	189
Amberjacks nei	*Seriola spp*	825	1 048	119	104	128	588	653
Atlantic bumper	*Chloroscombrus chrysurus*	952	1 035	4 065	3 519	1 573	5 304	5 757
Common dolphinfish	*Coryphaena hippurus*	2 500	4 028	4 117	2 848	4 359	4 299	4 818
Chub mackerel	*Scomber japonicus*	5 670	8 306	9 422	1 595	6 377	1 405	2 189
Barracudas nei	*Sphyraena spp*	11	12	828	217	527	378	371
Shortfin mako	*Isurus oxyrinchus*	83	190	...	100	120	...	...
Blue shark	*Prionace glauca*	743	1 103	...	500	580	...	2 000
Silky shark	*Carcharhinus falciformis*	502	279	...	70	80	...	...
Scalloped hammerhead	*Sphyrna lewini*	25	170	...	30	38	507	508
Angelsharks, sand devils nei	*Squatinidae*	1 587	...	...	...	...	...	...
Chola guitarfish	*Rhinobatos percellens*	404	...	...	...	...	...	...
Rays, stingrays, mantas nei	*Rajiformes*	3 104	3 010	4 673	4 277	4 867	4 860	6 294
Sharks, rays, skates, etc. nei	*Elasmobranchii*	8 446	10 189	12 596	13 576	15 900	14 626	12 032

English name / Nom anglais / Nombre inglés	Scientific name / Nom scientifique / Nombre científico	1996 t	1997 t	1998 t	1999 t	2000 t	2001 t	2002 t
Marine fishes nei	*Osteichthyes*	143 419	146 661	138 925	169 938	176 695	151 386	161 161
River prawns nei	*Macrobrachium spp*	2 677	2 157	1 506	3 235	2 437	...	...
Freshwater crustaceans nei	*Crustacea*	0	0	0	0	0	-	-
Dana swimcrab	*Callinectes danae*	2 020	2 600	3 014	1 626	1 597	1 062	1 310
Marine crabs nei	*Brachyura*	9 922	10 233	9 202	10 913	11 135	12 535	12 790
Caribbean spiny lobster	*Panulirus argus*	8 026	7 502	6 002	6 334	6 469	7 135	7 016
Redspotted shrimp	*Penaeus brasiliensis*	8 743	10 758	7 796	9 092	10 728	5 010	5 903
Penaeus shrimps nei	*Penaeus spp*	15 007	14 506	13 998	12 550	16 509	10 697	10 533
Atlantic seabob	*Xiphopenaeus kroyeri*	11 764	15 257	13 755	9 964	11 948	12 264	11 673
Marine crustaceans nei	*Crustacea*	288	310	244	266	227	162	247
Cupped oysters nei	*Crassostrea spp*	873	828	744	1 547	884	895	876
Sea mussels nei	*Mytilidae*	863	1 193	1 230	906	802	2 017	1 721
Triangular tivela	*Tivela mactroides*	126	196	664	1 684	273	280	835
Common squids nei	*Loligo spp*	950	1 486	758	1 696	1 187	1 186	2 137
Octopuses, etc. nei	*Octopodidae*	459	640	554	906	1 032	1 160	962
Marine molluscs nei	*Mollusca*	1 085	1 244	98	286	2 096	313	1 795
River and lake turtles nei	*Testudinata*	0	0	0	0	0	...	...
Marine turtles nei	*Testudinata*	0	0	0	0	0	-	-
	Country total	*715 482*	*744 585*	*706 789*	*703 941*	*766 846*	*806 672*	*822 159*

Chile

English name / Nom anglais / Nombre inglés	Scientific name / Nom scientifique / Nombre científico	1996 t	1997 t	1998 t	1999 t	2000 t	2001 t	2002 t
Common carp	*Cyprinus carpio*	-	-	4	-	-	-	-
Flatfishes nei	*Pleuronectiformes*	203	154	75	84	95	76	12
Blue antimora	*Antimora rostrata*	0	-	-	-	-	-	-
Tadpole codling	*Salilota australis*	481	647	352	245	372	641	360
Southern blue whiting	*Micromesistius australis*	25 445	32 875	40 857	36 506	27 459	28 755	29 409
Southern hake	*Merluccius australis*	23 788	24 666	22 458	24 656	29 402	28 806	23 431
South Pacific hake	*Merluccius gayi*	88 555	87 620	80 151	103 789	110 143	121 200	116 040
Patagonian grenadier	*Macruronus magellanicus*	379 015	71 479	354 184	309 904	91 333	162 082	133 418
Grenadiers nei	*Macrourus spp*	12	-	0	-	-	-	-
Mullets nei	*Mugilidae*	244	78	68	134	132	93	22
Groupers, seabasses nei	*Serranidae*	30	24	10	19	7	5	-
Cabinza grunt	*Isacia conceptionis*	93	142	54	156	42	28	149
Corvina	*Sciaena gilberti*	1 179	1 350	1 069	747	1 052	1 033	733
Peruvian weakfish	*Cynoscion analis*	11	12	18	19	14	14	30
Croakers nei	*Micropogonias spp*	128	101	9	0	0	0	0
Croakers, drums nei	*Sciaenidae*	27	25	50	39	25	15	5
Patagonian blennie	*Eleginops maclovinus*	287	133	103	179	164	109	240
Antarctic rockcods, noties nei	*Nototheniidae*	-	-	0	-	-	-	-
...A	*Normanichthys crockeri*	3 921	20 426	236	4 843	853	223	4 371
Pink cusk-eel	*Genypterus blacodes*	5 780	6 410	6 836	5 721	6 269	7 522	4 518
Red cusk-eel	*Genypterus chilensis*	982	745	584	415	608	730	213
Black cusk-eel	*Genypterus maculatus*	1 343	1 661	2 753	1 943	3 542	3 889	1 558
Alfonsinos nei	*Beryx spp*	-	-	144	706	4 366	5 182	8 166
Slimeheads nei	*Trachichthyidae*	-	-	-	779	1 482	1 868	1 514
Hapuku wreckfish	*Polyprion oxygeneios*	23	30	26	8	7	10	2
Tilefishes nei	*Branchiostegidae*	117	28	80	134	155	53	52
Peruvian morwong	*Cheilodactylus variegatus*	-	51	50	52	45	46	4
Antarctic toothfish	*Dissostichus mawsoni*	-	-	1	-	-	-	-
Patagonian toothfish	*Dissostichus eleginoides*	9 781	10 120	10 560	11 641	12 067	7 938	8 946
Mackerel icefish	*Champsocephalus gunnari*	-	-	6	-	715	365	-
Cardinal fishes nei	*Epigonus spp*	513	1 727	5 284	2 999	5 792	4 648	1 595
Snoek	*Thyrsites atun*	821	1 337	1 022	604	851	830	934
South Pacific breams nei	*Seriolella spp*	2 482	2 909	2 671	3 347	2 502	3 336	3 587
Scorpionfishes nei	*Scorpaenidae*	212	532	208	258	141	177	53
South American pilchard	*Sardinops sagax*	81 043	40 473	27 966	246 045	60 189	33 271	18 561
Pacific menhaden	*Ethmidium maculatum*	4 023	6 106	1 534	3 588	4 977	4 931	18 399
Araucanian herring	*Strangomera bentincki*	446 669	441 154	317 564	782 142	722 522	324 617	347 368
Anchoveta(=Peruvian anchovy)	*Engraulis ringens*	1 400 567	1 757 499	522 742	1 983 040	1 700 640	852 789	1 526 872
Eastern Pacific bonito	*Sarda chiliensis*	14	28	584	368	55	19	0
Skipjack tuna	*Katsuwonus pelamis*	-	53	47	9	0	57	1
Albacore	*Thunnus alalunga*	21	-	-	7	3	5	40
Yellowfin tuna	*Thunnus albacares*	32	57	78	48	77	66	15
Bigeye tuna	*Thunnus obesus*	16	6	29	6	20	5	7
Swordfish	*Xiphias gladius*	3 145	4 040	4 492	2 925	2 973	3 262	3 523
Atlantic saury	*Scomberesox saurus*	1	175	407	580	296	4 178	1 492
...A	*Odontesthes regia*	690	494	560	3 414	1 357	833	1 396
Chilean jack mackerel	*Trachurus murphyi*	3 883 326	2 917 064	1 612 912	1 219 689	1 234 299	1 649 933	1 518 994
Carangids nei	*Carangidae*	96	205	348	203	307	363	769
Atlantic pomfret	*Brama brama*	5 585	5 998	6 332	6 830	8 160	15 156	4 430
Common dolphinfish	*Coryphaena hippurus*	179	124	80	109	131	136	134
Chub mackerel	*Scomber japonicus*	146 649	211 649	71 769	120 123	95 789	365 031	343 371
Butterfishes, pomfrets nei	*Stromateidae*	9	6	11	12	10	11	141
Shortfin mako	*Isurus oxyrinchus*	320	888	830	379	592	964	425
Blue shark	*Prionace glauca*	11	114	10	7	262	445	605
Smooth-hounds nei	*Mustelus spp*	225	108	56	208	143	128	57
Rays, stingrays, mantas nei	*Rajiformes*	2 696	2 958	2 015	3 369	4 151	2 974	2 992
Elephantfishes nei	*Callorhinchus spp*	1 450	822	1 416	632	603	1 125	312
Marine fishes nei	*Osteichthyes*	2 253	1 014	935	2 469	6 414	5 861	8 581
Marine crabs nei	*Brachyura*	3 989	3 541	5 063	6 501	6 758	6 770	7 292
Juan Fernandez rock lobster	*Jasus frontalis*	36	32	21	22	17	21	9
Carrot squat lobster	*Pleuroncodes monodon*	7 726	8 939	12 602	12 710	11 129	1 754	2 499
Blue squat lobster	*Cervimunida johni*	6 402	10 322	9 426	7 273	5 069	2 178	925
Craylets, squat lobsters	*Galatheidae*	-	-	-	-	254	1	724
Southern king crab	*Lithodes antarcticus*	1 759	2 160	2 766	2 155	2 902	2 937	2 869
Softshell red crab	*Paralomis granulosa*	1 273	1 477	1 501	1 438	4 938	6 527	4 631
Chilean nylon shrimp	*Heterocarpus reedi*	10 535	10 239	7 301	7 951	5 448	4 863	4 112
Chilean knife shrimp	*Haliporoides diomedeae*	15	32	29	135	169	309	482

English name Nom anglais Nombre inglés	Scientific name Nom scientifique Nombre científico	1996 t	1997 t	1998 t	1999 t	2000 t	2001 t	2002 t
Giant barnacle	*Megabalanus psittacus*	879	579	683	620	620	685	199
Marine crustaceans nei	*Crustacea*	1	6	15	67	7	65	70
False abalone	*Concholepas concholepas*	2 541	3 154	2 564	2 294	1 274	828	1 622
Gastropods nei	*Gastropoda*	6 078	5 348	4 649	7 204	6 898	5 429	5 120
Chilean flat oyster	*Ostrea chilensis*	-	5	1	6	9	202	7
Chilean mussel	*Mytilus chilensis*	5 714	4 723	4 899	4 343	5 236	6 758	1 416
Choro mussel	*Choromytilus chorus*	323	266	127	155	217	166	91
Cholga mussel	*Aulacomya ater*	7 405	6 409	7 725	5 126	5 563	7 884	3 775
Peruvian calico scallop	*Argopecten purpuratus*	9	4	21	0	20	272	55
Scallops nei	*Pectinidae*	1 577	2 598	3 662	1 715	332	141	373
Gay's little venus	*Tawera gayi*	...	...	...	...	1	291	271
Taca clam	*Protothaca thaca*	20 016	12 475	24 254	16 429	16 303	26 483	5 360
Taquilla clams	*Mulinia spp*	999	2 757	2 549	1 536	1 491	1 699	7 034
Macha clam	*Mesodesma donacium*	6 144	6 770	6 464	1 728	1 249	1 396	1 303
Chilean semele	*Semele solida*	4 418	2 199	1 900	2 071	4 212	3 054	2 162
Clams, etc. nei	*Bivalvia*	18 720	17 880	11 900	20 804	16 551	16 775	12 723
Various squids nei	*Loliginidae, Ommastrephidae*	26	110	179	99	64	3 594	5 661
Octopuses, etc. nei	*Octopodidae*	3 477	4 404	4 877	3 168	1 682	2 008	1 354
Marine molluscs nei	*Mollusca*	9	69	72	35	28	17	-
Red sea squirt	*Pyura chilensis*	4 549	3 174	2 530	2 704	2 290	1 298	1 217
Chilean sea urchin	*Loxechinus albus*	51 437	45 560	44 843	55 654	54 096	46 794	60 166
Sea cucumbers nei	*Holothurioidea*	115	1	30	108	1 510	107	106
	Country total	*6 690 665*	*5 811 550*	*3 265 293*	*5 050 180*	*4 299 942*	*3 797 140*	*4 271 475*

Colombia

English name	Scientific name	1996	1997	1998	1999	2000	2001	2002
Characins nei	*Characidae*	5 504	3 730	6 427	8 667	6 600 F	6 600 F	6 600 F
Freshwater siluroids nei	*Siluroidei*	1 392	2 648	9 215	10 544	10 454 F	10 600 F	10 600 F
Freshwater fishes nei	*Osteichthyes*	16 165	14 232	6 031	9 577	7 800 F	7 800 F	7 800 F
Flatfishes nei	*Pleuronectiformes*	30	45	131	48	50 F	50 F	50 F
Panama hake	*Merluccius angustimanus*	...	267	165	143	250 F	391	380 F
Ladyfish	*Elops saurus*	143	20	4	11	20 F	20 F	20 F
Tarpon	*Megalops atlanticus*	135	2	2	-	-	-	-
Sea catfishes nei	*Ariidae*	449	370	138	106	120 F	120 F	120 F
Lebranche mullet	*Mugil liza*	39	-	2	-	-	-	-
Mullets nei	*Mugilidae*	103	60	63	44	60 F	58 F	58 F
Snooks(=Robalos) nei	*Centropomus spp*	452	78	96	62	70 F	70 F	65 F
Broomtail grouper	*Mycteroperca xenarcha*	180	125	155	82	80 F	210	200 F
Nassau grouper	*Epinephelus striatus*	11	5	3	0	...	...	...
Spotted grouper	*Epinephelus analogus*	43	48	32	39	30 F	28	28 F
Groupers nei	*Epinephelus spp*	180	125	155	82	70 F	57	55 F
Groupers, seabasses nei	*Serranidae*	86	17	37	26	30 F	30 F	30 F
Yellow snapper	*Lutjanus argentiventris*	43	40	100	82	80 F	80 F	80 F
Southern red snapper	*Lutjanus purpureus*	17	...	290	172	250 F	250 F	250 F
Lane snapper	*Lutjanus synagris*	236	35	12	11	20 F	20 F	20 F
Snappers nei	*Lutjanus spp*	460	43	452	515	600 F	802	780 F
Snappers, jobfishes nei	*Lutjanidae*	55	35	0	-	-	-	-
Grunts, sweetlips nei	*Haemulidae (=Pomadasyidae)*	224	12	6	1	5 F	5 F	5 F
Peruvian weakfish	*Cynoscion analis*	401	284	374	352	330	317	310 F
Croakers nei	*Micropogonias spp*	165	179	156	173	170 F	237	230 F
Croakers, drums nei	*Sciaenidae*	24	1	0	-	-	-	-
Mojarras, etc. nei	*Gerreidae*	125	6	23	0	5 F	5 F	5 F
Threadfins, tasselfishes nei	*Polynemidae*	16	4	3	16	10 F	10 F	10 F
South Pacific breams nei	*Seriolella spp*	...	...	...	...	...	8	8 F
Demersal percomorphs nei	*Perciformes*	202	236	165	211	220 F	220 F	210 F
Pacific anchoveta	*Cetengraulis mysticetus*	26 344	28 747	28 501	15 781	20 500 F	25 099	25 000 F
Clupeoids nei	*Clupeoidei*	180	555	469	61	250 F	250 F	250 F
Eastern Pacific bonito	*Sarda chiliensis*	6	5	8	9	10 F	13	10 F
Pacific sierra	*Scomberomorus sierra*	484	645	912	521	500 F	500 F	500 F
Seerfishes nei	*Scomberomorus spp*	539	22	30	55	50 F	50 F	50 F
Black skipjack	*Euthynnus lineatus*	70	...	...	...	79	...	...
Skipjack tuna	*Katsuwonus pelamis*	13 420	12 175	4 653	11 152	5 895	1 873	2 284
Yellowfin tuna	*Thunnus albacares*	10 379	9 610	15 718	13 448	17 443 F	25 572 F	30 836 F
Bigeye tuna	*Thunnus obesus*	7 368	3 289	561	1 430	230	103	159
Swordfish	*Xiphias gladius*	-	-	6	-	-	1	...
Tuna-like fishes nei	*Scombroidei*	12 498 F	13 299 F	23 259 F	6 881 F	5 141 F	5 120 F	5 100 F
Jacks, crevalles nei	*Caranx spp*	1 649	854	517	390	460 F	398 F	400 F
Amberjacks nei	*Seriola spp*	...	51	109	47	70 F	91	90 F
Smooth-hounds nei	*Mustelus spp*	1 007	435	361	388	360 F	311 F	310 F
Rays, stingrays, mantas nei	*Rajiformes*	3	2	2	1	1 F	1 F	1 F
Marine fishes nei	*Osteichthyes*	18 913 F	22 459 F	18 994 F	35 718 F	43 003 F	34 045 F	33 595 F
Marine crabs nei	*Brachyura*	49	166	44	11	120 F	120 F	120 F
Caribbean spiny lobster	*Panulirus argus*	185	108	319	175	250 F	250 F	250 F
Green spiny lobster	*Panulirus gracilis*	4	7	6	1	1 F	1 F	1 F
Western white shrimp	*Penaeus occidentalis*	1 091	1 445	1 211	2 686	2 000 F	1 219	1 200 F
Penaeus shrimps nei	*Penaeus spp*	710	1 745	377	775	3 039 F	3 000 F	3 000 F
Pacific seabob	*Xiphopenaeus riveti*	2 683	2 830	1 970	1 752	2 000 F	2 341	2 300 F
Kolibri shrimp	*Solenocera agassizii*	...	...	...	...	...	686	680 F
Natantian decapods nei	*Natantia*	456	559	345	555	570 F	591	580 F
Gastropods nei	*Gastropoda*	129	198	155	131	120 F	120 F	120 F
Mangrove cupped oyster	*Crassostrea rhizophorae*	-	-	8	-	-	-	-
Clams, etc. nei	*Bivalvia*	4	5	6	6	10 F	10 F	10 F
Various squids nei	*Loliginidae, Ommastrephidae*	324	203	43	38	60 F	87	85 F
Marine molluscs nei	*Mollusca*	454	857	87	19	158 F	160 F	155 F
	Country total	*125 829 F*	*122 918 F*	*122 908 F*	*122 995 F*	*129 644*	*130 000 F*	*135 000 F*

Ecuador

English name	Scientific name	1996	1997	1998	1999	2000	2001	2002
Freshwater fishes nei	*Osteichthyes*	300	400	400	400	400	400	400 F

English name Nom anglais Nombre inglés	Scientific name Nom scientifique Nombre científico	1996 t	1997 t	1998 t	1999 t	2000 t	2001 t	2002 t
Grunts, sweetlips nei	*Haemulidae (=Pomadasyidae)*	...	...	1 091	500	...	...	...
Croakers, drums nei	*Sciaenidae*	2 486	316	3 300	7 320	...	1 558	1 700 F
Hairtails, scabbardfishes nei	*Trichiuridae*	...	...	...	...	13 196	3 382	3 600 F
Gurnards, searobins nei	*Triglidae*	...	...	...	22 634	27 129	1 428	1 500 F
South American pilchard	*Sardinops sagax*	356 480	57 191	1 012	8 821	51 648	42 143	2 539
Red-eye round herring	*Etrumeus teres*	34 349	1 095	8 873	3 636	4 414	28	-
Pacific thread herring	*Opisthonema libertate*	41 041	43 145	40 530	22 253	20 519	19 886	10 951
Anchoveta(=Peruvian anchovy)	*Engraulis ringens*	-	-	-	-	-	2 071	71 013
Pacific anchoveta	*Cetengraulis mysticetus*	26 354	89 157	44 474	27 221	13 762	73 537	18 290
Frigate and bullet tunas	*Auxis thazard, A.rochei*	2 537	6 857	4 201	48 913	9 648	5 738	9 806
Black skipjack	*Euthynnus lineatus*	370	50	260	10	270	1 833	1 122
Skipjack tuna	*Katsuwonus pelamis*	37 468	67 400	67 453	126 992	105 146	68 217	77 791
Yellowfin tuna	*Thunnus albacares*	19 314	19 603	31 052	50 200	36 955	57 563	36 306
Bigeye tuna	*Thunnus obesus*	17 892	26 148	17 909	22 278	29 398	23 440	21 265
Marlins,sailfishes,etc. nei	*Istiophoridae*	-	-	3 727	6 962	...	...	...
Tuna-like fishes nei	*Scombroidei*	-	-	-	-	-	21 422	-
Chilean jack mackerel	*Trachurus murphyi*	56 781	30 302	25 900	19 072	7 144	134 011	604
Pacific bumper	*Chloroscombrus orqueta*	1 706	952	565	1 409	...	1 008	1 100 F
Carangids nei	*Carangidae*	65	45	839	1 632	...	68	70 F
Common dolphinfish	*Coryphaena hippurus*	119	34	15	75	88	158	44
Chub mackerel	*Scomber japonicus*	79 484	192 182	44 716	28 307	84 324	85 378	17 073
Butterfishes, pomfrets nei	*Stromateidae*	...	426	...	...	...	...	...
Marine fishes nei	*Osteichthyes*	20 307	8 262	9 495	97 162	186 431	39 897	39 730
Marine crabs nei	*Brachyura*	750	750	600	600	600	600	600 F
Green spiny lobster	*Panulirus gracilis*	50	50	50	50	50	50	50 F
Yellowleg shrimp	*Penaeus californiensis*	73	60	50	20	20	40	45 F
Whiteleg shrimp	*Penaeus vannamei*	4 497	4 000	3 000	1 100	1 100	2 250	2 450 F
Western white shrimp	*Penaeus occidentalis*	67	60	50	20	20	40	45 F
Penaeus shrimps nei	*Penaeus spp*	344	350	300	125	125	264	286 F
Stromboid conchs nei	*Strombus spp*	10	10	10	10	10	10	10 F
Pacific cupped oyster	*Crassostrea gigas*	5	5	5	5	5	5	5 F
Sea mussels nei	*Mytilidae*	3	3	5	5	5	5	5 F
Venus clams nei	*Veneridae*	0	0	0	0	0	0	-
Clams, etc. nei	*Bivalvia*	10	10	10	10	10	10	10 F
Common squids nei	*Loligo spp*	90	100	100	100	100	100	100 F
Octopuses, etc. nei	*Octopodidae*	0	0	5	5	5	5	5 F
Marine turtles nei	*Testudinata*	10	10	10	10	10	10	10 F
Sea urchins nei	*Strongylocentrotus spp*	0	0	0	0	0	0	-
Sea cucumbers nei	*Holothurioidea*	12	15	15	15	15	15	15 F
	Country total	*702 974*	*548 988*	*310 022*	*497 872*	*592 547*	*586 570*	*318 540*
Falkland Is								
Sea trout	*Salmo trutta*	1	1	1	1	1	1	1
Tadpole codling	*Salilota australis*	2 033	817	1 491	2 692	1 886	1 371	947
Southern blue whiting	*Micromesistius australis*	1 083	727	1 977	2 127	2 704	4 581	2 772
Argentine hake	*Merluccius hubbsi*	383	267	959	1 031	1 000	562	655
Patagonian grenadier	*Macruronus magellanicus*	2 569	1 829	4 246	5 109	3 404	5 452	9 768
Patagonian blennie	*Eleginops maclovinus*	4	4	4	4	10	61	47
Pink cusk-eel	*Genypterus blacodes*	297	154	253	451	304	347	333
Patagonian toothfish	*Dissostichus eleginoides*	50	178	570	1 113	927	1 460	1 321
Falkland sprat	*Sprattus fuegensis*	0	0	29	17	97	4	1
Silversides(=Sand smelts) nei	*Atherinidae*	2	2	2	2	4	1	2
Rays, stingrays, mantas nei	*Rajiformes*	184	204	216	314	353	417	466
Marine fishes nei	*Osteichthyes*	370	181	1 033	1 217	1 250	774	604
Softshell red crab	*Paralomis granulosa*	1	1	1	1	1	5	1
Chilean mussel	*Mytilus chilensis*	1	1	1	1	1	0	0
Patagonian squid	*Loligo gahi*	24 366	12 710	32 029	22 502	50 270	42 909	18 598
Argentine shortfin squid	*Illex argentinus*	196	37	804	2 582	716	1 879	140
Sevenstar flying squid	*Martialia hyadesi*	0	0	-	0	0	-	-
	Country total	*31 540*	*17 113*	*43 616*	*39 164*	*62 928*	*59 824*	*35 656*
Fr Guiana								
Freshwater fishes nei	*Osteichthyes*	0	0	0	0	0	0	0
Marine fishes nei	*Osteichthyes*	3 000 F	2 500	2 500	2 500 F	2 500 F	2 500 F	2 500 F
Penaeus shrimps nei	*Penaeus spp*	4 377	4 102	4 209	3 771	2 737	2 694	3 068
	Country total	*7 377 F*	*6 602*	*6 709*	*6 271 F*	*5 237 F*	*5 194 F*	*5 568 F*
Guyana								
Freshwater fishes nei	*Osteichthyes*	800	625	625	603	800	800	800
Southern red snapper	*Lutjanus purpureus*	...	...	...	...	570	524	423
King mackerel	*Scomberomorus cavalla*	-	270	440	398	214	239	267
Serra Spanish mackerel	*Scomberomorus brasiliensis*	211	571	625	1 143	308	329	441
Sharks, rays, skates, etc. nei	*Elasmobranchii*	765	1 892	...	2 175	...	...	953
Marine fishes nei	*Osteichthyes*	33 971	34 841	38 124	37 534	27 666	24 662	24 569
Penaeus shrimps nei	*Penaeus spp*	84	79	1 935	1 595	1 132	1 698	1 505
Atlantic seabob	*Xiphopenaeus kroyeri*	12 752	15 720	11 091	10 396	16 733	23 771	17 659
Whitebelly prawn	*Nematopalaemon schmitti*	-	-	-	-	1 464	1 382	1 400
	Country total	*48 583*	*53 998*	*52 840*	*53 844*	*48 887*	*53 405*	*48 017*
Paraguay								
Characins nei	*Characidae*	8 000 F	10 000 F	9 000 F	9 000 F	9 000 F	9 000 F	9 000 F
Freshwater siluroids nei	*Siluroidei*	10 000 F	13 000 F	12 000 F	12 000 F	12 000 F	12 000 F	12 000 F
Freshwater fishes nei	*Osteichthyes*	4 000 F	5 000 F	4 000 F	4 000 F	4 000 F	4 000 F	4 000 F
	Country total	*22 000*	*28 000*	*25 000 F*	*25 000 F*	*25 000 F*	*25 000 F*	*25 000 F*

English name Nom anglais Nombre inglés	Scientific name Nom scientifique Nombre científico	1996 t	1997 t	1998 t	1999 t	2000 t	2001 t	2002 t
Peru								
Velvety cichlids	*Astronotus spp*	308	177	141	251	188	183	...
Arapaima	*Arapaima gigas*	457	465	210	338	273	204	...
Cachama	*Colossoma macropomum*	693	762	254	1 016	963	904	...
Pirapatinga	*Piaractus brachypomus*	397	...	166	648	324	487	...
Netted prochilod	*Prochilodus reticulatus*	7 448	2 007	6 117	9 768	10 942	11 220	10 133
Freshwater fishes nei	*Osteichthyes*	19 524	27 941	28 047	24 018	19 387	22 464	19 453
Rainbow trout	*Oncorhynchus mykiss*	63	869	392	184	220	191	380
Flatfishes nei	*Pleuronectiformes*	528	212	230	263	177	313	256
South Pacific hake	*Merluccius gayi*	234 915	177 953	82 367	37 121	83 361	125 065	46 250
Mullets nei	*Mugilidae*	13 916	13 264	29 075	20 843	26 314	27 330	24 989
Groupers nei	*Epinephelus spp*	241	60	101	151	91	87	313
Peruvian rock seabass	*Paralabrax humeralis*	4 954	2 789	2 554	3 278	4 373	2 011	1 522
Snappers nei	*Lutjanus spp*	117	75	208	573	204	55	59
Cabinza grunt	*Isacia conceptionis*	1 955	1 892	2 079	2 791	3 251	3 293	5 506
Grunts, sweetlips nei	*Haemulidae (=Pomadasyidae)*	194	379	254	318	183	419	526
Corvina	*Sciaena gilberti*	7 920	2 215	5 085	5 695	3 692	3 295	5 318
Peruvian weakfish	*Cynoscion analis*	7 475	5 501	10 795	8 558	5 995	4 107	3 147
Croakers nei	*Micropogonias spp*	733	606	1 068	616	955	825	1 184
Peruvian banded croaker	*Paralonchurus peruanus*	4 263	2 737	4 363	6 063	5 729	4 167	1 886
Cusk-eels nei	*Genypterus spp*	1 121	439	425	196	557	552	1 029
Tilefishes nei	*Branchiostegidae*	892	292	119	146	117	1 485	2 507
Peruvian morwong	*Cheilodactylus variegatus*	283	411	90	236	335	260	356
South Pacific breams nei	*Seriolella spp*	3 704	388	505	1 589	1 473	3 192	2 192
South American pilchard	*Sardinops sagax*	1 056 413	625 143	908 291	187 924	226 294	60 298	6 853
Pacific menhaden	*Ethmidium maculatum*	5 769	7 135	39 311	25 848	19 014	9 085	8 938
Anchoveta(=Peruvian anchovy)	*Engraulis ringens*	7 463 147	5 927 599	1 206 322	6 740 225	9 575 717	6 358 217	8 104 729
Anchovies, etc. nei	*Engraulidae*	59 639	24 703	706 167	11 242	3 868	137 098	6 022
Eastern Pacific bonito	*Sarda chiliensis*	23 059	17 731	5 130	948	434	1 287	865
Pacific sierra	*Scomberomorus sierra*	439	98	861	2 712	930	181	288
Skipjack tuna	*Katsuwonus pelamis*	85	823	9 373	802	711	81	1 339
Yellowfin tuna	*Thunnus albacares*	953	908	12 747	2 784	2 548	4 175	5 967
Swordfish	*Xiphias gladius*	1	-	57	42	20	356	278
Flyingfishes nei	*Exocoetidae*	639	1 056	4 506	298 373	41 059	5 035	3 198
Silversides(=Sand smelts) nei	*Atherinidae*	3 802	5 184	45	6 692	11 215	7 528	11 220
Chilean jack mackerel	*Trachurus murphyi*	438 736	649 751	386 946	184 679	296 579	723 733	154 219
Jacks, crevalles nei	*Caranx spp*	129	16	69	607	35	11	62
Pompanos nei	*Trachinotus spp*	460	268	779	2 801	1 220	632	878
Amberjacks nei	*Seriola spp*	1 558	4 648	21 104	2 084	11 159	28 025	29 787
Chub mackerel	*Scomber japonicus*	49 221	206 183	401 903	527 729	73 263	176 202	32 698
Pelagic percomorphs nei	*Perciformes*	858	120	33	1 073	771	371	684
Smooth-hounds nei	*Mustelus spp*	3 230	3 166	8 038	2 892	4 042	4 648	7 015
Angelsharks, sand devils nei	*Squatinidae*	358	189	101	262	406	510	477
Pacific guitarfish	*Rhinobatos planiceps*	460	333	344	95	2 624	1 060	822
Rays, stingrays, mantas nei	*Rajiformes*	1 126	1 177	1 477	2 789	4 026	2 034	4 886
Sharks, rays, skates, etc. nei	*Elasmobranchii*	1 506	1 915	4 335	2 951	4 307	3 618	3 433
Marine fishes nei	*Osteichthyes*	45 788	83 203	375 784	164 855	91 905	122 447	63 296
Marine crabs nei	*Brachyura*	1 605	303	752	11 397	1 974	1 568	2 838
Green spiny lobster	*Panulirus gracilis*	52	12	669	496	278	62	20
Penaeus shrimps nei	*Penaeus spp*	9 245	15 648	17 752	7 255	1 852	2 075	1 537
Marine crustaceans nei	*Crustacea*	-	-	-	-	-	758	1 367
False abalone	*Concholepas concholepas*	2 728	4 366	830	2 289	1 250	544	686
Gastropods nei	*Gastropoda*	2 215	7 098	3 110	4 525	2 768	4 995	2 349
Cholga mussel	*Aulacomya ater*	6 023	9 669	15 106	14 612	13 370	14 700	15 658
Peruvian calico scallop	*Argopecten purpuratus*	2 086	4 009	23 525	30 141	11 810	6 272	2 031
Macha clam	*Mesodesma donacium*	1 060	1 061	578	-	10	0	85
Clams, etc. nei	*Bivalvia*	411	236	152	338	956	949	978
Common squids nei	*Loligo spp*	10 250	3 806	287	1 353	24 548	18 738	6 490
Jumbo flying squid	*Dosidicus gigas*	8 138	16 061	547	54 652	53 795	71 834	146 390
Octopuses, etc. nei	*Octopodidae*	760	1 856	5 123	1 593	819	635	1 415
Marine molluscs nei	*Mollusca*	537	2 538	1 546	3 676	2 312	2 116	7 940
Marine turtles nei	*Testudinata*	0	1	2	1	1	2	2
Echinoderms	*Echinodermata*	461	424	90	1 204	1 626	2 114	2 245
	Country total	*9 515 048*	*7 869 871*	*4 338 437*	*8 428 601*	*10 658 620*	*7 986 103*	*8 766 991*
Suriname								
Freshwater fishes nei	*Osteichthyes*	150 F	200 F	200 F	200 F	200 F	200	200 F
Marine fishes nei	*Osteichthyes*	10 400 F	11 777 F	11 845	9 800	10 500 F	11 300	11 180 F
Marine crabs nei	*Brachyura*	6 F	10 F	10 F	10 F	20	25	20 F
Penaeus shrimps nei	*Penaeus spp*	2 444	2 013	2 094	1 653	2 240 F	2 840	2 800 F
Atlantic seabob	*Xiphopenaeus kroyeri*	-	-	2 046	4 537	4 540 F	4 550	4 500 F
	Country total	*13 000 F*	*14 000 F*	*16 195*	*16 200*	*17 500 F*	*18 915*	*18 700 F*
Uruguay								
Characins nei	*Characidae*	338	1 924	1 687	1 762	1 690	107	260
Freshwater siluroids nei	*Siluroidei*	72	252	203	319	293	81	83
Freshwater fishes nei	*Osteichthyes*	188	40	41	342	312	254	39
Bastard halibuts nei	*Paralichthys spp*	483	502	496	413	256	307	400
Blue antimora	*Antimora rostrata*	-	-	-	0	-	-	-
Brazilian codling	*Urophycis brasiliensis*	223	272	344	281	235	361	382
Southern blue whiting	*Micromesistius australis*	-	-	-	-	9	0	5
Argentine hake	*Merluccius hubbsi*	57 937	48 367	49 111	32 045	27 710	27 618	32 231
Patagonian grenadier	*Macruronus magellanicus*	...	150	1 824	1 461	800	635	1 002
Grenadiers nei	*Macrourus spp*	-	-	0	3	-	8	180
Sea catfishes nei	*Ariidae*	18	15	24	78	51	42	96
Mullets nei	*Mugilidae*	346	214	272	57	194	219	282

English name Nom anglais Nombre inglés	Scientific name Nom scientifique Nombre científico	1996 t	1997 t	1998 t	1999 t	2000 t	2001 t	2002 t
Argentine seabass	*Acanthistius brasilianus*	85	51	42	19	88	9	2
Striped weakfish	*Cynoscion striatus*	12 654	15 186	15 286	8 481	13 440	10 890	8 958
Whitemouth croaker	*Micropogonias furnieri*	25 745	23 744	22 253	14 650	24 146	27 322	26 665
Kingcroakers nei	*Menticirrhus spp*	-	-	7	4	4	6	5
Argentine croaker	*Umbrina canosai*	1 210	547	1 112	1 401	1 071	1 416	1 546
King weakfish	*Macrodon ancylodon*	1 743	1 395	2 354	802	1 123	1 486	2 008
Black drum	*Pogonias cromis*	172	82	678	100	344	336	504
Red porgy	*Pagrus pagrus*	1	12	4	14	27	7	8
Brazilian flathead	*Percophis brasilianus*	-	-	-	1	-	0	0
Pink cusk-eel	*Genypterus blacodes*	43	41	86	206	368	756	569
Castaneta	*Cheilodactylus bergi*	2 938	4 149	9 681	3 105	1 351	1 269	349
Antarctic toothfish	*Dissostichus mawsoni*	-	-	-	-	-	23	-
Patagonian toothfish	*Dissostichus eleginoides*	-	-	1 607	1 532	3 468	7 766	6 544
Blackbelly rosefish	*Helicolenus dactylopterus*	2 204	3 404	4 389	2 581	2 111	1 830	1 544
Demersal percomorphs nei	*Perciformes*	3	57	-	334	51	34	71
Argentine menhaden	*Brevoortia pectinata*	84	219	415	86	66	103	47
Argentine anchovy	*Engraulis anchoita*	22	13	67	3 193	6	318	11
Atlantic bonito	*Sarda sarda*	-	-	-	-	1	23	15
Atlantic bluefin tuna	*Thunnus thynnus*	-	-	-	-	-	1	1
Albacore	*Thunnus alalunga*	75	56	110	69	90	40	111
Yellowfin tuna	*Thunnus albacares*	171	53	88	52	54	112	141
Bigeye tuna	*Thunnus obesus*	124	69	59	27	29	63	86
Blue marlin	*Makaira nigricans*	-	-	23	-	-	2	32
Atlantic white marlin	*Tetrapturus albidus*	2	50	22	-	-	0	2
Swordfish	*Xiphias gladius*	644	760	889	661	713	716	1 278
Tuna-like fishes nei	*Scombroidei*	2	50	94	136	95	34	60
Bluefish	*Pomatomus saltatrix*	11	11	84	18	48	93	56
Parona leatherjacket	*Parona signata*	439	431	465	360	371	618	446
Chub mackerel	*Scomber japonicus*	4	5	1	5	-	0	0
Pelagic percomorphs nei	*Perciformes*	2	243	150	10	18	-	-
Narrownose smooth-hound	*Mustelus schmitti*	204	174	2 156	3 212	1 037	1 153	1 121
Rays, stingrays, mantas nei	*Rajiformes*	2 614	2 342	398	1 576	1 004	989	2 000
Sharks, rays, skates, etc. nei	*Elasmobranchii*	1 760	2 367	444	1 901	991	890	1 145
Marine fishes nei	*Osteichthyes*	3 289	4 510	6 335	936	3 669	2 686	3 109
Red crab	*Geryon quinquedens*	1 513	3 432	2 682	3 382	5 259	2 089	2 004
Southern king crab	*Lithodes antarcticus*	0	0	0	0	0	0	-
Subantarctic stone crab	*Lithodes murrayi*	-	-	-	0	-	-	-
King crabs nei	*Lithodes spp*	-	-	-	0	-	-	-
Red stone crab	*Paralomis aculeata*	-	-	-	0	-	-	-
Sao Paulo shrimp	*Penaeus paulensis*	-	177	13	12	56	23	7
Argentine red shrimp	*Pleoticus muelleri*	-	-	-	40	0	0	2
Antarctic krill	*Euphausia superba*	-	-	-	3 444	6 477	-	-
Marine crustaceans nei	*Crustacea*	3	-	-	-	-	-	-
Angulate volute	*Zidona dufresnei*	...	...	...	...	990	825	900
River Plata mussel	*Mytilus platensis*	206	174	226	142	176	306	96
Patagonian scallop	*Zygochlamis patagonica*	...	...	...	...	890	3 638	215
Donax clams	*Donax spp*	0	0	0	0	0	0	-
Clams, etc. nei	*Bivalvia*	...	0	0	0	6	17	240
Argentine shortfin squid	*Illex argentinus*	5 669	20 857	13 175	13 679	12 144	7 373	11 811
Marine molluscs nei	*Mollusca*	89	557	1 310	3 651	-	-	91
Frogs	*Rana spp*	0	0	0	0	7	9	5
	Country total	*123 330*	*136 954*	*140 707*	*106 583*	*113 339*	*104 903*	*108 765*
Venezuela								
Common carp	*Cyprinus carpio*	218	1	3	5	1 390	0	0
Characins nei	*Characidae*	8 050	6 029	6 921	4 817	5 773	2 978	4 014
Prochilods nei	*Prochilodus spp*	18 740	10 927	15 670	9 320	9 613	6 743	14 061
Freshwater siluroids nei	*Siluroidei*	18 143	15 248	14 833	7 470	6 226	8 967	8 675
Freshwater fishes nei	*Osteichthyes*	11 452	11 977	11 011	17 334	7 222	12 189	13 443
Rainbow trout	*Oncorhynchus mykiss*	59	...	...	...	...	186	400
Diadromous clupeoids nei	*Clupeoidei*	552	355	467	403	287	200	183
Flatfishes nei	*Pleuronectiformes*	...	...	...	51	11	...	...
Sea catfishes nei	*Ariidae*	17 041	8 963	10 098	11 100	14 279	11 929	15 148
Flathead grey mullet	*Mugil cephalus*	8 839	6 465	6 284	5 151	5 467	5 008	4 866
Lebranche mullet	*Mugil liza*	3 248	1 550	2 874	2 855	3 186	2 537	2 294
Common snook	*Centropomus undecimalis*	2 177	1 560	2 022	1 704	1 850	1 903	1 437
Groupers nei	*Epinephelus spp*	2 185	2 235	1 990	1 591	1 408	811	898
Groupers, seabasses nei	*Serranidae*	265	322	384	328	140	123	137
Yellowtail snapper	*Ocyurus chrysurus*	338	335	272	220	291	158	213
Snappers, jobfishes nei	*Lutjanidae*	8 205	9 273	8 652	3 511	7 621	4 986	2 193
Grunts, sweetlips nei	*Haemulidae (=Pomadasyidae)*	5 466	6 336	7 685	4 108	4 191	9 132	3 314
Weakfishes nei	*Cynoscion spp*	13 879	10 503	13 481	6 303	12 716	10 734	10 803
Whitemouth croaker	*Micropogonias furnieri*	6 065	3 694	4 871	1 900	3 262	6 848	2 892
Porgies, seabreams nei	*Sparidae*	...	...	...	...	6	...	...
Mojarras, etc. nei	*Gerreidae*	0	0	0	0	0	0	-
Largehead hairtail	*Trichiurus lepturus*	4 609	5 050	5 408	4 017	3 716	8 793	6 484
Demersal percomorphs nei	*Perciformes*	3 532	4 760	3 616	4 118	1 322	1 687	2 084
Round sardinella	*Sardinella aurita*	153 782	140 571	186 060	126 468	73 534	71 168	158 125
Atlantic thread herring	*Opisthonema oglinum*	294	5 564	10 619	14 789	9 866	3 650	106
Atlantic anchoveta	*Cetengraulis edentulus*	8	0	119	0	0	2	0
Atlantic bonito	*Sarda sarda*	1 348	1 294	1 647	1 597	1 376	1 815	2 948
Wahoo	*Acanthocybium solandri*	479	498	349	448	150	297	275
King mackerel	*Scomberomorus cavalla*	2 140	3 530	340	2 424	1 498	1 861	2 324
Serra Spanish mackerel	*Scomberomorus brasiliensis*	3 609	3 670	3 651	1 686	3 624	3 062	2 337
Frigate and bullet tunas	*Auxis thazard, A.rochei*	2 758	2 525	1 926	1 524	1 410	1 342	1 068
Little tunny(=Atl.black skipj)	*Euthynnus alletteratus*	1 840	2 064	2 815	2 389	2 040	1 948	2 023
Black skipjack	*Euthynnus lineatus*	50	35	70	40	10	-	-
Skipjack tuna	*Katsuwonus pelamis*	6 994	11 114	10 224	17 031	8 133	9 050	6 269

English name Nom anglais Nombre inglés	Scientific name Nom scientifique Nombre científico	1996 t	1997 t	1998 t	1999 t	2000 t	2001 t	2002 t
Blackfin tuna	*Thunnus atlanticus*	758	498	1 034	1 192	589	1 902	1 210
Albacore	*Thunnus alalunga*	315	49	107	91	1 374	349	162
Yellowfin tuna	*Thunnus albacares*	77 452	74 001	76 605	68 927	80 059	128 621	128 497
Bigeye tuna	*Thunnus obesus*	619	519	462	150	436	708	1 028
Atlantic sailfish	*Istiophorus albicans*	165	185	258	179	93	126	159
Blue marlin	*Makaira nigricans*	137	130	205	220	28	72	76
Atlantic white marlin	*Tetrapturus albidus*	164	90	80	61	13	72	110
Longbill spearfish	*Tetrapturus pfluegeri*	1	0	1	0	-	4	0
Marlins,sailfishes,etc. nei	*Istiophoridae*	-	-	-	-	-	8	-
Swordfish	*Xiphias gladius*	85	20	37	30	30	21	34
Tuna-like fishes nei	*Scombroidei*	-	-	-	-	-	13	-
Bluefish	*Pomatomus saltatrix*	651	825	581	542	950	1 158	813
Jacks, crevalles nei	*Caranx spp*	3 732	2 823	2 898	2 642	3 462	2 115	2 161
Atlantic moonfish	*Selene setapinnis*	1 766	2 110	2 338	1 529	976	3 443	1 345
Pompanos nei	*Trachinotus spp*	307	285	435	146	132	117	174
Amberjacks nei	*Seriola spp*	338	355	336	450	610	399	538
Bigeye scad	*Selar crumenophthalmus*	2 353	2 425	2 653	3 765	1 704	1 774	2 267
Carangids nei	*Carangidae*	6	23	31	11	26	5	26
Common dolphinfish	*Coryphaena hippurus*	...	...	10	290	141	...	...
Chub mackerel	*Scomber japonicus*	416	549	753	631	399	514	664
Gulf butterfishes, etc. nei	*Peprilus spp*	1 889	568	1 175	1 004	809	742	872
Barracudas nei	*Sphyraena spp*	899	998	1 123	881	731	1 165	1 311
Pelagic percomorphs nei	*Perciformes*	759	442	381	219	182	109	74
Requiem sharks nei	*Carcharhinidae*	6 979	6 000	4 597	2 973	3 343	2 536	5 663
Rays, stingrays, mantas nei	*Rajiformes*	1 812	1 896	2 111	2 287	2 148	2 182	1 956
Marine fishes nei	*Osteichthyes*	33 049	36 793	16 376	-	464	11 304	29 592
Portunus swimcrabs nei	*Portunus spp*	1	289	224	4 041	4 994	4 001	2
Marine crabs nei	*Brachyura*	4 591	5 340	4 180	196	539	155	8 459
Caribbean spiny lobster	*Panulirus argus*	648	619	260	95	105	78	88
Penaeus shrimps nei	*Penaeus spp*	11 735	10 949	6 910	4 607	9 882	12 128	9 981
Mangrove cupped oyster	*Crassostrea rhizophorae*	2 695	1 705	2 594	2 037	1 590	3 754	800
South American rock mussel	*Perna perna*	223	295	3 802	451	316	1 081	1 272
Ark clams nei	*Arca spp*	31 925	39 128	27 981	38 646	44 709	43 675	44 725
Venus clams nei	*Veneridae*	536	533	295	378	353	388	262
Common squids nei	*Loligo spp*	771	719	787	102	463	1 261	637
Octopuses, etc. nei	*Octopodidae*	1 548	1 917	6 507	526	1 110	1 064	1 226
Marine molluscs nei	*Mollusca*	425	600	1 005	5 808	2 457	1 056	186
Green turtle	*Chelonia mydas*	0	0	0	-	-	-	-
Hawksbill turtle	*Eretmochelys imbricata*	0	0	0	-	-	-	-
Loggerhead turtle	*Caretta caretta*	0	0	0	-	-	-	-
River and lake turtles nei	*Testudinata*	0	0	0	-	-	-	-
Marine turtles nei	*Testudinata*	0	0	0	-	-	-	-
	Country total	*496 115*	*470 126*	*503 494*	*399 809*	*356 835*	*418 207*	*515 384*
Total		***19 789 286***	***17 230 905***	***10 702 894***	***16 534 825***	***17 992 410***	***14 929 607***	***16 021 401***

E-4 **Fish crustaceans, molluscs, etc** — **Capture production by countries or areas and species** — **Asia**
Poissons, crustacés, mollusques, etc — **Captures par pays ou zones et espèces** — **Asie**
Peces, crustáceos, moluscos, etc — **Capturas por países o áreas y especies** — **Asia**

English name Nom anglais Nombre inglés	Scientific name Nom scientifique Nombre científico	1996 t	1997 t	1998 t	1999 t	2000 t	2001 t	2002 t
Afghanistan								
Freshwater fishes nei	*Osteichthyes*	1 300 F	1 250 F	1 200 F	1 200 F	1 000 F	800 F	900 F
	Country total	*1 300 F*	*1 250 F*	*1 200 F*	*1 200 F*	*1 000 F*	*800 F*	*900 F*
Armenia								
Common carp	*Cyprinus carpio*	91	32	14	12	9	7	11
Crucian carp	*Carassius carassius*	28	23	42	26	38	32	54
Cyprinids nei	*Cyprinidae*	33	7	37	21	19	15	9
Freshwater fishes nei	*Osteichthyes*	0	0	0	0	0	-	22
Trouts nei	*Salmo spp*	0	6	0	163	186	180	167
Whitefishes nei	*Coregonus spp*	428	512	605	922	881	632	202
	Country total	*580*	*580*	*698*	*1 144*	*1 133*	*866*	*465*
Azerbaijan								
Freshwater bream	*Abramis brama*	402	346	314	52	55	127	112
Common carp	*Cyprinus carpio*	84	49	87	92	93	51	38
Tench	*Tinca tinca*	2	0	0	0	0	-	-
Crucian carp	*Carassius carassius*	...	...	...	6	4	17	9
Roach	*Rutilus rutilus*	74	89	62	81	8	64	39
Kutum	*Rutilus frisii*	...	...	...	...	...	...	16
Asp	*Aspius aspius*	9	6	4	2	1	5	4
Northern pike	*Esox lucius*	7	23	18	21	28	25	11
Wels(=Som)catfish	*Silurus glanis*	11	9	8	8	9	8	4
Pike-perch	*Stizostedion lucioperca*	22	11	7	5	5	19	35
Freshwater fishes nei	*Osteichthyes*	16	21	13	8	-	-	-
Sturgeons nei	*Acipenseridae*	69	63	61	69	70	76	70
Trouts nei	*Salmo spp*	-	-	-	-	-	5	9
Caspian shads	*Caspialosa spp*	75	42	82	60	1	52	14
Azov sea sprat	*Clupeonella cultriventris*	5 828	4 420	4 043	20 460	18 520	10 389	10 950
Mullets nei	*Mugilidae*	103	82	61	2	3	55	23
	Country total	*6 702*	*5 161*	*4 760*	*20 866*	*18 797*	*10 893*	*11 334*
Bahrain								
Freshwater fishes nei	*Osteichthyes*	0	0	0	0	-	-	-
Mullets nei	*Mugilidae*	99	60	73	10	59	39	29
Groupers nei	*Epinephelus spp*	532	300	331	525	670	794	725
Snappers nei	*Lutjanus spp*	133	207	100	294	157	103	142
Grunts, sweetlips nei	*Haemulidae (=Pomadasyidae)*	253	223	355	325	297	236	121
Emperors(=Scavengers) nei	*Lethrinidae*	1 506	892	944	1 227	1 403	1 377	1 675
Porgies, seabreams nei	*Sparidae*	496	493	757	582	591	401	488
Goatfishes	*Upeneus spp*	139	65	16	414	82	192	17
Mojarras(=Silver-biddies) nei	*Gerres spp*	219	350	347	261	334	372	172
Parrotfishes nei	*Scaridae*	56	46	63	66	32	21	27
Spinefeet(=Rabbitfishes) nei	*Siganus spp*	2 185	1 612	1 523	1 241	2 114	1 899	2 009
Narrow-barred Spanish mackerel	*Scomberomorus commerson*	158	47	85	44	66	109	121
Flyingfishes nei	*Exocoetidae*	92	85	145	299	52	112	28
Cobia	*Rachycentron canadum*	38	19	6	9	9	20	11
Carangids nei	*Carangidae*	668	498	359	524	414	450	470
Barracudas nei	*Sphyraena spp*	159	82	156	6	8	7	6
Marine fishes nei	*Osteichthyes*	1 185	1 029	852	938	859	1 077	803
Portunus swimcrabs nei	*Portunus spp*	1 047	1 289	1 017	2 179	2 380	2 556	2 828
Slipper lobsters nei	*Scyllaridae*	125	56	51	6	2	2	3
Penaeus shrimps nei	*Penaeus spp*	3 565	2 571	2 530	1 622	2 104	1 359	1 401
Cuttlefish,bobtail squids nei	*Sepiidae, Sepiolidae*	285	126	139	48	85	104	128
	Country total	*12 940*	*10 050*	*9 849*	*10 620*	*11 718*	*11 230*	*11 204*
Bangladesh								
Freshwater fishes nei	*Osteichthyes*	445 377	451 055	457 055	575 609	591 300	613 860	620 185
Hilsa shad	*Tenualosa ilisha*	225 598	214 434	205 739	214 519	219 532	229 714	220 593
Seerfishes nei	*Scomberomorus spp*	40	50	60	60	60	60	60
Marine fishes nei	*Osteichthyes*	119 484	135 860	145 260	137 285	161 977	193 746	231 041
Marine crustaceans nei	*Crustacea*	24 288	28 027	31 027	31 742	31 395	31 037	31 976
	Country total	*814 787*	*829 426*	*839 141*	*959 215*	*1 004 264*	*1 068 417*	*1 103 855*
Bhutan								
Freshwater fishes nei	*Osteichthyes*	300 F	300 F	300 F	300 F	300 F	300 F	300 F
	Country total	*300 F*	*300 F*	*300 F*	*300 F*	*300 F*	*300 F*	*300 F*
Brunei Darsm								
Freshwater fishes nei	*Osteichthyes*	1	0	0	0	0	0	1
Marine fishes nei	*Osteichthyes*	6 818	4 405	4 930	3 081	2 356	1 186	1 528
Giant river prawn	*Macrobrachium rosenbergii*	14	17	35	26	23	16	13
Natantian decapods nei	*Natantia*	275	0	0	0	0	0	0
Marine crustaceans nei	*Crustacea*	281	78	65	44	78	266	485
Marine molluscs nei	*Mollusca*	16	21	19	35	30	24	31
	Country total	*7 405*	*4 521*	*5 049*	*3 186*	*2 487*	*1 492*	*2 058*
Cambodia								
Freshwater fishes nei	*Osteichthyes*	63 440	72 900	75 600	230 700	245 300	384 500	359 800
Tuna-like fishes nei	*Scombroidei*	-	-	-	56	-	-	-
Marine fishes nei	*Osteichthyes*	22 958	21 928	23 701	28 070	26 585	30 980	33 830
Freshwater crustaceans nei	*Crustacea*	70	100	100	300	300	500	500

English name Nom anglais Nombre inglés	Scientific name Nom scientifique Nombre científico	1996 t	1997 t	1998 t	1999 t	2000 t	2001 t	2002 t
Marine crabs nei	*Brachyura*	2 350	2 240	2 420	2 850	3 593	4 200	4 600
Natantian decapods nei	*Natantia*	4 300	4 110	4 440	5 250	5 000	5 900	6 400
Argentine shortfin squid	*Illex argentinus*	-	-	-	-	2 768	1 200	32
Marine molluscs nei	*Mollusca*	1 550	1 480	1 600	1 900	801	900	1 000
Sea cucumbers nei	*Holothurioidea*	42	42	39	30	21	20	20
	Country total	*94 710*	*102 800*	*107 900*	*269 156*	*284 368*	*428 200*	*406 182*
China								
Freshwater fishes nei	*Osteichthyes*	994 971	1 032 861	1 218 152	1 394 610	1 222 955	1 033 302	924 203
Elongate ilisha	*Ilisha elongata*	51 339	77 532	84 290	110 359	107 689	101 342	94 509
Alaska pollock(=Walleye poll.)	*Theragra chalcogramma*	226 900	338 478	191 433	64 520	60 338	39 665	11 000
Hakes nei	*Merluccius spp*	-	-	0	-	-	-	-
Mullets nei	*Mugilidae*	67 423	85 967	99 023	113 454	107 243	119 246	113 879
Groupers nei	*Epinephelus spp*	23 241	30 201	36 098	40 245	41 513	44 975	45 448
Golden threadfin bream	*Nemipterus virgatus*	238 000	258 998	262 702	246 601	291 495	287 384	306 397
Large yellow croaker	*Larimichthys croceus*	80 072	69 950	70 935	65 806	122 920	76 289	83 087
Yellow croaker	*Larimichthys polyactis*	253 482	142 681	191 751	243 101	281 717	245 325	260 457
Croakers, drums nei	*Sciaenidae*	530	490	1 568	1 027	2 365	2 061	-
Porgies, seabreams nei	*Sparidae*	57 078	74 054	77 347	80 932	108 453	129 330	118 576
Filefishes nei	*Cantherhines(=Navodon) spp*	210 188	296 781	235 603	240 214	221 683	201 733	158 661
Daggertooth pike conger	*Muraenesox cinereus*	177 470	184 843	239 874	234 314	220 497	243 888	257 659
Alfonsinos nei	*Beryx spp*	-	-	-	-	-	-	159
Orange roughy	*Hoplostethus atlanticus*	-	-	-	-	623	710	585
Oreo dories nei	*Oreosomatidae*	-	-	-	-	-	180	107
Pelagic armourhead	*Pseudopentaceros richardsoni*	-	-	-	-	44	-	26
Cardinal fishes nei	*Epigonus spp*	-	-	-	-	-	-	3
Largehead hairtail	*Trichiurus lepturus*	1 071 914	1 014 598	1 223 360	1 222 454	1 285 469	1 282 698	1 287 798
Pacific herring	*Clupea pallasii*	1 665	15 780	21 796	17 936	15 258	51 950	48 693
Japanese pilchard	*Sardinops melanostictus*	92 918	124 844	121 120	147 125	153 944	160 825	186 281
European pilchard(=Sardine)	*Sardina pilchardus*	-	-	-	-	-	2	-
Japanese anchovy	*Engraulis japonicus*	671 376	1 201 964	1 373 328	1 096 916	1 142 884	1 260 712	1 173 196
Japanese Spanish mackerel	*Scomberomorus niphonius*	283 784	340 302	517 528	565 764	496 566	476 690	506 195
Skipjack tuna	*Katsuwonus pelamis*	...	...	4	...	1 050	3 880	9 295
Atlantic bluefin tuna	*Thunnus thynnus*	93	49	85	103	80	68	39
Albacore	*Thunnus alalunga*	28	2	1	3 722	3 403	4 968	4 514
Yellowfin tuna	*Thunnus albacares*	3 375	2 253	2 536	6 762	6 822	8 260	5 870
Bigeye tuna	*Thunnus obesus*	4 247	4 322	5 504	11 334	11 993	17 508	18 557
Atlantic sailfish	*Istiophorus albicans*	6	6	8	18	8	8	12
Blue marlin	*Makaira nigricans*	62	78	120	201	24	92	88
Atlantic white marlin	*Tetrapturus albidus*	9	11	15	30	2	20	23
Longbill spearfish	*Tetrapturus pfluegeri*	-	-	2	-	-	-	-
Marlins,sailfishes,etc. nei	*Istiophoridae*	18	-	-	401	1 000	799	971
Swordfish	*Xiphias gladius*	370	295	483	1 504	879	1 083	1 983
Tuna-like fishes nei	*Scombroidei*	15 935	14 600	14 087	20 524	30 543	1 130	882
Chilean jack mackerel	*Trachurus murphyi*	-	-	-	-	-	-	76 261
Jack and horse mackerels nei	*Trachurus spp*	-	-	1	-	0	3	-
Scads nei	*Decapterus spp*	607 686	505 991	532 986	502 590	502 289	544 728	603 015
Chub mackerel	*Scomber japonicus*	374 400	408 935	385 183	402 548	350 809	381 597	415 364
Silver pomfrets nei	*Pampus spp*	220 364	242 547	303 024	337 919	338 848	352 493	386 317
Shortfin mako	*Isurus oxyrinchus*	-	...	...	...	153	...	11
Sharks, rays, skates, etc. nei	*Elasmobranchii*	-	-	5	378	99	...	420
Marine fishes nei	*Osteichthyes*	4 399 261	4 159 651	4 294 630	4 404 719	3 996 124	3 535 190	3 409 290
Freshwater crustaceans nei	*Crustacea*	364 441	479 645	601 966	455 761	530 026	586 985	772 702
Blue swimming crab	*Portunus pelagicus*	51 288	45 749	48 190	52 577	56 092	59 416	61 981
Gazami crab	*Portunus trituberculatus*	283 394	237 960	266 630	270 280	335 078	333 556	347 103
Fleshy prawn	*Penaeus chinensis*	55 292	69 406	78 350	69 911	84 334	97 002	99 468
Penaeus shrimps nei	*Penaeus spp*	2 113	2 260	2 504	1 997	1 873	2 087	-
Southern rough shrimp	*Trachypenaeus curvirostris*	163 060	174 967	175 618	400 786	312 436	244 202	233 736
Akiami paste shrimp	*Acetes japonicus*	442 460	480 056	571 383	579 213	625 234	565 792	578 634
Marine crustaceans nei	*Crustacea*	914 672	1 086 501	1 231 473	1 131 643	1 213 725	1 264 976	1 214 762
Freshwater molluscs nei	*Mollusca*	403 448	374 461	460 126	434 993	480 249	529 645	551 021
Cuttlefish,bobtail squids nei	*Sepiidae, Sepiolidae*	173 142	243 431	227 520	216 303	260 906	318 367	296 836
Argentine shortfin squid	*Illex argentinus*	-	-	30 000	61 000	93 130	93 500	85 436
Jumbo flying squid	*Dosidicus gigas*	-	-	-	-	-	-	50 483
Various squids nei	*Loliginidae, Ommastrephidae*	229	66	117 200	132 237	124 188	87 290	87 056
Octopuses, etc. nei	*Octopodidae*	2 464	3 193	4 336	7 491	4 578	5 130	-
Marine molluscs nei	*Mollusca*	932 374	1 494 902	1 479 065	1 445 303	1 402 625	1 399 810	1 375 782
Sea urchins nei	*Strongylocentrotus spp*	200	200	200	200	200	200	200
Jellyfishes	*Rhopilema spp*	265 325	400 483	430 784	402 206	334 869	331 297	288 113
	Country total	*14 182 107*	*15 722 344*	*17 229 927*	*17 240 032*	*16 987 325*	*16 529 389*	*16 553 144*
China,H.Kong								
Cyprinids nei	*Cyprinidae*	0	0	0	0	0	0	0
Tonguefishes	*Cynoglossidae*	1 160	1 261	1 177	800 F	1 000 F	1 100 F	1 050 F
Lizardfishes nei	*Synodontidae*	9 034	8 070	6 619	4 700 F	5 800 F	6 400 F	6 250 F
Groupers, seabasses nei	*Serranidae*	1 349	1 318	1 240	900 F	1 100 F	1 200 F	1 150 F
Bigeyes nei	*Priacanthus spp*	5 391	4 962	4 149	3 000 F	3 600 F	4 000 F	3 900 F
Snappers, jobfishes nei	*Lutjanidae*	430	439	304	250 F	300 F	330 F	300 F
Threadfin breams nei	*Nemipterus spp*	19 568	20 024	19 449	14 000 F	17 000 F	18 800 F	18 300 F
Croakers, drums nei	*Sciaenidae*	3 776	4 881	3 992	2 800 F	3 500 F	3 900 F	3 800 F
Largeeye breams	*Gymnocranius spp*	208	170	143	100 F	120 F	130 F	120 F
Porgies, seabreams nei	*Sparidae*	925	883	852	600 F	700 F	780 F	750 F
Goatfishes	*Upeneus spp*	718	477	393	300 F	350 F	400 F	390 F
Pike-congers nei	*Muraenesox spp*	3 281	2 903	2 450	1 700 F	2 100 F	2 300 F	2 250 F
Tilefishes nei	*Branchiostegidae*	3 186	4 187	4 879	3 500 F	4 300 F	4 750 F	4 650 F
Hairtails, scabbardfishes nei	*Trichiuridae*	3 326	2 522	1 690	1 200 F	1 500 F	1 650 F	1 600 F
Indian driftfish	*Ariomma indica*	120	79	49	35 F	40 F	45 F	40 F
Pacific rudderfish	*Psenopsis anomala*	940	1 460	1 160	800 F	1 000 F	1 100 F	1 050 F

E-4

Fish crustaceans, molluscs, etc	Capture production by countries or areas and species	Asia
Poissons, crustacés, mollusques, etc	Captures par pays ou zones et espèces	Asie
Peces, crustáceos, moluscos, etc	Capturas por países o áreas y especies	Asia

English name Nom anglais Nombre inglés	Scientific name Nom scientifique Nombre científico	1996 t	1997 t	1998 t	1999 t	2000 t	2001 t	2002 t
Sardinellas nei	*Sardinella spp*	30	12	8	5 F	5 F	6 F	5 F
Slender rainbow sardine	*Dussumieria elopsoides*	48	19	17	10 F	15 F	17 F	15 F
Stolephorus anchovies	*Stolephorus spp*	31	20	17	10 F	15 F	17 F	15 F
Seerfishes nei	*Scomberomorus spp*	2 036	1 711	1 723	1 200 F	1 500 F	1 650 F	1 600 F
Tuna-like fishes nei	*Scombroidei*	18	1	4	0	0	0	0
Scads nei	*Decapterus spp*	5 368	4 143	2 962	2 100 F	2 600 F	2 900 F	2 800 F
Jacks, crevalles nei	*Caranx spp*	66	33	20	10 F	15 F	17 F	15 F
Butterfishes, pomfrets nei	*Stromateidae*	2 114	1 973	2 172	1 500 F	1 900 F	2 100 F	2 050 F
Sharks, rays, skates, etc. nei	*Elasmobranchii*	456	420	382	300 F	330 F	370 F	350 F
Marine fishes nei	*Osteichthyes*	101 269	101 062	108 206	76 460 F	94 272 F	104 550 F	102 250 F
Marine crabs nei	*Brachyura*	1 408	1 105	1 108	800 F	1 000 F	1 100 F	1 050 F
Tropical spiny lobsters nei	*Panulirus spp*	0	0	0	0	0	0	0
Natantian decapods nei	*Natantia*	6 589	6 225	5 285	3 800 F	4 600 F	5 100 F	5 000 F
Cuttlefish,bobtail squids nei	*Sepiidae, Sepiolidae*	991	841	510	400 F	450 F	500 F	490 F
Various squids nei	*Loliginidae, Ommastrephidae*	8 559	13 586	8 340	6 000 F	7 300 F	8 100 F	7 900 F
Marine molluscs nei	*Mollusca*	1 461	1 213	700	500 F	600 F	660 F	650 F
	Country total	*183 856*	*186 000*	*180 000*	*127 780*	*157 012*	*173 972*	*169 790*
China, Macao								
Freshwater fishes nei	*Osteichthyes*	0	0	0	0	0	0	0
Marine fishes nei	*Osteichthyes*	970	1 020 F	1 020 F	1 020 F	1 020 F	1 020 F	1 020 F
Natantian decapods nei	*Natantia*	218	230 F	230 F	230 F	230 F	230 F	230 F
Marine crustaceans nei	*Crustacea*	200	210 F	210 F	210 F	210 F	210 F	210 F
Marine molluscs nei	*Mollusca*	30	40 F	40 F	40 F	40 F	40 F	40 F
	Country total	*1 418*	*1 500 F*	*1 500 F*	*1 500 F*	*1 500 F*	*1 500 F*	*1 500 F*
China,Taiwan								
Common carp	*Cyprinus carpio*	26	19	22	45	49	50	54
Crucian carp	*Carassius carassius*	5	5	5	28	35	37	26
Mud carp	*Cirrhinus molitorella*	3	4	7	7	6	6	9
Grass carp(=White amur)	*Ctenopharyngodon idellus*	79	83	81	95	104	132	160
Silver carp	*Hypophthalmichthys molitrix*	29	35	34	11	9	11	18
Bighead carp	*Hypophthalmichthys nobilis*	67	72	99	241	204	202	183
Black carp	*Mylopharyngodon piceus*	15	16	22	27	34	36	25
Tilapias nei	*Oreochromis (=Tilapia) spp*	145	146	152	86	79	98	97
Torpedo-shaped catfishes nei	*Clarias spp*	-	-	6	-	-	-	-
Freshwater fishes nei	*Osteichthyes*	16	11	18	18	17	19	15
Chinese gizzard shad	*Clupanodon thrissa*	25	10	11	27	21	74	31
Milkfish	*Chanos chanos*	4	-	-	-	-	1	-
Barramundi(=Giant seaperch)	*Lates calcarifer*	14	9	26	47	32	80	73
Flatfishes nei	*Pleuronectiformes*	345	346	142	146	137	198	172
Alaska pollock(=Walleye poll.)	*Theragra chalcogramma*	12	7	9	9	9	-	-
Greater lizardfish	*Saurida tumbil*	6 561	3 208	3 049	3 075	3 612	2 248	1 520
Sea catfishes nei	*Ariidae*	122	134	208	257	725	435	529
Flathead grey mullet	*Mugil cephalus*	1 302	2 446	606	863	1 890	1 559	934
Groupers nei	*Epinephelus spp*	2 457	2 387	1 414	1 265	1 202	1 610	933
Red bigeye	*Priacanthus macracanthus*	7 386	4 824	3 670	2 756	3 986	3 079	2 142
Sillago-whitings	*Sillaginidae*	547	401	346	190	188	132	231
Moonfish	*Mene maculata*	725	902	1 010	1 747	1 496	1 582	1 110
Snappers, jobfishes nei	*Lutjanidae*	2 334	1 408	808	467	352	385	409
Golden threadfin bream	*Nemipterus virgatus*	4 261	4 401	3 681	3 990	4 824	3 536	2 995
Yellow croaker	*Larimichthys polyactis*	2 859	2 228	1 010	1 167	780	903	742
Blackmouth croaker	*Atrobucca nibe*	297	288	237	223	450	644	535
Silver croaker	*Pennahia argentata*	8 599	6 085	5 089	4 940	4 180	3 796	3 394
Croakers, drums nei	*Sciaenidae*	3 070	2 497	1 422	1 580	1 708	3 841	8 428
Silver seabream	*Pagrus auratus*	1 809	2 445	2 073	3 333	4 726	6 722	6 737
Blackhead seabream	*Acanthopagrus schlegeli*	207	277	404	604	717	878	784
Porgies, seabreams nei	*Sparidae*	5 401	6 334	6 440	7 312	6 288	9 953	7 370
Goatfishes	*Upeneus spp*	508	440	335	478	477	300	344
Parrotfishes nei	*Scaridae*	11	17	12	5	29	27	46
Fourfinger threadfin	*Eleutheronema tetradactylum*	3 157	2 831	3 248	2 414	6 917	1 715	1 236
Filefishes nei	*Cantherhines(=Navodon) spp*	871	446	586	995	829	959	1 322
Daggertooth pike conger	*Muraenesox cinereus*	2 846	3 246	5 733	9 001	5 874	5 084	2 827
Tilefishes nei	*Branchiostegidae*	1 227	626	372	496	448	512	306
Oilfish	*Ruvettus pretiosus*	2 634	2 622	3 043	2 661	2 584	3 678	4 190
Largehead hairtail	*Trichiurus lepturus*	11 830	8 955	7 991	9 375	7 271	8 834	8 390
Pacific rudderfish	*Psenopsis anomala*	8 834	6 001	8 258	5 075	4 506	5 646	5 154
Red-eye round herring	*Etrumeus teres*	1 462	2 152	1 248	893	1 118	1 306	1 553
Silver-stripe round herring	*Spratelloides gracilis*	517	650	886	521	639	505	1 033
Japanese anchovy	*Engraulis japonicus*	466	515	425	650	589	695	1 267
Dorab wolf-herring	*Chirocentrus dorab*	1	2	1	4	1	-	15
Clupeoids nei	*Clupeoidei*	3 873	4 222	4 338	3 721	4 221	4 393	4 581
Narrow-barred Spanish mackerel	*Scomberomorus commerson*	2 541	2 701	2 417	2 674	3 551	4 153	6 613
Indo-Pacific king mackerel	*Scomberomorus guttatus*	1 611	1 607	1 128	1 298	1 249	1 395	992
Japanese Spanish mackerel	*Scomberomorus niphonius*	7 546	11 761	8 579	4 516	5 955	11 497	12 565
Seerfishes nei	*Scomberomorus spp*	1 103	1 143	1 478	2 883	2 260	1 433	2 022
Frigate and bullet tunas	*Auxis thazard, A.rochei*	2 482	4 280	4 634	4 558	3 073	2 089	2 298
Kawakawa	*Euthynnus affinis*	2 616	2 577	2 307	2 065	1 977	2 136	1 722
Skipjack tuna	*Katsuwonus pelamis*	173 489	119 789	197 313	163 903	197 536	198 839	232 590
Atlantic bluefin tuna	*Thunnus thynnus*	472	504	456	250	407	397	352
Pacific bluefin tuna	*Thunnus orientalis*	956	1 813	1 910	3 089	2 690	2 084	1 841
Longtail tuna	*Thunnus tonggol*	8 900	6 762	6 466	9 206	14 429	18 621	4 311
Albacore	*Thunnus alalunga*	58 408	55 205	59 868	64 302	51 668	45 458	58 526
Southern bluefin tuna	*Thunnus maccoyii*	1 610	640	1 439	1 751	1 882	1 655	1 089
Yellowfin tuna	*Thunnus albacares*	82 891	101 379	122 484	95 004	99 279	109 574	97 197
Bigeye tuna	*Thunnus obesus*	64 498	74 770	76 250	76 761	82 484	81 244	104 260
Indo-Pacific sailfish	*Istiophorus platypterus*	2 296	3 350	3 975	3 610	2 527	2 216	2 989
Atlantic sailfish	*Istiophorus albicans*	102	46	153	99	135	-	198

English name Nom anglais Nombre inglés	Scientific name Nom scientifique Nombre científico	1996 t	1997 t	1998 t	1999 t	2000 t	2001 t	2002 t
Blue marlin	*Makaira nigricans*	14 167	11 312	13 135	13 904	14 325	14 401	14 233
Black marlin	*Makaira indica*	1 469	1 357	2 063	1 432	1 296	964	857
Striped marlin	*Tetrapturus audax*	3 416	3 539	4 116	2 696	2 995	2 017	-
Atlantic white marlin	*Tetrapturus albidus*	567	441	507	464	723	35	-
Marlins,sailfishes,etc. nei	*Istiophoridae*	24	5	24	0	0	807	-
Swordfish	*Xiphias gladius*	14 315	21 863	20 940	19 216	20 330	19 661	22 284
Pacific saury	*Cololabis saira*	8 236	21 887	12 794	12 541	27 868	39 764	51 295
Flyingfishes nei	*Exocoetidae*	662	2 497	1 077	618	287	366	502
Cobia	*Rachycentron canadum*	692	987	815	655	1 014	486	457
Japanese jack mackerel	*Trachurus japonicus*	4 218	4 711	7 121	2 661	6 003	3 903	7 458
Japanese scad	*Decapterus maruadsi*	5 059	19 667	12 090	20 532	4 387	6 224	7 755
Indian scad	*Decapterus russelli*	382	687	6 158	7 703	3 927	598	58
Scads nei	*Decapterus spp*	32 586	12 743	3 172	2 836	2 491	2 670	2 486
Jacks, crevalles nei	*Caranx spp*	7 065	11 433	8 228	4 882	6 355	3 273	4 614
Black pomfret	*Parastromateus niger*	3 427	2 534	1 744	1 307	1 671	2 129	1 779
Torpedo scad	*Megalaspis cordyla*	233	6 847	52	113	327	198	229
Common dolphinfish	*Coryphaena hippurus*	3 209	8 089	17 157	8 560	5 558	7 427	7 014
Chub mackerel	*Scomber japonicus*	53 906	47 279	35 527	45 339	28 635	24 298	34 950
Silver pomfret	*Pampus argenteus*	5 113	2 703	1 621	2 477	2 852	4 245	4 182
Butterfishes, pomfrets nei	*Stromateidae*	246	336	231	280	131	146	210
Barracudas nei	*Sphyraena spp*	547	761	771	519	362	439	311
Rays, stingrays, mantas nei	*Rajiformes*	2 457	1 367	246	235	351	851	1 143
Sharks, rays, skates, etc. nei	*Elasmobranchii*	38 701	38 722	39 779	42 698	45 572	41 504	43 269
Marine fishes nei	*Osteichthyes*	50 983	62 078	75 195	73 823	81 581	75 968	83 038
Blue swimming crab	*Portunus pelagicus*	1 391	1 141	966	511	1 966	1 703	1 030
Indo-Pacific swamp crab	*Scylla serrata*	935	180	215	269	299	230	337
Marine crabs nei	*Brachyura*	4 830	4 630	4 553	5 433	5 898	7 373	8 253
Longlegged spiny lobster	*Panulirus longipes*	14	14	14	12	11	23	13
Japanese fan lobster	*Ibacus ciliatus*	1 115	642	696	676	1 600	1 607	1 005
Kuruma prawn	*Penaeus japonicus*	1 561	2 665	1 906	4 071	8 168	4 179	1 590
Giant tiger prawn	*Penaeus monodon*	49	45	38	92	57	91	54
Redtail prawn	*Penaeus penicillatus*	5 020	2 473	647	316	308	312	509
Metapenaeus shrimps nei	*Metapenaeus spp*	626	369	299	210	202	241	227
Southern rough shrimp	*Trachypenaeus curvirostris*	4 663	5 288	3 926	2 241	532	593	366
Natantian decapods nei	*Natantia*	23 770	22 112	15 702	14 943	11 336	11 987	10 799
Freshwater molluscs nei	*Mollusca*	-	-	1	2	2	-	-
Abalones nei	*Haliotis spp*	8	6	2	2	42	1	-
Gastropods nei	*Gastropoda*	760	1 019	125	0	10	53	-
Pacific cupped oyster	*Crassostrea gigas*	-	-	-	-	-	4	-
Yesso scallop	*Patinopecten yessoensis*	76	63	114	112	90	47	25
Japanese hard clam	*Meretrix lusoria*	135	161	805	526	0	-	-
Cuttlefish,bobtail squids nei	*Sepiidae, Sepiolidae*	7 191	8 621	7 198	5 019	3 880	4 546	2 405
Common squids nei	*Loligo spp*	24 721	28 402	26 927	15 898	9 573	9 099	10 677
Argentine shortfin squid	*Illex argentinus*	101 328	185 775	163 180	264 089	238 334	146 783	110 887
Japanese flying squid	*Todarodes pacificus*	19 101	12 430	34 840	11 261	6 833	4 716	4 215
Wellington flying squid	*Nototodarus sloani*	14 747	6 620	3 974	761	-	-	-
Sevenstar flying squid	*Martialia hyadesi*	3 792	8 348	-	-	-	-	-
Octopuses, etc. nei	*Octopodidae*	967	615	941	500	287	374	481
Cephalopods nei	*Cephalopoda*	-	-	-	-	-	3	2
Marine molluscs nei	*Mollusca*	464	443	663	433	914	118	209
Sea urchins nei	*Strongylocentrotus spp*	59	61	39	33	41	50	63
	Country total	*967 483*	*1 038 048*	*1 091 768*	*1 099 715*	*1 093 889*	*1 005 199*	*1 042 756*
Cyprus								
Freshwater fishes nei	*Osteichthyes*	64	70	70	70	78	70 F	60 F
European hake	*Merluccius merluccius*	3	4	2	5	6	8	3
Hakes nei	*Merluccius spp*	...	...	40	164	189	115	-
Flathead grey mullet	*Mugil cephalus*	5	5	2	6	7	7	8
Groupers nei	*Epinephelus spp*	22	22	29	23	27	24	24
Groupers, seabasses nei	*Serranidae*	16	15	11	17	3	5	10
Seabasses nei	*Dicentrarchus spp*	...	...	...	...	...	3	3
Brown meagre	*Sciaena umbra*	2	5	3	7	4	3	-
Common pandora	*Pagellus erythrinus*	32	25	19	37	42	34	31
Axillary seabream	*Pagellus acarne*	22	25	23	50	25	20	19
Common dentex	*Dentex dentex*	38	18	18	22	18	19	16
Saddled seabream	*Oblada melanura*	4	4	3	7	8	11	14
Red porgy	*Pagrus pagrus*	35	27	25	31	23	21	15
Gilthead seabream	*Sparus aurata*	1	0	0	0	...	37	45
Bogue	*Boops boops*	285	230	233	259	354	216	162
Salema	*Sarpa salpa*	5	3	4	1	4	5	6
Porgies, seabreams nei	*Sparidae*	65	83	102	154	156	241	55
Picarels nei	*Spicara spp*	764	650	709	546	533	671	486
Surmullet	*Mullus surmuletus*	240	228	176	184	159	132	125
Red mullet	*Mullus barbatus*	108	119	115	145	103	91	84
Parrotfishes nei	*Scaridae*	32	37	78	58	33	42	55
Spinefeet(=Rabbitfishes) nei	*Siganus spp*	11	7	36	19	12	31	57
Largehead hairtail	*Trichiurus lepturus*	...	...	5 061	130	230	3 949	-
Scorpionfishes nei	*Scorpaenidae*	16	19	10	3	8	11	8
Sardinellas nei	*Sardinella spp*	...	...	2 172	4 492	15 043	9 455	-
European pilchard(=Sardine)	*Sardina pilchardus*	...	...	-	12	564	30	2
European anchovy	*Engraulis encrasicolus*	...	...	35	11 879	14 705	13 975	-
Atlantic bonito	*Sarda sarda*	-	-	-	-	14	13	10
Little tunny(=Atl.black skipj)	*Euthynnus alletteratus*	19	30	10	16	14	13	10
Skipjack tuna	*Katsuwonus pelamis*	3 210	5 200	300	-	-	-	-
Atlantic bluefin tuna	*Thunnus thynnus*	10	10	21	31	61	90	91
Albacore	*Thunnus alalunga*	-	-	-	-	6	-	12
Yellowfin tuna	*Thunnus albacares*	2 890	2 150	100	-	-	-	-
Bigeye tuna	*Thunnus obesus*	1 180	1 450	30	-	-	-	-
Swordfish	*Xiphias gladius*	40	51	61	92	82	135	104

E-4

Fish crustaceans, molluscs, etc — Capture production by countries or areas and species — Asia
Poissons, crustacés, mollusques, etc — Captures par pays ou zones et espèces — Asie
Peces, crustáceos, moluscos, etc — Capturas por países o áreas y especies — Asia

English name Nom anglais Nombre inglés	Scientific name Nom scientifique Nombre científico	1996 t	1997 t	1998 t	1999 t	2000 t	2001 t	2002 t
Tuna-like fishes nei	*Scombroidei*	...	...	39	72	180	268	8
Jack and horse mackerels nei	*Trachurus spp*	...	...	5 968	17 892	27 326	42 691	-
Greater amberjack	*Seriola dumerili*	28	21	8	17	25	14	15
Chub mackerel	*Scomber japonicus*	...	...	2 758	1 906	6 207	7 308	5
Sharks, rays, skates, etc. nei	*Elasmobranchii*	14	17	10	12	14	28	22
Marine fishes nei	*Osteichthyes*	3 057	14 003	635	944	921	995	183
Palinurid spiny lobsters nei	*Palinurus spp*	-	-	-	-	0	1	-
Lobsters nei	*Reptantia*	5	2	4	2	4	5	5
Deepwater rose shrimp	*Parapenaeus longirostris*	2	1	1	5	4	2	3
Common cuttlefish	*Sepia officinalis*	111	89	146	140	123	106	105
Various squids nei	*Loliginidae, Ommastrephidae*	-	-	-	9	1	-	-
Octopuses, etc. nei	*Octopodidae*	190	199	228	179	166	176	117
	Country total	*12 526*	*24 819*	*19 295*	*39 638*	*67 482*	*81 071*	*1 978 F*
Georgia								
Freshwater bream	*Abramis brama*	3	1	2	1	3	1	2
Common carp	*Cyprinus carpio*	0	0	2	11	12	5	5
Freshwater fishes nei	*Osteichthyes*	0	0	0	5	7	2	3
Sturgeons nei	*Acipenseridae*	...	...	...	3	4	3	7
Rainbow trout	*Oncorhynchus mykiss*	3	...	...	...	...	...	...
Flatfishes nei	*Pleuronectiformes*	-	-	-	5	9	11	11
Whiting	*Merlangius merlangus*	223	58	53	41	...	32	37
Mullets nei	*Mugilidae*	-	-	-	9	19	28	73
Surmullets(=Red mullets) nei	*Mullus spp*	-	14	11	8	3	22	67
Gobies nei	*Gobiidae*	...	...	...	2	3	5	32
European sprat	*Sprattus sprattus*	185	85	24	45	...	30	43
European anchovy	*Engraulis encrasicolus*	1 232	2 288	2 346	1 264	2 080	1 652	2 096
Jack and horse mackerels nei	*Trachurus spp*	-	18	13	...	35	7	19
Sharks, rays, skates, etc. nei	*Elasmobranchii*	71	1	550	18	21	27	65
Marine fishes nei	*Osteichthyes*	25	-	-	1	4	5	17
Sea snails	*Rapana spp*	711	118	-	-	-	-	60
	Country total	*2 453*	*2 583*	*3 001*	*1 413*	*2 200*	*1 830*	*2 537*
India								
Cyprinids nei	*Cyprinidae*	220 994	194 608	167 186	177 394	152 350	148 622	183 785
Freshwater siluroids nei	*Siluroidei*	...	...	63 724	71 152	64 179	57 999	67 621
Snakeheads(=Murrels) nei	*Channa spp*	94 944	105 496	159 065	50 349	58 669	58 856	45 584
Freshwater fishes nei	*Osteichthyes*	224 990	245 595	225 100	330 448	416 490	598 798	377 060
Kelee shad	*Hilsa kelee*	100 591	95 677	105 281	89 354	221 899	113 378	88 183
Diadromous clupeoids nei	*Clupeoidei*	1 481	2 497	1 517	1 143	917	2 090	2 090
Flatfishes nei	*Pleuronectiformes*	24 197	29 252	23 662	18 024	30 071	14 631	19 448
Unicorn cod	*Bregmaceros mcclellandi*	975	984	2 251	1 810	1 282	2 187	2 209
Bombay-duck	*Harpadon nehereus*	185 112	213 116	179 915	181 820	168 700	176 420	135 855
Lizardfishes nei	*Synodontidae*	19 670	13 528	12 765	12 914	5 070	3 728	4 509
Sea catfishes nei	*Ariidae*	73 125	78 150	72 891	86 096	80 817	92 077	87 609
Mullets nei	*Mugilidae*	24 265	26 871	19 457	19 872	28 837	30 750	18 649
Ponyfishes(=Slipmouths) nei	*Leiognathidae*	68 582	67 369	46 068	55 462	49 280	52 562	53 840
Croakers, drums nei	*Sciaenidae*	299 558	313 214	275 240	317 752	268 073	248 418	273 394
Goatfishes	*Upeneus spp*	18 129	15 870	17 260	17 892	10 536	16 736	17 530
Threadfins, tasselfishes nei	*Polynemidae*	7 052	7 095	7 764	10 058	5 776	5 752	6 427
Percoids nei	*Percoidei*	115 879	87 710	62 498	89 950	94 561	78 070	101 683
Lantern fish	*Lampanyctus achirus*	0	-	-	-	-	-	-
Pike-congers nei	*Muraenesox spp*	11 101	11 966	11 279	14 090	11 386	12 119	8 753
Hairtails, scabbardfishes nei	*Trichiuridae*	53 941	146 118	73 396	115 206	120 596	101 698	134 197
Indian oil sardine	*Sardinella longiceps*	136 713	212 060	173 351	147 715	327 425	343 076	334 163
Anchovies, etc. nei	*Engraulidae*	80 694	80 518	81 206	68 743	68 608	76 297	69 829
Wolf-herrings nei	*Chirocentrus spp*	19 510	25 153	15 127	17 511	16 074	17 182	21 254
Clupeoids nei	*Clupeoidei*	76 640	91 493	96 393	89 162	63 883	68 943	64 042
Wahoo	*Acanthocybium solandri*	23	21	23	29	84	152	...
Narrow-barred Spanish mackerel	*Scomberomorus commerson*	24 613	23 360	32 181	40 318	36 485	27 548	63 953
Indo-Pacific king mackerel	*Scomberomorus guttatus*	12 662	13 255	22 560	28 263	13 756	14 839	24 045
Streaked seerfish	*Scomberomorus lineolatus*	96	901	107	135	55	38	431
Seerfishes nei	*Scomberomorus spp*	1 216	3 335	-	-	-	-	
Frigate and bullet tunas	*Auxis thazard, A.rochei*	11 119	10 564	7 680	17 183	18 910	16 145	6 085
Kawakawa	*Euthynnus affinis*	14 778	23 425	15 451	21 641	23 617	20 782	8 139
Skipjack tuna	*Katsuwonus pelamis*	6 850	6 096	1 037	5 707	5 986	7 133	20 110
Longtail tuna	*Thunnus tonggol*	4 263	5 322	4 750	2 568	2 722	879	180
Yellowfin tuna	*Thunnus albacares*	7 327	4 106	3 618	2 073	2 157	4 750	60
Bigeye tuna	*Thunnus obesus*	-	-	4	-	-	2	-
Indo-Pacific sailfish	*Istiophorus platypterus*	13	12	14	9	9	-	-
Marlins,sailfishes,etc. nei	*Istiophoridae*	3 940	4 502	3 381	4 202	3 570	4 393	214
Swordfish	*Xiphias gladius*	62	58	25	15	-	30	...
Tuna-like fishes nei	*Scombroidei*	22	62	6 156	-	1	-	4 261
Halfbeaks nei	*Hemiramphus spp*	6 118	5 574	4 672	5 389	7 643	6 013	7 410
Flyingfishes nei	*Exocoetidae*	1 886	2 094	612	736	2 711	2 514	2 699
False trevally	*Lactarius lactarius*	6 930	7 942	9 528	6 827	6 061	6 984	5 194
Jacks, crevalles nei	*Caranx spp*	67 879	50 833	66 910	65 499	39 571	47 355	32 740
Pompanos nei	*Trachinotus spp*	6 784	3 697	2 738	2 329	12	11	12
Carangids nei	*Carangidae*	34 225	32 752	26 271	29 916	11 129	12 970	29 642
Indian mackerel	*Rastrelliger kanagurta*	293 526	190 372	184 419	179 378	78 821	46 158	53 074
Butterfishes, pomfrets nei	*Stromateidae*	37 496	36 731	33 319	31 185	25 391	24 657	35 753
Barracudas nei	*Sphyraena spp*	8 873	7 827	14 965	14 936	4 162	4 865	7 037
Sharks, rays, skates, etc. nei	*Elasmobranchii*	132 160	71 991	74 704	76 802	76 057	67 971	67 358
Marine fishes nei	*Osteichthyes*	455 702	503 234	499 863	473 445	561 368	706 594	769 876
Freshwater crustaceans nei	*Crustacea*	803	71	114	85	112	183	202
Marine crabs nei	*Brachyura*	1 547	2 237	1 789	1 693	23 650	18 005	19 169
Giant tiger prawn	*Penaeus monodon*	150 593	146 111	204 565	223 703	204 588	158 023	156 899
Penaeus shrimps nei	*Penaeus spp*	14 645	20 558	8 783	7 830	10 873	14 772	14 160

English name Nom anglais Nombre inglés	Scientific name Nom scientifique Nombre científico	1996 t	1997 t	1998 t	1999 t	2000 t	2001 t	2002 t
Natantian decapods nei	*Natantia*	155 181	134 352	121 044	120 664	128 399	156 146	198 677
Antarctic krill	*Euphausia superba*	6	-	-	-	-	-	-
Antarctic krill nei	*Euphausia spp*	0	-	-	-	-	-	-
Marine crustaceans nei	*Crustacea*	28 047	25 339	31 261	29 747	14 750	30 923	33 318
Cephalopods nei	*Cephalopoda*	85 120	117 624	95 834	93 709	96 408	53 271	90 076
Marine molluscs nei	*Mollusca*	15 299	4 750	2 718	2 217	1 796	2 597	2 424
Sea squirts nei	*Ascidiacea*	-	-	-	-	95	-	-
Salps	*Salpidae*	7	-	-	-	-	-	-
Jellyfishes	*Rhopilema spp*	0	-	-	-	-	-	-
	Country total	*3 447 954*	*3 523 448*	*3 373 492*	*3 472 150*	*3 666 428*	*3 777 092*	*3 770 912*
Indonesia								
Common carp	*Cyprinus carpio*	7 081	6 644	7 082	7 127	7 035	8 228	8 730
Hoven's carp	*Leptobarbus hoeveni*	6 892	5 836	3 241	4 608	3 382	2 362	2 270
Java barb	*Puntius javanicus*	19 622	19 469	20 189	17 939	17 791	17 425	17 840
Asian barbs nei	*Puntius spp*	13 253	11 976	12 131	11 263	12 406	11 649	11 840
Mozambique tilapia	*Oreochromis mossambicus*	16 943	17 715	17 865	21 172	19 831	20 210	20 980
Knifefishes	*Notopterus spp*	3 880	3 455	3 599	3 845	4 255	3 820	3 800
Glass catfishes	*Kryptopterus spp*	17 560	15 938	13 943	15 927	13 110	13 522	14 980
Torpedo-shaped catfishes nei	*Clarias spp*	8 385	7 257	7 535	6 413	5 496	7 609	7 110
Freshwater siluroids nei	*Siluroidei*	13 508	10 117	12 466	13 854	13 630	13 203	13 490
Gudgeons, sleepers nei	*Eleotridae*	1 109	1 290	1 078	1 243	1 125	2 604	3 080
Snakeskin gourami	*Trichogaster pectoralis*	30 407	21 375	20 936	23 265	21 733	20 857	20 440
Kissing gourami	*Helostoma temminckii*	12 611	18 376	16 598	23 320	18 771	19 429	20 140
Indonesian snakehead	*Channa micropeltes*	11 614	10 117	8 253	8 787	8 149	6 357	6 440
Snakeheads(=Murrels) nei	*Channa spp*	32 365	31 326	26 108	36 309	33 368	31 274	31 580
Freshwater fishes nei	*Osteichthyes*	115 323	103 812	96 597	110 347	112 554	107 609	107 690
River eels nei	*Anguilla spp*	4 339	657	787	1 212	4 553	1 907	3 790
Toli shad	*Tenualosa toli*	2 131	3 011	3 714	3 505	2 645	5 257	5 820
Barramundi(=Giant seaperch)	*Lates calcarifer*	48 312	55 942	65 193	65 173	68 788	63 485	68 720
Indian halibut	*Psettodes erumei*	7 397	15 075	9 939	12 071	11 143	11 373	13 510
Flatfishes nei	*Pleuronectiformes*	3 414	7 407	3 720	5 074	4 236	6 750	7 980
Bombay-duck	*Harpadon nehereus*	11 218	13 280	14 182	12 415	8 988	8 201	8 030
Lizardfishes nei	*Synodontidae*	7 226	15 158	11 998	12 944	13 383	11 954	13 570
Sea catfishes nei	*Ariidae*	62 412	78 578	66 298	69 646	70 266	74 443	78 450
Mullets nei	*Mugilidae*	35 450	35 478	35 582	35 437	36 077	33 595	34 460
Fusiliers	*Caesio spp*	32 714	38 358	34 142	37 944	33 712	38 312	42 120
Groupers nei	*Epinephelus spp*	38 286	42 164	43 766	43 472	48 422	48 516	52 320
Bigeyes nei	*Priacanthus spp*	3 679	4 448	4 612	4 982	6 203	5 344	5 760
Snappers nei	*Lutjanus spp*	60 338	69 585	66 280	66 492	62 306	67 773	69 850
Threadfin breams nei	*Nemipterus spp*	31 593	29 340	30 937	39 197	34 218	37 179	39 540
Ponyfishes(=Slipmouths) nei	*Leiognathidae*	71 401	89 403	79 532	91 219	69 512	87 757	95 010
Grunts, sweetlips nei	*Haemulidae (=Pomadasyidae)*	14 854	14 657	15 203	15 348	17 368	17 547	18 370
Croakers, drums nei	*Sciaenidae*	45 233	44 837	50 114	56 991	52 254	49 647	51 890
Emperors(=Scavengers) nei	*Lethrinidae*	34 559	31 908	31 706	32 557	30 657	29 575	31 200
Goatfishes	*Upeneus spp*	20 724	24 203	25 207	26 252	27 948	28 660	30 750
Threadfins, tasselfishes nei	*Polynemidae*	30 711	31 709	34 220	33 335	38 282	35 363	37 880
Hairtails, scabbardfishes nei	*Trichiuridae*	28 371	32 981	37 154	36 658	38 077	38 502	42 340
Goldstripe sardinella	*Sardinella gibbosa*	157 104	156 914	174 691	162 710	172 219	185 912	191 300
Bali sardinella	*Sardinella lemuru*	88 590	138 636	153 965	89 286	88 744	103 710	108 370
Rainbow sardine	*Dussumieria acuta*	21 906	22 829	22 321	24 185	23 761	20 401	21 100
Stolephorus anchovies	*Stolephorus spp*	161 781	183 591	166 808	163 117	173 944	190 182	197 810
Wolf-herrings nei	*Chirocentrus spp*	23 059	26 527	25 598	27 513	29 154	29 797	31 950
Wahoo	*Acanthocybium solandri*	12	15	14	16	12	12	12
Narrow-barred Spanish mackerel	*Scomberomorus commerson*	68 455	75 270	72 066	77 711	85 430	84 513	87 620
Indo-Pacific king mackerel	*Scomberomorus guttatus*	23 096	21 015	22 746	21 674	24 449	26 909	27 644
Frigate and bullet tunas	*Auxis thazard, A.rochei*	8 525	8 739	8 405	9 124	9 644	7 934	7 934
Kawakawa	*Euthynnus affinis*	373	383	368	400	423	348	348
Skipjack tuna	*Katsuwonus pelamis*	182 754	184 755	227 652	245 366	247 506	228 553	238 422
Albacore	*Thunnus alalunga*	1 300	1 561	1 461	1 707	2 659	2 865	2 865
Southern bluefin tuna	*Thunnus maccoyii*	1 349	1 888	1 245	2 351	1 068	1 472	1 472
Yellowfin tuna	*Thunnus albacares*	123 853	131 471	135 583	149 752	147 353	133 851	142 291
Bigeye tuna	*Thunnus obesus*	23 206	27 861	26 090	30 475	20 926	21 125	21 125
Indo-Pacific sailfish	*Istiophorus platypterus*	502	603	564	659	366	322	322
Blue marlin	*Makaira nigricans*	1 903	2 285	2 140	2 499	1 384	1 220	1 220
Black marlin	*Makaira indica*	391	470	440	514	286	252	252
Striped marlin	*Tetrapturus audax*	237	285	267	312	174	153	153
Marlins,sailfishes,etc. nei	*Istiophoridae*	753	772	742	806	851	701	701
Swordfish	*Xiphias gladius*	981	1 178	1 103	1 288	714	630	630
Tuna-like fishes nei	*Scombroidei*	211 622	220 442	225 031	229 787	253 985	229 288	238 503
Needlefishes nei	*Tylosurus spp*	26 688	29 797	29 741	29 409	32 870	27 320	27 980
Flyingfishes nei	*Exocoetidae*	16 534	16 126	18 961	17 822	19 980	16 467	17 490
Scads nei	*Decapterus spp*	251 289	276 924	277 593	261 138	255 375	258 393	267 050
Jacks, crevalles nei	*Caranx spp*	30 045	32 097	39 443	34 220	36 321	37 988	40 220
Black pomfret	*Parastromateus niger*	27 995	32 638	32 751	31 860	34 093	43 685	47 480
Rainbow runner	*Elagatis bipinnulata*	7 493	7 055	14 133	10 314	9 983	10 019	11 340
Torpedo scad	*Megalaspis cordyla*	14 915	14 937	17 561	19 457	20 485	21 596	22 680
Queenfishes	*Scomberoides spp*	15 623	13 573	15 872	15 979	14 552	17 285	18 110
Carangids nei	*Carangidae*	116 094	125 504	128 459	128 795	129 913	132 998	136 950
Indian mackerels nei	*Rastrelliger spp*	188 910	201 404	204 763	201 466	207 037	214 387	220 390
Silver pomfret	*Pampus argenteus*	19 932	20 461	21 308	21 340	25 492	30 285	32 630
Barracudas nei	*Sphyraena spp*	16 552	18 206	20 440	20 578	19 558	19 070	19 180
Rays, stingrays, mantas nei	*Rajiformes*	36 457	43 324	48 291	45 327	45 260	44 451	46 240
Sharks, rays, skates, etc. nei	*Elasmobranchii*	57 939	52 674	62 497	63 066	68 366	65 860	68 360
Marine fishes nei	*Osteichthyes*	433 431	413 080	456 546	470 576	508 966	568 594	607 320
Giant river prawn	*Macrobrachium rosenbergii*	6 126	5 208	5 362	5 937	5 449	5 185	5 240
Freshwater prawns, shrimps nei	*Palaemonidae*	3 356	3 242	3 871	3 900	4 210	3 898	4 000
Freshwater crustaceans nei	*Crustacea*	223	49	93	132	132	337	490
Blue swimming crab	*Portunus pelagicus*	11 558	15 388	12 370	14 276	14 053	22 040	26 130

English name Nom anglais Nombre inglés	Scientific name Nom scientifique Nombre científico	1996 t	1997 t	1998 t	1999 t	2000 t	2001 t	2002 t
Indo-Pacific swamp crab	*Scylla serrata*	7 342	8 298	8 161	8 707	8 774	11 752	12 590
Tropical spiny lobsters nei	*Panulirus spp*	2 464	4 021	2 394	3 244	3 596	4 490	5 800
Banana prawn	*Penaeus merguiensis*	53 914	53 924	62 192	64 179	66 644	65 269	68 720
Giant tiger prawn	*Penaeus monodon*	19 396	25 929	30 047	34 223	40 987	43 759	50 660
Metapenaeus shrimps nei	*Metapenaeus spp*	22 288	32 588	40 717	33 847	38 925	36 358	41 420
Natantian decapods nei	*Natantia*	95 780	102 120	94 481	111 277	106 358	120 882	127 190
Marine crustaceans nei	*Crustacea*	907	2 583	808	503	928	633	700
Freshwater molluscs nei	*Mollusca*	1 342	909	806	597	734	2 200	1 180
Cupped oysters nei	*Crassostrea spp*	1 596	1 678	2 029	2 025	2 281	1 194	1 780
Scallops nei	*Pectinidae*	1 142	424	837	839	578	419	480
Anadara clams nei	*Anadara spp*	46 183	41 919	31 591	33 910	34 695	64 308	68 050
Hard clams nei	*Meretrix spp*	13 481	14 027	17 146	14 767	14 177	13 375	14 600
Clams, etc. nei	*Bivalvia*	844	956	914	304	303	792	850
Cuttlefish,bobtail squids nei	*Sepiidae, Sepiolidae*	8 354	8 201	11 473	12 331	11 945	15 789	18 520
Common squids nei	*Loligo spp*	27 169	41 755	31 850	36 707	39 838	60 529	72 630
Octopuses, etc. nei	*Octopodidae*	645	1 269	3 519	2 337	1 985	3 227	3 350
Marine molluscs nei	*Mollusca*	1 148	382	674	617	524	2 845	3 040
Frogs	*Rana spp*	1 793	1 390	1 667	1 317	1 880	1 243	1 180
River and lake turtles nei	*Testudinata*	36	48	27	1	3	42	40
Marine turtles nei	*Testudinata*	721	638	781	706	744	350	390
Sea cucumbers nei	*Holothurioidea*	2 443	3 138	3 058	2 617	3 041	3 517	3 830
Jellyfishes	*Rhopilema spp*	6 740	17 719	3 861	32 652	29 516	21 465	23 300
Aquatic invertebrates nei	*Invertebrata*	1 240	1 685	709	2 649	2 849	1 918	2 080
	Country total	*3 604 795*	*3 857 660*	*3 961 004*	*4 044 543*	*4 120 126*	*4 273 662*	*4 505 474*
Iran								
Freshwater breams nei	*Abramis spp*	3	7	20	9	20	10	38
Kutum	*Rutilus frisii*	9 210	2 320	6 624	6 905	10 120	7 199	6 417
Roaches nei	*Rutilus spp*	878	203	607	626	1 515	1 316	787
Grass carp(=White amur)	*Ctenopharyngodon idellus*	4 205	4 517	5 039	4 600	2 500	580	1 301
Silver carp	*Hypophthalmichthys molitrix*	17 510	15 825	11 629	14 400	8 750	3 680	4 295
Bighead carp	*Hypophthalmichthys nobilis*	1 751	1 565	2 522	3 600	4 250	245	520
Cyprinids nei	*Cyprinidae*	12 364	12 870	19 573	10 800	10 000	2 680	3 645
Freshwater siluroids nei	*Siluroidei*	22	6	32	24	25	1	21
Pike-perch	*Stizostedion lucioperca*	6	4	105	19	20	26	32
Freshwater fishes nei	*Osteichthyes*	1 845	1 131	2 529	1 637	1 575	5 970	4 340
Sturgeons nei	*Acipenseridae*	1 600	1 300	1 200	1 000	1 000	870	643
Salmonoids nei	*Salmonoidei*	8	7	8	3	5	2	9
Caspian shads	*Caspialosa spp*	57 000	60 400	85 000	95 000	78 000	45 180	26 000
Diadromous clupeoids nei	*Clupeoidei*	330	566	817	697	595	623	932
Mullets nei	*Mugilidae*	2 554	3 074	4 558	4 080	5 125	5 263	6 873
Lanternfishes nei	*Myctophidae*	0	0	0	0	0	335	37
Clupeoids nei	*Clupeoidei*	10 000	10 000	9 708	13 030	14 955	17 453	12 865
Narrow-barred Spanish mackerel	*Scomberomorus commerson*	3 560	4 290	4 034	4 609	7 075	6 071	8 557
Indo-Pacific king mackerel	*Scomberomorus guttatus*	4 340	4 129	3 883	3 476	4 100	2 474	4 031
Frigate and bullet tunas	*Auxis thazard, A.rochei*	755	544	509	590	785	562	611
Kawakawa	*Euthynnus affinis*	5 665	8 451	7 947	10 858	13 500	12 474	16 361
Skipjack tuna	*Katsuwonus pelamis*	3 241	5 973	6 671	16 583	20 091	26 058	29 859
Longtail tuna	*Thunnus tonggol*	17 147	20 112	19 694	23 465	41 407	34 896	29 853
Albacore	*Thunnus alalunga*	10	16	9	-	-	-	-
Yellowfin tuna	*Thunnus albacares*	30 233	21 250	21 530	26 871	15 743	20 153	24 045
Bigeye tuna	*Thunnus obesus*	153	262	405	592	347	430	127
Indo-Pacific sailfish	*Istiophorus platypterus*	2 306	1 928	1 813	3 193	3 080	3 160	4 258
Black marlin	*Makaira indica*	3	3	5	1	1	-	3
Swordfish	*Xiphias gladius*	9	9	13	2	1	-	7
Tuna-like fishes nei	*Scombroidei*	4	5	9	-	-	-	-
Sharks, rays, skates, etc. nei	*Elasmobranchii*	...	...	1	...	...	...	8 071
Marine fishes nei	*Osteichthyes*	157 397	143 212	139 107	130 225	122 620	124 118	121 692
Tropical spiny lobsters nei	*Panulirus spp*	49	76	65	35	25	20	20
Natantian decapods nei	*Natantia*	5 837	7 620	5 774	4 570	9 850	6 940	5 726
Cephalopods nei	*Cephalopoda*	1 580	8 620	4 189	4 060	5 685	6 517	2 877
Marine molluscs nei	*Mollusca*	150	1 992	1 583	1 640	1 235	1 144	0
	Country total	*351 725*	*342 287*	*367 212*	*387 200*	*384 000*	*336 450*	*324 853*
Iraq								
Common carp	*Cyprinus carpio*	5 538	4 163	2 336	...	...	...	...
Cyprinids nei	*Cyprinidae*	7 694	4 782	1 703	5 723	4 013	3 800 F	3 800 F
Freshwater siluroids nei	*Siluroidei*	1 436	1 738	711	1 402	953	900 F	900 F
Freshwater fishes nei	*Osteichthyes*	4 381	9 836	4 361	2 205	3 412	3 300 F	3 300 F
Marine fishes nei	*Osteichthyes*	11 688	10 783	13 463	13 093	12 389	8 500 F	5 000 F
	Country total	*30 737*	*31 302*	*22 574*	*22 423*	*20 767*	*16 500 F*	*13 000 F*
Israel								
Common carp	*Cyprinus carpio*	35	38	150	165	189	97	98
Silver carp	*Hypophthalmichthys molitrix*	40	17	11	9	32	16	30
Kinneret bleak	*Acanthobrama terraesanctae*	1 155	626	1 171	1 048	1 052	811	1 170
Cyprinids nei	*Cyprinidae*	12	15	64	80	60	70	...
Blue tilapia	*Oreochromis aureus*	32	101	88	66	67	33	18
Mango tilapia	*Sarotherodon galilaeus*	350	462	391	405	262	110	93
Cichlids nei	*Cichlidae*	40	48	58	57	52	24	23
European hake	*Merluccius merluccius*	131	86	134	60	62	73	68
Brushtooth lizardfish	*Saurida undosquamis*	124	61	37	48	21	24	24
Flathead grey mullet	*Mugil cephalus*	439	550	514	599	310	300	294
Groupers nei	*Epinephelus spp*	406	384	394	381	263	188	226
Meagre	*Argyrosomus regius*	33	2	2	9	288	223	273
Common pandora	*Pagellus erythrinus*	100	65	47	67	517	...	...
Bogue	*Boops boops*	34	34	47	73	91	37	38

English name Nom anglais Nombre inglés	Scientific name Nom scientifique Nombre científico	1996 t	1997 t	1998 t	1999 t	2000 t	2001 t	2002 t
Porgies, seabreams nei	*Sparidae*	100	-	-	-	-	562	494
Surmullets(=Red mullets) nei	*Mullus spp*	227	185	276	350	340	401	368
Round sardinella	*Sardinella aurita*	242	435	557	306	88	117	143
Little tunny(=Atl.black skipj)	*Euthynnus alletteratus*	119	103	73	90	113	70	50
Atlantic bluefin tuna	*Thunnus thynnus*	14	-	-	-	-	-	-
Atlantic horse mackerel	*Trachurus trachurus*	65	226	172	178	175	90	129
Greater amberjack	*Seriola dumerili*	33	80	185	98	307	223	201
Chub mackerel	*Scomber japonicus*	71	47	54	44	8	4	...
Barracudas nei	*Sphyraena spp*	76	104	105	160	77	55	54
Sharks, rays, skates, etc. nei	*Elasmobranchii*	330	49	59	58	...	35	...
Marine fishes nei	*Osteichthyes*	721	1 083	1 410	1 227	1 143	1 078	848
Kuruma prawn	*Penaeus japonicus*	-	84	...	...	...	121	111
Deepwater rose shrimp	*Parapenaeus longirostris*	50	...	...	...	...	...	...
Natantian decapods nei	*Natantia*	200	221	225	186	184	116	-
Cephalopods nei	*Cephalopoda*	50	98	76	120	117	146	127
	Country total	*5 229*	*5 204*	*6 300*	*5 884*	*5 818*	*5 024*	*4 880*
Japan								
Common carp	*Cyprinus carpio*	4 771	4 607	4 477	4 259	4 079	3 558	3 359
Crucian carp	*Carassius carassius*	4 205	4 008	3 881	3 493	3 423	2 948	2 706
Freshwater fishes nei	*Osteichthyes*	13 344	13 163	12 073	11 020	10 341	9 610	9 160
Japanese eel	*Anguilla japonica*	901	860	860	817	765	677	610
Trouts nei	*Salmo spp*	1 222	1 207	1 187	1 122	1 136	1 053	1 043
Pink(=Humpback)salmon	*Oncorhynchus gorbuscha*	32 595	15 844	25 337	16 902	26 602	9 765	24 671
Chum(=Keta=Dog)salmon	*Oncorhynchus keta*	299 881	269 183	206 622	182 866	163 449	217 359	211 691 F
Masu(=Cherry) salmon	*Oncorhynchus masou*	2 507	1 800	2 570	1 979	1 811	1 605	1 879
Sockeye(=Red)salmon	*Oncorhynchus nerka*	5 723	9 246	2 768	2 750	2 147	2 740	3 222 F
Chinook(=Spring=King)salmon	*Oncorhynchus tshawytscha*	250	825	534	270	147	111	180
Coho(=Silver)salmon	*Oncorhynchus kisutch*	701	575	746	508	376	502	562
Rainbow trout	*Oncorhynchus mykiss*	728	672	618	562	536	484	447
Ayu sweetfish	*Plecoglossus altivelis*	12 732	12 619	11 386	11 380	11 172	11 148	10 663
Dotted gizzard shad	*Konosirus punctatus*	18 647	14 850	20 787	17 770	12 335	17 210	13 401
Atlantic halibut	*Hippoglossus hippoglossus*	4	6	7	5	3	36	14
Greenland halibut	*Reinhardtius hippoglossoides*	2 532	1 876	2 053	2 420	2 512	2 814	2 779
Witch flounder	*Glyptocephalus cynoglossus*	12	5	2	2	4	3	38
Amer. plaice(=Long rough dab)	*Hippoglossoides platessoides*	11	7	16	21	21	6	82
Bastard halibut	*Paralichthys olivaceus*	8 311	8 361	7 615	7 198	7 572	6 729	6 680
Flatfishes nei	*Pleuronectiformes*	80 425	76 690	72 850	68 847	68 518	60 979	63 812
Red codling	*Pseudophycis bachus*	24	14	15	27	70	26	20 F
Pacific cod	*Gadus macrocephalus*	57 576	58 477	57 243	55 292	51 052	43 550	29 516
Alaska pollock(=Walleye poll.)	*Theragra chalcogramma*	331 163	338 785	315 987	382 385	300 001	241 881	213 254
Southern blue whiting	*Micromesistius australis*	31 690	36 513	40 309	40 855	31 217	32 874	26 170 F
Argentine hake	*Merluccius hubbsi*	83	53	30	27	59	3	75
Patagonian grenadier	*Macruronus magellanicus*	544	644	844	400	1 889	866	1 612
Blue grenadier	*Macruronus novaezelandiae*	26 031	25 349	26 802	17 269	12 139	15 734	10 000 F
Roundnose grenadier	*Coryphaenoides rupestris*	35	40	37	40	27	146	-
Gadiformes nei	*Gadiformes*	4 131	3 186	4 098	3 415	2 519	2 356	1 500 F
Greater lizardfish	*Saurida tumbil*	8 144	7 638	7 860	7 716	7 591	6 094	5 652
Flathead grey mullet	*Mugil cephalus*	4 268	3 933	4 003	3 629	3 725	3 457	2 993
Mullets nei	*Mugilidae*	850	1 053	887	1 000	1 140	1 084	980
Groupers, seabasses nei	*Serranidae*	3	13	21	10	11	13	10 F
Japanese seabass	*Lateolabrax japonicus*	8 334	9 057	9 223	9 234	9 337	10 690	10 737
Grunts, sweetlips nei	*Haemulidae (=Pomadasyidae)*	5 303	4 801	5 703	5 282	5 136	4 995	5 606
Croakers, drums nei	*Sciaenidae*	7 062	5 998	5 430	4 850	4 791	4 362	4 112
Silver seabream	*Pagrus auratus*	16 468	15 611	15 375	15 731	15 041	14 633	16 527
Porgies, seabreams nei	*Sparidae*	11 605	11 256	11 284	10 675	9 066	9 545	10 000 F
Marbled rockcod	*Notothenia rossii*	-	-	-	-	-	-	0
Pacific sandlance	*Ammodytes personatus*	115 766	108 666	90 688	82 918	49 819	88 164	67 564
Okhotsk atka mackerel	*Pleurogrammus azonus*	181 513	206 763	240 971	169 481	165 118	161 160	154 736
Puffers nei	*Tetraodontidae*	8 238	7 103	8 329	9 647	10 989	7 820	7 869
Argentines	*Argentina spp*	1	-	-	-	-	-	-
Deepsea smelt	*Glossanodon semifasciatus*	8 134	7 431	7 142	6 312	5 970	5 414	4 926
Daggertooth pike conger	*Muraenesox cinereus*	1 989	2 060	2 081	2 298	2 400	2 738	2 843
Conger eels, etc. nei	*Congridae*	12 007	11 706	9 444	8 168	8 364	7 999	8 921
John dory	*Zeus faber*	16	19	17	13	2	46	19 F
Tilefishes nei	*Branchiostegidae*	3 648	2 994	2 284	1 949	1 678	1 781	1 804
Patagonian toothfish	*Dissostichus eleginoides*	264	411	...	...	...	...	3
Wolffishes(=Catfishes) nei	*Anarhichas spp*	20	17	26	21	15	53	30
Japanese sandfish	*Arctoscopus japonicus*	6 719	6 209	6 795	6 615	6 652	8 753	9 249
Snoek	*Thyrsites atun*	28	12	3	23	59	189	84
Largehead hairtail	*Trichiurus lepturus*	26 644	20 932	22 268	26 200	22 947	16 615	14 405
Pacific rudderfish	*Psenopsis anomala*	3 516	4 316	4 316	4 996	5 215	4 196	3 695
Pacific ocean perch	*Sebastes alutus*	1 730	767	896	766	458	668	967
Atlantic redfishes nei	*Sebastes spp*	1 585	596	922	460	140	187	165
Scorpionfishes nei	*Scorpaenidae*	2 317	2 082	1 638	1 314	1 244	1 132	1 323
Demersal percomorphs nei	*Perciformes*	4 511	4 812	3 767	3 097	2 895	6 489	1 300 F
Pacific herring	*Clupea pallasii*	2 021	1 926	2 531	2 579	2 260	2 385	1 366
Japanese pilchard	*Sardinops melanostictus*	319 354	284 054	167 073	351 207	149 616	178 423	50 313
Red-eye round herring	*Etrumeus teres*	49 752	55 043	48 441	28 712	23 833	31 525	26 355
Japanese anchovy	*Engraulis japonicus*	345 517	233 113	470 616	484 230	381 020	301 168	443 158
Clupeoids nei	*Clupeoidei*	58 232	59 619	52 427	79 405	74 581	58 286	62 694
Japanese Spanish mackerel	*Scomberomorus niphonius*	3 607	2 349	2 864	5 321	10 932	9 056	8 936
Seerfishes nei	*Scomberomorus spp*	7	-	-	-	-	-	-
Frigate and bullet tunas	*Auxis thazard, A.rochei*	20 374	32 574	21 613	29 517	27 139	37 247	37 247 F
Skipjack tuna	*Katsuwonus pelamis*	286 671	311 542	385 462	287 317	341 414	276 680	276 131 F
Atlantic bluefin tuna	*Thunnus thynnus*	4 561	3 496	4 262	3 822	3 140	2 815	3 501
Pacific bluefin tuna	*Thunnus orientalis*	6 668	6 487	3 667	11 188	10 128	5 832	5 832 F
Albacore	*Thunnus alalunga*	61 059	83 592	73 756	99 027	61 249	69 196	68 704 F
Southern bluefin tuna	*Thunnus maccoyii*	6 667	5 100	7 716	7 728	6 092	7 240	5 814 F

English name Nom anglais Nombre inglés	Scientific name Nom scientifique Nombre científico	1996 t	1997 t	1998 t	1999 t	2000 t	2001 t	2002 t
Yellowfin tuna	*Thunnus albacares*	80 135	114 156	98 430	96 434	103 368	98 466	98 713 F
Bigeye tuna	*Thunnus obesus*	101 591	105 580	115 571	92 673	96 067	104 247	102 059 F
Indo-Pacific sailfish	*Istiophorus platypterus*	1 095	844	1 583	1 360	1 109	896	868 F
Atlantic sailfish	*Istiophorus albicans*	80	60	91	71	73	33	15
Blue marlin	*Makaira nigricans*	8 820	8 999	9 593	7 298	7 272	6 551	6 573 F
Black marlin	*Makaira indica*	...	...	...	...	...	...	82
Striped marlin	*Tetrapturus audax*	7 613	6 537	8 136	5 951	4 545	4 783	4 799 F
Atlantic white marlin	*Tetrapturus albidus*	115	55	59	50	89	60	12
Shortbill spearfish	*Tetrapturus angustirostris*	...	...	...	...	...	...	34
Longbill spearfish	*Tetrapturus pfluegeri*	...	...	...	...	...	...	22
Swordfish	*Xiphias gladius*	15 652	12 595	16 898	13 585	13 181	12 818	13 072 F
Tuna-like fishes nei	*Scombroidei*	9 481	14 385	12 131	9 924	10 826	8 953	8 737 F
Pacific saury	*Cololabis saira*	229 227	290 812	144 983	141 011	216 471	269 797	205 282
Japanese flyingfish	*Cypselurus agoo*	8 501	7 486	8 933	6 738	9 615	8 286	6 909
Japanese jack mackerel	*Trachurus japonicus*	330 406	323 142	311 311	211 077	245 988	214 434	196 044
Chilean jack mackerel	*Trachurus murphyi*	-	-	-	7	-	-	-
Jack and horse mackerels nei	*Trachurus spp*	335	29	8	11	12	655	9
Japanese scad	*Decapterus maruadsi*	57 319	50 097	59 078	47 157	36 416	41 236	41 831
Japanese amberjack	*Seriola quinqueradiata*	246	...	...	...	...	...	...
Amberjacks nei	*Seriola spp*	50 333	47 211	45 484	54 918	77 461	66 925	51 194
Common dolphinfish	*Coryphaena hippurus*	8 527	10 229	14 944	9 280	8 959	9 142	9 000 F
Chub mackerel	*Scomber japonicus*	760 430	848 967	511 238	381 866	346 220	375 273	279 000
Butterfishes, pomfrets nei	*Stromateidae*	1 518	2 126	2 194	1 842	2 077	2 211	1 300 F
Whip stingray	*Dasyatis akajei*	4 029	3 959	4 329	4 407	5 388	4 312	4 512
Sharks, rays, skates, etc. nei	*Elasmobranchii*	20 177	25 438	29 336	28 703	26 500	23 609	20 038 F
Groundfishes nei	*Osteichthyes*	57	32	35	31	11	6	160
Marine fishes nei	*Osteichthyes*	292 617	292 089	269 472	261 613	275 442	251 517	334 640 F
Freshwater prawns, shrimps nei	*Palaemonidae*	2 222	2 413	1 872	1 942	1 676	1 158	1 002
Gazami crab	*Portunus trituberculatus*	4 022	3 112	3 528	2 752	3 131	3 596	3 485
Tanner crabs nei	*Chionoecetes spp*	3 447	4 870	4 677	4 892	5 640	5 355	5 016
Geryons nei	*Geryon spp*	3 250	1 937	1 562	1 730	2 980	1 650	1 650 F
Marine crabs nei	*Brachyura*	37 265	34 894	33 678	30 859	30 310	27 353	28 000 F
Longlegged spiny lobster	*Panulirus longipes*	1 092	1 068	1 084	1 154	1 244	1 486	1 378
King crabs	*Paralithodes spp*	322	154	132	117	89	181	129
Antarctic stone crab	*Paralomis spinosissima*	-	-	-	-	-	-	55
Globose king crab	*Paralomis formosa*	-	-	-	-	-	-	56
Kuruma prawn	*Penaeus japonicus*	2 262	2 144	2 069	1 523	1 447	1 271	1 123
Penaeus shrimps nei	*Penaeus spp*	192	41	-	-	-	-	-
Northern prawn	*Pandalus borealis*	-	-	-	-	-	130	100
Natantian decapods nei	*Natantia*	28 641	27 114	25 283	25 630	25 898	24 281	25 005 F
Antarctic krill	*Euphausia superba*	58 769	60 937	67 481	66 076	80 601	67 378	51 079
Marine crustaceans nei	*Crustacea*	58 892	63 028	67 945	49 783	74 072	49 619	50 000 F
Japanese corbicula	*Corbicula japonica*	26 714	21 822	19 932	20 009	19 295	17 295	17 779
Freshwater molluscs nei	*Mollusca*	1 305	1 361	1 132	900	628	583	338
Giant abalone	*Haliotis gigantea*	1 941	2 218	2 269	2 109	2 146	1 982	2 223
Horned turban	*Turbo cornutus*	10 119	12 132	12 556	11 000	9 839	10 241	9 504
Yesso scallop	*Patinopecten yessoensis*	271 124	261 164	287 802	299 628	304 286	290 974	306 666
Half-crenated ark	*Scapharca subcrenata*	16 328	14 133	10 120	10 413	7 308	4 899	9 828
Japanese hard clam	*Meretrix lusoria*	1 944	1 897	1 870	1 785	1 543	1 245	1 300
Japanese carpet shell	*Ruditapes philippinarum*	43 703	39 660	36 807	43 088	35 558	31 022	34 494
Imperial surf clam	*Pseudocardium sybillae*	8 738	7 541	8 227	8 808	8 883	8 914	8 861
Clams, etc. nei	*Bivalvia*	52 361	42 987	47 584	35 319	35 529	30 133	28 198
Cuttlefish,bobtail squids nei	*Sepiidae, Sepiolidae*	9 545	8 052	9 790	9 942	8 241	8 297	7 873
Common squids nei	*Loligo spp*	3 187	1 562	2 618	1 847	-	-	4 106
Neon flying squid	*Ommastrephes bartrami*	...	49 870	54 951	36 076	46 963	23 770	22 000 F
Northern shortfin squid	*Illex illecebrosus*	1	-	-	-	-	-	-
Argentine shortfin squid	*Illex argentinus*	73 896	126 504	77 452	154 129	119 966	70 652	7 800
Jumbo flying squid	*Dosidicus gigas*	14 297	23 179	-	3 305	56 825	72 412	72 201 F
Japanese flying squid	*Todarodes pacificus*	444 189	365 978	180 749	237 346	337 285	298 191	273 567 F
Wellington flying squid	*Nototodarus sloani*	11 342	5 182	3 734	1 853	1 853	1 396	988
Sevenstar flying squid	*Martialia hyadesi*	-	-	2	27	33	2	...
Various squids nei	*Loliginidae, Ommastrephidae*	121 718	63 116	56 371	56 575	49 607	46 451	30 565
Octopuses, etc. nei	*Octopodidae*	50 584	56 593	61 260	57 427	47 374	45 200	57 482
Sea urchins nei	*Strongylocentrotus spp*	12 996	14 297	13 653	13 530	12 455	11 208	12 733
Japanese sea cucumber	*Stichopus japonicus*	7 226	7 160	6 952	6 662	6 957	7 229	7 259
Aquatic invertebrates nei	*Invertebrata*	13 982	10 770	11 021	8 346	10 932	8 339	8 535 F
	Country total	*5 931 872*	*5 928 334*	*5 303 555*	*5 188 948*	*4 984 813*	*4 713 006*	*4 443 000*
Jordan								
Freshwater fishes nei	*Osteichthyes*	350	350	350	350	400	350	350
Fusiliers	*Caesio spp*	...	...	15	18	17	15	12
Emperors(=Scavengers) nei	*Lethrinidae*	...	...	2	2	1	2	1
Spinefeet(=Rabbitfishes) nei	*Siganus spp*	...	...	10	12	10	11	8
Narrow-barred Spanish mackerel	*Scomberomorus commerson*	...	...	0	1	0	1	1
Frigate and bullet tunas	*Auxis thazard, A.rochei*	...	...	2	4	2	5	5
Kawakawa	*Euthynnus affinis*	...	...	26	36	33	40	40
Skipjack tuna	*Katsuwonus pelamis*	...	...	35	44	45	52	52
Longtail tuna	*Thunnus tonggol*	...	...	2	2	2	3	3
Yellowfin tuna	*Thunnus albacares*	...	...	2	5	3	5	5
Scads nei	*Decapterus spp*	...	...	20	25	25	26	25
Marine fishes nei	*Osteichthyes*	90	100	6	11	12	10	24
	Country total	*440*	*450*	*470*	*510*	*550*	*520*	*526*
Kazakhstan								
Freshwater bream	*Abramis brama*	18 770 F	11 000 F	9 800 F	11 000 F	12 000 F	12 630	14 110
Common carp	*Cyprinus carpio*	440 F	320 F	230 F	500 F	650 F	710	1 371
Tench	*Tinca tinca*	-	-	-	-	-	-	183
Crucian carp	*Carassius carassius*	3 950 F	2 840 F	2 060 F	1 900 F	1 800 F	1 707	1 611

English name Nom anglais Nombre inglés	Scientific name Nom scientifique Nombre científico	1996 t	1997 t	1998 t	1999 t	2000 t	2001 t	2002 t
Roaches nei	*Rutilus spp*	2 420 F	1 740 F	1 260 F	1 000 F	850 F	579	1 682
Rudd	*Scardinius erythrophthalmus*	...	...	...	...	...	...	485
Grass carp(=White amur)	*Ctenopharyngodon idellus*	2 F	1 F	1 F	1 F	-	-	-
Silver carp	*Hypophthalmichthys molitrix*	240 F	170 F	120 F	70 F	30 F	-	-
Sichel	*Pelecus cultratus*	25 F	20 F	15 F	10 F	10 F	8	-
Asp	*Aspius aspius*	380 F	270 F	195 F	250 F	350 F	476	1 026
Northern pike	*Esox lucius*	580 F	420 F	300 F	550 F	700 F	819	920
Wels(=Som)catfish	*Silurus glanis*	1 380 F	990 F	700 F	750 F	750 F	780	608
European perch	*Perca fluviatilis*	670 F	480 F	350 F	450 F	500 F	547	422
Pike-perch	*Stizostedion lucioperca*	5 570 F	4 000 F	2 900 F	2 500 F	2 000 F	1 628	1 718
Snakeheads(=Murrels) nei	*Channa spp*	10 F	7 F	5 F	10 F	10 F	12	8
Freshwater fishes nei	*Osteichthyes*	269 F	358 F	364 F	10 809 F	13 715 F	1 434	532
Sturgeons nei	*Acipenseridae*	510 F	370 F	270 F	240 F	215 F	282	197
Whitefishes nei	*Coregonus spp*	27 F	20 F	15 F	20 F	30 F	35	37
Caspian shads	*Caspialosa spp*	0	0	0	0	0	-	-
Azov sea sprat	*Clupeonella cultriventris*	9 000 F	8 800	6 400	6 100	3 000	-	-
Mullets nei	*Mugilidae*	30 F	20 F	15 F	10 F	10 F	7	-
	Country total	*44 273*	*31 826*	*25 000 F*	*36 170*	*36 620*	*21 654*	*24 910*
Korea D P Rp								
Freshwater fishes nei	*Osteichthyes*	20 000	20 000	16 000 F	12 000 F	8 000 F	4 928	5 000 F
Flatfishes nei	*Pleuronectiformes*	1 966	6 972	4 000 F	4 000 F	3 800 F	3 800 F	3 800 F
Alaska pollock(=Walleye poll.)	*Theragra chalcogramma*	15 369	66 578	65 000 F	62 000 F	60 000 F	60 000 F	60 000 F
Croakers, drums nei	*Sciaenidae*	-	-	-	-	224	-	-
Porgies, seabreams nei	*Sparidae*	-	-	-	-	273	-	-
Okhotsk atka mackerel	*Pleurogrammus azonus*	6 487	3 535	3 500 F	3 500 F	3 000 F	3 000 F	3 000 F
Japanese pilchard	*Sardinops melanostictus*	5	0	...	...	...	...	...
Marine fishes nei	*Osteichthyes*	166 325	109 770	114 200 F	113 000 F	112 154 F	108 991 F	107 600 F
Natantian decapods nei	*Natantia*	-	-	-	-	52	-	-
Marine crustaceans nei	*Crustacea*	33 931	15 265	15 200 F	15 400 F	15 600 F	16 181	16 000 F
Cuttlefish,bobtail squids nei	*Sepiidae, Sepiolidae*	-	-	-	-	61	-	-
Various squids nei	*Loliginidae, Ommastrephidae*	8 892	14 192	10 000 F	10 000 F	9 500 F	9 500 F	9 500 F
Octopuses, etc. nei	*Octopodidae*	-	-	-	-	86	-	-
Sea urchins nei	*Strongylocentrotus spp*	150	150	100 F	100 F	100 F	100 F	100 F
	Country total	*253 125*	*236 462*	*228 000 F*	*220 000 F*	*212 850 F*	*206 500*	*205 000 F*
Korea Rep								
Common carp	*Cyprinus carpio*	1 979	842	874	438	...	...	...
Cyprinids nei	*Cyprinidae*	1 010	2 221	1 977	2 294	...	...	...
Freshwater siluroids nei	*Siluroidei*	279	251	...	136	...	...	...
Snakehead	*Channa argus*	169	120	...	28	...	...	...
Freshwater fishes nei	*Osteichthyes*	3 418	2 633	2 755	1 681	6 402	5 254	4 993
Japanese eel	*Anguilla japonica*	113	56	44	13	...	...	...
Ayu sweetfish	*Plecoglossus altivelis*	-	5	-	-	-	-	-
Salmonoids nei	*Salmonoidei*	1 463	652	1 122	276	28	19	...
Chinese gizzard shad	*Clupanodon thrissa*	4 908	13 836	11 349	9 511	6 366	9 120	4 717
Elongate ilisha	*Ilisha elongata*	630	349	381	265	134	40	19
Lefteye flounders nei	*Bothidae*	40	27	38	28	40	30	20
Yellow striped flounder	*Pseudopleuronectes herzenst.*	18 066	18 079	20 135	19 569	15 423	14 503	13 816
Southeast Atlantic soles nei	*Austroglossus spp*	480	1 197	773	926	965	650	492
Soles nei	*Soleidae*	102	-	-	256	1 016	-	-
Tonguefishes	*Cynoglossidae*	3 119	3 192	1 867	2 443	2 108	3 069	1 350
Bastard halibut	*Paralichthys olivaceus*	2 317	1 592	2 002	1 679	1 607	1 707	1 822
Flatfishes nei	*Pleuronectiformes*	377	323	97	415	218	3	12
Blue antimora	*Antimora rostrata*	0	1	-	-	-	-	-
Pacific cod	*Gadus macrocephalus*	2 740	3 984	6 249	6 509	10 098	13 110	9 240
Alaska pollock(=Walleye poll.)	*Theragra chalcogramma*	226 910	223 065	236 278	146 165	86 143	197 396	24 772
Southern hake	*Merluccius australis*	454	1 178	1 976	2 512	2 142	2 224	1 645
Benguela hake	*Merluccius polli*	-	-	4	-	-	-	-
Hakes nei	*Merluccius spp*	370	33	-	-	-	15	-
Patagonian grenadier	*Macruronus magellanicus*	33	-	31	282	1 076	1 553	1 584
Blue grenadier	*Macruronus novaezelandiae*	1 725	6 091	4 883	5 834	8 694	9 484	9 627
Grenadiers nei	*Macrourus spp*	5	10	-	-	-	-	-
Grenadiers, rattails nei	*Macrouridae*	-	-	-	-	140	-	-
Gadiformes nei	*Gadiformes*	1 694	278	834	2 243	1 718	3 079	412
Brushtooth lizardfish	*Saurida undosquamis*	152	207	215	35	795	758	33
Korean sandlance	*Hypoptychus dybowskii*	6 613	8 832	4 801	4 806	-	462	7
Flathead grey mullet	*Mugil cephalus*	4 200	5 564	4 962	9 678	8 730	9 784	8 887
Mullets nei	*Mugilidae*	-	-	-	133	-	-	-
Groupers nei	*Epinephelus spp*	389	174	120	1 054	353	259	230
Groupers, seabasses nei	*Serranidae*	436	760	1 121	964	1 483	991	748
Japanese seabass	*Lateolabrax japonicus*	1 178	1 501	2 957	2 933	1 201	1 566	1 479
Sillago-whitings	*Sillaginidae*	248	199	48	109	100	10	15
Snappers nei	*Lutjanus spp*	-	-	555	474	524	601	369
Snappers, jobfishes nei	*Lutjanidae*	-	-	14	-	-	-	-
Grunts, sweetlips nei	*Haemulidae (=Pomadasyidae)*	5 532	-	130	20	25	540	510
Brown meagre	*Sciaena umbra*	208	...	...	...	...	...	...
Honnibe croaker	*Nibea mitsukurii*	1 940	1 177	1 285	1 566	1 999	2 156	1 148
Large yellow croaker	*Larimichthys croceus*	22 225	1 565	294	270	354	426	325
Yellow croaker	*Larimichthys polyactis*	23 695	22 101	15 011	13 490	19 630	7 938	10 941
Croakers, drums nei	*Sciaenidae*	73 653	95 281	89 021	114 160	67 563	50 789	48 062
Emperors(=Scavengers) nei	*Lethrinidae*	-	-	93	165	98	459	-
Angolan dentex	*Dentex angolensis*	274	-	-	-	-	-	-
Dentex nei	*Dentex spp*	4	-	-	-	-	-	-
Silver seabream	*Pagrus auratus*	762	1 149	1 657	924	986	913	1 084
Pargo breams nei	*Pagrus spp*	10	-	-	-	-	-	-
Porgies, seabreams nei	*Sparidae*	8 009	7 648	7 152	8 316	8 505	6 767	10 390
Threadfins, tasselfishes nei	*Polynemidae*	239	-	1 241	-	-	-	-

E-4 Fish crustaceans, molluscs, etc / Poissons, crustacés, mollusques, etc / Peces, crustáceos, moluscos, etc

Capture production by countries or areas and species — Asia / Captures par pays ou zones et espèces — Asie / Capturas por países o áreas y especies — Asia

English name Nom anglais Nombre inglés	Scientific name Nom scientifique Nombre científico	1996 t	1997 t	1998 t	1999 t	2000 t	2001 t	2002 t
Pacific sandlance	*Ammodytes personatus*	-	-	-	-	16 293	4 803	5 145
Gobies nei	*Gobiidae*	1 598	1 722	1 299	897	788	703	388
Okhotsk atka mackerel	*Pleurogrammus azonus*	4 103	2 983	7 911	1 005	2 554	1 261	627
Bartail flathead	*Platycephalus indicus*	2 900	4 196	2 857	2 248	2 310	1 699	1 199
Purple puffer	*Takifugu vermicularis*	9 708	7 471	3 897	4 787	-	-	3 110
Puffers nei	*Tetraodontidae*	-	-	-	-	2 978	3 735	-
Filefishes nei	*Cantherhines(=Navodon) spp*	-	-	-	-	2 891	1 578	-
Threadsail filefish	*Stephanolepis cirrhifer*	1 772	16 318	9 364	2 999	-	-	933
Triggerfishes, durgons nei	*Balistidae*	306	-	-	16	-	-	-
Daggertooth pike conger	*Muraenesox cinereus*	1 411	2 518	1 506	190	1 862	1 080	833
Whitespotted conger	*Conger myriaster*	17 314	19 136	11 913	10 160	8 304	7 676	17 210
Conger eels, etc. nei	*Congridae*	69	63	87	159	122	78	6
Pink cusk-eel	*Genypterus blacodes*	920	1 378	1 242	1 894	1 684	1 604	1 864
Cusk-eels nei	*Genypterus spp*	-	-	-	-	57	-	-
Bearded brotula	*Brotula barbata*	-	-	-	-	16	-	-
Cusk-eels, brotulas nei	*Ophidiidae*	-	-	-	-	1	-	-
Alfonsinos nei	*Beryx spp*	-	-	77	-	4	-	-
Orange roughy	*Hoplostethus atlanticus*	-	-	-	230	-	47	-
John dory	*Zeus faber*	105	298	36	540	636	225	-
Tilefishes nei	*Branchiostegidae*	-	-	-	1 651	1 664	1 049	1 341
Patagonian toothfish	*Dissostichus eleginoides*	432	1 040	1 051	1 241	1 704	1 473	300
Antarctic toothfishes nei	*Dissostichus spp*	-	-	-	-	-	572	-
Blackfin icefish	*Chaenocephalus aceratus*	-	-	-	-	-	-	1
Mackerel icefish	*Champsocephalus gunnari*	-	-	-	-	-	-	602
South Georgia icefish	*Pseudochaenichthys georgianus*	-	-	-	-	-	-	1
Japanese sandfish	*Arctoscopus japonicus*	2 501	2 194	1 490	2 449	1 571	1 286	3 381
Largehead hairtail	*Trichiurus lepturus*	74 461	67 170	74 851	64 445	81 050	79 898	60 172
Hairtails, scabbardfishes nei	*Trichiuridae*	8 392	11 572	9 585	12 866	12 135	3 503	3 857
Ruffs, barrelfishes nei	*Centrolophidae*	-	-	-	493	317	209	636
Scorpionfishes nei	*Scorpaenidae*	5 070	5 725	4 705	5 036	5 098	5 354	2 097
Bluefin gurnard	*Chelidonichthys kumu*	75	86	146	43	79	140	260
Anglerfishes nei	*Lophiidae*	11 720	5 045	3 970	3 406	4 930	5 813	9 500
Pacific herring	*Clupea pallasii*	5 525	13 214	13 340	21 446	15 203	8 491	1 941
Japanese sardinella	*Sardinella zunasi*	10 663	5 593	1 973	6 674	4 603	766	796
Japanese pilchard	*Sardinops melanostictus*	18 560	9 041	7 595	17 142	2 207	129	8
Japanese anchovy	*Engraulis japonicus*	237 128	230 911	249 519	238 463	201 192	273 927	236 315
Clupeoids nei	*Clupeoidei*	-	-	-	1	-	-	-
Japanese Spanish mackerel	*Scomberomorus niphonius*	6 419	11 173	22 809	19 502	25 641	25 513	25 956
Seerfishes nei	*Scomberomorus spp*	1 830	3 112	2 935	1 242	1 609	2 097	1 337
Skipjack tuna	*Katsuwonus pelamis*	129 888	115 927	143 390	109 780	137 015	137 569	173 693
Atlantic bluefin tuna	*Thunnus thynnus*	683	613	-	-	-	-	-
Albacore	*Thunnus alalunga*	716	1 944	3 998	1 179	684	1 920	2 488
Southern bluefin tuna	*Thunnus maccoyii*	911	1 228	1 562	1 264	958	729	649
Yellowfin tuna	*Thunnus albacares*	35 267	61 244	73 354	43 255	49 461	58 957	48 778
Bigeye tuna	*Thunnus obesus*	28 418	31 191	33 042	26 558	30 079	31 335	31 962
Indo-Pacific sailfish	*Istiophorus platypterus*	250	1 302	389	496	185	33	128
Blue marlin	*Makaira nigricans*	24	405	805	300	301	350	396
Black marlin	*Makaira indica*	317	235	427	813	776	1 291	1 150
Striped marlin	*Tetrapturus audax*	451	1 172	795	589	637	349	275
Marlins,sailfishes,etc. nei	*Istiophoridae*	5 356	3 234	3 849	1 413	1 327	1 657	1 819
Swordfish	*Xiphias gladius*	129	967	573	474	665	1 031	1 151
Tuna-like fishes nei	*Scombroidei*	6 869	8 193	8 473	6 469	6 952	7 512	5 027
Pacific saury	*Cololabis saira*	28 368	68 853	18 531	29 538	44 340	26 205	27 187
Japanese halfbeak	*Hyporhamphus sajori*	990	1 193	1 160	913	956	613	670
Japanese jack mackerel	*Trachurus japonicus*	14 542	22 766	22 133	13 552	19 510	17 537	26 037
Jack and horse mackerels nei	*Trachurus spp*	1 202	2 133	1 983	2 182	259	313	14
Jacks, crevalles nei	*Caranx spp*	46	-	11	-	-	-	-
Amberjacks nei	*Seriola spp*	4 127	6 119	9 625	8 654	4 814	6 475	6 191
Chub mackerel	*Scomber japonicus*	415 003	160 479	172 925	177 609	145 945	203 743	142 068
Blue mackerel	*Scomber australasicus*	1	-	5	-	-	-	-
Silver pomfret	*Pampus argenteus*	4 806	919	1 029	3 374	-	-	2 247
Butterfishes, pomfrets nei	*Stromateidae*	9 484	10 770	13 249	15 235	7 838	6 819	6 191
Barracudas nei	*Sphyraena spp*	9	80	1	349	432	512	585
Rays, stingrays, mantas nei	*Rajiformes*	14 687	14 405	8 178	13 979	13 033	9 121	9 775
Sharks, rays, skates, etc. nei	*Elasmobranchii*	911	1 495	2 127	2 419	2 361	2 010	2 186
Marine fishes nei	*Osteichthyes*	183 016	165 430	140 162	109 914	106 683	128 975	105 062
Freshwater crustaceans nei	*Crustacea*	248	-	2	111	93	48	61
Gazami crab	*Portunus trituberculatus*	15 754	11 430	13 813	11 819	12 842	13 016	18 659
Tanner crabs nei	*Chionoecetes spp*	-	-	-	-	17 037	13 974	896
Marine crabs nei	*Brachyura*	61 513	58 855	43 999	35 439	9 564	9 031	6 758
Longlegged spiny lobster	*Panulirus longipes*	-	-	-	-	461	415	391
Antarctic stone crab	*Paralomis spinosissima*	0	0	-	-	-	-	-
King crabs, stone crabs nei	*Lithodidae*	-	0	-	-	-	-	-
Kuruma prawn	*Penaeus japonicus*	1 783	2 102	1 138	502	578	513	217
Fleshy prawn	*Penaeus chinensis*	1 242	1 911	1 245	814	1 211	581	222
Penaeus shrimps nei	*Penaeus spp*	560	143	136	502	-	-	-
Shiba shrimp	*Metapenaeus joyneri*	2 211	1 976	3 651	3 633	2 621	2 385	1 736
Southern rough shrimp	*Trachypenaeus curvirostris*	-	-	-	-	-	3 145	-
Akiami paste shrimp	*Acetes japonicus*	18 411	15 624	15 993	19 389	13 985	11 705	7 013
Natantian decapods nei	*Natantia*	16 296	17 768	24 588	18 698	17 640	12 471	20 446
Antarctic krill	*Euphausia superba*	-	-	2 850	27	7 233	7 525	14 353
Marine crustaceans nei	*Crustacea*	53	265	1 916	1 658	1 144	5 958	1 354
Japanese corbicula	*Corbicula japonica*	600	387	677	1 006	-	-	-
Freshwater molluscs nei	*Mollusca*	200	290	409	255	645	669	636
Abalones nei	*Haliotis spp*	188	214	71	79	113	104	75
Horned turban	*Turbo cornutus*	7 361	6 878	9 192	7 397	7 281	6 274	7 193
Gastropods nei	*Gastropoda*	2 975	1 617	1 325	1 670	818	730	1 555
Pacific cupped oyster	*Crassostrea gigas*	18 259	17 210	9 905	11 609	15 939	10 056	16 678
Korean mussel	*Mytilus coruscus*	2 191	3 211	1 469	1 414	1 133	1 085	946
Sea mussels nei	*Mytilidae*	837	4 098	6 456	6 771	5 795	3 828	3 500

English name Nom anglais Nombre inglés	Scientific name Nom scientifique Nombre científico	1996 t	1997 t	1998 t	1999 t	2000 t	2001 t	2002 t
Yesso scallop	*Patinopecten yessoensis*	122	196	42	6	0	11	-
Ark clams nei	*Arca spp*	782	667	1 311	534	318	153	228
Blood cockle	*Anadara granosa*	995	493	12 114	6 503	4 184	923	790
Japanese hard clam	*Meretrix lusoria*	2 315	2 075	3 472	1 799	1 430	1 044	1 704
Japanese carpet shell	*Ruditapes philippinarum*	12 392	16 854	14 585	13 963	20 982	20 004	14 758
Imperial surf clam	*Pseudocardium sybillae*	6 053	7 179	2 877	7 620	4 150	3 409	3 464
Cockles nei	*Cardiidae*	2 583	1 587	1 078	112	411	2 002	810
Clams, etc. nei	*Bivalvia*	33 446	8 851	15 665	14 728	12 277	15 705	11 453
Cuttlefish,bobtail squids nei	*Sepiidae, Sepiolidae*	2 832	4 946	4 457	8 309	2 107	2 966	3 019
Common squids nei	*Loligo spp*	-	-	-	1 629	885	898	-
Argentine shortfin squid	*Illex argentinus*	144 750	208 160	92 397	271 716	150 149	142 585	98 649
Jumbo flying squid	*Dosidicus gigas*	11 784	2 384	201	18 813	15 625	5 797	21 382
Japanese flying squid	*Todarodes pacificus*	252 618	224 959	163 016	249 280	226 309	225 616	226 656
Sevenstar flying squid	*Martialia hyadesi*	52	81	-	-	-	2	-
Various squids nei	*Loliginidae, Ommastrephidae*	15 609	18 399	23 333	18 666	11 999	14 710	21 321
Octopuses, etc. nei	*Octopodidae*	21 707	23 084	27 174	22 674	19 605	19 965	21 678
Marine molluscs nei	*Mollusca*	1 031	1 017	2 308	683	847	1 270	831
Sea squirts nei	*Ascidiacea*	16 747	2 780	891	1 171	1 443	1 053	1 025
Echinoderms	*Echinodermata*	2 802	2 771	1 410	1 182	1 461	1 454	1 459
Japanese sea cucumber	*Stichopus japonicus*	1 979	2 217	1 439	1 204	1 419	900	833
Aquatic invertebrates nei	*Invertebrata*	873	1 094	1 047	1 003	1 883	1 302	1 936
	Country total	*2 413 785*	*2 204 202*	*2 027 932*	*2 118 521*	*1 824 995*	*1 990 722*	*1 668 979*
Kuwait								
Freshwater fishes nei	*Osteichthyes*	0	0	0	0	0	0	0
Hilsa shad	*Tenualosa ilisha*	1 148	1 034	919	970	650 F	337	340 F
Brushtooth lizardfish	*Saurida undosquamis*	47	53	34	20	30 F	32	32 F
Flathead grey mullet	*Mugil cephalus*	39	65	69	89	60 F	34	35 F
Mullets nei	*Mugilidae*	540	628	1 068	965	700 F	422	430 F
Groupers nei	*Epinephelus spp*	287	241	264	237	250 F	268	265 F
Snappers nei	*Lutjanus spp*	84	78	64	51	40 F	27	30 F
Threadfin breams nei	*Nemipterus spp*	10	17	49	19	20 F	24	24 F
Grunts, sweetlips nei	*Haemulidae (=Pomadasyidae)*	180	163	174	245	210 F	191	200 F
Croakers, drums nei	*Sciaenidae*	974	1 127	1 211	1 385	1 100 F	853	860 F
Emperors(=Scavengers) nei	*Lethrinidae*	15	20	38	35	50 F	78	75 F
Yellowfin seabream	*Acanthopagrus latus*	234	249	280	464	350 F	271	280 F
Narrow-barred Spanish mackerel	*Scomberomorus commerson*	85	73	124	127	130 F	135	135 F
Indo-Pacific king mackerel	*Scomberomorus guttatus*	172	206	166	211	210 F	204	204 F
Cobia	*Rachycentron canadum*	...	...	...	...	...	38	40 F
Carangids nei	*Carangidae*	97	138	74	73	150 F	242	240 F
Silver pomfret	*Pampus argenteus*	862	560	501	259	200 F	133	140 F
Butterfishes, pomfrets nei	*Stromateidae*	230	186	111	170	-	-	-
Marine fishes nei	*Osteichthyes*	893	922	1 055	231	550 F	580	590 F
Natantian decapods nei	*Natantia*	2 358	2 066	1 598	720	1 300 F	1 977	1 980 F
	Country total	*8 255*	*7 826*	*7 799*	*6 271*	*6 000 F*	*5 846*	*5 900 F*
Kyrgyzstan								
Freshwater bream	*Abramis brama*	9 F	7 F	5 F	3 F	3 F	4	3
Common carp	*Cyprinus carpio*	31 F	23 F	15 F	9 F	10 F	11	9
Goldfish	*Carassius auratus*	2 F	1 F	1 F	1 F	1 F	2	1
Silver carp	*Hypophthalmichthys molitrix*	24 F	18 F	12 F	7 F	8 F	19	17
Cyprinids nei	*Cyprinidae*	35 F	26 F	18 F	11 F	12 F	8	7
Pike-perch	*Stizostedion lucioperca*	6 F	5 F	3 F	2 F	2 F	1	2
Whitefishes nei	*Coregonus spp*	53 F	40 F	26 F	15 F	16 F	12	9
	Country total	*160 F*	*120 F*	*80 F*	*48*	*52*	*57*	*48*
Laos								
Cyprinids nei	*Cyprinidae*	3 500 F	2 800 F	3 000 F	4 500 F	4 400 F	4 650 F	5 000 F
Freshwater fishes nei	*Osteichthyes*	19 500 F	16 057 F	16 642 F	25 541 F	24 850 F	26 350 F	28 440 F
	Country total	*23 000 F*	*18 857*	*19 642*	*30 041*	*29 250*	*31 000 F*	*33 440*
Lebanon								
Cyprinids nei	*Cyprinidae*	10	10	10	10	10	10	25
Freshwater fishes nei	*Osteichthyes*	10	10	10	10	10	10	272
Flatfishes nei	*Pleuronectiformes*	5	5	10	5	11	15	8
Gadiformes nei	*Gadiformes*	30	25	30	30	32	30	28
Mullets nei	*Mugilidae*	400	300	300	300	...	400	370
Groupers, seabasses nei	*Serranidae*	250	150	250	250	230	240	250
Porgies, seabreams nei	*Sparidae*	450	350	400	450	450	400	370
Picarels nei	*Spicara spp*	100	100	100	100	100	95	93
Surmullets(=Red mullets) nei	*Mullus spp*	200	150	200	200	250	200	200
European conger	*Conger conger*	5	5	10	5	8	10	9
Scorpionfishes nei	*Scorpaenidae*	100	100	100	150	150	100	125
Clupeoids nei	*Clupeoidei*	800	600	600	500	700	500	650
Tuna-like fishes nei	*Scombroidei*	500	700	400	400	500	450	400
Silversides(=Sand smelts) nei	*Atherinidae*	25	50	50	50	50	50	50
Carangids nei	*Carangidae*	450	350	400	350	450	400	400
Mackerels nei	*Scombridae*	500	300	300	300	350	350	350
Barracudas nei	*Sphyraena spp*	200	200	200	200	200	250	200
Sharks, rays, skates, etc. nei	*Elasmobranchii*	50	50	50	50	60	55	60
Marine crustaceans nei	*Crustacea*	25	150	50	125	55	55	60
Cuttlefish,bobtail squids nei	*Sepiidae, Sepiolidae*	25	50	25	50	25	25	25
Common octopus	*Octopus vulgaris*	...	...	25	25	25	25	25
	Country total	*4 135*	*3 655*	*3 520*	*3 560*	*3 666*	*3 670*	*3 970*

E-4

Fish crustaceans, molluscs, etc	Capture production by countries or areas and species	Asia
Poissons, crustacés, mollusques, etc	Captures par pays ou zones et espèces	Asie
Peces, crustáceos, moluscos, etc	Capturas por países o áreas y especies	Asia

English name Nom anglais Nombre inglés	Scientific name Nom scientifique Nombre científico	1996 t	1997 t	1998 t	1999 t	2000 t	2001 t	2002 t
Malaysia								
Freshwater fishes nei	*Osteichthyes*	3 683	3 949	4 626	3 366	3 549	3 446	3 185
Chacunda gizzard shad	*Anodontostoma chacunda*	3 991	3 286	2 659	4 373	4 016	4 037	3 632
Indian pellona	*Pellona ditchela*	15 285	13 374	9 921	9 188	12 705	12 664	12 058
Diadromous clupeoids nei	*Clupeoidei*	1 385	676	581	622	665	1 106	1 231
Barramundi(=Giant seaperch)	*Lates calcarifer*	1 322	874	1 880	1 397	1 701	1 518	1 440
Tonguefishes	*Cynoglossidae*	2 740	3 029	2 940	3 306	3 076	2 417	2 592
Flatfishes nei	*Pleuronectiformes*	2 233	2 193	2 557	3 026	1 641	1 712	1 592
Indo-Pacific tarpon	*Megalops cyprinoides*	99	514	338	226	133	104	62
Lizardfishes nei	*Synodontidae*	12 049	14 490	14 424	15 050	17 567	17 624	19 204
Sea catfishes nei	*Ariidae*	15 191	13 599	15 182	15 341	14 805	14 076	11 106
Eeltail catfishes	*Plotosus spp*	1 803	1 447	1 659	1 749	2 222	1 987	2 118
Mullets nei	*Mugilidae*	5 330	4 550	5 627	4 862	5 349	4 003	3 243
Fusiliers	*Caesio spp*	1 681	1 959	1 317	1 324	980	987	847
Groupers nei	*Epinephelus spp*	8 274	9 124	10 601	12 229	12 174	10 476	8 837
Sillago-whitings	*Sillaginidae*	1 895	1 934	1 809	2 014	1 812	2 057	1 607
Mangrove red snapper	*Lutjanus argentimaculatus*	11 486	10 848	6 162	12 934	11 439	11 044	11 262
Snappers nei	*Lutjanus spp*	3 247	3 210	4 486	4 666	4 270	4 332	3 931
Snappers, jobfishes nei	*Lutjanidae*	5 338	4 694	6 085	7 787	5 229	4 771	4 043
Threadfin breams nei	*Nemipterus spp*	29 534	30 102	40 327	39 694	32 510	28 910	30 519
Monocle breams	*Scolopsis spp*	1 244	950	829	1 036	1 701	1 642	1 442
Ponyfishes(=Slipmouths)	*Leiognathus spp*	2 539	2 362	3 090	3 049	2 461	2 285	2 340
Silver grunt	*Pomadasys argenteus*	1 644	1 156	2 844	2 235	1 500	1 593	1 469
Grunts, sweetlips nei	*Haemulidae (=Pomadasyidae)*	1 113	1 074	1 078	1 557	1 507	1 417	1 074
Croakers, drums nei	*Sciaenidae*	20 350	20 368	22 470	22 188	23 439	28 760	22 337
Goatfishes	*Upeneus spp*	8 107	10 302	7 650	8 606	10 845	12 960	13 532
Spotted sicklefish	*Drepane punctata*	875	583	516	843	678	689	789
Wrasses, hogfishes, etc. nei	*Labridae*	226	2 302	2 697	2 402	2 139	1 948	293
Threadfins, tasselfishes nei	*Polynemidae*	5 098	3 701	4 943	3 911	4 315	4 328	2 864
Spinefeet(=Rabbitfishes) nei	*Siganus spp*	1 152	1 335	1 464	1 208	1 393	1 494	1 732
Triggerfishes, durgons nei	*Balistidae*	1 606	1 941	1 793	2 826	1 428	2 195	1 926
Daggertooth pike conger	*Muraenesox cinereus*	4 533	6 844	7 335	5 830	5 418	5 019	5 373
Largehead hairtail	*Trichiurus lepturus*	6 368	11 273	23 951	18 009	11 574	9 711	5 423
Stolephorus anchovies	*Stolephorus spp*	24 361	23 772	25 651	23 045	22 516	17 723	23 683
Wolf-herrings nei	*Chirocentrus spp*	4 013	4 223	4 143	4 733	4 184	3 989	3 823
Clupeoids nei	*Clupeoidei*	44 525	38 110	46 315	45 517	33 613	40 750	40 611
Seerfishes nei	*Scomberomorus spp*	14 400	13 734	16 278	17 247	15 323	14 906	13 902
Kawakawa	*Euthynnus affinis*	33 944	48 498	49 409	57 281	58 132	53 217	59 383
Longtail tuna	*Thunnus tonggol*	2 045	2 125	3 106	2 824	3 586	2 773	4 753
Yellowfin tuna	*Thunnus albacares*	-	-	-	-	-	-	202
Bigeye tuna	*Thunnus obesus*	-	-	-	-	-	-	145
Marlins,sailfishes,etc. nei	*Istiophoridae*	274	362	324	2 046	161	155	219
Swordfish	*Xiphias gladius*	-	-	-	-	-	-	4
False trevally	*Lactarius lactarius*	387	561	575	681	422	439	363
Cobia	*Rachycentron canadum*	368	385	523	756	640	557	519
Indian scad	*Decapterus russelli*	59 733	71 184	53 426	70 160	84 203	77 394	90 301
Jacks, crevalles nei	*Caranx spp*	26 369	29 880	11 553	36 644	39 614	37 125	38 269
Black pomfret	*Parastromateus niger*	4 650	4 717	5 002	5 627	4 546	3 811	4 384
Rainbow runner	*Elagatis bipinnulata*	390	1 581	390	863	565	195	60
Torpedo scad	*Megalaspis cordyla*	14 912	13 662	18 785	19 895	17 332	13 952	22 505
Queenfishes	*Scomberoides spp*	3 240	2 872	3 253	3 811	3 482	3 527	2 696
Yellowstripe scad	*Selaroides leptolepis*	28 123	32 031	32 734	41 341	44 253	39 858	41 006
Indian mackerels nei	*Rastrelliger spp*	95 364	86 801	102 072	111 365	98 055	99 469	87 910
Silver pomfret	*Pampus argenteus*	5 020	5 093	5 401	4 371	3 531	4 114	3 700
Butterfishes, pomfrets nei	*Stromateidae*	460	467	495	570	461	377	488
Barracudas nei	*Sphyraena spp*	4 933	6 654	7 154	8 103	7 447	6 611	5 963
Rays, stingrays, mantas nei	*Rajiformes*	15 928	17 282	16 104	17 033	16 573	16 532	15 941
Sharks, rays, skates, etc. nei	*Elasmobranchii*	8 079	7 483	7 839	8 092	7 948	8 677	8 226
Marine fishes nei	*Osteichthyes*	352 837	331 727	382 806	374 189	411 228	413 420	446 300
Freshwater prawns, shrimps nei	*Palaemonidae*	-	-	-	-	-	-	265
Marine crabs nei	*Brachyura*	9 002	9 731	14 243	14 430	12 639	12 298	11 235
Tropical spiny lobsters nei	*Panulirus spp*	814	836	1 037	1 094	1 103	1 612	2 059
Sergestid shrimps nei	*Sergestidae*	20 827	15 240	11 336	16 480	10 448	8 367	8 560
Natantian decapods nei	*Natantia*	79 410	76 205	35 895	73 994	85 528	69 101	67 460
Clams, etc. nei	*Bivalvia*	18 606	25 500	6 001	1 677	6 231	3 949	3 254
Cuttlefish,bobtail squids nei	*Sepiidae, Sepiolidae*	18 776	20 591	26 182	22 068	26 314	21 467	23 601
Various squids nei	*Loliginidae, Ommastrephidae*	36 270	38 491	38 697	40 283	54 339	45 277	52 483
Octopuses, etc. nei	*Octopodidae*	989	1 270	1 347	1 512	1 551	1 480	1 531
Jellyfishes	*Rhopilema spp*	19 902	53 811	11 802	7 182	9 036	10 299	6 648
	Country total	*1 130 372*	*1 172 922*	*1 153 719*	*1 251 768*	*1 289 245*	*1 234 733*	*1 275 555*
Maldives								
Freshwater fishes nei	*Osteichthyes*	0	0	0	0	0	0	0
Dogtooth tuna	*Gymnosarda unicolor*	625	489	470	426	451	647	789
Frigate and bullet tunas	*Auxis thazard, A.rochei*	6 484	2 489	4 218	3 401	3 991	3 981	4 187
Kawakawa	*Euthynnus affinis*	3 789	2 089	3 624	1 692	1 898	2 150	2 242
Skipjack tuna	*Katsuwonus pelamis*	66 502	69 015	78 410	92 888	79 683	88 043	115 321
Yellowfin tuna	*Thunnus albacares*	11 811	12 489	13 566	13 664	11 713	15 088	20 771
Bigeye tuna	*Thunnus obesus*	630	540	606	604	472	536	958
Sharks, rays, skates, etc. nei	*Elasmobranchii*	11 856	10 643	10 887	6 883	13 523	11 935	11 498
Marine fishes nei	*Osteichthyes*	18 526	17 914	16 788	13 450	19 618	3 571	5 024
Tropical spiny lobsters nei	*Panulirus spp*	...	...	...	...	7	13	...
Marine molluscs nei	*Mollusca*	140	271	314	485	866	-	-
Sea cucumbers nei	*Holothurioidea*	145	318	85	54	205	226	191
	Country total	*120 508*	*116 257*	*128 968*	*133 547*	*132 427*	*126 190*	*160 981*
Mongolia								
Freshwater fishes nei	*Osteichthyes*	221	180	311	524	425	117	129

English name Nom anglais Nombre inglés	Scientific name Nom scientifique Nombre científico	1996 t	1997 t	1998 t	1999 t	2000 t	2001 t	2002 t
	Country total	*221*	*180*	*311*	*524*	*425*	*117*	*129*
Myanmar								
Freshwater fishes nei	*Osteichthyes*	146 494	149 069	149 279	159 746	189 708	235 376	304 529
Marine fishes nei	*Osteichthyes*	435 868	607 814	656 406	731 664	849 018	900 492	975 113
Natantian decapods nei	*Natantia*	16 000 F	22 000 F	24 000 F	27 000 F	30 000 F	30 000 F	32 000 F
Jellyfishes	*Rhopilema spp*	3 426	1 412	432	1 000 F	1 000 F	1 000 F	1 000 F
	Country total	*601 788*	*780 295*	*830 117*	*919 410*	*1 069 726*	*1 166 868*	*1 312 642*
Nepal								
Freshwater fishes nei	*Osteichthyes*	11 230	11 230	12 000	12 752	16 700	16 700	17 900
	Country total	*11 230*	*11 230*	*12 000*	*12 752*	*16 700*	*16 700*	*17 900*
Oman								
Freshwater fishes nei	*Osteichthyes*	0	0	0	0	0	0	0
Sea catfishes nei	*Ariidae*	372	679	1 024	1 161	1 306	2 238	1 163
Mullets nei	*Mugilidae*	79	139	176	158	123	543	424
Groupers nei	*Epinephelus spp*	3 409	3 366	5 345	4 829	5 013	3 799	3 292
Snappers, jobfishes nei	*Lutjanidae*	426	657	474	597	669	380	337
Threadfin and dwarf breams nei	*Nemipteridae*	233	2 068	466	1 166	317	1 206	2 865
Grunts, sweetlips nei	*Haemulidae (=Pomadasyidae)*	803	1 111	713	627	1 747	541	954
Croakers, drums nei	*Sciaenidae*	3 709	5 468	2 218	2 121	1 926	1 873	4 152
Emperors(=Scavengers) nei	*Lethrinidae*	4 087	5 767	6 630	6 954	7 664	6 526	7 179
Porgies, seabreams nei	*Sparidae*	4 692	4 322	4 016	6 098	4 419	3 976	8 776
Spinefeet(=Rabbitfishes) nei	*Siganus spp*	59	259	122	97	131	363	559
Hairtails, scabbardfishes nei	*Trichiuridae*	8 132	10 384	4 767	1 776	4 367	2 617	6 600
Demersal percomorphs nei	*Perciformes*	7 529	4 554	6 938	8 760	5 735	2 421	11 273
Indian oil sardine	*Sardinella longiceps*	26 741	16 765	20 650	21 710	40 044	58 960	37 931
Anchovies, etc. nei	*Engraulidae*	1 017	1 189	941	485	5 126	970	2 571
Striped bonito	*Sarda orientalis*	370	498	162	134	95	287	378
Narrow-barred Spanish mackerel	*Scomberomorus commerson*	5 243	5 944	3 145	3 390	2 559	2 785	2 088
Frigate and bullet tunas	*Auxis thazard, A.rochei*	613	846	611	583	488	638	170
Kawakawa	*Euthynnus affinis*	2 335	2 388	1 731	1 522	1 550	1 961	1 493
Skipjack tuna	*Katsuwonus pelamis*	408	730	227	320	293	1	2
Longtail tuna	*Thunnus tonggol*	5 316	5 020	4 379	4 798	5 318	6 011	6 854
Yellowfin tuna	*Thunnus albacares*	20 718	15 905	14 897	7 377	8 377	7 945	7 114
Indo-Pacific sailfish	*Istiophorus platypterus*	581	1 261	591	399	448	218	147
Tuna-like fishes nei	*Scombroidei*	184	366	124	99	521	188	7
Needlefishes nei	*Tylosurus spp*	201	252	116	209	131	130	541
Cobia	*Rachycentron canadum*	103	180	115	124	100	54	202
Jacks, crevalles nei	*Caranx spp*	2 927	3 896	2 957	2 552	1 806	1 218	1 666
Queenfishes	*Scomberoides spp*	823	703	528	693	408	528	420
Carangids nei	*Carangidae*	2 708	3 216	2 323	2 556	2 485	1 302	7 839
Indian mackerel	*Rastrelliger kanagurta*	1 712	2 207	1 994	2 024	2 426	3 223	3 459
Barracudas nei	*Sphyraena spp*	1 948	2 023	1 789	2 347	1 431	1 781	3 215
Pelagic percomorphs nei	*Perciformes*	2 045	2 949	6 055	10 790	5 353	3 391	5 468
Rays, stingrays, mantas nei	*Rajiformes*	372	359	189	289	240	198	199
Sharks, rays, skates, etc. nei	*Elasmobranchii*	5 870	6 342	4 805	4 020	3 651	3 632	3 803
Marine fishes nei	*Osteichthyes*	101	355	432	1	384	1 273	502
Tropical spiny lobsters nei	*Panulirus spp*	397	263	336	180	402	379	449
Natantian decapods nei	*Natantia*	276	376	65	356	432	627	467
Abalones nei	*Haliotis spp*	43	40	40	29	45	51	50
Cuttlefish,bobtail squids nei	*Sepiidae, Sepiolidae*	5 036	6 148	4 080	7 478	2 891	5 673	8 061
	Country total	*121 618*	*118 995*	*106 171*	*108 809*	*120 421*	*129 907*	*142 670*
Pakistan								
Freshwater fishes nei	*Osteichthyes*	142 092	167 530	163 524	179 865	176 468	180 100	181 000
Hilsa shad	*Tenualosa ilisha*	562	597	611	502	190	170	174
Barramundi(=Giant seaperch)	*Lates calcarifer*	214	209	196	204	-	-	-
Tonguefishes	*Cynoglossidae*	2 205	2 390	2 149	2 037	2 124	1 915	1 980
Bombay-duck	*Harpadon nehereus*	101	95	91	72	65	55	64
Greater lizardfish	*Saurida tumbil*	45	28	22	-	-	-	-
Sea catfishes nei	*Ariidae*	49 428	54 437	55 934	51 665	39 168	38 215	38 500
Mullets nei	*Mugilidae*	17 631	18 935	17 580	12 336	9 618	9 723	9 900
Groupers nei	*Epinephelus spp*	9 793	10 474	13 991	17 355	16 012	15 928	16 000
Sillago-whitings	*Sillaginidae*	289	266	218	201	194	204	210
Mangrove red snapper	*Lutjanus argentimaculatus*	2 002	2 394	3 192	3 195	3 003	2 900	3 000
Threadfin breams nei	*Nemipterus spp*	825	884	3 192	7 166	8 940	8 466	8 600
Grunts, sweetlips nei	*Haemulidae (=Pomadasyidae)*	5 268	6 010	6 221	8 147	9 961	9 752	9 850
Croakers, drums nei	*Sciaenidae*	19 934	20 428	19 625	24 665	21 976	21 725	22 886
Emperors(=Scavengers) nei	*Lethrinidae*	1 549	1 911	2 334	3 323	5 173	5 044	5 100
Porgies, seabreams nei	*Sparidae*	3 097	3 058	1 255	4 220	4 510	4 411	4 430
Fourfinger threadfin	*Eleutheronema tetradactylum*	516	1 783	969	...	63	55	60
Pike-congers nei	*Muraenesox spp*	4 904	5 637	5 627	8 377	5 937	5 834	5 940
Largehead hairtail	*Trichiurus lepturus*	9 073	11 583	12 337	31 623	28 754	27 355	28 440
Indian oil sardine	*Sardinella longiceps*	52 290	51 930	44 079	30 629	31 167	31 201	31 600
Anchovies, etc. nei	*Engraulidae*	14 091	16 113	13 165	15 154	15 191	15 001	15 400
Dorab wolf-herring	*Chirocentrus dorab*	1 580	1 931	2 051	2 266	2 775	2 604	2 720
Clupeoids nei	*Clupeoidei*	27 576	26 650	25 487	26 934	24 810	24 306	24 500
Narrow-barred Spanish mackerel	*Scomberomorus commerson*	10 108	12 009	12 232	11 734	9 366	8 455	7 922
Frigate and bullet tunas	*Auxis thazard, A.rochei*	49	54	56	59	42	52	31
Kawakawa	*Euthynnus affinis*	2 351	2 571	2 684	2 715	2 340	1 817	1 210
Skipjack tuna	*Katsuwonus pelamis*	4 140	4 480	4 372	4 505	4 308	3 405	3 102
Longtail tuna	*Thunnus tonggol*	4 121	5 360	5 220	5 600	5 315	4 735	4 466
Yellowfin tuna	*Thunnus albacares*	5 250	3 838	3 795	8 884	4 946	3 603	2 940
Indo-Pacific sailfish	*Istiophorus platypterus*	980	41	45	46	...	...	...

E-4 Fish crustaceans, molluscs, etc — Capture production by countries or areas and species — Asia
Poissons, crustacés, mollusques, etc — Captures par pays ou zones et espèces — Asie
Peces, crustáceos, moluscos, etc — Capturas por países o áreas y especies — Asia

English name Nom anglais Nombre inglés	Scientific name Nom scientifique Nombre científico	1996 t	1997 t	1998 t	1999 t	2000 t	2001 t	2002 t
Marlins,sailfishes,etc. nei	*Istiophoridae*	2 834	2 198	2 264	2 340	2 215	1 122	995
Tuna-like fishes nei	*Scombroidei*	1 990	1 500	1 592	4 610	4 240	2 775	2 480
False trevally	*Lactarius lactarius*	2	4	5	-	-	-	-
Cobia	*Rachycentron canadum*	1 574	1 449	1 254	1 136	2 896	2 797	2 808
Scads nei	*Decapterus spp*	1 010	1 225	3 505	4 661	4 600	4 355	4 400
Jacks, crevalles nei	*Caranx spp*	3 972	5 391	6 523	8 407	9 111	8 928	9 200
Black pomfret	*Parastromateus niger*	2 221	2 322	2 109	2 917	2 027	1 975	2 002
Torpedo scad	*Megalaspis cordyla*	3 000	2 100	1 100	1 450	2 017	1 825	1 950
Carangids nei	*Carangidae*	15 957	19 002	18 689	17 779	16 545	15 988	16 400
Common dolphinfish	*Coryphaena hippurus*	1 841	1 658	1 892	3 109	1 954	1 869	1 875
Butterfishes, pomfrets nei	*Stromateidae*	2 799	3 786	4 089	4 605	3 945	3 454	3 500
Barracudas nei	*Sphyraena spp*	2 878	2 683	2 664	3 520	3 981	3 889	3 900
Requiem sharks nei	*Carcharhinidae*	34 447	31 179	35 357	32 535	28 245	26 524	27 000
Guitarfishes, etc. nei	*Rhinobatidae*	1 422	1 481	1 564	1 643	2 185	1 944	2 004
Rays, stingrays, mantas nei	*Rajiformes*	15 563	15 769	17 576	20 780	20 740	20 801	20 900
Marine fishes nei	*Osteichthyes*	16 311	21 042	35 352	39 473	36 451	35 450	31 769
Marine crabs nei	*Brachyura*	3 200	3 989	5 680	5 109	5 187	5 099	6 060
Tropical spiny lobsters nei	*Panulirus spp*	724	765	782	1 077	807	756	802
Giant tiger prawn	*Penaeus monodon*	141	140	122	138	139	140	155
Penaeus shrimps nei	*Penaeus spp*	5 982	5 975	5 189	5 874	5 920	5 974	5 600
Metapenaeus shrimps nei	*Metapenaeus spp*	7 602	6 801	6 204	6 791	7 126	7 246	6 555
Parapenaeopsis shrimps nei	*Parapenaeopsis spp*	14 047	16 722	14 689	12 889	11 945	11 576	10 222
Cuttlefish,bobtail squids nei	*Sepiidae, Sepiolidae*	3 308	4 528	3 225	5 146	5 307	5 256	5 302
Various squids nei	*Loliginidae, Ommastrephidae*	2 600	4 460	3 300	5 062	4 070	4 024	3 200
	Country total	*537 489*	*589 795*	*596 980*	*654 530*	*614 069*	*600 798*	*599 104*
Palest, O.T.								
Freshwater fishes nei	*Osteichthyes*	0	0	0	0	0	0	0
Flatfishes nei	*Pleuronectiformes*	7	15	25	25 F	20 F	10 F	6
Lizardfishes nei	*Synodontidae*	16	32	32	30 F	20 F	15 F	9
Mullets nei	*Mugilidae*	32	19	35	35 F	30 F	20 F	17
Groupers nei	*Epinephelus spp*	46	45	44	45 F	40 F	30 F	23
Meagre	*Argyrosomus regius*	39	7	4	5 F	5 F	5 F	3
Sargo breams nei	*Diplodus spp*	4	17	24	20 F	20 F	20 F	16
Red porgy	*Pagrus pagrus*	35	47	35	35 F	25 F	10 F	4
Bogue	*Boops boops*	9	81	162	160 F	100 F	50 F	32
Surmullets(=Red mullets) nei	*Mullus spp*	56	57	85	85 F	70 F	50 F	35
Spinefeet(=Rabbitfishes) nei	*Siganus spp*	2	11	10	10 F	10 F	5 F	3
Sardinellas nei	*Sardinella spp*	978	2 483	1 780	1 750 F	1 450 F	1 300 F	1 299
Little tunny(=Atl.black skipj)	*Euthynnus alletteratus*	90	59	61	60 F	60 F	60 F	129
Tuna-like fishes nei	*Scombroidei*	50	102	92	100 F	80 F	60 F	54
Bluefish	*Pomatomus saltatrix*	36	8	31	30 F	20 F	10 F	5
Jack and horse mackerels nei	*Trachurus spp*	90	100	115	115 F	115 F	115 F	125
Chub mackerel	*Scomber japonicus*	130	120	337	340 F	280 F	220 F	160
Barracudas nei	*Sphyraena spp*	49	101	90	90 F	80 F	60 F	52
Smooth-hounds nei	*Mustelus spp*	24	11	9	10 F	10 F	5 F	2
Guitarfishes, etc. nei	*Rhinobatidae*	...	6	6	5 F	5 F	5 F	1
Rays, stingrays, mantas nei	*Rajiformes*	29	16	23	20 F	25 F	25 F	28
Marine fishes nei	*Osteichthyes*	559	140	194	200 F	170 F	140 F	131
Marine crabs nei	*Brachyura*	72	25	65	65 F	65 F	65 F	73
Natantian decapods nei	*Natantia*	55	161	161	160 F	120 F	80 F	60
Cuttlefish,bobtail squids nei	*Sepiidae, Sepiolidae*	62	98	144	145 F	120 F	80 F	55
Common squids nei	*Loligo spp*	23	30	61	60 F	60 F	60 F	56
	Country total	*2 493*	*3 791*	*3 625*	*3 600 F*	*3 000 F*	*2 500 F*	*2 378*
Philippines								
Cyprinids nei	*Cyprinidae*	6 497	5 717	6 453	4 677	5 032	5 562	9 127
Tilapias nei	*Oreochromis (=Tilapia) spp*	17 663	20 935	23 477	25 278	28 874	28 881	30 586
Torpedo-shaped catfishes nei	*Clarias spp*	2 696	2 396	1 628	2 058	2 200	2 366	2 601
Freshwater gobies nei	*Gobiidae*	3 585	4 300	3 803	4 027	4 563	4 280	5 920
Striped snakehead	*Channa striata*	5 457	4 547	4 856	5 789	6 386	6 698	7 388
Freshwater fishes nei	*Osteichthyes*	7 143	10 413	6 632	8 309	9 800	9 384	11 877
River eels nei	*Anguilla spp*	100	116	73	114	193	201	181
Chacunda gizzard shad	*Anodontostoma chacunda*	2 761	1 260	1 191	1 115	1 140	1 346	1 413
Indian pellona	*Pellona ditchela*	792	802	842	826	839	1 221	1 458
Milkfish	*Chanos chanos*	3 671	397	3 152	315	396	479	613
Barramundi(=Giant seaperch)	*Lates calcarifer*	3 049	553	655	784	642	751	822
Flatfishes nei	*Pleuronectiformes*	829	627	659	722	729	756	917
Indo-Pacific tarpon	*Megalops cyprinoides*	997	644	720	1 207	1 151	1 365	1 510
Lizardfishes nei	*Synodontidae*	8 435	6 671	7 345	7 649	5 539	5 751	7 031
Sea catfishes nei	*Ariidae*	10 497	9 025	7 797	8 312	8 396	9 044	9 786
Mullets nei	*Mugilidae*	15 409	15 039	14 204	15 660	14 951	16 003	15 070
Fusiliers	*Caesio spp*	13 788	16 754	15 931	14 952	13 516	13 677	12 162
Groupers, seabasses nei	*Serranidae*	12 776	12 197	13 160	13 675	12 492	11 339	13 913
Sillago-whitings	*Sillaginidae*	6 800	8 662	8 918	9 529	9 472	10 458	11 556
Moonfish	*Mene maculata*	4 810	11 933	11 516	12 372	11 703	13 268	16 945
Snappers, jobfishes nei	*Lutjanidae*	15 005	18 378	15 442	19 192	19 154	14 371	13 630
Threadfin breams nei	*Nemipterus spp*	32 884	29 839	30 511	29 301	29 487	27 079	49 257
Ponyfishes(=Slipmouths) nei	*Leiognathidae*	57 867	61 254	59 862	68 371	67 255	65 007	65 816
Goatfishes, red mullets nei	*Mullidae*	23 662	15 884	14 182	14 527	14 718	14 291	24 440
Mojarras(=Silver-biddies) nei	*Gerres spp*	5 909	4 285	4 854	5 008	5 117	5 525	6 230
Spotted sicklefish	*Drepane punctata*	64	174	96	133	114	135	138
Wrasses, hogfishes, etc. nei	*Labridae*	17 097	15 107	13 391	12 318	12 384	12 664	7 106
Threadfins, tasselfishes nei	*Polynemidae*	2 383	2 933	3 017	2 524	2 371	2 804	2 892
Glassfishes	*Ambassidae*	3 075	2 594	2 186	1 733	1 575	1 876	1 986
Percoids nei	*Percoidei*	36 825	20 212	20 036	20 717	19 912	21 207	23 813
Gobies nei	*Gobiidae*	8 404	7 542	6 756	8 021	7 719	9 032	9 291
Surgeonfishes nei	*Acanthuridae*	4 474	7 367	7 770	5 797	5 392	6 211	6 939

English name Nom anglais Nombre inglés	Scientific name Nom scientifique Nombre científico	1996 t	1997 t	1998 t	1999 t	2000 t	2001 t	2002 t
Batfishes	*Platax spp*	1 584	2 975	2 792	2 870	2 597	2 830	3 227
Scats	*Scatophagus spp*	4 204	2 152	2 496	2 541	2 698	3 217	3 235
Spinefeet(=Rabbitfishes) nei	*Siganus spp*	17 012	15 720	18 246	19 977	20 108	20 336	17 581
Conger eels, etc. nei	*Congridae*	2 687	2 053	2 540	2 459	2 349	2 663	2 775
Hairtails, scabbardfishes nei	*Trichiuridae*	12 266	9 412	9 877	10 363	8 641	8 315	9 187
Sardinellas nei	*Sardinella spp*	257 804	302 341	302 599	279 864	298 466	282 617	244 223
Rainbow sardine	*Dussumieria acuta*	19 441	15 216	20 639	24 080	12 993	16 902	8 501
Stolephorus anchovies	*Stolephorus spp*	71 456	78 678	77 049	78 087	79 630	100 899	74 095
Wolf-herrings nei	*Chirocentrus spp*	114	245	317	377	354	443	460
Clupeoids nei	*Clupeoidei*	266	325	663	634	602	736	772
Narrow-barred Spanish mackerel	*Scomberomorus commerson*	10 557	11 237	10 772	9 137	8 889	9 085	9 030
Frigate and bullet tunas	*Auxis thazard, A.rochei*	88 969	108 494	106 433	111 301	112 227	111 719	163 132
Kawakawa	*Euthynnus affinis*	24 345	26 573	24 424	25 406	27 963	27 280	34 681
Skipjack tuna	*Katsuwonus pelamis*	110 004	110 097	116 673	108 778	113 011	105 484	109 977
Albacore	*Thunnus alalunga*	-	-	506	198	101	68	-
Yellowfin tuna	*Thunnus albacares*	61 280	67 342	80 000	91 251	90 832	83 896	100 270
Bigeye tuna	*Thunnus obesus*	-	-	2 705	4 004	2 436	1 388	1 717
Blue marlin	*Makaira nigricans*	1 191	1 096	1 546	2 393	3 279	3 793	3 872
Striped marlin	*Tetrapturus audax*	-	-	121	54	32	13	16
Atlantic white marlin	*Tetrapturus albidus*	-	-	1	12	-	-	-
Marlins,sailfishes,etc. nei	*Istiophoridae*	4 317	4 799	5 338	5 082	4 969	6 037	6 135
Swordfish	*Xiphias gladius*	-	-	272	325	2 870	3 314	3 698
Tuna-like fishes nei	*Scombroidei*	4 002	5 554	4 804	3 914	3 621	3 809	-
Needlefishes nei	*Tylosurus spp*	9 454	9 967	9 532	10 323	10 084	11 819	12 139
Halfbeaks nei	*Hemiramphus spp*	3 231	1 655	1 949	2 281	2 309	2 820	2 990
Flyingfishes nei	*Exocoetidae*	17 300	27 801	36 134	38 280	36 050	33 212	34 650
Silversides(=Sand smelts) nei	*Atherinidae*	866	669	596	618	543	625	638
False trevally	*Lactarius lactarius*	26	202	213	235	253	285	346
Cobia	*Rachycentron canadum*	964	1 162	1 235	1 187	1 046	1 124	1 363
Scads nei	*Decapterus spp*	228 757	234 849	250 809	254 178	260 999	291 774	283 594
Rainbow runner	*Elagatis bipinnulata*	4 505	4 075	4 298	4 500	4 342	4 755	5 415
Torpedo scad	*Megalaspis cordyla*	5 864	11 429	14 817	16 033	16 274	19 590	21 747
Queenfishes	*Scomberoides spp*	2 965	2 631	2 830	4 404	4 665	5 242	5 625
Bigeye scad	*Selar crumenophthalmus*	43 660	54 167	61 999	65 776	71 365	80 858	100 786
Carangids nei	*Carangidae*	37 456	32 175	31 472	35 204	34 713	42 442	54 019
Common dolphinfish	*Coryphaena hippurus*	3 544	359	189	223	211	258	267
Chub mackerel	*Scomber japonicus*	2 991	2 025	1 507	1 485	1 422	1 664	1 866
Short mackerel	*Rastrelliger brachysoma*	25 224	22 978	23 350	25 713	26 771	28 091	32 657
Indian mackerel	*Rastrelliger kanagurta*	46 264	54 732	51 919	53 606	55 088	62 484	72 578
Butterfishes, pomfrets nei	*Stromateidae*	2 281	1 001	1 366	1 547	1 522	1 759	1 871
Barracudas nei	*Sphyraena spp*	10 003	6 707	7 722	8 976	7 778	9 128	10 295
Shortfin mako	*Isurus oxyrinchus*	-	-	-	3	-	-	-
Rays, stingrays, mantas nei	*Rajiformes*	4 756	2 125	2 174	2 299	2 248	2 733	2 848
Sharks, rays, skates, etc. nei	*Elasmobranchii*	3 839	1 690	2 119	2 188	2 080	2 571	2 682
Marine fishes nei	*Osteichthyes*	37 385	10 176	13 104	12 783	12 043	14 656	16 527
Freshwater prawns, shrimps nei	*Palaemonidae*	5 214	4 282	2 014	3 915	4 020	4 507	5 115
Freshwater crustaceans nei	*Crustacea*	55	34	32	1	5	67	69
Blue swimming crab	*Portunus pelagicus*	27 660	30 358	23 919	34 076	36 303	38 083	32 947
Indo-Pacific swamp crab	*Scylla serrata*	4 258	1 133	1 124	1 211	1 247	1 604	1 692
Tropical spiny lobsters nei	*Panulirus spp*	522	421	214	249	250	309	315
Slipper lobsters nei	*Scyllaridae*	334	8	65	89	90	121	123
Giant tiger prawn	*Penaeus monodon*	360	292	422	169	232	328	341
Penaeus shrimps nei	*Penaeus spp*	10 462	11 508	10 796	11 255	11 063	13 632	13 902
Endeavour shrimp	*Metapenaeus endeavouri*	225	94	45	116	136	174	192
Metapenaeus shrimps nei	*Metapenaeus spp*	5 431	5 420	6 545	6 419	5 987	7 059	7 248
Sergestid shrimps nei	*Sergestidae*	18 657	15 562	16 719	19 262	20 122	22 960	16 879
Squillids nei	*Squillidae*	695	2 133	2 219	2 068	2 141	2 612	3 344
Freshwater molluscs nei	*Mollusca*	122 636	101 841	90 154	87 259	85 575	68 958	52 571
Abalones nei	*Haliotis spp*	448	183	347	282	241	288	292
Slipper cupped oyster	*Crassostrea iredalei*	291	152	89	95	79	96	97
Green mussel	*Perna viridis*	334	47	22	20	17	20	20
Scallops nei	*Pectinidae*	139	61	40	62	53	54	55
Anadara clams nei	*Anadara spp*	6	3	4	4	3	3	3
Short neck clams nei	*Paphia spp*	31	2	1	1	2	2	2
Clams, etc. nei	*Bivalvia*	725	375	227	277	222	223	225
Cuttlefish,bobtail squids nei	*Sepiidae, Sepiolidae*	2 702	2 803	2 204	2 093	2 016	2 320	2 475
Common squids nei	*Loligo spp*	52 458	54 155	48 678	47 115	46 778	41 964	50 612
Octopuses, etc. nei	*Octopodidae*	9 025	7 991	5 235	5 813	5 502	6 945	7 179
Marine molluscs nei	*Mollusca*	62	-	-	-	-	-	-
Marine turtles nei	*Testudinata*	1	-	2	2	1	1	1
Sea urchins nei	*Strongylocentrotus spp*	452	296	161	143	125	127	129
Sea cucumbers nei	*Holothurioidea*	2 123	1 191	830	849	730	791	801
Jellyfishes	*Rhopilema spp*	57	20	10	12	12	12	12
	Country total	*1 783 601*	*1 805 806*	*1 833 380*	*1 872 818*	*1 896 638*	*1 949 026*	*2 030 542*

Qatar

English name	Scientific name	1996	1997	1998	1999	2000	2001	2002
Freshwater fishes nei	*Osteichthyes*	0	0	0	0	0	0	0
Greater lizardfish	*Saurida tumbil*	1	0	0	0	0	0	0
Sea catfishes nei	*Ariidae*	0	0	0	0	0	0	0
Mullets nei	*Mugilidae*	8	7	9	5	4	4	4
Groupers nei	*Epinephelus spp*	768	736	804	913	1 215	1 820	1 567
Snappers nei	*Lutjanus spp*	157	157	143	73	181	230	102
Threadfin and dwarf breams nei	*Nemipteridae*	13	15	3	0	0	0	0
Grunts, sweetlips nei	*Haemulidae (=Pomadasyidae)*	653	542	583	433	789	900	673
Emperors(=Scavengers) nei	*Lethrinidae*	1 031	1 172	1 326	798	1 442	1 820	1 512
King soldier bream	*Argyrops spinifer*	130	177	146	98	199	426	327
Porgies, seabreams nei	*Sparidae*	126	221	188	182	248	288	209
Goatfishes, red mullets nei	*Mullidae*	16	14	14	20	1	0	0
Mojarras(=Silver-biddies) nei	*Gerres spp*	50	56	92	77	78	82	52

English name / Nom anglais / Nombre inglés	Scientific name / Nom scientifique / Nombre científico	1996 t	1997 t	1998 t	1999 t	2000 t	2001 t	2002 t
Spinefeet(=Rabbitfishes) nei	*Siganus spp*	225	237	285	240	387	451	400
Sardinellas nei	*Sardinella spp*	0	-	0	0	0	0	0
Dorab wolf-herring	*Chirocentrus dorab*	0	-	0	0	0	0	0
Narrow-barred Spanish mackerel	*Scomberomorus commerson*	307	411	552	496	768	1 019	963
Tuna-like fishes nei	*Scombroidei*	0	-	0	0	0	0	0
Needlefishes nei	*Tylosurus spp*	14	25	12	20	1	2	1
Cobia	*Rachycentron canadum*	56	52	56	44	56	95	89
Scads nei	*Decapterus spp*	0	0	0	0	0	0	0
Jacks, crevalles nei	*Caranx spp*	222	247	308	289	409	388	318
Golden trevally	*Gnathanodon speciosus*	101	108	172	116	185	204	173
Queenfishes	*Scomberoides spp*	57	56	80	42	57	78	54
Carangids nei	*Carangidae*	308	308	326	287	420	552	364
Barracudas nei	*Sphyraena spp*	54	59	21	2	62	86	0
Requiem sharks nei	*Carcharhinidae*	0	-	0	0	0	0	0
Marine fishes nei	*Osteichthyes*	357	332	38	10	587	79	27
Marine crabs nei	*Brachyura*	47	47	69	39	26	21	17
Slipper lobsters nei	*Scyllaridae*	8	16	15	13	5	20	17
Green tiger prawn	*Penaeus semisulcatus*	0	-	0	0	0	0	0
Common squids nei	*Loligo spp*	30	37	37	10	22	41	11
	Country total	*4 739*	*5 032*	*5 279*	*4 207*	*7 142*	*8 606*	*6 880*
Saudi Arabia								
Freshwater fishes nei	*Osteichthyes*	0	0	0	0	0	0	0
Milkfish	*Chanos chanos*	130	70	82	81	64	73	93
Flatfishes nei	*Pleuronectiformes*	58	75	90	84	84	85	65
Lizardfishes nei	*Synodontidae*	172	188	215	214	169	199	196
Sea catfishes nei	*Ariidae*	302	366	315	309	564	534	485
Squirrelfishes nei	*Holocentridae*	40	50	34	53	46	40	97
Mullets nei	*Mugilidae*	320	369	510	317	326	397	414
Fusiliers	*Caesio spp*	-	-	-	-	-	-	19
Groupers, seabasses nei	*Serranidae*	4 207	4 403	5 053	5 430	5 402	5 273	6 838
Therapon pearch	*Terapon spp*	-	-	1	2	0	1	2
Bigeyes nei	*Priacanthus spp*	-	-	2	1	0	1	0
Snappers, jobfishes nei	*Lutjanidae*	1 704	2 148	2 302	2 022	1 647	2 092	1 988
Threadfin breams nei	*Nemipterus spp*	131	87	66	49	144	153	100
Grunts, sweetlips nei	*Haemulidae (=Pomadasyidae)*	1 063	1 014	779	846	1 350	1 206	1 390
Emperors(=Scavengers) nei	*Lethrinidae*	7 314	6 904	6 796	7 233	7 078	7 448	8 902
Porgies, seabreams nei	*Sparidae*	1 722	2 481	2 822	2 771	2 488	2 646	3 252
Goatfishes	*Upeneus spp*	83	73	107	123	72	122	88
Mojarras(=Silver-biddies) nei	*Gerres spp*	360	438	445	520	529	510	662
Blue sea chub	*Kyphosus cinerascens*	-	8	-	1	-	-	17
Wrasses, hogfishes, etc. nei	*Labridae*	108	105	66	155	91	79	153
Parrotfishes nei	*Scaridae*	777	906	702	655	756	758	849
Angelfishes nei	*Pomacanthidae*	-	-	7	14	1	3	9
Surgeonfishes nei	*Acanthuridae*	238	173	253	205	149	343	364
Spadefishes nei	*Ephippidae*	-	-	17	1	1	2	2
Spinefeet(=Rabbitfishes) nei	*Siganus spp*	2 779	2 173	1 761	1 691	1 832	1 822	1 998
Flatheads nei	*Platycephalidae*	-	-	18	7	5	6	329
Puffers nei	*Tetraodontidae*	-	-	-	0	0	18	0
Triggerfishes, durgons nei	*Balistidae*	-	26	8	8	5	7	24
Dorab wolf-herring	*Chirocentrus dorab*	-	-	8	7	10	9	23
Clupeoids nei	*Clupeoidei*	201	560	108	162	187	178	266
Narrow-barred Spanish mackerel	*Scomberomorus commerson*	5 276	5 511	6 722	6 032	6 057	5 532	6 399
Indo-Pacific king mackerel	*Scomberomorus guttatus*	-	114	300	303	303	276	320
Kawakawa	*Euthynnus affinis*	162	304	332	256	264	241	353
Longtail tuna	*Thunnus tonggol*	115	101	181	136	143	131	174
Indo-Pacific sailfish	*Istiophorus platypterus*	-	1	2	1	1	8	22
Tuna-like fishes nei	*Scombroidei*	804	861	773	797	783	709	1 197
Needlefishes nei	*Tylosurus spp*	96	111	131	140	107	126	136
Cobia	*Rachycentron canadum*	155	155	130	137	138	167	188
Snubnose pompano	*Trachinotus blochii*	-	-	4	0	31	-	-
Rainbow runner	*Elagatis bipinnulata*	-	415	98	132	5	52	4
Queenfishes	*Scomberoides spp*	349	385	572	456	572	543	580
Carangids nei	*Carangidae*	5 524	5 125	4 795	5 534	6 581	5 943	6 287
Indian mackerel	*Rastrelliger kanagurta*	1 549	1 990	2 072	1 979	1 525	1 803	1 921
Mackerels nei	*Scombridae*	...	...	12	16	11	10	3
Silver pomfret	*Pampus argenteus*	31	16	14	30	43	25	25
Barracudas nei	*Sphyraena spp*	1 251	1 246	1 489	1 267	1 563	1 562	1 492
Rays, stingrays, mantas nei	*Rajiformes*	-	-	4	8	4	6	8
Sharks, rays, skates, etc. nei	*Elasmobranchii*	398	543	697	497	649	651	731
Marine fishes nei	*Osteichthyes*	1 205	820	768	586	574	804	893
Marine crabs nei	*Brachyura*	1 060	1 371	1 062	959	1 021	1 002	1 209
Tropical spiny lobsters nei	*Panulirus spp*	13	18	13	19	8	14	19
Penaeus shrimps nei	*Penaeus spp*	7 423	7 011	7 812	3 612	5 639	4 761	3 996
Cuttlefish,bobtail squids nei	*Sepiidae, Sepiolidae*	578	598	656	760	739	796	748
Octopuses, etc. nei	*Octopodidae*	-	1	-	-	0	-	-
	Country total	*47 698*	*49 314*	*51 206*	*46 618*	*49 761*	*49 167*	*55 330*
Singapore								
Freshwater fishes nei	*Osteichthyes*	0	0	0	0	0	0	0
Barramundi(=Giant seaperch)	*Lates calcarifer*	39	58	39	29	41	52	51
Lizardfishes nei	*Synodontidae*	172	125	90	51	28	6	4
Sea catfishes nei	*Ariidae*	333	385	358	241	141	76	55
Mullets nei	*Mugilidae*	54	55	34	32	42	27	25
Fusiliers	*Caesio spp*	-	9	18	13	10	1	3
Groupers nei	*Epinephelus spp*	120	94	72	56	43	38	33
Sillago-whitings	*Sillaginidae*	48	45	41	24	10	14	8
Moonfish	*Mene maculata*	5	7	13	13	15	7	5

English name Nom anglais Nombre inglés	Scientific name Nom scientifique Nombre científico	1996 t	1997 t	1998 t	1999 t	2000 t	2001 t	2002 t
Snappers nei	*Lutjanus spp*	321	289	238	154	122	80	63
Snappers, jobfishes nei	*Lutjanidae*	61	66	50	31	32	16	11
Threadfin breams nei	*Nemipterus spp*	209	239	158	128	96	48	33
Ponyfishes(=Slipmouths)	*Leiognathus spp*	75	63	52	47	32	23	9
Grunts, sweetlips nei	*Haemulidae (=Pomadasyidae)*	53	45	27	25	18	18	16
Croakers, drums nei	*Sciaenidae*	123	180	160	114	68	45	56
Goatfishes	*Upeneus spp*	24	29	29	25	8	15	1
Threadfins, tasselfishes nei	*Polynemidae*	5	6	7	13	25	11	11
Spinefeet(=Rabbitfishes) nei	*Siganus spp*	14	21	24	28	5	8	12
Largehead hairtail	*Trichiurus lepturus*	144	170	140	127	83	50	61
Wolf-herrings nei	*Chirocentrus spp*	87	79	77	51	42	30	28
Clupeoids nei	*Clupeoidei*	379	351	329	206	78	73	63
Seerfishes nei	*Scomberomorus spp*	76	71	70	79	78	46	38
Skipjack tuna	*Katsuwonus pelamis*	5	47	12	23	2	10	6
Tuna-like fishes nei	*Scombroidei*	0	-	-	-	-	-	-
Scads nei	*Decapterus spp*	227	212	222	156	163	106	74
Jacks, crevalles nei	*Caranx spp*	312	313	234	175	139	66	69
Carangids nei	*Carangidae*	37	33	36	30	21	12	16
Indian mackerels nei	*Rastrelliger spp*	12	51	165	129	97	68	35
Butterfishes, pomfrets nei	*Stromateidae*	100	94	103	94	75	48	44
Barracudas nei	*Sphyraena spp*	170	134	149	105	79	86	65
Rays, stingrays, mantas nei	*Rajiformes*	327	308	336	250	261	187	162
Sharks, rays, skates, etc. nei	*Elasmobranchii*	94	93	80	59	43	32	30
Marine fishes nei	*Osteichthyes*	4 308	3 933	2 812	2 740	2 339	1 325	1 031
Indo-Pacific swamp crab	*Scylla serrata*	19	15	9	9	28	9	8
Marine crabs nei	*Brachyura*	176	203	264	175	189	203	173
Tropical spiny lobsters nei	*Panulirus spp*	-	5	11	8	8	7	2
Slipper lobsters nei	*Scyllaridae*	9	11	12	9	6	7	4
Natantian decapods nei	*Natantia*	857	706	621	522	422	250	222
Green mussel	*Perna viridis*	188	-	-	-	-	-	-
Cuttlefish,bobtail squids nei	*Sepiidae, Sepiolidae*	214	235	179	142	134	56	57
Common squids nei	*Loligo spp*	546	470	462	376	348	186	185
	Country total	*9 943*	*9 250*	*7 733*	*6 489*	*5 371*	*3 342*	*2 769*
Sri Lanka								
Tilapias nei	*Oreochromis (=Tilapia) spp*	22 250	27 250	29 900	28 520	33 220	27 230	24 900
Freshwater fishes nei	*Osteichthyes*	...	...	...	2 930	3 480	2 640	3 230
Demersal percomorphs nei	*Perciformes*	8 968	9 100	9 210	10 440	14 910	14 490	16 320
Clupeoids nei	*Clupeoidei*	48 221	47 200	50 800	51 370	53 250	49 270	52 310
Wahoo	*Acanthocybium solandri*	128	156	196	488	545	444	456
Narrow-barred Spanish mackerel	*Scomberomorus commerson*	817	999	1 246	856	766	1 015	589
Indo-Pacific king mackerel	*Scomberomorus guttatus*	0	-	-	-	-	-	-
Streaked seerfish	*Scomberomorus lineolatus*	0	-	-	-	-	-	-
Seerfishes nei	*Scomberomorus spp*	-	-	-	168	20	3	14
Frigate and bullet tunas	*Auxis thazard, A.rochei*	5 334	6 521	8 133	3 515	4 583	3 157	15 197
Kawakawa	*Euthynnus affinis*	2 262	2 765	3 449	2 167	2 167	3 434	4 011
Skipjack tuna	*Katsuwonus pelamis*	22 754	27 815	34 691	51 940	51 940	47 241	42 968
Longtail tuna	*Thunnus tonggol*	0	-	-	-	-	-	-
Southern bluefin tuna	*Thunnus maccoyii*	68	83	104	121	120	101	112
Yellowfin tuna	*Thunnus albacares*	12 889	15 756	19 651	27 538	22 091	15 050	17 765
Bigeye tuna	*Thunnus obesus*	491	600	749	462	348	336	338
Indo-Pacific sailfish	*Istiophorus platypterus*	5 360	6 552	8 172	6 979	8 878	7 574	6 827
Blue marlin	*Makaira nigricans*	0	-	-	-	-	-	-
Black marlin	*Makaira indica*	39	48	59	69	68	58	64
Striped marlin	*Tetrapturus audax*	0	0	0	-	-	-	-
Marlins,sailfishes,etc. nei	*Istiophoridae*	5 675	6 937	8 653	7 253	5 728	3 262	5 191
Swordfish	*Xiphias gladius*	2 591	3 167	3 950	2 132	5 545	4 757	2 467
Tuna-like fishes nei	*Scombroidei*	0	0	0	11	17	8	24
Carangids nei	*Carangidae*	6 088	6 900	8 500	8 680	10 450	9 950	10 760
Mackerels nei	*Scombridae*	17 700	20 000	20 900	21 350	22 180	16 760	17 250
Silky shark	*Carcharhinus falciformis*	21 000	15 000	20 875	20 700	14 130	15 870	18 510
Sharks, rays, skates, etc. nei	*Elasmobranchii*	6 954	11 920	7 625	8 660	13 884	8 240	6 830
Marine fishes nei	*Osteichthyes*	36 404	23 398	25 084	16 106	24 390	43 890	44 007
Marine crustaceans nei	*Crustacea*	2 502	2 360	880	3 080	230	4 450	7 530
Cuttlefish,bobtail squids nei	*Sepiidae, Sepiolidae*	300	300	300	365	310	290	380
Marine molluscs nei	*Mollusca*	0	-	-	10	15	30	60
Sea cucumbers nei	*Holothurioidea*	150	272	203	170	145	90	150
	Country total	*228 945*	*235 099*	*263 330*	*276 080*	*293 410*	*279 640*	*298 260*
Syria								
Freshwater fishes nei	*Osteichthyes*	3 103	3 557	4 347	5 338	3 991	5 969	6 355
European hake	*Merluccius merluccius*	250	300	125	110	87	52	63
Lizardfishes nei	*Synodontidae*	...	...	...	...	...	...	86
Mullets nei	*Mugilidae*	...	...	...	...	...	...	136
Groupers nei	*Epinephelus spp*	...	...	...	...	...	...	54
Pandoras nei	*Pagellus spp*	134	98	125	100	65	85	100
White seabream	*Diplodus sargus*	...	...	...	...	...	...	129
Pargo breams nei	*Pagrus spp*	50	80	90	62	74	77	112
Gilthead seabream	*Sparus aurata*	...	...	...	...	...	...	63
Bogue	*Boops boops*	...	...	...	...	...	...	141
Sand steenbras	*Lithognathus mormyrus*	...	...	...	...	...	...	25
Surmullets(=Red mullets) nei	*Mullus spp*	232	250	116	130	122	125	147
Spinefeet(=Rabbitfishes) nei	*Siganus spp*	...	...	...	...	...	...	118
Grey triggerfish	*Balistes carolinensis*	...	...	...	...	...	...	5
Scorpionfishes nei	*Scorpaenidae*	...	...	...	...	...	...	38
Gurnards, searobins nei	*Triglidae*	0	60	46	35	32	32	38
Demersal percomorphs nei	*Perciformes*	606	450	626	530	392	449	-
Sardinellas nei	*Sardinella spp*	550	300	292	338	251	197	334

English name Nom anglais Nombre inglés	Scientific name Nom scientifique Nombre científico	1996 t	1997 t	1998 t	1999 t	2000 t	2001 t	2002 t
Little tunny(=Atl.black skipj)	*Euthynnus alletteratus*	270	350	417	390	370	370	330
Atlantic horse mackerel	*Trachurus trachurus*	60	77	40	30	36	56	66
Jacks, crevalles nei	*Caranx spp*	...	...	...	...	...	...	36
Greater amberjack	*Seriola dumerili*	90	75	108	88	52	96	53
Atlantic mackerel	*Scomber scombrus*	116	210	274	245	384	187	264
Scomber mackerels nei	*Scomber spp*	...	...	...	...	...	...	41
Barracudas nei	*Sphyraena spp*	82	90	88	130	76	77	60
Smooth-hounds nei	*Mustelus spp*	50	-	-	-	-	-	-
Raja rays nei	*Raja spp*	...	...	...	...	...	...	182
Marine fishes nei	*Osteichthyes*	90	150	328	342	580	462	-
Marine crustaceans nei	*Crustacea*	90	84	75	70	60	57	66
Marine molluscs nei	*Mollusca*	...	...	...	...	...	...	136
	Country total	*5 773*	*6 131*	*7 097*	*7 938*	*6 572*	*8 291*	*9 178*
Tajikistan								
Freshwater bream	*Abramis brama*	...	...	...	...	...	37	25
Common carp	*Cyprinus carpio*	...	...	...	48	59	24	51
Crucian carp	*Carassius carassius*	...	...	...	...	...	...	17
Silver carp	*Hypophthalmichthys molitrix*	...	...	...	...	...	...	16
Sichel	*Pelecus cultratus*	...	...	...	...	...	8	4
Asp	*Aspius aspius*	...	...	...	...	...	6	5
Cyprinids nei	*Cyprinidae*	-	-	-	-	-	31	27
Wels(=Som)catfish	*Silurus glanis*	...	...	...	...	...	10	12
Pike-perch	*Stizostedion lucioperca*	...	...	...	...	...	21	24
Freshwater fishes nei	*Osteichthyes*	40 F	75 F	100 F	32 F	19 F	-	-
	Country total	*40 F*	*75 F*	*100 F*	*80 F*	*78 F*	*137*	*181*
Thailand								
Common carp	*Cyprinus carpio*	7 420	7 418	11 508	13 689	6 961	6 800	7 100
Asian barbs nei	*Puntius spp*	25 750	25 296	44 349	45 511	41 024	43 400	44 100
Nile tilapia	*Oreochromis niloticus*	29 253	28 727	40 173	49 840	40 037	43 100	44 700
Torpedo-shaped catfishes nei	*Clarias spp*	5 800	3 410	10 934	12 111	19 600	14 100	14 300
Pangas catfishes nei	*Pangasius spp*	541	522	917	1 061	1 274	1 100	1 200
Lai	*Monopterus albus*	-	-	-	-	162	100	100
Climbing perch	*Anabas testudineus*	3 905	3 754	4 637	6 340	6 730	5 800	6 300
Snakeskin gourami	*Trichogaster pectoralis*	385	353	1 486	511	723	500	500
Striped snakehead	*Channa striata*	25 509	24 099	16 664	17 995	20 453	18 300	18 600
Freshwater fishes nei	*Osteichthyes*	105 726	108 551	70 011	59 376	64 241	68 900	68 300
Barramundi(=Giant seaperch)	*Lates calcarifer*	589	169	1 049	85	833	563	562
Tonguefishes	*Cynoglossidae*	15 011	15 030	16 097	15 333	16 548	15 174	15 085
Indian halibut	*Psettodes erumei*	7 294	7 423	2 742	2 346	2 234	4 227	4 189
Lizardfishes nei	*Synodontidae*	61 684	71 315	76 467	73 308	69 791	71 510	71 776
Sea catfishes nei	*Ariidae*	5 684	5 704	7 541	12 598	11 217	8 627	8 639
Eeltail catfishes	*Plotosus spp*	841	763	1 201	631	1 311	1 026	1 034
Mullets nei	*Mugilidae*	5 087	5 397	5 549	5 498	8 374	5 907	5 902
Groupers, seabasses nei	*Serranidae*	9 349	8 944	9 310	8 050	7 627	8 645	8 646
Bigeyes nei	*Priacanthus spp*	82 122	77 825	80 234	82 735	75 688	81 754	82 278
Sillago-whitings	*Sillaginidae*	5 433	5 738	5 501	7 986	7 671	6 269	6 215
Snappers, jobfishes nei	*Lutjanidae*	14 653	12 828	14 747	13 241	7 959	12 807	12 832
Threadfin breams nei	*Nemipterus spp*	89 592	87 767	96 595	93 037	102 282	94 765	95 017
Monocle breams	*Scolopsis spp*	98	344	644	54	94	210	212
Croakers, drums nei	*Sciaenidae*	29 761	29 961	33 646	36 591	39 946	32 462	32 088
Threadfins, tasselfishes nei	*Polynemidae*	1 550	1 079	1 305	432	387	963	970
Daggertooth pike conger	*Muraenesox cinereus*	1 615	1 600	1 643	2 345	1 692	1 816	1 815
Largehead hairtail	*Trichiurus lepturus*	15 073	17 586	18 709	15 970	16 079	15 943	15 768
Sardinellas nei	*Sardinella spp*	214 857	202 792	185 858	182 813	164 014	192 556	193 197
Anchovies, etc. nei	*Engraulidae*	161 970	157 341	157 214	134 740	143 105	153 552	154 259
Dorab wolf-herring	*Chirocentrus dorab*	10 292	10 459	12 372	15 049	14 208	12 147	12 056
Seerfishes nei	*Scomberomorus spp*	15 242	14 500	15 249	14 640	12 779	13 866	13 817
Frigate and bullet tunas	*Auxis thazard, A.rochei*	21 839	19 723	19 997	23 119	19 434	20 635	20 918
Kawakawa	*Euthynnus affinis*	46 612	42 257	41 150	36 754	33 994	34 873	35 739
Skipjack tuna	*Katsuwonus pelamis*	-	-	-	-	1 110	460	...
Longtail tuna	*Thunnus tonggol*	54 383	49 162	52 404	51 617	58 245	46 550	47 399
Albacore	*Thunnus alalunga*	-	-	-	-	12	14	14
Yellowfin tuna	*Thunnus albacares*	-	-	-	-	478	231	26
Bigeye tuna	*Thunnus obesus*	-	-	-	-	280	188	90
Blue marlin	*Makaira nigricans*	-	-	-	-	...	1	1
Striped marlin	*Tetrapturus audax*	-	-	-	-	...	1	1
Marlins,sailfishes,etc. nei	*Istiophoridae*	-	-	-	-	16	-	-
Swordfish	*Xiphias gladius*	-	-	-	-	19	6	6
Indian scad	*Decapterus russelli*	85 205	78 156	86 163	84 574	93 233	85 806	85 910
Black pomfret	*Parastromateus niger*	6 383	6 253	2 918	5 879	4 009	5 076	5 080
Torpedo scad	*Megalaspis cordyla*	19 674	20 065	24 188	22 117	20 982	20 524	20 307
Bigeye scad	*Selar crumenophthalmus*	26 517	24 092	28 761	29 408	32 255	29 158	29 403
Blackbanded trevally	*Seriolina nigrofasciata*	7 269	7 096	6 042	6 045	5 415	6 238	6 218
Carangids nei	*Carangidae*	53 028	49 747	45 994	44 250	42 145	48 080	48 339
Indian mackerel	*Rastrelliger kanagurta*	45 129	42 676	43 682	47 885	35 203	42 341	42 187
Indian mackerels nei	*Rastrelliger spp*	140 826	138 621	151 010	164 110	152 884	151 181	151 638
Silver pomfret	*Pampus argenteus*	1 688	1 617	378	395	407	854	846
Butterfishes, pomfrets nei	*Stromateidae*	-	-	-	-	16	-	-
Barracudas nei	*Sphyraena spp*	14 389	14 245	14 025	16 860	16 389	14 613	14 460
Rays, stingrays, mantas nei	*Rajiformes*	9 978	10 353	8 289	12 279	13 650	10 831	10 809
Sharks, rays, skates, etc. nei	*Elasmobranchii*	7 775	7 616	7 737	10 118	11 039	8 762	8 732
Marine fishes nei	*Osteichthyes*	1 055 412	1 005 317	976 877	990 923	994 727	988 159	971 891
Freshwater prawns, shrimps nei	*Palaemonidae*	1 614	1 541	36	406	200	300	300
Blue swimming crab	*Portunus pelagicus*	41 915	40 089	46 678	41 250	43 871	45 707	46 616
Indo-Pacific swamp crab	*Scylla serrata*	4 243	4 031	3 732	5 736	6 921	5 414	5 585
Marine crabs nei	*Brachyura*	6 601	6 874	7 535	8 457	7 316	8 288	8 641

English name Nom anglais Nombre inglés	Scientific name Nom scientifique Nombre científico	1996 t	1997 t	1998 t	1999 t	2000 t	2001 t	2002 t
Flathead lobster	*Thenus orientalis*	3 015	2 962	3 538	1 797	2 312	2 540	2 464
Banana prawn	*Penaeus merguiensis*	13 645	12 633	12 498	12 780	15 001	12 077	11 685
Giant tiger prawn	*Penaeus monodon*	2 565	2 468	1 212	2 252	2 273	1 887	1 822
Green tiger prawn	*Penaeus semisulcatus*	3 469	3 517	3 019	2 821	2 298	2 703	2 609
Western king prawn	*Penaeus latisulcatus*	3 274	3 562	3 276	4 064	3 400	3 227	3 123
Penaeus shrimps nei	*Penaeus spp*	74 145	71 525	47 393	45 112	46 624	52 667	51 035
Metapenaeus shrimps nei	*Metapenaeus spp*	12 156	11 970	9 879	9 365	11 272	10 025	9 709
Sergestid shrimps nei	*Sergestidae*	22 127	17 488	15 372	7 660	3 757	12 363	11 983
Stomatopods nei	*Stomatopoda*	181	176	458	866	875	478	463
Horse mussels nei	*Modiolus spp*	-	-	-	-	94	20	20
Green mussel	*Perna viridis*	19 698	17 970	17 432	6 534	41 156	22 762	23 262
Scallops nei	*Pectinidae*	552	508	1 185	551	273	652	666
Blood cockle	*Anadara granosa*	-	-	441	1 757	972	948	969
Short neck clams nei	*Paphia spp*	52 889	35 852	49 661	69 978	49 094	54 566	55 014
Cuttlefish,bobtail squids nei	*Sepiidae, Sepiolidae*	70 525	71 588	63 340	66 247	67 294	67 797	67 733
Common squids nei	*Loligo spp*	79 235	78 948	92 908	83 135	86 203	83 080	83 030
Octopuses, etc. nei	*Octopodidae*	23 423	23 112	31 908	25 000	23 958	24 293	24 237
Marine molluscs nei	*Mollusca*	-	-	126	1 689	1 403	733	748
Jellyfishes	*Rhopilema spp*	30 496	42 393	64 760	84 568	139 622	66 627	67 931
Aquatic invertebrates nei	*Invertebrata*	-	-	-	34	219	1 749	-
	Country total	*3 013 961*	*2 902 898*	*2 930 354*	*2 952 308*	*2 997 394*	*2 932 374*	*2 921 216*
Timor-Leste								
Freshwater fishes nei	*Osteichthyes*	...	...	...	0	0	0	0
Yellowfin tuna	*Thunnus albacares*	...	...	...	1	3	3	3 F
Tuna-like fishes nei	*Scombroidei*	...	...	...	1	3	3	3 F
Marine fishes nei	*Osteichthyes*	...	...	...	400 F	350	344	338 F
Marine crabs nei	*Brachyura*	...	...	...	1 F	1	1	1 F
Tropical spiny lobsters nei	*Panulirus spp*	...	...	...	2 F	2	2	2 F
Natantian decapods nei	*Natantia*	...	...	...	1 F	1	1	1 F
Cephalopods nei	*Cephalopoda*	...	...	...	1 F	1	1	1 F
Marine turtles nei	*Testudinata*	...	...	...	1 F	1	1	1 F
	Country total	...	...	...	*408 F*	*362*	*356*	*350 F*
Turkey								
Freshwater bream	*Abramis brama*	...	...	...	259	200	151	198
Common carp	*Cyprinus carpio*	15 631	16 000	20 000	17 797	14 137	12 265	12 965
Tench	*Tinca tinca*	...	...	...	...	690	778	800
Common dace	*Leuciscus leuciscus*	215	250	300	176	104	91	73
Cyprinids nei	*Cyprinidae*	1 380	1 900	1 800	406	699	626	240
Northern pike	*Esox lucius*	225	350	200	276	224	192	217
Wels(=Som)catfish	*Silurus glanis*	705	1 000	1 000	958	1 019	813	987
North African catfish	*Clarias gariepinus*	...	...	...	216	576	520	495
Pike-perch	*Stizostedion lucioperca*	8 042	1 500	3 000	1 906	1 633	1 644	1 850
Freshwater gobies nei	*Gobiidae*	185	200	200	118	107	116	85
Freshwater fishes nei	*Osteichthyes*	4 621	1 800	1 700	2 434	1 697	3 290	3 478
European eel	*Anguilla anguilla*	342	400	300	99	176	122	147
Sea trout	*Salmo trutta*	395	200	200	263	277	364	352
Shads nei	*Alosa spp*	1 166	505	880	680	720	690	862
Turbot	*Psetta maxima*	2 035	980	1 860	1 870	2 700	2 455	459
Flatfishes nei	*Pleuronectiformes*	1 947	2 300	2 000	2 400	1 000	1 250	585
Greater forkbeard	*Phycis blennoides*	24	50	150	50	50	35	8
Blue whiting(=Poutassou)	*Micromesistius poutassou*	11 518	15 000	27 200	16 975	18 180	20 810	10 500
Whiting	*Merlangius merlangus*	21 450	15 500	13 150	14 110	18 000	10 000	8 808
European hake	*Merluccius merluccius*	150	25	15	5	10	-	-
Lizardfishes nei	*Synodontidae*	161	150	165	190	250	55	97
Mullets nei	*Mugilidae*	37 515	42 500	45 350	46 752	43 352	38 558	27 589
Dusky grouper	*Epinephelus marginatus*	700	463	640	124	85	79	71
Groupers, seabasses nei	*Serranidae*	790	1 137	700	361	400	411	100
Seabasses nei	*Dicentrarchus spp*	2 411	3 450	3 950	3 650	1 900	1 200	713
Brown meagre	*Sciaena umbra*	50	35	45	65	20	20	39
Shi drum	*Umbrina cirrosa*	116	65	150	155	45	55	70
Meagre	*Argyrosomus regius*	71	40	30	65	70	50	63
Sargo breams nei	*Diplodus spp*	2 345	1 750	2 210	2 110	1 420	655	751
Common dentex	*Dentex dentex*	442	330	370	220	100	60	79
Black seabream	*Spondyliosoma cantharus*	73	35	40	50	45	45	40
Saddled seabream	*Oblada melanura*	184	190	180	115	80	90	130
Red porgy	*Pagrus pagrus*	750	560	675	480	540	320	403
Gilthead seabream	*Sparus aurata*	1 340	1 200	1 400	1 665	830	1 070	700
Bogue	*Boops boops*	3 736	2 450	4 100	1 620	1 500	1 000	1 126
Salema	*Sarpa salpa*	331	340	300	155	200	160	243
Porgies, seabreams nei	*Sparidae*	412	220	180	240	110	135	137
Picarels nei	*Spicara spp*	1 525	1 650	3 700	1 680	1 500	2 250	1 105
Surmullet	*Mullus surmuletus*	3 962	2 950	2 050	2 100	2 300	1 570	1 450
Red mullet	*Mullus barbatus*	3 936	3 000	3 500	3 865	2 450	2 455	2 395
Gobies nei	*Gobiidae*	390	305	250	325	300	335	493
European conger	*Conger conger*	416	310	300	680	200	340	300
John dory	*Zeus faber*	73	50	120	135	100	130	55
Scorpionfishes nei	*Scorpaenidae*	585	435	515	315	360	640	296
Gurnards, searobins nei	*Triglidae*	2 319	750	700	710	260	200	271
European pilchard(=Sardine)	*Sardina pilchardus*	18 972	20 500	23 600	22 000	16 500	10 000	8 684
European anchovy	*Engraulis encrasicolus*	290 680	241 000	228 000	350 000	280 000	320 000	373 000
Atlantic bonito	*Sarda sarda*	10 284	7 810	24 000	17 900	12 000	13 460	6 286
Atlantic bluefin tuna	*Thunnus thynnus*	4 616	5 093	5 899	1 200	1 070	2 100	2 300
Swordfish	*Xiphias gladius*	320	350	450	230	373	360	370
Garfish	*Belone belone*	395	470	450	500	300	640	482
Silversides(=Sand smelts) nei	*Atherinidae*	974	2 240	2 300	2 755	2 083	2 260	2 251
Bluefish	*Pomatomus saltatrix*	4 117	3 050	3 350	2 995	4 250	13 060	25 000

English name Nom anglais Nombre inglés	Scientific name Nom scientifique Nombre científico	1996 t	1997 t	1998 t	1999 t	2000 t	2001 t	2002 t
Atlantic horse mackerel	*Trachurus trachurus*	7 559	5 100	4 500	4 000	7 200	10 635	6 982
Mediterranean horse mackerel	*Trachurus mediterraneus*	12 500	9 500	10 500	9 220	15 000	15 545	19 500
Leerfish	*Lichia amia*	1 544	1 650	1 950	2 780	320	255	330
Chub mackerel	*Scomber japonicus*	10 444	10 850	10 120	10 200	9 000	4 500	1 500
Atlantic mackerel	*Scomber scombrus*	592	300	650	850	900	550	486
Barracudas nei	*Sphyraena spp*	250	200	120	170	150	130	206
Smooth-hounds nei	*Mustelus spp*	2 158	1 720	1 450	1 625	2 880	1 000	686
Angelsharks, sand devils nei	*Squatinidae*	42	15	140	70	60	20	18
Rays, stingrays, mantas nei	*Rajiformes*	524	340	385	420	1 100	555	369
Marine fishes nei	*Osteichthyes*	1 855	1 095	1 861	1 375	7 365	1 230	2 380
Euro-American crayfishes nei	*Astacidae, Cambaridae*	850	1 100	1 500	1 372	1 681	1 634	1 894
Marine crabs nei	*Brachyura*	305	318	243	179	187	273	215
Common spiny lobster	*Palinurus elephas*	10	45	40	6	11	18	19
European lobster	*Homarus gammarus*	34	40	60	10	15	10	9
Natantian decapods nei	*Natantia*	1 100	1 380	1 400	890	2 000	3 000	4 000
Freshwater molluscs nei	*Mollusca*	1 500	2 000	1 500	1 585	1 592	1 601	1 937
European flat oyster	*Ostrea edulis*	1 140	1 495	1 050	840	150	10	70
Mediterranean mussel	*Mytilus galloprovincialis*	3 500	6 450	3 880	1 800	1 200	1 500	5 000
Great Mediterranean scallop	*Pecten jacobaeus*	52	95	50	68	570	150	470
Striped venus	*Chamelea gallina*	10 925	7 150	3 550	3 585	10 000	7 500	10 000
Common cuttlefish	*Sepia officinalis*	644	900	750	537	550	465	909
Common squids nei	*Loligo spp*	364	420	500	360	400	230	330
Common octopus	*Octopus vulgaris*	802	1 000	1 450	510	680	1 400	1 502
Marine molluscs nei	*Mollusca*	2 466	2 092	4 077	3 646	2 168	2 671	6 274
Frogs	*Rana spp*	740	160	100	118	77	873	898
Jellyfishes	*Rhopilema spp*	904	900	1 750	1 203	900	2 000	500
	Country total	*527 826*	*459 153*	*487 200*	*573 824*	*503 348*	*527 730*	*566 682*
Turkmenistan								
Freshwater bream	*Abramis brama*	75	100	142	147	153	126	107
Common carp	*Cyprinus carpio*	115	154	140	150	144	93	75
Crucian carp	*Carassius carassius*	85	1	225	233	228	154	141
Roach	*Rutilus rutilus*	71	69	48	39	41	1	2
Sichel	*Pelecus cultratus*	13	1	0	0	0	-	-
Asp	*Aspius aspius*	10	6	12	8	11	16	21
Wels(=Som)catfish	*Silurus glanis*	42	7	8	9	7	3	2
Pike-perch	*Stizostedion lucioperca*	30	28	109	87	96	42	34
Snakeheads(=Murrels) nei	*Channa spp*	0	1	0	0	0	1	-
Freshwater fishes nei	*Osteichthyes*	27	2	2	1	2	9	7
Sturgeons nei	*Acipenseridae*	...	...	...	11	3	3	5
Azov sea sprat	*Clupeonella cultriventris*	8 546	7 800	6 324	8 370	11 540	12 300	12 418
Mullets nei	*Mugilidae*	...	10	4	3	3	1	-
	Country total	*9 014*	*8 179*	*7 014*	*9 058*	*12 228*	*12 749*	*12 812*
Untd Arab Em								
Freshwater fishes nei	*Osteichthyes*	0	0	0	0	0	0	0
Milkfish	*Chanos chanos*	51	53	54	55	58	60 F	65
Sea catfishes nei	*Ariidae*	140	150	150	154	763	877	723
Mullets nei	*Mugilidae*	1 360	1 453	1 458	1 494	86	98	93
Groupers, seabasses nei	*Serranidae*	6 767	7 232	7 256	7 437	24 045	27 680	22 827
Snappers, jobfishes nei	*Lutjanidae*	3 474	3 713	3 725	3 818	1 718	1 977	1 187
Threadfin breams nei	*Nemipterus spp*	396	423	424	435	432	497	302
Ponyfishes(=Slipmouths) nei	*Leiognathidae*	731	781	784	804	780	...	...
Grunts, sweetlips nei	*Haemulidae (=Pomadasyidae)*	1 787	1 910	1 916	1 964	4 196	4 828	4 541
Emperors(=Scavengers) nei	*Lethrinidae*	11 455	12 242	12 283	12 590	19 647	22 619	21 056
King soldier bream	*Argyrops spinifer*	2 654	2 836	2 846	2 917	3 604	4 149	4 797
Yellowfin seabream	*Acanthopagrus latus*	...	...	...	...	...	313	122
Twobar seabream	*Acanthopagrus bifasciatus*	...	...	...	...	...	1 586	332
Goatfishes, red mullets nei	*Mullidae*	154	172	185	192	178	180 F	200
Mojarras(=Silver-biddies) nei	*Gerres spp*	1 235	1 320	1 324	1 351	1 651	1 900	371
Parrotfishes nei	*Scaridae*	...	...	...	...	333	383	712
Spinefeet(=Rabbitfishes) nei	*Siganus spp*	541	578	580	595	1 825	2 100	1 234
Hairtails, scabbardfishes nei	*Trichiuridae*	52	65	72	78	80	80 F	82
Indian oil sardine	*Sardinella longiceps*	3 491	4 077	4 085	4 144	...	...	...
Sardinellas nei	*Sardinella spp*	8 933	9 200	9 236	9 510	6 140	4 200	3 500
Stolephorus anchovies	*Stolephorus spp*	9 633	10 295	10 329	10 587	2 729	4 030	6 400
Wolf-herrings nei	*Chirocentrus spp*	72	77	78	80	75	...	...
Narrow-barred Spanish mackerel	*Scomberomorus commerson*	6 653	7 110	7 133	7 311	6 645	7 650	3 824
Seerfishes nei	*Scomberomorus spp*	1 594	2 500	2 886	3 117	2 876	2 870 F	...
Frigate and bullet tunas	*Auxis thazard, A.rochei*	578	618	620	636	376	380 F	320
Kawakawa	*Euthynnus affinis*	2 418	2 500	2 512	2 605	868	999	850
Longtail tuna	*Thunnus tonggol*	5 775	3 671	3 678	3 739	1 725	1 720 F	2 238
Marlins,sailfishes,etc. nei	*Istiophoridae*	239	230	233	233	250	250 F	...
Halfbeaks nei	*Hemiramphus spp*	36	42	58	61	55	50 F	55
Cobia	*Rachycentron canadum*	52	56	57	58	632	800 F	954
Scads nei	*Decapterus spp*	1 809	1 933	1 939	1 987	580	565 F	550
Jacks, crevalles nei	*Caranx spp*	6 905	7 379	7 403	7 588	3 107	2 762	4 759
Golden trevally	*Gnathanodon speciosus*	340	363	364	373	940	1 082	2 471
Torpedo scad	*Megalaspis cordyla*	954	1 019	1 022	1 048	1 100	...	...
Queenfishes	*Scomberoides spp*	2 002	2 140	2 147	2 201	609	701	453
Yellowstripe scad	*Selaroides leptolepis*	2 908	3 108	3 118	3 196	2 635	3 034	5 320
Carangids nei	*Carangidae*	5 861	6 205	6 810	7 020	3 568	4 107	1 164
Common dolphinfish	*Coryphaena hippurus*	56	60	61	63	55	...	...
Indian mackerel	*Rastrelliger kanagurta*	4 354	4 653	4 669	4 786	4 775	2 000	2 104
Barracudas nei	*Sphyraena spp*	2 116	2 261	2 269	2 326	543	625	487
Sharks, rays, skates, etc. nei	*Elasmobranchii*	1 902	1 832	1 881	1 945	1 530	1 762	2 541
Marine fishes nei	*Osteichthyes*	7 441	10 015	9 007	9 019	2 046	1 213	382
Marine crabs nei	*Brachyura*	56	60	60	62	1 710	1 969	353

English name Nom anglais Nombre inglés	Scientific name Nom scientifique Nombre científico	1996 t	1997 t	1998 t	1999 t	2000 t	2001 t	2002 t
Cuttlefish,bobtail squids nei	*Sepiidae, Sepiolidae*	25	26	27	28	491	465	205
	Country total	*107 000*	*114 358*	*114 739*	*117 607*	*105 456*	*112 561*	*97 574*
Uzbekistan								
Freshwater bream	*Abramis brama*	220 F	289	387	353	335	540	72
Common carp	*Cyprinus carpio*	193 F	843	804	826	617	906	148
Crucian carp	*Carassius carassius*	...	...	...	...	...	331	38
Goldfish	*Carassius auratus*	170 F	...	...	...	...	...	...
Roach	*Rutilus rutilus*	90 F	379	392	613	1 035	1 300	1 457
Grass carp(=White amur)	*Ctenopharyngodon idellus*	10 F	30	92	7	17	13	1
Silver carp	*Hypophthalmichthys molitrix*	481 F	893	249	322	586	544	75
Asp	*Aspius aspius*	20 F	...	...	...	...	...	...
Cyprinids nei	*Cyprinidae*	-	378	332	337	441	132	89
Northern pike	*Esox lucius*	20 F	7	10	60	23	24	1
Wels(=Som)catfish	*Silurus glanis*	10 F	9	14	16	8	16	0
Pike-perch	*Stizostedion lucioperca*	130 F	117	175	118	127	136	30
Snakeheads(=Murrels) nei	*Channa spp*	110 F	...	275	209	198	127	28
Freshwater fishes nei	*Osteichthyes*	40 F	130	69	10	-	1	70
	Country total	*1 494*	*3 075*	*2 799*	*2 871*	*3 387*	*4 070*	*2 009*
Viet Nam								
Freshwater fishes nei	*Osteichthyes*	163 936	176 589	137 800	168 107	169 000 F	132 280	148 200
Tuna-like fishes nei	*Scombroidei*	...	3 200 F	7 400 F	7 000 F	6 500 F	15 800 F	15 800 F
Marine fishes nei	*Osteichthyes*	808 226	832 118	849 310	922 690	927 205	1 033 623	1 025 800
Freshwater crustaceans nei	*Crustacea*	1 000 F	1 000 F	1 000 F	1 000 F	1 000 F	1 000 F	1 000 F
Marine crabs nei	*Brachyura*	28 300 F	26 400 F	48 000 F	35 800 F	40 000 F	36 900 F	46 500 F
Lobsters nei	*Reptantia*	...	2 000 F	1 000 F	200 F	500 F	700 F	700 F
Natantian decapods nei	*Natantia*	86 166	98 401	93 541	91 500	81 700 F	90 000 F	90 000 F
Cephalopods nei	*Cephalopoda*	92 000 F	92 500 F	103 000 F	110 000 F	180 000 F	130 000 F	130 000 F
Marine molluscs nei	*Mollusca*	44 016 F	44 117 F	52 903 F	50 003 F	44 685 F	50 000 F	50 000 F
	Country total	*1 223 644*	*1 276 325*	*1 293 954*	*1 386 300*	*1 450 590*	*1 490 303*	*1 508 000 F*
Yemen								
Indian halibut	*Psettodes erumei*	724	760 F	900 F	800 F	750 F	950 F	1 050 F
Sea catfishes nei	*Ariidae*	1 700	1 780 F	2 000 F	1 900 F	1 750 F	2 200 F	2 500 F
Mullets nei	*Mugilidae*	380	400 F	500 F	400 F	400 F	500 F	550 F
Groupers, seabasses nei	*Serranidae*	1 743	1 820 F	2 100 F	2 000 F	1 800 F	2 300 F	2 600 F
Cardinalfishes, etc. nei	*Apogonidae*	...	...	...	...	...	262	293
Snappers, jobfishes nei	*Lutjanidae*	1 460	1 530 F	1 700 F	1 700 F	1 500 F	1 900 F	2 150 F
Threadfin and dwarf breams nei	*Nemipteridae*	2 504	2 620 F	3 000 F	2 900 F	2 600 F	3 300 F	3 700 F
Grunts, sweetlips nei	*Haemulidae (=Pomadasyidae)*	1 318	1 380 F	1 600 F	1 500 F	1 350 F	1 700 F	1 900 F
Emperors(=Scavengers) nei	*Lethrinidae*	2 437	2 550 F	2 900 F	2 800 F	2 500 F	3 200 F	3 600 F
Demersal percomorphs nei	*Perciformes*	5 599	9 727 F	14 078 F	6 115 F	6 086 F	7 400	8 350 F
Indian oil sardine	*Sardinella longiceps*	4 120	4 310 F	4 900 F	4 700 F	4 300 F	5 500 F	6 200 F
Narrow-barred Spanish mackerel	*Scomberomorus commerson*	3 521	3 680	3 580	3 580 F	3 580 F	3 580 F	3 580 F
Seerfishes nei	*Scomberomorus spp*	500	520	510	510 F	510 F	510 F	510 F
Frigate and bullet tunas	*Auxis thazard, A.rochei*	20	20	20	20 F	20 F	20 F	20 F
Kawakawa	*Euthynnus affinis*	1 183	1 240	1 210	1 210 F	1 210 F	1 210 F	1 210 F
Skipjack tuna	*Katsuwonus pelamis*	88	90	90	90 F	90 F	90 F	90 F
Longtail tuna	*Thunnus tonggol*	1 887	1 970	1 920	1 920 F	1 920 F	1 920 F	1 920 F
Yellowfin tuna	*Thunnus albacares*	800	840	820	820 F	820 F	820 F	820 F
Tuna-like fishes nei	*Scombroidei*	300	310	300	300 F	300 F	300 F	300 F
Jack and horse mackerels nei	*Trachurus spp*	1 380	1 440 F	1 600 F	1 600 F	1 400 F	1 800 F	2 000 F
Jacks, crevalles nei	*Caranx spp*	413	430 F	500 F	500 F	400 F	500 F	550 F
Indian mackerel	*Rastrelliger kanagurta*	752	790	900 F	900 F	900 F	1 150 F	1 300 F
Barracudas nei	*Sphyraena spp*	1 813	1 900 F	2 100 F	2 100 F	1 900 F	2 400 F	2 700 F
Pelagic percomorphs nei	*Perciformes*	62 563	60 707 F	67 630 F	74 000 F	63 900 F	80 915 F	91 570 F
Rays, stingrays, mantas nei	*Rajiformes*	...	100 F	100 F	100 F	100 F	130 F	150 F
Sharks, rays, skates, etc. nei	*Elasmobranchii*	4 878	5 000 F	5 800 F	5 600 F	5 000 F	6 300 F	7 100 F
Marine crabs nei	*Brachyura*	...	...	...	...	...	37	34
Tropical spiny lobsters nei	*Panulirus spp*	323	482	828	334	178	202	227
Penaeus shrimps nei	*Penaeus spp*	665	547	904	686	526	1 600	1 700
Natantian decapods nei	*Natantia*	...	...	38	7	44	151	153
Cuttlefish,bobtail squids nei	*Sepiidae, Sepiolidae*	1 884	8 657	5 092	5 292	8 917	9 330	10 412
Octopuses, etc. nei	*Octopodidae*	...	...	...	...	...	21	9
Sea cucumbers nei	*Holothurioidea*	-	-	-	1	-	-	14
	Country total	*104 955*	*115 600*	*127 620*	*124 385*	*114 751*	*142 198*	*159 262*
Total		***42 002 404***	***43 904 775***	***44 804 664***	***45 777 965***	***45 621 911***	***45 495 225***	***45 495 299***

E-5

Fish crustaceans, molluscs, etc — Capture production by countries or areas and species — Europe
Poissons, crustacés, mollusques, etc — Captures par pays ou zones et espèces — Europe
Peces, crustáceos, moluscos, etc — Capturas por países o áreas y especies — Europa

English name Nom anglais Nombre inglés	Scientific name Nom scientifique Nombre científico	1996 t	1997 t	1998 t	1999 t	2000 t	2001 t	2002 t
Albania								
Common carp	*Cyprinus carpio*	45	38	230	216	230	300	260
Bleak	*Alburnus alburnus*	162	68	149	160	190	478	234
Crucian carp	*Carassius carassius*	...	...	77	62	65	326	260
Grass carp(=White amur)	*Ctenopharyngodon idellus*	1	0	3	5	45	10	14
Silver carp	*Hypophthalmichthys molitrix*	52	33	104	130	140	101	129
Wuchang bream	*Megalobrama amblycephala*	0	0	-	-	-	-	3
Pike-perch	*Stizostedion lucioperca*	10	4	-	-	-	45	30
Freshwater fishes nei	*Osteichthyes*	-	-	-	-	-	-	24
European eel	*Anguilla anguilla*	50	21	58	63	70	98	25
Rainbow trout	*Oncorhynchus mykiss*	-	-	-	-	-	-	67
Salmonoids nei	*Salmonoidei*	35	15	102	104	110	56	43
Shads nei	*Alosa spp*	2	1	-	-	-	2	23
European flounder	*Platichthys flesus*	3	9	42	41	45	15	3
Common sole	*Solea solea*	27	21	35	31	41	14	195
Megrim	*Lepidorhombus whiffiagonis*	1	0	-	-	-	1	6
Blue whiting(=Poutassou)	*Micromesistius poutassou*	2	0	-	-	-	-	6
European hake	*Merluccius merluccius*	293	185	340	341	330	380	200
Mullets nei	*Mugilidae*	105	42	136	140	150	180	128
Groupers nei	*Epinephelus spp*	2	1	-	-	-	-	3
European seabass	*Dicentrarchus labrax*	32	14	30	30	50	70	64
Brown meagre	*Sciaena umbra*	6	2	-	-	-	10	8
Shi drum	*Umbrina cirrosa*	...	...	...	...	...	...	10
Pandoras nei	*Pagellus spp*	27	25	33	35	34	15	26
Common dentex	*Dentex dentex*	...	...	...	...	...	26	80
Gilthead seabream	*Sparus aurata*	27	11	20	20	23	90	181
Bogue	*Boops boops*	104	65	220	220	220	120	150
Salema	*Sarpa salpa*	...	...	...	...	...	1	5
Porgies, seabreams nei	*Sparidae*	9	1	-	-	-	1	2
Picarels nei	*Spicara spp*	11	0	7	7	10	5	11
Surmullets(=Red mullets) nei	*Mullus spp*	64	61	143	145	140	170	97
Gobies nei	*Gobiidae*	10	4	-	-	-	5	7
European conger	*Conger conger*	1	2	-	-	-	-	10
John dory	*Zeus faber*	0	0	-	-	-	-	5
Wreckfish	*Polyprion americanus*	1	0	-	-	-	-	10
Silver scabbardfish	*Lepidopus caudatus*	0	0	18	19	18	0	0
Gurnards, searobins nei	*Triglidae*	15	0	-	-	-	34	72
Angler(=Monk)	*Lophius piscatorius*	46	0	42	48	44	...	14
European pilchard(=Sardine)	*Sardina pilchardus*	196	28	28	40	45	123	90
European anchovy	*Engraulis encrasicolus*	2	0	-	-	-	4	4
Atlantic bonito	*Sarda sarda*	2	0	12	30	25	30	24
Swordfish	*Xiphias gladius*	13	0	-	-	-	2	15
Silversides(=Sand smelts) nei	*Atherinidae*	20	8	11	15	20	10	16
Jack and horse mackerels nei	*Trachurus spp*	68	18	85	92	90	21	65
Greater amberjack	*Seriola dumerili*	2	1	-	-	-	2	19
Scomber mackerels nei	*Scomber spp*	10	5	4	4	4	11	11
Smooth-hounds nei	*Mustelus spp*	12	3	12	12	32	5	23
Dogfish sharks nei	*Squalidae*	64	13	-	-	-	10	77
Angelsharks, sand devils nei	*Squatinidae*	53	20	31	30	30	16	79
Guitarfishes, etc. nei	*Rhinobatidae*	1	0	-	-	-	-	2
Rays, stingrays, mantas nei	*Rajiformes*	23	24	86	78	85	14	28
Marine fishes nei	*Osteichthyes*	327	108	370	375	789	312	323
Freshwater crustaceans nei	*Crustacea*	0	-	-	-	-	-	-
Common spiny lobster	*Palinurus elephas*	...	...	...	...	...	1	2
Norway lobster	*Nephrops norvegicus*	3	0	-	-	-	10	5
Caramote prawn	*Penaeus kerathurus*	8	4	18	18	20	23	84
Deepwater rose shrimp	*Parapenaeus longirostris*	20	8	-	-	-	52	57
Blue and red shrimp	*Aristeus antennatus*	3	0	-	-	-	-	34
Mediterranean mussel	*Mytilus galloprovincialis*	...	24	-	-	-	-	350
Striped venus	*Chamelea gallina*	0	-	-	-	-	-	-
Common cuttlefish	*Sepia officinalis*	33	33	51	51	50	22	52
Common squids nei	*Loligo spp*	47	34	93	93	90	65	85
Common octopus	*Octopus vulgaris*	75	59	93	90	85	24	105
	Country total	*2 125*	*1 013*	*2 683*	*2 745*	*3 320*	*3 310*	*3 955*
Andorra								
Freshwater fishes nei	*Osteichthyes*	0	0	0	0	0	0	0
	Country total	*0*	*0*	*0*	*0*	*0*	*0*	*0*
Austria								
Freshwater fishes nei	*Osteichthyes*	450	465	451	432	439	362	350
	Country total	*450*	*465*	*451*	*432*	*439*	*362*	*350*
Belarus								
Freshwater bream	*Abramis brama*	244	182	130	98	27	198	403
Common carp	*Cyprinus carpio*	39	12	8	5	17	15	3 845
Tench	*Tinca tinca*	8	4	3	90	41	7	65
Crucian carp	*Carassius carassius*	69	35	106	138	154	188	239
Roaches nei	*Rutilus spp*	95	42	24	19	9	33	59
Orfe(=Ide)	*Leuciscus idus*	19	11	14	11	5	18	26
Cyprinids nei	*Cyprinidae*	92	61	47	27	19	77	195
Northern pike	*Esox lucius*	27	11	9	12	162	171	312
Burbot	*Lota lota*	17	13	18	20	7	29	38
European perch	*Perca fluviatilis*	101	75	49	38	34	86	105
Pike-perch	*Stizostedion lucioperca*	6	3	3	10	17	24	36

English name Nom anglais Nombre inglés	Scientific name Nom scientifique Nombre científico	1996 t	1997 t	1998 t	1999 t	2000 t	2001 t	2002 t
Freshwater fishes nei	*Osteichthyes*	-	-	-	-	-	-	147
European eel	*Anguilla anguilla*	20	15	18	16	14	25	10
Whitefishes nei	*Coregonus spp*	4	1	2	17	18	31	46
Three-spined stickleback	*Gasterosteus aculeatus*	80	34	26	13	29	41	49
	Country total	*821*	*499*	*457*	*514*	*553*	*943*	*5 575*
Belgium								
Freshwater bream	*Abramis brama*	60	60	60	60	60	60	60
Common carp	*Cyprinus carpio*	30	30	30	30	30	30	30
Tench	*Tinca tinca*	15	15	15	15	15	15	15
Roach	*Rutilus rutilus*	160	160	160	150	160	160	160
Cyprinids nei	*Cyprinidae*	50	50	50	50	50	50	50
Northern pike	*Esox lucius*	20	20	20	20	20	20	20
European perch	*Perca fluviatilis*	25	25	25	15	25	25	25
Pike-perch	*Stizostedion lucioperca*	15	15	15	10	15	15	15
European eel	*Anguilla anguilla*	30	30	30	30	30	30	30
Sea trout	*Salmo trutta*	100	100	100	150	100	100	100
Grayling	*Thymallus thymallus*	5	5	5	5	5	5	5
Whitefishes nei	*Coregonus spp*	1	1	1	1	1	1	1
Atlantic halibut	*Hippoglossus hippoglossus*	5	4	3	2	2	2	3
European plaice	*Pleuronectes platessa*	7 684	7 469	7 369	8 186	9 148	8 487	7 043
Common dab	*Limanda limanda*	690	789	961	980	865	850	752
Lemon sole	*Microstomus kitt*	1 094	975	1 256	1 020	1 057	1 076	1 089
European flounder	*Platichthys flesus*	278	146	307	363	322	316	230
Common sole	*Solea solea*	5 150	4 514	4 102	4 492	4 479	4 975	5 089
Megrim	*Lepidorhombus whiffiagonis*	208	188	142	143	132	87	80
Brill	*Scophthalmus rhombus*	422	349	341	365	423	491	448
Turbot	*Psetta maxima*	382	337	327	368	464	506	445
Atlantic cod	*Gadus morhua*	4 491	5 677	6 893	4 540	3 693	3 207	3 506
Ling	*Molva molva*	153	124	124	103	120	88	91
Haddock	*Melanogrammus aeglefinus*	394	746	976	569	512	840	737
Saithe(=Pollock)	*Pollachius virens*	161	264	256	208	126	30	120
Pollack	*Pollachius pollachius*	115	119	113	108	116	137	143
Pouting(=Bib)	*Trisopterus luscus*	377	336	323	364	468	561	577
Whiting	*Merlangius merlangus*	1 281	989	856	1 072	826	732	500
European hake	*Merluccius merluccius*	42	54	76	92	117	124	91
European conger	*Conger conger*	75	73	79	53	49	56	66
Atlantic wolffish	*Anarhichas lupus*	99	125	208	201	290	175	188
Atlantic redfishes nei	*Sebastes spp*	19	16	2	3	5	6	3
Red gurnard	*Chelidonichthys cuculus*	145	128	157	262	418	493	500
Grey gurnard	*Eutrigla gurnardus*	119	61	47	49	44	33	63
Gurnards, searobins nei	*Triglidae*	235	245	200	160	161	177	183
Angler(=Monk)	*Lophius piscatorius*	1 369	1 113	961	818	1 047	1 354	1 312
Atlantic herring	*Clupea harengus*	2	1	1	1	1	11	23
European sprat	*Sprattus sprattus*	3	7	4	2	2	2	1
Atlantic horse mackerel	*Trachurus trachurus*	28	19	19	21	19	20	30
Atlantic mackerel	*Scomber scombrus*	64	106	125	178	151	98	23
Picked dogfish	*Squalus acanthias*	16	15	17	10	11	13	23
Dogfishes and hounds nei	*Squalidae, Scyliorhinidae*	415	430	365	345	390	396	448
Raja rays nei	*Raja spp*	1 363	1 259	1 232	1 351	1 231	1 527	1 734
Various sharks nei	*Selachimorpha(Pleurotremata)*	19	18	11	14	15	18	12
Groundfishes nei	*Osteichthyes*	1 290	1 389	979	774	643	501	509
Pelagic fishes nei	*Osteichthyes*	4	2	1	1	2	4	30
Edible crab	*Cancer pagurus*	175	222	180	81	106	108	126
Norway lobster	*Nephrops norvegicus*	188	316	240	349	254	284	287
European lobster	*Homarus gammarus*	-	-	-	1	1	6	1
Common shrimp	*Crangon crangon*	903	743	378	1 053	616	988	538
Whelk	*Buccinum undatum*	226	166	117	83	101	119	92
Great Atlantic scallop	*Pecten maximus*	163	208	224	247	292	340	432
Common cuttlefish	*Sepia officinalis*	386	154	252	222	463	370	741
Common squids nei	*Loligo spp*	21	17	25	23	59	51	137
Octopuses, etc. nei	*Octopodidae*	28	45	49	41	44	29	70
Marine molluscs nei	*Mollusca*	30	31	26	22	4	10	1
	Country total	*30 823*	*30 500*	*30 835*	*29 876*	*29 800*	*30 209*	*29 028*
Bosnia Herzg								
Freshwater fishes nei	*Osteichthyes*	2 500 F	2 500 F	2 500 F	2 500 F	2 500 F	2 500 F	2 500 F
Marine fishes nei	*Osteichthyes*	0	0	0	0	0	0	0
	Country total	*2 500 F*	*2 500 F*	*2 500 F*	*2 500 F*	*2 500 F*	*2 500 F*	*2 500 F*
Bulgaria								
Freshwater bream	*Abramis brama*	91	90	82	71	25	4	6
Common carp	*Cyprinus carpio*	16	281	251	302	143	880	521
Tench	*Tinca tinca*	...	...	...	...	...	2	3
Bleak	*Alburnus alburnus*	6	62	67	92	24	3	19
Barbel	*Barbus barbus*	120	142	113	93	43	3	4
Common nase	*Chondrostoma nasus*	...	3	3	4	8	7	17
Crucian carp	*Carassius carassius*	...	328	392	360	179	138	247
Roach	*Rutilus rutilus*	...	...	10	40	...	...	14
Rudd	*Scardinius erythrophthalmus*	...	...	71	90	39	3	45
Orfe(=Ide)	*Leuciscus idus*	8	10	12	14	...	0	1
Chub	*Leuciscus cephalus*	...	92	121	150	43	4	17
Grass carp(=White amur)	*Ctenopharyngodon idellus*	8	8	17	20	12	18	23
Silver carp	*Hypophthalmichthys molitrix*	488	471	553	488	42	403	85
Bighead carp	*Hypophthalmichthys nobilis*	...	...	...	...	...	...	223
Vimba bream	*Vimba vimba*	...	84	15	50	4	1	2

E-5 Fish crustaceans, molluscs, etc — Capture production by countries or areas and species — Europe
Poissons, crustacés, mollusques, etc — Captures par pays ou zones et espèces — Europe
Peces, crustáceos, moluscos, etc — Capturas por países o áreas y especies — Europa

English name Nom anglais Nombre inglés	Scientific name Nom scientifique Nombre científico	1996 t	1997 t	1998 t	1999 t	2000 t	2001 t	2002 t
Asp	*Aspius aspius*	4	5	7	8	9	7	1
Cyprinids nei	*Cyprinidae*	...	1	251	250	102	-	14
Northern pike	*Esox lucius*	0	19	41	74	14	2	6
Wels(=Som)catfish	*Silurus glanis*	27	34	44	106	59	4	34
Burbot	*Lota lota*	-	-	-	-	1	2	5
European perch	*Perca fluviatilis*	...	63	81	102	26	6	23
Pike-perch	*Stizostedion lucioperca*	26	45	54	68	27	5	18
Freshwater fishes nei	*Osteichthyes*	182	0	1	-	-	-	22
Danube sturgeon(=Osetr)	*Acipenser gueldenstaedtii*	2	6	7	6	1	1	3
Sterlet sturgeon	*Acipenser ruthenus*	1	1	1	2	2	1	3
Starry sturgeon	*Acipenser stellatus*	0	0	4	6	1	1	2
Beluga	*Huso huso*	29	42	43	37	19	7	14
Sea trout	*Salmo trutta*	-	8	11	11	4	1	16
Rainbow trout	*Oncorhynchus mykiss*	-	12	10	10	5	17	21
Brook trout	*Salvelinus fontinalis*	-	1	1	1	1	3	6
European whitefish	*Coregonus lavaretus*	-	0	0	0	-	-	-
Houting	*Coregonus oxyrinchus*	-	0	0	0	-	-	-
Pontic shad	*Alosa pontica*	233	165	171	73	39	32	141
European flounder	*Platichthys flesus*	-	-	-	-	-	-	9
Common sole	*Solea solea*	-	-	-	-	-	-	10
Turbot	*Psetta maxima*	62	60	64	54	55	57	136
Rocklings nei	*Gaidropsarus spp*	-	-	-	-	-	-	20
Whiting	*Merlangius merlangus*	-	-	-	-	9	8	16
Senegalese hake	*Merluccius senegalensis*	-	-	10	-	-	-	-
Flathead grey mullet	*Mugil cephalus*	26	28	11	14	15	47	71
Leaping mullet	*Liza saliens*	3	2	2	2	0	10	25
Sargo breams nei	*Diplodus spp*	-	-	-	-	-	90	-
Porgies, seabreams nei	*Sparidae*	-	-	35	-	-	-	-
Red mullet	*Mullus barbatus*	-	-	-	-	5	26	33
Gobies nei	*Gobiidae*	477	424	381	437	145	142	142
Largehead hairtail	*Trichiurus lepturus*	-	-	1 383	-	-	-	-
Sardinellas nei	*Sardinella spp*	-	-	3 216	-	-	-	-
European pilchard(=Sardine)	*Sardina pilchardus*	-	-	48	-	-	-	-
European sprat	*Sprattus sprattus*	3 535	3 646	3 275	3 595	1 737	695	11 595
European anchovy	*Engraulis encrasicolus*	23	44	896	36	64	102	237
Atlantic bonito	*Sarda sarda*	33	16	51	20	35	49	-
Tuna-like fishes nei	*Scombroidei*	-	-	225	-	-	-	-
Garfish	*Belone belone*	2	2	4	4	9	16	37
Big-scale sand smelt	*Atherina boyeri*	-	-	0	1	21	2	0
Bluefish	*Pomatomus saltatrix*	10	12	10	8	18	2	102
Mediterranean horse mackerel	*Trachurus mediterraneus*	68	36	40	30	111	130	142
Jack and horse mackerels nei	*Trachurus spp*	-	-	1 669	-	-	-	-
Chub mackerel	*Scomber japonicus*	-	-	486	-	-	-	-
Picked dogfish	*Squalus acanthias*	64	40	28	25	102	126	100
Marine fishes nei	*Osteichthyes*	49	51	376	-	-	2	15
Euro-American crayfishes nei	*Astacidae, Cambaridae*	-	-	-	-	-	1	8
Natantian decapods nei	*Natantia*	1	3	2	2	-	-	-
Sea snails	*Rapana spp*	3 260	4 900	4 300	3 800	3 800	3 353	698
Mediterranean mussel	*Mytilus galloprovincialis*	-	-	-	-	-	7	55
	Country total	*8 854*	*11 237*	*18 946*	*10 556*	*6 998*	*6 420*	*15 007*
Channel Is								
Freshwater fishes nei	*Osteichthyes*	0	0	0	0	0	0	0
European plaice	*Pleuronectes platessa*	26	19	17	23	18	13	11
Lemon sole	*Microstomus kitt*	0	0	1	1	3	1	1
Common sole	*Solea solea*	9	26	21	21	26	25	20
Sand sole	*Solea lascaris*	2	3	3	1	1	1	1
Megrim	*Lepidorhombus whiffiagonis*	-	-	-	-	-	1	0
Brill	*Scophthalmus rhombus*	10	10	10	6	17	17	16
Turbot	*Psetta maxima*	5	5	3	4	6	9	6
Atlantic cod	*Gadus morhua*	6	14	19	21	11	6	9
Ling	*Molva molva*	20	37	25	19	13	3	2
Haddock	*Melanogrammus aeglefinus*	44	0	0	-	-	-	0
Saithe(=Pollock)	*Pollachius virens*	2	4	0	2	-	-	-
Pollack	*Pollachius pollachius*	27	35	52	75	97	57	40
Pouting(=Bib)	*Trisopterus luscus*	1	1	0	1	5	6	14
Blue whiting(=Poutassou)	*Micromesistius poutassou*	-	1	1	1	-	-	-
Whiting	*Merlangius merlangus*	1	0	3	2	3	3	1
European hake	*Merluccius merluccius*	0	0	0	0	-	-	-
Gadiformes nei	*Gadiformes*	-	-	-	1	-	-	-
Mullets nei	*Mugilidae*	23	14	11	13	12	9	7
European seabass	*Dicentrarchus labrax*	56	74	79	107	129	80	73
Gilthead seabream	*Sparus aurata*	...	...	...	15	-	-	-
Porgies, seabreams nei	*Sparidae*	9	48	126	132	105	108	125
Surmullet	*Mullus surmuletus*	4	11	17	22	23	19	15
Ballan wrasse	*Labrus bergylta*	-	-	-	2	1	-	-
Sandeels(=Sandlances) nei	*Ammodytes spp*	2	63	61	41	41	41	40
European conger	*Conger conger*	57	17	32	30	24	25	17
John dory	*Zeus faber*	0	0	1	1	2	1	1
Wreckfish	*Polyprion americanus*	-	2	1	0	-	-	-
Red gurnard	*Chelidonichthys cuculus*	10	8	6	10	7	-	15
Gurnards, searobins nei	*Triglidae*	...	10	31	22	10	41	7
Angler(=Monk)	*Lophius piscatorius*	2	3	3	2	3	5	2
Garfish	*Belone belone*	0	1	2	0	2	2	1
Atlantic horse mackerel	*Trachurus trachurus*	5	7	11	10	8	8	9
Atlantic mackerel	*Scomber scombrus*	9	9	23	18	16	14	12
Porbeagle	*Lamna nasus*	...	...	1	0	2	2	2
Blue shark	*Prionace glauca*	...	...	1	0	-	-	-

E-5

Fish crustaceans, molluscs, etc	Capture production by countries or areas and species	Europe
Poissons, crustacés, mollusques, etc	Captures par pays ou zones et espèces	Europe
Peces, crustáceos, moluscos, etc	Capturas por países o áreas y especies	Europa

English name Nom anglais Nombre inglés	Scientific name Nom scientifique Nombre científico	1996 t	1997 t	1998 t	1999 t	2000 t	2001 t	2002 t
Dogfishes and hounds nei	*Squalidae, Scyliorhinidae*	48	30	22	15	33	51	53
Raja rays nei	*Raja spp*	180	33	223	262	181	241	235
Various sharks nei	*Selachimorpha(Pleurotremata)*	2	3	3	7	1	-	-
Marine fishes nei	*Osteichthyes*	22	203	27	40	-	-	-
Edible crab	*Cancer pagurus*	2 043	2 268	2 214	1 655	1 440	1 560	1 423
Spinous spider crab	*Maja squinado*	868	445	460	553	522	428	406
Marine crabs nei	*Brachyura*	-	-	2	2	2	-	0
Palinurid spiny lobsters nei	*Palinurus spp*	1	0	1	1	1	3	2
European lobster	*Homarus gammarus*	260	221	214	212	153	178	200
Tuberculate abalone	*Haliotis tuberculata*	-	-	-	-	3	2	3
Whelk	*Buccinum undatum*	550	438	135	8	338	519	204
Queen scallop	*Aequipecten opercularis*	-	-	-	7	-	-	-
Great Atlantic scallop	*Pecten maximus*	29	109	155	203	295	439	463
Scallops nei	*Pectinidae*	-	50	80	-	-	-	-
Common cuttlefish	*Sepia officinalis*	12	10	15	22	26	8	11
Common squids nei	*Loligo spp*	1	6	5	11	9	1	2
	Country total	*4 346*	*4 238*	*4 117*	*3 601*	*3 589*	*3 927*	*3 449*
Croatia								
Common carp	*Cyprinus carpio*	143	126	1 F	1	3	1	2
Tench	*Tinca tinca*	3	0	-	-	-	-	-
Northern pike	*Esox lucius*	30	30	1 F	1	1	1	1
Wels(=Som)catfish	*Silurus glanis*	24	20	1 F	1	1	1	2
Pike-perch	*Stizostedion lucioperca*	12	14	1 F	1	1	1	1
Freshwater fishes nei	*Osteichthyes*	217	213	5 F	5	8	12	17
European eel	*Anguilla anguilla*	6	7	-	-	-	-	-
Rainbow trout	*Oncorhynchus mykiss*	5	5	1 F	1	3	18	2
Flatfishes nei	*Pleuronectiformes*	130	133	150	65	113	111	106
European hake	*Merluccius merluccius*	929	828	935	650	583	557	624
Flathead grey mullet	*Mugil cephalus*	105	105	195	120	68	14	7
European seabass	*Dicentrarchus labrax*	...	...	31	20	22	13	2
Common dentex	*Dentex dentex*	56	76	70	50	55	10	7
Black seabream	*Spondyliosoma cantharus*	51	56	60	23	19	5	3
Saddled seabream	*Oblada melanura*	183	169	185	130	120	55	30
Gilthead seabream	*Sparus aurata*	13	44	84	27	25	11	6
Bogue	*Boops boops*	303	342	335	140	147	109	55
Salema	*Sarpa salpa*	127	132	125	42	24	28	14
Picarels nei	*Spicara spp*	290	340	255	245	231	189	72
Surmullet	*Mullus surmuletus*	311	280	285	200	277	480	294
European conger	*Conger conger*	122	114	130	49	38	25	9
Scorpionfishes nei	*Scorpaenidae*	137	133	185	46	40	39	22
Sardinellas nei	*Sardinella spp*	163	344	170	25	35	37	42
European pilchard(=Sardine)	*Sardina pilchardus*	9 199	6 996	12 500	10 500	11 226	9 097	12 626
European anchovy	*Engraulis encrasicolus*	220	545	990	3 000	3 735	2 850	3 187
Atlantic bonito	*Sarda sarda*	159	171	158	120	120	54	28
Frigate and bullet tunas	*Auxis thazard, A.rochei*	26	16	12	0	-	-	-
Little tunny(=Atl.black skipj)	*Euthynnus alletteratus*	9	9	16	0	-	-	-
Atlantic bluefin tuna	*Thunnus thynnus*	1 360	1 105	906	970	930	903	977
Swordfish	*Xiphias gladius*	...	...	10	20	...	...	...
Garfish	*Belone belone*	63	85	60	10	5	11	4
Silversides(=Sand smelts) nei	*Atherinidae*	148	171	260	50	44	20	8
Mediterranean horse mackerel	*Trachurus mediterraneus*	361	336	200	90	75	192	84
Greater amberjack	*Seriola dumerili*	80	64	62	30	25	19	38
Scomber mackerels nei	*Scomber spp*	650	998	585	175	377	490	420
Dogfish sharks nei	*Squalidae*	260	239	105	53	50	74	34
Rays, stingrays, mantas nei	*Rajiformes*	141	119	120	68	57	42	34
Marine fishes nei	*Osteichthyes*	619	613	590	619	1 080	863	1 625
Spinous spider crab	*Maja squinado*	25	54	50	15	14	21	61
Norway lobster	*Nephrops norvegicus*	486	473	500	240	250	274	140
European lobster	*Homarus gammarus*	31	43	40	18	18	13	3
European flat oyster	*Ostrea edulis*	...	...	35	14	12	24	5
Mediterranean mussel	*Mytilus galloprovincialis*	...	...	200	60	57	32	45
Cuttlefish,bobtail squids nei	*Sepiidae, Sepiolidae*	102	151	215	145	138	85	45
Common squids nei	*Loligo spp*	233	287	290	170	127	162	53
Common octopus	*Octopus vulgaris*	180	293	435	...	...	...	...
Cephalopods nei	*Cephalopoda*	483	443	265	646	873	885	403
Marine molluscs nei	*Mollusca*	38	313	125	40	27	69	33
Aquatic invertebrates nei	*Invertebrata*	-	-	4	5	8	592	59
	Country total	*18 233*	*17 035*	*21 938*	*18 900*	*21 062*	*18 489*	*21 230*
Czech Rep								
Freshwater bream	*Abramis brama*	247	232	253	297	261	247	243
Common carp	*Cyprinus carpio*	2 522	2 312	2 899	3 006	3 558	3 560	3 909
Tench	*Tinca tinca*	23	21	29	30	27	24	24
Goldfish	*Carassius auratus*	49	40	40	43	35	37	33
Orfe(=Ide)	*Leuciscus idus*	27	0	0	0	0	0	1
Grass carp(=White amur)	*Ctenopharyngodon idellus*	47	49	53	70	60	60	69
Bighead carp	*Hypophthalmichthys nobilis*	3	5	6	8	10	12	12
Asp	*Aspius aspius*	10	15	16	16	13	17	18
Northern pike	*Esox lucius*	163	159	168	183	180	176	172
Wels(=Som)catfish	*Silurus glanis*	36	44	49	52	53	57	61
European perch	*Perca fluviatilis*	33	34	36	37	34	34	30
Pike-perch	*Stizostedion lucioperca*	130	157	125	130	134	139	144
Freshwater fishes nei	*Osteichthyes*	102	121	134	168	151	131	126
European eel	*Anguilla anguilla*	28	27	28	28	24	29	28
Trouts nei	*Salmo spp*	53	57	70	64	55	56	50
Rainbow trout	*Oncorhynchus mykiss*	28	29	30	38	39	48	44

English name Nom anglais Nombre inglés	Scientific name Nom scientifique Nombre científico	1996 t	1997 t	1998 t	1999 t	2000 t	2001 t	2002 t
Brook trout	*Salvelinus fontinalis*	2	2	2	3	3	3	5
Grayling	*Thymallus thymallus*	19	15	13	16	16	15	13
European whitefish	*Coregonus lavaretus*	2	2	1	1	1	1	1
	Country total	*3 524*	*3 321*	*3 952*	*4 190*	*4 654*	*4 646*	*4 983*
Denmark								
Freshwater bream	*Abramis brama*	66	94	144	80	48	28	26
Common carp	*Cyprinus carpio*	1	1	1	0	0	0	10
Tench	*Tinca tinca*	0	0	0	0	1	0	0
Crucian carp	*Carassius carassius*	-	0	-	0	0	0	0
Roach	*Rutilus rutilus*	43	38	86	39	43	26	19
Rudd	*Scardinius erythrophthalmus*	-	-	-	0	2	1	0
Northern pike	*Esox lucius*	11	14	16	11	9	10	10
Burbot	*Lota lota*	0	0	0	0	0	0	0
European perch	*Perca fluviatilis*	63	83	100	113	100	73	79
Pike-perch	*Stizostedion lucioperca*	12	51	53	15	28	9	9
Ruffe	*Gymnocephalus cernuus*	13	5	9	3	2	1	1
Freshwater fishes nei	*Osteichthyes*	3	-	-	0	0	1	0
Sturgeons nei	*Acipenseridae*	-	-	0	-	-	-	-
European eel	*Anguilla anguilla*	734	796	600	716	620	658	569
Atlantic salmon	*Salmo salar*	528	493	487	389	413	434	321
Sea trout	*Salmo trutta*	77	52	59	101	70	52	35
Rainbow trout	*Oncorhynchus mykiss*	1	1	-	6	-	0	0
European smelt	*Osmerus eperlanus*	53	34	19	20	29	25	10
European whitefish	*Coregonus lavaretus*	30	31	29	25	7	77	51
Houting	*Coregonus oxyrinchus*	-	-	1	-	22	-	-
Three-spined stickleback	*Gasterosteus aculeatus*	-	4	-	-	-	-	-
Atlantic halibut	*Hippoglossus hippoglossus*	71	70	62	51	53	53	82
European plaice	*Pleuronectes platessa*	22 571	24 621	18 927	23 123	23 902	26 849	22 998
Greenland halibut	*Reinhardtius hippoglossoides*	3	2	1	1	1	1	1
Witch flounder	*Glyptocephalus cynoglossus*	999	1 563	1 906	2 116	2 338	2 049	1 907
Amer. plaice(=Long rough dab)	*Hippoglossoides platessoides*	0	5	8	31	5	31	0
Common dab	*Limanda limanda*	3 952	3 211	2 646	2 514	2 113	2 300	2 648
Lemon sole	*Microstomus kitt*	1 108	1 172	1 591	1 812	2 037	1 820	1 447
European flounder	*Platichthys flesus*	5 469	5 378	4 725	3 528	4 604	6 066	5 143
Common sole	*Solea solea*	2 083	1 478	1 050	1 433	1 804	1 324	1 210
Megrim	*Lepidorhombus whiffiagonis*	7	6	26	21	30	55	8
Brill	*Scophthalmus rhombus*	220	148	189	244	237	163	148
Turbot	*Psetta maxima*	1 117	908	770	727	809	872	994
Tusk(=Cusk)	*Brosme brosme*	130	146	105	177	225	274	221
Atlantic cod	*Gadus morhua*	90 741	80 491	69 025	70 547	57 018	46 185	37 859
Ling	*Molva molva*	868	969	823	837	741	910	829
Blue ling	*Molva dypterygia*	8	14	4	7	15	27	13
Haddock	*Melanogrammus aeglefinus*	5 050	5 227	5 786	3 130	2 707	4 001	8 920
Saithe(=Pollock)	*Pollachius virens*	4 708	4 517	3 973	4 501	3 536	3 592	5 686
Pollack	*Pollachius pollachius*	1 049	637	564	480	490	358	464
Norway pout	*Trisopterus esmarkii*	162 943	153 047	63 678	57 441	150 040	62 913	78 243
Pouting(=Bib)	*Trisopterus luscus*	1	2	2	1	1	3	2
Blue whiting(=Poutassou)	*Micromesistius poutassou*	52 699	33 486	69 305	79 810	62 074	65 058	51 040
Whiting	*Merlangius merlangus*	391	196	144	175	326	326	381
European hake	*Merluccius merluccius*	868	670	591	846	811	1 043	1 163
Roundnose grenadier	*Coryphaenoides rupestris*	1 174	2 124	4 429	2 521	1 981	2 229	3 150
Mullets nei	*Mugilidae*	22	29	24	26	22	29	17
European seabass	*Dicentrarchus labrax*	1	1	2	1	5	2	1
Surmullet	*Mullus surmuletus*	1	1	1	3	2	5	12
Eelpout	*Zoarces viviparus*	3	7	2	7	3	5	4
Sandeels(=Sandlances) nei	*Ammodytes spp*	669 035	840 774	646 905	528 551	567 350	666 295	662 402
Greater weever	*Trachinus draco*	45	50	36	52	39	48	98
Argentines	*Argentina spp*	1 446	1 455	748	1 420	1 039	907	614
Atlantic wolffish	*Anarhichas lupus*	195	220	273	298	294	223	248
Golden redfish	*Sebastes marinus*	4	4	7	5	8	11	1
Beaked redfish	*Sebastes mentella*	17	19	20	48	35	89	42
Tub gurnard	*Chelidonichthys lucerna*	5	4	5	19	15	12	63
Grey gurnard	*Eutrigla gurnardus*	70	36	56	85	96	289	65
Lumpfish(=Lumpsucker)	*Cyclopterus lumpus*	1 973	1 553	188	835	425	422	592
Angler(=Monk)	*Lophius piscatorius*	1 835	2 040	1 873	1 858	1 724	1 917	1 932
Atlantic herring	*Clupea harengus*	153 009	125 300	139 710	137 578	153 899	141 263	112 476
European pilchard(=Sardine)	*Sardina pilchardus*	13 704	1 740	17 337	17 676	7 893	1 399	3 523
European sprat	*Sprattus sprattus*	226 135	284 489	270 439	282 299	276 878	256 517	237 466
European anchovy	*Engraulis encrasicolus*	-	-	-	-	-	0	-
Atlantic bluefin tuna	*Thunnus thynnus*	0	-	1	-	-	-	-
Capelin	*Mallotus villosus*	60 898	48 524	40 349	3 837	20 807	17 588	23 165
Garfish	*Belone belone*	708	994	648	571	714	558	580
Atlantic horse mackerel	*Trachurus trachurus*	63 929	63 430	32 597	32 047	25 083	23 631	11 611
Atlantic pomfret	*Brama brama*	-	-	-	-	-	0	0
Atlantic mackerel	*Scomber scombrus*	26 238	24 054	27 415	29 705	31 642	31 370	33 046
Porbeagle	*Lamna nasus*	71	69	85	107	73	76	42
Blue shark	*Prionace glauca*	3	1	1	1	2	1	13
Tope shark	*Galeorhinus galeus*	2	2	3	4	7	4	4
Picked dogfish	*Squalus acanthias*	142	196	126	131	146	156	256
Blue skate	*Raja batis*	32	9	7	11	47	0	0
Raja rays nei	*Raja spp*	44	40	20	46	87	122	60
Rabbit fish	*Chimaera monstrosa*	-	-	-	-	-	1	0
Groundfishes nei	*Osteichthyes*	0	0	0	-	-	-	-
Finfishes nei	*Osteichthyes*	118	103	116	78	142	107	209
Euro-American crayfishes nei	*Astacidae, Cambaridae*	0	0	0	0	0	0	0
Edible crab	*Cancer pagurus*	15	14	21	25	33	53	67
Spinous spider crab	*Maja squinado*	0	1	1	1	1	2	2

English name Nom anglais Nombre inglés	Scientific name Nom scientifique Nombre científico	1996 t	1997 t	1998 t	1999 t	2000 t	2001 t	2002 t
Marine crabs nei	*Brachyura*	171	193	220	193	205	189	245
Norway lobster	*Nephrops norvegicus*	4 176	4 282	4 982	5 455	5 083	4 813	5 437
European lobster	*Homarus gammarus*	44	39	18	11	11	11	11
Northern prawn	*Pandalus borealis*	9 071	8 552	8 100	4 880	5 721	5 331	5 515
Common prawn	*Palaemon serratus*	144	182	122	141	151	141	144
Common shrimp	*Crangon crangon*	2 207	3 250	2 509	2 911	2 322	1 826	3 197
Norwegian krill	*Meganyctiphanes norvegica*	-	-	88	-	-	36	-
Marine crustaceans nei	*Crustacea*	117	0	-	-	-	-	-
Periwinkles nei	*Littorina spp*	4	1	2	-	2	0	0
Whelk	*Buccinum undatum*	-	-	-	1	-	1	1
European flat oyster	*Ostrea edulis*	9	24	6	8	9	23	528
Blue mussel	*Mytilus edulis*	86 002	90 765	108 329	96 215	110 618	122 480	110 873
Scallops nei	*Pectinidae*	0	0	0	0	0	0	0
Solid surf clam	*Spisula solida*	-	-	-	-	55	214	1 709
Common edible cockle	*Cerastoderma edule*	5	2 603	1 993	246	2 089	2 392	78
Cuttlefish,bobtail squids nei	*Sepiidae, Sepiolidae*	2	0	-	26	20	5	22
Northern shortfin squid	*Illex illecebrosus*	-	17	16	-	-	-	-
Starfishes nei	*Asteroidea*	-	-	-	0	0	0	0
European edible sea urchin	*Echinus esculentus*	0	0	0	1	0	0	0
Aquatic invertebrates nei	*Invertebrata*	0	-	-	-	-	-	-
	Country total	*1 681 517*	*1 826 852*	*1 557 335*	*1 405 005*	*1 534 089*	*1 510 461*	*1 442 068*
Estonia								
Freshwater bream	*Abramis brama*	175	230	247	194	204	328	401
Roach	*Rutilus rutilus*	502	492	449	324	478	503	767
Orfe(=Ide)	*Leuciscus idus*	132	90	70	52	65	40	30
Vimba bream	*Vimba vimba*	165	185	165	122	101	83	115
Northern pike	*Esox lucius*	139	105	131	152	177	199	218
Burbot	*Lota lota*	36	31	21	53	40	38	42
European perch	*Perca fluviatilis*	1 030	1 202	1 052	976	841	686	824
Pike-perch	*Stizostedion lucioperca*	726	460	891	771	677	517	977
Freshwater fishes nei	*Osteichthyes*	123	160	134	168	170	209	28
European eel	*Anguilla anguilla*	55	56	44	60	67	67	49
Atlantic salmon	*Salmo salar*	9	10	7	14	21	14	16
Sea trout	*Salmo trutta*	15	11	8	10	13	13	16
European smelt	*Osmerus eperlanus*	484	415	1 431	1 008	1 194	762	2 320
Vendace	*Coregonus albula*	127	153	159	47	1	-	-
European whitefish	*Coregonus lavaretus*	21	20	20	28	33	33	47
Houting	*Coregonus oxyrinchus*	63	30	60	35	9	9	11
Sea lamprey	*Petromyzon marinus*	18	3	4	7	8	3	2
River lamprey	*Lampetra fluviatilis*	0	7	16	9	26	25	23
Lefteye flounders nei	*Bothidae*	-	-	-	-	-	-	1
Atlantic halibut	*Hippoglossus hippoglossus*	-	-	-	-	-	0	-
Greenland halibut	*Reinhardtius hippoglossoides*	-	-	-	-	181	1 100	1 125
Witch flounder	*Glyptocephalus cynoglossus*	-	-	-	-	5	3	2
Amer. plaice(=Long rough dab)	*Hippoglossoides platessoides*	-	-	-	-	94	54	24
Yellowtail flounder	*Limanda ferruginea*	-	-	-	-	53	47	14
European flounder	*Platichthys flesus*	297	333	355	416	419	482	515
Winter flounder	*Pseudopleuronectes americanus*	-	-	-	-	25	-	-
Flatfishes nei	*Pleuronectiformes*	-	-	-	-	1	-	-
Atlantic cod	*Gadus morhua*	1 392	1 174	1 070	1 059	520	799	59
Blue ling	*Molva dypterygia*	-	-	-	-	-	85	22
White hake	*Urophycis tenuis*	-	-	-	-	3	2	6
Saithe(=Pollock)	*Pollachius virens*	-	16	-	-	-	-	-
Blue whiting(=Poutassou)	*Micromesistius poutassou*	10 982	5 678	6 321	0	-	-	-
Patagonian grenadier	*Macruronus magellanicus*	113	-	-	-	-	108	-
Roughhead grenadier	*Macrourus berglax*	-	-	-	-	1	-	-
Roundnose grenadier	*Coryphaenoides rupestris*	-	-	-	-	20	701	848
Sea chubs nei	*Kyphosus spp*	-	-	-	-	-	-	37
Patagonian blennie	*Eleginops maclovinus*	-	-	-	14	-	-	-
Eelpout	*Zoarces viviparus*	2	8	9	2	1	1	1
Baird's slickhead	*Alepocephalus bairdii*	-	-	-	-	-	153	259
Atlantic wolffish	*Anarhichas lupus*	-	-	-	-	-	-	1
Wolffishes(=Catfishes) nei	*Anarhichas spp*	-	-	-	-	6	5	1
Black scabbardfish	*Aphanopus carbo*	-	-	-	-	-	224	-
Pacific ocean perch	*Sebastes alutus*	-	-	-	8	-	-	-
Atlantic redfishes nei	*Sebastes spp*	7 091	3 720	3 968	2 108	8 652	785	41
Atlantic herring	*Clupea harengus*	45 296	52 435	42 721	44 038	41 735	41 738	36 250
Sardinellas nei	*Sardinella spp*	252	415	3 171	2 474	-	-	-
European pilchard(=Sardine)	*Sardina pilchardus*	274	480	-	-	-	-	-
European sprat	*Sprattus sprattus*	22 493	39 693	32 165	36 407	41 394	40 777	40 717
Garfish	*Belone belone*	405	400	167	122	135	111	148
Atlantic horse mackerel	*Trachurus trachurus*	80	203	34	0	-	-	-
Jack and horse mackerels nei	*Trachurus spp*	1 903	2 080	1 550	-	-	-	-
Jacks, crevalles nei	*Caranx spp*	-	-	-	1 622	-	-	-
Pompanos nei	*Trachinotus spp*	-	-	-	-	-	-	3
Chub mackerel	*Scomber japonicus*	4 486	1 980	7 215	3 416	-	-	-
Atlantic mackerel	*Scomber scombrus*	3 741	6 324	7 356	3 595	2 673	218	-
Barracudas nei	*Sphyraena spp*	-	-	-	-	-	-	1
Shortfin mako	*Isurus oxyrinchus*	-	-	-	2	-	-	-
Angelshark	*Squatina squatina*	-	-	-	-	-	-	3
Raja rays nei	*Raja spp*	-	-	-	-	240	1 079	344
Various sharks nei	*Selachimorpha(Pleurotremata)*	-	-	-	-	-	-	53
Sharks, rays, skates, etc. nei	*Elasmobranchii*	-	-	-	-	-	-	188
Finfishes nei	*Osteichthyes*	26	23	28	32	41	91	125
Marine fishes nei	*Osteichthyes*	148	-	469	-	0	-	4
Euro-American crayfishes nei	*Astacidae, Cambaridae*	0	...	...	...	1	1	1
Deepwater rose shrimp	*Parapenaeus longirostris*	-	-	-	-	3	6	4

English name Nom anglais Nombre inglés	Scientific name Nom scientifique Nombre científico	1996 t	1997 t	1998 t	1999 t	2000 t	2001 t	2002 t
Northern prawn	*Pandalus borealis*	3 220	4 991	7 206	12 448	12 816	11 235	14 236
Argentine shortfin squid	*Illex argentinus*	-	-	-	-	-	1 833	533
Various squids nei	*Loliginidae, Ommastrephidae*	2 425	-	-	-	-	-	-
Marine molluscs nei	*Mollusca*	-	-	-	-	2	-	-
	Country total	*108 446*	*123 613*	*118 714*	*111 793*	*113 146*	*105 167*	*101 452*
Faeroe Is								
Freshwater fishes nei	*Osteichthyes*	0	0	0	0	0	0	-
Atlantic salmon	*Salmo salar*	-	-	5	-	0	-	-
Atlantic halibut	*Hippoglossus hippoglossus*	459	479	391	432	318	205	360
European plaice	*Pleuronectes platessa*	443	506	443	325	259	250	419
Greenland halibut	*Reinhardtius hippoglossoides*	6 241	5 124	3 787	4 111	5 584	4 128	2 362
Witch flounder	*Glyptocephalus cynoglossus*	13	2	3	7	...	...	...
Amer. plaice(=Long rough dab)	*Hippoglossoides platessoides*	8	-	-	3	2	-	-
Common dab	*Limanda limanda*	37	39	39	29	...	...	...
Lemon sole	*Microstomus kitt*	236	332	464	433	389	694	1 175
Turbot	*Psetta maxima*	2	-	-	-	-	-	-
Flatfishes nei	*Pleuronectiformes*	2	757	1 059	292	-	-	16
Tusk(=Cusk)	*Brosme brosme*	2 562	2 593	2 389	2 714	2 585	2 922	2 177
Atlantic cod	*Gadus morhua*	60 504	57 921	39 689	33 725	32 601	38 706	38 117
Ling	*Molva molva*	3 132	4 056	3 547	2 998	2 358	2 558	2 139
Blue ling	*Molva dypterygia*	1 624	1 172	1 274	2 136	1 756	2 454	801
Haddock	*Melanogrammus aeglefinus*	13 896	21 409	22 598	19 697	16 212	16 061	23 714
Saithe(=Pollock)	*Pollachius virens*	20 398	22 598	26 751	34 423	35 997	45 792	50 232
Pollack	*Pollachius pollachius*	-	1	-	-	-	-	-
Norway pout	*Trisopterus esmarkii*	9 133	11 215	6 222	4 045	1 754	2 429	2 429 F
Blue whiting(=Poutassou)	*Micromesistius poutassou*	21 483	28 773	71 217	105 106	152 687	259 761	235 000 F
Whiting	*Merlangius merlangus*	1 042	1 018	1 724	1 756	1 593	1 289	1 026
European hake	*Merluccius merluccius*	1	6	5	5	...	...	...
Roundnose grenadier	*Coryphaenoides rupestris*	234	200	84	118	76	91	159
Gadiformes nei	*Gadiformes*	449	898	57	338	-	-	51
Sandeels(=Sandlances) nei	*Ammodytes spp*	5 023	11 221	11 071	7 487	8 513	6 030	6 030 F
...A	*Normanichthys crockeri*	3	-	-	-	-	-	-
Argentines	*Argentina spp*	9 496	8 433	17 167	8 186	6 388	9 952	7 015
Alfonsinos nei	*Beryx spp*	-	5	-	-	-	-	-
Orange roughy	*Hoplostethus atlanticus*	950	854	747	349	155	1	29
Black cardinal fish	*Epigonus telescopus*	31	129	94	4	...	...	...
Atlantic wolffish	*Anarhichas lupus*	146	196	264	291	...	...	...
Wolffishes(=Catfishes) nei	*Anarhichas spp*	31	64	60	118	154	155	258
Black scabbardfish	*Aphanopus carbo*	256	126	89	45	116	412	1 094
Golden redfish	*Sebastes marinus*	11 028	11 402	7 951	7 129	13 324	13 572	3 500
Atlantic redfishes nei	*Sebastes spp*	1	-	-	-	-	-	64
Gurnards, searobins nei	*Triglidae*	-	-	-	1	...	...	...
Angler(=Monk)	*Lophius piscatorius*	1 610	1 765	1 885	2 578	2 216	2 006	1 781
Atlantic herring	*Clupea harengus*	57 853	65 949	70 214	56 476	65 270	35 172	35 172 F
European sprat	*Sprattus sprattus*	100	-	753	1 719	...	...	...
Atlantic bluefin tuna	*Thunnus thynnus*	-	-	67	104	118	...	...
Yellowfin tuna	*Thunnus albacares*	-	-	-	-	1	...	...
Bigeye tuna	*Thunnus obesus*	-	-	-	11	8	...	...
Swordfish	*Xiphias gladius*	-	-	-	5	4	...	...
Capelin	*Mallotus villosus*	39 777	43 466	41 966	24 275	59 855	32 110	35 000 F
Opah	*Lampris guttatus*	-	-	-	1	...	...	...
Atlantic horse mackerel	*Trachurus trachurus*	863	1 005	216	3 643	2 014	180	...
Atlantic mackerel	*Scomber scombrus*	19 530	8 401	10 654	11 334	21 022	24 005	24 005 F
Porbeagle	*Lamna nasus*	7	9	7	10	13	8	10
Picked dogfish	*Squalus acanthias*	51	212	356	484	354	613	337
Leafscale gulper shark	*Centrophorus squamosus*	53	58	129	11	...	...	...
Portuguese dogfish	*Centroscymnus coelolepis*	282	228	80	35	...	...	...
Dogfish sharks nei	*Squalidae*	7	8	8	3	...	...	...
Raja rays nei	*Raja spp*	169	187	151	183	125	90	33
Groundfishes nei	*Osteichthyes*	-	474	133	-	-	-	-
Marine fishes nei	*Osteichthyes*	1 201	1 362	858	247	3 899	2 911	2 932 F
Marine crabs nei	*Brachyura*	3	1	4	1	6	1	...
Norway lobster	*Nephrops norvegicus*	66	40	57	80	73	51	29
Northern prawn	*Pandalus borealis*	10 592	10 868	12 985	14 843	12 611	16 175	12 534 F
Queen scallop	*Aequipecten opercularis*	3 559	3 581	4 751	5 993	3 989	4 053	...
Northern shortfin squid	*Illex illecebrosus*	-	2	32	23	...	...	...
	Country total	*304 587*	*329 145*	*364 497*	*358 362*	*454 399*	*524 837*	*490 000 F*
Finland								
Freshwater bream	*Abramis brama*	3 143	3 275	2 854	2 845	2 546	2 647	2 609
Roach	*Rutilus rutilus*	2 172	2 271	1 465	1 465	1 412	1 459	1 501
Roaches nei	*Rutilus spp*	6 775	6 775	5 362	5 362	3 999	3 999	3 999
Orfe(=Ide)	*Leuciscus idus*	492	491	400	396	468	471	465
Cyprinids nei	*Cyprinidae*	23	23	-	-	-	-	-
Northern pike	*Esox lucius*	12 345	12 377	11 649	11 663	10 449	10 428	10 458
Burbot	*Lota lota*	1 647	1 663	1 147	1 154	1 176	1 168	1 161
European perch	*Perca fluviatilis*	17 647	17 860	14 721	14 694	13 374	13 395	13 476
Pike-perch	*Stizostedion lucioperca*	2 476	2 630	3 241	3 188	1 824	1 786	1 981
Freshwater fishes nei	*Osteichthyes*	957	928	673	736	572	602	647
European eel	*Anguilla anguilla*	22	22	...	...	...	...	...
Atlantic salmon	*Salmo salar*	1 714	1 790	1 020	912	964	817	814
Sea trout	*Salmo trutta*	672	661	460	441	437	416	396
Trouts nei	*Salmo spp*	1 286	1 286	879	879	610	610	610
Rainbow trout	*Oncorhynchus mykiss*	917	918	1 185	1 146	737	788	739
European smelt	*Osmerus eperlanus*	2 223	1 898	1 209	1 330	779	899	1 029
Vendace	*Coregonus albula*	5 988	5 975	5 115	5 125	5 022	5 024	5 003

English name Nom anglais Nombre inglés	Scientific name Nom scientifique Nombre científico	1996 t	1997 t	1998 t	1999 t	2000 t	2001 t	2002 t
European whitefish	*Coregonus lavaretus*	5 126	5 003	4 882	4 703	4 541	4 247	4 175
Salmonoids nei	*Salmonoidei*	472	472	329	329	288	282	282
European flounder	*Platichthys flesus*	715	702	555	558	449	500	449
Turbot	*Psetta maxima*	...	...	...	...	6	4	3
Atlantic cod	*Gadus morhua*	3 139	1 543	1 037	1 572	1 824	1 723	1 051
Atlantic herring	*Clupea harengus*	94 548	91 544	86 350	83 042	81 648	82 867	76 531
European sprat	*Sprattus sprattus*	14 351	19 851	27 014	18 886	23 242	15 850	17 353
Euro-American crayfishes nei	*Astacidae, Cambaridae*	227	227	134	134	134	114	76
	Country total	*179 077*	*180 185*	*171 681*	*160 560*	*156 501*	*150 096*	*144 808*
France								
Freshwater fishes nei	*Osteichthyes*	4 500	4 500	4 460	2 000 F	2 000 F	2 000 F	2 000 F
Sturgeons nei	*Acipenseridae*	0	0	0	1	-	-	-
European eel	*Anguilla anguilla*	403	1 782	449	289	399	415 F	392 F
Atlantic salmon	*Salmo salar*	0	0	0	2	10	12	5
Sea trout	*Salmo trutta*	0	0	0	0	-	1	1
European smelt	*Osmerus eperlanus*	95	106	69	75	103	96	64
Salmonoids nei	*Salmonoidei*	-	-	-	30	16	1	-
Allis shad	*Alosa alosa*	59	56	2	...	38	45	94
Twaite shad	*Alosa fallax*	2	1	...	...	11	11	9
Shads nei	*Alosa spp*	2	3	4	49	25	14	18
Sea lamprey	*Petromyzon marinus*	13	15	11	27	22	23	35
Atlantic halibut	*Hippoglossus hippoglossus*	34	16	16	21	29	44	19
European plaice	*Pleuronectes platessa*	3 843	4 326	4 667	5 642	4 709	4 254	4 708
Greenland halibut	*Reinhardtius hippoglossoides*	603	647	282	268	158	230	351
Witch flounder	*Glyptocephalus cynoglossus*	719	816	602	484	587	582	631
Amer. plaice(=Long rough dab)	*Hippoglossoides platessoides*	13	14	11	14	10	11	-
Common dab	*Limanda limanda*	1 120	1 446	1 576	1 194	1 106	950	1 182
Lemon sole	*Microstomus kitt*	2 396	1 782	1 522	1 349	1 308	1 366	1 444
European flounder	*Platichthys flesus*	238	223	163	204	221	243	287
Common sole	*Solea solea*	7 338	7 281	7 073	8 402	8 447	7 828	7 303
Sand sole	*Solea lascaris*	127	150	137	118	131	153	180
Wedge sole	*Dicologlossa cuneata*	798	647	488	602	686	579	606
Thickback soles	*Microchirus spp*	50	57	77	71	71	80	139
Soles nei	*Soleidae*	-	-	-	31	65	26	45
Megrim	*Lepidorhombus whiffiagonis*	4 267	3 987	3 601	3 529	4 154	4 125	3 311
Brill	*Scophthalmus rhombus*	417	357	351	397	512	490	539
Turbot	*Psetta maxima*	810	646	629	553	650	639	659
Flatfishes nei	*Pleuronectiformes*	14	17	24	20	3 075	1 855	3 143
Common mora	*Mora moro*	-	-	-	-	-	1	39
Tadpole codling	*Salilota australis*	31	25	11	5	29	-	-
Moras nei	*Moridae*	-	-	75	67	60	73	12
Tusk(=Cusk)	*Brosme brosme*	439	439	453	428	335	282	184
Atlantic cod	*Gadus morhua*	21 500	25 123	20 640	14 431	11 886	11 336	11 660
Ling	*Molva molva*	5 738	5 463	5 510	5 112	3 099	2 987	2 546
Blue ling	*Molva dypterygia*	4 127	4 736	5 955	4 794	5 571	3 666	3 223
Greater forkbeard	*Phycis blennoides*	562	605	476	528	733	746	664
Haddock	*Melanogrammus aeglefinus*	6 027	8 060	4 983	3 582	4 379	5 970	5 883
Saithe(=Pollock)	*Pollachius virens*	19 598	17 802	18 218	24 638	26 941	28 533	30 148
Pollack	*Pollachius pollachius*	3 881	3 626	3 359	2 935	3 775	3 649	3 924
Rocklings nei	*Gaidropsarus spp*	...	...	...	24	43	20	19
Poor cod	*Trisopterus minutus*	603	428	428	637	888	754	859
Pouting(=Bib)	*Trisopterus luscus*	6 119	6 283	6 108	6 333	7 619	7 293	7 081
Blue whiting(=Poutassou)	*Micromesistius poutassou*	6 463	12 467	8 013	6 372	16 076	19 078	14 785
Southern blue whiting	*Micromesistius australis*	67	-	-	-	-	-	-
Whiting	*Merlangius merlangus*	21 315	21 590	20 074	21 572	19 028	19 416	18 385
European hake	*Merluccius merluccius*	10 322	10 015	6 588	9 160	11 635	10 029	13 582
Argentine hake	*Merluccius hubbsi*	17	4	3	3	0	-	-
Hakes nei	*Merluccius spp*	-	...	44	5	-	-	-
Patagonian grenadier	*Macruronus magellanicus*	29	-	-	2	0	-	-
Roughhead grenadier	*Macrourus berglax*	21	25	29	116	4	4	7
Grenadiers nei	*Macrourus spp*	-	11	14	68	143	158	371
Roundnose grenadier	*Coryphaenoides rupestris*	7 650	7 389	6 820	7 851	9 968	8 494	8 635
Gadiformes nei	*Gadiformes*	24	442	861	10	2 436	3 092	4 154
Mullets nei	*Mugilidae*	1 086	1 292	1 299	1 458	1 272	1 485	1 238
Dusky grouper	*Epinephelus marginatus*	-	-	-	1	-	-	-
Spotted seabass	*Dicentrarchus punctatus*	68	42	75	68	71	79	57
European seabass	*Dicentrarchus labrax*	3 330	3 012	2 793	3 503	4 152	4 208	3 888
Cardinalfishes, etc. nei	*Apogonidae*	-	-	-	294	-	-	-
Canary drum (=Baardman)	*Umbrina canariensis*	1	-	-	3	7	7	7
Meagre	*Argyrosomus regius*	250	409	457	349	189	162	153
Blackspot(=red) seabream	*Pagellus bogaraveo*	27	39	40	69	60	51	77
Common pandora	*Pagellus erythrinus*	78	77	78	167	126	97	109
Axillary seabream	*Pagellus acarne*	80	39	36	318	282	198	253
White seabream	*Diplodus sargus*	54	41	35	55	64	51	78
Sargo breams nei	*Diplodus spp*	80	109	109	82	79	83	84
Common dentex	*Dentex dentex*	10	9	9	5	2	1	0
Black seabream	*Spondyliosoma cantharus*	2 803	2 885	3 116	3 031	3 029	2 799	3 139
Saddled seabream	*Oblada melanura*	13	14	14	10	14	8	8
Red porgy	*Pagrus pagrus*	3	1	1	2	1	1	0
Gilthead seabream	*Sparus aurata*	287	221	213	378	376	369	455
Bogue	*Boops boops*	189	203	219	217	299	329	300
Sand steenbras	*Lithognathus mormyrus*	60	121	110	112	118	97	89
Salema	*Sarpa salpa*	30	82	79	64	66	70	68
Porgies, seabreams nei	*Sparidae*	-	...	77	55	-	3	-
Picarels nei	*Spicara spp*	8	9	9	10	9	5	6
Surmullet	*Mullus surmuletus*	2 936	1 861	3 666	2 592	4 054	2 969	2 636
Surmullets(=Red mullets) nei	*Mullus spp*	149	149	149	266	270	206	188

English name Nom anglais Nombre inglés	Scientific name Nom scientifique Nombre científico	1996 t	1997 t	1998 t	1999 t	2000 t	2001 t	2002 t
Wrasses, hogfishes, etc. nei	*Labridae*	...	...	...	222	250	250	221
Marbled rockcod	*Notothenia rossii*	-	1	-	1	-	-	-
Humped rockcod	*Notothenia gibberifrons*	-	-	-	-	-	0	-
Grey rockcod	*Notothenia squamifrons*	15	-	-	-	-	-	-
Sandeels(=Sandlances) nei	*Ammodytes spp*	95	159	70	93	90	51	77
Greater weever	*Trachinus draco*	129	169	184	218	224	206	215
Gobies nei	*Gobiidae*	4	2	0	4	3	1	3
Grey triggerfish	*Balistes carolinensis*	-	1	1	2	-	4	-
Argentines	*Argentina spp*	-	-	-	121	59	45	8
European conger	*Conger conger*	4 657	4 110	4 574	4 612	5 470	5 225	5 201
Pink cusk-eel	*Genypterus blacodes*	2	1	0	0	-	-	-
Alfonsinos nei	*Beryx spp*	0	3	27	75	48	52	43
Orange roughy	*Hoplostethus atlanticus*	1 067	1 012	1 110	1 330	1 048	1 254	484
John dory	*Zeus faber*	712	690	787	906	1 275	1 364	1 230
Wreckfish	*Polyprion americanus*	4	10	13	42	52	22	15
Patagonian toothfish	*Dissostichus eleginoides*	3 484	4 090	4 619	5 018	7 156	5 838	3 569
Mackerel icefish	*Champsocephalus gunnari*	5	0	-	-	-	386	-
South Georgia icefish	*Pseudochaenichthys georgianus*	-	-	-	-	-	0	-
Unicorn icefish	*Channichthys rhinoceratus*	-	5	1	1	-	-	-
Icefishes nei	*Channichthyidae*	-	-	-	-	-	0	-
Black cardinal fish	*Epigonus telescopus*	52	52	232	...	197	153	63
Atlantic wolffish	*Anarhichas lupus*	3	1	2	14	9	7	8
Largehead hairtail	*Trichiurus lepturus*	-	...	3 400	509	-	-	-
Silver scabbardfish	*Lepidopus caudatus*	41	11	11	255	16	31	35
Black scabbardfish	*Aphanopus carbo*	2 868	2 118	1 710	1 833	3 707	5 070	4 628
Atlantic redfishes nei	*Sebastes spp*	2 244	2 567	1 604	1 254	943	1 076	569
Blackbelly rosefish	*Helicolenus dactylopterus*	104	115	140	129	123	125	49
Scorpionfishes nei	*Scorpaenidae*	56	143	81	38	43	42	74
Tub gurnard	*Chelidonichthys lucerna*	...	...	...	...	1 288	1 202	1 261
Red gurnard	*Chelidonichthys cuculus*	...	...	...	...	4 385	4 869	5 316
Streaked gurnard	*Chelidonichthys lastoviza*	...	...	...	...	56	51	68
Grey gurnard	*Eutrigla gurnardus*	...	...	...	...	236	222	196
Gurnards, searobins nei	*Triglidae*	5 895	5 683	5 444	5 851	253	192	90
Angler(=Monk)	*Lophius piscatorius*	18 701	17 370	14 278	10 838	13 055	13 543	15 373
Demersal percomorphs nei	*Perciformes*	3	3	3	19	43	26	26
Atlantic herring	*Clupea harengus*	11 742	21 222	23 398	25 531	25 398	27 577	26 627
Sardinellas nei	*Sardinella spp*	-	...	7 228	624	5 517	1 220	-
European pilchard(=Sardine)	*Sardina pilchardus*	16 552	20 973	18 589	38 861	29 209	29 075	34 363
European sprat	*Sprattus sprattus*	597	68	0	83	97	9	1
European anchovy	*Engraulis encrasicolus*	19 499	16 223	25 477	24 677	26 479	23 699	21 023
Clupeoids nei	*Clupeoidei*	-	-	-	837	180	1 103	1 865
Atlantic bonito	*Sarda sarda*	-	-	-	24	32	70	45
Frigate and bullet tunas	*Auxis thazard, A.rochei*	5 430	4 605	4 648	5 772	6 859	142	136
Little tunny(=Atl.black skipj)	*Euthynnus alletteratus*	2 109	1 981	1 731	2 438	2 702	...	3
Skipjack tuna	*Katsuwonus pelamis*	63 164	48 299	48 721	63 009	58 118	48 616	70 608
Atlantic bluefin tuna	*Thunnus thynnus*	9 690	8 470	7 713	6 741	7 322	6 748	6 565
Albacore	*Thunnus alalunga*	5 275	5 200	4 216	7 356	6 392	7 007	4 502
Yellowfin tuna	*Thunnus albacares*	69 397	61 194	53 120	62 046	67 483	63 588	67 864
Bigeye tuna	*Thunnus obesus*	16 079	13 804	12 013	14 045	12 622	10 066	11 691
Atlantic sailfish	*Istiophorus albicans*	97	110	138	131	98	-	-
Blue marlin	*Makaira nigricans*	96	82	80	83	79	-	-
Atlantic white marlin	*Tetrapturus albidus*	7	7	9	8	7	-	-
Longbill spearfish	*Tetrapturus pfluegeri*	59	68	86	81	60	-	-
Marlins,sailfishes,etc. nei	*Istiophoridae*	-	-	-	-	66	-	-
Swordfish	*Xiphias gladius*	97	164	110	104	126	113	101
Tuna-like fishes nei	*Scombroidei*	-	-	-	-	-	-	53
Capelin	*Mallotus villosus*	-	-	-	-	1	1	4
Garfish	*Belone belone*	59	70	43	55	60	55	104
Silversides(=Sand smelts) nei	*Atherinidae*	99	127	120	135	83	96	66
Atlantic horse mackerel	*Trachurus trachurus*	25 396	26 862	28 103	27 087	21 628	19 852	21 241
Jack and horse mackerels nei	*Trachurus spp*	422	2 148	6 771	1 887	915	1 038	705
Atlantic pomfret	*Brama brama*	1	0	2	7	7	5	5
Chub mackerel	*Scomber japonicus*	184	98	1 663	362	470	187	473
Atlantic mackerel	*Scomber scombrus*	13 908	16 567	20 963	18 256	22 640	22 300	23 050
Mackerels nei	*Scombridae*	-	120	25	13	4	2 649	-
Basking shark	*Cetorhinus maximus*	0	1	0	3	-	-	-
Thresher	*Alopias vulpinus*	7	13	7	35	128	132	24
Porbeagle	*Lamna nasus*	267	315	219	318	410	368	461
Small-spotted catshark	*Scyliorhinus canicula*	4 871	5 338	5 480	5 707	5 944	6 284	5 739
Nursehound	*Scyliorhinus stellaris*	197	292	181	135	159	180	156
Catsharks, nursehounds nei	*Scyliorhinus spp*	18	6	51	13	68	7	7
Blue shark	*Prionace glauca*	278	213	163	233	396	207	109
Smooth-hounds nei	*Mustelus spp*	578	624	749	824	1 050	1 249	1 586
Tope shark	*Galeorhinus galeus*	403	454	369	386	450	469	349
Picked dogfish	*Squalus acanthias*	1 726	1 715	1 417	1 197	1 100	1 335	1 128
Leafscale gulper shark	*Centrophorus squamosus*	-	-	-	-	-	-	48
Portuguese dogfish	*Centroscymnus coelolepis*	-	-	-	-	-	-	456
Longnose velvet dogfish	*Centroscymnus crepidater*	-	-	-	-	-	-	12
Black dogfish	*Centroscyllium fabricii*	-	-	-	-	269	271	28
Dogfish sharks nei	*Squalidae*	3 135	2 811	2 288	3 077	4 236	4 211	1 966
Angelsharks, sand devils nei	*Squatinidae*	1	0	0	1	1	1	-
Blue skate	*Raja batis*	295	314	296	467	653	667	447
Thornback ray	*Raja clavata*	1 756	1 579	1 343	1 335	1 251	1 231	1 178
Spotted ray	*Raja montagui*	977	1 163	1 179	1 260	1 341	1 563	1 451
Sandy ray	*Raja circularis*	438	438	410	435	369	330	301
Shagreen ray	*Raja fullonica*	46	39	38	65	38	68	71
Small-eyed ray	*Raja microocellata*	-	-	1	11	-	-	-
Cuckoo ray	*Raja naevus*	4 077	4 721	4 015	3 638	3 064	2 885	2 735
Longnosed skate	*Raja oxyrinchus*	346	311	327	194	140	89	211

English name Nom anglais Nombre inglés	Scientific name Nom scientifique Nombre científico	1996 t	1997 t	1998 t	1999 t	2000 t	2001 t	2002 t
Raja rays nei	*Raja spp*	2 902	3 196	2 870	3 408	3 111	3 195	3 126
Stingrays nei	*Dasyatis spp*	1	2	5	6	10	7	10
Eagle rays	*Myliobatidae*	0	0	0	2	2	2	-
Rays, stingrays, mantas nei	*Rajiformes*	84	82	97	76	157	183	417
Torpedo rays	*Torpedo spp*	16	18	19	34	32	43	34
Ratfishes nei	*Hydrolagus spp*	-	-	-	38	573	822	627
Sharks, rays, skates, etc. nei	*Elasmobranchii*	28	...	...	43	-	-	-
Groundfishes nei	*Osteichthyes*	228	-	-	-	-	-	-
Finfishes nei	*Osteichthyes*	3 771	3 732	2 344	2 179	8 759	12 661	5 844
Marine fishes nei	*Osteichthyes*	14 034	13 920	2 583	931	751	1 749	745
Edible crab	*Cancer pagurus*	5 927	6 554	5 598	6 503	6 549	6 604	5 877
Portunus swimcrabs nei	*Portunus spp*	225	261	155	163	242	267	308
Green crab	*Carcinus maenas*	350	359	204	465	558	541	471
Mediterranean shore crab	*Carcinus aestuarii*	16	16	16	7	14	9	9
Spinous spider crab	*Maja squinado*	3 171	3 435	3 126	3 380	4 428	5 440	4 143
Marine crabs nei	*Brachyura*	160	5	5	-	13	10	6
Pink spiny lobster	*Palinurus mauritanicus*	1	0	25	11	9	5	1
Common spiny lobster	*Palinurus elephas*	113	99	53	61	62	68	49
Palinurid spiny lobsters nei	*Palinurus spp*	-	-	-	1	-	-	-
Norway lobster	*Nephrops norvegicus*	8 623	7 125	6 611	5 843	6 639	6 992	6 870
European lobster	*Homarus gammarus*	267	327	219	308	332	329	359
Lobsters nei	*Reptantia*	-	-	-	0	2	2	-
Craylets, squat lobsters	*Galatheidae*	90	94	81	102	98	84	84
Caramote prawn	*Penaeus kerathurus*	...	...	...	...	4	6	7
Deepwater rose shrimp	*Parapenaeus longirostris*	7	17	17	1	1	1	4
Delta prawn	*Palaemon longirostris*	...	...	...	10	19	26	18
Common prawn	*Palaemon serratus*	311	288	213	296	307	326	247
Common shrimp	*Crangon crangon*	309	237	272	399	472	391	287
Natantian decapods nei	*Natantia*	-	-	-	5	2	1	1
Goose barnacles	*Lepas spp*	2	-	-	-	-	-	-
Spottail mantis squillid	*Squilla mantis*	15	10	10	34	44	33	43
Marine crustaceans nei	*Crustacea*	887	1 156	678	701	686	689	1 388
Periwinkles nei	*Littorina spp*	2	-	-	-	-	-	-
Murex	*Murex spp*	37	41	41	35	51	52	60
Tuberculate abalone	*Haliotis tuberculata*	62	75	36	37	61	60	31
Whelk	*Buccinum undatum*	4 406	12 887	6 266	7 691	12 724	11 030	10 532
Gastropods nei	*Gastropoda*	-	-	-	11	15	19	26
European flat oyster	*Ostrea edulis*	49	18	14	12	9	132	197
Pacific cupped oyster	*Crassostrea gigas*	54	14	14	50	108	63	10
Blue mussel	*Mytilus edulis*	197	6 972	1 355	9 564	8 660	8 015	4 782
Mediterranean mussel	*Mytilus galloprovincialis*	500	1 078	1 078	23	24	12	4
Sea mussels nei	*Mytilidae*	1 077	0	0	0	1 792	1 110	1 001
Queen scallop	*Aequipecten opercularis*	311	595	637	2 574	3 475	5 989	4 243
Great Atlantic scallop	*Pecten maximus*	12 169	14 451	12 866	13 719	13 745	16 458	19 389
Scallops nei	*Pectinidae*	268	185	149	328	313	430	262
Striped venus	*Chamelea gallina*	706	810	715	790	696	627	798
Pullet carpet shell	*Venerupis pullastra*	934	1 493	2 216	2 306	1 311	1 595	1 711
Grooved carpet shell	*Ruditapes decussatus*	106	71	510	490	101	304	55
Venus clams nei	*Veneridae*	1	99	-	207	252	311	367
Solid surf clam	*Spisula solida*	...	...	...	...	127	101	101
Donax clams	*Donax spp*	-	-	-	34	94	88	68
Razor clams nei	*Solen spp*	...	...	...	5	5	5	5
Common edible cockle	*Cerastoderma edule*	664	731	418	481	81	11	18
Clams, etc. nei	*Bivalvia*	3 746	2 893	5 822	4 638	4 280	4 696	5 109
Common cuttlefish	*Sepia officinalis*	85	85	85	88	106	83	96
Cuttlefish,bobtail squids nei	*Sepiidae, Sepiolidae*	14 532	11 982	13 015	14 465	18 939	13 814	16 105
Patagonian squid	*Loligo gahi*	4 394	1 512	4 146	2 309	2 024	-	-
Common squids nei	*Loligo spp*	135	465	465	241	264	177	177
Broadtail shortfin squid	*Illex coindetii*	395	411	216	338	402	250	527
Argentine shortfin squid	*Illex argentinus*	28	0	0	56	0	-	-
European flying squid	*Todarodes sagittatus*	87	87	87	-	-	-	-
Various squids nei	*Loliginidae, Ommastrephidae*	3 895	4 456	4 332	5 946	5 399	4 690	6 358
Common octopus	*Octopus vulgaris*	706	439	439	1 493	1 643	1 406	2 100
Octopuses, etc. nei	*Octopodidae*	67	75	90	246	484	388	135
Cephalopods nei	*Cephalopoda*	-	-	-	55	49	48	55
Grooved sea squirt	*Microcosmus sulcatus*	28	22	22	30	30	76	78
Stony sea urchin	*Paracentrotus lividus*	63	48	59	84	198	101	164
	Country total	*564 694*	*574 076*	*549 545*	*593 479*	*634 819*	*605 629*	*620 078*
Germany								
Freshwater breams nei	*Abramis spp*	162	275	144	249	124	174	107
Common carp	*Cyprinus carpio*	386	386	386	387	386	387	387
Tench	*Tinca tinca*	36	36	36	37	36	37	37
Roach	*Rutilus rutilus*	176	242	347	462	269	320	347
Grass carp(=White amur)	*Ctenopharyngodon idellus*	10	10	10	5	5	5	3
Silver carp	*Hypophthalmichthys molitrix*	76	76	76	39	39	39	23
Bighead carp	*Hypophthalmichthys nobilis*	12	12	12	6	6	6	4
Cyprinids nei	*Cyprinidae*	1 653	1 653	1 653	1 653	1 653	1 653	1 653
Northern pike	*Esox lucius*	288	306	304	336	322	309	309
Wels(=Som)catfish	*Silurus glanis*	12	12	12	12	12	12	12
Burbot	*Lota lota*	2	3	3	3	3	4	4
European perch	*Perca fluviatilis*	764	782	822	714	501	541	478
Pike-perch	*Stizostedion lucioperca*	546	560	491	524	508	494	524
Freshwater fishes nei	*Osteichthyes*	19 079	19 011	19 009	19 010	19 011	19 009	19 011
Sturgeons nei	*Acipenseridae*	0	0	0	0	0	0	0
European eel	*Anguilla anguilla*	696	746	717	747	686	638	635
Atlantic salmon	*Salmo salar*	27	35	42	30	45	39	29
Trouts nei	*Salmo spp*	7	8	6	9	12	11	13

English name Nom anglais Nombre inglés	Scientific name Nom scientifique Nombre científico	1996 t	1997 t	1998 t	1999 t	2000 t	2001 t	2002 t
European smelt	*Osmerus eperlanus*	29	87	32	46	4	6	13
European whitefish	*Coregonus lavaretus*	-	5	8	21	47	63	34
Whitefishes nei	*Coregonus spp*	459	459	459	459	459	459	459
Salmonoids nei	*Salmonoidei*	29	29	29	29	29	29	29
Lefteye flounders nei	*Bothidae*	...	...	...	293	370	290	292
Atlantic halibut	*Hippoglossus hippoglossus*	43	23	28	42	23	20	22
European plaice	*Pleuronectes platessa*	4 935	4 304	3 050	3 462	4 501	4 842	4 148
Greenland halibut	*Reinhardtius hippoglossoides*	3 924	3 846	3 870	3 532	3 742	3 402	2 629
Witch flounder	*Glyptocephalus cynoglossus*	7	9	13	8	13	8	5
Amer. plaice(=Long rough dab)	*Hippoglossoides platessoides*	0	-	-	-	0	-	3
Common dab	*Limanda limanda*	1 880	1 384	1 129	1 104	1 124	1 074	762
Lemon sole	*Microstomus kitt*	67	78	151	68	74	77	121
European flounder	*Platichthys flesus*	1 637	2 449	2 159	2 347	2 782	2 349	2 385
Common sole	*Solea solea*	685	513	786	1 462	1 291	959	771
Megrim	*Lepidorhombus whiffiagonis*	1	2	3	1	3	1	0
Brill	*Scophthalmus rhombus*	47	48	60	54	80	65	60
Turbot	*Psetta maxima*	256	330	267	309	454	363	343
Flatfishes nei	*Pleuronectiformes*	404	508	279	-	-	-	3
Tusk(=Cusk)	*Brosme brosme*	59	26	19	16	13	10	10
Atlantic cod	*Gadus morhua*	37 629	26 491	23 075	21 990	18 414	19 222	15 412
Ling	*Molva molva*	1 409	965	308	247	215	110	114
Blue ling	*Molva dypterygia*	119	11	16	15	110	26	10
Greater forkbeard	*Phycis blennoides*	-	-	-	1	8	12	11
Haddock	*Melanogrammus aeglefinus*	2 718	2 441	1 712	1 039	1 225	1 368	1 786
Saithe(=Pollock)	*Pollachius virens*	15 197	15 993	13 562	13 307	12 385	13 320	14 542
Pollack	*Pollachius pollachius*	102	117	43	63	39	41	116
Norway pout	*Trisopterus esmarkii*	-	-	-	-	2	-	-
Blue whiting(=Poutassou)	*Micromesistius poutassou*	6 865	4 722	17 970	3 170	12 654	19 059	17 052
Whiting	*Merlangius merlangus*	711	276	191	371	754	680	568
European hake	*Merluccius merluccius*	83	76	69	68	46	73	71
Roundnose grenadier	*Coryphaenoides rupestris*	14	38	121	101	42	23	25
Mullets nei	*Mugilidae*	1	3	6	21	13	23	40
Groupers, seabasses nei	*Serranidae*	-	-	-	-	-	1	-
Cardinalfishes, etc. nei	*Apogonidae*	-	-	-	-	-	10	-
Porgies, seabreams nei	*Sparidae*	-	-	-	-	-	-	21
Surmullets(=Red mullets) nei	*Mullus spp*	-	-	-	-	13	10	9
Eelpout	*Zoarces viviparus*	3	2	2	2	1	5	1
Argentines	*Argentina spp*	1 394	1 498	633	24	483	189	150
Baird's slickhead	*Alepocephalus bairdii*	-	-	-	-	12	1	-
Black cardinal fish	*Epigonus telescopus*	-	-	-	-	50	-	-
Atlantic wolffish	*Anarhichas lupus*	94	44	88	67	86	66	96
Silver scabbardfish	*Lepidopus caudatus*	-	-	-	64	4	-	-
Black scabbardfish	*Aphanopus carbo*	2	-	-	-	-	-	-
Atlantic redfishes nei	*Sebastes spp*	22 512	21 096	20 342	18 417	14 278	12 790	15 554
Gurnards, searobins nei	*Triglidae*	94	136	141	187	180	150	198
Lumpfish(=Lumpsucker)	*Cyclopterus lumpus*	2	2	0	4	1	3	2
Angler(=Monk)	*Lophius piscatorius*	542	1 137	1 446	847	568	364	362
Atlantic herring	*Clupea harengus*	42 153	42 749	47 028	50 857	47 048	50 680	56 701
Sardinellas nei	*Sardinella spp*	10 115	23 866	17 448	24 150	-	1 683	1 518
European pilchard(=Sardine)	*Sardina pilchardus*	50	3 295	4 781	1 446	307	500	3 051
European sprat	*Sprattus sprattus*	161	427	4 551	183	22	791	950
European anchovy	*Engraulis encrasicolus*	0	-	16	-	-	42	-
Atlantic bonito	*Sarda sarda*	714	417	42	143	-	51	38
Skipjack tuna	*Katsuwonus pelamis*	3	-	-	-	-	-	-
Capelin	*Mallotus villosus*	-	-	5 001	-	-	-	95
Garfish	*Belone belone*	168	130	58	125	82	73	113
Atlantic horse mackerel	*Trachurus trachurus*	22 028	38 077	34 376	24 377	16 778	12 464	15 926
Jack and horse mackerels nei	*Trachurus spp*	-	-	-	-	-	708	278
Chub mackerel	*Scomber japonicus*	87	645	315	334	-	20	318
Atlantic mackerel	*Scomber scombrus*	16 229	15 864	21 490	19 960	22 980	25 325	26 536
Porbeagle	*Lamna nasus*	-	-	-	0	17	1	3
Picked dogfish	*Squalus acanthias*	-	-	-	45	188	303	119
Dogfish sharks nei	*Squalidae*	19	12	16	235	271	433	519
Raja rays nei	*Raja spp*	65	74	81	102	130	27	26
Various sharks nei	*Selachimorpha(Pleurotremata)*	309	139	110	-	-	-	-
Finfishes nei	*Osteichthyes*	270	291	164	150	66	37	36
Marine fishes nei	*Osteichthyes*	40	45	53	110	-	132	77
Freshwater crustaceans nei	*Crustacea*	12	12	12	12	12	12	12
Edible crab	*Cancer pagurus*	-	38	44	57	64	44	74
Marine crabs nei	*Brachyura*	13	6	4	-	-	-	141
Norway lobster	*Nephrops norvegicus*	77	70	70	110	86	141	132
European lobster	*Homarus gammarus*	17	0	0	0	-	0	0
Northern prawn	*Pandalus borealis*	-	-	-	1 585	-	-	-
Common shrimp	*Crangon crangon*	15 994	19 890	14 814	17 457	17 423	12 571	15 966
Various squids nei	*Loliginidae, Ommastrephidae*	2	4	11	6	5	3	17
	Country total	*236 411*	*259 352*	*266 622*	*238 925*	*205 689*	*211 282*	*224 451*
Gibraltar								
Marine fishes nei	*Osteichthyes*	0	0	0	0	0	0	0
	Country total	*0*	*0*	*0*	*0*	*0*	*0*	*0*
Greece								
Common carp	*Cyprinus carpio*	208	256	279	247	220	198	263
Goldfish	*Carassius auratus*	530	296	235	548	448	415	350
Roaches nei	*Rutilus spp*	164	81	252	309	345	324	334
Rudd	*Scardinius erythrophthalmus*	3	70	69	30	50	45	50
Cyprinids nei	*Cyprinidae*	173	37	38	44	40	35	32

English name Nom anglais Nombre inglés	Scientific name Nom scientifique Nombre científico	1996 t	1997 t	1998 t	1999 t	2000 t	2001 t	2002 t
Northern pike	*Esox lucius*	12	9	5	8	10	9	10
Wels(=Som)catfish	*Silurus glanis*	11	13	20	14	15	18	20
European perch	*Perca fluviatilis*	24	15	19	24	15	23	20
Freshwater fishes nei	*Osteichthyes*	1 730	1 624	1 410	1 592	1 803	1 663	1 451
European eel	*Anguilla anguilla*	31	31	43	27	34	32	25 F
Salmonoids nei	*Salmonoidei*	1	6	38	11	50	42	40
Shads nei	*Alosa spp*	1 010	1 284	1 784	1 898	1 283	2 217	2 008 F
Common sole	*Solea solea*	1 235	1 169	745	619	687	562	645 F
Turbot	*Psetta maxima*	60	60	47	65	63	77	70 F
Blue whiting(=Poutassou)	*Micromesistius poutassou*	1 227	1 558	846	630	566	471	450 F
European hake	*Merluccius merluccius*	4 649	4 257	3 052	3 128	2 969	2 753	2 999 F
Flathead grey mullet	*Mugil cephalus*	3 824	3 077	1 760	2 023	1 748	1 403	1 308 F
Dusky grouper	*Epinephelus marginatus*	112	94	57	64	90	99	91 F
White grouper	*Epinephelus aeneus*	306	125	123	148	137	161	109
Groupers nei	*Epinephelus spp*	54	91	65	85	63	142	130 F
Groupers, seabasses nei	*Serranidae*	410	619	330	336	286	463	250 F
European seabass	*Dicentrarchus labrax*	455	380	258	289	345	300	281 F
Shi drum	*Umbrina cirrosa*	160	78	109	136	151	103	99 F
Pandoras nei	*Pagellus spp*	1 443	1 544	839	1 000	1 050	937	506 F
White seabream	*Diplodus sargus*	409	509	327	503	581	347	330 F
Sargo breams nei	*Diplodus spp*	19	7	22	18	117	88	50
Large-eye dentex	*Dentex macrophthalmus*	653	1 026	544	516	398	330	314 F
Common dentex	*Dentex dentex*	744	448	522	474	364	480	594 F
Black seabream	*Spondyliosoma cantharus*	391	318	219	330	176	138	125 F
Saddled seabream	*Oblada melanura*	607	399	332	239	443	287	270 F
Red porgy	*Pagrus pagrus*	1 280	933	853	887	913	1 139	1 050 F
Pargo breams nei	*Pagrus spp*	531	465	371	435	386	326	317 F
Gilthead seabream	*Sparus aurata*	199	138	125	142	248	176	170 F
Bogue	*Boops boops*	6 744	5 973	4 228	4 658	4 096	3 674	3 500 F
Salema	*Sarpa salpa*	687	485	369	404	408	346	330 F
Picarels nei	*Spicara spp*	8 307	8 371	4 549	4 663	4 029	4 031	3 866 F
Surmullet	*Mullus surmuletus*	2 004	1 944	1 095	1 274	1 368	1 238	1 200 F
Surmullets(=Red mullets) nei	*Mullus spp*	2 495	2 426	1 744	1 735	1 810	1 541	1 500 F
West African goatfish	*Pseudupeneus prayensis*	1 065	692	805	838	715	1 081	675
European conger	*Conger conger*	1 160	1 922	1 293	1 211	1 019	1 062	1 000 F
John dory	*Zeus faber*	487	309	274	201	196	296	247 F
Scorpionfishes nei	*Scorpaenidae*	841	783	518	634	640	726	651 F
Gurnards, searobins nei	*Triglidae*	376	380	237	229	207	169	160 F
Angler(=Monk)	*Lophius piscatorius*	942	912	757	739	882	694	650 F
Blackbellied angler	*Lophius budegassa*	-	4	7	4	27	-	18
European pilchard(=Sardine)	*Sardina pilchardus*	18 896	20 561	17 734	15 214	16 026	14 395	14 000 F
European sprat	*Sprattus sprattus*	262	279	216	110	266	474	400 F
European anchovy	*Engraulis encrasicolus*	15 073	14 583	17 099	16 456	9 863	10 770	10 500 F
Atlantic bonito	*Sarda sarda*	1 752	1 559	945	2 135	1 914	1 550	1 420
Frigate and bullet tunas	*Auxis thazard, A.rochei*	1 426	1 426	...	...	196	125	120
Little tunny(=Atl.black skipj)	*Euthynnus alletteratus*	-	-	-	-	-	-	132
Atlantic bluefin tuna	*Thunnus thynnus*	878	1 218	287	248	622	361	438
Albacore	*Thunnus alalunga*	952	741	1 152	2 005	1 786	1 840	1 352
Swordfish	*Xiphias gladius*	1 238	750	1 650	1 520	1 960	1 730	978
Tuna-like fishes nei	*Scombroidei*	116	145	300	...	195	128	-
Garfish	*Belone belone*	331	253	188	83	184	126	120 F
Big-scale sand smelt	*Atherina boyeri*	8	110	350	350	350	326	390
Bluefish	*Pomatomus saltatrix*	128	111	144	259	265	511	450 F
Atlantic horse mackerel	*Trachurus trachurus*	2 484	2 956	1 360	942	774	747	700 F
Mediterranean horse mackerel	*Trachurus mediterraneus*	8 039	7 169	4 350	3 534	3 902	3 408	3 250 F
Jack and horse mackerels nei	*Trachurus spp*	-	-	-	-	1	-	2
Greater amberjack	*Seriola dumerili*	697	336	252	205	176	396	350 F
Carangids nei	*Carangidae*	-	-	11	-	4	1	1
Chub mackerel	*Scomber japonicus*	6 520	5 640	2 173	1 850	2 286	1 829	1 750 F
Atlantic mackerel	*Scomber scombrus*	886	305	141	200	443	155	150 F
Smooth-hounds nei	*Mustelus spp*	440	517	359	576	617	372	359 F
Dogfish sharks nei	*Squalidae*	290	243	290	258	270	224	224 F
Guitarfishes, etc. nei	*Rhinobatidae*	112	63	87	73	94	89	86 F
Raja rays nei	*Raja spp*	1 002	900	715	718	746	579	550 F
Marine fishes nei	*Osteichthyes*	11 086	12 433	9 083	9 606	12 595	11 353	10 397 F
Euro-American crayfishes nei	*Astacidae, Cambaridae*	...	15	23	28	23	27	25
Mediterranean shore crab	*Carcinus aestuarii*	49	28	50	37	16	17	17 F
Marine crabs nei	*Brachyura*	60	84	90	74	59	54	63
Norway lobster	*Nephrops norvegicus*	487	351	455	243	268	242	230 F
European lobster	*Homarus gammarus*	212	374	155	170	201	233	225 F
Caramote prawn	*Penaeus kerathurus*	1 261	1 654	1 499	1 116	1 448	1 503	1 441 F
Natantian decapods nei	*Natantia*	2 278	2 470	1 808	1 812	1 638	1 355	1 317 F
European flat oyster	*Ostrea edulis*	1 003	344	80	47	105	120	105 F
Mediterranean mussel	*Mytilus galloprovincialis*	12 625	24 139	6 389	15 860	469	254	250 F
Common cuttlefish	*Sepia officinalis*	2 473	2 969	2 279	3 123	1 766	2 238	2 760 F
Common squids nei	*Loligo spp*	856	623	426	397	560	405	380 F
Various squids nei	*Loliginidae, Ommastrephidae*	843	513	567	686	886	750	713 F
Common octopus	*Octopus vulgaris*	2 919	2 952	1 673	1 910	2 193	2 058	1 950 F
Octopuses, etc. nei	*Octopodidae*	904	815	782	1 781	934	813	830 F
Marine molluscs nei	*Mollusca*	3 076	2 241	1 789	1 736	1 688	2 169	2 000 F
	Country total	*149 435*	*157 088*	*108 580*	*118 771*	*99 280*	*94 388*	*88 983 F*
Hungary								
Common carp	*Cyprinus carpio*	2 717	2 255	3 373	3 279	3 212	2 470	2 787
Tench	*Tinca tinca*	-	5	-	-	-	-	-
Barbel	*Barbus barbus*	37	64	46	50	30	52	41
Grass carp(=White amur)	*Ctenopharyngodon idellus*	346	305	301	318	356	309	400
Silver carp	*Hypophthalmichthys molitrix*	862	1 483	731	676	365	997	525

E-5

Fish crustaceans, molluscs, etc — Capture production by countries or areas and species — Europe
Poissons, crustacés, mollusques, etc — Captures par pays ou zones et espèces — Europe
Peces, crustáceos, moluscos, etc — Capturas por países o áreas y especies — Europa

English name Nom anglais Nombre inglés	Scientific name Nom scientifique Nombre científico	1996 t	1997 t	1998 t	1999 t	2000 t	2001 t	2002 t
Bighead carp	*Hypophthalmichthys nobilis*	74	83	-	-	-	-	-
Asp	*Aspius aspius*	22	44	38	42	38	21	20
Cyprinids nei	*Cyprinidae*	1 731	1 730	1 510	1 666	1 710	2 155	2 418
Northern pike	*Esox lucius*	46	203	158	241	280	191	190
Wels(=Som)catfish	*Silurus glanis*	201	121	113	145	104	120	134
Pike-perch	*Stizostedion lucioperca*	224	199	156	169	200	196	190
Freshwater fishes nei	*Osteichthyes*	733	776	648	714	718	89	15
Sterlet sturgeon	*Acipenser ruthenus*	34	14	9	35	12	11	12
European eel	*Anguilla anguilla*	579	124	182	179	76	27	18
	Country total	*7 606*	*7 406*	*7 265*	*7 514*	*7 101*	*6 638*	*6 750*
Iceland								
Freshwater fishes nei	*Osteichthyes*	0	0	0	0	0	0	-
Atlantic salmon	*Salmo salar*	597	202	202	142	87	88	88 F
Trouts nei	*Salmo spp*	250	250	250	250	91	72	72 F
Atlantic halibut	*Hippoglossus hippoglossus*	837	677	501	567	493	589	683
European plaice	*Pleuronectes platessa*	11 070	10 557	7 111	7 064	5 218	4 905	5 126
Greenland halibut	*Reinhardtius hippoglossoides*	22 125	18 631	10 751	11 187	15 060	16 642	19 229
Witch flounder	*Glyptocephalus cynoglossus*	1 486	1 272	947	1 408	1 098	1 132	1 147
Amer. plaice(=Long rough dab)	*Hippoglossoides platessoides*	7 027	6 468	3 329	3 833	3 176	3 473	3 579
Common dab	*Limanda limanda*	7 954	7 891	5 061	3 981	3 015	4 373	4 358
Lemon sole	*Microstomus kitt*	984	1 135	1 432	1 860	1 438	1 371	950
Megrim	*Lepidorhombus whiffiagonis*	419	281	221	124	97	96	78
Turbot	*Psetta maxima*	0	0	0	0	0	0	0
Flatfishes nei	*Pleuronectiformes*	11	13	3	5	-	2	1
Blue antimora	*Antimora rostrata*	2	-	-	-	-	-	-
Tusk(=Cusk)	*Brosme brosme*	5 226	4 847	4 118	5 796	4 741	3 425	3 935
Atlantic cod	*Gadus morhua*	204 058	208 636	242 968	260 643	238 324	240 002	213 417
Ling	*Molva molva*	3 670	3 634	3 603	3 976	3 223	2 864	2 833
Blue ling	*Molva dypterygia*	1 284	1 320	1 208	2 321	1 623	765	1 274
Haddock	*Melanogrammus aeglefinus*	56 223	43 256	40 712	44 729	41 698	39 825	49 951
Saithe(=Pollock)	*Pollachius virens*	39 297	36 548	30 532	30 729	32 947	31 941	41 839
Norway pout	*Trisopterus esmarkii*	0	-	-	-	-	160	253
Blue whiting(=Poutassou)	*Micromesistius poutassou*	513	10 480	68 514	160 424	259 157	365 101	286 381
Whiting	*Merlangius merlangus*	430	443	531	931	1 349	1 179	1 295
Cape hakes	*Merluccius capensis,M.paradox.*	-	352	206	-	-	-	-
Roughhead grenadier	*Macrourus berglax*	15	4	1	-	5	3	11
Roundnose grenadier	*Coryphaenoides rupestris*	216	207	120	146	70	57	70
Sandeels(=Sandlances) nei	*Ammodytes spp*	-	-	-	-	-	8	-
Argentines	*Argentina spp*	808	3 367	13 387	5 495	4 595	2 478	4 357
Baird's slickhead	*Alepocephalus bairdii*	0	0	0	0	0	2	-
Kingklip	*Genypterus capensis*	-	31	2	-	-	-	-
Alfonsinos nei	*Beryx spp*	7	466	126	-	-	-	-
Orange roughy	*Hoplostethus atlanticus*	43	79	28	14	68	18	10
John dory	*Zeus faber*	-	7	-	-	-	-	-
Atlantic wolffish	*Anarhichas lupus*	14 638	11 685	11 844	13 769	15 043	17 953	14 303
Spotted wolffish	*Anarhichas minor*	1 109	1 180	1 599	1 545	1 896	2 126	2 128
Snoek	*Thyrsites atun*	-	1	-	-	-	-	-
Black scabbardfish	*Aphanopus carbo*	17	1	0	9	18	8	15
Golden redfish	*Sebastes marinus*	-	-	-	45 406	44 829	38 762	54 766
Beaked redfish	*Sebastes mentella*	-	-	-	21 463	45 231	42 440	44 504
Atlantic redfishes nei	*Sebastes spp*	120 751	111 652	116 132	43 475	26 244	11 325	11 606
Scorpionfishes nei	*Scorpaenidae*	-	5	-	-	-	-	-
Grey gurnard	*Eutrigla gurnardus*	1	0	0	0	0	-	0
Lumpfish(=Lumpsucker)	*Cyclopterus lumpus*	4 201	6 520	3 165	3 373	2 458	412	206
Angler(=Monk)	*Lophius piscatorius*	669	787	850	977	1 570	1 353	965
Devil anglerfish	*Lophius vomerinus*	-	5	2	-	-	-	-
Atlantic herring	*Clupea harengus*	265 413	291 117	277 461	298 435	287 663	178 950	223 842
Atlantic bluefin tuna	*Thunnus thynnus*	-	1	2	33	29	-	1
Bigeye tuna	*Thunnus obesus*	-	-	-	-	5	-	-
Swordfish	*Xiphias gladius*	-	-	-	2	2	-	-
Capelin	*Mallotus villosus*	1 179 051	1 319 191	750 065	703 694	892 405	918 417	1 078 818
Chub mackerel	*Scomber japonicus*	-	28	-	-	-	-	-
Atlantic mackerel	*Scomber scombrus*	92	927	357	144	0	1	53
Mackerels nei	*Scombridae*	-	11	-	-	-	-	-
Porbeagle	*Lamna nasus*	5	3	4	2	2	3	2
Greenland shark	*Somniosus microcephalus*	61	73	87	51	45	57	56
Picked dogfish	*Squalus acanthias*	157	106	78	57	109	136	276
Portuguese dogfish	*Centroscymnus coelolepis*	-	-	5	0	0	-	-
Black dogfish	*Centroscyllium fabricii*	4	0	-	-	2	-	-
Blue skate	*Raja batis*	181	118	108	80	94	85	59
Starry ray	*Raja radiata*	1 493	1 431	1 252	996	1 076	1 211	1 781
Shagreen ray	*Raja fullonica*	19	16	12	21	27	37	32
Raja rays nei	*Raja spp*	-	14	-	-	-	-	-
Rabbit fish	*Chimaera monstrosa*	21	15	29	11	5	1	-
Chimaeras, etc. nei	*Chimaeriformes*	1	0	-	-	-	-	-
Groundfishes nei	*Osteichthyes*	237	179	2	82	...	45	55
Marine fishes nei	*Osteichthyes*	-	-	4	-	-	-	-
Marine crabs nei	*Brachyura*	0	0	-	-	-	1	0
Norway lobster	*Nephrops norvegicus*	1 623	1 215	1 411	1 389	1 230	1 420	1 548
Northern prawn	*Pandalus borealis*	89 633	82 627	62 727	42 958	33 539	30 790	36 157
Whelk	*Buccinum undatum*	520	1 199	13	298	770	678	-
Iceland scallop	*Chlamys islandica*	8 978	10 403	10 098	8 858	9 074	6 499	5 192
Ocean quahog	*Arctica islandica*	6 315	4 351	8 776	3 501	1 584	7 434	12 353
European flying squid	*Todarodes sagittatus*	3	5	4	3	1	-	0
Various squids nei	*Loliginidae, Ommastrephidae*	-	4	0	-	-	-	-
European edible sea urchin	*Echinus esculentus*	423	20	-	10	-	-	-
	Country total	*2 060 168*	*2 205 944*	*1 681 951*	*1 736 267*	*1 982 524*	*1 980 715*	*2 129 655*

English name Nom anglais Nombre inglés	Scientific name Nom scientifique Nombre científico	1996 t	1997 t	1998 t	1999 t	2000 t	2001 t	2002 t
Ireland								
Northern pike	*Esox lucius*	2 000	2 000	2 000	...	...	...	...
European perch	*Perca fluviatilis*	...	...	200	...	...	...	...
Freshwater fishes nei	*Osteichthyes*	0	0	0	...	...	...	9
European eel	*Anguilla anguilla*	550	550	650	500	250	110	104
Atlantic salmon	*Salmo salar*	816	675	725	525	621	792	674
Sea trout	*Salmo trutta*	1 200	1 200	1 200	547	10	...	2
Rainbow trout	*Oncorhynchus mykiss*	75	99	75	...	...	...	...
Atlantic halibut	*Hippoglossus hippoglossus*	8	4	9	11	1	17	23
European plaice	*Pleuronectes platessa*	1 679	1 699	1 731	1 424	1 028	841	801
Greenland halibut	*Reinhardtius hippoglossoides*	2	2	21	78	22	71	84
Witch flounder	*Glyptocephalus cynoglossus*	615	605	657	713	551	915	831
Common dab	*Limanda limanda*	76	113	109	66	39	34	32
Lemon sole	*Microstomus kitt*	581	667	527	531	468	440	482
European flounder	*Platichthys flesus*	13	13	13	13	12	18	19
Common sole	*Solea solea*	463	483	526	492	376	375	334
Sand sole	*Solea lascaris*	13	12	15	1	2	1	2
Megrim	*Lepidorhombus whiffiagonis*	3 507	3 063	3 383	3 162	3 364	3 713	2 848
Brill	*Scophthalmus rhombus*	126	181	141	126	119	95	99
Turbot	*Psetta maxima*	261	257	234	261	236	185	183
Flatfishes nei	*Pleuronectiformes*	141	210	184	37	23	15	14
Common mora	*Mora moro*	-	-	-	-	-	32	44
Tusk(=Cusk)	*Brosme brosme*	64	45	43	43	113	122	109
Atlantic cod	*Gadus morhua*	7 258	5 702	5 294	3 860	2 928	2 653	2 503
Ling	*Molva molva*	1 379	1 305	1 272	1 138	1 089	1 463	1 303
Blue ling	*Molva dypterygia*	-	1	22	43	91	827	583
Greater forkbeard	*Phycis blennoides*	154	228	318	379	399	679	720
Haddock	*Melanogrammus aeglefinus*	4 421	6 234	6 572	4 898	5 812	5 404	3 509
Saithe(=Pollock)	*Pollachius virens*	2 514	1 841	1 687	1 704	1 743	2 048	1 354
Pollack	*Pollachius pollachius*	1 288	1 052	946	1 049	1 131	1 382	1 334
Norway pout	*Trisopterus esmarkii*	-	-	-	-	1	-	-
Pouting(=Bib)	*Trisopterus luscus*	2	12	1	21	10	28	12
Blue whiting(=Poutassou)	*Micromesistius poutassou*	1 709	25 987	45 538	35 880	26 067	29 910	17 825
Whiting	*Merlangius merlangus*	10 326	9 394	7 762	7 643	6 505	6 621	6 669
European hake	*Merluccius merluccius*	1 741	2 270	1 971	2 090	2 037	1 124	698
Roundnose grenadier	*Coryphaenoides rupestris*	1	4	-	1	45	-	-
Gadiformes nei	*Gadiformes*	105	190	-	279	-	55	8
Mullets nei	*Mugilidae*	40	33	15	29	11	3	8
Cardinalfishes, etc. nei	*Apogonidae*	-	-	-	-	-	5	55
Grunts, sweetlips nei	*Haemulidae (=Pomadasyidae)*	-	-	-	-	-	5	7
Blackspot(=red) seabream	*Pagellus bogaraveo*	8	8	6	1	...	11	7
Surmullet	*Mullus surmuletus*	40	-	38	-	-	-	-
Sandeels(=Sandlances) nei	*Ammodytes spp*	-	-	-	389	-	-	10
Greater weever	*Trachinus draco*	-	-	-	-	-	691	616
Argentines	*Argentina spp*	295	1 089	405	396	4 709	7 505	7 592
European conger	*Conger conger*	142	202	374	295	279	253	277
Orange roughy	*Hoplostethus atlanticus*	-	-	-	-	3	2 759	4 647
John dory	*Zeus faber*	125	112	98	145	174	169	153
Boarfishes nei	*Caproidae*	-	-	-	-	-	120	91
Wreckfish	*Polyprion americanus*	-	-	5	-	1	1	-
Atlantic wolffish	*Anarhichas lupus*	39	22	39	35	66	27	50
Largehead hairtail	*Trichiurus lepturus*	-	-	-	-	-	776	906
Black scabbardfish	*Aphanopus carbo*	0	1	-	1	12	299	259
Atlantic redfishes nei	*Sebastes spp*	15	48	71	171	186	433	297
Red gurnard	*Chelidonichthys cuculus*	-	-	25	47	...	...	...
Grey gurnard	*Eutrigla gurnardus*	-	-	38	71	...	...	...
Gurnards, searobins nei	*Triglidae*	77	82	-	-	79	97	101
Angler(=Monk)	*Lophius piscatorius*	3 348	3 880	4 251	4 298	3 839	3 112	2 523
Atlantic herring	*Clupea harengus*	71 953	57 155	58 248	45 334	42 114	40 640	30 606
Sardinellas nei	*Sardinella spp*	-	-	-	-	-	52 980	24 552
European pilchard(=Sardine)	*Sardina pilchardus*	-	-	-	3 195	2 592	7 855	12 159
European sprat	*Sprattus sprattus*	4 214	2 085	1 578	5 826	6 032	455	1 729
Atlantic bluefin tuna	*Thunnus thynnus*	-	14	21	52	22	8	15
Albacore	*Thunnus alalunga*	874	1 913	3 750	4 858	3 464	2 093	1 324
Yellowfin tuna	*Thunnus albacares*	-	-	-	-	-	3	-
Bigeye tuna	*Thunnus obesus*	-	-	-	-	-	10	-
Swordfish	*Xiphias gladius*	15	15	132	81	35	17	4
Capelin	*Mallotus villosus*	-	0	1	-	-	-	-
Atlantic horse mackerel	*Trachurus trachurus*	127 876	75 002	74 253	58 201	55 438	54 975	33 072
Jack and horse mackerels nei	*Trachurus spp*	-	-	-	-	-	8 522	3 411
Atlantic pomfret	*Brama brama*	-	-	-	-	1	184	404
Atlantic mackerel	*Scomber scombrus*	49 966	53 094	67 310	59 609	70 184	76 586	76 662
Ocean sunfish	*Mola mola*	-	-	-	-	-	13	-
Porbeagle	*Lamna nasus*	-	-	-	8	1	6	3
Small-spotted catshark	*Scyliorhinus canicula*	...	...	...	...	...	633	564
Blue shark	*Prionace glauca*	-	-	-	67	23	66	11
Tope shark	*Galeorhinus galeus*	...	...	...	...	...	4	1
Picked dogfish	*Squalus acanthias*	2 095	1 407	1 259	962	880	1 301	1 293
Portuguese dogfish	*Centroscymnus coelolepis*	...	...	...	...	...	216	341
Dogfish sharks nei	*Squalidae*	1 170	917	1 144	683	4	30	14
Raja rays nei	*Raja spp*	2 212	2 715	2 120	2 283	2 078	2 140	2 501
Rabbit fish	*Chimaera monstrosa*	...	...	2	-	5	15	16
Ratfishes nei	*Hydrolagus spp*	...	...	...	...	...	5	-
Various sharks nei	*Selachimorpha(Pleurotremata)*	23	32	169	90	175	455	496
Marine fishes nei	*Osteichthyes*	-	-	575	177	1	709	1 080
Edible crab	*Cancer pagurus*	5 649	7 572	7 392	7 772	9 598	9 738	10 099
Portunus swimcrabs nei	*Portunus spp*	...	...	314	463	...	214	...

English name Nom anglais Nombre inglés	Scientific name Nom scientifique Nombre científico	1996 t	1997 t	1998 t	1999 t	2000 t	2001 t	2002 t
Green crab	*Carcinus maenas*	...	...	79	16	...	68	...
Spinous spider crab	*Maja squinado*	192	153	185	299	163	264	330
Marine crabs nei	*Brachyura*	312	272	-	1	268	-	-
Palinurid spiny lobsters nei	*Palinurus spp*	62	48	46	35	42	35	36
Norway lobster	*Nephrops norvegicus*	5 171	7 020	6 950	8 492	7 709	7 074	6 983
European lobster	*Homarus gammarus*	567	513	611	597	606	781	740
Palaemonid shrimps nei	*Palaemonidae*	399	358	505	551	450	268	208
Marine crustaceans nei	*Crustacea*	-	-	-	-	-	401	595
Common periwinkle	*Littorina littorea*	2 814	3 152	2 636	3 018	2 641	2 781	2 287
Whelk	*Buccinum undatum*	6 575	3 852	3 667	4 561	4 942	6 364	7 901
European flat oyster	*Ostrea edulis*	415	773	...	...	...	...	...
Blue mussel	*Mytilus edulis*	1 372	1 963	955	503	-	-	-
Queen scallop	*Aequipecten opercularis*	3	7	5	29	3	13	58
Great Atlantic scallop	*Pecten maximus*	560	633	693	1 497	1 579	1 413	1 140
Grooved carpet shell	*Ruditapes decussatus*	21	23	3	3	3	130	-
Razor clams nei	*Solen spp*	-	28	316	407	334	201	167
Common edible cockle	*Cerastoderma edule*	10	64	296	1	8	6	112
Clams, etc. nei	*Bivalvia*	-	-	-	-	301	126	83
Northern shortfin squid	*Illex illecebrosus*	...	...	...	...	...	121	114
European flying squid	*Todarodes sagittatus*	...	...	...	...	...	14	40
Various squids nei	*Loliginidae, Ommastrephidae*	481	442	610	282	135	242	356
Octopuses, etc. nei	*Octopodidae*	13	7	3	10	8	14	13
European edible sea urchin	*Echinus esculentus*	2	5	1	2	1	5	-
	Country total	*332 233*	*292 809*	*325 020*	*283 358*	*276 292*	*356 430*	*282 331*
Isle of Man								
Freshwater fishes nei	*Osteichthyes*	0	0	0	0	0	0	0
European plaice	*Pleuronectes platessa*	16	11	14	5	6	1	0
Witch flounder	*Glyptocephalus cynoglossus*	0	1	0	0	0	-	-
Common dab	*Limanda limanda*	0	0	-	-	-	-	-
Lemon sole	*Microstomus kitt*	4	0	4	3	3	1	0
Common sole	*Solea solea*	4	5	3	1	1	1	0
Megrim	*Lepidorhombus whiffiagonis*	-	3	2	...	...	-	-
Brill	*Scophthalmus rhombus*	1	0	0	1	1	0	0
Turbot	*Psetta maxima*	1	1	0	0	-	0	-
Atlantic cod	*Gadus morhua*	27	19	34	9	11	1	7
Ling	*Molva molva*	3	2	1	1	1	0	-
Haddock	*Melanogrammus aeglefinus*	38	9	13	7	19	1	0
Saithe(=Pollock)	*Pollachius virens*	11	9	7	2	1	0	4
Pollack	*Pollachius pollachius*	16	11	11	2	1	-	3
Whiting	*Merlangius merlangus*	28	24	33	5	2	1	1
European hake	*Merluccius merluccius*	18	28	30	3	3	1	-
Surmullet	*Mullus surmuletus*	-	-	-	-	-	4	-
European conger	*Conger conger*	0	0	...	-	-	-	-
Gurnards, searobins nei	*Triglidae*	2	1	1	1	...	1	-
Angler(=Monk)	*Lophius piscatorius*	34	27	28	9	5	2	1
Atlantic herring	*Clupea harengus*	693	821	0	1	...	35	-
Atlantic mackerel	*Scomber scombrus*	0	0	0	4	0	8	6
Dogfish sharks nei	*Squalidae*	25	25	12	19	11	3	1
Raja rays nei	*Raja spp*	10	6	6	3	5	1	0
Finfishes nei	*Osteichthyes*	-	-	0	-	-	-	-
Edible crab	*Cancer pagurus*	94	478	274	231	142	170	387
Norway lobster	*Nephrops norvegicus*	20	24	17	10	3	2	-
European lobster	*Homarus gammarus*	0	26	25	14	8	12	23
Whelk	*Buccinum undatum*	296	193	...	227	89	2	62
Queen scallop	*Aequipecten opercularis*	1 129	1 630	991	1 255	2 275	1 749	1 655
Great Atlantic scallop	*Pecten maximus*	1 064	933	706	794	965	1 115	977
Common squids nei	*Loligo spp*	3	2	2	2	-	1	0
	Country total	*3 537*	*4 289*	*2 214*	*2 609*	*3 552*	*3 112*	*3 127*
Italy								
Cyprinids nei	*Cyprinidae*	1 146	2 378	1 155	1 900	725	1 821	799
Freshwater fishes nei	*Osteichthyes*	3 647	2 895	2 346	2 316	2 819	2 643	2 341
European eel	*Anguilla anguilla*	883	1 010	682	645	549	446	402
Salmonoids nei	*Salmonoidei*	1 625	1 091	897	937	692	846	945
Common sole	*Solea solea*	3 597	3 085	2 638	2 252	2 165	2 966	2 866
Tonguefishes	*Cynoglossidae*	66	-	-	-	-	-	-
Turbots nei	*Scophthalmidae*	1 377	964	528	478	643	622	482
Flatfishes nei	*Pleuronectiformes*	-	261	244	96	107	117	115
Blue whiting(=Poutassou)	*Micromesistius poutassou*	1 546	1 300	1 449	1 451	1 261	1 167	755
European hake	*Merluccius merluccius*	30 707	17 971	13 166	9 754	9 220	9 304	9 918
Mullets nei	*Mugilidae*	5 172	5 281	5 344	4 799	4 095	5 023	5 716
Dusky grouper	*Epinephelus marginatus*	558	640	124	89	97	252	229
European seabass	*Dicentrarchus labrax*	2 481	2 030	1 889	1 881	2 195	2 735	3 428
Shi drum	*Umbrina cirrosa*	495	351	138	138	158	259	156
Pandoras nei	*Pagellus spp*	1 152	1 445	836	751	1 171	949	800
Sargo breams nei	*Diplodus spp*	1 069	706	382	340	321	462	475
Common dentex	*Dentex dentex*	1 253	389	190	205	309	201	207
Gilthead seabream	*Sparus aurata*	1 743	1 859	1 717	1 754	1 939	2 675	3 004
Bogue	*Boops boops*	5 281	4 178	4 074	3 105	3 541	3 537	3 129
Picarels nei	*Spicara spp*	953	647	545	547	385	313	287
Surmullets(=Red mullets) nei	*Mullus spp*	11 325	7 499	7 491	8 771	9 044	7 121	6 111
Gobies nei	*Gobiidae*	1 311	1 085	991	800	712	665	695
Scorpionfishes nei	*Scorpaenidae*	-	-	-	-	-	14	-
Gurnards, searobins nei	*Triglidae*	3 908	3 473	3 293	2 668	2 168	2 264	1 941
Angler(=Monk)	*Lophius piscatorius*	2 345	6 672	2 845	1 705	1 269	1 244	1 083
European pilchard(=Sardine)	*Sardina pilchardus*	42 129	38 174	36 387	28 876	25 805	23 980	18 049

English name Nom anglais Nombre inglés	Scientific name Nom scientifique Nombre científico	1996 t	1997 t	1998 t	1999 t	2000 t	2001 t	2002 t
European anchovy	*Engraulis encrasicolus*	40 541	53 439	44 429	39 783	50 728	53 047	42 068
Atlantic bonito	*Sarda sarda*	2 233	4 580	2 121	1 614	1 116	1 006	944
Frigate and bullet tunas	*Auxis thazard, A.rochei*	229	499	254	439	215	375	251
Skipjack tuna	*Katsuwonus pelamis*	-	1 754	3 024	3 416	...	1 681	2 660
Atlantic bluefin tuna	*Thunnus thynnus*	10 006	9 548	4 059	3 279	3 845	4 377	4 628
Albacore	*Thunnus alalunga*	1 769	1 426	1 472	2 561	3 630	2 882	4 071
Yellowfin tuna	*Thunnus albacares*	-	1 340	2 299	2 626	...	1 332	1 746
Bigeye tuna	*Thunnus obesus*	-	457	612	848	...	57	315
Swordfish	*Xiphias gladius*	5 286	6 104	6 104	6 312	7 515	6 388	3 539
Tuna-like fishes nei	*Scombroidei*	-	1	-	-	-	8	-
Garfish	*Belone belone*	243	216	238	209	134	139	139
Silversides(=Sand smelts) nei	*Atherinidae*	1 112	1 101	883	851	725	736	772
Jack and horse mackerels nei	*Trachurus spp*	6 790	5 168	6 314	4 315	3 428	3 927	2 913
Leerfish	*Lichia amia*	643	400	197	249	185	183	223
Scomber mackerels nei	*Scomber spp*	8 012	7 866	7 277	5 748	5 522	6 033	5 041
Smooth-hounds nei	*Mustelus spp*	2 659	621	636	440	462	369	325
Rays, stingrays, mantas nei	*Rajiformes*	2 309	5 325	2 807	1 117	507	555	521
Marine fishes nei	*Osteichthyes*	33 212	28 258	27 773	23 355	24 131	27 129	21 595
Marine crabs nei	*Brachyura*	-	-	-	-	8	-	-
Common spiny lobster	*Palinurus elephas*	312	331	174	161	123	166	152
Norway lobster	*Nephrops norvegicus*	5 101	4 834	2 582	3 033	2 485	2 287	2 051
Caramote prawn	*Penaeus kerathurus*	3	-	-	-	-	-	-
Penaeus shrimps nei	*Penaeus spp*	30	-	-	-	-	-	-
Deepwater rose shrimp	*Parapenaeus longirostris*	7 065	7 019	4 410	4 631	7 500	6 980	6 378
Aristeid shrimps nei	*Aristeidae*	2 258	2 406	1 231	2 128	4 463	1 833	1 768
Natantian decapods nei	*Natantia*	306	345	456	493	370	686	473
Spottail mantis squillid	*Squilla mantis*	5 431	4 497	3 670	4 767	5 244	5 570	5 374
Marine crustaceans nei	*Crustacea*	3 657	3 681	3 245	2 294	2 130	1 849	1 695
Mediterranean mussel	*Mytilus galloprovincialis*	22 174	21 430	27 270 F	26 510 F	44 200 F	44 160 F	46 030
Striped venus	*Chamelea gallina*	36 707	28 604	28 830	36 462	34 191	34 916	25 971
Cuttlefish,bobtail squids nei	*Sepiidae, Sepiolidae*	8 937	9 398	8 330	7 080	6 325	7 370	4 995
Common squids nei	*Loligo spp*	5 368	4 152	2 237	1 909	1 890	2 352	1 785
European flying squid	*Todarodes sagittatus*	4 672	2 614	3 995	2 056	2 516	2 346	2 034
Various squids nei	*Loliginidae, Ommastrephidae*	411	36	54	101	95	11	105
Common octopus	*Octopus vulgaris*	9 301	9 519	10 411	8 844	9 173	10 036	8 157
Horned and musky octopuses	*Eledone spp*	1 939	1 590	1 506	1 041	1 621	1 661	1 309
Octopuses, etc. nei	*Octopodidae*	176	-	-	-	-	-	-
Marine molluscs nei	*Mollusca*	11 238	9 750	7 845	7 860	6 282	6 324	5 885
	Country total	*365 899*	*343 693*	*306 096*	*282 790*	*302 149*	*310 397*	*269 846*
Latvia								
Freshwater bream	*Abramis brama*	242	241	191	235	218	246	252
Common carp	*Cyprinus carpio*	3	3	6	5	3	5	5
Tench	*Tinca tinca*	15	24	22	41	29	35	40
Crucian carp	*Carassius carassius*	7	11	10	14	15	24	36
Roach	*Rutilus rutilus*	46	64	56	67	70	60	59
Orfe(=Ide)	*Leuciscus idus*	2	2	2	3	2	3	2
Vimba bream	*Vimba vimba*	58	57	92	119	92	109	99
Cyprinids nei	*Cyprinidae*	34	42	26	27	-	-	-
Northern pike	*Esox lucius*	56	47	55	72	72	73	88
Wels(=Som)catfish	*Silurus glanis*	1	2	1	2	-	-	-
European perch	*Perca fluviatilis*	56	56	55	87	58	83	95
Pike-perch	*Stizostedion lucioperca*	73	40	38	58	48	59	78
Ruffe	*Gymnocephalus cernuus*	5	5	-	1	-	-	-
Freshwater fishes nei	*Osteichthyes*	20	13	13	20	63	50	62
European eel	*Anguilla anguilla*	26	29	27	17	15	19	11
Atlantic salmon	*Salmo salar*	151	169	125	166	151	138	111
Sea trout	*Salmo trutta*	10	7	7	10	14	11	13
European smelt	*Osmerus eperlanus*	386	335	218	180	261	128	14
Vendace	*Coregonus albula*	5	5	6	7	4	5	3
Whitefishes nei	*Coregonus spp*	2	1	1	-	-	-	-
River lamprey	*Lampetra fluviatilis*	140	80	79	120	135	88	79
Three-spined stickleback	*Gasterosteus aculeatus*	34	110	82	-	-	-	-
Greenland halibut	*Reinhardtius hippoglossoides*	-	-	-	-	215	291	11
European flounder	*Platichthys flesus*	294	367	364	509	418	613	599
Turbot	*Psetta maxima*	42	46	36	54	16	6	9
Atlantic cod	*Gadus morhua*	8 741	6 187	7 778	6 914	6 280	6 298	4 867
Southern blue whiting	*Micromesistius australis*	2	-	-	-	-	-	-
Senegalese hake	*Merluccius senegalensis*	68	27	16	320	280	126	105
Patagonian grenadier	*Macruronus magellanicus*	15	-	-	-	-	-	-
Mullets nei	*Mugilidae*	21	1	20	44	19	-	-
Grunts, sweetlips nei	*Haemulidae (=Pomadasyidae)*	-	2	-	-	-	-	-
Sargo breams nei	*Diplodus spp*	19	20	13	81	90	176	135
Large-eye dentex	*Dentex macrophthalmus*	-	24	57	91	190	71	29
Common dentex	*Dentex dentex*	436	8	...	...	...	19	29
Porgies, seabreams nei	*Sparidae*	-	-	48	80	53	-	-
Eelpout	*Zoarces viviparus*	139	80	41	32	23	26	37
Largehead hairtail	*Trichiurus lepturus*	12	-	1 232	1 502	544	13	46
Atlantic redfishes nei	*Sebastes spp*	1 084	-	-	-	13	11	1 841
Atlantic herring	*Clupea harengus*	27 523	29 330	24 417	27 163	26 768	26 652	25 284
Sardinellas nei	*Sardinella spp*	24 209	6 497	6 064	15 031	7 886	7 689	6 132
European pilchard(=Sardine)	*Sardina pilchardus*	474	-	7	23	633	54	362
European sprat	*Sprattus sprattus*	34 211	49 314	44 858	42 834	46 186	42 769	47 540
European anchovy	*Engraulis encrasicolus*	-	-	1 978	4 876	10 142	9 143	6 872
Atlantic bonito	*Sarda sarda*	301	887	318	510	416	396	639
Yellowfin tuna	*Thunnus albacares*	151	223	97	25	36	72	334
Tuna-like fishes nei	*Scombroidei*	-	-	147	27	-	-	-
Garfish	*Belone belone*	-	-	-	-	-	11	-

English name Nom anglais Nombre inglés	Scientific name Nom scientifique Nombre científico	1996 t	1997 t	1998 t	1999 t	2000 t	2001 t	2002 t
Bluefish	*Pomatomus saltatrix*	155	116	31	116	144	17	48
Jack and horse mackerels nei	*Trachurus spp*	14 818	4 881	8 710	14 284	22 591	18 522	7 768
Jacks, crevalles nei	*Caranx spp*	-	11	-	-	-	-	-
Leerfish	*Lichia amia*	152	236	127	172	274	96	454
Carangids nei	*Carangidae*	-	36	-	-	-	-	-
Chub mackerel	*Scomber japonicus*	3 765	1 931	2 562	3 123	7 151	9 924	7 079
Atlantic mackerel	*Scomber scombrus*	233	-	-	-	-	-	-
Marine fishes nei	*Osteichthyes*	19 198	3 118	1 107	3 247	1 616	996	407
Palinurid spiny lobsters nei	*Palinurus spp*	-	-	-	-	-	-	10
Northern prawn	*Pandalus borealis*	1 253	997	1 191	3 080	3 169	3 028	1 951
Natantian decapods nei	*Natantia*	-	-	-	-	-	19	7
Cuttlefish,bobtail squids nei	*Sepiidae, Sepiolidae*	-	-	-	-	-	-	17
Various squids nei	*Loliginidae, Ommastrephidae*	3 956	-	-	-	-	-	12
Octopuses, etc. nei	*Octopodidae*	-	-	-	-	-	2	6
	Country total	*142 644*	*105 682*	*102 331*	*125 389*	*136 403*	*128 176*	*113 677*
Liechtensten								
Freshwater fishes nei	*Osteichthyes*	0	0	0	0	0	0	0
	Country total	*0*	*0*	*0*	*0*	*0*	*0*	*0*
Lithuania								
Freshwater bream	*Abramis brama*	397	448	454	466	467	470	512
Freshwater breams nei	*Abramis spp*	7	10	12	11	13	26	32
Common carp	*Cyprinus carpio*	31	14	13	12	16	16	18
Tench	*Tinca tinca*	9	12	9	11	13	15	16
Bleak	*Alburnus alburnus*	-	-	-	-	3	4	6
Crucian carp	*Carassius carassius*	-	-	-	-	-	-	15
Goldfish	*Carassius auratus*	30	30	38	32	45	38	33
Roach	*Rutilus rutilus*	431	593	645	647	635	643	746
Rudd	*Scardinius erythrophthalmus*	14	18	14	13	17	20	25
Orfe(=Ide)	*Leuciscus idus*	1	1	-	-	1	1	1
Chub	*Leuciscus cephalus*	-	-	-	-	3	3	10
Vimba bream	*Vimba vimba*	2	3	3	11	48	40	78
Sichel	*Pelecus cultratus*	...	...	4	3	8	12	4
Asp	*Aspius aspius*	4	6	9	6	6	5	10
Cyprinids nei	*Cyprinidae*	1	2	-	-	-	19	0
Northern pike	*Esox lucius*	66	71	56	62	71	68	74
Wels(=Som)catfish	*Silurus glanis*	0	-	-	-	0	0	0
Burbot	*Lota lota*	5	14	10	13	13	9	15
European perch	*Perca fluviatilis*	83	114	104	116	115	115	132
Pike-perch	*Stizostedion lucioperca*	56	65	51	60	78	115	173
Ruffe	*Gymnocephalus cernuus*	...	...	...	52	97	64	31
Freshwater fishes nei	*Osteichthyes*	45	90	88	10	6	2	1
European eel	*Anguilla anguilla*	12	11	17	18	11	12	13
Atlantic salmon	*Salmo salar*	10	4	5	6	6	4	11
Sea trout	*Salmo trutta*	-	2	3	4	5	3	3
European smelt	*Osmerus eperlanus*	81	190	334	365	214	360	287
Vendace	*Coregonus albula*	...	...	...	...	7	6	4
European whitefish	*Coregonus lavaretus*	...	...	...	...	1	4	2
Whitefishes nei	*Coregonus spp*	18	20	10	11	-	-	-
Shads nei	*Alosa spp*	-	-	-	-	-	-	1
River lamprey	*Lampetra fluviatilis*	1	-	-	-	-	3	2
Three-spined stickleback	*Gasterosteus aculeatus*	-	-	-	13	22	18	6
Greenland halibut	*Reinhardtius hippoglossoides*	-	-	-	-	21	395	346
Witch flounder	*Glyptocephalus cynoglossus*	-	-	-	-	-	3	2
Amer. plaice(=Long rough dab)	*Hippoglossoides platessoides*	-	-	-	-	-	3	26
Yellowtail flounder	*Limanda ferruginea*	-	-	-	-	-	1	-
European flounder	*Platichthys flesus*	330	624	736	571	618	...	...
Turbot	*Psetta maxima*	-	-	62	58	23	18	18
Flatfishes nei	*Pleuronectiformes*	-	-	-	-	-	1 137	1 082
Atlantic cod	*Gadus morhua*	5 520	4 694	3 296	4 371	4 721	3 852	2 964
Blue ling	*Molva dypterygia*	-	-	-	-	-	16	29
White hake	*Urophycis tenuis*	-	-	-	-	-	-	2
Blue whiting(=Poutassou)	*Micromesistius poutassou*	651	-	-	1 231	-	-	-
Senegalese hake	*Merluccius senegalensis*	...	...	180	307	180	43	189
Roughhead grenadier	*Macrourus berglax*	-	-	-	-	1	28	3
Roundnose grenadier	*Coryphaenoides rupestris*	-	-	-	-	-	137	1 849
Sea catfishes nei	*Ariidae*	-	-	-	-	-	-	113
Mullets nei	*Mugilidae*	-	-	-	-	-	-	102
Bigeye grunt	*Brachydeuterus auritus*	-	-	-	-	-	-	78
Croakers, drums nei	*Sciaenidae*	-	-	-	-	-	-	61
Large-eye dentex	*Dentex macrophthalmus*	-	-	-	-	-	-	121
Porgies, seabreams nei	*Sparidae*	...	...	192	157	155	32	137
Baird's slickhead	*Alepocephalus bairdii*	-	-	-	-	-	460	-
Alfonsinos nei	*Beryx spp*	-	-	-	-	-	-	98
John dory	*Zeus faber*	-	-	-	-	-	-	1
Bonnetmouths, rubyfishes nei	*Emmelichthyidae*	-	-	-	-	-	-	8
Largehead hairtail	*Trichiurus lepturus*	...	...	9 708	13	32	167	8 120
Black scabbardfish	*Aphanopus carbo*	-	-	-	-	-	3	9
Ruffs, barrelfishes nei	*Centrolophidae*	-	-	-	-	-	-	51
Atlantic redfishes nei	*Sebastes spp*	10 649	-	1 769	3 884	6 687	20 182	21 853
Atlantic herring	*Clupea harengus*	4 257	3 330	2 368	1 313	1 198	1 639	1 539
Round sardinella	*Sardinella aurita*	...	...	10 575	8 680	6 324	4 309	22 205
European pilchard(=Sardine)	*Sardina pilchardus*	...	...	20	6	292	22	206
European sprat	*Sprattus sprattus*	10 165	6 018	4 460	3 117	1 682	3 135	2 800
European anchovy	*Engraulis encrasicolus*	...	...	3 612	13 774	16 137	9 492	14 064
Clupeoids nei	*Clupeoidei*	2 400	-	-	-	-	-	-

English name Nom anglais Nombre inglés	Scientific name Nom scientifique Nombre científico	1996 t	1997 t	1998 t	1999 t	2000 t	2001 t	2002 t
Atlantic bonito	*Sarda sarda*	-	-	-	-	-	-	793
West African Spanish mackerel	*Scomberomorus tritor*	-	-	-	-	-	-	298
Tuna-like fishes nei	*Scombroidei*	...	...	467	110	80	154	216
Bluefish	*Pomatomus saltatrix*	-	-	-	-	-	-	742
Atlantic horse mackerel	*Trachurus trachurus*	7 400	-	-	421	5	344	-
Jack and horse mackerels nei	*Trachurus spp*	...	...	11 902	20 657	25 464	17 672	46 282
Leerfish	*Lichia amia*	-	-	-	-	-	-	386
Chub mackerel	*Scomber japonicus*	...	...	5 420	2 105	3 871	3 074	12 119
Atlantic mackerel	*Scomber scombrus*	7 334	-	2 823	4 936	2 085	1 949	1 600
Blue butterfish	*Stromateus fiatola*	-	-	-	-	-	-	4
Barracudas nei	*Sphyraena spp*	-	-	-	-	-	-	40
Dogfish sharks nei	*Squalidae*	-	-	-	-	-	14	40
Raja rays nei	*Raja spp*	-	-	-	-	-	4	18
Finfishes nei	*Osteichthyes*	188	152	41	40	204	3	564
Marine fishes nei	*Osteichthyes*	33 330	25 680	3 495	1 099	899	76 035	-
Euro-American crayfishes nei	*Astacidae, Cambaridae*	1	1	0	1	1	0	0
Northern prawn	*Pandalus borealis*	1 585	1 785	3 340	4 167	6 376	5 413	6 707
Natantian decapods nei	*Natantia*	-	-	-	-	11	-	-
Various squids nei	*Loliginidae, Ommastrephidae*	3 400	...	233	2	-	-	-
	Country total	*88 514*	*44 002*	*66 578*	*72 962*	*78 988*	*151 831*	*150 146*
Luxembourg								
Freshwater fishes nei	*Osteichthyes*	0	0	0	0	0	0	0
	Country total	*0*	*0*	*0*	*0*	*0*	*0*	*0*
Macedonia								
Common carp	*Cyprinus carpio*	10	9	...	...	25	6	22
Freshwater fishes nei	*Osteichthyes*	31	68	107	113	52	7	6
Sturgeons nei	*Acipenseridae*	0	2	6	-	-	-	-
Trouts nei	*Salmo spp*	37	51	18	22	131	115	120
	Country total	*78*	*130*	*131*	*135*	*208*	*128*	*148*
Malta								
Freshwater fishes nei	*Osteichthyes*	0	0	0	0	0	0	-
Common sole	*Solea solea*	0	0	0	0	0	0	-
Greater forkbeard	*Phycis blennoides*	-	2	3	4	5	0	-
European hake	*Merluccius merluccius*	2	4	5	6	6	0	-
Mullets nei	*Mugilidae*	0	0	0	0	0	0	-
Groupers nei	*Epinephelus spp*	19	27	15	0	15	15	31
Groupers, seabasses nei	*Serranidae*	1	1	1	2	2	2	1
European seabass	*Dicentrarchus labrax*	0	0	0	15	0	0	-
Common pandora	*Pagellus erythrinus*	5	6	6	6	5	2	2
Axillary seabream	*Pagellus acarne*	5	4	3	3	2	-	-
Sargo breams nei	*Diplodus spp*	2	0	4	4	2	2	3
Common dentex	*Dentex dentex*	0	1	0	1	1	1	2
Saddled seabream	*Oblada melanura*	1	2	1	2	2	1	1
Red porgy	*Pagrus pagrus*	8	9	8	6	6	4	4
Bogue	*Boops boops*	17	16	15	12	21	27	16
Salema	*Sarpa salpa*	0	0	1	0	0	0	-
Picarels nei	*Spicara spp*	7	7	7	8	9	6	3
Surmullets(=Red mullets) nei	*Mullus spp*	7	7	8	12	7	5	4
Greater weever	*Trachinus draco*	2	2	0	3	0	0	-
Gobies nei	*Gobiidae*	0	0	0	0	0	-	-
European conger	*Conger conger*	2	4	3	2	2	3	3
John dory	*Zeus faber*	0	1	1	2	1	0	-
Wreckfish	*Polyprion americanus*	9	14	8	8	8	8	16
Scorpionfishes nei	*Scorpaenidae*	8	3	8	12	11	0	-
Gurnards, searobins nei	*Triglidae*	2	0	2	4	2	1	1
Angler(=Monk)	*Lophius piscatorius*	0	1	0	2	0	-	-
Clupeoids nei	*Clupeoidei*	0	0	0	0	-	-	-
Atlantic bonito	*Sarda sarda*	2	7	2	2	1	-	-
Frigate and bullet tunas	*Auxis thazard, A.rochei*	3	6	6	3	1	-	-
Little tunny(=Atl.black skipj)	*Euthynnus alletteratus*	3	3	0	0	0	5	4
Atlantic bluefin tuna	*Thunnus thynnus*	399	393	407	447	376	219	176
Albacore	*Thunnus alalunga*	-	1	1	1	4	-	-
Marlins,sailfishes,etc. nei	*Istiophoridae*	1	1	-	-	-	-	-
Swordfish	*Xiphias gladius*	72	100	153	187	175	102	253
Tuna-like fishes nei	*Scombroidei*	0	0	0	-	-	-	-
Mediterranean horse mackerel	*Trachurus mediterraneus*	5	4	2	4	0	0	-
Greater amberjack	*Seriola dumerili*	9	6	6	6	3	2	3
Carangids nei	*Carangidae*	7	4	13	23	28	8	8
Common dolphinfish	*Coryphaena hippurus*	307	295	363	349	234	303	347
Chub mackerel	*Scomber japonicus*	23	29	40	19	34	32	3
Porbeagle	*Lamna nasus*	1	0	0	0	0	0	-
Picked dogfish	*Squalus acanthias*	28	28	23	18	19	17	24
Dogfish sharks nei	*Squalidae*	4	5	3	1	2	3	2
Angelsharks, sand devils nei	*Squatinidae*	0	0	0	0	0	0	-
Rays, stingrays, mantas nei	*Rajiformes*	7	8	5	6	7	0	-
Sharks, rays, skates, etc. nei	*Elasmobranchii*	3	2	11	4	13	0	-
Marine fishes nei	*Osteichthyes*	8 199	1	12	16	29	82	59
Natantian decapods nei	*Natantia*	9	16	18	24	23	36	29
Cuttlefish,bobtail squids nei	*Sepiidae, Sepiolidae*	5	3	3	5	4	0	-
Common squids nei	*Loligo spp*	2	2	2	2	3	2	2
European flying squid	*Todarodes sagittatus*	-	-	2	2	2	0	-
Octopuses, etc. nei	*Octopodidae*	11	11	9	11	9	5	7
	Country total	*9 197*	*1 036*	*1 180*	*1 244*	*1 074*	*893*	*1 004*

E-5 Fish crustaceans, molluscs, etc — Capture production by countries or areas and species — Europe
Poissons, crustacés, mollusques, etc — Captures par pays ou zones et espèces — Europe
Peces, crustáceos, moluscos, etc — Capturas por países o áreas y especies — Europa

English name Nom anglais Nombre inglés	Scientific name Nom scientifique Nombre científico	1996 t	1997 t	1998 t	1999 t	2000 t	2001 t	2002 t
Moldova Rep								
Common carp	*Cyprinus carpio*	408	349	280	178	192	212	274
Crucian carp	*Carassius carassius*	128	166	159	104	132	127	183
Northern pike	*Esox lucius*	53	39	38	25	19	36	62
Pike-perch	*Stizostedion lucioperca*	5	4	1	2	1	12	19
Freshwater fishes nei	*Osteichthyes*	9	11	13	-	-	-	27
	Country total	*603*	*569*	*491*	*309*	*344*	*387*	*565*
Monaco								
Marine fishes nei	*Osteichthyes*	3 F	3 F	3 F	3 F	3 F	3 F	3 F
	Country total	*3 F*	*3 F*	*3 F*	*3 F*	*3 F*	*3 F*	*3 F*
Netherlands								
Freshwater bream	*Abramis brama*	75	65	399	355	350 F	300 F	282
Roaches nei	*Rutilus spp*	100	123	107	100	90 F	80 F	64
European perch	*Perca fluviatilis*	376	336	155	177	170 F	150 F	131
Pike-perch	*Stizostedion lucioperca*	100	89	61	104	161 F	254 F	302
Freshwater fishes nei	*Osteichthyes*	350	362	176	154	150 F	150 F	134
European eel	*Anguilla anguilla*	336	315	345	372	351 F	374 F	373
Atlantic salmon	*Salmo salar*	2	1	1	1	-	0	0
European smelt	*Osmerus eperlanus*	856	1 033	327	1 097	1 074 F	1 041 F	1 126
Twaite shad	*Alosa fallax*	-	-	-	-	1	5	8
Atlantic halibut	*Hippoglossus hippoglossus*	3	5	4	2	1	-	0
European plaice	*Pleuronectes platessa*	35 539	34 272	30 592	37 543	35 079	33 835	29 083
Witch flounder	*Glyptocephalus cynoglossus*	0	1	4	9	7	1	-
Common dab	*Limanda limanda*	...	...	7 983	8 656	6 544	5 969	4 955
Lemon sole	*Microstomus kitt*	...	...	839	681	492	456	402
European flounder	*Platichthys flesus*	...	...	4 942	3 159	2 658	2 621	3 530
Common sole	*Solea solea*	15 563	10 370	15 308	16 329	15 343	13 737	12 120
Megrim	*Lepidorhombus whiffiagonis*	11	23	31	28	...	11	8
Brill	*Scophthalmus rhombus*	736	598	811	809	1 005	1 093	908
Turbot	*Psetta maxima*	1 780	1 866	1 700	1 812	2 287	2 277	1 899
Atlantic cod	*Gadus morhua*	9 307	11 838	14 724	9 075	6 000	3 656	4 714
Ling	*Molva molva*	-	-	-	-	5	4	3
Haddock	*Melanogrammus aeglefinus*	111	494	289	115	121	295	360
Saithe(=Pollock)	*Pollachius virens*	19	42	8	7	11	19	5
Pollack	*Pollachius pollachius*	19	15	7	5	5	1	1
Norway pout	*Trisopterus esmarkii*	13	85	3	1	3	-	2
Pouting(=Bib)	*Trisopterus luscus*	-	-	-	-	612	645	735
Blue whiting(=Poutassou)	*Micromesistius poutassou*	16 407	24 132	27 693	32 889	43 145	63 625	35 624
Whiting	*Merlangius merlangus*	3 411	2 554	1 981	1 806	1 899	2 619	2 448
European hake	*Merluccius merluccius*	111	62	75	98	43	72	18
Mullets nei	*Mugilidae*	-	0	-	17	36	184	113
Groupers, seabasses nei	*Serranidae*	-	-	17	14	-	-	-
European seabass	*Dicentrarchus labrax*	8	1	49	32	60	79	96
Blackspot(=red) seabream	*Pagellus bogaraveo*	38	-	-	28	71	2	11
Canary dentex	*Dentex canariensis*	-	-	-	-	20	-	-
Black seabream	*Spondyliosoma cantharus*	-	-	-	-	-	-	1
Surmullet	*Mullus surmuletus*	1	0	-	-	235	560	337
Greater weever	*Trachinus draco*	-	-	-	-	-	6	0
Argentines	*Argentina spp*	3 953	4 696	4 964	8 033	3 636	3 659	4 213
European conger	*Conger conger*	-	-	-	-	-	1	0
Wolffishes(=Catfishes) nei	*Anarhichas spp*	6	16	36	21	10	2	3
Largehead hairtail	*Trichiurus lepturus*	33	-	103	401	115	697	547
Black scabbardfish	*Aphanopus carbo*	-	-	-	11	7	-	21
Atlantic redfishes nei	*Sebastes spp*	41	53	20	16	19	8	15
Tub gurnard	*Chelidonichthys lucerna*	-	-	-	-	1 164	-	1 438
Red gurnard	*Chelidonichthys cuculus*	-	-	-	-	46	1 724	53
Grey gurnard	*Eutrigla gurnardus*	-	-	-	-	459	295	286
Angler(=Monk)	*Lophius piscatorius*	227	319	259	169	170	168	86
Atlantic herring	*Clupea harengus*	77 605	65 448	77 090	78 741	75 221	66 357	78 557
Round sardinella	*Sardinella aurita*	...	...	...	...	...	134 490	93 032
Sardinellas nei	*Sardinella spp*	41 488	86 635	110 091	115 753	122 783	-	5 131
European pilchard(=Sardine)	*Sardina pilchardus*	1 242	6 488	8 998	7 624	17 862	11 786	27 581
European sprat	*Sprattus sprattus*	293	806	54	264	307	136	169
European anchovy	*Engraulis encrasicolus*	6	1	16	3	-	3	4
Atlantic bonito	*Sarda sarda*	1 694	1 625	2 171	966	1 507	1 791	1 793
Garfish	*Belone belone*	-	-	-	-	-	2	2
Atlantic horse mackerel	*Trachurus trachurus*	135 965	122 683	103 248	84 891	65 994	84 011	56 575
Jack and horse mackerels nei	*Trachurus spp*	1 938	3 245	3 163	2 847	9 053	14 476	11 003
Chub mackerel	*Scomber japonicus*	857	3 202	1 836	1 561	12 005	12 708	23 254
Atlantic mackerel	*Scomber scombrus*	24 246	23 702	30 163	27 816	32 403	33 109	43 460
Tiger shark	*Galeocerdo cuvier*	-	-	-	-	-	-	13
Picked dogfish	*Squalus acanthias*	-	-	-	-	28	39	27
Raja rays nei	*Raja spp*	-	-	550	480	631	748	793
Various sharks nei	*Selachimorpha(Pleurotremata)*	-	-	-	-	-	3	-
Marine fishes nei	*Osteichthyes*	16 914	19 406	4 127	3 897	1 201	1 781	1 450
Edible crab	*Cancer pagurus*	-	-	-	-	146	300	506
Norway lobster	*Nephrops norvegicus*	423	627	694	662	572	853	966
European lobster	*Homarus gammarus*	-	-	-	13	12	33	11
Penaeus shrimps nei	*Penaeus spp*	-	-	-	1	1	3	5
Common shrimp	*Crangon crangon*	12 067	13 054	11 871	13 772	11 496	14 081	11 453
Freshwater molluscs nei	*Mollusca*	-	-	-	-	-	-	451
Whelk	*Buccinum undatum*	-	-	-	-	121	163	178
Great Atlantic scallop	*Pecten maximus*	228	188	408	306	249	274	473

English name Nom anglais Nombre inglés	Scientific name Nom scientifique Nombre científico	1996 t	1997 t	1998 t	1999 t	2000 t	2001 t	2002 t
Common edible cockle	*Cerastoderma edule*	6 300	10 923	68 133	50 888	19 633	-	-
Cuttlefish,bobtail squids nei	*Sepiidae, Sepiolidae*	-	-	-	-	101	162	381
Various squids nei	*Loliginidae, Ommastrephidae*	-	-	-	-	773	171	312
Octopuses, etc. nei	*Octopodidae*	-	-	-	-	-	7	-
	Country total	*410 798*	*451 799*	*536 626*	*514 611*	*495 754*	*518 162*	*464 035*
Norway								
Northern pike	*Esox lucius*	...	13	7	...	...	...	...
Burbot	*Lota lota*	...	1	1	...	...	...	...
European perch	*Perca fluviatilis*	...	9	3	...	...	...	...
Pike-perch	*Stizostedion lucioperca*	...	5	3	...	...	...	...
European eel	*Anguilla anguilla*	352	497	363	475	281	304	310
Atlantic salmon	*Salmo salar*	793	638	753	827	1 054	1 125 F	1 022 F
Sea trout	*Salmo trutta*	-	-	-	-	12	0	11
Trouts nei	*Salmo spp*	-	-	-	-	4	11	0
Chars nei	*Salvelinus spp*	81	78	94	93	103	100 F	100 F
Grayling	*Thymallus thymallus*	...	1	...	...	...	...	...
Vendace	*Coregonus albula*	...	10	4	10	6	5 F	5 F
European whitefish	*Coregonus lavaretus*	...	57	52	54	47	45 F	45 F
Atlantic halibut	*Hippoglossus hippoglossus*	678	879	672	696	1 039	868	676
European plaice	*Pleuronectes platessa*	1 731	2 857	1 872	1 816	1 944	2 773	2 945
Greenland halibut	*Reinhardtius hippoglossoides*	17 073	12 343	11 948	19 704	13 022	15 152	11 535
Witch flounder	*Glyptocephalus cynoglossus*	80	86	140	135	97	88	82
Amer. plaice(=Long rough dab)	*Hippoglossoides platessoides*	-	119	24	15	0	15	0
Common dab	*Limanda limanda*	-	-	-	-	49	54	55
Lemon sole	*Microstomus kitt*	47	63	59	59	60	53	60
European flounder	*Platichthys flesus*	-	-	-	-	5	3	3
Common sole	*Solea solea*	-	-	-	-	198	88	53
Soles nei	*Soleidae*	...	...	...	...	17	7	40
Brill	*Scophthalmus rhombus*	21	26	26	30	27	26	23
Turbot	*Psetta maxima*	54	57	45	48	69	94	99
Flatfishes nei	*Pleuronectiformes*	376	477	389	475	76	153	117
Moras nei	*Moridae*	-	-	-	-	68	277	97
Tusk(=Cusk)	*Brosme brosme*	19 483	13 797	21 032	23 274	21 915	18 778	18 171
Atlantic cod	*Gadus morhua*	358 395	401 277	321 428	256 555	219 192	208 856	228 672
Ling	*Molva molva*	18 931	15 295	22 719	19 217	16 899	13 562	15 358
Blue ling	*Molva dypterygia*	530	497	420	544	834	1 020	902
Greater forkbeard	*Phycis blennoides*	-	-	-	-	709	1 340	1 334
Haddock	*Melanogrammus aeglefinus*	97 115	106 155	79 008	53 243	45 934	51 648	55 233
Saithe(=Pollock)	*Pollachius virens*	221 638	183 451	194 452	198 387	169 746	169 506	203 963
Pollack	*Pollachius pollachius*	2 318	2 230	2 247	2 928	3 385	2 888	3 465
Norway pout	*Trisopterus esmarkii*	103 126	47 032	27 575	51 124	52 912	27 123	25 996
Blue whiting(=Poutassou)	*Micromesistius poutassou*	356 054	348 268	570 665	534 570	553 478	573 686	558 070
Whiting	*Merlangius merlangus*	210	140	116	143	145	237	171
European hake	*Merluccius merluccius*	938	981	825	609	693	635	534
Southern hake	*Merluccius australis*	210	117	16	-	-	-	-
Blue grenadier	*Macruronus novaezelandiae*	6 614	5 576	4 633	-	-	-	-
Roughhead grenadier	*Macrourus berglax*	-	-	-	-	63	148	97
Roundnose grenadier	*Coryphaenoides rupestris*	-	-	-	-	31	78	59
Gadiformes nei	*Gadiformes*	217	262	940	438	7	-	-
Wrasses, hogfishes, etc. nei	*Labridae*	-	-	-	-	10	6	4
Sandeels(=Sandlances) nei	*Ammodytes spp*	160 702	350 672	343 625	187 589	119 015	187 459	175 985
Argentines	*Argentina spp*	6 817	5 167	8 654	7 823	6 107	14 668	7 406
European conger	*Conger conger*	0	0	0	1	0	0	1
Alfonsinos nei	*Beryx spp*	-	836	1 066	-	324	-	-
Orange roughy	*Hoplostethus atlanticus*	5	34	15	-	642	-	-
Oreo dories nei	*Oreosomatidae*	1	-	7	-	175	-	-
Atlantic wolffish	*Anarhichas lupus*	...	...	...	...	917	1 111	870
Spotted wolffish	*Anarhichas minor*	...	...	...	...	1 091	1 111	1 394
Wolffishes(=Catfishes) nei	*Anarhichas spp*	6 819	12 769	16 332	6 398	4 370	12 205	3 435
Giant stargazer	*Kathetostoma giganteum*	1	1	1	-	-	-	-
Oilfish	*Ruvettus pretiosus*	-	5	-	-	-	-	-
South Pacific breams nei	*Seriolella spp*	121	70	4	-	-	-	-
Golden redfish	*Sebastes marinus*	-	-	-	-	15 996	13 698	9 438
Beaked redfish	*Sebastes mentella*	-	-	-	-	9 484	14 958	6 883
Atlantic redfishes nei	*Sebastes spp*	28 679	22 687	28 560	30 856	114	1	9
Lumpfish(=Lumpsucker)	*Cyclopterus lumpus*	4 355	5 652	1 365	2 059	2 374	5 184	5 936
Angler(=Monk)	*Lophius piscatorius*	2 071	1 447	2 646	3 239	4 357	4 974	3 172
Atlantic herring	*Clupea harengus*	763 073	923 165	831 844	829 008	800 059	581 161	573 965
European pilchard(=Sardine)	*Sardina pilchardus*	-	-	3 421	-	-	-	-
European sprat	*Sprattus sprattus*	59 115	7 051	35 166	22 214	6 353	12 465	2 609
Atlantic bluefin tuna	*Thunnus thynnus*	-	-	-	5	0	-	-
Swordfish	*Xiphias gladius*	-	1	-	-	-	-	-
Capelin	*Mallotus villosus*	207 706	157 889	88 226	91 813	370 769	482 835	522 349
Garfish	*Belone belone*	1	2	1	1	0	0	1
Atlantic horse mackerel	*Trachurus trachurus*	15 556	46 491	13 366	46 657	2 084	7 988	36 686
Jack and horse mackerels nei	*Trachurus spp*	0	1	-	-	-	-	-
Atlantic mackerel	*Scomber scombrus*	136 699	137 256	158 340	161 046	174 228	180 603	184 371
Basking shark	*Cetorhinus maximus*	1 979	1 159	137	77	293	200	135
Porbeagle	*Lamna nasus*	28	17	28	33	22	17	19
Picked dogfish	*Squalus acanthias*	2 749	1 567	1 293	1 461	1 644	1 424	1 126
Leafscale gulper shark	*Centrophorus squamosus*	-	-	-	-	-	1	-
Portuguese dogfish	*Centroscymnus coelolepis*	-	-	-	-	-	13	-
Dogfish sharks nei	*Squalidae*	-	-	-	-	118	313	10
Blue skate	*Raja batis*	...	...	...	...	72	65	53
Thornback ray	*Raja clavata*	-	-	-	-	2	-	-
Raja rays nei	*Raja spp*	798	591	752	791	704	725	484
Rabbit fish	*Chimaera monstrosa*	-	-	-	-	1	70	46

E-5

Fish crustaceans, molluscs, etc	Capture production by countries or areas and species	Europe
Poissons, crustacés, mollusques, etc	Captures par pays ou zones et espèces	Europe
Peces, crustáceos, moluscos, etc	Capturas por países o áreas y especies	Europa

English name Nom anglais Nombre inglés	Scientific name Nom scientifique Nombre científico	1996 t	1997 t	1998 t	1999 t	2000 t	2001 t	2002 t
Ratfishes nei	*Hydrolagus spp*	-	-	-	-	-	6	7
Various sharks nei	*Selachimorpha(Pleurotremata)*	0	1	0	13	1	72	1
Finfishes nei	*Osteichthyes*	158	322	2 791	3 389	3 199	1 909	1 179
Marine fishes nei	*Osteichthyes*	235	155	187	-	42	-	-
Noble crayfish	*Astacus astacus*	...	10	10	10	...	...	...
Edible crab	*Cancer pagurus*	1 889	2 204	2 984	2 837	2 889	3 476	4 345
Marine crabs nei	*Brachyura*	-	-	-	-	-	2	0
Norway lobster	*Nephrops norvegicus*	188	187	293	383	346	281	280
European lobster	*Homarus gammarus*	30	35	45	59	52	40	42
Red king crab	*Paralithodes camtschaticus*	70	71	124	202	211	434	414
Northern prawn	*Pandalus borealis*	41 505	41 961	57 141	63 538	66 578	66 336	70 524
Cupped oysters nei	*Crassostrea spp*	-	-	-	-	-	5	-
Blue mussel	*Mytilus edulis*	4	0	-	1	10	-	-
Horse mussels nei	*Modiolus spp*	20	30	20	7	2	2	12
Great Atlantic scallop	*Pecten maximus*	14	39	114	425	571	670	575
Iceland scallop	*Chlamys islandica*	-	-	-	-	14	14	16
Scallops nei	*Pectinidae*	3	16	21	12	4	13	37
Common edible cockle	*Cerastoderma edule*	-	-	-	-	38	33	19
European flying squid	*Todarodes sagittatus*	0	190	2	0	0	-	-
Marine molluscs nei	*Mollusca*	1	14	111	118	17	9	43
Sea urchins nei	*Strongylocentrotus spp*	-	-	-	-	1	0	0
	Country total	*2 648 457*	*2 863 059*	*2 861 223*	*2 627 534*	*2 699 421*	*2 687 303*	*2 743 184*
Poland								
Freshwater bream	*Abramis brama*	3 622	2 367	2 057	2 321	2 945	2 576	2 626
Freshwater breams nei	*Abramis spp*	373	...	...	235	490	400	398
Common carp	*Cyprinus carpio*	77	82	78	37	45	50	54
Tench	*Tinca tinca*	97	101	91	102	160	113	141
Crucian carp	*Carassius carassius*	46	50	47	49	95	87	115
Roach	*Rutilus rutilus*	3 095	1 834	1 662	1 962	2 188	2 175	2 344
Orfe(=Ide)	*Leuciscus idus*	0	-	-	-	9	7	7
Grass carp(=White amur)	*Ctenopharyngodon idellus*	4	11	4	2	4	4	5
Silver carp	*Hypophthalmichthys molitrix*	211	180	106	136	185	120	198
Vimba bream	*Vimba vimba*	...	...	...	...	5	2	1
Asp	*Aspius aspius*	...	...	...	...	6	5	6
Cyprinids nei	*Cyprinidae*	0	395	330	-	2	2	3
Northern pike	*Esox lucius*	1 076	280	226	262	363	311	354
Wels(=Som)catfish	*Silurus glanis*	0	0	1	1	7	4	4
Burbot	*Lota lota*	0	9	10	12	18	26	38
European perch	*Perca fluviatilis*	1 169	1 584	1 172	1 124	922	1 119	1 057
Pike-perch	*Stizostedion lucioperca*	512	480	381	537	546	478	509
Freshwater fishes nei	*Osteichthyes*	13 845	10 074	10 327	10 417	12 508	13 276	13 295
European eel	*Anguilla anguilla*	639	489	454	474	429	426	362
Atlantic salmon	*Salmo salar*	125	110	114	118	145	161	198
Trouts nei	*Salmo spp*	190	200	329	385	718	577	866
Rainbow trout	*Oncorhynchus mykiss*	...	35	27	14	12	12	13
European smelt	*Osmerus eperlanus*	-	38	53	212	19	20	57
Vendace	*Coregonus albula*	234	297	217	286	275	227	207
European whitefish	*Coregonus lavaretus*	33	24	14	21	29	13	11
Lampreys nei	*Petromyzontidae*	0	2	2	2	6	5	3
Atlantic halibut	*Hippoglossus hippoglossus*	-	-	-	-	-	488	-
Pacific halibut	*Hippoglossus stenolepis*	-	-	-	-	-	4	-
Greenland halibut	*Reinhardtius hippoglossoides*	-	12	31	8	4	4	22
Amer. plaice(=Long rough dab)	*Hippoglossoides platessoides*	-	-	-	-	-	1	-
Flatfishes nei	*Pleuronectiformes*	8 836	6 168	5 835	5 779	5 601	6 725	9 232
Atlantic cod	*Gadus morhua*	35 968	34 295	27 705	28 056	23 340	23 310	17 209
Blue ling	*Molva dypterygia*	-	-	-	-	-	19	8
Haddock	*Melanogrammus aeglefinus*	18	35	27	24	16	96	52
Saithe(=Pollock)	*Pollachius virens*	365	822	813	862	747	727	752
Alaska pollock(=Walleye poll.)	*Theragra chalcogramma*	116 257	125 413	81 889	65 508	33 192	16 590	-
Blue whiting(=Poutassou)	*Micromesistius poutassou*	-	-	-	-	-	-	38
Southern blue whiting	*Micromesistius australis*	3 402	-	-	-	-	-	-
Whiting	*Merlangius merlangus*	-	-	1	-	-	-	-
Senegalese hake	*Merluccius senegalensis*	64	-	-	-	-	87	-
Argentine hake	*Merluccius hubbsi*	-	-	-	35	-	-	-
North Pacific hake	*Merluccius productus*	-	-	-	-	977	-	-
Cape hakes	*Merluccius capensis,M.paradox.*	3	-	-	-	-	-	-
Patagonian grenadier	*Macruronus magellanicus*	146	-	-	86	-	73	-
Grenadiers nei	*Macrourus spp*	-	-	-	13	-	3	-
Roundnose grenadier	*Coryphaenoides rupestris*	-	5 867	6 769	546	-	179	942
Porgies, seabreams nei	*Sparidae*	7	5	-	-	-	-	-
Antarctic rockcods, noties nei	*Nototheniidae*	-	-	-	207	-	-	-
Alfonsinos nei	*Beryx spp*	-	1 964	-	-	-	-	-
Mackerel icefish	*Champsocephalus gunnari*	-	-	-	-	-	-	296
Atlantic wolffish	*Anarhichas lupus*	-	19	40	6	18	8	12
Black scabbardfish	*Aphanopus carbo*	-	-	-	-	-	-	2
Pacific ocean perch	*Sebastes alutus*	-	-	-	-	21	-	-
Beaked redfish	*Sebastes mentella*	-	777	12	6	2	9	437
Atlantic herring	*Clupea harengus*	31 246	28 939	21 873	19 229	24 516	37 611	36 778
Sardinellas nei	*Sardinella spp*	7 166	2 553	-	-	-	3 463	4 824
European pilchard(=Sardine)	*Sardina pilchardus*	2 439	1 269	-	-	-	-	14 244
European sprat	*Sprattus sprattus*	77 472	105 298	59 090	71 706	84 324	85 757	81 243
Atlantic bonito	*Sarda sarda*	225	39	-	-	-	521	79
Cape horse mackerel	*Trachurus capensis*	1 700	-	-	-	-	3 098	3 557
Jack and horse mackerels nei	*Trachurus spp*	3 583	281	-	-	-	1 449	1 638
Scads nei	*Decapterus spp*	54	-	-	-	-	-	-
Chub mackerel	*Scomber japonicus*	4 086	480	-	-	-	1 666	6 633
Atlantic mackerel	*Scomber scombrus*	-	22	-	-	-	-	-

English name Nom anglais Nombre inglés	Scientific name Nom scientifique Nombre científico	1996 t	1997 t	1998 t	1999 t	2000 t	2001 t	2002 t
Raja rays nei	*Raja spp*	-	-	-	-	-	2	-
Sharks, rays, skates, etc. nei	*Elasmobranchii*	-	-	-	-	-	11	8
Groundfishes nei	*Osteichthyes*	36	39	41	-	-	-	-
Pelagic fishes nei	*Osteichthyes*	60	66	61	-	-	-	-
Finfishes nei	*Osteichthyes*	23	34	23	-	17	307	339
Marine fishes nei	*Osteichthyes*	2 184	642	-	8	25	6 001	2 056
Natantian decapods nei	*Natantia*	-	824	691	894	1 732	263	-
Antarctic krill	*Euphausia superba*	22 104	14 408	19 388	19 167	20 049	13 696	16 365
Patagonian squid	*Loligo gahi*	-	-	19	4 875	-	-	-
Argentine shortfin squid	*Illex argentinus*	1	-	-	-	970	701	2 803
	Country total	*342 793*	*348 913*	*242 010*	*235 724*	*217 682*	*225 065*	*222 441*
Portugal								
Common carp	*Cyprinus carpio*	0	0	0	0	0	0	-
Crucian carp	*Carassius carassius*	0	0	0	0	0	0	0
Northern pike	*Esox lucius*	0	0	-	-	0	0	-
Freshwater fishes nei	*Osteichthyes*	0	0	0	0	0	1	12
Sturgeons nei	*Acipenseridae*	0	-	-	-	-	-	-
European eel	*Anguilla anguilla*	...	...	...	30	29	37	36
Atlantic salmon	*Salmo salar*	-	-	-	0	0	-	-
Sea trout	*Salmo trutta*	0	-	-	1	1	1	8
Shads nei	*Alosa spp*	18	17	21	17	20	22	21
West African ilisha	*Ilisha africana*	-	-	-	1	-	-	-
Diadromous clupeoids nei	*Clupeoidei*	35	36	30	-	-	-	-
Sea lamprey	*Petromyzon marinus*	3	2	2	4	6	6	13
Lefteye flounders nei	*Bothidae*	22	2	...	89	119	119	117
Atlantic halibut	*Hippoglossus hippoglossus*	14	17	31	51	30	45	55
European plaice	*Pleuronectes platessa*	137	89	115	95	124	145	184
Greenland halibut	*Reinhardtius hippoglossoides*	3 395	3 393	3 341	4 044	4 725	5 066	4 464
Witch flounder	*Glyptocephalus cynoglossus*	270	380	403	535	254	599	454
Amer. plaice(=Long rough dab)	*Hippoglossoides platessoides*	376	446	645	789	510	704	723
Yellowtail flounder	*Limanda ferruginea*	-	-	85	426	153	351	122
European flounder	*Platichthys flesus*	0	-	-	-	-	-	-
Common sole	*Solea solea*	167	151	113	121	152	201	115
Sand sole	*Solea lascaris*	77	95	118	90	116	142	97
Wedge sole	*Dicologlossa cuneata*	87	8	0	111	119	121	97
Soles nei	*Soleidae*	248	124	60	854	1 124	904	794
Megrim	*Lepidorhombus whiffiagonis*	58	26	47	54	47	22	26
Brill	*Scophthalmus rhombus*	48	39	33	39	46	57	47
Turbot	*Psetta maxima*	40	28	27	34	63	83	69
Flatfishes nei	*Pleuronectiformes*	1 304	1 083	964	11	11	6	132
Tadpole codling	*Salilota australis*	-	-	-	-	12	-	-
Tusk(=Cusk)	*Brosme brosme*	0	-	-	-	-	-	-
Atlantic cod	*Gadus morhua*	8 083	9 079	6 042	4 212	3 778	4 384	4 118
Blue ling	*Molva dypterygia*	25	21	14	10	14	9	13
Greater forkbeard	*Phycis blennoides*	113	46	45	54	98	92	63
Brazilian codling	*Urophycis brasiliensis*	-	-	-	-	-	3	8
Red hake	*Urophycis chuss*	125	56	18	77	42	273	1 969
White hake	*Urophycis tenuis*	-	-	-	-	-	-	1 678
Haddock	*Melanogrammus aeglefinus*	208	207	55	48	144	128	231
Saithe(=Pollock)	*Pollachius virens*	24	13	49	37	64	86	132
Pollack	*Pollachius pollachius*	2	2	1	1	15	41	45
Pouting(=Bib)	*Trisopterus luscus*	2 491	2 051	2 254	2 792	3 299	4 511	3 103
Blue whiting(=Poutassou)	*Micromesistius poutassou*	3 565	2 448	1 900	2 676	2 169	1 763	1 698
Southern blue whiting	*Micromesistius australis*	-	-	-	-	1	-	-
Whiting	*Merlangius merlangus*	184	139	115	76	77	38	42
European hake	*Merluccius merluccius*	3 622	2 578	2 563	3 217	3 061	3 032	2 880
Senegalese hake	*Merluccius senegalensis*	223	102	42	17	...	...	...
Silver hake	*Merluccius bilinearis*	-	-	-	-	-	-	29
Argentine hake	*Merluccius hubbsi*	4 253	603	310	-	3	-	-
Hakes nei	*Merluccius spp*	1 515	914	1 027	474	-	365	186
Cape hakes	*Merluccius capensis,M.paradox.*	-	-	-	-	-	1	-
Patagonian grenadier	*Macruronus magellanicus*	-	-	-	-	32	-	-
Roughhead grenadier	*Macrourus berglax*	787	762	1 090	1 299	395	610	508
Grenadiers nei	*Macrourus spp*	80	-	-	-	-	1	19
Roundnose grenadier	*Coryphaenoides rupestris*	0	-	-	-	-	-	-
Gadiformes nei	*Gadiformes*	724	724	726	596	514	492	597
Morays	*Muraenidae*	...	...	...	193	169	145	130
Mullets nei	*Mugilidae*	297	303	336	324	336	376	321
Dusky grouper	*Epinephelus marginatus*	123	13	27	8	2	2	-
White grouper	*Epinephelus aeneus*	5	0	-	-	-	1	0
Groupers nei	*Epinephelus spp*	203	36	11	9	2	5	11
Groupers, seabasses nei	*Serranidae*	461	298	338	336	292	180	166
European seabass	*Dicentrarchus labrax*	57	40	38	37	49	43	43
Seabasses nei	*Dicentrarchus spp*	-	5	6	336	405	378	357
Cardinalfishes, etc. nei	*Apogonidae*	11	0	-	-	0	-	-
Rubberlip grunt	*Plectorhinchus mediterraneus*	1 436	544	115	267	87	12	183
Bastard grunt	*Pomadasys incisus*	1	0	-	-	0	0	-
Grunts, sweetlips nei	*Haemulidae (=Pomadasyidae)*	186	87	69	47	35	27	11
Meagre	*Argyrosomus regius*	0	0	3	3	4	30	36
Southern meagre(=Mulloway)	*Argyrosomus hololepidotus*	-	-	-	-	-	1	-
Boe drum	*Pteroscion peli*	582	136	101	18	2	0	17
West African croakers nei	*Pseudotolithus spp*	157	36	15	49	635	25	31
Croakers, drums nei	*Sciaenidae*	110	52	30	239	138	175	376
Blackspot(=red) seabream	*Pagellus bogaraveo*	1 446	1 350	1 511	1 490	1 033	1 128	1 312
Common pandora	*Pagellus erythrinus*	122	158	134	105	151	128	142
Axillary seabream	*Pagellus acarne*	1 198	970	883	997	1 298	1 202	1 130
Red pandora	*Pagellus bellottii*	1	-	-	-	-	-	-

E-5

Fish crustaceans, molluscs, etc — **Capture production by countries or areas and species** — **Europe**
Poissons, crustacés, mollusques, etc — **Captures par pays ou zones et espèces** — **Europe**
Peces, crustáceos, moluscos, etc — **Capturas por países o áreas y especies** — **Europa**

English name Nom anglais Nombre inglés	Scientific name Nom scientifique Nombre científico	1996 t	1997 t	1998 t	1999 t	2000 t	2001 t	2002 t
Sargo breams nei	*Diplodus spp*	65	18	4	1 051	979	799	899
Large-eye dentex	*Dentex macrophthalmus*	71	74	78	66	0	0	1
Common dentex	*Dentex dentex*	20	21	38	25	13	16	10
Dentex nei	*Dentex spp*	15	2	63	28	6	2	3
Black seabream	*Spondyliosoma cantharus*	23	5	4	165	177	158	225
Saddled seabream	*Oblada melanura*	1	0	1	-	0	0	0
Red porgy	*Pagrus pagrus*	171	274	635	594	476	269	135
Pargo breams nei	*Pagrus spp*	196	31	11	4	41	145	11
Gilthead seabream	*Sparus aurata*	213	189	173	151	183	213	268
Bogue	*Boops boops*	417	420	358	426	670	958	849
Sand steenbras	*Lithognathus mormyrus*	...	...	...	178	158	109	145
Salema	*Sarpa salpa*	...	...	...	336	246	320	239
Porgies, seabreams nei	*Sparidae*	2 477	2 044	1 859	6	21	3	4
Picarels nei	*Spicara spp*	...	...	...	43	22	25	36
Surmullet	*Mullus surmuletus*	52	69	66	180	155	191	166
Surmullets(=Red mullets) nei	*Mullus spp*	2	0	0	-	-	-	-
West African goatfish	*Pseudupeneus prayensis*	1	-	1	0	-	-	-
Wrasses, hogfishes, etc. nei	*Labridae*	...	...	...	44	37	40	55
Parrotfish	*Sparisoma cretense*	-	-	-	56	89	162	153
Sandeels(=Sandlances) nei	*Ammodytes spp*	41	18	9	13	29	40	34
Greater weever	*Trachinus draco*	0	0	2	7	12	11	22
Gobies nei	*Gobiidae*	0	0	-	0	-	-	-
Triggerfishes, durgons nei	*Balistidae*	9	4	1	44	38	21	32
Toadfishes, etc. nei	*Batrachoididae*	-	-	-	70	91	75	86
European conger	*Conger conger*	3 355	2 867	2 861	2 473	2 002	1 896	1 527
Conger eels, etc. nei	*Congridae*	-	-	-	-	-	1	-
Longspine snipefish	*Macroramphosus scolopax*	-	-	-	-	-	-	0
Pink cusk-eel	*Genypterus blacodes*	-	-	-	-	13	89	98
Alfonsinos nei	*Beryx spp*	126	58	51	132	89	67	81
Orange roughy	*Hoplostethus atlanticus*	-	-	-	117	157	161	122
Slimeheads nei	*Trachichthyidae*	-	-	-	-	-	470	-
John dory	*Zeus faber*	166	179	317	366	431	457	417
Silvery John dory	*Zenopsis conchifer*	9	6	6	3	-	0	-
Wreckfish	*Polyprion americanus*	644	416	389	321	348	310	386
Patagonian toothfish	*Dissostichus eleginoides*	-	-	-	-	3	-	-
Spotted wolffish	*Anarhichas minor*	...	...	...	...	...	20	-
Wolffishes(=Catfishes) nei	*Anarhichas spp*	292	410	617	645	229	304	714
Stargazer	*Uranoscopus scaber*	...	...	...	15	50	46	104
Oilfish	*Ruvettus pretiosus*	-	-	-	14	14	11	17
Roudi escolar	*Promethichthys prometheus*	-	-	-	-	-	6	7
Largehead hairtail	*Trichiurus lepturus*	0	0	0	3	0	0	0
Silver scabbardfish	*Lepidopus caudatus*	10 996	7 584	5 512	3 285	66	86	82
Black scabbardfish	*Aphanopus carbo*	10 434	7 576	7 583	7 181	7 070	6 753	6 565
Golden redfish	*Sebastes marinus*	2	7	10	2	269	2	5
Atlantic redfishes nei	*Sebastes spp*	5 057	5 336	6 628	10 451	9 699	8 319	9 718
Blackbelly rosefish	*Helicolenus dactylopterus*	...	...	...	334	436	313	297
Scorpionfishes nei	*Scorpaenidae*	730	658	685	298	273	372	225
Tub gurnard	*Chelidonichthys lucerna*	-	-	-	3	5	3	8
Gurnards, searobins nei	*Triglidae*	641	615	504	503	650	616	644
Angler(=Monk)	*Lophius piscatorius*	224	323	179	1 471	880	617	512
Blackbellied angler	*Lophius budegassa*	127	55	26	16	18	2	6
American angler	*Lophius americanus*	-	-	-	-	-	-	134
Atlantic herring	*Clupea harengus*	2	0	0	1	0	2	0
Madeiran sardinella	*Sardinella maderensis*	0	1	5	11	2	3	1
European pilchard(=Sardine)	*Sardina pilchardus*	86 855	81 477	82 992	71 972	66 319	71 947	68 763
European sprat	*Sprattus sprattus*	0	-	-	-	-	-	-
European anchovy	*Engraulis encrasicolus*	2 775	633	1 657	1 408	310	855	934
Atlantic bonito	*Sarda sarda*	83	49	98	98	161	47	61
Plain bonito	*Orcynopsis unicolor*	-	-	-	-	-	-	3
Frigate and bullet tunas	*Auxis thazard, A.rochei*	-	1	59	268	503	236	176
Little tunny(=Atl.black skipj)	*Euthynnus alletteratus*	218	320	171	14	50	-	2
Skipjack tuna	*Katsuwonus pelamis*	8 297	4 399	4 544	1 810	1 307	2 168	2 959
Atlantic bluefin tuna	*Thunnus thynnus*	473	749	377	487	502	468	179
Albacore	*Thunnus alalunga*	2 128	651	215	556	764	1 217	2 387
Yellowfin tuna	*Thunnus albacares*	288	176	267	177	204	23	35
Bigeye tuna	*Thunnus obesus*	5 810	5 437	6 334	3 313	1 498	1 606	2 590
Atlantic sailfish	*Istiophorus albicans*	-	-	-	53	18	3	10
Blue marlin	*Makaira nigricans*	7	3	47	8	17	18	7
Atlantic white marlin	*Tetrapturus albidus*	-	-	-	-	-	-	1
Marlins,sailfishes,etc. nei	*Istiophoridae*	0	-	18	13	75	470	691
Swordfish	*Xiphias gladius*	2 092	1 344	1 262	1 388	1 335	1 809	1 938
Tuna-like fishes nei	*Scombroidei*	252	164	291	141	220	178	180
Garfish	*Belone belone*	41	35	43	54	55	57	30
Needlefishes, etc. nei	*Belonidae*	1	-	19	-	-	-	-
Atlantic saury	*Scomberesox saurus*	0	0	54	0	1	1	-
Dealfishes	*Trachipterus spp*	...	...	...	29	20	25	16
Silversides(=Sand smelts) nei	*Atherinidae*	3	6	22	6	3	0	1
Bluefish	*Pomatomus saltatrix*	62	48	37	22	20	15	15
Atlantic horse mackerel	*Trachurus trachurus*	14 065	18 739	21 404	15 535	15 471	15 305	20 683
Jack and horse mackerels nei	*Trachurus spp*	599	764	660	495	562	386	358
Jacks, crevalles nei	*Caranx spp*	134	114	95	32	1	1	0
Pompanos nei	*Trachinotus spp*	0	1	2	3	-	-	0
Greater amberjack	*Seriola dumerili*	...	...	...	10	21	20	37
Amberjacks nei	*Seriola spp*	33	45	101	32	8	7	8
Leerfish	*Lichia amia*	0	0	-	0	0	0	-
Carangids nei	*Carangidae*	...	...	...	45	29	30	20
Atlantic pomfret	*Brama brama*	34	91	9	9	8	10	80
Common dolphinfish	*Coryphaena hippurus*	1	0	1	1	1	4	4
Chub mackerel	*Scomber japonicus*	7 658	7 781	7 771	14 962	11 593	4 938	5 805

English name Nom anglais Nombre inglés	Scientific name Nom scientifique Nombre científico	1996 t	1997 t	1998 t	1999 t	2000 t	2001 t	2002 t
Atlantic mackerel	*Scomber scombrus*	3 009	2 083	2 898	2 035	2 254	3 121	3 090
Mackerels nei	*Scombridae*	237	340	247	35	0	-	-
Blue butterfish	*Stromateus fiatola*	87	72	24	17	29	-	70
Butterfishes, pomfrets nei	*Stromateidae*	-	-	5	94	35	94	61
Barracudas nei	*Sphyraena spp*	20	9	10	43	43	42	39
Ocean sunfish	*Mola mola*	-	-	-	12	-	-	-
Bluntnose sixgill shark	*Hexanchus griseus*	-	-	-	-	-	1	7
Basking shark	*Cetorhinus maximus*	1	1	-	1	1	3	1
Thresher	*Alopias vulpinus*	...	...	...	13	20	39	23
Shortfin mako	*Isurus oxyrinchus*	...	...	22	163	659	513	263
Porbeagle	*Lamna nasus*	0	0	0	0	16	5	16
Blue shark	*Prionace glauca*	...	...	85	905	3 083	4 663	4 575
Smooth hammerhead	*Sphyrna zygaena*	...	...	...	8	22	10	21
Smooth-hounds nei	*Mustelus spp*	187	27	14	81	41	43	48
Tope shark	*Galeorhinus galeus*	...	...	...	...	2	1	2
Greenland shark	*Somniosus microcephalus*	0	0	-	0	0	1	0
Picked dogfish	*Squalus acanthias*	2	2	2	21	2	3	4
Gulper shark	*Centrophorus granulosus*	...	...	...	73	54	93	152
Leafscale gulper shark	*Centrophorus squamosus*	...	...	...	440	506	537	659
Birdbeak dogfish	*Deania calcea*	...	...	...	...	18	50	90
Portuguese dogfish	*Centroscymnus coelolepis*	...	...	...	607	640	643	598
Kitefin shark	*Dalatias licha*	...	...	...	45	311	189	40
Dogfish sharks nei	*Squalidae*	977	999	905	-	-	-	-
Dogfishes and hounds nei	*Squalidae, Scyliorhinidae*	1 341	1 376	1 266	754	803	810	762
Angular roughshark	*Oxynotus centrina*	...	...	...	81	33	63	86
Raja rays nei	*Raja spp*	2 459	2 615	2 807	3 728	2 325	2 563	2 999
Eagle rays	*Myliobatidae*	...	...	...	11	8	9	12
Rays, stingrays, mantas nei	*Rajiformes*	261	257	189	74	0	82	54
Various sharks nei	*Selachimorpha(Pleurotremata)*	1 659	1 874	1 882	352	297	216	345
Sharks, rays, skates, etc. nei	*Elasmobranchii*	2 366	1 241	1 214	1 933	570	1 069	451
Finfishes nei	*Osteichthyes*	9 352	9 407	10 767	5 072	2 514	1 572	731
Marine fishes nei	*Osteichthyes*	741	486	410	299	90	149	103
Edible crab	*Cancer pagurus*	4	15	13	21	14	13	12
Portunus swimcrabs nei	*Portunus spp*	29	16	19	0	59	67	123
Green crab	*Carcinus maenas*	200	125	156	77	111	125	335
Spinous spider crab	*Maja squinado*	40	47	58	60	59	51	49
Marine crabs nei	*Brachyura*	33	61	36	36	120	243	44
Palinurid spiny lobsters nei	*Palinurus spp*	139	45	27	20	20	12	5
Norway lobster	*Nephrops norvegicus*	185	162	187	258	289	370	425
European lobster	*Homarus gammarus*	3	3	2	1	2	2	2
Lobsters nei	*Reptantia*	9	25	17	89	83	8	11
Penaeus shrimps nei	*Penaeus spp*	354	497	808	-	-	-	-
Deepwater rose shrimp	*Parapenaeus longirostris*	534	340	646	2 345	1 756	1 504	724
Blue and red shrimp	*Aristeus antennatus*	...	...	...	194	269	182	97
Northern prawn	*Pandalus borealis*	1	241	374	1 062	555	640	16
Palaemonid shrimps nei	*Palaemonidae*	1	15	5	38	63	13	7
Common shrimp	*Crangon crangon*	0	0	-	1	3	0	-
Natantian decapods nei	*Natantia*	1 273	679	803	211	379	1 074	416
Marine crustaceans nei	*Crustacea*	66	182	123	189	15	28	21
Periwinkles nei	*Littorina spp*	0	-	-	-	-	-	-
Gastropods nei	*Gastropoda*	48	41	41	189	200	188	180
Blue mussel	*Mytilus edulis*	35	46	24	87	48	74	165
Queen scallop	*Aequipecten opercularis*	0	-	-	-	-	-	1
Great Atlantic scallop	*Pecten maximus*	1	2	0	0	0	1	-
Striped venus	*Chamelea gallina*	7	17	185	129	156	48	865
Pullet carpet shell	*Venerupis pullastra*	153	419	240	212	87	81	73
Grooved carpet shell	*Ruditapes decussatus*	185	27	33	75	16	21	18
Solid surf clam	*Spisula solida*	-	-	-	765	1 153	1 125	1 220
Donax clams	*Donax spp*	...	...	...	456	401	540	347
Razor clams nei	*Solen spp*	1 729	124	15	4	12	214	195
Common edible cockle	*Cerastoderma edule*	3 522	1 285	1 264	1 409	1 292	683	3 518
Clams, etc. nei	*Bivalvia*	1 482	741	1 007	493	395	398	322
Common cuttlefish	*Sepia officinalis*	2 076	1 640	1 865	1 319	1 432	1 463	1 478
Common squids nei	*Loligo spp*	4 776	1 567	1 499	704	1 006	1 274	1 092
Northern shortfin squid	*Illex illecebrosus*	4	-	1	-	-	-	-
Argentine shortfin squid	*Illex argentinus*	640	712	1 531	-	-	1 049	2 355
Various squids nei	*Loliginidae, Ommastrephidae*	0	0	2	0	10	5	-
Common octopus	*Octopus vulgaris*	-	-	-	12	624	823	659
Octopuses, etc. nei	*Octopodidae*	11 785	9 333	6 515	9 509	9 220	7 332	7 712
Marine molluscs nei	*Mollusca*	626	687	799	32	68	121	60
Echinoderms	*Echinodermata*	-	-	-	-	-	15	1
	Country total	*259 846*	*221 879*	*224 171*	*208 407*	*187 521*	*191 017*	*200 037*
Romania								
Freshwater bream	*Abramis brama*	827	328	951	1 052	936	800	911
Freshwater breams nei	*Abramis spp*	-	-	3	1	-	-	7
Common carp	*Cyprinus carpio*	441	173	147	310	458	566	486
Tench	*Tinca tinca*	-	-	2	7	4	31	19
Bleak	*Alburnus alburnus*	-	-	24	16	30	60	27
Barbel	*Barbus barbus*	-	-	32	11	17	34	12
Crucian carp	*Carassius carassius*	-	33	2	23	128	97	87
Goldfish	*Carassius auratus*	1 954	817	1 113	1 199	1 212	1 149	1 097
Roaches nei	*Rutilus spp*	301	125	292	234	174	278	276
Orfe(=Ide)	*Leuciscus idus*	-	1	5	1	-	-	-
Chub	*Leuciscus cephalus*	-	-	45	17	24	28	9
Grass carp(=White amur)	*Ctenopharyngodon idellus*	124	12	46	21	83	73	10
Silver carp	*Hypophthalmichthys molitrix*	1 272	1 940	428	1 308	634	644	411
Bighead carp	*Hypophthalmichthys nobilis*	449	299	369	396	356	313	278

E-5

Fish crustaceans, molluscs, etc	Capture production by countries or areas and species	Europe
Poissons, crustacés, mollusques, etc	Captures par pays ou zones et espèces	Europe
Peces, crustáceos, moluscos, etc	Capturas por países o áreas y especies	Europa

English name Nom anglais Nombre inglés	Scientific name Nom scientifique Nombre científico	1996 t	1997 t	1998 t	1999 t	2000 t	2001 t	2002 t
Vimba bream	*Vimba vimba*	-	26	10	31	3	2	2
Asp	*Aspius aspius*	-	-	11	7	17	4	7
Cyprinids nei	*Cyprinidae*	112	16	396	126	39	298	282
Northern pike	*Esox lucius*	8	4	40	47	47	95	83
Wels(=Som)catfish	*Silurus glanis*	22	9	58	87	73	116	120
European perch	*Perca fluviatilis*	38	8	40	60	28	42	32
Pike-perch	*Stizostedion lucioperca*	94	30	78	154	155	92	118
Freshwater fishes nei	*Osteichthyes*	67	5	-	-	169	188	230
Danube sturgeon(=Osetr)	*Acipenser gueldenstaedtii*	-	-	5	10	19	17	4
Starry sturgeon	*Acipenser stellatus*	-	5	2	11	22	20	13
Beluga	*Huso huso*	-	2	7	7	32	20	22
Sturgeons nei	*Acipenseridae*	7	7	7	4	1	2	2
European eel	*Anguilla anguilla*	-	1	1	0	26	-	-
Sea trout	*Salmo trutta*	39	28	10	63	18	48	15
Rainbow trout	*Oncorhynchus mykiss*	...	3	25	25	32	65	12
Grayling	*Thymallus thymallus*	-	-	-	2	-	-	-
Pontic shad	*Alosa pontica*	400	487	446	20	106	128	266
Shads nei	*Alosa spp*	101	43	114	60	77	22	3
Azov sea sprat	*Clupeonella cultriventris*	4	2	52	4	5	11	4
Common sole	*Solea solea*	-	-	4	5	6	9	6
Turbot	*Psetta maxima*	6	1	-	2	2	13	17
Whiting	*Merlangius merlangus*	372	441	640	272	275	306	85
Mullets nei	*Mugilidae*	-	-	-	-	-	-	6
Surmullets(=Red mullets) nei	*Mullus spp*	1	3	3	1	2	3	2
Gobies nei	*Gobiidae*	8	2	6	30	42	24	46
Atlantic herring	*Clupea harengus*	1 794	-	-	-	-	-	-
European pilchard(=Sardine)	*Sardina pilchardus*	2	-	-	-	-	-	-
European sprat	*Sprattus sprattus*	2 014	3 318	3 293	1 933	1 803	1 792	1 617
European anchovy	*Engraulis encrasicolus*	138	45	146	155	204	186	296
Silversides(=Sand smelts) nei	*Atherinidae*	3	10	73	33	42	29	8
Bluefish	*Pomatomus saltatrix*	-	-	12	3	4	10	2
Atlantic horse mackerel	*Trachurus trachurus*	360	-	-	-	-	-	-
Mediterranean horse mackerel	*Trachurus mediterraneus*	13	1	15	3	8	17	21
Atlantic mackerel	*Scomber scombrus*	7 265	...	...	...	...	...	...
Marine fishes nei	*Osteichthyes*	23	2	2	1	1	5	6
Euro-American crayfishes nei	*Astacidae, Cambaridae*	-	181	65	56	32	-	3
Frogs	*Rana spp*	...	38	41	35	26	-	29
	Country total	*18 259*	*8 446*	*9 061*	*7 843*	*7 372*	*7 637*	*6 989*
Russian Fed								
Freshwater bream	*Abramis brama*	29 374	30 143	32 189	25 687	25 473	28 754	29 144
Freshwater breams nei	*Abramis spp*	1 664	1 046	2 058	1 951	1 593	1 751	1 886
Common carp	*Cyprinus carpio*	3 935	2 716	3 215	3 508	4 007	2 699	2 614
Tench	*Tinca tinca*	936	1 647	990	1 071	1 307	1 409	1 456
Roaches nei	*Rutilus spp*	21 968	15 971	15 963	10 433	14 547	15 312	16 957
Orfe(=Ide)	*Leuciscus idus*	2 914	3 037	3 280	2 604	2 378	2 257	2 492
Silver carp	*Hypophthalmichthys molitrix*	...	...	31	11	8	11	8
Sichel	*Pelecus cultratus*	451	342	1 724	1 786	1 201	1 548	1 297
Asp	*Aspius aspius*	84	36	45	53	57	65	148
Cyprinids nei	*Cyprinidae*	9 962	17 169	23 453	20 387	20 148	18 307	20 258
Northern pike	*Esox lucius*	5 688	4 646	5 567	6 118	8 864	8 444	9 759
Wels(=Som)catfish	*Silurus glanis*	5 864	7 026	9 305	7 031	7 199	6 540	5 978
Burbot	*Lota lota*	1 531	1 533	2 121	1 763	1 778	1 644	1 697
European perch	*Perca fluviatilis*	2 628	3 527	4 459	3 455	3 726	3 986	6 719
Pike-perch	*Stizostedion lucioperca*	6 215	6 104	5 390	5 320	6 128	5 711	6 041
Freshwater fishes nei	*Osteichthyes*	9 648	6 138	4 040	4 160	15 849	4 457	5 213
Danube sturgeon(=Osetr)	*Acipenser gueldenstaedtii*	482	721	646	359	250	251	219
Sterlet sturgeon	*Acipenser ruthenus*	-	2	1	1	-	1	2
Starry sturgeon	*Acipenser stellatus*	681	448	336	234	176	172	136
Beluga	*Huso huso*	105	127	78	40	44	40	32
Sturgeons nei	*Acipenseridae*	430	441	372	297	178	107	84
European eel	*Anguilla anguilla*	46	47	49	23	46	56	55
Atlantic salmon	*Salmo salar*	173	148	164	127	154	161	148
Sea trout	*Salmo trutta*	-	-	-	1	1	4	10
Pink(=Humpback)salmon	*Oncorhynchus gorbuscha*	113 181	187 667	191 439	187 181	157 138	167 566	117 584
Chum(=Keta=Dog)salmon	*Oncorhynchus keta*	23 083	22 898	26 046	28 162	36 490	32 067	36 562
Masu(=Cherry) salmon	*Oncorhynchus masou*	45	8	7	10	3	5	3
Sockeye(=Red)salmon	*Oncorhynchus nerka*	22 891	10 177	12 767	14 889	19 548	22 475	28 372
Chinook(=Spring=King)salmon	*Oncorhynchus tshawytscha*	521	636	556	793	479	499	665
Coho(=Silver)salmon	*Oncorhynchus kisutch*	1 976	1 310	2 319	1 668	2 278	2 034	2 260
European smelt	*Osmerus eperlanus*	1 022	760	835	409	844	976	730
Rainbow smelt	*Osmerus mordax*	640	1 113	1 758	1 340	609	725	1 440
Smelts nei	*Osmerus spp, Hypomesus spp*	2 049	4 416	3 424	3 130	4 560	4 180	4 997
Whitefishes nei	*Coregonus spp*	9 313	9 355	8 786	8 871	9 161	9 061	13 177
Salmonoids nei	*Salmonoidei*	1 338	1 565	2 103	1 446	2 728	1 511	1 922
Pontic shad	*Alosa pontica*	9	2	1	-	-	-	2
Twaite shad	*Alosa fallax*	-	-	-	-	-	-	5
Caspian shads	*Caspialosa spp*	1 882	2 304	3 284	4 654	1 273	176	239
Azov sea sprat	*Clupeonella cultriventris*	92 190	81 911	116 414	152 934	122 811	55 717	46 747
Lampreys nei	*Petromyzontidae*	76	31	37	67	39	49	117
Three-spined stickleback	*Gasterosteus aculeatus*	271	314	884	1 021	1 630	1 507	2 418
Atlantic halibut	*Hippoglossus hippoglossus*	-	-	-	6	5	2	13
European plaice	*Pleuronectes platessa*	2 161	3 531	3 729	3 911	3 114	1 250	1 133
Greenland halibut	*Reinhardtius hippoglossoides*	2 565	1 075	5 144	7 630	8 879	9 568	10 209
Kamchatka flounder	*Atheresthes evermanni*	9 739	8 476	9 567	10 743	23 473	19 049	17 609
Witch flounder	*Glyptocephalus cynoglossus*	-	-	52	110	114	65	135
Amer. plaice(=Long rough dab)	*Hippoglossoides platessoides*	778	1 482	6 569	7 878	3 207	3 754	3 059
Yellowtail flounder	*Limanda ferruginea*	-	-	-	96	212	148	103

E-5 | Fish crustaceans, molluscs, etc / Poissons, crustacés, mollusques, etc / Peces, crustáceos, moluscos, etc | Capture production by countries or areas and species / Captures par pays ou zones et espèces / Capturas por países o áreas y especies | Europe / Europe / Europa

English name Nom anglais Nombre inglés	Scientific name Nom scientifique Nombre científico	1996 t	1997 t	1998 t	1999 t	2000 t	2001 t	2002 t
European flounder	*Platichthys flesus*	-	-	-	-	1 392	1 351	1 327
Common sole	*Solea solea*	-	-	-	-	-	-	190
Turbot	*Psetta maxima*	-	-	-	-	53	69	50
Flatfishes nei	*Pleuronectiformes*	57 223	63 812	79 723	97 050	103 108	95 131	79 831
Blue antimora	*Antimora rostrata*	-	-	-	-	-	0	-
Moras nei	*Moridae*	-	152	11 115	31 295	39 316	32 392	29 323
Tusk(=Cusk)	*Brosme brosme*	-	-	-	-	46	83	54
Atlantic cod	*Gadus morhua*	309 391	316 147	248 719	215 616	171 018	188 884	188 213
Pacific cod	*Gadus macrocephalus*	93 890	79 927	94 282	101 929	68 415	59 783	60 625
Ling	*Molva molva*	-	-	-	-	8	2	11
Blue ling	*Molva dypterygia*	-	-	-	-	-	-	3
Greater forkbeard	*Phycis blennoides*	-	-	-	-	2	11	-
Brazilian codling	*Urophycis brasiliensis*	-	-	-	-	-	-	70
Red hake	*Urophycis chuss*	-	-	4	2	120	118	1 284
Haddock	*Melanogrammus aeglefinus*	73 857	41 228	20 560	30 978	24 894	34 970	38 817
Navaga(=Wachna cod)	*Eleginus navaga*	659	1 152	1 185	480	674	1 166	962
Saffron cod	*Eleginus gracilis*	21 110	27 803	40 426	47 032	35 763	33 753	32 591
Saithe(=Pollock)	*Pollachius virens*	1 177	1 802	3 837	3 932	4 564	4 953	5 413
Alaska pollock(=Walleye poll.)	*Theragra chalcogramma*	2 439 980	2 252 742	1 930 650	1 500 450	1 215 065	1 145 016	826 707
Polar cod	*Boreogadus saida*	20 784	6 826	3 592	22 005	40 743	39 445	37 191
Blue whiting(=Poutassou)	*Micromesistius poutassou*	87 310	118 656	130 042	182 637	241 905	315 586	298 367
Southern blue whiting	*Micromesistius australis*	377	610	-	-	-	30	4
Whiting	*Merlangius merlangus*	11	10	119	184	341	642	656
Senegalese hake	*Merluccius senegalensis*	1 112	1 081	1 171	1 230	604	207	...
Southern hake	*Merluccius australis*	27	202	-	-	-	-	-
Silver hake	*Merluccius bilinearis*	639	-	163	-	1 679	2 055	2 525
Argentine hake	*Merluccius hubbsi*	-	-	-	-	-	197	86
Benguela hake	*Merluccius polli*	-	-	-	-	-	-	1
Hakes nei	*Merluccius spp*	-	-	-	-	-	-	647
Cape hakes	*Merluccius capensis,M.paradox.*	98	252	321	67	41	49	2
Patagonian grenadier	*Macruronus magellanicus*	-	-	-	-	-	173	2
Blue grenadier	*Macruronus novaezelandiae*	3 905	3 191	-	-	-	-	-
Whitson's grenadier	*Macrourus whitsoni*	-	-	-	-	-	0	-
Grenadiers nei	*Macrourus spp*	-	-	-	-	-	1	524
Roundnose grenadier	*Coryphaenoides rupestris*	267	1 041	921	641	2 627	1 992	1 002
Grenadiers, rattails nei	*Macrouridae*	310	1 044	62	52	579	2 232	10 853
Gadiformes nei	*Gadiformes*	10	-	2	6	6	1	13
Bonefish	*Albula vulpes*	164	-	-	-	-	-	-
Sea catfishes nei	*Ariidae*	-	-	-	2	2	-	20
Mullets nei	*Mugilidae*	735	1 270	1 796	2 618	2 693	1 677	1 869
Bigeye grunt	*Brachydeuterus auritus*	-	-	36	-	-	5	561
Grunts, sweetlips nei	*Haemulidae (=Pomadasyidae)*	-	-	2	12	1	8	1
Meagre	*Argyrosomus regius*	-	-	-	-	5	-	-
West African croakers nei	*Pseudotolithus spp*	-	15	33	13	13	37	25
Croakers, drums nei	*Sciaenidae*	58	-	-	-	-	59	4
Large-eye dentex	*Dentex macrophthalmus*	1 210	1 566	2 242	4 003	2 385	1 181	965
Salema	*Sarpa salpa*	-	-	-	-	-	1	-
Porgies, seabreams nei	*Sparidae*	902	623	1 341	1 492	1 361	868	2 102
Surmullets(=Red mullets) nei	*Mullus spp*	76	68	119	92	127	119	47
Goatfishes, red mullets nei	*Mullidae*	-	-	-	-	-	-	28
Pacific sandlance	*Ammodytes personatus*	-	-	-	-	17	-	192
Gobies nei	*Gobiidae*	14 524	16 177	20 554	30 177	32 745	27 346	22 143
Okhotsk atka mackerel	*Pleurogrammus azonus*	21 260	36 075	40 889	40 283	52 784	49 171	55 570
Triggerfishes, durgons nei	*Balistidae*	-	154	-	-	-	-	-
Argentines	*Argentina spp*	-	-	-	-	1 219	496	293
Lanternfishes nei	*Myctophidae*	-	-	-	5	67	-	-
Conger eels, etc. nei	*Congridae*	-	-	-	-	-	-	1 192
Longspine snipefish	*Macroramphosus scolopax*	-	89	443	-	-	44	-
Kingklip	*Genypterus capensis*	-	-	-	-	-	5	1
Cusk-eels, brotulas nei	*Ophidiidae*	-	-	-	-	-	18	-
Alfonsinos nei	*Beryx spp*	162	48	83	59	35	231	816
Orange roughy	*Hoplostethus atlanticus*	-	-	-	-	14	-	-
John dory	*Zeus faber*	-	-	-	-	2	-	-
Boarfishes nei	*Caproidae*	-	-	5	-	-	-	-
Oreo dories nei	*Oreosomatidae*	5	-	-	2	-	14	8
Cape bonnetmouth	*Emmelichthys nitidus*	-	70	7	-	-	-	-
Bonnetmouths, rubyfishes nei	*Emmelichthyidae*	-	-	-	-	-	6	-
Patagonian toothfish	*Dissostichus eleginoides*	103	-	-	-	-	225	316
Patagonian rockcod	*Patagonotothen brevicauda*	-	-	-	3	0	-	4
Blackfin icefish	*Chaenocephalus aceratus*	-	-	-	0	-	-	-
Mackerel icefish	*Champsocephalus gunnari*	-	-	-	265	3 395	0	1 373
South Georgia icefish	*Pseudochaenichthys georgianus*	-	-	-	0	-	-	-
Atlantic wolffish	*Anarhichas lupus*	10 523	18 247	23 730	23 794	22 792	19 120	20 212
Wolffishes(=Catfishes) nei	*Anarhichas spp*	-	-	38	-	7	23	56
Eelpouts	*Lycodes spp*	18	-	2	1	28	48	62
Snoek	*Thyrsites atun*	15	11	-	-	-	-	-
Largehead hairtail	*Trichiurus lepturus*	51 778	33 465	7 422	7 217	4 846	2 326	274
Silver scabbardfish	*Lepidopus caudatus*	-	-	-	-	-	2	-
South Pacific breams nei	*Seriolella spp*	185	352	34	28	-	-	-
Golden redfish	*Sebastes marinus*	-	-	-	-	1 066	978	832
Pacific ocean perch	*Sebastes alutus*	2 615	2 095	1 739	2 478	1 023	793	1 232
Beaked redfish	*Sebastes mentella*	-	-	-	-	14 715	32 208	35 303
Atlantic redfishes nei	*Sebastes spp*	47 660	41 988	31 215	29 615	23 356	14 152	20 416
Scorpionfishes nei	*Scorpaenidae*	5	-	-	-	46	56	61
Gurnards, searobins nei	*Triglidae*	-	-	628	2 426	26 081	3 155	60
Sablefish	*Anoplopoma fimbria*	502	-	-	-	6	6	50
Lumpfish(=Lumpsucker)	*Cyclopterus lumpus*	-	-	-	-	20	28	28
Angler(=Monk)	*Lophius piscatorius*	-	-	-	-	-	1	-
Atlantic herring	*Clupea harengus*	134 380	181 179	139 566	170 601	174 204	127 420	128 129

English name Nom anglais Nombre inglés	Scientific name Nom scientifique Nombre científico	1996 t	1997 t	1998 t	1999 t	2000 t	2001 t	2002 t
Pacific herring	*Clupea pallasii*	171 810	313 397	395 595	359 194	361 241	278 511	203 411
Round sardinella	*Sardinella aurita*	116 705	101 457	106 318	109 445	42 400	22 557	18 800
Sardinellas nei	*Sardinella spp*	229	2 739	9 537	22 169	5 645	443	1 513
Japanese pilchard	*Sardinops melanostictus*	-	-	-	3	-	-	-
European pilchard(=Sardine)	*Sardina pilchardus*	56 168	24 864	5 100	5 504	11 200	1 829	6 450
European sprat	*Sprattus sprattus*	19 824	22 900	22 321	36 100	35 912	43 081	44 072
European anchovy	*Engraulis encrasicolus*	3 503	21 581	47 358	33 407	34 140	23 325	31 036
Anchovies, etc. nei	*Engraulidae*	22	5	23	34	69	114	260
Atlantic bonito	*Sarda sarda*	175	1 937	4 960	2 156	878	574	1 441
West African Spanish mackerel	*Scomberomorus tritor*	44	-	14	19	7	4	-
Frigate and bullet tunas	*Auxis thazard, A.rochei*	46	500	2 433	460	420	1 053	768
Little tunny(=Atl.black skipj)	*Euthynnus alletteratus*	49	-	88	-	-	-	74
Skipjack tuna	*Katsuwonus pelamis*	381	1 146	2 086	1 426	374	33	-
Yellowfin tuna	*Thunnus albacares*	2 696	4 275	4 931	4 359	737	-	-
Bigeye tuna	*Thunnus obesus*	13	38	4	8	91	-	-
Swordfish	*Xiphias gladius*	-	-	-	-	-	-	2
Tuna-like fishes nei	*Scombroidei*	-	-	6	11	22	18	3
Capelin	*Mallotus villosus*	180	160	405	32 555	94 984	181 566	250 921
Garfish	*Belone belone*	-	-	-	15	83	1	1
Pacific saury	*Cololabis saira*	10 280	7 091	4 665	4 808	17 390	40 407	51 709
Big-scale sand smelt	*Atherina boyeri*	31	13	25	30	63	95	99
Bluefish	*Pomatomus saltatrix*	96	406	283	536	226	157	168
Atlantic horse mackerel	*Trachurus trachurus*	804	554	345	121	86	16	3
Pacific jack mackerel	*Trachurus symmetricus*	-	-	230	10	-	-	-
Cape horse mackerel	*Trachurus capensis*	78 429	69 248	104 935	55 642	50 456	28 215	1 738
Cunene horse mackerel	*Trachurus trecae*	58 466	51 583	6 265	33 973	16 387	5 934	1 934
Greenback horse mackerel	*Trachurus declivis*	2 280	886	52	223	-	-	-
Jack and horse mackerels nei	*Trachurus spp*	73 446	56 008	84 736	71 262	70 949	56 235	41 423
Jacks, crevalles nei	*Caranx spp*	15	24	-	32	2	1	-
African moonfish	*Selene dorsalis*	15	12	1	-	-	-	-
Yellowtail amberjack	*Seriola lalandi*	-	-	-	-	-	-	3
Amberjacks nei	*Seriola spp*	82	232	-	213	-	-	-
Leerfish	*Lichia amia*	84	393	736	622	200	84	354
Atlantic bumper	*Chloroscombrus chrysurus*	-	2	49	-	-	-	-
Carangids nei	*Carangidae*	40	227	81	-	38	805	583
Atlantic pomfret	*Brama brama*	103	56	-	112	21	-	29
Chub mackerel	*Scomber japonicus*	72 940	64 748	68 368	48 365	45 836	31 710	37 501
Atlantic mackerel	*Scomber scombrus*	43 046	53 732	67 838	51 348	50 772	41 571	45 811
Blue butterfish	*Stromateus fiatola*	-	-	-	4	-	-	13
Barracudas nei	*Sphyraena spp*	1 176	399	125	844	-	1	5
Raja rays nei	*Raja spp*	7	476	932	975	4 485	2 914	3 379
Eaton's skate	*Bathyraja eatonii*	-	-	-	-	-	0	-
Rays, stingrays, mantas nei	*Rajiformes*	28	25	34	344	1 440	1 835	1 372
Sharks, rays, skates, etc. nei	*Elasmobranchii*	19	9	107	30	12	127	19
Pelagic fishes nei	*Osteichthyes*	8	-	-	-	-	-	-
Finfishes nei	*Osteichthyes*	801	520	603	845	22 659	698	3 386
Marine fishes nei	*Osteichthyes*	7 439	8 150	11 363	8 997	76 914	45 431	16 331
Freshwater crustaceans nei	*Crustacea*	28	34	33	41	957	1 733	1 754
Tanner crabs nei	*Chionoecetes spp*	22 769	27 112	20 848	21 234	21 848	24 493	23 759
Hair crab	*Erimacrus isenbeckii*	789	612	409	440	198	162	117
Marine crabs nei	*Brachyura*	-	49	-	-	-	194	856
Red king crab	*Paralithodes camtschaticus*	34 300	23 262	32 557	37 072	28 632	16 316	10 908
Blue king crab	*Paralithodes platypus*	8 762	10 268	4 508	5 455	5 233	4 500	4 598
Brown king crab	*Paralithodes brevipes*	204	418	194	256	347	254	350
Golden king crab	*Lithodes aequispina*	4 666	4 917	3 897	2 746	1 797	2 245	2 291
Coonstripe shrimp	*Pandalus hypsinotus*	...	467	388	288	275	359	231
Northern prawn	*Pandalus borealis*	10 191	4 369	8 022	16 964	35 253	20 057	12 629
Humpy shrimp	*Pandalus goniurus*	...	-	1 199	330	1 200	247	32
Hokkai shrimp	*Pandalus kessleri*	...	123	55	97	75	94	81
Pandalus shrimps nei	*Pandalus spp*	3 000	-	-	-	-	-	-
Morotoge shrimp	*Pandalopsis japonica*	...	12	86	35	11	36	22
Sculptured shrimps nei	*Sclerocrangon spp*	...	38	59	82	20	45	41
Natantian decapods nei	*Natantia*	-	2	-	16	92	83	263
Japanese corbicula	*Corbicula japonica*	-	-	-	-	-	74	209
Cupped oysters nei	*Crassostrea spp*	-	-	-	1	4	32	1
Mediterranean mussel	*Mytilus galloprovincialis*	-	-	-	4	-	-	-
Sea mussels nei	*Mytilidae*	-	-	9	80	63	68	35
Yesso scallop	*Patinopecten yessoensis*	5 084	5 534	6 253	5 764	5 728	2 236	3 612
Scallops nei	*Pectinidae*	7 634	13 878	12 887	11 948	12 717	13 598	5 751
Blood cockle	*Anadara granosa*	-	-	-	-	-	100	650
Clams, etc. nei	*Bivalvia*	-	-	-	-	216	317	200
Cuttlefish,bobtail squids nei	*Sepiidae, Sepiolidae*	-	-	-	-	-	5	-
Neon flying squid	*Ommastrephes bartrami*	-	-	-	-	405	100	483
Northern shortfin squid	*Illex illecebrosus*	-	-	29	-	12	-	-
Argentine shortfin squid	*Illex argentinus*	20 254	884	-	-	3 404	2 578	5 453
Various squids nei	*Loliginidae, Ommastrephidae*	41 678	60 722	52 464	56 108	69 835	44 249	72 538
Octopuses, etc. nei	*Octopodidae*	235	210	34	24	29	64	77
Marine molluscs nei	*Mollusca*	3 530	4 780	6 301	10 586	5 643	4 780	4 679
Sea urchins nei	*Strongylocentrotus spp*	1 608	1 153	1 560	1 245	1 677	1 763	2 620
Jellyfishes	*Rhopilema spp*	-	-	-	-	-	142	-
	Country total	*4 675 738*	*4 661 853*	*4 454 759*	*4 141 158*	*3 973 535*	*3 628 459*	*3 232 295*
Serbia-Monte								
Northern pike	*Esox lucius*	16	33	25	19	22	16	18
Wels(=Som)catfish	*Silurus glanis*	60	66	49	47	47	33	52
Freshwater fishes nei	*Osteichthyes*	1 156	964	792	762	603	521	867
European eel	*Anguilla anguilla*	3	3	3	...	...	...	2
Flatfishes nei	*Pleuronectiformes*	10	8	10	9	9	11	10

English name Nom anglais Nombre inglés	Scientific name Nom scientifique Nombre científico	1996 t	1997 t	1998 t	1999 t	2000 t	2001 t	2002 t
European hake	*Merluccius merluccius*	22	20	21	19	17	18	18
Flathead grey mullet	*Mugil cephalus*	25	24	26	25	26	25	27
Seabasses nei	*Dicentrarchus spp*	5	4	7	6	7	7	7
Common dentex	*Dentex dentex*	7	7	9	9	11	11	10
Black seabream	*Spondyliosoma cantharus*	5	3	5	3	4	4	4
Saddled seabream	*Oblada melanura*	4	4	6	7	5	6	6
Gilthead seabream	*Sparus aurata*	4	4	4	6	6	7	7
Bogue	*Boops boops*	21	20	20	22	24	25	33
Salema	*Sarpa salpa*	7	7	9	10	10	10	10
Picarels nei	*Spicara spp*	11	12	13	16	21	27	38
Surmullet	*Mullus surmuletus*	11	10	11	9	10	11	11
European conger	*Conger conger*	20	19	19	17	16	14	12
Scorpionfishes nei	*Scorpaenidae*	7	7	7	7	8	7	7
Sardinellas nei	*Sardinella spp*	11	11	13	17	13	16	13
European pilchard(=Sardine)	*Sardina pilchardus*	42	45	49	49	36	28	25
European anchovy	*Engraulis encrasicolus*	0	1	2	4	5	10	9
Atlantic bonito	*Sarda sarda*	10	12	12	14	17	17	16
Frigate and bullet tunas	*Auxis thazard, A.rochei*	6	6	6	7	8	9	8
Little tunny(=Atl.black skipj)	*Euthynnus alletteratus*	22	18	20	18	16	16	16
Atlantic bluefin tuna	*Thunnus thynnus*	4	4	6	7	4	5	6
Garfish	*Belone belone*	1	1	2	5	5	6	6
Silversides(=Sand smelts) nei	*Atherinidae*	6	7	9	9	13	14	22
Mediterranean horse mackerel	*Trachurus mediterraneus*	16	14	15	17	14	15	16
Greater amberjack	*Seriola dumerili*	11	9	11	9	12	13	12
Scomber mackerels nei	*Scomber spp*	8	9	13	14	13	13	13
Dogfish sharks nei	*Squalidae*	10	10	8	9	9	7	8
Rays, stingrays, mantas nei	*Rajiformes*	12	12	12	12	11	11	10
Marine fishes nei	*Osteichthyes*	19	25	20	25	27	-	22
Spinous spider crab	*Maja squinado*	0	0	0	0	0	0	1
Norway lobster	*Nephrops norvegicus*	5	5	7	5	7	9	10
European lobster	*Homarus gammarus*	5	5	6	5	6	7	5
European flat oyster	*Ostrea edulis*	0	0	0	1	1	0	0
Cuttlefish,bobtail squids nei	*Sepiidae, Sepiolidae*	10	9	10	10	10	10	10
Common squids nei	*Loligo spp*	12	13	13	12	13	14	13
Common octopus	*Octopus vulgaris*	8	8	9	9	10	12	11
Marine molluscs nei	*Mollusca*	0	0	0	0	-	1	-
	Country total	*1 612*	*1 439*	*1 279*	*1 251*	*1 096*	*986*	*1 391*
Slovakia								
Freshwater breams nei	*Abramis spp*	111	102	99	98	94	95	103
Common carp	*Cyprinus carpio*	778	746	778	822	854	967	1 166
Tench	*Tinca tinca*	8	9	8	8	8	5	5
Barbel	*Barbus barbus*	19	21	15	10	17	19	15
Common nase	*Chondrostoma nasus*	24	23	25	24	24	20	15
Crucian carp	*Carassius carassius*	-	-	2	1	0	...	...
Goldfish	*Carassius auratus*	76	53	54	62	0	73	63
Chubs nei	*Leuciscus spp*	47	39	32	38	33	31	24
Grass carp(=White amur)	*Ctenopharyngodon idellus*	15	21	16	17	15	9	19
Silver carp	*Hypophthalmichthys molitrix*	-	-	4	5	7	8	8
Vimba bream	*Vimba vimba*	20	25	27	14	11	10	11
Asp	*Aspius aspius*	12	9	9	9	8	8	9
Cyprinids nei	*Cyprinidae*	8	44	37	38	53	12	29
Northern pike	*Esox lucius*	103	85	68	69	76	73	64
Wels(=Som)catfish	*Silurus glanis*	21	21	20	20	22	28	29
Burbot	*Lota lota*	1	3	4	4	2	4	4
European perch	*Perca fluviatilis*	18	11	13	13	13	14	13
Pike-perch	*Stizostedion lucioperca*	76	70	65	64	56	62	79
Sterlet sturgeon	*Acipenser ruthenus*	1	0	0	0	1	-	0
European eel	*Anguilla anguilla*	7	8	8	8	4	6	7
Sea trout	*Salmo trutta*	34	40	42	41	37	38	33
Rainbow trout	*Oncorhynchus mykiss*	16	16	17	16	19	30	31
Brook trout	*Salvelinus fontinalis*	0	1	1	0	0	1	0
Huchen	*Hucho hucho*	1	1	1	1	1	1	1
Grayling	*Thymallus thymallus*	18	16	16	14	13	17	18
	Country total	*1 414*	*1 364*	*1 361*	*1 396*	*1 368*	*1 531*	*1 746*
Slovenia								
Freshwater bream	*Abramis brama*	10	11	11	10	11	10	9
Freshwater breams nei	*Abramis spp*	5	5	5	4	0	1	-
Common carp	*Cyprinus carpio*	86	90	94	78	71	75	83
Tench	*Tinca tinca*	1	2	2	3	2	1	1
Bleak	*Alburnus alburnus*	-	-	-	-	-	-	3
Barbel	*Barbus barbus*	10	12	9	12	11	11	11
Crucian carp	*Carassius carassius*	-	-	0	1	0	0	2
Goldfish	*Carassius auratus*	-	-	0	1	1	1	-
Roach	*Rutilus rutilus*	2	2	2	2	3	4	4
Roaches nei	*Rutilus spp*	6	8	9	10	11	7	8
Rudd	*Scardinius erythrophthalmus*	7	6	4	4	1	0	0
Orfe(=Ide)	*Leuciscus idus*	17	17	16	14	15	14	14
Common dace	*Leuciscus leuciscus*	5	5	5	4	0	0	0
Grass carp(=White amur)	*Ctenopharyngodon idellus*	-	-	2	2	3	2	2
Silver carp	*Hypophthalmichthys molitrix*	7	6	4	4	-	0	0
Vimba bream	*Vimba vimba*	-	-	1	1	2	2	1
Asp	*Aspius aspius*	-	-	0	-	0	0	-
Cyprinids nei	*Cyprinidae*	54	40	42	35	35	25	24
Northern pike	*Esox lucius*	11	12	9	10	11	9	9
Wels(=Som)catfish	*Silurus glanis*	4	7	7	5	7	6	8

E-5

Fish crustaceans, molluscs, etc — **Capture production by countries or areas and species** — **Europe**
Poissons, crustacés, mollusques, etc — **Captures par pays ou zones et espèces** — **Europe**
Peces, crustáceos, moluscos, etc — **Capturas por países o áreas y especies** — **Europa**

English name Nom anglais Nombre inglés	Scientific name Nom scientifique Nombre científico	1996 t	1997 t	1998 t	1999 t	2000 t	2001 t	2002 t
European perch	*Perca fluviatilis*	1	1	1	0	0	1	1
Pike-perch	*Stizostedion lucioperca*	5	5	5	4	5	4	4
Freshwater fishes nei	*Osteichthyes*	0	17	0	3	-	-	2
Beluga	*Huso huso*	1	1	1	1	0	1	0
European eel	*Anguilla anguilla*	-	-	-	-	-	-	2
Sea trout	*Salmo trutta*	14	13	11	10	9	7	9
Trouts nei	*Salmo spp*	3	2	2	2	2	1	0
Rainbow trout	*Oncorhynchus mykiss*	32	33	23	19	21	21	25
Brook trout	*Salvelinus fontinalis*	1	0	0	0	1	0	0
Arctic char	*Salvelinus alpinus*	0	1	0	0	1	1	2
Grayling	*Thymallus thymallus*	7	6	4	4	3	2	2
European flounder	*Platichthys flesus*	-	-	-	-	1	3	2
Common sole	*Solea solea*	1	1	1	1	2	3	4
Poor cod	*Trisopterus minutus*	-	-	-	-	2	1	1
Whiting	*Merlangius merlangus*	-	-	13	16	14	37	19
European hake	*Merluccius merluccius*	4	4	2	1	0	2	2
Mullets nei	*Mugilidae*	12	2	27	13	22	49	19
European seabass	*Dicentrarchus labrax*	-	-	-	1	1	5	4
Shi drum	*Umbrina cirrosa*	-	-	-	-	-	-	1
Common pandora	*Pagellus erythrinus*	-	-	1	1	3	7	8
Sargo breams nei	*Diplodus spp*	-	-	-	-	1	2	3
Gilthead seabream	*Sparus aurata*	-	-	-	1	1	4	4
Bogue	*Boops boops*	-	-	2	1	3	2	1
Sand steenbras	*Lithognathus mormyrus*	-	-	-	-	1	2	2
Salema	*Sarpa salpa*	-	-	-	-	-	-	2
Picarels nei	*Spicara spp*	-	-	3	3	4	5	6
Surmullets(=Red mullets) nei	*Mullus spp*	-	-	1	1	1	4	1
Black goby	*Gobius niger*	-	-	-	-	1	1	1
European conger	*Conger conger*	-	-	-	-	0	1	0
Gurnards, searobins nei	*Triglidae*	-	-	-	-	1	1	1
European pilchard(=Sardine)	*Sardina pilchardus*	1 982	1 973	1 788	1 614	1 415	1 219	1 223
European sprat	*Sprattus sprattus*	7	1	-	2	3	8	14
European anchovy	*Engraulis encrasicolus*	24	33	51	75	96	97	72
Atlantic bonito	*Sarda sarda*	-	-	-	-	-	-	1
Garfish	*Belone belone*	-	-	-	-	1	1	0
Silversides(=Sand smelts) nei	*Atherinidae*	-	-	-	-	1	2	5
Atlantic horse mackerel	*Trachurus trachurus*	9	8	4	5	4	4	2
Chub mackerel	*Scomber japonicus*	10	4	17	3	4	3	1
Atlantic mackerel	*Scomber scombrus*	5	5	10	13	14	16	4
Smooth-hounds nei	*Mustelus spp*	-	-	-	-	2	4	2
Picked dogfish	*Squalus acanthias*	0	0	1	1	-	-	-
Marine fishes nei	*Osteichthyes*	8	7	5	4	-	-	-
Spottail mantis squillid	*Squilla mantis*	-	-	-	-	0	3	4
European flat oyster	*Ostrea edulis*	0	0	-	-	0	0	0
Mediterranean mussel	*Mytilus galloprovincialis*	1	1	-	-	-	0	0
Striped venus	*Chamelea gallina*	-	-	-	-	0	1	1
Common cuttlefish	*Sepia officinalis*	6	5	18	18	11	72	22
Common squids nei	*Loligo spp*	2	4	3	2	3	8	3
Common octopus	*Octopus vulgaris*	2	7	-	-	-	-	-
Marine molluscs nei	*Mollusca*	5	10	12	8	18	54	25
	Country total	*2 367*	*2 367*	*2 228*	*2 027*	*1 856*	*1 827*	*1 686*
Spain								
Freshwater fishes nei	*Osteichthyes*	4 000 F	4 000 F	4 000 F	4 000 F	4 000 F	4 000 F	4 000 F
Sturgeons nei	*Acipenseridae*	-	-	1	1	-	-	-
European eel	*Anguilla anguilla*	68	72	23	39	70	62	93
Atlantic salmon	*Salmo salar*	10 F	10 F	10 F	10 F	10 F	10 F	10 F
Sea trout	*Salmo trutta*	2 200 F	2 200 F	2 200 F	2 200 F	2 200 F	2 200 F	2 200 F
Salmonoids nei	*Salmonoidei*	18	24	26	8	33	13	1 263
Alewife	*Alosa pseudoharengus*	...	...	...	...	61	-	-
Shads nei	*Alosa spp*	-	-	-	-	-	-	382
Atlantic halibut	*Hippoglossus hippoglossus*	58	46	69	103	112	152	159
European plaice	*Pleuronectes platessa*	14	3	6	3	39	41	13
Greenland halibut	*Reinhardtius hippoglossoides*	7 525	8 200	7 484	9 611	10 401	13 337	12 833
Witch flounder	*Glyptocephalus cynoglossus*	3 387	5 527	6 154	5 328	4 443	4 225	4 118
Amer. plaice(=Long rough dab)	*Hippoglossoides platessoides*	878	1 118	1 346	1 522	1 622	1 412	915
Yellowtail flounder	*Limanda ferruginea*	259	656	562	752	775	622	216
Common dab	*Limanda limanda*	-	0	130	129	29	24	70
Lemon sole	*Microstomus kitt*	-	235	1 197	1 282	2 207	4 040	408
European flounder	*Platichthys flesus*	74	319	873	323	88	139	21
Common sole	*Solea solea*	574	553	5 783	248	1 093	4 805	470
West coast sole	*Austroglossus microlepis*	54	-	-	-	-	-	-
Soles nei	*Soleidae*	6 846	6 208	7 519	1 781	227	979	829
Megrim	*Lepidorhombus whiffiagonis*	175	185	44	-	101	131	73
Megrims nei	*Lepidorhombus spp*	6 405	7 006	7 446	6 969	8 361	8 161	6 277
Brill	*Scophthalmus rhombus*	478	429	60	36	29	23	9
Turbot	*Psetta maxima*	282	339	231	252	124	122	43
Flatfishes nei	*Pleuronectiformes*	10 253	18 234	15 696	15 850	11 066	8 123	10 540
Blue antimora	*Antimora rostrata*	16	0	26	24	21	-	0
Tadpole codling	*Salilota australis*	3 484	2 505	6 140	5 935	3 914	2 250	545
Tusk(=Cusk)	*Brosme brosme*	62	98	106	150	249	72	151
Atlantic cod	*Gadus morhua*	16 356	17 228	14 392	10 159	8 770	20 283	8 409
Ling	*Molva molva*	5 301	6 153	9 256	8 907	6 259	4 276	6 506
Blue ling	*Molva dypterygia*	299	1 166	1 405	1 710	3 113	4 472	2 041
Greater forkbeard	*Phycis blennoides*	1 741	5 209	4 671	3 675	2 369	1 181	1 757
Red hake	*Urophycis chuss*	893	958	1 200	1 350	1 551	1 755	1 164
White hake	*Urophycis tenuis*	187	304	491	426	802	689	3 258
Haddock	*Melanogrammus aeglefinus*	718	505	541	780	669	2 217	156

English name Nom anglais Nombre inglés	Scientific name Nom scientifique Nombre científico	1996 t	1997 t	1998 t	1999 t	2000 t	2001 t	2002 t
Saithe(=Pollock)	*Pollachius virens*	33	83	397	82	158	152	54
Pollack	*Pollachius pollachius*	185	213	218	175	175	436	222
Pouting(=Bib)	*Trisopterus luscus*	2 823	3 637	2 269	2 736	3 635	3 240	4 089
Blue whiting(=Poutassou)	*Micromesistius poutassou*	35 113	41 054	33 359	35 371	34 276	34 283	28 510
Southern blue whiting	*Micromesistius australis*	2 471	1 591	3 435	3 128	3 346	5 243	3 160
Whiting	*Merlangius merlangus*	44	72	187	233	353	299	248
European hake	*Merluccius merluccius*	27 012	26 304	27 200	29 354	31 183	19 159	22 105
Senegalese hake	*Merluccius senegalensis*	15 001	14 478	13 594	15 678	11 211	16 716	6 978
Silver hake	*Merluccius bilinearis*	-	-	-	-	4	9	27
Argentine hake	*Merluccius hubbsi*	21 697	14 816	18 298	26 810	22 196	25 302	13 572
Benguela hake	*Merluccius polli*	-	-	-	18	-	-	-
Cape hakes	*Merluccius capensis,M.paradox.*	1 724	1 592	3	2 154	2 877	3 349	338
Patagonian grenadier	*Macruronus magellanicus*	7 733	7 422	16 104	11 132	9 794	13 599	2 280
Roughhead grenadier	*Macrourus berglax*	257	3 740	6 050	5 705	8 097	1 103	905
Grenadiers nei	*Macrourus spp*	-	2	-	-	-	-	-
Roundnose grenadier	*Coryphaenoides rupestris*	4 066	2 476	3 935	6 224	15 465	38 225	10 232
Gadiformes nei	*Gadiformes*	35 932	22 256	12 029	14 338	3 724	29 664	8 656
Mullets nei	*Mugilidae*	489	456	660	210	117	75	42
Dusky grouper	*Epinephelus marginatus*	222	204	256	123	43	71	134
Groupers nei	*Epinephelus spp*	119	32	21	28	42	36	5
Groupers, seabasses nei	*Serranidae*	214	467	719	693	663	624	700
Spotted seabass	*Dicentrarchus punctatus*	-	-	-	-	-	-	27
European seabass	*Dicentrarchus labrax*	534	474	457	383	473	326	303
Seabasses nei	*Dicentrarchus spp*	-	-	-	158	197	259	254
Rubberlip grunt	*Plectorhinchus mediterraneus*	257	239	174	453	94	135	74
Bastard grunt	*Pomadasys incisus*	0	5	-	-	-	-	-
Grunts, sweetlips nei	*Haemulidae (=Pomadasyidae)*	476	205	261	100	198	163	67
Canary drum (=Baardman)	*Umbrina canariensis*	-	-	-	-	-	4	-
Croakers, drums nei	*Sciaenidae*	117	-	184	275	174	58	425
Blackspot(=red) seabream	*Pagellus bogaraveo*	739	3 075	1 395	947	699	94	246
Common pandora	*Pagellus erythrinus*	-	247	322	277	305	229	514
Axillary seabream	*Pagellus acarne*	-	424	373	-	-	-	-
Red pandora	*Pagellus bellottii*	-	43	44	42	-	-	-
Pandoras nei	*Pagellus spp*	3 083	763	708	2 122	1 609	2 056	1 969
White seabream	*Diplodus sargus*	-	131	106	137	114	182	548
Sargo breams nei	*Diplodus spp*	151	117	111	77	74	98	7
Common dentex	*Dentex dentex*	60	62	61	149	265	56	44
Dentex nei	*Dentex spp*	335	338	494	649	859	869	1 168
Black seabream	*Spondyliosoma cantharus*	340	360	325	7	18	23	176
Saddled seabream	*Oblada melanura*	-	30	44	78	66	81	56
Red porgy	*Pagrus pagrus*	142	425	841	357	222	181	234
Pargo breams nei	*Pagrus spp*	100	106	223	313	177	141	5
Gilthead seabream	*Sparus aurata*	681	546	508	956	1 229	2 164	1 174
Bogue	*Boops boops*	1 356	1 642	1 919	3 898	1 943	1 819	1 977
Sand steenbras	*Lithognathus mormyrus*	-	270	323	255	290	246	203
Salema	*Sarpa salpa*	-	242	199	167	147	164	253
Porgies, seabreams nei	*Sparidae*	3 144	2 770	3 085	2 237	3 341	3 566	791
Picarels nei	*Spicara spp*	1	-	-	494	712	759	346
Surmullet	*Mullus surmuletus*	292	300	132	550	819	825	936
Red mullet	*Mullus barbatus*	...	...	...	328	263	262	560
Surmullets(=Red mullets) nei	*Mullus spp*	754	527	547	1 927	1 688	1 778	1 199
Threadfins, tasselfishes nei	*Polynemidae*	-	2	-	-	-	-	-
Antarctic rockcods, noties nei	*Nototheniidae*	-	0	-	-	-	-	-
Sandeels(=Sandlances) nei	*Ammodytes spp*	616	656	466	757	1 117	689	708
Greater weever	*Trachinus draco*	60	69	67	-	30	57	67
Gobies nei	*Gobiidae*	-	-	-	478	311	378	294
Argentines	*Argentina spp*	-	-	-	23	72	99	53
European conger	*Conger conger*	5 219	5 349	5 193	4 750	3 668	3 311	3 193
Pink cusk-eel	*Genypterus blacodes*	706	779	1 800	1 901	1 392	1 408	388
Bearded brotula	*Brotula barbata*	-	-	-	-	-	61	-
Alfonsinos nei	*Beryx spp*	24	233	420	-	79	247	435
Orange roughy	*Hoplostethus atlanticus*	22	26	26	38	20	16	55
Slimeheads nei	*Trachichthyidae*	833	1 052	33	25	3	6	2
John dory	*Zeus faber*	284	510	621	615	541	949	846
Boarfishes nei	*Caproidae*	-	-	-	-	-	7	-
Wreckfish	*Polyprion americanus*	166	280	162	186	79	133	81
Red bandfish	*Cepola rubescens*	-	-	-	-	-	-	17
Patagonian toothfish	*Dissostichus eleginoides*	79	596	355	756	846	931	1 047
Patagonian rockcod	*Patagonotothen brevicauda*	-	-	-	-	-	-	5
Longtail Southern cod	*Patagonotothen ramsayi*	-	-	-	-	-	-	10
Black cardinal fish	*Epigonus telescopus*	-	-	-	-	-	-	70
Atlantic wolffish	*Anarhichas lupus*	7	23	7	-	2	7	0
Wolffishes(=Catfishes) nei	*Anarhichas spp*	695	555	510	455	584	883	641
Largehead hairtail	*Trichiurus lepturus*	-	-	-	-	-	5	36
Silver scabbardfish	*Lepidopus caudatus*	2 197	2 000	3 004	2 747	284	3 287	2 642
Black scabbardfish	*Aphanopus carbo*	41	106	137	117	1 029	1 323	992
Hairtails, scabbardfishes nei	*Trichiuridae*	-	-	-	-	-	13	-
Ruffs, barrelfishes nei	*Centrolophidae*	-	-	-	-	36	-	-
Golden redfish	*Sebastes marinus*	-	-	-	50	23	8	3
Beaked redfish	*Sebastes mentella*	4 307	4 438	4 587	3 526	2 530	5 575	1 652
Atlantic redfishes nei	*Sebastes spp*	968	5 965	5 218	8 185	5 718	4 644	4 104
Blackbelly rosefish	*Helicolenus dactylopterus*	-	-	-	8	8	7	-
Scorpionfishes nei	*Scorpaenidae*	1 122	1 881	1 147	1 243	1 646	985	2 090
Gurnards, searobins nei	*Triglidae*	1 356	1 525	868	865	1 114	852	1 083
Angler(=Monk)	*Lophius piscatorius*	3 534	5 161	6 301	7 935	6 890	6 737	5 746
Blackbellied angler	*Lophius budegassa*	524	336	524	126	208	369	491
American angler	*Lophius americanus*	-	1	2	0	3	9	93
Devil anglerfish	*Lophius vomerinus*	58	...	123	32	44	52	55
Anglerfishes nei	*Lophiidae*	-	-	-	-	-	3	-

E-5 Fish crustaceans, molluscs, etc — Capture production by countries or areas and species — Europe
Poissons, crustacés, mollusques, etc — Captures par pays ou zones et espèces — Europe
Peces, crustáceos, moluscos, etc — Capturas por países o áreas y especies — Europa

English name Nom anglais Nombre inglés	Scientific name Nom scientifique Nombre científico	1996 t	1997 t	1998 t	1999 t	2000 t	2001 t	2002 t
Demersal percomorphs nei	*Perciformes*	...	...	...	...	841	758	3 954
Atlantic herring	*Clupea harengus*	-	-	-	-	232	232	266
Round sardinella	*Sardinella aurita*	2 224	3 210	3 411	17	41	49	5
Sardinellas nei	*Sardinella spp*	6 879	7 156	7 996	9 785	7 579	13 556	7 368
European pilchard(=Sardine)	*Sardina pilchardus*	221 956	192 443	200 420	128 231	81 028	71 144	54 370
European sprat	*Sprattus sprattus*	5	17	-	6	17	17	41
European anchovy	*Engraulis encrasicolus*	29 823	26 220	24 878	31 070	28 109	37 062	28 357
Clupeoids nei	*Clupeoidei*	7 873	6 542	15 572	19 461	10 945	20 242	9 481
Atlantic bonito	*Sarda sarda*	692	629	333	445	354	359	466
Wahoo	*Acanthocybium solandri*	25	25	29	32	32	38	46
Frigate and bullet tunas	*Auxis thazard, A.rochei*	4 470	1 393	1 057	692	1 356	1 784	1 337
Little tunny(=Atl.black skipj)	*Euthynnus alletteratus*	73	36	125	6	10	104	40
Skipjack tuna	*Katsuwonus pelamis*	108 604	111 076	114 784	159 685	137 211	122 139	140 436
Atlantic bluefin tuna	*Thunnus thynnus*	8 762	8 047	5 800	5 363	6 246	5 867	6 305
Albacore	*Thunnus alalunga*	17 737	18 961	13 877	16 950	16 932	9 947	10 650
Yellowfin tuna	*Thunnus albacares*	98 306	88 490	75 711	82 532	85 876	90 425	89 932
Bigeye tuna	*Thunnus obesus*	29 307	31 622	23 724	41 898	43 191	25 626	30 130
Indo-Pacific sailfish	*Istiophorus platypterus*	-	-	9	7	1	1	2
Atlantic sailfish	*Istiophorus albicans*	28	41	38	25	23	425	658
Blue marlin	*Makaira nigricans*	110	76	88	128	142	97	33
Black marlin	*Makaira indica*	0	-	-	1	0	0	0
Striped marlin	*Tetrapturus audax*	-	-	-	1	0	0	0
Atlantic white marlin	*Tetrapturus albidus*	107	74	69	133	186	62	6
Shortbill spearfish	*Tetrapturus angustirostris*	-	1	2	2	0	0	1
Longbill spearfish	*Tetrapturus pfluegeri*	1	1	1	52	64	31	12
Marlins,sailfishes,etc. nei	*Istiophoridae*	17	95	127	-	-	1	7
Swordfish	*Xiphias gladius*	17 156	17 412	14 125	13 791	15 209	13 930	21 020
Tuna-like fishes nei	*Scombroidei*	9	-	2 830	20	789	599	744
Garfish	*Belone belone*	455	509	998	600	270	380	1 213
Needlefishes, etc. nei	*Belonidae*	-	-	-	-	-	-	3
Atlantic saury	*Scomberesox saurus*	483	639	907	574	184	992	589
Silversides(=Sand smelts) nei	*Atherinidae*	297	313	242	244	254	212	340
Bluefish	*Pomatomus saltatrix*	3 885	2 114	1 169	958	261	206	132
Cape horse mackerel	*Trachurus capensis*	-	2	-	-	-	-	-
Jack and horse mackerels nei	*Trachurus spp*	32 628	39 970	36 946	43 750	44 468	46 996	38 261
Pompanos nei	*Trachinotus spp*	-	-	-	4	-	7	85
Greater amberjack	*Seriola dumerili*	-	-	-	704	375	430	283
Amberjacks nei	*Seriola spp*	114	142	161	71	88	116	26
Leerfish	*Lichia amia*	24	32	38	28	43	45	132
Carangids nei	*Carangidae*	-	-	-	39	34	393	14
Atlantic pomfret	*Brama brama*	5 339	3 491	2 908	2 346	586	348	77
Common dolphinfish	*Coryphaena hippurus*	-	-	-	92	137	70	170
Chub mackerel	*Scomber japonicus*	894	100	378	302	3 697	1 138	86
Atlantic mackerel	*Scomber scombrus*	17 982	21 256	28 212	25 636	28 190	27 340	29 772
Scomber mackerels nei	*Scomber spp*	1 489	-	-	23	-	-	557
Mackerels nei	*Scombridae*	26 414	34 628	36 547	19 200	13 609	22 949	19 936
Blue butterfish	*Stromateus fiatola*	...	...	...	...	20	58	50
Butterfishes, pomfrets nei	*Stromateidae*	-	-	-	-	-	7	24
Barracudas nei	*Sphyraena spp*	24	31	19	57	125	167	169
Pelagic percomorphs nei	*Perciformes*	...	...	...	...	1	2	...
Basking shark	*Cetorhinus maximus*	2	6	6	-	-	-	4
Thresher	*Alopias vulpinus*	...	30	45	...	...	...	...
Bigeye thresher	*Alopias superciliosus*	...	149	114	...	...	...	...
Thresher sharks nei	*Alopias spp*	...	34	55	66	...	...	...
Shortfin mako	*Isurus oxyrinchus*	-	990	1 264	335	264	228	2 163
Porbeagle	*Lamna nasus*	31	124	686	1 001	1 184	1 009	65
Blackmouth catshark	*Galeus melastomus*	...	...	...	...	...	...	288
Small-spotted catshark	*Scyliorhinus canicula*	...	...	...	...	...	...	6
Catsharks, nursehounds nei	*Scyliorhinus spp*	-	-	-	213	331	379	185
Blue shark	*Prionace glauca*	-	5 260	2 695	2 233	2 803	2 795	10 976
Silky shark	*Carcharhinus falciformis*	-	-	-	-	-	-	31
Requiem sharks nei	*Carcharhinidae*	-	43	810	-	9	8	57
Smooth hammerhead	*Sphyrna zygaena*	-	220	103	-	-	-	9
Scalloped hammerhead	*Sphyrna lewini*	-	-	-	-	-	-	290
Hammerhead sharks, etc. nei	*Sphyrnidae*	...	808	746	...	...	...	2
Smooth-hounds nei	*Mustelus spp*	118	-	-	21	15	19	12
Tope shark	*Galeorhinus galeus*	-	-	-	-	-	37	6
Little sleeper shark	*Somniosus rostratus*	...	...	...	...	...	...	2
Picked dogfish	*Squalus acanthias*	63	0	27	94	381	372	367
Gulper shark	*Centrophorus granulosus*	...	...	...	...	...	...	9
Leafscale gulper shark	*Centrophorus squamosus*	-	-	-	-	-	-	495
Lanternsharks nei	*Etmopterus spp*	...	...	...	573	...	...	102
Birdbeak dogfish	*Deania calcea*	-	-	-	-	-	-	12
Portuguese dogfish	*Centroscymnus coelolepis*	-	-	-	-	-	-	135
Dogfish sharks nei	*Squalidae*	676	1 442	1 235	939	923	1 077	665
Sailfin roughshark	*Oxynotus paradoxus*	...	...	...	...	...	...	1
Raja rays nei	*Raja spp*	8 550	19 355	20 018	24 387	25 233	19 745	9 864
Rays, stingrays, mantas nei	*Rajiformes*	451	291	307	511	536	375	835
Various sharks nei	*Selachimorpha(Pleurotremata)*	3 642	30 377	20 450	23 662	31 793	30 815	17 771
Sharks, rays, skates, etc. nei	*Elasmobranchii*	5 529	40 512	18 757	16 765	18 877	20 244	18 644
Groundfishes nei	*Osteichthyes*	12 354	2 506	5 411	3 506	7 687	828	743
Pelagic fishes nei	*Osteichthyes*	25 609	8 169	15 685	4 426	2 947	100	4 701
Finfishes nei	*Osteichthyes*	10 721	19 532	33 711	22 091	17 492	12 646	14 107
Marine fishes nei	*Osteichthyes*	32 897	35 699	37 203	21 280	17 454	16 836	13 670
White-clawed crayfish	*Austropotamobius pallipes*	0	0	0	0	0	0	0
Red swamp crawfish	*Procambarus clarkii*	2 500 F	2 500 F	2 500 F	2 500 F	2 500 F	2 500 F	2 500 F
Edible crab	*Cancer pagurus*	120	51	23	49	48	35	64
Portunus swimcrabs nei	*Portunus spp*	0	6	20	18	-	-	-
Green crab	*Carcinus maenas*	1	2	8	8	10	12	11

English name Nom anglais Nombre inglés	Scientific name Nom scientifique Nombre científico	1996 t	1997 t	1998 t	1999 t	2000 t	2001 t	2002 t
Mediterranean shore crab	*Carcinus aestuarii*	-	-	-	-	-	-	36
Spinous spider crab	*Maja squinado*	185	209	196	210	199	113	117
Geryons nei	*Geryon spp*	370	-	300	135	156	168	-
Marine crabs nei	*Brachyura*	3 883	11 608	5 842	1 933	1 674	2 063	1 672
Palinurid spiny lobsters nei	*Palinurus spp*	149	104	103	96	46	111	163
Norway lobster	*Nephrops norvegicus*	1 735	1 928	1 647	2 542	1 579	1 540	1 497
European lobster	*Homarus gammarus*	30	17	15	13	10	11	7
Lobsters nei	*Reptantia*	11	-	14	1	-	-	2
Antarctic stone crab	*Paralomis spinosissima*	-	0	-	-	-	-	-
Caramote prawn	*Penaeus kerathurus*	213	53	196	103	63	64	250
Southern pink shrimp	*Penaeus notialis*	707	-	-	-	-	-	-
Penaeus shrimps nei	*Penaeus spp*	373	368	498	336	318	433	256
Deepwater rose shrimp	*Parapenaeus longirostris*	4 993	5 217	9 274	6 489	5 749	9 346	737
Scarlet shrimp	*Plesiopenaeus edwardsianus*	352	479	337	605	54	39	34
Blue and red shrimp	*Aristeus antennatus*	-	-	-	414	772	1 418	1 033
Striped red shrimp	*Aristeus varidens*	764	780	1 932	1 007	1 870	2 205	546
Northern prawn	*Pandalus borealis*	940	1 429	1 366	1 296	2 294	1 683	938
Common prawn	*Palaemon serratus*	80	91	79	66	47	41	110
Common shrimp	*Crangon crangon*	222	150	100	87	...	1	138
Natantian decapods nei	*Natantia*	16 437	19 065	41 445	14 598	10 341	11 875	13 170
Spottail mantis squillid	*Squilla mantis*	-	-	-	1 222	1 043	1 000	810
Marine crustaceans nei	*Crustacea*	1 600	1 404	2 376	2 146	2 153	2 271	964
Common periwinkle	*Littorina littorea*	132	147	117	126	139	...	28
Periwinkles nei	*Littorina spp*	8	8	15	8	3	5	3
Whelk	*Buccinum undatum*	-	-	-	-	-	-	4
Gastropods nei	*Gastropoda*	-	-	-	955	799	851	889
European flat oyster	*Ostrea edulis*	253	388	316	443	152	37	59
Pacific cupped oyster	*Crassostrea gigas*	9	-	-	1	9	9	-
Blue mussel	*Mytilus edulis*	18	176	84	27	-	33	66
Mediterranean mussel	*Mytilus galloprovincialis*	0	29	0	19	18	24	11
Great Atlantic scallop	*Pecten maximus*	675	391	299	86	508	84	223
Scallops nei	*Pectinidae*	94	78	50	141	63	35	32
Striped venus	*Chamelea gallina*	2 156	2 213	3 518	3 639	4 779	5 121	2 894
Pullet carpet shell	*Venerupis pullastra*	1 739	2 146	2 710	2 451	990	744	1 037
Grooved carpet shell	*Ruditapes decussatus*	761	1 088	1 340	1 283	612	289	645
Donax clams	*Donax spp*	-	-	-	-	-	-	12
Razor clams nei	*Solen spp*	88	209	167	171	50	129	-
Common edible cockle	*Cerastoderma edule*	2 358	2 964	2 472	3 105	3 740	1 486	737
Clams, etc. nei	*Bivalvia*	5 836	5 422	6 494	5 676	3 182	2 842	3 796
Common cuttlefish	*Sepia officinalis*	176	336	1 078	1 411	1 895	833	712
Cuttlefish,bobtail squids nei	*Sepiidae, Sepiolidae*	3 031	3 499	2 869	6 514	2 901	3 994	1 841
Patagonian squid	*Loligo gahi*	35 692	3 967	8 933	8 185	9 392	9 011	3 673
Cape Hope squid	*Loligo reynaudi*	58	-	-	0	-	-	1
Common squids nei	*Loligo spp*	1 520	2 982	2 595	2 378	2 062	2 279	1 405
Northern shortfin squid	*Illex illecebrosus*	2 601	2 427	765	1 932	1 831	1 503	2 406
Argentine shortfin squid	*Illex argentinus*	17 091	25 374	23 829	30 694	26 140	41 318	7 750
European flying squid	*Todarodes sagittatus*	1 113	2 657	2 256	2 625	2 204	1 708	2 952
Sevenstar flying squid	*Martialia hyadesi*	1	0	0	0	-	-	-
Various squids nei	*Loliginidae, Ommastrephidae*	9 206	4 317	9 268	14 724	3 449	3 158	2 334
Common octopus	*Octopus vulgaris*	26 874	22 545	38 235	21 692	11 379	14 913	8 483
Octopuses, etc. nei	*Octopodidae*	11 526	11 190	11 902	10 065	14 275	9 549	9 067
Cephalopods nei	*Cephalopoda*	6 240	5 636	8 541	242	401	190	335
Marine molluscs nei	*Mollusca*	1 301	1 543	11 230	404	294	287	96
Echinoderms	*Echinodermata*	487	590	560	621	309	306	603
Sea cucumbers nei	*Holothurioidea*	4	4	4	1	9	4	10
Aquatic invertebrates nei	*Invertebrata*	5	4	10	529	-	-	30
	Country total	*1 174 683*	*1 205 401*	*1 263 016*	*1 172 995*	*1 045 616*	*1 092 525*	*882 633*
Svalbard Is								
Marine fishes nei	*Osteichthyes*	0	0	0	0	0	0	0
	Country total	*0*	*0*	*0*	*0*	*0*	*0*	*0*
Sweden								
Freshwater bream	*Abramis brama*	25	18	22	20	6	4	5
Tench	*Tinca tinca*	0	0	1	1	-	-	-
Roach	*Rutilus rutilus*	0	1	-	-	-	-	-
Roaches nei	*Rutilus spp*	0	1	-	-	-	-	-
Cyprinids nei	*Cyprinidae*	0	0	-	-	-	-	-
Northern pike	*Esox lucius*	360	338	302	322	290	242	234
Burbot	*Lota lota*	75	66	68	65	4	3	2
European perch	*Perca fluviatilis*	267	386	371	301	231	201	250
Pike-perch	*Stizostedion lucioperca*	312	366	334	346	327	267	398
Freshwater fishes nei	*Osteichthyes*	131	230	134	157	197	155	179
European eel	*Anguilla anguilla*	1 042	1 073	645	734	560	580	633
Atlantic salmon	*Salmo salar*	805	715	657	433	514	412	331
Trouts nei	*Salmo spp*	131	118	138	94	88	63	56
Rainbow trout	*Oncorhynchus mykiss*	-	-	-	0	-	-	-
Arctic char	*Salvelinus alpinus*	25	24	28	21	24	18	17
Grayling	*Thymallus thymallus*	3	0	0	0	-	-	-
European smelt	*Osmerus eperlanus*	9	10	8	9	-	-	-
Vendace	*Coregonus albula*	1 321	1 208	676	573	830	820	1 127
European whitefish	*Coregonus lavaretus*	530	497	489	460	420	336	416
Salmonoids nei	*Salmonoidei*	0	1	0	0	-	-	-
Atlantic halibut	*Hippoglossus hippoglossus*	8	10	8	8	10	8	8
European plaice	*Pleuronectes platessa*	531	558	431	405	449	416	387
Witch flounder	*Glyptocephalus cynoglossus*	299	355	448	501	578	576	584
Amer. plaice(=Long rough dab)	*Hippoglossoides platessoides*	22	6	0	-	-	-	-

E-5 Fish crustaceans, molluscs, etc — Capture production by countries or areas and species — Europe
Poissons, crustacés, mollusques, etc — Captures par pays ou zones et espèces — Europe
Peces, crustáceos, moluscos, etc — Capturas por países o áreas y especies — Europa

English name Nom anglais Nombre inglés	Scientific name Nom scientifique Nombre científico	1996 t	1997 t	1998 t	1999 t	2000 t	2001 t	2002 t
Common dab	*Limanda limanda*	37	46	33	16	10	14	10
Lemon sole	*Microstomus kitt*	117	121	105	94	71	61	48
European flounder	*Platichthys flesus*	1 262	1 073	526	274	341	467	357
Common sole	*Solea solea*	61	52	41	43	30	20	15
Brill	*Scophthalmus rhombus*	7	11	13	18	17	13	12
Turbot	*Psetta maxima*	296	294	188	159	106	64	55
Flatfishes nei	*Pleuronectiformes*	0	1	0	0	-	-	-
Tusk(=Cusk)	*Brosme brosme*	7	3	3	3	8	6	4
Atlantic cod	*Gadus morhua*	41 827	34 797	22 475	22 597	23 174	24 111	17 383
Ling	*Molva molva*	73	61	44	44	46	47	47
Blue ling	*Molva dypterygia*	0	2	0	-	-	-	-
Haddock	*Melanogrammus aeglefinus*	1 226	1 519	1 013	895	964	1 087	965
Saithe(=Pollock)	*Pollachius virens*	1 773	1 649	1 857	1 929	1 468	1 628	1 868
Pollack	*Pollachius pollachius*	355	261	180	160	124	108	112
Norway pout	*Trisopterus esmarkii*	237	2	-	-	133	780	-
Blue whiting(=Poutassou)	*Micromesistius poutassou*	4 038	4 568	6 034	15 511	3 362	2 058	18 483
Whiting	*Merlangius merlangus*	374	101	90	128	177	153	178
European hake	*Merluccius merluccius*	45	33	26	27	34	63	51
Roundnose grenadier	*Coryphaenoides rupestris*	0	42	0	-	-	258	262
Gadiformes nei	*Gadiformes*	0	-	-	-	57	160	-
Sandeels(=Sandlances) nei	*Ammodytes spp*	-	1	8 585	23 225	28 165	50 559	55 953
Greater weever	*Trachinus draco*	10	1	3	12	7	26	9
Argentines	*Argentina spp*	...	541	428	0	273	1 010	484
Atlantic wolffish	*Anarhichas lupus*	117	174	157	163	154	95	74
Atlantic redfishes nei	*Sebastes spp*	0	-	0	1	-	-	-
Grey gurnard	*Eutrigla gurnardus*	4	5	8	133	5	5	2
Lumpfish(=Lumpsucker)	*Cyclopterus lumpus*	147	129	20	72	47	83	213
Angler(=Monk)	*Lophius piscatorius*	38	44	44	48	107	92	119
Atlantic herring	*Clupea harengus*	132 153	166 311	201 738	157 541	174 081	125 749	97 625
European pilchard(=Sardine)	*Sardina pilchardus*	-	-	-	-	-	1 031	142
European sprat	*Sprattus sprattus*	168 582	126 361	149 664	112 452	91 164	88 562	78 365
Capelin	*Mallotus villosus*	-	-	-	-	-	-	7 570
Garfish	*Belone belone*	14	12	27	30	7	9	8
Atlantic horse mackerel	*Trachurus trachurus*	166	1 761	3 418	2 004	1 162	119	575
Atlantic mackerel	*Scomber scombrus*	5 387	4 390	5 161	5 003	4 500	5 098	5 232
Porbeagle	*Lamna nasus*	1	1	1	1	1	1	0
Picked dogfish	*Squalus acanthias*	154	197	140	114	124	238	270
Raja rays nei	*Raja spp*	9	8	2	3	3	12	8
Sharks, rays, skates, etc. nei	*Elasmobranchii*	0	0	0	0	0	-	-
Finfishes nei	*Osteichthyes*	3 065	3 003	338	382	540	508	300
Euro-American crayfishes nei	*Astacidae, Cambaridae*	6	9	6	8	17	37	49
Edible crab	*Cancer pagurus*	87	79	93	122	128	133	146
Norway lobster	*Nephrops norvegicus*	1 105	1 130	1 317	1 263	1 232	1 067	1 046
European lobster	*Homarus gammarus*	26	27	26	25	20	18	19
Northern prawn	*Pandalus borealis*	2 176	2 598	2 283	2 297	2 073	2 113	2 188
Marine crustaceans nei	*Crustacea*	-	-	-	2	1	-	-
European flat oyster	*Ostrea edulis*	3	3	2	4	2	1	2
Blue mussel	*Mytilus edulis*	-	3	36	0	70	51	86
Various squids nei	*Loliginidae, Ommastrephidae*	0	1	1	1	-	0	1
Marine molluscs nei	*Mollusca*	-	-	-	-	1	-	-
	Country total	*370 881*	*357 406*	*410 886*	*351 254*	*338 534*	*311 816*	*294 963*
Switzerland								
Freshwater bream	*Abramis brama*	24	20	16	16	13	9	10
Common carp	*Cyprinus carpio*	1	2	1	1	1	1	2
Tench	*Tinca tinca*	4	3	3	3	3	3	4
Roach	*Rutilus rutilus*	146	164	168	154	136	137	159
Cyprinids nei	*Cyprinidae*	42	20	10	6	8	13	13
Northern pike	*Esox lucius*	36	34	33	34	48	48	46
Wels(=Som)catfish	*Silurus glanis*	1	1	0	1	1	1	1
Burbot	*Lota lota*	6	9	6	7	8	8	10
European perch	*Perca fluviatilis*	475	364	422	491	395	262	288
Pike-perch	*Stizostedion lucioperca*	10	6	7	6	5	9	9
Freshwater fishes nei	*Osteichthyes*	0	0	3	0	4	1	2
European eel	*Anguilla anguilla*	3	2	3	3	2	2	2
Sea trout	*Salmo trutta*	8	13	12	11	13	13	13
Rainbow trout	*Oncorhynchus mykiss*	0	0	0	0	0	0	0
Arctic char	*Salvelinus alpinus*	28	27	21	22	22	16	16
Whitefishes nei	*Coregonus spp*	1 050	1 182	1 098	1 074	980	1 174	946
Shads nei	*Alosa spp*	7	12	6	11	20	18	23
	Country total	*1 841*	*1 859*	*1 809*	*1 840*	*1 659*	*1 715*	*1 544*
Ukraine								
Freshwater bream	*Abramis brama*	1 008	832	742	840	881	942	502
Freshwater breams nei	*Abramis spp*	101	62	86	122	80	137	107
Common carp	*Cyprinus carpio*	598	27	33	41	44	40	43
Bleak	*Alburnus alburnus*	-	-	-	-	-	-	2
Goldfish	*Carassius auratus*	638	461	452	638	628	747	773
Roach	*Rutilus rutilus*	799	795	1 141	667	451	526	325
Rudd	*Scardinius erythrophthalmus*	15	19	34	38	38	33	32
Grass carp(=White amur)	*Ctenopharyngodon idellus*	2	-	-	-	0	-	1
Silver carp	*Hypophthalmichthys molitrix*	3 188	1 380	1 008	442	444	267	152
Vimba bream	*Vimba vimba*	25	10	18	13	13	20	15
Sichel	*Pelecus cultratus*	9	40	37	27	16	14	11
Asp	*Aspius aspius*	25	19	7	15	26	7	7
Northern pike	*Esox lucius*	26	23	15	18	15	14	5
Wels(=Som)catfish	*Silurus glanis*	2	1	1	1	3	3	2

E-5 Fish crustaceans, molluscs, etc — Capture production by countries or areas and species — Europe
Poissons, crustacés, mollusques, etc — Captures par pays ou zones et espèces — Europe
Peces, crustáceos, moluscos, etc — Capturas por países o áreas y especies — Europa

English name Nom anglais Nombre inglés	Scientific name Nom scientifique Nombre científico	1996 t	1997 t	1998 t	1999 t	2000 t	2001 t	2002 t
European perch	*Perca fluviatilis*	126	80	107	79	155	74	64
Percarina	*Percarina demidoffi*	-	-	-	-	-	-	18
Pike-perch	*Stizostedion lucioperca*	811	955	965	986	1 171	1 722	1 963
Freshwater gobies nei	*Gobiidae*	...	...	60	16	...	...	70
Freshwater fishes nei	*Osteichthyes*	213	15	1	2	...	7	-
Danube sturgeon(=Osetr)	*Acipenser gueldenstaedtii*	130	134	113	34	22	8	8
Starry sturgeon	*Acipenser stellatus*	18	49	13	11	5	3	2
European eel	*Anguilla anguilla*	-	-	-	-	-	-	5
Pontic shad	*Alosa pontica*	283	329	213	24	83	148	210
Shads nei	*Alosa spp*	23	57	29	5	0	-	-
Azov sea sprat	*Clupeonella cultriventris*	3 289	3 218	3 324	9 992	8 446	18 989	13 352
European flounder	*Platichthys flesus*	30	30	16	15	8	10	5
Turbot	*Psetta maxima*	120	82	63	110	118	171	157
Common mora	*Mora moro*	-	-	-	-	3	2	-
Red codling	*Pseudophycis bachus*	...	...	51	...	63	50	9
Southern blue whiting	*Micromesistius australis*	3 713	3 607	7 730	8 306	3 502	267	-
Whiting	*Merlangius merlangus*	3	29	55	18	20	18	9
Senegalese hake	*Merluccius senegalensis*	10	580	2 361	300	1 127	648	23
Southern hake	*Merluccius australis*	111	181	578	1 422	1 100	8	13
Argentine hake	*Merluccius hubbsi*	1	-	-	-	-	-	-
Cape hakes	*Merluccius capensis,M.paradox.*	-	-	-	6	22	4	-
Blue grenadier	*Macruronus novaezelandiae*	14 754	12 632	16 064	15 141	19 333	1 300	28
Grenadiers, rattails nei	*Macrouridae*	-	-	-	-	-	-	11
Gadiformes nei	*Gadiformes*	-	-	-	-	-	997	-
Sea catfishes nei	*Ariidae*	-	-	-	-	1	-	-
So-iuy mullet	*Mugil soiuy*	1 039	2 718	3 674	5 364	5 575	2 810	2 601
Mullets nei	*Mugilidae*	3	0	5	90	121	78	194
Grunts, sweetlips nei	*Haemulidae (=Pomadasyidae)*	-	-	-	-	132	6	166
West African croakers nei	*Pseudotolithus spp*	...	...	28	12	87	5	23
Large-eye dentex	*Dentex macrophthalmus*	52	4	207	272	924	185	49
Dentex nei	*Dentex spp*	-	-	2	-	-	-	-
Silver seabream	*Pagrus auratus*	-	-	-	-	-	-	3
Porgies, seabreams nei	*Sparidae*	153	1 245	757	1 264	1 966	363	214
Picarels nei	*Spicara spp*	-	-	-	-	3	-	-
Red mullet	*Mullus barbatus*	...	18	35	30	12	201	42
Gobies nei	*Gobiidae*	72	96	286	601	825	1 510	3 582
Argentines	*Argentina spp*	-	-	-	-	-	-	1
Pink cusk-eel	*Genypterus blacodes*	...	...	21	35	258	35	-
Alfonsinos nei	*Beryx spp*	3 826	1 423	859	1 964	1 578	380	-
Orange roughy	*Hoplostethus atlanticus*	-	-	-	-	102	195	-
John dory	*Zeus faber*	...	...	99	7	59	9	-
Dories nei	*Zeidae*	-	-	-	-	-	2	74
Bonnetmouths, rubyfishes nei	*Emmelichthyidae*	28	7	275	181	-	86	1 410
Pelagic armourhead	*Pseudopentaceros richardsoni*	298	51	78	108	77	12	-
Patagonian toothfish	*Dissostichus eleginoides*	969	1 048	885	593	220	157	-
Snoek	*Thyrsites atun*	2 244	3 898	3 881	7 922	6 113	2 970	3 846
Largehead hairtail	*Trichiurus lepturus*	-	-	2 490	1 314	1 423	2 096	59
Silver scabbardfish	*Lepidopus caudatus*	-	-	-	-	83	11	40
Common warehou	*Seriolella brama*	-	-	-	-	-	-	169
Silver warehou	*Seriolella punctata*	-	-	-	-	-	-	397
South Pacific breams nei	*Seriolella spp*	484	1 183	320	328	1 558	666	34
Ruffs, barrelfishes nei	*Centrolophidae*	290	440	395	753	360	299	-
Beaked redfish	*Sebastes mentella*	518	-	-	-	-	-	-
Bluefin gurnard	*Chelidonichthys kumu*	-	-	-	-	-	-	2
Sardinellas nei	*Sardinella spp*	83 333	97 009	161 821	74 729	38 120	10 613	13 714
Australian pilchard	*Sardinops neopilchardus*	-	-	-	-	-	-	72
European pilchard(=Sardine)	*Sardina pilchardus*	44 221	10 126	12 828	49 136	40 902	27 939	19 808
European sprat	*Sprattus sprattus*	20 720	20 208	30 282	29 238	32 655	49 004	45 503
European anchovy	*Engraulis encrasicolus*	4 398	9 444	3 914	5 527	16 390	16 668	12 941
Atlantic bonito	*Sarda sarda*	342	2 786	1 918	1 114	399	231	656
West African Spanish mackerel	*Scomberomorus tritor*	-	-	-	-	21	...	42
Frigate and bullet tunas	*Auxis thazard, A.rochei*	-	-	-	36	48	...	43
Tuna-like fishes nei	*Scombroidei*	...	...	303	-	28	213	54
Garfish	*Belone belone*	0	0	1	1	0	-	-
Dealfishes	*Trachipterus spp*	-	-	-	-	-	-	1
Silversides(=Sand smelts) nei	*Atherinidae*	326	396	632	388	653	332	540
Bluefish	*Pomatomus saltatrix*	-	209	13	238	97	29	325
Mediterranean horse mackerel	*Trachurus mediterraneus*	...	5	-	-	1	1	34
Cape horse mackerel	*Trachurus capensis*	27 121	5 179	18 345	4 592	13 837	3 693	-
Cunene horse mackerel	*Trachurus trecae*	-	-	-	-	-	-	501
Greenback horse mackerel	*Trachurus declivis*	13 093	9 740	9 309	15 306	12 213	7 577	5 667
Jack and horse mackerels nei	*Trachurus spp*	42 471	38 311	64 237	53 333	66 266	37 240	12 995
White trevally	*Pseudocaranx dentex*	-	-	-	-	-	-	5
Jacks, crevalles nei	*Caranx spp*	...	...	-	2	-	-	-
African moonfish	*Selene dorsalis*	-	-	-	-	-	-	3
Amberjacks nei	*Seriola spp*	-	-	-	15	...	...	17
Leerfish	*Lichia amia*	...	428	-	266	109	31	389
Atlantic bumper	*Chloroscombrus chrysurus*	-	-	-	-	2	44	-
Atlantic pomfret	*Brama brama*	-	-	-	-	9	-	-
Pomfrets, ocean breams nei	*Bramidae*	-	-	-	-	2	-	3
Chub mackerel	*Scomber japonicus*	98 944	120 272	72 959	42 283	44 961	18 988	8 015
Blue mackerel	*Scomber australasicus*	156	9	214	3 457	1 677	2 040	1 849
Blue butterfish	*Stromateus fiatola*	-	-	-	-	1	-	8
Barracudas nei	*Sphyraena spp*	-	-	-	-	-	-	3
Thresher	*Alopias vulpinus*	-	-	-	-	-	-	1
Shortfin mako	*Isurus oxyrinchus*	-	-	-	-	-	-	1
Porbeagle	*Lamna nasus*	-	-	-	-	-	-	8
Tope shark	*Galeorhinus galeus*	-	-	-	-	-	-	1
Picked dogfish	*Squalus acanthias*	44	20	38	94	71	134	259

English name Nom anglais Nombre inglés	Scientific name Nom scientifique Nombre científico	1996 t	1997 t	1998 t	1999 t	2000 t	2001 t	2002 t
Thornback ray	*Raja clavata*	17	9	24	31	24	65	79
Common stingray	*Dasyatis pastinaca*	...	...	-	-	4	11	-
Rays, stingrays, mantas nei	*Rajiformes*	0	1	-	-	-	-	-
Ghost shark	*Callorhinchus milii*	-	-	-	-	-	-	1
Sharks, rays, skates, etc. nei	*Elasmobranchii*	...	...	...	...	...	1	-
Finfishes nei	*Osteichthyes*	32	-	-	-	-	-	-
Marine fishes nei	*Osteichthyes*	16 663	8 279	29 968	59 791	60 146	123 281	67 794
Freshwater crustaceans nei	*Crustacea*	...	...	...	...	0	2	3
Marine crabs nei	*Brachyura*	-	-	-	-	-	-	6
Tropical spiny lobsters nei	*Panulirus spp*	-	-	1	-	-	-	-
Northern prawn	*Pandalus borealis*	-	-	-	-	-	405	-
Natantian decapods nei	*Natantia*	1	1	1	1	1	...	4
Antarctic krill	*Euphausia superba*	10 277	-	-	6 719	-	14 023	32 015
Sea snails	*Rapana spp*	378	476	371	619	913	400	93
Mediterranean mussel	*Mytilus galloprovincialis*	246	159	82	163	115	71	75
Wellington flying squid	*Nototodarus sloani*	4 136	7 955	5 321	1 462	2 872	8 623	11 230
Various squids nei	*Loliginidae, Ommastrephidae*	339	-	-	3	-	-	-
Marine molluscs nei	*Mollusca*	-	-	-	-	-	-	2
Aquatic invertebrates nei	*Invertebrata*	-	-	-	-	-	3	9
	Country total	*407 305*	*368 800*	*462 196*	*408 711*	*391 831*	*360 914*	*265 599*
UK								
Freshwater bream	*Abramis brama*	1	9	8	8	2	0	2
Roach	*Rutilus rutilus*	-	4	5	5	-	-	-
Northern pike	*Esox lucius*	5	2	4	4	6	5	4
Pike-perch	*Stizostedion lucioperca*	1	0	0	2	1	1	-
European eel	*Anguilla anguilla*	895	812	741	697	796	595	571
Atlantic salmon	*Salmo salar*	568	457	419	403	480	519	512
Sea trout	*Salmo trutta*	198	141	572	489	150	571	148
Trouts nei	*Salmo spp*	-	-	-	-	-	254	426
Rainbow trout	*Oncorhynchus mykiss*	300	100	2 831	3 224	1 267	1 174	1 765
Arctic char	*Salvelinus alpinus*	-	-	-	-	-	1	1
Grayling	*Thymallus thymallus*	-	-	-	-	0	0	-
European smelt	*Osmerus eperlanus*	2 677	2 635	1 217	888	5 316	21	32
Shads nei	*Alosa spp*	6	0	1	1	3	8	5
River lamprey	*Lampetra fluviatilis*	-	-	-	-	2	2	2
Lefteye flounders nei	*Bothidae*	-	-	-	-	0	-	-
Atlantic halibut	*Hippoglossus hippoglossus*	392	367	202	254	208	159	222
European plaice	*Pleuronectes platessa*	25 319	26 599	23 510	20 394	23 701	21 936	19 647
Greenland halibut	*Reinhardtius hippoglossoides*	2 358	1 515	1 777	1 611	1 746	1 690	1 346
Witch flounder	*Glyptocephalus cynoglossus*	3 479	3 748	3 889	3 661	4 102	4 535	3 896
Amer. plaice(=Long rough dab)	*Hippoglossoides platessoides*	1	0	0	0	0	15	-
Common dab	*Limanda limanda*	2 221	2 590	2 467	2 248	1 705	1 530	1 355
Lemon sole	*Microstomus kitt*	5 444	5 723	5 230	5 010	4 367	4 005	2 719
European flounder	*Platichthys flesus*	357	379	293	149	201	148	172
Common sole	*Solea solea*	3 022	2 671	2 561	2 808	2 443	2 720	2 574
Sand sole	*Solea lascaris*	27	28	32	12	47	42	55
Wedge sole	*Dicologlossa cuneata*	-	1	0	0	-	-	-
Megrim	*Lepidorhombus whiffiagonis*	7 212	7 479	6 520	5 606	5 513	4 672	4 418
Brill	*Scophthalmus rhombus*	688	568	513	410	470	535	475
Turbot	*Psetta maxima*	1 270	1 148	974	851	877	1 001	1 067
Antarctic armless flounder	*Mancopsetta maculata*	-	-	-	-	0	-	-
Flatfishes nei	*Pleuronectiformes*	172	172	-	-	158	171	143
Smalleye moray cod	*Muraenolepis microps*	-	-	-	-	0	-	0
Moray cods nei	*Muraenolepis spp*	-	-	-	0	0	-	-
Blue antimora	*Antimora rostrata*	-	1	-	-	-	-	-
Tadpole codling	*Salilota australis*	18	39	24	188	30	17	15
Moras nei	*Moridae*	-	415	0	-	2	-	-
Tusk(=Cusk)	*Brosme brosme*	391	507	723	852	1 016	872	672
Atlantic cod	*Gadus morhua*	78 364	74 637	77 182	51 695	41 750	32 840	31 548
Ling	*Molva molva*	13 942	12 924	13 594	11 350	9 244	8 095	8 465
Blue ling	*Molva dypterygia*	1 772	2 811	2 450	4 119	3 019	5 980	3 166
Greater forkbeard	*Phycis blennoides*	2 022	2 054	1 887	1 495	1 563	1 204	985
Haddock	*Melanogrammus aeglefinus*	87 420	83 388	83 436	72 001	50 644	42 865	52 869
Saithe(=Pollock)	*Pollachius virens*	15 413	14 609	12 261	12 442	10 924	10 585	12 310
Pollack	*Pollachius pollachius*	3 631	3 845	3 211	2 398	2 706	2 839	2 765
Norway pout	*Trisopterus esmarkii*	215	13	-	2	0	0	1
Pouting(=Bib)	*Trisopterus luscus*	962	813	554	458	885	899	724
Blue whiting(=Poutassou)	*Micromesistius poutassou*	14 326	33 701	98 936	106 491	45 048	51 889	28 679
Southern blue whiting	*Micromesistius australis*	108	20	48	85	22	30	181
Whiting	*Merlangius merlangus*	36 788	35 189	27 243	25 561	23 458	15 287	11 538
European hake	*Merluccius merluccius*	5 380	5 716	5 168	5 537	4 519	2 775	2 663
Argentine hake	*Merluccius hubbsi*	38	104	67	53	30	83	233
Patagonian grenadier	*Macruronus magellanicus*	86	166	2	347	42	30	52
Roughhead grenadier	*Macrourus berglax*	-	14	-	1 038	8	16	7
Whitson's grenadier	*Macrourus whitsoni*	-	-	-	-	1	-	-
Grenadiers nei	*Macrourus spp*	-	7	4	10	2	2	-
Roundnose grenadier	*Coryphaenoides rupestris*	184	228	-	-	587	1 030	759
Gadiformes nei	*Gadiformes*	-	-	-	-	-	3	5
Mullets nei	*Mugilidae*	63	126	89	115	67	65	35
Dusky grouper	*Epinephelus marginatus*	-	12	1	-	-	-	-
European seabass	*Dicentrarchus labrax*	582	572	501	687	406	457	640
Blackspot(=red) seabream	*Pagellus bogaraveo*	1	36	5	15	13	38	41
Black seabream	*Spondyliosoma cantharus*	-	-	5	260	240	202	294
Porgies, seabreams nei	*Sparidae*	482	567	496	134	146	111	234
Surmullet	*Mullus surmuletus*	192	173	180	147	220	310	217
Wrasses, hogfishes, etc. nei	*Labridae*	17	16	20	20	22	25	20
Patagonian blennie	*Eleginops maclovinus*	-	-	13	-	-	-	-

English name Nom anglais Nombre inglés	Scientific name Nom scientifique Nombre científico	1996 t	1997 t	1998 t	1999 t	2000 t	2001 t	2002 t
Marbled rockcod	*Notothenia rossii*	-	-	-	-	0	-	5
Humped rockcod	*Notothenia gibberifrons*	-	-	-	0	1	0	1
Grey rockcod	*Notothenia squamifrons*	-	-	-	-	5	-	0
Painted notie	*Nototheniops larseni*	-	-	-	-	0	-	0
Yellowfin notie	*Nototheniops nudifrons*	-	-	-	-	0	-	-
Striped rockcod	*Pagothenia hansoni*	-	-	-	-	0	-	-
Antarctic rockcods, noties nei	*Nototheniidae*	-	-	-	-	-	0	10
Sandeels(=Sandlances) nei	*Ammodytes spp*	22 936	39 271	29 080	14 102	16 530	1 264	3 691
Argentines	*Argentina spp*	-	-	-	28	-	7 955	4 862
Nichol's lanternfish	*Gymnoscopelus nicholsi*	-	-	-	-	0	-	0
European conger	*Conger conger*	980	957	1 044	1 081	1 157	1 255	1 301
Pink cusk-eel	*Genypterus blacodes*	6	11	7	32	7	9	7
Alfonsinos nei	*Beryx spp*	-	4	-	-	7	16	29
Orange roughy	*Hoplostethus atlanticus*	0	0	0	12	2	35	70
John dory	*Zeus faber*	220	159	136	181	296	267	269
Wreckfish	*Polyprion americanus*	8	0	0	0	0	1	-
Patagonian toothfish	*Dissostichus eleginoides*	1	664	734	1 079	1 443	927	1 736
Patagonian rockcod	*Patagonotothen brevicauda*	-	-	-	-	0	-	0
...A	*Parachaenichthys georgianus*	-	-	-	-	0	-	-
Blackfin icefish	*Chaenocephalus aceratus*	-	-	-	-	0	-	4
Mackerel icefish	*Champsocephalus gunnari*	-	-	-	-	4	208	396
South Georgia icefish	*Pseudochaenichthys georgianus*	-	-	-	-	0	6	5
Icefishes nei	*Channichthyidae*	-	-	-	-	-	-	0
Black cardinal fish	*Epigonus telescopus*	-	-	-	-	1	22	-
Atlantic wolffish	*Anarhichas lupus*	1 128	199	893	928	917	140	151
Wolffishes(=Catfishes) nei	*Anarhichas spp*	0	...	...	...	...	600	437
Silver scabbardfish	*Lepidopus caudatus*	-	-	-	-	12	5	1
Black scabbardfish	*Aphanopus carbo*	40	2	159	201	428	742	1 065
Atlantic redfishes nei	*Sebastes spp*	1 776	1 507	1 554	2 672	2 495	2 875	2 117
Red scorpionfish	*Scorpaena scrofa*	...	...	...	...	1	2	5
Blackbelly rosefish	*Helicolenus dactylopterus*	15	3	116	253	184	186	106
Red gurnard	*Chelidonichthys cuculus*	59	17	...	...	4	207	281
Grey gurnard	*Eutrigla gurnardus*	56	59	...	...	...	46	41
Gurnards, searobins nei	*Triglidae*	583	484	629	631	953	1 131	1 154
Lumpfish(=Lumpsucker)	*Cyclopterus lumpus*	0	0	0	0	1	-	-
Angler(=Monk)	*Lophius piscatorius*	31 451	29 783	21 395	16 989	15 955	16 249	14 289
Anglerfishes nei	*Lophiidae*	-	-	-	-	-	2 105	1 199
Atlantic herring	*Clupea harengus*	120 935	103 405	104 627	104 752	82 658	81 363	72 893
European pilchard(=Sardine)	*Sardina pilchardus*	7 304	7 400	6 873	4 815	4 358	10 427	9 401
European sprat	*Sprattus sprattus*	7 172	8 317	7 021	15 168	8 336	5 091	5 792
European anchovy	*Engraulis encrasicolus*	...	...	79	3	0	274	0
Clupeoids nei	*Clupeoidei*	4	17	4	-	-	-	-
Kawakawa	*Euthynnus affinis*	-	-	-	-	-	-	1
Skipjack tuna	*Katsuwonus pelamis*	-	-	-	-	-	-	1
Atlantic bluefin tuna	*Thunnus thynnus*	0	1	1	12	0	-	-
Albacore	*Thunnus alalunga*	49	33	117	343	15	2	-
Yellowfin tuna	*Thunnus albacares*	-	-	-	-	-	-	26
Swordfish	*Xiphias gladius*	5	11	0	2	1	-	-
Capelin	*Mallotus villosus*	-	-	1 115	79	-	-	-
Garfish	*Belone belone*	0	0	1	4	3	2	3
Silversides(=Sand smelts) nei	*Atherinidae*	-	-	-	-	-	1	-
Atlantic horse mackerel	*Trachurus trachurus*	49 927	51 909	32 832	21 025	17 100	19 581	12 332
Atlantic mackerel	*Scomber scombrus*	144 964	149 448	179 711	166 658	193 638	198 953	200 405
Shortfin mako	*Isurus oxyrinchus*	-	-	0	2	3	2	1
Porbeagle	*Lamna nasus*	-	-	-	6	6	10	7
Small-spotted catshark	*Scyliorhinus canicula*	273	275	260	111	238	155	162
Nursehound	*Scyliorhinus stellaris*	109	86	77	139	115	84	39
Dusky catshark	*Halaelurus canescens*	-	-	-	-	-	-	0
Blue shark	*Prionace glauca*	-	-	-	-	12	9	6
Smooth-hound	*Mustelus mustelus*	-	-	-	-	15	-	56
Tope shark	*Galeorhinus galeus*	53	55	55	74	110	82	73
Picked dogfish	*Squalus acanthias*	9 423	8 691	8 926	7 527	7 138	7 306	5 170
Birdbeak dogfish	*Deania calcea*	-	-	-	-	-	1	-
Portuguese dogfish	*Centroscymnus coelolepis*	54	52	147	75	514	1 663	1 456
Longnose velvet dogfish	*Centroscymnus crepidater*	-	-	3	-	-	-	-
Dogfish sharks nei	*Squalidae*	-	-	4	-	-	477	752
Dogfishes and hounds nei	*Squalidae, Scyliorhinidae*	218	247	358	999	1 806	1 443	1 795
Angelsharks, sand devils nei	*Squatinidae*	0	47	-	-	-	-	-
Antarctic starry skate	*Raja georgiana*	-	-	-	-	0	-	-
Raja rays nei	*Raja spp*	9 157	8 088	7 537	6 233	6 457	6 392	6 059
...A	*Bathyraja meridionalis*	-	-	-	-	0	-	-
Rays, stingrays, mantas nei	*Rajiformes*	8	22	13	51	17	38	24
Rabbit fish	*Chimaera monstrosa*	-	0	1	1	4	36	7
Ratfishes nei	*Hydrolagus spp*	-	-	-	-	-	2	-
Straightnose rabbitfish	*Rhinochimaera atlantica*	-	-	-	-	-	2	1
Various sharks nei	*Selachimorpha(Pleurotremata)*	2 040	3 865	2 669	2 342	954	1 640	1 198
Sharks, rays, skates, etc. nei	*Elasmobranchii*	-	16	31	-	-	4	26
Groundfishes nei	*Osteichthyes*	4 131	3 689	3 684	2 498	911	977	893
Pelagic fishes nei	*Osteichthyes*	-	10	-	12	-	-	-
Marine fishes nei	*Osteichthyes*	48	21	85	0	13	19	23
Signal crayfish	*Pacifastacus leniusculus*	-	-	-	10	81	80	50
Edible crab	*Cancer pagurus*	13 182	18 516	21 799	19 988	21 607	21 909	20 760
Portunus swimcrabs nei	*Portunus spp*	1 459	2 929	2 513	2 282	1 672	2 418	2 397
Green crab	*Carcinus maenas*	286	411	462	333	194	331	402
Spinous spider crab	*Maja squinado*	1 224	1 849	1 843	1 310	990	1 199	1 171
Red crab	*Geryon quinquedens*	-	-	61	-	-	627	417
Geryons nei	*Geryon spp*	1 477	325	587	1 015	763	382	1 041
Marine crabs nei	*Brachyura*	639	205	57	90	17	96	492
Palinurid spiny lobsters nei	*Palinurus spp*	7	42	19	15	15	15	15

English name Nom anglais Nombre inglés	Scientific name Nom scientifique Nombre científico	1996 t	1997 t	1998 t	1999 t	2000 t	2001 t	2002 t
Norway lobster	*Nephrops norvegicus*	29 218	31 713	29 212	31 312	28 422	28 536	28 502
European lobster	*Homarus gammarus*	1 043	1 534	1 482	1 819	1 141	1 086	1 188
Craylets, squat lobsters	*Galatheidae*	15	12	-	28	-	-	-
Red king crab	*Paralithodes camtschaticus*	-	-	37	139	89	44	87
King crabs	*Paralithodes spp*	-	-	-	0	-	-	-
Antarctic stone crab	*Paralomis spinosissima*	-	-	-	0	0	4	0
Globose king crab	*Paralomis formosa*	-	-	-	2	2	11	-
King crabs, stone crabs nei	*Lithodidae*	-	1	0	0	0	4	2
Northern prawn	*Pandalus borealis*	1 996	417	595	1 815	539	996	68
Pandalus shrimps nei	*Pandalus spp*	39	48	1 421	46	460	74	290
Common prawn	*Palaemon serratus*	8	15	6	22	15	24	13
Common shrimp	*Crangon crangon*	742	598	739	1 453	1 129	2 290	1 430
Natantian decapods nei	*Natantia*	2	-	-	-	-	-	-
Antarctic krill	*Euphausia superba*	-	308	634	-	-	-	-
Marine crustaceans nei	*Crustacea*	2	-	-	-	-	12	7
Periwinkles nei	*Littorina spp*	1 756	2 870	1 925	1 336	1 059	760	271
Whelk	*Buccinum undatum*	12 150	8 892	3 525	4 925	10 733	11 270	8 669
European flat oyster	*Ostrea edulis*	584	553	1 047	407	439	611	586
Pacific cupped oyster	*Crassostrea gigas*	125	98	59	6	5	20	5
Cupped oysters nei	*Crassostrea spp*	2	5	-	-	-	-	-
Blue mussel	*Mytilus edulis*	12 337	18 994	11 434	7 972	7 468	14 905	16 738
Queen scallop	*Aequipecten opercularis*	2 181	5 630	8 102	5 888	5 149	8 660	10 774
Great Atlantic scallop	*Pecten maximus*	9 195	9 706	9 707	...	...	9 796	9 171
Scallops nei	*Pectinidae*	7 421	8 995	10 371	19 108	19 507	9 722	9 553
Grooved carpet shell	*Ruditapes decussatus*	-	2	-	-	-	29	-
Northern quahog(=Hard clam)	*Mercenaria mercenaria*	-	0	0	-	175	-	-
Razor clams nei	*Solen spp*	56	220	134	114	67	59	37
Common edible cockle	*Cerastoderma edule*	24 176	19 493	12 035	14 123	20 306	19 048	14 268
Clams, etc. nei	*Bivalvia*	49	142	13	67	15	91	43
Cuttlefish,bobtail squids nei	*Sepiidae, Sepiolidae*	4 607	2 202	2 760	2 260	3 076	2 705	3 535
Patagonian squid	*Loligo gahi*	4 043	2 334	3 337	2 148	5 328	4 015	2 705
Common squids nei	*Loligo spp*	3 264	3 004	3 039	2 611	1 757	850	1 002
Argentine shortfin squid	*Illex argentinus*	-	-	-	336	6	21	-
European flying squid	*Todarodes sagittatus*	-	18	293	204	186	193	171
Various squids nei	*Loliginidae, Ommastrephidae*	13	8	8	4	3	815	1 346
Antarctic octopuses	*Pareledone spp*	-	-	-	-	0	-	-
Octopuses, etc. nei	*Octopodidae*	229	148	111	63	135	164	179
Marine molluscs nei	*Mollusca*	15	23	8	113	16	1	21
Sea urchins nei	*Strongylocentrotus spp*	1	0	-	-	-	-	-
Jellyfishes	*Rhopilema spp*	-	-	-	-	5	-	0
Aquatic invertebrates nei	*Invertebrata*	-	-	-	-	-	24	4
	Country total	*867 686*	*890 936*	*923 219*	*840 719*	*747 571*	*741 045*	*689 919*
Total		***17 490 005***	***17 912 203***	***17 109 957***	***16 088 259***	***16 170 292***	***15 981 378***	***15 163 631***

English name Nom anglais Nombre inglés	Scientific name Nom scientifique Nombre científico	1996 t	1997 t	1998 t	1999 t	2000 t	2001 t	2002 t
Amer Samoa								
Freshwater fishes nei	*Osteichthyes*	0	0	0	0	0	0	-
Squirrelfishes nei	*Holocentridae*	1	1	0	1	1	0	1
Groupers nei	*Epinephelus spp*	3	2	2	1	1	1	1
Snappers, jobfishes nei	*Lutjanidae*	8	7	4	5	5	9	10
Emperors(=Scavengers) nei	*Lethrinidae*	5	2	0	1	4	6	6
Parrotfishes nei	*Scaridae*	8	3	10	7	6	3	1
Surgeonfishes nei	*Acanthuridae*	8	-	13	16	10	3	2
Wahoo	*Acanthocybium solandri*	5	7	12	17	20	47	108
Skipjack tuna	*Katsuwonus pelamis*	32	16	8	20	14	56	171
Albacore	*Thunnus alalunga*	86	309	446	338	624	3 253	5 944
Yellowfin tuna	*Thunnus albacares*	12	22	42	64	86	183	484
Bigeye tuna	*Thunnus obesus*	4	4	10	9	21	74	196
Indo-Pacific sailfish	*Istiophorus platypterus*	2	3	2	3	1	1	1
Blue marlin	*Makaira nigricans*	14	18	16	13	17	5	20
Swordfish	*Xiphias gladius*	-	-	2	0	1	1	1
Tuna-like fishes nei	*Scombroidei*	3	0	0	0	0	1	0
Opah	*Lampris guttatus*	-	-	-	1	1	1	-
Carangids nei	*Carangidae*	1	2	1	1	2	1	1
Pomfrets, ocean breams nei	*Bramidae*	-	-	-	-	-	1	0
Common dolphinfish	*Coryphaena hippurus*	5	17	10	13	15	16	14
Barracudas nei	*Sphyraena spp*	2	4	1	0	0	0	1
Sharks, rays, skates, etc. nei	*Elasmobranchii*	-	4	-	-	-	-	-
Marine fishes nei	*Osteichthyes*	10	9	5	6	0	0	1
Tropical spiny lobsters nei	*Panulirus spp*	1	1	2	1	1	1	0
Octopuses, etc. nei	*Octopodidae*	-	-	0	1	0	0	0
	Country total	*210*	*431*	*586*	*518*	*830*	*3 663*	*6 963*
Australia								
Freshwater fishes nei	*Osteichthyes*	252	194	200	154	166	158	127
Short-finned eel	*Anguilla australis*	621	523	453	425	470 F	512 F	497 F
River eels nei	*Anguilla spp*	40	98	100	178	176	182	150 F
Barramundi(=Giant seaperch)	*Lates calcarifer*	1 272	1 178	1 306	1 690	1 747	1 915	1 783
Sand flounders	*Rhombosolea spp*	27	27	28	57	1	...	...
Flatfishes nei	*Pleuronectiformes*	7	6	6	12	15	12	8
Blue antimora	*Antimora rostrata*	-	0	-	0	-	-	-
Tadpole codling	*Salilota australis*	-	-	85	60	-	-	-
Southern blue whiting	*Micromesistius australis*	-	-	23	165	-	-	-
Argentine hake	*Merluccius hubbsi*	-	-	3	10	-	-	-
Patagonian grenadier	*Macruronus magellanicus*	-	-	31	377	-	-	-
Blue grenadier	*Macruronus novaezelandiae*	2 680	2 851	4 752	6 209	9 511	7 580	9 178
Whitson's grenadier	*Macrourus whitsoni*	-	-	-	-	0	-	-
Bigeye grenadier	*Macrourus holotrachys*	-	1	-	-	-	-	-
Grenadiers nei	*Macrourus spp*	-	0	0	1	3	-	-
Gadiformes nei	*Gadiformes*	24	21	33	26	74	68	53
Sea catfishes nei	*Ariidae*	256	39	21	32	-	-	-
Mullets nei	*Mugilidae*	6 492	6 610	6 614	6 396	5 549	6 665	6 039
Groupers, seabasses nei	*Serranidae*	860	155	179	71	99	92	113
Sillago-whitings	*Sillaginidae*	4 389	3 945	3 654	3 961	4 078	3 852	4 120
Ruff	*Arripis georgianus*	1 302	1 287	1 008	1 066	1 143	992	860
Australian salmon	*Arripis trutta*	4 669	4 773	3 919	3 526	4 681	4 734	5 162
Snappers, jobfishes nei	*Lutjanidae*	1 120	1 686	1 790	2 857	4 411	4 973	5 197
Southern meagre(=Mulloway)	*Argyrosomus hololepidotus*	183	177	183	239	458	571	553
Geelbek croaker	*Atractoscion aequidens*	20	27	27	29	32	36	20
Silver seabream	*Pagrus auratus*	4 261	4 180	3 631	4 601	3 797	4 590	4 374
Porgies, seabreams nei	*Sparidae*	940	780	755	859	1 309	1 340	1 101
Parore	*Girella tricuspidata*	501	507	496	503	541	513	470
Wrasses, hogfishes, etc. nei	*Labridae*	77	77	77	99	98	101	101
Threadfins, tasselfishes nei	*Polynemidae*	906	216	272	914	1 013	1 094	1 009
Marbled rockcod	*Notothenia rossii*	-	0	-	0	-	-	0
Grey rockcod	*Notothenia squamifrons*	-	4	3	10	363	329	330
Antarctic rockcods, noties nei	*Nototheniidae*	-	-	3	-	-	0	-
Percoids nei	*Percoidei*	497	131	131	-	69	41	58
Flatheads nei	*Platycephalidae*	4 388	4 575	3 285	4 738	4 817	4 133	4 410
Puffers nei	*Tetraodontidae*	228	219	184	3	662	607	873
Pink cusk-eel	*Genypterus blacodes*	1 397	1 923	1 832	1 891	2 089	1 714	1 646
Redfish	*Centroberyx affinis*	1 284	1 357	1 812	1 555	1 407	1 213	1 158
Orange roughy	*Hoplostethus atlanticus*	4 883	3 479	8 064	7 581	5 717	6 086	4 606
John dory	*Zeus faber*	122	118	103	175	199	178	165
Mirror dory	*Zenopsis nebulosus*	352	429	544	378	315	257	416
Morwongs	*Nemadactylus spp*	1 185	1 513	1 178	1 334	1 116	1 197	1 055
Trumpeters nei	*Latridae*	20	15	34	161	148	90	78
Patagonian toothfish	*Dissostichus eleginoides*	-	1 868	3 506	3 410	3 048	2 640	2 567
Mackerel icefish	*Champsocephalus gunnari*	-	227	128	2	87	1 073	966
Ocellated icefish	*Chionodraco rastrospinosus*	-	1	-	-	-	-	-
Unicorn icefish	*Channichthys rhinoceratus*	-	4	5	1	2	1	3
Spiny icefish	*Chaenodraco wilsoni*	-	-	-	-	-	11	-
Snoek	*Thyrsites atun*	850 F	300	200	102	141	183	170
Silver gemfish	*Rexea solandri*	3 000 F	339	598	458	667	569	307
South Pacific breams nei	*Seriolella spp*	3 406	4 020	3 356	3 327	3 449	4 192	4 009
Scorpionfishes nei	*Scorpaenidae*	316	349	371	356	67	36	56
Bluefin gurnard	*Chelidonichthys kumu*	60	350	372	0	412	440	376
Latchet(=Sharpbeak gurnard)	*Pterygotrigla polyommata*	58	136	92	94	120	149	140
Antarctic horsefish	*Zanclorhynchus spinifer*	-	0	-	-	-	-	-
Demersal percomorphs nei	*Perciformes*	2 600 F	-	-	-	-	-	-
Anchovies, etc. nei	*Engraulidae*	787	19	19	2	39	22	116
Clupeoids nei	*Clupeoidei*	13 030	13 445	8 218	4 377	5 564	8 563	14 474

English name Nom anglais Nombre inglés	Scientific name Nom scientifique Nombre científico	1996 t	1997 t	1998 t	1999 t	2000 t	2001 t	2002 t
Wahoo	*Acanthocybium solandri*	1	4	4	8	17	19	22
Narrow-barred Spanish mackerel	*Scomberomorus commerson*	1 160	1 475	1 617	336	558	741	468
Seerfishes nei	*Scomberomorus spp*	626	171	-	-	22	55	65
Kawakawa	*Euthynnus affinis*	0	1	-	-	-	0	-
Skipjack tuna	*Katsuwonus pelamis*	2 979	5 598	3 098	5 416	4 914	3 172	1 372
Longtail tuna	*Thunnus tonggol*	13	0	0	34	13	71	33
Albacore	*Thunnus alalunga*	472	340	444	435	387	532	753
Southern bluefin tuna	*Thunnus maccoyii*	5 355	5 940	4 791	5 655	5 267	5 279	5 296
Yellowfin tuna	*Thunnus albacares*	1 850	2 041	2 111	2 662	1 695	2 943	2 838
Bigeye tuna	*Thunnus obesus*	318	1 039	2 056	1 379	1 140	1 548	1 437
Indo-Pacific sailfish	*Istiophorus platypterus*	...	...	3	1	...	...	7
Blue marlin	*Makaira nigricans*	0	2	-	-	-	-	0
Black marlin	*Makaira indica*	-	4	-	-	-	-	-
Striped marlin	*Tetrapturus audax*	3	17	12	62	519	720	770
Shortbill spearfish	*Tetrapturus angustirostris*	...	...	...	1	...	0	16
Marlins,sailfishes,etc. nei	*Istiophoridae*	199	1 084	1 929	2 185	3	4	-
Swordfish	*Xiphias gladius*	22	43	337	1 360	3 875	4 753	4 342
Tuna-like fishes nei	*Scombroidei*	305	290	444	34	237	176	163
Needlefishes, etc. nei	*Belonidae*	1	0	0	-	-	-	-
Halfbeaks nei	*Hemiramphus spp*	237	680	695	530	838	948	865
Southern opah	*Lampris immaculatus*	-	-	-	-	-	-	0
Bluefish	*Pomatomus saltatrix*	260	243	192	250	209	351	309
Greenback horse mackerel	*Trachurus declivis*	68	30	18	16	452	152	636
White trevally	*Pseudocaranx dentex*	964	872	558	479	418	415	491
Amberjacks nei	*Seriola spp*	294	94	86	2	3	2	2
Mackerels nei	*Scombridae*	987	89	91	3 728	3 311	2 959	3 886
Pelagic percomorphs nei	*Perciformes*	4 020 F	-	-	-	-	-	-
Porbeagle	*Lamna nasus*	-	2	-	-	-	-	-
Smooth-hounds nei	*Mustelus spp*	3 878	4 169	2 858	2 462	1 896	2 572	2 524
Tope shark	*Galeorhinus galeus*	...	...	...	...	498	325	350
Pacific sleeper shark	*Somniosus pacificus*	-	-	-	1	-	-	3
Sawsharks nei	*Pristiophorus spp*	...	...	...	...	270	423	371
Angelsharks, sand devils nei	*Squatinidae*	102	129	120	102	98	71	118
Rays, stingrays, mantas nei	*Rajiformes*	125	75	73	76	56	64	71
Ghost shark	*Callorhinchus milii*	...	49	21	14	82	105	102
Sharks, rays, skates, etc. nei	*Elasmobranchii*	4 613	3 894	3 426	3 673	4 643	4 966	5 416
Marine fishes nei	*Osteichthyes*	29 654 F	30 684	35 748	31 158	12 948	11 631	17 668
Australian crayfish	*Euastacus armatus*	-	-	-	66	24	23	69
Blue swimming crab	*Portunus pelagicus*	6 138	6 008	6 000	5 077	5 938	6 778	5 942
Australian spiny lobster	*Panulirus cygnus*	9 902	9 896	10 400	13 065	14 605	11 353	9 050
Tropical spiny lobsters nei	*Panulirus spp*	205	233	551	520	359	274	330
Green rock lobster	*Jasus verreauxi*	97	104	108	110	117	104	103
Southern rock lobster	*Jasus novaehollandiae*	4 856	4 888	4 615	4 655	4 756	4 677	4 386
Slipper lobsters nei	*Scyllaridae*	622	875	945	7	0	0	...
Metanephrops nei	*Metanephrops spp*	-	-	-	-	39	105	97
King crabs, stone crabs nei	*Lithodidae*	-	0	-	-	-	-	-
Banana prawn	*Penaeus merguiensis*	4 347	4 546	3 711	4 125	2 586	6 646	5 794
Giant tiger prawn	*Penaeus monodon*	3 841	3 055	3 686	3 477	4 601	4 272	4 194
Western king prawn	*Penaeus latisulcatus*	76	67	110	81	92	71	169
Penaeus shrimps nei	*Penaeus spp*	17 034	16 149	16 870	15 454	12 999	12 854	12 235
Endeavour shrimp	*Metapenaeus endeavouri*	2 175	2 245	2 885	2 575	3 495	3 351	2 954
Marine crustaceans nei	*Crustacea*	1 352	5 474	5 328	1 886	2 107	1 969	1 883
Blacklip abalone	*Haliotis rubra*	5 425	5 240	5 247	5 620	5 532	5 609	6 126
Australian mussel	*Mytilus planulatus*	75	1	1	1	1	1	0
Scallops nei	*Pectinidae*	12 320	8 589	9 874	11 574	12 014	9 198	5 658
Pipi wedge clam	*Paphies australis*	702	1 295	1 496	1 489	1 713	1 875	1 560
Cuttlefish,bobtail squids nei	*Sepiidae, Sepiolidae*	438	436	313	451	302	292	297
Patagonian squid	*Loligo gahi*	-	-	3 198	2 486	-	-	-
Argentine shortfin squid	*Illex argentinus*	-	-	-	167	-	-	-
Various squids nei	*Loliginidae, Ommastrephidae*	1 884	3 351	1 617	3 641	2 770	4 245	2 295
Octopuses, etc. nei	*Octopodidae*	74	590	783	560	446	721	545
Cephalopods nei	*Cephalopoda*	-	1	-	-	-	-	-
Marine molluscs nei	*Mollusca*	818	381	1 046	7 558	1 245	1 123	1 097
Jellyfishes	*Rhopilema spp*	-	11	2	-	-	-	-
Aquatic invertebrates nei	*Invertebrata*	-	1	-	-	110	169	106
	Country total	*200 249*	*196 674*	*203 202*	*211 176*	*190 090*	*194 411*	*194 178*
Christmas Is								
Freshwater fishes nei	*Osteichthyes*	0	0	0	0	0	0	0
Marine fishes nei	*Osteichthyes*	0	0	0	0	0	0	0
	Country total	*0*	*0*	*0*	*0*	*0*	*0*	*0*
Cocos Is								
Freshwater fishes nei	*Osteichthyes*	0	0	0	0	0	0	0
Marine fishes nei	*Osteichthyes*	0	0	0	0	0	0	0
	Country total	*0*	*0*	*0*	*0*	*0*	*0*	*0*
Cook Is								
Freshwater fishes nei	*Osteichthyes*	0	0	0	0	0	0	0
Mullets nei	*Mugilidae*	5 F	5 F	5 F	5 F	5 F	5 F	5 F
Groupers nei	*Epinephelus spp*	110 F	100 F	90 F	80 F	60 F	60 F	60 F
Snappers nei	*Lutjanus spp*	20 F	20 F	20 F	20 F	20 F	20 F	20 F
Albacore	*Thunnus alalunga*	5	5 F	5 F	5 F	5 F	2	879
Yellowfin tuna	*Thunnus albacares*	8	5 F	5 F	5 F	5 F	1	49
Bigeye tuna	*Thunnus obesus*	3	3 F	3 F	3 F	3 F	1	66
Marlins,sailfishes,etc. nei	*Istiophoridae*	40 F	17 F	15 F	10 F	5 F	1 F	62 F

English name Nom anglais Nombre inglés	Scientific name Nom scientifique Nombre científico	1996 t	1997 t	1998 t	1999 t	2000 t	2001 t	2002 t
Swordfish	*Xiphias gladius*	41 F	17 F	15 F	10 F	5 F	2 F	62 F
Flyingfishes nei	*Exocoetidae*	30 F	30 F	30 F	30 F	30 F	30 F	30 F
Jacks, crevalles nei	*Caranx spp*	40 F	40 F	40 F	40 F	40 F	40 F	40 F
Sharks, rays, skates, etc. nei	*Elasmobranchii*	20 F	20 F	20 F	20 F	20 F	20 F	20 F
Marine fishes nei	*Osteichthyes*	333 F	343 F	337 F	347 F	347 F	343 F	300 F
Marine crabs nei	*Brachyura*	5 F	5 F	5 F	5 F	5 F	5 F	5 F
Octopuses, etc. nei	*Octopodidae*	60 F	50 F	40 F	30 F	30 F	30 F	30 F
Marine molluscs nei	*Mollusca*	160 F	140 F	120 F	120 F	120 F	120 F	120 F
Sea urchins nei	*Strongylocentrotus spp*	20 F	20 F	20 F	20 F	20 F	20 F	20 F
	Country total	*900 F*	*820 F*	*770 F*	*750 F*	*720 F*	*700 F*	*1 768 F*
Fiji Islands								
Nile tilapia	*Oreochromis niloticus*	37	40	288	290	280 F	278	280 F
Milkfish	*Chanos chanos*	34	26	21	30	30 F	32	30 F
Mullets nei	*Mugilidae*	759	860	1 195	3 067	2 800 F	2 915	2 900 F
Groupers nei	*Epinephelus spp*	1 017	1 060	1 160	1 600	1 450 F	1 544	1 530 F
Cardinalfishes, etc. nei	*Apogonidae*	200	180	60	71	60 F	67	65 F
Snappers nei	*Lutjanus spp*	1 347	1 315	1 155	1 710	1 600 F	1 728	1 700 F
Ponyfishes(=Slipmouths)	*Leiognathus spp*	76	75	69	84	80 F	87	85 F
Largeeye breams	*Gymnocranius spp*	20	22	25	28	25 F	30	30 F
Emperors(=Scavengers) nei	*Lethrinidae*	1 230	1 780	1 731	2 990	2 700 F	2 883	2 850 F
Goatfishes	*Upeneus spp*	108	190	155	160	140 F	157	150 F
Mojarras(=Silver-biddies) nei	*Gerres spp*	6	10	13	11	10 F	11	10 F
Wrasses, hogfishes, etc. nei	*Labridae*	204	375	180	399	360 F	400	390 F
Surgeonfishes nei	*Acanthuridae*	125	105	180	206	180 F	196	195 F
Spinefeet(=Rabbitfishes) nei	*Siganus spp*	73	80	93	100	90 F	96	95 F
Triggerfishes, durgons nei	*Balistidae*	4	8	10	14	10 F	...	...
Demersal percomorphs nei	*Perciformes*	380	250	-	-	-	-	-
Sardinellas nei	*Sardinella spp*	120	140	30	47	35 F	...	35 F
Silver-stripe round herring	*Spratelloides gracilis*	140	135	10	40	30 F	...	30 F
Bluestripe herring	*Herklotsichthys quadrimaculat.*	140	120	22	20	20 F	...	20 F
Wahoo	*Acanthocybium solandri*	130	145	148	160	150 F	167	160 F
Narrow-barred Spanish mackerel	*Scomberomorus commerson*	1 247	1 025	1 455	2 296	2 000 F	2 120	2 100 F
Skipjack tuna	*Katsuwonus pelamis*	3 124	987	459	507	343	431	420 F
Albacore	*Thunnus alalunga*	1 446	1 842	2 121	2 279	6 065	7 971	8 026
Yellowfin tuna	*Thunnus albacares*	1 540	1 016	869	725	2 467	2 126	2 027
Bigeye tuna	*Thunnus obesus*	593	409	460	462	687	662	853
Halfbeaks nei	*Hemiramphus spp*	70	62	49	61	50 F	56	55 F
Silversides(=Sand smelts) nei	*Atherinidae*	31	25	80	91	80 F	86	85 F
Jacks, crevalles nei	*Caranx spp*	384	695	647	730	650 F	709	700 F
Indian mackerel	*Rastrelliger kanagurta*	400	312	224	721	650 F	722	700 F
Barracudas nei	*Sphyraena spp*	567	1 562	1 626	2 979	2 700 F	2 809	2 800 F
Marine fishes nei	*Osteichthyes*	1 800	1 035	1 358	1 218	1 200 F	1 113	1 070 F
Freshwater crustaceans nei	*Crustacea*	327	315	323	332	340 F	343	340 F
Indo-Pacific swamp crab	*Scylla serrata*	208	290	270	281	250 F	268	265 F
Marine crabs nei	*Brachyura*	200	280	255	300	270 F	290	280 F
Tropical spiny lobsters nei	*Panulirus spp*	105	130	211	220	200 F	224	215 F
Marine crustaceans nei	*Crustacea*	85	78	85	91	80 F	87	85 F
Freshwater molluscs nei	*Mollusca*	2 670	3 970	4 500	5 000	5 080 F	5 300	5 180 F
Anadara clams nei	*Anadara spp*	1 044	1 950	2 800	2 990	2 750 F	2 884	2 870 F
Clams, etc. nei	*Bivalvia*	54	45	52	50	40 F	43	45 F
Octopuses, etc. nei	*Octopodidae*	48	35	62	61	50 F	57	55 F
Marine molluscs nei	*Mollusca*	2 000	3 880	3 200	3 302	3 100 F	3 150	3 150 F
Green turtle	*Chelonia mydas*	22	4	2	2	2 F	3	2 F
Marine turtles nei	*Testudinata*	24	7	2	8	6 F	7	7 F
Sea urchins nei	*Strongylocentrotus spp*	40	95	103	100	90 F	96	95 F
Sea cucumbers nei	*Holothurioidea*	850	790	400	880	800 F	824	820 F
	Country total	*25 029*	*27 755*	*28 158*	*36 713*	*40 000 F*	*42 972*	*42 800 F*
Fr Polynesia								
Freshwater fishes nei	*Osteichthyes*	0	50	50	50	50	50	50
Snappers, jobfishes nei	*Lutjanidae*	...	100	81	109	127	112	80
Wahoo	*Acanthocybium solandri*	...	119	188	269	229	259	291
Skipjack tuna	*Katsuwonus pelamis*	1 400	1 126	1 560	1 386	1 189	1 557	1 470
Albacore	*Thunnus alalunga*	1 750	2 717	3 235	2 642	3 580	4 432	4 678
Yellowfin tuna	*Thunnus albacares*	811	860	843	1 225	1 762	1 514	1 163
Bigeye tuna	*Thunnus obesus*	186	310	403	278	712	746	651
Marlins,sailfishes,etc. nei	*Istiophoridae*	587	598	518	703	566	551	527
Swordfish	*Xiphias gladius*	84	56	58	66	47	79	70
Tuna-like fishes nei	*Scombroidei*	...	...	...	2	0	0	4
Flyingfishes nei	*Exocoetidae*	...	55	92	...	88	82	79
Opah	*Lampris guttatus*	96	...	...	137	124	148	140
Common dolphinfish	*Coryphaena hippurus*	257	427	437	429	446	651	610
Shortfin mako	*Isurus oxyrinchus*	...	...	...	...	27	53	41
Sharks, rays, skates, etc. nei	*Elasmobranchii*	387	367	347	427	582	705	1 063
Marine fishes nei	*Osteichthyes*	4 332	4 820	4 585	4 538	4 274	4 362	4 526
Giant river prawn	*Macrobrachium rosenbergii*	0	3	3	3	3	3	3
Indo-Pacific swamp crab	*Scylla serrata*	-	2	2	2	2	2	3
Tropical spiny lobsters nei	*Panulirus spp*	20	40	51	50	51	58	54
Marine crustaceans nei	*Crustacea*	0	0	0	0	0	-	-
Marine molluscs nei	*Mollusca*	0	10	10	10	25	25	25
Echinoderms	*Echinodermata*	0	10	10	10	15	15	15
	Country total	*9 910*	*11 670*	*12 473*	*12 336*	*13 899*	*15 404*	*15 543*
Guam								
Tilapias nei	*Oreochromis (=Tilapia) spp*	-	-	6	-	-	-	-

E-6 Fish crustaceans, molluscs, etc — Capture production by countries or areas and species — Oceania
Poissons, crustacés, mollusques, etc — Captures par pays ou zones et espèces — Océanie
Peces, crustáceos, moluscos, etc — Capturas por países o áreas y especies — Oceanía

English name Nom anglais Nombre inglés	Scientific name Nom scientifique Nombre científico	1996 t	1997 t	1998 t	1999 t	2000 t	2001 t	2002 t
Freshwater fishes nei	*Osteichthyes*	0	-	-	-	-	-	-
Groupers nei	*Epinephelus spp*	0	1	0	1	1	1	1
Snappers, jobfishes nei	*Lutjanidae*	1	2	3	10	6	5	3
Emperors(=Scavengers) nei	*Lethrinidae*	1	1	1	1	1	3	3
Wrasses, hogfishes, etc. nei	*Labridae*	-	-	1	2	0	1	0
Parrotfishes nei	*Scaridae*	1	1	4	1	0	0	0
Surgeonfishes nei	*Acanthuridae*	1	...	3	0	-	0	5
Spinefeet(=Rabbitfishes) nei	*Siganus spp*	0	...	1	2	0	0	1
Wahoo	*Acanthocybium solandri*	19	20	30	19	20	23	22
Skipjack tuna	*Katsuwonus pelamis*	17	24	28	19	61	60	47
Yellowfin tuna	*Thunnus albacares*	15	17	25	15	18	10	13
Indo-Pacific sailfish	*Istiophorus platypterus*	0	0	1	0	1	1	1
Blue marlin	*Makaira nigricans*	15	24	13	14	30	15	13
Tuna-like fishes nei	*Scombroidei*	1	2	1	1	0	2	2
Scads nei	*Decapterus spp*	0	3	2	5	4	5	2
Rainbow runner	*Elagatis bipinnulata*	-	-	1	2	1	2	3
Carangids nei	*Carangidae*	1	1	1	2	1	1	2
Common dolphinfish	*Coryphaena hippurus*	35	43	79	40	34	53	36
Barracudas nei	*Sphyraena spp*	1	1	1	1	2	4	2
Sharks, rays, skates, etc. nei	*Elasmobranchii*	0	0	-	0	0	0	0
Marine fishes nei	*Osteichthyes*	13	18	49	85	92	89	72
Tropical spiny lobsters nei	*Panulirus spp*	0	0	1	1	2	1	1
Natantian decapods nei	*Natantia*	-	-	1	0	0	0	0
Octopuses, etc. nei	*Octopodidae*	0	-	1	2	1	2	2
	Country total	*121*	*158*	*253*	*223*	*275*	*278*	*231*
Kiribati								
Freshwater fishes nei	*Osteichthyes*	0	0	0	0	0	0	-
Milkfish	*Chanos chanos*	290	290	...	80	2 175	58	57
Mullets nei	*Mugilidae*	450	500	149	994	611	1 300	2 534
Snappers, jobfishes nei	*Lutjanidae*	1 950	1 940	1 047	1 505	2 141	2 794	947
Emperors(=Scavengers) nei	*Lethrinidae*	1 970	1 960	675	1 299	2 137	3 930	1 363
Goatfishes	*Upeneus spp*	450	450	149	582	639	1 609	404
Mojarras(=Silver-biddies) nei	*Gerres spp*	1 720	1 730	566	890	674	1 455	847
Percoids nei	*Percoidei*	1 590	1 580	10 713	24 588	632	1 069	1 198
Clupeoids nei	*Clupeoidei*	3 340	3 320	613	2 638	2 725	2 079	3 619
Skipjack tuna	*Katsuwonus pelamis*	4 111	2 855	5 544	4 493	3 701	3 286	3 793
Yellowfin tuna	*Thunnus albacares*	651	2 223	2 076	1 423	1 209	1 220	793
Bigeye tuna	*Thunnus obesus*	69	130	99	157	63	113	74
Flyingfishes nei	*Exocoetidae*	1 780	1 770	2 525	2 547	836	1 594	1 617
Jacks, crevalles nei	*Caranx spp*	500	510	2 530	1 910	1 108	2 858	3 702
Barracudas nei	*Sphyraena spp*	2 160	450	987	855	420	732	1 360
Sharks, rays, skates, etc. nei	*Elasmobranchii*	1 840	1 830	2 381	3 012	1 581	1 273	2 769
Marine fishes nei	*Osteichthyes*	2 400	2 380	3 875	4 216	2 396	3 198	3 130
Marine crustaceans nei	*Crustacea*	220	4	...	...	131	418	174
Octopuses, etc. nei	*Octopodidae*	2 230	1 874	650	687	373	69	23
Marine molluscs nei	*Mollusca*	4 150	4 120	571	776	1 947	3 260	2 337
Sea cucumbers nei	*Holothurioidea*	...	136	154	89	64	60	260
	Country total	*31 871*	*30 052*	*35 304*	*52 741*	*25 563*	*32 375*	*31 001*
Marshall Is								
Freshwater fishes nei	*Osteichthyes*	0	0	0	0	0	0	0
Skipjack tuna	*Katsuwonus pelamis*	-	-	-	-	6 625	31 983	37 057
Yellowfin tuna	*Thunnus albacares*	-	-	-	-	900	2 927	1 101
Bigeye tuna	*Thunnus obesus*	-	-	-	-	35	134	84
Marine fishes nei	*Osteichthyes*	2 772	400 F	500 F	500 F	500 F	500 F	500 F
Indo-Pacific swamp crab	*Scylla serrata*	0	0	0	0	0	0	0
Tropical spiny lobsters nei	*Panulirus spp*	0	0	0	0	-	-	-
Natantian decapods nei	*Natantia*	0	0	0	0	-	-	-
	Country total	*2 772*	*400 F*	*500 F*	*500 F*	*8 060*	*35 544*	*38 742*
Micronesia								
Freshwater fishes nei	*Osteichthyes*	5 F	5 F	5 F	5 F	5 F	5 F	5 F
Skipjack tuna	*Katsuwonus pelamis*	6 745	5 501	11 314	6 972	15 843	11 267	13 957
Albacore	*Thunnus alalunga*	-	1	-	2	5	4	-
Yellowfin tuna	*Thunnus albacares*	891	2 845	3 212	3 403	5 411	5 848	3 922
Bigeye tuna	*Thunnus obesus*	183	430	705	1 016	1 175	984	958
Marine fishes nei	*Osteichthyes*	1 200 F	1 300 F	1 300 F	1 400 F	1 400 F	1 500 F	1 500 F
Indo-Pacific swamp crab	*Scylla serrata*	5 F	5 F	5 F	5 F	5 F	5 F	5 F
Tropical spiny lobsters nei	*Panulirus spp*	20 F	20 F	20 F	20 F	20 F	20 F	20 F
Natantian decapods nei	*Natantia*	0	0	0	0	0	0	0
Octopuses, etc. nei	*Octopodidae*	20 F	20 F	20 F	20 F	20 F	20 F	20 F
Marine turtles nei	*Testudinata*	0	0	0	0	0	0	0
	Country total	*9 069 F*	*10 127 F*	*16 581 F*	*12 843 F*	*23 884 F*	*19 653 F*	*20 387 F*
Nauru								
Skipjack tuna	*Katsuwonus pelamis*	...	...	...	...	...	...	2
Yellowfin tuna	*Thunnus albacares*	...	...	...	...	...	...	10
Bigeye tuna	*Thunnus obesus*	...	...	...	...	...	...	2
Marine fishes nei	*Osteichthyes*	300 F	250 F	200 F	150 F	100 F	50 F	7
	Country total	*300 F*	*250 F*	*200 F*	*150 F*	*100 F*	*50 F*	*21*
NewCaledonia								
Freshwater fishes nei	*Osteichthyes*	0	0	0	0	0	0	0

English name Nom anglais Nombre inglés	Scientific name Nom scientifique Nombre científico	1996 t	1997 t	1998 t	1999 t	2000 t	2001 t	2002 t
Mullets nei	*Mugilidae*	20	61	64	63	75	88	88 F
Snappers nei	*Lutjanus spp*	45	36	43	22	24	23	23 F
Alfonsinos nei	*Beryx spp*	0	0	0	0	0	0	0
Ruffs, barrelfishes nei	*Centrolophidae*	0	0	0	0	0	0	0
Seerfishes nei	*Scomberomorus spp*	19	3	16	41	4	1	1 F
Skipjack tuna	*Katsuwonus pelamis*	0	1	1	0	0	0	0
Albacore	*Thunnus alalunga*	414	277	860	690	895	1 020	1 165
Yellowfin tuna	*Thunnus albacares*	554	466	185	373	250	570	572
Bigeye tuna	*Thunnus obesus*	233	234	498	553	517	128	189
Tuna-like fishes nei	*Scombroidei*	244	169	285	236	294	310	265
Mackerels nei	*Scombridae*	119	102	104	161	161	30	30 F
Marine fishes nei	*Osteichthyes*	415	385	458	415	509	618	602 F
Marine crabs nei	*Brachyura*	9	29	20	58	22	14	14 F
Tropical spiny lobsters nei	*Panulirus spp*	17	14	17	17	13	9	9 F
Penaeus shrimps nei	*Penaeus spp*	-	-	-	0	0	0	0
Marine crustaceans nei	*Crustacea*	0	0	-	0	0	0	0
Cuttlefish,bobtail squids nei	*Sepiidae, Sepiolidae*	3	1	2	3	1	0	0
Marine molluscs nei	*Mollusca*	130	95	150	27	6	9	9 F
Sea cucumbers nei	*Holothurioidea*	776	565	402	493	615	489	450 F
	Country total	*2 998*	*2 438*	*3 105*	*3 152*	*3 386*	*3 309*	*3 417 F*
New Zealand								
Common carp	*Cyprinus carpio*	0	28	2	3	4	6	7
Brown bullhead	*Ameiurus nebulosus*	1	5	12	14	10	8	7
Freshwater fishes nei	*Osteichthyes*	600	600	400	400	400	300	300
River eels nei	*Anguilla spp*	1 002	980	1 293	1 151	1 055	1 075	923
Chinook(=Spring=King)salmon	*Oncorhynchus tshawytscha*	1	1	3	1	1	1	0
Sand flounders	*Rhombosolea spp*	224	251	...	...	...	37	204
Flatfishes nei	*Pleuronectiformes*	4 284	7 747	4 270	3 505	2 939	3 220	2 810
Smalleye moray cod	*Muraenolepis microps*	-	-	-	4	5	-	0
Moray cods nei	*Muraenolepis spp*	-	-	0	1	2	3	5
Common mora	*Mora moro*	694	1 410	1 324	1 122	1 355	1 209	1 308
Red codling	*Pseudophycis bachus*	10 572	11 073	16 429	12 528	5 232	4 454	4 414
Blue antimora	*Antimora rostrata*	-	-	0	0	0	3	3
Southern blue whiting	*Micromesistius australis*	2 753	10 234	35 059	39 012	23 000	29 789	42 086
Southern hake	*Merluccius australis*	8 317	9 692	15 047	15 499	12 799	12 956	12 176
Blue grenadier	*Macruronus novaezelandiae*	145 308	229 890	267 616	236 652	234 029	223 703	192 482
Whitson's grenadier	*Macrourus whitsoni*	-	-	-	1	5	48	158
Ridge scaled rattail	*Macrourus carinatus*	-	-	-	20	65	-	0
Grenadiers nei	*Macrourus spp*	-	-	9	1	0	-	-
Grenadiers, whiptails nei	*Coryphaenoides spp*	-	-	-	-	-	-	0
Thorntooth grenadier	*Lepidorhynchus denticulatus*	670	2 361	4 627	3 678	3 833	4 783	5 349
Grenadiers, rattails nei	*Macrouridae*	1 579	5 197	4 350	3 670	2 394	3 094	3 866
Gadiformes nei	*Gadiformes*	36	22	34	11	14	8	11
Mullets nei	*Mugilidae*	897	872	720	846	748	910	748
Orange perch	*Lepidoperca pulchella*	...	...	193	32	...	46	97
Australian salmon	*Arripis trutta*	3 496	2 777	2 928	4 780	3 825	2 934	3 637
Emperors(=Scavengers) nei	*Lethrinidae*	-	4	-	-	-	-	-
Porgies	*Calamus spp*	-	-	0	-	-	-	-
Silver seabream	*Pagrus auratus*	5 814	6 233	6 278	6 876	6 852	6 209	6 571
Parore	*Girella tricuspidata*	60	81	80	76	94	71	78
Wrasses, hogfishes, etc. nei	*Labridae*	0	24	2	3	4	2	3
Antarctic rockcods, noties nei	*Nototheniidae*	16	34	3	19	3	16	42
New Zealand blue cod	*Parapercis colias*	2 115	2 227	2 313	2 286	2 130	2 441	2 376
Flatheads nei	*Platycephalidae*	1	1	2	6	7	19	51
Velvet leatherjacket	*Parika scaber*	442	1 095	312	738	1 279	1 142	1 012
Argentines	*Argentina spp*	8	41	68	63	42	56	101
Conger eels, etc. nei	*Congridae*	109	113	97	88	96	106	144
Banded yellowfish	*Centriscops humerosus*	1	2	11	39	...	52	24
Pink cusk-eel	*Genypterus blacodes*	12 454	22 594	22 215	21 424	21 617	18 620	20 295
Alfonsinos nei	*Beryx spp*	2 159	2 617	3 516	2 579	2 880	3 044	2 888
Redfish	*Centroberyx affinis*	123	194	172	134	176	202	117
Orange roughy	*Hoplostethus atlanticus*	28 639	20 545	21 485	23 780	17 879	14 044	17 954
Slimeheads nei	*Trachichthyidae*	-	-	7	3	4	2	12
John dory	*Zeus faber*	729	800	828	882	841	914	1 141
Dories nei	*Zeidae*	288	641	754	763	778	685	1 012
Oreo dories nei	*Oreosomatidae*	18 776	21 850	21 095	22 646	22 775	24 165	17 625
Hapuku wreckfish	*Polyprion oxygeneios*	1 155	1 657	1 571	1 547	1 497	1 579	1 610
Cape bonnetmouth	*Emmelichthys nitidus*	1 845	1 635	2 064	2 846	2 825	1 881	2 825
Bonnetmouths, rubyfishes nei	*Emmelichthyidae*	596	431	378	271	582	434	403
Giant boarfish	*Paristiopterus labiosus*	19	27	75	6	9	3	9
Pelagic armourhead	*Pseudopentaceros richardsoni*	7	2	78	13	6	7	37
Tarakihi	*Nemadactylus macropterus*	4 366	5 441	5 239	5 589	5 739	6 129	6 149
Morwongs	*Nemadactylus spp*	92	107	109	78	99	92	91
Trumpeters nei	*Latridae*	727	761	567	574	476	505	493
Antarctic toothfish	*Dissostichus mawsoni*	-	-	41	296	751	582	1 354
Patagonian toothfish	*Dissostichus eleginoides*	1 061	5	44	2	0	44	24
Blackfin icefish	*Chaenocephalus aceratus*	-	-	-	-	0	-	-
Icefishes nei	*Channichthyidae*	-	-	0	0	0	2	2
Black cardinal fish	*Epigonus telescopus*	3 002	4 334	2 568	2 869	4 095	1 957	2 741
Plunderfish	*Pogonophryne permitini*	-	-	-	-	0	0	0
Giant stargazer	*Kathetostoma giganteum*	2 122	3 990	2 195	3 370	3 638	4 233	3 272
Snoek	*Thyrsites atun*	15 849	22 047	25 972	20 642	21 905	25 222	23 121
Escolar	*Lepidocybium flavobrunneum*	0	1	1	22	47	87	76
Oilfish	*Ruvettus pretiosus*	53	34	57	53	61	86	131
Silver gemfish	*Rexea solandri*	1 344	1 914	1 301	902	1 029	827	688
Frostfishes	*Benthodesmus spp*	-	-	-	-	-	1	80
Silver scabbardfish	*Lepidopus caudatus*	941	2 342	3 344	2 638	1 536	2 876	2 435

English name Nom anglais Nombre inglés	Scientific name Nom scientifique Nombre científico	1996 t	1997 t	1998 t	1999 t	2000 t	2001 t	2002 t
Common warehou	*Seriolella brama*	1 760	4 108	3 101	3 881	4 259	4 101	4 005
Silver warehou	*Seriolella punctata*	4 926	11 253	10 993	9 029	11 218	11 268	8 778
White warehou	*Seriolella caerulea*	1 467	2 432	2 296	2 366	2 407	1 962	1 975
Bluenose warehou	*Hyperoglyphe antarctica*	2 432	2 974	2 630	2 755	2 793	2 954	3 001
Ruffs, barrelfishes nei	*Centrolophidae*	169	214	202	249	274	211	169
Scorpionfishes nei	*Scorpaenidae*	1 133	1 643	1 843	2 088	1 819	1 897	2 306
Bluefin gurnard	*Chelidonichthys kumu*	2 260	2 625	2 454	2 199	2 663	3 670	3 447
Spotted gurnard	*Pterygotrigla picta*	87	71	87	74	55	65	49
Australian pilchard	*Sardinops neopilchardus*	169	385	519	894	1 253	1 399	920
Clupeoids nei	*Clupeoidei*	11	8	1	12	11	11	3
Wahoo	*Acanthocybium solandri*	-	6	-	-	-	-	-
Frigate and bullet tunas	*Auxis thazard, A.rochei*	0	2	4	-	5	5	5
Skipjack tuna	*Katsuwonus pelamis*	3 631	4 792	8 156	5 688	9 699	3 691	3 344
Pacific bluefin tuna	*Thunnus orientalis*	5	12	20	21	21	50	55
Albacore	*Thunnus alalunga*	7 150	3 220	6 525	3 903	4 500	5 353	5 645
Southern bluefin tuna	*Thunnus maccoyii*	81	138	337	460	380	358	450
Yellowfin tuna	*Thunnus albacares*	181	118	127	153	107	137	25
Bigeye tuna	*Thunnus obesus*	86	140	388	420	421	480	200
Striped marlin	*Tetrapturus audax*	0	1	-	-	-	-	-
Shortbill spearfish	*Tetrapturus angustirostris*	42	18	-	-	-	-	0
Swordfish	*Xiphias gladius*	152	170	564	1 004	975	1 029	929
Tuna-like fishes nei	*Scombroidei*	36	83	66	101	69	88	96
Halfbeaks nei	*Hemiramphus spp*	16	17	28	24	18	13	11
Flyingfishes nei	*Exocoetidae*	4	2	3	1	2	1	7
Opah	*Lampris guttatus*	76	126	254	335	283	340	329
Dealfishes	*Trachipterus spp*	49	60	74	65	67	103	63
King of herrings	*Regalecus glesne*	10	64	60	34	20	1	0
Jack and horse mackerels nei	*Trachurus spp*	29 085	34 057	36 059	34 003	22 544	28 507	32 285
White trevally	*Pseudocaranx dentex*	2 869	2 941	3 608	4 192	3 603	3 116	3 280
Scads nei	*Decapterus spp*	23	11	63	29	82	87	70
Amberjacks nei	*Seriola spp*	381	349	327	317	296	278	242
Atlantic pomfret	*Brama brama*	480	413	488	465	401	903	552
Common dolphinfish	*Coryphaena hippurus*	0	0	0	1	0	15	7
Blue mackerel	*Scomber australasicus*	2 837	8 768	7 041	12 417	10 431	9 761	13 287
Broadnose sevengill shark	*Notorynchus cepedianus*	...	...	2	3	4	5	4
Basking shark	*Cetorhinus maximus*	2	2	49	129	95	84	40
Thresher	*Alopias vulpinus*	13	24	21	32	51	57	53
Shortfin mako	*Isurus oxyrinchus*	52	40	74	110	208	327	238
Porbeagle	*Lamna nasus*	16	21	164	246	188	127	130
Blue shark	*Prionace glauca*	246	120	540	593	1 169	1 328	1 186
Copper shark	*Carcharhinus brachyurus*	...	...	15	14	25	38	38
Smooth hammerhead	*Sphyrna zygaena*	10	3	6	11	13	17	10
Spotted estuary smooth-hound	*Mustelus lenticulatus*	1 350	3 464	1 707	1 662	1 643	1 563	1 403
Tope shark	*Galeorhinus galeus*	3 044	2 864	3 083	3 633	3 100	3 091	3 315
Picked dogfish	*Squalus acanthias*	2 477	7 232	3 064	4 409	3 362	4 192	6 024
Leafscale gulper shark	*Centrophorus squamosus*	...	...	4	1	0	0	1
Lanternsharks nei	*Etmopterus spp*	0	2	-	-	-	4	25
Birdbeak dogfish	*Deania calcea*	...	...	36	17	28	66	86
Kitefin shark	*Dalatias licha*	175	352	434	328	317	375	520
Dogfish sharks nei	*Squalidae*	693	1 705	701	1 010	770	705	325
Antarctic starry skate	*Raja georgiana*	-	-	-	11	36	7	24
Eaton's skate	*Bathyraja eatonii*	-	-	-	1	5	0	1
Bathyraja rays nei	*Bathyraja spp*	-	-	-	1	-	-	-
Rays and skates nei	*Rajidae*	-	-	-	6	-	-	-
Eagle rays	*Myliobatidae*	0	1	1	2	2	5	9
Rays, stingrays, mantas nei	*Rajiformes*	1 582	2 227	2 318	2 821	2 634	2 784	2 648
Dark ghost shark	*Hydrolagus novaezealandiae*	1 614	2 064	1 956	1 975	1 819	1 572	2 055
Ratfishes nei	*Hydrolagus spp*	...	0	36	453	975	2 184	1 901
Ghost shark	*Callorhinchus milii*	595	913	951	1 260	1 228	1 189	1 085
Chimaeras, etc. nei	*Chimaeriformes*	49	5	5	21	40	76	103
Sharks, rays, skates, etc. nei	*Elasmobranchii*	2 375	1 580	673	1 062	6	-	13
Marine fishes nei	*Osteichthyes*	26 526	26 264	54	74	122	68	125
Marine crabs nei	*Brachyura*	294	359	393	407	355	360	292
Green rock lobster	*Jasus verreauxi*	64	54	16	13	35	10	8
Red rock lobster	*Jasus edwardsii*	3 121	5 009	2 707	2 818	2 789	2 551	2 485
Slipper lobsters nei	*Scyllaridae*	0	0	4	0	1	4	2
New Zealand lobster	*Metanephrops challengeri*	670	1 093	989	925	1 034	1 093	1 020
Subantarctic stone crab	*Lithodes murrayi*	-	-	-	-	-	-	0
Red stone crab	*Paralomis aculeata*	-	-	0	-	-	-	-
King crabs, stone crabs nei	*Lithodidae*	-	-	0	-	-	-	-
Natantian decapods nei	*Natantia*	2	23	1	-	-	0	-
Abalones nei	*Haliotis spp*	1 020	1 180	1 300	1 170	1 265	1 064	1 090
New Zealand dredge oyster	*Ostrea lutaria*	1 931	2 172	1 000	995	766	832	816
Pacific cupped oyster	*Crassostrea gigas*	2	2	0	62	0	0	-
Sea mussels nei	*Mytilidae*	450	0	664	2 977	4 467	2 270	1 725
New Zealand scallop	*Pecten novaezelandiae*	5 080	18 848	4 592	6 152	2 912	6 792	4 408
Delicate scallop	*Zygochlamis delicatula*	124	201	91	128	0	222	118
Stutchbury's venus	*Chione stutchburyi*	815	541	1 325	1 396	1 789	1 748	1 687
Short neck clams nei	*Paphia spp*	211	114	204	181	131	202	212
Wellington flying squid	*Nototodarus sloani*	23 474	44 845	42 541	27 282	20 878	35 100	50 016
Various squids nei	*Loliginidae, Ommastrephidae*	7	17	27	48	74	45	41
Octopuses, etc. nei	*Octopodidae*	227	267	167	148	119	140	100
Marine molluscs nei	*Mollusca*	16	4	-	-	2	1	15
Echinoderms	*Echinodermata*	277	627	832	643	712	853	738
Starfishes nei	*Asteroidea*	4	0	13	2	6	11	39
Sea cucumbers nei	*Holothurioidea*	1	0	-	-	-	0	1
Aquatic invertebrates nei	*Invertebrata*	-	-	-	-	-	-	1
	Country total	*425 564*	*612 424*	*640 236*	*599 466*	*549 318*	*561 940*	*559 289*

E-6

Fish crustaceans, molluscs, etc — **Capture production by countries or areas and species** — **Oceania**
Poissons, crustacés, mollusques, etc — **Captures par pays ou zones et espèces** — **Océanie**
Peces, crustáceos, moluscos, etc — **Capturas por países o áreas y especies** — **Oceanía**

English name Nom anglais Nombre inglés	Scientific name Nom scientifique Nombre científico	1996 t	1997 t	1998 t	1999 t	2000 t	2001 t	2002 t
Niue								
Freshwater fishes nei	*Osteichthyes*	0	0	0	0	0	0	0
Marine fishes nei	*Osteichthyes*	200 F	200 F	200 F	200 F	200 F	200 F	200 F
	Country total	*200 F*	*200 F*	*200 F*	*200 F*	*200 F*	*200 F*	*200 F*
Norfolk Is								
Freshwater fishes nei	*Osteichthyes*	0	0	0	0	0	0	0
Marine fishes nei	*Osteichthyes*	0	0	0	0	0	0	0
	Country total	*0*	*0*	*0*	*0*	*0*	*0*	*0*
N Marianas								
Freshwater fishes nei	*Osteichthyes*	0	0	0	0	0	0	-
Groupers nei	*Epinephelus spp*	3	5	3	2	2	3	2
Snappers, jobfishes nei	*Lutjanidae*	10	15	7	16	5	11	10
Emperors(=Scavengers) nei	*Lethrinidae*	5	13	50	4	4	8	3
Goatfishes, red mullets nei	*Mullidae*	12	9	1	1	1	1	0
Parrotfishes nei	*Scaridae*	3	7	1	2	4	13	2
Surgeonfishes nei	*Acanthuridae*	3	-	0	3	3	10	2
Spinefeet(=Rabbitfishes) nei	*Siganus spp*	...	...	2	3	5	4	4
Wahoo	*Acanthocybium solandri*	5	4	2	4	2	2	4
Skipjack tuna	*Katsuwonus pelamis*	75	64	61	48	56	61	73
Yellowfin tuna	*Thunnus albacares*	17	11	5	11	7	7	12
Blue marlin	*Makaira nigricans*	3	4	2	2	2	1	1
Tuna-like fishes nei	*Scombroidei*	7	7	8	8	7	3	6
Scads nei	*Decapterus spp*	2	4	0	5	10	13	10
Carangids nei	*Carangidae*	1	2	2	2	1	3	2
Common dolphinfish	*Coryphaena hippurus*	16	17	9	6	3	6	9
Marine fishes nei	*Osteichthyes*	61	88	80	75	75	49	55
Tropical spiny lobsters nei	*Panulirus spp*	2	0	2	1	2	2	2
Natantian decapods nei	*Natantia*	-	-	-	-	-	-	0
Octopuses, etc. nei	*Octopodidae*	0	-	0	0	-	0	1
Marine turtles nei	*Testudinata*	0	-	-	-	-	-	-
	Country total	*225*	*250*	*235*	*193*	*189*	*197*	*198*
Palau								
Freshwater fishes nei	*Osteichthyes*	0	0	0	0	0	0	0
Milkfish	*Chanos chanos*	1	0	1	0	1	1	0
Mullets nei	*Mugilidae*	1	1	2	0	1	0	0
Groupers nei	*Epinephelus spp*	7	1	2	4	2	10	5
Snappers, jobfishes nei	*Lutjanidae*	15	4	11	6	3	3	4
Emperors(=Scavengers) nei	*Lethrinidae*	12	6	17	8	6	9	3
Goatfishes, red mullets nei	*Mullidae*	2	1	2	1	1	1	0
Mojarras(=Silver-biddies) nei	*Gerres spp*	1	1	1	0	1	0	0
Sea chubs nei	*Kyphosus spp*	3	2	2	1	1	1	1
Wrasses, hogfishes, etc. nei	*Labridae*	0	0	0	2	5	7	11
Parrotfishes nei	*Scaridae*	25	18	12	12	12	26	12
Surgeonfishes nei	*Acanthuridae*	42	10	7	13	13	46	36
Spinefeet(=Rabbitfishes) nei	*Siganus spp*	14	5	4	7	14	11	6
Wahoo	*Acanthocybium solandri*	2	1	1	1	0	2	1
Narrow-barred Spanish mackerel	*Scomberomorus commerson*	1	0	1	0	0	1	0
Kawakawa	*Euthynnus affinis*	1	1	3	1	2	1	0
Yellowfin tuna	*Thunnus albacares*	2	0	1	1	42	34	2
Bigeye tuna	*Thunnus obesus*	...	...	...	...	49	19	...
Marlins,sailfishes,etc. nei	*Istiophoridae*	1	0	0	0	1	4	0
Swordfish	*Xiphias gladius*	...	...	...	...	1	2	...
Tuna-like fishes nei	*Scombroidei*	3	1	1	1	1	2	51 F
Carangids nei	*Carangidae*	14	4	10	4	6	3	2
Common dolphinfish	*Coryphaena hippurus*	0	0	0	0	0	0	0
Indian mackerel	*Rastrelliger kanagurta*	1	0	1	1	0	0	0
Barracudas nei	*Sphyraena spp*	3	1	1	1	0	0	0
Marine fishes nei	*Osteichthyes*	846	845	868	886	878	883	868
Indo-Pacific swamp crab	*Scylla serrata*	1	1	2	8	6	8	5
Tropical spiny lobsters nei	*Panulirus spp*	1	1	1	1	1	1	1
Sea cucumbers nei	*Holothurioidea*	0	0	0	0	0	0	0
Aquatic invertebrates nei	*Invertebrata*	0	0	0	1	0	0	0
	Country total	*999*	*904*	*951*	*960*	*1 047*	*1 075*	*1 008 F*
Papua N Guin								
Mozambique tilapia	*Oreochromis mossambicus*	2 310 F	2 310 F	2 310 F	2 310 F	2 310 F	2 310 F	2 310 F
Gudgeons, sleepers nei	*Eleotridae*	1 850 F	1 850 F	1 850 F	1 850 F	1 850 F	1 850 F	1 850 F
Freshwater fishes nei	*Osteichthyes*	6 649 F	6 649 F	6 641 F	6 649 F	6 655 F	6 654 F	6 654 F
Diadromous clupeoids nei	*Clupeoidei*	480 F	480 F	480 F	480 F	480 F	480 F	480 F
Barramundi(=Giant seaperch)	*Lates calcarifer*	356 F	393 F	395 F	508 F	423 F	499 F	480 F
Sea catfishes nei	*Ariidae*	1 850 F	1 850 F	1 850 F	1 850 F	1 850 F	1 850 F	1 850 F
Kawakawa	*Euthynnus affinis*	0	-	-	-	-	-	-
Skipjack tuna	*Katsuwonus pelamis*	9 512	11 270	37 214	29 949	52 289	64 355	89 948
Longtail tuna	*Thunnus tonggol*	0	-	-	-	-	-	-
Albacore	*Thunnus alalunga*	38	101	104	129	159	123	136
Yellowfin tuna	*Thunnus albacares*	971	6 968	11 924	7 875	14 659	25 779	27 912
Bigeye tuna	*Thunnus obesus*	50	1 060	1 486	1 028	1 539	5 004	3 583
Marlins,sailfishes,etc. nei	*Istiophoridae*	16	6	87	230	418	368	340 F
Marine fishes nei	*Osteichthyes*	10 500 F	10 500 F	10 000 F	10 000 F	10 000 F	10 000 F	10 000 F
River prawns nei	*Macrobrachium spp*	5 F	5 F	5 F	5 F	5 F	5 F	5 F
Oceanian crayfishes nei	*Parastacidae*	6	6	14	6	0	1	1 F

English name Nom anglais Nombre inglés	Scientific name Nom scientifique Nombre científico	1996 t	1997 t	1998 t	1999 t	2000 t	2001 t	2002 t
Indo-Pacific swamp crab	*Scylla serrata*	25 F	24 F	23 F	22 F	24 F	24 F	24 F
Tropical spiny lobsters nei	*Panulirus spp*	165	205	217	203	197	131	130 F
Banana prawn	*Penaeus merguiensis*	820	676	1 233	949	1 136	1 017	1 000 F
Giant tiger prawn	*Penaeus monodon*	109	99	209	164	126	117	117 F
Metapenaeus shrimps nei	*Metapenaeus spp*	219	159	187	279	328	346	320 F
Natantian decapods nei	*Natantia*	104	59	50	99	135	117	110 F
Sea cucumbers nei	*Holothurioidea*	1 788	1 515	2 037	1 185	1 824	1 453	1 450 F
	Country total	*37 823 F*	*46 185 F*	*78 316 F*	*65 770 F*	*96 407 F*	*122 483 F*	*148 700 F*
Pitcairn Is								
Freshwater fishes nei	*Osteichthyes*	0	0	0	0	0	0	0
Marine fishes nei	*Osteichthyes*	8 F	8 F	8 F	8 F	8 F	8 F	8 F
	Country total	*8 F*	*8 F*	*8 F*	*8 F*	*8 F*	*8 F*	*8 F*
Samoa								
Freshwater fishes nei	*Osteichthyes*	0	0	0	0	1	1 F	1 F
Albacore	*Thunnus alalunga*	1 775	4 108	4 742	4 027	4 067	4 820	4 360
Yellowfin tuna	*Thunnus albacares*	573	1 327	801	681	1 120	470	388
Bigeye tuna	*Thunnus obesus*	27	63	334	283	177	185	153
Tuna-like fishes nei	*Scombroidei*	-	-	-	-	479	470 F	470 F
Sharks, rays, skates, etc. nei	*Elasmobranchii*	...	...	...	...	20	20 F	20 F
Marine fishes nei	*Osteichthyes*	322	1 513	1 640	5 163	5 289	5 200 F	5 200 F
Marine crustaceans nei	*Crustacea*	10	10	10	30	207	200 F	200 F
Marine molluscs nei	*Mollusca*	20	20	20	20	1 644	1 600 F	1 600 F
	Country total	*2 727*	*7 041*	*7 547*	*10 204*	*13 004*	*12 966 F*	*12 392 F*
Solomon Is								
Freshwater fishes nei	*Osteichthyes*	0	0	0	0	0	0	0
Skipjack tuna	*Katsuwonus pelamis*	26 485	36 311	38 662	35 613	8 368	12 530	13 897
Albacore	*Thunnus alalunga*	100	109	370	136	224	54	115
Yellowfin tuna	*Thunnus albacares*	11 003	9 588	8 114	8 843	3 334	4 692	3 913
Bigeye tuna	*Thunnus obesus*	1 109	1 434	1 232	1 070	577	576	653
Marlins,sailfishes,etc. nei	*Istiophoridae*	50 F	50 F	50 F	50 F	50 F	50 F	50 F
Sharks, rays, skates, etc. nei	*Elasmobranchii*	50 F	4 000 F	600 F	310 F	300 F	300 F	300 F
Marine fishes nei	*Osteichthyes*	12 000 F	12 000 F	12 000 F	12 000 F	12 000 F	12 000 F	12 000 F
Banana prawn	*Penaeus merguiensis*	20 F	30 F	40 F	20 F	20 F	20 F	20 F
Abalones nei	*Haliotis spp*	-	-	1	-	-	-	-
Gastropods nei	*Gastropoda*	-	-	0	0	-	-	-
Clams, etc. nei	*Bivalvia*	60	280 F	10 F	10 F	5 F	5 F	5 F
Sea cucumbers nei	*Holothurioidea*	113	203	253	376	48	50 F	50 F
	Country total	*50 990 F*	*64 005 F*	*61 332 F*	*58 428 F*	*24 926 F*	*30 277 F*	*31 003 F*
Tokelau								
Freshwater fishes nei	*Osteichthyes*	0	0	0	0	0	0	0
Marine fishes nei	*Osteichthyes*	200 F	200 F	200 F	200 F	200 F	200 F	200 F
	Country total	*200 F*	*200 F*	*200 F*	*200 F*	*200 F*	*200 F*	*200 F*
Tonga								
Freshwater fishes nei	*Osteichthyes*	0	0	0	0	0	0	0
Skipjack tuna	*Katsuwonus pelamis*	2	4	7	3	2	12	23
Albacore	*Thunnus alalunga*	431	493	616	801	862	1 268	1 199
Yellowfin tuna	*Thunnus albacares*	88	100	125	163	175	259	262
Bigeye tuna	*Thunnus obesus*	60	69	86	112	120	191	219
Black marlin	*Makaira indica*	20	10	14	6	13	10	37
Swordfish	*Xiphias gladius*	7	6	8	5	53	8	42
Tuna-like fishes nei	*Scombroidei*	1	8	14	34	42	85	35
Marine fishes nei	*Osteichthyes*	2 100	2 000	2 936	2 890	2 305	2 548	2 612
Marine crustaceans nei	*Crustacea*	120	100	177	200	175	270	337
Marine molluscs nei	*Mollusca*	...	1	3	7	13	22	38
Sea cucumbers nei	*Holothurioidea*	86	80	90	0	0	0	0
	Country total	*2 915*	*2 871*	*4 076*	*4 221*	*3 760*	*4 673*	*4 804*
Tuvalu								
Freshwater fishes nei	*Osteichthyes*	0	0	0	0	0	0	0
Skipjack tuna	*Katsuwonus pelamis*	260 F	300 F	300 F	300 F	300 F	300 F	300 F
Yellowfin tuna	*Thunnus albacares*	15 F	20 F	20 F	20 F	20 F	20 F	20 F
Tuna-like fishes nei	*Scombroidei*	15 F	20 F	20 F	20 F	20 F	20 F	20 F
Marine fishes nei	*Osteichthyes*	110 F	160 F	160 F	160 F	160 F	160 F	160 F
	Country total	*400 F*	*500 F*	*500 F*	*500 F*	*500 F*	*500 F*	*500 F*
US Minor Is								
Marine fishes nei	*Osteichthyes*	0	0	0	0	0	0	0
	Country total	*0*	*0*	*0*	*0*	*0*	*0*	*0*
Vanuatu								
Freshwater fishes nei	*Osteichthyes*	0	0	0	0	0	0	0
Porgies, seabreams nei	*Sparidae*	0	-	0	-	-	-	-
Black skipjack	*Euthynnus lineatus*	-	-	10	-	-	-	19
Skipjack tuna	*Katsuwonus pelamis*	19 600	30 860	40 112	57 151	43 032	6 901	6 230
Albacore	*Thunnus alalunga*	192	95	10	...	...	...	225
Yellowfin tuna	*Thunnus albacares*	13 182	31 267	28 243	28 477	17 586	10 936	5 808

English name Nom anglais Nombre inglés	Scientific name Nom scientifique Nombre científico	1996 t	1997 t	1998 t	1999 t	2000 t	2001 t	2002 t
Bigeye tuna	*Thunnus obesus*	10 231	6 107	4 143	5 099	5 978	4 284	2 457
Marlins,sailfishes,etc. nei	*Istiophoridae*	98	97	99	14	...	...	...
Marine fishes nei	*Osteichthyes*	1 209 F	1 327 F	1 310 F	1 400 F	1 400 F	1 500 F	1 500 F
Penaeus shrimps nei	*Penaeus spp*	170	150	70	-	-	-	-
Natantian decapods nei	*Natantia*	-	-	0	-	-	-	-
Marine crustaceans nei	*Crustacea*	250 F	250 F	250 F	250 F	250 F	250 F	250 F
Cuttlefish,bobtail squids nei	*Sepiidae, Sepiolidae*	33	19	11	-	-	-	-
Octopuses, etc. nei	*Octopodidae*	-	-	0	-	-	-	-
Marine molluscs nei	*Mollusca*	600 F	600 F	600 F	600 F	600 F	600 F	600 F
Sea cucumbers nei	*Holothurioidea*	50 F	50 F	50 F	50 F	50 F	50 F	50 F
	Country total	*45 615 F*	*70 822 F*	*74 908 F*	*93 041 F*	*68 896 F*	*24 521 F*	*17 139 F*
Wallis Fut I								
Freshwater fishes nei	*Osteichthyes*	0	0	0	0	0	0	0
Marine fishes nei	*Osteichthyes*	174	170	294	294 F	294 F	294 F	294 F
Marine crabs nei	*Brachyura*	1	1	1	1 F	1 F	1 F	1 F
Tropical spiny lobsters nei	*Panulirus spp*	2	2	2	2 F	2 F	2 F	2 F
Octopuses, etc. nei	*Octopodidae*	1	1	1	1 F	1 F	1 F	1 F
Marine turtles nei	*Testudinata*	2	2	2	2 F	2 F	2 F	2 F
	Country total	*180*	*176*	*300*	*300 F*	*300 F*	*300 F*	*300 F*
Total		***851 275***	***1 086 361***	***1 169 941***	***1 164 593***	***1 065 562***	***1 107 699***	***1 130 792***

E-9 Fish crustaceans, molluscs, etc / Poissons, crustacés, mollusques, etc / Peces, crustáceos, moluscos, etc

Capture production by countries or areas and species / Captures par pays ou zones et espèces / Capturas por países o áreas y especies

Other nei / Autres nca / Otros nep

English name Nom anglais Nombre inglés	Scientific name Nom scientifique Nombre científico	1996 t	1997 t	1998 t	1999 t	2000 t	2001 t	2002 t
Other nei								
Benguela hake	*Merluccius polli*	-	-	-	8	44	139	504
Cape hakes	*Merluccius capensis,M.paradox.*	1	-	-	-	-	-	-
Croakers, drums nei	*Sciaenidae*	-	-	-	-	-	12	...
Dentex nei	*Dentex spp*	462	-	-	-	-	-	-
Porgies, seabreams nei	*Sparidae*	-	-	-	63	49	341	114
Largehead hairtail	*Trichiurus lepturus*	-	-	-	147	948	660	301
Sardinellas nei	*Sardinella spp*	-	-	-	5 812	2 306	38 577	31 211
European pilchard(=Sardine)	*Sardina pilchardus*	-	-	-	226	1 876	3 949	7 677
European anchovy	*Engraulis encrasicolus*	-	-	-	3 560	4 419	4 068	9 931
Atlantic bonito	*Sarda sarda*	300	75	-	-	-	-	-
Narrow-barred Spanish mackerel	*Scomberomorus commerson*	3 060	3 060	-	-	-	-	-
Seerfishes nei	*Scomberomorus spp*	431	431	-	-	-	-	-
Frigate tuna	*Auxis thazard*	80	120	309	491	291	420	186
Frigate and bullet tunas	*Auxis thazard, A.rochei*	100	100	-	18	367	110	513
Little tunny(=Atl.black skipj)	*Euthynnus alletteratus*	200	200	200	200	-	-	33
Black skipjack	*Euthynnus lineatus*	-	-	-	-	-	32	156
Kawakawa	*Euthynnus affinis*	140	140	-	-	-	-	-
Skipjack tuna	*Katsuwonus pelamis*	45 050	43 096	50 484	60 271	69 316	63 497	83 795
Atlantic bluefin tuna	*Thunnus thynnus*	244	1 375	1 921	2 564	396	620	508
Longtail tuna	*Thunnus tonggol*	351	351	-	-	85	-	-
Albacore	*Thunnus alalunga*	8 766	5 822	10 954	11 034	11 589	6 718	6 689
Southern bluefin tuna	*Thunnus maccoyii*	295	333	476	483	49	80	80
Yellowfin tuna	*Thunnus albacares*	57 783	51 254	54 640	59 061	65 787	77 463	89 154
Bigeye tuna	*Thunnus obesus*	34 733	38 131	52 114	54 683	43 566	31 773	27 804
Atlantic sailfish	*Istiophorus albicans*	40	40	-	-	-	-	-
Blue marlin	*Makaira nigricans*	1 257	1 240	1 787	1 373	1 446	538	538
Black marlin	*Makaira indica*	208	113	220	173	155	62	62
Striped marlin	*Tetrapturus audax*	1 443	743	1 089	836	838	323	323
Atlantic white marlin	*Tetrapturus albidus*	100	100	-	-	-	-	-
Marlins,sailfishes,etc. nei	*Istiophoridae*	143	344	357	294	310	149	149
Swordfish	*Xiphias gladius*	8 549	6 064	8 104	7 084	7 833	3 065	3 065
Tuna-like fishes nei	*Scombroidei*	80	20	41	40	751	2	6
Cape horse mackerel	*Trachurus capensis*	9 805	-	-	-	-	-	-
Jack and horse mackerels nei	*Trachurus spp*	-	-	-	3 375	8 994	12 012	58 030
Chub mackerel	*Scomber japonicus*	-	-	-	781	4 002	4 845	15 193
Sharks, rays, skates, etc. nei	*Elasmobranchii*	489	359	288	194	231	560	560
Marine fishes nei	*Osteichthyes*	13	202	-	740	675	648	1 012
Penaeus shrimps nei	*Penaeus spp*	-	-	-	19	31	-	-
Natantian decapods nei	*Natantia*	-	-	-	-	17	-	-
Cuttlefish,bobtail squids nei	*Sepiidae, Sepiolidae*	-	-	-	138	154	148	-
Various squids nei	*Loliginidae, Ommastrephidae*	-	-	-	79	91	77	-
Octopuses, etc. nei	*Octopodidae*	-	-	-	196	245	264	-
	Country total	*174 123*	*153 713*	*182 984*	*213 943*	*226 861*	*251 152*	*337 594*
Total		***174 123***	***153 713***	***182 984***	***213 943***	***226 861***	***251 152***	***337 594***

Notes on individual countries or areas

Notes sur divers pays ou zones

Notas sobre los distintos países o áreas

Notes on individual countries or areas

ALGERIA

Data concerning tuna catches are reviewed in collaboration with the International Commission for the Conservation of Atlantic Tunas (ICCAT), the regional agency concerned with tuna statistics.

AMERICAN SAMOA

Data concerning tuna catches are reviewed in collaboration with the Secretariat of the Pacific Community (SPC), the regional agency concerned with tuna statistics.

ANGOLA

Capture production in Angolan waters by chartered foreign vessels have not been included pending further information on the nationality to which they should be attributed.

Data concerning tuna catches are reviewed in collaboration with ICCAT, the regional agency concerned with tuna statistics.

ANTIGUA AND BARBUDA

Data for 'Stromboid conchs nei' since 1993 have been converted to live weight equivalents by using the conversion factor '7.5'.

AUSTRALIA

Data refer to a split-year (1 July – 30 June) shown under the calendar year in which the split-year ends.

Catch data, excluding those for tuna species, for the Southwest Atlantic area are provided by the Fisheries Department, Falkland Islands Government.

Data concerning tuna catches in the Indian Ocean are reviewed in collaboration with IOTC, the regional agency concerned with tuna statistics.

AUSTRIA

Excludes recreational fisheries.

BAHAMAS

Data for Caribbean spiny lobster and for groupers have been converted to live weight equivalents by using the conversion factors '3' and '2.5' respectively.

BANGLADESH

Data refer to a split-year (1 July – 30 June) shown under the calendar year in which the split-year ends.

Data concerning tuna catches are reviewed in collaboration with the Indian Ocean Tuna Commission (IOTC), the regional agency concerned with tuna statistics.

Notes sur divers pays ou zones

ALGÉRIE

Les données concernant les captures des thonidés sont révisées en collaboration avec la Commission internationale pour la conservation des thonidés de l'Atlantique (CICTA), l'organisation régionale chargée des statistiques des thonidés.

SAMOA AMÉRICAINES

Les données concernant les captures des thonidés sont révisées en collaboration avec le Secrétariat général de la Communauté du Pacifique (CPS), l'organisation régionale chargée des statistiques des thonidés.

ANGOLA

Les captures dans les eaux angolaises des bateaux de pêche étrangers n'ont pas été incluses dans l'attente d'un complément d'information sur les pays auxquels elles doivent être attribuées.

Les données concernant les captures des thonidés sont révisées en collaboration avec la CICTA, l'organisation régionale chargée des statistiques des thonidés.

ANTIGUA-ET-BARBUDA

Les données relatives aux 'Strombes nca' depuis 1993 ont été converties en équivalents poids vif en utilisant le facteur de conversion '7,5'.

AUSTRALIE

Les données se réfèrent à une année fractionnée (1er juillet - 30 juin) et figurent sous l'année civile durant laquelle se termine l'année fractionnée.

Les données sur les captures, à l'exclusion de celles concernant le thon, pour l'Atlantique sud-ouest, sont fournies par le Département des pêches du Gouvernement des Îles Falkland.

Les données concernant les captures des thonidés dans l'océan Indien sont révisées en collaboration avec la CTOI, l'organisation régionale chargée des statistiques des thonidés.

AUTRICHE

Non compris la pêche récréative.

BAHAMAS

Les données relatives aux langoustes des Caraïbes et aux mérous ont été converties en équivalents poids vif en utilisant respectivement les facteurs de conversion '3' et '2,5'.

BANGLADESH

Les données se réfèrent à une année fractionnée (1er juillet - 30 juin) et figurent sous l'année civile durant laquelle se termine l'année fractionnée.

Les données concernant les captures des thonidés sont révisées en collaboration avec la Commission des thons de l'océan Indien (CTOI), l'organisation régionale chargée des statistiques des thonidés.

Notas sobre los distintos países o áreas

ARGELIA

Los datos relativos a las capturas de atún se revisan en colaboración con la Comisión Internacional para la Conservación del Atún Atlántico (CICAA), el organismo regional encargado de las estadísticas atuneras.

SAMOA AMERICANA

Los datos relativos a las capturas de atún se revisan en colaboración con la Secretaría general de la Comunidad del Pacífico (SCP), el organismo regional encargado de las estadísticas atuneras.

ANGOLA

No se han incluido capturas en aguas de Angola por buques fletados extranjeros, en espera de más información sobre la nacionalidad que se les debe atribuir.

Los datos relativos a las capturas de atún se revisan en colaboración con la CICAA, el organismo regional encargado de las estadísticas atuneras.

ANTIGUA Y BARBUDA

Los datos de 'Cobos nep' desde 1993 se han convertido en sus equivalentes de peso en vivo aplicando el factor de conversión '7,5'.

AUSTRALIA

Los datos se refieren a un año emergente (1 de julio - 30 de junio) que se indica como el año civil en que finaliza el año emergente.

Los datos relativos a las capturas en la zona del Atlántico Sudoccidental, con exclusión de las de atunes, fueron proporcionados por el Departamento de Pesca del Gobierno de las Islas Malvinas (Falkland).

Los datos relativos a las capturas de atún en el océano Índico se revisan en colaboración con la CAOI, el organismo regional encargado de las estadísticas atuneras.

AUSTRIA

Se excluye la pesca recreativa.

BAHAMAS

Los datos de la langosta del Caribe y de los meros se han convertido en sus equivalentes de peso en vivo aplicando respectivamente los factores de conversión '3' y '2,5'.

BANGLADESH

Los datos se refieren a un año emergente (1 de julio - 30 de junio) que se indica como el año civil en que finaliza el año emergente.

Los datos relativos a las capturas de atún se revisan en colaboración con la Comisión del Atún para el Océano Índico (CAOI), el organismo regional encargado de las estadísticas atuneras.

Notes on individual countries or areas

BELIZE

Data concerning tuna catches are reviewed in collaboration with the Inter-American Tropical Tuna Commission (IATTC) and ICCAT, the regional agencies concerned with tuna statistics.

Catch data, excluding those for tuna species, for the Eastern Central and Southeast Atlantic areas are provided by the Las Palmas Survey (LPS) and the "Bulletin Statistique" published by IMROP, Mauritania.

Catch data, excluding those for tuna species, for the Southwest Atlantic area are provided by the Fisheries Department, Falkland Islands Government.

Data reported since 1980 as "fish fillet", "lobster tail", "crab claws", and "conch" have been converted to live weight equivalents by using the conversion factors of '2', '2', '4' and '7.5' respectively.

BENIN

Data concerning tuna catches are reviewed in collaboration with ICCAT, the regional agency concerned with tuna statistics.

BHUTAN

Country does not submit returns. Data estimated on the basis of reports of visiting missions and other documentation.

BOTSWANA

Submission of capture data resumed with 2000 statistics after twelve years of not reporting. Data before 1996 were probably overestimated.

BRAZIL

Since 1995, 100,000 tonnes of estimated subsistence catches have been included under 'Marine fishes nei'.

Data concerning tuna catches are reviewed in collaboration with ICCAT, the regional agency concerned with tuna statistics.

BRITISH VIRGIN ISLANDS

Include Anegada, Jost Van Dyke, Tortola and Virgin Gorda.

In 1999, data reported since 1976 in thousands of pounds instead of tonnes have been revised.

BULGARIA

Inland water 1994-96 catches refer only to Danube River.

Catch data, excluding those for tuna species, for the Eastern Central Atlantic area have also been taken from the "Bulletin Statistique" published by IMROP, Mauritania.

BURUNDI

Decreased catch for 1996 due to limited period of fishing in the Lake Tanganyka (only two months).

Notes sur divers pays ou zones

BELIZE

Les données concernant les captures des thonidés sont révisées en collaboration avec la Commission interaméricaine du thon tropical (CITT) et la CICTA, les organisations régionales chargées des statistiques des thonidés.

Les données sur les captures, à l'exclusion de celles concernant le thon, pour l'Atlantique du centre-est et du sud-est sont tirées de l'Enquête Las Palmas et du "Bulletin statistique" publié par l'IMROP, Mauritanie.

Les données sur les captures, à l'exclusion de celles concernant le thon, pour l'Atlantique sud-ouest, sont fournies par le Département des pêches du Gouvernement des Îles Falkland.

Les données rapporté depuis 1980 comme "fish fillet", "lobster tail", "crab claws" et "conch" ont été converties en équivalents poids vif en utilisant respectivement les facteurs de conversion '2' '2', '4' et '7.5'.

BÉNIN

Les données concernant les captures des thonidés sont révisées en collaboration avec la CICTA, l'organisation régionale chargée des statistiques des thonidés.

BHOUTAN

Pays non déclarant. Données estimées à partir de rapports de missions et d'autres documents.

BOTSWANA

L'envoi des données de captures a repris avec les statistiques de 2000, après douze ans de non-rapport. Les données avant 1996 ont été probablement surestimées.

BRÉSIL

Depuis 1995, 100 000 tonnes des estimations de captures de la pêche de subsistance sont incluses dans 'Poissons marins nca'.

Les données concernant les captures des thonidés sont révisées en collaboration avec la CICTA, l'organisation régionale chargée des statistiques des thonidés.

ÎLES VIERGES BRITANNIQ.

Comprend Anegada, Jost Van Dyke, Tortola et Virgin Gorda.

En 1999 il a été procédé à une révision des séries de données commençant en 1976, car celles-ci étaient auparavant présentées en milliers de livres plutôt qu'en tonnes.

BULGARIE

Les données de 1994-96 concernant les captures d'eau douce se rapportent uniquement au fleuve Danube.

Les données sur les captures, à l'exclusion de celles concernant le thon, pour l'Atlantique centre-est ont également été tirées du "Bulletin statistique" publié par l'IMROP, Mauritanie.

BURUNDI

La diminution des captures enregistrées en 1996 est due à la durée limitée de pêche dans le lac Tanganyika (la pêche n'a été ouverte que pendant deux mois).

Notas sobre los distintos países o áreas

BELICE

Los datos relativos a las capturas de atún se revisan en colaboración con la Comisión Interamericana del Atún Tropical (CIAT) y la CICAA, los organismos regionales encargados de las estadísticas atuneras.

Los datos relativos a las capturas en las zonas del Atlántico centro-oriental y Sudoriental, con exclusión de las de atunes, fueron proporcionados por Las Palmas Survey (LPS) y por la publicación "Bulletin Statistique" del IMROP, Mauritania.

Los datos relativos a las capturas en la zona del Atlántico Sudoccidental, con exclusión de las de atunes, fueron proporcionados por el Departamento de Pesca del Gobierno de las Islas Malvinas (Falkland).

Los datos proporcionados desde 1980 como "fish fillet", "lobster tail", "crab claws" y "conch" se han convertido en sus equivalentes de peso en vivo aplicando respectivamente los factores de conversión '2' '2', '4' et '7.5'.

BENIN

Los datos relativos a las capturas de atún se revisan en colaboración con la CICAA, el organismo regional encargado de las estadísticas atuneras.

BHUTÁN

El país no proporciona información. Datos estimados sobre la base de informes de misiones al país y de otro tipo de documentación.

BOTSWANA

El envío de datos de capturas empezó de nuevo en 2000 después de doce años de no envíos. Los datos antes de 1996 son en exceso.

BRASIL

Desde 1995, 100 000 toneladas métricas de capturas de la pesca de subsistencia se incluyen en 'Peces marinos nep'.

Los datos relativos a las capturas de atún se revisan en colaboración con la CICAA, el organismo regional encargado de las estadísticas atuneras.

ISLAS VÍRGENES BRITÁN.

Incluye Anegada, Jost Van Dyke, Tortola y Virgin Gorda.

En 1999 se revisaron los datos a partir de 1976, notificados anteriormente a la FAO en miles de libras en lugar de toneladas.

BULGARIA

Los datos de capturas de 1994-96 relativos a la pesca en aguas continentales se refieren sólo al río Danubio.

Los datos sobre las capturas de la zona del Atlántico centro-oriental, con exclusión de las de atunes, también se han extraído de la publicación "Bulletin Statistique" del IMROP, Mauritania.

BURUNDI

La disminución de captura registrada en 1996 se debe a la duración limitada de la pesca en el lago Tanganica (la pesca se permitió solamente por dos meses).

Notes on individual countries or areas

Notes sur divers pays ou zones

Notas sobre los distintos países o áreas

CAMBODIA

Due to the introduction of a new methodology to estimate production of semi-commercial and subsistence fisheries, inland water statistics since 1999 are not comparable with those of previous years.

Catch data for the Southwest Atlantic area are provided by the Fisheries Department, Falkland Islands Government and tuna catches are reviewed in collaboration with ICCAT.

CAMBODGE

Du fait de l'introduction d'une nouvelle méthodologie d'estimation de la production des pêches semi-commerciales et de subsistance, les statistiques des pêches dans les eaux intérieures depuis 1999 ne sont plus comparables avec celles des années précédentes.

Les données sur les captures dans l'Atlantique sud-ouest sont fournies par le Département des pêches du Gouvernement des Îles Falkland et les captures des thonidés sont révisées en collaboration avec la CICTA.

CAMBOYA

Debido a la introducción de una nueva metodología para calcular la producción de las pesquerías semicomerciales y de subsistencia, las estadísticas de la pesca continental desde el año 1999 no son comparables con las de años anteriores.

Los datos relativos a las capturas en la zona del Atlántico Sudoccidental fueron proporcionados por el Departamento de Pesca del Gobierno de las Islas Malvinas (Falkland) y las capturas de atún se revisaron en colaboración con la CICAA.

CANADA

Data concerning tuna catches are reviewed in collaboration with ICCAT and SPC, the regional agencies concerned with tuna statistics.

CANADA

Les données concernant les captures des thonidés sont révisées en collaboration avec la CICTA et la CPS, les organisations régionales chargées des statistiques des thonidés.

CANADÁ

Los datos relativos a las capturas de atún se revisan en colaboración con la CICAA y la SCP, los organismos regionales encargados de las estadísticas atuneras.

CAPE VERDE

Data concerning tuna catches are reviewed in collaboration with ICCAT, the regional agency concerned with tuna statistics.

CAP-VERT

Les données concernant les captures des thonidés sont révisées en collaboration avec la CICTA, l'organisation régionale chargée des statistiques des thonidés.

CABO VERDE

Los datos relativos a las capturas de atún se revisan en colaboración con la CICAA, el organismo regional encargado de las estadísticas atuneras.

CAYMAN IS

Data for the Eastern Central Atlantic refer to catches of vessels fishing with flag of convenience.

ÎLES CAÏMANES

Les données dans l'Atlantique centre-est se rapportent aux captures effectuées sous pavillon de complaisance.

IS CAIMÁN

Los datos en el Atlántico centro-oriental se refieren a capturas de pabellones de conveniencia.

CHANNEL ISLANDS

The 1970-82 catch data refer to the Bailiwick of Guernsey only. The 1983-2002 catch data include also statistics relating to the Bailiwick of Jersey.

ÎLES ANGLO-NORMANDES

Les données relatives aux captures de 1970-82 se rapportent uniquement au bailliage de Guernsey. Les données relatives aux captures de 1983-2002 comprennent aussi les statistiques relatives au bailliage de Jersey.

ISLAS ANGLONORMANDAS

Los datos de capturas de 1970-82 se refieren sólo a la bailía de Guernsey. Los datos de capturas de 1983-2002 incluyen también las estadísticas relativas a la bailía de Jersey.

CHINA

For statistical purposes, the data for China do not include Hong Kong Special Administrative Region (Hong Kong SAR), Macao SAR and Taiwan Province of China.

Catch data, considered to be overstated since the early 1990s, under review and subject to possible downward revisions.

Data concerning tuna catches are reviewed in collaboration with ICCAT, IOTC and SPC, the regional agencies concerned with tuna statistics.

Catch data, excluding those for tuna species, for the Eastern Central and Southeast Atlantic areas are provided by the Las Palmas Survey (LPS).

Catch data, excluding those for tuna species, for the Southwest Atlantic area are provided also by the Fisheries Department, Falkland Islands Government.

Since 1970 data for "Aquatic plants nei", recorded on a dry-weight basis have been converted to wet-weight equivalents by using the conversion factor '10'.

Data for "Freshwater molluscs nei" and "Marine molluscs nei" prior to 1997 have been converted to live weight equivalents by using the conversion factor '2.13'.

CHINE

Les données statistiques relatives à la Chine ne comprennent pas celles qui concernent la Région administrative spéciale de Hong Kong (la RAS de Hong-Kong), la RAS de Macao et à Taïwan Province de Chine.

On considère les données sur les captures exagérées à partir du début des années 90. Elles sont actuellement examinées et seront éventuellement révisées à la baisse.

Les données concernant les captures des thonidés sont révisées en collaboration avec la CICTA, la CTOI et la CPS, les organisations régionales chargées des statistiques des thonidés.

Les données sur les captures, à l'exclusion de celles concernant le thon, pour l'Atlantique du centre-est et du sud-est sont tirées de l'Enquête Las Palmas.

Les données sur les captures, à l'exclusion de celles concernant le thon, pour l'Atlantique sud-ouest, sont fournies également par le Département des pêches du Gouvernement des Îles Falkland.

Depuis 1970 les données relatives aux "Plantes aquatiques nca" enregistrées en poids sec, ont été converties en leurs équivalents poids vert en utilisant le facteur de conversion '10'.

Les données relatives aux "Mollusques d'eau douce nca" et "Mollusques marins nca" avant 1997 ont été converties en équivalents poids vif en utilisant le facteur de conversion '2,13'.

CHINA

Los datos estadísticos relativos a China excluyen los datos correspondientes a la Región Administrativa Especial de Hong Kong (la RAE de Hong Kong), a la RAE de Macao y a Taiwán Provincia de China.

Los datos relativos a las capturas se consideran exagerados desde el principio de los años noventa y están siendo examinados para una posible rectificación.

Los datos relativos a las capturas de atún se revisan en colaboración con la CICAA, la CAOI y la SCP, los organismos regionales encargados de las estadísticas atuneras.

Los datos relativos a las capturas en las zonas del Atlántico centro-oriental y Sudoriental, con exclusión de las de atunes, fueron proporcionados por Las Palmas Survey (LPS).

Los datos relativos a las capturas en la zona del Atlántico Sudoccidental, con exclusión de las de atunes, fueron proporcionados por el Departamento de Pesca del Gobierno de las Islas Malvinas (Falkland).

Desde 1970 los datos relativos a las "Plantas acuáticas nep" registrados en peso en seco se han convertido en sus equivalentes de peso húmedo aplicando el factor de conversión '10'.

Los datos de "Moluscos de agua dulce nep" y de "Moluscos marinos nep" antes de 1997 se han convertido en sus equivalentes de peso en vivo aplicando el factor de conversión '2,13'.

Notes on individual countries or areas

CHINA, HONG KONG SAR

Since 1999 only a total of marine capture production has been made available to FAO. Species breakdown is estimated.

CHINA, MACAO SAR

Fishery surveys have been suspended since July 1996. Data for 1997 onwards are FAO estimates.

COLOMBIA

Data concerning tuna catches are reviewed in collaboration with IATTC and ICCAT, the regional agency concerned with tuna statistics.

CONGO DEM. REP.

Formerly Zaire.

COOK ISLANDS

Data concerning tuna catches are reviewed in collaboration with SPC, the regional agency concerned with tuna statistics.

COSTA RICA

Data for shrimps have been converted to live weight equivalents by using the conversion factor '2'.

Since 1992 data for "Elasmobranchii – Sharks, rays, skates, etc. nei" include also production of shark fins converted to live weight equivalents by using the conversion factor '20'.

CÔTE D'IVOIRE

Data concerning tuna catches are reviewed in collaboration with ICCAT, the regional agency concerned with tuna statistics.

CROATIA

Catches in inland waters decreased since 1998 as statistics on recreational fishery are no longer collected.

Data concerning tuna catches are reviewed in collaboration with ICCAT, the regional agency concerned with tuna statistics.

CUBA

Total production of blue tilapia (*Oreochromis aureus*) is estimated as 80 percent capture production and 20 percent aquaculture production.

Notes sur divers pays ou zones

CHINE, RAS DE HONG-KONG

Depuis 1999 la FAO n'a reçu que les données des captures marines totales. La répartition entre les espèces correspond à une estimation.

CHINE, RAS DE MACAO

Les prospections des pêches ont été suspendues à partir de juillet 1996. Les données à partir de 1997 sont des estimations de la FAO.

COLOMBIE

Les données concernant les captures des thonidés sont révisées en collaboration avec la CITT et la CICTA, les organisations régionales chargées des statistiques des thonidés.

REP. DÉM DU CONGO

Anciennement Zaïre.

ÎLES COOK

Les données concernant les captures des thonidés sont révisées en collaboration avec la CPS, l'organisation régionale chargée des statistiques des thonidés.

COSTA RICA

Les données concernant les crevettes ont été converties en équivalents poids vif en utilisant le facteur de conversion '2'.

Depuis 1992 les données sous "Elasmobranchii – Requins, raies, etc. nca" comprennent également la production d'ailerons de requins, converties en équivalents poids vif en utilisant le facteur de conversion '20'.

CÔTE D'IVOIRE

Les données concernant les captures des thonidés sont révisées en collaboration avec la CICTA, l'organisation régionale chargée des statistiques des thonidés.

CROATIE

La diminution des captures dans les eaux continentales depuis 1998 est due à ce que les statistiques concernant la pêche récréative ne sont plus recueillies.

Les données concernant les captures des thonidés sont révisées en collaboration avec la CICTA, l'organisation régionale chargée des statistiques des thonidés.

CUBA

La production totale de la tilapia *Oreochromis aureus* est estimée à 80 pour cent de production de capture et 20 pour cent de production d'aquaculture.

Notas sobre los distintos países o áreas

CHINA, RAE DE HONG KONG

Desde 1999 sólo se proporcionaron a la FAO datos de la producción total de capturas marinas. El desglose de especies constituye una estimación.

CHINA, RAE DE MACAO

Las encuestas pesqueras se suspendieron desde julio de 1996. Los datos relativos a 1997 y períodos sucesivos constituyen estimaciones de la FAO.

COLOMBIA

Los datos relativos a las capturas de atún se revisan en colaboración con la CIAT y la CICAA, los organismos regionales encargados de las estadísticas atuneras.

REP. DEM. DEL CONGO

Antes Zaire.

ISLAS COOK

Los datos relativos a las capturas de atún se revisan en colaboración con la SCP, el organismo regional encargado de las estadísticas atuneras.

COSTA RICA

Los datos relativos a los camarones se han convertido en sus equivalentes de peso en vivo aplicando el factor de conversión '2'.

Desde 1992 los datos incluidos en "Elasmobranchii – Tiburones, rayas, etc. nep" comprenden también la producción de aletas de tiburón, convertida en sus equivalentes de peso en vivo aplicando el factor de conversión '20'.

CÔTE D'IVOIRE

Los datos relativos a las capturas de atún se revisan en colaboración con la CICAA, el organismo regional encargado de las estadísticas atuneras.

CROACIA

La disminución de las capturas en aguas continentales desde 1998 se debe a que ya no se recogen datos estadísticos sobre la pesca recreativa.

Los datos relativos a las capturas de atún se revisan en colaboración con la CICAA, el organismo regional encargado de las estadísticas atuneras.

CUBA

La producción de la tilapia *Oreochromis aureus* ha sido estimada como 80 por ciento de producción de captura y 20 por ciento de producción de acuicultura.

Notes on individual countries or areas

CYPRUS

Data refer to government-controlled area only.

Data concerning tuna catches are reviewed in collaboration with IATTC and ICCAT the regional agencies concerned with tuna statistics.

Catch data, excluding those for tuna species, for the Eastern Central Atlantic area have also been taken from the "Bulletin Statistique" published by IMROP, Mauritania.

ECUADOR

Data concerning tuna catches are reviewed in collaboration with IATTC, the regional agency concerned with tuna statistics.

EGYPT

Data concerning tuna catches in the Indian Ocean are reviewed in collaboration with IOTC, the regional agency concerned with tuna statistics.

Catch data, excluding those for tuna species, for the Eastern Central Atlantic area have also been taken from the "Bulletin Statistique" published by IMROP, Mauritania.

EL SALVADOR

Large increases in artisanal catch data since 2002 are probably due to an improved data collection system.

ERITREA

Formerly part of Ethiopia.

Data concerning tuna catches are reviewed in collaboration with IOTC, the regional agency concerned with tuna statistics.

ESTONIA

Catch data, excluding those for tuna species, for the Eastern Central Atlantic area have also been taken from the "Bulletin Statistique" published by IMROP, Mauritania.

FIJI ISLANDS

Includes Viti Levu, Vanua Levu, and Rotuma islands.

Data concerning tuna catches are reviewed in collaboration with SPC, the regional agency concerned with tuna statistics.

Data for red and green seaweeds recorded on a dry-weight basis have been converted to wet-weight equivalents by using the conversion factor '8'.

FINLAND

Increase of catches in inland waters since 1988 due to changes in the statistical methods and to inclusion of recreational fishing.

Notes sur divers pays ou zones

CHYPRE

Les données ne concernent que la partie du territoire sous contrôle du gouvernement.

Les données concernant les captures des thonidés sont révisées en collaboration avec la CITT et la CICTA, les organisations régionales chargées des statistiques des thonidés.

Les données sur les captures, à l'exclusion de celles concernant le thon, pour l'Atlantique centre-est ont également été tirées du "Bulletin statistique" publié par l'IMROP, Mauritanie.

ÉQUATEUR

Les données concernant les captures des thonidés sont révisées en collaboration avec la CITT, l'organisation régionale chargée des statistiques des thonidés.

ÉGYPTE

Les données concernant les captures des thonidés dans l'océan Indien sont révisées en collaboration avec la CTOI, l'organisation régionale chargée des statistiques des thonidés.

Les données sur les captures, à l'exclusion de celles concernant le thon, pour l'Atlantique centre-est ont également été tirées du "Bulletin statistique" publié par l'IMROP, Mauritanie.

EL SALVADOR

L'augmentation des captures artisanales enregistrées depuis 2002 est probablement due en grande partie à une meilleure couverture.

ÉRYTHRÉE

Anciennement partie de l'Ethiopie.

Les données concernant les captures des thonidés sont révisées en collaboration avec la CTOI, l'organisation régionale chargée des statistiques des thonidés.

ESTONIE

Les données sur les captures, à l'exclusion de celles concernant le thon, pour l'Atlantique centre-est ont également été tirées du "Bulletin statistique" publié par l'IMROP, Mauritanie.

ÎLES FIDJI

Comprend les îles Viti Levu, Vanua Levu et Rotuma.

Les données concernant les captures des thonidés sont révisées en collaboration avec la CPS, l'organisation régionale chargée des statistiques des thonidés.

Les données relatives aux algues rouges et vertes enregistrées en poids sec, ont été converties en leurs équivalents poids vert en utilisant le facteur de conversion '8'.

FINLANDE

L'augmentation des captures dans les eaux continentales depuis 1988 s'explique par une modification des méthodes statistiques et l'inclusion de la pêche récréative.

Notas sobre los distintos países o áreas

CHIPRE

Los datos se refieren solamente a la zona controlada por el Gobierno.

Los datos relativos a las capturas de atún se revisan en colaboración con la CIAT y la CICAA, los organismos regionales encargados de las estadísticas atuneras.

Los datos sobre las capturas de la zona del Atlántico centro-oriental, con exclusión de las de atunes, también se han extraído de la publicación "Bulletin Statistique" del IMROP, Mauritania.

ECUADOR

Los datos relativos a las capturas de atún se revisan en colaboración con la CIAT, el organismo regional encargado de las estadísticas atuneras.

EGIPTO

Los datos relativos a las capturas de atún en el océano Índico se revisan en colaboración con la CAOI, el organismo regional encargado de las estadísticas atuneras.

Los datos sobre las capturas de la zona del Atlántico centro-oriental, con exclusión de las de atunes, también se han extraído de la publicación "Bulletin Statistique" del IMROP, Mauritania.

EL SALVADOR

Las mayores capturas artesanales registradas desde 2002 se deben probablemente, en gran parte, a una mejor cobertura.

ERITREA

Antes parte de Etiopía.

Los datos relativos a las capturas de atún se revisan en colaboración con la CAOI, el organismo regional encargado de las estadísticas atuneras.

ESTONIA

Los datos sobre las capturas de la zona del Atlántico centro-oriental, con exclusión de las de atunes, también se han extraído de la publicación "Bulletin Statistique" del IMROP, Mauritania.

ISLAS FIJI

Incluyen las islas Viti Levu, Vanua Levu y Rotuma.

Los datos relativos a las capturas de atún se revisan en colaboración con la SCP, el organismo regional encargado de las estadísticas atuneras.

Los datos relativos a las algas rojas y verdes registrados en peso en seco se han convertido en sus equivalentes de peso húmedo aplicando el factor de conversión '8'

FINLANDIA

El incremento de las capturas en aguas continentales desde 1988 se debe a cambios en los métodos estadísticos y a la inclusión de la pesca recreativa.

Notes on individual countries or areas

FRANCE

Data concerning tuna catches are reviewed in collaboration with ICCAT and IOTC, the regional agencies concerned with tuna statistics.

Catch data, excluding those for tuna species, for the Eastern Central Atlantic area have also been taken from the "Bulletin Statistique" published by IMROP, Mauritania.

Catch data, excluding those for tuna species, for the Southwest Atlantic area are provided by the Fisheries Department, Falkland Islands Government.

FRENCH POLYNESIA

Since 1997, data for 'Trochus shells', 'Pearl oyster shells nei' and 'Sponges' are derived from export statistics. Data for 'Pearl oyster shells nei' include also production of pearls.

FRENCH SOUTHERN TERR.

Include Amsterdam and St. Paul Islands in area 51 (Western Indian Ocean) and Kerguelen and Crozet Islands in area 58 (Antarctic Indian Ocean). Catches reported for area 58 are included with those of France.

GERMANY

Increased catches of 'Freshwater fishes nei' since 1995 due to the inclusion of estimated catches (about 18,800 mt) from recreational fisheries.

GHANA

Data concerning tuna catches are reviewed in collaboration with ICCAT, the regional agency concerned with tuna statistics.

GREECE

Since 1988, the data for catches in inland waters are those provided by the Ministry of Agriculture while data prior to 1988 were those provided by the Ministry of National Economy.

Data concerning tuna catches are reviewed in collaboration with ICCAT, the regional agency concerned with tuna statistics.

GREENLAND

Starting with 1993, data reported on marine mammals include also harvest for subsistence. Data for 2002 cover only the period between January and September.

GUATEMALA

Data concerning tuna catches are reviewed in collaboration with IATTC, the regional agency concerned with tuna statistics.

Data exclude significant quantities of unrecorded subsistence catches.

Notes sur divers pays ou zones

FRANCE

Les données concernant les captures des thonidés sont révisées en collaboration avec la CICTA et la CTOI, les organisations régionales chargées des statistiques des thonidés.

Les données sur les captures, à l'exclusion de celles concernant le thon, pour l'Atlantique centre-est ont également été tirées du "Bulletin statistique" publié par l'IMROP, Mauritanie.

Les données sur les captures, à l'exclusion de celles concernant le thon, pour l'Atlantique sud-ouest, sont fournies par le Département des pêches du Gouvernement des Îles Falkland.

POLYNÉSIE FRANÇAISE

Depuis 1997 les données relatives aux 'Troques', 'Coquilles d'huîtres perl. nca' et 'Eponges' proviennent des statistiques d'exportation. Les données de 'Coquilles d'huîtres perl. nca' comprennent également la production des perles.

TERRES AUSTRALES FR.

Comprennent les îles d'Amsterdam et de Saint-Paul dans la zone de pêche 51 (océan Indien, ouest) et les îles Kerguelen et Crozet dans la zone de pêche 58 (océan Indien, antarctique). Les données relatives à la zone de pêche 58 sont incluses avec celles de la France.

ALLEMAGNE

L'augmentation des captures de 'Poissons d'eau nca' douce depuis 1995 est imputable à l'inclusion de l'estimation (approximativement 18 800 tonnes) de pêche récréative.

GHANA

Les données concernant les captures des thonidés sont révisées en collaboration avec la CICTA, l'organisation régionale chargée des statistiques des thonidés.

GRÈCE

Depuis 1988 les données sur les captures dans les eaux continentales sont fournies par le Ministère de l'agriculture tandis que les données avant 1988 étaient fournies par le Ministère de l'économie nationale.

Les données concernant les captures des thonidés sont révisées en collaboration avec la CICTA, l'organisation régionale chargée des statistiques des thonidés.

GROENLAND

A partir de 1993, les données reportées officiellement concernant les mammifères aquatiques comprennent également la pêche destinée à la subsistance. Les données pour 2002 se réfèrent exclusivement à la période entre janvier et septembre.

GUATEMALA

Les données concernant les captures des thonidés sont révisées en collaboration avec la CITT, l'organisation régionale chargée des statistiques des thonidés.

Non compris des captures importantes non enregistrées concernant la pêche de subsistance.

Notas sobre los distintos países o áreas

FRANCIA

Los datos relativos a las capturas de atún se revisan en colaboración con la CICAA y la CAOI, los organismos regionales encargados de las estadísticas atuneras.

Los datos sobre las capturas de la zona del Atlántico centro-oriental, con exclusión de las de atunes, también se han extraído de la publicación "Bulletin Statistique" del IMROP, Mauritania.

Los datos relativos a las capturas en la zona del Atlántico Sudoccidental, con exclusión de las de atunes, fueron proporcionados por el Departamento de Pesca del Gobierno de las Islas Malvinas (Falkland).

POLINESIA FRANCESA

Desde 1997 los datos relativos a 'Tróquidos', 'Conchas de ostras perleras nep' y 'Esponjas' han sido obtenidos de las estadísticas de exportación. Datos de 'Conchas de ostras perleras nep' incluyen también la producción de perlas.

TIERRAS AUSTRALES FR.

Incluyen las islas Amsterdam y San Pablo en área de pesca 51 (océano Índico occidental) y las islas Kerguélen y Crozet en área de pesca 58 (océano Índico Antártico). Datos relativos a la área de pesca 58 se incluyen en los datos de la Francia.

ALEMANIA

El incremento de capturas de 'Peces de agua dulce nep' desde 1995 se debe a la inclusión de la estimación (aproximadamente 18 800 toneladas) de la pesca recreativa.

GHANA

Los datos relativos a las capturas de atún se revisan en colaboración con la CICAA, el organismo regional encargado de las estadísticas atuneras.

GRECIA

Desde 1988 las cifras de las capturas en aguas continentales son aquellas proporcionadas por el Ministerio de la Agricultura mientras las cifras antecedentes fueron proporcionadas por el Ministerio de la Economía Nacional.

Los datos relativos a las capturas de atún se revisan en colaboración con la CICAA, el organismo regional encargado de las estadísticas atuneras.

GROENLANDIA

A partir de 1993, los datos proporcionados oficialmente incluyen también la recogida hecha con fin de subsistencia. Datos sobre 2002 se refieren solamente al período entre enero y septiembre.

GUATEMALA

Los datos relativos a las capturas de atún se revisan en colaboración con la CIAT, el organismo regional encargado de las estadísticas atuneras.

Se excluyen importantes cantidades de capturas no registradas relativas a la pesca de subsistencia.

Notes on individual countries or areas

HONDURAS

Data concerning tuna catches are reviewed in collaboration with IATTC and ICCAT, the regional agencies concerned with tuna statistics. A portion of the tunas caught in the Indian Ocean by vessels flying the Honduran flag may be included under "Other nei".

Catch data, excluding those for tuna species, for the Eastern Central and Southeast Atlantic areas are provided by the Las Palmas Survey (LPS).

INDIA

Data concerning tuna catches are reviewed in collaboration with IOTC, the regional agency concerned with tuna statistics.

INDONESIA

Starting with the 2001 data and backwards to 1975, officially provided capture statistics are in accordance with the boundary change between fishing areas 57 and 71 in the Australian-Indonesian region.

Data concerning tuna catches in the Indian Ocean are reviewed in collaboration with IOTC, the regional agency concerned with tuna statistics.

IRAN (ISLAMIC REP. OF)

Data concerning tuna catches are reviewed in collaboration with IOTC, the regional agency concerned with tuna statistics.

IRELAND

Since 1999 catches in inland waters are no longer collected except those for European eel, Atlantic salmon and sea trout.

Data concerning tuna catches are reviewed in collaboration with ICCAT, the regional agency concerned with tuna statistics.

ISRAEL

Up to 1994 includes Palestine, Occupied Tr.

ITALY

Data concerning tuna catches are reviewed in collaboration with ICCAT, the regional agency concerned with tuna statistics.

Catch data for 'Mediterranean mussel' are subject to revisions; increased catches since 2000 due to improved coverage.

JAMAICA

Data for 'Stromboid conchs nei' have been converted to live weight equivalents by using the conversion factor '7.5'.

Notes sur divers pays ou zones

HONDURAS

Les données concernant les captures des thonidés sont révisées en collaboration avec la CITT et la CICTA, les organisations régionales chargées des statistiques des thonidés. Une partie des données de thonidés attrapés dans l'océan Indien par des navires battant pavillon hondurien peut être incluse avec celles de "Autres nca".

Les données sur les captures, à l'exclusion de celles concernant le thon, pour l'Atlantique du centre-est et du sud-est sont tirées de l'Enquête Las Palmas.

INDE

Les données concernant les captures des thonidés sont révisées en collaboration avec la CTOI, l'organisation régionale chargée des statistiques des thonidés.

INDONÉSIE

A partir des données de 2001 et précédemment à 1975, les statistiques des captures communiquées par le service national sont conformes au changement de la limite entre les zones de pêche 57 et 71 dans la région australienne-indonésienne.

Les données concernant les captures des thonidés dans l'océan Indien sont révisées en collaboration avec la CTOI, l'organisation régionale chargée des statistiques des thonidés.

IRAN (RÉP. ISLAMIQUE D')

Les données concernant les captures des thonidés sont révisées en collaboration avec la CTOI, l'organisation régionale chargée des statistiques des thonidés.

IRLANDE

Depuis 1999 les données de captures dans les eaux continentales ne sont plus rassemblées à l'exception de celles relatives à anguille d'Europe, saumon de l'Atlantique et truite de mer.

Les données concernant les captures des thonidés sont révisées en collaboration avec la CICTA, l'organisation régionale chargée des statistiques des thonidés.

ISRAËL

Jusqu'en 1994 comprend la Palestine, terr. occupés.

ITALIE

Les données concernant les captures des thonidés sont révisées en collaboration avec la CICTA, l'organisation régionale chargée des statistiques des thonidés.

Les données relatives aux captures de 'Moule commune' sont sujettes à révisions; l'augmentation de captures depuis 2000 s'explique par une meilleure couverture.

JAMAÏQUE

Les données relatives aux 'Strombes nca' ont été converties en équivalents poids vif en utilisant le facteur de conversion '7,5'.

Notas sobre los distintos países o áreas

HONDURAS

Los datos relativos a las capturas de atún se revisan en colaboración con la CIAT y la CICAA, los organismos regionales encargados de las estadísticas atuneras. Una porción de las capturas de atún en el océano Índico por embarcaciones con pabellón hondureño puede ser incluida en los datos de "Otros nep".

Los datos relativos a las capturas en las zonas del Atlántico centro-oriental y Sudoriental, con exclusión de las de atunes, fueron proporcionados por Las Palmas Survey (LPS).

INDIA

Los datos relativos a las capturas de atún se revisan en colaboración con la CAOI, el organismo regional encargado de las estadísticas atuneras.

INDONESIA

A partir de los datos de 2001, y hasta 1975 retrocediendo en el tiempo, estadísticas de capturas proporcionadas por la oficina nacional son en conformidad del cambio del límite entre las áreas de pesca 57 y 71 en la región australiana-indonesia.

Los datos relativos a las capturas de atún en el océano Índico se revisan en colaboración con la CAOI, el organismo regional encargado de las estadísticas atuneras.

IRÁN (REP. ISLÁMICA DEL)

Los datos relativos a las capturas de atún se revisan en colaboración con la CAOI, el organismo regional encargado de las estadísticas atuneras.

IRLANDA

Desde 1999 las capturas en aguas continentales ya no se recogen a excepión de datos por anguila europea, salmón del Atlántico y trucha marina.

Los datos relativos a las capturas de atún se revisan en colaboración con la CICAA, el organismo regional encargado de las estadísticas atuneras.

ISRAEL

Hasta el 1994 incluye la Palestina, Terri. Ocupado.

ITALIA

Los datos relativos a las capturas de atún se revisan en colaboración con la CICAA, el organismo regional encargado de las estadísticas atuneras.

Los datos de capturas relativos al 'Mejillón común' están sujetos a revisiones; el incremento de capturas desde 2000 se debe a una mejor cobertura.

JAMAICA

Los datos de 'Cobos nep' se han convertido en sus equivalentes de peso en vivo aplicando el factor de conversión '7,5'.

Notes on individual countries or areas

Notes sur divers pays ou zones

Notas sobre los distintos países o áreas

JAPAN

Data for tuna species are not comparable with those of IATTC, ICCAT, IOTC and SPC, with the exception of 2002 data.

KIRIBATI

Includes Fanning Island, Washington Island and Christmas Island in the Line Islands; Ocean Island, Phoenix Islands (Birnie, Gardner, Hull, McKean, Phoenix, Sydney, Canton and Enderbury).

Data concerning tuna catches are reviewed in collaboration with the SPC, the regional agency concerned with tuna statistics.

KOREA DEM. PEOPLE'S REP.

The 1961-93 capture data have been extensively revised on the basis of the information in a report to FAO on anomalies in global capture statistics.

Catch data, excluding those for tuna species, for the Eastern Central Atlantic areas are provided by the Las Palmas Survey (LPS).

KOREA REPUBLIC

Data for tuna species not comparable with those of IATTC, ICCAT, IOTC and SPC.

LATVIA

Catch data, excluding those for tuna species, for the Eastern Central Atlantic area have also been taken from the "Bulletin Statistique" published by IMROP, Mauritania.

LIBYAN ARAB JAMAHIRIYA

Data concerning tuna catches are reviewed in collaboration with ICCAT, the regional agency concerned with tuna statistics.

LITHUANIA

Since 2002, data for the Curronian lagoon have been reported as captures in area 27 (Northeast Atlantic) instead of area 05 (Europe – Inland waters) causing a displacement of catches between the two areas.

Catch data, excluding those for tuna species, for the Eastern Central Atlantic area have also been taken from the "Bulletin Statistique" published by IMROP, Mauritania.

MADAGASCAR

Data concerning tuna catches are reviewed in collaboration with IOTC, the regional agency concerned with tuna statistics.

Data reported since 1964 for "Sea cucumber nei" recorded on a dry-weight basis have been converted to wet-weight equivalents by using the conversion factor '3'.

JAPON

Les données concernant les thonidés ne sont pas comparables avec les renseignements fournis par la CITT, la CICTA, la CTOI et la CPS sauf pour les captures relatives à 2002.

KIRIBATI

Comprend Fanning, Washington, l'île Christmas dans les îles de la Ligne; l'île Océan, les îles Phoenix (Birnie, Gardner, Hull, McKean, Phoenix, Sidney, Canton et Enderbury).

Les données concernant les captures des thonidés sont révisées en collaboration avec la CPS, l'organisation régionale chargée des statistiques des thonidés.

RÉP. POP. DÉM. DE CORÉE

Les données sur les captures de la période 1961-93 ont été largement révisées en fonction des informations par un rapport à la FAO sur les anomalies dans les statistiques globales des captures.

Les données sur les captures, à l'exclusion de celles concernant le thon, pour l'Atlantique du centre-est sont tirées de l'Enquête Las Palmas.

RÉPUBLIQUE DE CORÉE

Les données concernant les thonidés ne sont pas comparables avec les renseignements fournis par la CITT, la CICTA, la CTOI et la CPS.

LETTONIE

Les données sur les captures, à l'exclusion de celles concernant le thon, pour l'Atlantique centre-est ont également été tirées du "Bulletin statistique" publié par l'IMROP, Mauritanie.

JAMAHIRIYA ARABE LIBYEN.

Les données concernant les captures des thonidés sont révisées en collaboration avec la CICTA, l'organisation régionale chargée des statistiques des thonidés.

LITUANIE

Depuis 2002 les données concernant la lagune de Curronian ont été rapportées comme captures dans la zone 27 (Atlantique nord-est) à la place de la zone 05 (Europe - Eaux continentales) causant un déplacement des captures entre les deux zones.

Les données sur les captures, à l'exclusion de celles concernant le thon, pour l'Atlantique centre-est ont également été tirées du "Bulletin statistique" publié par l'IMROP, Mauritanie.

MADAGASCAR

Les données concernant les captures des thonidés, poissons type thon nca et poissons marins nca sont révisées en collaboration avec la CTOI, l'organisation régionale chargée des statistiques des thonidés.

Les données rapporté depuis 1964 comme "Bèches-de-mer nca" enregistrées en poids sec, ont été converties en leurs équivalents poids vert en utilisant le facteur de conversion '3'.

JAPÓN

Los datos sobre las especies de atún no son comparables con la información de la CIAT, la CICAA, la CAOI y la SCP con la excepción de las capturas de 2002.

KIRIBATI

Incluye las islas Fanning, Washington, y Christmas en las islas de la Línea; la isla Océano, y las islas Phoenix (Birnie, Gardner, Hull, McKean, Phoenix, Sidney, Canton y Enderbury).

Los datos relativos a las capturas de atún se revisan en colaboración con la SCP, el organismo regional encargado de las estadísticas atuneras.

REP. POP. DEM. DE COREA

Los datos de capturas para el período 1990-98 han sido revisados ampliamente para tener en cuenta la información en un informe a la FAO sobre anomalías en las estadísticas globales de capturas.

Los datos relativos a las capturas en las zonas del Atlántico centro-oriental, con exclusión de las de atunes, fueron proporcionados por Las Palmas Survey (LPS).

REPÚBLICA DE COREA

Los datos sobre las especies de atún no son comparables con la información de la CIAT, la CICAA, la CAOI y la SCP.

LETONIA

Los datos sobre las capturas de la zona del Atlántico centro-oriental, con exclusión de las de atunes, también se han extraído de la publicación "Bulletin Statistique" del IMROP, Mauritania.

JAMAHIRIYA ÁRABE LIBIA

Los datos relativos a las capturas de atún se revisan en colaboración con la CICAA, el organismo regional encargado de las estadísticas atuneras.

LITUANIA

Desde 2002 los datos sobre la laguna de Curronian han sido proporcionados como capturas en el área 27 (Atlántico nordeste) en lugar del área 05 (Europa - Aguas continentales) causando una dislocación de capturas entre las dos áreas.

Los datos sobre las capturas de la zona del Atlántico centro-oriental, con exclusión de las de atunes, también se han extraído de la publicación "Bulletin Statistique" del IMROP, Mauritania.

MADAGASCAR

Los datos relativos a la captura nominal de atunes, peces parecidos a los atunes nep y peces marinos nep se revisan en colaboración con la CAOI, el organismo regional encargado de las estadísticas atuneras.

Los datos proporcionados desde 1964 como "Cohombros de mar nep" registrados en peso en seco se han convertido en sus equivalentes de peso húmedo aplicando el factor de conversión'3'.

Notes on individual countries or areas

MALAYSIA

Includes Peninsular Malaysia, Sabah and Sarawak.

Data concerning tuna catches in the Indian Ocean are reviewed in collaboration with IOTC, the regional agency concerned with tuna statistics.

MALDIVES

Data concerning tuna catches are reviewed in collaboration with IOTC, the regional agency concerned with tuna statistics.

MALTA

Data concerning tuna catches are reviewed in collaboration with ICCAT, the regional agency concerned with tuna statistics.

Catch data, excluding those for tuna species, for the Eastern Central Atlantic area have also been taken from the "Bulletin Statistique" published by IMROP, Mauritania.

MARSHALL ISLANDS

Data concerning tuna catches are reviewed in collaboration with the SPC, the regional agency concerned with tuna statistics. Increase of tuna catches since 2001 due to vessels, previously flagged in Vanuatu, re-flagged in the Marshall Islands.

Catch data, excluding those for tuna species, for the Eastern Central Atlantic area have also been taken from the "Bulletin Statistique" published by IMROP, Mauritania.

MAURITIUS

Data concerning tuna catches are reviewed in collaboration with IOTC, the regional agency concerned with tuna statistics.

MAYOTTE

Includes Grand-Terre and Pamandzi.

Data concerning tuna catches are reviewed in collaboration with IOTC, the regional agency concerned with tuna statistics.

MICRONESIA

Includes Yap, Truk, Pohnpei and Kosrae.

Data concerning tuna catches are reviewed in collaboration with the SPC, the regional agency concerned with tuna statistics.

MOROCCO

Data concerning tuna catches are reviewed in collaboration with ICCAT, the regional agency concerned with tuna statistics.

Notes sur divers pays ou zones

MALAISIE

Comprend la Malaisie Péninsulaire, Sabah et Sarawak.

Les données concernant les captures des thonidés dans l'océan Indien sont révisées en collaboration avec la CTOI, l'organisation régionale chargée des statistiques des thonidés.

MALDIVES

Les données concernant les captures des thonidés, poissons type thon nca et poissons marins nca sont révisées en collaboration avec la CTOI, l'organisation régionale chargée des statistiques des thonidés.

MALTE

Les données concernant les captures des thonidés sont révisées en collaboration avec la CICTA, l'organisation régionale chargée des statistiques des thonidés.

Les données sur les captures, à l'exclusion de celles concernant le thon, pour l'Atlantique centre-est ont également été tirées du "Bulletin statistique" publié par l'IMROP, Mauritanie.

ÎLES MARSHALL

Les données concernant les captures des thonidés sont révisées en collaboration avec la CPS, l'organisation régionale chargée des statistiques des thonidés. L'augmentation depuis 2001 des captures pour les thonidés s'explique par les captures effectuées par bateaux, précédemment battant pavillon de Vanuatu, transférés en pavillon des Îles Marshall.

Les données sur les captures, à l'exclusion de celles concernant le thon, pour l'Atlantique centre-est ont également été tirées du "Bulletin statistique" publié par l'IMROP, Mauritanie.

MAURICE

Les données concernant les captures des thonidés sont révisées en collaboration avec la CTOI, l'organisation régionale chargée des statistiques des thonidés

MAYOTTE

Comprend Grande-Terre et Pamandzi.

Les données concernant les captures des thonidés, poissons type thon nca et poissons marins nca sont révisées en collaboration avec la CTOI, l'organisation régionale chargée des statistiques des thonidés.

MICRONÉSIE

Comprend Yap, Truk, Pohnpei et Kosrae.

Les données concernant les captures des thonidés sont révisées en collaboration avec la CPS, l'organisation régionale chargée des statistiques des thonidés.

MAROC

Les données concernant les captures des thonidés sont révisées en collaboration avec la CICTA, l'organisation régionale chargée des statistiques des thonidés.

Notas sobre los distintos países o áreas

MALASIA

Incluye la Malasia Peninsular, Sabah y Sarawak.

Los datos relativos a las capturas de atún en el océano Índico se revisan en colaboración con la CAOI, el organismo regional encargado de las estadísticas atuneras.

MALDIVAS

Los datos relativos a la captura nominal de atunes, peces parecidos a los atunes nep y peces marinos nep se revisan en colaboración con la CAOI, el organismo regional encargado de las estadísticas atuneras.

MALTA

Los datos relativos a las capturas de atún se revisan en colaboración con la CICAA, el organismo regional encargado de las estadísticas atuneras.

Los datos sobre las capturas de la zona del Atlántico centro-oriental, con exclusión de las de atunes, también se han extraído de la publicación "Bulletin Statistique" del IMROP, Mauritania.

ISLAS MARSHALL

Los datos relativos a las capturas de atún se revisan en colaboración con la SCP, el organismo regional encargado de las estadísticas atuneras. El incremento desde 2001 de las capturas de atún se debe a embarcaciones, previamente con pabellón de Vanuatu, que cambiaron de pabellón en las Islas Marshall.

Los datos sobre las capturas de la zona del Atlántico centro-oriental, con exclusión de las de atunes, también se han extraído de la publicación "Bulletin Statistique" del IMROP, Mauritania.

MAURICIO

Los datos relativos a las capturas de atún se revisan en colaboración con la CAOI, el organismo regional encargado de las estadísticas atuneras.

MAYOTTE

Incluye Grande-Terre y Pamandzi.

Los datos relativos a la captura nominal de atunes, peces parecidos a los atunes nep y peces marinos nep se revisan en colaboración con la CAOI, el organismo regional encargado de las estadísticas atuneras.

MICRONESIA

Incluye Yap, Truk, Pohnpei y Kosrae.

Los datos relativos a las capturas de atún se revisan en colaboración con la SCP, el organismo regional encargado de las estadísticas atuneras.

MARRUECOS

Los datos relativos a las capturas de atún se revisan en colaboración con la CICAA, el organismo regional encargado de las estadísticas atuneras.

Notes on individual countries or areas

MOZAMBIQUE

Data collection system considered to largely underestimate capture production, particularly for the artisanal sector.

MYANMAR

Data refer to a fiscal year period (1 April – 31 March) shown under the calendar year in which the fiscal year ends.

Data for jellyfishes expressed in dry-weight.

NAMIBIA

A remarkable amount of 'Cape horse mackerel' (*Trachurus capensis*) catches has been added to Namibian capture production for the 1992-2000 period to take account of information provided by the national reporting office.

Data concerning tuna catches are reviewed in collaboration with ICCAT, the regional agency concerned with tuna statistics.

Catch data, excluding those for tuna species, for the Southwest Atlantic area are provided by the Fisheries Department, Falkland Islands Government.

Data for 'Gracilaria seaweeds' recorded on a dry-weight basis have been converted to wet-weight equivalents by using the conversion factor '8'.

NEPAL

Data refer to a split-year (16 July - 15 July) shown under the calendar year in which the split-year ends.

NETHERLANDS ANTILLES

Include Curaçao, Bonaire, Saba, St. Eustatius and St. Martin.

Data concerning tuna catches are reviewed in collaboration with ICCAT, the regional agency concerned with tuna statistics.

NEW CALEDONIA

Data concerning tuna catches are reviewed in collaboration with the SPC, the regional agency concerned with tuna statistics.

NEW ZEALAND

For the period 1995-2000, data for "Elasmobranchii – Sharks, rays, skates, etc. nei" include also production of shark fins converted to live weight equivalents by using the conversion factor '20'.

Data for 'New Zealand scallop' since 1987 have been converted to live weight equivalents by using the conversion factor '8'

NICARAGUA

Data concerning tuna catches are reviewed in collaboration with IATTC, the regional agency concerned with tuna statistics.

Notes sur divers pays ou zones

MOZAMBIQUE

On considère que le système de collecte des données sous-estime partie des captures, en particulier pour la pêche artisanale.

MYANMAR

Les données se réfèrent à l'exercice financier (1er avril - 31 mars) et figurent sous l'année civile durant laquelle se termine l'exercice financier.

Les données sur les méduses sont exprimées en poids sec.

NAMIBIE

Une quantité remarquable de captures de 'Chinchard du Cap' (*Trachurus capensis*) a été ajoutée aux captures de Namibie pour la période 1992-2000 en fonction des informations fournies par le service national déclarant.

Les données concernant les captures des thonidés sont révisées en collaboration avec la CICTA, l'organisation régionale chargée des statistiques des thonidés.

Les données sur les captures, à l'exclusion de celles concernant le thon, pour l'Atlantique sud-ouest, sont fournies par le Département des pêches du Gouvernement des Îles Falkland.

Les données relatives aux 'Algues gracilaires' enregistrées en poids sec, ont été converties en leurs équivalents poids vert en utilisant le facteur de conversion '8'.

NÉPAL

Les données se réfèrent à une année fractionnée (16 juillet - 15 juillet) et figurent sous l'année civile durant laquelle se termine l'année fractionnée.

ANTILLES NÉERLANDAISES

Comprennent Curaçao, Bonaire, Saba, Saint-Eustache et Saint-Martin.

Les données concernant les captures des thonidés sont révisées en collaboration avec la CICTA, l'organisation régionale chargée des statistiques des thonidés.

NOUVELLE-CALÉDONIE

Les données concernant les captures des thonidés sont révisées en collaboration avec la CPS, l'organisation régionale chargée des statistiques des thonidés.

NOUVELLE-ZÉLANDE

Pour la période 1995-2000 les données sous "Elasmobranchii – Requins, raies, etc. nca" comprennent également la production d'ailerons de requins, converties en équivalents poids vif en utilisant le facteur de conversion '20'.

Les données relatives aux 'Pecten de Nouvelle-Zélande' depuis 1987 ont été converties en équivalents poids vif en utilisant le facteur de conversion '8'.

NICARAGUA

Les données concernant les captures des thonidés sont révisées en collaboration avec la CITT, l'organisation régionale chargée des statistiques des thonidés.

Notas sobre los distintos países o áreas

MOZAMBIQUE

Se considera que el sistema de colección de datos subestima demasiado la producción de capturas, especialmente las de la pesca artesanal.

MYANMAR

Los datos se refieren a un año fiscal (1 de abril - 31 de marzo) que se indica como el año civil en que finaliza el año fiscal.

Los datos relativos a las medusas se han expresado en peso seco.

NAMIBIA

Una cantidad notable de capturas de 'Jurel del Cabo' (*Trachurus capensis*) se ha agregado a la producción des capturas de Namibia para el período 1992-2000 para tener en cuenta la información notificada por parte de la oficina nacional.

Los datos relativos a las capturas de atún se revisan en colaboración con la CICAA, el organismo regional encargado de las estadísticas atuneras.

Los datos relativos a las capturas en la zona del Atlántico Sudoccidental, con exclusión de las de atunes, fueron proporcionados por el Departamento de Pesca del Gobierno de las Islas Malvinas (Falkland).

Los datos relativos a las algas 'Gracilarias' registrados en peso en seco se han convertido en sus equivalentes de peso húmedo aplicando el factor de conversión '8'.

NEPAL

Los datos se refieren a un año emergente (16 de julio - 15 de julio) que se indica como el año civil en que finaliza el año emergente.

ANTILLAS NEERLANDESAS

Incluye Curaçao, Bonaire, Saba, San Eustaquio y San Martín.

Los datos relativos a las capturas de atún se revisan en colaboración con la CICAA, el organismo regional encargado de las estadísticas atuneras.

NUEVA CALEDONIA

Los datos relativos a las capturas de atún se revisan en colaboración con la SCP, el organismo regional encargado de las estadísticas atuneras.

NUEVA ZELANDIA

Para el período 1995-2000 los datos incluidos en "Elasmobranchii – Tiburones, rayas, etc. nep" comprenden también la producción de aletas de tiburón, convertida en sus equivalentes de peso en vivo aplicando el factor de conversión '20'.

Los datos de 'Vieira de Nueva Zelandia' desde 1987 se han convertido en sus equivalentes de peso en vivo aplicando el factor de conversión '8'.

NICARAGUA

Los datos relativos a las capturas de atún se revisan en colaboración con la CIAT, el organismo regional encargado de las estadísticas atuneras.

Notes on individual countries or areas

NORTHERN MARIANA IS.

Includes Marianas Islands except Guam.

NORWAY

Up to 1995, reported weight of the basking shark (*Cetorhinus maximus*) liver was converted to live weight equivalents by using the conversion factor '10'. Since 1996, basking shark fins have been converted to live weight equivalents by using the conversion factor '100'.

OTHER NEI

Data refer to catches, mainly of tuna species, reported by IATTC, ICCAT, IOTC, Las Palmas Survey (LPS) and the "Bulletin Statistique" published by IMROP, Mauritania, as caught by unidentified countries.

PAKISTAN

Data concerning tuna catches are reviewed in collaboration with IOTC, the regional agency concerned with tuna statistics.

PALESTINE, OCCUPIED TR.

Formerly Gaza Strip.

PANAMA

Data concerning tuna catches are reviewed in collaboration with IATTC and ICCAT, the regional agencies concerned with tuna statistics.

Catch data, excluding those for tuna species, for the Eastern Central and Southeast Atlantic areas are provided by the Las Palmas Survey (LPS) and the "Bulletin Statistique" published by IMROP, Mauritania.

PAPUA NEW GUINEA

Data concerning tuna catches are reviewed in collaboration with SPC, the regional agency concerned with tuna statistics.

PHILIPPINES

Data concerning tuna catches in the Atlantic and in the Indian Oceans are reviewed in collaboration with ICCAT and IOTC, the regional agencies concerned with tuna statistics.

POLAND

Estimated catches (13,000 mt) from recreational fisheries included in 'Freshwater fishes nei'.

PORTUGAL

Data concerning tuna catches are reviewed in collaboration with ICCAT and IOTC, the regional agencies concerned with tuna statistics.

Catch data, excluding those for tuna species, for the Eastern Central Atlantic area have also been taken from the "Bulletin Statistique" published by IMROP, Mauritania.

Notes sur divers pays ou zones

ÎLES MARIANNES SEPTENTR.

Comprend les îles Mariannes sauf Guam.

NORVÈGE

Jusqu'en 1995, le poids du foie du requin pèlerin (*Cetorhinus maximus*) était converti en équivalent poids vif en utilisant le facteur de conversion '10'. Depuis 1996, les ailerons de requins pèlerins ont été converties en équivalents poids vif en utilisant le facteur de conversion '100'.

AUTRES NCA

Les données se rapportent aux captures, principalement des thonidés, indiqués par la CITT, la CICTA, la CTOI, l'Enquête Las Palmas et du "Bulletin statistique" publié par l'IMROP, Mauritanie, comme captures de pays non identifiables.

PAKISTAN

Les données concernant les captures des thonidés sont révisées en collaboration avec la CTOI, l'organisation régionale chargée des statistiques des thonidés.

PALESTINE, TERR. OCCUPÉS

Anciennement Bande de Gaza.

PANAMA

Les données concernant les captures des thonidés sont révisées en collaboration avec la CITT et la CICTA, les organisations régionales chargées des statistiques des thonidés.

Les données sur les captures, à l'exclusion de celles concernant le thon, pour l'Atlantique du centre-est et du sud-est sont tirées de l'Enquête Las Palmas et du "Bulletin statistique" publié par l'IMROP, Mauritanie.

PAPOUASIE-NLLE-GUINÉE

Les données concernant les captures des thonidés sont révisées en collaboration avec la CPS, l'organisation régionale chargée des statistiques des thonidés.

PHILIPPINES

Les données concernant les captures des thonidés dans les océans Atlantique et Indien sont révisées en collaboration avec la CICTA et la CTOI, les organisations régionales chargées des statistiques des thonidés.

POLOGNE

Estimation (13 000 tonnes) de pêche récréative comprises en 'Poissons d'eau douce nca'.

PORTUGAL

Les données concernant les captures des thonidés sont révisées en collaboration avec la CICTA et la CTOI, les organisations régionales chargées des statistiques des thonidés.

Les données sur les captures, à l'exclusion de celles concernant le thon, pour l'Atlantique centre-est ont également été tirées du "Bulletin statistique" publié par l'IMROP, Mauritanie.

Notas sobre los distintos países o áreas

ISLAS MARIANAS SEPTENT.

Incluye las islas Marianas excepto Guam.

NORUEGA

Hasta 1995, el peso notificado de hígado de peregrino (*Cetorhinus maximus*) se convirtió en sus equivalentes de peso en vivo aplicando el factor de conversión '10'. Desde 1996, las aletas de peregrinos se han convertido en sus equivalentes de peso en vivo aplicando el factor de conversión '100'.

OTROS NEP

Los datos abarcan capturas, principalmente de atún, indicado por la CIAT, la CICAA, la CAOI, Las Palmas Survey (LPS) y por la publicación "Bulletin Statistique" del IMROP, Mauritania, como capturado por países desconocidos.

PAKISTÁN

Los datos relativos a las capturas de atún se revisan en colaboración con la CAOI, el organismo regional encargado de las estadísticas atuneras.

PALESTINA, TERRI. OCUPADO

Antes Faja de Gaza.

PANAMÁ

Los datos relativos a las capturas de atún se revisan en colaboración con la CIAT y la CICAA, los organismos regionales encargados de las estadísticas atuneras.

Los datos relativos a las capturas en las zonas del Atlántico centro-oriental y Sudoriental, con exclusión de las de atunes, fueron proporcionados por Las Palmas Survey (LPS) y por la publicación "Bulletin Statistique" del IMROP, Mauritania.

PAPUA NUEVA GUINEA

Los datos relativos a las capturas de atún se revisan en colaboración con la SCP, el organismo regional encargado de las estadísticas atuneras.

FILIPINAS

Los datos relativos a las capturas de atún en los océanos Atlántico e Índico se revisan en colaboración con la CICAA y la CAOI, los organismos regionales encargados de las estadísticas atuneras.

POLONIA

Estimación (13 000 toneladas) de la pesca recreativa incluida en 'Peces de agua dulce nep'.

PORTUGAL

Los datos relativos a las capturas de atún en el Atlántico se revisan en colaboración con la CICAA y la CAOI, los organismos regionales encargados de las estadísticas atuneras.

Los datos sobre las capturas de la zona del Atlántico centro-oriental, con exclusión de las de atunes, también se han extraído de la publicación "Bulletin Statistique" del IMROP, Mauritania.

Notes on individual countries or areas

PUERTO RICO

Data for 'Stromboid conchs nei' have been converted to live weight equivalents by using the conversion factor '7.5'.

RÉUNION

Data concerning tuna catches are reviewed in collaboration with IOTC, the regional agency concerned with tuna statistics.

RUSSIAN FEDERATION

Data for tuna species not comparable with those of IATTC, ICCAT, IOTC and SPC.

SAINT HELENA

Includes the islands of Ascensión and Tristan da Cunha.

Data for 'Tristan da Cunha rock lobster' (*Jasus tristani*), octopuses and, since 1994, for most of the quantities included under 'Marine fishes nei' refer to catches from the Tristan da Cunha area. A portion of these catches are taken by foreign fleets.

SAINT VINCENT/GRENADINES

Data concerning tuna catches are reviewed in collaboration with ICCAT, the regional agency concerned with tuna statistics.

Catch data, excluding those for tuna species, for the Eastern Central and Southeast Atlantic areas are provided by the Las Palmas Survey (LPS) and the "Bulletin Statistique" published by IMROP, Mauritania.

Catch data, excluding those for tuna species, for the Southwest Atlantic area are provided by the Fisheries Department, Falkland Islands Government.

SAMOA

Since 1993, data concerning tuna catches are reviewed in collaboration with the SPC, the regional agency concerned with tuna statistics.

SAUDI ARABIA

Data concerning tuna catches are reviewed in collaboration with IOTC, the regional agency concerned with tuna statistics.

SENEGAL

Inland water statistics for the period 1994-99 have been revised to exclude quantities probably referring to marine species already included in marine capture statistics.

SERBIA AND MONTENEGRO

Formerly Yugoslavia, Fed. Rep. of.

Notes sur divers pays ou zones

PORTO RICO

Les données relatives aux 'Strombes nca' ont été converties en équivalents poids vif en utilisant le facteur de conversion '7,5'.

RÉUNION

Les données concernant les captures des thonidés sont révisées en collaboration avec la CTOI, l'organisation régionale chargée des statistiques des thonidés.

FÉDÉRATION DE RUSSIE

Les données concernant les thonidés ne sont pas comparables avec les renseignements fournis par la CITT, la CICTA, la CTOI et la CPS.

SAINTE-HÉLÈNE

Comprend les îles de l'Ascension et Tristan da Cunha.

Les données concernant la 'Langouste de Tristan da Cunha' (*Jasus tristani*), les poulpes et, depuis 1994, pour la plupart des quantités incluses sous 'Poissons marins nca' se réfèrent aux captures effectuées dans la région de Tristan da Cunha. Une partie de ces quantités sont capturées par les flottes étrangères.

SAINT-VINCENT/GRENADINES

Les données concernant les captures des thonidés sont révisées en collaboration avec la CICTA, l'organisation régionale chargée des statistiques des thonidés.

Les données sur les captures, à l'exclusion de celles concernant le thon, pour l'Atlantique du centre-est et du sud-est sont tirées de l'Enquête Las Palmas et du "Bulletin statistique" publié par l'IMROP, Mauritanie.

Les données sur les captures, à l'exclusion de celles concernant le thon, pour l'Atlantique sud-ouest, sont fournies par le Département des pêches du Gouvernement des Îles Falkland.

SAMOA

Depuis 1993 les données concernant les captures des thonidés sont révisées en collaboration avec la CPS, l'organisation régionale chargée des statistiques des thonidés.

ARABIE SAOUDITE

Les données concernant les captures des thonidés sont révisées en collaboration avec la CTOI, l'organisation régionale chargée des statistiques des thonidés.

SÉNÉGAL

Les statistiques des pêches dans les eaux intérieures pour la période 1994-99, ont été révisées afin d'exclure des quantités qui se référaient probablement aux espèces marines déjà comprises dans les statistiques de pêche maritime.

SERBIE-ET-MONTÉNÉGRO

Anciennement Rép. féd. de Yougoslavie.

Notas sobre los distintos países o áreas

PUERTO RICO

Los datos de 'Cobos nep' se han convertido en sus equivalentes de peso en vivo aplicando el factor de conversión '7,5'.

REUNIÓN

Los datos relativos a las capturas de atún se revisan en colaboración con la CAOI, el organismo regional encargado de las estadísticas atuneras.

FEDERACIÓN DE RUSIA

Los datos sobre las especies de atún no son comparables con la información de la CIAT, la CICAA, la CAOI y la SCP.

SANTA ELENA

Incluye las islas de Ascensión y Tristán de Cunha.

Los datos relativos a la 'Langosta de Tristán da Cunha' (*Jasus tristani*), los pulpos y, desde 1994, para la mayoría de las cantidades incluidas en 'Peces marinos nep' se refieren a capturas desde el área de Tristan da Cunha. Una porción de estas cantidades se captura por las flotas extranjeras.

SAN VICENTE/GRENADINAS

Los datos relativos a las capturas de atún en el Atlántico se revisan en colaboración con la CICAA, el organismo regional encargado de las estadísticas atuneras.

Los datos relativos a las capturas en las zonas del Atlántico centro-oriental y Sudoriental, con exclusión de las de atunes, fueron proporcionados por Las Palmas Survey (LPS) y por la publicación "Bulletin Statistique" del IMROP, Mauritania.

Los datos relativos a las capturas en la zona del Atlántico Sudoccidental, con exclusión de las de atunes, fueron proporcionados por el Departamento de Pesca del Gobierno de las Islas Malvinas (Falkland).

SAMOA

Desde 1993 los datos relativos a las capturas de atún se revisan en colaboración con la SCP, el organismo regional encargado de las estadísticas atuneras.

ARABIA SAUDITA

Los datos relativos a las capturas de atún se revisan en colaboración con la CAOI, el organismo regional encargado de las estadísticas atuneras.

SENEGAL

Las estadísticas de la pesca continental para el período 1994-99 han sido revisadas para excluir cantidades que probablemente se referían a especies marinas ya incluidas en las estadísticas de pesca marina.

SERBIA Y MONTENEGRO

Antes Rep. Fed. de Yugoslavia.

Notes on individual countries or areas

SEYCHELLES

Data concerning tuna catches are reviewed in collaboration with ICCAT and IOTC, the regional agencies concerned with tuna statistics.

Increased catches of tuna species since 1997 due to fishing activities of purse seiners re-flagged under Seychelles flag.

SOLOMON ISLANDS

Data concerning tuna catches are reviewed in collaboration with SPC, the regional agency concerned with tuna statistics.

SOMALIA

Since 1991, the only data available have been estimates of the total marine capture production in 1994 provided by the FAO Representative in Somalia. Approximately half of the total capture production for 1994 was made of shark catches but this datum is included under "Marine fishes nei", as no shark statistics are available for other years.

SOUTH AFRICA

Data concerning tuna catches are reviewed in collaboration with ICCAT and IOTC, the regional agencies concerned with tuna statistics.

Data for brown and red seaweeds recorded on a dry-weight basis have been converted to wet-weight equivalents by using the conversion factor '8'.

SPAIN

Data concerning tuna catches are reviewed in collaboration with IATTC, ICCAT, IOTC and SPC, the regional agencies concerned with tuna statistics.

Catch data, excluding those for tuna species, for the Eastern Central Atlantic area have also been taken from the "Bulletin Statistique" published by IMROP, Mauritania.

Catch data, excluding those for tuna species, for the Southwest Atlantic area are provided also by the Fisheries Department, Falkland Islands Government.

SRI LANKA

In 1999, fishery statistics formerly assigned to area 51 (Western Indian Ocean) have been moved to area 57 (Eastern Indian Ocean) following a modification of the boundary between the two areas.

Data concerning tuna catches are reviewed in collaboration with IOTC, the regional agency concerned with tuna statistics.

Up to 1996, data for 'Silky shark' estimated as 75% of the officially reported total catch data for 'Sharks, rays, skates, etc.'.

TAIWAN PROVINCE OF CHINA

Data for tuna species not comparable with those of IATTC, ICCAT, IOTC and SPC.

Notes sur divers pays ou zones

SEYCHELLES

Les données concernant les captures des thonidés sont révisées en collaboration avec la CICTA et la CTOI, les organisations régionales chargées des statistiques des thonidés.

L'augmentation de captures concernant le thon depuis 1997 s'explique par les activités de pêche de quatre senneurs à senne coulissante battant pavillon des Seychelles.

ÎLES SALOMON

Les données concernant les captures des thonidés sont révisées en collaboration avec la CPS, l'organisation régionale chargée des statistiques des thonidés.

SOMALIE

Depuis 1991, les seules données disponibles sont des estimations des captures marines totales en 1994 fournies par le Représentant de la FAO en Somalie. Environ la moitié des captures totales pour 1994 était composée de requins, mais ces données sont comprises sous "Poissons marins nca" car on ne dispose pas de statistiques sur les requins pour les autres années.

AFRIQUE DU SUD

Les données concernant les captures des thonidés sont révisées en collaboration avec la CICTA et la CTOI, les organisations régionales chargées des statistiques des thonidés.

Les données relatives aux algues brunes et rouges enregistrées en poids sec, ont été converties en leurs équivalents poids vert en utilisant le facteur de conversion '8'.

ESPAGNE

Les données concernant les captures des thonidés sont révisées en collaboration avec la CITT, la CICTA, la CTOI et la CPS, les organisations régionales chargées des statistiques des thonidés.

Les données sur les captures, à l'exclusion de celles concernant le thon, pour l'Atlantique centre-est ont également été tirées du "Bulletin statistique" publié par l'IMROP, Mauritanie.

Les données sur les captures, à l'exclusion de celles concernant le thon, pour l'Atlantique sud-ouest, sont fournies également par le Département des pêches du Gouvernement des Îles Falkland.

SRI LANKA

En 1999 les statistiques des pêches ont été déplacées de la zone 51 (l'océan Indien ouest) à la zone 57 (l'océan Indien est) à la suite d'une modification des limites entre les deux zones.

Les données concernant les captures des thonidés sont révisées en collaboration avec la CTOI, l'organisation régionale chargée des statistiques des thonidés.

Jusqu'à l'année 1996 les captures de 'Requin soyeux' sont des estimations calculées comme 75% des quantités totales relatives aux 'Requins, raies, etc.' communiquées par le service national.

TAÏWAN PROVINCE DE CHINE

Les données concernant les thonidés ne sont pas comparables avec les renseignements fournis par la CITT, la CICTA, la CTOI et la CPS.

Notas sobre los distintos países o áreas

SEYCHELLES

Los datos relativos a las capturas de atún se revisan en colaboración con la CICAA y la CAOI, los organismos regionales encargados de las estadísticas atuneras.

El incremento desde 1997 de las capturas para atunes se debe a los desembarques de cuatro cerqueros con jareta con pabellón de Seychelles.

ISLAS SALOMÓN

Los datos relativos a las capturas de atún se revisan en colaboración con la SCP, el organismo regional encargado de las estadísticas atuneras.

SOMALIA

Desde 1991 los únicos datos disponibles han sido las estimaciones de la producción total de capturas marinas en 1994 proporcionadas por el Representante de la FAO en Somalia. Aproximadamente la mitad de la producción total de capturas en 1994 correspondía a tiburones, pero este dato se ha incluido en "Peces marinos nep" por no disponerse de estadísticas sobre tiburones relativas a otros años.

SUDÁFRICA

Los datos relativos a las capturas de atún se revisan en colaboración con la CICAA y la CAOI, los organismos regionales encargados de las estadísticas atuneras.

Los datos relativos a las algas pardas y rojas registrados en peso en seco se han convertido en sus equivalentes de peso húmedo aplicando el factor de conversión '8'.

ESPAÑA

Los datos relativos a las capturas de atún se revisan en colaboración con la CIAT, la CICAA, la CAOI y la SCP, los organismos regionales encargados de las estadísticas atuneras.

Los datos sobre las capturas de la zona del Atlántico centro-oriental, con exclusión de las de atunes, también se han extraído de la publicación "Bulletin Statistique" del IMROP, Mauritania.

Los datos relativos a las capturas en la zona del Atlántico Sudoccidental, con exclusión de las de atunes, fueron proporcionados por el Departamento de Pesca del Gobierno de las Islas Malvinas (Falkland).

SRI LANKA

En 1999 las estadísticas pesqueras se desplazaron de la zona 51 (océano Índico occidental) a la zona 57 (océano Índico oriental) tras una modificación del límite entre ambas.

Los datos relativos a las capturas de atún se revisan en colaboración con la CAOI, el organismo regional encargado de las estadísticas atuneras.

Hasta el 1996 la captura de 'Tiburón jaquetón' ha sido estimada según el 75% del total de 'Tiburones, rayas, etc.' proporcionado por la oficina nacional.

TAIWÁN PROVINCIA DE CHINA

Los datos sobre las especies de atún no son comparables con la información de la CIAT, la CICAA, la CAOI y la SCP.

Notes on individual countries or areas

TANZANIA, UNITED REP. OF

Data refer only to capture production of mainland Tanzania and do not include captures of the Zanzibar state.

Data concerning tuna catches are reviewed in collaboration with IOTC, the regional agency concerned with tuna statistics.

THAILAND

Data concerning tuna catches in the Indian Ocean are reviewed in collaboration with IOTC, the regional agency concerned with tuna statistics.

TIMOR-LESTE

Formerly part of Indonesia.

In 2001, fishery statistics formerly assigned to area 71 (Western Central Pacific) have been moved to area 57 (Eastern Indian Ocean) following a modification of the boundary between the two areas.

Data concerning tuna catches in the Indian Ocean are reviewed in collaboration with IOTC, the regional agency concerned with tuna statistics.

TONGA

Data concerning tuna catches are reviewed in collaboration with the SPC, the regional agency concerned with tuna statistics.

TRINIDAD AND TOBAGO

Data refer only to industrial catches of tunas and shrimps and to artisanal catches from the island of Trinidad.

Data concerning tuna catches are reviewed in collaboration with ICCAT, the regional agency concerned with tuna statistics.

TUNISIA

Data concerning tuna catches are reviewed in collaboration with ICCAT, the regional agency concerned with tuna statistics.

TURKEY

Data concerning tuna catches are reviewed in collaboration with ICCAT, the regional agency concerned with tuna statistics.

TURKS AND CAICOS IS.

Data refer to a split-year fishing season (1 August – 31 July) shown under the calendar year in which the fishing season ends.

Data for 'Stromboid conchs nei' have been converted to live weight equivalents by using the conversion factor '7.5'.

Notes sur divers pays ou zones

RÉP.-UNIE DE TANZANIE

Les données se réfèrent seulement aux captures de la Tanzanie continentale et n'incluent pas les captures de l'état de Zanzibar.

Les données concernant les captures des thonidés sont révisées en collaboration avec la CTOI, l'organisation régionale chargée des statistiques des thonidés.

THAÏLANDE

Les données concernant les captures des thonidés dans l'océan Indien sont révisées en collaboration avec la CTOI, l'organisation régionale chargée des statistiques des thonidés.

TIMOR-LESTE

Anciennement partie de l'Indonésie.

En 2001 les statistiques des pêches ont été déplacées de la zone 71 (Pacifique centre-ouest) à la zone 57 (Océan Indien est) à la suite d'une modification des limites entre les deux zones.

Les données concernant les captures des thonidés dans l'océan Indien sont révisées en collaboration avec la CTOI, l'organisation régionale chargée des statistiques des thonidés.

TONGA

Les données concernant les captures des thonidés sont révisées en collaboration avec la CPS, l'organisation régionale chargée des statistiques des thonidés.

TRINITÉ-ET-TOBAGO

Les données se réfèrent seulement aux captures industrielles des thons et des crevettes et aux captures artisanales de l'île de Trinité.

Les données concernant les captures des thonidés sont révisées en collaboration avec la CICTA, l'organisation régionale chargée des statistiques des thonidés.

TUNISIE

Les données concernant les captures des thonidés sont révisées en collaboration avec la CICTA, l'organisation régionale chargée des statistiques des thonidés.

TURQUIE

Les données concernant les captures des thonidés sont révisées en collaboration avec la CICTA, l'organisation régionale chargée des statistiques des thonidés.

ÎLES TURQUES ET CAÏQUES

Les données se réfèrent à une période de pêche fractionnée (1er août – 31 juillet) et figurent sous l'année civile durant laquelle se termine la période de pêche.

Les données relatives aux 'Strombes nca' ont été converties en équivalents poids vif en utilisant le facteur de conversion '7,5'.

Notas sobre los distintos países o áreas

REP. UNIDA DE TANZANÍA

Los datos se refieren solamente a las capturas de la Tanzania continental y no incluyen las capturas del estado de Zanzibar.

Los datos relativos a las capturas de atún se revisan en colaboración con la CAOI, el organismo regional encargado de las estadísticas atuneras.

TAILANDIA

Los datos relativos a las capturas de atún en el océano Índico se revisan en colaboración con la CAOI, el organismo regional encargado de las estadísticas atuneras.

TIMOR-LESTE

Antes parte de Indonesia.

En 2001 las estadísticas pesqueras se desplazaron de la zona 71 (Pacífico centro-occidental) a la zona 57 (océano Índico oriental) tras una modificación del límite entre ambas.

Los datos relativos a las capturas de atún en el océano Índico se revisan en colaboración con la CAOI, el organismo regional encargado de las estadísticas atuneras.

TONGA

Los datos relativos a las capturas de atún se revisan en colaboración con la SCP, el organismo regional encargado de las estadísticas atuneras.

TRINIDAD Y TABAGO

Los datos se refieren solamente a las capturas industriales de atunes y camarones y a las capturas artesanales de la isla de Trinidad.

Los datos relativos a las capturas de atún en el Atlántico se revisan en colaboración con la CICAA, el organismo regional encargado de las estadísticas atuneras.

TÚNEZ

Los datos relativos a las capturas de atún en el Atlántico se revisan en colaboración con la CICAA, el organismo regional encargado de las estadísticas atuneras.

TURQUÍA

Los datos relativos a las capturas de atún en el Atlántico se revisan en colaboración con la CICAA, el organismo regional encargado de las estadísticas atuneras.

ISLAS TURCAS Y CAICOS

Los datos se refieren a una temporada de pesca (1 de agosto – 31 de julio) que se indica como el año civil en que finaliza la temporada.

Los datos de 'Cobos nep' se han convertido en sus equivalentes de peso en vivo aplicando el factor de conversión '7,5'.

Notes on individual countries or areas

Notes sur divers pays ou zones

Notas sobre los distintos países o áreas

UK

Data for England and Wales, Scotland and Northern Ireland have been combined.

Increase of catches in inland waters since 1998 due to inclusion of significant quantities from recreational fisheries.

Catch data, excluding those for tuna species, for the Southwest Atlantic area are provided by the Fisheries Department, Falkland Islands Government.

Data concerning tuna catches catches in the Indian Ocean are reviewed in collaboration with IOTC, the regional agency concerned with tuna statistics.

ROYAUME-UNI

Les chiffres concernent l'Angleterre et le pays de Galles, l'Ecosse et l'Irlande du Nord.

L'augmentation des captures dans les eaux continentales depuis 1998 s'explique par l'inclusion de quantités importantes de pêche récréative.

Les données sur les captures, à l'exclusion de celles concernant le thon, pour l'Atlantique sud-ouest, sont fournies par le Département des pêches du Gouvernement des Îles Falkland.

Les données concernant les captures des thonidés dans l'océan Indien sont révisées en collaboration avec la CTOI, l'organisation régionale chargée des statistiques des thonidés.

REINO UNIDO

Se han agrupado los datos sobre Inglaterra y Gales, Escocia e Irlanda del Norte.

El incremento de las capturas en aguas continentales desde 1998 se debe a la inclusión de importantes cantidades de pesca recreativa.

Los datos relativos a las capturas en la zona del Atlántico Sudoccidental, con exclusión de las de atunes, fueron proporcionados por el Departamento de Pesca del Gobierno de las Islas Malvinas (Falkland).

Los datos relativos a las capturas de atún en el océano Índico se revisan en colaboración con la CAOI, el organismo regional encargado de las estadísticas atuneras.

UKRAINE

Catch data, excluding those for tuna species, for the Eastern Central Atlantic area have also been taken from the "Bulletin Statistique" published by IMROP, Mauritania.

Catch data for inland waters are considered underestimated in recent years.

UKRAINE

Les données sur les captures, à l'exclusion de celles concernant le thon, pour l'Atlantique centre-est ont également été tirées du "Bulletin statistique" publié par l'IMROP, Mauritanie.

Les captures en eaux continentales sont considérées sous-estimées ces dernières années.

UCRANIA

Los datos sobre las capturas de la zona del Atlántico centro-oriental, con exclusión de las de atunes, también se han extraído de la publicación "Bulletin Statistique" del IMROP, Mauritania.

Las capturas en aguas continentales se consideran subestimadas en años recientes.

URUGUAY

Data concerning tuna catches are reviewed in collaboration with ICCAT, the regional agency concerned with tuna statistics.

URUGUAY

Les données concernant les captures des thonidés sont révisées en collaboration avec la CICTA, l'organisation régionale chargée des statistiques des thonidés.

URUGUAY

Los datos relativos a las capturas de atún en el Atlántico se revisan en colaboración con la CICAA, el organismo regional encargado de las estadísticas atuneras.

USA

Capture data for recreational fisheries are not included.

Data for tuna species not comparable with those of IATTC, ICCAT and SPC.

Catch data officially reported for selected clams, mussels and oysters, have been revised to exclude quantities derived from aquaculture.

Due to confidentiality reasons, data for calico scallop (*Argopecten gibbus*) are not available since 1996.

Since 1985, data for seals refer exclusively to harvest for subsistence.

ÉTATS-UNIS D'AMÉRIQUE

Les données relatives aux captures ne comprennent pas la pêche récréative.

Les données concernant les thonidés ne sont pas comparables avec les renseignements fournis par la CITT, la CICTA et la CPS.

Les données relatives aux captures de certaines clams, moules et huîtres, reportées officiellement, ont été révisées afin d'exclure les quantités provenant de l'aquaculture.

Le manque de données concernant le peigne calicot (*Argopecten gibbus*) depuis 1996 s'explique en raison de leur caractère confidentiel.

Depuis 1985 les données concernant les otaries se réfèrent exclusivement à la pêche destinée à la subsistance.

ESTADOS UNIDOS DE AMÉRICA

Los datos de capturas no incluyen la pesca recreativa.

Los datos sobre las especies de atún no son comparables con la información de la CIAT, la CICAA y la SCP.

Las capturas relativas a algunas especies de almejas, mejillones y ostras notificadas por parte de la oficina nacional, se han revisado para que excluyan cantidades procedentes de la acuicultura.

La falta de los datos para peine percal (*Argopecten gibbus*) desde 1996 se debe al carácter confidencial de la información.

Desde 1985 los datos sobre los lobos se refieren exclusivamente a la recogida hecha con fin de subsistencia.

US VIRGIN ISLANDS

Includes Saint Croix, Saint John and Saint Thomas.

Data refer to a split-year (1 July – 30 June) shown under the calendar year in which the split-year ends.

ÎLES VIERGES AMÉRICAINES

Comprend Sainte-Croix, Saint-John et Saint-Thomas.

Les données se réfèrent à une année fractionnée (1er juillet - 30 juin) et figurent sous l'année civile durant laquelle se termine l'année fractionnée.

ISLAS VÍRGENES DE LOS EU

Incluye Sainte Croix, Saint John y Saint Thomas.

Los datos se refieren a un año emergente (1 de julio - 30 de junio) que se indica como el año civil en que finaliza el año emergente.

VANUATU

Data concerning tuna catches are reviewed in collaboration with IATTC and SPC, the regional agencies concerned with tuna statistics. Increase of tuna catches in the Western Central Pacific since 1995 due to vessels fishing with flag of convenience. In 2001, these vessels were re-flagged in the Marshall Islands.

Catch data, excluding those for tuna species, for the Eastern Central and Southeast Atlantic areas are provided by the Las Palmas Survey (LPS).

VANUATU

Les données concernant les captures des thonidés sont révisées en collaboration avec la CITT et la CPS, les organisations régionales chargées des statistiques des thonidés. A partir de 1995, l'augmentation des captures pour les thonidés dans le Pacifique centre-ouest s'explique par les captures effectuées sous pavillon de complaisance. En 2001 ces bateaux ont été transférés en pavillon des Îles Marshall.

Les données sur les captures, à l'exclusion de celles concernant le thon, pour l'Atlantique du centre-est et du sud-est sont tirées de l'Enquête Las Palmas.

VANUATU

Los datos relativos a las capturas de atún se revisan en colaboración con la CIAT y la SCP, los organismos regionales encargados de las estadísticas atuneras. Desde 1995 el incremento de las capturas de atún en el Pacífico centro-occidental se debe a las capturas de pabellones de conveniencia. En 2001 estas embarcaciones cambiaron de pabellón en las Islas Marshall.

Los datos relativos a las capturas en las zonas del Atlántico centro-oriental y Sudoriental, con exclusión de las de atunes, fueron proporcionados por Las Palmas Survey (LPS).

Notes on individual countries or areas

VENEZUELA

Data concerning tuna catches are reviewed in collaboration with IATTC and ICCAT, the regional agencies concerned with tuna statistics.

VIET NAM

The 1986-99 catch data have been extensively revised on the basis of the publication "Statistical data of Vietnam agriculture, forestry and fishery 1975-2000" published by the General Statistical Office, and of trade data.

YEMEN

Includes Kamaran Islands, Perim and Socotra.

The 1982-95 catch data for 'Pelagic Percomorphs' include quantities of 'Indian oil sardine'.

ZAMBIA

Data for 'Stolothrissa, Limnothrissa spp' refer only to the capture fisheries of Lake Kariba.

ZIMBABWE

Data refer only to the capture fisheries of Lake Kariba.

Notes sur divers pays ou zones

VENEZUELA

Les données concernant les captures des thonidés sont révisées en collaboration avec la CITT et la CICTA, les organisations régionales chargées des statistiques des thonidés.

VIET NAM

Les données sur les captures de la période 1986-99 ont été largement révisées en fonction de la publication "Statistical data of Vietnam agriculture, forestry and fishery 1975-2000" publié par le Bureau Général de Statistique, et des statistiques du commerce.

YÉMEN

Comprend les îles Kamaran, Perim et Socotra.

Les données de 1982-95 concernant les captures de 'Percomorphes pélagiques' comprennent des quantités de 'Sardinelle indienne'.

ZAMBIE

Les données relatives aux 'Stolothrissa, Limnothrissa spp' se réfèrent exclusivement aux pêches de capture du lac Kariba.

ZIMBABWE

Les données se réfèrent exclusivement aux pêches de capture du lac Kariba.

Notas sobre los distintos países o áreas

VENEZUELA

Los datos relativos a las capturas de atún se revisan en colaboración con la CIAT y la CICAA, los organismos regionales encargados de las estadísticas atuneras.

VIET NAM

Los datos de las capturas de 1986-99 han sido revisados ampliamente para tener en cuenta la publicación "Statistical data of Vietnam agriculture, forestry and fishery 1975-2000" publicada por la Dirección General de Estadística, y de las estadísticas de comercio.

YEMEN

Incluyen las islas Kamaran, Perim y Socotra.

Los datos de 1982-95 relativos a las capturas de de 'Percomorfos pelágicos' incluyen cantidades de 'Sardinella aceitera'.

ZAMBIA

Los datos de 'Stolothrissa, Limnothrissa spp' se refieren únicamente a las pesquerías de capturas del lago Kariba.

ZIMBABWE

Los datos se refieren únicamente a las pesquerías de capturas del lago Kariba.

Index of FAO English, French, Spanish and scientific names

Index des noms scientifiques et des noms FAO en anglais, français et espagnol

Índice de nombres científicos y de nombres FAO en inglés, francés y español

Index of FAO English, French, Spanish and scientific names

Index des noms scientifiques et des noms FAO en anglais, français et espagnol

Índice de nombres científicos y de nombres FAO en inglés, francés y español

Index of FAO English, French, Spanish and scientific names

Index des noms scientifiques et des noms FAO en anglais, français et espagnol

Índice de nombres científicos y de nombres FAO en inglés, francés y español

Index of FAO English, French, Spanish and scientific names | **Index des noms scientifiques et des noms FAO en anglais, français et espagnol** | **Índice de nombres científicos y de nombres FAO en inglés, francés y español**

Index of FAO English, French, Spanish and scientific names

Index des noms scientifiques et des noms FAO en anglais, français et espagnol

Índice de nombres científicos y de nombres FAO en inglés, francés y español

Index of FAO English, French, Spanish and scientific names

Index des noms scientifiques et des noms FAO en anglais, français et espagnol

Índice de nombres científicos y de nombres FAO en inglés, francés y español

Index of FAO English, French, Spanish and scientific names

Index des noms scientifiques et des noms FAO en anglais, français et espagnol

Índice de nombres científicos y de nombres FAO en inglés, francés y español

Index of FAO English, French, Spanish and scientific names | **Index des noms scientifiques et des noms FAO en anglais, français et espagnol** | **Índice de nombres científicos y de nombres FAO en inglés, francés y español**

Index of FAO English, French, Spanish and scientific names

Index des noms scientifiques et des noms FAO en anglais, français et espagnol

Índice de nombres científicos y de nombres FAO en inglés, francés y español

Index of FAO English, French, Spanish and scientific names

Index des noms scientifiques et des noms FAO en anglais, français et espagnol

Índice de nombres científicos y de nombres FAO en inglés, francés y español

Index of FAO English, French, Spanish and scientific names

Index des noms scientifiques et des noms FAO en anglais, français et espagnol

Índice de nombres científicos y de nombres FAO en inglés, francés y español

Index of FAO English, French, Spanish and scientific names

Index des noms scientifiques et des noms FAO en anglais, français et espagnol

Índice de nombres científicos y de nombres FAO en inglés, francés y español

Index of FAO English, French, Spanish and scientific names

Index des noms scientifiques et des noms FAO en anglais, français et espagnol

Índice de nombres científicos y de nombres FAO en inglés, francés y español

Index of FAO English, French, Spanish and scientific names

Index des noms scientifiques et des noms FAO en anglais, français et espagnol

Índice de nombres científicos y de nombres FAO en inglés, francés y español

Index of FAO English, French, Spanish and scientific names

Index des noms scientifiques et des noms FAO en anglais, français et espagnol

Índice de nombres científicos y de nombres FAO en inglés, francés y español

Index of FAO English, French, Spanish and scientific names

Index des noms scientifiques et des noms FAO en anglais, français et espagnol

Índice de nombres científicos y de nombres FAO en inglés, francés y español

Index of FAO English, French, Spanish and scientific names | **Index des noms scientifiques et des noms FAO en anglais, français et espagnol** | **Índice de nombres científicos y de nombres FAO en inglés, francés y español**

Index of FAO English, French, Spanish and scientific names

Index des noms scientifiques et des noms FAO en anglais, français et espagnol

Índice de nombres científicos y de nombres FAO en inglés, francés y español

Index of FAO English, French, Spanish and scientific names

Index des noms scientifiques et des noms FAO en anglais, français et espagnol

Índice de nombres científicos y de nombres FAO en inglés, francés y español

Index of FAO English, French, Spanish and scientific names

Index des noms scientifiques et des noms FAO en anglais, français et espagnol

Índice de nombres científicos y de nombres FAO en inglés, francés y español

Index of FAO English, French, Spanish and scientific names

Index des noms scientifiques et des noms FAO en anglais, français et espagnol

Índice de nombres científicos y de nombres FAO en inglés, francés y español

Index of FAO English, French, Spanish and scientific names

Index des noms scientifiques et des noms FAO en anglais, français et espagnol

Índice de nombres científicos y de nombres FAO en inglés, francés y español

Index of FAO English, French, Spanish and scientific names

Index des noms scientifiques et des noms FAO en anglais, français et espagnol

Índice de nombres científicos y de nombres FAO en inglés, francés y español

Index of FAO English, French, Spanish and scientific names

Index des noms scientifiques et des noms FAO en anglais, français et espagnol

Índice de nombres científicos y de nombres FAO en inglés, francés y español

Index of FAO English, French, Spanish and scientific names

Index des noms scientifiques et des noms FAO en anglais, français et espagnol

Índice de nombres científicos y de nombres FAO en inglés, francés y español

Index of FAO English, French, Spanish and scientific names

Index des noms scientifiques et des noms FAO en anglais, français et espagnol

Índice de nombres científicos y de nombres FAO en inglés, francés y español

Index of FAO English, French, Spanish and scientific names

Index des noms scientifiques et des noms FAO en anglais, français et espagnol

Índice de nombres científicos y de nombres FAO en inglés, francés y español

Index of FAO English, French, Spanish and scientific names

Index des noms scientifiques et des noms FAO en anglais, français et espagnol

Índice de nombres científicos y de nombres FAO en inglés, francés y español

Index of FAO English, French, Spanish and scientific names

Index des noms scientifiques et des noms FAO en anglais, français et espagnol

Índice de nombres científicos y de nombres FAO en inglés, francés y español

List of yearbooks of fishery statistics

Liste des annuaires statistiques des pêches

Lista de los anuarios estadísticos de pesca

Volumes published in 1948-1963 | Volumes publiés en 1948-1963 | Volúmenes publicados en 1948-1963

Production / Production / Producción		*Production and fishing craft / Production et bateaux de pêche / Producción y embarcaciones de pesca*			*International Trade / Commerce international / Comercio internacional*		
		‣ Vol. I	*(1947)*				
		‣ Vol.II	*(1948-49)*				
		‣ Vol.III	*(1950-51)*				
		‣ Vol.IV	*(1952-53)*	Part 1	‣ Vol.IV	*(1952-53)*	Part 2
		‣ Vol.V	*(1954-55)*				
		‣ Vol.VI	*(1955-56)*				
‣ Vol.VII	*(1957)*				‣ Vol.VIII	*(1957)*	
		‣ Vol.IX	*(1958)*		‣ Vol.X	*(1958-59)*	
‣ Vol.XI	*(1959)*				‣ Vol.XIII	*(1960-61)*	
‣ Vol.XIV	*(1961)*	‣ Vol.XII	*(1960)*				
		‣ Vol.XV	*(1962)*				

Volumes published in 1964-1997 | Volumes publiés en 1964-1997 | Volúmenes publicados en 1964-1997

'Catches and landings' / 'Captures et quantités débarquées' / 'Capturas y desembarques'			*'Fishery Commodities' / 'Produits des pêches' / 'Productos pesqueros'*		
‣ Vol.16	('----' *1963*)	Dec. 1964	‣ Vol.17	('----' *1963*)	Jan. 1965
‣ Vol.18	('----' *1964*)	Oct. 1965	‣ Vol.19	('----' *1964*)	Dec. 1965
‣ Vol.20	('----' *1965*)	Oct. 1966	‣ Vol.21	('----' *1965*)	Dec. 1966
‣ Vol.22	('----' *1966*)	Oct. 1967	‣ Vol.23	('----' *1966*)	Dec. 1967
‣ Vol.24	('----' *1967*)	Oct. 1968	‣ Vol.25	('----' *1967*)	Dec. 1968
‣ Vol.26	('----' *1968*)	Oct. 1969	‣ Vol.27	('----' *1968*)	Dec. 1969
‣ Vol.28	('----' *1969*)	Oct. 1970	‣ Vol.29	('----' *1969*)	Dec. 1970
‣ Vol.30	('----' *1970*)	Nov. 1971	‣ Vol.31	('----' *1970*)	Dec. 1971
‣ Vol.32	('----' *1971*)	Nov. 1972	‣ Vol.33	('----' *1971*)	Dec. 1972
‣ Vol.34	('----' *1972*)	Nov. 1973	‣ Vol.35	('----' *1972*)	Dec. 1973
‣ Vol.36	('----' *1973*)	Nov. 1974	‣ Vol.37	('----' *1973*)	Dec. 1974
‣ Vol.38	('----' *1974*)	Dec. 1975	‣ Vol.39	('----' *1974*)	Dec. 1975
‣ Vol.40	('----' *1975*)	Dec. 1976	‣ Vol.41	('----' *1975*)	Dec. 1976
‣ Vol.42	('----' *1976*)	Nov. 1977	‣ Vol.43	('----' *1976*)	Dec. 1977
‣ Vol.44	('----' *1977*)	Nov. 1978	‣ Vol.45	('----' *1977*)	Dec. 1978
‣ Vol.46	('----' *1978*)	Nov. 1979	‣ Vol.47	('----' *1978*)	Dec. 1979
‣ Vol.48	('----' *1979*)	Dec. 1980	‣ Vol.49	('----' *1979*)	Dec. 1980
‣ Vol.50	('----' *1980*)	Dec. 1981	‣ Vol.51	('----' *1980*)	Dec. 1981
‣ Vol.52	('----' *1981*)	Jan. 1983	‣ Vol.53	('----' *1981*)	Feb. 1983
‣ Vol.54	('----' *1982*)	Jan. 1984	‣ Vol.55	('----' *1982*)	Jan. 1984
‣ Vol.56	('----' *1983*)	Dec. 1984	‣ Vol.57	('----' *1983*)	Dec. 1984
‣ Vol.58	('----' *1984*)	Jun. 1986	‣ Vol.59	('----' *1984*)	Jun. 1986
Vol.60	('----' *1985*)	May 1987	‣ Vol.61	('----' *1985*)	May 1987
‣ Vol.62	('----' *1986*)	Mar. 1988	‣ Vol.63	('----' *1986*)	Mar. 1988
‣ Vol.64	('----' *1987*)	Mar. 1989	‣ Vol.65	('----' *1987*)	Mar. 1989
‣ Vol.66	('----' *1988*)	Apr. 1990	‣ Vol.67	('----' *1988*)	May 1990
‣ Vol.68	('----' *1989*)	Apr. 1991	Vol.69	('----' *1989*)	May 1991
Vol.70	('----' *1990*)	Apr. 1992	Vol.71	('----' *1990*)	May 1992
Vol.72	('----' *1991*)	Apr. 1993	Vol.73	('----' *1991*)	May 1993
‣ Vol.74	('----' *1992*)	May 1994	Vol.75	('----' *1992*)	May 1994
Vol.76	('----' *1993*)	Apr. 1995	Vol.77	('----' *1993*)	May 1995
Vol.78	('----' *1994*)	Apr. 1996	Vol.79	('----' *1994*)	May 1996
Vol.80	('----' *1995*)	Apr. 1997	Vol.81	('----' *1995*)	May 1997

‣ Out of Print | ‣ Epuisé | ‣ Agotado

List of yearbooks of fishery statistics

Liste des annuaires statistiques des pêches

Lista de los anuarios estadísticos de pesca

Volumes published since 1998 — Volumes publiés à partir de 1998 — Volúmenes publicados a partir de 1998

'Capture production' *'Captures'* *'Capturas'*			*'Aquaculture production'a/* *'Production de l'aquaculture'a/* *'Producción de acuicultura'a/*			*'Fishery Commodities'* *'Produits des pêches'* *'Productos pesqueros'*		
Vol.82	('----' *1996*)	Apr. 1998				Vol.83	('----' *1996*)	Apr. 1998
Vol.84	('----' *1997*)	Apr. 1999				Vol.85	('----' *1997*)	Apr. 1999
Vol.86/1	('----' *1998*)	Apr. 2000	Vol.86/2	('----' *1998*)	Apr. 2000	Vol.87	('----' *1998*)	Apr. 2000
Vol.88/1	('----' *1999*)	Apr. 2001	Vol.88/2	('----' *1999*)	Apr. 2001	Vol.89	('----' *1999*)	Apr. 2001
Vol.90/1	('----' *2000*)	Apr. 2002	Vol.90/2	('----' 2000)	Apr. 2002	Vol.91	('----' *2000*)	Apr. 2002
Vol.92/1	('----' *2001*)	Apr. 2003	Vol.92/2	('----' 2001)	Apr. 2003	Vol.93	('----' *2001*)	Apr. 2003
Vol.94/1	('----' *2002*)	Apr. 2004	Vol.94/2	('----' 2002)	Apr. 2004	Vol.95	('----' *2002*)	Apr. 2004

a/ Aquaculture production statistics were combined with those of capture fisheries and published jointly in the "FAO yearbook. Fishery statistics. Catches and landings" until Volume 80. Since 1989, they were also published separately as "FAO Fisheries Circular No. 815: Aquaculture production statistics".

a/ Les statistiques de production de l'aquaculture étaient confondues avec celles de la production de pêche et publiées dans l'"Annuaire de la FAO. Statistiques des pêches. Captures et quantités débarquées" jusqu'au Volume 80. A partir de 1989, les données étaient aussi publiées séparément dans la "Circulaire de la FAO sur les pêches n° 815: Statistiques de la production de l'aquaculture".

a/ Las estadísticas de producción de acuicultura estaban incluidas en la producción de pesca y publicadas juntas en el "Anuario de la FAO. Estadísticas de pesca. Capturas y desembarques", hasta el Volumen 80. Desde 1989, los datos han sido publicados también por separado en la "Circular de Pesca de la FAO N° 815: Estadísticas de la producción de acuicultura".

Tel.: (+81) 3 3265 7531
Fax: (+81) 3 3265 4656
Maruzen Company Ltd
5-7-1 Heiwajima, Ohta-Ku
Tokyo 143-0006
Tel.: (+81) 3 3763 2259
Fax: (+81) 3 3763 2830
E-mail: o_miyakawa@maruzen.co.jp

• KENYA
Text Book Centre Ltd
Kijabe Street
PO Box 47540, Nairobi
Tel.: +254 2 330 342
Fax: +254 2 22 57 79
Legacy Books
Mezzanine 1, Loita House, Loita Street
Nairobi, PO Box 68077
Tel.: (+254) 2 303853
Fax: (+254) 2 330854
E-mail: info@legacybookshop.com

• LUXEMBOURG
M.J. De Lannoy
202, avenue du Roi
B-1060, Bruxelles (Belgique)
Courriel: jean.de.lannoy@infoboard.be

• MADAGASCAR
Centre d'Information et de Documentation Scientifique et Technique
Ministère de la recherche appliquée au développement
B.P. 6224, Tsimbazaza, Antananarivo

• MALAYSIA
MDC Publishers Printers Sdn Bhd
MDC Building
2717 & 2718, Jalan Parmata Empat
Taman Permata, Ulu Kelang
53300 Kuala Lumpur
Tel.: (+60) 3 41086600
Fax: (+60) 3 41081506
E-mail: inquiries@mdcbd.com.my
Web site: www.mdcppd.com.my

• MAROC
La Librairie Internationale
70, rue T'ssoule
B.P. 302 (RP), Rabat
Tél.: (+212) 37 75 0183
Fax: (+212) 37 75 8661

• MÉXICO
Librería, Universidad Autónoma de Chapingo
56230 Chapingo
Libros y Editoriales S.A.
Av. Progreso N° 202-1° Piso A
Apartado Postal 18922
Col. Escandón, 11800 México D.F.
Correo electrónico: lyesa99@mail.com/ventas@lyesa.com
Mundi Prensa Mexico, S.A.
Río Pánuco, 141 Col. Cuauhtémoc
C.P. 06500, México, DF
Tel.: (+52) 5 533 56 58
Fax: (+52) 5 514 67 99
Correo electrónico:
resavbp@data.net.mx

• NETHERLANDS
Roodveldt Import b.v.
Brouwersgracht 288
1013 HG Amsterdam
Tel.: (+31) 20 622 80 35
Fax: (+31) 20 625 54 93
E-mail: roodboek@euronet.nl
Swets & Zeitlinger b.v.
PO Box 830, 2160 Lisse
Heereweg 347 B, 2161 CA Lisse
E-mail: infono@swets.nl
Web site: www.swets.nl

• NEW ZEALAND
Legislation Direct
c/o Securacopy, PO Box 12 418
1st oor, 242 Thorndon Quay,
Wellington
Tel.: (+64) 4 496 56 94
Fax: (+64) 4 496 56 98
E-mail: Jeanette@legislationdirect.co.nz
Web site: www.gplegislation.co.nz

• NICARAGUA
Librería HISPAMER
Costado Este Univ. Centroamericana
Apartado Postal A-221, Managua
Correo electrónico:
hispamer@munditel.com.ni

• NIGERIA
University Bookshop (Nigeria) Ltd
University of Ibadan, Ibadan

• PAKISTAN
Mirza Book Agency
65 Shahrah-e-Quaid-e-Azam
PO Box 729, Lahore 3

• PARAGUAY
Librería Intercontinental Editora e Impresora S.R.L.
Caballero 270 c/Mcal Estigarribia
Asunción

• PERU
Librería de la Biblioteca Agrícola Nacional - Universidad Nacional Agraria
Av. La Universidad s/n
La Molina, Lima
Tel. : (+51) 1 3493910; Fax: (+51) 1 3493910
Correo electrónico:
ban@lamolina.edu.pe
Web site:
http//tumi.lamolina.edu.pe/ban.htm

• PHILIPPINES
International Booksource Center, Inc.
1127-A Antipolo St, Barangay Valenzuela
Makati City
Tel.: (+63) 2 8966501/8966505/8966507
Fax: (+63) 2 8966497
E-mail: ibcdina@pacific.net.ph

• POLAND
Ars Polona S.A.
ul. Obronców 25
03-933 Warsaw
Tel.: (+48) 22 8261201
Fax: (+48) 22 8266240
E-mail: arspolona@arspolona.com.pl
Web site: http://www.arspolona.com.pl

• PORTUGAL
Livraria Portugal, Dias e Andrade Ltda.
Rua do Carmo, 70-74
Apartado 2681, 1200 Lisboa Codex
Correo electrónico: liv.portugal@mail.telepac.pt

• REPÚBLICA DOMINICANA
CEDAF - Centro para el Desarrollo Agropecuario y Forestal, Inc.
Calle José Amado Soler, 50 - Urban. Paraíso
Apartado Postal, 567-2, Santo Domingo
Tel.: (+001) 809 5440616/5440634/5655603
Fax: (+001) 809 5444727/5676989
Correo electrónico: fda@Codetel.net.do
Web site: www.cedaf.org.do

• RUSSIAN FEDERATION
tsdatelstovo VES MIR
9a, Kolpachniy pereulok
101831 Moscow
Tel.: (+7) 095 9236839/9238568
Fax: (+7) 095 9254269
E-mail: orders@vesmirbooks,ru
Web site: www.vesmirbooks.ru

• SERBIA AND MONTENEGRO
Jugoslovenska Knjiga DD
Terazije 27
POB 36, 11000 Beograd
Tel.: (+381) 11 3340 025
Fax: (+381) 11 3231 079
E-mail: juknjiga@eunet.yu
or babicmius@yahoo.com

• SINGAPORE
Select Books Pte Ltd
Tanglin Shopping Centre
19 Tanglin Road, #03-15,
Singapore 247909
Tel.: (+65) 732 1515
Fax: (+65) 736 0855
E-mail: info@selectbooks.com.sg
Web site: www.selectbooks.com.sg

• SLOVAK REPUBLIC
Institute of Scientific and Technical Information for Agriculture
Samova 9, 950 10 Nitra
Tel.: (+421) 87 522 185
Fax: (+421) 87 525 275
E-mail: uvtip@nr.sanet.sk

• SOUTH AFRICA
Preasidium Books (Pty) Ltd
810 - 4th Street, Wynberg 2090
Tel.: (+27) 11 88 75994
Fax: (+27) 11 88 78138
E-mail: pbooks@global.co.za

• SUISSE
UN Bookshop
Palais des Nations
CH-1211 Genève 1
Site Web: www.un.org
Adeco - Editions Van Diermen
Chemin du Lacuez, 41
CH-1807 Blonay
Tel.: (+41) (0) 21 943 2673
Fax: (+41) (0) 21 943 3605
E-mail: mvandier@ip-worldcom.ch
Münstergass Buchhandlung
Docudisp, PO Box 584
CH-3000 Berne 8
Tel.: (+41) 31 310 2321
Fax: (+41) 31 310 2324
E-mail: docudisp@muenstergass.ch
Web site: www.docudisp.ch

• SURINAME
Vaco n.v. in Suriname
Domineestraat 26, PO Box 1841
Paramaribo

• SWEDEN
Swets Blackwell AB
PO Box 1305, S-171 25 Solna
Tel.: (+46) 8 705 9750
Fax: (+46) 8 27 00 71
E-mail: awahlquist@se.swetsblackwell.com
Web site: www.swetsblackwell.com/se/
Bokdistributören
c/o Longus Books Import
PO Box 610, S-151 27 Södertälje
Tel.: (+46) 8 55 09 49 70
Fax: (+46) 8 55 01 76 10; E-mail:
lis.ledin@hk.akademibokhandeln.se

• THAILAND
Suksapan Panit
Mansion 9, Rajdamnern Avenue,
Bangkok

• TOGO
Librairie du Bon Pasteur
B.P. 1164, Lomé

• TRINIDAD AND TOBAGO
Systematics Studies Limited
St Augustine Shopping Centre
Eastern Main Road, St Augustine
Tel.: (+001) 868 645 8466
Fax: (+001) 868 645 8467
E-mail: tobe@trinidad.net

• TURKEY
DUNYA ACTUEL A.S.
"Globus" Dunya Basinevi
100. Yil Mahallesi
34440 Bagcilar, Istanbul
Tel.: (+90) 212 629 0808
Fax: (+90) 212 629 4689
E-mail: aktuel.info@dunya.comr
Web site: www.dunyagazetesi.com.tr/

• UNITED ARAB EMIRATES
Al Rawdha Bookshop
PO Box 5027, Sharjah
Tel.: (+971) 6 538 7933
Fax: (+971) 6 538 4473
E-mail: alrawdha@hotmail.com

• UNITED KINGDOM
The Stationery Office
51 Nine Elms Lane
London SW8 5DR
Tel.: (+44) (0) 870 600 5522 (orders)
(+44) (0) 207 873 8372 (enquiries)
Fax: (+44) (0) 870 600 5533 (orders)
(+44) (0) 207 873 8247 (enquiries)
E-mail: ipa.enquiries@theso.co.uk
Web site: www.clicktso.com
and through The Stationery Office Bookshops
E-mail: postmaster@theso.co.uk
Web site: www.the-stationery-office.co.uk
Intermediate Technology Bookshop
103-105 Southampton Row
London WC1B 4HH
Tel.: (+44) 207 436 9761
Fax: (+44) 207 436 2013
E-mail: orders@itpubs.org.uk
Web site: www.developmentbookshop.com

• UNITED STATES
Publications:
BERNAN Associates (ex UNIPUB)
4611/F Assembly Drive
Lanham, MD 20706-4391
Toll-free: (+1) 800 274 4447
Fax: (+1) 800 865 3450
E-mail: query@bernan.com
Web site: www.bernan.com
United Nations Publications
Two UN Plaza, Room DC2-853
New York, NY 10017
Tel.: (+1) 212 963 8302/800 253 9646
Fax: (+1) 212 963 3489
E-mail: publications@un.org
Web site: www.unog.ch
UN Bookshop (direct sales)
The United Nations Bookshop
General Assembly Building Room 32
New York, NY 10017
Tel.: (+1) 212 963 7680
Fax: (+1) 212 963 4910
E-mail: bookshop@un.org
Web site: www.un.org
Periodicals:
Ebsco Subscription Services
PO Box 1943
Birmingham, AL 35201-1943
Tel.: (+1) 205 991 6600
Fax: (+1) 205 991 1449
The Faxon Company Inc.
15 Southwest Park
Westwood, MA 02090
Tel.: (+1) 617 329 3350
Telex: 95 1980
Cable: FW Faxon Wood

• URUGUAY
Librería Agropecuaria S.R.L.
Buenos Aires 335, Casilla 1755
Montevideo C.P. 11000

• VENEZUELA
Tecni-Ciencia Libros
CCCT Nivel C-2
Caracas
Tel.: (+58) 2 959 4747
Fax: (+58) 2 959 5636
Correo electrónico:
tclibros@attglobal.net
Fudeco, Librería
Avenida Libertador-Este
Ed. Fudeco, Apartado 254
Barquisimeto C.P. 3002, Ed. Lara
Tel.: (+58) 51 538 022
Fax: (+58) 51 544 394
Librería FAGRO
Universidad Central de Venezuela (UCV)
Maracay

• YUGOSLAVIA
See Serbia and Montenegro

• ZIMBABWE
Prestige Books
The Book Café
Fife Avenue Shops
Harare
Tel.: (+263) 4 336298/336301
Fax: (+263) 4 335105
E-mail: books@prestigebooks.co.zw

06/04

Sales and Marketing Group, Information Division, FAO
Viale delle Terme di Caracalla, 00100 Rome, Italy
Tel.: (+39) 06 57051 – Fax: (+39) 06 57053360
E-mail: publications-sales@fao.org
www.fao.org/catalog/giphome.htm

أماكن بيع مطبوعات المنظمة
当地何处可以购买粮农组织出版物

WHERE TO PURCHASE FAO PUBLICATIONS LOCALLY
POINTS DE VENTE DES PUBLICATIONS DE LA FAO
PUNTOS DE VENTA DE PUBLICACIONES DE LA FAO

• ANGOLA
Empresa Nacional do Disco e de Publicações, ENDIPU-U.E.E.
Rua Cirilo da Conceição Silva, Nº 7
C.P. Nº 1314-C, Luanda

• ARGENTINA
Librería Hemisferio Sur
Pasteur 743, 1028 Buenos Aires
Correo electrónico: adolfop@hemisferiosur.com.ar
World Publications S.A.
Av. Córdoba 1877, 1120 Buenos Aires
Tel./Fax: (+54) 11 48158156

• AUSTRALIA
Hunter Publications (Tek Imaging Pty. Ltd)
PO Box 404, Abbotsford, Vic. 3067
Tel.: (+61) 3 9417 5361
Fax: (+61) 3 9419 7154
E-mail: admin@tekimaging.com.au

• AUSTRIA
Uno Verlag
Am Hofgarten, 10
D-53113 Bonn, Germany
Tel.: (+49) 228 949020
Fax: (+49) 228 9490222
E-mail: info@uno-verlag.de
Web site: www.uno-verlag.de

• BELGIQUE
M.J. De Lannoy
202, avenue du Roi, B-1060 Bruxelles
CCP: 000-0808993-13
Courriel: jean.de.lannoy@infoboard.be

• BOLIVIA
Los Amigos del Libro
Av. Heroínas 311, Casilla 450
Cochabamba;
Mercado 1315, La Paz
Correo electrónico:
gutten@amigol.bo.net

• BOTSWANA
Botsalo Books (Pty) Ltd
PO Box 1532, Gaborone
Tel.: (+267) 312576
Fax: (+ 267) 372608
E-mail: botsalo@botsnet.bw

• BRAZIL
Fundação Getúlio Vargas
Praia do Botafogo 190, C.P. 9052
Rio de Janeiro
Correo eléctronico: livraria@fgv.br

• CANADA
Renouf Publishing
5369 chemin Canotek Road, Unit 1
Ottawa, Ontario K1J 9J3
Tel.: (+1) 613 745 2665
Fax: (+1) 613 745 7660
E-mail: order.dept@renoufbooks.com
Web site: www.renoufbooks.com

• CHILE
Librería - Marta Caballero
c/o FAO, Oficina Regional para América Latina y el Caribe (RLC)
Avda. Dag Hammarskjold, 3241
Vitacura, Santiago
Tel.: (+56) 2 33 72 314
Correo electrónico:
german.rojas@field.fao.org
Correo electrónico:
caballerocastillo@hotmail.com

• CHINA
China National Publications Import & Export Corporation
16 Gongti East Road, Beijing 100020
Tel.: (+86) 10 6506 3070
Fax: (+86) 10 6506 3101
E-mail: serials@cnpiec.com.cn

• COLOMBIA
INFOENLACE LTDA
Cra. 15 No. 86A–31
Santafé de Bogotá
Tel.: (+57) 1 6009474–6009480
Fax: (+57) 1 6180195
Correo electrónico:
servicliente@infoenlace.com.co
Sitio Web: www.infoenlace.com.co

• CONGO
Office national des librairies populaires
B.P. 577, Brazzaville

• COSTA RICA
Librería Lehmann S.A.
Av. Central, Apartado 10011
1000 San José
Correo eléctronico:
llehmann@solracsa.co.cr

• CÔTE D'IVOIRE
CEDA
04 B.P. 541, Abidjan 04
Tél.: (+225) 22 20 55
Télécopie: (+225) 21 72 62

• CUBA
Ediciones Cubanas
Empresa de Comercio Exterior de Publicaciones
Obispo 461, Apartado 605, La Habana

• CZECH REPUBLIC
Myris Trade Ltd
V Stinhlach 1311/3, PO Box 2
142 01 Prague 4
Tel.: (+420) 2 34035200
Fax: (+420) 2 34035207
E-mail: myris@myris.cz
Web site: www.myris.cz

• DENMARK
Gad Import Booksellers
c/o Gad Direct
31-33 Fiolstraede
DK-1171 Copenhagen K
Tel.: (+45) 3313 7233
Fax: (+45) 3254 2368
E-mail: info@gaddirect.dk

• ECUADOR
Libri Mundi, Librería Internacional
Juan León Mera 851
Apartado Postal 3029, Quito
Correo electrónico:
librimu1@librimundi.com.ec
Web site: www.librimundi.com
Universidad Agraria del Ecuador
Centro de Información Agraria
Av. 23 de julio, Apartado 09-01-1248
Guayaquil
Librería Española
Murgeón 364 y Ulloa, Quito

• EGYPT
MERIC The Middle East Readers' Information Centre
2 Baghat Aly Street, Appt. 24
El Masry Tower D
Cairo/Zamalek
Tel.: (+20) 2 3413824/34038818
Fax: (+20) 2 3419355
E-mail: info@mericonline.com

• ESPAÑA
Librería Agrícola
Fernando VI 2, 28004 Madrid
Librería de la Generalitat de Catalunya
Rambla dels Estudis 118 (Palau Moja)
08002 Barcelona
Tel.: (+34) 93 302 6462
Fax: (+34) 93 302 1299
Mundi Prensa Libros S.A.
Castelló 37, 28001 Madrid
Tel.: +34 91 436 37 00
Fax: +34 91 575 39 98
Sitio Web: www.mundiprensa.com
Correo electrónico:
libreria@mundiprensa.es
Mundi Prensa - Barcelona
Consejo de Ciento 391
08009 Barcelona
Tel.: (+34) 93 488 34 92
Fax: (+34) 93 487 76 59

• FINLAND
Akateeminen Kirjakauppa
PL 23, 00381 Helsinki
(Myymälä/Shop: Keskuskatu 1
00100 Helsinki)
Tel.: (+358) 9 121 4385
Fax: (+358) 9 121 4450
E-mail: akatilaus@akateeminen.com
Web site: www.akateeminen.com/suurasiakkaat/palvelut.htm

• FRANCE
Lavoisier Tec & Doc
14, rue de Provigny
94236 Cachan Cedex
Courriel: livres@lavoisier.fr
Site Web: www.lavoisier.fr
Librairie du commerce international
10, avenue d'Iéna
75783 Paris Cedex 16
Courriel: librarie@cfce.fr
Site Web: www.planetexport.fr

• GERMANY
Alexander Horn Internationale Buchhandlung
Friedrichstrasse 34
D-65185 Wiesbaden
Tel.: (+49) 611 9923540/9923541
Fax: (+49) 611 9923543
E-mail: alexhorn1@aol.com
TRIOPS - Tropical Scientific Books
S. Toeche-Mittler Versandbuchhandlung GmbH
Hindenburstr. 33
D-64295 Darmstadt
Tel.: (+49) 6151 336 65
Fax: (+49) 6151 314 048
E-mail for orders: orders@net-library.de
E-mail for info.: info@net-library.de / triops@triops.de
Web site: www.net-library.de / www.triops.de
Uno Verlag
Am Hofgarten, 10
D-53113 Bonn
Tel.: (+49) 228 94 90 20
Fax: (+49) 228 94 90 222
E-mail: info@uno-verlag.de
Web site: www.uno-verlag.de

• GHANA
Readwide Bookshop Ltd
PO Box 0600 Osu, Accra
Tel.: (+233) 21 22 1387
Fax: (+233) 21 66 3347
E-mail: readwide@africaonline.cpm.gh

• GREECE
Librairie Kauffmann SA
28, rue Stadiou, 10564 Athens
Tel.: (+30) 1 3236817
Fax: (+30) 1 3230320
E-mail: ord@otenet.gr

• GUYANA
Guyana National Trading Corporation Ltd
45-47 Water Street, PO Box 308
Georgetown

• HONDURAS
Escuela Agrícola Panamericana
Librería RTAC
El Zamorano, Apartado 93, Tegucigalpa
Correo electrónico:
libreriazam@zamorano.edu.hn

• HUNGARY
Librotrade Kft.
PO Box 126, H-1656 Budapest
Tel.: (+36) 1 256 1672
Fax: (+36) 1 256 8727

• INDIA
Allied Publisher Ltd
751 Mount Road
Chennai 600 002
Tel.: (+91) 44 8523938/8523984
Fax: (+91) 44 8520649
E-mail:
allied.mds@smb.sprintrpg.ems.vsnl.net.in
Bookwell
Head Office:
2/72, Nirankari Colony, New Delhi - 110009
Tel.: (+91) 11 725 1283
Fax: (+91) 11 328 13 15
Sales Office:
24/4800, Ansari Road
Darya Ganj, New Delhi - 110002
Tel.: (+91) 11 326 8786
E-mail: bkwell@nde.vsnl.net.in
EWP Affiliated East-West Press PVT, Ltd
G-I/16, Ansari Road, Darya Gany
New Delhi 110 002
Tel.: (+91) 11 3264 180
Fax: (+91) 11 3260 358
E-mail: affiliat@nda.vsnl.net.in
Monitor Information Services
203, Moghal Marc Ratan Complex
Narayanguda
Hyderabad – 500029, Andhra Pradesh
Tel.: (+91) 40 55787065
Fax: (+91) 40 27552390
E-mail: helpdesk@monitorinfo.com
Web site: www.monitorinfo.com
M/S ResearchCo Book Centre
25-B/2, New Rohtak Road (near Liberty Cinema), New Delhi 110 005
Tel.: (+91) 11 551 50445
Fax: (+91) 11 287 16134
E-mail: researchco@dishnetdsl.net
Oxford Book and Stationery Co.
Scindia House
New Delhi 110001
Tel.: (+91) 11 3315310
Fax: (+91) 11 3713275
E-mail: oxford@vsnl.com
Periodical Expert Book Agency
G-56, 2nd Floor, Laxmi Nagar
Vikas Marg, Delhi 110092
Tel.: (+91) 11 2215045/2150534
Fax: (+91) 11 2418599
E-mail: pebe@vsnl.net.in

• INDONESIA
P.F. Book
Jl. Setia Budhi No. 274, Bandung 40143
Tel.: (+62) 22 201 1149
Fax: (+62) 22 201 2840
E-mail: pfbook@bandung.wasantara.net.id

• IRAN
The FAO Bureau, International and Regional Specialized Organizations Affairs
Ministry of Agriculture of the Islamic Republic of Iran
Keshavarz Bld, M.O.A., 17th oor
Teheran

• ITALY
FAO Bookshop
Viale delle Terme di Caracalla
00100 Roma
Tel.: (+39) 06 57053597
Fax: (+39) 06 57053360
E-mail: publications-sales@fao.org
Il Mare Libreria Internationale
Via di Ripetta 239
00186 Roma
Tel.: (+39) 06 3612155
Fax: (+39) 06 3612091
E-mail: ilmare@ilmare.com
Web site: www.iimare.com
Libreria Commissionaria Sansoni S.p.A. - Licosa
Via Duca di Calabria 1/1
50125 Firenze
Tel.: (+39) 55 64831
Fax: (+39) 55 64 2 57
E-mail: licosa@ftbcc.it
Libreria Scientifica "AEIOU"
dott. Lucio de Biasio
Via Coronelli 6, 20146 Milano
Tel.: (+39) 02 48954552
Fax: (+39) 02 48954548
E-mail: in@aeioulib.com or commerciale@aeioulib.com

• JAPAN
Far Eastern Booksellers (Kyokuto Shoten Ltd)
12 Kanda-Jimbocho 2 chome
Chiyoda-ku - PO Box 72, Tokyo 101-91